Microbiology

Principles and Explorations

Jacquelyn G. Black

Seventh Edition

Cover image: © Russell Kightley/Photoresearchers, Inc.
Author photos: Paul D. Robertson

To order books or for customer service, please call 1(800)-CALL-WILEY (225-5945).

Printed in the United States of America.

ISBN 978-1-118-10548-1 .

Printed and bound by IPAK.

10 9 8 7 6 5 4 3 2 1

Diseases and the Organisms that Cause Them

BACTERIAL DISEASES—ALSO SEE APPENDIX B

Disease	Organism	Type*	Page	Disease	Organism	Type*	Page
acne	*Propionibacterium acnes*	R, +	580	ornithosis (psittacosis)	*Chlamydia psittaci*	coccoid, NA	659
actinomycosis	*Actinomyces israelii*	I, +	591	Oroyo fever (Carrion's	*Bartonella*	coccoid, −	737–738
anthrax	*Bacillus anthracis*	R, +	97, 724–727	disease, bartonellosis)	*bacilliformis*		
bacterial meningitis	*Haemophilus influenzae*	R, −	755	peptic ulcer	*Helicobacter pylori*	R, −	692–694
	Neisseria meningitidis	C, −	444, 756	periodontal disease	*Porphyromonas gingivalis* and others	R, −	681–682
	Streptococcus pneumoniae	C, +	757	pharyngitis (strep throat)	*Streptococcus pyogenes*	C, +	643–644
	Listeria monocytogenes	R, −	757	plague (black death)	*Yersinia pestis*	R, −	334, 727–729
bacterial vaginitis	*Gardnerella vaginalis*	R, −	613	bubonic plague pneumonic plague			
botulism	*Clostridium botulinum*	R, +	413, 684, 767–768				
brucellosis (undulant fever, Malta fever)	*Brucella* sp.†	CB, −	730–731	pneumonia	*Streptococcus pneumoniae*	C, +	652–653
cat scratch fever	*Afipia felis,* *Bartonella henselae*	R, − CB, NA	597		*Klebsiella pneumoniae*	R, −	128, 172, 652, 670
chancroid	*Haemophilus ducreyi*	R, −	623	pneumonia, atypical (walking pneumonia)	*Mycoplasma pneumoniae*	I, NA	653
cholera (Asiatic cholera)	*Vibrio cholerae*	vibrio, −	413, 688–690	pseudomembranous colitis	*Clostridium difficile*	R, +	694–695
conjunctivitis	*Haemophilus aegyptius*	CB, −	593	puerperal fever (childbed fever)	*Streptococcus pyogenes*	C, +	719
dental caries	*Streptococcus mutans*	C, +	679–681	Q fever	*Coxiella burnetti*	CB, NA	334, 659–660
diptheria	*Corynebacterium diptheriae*	R, +	645–646	rat bite fever	*Spirillum minor*	S, −	597
ehrlichiosis	*Ehrlichia* sp.	R, NA	738		*Streptobacillus moniliformis*	R, −	597
endocarditis	*Enterococcus faecalis*	C, +	720–721	relapsing fever	*Borrelia* sp.	S, −	731–733
food poisoning	*Staphylococcus aureus*	C, +	413, 684	rheumatic fever	*Streptococcus pyogenes*	C, +	720
	Streptococcus pyogenes	C, +	720	rickettsialpox	*Rickettsia akari*	CB, NA	737
	Clostridium perfringens	R, +	413, 684	Rocky Mountain spotted fever	*Rickettsia rickettsii*	CB, NA	736–737
	Clostridium botulinum	R, +	685	salmonellosis	*Salmonella* sp.	R, −	685–686
	Bacillus cereus	R, +	685	shigellosis (bacillary dysentery)	*Shigella* sp.	R, −	687–688
	Listeria monocytogenes	R, +	757	skin and wound infections (scalded	*Staphylococcus aureus*	C, +	578
	Campylobacter sp.	R, −	373, 684–685, 692	skin syndrome,	*Staphylococcus epidermidis*	C, +	579
	Shigella sp.	R, −	413, 687–688	scarlet fever,	*Streptococcus* sp.	C, +	579
	Salmonella sp.	R, −	334, 685–686	erysipelas, impetigo,	*Providencia stuartii*	R, −	580
	Vibrio parahaemolyticus	R, −	688	etc.)	*Pseudomonas aeruginosa*	R, −	580–581
gas gangrene	*Clostridium perfringens* and others	R, −	595–596		*Serratia marcescens*	R, −	197, 580
gonorrhea	*Neisseria gonorrhoeae*	C, −	616–620	syphillis	*Treponema pallidum*	S, −	620–624
granuloma inguinale (donovanosis)	*Calymmatobacterium granulomatis*	R, −	627	tetanus	*Clostridium tetani*	R, +	765–767
Hansen's disease (leprosy)	*Mycobacterium leprae*	R, A-F	407, 763–765	toxic shock syndrome	*Staphylococcus aureus*	C, +	614–615
Legionnaires' disease (legionellosis)	*Legionella pneumophilia*	R, −	653–654	trachoma	*Chlamydia trachomatis*	coccoid, NA	593
leptospirosis	*Leptospira interrogans*	S, −	612–613	trench fever	*Rochalimaea quintana*	CB, NA	334, 737
listeriosis	*Listeria monocytogenes*	R, +	757	tuberculosis	*Mycobacterium tuberculosis*	R, A-F	654–658
Lyme disease	*Borrelia burgdorferi*	S, −	334, 733–734	tuberculosis, avian	*Mycobacterium avium*	R, A-F	655
lymphogranuloma venereum	*Chlamydia trachomatis*	coccoid, NA	626–627	tularemia	*Francisella tularensis*	R, −	334, 729–730
Madura foot (maduromycosis)	*Actinomadura, Streptomyces, Nocardia*	I, +, some A-F	591	typhoid fever	*Salmonella typhi*	R, −	686–687
nongonococcal urethritis (NGU)	*Chlamydia trachomatis*	R, VAR	625–626	typhus, endemic (murine typhus)	*Rickettsia typhi*	CB, NA	736
	Ureaplasma urealyticum	I, NA	626	typhus, epidemic	*Rickettsia prowazekii*	CB, NA	735
				typhus, recrudescent (Brill-Zinsser disease)	*Rickettsia prowazekii*	CB, NA	735
				typhus, scrub (tsutsugamushi disease)	*Rickettsia tsutsugamushi*	CB, NA	736

Diseases and the Organisms that Cause Them (*Countinued*)

BACTERIAL DISEASES—ALSO SEE APPENDIX B

Disease	Organism	Type*	Page
verruga peruana (bartonellosis)	*Bartonella bacilliformis*	coccoid, −	737
vibriosis	*Vibrio parahaemolyticus*	R, −	690
whooping cough (pertussis)	*Bordetella pertussis*	CB, −	649–651
yersiniosis	*Yersinia enterocolitica*	R, −	692

*Key to types:
C = coccus	I = irregular	VAR = Gram-variable
CB = coccobacillus	− = Gram-negative	A-F = acid-fast
R = rod	+ = Gram-positive	NA = not applicable
S = spiral		

†Species

VIRAL DISEASES

Disease	Virus	Reservoir	Page
aplastic crisis in sickle cell anemia	erythrovirus (B19)	humans	743
avian (bird) flu	influenza	birds	660–663
bronchitis, rhinitis	parainfluenza	humans, some other mammals	648–649
Burkitt's lymphoma	Epstein-Barr	humans	740–741
cervical cancer	human papillomavirus	humans	271, 587 631
chickenpox	varicella-zoster	humans	277–282 583–584
coryza (common cold)	rhinovirus	humans	277, 647–648
	coronavirus	humans	648
cytomegalic inclusion disease	cytomegalovirus	humans	632
Dengue fever	Dengue	humans	334, 739
encephalitis	Colorado tick fever	mammals	334, 743
	Eastern equine encephalitis	birds	277, 428, 761
	St. Louis encephalitis	birds	761
	Venezuelan equine encephalitis	rodents	280, 761
	Western equine encephalitis	birds	280, 335, 428, 761
epidemic keratoconjunctivitis	adenovirus	humans	593–594
fifth disease (erythema infectiosum)	erythrovirus (B19)	humans	277, 743
hantavirus pulmonary syndrome	bunyavirus	rodents	277, 666
hemorrhagic fever	Ebola virus (filovirus)	humans (?)	277, 742
	Marburg virus (filovirus)	humans (?)	277, 742
hemorrhagic fever, Bolivian	arenavirus	rodents and humans	743
hemorrhagic fever, Korean	bunyavirus (Hantaan)	rodents	277, 742
hepatitis A (infectious hepatitis)	hepatitis A	humans	277, 696–698
hepatitis B (serum hepatitis)	hepatitis B	humans	277, 699
hepatitis C (non-A, non-B)	hepatitis C	humans	699
hepatitis D (delta hepatitis)	hepatitis D	humans	700
hepatitis E (enterically transmitted non-A, non-B, non-C)	hepatitis E	humans	700
herpes, genital	usually herpes simplex type 2, sometimes type 1	humans	277, 629–631
herpes, oral	usually herpes simplex type 1, sometimes type 2	humans	277, 628
HIV disease, AIDS	human immunodeficiency virus (HIV)	humans	277, 555–560
infectious mononucleosis	Epstein-Barr	humans	740
influenza	influenza	swine, humans (type A)	277, 280, 513 660–664
		humans (type B)	277, 280, 515, 660– 664, 757
		humans (type C)	660–664
Lassa fever	arenavirus	rodents	743
measles (rubeola)	measles	humans	277, 581–582
meningoencephalitis	herpes	humans	630, 762
molluscum contagiosum	poxvirus group	humans	586
monkeypox	orthopoxvirus	humans, monkeys	586
mumps	paramyxovirus	humans	682–683
pneumonia	adenoviruses, respiratory syncytial virus	humans	652–653
poliomyelitis	poliovirus	humans	277, 768–771
rabies	rabies	all warm-blooded animals	758–761
respiratory infections	adenovirus	humans	667
	polyomavirus	none	762
Rift Valley fever	bunyavirus (phlebovirus)	humans sheep, cattle	742
roseola	human herpes virus-6	humans	583
rubella (German measles)	rubella	humans	277, 581–582
SARS (sudden acute respiratory syndrome)	coronavirus	animal	665–666
shingles	varicella-zoster	humans	277, 583–585
smallpox	variola (major and minor)	humans	277, 585–586
viral enteritis	rotavirus	humans	696
warts, common (papillomas)	human papillomavirus	humans	277, 586–588
warts, genital (condylomas)	human papillomavirus	humans	277, 586–588, 631–633
West Nile	West Nile	birds	761
yellow fever	yellow fever	monkeys, humans, mosquitoes	277, 280, 334, 739

The tables of fungal and parasitic diseases appear on the following page.

Diseases and the Organisms that Cause Them (*Concluded*)

UNCONVENTIONAL AGENTS

Disease	Agent	Reservior	Page	Disease	Agent	Reservior	Page
chronic wasting disease	prion	elk, deer	773	mad cow disease	prion	cattle	772–773
Creutzfeldt-Jacob disease	prion	humans	769–770	(bovine spongiform encephalopathy)			
kuru	prion	humans	770	scrapie	prion	sheep	771
				tomato stunt	viroid	plants	

FUNGAL DISEASES

Disease	Organism	Page	Disease	Organism	Page
aspergillosis	*Aspergillus* sp	590, 669	histoplasmosis	*Histoplasma capsulatum*	668
blastomycosis	*Blastomyces dermatitidis*	589–590	*Pneumocystis* pneumonia	*Pneumocystis carinii*	669
candidiasis	*Candida albicans*	590	ringworm (tinea)	various species of *Epidermophyton, Trichophyton, Microsporum*	588–589
coccidioidomycosis (San Joaquin valley fever)	*Coccidioides immitis*	667–668			
cryptococcosis	*Filobasidiella neoformans*	668–669	sporotrichosis	*Sporothrix schenckii*	589
ergot poisoning	*Claviceps purpurea*	816	zygomycosis	*Rhizopus* sp., *Mucor* sp	590–591

PARASITIC DISEASES

Disease	Organism	Type	Page	Disease	Organism	Type	Page
Acanthamoeba keratitis	*Acanthamoeba culbertsoni*	protozoan	439	malaria	*Plasmodium* sp.	protozoan	317–318, 443, 745–747
African sleeping sickness (trypanosomiasis)	*Trypanosoma brucei gambiense* and *T. brucei rhodesiense*	protozoan	334, 773–775	pediculosis (lice infestation)	*Pediculus humanus*	louse	599
amoebic dysentery	*Entamoeba histolytica*	protozoan	701	pinworm	*Enterobius vermicularis*	roundworm	711
ascariasis	*Ascaris lumbricoides*	roundworm	708	river blindness (onchocerciasis)	*Onchocerca volvulus*	roundworm	594–595
babesiosis	*Babesia microti*	protozoan	749				
balantidiasis	*Balantidium coli*	protozoan	701–702	scabies (sarcoptic mange)	*Sarcoptes scabiei*	mite	598
Chagas' disease	*Trypanosoma cruzi*	protozoan	334, 775–776	schistosomiasis	*Schistosoma* sp.	flatworm	328, 721–723
chigger dermatitis	*Trombicula* sp.	mite	598	sheep liver fluke (fascioliasis)	*Fasciola hepatica*	flatworm	704
chigger infestation	*Tunga penetrans*	sandflea	598				
Chinese liver fluke	*Clonorchis sinensis*	flatworm	704	strongyloidiasis	*Strongyloides stercoralis*	roundworm	709–711
crab louse	*Phthirus pubis*	louse	599	swimmer's itch	*Schistosoma* sp.	flatworm	591
cryptosporidiosis	*Cryptosporidium* sp.	protozoan	702	tapeworm infestation (taeniasis)	*Hymenolepis nana* (dwarf tapeworm)	flatworm	705–707
dracunculiasis (Guinea worm)	*Dracunculus medinensis*	roundworm	330, 591		*Taenia saginata* (beef tapeworm)	flatworm	326–327, 705–707
elephantiasis (filariasis)	*Wuchereria bancrofti*	roundworm	331–332, 723		*Taenia solium* (pork tapeworm)	flatworm	705–707
fasciiolopsiasis	*Fasciolopsis buski*	flatworm	705		*Diphyllobothrium latum* (fish tapeworm)	flatworm	705–707
giardiasis	*Giardia intestinalis*	protozoan	700				
heartworm disease	*Dirofilaria immitis*	roundworm	312, 719		*Echinococcus granulosus* (dog tapeworm)	flatworm	705–707
hookworm	*Ancylostoma duodenale* (Old World hookworm)	roundworm	707				
	Necator americanus (New World hookworm)	roundworm	707	toxoplasmosis	*Toxoplasma gondii*	protozoan	747–749
leishmaniasis	*Leishmania braziliensis*	protozoan	334, 744	trichinosis	*Trichinella spiralis*	roundworm	330, 707
kala azar	*L. donovani*			trichomoniasis	*Trichomonas vaginalis*	protozoan	615
oriental sore	*L. tropica*			trichuriasis (whipworm)	*Trichuris trichiura*	roundworm	709
liver/lung fluke (paragonimiasis)	*Paragonimus westermani*	flatworm	327, 669–670	visceral larva migrans	*Toxocara* sp.	roundworm	709
loaiasis	*Loa loa*	roundworm	330, 595				

Achieve Positive Learning Outcomes with WileyPLUS...

Every one of your students has the potential to make a difference. And realizing that potential starts right here, in your course.

When students succeed in your course—when they stay on-task and make the breakthrough that turns confusion into confidence—they are empowered to realize the possibilities for greatness that lie within each of them. We know your goal is to create an environment where students reach their full potential and experience the exhilaration of academic success that will last them a lifetime. *WileyPLUS* can help you reach that goal.

Wiley**PLUS** is an online suite of resources—including the complete text—that will help your students:

*   come to class better prepared for your lectures
*   get immediate feedback and context-sensitive help on assignments and quizzes
*   track their progress throughout the course

"I just wanted to say how much this program helped me in studying... I was able to actually see my mistakes and correct them. ... I really think that other students should have the chance to use *WileyPLUS*."

Ashlee Krisko, *Oakland University*

www.wileyplus.com

88% of students surveyed said it improved their understanding of the material.*

MICROBIOLOGY 7e

PRINCIPLES AND EXPLORATIONS

JACQUELYN G. BLACK

Marymount University, Arlington, Virginia

Contributor: **LAURA J. BLACK**

Laura Black has been working on this book since she was ten years old. She has now been brought on as a contributing author for the seventh edition.

Jacquelyn and Laura Black

WILEY

JOHN WILEY & SONS, INC.

To Laura . . .
for sharing her mother and much of her childhood
with that greedy sibling "the book."

Senior Acquisitions Editor Kevin Witt
Associate Editor Merillat Staat
Senior Production Editor Elizabeth Swain
Executive Marketing Manager Clay Stone
Text Designer Madelyn Lesure
Cover Designers Madelyn Lesure and Merillat Staat
Senior Illustration Editor Anna Melhorn
Photo Editor Hilary Newman
Photo Researcher Mary Ann Price
Senior Media Editor Linda Muriello
Editorial Assistant Alissa Rufino

Cover Image: ©Russell Kightley/Photo Researchers, Inc.
Author photos: Paul D. Robertson

This book was set in 10/12 Times Ten by GGS Book Services, Atlantic Highlands and printed and bound by R. R. Donnelley, Jefferson City. The cover was printed by R. R. Donnelley, Jefferson City.

This book is printed on acid-free paper.

To order books or for customer service please, call 1-800-CALL WILEY (225-5945).

Library of Congress Cataloging-in-Publication Data
Black, Jacquelyn G.
  Microbiology : principles and explorations / Jacquelyn G. Black. – 7th ed.
     p. cm.
  Includes index.
  ISBN 978-0-470-10748-5
  1. Microbiology. I. Title.
  QR41.2.B58 2008
  616.9'041–dc22
                                   2007033363

Printed in the United States of America
10 9 8 7 6 5 4 3

The development of microbiology—from Leeuwenhoek's astonished observations of "animalcules," to Pasteur's first use of rabies vaccine on a human, to Fleming's discovery of penicillin, to today's search to stop the spread of SARS and race to develop an AIDS vaccine is one of the most dramatic stories in the history of science. To understand the roles microbes play in our lives, including the interplay between microorganisms and humans, we must examine, learn about, and study their world—the world of microbiology.

Microorganisms are everywhere. They exist in a range of environments from mountains and volcanoes to deepseas vents and hot springs. Microorganisms can be found in the air we breathe, in the food we eat, and even within our own body. In fact, we come in contact with countless numbers of microorganisms every day. Although some microbes can cause disease, most are not disease producers; rather they play a critical role in the processes that provide energy and make life possible. Some even prevent disease, and others are used in attempts to cure disease.

Because microorganisms play diverse roles in the world, microbiology continues to be an exciting and critical discipline of study. And because microbes affect our everyday lives, microbiology provides many challenges and offers many rewards. Look at your local newspaper, and you will find items concerning microbiology: to mention a few, reports on diseases such as AIDS, tuberculosis, and cancer; the resurgence of malaria and dengue fever, or "new" diseases such as Avian Flu, HPV, West Nile fever, Monkey Pox, SARS, and those caused by the Ebola virus and the hantavirus; a bacterium that can cause ulcers and stomach cancer; technologies designed to increase food production; bacteriophages used to circumvent antibiotic resistance; microorganisms used to clean up toxic wastes and oil spills; and the Human Genome Project that identifies the complete set of genetic instructions within the cells of the human body.

THEME

The theme that permeates this book is that microbiology is a current, relevant, exciting central science that affects all of us. I would like to share this excitement with you. Come with me as I take you, and your students, on a journey through the relevancy of microbiology. In countless areas—from agriculture to evolution, from ecology to dentistry—microbiology is contributing to scientific knowledge as well as solving human problems. Accordingly, a goal of this text is to offer a sense of the history of this science, its methodology, its many contributions to humanity, and the many ways in which it continues to be on the cutting edge of scientific advancement.

AUDIENCE AND ORGANIZATION

This book meets the needs of students in the health sciences as well as biology majors and students enrolled in other science programs who need a solid foundation in microbiology. It is designed to serve both audiences—in part by using an abundance of clinically important information to illustrate the general principles of microbiology and in part by offering a wide variety of additional applications.

In this edition, boxed essays have been newly organized to help students readily recognize the type of application being presented. Each application is identified with an appropriate icon.

BIOTECHNOLOGY

Whose DNA Do You Have?

Eukaryotic nuclei can be removed from cells of one species (such as human) and implanted into the egg cytoplasm of another species (such as a cow) from which the original nucleus has been removed. However, the cytoplasmic mitoch...

PUBLIC HEALTH

Can a Caterpillar Make Your Next Flu Vaccine?

Allergic to vaccines made in eggs? Cheer up, a faster new method using insect cell culture may soon replace the egg. An...

CLOSE UP

Fungi and Orchids

When explorers first brought orchids from South America back to England during the nineteenth century, the English were delighted to have such handsome specimens in their conservatories. However, they suffered...

TRY IT

Find Your Own Killer Phage

Obtain a sample likely to contain bacteriophages. The original samples used when phages were discovered were sewage-contaminated river waters. Manure is a rich source. Centrifuge to remove large particles and macroscopic organ...

APPLICATIONS

Plant Viruses

Besides the specificity shown by some viruses for bacteria and humans, other viruses are specific to and infect plants. Most viruses enter plant cells through dam...

The organization of the seventh edition continues to combine logic with flexibility. The chapters are grouped in units from the fundamentals of chemistry, cells, and microscopy; to metabolism, growth, and genetics; to taxonomy of microbes and multicellular parasites; to control of microorganisms; to host-microbe interactions; to infectious diseases of humans; and finally to environmental and applied microbiology. The chapter sequence will be useful in most microbiology courses as they are usually taught. However, it is not essential that chapters be assigned in their present order; it is possible to use this book in courses organized along different lines.

▮▮▮ STYLE AND CURRENCY

▮ In a field that changes so quickly—with new research, new drugs, and even new diseases—it is essential that a text be as up-to-date as possible. This book incorporates the latest information on all aspects of microbiology, including geomicrobiology, phage therapy, deep hot biosphere vents, and clinical practice. Special attention has been paid to such important, rapidly evolving topics as genetic engineering, taxonomy, lateral gene transfer, cervical cancer, mad cow disease, and immunology.

The rapid advances being made in microbiology make teaching about—and learning about—microorganisms challenging. Therefore, every effort has been made in the seventh edition of *Microbiology: Principles and Explorations* to ensure that the writing is simple, straightforward, and functional; that microbiological concepts and methodologies are clearly and thoroughly described; and that the information presented is as accessible as possible to students. Students who enjoy a course are likely to retain far more of its content for a longer period of time than those who take the course like a dose of medicine. There is no reason for a text to be any less interesting than the subject it describes. So, in addition to a narrative that is direct and authoritative, students will find injections of humor, engaging stories, and personal reflections that I hope impart a sense of discovery and wonder and a bit of my passion for microbial life. Because students find courses most interesting when they can relate topics to their everyday life or to career goals, I have emphasized the connection between microbiological knowledge and student experiences. One way that this connection is made is through the many boxed essays described previously. Another is through the use of factoids, post-it type notes that are tidbits of information relating to the running text and that add an extra dimension of flavor to the discussion at hand.

Rhinoviruses, causes of the common cold, can survive on household objects for up to 3 days.

TABLE 15.6—See Table 15.6 Key on p. 458

Agents of Bioterrorism

Agent	Incubation	Signs/Symptoms	Diagnostic Tests	Precautions
Anthrax	1–5 d	Fever, malaise, fatigue, cough, mild chest discomfort, cold/flu-like symptoms. Improvement 2–3 days. Abrupt onset resp. distress, shock, CXR: widened mediastinum.	Nasal, resp. culture, FA, PCR, blood–Gram stain, culture, PCR, serum–Ag ELISA	Standard
Botulism	1–5 d	Cranial nerve palsy–ptosis, blurred vision, dry mouth, dysphagia, dysphonia, descending flaccid	Nasal swab & resp.: PCR, Ag ELISA, serum toxin assays, blood & stool	Standard
Brucell				

Chemotherapy	Chemoprophylaxis	Vaccine	Comments/(Human to Human Trans.?)
• Cipro 400 mg IV Q 8–12 h • Doxycycline 200 mg IV, then, 100 mg IV Q 12 h • PCN 2 million units IV Q 2 h • Streptomycin 30 mg/kg IM QD OR Gent. Child/Preg: Cipro, PCN, Doxycycline 3rd choice	• Cipro 500 mg po bid × 4 wk. If not vaccinated give vaccine. • Doxycycline 100 mg po bid + vaccine	Bioport vaccine 0.5 ml SC Q 1, 2, 4 wk, 6, 12, 18 mo. & annually	(No)
• DOD Heptavalent Equine—Despericiated antitoxin for serotypes (A–G IND) 1 vial IV	Pentavalent toxid vaccine Types (A–E)	DOD Heptavalent Equine Toxoid for Serotypes A–E (IND) 0.5 ml deep SC @ 0, 2, & 12 wk & annually	Skin test before vaccine. (NO)
• Doxycycline 200 mg/d po + Rifampin 600 mg, 900 mg/d po × 6 wks.	Doxycycline & Rifampin × 3 wk	No vaccine available	(NO)

DESIGN AND ILLUSTRATIONS

The seventh edition of *Microbiology: Principles and Explorations* has been completely redesigned with an eye toward increasing the readability, enhancing the presentation of illustrations and photographs, and making the pedagogical features more effective for use. The use of clear, attractive drawings and carefully chosen photographs can significantly contribute to the student's understanding of a scientific subject. Throughout, color has been used not just decoratively but for its pedagogic value. For example, every effort has been made to color similar molecules and structures the same way each time they appear, making them easier to recognize.

Illustrations have been carefully developed to amplify and enhance the narrative. The line art in this text is sometimes as simple as a flow diagram or just as often a complex illustration of a structure drawn by some of the best medical illustrators working today. Photographs, many of which I have personally taken or researched, also richly enhance the text. The diversity of the photo program encompasses numerous micrographs, photographs of clinical conditions, microbiologists at work, and some laboratory techniques and results. Often, you will find a photograph accompanied by a line drawing aiding in the understanding of an unfamiliar subject.

Black smoking vents

The fluid-mosaic model of the cell membrane

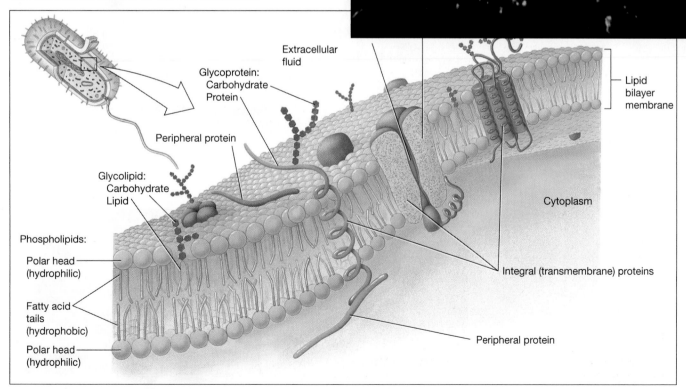

Extracellular fluid

Glycoprotein:
Carbohydrate
Protein

Peripheral protein

Lipid bilayer membrane

Glycolipid:
Carbohydrate
Lipid

Cytoplasm

Phospholipids:

Polar head (hydrophilic)

Fatty acid tails (hydrophobic)

Polar head (hydrophilic)

Integral (transmembrane) proteins

Peripheral protein

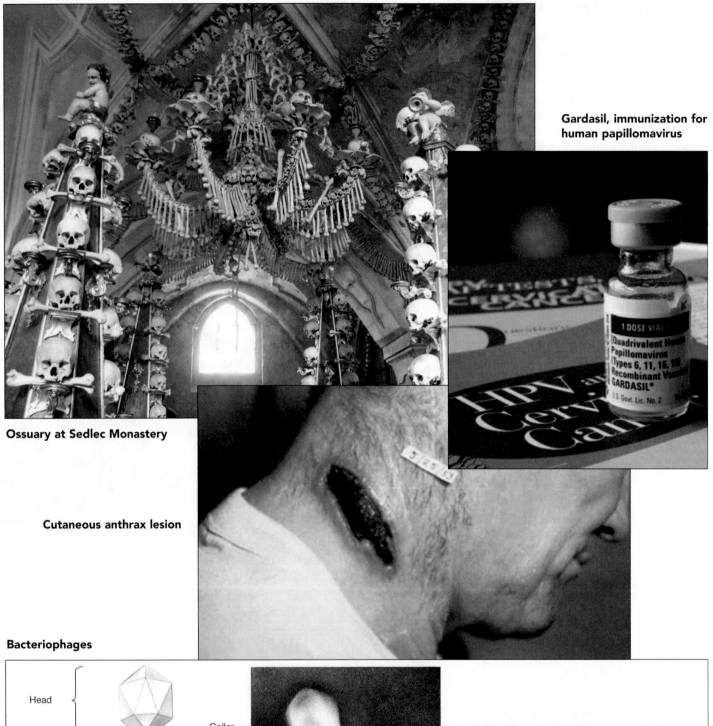

Gardasil, immunization for human papillomavirus

Ossuary at Sedlec Monastery

Cutaneous anthrax lesion

Bacteriophages

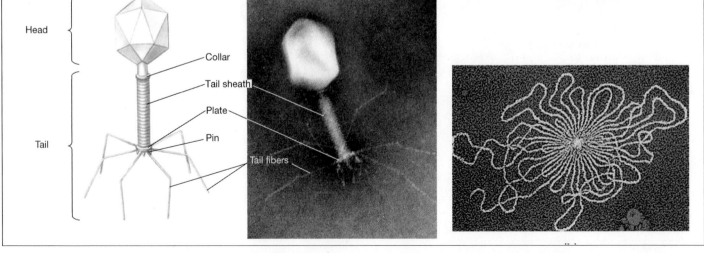

Head

Collar

Tail sheath

Plate

Tail

Pin

Tail fibers

▌ SPECIAL PEDAGOGICAL FEATURES

The distinctive learning aids developed for this text help students to study and comprehend microbiology more efficiently. Learning microbiology requires learning a new vocabulary, understanding basic concepts and relating them to other concepts, and applying those concepts to the world around us. The pedagogical structure of *Microbiology: Principles and Explorations* was designed with these goals in mind. These features include Chapter Opening "Come With Me" Vignettes with Video Clips, Checklists, Concept Links, Retracing Our Steps, Take Another Look, Terminology Checks, Clinical Case Studies, Critical Thinking Questions, Self-Quizzes, and Explorations on the Web. All of these are fully described and illustrated in "A Student's Guide to Success" that follows this preface.

▌ FEATURES

Carefully crafted ancillaries, developed to create a successful teaching and learning environment, support the seventh edition of *Microbiology: Principles and Explorations*. These include the following.

NEW AND DYNAMIC MEDIA

The study of microbiology is a fascinating, but challenging pursuit. With that in mind, we are pleased to offer a series of animated tutorials and video components.

COME WITH ME

An exciting new resource, Come With Me features a video to accompany each visually stimulating chapter opener in which Jackie Black invites the student to accompany her into the exciting world of microbiology. Access Come with Me videos within WileyPLUS.

TAKE ANOTHER LOOK

This feature provides additional information on key topics of interest to students. Comprised of an eclectic mix of animation, video, photos, and textual information students can explore different aspects of featured concepts. Icons placed in the text indicate when a Take Another Look may be found within WileyPLUS.

ANIMATIONS

Expanded animation offerings, listed below, continue to bring to life illustrations in *Microbiology: Principles and Explorations*. An animation icon accompanying an illustration indicates when students may access animations within WileyPLUS.

CHAPTER 2
Acids and Bases
Chemical Bonding
Polarity and Solubility
Types of Reactions and Equilibrium

CHAPTER 3
Staining Bacteria: The Gram Stain
Wavelength Analogy

CHAPTER 4
Eukaryotic Cell Structure and Function
Simple Diffusion
Prokaryotic Cell Structure and Function
Endocytosis and Exocytosis
Mitosis and Meiosis Compared
Osmosis
Peptidoglycan
Lipopolysaccharide

CHAPTER 5
Catabolism of Fats and Proteins
Competitive and Noncompetitive Inhibition of Enzymes
Metabolism, the Sum of Catabolism and Anabolism
Functions of Enzymes and Uses of ATP
Nonspecific Disease-Resistance Mechanisms
Cell Respiration

CHAPTER 6
Binary Fission
Endospore Formation
Budding
Streak Plate Method
Enterotube

CHAPTER 7
End Product Inhibition
Enzyme Induction; The lac Operon
Eukaryotic Genes Contain Introns
Mutations
The Polymerase Chain Reaction
DNA Replication in a Prokaryote
Thymine Dimer Repair
Protein Synthesis

CHAPTER 8
Gene Transfer: Transformation
Transduction
Conjugation
Recombinant DNA

CHAPTER 9
Five-Kingdom System
Shrub of Life
Lateral Gene Transfer
DNA Hybridization
Three-Domain System

CHAPTER 10
Viruses
Prion Proteins

CHAPTER 13
Antibiotic Treatment
Antibiotic Actions
Antiviral Base Analogs

CHAPTER 16
Innate Host Defenses
Inflammation

CHAPTER 17
Introduction to Specific Immunity
Production of Monoclonal Antibodies
Antibody Mediated Immunity
Cell Mediated Immunity

CHAPTER 18
Myasthenia Gravis
Four Types of Hypersensitivity

FOR THE STUDENT

Book Companion Website is located at www.wiley.com/college/black. Quizzes for student self testing, the *Biology NewsFinder*, anatomy overviews, Insights and Explorations, and Flashcards are available.

Laboratory Exercises in Microbiology, 3e (978-0-470-13392-7) Robert A. Pollack, Lorraine Findlay, Walter Mondschein, and R. Ronald Modesto is a new publication that dovetails perfectly with this textbook. This hands-on laboratory manual contains a variety of interactive activities and experiments that teach students the basic concepts of microbiology. It also covers methods that allow the safe movement or transfer of microbial cells from one type of growth environment to another, classification and identification of microbes, and microbial biochemistry.

FOR THE INSTRUCTOR

Book Companion Website is easily accessed at www.wiley.com/college/black and provides instructors with a host of instructional support materials including:

Instructor's Resource Manual, written by Dubear Kroening, University of Wisconsin–Fox Valley, includes lecture outlines, teaching suggestions for each chapter in the text, in-class activities, and helpful tips on teaching with technology in the classroom.

Testbank and Computerized Testbank, written by Elisa Margolis of Emory University, include multiple-choice and essay questions to test a variety of concepts and applications. The computerized testbank is an easy-to-use test-generation program that fully support graphics, print tests, student answer sheets, and answer keys.

The software's advanced features allow you to create an exam to your exact specifications.

Personal Response System Questions by Anne Hemsley, Antelope Valley College are specifically designed to foster student discussion and debate in class.

Animations by Richard Anderson of Modesto Junior College bring to life key pieces of art from the text and are available in flash files as well as embedded into PowerPoint slides for ease in incorporating them into your lecture presentations.

All Line Illustrations and Photos from *Microbiology: Principles and Explorations,* 7e are available in jpeg format on the book website.

PowerPoint Presentations by Anne Hemsley, Antelope Valley College are tailored to *Microbiology: Principles and Explorations* 7e's topical coverage and learning objectives. These presentations are designed to convey key text concepts with embedded illustrations and photos.

Transparencies on Demand is a link on the instructor's book companion site that offers *Microbiology: Principles and Explorations* users the ability to build their own custom transparency set. Search by text chapter or key word to select from all illustrations in *Microbiology: Principles and Explorations* 7e as well as a host of other Wiley life science illustrations.

WileyPLUS

WileyPLUS provides an integrated suite of teaching and learning resources, including an online version of the text, in one easy-to-use website. Organized around the essential activities you perform in class, *WileyPLUS* helps you:

- **Prepare and Present.** Create class presentations using a wealth of Wiley-provided resources, including an online version of the textbook, PowerPoint slides, animations, and more—making your preparation time more efficient. You may easily adapt, customize, and add to this content to meet the needs of your course.

- **Create Assignments.** Automate the assigning and grading of homework or quizzes by using Wiley-provided question banks or by writing your own. Student results will be automatically graded and recorded in your gradebook. *WileyPLUS* can link homework problems to the relevant section of the online text, providing students with context-sensitive help.

- **Track Student Progress.** Keep track of your students' progress via an instructor's gradebook, which allows you to analyze individual and overall class results to determine student progress and level of understanding.

- **Administer Your Course.** WileyPLUS can easily be integrated with other course management systems, gradebooks, or other resources you are using in your class, providing you with the flexibility to build your course in your own way.

▌▌▌ ACKNOWLEDGMENTS

Thanks really must go to the many people who have helped this seventh edition become a reality. Critical team members include Kevin Witt, Senior Acquisitions Editor; Merillat Staat, Associate Editor; Elizabeth Swain, Senior Production Editor; Madelyn Lesure, Senior Designer; Clay Stone, Executive Marketing Manager; Anna Melhorn, Senior Illustration Editor; Hilary Newman, Photo Manager; Mary Ann Price, Photo Researcher; Alissa Rufino, Editorial Assistant; and Lisa Vecchio, Senior Marketing Assistant.

My thanks and appreciation go to Michael Chase of Marymount University for updating and revising the Self-Quiz and Critical Thinking questions at the end of each chapter. Anne Hemsley of Antelope Valley College is owed great thanks for her insightful advice and many comments on the revision. Special thanks go to Martha Roberts for all her help.

Most importantly I would like to thank the many reviewers who have taken the time to share their comments and suggestions for enhancing each edition of this text. Your input makes a considerable difference.

REVIEWERS FOR THE SEVENTH EDITION

Sally McLaughlin Bauer, Hudson Valley Community College
Gregory Bertoni, Columbus State Community College
Margaret Beucher, University of Pittsburgh
Judith K. Davis, Florida Community College at Jacksonville
Nwadiuto Esiobu, Florida Atlantic University
Sara K. Germain, Southwest Tennessee Community College
Anne Hemsley, Antelope Valley College
Dale R. Horeth, Tidewater Community College
Karen Kendall-Fite, Columbia State Community College
Marty Lowe, Bergen Community College
Rebecca Nelson, Pulaski Technical College
Robert A. Pollack, Nassau Community College
Madhura Pradhan, Ohio State University
Mary V. Mawn, Hudson Valley Community College
Eric Raymond Paul, Texas Tech University
Karl J. Roberts, Prince George's Community College
Victoria C. Sharpe, Blinn College
Jia Shi, De Anza College
Kent R. Thomas, Wichita State University
Winfred E. Watkins, McLennan Community College
Valerie A. Watson, West Virginia University
Mark Zelman, Aurora University

REVIEWERS FOR PREVIOUS EDITIONS

Ronald W. Alexander, Tompkins Cortland Community College
D.Andy Anderson, Utah State University
Richard Anderson, Modesto Community College
Rod Anderson, Ohio Northern University
Oswald G. Baca, University of New Mexico
David L. Balkwill, Florida State University
Keith Bancroft, Southeastern Louisiana University
James M. Barbaree, Auburn University
Jeanne K. Barnett, University of Southern Indiana
Rebekah Bell, University of Tennessee at Chattanooga
R. L. Bernstein, San Francisco State University
David L. Berryhill, North Dakota State University
Steven Blanke, University of Houston
Alexandra Blinkova, University of Texas
Richard D. Bliss, Yuba College
Kathleen A. Bobbitt, Wagner College
Katherine Boettcher, University of Maine
Clifford Bond, Montana State University
Edward A. Botan, New Hampshire Technical College
Benita Brink, Adams State College
Kathryn H. Brooks, Michigan State University
Burke L. Brown, University of South Alabama
Daniel Brown, Santa Fe Community College
Linda Brushlind, Oregon State University
Barry Chess, Pasadena Community College
Kotesward Chintalacharuvu, UCLA
Richard Coico, City University of New York Medical School
William H. Coleman, University of Hartford
Iris Cook, Westchester Community College
Thomas R. Corner, Michigan State University
Christina Costa, Mercy College
Mark Davis, University of Evansville
Dan C. DeBorde, University of Montana
Sally DeGroot, St. Petersburg Junior College
Michael Dennis, Montana State University at Billings
Monica A. Devanas, Rutgers University
Von Dunn, Tarrant County Junior College
John G. Dziak, Community College of Allegheny County
Susan Elrod, California Polytechnic State University
Mark Farinha, University of North Texas
David L.Filmer, Purdue University
Eugene Flaumenhaft, University of Akron
Pamela B.Fouche,Walters State Community College
Christine L.Frazier, Southeast Missouri State University
Denise Y.Friedman, Hudson Valley Community College
Ron Froehlich, Mt. Hood Community College
David E. Fulford, Edinboro University of Pennsylvania
William R. Gibbons, South Dakota State University
Eric Gillock, Fort Hays State University
Mike Griffin, Angelo State University
Van H. Grosse, Columbus State University
Richard Hanke, Rose State College
Pamela L. Hanratty, Indiana University
Janet Hearing, State University of New York, Stony Brook
Ali Hekmati, Mott Community College
Donald Hicks, Los Angeles Community College
Lawrence W. Hinck, Arkansas State University
Elizabeth A. Hoffman, Ashland Community College
Clifford Houston, University of Texas
Ronald E. Hurlbert, Washington State University

Michael Hyman, North Carolina State University
John J. Iandolo, Kansas State University
Robert J. Janssen, University of Arizona
Thomas R. Jewell, University of Wisconsin—Eau Claire
Wallace L. Jones, De Kalb College
Ralph Judd, University of Montana
John W. Kimball, Harvard University
Karen Kirk, Lake Forest College
Timothy A. Kral, University of Arkansas
Helen Kreuzer, University of Utah
Michael Lawson, Montana Southern State College
Donald G. Lehman, Wright State University
Jeff Leid, Northern Arizona University
Harvey Liftin, Broward Community College
Roger Lightner, University of Arkansa, Fort Smith
Tammy Liles, Lexington Community College
Jeff Lodge, Rochester Institute of Technology
William Lorowitz, Weber State University
Caleb Makukutu, Kingwood College
Stanley Maloy, Sand Diego State University
Alesandria Manrov, Tidewater Community College,
    Virginia Beach
Judy D. Marsh, Emporia State University
Rosemarie Marshall, California State University, Los Angeles
John Martinko, Southern Illinois University
Anne Mason, Yavapai College
William C. Matthai, Tarrant County Junior College
Pam McLaughlin, Madisonville Community College
Robert McLean, Southwest Texas State University
Karen Messley, Rock Valley College
Chris H. Miller, Indiana University
Rajeev Misra, Arizona State University
Barry More, Florida Community College at Jacksonville
Timothy Nealon, St. Philip's College
Russell Nordeen, University of Arkansa, Monticello
Russell A. Normand, Northeast Louisiana University
Christian C.Nwamba, Wayne State University
Douglas Oba, Brigham Young University—Hawaii
Roselie Ocamp-Friedmann, University of Florida
Cathy Oliver, Manatee Community College
Raymond B. Otero, Eastern Kentucky University
Curtis Pantle, Community College of Southern Nevada
C.O. Patterson, Texas A&M University
Kimberley Pearlstein, Adelphi University
Roberta Petriess, Witchita State University
Robin K. Pettit, State University of New York, Potsdam
Robert W. Phelps, San Diego Mesa College
Holly Pinkart, Central Washington University
Robert A. Pollack, Nassau Community College
Jeff Pommerville, Glendale Community College

Leodocia Pope, University of Texas at Austin
Jennifer Punt, Haverford College
Ben Rains, Pulaski Technical College
Jane Repko, Lansing Community College
Quentin Reuer, University of Alaska, Anchorage
Kathleen Richardson, Portland Community College
Robert C. Rickert, University of California, San Diego
Russell Robbins, Drury College
Richard A. Robison, Brigham Young University
Dennis J. Russell, Seattle Pacific University
Frances Sailer, University of North Dakota
Gordon D. Schrank, St. Cloud State University
Alan J. Sexstone, West Virginia University
Deborah Simon-Eaton, Santa Fe Community College
K.T. Shanmugam, University of Florida
Pocahontas Shearin Jones, Halifax Community College
Brian R. Shmaefsky, Kingwood College
Sara Silverstone, State University of New York, Brockport
Robert E. Sjogren, University of Vermont
Ralph Smith, Colorado State University
D. Peter Snustad, University of Minnesota
Larry Snyder, Michigan State University
Joseph M. Sobek, University of Southwestern Louisiana
J. Glenn Songer, University of Arizona
Jay Sperry, University of Rhode Island
Paul M. Steldt, St. Philips College
Bernice C. Stewart, Prince George's Community College
Gerald Stine, University of Florida
Larry Streans, Central Piedmont Community College
Paul E.Thomas, Rutgers College of Pharmacy
Teresa Thomas, Southwestern College
Grace Thornhill, University of Wisconsin—River Falls
Jack Turner, University of Southern California—Spartanburg
James E. Urban, Kansas State University
Manuel Varella, Eastern New Mexico University
Phylis K.Williams, Sinclair Community College
George A. Wistreich, East Los Angeles College
Shawn Wright, Albuquerque Technical-Vocational Institute
Michael R. Yeaman, University of New Mexico
John Zak, Texas Tech University
Thomas E. Zettle, Illinois Central College

Comments and suggestions about the book are most welcome. You can contact me through my editors at John Wiley and Sons.

Jacquelyn Black
Arlington, Virginia

While writing this text, I developed and incorporated a variety of special features to help you learn microbiology more efficiently and confidently. Some students like to race through their texts, highlighting things that they feel might be important, but they never really assimilate the information into their long-term memory. Learning microbiology requires a number of steps to grasp the basic concepts, their interrelationships with one another, and their applications to the world around you. Stepping your way through each chapter using these features will help you optimize your study time.

The features of this text reflect the feedback from students—like you—who have used previous editions of the text and who—like you—want to make the most of their exploration of this discipline. Knowing how to use this book will help you enjoy your discoveries about the world of microbes. Below are some hints for using these helpful tools. A review of the preface will give you further insights into all of the text's distinctive features. Begin your study by anticipating what is to be learned from each chapter and by making the connections to concepts previously learned that have relevance to this new material.

Opening Vignettes, accompanied by a photograph, introduce you to the upcoming subject with a story or discussion that shows how the chapter concepts are relevant to the real world. Sometimes they are even personal stories, such as the one in this example where I describe my scalded hand and wounded pride!

The **Chapter Outline** will give you a brief overview of the topics to be covered in each chapter and will help you focus on what is important as you read it. As you read the chapter, a number of features are implemented to help you navigate through the material. The beautiful illustrations and photographs are an important part of your learning experience. So take time when a **figure reference** is mentioned in the narrative to stop and fully investigate the accompanying art. To make it easier for you to toggle back and forth between an illustration and the running narrative, figure call-outs in the text are colored red. Because connecting concepts is so important to success in this course, numerous **Concept Links** are included in each chapter. Red arrowheads with chapter and page references link to relevant concepts covered elsewhere in the text.

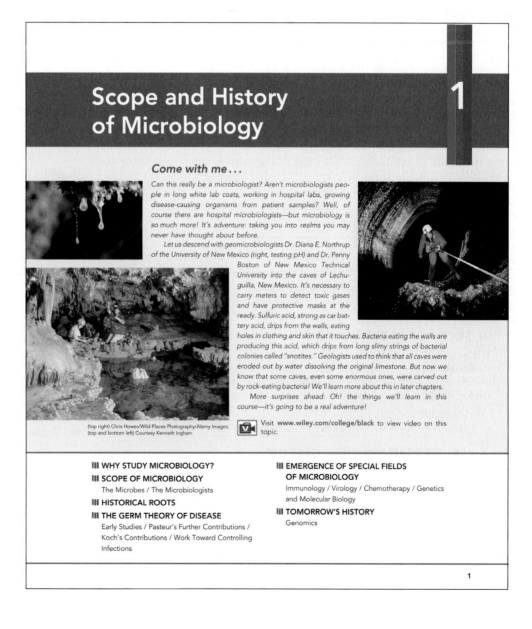

Scope and History of Microbiology

1

Come with me...

Can this really be a microbiologist? Aren't microbiologists people in long white lab coats, working in hospital labs, growing disease-causing organisms from patient samples? Well, of course there are hospital microbiologists—but microbiology is so much more! It's adventure: taking you into realms you may never have thought about before.

Let us descend with geomicrobiologists Dr. Diana E. Northrup of the University of New Mexico (right, testing pH) and Dr. Penny Boston of New Mexico Technical University into the caves of Lechuguilla, New Mexico. It's necessary to carry meters to detect toxic gases and have protective masks at the ready. Sulfuric acid, strong as car battery acid, drips from the walls, eating holes in clothing and skin that it touches. Bacteria eating the walls are producing this acid, which drips from long slimy strings of bacterial colonies called "snotites." Geologists used to think that all caves were eroded out by water dissolving the original limestone. But now we know that some caves, even some enormous ones, were carved out by rock-eating bacteria! We'll learn more about this in later chapters.

More surprises ahead: Oh! the things we'll learn in this course—it's going to be a real adventure!

(top right) Chris Howes/Wild Places Photography/Alamy Images; (top and bottom left) Courtesy Kenneth Ingham

Visit www.wiley.com/college/black to view video on this topic.

III WHY STUDY MICROBIOLOGY?
III SCOPE OF MICROBIOLOGY
   The Microbes / The Microbiologists
III HISTORICAL ROOTS
III THE GERM THEORY OF DISEASE
   Early Studies / Pasteur's Further Contributions / Koch's Contributions / Work Toward Controlling Infections

III EMERGENCE OF SPECIAL FIELDS OF MICROBIOLOGY
   Immunology / Virology / Chemotherapy / Genetics and Molecular Biology
III TOMORROW'S HISTORY
   Genomics

1

> ✓ **CHECKLIST**
>
> 1. List three reasons to study microbiology.
> 2. What is the difference between microbiology and bacteriology?
> 3. What is the difference between etiology and epidemiology?
> 4. List five bacterial diseases and five viral diseases.

When you reach the end of major sections, you will find numbered **Checklists** marked with a red check mark. Stop and check your understanding of the material you've just read. Some of the questions ask you to recall basic facts or concepts, while others ask you to apply what you've learned. Try and answer all of the questions. If you can, you are ready to move on to the next section. At the end of each chapter are other resources that you will find useful. **Retracing Our Steps** is the title signaling a chapter summary. It serves as a road map detailing the concepts and facts visited in the chapter. It is organized in an outline format that helps highlight and connect the main points of the chapter. Review it carefully and ask yourself questions about the information being summarized. Try to recall what you read, and mentally connect the points of the summary outline to the things your instructor said in lecture. You'll find yourself remembering a lot more than you thought possible.

Being able to speak the language of microbiology is half the battle of mastering it. Caution: Without knowing the vocabulary, you can't expect to get a good grade. Use the **Terminology Check** to review the key vocabulary from each chapter. Say the word out loud along with your best shot at its definition. Try to recall how the terms fit into the fabric of the whole chapter. If you have trouble recalling the word, use the page reference and go back to where it was first introduced in the chapter.

The **Glossary** can also be used to quickly find a definition. **Appendix C** contains a list of word roots that will make it easier to remember or figure out the meaning of the words. On the companion web site we have an audio glossary where you can actually hear the words pronounced correctly.

A NOTE ON PRONUNCIATION
The study of microbiology requires learning a new vocabulary that sometimes challenges pronunciation skills. For the more difficult terms, a pronunciation key is included with the new term. The scheme used for pronunciation is simple:

- A single accent mark (') is used for the main accent in a word;
- A double accent mark (") is used for secondary accents, if any;
- Any vowel not followed by a consonant is assumed to be long;
- Syllables are separated by either a hyphen or an accent mark.

▮ RETRACING OUR STEPS

▮▮▮ WHY STUDY MICROBIOLOGY?

- Microorganisms are part of the human environment and are therefore important to human health and activities.
- The study of microorganisms provides insight into life processes in all forms of life.

▮▮▮ SCOPE OF MICROBIOLOGY

THE MICROBES

- **Microbiology** is the study of all **microorganisms (microbes)** in the microscopic range. These include **bacteria, algae, fungi, viruses, viroids, prions,** and **protozoa.**

THE MICROBIOLOGISTS

- Immunology, virology, chemotherapy, and genetics are espe...

- The development of high-quality lenses by Leeuwenhoek made it possible to observe microorganisms and later to formulate the **cell theory**.

▮▮▮ THE GERM THEORY OF DISEASE

- The **germ theory of disease** states that microorganisms (germs) can invade other organisms and cause disease.

EARLY STUDIES

- Progress in microbiology and acceptance of the germ theory of disease required that the idea of **spontaneous generation** be refuted. Redi and Spallanzani demonstrated that organisms did not arise from nonliving material. Pasteur, with his swan-necked flasks, and Tyndall, with his dust-free air, fi-

---

▮ TERMINOLOGY CHECK

algae *(p. 4)*	control variable *(p. 19)*	Koch's Postulates *(p. 13)*	protozoa *(p. 5)*
antibiotics *(p. 17)*	experimental variable *(p. 19)*	microbe *(p. 4)*	spontaneous generation
bacteria *(p. 4)*	fungi *(p. 5)*	microbiology *(p. 4)*	*(p. 9)*
bacteriophage *(p. 22)*	germ theory of disease	microorganism *(p. 2)*	variable *(p. 19)*
cell theory *(p. 9)*	*(p. 9)*	prediction *(p. 19)*	viroid *(p. 5)*
conclusions *(p. 19)*	hypothesis *(p. 19)*	prion *(p. 5)*	viruses *(p. 5)*

The following end-of-chapter study aids will help you double-check your newly gained knowledge. We have included new **Clinical Case Study Questions** in the seventh edition. These are word problems that ask you to diagnose and prescribe treatment for medical conditions based on your chapter study. For additional, interactive case studies, check out the companion web site. **Critical Thinking Questions** go further than simple recall and test your understanding of the basic concepts, how they are connected, and how they are applied. The **Self-Quiz** offers you the opportunity to test your knowledge before an actual in-class test. These consist of multiple choice, true-false, matching, and art questions.

Before you conclude your study of each chapter, I hope you will go to **Explorations on the Web**. The World Wide Web is a great way to enhance your study of microbiology. The companion website for the text (http://www.wiley.com/college/black) is designed to offer you a variety of activities to expand your exploration of the chapter material, including self-tests, interactive case studies, an audio glossary with tools for creating flashcards and terminology quizzing, focused web searches, exploration activities, expanded interviews with microbiologists, web-links to related content, access to animations, and more. At the end of each chapter you will find several questions or statements that will give you a clue to some of the activities you will find on the web for that chapter.

And, finally, for the most important advice that I can give you: enjoy your study of microbiology. I hope that I have been able to transmit in the pages of this book a sense of the excitement that is microbiology. The longer I am a microbiologist, the more excited I become. It never grows old or boring—it is constantly new and vital. For me, it is all about having a passion for life. I am delighted to be able to share it with you. Have a successful semester!

▮ CLINICAL CASE STUDY

When doctors first began noticing cases of AIDS in the U.S., they did not know much about it. They had many questions to answer. What did they find to be its etiology? What were some observations about its epidemiology?

▮ CRITICAL THINKING QUESTIONS

1. Edward Jenner, in eighteenth-century England, first injected a child with a totally untested smallpox vaccine and then, after a time, injected that child with living smallpox virus. What would be the likely reaction to someone performing a similar experiment today? How do you think a scientist of today would test a potential new vaccine?

2. Can you think of some reasons why it might be hard to fulfill Koch's Postulates in order to support the "Germ Theory" of disease?

3. As often happens in science, one observation or experiment that is used to look at one aspect or subdiscipline of science can lead to profound explanations or solutions in

▮ SELF-QUIZ

1. Less than 1% of microorganisms are harmful and cause disease. True or false?

2. Life on earth would be much better if all microbes were eradicated. True or false?

3. Which of the following is not true?
  (a) A single bacterium weighs approximately 1×10^{-11} grams.
  (b) On average there are 100 trillion microorganisms on any given human.
  (c) Microbes can only be found where man naturally habituates.
  (d)

8. Animals such as worms and ticks are too large to be included in a microbiology course. True or False? Explain.

9. What is the difference between etiology and epidemiology?

10. The epidemic that infected Europe, North Africa, and the Middle East and killed tens of millions was known as the Black Death. The disease was caused by:
  (a) Smallpox
  (b) Bubonic plague
  (c) Breathing of foul air
  (d) Anthrax
  (e) Swine flu

▮ EXPLORATIONS ON THE WEB   http://www.wiley.com/college/black

If you think you've mastered this chapter, there's more to challenge you on the web. Go to the companion web site to fine-tune your understanding of the chapter concepts and discover answers to the questions posed below.

1. Did you know that there is a museum in Vienna, Austria where you can see one of the washbasins used by Semmelweis in hospitals there? Find out more about Semmelweis and how he decreased the incidence of childbirth fever.

2. Check out the web site to find out what procedures were developed to kill microorganisms affecting the wine industry without altering the flavor of the wine itself.

Brief Contents

Contents

6 ▌ Growth and Culturing of Bacteria   147

7 ▌ Microbial Genetics   178

8 ▌ Gene Transfer and Genetic Engineering   211

9 ▌ An Introduction to Taxonomy: The Bacteria   240

10 ▊ Viruses   271

11 ▊ Eukaryotic Microorganisms and Parasites   310

12 ▊ Sterilization and Disinfection   341

13 ▊ Antimicrobial Therapy   366

23 ▌▌▌ Cardiovascular, Lymphatic, and Systemic Diseases   717

24 ▌▌▌ Diseases of the Nervous System   754

25 ▌▌▌ Environmental Microbiology   782

26 ▌▌▌ Applied Microbiology   815

List of Boxes

PUBLIC HEALTH

CLOSEUP

Scope and History of Microbiology

Come with me . . .

Can this really be a microbiologist? Aren't microbiologists people in long white lab coats, working in hospital labs, growing disease-causing organisms from patient samples? Well, of course there are hospital microbiologists—but microbiology is so much more! It's adventure: taking you into realms you may never have thought about before.

Let us descend with geomicrobiologists Dr. Diana E. Northrup of the University of New Mexico (right, testing pH) and Dr. Penny Boston of New Mexico Technical University into the caves of Lechuguilla, New Mexico. It's necessary to carry meters to detect toxic gases and have protective masks at the ready. Sulfuric acid, strong as car battery acid, drips from the walls, eating holes in clothing and skin that it touches. Bacteria eating the walls are producing this acid, which drips from long slimy strings of bacterial colonies called "snotites." Geologists used to think that all caves were eroded out by water dissolving the original limestone. But now we know that some caves, even some enormous ones, were carved out by rock-eating bacteria! We'll learn more about this in later chapters.

More surprises ahead: Oh! the things we'll learn in this course—it's going to be a real adventure!

(top right) Chris Howes/Wild Places Photography/Alamy Images;
(top and bottom left) Courtesy Kenneth Ingham

 Video related to this topic is available within WileyPLUS.

||| WHY STUDY MICROBIOLOGY?

||| SCOPE OF MICROBIOLOGY
The Microbes / The Microbiologists

||| HISTORICAL ROOTS

||| THE GERM THEORY OF DISEASE
Early Studies / Pasteur's Further Contributions / Koch's Contributions / Work Toward Controlling Infections

||| EMERGENCE OF SPECIAL FIELDS OF MICROBIOLOGY
Immunology / Virology / Chemotherapy / Genetics and Molecular Biology

||| TOMORROW'S HISTORY
Genomics

Almost one-half of children under the age of 10 died of infectious disease prior to the last century.

"It's just some 'bug' going around." You have heard that from others or said it yourself when you have been ill for a day or two. Indeed, the little unidentified illnesses we all have from time to time and attribute to a "bug" are probably caused by viruses, the tiniest of all *microbes*. Other groups of **microorganisms**—bacteria, fungi, protozoa, and some algae—also have disease-causing members. Before studying microbiology, therefore, we are likely to think of microbes as germs that cause disease. Health scientists are concerned with just such microbes and with treating and preventing the diseases they cause. Yet less than 1% of known microorganisms cause disease, so focusing our study of microbes exclusively on disease gives us too narrow a view of microbiology.

WHY STUDY MICROBIOLOGY?

If you were to dust your desk and shake your dust cloth over the surface of a medium designed for growing microorganisms, after a day or so you would find a variety of organisms growing on that medium. If you were to cough onto such a medium or make fingerprints on it, you would later find a different assortment of microorganisms growing on the medium. When you have a sore throat and your physician orders a throat culture, a variety of organisms will be present in the culture—perhaps including the one that is causing your sore throat. Thus, microorganisms have a close association with humans. They are in us, on us, and nearly everywhere around us **(Figure 1.1)**. One reason for studying

Figure 1.1 A simple experiment shows that microorganisms are almost everywhere in our environment. Soil was added to nutrient agar, a culture medium (dish on top); another dish with agar was exposed to air (bottom left); and a tongue print was made on an agar surface (bottom right). After 3 days of incubation under favorable conditions, abundant microbial growth is easily visible in all three dishes. (*Courtesy Jacquelyn G. Black*)

microbiology is that *microorganisms are part of the human environment and are therefore important to human health.*

Microorganisms are essential to the web of life in every environment. Many microorganisms in the ocean and in bodies of fresh water capture energy from sunlight and store it in molecules that other organisms use as food. Microorganisms decompose dead organisms, waste material from living organisms, and even some kinds of industrial wastes. They make nitrogen available to plants.

These are only a few of the many examples of how microorganisms interact with other organisms and help maintain the balance of nature. The vast majority of microorganisms are directly or indirectly beneficial, not only to other organisms, but also to humans. They form essential links in many food chains that produce plants and animals that humans eat. Aquatic microbes serve as food for small macroscopic animals that, in turn, serve as food for fish and shellfish that humans eat. Certain microorganisms live in the digestive tracts of grazing animals such as cattle and sheep and aid in their digestive processes. Without these microbes, cows could not digest grass, and horses would get no nourishment from hay. Humans occasionally eat microbes, such as some algae and fungi, directly. Mushrooms, for instance, are the macroscopic reproductive bodies of masses of microscopic fungi. Biochemical reactions carried out by microbes also are used by the food industry to make pickles, sauerkraut, yogurt and other dairy products, fructose used in soft drinks, and the artificial sweetener aspartame. Fermentation reactions in microorganisms are used in the brewing industry to make beer and wine, and in baking to leaven dough.

One of the most significant benefits that microorganisms provide is their ability to synthesize *antibiotics*, substances derived from one microorganism that kill or restrict the growth of other microorganisms. Therefore, microorganisms can be used to cure diseases as well as cause them. Finally, microorganisms are the major tools of genetic engineering. Several products important to humans, such as interferon and growth hormones, can now be produced economically by microbes because of genetic engineering.

CLOSE UP

We Are Not Alone

"We are outnumbered. The average human contains about 10 trillion cells. On that average human are about 10 times as many microorganisms, or 100 trillion microscopic beings. ... As long as they stay in balance and where they belong, [they] do us no harm. ... In fact, many of them provide some important services to us. [But] most are opportunists, who if given the opportunity of increasing growth or invading new territory, will cause infection."

—Robert J. Sullivan, 1989

New organisms are being engineered to degrade oil spills, to remove toxic materials from soil, and to digest explosives that are too dangerous to handle. They will be major tools in cleaning up our environment. Other organisms will be designed to turn waste products into energy. Still other organisms will receive desirable genes from other types of organisms—for example, crop plants will be given bacterial genes that produce nitrogen-containing compounds needed for plant growth. The citizen of today, and even more so of tomorrow, must be scientifically literate, understanding many microbial products and processes.

Although only certain microbes cause disease, learning how such diseases are transmitted and how to diagnose, treat, and prevent them is of great importance in a health-science career. Such knowledge will help those of you who pursue such a career to care for patients and avoid becoming infected yourself.

Another reason for studying microbiology is that such study *provides insight into life processes in all life-forms.* Biologists in many different disciplines use ideas from microbiology and use the organisms themselves. Ecologists draw on principles of microbiology to understand how matter is decomposed and made available for continuous recycling. Biochemists use microbes to study metabolic pathways—sequences of chemical reactions in living organisms. Geneticists use microbes to study how hereditary information is transferred and how such information controls the structure and functions of organisms.

A bacterium may weigh approximately 0.00000000001 grams, yet collectively, microbes constitute about 60% of the earth's biomass.

Microorganisms are especially useful in research for at least three reasons:

1. Compared to other organisms, microbes have relatively simple structures. It is easier to study most life processes in simple unicellular organisms than in complex multicellular ones.

2. Large numbers of microorganisms can be used in an experiment to obtain statistically reliable results at a reasonable cost. Growing a billion bacteria costs less than maintaining 10 rats. Experiments with large numbers of microorganisms give more reliable results than do those with small numbers of organisms with individual variations.

3. Because microorganisms reproduce very quickly, they are especially useful for studies involving the transfer of genetic information. Some bacteria can undergo three cell divisions in an hour, so the effects of gene transfer can quickly be followed through many generations.

By studying microbes, scientists have achieved remarkable success in understanding life processes and disease control. For example, within the last few decades, vaccines have nearly eradicated several dreaded childhood diseases—including measles, polio, German measles, mumps, and chickenpox. Smallpox, which once accounted for 1 out of every 10 deaths in Europe, has not been reported anywhere on the planet since 1978. Much has also been learned about genetic changes that lead to antibiotic resistance and about how to manipulate genetic information in bacteria. Much more remains to be learned. For example, how can vaccines be made available on a worldwide basis? How can the development of new antibiotics keep pace with genetic changes in microorganisms? How will increased world travel continue to affect the spread of infections? Will the continued encroachment of humans into virgin jungles result in new, emerging diseases? Can a vaccine or effective treatment for acquired immunodeficiency syndrome (AIDS) be developed? Therein lie some of the challenges for the next generation of biologists and health scientists.

The full extent and importance of bacteria to our planet is just now being revealed. Deep drilling projects have discovered bacteria living at depths which no one had believed possible. At first their presence was attributed to contaminated drilling materials from the surface. But now several careful studies have confirmed populations of bacteria truly native to depths such as 1.6 km in France, 4.2 km in Alaska, and 5.2 km in Sweden. It seems that no matter how far down we drill, we always find bacteria living there. But, as we approach the hot interior of the Earth, temperature increases with depth. The Alaskan bacteria were living at 110°C! Evidence is accumulating that there is a "deep, hot biosphere," as named by American scientist Thomas Gold. This region of microbial life may extend down as far as 10 km below our "surface biosphere." At places along the border between these two biospheres, materials such as oil, hydrogen sulfide (H_2S), and methane (CH_4) are upwelling, carrying along with them bacteria from deep inside our planet. Scientists now speak of a "continuous subcrustal culture" of bacteria filling a deep hot zone lying beneath the entire Earth's surface. The mass of bacteria in the surface biosphere by far exceeds the total weight of all other living things. Add to this the weight of all the bacteria living inside the deep hot biosphere, and it is apparent that our Earth is truly "the planet of bacteria."

The cave shown in the photo at the beginning of this chapter is one of those places along the border between the two biospheres where gases are rising up from the deep hot interior. In this book we will examine bacteria at other borderland sites (for example, black hot smoking vents located deep at the ocean bottom; cold seeps higher up in the ocean on the continental shelves; and boiling mud pots such as those at Yellowstone National Park in the United States, and at the Kamchatka peninsula of Russia). And, of course, we will take a closer look at those fascinating caves shown at the beginning of this chapter.

▓ SCOPE OF MICROBIOLOGY

▓ Microbiology is the study of **microbes**, organisms so small that a microscope is needed to study them. We consider two dimensions of the scope of microbiology: (1) the variety of kinds of microbes and (2) the kinds of work microbiologists do.

THE MICROBES

The major groups of organisms studied in microbiology are bacteria, algae, fungi, viruses, and protozoa **(Figure 1.2a–e)**. All are widely distributed in nature. For example, a recent study of bee bread (a pollen-derived nutrient eaten by worker bees) showed it to contain 188 kinds of fungi and 29 kinds of bacteria. Most microbes consist of a single cell. (Cells are the basic units of structure and function in living things; they are discussed in ◄Chapter 4.) Viruses, tiny acellular entities on the borderline between the living and the nonliving, behave like living organisms when they gain entry to cells. They, too, are studied in microbiology. Microbes range in size from small viruses

500 bacteria, each 1 μm (1/1000 of a millimeter) long, would fit end-to-end across the dot above the letter "i."

20 nm in diameter to large protozoans 5 mm or more in diameter. In other words, the largest microbes are as much as 250,000 times the size of the smallest ones! (Refer to ◄Appendix A for a review of metric units.)

Among the great variety of microorganisms that have been identified, bacteria probably have been the most thoroughly studied. The majority of **bacteria** (singular: *bacterium*) are single-celled organisms with spherical, rod, or spiral shapes, but a few types form filaments. Most are so small they can be seen with a light microscope only under the highest magnification. Although bacteria are cellular, they do not have a cell nucleus, and they lack the membrane-enclosed intracellular structures found in most other cells. Many bacteria absorb nutrients from their environment, but some make their own nutrients by photosynthesis or other synthetic processes. Some are stationary, and others move about. Bacteria are widely distributed in nature, for example, in aquatic environments and in decaying matter. And some occasionally cause diseases.

In contrast to bacteria, several groups of microorganisms consist of larger, more complex cells that have a cell nucleus. They include algae, fungi, and protozoa, all of which can easily be seen with a light microscope.

Many **algae** (al'je; singular: *alga*) are single-celled microscopic organisms, but some marine algae are

Figure 1.2 Typical microorganisms (enhanced with artificial coloring). (a) Several *Klebsiella pneumoniae* cells, a bacterium that can cause pneumonia in humans (5, 821X). *(CNRI/Photo Researchers, Inc.)* **(b)** *Micrasterias*, a type of green algae that lives in fresh water (334X). *(R. B. Taylor/Photo Researchers, Inc.)* **(c)** Fruiting bodies of the fungus *Philobolus crystallinus* with black spore packets on top that will be shot into the air to colonize new areas (50X). *(CBC/Phototake)* **(d)** Bacteriophages (viruses that infect bacteria; 35,500X). *(Thomas Broker/Phototake)* **(e)** *Amoeba*, a protozoan (183X). *(Michael Abbey/Visuals Unlimited)* **(f)** Head of the tapeworm *Acanthrocirrus retrirostris* (189X). At the top of the head are hooks and suckers that the worm uses to attach to a host's intestinal tissues. *(Cath Ellis/Photo Researchers, Inc.)*

TABLE 1.1

Reportable Diseases Caused by Microorganisms and Parasites[a]

Bacterial Diseases	Bacterial Diseases	Bacterial Diseases	Viral Diseases	Algal Diseases
Anthrax	Listeriosis	Streptococcal toxic-shock syndrome	Hantavirus pulmonary syndrome	None
Bacterial meningitis	Lyme disease			
Botulism	Meningitis	Syphilis	Hepatitis A, B, and C	
Brucellosis	Pertussis (whooping cough)	Tetanus	Hepatitis (unspecified)	**Fungal Diseases**
Chancroid	Plague	Toxic-shock syndrome	HIV infection, adult	Coccidiomycosis
Chlamydial genital infections	Psittacosis	(other than streptococcal)	HIV infection, pediatric	
Cholera	Q Fever	Tuberculosis	Influenza	
Diphtheria	Rocky Mountain spotted fever	Tularemia	Measles (rubeola)	**Protozoan Disease**
Ehrlichiosis	Salmonellosis	Typhoid fever	Mumps	Cryptosporidiosis
Escherichia coli O157:H7	Shiga-toxin-producing *E. coli*	Vancomycin-resistant *Staphylococcus aureus*	Poliomyelitis (paralytic)	Cyclosporiasis
Food poisoning			Rabies (animal and human)	Giardiasis
Gonorrhea	Shigellosis	**Viral Diseases**	Rubella (German measles)	Malaria
Haemophilus influenzae infections (invasive)	Streptococcal disease, invasive, Group A	AIDS (symptomatic cases)	SARS (severe acute respiratory syndrome)	
Hansen's disease (leprosy)	Streptococcal pneumonia, drug-resistant invasive disease	Arbovirus infection	Smallpox	**Helminth Disease**
		Encephalitis: eastern equine, St. Louis, West Nile, western equine	Varicella (chickenpox, shingles)	Trichinosis
Legionnaires' disease			Yellow fever	

[a]Infectious disease reporting varies by state. This table lists most of the diseases commonly reported to the U.S. Centers for Disease Control and Prevention (CDC).

large, relatively complex, multicellular organisms. Unlike bacteria, algae have a clearly defined cell nucleus and numerous membrane-enclosed intracellular structures. All algae photosynthesize their own food as plants do, and many can move about. Algae are widely distributed in both fresh water and oceans. Because they are so numerous and because they capture energy from sunlight in the food they make, algae are an important source of food for other organisms. Algae are of little medical importance; only one species, *Prototheca*, has been found to cause disease in humans. Having lost its chlorophyll, and therefore the ability to produce its own food, it now makes meals of humans.

Like algae, many **fungi** (fun′ji; singular: *fungus*), such as yeasts and some molds, are single-celled microscopic organisms. Some, such as mushrooms, are multicellular, macroscopic organisms. Fungi also have a cell nucleus and intracellular structures. All fungi absorb ready-made nutrients from their environment. Some fungi form extensive networks of branching filaments, but the organisms themselves generally do not move. Fungi are widely distributed in water and soil as decomposers of dead organisms. Some are important in medicine either as agents of diseases such as ringworm and vaginal yeast infections or as sources of antibiotics.

Viruses are acellular entities too small to be seen with a light microscope. They are composed of specific chemical substances—a nucleic acid and a few proteins ◀(Chapter 2). Indeed, some viruses can be crystallized and stored in a container on a shelf for years, but they retain the ability to invade cells. Viruses replicate themselves and display other properties of living organisms only when they have invaded cells. Many viruses can invade human cells and cause disease. Even smaller acellular agents of disease are **viroids** (nucleic acid without a protein coating), and **prions** (protein without any nucleic acid). Viroids have been shown to cause various

plant diseases, whereas prions cause mad cow disease and related disorders.

Protozoa (pro-to-zo′ah; singular: *protozoan*) also are single-celled, microscopic organisms with at least one nucleus and numerous intracellular structures. A few species of amoebae are large enough to be seen with the naked eye, but we can study their structure only with a microscope. Many protozoa obtain food by engulfing or ingesting smaller microorganisms. Most protozoa can move, but a few, especially those that cause human disease, cannot. Protozoa are found in a variety of water and soil environments, as well as in animals such as malaria-carrying mosquitoes.

In addition to organisms properly in the domain of microbiology, in this text we consider some macroscopic *helminths* (worms) **(Figure 1.2f)** and *arthropods* (insects and similar organisms). The helminths have microscopic stages in their life cycles that can cause disease, and the arthropods can transmit these stages, as well as other disease-causing microbes.

We will learn more about the classification of microorganisms in ◀Chapter 9. For now it is important to know only that cellular organisms are referred to by two names: their *genus* and *species* names. For example, a bacterial species commonly found in the human gut is called *Escherichia coli*, and a protozoan species that can cause severe diarrhea is called *Giardia intestinalis*. The naming of viruses is less precise. Some viruses, such as herpesviruses, are named for the group to which they belong. Others, such as polioviruses, are named for the disease they cause.

Disease-causing organisms and the diseases they cause in humans are discussed in detail in ◀Chapters 19–24. Hundreds of infectious diseases are known to medical science. Some of the most important—those diseases that physicians should report to the U.S. Centers for Disease Control and Prevention (CDC)—are listed in **Table 1.1**. The CDC is a federal agency that

TABLE 1.2

Fields of Microbiology	
Field (Pronunciation)	**Examples of What Is Studied**
Microbial taxonomy	Classification of microorganisms
Fields According to Organisms Studied	
Bacteriology (bak″ter-e-ol′o-je)	Bacteria
Phycology (fi-kol′o-je)	Algae (*phyco*, "seaweed")
Mycology (mi-kol′o-je)	Fungi (*myco*, "a fungus")
Protozoology (pro″to-zo-ol′o-je)	Protozoa (*proto*, "first"; *zoo*, "animal")
Parasitology (par″a-si-tol′o-je)	Parasites
Virology (vi-rol′o-je)	Viruses
Fields According to Processes or Functions Studied	
Microbial metabolism	Chemical reactions that occur in microbes
Microbial genetics	Transmission and action of genetic information in microorganisms
Microbial ecology	Relationships of microbes with each other and with the environment
Health-Related Fields	
Immunology (im″u-nol′o-je)	How host organisms defend themselves against microbial infection
Epidemiology (epi-i-de-me-ol′o-je)	Frequency and distribution of diseases
Etiology (e-te-ol′-o-je)	Causes of disease
Infection control	How to control the spread of nosocomial (nos-o-ko′me-al), or hospital-acquired, infections
Chemotherapy	The development and use of chemical substances to treat diseases
Fields According to Applications of Knowledge	
Food and beverage technology	How to protect humans from disease organisms in fresh and pre-served foods
Environmental microbiology	How to maintain safe drinking water, dispose of wastes, and control environmental pollution
Industrial microbiology	How to apply knowledge of microorganisms to the manufacture of fermented foods and other products of microorganisms
Pharmaceutical microbiology	How to manufacture antibiotics, vaccines, and other health products
Genetic engineering	How to use microorganisms to synthesize products useful to humans

Figure 1.3 Microbiology is used in diverse careers. These careers include such activities as **(a)** using genetically engineered bacteria to investigate how diet influences the risk of developing cancer *(Courtesy United States Department of Agriculture)*; **(b)** inspecting plastics made with as much as 40% starch (pieces inside baskets) for signs that aquatic microbes are degrading them *(Courtesy United States Department of Agriculture)*; **(c)** using bacteria to decontaminate toxic wastes *(David Parker/Photo Researchers, Inc.)*; **(d)** using beating nets to survey for ticks that can spread disease to livestock and humans *(Courtesy United States Department of Agriculture)*; **(e)** keeping pets and domestic animals healthy, as well as improving their productivity, by means of advances in veterinary science. *(Courtesy United States Department of Agriculture)*

collects data about diseases and about developing ways to control them.

THE MICROBIOLOGISTS

Microbiologists study many kinds of problems that involve microbes. Some study microbes mainly to find out more about a particular type of organism—the life stages of a particular fungus, for example. Other microbiologists are interested in a particular kind of function, such as the metabolism of a certain sugar or the action of a specific gene. Still others focus directly on practical problems, such as how to purify or synthesize a new antibiotic or how to make a vaccine against a particular disease. Quite often the findings from one project are useful in another, as when agricultural scientists use information from microbiologists to control pests and improve crop yields, or when environmentalists attempt to maintain natural food chains and prevent damage to the environment. Some fields of microbiology are described in **Table 1.2**.

Microbiologists work in a variety of settings **(Figure 1.3)**. Some work in universities, where they are likely to teach, do research, and train students to do research. Microbiologists in both university and commercial laboratories are helping to develop the microorganisms used in genetic engineering. Some law firms are hiring microbiologists to help with the complexities of patenting new genetically engineered organisms. These organisms can be used in such important ways as cleaning up the environment *(bioremediation)*, controlling insect pests, improving foods, and fighting disease. Many microbiologists work in health-related positions. Some work in clinical laboratories, performing tests to diagnose diseases or determining which antibiotics will cure a particular disease. A few microbiologists develop new clinical tests. Others work in industrial laboratories to develop or manufacture antibiotics, vaccines, or similar biological products. Still others, concerned with controlling the spread of infections and related public health matters, work in hospitals or government labs. Please go to the website for this chapter, http://www.wiley.com/college/black, to read an interview with a keeper and a veterinarian from the Smithsonian's National Zoo in Washington, D.C. See how important microbiology is there!

From the point of view of health scientists, today's research is the source of tomorrow's new technologies. Research in *immunology* is greatly increasing our knowledge of how microbes trigger host responses and how the microbes escape these responses. It also is contributing to the development of new vaccines and to the treatment of immunologic disorders. Research in *virology* is improving our understanding of how viruses cause infections and how they are involved in cancer. Research in *chemotherapy* is increasing the number of drugs available to treat infections and is also improving our knowledge of how these drugs work. Finally, research in *genetics* is providing new information about the transfer of genetic information and, especially, about how genetic information acts at the molecular level.

✓ CHECKLIST

1. List three reasons to study microbiology.
2. What is the difference between microbiology and bacteriology?
3. What is the difference between etiology and epidemiology?
4. List five bacterial diseases and five viral diseases.

▌▌▌ HISTORICAL ROOTS

Many of the ancient Mosaic laws found in the Bible about basic sanitation have been used through the centuries and still contribute to our practices of preventive medicine. In Deuteronomy, Chapter 13, Moses instructed the soldiers to carry spades and bury solid waste matter. The Bible also refers to leprosy and to the isolation of lepers. Although in those days the term *leprosy* probably included other infectious and noninfectious diseases, isolation did limit the spread of the infectious diseases.

The Greeks anticipated microbiology, as they did so many things. The Greek physician Hippocrates, who lived around 400 B.C., set forth ethical standards for the practice of medicine that are still in use today. Hippocrates was wise in human relations and also a shrewd observer. He associated particular signs and symptoms with certain illnesses and realized that diseases could be transmitted from one person to another by clothing or other objects. At about the same time, the Greek historian Thucydides observed that people who had recovered from the plague could take care of plague victims without danger of getting the disease again.

The Romans also contributed to microbiology, as early as the first century B.C. The scholar and writer Varro proposed that tiny invisible animals entered the body through the mouth and nose to cause disease. Lucretius, a philosophical poet, cited "seeds" of disease in his *De Rerum Natura (On the Nature of Things)*.

Bubonic plague, also called the Black Death, appeared in the Mediterranean region around 542 A.D., where it reached epidemic proportions and killed millions. In 1347 the plague invaded Europe along the caravan routes and sea lanes from central Asia, affecting Italy first, then France, England, and finally northern Europe. Although no accurate records were kept at that time, it is estimated that tens of millions of people in Europe died during this and successive waves of plague over the next 300 years. The Black Death was a great leveler—it killed rich and poor alike **(Figure 1.4)**. The wealthy fled to isolated summer homes but carried plague-infected fleas with them in unwashed hair and clothing. In the mid-fourteenth century (1347–1351) plague alone wiped out 25 million people—one-fourth of the population of Europe and neighboring regions—in just 5 years. In the vicinity of the Sedlec Monastery near Prague, over 30,000 people died of plague in one year. Their bones are now displayed in an ossuary **(Figure 1.5)**.

Take another look

Until the seventeenth century, the advance of microbiology was hampered by the lack of appropriate tools to observe microbes. Around 1665, the English scientist Robert Hooke built a compound microscope (one in which light passes through two lenses) and used it to observe thin slices of cork. He coined the term *cell* to describe the orderly arrangement of small boxes that he saw because they reminded him of the cells (small, bare rooms) of monks. However, it was Anton van Leeuwenhoek **(Figure 1.6)**, a Dutch cloth merchant and amateur lens grinder, who first made and used lenses to observe living microorganisms. The lenses Leeuwenhoek made were of excellent quality; some gave magnifications up to 300X and were remarkably free of distortion. Making these lenses and looking through them were the passions of his life. Everywhere he looked he found what he called "animalcules." He found them in stagnant water, in sick people, and even in his own mouth.

Over the years Leeuwenhoek observed all the major kinds of microorganisms—protozoa, algae, yeast, fungi, and bacteria in spherical, rod, and spiral forms. He once wrote, "For my part I judge, from myself (howbeit I clean my mouth like I've already said), that all the people living in our United Netherlands are not as many as the living animals that I carry in my own mouth this very day." Starting in the 1670s he wrote numerous letters to the Royal Society in London and pursued his studies until his death in 1723 at the age of 91. Leeuwenhoek refused to sell his microscopes to others and so failed to foster the development of microbiology as much as he could have.

After Leeuwenhoek's death, microbiology did not advance for more than a century. Eventually microscopes became more widely available, and progress resumed. Several workers discovered ways to stain microorganisms with dyes to make them more visible.

Figure 1.4 *The Triumph of Death* **by Pieter Brueghel the Elder.** The picture, painted in the mid-sixteenth century, a time when outbreaks of plague were still common in many parts of Europe, dramatizes the swiftness and inescapability of death for people of all social and economic classes. *(Pieter Brueghel the Elder (1528–1569), Flemish, Trionfolo della Morte, Painting, Prado, Madrid, Spain/Scala/ Art Resource, NY)*

The Swedish botanist Carolus Linnaeus developed a general classification system for all living organisms. The German botanist Matthias Schleiden and the German zoologist Theodor Schwann formulated the **cell theory**, which states that cells are the fundamental units of life and carry out all the basic functions of living things. Today this theory still applies to all cellular organisms, but not to viruses.

▋▋▋ THE GERM THEORY OF DISEASE

▋▋▋ The **germ theory of disease** states that microorganisms (germs) can invade other organisms and cause disease. Although this is a simple idea and is generally accepted today, it was not widely accepted when formulated in the mid-nineteenth century. Many people believed that broth, left standing, turned cloudy because of something about the broth itself. Even after it was shown that microorganisms in the broth caused it to turn cloudy, people believed that the microorganisms, like the "worms" (fly larvae, or maggots) in rotting meat, arose from nonliving things, a concept known as **spontaneous generation**. Widespread belief in spontaneous generation, even among scientists, hampered further development of the science of microbiology and the acceptance of the germ theory of disease. As long as they believed that microorganisms could arise from nonliving substances, scientists saw no purpose in considering how diseases were transmitted or how they could be controlled. Dispelling the belief in spontaneous generation took years of painstaking effort.

EARLY STUDIES

For as long as humans have existed, some probably have believed that living things somehow originated spontaneously from nonliving matter. Aristotle's theories about his four "elements"—fire, earth, air, and water—seem to have suggested that nonliving forces

Figure 1.5   Ossuary (bone display) at the Sidlec Monastery, located near Prague. Most of the bones are from victims of the plague outbreak of 1347–1351, in which over 30,000 people died. *(Photo by J. Frisco Arenas Ramirez)*

Figure 1.6   Anton van Leeuwenhoek (1632–1723). He is shown holding one of his simple microscopes. *(© Corbis)*

somehow contributed to the generation of life. Even some naturalists believed that rodents arose from moist grain, beetles from dust, and worms and frogs from mud. As late as the nineteenth century, it seemed obvious to most people that rotting meat gave rise to "worms."

In the late seventeenth century, the Italian physician Francesco Redi devised a set of experiments to demonstrate that if pieces of meat were covered with gauze so that flies could not reach them, no "worms" appeared in the meat, no matter how rotten it was **(Figure 1.7)**. Maggots did, however, hatch from fly eggs laid on top of the gauze. Despite the proof that maggots did not arise spontaneously, some scientists still believed in spontaneous generation—at least that of microorganisms. Lazzaro Spallanzani, an Italian cleric and scientist, was more skeptical. He boiled broth infusions containing organic (living or previously living) matter and sealed the flasks to demonstrate that no organisms would develop spontaneously in them. Critics did not accept this as disproof of spontaneous generation. They

Figure 1.7 Redi's experiments refuting the spontaneous generation of maggots in meat. When meat is exposed in an open jar, flies lay their eggs on it, and the eggs hatch into maggots (fly larvae). In a sealed jar, however, no maggots appear. If the jar is covered with gauze, maggots hatch from eggs that the flies lay on top of the gauze but still no maggots appear on the meat.

argued that boiling drove off oxygen (which they thought all organisms required) and that sealing the flasks prevented its return.

Several scientists tried different ways of introducing air to counter this criticism. Schwann heated air before introducing it into flasks, and other scientists filtered air through chemicals or cotton plugs. All these methods prevented the growth of microorganisms in the flasks. But the critics still argued that altering the air prevented spontaneous generation.

Figure 1.8 A "swan-necked" flask that Pasteur used in refuting the theory of spontaneous generation. Although air could enter the flasks, microbes became trapped in the curved necks and never reached the contents. The contents, therefore, remained sterile—and still are today, in the museum—despite their exposure to the air. (© Charles O'Rear/Corbis)

Even nineteenth-century scientists of some stature continued to argue vociferously in favor of spontaneous generation. They believed that an organic compound previously formed by living organisms contained a "vital force" from which life sprang. The force, of course, required air, and they believed that all the methods of introducing air somehow changed it so that it could not interact with the force.

The proponents of spontaneous generation were finally defeated, mainly by the work of the French chemist Louis Pasteur and the English physicist John Tyndall. When the French Academy of Science sponsored a competition in 1859 "to try by well-performed experiment to throw new light on the question of spontaneous generation," Pasteur entered the competition.

During the years Pasteur worked in the wine industry, he had established that alcohol was produced in wine only if yeast was present, and he learned a lot about the growth of microorganisms. Pasteur's experiment for the competition involved his famous "swan-necked" flasks (**Figure 1.8**). He boiled *infusions* (broths of foodstuffs) in flasks, heated the glass necks, and drew them out into long, curved tubes open at the end. Air could enter the flasks without being subjected to any of the treatments that critics had claimed destroyed its effectiveness. Airborne microorganisms could also enter the necks of the flasks, but they became trapped in the curves of the neck and never reached the infusions. The infusions from Pasteur's experiments remained sterile unless a flask was tipped so that the infusion flowed into the neck and then back into the flask. This manipulation allowed microorganisms trapped in the neck to wash into the infusion, where they could grow and cause the infusion to become cloudy. In another experiment Pasteur filtered air through three cotton plugs. He then immersed the plugs in sterile infusions, demonstrating that growth occurred in the infusions from organisms trapped in the plugs.

Tyndall delivered another blow to the idea of spontaneous generation when he arranged sealed flasks of boiled infusion in an airtight box. After allowing time for all dust particles to settle to the bottom of the box, he carefully removed the covers from the flasks. These flasks, too, remained sterile. Tyndall had shown that air could be sterilized by settling, without any treatment that would prevent the "vital force" from acting.

Both Pasteur and Tyndall were fortunate that the organisms present in their infusions at the time of boiling were destroyed by heat. Others who tried the same experiments observed that the infusions became cloudy from growth of microorganisms. We now know that the infusions in which growth occurred contained heat-resistant or spore-forming microorganisms, but at the time, the growth of such organisms was seen as evidence of spontaneous generation. Still, the works of Pasteur and Tyndall successfully disproved spontaneous generation to most scientists of the time. Recognition that microbes must be introduced into a medium before their growth can be observed paved the way for further development of microbiology—especially for development of the germ theory of disease.

PASTEUR'S FURTHER CONTRIBUTIONS

Louis Pasteur (**Figure 1.9**) was such a giant among nineteenth-century scientists working in microbiology that we must consider some of his many contributions. Born in 1822, the son of a sergeant in Napoleon's army, Pasteur worked as a portrait painter and a teacher before he began to study chemistry in his spare time. Those studies led to posts in several French universities as professor of chemistry and to significant contributions to the wine and silkworm industries. He discovered that carefully selected yeasts made good wine, but that mixtures of other microorganisms competed with the yeast for sugar and made wine taste oily or sour. To combat this problem, Pasteur developed the technique of pasteurization (heating wine to 56°C in the absence of oxygen for 30 minutes) to kill unwanted organisms. While studying silkworms, he identified three different microorganisms, each of which caused a different disease. His association of specific organisms with particular diseases, even though in silkworms rather than in humans, was an important first step in proving the germ theory of disease.

Despite personal tragedy—the deaths of three daughters and a cerebral hemorrhage that left him with permanent paralysis—Pasteur went on to contribute to the development of vaccines. The best known of his vaccines is the rabies vaccine, made of dried spinal cord from rabbits infected with rabies, which was tested in animals. When a 9-year-old boy who had been severely bitten by a rabid dog was brought to him, Pasteur administered the vaccine, but only after a long night of soul searching. He was not a physician, and had never

Figure 1.9 Louis Pasteur in his laboratory. The first rabies vaccine, developed by Pasteur, was made from the dried spinal cords of infected rabbits. *(Granger Collection)*

**Figure 1.10
Robert Koch and his wife.** Koch formulated four postulates for linking a given organism to a specific disease. *(National Library of Medicinea/SPL/Photo Researchers, Inc.)*

practiced medicine before. The boy, who had been doomed to die, survived and became the first person to be immunized against rabies. Later, during World War II, the then grown-up boy was killed by German soldiers for refusing to give them access to Pasteur's tomb so that they could desecrate his bones.

In 1894 Pasteur became director of the Pasteur Institute, which was built for him in Paris. Until his death in 1895, he guided the training and work of other scientists at the institute. Today the Pasteur Institute is a thriving research center—an appropriate memorial to its founder.

KOCH'S CONTRIBUTIONS

Robert Koch (**Figure 1.10**), a contemporary of Pasteur, finished his medical training in 1872 and worked as a physician in Germany throughout most of his career. After he bought a microscope and photographic equipment, he spent most of his time studying bacteria, especially those that cause disease. Koch identified the bacterium that causes anthrax, a highly contagious and lethal disease in cattle and sometimes in humans. He recognized both actively dividing cells and dormant cells (spores) and developed techniques for studying them *in vitro* (outside a living organism).

Koch also found a way to grow bacteria in *pure cultures*—cultures that contained only one kind of organism. He tried streaking bacterial suspensions on potato slices and then on solidified gelatin. But gelatin melts at incubator (body) temperature; even at room temperature, some microbes liquefy it. Finally, Angelina Hesse (**Figure 1.11**), the American wife of one of Koch's colleagues, suggested that Koch add agar (a thickener used in cooking) to his bacteriological media. This created a firm surface over which microorganisms could be spread very thinly—so thinly that some individual organisms were separated from all

Figure 1.11 Angelina and Walther Hesse. The American wife of Koch's assistant suggested solidifying broths with agar as an aid to obtaining pure cultures. She had used it to solidify broths in her kitchen, and we still use it in our labs today. *(From ASM News 47(7) 392, 1961. Reproduced with permission of American Society for Microbiology.)*

others. Each individual organism then multiplied to make a colony of thousands of descendants. Koch's technique of preparing pure cultures is still used today.

Koch's outstanding achievement was the formulation of four postulates to associate a particular organism with a specific disease. **Koch's Postulates**, which provided scientists with a method of establishing the germ theory of disease, are as follows:

1. The specific causative agent must be found in every case of the disease.
2. The disease organism must be isolated in pure culture.
3. Inoculation of a sample of the culture into a healthy, susceptible animal must produce the same disease.
4. The disease organism must be recovered from the inoculated animal.

Implied in Koch's postulates is his one organism–one disease concept. The postulates assume that an infectious disease is caused by a single organism, and they are directed toward establishing that fact. This concept also was an important advance in the development of the germ theory of disease.

After obtaining a laboratory post at Bonn University in 1880, Koch was able to devote his full time to studying microorganisms. He identified the bacterium that causes tuberculosis, developed a complex method of staining this organism, and disproved the idea that tuberculosis was inherited. He also guided the research that led to the isolation of *Vibrio cholerae*, the bacterium that causes cholera.

In a few years Koch became professor of hygiene at the University of Berlin, where he taught a microbiology course believed to be the first ever offered. He also developed *tuberculin*, which he hoped would be a vaccine against tuberculosis. Because he underestimated the difficulty of killing the organism that causes tuberculosis, the use of tuberculin resulted in several deaths from that disease. Although tuberculin was unacceptable as a vaccine, its development laid the groundwork for a skin test to diagnose tuberculosis. After the vaccine disaster, Koch left Germany. He made several visits to Africa, at least two visits to Asia, and one to the United States.

In the remaining 15 years of his life, his accomplishments were many and varied. He conducted research on malaria, typhoid fever, sleeping sickness, and several other diseases. His studies of tuberculosis won him the Nobel Prize for Physiology and Medicine in 1905, and his work in Africa and Asia won him great respect on those continents.

WORK TOWARD CONTROLLING INFECTIONS

Like Koch and Pasteur, two nineteenth-century physicians, Ignaz Philipp Semmelweis of Austria and Joseph Lister of England, were convinced that microorganisms caused infections **(Figure 1.12)**. Semmelweis recognized a connection between autopsies and puerperal (childbed) fever. Many physicians went directly from performing autopsies to examining women in labor without so much as washing their hands. When Semmelweis attempted to encourage more sanitary practices, he was ridiculed and harassed until he had a nervous breakdown and was sent to an asylum. Ultimately, he suffered the curious irony of succumbing to an infection caused by the same organism that produces puerperal

CLOSE UP

What's in the Last Drop?

During the nineteenth century, French and German scientists were fiercely competitive. One area of competition was the preparation of pure cultures. Koch's reliable method of preparing pure cultures from colonies on solid media allowed German microbiologists to forge ahead. The French microbiologists' method of broth dilution, though now often used to count organisms ◄(Chapter 6), hampered their progress. They added a few drops of a culture to fresh broth, mixed it, and added a few drops of the mixture to more fresh broth. After several successive dilutions, they assumed that the last broth that showed growth of microbes had contained a single organism. Unfortunately, the final dilution often contained more than one organism, and sometimes the organisms were of different kinds. This faulty technique led to various fiascos, such as inoculating animals with deadly organisms instead of vaccinating them.

(a)   (b)

Figure 1.12 **Two nineteenth-century pioneers in the control of infections.** **(a)** Ignaz Philipp Semmelweis, who died in an asylum before his innovations were widely accepted, depicted on a 1965 Austrian postage stamp *(Granger Collection)*; **(b)** Joseph Lister, performing surgery using a spray of carbolic acid above the surgical field, successfully carried on Semmelweis's work toward achieving aseptic techniques. *(© Corbis)*

fever. In 1865, Lister, who had read of Pasteur's work on pasteurization and Semmelweis's work on improving sanitation, initiated the use of dilute carbolic acid on bandages and instruments to reduce infection. Lister, too, was ridiculed, but with his imperturbable temperament, resolute will, and tolerance of hostile criticism, he was able to continue his work. His methods, the first *aseptic techniques*, were proven effective by the decrease in surgical wound infections in his surgical wards. At age 75, some 37 years after he had introduced the use of carbolic acid, Lister was awarded the Order of Merit for his work in preventing the spread of infection. He is considered the father of antiseptic surgery.

✓ CHECKLIST

1. What similarities and differences do you see when comparing past epidemics of plague with today's AIDS epidemic?

2. State the germ theory of disease. Try to think of an explanation of disease causation that would be contrary to the germ theory.

3. How did Pasteur's experiment with "swan-necked" flasks disprove the theory of spontaneous generation?

4. Why was the French microbiologists' method of broth dilution inadequate for obtaining pure cultures of organisms?

▌▌▌ EMERGENCE OF SPECIAL FIELDS OF MICROBIOLOGY

▌▌▌ Pasteur, Koch, and most other microbiologists considered to this point were generalists interested in a wide variety of problems. Certain other contributors to

microbiology had more specialized interests, but their achievements were no less valuable. In fact, those achievements helped establish the special fields of immunology, virology, chemotherapy, and microbial genetics—fields that are today prolific research areas. Selected fields of microbiology are defined in Table 1.2.

IMMUNOLOGY

Disease depends not only on microorganisms invading a host but also on the host's response to that invasion. Today, we know that the host's response is in part a response of the immune system.

The ancient Chinese knew that a person scarred by smallpox would not again get the disease. They took dried scabs from lesions of people who were recovering from the disease and ground them into a powder that they sniffed. As a result of inhaling weakened organisms, they acquired a mild case of smallpox but were protected against subsequent infection.

Smallpox was unknown in Europe until the Crusaders carried it back from the Near East in the twelfth century. By the seventeenth century, it was widespread. In 1717 Lady Mary Ashley Montagu, wife of the British ambassador to Turkey, introduced a kind of immunization to England. A thread was soaked in fluid from a smallpox vesicle (blister) and drawn through a small incision in the arm. This

Pocahontas died of smallpox in 1617, in England

technique, called *variolation*, was used at first by only a few prominent people, but eventually it became widespread.

In the late eighteenth century, Edward Jenner realized that milkmaids who got cowpox did not get smallpox, and he inoculated his own son with fluid from a cowpox blister. He later similarly inoculated an 8-year-old and subsequently inoculated the same child with smallpox. The child remained healthy. The word *vaccinia* (*vacca*, the Latin name for "cow") gave rise both to the name of the virus that causes cowpox and to the word *vaccine*. In the early 1800s Jenner received grants amounting to a total of 30,000 British pounds to extend his work on vaccination. Today, those grants would be worth more than $1 million. They may have been the first grants for medical research.

Calves—shaved, inoculated, and covered with cowpox lesions—were led house to house by entrepreneurs who offered vaccination during American colonial times.

Pasteur contributed significantly to the emergence of immunology with his work on vaccines for rabies and cholera. In 1879, when Pasteur was studying chicken cholera, his assistant accidentally used an old chicken cholera culture to inoculate some chickens. The chickens did not develop disease symptoms. When the assistant later inoculated the same chickens with a fresh chicken cholera culture, they remained healthy. Although he had not planned to use the old culture first, Pasteur did realize that the chickens had been immunized against chicken cholera. He reasoned that the organisms must have lost their ability to produce disease but retained their ability to produce immunity. This finding led Pasteur to look for techniques that would have the same effect on other organisms. His development of the rabies vaccine was a successful attempt.

Along with Jenner and Pasteur, the nineteenth-century Russian zoologist Elie Metchnikoff was a pioneer in immunology (**Figure 1.13**). In the 1880s many scientists believed that immunity was due to noncellular substances in the blood. Metchnikoff discovered that certain cells in the body would ingest microbes. He named those cells *phagocytes*, which literally means "cell eating." The identification of phagocytes as cells that defend the body against invading microorganisms was a first step in understanding immunity. Metchnikoff also developed several vaccines. Some were successful, but unfortunately some infected the recipients with the microorganisms against which they were supposedly being immunized. A few of his subjects acquired gonorrhea and syphilis from his vaccines. Metchnikoff had used the French method of obtaining supposedly "pure" cultures.

VIROLOGY

The science of virology emerged after that of bacteriology because viruses could not be recognized until certain techniques for studying and isolating larger

Figure 1.13 Elie Metchnikoff. Metchnikoff was one of the first scientists to study the body's defenses against invading microorganisms. *(Stock Montage, Inc./Historical Pictures Collection)*

particles such as bacteria had been developed. When Pasteur's collaborator Charles Chamberland developed a porcelain filter to remove bacteria from water in 1884, he had no idea that any kind of infectious agent could pass through the filter. But researchers soon realized that some filtrates (materials that passed through the filters) remained infectious even after the bacteria were filtered out. The Dutch microbiologist Martinus Beijerinck determined why such filtrates were infectious and was thus the first to characterize viruses. The term *virus* had been used earlier to refer to poisons and to

CLOSE UP

A "Thorny" Problem

Metchnikoff's personal life played a role in his discovery of phagocytes. A widower, when he remarried he took in his 16-year-old wife's dozen younger brothers and sisters as part of the marriage agreement. On one occasion, Metchnikoff left under his microscope a starfish he was studying before he went to lunch with his wife. While the starfish was unattended, the mischievous children poked thorns into it. Metchnikoff was enraged to discover his mutilated starfish, but he looked through the microscope before he discarded his ruined specimen. He was amazed to discover that cells of the starfish had gathered around the thorns. After further study, he identified those cells as *leukocytes* (white blood cells) and found that they could devour foreign particles. He coined the word *phagocytosis* (*phago*: "to eat"; *cyte*: "cell") to describe this process. He realized that this process was a major mechanism of body defense—for which he received the Nobel Prize in 1908, and has become known as the father of immunology. Incidentally, he never did have any children of his own.

infectious agents in general. Beijerinck used the term to refer to specific *pathogenic* (disease-causing) molecules incorporated into cells. He also believed that these molecules could borrow for their own use existing metabolic and replicative mechanisms of the infected cells, known as *host cells*.

Further progress in virology required development of techniques for isolating, propagating, and analyzing viruses. The American scientist Wendell Stanley crystallized tobacco mosaic virus in 1935, showing that an agent with properties of a living organism also behaved as a chemical substance **(Figure 1.14)**. The crystals consisted of protein and ribonucleic acid (RNA). The nucleic acid was soon shown to be important in the infectivity of viruses. Viruses were first observed with an electron microscope in 1939. From that time both chemical and microscopic studies were used to investigate viruses.

By 1952 the American biologists Alfred Hershey and Martha Chase had demonstrated that the genetic material of some viruses is another nucleic acid, deoxyribonucleic acid (DNA). In 1953 the American postdoctoral student James Watson and the English biophysicist Francis Crick determined the structure of DNA. The stage was set for rapid advances in understanding how DNA functions as genetic material both in viruses and in cellular organisms. Since the 1950s hundreds of viruses have been isolated and characterized. Although much remains to be learned about viruses, tremendous progress has been made in understanding their structure and how they function.

CHEMOTHERAPY

The Greek physician Dioscorides compiled *Materia Medica* in the first century A.D. This five-volume work listed a number of substances derived from medicinal plants still in use today—digitalis, curare, ephedrine, and morphine—along with a number of herbal medications. Credit for bringing herbal medicine to the United States is given to many groups of settlers, but Native Americans used many medicinal plants before the arrival of Europeans in the Americas. Many so-called primitive peoples still use herbs extensively, and some pharmaceutical companies finance expeditions into the Amazon Basin and other remote areas to investigate the uses the natives make of the plants around them.

During the Middle Ages virtually no advances were made in the use of chemical substances to treat diseases. Early in the sixteenth century the Swiss physician Aureolus Paracelsus used metallic chemical elements to treat diseases—antimony for general infections and mercury for syphilis. In the mid-seventeenth century Thomas Sydenham, an English physician, introduced cinchona tree bark to treat malaria. This bark, which we now know contains quinine, had been used to treat fevers in Spain and South America. In the nineteenth century morphine was extracted from the opium poppy and used medicinally to alleviate pain.

Paul Ehrlich, the first serious researcher in the field of chemotherapy **(Figure 1.15)**, received his doctoral degree from the University of Leipzig, Germany, in 1878. His discovery that certain dyes stained microorganisms but not animal cells suggested that the dyes or other chemicals might selectively kill microbial cells. This led him to search for the "magic bullet," a chemical that would destroy specific bacteria without damaging surrounding tissues. Ehrlich coined the term *chemotherapy* and headed the world's first institute concerned with the development of drugs to treat disease.

Early in the twentieth century the search for the magic bullet continued, especially among scientists at

(a)   EM

— RNA

▷ Protein subunits

(b)

Figure 1.14 The tobacco mosaic virus. (a) Electron micrograph of tobacco mosaic virus (magnification approx. 617,000X). *(Omikron/Photo Researchers, Inc.)* **(b)** The structure of the tobacco mosaic virus. A helical core of RNA is surrounded by a coat that consists of repeating protein units. The structure of the particles is so regular that the viruses can be crystallized.

Ehrlich's institute. After testing hundreds of compounds (and numbering each compound), Ehrlich found compound 418 (arsenophenylglycine) to be effective against sleeping sickness and compound 606 (Salvarsan) to be effective against syphilis. For 40 years Salvarsan remained the best available treatment for this disease. In 1922 Alexander Fleming, a Scottish physician, discovered that lysozyme, an enzyme found in tears, saliva, and sweat, could kill bacteria. Lysozyme was the first body secretion shown to have chemotherapeutic properties.

The development of **antibiotics** began in 1917 with the observation that certain bacteria (actinomycetes) stopped the growth of other bacteria. In 1928 Fleming **(Figure 1.16)** observed that a colony of *Penicillium* mold contaminating a culture of *Staphylococcus* bacteria had prevented growth of bacteria adjacent to itself. Although Fleming was not the first to observe this phenomenon, he was the first to recognize its potential for countering infections. However, purification of sufficient quantities of the substance he called *penicillin* proved to be very difficult. The great need for such a drug during World War II, money from the Rockefeller Institute, and the hard work of the German biochemist Ernst Chain, the Australian pathologist Howard Florey, and researchers at Oxford University accomplished the task. Penicillin became available as a safe and versatile chemotherapeutic agent for use in humans.

While this work was going on, sulfa drugs were being developed. In 1935, prontosil rubrum, a reddish dye containing a sulfonamide chemical group, was used in treating streptococcal infections. Further study showed that sulfonamides were converted in the body to sulfanilamides; much subsequent work was devoted to developing drugs containing sulfanilamide. The German chemist Gerhard Domagk played an important role in this work, and one of the drugs, prontosil, saved the life of his daughter. In 1939 he was awarded a Nobel Prize for his work, but Hitler refused to allow him to make the trip to receive it. Extensions of Domagk's work led to the development of isoniazid, an effective agent against tuberculosis. Both sulfa drugs and isoniazid are still used today.

The development of antibiotics resumed with the work of Selman Waksman, who was born in Ukraine and moved to the United States in 1910. Inspired by the 1939 discovery, by the French microbiologist Rene

Figure 1.15   Paul Ehrlich. Ehrlich was a pioneer in the development of chemotherapy for infectious disease. *(© Bettmann/ Corbis)*

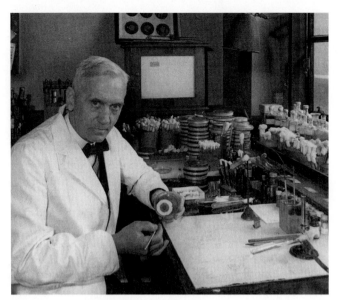

Figure 1.16 Alexander Fleming. Fleming discovered the antibacterial properties of penicillin. (© UPI/Bettmann/Corbis)

Dubos, of tyrothricin, an antibiotic produced by soil bacteria, Waksman examined soil samples from all over the world for growth-inhibiting microorganisms or their products. He coined the term *antibiotic* in 1941 to describe actinomycin and other products he isolated. Both tyrothricin and actinomycin proved to be too toxic for general use as antibiotics. After repeated efforts, Waksman isolated the less toxic drug streptomycin in 1943. Streptomycin constituted a major breakthrough in the treatment of tuberculosis. In the same decade Waksman and others isolated neomycin, chloramphenicol, and chlortetracycline.

Examining soil samples proved to be a good way to find antibiotics, and explorers and scientists still collect soil samples for analysis. The more common antibiotic-producing organisms are rediscovered repeatedly, but the possibility of finding a new one always remains. Even the sea has yielded antibiotics, especially from the fungus *Cephalosporium acremonium*. The Italian microbiologist Giuseppe Brotzu noted the absence of disease organisms in seawater where sewage entered, and he determined that an antibiotic must be present. Cephalosporin was subsequently purified, and a variety of cephalosporin derivatives are now available for treating human diseases.

The fact that many antibiotics have been discovered does not stop the search for more. As long as there are untreatable infectious diseases, the search will continue. Even when effective treatment becomes available, it is always possible that a better, less toxic, or cheaper treatment can be found. Of the many chemotherapeutic agents currently available, none can cure viral infections. Consequently, much of today's drug research is focused on developing effective antiviral agents.

GENETICS AND MOLECULAR BIOLOGY

Modern genetics began with the rediscovery in 1900 of Gregor Mendel's principles of genetics. Even after this significant event, for nearly three decades little progress was made in understanding how microbial characteristics are inherited. For this reason, microbial genetics is the youngest branch of microbiology. In 1928 the British scientist Frederick Griffith discovered that previously harmless bacteria could change their own nature and become capable of causing disease. The remarkable thing about this discovery is that live bacteria were shown to acquire heritable traits from dead ones. During the early 1940s, Oswald Avery, Maclyn McCarty, and Colin MacLeod of the Rockefeller Institute in New York City demonstrated that the change was produced by DNA. After that finding came the crucial discovery of the structure of DNA by James Watson and Francis Crick. This breakthrough ushered in the modern era of molecular genetics.

About the same time, the American geneticists Edward Tatum and George Beadle used genetic variations in the mold *Neurospora* to demonstrate how genetic information controls metabolism. In the early 1950s the American geneticist Barbara McClintock discovered that some genes (units of inherited information) can move from one location to another on a chromosome. Before McClintock's work, genes were thought to remain stationary. Her revolutionary discovery forced geneticists to revise their thinking about genes.

More recently, scientists have discovered the genetic basis that underlies the human body's ability to make an enormous diversity of *antibodies*, molecules that the immune system produces to combat invading microbes and their toxic products. Within cells of the immune system, genes are shuffled about and spliced together in various combinations, allowing the body to make millions of different antibodies, including some that can protect us from threats that the body has never previously encountered.

▋▋▋ TOMORROW'S HISTORY

Today's discovery is tomorrow's history. In an active research field such as microbiology, it is impossible to present a complete history. Some of the microbiologists omitted from this discussion are listed in **Table 1.3**. The period represented there, 1874–1917, is called the Golden Age of Microbiology. You may find that many of the terms used to describe these scientists' accomplishments are unfamiliar, but you will become familiar with them as you pursue the study of microbiology. Since 1900, Nobel Prizes have been awarded annually to outstanding scientists, many of whom were in the fields of physiology or medicine (**Table 1.4**). In some years the prize has been shared by several

CLOSE UP

How Microbiologists Investigate Problems

Like other scientists, microbiologists investigate problems by designing and carrying out experiments. Such experiments provide information that health scientists use to solve medical problems. Much of this text is devoted to presenting information obtained from experiments and to showing how that information is used in understanding infectious diseases. How do scientists investigate problems?

First, a scientific problem must concern some aspect of the natural world because scientific methods can deal only with natural conditions and events. Microbiological problems deal with natural conditions and events involving microbes. Second, scientific problems must be clearly defined and sufficiently limited in scope so that a hypothesis and a prediction can be formulated. A **hypothesis** is a tentative explanation to account for an observed condition or event. The hypothesis in a particular experiment must be (1) an explanation for the defined problem and (2) testable. A testable hypothesis is one for which evidence can be collected to support or refute the hypothesis. A **prediction** is an outcome or consequence that will result if the hypothesis is true. Before beginning a scientific experiment, one must define the problem and make a hypothesis and a prediction.

A good hypothesis is one that offers the most reasonable explanation and the simplest solution to a problem. The purpose of scientific experiments is to test hypotheses by determining the correctness of predictions derived from the hypotheses. Scientific progress is made by making and testing hypotheses.

For example, suppose that a microbiologist has isolated an organism in pure culture and wants to know the effects of temperature on its growth. On the basis of information from prior research, he or she might (1) hypothesize that the organism's growth rate increases with temperature and (2) predict that the rate of increase in the number of organisms in a culture is proportional to the increase in temperature. After making the hypothesis and a prediction, the investigator designs an experiment to test the hypothesis. The experiment must be designed specifically to test the hypothesis and to collect evidence to determine whether the prediction is true.

To design a good experiment, an investigator must consider all variables that might affect the outcome. A **variable** is anything that can change for the purposes of an experiment. An experiment should have only one **experimental variable**, the factor that is purposely changed for the experiment. For example, in the study of the effects of temperature on the growth of an organism, temperature is the experimental variable. The hypothesis and prediction are related to the experimental variable. All other variables are **control variables**, factors that *can* change but that are prevented from changing for the duration of the experiment. In our example, the control variables include the number and characteristics of the organism, the quantity and properties of the medium, and all environmental factors except temperature.

When all variables have been identified, the investigator establishes the procedures for carrying out the experiment. Once the experiment has been designed, it must be carried out exactly as planned. All observations must be made and recorded accurately and precisely. If problems or unusual situations are encountered, they must be noted carefully. For example, should an incubator fail to maintain certain cultures at the proper temperature for the appropriate length of time, this failure should be noted and taken into consideration in interpreting the experiment. When the experiment is completed, the researcher analyzes and interprets the results in light of the hypothesis and prediction. The analysis of the results of an experiment often involves preparation of tables and graphs and usually compares results obtained under experimental and control conditions.

The goal of an experiment is to draw **conclusions** as to whether the prediction is true. If the experimental results, when analyzed, do not support the hypothesis, they may nevertheless suggest a better alternative hypothesis. The experimenter might wish to design further experiments to test this new hypothesis. Often, it is unexpected experimental results that lead to the most interesting discoveries.

In this textbook, you will find some boxes, such as A Winter Dilemma (p. 35) and What Grows in Your Health and Beauty Aids? (p. 593), that suggest projects you might try yourself. Even if you are not able to carry out these projects, at least make a mental plan of what hypotheses and steps you could use to investigate such a problem. Perhaps you can discuss your experimental designs during a lecture or in a lab. What other questions occur to you? How would you go about forming and testing hypotheses for them? In addition to performing experiments, scientists also should report their results so that other scientists can verify and use the information. Scientific knowledge increases by the sharing of information. This allows other scientists to repeat experiments and determine whether the results are reproducible. It also allows them to develop new experiments that build on existing information.

scientists, although the scientists may have made independent contributions. Refer to Tables 1.3 and 1.4 as you begin to study each new area of microbiology.

You can see from Table 1.4 that microbiology has been in the forefront of research in medicine and biology for several decades and probably never more so than today. One reason is the renewed focus on infectious disease brought about by the advent of AIDS. Another is the dramatic progress in genetic engineering that has been made in the past two decades. Microorganisms have been and continue to be an essential part of the genetic engineering revolution. Most of the key discoveries that led to our present understanding of genetics emerged from research with microbes. Today scientists

TABLE 1.3

	The Golden Age of Microbiology: Early Microbiologists and Their Achievements	
Year	Investigator	Achievement
1874	Billroth	Discovery of round bacteria in chains
1876	Koch	Identification of *Bacillus anthracis* as causative agent of anthrax
1878	Koch	Differentiation of staphylococci
1879	Hansen	Discovery of *Mycobacterium leprae* as causative agent of leprosy
1880	Neisser	Discovery of *Neisseria gonorrhoeae* as causative agent of gonorrhea
1880	Laveran and Ross	Identification of life cycle of malarial parasites in red blood cells of infected humans
1880	Eberth	Discovery of *Salmonella typhi* as causative agent of typhoid fever
1880	Pasteur and Sternberg	Isolation and culturing of pneumonia cocci from saliva
1881	Koch	Animal immunization with attenuated anthrax bacilli
1882	Leistikow and Loeffler	Cultivation of *Neisseria gonorrhoeae*
1882	Koch	Discovery of *Mycobacterium tuberculosis* as causative agent of tuberculosis
1882	Loeffler and Schutz	Identification of actinobacillus that causes the animal disease glanders
1883	Koch	Identification of *Vibrio cholerae* as causative agent of cholera
1883	Klebs	Identification of *Corynebacterium diphtheriae* and toxin as causative agent of diphtheria
1884	Loeffler	Culturing of *Corynebacterium diphtheriae*
1884	Rosenbach	Pure culturing of streptococci and staphylococci
1885	Escherich	Identification of *Escherichia coli* as a natural inhabitant of the human gut
1885	Bumm	Pure culturing of *Neisseria gonorrhoeae*
1886	Flugge	Staining to differentiate bacteria
1886	Fraenckel	*Streptococcus pneumoniae* related to pneumonia
1887	Weichselbaum	*Neisseria meningitidis* related to meningitis
1887	Bruce	Identification of *Brucella melitensis* as causative agent of brucellosis in cattle
1887	Petri	Invention of culture dish
1888	Roux and Yersin	Discovery of action of diphtheria toxin
1889	Charrin and Roger	Discovery of agglutination of bacteria in immune serum
1889	Kitasato	Discovery that *Clostridium tetani* produces tetanus toxin
1890	Pfeiffer	Identification of Pfeiffer bacillus, *Haemophilus influenzae*
1890	von Behring and Kitasato	Immunization of animals with diphtheria toxin
1892	Ivanovski	Discovery of filterability of tobacco mosaic virus
1894	Roux and Kitasato	Identification of *Yersinia pestis* as causative agent of bubonic plague
1894	Pfeiffer	Discovery of bacteriolysis in immune serum
1895	Bordet	Discovery of alexin (complement) and hemolysis
1896	Widal and Grunbaum	Development of diagnostic test based on agglutination of typhoid bacilli by immune serum
1897	van Ermengem	Discovery of *Clostridium botulinum* as causative agent of botulism
1897	Kraus	Discovery of precipitins
1897	Ehrlich	Formulation of side-chain theory of antibody formation
1898	Shiga	Discovery of *Shigella dysenteriae* as causative agent of dysentery
1898	Loeffler and Frosch	Discovery of filterability of virus that causes foot-and-mouth disease
1899	Beijerinck	Discovery of intracellular reproduction of tobacco mosaic virus
1901	Bordet and Gengou	Identification of *Bordetella pertussis* as causative agent of whooping cough; development of complement fixation test
1901	Reed and colleagues	Identification of virus that causes yellow fever
1902	Portier and Richet	Work on anaphylaxis
1903	Remlinger and Riffat-Bey	Identification of virus that causes rabies
1905	Schaudinn and Hoffmann	Identification of *Treponema pallidum* as causative agent of syphilis
1906	Wasserman, Neisser, and Bruck	Development of Wasserman reaction for syphilis antibodies
1907	Asburn and Craig	Identification of virus that causes dengue fever
1909	Flexner and Lewis	Identification of virus that causes poliomyelitis
1915	Twort	Discovery of viruses that infect bacteria
1917	d'Herelle	Independent rediscovery of viruses that infect bacteria (bacteriophages)

TABLE 1.4

Nobel Prize Awards for Research Involving Microbiology		
Year of Prize	**Prize Winner**	**Topic Studied**
1901	von Behring	Serum therapy against diphtheria
1902	Ross	Malaria
1905	Koch	Tuberculosis
1907	Laveran	Protozoa and the generation of disease
1908	Ehrlich and Metchnikoff	Immunity
1913	Richet	Anaphylaxis
1919	Bordet	Immunity
1928	Nicolle	Typhus exanthematicus
1939	Domagk	Antibacterial effect of prontosil
1945	Fleming, Chain, and Florey	Penicillin
1951	Theiler	Vaccine for yellow fever
1952	Waksman	Streptomycin
1954	Enders, Weller, and Robbins	Cultivation of polio virus
1958	Lederberg	Genetic mechanisms
	Beadle and Tatum	Transmission of hereditary characteristics
1959	Ochoa and Kornberg	Chemical substances in chromosomes that play a role in heredity
1960	Burnet and Medawar	Acquired immunological tolerance
1962	Watson, Crick, and Wilkins	Structure of DNA
1965	Jacob, Lwoff, and Monod	Regulatory mechanisms in microbial genes
1966	Rous	Viruses and cancer
1968	Holley, Khorana, and Nirenberg	Genetic code
1969	Delbruck, Hershey, and Luria	Mechanism of virus infection in living cells
1972	Edelman and Porter	Structure and chemical nature of antibodies
1975	Baltimore, Temin, and Dulbecco	Interactions between tumor viruses and genetic material of the cell
1976	Blumberg and Gajdusek	New mechanisms for the origin and dissemination of infectious diseases
1978	Smith, Nathans, and Arber	Restriction enzymes for cutting DNA
1980	Benacerraf, Snell, and Dausset	Immunological factors in organ transplants
1980	Berg	Recombinant DNA
1984	Milstein, Köhler, and Jerne	Immunology
1987	Tonegawa	Genetics of antibody diversity
1988	Black, Elion, and Hitchings	Principles of drug therapy
1989	Bishop and Varmus	Genetic basis of cancer
1990	Murray, Thomas, and Corey	Transplant techniques and drugs
1993	Mullis	Polymerase chain reaction method to amplify (copy) DNA
1993	Smith	Method to splice foreign components into DNA
1993	Sharp and Roberts	Genes can be discontinuous
1996	Doherty and Zinkernagel	Recognition of virus-infected cells by immune defenses
1997	Prusiner	Prions
2005	Marshall and Warren	*Helicobacter pylori* causes stomach ulcers

are attempting to redesign microorganisms for a variety of purposes (as we will see in ◀Chapter 8). Bacteria have been converted into factories that produce drugs, hormones, vaccines, and a variety of biologically important compounds. And microbes, viruses in particular, are often the vehicle by which scientists insert new genes into other organisms. Such techniques are beginning to enable us to produce improved varieties of plants and animals such as pest-resistant crops and may even enable us to correct genetic defects in human beings.

In September 1990, a 4-year-old girl became the first gene-therapy patient. She had inherited a defective gene that crippled her immune system. Doctors at the National Institutes of Health (NIH) inserted a normal copy of the gene into some of her white blood cells in the laboratory and then injected these gene-treated cells back into her body, where, it was hoped, they would restore her immune system. Critics were worried that a new gene randomly inserted into the girl's white blood cells could damage other genes and cause cancer. The experiment was a success and she enjoys good health today.

New information is constantly being discovered and sometimes supersedes earlier findings. Occasionally, new discoveries lead almost immediately to the development of medical applications, as occurred with penicillin and as will most certainly occur when a cure or vaccine for AIDS is discovered. However, old ideas such as spontaneous generation and old practices such as unsanitary measures in medicine can take years to replace. Many new bioethics problems will require considerable thought. Decisions about AIDS testing and reporting, transplants, cloning, environmental cleanup, and related issues will not come easily or quickly. Because of the wealth of prior knowledge, it is likely that you will learn more about microbiology in a single course than many pioneers learned in a lifetime. Yet, those pioneers deserve great credit because they worked with the unknown and had few people to teach them.

The future of microbiology holds exciting developments. One area involves use of **bacteriophages**, viruses that attack and kill specific kinds of bacteria. Yes, "the big fleas do have little fleas to bite them, and so on ad infinitum." Using phages to treat bacterial infections was developed back in the 1920s and 1930s in Eastern Europe and the Soviet Union. With the discovery of antibiotics in the 1940s, use of phages fell into disfavor and never really made it into Western medical practice. However, in Soviet countries phage therapy has been preferred over use of antibiotics even through today. Soviet troops carried color-coded packets of various phages with them, specific for the bacterial diseases they were likely to encounter. When the first few men came down with disease "A," everyone was instructed to open the "red" packet and consume it; for disease "B", the "blue packet", etc., and thus epidemic outbreaks were prevented. Schoolchildren were also given such packets. Today, in the West, we are struggling against bacteria that have developed resistance to antibiotics—some against every known antibiotic—leading scientists to reexamine the usefulness of phages.

Agricultural and food problems have recently been shown to have phage-mediated solutions. *Listeria monocytogenes*, a foodborne pathogen that can live and grow at refrigerator temperatures, causes bacterial diarrhea which is fatal in 20% of its cases. A phage has been shown to control growth of *Listeria* on cut apples and melons better than chemical sanitizers, or washing.

Herds of animals can also be protected with phages. The U.S. Department of Agriculture, in January of 2007, approved use of a spray or wash containing phages targeted against *E. coli* O157: H7, to be applied to live animals prior to slaughter. This bacterium causes an often-fatal hemorrhagic dysentery. Removing it from hides before slaughter will help prevent its getting into products such as ground beef and will help to keep our food supply safe. Tests of effectiveness of phages on human diseases are underway now. Hopefully we will soon have replacements for antibiotics.

Unfortunately another thing the future holds for us is the threat of bioterrorism. Perhaps phages will be able to help us there, too, as we have already isolated phages that can destroy many strains of anthrax, for example. Other terrorist diseases and methods of their use and control will be discussed in ◄Chapter 15, as well as in relevant parts of the chapters on disease. The words of Dr. Ken Alibek, former head of the Soviet secret germ warfare program, in his book, *Biohazard* (1999), should send shivers through us, "Our factory could turn out two tons of anthrax a day in a process as reliable and efficient as producing tanks, trucks, cars, or Coca-Cola." He also explains that, "It would take only five kilograms of the Anthrax 836 developed at the Kazakhstan base to infect half the people living in a square kilometer of territory."

Farmers fighting weeds in their fields may soon have help from the U.S. Department of Agriculture. Scientists there **(Figure 1.17)** are searching for specialist microbes that will selectively attack weed seeds in soil, causing them to rot and die, without the use of chemical sprays.

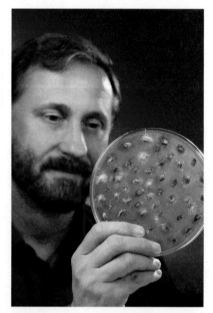

Figure 1.17   Let microbes kill weed seeds.
U.S. Department of agriculture scientist holds a culture of giant ragweed (*Ambrosia trifida*) seeds embedded in agar, some overgrown with soil microorganisms, He is investigating how and why some weed seeds escape decay by these organisms. *(Peggy Greb/U.S. Department of Agriculture)*

GENOMICS

Microbial genetic techniques have made possible a colossal scientific undertaking: the Human Genome Project. Its purpose is to identify the location and chemical sequence of all the genes in the human genome—that is, all the genetic material in the human species. Begun in 1990, it was to be completed by 2005, at a cost of approximately $3 billion. Amazingly, it was finished in May 2000, ahead of schedule and under budget! Another surprise was the finding that humans have just over 25,000 genes, instead of estimates that ranged up to 142,000 genes. In February 2001 reports were published in separate scientific journals by the two rival groups that had completed the project: Dr. J. Craig Venter, then president of Celera Genomics (Rockville, Maryland) in *Science*, and in *Nature* by Dr. Eric Lander, representing the International Human Genome Sequencing Consortium, a group of academic centers funded mainly by the NIH and the Wellcome Trust of London. The 3 billion base pairs in the human genome do not all code for useful genes. An estimated 75% of them code for "junk DNA." However, many scientists believe that we may eventually discover uses for what we now consider "junk."

Work on the Human Genome Project was based on techniques that were first developed for sequencing microbial genomes, which are smaller and easier to work with. Over 100 microbial genomes have been sequenced so far. A great surprise was finding that some bacteria have two or three chromosomes instead of the single one that was thought to be all any bacterium could have. It is interesting that 113 genes, and possibly scores more, have come to the human genome directly from bacteria. Venter has sequenced the mouse genome and reports that humans have only 300 genes not found in the mouse. The functions of 41.7% of human genes are still unknown. Says Venter, "The secrets of life are all spelled out for us in the genome, we just have to learn how to read it." Help reading the human genome is coming in great part from experiments using microbes.

✓ CHECKLIST

1. What were the scientific contributions of Jenner, Metchnikoff, Ehrlich, Fleming, McClintock, and Venter?

2. When was the Golden Age of Microbiology? What types of discoveries were mostly made during this period?

3. What is the Human Genome Project? How has microbiology been associated with it?

▌ RETRACING OUR STEPS

❚❚❚ WHY STUDY MICROBIOLOGY?

• Microorganisms are part of the human environment and are therefore important to human health and activities.

• The study of microorganisms provides insight into life processes in all forms of life.

❚❚❚ SCOPE OF MICROBIOLOGY

THE MICROBES

• **Microbiology** is the study of all **microorganisms (microbes)** in the microscopic range. These include **bacteria**, **algae**, **fungi**, **viruses**, **viroids**, **prions**, and **protozoa**.

THE MICROBIOLOGISTS

• Immunology, virology, chemotherapy, and genetics are especially active research fields of microbiology.

• Microbiologists work as researchers or teachers in university, clinical, and industrial settings. They do basic research in the biological sciences; help to perform or devise diagnostic tests; develop and test antibiotics and vaccines; work to control infection, protect public health, and safeguard the environment; and play important roles in the food and beverage industries.

❚❚❚ HISTORICAL ROOTS

• The ancient Greeks, Romans, and Jews all contributed to early understandings of the spread of disease.

• Diseases such as bubonic plague and syphilis caused millions of deaths because of the lack of understanding of how to control or treat the infections.

• The development of high-quality lenses by Leeuwenhoek made it possible to observe microorganisms and later to formulate the **cell theory**.

❚❚❚ THE GERM THEORY OF DISEASE

• The **germ theory of disease** states that microorganisms (germs) can invade other organisms and cause disease.

EARLY STUDIES

• Progress in microbiology and acceptance of the germ theory of disease required that the idea of **spontaneous generation** be refuted. Redi and Spallanzani demonstrated that organisms did not arise from nonliving material. Pasteur, with his swan-necked flasks, and Tyndall, with his dust-free air, finally dispelled the idea of spontaneous generation.

PASTEUR'S FURTHER CONTRIBUTIONS

• Pasteur also studied wine making and disease in silkworms and developed the first rabies vaccine. His association of particular microbes with certain diseases furthered the establishment of the germ theory.

KOCH'S CONTRIBUTIONS

• Koch developed four postulates that aided in the definitive establishment of the germ theory of disease. **Koch's Postulates** are as follows:

1. The specific causative agent must be found in every case of the disease.

2. The disease organism must be isolated in pure culture.

3. Inoculation of a sample of the culture into a healthy, susceptible animal must produce the same disease.

4. The disease organism must be recovered from the inoculated animal.

• Koch also developed techniques for isolating organisms, identified the bacillus that causes tuberculosis, developed tuberculin, and studied various diseases in Africa and Asia.

WORK TOWARD CONTROLLING INFECTIONS

• Lister and Semmelweis contributed to improved sanitation in medicine by applying the germ theory and using aseptic technique.

III EMERGENCE OF SPECIAL FIELDS OF MICROBIOLOGY

IMMUNOLOGY

• Immunization was first used against smallpox; Jenner used fluid from cowpox blisters to immunize against it.

• Pasteur developed techniques to weaken organisms so they would produce immunity without producing disease.

VIROLOGY

• Beijerinck characterized viruses as pathogenic molecules that could take over a host cell's mechanisms for their own use.

• Reed demonstrated that mosquitoes carry the yellow fever agent, and several other investigators identified viruses in the early twentieth century. The structure of DNA—the genetic material in many viruses and in all cellular organisms—was discovered by Watson and Crick.

• Techniques for isolating, propagating, and analyzing viruses were developed. Viruses could then be observed and in many cases crystallized, and their nucleic acids could be studied.

CHEMOTHERAPY

• Substances derived from medicinal plants were virtually the only source of chemotherapeutic agents until Ehrlich began a systematic search for chemically defined substances that would kill bacteria.

• Fleming and his colleagues developed penicillin, and Domagk and others developed sulfa drugs.

• Waksman and others developed streptomycin and other antibiotics derived from soil organisms.

GENETICS AND MOLECULAR BIOLOGY

• Griffith discovered that previously harmless bacteria could change their nature and become capable of causing disease. This genetic change was shown by Avery, McCarty, and MacLeod to be due to DNA. Tatum and Beadle studied biochemical mutants of *Neurospora* to show how genetic information controls metabolism.

III TOMORROW'S HISTORY

• Microbiology has been at the forefront of research in medicine and biology, and microorganisms continue to play a critical role in genetic engineering and gene therapy.

• Bacteriophage viruses may be able to cure diseases, and help ensure food safety.

GENOMICS

• The Human Genome Project has identified the location and sequence of all bases in the human genome. Microbes and microbiological techniques have contributed to this work.

• Over 100 bacterial genomes have been sequenced completely. A few have two instead of one chromosome.

■ TERMINOLOGY CHECK

algae *(p. 4)*
antibiotics *(p. 17)*
bacteria *(p. 4)*
bacteriophage *(p. 22)*
cell theory *(p. 9)*
conclusions *(p. 19)*

control variable *(p. 19)*
experimental variable *(p. 19)*
fungi *(p. 5)*
germ theory of disease *(p. 9)*
hypothesis *(p. 19)*

Koch's Postulates *(p. 13)*
microbe *(p. 4)*
microbiology *(p. 4)*
microorganism *(p. 2)*
prediction *(p. 19)*
prion *(p. 5)*

protozoa *(p. 5)*
spontaneous generation *(p. 9)*
variable *(p. 19)*
viroid *(p. 5)*
viruses *(p. 5)*

■ CLINICAL CASE STUDY

When doctors first began noticing cases of AIDS in the U.S., they did not know much about it. They had many questions to answer. What did they find to be its etiology? What were some observations about its epidemiology?

■ CRITICAL THINKING QUESTIONS

1. Edward Jenner, in eighteenth-century England, first injected a child with a totally untested smallpox vaccine and then, after a time, injected that child with living smallpox virus. What would be the likely reaction to someone performing a similar experiment today? How do you think a scientist of today would test a potential new vaccine?

2. Can you think of some reasons why it might be hard to fulfill Koch's Postulates in order to support the "Germ Theory" of disease?

3. As often happens in science, one observation or experiment that is used to look at one aspect or subdiscipline of science can lead to profound explanations or solutions in another aspect or discipline of science. Serendipity and experimental mistakes also play a role in this. Explain how this might apply to Angelina Hesse and the success of Robert Koch's pure cultures, Louis Pasteur's assistant and the success of the immunizations of chickens against chicken cholera, plus Alexander Fleming and *Penicillium*.

4. It is likely that others beside Anton van Leeuwenhoek were using lenses to look at microorganisms. After all, Robert Hooke had developed and used the compound microscope in about 1665, and the first letter from van Leeuwenhoek to the Royal Society of London was written in 1673. Why is it that we know about the observations of van Leeuwenhoek and not others? What reasons can you give for why it is important for scientists to publish the results of their research?

5. The completion of chromosomal mapping and sequencing of genes in the Human Genome Project has been one of humanity's greatest accomplishments, yet it really serves as just the beginning for a new era of genomic science.
   (a) Can you think of new burgeoning fields of science that will result from the newly acquired sequencing data?
   (b) Antibiotics are effective because of their selective killing of bacterial and not animal cells. Can you think of ways in which comparing bacerial genomes to the human genome might result in additional cures of bacterial infections/diseases?

6. Which of the following factors in today's world make it difficult to keep disease-causing microorganisms in check from a health point of view?
   (a) Lack of a balanced distribution of wealth
   (b) Increased and quick world travel
   (c) Encroachment of humans into virgin jungles
   (d) Antibiotic-resistant bacteria
   (e) All of the above

| SELF-QUIZ

1. Less than 1% of microorganisms are harmful and cause disease. True or false?

2. Life on earth would be much better if all microbes were eradicated. True or false?

3. Which of the following is not true?
   (a) A single bacterium weighs approximately 1×10^{-11} grams.
   (b) On average there are 100 trillion microorganisms on any given human.
   (c) Microbes can only be found where man naturally habituates.
   (d) There are more microbes in your mouth than all the people who have lived in the history of man.

4. Which of the following is not a reason for microoranisms being useful in research?
   (a) Microbes have high reproduction rates.
   (b) Microbes are easily controlled.
   (c) Microbes have relatively simple structures.
   (d) Microbes make it easier to prove the statistical significance of an experiment because of their ease and cost effectiveness resulting from their large numbers.

5. Why are microbes important to study and how are they directly useful to man? important to environment

6. People in central Asia are still suffering from smallpox infections. True or false?

7. Match the following microorganisms with the description that best applies:

   C Algae
   F Bacteria
   A Fungi
   E Protozoa
   B Viruses
   D Helminthes

   (a) Multicellular nucleated microorganism that has branching filaments
   (b) Acellular entities that require a host for multiplication
   (c) Photosynthetic large cells that rarely cause human disease
   (d) Parasitic worms
   (e) Large, single-celled nucleated microorganisms
   (f) Single-celled non-nucleated microorganisms

8. Animals such as worms and ticks are too large to be included in a microbiology course. True or False? Explain.

9. What is the difference between etiology and epidemiology? etiology is cause of disease   epidemiology – distributi

10. The epidemic that infected Europe, North Africa, and the Middle East and killed tens of millions was known as the Black Death. The disease was caused by:
    (a) Smallpox   (d) Anthrax
    (b) Bubonic plague   (e) Swine flu
    (c) Breathing of foul air

11. Which of the following people in history did not make a contribution toward preventive medicine by stricter sanitation practices?
    (a) Angelina Hesse
    (b) Moses and the mosaic laws found in the bible
    (c) Joseph Lister
    (d) Ignaz Philipp Semmelweis

12. The event that triggered the development and establishment of microbiology as a science is the:
    (a) Spontaneous generation
    (b) Use of disinfectants
    (c) Vaccinations
    (d) Germ theory of disease
    (e) Development of the microscope

13. What was Leeuwenhoek's contribution to microbiology? development of highquality lens in microscope

14. Which scientist first disproved spontaneous generation by showing that maggots only appear on decaying meat that has been exposed to flies?
    (a) Lister   (d) Redi
    (b) Pasteur   (e) Koch
    (c) Hooke

15. Which of the following experiments was not useful in disproving "spontaneous generation"?
    (a) John Tyndall's "dust settlement" experiment with sealed and open flasks of boiled infusions in airtight boxes.

(b) Louis Pasteur's experiment with boiled infusions in long swan-necked flasks.

(c) Louis Pasteur's experiment with filtered air through cotton plugs and sterile infusions.

(d) Louis Pasteur's "pasteurization process" in wine making.

16. The biggest obstacle in the acceptance and development of the science of microbiology was:
    (a) Lack of effective vaccines
    (b) Lack of sterile containers
    (c) Theory of spontaneous generation
    (d) Absence of debilitating diseases before the seventeenth century
    (e) Use of aseptic technique

17. Besides providing strong evidence toward the disproof of spontaneous generation, Louis Pasteur made many other contributions toward the advancement of microbiology. Which of the following is not one of Pasteur's contributions?
    (a) Provided evidence for the germ theory with his association of specific microbes with certain diseases in silkworms
    (b) Developed the first rabies vaccine
    (c) Developed the technique of pasteurization to cure sour wine
    (d) Developed a cowpox vaccine for smallpox
    (e) Contributed to the emerging science of immunology with the study of chicken cholera in chickens

18. The germ theory of disease states that:
    (a) Microorganisms that invade other organisms can cause disease in those organisms
    (b) Microorganisms can spontaneously arise in debilitated hosts
    (c) Microorganisms do not cause infectious diseases
    (d) Not all microorganisms are harmful
    (e) Malaria is caused by bad air ("Mal"—"Aria")

19. Put Koch's postulates in order.
    2 (a) The disease organism must be isolated in pure culture.
    4 (b) The disease organism must be recovered from the inoculated animal.
    1 (c) The specific causative agent must be found in every case of the disease.
    3 (d) Inoculation of a sample of the culture into a healthy, susceptible animal must produce the same disease.

20. What did Semmelweis and Lister contribute to microbiology?
    to improve sanitation in medicine

21. Match the following scientists who emerged in specialized fields of microbiology to their famous contributions and specialized field:

    I. Metchinikoff   A. Virology (4)
    II. Beijerinck   B. Chemotherapy (2)
    III. McClintock   C. Immunology (3)
    IV. Ehrlich   D. Genetics (1)

    1. Mobile ("jumping") genes
    2. Salvarsan against syphilis
    3. Cellular immunity (phagocytes)
    4. Infectious filtrates contain viruses

22. Describe the contributions of the following scientists to the field of microbiology: Beijerinck, Fleming, and Metchnikoff.

23. (a) How do bacteria differ from viruses? (b) Are there ways to fight infectious diseases caused by bacteria other than through the use of antibiotics?

24. Scientists have found many antibiotics by examining microorganisms in soil. True or false?

25. Some of the most important emerging fields of microbiology are virology, chemotherapy (antibiotics), and genetics molecular biology. Match the following scientists to their contribution in these emerging fields:

    D James Watson & Francis Crick
    E Reed and colleagues
    F Avery, McCarty, & McLeod
    B Beadle & Tatum
    C Selman Waksman
    A Frederick Griffith

    (a) Discovered that live, harmless bacteria could become disease causing by acquiring heritable traits from dead ones
    (b) Used mold to demonstrate how genetic information controls metabolism
    (c) Discovered and isolated the antibiotic streptomycin in 1943
    (d) Discovered the alpha helical structure of DNA
    (e) Identified the virus that caused yellow fever
    (f) Proved that DNA was responsible for acquiring new changes

26. Use the following diagram to explain how Pasteur's swan-necked flasks prevent contamination of sterile broth in the flasks. Describe what happens to the sterile broth in (a) after it has been allowed to cool as in (b). What happens to the broth after the flask has been tipped enough to let the broth come in contact with the dust and microorganisms and is tipped back as in (c)?

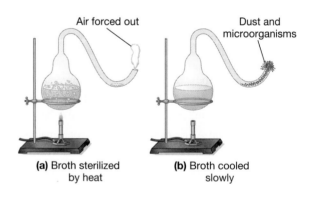

(a) Broth sterilized by heat   (b) Broth cooled slowly

(c) Flask tipped

▌ EXPLORATIONS ON THE WEB

If you think you've mastered this chapter, there's more to challenge you on the web. Go to the companion web site to fine-tune your understanding of the chapter concepts and discover answers to the questions posed below.

1. Did you know that there is a museum in Vienna, Austria where you can see one of the washbasins used by Semmelweis in hospitals there? Find out more about Semmelweis and how he decreased the incidence of childbirth fever.

2. Check out the web site to find out what procedures were developed to kill microorganisms affecting the wine industry without altering the flavor of the wine itself.

2 Fundamentals of Chemistry

Come with me...

Free-falling and weightless, you climb toward the others. You link hands with your skydiving partners to form the pattern you discussed just moments ago in the safety of the airplane. Suspended between heaven and earth, you consider the significance of the connections that bind you.

Just as the hands of these skydivers hold them in a complex pattern, chemical bonds between atoms hold together complex molecular patterns. The shapes of molecules are very important— when shapes change, the properties of the molecules themselves change. In an organism, this change can mean the difference between life and death.

© Michael McGowan Fun Air Productions

 Video related to this topic is available within WileyPLUS.

▌▌▌ WHY STUDY CHEMISTRY?

▌▌▌ CHEMICAL BUILDING BLOCKS AND CHEMICAL BONDS

Chemical Building Blocks / The Structure of Atoms / Chemical Bonds / Chemical Reactions

▌▌▌ WATER AND SOLUTIONS

Water / Solutions and Colloids / Acids, Bases, and pH

▌▌▌ COMPLEX ORGANIC MOLECULES

Carbohydrates / Lipids / Proteins / Nucleotides and Nucleic Acids

All living and nonliving things, including microbes, are composed of matter. Thus, it is not surprising that all properties of microorganisms are determined by the properties of matter.

WHY STUDY CHEMISTRY?

Chemistry is the science that deals with the basic properties of matter. Therefore, we need to know some chemistry to begin to understand microorganisms. Chemical substances undergo changes and interact with one another in *chemical reactions*. Metabolism, the use of nutrients for energy or for making the substance of cells, consists of many different chemical reactions. This is true regardless of whether the organism is a human or a microorganism. Thus, understanding the basic principles of chemistry is essential to understanding metabolic processes in living things. A microbiologist uses chemistry to understand the structure and function of microorganisms themselves and to understand how they affect humans in disease processes, as well as how they affect all life on earth.

CHEMICAL BUILDING BLOCKS AND CHEMICAL BONDS

CHEMICAL BUILDING BLOCKS

Matter is composed of very small particles that form the basic chemical building blocks. Over the years, chemists have observed matter and deduced the characteristics of these particles. Just as the alphabet can be used to make thousands of words, the chemical building blocks can be used to make thousands of different substances. The complexity of chemical substances greatly exceeds the complexity of words. Words rarely contain more than 20 letters, whereas some complex chemical substances contain as many as 20,000 building blocks!

The smallest chemical unit of matter is the **atom**. Many different kinds of atoms exist. Matter composed of one kind of atom is called an **element**. Each element has specific properties that distinguish it from other elements. Carbon is an element; a pure sample of carbon consists of a vast number of carbon atoms. Oxygen and nitrogen also are elements; they are found as gases in the earth's atmosphere. Chemists use one- or two-letter symbols to designate elements—such as C for carbon, O for oxygen, N for nitrogen, and Na for sodium (from its Latin name, *natrium*).

Atoms combine chemically in various ways. Sometimes atoms of a single element combine with each other. For example, carbon atoms form long chains that are important in the structure of living things. Both oxygen and nitrogen form paired atoms, O_2 and N_2. More often, atoms of one element combine with atoms of other elements. Carbon dioxide (CO_2) contains one atom of carbon and two atoms of oxygen; water (H_2O) contains two atoms of hydrogen and one atom of oxygen. (The subscripts in these formulas indicate how many atoms of each element are present.)

When two or more atoms combine chemically, they form a **molecule**. Molecules can consist of atoms of the same element, such as N_2, or atoms of different elements, such as CO_2. Molecules made up of atoms of two or more elements are called **compounds**. Thus, CO_2 is a compound, but N_2 is not. The properties of compounds are different from those of their component elements. For example, in their elemental state, both hydrogen and oxygen are gases at ordinary temperatures. They can combine to form water, however, which is a liquid at ordinary temperatures.

Living things consist of atoms of relatively few elements, principally carbon, hydrogen, oxygen, and nitrogen, but these are combined into highly complex compounds. A simple sugar molecule, $C_6H_{12}O_6$, contains 24 atoms. Many molecules found in living organisms contain thousands of atoms.

THE STRUCTURE OF ATOMS

Although the atom is the smallest unit of any element that retains the properties of that element, atoms do contain even smaller particles that together account for those properties. Physicists study many such subatomic particles, but we discuss only **protons**, **neutrons**, and **electrons**. Three important properties of these particles are atomic mass, electrical charge, and location in the atom (Table 2.1). *Atomic mass* is measured in terms of *atomic mass units (AMU)*. The mass of a proton or a neutron is almost exactly equal to 1 AMU; electrons have a much smaller mass. With respect to electrical charge, electrons are negatively ($-$) charged, and protons are positively ($+$) charged. Neutrons are neutral, with no charge. Atoms normally have an equal number of protons and electrons and so are electrically neutral. The heavy protons and neutrons are packed into the tiny, central *nucleus* of the atom, whereas the lighter electrons move around the nucleus in what have commonly been described as orbits.

Nitrogen is called "stickstoff" in German and "azoto" in Italian, but its symbol is N in every country of the world.

TABLE 2.1

Properties of Atomic Particles			
Particle	Atomic Mass	Electrical Charge	Location
Proton	1	+	Nucleus
Neutron	1	None	Nucleus
Electron	1/1,836	−	Orbiting the nucleus

The atoms of a particular element always have the same number of protons; that number of protons is the **atomic number** of the element. Atomic numbers range from 1 to over 100. The numbers of neutrons and electrons in the atoms of many elements can change, but the number of protons—and therefore the atomic number—remains the same for all atoms of a given element.

Protons and electrons are oppositely charged. Consequently, they attract each other. This attraction keeps the electrons near the nucleus of an atom. The electrons are in constant, rapid motion, forming an electron cloud around the nucleus. Because some electrons have more energy than others, chemists use a model with concentric circles, or *electron shells*, to suggest different energy levels. Electrons with the least energy are located nearest the nucleus, and those with more energy are farther from the nucleus. Each energy level corresponds to an electron shell (**Figure 2.1**).

An atom of hydrogen has only one electron, which is located in the innermost shell. An atom of helium has two electrons in that shell; two is the maximum number of electrons that can be found in the innermost shell. Atoms with more than two electrons always have two electrons in the inner shell and up to eight additional electrons in the second shell. The inner shell is filled before any electrons occupy the second shell; the second shell is filled before any electrons occupy the third shell, and so on. Very large atoms have several more electron

Figure 2.2
The formation of ions, or electrically charged atoms.
(a) When a neutral sodium atom loses the single electron in its outermost shell, it becomes a sodium ion, Na$^+$. When a neutral chlorine atom gains an extra electron in its outer shell, it becomes a chloride ion, Cl$^-$.
(b) Oppositely charged ions attract one another. Such attraction creates an ionic bond and results in the formation of an ionic compound, in this case sodium chloride (NaCl).

Figure 2.1   **The structure of four biologically important atoms.** Hydrogen, the simplest element, has an atom whose nucleus is made up of a single proton and a single electron in the first shell. In carbon, nitrogen, and oxygen, the first shell is filled with two electrons and the second shell is partly filled. Carbon, with six protons in its nucleus, has six electrons, four of them in the second shell. Nitrogen has five electrons, and oxygen six electrons in the second shell. It is the electrons in the outermost shell that take part in chemical bonding.

shells of larger capacity, but in elements found in living things, the outer shell is chemically stable if it contains eight electrons. This principle, known as the **rule of octets**, is important for understanding chemical bonding, which we will discuss shortly.

Atoms whose outer electron shells are nearly full (containing six or seven electrons) or nearly empty (containing one or two electrons) tend to form ions. An **ion** is a charged atom produced when an atom gains or loses one or more electrons (**Figure 2.2a**). When an atom of sodium (atomic number 11) loses the one electron in its outer shell without losing a proton, it becomes a positively charged ion, called a **cation** (kat′i-on). When an atom of chlorine (atomic number 17) gains an electron to fill its outer shell, it becomes a negatively charged ion, called an **anion** (an′i-on). In the ionized state, chlorine is referred to as chloride. Ions of elements such as sodium or chlorine are chemically more *stable* than atoms of these same elements because the ions' outer electron shells are full. Many elements are found in microorganisms or their environments as ions (**Table 2.2**). Those with one or two electrons in their outer shell tend to lose electrons and form ions with +1 or +2 charges, respectively; those with seven electrons in their outer shell tend to gain an electron and form ions with a charge of −1. Some ions, such as the hydroxyl (hi-drok′sil) ion (OH$^-$), are compounds—they contain more than one element.

Take another look

TABLE 2.2

Some Common Ions		
Ion	**Name**	**Brief Description**
Na^+	Sodium	Contributes to salinity of natural bodies of water and body fluids of multicellular organisms.
K^+	Potassium	Important ion that maintains cell turgor.
H^+	Hydrogen	Responsible for the acidity of solutions and commonly regulates motility.
Ca^{2+}	Calcium	Often acts as a chemical messenger.
Mg^{2+}	Magnesium	Commonly required for chemical reactions to occur.
Fe^{2+}	Ferrous iron	Carries electrons to oxygen during some chemical reactions that produce energy. Can prevent growth of some microbes that cause human disease.
NH_4^+	Ammonium	Found in animal wastes and degraded by some bacteria.
Cl^-	Chloride	Often found with a positively charged ion, where it usually neutralizes charge.
OH^-	Hydroxyl	Usually present in excess in basic solutions where H^+ is depleted.
HCO_3^-	Bicarbonate	Often neutralizes acidity of bodies of water and body fluids.
NO_3^-	Nitrate	A product of the action of certain bacteria that convert nitrite into a form plants can use.
SO_4^{2-}	Sulfate	Component of sulfuric acid in atmospheric pollutants and acid rain.
PO_4^{3-}	Phosphate	Can be combined with certain other molecules to form high-energy bonds, where energy is stored in form living things can use.

Although all atoms of the same element have the same atomic number, they may not have the same atomic weight. **Atomic weight** is the sum of the number of protons and neurons in an atom. Many elements consist of atoms with differing atomic weights. For example, carbon usually has six protons and six neutrons; it has an atomic weight of 12. But some naturally occurring carbon atoms have one or two extra neutrons, giving these atoms an atomic weight of 13 or 14. In addition, laboratory techniques are available to create atoms with different numbers of neutrons. Atoms of a particular element that contain different numbers of neutrons are called **isotopes**. The superscript to the left of the symbol for the element indicates the atomic weight of the particular isotope. For example, carbon with an atomic weight of 14, which is often used to date fossils, is written ^{14}C. The atomic weight of an element that has naturally occurring isotopes is the average atomic weight of the natural mixture of isotopes. Thus, atomic weights are not always whole numbers, even though any particular atom contains a specific number of whole neutrons and protons. Table 2.3 gives the atomic weights of some elements found in living things, as well as some other properties.

A **gram molecular weight**, or **mole**, is the weight of a substance in grams (g) equal to the sum of the atomic weights of the atoms in a molecule of the substance. For example, a mole of glucose, $C_6H_{12}O_6$, weighs 180 grams: [6 carbon atoms × 12 (atomic weight)] + [12 hydrogen atoms × 1 (atomic weight)] + [6 oxygen atoms × 16 (atomic weight)] = 180 grams. The mole is defined so that 1 mole of any substance always contains 6.023 × 10^{23} particles.

Some isotopes are stable, and others are not. The nuclei of unstable isotopes tend to emit subatomic particles and radiation. Such isotopes are said to be *radioactive* and are called **radioisotopes**. Emissions from radioactive nuclei can be detected by radiation counters. Such emissions can be useful in studying chemical processes, but they also can harm living things.

CHEMICAL BONDS

Chemical bonds form between atoms through interactions of electrons in their outer shells. Energy associated with these bonding electrons holds the atoms together, forming molecules. Three kinds of chemical bonds commonly found in living organisms are ionic, covalent, and hydrogen bonds.

Ionic bonds result from the attraction between ions that have opposite charges. For example, sodium ions, with a positive charge (Na^+) combine with chloride ions, with a negative charge (Cl^-) **(Figure 2.2b)**.

Many compounds, especially those that contain carbon, are held together by **covalent bonds**. Instead of gaining or losing electrons, as in ionic bonding, carbon and some other atoms in covalent bonds share pairs of electrons **(Figure 2.3)**. One carbon atom, which has four electrons in its outer shell, can share an electron with each of four hydrogen atoms. At the same time, each of the four hydrogen atoms shares an electron with the carbon atom. Four pairs of electrons are shared, each pair consisting of one electron from carbon and one electron from hydrogen. Such mutual sharing makes a carbon atom stable with eight electrons in its outer shell, and a hydrogen atom is stable with two electrons in its outer shell. Equal sharing produces *nonpolar compounds*—compounds with no charged regions. Sometimes a carbon atom and an atom such as an oxygen atom share

TABLE 2.3

				Electrons in Outer	
Element	Symbol	Atomic Number	Atomic Weight	Shell	Biological Occurrence
Oxygen	O	8	16.0	6	Component of biological molecules; required for aerobic metabolism
Carbon	C	6	12.0	4	Essential atom of all organic compounds
Hydrogen	H	1	1.0	1	Component of biological molecules; H^+ released by acids
Nitrogen	N	7	14.0	5	Component of proteins and nucleic acids
Calcium	Ca	20	40.1	2	Found in bones and teeth; regulator of many cellular processes
Phosphorus	P	15	31.0	5	Found in nucleic acids, ATP, and some lipids
Sulfur	S	16	32.0	6	Found in proteins; metabolized by some bacteria
Iron	Fe	26	55.8	2	Carries oxygen; metabolized by some bacteria
Potassium	K	19	39.1	1	Important intracellular ion
Sodium	Na	11	23.0	1	Important extracellular ion
Chlorine	Cl	17	35.4	7	Important extracellular ion
Magnesium	Mg	12	24.3	2	Needed by many enzymes
Copper	Cu	29	63.6	1	Needed by some enzymes; inhibits growth of some microorganisms
Iodine	I	53	126.9	7	Component of thyroid hormones
Fluorine	F	9	19.0	7	Inhibits microbial growth
Manganese	Mn	25	54.9	2	Needed by some enzymes
Zinc	Zn	30	65.4	2	Needed by some enzymes; inhibits microbial growth

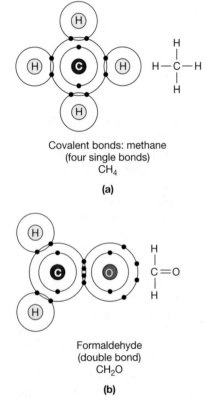

Covalent bonds: methane
(four single bonds)
CH_4

(a)

Formaldehyde
(double bond)
CH_2O

(b)

Figure 2.3 Covalent bonds are formed by sharing electrons. (a) In methane, a carbon atom, with four electrons in its outermost shell, shares pairs of electrons with four hydrogen atoms. In this way all five atoms acquire stable, filled outer shells. Each shared electron pair constitutes a single covalent bond. (b) In formaldehyde, a carbon atom shares pairs of electrons with two hydrogen atoms and also shares two pairs of electrons with an oxygen atom, forming a double covalent bond.

two pairs of electrons to form a double bond. The octet rule still applies, and each atom has eight electrons in its outer shell and is therefore stable. In structural formulas, chemists use a single line to represent a single pair of shared electrons and a double line to represent two pairs of shared electrons (Figure 2.3).

Atoms of four elements—carbon, hydrogen, oxygen, and nitrogen—commonly form covalent bonds that fill their outer electron shells. Carbon shares four electrons, hydrogen one electron, oxygen two electrons, and nitrogen three electrons. Unlike many ionic bonds, covalent bonds are stable and thus are important in molecules that form biological structures.

Hydrogen bonds, though weaker than ionic and covalent bonds, are important in biological structures and are typically present in large numbers. The atomic nuclei of oxygen and nitrogen attract electrons very strongly. When hydrogen is covalently bonded to oxygen or nitrogen, the electrons of the covalent bond are shared unevenly—they are held closer to the oxygen or nitrogen than to the hydrogen. The hydrogen atom then has a partial positive charge, and the other atom has a partial negative charge. In this case of unequal sharing, the molecule is called a **polar compound** because of its oppositely charged regions. The weak attraction between such partial charges is called a hydrogen bond.

Polar compounds such as water often contain hydrogen bonds. In a water molecule, electrons from the hydrogen atoms stay closer to the oxygen atom, and the

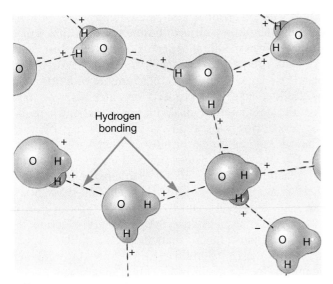

Figure 2.4 **Polar compounds and hydrogen bonding.** Water molecules are polar—they have a region with a partial positive charge (the hydrogen atoms) and a region with a partial negative charge (the oxygen atom). Hydrogen bonds, created by the attraction between oppositely charged regions of different molecules, hold the water molecules together in clusters.

hydrogen atoms lie to one side of the oxygen atom (Figure 2.4). Thus, water molecules are polar molecules that have a positive hydrogen region and a negative oxygen region. Covalent bonds between the hydrogen and oxygen atoms hold the atoms together. Hydrogen bonds between the hydrogen and oxygen regions of different water molecules hold the molecules in clusters.

Hydrogen bonds also contribute to the structure of large molecules such as proteins and nucleic acids, which contain long chains of atoms. The chains are coiled or folded into a three-dimensional configuration that is held together in part by hydrogen bonds.

✓ CHECKLIST

1. Which number tells us the identity of an atom?

2. Is it possible to have a molecule of an element? A molecule of a compound? Give examples.

3. If isotopes can be thought of as "twins," "triplets," "sextuplets," and so on, in what ways are they identical? Different?

4. What type of bonding is produced by the equal sharing of an electron pair between two atoms? By unequal sharing?

CHEMICAL REACTIONS

Chemical reactions in living organisms typically involve the use of energy to form chemical bonds and the release of energy as chemical bonds are broken. For example, the food we eat consists of molecules that have much energy stored in their chemical bonds. During

catabolism (ka-tab′o-lizm), the breakdown of substances, food is degraded and some of that stored energy is released. Microorganisms use nutrients in the same general way. A catabolic reaction can be symbolized as follows:

$$X—Y \rightarrow X + Y + energy$$

where X—Y represents a nutrient molecule and where energy was originally stored in the bond between X and Y.

Catabolic reactions are **exergonic**—that is, they release energy. Conversely, energy is used to form chemical bonds in the synthesis of new compounds. In **anabolism** (a-nab′o-lizm), the buildup, or *synthesis*, of substances, energy is used to create bonds. An anabolic reaction can be symbolized as follows:

$$X + Y + energy \rightarrow X—Y$$

where energy is stored in the new substance X—Y. Anabolic reactions occur in living cells when small molecules are used to synthesize large molecules. Cells can store small amounts of energy for later use or can expend energy to make new molecules. Most anabolic reactions are **endergonic**—that is, they require energy.

▓ WATER AND SOLUTIONS

Water, one of the simplest of chemical compounds, is also one of the most important to living things. It takes part directly in many chemical reactions. Numerous substances dissolve in water or form mixtures called colloidal dispersions. Acids and bases exist and function principally in water mixtures.

WATER

Water is so essential to life that humans can live only a few days without it. Many microorganisms die almost immediately if removed from their normal aqueous environments, such as lakes, ponds, oceans, and moist soil. Yet, others can survive for several hours or days without water, and spores formed by a few microorganisms survive for many years away from water. Several kinds of bacteria find the moist, nutrient-rich secretions of human skin glands to be an ideal environment.

Water has several properties that make it important to living things. Because water is a polar compound and forms hydrogen bonds, it can form thin layers on surfaces and can act as a *solvent*, or dissolving medium. Water is a good solvent for ions because the polar water molecules surround the ions. The positive region of water molecules is attracted to negative ions, and the negative region of water molecules is attracted to positive ions. Many different kinds of ions can therefore be distributed evenly through a water medium, forming a *solution* (Figure 2.5).

Water forms thin layers because it has a high surface tension. **Surface tension** is a phenomenon in which the

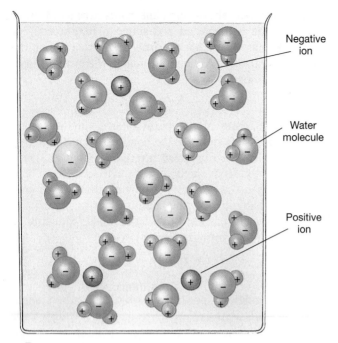

Figure 2.5 Polarity and water molecules. Polarity enables water to dissolve many ionic compounds. The positive regions of the water molecules surround negative ions, and the negative regions of the water molecules surround positive ions, holding the ions in solution.

surface of water acts as a thin, invisible, elastic membrane **(Figure 2.6)**. The polarity of water molecules gives them a strong attraction for one another but no attraction for gas molecules in air at the water's surface. Therefore, surface water molecules cling together, forming hydrogen

(a)

(b)

Figure 2.6 Surface Tension. (a) Hydrogen bonding between water molecules creates surface tension, which causes the surface of water to behave like an elastic membrane. **(b)** The surface tension of water is strong enough to support the weight of the insects known as water striders. *(Biophoto Associates/ Photo Researchers, Inc.)*

bonds with other molecules below the surface. In living cells this feature of surface tension allows a thin film of water to cover membranes and to keep them moist.

Water has a high *specific heat*; that is, it can absorb or release large quantities of heat energy with little temperature change. This property of water helps to stabilize the temperature of living organisms, which are composed mostly of water, as well as bodies of water where many microorganisms live.

Finally, water provides the medium for most chemical reactions in cells, and it participates in many of these reactions. Suppose, for example, that substance X can gain or lose H^+ and that substance Y can gain or lose OH^-. The substances that enter a reaction are called **reactants**. In an anabolic reaction, the components of water (H^+ and OH^-) are removed from the reactants to form a larger product molecule:

$$X{-}H + HO{-}Y \rightarrow X{-}Y + H_2O$$

This kind of reaction, called **dehydration synthesis**, is involved in the synthesis of complex carbohydrates, some lipids (fats), and proteins. Conversely, in many catabolic reactions, water is added to a reactant to form simple products:

$$X{-}Y + H_2O \rightarrow X{-}H + HO{-}Y$$

This kind of reaction, called **hydrolysis**, occurs in the breakdown of large nutrient molecules to release simple sugars, fatty acids, and amino acids.

SOLUTIONS AND COLLOIDS

Solutions and colloidal dispersions are examples of mixtures. Unlike a chemical compound, which consists of molecules whose atoms are present in specific proportions, a **mixture** consists of two or more substances that are combined in any proportion and are not chemically bound. Each substance in a mixture contributes its properties to the mixture. For example, a mixture of sugar and salt could be made by using any proportions of the two ingredients. The degree of sweetness or saltiness of the mixture would depend on the relative amounts of sugar and salt present, but both sweetness and saltiness would be detectable.

A **solution** is a mixture of two or more substances in which the molecules of the substances are evenly distributed and ordinarily will not separate out upon standing. In a solution the medium in which substances are dissolved is the **solvent**. The substance dissolved in the solvent is the **solute**. Solutes can consist of atoms, ions, or molecules. In cells and in the medium in which cells live, water is the solvent in nearly all solutions. Typical solutes include the sugar glucose, the gases carbon dioxide and oxygen, and many different kinds of ions. Many smaller proteins also can act as solutes in true solutions.

Few living things can survive in highly concentrated solutions. We make use of this fact in preserving several kinds of foods. Can you think of foods that are often kept unrefrigerated and unsealed for long periods of time? Jellies, jams, and candies do not readily spoil because most microorganisms cannot tolerate the high concentration of sugar. Salt-cured meats are too salty to allow growth of most microorganisms, and pickles are too acidic for most microbes.

Particles too large to form true solutions can sometimes form *colloidal dispersions*, or **colloids**. Gelatin dessert is a colloid in which the protein gelatin is dispersed in a watery medium. Similarly, colloidal dispersions in cells usually are formed from large protein molecules dispersed in water. The fluid or semifluid substance inside living cells is a complex colloidal system.

Large particles are suspended by opposing electrical charges, layers of water molecules around them, and other forces. Media for growing microorganisms sometimes are solidified with agar; these media are colloidal dispersions. Some colloidal systems have the ability to change from a semisolid state, such as gelatin that has "set," to a more fluid state, such as gelatin that has melted. Amoebae seem to move, in part, by the ability of the colloidal material within them to change back and forth between semisolid and fluid states.

ACIDS, BASES, AND pH

In chemical terms, most living things exist in relatively neutral environments, but some microorganisms live in environments that are *acidic* or *basic* (*alkaline*). Understanding acids and bases is important in studying microorganisms and their effects on human cells. An **acid** is a hydrogen ion (H^+) donor. (A hydrogen ion is a proton.) An acid donates H^+ to a solution. The acids found in living organisms usually are weak acids such as acetic acid (vinegar), although some are strong acids such as hydrochloric acid. Acids release H^+ when carboxyl groups (—COOH) ionize to COO^- and H^+. A **base** is a proton acceptor, or a hydroxyl ion donor. It accepts H^+ from, or donates OH^- (hydroxyl ion) to, the solution. The bases found in living organisms usually are weak bases such as the amino (NH_2) group, which accepts H^+ to form NH_3^+.

Chemists have devised the concept of pH to specify the acidity or alkalinity of a solution. The **pH** scale **(Figure 2.7)**, which relates proton concentrations to pH, is a logarithmic scale. This means that the concentration of hydrogen ions (protons) changes by a factor of 10 for each unit of the scale. The practical range of the pH

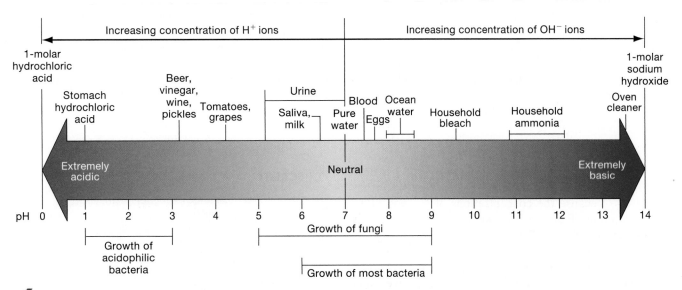

Figure 2.7 The pH values of some common substances. Each unit of the pH scale represents a 10-fold increase or decrease in the concentration of hydrogen ions. Thus, vinegar, for example, is 10,000 times more acidic than pure water.

APPLICATIONS

Bacterial Acids Are Eating The Last Supper

Past civilizations built their temples, tombs, and monuments out of stone to last forever. Indeed, structures such as the Egyptian pyramids have lasted for thousands of years. But recently, the ancient stone has begun to crumble into dust. Experts first believed that the destruction of stone was due to the gaseous pollutants spewed into the air from automobiles and industrial chimneys. Now, however, experts have found that a whole new world—a microsystem of microbes—is living in the stone itself, interacting with the pollutant gases and wreaking chemical havoc.

The prime villain devouring ancient marble is *Thiobacillus thioparis*, a bacterium that converts the pollutant sulfur dioxide (SO_2) gas into sulfuric acid (H_2SO_4). The sulfuric acid acts on the calcium carbonate ($CaCO_3$) in marble, forming carbon dioxide (CO_2) and the salt calcium sulfate ($CaSO_4$), which is a form of plaster. The bacteria use the CO_2 as their source of carbon. The plaster produced by this process is soft and washes away by rain or just crumbles and falls off. On many buildings, such as the Parthenon in Greece, this bacterial "epidemic" has already eroded a 2-in.-thick layer of marble surface into plaster. In fact, more marble has been destroyed in the last 35 years than had been destroyed in the previous 300 years.

Not all the damage, however, is caused by chemical processes. Some is due to the physical activities of fungi that are pushing their threadlike growth (hyphae) into the rock, splitting and reducing it to powder. Dozens of kinds of bacteria, yeast, filamentous fungi, and even algae produce acids that eventually eat into the stone. Even paintings aren't immune to the effects of microbial growth. Frescoes painted on walls, such as Leonardo da Vinci's *Last Supper*, are also being consumed, pieces flaking off and brilliant colors being dulled.

Is there a cure for these microbial invasions? A treatment plan identifies which microorganisms are attacking the stone and then determines which antibiotics will kill them most effectively. Administering antibiotics to a building is a tricky business. A spray gun must be used instead of a hypodermic needle or pills. But some antibiotics cannot be applied by spraying, so it is difficult to find a suitable one. Sometimes

A close-up of Leonardo da Vinci's *The Last Supper* shows deterioration due to microbial action. Santa Maria della Grazie, Milano. *(Scala/ Art Resource)*

disinfectants such as isothiazolinone chloride are used, but such chemicals must be used carefully. In Angkor, Cambodia, for instance, unskilled Army workers armed with hard brushes and fungicidal solutions have scrubbed away ancient temple paintings in their zeal to remove fungal growth.

Unfortunately, antibiotic treatment does not reverse damage already done. Statues cannot grow a new skin, as you might do after recovering from an infection. Therefore, researchers are working on a process to harden the plaster layer formed by microbes by baking the affected pieces at high temperatures. This method, however, is hardly feasible for an entire statue or a cathedral. And how do we protect these vulnerable structures as long as the pollutants remain in the atmosphere? Stopgap measures have included taking some statues inside and building protective domes over others. The real—but elusive—answer lies in cleaning up our environment.

scale is from 0 to 14. A solution with a pH of 7 is **neutral**—neither acidic nor **alkaline** (basic). Pure water has a pH of 7 because the concentrations of H^+ and OH^- in it are equal. Figure 2.7 shows the pH of some body fluids, selected foods, and other substances. The hydrochloric acid in your stomach digests your meal as well as most bacteria that may be on or in the food. People who lack stomach acid get far more digestive tract infections.

✓ CHECKLIST

1. What is the difference between endergonic and exergonic chemical reactions?

2. What properties of a water molecule enable it to act as a good solvent for ionic molecules?

3. Why does a higher number, such as pH 11, indicate a stronger base than pH 9, whereas a higher number, such as pH 5, indicates a weaker acid than pH 3?

Figure 2.8 Four classes of organic compounds that contain oxygen. Alcohols contain one or more hydroxyl groups (—OH), aldehydes and ketones contain carbonyl groups (—C=O), and organic acids contain carboxyl groups (—COOH).

COMPLEX ORGANIC MOLECULES

The basic principles of general chemistry also apply to **organic chemistry**, the study of compounds that contain carbon. The study of the chemical reactions that occur in living systems is the branch of organic chemistry known as **biochemistry**. Early in the 1800s it was believed that molecules from living things were filled with a supernatural "vital force" and therefore could not be explained by the laws of chemistry and physics. It was considered impossible to make *organic compounds* outside of living systems. That idea was disproved in 1828 when the German scientist Friedrich Wohler synthesized the organic compound urea, a small molecule excreted as a waste material by many animals. Since that time thousands of organic compounds—plastics, fertilizers, and medicines—have been made in the laboratory. Organic compounds such as carbohydrates, lipids, proteins, and nucleic acids occur naturally in living things and in the products or remains of living things. The ability of carbon atoms to form covalent bonds and to link up in long chains makes possible the formation of an almost infinite number of organic compounds.

The simplest carbon compounds are the *hydrocarbons*, chains of carbon atoms with their associated hydrogen atoms. The structure of the hydrocarbon propane, C_3H_8, for example, is as follows:

Carbon chains can have not only hydrogen but also other atoms such as oxygen and nitrogen bound to them. Some of these atoms form functional groups. A **functional group** is a part of a molecule that generally participates in chemical reactions as a unit and that gives the molecule some of its chemical properties.

Four significant groups of compounds—alcohols, aldehydes, ketones, and organic acids—have functional groups that contain oxygen (**Figure 2.8**). An alcohol has one or more hydroxyl groups (—OH). An aldehyde has a carbonyl group (—CO) at the end of the carbon chain; a ketone has a carbonyl group within the chain. An organic acid has one or more carboxyl groups (—COOH). One key functional group that does not contain oxygen is the amino group (—NH_2). Found in amino acids, amino groups account for the nitrogen in proteins.

The relative amount of oxygen in different functional groups is significant. Groups with little oxygen, such as alcohol groups, are said to be *reduced*; groups with relatively more oxygen, such as carboxyl groups, are said to be *oxidized* (Figure 2.8). As we shall see in ◄Chapter 5, *oxidation* is the addition of oxygen or the removal of hydrogen or electrons from a substance. Burning is an example of oxidation. *Reduction* is the removal of oxygen or the addition of hydrogen or electrons to a substance. In general, the more reduced a molecule, the more energy it contains. Hydrocarbons, such as gasoline, have no oxygen and thus represent the extreme in energy-rich, reduced molecules. They make good fuels because they contain so much energy. Conversely, the more oxidized a molecule, the less energy it contains. Carbon dioxide (CO_2) represents the extreme in an oxidized molecule because no more than two oxygen atoms can bond to a single carbon atom. As we shall see, oxidation releases energy from molecules.

Let us now consider the major classes of large, complex biochemical molecules of which all living things, including microbes, are composed.

Figure 2.9 **Isomers.** Glucose and fructose are isomers: They contain the same atoms, but they differ in structure.

Glucose
($C_6H_{12}O_6$)

Fructose
($C_6H_{12}O_6$)

Figure 2.10 **Three ways of representing the glucose molecule.** **(a)** In solution, the straight-chain form is rarely found. **(b)** Instead, the molecule bonds to itself, forming a six-membered ring. The ring is conventionally depicted as a flat hexagon. **(c)** The actual three-dimensional structure is more complex. The spheres in this depiction represent carbon atoms.

(a)   (b)   (c)

CARBOHYDRATES

Carbohydrates serve as the main source of energy for most living things. Plants make carbohydrates, including structural carbohydrates such as cellulose, and energy-storage carbohydrates such as starch. Animals use carbohydrates as food, and many, including humans, store energy in a carbohydrate called *glycogen*. Many microorganisms use carbohydrates from their environment for energy and also make a variety of carbohydrates. Carbohydrates in the membranes of cells can act as markers that make a cell chemically recognizable. Chemical recognition is important in immunological reactions and other processes in living things.

All carbohydrates contain the elements carbon, hydrogen, and oxygen, generally in the proportion of two hydrogen atoms to each carbon and oxygen. There are three groups of carbohydrates: monosaccharides, disaccharides, and polysaccharides. **Monosaccharides** consist of a carbon chain or ring with several alcohol groups

and one other functional group, either an aldehyde group or a ketone group. Several monosaccharides, such as glucose and fructose, are **isomers**—they have the same molecular formula, $C_6H_{12}O_6$, but different structures and different properties **(Figure 2.9)**. Thus, even at the chemical level we can see that structure and function are related.

Glucose, the most abundant monosaccharide, can be represented schematically as a straight chain, a ring, or a three-dimensional structure. The chain structure, in **Figure 2.10a**, clearly shows a carbonyl group at carbon 1 (the first carbon in the chain, at the top in this orientation) and alcohol groups on all the other carbons. **Figure 2.10b** shows how a glucose molecule in solution rearranges and bonds to itself to form a closed ring. The three-dimensional projection in **Figure 2.10c** more closely approximates the actual shape of the molecule. In studying structural formulas, it is important to imagine each molecule as a three-dimensional object.

Monosaccharides can be reduced to form deoxy sugars and sugar alcohols **(Figure 2.11)**. The deoxy sugar

(a) Deoxyribose   **(b)** Ribose   **(c)** Glycerol   **(d)** Mannitol

Figure 2.11 **Deoxy sugars and sugar alcohols.** **(a)** "Deoxy" indicates one less oxygen atom—one of the carbon atoms of the deoxy sugar deoxyribose lacks a hydroxyl group that **(b)** ribose has. **(c)** Glycerol is a three-carbon sugar alcohol that is a component of fats. **(d)** Mannitol is a sugar alcohol used in diagnostic tests for certain microbes.

deoxyribose, which has a hydrogen atom instead of an —OH group on one of its carbons, is a component of DNA. Certain sugar alcohols, which have an additional alcohol group instead of an aldehyde or ketone group, can be metabolized by particular microorganisms. Mannitol and other sugar alcohols are used to identify some microorganisms in diagnostic tests.

Disaccharides are formed when two monosaccharides are connected by the removal of water and the formation of a **glycosidic bond**, a sugar alcohol/sugar linkage (**Figure 2.12a**). Sucrose, common table sugar, is a disaccharide made of glucose and fructose. **Polysaccharides** are formed when many monosaccharides are linked by glycosidic bonds (**Figure 2.12b**). Polysaccharides such as starch, glycogen, and cellulose are **polymers**—long chains of repeating units—of glucose. However, the glycosidic bonds in each polymer are arranged differently. Plants and most algae make starch and cellulose. Starch serves as a way to store energy, and cellulose is a structural component of cell walls. Animals make and store glycogen, which they can break down to glucose as energy is needed. Microorganisms contain several other important polysaccharides, as we shall see in later chapters.

Table 2.4 summarizes the types of carbohydrates.

LIPIDS

Lipids constitute a chemically diverse group of substances that includes fats, phospholipids, and steroids. They are relatively insoluble in water but are soluble in nonpolar solvents such as ether and benzene. Lipids form part of the structure of cells, especially cell membranes, and many can be used for energy. Generally, lipids contain relatively more hydrogen and less oxygen than carbohydrates and therefore contain more energy than carbohydrates.

Fats contain the three-carbon alcohol glycerol and one or more fatty acids. A **fatty acid** consists of a long chain of carbon atoms with associated hydrogen atoms and a carboxyl group at one end of the chain. The

The higher the fat content of food, the slower the movement of feces through the bowels, where bacteria convert the undigested fats into cancer-causing compounds.

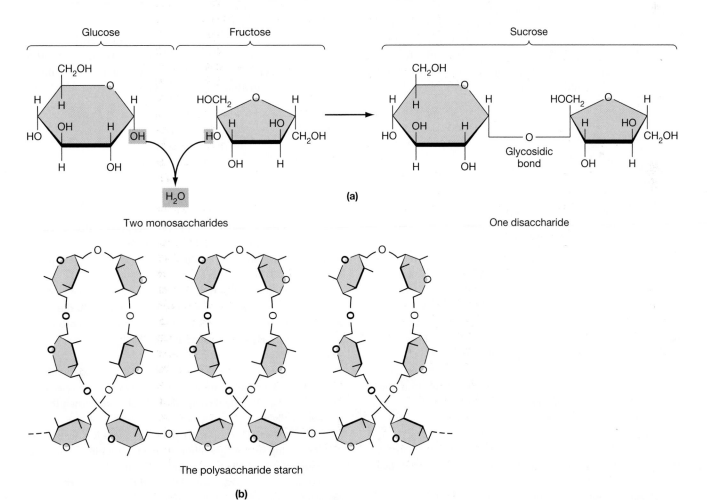

(a)

Two monosaccharides   One disaccharide

The polysaccharide starch

(b)

Figure 2.12 Disaccharides and polysaccharides. (a) Two monosaccharides are joined to form a disaccharide by dehydration synthesis and the formation of a glycosidic bond. **(b)** Polysaccharides such as starch are formed by similar reactions that link many monosaccharides into long chains.

TABLE 2.4

Types of Carbohydrates		
Class of Carbohydrates	**Examples**	**Description and Occurrence**
Monosaccharides	Glucose	Sugar found in most organisms
	Fructose	Sugar found in fruit
	Galactose	Sugar found in milk
	Ribose	Sugar found in RNA
	Deoxyribose	Sugar found in DNA
Disaccharides	Sucrose	Glucose and fructose; table sugar
	Lactose	Glucose and galactose; milk sugar
	Maltose	Two glucose units; product of starch digestion
Polysaccharides	Starch	Polymer of glucose stored in plants, digestible by humans
	Glycogen	Polymer of glucose stored in animal liver and skeletal muscles
	Cellulose	Polymer of glucose found in plants, not digestible by humans; digested by some microbes

Ticks' body walls are impervious to most pesticides thanks to chitin, a polysaccharide, also found in fungal cell walls.

synthesis of a fat from glycerol and fatty acids involves removing water and forming an ester bond between the carboxyl group of the fatty acid and an alcohol group of glycerol (**Figure 2.13a**). A **triacylglycerol**, formerly called a *triglyceride*, is a fat formed when three fatty acids are bonded to glycerol. *Monacylglycerols* (monoglycerides) and *diacylglycerols* (diglycerides) contain one and two fatty acids, respectively, and usually are formed from the digestion of triacylglycerols.

Fatty acids can be saturated or unsaturated. A **saturated fatty acid** contains all the hydrogen it can have; that is, it is saturated with hydrogen (**Figure 2.13b**). An

APPLICATIONS

Can a Cow Actually Explode?

Cows can derive a good deal of nourishment from grass, hay, and other fibrous vegetable matter that are inedible to humans. We can't digest cellulose, the chief component of plants. If you had to live on hay, you would probably starve to death. How, then, do cows and other hooved animals manage on such a diet?

Oddly enough, cows can't digest cellulose either. But they don't need to—it's done for them. Cows and their relatives harbor in their stomachs large populations of microorganisms that do the work of breaking down cellulose into sugars that the animal can use. The same is true of termites: If it weren't for microbes in their guts that help them digest cellulose, they couldn't dine on the wooden beams in your house.

Cellulose is very similar to starch—both consist of long chains of glucose molecules. The bonds between these molecules, however, are slightly different in their geometry in the two substances. As a result, the enzymes that animals use to break down a starch molecule into its component glucose units have no effect on cellulose. In fact, very few organisms produce enzymes that can attack cellulose. Even the protists (unicellular organisms with a nucleus) that live in the stomachs

of cows and termites cannot always do it by themselves. Just as cows and termites depend on the protists in their stomachs, the protists frequently depend on certain bacteria that reside permanently within them. It is these bacteria that actually make the essential digestive enzymes.

The activities of the intestinal microorganisms that perform these digestive services are a mixed blessing, both to the cows and to the humans who keep them. The bacteria also produce methane gas, CH_4—as much as 190 to 380 liters per day from a single cow. Methane production can be so rapid that a cow's stomach may rupture if the cow can't burp. Some ingenious inventors have actually patented cow safety valves to release the gas buildup directly through the animal's side. When this gas eventually makes its way out of the cow by one route or another, it rises to the upper atmosphere. There it is suspected of contributing to the "greenhouse effect," trapping solar heat and causing an overall warming of the Earth's climate ◄(Chapter 25). Scientists have estimated that the world's cows release 50 million metric tons of methane annually—and that's not counting what the sheep, goats, antelope, water buffalo, and other grass eaters release!

Figure 2.13 **The structure of fats.** **(a)** Three fatty acids combine with glycerol to form a molecule of triacylglycerol, a type of fat. The group designated R is a long hydrocarbon chain that varies in length in different fatty acids. It may be saturated or unsaturated. **(b)** Saturated fatty acids have only single covalent bonds between carbon atoms in their carbon chains and can therefore accommodate the maximum possible number of hydrogens. **(c)** Unsaturated fatty acids, such as oleic acid, have one or more double bonds between carbons and thus contain fewer hydrogens. The double bond causes a bend in the carbon chain. In (b) and (c), both structural formulas and space-filling models are shown.

unsaturated fatty acid has lost at least two hydrogen atoms and contains a double bond between the two carbons that have lost hydrogen atoms **(Figure 2.13c)**. "Unsaturated" thus means not completely saturated with hydrogen. Oleic acid is an unsaturated fatty acid. *Polyunsaturated fats*, many of which are vegetable oils that remain liquid at room temperature, contain many unsaturated fatty acids.

Some lipids contain one or more other molecules in addition to fatty acids and glycerol. For example, **phospholipids**, which are found in all cell membranes, differ from fats by the substitution of phosphoric acid (H_3PO_4) for one of the fatty acids **(Figure 2.14a)**. The charged phosphate group ($-HPO_4^-$) is typically attached to another charged group. Both can mix with water, but the fatty acid end cannot **(Figure 2.14b)**.

Such properties of phospholipids are important in determining the characteristics of cell membranes ◀(Chapter 4).

Steroids have a four-ring structure **(Figure 2.15a)** and are quite different from other lipids. They include cholesterol, steroid hormones, and vitamin D. Cholesterol **(Figure 2.15b)** is insoluble in water and is found in the cell membranes of animal cells and the group of bacteria called mycoplasmas. Steroid hormones and vitamin D are important in many animals.

PROTEINS

Properties of Proteins and Amino Acids

Among the molecules found in living things, proteins have the greatest diversity of structure and function.

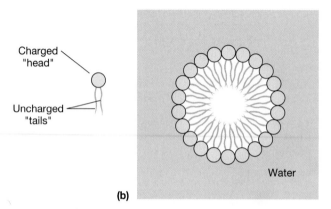

$$CH_3-CH_2-CH_2-CH_2-CH_2-CH_2-CH_2-CH_2-CH_2-CH_2-CH_2-CH_2-CH_2-C$$
$$CH_3-CH_2-CH_2-CH_2-CH_2-CH_2-CH_2-CH_2-CH=CH-CH_2-CH_2-CH_2-CH_2-CH_2-C$$

Uncharged fatty acid chains

Other charged group

Charged phosphate group

Glycerol portion

(a)

Charged "head"

Uncharged "tails"

Water

(b)

Figure 2.14   Phospholipids. (a) In phospholipids, one of the fatty acid chains of a fat molecule is replaced by phosphoric acid. The charged phosphate group and another attached group can interact with water molecules, which are polar, but the two long, uncharged fatty acid tails cannot. **(b)** As a result, phospholipid molecules in water tend to form globular structures with the phosphate groups facing outward and the fatty acids in the interior.

Proteins are composed of building blocks called **amino acids**, which have at least one amino ($-NH_2$) group and one acidic carboxyl ($-COOH$) group. The general structures of an amino acid and some of the 20 amino acids found in proteins are shown in **Figure 2.16**. Each amino acid is distinguishable by a different chemical group, called an **R group**, attached to the central carbon atom. Because all amino acids contain carbon, hydrogen, oxygen, and nitrogen, and some contain sulfur, proteins also contain these elements.

A protein is a polymer of amino acids joined by **peptide bonds**—that is, covalent bonds that link an amino group of one amino acid and a carboxyl group of another amino acid **(Figure 2.17)**. Two amino acids

Steroid

Cholesterol

Side chain

(a)   **(b)**

Figure 2.15   Steroids. (a) Steroids are lipids with a characteristic four-ring structure. The specific chemical groups attached to the rings determine the properties of different steroids. **(b)** One of the most biologically important steroids is cholesterol, a component of the membranes of animal cells and one group of bacteria.

APPLICATIONS

Down the Drain

You have washed enough dishes to know that soaps get oily substances off the dishes and into the water. Although you probably don't think about how soap works, soap chemistry is an application of something you have learned in this chapter. Because water is a polar molecule, it has a high surface tension and beads on clean surfaces. To make water "wetter," it is necessary to reduce the surface tension by adding surfactants. Soaps are anionic surfactants made from fats and oils treated with strong alkali. This process produces a complex molecule with a charged carboxyl group at one end and a nonionized saturated hydrocarbon at the other. The saturated hydrocarbon portion of the soap molecule mixes with the fats on dishes while the charged carboxyl group mixes with the dishwater. The chemical interaction between the grease, soap, and water loosens the food particles off your dishes and carries them down the drain.

linked together make a *dipeptide*, three make a *tripeptide*, and many make a **polypeptide**. In addition to the amino and carboxyl groups, some amino acids have an R group called a *sulfhydryl group* (—SH). Sulfhydryl groups in adjacent chains of amino acids can lose hydrogen and form *disulfide linkages* (—S—S—) from one chain to the other.

The Structure of Proteins

Proteins have several levels of structure. The **primary structure** of a protein consists of the specific sequence of amino acids in a polypeptide chain (**Figure 2.18a**). The **secondary structure** of a protein consists of the folding or coiling of amino acid chains into a particular pattern, such as a helix or pleated sheet (**Figure 2.18b**). Hydrogen bonds are responsible for such patterns. Further bending and folding of the protein molecule into globular (irregular spherical) shapes or fibrous thread-like strands produces the **tertiary structure** (**Figure 2.18c**). Some large proteins such as hemoglobin have **quaternary structure**, formed by the association of several tertiary-structured polypeptide chains (**Figure 2.19**). Tertiary and quaternary structures are maintained by disulfide linkages, hydrogen bonds, and other forces between R groups of amino acids. The three-dimensional shapes of protein molecules and the nature of sites at which other molecules can bind to them are extremely important in determining how proteins function in living organisms.

Several conditions can disrupt hydrogen bonds and other weak forces that maintain protein structure. They include highly acidic or basic conditions and temperatures above 50°C. Such disruption of secondary, tertiary,

Figure 2.16 Amino acids. (a) The general structure of an amino acid, and **(b)** six representative examples. All amino acids have four groups that are attached to the central carbon atom: an amino (—NH$_2$) group, a carboxyl (—COOH) group, a hydrogen atom, and a group designated R that is different in each amino acid. The R group determines many of the chemical properties of the molecule—for example, whether it is nonpolar, polar, acidic, or basic.

Figure 2.17 **Peptide linkage.** Two amino acids are joined by the removal of a water molecule (dehydration synthesis) and the formation of a peptide bond between the —COOH group of one and the —NH₂ group of the other.

Figure 2.18 **Three levels of protein structure.** **(a)** Primary structure is the sequence of amino acids (A1, A2, etc.) in a polypeptide chain. Imagine it as a straight telephone cord. **(b)** Polypeptide chains, especially those of structural proteins, tend to coil or fold into a few simple, regular, three-dimensional patterns called secondary structure. Imagine the telephone cord as a coiled cord. **(c)** Polypeptide chains of enzymes and other soluble proteins may also exhibit secondary structure. In addition, the chains tend to fold up into complex, globular shapes that constitute the protein's tertiary structure. Imagine the knot formed when a coiled telephone cord tangles. *(Courtesy Jacquelyn G. Black)*

Figure 2.19
Quaternary protein structure. **(a)** Many large proteins such as hemoglobin, which carries oxygen in human red blood cells, are made up of several polypeptide chains. The arrangement of these chains makes up the protein's quaternary structure. **(b)** Keratin, a component of human skin and hair, also consists of several polypeptide chains and so has quaternary structure.

and quaternary structures is called **denaturation**. Sterilization and disinfection procedures often make use of heat or chemicals that kill microorganisms by denaturing their proteins. Also, the cooking of meat tenderizes it by denaturing proteins. Therefore, microbes and cells of larger organisms must be maintained within fairly narrow ranges of pH and temperature to prevent disruption of protein structure.

Classification of Proteins

Most proteins can be classified by their major functions as either structural proteins or enzymes. **Structural proteins**, as the name implies, contribute to the three-dimensional structure of cells, cell parts, and membranes. Certain proteins, called *motile proteins*, contribute both to structure and to movemnt. They account for the contraction of animal muscle cells and for some kinds of movement in microbes. **Enzymes** are protein *catalysts*—substances that control the rate of chemical reactions in cells. A few proteins are neither structural proteins nor enzymes. They include proteins that form receptors for certain substances on cell membranes and antibodies that participate in the body's immune responses ◀(Chapter 17).

Enzymes

Enzymes increase the rate at which chemical reactions take place within living organisms in the temperature range compatible with life. We discuss enzymes in more detail in ◀Chapter 5 but summarize their properties here. In general, enzymes speed up reactions by decreasing the energy required to start reactions. They also hold reactant molecules close together in the proper orientation for reactions to occur. Each enzyme has an **active site**, which is the site at which it combines with its **substrate**, the substance on which an enzyme acts. Enzymes have **specificity**—that is, each enzyme acts on a particular substrate or on a certain kind of chemical bond.

Like catalysts in inorganic chemical reactions, enzymes are not permanently affected or used up in the reactions they initiate. Enzyme molecules can be used over and over again to catalyze a reaction, although not indefinitely. Because enzymes are proteins, they are denatured by extremes of temperature and pH. However, some microbes that live in extreme conditions of high temperature or very acidic environments, have enzymes that can resist these conditions.

NUCLEOTIDES AND NUCLEIC ACIDS

The chemical properties of *nucleotides* allow these compounds to perform several essential functions. One key function is storage of energy in **high-energy bonds**—bonds that, when broken, release more energy than do most covalent bonds. Nucleotides joined to form *nucleic acids* are, perhaps, the most remarkable of all biochemical substances. They store information that directs protein synthesis and that can be transferred from parent to progeny.

A **nucleotide** consists of three parts: (1) a nitrogenous base, so named because it contains nitrogen and has alkaline properties; (2) a five-carbon sugar; and (3) one or more phosphate groups, as **Figure 2.20a**

Figure 2.20 Nucleotides. (a) A nucleotide consists of a nitrogenous base, a five-carbon sugar, and one or more phosphate groups. **(b)** A nucleoside is comprised of the sugar and base without the phosphates. **(c)** The nucleotide adenosine triphosphate (ATP) is the immediate source of energy for most activities of living cells. In ATP, the base is adenine, and the sugar is ribose. Adding a phosphate group to adenosine diphosphate greatly increases the energy of the molecule; removal of the third phosphate group releases energy that can be used by the cell.

shows for the nucleotide *adenosine triphosphate (ATP)*. The sugar and base alone make up a *nucleoside* (**Figure 2.20b**).

The nucleotide ATP is the main source of energy in cells because it stores chemical energy in a form cells can use. The bonds between phosphates in ATP that are high-energy bonds are designated by wavy lines (**Figure 2.20c**). They contain more energy than most covalent bonds, in that more energy is released when they are broken. Enzymes control the forming and breaking of high-energy bonds so that energy is released as needed within cells. The capture, storage, and use of energy is an important component of cellular metabolism ◄(Chapter 5).

Nucleic acids consist of long polymers of nucleotides, called **polynucleotides**. They contain genetic information that determines all the heritable characteristics of a living organism, be it a microbe or a human. Such information is passed from generation to generation and directs protein synthesis in each organism. By directing protein

synthesis, nucleic acids determine which structural proteins and enzymes an organism will have. The enzymes determine what other substances the organism can make and what reactions it can carry out.

The two nucleic acids found in living organisms are **ribonucleic acid (RNA)** and **deoxyribonucleic acid (DNA)**. Except in a few viruses, RNA is a single polynucleotide chain, and DNA is a double chain of polynucleotides arranged as a double helix. In both nucleic acids, the phosphate and sugar molecules form a sturdy but inert "backbone" from which nitrogenous bases protrude. In DNA each chain is connected by hydrogen bonds between the bases, so the whole molecule resembles a ladder with many rungs (**Figure 2.21a**).

DNA and RNA contain somewhat different building blocks (**Table 2.5**). RNA contains the sugar ribose, whereas DNA contains deoxyribose, which has one less oxygen atom than ribose. Three nitrogenous bases, adenine, cytosine, and guanine, are found in both DNA and RNA. In addition, DNA contains the base

Figure 2.21 **Nucleic acid structure.** Nucleic acids consist of a backbone of alternating sugar and phosphate groups to which nitrogenous bases are attached. **(a)** RNA is usually single-stranded. DNA molecules typically consist of two chains held together by hydrogen bonds between bases. **(b)** The complementary base pairs in DNA, showing how hydrogen bonds are formed.

TABLE 2.5

Components of DNA and RNA			
Component		**DNA**	**RNA**
Sugars	Phosphoric acid	X	X
	Ribose		X
	Deoxyribose	X	
Bases	Adenine	X	X
	Guanine	X	X
	Cytosine	X	X
	Thymine	X	
	Uracil		X

thymine, and RNA contains the base uracil. Of these bases, adenine and guanine are **purines**, nitrogenous base molecules that contain double-ring structures, and thymine, cytosine, and uracil are **pyrimidines**, nitrogenous base molecules that contain a single-ring structure **(Figure 2.22)**. All cellular organisms have both DNA and RNA. Viruses have either DNA or RNA but not both.

The two nucleotide chains of DNA are held together by hydrogen bonds between the bases and by other forces. The hydrogen bonds always connect adenine to thymine and cytosine to guanine, as shown in **Figure 2.21b**. This linking of specific bases is called **complementary base pairing**. It is determined by the sizes and shapes of the bases. The same kind of complementary base pairing also occurs when information is transmitted from DNA to RNA at the beginning of protein synthesis ◄(Chapter 7). In that situation, adenine in DNA base pairs with uracil in RNA.

DNA and RNA chains contain hundreds or thousands of nucleotides with bases arranged in a particular sequence. This sequence of nucleotides, like the sequence of letters in words and sentences, contains information that determines what proteins an organism will have. As noted earlier, an organism's structural proteins and enzymes, in turn, determine what the organism is and what it can do. Like changing a letter in a word, changing a nucleotide in a sequence can change the information it carries. The number of different possible sequences of bases is almost infinite, so DNA and RNA can contain a great many different pieces of information.

The functions of DNA and RNA are related to their ability to convey information. DNA is transmitted from one generation to the next. It determines the heritable characteristics of the new individual by supplying the information for the proteins its cells will contain. In contrast, RNA carries information from the DNA to the sites where proteins are manufactured in cells. There it directs and participates in the actual assembly of proteins. The functions of nucleic acids are discussed in more detail in ◄Chapters 7 and 8.

✔**CHECKLIST**

1. Why are starch, DNA, and RNA all considered to be polymers?
2. Distinguish among primary, secondary, tertiary, and quaternary levels of protein structure.
3. How could you distinguish between carbohydrates, lipids, proteins, and nucleic acids?

Purines   **Pyrimidines**

Adenine

Thymine
(in DNA only)

Cytosine

Guanine

Uracil
(in RNA only)

Figure 2.22 The five bases found in nucleic acids. DNA contains the purines adenine and guanine and the pyrimidines cytosine and thymine. In RNA, thymine is replaced by the pyrimidine uracil.

▍ RETRACING OUR STEPS

III WHY STUDY CHEMISTRY?

- A knowledge of basic chemistry is needed to understand how microorganisms function and how they affect humans and our environment.

III CHEMICAL BUILDING BLOCKS AND CHEMICAL BONDS

CHEMICAL BUILDING BLOCKS

- The smallest chemical unit of matter is an **atom**. An **element** is a fundamental kind of matter, and the smallest unit of an element is an atom. An element is composed of only one type of atom. A **molecule** consists of two or more atoms of the same or different kinds that are chemically combined. A molecule of a **compound** consists of two or more different kinds of atoms that are chemically combined.
- The most common elements in all forms of life are carbon (C), hydrogen (H), oxygen (O), and nitrogen (N).

THE STRUCTURE OF ATOMS

- Atoms consist of positively charged **protons** and neutral **neutrons** in the atomic nucleus and very small, negatively charged **electrons** orbiting the nucleus.
- The number of protons in an atom is equal to its **atomic number**. The total number of protons and neutrons determines the element's **atomic weight**.
- **Ions** are atoms or molecules that have gained or lost one or more electrons.
- **Isotopes** are atoms of the same element that contain different numbers of neutrons. **Radioisotopes** are unstable isotopes that emit subatomic particles and radiation.

CHEMICAL BONDS

- Atoms of molecules are held together by **chemical bonds**.
- **Ionic bonds** form from the attraction of oppositely charged ions. In **covalent bonds**, atoms share pairs of electrons. **Hydrogen bonds** are weak attractions between polar regions of hydrogen atoms and oxygen or nitrogen atoms.

CHEMICAL REACTIONS

- Chemical reactions break or form chemical bonds and release or use energy.
- **Catabolism**, the breaking down of molecules, releases energy. **Anabolism**, the synthesis of larger molecules, requires energy.

III WATER AND SOLUTIONS

WATER

- Water is a **polar compound**, acts as a solvent, and forms thin layers because it has high **surface tension**.
- Water also has high specific heat, and it serves as a medium for, and participates in, many chemical reactions.

SOLUTIONS AND COLLOIDS

- **Solutions** are **mixtures** with one or more **solutes** evenly distributed throughout a **solvent**.
- **Colloids** contain particles too large to form true solutions.

ACIDS, BASES, AND pH

- In most solutions containing acids or bases, **acids** release H^+ ions, and **bases** accept H^+ ions (or release OH^- ions).
- The **pH** of a solution is a measure of its acidity or alkalinity. A pH of 7 is **neutral**, below 7 is acidic, and above 7 is basic, or **alkaline**.

III COMPLEX ORGANIC MOLECULES

- **Organic chemistry** is the study of carbon-containing compounds.
- Organic compounds such as alcohols, aldehydes, ketones, organic acids, and amino acids can be identified by their **functional groups**.

CARBOHYDRATES

Carbohydrates consist of carbon chains in which most of the carbon atoms have an associated alcohol group and one carbon has either an aldehyde or a ketone group.
- The simplest carbohydrates are **monosaccharides**, which can combine to form **disaccharides** and **polysaccharides**. Long chains of repeating units are called **polymers**.
- The body uses carbohydrates primarily for energy.

LIPIDS

- All **lipids** are insoluble in water but soluble in nonpolar solvents.
- **Fats** consist of glycerol and **fatty acids**.
- **Phospholipids** contain a phosphate group in place of a fatty acid.
- **Steroids** have a complex four-ring structure.

PROTEINS

- **Proteins** consist of chains of **amino acids** linked by **peptide bonds**.
- Proteins form part of the structure of cells, act as enzymes, and contribute to other functions such as motility, transport, and regulation.
- **Enzymes** are biological catalysts of great **specificity** that increase the rate of chemical reactions in living organisms. Each enzyme has an **active site** to which its **substrate** binds.

NUCLEOTIDES AND NUCLEIC ACIDS

A **nucleotide** consists of a nitrogenous base, a sugar, and one or more phosphates.
- Some nucleotides contain **high-energy bonds**.
- **Nucleic acids** are important information-containing molecules that consist of chains of nucleotides. The nucleic acids that occur in living organisms are **ribonucleic acid (RNA)** and **deoxyribonucleic acid (DNA)**.

▍ TERMINOLOGY CHECK

acid *(p. 35)*	alkaline *(p. 36)*	anabolism *(p. 33)*	atom *(p. 29)*
active site *(p. 45)*	amino acid *(p. 42)*	anion *(p. 30)*	atomic number *(p. 30)*

▌ CLINICAL CASE STUDY

Certain bacteria, such as those that cause tuberculosis and leprosy, have a lot of waxy lipid material in their cell walls. They are much more difficult to stain than are regular bacteria. You must expose them to steam heat in order to drive the stain in. Also, most antibiotics do not affect them. What chemical explanation can you suggest to explain these properties?

▌ CRITICAL THINKING QUESTIONS

1. You may have noticed that in space exploration, attention is often focused on the presence or absence of water on other planets. What characteristics of water make it so essential to life as we know it on Earth? Do you think that a living system could develop on a water-free planet?

2. What properties of carbon have led to its being the "central" element in most of the essential chemicals within living organisms? Could some other element have assumed the role that carbon plays in all living things on Earth? Here? Elsewhere in the universe?

3. Hair follicles contain a high content of α-keratin protein. Knowing some of the properties of proteins, how do chemical treatments and heat (curlers and hair dryers) contribute toward styling hair into a new or different pattern?

4. Because bacteria are able to replicate at high rates, they are subjected to mistakes or mutations being made in their nucleotide sequence. For example, the nucleotide sequence in a gene coding for a protein might be AATTGGCCA, but because of mutation might become GGTTGGCCA. How may such a mutation become beneficial to the bacterium? How about harmful?

5. Mad cow disease is caused by a protein particle called a prion. It contains no nucleic acid. What activity associated with proteins might be involved in producing more prion particles? Use the index of your textbook to help you consider how a prion can make more prions.

▌ SELF-QUIZ

1. The smallest unit of an element that retains all the properties of that element is:

(a) Molecule
(b) Compound
(c) Atom
(d) Electron
(e) Cation
(f) Anion

2. Define atom (A), element (E), molecule (M), and compound (C). Place the appropriate letter(s) next to each of the following:

A H₂O
A O₂
C Salt

E Sulfur
C CH₄
E Sodium

I Glucose
A H₂
E Chlorine

3. How do ions differ from atoms?

4. Atoms consist of _pYotoNS_ that are positively charged, neutrons which carry a _Neutron_ charge, and _electros_ which carry a negative charge. The protons and neutrons together form the atomic _number_, while the electrons _orbit_ the nucleus. Atomic weight refers to the number of _protons_ and _neutrons_ found in an element. Atomic _mass_ refers to the number of protons found in an atom of a given element.

5. How do isotopes of an element differ from each other?

6. Match the following terms:

_C_ Solute
_D_ Mixture
_A_ Solution
_B_ Solvent

(a) A mixture of two or more substances in which the molecules of the substances are evenly distributed
(b) The medium in which substances are dissolved
(c) The substance dissolved in the solvent
(d) Two or more substances that combine in any proportion and are not chemically bound

7. Which of the following are properties of water that make it important for living cells?

(a) It is a polar molecule that can form solutions
(b) It has high surface tension
(c) It has a high specific heat
(d) It can participate in dehydration and hydrolysis reactions
(e) All of the above

8. (a) How is the pH scale used to measure acidity? Among the pH values 3, 5, 7, 11, and 13, which represents (b) the strongest acid and (c) the strongest base? (d) At what pH do you think the bacterium *Thiobacillus thioparis* can damage ancient marbic stone?

9. Organic compounds are present in all living cells. They all share the following characteristic:

(a) Are used in protein synthesis
(b) Are biological catalysts
(c) Are composed of carbon atom backbone surrounded by chloride atoms
(d) Are hydrophobic
(e) Are composed of carbon atom backbone surrounded by hydrogen atoms

10. (a) What is the basic structure of a monosaccharide? (b) How are disaccharides and polysaccharides different from monosaccharides?

11. The most immediate source of energy for living cells is generally in the form of:

(a) Lipids   (d) Protein
(b) Carbohydrates   (e) Vitamins
(c) Ketones

12. In what ways are carbohydrates used in living organisms?

13. Describe the structure and uses in the body of simple lipids, phospholipids, and steroids?

14. Lipids are generally:

(a) Hydrophobic
(b) Present in cell membranes
(c) Composed of fatty acids
(d) A high energy source
(e) All of the above

15. A peptide bond is formed between two amino acids by the reaction of the _____ of one amino acid with the _____ of the other.

(a) R group/R group
(b) R group/carboxyl group
(c) —OH of the amino group/—C=O of the carboxyl group
(d) —H of the amino group/—OH of the carboxyl group
(e) —NH$_3$ group/central carbon atom

16. Describe the four levels of protein structure. How is each maintained?

17. Enzymes are biological catalysts. They all share which of the following characteristics?

(a) Are not consumed in chemical reactions
(b) Lower the activation energy of reactions
(c) Increase the rate of reactions
(d) Allow reactions to occur that would otherwise require a higher temperature
(e) All the above

18. At high temperature and pH extremes, enzymes are generally denatured and lose their functionality. True or false?

19. Energy inside all living cells is rapidly consumed and generated in metabolic reactions. In what form is energy traded inside these cells?

(a) Glucose   (d) Metabolic enzymes
(b) Lactose   (e) DNA
(c) ATP

20. Nucleotides are:

(a) Building blocks of DNA
(b) Small enclosures inside the nucleus
(c) Sources of immediate energy
(d) Building blocks of proteins
(e) Readily present in cell membranes

21. Match the following:

___Adenine
___Thymine
___Phosphate
___Ribose
___Nucleotide
___Deoxynucleotide
___Uracil
___Guanine
___Nitrogenous base

(a) Present in only DNA
(b) Present in only RNA
(c) Present in both DNA and RNA
(d) Present in DNA, RNA, and ATP
(e) Present in RNA and ATP

22. Indicate which base is (or bases are) the complementary paired base(s) for each of the following:

Adenine ___
Cytosine ___
Guanine ___
Uracil   ___
Thymine ___

23. Match the following macromolecules:

D ___ Polysaccharides   (a) Protein
E ___ Polypeptide   (b) Chromosomes
C ___ Fat   (c) Lipids
B ___ DNA   (d) Carbohydrate
A ___ Steroids   (e) Protein synthesis

24. Which of the following statements is true about anabolism and catabolism?

   (a) The breakdown of a compound is known as anabolism and consumes energy, while the synthesis of a compound is known as catabolism and releases energy.
   (b) The breakdown of a compound is known as catabolism and consumes energy, while the synthesis of a compound is known as anabolism and releases energy.
   (c) The synthesis of a compound is known as anabolism and is an endergonic reaction, while the breakdown of a compound is known as catabolism and is an endergonic reaction.
   (d) The synthesis of a compound is known as anabolism and is an endergonic reaction, while the breakdown of

a compound is known as catabolism and is an exergonic reaction.
   (e) None of the above.

25. Which of the following is not a property of water?

   (a) Because water is a polar compound, it serves as a good solvent for ions.
   (b) Water's polarity and hydrogen bonding are responsible for its high surface tension.
   (c) Water's high specific heat helps stabilize the temperature of living organisms.
   (d) Water provides the medium for most cellular reactions but does not participate in them.
   (e) All the above are properties of water.

26. What is the chemical nature of this compound? Identify each of the circled parts of the molecule?

a. ___
b. ___
c. ___
d. ___

∎ EXPLORATIONS ON THE WEB

http://www.wiley.com/college/black

If you think you've mastered this chapter, there's more to challenge you on the web. Go to the companion web site to fine-tune your understanding of the chapter concept and discover answers to the questions posed below.

   1. Life on our planet is based on the element carbon. Science fiction writers have for years suggested that other worlds could exist where silicon is the element on which all life is on based. What do you think such life would be like?

   2. In February 1953, Watson and Crick announced to the patrons of the Eagle pub in Cambridge, England, that they had discovered the secret of life. What did they mean?

Microscopy and Staining

Come with me...

Very rare surviving simple microscope, c.1700. Ivary, 3.5" high, (Courtesy, George Black)

Come closer to an ordinary, "clean" household pin. What do you see? Probably not a lot more than you saw from farther away. If you enjoy normal vision, you can see dust particles as small as 20 micrometers (1 micrometer = 1/1000 of a millimeter) glimmering in a shaft of sunlight. But what's amazing is how much you aren't seeing! Most bacteria are only 1/2 to 2 micrometers long and therefore remain invisible without the help of a microscope. With just a 10-fold increase in visual acuity, you could begin to see this busy, squirming microbial world. How might such enhanced vision affect your desire to lift that "clean" fork to your mouth or your appetite for those "clean" vegetable sticks from the salad bar?

How many bacteria can fit on a pinpoint? Look at the successive close-up pictures here, and start counting. Each gold rod is an artificially colored bacterium. If you were to stick yourself with a "clean" pin, how many bacteria might be left behind in your finger? How many are necessary to start an infection? Sometimes as few as 10 bacteria are enough.

Today's microscopes allow us to see all this detail. But what if you were a microbiology student many years ago? Come with me to see what biologists before you used. If you think that the microscope you use in the lab is difficult to use, do you think any of these would be easier?

 Video related to this topic is available within WileyPLUS.

SEM (91X)   SEM (455X)

SEM (2,2764X)   SEM (12,548X)

Dr. Tony Brian & David Parker/Photo Researchers, Inc.

▌▌ HISTORICAL MICROSCOPY
▌▌ PRINCIPLES OF MICROSCOPY
Metric Units / Properties of Light: Wavelength and Resolution / Properties of Light: Light and Objects

▌▌ LIGHT MICROSCOPY
The Compound Light Microscope / Dark-Field Microscopy / Phase-Contrast Microscopy / Nomarski (Differential Interference Contrast)

Microscopy / Fluorescence Microscopy / Confocal Microscopy / Digital Microscopy

▌▌ ELECTRON MICROSCOPY
Transmission Electron Microscopy / Scanning Electron Microscopy / Scanning Tunneling Microscopy

▌▌ TECHNIQUES OF LIGHT MICROSCOPY
Preparation of Specimens for the Light Microscope / Principles of Staining

HISTORICAL MICROSCOPY

Anton van Leeuwenhoek (1632–1723), living in Delft, Holland, was almost certainly the first person to see individual microorganisms. He constructed simple microscopes capable of magnifying objects 100 to 300 times. These instruments were unlike what we commonly think of as microscopes today. Consisting of a single tiny lens, painstakingly ground, they were actually very powerful magnifying glasses (Figure 3.1). It was so difficult to focus one of Leeuwenhoek's microscopes that instead of changing specimens, he built a new microscope for each specimen, leaving the previous specimen and microscope together. When foreign investigators came to Leeuwenhoek's laboratory to look through his microscopes, he made them keep their hands behind their backs to prevent them from touching the focusing apparatus!

In a letter to the Royal Society of London in 1676, Leeuwenhoek described his first observations of bacteria and protozoa in water. He kept his techniques secret, however. Even today we are not sure of his methods of illumination, although it is likely that he used indirect lighting, with light bouncing off the sides of specimens rather than passing through them. Leeuwenhoek was also unwilling to part with any of the 419 microscopes he made. It was only near the time of his death that his daughter, at his direction, sent 100 of them to the Royal Society.

Following Leeuwenhoek's death in 1723, no one came forward to continue the work of perfecting the design and construction of microscopes, and the progress of microbiology slowed. Still, he had taken the first steps. Through Leeuwenhoek's letters to the Royal Society in the mid-1670s, the existence of microbes was revealed to the scientific community. And in 1683 he described bacteria taken from his own mouth. However, Leeuwenhoek could see very little detail of their structure. Further study required the development of more complex microscopes, as we shall soon see.

PRINCIPLES OF MICROSCOPY

METRIC UNITS

Microscopy is the technology of making very small things visible to the human eye. Because microorganisms are so small, the units used to measure them are likely to be unfamiliar to beginning students used to dealing with a macroscopic world (Table 3.1).

The **micrometer** (μm), formerly called a micron (μ), is equal to 0.000001 m. A micrometer can also be expressed as 10^{-6} m. The second unit, the **nanometer** (nm), formerly called a millimicron (mμ), is equal to 0.000000001 m. It also is expressed as 10^{-9} m. A third unit, the **angstrom** ($\overset{\circ}{\text{A}}$), is found in much of the current and older literature but no longer has any official recognition. It is equivalent to 0.0000000001 m, 0.1 nm, or 10^{-10} m. **Figure 3.2** shows a scale that summarizes the metric system unit equivalents, the ranges of sizes that

The Secret Kingdom Uncovered
"I have had several gentlewomen in my house, who were keen on seeing the little eels in vinegar; but some of 'em were so disgusted at the spectacle that they vowed they'd never use vinegar again. But what if one should tell such people in [the] future that there are more animals living in the scum on the teeth in a man's mouth, than there are men in a whole kingdom?"
—Anton van Leeuwenkoek, 1683

(a) Single glass lens — Specimen mounted at tip of needle — Positioning screw — Focusing screw — Handle

(b)

(c)

Figure 3.1 Leeuwenhoek's investigations. **(a)** A replica of one of Leeuwenhoek's microscopes. This simple microscope, really a very powerful magnifying glass, uses a single, tiny, almost spherical lens set into the metal plate. The specimen is mounted on the needlelike end of the vertical shaft and examined through the lens from the opposite side. The various screws are used to position the specimen and bring it into focus—a very difficult process. **(b)** The proper way of looking through Leeuwenhoek's microscope. **(c)** An excerpt from Leeuwenhoek's writing, and the vinegar eels (nematodes) that so upset Leeuwenhoek's friends (80X). *(a, b: Kathy Talaro/Visuals Unlimited; c: John D. Cunningham/Visuals Unlimited)*

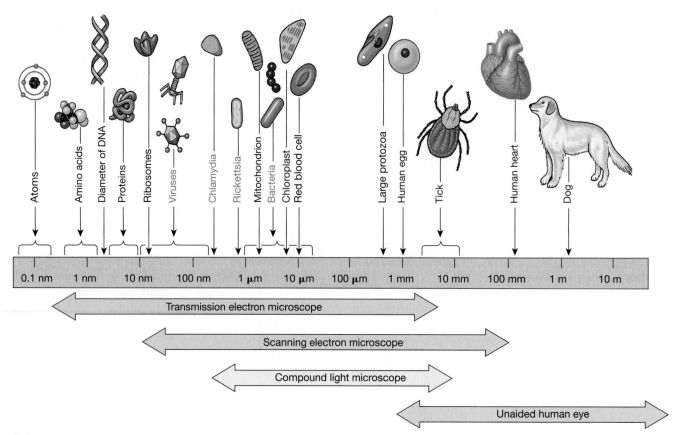

Figure 3.2 **Relative sizes of objects.** Sizes are shown relative to a metric scale; names in red are organisms studied in microbiology. Chlamydia and Rickettsia are groups of bacteria that are much smaller in size than other bacteria. The range of effective use for various instruments is also depicted.

can be detected by the unaided human eye and by various types of microscopy, and examples of where various organisms fall on this scale.

PROPERTIES OF LIGHT: WAVELENGTH AND RESOLUTION

Light has a number of properties that affect our ability to visualize objects, both with the unaided eye and (more crucially) with the microscope. Understanding these properties will allow you to improve your practice of microscopy.

One of the most important properties of light is its **wavelength**, or the length of a light ray **(Figure 3.3)**.

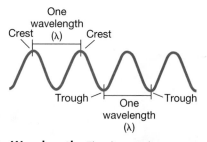

Figure 3.3 **Wavelength.** The distance between two adjacent crests or two adjacent troughs of any wave is defined as 1 wavelength, designated by the Greek letter lambda (λ).

TABLE 3.1

Some Commonly Used Units of Length			
Unit (Abbreviation)	**Prefix**	**Metric Equivalent**	**English Equivalent**
meter (m)			3.28 ft
centimeter (cm)	*centi* = one hundredth	$0.01 \text{ m} = 10^{-2} \text{ m}$	0.39 in.
millimeter (mm)	*milli* = one thousandth	$0.001 \text{ m} = 10^{-3} \text{ m}$	0.039 in.
micrometer (μm)	*micro* = one millionth	$0.000001 \text{ m} = 10^{-6} \text{ m}$	0.000039 in.
nanometer (nm)	*nano* = one billionth	$0.000000001 \text{ m} = 10^{-9} \text{ m}$	0.000000039 in.
angstrom (Å)		$0.0000000001 \text{ m} = 10^{-10} \text{ m}$	0.0000000039 in.

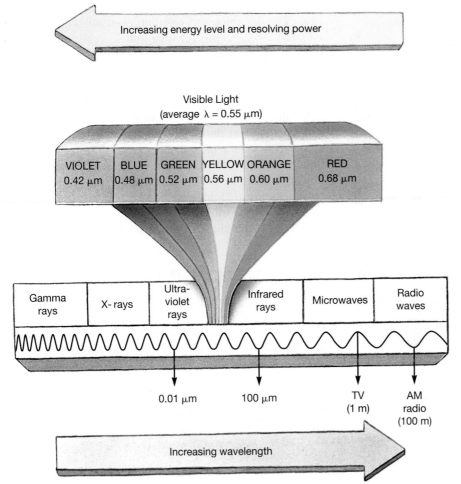

Figure 3.4 **The electromagnetic spectrum.** Only a narrow range of wavelengths—those of visible and ultraviolet light—are used in light microscopy. The shorter the wavelength used, the greater the resolution that can be attained. White light is the combination of all colors of visible light.

Represented by the Greek letter lambda (λ), wavelength is equal to the distance between two adjacent crests or two adjacent troughs of a wave. The sun produces a continuous *spectrum* of electromagnetic radiation with waves of various lengths **(Figure 3.4)**. Visible light rays, as well as ultraviolet and infrared rays, constitute particular parts of this spectrum. White light is the combination of all colors of visible light. Black is the absence of visible light.

The wavelength used for observation is crucially related to the resolution that can be obtained. **Resolution** refers to the ability to see two items as separate and discrete units **(Figure 3.5a)** rather than as a fuzzy, overlapped single image **(Figure 3.5b)**. We can magnify objects, but if the objects cannot be resolved, the

magnification is useless. Light must pass between two objects for them to be seen as separate things. If the wavelength of the light by which we see the objects is too long to pass between them, they will appear as one. The key to resolution is to get light of a short-enough wavelength to fit between the objects we want to see separately. Cell structures less than one-half a wavelength long will not be visible.

To visualize this phenomenon, imagine a target with a foot-high letter E hanging in front of a white background. Suppose that you throw at the target ink-covered objects with diameters corresponding to various wavelengths **(Figure 3.6)**. If one object has a diameter smaller than the distance between the "arms" of the letter E, the object will pass between the arms and the arms will be distinguishable as separate structures. First, imagine tossing basketballs. Because they cannot fit between the arms, light rays of that size would give poor resolution. Next toss tennis balls at the target. The resolution will improve. Then try jelly beans and, finally, tiny beads. With each decrease in the diameter of the object thrown, the number of such objects that can pass between the arms of the E increases. Resolution

Figure 3.5 Resolution. (a) The two dots are resolved—that is, they can clearly be seen as separate structures. **(b)** These two dots are not resolved—they appear to be fused.

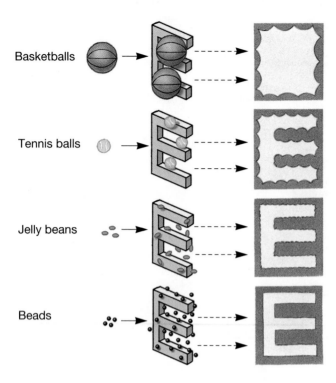

Basketballs

Tennis balls

Jelly beans

Beads

Figure 3.6 An analogy for the effect of wavelength on resolution. Smaller objects (corresponding to shorter wavelengths) can pass more easily between the arms of the letter E, defining it more clearly and producing a sharper image.

improves, and the shape of the letter is revealed with greater and greater precision.

Microscopists use shorter and shorter wavelengths of electromagnetic radiation to improve resolution. Visible light, which has an average wavelength of 550 nm, cannot resolve objects separated by less than 220 nm. Ultraviolet light, which has a wavelength of 100 to 400 nm, can resolve separations as small as 110 nm. Thus, microscopes that used ultraviolet light instead of visible light allowed researchers to find out more about the details of cellular structures. But the invention of the electron microscope, which uses electrons rather than light, was the major step in increasing the ability to resolve objects. Electrons behave both as particles and as waves. Their wavelength is about 0.005 nm, which allows resolution of separations as small as 0.2 nm.

Galileo (an astronomer) was the first person to record microscopic observations of biological nature—an insect's eye.

The **resolving power (RP)** of a lens is a numerical measure of the resolution that can be obtained with that lens. The smaller the distance between objects that can be distinguished, the greater the resolving power of the lens. We can calculate the RP of a lens if we know its **numerical aperture (NA)**, a mathematical expression relating to the extent that light is concentrated by the condenser lens and collected by the objective. The formula for calculating resolving power is RP = λ/2NA. As this formula indicates, the smaller the value of λ and the larger the value of NA, the greater the resolving power of the lens.

The NA values of lenses differ in accordance with the power of magnification and other properties. The NA value is engraved on the side of each objective lens (the lens nearest the stage) of a light microscope. Look at the NA values on the microscope you use the next time you are in the laboratory. Typical values for the objective lenses commonly found on modern light microscopes are 0.25 for low power, 0.65 for high power, and 1.25 for the oil immersion lens. The higher the NA value, the better the resolution that can be obtained.

PROPERTIES OF LIGHT: LIGHT AND OBJECTS

Various things can happen to light as it travels through a medium such as air or water and strikes an object (**Figure 3.7**). Let us look at some of those things now and consider how they can affect what we see through a microscope.

Reflection

If the light strikes an object and bounces back (giving the object color), we say that **reflection** has occurred.

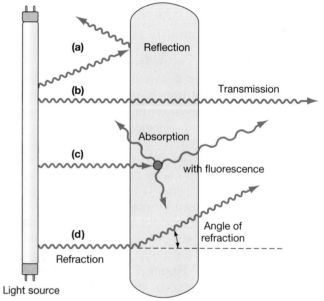

Figure 3.7 Various interactions of light with an object it strikes. (a) Light may be reflected back from the object. The particular wavelengths reflected back to the eye determine the perceived color of the object. **(b)** Light may be transmitted directly through the object. **(c)** Light may be absorbed, or taken up, by the object. In some cases, the absorbed light rays are reemitted as longer wavelengths, a phenomenon known as fluorescence. **(d)** Light passing through the object may be refracted, or bent, by it.

For example, light rays in the green range of the spectrum are reflected off the surfaces of the leaves of plants. Those reflected rays are responsible for our seeing the leaves as green.

Transmission

Transmission refers to the passage of light through an object. You cannot see through a rock because light cannot pass through it, as it does through a glass window. In order for you to see objects through a microscope, light must either be reflected from the objects or transmitted through them. Most of your observations of microorganisms will make use of transmitted light.

Absorption

If light rays neither pass through nor bounce off an object but are taken up by the object, **absorption** has occurred. Energy in absorbed light rays can be used in various ways. For example, all wavelengths of the sun's light rays except those in the green range are absorbed by a leaf. Some of the energy in these other light rays is captured in photosynthesis and used by the plant to make food. Energy from absorbed light can also raise the temperature of an object. A black object, which reflects no light, will gain heat much faster than a white object, which reflects all light rays.

In some cases, absorbed light rays, especially ultraviolet light rays, are changed into longer wavelengths and reemitted. This phenomenon is known as **luminescence**. If luminescence occurs only during irradiation (when light rays are striking an object), the object is said to **fluoresce**. Many fluorescent dyes are important in microbiology, especially in the field of immunology, because they help us visualize immune reactions and internal processes in microorganisms. If an object continues to emit light when light rays no longer strike it, the object is **phosphorescent**. Some bacteria that live deep in the ocean are phosphorescent.

Refraction

Refraction is the bending of light as it passes from one medium to another of different density. The bending of the light rays gives rise to an *angle of refraction*, the degree of bending (Figure 3.7d). You have probably seen how the underwater portion of a pole that is sticking out of water or a drinking straw in a glass of water seems to bend **(Figure 3.8)**. When you remove the object from the water, it is clearly straight. It looks bent because light rays deviate, or bend, when they pass from the water into the air as their speed changes across the water-air interface. The **index of refraction** of a material is a measure of the speed at which light passes through the material. When two substances have different indices of refraction, light will bend as it passes from one material into the other.

Light passing through a glass microscope slide, through air, and then through a glass lens is refracted

Figure 3.8 Refraction. The refraction of light rays passing from water into air causes the pencil to appear bent. *(Southern Illinois University Niomed/Custom Medical Stock Photo, Inc.)*

each time it goes from one medium to another. This causes loss of light and a blurred image. To avoid this problem, microscopists use **immersion oil**, which has the

TRY IT

A Life of Crime

If you want to take some diamonds through customs without declaring them, here's how to do it. Obtain an oil with the same refractive index as the diamonds. Pour it into a clear glass bottle labeled "Baby Oil," and drop the diamonds in. The diamonds are invisible if their surfaces are clean. This trick works because light is not bent when it passes from one medium to another with the same index of refraction. Thus, the boundary between the diamonds and the oil is not apparent.

If larceny is not in your heart or the price of diamonds is not in your pocket, try this entertaining, but legal, activity. Clean a glass rod and dip it in and out of a bottle of immersion oil. Watch it disappear and reappear. This experiment will help you understand what is happening when you use the oil immersion lens.

Immersion oil · · · · · · Water

(Richard Megna/Fundamental Photographs)

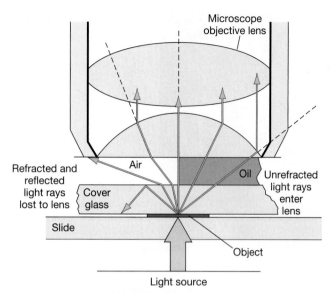

Figure 3.9 Immersion oil. Immersion oil is used to prevent the loss of light that results from refraction. The focusing of as much light as possible adds to the clarity of the image. Immersion oil may also be added between the top of the condenser and the bottom of the slide to eliminate another site for refraction.

Immersion oil is nothing new. Robert Hooke, an early microscopist, first mentioned using a form of it back in 1678.

same index of refraction as glass, to replace the air. The slide and the lens are joined by a layer of oil; there is no refraction to cause the image to blur **(Figure 3.9)**. If you forget to use oil with the oil immersion lens of a microscope, it will be impossible to focus clearly on a specimen. Staining (dyeing) a specimen increases differences in indices of refraction, making it easier to observe details.

Diffraction

As light passes through a small opening, such as a hole, slit, or space between two adjacent cellular structures, the light waves are bent around the opening. This phenomenon is **diffraction**. **Figure 3.10** shows diffraction patterns formed when light passes through a small aperture or around the edge of an object. Similar patterns occur when water passes through an opening in, or around the back of, a breakwater. Look for these patterns the next time you are flying over water.

Take another look

Diffraction is a problem for microscopists because the lens acts as a small aperture through which the light must pass. A blurry image results. The higher the magnifying power of a lens, the smaller the lens must be, and therefore the greater the diffraction and blurring it causes. The oil immersion (100X) lens, with its total magnification capacity of about 1,000X (when combined with a 10X ocular lens), represents the limit of useful magnification with the light microscope. The small size of higher-power lenses causes such severe diffraction that resolution is impossible.

✔ CHECKLIST

1. Which color of light would give you better resolution when using a microscope: red (wavelength 0.68 μm) or blue (wavelength 0.48 μm)? Why?

2. If you built a light microscope having a total magnification of 5,000X, would it give you better, worse, or the same resolution as one that has a magnification of 1,000X? Why?

3. Why would radio waves and microwaves be unsuitable for examining microbes?

4. Would standard immersion oil placed on a plastic slide prevent refraction? Why or why not?

Figure 3.10 Diffraction. Light waves are diffracted as they pass **(a)** around the edge of an object and **(b)** through a small aperture. **(c)** Water waves being diffracted as they pass through an opening in a breakwater. *(Runk Schoenberger/Grant Heilman Photography)*

LIGHT MICROSCOPY

Light microscopy refers to the use of any kind of microscope that uses visible light to make specimens observable. The modern light microscope is a descendant not of Leeuwenhoek's single lenses but of Hooke's compound microscope—a microscope with more than one lens. (◄Chapter 1, p. 9) Single lenses produce two problems: They cannot bring the entire field into focus simultaneously, and there are colored rings around objects in the field. Both problems are solved today by the use of multiple correcting lenses placed next to the primary magnifying lens **(Figure 3.11)**. Used in the objectives and eyepieces of modern compound microscopes, correcting lenses give us nearly distortion-free images.

Over the years, several kinds of light microscopes have been developed, each adapted for making certain kinds of observations. We look first at the standard light microscope and then at some special kinds of microscopes.

THE COMPOUND LIGHT MICROSCOPE

The **optical microscope**, or *light microscope*, has undergone various improvements since Leeuwenhoek's time and essentially reached its current form shortly before the beginning of the twentieth century. This microscope is a **compound light microscope**—that is, it has more than one lens. The parts of a modern compound microscope and the path light takes through it are shown in **Figure 3.12**. A compound microscope with a single eyepiece (*ocular*) is said to be **monocular**; one with two eyepieces is said to be **binocular**.

Light enters the microscope from a source in the **base** and often passes through a blue filter, which filters out the long wavelengths of light, leaving the shorter wavelengths and improving resolution. It then goes through a **condenser**, which converges the light beams so that they pass through the specimen. The **iris diaphragm** controls the amount of light that passes through the specimen and into the objective lens. The higher the

Figure 3.12 The compound light microscope. Yellow indicates the path of light through the microscope.

magnification, the greater the amount of light needed to see the specimen clearly. The **objective lens** magnifies the image before it passes through the **body tube** to the ocular lens in the eyepiece. The **ocular lens** further magnifies the image. A **mechanical stage** allows precise control of moving the slide, which is especially useful in the study of microbes.

The focusing mechanism consists of a **coarse adjustment** knob, which changes the distance between the objective lens and the specimen fairly rapidly, and a **fine adjustment** knob, which changes the distance very slowly. The coarse adjustment knob is used to locate the specimen. The fine adjustment knob is used to bring it into sharp focus.

Compound microscopes have up to six interchangeable objective lenses that have different powers of magnification.

The **total magnification** of a light microscope is calculated by multiplying the magnifying power of the objective lens (the lens used to view your specimen) by the magnifying power of the ocular lens (the lens nearest your eye). Typical values for a microscope with a 10X ocular lens are:

- scanning (3X) × (10X) = 30X magnification
- low power (10X) × (10X) = 100X magnification
- high "dry" (40X) × (10X) = 400X magnification
- oil immersion (100X) × (10X) = 1,000X magnification

Most microscopes are designed so that when the microscopist increases or decreases the magnification by changing from one objective lens to another, the

Figure 3.11 Cutaway view of a modern microscope objective. What we refer to as a single objective lens is really a series of several lenses, which are necessary to correct aberrations of color and focus. The best objectives may have as many as a dozen or more elements.

(a)

Figure 3.13   A comparison of the illumination in bright-field and dark-field microscopy. **(a)** The condenser of the bright-field microscope concentrates and transmits light directly through the specimen. **(b)** The dark-field condenser deflects light rays so that they reflect off the specimen at an angle before they are collected and focused into an image.

(b)

Figure 3.14   A comparison of bright-field and dark-field images. **(a)** Bright-field and **(b)** dark-field microscope views of *Saccharomyces cerevisiae* (brewer's yeast, magnified 975X). Dark-field illumination provides an enormous increase in contrast. *(Jim Solliday/Biological Photo Service)*

specimen will remain very nearly in focus. Such microscopes are said to be **parfocal** (*par* means "equal"). The development of parfocal microscopes greatly improved the efficiency of microscopes and reduced the amount of damage to slides and objective lenses. Most student-grade microscopes are parfocal today.

Some microscopes are equipped with an **ocular micrometer** for measuring objects viewed. This is a glass disc with a scale marked on it that is placed inside the eyepiece between its lenses. This scale must first be calibrated with a stage micrometer, which has metric units engraved on it. When these units are viewed through the microscope at various magnifications, the microscopist can determine the corresponding metric values of the divisions on the ocular micrometer for each objective lens. Thereafter, he or she needs only to count the number of divisions covered by the observed object and multiply by the calibration factor for that lens in order to determine the actual size of the object.

DARK-FIELD MICROSCOPY

The condenser used in an ordinary light microscope causes light to be concentrated and transmitted directly through the specimen, as shown in **Figure 3.13a**. This gives **bright-field illumination (Figure 3.14a)**. In some cases, however, it is more useful, especially with light-sensitive organisms, to examine specimens that would lack contrast with their background in a bright field under other illumination. Live spirochetes (spi′ro-kets), spiral-shaped bacteria that cause syphilis and other diseases, are just such organisms. In this situation **dark-field illumination** is used. A microscope adapted for dark-field illumination has a condenser that prevents light from being transmitted through the specimen but

**Figure 3.15
A phase-contrast image.** *Amoeba,* a protozoan (160X). *(Biophoto Associates/ Photo Researchers, Inc.)*

Figure 3.16 A Nomarski image. The protozoan *Paracineta* is attached by a long stalk to the green alga *Spongomorpha* (magnified 400X). *(Biological Photo Service)*

instead causes the light to reflect off the specimen at an angle (**Figure 3.13b**). When these rays are gathered and focused into an image, a light object is seen on a dark background (**Figure 3.14b**).

PHASE-CONTRAST MICROSCOPY

Most living microorganisms are difficult to examine because they cannot be stained by coloring them with dyes—stains usually kill the organisms. To observe them alive and unstained requires the use of **phase-contrast microscopy**. A phase-contrast microscope has a special condenser and objective lenses that accentuate small differences in the refractive index of various structures within the organism. Light passing through objects of different refractive indices is slowed down and diffracted. The changes in the speed of light are seen as different degrees of brightness (**Figure 3.15**).

NOMARSKI (DIFFERENTIAL INTERFERENCE CONTRAST) MICROSCOPY

Nomarski microscopy, like phase-contrast microscopy, makes use of differences in refractive index to visualize unstained cells and structures. However, the microscope used, a *differential interference contrast microscope*, produces much higher resolution than the standard phase-contrast microscope. It has a very short *depth of field* (the thickness of specimen that is in focus at any one time) and can produce a nearly three-dimensional image (**Figure 3.16**).

FLUORESCENCE MICROSCOPY

In **fluorescence microscopy**, ultraviolet light is used to excite molecules so that they release light of a longer

wavelength than that originally striking them (see Figure 3.7c). The different wavelengths produced are often seen as brilliant shades of orange, yellow, or yellow green. Some organisms, such as *Pseudomonas*, fluoresce naturally when irradiated with ultraviolet light. Other organisms, such as *Mycobacterium tuberculosis* and *Treponema pallidum* (the cause of syphilis), must be treated with a fluorescent dye called a *fluorochrome*. They then stand out sharply against a dark background (**Figure 3.17**). Acridine orange is a fluorochrome that binds to nucleic acids, and it colors bright green, orange green, or yellow, depending on the filter system used with the fluorescence microscope. It is sometimes used to screen samples for microbial growth, with live cells showing up in bright orange or green.

Fluorescent antibody staining is now widely used in diagnostic procedures to determine whether an *antigen* (a foreign substance such as a microbe) is present. *Antibodies*—molecules produced by the body as an immune response to an invading antigen—are found in many clinical specimens such as blood and serum. If a patient's specimen contains a particular antigen, that antigen and the antibodies specifically made against it will clump together. However, this reaction is ordinarily not visible. Therefore, fluorescent dye molecules are attached to the antibody molecules. If the dye molecules are retained by the specimen, the antigen is presumed to be present, and a positive diagnosis can be made. Thus, if fluorescent dye-tagged antibodies against syphilis organisms are added to a specimen containing spirochetes and are seen to bind to the tagged organisms, those organisms can be identified as the cause of syphilis. This technique is especially important in *immunology*, in which the reactions of antigens and antibodies are studied in great detail (see ◀Chapters 17

LM

Figure 3.17 Fluorescent antibody staining. The fluorescent dye-tagged antibodies clearly show live bacterial cells (green) and dead (red) cells (854X). *(David Phillips/Visuals Unlimited)*

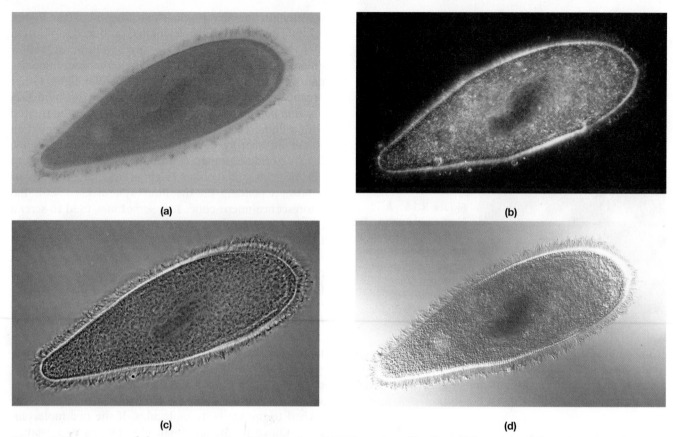

Figure 3.18 **Images of the same organism (*Paramecium*, 600X) produced by four different techniques.** **(a)** Bright-field microscopy, **(b)** dark-field microscopy, **(c)** phase-contrast microscopy, **(d)** Nomarski microscopy. One microscope can have the optics for all four techniques. *(David M. Phillips/Visuals Unlimited)*

and 18, especially Figure 18.35 on the technique of fluorescent antibody staining). Diagnoses can often be made in minutes rather than the hours or days it would take to isolate, culture, and identify organisms.

Figure 3.18 shows the images produced by four different microscopic techniques.

CONFOCAL MICROSCOPY

Confocal systems use beams of ultraviolet laser light to excite fluorescent chemical dye molecules into emitting (returning) light **(Figure 3.19)**. The exciting light beam is focused onto the specimen (usually nonliving) either through a thin optical fiber, or by passing through a

Figure 3.19 **(a)** A confocal microscope system manufactured by Olympus. Cell with microtubular fragments shown using **(b)** standard fluorescent microscopy, and **(c)** confocal microscopy. *(Courtesy Olympus Corporation, Scientific Equipment Division)*

(a)

(b)

Figure 3.20 **(a)** A digital microscope system manufactured by Nikon Instruments, Inc. *(Courtesy of Nikon Instruments Inc.)* **(b)** Cyanobacterium, *Chroococcus*, viewed by digital microscopy. *(© Wim van Egmond)*

small aperture shaped as a pinhole or a slit. Resultant fluorescent emissions are focused on a detector which also has a small aperture or slit in front of it. The smaller the apertures used at both sites, the greater the amounts of out-of-focus light blocked from the detector. A computer reconstructs an image from the emitted light with resolution that can be up to 40% better than with other types of light microscopy. Because of the sharpness of focus, the image is like a very thin knife-blade cut through the specimen. For thick specimens, a whole series of successive focal plane cuts can be recorded, and assembled into a three-dimensional model. This is very helpful in studying communities of microbes without disturbing them, as in examining living biofilms. Time-lapse images can also be collected.

DIGITAL MICROSCOPY

Have you had frustrating moments in the lab when you just couldn't get a slide into focus? Then you would like the auto-focus, auto-aperture, auto-light, motorized stage and magnification changers of a **digital microscope** (Figure 3.20). Not only that, but these microscopes also come with a built-in digital camera and preloaded software. All you do is plug in the unit, turn on the power, and use the mouse to view live or stained specimens on a monitor, or in a group situation on a screen through use of a projector. It can also be integrated into a local or wide area network for distance learning or teleconferencing. Imagine being able to show your cousin in Kansas live images of what's swimming in your sample of pond water. There are, however, some limitations: maximum magnification is quite limited, and price is high.

▌▌▌ ELECTRON MICROSCOPY

The light microscope opened doors to the world of microbes. However, because it could not resolve objects

(a)

Figure 3.21 **Electron microscopy.** **(a)** A cross-sectional diagram of an electron microscope, showing the pathways of the electron beam as it is focused by electromagnetic lenses. **(b)** A modern scanning electron microscope in use. *(Pascal Goetgheluck/ Photo Researchers, Inc.)*

Electron source
Condenser lens
Specimen
Objective lens
Intermediate image
Projector lens
Viewing lenses
Fluorescent screen or photographic film (final image)

(b)

(a)

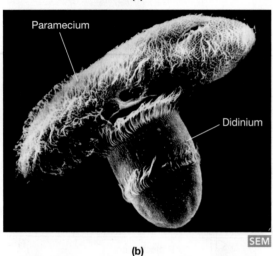

(b)

Figure 3.22 Light and electron microscopy images compared. (a) Light (160X) and **(b)** electron (425X) microscope images of a *Didinium* eating a *Paramecium*. Notice how much more detail is revealed by the scanning electron micrograph. *(top: Eric V. Grave/Photo Researchers, Inc.; bottom: Biophoto Associates/Photo Researchers, Inc.)*

separated by less than 0.2 μm, the view was limited to observations at the level of whole cells and their arrangements. Few subcellular structures could be seen; neither could viruses. The advent of the **electron microscope (EM)** allowed these small structures to be visualized and studied. The EM was developed in 1932 and was in use in many laboratories by the early 1940s.

The EM uses a beam of electrons instead of a beam of light, and electromagnets rather than glass lenses are used to focus the beam **(Figure 3.21)**. The electrons must travel through a vacuum because collisions with air molecules would scatter the electrons and result in a distorted image. Electron microscopes are much

Germans invented the EM. At the end of World War II, the U.S. confiscated Hitler's personal physician's EM. U.S. scientists studied it to learn to make American EMs.

more expensive than light microscopes. They also take up much more space and require additional rooms for preparation of specimens and for processing of photographs. Photographs taken on any microscope are called *micrographs*; those taken on an electron microscope are called **electron micrographs**. Nothing else can show the great detail of minute biological structures that EMs can **(Figure 3.22)**.

The two most common types of electron microscope are the transmission electron microscope and the scanning electron microscope. Both are used to study various life forms, including microbes. The more advanced scanning tunneling microscope and atomic force microscope let us see actual molecules and even individual atoms.

TRANSMISSION ELECTRON MICROSCOPY

The **transmission electron microscope (TEM)** gives a better view of the internal structure of microbes than do other types of microscopes. Because of the very short wavelength of illumination (electrons) on which the TEM operates, it can resolve objects as close as 1 nm and magnify microbes (and other objects) up to 500,000X. To prepare specimens for transmission electron microscopy, a specimen may be embedded in a block of plastic and cut with a glass or diamond knife to produce very thin slices (*sections*). These sections are placed on thin wire grids for viewing so that a beam of electrons will pass directly through the section. The section must be exceedingly thin (70–90 nm) because electrons cannot penetrate very far into materials. The specimens can also be treated with special preparations that contain heavy metal elements. The heavy metals scatter electrons and contribute to forming an image.

Figure 3.23 Shadow casting. Spraying a heavy metal (such as gold or platinum) at an angle over a specimen leaves a "shadow," or darkened area, where metal is not deposited. This technique, known as *shadow casting*, produces images with a three-dimensional appearance, as in this photograph of polio viruses (magnified 330,480X). You can calculate the height of the organisms from the length of their shadows if you know the angle of the metal spray. *(John J. Cardamone Jr. & B.A. Phillips/Biological Photo Service)*

Figure 3.24 Freeze-fracturing and freeze-etching. (a) In freeze-fracturing, a specimen is frozen in a block of ice and broken apart with a very sharp knife. The fracture reveals the interiors of cellular structures and typically passes through the center of membrane bilayers, exposing their inner faces. **(b)** In freeze-etching, water is evaporated directly from the ice and frozen cytoplasm of the fractured specimen, uncovering additional surfaces for observation.

Very small specimens, such as molecules or viruses, can be placed directly on plastic-coated grids. Then a heavy metal such as gold or platinum is sprayed at an angle onto the specimen, a technique known as **shadow casting**. A thin layer of the metal is deposited. Areas behind the specimen that did not receive a coating of metal appear as "shadows," which can give a three-dimensional effect to the image **(Figure 3.23)**. Electron beams are deflected by the densely coated parts of the specimen but, for the most part, pass through the shadows.

It is also possible to view the interior of a cell with a TEM by a technique called **freeze-fracturing**. In this technique the cell is frozen and then fractured with a knife. The cleaving of a specimen reveals the surfaces of structures inside the cell **(Figure 3.24a)**. **Freeze-etching**, which involves the evaporation of water from the frozen and fractured specimen, can then expose additional surfaces for examination **(Figure 3.24b)**. These surfaces must also be coated with a heavy metal layer that produces shadows. This layer, called a *replica*, is viewed by TEM **(Figure 3.25)**.

The image formed by the electron beam is made visible as a light image on a fluorescent screen, or monitor. (The actual image made by the electron beam is not visible and would burn your eyes if you tried to view it directly.) The electrons are used to excite the phosphors (light-generating compounds) coating the screen. However, the electron beam will eventually burn through the specimen. Therefore, before this happens, electron micrographs are made, either by photographing the image on the video screen or by replacing the screen itself with a photographic plate **(Figure 3.26a)**. Electron micrographs can be enlarged, just as you would enlarge any photograph, to obtain an image magnified 20 million times! The micrographs are permanent records of specimens observed and can be studied at leisure. The study of electron micrographs has provided much of our knowledge of the internal structure of microbes. The "M" in TEM and SEM (next page) can refer to either microscope or micrograph.

Figure 3.25 A freeze-etch preparation. The toxic cyanobacterium *Microcystis aeruginosa* (magnified 18,000X), showing details of large spherical gas vesicles. *(Biological Photo Service)*

TEM

SEM

(a)   (b)

Figure 3.26   TEM and SEM compared. Colorized electron micrographs of *Escherichia coli* produced by **(a)** transmission electron microscopy (66,952X) and **(b)** scanning electron microscopy (39,487X). *(a: Dennis Kunkel/Phototake; b: David M. Phillips/Visuals Unlimited)*

SCANNING ELECTRON MICROSCOPY

The **scanning electron microscope (SEM)** is a more recent invention than the TEM and is used to create images of the surfaces of specimens. The SEM can resolve objects as close as 20 nm, giving magnifications up to approximately 50,000X. The SEM gives us wonderful three-dimensional views of the exterior of cells (**Figure 3.26b**).

Preparing a specimen for the SEM involves coating it with a thin layer of a heavy metal, such as gold or palladium. The SEM is operated by scanning, or sweeping, a very narrow beam of electrons (an electron probe) back and forth across a metal-coated specimen. Secondary, or backscattered, electrons leaving the specimen surface are collected, the current is increased, and the resulting image is displayed on a screen. Photographs of the image can be made and enlarged for further study.

Views of the three-dimensional world of microbes, as shown in **Figure 3.27**, are breathtakingly beautiful.

SCANNING TUNNELING MICROSCOPY

In 1980, Gerd Binnig and Heinrich Rohrer invented the first of a series of rapidly improving **scanning tunneling microscopes (STMs)**, also called scanning probe microscopes. Five years later, they received the Nobel Prize for their discovery.

A thin wire probe made of platinum and viridium is used to trace the surface of a substance, much as you

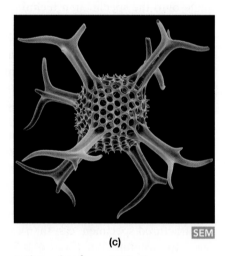

SEM

SEM

SEM

(a)   (b)   (c)

SEM

(d)

Figure 3.27   Colorized SEM photos of representative microbes. (a) The fungus *Aspergillus*, a cause of human respiratory disease (10,506X); **(b)** *Actinomyces*, a branching bacterium (5670X); **(c)** a radiolarian from the Indian Ocean (1761X); **(d)** the diatom *Cyclotella meneghiniana*, one of many that carry on photosynthesis and form the base of many aquatic food chains (1584X). *(a: Visuals Unlimited; b: David M. Phillips/Photo Researchers, Inc.; c: Manfred Kage/Peter Arnold, Inc.; d: Dr. Anne Smith/Photo Researchers, Inc.)*

would use your finger to feel the bumps while reading Braille. Electron clouds (regions of electron movement) from the surfaces of the probe and the specimen overlap, producing a kind of pathway through which electrons can "tunnel" into one another's clouds. This tunneling sets up an observable current. The stronger the current, the closer the top of the atom is to the probe. Running the probe across in a straight line reveals the highs and lows of individual molecules or atoms in a surface (**Figure 3.28**).

Even movies can be made using this technique. The first one ever produced showed individual fibrin molecules coming together to form a blood clot. Scanning tunneling microscopy also works well under water, and it can be used to examine *live* specimens, such as virus-infected cells exploding and releasing newly formed viruses.

The **atomic force microscope (AFM)**, a more advanced member of this family of microscopes, allows three-dimensional imaging and measurement of structures from atomic size to about 1 μm. The AFM has been very useful in studying DNA because it enables investigators to distinguish between bases, such as adenine and guanine, from differences in their electron density states. Atomic force microscopy has also been used underwater to study chemical reactions at living cell surfaces (**Figure 3.29**), which help to confirm chemical analyses of the cell wall material.

In addition to producing images, AFM can measure forces, for example, the force needed to unfold a protein located in a membrane. One can also determine the flexibility of a polysaccharide molecule, that is, to know how far it can elongate before it ruptures. This information is important in studying attachment of adjacent cells to form aggregations such as colonies and films, or for the ability to attach to host cells.

The various types of microscopy and their uses are summarized in **Table 3.2**.

(a)

100 nm   200 nm

Figure 3.29 **(a)** An atomic force microscope. **(b)** Surface of a dormant spore of the fungus *Aspergillus oryzae* shows a covering of protein rodlets. **(c)** A few hours later, the rodlets have disintegrated into a layer of soft material, beginning to reveal inner spore walls composed of polysaccharides. The entire process can be watched live, underwater, and filmed with the atomic force microscope. *(Michael L Abramson/Time Life Picture/Getty Images, From B. Jean and H. Hörben, Force Microscopy, Wiley Liss, 2006, Fig. 5-4cd, page 77)*

Figure 3.28 Scanning tunneling microscopy. Individual atoms of the element xenon can be clearly distinguished by using a scanning tunneling microscope (STM). This chain of seven xenon atoms was built by IBM scientists moving each atom into position, one at a time. The atoms are 20 billionths of an inch, or 0.5 nm, apart. The atoms are bonded together; moving an end atom will relocate up to three of them at a time. *(Courtesy of International Business Machines Corporation. Unauthorized use not permitted.)*

✓**CHECKLIST**

1. What does the electron microscope use instead of light beams and glass lenses? What does the fluorescence microscope use?

2. How can you distinguish between TEM and SEM micrographs? Which type is Figure 3.25?

3. Why are colored photos of electron micrographs referred to as "colorized" or "false color"?

4. Rank the following types of microscopes according to the wavelength of illuminating beam they use, beginning with the longest wavelength: fluorescent, TEM, bright-field. What effect does wavelength have on the resolving power of these microscopes?

TABLE 3.2

Comparison of Types of Microscopy

Type	Special Features		Appearance	Uses
Bright-Field	Uses visible light; simple to use, least expensive	a	Colored or clear specimen on light background	Observation of dead stained organisms or live ones with sufficient natural color contrast
Dark-Field	Uses visible light with a special condenser that causes light rays to reflect off specimen at an angle	b	Bright specimen on dark background	Observation of unstained living or difficult-to-stain organisms Allows one to see motion
Phase-Contrast	Uses visible light plus phase-shifting plate in objective with a special condenser that causes some light rays to strike specimen out of phase with each other	c	Specimen has different degrees of brightness and darkness	Detailed observation of internal structure of living unstained organisms
Nomarski	Uses visible light out of phase; has higher resolution than standard phase-contrast microscope	d	Produces a nearly three-dimensional image	Observation of finer details of internal structure of living unstained organisms
Fluorescence	Uses ultraviolet light to excite molecules to emit light of different wavelengths, often brilliant colors, because UV can burn eyes, special lens materials are used	e	Bright, fluorescent, colored specimen on dark background	Diagnostic tool for detection of organisms or antibodies in clinical specimens or for immunologic studies
Confocal	Uses laser light to obtain thin focal-level sections through a specimen, with 40 times greater resolution and less out-of-focus light	f	Non-fuzzy thin-section image	Observation of very specific levels of specimen
Digital	Uses computer technology to automatically focus, adjust light, and take photographs of specimens; can put directly online	g	Standard image	Ease of operation plus online use
Transmission Electron	Uses electron beam instead of light rays and electromagnetic lenses instead of glass lenses; image is projected on a video screen; very expensive; preparation requires considerable time and practice	h	Highly magnified, detailed image; not three-dimensional except with shadow casting	Examination of thin sections of cells for details of internal structure, exterior of cells, and viruses, or surfaces when freeze-fracturing is used
Scanning Electron	Uses electron beam and electromagnetic lenses; expensive; preparation requires considerable time and practice	i	Three-dimensional view of surfaces	Observation of exterior surfaces of cells or internal surfaces
Scanning Tunneling	Uses a wire probe to trace over surfaces, allowing electrons to move (tunnel), thereby generating electric currents that reveal highs and lows of specimen's surface	j	Three-dimensional view of surfaces	Observation of exterior surfaces of atoms or molecules

TECHNIQUES OF LIGHT MICROSCOPY

Microscopes are of little use unless the specimens for viewing are prepared properly. Here we explain some important techniques used in light microscopy.

Although resolution and magnification are important in microscopy, the degree of contrast between structures to be observed and their backgrounds is equally important. Nothing can be seen without contrast, so special techniques have been developed to enhance contrast.

PREPARATION OF SPECIMENS FOR THE LIGHT MICROSCOPE

Wet Mounts

Wet mounts, in which a drop of medium containing the organisms is placed on a microscope slide, can be used to view living microorganisms. The addition of a 2% solution of carboxymethylcellulose, a thick, syrupy solution, helps to slow fast-moving organisms so they can be studied. A special version of the wet mount, called a **hanging drop**, often is used with dark-field illumination (Figure 3.30). A drop of culture is placed on a coverslip that is encircled with petroleum jelly. The coverslip and drop are then inverted over the well of a depression slide. The drop hangs from the coverslip, and the petroleum jelly forms a seal that prevents evaporation. This preparation gives good views of microbial motility.

Smears

Smears, in which microorganisms from a loopful of medium are spread onto the surface of a glass slide, can be used to view killed organisms. Although they are living when placed on the slide, the organisms are killed by the techniques used to fix (attach) them to the slide. Smear preparation often is difficult for beginners. If you make smears too thick, you will have trouble seeing individual cells; if you make them too thin, you may find no organisms. If you stir the drop of medium too much as you spread it on the slide, you will disrupt cell arrangements. You may see organisms that normally appear in tetrads (groups of four) as single or double organisms. Such variations lead some beginners to imagine that they see more than one kind of organism when, in fact, the organisms are all of the same species.

After a smear is made, it is allowed to air-dry completely. Then it is quickly passed three or four times through an open flame. This process is called **heat fixation**. Heat fixation accomplishes three things: (1) It kills the organisms, (2) it causes the organisms to adhere to the slide, and (3) it alters the organisms so that they more readily accept stains (dyes). If the slide is not completely dry when you pass it through the flame, the organisms will be boiled and destroyed. If you heat-fix

Depression (well) slide — Coverslip — Vaseline jelly (seal)

Drop of liquid with specimen (hanging upside down from undersurface of coverslip)

(a)

(b) LM

Figure 3.30 The hanging-drop technique. (a) A drop of culture is placed on a coverslip, ringed with petroleum jelly and then inverted and placed over the well in a depression slide. The petroleum jelly forms a seal to prevent evaporation. **(b)** Dark-field micrograph of a hanging-drop preparation (2500X) showing the spiral bacterium *Treponema pallidum*, the cause of syphilis. *(A. M. Siegelman/Visuals Unlimited)*

too little, the organisms may not stick and will wash off the slide in subsequent steps. Any cells remaining alive will stain poorly. If you heat-fix too much, the organisms may be incinerated, and you will see distorted cells and cellular remains. Certain structures, such as the capsules found on some microbes, are destroyed by heat-fixing, so this step is omitted and these microbes are affixed to the slide just by air-drying.

PRINCIPLES OF STAINING

A **stain**, or dye, is a molecule that can bind to a cellular structure and give it color. Staining techniques make the microorganisms stand out against their backgrounds. They are also used to help investigators group major categories of microorganisms, examine the structural and chemical differences in cellular structures, and look at the parts of the cell.

In microbiology the most commonly used dyes are **cationic** (positively charged), or **basic, dyes**, such as methylene blue, crystal violet, safranin, and malachite green. These dyes are attracted to any negatively charged cell components. The cell membranes of most bacteria have negatively charged surfaces and thus

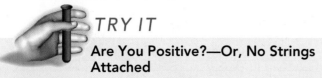

TRY IT

Are You Positive?—Or, No Strings Attached

The Gram stain is not foolproof. Some anaerobic Gram-positive organisms decolorize easily and may falsely appear to be Gram-negative. Gram-negative organisms such as *Streptobacillus moniliformis* can stain Gram-positive. Is there some way to be sure of the organism's correct Gram reaction? There are several ways. One is the potassium hydroxide (KOH) test.

Place 2 drops of a 3% solution of KOH on a slide. Remove an inoculating loopful of the organism in question from a pure colony. Add it to the KOH on the slide, and mix continuously for 30 seconds. As you stir, occasionally lift the loop up 1 or 2 cm from the surface to see if "strings" of gooey material hang down. If the organism is truly Gram-negative, the KOH will break down its cell walls, releasing its DNA and forming strings. Gram-positive organisms will not form strings.

attract the positively charged basic dyes. Other stains, such as eosin and picric acid, are **anionic** (negatively charged), or **acidic**, **dyes**. They are attracted to any positively charged cell materials.

Two main types of stains, simple stains and differential stains, are used in microbiology. They are compared in **Table 3.3**. A **simple stain** makes use of a single dye and reveals basic cell shapes and cell arrangements. Methylene blue, safranin, carbolfuchsin, and crystal violet are commonly used simple stains. A **differential stain** makes use of two or more dyes and distinguishes between two kinds of organisms or between two different parts of an organism. Common differential stains are the Gram stain, the Ziehl-Neelsen acid-fast stain, and the Schaeffer-Fulton spore stain.

The Gram Stain

The **Gram stain**, probably the most frequently used differential stain, was devised by a Danish physician, Hans Christian Gram, in 1884. Gram was testing new methods of staining biopsy and autopsy materials, and he noticed that with certain methods some bacteria were stained differently than the surrounding tissues. As a result of his experiments with stains, the highly useful Gram stain was developed. In Gram staining, bacterial cells take up crystal violet. Iodine is then added; it acts as a **mordant**, a chemical that helps retain the stain in certain cells. Those structures that cannot retain crystal violet are decolorized with 95% ethanol or an ethanol-acetone solution, rinsed, and subsequently stained (counterstained) with safranin. The steps in the Gram-staining procedure are shown in **Figure 3.31**.

Four groups of organisms can be distinguished with the Gram stain: (1) *Gram-positive* organisms, whose cell walls retain crystal violet stain; (2) *Gram-negative* organisms, whose cell walls do not retain crystal violet

stain; (3) *Gram-nonreactive* organisms, which do not stain or which stain poorly; and (4) *Gram-variable* organisms, which stain unevenly. The differentiation between Gram-positive and Gram-negative organisms reveals a fundamental difference in the nature of the cell walls of bacteria, as is explained in ◀Chapter 4. Furthermore, the reactions of bacteria to the Gram stain have helped in distinguishing Gram-positive, Gram-negative, and Gram-nonreactive groups that belong to radically different taxonomic groups ◀(Chapter 9).

Gram-variable organisms have somehow lost their ability to react distinctively to the Gram stain. Organisms from cultures over 48 hours old (and sometimes only 24 hours old) are often Gram-variable, probably because of changes in the cell wall with aging. Therefore, to determine the reaction of an organism to the Gram stain, you should use organisms from cultures 18–24 hours old.

The Ziehl-Neelsen Acid-Fast Stain

The **Ziehl-Neelsen acid-fast stain** is a modification of a staining method developed by Paul Ehrlich in 1882. It can be used to detect tuberculosis- and leprosy-causing organisms of the genus *Mycobacterium* (**Figure 3.32**). Slides of organisms are covered with carbolfuchsin and are heated, rinsed, and decolorized with 3% hydrochloric acid (HCl) in 95% ethanol, rinsed again, and then stained with Loeffler's methylene blue. Most genera of bacteria will lose the red carbolfuchsin stain when decolorized. However, those that are "*acid-fast*" retain the bright red color. The lipid components of their walls, which are responsible for this characteristic, are discussed in ◀Chapter 4. Bacteria that are not acid-fast lose the red color and can therefore be stained blue with the Loeffler's methylene blue counterstain.

Special Staining Procedures

Negative Staining. Negative stains are used when a specimen—or a part of it, such as the capsule—resists taking up a stain. The *capsule* is a layer of polysaccharide material that surrounds many bacterial cells and can act as a barrier to host defense mechanisms. It also repels stains. In negative staining, the background around the organisms is filled with a stain, such as India ink, or an acidic dye, such as nigrosin. This process leaves the organisms themselves as clear, unstained objects that stand out against the dark background. A second simple or differential stain can be used to demonstrate the presence of the cell inside the capsule. Thus, a typical slide will show a dark background and clear, unstained areas of capsular material, inside of which are purple cells stained with crystal violet (**Figure 3.33**) or blue cells stained with methylene blue.

Flagellar Staining. *Flagella*, appendages that some cells have and use for locomotion, are too thin to be seen easily with the light microscope. When it is necessary to determine their presence or arrangement,

TABLE 3.3

Comparison of Staining Techniques				
Type	**Examples**		**Result**	**Uses**
Simple Stains				
Use a single dye; do not distinguish organisms or structures by different staining reactions	Methylene blue Safranin Crystal violet ⟶	a	Uniform blue stain Uniform red stain Uniform purple stain	Shows sizes, shapes, and arrangements of cells
Differential Stains				
Use two or more dyes that react differently with various kinds or parts of bacteria, allowing them to be distinguished	Gram stain	b	Gram +: purple with crystal violet Gram −: red with safranin counterstain Gram-variable: intermediate or mixed colors (some stain + and some − on same slide) Gram-nonreactive: stain poorly or not at all	Distinguishes Gram +, Gram −, Gram-variable, and Gram nonreactive organisms
	Ziehl-Neelsen acid-fast stain	c	Acid-fast bacteria retain carbolfuchsin and appear red. Non–acid-fast bacteria accept the methylene blue counterstain and appear blue	Distinguishes members of the genera *Mycobacterium* and *Nocardia* from other bacteria
	Negative stain	d	Capsules appear clear against a dark background	Allows visualization of organisms with structures that will not accept most stains, such as capsules
Special Stains				
Identify various specialized structures	Flagellar stain		Flagella appear as dark lines with silver, or red with carbolfuchsin	Indicates presence of flagella by building up layers of stain on their surface
	Schaeffer-Fulton spore stain	e	Endospores retain malachite green stain. Vegetative cells accept safranin counterstain and appear red	Allows visualization of hard-to-stain bacterial endospores such as members of genera *Clostridium* and *Bacilllus*

(a: Michael Abbey/Visuals Unlimited; b: Michael Abbey/Visuals Unlimited; c: John D. Cunningham/Visuals Unlimited; d: Jack Bostrack/Visuals Unlimited; e: CDC/Courtesy of Larry Stauffer/Oregon State Public Health Lab)

flagellar stains are painstakingly prepared to coat the surfaces of the flagella with dye or a metal such as silver. These techniques are very difficult and time consuming and so are usually omitted from the beginning course in microbiology. (See ◀Figure 4.12 for some examples of stained flagella.)

Endospore Staining. A few types of bacteria produce resistant cells called *endospores*. Endospore walls are very resistant to penetration of ordinary stains. When a simple stain is used, the spores will be seen as clear, glassy, easily recognizable areas within the bacterial cell. Thus, strictly speaking, it is not absolutely necessary to perform an endospore stain to see the spores. However, the differential **Schaeffer-Fulton spore stain** makes spores easier to visualize **(Figure 3.34)**. Heat-fixed smears are covered with malachite green and then gently heated until they steam. Approximately

(a) Crystal violet (1 minute)

All purple

Drain, rinse

(b) Iodine (1 minute)

All purple;
iodine acts as mordant
to set stain

Drain, rinse

(c) Decolorize with alcohol
(one quick rinse);
immediately after, rinse with water

Gram + cocci = purple
Gram − rods = clear

(d) Safranin (30–60 seconds)

Gram + cocci = purple
Gram − rods = red (pink)

Drain, rinse, blot

(a)

(b)

(c)

(d)

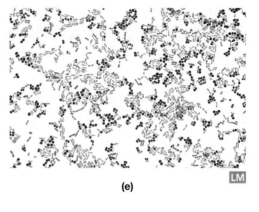

(e)

Figure 3.31 The Gram stain. (a)–(d) Steps in Gram staining. **(e)** Gram-positive cells retain the purple color of crystal violet, whereas Gram-negative cells are decolorized with alcohol and subsequently pick up the red color of the safranin counterstain. *(Michael Abbey/Visuals Unlimited)*

Figure 3.32 The Ziehl-Neelsen acid-fast stain. This stain produces vivid red color in acid-fast organisms such as *Mycobacterium leprae* (magnified 3844X), the cause of leprosy. *(John D. Cunningham/Visuals Unlimited)*

Figure 3.33 Negative staining. Negative staining for capsules reveals a clear area (the capsule, which does not accept stain) in a dark pink background of India ink and crystal violet counterstain. The cells themselves are stained deep purple with the counterstain. The bacteria are *Streptococcus pneumoniae* (3399X), which are arranged in pairs. *(Jack Bostrack/Visuals Unlimited)*

Figure 3.34 The Schaeffer-Fulton spore stain. Endospores of *Bacillus megaterium* (2,335X) are visible as green, oval structures inside and outside the rod-shaped cells. Vegetative cells, which represent a non–spore-forming stage, and cellular regions without spores stain red. *(CDC/Courtesy of Larry Stauffer /Oregon State Public Health Lab)*

TRY IT

Take a Stab at It

Want to find spores in the real world around you? Take a straight inoculating needle (no loop at the end), rub it around in some soil, and then "stab" it with a single stroke down to the bottom of a deep test tube full of nutrient agar. Withdraw the needle and incubate the tube for 24 to 48 hours. Growth should become obvious along the pathway of the stab. Remove some of the growth and stain it by the Schaeffer-Fulton method. You should see lots of cells containing endospores, probably members of the genus *Clostridium*, famous for causing diseases such as tetanus, botulism, and gas gangrene.

What do you think the oxygen conditions are deep in the agar? What might this have to do with endospore formation?

5 minutes of such steaming causes the endospore walls to become more permeable to the dye. However, newer stains are now available that do not require steaming. The slide is then rinsed with water for 30 seconds to remove the green dye from all parts of the cell except for the endospores, which retain it. Then a counterstain of safranin is placed on the slide to stain the non–spore-forming, or vegetative, areas of the cells. Cells of cultures without endospores appear red; those with endospores have green spores and red vegetative cells.

Although microscopy and staining techniques can offer valuable information about microorganisms, these methods are usually not enough to permit identification of most microbes. Many species look identical under the microscope—after all, there are only a limited number of basic shapes, arrangements, and staining reactions but thousands of kinds of bacteria. This means that biochemical and genetic characteristics usually must be determined before an identification can be made ◄(Chapter 9).

✓CHECKLIST

1. What can be observed in a wet-mount or hanging-drop preparation that cannot be oserved in heat-fixed slides?

2. What is the difference between simple stains and differential stains?

3. What color are Gram-negative organisms after Gram staining? What color are Gram-positive organisms?

▌ RETRACING OUR STEPS

▌▌▌ HISTORICAL MICROSCOPY

• The existence of microorganisms was unknown until the invention of the microscope. Leeuwenhoek, probably the first to see microorganisms (in the 1600s), set the stage for **microscopy**, the technology of making very small things visible to the human eye.

• Leeuwenhoek's simple microscopes could reveal little detail of specimens. Today, multiple-lens, compound microscopes give us nearly distortion-free images, enabling us to delve further into the study of microbes.

▌▌▌ PRINCIPLES OF MICROSCOPY

METRIC UNITS

• The three units most used to describe microbes are the **micrometer** (μm), formerly called a micron, which is equal to 0.000001 m, also written as 10^{-6} m; the **nanometer** (nm), formerly called a millimicron (mμ), which is equal to 0.000000001 m, or 10^{-9} m; and the **angstrom** (Å), which is equal to 0.0000000001 m, 0.1 nm, or 10^{-10} m, but is no longer officially recognized.

PROPERTIES OF LIGHT: WAVELENGTH AND RESOLUTION

• The **wavelength**, or the length of light rays, is the limiting factor in resolution.

• **Resolution** is the ability to see two objects as separate, discrete entities. Light wavelengths must be small enough to fit between two objects for them to be resolved.

• **Resolving power** can be defined as RP = λ/2NA, where λ = wavelength of light. The smaller the value of λ and the larger the value of NA, the greater the resolving power of the lens.

• **Numerical aperture** (NA) relates to the extent to which light is concentrated by the condenser and collected by the objective. Its value is engraved on the side of each objective lens.

PROPERTIES OF LIGHT: LIGHT AND OBJECTS

- If light strikes an object and bounces back, **reflection** (which gives an object its color) has occurred.
- **Transmission** is the passage of light through an object. Light must either be reflected from or transmitted through an object for it to be seen with a light microscope.
- **Absorption** of light rays occurs when they neither bounce off nor pass through an object but are taken up by that object. Absorbed light energy is used in performing photosynthesis or in raising the temperature of the irradiated body.
- Reemission of absorbed light as light of longer wavelengths is known as **luminescence**. If reemission occurs only during irradiation, the object is said to **fluoresce**. If reemission continues after irradiation ceases, the object is said to be **phosphorescent**.
- **Refraction** is the bending of light as it passes from one medium to another of different density. **Immersion oil**, which has the same **index of refraction** as glass, is used to replace air and to prevent refraction at a glass-air interface.
- **Diffraction** is the bending of light waves as they pass through a small opening, such as a hole, a slit, a space between two adjacent cellular structures, or a small, high-powered, magnifying lens in a microscope. The bent light rays distort the image obtained and limit the usefulness of the light microscope.

▌ LIGHT MICROSCOPY

THE COMPOUND LIGHT MICROSCOPE

The major parts of a compound light microscope and their functions are as follows:
- **Base** Supporting structure that generally contains the light source.
- **Condenser** Converges light beams to pass through the specimen.
- **Iris diaphragm** Controls the amount of light passing through the specimen.
- **Objective lens** Magnifies image.
- **Body tube** Conveys light to the ocular lens.
- **Ocular lens** Magnifies the image from the objective. A microscope with one ocular lens (eyepiece) is **monocular**; a microscope with two oculars is **binocular**.
- **Mechanical stage** Allows precise control in moving the slide.
- **Coarse adjustment** Knob used to locate specimen.
- **Fine adjustment** Knob used to bring specimen into sharp focus.
- The **total magnification** of a light microscope is calculated by multiplying the magnifying power of the objective lens by the magnifying power of the ocular lens. Increased magnification is of no value unless good resolution can also be maintained.

DARK-FIELD MICROSCOPY

- **Bright-field illumination** is used in the ordinary light microscope, with light passing directly through the specimen.
- **Dark-field illumination** uses a special condenser that causes light to reflect off the specimen at an angle rather than pass directly through it.

PHASE-CONTRAST MICROSCOPY

- **Phase-contrast microscopy** uses microscopes with special condensers that accentuate small differences in the refractive index of structures within the cell, allowing live, unstained organisms to be examined.

NOMARSKI (DIFFERENTIAL INTERFERENCE CONTRAST) MICROSCOPY

- **Nomarski microscopy** uses microscopes that operate essentially like phase-contrast microscopes but with a much greater resolution and a very short depth of field. They produce a nearly three-dimensional image.

FLUORESCENCE MICROSCOPY

- **Fluorescence microscopy** uses ultraviolet light instead of white light to excite molecules within the specimen or dye molecules attached to the specimen. These molecules emit different wavelengths, often of brilliant colors.

CONFORAL MICROSCOPY

- **Confocal microscopy** uses laser light to obtain thin, focal-level sections through a specimen, with 40 times greater resolution, and less out-of-focus light.

DIGITAL MICROSCOPY

- **Digital microscopy** uses computer technology to automatically focus, adjust light, and take photographs of specimens. These can be directly uploaded and viewed online.

▌ ELECTRON MICROSCOPY

- The **electron microscope (EM)** uses a beam of electrons instead of a beam of light and electromagnets instead of glass lenses for focusing. They are much more expensive and difficult to use but give magnifications of up to 500,000X and a resolving power of less than 1 nm. Viruses can be seen only by using EMs.
- Advanced types of EMs can visualize actual molecules and individual atoms.

TRANSMISSION ELECTRON MICROSCOPY

- For the **transmission electron microscope (TEM)**, very thin slices (sections) of a specimen are used, revealing the internal structure of microbial and other cells.

SCANNING ELECTRON MICROSCOPY

- For the **scanning electron microscope (SEM)**, a specimen is coated with a metal. The electron beam is scanned, or swept, over this coating to form a three-dimensional image.

SCANNING TUNNELING MICROSCOPY

- The **scanning tunneling microscope (STM)** can produce three-dimensional images of individual molecules and atoms, as well as movies. The atomic force microscope also can show molecular changes at cell surfaces.

▌ TECHNIQUES OF LIGHT MICROSCOPY

PREPARATION OF SPECIMENS FOR THE LIGHT MICROSCOPE

- **Wet mounts** are used to view living organisms. The **hanging-drop** technique is a special type of wet mount, often used to determine whether organisms are motile.
- **Smears** of appropriate thickness are allowed to air-dry completely and are then passed through an open flame. This process, called **heat fixation**, kills the organisms, causing them to adhere to the slide and more readily accept stains.

PRINCIPLES OF STAINING

• A **stain**, or dye, is a molecule that can bind to a structure and give it color.

• Most microbial stains are **cationic** (positively charged), or **basic**, **dyes**, such as methylene blue. Because most bacterial surfaces are negatively charged, these dyes are attracted to them.

• **Simple stains** use one dye and reveal basic cell shapes and arrangements. **Differential stains** use two or more dyes and distinguish various properties of organisms. The **Gram stain**, the **Schaeffer-Fulton spore stain**, and the **Ziehl-Neelsen acid-fast stain** are examples.

• **Negative stains** color the background around cells and their parts, which resist taking up stain.

• **Flagellar stains** add layers of dye or metal to the surface of flagella to make those surfaces visible.

• In the Schaeffer-Fulton spore stain, endospores stain green due to the uptake of malachite green, whereas vegetative cells stain red due to safranin uptake.

▌ TERMINOLOGY CHECK

absorption *(p. 57)*
angstrom *(p. 53)*
anionic (acidic) dye *(p. 70)*
atomic force microscope (AFM) *(p. 67)*
base *(p. 59)*
binocular *(p. 59)*
body tube *(p. 59)*
bright-field illumination *(p. 60)*
cationic (basic) dye *(p. 69)*
coarse adjustment *(p. 59)*
compound light microscope *(p. 59)*
condenser *(p. 59)*
confocal microscopy *(p. 62)*
dark-field illumination *(p. 60)*
differential stain *(p. 70)*
diffraction *(p. 58)*
digital microscope *(p. 63)*

electron micrograph *(p. 64)*
electron microscope (EM) *(p. 64)*
fine adjustment *(p. 59)*
flagellar stain *(p. 71)*
fluoresce *(p. 57)*
fluorescence microscopy *(p. 61)*
fluorescent antibody staining *(p. 61)*
freeze-etching *(p. 65)*
freeze-fracturing *(p. 65)*
Gram stain *(p. 70)*
hanging drop *(p. 69)*
heat fixation *(p. 69)*
immersion oil *(p. 57)*
index of refraction *(p. 57)*
iris diaphragm *(p. 59)*
light microscopy *(p. 59)*
luminescence *(p. 57)*

mechanical stage *(p. 59)*
micrometer *(p. 53)*
microscopy *(p. 53)*
monocular *(p. 59)*
mordant *(p. 70)*
nanometer *(p. 53)*
negative stain *(p. 70)*
Nomarski microscopy *(p. 61)*
numerical aperture (NA) *(p. 56)*
objective lens *(p. 59)*
ocular lens *(p. 59)*
ocular micrometer *(p. 60)*
optical microscope *(p. 59)*
parfocal *(p. 60)*
phase-contrast microscopy *(p. 61)*
phosphorescent *(p. 57)*
reflection *(p. 56)*
refraction *(p. 57)*

resolution *(p. 55)*
resolving power (RP) *(p. 56)*
scanning electron microscope (SEM) *(p. 66)*
scanning tunneling microscope (STM) *(p. 66)*
Schaeffer-Fulton spore stain *(p. 71)*
shadow casting *(p. 65)*
simple stain *(p. 70)*
smear *(p. 69)*
stain *(p. 69)*
total magnification *(p. 59)*
transmission *(p. 57)*
transmission electron microscope (TEM) *(p. 64)*
wavelength *(p. 54)*
wet mount *(p. 69)*
Ziehl-Neelsen acid-fast stain *(p. 70)*

▌ CLINICAL CASE STUDY

Mr. Jones comes into the hospital with symptoms that are indicative of any one of three diseases. The treatments and prognoses for these three diseases are all very different. It is important to correctly identify which disease Mr. Jones has, and quickly, because he is very ill. The technician up in the lab has bottles of antibodies labeled with fluorescent dye molecules. How can he determine if the bacterium he cultured from the patient's specimen is one of the 3 suspect organisms? What kind of microscope should he use?

▌ CRITICAL THINKING QUESTIONS

1. Katy started the primary crystal violet of her Gram stain on her "unknown" culture, but ran out of time. She left the slide in her drawer, with plans to complete the washes and staining next week. What has she overlooked that may prevent her from obtaining proper results? If Katy were to initiate a new Gram stain on a smear she made from last week's bacterial culture, would she get good, reliable results? Why or why not?

2. Craig performed a side-by-side Gram stain of a known Gram-positive and a known Gram-negative organism. Unfortunately he forgot to do the iodine step. When he observed his completed slide under oil immersion, what do you think he saw? Why?

3. What are the advantages and disadvantages of observing living-specimen slides and heat-fixed-specimen slides of microorganisms?

▌ SELF-QUIZ

1. If a bacterium is 0.5 μm wide and 15 μm long, how many nanometers wide is it, and how many millimeters long is it?

2. The compound light microscope can be used to observe:
   (a) Atoms, proteins, viruses, and bacteria
   (b) Viruses, bacteria, cell organelles, and red blood cells
   (c) Amino acids, bacteria, and red blood cells
   (d) Ribosomes, bacteria, cell organelles, and red blood cells
   (e) Bacteria, cell organelles, and red blood cells

3. The average wavelength of light visible to our eyes is:
   (a) 800 nm   (d) 100 nm
   (b) 200 nm   (e) 420 nm
   (c) 550 nm

4. What is resolution, and why is it important in microscopy?

5. Define and contrast absorption, reflection, transmission, and refraction. Identify each on the following diagram.

Light source

(a) _____
(b) _____
(c) _____
(d) _____

6. In light microscopes, what function does a condenser serve?
   (a) Focuses the light rays onto our eyes
   (b) Magnifies the light rays after their passage through the sample
   (c) Focuses the light rays on the sample
   (d) Increases light intensity
   (e) Reduces glare

7. Match the following:
   ___Fluorescence   (a) Prevents refraction and blurry images
   ___Diffraction
   ___Immersion oil   (b) When absorbed light rays are changed into longer wavelengths and reemitted
   ___Phosphorescence
   ___Luminescence   (c) Luminescence occurs only when light rays are striking an object
                      (d) Object continues to emit light even after light rays no longer strike it
                      (e) Bending of light rays around an opening

8. The total magnification of a microscope is calculated by:
   (a) Addition of the objective lens and ocular lens magnification powers
   (b) Multiplication of the objective lens and ocular lens magnification powers
   (c) Multiplication of the objective lens and condenser lens magnification powers
   (d) The objective lens power squared
   (e) None of the above

9. Match the following types of microscope to their description:
   ___Phase contrast   (a) Uses visible light out of phase
   ___Dark field   (b) Uses laser light to get thin focal-level sections through specimen, resulting in 40X greater resolution and less out-of-focus light
   ___Bright field
   ___Transmission electron
   ___Confocal   (c) Uses UV light to excite molecules to emit light of different wavelengths
   ___Scanning electron
   ___Fluorescence   (d) Uses visible light, but causes some light rays to strike the specimen out of phase with each other
   ___Nomarski
                            (e) Uses electron beam instead of light rays and electromagnetic lenses instead of glass lenses; useful for viewing surface images of specimen
                            (f) Uses visible light only, with light passing directly through specimen.
                            (g) Uses visible light rays but causes them to reflect off specimen at an angle
                            (h) Uses electron beam and electromagnetic lenses; useful for viewing internal structures of cells

10. What is the difference between a simple and a differential stain?

11. Which of the following stains is used frequently to identify *Mycobacterium* and other bacteria whose cell walls contain high amounts of lipids?
    (a) Gram stain   (d) Lipidialar stain
    (b) Schaeffer-Fulton stain (e) Spore stain
    (c) Acid-fast stain

12. Which of the following stains is used to classify microorganisms based on their cell wall content?
    (a) Capsular stain   (d) Negative stain
    (b) Gram stain   (e) Methylene blue
    (c) Spore stain

13. Which of the following groups of organisms can the Gram stain not distinguish?
    (a) Gram-positive organisms, whose cell walls retain the primary crystal violet stain
    (b) Gram-variable organisms, which stain unevenly
    (c) Gram-negative organisms, whose cell walls retain the mordant iodine
    (d) Gram-nonreactive organisms, which do not stain or stain poorly
    (e) Gram-negative organisms, whose cell walls do not retain the primary crystal violet stain

14. Commonly used dyes in microbiology are _____ (positively charged) or basic dyes that are attracted to negative cell components (such as in most bacterial cell walls) and anionic (negatively charged) or _____ dyes which are attracted to _____ charged cell material.

15. Which of the following can give you ambiguous results for the Gram stain?
(a) Too much decolorizing
(b) Improper heat-fixing
(c) Concentration and freshness of the Gram-staining reagents
(d) Cell density of the smear
(e) a, b, and c
(f) All of the above

16. The order of reagents used in the Gram stain are:
(a) Crystal violet, iodine, safranin, alcohol
(b) Alcohol, crystal violet, iodine, safranin
(c) Iodine, crystal violet, safranin, alcohol
(d) Crystal violet, iodine, alcohol, safranin
(e) Crystal violet, safranin, alcohol, iodine

17. Which of the following is/are true about fluorescent microscopy?
(a) Fluorescent microscopes use an infrared light source
(b) Fluorochromes are sometimes necessary to visualize cellular structures or cells
(c) Antibodies can be "tagged" with fluorescent molecules to help visualize and prove the presence of their corresponding antigen or foreign substance, such as a microbe in a blood sample
(d) a, b, and c
(e) b and c
(f) None of the above

18. All of the following are examples of special stains except:
(a) Endospore stain
(b) Flagellar stain
(c) Ziehl-Neelsen acid-fast stain
(d) Negative stains
(e) All of the above are special stains

19. The presence of a capsule around bacterial cells usually indicates their increased disease-causing potential and resistance to disinfection. Capsules are generally viewed by:
(a) Spore staining
(b) Scanning electron microscopy
(c) Gram staining
(d) Ziehl-Neelsen staining
(e) Negative staining

20. Special stains can definitively help in specific identification of microbes. True or False?

21. Which of the following microscopic techniques provide three-dimensional images of a bacterial cell?
(a) Transmission electron microscopy
(b) Scanning electron microscopy
(c) Negative-staining microscopy
(d) Dark-field microscopy
(e) Fluorescent microscopy

22. Describe the differences between scanning electron microscopy and transmission electron microscopy.

23. The transmission electron microscope has the greatest resolving power because it uses an electron beam to view the sample instead of a light beam. The electron beam is used because:
(a) Electrons have longer wavelengths than light waves
(b) Electrons do not penetrate the sample
(c) Light waves are less visible
(d) Electrons have shorter wavelengths than light waves
(e) Electrons are less invasive

24. Describe the process and advantages of using differential interference contrast (Nomarski) microscopy.

25. Compound Microscope
Label the parts of the microscope (a) through (g) and indicate their function.

(a) _____
(b) _____
(c) _____
(d) _____
(e) _____
(f) _____
(g) _____

26. Match the following microscopic techniques with their respective functions
___Phase-contrast microscopy
___Fluorescent microscopy
___Transmission electron microscopy
___Bright-field microscopy
___Dark-field microscopy
___Scanning electron microscopy

(a) View Gram-stained microbial cells
(b) View nano-sized details of sections through interior of bacteria
(c) View internal structures of unstained, living cells
(d) View antibody-tagged cells
(e) View translucent microbes
(f) View nano-sized details of exterior of bacteria

27. Which of the following are true about wet mounts and the hanging-drop technique?
    (a) Wet mounts are useful for staining microorganisms while the hanging-drop technique is a variation of the wet mount allowing one to use the electron microscope
    (b) Wet mounts are useful for viewing living microorganisms while the hanging-drop technique is a variation of the wet mount allowing one to view their motility
    (c) Wet mounts are useful for preparing smears of microorganisms while the hanging-drop technique is a variation of the wet mount and must be performed upside down
    (d) a and b
    (e) b and c

▌ EXPLORATIONS ON THE WEB

http://www.wiley.com/college/black

If you think you've mastered this chapter, there's more to challenge you on the web. Go to the companion web site to fine-tune your understanding of the chapter concepts and discover answers to the questions posed below.

1. If Galileo and Leeuwenhoek didn't invent the microscope, then who did?

2. In 1660 Marcello Malpighi used a microscope to see capillaries, proving William Harvey's theory on blood circulation.

3. Did you know that in one second, bacteria can move ten times the length of their bodies. In the same time period, Michael Jordan can only move 5.4 times the length of his body.

Characteristics of Prokaryotic and Eukaryotic Cells

Courtesy Jill Banfield, University of California, Berkeley

Come with me . . .

The smallest known bacteria just keep getting smaller. Since 1996, Jill Banfield, at the University of California, Berkeley, has studied the bacteria living in slime that drains from the Richmond Mine at Iron Mountain, California. In December of 2006, she announced that three new species of bacteria, the size of viruses, were found in a pink slick layer which floated on top of green water draining from the mine. This water was as acidic as battery acid, hotter than body temperature at 108°F, and rich in the poisonous metals: arsenic, copper, zinc, and iron. This runoff is so toxic that the Environmental Protection Agency (EPA) catches and treats it rather than letting it pollute local waters. And these microbes create the extreme conditions they live in! Come with me to the website to learn how they do this. And consider whether similar tiny bacteria might be living in soil on Mars, or in kidney stones found in humans?

 Video related to this topic is available within WileyPLUS.

BASIC CELL TYPES

All living cells can be classified as either prokaryotic, from the Greek words *pro* (before) and *karyon* (nucleus), or eukaryotic, from *eu* (true) and *karyon* (nucleus). **Prokaryotic** (pro-kar″e-ot′ik) **cells** lack a nucleus and other membrane-enclosed structures, whereas **eukaryotic** (u-kar″e-ot′ik) **cells** have such structures.

All prokaryotes are single-celled organisms, and all are bacteria. Most of this book will be devoted to the study of prokaryotes. Eukaryotes include all plants, animals, fungi, and protists (organisms such as *Amoeba, Paramecium*, and the malaria parasite). We will also spend some time studying eukaryotes, especially the fungi and various parasites, plus the interactions of eukaryotic cells and prokaryotes.

Prokaryotic and eukaryotic cells are *similar* in several ways. Both are surrounded by a *cell membrane*, or *plasma membrane*. Although some cells have structures that extend beyond this membrane or surround it, the membrane defines the boundaries of the living cell. Both prokaryotic and eukaryotic cells also encode genetic information in DNA molecules.

These two types of cells are *different* in other, important ways. In eukaryotic cells, DNA is in a nucleus surrounded by a membranous *nuclear envelope*, but in prokaryotic cells, DNA is in a nuclear region not surrounded by a membrane. Eukaryotic cells also have a variety of internal structures called **organelles** (or-ga-nelz′), or "little organs," that are surrounded by one or more membranes. Prokaryotic cells generally lack organelles that are membrane-enclosed. We take advantage of some of the differences between eukaryotic human cells and prokaryotic bacterial cells when we try to control disease-causing bacteria without harming the human host.

In this chapter we examine the similarities and differences of prokaryotic cells and eukaryotic cells, as summarized in Table 4.1. (Refer to this table each time you learn

TABLE 4.1

Similarities and Differences Between Prokaryotic and Eukaryotic Cells		
Characteristic	**Prokaryotic Cells**	**Eukaryotic Cells**
Genetic Structures		
Genetic material (DNA)	Usually found in single circular chromosome	Typically found in paired chromosomes
Location of genetic information	Nuclear region (nucleoid)	Membrane-enclosed nucleus
Nucleolus	Absent	Present
Histones	Absent	Present
Extrachromosomal DNA	In plasmids	In organelles, such as mitochondria and chloroplasts, and in plasmids
Intracellular Structures		
Mitotic spindle	Absent	Present during cell division
Plasma membrane	Fluid-mosaic structure lacking sterols	Fluid-mosaic structure containing sterols
Internal membranes	Only in photosynthetic organisms	Numerous membrane-enclosed organelles
Endoplasmic reticulum	Absent	Present
Respiratory enzymes	Cell membrane	Mitochondria
Chromatophores	Present in photosynthetic bacteria	Absent
Chloroplasts	Absent	Present in some
Golgi apparatus	Absent	Present
Lysosomes	Absent	Present
Peroxisomes	Absent	Present
Ribosomes	70S	80S in cytoplasm and on endoplasmic reticulum, 70S in organelles
Cytoskeleton	Absent	Present
Extracellular Structures		
Cell wall	Peptidoglycan found on most cells	Cellulose, chitin, or both found on plant and fungal cells
External layer	Capsule or slime layer	Pellicle, test, or shell in certain protists
Flagella	When present, consist of fibrils of flagellin	When present, consist of complex membrane-enclosed structure with "9 + 2" microtubule arrangement
Cilia	Absent	Present as structures shorter than, but similar to, flagella in some eukaryotic cells
Pili	Present as attachment or conjugation pili in some prokaryotic cells	Absent
Reproductive Process		
Cell division	Binary fission	Mitosis and/or meiosis
Sexual exchange of genetic material	Not part of reproduction	Meiosis
Sexual or asexual reproduction	Only asexual reproduction	Sexual or asexual reproduction

about a new cellular structure.) Viruses do not fit in either category, as they are acellular. However, some viruses infect prokaryotic cells, while other viruses infect eukaryotes. ◄Chapter 10 will examine viruses in detail.

▌▌▌ PROKARYOTIC CELLS

▌▌▌ Detailed studies of cells have revealed that prokaryotes differ enough to be split into two large groups called *domains*. A relatively new concept in biological classification, domain is the highest category, higher even than kingdom. Three domains exist: two prokaryotic and one eukaryotic:

- Archaea (archaeobacteria) (from *archae*, ancient)
- Bacteria (eubacteria)
- Eukarya

All members of Archaea and Bacteria are prokaryotes and have traditionally been called types of bacteria. A problem of terminology arises over the use of a capital versus a lowercase b in the word *bacteria*. All bacteria (lowercase b) are prokaryotes, but not all prokaryotes belong to the domain Bacteria (capital B). The differences between Archaea and Bacteria are not so much structural as molecular. Therefore, most of what we have to say about "bacteria" in this chapter applies to both Archaea and Bacteria. (We will discuss Archaea further in ◄Chapter 11.)

Most bacteria on this planet, both in the environment and living in and on humans, are members of the domain Bacteria. As yet, we know of no disease-causing Archaea, but they may be involved, in disease of the gums. However, they are very important in the ecology of our planet, especially in extreme environments, such as in deep-sea hydrothermal vents, where sulfur-laden water, at temperatures exceeding the boiling point of water, gushes out from openings in the ocean floor.

SIZE, SHAPE, AND ARRANGEMENT

Size

Prokaryotes are among the smallest of all organisms. Most prokaryotes range from 0.5 to 2.0 μm in diameter. For comparison, a human red blood cell is about 7.5 μm in diameter. Keep in mind, however, that although we often use diameter to specify cell size, many cells are not spherical in shape. Some spiral bacteria have a much larger diameter, and some cyanobacteria (formerly called blue-green algae) are 60 μm long. Because of their small size, bacteria have a large surface-to-volume ratio. For example, spherical bacteria with a diameter of 2 μm have a surface area of about 12 μm^2 and a volume of about 4 μm^3. Their surface-to-volume ratio is 12:4, or 3:1. In contrast, eukaryotic cells with a diameter of 20 μm have a surface area of about 1,200 μm^2 and a volume of about 4,000 μm^3. Their surface-to-volume ratio is 1,200:4,000,

or 0.3:1—only one-tenth as great. The large surface-to-volume ratio of bacteria means that no internal part of the cell is very far from the surface and that nutrients can easily and quickly reach all parts of the cell.

Shape

Typically bacteria come in three basic shapes—spherical, rodlike, and spiral (**Figure 4.1**)—but variations abound. A spherical bacterium is called a **coccus** (kok′us; plural: *cocci* [kok′se]), and a rodlike bacterium is called a **bacillus** (ba-sil′us; plural: *bacilli* [bas-il′e]). Some bacteria, called *coccobacilli*, are short rods intermediate in shape between cocci and bacilli. Spiral bacteria have a variety of curved shapes. A comma-shaped bacterium is called a **vibrio** (vib′re-o); a rigid, wavy-shaped one, a **spirillum** (spiril′um; plural: *spirilla*); and a corkscrew-shaped one, a **spirochete** (spi′ro-ket). Some bacteria do not fit any of the preceding categories but rather have spindle shapes or irregular, lobed shapes. Square bacteria were discovered on the shores of the Red Sea in 1981. They are 2 to 4 μm on a side and sometimes aggregate in wafflelike sheets. Triangular bacteria were not discovered until 1986.

Even bacteria of the same kind sometimes vary in size and shape. When nutrients are abundant in the environment and cell division is rapid, rods are often twice as large as those in an environment with only a moderate supply of nutrients. Although variations in shape within a single species of bacteria are generally small, there are exceptions. Some bacteria vary widely in form even within a single culture, a phenomenon known as **pleomorphism**. Moreover, in aging cultures where organisms have used up most of the nutrients and have deposited wastes, cells not only are generally smaller, but they often display a great diversity of unusual shapes.

Arrangement

In addition to characteristic shapes, many bacteria also are found in distinctive arrangements of groups of cells (**Figure 4.2**). Such groups form when cells divide without separating. Cocci can divide in one or more planes, or randomly. Division in one plane produces cells in pairs (indicated by the prefix **diplo-**) or in chains (**strepto-**). Division in two planes produces cells in **tetrads** (four cells arranged in a cube). Division in three planes produces **sarcinae** (singular: *sarcina*; eight cells arranged in a cube). Random division planes produce grapelike clusters (**staphylo-**). Bacilli divide in only one

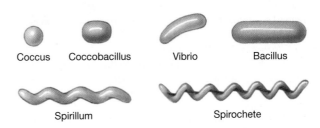

Coccus   Coccobacillus   Vibrio   Bacillus

Spirillum   Spirochete

Figure 4.1   The most common bacterial shapes.

Figure 4.2 Arrangements of bacteria. (a) Cocci arranged in pairs (diplococci of *Neisseria*) and in chains (*Streptococcus*), formed by division in one plane (top, 22,578X; bottom, 9,605X). *(Kwangshin Kim/Photo Researchers, Inc., David M. Philips/Visuals Unlimited)* **(b)** Cocci arranged in a tetrad (*Merisopedia*, 100X), formed by division in two planes. *(Science Vu/Visuals Unlimited)* **(c)** Cocci arranged in a sarcina (*Sarcina lutea*, 16,000X), formed by division in three planes. *(R. Kessel & G. Shih/Visuals Unlimited)* **(d)** Cocci arranged randomly in a cluster (*Staphylococcus*, 5,400X), formed by division in many planes. *(Dr. Tony Brain/Photo Researchers, Inc.)* **(e)** Bacilli arranged in chains are called streptobacilli (*Bacillus megaterium*, 6,017X) *(David Scharf/Science Faction)* **(f)** Bacillus arranged in a rosette (*Caulobacter*, 2,400X), attached by stalks to a substrate. *(Courtesy James T. Staley, University of Washington)* **(g)** Star-shaped bacteria (*Stella*). *(Courtesy Dr. Heinz Schlesner, University of Kiel, Germany)* **(h)** Square-shaped bacterium, *Haloarcula*, a salt-loving member of the Archaea. *(Courtesy Dr. Mike Dyall-Smith, University of Melbourne, Australia)*

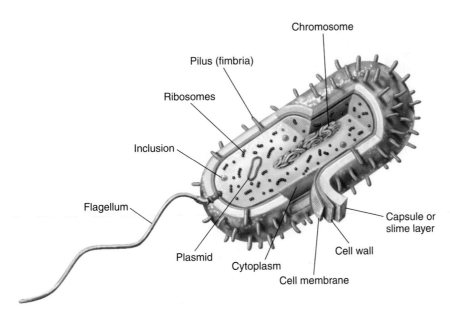

Chromosome

Pilus (fimbria)

Ribosomes

Inclusion

Flagellum

Plasmid

Cytoplasm

Cell membrane

Cell wall

Capsule or slime layer

Figure 4.3 A typical prokaryotic cell. The cell depicted is a bacillus with a polar flagellum (a flagellum at one end).

plane, but they can produce cells connected end-to-end (like train cars) or side-by-side. Spiral bacteria are not generally grouped together.

Prokaryotes divide by binary fission, rather than by mitosis or meiosis. New cell wall material grows, and the cell pinches in half through this area. Inside, the chromosome has duplicated, and one is found in each daughter cell.

✓CHECKLIST

1. Are viruses prokaryotes? Eukaryotes? Why or why not?

2. Compare surface-to-volume ratios of prokaryotes and eukaryotes. Of what importance is this difference?

3. Explain how sex is *not* related to reproduction in prokaryotes. What then *is* its purpose?

4. Prokaryotes lack mitochondria. What structure *does* perform the functions of mitochondria in prokaryotes?

AN OVERVIEW OF STRUCTURE

Structurally, bacterial cells **(Figure 4.3)** consist of the following:

1. A cell membrane, usually surrounded by a cell wall and sometimes by an additional outer layer.
2. An internal cytoplasm with ribosomes, a nuclear region, and in some cases granules and/or vesicles.
3. A variety of external structures, such as capsules, flagella, and pili.

Let us look at each of these kinds of structures in some detail.

THE CELL WALL

The semirigid **cell wall** lies outside the cell membrane in nearly all bacteria. It performs two important functions. First, it maintains the characteristic shape of the cell. If the cell wall is digested away by enzymes, the cell takes on a spherical shape. Second, it prevents the cell from

bursting when fluids flow into the cell by *osmosis* (described later in this chapter). Although the cell wall surrounds the cell membrane, in many cases it is extremely porous and does not play a major role in regulating the entry of materials into the cell.

Components of Cell Walls

Peptidoglycan.   **Peptidoglycan** (pep″ti-do-gly′-kan), also called *murein* (from *murus*, wall), is the single most important component of the bacterial cell wall. It is a polymer so large that it can be thought of as one immense, covalently linked molecule. It forms a supporting net around a bacterium that resembles multiple layers of chain-link fence **(Figure 4.4)**. Gram-positive cells may have as many as 40 such layers. In the peptidoglycan polymer, molecules of *N*-acetylglucosamine (gluNAc) alternate with molecules of *N*-acetylmuramic acid

CLOSE UP

Having Trouble Visualizing Cell Wall Structure?

Yes, it's hard to envision something in three-dimensions when all you've got is a drawing on a flat two-dimensional page. So, try thinking of a bacterial cell wall in this way: imagine two chain-link fences running parallel to each other about a foot or two apart. Add lots of strong metal bars reaching across the space between them, locking them firmly to one another. Now you have a nice sturdy structure. Next, add many more rows of chain-link fences and link them all to each other with lots of those metal crossbars. Do you think it would be easy to break through such a massive fence? No, and it's not easy to break through the bacterial cell wall, either. But there are guys outside the fence with strong "wire cutters" (e.g., antibiotics and enzymes) who can sometimes get in.

(a)

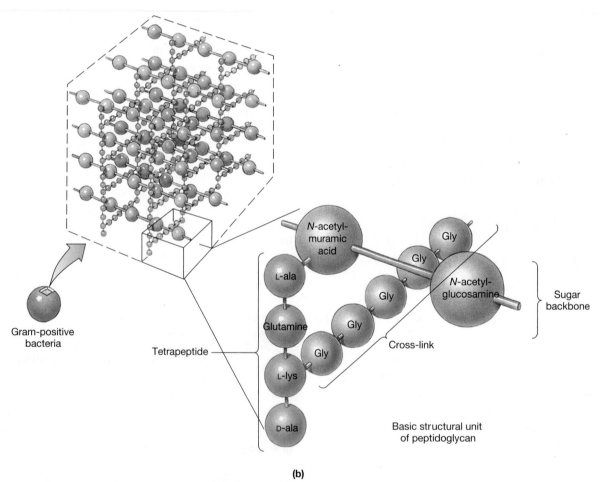

(b)

Figure 4.4 Peptidoglycan. **(a)** A two-dimensional view of the peptidoglycan of the Gram-negative bacterium *Escherichia coli*; a polymer of two alternating sugar units (purple), *N*-acetylglucosamine and *N*-acetylmuramic acid, both of which are derivatives of glucose. The sugars are joined by short peptide chains (tetrapeptides) that consist of four amino acids (red). The sugars and tetrapeptides are cross-linked by a simple peptide bond. **(b)** A three-dimensional view of peptidoglycan for the Gram-positive bacterium *Staphylococcus aureus*. Amino acids are shown in red. Compare the components with those in **(a)**. Different organisms can have different amino acids in the tetrapeptide chain, as well as different cross-links.

Figure 4.5 Lipopolysaccharide. (a) Lipopolysaccharide (LPS), also called endotoxin, is an important component of the outer membrane in Gram-negative cell walls. The lipid A portion of the molecule consists of a backbone of alternating pyrophosphate units (POP, linked phosphate groups) and glucosamine (G, a glucose derivative), to which long fatty acid side chains are attached. Lipid A is a toxic substance that contributes to the danger of infection by Gram-negative bacteria. Polysaccharide side chains extending outward from the glucosamine units make up the remainder of the molecule. (b) Waiting too long before beginning antibiotic therapy for a Gram-negative infection is dangerous. Killing a large population of Gram-negative cells can lead to a huge release of LPS (endotoxin) as their cell walls disintegrate. Hospitalization and even death can result. (*Arthur Tilley/Taxi/Getty Images*)

(murNAc). These molecules are cross-linked by tetrapeptides, chains of four amino acids. In most Gram-positive organisms, the third amino acid is lysine; in most Gram-negative organisms, it is diaminopimelic acid. Amino acids, like many other organic compounds, have *stereoisomers*—structures that are mirror images of each other, just as a left hand is a mirror image of a right hand. Some of the amino acids in the tetrapeptide chains are mirror images of those amino acids most commonly found in living things. Those chains are not readily broken down because most organisms lack enzymes that can digest the stereoisomeric forms.

Cell walls of Gram-positive organisms have an additional molecule, teichoic acid. **Teichoic** (tie-ko′ik) **acid**, which consists of glycerol, phosphates, and the sugar alcohol ribitol, occurs in polymers up to 30 units long. These polymers extend beyond the rest of the cell wall, even beyond the capsule in encapsulated bacteria. Although its exact function is unclear, teichoic acid furnishes attachment sites for bacteriophages (viruses that infect bacteria) and probably serves as a passageway for movement of ions into and out of the cell.

Outer Membrane. The **outer membrane**, found primarily in Gram-negative bacteria, is a bilayer membrane (discussed later in this chapter). It forms the outermost layer of the cell wall and is attached to the peptidoglycan by an almost continuous layer of small lipoprotein molecules (proteins combined with a lipid). The lipoproteins are embedded in the outer membrane and covalently bonded to the peptidoglycan. The outer membrane acts as a coarse sieve and exerts little control over the movement of substances into and out of the cell. However, it does control the transport of certain proteins from the environment. Proteins called porins form channels through the outer membrane. Gram-negatives are less sensitive to penicillin than are Gram-positives, in part because the outer membrane inhibits entrance of penicillin into the cell. The outer surface of the outer membrane has surface antigens and receptors. Certain viruses can bind to some receptors as the first step in infecting the bacterium.

Lipopolysaccharide (LPS), also called **endotoxin**, is an important part of the outer membrane and can be used to identify Gram-negative bacteria. It is an integral part of the cell wall and is not released until the cell walls of dead bacteria are broken down. LPS consists of polysaccharides and **lipid A (Figure 4.5)**. The polysaccharides are found in repeating side chains that extend outward from the organism. It is these repeating units that are used to identify different Gram-negative bacteria. The lipid A portion is responsible for the toxic properties that make any Gram-negative infection a potentially serious medical problem. It causes fever and dilates blood vessels, so the blood pressure drops precipitously. Because bacteria release endotoxin mainly when they are dying, killing them may increase the concentration of this very toxic substance. Thus, antibiotics given late in an infection may cause a worsening of symptoms, or even death of the patient.

Periplasmic Space. Another distinguishing characteristic of many bacteria is the presence of a gap between the cell membrane and the cell wall. The gap is most easily observed by electron microscopy of Gram-negative bacteria. In these organisms the gap is called the

CLOSE UP

The Big and Bigger Bacteria

An exception to the rule of prokaryotic cells being very tiny, found in 1985, lives symbiotically inside the intestines of sturgeon fish caught in the Red Sea and on the Great Barrier Reef of Australia. The bacterium, *Epulopiscium fishelsoni*, can be seen with the naked eye. At 600 μm long by 80 μm in diameter, it is several times the length of single-celled eukaryotes such as *Paramecium caudatum*, and a million times longer than the bacterium *Escherichia coli*. You can see it with the naked eye—no microscope needed! But that's only one of its unique features. These bacteria reproduce not by binary fission, but by a strange kind of "live birth." Inside the cytoplasm of the parent cell, two miniature "baby bacteria" develop. These are released ("born") through a slit-like opening at one end of the parent cell. It is thought that this mechanism may be related to ancestors who were spore formers. Another unique feature of *Epulopiscium* is that, because of its great size, it may be possible to carefully insert intracellular probes without significantly damaging the cell. All sorts of cellular functions could then be studied. Obviously, regular bacteria are too tiny for such experiments.

Then in 1997, a German marine microbiology student, Heide Shultz, discovered the largest bacterium yet known, a huge coccus that can reach up to 750 μm in diameter. It lives in ocean sediments off the coast of Namibia, Africa, growing in chains of up to 50 cells, which glisten white from hundreds of sulfur granules inside its cytoplasm, making it look like a strand of pearls. It was given the very appropriate name of *Thiomargarita namibia* (*thio*, sulfur; *margarita*, pearl). Their giant size allows them room to store enough food to last 3 months without receiving additional nutrients.

Epulopiscium fishelsoni. (Esther R. Angert/Phototake) SEM

Thiomargarita namibia, giant coccus, containing SEM
hundreds of sulfur granules. *(Courtesy Heide Schulz, Max Planck Institute for Marine Mikrobiology, Bremen, Germany)*

periplasmic (per′e-plaz″mik) **space**. It represents a very active area of cell metabolism. This space contains not only the cell wall peptidoglycan but also many digestive enzymes and transport proteins that destroy potentially harmful substances and transport metabolites into the bacterial cytoplasm, respectively. The *periplasm* consists of the peptidoglycan, protein constituents, and metabolites found in the periplasmic space.

Periplasmic spaces are rarely observed in Gram-positive bacteria. However, such bacteria must accomplish many of the same metabolic and transport functions that Gram-negative bacteria do. At present most Gram-positive bacteria are thought to have only periplasms—not periplasmic spaces—where metabolic digestion occurs and new cell wall peptidoglycan is attached. The periplasm in Gram-positive cells is thus part of the cell wall.

TABLE 4.2

Characteristics of the Cell Walls of Gram-Positive, Gram-Negative, and Acid-Fast Bacteria			
Characteristic	**Gram-Positive Bacteria**	**Gram-Negative Bacteria**	**Acid-Fast Bacteria**
Peptidoglycan	Thick layer	Thin layer	Relatively small amount
Teichoic acid	Often present	Absent	Absent
Lipids	Very little present	Lipopolysaccharide	Mycolic acid and other waxes and glycolipids
Outer membrane	Absent	Present	Absent
Periplasmic space	Absent	Present	Absent
Cell shape	Always rigid	Rigid or flexible	Rigid or flexible
Results of enzyme digestion	Protoplast	Spheroplast	Difficult to digest
Sensitivity to dyes and antibiotics	Most sensitive	Moderately sensitive	Least sensitive

Figure 4.6 The bacterial cell wall. Schematic drawings, paired with TEM photos of representative bacteria. **(a)** Gram-positive (*Bacillus fastidosus*), magnification unknown *(Biological Photo Service)*, **(b)** Gram-negative (*Azomonas insignis*) (280,148X), *(Dr. T. J. Beveridge/Biological Photo Service)*, **(c)** acid-fast (*Mycobacterium phlei*) (24,013X). *(Terrance J. Beveridge & T. Paul/Visuals Unlimited)*

Distinguishing Bacteria by Cell Walls

Certain properties of cell walls produce different staining reactions. Gram-positive, Gram-negative, and acid-fast bacteria can be distinguished on the basis of these reactions (Table 4.2 and Figure 4.6).

Gram-Positive Bacteria. The cell wall in Gram-positive bacteria has a relatively thick layer of peptidoglycan, 20 to 80 nm across. The peptidoglycan layer is closely attached to the outer surface of the cell membrane. Chemical analysis shows that 60 to 90% of the cell wall of a Gram-positive bacterium is peptidoglycan. Except for those of streptococci, most Gram-positive cell walls contain very little protein. If peptidoglycan is digested from their cell walls, Gram-positive bacteria become **protoplasts**, or cells with a cell membrane but no cell wall. Protoplasts shrivel or burst unless they are kept in an *isotonic* solution—a solution that has the same pressure as that inside the cell.

The thick cell walls of Gram-positive bacteria retain such stains as the crystal violet-iodine dye in the cytoplasm, but yeast cells, many of which have thick walls, but no peptidoglycan, also retain these stains. Thus, retention of Gram stain seems to be directly related to wall thickness and not to peptidoglycan. Physiological damage or aging can make a Gram-positive cell wall leaky, so the dye complex escapes. Such organisms can become Gram-variable or even Gram-negative as they age. Therefore, Gram staining must be performed on cultures less than 24 hours old.

Gram-positive bacteria lack both an outer membrane and a periplasmic space. Thus, digestive enzymes not retained in the periplasm are released into the environment, where they sometimes become so diluted that the organisms derive no benefit from them.

Gram-Negative Bacteria. The cell wall of a Gram-negative bacterium is thinner but more complex than that of a Gram-positive bacterium. Only 10 to 20% of the cell wall is peptidoglycan; the remainder consists of various polysaccharides, proteins, and lipids. The cell wall contains an outer membrane, which constitutes the outer surface of the wall, leaving only a very narrow periplasmic space. The inner surface of the wall is separated from the cell membrane by a wider periplasmic space. Toxins and enzymes remain in the periplasmic space in sufficient concentrations to help destroy substances that might harm the bacterium, but they do not harm the organism that produced them. If the cell wall is digested away, Gram-negative bacteria become **spheroplasts**, which have both a cell membrane and most of the outer membrane. Gram-negative bacteria fail to retain the crystal violet-iodine dye during the decolorizing procedure partly because of their thin cell walls and partly because of the relatively large quantities of lipoproteins and lipopolysaccharides in the walls.

Acid-Fast Bacteria. Although the cell wall of *acid-fast bacteria*, the mycobacteria, is thick, like that of Gram-positive bacteria, it is approximately 60% lipid and contains much less peptidoglycan. In the acid-fast staining process, carbolfuchsin binds to cytoplasm and resists removal by an acid-alcohol mixture (◄Chapter 3, p. 70). The lipids make acid-fast organisms impermeable to most other stains and protect them from acids and alkalis. The organisms grow slowly because the lipids impede entry of nutrients into cells, and the cells must expend large quantities of energy to synthesize lipids. Acid-fast cells can be stained by the Gram stain method; they stain as Gram-positive.

Controlling Bacteria by Damaging Cell Walls. Some methods of controlling bacteria are based on properties of the cell wall. For example, the antibiotic penicillin blocks the final stages of peptidoglycan synthesis. If penicillin is present when bacterial cells are dividing, the cells cannot form complete walls, and they die. Similarly, the enzyme lysozyme, found in tears and other human body secretions, digests peptidoglycan. This enzyme helps prevent bacteria from entering the body and is the body's main defense against eye infections (see ◄Figure 19.4).

Wall-Deficient Organisms

Bacteria that belong to the genus *Mycoplasma* have no cell walls. They are protected from osmotic swelling and bursting by a strengthened cell membrane that contains sterols. These are molecules typical of eukaryotes and are rarely found in prokaryotes. However, the protection is not complete, and often mycoplasmas must be grown in special media. Without a rigid cell wall, they vary widely in shape, often forming slender, branched filaments and exhibiting extreme pleomorphism.

Other genera of bacteria may normally have a cell wall but can suddenly lose their ability to form cell walls. These wall-deficient strains are called **L-forms**, named after the Lister Institute, where they were discovered over 70 years ago. The loss may occur naturally or be caused by chemical treatment. L-forms may play a role in chronic or recurrent diseases. Treatment with antibiotics that affect cell wall synthesis will kill most of the bacteria in some infections, but it leaves a few alive as L-forms. When treatment is discontinued, the L-forms can revert to walled forms and regrow an infecting population. An example of this is found in the association of the bacterium *Mycobacterium paratuberculosis* with Crohn's disease, a chronic disorder of the intestine.

Some Archaea may entirely lack cell walls, while others have unusual walls of polysaccharides, or of proteins, but lack true peptidoglycan. Instead they have a similar compound called pseudomurein.

(a)

Figure 4.7 The fluid-mosaic model of the cell membrane. **(a)** The basic structural component of the membrane is the phospholipid molecule. A phospholipid has two long fatty acid "tails" of hydrocarbon. The tails are very hydrophobic—they do not interact with water and form an oily barrier to most water-soluble substances. The "head" of the molecule consists of a charged phosphate group, usually joined to a charged nitrogen-containing group. The head is very hydrophilic—it interacts with water. **(b)** The fluid-mosaic model of membrane structure. The phospholipids form a bilayer in which the hydrophobic tails form the central core and the hydrophilic heads form the surfaces that face both the interior of the cell and the outside environment. In this fluid bilayer, proteins float like icebergs. Some extend through the bilayer; others are anchored to the inner or outer surface. Proteins and membrane lipids to which carbohydrate chains are attached are called *glycoproteins* and *glycolipids*, respectively. A few bacteria, such as mycoplasmas, have cholesterol molecules in their cell membranes, as do most eukaryotes. Mycoplasmas lack cell walls; cholesterol molecules add rigidity to the cell membrane.

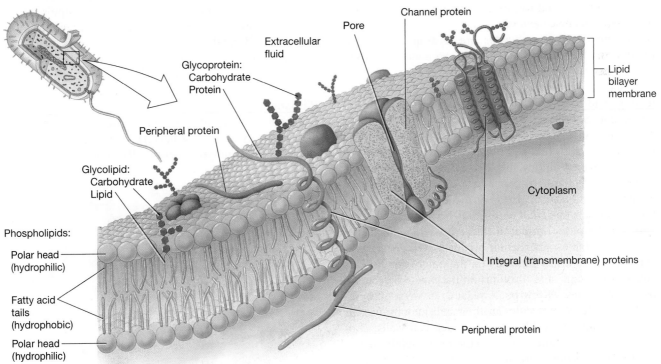

(b)

THE CELL MEMBRANE

The **cell membrane**, or *plasma membrane*, is a living membrane that forms the boundary between a cell and its environment. Also known as the cytoplasmic membrane, this dynamic, constantly changing membrane is not to be confused with the cell wall. The latter is a more static structure external to the cell membrane.

Bacterial cell membranes have the same general structure as the membranes of all other cells. Such membranes, formerly called *unit membranes*, consist mainly of phospholipids and proteins. The **fluid-mosaic model (Figure 4.7)** represents the current understanding of the structure of such a membrane. The model's name is derived from the fact that phospholipids in the membrane are in a fluid state and that proteins are dispersed among the lipid molecules in the membrane, forming a mosaic pattern.

Membrane phospholipids form a *bilayer*, or two adjacent layers. In each layer, the phosphate ends of the lipid molecules extend toward the membrane surface, and the fatty acid ends extend inward. The charged phosphate ends of the molecules are **hydrophilic** (water-loving) and thus can interact with the watery environment (Figure 4.7a). The fatty acid ends, consisting largely of nonpolar hydrocarbon chains, are **hydrophobic** (water-fearing) and form a barrier between the cell and its environment. Some membranes also contain other lipids. The membranes of mycoplasmas, bacteria that lack a cell wall, include lipids that add rigidity and are called *sterols*.

Interspersed among the lipid molecules are protein molecules (Figure 4.7b). Some extend through the entire membrane and act as carriers or form pores or channels through which materials enter and leave the cell. Proteins on the outer surface include those that make the cell identifiable as a particular organism. Others are embedded in, or loosely attached to, the inner or outer surface of the membrane. Proteins on the inner surface are usually enzymes. A few bacteria, such as mycoplasmas, have cholesterol molecules in their cell membranes, as do most eukaryotes. Mycoplasmas lack cell walls; cholesterol molecules add rigidity to their cell membranes.

Cell membranes are dynamic, constantly changing entities. Materials constantly move through pores and through the lipids themselves, although selectively. Also, both the lipids and the proteins in membranes are continuously changing positions. Some antibiotics and disinfectants kill bacteria by causing their cell membranes to leak, as will be discussed in ◄Chapter 13.

The main function of the cell membrane is to regulate the movement of materials into and out of a cell by transport mechanisms, which are discussed in this chapter. In bacteria this membrane also performs some functions carried out by other structures in eukaryotic cells. It synthesizes cell wall components, assists with DNA replication, secretes proteins, carries on respiration, and captures energy as ATP. It also contains bases of appendages called *flagella*; the actions of the bases cause the flagella to move. Finally, some proteins in the bacterial cell membrane respond to chemical substances in the environment.

INTERNAL STRUCTURE

Bacterial cells typically contain *ribosomes*, a *nucleoid*, and a variety of *vacuoles* within their *cytoplasm*. Figure 4.3 shows the locations of these structures in a generalized prokaryotic cell. Certain bacteria sometimes contain *endospores* as well.

Cytoplasm

The **cytoplasm** of prokaryotic cells is the semifluid substance inside the cell membrane. Because these cells typically have only a few clearly defined structures, such as one, two, or three chromosomes and some ribosomes, they consist mainly of cytoplasm. Cytoplasm is about four-fifths water and one-fifth substances dissolved or suspended in the water. These substances include enzymes and other proteins, carbohydrates, lipids, and a variety of inorganic ions. Many chemical reactions, both anabolic and catabolic, occur in the cytoplasm. Unlike eukaryotic cytoplasm, that of prokaryotes does not carry out the movement known as "streaming."

Ribosomes

Ribosomes consist of RNA and protein. They are abundant in the cytoplasm of bacteria, often grouped in long chains called **polyribosomes**. Ribosomes are nearly

Figure 4.8   A centrifuge. Suspended particles in tubes of liquid are whirled around at high speeds, causing them to settle to the bottom of the tubes or to form bands at different levels. The rate of settling or the locations of the bands can be used to determine the size, weight, and shape of the particles.

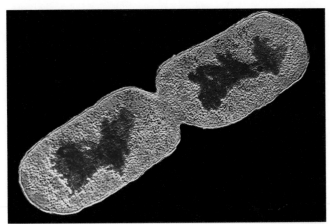

Figure 4.9 The bacterial nuclear region. A colorized TEM of a thin section of *Escherichia coli* with the DNA shown in red (42,382X). *(CNRI/Custom Medical Stock Photo, Inc.)*

spherical, stain densely, and contain a large subunit and a small subunit. Ribosomes serve as sites for protein synthesis ◄(Chapter 7).

The relative sizes of ribosomes and their subunits can be determined by measuring their *sedimentation rates*—the rates at which they move toward the bottom of a tube when the tube is rapidly spun in an instrument called a *centrifuge* (Figure 4.8). Sedimentation rates, which generally vary with molecular size, are expressed in terms of *Svedberg* (S) *units*. Whole bacterial ribosomes, which are smaller than eukaryotic ribosomes, have a rate of 70S; their subunits have rates of 30S and 50S. Certain antibiotics, such as streptomycin and erythromycin, bind specifically to 70S ribosomes and disrupt bacterial protein synthesis. Because those antibiotics do not affect the larger 80S ribosomes found in eukaryotic cells, they kill bacteria without harming host cells.

Nuclear Region

One of the key features differentiating prokaryotic cells from eukaryotic cells is the absence of a nucleus bounded by a nuclear membrane. Instead of a nucleus, bacteria have a **nuclear region**, or **nucleoid** (Figure 4.9). The centrally located nuclear region consists mainly of DNA, but has some RNA and protein associated with it. It was long believed that the DNA was always arranged in one large, circular chromosome. Then in 1989, two circular chromosomes were found in the aquatic photosynthetic bacterium *Rhodobacter sphaeroides*. *Agrobacterium rhizogenes* likewise has two circular chromosomes, but its close relative *Agrobacterium tumefaciens*, which causes tumors in plants, has one circular chromosome and a second chromosome that is linear. *Brucella suis*, a pathogen of pigs, is unusual in that some of its strains have two chromosomes, whereas other strains in the same species have only one. The cholera-causing bacterium, *Vibrio cholerae*, has two circular chromosomes: one large, the other about one-fourth the size of the first. Both are essential for reproduction. Some bacteria also contain smaller circular molecules of DNA called

plasmids. Genetic information in plasmids supplements information in the chromosome ◄(Chapter 8). Questions regarding the evolution of chromosome number in bacteria will be discussed in ◄Chapter 9.

Internal Membrane Systems

Photosynthetic bacteria and cyanobacteria contain internal membrane systems, sometimes known as **chromatophores** (Figure 4.10). The membranes of the chromatophores, derived from the cell membrane, contain the pigments used to capture light energy for the synthesis of sugars. Nitrifying bacteria, soil organisms that convert nitrogen compounds into forms usable by green plants, also have internal membranes. They house the enzymes used in deriving energy from the oxidation of nitrogen compounds ◄(Chapter 5).

Electron micrographs of bacterial cells often show large infoldings of the cell membrane called *mesosomes*. Although these were originally thought to be structures present in living cells, they have now been proven to be artifacts: That is, they were created by the processes used to prepare specimens for electron microscopy.

Inclusions

Bacteria can have within their cytoplasm a variety of small bodies collectively referred to as **inclusions**. Some are called *granules*; others are called *vesicles*.

Granules, although not bounded by membrane, contain substances so densely compacted that they do not easily dissolve in cytoplasm. Each granule contains a specific substance, such as glycogen or polyphosphate. *Glycogen*, a glucose polymer, is used for energy. *Polyphosphate*, a phosphate polymer, supplies phosphate for

photo-
synthetic
membrane

carbohydrate
granules

0.5 μm

Figure 4.10 Internal membrane systems. TEM of the cyanobacterium *Synechocystis* showing chromatophores. The cell's outer regions are filled with photosynthetic membranes. The dark spots between the membranes are granules in which carbohydrates produced by photosynthesis are stored. *(Courtesy Dirk Paul Stephan, University of Bielefeld, Germany)* a variety of metabolic processes. Polyphosphate granules are called **volutin** (vo-lu′tin), or **metachromatic granules**, because they display **metachromasia**. That is, although most substances stained with a simple stain such as methylene blue take on a uniform, solid color, metachromatic granules exhibit different intensities of color. Although quite numerous in some bacteria, these granules become depleted during starvation. Bacteria that obtain energy by the metabolism of sulfur may contain reserve granules of sulfur in their cytoplastm.

Certain bacteria have specialized membrane-enclosed structures called **vesicles** (or *vacuoles*). Some aquatic photosynthetic bacteria and cyanobacteria have rigid gas-filled vacuoles (shown in ◀Figure 3.25). These organisms regulate the amount of gas in vacuoles and, therefore, the depth at which they float to obtain optimum light for photosynthesis. Another type of vesicle, found only in bacteria, contains deposits of poly-β-hydroxybutyrate. These lipid deposits serve as storehouses of energy and as sources of carbon for building new molecules. See the "Living Magnets" box on page 97 for a description of iron-containing vesicles called **magnetosomes**.

Endospores

The properties of bacterial cells just described pertain to **vegetative cells**, or cells that are metabolizing nutrients. However, vegetative cells of some bacteria, such as *Bacillus* and *Clostridium*, produce resting stages called **endospores**. Although bacterial endospores are commonly referred to simply as *spores*, do not confuse them with fungal spores. A bacterium produces a single endospore, which merely helps that organism survive and is not a means of reproduction. A fungus produces numerous spores, which help the organism survive and provide a means of reproduction.

Endospores, which are formed within cells, contain very little water and are highly resistant to heat, drying, acids, bases, certain disinfectants, and even radiation. The depletion of a nutrient will usually induce a large number of cells to produce spores. However, many investigators believe that spores are part of the normal life cycle and that a few are formed even when nutrients are adequate and environmental conditions are favorable. Thus, *sporulation*, or endospore formation, seems to be a means by which some bacteria prepare for the possibility of future adverse conditions, in much the same way as countries keep "standing armies" ready in case of war.

In 1877, F. Cohn first demonstrated bacterial spores.

Structurally, an endospore consists of a *core*, surrounded by a *cortex*, a *spore coat*, and in some species a delicately thin layer called the *exosporium* (Figure 4.11). The core has an outer core wall, a cell membrane, nuclear region, and other cell components. Unlike vegetative cells, endospores contain *dipicolinic acid* and a large quantity of calcium ions (Ca^{2+}). These

Figure 4.11 Endospores. A colorized electron micrograph of an endospore within a *Clostridium perfringens* cell (29,349X). *(Institut Pasteur/Phototake)*

materials, which are probably stored in the core, appear to contribute to the heat resistance of endospores, as does their very low water content.

Endospores are capable of surviving adverse environmental conditions for long periods of time, some for over 10,000 years (there are claims that endospores sealed in amber have survived for over 25 million years). Spores of Antarctic bacteria can remain dormant for at least 10,000 years at a temperature of −14°C in ice 430 meters deep. Some withstand hours of boiling. When conditions become more favorable, endospores *germinate*, or begin to develop into functional vegetative cells. (The processes of spore formation and germination are discussed in ◀Chapter 6 and atomic force photos of changes in the surface of a germinating fungal spore are shown in ◀Chapter 3, p. 67.) Because endospores are so resistant, special methods must be used to kill them during sterilization. Otherwise, they germinate and grow in media thought to be sterile. Methods to ensure that endospores are killed when culture media or foods are sterilized are described in ◀Chapter 13. You will also find in lab that endospores can be difficult to stain. Killing anthrax spores in United States government buildings contaminated by terrorist activities has proven to be difficult and costly.

EXTERNAL STRUCTURE

In addition to cell walls, many bacteria have structures that extend beyond or surround the cell wall. *Flagella* and *pili* extend from the cell membrane through the cell wall and beyond it. *Capsules* and *slime layers* surround the cell wall.

Flagella

About half of all known bacteria are *motile*, or capable of movement. They often move with speed and apparent purpose, and they usually move by means of long, thin, helical appendages called **flagella** (singular:

(a) LM

(b) LM

(c) LM

(d) LM

(e) SEM

Figure 4.12 Arrangements of bacterial flagella. **(a)** Polar, monotrichous (single flagellum at one end) *Pseudomonas* (3,300X). *(E. C. S. Chan/Visuals Unlimited)* **(b)** Polar, amphitrichous (single flagellum at each end) *Spirillum* (694X). *(Jack M. Bostrack/Visuals Unlimited)* **(c)** Lophotrichous (with tuft of flagella at one or both ends) *Spirillum* (1,214X). *(E. C. S. Chan/Visuals Unlimited)* **(d)** Peritrichous (flagella distributed all over) *Salmonella* (1,200X). *(John D. Cunningham/Visuals Unlimited)* **(e)** SEM of peritrichous flagella, *Proteus* (29,400X). *(Fred Hossler/Visuals Unlimited)*

flagellum). A bacterium can have one flagellum or two or many flagella. Bacteria with a single *polar* flagellum located at one end, or pole, are said to be **monotrichous** (mon-o-trik′-us; **Figure 4.12a**); bacteria with two flagella, one at each end, are **amphitrichous** (am-fe-trik′-us; **Figure 4.12b**); both types are said to be *polar*. Bacteria with two or more flagella at one or both ends are **lophotrichous** (lo-fo-trik′us; **Figure 4.12c**); and those with flagella all over the surface are **peritrichous** (pe-ri-trik′us; **Figure 4.12d** and **e**). Bacteria without flagella are **atrichous** (a-trik′us). Cocci rarely have flagella.

The diameter of a prokaryote's flagellum is about one-tenth that of a eukaryote's flagellum. It is made of protein subunits called *flagellin*. Each flagellum is attached to the cell membrane by a basal region consisting of a protein other than flagellin (**Figure 4.13**). The basal region has a hooklike structure and a complex *basal body*. The basal body consists of a central rod or shaft surrounded by a set of rings. Gram-negative bacteria have a pair of rings embedded in the cell membrane and another pair of rings associated with the peptidoglycan and lipopolysaccharide layers of the cell wall. Gram-positive bacteria have one ring embedded in the cell membrane and another in the cell wall.

Most flagella rotate like twirling L-shaped hooks, such as a dough hook on a kitchen mixer or the rotating string on a hand-carried grass trimmer. Motion is thought to occur as energy is used to make one of the rings in the cell membrane rotate with respect to the other. When flagella bundle together (**Figure 4.14a**) they rotate counterclockwise, and the bacteria *run*, or move in a straight line. When the flagella rotate clockwise, the flagellar bundle comes apart, causing the bacterium to *tumble* randomly (**Figure 4.14b**). Both runs and tumbles are generally random movements; that is, no one direction of movement is more likely than any other direction. Runs last an

average of 1.0 second, during which the bacterium swims about 10 to 20 times the length of its body. Tumbles last about 0.1 second, and no forward progress is made. "Cruising speed" for bacteria is about 10 body lengths/ second, which would be "flying speed" for humans!

Chemotaxis. Sometimes bacteria move toward or away from substances in their environment by a nonrandom process called **chemotaxis** (**Figure 4.14c**). Concentrations of most substances in the environment vary along a gradient—that is, from high to low concentration. When a bacterium is moving in the direction of increasing concentration of an attractant (such as a nutrient), it tends to lengthen its runs and to reduce the frequency of its tumbles. When it is moving away from the attractant, it shortens its runs and increases the frequency of its tumbles. Even though the direction of the individual runs is still random, the net result is movement toward the attractant, or *positive chemotaxis*. Movement away from the repellent, or *negative chemotaxis*, results from the opposite responses: long runs and few tumbles while the bacterium moves in the direction of lower concentration of the harmful substances, short runs and many tumbles while it moves in the direction of higher concentration. The exact mechanism that produces these behaviors is not fully understood, but certain structures on bacterial cell surfaces can detect changes in concentration over time. *Escherichia coli* cells have at least four different types of receptors (called *transducers*) that extend through the cell membrane and detect chemicals and signal the cells to respond.

We also know that bacteria use helical motion to orient to external signals. They do not swim in a straight line, but in a helical pathway which has a net trajectory in one direction. Organisms can change the handedness of the helix. Right-handedness gives a

(a)

Cell wall

Outer membrane   Peptidoglycan

Cell membrane

Cytoplasm

Filament

Hook

Outer rings   Rod   Inner rings

Basal body

(b)

EM

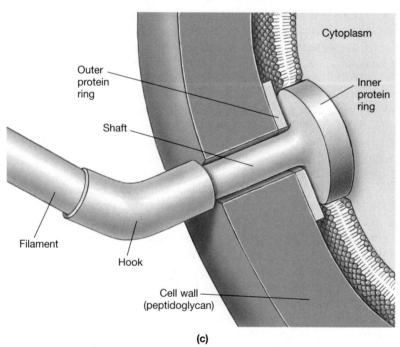

Cytoplasm

Outer protein ring

Inner protein ring

Shaft

Filament

Hook

Cell wall (peptidoglycan)

(c)

Figure 4.13 Structure of two different bacterial flagella. **(a)** Drawing and **(b)** electron micrograph of the basal region of the flagellum of a Gram-negative bacterium (1,419,748X). The flagellum has three main parts: a filament, a hook, and a basal body consisting of a rod surrounded by four rings. *(Julius Adler/Visuals Unlimited)* **(c)** Gram-positive bacteria have only two rings, one attached to the peptidoglycan of the cell wall and one to the cell membrane.

Figure 4.14 **Chemotaxis.** **(a)** When all the flagella of a bacterium rotate in a counterclockwise direction, the flagella bundle together and push the bacterium in a fairly straight, forward movement called a *run*. When the flagella reverse and rotate in a clockwise direction, the bundle comes apart, each flagellum acts independently, and the cells tumble about in random directions, a movement called a *tumble*. **(b)** Peritrichous and lophotrichous flagellated bacteria doing runs and tumbles. Note that the cell swims forward (a run) only when flagella are bundled and that the bacterium changes direction following a tumble. **(c)** When nothing attracts or repels a bacterium, it has frequent tumbles and short runs, resulting in random movement.

positive taxis, whereas left-handedness gives a negative taxis. The U.S. Navy has expanded this microbial study to build a tiny underwater autonomous robot ("Micro Hunter," 17 cm, 70 g) that uses helical paths to search for "lost asset recovery."

Phototaxis. Some bacteria can move toward or away from light; this response is called **phototaxis**. Bacteria that move toward light exhibit *positive phototaxis*, whereas those that move away from light exhibit *negative phototaxis*. The movement may be accomplished by means of flagella. Or, in the case of some photosynthetic aquatic bacteria, oil droplet inclusions in their cytoplasm may give them the buoyancy to rise toward the water surface, where light is more available.

Axial Filaments
Spirochetes have **axial filaments**, or **endoflagella**, instead of flagella that extend beyond the cell wall **(Figure 4.15)**. Each filament is attached at one of its ends to an end of the cytoplasmic cylinder that forms the body of the spirochete. Because the axial filaments lie between the outer

Axial
filaments

(a)   SEM

TEM

Figure 4.15   Axial filaments, or endoflagella. **(a)** Axial filaments made visible by false coloring are clearly seen as spiraling yellow ribbons running inside the cell wall along the body of the spirochete *Leptospira interrogans* (50,000X). *(CNRI/Photo Researchers, Inc.)* **(b)** TEM (cross-section) of a spirochete, showing numerous axial filaments (dark circles). Axial filaments lie between the outer sheath and the cell wall. *(Courtesy Dr. Max Listgarten, School of Dental Medicine, University of Pennsylvania, as published in Journal of Bacteriology 88:1087–1103.)*

Sex does not result in bacterial reproduction. Sex does not get a female bacterium pregnant. Instead, it leaves her the older but wiser girl, having acquired new genetic information from her male partner.

sheath and the cell wall, their twisting causes the rigid spirochete body to rotate like a corkscrew.

Pili

Pili (singular: *pilus*) are tiny, hollow projections. They are used to attach bacteria to surfaces and are not involved in movement. A pilus is composed of subunits of the protein *pilin*. Bacteria can have two kinds of pili **(Figure 4.16)**: (1) long *conjugation pili*, or *F pili* (also called *sex pili*), and (2) short *attachment pili*, or *fimbriae* (fim'-bre-e; singular: *fimbria*).

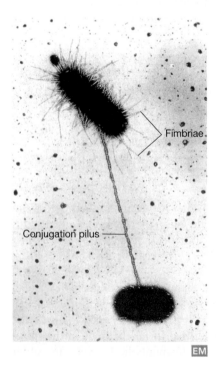

Fimbriae

Conjugation pilus

EM

Figure 4.16   Pili. An *Escherichia coli* cell (14,300X), showing two kinds of pili. The shorter ones are fimbriae, used for attachment to surfaces. The long tube reaching to another cell is a conjugation pilus, perhaps used to transfer DNA. *(Courtesy Charles C. Brinton, Jr., and Judith Carnahan)*

Conjugation Pili.   Conjugation pili (or *sex pili*), found only in certain groups of bacteria, attach two cells and may furnish a pathway for the transfer of the genetic material DNA. This transfer process is called *conjugation* (Figures 4.16 and 8.7). Transfer of DNA furnishes genetic variety for bacteria, as sexual reproduction does for many other life forms. Such transfers among bacteria cause problems for humans because antibiotic resistance can be passed on with the DNA transfer. Consequently, more and more bacteria acquire resistance, and humans must look for new ways to control the growth of these bacteria.

Attachment Pili.   Attachment pili, or **fimbriae,** help bacteria adhere to surfaces, such as cell surfaces and the interface of water and air. They contribute to the *pathogenicity* of certain bacteria—their ability to produce disease—by enhancing colonization (the development of colonies) on the surfaces of the cells of other organisms. For example, some bacteria adhere to red blood cells by attachment pili and cause the blood cells to clump, a process called *he-magglutination.* In certain species of bacteria, some individuals have attachment pili and others lack them. In *Neisseria gonorrhoeae,* strains without pili are rarely able to cause gonorrhea, but those with pili are highly infectious because they attach to epithelial cells of the urogenital system. Such pili also allow them to attach to sperm cells and thereby spread to the next individual.

A chemical in cranberry juice helps prevent urinary tract infections by preventing formation of attachment pili.

Some aerobic bacteria form a shiny or fuzzy, thin layer at the air-water interface of a broth culture. This layer, called a **pellicle,** consists of many bacteria that adhere to the surface by their attachment pili. Thus, attachment pili allow the organisms to remain in the broth, from which they take nutrients, while they congregate near air, where the oxygen concentration is greatest.

TRY IT

Living Magnets

Magnetotactic bacteria synthesize magnetite (Fe_3O_4), or lodestone, and store it in membranous vesicles called *magnetosomes*. (Lodestone was the first substance with magnetic properties to be discovered.) The presence of these magnetic inclusions enables these bacteria to respond to magnetic fields. In the Northern Hemisphere, magnetotactic bacteria swim toward the North Pole; in the Southern Hemisphere, they swim toward the South Pole; and near the equator, some swim north and others south. However, the bacteria also swim downward in water because the magnetic force from the Earth's poles is deflected through the Earth and not over its horizon. The swimming toward a magnetic pole is called *magnetotaxis*. Their downward magnetotactic response appears to help these anaerobic bacteria move down toward sediments where their food (iron oxide) is abundant and where oxygen, which they cannot tolerate, is deficient.

Magnetotactic bacteria live in mud and brackish waters, and more than a dozen species have been identified. Most have a single flagellum, but *Aquaspirillum magnetotacticum* has two, one at each end, so it can swim forward or backward. When magnetotactic bacteria with one flagellum are placed in the field of an electromagnet, they make U-turns as the poles of the magnet are reversed.

Magnetosomes are nearly constant in size and are oriented in parallel chains like a string of tiny magnets. Experiments are currently under way to use these bacteria in the manufacture of magnets, audiotapes, and videotapes.

Magnetosomes have been found inside meteorites from Mars. They are associated with bacillus-shaped structures, which some scientists believe to be Martian bacterial life forms.

TRY IT: You can easily find your own magnetotactic bacteria. They are very common organisms. Bring back a bucket with a few inches of mud from a pond and enough pondwater to fill the rest of the bucket. Try mud from different locations: fresh, salt, and brackish (mixed fresh and salt) waters. Cover the bucket and store it in the dark (you don't want to grow algae) for about one month.

(a)

(b)

(a) Electron micrograph of the magnetotactic bacterium *Aquaspirillum magnetotacticum*. The numerous dark, square inclusions, called magnetosomes, are composed of iron oxide (Fe_3O_4). **(b)** The magnetosomes enable these organisms to orient themselves in a magnetic field.

(**a**: D. Balkwill & D. Maratea/Visuals Unlimited; **b**: Courtesy Richard Blakemore, University of New Hampshire)

Remove a large beaker of the water, and place a magnet against the outside of the glass. Allow this to stand for a day or two, until you see a whitish spot in the water near the magnet's end. Which attracts more organisms, the north or south end of the magnet? With a pipette, remove a sample of the cloudy liquid and examine it under the microscope. When you lay a magnet on the stage, any magnetotactic bacteria that are present will orient themselves to the field. They can be purified by streaking on agar and then incubating them in the absence of oxygen, processes described in ◀Chapter 6. Microbiology doesn't exist only in books or labs. It's everywhere around you in the real world—just look for it!

Glycocalyx

Glycocalyx is the currently accepted term used to refer to all polysaccharide-containing substances found external to the cell wall, from the thickest *capsules* to the thinnest *slime layers*. All bacteria have at least a thin slime layer.

Capsule. A **capsule** is a protective structure outside the cell wall of the organism that secretes it. Only certain bacteria are capable of forming capsules, and not all members of a species have capsules. For example, the bacterium that causes anthrax, a disease naturally found mainly in cattle, does not produce a capsule when it grows outside an organism but does when it infects an animal. Capsules typically consist of complex polysaccharide molecules arranged in a loose gel. However, the chemical composition of each capsule is unique to the strain of bacteria that secreted it. Anthrax bacteria have a capsule composed of protein. When encapsulated bacteria invade a host, the capsule prevents host defense mechanisms, such as phagocytosis, from destroying the bacteria. If bacteria lose their capsules, they become less likely to cause disease and more vulnerable to destruction.

Slime Layer. A **slime layer** is less tightly bound to the cell wall and is usually thinner than a capsule. When present, it protects the cell against drying, helps trap nutrients near the cell, and sometimes binds cells together. Slime layers allow bacteria to adhere to objects in their environments, such as rock surfaces or the root hairs of plants, so that they can remain near sources of nutrients or oxygen. This "biofilm" protects bacteria on the bottom of the layers from environmental or man-made chemicals. Some

Figure 4.17 **The slime layer.** Bacteria growing [SEM]
on tooth enamel, to which they initially adhere by means
of their slime layer (9088X). This constitutes a "biofilm,"
and protects the bacteria at the bottom of the layer from toothpaste
and mouthwashes. *(Dr. Tony Brain/Photo
Researchers, Inc.)*

oral bacteria, for example, adhere by their slime layers
and form dental plaque **(Figure 4.17)**. The slime layer
keeps the bacteria in close proximity to the tooth surface,
where they can cause dental caries. Plaque is extremely
tightly bound to tooth surfaces. If not removed regularly
by brushing, it can be removed only by a dental professional in a procedure called *scaling*.

✓ CHECKLIST

1. How do bacteria move in chemotaxis? Distinguish
between runs and tumbles, and their frequencies.

2. What risk would a spore-forming bacterial species run if
it did not produce any spores until conditions became
adverse?

3. What are the functions of pili?

4. Distinguish between nucleus and nucleoid.

EUKARYOTIC CELLS

AN OVERVIEW OF STRUCTURE

Eukaryotic cells are larger and more complex than
prokaryotic cells. Most eukaryotic cells have a diameter
of more than 10 μm, and many are much larger. They
also contain a variety of highly differentiated structures.
These cells are the basic structural unit of all organisms

Figure 4.18 **A generalized eukaryotic cell.** Most of the features shown are present in nearly all eukaryotic
cells, but some (the centrioles, microvilli, and lysosomes) occur only in animal cells, and others (the chloroplast) are
found only in cells capable of carrying out photosynthesis.

Figure 4.19 Pores through the cell nucleus. (a) The dark, granular material is chromatin. The pores in the nuclear membrane allow for entry and exit of materials (120,000X). *(Don Fawcett/Photo Researchers, Inc.)* **(b)** Freeze-fracture of a nucleus (compare with ◀Figure 3.23). The many circular structures are nuclear pores (264,139X). *(Don Fawcett/Visuals Unlimited)*

in the kingdoms Protista, Plantae, Fungi, and Animalia ◀(Chapter 9). Eukaryotic organisms include microscopic protozoa, algae, and fungi and are thus appropriately considered in microbiology. The general structure of the eukaryotic cell is shown diagrammatically in **Figure 4.18**.

THE PLASMA MEMBRANE

The cell membrane, or **plasma membrane**, of a eukaryotic cell has the same fluid-mosaic structure as that of a prokaryotic cell. In addition, eukaryotes also contain several organelles enclosed by membranes that have a similar membrane structure.

Eukaryotic membranes differ from prokaryotic membranes in some respects, especially in the greater variety of lipids they contain. Eukaryotic membranes contain sterols, found among prokaryotes only in the mycoplasmas. Sterols add rigidity to a membrane, and this may be important in keeping membranes intact in eukaryotic cells. Because of their larger size, eukaryotic cells have a much lower surface-to-volume ratio than prokaryotic cells. As the volume of cytoplasm enclosed by a membrane increases, the membrane is placed under greater stress. The sterols in the membrane may help it withstand the stress.

Functionally, eukaryotic plasma membranes are less versatile than prokaryotic ones. They do not have respiratory enzymes that capture metabolic energy and store it in ATP; in the course of evolution, that function has been taken over by mitochondria in all but a few eukaryotes.

INTERNAL STRUCTURE

The internal structure of eukaryotic cells is exceedingly more complex than that of prokaryotic cells. It is also much more highly organized and contains numerous organelles.

Cytoplasm

The cytoplasm makes up a relatively smaller portion of eukaryotic cells than of prokaryotic cells because the *nucleus* and many organelles fill much of the space in a eukaryotic cell. Like the cytoplasm of prokaryotic cells, the cytoplasm of eukaryotic cells is a semifluid substance consisting mainly of water with the same substances dissolved in it. In addition, this cytoplasm contains elements of a *cytoskeleton*, a fibrous network that gives these larger cells shape and support.

Cell Nucleus

The most obvious difference between eukaryotic and prokaryotic cells is the presence of a nucleus in the eukaryotic cells. The **cell nucleus (Figure 4.19)** is a distinct

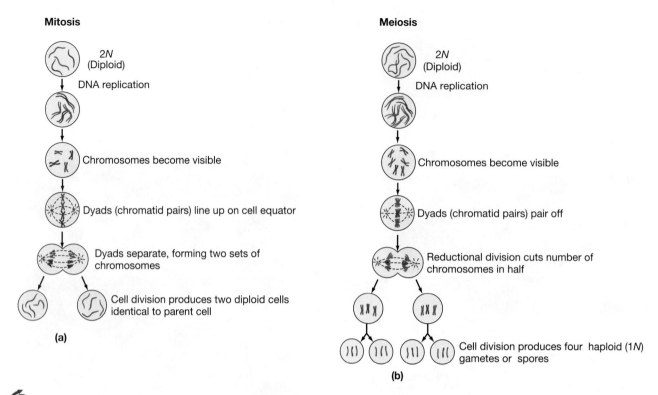

Figure 4.20 Mitosis and meiosis compared. Both processes are preceded by duplication of DNA; soon after, the chromosomes become visible. **(a)** Mitosis produces two identical daughter cells with the same number and kinds of chromosomes. **(b)** In meiosis, two divisions give rise to four cells, each with half the number of chromosomes as the original parent cell. For this reason, meiosis is sometimes called *reduction division.*

organelle enclosed by a nuclear envelope and contains nucleoplasm, nucleoli, and (typically paired) chromosomes. The **nuclear envelope** consists of a double membrane, each layer of which is structurally like the plasma membrane. **Nuclear pores** in the envelope allow RNA molecules to leave the semifluid portion of the nucleus, known as **nucleoplasm**, and to participate in protein synthesis. Each nucleus has one or more **nucleoli** (singular: *nucleolus*), which contain a significant amount of RNA and serve as sites for the assembly of ribosomes.

Also present in the nucleus of most eukaryotic organisms are paired **chromosomes**, each of which contains DNA and proteins called **histones**. Histones contribute directly to the structure of chromosomes, and other proteins probably regulate the chromosomes' function. During cell division, the chromosomes are extensively coiled and folded into compact structures. Between divisions, however, the chromosomes are uncoiled and visible only as a tangle of fine threads called **chromatin** that give the nucleus a granular appearance.

The nuclei of eukaryotic cells divide by the process of **mitosis (Figure 4.20a)**. Prior to the actual division of the nucleus, the chromosomes replicate but remain attached, forming **dyads**. In most eukaryotic cells, the nuclear envelope breaks apart during mitosis, and a system of tiny fibers called the **spindle apparatus** guides the movement of chromosomes. Dyads aggregate in the center of the spindle and separate into single chromosomes

as they move along fibers to the poles of the spindle. Each new cell receives one copy of each chromosome that was present in the parent cell. Because the parent cell contained paired chromosomes, the progeny likewise contain paired chromosomes. Cells with paired chromosomes are said to be **diploid** (2*N*) cells.

During sexual reproduction, the nuclei of sex cells divide by a process called **meiosis (Figure 4.20b)**. After the chromosomes replicate, forming dyads, pairs of dyads come together. During the course of two cell divisions, the dyads are distributed to four new cells. Thus, each cell receives only one chromosome from each pair. Such cells are said to be **haploid** (1*N*) cells. Haploid cells can become gametes or spores. **Gametes** are haploid cells that participate in sexual reproduction; gametes from each of two parent organisms unite to form a diploid **zygote**, the first cell of a new individual. Some **spores** become dormant, whereas others reproduce by mitosis as haploid vegetative cells. Dormant spores allow for survival during adverse environmental conditions. When conditions improve, the spores germinate and begin to divide. Eventually some of these cells produce gametes, which can unite to form zygotes. Thus, the organism alternates between haploid and diploid generations.

Mitochondria and Chloroplasts

Mitochondria (singular: *mitochondrion*), known as the powerhouses of eukaryotic cells, are exceedingly

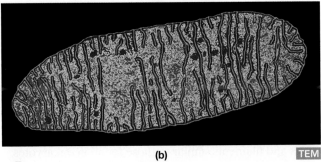

Figure 4.21 Mitochondria. (a) Respiratory enzymes that make ATP are located on the surfaces of the inner membrane and the cristae, which are infoldings of the inner membrane. **(b)** TEM of mitochondrion in longitudinal section (45,000X). *(Dr. Donald Fawcett & Dr. Porter/Visuals Unlimited)*

important organelles. They are quite numerous in some cells and can account for up to 20% of the cell volume. Mitochondria are complex structures about 1 μm in diameter, with an outer membrane, an inner membrane, and a fluid-filled **matrix** inside the inner membrane (**Figure 4.21**). The inner membrane is extensively folded to form **cristae**, which extend into the matrix. Mitochondria carry out the oxidative reactions that capture energy in ATP. Energy in ATP is in a form usable by cells for their activities.

Eukaryotic cells capable of carrying out photosynthesis contain **chloroplasts (Figure 4.22)**. They, too, have an outer and an inner membrane. The inner **stroma** of these organelles corresponds structurally with the matrix of mitochondria. Unlike mitochondria, chloroplasts have separate internal membranes, called **thylakoids**, that contain the pigment *chlorophyll*, which captures energy from light during photosynthesis. Both mitochondria and chloroplasts contain DNA and can replicate independently of the cell in which they function. This and other evidence has led many biologists to speculate that these organelles may have originated as free-living organisms.

Ribosomes

Ribosomes of eukaryotic cells, which are larger than those of prokaryotic cells, are about 60% RNA and 40% protein. They have a sedimentation rate of 80S, and their subunits have sedimentation rates of 60S and 40S. Ribosomes are assembled in the nucleoli of the nucleus. All ribosomes provide sites for protein synthesis, and some are arranged in chains as polyribosomes. Those that are attached to an organelle called the *endoplasmic reticulum* usually make proteins for secretion from the cell; those that are free in the cytoplasm usually make proteins for use in the cell.

Endoplasmic Reticulum

The **endoplasmic reticulum** (Figure 4.18) (ER) is an extensive system of membranes that forms numerous tubes and plates in the cytoplasm. Endoplasmic reticulum can be smooth or rough-textured. *Smooth*

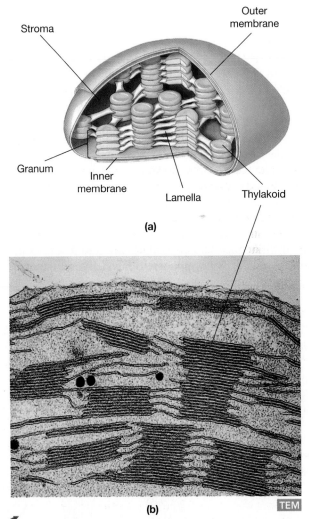

Figure 4.22 Chloroplasts. (a) The thylakoid membranes contain chlorophyll and other pigments and enzymes needed for photosynthesis. Thylakoids occur in stacks called *grana*; grana are joined by membranous flat sheets called *lamellae*. **(b)** A colorized TEM of a chloroplast (magnified 61,680X) from a leaf of corn. *(Dr. Kenneth R. Miller/Photo Researchers)*

endoplasmic reticulum contains enzymes that synthesize lipids, especially those to be used in making membranes. *Rough endoplasmic reticulum* has ribosomes bound to its surface, which give it a rough texture. Its function is, along with the ribosomes, to manufacture proteins. Vesicles from this membrane system transport to the Golgi apparatus the lipids and proteins synthesized in or on the endoplasmic reticulum membrane.

Golgi Apparatus

The **Golgi** (gol′je) **apparatus** (Figure 4.18) consists of a stack of flattened membranous sacs. The Golgi apparatus receives substances transported from the endoplasmic reticulum, stores the substances, and typically alters their chemical structure. It packages these substances in small segments of membrane called **secretory vesicles**. The secretory vesicles fuse with the plasma membrane and release secretions to the exterior of the cell. The Golgi apparatus also helps to form the plasma membrane and membranes of the lysosomes.

Lysosomes

Lysosomes (Figure 4.18) are extremely small, membrane-covered organelles made by the Golgi apparatus in animal cells. They contain multiple kinds of digestive enzymes that could destroy a cell if those enzymes were released into the cytoplasm. Lysosomes fuse with *vacuoles* that form as a cell ingests substances, and they release enzymes that digest the substances in the vacuoles. Many bacteria that enter cells, especially ones that were engulfed by white blood cells, are killed by lysosomal enzymes.

Peroxisomes

Peroxisomes are small, membrane-enclosed organelles filled with enzymes. Peroxisomes are found in both plant and animal cells but appear to have different functions in the two kinds of cells. In animal cells their enzymes oxidize amino acids, whereas in plant cells they typically oxidize fats. Peroxisomes are so named because their enzymes convert hydrogen peroxide to water in both plant and animal cells. If hydrogen peroxide were to accumulate in cells, it would kill them, just as it kills bacteria when humans use it as an antiseptic.

Vacuoles

In eukaryotic cells **vacuoles** are membrane-enclosed structures that store materials such as starch, glycogen, or fat to be used for energy. Some vacuoles form when cells engulf food particles. As we have already noted, the contents of these vacuoles are eventually digested by lysosomal enzymes. Water-filled vacuoles add rigidity to plant cells. Loss of this water causes wilting of plant structures.

Cytoskeleton

The **cytoskeleton** is a network of protein fibers made of **microtubules** (which are hollow tubes) and **microfilaments** (which are filamentous fibers). The cytoskeleton supports and gives rigidity and shape to a cell. It also is involved in cell movements, such as those that occur when cells engulf substances or when they make *amoeboid movements* (which are explained later in the chapter). Recent studies indicate that some bacteria may have tubules and filaments similar to those found in eukaryotes.

EXTERNAL STRUCTURE

Like external structures of prokaryotic cells, external structures of eukaryotic cells either assist with movement or provide a protective covering for the plasma membrane. These structures include flagella, cilia, and cell walls and other coverings. Although *pseudopodia* are not, strictly speaking, external structures, they do

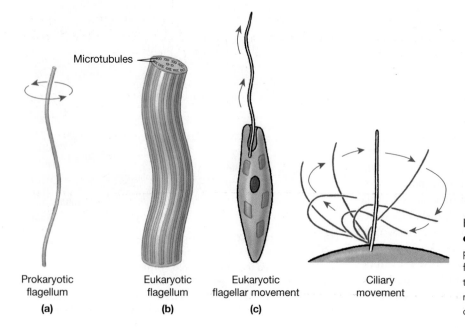

Microtubules

Prokaryotic flagellum
(a)

Eukaryotic flagellum
(b)

Eukaryotic flagellar movement
(c)

Ciliary movement

Figure 4.23 Prokaryotic and eukaryotic flagella compared. (a) A prokaryotic flagellum; **(b)** a eukaryotic flagellum. Notice the substantial difference in the diameter of these two structures. **(c)** The movement of a eukaryotic flagellum and cilium.

achieve movement and so are discussed here. Cells of algae and macroscopic green plants have cell walls, and some protozoa have special cell coverings.

Flagella

Flagella in eukaryotes, which are larger and more complex than those in prokaryotes (**Figure 4.23a**), consist of two central microtubules and nine pairs of peripheral microtubules (a 9 + 2 arrangement) surrounded by a membrane (**Figure 4.23b**). Each fiber is a microtubule made of the protein *tubulin*. One of these microtubules is about the same size as an entire prokaryotic flagellum. Associated with each pair of peripheral microtubules are small molecules of the protein *dynein*. Eukaryotic flagella move like a whip (**Figure 4.23c**), whereas prokaryotic flagella move like a rotating hook. One mechanism of eukaryotic flagellar movement is a cross-bridging among dynein and other flagellar proteins. Through ATP hydrolysis, dynein plays a role in converting chemical energy in ATP to mechanical energy, which makes the flagellum move. Microtubules in the flagellum are thought to slide toward or away from the base of the cell in a wavelike manner and thereby cause the whole flagellum to move.

Flagella are most common among protozoa but are found among algae as well. Most flagellated eukaryotes have one flagellum, but some have two or more. The only flagellated human cells are spermatozoa.

Cilia

Cilia are shorter and more numerous than flagella, but they have the same chemical composition and basic arrangement of microtubules. Cilia are found mainly among ciliated protozoa, which have 10,000 or more cilia distributed over their cell surface (**Figure 4.24**). Each cilium passes through a stroke-and-recovery cycle as it beats. Together the cilia of an organism beat in a coordinated pattern, which creates a wave that passes from one end of the organism to the other. The large number of cilia and their coordinated beating allow ciliated organisms, such as paramecia, to move much more rapidly than those with flagella. Cilia on some cells can also propel fluids, dissolved particles, bacteria, mucus, and so on past the cell. This function can be of

great importance in host defenses against diseases, particularly in the respiratory tract, where it is known in humans as the mucociliary escalator.

Pseudopodia

Pseudopodia (su″do-po′de-a; singular: *pseudopodium*), or "false feet," are temporary projections of cytoplasm associated with **amoeboid movement**. This kind of movement occurs only in cells without walls, such as amoebas and some white blood cells, and only when the cell is resting on a solid surface. When an amoeba first extends a portion of its body to form a

> *Amoeboid movement results from interaction between actin and myosin filaments, similar to those present in our own muscles.*

(a) SEM

(b)

Stroke Recovery Stroke

Movement →

Direction of wave →

Stroke Recovery Stroke

Direction of cell movement →

(c)

Figure 4.24 Cilia. (a) The ciliated protozoan *Oxytricha* (918X). *(Manfred Kage /Peter Arnold, Inc.)* **(b)** The stroke-and-recovery motion of a cilium. **(c)** Cilia on an organism move in a synchronized fashion, creating a wave that propels the organism forward.

Resting amoeba with cytoplasm distributed evenly.

Newly formed pseudopodium with less dense cytoplasm.

(a)

Figure 4.25 Pseudopodia. (a) The formation of a pseudopodium, a cytoplasmic extension that allows organisms such as amoebas to move and capture food. **(b)** Micrographs of an amoeba engulfing food particles (132X). *(Michael Abbey/Visuals Unlimited)*

pseudopodium, the cytoplasm is much less dense in the pseudopodium than in other areas of the cell (**Figure 4.25**). As a result, cytoplasm from elsewhere in the organism flows into the pseudopodium by **cytoplasmic streaming**. Amoeboid movement is a slow, inching-along process.

Cell Walls

A number of unicellular eukaryotic organisms have cell walls, none of which contains the peptidoglycan that is characteristic of bacteria. Algal cell walls consist mainly of cellulose, but some contain other polysaccharides.

Cell walls of fungi consist of cellulose or chitin, or both. *Chitin* is a structural polysaccharide that is also common in the exoskeletons of arthropods such as insects and crustacea. Protozoans have flexible external coverings called pellicles. Regardless of composition, cell walls give cells rigidity and protect them from bursting when water moves into them from the environment.

▌▌▌ EVOLUTION BY ENDOSYMBIOSIS

Biologists believe that life arose on Earth (or perhaps was "seeded in" by meteorites) about 4 billion years ago, in the form of simple organisms much like the prokaryotic organisms of today. However, fossil evidence suggests that eukaryotic organisms arose only about 1 billion years ago. How development from prokaryote to eukaryote took place is unknown, but the **endosymbiotic theory** offers a plausible explanation. As we have seen, the major difference between prokaryotes and eukaryotes is that eukaryotes possess specialized membrane-enclosed organelles, including a true nucleus. According to the endosymbiotic theory, the organelles of eukaryotic cells arose from prokaryotic cells that had developed a *symbiotic* relationship with the eukaryote-to-be. Symbiosis is a relationship between two different kinds of organisms that live in close contact. If one lives inside the other, the relationship is known as *endosymbiosis*.

It is suggested that the first eukaryotic cell was an amoeba-like cell that somehow had developed a nucleus. Knowing the ease with which bits of cell membrane pinch off to form vesicles, it is fairly easy to imagine that a primitive chromosome might have become surrounded by membrane, thereby creating a rudimentary nucleus. This primitive eukaryote was probably a phagocytic cell, that is, one that obtains its nutrients by engulfing material from its environment, including, presumably, other cells. Although most engulfed prokaryotic cells were probably digested and used to nourish the phagocyte, some apparently survived and became permanent residents within the cytoplasm, eventually becoming incorporated as

(a)

(b)

Figure 4.26 Mitochondria, as befits formerly independent organisms, have their own DNA. (a) DNA filaments seen inside the mitochondrion of a frog cell (390,206X) *(D.W. Fawcett/Visuals Unlimited)* **(b)** DNA filaments (5–6 μm in length) isolated from the mitochondrion. *(I. B. David, D. R. Wolstenholme, D. W. Fawcett/Visuals Unlimited)*

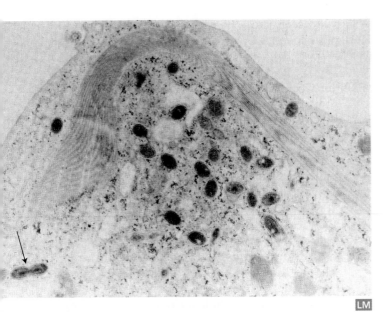

LM

Figure 4.27 **Endosymbiosis.** In the cytoplasm of *Pyrsonympha*, a protist that lives symbiotically in the hindgut of termites, bacteria (dark ovals) act as mitochondria for the protist. At lower left, one of the bacteria is dividing (arrow). *(Micrograph by David G. Chase from Early Life by Lynn Margulies ((c)1984 by Jones & Bartlett Publishers, Inc., Fig. 4-8, p. 90))*

organelles. Both organisms benefited from this arrangement. The engulfed prokaryotes were protected by the eukaryote, and the eukaryote acquired some new capabilities through the presence of its symbionts.

Evidence supporting this theory comes from a comparison of the characteristics of eukaryotic organelles with those of prokaryotic organisms:

- Mitochondria and chloroplasts are approximately the same size as prokaryotic cells.
- Unlike other organelles, mitochondria and chloroplasts have their own DNA. Organelle DNA is present in the form of a single circular loop, like the chromosome of a prokaryote **(Figure 4.26)**.
- Organelles have their own 70S ribosomes, which are like prokaryotic ribosomes, in contrast to the 80S ribosomes of eukaryotes.
- Organelle DNA and ribosomes carry out protein synthesis as it occurs in bacteria, rather than as it occurs when directed by nuclear DNA of modern eukaryotes.
- Antibiotics inhibiting protein synthesis by bacterial ribosomes do likewise on the ribosomes of chloroplasts and mitochondria.
- Mitochondria and chloroplasts divide independently of the eukaryotic cell cycle, by means of binary fission.
- The double-membrane structures of mitochondria and chloroplasts strongly resemble the cell membranes of Gram-negative bacteria, even having the same type of pores.

- Chloroplasts strongly resemble the structure of photosynthetic, chlorophyll-containing prokaryotic cyanobacteria.
- Mitochondrial DNA most closely matches the DNA of the bacterium *Rickettsia prowazekii*.

Furthermore, the notion of prokaryotic endosymbionts living in eukaryotes is not mere speculation. Examples of such relationships abound in nature. Certain eukaryotes living in low-oxygen environments lack mitochondria, yet they get along quite well, thanks to bacteria that live inside them and serve as "surrogate mitochondria." Protists living symbiotically in the hindgut of termites are, in turn, colonized by symbiotic bacteria similar in size and distribution to mitochondria **(Figure 4.27)**. The bacteria function better in this low-oxygen condition than mitochondria would. They oxidize food and provide energy in the form of ATP for their protist partner. Some primitive eukaryotes today still lack mitochondria. *Giardia*, a parasitic protist that causes diarrhea, is an example of a eukaryote that probably never acquired any mitochondria.

Living in the mud at the bottom of ponds is a giant amoeba, *Pelomyxa palustris*. It also lacks mitochondria and has at least two kinds of endosymbiotic bacteria. Killing just the bacteria with antibiotics allows lactic acid to accumulate. This suggests that the bacteria oxidize the end-products of glucose fermentation, a function that mitochondria ordinarily perform. What else do mitochondria do? They must perform some task necessary for the formation or functioning of the Golgi apparati. This group includes all prokaryotes. Perhaps integration of bacterial endosymbionts into a cell led to the development of mitochondria and Golgi apparati.

Dr. Lynn Margulis proposes that eukaryotic flagellae and cilia (she calls them "undulipodia") originated from symbiotic associations of motile bacteria, called spirochetes, with nonphotosynthetic protists. Such associations of present-day species are well known. *Mixotricha paradoxa*, a protist endosymbiont found in the hindgut of the Australian termite *Mastotermes darwiniensis*, uses the four flagellae at its front end to steer but depends on the half-million spirochetes covering its surface for driving power. These spirochetes have a natural tendency to coat living or dead surfaces. Dramatic motion pictures show that once attached, they coordinate their undulations and beat in unison, propelling their host particle along. Margulis hypothesizes that some ancient spirochetes integrated into their host cells to become cilia and flagella. She further suggests that other spirochetes were drawn down inside the cell (a process that can be observed in modern species) and eventually transformed into microtubules.

The spirochetes would have obtained nutrients that leaked from the eukaryote, while giving the eukaryote motility. Giant tube worms (6 feet long) living near hydrothermal vents deep in the ocean lack mouths, anuses, and digestive tracts. What keeps them alive? Prokaryotic

Take another look

endosymbiont bacteria colonize their internal tissues. The bacteria generate energy by metabolizing the hydrogen sulfide spewing forth from the hot vents. Excess energy is transferred to the tube worms. A similar relationship exists between endosymbiont bacteria and giant clams that live at the vents. Endosymbiosis is a common pattern of life.

THE MOVEMENT OF SUBSTANCES ACROSS MEMBRANES

A living cell, either prokaryotic or eukaryotic, is a dynamic entity. A cell is separated from its environment by a membrane, across which substances constantly move in a carefully controlled manner. Understanding how these movements occur is essential to understanding how a cell functions. Very small polar substances, such as water, small ions, and small water-soluble molecules, probably pass through pores in the membrane. Nonpolar substances, such as lipids and other uncharged particles (molecules or ions), dissolve in and pass through the membrane lipids. Still other substances are moved through the membrane by carrier molecules. Most large molecules are unable to enter cells without the aid of specific carriers.

The mechanisms by which substances move across membranes can be passive or active. In passive transport, the cell expends no energy to move substances down a *concentration gradient*, that is, from higher to lower concentration. Passive processes include *simple diffusion*, *facilitated diffusion*, and *osmosis*. In active processes, the cell expends energy from ATP, enabling it to transport substances against a concentration gradient. These processes include *active transport*. The processes *endocytosis* and *exocytosis*, which occur only in

eukaryotic cells, are separate mechanisms for moving substances across the plasma membrane.

SIMPLE DIFFUSION

All molecules have kinetic energy; that is, they are constantly in motion and are continuously redistributed. **Simple diffusion** is the net movement of particles from a region of higher to lower concentration (**Figure 4.28**). Suppose, for example, that you drop a lump of sugar into a cup of coffee. At first a concentration gradient exists, with the sugar concentration greatest at the lump and least at the rim of the cup. Eventually, though, sugar molecules become evenly distributed throughout the coffee (they reach *equilibrium*) even without stirring.

Diffusion occurs because of random movement of particles. Although particles move at high velocity, they do not travel far in a straight line before they collide with other randomly moving particles. Even so, some particles from a region of high concentration eventually move toward a region of lower concentration. Fewer particles move in the opposite direction, for two reasons: (1) There are fewer of them in regions of low concentration to begin with, and (2) they are likely to be repelled by collision with particles from a region of high concentration.

The length of time required for particles to diffuse across a cell increases with cell diameter. Materials can diffuse throughout small prokaryotic cells very quickly and throughout larger eukaryotic cells fast enough to supply nutrients and to remove wastes fairly efficiently. If cells were much larger, diffusion throughout the cell would be too slow to sustain life, so diffusion rates may be responsible in part for limiting the size of cells.

CLOSE UP

A Bug in a Bug

If the mitochondrion was originally a bacterium, what's another bacterium doing inside a mitochondrion? Eating it, is what! Investigators studying the tick, *Ixodes ricinus*, the main vector of Lyme disease, found strange DNA in their specimens. They tracked it down to a new species of bacterium living inside the egg cells of the tick's ovaries. But the new "bugs" were inside the mitochondria, not out in the cytoplasm. Somehow they get inside, between the inner and outer mitochondrial membranes, and eat the entire contents of the mitochondrion, leaving just the outer membrane, the ticks don't seem to be harmed by this. Perhaps it is because only about half the mitochondria get eaten.

Amazed at the first ever known case of a bacterium infecting a mitochondrion, the investigators obtained other ticks of the same species from all around the world. And, sure enough, 100% of the female specimens had the same bacterium infecting their egg cells. In 2006, the bacterium officially received the name of *Midichloria mitochondrii*.

Symbiotic bacteria in the mitochondria of a tick. Note that these mitochondria are large enough to readily accommodate bacteria. (*Courtesy Luciano Sacchi, The University of Pavia, Italy*)

Figure 4.28 Simple diffusion. The random movements of molecules cause them to spread out (diffuse) from an area of high concentration to areas of lower concentration until eventually they are equally distributed throughout the available space (that is, they reach equilibrium).

Any membrane severely limits diffusion, but many substances diffuse through the lipids of membranes. Diffusion through the phospholipid bilayer is affected by several factors: (1) the solubility of the diffusing substance in lipid, (2) the temperature, and (3) the difference between the highest and lowest concentration of the diffusing substance. Nonpolar substances such as steroids and gases (CO_2, O_2) cross the membrane rapidly by dissolving in the nonpolar fatty acid tails of the membrane phospholipids.

A few substances also diffuse through pores. Such diffusion is affected by the size and charge of the diffusing particles and the charges on the pore surface. Pores probably have a diameter of less than 0.8 nm, so only water, small water-soluble molecules, and ions such as H^+, K^+, Na^+, and Cl^- pass through them. This is one reason a membrane is said to be **selectively permeable** (*semipermeable*).

FACILITATED DIFFUSION

Facilitated diffusion is diffusion down a concentration gradient and across a membrane with the assistance of special pores or carrier molecules. In fact, membranes contain protein-lined pores for specific ions. These pores have an arrangement of charges that allows rapid passage of a particular ion. The carrier molecules are proteins, embedded in the membrane, that bind to one or to a few specific molecules and assist in their movement. By one possible mechanism for facilitated diffusion, a carrier acts like a revolving door or shuttle that provides a convenient one-way channel for the movement of substances across a membrane **(Figure 4.29)**. Carrier molecules can become saturated, and similar molecules sometimes compete for the same carrier. Saturation occurs when all the carrier molecules are moving the diffusing substance as fast as they can. Under these conditions the rate of diffusion reaches a maximum and cannot increase further. When a carrier molecule can transport more than one substance, the substances compete for the carrier in proportion to

their concentrations. For example, if there is twice as much of substance A as substance B, substance A will move across the membrane twice as fast as substance B.

OSMOSIS

Osmosis is a special case of diffusion in which water molecules diffuse across a selectively permeable

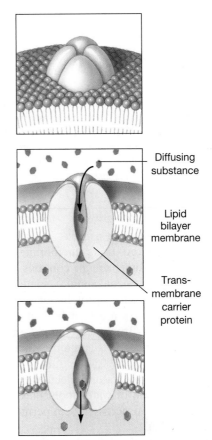

Diffusing substance

Lipid bilayer membrane

Trans-membrane carrier protein

Figure 4.29 Facilitated diffusion. Carrier protein molecules aid in the movement of substances through the cell membrane, but only down their concentration gradient (from a region where their concentration is high to one where their concentration is low). This process does not require the expenditure of any energy (ATP) by the cell.

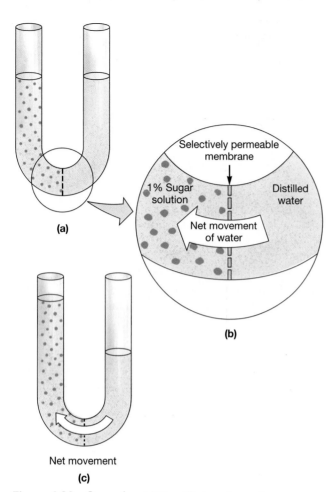

Figure 4.30   Osmosis.   (a) The diffusion of water from an area of higher water concentration (the right side) to an area of lower water concentration (the left side) through a semipermeable membrane. **(b)** Here the net movement of water is into the sugar solution because the concentration of water there is slightly lower than on the other side of the membrane. **(c)** As a result of the net movement of water, the column rises on the left.

membrane. To demonstrate osmosis, we start with two compartments separated by a membrane permeable only to water. One compartment contains pure water, and the other compartment contains some large, non-diffusible molecules, such as proteins or sugars **(Figure 4.30a)**. Water molecules move in both directions, but their net movement is from pure water (concentration 100%) toward the water that contains other molecules (concentration less than 100%; **Figure 4.30b**). Thus, osmosis is the net flow of water molecules from a region of higher concentration of water molecules to a region of lower concentration across a semipermeable membrane **(Figure 4.30c)**.

Osmotic pressure is defined as the pressure required to *prevent* the net flow of water by osmosis. The least amount of hydrostatic pressure required to prevent the movement of water from a given solution into pure water is the osmotic pressure of the solution. The osmotic pressure of a solution is proportional to the number

of particles dissolved in a given volume of that solution. Thus, NaCl and other salts that form two ions per molecule exert twice as much osmotic pressure as glucose and other substances that do not ionize, provided each compound is present at the same concentration.

The important thing for a microbiologist to know about osmosis and osmotic pressure is how particles dissolved in fluid environments affect microorganisms in those environments **(Figure 4.31)**. For this purpose, tonicity is a useful concept. *Tonicity* describes the behavior of cells in a fluid environment. The cells are the reference point, and the fluid environments are compared to the cells. The fluid surrounding cells is **isotonic** to the cells when no change in the cell volume occurs (Figure 4.31a). The fluid is **hypotonic** to the cells if the cells swell or burst as water moves from the environment into the cells (Figure 4.31b); it is **hypertonic** to the cells if the cells shrivel or shrink as water moves out of them into the fluid environment (Figure 4.31c). Although bacteria become dehydrated and their cytoplasm shrinks away from the cell wall in a hypertonic environment, their cell walls usually prevent them from swelling or bursting in the hypotonic environments they typically inhabit. The high concentration of sugar in jams and jellies is an example of tonicity at work, preventing growth of bacteria.

ACTIVE TRANSPORT

In contrast to passive processes, **active transport** moves molecules and ions against concentration gradients from regions of lower concentration to ones of higher concentration. This process is analogous to rolling something uphill, and it does require the cell to expend energy from ATP. Active transport is important in microorganisms for moving nutrients that are present in low concentrations in the environment of the cells. It requires membrane proteins that act as both carriers and enzymes **(Figure 4.32)**. These proteins display specificity in that each carrier transports a single substance or a few closely related substances. The results of active transport are to concentrate a substance on one side of a membrane and to maintain that concentration against a gradient. As with facilitated diffusion, active transport carriers also are subject to saturation and competition for binding sites by similar molecules.

Group translocation reactions move a substance from the outside of a bacterial cell to the inside while chemically modifying the substance so that it cannot diffuse out. This process allows molecules such as glucose to be accumulated against a concentration gradient. Because the modified molecule inside the cell is different from those outside, no actual concentration gradient exists. Energy for this process is supplied by phosphoenolpyruvate (PEP), a high-energy phosphate compound. Many eukaryotic cells have a similar active transport mechanism for preventing diffusion.

Situation	Isotonic	Hypotonic	Hypertonic
A bag, permeable to water but not salt, is placed in a beaker containing one of three different salt solutions.	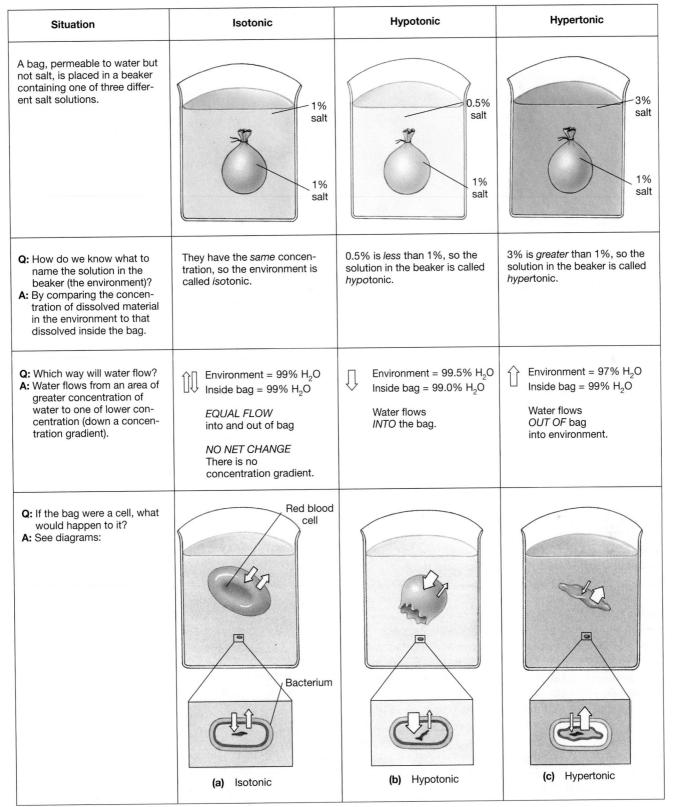 1% salt / 1% salt	0.5% salt / 1% salt	3% salt / 1% salt
Q: How do we know what to name the solution in the beaker (the environment)? **A:** By comparing the concentration of dissolved material in the environment to that dissolved inside the bag.	They have the *same* concentration, so the environment is called *iso*tonic.	0.5% is *less* than 1%, so the solution in the beaker is called *hypo*tonic.	3% is *greater* than 1%, so the solution in the beaker is called *hyper*tonic.
Q: Which way will water flow? **A:** Water flows from an area of greater concentration of water to one of lower concentration (down a concentration gradient).	Environment = 99% H$_2$O Inside bag = 99% H$_2$O *EQUAL FLOW* into and out of bag *NO NET CHANGE* There is no concentration gradient.	Environment = 99.5% H$_2$O Inside bag = 99.0% H$_2$O Water flows *INTO* the bag.	Environment = 97% H$_2$O Inside bag = 99% H$_2$O Water flows *OUT OF* bag into environment.
Q: If the bag were a cell, what would happen to it? **A:** See diagrams:	Red blood cell Bacterium **(a)** Isotonic	**(b)** Hypotonic	**(c)** Hypertonic

Figure 4.31 Experiments that examine the effects of tonicity on osmosis. (a) A cell in an isotonic environment—one that has the same concentration of dissolved material as the interior of the cell—will experience no net gain or loss of water and will retain its original shape. **(b)** A cell in a hypotonic environment—one with a lower concentration of dissolved material than the interior of the cell—will gain water and swell. Unlike a bacterial cell, a red blood cell will burst because it lacks a cell wall. **(c)** A cell in a hypertonic environment—one with a higher concentration of dissolved material than the interior of the cell—will lose water and shrink.

Nutrient

Accessory protein

Carrier protein

Lipid bilayer membrane

ATP

Cytoplasm

ADP + P$_i$

Figure 4.32 Active transport. Carrier protein molecules aid in movement of molecules through the membrane. This process can take place against a concentration gradient and so requires the use of energy (in the form of ATP) by the cell. The accessory protein participates in the carrier protein's function. (P$_i$ is inorganic phosphate, HPO$_4^-$.)

ENDOCYTOSIS AND EXOCYTOSIS

In addition to the processes that move substances directly across membranes, eukaryotic cells move substances by forming membrane-enclosed vesicles. Such vesicles are made from portions of the plasma membrane. If they form by invagination (poking in) and surround substances from outside the cell, the process is called **endocytosis**. These vesicles pinch off from the plasma membrane and enter the cell. If vesicles inside the cell fuse with the plasma membrane and extrude their contents from the cell, the process is called **exocytosis**. Both endocytosis and exocytosis require energy, probably to allow contractile proteins of the cell's cytoskeleton to move vesicles.

Endocytosis

There are several types of endocytosis. In one type, known as *receptor-mediated endocytosis*, a substance out-

Using endocytosis, in just 5 minutes, an amoeba can take in 50 times as much protein as its original protein content.

side the cell binds to the plasma membrane, which invaginates and surrounds the substance. The exact mechanisms that trigger binding and invagination depend on specific receptor sites on the plasma membrane. Once the substance is completely surrounded by plasma membrane to form a vesicle, the vesicle pinches off from the plasma membrane.

Of all the types of endocytosis, only phagocytosis is of special interest to microbiologists. In **phagocytosis**, large vacuoles called *phagosomes* form around microorganisms and debris from tissue injury. These vacuoles enter the cell, taking with them large amounts of the plasma membrane **(Figure 4.33)**. The vacuole membrane fuses with lysosomes, which release their enzymes into

the vacuoles. The enzymes digest the contents of the vacuoles (*phagolysosomes*), and release small molecules into the cytoplasm. Often, undigested particles in residual bodies are returned to and fuse with the plasma membrane. The particles are released from the cell by exocytosis. Certain white blood cells are especially adept at phagocytosis and play an important role in defending the body against infection by microorganisms.

Exocytosis

Exocytosis, the mechanism by which cells release secretions, can be thought of as the opposite of endocytosis. Most secretory products are synthesized on ribosomes or smooth endoplasmic reticulum. They are transported through the membrane of the endoplasmic reticulum; packaged in vesicles; and moved to the Golgi apparatus, where their contents are processed to form the final secretory product. Once secretory vesicles form, they move toward the plasma membrane and fuse with it (Figure 4.33). The contents of the vesicles are then released from the cell.

✓**CHECKLIST**

1. Most prokaryotes lack sterols in their plasma membranes. What functions do sterols, such as cholesterol, fulfill in eukaryotic plasma membranes?

2. Compare the number and structure of chromosomes in prokaryotes and eukaryotes.

3. Give two specific arguments that support the idea that prokaryotes were involved in the evolution of eukaryotes by means of endosymbiosis.

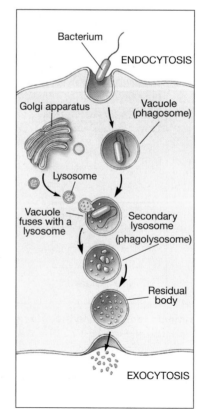

Figure 4.33 Endocytosis and exocytosis. Endocytosis is the process of taking materials into the cell; exocytosis is the process of releasing materials from the cell. Material taken in by the form of endocytosis called *phagocytosis* is enclosed in vacuoles known as *phagosomes*. The phagosomes fuse with lysosomes, which release powerful enzymes that degrade the vacuolar contents. Reusable components are absorbed into the cell, and debris is released by exocytosis.

▌ RETRACING OUR STEPS

▌▌▌ BASIC CELL TYPES

- Both **prokaryotic cells** and **eukaryotic cells** have membranes that define the bounds of the living cell, and both contain genetic information stored in DNA.
- Prokaryotic cells differ from eukaryotic cells in that they lack a defined nucleus and membrane-enclosed **organelles** (except for a few simple membrane-covered bodies in certain types of prokaryotes).

▌▌▌ PROKARYOTIC CELLS

- All prokaryotes are classified in either the domain Archaea or the domain Bacteria.

SIZE, SHAPE, AND ARRANGEMENT

- Prokaryotes are the smallest living organisms.
- Bacteria are grouped by shape: **cocci** (spherical), **bacilli** (rod-shaped), **spirilli** (rigid, wavy), **vibrios** (comma-shaped), and **spirochetes** (corkscrew-shaped).
- Arrangements of bacteria include groupings such as pairs, **tetrads**, **sarcinae**, grapelike clusters (**staphylo-**), and long chains (**strepto-**).

AN OVERVIEW OF STRUCTURE

- Bacterial cells have a cell membrane, cytoplasm, ribosomes, a nuclear region, and external structures.

THE CELL WALL

- The rigid **cell wall** outside the cell membrane is composed mainly of the polymer **peptidoglycan**.
- Cell walls differ in composition and structure. In Gram-positive bacteria, the cell wall consists of a thick, dense layer of peptidoglycan, with **teichoic acid** in it. In Gram-negative bacteria, the cell wall has a thin layer of peptidoglycan, separated from the cytoplasmic membrane by the **periplasmic space** and enclosed by an **outer membrane** made of **lipopolysaccharide**, or **endotoxin**. In acid-fast bacteria, the cell wall consists mainly of lipids, some of which are true waxes, and some of which are glycolipids.
- Some bacterial cell walls are damaged by penicillin and lysozyme.

THE CELL MEMBRANE

- The **cell membrane** has a **fluid-mosaic** structure with phospholipids forming a bilayer and proteins interspersed in a mosaic pattern.
- The main function of the cell membrane is to regulate the movement of materials into and out of cells.
- Bacterial cell membranes also perform functions usually carried out by organelles of eukaryotic cells.

INTERNAL STRUCTURE

- The **cytoplasm** is the semifluid substance inside the cell membrane.
- **Ribosomes**, which consist of RNA and protein, serve as sites for protein synthesis.
- The **nuclear region** usually includes just one, large, circular chromosome but can have two or three, some of which may be linear in shape, and which contain the prokaryotic cell's DNA and some RNA and protein.

- Bacteria contain a variety of **inclusions**, including **granules** that store glycogen or other substances and **vesicles** filled with gas or iron compounds (**magnetosomes**).
- Some bacteria form resistant **endospores**. The core of an endospore contains living material and is surrounded by a cortex, spore coat, and exosporium.

EXTERNAL STRUCTURE

- Motile bacteria have one or more **flagella**, which propel the cell by the action of rings in their basal body.
- Much bacterial movement is random, but some bacteria exhibit **chemotaxis** (movement toward attractants and away from repellents) and/or **phototaxis** (movement toward or away from light).
- Some bacteria have **pili**: **Conjugation pili** allow exchange of DNA, whereas **attachment pili** (**fimbriae**) help bacteria adhere to surfaces.
- The **glycocalyx** includes all polysaccharides external to a bacterial cell wall. **Capsules** prevent host cells from destroying a bacterium; capsules of any species of bacteria have a specific chemical composition. **Slime layers** protect bacterial cells from drying, trap nutrients, and sometimes bind cells together, as in dental plaque.

▌▌▌ EUKARYOTIC CELLS

AN OVERVIEW OF STRUCTURE

- Eukaryotic cells, which are generally larger and more complex than prokaryotic cells, are the basic structural unit of microscopic and macroscopic organisms of the kingdoms Protista, Plantae, Fungi, and Animalia.

THE PLASMA MEMBRANE

- **Plasma membranes** of eukaryotic cells are almost identical to those of prokaryotic cells, except that they contain sterols. The function of eukaryotic plasma membranes, however, is limited primarily to regulating movement of substances into and out of cells.

INTERNAL STRUCTURE

- Eukaryotic cells are characterized by the presence of a membrane-enclosed **cell nucleus**, with a **nuclear envelope**, **nucleoplasm**, **nucleoli**, and **chromosomes** (typically paired) that contain DNA and proteins called **histones**.
- In cell division by **mitosis**, each cell receives one of each chromosome found in parent cells. In cell division by **meiosis**, each cell receives one member of each pair of chromosomes, and the progeny can be **gametes** or **spores**.
- **Mitochondria**, the powerhouses of eukaryotic cells, carry out the oxidative reactions that capture energy in ATP.
- Photosynthetic cells contain **chloroplasts**, which capture energy from light.
- Eukaryotic ribosomes are larger than those of prokaryotes and can be free or attached to endoplasmic reticulum. Free ribosomes make protein to be used in the cell; those that are attached to endoplasmic reticulum make proteins to be secreted.
- The **endoplasmic reticulum** is an extensive membrane network. Without ribosomes (smooth ER), the endoplasmic reticulum synthesizes lipids; when combined with ribosomes (rough ER), it produces proteins.

- The **Golgi apparatus** is a set of stacked membranes that receive, modify, and package proteins into **secretory vesicles**.
- **Lysosomes**, in animal cells, are organelles that contain digestive enzymes, which destroy dead cells and digest contents of vacuoles.
- **Peroxisomes** are membrane-enclosed organelles that convert peroxides to water and oxygen and sometimes oxidize amino acids and fats.
- **Vacuoles** contain various stored substances and materials engulfed by phagocytosis.
- The **cytoskeleton** is a network of **microfilaments** and **microtubules** that support and give rigidity to cells and provide for cell movements.

EXTERNAL STRUCTURE

- Most external components of eukaryotic cells are concerned with movement. Eukaryotic flagella are composed of microtubules; sliding of proteins at their bases causes them to move.
- **Cilia** are smaller than flagella and beat in coordinated waves.
- **Pseudopodia** are projections into which cytoplasm flows, causing a creeping movement.
- Eukaryotic cells of the plant and fungi kingdoms have cell walls, as do the algal protists.

III EVOLUTION BY ENDOSYMBIOSIS

- The **endosymbiont theory** holds that organelles of eukaryotic cells arose from prokaryotes that had been engulfed and survived to develop a symbiotic relationship by living inside the larger cell.
- Mitochondria, chloroplasts, flagella, and microtubules are believed to have originated from endosymbiont prokaryotes.
- Many examples exist of modern prokaryotes living endosymbiotically inside eukaryotes.

III THE MOVEMENT OF SUBSTANCES ACROSS MEMBRANES

- In all passive processes of movement of substances across membranes the net movement is from a region of higher concentration to a region of lower concentration. These processes do not require expenditure of energy by the cell.

SIMPLE DIFFUSION

- **Simple diffusion** results from the molecular kinetic energy and random movement of particles. The role of diffusion in living cells depends on the size of particles, nature of membranes, and distances substances must move inside cells.

FACILITATED DIFFUSION

- **Facilitated diffusion** uses protein carrier molecules or protein-lined pores in membranes in moving ions or molecules from high to low concentrations.

OSMOSIS

- **Osmosis** is the net movement of water molecules through a **selectively permeable** membrane from a region of higher concentration of water to a region of lower concentration. The **osmotic pressure** of a solution is the pressure required to prevent such a flow.

ACTIVE TRANSPORT

- Active processes that move substances across membranes generally result in movement from regions of lower concentration of the substances to regions of higher concentration and require the cell to expend energy.
- **Active transport** requires a protein carrier molecule in a membrane, a source of ATP, and an enzyme that releases energy from ATP.
- Active transport is important in cell functions because it allows cells to take up substances that are in low concentration in the environment and to concentrate those substances within the cell.

ENDOCYTOSIS AND EXOCYTOSIS

- **Endocytosis** and **exocytosis**, which occur only in eukaryotic cells, involve formation of vesicles from fragments of plasma membrane and fusion of vesicles with the plasma membrane, respectively.
- In endocytosis the vesicle enters the cell, as in **phagocytosis**.
- In exocytosis the vesicle leaves the cell, as in secretion.
- Endocytosis and exocytosis are important because they allow the movement of relatively large quantities of materials across plasma membranes.

TERMINOLOGY CHECK

active transport (*p. 108*)
amoeboid movement
   (*p. 103*)
amphitrichous (*p. 93*)
atrichous (*p. 93*)
attachment pilus (*p. 96*)
axial filament (*p. 95*)
bacillus (*p. 81*)
capsule (*p. 97*)
cell membrane (*p. 90*)
cell nucleus (*p. 99*)
cell wall (*p. 83*)
chemotaxis (*p. 93*)

chloroplast (*p. 101*)
chromatin (*p. 100*)
chromatophore (*p. 91*)
chromosome (*p. 100*)
cilium (*p. 103*)
coccus (*p. 81*)
conjugation pilus (*p. 96*)
crista (*p. 101*)
cytoplasm (*p. 90*)
cytoplasmic streaming
   (*p. 104*)
cytoskeleton (*p. 102*)
diplo- (*p. 81*)

diploid (*p. 100*)
dyad (*p. 100*)
endocytosis (*p. 110*)
endoflagellum (*p. 95*)
endoplasmic reticulum
   (*p. 101*)
endospore (*p. 92*)
endosymbiotic theory (*p. 104*)
endotoxin (*p. 85*)
eukaryotic cell (*p. 80*)
exocytosis (*p. 110*)
facilitated diffusion (*p. 107*)
fimbria (*p. 96*)

flagellum (*p. 92*)
fluid-mosaic model (*p. 90*)
gamete (*p. 100*)
glycocalyx (*p. 97*)
Golgi apparatus (*p. 102*)
granule (*p. 91*)
group translocation reaction
   (*p. 108*)
haploid (*p. 100*)
histone (*p. 100*)
hydrophilic (*p. 90*)
hydrophobic (*p. 90*)
hypertonic (*p. 108*)

hypotonic (*p. 108*)
inclusion (*p. 91*)
isotonic (*p. 108*)
L-form (*p. 88*)
lipid A (*p. 85*)
lipopolysaccharide (LPS)
    (*p. 85*)
lophotrichous (*p. 93*)
lysosome (*p. 102*)
magnetosome (*p. 92*)
matrix (*p. 101*)
meiosis (*p. 100*)
metachromasia (*p. 92*)
metachromatic granule
    (*p. 92*)
microfilament (*p. 102*)
microtubule (*p. 102*)

mitochondrion (*p. 100*)
mitosis (*p. 100*)
monotrichous (*p. 93*)
nuclear envelope (*p. 100*)
nuclear pore (*p. 100*)
nuclear region (*p. 91*)
nucleoid (*p. 91*)
nucleolus (*p. 100*)
nucleoplasm (*p. 100*)
organelle (*p. 80*)
osmosis (*p. 107*)
osmotic pressure (*p. 108*)
outer membrane (*p. 85*)
pellicle (*p. 96*)
peptidoglycan (*p. 83*)
periplasmic space (*p. 86*)
peritrichous (*p. 93*)

peroxisome (*p. 102*)
phagocytosis (*p. 110*)
phototaxis (*p. 95*)
pilus (*p. 96*)
plasma membrane (*p. 99*)
pleomorphism (*p. 81*)
polyribosome (*p. 90*)
prokaryotic cell (*p. 80*)
protoplast (*p. 88*)
pseudopodium (*p. 103*)
ribosome (*p. 90*)
sarcina (*p. 81*)
secretory vesicle (*p. 102*)
selectively permeable
    (*p. 107*)
simple diffusion (*p. 106*)
slime layer (*p. 97*)

spheroplast (*p. 88*)
spindle apparatus (*p. 100*)
spirillum (*p. 81*)
spirochete (*p. 81*)
spore (*p. 100*)
staphylo- (*p. 81*)
strepto- (*p. 81*)
stroma (*p. 101*)
teichoic acid (*p. 85*)
tetrad (*p. 81*)
thylakoid (*p. 101*)
vacuole (*p. 102*)
vegetative cell (*p. 92*)
vesicle (*p. 92*)
vibrio (*p. 81*)
volutin (*p. 92*)
zygote (*p. 100*)

▌ CLINICAL CASE STUDY

Luis had a very painful coldsore on his upper lip. Earlier that semester he had used a topical cream containing polymyxin to treat an infected cut on his hand. At that time, he had read in his microbiology textbook that polymyxin is an antibiotic that disrupts cell membranes. He applied some polymyxin cream to his coldsore daily, but obtained no relief. Why did this treatment fail?

▌ CRITICAL THINKING QUESTIONS

1. Your roommate has noticed that you now spend most of your time studying microbiology and has become curious about the subject. She asks you to explain in the simplest possible way how prokaryotes differ from eukaryotes. What do you tell her?

2. Many of today's antibacterial drugs work by interfering with the growth of cell walls. Why do these drugs tend to have little toxicity to human cells?

3. According to the **endosymbiotic theory**, the organelles of eukaryotic cells arose from prokaryotic cells that had developed a symbiotic relationship with the eukaryote-to-be. Strong evidence supporting the theory comes from a comparison of prokaryotic organisms to eukaryotic organelles (especially mitochondria, chloroplasts, and ribosomes). Can you think of any other evidence supporting the theory?

▌ SELF-QUIZ

1. Most prokaryotes range in size from 0.5 to 2.0 μm yet have large surface-to-volume ratios. This large surface-to-volume ratio allow prokaryotes to:
(a) Ward off invaders
(b) Resist antibiotics
(c) Get nutrients easily to all parts of the cell
(d) Undergo meiosis
(e) b and d

2. Attribute each of the following to either P, prokaryotes only; E, eukaryotes only; B, both; N, neither:
(a) P Single chromosome
(b) E Membrane-bound nucleus
(c) B Fluid-mosaic membrane
(d) N Viruses
(e) P 70S ribosomes
(f) E Endoplasmic reticulum
(g) E Respiratory enzymes in mitochondria

(h) E Mitosis
(i) P Peptidoglycan in cell wall
(j) E Cilia
(k) E 80S ribosomes
(l) E Chloroplasts
(m) E "9+2" microtubule arrangement in flagella
(n) P Bacteria
(o) B Can have extrachromosomal DNA
(p) E Meiosis

3. Match the following bacterial morphology designations with their description:
C Coccus
A Bacillus
D Spirillum
F Vibrio
B Staph
E Tetrad

(a) Rod-shaped
(b) Grapelike clusters
(c) Round spheres
(d) Corkscrew shaped
(e) Cube of 4 cells
(f) Curved rods

4. Members of the Archaea and Bacteria domains are both prokaryotic and have similar structure, but differ molecularly. True or false?

5. Describe the properties of the cell membrane, and relate them to the fluid-mosaic model.

6. Which of the following are characteristics of the Gram-positive cell wall?
   (a) Lacks outer membrane
   (b) Lacks teichoic acid
   (c) Lacks a periplasmic space
   (d) Lacks lipopolysaccharide
   (e) Contains a thick peptidoglycan layer
   (f) Two of the above

7. The association of endotoxin in Gram-negative bacteria is a result of the presence of:
   (a) Peptidoglycan   (d) Steroids
   (b) Lipopolysaccharide   (e) Calcified proteins
   (c) Polypeptide

8. The reason that bacterial cells are more resistant to osmotic shock than are eukaryotic cells is that they:
   (a) Contain a cell wall composed of cellulose
   (b) Contain osmo-regulating porins
   (c) Actively block water molecules from entering the cell
   (d) Are selectively permeable
   (e) Contain a cell wall composed of peptidoglycan

9. Which of the following describes prokaryotic cell membranes?
   (a) Selectively permeable
   (b) Regulates passage of materials into and out of the cell
   (c) Contains proteins and phospholipids
   (d) Contains metabolic enzymes
   (e) All of these

10. Match the following bacterial locomotion and external structure terms to their descriptions:
    ___Phototaxis   (a) Spirochete endoflagella causing
    ___Flagellum   corkscrew motion
    ___Conjugation pilus (b) Tiny, hollow projection that at-
    ___Slime layer   taches 2 cells, providing a conduit
    ___Chemotaxis   for exchange of genetic material
    ___Glycocalyx   (c) Term used to describe all poly-
    ___Axial filaments   saccharide-containing substances
    ___Capsule   external to the cell wall
                         (d) A response of some bacteria to
                             move toward or away from light
                         (e) A thick, protective poly-
                             saccharide-containing structure
                             located outside of the cell wall
                         (f) Long, thin, helical appendage
                             used for movement
                         (g) Thin glycocalyx that prevents
                             dehydration, traps nutrients, and
                             allows for attachment to other
                             cells and objects in the
                             environment
                         (h) Nonrandom response of move-
                             ment toward or away from
                             chemical concentration gradients
                             in the environment

11. Draw a diagram describing what will happen to a bacterial cell when it is placed in:
    (a) Hypotonic solution
    (b) Isotonic solution
    (c) Hypertonic solution

12. Which of the following structures are essential for survival of most bacteria?
    (a) Cell wall   (d) a and b
    (b) Plasma membrane   (e) a, b, and c
    (c) Capsule

13. Bacterial fimbriae present on the outer cell surface are used for:
    (a) Cellular motility   (d) Adherence to surfaces
    (b) Sexual reproduction   (e) Adherence and exchange
    (c) Cell wall synthesis   of genetic information

14. Describe the structure and function of cilia and pseudopodia. What types of organisms have flagella, cilia, and pseudopodia?

15. Usually, bacteria form more endospores in response to:
    (a) Need for reproduction
    (b) Colony formation
    (c) Adverse environmental stress
    (d) Nutrient surplus
    (e) Increased aeration

16. Match the following bacterial structures with the type of bacterium on which they are found:
    ___Cell wall   (a) Gram-positive bacterium
    ___Lipopolysaccharide (b) Gram-negative bacterium
    ___Flagellum   (c) Both Gram-positive and
    ___Cilia   Gram-negative bacteria
    ___Teichoic acid   (d) Not found associated with
                               bacteria

17. The use of antibiotics that inhibit or inactivate cellular ribosomes will result directly in the loss of which of the following functions:
    (a) ATP production   (d) Protein synthesis
    (b) DNA replication   (e) Cell division
    (c) Phagocytosis

18. Plasmids are small extrachromosomal DNA molecules that contain nonessential genes that give bacteria competitive characteristics in their environments. True or False?

19. The site of ATP synthesis in prokaryotic cells is the _____ and in the eurkaryotic cells is the _____.

20. List the characteristics of a chloroplast and describe its function in photosynthesis.

21. Peptidoglycan digested from Gram- _____ bacteria retain their cell membrane but lose their cell walls, making them protoplasts, whereas cell wall digests of Gram-negative bacteria retain their cell and outer membranes intact, making them _____. Other genera of bacteria normally have cell walls but can lose their ability to form cell walls; such bacteria are called _____.

22. Match each following organelle with its function:
___Cytoskeleton
___Lysosomes
___Smooth endoplasmic reticulum
___Rough endoplasmic reticulum
___Nucleus
(a) Contains enzymes for lipid synthesis
(b) Vacuole that contains digestive enzymes
(c) Has sites for protein synthesis
(d) Site of ribosome synthesis
(e) Network of microtubules and microfilaments

23. Which of the following is not true about phagocytosis?
(a) It is a form of exocytosis.
(b) It is energy dependent.
(c) It only occurs in eukaryotes.
(d) A larger cell engulfs a smaller cell that will eventually be present in an internal vacuole.
(e) It requires fusion of internal lysosomes to engulfed vacuole for contents to be digested.

24. Match the following mechanisms by which substances move across membranes to their descriptive terms:
___Facilitated diffusion
___Osmosis
___Simple diffusion
___Passive transport
___Active transport
___Hypotonic solution
___Isotonic solution
___Hypertonic solution

(a) Diffusion in which water molecules diffuse across a selectively permeable membrane
(b) Movement of substances down a concentration gradient with no expenditure of energy
(c) Passive diffusion down a concentration gradient and across a membrane with the aid of special pores or carrier molecules
(d) Fluid environment surrounding cells that contains a higher concentration of a dissolved substance, causing cells to shrink
(e) Fluid environment surrounding cells that contains a lower concentration of a dissolved substance, causing cells to burst
(f) Requires ATP energy to move molecules and ions against their concentration gradient
(g) Net movement of particles from a region of higher to lower concentration

(h) Fluid environment surrounding cells that contains an equal concentration of a dissolved substance, causing no change in cell volume

25. Mitosis differs from meiosis in the following ways EXCEPT:
(a) Mitosis results in a full complement of chromosomes in two cells whereas meiosis results in four cells having half the number of chromosomes.
(b) In mitosis, all chromosomes are replicated whereas in meiosis, only half are replicated.
(c) Meiosis only occurs in somatic or body cells whereas meiosis occurs in production of gametes or sex cells.
(d) None of the above.
(e) a and c only.

26. For each of the lettered regions identified on this figure, give its name and function.

(a) _____
(b) _____
(c) _____
(d) _____
(e) _____
(f) _____
(g) _____
(h) _____
(i) _____
(j) _____

▌ EXPLORATIONS ON THE WEB

http://www.wiley.com/college/black

If you think you've mastered this chapter, there's more to challenge you on the web. Go to the companion web site to fine-tune your understanding of the chapter concepts and discover answers to the questions posed below.

1. Everyone knows that the nucleus contains DNA, but did you also know that the mitochondria and chloroplast have their own DNA? And that the mitochondria and chloroplast probably originated from bacteria that were engulfed by larger bacteria?

2. Did you know that the components of the cell membrane can move side to side and flip-flop?

5

Essential Concepts of Metabolism

Come with me . . .

You're right! I do have my arm inside the cow. It's warm and squishy there inside her rumen, one of the cow's four stomach-like compartments. The grass and hay she's eaten are being digested there. Why can't we digest such low-cost meals? We lack the enzymes needed for the metabolic pathways that digest grass—as does the cow. However, she has billions of microbes, a different mix in each of the four "stomachs," that do metabolize the grass for her. Without them she would starve. I'm going to remove a sample of rumen contents. I'll squeeze out the juice, and examine it under the microscope, plus try to grow some of the fascinating microbes in culture. Microbes are able to do far more types of metabolism than are humans. And remember, most of the methane in the earth's atmosphere comes from rumen microbes. Microbial metabolism keeps our world running. So, come with me now to visit this very cooperative cow.

Courtesy Jacquelyn G. Black

 Video related to this topic is available within WileyPLUS.

Until the middle of the nineteenth century, people didn't know what caused a fruit juice to become wine, or milk to sour. Then, in 1857, Louis Pasteur proved that alcoholic fermentation was caused by microorganisms. A few years later he identified specific organisms from samples of fermenting juices and souring milk. Pasteur was one of the first to study chemical processes in a living organism. Since his time, much has been learned about such processes.

METABOLISM: AN OVERVIEW

Metabolism is the sum of all the chemical processes carried out by living organisms **(Figure 5.1)**. It includes **anabolism**, reactions that require energy to synthesize complex molecules from simpler ones, and **catabolism**, reactions that release energy by breaking complex molecules into simpler ones that can then be reused as building blocks. Anabolism is needed for growth, reproduction, and repair of cellular structures. Catabolism provides an organism with energy for its life processes, including movement, transport, and the synthesis of complex molecules—that is, anabolism.

All catabolic reactions involve *electron transfer*, which allows energy to be captured in high-energy bonds in ATP and similar molecules (see ◄Appendix E). Electron transfer is directly related to oxidation and reduction **(Table 5.1)**. **Oxidation** can be defined as the loss or removal of electrons. Although many substances combine with oxygen and transfer electrons to oxygen, oxygen need not be present if another electron acceptor is available. **Reduction** can be defined as the gain of electrons. When a substance loses electrons, or is oxidized, energy is released, but another substance must gain the electrons, or be reduced, at the same time. For example, during the oxidation of organic molecules, hydrogen atoms are removed and used to reduce oxygen to form water:

$$2H_2 + O_2 \rightarrow 2H_2O$$
hydrogen   oxygen   water

In this reaction, hydrogen is an **electron donor**, or *reducing agent*, and oxygen is an **electron acceptor**, or

Figure 5.1   Metabolism, the sum of catabolism and anabolism. Large, complex molecules are generally richer in energy than are small, simple ones. Catabolic reactions break down large molecules into smaller ones, releasing energy. Organisms capture some of this energy for their life processes. Anabolic reactions use energy to build larger molecules from smaller components. The molecules synthesized in this way are used for growth, reproduction, and repair.

TABLE 5.1

Comparison of Oxidation and Reduction	
Oxidation	**Reduction**
Loss of electrons (A)	Gain of electrons (B)
Gain of oxygen	Loss of oxygen
Loss of hydrogen	Gain of hydrogen
Loss of energy (liberates energy)	Gain of energy (stores energy in the reduced compound)
Exothermic; exergonic (gives off heat energy)	Endothermic; endergonic (requires energy, such as heat)

oxidizing agent. Because oxidation and reduction must occur simultaneously, the reactions in which they occur are sometimes called *redox reactions.*

Among all living things, microorganisms are particularly versatile in the ways in which they obtain energy. The ways different microorganisms capture energy, and obtain carbon, can be classified as **autotrophy** (aw-to-trof′-e)—"self-feeding"—or **heterotrophy** (het″er-o-trof′e)—"other-feeding" **(Figure 5.2)**. **Autotrophs** use carbon dioxide (an inorganic substance) to synthesize organic molecules. They include **photoautotrophs**, which obtain energy from light, and **chemoautotrophs**, which obtain energy from oxidizing simple inorganic substances such as sulfides and nitrites. **Heterotrophs** get their carbon from ready-made organic molecules, which they obtain from other organisms, living or dead. There are **photoheterotrophs**, which obtain chemical energy from light, and **chemoheterotrophs**, which obtain chemical energy from breaking down ready-made organic compounds.

Eyeliner is a favorite food for the follicle eye mite because it contains all the nutrients they need to survive.

Autotrophic metabolism (especially photosynthesis) is important as a means of energy capture in many free-living microorganisms. However, such microorganisms do not usually cause disease. We emphasize metabolic processes that occur in chemoheterotrophs because many microorganisms, including nearly all infectious ones, are chemoheterotrophs. These processes include *glycolysis* (oxidation of glucose to pyruvic acid), *fermentation* (conversion of pyruvic acid to ethyl alcohol, lactic acid, or other organic compounds), and *aerobic respiration* (oxidation of pyruvic acid to carbon dioxide and

Figure 5.2 **The main types of energy-capturing metabolism.**

water). Glycolysis and fermentation (anaerobic processes) do not require oxygen, and only a small amount of the energy in a glucose molecule is captured in ATP. Aerobic respiration does require oxygen as an electron acceptor and captures a relatively large amount of the energy in a glucose molecule in ATP. Complete oxidation of glucose by glycolysis and aerobic respiration is summarized in the following equation:

$$C_6H_{12}O_6 + 6O_2 \rightarrow 6CO_2 + 6H_2O + energy$$
glucose   oxygen   carbon   water
dioxide

A large number of microorganisms obtain energy by *photosynthesis*, the use of light energy and hydrogen from water or other compounds to reduce carbon dioxide to an organic substance that contains more energy. The overall synthesis of glucose by photosynthesis in cyanobacteria and algae (and green plants) is summarized in the following equation:

$$6CO_2 + 6H_2O \xrightarrow[\text{chlorophyll}]{\text{light energy}} C_6H_{12}O_6 + 6O_2$$
carbon   water   glucose   oxygen
dioxide

(Other photosynthetic bacteria, as we shall see later, use a different version of this process.) Photosynthetic organisms then use the glucose or other carbohydrates made in this way for energy. Notice that the two equations above are the reverse of each other. **Figure 5.3** shows the relationship between respiration and photosynthesis.

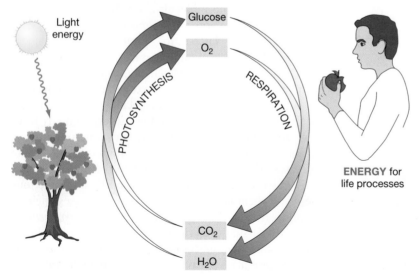

ENERGY for life processes

Figure 5.3 Photosynthesis and respiration form a cycle. In photosynthesis, light energy is used to reduce carbon dioxide, forming energy-rich compounds such as glucose and other carbohydrates. In aerobic respiration, energy-rich compounds are oxidized to carbon dioxide and water, and some of the energy released is captured for use in life processes. (The form of photosynthesis depicted here is carried out by cyanobacteria, algae, and green plants. Green and purple bacteria use compounds other than water as a source of hydrogen atoms to reduce CO_2.)

Like nearly all other chemical processes in living organisms, glycolysis, fermentation, aerobic respiration, and photosynthesis each consist of a series of chemical reactions in which the *product* of one reaction serves as the *substrate* (reacting material) for the next: A → B → C → D → E, and so on. Such a chain of reactions is called a **metabolic pathway**. Each reaction in a pathway is controlled by a particular enzyme. In this pathway, A is the initial *substrate*, E is the final product, and B, C, and D are *intermediates*.

Metabolic pathways can be catabolic or anabolic (biosynthetic). **Catabolic pathways** capture energy in a form cells can use. **Anabolic pathways** make the complex molecules that form the structure of cells, enzymes, and other molecules that control cells. These pathways use building blocks such as sugars, glycerol, fatty acids, amino acids, nucleotides, and other molecules to make carbohydrates, lipids, proteins, nucleic acids, or combinations such as glycolipids (made from carbohydrates and lipids), glycoproteins (from carbohydrates and proteins), lipoproteins (from lipids and proteins), and nucleoproteins (from nucleic acids and proteins). ATP molecules are the links that couple catabolic and anabolic pathways. Energy released in catabolic reactions is captured and stored in the form of ATP molecules, which are later broken down to provide the energy needed to build up new molecules in biosynthetic pathways. Bacteria transfer approximately 40% of the energy in a glucose molecule to ATP during aerobic metabolism and 5% during anaerobic fermentation processes. Yields are higher in aerobic processes because their end products are highly oxidized, whereas end products of anaerobic processes are only partially oxidized.

✓ CHECKLIST

1. How are photosynthesis and respiration related to each other?

2. What is the main difference between chemoautotrophs and chemoheterotrophs? Which of these groups include bacteria? Which include organisms that cause human disease?

▌▌▌ ENZYMES

Enzymes are a special category of proteins found in all living organisms. In fact, most cells contain hundreds of enzymes, and cells are constantly synthesizing proteins, many of which are enzymes. Enzymes act as *catalysts*—substances that remain unchanged while they speed up reactions to as much as a million times the uncatalyzed rate, which is ordinarily not sufficient to sustain life. The only other way to speed up the reaction rate would be to increase

Ecologically friendly drain cleaners contain bacteria or bacterial enzymes, which break down hair and grease.

the temperature: In general, a 10-degree increase in temperature results in a doubling of the reaction rate. However, most cells would die when exposed to such a rise in temperature. Thus, enzymes are necessary for life at temperatures that cells can withstand. To explain how enzymes do these things, we must consider their properties (◀Chapter 2, p. 45).

PROPERTIES OF ENZYMES

In general, chemical reactions that release energy can occur without input of energy from the surroundings. Nevertheless, such reactions often occur at unmeasurably low rates because the molecules lack the energy to start the reaction. For example, although the oxidation of glucose releases energy, that reaction does not occur unless energy to start it is available. The energy required to start such a reaction is called **activation energy (Figure 5.4)**. Activation energy can be thought of as a hurdle over which molecules must be raised to get a reaction started. By analogy, a rock resting in a depression at the top of a hill would easily roll down the hill if pushed out of the depression. Activation energy is like the energy required to lift the rock out of the depression.

A common way to activate a reaction is to raise the temperature, thereby increasing molecular movement, as you do when you strike a match. Matches ordinarily do not burst into a flame spontaneously. If the energy from friction (striking) is added to the reactants on the match head, the temperature increases, and the match bursts into flame. Such a reaction in cells would raise the temperature enough to denature proteins and evaporate liquids. Enzymes lower the activation energy so reactions can occur at mild temperatures in living cells.

Enzymes also provide a surface on which reactions take place. Each enzyme has a certain area on its surface called the **active site**, a binding site. The active site is the region at which the enzyme forms a loose

Figure 5.4 The effect of enzymes on activation energy. A chemical reaction cannot take place unless a certain amount of activation energy is available to start it. Enzymes lower the amount of activation energy needed to initiate a reaction. They thus make it possible for biologically important reactions to occur at the relatively low temperatures that living organisms can tolerate.

Figure 5.5 The action of enzymes on substrates to yield products. (a) A computer-generated model of an enzyme (blue and purple) with a substrate molecule (yellow) bound to the active site. The active site of the enzyme is a cleft or pocket with a shape and chemical composition that enable it to bind a particular substrate—the molecule on which the enzyme acts. *(Clive Freeman, The Royal Institution/Photo Researchers, Inc.)* **(b)** Each substrate binds to an active site, producing an enzyme-substrate complex. The enzyme helps a chemical reaction occur, and one or more products are formed. In this example, the reaction is one that joins two substrate molecules. Other enzyme-catalyzed reactions can involve the splitting of one substrate molecule into two or more parts or the chemical modification of a substrate.

(a)

(b)

association with its **substrate**, the substance on which the enzyme acts **(Figure 5.5a)**. Like all molecules, a substrate molecule has kinetic energy, and it collides with various molecules within a cell. When it collides with the active site of its enzyme, an **enzyme-substrate complex** forms **(Figure 5.5b)**. As a result of binding to the enzyme, some of the chemical bonds in the substrate are weakened. The substrate then undergoes chemical change, the product or products are formed, and the enzyme detaches.

Enzymes generally have a high degree of **specificity**; they catalyze only one type of reaction, and most act on only one particular substrate. An enzyme's shape (its tertiary structure; ◄Chapter 2, p. 45), especially the shape and electrical charges at its active site, accounts for its specificity. When an enzyme acts on more than one substrate, it usually acts on substrates with the same functional group or the same kind of chemical bond. For example, *proteolytic*, or protein-splitting, enzymes act on different proteins but always act on the peptide bonds in those proteins.

Enzymes are usually named by adding the suffix *-ase* to the name of the substrate on which they act. For example, phosphatases act on phosphates, sucrase breaks down the sugar sucrose, lipases break down lipids, and peptidases break peptide bonds. Enzymes can be divided into two categories on the basis of where they act. **Endoenzymes**, or intracellular enzymes, act within the cell that produced them. **Exoenzymes**, including extracellular enzymes, are synthesized in a cell but cross the cell membrane to act in the periplasmic space or in the cell's immediate environment.

PROPERTIES OF COENZYMES AND COFACTORS

Many enzymes can catalyze a reaction only if substances called *coenzymes*, or *cofactors*, are present. Such enzymes consist of a protein portion, called the **apoenzyme**, that must combine with a nonprotein coenzyme or cofactor to form an active **holoenzyme** **(Figure 5.6)**. A **coenzyme** is a nonprotein organic molecule bound to or loosely associated with an enzyme.

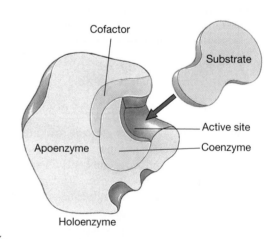

Figure 5.6 The parts of an enzyme. Many enzymes consist of a protein apoenzyme that must combine with a nonprotein coenzyme (an organic molecule) or cofactor (an inorganic ion) or both to form the functional holoenzyme. Can you explain why vitamins and minerals are important in your diet?

TRY IT

Spaghetti or Macaroni?

Some bacteria, for example, *Bacillus* spp., produce an exoenzyme that digests the protein gelatin. You will probably test for its presence or absence in the lab using test tubes of solidified gelatin. But they only remain solid at cold temperatures—try taking a Jell-oT mold to a picnic on a hot day and watch it melt all over the table. Once you have inoculated test microbes into your tube of solid gelatin, you need to put it into an incubator—the temperature there melts it. So, when you take it out you'll need to refrigerate it to see if it can harden again. If not, the bacteria must have released the exoenzyme gelatinase, which digested the gelatin and you have a positive test result. So what does this mean? Think of it this way: imagine the long protein molecules of gelatin as being long spaghetti strands. Trying to pull out just one strand from a mass of spaghetti is difficult because all the strands are tangled up into a big ball. This is like when the gelatin is set or hardened. Now think of the exoenzyme gelatinase as being a pair of scissors. It will cut (hydrolyze) the spaghetti strands into short little pieces, like macaroni. How easy is it to pick up a mass of macaroni? They slip through your fingers and don't stay together. This is like the liquid state of a positive gelatinase test. The tricky part of a gelatinase test is reading the results at the correct temperature: too warm and you may get a false-positive; too cold (frozen) and you may get a false-negative. So, is there an easier way? Try this. Cut a strip of exposed, but not developed, X-ray film small enough to fit into a test tube. Add enough nutrient broth to cover the bottom half of the strip. Inoculate your test organism into the broth and incubate. If the bacterium produces gelatinase, the exoenzyme will digest (hydrolyze) away the coating of gelatin that holds the light-sensitive silver grains to the plastic film backing. You will see only the clear plastic. If it stays dark, the gelatin was not digested. No more worrying about proper temperatures!

The clear plastic film is revealed due to the action of gelatinase *(Reprinted from L. De la Maza, M.Pezzlo & E.J. Baron,* Color Atlas of Diagnostic Microbiology, *Mosby, 1997, page 47. Reproduced with Permission by Elsevier.)*

Many coenzymes are synthesized from *vitamins*, which are essential nutrients precisely because they are required to make coenzymes. For example, *coenzyme A* is made from the vitamin pantothenic acid, and **NAD** (*nicotinamide adenine dinucleotide*) is made from the vitamin niacin. A **cofactor** is usually an inorganic ion, such as magnesium, zinc, or manganese. Cofactors often improve the fit of an enzyme with its substrate, and their presence can be essential in allowing the reaction to proceed.

Carrier molecules such as cytochromes and coenzymes carry hydrogen atoms or electrons in many oxidative reactions (**Figure 5.7**). When a coenzyme receives hydrogen atoms or electrons, it is reduced;

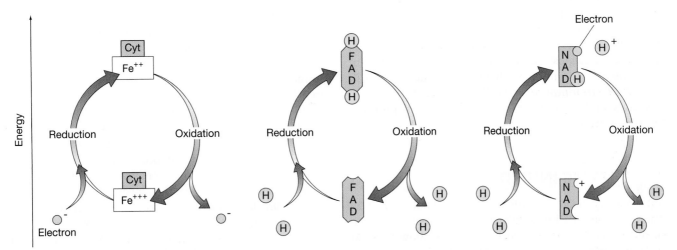

Figure 5.7 Energy transfer by carrier molecules. Carrier molecules such as cytochromes (cyt) and some coenzymes carry energy in the form of electrons in many biochemical reactions. Coenzymes such as FAD carry whole hydrogen atoms (electrons together with protons); NAD carries one hydrogen atom and one "naked" electron. When coenzymes are reduced (gain electrons), they increase in energy; when they are oxidized (lose electrons), they decrease in energy.

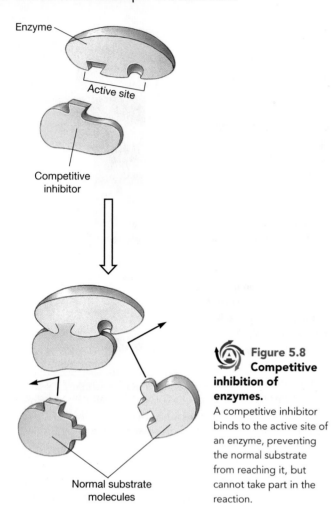

Figure 5.8 Competitive inhibition of enzymes.

A competitive inhibitor binds to the active site of an enzyme, preventing the normal substrate from reaching it, but cannot take part in the reaction.

when it releases them, it is oxidized. The coenzyme **FAD** (*flavin adenine dinucleotide*), for example, receives two hydrogen atoms to become FADH$_2$ (reduced FAD). The coenzyme NAD has a positive charge in its oxidized state (NAD$^+$). In its reduced state, NADH, it carries a hydrogen atom and an electron from another hydrogen atom, the proton of which remains in the cellular fluids. In all such oxidation-reduction reactions, the electron carries the energy that is transferred from one molecule to another. Thus, for simplicity, we will refer to *electron transfer* regardless of whether "naked" electrons or hydrogen atoms (electrons with protons) are transferred.

▌▌▌ ENZYME INHIBITION

No organism can afford to allow continual maximum activity of all its enzymes. Not only is this a waste of materials and energy, but it also may allow harmful quantities of compounds to accumulate, while others are lacking. Therefore, there must be ways to inhibit enzyme activity in order to slow or even stop its rate. How, then, are enzymes inhibited? Knowing the answers can help

us to control the rates of microbial growth, or the production of certain products that they form.

A molecule similar in structure to a substrate can sometimes bind to an enzyme's active site even though the molecule is unable to react. This nonsubstrate molecule is said to act as a **competitive inhibitor** of the reaction because it competes with the substrate for the active site **(Figure 5.8)**. When the inhibitor binds to an active site, it prevents the substrate from binding and thereby inhibits the reaction.

AZT, a drug used in the treatment of AIDS, is a competitive inhibitor of the reverse transcriptase enzyme, necessary for viral replication.

Because the attachment of such a competitive inhibitor is reversible, the degree of inhibition depends on the relative concentrations of substrate and inhibitor. When the concentration of the substrate is high and that of the inhibitor is low, the active sites of only a few enzyme molecules are occupied by the inhibitor, and the rate of the reaction is only slightly reduced. When the concentration of the substrate is low and that of the inhibitor is high, the active sites of many enzyme molecules are occupied by the inhibitor, and the rate of the reaction is greatly reduced.

The *sulfa drugs* ◀(Chapter 13) are competitive inhibitors. Normally, bacterial cells have the enzymes to convert *para-aminobenzoic acid* (*PABA*) to folic acid, an essential vitamin. If sulfa drugs are present, they compete with PABA for the enzymes' active sites. The greater the sulfa drug concentration, the greater the inhibition of folic acid synthesis.

Enzymes also can be inhibited by substances called **noncompetitive inhibitors**. Some noncompetitive inhibitors attach to the enzyme at an **allosteric site**, which is a site other than the active site **(Figure 5.9)**. Such inhibitors distort the tertiary protein structure and alter the shape of the active site. Any enzyme molecule thus affected no longer can bind substrate, so it cannot catalyze a reaction. Although some noncompetitive inhibitors bind reversibly, others bind irreversibly and permanently inactivate enzyme molecules, thereby greatly decreasing the reaction rate. In noncompetitive inhibition, increasing the substrate concentration does not increase the reaction rate as it does in the presence of a competitive inhibitor. Lead, mercury, and other heavy metals, although not noncompetitive inhibitors, can bind to other sites on the enzyme molecule and permanently change its shape, thus inactivating it.

Feedback inhibition, a kind of reversible noncompetitive inhibition, regulates the rate of many metabolic pathways. For example, when an end product of a pathway accumulates, the product often binds to and inactivates the enzyme that catalyzes the first reaction in the pathway. Feedback inhibition is discussed in more detail in ◀Chapter 7.

Figure 5.9 **Noncompetitive (allosteric) inhibition of enzymes.** A noncompetitive (allosteric) inhibitor usually binds at a site other than the active site (that is, at an allosteric site). Its presence changes the shape of the enzyme enough to interfere with binding of the normal substrate. Some noncompetitive inhibitors are used in the regulation of metabolic pathways, but others are poisons.

✔CHECKLIST

1. Distinguish between coenzyme and cofactor. How are vitamins related to these?

2. Why are inhibitors attaching to the allosteric site called noncompetitive inhibitors?

FACTORS THAT AFFECT ENZYME REACTIONS

Factors that affect the rate of enzyme reactions include:

- Temperature
- pH
- Concentrations of substrate, product, and enzyme

Temperature and pH

Like other proteins, enzymes are affected by heat and by extremes of pH. Even small pH changes can alter the electrical charge on various chemical groups in enzyme molecules, thereby altering the enzyme's ability to bind its substrate and catalyze a reaction.

Most human enzymes have an *optimum temperature*, near normal body temperature, and an *optimum pH*, near neutral, at which they catalyze a reaction most rapidly. Microbial enzymes likewise function best at optimum temperature and pHs, which are related to an organism's normal environment. The enzymes of microbes that infect humans have approximately the same optimum temperature and pH requirements as human enzymes.

Changes in *enzyme activity*, the rate at which an enzyme catalyzes a reaction, are shown in **Figure 5.10**. In the first graph, enzyme activity increases with temperature up to the enzyme's optimum temperature. Above 40°C, however, the enzyme is rapidly denatured, and its activity decreases accordingly (◀Chapter 2, p. 45). The second graph shows that activity is maximal at an enzyme's optimum pH and decreases as the pH rises or drops from the

APPLICATIONS

How to Ruin an Enzyme

If an enzyme has a vital metabolic function, its inhibitor acts as a poison. A competitive inhibitor temporarily poisons enzyme molecules and slows the reaction. If it binds to the active site of all molecules at one time, it can stop the reaction. The enzyme itself is unharmed, however, and resumes function if the poison is removed. If the poison forms a covalent bond to the enzyme, it is a permanent poison, and the enzyme's ability to function is irreversibly destroyed.

Enzyme inhibition has wide-ranging applications. Some antibiotics, such as penicillin, kill bacteria by damaging their cell wall and causing lysis. The antibiotic binds to and inactivates enzymes that are needed to reseal breaks in the peptidoglycan molecules of the cell wall during cell growth. When the cell becomes sufficiently weakened, lysis occurs. Fluoride, which prevents tooth decay, hardens enamel and poisons enzymes. In low concentrations it kills bacteria in the mouth without damaging human cells, but if the concentration is high enough, it can kill human cells, too. Many pesticides and herbicides exert their effects through competitive inhibition. Certain chemotherapeutic agents used to treat cancer inhibit enzymes that are most active in rapidly dividing cells, including malignant cells. Finally, heavy metals inactivate enzymes noncompetitively and permanently and so function as active ingredients in many disinfectants.

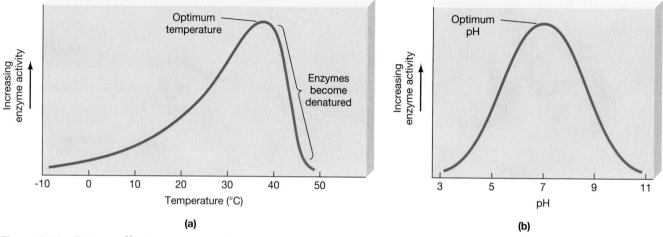

Figure 5.10 Factors affecting enzyme activity. (a) Enzymes become more active as the temperature rises. Above about 40°C, however, most enzymes become denatured, and their activity falls off sharply. **(b)** Most enzymes also have an optimal pH at which they function most effectively. How might adding vinegar (an acid) to food retard spoilage by microbes?

optimum. Like high temperatures, extremely acidic or alkaline conditions also denature enzymes. Such conditions are used to kill or control the growth of microorganisms.

Concentration

To understand the effects of concentrations of substrates and products on enzyme-catalyzed reactions, we must first note that all chemical reactions are, in theory, reversible. Enzymes can catalyze a reaction to go in either direction: AB → A + B or A + B → AB. The concentrations of substrates and products are among several factors that determine the direction of a reaction. A high concentration of AB drives the reaction toward formation of A and B. Use of A and B in other reactions as fast as these products are formed also drives the reaction toward the formation of more A and B. Conversely, use of AB in another reaction so that its concentration remains low drives the reaction toward the formation of AB. When neither AB nor A and B are removed from the system, the reaction will ultimately reach a steady state known as **chemical equilibrium**. At equilibrium, no net change in the concentrations of AB, A, or B occurs.

The quantity of enzyme available usually controls the rate of a metabolic reaction. A single enzyme molecule can catalyze only a specific number of reactions per second, that is, can act on only a specific number of substrate molecules. The reaction rate increases with the number of enzyme molecules and reaches a maximum when all available enzyme molecules are working at full capacity. However, if the substrate concentration is too low to keep all enzyme molecules working at capacity, the substrate concentration will determine the rate of the reaction.

With an overview of metabolic processes and an understanding of enzymes and how they work, we are ready to look at metabolic processes in more detail. We begin with glycolysis, fermentation, and aerobic respiration, the processes used by most microorganisms to capture energy.

ANAEROBIC METABOLISM: GLYCOLYSIS AND FERMENTATION

GLYCOLYSIS

Glycolysis (gli-kol′i-sis), also called the Embden-Meyerhof pathway, is the metabolic pathway used by most autotrophic and heterotrophic organisms, both aerobes and anaerobes, to begin to break down glucose. The name glycolysis literally means splitting (*lysis*) of sugar (*glyco-*). It does not require oxygen, but it can occur in either the presence or absence of oxygen. **Figure 5.11** shows the 10 steps of the glycolytic pathway, within which four important events occur:

1. Substrate-level phosphorylation (the transfer of phosphate groups from ATPs to glucose)
2. Breaking of a six-carbon molecule (glucose) into two three-carbon molecules
3. The transfer of two electrons to the coenzyme NAD
4. The capture of energy in ATP

Phosphorylation (fos″for-i-la′shun) is the addition of a phosphate group to a molecule, often from ATP. This addition generally increases the molecule's energy. Thus, phosphate groups commonly serve as energy carriers in biochemical reactions. Early in glycolysis, phosphate groups from two molecules of ATP are added to glucose. This expenditure of two ATPs raises the energy level of glucose. It can then participate in subsequent reactions (like the rock pushed out of the depression atop the hill) and glucose is rendered incapable of leaving the cell. The phosphorylated molecule drives the cell's metabolic reactions.

After phosphorylation, glucose is broken into two three-carbon molecules, and each molecule is oxidized as two electrons are transferred from it to NAD. The

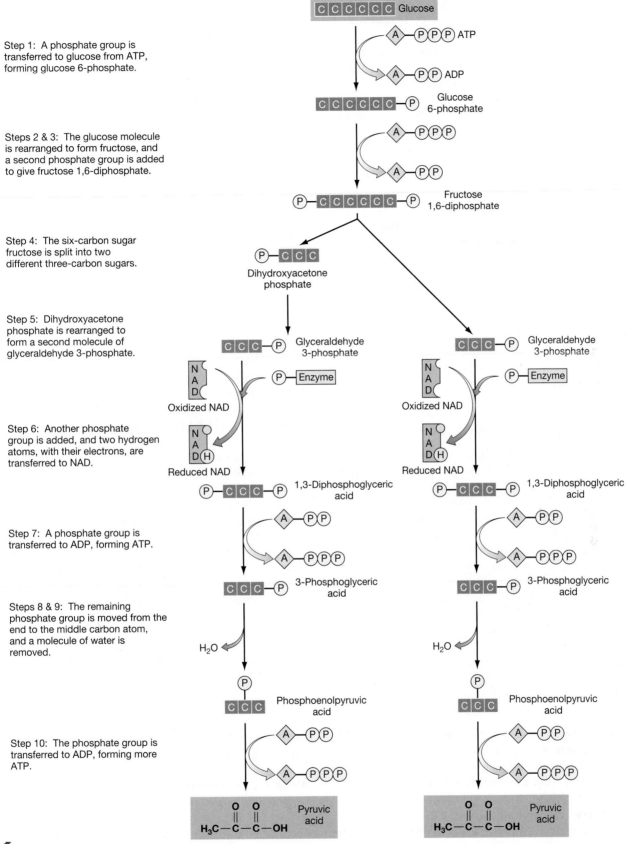

Step 1: A phosphate group is transferred to glucose from ATP, forming glucose 6-phosphate.

Steps 2 & 3: The glucose molecule is rearranged to form fructose, and a second phosphate group is added to give fructose 1,6-diphosphate.

Step 4: The six-carbon sugar fructose is split into two different three-carbon sugars.

Step 5: Dihydroxyacetone phosphate is rearranged to form a second molecule of glyceraldehyde 3-phosphate.

Step 6: Another phosphate group is added, and two hydrogen atoms, with their electrons, are transferred to NAD.

Step 7: A phosphate group is transferred to ADP, forming ATP.

Steps 8 & 9: The remaining phosphate group is moved from the end to the middle carbon atom, and a molecule of water is removed.

Step 10: The phosphate group is transferred to ADP, forming more ATP.

Figure 5.11 **The reactions of glycolysis.** Note that in steps 1 and 3, two molecules of ATP are used (A = adenosine). In steps 7 and 10, two molecules of ATP are formed. Because each glucose molecule yields two of the three-carbon sugars that undergo reactions 7 and 10, four molecules of ATP are actually formed, giving a net yield of two ATP per glucose.

end products are two molecules of pyruvic acid (called pyruvate in its ionized form) and two molecules of reduced NAD (NADH).

Energy is captured in ATP at the substrate level—that is, in the direct course of glycolysis—in two separate reactions late in the process. With *adenosine diphosphate (ADP)* and *inorganic phosphate* (P_i) available in the cytoplasm, the energy released from substrate molecules is used to form high-energy bonds between ADP and P_i:

$$ADP + P_i + energy \rightarrow ATP$$

Glycolysis provides cells with a relatively small amount of energy. Energy is captured in two molecules of ATP during the metabolism of each three-carbon molecule, and a total of four ATPs are formed as one six-carbon glucose molecule is metabolized by glycolysis to two molecules of pyruvic acid. (See ◄Appendix E for a more detailed account of glycolysis.) Because energy from two ATPs was used in the initial phosphorylations, glycolysis results in a net energy capture of only two ATPs per glucose molecule. When atmospheric oxygen is present and the organism has the enzymes to carry out aerobic respiration, electrons from reduced NAD are transferred to oxygen during biological oxidation, as is explained later.

ALTERNATIVES TO GLYCOLYSIS

Besides glycolysis, many microorganisms have one or two other metabolic pathways for glucose oxidation. For example, many bacteria, including *Escherichia coli* and *Bacillus subtilis*, have a *pentose phosphate pathway*. This pathway, which can function at the same time as glycolysis, breaks down not only glucose but also five-carbon sugars (pentoses). (◄Appendix E outlines this pathway.) In a few species of bacteria, including *Pseudomonas*, enzymes carry out the *Entner-Doudoroff pathway*, which replaces the glycolytic and pentose phosphate pathways. In this pathway, glucose goes through a short series of reactions, one intermediate (glyceraldehyde 3-phosphate) of which goes through the last five steps of a typical glycolysis and produces two ATP molecules in the process of forming pyruvic acid.

Two additional features illustrate principles that apply to metabolic pathways in general:

1. Each reaction is catalyzed by a specific enzyme. Although enzyme names have been omitted in our discussion for simplicity, it is important to remember that each reaction in a metabolic pathway is catalyzed by an enzyme.

2. When electrons are removed from intermediates in metabolic pathways, they are transferred to one of two coenzymes–NAD or NADP (*nicotinamide adenine dinucleotide phosphate*). In a reduced form (NADH or NADPH), these coenzymes store a cell's *reducing power*. For example, in glycolysis,

oxidized NAD^+ becomes reduced NAD (NADH). As explained next, electrons are removed from reduced NAD during fermentation, freeing it to remove more electrons from glucose and keep glycolysis operating. Because cells contain limited quantities of both enzymes and coenzymes, the rate at which the reactions of glycolysis and other pathways occur is limited by the availability of these important molecules.

Although glucose is the main nutrient of some microorganisms, other microbes can obtain energy from other sugars. Such organisms usually have specific enzymes to convert a sugar into an intermediate in the glycolytic pathway. Once the sugar has entered glycolysis, it is metabolized to pyruvic acid and then fermented or metabolized aerobically by processes to be described later.

FERMENTATION

The metabolism of glucose or another sugar by glycolysis is a process carried out by nearly all cells. One process by which pyruvic acid is subsequently metabolized in the absence of oxygen is **fermentation**. Fermentation is the result of the need to recycle the limited amount of NAD by passing the electrons of reduced NAD off to other molecules. It occurs by many different pathways (**Figure 5.12**). Two of the most important and commonly occurring pathways are homolactic acid fermentation and alcoholic fermentation. Neither captures energy in ATP from the metabolism of pyruvic acid, but both pathways remove electrons from reduced NAD so that it can continue to act as an electron acceptor. Thus, they indirectly foster energy capture by keeping glycolysis going.

Yeast and bacteria on berries can ferment their sugars to alcohol. Birds eating these berries can become quite drunk as is obvious from their flight patterns.

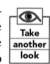

Take another look

Homolactic Acid Fermentation

The simplest pathway for pyruvic acid metabolism is **homolactic acid fermentation**, in which only (homo-) lactic acid is made (**Figure 5.13**). Pyruvic acid is converted directly to lactic acid, using electrons from reduced NAD. Unlike other fermentations, this type produces no gas. It occurs in some types of the bacteria called lactobacilli, in streptococci, and in mammalian muscle cells. This pathway in lactobacilli is used in making some cheeses.

Alcoholic Fermentation

In **alcoholic fermentation** (**Figure 5.14**), carbon dioxide is released from pyruvic acid to form the intermediate acetaldehyde, which is quickly reduced to ethyl alcohol by electrons from reduced NAD. Alcoholic fermentation, although rare in bacteria, is common in yeasts and is used in making bread and wine. ◄Chapter 26 deals extensively with these topics.

Feet darkened by gangrene

Cheese

Wine

Figure 5.12 Fermentation pathways. Many different fermentation pathways are found among microorganisms. Would two different microbes, each fermenting a quantity of the same material, necessarily produce the same products or byproducts? What about flavor? *(top left and center, SUPERSTOCK; top right, Mike Pares/Custom Medical Stock Photo, Inc.)*

Other Kinds of Fermentation

The other kinds of fermentation summarized in Figure 5.12 are performed by a great variety of microorganisms. One of the most important things about these processes is that they occur in certain infectious organisms, and their products are used in diagnosis. For example,

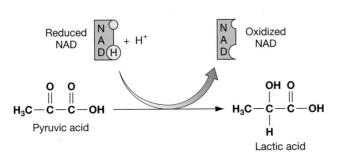

Figure 5.13 Homolactic acid fermentation. Pyruvic acid is reduced to lactic acid by the NAD from step 6 of glycolysis (Figure 5.11).

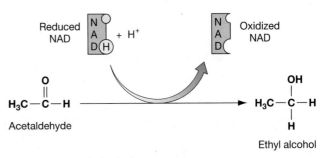

Figure 5.14 Alcoholic fermentation. In this two-step process, a molecule of carbon dioxide is first removed from pyruvic acid to form acetaldehyde. Acetaldehyde is then reduced to ethyl alcohol by NAD.

the Voges-Proskauer test for acetoin, an intermediate in butanediol fermentation, helps detect the bacterium *Klebsiella pneumoniae*, which can cause pneumonia. Anaerobic butyric-butylic fermentation occurs in *Clostridium* species that cause tetanus and botulism. The production of butyric acid by *Clostridium perfringens* is an important cause of the severe tissue damage of gangrene. This fermentation also produces the unpleasant odors of rancid butter and cheese.

The ability to ferment sugars other than glucose forms the basis of other diagnostic tests. One such test (**Figure 5.15**) uses the sugar mannitol and the pH indicator phenol red. The pathogenic bacterium *Staphylococcus aureus* ferments mannitol and produces acid, which causes the phenol

Figure 5.15 A positive (yellow) mannitol-fermentation test. This test distinguishes the pathogenic *Staphylococcus aureus* (right) from most nonpathogenic *Staphylococcus* species. *S. aureus* ferments mannitol, producing acid that turns the pH indicator (phenol red) in the medium to yellow. The medium before inoculation (left) is light red. *(Courtesy George A. Wistreich, East Los Angeles College)*

red in the medium to turn yellow. The nonpathogenic bacterium *Staphylococcus epidermidis* fails to ferment mannitol and does not change the color of the medium.

Many of the products formed by these and other fermentations, such as acetic acid, acetone, and glycerol, are of commercial value. ◄Chapter 26 discusses industrial, pharmaceutical, and food products produced by microbial fermentation. Some of these may allow us to reduce our dependence on costly petrochemicals.

AEROBIC METABOLISM: RESPIRATION

As we have noted, most organisms obtain some energy by metabolizing glucose to pyruvate by glycolysis. Among microorganisms, both anaerobes and aerobes carry out these reactions. **Anaerobes** are organisms that do not use oxygen; they include some that are killed by exposure to oxygen. **Aerobes** are organisms that *do* use oxygen; they include some that must have oxygen. In addition, a significant number of species of microorganisms are *facultative anaerobes* ◄(Chapter 6), which use oxygen if it is available but can function without it. Although aerobes obtain some of their energy from glycolysis, they use glycolysis chiefly as a prelude to a much more productive process, one that allows them to obtain far more of the energy potentially available in glucose. This process is **aerobic respiration** via the *Krebs cycle* and *oxidative phosphorylation*.

THE KREBS CYCLE

The **Krebs cycle**, named for the German biochemist Hans Krebs, who identified its steps in the late 1930s, metabolizes two-carbon units called *acetyl groups* to

APPLICATIONS

Light of My Death

The culture surrounding death and funeral traditions has undergone many changes. In early America, and through the early 1900s, the corpse was often kept at home for several days before burial. Later, when mortuaries and funeral homes became popular, the body was displayed for an extended time. Before bodies were stored in refrigerated mortuaries, gas build-up and bloating from fermentation was a practical and cosmetic problem that undertakers had to control. To prevent distortion of the corpse due to the accumulation of bacterial gases, undertakers would prick tiny holes in the body and briefly hold a candle to the openings. Long blue flames would appear, fed by the gases escaping from the corpse. These flames would last 3 or 4 days until all the gas had been consumed. How does this evidence explain the observation that 19 times as much energy is captured in aerobic metabolism as in fermentation?

CO_2 and H_2O. It also is called the **tricarboxylic acid (TCA) cycle**, because some molecules in the cycle have three carboxyl (COOH) groups, or the **citric acid cycle**, because citric acid is an important intermediate.

Before pyruvic acid (the product of glycolysis) can enter the Krebs cycle, it must first be converted to *acetyl-CoA*. This complex reaction involves the removal of one molecule of CO_2, transfer of electrons to NAD, and addition of coenzyme A (CoA) (**Figure 5.16**). In prokaryotes, these reactions occur in the cytoplasm; in eukaryotes, they occur in the matrix of mitochondria.

The Krebs cycle is a sequence of reactions in which acetyl groups are oxidized to carbon dioxide. Hydrogen atoms are also removed, and their electrons are transferred to coenzymes that serve as electron carriers (**Figure 5.17**). (The hydrogens, as we will see, are eventually combined with oxygen to form water.) Each reaction in the Krebs cycle is controlled by a specific enzyme, and the molecules are passed from one enzyme to the next as they go through the cycle. The reactions form a cycle because oxaloacetic acid (oxaloacetate), a first reactant, is regenerated at the end of the cycle. As one acetyl group is metabolized, oxaloacetate combines with another to form citric acid and goes through the cycle again. (◄Appendix E gives more detail on the Krebs cycle.)

Certain events in the Krebs cycle are of special significance:

- The oxidation of carbon
- The transfer of electrons to coenzymes
- Substrate-level energy capture

As each acetyl group goes through the cycle, two molecules of carbon dioxide arise from the complete oxidation of its two carbons. Four pairs of electrons are transferred to coenzymes: three pairs to NAD and one pair to FAD. Much energy is derived from these electrons in the next phase of aerobic respiration, as we will soon see. Finally, some energy is captured in a high-energy bond in guanosine triphosphate (GTP). This reaction takes place at the substrate level; that is, it occurs directly in the course of a reaction of the Krebs cycle. Energy in GTP is easily transferred to ATP. Note that because each glucose molecule produces two molecules of acetyl-CoA, the quantities of the products just given must be doubled to represent the yield from the metabolism of a single glucose molecule.

BIOTECHNOLOGY

Putting Microbes to Work

What can you get from *Klebsiella pneumoniae* besides pneumonia? Acrylic plastics, clothing, pharmaceuticals, paints—all products of the parent compound 3-hydroxy-propionaldehyde, which *K. pneumoniae* produces by fermentation of glycerol. Glycerol is a common byproduct of the processing of animal fats and vegetable oils, such as those from soybeans. Thus, our nation's surplus farm commodities could eventually help replace costly imported petrochemicals, thanks to microbial fermentations.

Scientists are also exploring the possibility of modifying other microbes such as *Saccharomyces cerevisiae* (baker's yeast) and *S. carlsbergensis* (brewer's yeast) to make them more useful to us. These scientists hope to obtain from yeast a fast-acting enzyme that, when given intravenously, will dissolve blood clots and thus lessen the damage caused by heart attacks and strokes. The biochemical versatility of yeast may also help clear our air of petrochemical pollutants. The yeast *Pachysolen tannaphilus* converts xylose, a sugar found in woody plant parts such as corn stalks, directly into ethyl alcohol. The U.S. Department of Agriculture estimates that such yeasts could produce 4 billion gallons of clean-burning fuel alcohol from agricultural wastes per year.

Figure 5.16 The doorway into the Krebs cycle.
Pyruvic acid loses a molecule of CO_2 and is oxidized by NAD. The resulting two-carbon acetyl group is attached to coenzyme A, forming acetyl-CoA.

Figure 5.17 The reactions of the Krebs cycle. The intermediates are simplified to show only the number of carbon atoms and carboxyl groups for each. A two-carbon acetyl group enters the cycle as acetyl-CoA in step 1, and two carbon atoms leave the cycle as molecules of CO_2 in steps 3 and 4. Energy is captured in guanosine triphosphate (GTP) in step 5 and will eventually be transferred to ATP. In addition, electrons are removed by coenzymes in steps 3, 4, 6, and 8. More energy will be extracted from these electrons when they are subsequently fed into the electron transport chain.

ELECTRON TRANSPORT AND OXIDATIVE PHOSPHORYLATION

Electron transport and oxidative phosphorylation can be likened to a series of waterfalls in which the water makes many small descents and three larger ones (Figure 5.18). In most electron transfers (the small descents), only small amounts of energy are released. At three points (the larger descents), more energy is released, some of which is used to form ATP by the addition of P_i to ADP.

Electron transport, the process leading to the transfer of electrons from substrate to O_2, begins during one of the energy-releasing dehydrogenation reactions of catabolism. Two hydrogen atoms (each consisting of one electron and one proton) are transferred to NAD, forming reduced NAD. The resulting compound in turn transfers the pairs of atoms to one of a series of other carrier compounds embedded in the cell membrane of bacteria or in the inner membrane of mitochondria. These carrier compounds form an **electron transport chain**, which is often called the *respiratory chain* (**Figure 5.19**). Through a series of oxidation-reduction reactions, the electron transport chain performs two basic functions: (1) accepting electrons from an electron donor and transferring them to an electron acceptor, and (2) conserving for ATP synthesis some of the energy released during the electron transfer. The large amounts of energy obtained from aerobic respiration result from

Figure 5.18 The electron transport chain modeled as a waterfall. As the electrons are passed from carrier to carrier in the chain, they decrease in energy, and some of the energy they lose is harnessed to make ATP.

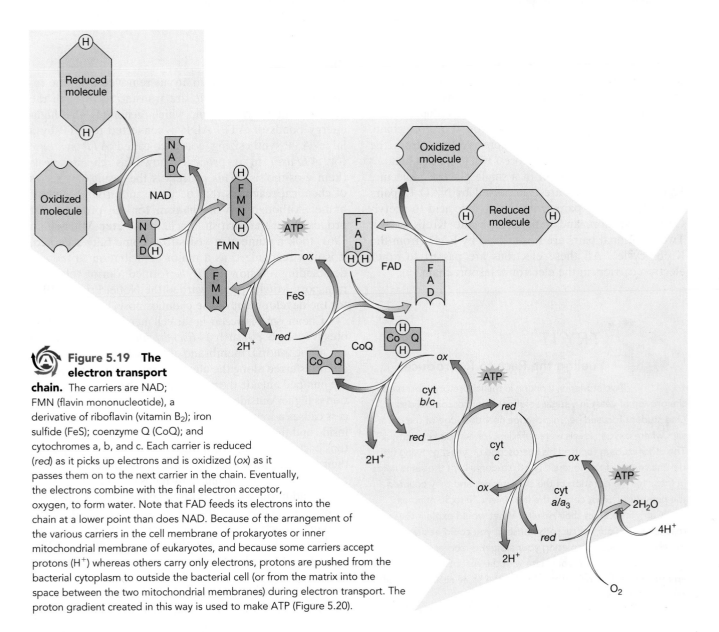

Figure 5.19 The electron transport chain. The carriers are NAD; FMN (flavin mononucleotide), a derivative of riboflavin (vitamin B_2); iron sulfide (FeS); coenzyme Q (CoQ); and cytochromes a, b, and c. Each carrier is reduced (*red*) as it picks up electrons and is oxidized (*ox*) as it passes them on to the next carrier in the chain. Eventually, the electrons combine with the final electron acceptor, oxygen, to form water. Note that FAD feeds its electrons into the chain at a lower point than does NAD. Because of the arrangement of the various carriers in the cell membrane of prokaryotes or inner mitochondrial membrane of eukaryotes, and because some carriers accept protons (H^+) whereas others carry only electrons, protons are pushed from the bacterial cytoplasm to outside the bacterial cell (or from the matrix into the space between the two mitochondrial membranes) during electron transport. The proton gradient created in this way is used to make ATP (Figure 5.20).

the transfer of electrons through the electron transport chain, from a level of high energy to one of low energy, with the formation of ATP (Figure 5.19). Energy is captured in high-energy bonds as P_i combines with ADP to form ATP. This process is known as **oxidative phosphorylation**. Each member of the chain becomes reduced as it picks up electrons; then upon giving up electrons to the next member in line, it is oxidized. In aerobic respiration, oxygen is the final electron acceptor and becomes reduced to water (Figure 5.19).

Several types of enzyme complexes are involved in electron transport. These include NADH dehydrogenase, cytochrome reductase, and cytochrome oxidase. The electron carriers include **flavoproteins** (such as FAD and flavin mononucleotide, FMN), iron-sulfur (FeS) proteins, and **cytochromes**, proteins with an iron-containing ring called *heme*. A group of nonprotein, lipid-soluble electron carriers known as the **quinones**, or *coenzymes Q*, are also found in electron transport systems.

All electron transport chains are not alike; they differ from organism to organism, and sometimes a given organism may have more than one kind. However, they all have compounds such as flavoproteins and quinones, which accept only hydrogen atoms and compounds such as cytochromes, which accept only electrons. Unless electrons are continuously transferred from reduced NAD and FAD to oxygen via the electron transport chain, these enzymes cannot accept more electrons from the Krebs cycle, and the entire process will grind to a halt.

From the metabolism of a single glucose molecule, 10 pairs of electrons are transported by NAD (2 pairs from glycolysis, 2 pairs from the pyruvic acid to acetyl-CoA conversion, and 6 pairs from the Krebs cycle). Two additional pairs are transported by FAD (from the Krebs cycle). All these electrons are passed to other electron carriers in the electron transport chain.

TRY IT

Fueling the Race to Reproduce

Two students, working in the laboratory, observed the growth of yeast in a sugar solution on microscopic slides. One student focused her microscope near the edge of the coverslip, where oxygen levels were sufficient for aerobic respiration. The other student focused his microscope on yeast growing under the center of the coverslip. Both students kept the same fields in focus for the duration of the class period, and they counted the number of yeast cells in the field every 20 to 30 minutes.

Can you predict their results? What would explain these results? With your instructor's consent, you could easily try this experiment while completing your assigned laboratory exercises and then share your data with the rest of the class. Do you think this experiment would give the same results with all types of organisms? Why?

In our waterfall analogy, we can think of water entering the falls at two sites, one higher up the mountain than the other. Water from the higher site falls farther than water entering lower down the mountain. In bacteria, electrons entering the electron transport chain at NAD start at the top, and their descent releases enough energy to make three ATPs. Electrons entering at FAD start partway down the chain and contribute only enough energy to make two ATPs. Thus, during aerobic metabolism of a glucose molecule, the 10 pairs of electrons from NAD produce 30 ATPs, and 2 pairs from FAD produce 4 ATPs, for a total of 34 ATPs. Including the 2 ATP molecules from glycolysis and the 2 GTP molecules (=2 ATPs) from the Krebs cycle gives a total yield of 38 ATPs per glucose molecule.

Oxidative phosphorylation, when compared with fermentation, generates the greater amount of energy from glucose. Fermentation, through the substrate-level production of ATP during glycolysis, yields only about 5% as much. The net gain of ATP molecules from fermentation is 2.

Chemiosmosis

Electrons for the hydrogen atoms removed from the reactions of the Krebs cycle are transferred through the electron transport system, which generates the high-energy bonds of ATP. ADP is converted to ATP by a large ATP-synthesizing complex called *ATP synthase* (or *ATPase*) in a process known as **chemiosmosis** (kem″e-os-mo′sis). This process is the result of a series of chemical reactions that occur in and around a membrane. Although the mechanism for the process, first proposed by the British biochemist Peter Mitchell in 1961, took a number of years to become fully accepted, it is now recognized as a major contribution to the understanding of how ATP is formed during electron transport. Mitchell was awarded the Nobel Prize in 1978 for the development of the chemiosmosis hypothesis.

Chemiosmosis occurs in the cell membrane of prokaryotes (**Figure 5.20**) such as *Escherichia coli* and in the inner mitochondrial membrane of eukaryotes. As electrons are transferred along the electron transport chain, protons are pumped outside the membrane, so the ions' concentration is higher outside the membrane than inside. This process causes a lowering of the proton concentration on the inside and the development of a force that drives the protons back into the cell or mitochondrial matrix to equalize their concentration on both sides of the membrane. Any concentration gradient naturally tends to equalize itself.

In addition to the proton concentration gradient across the membrane, there is also an electrochemical gradient, which makes the membrane a type of biological battery that can power the formation of ATP. The H^+ excess on one side of the membrane gives that side a positive charge compared with the other side. The power generated by the gradient is called the *proton motive force*. The protons flow through special channels within

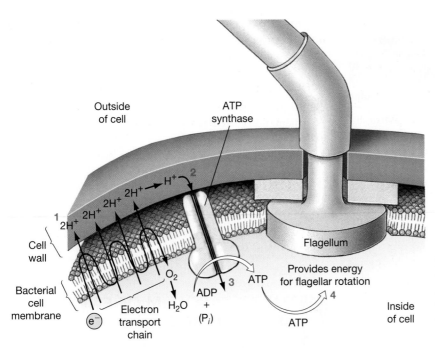

Figure 5.20 Chemiosmosis.
Energy capture by chemiosmosis in a bacterial cell membrane is shown here. (1) Protons "pumped" out during electron transport (2) pass back in through channels in the ATP synthases that (3) phosphorylate ADP to make ATP. The flow of protons provides the energy to drive this reaction (4) and to power rotation of flagella.

the synthase complex. Energy is thus released and used to form ATP from ADP and inorganic phosphate (P_i).

Anaerobic Respiration—A Bacterial Alternative

Some bacteria use only parts of the Krebs cycle and the electron transport chain. They are anaerobes who do not use free O_2 as their final electron acceptor. Instead, in a process called **anaerobic respiration**, they use inorganic oxygen-containing molecules such as nitrate (NO_3^-), nitrite (NO_2^-), and sulfate (SO_4^{2-}) **(Figure 5.21)**. Because anaerobes use less of the metabolic pathways, they produce fewer ATP molecules than do aerobic organisms.

One of the reactions of anaerobic respiration commonly tested for in urinalysis is removal of one oxygen atom from nitrate to form nitrite. A positive nitrite test indicates the presence of bacteria such as *E. coli*. Other bacteria can further reduce nitrites to compounds such as ammonia (NH_3) or even free nitrogen gas (N_2). These are important reactions in the nitrogen cycle and will be further discussed in ◀Chapter 25.

THE SIGNIFICANCE OF ENERGY CAPTURE

In glycolysis and fermentation, as we noted earlier, a net of 2 ATPs is usually produced for every glucose metabolized anaerobically. When glycolysis is followed by aerobic respiration, each glucose molecule produces an additional 2 ATPs at the substrate level in the Krebs cycle and 34 ATPs by oxidative phosphorylation. Thus, a glucose molecule yields 38 ATPs by aerobic metabolism

Figure 5.21 Final electron acceptors.
Aerobic respiration, anaerobic respiration, and fermentation have different final electron acceptors.

TABLE 5.2

Energy Captured in ATP Molecules from a Glucose Molecule by Anaerobic and Aerobic Metabolism in Prokaryotes		
	Number of ATP Molecules	
Prokaryotic Metabolic Process	**Anaerobic Conditions**	**Aerobic Conditions**
Glycolysis		
Substrate level	4	4
Hydrogen to NAD	0	6
Pyruvate to Acetyl-CoA		
Hydrogen to NAD	0	6
Krebs Cycle		
Substrate level	0	2
Hydrogen to NAD	0	18
Hydrogen to FAD	0	4
Less Energy for Phosphorylation	−2	−2
Total	2	38

but only 2 ATPs by glycolysis and fermentation (Table 5.2). Thus, 19 times as much energy is captured in aerobic metabolism as in fermentation! Therefore, aerobic microorganisms in environments with ample oxygen generally grow more rapidly than anaerobes. But aerobes will die if oxygen is depleted, unless they can switch to fermentation. **Table 5.3** summarizes the metabolic processes we have studied thus far.

✓**CHECKLIST**

1. If oxygen is present, will it stop glycolysis? Fermentation? The Krebs cycle? The electron transport chain?

2. If four actual molecules of ATP are produced for each glucose molecule going through glycolysis, why do we say glycolysis yields only two ATPs?

3. What are the functions of NAD (NADH) and FAD (FADH)?

4. Where does the electron transport chain function in prokaryotes? In eukaryotes?

▌▌▌ THE METABOLISM OF FATS AND PROTEINS

For most organisms, including microorganisms, glucose is a major source of energy. However, for almost any organic substance, we can find a type of microorganism that can degrade that substance for energy. This attribute of microorganisms, and the fact that microbes are found almost everywhere on our planet, accounts for their ability to degrade dead and decaying remains and wastes of all organisms.

TABLE 5.3

A Comparison of Metabolic Processes				
	Glycolysis	**Fermentation**	**Krebs Cycle**[a]	**Electron Transport Chain**
Location	In cytoplasm	In cytoplasm	Prokaryotes: in cytoplasm Eukaryotes: in the mitochondrial matrix	Prokaryotes: in cell membrane Eukaryotes: in inner mitochondrial membranes
Oxygen Conditions	Anaerobic; oxygen is not required; does not stop, however, if oxygen is present	Without O_2; presence of oxygen will cause it to stop	Aerobic	Aerobic
Starting Molecule(s)	1 glucose (6C)	Various substrate molecules go through glycolysis, yielding 2 pyruvic acid	2 pyruvic acid	6 O_2
Ending Molecules	2 pyruvic acid (3C) 2 NADH	Various, depending on which form of fermentation occurs, e.g., ethanol, lactic acid, CO_2, acetic acid	6 CO_2 8 NADH 2 FADH	6 H_2O
Amount of ATP Produced	4 ATP (net 2 ATP)	Various, depending on which form of fermentation occurs, usually 2 or 3 ATP; always far less than is produced in aerobic respiration	2 GTP (=2 ATP)	34 ATP

[a]Includes the pyruvic acid → acetyl-CoA step.

BIOTECHNOLOGY

Microbial Clean-Up

A few species of bacteria, such as some members of the genus *Pseudomonas*, can use crude oil for energy. They can grow in seawater with only oil, potassium phosphate, and urea (a nitrogen source) as nutrients. These organisms can clean up oil spills in the ocean, acting as "bioremediators." They have also proven useful in degrading oil that remains in the water carried by tankers as ballast after unloading their oil. Then the water pumped from the tankers into the sea in preparation for a new cargo of oil does not pollute. A detergent-like substance has been isolated recently from these organisms. When the detergent is added to a quantity of oil sludge, it converts 90% of the sludge into usable petroleum in about 4 days, thereby reducing waste and providing a convenient means of cleaning oil-fouled tanks.

(a)

(b)

(c)

(a) The 1989 Exxon *Valdez* oil spill left great quantities of pooled oil on sites in the Gulf of Alaska, such as on Green Island. Bioremediation in 1989 **(b)** by the application of nutrients (nitrogen and phosphorus) to the shoreline accelerated the bacterial biodegradation of the oil into carbon dioxide and water. **(c)** In 1991, the area was surveyed and found to be mostly cleared of oil, with no further treatment recommended. *(Courtesy Exxon Corporation)*

FAT METABOLISM

Most microorganisms, like most animals, can obtain energy from lipids. The following examples give a general idea of how such processes occur. Fats are hydrolyzed to glycerol and three fatty acids. The glycerol is metabolized by glycolysis. The fatty acids, which usually have an even number of carbons (16, 18, or 20), are broken down into 2-carbon pieces by a metabolic pathway called **beta oxidation**. In this process a fatty acid first combines with coenzyme A. Oxidation of the beta carbon (second carbon from the carboxyl group) of the fatty acid results in the release of acetyl-CoA and the formation of a fatty acid shorter by 2 carbon atoms. The process is then repeated, and another acetyl-CoA molecule is released. The newly formed acetyl Co-A is then oxidized via the Krebs cycle to obtain additional energy **(Figure 5.22)**.

PROTEIN METABOLISM

Proteins also can be metabolized for energy **(Figure 5.23)**. They are first hydrolyzed into individual amino

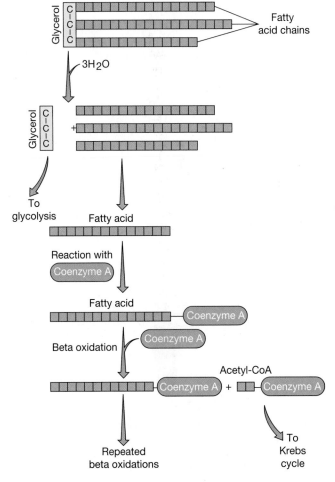

Figure 5.22 The catabolism of fats.
Triglycerides are hydrolyzed into glycerol and fatty acids. The glycerol is broken down via glycolysis. The fatty acids are broken down into 2-carbon units and fed into the Krebs cycle, where they are metabolized to produce additional energy.

Figure 5.23 The catabolism of proteins.
Polypeptides are hydrolyzed to amino acids. The amino acids are deaminated, and the resulting molecules enter pathways leading to the Krebs cycle.

acids by *proteolytic* (protein-digesting) *enzymes*. Then the amino acids are *deaminated*—that is, their amino groups are removed. The resulting deaminated molecules enter glycolysis, fermentation, or the Krebs cycle. The metabolism of all major nutrients (fats, carbohydrates, and proteins) for energy is summarized in **Figure 5.24**.

▌▌▌ OTHER METABOLIC PROCESSES

Having considered energy capture in chemoheterotrophs, we will now briefly consider energy capture in photoautotrophs, photoheterotrophs, and chemoautotrophs.

PHOTOAUTOTROPHY

Organisms called photoautotrophs carry out **photosynthesis**, the capture of energy from light and the use of this energy to manufacture carbohydrates from carbon dioxide. Photosynthesis occurs in green and purple bacteria, in cyanobacteria, in algae, and in higher plants. Photosynthetic bacteria, which probably evolved early in the evolution of living organisms, perform their own version of photosynthesis in the absence of O_2. However, algae and green plants make much more of the world's carbohydrate supply, so we will consider the process in those organisms first and then see how it differs in green and purple bacteria.

In green plants, algae, and cyanobacteria, photosynthesis occurs in two parts—the "photo" part, or the *light reactions*, in which light energy is converted to chemical energy, and the "synthesis" part, or the *dark reactions*, in which chemical energy is used to make organic molecules. Each part involves a series of steps.

In the **light-dependent (light) reactions**, light strikes the green pigment chlorophyll *a* in thylakoids of chloroplasts (◀Chapter 4, p. 101). Electrons in the chlorophyll become excited—that is, raised to a higher energy level. These electrons participate in generating ATP in cyclic photophosphorylation and in noncyclic photoreduction

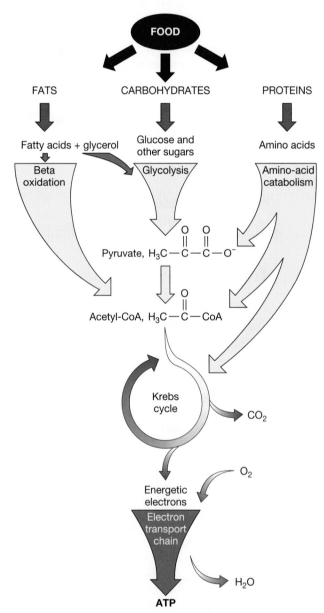

Figure 5.24 Metabolism of the major classes of biomolecules: a summary. This can be thought of as a giant funnel that eventually channels all three types of nutrients into the Krebs cycle.

(Figure 5.25). In **cyclic photophosphorylation**, excited electrons from chlorophyll are passed down an electron transport chain. As they are transferred, energy is captured in ATP by chemiosmosis (as described previously in connection with oxidative phosphorylation). When the electrons return to the chlorophyll, they can be excited over and over again, so the process is said to be cyclic.

In **noncyclic photoreduction**, energy is also captured by chemiosmosis. In addition, membrane proteins and energy from light are used to split water molecules into protons, electrons, and oxygen molecules, a process called **photolysis** (fo-tol′eh-sis). The electrons replace those lost from chlorophyll, which are thus freed to reduce the coenzyme NADP. ATP and reduced NADP

Figure 5.26 The relation between the light and dark reactions. In the dark reactions, ATP and NADPH (the products of the light reactions) are used to reduce carbon dioxide, forming carbohydrates such as glucose. The dark reactions do not require darkness; they are so named because they can take place in the dark, as long as the products of the light reactions are available.

(a)

(b)

Figure 5.25 The light reactions of photosynthesis, performed by cyanobacteria, algae, and green plants. Electrons in chlorophyll receive a boost in energy from light, and their extra energy is used to make ATP. In the pathway of cyclic photophosphorylation **(a)** the electrons return to chlorophyll and thus can be used over and over again. In noncyclic photoreduction **(b)** the electrons receive a second boost that gives them enough energy to reduce NADP. The electrons are replaced by the splitting of water.

(NADPH)—the products of the light reaction—and atmospheric CO_2 subsequently participate in the dark reactions.

The **light-independent (dark) reactions**, or *carbon fixation*, occur in the stroma of chloroplasts. Carbon dioxide is reduced by electrons from NADPH in a process known as the *Calvin-Benson cycle* (see ◀Appendix E). Energy from ATP and electrons from NADPH are required in this synthetic process. Various carbohydrates, chiefly glucose, are the products of the dark reactions **(Figure 5.26)**.

Photosynthesis in green and purple sulfur bacteria differs from that in green plants, algae, and cyanobacteria in ways related to the evolution of organisms. The first photosynthetic organisms probably were purple and green bacteria, which evolved in an atmosphere containing much hydrogen but no oxygen. They differ from green plants, algae, and cyanobacteria as follows:

1. Bacterial chlorophyll absorbs slightly longer wavelengths of light than does chlorophyll *a*.
2. They use hydrogen compounds such as hydrogen sulfide (H_2S), rather than water (H_2O), for reducing carbon dioxide. Electrons from their pigments reach an energy level high enough to split H_2S (but not high enough to split H_2O) and to generate an H^+ gradient for ATP synthesis. (Some purple and green bacteria produce elemental sulfur as a byproduct; a few produce strong sulfuric acid.)

3. They are usually strict anaerobes and can live only in the absence of oxygen. They do not release oxygen as a product of photosynthesis, as green plants do.

Characteristics of the groups of bacteria that carry out this primitive form of photosynthesis are summarized in Table 5.4.

The cyanobacteria also are photosynthetic, but they probably evolved after the purple and green bacteria. Although prokaryotic, the cyanobacteria release oxygen during photosynthesis, as do green plants and algae. In fact, cyanobacteria are probably responsible for the addition of oxygen to the primitive atmosphere.

PHOTOHETEROTROPHY

Photoheterotrophs are a small group of bacteria that can use energy from light but require organic substances such as alcohols, fatty acids, or carbohydrates as carbon sources. These organisms include the nonsulfur, purple or green bacteria.

CHEMOAUTOTROPHY

Chemoautotrophic bacteria (also called *chemolithotrophs*) are unable to carry out photosynthesis but can oxidize inorganic substances for energy. With this energy and carbon dioxide as a carbon source, these bacteria can synthesize a great variety of substances, including carbohydrates, fats, proteins, nucleic acids, and substances that are required as vitamins by many organisms.

The ability to oxidize, and therefore extract energy from, inorganic substances is probably the most outstanding characteristic of chemoautotrophs, but these bacteria have other noteworthy attributes. The nitrifying bacteria are especially important because they increase the quantity of usable nitrogen compounds available to plants and replace nitrogen that plants remove from the soil. *Thiobacillus* and some other sulfur bacteria produce sulfuric acid by oxidizing elemental sulfur or hydrogen sulfide. Acidity lower than pH 1 has been produced by sulfur bacteria. Sulfur is sometimes

TABLE 5.4

Characteristics of Photosynthetic Bacteria		
Group	Family and Representative Genus	Pigments
Green sulfur bacteria	Chlorobiaceae *Chlorobium*	Bacterial chlorophyll
Purple sulfur bacteria	Chromaticeae *Chromatium*	Bacterial chlorophyll and red and purple carotenoid pigments

TABLE 5.5

Characteristics of Chemoautotrophic Bacteria		
Group and Representative Genus/Genera	Source of Energy	Products after Oxidizing Reaction
Nitrifying bacteria		
Nitrobacter	HNO_2	HNO_3
Nitrosomonas	NH_3	$HNO_2 + H_2O$
Nonphotosynthetic sulfur bacteria		
Thiothrix	H_2S	$H_2O + 2S$
Thiobacillus	S	H_2SO_4
Iron bacteria		
Siderocapsa	Fe^{2+}	$Fe^{3+} + OH^-$
Hydrogen bacteria		
Alcaligenes	H_2	H_2O

added to alkaline soil to acidify it, a practice that works because of the numerous thiobacilli present in most soils. Finally, some chemoautotrophic Archaeobacteria have been found near volcanic vents in the ocean floor, where they grow at extremely high temperatures and sometimes under very acidic conditions. Characteristics of chemoautotrophs are summarized in Table 5.5.

✓CHECKLIST

1. Lipids are broken down into glycerol and fatty acids. How are each of these then further metabolized?

2. What is returned to chlorophyll in cyclic photophosphorylation that is not returned in noncyclic photoreduction?

3. What were probably the first kind of photosynthetic organisms on our planet? Were they aerobes or anaerobes? Did they release oxygen gas as a product of photosynthesis?

4. What type of metabolism is characteristic of nitrifying bacteria? Why are they important organisms?

▌▌▌ THE USES OF ENERGY

Microorganisms use energy for such processes as biosynthesis, membrane transport, movement, and growth. Here we will summarize some biosynthetic activities and some mechanisms for membrane transport and movement. We will consider microbial growth in ◄Chapter 6.

BIOSYNTHETIC ACTIVITIES

Microorganisms share many biochemical characteristics with other organisms. All organisms require the same building blocks to make proteins and nucleic acids. Many of these building blocks (amino acids, purines, pyrimidines, and ribose) can be derived from intermediate products of energy-yielding pathways (Figure 5.27).

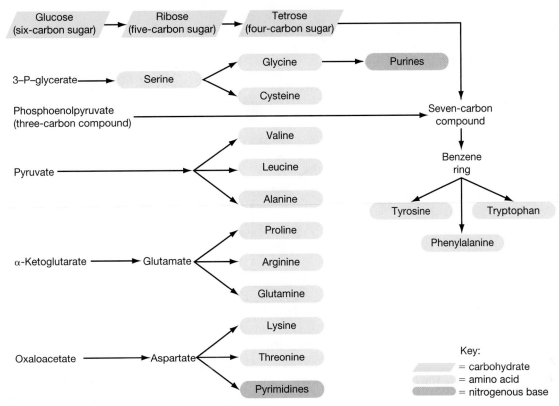

Figure 5.27 Some biosynthetic pathways. This flow diagram shows how amino acids, nucleic acid bases, and ribose are made from intermediates in glycolysis and from the Krebs cycle.

When the energy-yielding pathways were first discovered, they were thought to be purely catabolic. Now that many of their intermediates are known to be involved in biosynthesis, they are more properly called **amphibolic** (am-fe-bol′ik) **pathways** (*amphi-*, either) because they can yield either energy or building blocks for synthetic reactions.

Some biosynthetic pathways are quite complex. For example, synthesis of amino acids in those organisms that can make them often requires many reactions, with an enzyme for each reaction. Tyrosine synthesis requires no fewer than 10 enzymes, and tryptophan synthesis needs at least 13. The synthetic pathways for making purines and pyrimidines also are complex. The absence of a single enzyme in a synthetic pathway can prevent the synthesis of a substance. Any essential substance that an organism cannot synthesize must be accessible in the environment, or the organism will die. Missing enzymes thus increase the nutritional needs of organisms.

Microorganisms of many different types also synthesize a variety of carbohydrates and lipids. The rate at which they are synthesized varies and depends on the availability and activity of enzymes. Some organisms, such as the aerobe *Acetobacter*, synthesize cellulose, which is ordinarily found in plants. As strands of cellulose reach the cell surface, they form a mat that traps carbon dioxide bubbles and keeps the cell afloat. Because these organisms must have oxygen, the mat contributes to their survival by keeping them near the surface, where oxygen is plentiful.

Many bacteria synthesize peptidoglycan, lipopolysaccharide, and other polymers associated with cell walls (◄Chapter 4, p. 83). Some bacteria form capsules, especially in media that contain serum or large amounts of sugar. Capsules usually consist of polymers of one or more monosaccharides. However, in *Bacillus anthracis*, the bacterium that causes anthrax, the capsule is a

CLOSE UP

The Swamp of Eternal Stench

It is possible to convert a lovely lake into a "Swamp of Eternal Stench" because such a delicate environmental balance exists between aerobic respiration and fermentation. When excess nutrients such as carbon, nitrogen, or phosphorus are present, there is an increase in both plant and bacterial growth. The aerobic bacteria will deplete the lake of dissolved oxygen. As more fish, plants, and animals die, the anaerobic bacteria get the ecological advantage. Fermentation rules! Our bucolic lake is now a bubbling stew, emitting methane and hydrogen sulfide gases that are spiced with putrescine and cadaverine—the odors of death.

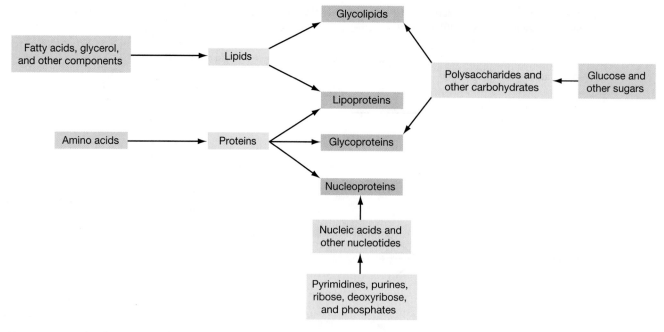

Figure 5.28 **The formation of complex biomolecules from simpler components.**

polypeptide of glutamic acid. The biosynthetic processes in microorganisms are summarized in **Figure 5.28**.

MEMBRANE TRANSPORT AND MOVEMENT

In addition to using energy for biosynthetic processes, microorganisms also use energy for transporting substances across membranes and for their own movement. These energy uses are as important to the survival of the organisms as are their biosynthetic activities.

Membrane Transport

Microbes use energy to move most ions and metabolites across cell membranes against concentration gradients. For example, bacteria can transport a sugar or an amino acid from a region of low concentration outside the cell to a region of higher concentration inside the cell. This means that they accumulate nutrients within cells in concentrations a hundred to a thousand times the concentration outside the cell. They also concentrate certain inorganic ions by the same means.

Two mechanisms exist in bacteria for concentrating substances inside cells, and both require energy. One active transport mechanism is specific to Gram-negative bacteria, such as *E. coli*. Such bacteria have two membranes—the cell membrane, which surrounds the cell's cytoplasm, and the outer membrane, which forms part of the cell wall (◄Chapter 4, p. 85). Transmembrane carrier proteins called **porins** form channels through the outer membrane. Porins allow entry of ions and small hydrophilic metabolites via *facilitated diffusion* (◄Chapter 4, p. 107). After entering the periplasmic space, a specific periplasmic protein combines with one of the diffusing ions or metabolites. The periplasmic

protein then facilitates the transport of the substance into the cytoplasm via a specific carrier protein in the cell membrane. Such substances generally gain entry by active transport. Through ATP hydrolysis, the carrier protein changes shape, allowing the metabolite into the cytoplasm (see ◄Figure 4.32).

Another mechanism, present in all bacteria, is called the **phosphotransferase system (PTS)**. It consists of sugar-specific enzyme complexes called **permeases** (per'me-a-sez), which form a transport system through the cell membrane. The PTS uses energy from the high-energy phosphate molecule phosphoenolpyruvate (PEP). When PEP is present in the cytoplasm, it can provide energy and a phosphate group to a permease in the membrane. Then the permease transfers the phosphate to a sugar molecule and at the same time moves the sugar across the membrane. A phosphorylated sugar is thus transported inside the cell and is prepared to undergo metabolism. This *group translocation* was discussed in ◄Chapter 4 (p. 108).

Movement

Most motile bacteria move by means of flagella, but some move by gliding or creeping or in a corkscrew motion. Flagellated bacteria move by rotating their flagella (◄Chapter 4, p. 95). The mechanism for rotation, though not fully understood, appears to involve a proton gradient, as in chemiosmosis. As the protons move down the gradient, they drive the rotation. Gliding bacteria move only when in contact with a solid surface, such as decaying organic matter. Rotation of the cell on its own axis often occurs with gliding. A number of mechanisms have been proposed to explain gliding, but the mechanism that propels the gliding bacterium

Myxococcus is best understood. This organism uses energy to secrete a substance called a **surfactant** (ser-fak'-tant), which lowers surface tension at the bacterium's posterior end. The difference in surface tension between the anterior and posterior ends (a passive phenomenon) causes *Myxococcus* to glide.

Spirochetes expend energy for both creeping and thrashing motions. On a solid surface they creep along like an inchworm by alternately attaching front and rear ends. Suspended in a liquid medium, they thrash (twist and turn). Both creeping and thrashing motions probably occur by waves of contraction within the cell substance that exert force against axial filaments.

BIOLUMINESCENCE

Bioluminescence, the ability of an organism to emit light, appears to have evolved as a by-product of aerobic metabolism. Bacteria of the genera *Photobacterium* and *Achromobacter*, fireflies, glowworms, and certain marine organisms living at great depths in the ocean exhibit bioluminescence (**Figure 5.29**). Many light-emitting organisms have the enzyme *luciferase* (lu-sif'er-ace), along with other components of the electron transport system. (Luciferase derives its name from Lucifer, which means "morning star.") Luciferase catalyzes a complex reaction in which molecular oxygen is used to oxidize a long-chain aldehyde or ketone to a carboxylic acid. At the same time, $FMNH_2$ from the electron transport chain is oxidized to an excited form of *flavin mononucleotide* (FMN), a carrier molecule derived from riboflavin (vitamin B_2), that emits light as it returns to its unexcited state. In this process, phosphorylation reactions are bypassed, and no ATP is generated. Instead, energy is released as light.

Luminescent microorganisms often live on the surface of marine organisms such as some squids and fish. More than 300 years ago, the Irish chemist Robert Boyle observed that the familiar glow of the skin of dead fish lasted only as long as oxygen was available. At that time the electron transport system, and the role of oxygen in it, were not understood.

Bioluminescence exhibited by larger organisms has survival value. It is the sole light source for marine

(a)

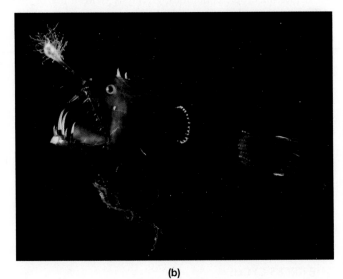

(b)

Figure 5.29 Microbial bioluminescence. **(a)** Bioluminescent bacteria in the Petri dish produce enough light to read by. *(John D. Cunningham/Visuals Unlimited)* **(b)** Angler fish lights up the dark, deep-ocean depths with bioluminescent bacteria that live symbiotically in its long "lure," which attracts prey to within reach of its jaws. *(Peter David/Taxi/Getty Images)*

APPLICATIONS

Something for Everyone

Environmental scientists observe that hazardous-waste sites that support the growth of highly diverse microbial populations are much easier to bioremediate than sites inhabited by a few microbial species. Sites inhabited by diverse populations remediate faster and with fewer problems associated with the production and accumulation of toxic metabolic by-products. The reason for this is co-metabolism—an advantageous situation in which an organism, in the process of oxidizing a particular substrate, oxidizes a second substrate as well. The second metabolite is not a source of nutrients or energy for the oxidizing organism but is a nutrient for the second organism. Hazardous-waste sites where co-metabolism is encouraged can efficiently mineralize organic wastes to carbon dioxide and water.

creatures that live at great depths, and it helps land organisms such as fireflies find mates. How bioluminescence came to be established among microorganisms is less clear. One hypothesis is that early in the evolution of living things, bioluminescence served to remove oxygen from the atmosphere as it was produced by some of the first photosynthetic organisms. Although this is not an advantage to aerobes, it is an advantage to strict anaerobes. Because most of the microorganisms in existence at that time were anaerobes susceptible to the toxic effects of oxygen, bioluminescence would have been beneficial to them. Today, many bioluminescent microbes are beneficiaries of symbiotic relationships with their hosts. They provide light in return for shelter and nutrients.

Scientists have found a way to put bioluminescent bacteria to work. In the Microtox Acute Toxicity Test, bioluminescent bacteria are exposed to a water sample to determine if the sample is toxic. Any change—positive or negative—in the bacteria's growth is observable as a change in their light output. The sample's toxicity is calculated by comparing before-and-after readings of the light levels. The brainchild of Microbics Corp. of Carlsbad, California, this toxicity test takes only minutes to perform.

The Microtox Acute Toxicity Test is useful for testing the quality of drinking water and for numerous other industrial applications. For example, waste-water treatment plants use it to determine quickly whether their treated effluent will be able to pass government toxicity compliance tests. Paper mills use the test to determine how much disinfectant is needed to rid their equipment of the microbial growth that slows down the manufacturing process and affects product quality. Makers of household cleansers, shampoos, or cosmetics use the test in place of controversial animal testing in which drops of the products are put into the eyes of rabbits to determine the products' irritancy levels. And unlike cell-culturing techniques, the test requires little skill to perform and to interpret. Bioluminescence could prove to be a very important process to industry in the future.

∎ RETRACING OUR STEPS

∎∎∎ METABOLISM: AN OVERVIEW

* **Metabolism** is the sum of all the chemical processes in a living organism. It consists of **anabolism**, reactions that require energy to synthesize complex molecules from simpler ones, and **catabolism**, reactions that release energy by breaking complex molecules into simpler ones.
* **Autotrophs**, which use carbon dioxide to synthesize organic molecules, include **photoautotrophs** (which carry on photosynthesis) and **chemoautotrophs**.
* **Heterotrophs**, which use organic molecules made by other organisms, include **chemoheterotrophs** and **photoheterotrophs**.
* For growth, movement, and other activities, **metabolic pathways** use energy captured in the **catabolic pathways**.

∎∎∎ ENZYMES

PROPERTIES OF ENZYMES

* **Enzymes** are proteins that catalyze chemical reactions in living organisms by lowering the **activation energy** needed for a reaction to occur.
* Enzymes have an **active site**, the binding site to which the **substrate** (the substance on which the enzyme acts) attaches to form an **enzyme-substrate complex**. Enzymes typically exhibit a high degree of **specificity** in the reactions they catalyze.

PROPERTIES OF COENZYMES AND COFACTORS

* Some enzymes require **coenzymes**, nonprotein organic molecules that can combine with the **apoenzyme**, the protein portion of the enzyme, to form a **holoenzyme**. Some enzymes also require inorganic ions as **cofactors**.

∎∎∎ ENZYME INHIBITION

* Enzyme activity can be reduced by **competitive inhibitors**, molecules that compete with the substrate for the enzyme's active site, or by **noncompetitive inhibitors**, molecules that bind to an **allosteric site**, a site other than the active site.

FACTORS THAT AFFECT ENZYME REACTIONS

* Factors that affect the rate of enzyme reactions include temperature, pH, and concentrations of substrate, product, and enzyme.

∎∎∎ ANAEROBIC METABOLISM: GLYCOLYSIS AND FERMENTATION

GLYCOLYSIS

* **Glycolysis** is a metabolic pathway by which glucose is oxidized to pyruvic acid.
* Under anaerobic conditions, glycolysis yields a net of two ATPs per molecule of glucose.

ALTERNATIVES TO GLYCOLYSIS

* Some organisms use the pentose phosphate pathway, or else use the Entner-Doudoroff pathway instead of glycolysis.

FERMENTATION

* **Fermentation** refers to the reactions of metabolic pathways by which NADH is oxidized to NAD. An organic molecule is the final electron acceptor.
* Six pathways of fermentation are summarized in Figure 5.12. **Homolactic acid** and **alcoholic fermentations** are two of the most important and commonly occurring fermentation pathways.

∎∎∎ AEROBIC METABOLISM: RESPIRATION

* **Anaerobes** do not use oxygen; **aerobes** use oxygen and obtain energy chiefly via **aerobic respiration**.

THE KREBS CYCLE

* The **Krebs cycle** metabolizes two-carbon compounds to CO_2 and H_2O, produces one ATP directly from each acetyl group, and transfers hydrogen atoms to the electron transport system.
* In energy production the Krebs cycle processes acetyl-CoA, so that (in the electron transport chain) hydrogen atoms can be oxidized for energy.

ELECTRON TRANSPORT AND OXIDATIVE PHOSPHORYLATION

- **Electron transport** is the transfer of electrons to oxygen (the final electron acceptor).
- **Oxidative phosphorylation** involves the **electron transport chain** for ATP synthesis and is a membrane-regulated process not directly related to the metabolism of specific substrates.
- The theory of **chemiosmosis** explains how energy is used to synthesize ATP.
- **Anaerobic respiration** occurs in the absence of free oxygen, and does not use all parts of the Krebs cycle or electron transport chain, thus making less ATP. Inorganic molecules are the final electron acceptors.

THE SIGNIFICANCE OF ENERGY CAPTURE

- In prokaryotes, aerobic (oxidative) metabolism captures 19 times as much energy as does anaerobic metabolism.

III THE METABOLISM OF FATS AND PROTEINS

- Most organisms get energy mainly from glucose. But for almost any organic substance, there is some microorganism that can metabolize it.

FAT METABOLISM

- Fat metabolism involves hydrolysis and the enzymatic formation of glycerol and free fatty acids. Fatty acids are in turn oxidized by **beta oxidation**, which results in the release of acetyl-CoA. Acetyl-CoA then enters the Krebs cycle.

PROTEIN METABOLISM

- The metabolism of proteins involves the breakdown of proteins to amino acids, the deamination of the amino acids, and their subsequent metabolism in glycolysis, fermentation, or the Krebs cycle.

III OTHER METABOLIC PROCESSES

PHOTOAUTOTROPHY

- **Photosynthesis** is the use of light energy to synthesize carbohydrates: (1) The **light-dependent (light) reactions** can include **cyclic photophosphorylation** or **photolysis** accompanied by **noncyclic photoreduction** of NADP; (2) the **light-independent (dark) reactions** involve the reduction of CO_2 to carbohydrate.

- Photosynthesis in cyanobacteria and algae provides a means of making nutrients, as it does in green plants; however, photosynthetic bacteria generally use some substances besides water to reduce carbon dioxide.

PHOTOHETEROTROPHY

- **Photoheterotrophy** is the use of light as a source of energy. It requires organic compounds as sources of carbon.

CHEMOAUTOTROPHY

- **Chemoautotrophs**, or **chemolithotrophs**, oxidize inorganic substances to obtain energy. Chemolithotrophs require only carbon dioxide as a carbon source.

III THE USES OF ENERGY

BIOSYNTHETIC ACTIVITIES

- An **amphibolic pathway** is a metabolic pathway that can capture energy or synthesize substances needed by the cell.
- Figure 5.27 summarizes the intermediate products of energy-yielding metabolism and some of the building blocks for synthetic reactions that can be made from them.
- Bacteria synthesize a variety of cell wall polymers.

MEMBRANE TRANSPORT AND MOVEMENT

- Membrane transport uses energy derived from the ATP-producing electron transport system in the membrane to concentrate substances against a gradient. It occurs by active transport and by the **phosphotransferase system**.
- Movement in bacteria can be by flagella, by gliding or creeping, or by axial filaments.

BIOLUMINESCENCE

- The ability of an organism to emit light may have evolved as a way to remove oxygen from the surroundings of primitive anaerobic microbes early in the Earth's history. Today it often functions in symbiotic relationships with larger organisms.

▋ TERMINOLOGY CHECK

activation energy (*p. 119*)
active site (*p. 119*)
aerobe (*p. 128*)
aerobic respiration (*p. 128*)
alcoholic fermentation (*p. 126*)
allosteric site (*p. 122*)
amphibolic pathway (*p. 139*)
anabolic pathway (*p. 119*)
anabolism (*p. 117*)
anaerobe (*p. 128*)
anaerobic respiration (*p. 133*)

apoenzyme (*p. 120*)
autotroph (*p. 117*)
autotrophy (*p. 117*)
beta oxidation (*p. 135*)
catabolic pathway (*p. 119*)
catabolism (*p. 117*)
chemical equilibrium (*p. 124*)
chemiosmosis (*p. 132*)
chemoautotroph (*p. 117*)
chemoheterotroph (*p. 117*)
citric acid cycle (*p. 129*)
coenzyme (*p. 120*)
cofactor (*p. 121*)
competitive inhibitor (*p. 122*)

cyclic photophosphorylation (*p. 137*)
cytochrome (*p. 132*)
electron acceptor (*p. 117*)
electron donor (*p. 117*)
electron transport (*p. 131*)
electron transport chain (*p. 131*)
endoenzyme (*p. 120*)
enzyme (*p. 119*)
enzyme-substrate complex (*p. 120*)
exoenzyme (*p. 120*)
FAD (*p. 122*)

feedback inhibition (*p. 122*)
fermentation (*p. 126*)
flavoprotein (*p. 132*)
glycolysis (*p. 124*)
heterotroph (*p. 117*)
heterotrophy (*p. 117*)
holoenzyme (*p. 120*)
homolactic acid fermentation (*p. 126*)
Krebs cycle (*p. 128*)
light-dependent (light) reactions (*p. 136*)
light-independent (dark) reactions (*p. 137*)

metabolic pathway **(p. 119)** oxidation **(p. 117)** photoautotroph **(p. 117)** specificity **(p. 120)**
metabolism **(p. 117)** oxidative photoheterotroph **(p. 117)** substrate **(p. 119)**
NAD **(p. 121)** phosphorylation **(p. 132)** photolysis **(p. 137)** surfactant **(p. 141)**
noncompetitive inhibitor permease **(p. 140)** photosynthesis **(p. 136)** tricarboxylic acid **(TCA)**
(p. 122) phosphorylation **(p. 124)** porin **(p. 140)** **cycle(p. 129)**
noncyclic photoreduction phosphotransferase system quinone **(p. 132)**
(p. 137) **(PTS)(p. 140)** reduction **(p. 117)**

∎ CLINICAL CASE STUDY

Kim, a new mother, had heard that babies can suddenly develop high fevers, which are dangerous. Her 6-month-old son had symptoms of a mild respiratory infection but was resting peacefully in his crib. Imagine her surprise when, upon taking his temperature, she found it to be 106°F, or 41.2°C. She wondered if she should call the pediatrician, given that her son did not seem to be in any distress. Based on the information in this chapter, what could you tell Kim about the dangers of fevers over 40°C?

∎ CRITICAL THINKING QUESTIONS

1. Suppose that you had a culture known to contain *Klebsiella pneumoniae* and *Staphylococcus aureus*. Devise a way to separate and identify the organisms.

2. In what sequence might the different kinds of metabolism mentioned in this chapter have evolved? Why didn't just one type of metabolism evolve? Give reasons.

3. Many of the drugs we use to combat infections by microorganisms act as enzyme inhibitors. If you were engaged in developing new antimicrobial drugs, how might you proceed to develop new enzyme-inhibiting drugs?

∎ SELF-QUIZ

1. Which of the following is not true about photoautotrophs?
(a) They require CO_2 and light.
(b) They synthesize organic molecules from inorganic molecules.
(c) They are a subdivision of heterotrophs.
(d) They are a subdivision of autotrophs.
(e) All the above are true.

2. Match the following:
___ Photoautotrophs (a) Use inorganic chemical reactions for energy production
___ Chemoautotrophs
___ Photoheterotrophs (b) Use sunlight as a source of energy, and organic compounds as a carbon source
___ Chemoheterotrophs
(c) Use sunlight and carbon dioxide
(d) Use organic compounds for energy production

3. Match the following characteristics to either (a) autotrophs or (b) chemoheterotrophs:
___ Many microorganisms in this group are infectious.
___ Many microorganisms in this group can carry out photosynthesis.
___ Members of this group usually do not cause disease.
___ Members of this group carry out the same metabolic processes as man.
___ Members of this group break down organic compounds to obtain energy.
___ Members of this group synthesize organic compounds to obtain energy.

4. Match the following chemical processes:
___ Oxidation (a) Breakdown of nutrients
___ Catabolic reaction (b) Addition of phosphate group
___ Anabolic reaction (c) Loss of electrons
___ Reduction (d) Formation of macromolecules
___ Phosphorylation (e) Gain of electrons

5. All of the following statements are true about activation energy (AE) EXCEPT:
(a) Chemical reactions could not proceed inside of cells without AE.
(b) Cellular enzymes increase the AE of a reaction.
(c) AE is the energy needed to start a reaction.
(d) Uncontrolled AE in cellular reactions could denature proteins and evaporate liquids.
(e) All of the above are true.

6. Define metabolism and distinguish between anabolism and catabolism.

7. Match the following terms for microorganisms' "uses of energy" to their corresponding descriptions:
___ Membrane transport
___ Bioluminescence
___ Movement
___ Biosynthesis

(a) Amphibolic pathways
(b) Building a proton gradient for flagellar rotation
(c) Production of the amino acid tryptophan
(d) Gliding and creeping microorganisms
(e) Transmembrane carrier proteins (porins)
(f) Ability of microorganisms to emit light
(g) Phosphotransferase system and permeases

8. Metabolic pathways rely on many enzymes to synthesize or catabolize substrates to an end product. Within a given metabolic pathway, a product can become another enzyme's substrate. True or false?

9. Which of the following statements about enzyme characteristics is true?
(a) Enzymes generally exhibit a high degree of specificity for one particular substrate.
(b) Enzyme-substrate complexes occur when a substrate molecule collides with the allosteric site of an enzyme.
(c) Chemical bonds within a substrate are strengthened when this substrate forms an enzyme-substrate complex.
(d) Enzymes have a region called the active site which provides an area where it can form a loose association with its substrate.
(e) a and d.

10. Enzyme cofactors are usually inorganic ions that enhance enzymatic activity by improving the "fit" between an enzyme and its substrate. True or false?

11. All of the following statements about competitive and noncompetitive inhibitors and true EXCEPT:
(a) Competitive inhibitors are structurally similar to an enzyme's substrate and bind to the enzyme's allosteric site.
(b) Competitive inhibitors work by competing with a substrate for binding to an enzyme's active site.
(c) Noncompetitive inhibitors can bind at sites other than the active site of an enzyme, distorting the tertiary protein structure, which alters the shape of the active site, rendering it ineffective for substrate binding.
(d) Some noncompetitive inhibitors bind reversibly while some bird irreversibly to their enzyme.
(e) b and d.

12. Which of the following would influence the rate of an enzyme reaction?
(a) Temperature
(b) pH
(c) Concentration of substrate molecules
(d) Concentration of product molecules
(e) All of these

13. What is feedback inhibition?
(a) When the end product competitively inhibits the enzyme that produced it.
(b) When the first enzyme in line shuts down because of a buildup in its substrate.
(c) When an end product accumulates, it often binds to and inactivates the first enzyme that catalyzes the first reaction in the pathway.
(d) It is a reversible noncompetitive inhibition that regulates the rate of many metabolic pathways.
(e) c and d.

14. The principal energy-exchange molecule in living cells is:
(a) Glucose
(b) ATP
(c) RNA
(d) AMP
(e) Enertran

15. Which of the following statements about glycolysis is not true?
(a) Glycolysis, like fermentation, is an aerobic metabolic pathway that reduces glucose, transferring the electrons to the coenzyme ATP, which in turn passes these electrons to the final electron acceptor, an organic molecule.
(b) Glycolysis can occur under aerobic or anaerobic conditions and is a metabolic pathway by which glucose is oxidized to pyruvic acid.
(c) Glycolysis depends on the expenditure of two ATPs in substrate-level phosphorylations of the glucose molecule to initiate the metabolic pathway.
(d) Energy from glycolysis is captured in the form of ATP at the substrate level when released energy from substrate molecules (late in the process) is used to form high-energy bonds between ADP and P_i.
(e) Glycolysis is a metabolic process that splits glucose (a 6 carbon molecule) into two three-carbon molecules of pyruvic acid that captures a relatively small amount of energy in the form of ATP compared to electron transport and oxidative phosphorylation.

16. Which of the following is a characteristic of fermentation:
(a) Produces acids, gases, and alcohol
(b) Occurs in the absence of oxygen
(c) Starts with the breakdown of pyruvic acid
(d) Occurs following glycolysis and produces NAD
(e) All the above

17. During aerobic cell respiration most of the energy is produced during:
(a) Krebs cycle
(b) Glycolysis
(c) Fermentation
(d) ATP → ADP
(e) Electron transport chain reactions

18. The typical end products of complete aerobic cell respiration are carbon dioxide, water, and:
(a) ATP
(b) Glucose
(c) Citric acid
(d) Lactic acid
(e) Pyruvate

19. Is glycolysis the only metabolic pathway microorganisms have for oxidizing glucose? If not, then what are they?

20. How are fats and proteins used for energy?

21. Match the following electron transport and oxidative phosphorylation terms to their description:

___ Oxidative phosphorylation

___ Chemiosmosis

___ Flavoproteins, cytochromes, and quinones

___ Electron transport

(a) Transfer of electrons to final electron acceptor (oxygen)

(b) Energy capture in the form of ATP harnessed from a series of redox reactions, with oxygen being the final electron acceptor

(c) Electron carriers

(d) ATP production from a proton gradient across the plasma membrane

22. The end products of photosynthesis in cyanobacteria and plant cells are:
(a) Water and oxygen
(b) Glucose and water
(c) Glucose and oxygen
(d) Water and carbon dioxide
(e) Glucose and carbon dioxide

23. The energy source that drives the photosynthetic reactions in cyanobacteria is:
(a) Heat
(b) Light
(c) Complex sugars
(d) ATP
(e) Oxygen

24. In the photosynthetic reactions, which of the following is NOT true?
(a) Carbon dioxide is required in the dark reactions.
(b) Energy is produced in the dark reactions.

(c) Light reactions require light energy.
(d) Occur in the thylakoids of the eukaryotic cells.
(e) Generally result in the formation of glucose.

25. Match the following:

___ Chemiosmosis

___ Glycolysis

___ Electron transport chain

___ Fermentation

___ Photosynthesis

___ Krebs cycle

(a) Pathway that begins the breakdown of glucose

(b) ATP production from a proton gradient across the plasma membrane

(c) Anaerobic pathway that uses an organic final electron acceptor

(d) Pathway that uses carbon dioxide, light, and chlorophyll to produce carbohydrates

(e) Also is known as the tricarboxylic acid cycle (TCA) or as the citric acid cycle

(f) Flavoproteins, cytochromes, and quinones

26. Label parts (a) through (g) of this enzyme.

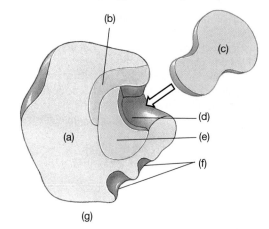

▌ EXPLORATIONS ON THE WEB

http://www.wiley.com/college/black

If you think you've mastered this chapter, there's more to challenge you on the web. Go to the companion web site to fine-tune your understanding of the chapter concepts and discover answers to the questions posed below.

1. How do you supercharge a chemical reaction?

2. Did you know that without enzymes most chemical reactions that maintain a viable organism would not occur below 90°C (or 200°F)?

3. Find out how one chemical cycle has three different names: tricarboxylic acid cycle, citric acid cycle, or the Krebs cycle.

Growth and Culturing of Bacteria

Courtesy Jacquelyn G. Black

Come with me...

I stood in awe near the geyser in Iceland, completely oblivious to the grandeur of my surroundings. The entire reason for my expedition undulated gently in the current of the runoff stream before me. The long, waving filaments of sulfur bacteria looked like long blond hair blowing in a light breeze. They were magnificent! Finally I was able to see with my own eyes, bacteria I had been reading about for years.

My excitement overcame me, and despite the warning steam billowing off the water, I plunged my hand into the water. I just wanted to find out what the strands felt like, but I guess I will never know. The near boiling water scalded my hand immediately. Later, as I nursed my blisters and wounded pride, I pondered the phenomena that allow the bacteria to thrive in an environment that is so hostile to most life forms (including me).

 Video related to this topic is available within WileyPLUS.

||| GROWTH AND CELL DIVISION
Microbial Growth Defined / Cell Division / Phases of Growth / Measuring Bacterial Growth

||| FACTORS AFFECTING BACTERIAL GROWTH
Physical Factors / Nutritional Factors

||| SPORULATION
Other Sporelike Bacterial Structures

||| CULTURING BACTERIA
Methods of Obtaining Pure Cultures / Culture Media / Methods of Performing Multiple Diagnostic Tests

||| LIVING, BUT NONCULTURABLE ORGANISMS

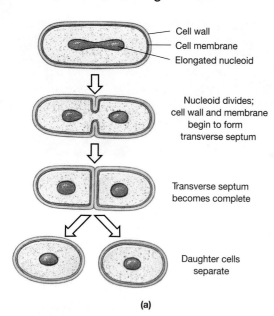

Cell wall
Cell membrane
Elongated nucleoid

Nucleoid divides;
cell wall and membrane
begin to form
transverse septum

Transverse septum
becomes complete

Daughter cells
separate

(a)

Figure 6.1 Binary fission.
(a) The stages of binary fission in a bacterial cell. **(b)** A thin section of the bacterium *Staphylococcus*, which is undergoing binary fission (51,027X). *(George Musil/Visuals Unlimited)* **(c)** The nucleoid of a bacterial cell (77,307X). *(Ralph A. Slepecky/Visuals Unlimited)*

(b) TEM

nucleoid

(c) TEM

In this chapter, we will use what we learned in Chapter 5 about energy in microorganisms to study how to grow them in the laboratory. Bacterial growth, which has been more thoroughly studied than growth in other microorganisms, is affected by a variety of physical and nutritional factors. Knowing how these factors influence growth is useful in culturing organisms in the laboratory and in preventing their growth in undesirable places. Furthermore, growing the microbes in pure cultures is essential in performing diagnostic tests that are used to identify a number of disease-causing organisms.

GROWTH AND CELL DIVISION

MICROBIAL GROWTH DEFINED

In everyday language, growth refers to an increase in size. We are accustomed to seeing children, other animals, and plants grow. Unicellular organisms also grow, but as soon as a cell, called the **mother** (or *parent*) **cell**, has approximately doubled in size and duplicated its contents, it divides into two **daughter cells**. Then the daughter cells grow, and subsequently they also divide. Because individual cells grow larger only to divide into two new individuals, **microbial growth** is defined not in terms of cell size but as the increase in the number of cells, which occurs by cell division.

CELL DIVISION

Cell division in bacteria, unlike cell division in eukaryotes, usually occurs by *binary fission* or sometimes by *budding*. In **binary fission**, a cell duplicates its components and divides into two cells **(Figure 6.1a)**. The daughter cells become independent when a *septum* (partition) grows between them and they separate **(Figure 6.1b)**. Unlike eukaryotic cells, prokaryotic cells do not have a cell cycle with a specific period of DNA synthesis. Instead, in continuously dividing cells, DNA synthesis also is continuous and replicates the bacterial chromosome shortly before the cell divides. The chromosome is attached to the cell membrane, which grows and separates the replicated chromosomes. Replication of the chromosome is completed before cell division, when the cell may temporarily contain two or more nucleoids. In some species, incomplete separation of the cells produces linear chains (linked bacilli), **tetrads** (cuboidal groups of four cocci), **sarcinae** (singular: *sarcina*; groups of eight cocci in a cubical packet), or grapelike clusters (staphylococci) (Figure 4.2). Some bacilli always form chains or filaments; others form them only under unfavorable growth conditions. Streptococci form chains when grown on artificial media but exist as single or paired cells when isolated from a rapidly growing lesion in an infected human host.

Cell division in yeast and a few bacteria occurs through **budding**. In that process, a small, new cell develops from the surface of an existing cell and subsequently separates from the parent cell **(Figure 6.2)**.

Figure 6.2 Budding in yeast (12,000X). *(SPL/Photo Researchers, Inc.)*

PHASES OF GROWTH

Consider a population of organisms introduced into a fresh, nutrient-rich **medium** (plural: media), a mixture of substances on or in which microorganisms grow. Such organisms display four major phases of growth: (1) the lag phase, (2) the log (logarithmic) phase, (3) the stationary phase, and (4) the decline phase, or death phase. These phases form the **standard bacterial growth curve (Figure 6.3)**.

Each square centimeter of skin hosts an average of 100,000 organisms. Bacteria reproduce so quickly, their population is restored within hours after washing.

The Lag Phase

In the **lag phase**, the organisms do not increase significantly in number, but they are metabolically active—growing in size, synthesizing enzymes, and incorporating various molecules from the medium. During this phase the individual organisms increase in size, and they produce large quantities of energy in the form of ATP.

The length of the lag phase is determined in part by characteristics of the bacterial species and in part by conditions in the media—both the medium from which the organisms are taken and the one to which they are transferred. Some species adapt to the new medium in an hour or two; others take several days. Organisms from old cultures, adapted to limited nutrients and large accumulations of wastes, take longer to adjust to a new medium than do those transferred from a relatively fresh, nutrient-rich medium.

The Log Phase

Once organisms have adapted to a medium, population growth occurs at an **exponential**, or **logarithmic** (log), **rate**. When the scale of the vertical axis is logarithmic, growth in this **log phase** appears on a graph as a straight diagonal line, which represents the size of the bacterial population. (On the base-10 logarithmic scale, each successive unit represents a 10-fold increase in the number of organisms; see ◀Appendix A.) During the log phase, the

Under ideal conditions, one bacterium can multiply to 2,097,152 within 7 hours.

organisms divide at their most rapid rate—a regular, genetically determined interval called the **generation time**. The population of organisms doubles in each generation time. For example, a culture containing 1,000 organisms per milliliter with a generation time of 20 minutes would contain 2,000 organisms per milliliter after 20 minutes, 4,000 organisms after 40 minutes, 8,000 after 1 hour, 64,000 after 2 hours, and 512,000 after 3 hours. Such growth is said to be *exponential*, or *logarithmic*.

The generation time for most bacteria is between 20 minutes and 20 hours, and is typically less than 1 hour. Some bacteria, such as those that cause tuberculosis and leprosy, have much longer generation times. Some individual cells take slightly longer than others to go from the lag phase to the log phase, and they do not all divide precisely together. If they divided together and the generation time was exactly 20 minutes, the number of cells in a culture would increase in a stair-step pattern, exactly doubling every 20 minutes—a hypothetical situation called **synchronous growth**. In an actual culture, each cell divides sometime during the 20-minute generation time, with about 1/20 of the cells dividing each minute—a natural situation called **nonsynchronous growth**. Nonsynchronous growth appears as a smooth line, not as steps, on a graph **(Figure 6.4)**.

Organisms in a tube of culture medium can maintain logarithmic growth for only a limited time. As the number of organisms increases, nutrients are used up, metabolic wastes accumulate, living space may become limited, and aerobes suffer from oxygen depletion. Generally, the limiting factor for logarithmic growth seems to be the rate at which energy can be produced in the form of ATP. As the availability of nutrients decreases, the cells become less able to generate ATP, and their growth rate decreases. The decrease in growth rate is shown in Figure 6.3 by a gradual leveling off of the growth curve (the curved segment to the right of the log phase).

Leveling off of growth is followed by the stationary phase unless fresh medium is added or organisms are transferred to fresh medium. Logarithmic growth can be maintained by a device, much like a thermostat, called a **chemostat (Figure 6.5)**, which has a growth chamber and a reservoir from which fresh medium is continuously added to the growth chamber as old medium is withdrawn. Alternatively, organisms from a culture in the stationary phase can be transferred to a fresh medium. After a brief lag phase, such organisms quickly reenter the log phase of growth.

The Stationary Phase

When cell division decreases to the point that new cells are produced at the same rate as old cells die, the number of live cells stays constant. The culture is then in the **stationary phase**, represented by a horizontal straight line in Figure 6.3. The medium contains a limited amount of nutrients and may contain toxic quantities of waste materials. Also, the oxygen supply may become inadequate for aerobic organisms, and damaging pH changes may occur.

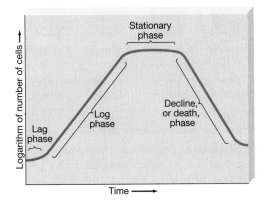

Figure 6.3 A standard bacterial growth curve.

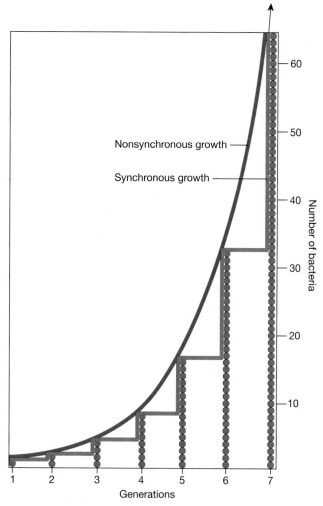

Figure 6.4 Synchronous versus nonsynchronous growth.
A growth curve for an exponentially increasing population plotted for a synchronously dividing population (red line) and for a nonsynchronously dividing population (green line). The blue spheres represent the number of bacteria present in each generation, after beginning with a single cell.

The Decline (Death) Phase

As conditions in the medium become less and less supportive of cell division, many cells lose their ability to divide, and thus the cells die. In this **decline phase**, or **death phase**, the number of live cells decreases at a logarithmic rate, as indicated by the straight, downward-sloping diagonal line in Figure 6.3. During the decline phase, many cells undergo *involution*—that is, they assume a variety of unusual shapes, which makes them difficult to identify. In cultures of spore-forming organisms, more spores than vegetative (metabolically active) cells survive. The duration of this phase is as highly variable as the duration of the logarithmic growth phase. Both depend primarily on the genetic characteristics of the organism. Cultures of some bacteria go through all growth phases and die in a few days; others contain a few live organisms after months or even years.

Growth in Colonies

Growth phases are displayed in different ways in colonies growing on a solid medium. Typically, a cell divides

Figure 6.5 The BIOSTAT® is a compact, autoclavable bioreactor fermentor system also known as a chemostat. The data are transferred to a standard Notebook-PC. Constantly renewing nutrients in a culture makes it possible to grow organisms continuously in the log phase. *(Courtesy Sartorius BBI Systems, Inc.)*

exponentially, forming a small **colony**—all the descendants of the original cell. The colony grows rapidly at its edges; cells nearer the center grow more slowly or begin to die because they have smaller quantities of available nutrients and are exposed to more toxic waste products. All phases of the growth curve occur simultaneously in a colony—that is, growth is nonsynchronous.

MEASURING BACTERIAL GROWTH

Bacterial growth is measured by estimating the number of cells that have arisen by binary fission during a growth phase. This measurement is expressed as the number of *viable* (living) organisms per milliliter of culture. Several methods of measuring bacterial growth are available.

Figure 6.6 Serial dilution. One milliliter is taken from a broth culture and added to 9 ml of sterile water, thereby diluting the culture by a factor of 10. This procedure is repeated until the desired concentration is reached.

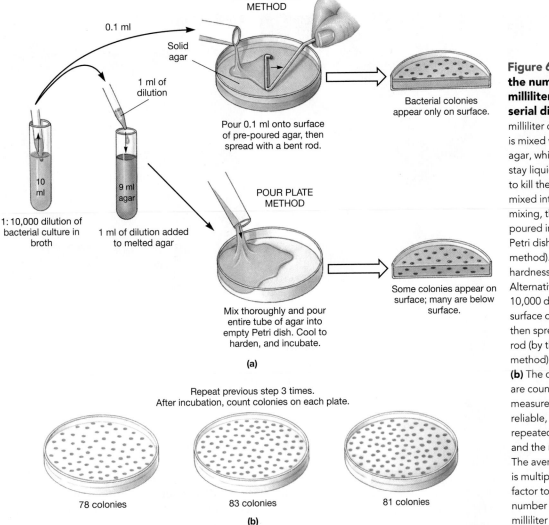

SPREAD PLATE METHOD

0.1 ml

Solid agar

Pour 0.1 ml onto surface of pre-poured agar, then spread with a bent rod.

Bacterial colonies appear only on surface.

1 ml of dilution

10 ml

9 ml agar

1 : 10,000 dilution of bacterial culture in broth

1 ml of dilution added to melted agar

POUR PLATE METHOD

Mix thoroughly and pour entire tube of agar into empty Petri dish. Cool to harden, and incubate.

Some colonies appear on surface; many are below surface.

(a)

Repeat previous step 3 times.
After incubation, count colonies on each plate.

78 colonies   83 colonies   81 colonies

(b)

Figure 6.7 Calculation of the number of bacteria per milliliter of culture using serial dilution. (a) One milliliter of a 1 : 10,000 dilution is mixed with 9 ml of melted agar, which is warm enough to stay liquid but not hot enough to kill the organisms being mixed into it. After thorough mixing, the warm agar is quickly poured into an empty, sterile Petri dish (by the pour plate method). When cooled to hardness, it is incubated. Alternatively, 0.1 ml of a 1 : 10,000 dilution is poured onto a surface of pre-poured agar and then spread with a bent sterile rod (by the spread plate method). Next it is incubated. **(b)** The colonies that develop are counted. A single measurement is not very reliable, so the procedure is repeated at least three times, and the results are averaged. The average number of colonies is multiplied by the dilution factor to ascertain the total number of organisms per milliliter of the original culture.

Serial Dilution and Standard Plate Counts

One method of measuring bacterial growth is the *standard plate count*. This technique relies on the fact that under proper conditions, only a living bacterium will divide and form a visible colony on an agar plate. An *agar plate* is a Petri dish containing a nutrient medium solidified with **agar**, a complex polysaccharide extracted from certain marine algae. Because it is difficult to count more than 300 colonies on one agar plate, it is usually necessary to dilute the original bacterial culture before you plate (transfer) a known volume of the culture onto the solid plate. *Serial dilutions* accomplish this purpose.

To make **serial dilutions (Figure 6.6)**, you start with organisms in liquid medium. Adding 1 ml of this medium to 9 ml of sterile water makes a 1 : 10 dilution; adding 1 ml of the 1 : 10 dilution to 9 ml of sterile water makes a 1 : 100 dilution; and so on. The number of bacteria per milliliter of fluid is reduced by 9/10 in each dilution. Subsequent dilutions are made in ratios of 1 : 1,000, 1 : 10,000, 1 : 100,000, 1 : 1,000,000, or even 1 : 10,000,000 if the original culture contained an extremely large number of organisms.

From each dilution, usually beginning with the 1 : 100, 0.1 ml of the culture is transferred to an agar plate. (One-tenth milliliter of the 1 : 10 dilution typically contains too many organisms to yield countable colonies when transferred to a Petri plate.) The transfer can be done by either the pour plate method or the spread plate method **(Figure 6.7)**. A **pour plate** is made by first adding 1.0 ml of a diluted culture from a serial dilution to 9 ml of melted nutrient agar. After the medium is mixed, it is poured into an empty Petri plate. Once the agar medium cools, solidifies, and is incubated, colonies will develop both within the medium and on its surface. Cells suspended in the melted agar during preparation may be heat-damaged, and then they will not form colonies. Those that do grow inside the agar will form smaller colonies than those growing on the surface. The **spread plate method** eliminates such problems because all cells remain on the surface of the solid medium. The diluted sample is first placed on the center of a solid, cooled agar medium. The sample is then spread evenly over the medium's surface with a sterile, bent glass rod. After incubation, colonies develop on the agar surface.

Wherever a single living bacterium is deposited on an agar plate, it will divide to form a colony. Each bacterium represents a **colony-forming unit (CFU)**. One or

(a)

(b)

(c)

Figure 6.8
Counting colonies.
(a) Using a bacterial colony counter. *(SIM/ Visuals Unlimited)*
(b) Bacterial colonies viewed through the magnifying glass against a colony-counting grid. The plate was produced by the pour plate method. How many different colony types can you identify on this plate? *(Biological Photo Service)* **(c)** Which of these plates would be the correct one to count? Why? *(M. Gabridge/Visuals Unlimited)*

number of colonies found on a plate by the *dilution factor*; if it is a fraction, use the denominator. A dilution factor of 1,000 would be expressed as 1 : 1,000 or 1/1,000, and a dilution factor of 10,000 would be expressed as 1 : 10,000. A typical calculation for an average colony count of 81 produced by plating a 1/100,000 dilution (dilution factor = 100,000) would be as follows:

$$81 \times 100,00 = 8,100,000 \quad \text{or} \quad 8.1 \times 10^6 \text{ CFU/ml}$$

The accuracy of the serial dilution and plate count method depends on homogeneous dispersal of organisms in each dilution. Error can be minimized by shaking each culture before sampling and making several plates from each dilution. Accuracy is also affected by the death of cells. Because the number of colonies counted represents the number of living organisms, it does not include organisms that may have died by the time plating was done; nor does it include organisms that cannot grow on the chosen medium. Using young cultures in the log phase of growth minimizes this kind of error.

Direct Microscopic Counts

Bacterial growth can be measured by **direct microscopic counts**. In this method a known volume of medium is introduced into a specially calibrated, etched glass slide called a *Petroff-Hausser counting chamber* (**Figure 6.9**), also known as a hemocytometer. A bacterial suspension is introduced onto the chamber with a calibrated pipette. After the bacteria settle and the liquid currents have slowed, the microorganisms are counted in specific calibrated areas. Their number per unit volume of the original suspension is calculated by using an appropriate formula. The number of bacteria per milliliter of medium can be estimated with a reasonable degree of accuracy. The accuracy of direct microscopic counts depends on the presence of more than 10 million bacteria per milliliter of

more plates should have a small enough number of colonies such that one is clearly distinguishable and can be counted. If you have made dilutions properly, you should get plates with a **countable number** of colonies (30 to 300 per plate).

To count the actual number of colonies present, you would place the plate under the magnifying lens of a *colony counter* (**Figure 6.8**), and colonies on the entire plate are counted. To determine the number of colony-forming units in the original culture, multiply the

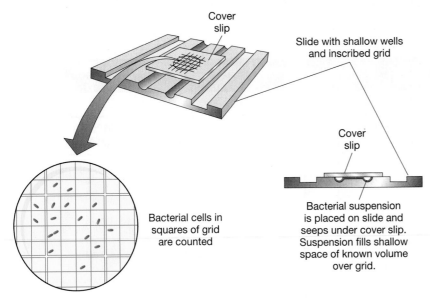

Cover slip

Slide with shallow wells and inscribed grid

Cover slip

Bacterial cells in squares of grid are counted

Bacterial suspension is placed on slide and seeps under cover slip. Suspension fills shallow space of known volume over grid.

Figure 6.9 **The Petroff-Hausser counting chamber (hemocytometer).** The volume of suspension filling the narrow space between the grid and the cover slide is known, so the number of bacteria per unit of volume can be calculated.

culture. This is because counting chambers are designed to allow accurate counts only when large numbers of cells are present. An accurate count also requires that the bacteria be homogeneously distributed throughout the culture. This technique has the disadvantage of generally not distinguishing between living and dead cells.

Most Probable Number

When samples contain too few organisms to give reliable measures of population size by the standard plate count method, as in food and water sanitation studies, or when organisms will not grow on agar, the **most probable number (MPN)** method is used. With this method, the technician observes the sample, estimates the number of cells in it, and makes a series of progressively greater dilutions. As the dilution factor increases, a point will be reached at which some tubes will contain a single organism and others, none. A typical MPN test consists of five tubes of each of three volumes (using 10, 1, and 0.1 ml) of a dilution (**Figure 6.10**). Those that contain an organism will

Volume of Dilution Added	Culture Results	Number of Positive Tubes
10 ml	+ + + + +	5
1 ml	+ − − − +	2
0.1 ml	− − − − −	0

(a)

(b) (1) (2) (3)

Figure 6.10 **A most probable number (MPN) test.** **(a)** Those tubes in which gas bubbles are visible (labeled +) contain organisms. The gas they have produced by fermenting the medium has risen and been trapped as bubbles in the tops of the small inverted tubes (Durham tubes). **(b)** An enlarged view of a (1) positive carbohydrate fermentation test showing CO_2 gas trapped inside a Durham tube; (2) a positive test where acid but not gas was produced; and (3) a negative test where neither acid nor gas was produced. The pH indicator in the broth remains red instead of turning yellow as it does in the presence of acid. *(Courtesy Jacquelyn G. Black)*

TABLE 6.1

Most Probable Number (/) Index for Combinations of Positive and Negative Results When Five Tubes Are Used per Dilution (Five Each of 10 ml, 1 ml, and 0.1 ml)							
Number of Tubes with Positive Results							
10 ml	1 ml	0.1 ml	MPN Index/100 ml	10 ml	1 ml	0.1 ml	MPN Index/100 ml
0	0	0	<2	4	3	1	33
0	0	1	2	4	4	0	34
0	1	0	2	5	0	0	23
0	2	0	4	5	0	1	30
1	0	0	2	5	0	2	40
1	0	1	4	5	1	0	30
1	1	0	4	5	1	1	50
1	1	1	6	5	1	2	60
1	2	0	6	5	2	0	50
2	0	0	4	5	2	1	70
2	0	1	7	5	2	2	90
2	1	0	7	5	3	0	80
2	1	1	9	5	3	1	110
2	2	0	9	5	3	2	140
2	3	0	12	5	3	3	170
3	0	0	8	5	4	0	130
3	0	1	11	5	4	1	170
3	1	0	11	5	4	2	220
3	1	1	14	5	4	3	280
3	2	0	14	5	4	4	350
3	2	1	17	5	5	0	240
4	0	0	13	5	5	1	300
4	0	1	17	5	5	2	500
4	1	0	17	5	5	3	900
4	1	1	21	5	5	4	1600
4	1	2	26	5	5	5	≥1600
4	2	0	22				
4	2	1	26				
4	3	0	27				

Source: A. E. Greenberg, L. S. Clesceri, and A. D. Eaton, Eds. *Standard Methods for the Examination of Water and Wastewater.* 18th ed. Washington, DC: American Public Health Association, 1992.

display growth by producing gas bubbles and/or by becoming cloudy, when incubated. The number of organisms in the original culture is estimated from a most probable number table. The values in the table, which are based on statistical probabilities, specify that the number of organisms in the original culture has a 95% chance of falling within a particular range. A complete MPN table is given in **Table 6.1**. The more tubes that show growth, especially at greater dilutions, the more organisms were present in the sample. To use the MPN table, match the number of positive tubes for each dilution (5, 2, and 0 in Figure 6.10a) with the value in the MPN Index/100 ml column (50 organisms/100 ml in this example).

One of the most useful applications of the MPN method is in testing water purity. See Chapter 25, p. 803 for an explanation of the multiple-tube fermentation method which provides an estimate of the number of coliforms (bacteria of fecal origin).

Filtration

Another method of estimating the size of small bacterial populations uses **filtration**. A known volume of water or air is drawn through a filter with pores too small to allow passage of bacteria. When the filter is then placed on a solid medium, each colony that grows represents originally one organism collected by the filter. Thus, the number of organisms per liter of water or air can be calculated. (Figure 25.19 shows the filtration process and colonies grown on a filter pad.)

Other Methods

Several other methods of monitoring bacterial growth are available. They include simple observation with or without special instruments, measurement of metabolic products by the detection of gas or acid production, and determination of dry weight of cells.

Figure 6.11 Turbidity. Turbidity, or a cloudy appearance, is an indicator of bacterial growth in urine in the tube on the left. *(Richard Megna/Fundamental Photographs)*

Turbidity (a cloudy appearance) in a culture tube indicates the presence of organisms **(Figure 6.11)**. Fairly accurate estimates of growth can be obtained by measuring turbidity with a photoelectric device, such as a *colorimeter* or a *spectrophotometer* **(Figure 6.12)**. This method is particularly useful in monitoring the rate of growth without disturbing the culture. Samples with very high cell densities, however, must be diluted to ensure accurate readings. Measures of bacterial growth based on turbidity are likewise especially subject to error when cultures contain fewer than 1 million cells per milliliter. Such cultures can display little or no turbidity even when growth is occurring. Conversely, turbidity can be produced by a high concentration of dead cells in a culture.

Figure 6.12 A spectrophotometer. This instrument can be used to measure bacterial growth by determining the degree of light transmission through the culture. Samples of culture in special optically clear tubes are placed inside the spectrophotometer (under the square lid at left of machine) and are measured against standards. *(Courtesy Thermo Electron Corporation)*

Measuring the metabolic products of a population can be used to estimate bacterial growth indirectly. The rate at which metabolic products such as gases and/or acids are formed by a culture reflects the mass of bacteria present. Gas production can be detected (rather than measured) by capturing the gas in small inverted tubes placed inside larger tubes of liquid medium containing bacteria. Acid production can be detected by incorporating *pH indicators*—chemical substances that change color with changes in pH—in a liquid medium containing metabolically active bacteria (Figure 6.10b).

The rate at which a substrate such as glucose or oxygen is used up also reflects cell mass. For example, one method for estimating bacterial mass is the *dye reduction test*, which measures the direct or indirect uptake of oxygen. In this test, a dye such as methylene blue is incorporated into a medium containing milk. Bacteria inoculated into the medium use oxygen as they metabolize the milk. Methylene blue is blue in the presence of oxygen and turns colorless in its absence. Thus, the faster the medium loses color, the faster the oxygen is being used up, and the more bacteria are presumed to be present. The rate at which the dye is decolorized (dye reduction) is a highly indirect approach; it is not an accurate measure of bacterial mass.

Finally, the number of cells in a culture can be determined by *dry weight measurements*. To calculate the dry weight of cells, they must be separated from the medium by some physical means such as filtration or centrifugation. The cells are then dried, and the resulting mass is weighed.

✓ CHECKLIST

1. What are the differences between the lag phase and the log phase of a bacterial growth curve?

2. How does logarithmic rate of increase differ from arithmetic rate of increase? The following sequence of numbers of cells is an example of which type of rate of increase? 1, 2, 4, 8, 16, 32.

3. If a broth culture of bacteria, initially containing 37,000 bacteria/ml, is diluted to a 1 : 1,000 dilution, how many bacteria/ml of the diluted broth would be present on average?

4. Why does a direct microscopic count of bacteria using a Petroff-Hausser chamber not give you a viable count? How does this differ from the spread plate and pour plate methods?

▌▌▌ FACTORS AFFECTING BACTERIAL GROWTH

Microorganisms are found in nearly every environment on Earth, including environments in which no other life forms can survive. Microbes can exist in a great many environments because they are small and easily dispersed, occupy little space, need only small quantities of nutrients, and are remarkably diverse in their nutritional requirements. They also have great capacity for adapting to environmental changes. For almost any substance, there is some microbe that can metabolize it as a nutrient; for almost any environmental change, there is some microbe that can survive the change.

As warmblooded, air-breathing, land-dwelling mammals, we tend to forget that 72% of our planet's surface is water, that 90% of that water is salt water, and that environments containing living organisms have an average temperature of about 5°C. Unlike humans, microorganisms live mostly in water, and many are adapted to temperatures above or below those we consider optimum. The organisms of particular interest in the health sciences account for only a fraction of all microorganisms—those that have adapted to conditions found in or on the human body.

Different species of microorganisms can grow in a wide range of environments—from highly acidic to somewhat alkaline conditions, from Antarctic ice to hot springs, in pure spring water or in salty marshes, in oceans with or without oxygen, and even under great pressure and in boiling stream vents on the ocean floor. Microorganisms use a variety of substances to obtain energy, and some require special nutrients.

The total weight of the number of bacteria that live in the soil and underground is estimated to be 10,034 trillion tons.

The kinds of organisms found in a given environment and the rates at which they grow can be influenced by a variety of factors, both physical and biochemical. **Physical factors** include pH, temperature, oxygen concentration, moisture, hydrostatic pressure, osmotic pressure, and radiation. **Nutritional** (*biochemical*) **factors** include availability of carbon, nitrogen, sulfur, phosphorus, trace elements, and, in some cases, vitamins.

PHYSICAL FACTORS

pH

Remember that the acidity or alkalinity of a medium is expressed in terms of pH (◀Chapter 2, p. 35). Although the pH scale is now widely used in chemistry, it was invented by the Danish chemist Søren Sørenson to describe the limits of growth of microorganisms in various media. Microorganisms have an **optimum pH**—the pH at which they grow best. Their optimum pH is usually near neutrality (pH 7). Most microbes do not grow at a pH more than 1 pH unit above or below their optimum pH.

According to their tolerance for acidity or alkalinity, bacteria are classified as:

- acidophiles,
- neutrophiles, or
- alkaliphiles.

However, no single species can tolerate the full pH range of any of these categories, and many tolerate a pH range that overlaps two categories. **Acidophiles** (a-sid′o-

filz), or acid-loving organisms, grow best at a pH of 0.1 to 5.4. *Lactobacillus*, which produces lactic acid, is an acidophile, but it tolerates only mild acidity. Some bacteria that oxidize sulfur to sulfuric acid, however, can create and tolerate conditions as low as pH 1.0. Bacteria producing sulfuric acid strong enough to eat through your clothing are now known to have eaten out some huge caves from limestone (e.g., Carlsbad Caverns in the American Southwest). Acid drips from long, hanging colonies of bacteria which have the consistency of strings of mucus, leading to their name of "snotites," as shown in one of the opening photos for Chapter 1, p. 1. **Neutrophiles** (nu′tro-filz) exist from pH 5.4 to 8.0. Most of the bacteria that cause disease in humans are neutrophiles. **Alkaliphiles** (al′kah-li-filz), or alkali-loving (base-loving) organisms, exist from pH 7.0 to 11.5. *Vibrio cholerae*, the causative agent of the disease Asiatic cholera, grows best at a pH of about 9.0. *Alcaligenes faecalis*, which sometimes infects humans already weakened by another disease, can create and tolerate alkaline conditions of pH 9.0 or higher. The soil bacterium *Agrobacterium* grows in alkaline soil of pH 12.0.

The effects of pH on organisms can, in part, be related to the concentration of organic acids in the medium and to the protection that bacterial cell walls sometimes provide. *Lactobacillus* and other organisms that produce organic acids during fermentation inhibit their own growth by producing acids such as lactic acid and pyruvic acid, which accumulate in the medium. It appears that the acids themselves, rather than the hydrogen ions per se, inhibit growth. Changes in pH can lead to denaturing of enzymes and other proteins and can interfere with pumping of ions at the cell membrane. Other organisms have relatively impervious cell walls that prevent the cell membrane from being exposed to an extreme pH in the medium. These organisms appear to tolerate environmental acidity or alkalinity because the cell itself is maintained at a nearly neutral pH.

Many bacteria often produce sufficient quantities of acids as metabolic by-products that eventually interfere with their own growth. To prevent this situation in the laboratory cultivation of bacteria, *buffers* are incorporated into growth media to maintain the proper pH levels. Phosphate salts are commonly used for this purpose.

Temperature

Most species of bacteria can grow over a 30°C temperature range, but the minimum and maximum temperatures for different species vary considerably. Seawater remains liquid below 0°C, and organisms living in cold ocean waters can tolerate below-freezing temperatures. According to their growth temperature range, bacteria can be classified as:

- psychrophiles,
- mesophiles, or
- thermophiles.

CLOSE UP

Every Nook and Cranny

Bacteria can effectively inhabit every place suitable for the existence of life—our ecologically complex gut, frozen Antarctic glaciers, and under the extreme barometric pressures and temperatures encountered in ocean floor vents. At temperatures greater than 160°F, all life on Earth is bacterial. The bacterium *Thermophila acidophilum* thrives at 140°F at a pH of 1 or 2. This organism, found on the surface of burning coal and in thermal hot springs, "freezes" to death at 100°F. Recently, microbial communities have been found living 3,000 feet below the Earth's surface in Columbia River basalt. These bacteria are anaerobic, getting their energy from the hydrogen reaction produced between the minerals in the basalt and groundwater seeping through the rocks. The characterization of bacteria living in extreme environments provides insight into the diversity of life strategies, as well as an opportunity to produce and use biological molecules with unique capabilities.

Most bacteria, however, do not tolerate the whole temperature range of a category, and some tolerate a range that overlaps categories. Within these groups, bacteria are further classified as obligate or facultative. **Obligate** means that the organism *must* have the specified environmental condition. **Facultative** means that the organism is *able* to adjust to and tolerate the environmental condition, but it can also live in other conditions.

Psychrophiles (si'kro-filz), or cold-loving organisms, grow best at temperatures of 15° to 20°C, although some live quite well at 0°C. They can be further divided into **obligate psychrophiles**, such as *Bacillus globisporus*, which cannot grow above 20°C, and **facultative psychrophiles**, such as *Xanthomonas pharmicola*, which grows best below 20°C but also can grow above that temperature. Psychrophiles live mostly in cold water and soil. None can live in the human body, but some, such as *Listeria monocytogenes*, are known to cause spoilage of refrigerated foods, and subsequent disease in humans, sometimes fatal.

Bird flu viruses can survive decades in frozen lake ice. Global warming may release them to infect migratory birds who will spread it.

Mesophiles (mes'o-filz), which include most bacteria, grow best at temperatures between 25° and 40°C. Human pathogens are included in this category, and most of them grow best near human body temperature (37°C). *Thermoduric* organisms ordinarily live as mesophiles but can withstand short periods of exposure to high temperatures. Inadequate heating during canning or in pasteurization may leave such organisms alive and therefore able to spoil food.

Thermophiles (therm'o-filz), or heat-loving organisms, grow best at temperatures from 50° to 60°C. Many are found in compost heaps, and a few tolerate temperatures as high as 110°C in boiling hot springs. They can be further classified as **obligate thermophiles**, which can grow only at temperatures above 37°C, or **facultative thermophiles**, which can grow both above and below 37°C. *Bacillus stearothermophilus*, which usually is considered an obligate thermophile, grows at its maximum rate at 65° to 75°C but can display minimal growth and cause food spoilage at temperatures as low as 30°C. Thermophilic sulfur bacteria find zones of optimum growth temperatures in the runoff troughs of geysers (**Figure 6.13**). Different species collect at various locations along the sides of the trough. The most heat-tolerant are near the geyser, and those with lesser heat tolerance are distributed in regions where the water has cooled to their optimum temperature. In deep channels, the most heat-tolerant species are found at the greatest depths and the least heat-tolerant near the surface, where the water has cooled. Under laboratory conditions that use high pressure to increase water temperature above 100°C, archaeobacteria from deep-sea vents have grown at 115°C (238°F). (Chapter 9 provides more information about these remarkable organisms.)

Tomato juice agar, which is actually made with tomato juice, is used for the cultivation of lactobacilli.

The temperature range over which an organism grows is determined largely by the temperatures at which its enzymes function. Within this temperature range, three critical temperatures can be identified:

1. The *minimum growth temperature*, the lowest temperature at which cells can divide.
2. The *maximum growth temperature*, the highest temperature at which cells can divide.
3. The *optimum growth temperature*, the temperature at which cells divide most rapidly—that is, have the shortest generation time.

Figure 6.13 **Thermophiles.** Geyser Hot Springs, Black Rock Desert, Nevada. Thermophilic sulfur bacteria can live and grow in the runoff waters from such geysers despite the near-boiling temperatures. *(Stephen Trimble/DRK Photo)*

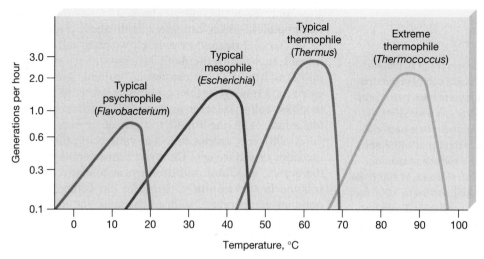

Figure 6.14 Growth rates of psychrophilic, mesophilic, and thermophilic bacteria. Notice the overlap of temperature ranges at which these organisms can survive. Growth rates are much lower at the extreme ends of the ranges.

Regardless of the type of bacteria, growth gradually increases from the minimum to the optimum temperature and decreases very sharply from the optimum to the maximum temperature. Furthermore, the optimum temperature is often very near the maximum temperature **(Figure 6.14)**. These growth properties are due to changes in enzyme activity (◀Chapter 5, p. 123). Enzyme activity generally doubles for every 10°C rise in temperature until the high temperature begins to denature all proteins, including enzymes. The sharp decrease in enzyme activity at a temperature only slightly higher than the optimum temperature occurs as enzyme molecules become so distorted by denaturation that they cannot catalyze reactions.

Temperature is important not only in providing conditions for microbial growth, but also in preventing such growth. The refrigeration of food, usually at 4°C, reduces the growth of psychrophiles and prevents the growth of most other bacteria. However, food and other materials, such as blood, can support growth of some bacteria even when refrigerated. For this reason, perishable materials that can withstand freezing are stored at temperatures of −30°C if they are to be kept for long periods of time. High temperatures also can be used to prevent bacterial growth (Chapter 12). Laboratory equipment and media are generally sterilized with heat, and food is frequently preserved by heating and storing in closed containers. Bacteria are more apt to survive extremes of cold than extremes of heat; enzymes are not denatured by chilling but can be permanently denatured by heat.

Cold temperatures may have helped to preserve *Exiguobacterium* sp., a bacterium isolated from 2- to 3-million-year-old Siberian permafrost soil. It grows well at −2.5°C, and is associated with human infections. And, contrary to expectations, soil-dwelling fungi in mountainous parts of the United States have been found to increase in numbers and biomass beneath the ice and snow of winter, compared to their abundance in summer.

Oxygen

Bacteria, especially heterotrophs, can be divided into aerobes, which require oxygen to grow, and anaerobes, which do not require it (◀Chapter 5, p. 128). Among the aerobes, cultures of rapidly dividing cells require more oxygen than do cultures of slowly dividing cells. **Obligate aerobes**, such as *Pseudomonas*, which is a common cause of hospital-acquired infections, must have free oxygen for aerobic respiration, whereas **obligate anaerobes**, such as *Clostridium botulinum*, *C. tetani*, and *Bacteroides*, are killed by free oxygen. In a culture tube containing nutrient broth, obligate aerobes grow near the surface, where atmospheric oxygen diffuses into the medium; obligate anaerobes grow near the bottom of the tube, where little or no free oxygen reaches them **(Figure 6.15)**.

For aerobes, oxygen is often the environmental factor that limits growth rate. Oxygen is poorly soluble in water, and various methods are sometimes employed to maintain a high O_2 concentration in cultures, including vigorous mixing or forced aeration by bubbling air through a culture, as is done in a fish tank. This is especially important in such commercial processes as the production of antibiotics and in sewage treatment.

Figure 6.15 Patterns of oxygen use. Different organisms incubated for 24 hours in tubes of a nutrient broth accumulate in different regions depending on their need for, or sensitivity to, oxygen.

When the Going Gets Tough, Hide Inside a Rock

A few bacteria live in bitterly cold, dry valleys of Antarctica, where very few other organisms can survive. The relative humidity is so low that water passes directly from the frozen to the vapor state and is rarely found as a liquid. Organisms that live there nevertheless manage to carry out their metabolic activities, either by using water vapor or by melting tiny amounts of ice with their metabolic heat. But they cannot survive the harsh conditions of the Antarctic atmosphere. The bacteria must hide inside translucent rocks (such as quartz, feldspar, and certain marbles), which allow the sun's rays to penetrate so these bacteria can carry out photosynthesis. Because they do not produce mineral-dissolving chemicals, these endolithic organisms must colonize only porous rocks. They are usually able to invade several millimeters into the rock, where they find safe refuge until the rock is eroded by wind.

Mars was originally a warm planet, but it cooled down when it lost its atmosphere. If life had evolved on Mars during its warm phase, would that life have sought shelter inside surface rocks? Examination of Martian meteorites reveals evidence of possible early life forms that resemble bacteria entombed in what may have been their last refuge.

Between the extremes of obligate aerobes and obligate anaerobes are the *microaerophiles*, the *facultative anaerobes*, and the *aerotolerant anaerobes*. **Microaerophiles** (mi″kro-aer′o-filz) appear to grow best in the presence of a small amount of free oxygen. They grow below the surface of the medium in a culture tube at the level where oxygen availability matches their needs. Microaerophiles such as *Campylobacter*, which can cause intestinal disorders, also are **capnophiles**, or carbon dioxide–loving organisms. They thrive under conditions of low oxygen and high carbon dioxide concentration. **Facultative anaerobes** ordinarily carry on aerobic metabolism when oxygen is present, but they shift to anaerobic metabolism when oxygen is absent. *Staphylococcus* and *Escherichia coli* are facultative anaerobes; they often are found in the intestinal and urinary tracts, where only a small amount of oxygen is available. **Aerotolerant anaerobes** can survive in the presence of oxygen but do not use it in their metabolism. *Lactobacillus*, for example, always captures energy by fermentation, regardless of whether the environment contains oxygen.

Compared with other groups of organisms defined according to oxygen requirements, facultative anaerobes have the most complex enzyme systems. They have one set of enzymes that enables them to use oxygen as an electron acceptor and another set that enables them to use another electron acceptor when oxygen is not available. In contrast, the enzymes of the other groups defined here are limited to either aerobic or anaerobic respiration.

Obligate anaerobes are killed not by gaseous oxygen but by a highly reactive and toxic form of oxygen called **superoxide** (O_2^-). Superoxide is formed by certain oxidative enzymes and is converted to molecular oxygen (O_2) and toxic hydrogen peroxide (H_2O_2) by an enzyme called **superoxide dismutase**. Hydrogen peroxide is converted to water and molecular oxygen by the enzyme **catalase**. Obligate aerobes and most facultative anaerobes have both enzymes. Some facultative and aerotolerant anaerobes have superoxide dismutase but lack catalase. Most obligate anaerobes lack both enzymes and succumb to the toxic effects of superoxide and hydrogen peroxide.

Moisture

All actively metabolizing cells generally require a water environment. Unlike larger organisms that have protective coverings and internal fluid environments, single-celled organisms are exposed directly to their environment. Most vegetative cells can live only a few hours without moisture; only the spores of spore-forming organisms can exist in a dormant state in a dry environment.

Hydrostatic Pressure

Water in oceans and lakes exerts **hydrostatic pressure**, pressure exerted by standing water, in proportion to its depth. Such pressure doubles with every 10 m increase in depth. For example, in a lake 50 m deep, the pressure is 32 times the atmospheric pressure. Some ocean valleys have depths in excess of 7,000 m, and certain bacteria are the only organisms known to survive the extreme pressure at such depths. Bacteria that live at high pressures, but die if left in the laboratory for only a few hours at standard atmospheric pressure, are called **barophiles**. It appears that their membranes and enzymes do not simply tolerate pressure but require pressure to function properly. The high pressure is necessary to keep their enzyme molecules in the proper three-dimensional configuration. Without it, the enzymes lose their shape and denature, and the organisms die.

Osmotic Pressure

We saw in Chapter 4 that the membranes of all microorganisms are selectively permeable. The cell membrane allows water to move by osmosis between the cytoplasm and the environment (Figure 4.31). Environments that contain dissolved substances exert osmotic pressure, and the pressure can exceed that exerted by dissolved substances in cells. Cells in such *hyperosmotic* environments lose water and undergo **plasmolysis** (plas-mol′e-sis), or shrinking of the cell. In microorganisms with a cell wall, the cell or plasma membrane separates from the cell wall. Conversely, cells in distilled water have a higher osmotic pressure than their environment and, therefore, gain water. In bacteria, the rigid cell wall prevents cells from swelling and bursting, but the cells fill with water and become *turgid* (distended).

Most bacterial cells can tolerate a fairly wide range of concentrations of dissolved substances. Their cell

Figure 6.16 Responses to salt. **(a)** Growth rates of halophilic (salt-loving) and nonhalophilic organisms are related to sodium ion concentration. **(b)** The Great Salt Lake in Utah, an example of an environment in which halophilic organisms thrive. Note the white areas of dried salt around the edges of the lake. *(© Tony Hamblin/Corbis)*

membranes contain transport systems that regulate the movement of dissolved substances across the membrane (◄Chapter 5, p. 140). Yet, if concentrations outside the cells become too high, water loss can inhibit growth or even kill the cells.

CLOSE UP

In the Pink

The Great Salt Lake in Utah is an extremely salty environment that supports bacterial growth. The Great Salt Lake, nearly 10 times saltier than ocean water, supports many varieties of halobacteria. Because all halobacteria lack peptidoglycan in their cell wall, they are Gram-negative. In addition, they are insensitive to most antibiotics, contain unusually large plasmids, and are obligate aerobes. The halobacteria require large amounts of sodium for growth, a requirement that is not satisfied when a similar ion is used in its place. Certain species of the extreme halophiles use a light-mediated mechanism to produce ATP. Unlike green plants, the pigments they use for light-dependent ATP synthesis are the red-orange carotinoids and the red-purple bacterioruberins and bacteriorhodopsins. The brilliant color of these bacteria can be seen when high-salinity lakes and holding ponds are photographed from the air—giving the distinct appearance of a pink patchwork quilt.

(Aerial Archives)

The use of salt as a preservative in curing hams and bacon and in making pickles is based on the fact that high concentrations of dissolved substances exert sufficient osmotic pressure to kill or inhibit microbial growth. The use of sugar as a preservative in making jellies and jams is based on the same principle.

Bacteria called **halophiles** (hal′o-filz), or salt-loving organisms, require moderate to large quantities of salt (sodium chloride). Their membrane transport systems actively transport sodium ions out of the cells and concentrate potassium ions inside them. Two possible explanations for why halophiles require sodium have been proposed. One is that the cells need sodium to maintain a high intracellular potassium concentration so that their enzymes will function. The other is that they need sodium to maintain the integrity of their cell walls.

Halophiles are typically found in the ocean, where the salt concentration (3.5%) is optimum for their growth. Extreme halophiles require salt concentrations of 20% to 30% **(Figure 6.16)**. They are found in exceptionally salty bodies of water, such as the Dead Sea, and sometimes even in brine vats, where they cause spoilage of pickles being made there.

Radiation

Radiant energy, such as gamma rays and ultraviolet light, can cause mutations (changes in DNA) and even kill organisms. However, some microorganisms have pigments that screen radiation and help to prevent DNA damage. Others have enzyme systems that can repair certain kinds of DNA damage.

The bacterium, *Deinococcus radiodurans* can survive 10,000 Grays (Gy) of radiation. The Gy is a unit of measurement for absorbed dose of radiation. 5 Gy will kill a human, and 1,000 Gy will sterilize a culture of *E. coli*. Bacteria which can withstand high levels of radiation may be valuable for use in cleaning up contaminated sites.

NUTRITIONAL FACTORS

The growth of microorganisms is affected by nutritional factors, as well as by physical factors. Nutrients needed by microorganisms include carbon, nitrogen, sulfur, phosphorus, certain trace elements, and vitamins. Although we are concerned with ways in which microorganisms satisfy their own nutritional needs, we can note that in satisfying such needs they also help to recycle elements in the environment. Activities of microbes in the carbon, nitrogen, sulfur, and phosphorus cycles are described in Chapter 26. A few microbes are **fastidious**—that is, they have special nutritional needs that can be difficult to meet in the laboratory. Some fastidious organisms, including those that cause gonorrhea, grow quite well in the human body but still cannot be easily grown in the laboratory on nutrient media.

Carbon Sources

Most bacteria use some carbon-containing compound as an energy source, and many use carbon-containing compounds as building blocks to synthesize cell components. Photoautotrophic organisms reduce carbon dioxide to glucose and other organic molecules. Both autotrophic and heterotrophic organisms can obtain energy from glucose by glycolysis, fermentation, and the Krebs cycle.

They also synthesize some cell components from intermediates in these pathways.

Nitrogen Sources

All organisms, including microorganisms, need nitrogen to synthesize enzymes, other proteins, and nucleic acids. Some microorganisms obtain nitrogen from inorganic sources, and a few even obtain energy by metabolizing inorganic nitrogen-containing substances. Many microorganisms reduce nitrate ions (NO_3^-) to amino groups (NH_2) and use the amino groups to make amino acids. Some can synthesize all 20 amino acids found in proteins, whereas others must have one or a few amino acids provided in their medium. Certain fastidious organisms require all 20 amino acids and other building blocks in their medium. Many disease-causing organisms obtain amino acids for making proteins and other nitrogenous molecules from the cells of humans and other organisms they invade.

Once amino acids are synthesized or obtained from the medium, they can be used in protein synthesis. Similarly, purines and pyrimidines can be used to make DNA and RNA. The processes by which proteins and nucleic acids are synthesized are directly related to the genetic information contained in a cell. Thus, the synthesis of proteins and of nucleic acids will be discussed in Chapters 7 and 8.

CLOSE UP

Picky Eaters

Spiroplasma species, tiny spiral bacteria that lack cell walls, are among the most nutritionally fastidious organisms known. Recently, Kevin Hackett, a U.S. Department of Agriculture (USDA) scientist, devised an exact formula of 80 ingredients, including lipids, carbohydrates, amino acids, salts, vitamins, organic acids, and penicillin (to suppress potential competitors) to meet their needs. In his laboratory, he uses this medium to keep more than 30 species of spiroplasmas alive and well, making it possible for researchers to study them outside

Spiroplasma (magnified 55,100X). (David M. Phillips/Visuals Unlimited)

the more than 100 insect, tick, and plant species that they normally inhabit. Until now, most *Spiroplasma* species have been impossible to keep alive outside their hosts.

The spiroplasmas are responsible for hundreds of crop and animal diseases. Medical researchers are particularly interested in one species that experimentally causes tumors in animals. Another species kills honey bees, and a third lives harmlessly in the Colorado potato beetle, an insect that damages potato, eggplant, and tomato plants. Scientists hope to alter this last species genetically so that it will kill its potato beetle host.

USDA scientists are now trying to formulate complex media to grow mycoplasma-like organisms, a related group of bacteria that also lack cell walls. These bacteria cause hundreds of crop diseases and millions of dollars in economic losses each year. They are spread from plant to plant by infected insects. Another medium being designed would grow the bacterium *Mycoplasma pneumoniae*, which is the cause of a form of "walking pneumonia" in humans.

USDA scientist Kevin Hackett working on his microbial "witch's brew"—a mix of some 80 ingredients that will support the growth of nutritionally fastidious spiroplasmas outside their hosts. (*Courtesy Agricultural Research Service, USDA*)

Sulfur and Phosphorus

In addition to carbon and nitrogen, microorganisms need a supply of certain minerals, especially sulfur and phosphorus, which are important cell components. Microorganisms obtain sulfur from inorganic sulfate salts and from sulfur-containing amino acids. They use sulfur and sulfur-containing amino acids to make proteins, coenzymes, and other cell components. Some organisms can synthesize sulfur-containing amino acids from inorganic sulfur and other amino acids. Microorganisms obtain phosphorus mainly from inorganic phosphate ions (PO_4^{3-}) They use phosphorus (as phosphate) to synthesize ATP, phospholipids, and nucleic acids.

Trace Elements

Many microorganisms require a variety of **trace elements**, tiny amounts of minerals such as copper, iron, zinc, and cobalt, usually in the form of ions. Trace elements often serve as cofactors in enzymatic reactions. All organisms require some sodium and chloride, and halophiles require large amounts of these ions. Potassium, zinc, magnesium, and manganese are used to activate certain enzymes. Cobalt is required by organisms that can synthesize vitamin B_{12}. Iron is required for the synthesis of heme-containing compounds (such as the cytochromes of the electron transport system) and for certain enzymes. Although little iron is required, a shortage severely retards growth. Calcium is required by Gram-positive bacteria for synthesis of cell walls and by spore-forming organisms for synthesis of spores.

Vitamins

A **vitamin** is an organic substance that an organism requires in small amounts and that is typically used as a coenzyme. Many microorganisms make their own vitamins from simpler substances. Other microorganisms require several vitamins in their media because they lack the enzymes to synthesize them. Vitamins required by some microorganisms include folic acid, vitamin B_{12}, and vitamin K. Human pathogens often require a variety of vitamins and thus are able to grow well only when they can obtain these substances from the host organism. Growing such organisms in the laboratory requires a complex medium that contains all the nutrients they normally obtain from their hosts. Microbes living in the human intestine manufacture vitamin K, which is necessary for blood clotting, and some of the B vitamins, thus benefiting their host.

Nutritional Complexity

An organism's **nutritional complexity**, the number of nutrients it must obtain to grow, is determined by the kind and number of its enzymes. The absence of a single enzyme can render an organism incapable of synthesizing a specific substance. The organism therefore must obtain the substance as a nutrient from its environment. Microorganisms vary in the number of enzymes they possess. Those with many enzymes have simple nutritional needs because they can synthesize nearly all the substances they need. Those with fewer enzymes have complex nutritional requirements because they lack the ability to synthesize many of the substances they need for growth. Thus, nutritional complexity reflects a deficiency in biosynthetic enzymes.

Locations of Enzymes

Most microorganisms move a variety of small molecules across their cell or plasma membranes and metabolize them. These substances include glucose, amino acids, small peptides, nucleosides, and phosphates as well as various inorganic ions. In addition to the endoenzymes that are produced for use within the cell (Chapter 5, p. 120), many bacteria (and fungi) produce *exoenzymes* and release them through the cell or plasma membrane. These enzymes include **extracellular enzymes**, usually produced by Gram-positive rods, which act in the medium around the organism, and **periplasmic enzymes**, usually produced by Gram-negative organisms, which act in the periplasmic space. Most exoenzymes are hydrolases; they add water as they split large molecules of carbohydrate, lipid, or protein into smaller ones that can be absorbed **(Table 6.2)**. Although microbes cannot move large molecules across membranes, in nature they use large molecules from other organisms by digesting those molecules with exoenzymes before absorbing them.

TABLE 6.2

Examples of Exoenzymes	
Enzyme	**Action**
Enzymes That Act on Complex Carbohydrates	
Carbohydrases	Break down large carbohydrate molecules into smaller ones
Amylase	Breaks down starch to maltose
Cellulase	Breaks down cellulose to cellobiose
Enzymes That Act on Sugars	
Sucrase	Breaks down sucrose to glucose and fructose
Lactase	Breaks down lactose to glucose and galactose
Maltase	Breaks down maltose to two glucose molecules
Enzymes That Act on Lipids	
Lipase	Breaks down fats to glycerol and fatty acids
Enzymes That Act on Proteins	
Proteases	Break down proteins to peptides and amino acids
Caseinase	Breaks down milk protein to amino acids and peptides
Gelatinase	Breaks down gelatin to amino acids and peptides

Figure 6.17
The vegetative and sporulation cycles in bacteria capable of sporulation.

Adaptation to Limited Nutrients

Microorganisms adapt to limited nutrients in several ways:

1. Some synthesize increased amounts of enzymes for uptake and metabolism of limited nutrients. This allows the organisms to obtain and use a larger proportion of the few nutrient molecules that are available.

2. Others have the ability to synthesize enzymes needed to use a different nutrient. For example, if glucose is in short supply, some microorganisms can make enzymes to take up and use a more plentiful nutrient such as lactose.

3. Many organisms adjust the rate at which they metabolize nutrients and the rate at which they synthesize molecules required for growth to fit the availability of the least plentiful nutrient. Both metabolism and growth are slowed, but no energy is wasted on synthesizing products that cannot be used. Growth is as rapid as conditions will allow.

✓ CHECKLIST

1. What does the ending *-phile* mean? Distinguish between the terms *obligate* and *facultative*.

2. What enzymes do most obligate anaerobes lack? How does this cause them to die in the presence of oxygen?

3. Do fastidious microorganisms have greater or lesser numbers of enzyme types than microbes with simpler nutritional needs? Why?

▌▌▌ SPORULATION

Sporulation, the formation of endospores, occurs in *Bacillus*, *Clostridium*, and a few other Gram-positive genera but has been studied most carefully in *B. subtilis* and *B. megaterium*. Do not confuse bacterial endospores, a single one of which forms inside the bacterial cell, with fungal spores. Fungal spores are produced in great numbers, and are form of reproduction (Chapter 11, p. 320). Bacteria that form endospores generally do so during the stationary phase in response to environmental, metabolic, and cell cycle signals.

When nutrients such as carbon or nitrogen become limiting, highly resistant endospores form inside mother cells. (With very low frequency, some bacteria form endospores even when nutrients are available.) Although endospores are not metabolically active, they can survive long periods of drought and are resistant to killing by extreme temperatures, radiation, and some toxic chemicals. Some endospores can withstand much higher temperatures than vegetative cells can. The endospore itself cannot divide, and the parent cell can produce only one endospore, so sporulation is a protective or survival mechanism, not a means of reproduction.

As endospore formation begins, DNA is replicated and forms a long, compact, *axial nucleoid* **(Figure 6.17)**. The two chromosomes formed by replication separate and move to different locations in the cell. In some bacteria the endospore forms near the middle of the cell, and in others it forms at one end **(Figure 6.18)**. The DNA where the endospore will form directs endospore formation. Most of the cell's RNA and some cytoplasmic protein molecules gather around the DNA to make the **core**, or living part, of the endospore. The core contains **dipicolinic** (di-pik-o-lin'ik) **acid** and calcium ions, which probably contribute to an endospore's heat resistance by stabilizing protein structure. An **endospore septum**, consisting of a cell membrane but lacking a cell wall, grows around the core, enclosing it in a double thickness of cell membrane (Figure 6.17). Both layers of this membrane synthesize peptidoglycan and

(a) [LM]

(b) [TEM]

Figure 6.18 **Bacterial endospores in two *Clostridium* species.** **(a)** Cells with centrally located endospores (2,121X). *(Soad Tabaqchali/ Visuals Unlimited)* **(b)** Cells with terminally located endospores, which give the organisms a club-shaped appearance (17,976X). *(Alfred Pasieka/Photo Researchers, Inc.)*

release it into the space between the membranes. Thus, a laminated layer called the **cortex** is formed. The cortex protects the core against changes in osmotic pressure, such as those that result from drying. A **spore coat** of keratin-like protein, which is impervious to many chemicals, is laid down around the cortex by the mother cell. Finally, in some endospores an **exosporium**, a lipid-protein membrane, is formed outside the coat by the mother cell. The function of the exosporium is unknown. Under laboratory conditions, sporulation takes about 7 hours.

Once favorable conditions return, an endospore develops into a vegetative cell, which lacks the endospore's resistant properties. **Germination**, in which a spore returns to its vegetative state, occurs in three stages. The first stage, *activation*, usually requires some traumatic agent such as low pH or heat, which damages the coat. Without such damage, some endospores germinate slowly, if at all. The second state, *germination proper*, requires water and a germination agent (such as the amino acid alanine or certain inorganic ions) that penetrates the damaged coat. During this process, much of the cortical peptidoglycan is broken down, and its fragments are released into the medium. The living cell (which occupied the core) now takes in large quantities of water and loses its resistance to heat and staining, as well as its *refractility* (ability to bend light rays). Finally, *outgrowth* occurs in a medium with adequate nutrients. Proteins and RNA are synthesized, and in about an hour, DNA synthesis begins. The cell is now a vegetative cell and undergoes binary fission.

Endospores trapped in amber for 25 million years germinate when placed in nutrient media.

Thus, bacterial cells capable of sporulation display two cycles—the *vegetative cycle* and the *sporulation*

cycle (Figure 6.17). The vegetative cycle is repeated at intervals of 20 minutes or more, and the sporulation cycle is initiated periodically. Endospores known to be 300 or more years old have been observed to undergo germination when placed in a favorable medium. They are a very good form of insurance against extinction.

An unusual relative of *Epulopiscium fishelsoni* is *Metabacterium polyspora*, a bacterium found in the digestive tract of guinea pigs. Both species are relatives of the spore-forming *Clostridium*. *M. polyspora* produces multiple endospores at each end of its cell, by divisions of the endospores early in their development. This is an exception to the rule that endospore formation is not reproductive in bacteria.

OTHER SPORELIKE BACTERIAL STRUCTURES

Certain bacteria, such as *Azotobacter*, form resistant **cysts**, or spherical, thick-walled cells, that resemble endospores. Like endospores, cysts are metabolically inactive and resist drying. Unlike endospores, they lack dipicolinic acid and have only limited resistance to high temperatures. Cysts germinate into single cells and therefore are not a means of reproduction.

Some filamentous bacteria, such as *Micromonospora* and *Streptomyces*, form asexually reproduced **conidia** (ko-nid′e-ah), or chains of aerial spores with thick outer walls. These spores are temporarily dormant but are not especially resistant to heat or drying. When the spores, which are produced in large numbers, are dispersed to a suitable environment, they form new filaments. Unlike endospores, these spores do contribute to reproduction of the species.

CULTURING BACTERIA

Culturing of bacteria in the laboratory presents two problems. First, a pure culture of a single species is needed to study an organism's characteristics. Second, a medium must be found that will support the growth of the desired organism. Let's look at some of the ways in which these problems are solved.

METHODS OF OBTAINING PURE CULTURES

To study bacteria in the laboratory, it is important to obtain a **pure culture**, a culture that contains only a single species of organism. Prior to the development of pure culture techniques, scientists studied *mixed cultures*, or cultures containing several different kinds of organisms. Researchers could make observations of different shapes and sizes of organisms, but they could find out little about the nutritional needs or growth characteristics of individual species. Today, pure cultures are obtained by isolating the progeny of a single cell.

Simple as it seems now, the technique of isolating pure cultures was difficult to develop. Attempts to isolate single cells by serial dilution were often unsuccessful because two or more organisms of different species were often present in the highest dilutions. Koch's technique of spreading bacteria thinly over a solid surface was more effective because it deposited a single bacterium at some sites. However, he tried several different solid substances. Using the discovery of Angelina Hesse, the wife of an associate, he settled on agar as the ideal solidifying agent. Only a very few organisms digest it, and in 1.5% solution it does not melt below 95°C. Furthermore, after being melted, agar remains in the liquid state until it has cooled to about 40°C, a temperature cool enough to allow the addition of nutrients and living organisms that might be destroyed by heat.

The Streak Plate Method

Today, the accepted way to prepare pure cultures is the **streak plate method**, which uses agar plates. Bacteria are picked up on a sterile wire loop, and the wire is moved lightly along the agar surface, depositing streaks of bacteria on the surface. The inoculating loop is flamed, and a few bacteria are picked up from the region already deposited and streaked onto a new region, as shown in **Figure 6.19**. Fewer and fewer bacteria are deposited as the streaking continues, and the loop is flamed after each streaking. Individual organisms are deposited in the region streaked last. After the plate is incubated at a suitable growth temperature for the organism, small colonies—each derived from a single bacterial cell—appear. The wire loop is used to pick up a portion of an isolated colony and transfer it to any appropriate sterile medium for further study. The use of sterile (aseptic) technique ensures that the new medium will contain organisms of only a single species.

(a)

(b)

Figure 6.19   The streak plate method of obtaining pure cultures.   (a) A drop of culture on a wire inoculating loop is lightly streaked across the top of the agar in region 1. The loop is flamed, the plate is rotated, and a few organisms are picked up from region 1 and streaked out into region 2. The loop is flamed again, and the process is repeated in region 3. The plate is then incubated. **(b)** A streak plate of *Serratia marcescens* after incubation. Note the greatly reduced numbers of colonies in each successive region. *(Christine Case/Visuals Unlimited)*

The Pour Plate Method

Another way to obtain pure cultures, the **pour plate method**, makes use of serial dilutions (Figure 6.7a). A series of dilutions are made such that the final dilution contains about 1,000 organisms. Then 1 ml of liquid medium from the final dilution is placed in 9 ml of melted agar medium (45°C), and the medium is quickly poured into a sterile plate. The resulting pour plate will contain a small number of bacteria, some of which will form isolated colonies on the agar. Because this method embeds some organisms in the medium, it is particularly useful for growing microaerophiles that cannot tolerate exposure to oxygen in the air at the surface of the medium.

CULTURE MEDIA

In nature, many species of bacteria and other microorganisms are found growing together in oceans, lakes,

TABLE 6.3

A Defined Synthetic Medium for Growing *Proteus vulgaris*			
Ingredient	**Amount**	**Ingredient**	**Amount**
Water	1 liter	K_2HPO_4	1 g
$MgSO_4 \cdot 7H_2O$	200 mg	$FeSO_4 \cdot 7H_2O$	10 mg
$CaCl_2$	10 mg	Glucose	5 g
NH_4Cl	1 g	Nicotinic acid	0.1 mg
Trace elements (Mn, Mo, Cu, Co, Zn as inorganic salts, known quantities of 0.02–0.5 mg each)			

Source: Adapted from R. Y. Stanier et al. *The Microbial World.* 5th ed. Upper Saddle River, NJ: Prentice-Hall, 1986.

and soil and on living or dead organic matter. These materials might be thought of as *natural media.* Although soil and water samples are often brought into the laboratory, organisms from them are typically isolated and pure cultures are prepared for study.

Growing bacteria in the laboratory requires knowledge of their nutritional needs and the ability to provide the needed substances in a medium. Through years of experience in culturing bacteria in the laboratory, microbiologists have learned what nutrients must be supplied to each of many different organisms. Certain organisms, such as those that cause syphilis and leprosy, still cannot be cultured in laboratory media. They must be grown in cultures that contain living human or other animal cells. Many other organisms whose nutritional needs are reasonably well known can be grown in one or more types of media.

Types of Media

Laboratory media are generally synthetic media, as opposed to the natural media mentioned previously. A **synthetic medium** is a medium prepared in the laboratory from materials of precise or reasonably well-defined composition. A **defined synthetic medium** is one that contains known specific kinds and amounts of chemical substances. Examples of defined synthetic media are given in Tables 6.3 and 6.4. A **complex medium**, or **chemically nondefined medium**, is one that contains reasonably familiar materials but varies slightly in chemical composition from batch to batch. Such media contain blood or extracts from beef, yeasts, soybeans, and other organisms. A common ingredient is **peptone**, a product of enzyme digestion of proteins. It provides small peptides that microorganisms can use. Although the exact concentrations are not known, trace elements and vitamins are present in sufficient quantities in complex media to support the growth of many organisms. Both liquid nutrient broth and solidified agar medium used to culture many organisms are complex media. An example of a complex medium is given in Table 6.5.

Commonly Used Media

Most routine laboratory cultures use media containing peptone from meat or fish in nutrient broth or solid agar

TABLE 6.4

A Defined Synthetic Medium for Growing the Fastidious Bacterium *Leuconostoc mesenteroides*			
Ingredient	**Amount**	**Ingredient**	**Amount**
Water	1 liter		
Energy Source			
Glucose	25 g		
Nitrogen Source			
NH_4Cl	3 g		
Minerals			
KH_2PO_4	600 mg	$FeSO_4 \cdot 7H_2O$	10 mg
K_2HPO_4	600 mg	$MnSO_4 \cdot 4H_2O$	20 mg
$MgSO_4 \cdot 7H_2O$	200 mg	NaCl	10 mg
Organic Acid			
Sodium acetate	20 g		
Amino Acids			
DL-α-Alanine	200 mg	L-Lysine · HCl	250 mg
L-Arginine	242 mg	DL-Methionine	100 mg
L-Asparagine	400 mg	DL-Phenylalanine	100 mg
L-Aspartic acid	100 mg	L-Proline	100 mg
L-Cysteine	50 mg	DL-Serine	50 mg
L-Glutamic acid	300 mg	DL-Threonine	200 mg
Glycine	100 mg	DL-Tryptophan	40 mg
L-Histidine · HCl	62 mg	L-Tyrosine	100 mg
DL-Isoleucine	250 mg	DL-Valine	250 mg
DL-Leucine	250 mg		
Purines and Pyrimidines			
Adenine sulfate · H_2O	10 mg	Uracil	10 mg
Guanine · HCl · $2H_2O$	10 mg	Xanthine · HCl	10 mg
Vitamins			
Thiamine · HCl	0.5 mg	Riboflavin	0.5 mg
Pyridoxine · HCl	1.0 mg	Nicotinic acid	1.0 mg
Pyridoxamine · HCl	0.3 mg	p-Aminobenzoic acid	0.1 mg
Pyridoxal · HCl	0.3 mg	Biotin	0.001 mg
Calcium pantothenate	0.5 mg	Folic acid	0.01 mg

Source: H. E. Sauberlich and C. A. Baumann. "A factor required for the growth of *Leuconostoc citrovorum.*" *J. Biol. Chem.* 176(1948):166.

TABLE 6.5

A Complex Medium Suitable for Many Heterotrophic Organisms	
Nutrient Broth Ingredient	**Amount**
Water	1 liter
Peptone	5 g
Beef extract	3 g
NaCl	8 g
Solidified Medium	
Agar	15 g
Above ingredients in amounts specified	

medium. Such media are sometimes enriched with **yeast extract**, which contains a number of vitamins, coenzymes, and nucleosides. **Casein hydrolysate**, made from milk protein, contains many amino acids and is used to enrich certain media. Because blood contains many nutrients needed by fastidious pathogens, **serum** (the liquid part of the blood after clotting factors have been removed), whole blood, and heated whole blood can be useful in enriching media. **Blood agar** is useful in identifying organisms that can cause hemolysis, or breakdown of red blood cells. Sheep's blood is used because its hemolysis is more clearly defined than when human blood is used in the agar medium.

'Chocolate agar' sounds yummy, but contains no actual chocolate. The color comes from cooked blood! It is used to culture fastidious organisms.

Selective, Differential, and Enrichment Media

To isolate and identify particular microorganisms, especially those from patients with infectious diseases, *selective*, *differential*, or *enrichment media* are often used. Such special media are an essential part of modern diagnostic microbiology. **Table 6.6** shows some examples of special diagnostic media. Additional photos and descriptions of other examples of media can be found on the website for this chapter.

A **selective medium** is one that encourages the growth of some organisms but suppresses the growth of others. For example, to identify *Clostridium botulinum* in food samples suspected of being agents of food poisoning, the antibiotics sulfadiazine and polymyxin sulfate (SPS) are added to anaerobic cultures of *Clostridium* species. This culture medium is called *SPS agar*. It allows growth of *Clostridium botulinum* while inhibiting growth of most other *Clostridium* species.

A **differential medium** has a constituent that causes an observable change (a color change or a change in pH) in the medium when a particular biochemical reaction occurs. This change allows microbiologists to distinguish a certain type of colony from others growing on the same plate **(Figure 6.20)**. SPS agar also serves as a differential medium. Colonies of *Clostridium botulinum* formed on this medium are black because of hydrogen sulfide made by the organisms from the sulfur-containing additives.

Many media, such as SPS agar and MacConkey agar, are both selective and differential. *MacConkey agar* contains crystal violet and bile salts, which inhibit growth of Gram-positive bacteria while allowing growth of Gram-negative bacteria. MacConkey agar also contains the sugar lactose plus a pH indicator that turns colonies of lactose fermenters red and leaves colonies of nonfermenters colorless and translucent. Although there are some exceptions, most organisms that are normally found in the human intestines ferment lactose, whereas most pathogens (disease-causing microorganisms) do not.

An **enrichment medium** contains special nutrients that allow growth of a particular organism that might not otherwise be present in sufficient numbers to allow it to be isolated and identified. Unlike a selective medium, an enrichment medium does not suppress others. For example, because *Salmonella typhi* organisms may not be sufficiently numerous in a fecal sample to allow positive identification, they are cultured on a medium containing the trace element selenium, which supports growth of the organism. After incubation in the enrichment medium, the greater numbers of the organisms increase the likelihood of a positive identification.

Controlling Oxygen Content of Media

Obligate aerobes, microaerophiles, and obligate anaerobes require special attention to maintain oxygen concentrations suitable for growth. Most obligate aerobes obtain sufficient oxygen from nutrient broth or on the surface of solidified agar medium, but some need more. Oxygen gas is bubbled through the medium or into the incubation environment with filters between the gas source and the medium to prevent contamination of the

Figure 6.20 Differential media. (a) Identification of urinary tract bacterial pathogens is made easy with this special agar produced by CHROMagarE. *(Courtesy CHROMAGAR/DRG International, Inc)* **(b)** Three species of the fungal genus *Candida* can be differentiated in mixed culture when grown on CHROMagar Candida plates. *(Courtesy CHROMAGAR/DRG International, Inc)*

TABLE 6.6

Selected Examples of Diagnostic Media

Medium		Organism(s) Identified	Selectivity and/or Differentiation Achieved
Brilliant green agar	a	*Salmonella*	**Selective** Brilliant green dye inhibits Gram-positive bacteria and thus selects Gram-negative ones. **Differential** Differentiates *Shigella* colonies (which do not ferment lactose or sucrose and are red to white) from other organisms that do ferment one of those sugars and are yellow to green.
Eosin methylene blue agar (EMB)	b	Gram-negative enterics (Enterobacteriaceae)	**Selective** Medium partially inhibits Gram-positive bacteria. **Differential** Eosin and methylene blue differentiate among organisms: *Escherichia coli* colonies are purple and typically have a metallic green sheen; *Enterobacter aerogenes* colonies are pink, indicating that they ferment lactose; and colonies of other organisms are colorless, indicating they do not ferment lactose.
MacConkey agar	c	Gram-negative enterics	**Selective** Crystal violet and bile salts inhibit Gram-positive bacteria. **Differential** Lactose and the pH indicator neutral red (red when acidic) identify lactose fermenters as red colonies and nonfermenters as light pink. Most intestinal pathogens are nonfermenters and hence do not produce acid.
Triple sugar-iron agar (TSI)	① d ② e ③ f ④ g	Gram-negative enterics	**Not Selective** **Differential** Used in agar slants (tubes cooled in slanted position), where differentiation is based on both aerobic surface growth (slant) and anaerobic growth in agar in base of tube (butt). Medium contains specific amounts of glucose, sucrose, and lactose, sulfur-containing amino acids, iron, and a pH indicator, so relative use of each sugar and H_2S formation can be detected. 1. Uninoculated tube of TSI. 2. Inoculated: red slant and red butt = no change; no sugar fermented. 3. Yellow slant and yellow butt = lactose and glucose fermented to acid; trapped bubbles in butt indicate fermentation to acid and gas. 4. Red slant (lactose not fermented) and yellow butt (glucose fermented to acid); black precipitate = H_2S produced; sometimes obscures yellow butt. Almost all enteric pathogens produce red slant and yellow butt, with or without H_2S and/or gas.

Differentiation of Intestinal Bacilli Based on TSI

red slant red butt	yellow slant yellow butt no H_2S	yellow slant yellow butt H_2S produced	red slant yellow butt no H_2S	red slant yellow butt H_2S produced
Pseudomonas *Acinetobacter* *Alcaligenes*	*Escherichia* *Enterobacter* *Klebsiella*	*Citrobacter* *Arizona* some *Proteus* sp.	*Shigella* some *Proteus* sp.	most *Salmonella* *Citrobacter* *Arizona*

(a: Courtesy George A. Wistreich, East Los Angeles College; b: Runk Schoenberger/Grant Heilman Photography; c: Cytographics, Inc./ Visuals Unlimited; d: LeBeau/Custom Medical Stock Photo; e: (LeBeau/Custom Medical Stock Photo); f: Raymond B. Otero/Visuals Unlimited; g: SUPERSTOCK)

APPLICATIONS

Don't Leave Home Without Your CO₂!

Transporting specimens from patients to laboratories sometimes presents problems. The organisms must not be subjected to drying conditions or to too much or too little oxygen. Of course, specimen handlers must be protected from infection. Cultures that may contain *Neisseria gonorrhoeae* from patients with gonorrhea pose one such problem—providing an atmosphere relatively high in carbon dioxide. Commercial systems, such as the JEMBEC (John E. Martin Biological Environment Chamber), are available. This system consists of a small plastic plate of selective medium and a tablet of sodium bicarbonate and citric acid. The plate is inoculated, and the tablet is placed in it. The plate is then placed in a plastic bag and sealed. Moisture from the medium causes the tablet to release carbon dioxide; an appropriate concentration of 5% to 10% is obtained. The culture is incubated to allow growth to begin before shipment to the laboratory.

culture. Microaerophiles can be incubated in tubes of a nutrient medium or agar plates in a jar in which a candle is lit before the jar is sealed (**Figure 6.21**). (Scented candles should not be used because oils from them inhibit bacterial growth.) The burning candle uses the oxygen in the jar and adds carbon dioxide to it. When the carbon dioxide extinguishes the flame, conditions are optimum for the growth of microorganisms that require small amounts of carbon dioxide, such as the bacterium *Neisseria gonorrhoeae*, which causes gonorrhea.

To culture obligate anaerobes, all molecular oxygen must be removed and kept out of the medium. Addition

of oxygen-binding agents such as thioglycollate, the amino acid cysteine, or sodium sulfide to the medium prevents oxygen from exerting toxic effects on anaerobes. Media can be dispensed in sealed, screw-cap tubes, completely filled to exclude air, or in Petri plates. When culture must be grown in plates so that colonial growth can be studied, special jars are used that can hold both plates and tubes. Agar plates are incubated in sealed jars containing chemical substances that remove oxygen from the air and generate carbon dioxide (**Figure 6.22**). *Stab cultures* can be made by stabbing a straight inoculating wire coated with organisms into a tube of agar-solidified medium. In laboratories where anaerobes are regularly handled, an *anaerobic transfer chamber* is often used (**Figure 6.23**). Equipment and cultures are introduced through an air lock, and the technician uses glove ports to manipulate the cultures.

Maintaining Cultures

Once an organism has been isolated, it can be maintained indefinitely in a pure culture called a **stock culture**. When needed for study, a sample from a stock culture is inoculated into fresh medium. The stock culture itself is never used for laboratory studies. However, organisms in stock cultures go through growth phases, deplete nutrients, and accumulate wastes just as those in any culture do. As the culture ages, the organisms may acquire odd shapes or other altered characteristics. Stock cultures are maintained by making subcultures in fresh medium at frequent intervals to keep the organisms growing.

The use of careful aseptic techniques is important in all manipulations of cultures. **Aseptic techniques** minimize the chances that cultures will be contaminated by

Figure 6.21 Candle jar culture of anaerobes and microaerophiles. Microaerophiles are growing in culture tubes and on Petri plates in a sealed jar in which a candle burned until it was extinguished by carbon dioxide accumulation in the atmosphere of the jar. A small amount of oxygen remains. Although this method is not used much anymore, years ago it was the main means of growing anaerobes.

Plugs

Broth cultures

Candle

Agar media in inverted Petri plates

Figure 6.22 CO₂ incubator. When activated, these chemicals remove oxygen and are enclosed with cultures in a sealed jar to create an anaerobic chamber. These are useful for the small laboratory that has only a few plates needing anaerobic incubation. *(Jack M. Bostrack/Visuals Unlimited)*

Figure 6.23 Anaerobic transfer. A large transfer chamber with an air lock for introducing equipment and cultures and with ports to allow manipulation of the cultures. *(Courtesy Shellab)*

organisms from the environment or that organisms, especially pathogens, will escape into the environment. Such techniques are especially important in making subcultures from stock cultures. Otherwise an undesirable organism might be introduced, and the stock organism would have to be reisolated. Even with regular transfers of organisms from stock cultures to fresh media, organisms can undergo mutations (changes in DNA) and develop altered characteristics.

Preserved Cultures

To avoid the risk of contamination and to reduce the mutation rate, stock culture organisms also should be kept in a **preserved culture**, a culture in which organisms are maintained in a dormant state. The most commonly used technique for preserving cultures is *lyophilization* (freeze-drying), in which cells are quickly frozen, dehydrated while frozen, and sealed in vials under vacuum (Chapter 12). Such cultures can be kept indefinitely at room temperature.

Because microorganisms frequently undergo genetic changes, reference cultures are maintained. A **reference culture** is a preserved culture that maintains the organisms with the characteristics as originally defined. Reference cultures of all known species and strains of bacteria and many other microorganisms are maintained in the American Type Culture Collection, and many also are maintained in universities and research centers. Then if stock cultures in a particular laboratory undergo change or if other laboratories wish to obtain certain organisms for study, reference cultures are always available.

✔**CHECKLIST**

1. Why are endospores not considered to be a form of reproduction? Why would waiting until unfavorable conditions occurred before beginning spore formulation place an organism at greater risk than it would be if it continuously formed spores?

2. Distinguish between the various kinds of media: synthetic, complex, defined, selective, differential, enrichment, and transport. Can a given medium be more than one of these types?

3. What is the purpose of a stock culture? Why is it never used itself for laboratory tests?

METHODS OF PERFORMING MULTIPLE DIAGNOSTIC TESTS

Many diagnostic laboratories use culture systems that contain a large number of differential and selective media, such as the Enterotube Multitest System® or the Analytical Profile Index (API). These systems allow simultaneous determination of an organism's reaction to a variety of carefully chosen diagnostic media from a single inoculation. The advantages of these systems are that they use small quantities of media, occupy little space in an incubator, and provide an efficient and reliable means of making positive identification of infectious organisms.

The Enterotube System® is used to identify enteric pathogens, or organisms that cause intestinal diseases such as typhoid and paratyphoid fevers, shigellosis, gastroenteritis, and some kinds of food poisoning. The causative organisms are all Gram-negative rods indistinguishable from one another without biochemical tests. The Enterotube System® consists of a tube with compartments, each of which contains one or more different media, and a sterile inoculating rod (**Figure 6.24a**). Each compartment is inoculated when the tip of the rod is touched to a colony and the rod is drawn through the tube. After the tube has been incubated for 24 hours at 37°C, the results of 15 biochemical tests can be obtained by observing (1) whether gas was produced and (2) the color of the medium in each compartment. Tests are grouped in sets of three; within each group, tests are assigned a number 1, 2, or 4 (**Figure 6.24b**). The sum of the numbers of positive tests in each group indicates which tests are positive. The sum 3 means that tests 1 and 2 were positive; the sum 5 means that tests 1 and 4 were positive; the sum 6, that tests 2 and 4 were positive; and the sum 7, that all tests were positive. The single-digit sums for each of the five sets of tests are combined to form a five-digit identification number for a particular organism. For example, 36601 is *Escherchia coli*, and 34363 is *Klebsiella pneumoniae*. A list of identification numbers and the corresponding organisms is provided with the system.

(a)

BBL® Enterotube™ II
4343128

(b)

 Figure 6.24   The enterotube multitest system®.   (a) Removal of a sterile end cap allows you to "pick" a colony from the surface of a plate. The wire is then drawn through all compartments, thereby inoculating each of them. *(Courtesy BD DiagnosticSystems)* **(b)** After inoculation and incubation, compartments with positive test results are assigned a number. The numbers are summed within zones to get a definitive index number that identifies an organism on the list in the coding manual. Any necessary confirmatory tests are also noted there. By numbering each test in a zone with a digit equal to a power of 2 (1, 2, 4, 8, and so on), the sum of any set of positive reactions results in a unique number. A given species may, however, be coded for by many different numbers, as individual strains of that species vary somewhat in their characteristics. *(Courtesy Becton Dickinson Microbiology Systems)*

The API consists of a plastic tray with 20 microtubes called *cupules*, each containing a different kind of dehydrated medium **(Figure 6.25)**. Each cupule medium is rehydrated and inoculated with a suspension of bacteria from an isolated colony. As with Enterotubes, the tray is incubated, test results are determined, and the values 1, 2, and 4 are summed for sets of three tests. The seven-digit profile number identifies the organism.

In this brief discussion of diagnostic systems, we have considered only the tip of the iceberg. Of the many other available tests, a large number are based on immunological properties of organisms. We will consider some of them when we discuss immunology or particular infectious agents. Also, much is known about which organisms are likely to infect certain human organs and tissues, and many diagnostic tests are designed to distinguish among organisms found in respiratory secretions, fecal samples, blood, other tissues, and body fluids.

LIVING, BUT NONCULTURABLE, ORGANISMS

We have spent an entire chapter discussing how to culture microbes. Perhaps it will surprise you to learn that *most* microbes cannot be cultured in the laboratory, and have never even been identified. We can see them under the microscope and can retrieve their DNA, but we can't grow them, or understand their activities and place in our environment. In the next two chapters (on genetics), we will learn about identifying microbes from samples of their DNA. The hospital and environmental labs of the future will not be filled with the great number of racks of cultures that we see in today's labs. No longer will we wait days or even weeks for cultures to grow. We'll just see whose DNA is present—a matter of a few minutes or hours. But someone is going to have to figure out how to grow those nonculturable organisms so that we can study them, not just identify them. Could that someone be you?

Take another look

Klebsiella pneumoniae pneumoniae ATCC 35657

ONPG	ADH	LDC	ODC	CIT	H2S	URE	TDA	IND	VP		GEL	GLU	MAN	INO	SOR	RHA	SAC	MEL	AMY	ARA
+	−	+	−	+	−	+*	−	−	+*		−	+	+	+	+	+	+	+	+	+

Enterobacter cloacae ATCC 13407

ONPG	ADH	LDC	ODC	CIT	H2S	URE	TDA	IND	VP		GEL	GLU	MAN	INO	SOR	RHA	SAC	MEL	AMY	ARA
+	+	−	+	+	−	−	−	−	+		−	+	+	−	+	+	+	+	+	+

Proteus mirabilis ATCC 35659

ONPG	ADH	LDC	ODC	CIT	H2S	URE	TDA	IND	VP		GEL	GLU	MAN	INO	SOR	RHA	SAC	MEL	AMY	ARA
−	−	−	+	±*	+	+	+	−	−		−*	+	−	−	−	−	−•	−	−	−

Escherichia coli ATCC 25922

ONPG	ADH	LDC	ODC	CIT	H2S	URE	TDA	IND	VP		GEL	GLU	MAN	INO	SOR	RHA	SAC	MEL	AMY	ARA
+	−	+	+	−	−	−	−	+	−		−	+	+	−	+	+	−	+	−	+

Figure 6.25 The analytical profile index (API) 20E system. Various species of Enterobacteriaceae are shown here with the differences in reactions that enable them to be identified. This system allows identification to species level of 125 Gram-negative intestinal bacilli. *(Courtesy API/CounterPart Diagnostics/bioMerieux Vitek, Inc.)*

▌ RETRACING OUR STEPS

▐▐▐ GROWTH AND CELL DIVISION

MICROBIAL GROWTH DEFINED

• Microbial growth can be defined as the orderly increase in quantity of all cell components and in the number of cells of an organism.

• Because of limited increase in cell size and the frequency of cell division, growth in microorganisms is measured by increase in cell number.

CELL DIVISION

• Most cell divisions in bacteria occur by **binary fission**, in which the nuclear body divides and the cell forms a transverse septum that separates the original cell into two cells.

• Yeast cells and some bacteria divide by **budding**, in which a small, new cell develops from the surface of an existing cell.

PHASES OF GROWTH

• In a nutrient-rich **medium** (a mixture of substances on or in which microorganisms grow), bacteria divide rapidly. The length of time required for one division is called the **generation time**. Such growth is said to occur at an **exponential**, or **logarithmic, rate**.

• Bacteria introduced into a fresh, nutrient-rich medium display four major phases of growth: (1) In the **lag phase**, the organisms are metabolically active—growing and synthesizing various substances but not increasing in number. (2) In the **log phase**, organisms divide at an exponential, or logarithmic, rate and with a constant generation time. These properties of growth in the log phase can be used to calculate both the number of generations and the generation time. Cultures can be maintained by the use of a **chemostat**, which allows continuous addition of fresh medium. (3) In the **stationary phase**, the number of new cells produced equals the number of cells dying. The medium contains limited amounts of nutrients and may contain toxic quantities of waste materials. (4) In the **decline phase**, or **death phase**, many cells lose their ability to divide and eventually die. A logarithmic decrease in the number of cells results.

• Growth in colonies parallels that in liquid medium, except that most growth occurs at the edge of the colony, and all phases of growth occur simultaneously somewhere in the colony.

MEASURING BACTERIAL GROWTH

• Growth can be measured by **serial dilution**, in which successive 1:10 dilutions of a liquid culture of bacteria are made and transferred onto an **agar plate**; the colonies that arise are counted. Each colony represents one live cell from the original sample.

• Growth also can be measured by **direct microscopic counts**, the **most probable number (MPN)** technique, **filtration**, observing or measuring **turbidity**, measuring products of metabolism, and obtaining the dry weight of cells.

III FACTORS AFFECTING BACTERIAL GROWTH

PHYSICAL FACTORS

• Acidity and alkalinity of the medium affect growth, and most organisms have an **optimum pH** range of no more than one pH unit.

• Temperature affects bacterial growth. (1) Most bacteria can grow over a 30°C temperature range. (2) Bacteria can be classified according to growth temperature into three categories: **psychrophiles**, which grow at low temperatures (below 25°C); **mesophiles**, which grow best at temperatures between 25°C and 40°C; and **thermophiles**, which grow at high temperatures (above 40°C). (3) The temperature range of an organism is closely related to the temperature at which its enzymes function best.

• The quantity of oxygen in the environment affects the growth of bacteria. (1) **Obligate aerobes** require relatively large amounts of free molecular oxygen to grow. (2) **Obligate anaerobes** are killed by free oxygen and must be grown in the absence of free oxygen. (3) **Facultative anaerobes** can metabolize substances aerobically if oxygen is available or anaerobically if it is absent. (4) **Aerotolerant anaerobes** metabolize substances anaerobically but are not harmed by free oxygen. (5) **Microaerophiles** must have only small amounts of oxygen to grow.

• Actively metabolizing bacteria require some water in their environment.

• Some bacteria, but no other living things, can withstand extreme **hydrostatic pressures** in deep valleys in the ocean.

• Osmotic pressure affects bacterial growth, and water can be drawn into or out of cells according to the relative osmotic pressure created by dissolved substances in the cell and the environment. (1) Active transport minimizes the effects of high osmotic pressure in the environment. (2) Bacteria called **halophiles** require moderate to large amounts of salt and are found in the ocean and in exceptionally salty bodies of water.

NUTRITIONAL FACTORS

• All organisms require a carbon source: (1) Autotrophs use CO_2 as their carbon source and synthesize other substances they need. (2) Heterotrophs require glucose or another organic carbon source from which they obtain energy and intermediates for synthetic processes.

• Microorganisms require an organic or inorganic nitrogen source from which to synthesize proteins and nucleic acids. They also require a source of other elements found within them, including sulfur, phosphorus, potassium, iron, and many **trace elements**.

• Microorganisms that lack the enzymes to synthesize particular **vitamins** must obtain those vitamins from their environment.

• The nutritional requirements of an organism are determined by the kind and number of its enzymes. **Nutritional complexity** reflects a deficiency in biosynthetic enzymes.

• Bioassay techniques use metabolic properties of organisms to determine quantities of vitamins and other compounds in foods and other materials.

• Most microorganisms move substances of low molecular weight across their cell membranes and metabolize them internally. Some bacteria (and fungi) also produce exoenzymes that digest large molecules outside the cell membrane of the organism.

• Microorganisms adjust to limited nutrient supplies by increasing the quantities of enzymes they produce, by making enzymes to metabolize another available nutrient, or by adjusting their metabolic activities to grow at a rate consistent with availability of nutrients.

III SPORULATION

• **Sporulation**, which occurs in *Bacillus*, *Clostridium*, and a few other Gram-positive genera, involves the steps summarized in Figure 6.17.

• Sporulation lets the bacterium withstand long periods of dry conditions and extreme temperatures.

• When more favorable conditions are restored, **germination** occurs—endospores begin to develop into vegetative cells.

III CULTURING BACTERIA

METHODS OF OBTAINING PURE CULTURES

• The **streak plate method** of obtaining a **pure culture** involves spreading bacteria across a sterile, solid surface such as an agar plate so that the progeny of a single cell can be picked up from the surface and transferred to a sterile medium.

• The **pour plate method** of obtaining a pure culture involves serial dilution, transferring to melted agar a specific volume of the dilution, which contains a few organisms, and picking up cells from a colony on the agar.

CULTURE MEDIA

- In nature, microorganisms grow on natural media, or the nutrients available in water, soil, and living or dead organic material.
- In the laboratory, microorganisms are grown in **synthetic media**: (1) **Defined synthetic media** consist of known quantities of specific nutrients. (2) **Complex media** consist of nutrients of reasonably well-known composition that vary in composition from batch to batch.
- Most routine laboratory cultures make use of **peptones**, or digested meat or fish proteins. Other substances such as **yeast extract**, **casein hydrolysate**, **serum**, whole blood, or heated whole blood are sometimes added.
- Diagnostic media are (1) **selective media** if they encourage growth of some organisms and inhibit growth of others, (2) **differential media** if they allow different kinds of colonies on the same plate to be distinguished from one another, or (3) **enrichment media** if they provide a nutrient that fosters growth of a particular organism.

- Cultures are maintained as **stock cultures** for routine work, as **preserved cultures** to prevent risk of contamination or change in characteristics, and as **reference cultures** to preserve specific characteristics of species and strains.

METHODS OF PERFORMING MULTIPLE DIAGNOSTIC TESTS

- These systems allow simultaneous reaction of an organism to a large number of differential and selective media. Analysis of results allows rapid identification of the organism.
- The two most common used systems are the Enterotube Multitest System® and the Analytic Profile Index (API).

III LIVING, BUT NONCULTURABLE ORGANISMS

- Most organisms cannot be cultured. Eventually they may be identified by their DNA.

▌ TERMINOLOGY CHECK

acidophile *(p. 156)*
aerotolerant anaerobe *(p. 159)*
agar *(p. 151)*
alkaliphile *(p. 156)*
aseptic technique *(p. 169)*
barophile *(p. 159)*
binary fission *(p. 148)*
blood agar *(p. 167)*
budding *(p. 148)*
capnophile *(p. 159)*
casein hydrolysate *(p. 167)*
catalase *(p. 159)*
chemically nondefined medium *(p. 166)*
chemostat *(p. 149)*
colony *(p. 150)*
colony-forming unit (CFU) *(p. 151)*
complex medium *(p. 166)*
conidium *(p. 164)*
core *(p. 163)*
cortex *(p. 164)*
countable number *(p. 152)*
cyst *(p. 164)*
daughter cell *(p. 148)*
death phase *(p. 150)*

decline phase *(p. 150)*
defined synthetic medium *(p. 166)*
differential medium *(p. 167)*
dipicolinic acid *(p. 163)*
direct microscopic count *(p. 152)*
endospore septum *(p. 163)*
enrichment medium *(p. 167)*
exosporium *(p. 164)*
exponential rate *(p. 149)*
extracellular enzyme *(p. 162)*
facultative *(p. 157)*
facultative anaerobe *(p. 159)*
facultative psychrophile *(p. 157)*
facultative thermophile *(p. 157)*
fastidious *(p. 161)*
filtration *(p. 154)*
generation time *(p. 149)*
germination *(p. 164)*
halophile *(p. 160)*
hydrostatic pressure *(p. 159)*
lag phase *(p. 149)*
logarithmic rate *(p. 149)*
log phase *(p. 149)*

medium *(p. 149)*
mesophile *(p. 157)*
microaerophile *(p. 159)*
microbial growth *(p. 148)*
most probable number (MPN) *(p. 153)*
mother cell *(p. 148)*
neutrophile *(p. 156)*
nonsynchronous growth *(p. 149)*
nutritional complexity *(p. 162)*
nutritional factor *(p. 156)*
obligate *(p. 157)*
obligate aerobe *(p. 158)*
obligate anaerobe *(p. 158)*
obligate psychrophile *(p. 157)*
obligate thermophile *(p. 157)*
optimum pH *(p. 156)*
peptone *(p. 166)*
periplasmic enzyme *(p. 162)*
physical factor *(p. 156)*
plasmolysis *(p. 159)*
pour plate *(p. 151)*
pour plate method *(p. 165)*
preserved culture *(p. 170)*

psychrophile *(p. 157)*
pure culture *(p. 165)*
reference culture *(p. 170)*
sarcina *(p. 148)*
selective medium *(p. 167)*
serial dilution *(p. 151)*
serum *(p. 167)*
spore coat *(p. 164)*
sporulation *(p. 163)*
spread plate method *(p. 151)*
standard bacterial growth curve *(p. 149)*
stationary phase *(p. 149)*
stock culture *(p. 169)*
streak plate method *(p. 165)*
superoxide *(p. 159)*
superoxide dismutase *(p. 159)*
synchronous growth *(p. 149)*
synthetic medium *(p. 166)*
tetrad *(p. 148)*
thermophile *(p. 157)*
trace element *(p. 162)*
turbidity *(p. 155)*
vitamin *(p. 162)*
yeast extract *(p. 167)*

▌ CLINICAL CASE STUDY

Helicobacter pylori is the bacterium that causes most stomach ulcers. This bacterium has found a way to live in the midst of the extremely low pH of the stomach. Scientists have found that it makes an enzyme called urease that produces ammonia from urea. How could this protect the bacterium from the acid pH of the stomach?

∎ CRITICAL THINKING QUESTIONS

1. Exactly 100 bacteria with a generation time of 30 minutes are introduced into fresh sterile broth at 8:00 A.M. and maintained at an optimum incubation temperature throughout the day. How many bacteria are present at 3:00 P.M.? How many generations will take place by 5:00 P.M.?

2. In the above example, do you think that the number of bacteria could continue to double every 30 minutes indefinitely? Why do you say that?

3. An attempt to transfer bacteria to new media during the death phase of a culture resulted in actual growth of the organisms. What is the most likely explanation for this phenomenon?

∎ SELF-QUIZ

1. The chemical nutrients that are needed for the maintenance and growth of bacteria are collectively termed CHNOPS. True or false?

2. Most cell divisions in bacteria occur by budding, in which a small new cell develops from the surface of an existing cell, whereas yeasts and some bacteria divide by binary fission, in which the nuclear body divides and the cell forms a transverse septum that separates the original cell into two cells. True or false?

3. Microbial growth is measured by what parameter?
(a) Increased cell size
(b) Increased size of cellular components
(c) Increase in total number of cells
(d) b and c
(e) a and b

4. Match the following growth phase terms to their definitions:
___ Decline/death phase
___ Stationary phase
___ Lag phase
___ Chemostat
___ Log phase
___ Medium

(a) A nutrient-rich mix of substances on or in which microorganisms grow
(b) Cells lose their ability to divide and die
(c) The number of new cells produced equals number of cells dying
(d) Organisms divide at an exponential rate with constant generation time
(e) Organisms are metabolically active but not increasing in cell number
(f) Continuous addition of fresh medium which allows cultures to be maintained

5. Which of the following is the best definition of generation time?
(a) The length of time it takes for lag phase to occur
(b) The length of time it takes a population of cells to double
(c) The minimum length of time it takes a cell to divide
(d) The length of time a culture stays in stationary phase
(e) The length of time it takes log phase to occur

6. (a) If a bacterial cell that has a generation time of 30 minutes is placed in a suitable sterile nutrient broth at time 0 and allowed to get to a logarithmic phase, after 3 hours how many cells would there be in the culture?

(b) If you froze that culture for 1 hour at $-80°C$, then took the culture out of the freezer and placed it at room temperature and then measured cell number after 30 minutes, how many cells would be in the culture now?

7. Optimum growth conditions for bacteria are determined by the conditions that optimize bacterial:
(a) Membrane fluidity
(b) Enzymes
(c) DNA structure
(d) RNAse activity
(e) DNAse activity

8. All of the following are ways in which bacterial growth can be measured EXCEPT:
(a) Direct microscope counts
(b) Serial dilution with transference to agar plate
(c) Turbidity measurements
(d) Most probable number (MPN) technique
(e) Automatic pipetting
(f) Filtration

9. Bacteria make exoenzymes in order to:
(a) Make structures within the bacterium
(b) Synthesize large molecules of carbohydrates, proteins, and nucleic acids
(c) Hydrolyze large molecules to smaller ones so that they can be transported into the bacterium
(d) Synthesize molecules needed by fastidious organisms
(e) None of the above, bacteria do not make exoenzymes

10. Why do foods containing a high concentration of salt or sugar usually not require refrigeration to prevent their spoilage?

11. Some bacteria have complex nutritional requirements because they:
(a) Are composed of a large number of different types of molecules
(b) Can make a great many of the molecules found in the cell from simple precursors
(c) Have many different enzymes and therefore can make many molecules
(d) Contain unique molecules not normally found in bacterial cells
(e) Lack many enzymes and must therefore be provided with many of the molecules they need for growth

12. The loss of water and shrinkage of a cell in a hyperosmotic environment is known as _____ . The enzyme _____ is important to many bacteria because it

_____ down toxic hydrogen peroxide. Some bacteria are barophiles and can only tolerate living under deep water because the water exerts extreme _____ pressure.

13. Bacteria that customarily grow near the surface of bodies of water are usually:
    (a) Facultative anaerobes   (d) Obligate aerobes
    (b) Obligate anaerobes   (e) Capnophiles
    (c) Microaerophiles

14. Match the following microbial oxygen growth requirements with their descriptions:
    ___ Aerotolerant anaerobe   (a) Killed by oxygen
    ___ Obligate aerobe   (b) Must have abundant
    ___ Capnophile   oxygen
    ___ Microaerophile   (c) Likes carbon dioxide
    ___ Facultative anaerobe   (d) Needs a small amount of
    ___ Obligate anaerobe   oxygen
                                 (e) Grows with or without
                                     oxygen

15. Bacteria that require moderate to large amounts of salt for their survival are known as:
    (a) Capnophiles   (d) Acidophiles
    (b) Barophiles   (e) Halophiles
    (c) Mesophiles

16. Which of the following statements about endospores is true?
    (a) Endospore formation in some bacteria occurs because of environmental stressors such as a limiting nutrient or extremes in pH.
    (b) Endospore formation in bacteria is a means of reproduction.
    (c) Endospore formation occurs in *Bacillus, Clostridium,* and a few other Gram-positive genera.
    (d) When favorable conditions are restored, endospores undergo germination or development into a vegetative cell.
    (e) a, c, and d.

17. Bacteriological media that are composed of ingredients whose exact chemical composition are known are called:
    (a) Designated   (d) Selective
    (b) Exact   (e) Aesthetic
    (c) Defined

18. Blood agar is often used to observe changes in the appearance of the agar around the colonies growing on this medium. This medium could then be called:
    (a) Selective
    (b) Designated
    (c) Differential
    (d) Defined
    (e) Exact

19. A bacterial medium that contains 20 grams of beef extract and 10 grams of sodium chloride dissolved in 1 liter of water is a defined medium. True or false?

20. MacConkey agar contains the dye, crystal violet, that inhibits the growth of Gram-positive bacteria and also contains lactose and a pH indicator that allow the detection of lactose-fermenting bacteria. MacConkey agar is classified as:
    (a) Differential, selective
    (b) Complex, selective
    (c) Defined, selective, differential

    (d) Differential
    (e) Regulatory, selective

21. Which of the following bacterial groups would you expect to be MOST likely associated with human infections?
    (a) Thermophiles
    (b) Lactophiles
    (c) Psychrophiles
    (d) Pedophiles
    (e) Mesophiles

22. What are the purposes of carrying out the streak plate and pour plate techniques in microbiology?

23. The bacteria that multiply in improperly treated sealed canned food are most likely to be:
    (a) Aerobes   (d) Anaerobes
    (b) Carnivores   (e) Facultative anaerobes
    (c) Omnivores

24. Many diagnostic laboratories use culture systems that contain a large number of differential and selective media. The advantages of such systems are that they use small quantities of media, occupy little incubator space, and provide efficient and reliable means of making positive identification of infectious organisms. Which of the following is a common culture system in use today?
    (a) Analytical Profile Index (API)
    (b) Kirby-Bauer plate
    (c) Brewer's jar
    (d) Enterotube Multitest System
    (e) a and d

25. Which of the following methods allows determination of the specific number of viable cells in a specimen?
    (a) Turbidity measurement
    (b) Dry weight measurement
    (c) Total plate count
    (d) Petroff-Hausser bacterial counter
    (e) Total nitrogen measurement

26. Identify the position of each of the following on the accompanying graph:
    ___ Organisms divide at their most rapid rate
    ___ New cells are produced at same rate as old cells die
    ___ Lag phase
    ___ Log phase
    ___ Many cells undergo involution and death

▍ EXPLORATIONS ON THE WEB

If you think you've mastered this chapter, there's more to challenge you on the web. Go to the companion web site to fine-tune your understanding of the chapter concepts and discover answers to the questions posed below.

1. Did you know that some bacteria grow in environments hotter than 100°C or that other bacteria populate the Antarctic Ocean?

2. Iron-eating bacteria, responsible for turning water yellow, red, or orange, and leaving a rusty slime deposit in toilet tanks, may have been the first living organisms on this planet.

Microbial Genetics

Come with me...

As I pulled into a visitor's parking place outside TIGR (pronounced like "tiger"). The Institute for Genomic Research, I was excited to be visiting this world-famous facility. Housed in several modern buildings on 17 acres, located in Rockville, Maryland, TIGR looks like a small college campus. I had met Dr. Karen E. Nelson earlier at conferences, and she had graciously agreed to take me on a tour of TIGR and to talk with me about her research and career.

(Courtesy Karen Nelson/TIGR)

Settling into her office, I asked her to first fill me in on the background of TIGR. Genomics is the study of all the genes (DNA) of an organism. The TIGR Institute was founded in 1992 by J. Craig Venter, now famous for his completion of the Human Genome Project.

Since TIGR is a nonprofit organization, I asked where they got the money to run such extensive and expensive research projects. Karen explained that much of their financing comes from grants from foundations and government agencies such as the National Institutes of Health (NIH), Department of Energy, Department of Defense, National Science Foundation (NSF), U.S. Department of Agriculture, Merck Genome Research Institute, Office of Naval Research, Amgen, Burroughs Wellcome Fund, and Chiron Corporation. Being a microbiologist, I was familiar with the great successes that TIGR has had with microbes. What I hadn't realized was how much work they've been doing with other kinds of organisms such as rice, tomatoes, and potatoes. They also worked on chromosome 16 of the Human Genome Project.

In 1995 TIGR scientists led by Dr. Venter and Dr. Hamilton O. Smith achieved a marvelous first: the first ever sequencing of the entire genome of a microbe, namely Haemophilus influenzae, a major cause of childhood ear infections and meningitis. This was quickly followed that same year by a team led by Claire Fraser with the complete sequencing of the DNA of a second microbe, Mycoplasma genitalium, having the smallest microbial genome known, and in 1996 the sequence of Methanococcus jannaschii, an archaeal microbe that lives in deep sea vents at very high temperatures and pressure, metabolizing hydrogen and carbon dioxide to produce methane. Since then, genome after genome has been sequenced at TIGR. Karen sequenced the very first bacterium found to have two chromosomes, rather than one.

 Video related to this topic is available within WileyPLUS.

We have considered many aspects of metabolism and growth, but we have yet to consider the synthesis of nucleic acids and proteins. Synthesis of these complex molecules is the basis of **genetics**, the study of heredity. The genetics of microorganisms is an exciting and active research area, and it is also a rewarding area for microbiologists. Since the inception of the annual Nobel Prize in physiology or medicine in 1900, more than 30 prizes have been awarded in microbiology-related fields, especially microbial genetics. Because of this intensive investigation, much is now known about microbial genetics. We will begin our study of genetics by seeing how bacteria synthesize nucleic acids—DNA and RNA—and how the nucleic acids are involved in the synthesis of proteins. We will also see how *genes* (specific segments of DNA) act, how they are regulated, and how they are altered by mutation. In the next chapter we will discuss the mechanisms by which genetic information is transferred among microorganisms.

AN OVERVIEW OF GENETIC PROCESSES

THE BASIS OF HEREDITY

All information necessary for life is stored in an organism's genetic material, DNA, or, for many viruses, RNA. To explain **heredity**—the transmission of this information from an organism to its progeny (offspring)—we must consider the nature of chromosomes and genes.

A **chromosome** is typically a circular (in prokaryotes) or linear (in eukaryotes), threadlike molecule of DNA. Recall that DNA consists of a double chain of nucleotides with each nucleotide made up of a sugar, a phosphate, and a base (adenine, thymine, guanine, or cytosine). The nucleotides are arranged in a helix, with the nucleotide base pairs held together by hydrogen bonds (**Figure 7.1**; ◀Chapter 2, p. 46). The specific sequence of nucleotides in the DNA can be copied to make another molecule of DNA, or used to make RNA which then does protein synthesis.

A typical prokaryotic cell contains a single circular chromosome, composed primarily of a single DNA molecule about 1 mm long when fully stretched out—some 1,000 times longer than the cell itself. This immense molecule fits compactly into the cell, where it forms the nucleoid (◀Chapter 4, p. 91) by twisting tightly around itself, a process known as *supercoiling*. When a prokaryotic cell reproduces by binary fission, the chromosome reproduces, or *replicates*, itself, and each daughter cell receives one of the chromosomes. This mechanism provides for the orderly transmission of genetic information from parent cell to daughter cells.

(a)

(b)

Figure 7.1 The structure of DNA. **(a)** The two upright strands, composed of the sugar deoxyribose (D) and phosphate groups (P), are held together by hydrogen bonding between complementary bases. Adenine (A) always pairs with thymine (T), and guanine (G) always pairs with cytosine (C). Each strand can thus provide the information needed for the formation of a new DNA molecule. **(b)** The DNA molecule is twisted into a double helix. The two sugar-phosphate strands run in opposite (antiparallel) directions. Each new strand grows from the 5′ end toward the 3′ end.

It seems like there are always exceptions in microbiology. Recall that microbiologists have discovered bacteria so large that they do not require a microscope to be seen, and that bacteria in the genus *Mycoplasma* have no cell wall. So, it is no surprise that scientists have found exceptions for bacterial chromosomes. First, some bacteria were discovered to have a linear rather than circular chromosome. In the last few years, scientists such as Karen E. Nelson (featured in the chapter opening vignette) have discovered that at least a couple dozen bacterial species have two chromosomes (or even 3!) and that sometimes one of these chromosomes is linear! *Vibrio cholerae* has two circular chromosomes, one large and one small. Why isn't the smaller one considered to be just a large plasmid? The answer lies in the definition of a chromosome: To be a chromosome, a DNA molecule must contain genetic information essential for the continuous survival of the organisms. Plasmids contain only genetic information that may be helpful to organisms, but that they could survive without. In contrast, *V. cholerae*'s small chromosome contains "essential" genes. Several important metabolic pathways have the information for some steps controlled by genes on one chromosome, while other steps are controlled by genes on the other chromosome. Cells having just the large chromosome can stay alive for awhile, but cannot reproduce.

In the absence of mitosis, it is not yet understood how daughter cells get one copy of each kind of chromosome. Evidently mistakes happen frequently, as populations of cells with just the large chromosome accumulate in biofilms (thin layers of bacteria that are growing on a surface). It is thought that while they live, they may be pumping out molecules helpful to the nearby two-chromosome cells, which then can grow and reproduce faster. Cells that receive other numbers and combinations of the two chromosomes have yet to be studied—another adventure and more discoveries awaiting microbiologists! Likewise they will be looking closely at *Deinococcus radiodurans*, a bacterium exceedingly resistant to radiation that also has two chromosomes.

A **gene**, the basic unit of heredity, is a linear sequence of nucleotides of DNA that form a functional unit of a chromosome or of a plasmid. All information for the structure and function of an organism is coded in its genes. In many cases, a gene determines a single characteristic. However, the information in a specific gene, found at a particular **locus** (location) on the chromosome or plasmid, is not always the same. Genes with different information at the same locus are called **alleles** (al-eelz′). Because prokaryotes have a single chromosome, they generally have only one version, or allele, of each gene. (In ◄Chapter 8, we will discover exceptions to this rule.) Many (but not all) eukaryotes have two sets of chromosomes and thus two alleles of each gene, which may be the same or different. For example, in human blood types, any one of three genes—A, B, or O—can occupy a certain locus. Allele A causes red blood cells to have a certain glycoprotein, which

we will designate as molecule A, on their surfaces. Allele B causes them to have molecule B, and allele O does not cause them to have any glycoprotein molecule on the cell surfaces. People with type AB blood produce both molecules A and B because they have both alleles A and B.

Heritable variations in the characteristics of progeny can arise from mutations. A **mutation** is a permanent alteration in DNA. Mutations usually change the sequence of nucleotides in DNA and thereby change the information in the DNA. When the mutated DNA is transmitted to a daughter cell, the daughter cell can differ from the parent cell in one or more characteristics. We will see in ◄Chapter 8 that heritable variations in the characteristics of prokaryotic organisms can arise by a variety of mechanisms.

NUCLEIC ACIDS IN INFORMATION STORAGE AND TRANSFER

Information Storage

All the information for the structure and functioning of a cell is stored in DNA. For example, in the chromosome of the bacterium *Escherichia coli*, each of the paired strands of DNA contains about 5 million bases arranged in a particular linear sequence. The information in those bases is divided into units of several hundred bases each. Each of these units is a gene. Some of the genes and their locations on the chromosome of *E. coli* are shown in **Figure 7.2**.

We might think of a gene as a sentence in the language of nucleic acids. Each sentence in this language is constructed from a four-letter alphabet corresponding to the four nitrogenous bases in DNA: adenine (A), thymine (T), cytosine (C), and guanine (G). When these four letters combine to make "sentences" several hundred letters long, the number of possible sentences becomes almost infinite. Likewise, an almost infinite number of possible genes exists. If each gene contained 500 bases, a chromosome containing 5 million bases could contain 10,000 different genes, Thus, the information storage capacity of DNA is exceedingly large!

Haemophilus influenzae is the first microbe to have its genome (1.83 kilobases; 1 kilobase (kb) = 1,000 base pairs) completely sequenced. Its sequence was published in *Science*, July 28, 1995. Since then over 350 more microbial genomes have been sequenced, including one that was completed in a single day by five different labs, each working on a different part. Today it takes about 13 hours to completely sequence a microbial genome, with the help of automated equipment.

Microbial genomes are small and easier to study, usually no more than 10 million DNA base pairs, compared with about 3 billion in the human and mouse genomes.

The E. coli genome consists of 4,639,221 base pairs, which code for at least 4,288 proteins.

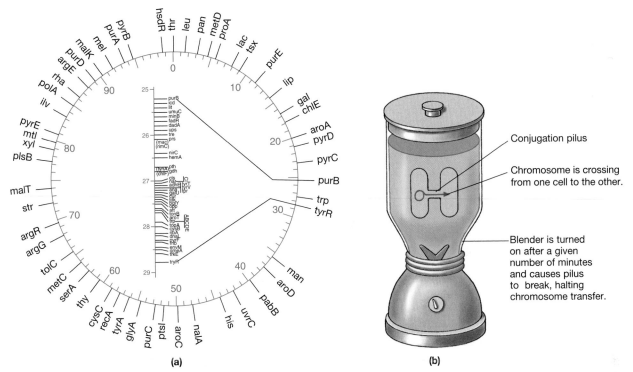

Figure 7.2 A partial chromosome map of E. coli. (a) The entire genome of *E. coli* consists of approximately 4,000 genes. The outer circle is a simplified representation of the chromosome, with some commonly studied genes marked on it. It takes about 100 minutes to transfer the entire chromosome from a donor to a recipient cell in conjugation ◄(Chapter 8), a mechanism by which genes are transferred between bacteria. The numbers marked inside the circle represent the number of minutes of transfer required to reach that point on the chromosome. The insert is a small segment of the *E. coli* map, enlarged to show some of the additional genes that have been located within that region (after Bachman). (b) These times are determined by allowing two different strains of bacteria to conjugate inside a blender that is turned on after a given number of minutes, causing the pili to break, thus halting chromosome transfer. Recipient cells are then grown and examined to determine which genes have been transferred to them. By varying the times used, researchers discover the order of the genes on the chromosome.

CLOSE UP

Smallest Known Bacterial Genome—On the Way to Becoming an Organelle?

October 2006 found *Nanoarchaeum equitans* dethroned as the holder of the record for smallest known bacterial genome. It has 491,000 bases pairs of DNA. The new champion, *Carsonella ruddii*, has only one-third that number, 159,662, which provide only 182 protein-coding genes.

C. ruddii is an endosymbiont living inside special cells (bacteriocytes) within specialized structures called bacteriomes (see photo), located inside a sap-sucking psyllid insect which spreads "greening" disease among citrus plants. Plant sap is poor in nutrients, so some sap-sucking insects rely on endosymbiont bacteria to manufacture necessities such as amino acids which are then shared. In some cases, the insect and bacteria have evolved together for so long that neither one can survive without the other. Inside their insect host, the bacteria need fewer genes to survive, since the host supplies many of their needs. Perhaps some of the bacterial genes were transferred into the insect's genome, which then took on the job of providing for the bacteria. Later, if the bacteria lost these genes, it was no problem, it just allowed reduction of the bacterial genome. *C. ruddii* has lost genes that are considered absolutely necessary for life. These reduced genome bacteria may be on their way to becoming organelles inside the host cell. Remember "evolution by endosymbiosis" back in ◄Chapter 4, p. 104!

The bright yellow structure (bacteriome) inside this insect contains endosymbiotic bacteria, *Carsonella ruddii*, which have the smallest genome yet (as of October, 2006) found among bacteria. (*Courtesy Nancy Moran, University of Arizona*)

APPLICATIONS

Enemy DNA

How do cells of your immune system identify foreign invaders, such as bacteria and viruses? Until now, immunologists have believed that proteins on the outer surface of invading cells and viruses were the triggers that alerted the body's defense system. Now, however, preliminary studies by Dr. Arthur M. Krieg of the University of Iowa College of Medicine indicate that even before your body responds to these surface proteins, it recognizes bacterial and viral DNA and begins an assault on the invaders. It detects a pattern of bases unique to bacteria and viruses—the frequent occurrence of C–G sequences. This combination of bases is uncommon in mammalian DNA. When it does occur, attached to the cytosine is a methyl group—a group that bacterial and viral C–G sequences do not have.

There is some evidence that people with systemic lupus erythematosus, an autoimmune disease in which the immune system attacks the body's own DNA, may not have the normal ability to add methyl groups to their DNA. Thus their DNA may look like foreign DNA to the immune system. Perhaps the cure for this disease lies in increasing the patient's ability to add methyl groups. Another clinical application may lie in administering artificially produced C–G sequences to patients whose immune systems need stimulation. In the laboratory, the addition of such C–G sequences to flasks of B cells (immune cells that make antibodies, or proteins that respond to foreign invaders) causes 95% of the cells to begin multiplying within a half hour. Further research is needed to see if the same effect occurs in whole organisms.

Information Transfer

Information stored in DNA is used both to guide the replication of DNA in preparation for cell division and to direct protein synthesis. The three ways in which this information is transferred are as follows:

1. *Replication:* DNA makes new DNA.
2. *Transcription:* DNA makes RNA as the first step in protein synthesis.
3. *Translation:* RNA links amino acids together to form proteins.

In both DNA replication and transcription, DNA serves as a **template** (much like a sewing pattern) for the synthesis of a new nucleotide polymer. The sequence of bases in each new polymer is complementary to that in the original DNA. Such an arrangement is accomplished by base pairing. Recall from ◄Chapter 2 that in complementary base pairing in DNA, adenine always pairs with thymine (A–T), and guanine always pairs with cytosine (G–C). Recall also that when DNA serves as a template for synthesis of RNA, the pairing is different: In RNA, thymine is replaced by uracil (U), which pairs with adenine.

In **DNA replication**, the new polymer is also DNA. In protein synthesis, the new polymer is a particular type of RNA called *messenger RNA (mRNA)*, which then serves as a second template that dictates the arrangement of amino acids in a protein. Some proteins form the structure of a cell, others (enzymes) regulate its metabolism, and still others transport substances across a membrane.

In the overall process of protein synthesis, the synthesis of mRNA from a DNA template is called **transcription**, and the synthesis of protein from information in mRNA is called **translation**. By analogy, transcription transfers information from one nucleic acid to another as you might transcribe handwritten sentences to typewritten sentences in the same language.[1] Translation transfers information from the language of nucleic acids to the language of amino acids as you might translate English sentences into another language. There are even "proofreading" enzymes that try to eliminate any errors that occur, ensuring that a correct copy is passed on.

In the case of viruses that have RNA as their genetic material, scientists were initially unable to understand how these viruses could make more RNA. Then, the discovery of enzymes for **reverse transcription** revealed a process whereby RNA can make DNA. This DNA can then make more RNA. Such viruses are known as *retroviruses* because of this reverse process. (We will study them in greater detail in ◄Chapter 10.) HIV, the virus that causes AIDS, is a retrovirus. Reverse transcription is a less accurate process than regular transcription. Uncorrected errors are passed on as mutations, or permanent changes in the genes of an organism. HIV has a mutation rate 500 times higher than that of most organisms, an unfortunate fact for would-be vaccine makers.

DNA replication, transcription, and translation all transfer information from one molecule to another (Figure 7.3). These processes allow information in DNA to be transferred to each new generation of cells and to be used to control the functioning of cells through protein synthesis.

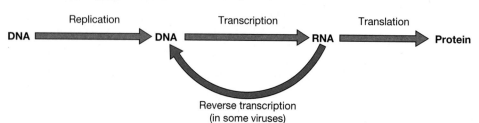

Figure 7.3 The transfer of information from DNA to protein. As we shall see later, certain viruses, such as the one that causes AIDS, can direct synthesis of DNA from their RNA (reverse transcription).

✔CHECKLIST

1. Compare and contrast chromosomes in prokaryotes and eukaryotes.

2. DNA is not always the genetic material. What are the exceptions?

3. How could mutations give rise to new alleles of a gene?

4. How do bacterial DNA and viral DNA differ from human DNA? What disease may be related to this difference?

5. How does translation differ from transcription?

DNA REPLICATION

To understand DNA replication, we need to recall from ◄Chapter 2 that pairs of helical DNA strands are held together by base pairing of adenine with thymine and cytosine with guanine. We also need to know that the ends of each strand are different. At one end, called the $3'$ (3 prime) end, carbon 3 of deoxyribose is free to bind to other molecules. At the other end, the $5'$ (5 prime) end, carbon 5 of deoxyribose is attached to a phosphate (Figure 7.1). This structure is somewhat analogous to that of a freight train, with the $3'$ end the engine and the $5'$ end the caboose. When the two strands of a double helix combine by base pairing, they do so in a head-to-tail, or **antiparallel**, fashion. The arrangement of the strands is somewhat like two trains pointed in opposite directions, and base pairing is like passengers in the two trains shaking hands.

DNA replication begins at a specific location (the origin) in the circular chromosome of a prokaryotic cell and usually proceeds simultaneously away from the origin in both directions. This creates two moving **replication forks**, the points at which the two strands of DNA separate to allow replication of DNA **(Figure 7.4)**. Various enzymes (helicases) break the hydrogen bonds between the bases in the two DNA strands, unwind the strands from each other, and stabilize the exposed single strands, preventing them from joining back together. Molecules of the enzyme **DNA polymerase** then move along behind each replication fork, synthesizing new DNA strands complementary to the original ones at a speed of approximately 1,000 nucleotides per second. DNA polymerase also "*proofreads*" the growing strand, correcting errors such as mismatched bases. Even at such high speeds, proofreading usually leaves only one in 10 base pairs with an error.

The enzyme DNA polymerase can add nucleotides only to the $3'$ end of a growing DNA strand. Consequently, only one strand of original DNA can serve as a template for the synthesis of a continuous new strand, the **leading strand**, going in the $5'$ to $3'$ direction. Along the other strand, which runs in the $3'$ to $5'$ direction, the synthesis of new DNA, the **lagging strand**, must be *discontinuous*; that is, the polymerase must continually

jump ahead and work backward, making a series of short DNA segments called **Okazaki fragments** which consist of 100 to 1,000 base pairs. Each fragment must have a short piece of RNA called an **RNA primer** attached to the parent DNA in order to start synthesis of new DNA. Later DNA polymerase will digest the RNA primer and replace it with DNA. The fragments are then joined together by another enzyme called a **ligase**. Formation of leading and lagging strands goes on simultaneously. But because the DNA polymerase producing Okazaki fragments must wait until enough DNA has been opened up at the replication fork for an RNA primer to form, it is said to be "lagging." Ultimately, two separate chromosomes are formed (Figure 7.4), each double helix consisting of one strand of old, or parent, DNA and one strand of new DNA. Such replication is called **semiconservative replication** because one strand is always conserved.

PROTEIN SYNTHESIS

TRANSCRIPTION

All cells must constantly synthesize proteins to carry out their life processes: reproduction, growth, repair, and regulation of metabolism. This synthesis involves the accurate transfer of linear information of the DNA strands (genes) into a linear sequence of amino acids in proteins.

To set the stage for protein synthesis, hydrogen bonds between bases in DNA strands are broken enzymatically in certain regions so that the strands separate. Short sequences of unpaired DNA bases are thus exposed to serve as templates in transcription. Only one strand directs the synthesis of mRNA for any one gene; the complementary strand is used as a template during DNA replication or during the transcription of some other gene. Recall that RNA contains the base uracil instead of thymine (◄Chapter 2, p. 46). Thus, when mRNA is transcribed

APPLICATIONS

If DNA Makes Only Proteins, What Makes Carbohydrates and Lipids?

If genetic information in DNA is used specifically to determine the structure of proteins, how are the structures of carbohydrates and lipids determined? Stop and think of the kinds of proteins a cell has. Many are enzymes, and, of course, some of those enzymes direct the synthesis of carbohydrates and lipids. The entire cell is controlled by DNA—either directly, in DNA replication and synthesis of structural proteins, or indirectly, by the synthesis of enzymes that in turn control the synthesis of carbohydrates and lipids.

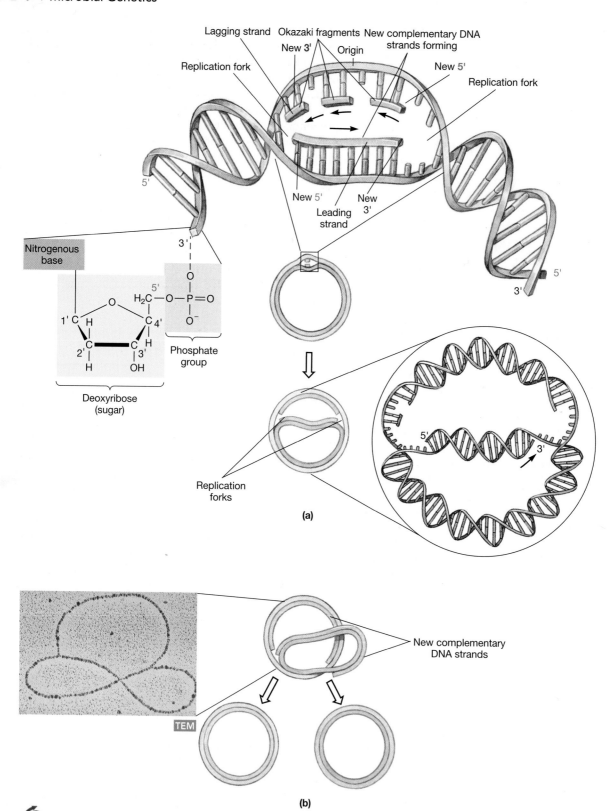

(a)

(b)

Figure 7.4 **DNA replication in a prokaryote.** **(a)** The DNA strands separate, and replication begins at a replication fork on each strand. As synthesis proceeds, each strand of DNA serves as a template for the replication of its partner. Note the antiparallel arrangement of the complementary strands of the DNA double helix. Because synthesis of new DNA can take place in only one direction, the process must be discontinuous along one strand. Short segments are formed and then spliced together as the arrows indicate. **(b)** As the 2 replication forks meet, the 2 new chromosomes come apart, each made of 1 old and 1 new strand, i.e., examples of semiconservative replication. Each new cell can undergo subsequent replications. (21,011X). *(NIH/Kakefuda/Photo Researchers, Inc.)*

from DNA, uracil pairs with adenine; otherwise, the bases pair just as they do in DNA replication. Messenger RNA is formed in the 5′ to 3′ direction.

To transcribe its DNA, a cell must have sufficient quantities of nucleotides that contain high-energy phosphate bonds, which provide energy for the nucleotides to participate in subsequent reactions. After separating the DNA strands, the enzyme **RNA polymerase** binds to one strand of exposed DNA recognizing a sequence of nucleotide bases in the DNA that indicates this is the start of a gene (*promoter sequence*). As shown in **Figure 7.5**, after an enzyme binds to the first base in DNA (adenine, in this case), the appropriate nucleotide joins the DNA base-enzyme complex. The new base then attaches by base pairing to the template base of DNA. The enzyme moves to the next DNA base, and the appropriate phosphorylated nucleotide joins the complex. The phosphate of the second nucleotide is linked to the ribose of the first nucleotide, and *pyrophosphate* (two attached molecules of phosphate) is released. This forms the first link in a new polymer of RNA. Energy to form this link comes from the hydrolysis of ATP and the release of two more phosphate groups. This process is repeated until the RNA molecule is completed.

In prokaryotes, transcription and translation both take place in the cytoplasm, whereas in eukaryotes, transcription takes place in the cell nucleus. The mRNA of eukaryotic transcription must be completely formed and transported through the nuclear envelope to the cytoplasm before translation can begin. Moreover, the mRNA molecule undergoes additional processing before it is ready to leave the nucleus. In eukaryotic cells, as well as in certain types of bacteria known as Archaea (to be discussed in ◀Chapter 9), the regions of genes that code for proteins are called **exons**. Exons are typically separated within a gene by DNA segments that do not code for proteins. Such noncoding *intervening*

regions are called **introns**. In the nucleus, RNA polymerase first forms mRNA from the entire gene, including all exons and introns. The newly formed, long mRNA molecule is streamlined by other enzymes, which remove the introns and splice together the exons. The resulting mRNA is ready to direct protein synthesis and to leave the nucleus **(Figure 7.6)**.

KINDS OF RNA

Three kinds of RNA—*ribosomal RNA*, *messenger RNA*, and *transfer RNA*—participate in protein synthesis. Each RNA consists of a single strand of nucleotides and is synthesized by transcription, using DNA as a template. To complete the story of protein synthesis, we will need more information about these types of RNA.

Ribosomal RNA (rRNA) binds closely to certain proteins to form two kinds of ribosome subunits. A subunit of each kind combines to form a ribosome. Recall that ribosomes are sites of protein synthesis in a cell (◀Chapter 4, p. 91). They serve as binding sites for transfer RNA, and some of their proteins act as enzymes that control protein synthesis. Prokaryotic ribosomes are made of a small (30S) and a large (50S) subunit. (Eukaryotic ribosomes are formed from a 40S and a 60S subunit.) After the two subunits join together around the strand of mRNA **(Figure 7.7)**, the synthesis of a peptide begins. The newly formed polypeptide chain grows out through a tunnel in the 50S subunit.

Messenger RNA (mRNA) is synthesized in units that contain sufficient information to direct the synthesis of one or more polypeptide chains. One mRNA molecule corresponds to one or more genes, the functional units of DNA. Each mRNA molecule becomes associated with one or more ribosomes. At the ribosome, the information coded in mRNA acts during translation to dictate the sequence of amino acids in the protein.

Figure 7.5 The transcription of RNA from template DNA. The PPP represents a triphosphate, and PP represents pyrophosphate. In RNA, U (rather than T) pairs with A.

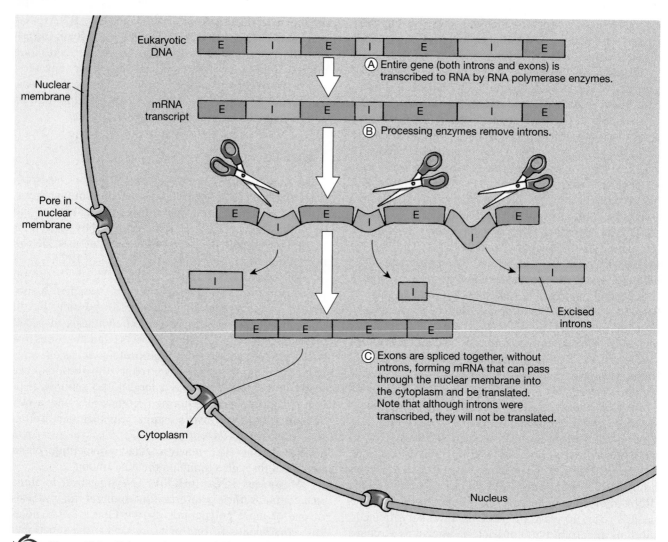

Figure 7.6 Eukaryotic genes differ in complexity from prokaryotic genes. Coding sequences of DNA called exons (E) alternate with noncoding sequences called introns (I). After both are transcribed into RNA, the introns are removed, leaving only the spliced-together exons to enter the cytoplasm, where they will be translated. All prokaryotes except for Archaea lack introns.

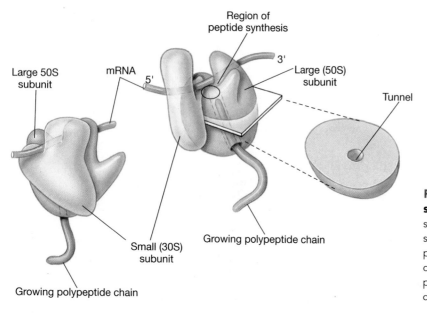

Figure 7.7 Prokaryotic ribosomal structure. The small (30S) and large (50S) subunits are shown from two different angles. The subunits enfold the mRNA strand. The region of peptide synthesis is the junction of these three components. The growing polypeptide chain passes through a tunnel in the 50S subunit, which can be seen in cross section.

In translation, each triplet (sequence of three bases) in mRNA constitutes a **codon** (ko'don). Codons are the "words" in the language of nucleic acids. Each codon specifies a particular amino acid or acts as a terminator codon. The first codon in a molecule of mRNA acts as a **start codon**. It always codes for the amino acid methionine, even though the methionine may be removed from the protein later. The last codon to be translated in a molecule of mRNA is a **terminator**, or **stop codon**. It acts as a kind of punctuation mark to indicate the end of a protein molecule. Using a sentence as an analogy, the methionine codon is the capital letter at the beginning of the sentence, and the terminator codon is the punctuation mark at the end.

At least one codon exists for each of the 20 amino acids found in proteins. Several codons exist for some amino acids; for example, six different codons code for leucine (Leu). Find these in **Figure 7.8**. The relationship between each codon and a specific amino acid constitutes the **genetic code** (Figure 7.8). Those codons that code for an amino acid are called **sense codons**. Early in the study of the genetic code, investigators found a few codons that did not code for any amino acid. Those codons were therefore named **nonsense codons**. It was later found that they were stop codons. Although genetic information is stored in DNA, the genetic code is written in codons of mRNA. Of course, the information in the codons is derived *directly* from DNA by complementary base pairing during transcription.

Comparisons of the codons among different organisms have shown them to be nearly the same in all organisms, from bacteria to humans. This universality of the genetic code allows research on other organisms to be applied to the understanding of information transmission in human cells. Much of what is known about how the genetic code operates has been learned from research on bacteria.

The function of **transfer RNA (tRNA)** is to transfer amino acids from the cytoplasm to the ribosomes for placement in a protein molecule. Many different kinds of tRNAs have been isolated from the cytoplasm of cells. A tRNA molecule consists of 75 to 80 nucleotides and is folded back on itself to form several loops that are stabilized by complementary base pairing (**Figure 7.9**). Each tRNA has a three-base **anticodon** (an'ti-ko''don) that is complementary to a particular mRNA codon. It also has a binding site for an amino acid—the particular amino acid specified by the mRNA codon. (The mRNA codon, of course, got its information directly from DNA.) Thus, the tRNAs are the link between the codons and the corresponding amino acids. Amino acid attachment to specific

The 2006 Nobel Prize was awarded to Andrew Fire and Craig Mello for discovery of another type of RNA, interference RNA (RNAi) which silences a gene by blocking the expression of a specific messenger RNA. This may reveal the function of that gene, or how it interacts with other genes.

First position	Second position				Third position
	U	C	A	G	
U	UUU Phe UUC Phe UUA Leu UUG Leu	UCU Ser UCC Ser UCA Ser UCG Ser	UAU Tyr UAC Tyr UAA Stop UAG Stop	UGU Cys UGC Cys UGA Stop UGG Trp	U C A G
C	CUU Leu CUC Leu CUA Leu CUG Leu	CCU Pro CCC Pro CCA Pro CCG Pro	CAU His CAC His CAA Gln CAG Gln	CGU Arg CGC Arg CGA Arg CGG Arg	U C A G
A	AUU Ile AUC Ile AUA Ile AUG Met	ACU Thr ACC Thr ACA Thr ACG Thr	AAU Asn AAC Asn AAA Lys AAG Lys	AGU Ser AGC Ser AGA Arg AGG Arg	U C A G
G	GUU Val GUC Val GUA Val GUG Val	GCU Ala GCC Ala GCA Ala GCG Ala	GAU Asp GAC Asp GAA Glu GAG Glu	GGU Gly GGC Gly GGA Gly GGG Gly	U C A G

Figure 7.8 The genetic code, with standard three-letter abbreviations for amino acids. To find the amino acid for which the mRNA codon AGU codes, go down the left column to the block labeled A, move across to the fourth square labeled G at the top of the figure, and find the first line in the square labeled U on the right side of the figure. There you will find Ser, the abbreviation for serine. *Stop* designates a terminator codon of which there are three. The *Start* codon is AUG, which also codes for methionine. Therefore, protein synthesis always begins with methionine. The methionine is usually removed later, however, so not all proteins actually start with methionine. When found in the middle of an mRNA strand, AUG codes for methionine.

TABLE 7.1

Properties of the Different Kinds of RNA	
Kind of RNA	**Properties**
Ribosomal	Combines with specific proteins to form ribosomes.
	Serves as a site for protein synthesis.
	Associated enzymes function in controlling protein synthesis.
Messenger	Carries information from DNA for synthesis of a protein.
	Molecules correspond in length to one or more genes in DNA.
	Has base triplets called codons that constitute the genetic code.
	Attaches to one or more ribosomes.
Transfer	Found in the cytoplasm, where they pick up amino acids and transfer them to mRNA.
	Molecules have a cloverleaf shape with an attachment site for a specific amino acid.
	Each has a single triplet of bases called an anticodon, which pairs complementarily the corresponding codon in mRNA.

Figure 7.9 Transfer RNA. **(a)** The two-dimensional structure of the tryptophan transfer RNA. The anticodon end will pair up with a codon on a strand of messenger RNA and deliver the desired amino acid (tryptophan), which is bonded to the acceptor arm at its opposite end. The molecule is maintained in its cloverleaf pattern by hydrogen bonding between strands that form the arms (dashed lines). **(b)** A tRNA molecule folded into its complex three-dimensional shape, in diagram form and as a computer-generated model. *(Courtesy Tripos, Inc.)*

tRNA molecules is achieved by the action of amino-acid-activating enzymes and energy derived from ATP.

The anticodon attaches by complementary base pairing to the appropriate mRNA codon so that its amino acid is aligned for incorporation into a protein. The accuracy of amino acid placement in protein synthesis depends on this precise pairing of codons and anticodons. The properties of the three types of RNA are summarized in **Table 7.1**.

TRANSLATION

Protein synthesis, an important process in bacterial growth, uses 80 to 90% of a bacterial cell's energy. Generally, during protein synthesis, the various RNAs and amino acids are available in sufficient quantities. The RNAs can be reused many times before they lose their ability to function. Of the types of RNA, mRNA is produced in the most

precise quantity in accordance with the cell's need for a particular protein. **Figure 7.10** shows the three types of RNA and how they function in protein synthesis.

Once an mRNA molecule has been transcribed and has combined with a ribosome, the ribosome initiates protein synthesis and provides the site for protein assembly. Each ribosome attaches first to the end of the mRNA that corresponds to the beginning of a protein. The length of each polypeptide chain extending from a ribosome corresponds to the amount of mRNA the ribosome has "read." Several ribosomes can be attached at different points along an mRNA molecule to form a **polyribosome** (or *polysome*) **(Figure 7.11)**.

In prokaryotes (unlike eukaryotes), transcription and translation take place in the cytoplasm, where all necessary enzymes and ribosomes are present. In eukaryotes, the mRNA formed in the nucleus must pass through the nuclear membrane before it is available to the ribosomes, which carry out protein synthesis.

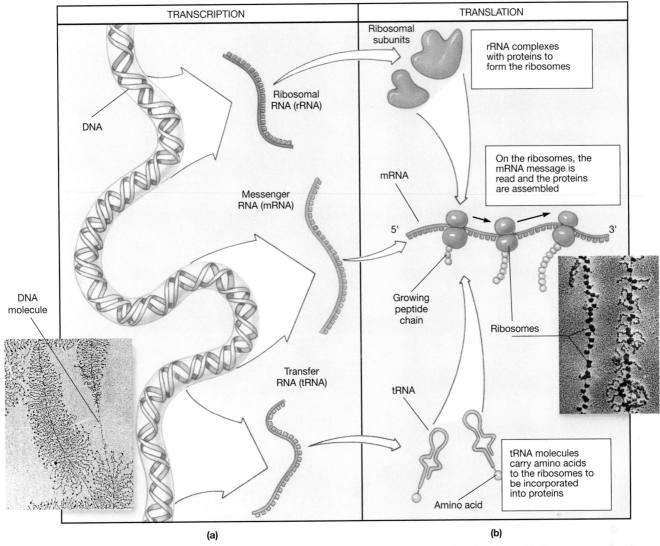

Figure 7.10 Transcription and translation. (a) Transcription from DNA to RNA. *(Fron O. L. Miller, Jr., and B. R. Beatty, Journal of Cellular Physiology 74 (1969))* **(b)** Translation from RNA to protein. Many ribosomes that are connected to and read the same piece of mRNA are called a polyribosome. *(E. Kiselva & D. Fawcett/Visuals Unlimited)*

The main steps in protein synthesis **(Figure 7.12)** can be summarized as follows: The process begins when a molecule of mRNA becomes properly oriented on a ribosome. As each codon of the mRNA is "read," the appropriate tRNA combines with it and thereby delivers a particular amino acid to the protein assembly site. The location on the ribosome where the first tRNA pairs is called the *P site*. The second codon of the mRNA then pairs with a tRNA that transports the second amino acid to the *A site*, which is next to the P site. Matching of codon and anticodon by base pairing allows coded information in mRNA to specify the sequence of amino acids in a protein. Any tRNAs with nonmatching anticodons simply do not bind to the ribosome. As amino acids are delivered one after another and peptide bonds form between them, the length of the polypeptide chain increases. This

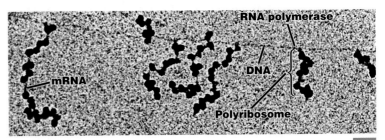

Figure 7.11 Concurrent transcription and translation in prokaryotes. A portion of *E. coli* DNA runs horizontally across this electron micrograph (24,013X). Ribosomes have attached to the pieces of mRNA and are synthesizing proteins which can be seen increasing in length from right to left, indicating the direction of transcription. The presence of many ribosomes all "riding" simultaneously along one piece of mRNA, gives this the name of *polyribosome* (or *polysome*). *(O.L. Miller Jr./Visuals Unlimited)*

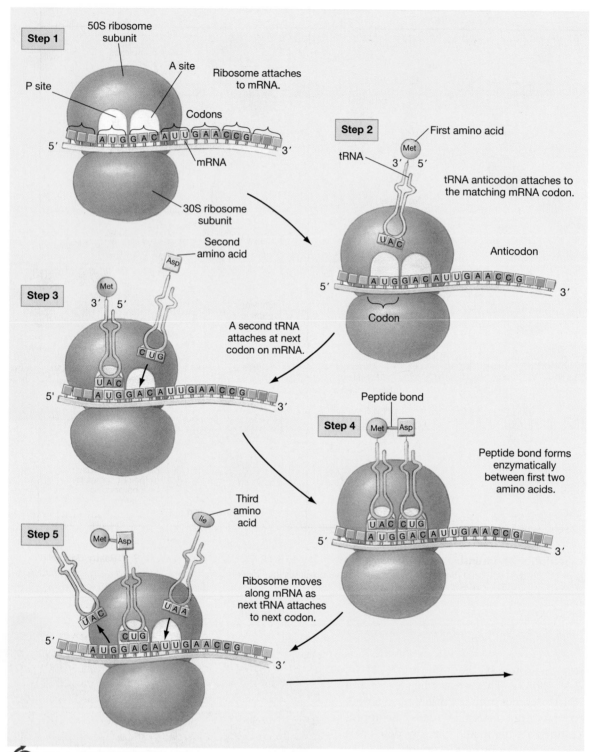

Figure 7.12 **Protein synthesis.** Steps 1–5: The main steps in protein synthesis. Steps 6 and 7: Many ribosomes can "read" the same strand of mRNA simultaneously. The ribosomes are shown moving from left to right.

process continues until the ribosome recognizes a stop codon. When the ribosome "reads" a stop codon at the A site, it releases the finished protein from the P site.

Any mRNA molecule can direct simultaneous synthesis of many identical protein molecules—one for each ribosome passing along it. Ribosomes, mRNAs, and tRNAs are reusable. The tRNAs shuttle back and forth picking up amino acids in the cytoplasm, and bringing them to the ribosome, where the amino acids are incorporated into protein.

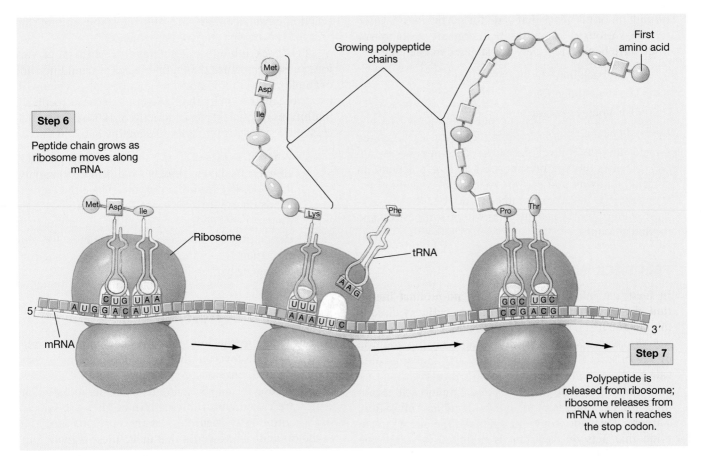

Growing polypeptide chains

First amino acid

Step 6

Peptide chain grows as ribosome moves along mRNA.

Ribosome

tRNA

mRNA

Step 7

Polypeptide is released from ribosome; ribosome releases from mRNA when it reaches the stop codon.

Figure 7.12 *(Continued)*

✓ CHECKLIST

1. Distinguish between leading and lagging strands.
2. What do 5′ and 3′ refer to? How do they determine the direction of synthesis of new DNA?
3. Does the genetic code have synonyms?
4. Contrast the three kinds of RNA. Does DNA make all three kinds?

THE REGULATION OF METABOLISM

THE SIGNIFICANCE OF REGULATORY MECHANISMS

Bacteria use most of their energy to synthesize substances needed for growth. These substances include structural proteins, which form cell parts, and enzymes, which control both energy production and synthetic reactions. The survival of bacteria depends on their ability to grow even when conditions are less than ideal—for example, when nutrients are in short supply. In their evolution, cells of bacteria (and all other organisms) have developed mechanisms to turn reactions on and off in accordance with their needs. Energy and materials are too valuable to waste. Also, the cell has a limited amount of space for storing excesses of materials it synthesizes. Thus, cells use energy to synthesize substances in the amounts needed and shut off synthesis before wasteful excesses are produced.

All living organisms are presumed to have control mechanisms that regulate their metabolic activities. However, more research on control mechanisms has been done on bacteria than on all other organisms. Bacteria are ideal for such studies for several reasons:

1. They can be grown in large numbers relatively inexpensively under a variety of controlled environmental conditions.
2. They produce many new generations quickly.
3. Because they reproduce so rapidly, a variety of mutations can be observed in a relatively short time.

Mutant organisms that have an alteration in their control mechanisms can be isolated and studied along with nonmutated organisms to better understand the operation of control mechanisms.

CATEGORIES OF REGULATORY MECHANISMS

The mechanisms that control metabolism either regulate enzyme activity directly or regulate enzyme synthesis by turning on or off genes that code for particular enzymes. Of the various mechanisms that regulate metabolism, three have been extensively investigated in bacteria:

- Feedback inhibition
- Enzyme induction
- Enzyme repression

In *feedback inhibition*, enzyme activity is regulated directly, and the control mechanism determines how rapidly enzymes already present will catalyze reactions. In *enzyme induction* and *enzyme repression*, regulation occurs indirectly by enzyme synthesis, and the control mechanism determines which enzymes will be synthesized and in what amounts.

FEEDBACK INHIBITION

In **feedback inhibition**, also called **end-product inhibition**, the end product of a biosynthetic pathway directly inhibits the first enzyme in the pathway. This mechanism was discovered when it was observed that the addition of one of several amino acids to a growth medium could cause a bacterium suddenly to stop synthesizing that particular amino acid. Synthesis of the amino acid threonine, for example, is regulated by feedback inhibition. Threonine is made from aspartate, and the allosteric enzyme that acts on aspartate is inhibited by threonine **(Figure 7.13)**. (Aspartate is derived from the oxaloacetate formed in the Krebs cycle.) When an inhibitor (threonine) attaches to the allosteric site, it alters the enzyme's shape so the substrate (aspartate) cannot attach to the active site (◀Chapter 5, p. 122). Thus, feedback inhibition occurs when the end product of a reaction sequence binds to the allosteric site of the enzyme for the first step in the sequence.

Feedback inhibition regulates the synthesis of various substances other than amino acids (pyrimidines, for example). This regulatory mechanism also occurs in many organisms other than bacteria. Because feedback inhibition acts quickly and directly on a metabolic process, it allows the cell to conserve energy in two ways:

1. When it is plentiful, the inhibitor (end product) attaches to the enzyme; when it is in short supply, it is released from the enzyme. Thus, the cell expends energy to synthesize the end product only when it is needed.
2. Regulation of enzyme activity requires less energy than the more complex processes that regulate gene expression.

ENZYME INDUCTION

At one point in the investigation of metabolic regulation, it was discovered that certain organisms always contain active enzymes for glucose metabolism even when glucose is not present in the medium. Such enzymes are called **constitutive enzymes**; they are synthesized continuously regardless of the nutrients available to the organism. The genes that make these enzymes are always active. In contrast, enzymes that are synthesized by genes that are sometimes active and sometimes inactive, depending on the presence or absence of substrate, are called **inducible enzymes**.

When bacteria such as *E. coli* are grown on a nutrient medium that contains no lactose, the cells do not

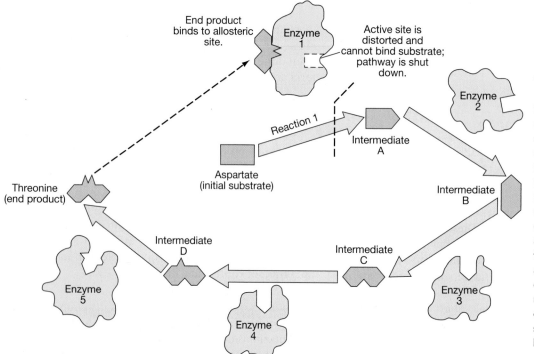

Figure 7.13
Feedback inhibition. The synthesis of threonine has five enzymatically controlled reactions (arrows) and four intermediate products (A, B, C, and D). Threonine (the end product) inhibits an allosteric enzyme 1 that catalyzes Reaction 1. The allosteric enzyme is functional when its allosteric site is not occupied and is nonfunctional when the end product of the sequence of reactions is bound to that site.

make any of the enzymes that they would need to utilize lactose as an energy source. When lactose is present, however, the cells synthesize the enzymes needed for its metabolism. This phenomenon is an example of **enzyme induction**. Enzyme induction controls the breakdown of nutrients as they become available in the growth medium. Such a system is turned on when a nutrient is available and turned off when it is depleted. The nutrient itself acts as an **inducer** of enzyme production.

The *operon* (op′er-on) *theory*, a model that explains the regulation of some protein synthesis in bacteria, was proposed in 1961 by French scientists François Jacob and Jacques Monod, who received a Nobel Prize in 1965 for their work. Although the model applies to several operons, we will illustrate it with the *lac* operon, which regulates lactose metabolism. An **operon** is a sequence of closely associated genes that regulate enzyme production. An operon includes one or more **structural genes**, which carry information for the synthesis of specific proteins such as enzyme molecules, and **regulatory sites**, which control the expression of the structural genes. A **regulator** (*i*) **gene** works in conjunction with the operon but may be located some distance from it. In prokaryotes, several structural genes are controlled by one operon—a more efficient method than that of eukaryotes, in which each gene is controlled by its own regulatory site. The operon seems almost totally limited to prokaryotes. Thus far, the only eukaryotes found to have operons are the nematodes (roundworms), such as *Caenorhabditis elegans*.

The *lac* operon **(Figure 7.14)** consists of regulatory sites, called a *promoter* and an *operator*, and three structural genes, *Z*, *Y*, and *A*, which direct synthesis of specific enzymes. An RNA polymerase molecule must bind to the promoter before transcription can begin. The separate *i* gene directs synthesis of a substance called the *lac repressor*. The **repressor** is a protein that binds to the

operator and prevents transcription of the adjacent *Z*, *Y*, and *A* genes. Consequently, the enzymes that metabolize lactose are not synthesized. The *i* gene is an example of a *constitutive gene*—it is always undergoing protein synthesis to produce more repressor protein and is not controlled by the promoter.

When present in the medium, lactose acts as the inducer by binding to and inactivating the *lac* repressor. The repressor then no longer blocks the operator. The RNA polymerase then binds to the promoter, causing the operator to initiate transcription of the *Z*, *Y*, and *A* genes as a single long strand of mRNA. This mRNA becomes associated with ribosomes and directs synthesis of three enzymes: β-galactosidase (*Z* gene), permease (*Y* gene), and transacetylase (*A* gene). Discovery of the operon led to the realization that a single mRNA molecule could code for the production of more than one protein, e.g., the 3 enzymes in the *lac* operon. Permease transports lactose into cells, and β-galactosidase breaks down lactose into glucose and galactose (Figure 7.14b). Although the role of transacetylase is not clear, it is thought to facilitate the escape of galactosides. When the available lactose has been broken down, there is none left to bind to the repressor. The active repressor again binds to the operator, and the operon is turned off.

ENZYME REPRESSION

In contrast to enzyme induction, which typically regulates catabolism, **enzyme repression** typically regulates anabolism. It controls processes in which substances needed for growth are synthesized. Synthesis of the amino acid tryptophan, for example, is regulated by enzyme repression through actions of the *trp* operon, which consists of five structural genes.

When tryptophan is available to a bacterial cell, the amino acid binds to an inactive repressor. Binding activates the repressor protein, which can then bind to the promoter and repress synthesis of the enzymes needed to make tryptophan. When tryptophan is not available, the repressor protein remains inactive, and repression does not occur. Structural genes are transcribed, and tryptophan is synthesized. When tryptophan becomes plentiful, it again represses the operon. An even finer control mechanism, called **attenuation**, allows transcription of the *trp* operon to begin but terminates it prematurely by a complex process when sufficient amounts of tryptophan are already present in the cell. Several operons, especially those for amino acid synthesis, have attenuation mechanisms.

Although typical enzyme repression regulates anabolic pathways, there is a slightly different kind of repression that operates in connection with some catabolic pathways. When certain bacteria (*E. coli*, for example) are grown on a nutrient medium containing both glucose and lactose, they grow at a logarithmic rate as long as glucose is available. When the glucose is depleted, they

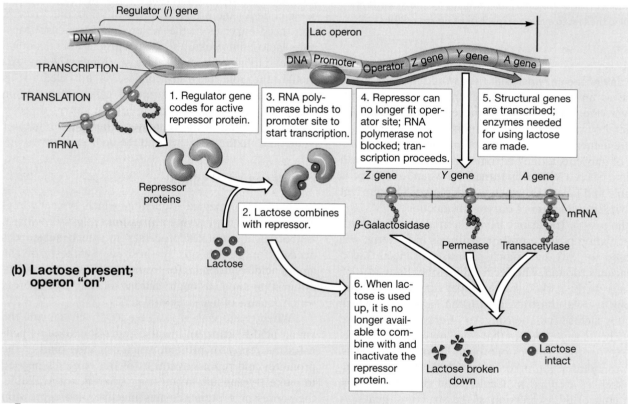

Figure 7.14 **Enzyme Induction.** The mechanisms of operation of the *lac* operon. **(a)** In the absence of lactose, the repressor binds to the operator, preventing transcription of the genes coding for enzymes used to metabolize lactose. **(b)** When lactose is present, it binds to the repressor and inactivates it. The structural genes of the operon are transcribed, and enzymes for metabolizing lactose are synthesized. The regulator (*i*) gene may be some distance away from the operon.

enter a stationary phase but soon begin to grow again at a logarithmic rate, though not quite as rapidly (**Figure 7.15**). This time the logarithmic growth rate results from the metabolism of lactose. The stationary phase is the period during which the enzymes needed to utilize lactose are being synthesized.

Why was the synthesis of these enzymes not induced before the glucose was depleted, since lactose was present in the medium from the start? The answer is that bacteria use glucose as a nutrient with high efficiency.

The enzymes for metabolizing glucose, being constitutive, are always present in the cell. Thus, when glucose is abundant, there is no advantage in making enzymes for metabolizing lactose even if lactose is also available. Consequently, the *lac* operon that we described previously is repressed when glucose is present in adequate quantities, an effect known as **catabolite repression**. In this way the cell saves energy by not making enzymes it doesn't need. When glucose supplies fall, the repression is lifted, the *lac* operon genes are transcribed,

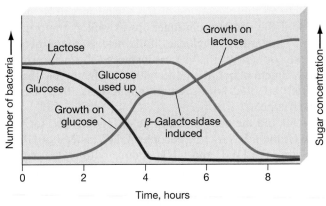

Figure 7.15 Catabolite repression. The growth curve for bacteria in a medium initially containing both glucose and lactose. When glucose is used up, growth stops temporarily but begins again at a slower rate, using lactose as an energy source.

and the cell is ready to switch over to using lactose. In short, the transcription of the *lac* operon requires both that lactose be present and that glucose *not* be present.

Both enzyme induction and enzyme repression are regulatory mechanisms that control enzyme production by altering gene expression. Although these two mechanisms have different effects, they actually represent two examples of the operation of a single mechanism for turning genes on and off **(Table 7.2)**.

✓ CHECKLIST

1. What "feeds back" in feedback inhibition? What does it inhibit? How does it do this?
2. What is the inducer for the *lac* operon?
3. Compare enzyme induction and enzyme repression.

▌▌▌ MUTATIONS

Mutations, or changes in DNA, can now be defined more precisely as heritable changes in the sequence of nucleotides in DNA. Mutations account for evolutionary changes in microorganisms (and larger organisms) and for alterations that produce different strains within species. Here we will consider how DNA changes during mutations and how these changes affect the organisms.

TYPES OF MUTATIONS AND THEIR EFFECTS

Before we can consider mutations and their effects, we need to distinguish between an organism's genotype and its phenotype. **Genotype** refers to the genetic information contained in the DNA of the organism. **Phenotype** refers to the specific characteristics displayed by the organism. Mutations always change the genotype. Such a change may or may not be expressed in the phenotype, depending on the nature of the mutation.

Two important kinds of mutation are *point mutations*, which affect a single base, and *frameshift mutations*, which can affect more than one base in DNA. Mutations often make an organism unable to synthesize one or more proteins. The absence of a protein often leads to changes in the organism's structure or in its ability to metabolize a particular substance.

A third type of mutation does not involve a change as to which bases are present, as is the case in point and frameshift mutations. Instead, a portion of the chromosome changes its position, perhaps even breaking off and jumping to another part of the same or a different chromosome (*transposons*). Or it may reinsert itself in the same location, but upside down (*inversions*). As you think back to the *lac* operon, you can see why it is important for genes to retain their correct order on a chromosome. Imagine what would happen if a piece of

TABLE 7.2

Effects of Regulatory Systems Involving an Operon			
Regulatory Mechanism (Example)	Type of Pathway Regulated	Regulating Substance	Condition That Leads to Gene Expression
Enzyme induction (*lac* operon)	Catabolic (degradational) and releases energy	Nutrient (lactose)	Presence of nutrient (lactose)
Enzyme repression (*trp* operon)	Anabolic (biosynthetic) and uses energy	End product (tryptophan)	Absence of end product (tryptophan)

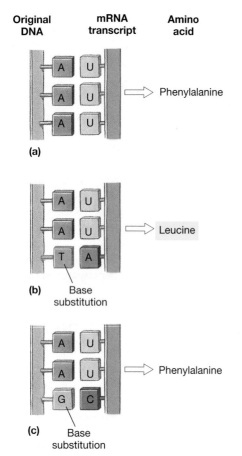

Original DNA	mRNA transcript	Amino acid

(a) → Phenylalanine

(b) Base substitution → Leucine

(c) Base substitution → Phenylalanine

Take another look

Figure 7.16 The effects of base substitution (a point mutation). The resulting protein may or may not be significantly affected depending on whether the new codon specifies the same amino acid, one with similar properties, or an entirely different one. In this illustration, change in a single base turns a codon coding for phenylalanine **(a)** into one coding for leucine or **(b)** into another codon which is a synonym for phenylalanine **(c)** , i.e., a "silent" mutation.

chromosome suddenly inserted itself into the middle of an operon! In ◄Chapter 8 we will examine how genetic engineers who insert genes into chromosomes must take this into consideration.

A **point mutation** is a base substitution, or nucleotide replacement, in which one base is substituted for another at a specific location in a gene. The mutation changes a single codon in mRNA, and it may or may not change the amino acid sequence in a protein. Let's look at some examples **(Figure 7.16)**.

Suppose a three-base sequence of DNA is changed from AAA to AAT. During transcription the mRNA codon will change from UUU to UUA. (Recall that uracil in RNA pairs with adenine in DNA; ◄Chapter 2, p. 47.) When the information in the mRNA is used to synthesize protein, the amino acid leucine will be substituted for phenylalanine in the protein. (To verify this for yourself, refer to the genetic code in Figure 7.8.) Because of the single amino acid substitution, the new protein will be different from the normal protein. The

effects on the phenotype of the organism will be negligible if the new protein functions as well as the original one. They will be significant if the new protein functions poorly or not at all. In rare instances the new protein may function better and produce a phenotype that is better adapted to its environment than the original phenotype.

Should the code in DNA be changed from AAA to AAG, the mRNA code becomes UUC instead of UUU. Because the UUC and UUU codons both code for phenylalanine, the mutation has no effect on the protein being synthesized. In this case, although the genotype has changed, the phenotype is unaffected.

Sometimes the substitution of a single base in DNA produces a terminator codon in mRNA. If the terminator codon is introduced in the middle of a molecule of mRNA destined to produce a single protein, synthesis will be terminated part way through the molecule. A polypeptide that will most likely be unable to function in the cell will be released, and the appropriate protein will not be synthesized. If the missing protein is essential to cell structure or function, the effect can be lethal.

A **frameshift mutation** is a mutation in which there is a **deletion** or an **insertion** of one or more bases **(Figure 7.17)**. Such mutations alter all the three-base sequences beyond the deletion or insertion. When mRNA transcribed from such altered DNA is used to synthesize a protein, many amino acids in the sequence may be altered. (Remember, a ribosome reads an mRNA in codons, sets of three bases.) Such mutations also commonly introduce terminator codons and cause protein synthesis to stop when only a short polypeptide has been made. Frameshift mutations usually prevent synthesis of a particular protein, and they change both the genotype and the phenotype. Their effect on the organism depends on the role of the missing protein in the organism's function. Point and frameshift mutations and their effects are summarized in **Table 7.3**. If three bases, or a multiple of three bases, were inserted or lost, what would happen? One or more amino acids would be gained or lost.

PHENOTYPIC VARIATION

Phenotypic variations frequently seen in mutated bacteria include alterations in colony morphology, colony color, or nutritional requirements. For example, instead of being a normal smooth, glossy, raised colony, a colony with mutated DNA may have a flat, rough appearance. The mutation has impaired synthesis of certain cell surface substances. In organisms that typically form capsules, a mutation can prevent synthesis of capsular polysaccharides. Mutations that alter nutritional requirements generally increase the nutritional needs of an organism, usually by impairing the organism's ability to synthesize one or more enzymes. As a result, the organism may require certain amino acids or vitamins in its medium because it can no longer make them itself.

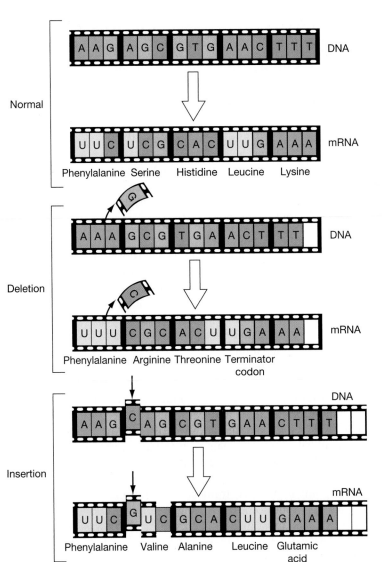

Figure 7.17 **The effects of frameshift mutations.** The addition or deletion of one or more nucleotides changes the amino acid sequence coded for by the entire gene from that point on. (The addition or deletion of three nucleotides might not affect the resulting protein very much. Can you see why?)

TABLE 7.3

Types of Mutations and Their Effects on Organisms	
Types of Mutations	**Effects on Organisms**
Point Mutation	
Single base change in DNA with no change in the amino acid specified by the mRNA codon.	No effect on protein; a "silent" mutation.
Change in DNA with change in the amino acid sequence specified by the mRNA codon.	Change in protein by substitution of one amino acid for another; can significantly alter function of protein.
Change in DNA that creates a terminator codon in mRNA.	Produces polypeptide of no use to organism and prevents synthesis of normal protein.
Frameshift Mutation	
Deletion or insertion of one or more bases in DNA.	Changes entire sequence of codons and greatly alters amino acid sequence; can introduce terminator codon and produce useless polypeptides instead of normal proteins.

Studies of bacteria that have lost the ability to synthesize a particular enzyme have played an important role in our understanding of metabolic pathways. Such nutritionally deficient mutants are called **auxotrophs** (awks′o-trofs″; *auxo*, "increase," and *trophos*, "food"); they require special substances in their medium to maintain growth. In contrast to auxotrophs, normal, nonmutant forms are called **prototrophs** (pro′to-trofs), or *wild types*. Comparisons of characteristics of auxotrophs and prototrophs show the effects of a mutation on metabolism. By observing which metabolites accumulate and which nutrients must be added to the medium of auxotrophs, microbiologists have determined the specific steps in the metabolism of certain substances.

Still another type of phenotypic variation of genetic origin is temperature sensitivity. For example, suppose that an organism at one time could grow over a wide range of environmental temperatures. As a result of a mutation, it loses the ability to grow at the higher temperatures of its former range. It can still grow at 25°C, but it can no longer grow at 40°C. This phenomenon may be due to a point mutation that changed a single amino acid in an enzyme. The slightly altered enzyme may function at moderate temperatures but may be easily denatured and inactivated at higher temperatures.

Some phenotypic variations are caused by environmental factors and occur without any change in the genotype (alteration in DNA). For example, large amounts of sugar or irritants in the medium can cause some organisms to form a larger-than-normal capsule. Some organisms, such as the anthrax bacterium, form spores in open air, in spilled blood, or on tissue surfaces but not inside tissues. Variations in environmental temperature can affect pigment synthesis. *Serratia marcescens* usually produces pigment at room temperature but may not do so at higher temperatures. It has the gene for pigment production, but the gene is expressed only at certain temperatures.

SPONTANEOUS AND INDUCED MUTATIONS

Mutations appear to be random or chance events; it is usually impossible to predict when a mutation will occur or which genes will be altered. Although all mutations

When starved, E. coli increase their mutation rate, thereby increasing their chances of survival due to a helpful mutation.

result from permanent changes in DNA, they can be spontaneous or induced. **Spontaneous mutations** occur in the absence of any agent known to cause changes in DNA. They arise during the replication of DNA and appear to be due to errors in the base pairing of nucleotides in the old and new strands of DNA. Various genes in the DNA of bacteria have different spontaneous *mutation rates*, ranging from 10^{-3} to 10^{-9} per cell division. In other words, one gene might undergo a spontaneous mutation once in every thousand $(1/10^3)$ cell divisions, whereas another gene might undergo a spontaneous mutation only once in every billion $(1/10^9)$ cell divisions. **Induced mutations** are produced by agents called **mutagens**, which increase the mutation rate above the spontaneous mutation rate. Mutagens include chemical agents and radiation **(Table 7.4)**.

CHEMICAL MUTAGENS

Chemical mutagens act at the molecular level to alter the sequence of bases in DNA. They include base analogs, alkylating agents, deaminating agents, and acridine derivatives.

A **base analog** is a molecule quite similar in structure to one of the nitrogenous bases normally found in DNA. A cell may incorporate a base analog into its DNA in place of the normal base. For example, 5-bromouracil can be inserted in DNA instead of thymine **(Figure 7.18)**. When DNA containing 5-bromouracil is replicated, the analog can cause an error in base pairing. The 5-bromouracil that replaced thymine may pair with guanine instead of with adenine, which normally pairs with thymine. When DNA replicates in the presence of a significant quantity

Figure 7.18 Base analogs. The similarity between the structure of the base analog 5-bromouracil and the structure of the normal base thymine allows it in some cases to be taken up in place of thymine. The bromine (Br) group occupies an area about the same size as the methyl (CH_3) group.

of 5-bromouracil, the analog can be incorporated at many sites in the DNA molecule. A mutation occurs wherever the analog causes the insertion of guanine instead of adenine in the subsequent replication. Another purine base analog, caffeine, can cause mutations in an unborn child. For this reason pregnant women are advised to avoid or limit their caffeine intake.

Alkylating agents are substances that add alkyl groups (such as a methyl group, $—CH_3$) to other molecules. Adding an alkyl group to a nitrogenous base alters the shape of the base and can cause an error in base pairing. For example, the addition of a methyl group to guanine can cause it to pair with thymine instead of cytosine. Such a change can give rise to a point mutation. Some alkylating agents can cause several types of mutations: point mutations; frameshift mutations; and even breaks in chromosomes, resulting in very severe damage or death. The most infamous alkylating agent was probably mustard gas, used in World War I trench warfare, where it killed thousands of soldiers.

Deaminating agents such as nitrous acid (HNO_2) remove an amino group ($—NH_2$) from a nitrogenous

TABLE 7.4

Some Mutagens and Their Effects	
Mutagen	**Effects**
Chemical Agents	
Base analog *Examples:* caffeine, 5-bromouracil	Substitutes "look-alike" molecule for the normal nitrogenous base during DNA replication → point mutation.
Alkylating agent *Example:* nitrosoguanidine	Adds an alkyl group, such as methyl group ($—CH_3$), to nitrogenous base, resulting in incorrect pairing → point mutation.
Deaminating agent *Example:* nitrous acid, nitrates, nitrites	Removes an amino group ($—NH_2$) from a nitrogenous base → point mutation.
Acridine derivative *Example:* acridine dyes, quinacrine	Inserts into DNA ladder between backbones to form a new rung, distorting the helix → frameshift mutation.
Radiation	
Ultraviolet	Links adjacent pyrimidines to each other, as in thymine dimer formation, and thereby impairs replication.
X-ray and gamma ray	Ionize and break molecules in cells to form free radicals, which in turn break DNA.

base. Removing an amino group from adenine causes it to resemble guanine, and the deaminated base pairs with cytosine instead of thymine. Nitrates (NO_3^-) and nitrites (NO_2^-) are sometimes added to foods such as hot dogs and cold cuts for coloring, flavoring, or antibacterial action. The hazard of such additives is that, in the body, they form nitrosamines—deaminating agents known to cause birth defects, cancer, and other mutations in laboratory animals.

In contrast to these alterations, which cause point mutations, **acridine derivatives** cause frameshift mutations. The acridine molecule contains one pyrimidine ring and two benzene rings (**Figure 7.19**). This molecule or one of its derivatives can become inserted in the DNA double helix, displacing both members of a base pair. Such a modification distorts the helix and causes partial unwinding of the DNA strands. The distortion allows one or more bases to be added or deleted, and a frameshift mutation results. The drug quinacrine (Atabrine) is an acridine derivative that was used to treat malaria until other drugs with less unpleasant side effects were developed. It causes mutations in the malarial parasite and possibly in the human host that receives the drug.

RADIATION AS A MUTAGEN

Radiation such as X-rays and ultraviolet rays can act as a mutagen. Ultraviolet rays affect only the skin of humans because the rays lack energy for deeper penetration, but they have significant effects on microorganisms, which they penetrate easily. Ultraviolet lights are sometimes mounted in hospitals and laboratories to kill airborne bacteria. When ultraviolet rays strike DNA, they can cause adjacent pyrimidine bases to bond to each other, thereby creating a pyrimidine dimer. A **dimer** (di′mer) consists of two adjacent pyrimidines (two thymines, two cytosines, or thymine and cytosine) bonded together in a DNA strand (**Figure 7.20**). Binding of pyrimidines to each other prevents base pairing during replication of the adjacent DNA strand, so a gap is produced in the

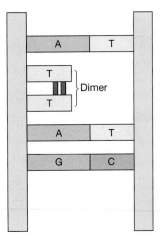

Figure 7.20 Thymine dimers caused by radiation. The formation of a dimer prevents the affected bases from pairing with bases in the complementary chain of DNA, impairing replication and preventing transcription.

replicated DNA. Transcription of mRNA stops at the gap, and the affected gene fails to transmit information.

X-rays and gamma rays, which are more energetic than ultraviolet radiation, easily break chemical bonds in molecules (◄Chapter 3, p. 55). The product is often a *free radical*, a highly reactive atom, molecule, or ion that in turn attacks other cell molecules, including DNA.

Until recently, microbiologists had no control over which genes underwent mutation when organisms were treated with mutagens. Now certain enzymes are available that greatly facilitate such studies. **Restriction endonucleases** cut DNA at precise base sequences, and **exonucleases** remove segments of DNA. These enzymes allow individual genes to be isolated and mutated at predetermined sites. The mutated gene can be inserted into a host's chromosome, and the effect of the specific mutation studied.

THE REPAIR OF DNA DAMAGE

Many bacteria, and other organisms as well, have enzymes that can repair certain kinds of damage to DNA. Two mechanisms, *light repair* and *dark repair*, are known to repair damage caused by dimers.

Light repair, or **photoreactivation**, occurs in the presence of visible light in bacteria previously exposed to ultraviolet light. When organisms containing dimers are kept in visible light, the light activates an enzyme that breaks the bonds between the pyrimidines of a dimer (**Figure 7.21a**). Thus, mutations that might have been passed along to daughter cells are corrected, and the DNA is returned to its normal state. This mechanism contributes to the survival of the bacteria but creates a problem for microbiologists. Cultures that are irradiated with ultraviolet light to induce mutations must be kept in the dark for the mutations to be retained.

Dark repair, which occurs in some bacteria, and can take place in the presence or absence of light, requires several enzyme-controlled reactions (**Figure 7.21b**). First, an endonuclease breaks the defective DNA strand near the dimer. Second, a DNA polymerase synthesizes

Backbone of DNA (deoxyribose and phosphate groups)

Acridine

Base pairs

Figure 7.19 Acridine, a chemical mutagen. Insertion of acridine into a DNA helix can produce a frameshift mutation.

Figure 7.21   Thymine dimer repairs.   (a) Light repair of DNA (photoreactivation) removes dimers. **(b)** In dark repair, a defective segment of DNA is cut out and replaced.

new DNA to replace the defective segment, using the normal complementary strand as a template. Third, an exonuclease removes the defective DNA segment. Finally, a ligase connects the repaired segment to the remainder of the DNA strand. These reactions were identified in *E. coli* but are now known to occur in many other bacteria. Human cells have similar mechanisms; some human skin cancers, such as xeroderma pigmentosum **(Figure 7.22)**, are caused by a defect in the cellular DNA repair mechanism.

THE STUDY OF MUTATIONS

Microorganisms are especially useful in studying mutations because of their short generation time and the relatively small expense of maintaining large populations of mutant organisms for study. Comparisons of normal and mutant organisms have led to important advances in the understanding of both genetic mechanisms and metabolic pathways. Microorganisms continue to be important to

researchers who are attempting to further our knowledge of these processes. However, the study of mutations is not without its problems. Two common problems are (1) distinguishing between spontaneous and induced mutations and (2) isolating particular mutants from a culture containing both mutated and normal organisms. The *fluctuation test* and the technique of *replica plating* are used to distinguish between spontaneous and induced mutations; replica plating also is used to isolate mutants.

Why is it important to differentiate between spontaneous and induced mutations? Making this distinction helps us understand mechanisms in the evolution of microorganisms and presumably other organisms as well. For example, some organisms are penicillin resistant—they grow in the presence of penicillin despite its antibiotic properties. Theoretically, there are two ways in which organisms can acquire such resistance: Either the penicillin *induces* a change in the organism that enables it to grow in the presence of penicillin, or a mutation occurs *spontaneously* that will allow the organism to grow

(a)

(b)

Figure 7.22 The inability to repair UV-caused dimers. (a) Sunbathers acquire dimers caused by UV radiation, which can cause skin cancer if they are not repaired. *(Steven Frame/Stock Boston)* **(b)** Xeroderma pigmentosum is a genetic disease in which the enzymes that normally repair UV damage to DNA are defective, and exposure to sunlight results in multiple skin cancers. *(Dr. Ken Greer/Visuals Unlimited)*

if it is later exposed to penicillin. In the latter case, penicillin will kill nonresistant organisms, thereby *selecting* for the resistant mutant. Various experiments, two of which are described next, have shown that the second mechanism, the selection of spontaneous mutants, is the primary means of evolution in microorganisms.

The **fluctuation test**, designed by Salvador Luria and Max Delbruck in 1943, is based on the following hypothesis: If mutations that confer resistance occur spontaneously and at random, we would expect great fluctuation in the number of resistant organisms per culture among a large number of cultures. This fluctuation would occur regardless of whether the substance to which resistance develops is present. A mutation might occur

early in the incubation period, late in that period, or not at all. Cultures with early mutations would contain many mutated progeny. Those with late mutations would have few mutated progeny, and those without a mutation would have none. An alternative hypothesis is that mutations conferring resistance to a substance occur only in the presence of the substance. Then cultures containing the substance would be expected to have approximately equal numbers of resistant organisms, whereas cultures lacking the substance would have no resistant organisms.

To test these hypotheses, Luria and Delbruck inoculated a large flask of liquid medium with a type of bacterium that was sensitive to the antibiotic streptomycin. At the same time they inoculated 100 small tubes of

BIOTECHNOLOGY

Ozone Biosensors

Because ozone (O_3) filters out harmful ultraviolet radiation, the discovery of holes in the ozone layer of the earth's atmosphere has raised concern about how much ultraviolet light reaches the earth's surface. Of particular concern are the questions of how deeply into seawater ultraviolet radiation penetrates and how it affects marine organisms, especially plankton (floating microorganisms) and viruses that attack plankton. Plankton form the base of the marine food chains and are believed to affect our planet's temperature and weather by their uptake of CO_2 for photosynthesis.

Deneb Karentz, a researcher at the Laboratory of Radiobiology and Environmental Health (University of California, San Francisco), has devised a simple method for measuring ultraviolet penetration and intensity. Working in the Antarctic Ocean, she submerged to various depths thin plastic bags containing special strains of *E. coli* that are almost totally unable to repair ultraviolet (UV) radiation damage to their DNA. Bacterial death rates in these bags were compared with rates in unexposed control bags of the same organism. The bacterial "biosensors" revealed constant,

significant ultraviolet damage at depths of 10 m and frequently at 20 and 30 m. Karentz plans additional studies of how ultraviolet may affect seasonal plankton blooms (growth spurts) in the oceans.

Collecting *E. coli* samples in the Antarctic Ocean to measure penetration and intensity. *(Courtesy Deneb Karentz, University of San Francisco)*

Figure 7.23 Fluctuation testing. Luria and Delbruck's fluctuation test proves that mutations that confer antibiotic resistance are random—they are not induced by exposure to the antibiotic.

liquid medium with the same bacteria. No streptomycin was present in either the flask or the tubes. Both the flask and the tubes were allowed to reach maximum growth (10^9 organisms per milliliter). One-milliliter samples were then used to inoculate agar plates containing streptomycin; a plate was made from each tube, and many plates were made from the flask. After 24 hours, the colonies on each plate were counted. Each colony represented a resistant mutant that could grow in the presence of streptomycin. There was far greater fluctuation in the number of colonies among the plates inoculated from the tubes than among the plates inoculated from the flask (**Figure 7.23**). Therefore, mutations must have occurred at different times or not at all in the various tubes. Mutations also must have occurred at different times in the flask, but progeny of mutated organisms became distributed through the medium, so that the number of mutants in each sample did not vary greatly. Luria and Delbruck concluded that resistance was conferred from random mutations occurring at different times among the organisms in the tubes and not from exposure to streptomycin. (Can you predict what results would have been obtained if resistance arose only from exposure to streptomycin?)

The technique of **replica plating**, devised by Joshua and Esther Lederberg in 1952, is also used to study mutations. Based on the same reasoning as the fluctuation test, it hypothesizes that resistance to a substance arises spontaneously and at random without the need for exposure to the substance. In the original replica plating studies (**Figure 7.24**), bacteria from a liquid culture were evenly spread on a master agar plate and allowed to grow for 4 to 5 hours. A sterile velveteen pad was then gently pressed

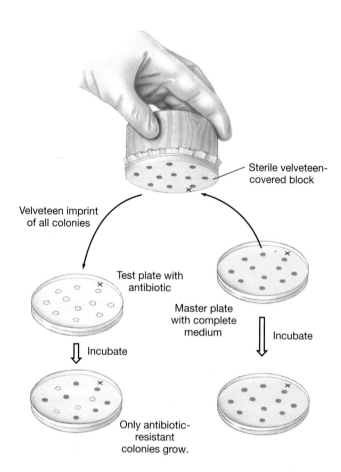

Figure 7.24 Replica plating. This technique allows detection of antibiotic-resistant organisms. The X on the side of the plate provides a reference for identifying colonies from the same organism.

APPLICATIONS

Polymerase Chain Reaction (PCR)—Key to Past and Future Worlds of DNA

Is there privacy in the grave? Not anymore. Ancient DNA, sometimes as old as 17 million years, is being recovered and its bases sequenced. Which genes did past organisms have? How many were handed down to us, and how many mutant versions do we have? DNA has been extracted from the brains of 91 prehistoric Native Americans buried and mummified in Florida peat bogs 7,500 years ago. That DNA is now undergoing analysis, thanks to **polymerase chain reaction (PCR)**, a technique that first became available in 1985. PCR allows us to produce rapidly (amplify) a billion copies of DNA without needing a living cell. These large quantities are then easily analyzed.

How does PCR amplification of DNA work? To replicate or make more copies of a piece of DNA you need to know a short sequence of nucleotides at the ends of the piece of DNA you wish to copy. Copies of this short sequence can be made in less than 24 hours in automated synthesizing equipment. These short sequences are called oligonucleotides (*oligos*, "few"). These oligonucleotides will act as a primer molecule by binding to the target DNA and providing a starting point for DNA synthesis in the PCR reaction. If the target DNA to be replicated is very long it may be cut into smaller pieces with enzymes called *restriction endonucleases* that cut at specific nucleotide sequences in DNA.

The sequence of events in PCR amplification is shown in the figure. The heating process (thermal cycling), which converts newly formed DNA into single strands, is repeated until it has produced billions of copies of the desired piece of DNA. The DNA is then easily detected (as in a clinical diagnostic test) or analyzed for total base sequence. Cutting a large piece of DNA into smaller pieces, sequencing the PCR-amplified quantities of the pieces, and then looking for overlaps at the ends allows us finally to determine the sequence of the entire original DNA piece.

Scientists have applied this tool to many questions concerning the past. For example, they have extracted DNA from fossil flies embedded in amber—and, ironically, did so shortly after the book *Jurassic Park* was written. They have also studied DNA from fossil leaves embedded in Idaho shale 17 million years ago (the DNA is very similar to that of modern magnolias) and from the bloodstains, hair, and bone chips preserved by doctors attending Abraham Lincoln at the time of his assassination. Lincoln is suspected of having had the hereditary disease Marfan's syndrome, which causes weakened arteries that can rupture and cause death. Most people with Marfan's syndrome die before they reach Lincoln's age. Would Lincoln have died soon had he not been assassinated? We can now create a library of Lincoln's DNA. As the Human Genome Project identifies the sequence of various genes (including those for Marfan's), we can match them to Lincoln's DNA and know with certainty which genes he had.

A modern forensic problem has brought the term *PCR* to the lips of the average American. Attorneys in the O. J. Simpson trial argued about DNA analysis and about the reliability of the PCR techniques in front of millions worldwide. Seldom has a scientific advance entered public awareness so rapidly. Prisoners already incarcerated for many years began asking for DNA

(a) *Enterobacter cloacae*. These bacteria were isolated and cultured from the remains of an 11,000-year-old mastodon fossil found in Newark, Ohio, in 1989. *(Courtesy Dr. Gerald Goldstein, Licking County Archaeological and Landmarks Society)* **(b)** The mastodon's remains. Its digestive tract could be identified as a darker-colored, discrete cylindrical mass bent into the shape of intestinal loops. No organisms were found in areas sampled adjacent to the intestine. *(Courtesy Licking County Archaeological and Landmarks Society)*

analysis of evidence from their trials. PCR made it possible to bring forth evidence that was not available earlier.

DNA analysis can be used to protect and free the innocent, as well as to convict the guilty. After PCR and DNA analysis were conducted on semen samples, one man convicted of rape was shown not to have been the rapist. His family had never lost faith in his innocence throughout the 10 years he had been locked up. (Analysis of semen in sexual assault cases has led to the freeing of 30% of initial suspects.)

These techniques are powerful forensic tools. Enough DNA can be recovered, and amplified by PCR, from the sweatband of a baseball cap to identify its wearer with extremely high certainty. DNA analysis of saliva on the back of a stamp on an envelope can identify the person who licked the stamp.

Rising from an 11,000-year-old grave in Ohio are cultures of two strains of *Enterobacter cloacae*, recovered alive but frozen in a state of suspended animation, from the intestine of a 4-ton mastodon that had been killed by prehistoric hunters, butchered, and then sunk into a peat bog (a primitive form of food preservation) (see the photos). And preserved they are! No mutations have occurred over the past 11,000 years. PCR analysis will reveal how these ancient organisms differ from today's strains of *E. cloacae*. Botanists awaited the tantalizing analysis of DNA from the mastodon's last meal: pollen, grains stuck in his teeth, swamp grass, mosses, leaves, and water lily. Chapters of evolutionary history are being rewritten.

For the living, PCR-amplified DNA analysis can reveal the presence of organisms that are difficult, dangerous, slow-growing, or require extra skill to culture in standard clinical laboratories. Tuberculosis cultures require 8 weeks to grow; PCR techniques will confirm the presence of the DNA of tuberculosis organisms in just hours. Medical-technology programs will need to train students in these techniques of the future, and current personnel will need to be retrained.

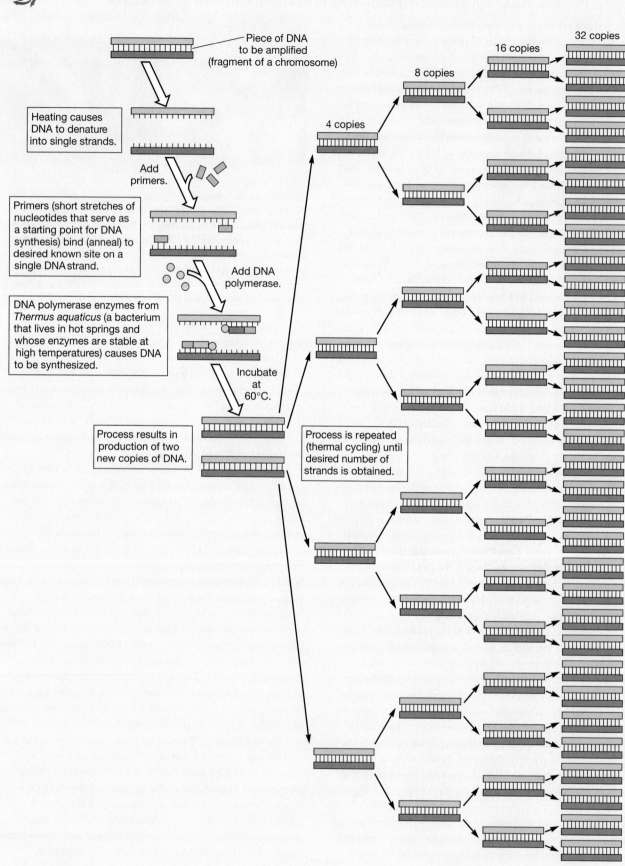

Piece of DNA to be amplified (fragment of a chromosome)

32 copies

16 copies

8 copies

4 copies

Heating causes DNA to denature into single strands.

Add primers.

Primers (short stretches of nucleotides that serve as a starting point for DNA synthesis) bind (anneal) to desired known site on a single DNA strand.

Add DNA polymerase.

DNA polymerase enzymes from *Thermus aquaticus* (a bacterium that lives in hot springs and whose enzymes are stable at high temperatures) causes DNA to be synthesized.

Incubate at 60°C.

Process results in production of two new copies of DNA.

Process is repeated (thermal cycling) until desired number of strands is obtained.

against the surface of the master plate to pick up organisms from each colony. The tiny fibers of velveteen acted like hundreds of tiny inoculating needles. The pad was carefully kept in the same orientation and used to inoculate an agar plate containing a substance such as penicillin, to which bacteria might be resistant. After incubation, the exact positions of corresponding colonies on the two plates were noted. The bacteria in colonies found on the penicillin plate had resistance to penicillin without ever having been exposed to it.

Replica plating not only demonstrates spontaneity of mutations that confer resistance, it also provides a means of isolating resistant organisms without exposing them to a substance. By keeping the velveteen pad in perfect alignment during the transfer process, colonies on the original plate that contain resistant organisms can be identified by their location relative to colonies on the master penicillin plate.

Replica plating is now widely used to study changes in the characteristics of many bacteria. The velveteen pads have been replaced by other materials that are easier to sterilize and manipulate. The technique is especially useful for identifying mutants whose nutritional needs have changed. Replicas can be transferred to a variety of different media, each deficient in a particular nutrient. Failure of particular colonies to grow on the deficient medium indicates that a mutation prevented the organism from synthesizing that nutrient.

THE AMES TEST

Human cancers can be induced by environmental substances that act by altering DNA. Much research effort is now being devoted to determining which substances are **carcinogens** (cancer-producing compounds). Carcinogens tend to be mutagenic, so determining whether a substance is mutagenic is often a first step in identifying it as a carcinogen. Bacteria, being subject to mutation and being easier and cheaper to study than larger organisms, are ideal organisms to use in screening substances for mutagenic properties. Proving that a substance causes mutations in bacteria does not prove that it does so in human cells. Even proving that a substance causes mutations in human cells does not prove that the mutations will lead to cancer. Additional tests, including tests in animals, are necessary to identify carcinogens, but initial screening with bacteria can eliminate some substances from further study. If a substance induces no mutations in a large population of bacteria, most researchers, including the U.S. Food and Drug Administration, believe that it is not likely to be a carcinogen.

The **Ames test** (Figure 7.25a), devised by the American microbiologist Bruce Ames, is used to test whether substances induce mutations in certain strains of *Salmonella* (auxotrophs) that have lost their ability to synthesize the amino acid histidine. These strains easily undergo another mutation that restores their ability to

(a)

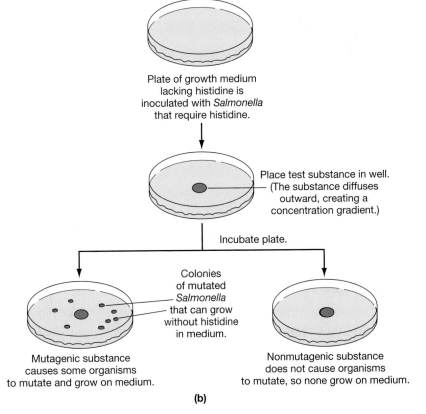

Plate of growth medium lacking histidine is inoculated with *Salmonella* that require histidine.

Place test substance in well. (The substance diffuses outward, creating a concentration gradient.)

Incubate plate.

Colonies of mutated *Salmonella* that can grow without histidine in medium.

Mutagenic substance causes some organisms to mutate and grow on medium.

Nonmutagenic substance does not cause organisms to mutate, so none grow on medium.

(b)

Figure 7.25 The Ames test for mutagenic properties of chemicals. **(a)** Plates used in the Ames test. *(Courtesy Bruce N. Ames, University of California at Berkeley).* **(b)** The test is used to determine whether a substance is a mutagen and therefore a potential carcinogen.

synthesize histidine. The Ames test is based on the hypothesis that if a substance is a mutagen, it will increase the rate at which these organisms revert to being histidine synthesizers (**Figure 7.25b**). Furthermore, the more powerful a substance's mutagenic capacity, the greater the number of reverted organisms it causes to appear. In practice, the organism is grown in the presence of a test substance. If any organisms regain the ability to synthesize histidine, the substance is suspected of being a mutagen. The larger the number of organisms that regain the synthetic ability, the stronger the substance's mutagenic capacity is likely to be.

✓ CHECKLIST

1. If you had to be exposed to a mutagen, would you rather it be one that causes point mutations or one that causes frameshift mutations? Why?

2. How can a change in genotype fail to produce a change in phenotype? Does a change in phenotype always require a change in genotype?

3. Why are nitrates and nitrites, which have been added to sausages to retard bacterial growth and thereby prevent food poisoning, a danger to humans?

4. Does exposure to antibiotics cause bacteria to mutate so as to be resistant to these antibiotics? How was this proven?

❚ RETRACING OUR STEPS

❙❙❙ AN OVERVIEW OF GENETIC PROCESSES

THE BASIS OF HEREDITY

- **Heredity** involves the transmission of information from an organism to its progeny.
- **Genes** are linear sequences of DNA that carry coded information for the structure and function of an organism.
- Prokaryotic chromosomes are threadlike circular structures made of DNA. Transmission of information in prokaryotes typically occurs during asexual reproduction in which the chromosome is reproduced (replicates), and each daughter cell receives a chromosome like the one in the parent cell. A very few bacteria contain two, and more rarely three, chromosomes per cell.
- **Mutations** (alterations in DNA) transmitted to progeny account for much of the variation in organisms.

NUCLEIC ACIDS IN INFORMATION STORAGE AND TRANSFER

- All information for the functioning of a cell is stored in DNA in a specific sequence of the nitrogenous bases: adenine, thymine, cytosine, and guanine.
- Information stored in DNA is used for two purposes: (1) to replicate DNA in preparation for cell division and (2) to provide information for protein synthesis. In both of these processes, information is transferred by base pairing.

❙❙❙ DNA REPLICATION

- **Replication** of bacterial DNA begins at a specific point in the circular chromosome and usually proceeds in both directions simultaneously. The main steps in DNA replication are summarized in Figure 7.4.
- DNA replication is **semiconservative**—each chromosome consists of one strand of old (parent) DNA and one of newly synthesized DNA.

❙❙❙ PROTEIN SYNTHESIS

TRANSCRIPTION

In **transcription**, **messenger RNA (mRNA)** is transcribed from DNA, as summarized in Figure 7.5, and serves as a **template** for protein synthesis.

KINDS OF RNA

- Besides mRNA, two other kinds of RNA are similarly produced and then used for protein synthesis: (1) **ribosomal RNA (rRNA)**, which combines with specific proteins to form ribosomes, the sites for protein assembly, and (2) **transfer RNA (tRNA)**, which carries amino acids to the assembly site.

TRANSLATION

- In the process of **translation**, three-base sequences in mRNA act as **codons** and are matched by base pairing with **anticodons** of tRNA. The mRNA codons constitute the **genetic code**—a code that is essentially the same for all living organisms, and determines the order in which specific amino acids are linked together to eventually form a protein.
- Once the mRNA and ribosomes are aligned, the process of protein synthesis proceeds as summarized in Figure 7.12 until a **terminator**, or **stop codon**, is reached.

❙❙❙ THE REGULATION OF METABOLISM

THE SIGNIFICANCE OF REGULATORY MECHANISMS

- Mechanisms that regulate metabolism turn reactions on and off in accordance with the needs of cells, allowing the cells to use various enegy sources and to limit synthesis of substances to the amounts needed.

CATEGORIES OF REGULATORY MECHANISMS

- The two basic categories of regulatory mechanisms are: (1) mechanisms that regulate the activity of enzymes already available in the cell and (2) mechanisms that regulate the action of genes, which determine what enzymes and other proteins will be available.

FEEDBACK INHIBITION

- In **feedback inhibition**, the end product of a biochemical pathway directly inhibits the first enzyme in the pathway (Figure 7.13).
- Enzymes subject to such regulation are generally allosteric.
- Feedback inhibition regulates the activity of existing enzymes and is a quick-acting control mechanism.

ENZYME INDUCTION

- In **enzyme induction** (Figure 7.14), the presence of a substrate activates an **operon**, a sequence of closely associated genes that include **structural genes** and **regulatory sites**: (1) In the absence of lactose, a **repressor**—a product of the **regulator** (*i*) **gene**—attaches to the operator and prevents transcription of the genes of the *lac* operon. (2) When lactose is present, it inactivates the repressor and allows transcription of the genes of the *lac* operon.

ENZYME REPRESSION

• In **enzyme repression**, the presence of a synthetic product inhibits its further synthesis by inactivating an operon: (1) When tryptophan is present, it attaches to the repressor protein and represses genes of the *trp* operon. (2) In the absence of tryptophan, the repressor is not activated, and genes of the *trp* operon are transcribed.

• In **catabolite repression**, the presence of a preferred nutrient (often glucose) represses the synthesis of enzymes that would be used to metabolize some alternative substance.

• Both enzyme induction and enzyme repression regulate by altering gene expression. The effect on enzyme synthesis in both cases depends on the presence or absence of the regulatory substance—lactose, tryptophan, or glucose in the preceding examples.

III MUTATIONS

TYPES OF MUTATIONS AND THEIR EFFECTS

• The genetic makeup of an organism is its **genotype**; the physical expression of the genotype is the **phenotype**.

• Mutations cause a change in the genotype of an organism; the change may or may not be expressed in the phenotype.

• Two major classes of mutations are (Table 7.3): (1) **point mutations**, which consist of changes in a single nucleotide, and (2) **frameshift mutations**, which consist of the **insertion** or **deletion** of one or more nucleotides.

• A third class of mutations involves movements of parts of chromosomes.

PHENOTYPIC VARIATION

• Phenotypic variations produced by mutations can be alterations in colony morphology, nutritional requirements, or temperature sensitivity.

SPONTANEOUS AND INDUCED MUTATIONS

• **Spontaneous mutations** occur in the absence of any known mutagen and appear to be due to errors in base pairing during DNA replication. Various genes have different rates of mutation.

• **Induced mutations** are mutations produced by agents called **mutagens**. Mutagens increase the mutation rate.

CHEMICAL MUTAGENS

• Chemical mutagens include **base analogs**, **alkylating agents**, **deaminating agents**, and **acridine derivatives**.

RADIATION AS A MUTAGEN

• **Radiation** often causes the formation of **dimers**—such as two adjacent pyrimidine bases that are bound to each other forming a thymine dimer which interferes with DNA replication.

THE REPAIR OF DNA DAMAGE

• Many bacteria have enzymes that can repair certain damages to DNA (Figure 7.20). (1) **Light repair** uses an enzyme that is activated by visible light and that breaks bonds between pyrimidines of a dimer. (2) **Dark repair** uses several enzymes that do not require light for activation; they excise defective DNA and replace it with DNA complementary to the normal DNA strand.

THE STUDY OF MUTATIONS

• Microorganisms are useful in studying mutations because many generations can be produced quickly and inexpensively.

• The **fluctuation test** demonstrates that resistance to chemical substances occurs spontaneously rather than being induced.

• **Replica plating** likewise demonstrates the spontaneous nature of mutations; it also can be used for isolating mutants without exposing them to the substance to which they are resistant.

THE AMES TEST

• The **Ames test** is based on the ability of **auxotrophic** bacteria to mutate by reverting to their original synthetic ability. It is used for screening chemicals for mutagenic properties, which indicate potential **carcinogens**.

▮ TERMINOLOGY CHECK

acridine derivative **(p. 200)**
alkylating agent **(p. 199)**
allele **(p. 181)**
Ames test **(p. 206)**
anticodon **(p. 188)**
antiparallel **(p. 184)**
attenuation **(p. 194)**
auxotroph **(p. 198)**
base analog **(p. 199)**
carcinogen **(p. 206)**
catabolite repression
    (p. 195)
chromosome **(p. 180)**
codon **(p. 188)**
constitutive enzyme
    (p. 192)

dark repair **(p. 200)**
deaminating agent **(p. 199)**
deletion **(p. 197)**
dimer **(p. 200)**
DNA polymerase **(p. 184)**
DNA replication **(p. 183)**
end-product inhibition
    (p. 191)
enzyme induction **(p. 194)**
enzyme repression **(p. 194)**
exon **(p. 186)**
exonuclease **(p. 200)**
feedback inhibition **(p. 191)**
fluctuation test **(p. 202)**
frameshift mutation **(p. 197)**
gene **(p. 181)**

genetic code **(p. 188)**
genetics **(p. 180)**
genotype **(p. 196)**
heredity **(p. 180)**
induced mutation **(p. 199)**
inducer **(p. 194)**
inducible enzyme **(p. 193)**
insertion **(p. 197)**
intron **(p. 186)**
lagging strand **(p. 184)**
leading strand **(p. 184)**
ligase **(p. 184)**
light repair **(p. 200)**
locus **(p. 181)**
messenger RNA (mRNA)
    (p. 186)

mutagen **(p. 199)**
mutation **(p. 181)**
nonsense codon **(p. 188)**
Okazaki fragment **(p. 184)**
operon **(p. 194)**
phenotype **(p. 196)**
photoreactivation **(p. 200)**
point mutation **(p. 197)**
polymerase chain reaction
    (p. 204)
polyribosome **(p. 189)**
prototroph **(p. 198)**
radiation **(p. 200)**
regulator gene **(p. 194)**
regulatory site **(p. 194)**
replica plating **(p. 203)**

replication fork *(p. 184)*
repressor *(p. 194)*
restriction endonuclease
    (p. 200)
reverse transcription
    (p. 183)

ribosomal RNA (rRNA)
    (p. 186)
RNA polymerase *(p. 186)*
RNA primer *(p. 184)*
semiconservative replication
    (p. 184)

sense codon *(p. 188)*
spontaneous mutation
    (p. 199)
start codon *(p. 188)*
stop codon *(p. 188)*
structural gene *(p. 194)*

template *(p. 183)*
terminator *(p. 188)*
transcription *(p. 183)*
transfer RNA (tRNA)
    (p. 188)
translation *(p. 183)*

▌ CLINICAL CASE STUDY

Cathy develops a painful cluster of blister-like lesions on her genitals. The doctor tells her that she has an infection with herpes virus, and gives her a prescription for acyclovir, an antiviral drug. Cathy goes home, uses the acyclovir faithfully according to directions, but gets no relief. Her good friend, Mary, also has had herpes for quite some time, and always gets relief when she uses acyclovir. Based on a conversation about this with Mary, Cathy wonders whether she really has herpes; maybe the doctor was wrong. She visits the doctor again, and he tells her that she probably has a mutant strain of herpes virus which is not affected by acyclovir, and gives her a different antiviral prescription. On the way home, Cathy wonders whether viruses can really mutate, and if the doctor really knows what he's talking about. What could you tell her?

▌ CRITICAL THINKING QUESTIONS

1. Prokaryotes usually have just one chromosome and carry just one gene for each trait. Human body cells, in contrast, carry duplicate genetic material: two genes for each trait. How does this affect the expression of genes in bacteria as opposed to human genes?

2. During the early stages of development of the earth's atmosphere, the planet was exposed to greater amounts of ultraviolet radiation than it is today. What do you suppose were the effects of this radiation on the longevity of individual organisms and on the rate of evolution of life forms?

3. Suppose you are a medical technician working in a hospital laboratory and you collected a sputum sample from a critically ill patient suspected of having tuberculosis. You of course go back to the lab and attempt to culture and identify this dangerous, fastidious, and slow-growing microorganism, a process that can take 8 weeks. However, a quick verification of the tuberculosis diagnosis is needed. Can you think of a possible method or technology that can help speed the process? If so, what is it and how does it work?

▌ SELF-QUIZ

1. Match the following with their description:

C Heredity
A Chromosome
E Phenotype
F Gene
G Alleles
B Mutation
D Genotype

(a) Threadlike molecule of DNA, typically circular in prokaryotes
(b) Permanent alteration in DNA
(c) Involves the transmission of information from an organism to its progeny
(d) Refers to the genetic information contained in the DNA of an organism (what it actually is)
(e) The specific characteristics displayed by the organism (what it appears to be)
(f) Linear sequence of DNA that carries coded instructions for structure and function of an organism
(g) Different forms of a gene found at a single location (locus)

2. All of the following are true about the information stored in DNA EXCEPT:
(a) All information for the functioning of a cell is stored in DNA in a specific sequence of the nitrogenous bases: adenine, thymine, guanine, and uracil.
(b) Information storage in DNA provides instructions for protein synthesis.
(c) Information storage in DNA is used to replicate DNA in preparation for cell division.
(d) DNA's information is transferred via base pairing.
(e) c and d.

3. How many chromosomes are found in a typical bacterial cell?
(a) 2
(b) 1
(c) 4
(d) 23
(e) 16

4. Without the action of DNA ligase, cells would not be able to complete their replication. What is the function of DNA ligase?
(a) Unzips the DNA double helix
(b) Stabilizes single-stranded DNA
(c) Binds DNA sequences together to generate a continuous strand
(d) Proofreads the replication process
(e) Creates a RNA copy of the DNA

5. Match the following terms with their respective description:

_C_ Semiconservative replication
_D_ Anticodon
_F_ Translation
_A_ Replication fork
_B_ Transcription
_E_ Okazaki fragment

(a) Point where the helix separates during DNA replication
(b) mRNA synthesized from a DNA template
(c) Each chromosome consists of one strand of old (parental) DNA and one of newly synthesized DNA
(d) Three bases that are complementary to a particular mRNA codon
(e) RNA-primed, short, discontinuously synthesized DNA fragment known as the lagging strand
(f) Production of polypeptide chain from the RNA template

6. From the DNA template sequence 3′-ATGCAGTAG-5′, what is the complementary messenger RNA sequence, transfer RNA anticodon sequences, and corresponding amino acids? Is there a terminator (nonsense) codon in the sequence? If so, what is it?

7. What type of RNA is involved in protein synthesis?
   D (mRNA & tRNA)

8. What type of RNA carries and transfers amino acids from the cytoplasm to the ribosome for placement in the synthesis of a polypeptide chain?
   (a) Messenger RNA
   (b) Transfer RNA
   (c) Ribosomal RNA
   (d) a and b
   (e) All of these

9. What type of RNA carries the genetic information required for protein synthesis?
   (a) Transfer RNA
   (b) Messenger RNA
   (c) Ribosomal RNA
   (d) All of these
   (e) None of these

10. What is the significance of the presence of mechanisms to regulate metabolism?

11. What factors distinguish the two basic regulatory mechanisms?

12. Match the following metabolic regulation terms with their descriptions:

_E_ Enzyme repression
_F_ Feedback inhibition
_A_ Catabolite repression
_C_ Enzyme induction
_D_ Repressor
_B_ Operon

(a) Presence of preferred nutrient represses synthesis of enzymes that would be used to metabolize an alternative substance
(b) Sequence of closely associated genes and regulatory sites that regulate enzyme production
(c) Presence of a substrate induces the activation of a gene which produces the corresponding enzyme

needed for the catabolism of this specific substrate
(d) A protein that binds to the operator preventing transcription of adjacent genes
(e) Presence of a synthetic product inhibits its further synthesis by inactivating its operon
(f) End product of a biochemical pathway directly inhibits the first enzyme in the pathway

13. For the *lac* operon, match the following:

_D_ Inducer
_E_ Place where repressor binds to shut off operon
_F_ Substance that binds to promoter site to start transcription
_D_ Combines with repressor to keep operon "on"
_C_ Z, Y, A
_A_ May be located some distance from the operon and is not under control of the promoter
_G_ Protein that binds to operator preventing transcription of structural genes

(a) Regulator gene
(b) Promoter
(c) Structural genes
(d) Lactose
(e) Operator
(f) RNA polymerase
(g) Repressor

14. Bacteria typically use repression to control:
   (a) Catabolic pathways
   (b) Anabolic pathways
   (c) Amphibolic pathways
   (d) Protein synthesis
   (e) DNA synthesis

15. Catabolite repression is regulated by the concentration of:
   (a) Lactose
   (b) Glucose
   (c) Messenger RNA
   (d) Amino acids
   (e) Active ribosomes

16. Two daughter cells would inherit which of the following changes from the parent cell:
   (a) A change in a protein
   (b) A change in a tRNA
   (c) A change in a rRNA
   (d) A change in a mRNA
   (e) A change in chromosomal DNA

17. A frameshift mutation occurs following the:
   (a) Insertion of one base
   (b) Insertion of more than one base
   (c) Deletion of one base
   (d) Deletion of more than one base
   (e) All of these

18. Radiation causes damage by causing the formation of dimers of:
   (a) Guanidine and cytosine
   (b) Cytosine and thymidine
   (c) Adenine and cytosine
   (d) Guanidine and adenine
   (e) Thymidine and adenine

19. The antibiotic streptomycin inhibits bacterial growth by binding to a protein in the 30S (subunit) of the ribosome. Based on this information, streptomycin inhibits:
(a) DNA synthesis
(b) Transcription in eukaryotes
(c) Translation in prokaryotes
(d) Translation in eukaroytes
(e) Transcription in prokaryotes

20. Suppose a point mutation occurred in the third position of a codon in a DNA template coding for a protein, changing it from TTT to TTC. What would be the consequences of this mutation?
(a) It would cause a frameshift mutation downstream of the point mutation, resulting in a different protein.
(b) It would cause a different amino acid to be placed at this position in the polypeptide, making it a mutant protein
(c) There would be no change in the amino acid at this position because of the redundancy of the genetic code. Such point mutations are known as "silent mutations" because although the genotype is different, the phenotype remains the same.
(d) All of the above.
(e) None of the above.

21. The sequence of bases in DNA can be altered by chemical mutagens. These cause changes by:
(a) Acting as base analogs and being incorporated into DNA
(b) Adding methyl groups to bases, leading to errors in base pairing
(c) Removing an amino group from a base
(d) Being inserted into double-stranded DNA
(e) All of these

22. (a) Why are bacteria useful in the study of mutations?
(b) How do spontaneous and induced mutations differ?
(c) What two tests are used to distinguish between the spontaneity or induction of a mutation and how do they work?
(d) What is the Ames test and what is it used for?

23. DNA damage in the form of dimers induced by UV light can be repaired by many bacteria using ___ ___ or

①light repair *② enzyme control*

photoreactivation, which occurs when the bacteria get back into the presence of visible light, or by dark repair, which can occur in the presence or absence of visible light and requires several ___ -controlled reactions.

24. Which of the following is NOT associated with prokaryotic cells?
(a) Semiconservative replication
(b) Inducible operons
(c) Lagging and leading strands
(d) Introns and exons
(e) Orgin of replication

25. The process used in the laboratory to produce millions of copies of DNA is:
(a) Ames assay
(b) *In situ* polymerization
(c) Fluctuation test
(d) Polymerase chain reaction (PCR)
(e) Reverse transcriptase

26. Name and describe the effects of the following mutations (read from left to right):

(a) Frameshift deletion
(b) Point base substitution
(c) Frameshift insertion of two bases

▌ EXPLORATIONS ON THE WEB

http://www.wiley.com/college/black

If you think you've mastered this chapter, there's more to challenge you on the web. Go to the companion web site to fine-tune your understanding of the chapter concepts and discover answers to the questions posed below.

1. Why do bacteria have a single, circular chromosome while eukaryotic cells have linear chromosomes?

2. Find out why DNA contains the information for **everything** that occurs within a cell—every action, every substance made, every event, every response.

3. What changes occur in just three genes in *Yersinia pestis* (plague) to transform it from a harmless, long-term inhabitant in the flea midgut to one that leads to a blood-feeding frenzy during which the flea regurgitates a mass of bacteria and thereby efficiently transmits the plague?

8

Gene Transfer and Genetic Engineering

Come with me...

Can man design and create new life forms, beginning with genes he has chosen, or even constructed? Well here's the man who is trying to do exactly that: Dr. J. Craig Venter. Time magazine has included him on its 2007 list of the 100 most influential people in the world. Already famous for his role in completing the Human Genome Project, he has now embarked on a new project, "synthetic biology."

In January 2006, Venter finished a 3-year-long voyage on his floating laboratory yacht, Sorcerer II, The map shows its route, circumnavigating the globe, retracing much of Charles Darwin's epic voyage on H. M. S. Beagle. Every 200 miles, samples of seawater were taken and filtered; the filter papers were then frozen and sent back to his labs in Rockville, Maryland where the DNA was extracted and sequenced. In an analysis of the material collected from the Sargasso Sea alone, he has found at least 1,800 new species and 1.2 million never before discovered genes. Just how many new species and genes will be found, when all the material has been sorted through, is mind-boggling! And, all this information is being posted free on the Internet, for use by any investigator.

© Ian White Photography

But what is Venter planning to do with the information? He has already made a totally new, synthetic bacteriophage virus, in one week, by stringing together genes in his laboratory. But most biologists do not consider viruses to be alive. His current aim is to use some of these new genes, along with other known ones, to create and patent a totally new bacterium, possibly to be named Mycobacterium laboratorium. He will have created life. His new company, Synthetic Genomics is dedicated to constructing such microbes to produce ethanol and hydrogen as alternative fuels. To quote him, "We're moving from reading the genetic code to writing it."

Wired © 2004, Conde Nast Publications

 Video related to this topic is available within WileyPLUS.

The transfer of genetic material from one organism to another can have far-reaching consequences. In microbes, it provides ways for viruses to introduce genetic information into bacteria and mechanisms for bacteria to increase their disease-causing capabilities or to become resistant to antibiotics. Information obtained from studying the transfer of genetic material between microorganisms can be applied to agricultural, industrial, and medical problems and to the unique problems of the prevention and treatment of infectious diseases. In this chapter we will discuss the mechanisms by which genetic transfers occur, and the significance of such transfers.

THE TYPES AND SIGNIFICANCE OF GENE TRANSFER

Gene transfer refers to the movement of genetic information between organisms. In most eukaryotes, it is an essential part of the organism's life cycle and usually occurs by sexual reproduction. Male and female parents produce *gametes* (sex cells), which unite to form a zygote, the first cell of a new individual. Because each parent produces many genetically different gametes, many different combinations of genetic material can be transferred to offspring. In bacteria, gene transfer is not an essential part of the life cycle. When it does take place, usually only some of the genes of the *donor* cell are transferred to the other participating, or *recipient*, cell. This combining of genes (DNA) from two different cells is called **recombination**, and the resulting cell is referred to as a *recombinant*.

When genes pass from parents to offspring, this is called **vertical gene transfer**. Sexual reproduction of plants and animals is what we usually think of as vertical gene transfer. In contrast, bacteria do vertical gene transfer when they reproduce asexually by binary fission. Furthermore, bacteria can also do horizontal, or **lateral**, **gene transfer**, when they pass genes to other microbes of their same generation. Before the 1920s, bacteria were thought to reproduce only by binary fission and to have no means of genetic transfer comparable to that achieved through sexual reproduction in eukaryotes. Since then, three mechanisms of lateral gene transfer in bacteria have been discovered, none of which is associated with reproduction. We will discuss each mechanism—*transformation*, *transduction*, and *conjugation*—in this chapter.

Gene transfer is significant because it greatly increases the genetic diversity of organisms. As noted in ◀Chapter 7, p. 196, mutations account for some genetic diversity, but gene transfer between organisms accounts for even more. When organisms are subjected to changing environmental conditions, genetic diversity increases the likelihood that some organisms will adapt to any particular condition. Such diversity leads to evolutionary changes. Organisms with genes that allow them to adapt

to an environment survive and reproduce, whereas organisms lacking those genes perish. If all organisms were genetically identical, all would survive and reproduce, or all would die. In ◀Chapter 9, we will discuss the newly found evidence that lateral gene transfer has been far more common in evolutionary history than was ever suspected. This discovery has led to major changes in how we now view evolutionary relationships.

In *recombinant DNA technology*, genes from one kind of organism are introduced by lateral transfer into the genome of another kind of organism (for example, when human genes are inserted into the cells of a pig). Genetic engineers have learned to artificially manipulate the three natural means of lateral gene transfer to create desired recombinant DNA and organisms. Let us now examine the three basic mechanisms.

TRANSFORMATION

THE DISCOVERY OF TRANSFORMATION

Bacterial **transformation**, a change in an organism's characteristics because of the transfer of genetic information, was discovered in 1928 by Frederick Griffith, an English military physician, while he was studying pneumococcal infections in mice. Pneumococci with capsules (◀Chapter 4, p. 97) produce smooth (S-type), glistening colonies. Those lacking capsules produce rough (R-type) colonies with a coarse, nonglistening appearance. Only the capsule-producing (encapsulated) pneumococci inoculated into mice were *pathogens*—that is, they had the power to cause disease (pneumonia). One such organism can multiply rapidly and kill a mouse! Mice are said to be "exquisitely" sensitive to pneumococci; therefore, they make excellent test animals. Capsules help prevent molecules produced by the mouse's immune system from reaching the surface of the bacterium. They also make it difficult for white blood cells to engulf the invading bacteria. In other words, the capsule protects the bacteria from the mouse's immune system.

Griffith injected one group of mice with heat-killed smooth pneumococci, a second group with live smooth pneumococci, a third group with live rough pneumococci, and a fourth group with a mixture of live rough and heat-killed smooth pneumococci (Figure 8.1). As expected, mice that received live smooth pneumococci developed pneumonia and died, whereas those that received either heat-killed smooth pneumococci or live rough pneumococci did not develop pneumonia and survived. Surprisingly, those that received the mixture also died of pneumonia. Imagine Griffith's shock when he isolated live smooth organisms from these mice. He had no way of knowing exactly what had happened, but he realized that some cells had been "transformed" from type R to type S. Moreover, the change was heritable. We now know that the R-type bacteria picked up naked DNA, liberated from

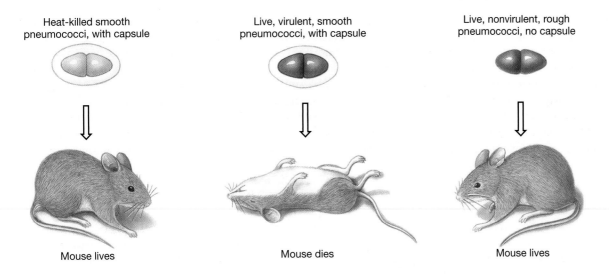

Heat-killed smooth pneumococci, with capsule

Live, virulent, smooth pneumococci, with capsule

Live, nonvirulent, rough pneumococci, no capsule

Mouse lives

Mouse dies

Mouse lives

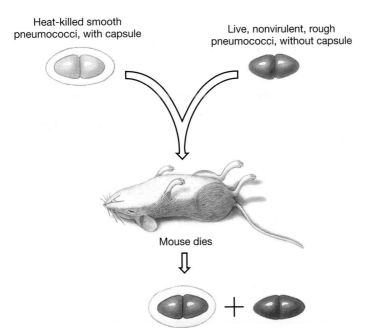

Heat-killed smooth pneumococci, with capsule

Live, nonvirulent, rough pneumococci, without capsule

Mouse dies

Live smooth pneumococci (with capsule) plus live rough pneumococci (without capsule) isolated from dead mouse

Figure 8.1 The discovery of transformation: Griffith's experiment with pneumococcal infections in mice. When S-type pneumococci (which produce smooth-appearing colonies, due to the presence of capsules) are injected into mice, the mice die of pneumonia. The mice survive when R-type pneumococci (which produce rough-appearing colonies, due to the lack of capsules) or heat-killed S-type pneumococci are injected. But when a mixture of live R-type and heat-killed S-type pneumococci—neither of which is lethal by itself—is injected, the mice die, and live S-type organisms as well as R-types are recovered from the dead animals.

disintegrated dead S-type bacteria, and incorporated it into their own DNA. R-type bacteria that picked up DNA that had genes for capsule production were genetically transformed into S-type organisms.

In subsequent studies of transformation, Oswald Avery discovered that a capsular polysaccharide was responsible for the virulence of pneumococci. In 1944, Avery, Colin MacLeod, and Maclyn McCarty isolated the substance responsible for the transformation of pneumococci and determined that it was DNA. In retrospect, this discovery marked the "birth" of molecular genetics, but at the time DNA was not known to carry genetic information. Researchers working with plant and animal chromosomes had isolated both DNA and protein from them, but they thought the genetic information was in the protein. Only when James Watson and Francis Crick determined

the structure of DNA did it become clear how DNA encodes genetic information. After this original work with pneumococci (now called *Streptococcus pneumoniae*), natural transformation was observed in organisms from a wide variety of genera, including *Acinetobacter, Bacillus, Haemophilus, Neisseria,* and *Staphylococcus,* and in the yeast *Saccharomyces cerevisiae.* In addition to natural transformation, scientists have discovered ways to artificially transform bacteria in the laboratory.

THE MECHANISM OF TRANSFORMATION

To study the mechanism of transformation, scientists extract DNA from donor organisms by a complex biochemical process that yields hundreds of naked DNA fragments from each bacterial chromosome. (*Naked DNA* is DNA that has been released from an organism, often after the cell is lysed, and the DNA is no longer incorporated into chromosomes or other structures.) When the extracted, naked DNA is placed in a medium with organisms capable of incorporating it, most organisms can take up a maximum of about 10 fragments, which is less than 5 percent of the amount of DNA normally present in the organism.

Uptake of DNA occurs only at a certain stage in a cell's growth cycle, in response to high cell density and depletion of nutrients. In this stage, a protein called **competence factor** is released into the medium and apparently facilitates the entry of DNA. When competence factor from one culture is used to treat a culture that lacks it, cells in the treated culture become *competent* to receive DNA—they can now take up DNA fragments. However, not all bacteria can become competent; thus not all bacteria can be transformed. DNA entry depends on such factors as modifications of the cell wall and the formation of specific receptor sites on the plasma membrane that can bind DNA. DNA transport proteins (proteins that bring DNA into the cell), and a DNA exonuclease (an enzyme that cuts up DNA) are also needed. Most naturally transformable bacteria will take up DNA from any source, exceptions are *Neisseria gonorrhoeae* and *Haemophilus influenzae* which only take up DNA from their own species. Specific nucleotide sequences in the DNA from these two species are recognized by the receptor protein on the surface of the competent bacteria in the same species (**Figure 8.2**).

Once DNA reaches the entry sites, endonucleases cut double-stranded DNA into units of 7,000 to 10,000 nucleotides. The strands separate, and only one strand enters the cell. Single-stranded DNA is vulnerable to attack by various nucleases and can enter a cell only if the nucleases on the cell surface somehow have been inactivated. Inside the cell, the donor single-stranded DNA must combine by base pairing with a portion of the recipient chromosome immediately or else be destroyed. In transformation, as well as in other mechanisms of gene transfer, the donor single-stranded DNA is positioned alongside the recipient DNA so that identical loci are next to one another. Splicing of a DNA strand involves breaking the strand, removing a segment, inserting a new segment, and attaching the ends. This process is called homologous recombination. Enzymes in the recipient cell excise (cut out) a portion of the recipient's DNA and replace (recombine) it with the donor DNA, which now becomes a permanent part of the recipient's chromosome. The leftover recipient DNA is subsequently broken down, so the number of nucleotides in the cell's DNA remains constant.

THE SIGNIFICANCE OF TRANSFORMATION

Although transformation has been observed mainly in the laboratory, it occurs in nature. It probably follows the breakdown of dead organisms in an environment where live ones of the same or a closely related species are present. However, the degree to which transformation contributes to the genetic diversity of organisms in nature is not fully known. In the laboratory, researchers induce transformation artificially, using chemicals, heat, cold, or a strong electric field, in order to study the effects of DNA

1. Naked DNA fragments from disintegrated cells in the area of a potential recipient cell. This cell must be of the correct genus and be in a state of competence, a proper physiologic condition, to permit entry of the DNA fragments.

2. Entry of naked DNA into competent cell

3. Recombination

Some DNA fragments replace (recombine with) original host cell DNA. The resultant recombinant cell is said to have been genetically transformed and will now express the foreign genes it has received and pass them on to all its offspring.

DNA that has not recombined is broken down by enzymes.

 Figure 8.2 **The mechanism of bacterial transformation.**

that differs from the DNA that the organism already has. Transformation also can be used to study the locations of genes on a chromosome and to insert DNA from one species into that of another species, thereby producing *recombinant* DNA.

TRANSDUCTION

THE DISCOVERY OF TRANSDUCTION

Transduction, like transformation, is a method of transferring genetic material from one bacterium to another (*trans*, "across," *ductio*, "to pull"; viruses carry or pull genes from one cell to another). Unlike transformation, in which naked DNA is transferred, in transduction DNA is carried by a **bacteriophage** (bak-ter'-e-o-faj)—a virus that can infect bacteria. The phenomenon of transduction was originally discovered in

Human beings are not the only ones infected by viruses. There are viruses that specifically infect bacteria, plants, other animals, fungi, algae, and even protozoa.

Salmonella in 1952 by Joshua Lederberg and Norton Zinder and has since been observed in many different genera of bacteria.

THE MECHANISMS OF TRANSDUCTION

To understand the mechanisms of transduction, we need to describe briefly the properties of bacteriophages, also called **phages** (faj'ez). Phages, which are described in more detail in ◄Chapter 10, are composed of a core of nucleic acid covered by a protein coat. They infect bacterial cells (*hosts*) and reproduce within them, as shown in **Figure 8.3**. A phage capable of infecting a bacterium attaches to a receptor site on the cell wall. The phage nucleic acid enters the bacterial cell after a phage enzyme weakens the cell wall. The protein coat remains outside, attached to the cell wall. Once the nucleic acid is in the cell further events follow one of two pathways, depending on whether the phage is *virulent* or *temperate*.

A **virulent phage** is capable of causing infection and, eventually, the destruction and death of a bacterial cell. Once the phage nucleic acid enters the cell, phage genes

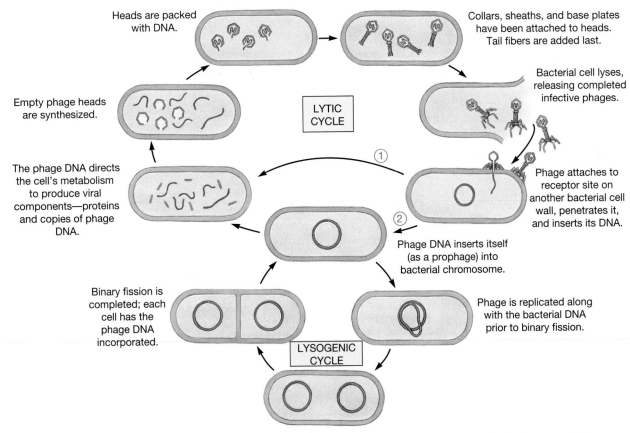

Figure 8.3 **Bacteriophage life cycles.** When a bacteriophage injects its viral DNA into a host bacterial cell, at least two different outcomes are possible. In the lytic cycle, characteristic of virulent phages, the phage DNA takes control of the cell and ① causes it to synthesize new viral components, which are assembled into whole viral particles. The cell lyses, releasing the infective viruses, which can then enter new host cells. In the lysogenic cycle, the DNA of a temperate phage enters the host cell, ② becomes incorporated into the bacterial chromosome as a prophage, and replicates along with the chromosome through many cell divisions. However, a lysogenic phage can suddenly revert to the lytic life cycle. A prophage is thus a sort of "time bomb" sitting inside the infected cell.

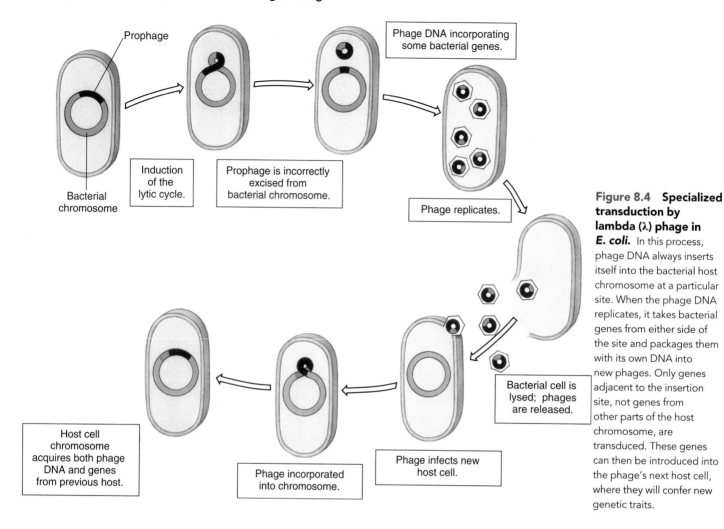

Prophage

Bacterial chromosome

Induction of the lytic cycle.

Prophage is incorrectly excised from bacterial chromosome.

Phage DNA incorporating some bacterial genes.

Phage replicates.

Bacterial cell is lysed; phages are released.

Host cell chromosome acquires both phage DNA and genes from previous host.

Phage incorporated into chromosome.

Phage infects new host cell.

Figure 8.4 Specialized transduction by lambda (λ) phage in E. coli. In this process, phage DNA always inserts itself into the bacterial host chromosome at a particular site. When the phage DNA replicates, it takes bacterial genes from either side of the site and packages them with its own DNA into new phages. Only genes adjacent to the insertion site, not genes from other parts of the host chromosome, are transduced. These genes can then be introduced into the phage's next host cell, where they will confer new genetic traits.

direct the cell to synthesize phage-specific nucleic acids and proteins. Some of the proteins destroy the host cell's DNA, whereas other proteins and the nucleic acids assemble into complete phages. When the cell becomes filled with a hundred or more phages, phage enzymes rupture the cell, releasing newly formed phages, which can then infect other cells. Because this cycle results in **lysis** (li′sis), or rupture, of the infected (host) cell, it is called a **lytic** (lit′ik) **cycle**.

A **temperate phage** ordinarily does not cause a disruptive infection. Instead the phage DNA is incorporated into a bacterium's DNA and is replicated with it. This phage also produces a repressor substance that prevents the destruction of bacterial DNA, and the phage's DNA does not direct the synthesis of phage particles. Phage DNA that is incorporated into the host bacterium's DNA is called a **prophage** (pro′faj). Persistence of a prophage without phage replication and destruction of the bacterial cell is called **lysogeny** (li-soj′e-ne), and cells containing a prophage are said to be **lysogenic** (li0so-jen′ik). Several ways to induce such cells to enter the lytic cycle are known, and most involve inactivation of the repressor substance.

Temperate phages can replicate themselves either as a prophage in a bacterial chromosome or independently by assembling into new phages. Transduction happens when, instead of only phage DNA being packed into newly forming phage heads, some bacterial DNA is also packed into the heads. Temperate phages can carry out both generalized and specialized forms of transduction. In *generalized transduction*, any bacterial gene can be transferred by the phage; in *specialized transduction*, only specific genes are transferred.

Specialized Transduction

Several lysogenic phages are known to carry out specialized transduction, but lambda (λ) phage in *Escherichia coli* has been extensively studied. Phages usually insert at a specific location when they integrate with a chromosome. Lambda phage inserts into the *E. coli* chromosome between the *gal* gene, which controls galactose use, and the *bio* gene, which controls biotin synthesis. The *gal* gene and *bio* gene are part of operons (◄Chapter 7, p. 194). When cells containing lambda phage are induced to enter the lytic cycle, genes of the phage form a loop and are excised from the bacterial chromosome **(Figure 8.4)**.

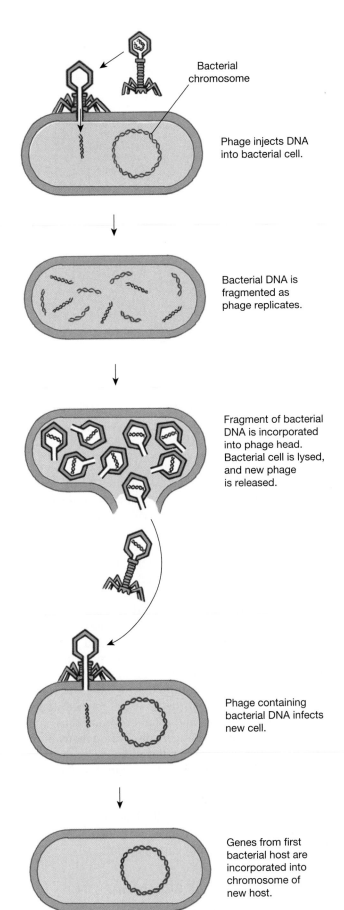

Phage injects DNA into bacterial cell.

Bacterial chromosome

Bacterial DNA is fragmented as phage replicates.

Fragment of bacterial DNA is incorporated into phage head. Bacterial cell is lysed, and new phage is released.

Phage containing bacterial DNA infects new cell.

Genes from first bacterial host are incorporated into chromosome of new host.

Lambda phage DNA then directs the synthesis and assembly of new phage particles, and the cell lyses.

In most cases, the new phage particles released contain only phage genes. Rarely (about one excision in a million) the phage contains one or more bacterial genes that were adjacent to the phage DNA when it was part of the bacterial chromosome. For example, the *gal* gene might be incorporated into a phage particle. When it infects another bacterial cell, the particle transfers not only the phage genes but also the *gal* gene. This process, in which a phage particle transduces (transfers) specific genes from one bacterial cell to another, is called **specialized transduction**. In specialized transduction, the bacterial DNA transduced is limited to one or a few genes lying adjacent to the prophage.

Generalized Transduction

When bacterial cells containing phage DNA enter the lytic cycle, phage enzymes break host cell DNA into many small segments (Figure 8.5). As the phage directs synthesis and assembly of new phage particles, it packages DNA by the "headful" (enough DNA to fill the head of a virus). This allows a bacterial DNA fragment occasionally to be incorporated into a phage particle. Likewise, DNA from plasmids, or other viruses infecting the cell can be packaged into the phage head. Once this phage particle, with its newly acquired bacterial DNA, leaves the infected host, it may infect another susceptible bacterium, thereby transferring the bacterial genes through the process of **generalized transduction**. Each bacterial fragment from the host cell chromosome has an equal chance of accidentally becoming a part of phage particles during the phage's replication cycle.

THE SIGNIFICANCE OF TRANSDUCTION

Transduction is significant for several reasons. First, it transfers genetic material from one bacterial cell to another and alters the genetic characteristics of the recipient cell. As demonstrated by the specialized transduction of the *gal* genes, a cell lacking the ability to metabolize galactose could acquire that ability. Other characteristics can also be transferred either by specialized or generalized transduction.

Second, the incorporation of phage DNA into a bacterial chromosome demonstrates a close evolutionary relationship between the prophage and the host bacterial cell. The DNA of the prophage and that of the host

Figure 8.5 Generalized transduction.
Bacteriophage infection of a host bacterium initiates the lytic cycle. The bacterial chromosome is broken into many fragments, any of which can be picked up and packaged along with phage DNA into new phage particles. When those particles are released and infect another bacterial cell, the new host acquires the genes that were brought along (transduced) from the previous bacterial host cell.

chromosome must have regions of quite similar base sequences. Otherwise, the prophage would not bind to the bacterial chromosome.

Third, the discovery that a prophage can exist in a cell for a long period of time suggests a similar possible mechanism for the viral origin of cancer. If a prophage can exist in a bacterial cell and at some point alter the expression of the cell's DNA, this could explain how animal viruses cause malignant changes. For example, viral genes inserted into a human chromosome could disrupt regulation of some genes, allowing structural genes to be active at the wrong times, continuously, or maybe even not at all. Fetal genes cause rapid proliferation of cells during early development, but growth soon slows down and eventually stops in adulthood. If these fetal genes were turned on again later in life in a group of cells, they could rapidly grow into a tumor. (Viruses and cancer are discussed in Chapter 10.)

Fourth, an interesting thought is that some animal viruses probably brought along genes from their previous host(s) when they infected new human hosts (such as you). These previous hosts were not necessarily humans. In that sense, you may not be entirely human now! You may be "transgenic."

Finally, and most importantly to molecular geneticists, transduction provides a way to study gene linkage. Genes are said to be *linked* when they are so close together on a DNA segment that they are likely to be transferred together. Different phages can be incorporated into a bacterial chromosome, each kind usually entering at a specific site. By studying many different phage transductions, scientists can determine where they were inserted on the chromosome and which adjacent genes they are capable of transferring. The combined findings of many such studies eventually allow identification of the sequence of genes in a chromosome. This procedure is called a **chromosome mapping**.

✓CHECKLIST

1. How does transformation differ from transduction?
2. How does gene transfer in prokaryotes differ from that in eukaryotes?
3. Which genes are transferred in generalized transduction? In specialized transduction?

▌▌▌ CONJUGATION

THE DISCOVERY OF CONJUGATION

In **conjugation**, like transformation and transduction, genetic information is transferred from one bacterial cell to another. Conjugation differs from those other mechanisms in two ways: (1) It requires contact between donor and recipient cells, and (2) it transfers much larger quantities of DNA (occasionally whole chromosomes).

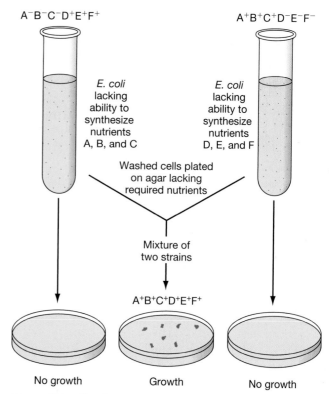

Figure 8.6 **The discovery of conjugation: Lederberg's experiment.**

Conjugation was first discovered in 1946 by Joshua Lederberg, who at that time was still a medical student. In his experiments, Lederberg used mutated strains of *E. coli* that were unable to synthesize certain substances. He selected two strains, each defective in a different synthetic pathway, and grew them in a nutrient-rich medium **(Figure 8.6)**. He removed cells from each culture and washed them to remove the residue of the nutrient medium. He then attempted to culture cells of each strain on agar plates that lacked the special nutrients needed by the strain. He also mixed cells from the two strains and plated them on the same medium. Whereas cells from the original cultures failed to grow, some from the mixed cultures did grow. The latter must have acquired the ability to synthesize all the substances they needed. Lederberg and others continued to study this phenomenon and eventually discovered many of the details of the mechanism of conjugation.

Lederberg was indeed fortunate in his choice of organisms, because similar studies of other strains of *E. coli* failed to demonstrate conjugation. In addition to the mutations that led to synthetic deficiencies in Lederberg's organisms, he also happened to use two *E. coli* cell types that were capable of conjugation.

THE MECHANISMS OF CONJUGATION

The mechanisms involved in conjugation were clarified through several important experiments, each of which built on the findings of the preceding one. Of those experiments, we will consider three: transfer of F plasmids, high-frequency recombinations, and transfer of F′ plasmids. Recall from ◄Chapter 4, p. 91, that **plasmids** are small *extrachromosomal DNA* molecules. Bacterial cells often contain several different plasmids that carry genetic information for various nonessential cell functions.

The Transfer of F Plasmids

After Lederberg's initial experiment, an important discovery about the mechanism of conjugation was made.

F plasmids are circular, double-stranded DNA molecules containing about 100,000 nucleotide pairs (that's the same as about 2 percent of a bacterial chromosome).

Two types of cells, called F⁺ and F⁻, were found to exist in any population of *Escherichia coli* capable of conjugating. **F⁺ cells** contain extrachromosomal DNA called **F (fertility) plasmids**; **F⁻ cells** lack F plasmids. (Lederberg coined the term *plasmid* in the 1950s to describe these fragments of DNA.)

Among the genetic information carried on the F plasmid is information for the synthesis of proteins that form F pili. The F⁺ cell makes an **F pilus** (or *sex pilus* or *conjugation pilus*), a bridge by which it attaches to the F⁻ cell when F⁺ and F⁻ cells conjugate **(Figure 8.7)** (◄Chapter 4, p. 96). A copy of the F plasmid

Figure 8.7 **TEM of F pili of *E. coli*.** (1 F⁺ and 2 F⁻; 18,000X.) Phages along the pili make them visible. Unlike the shorter attachment pili (fimbriae), this long type of pilus is used for transfer of genes in conjugation and is often called a sex pilus. (*Dr. L. Caro/Photo Researchers, Inc.*)

is then transferred from the F⁺ cell to the F⁻ cell **(Figure 8.8)**. F⁺ cells are called *donor* or *male* cells, and F⁻ cells are called *recipient* or *female* cells.

Although the exact transfer process remains unknown, the DNA is transferred as a single strand via a *conjugation bridge* (mating channel). Because the sex pilus contains a hole that would permit the passage of single-stranded DNA, it is possible, but uncertain, that DNA enters the recipient through this channel. However, there is also evidence to suggest that mating cells temporarily fuse, during which time the DNA is transferred. The pilus makes contact with a receptor site on the surface of the F⁻ (recipient) cell. A pore forms at this site. Inside the F⁻ cell, the pilus is pulled in and dismantled. This draws the two cells closer together. DNA from the F⁺ cell enters the F⁻ cell at this site. Each cell then synthesizes the complementary strand of DNA, so both have a complete F plasmid. Because all F⁻ cells in a mixed culture of F⁺ and F⁻ cells receive the F plasmid, the entire population quickly becomes F⁺; but in a culture of only F⁻ cells, no transfer occurs, and cells remain F⁻ cells.

High-Frequency Recombinations

The mechanisms of conjugation were further clarified when the Italian scientist L. L. Cavalli-Sforza isolated a **clone**, a group of identical cells descended from a single parent cell, from an F⁺ strain that could induce more than a thousand times the number of genetic recombinations seen in the F⁺ and F⁻ conjugations. Such a donor strain is called a **high frequency of recombination (Hfr) strain**.

Hfr strains arise from F⁺ strains when the F plasmid is incorporated into the bacterial chromosome at one of several possible sites **(Figure 8.9a)**. When an Hfr cell serves as a donor in conjugation, the F plasmid initiates transfer of chromosomal DNA. Usually, only part of the F plasmid, called the **initiating segment**, is transferred, along with some adjacent chromosomal genes **(Figure 8.9b)**. The recipient cell does not become an F⁺ donor cell, as only a part of the F plasmid is transferred.

In the 1950s the French scientist Elie Wollman and François Jacob studied this Hfr process in a series of interrupted mating experiments. They combined cells of an Hfr strain with cells of an F⁻ strain and removed samples of cells at short intervals. Each cell sample was subjected to mechanical agitation through vibration or whirling in a blender to disrupt the conjugation process. Cells from each sample were plated on a variety of media, each of which lacked a particular nutrient, to determine their nutrient requirements. By careful observation of the genetic characteristics of cells from many experiments, the investigators determined that transfer of DNA in conjugation occurred in a linear fashion and according to a precise time schedule. When conjugation was disrupted after 8 minutes, most recipient cells had received one gene. When it was disrupted after 120 minutes, recipient cells had received much more DNA, sometimes an entire chromosome. At intermediate intervals, the number of donor

Figure 8.9 High-frequency recombinations. **(a)** Conversion of F⁺ cells to the Hfr condition. Hfr cells arise from F⁺ cells when their F plasmid is incorporated into a bacterial chromosome at one of several possible sites. **(b)** During conjugation, the (pink) initiating site of the F plasmid and adjacent genes are transferred to a recipient cell. Genes are transferred in linear sequence, and the number of genes transferred depends on the duration of conjugation and whether the DNA strand breaks or remains intact.

Figure 8.8 An F⁺ × F⁻ mating. The F⁺ cell transfers one strand of DNA from its F plasmid to the F⁻ cell via the conjugation bridge. As this occurs, the complementary strands of F plasmid DNA are synthesized. Thus, the recipient cell gets a complete copy of the F plasmid, and the donor cell retains a complete copy.

genes transferred was proportional to the length of time conjugation was allowed to proceed. However, because of a tendency of chromosomes to break during transfer, some cells received fewer genes than would have been predicted by the time allowed. Whatever the number of genes transferred, they were always transferred in linear sequence from the initiation site created by the incorporation of the F plasmid.

The Transfer of F′ Plasmids

The process of incorporating an F plasmid into a bacterial chromosome is reversible. In other words, DNA incorporated into a chromosome can separate from it and again become an F plasmid. In some cases this separation occurs imprecisely, and a fragment of the chromosome is carried with the F plasmid, creating what is

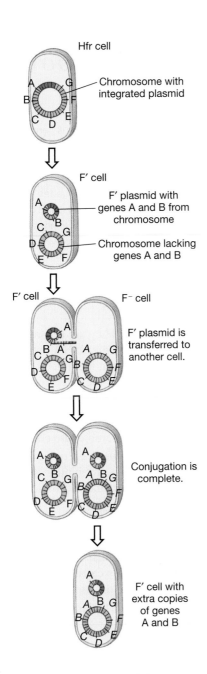

Figure 8.10 **The formation and transfer of F′ plasmids.**
When the F plasmid in an Hfr cell separates from the bacterial chromosome, it may carry some chromosomal DNA with it. Such an F′ plasmid may then be transferred by conjugation to an F⁻ cell. The recipient cell will then have two copies of some genes—one on its chromosome and one on the plasmid.

called an **F′ (F prime) plasmid** (Figure 8.10). Cells containing such plasmids are called *F′ strains*. When F′ cells conjugate with F⁻ cells, the whole plasmid (including the genes from the chromosome) is transferred. Hence, recipient cells have two of some chromosomal genes—one on the chromosomes and one associated with the plasmid. F′ plasmids generally do not become part of the recipient cell's chromosome.

In the transfer of F⁺ and F′ plasmids, as in all other transfers during conjugation, the donor cell retains all the genes it had prior to the transfer, including copies of the F plasmid. Single-stranded DNA is transferred, and both donor and recipient cells synthesize a complementary strand for any single-stranded DNA they contain.

The results of conjugation with F⁺, Hfr, and F′ transfers are summarized in **Table 8.1**.

THE SIGNIFICANCE OF CONJUGATION

Like other mechanisms for gene transfer, conjugation is significant because it contributes to genetic variation. Larger amounts of DNA are transferred in conjugation than in other transfers, so conjugation is especially important in increasing genetic diversity. In fact, conjugation may represent an evolutionary stage between the asexual processes of transduction and transformation and the actual fusion of whole cells (the gametes) that occur during sexual reproduction in eukaryotes. For the microbial geneticist, conjugation is of special significance because precise linear transfer of genes is useful in chromosome mapping.

Plasmids that are self-transmissible—that is, have genes for the formation of an F pilus—can sometimes transfer into species other than their own kind. Those that can are said to be **promiscuous**. Sometimes the species are only distantly related; in other cases, transfer even occurs into eukaryotic cells! Obviously this has important implications for health and evolution.

Some Gram-positive bacteria have self-transmissible plasmids that do not form F pili. Instead, bacteria lacking these plasmids secrete peptide compounds, which stimulate nearby bacteria that do contain the plasmids to mate with them. Once a bacterium acquires the plasmid, it stops producing the attracting peptide for it. This is a neat conservation of energy. However, these cells will still secrete other peptides that will serve as mating lures for other plasmids they have not yet acquired.

Take another look

TABLE 8.1

Results of Selected Conjugations			
Donor	Recipient	Molecule(s) Transferred	Product
F⁺	F⁻	F plasmid	F⁺ cells
Hfr	F⁻	Initiating segment of F plasmid and variable quantity of chromosomal DNA	F⁻ with variable quantity of chromosomal DNA
F′	F⁻	F′ plasmid and some chromosomal genes it carries with it	F′ cell with some duplicate gene pairs: one on chromosome, one on plasmid

GENE TRANSFER MECHANISMS COMPARED

The most fundamental differences among the major types of transfers of genetic information concern the quantity of DNA transferred and the mechanism by which the transfer takes place. In transformation, less than 1 percent of the DNA in one bacterial cell is transferred to another, and the transfer involves only chromosomal DNA.

In transduction, the quantity of DNA transferred varies from a few genes to large fragments of the chromosome, and a bacteriophage is always involved in the transfer. In specialized transduction, the phage inserts into a bacterial chromosome and carries a few host genes with it when it separates. In generalized transduction, the phage causes fragmentation of the bacterial chromosome; some of the fragments are accidentally packed into viruses as they are assembled.

In conjugation, the quantity of DNA transferred is highly variable, depending on the mechanism. A plasmid is always involved in the transfer. An F plasmid itself can be transferred, as occurs in F^+ and F^- conjugation. An initiating segment of a plasmid and any quantity of chromosomal DNA—from a few genes to the whole chromosome—is transferred in Hfr conjugation. A plasmid and whatever chromosomal genes it has carried with it from the chromosome are transferred in F' conjugation. These characteristics are summarized in **Table 8.2**.

PLASMIDS

CHARACTERISTICS OF PLASMIDS

The F plasmid just described was the first plasmid to be discovered. Since its discovery, many other plasmids have been identified. Most are circular, double-stranded,

extrachromosomal DNA. They are self-replicating by the same mechanism that any other DNA uses to replicate itself. Most plasmids have been identified by virtue of some recognizable function that they serve in a bacterium. These functions include the following:

1. F plasmids (fertility factors) direct the synthesis of proteins that self-assemble into conjugation pili.
2. *Resistance (R) plasmids* carry genes that provide resistance to various antibiotics such as chloramphenicol and tetracycline and to heavy metals such as arsenic and mercury.
3. Other plasmids direct the synthesis of bacteriocidal (bacteria-killing) proteins called *bacteriocins*.
4. Virulence plasmids, such as those in *Salmonella*, or the neurotoxin genes carried on plasmids in *Clostridium tetani*, cause disease signs and symptoms.
5. Tumor-inducing (Ti) plasmids can cause tumor formation in plants.

TABLE 8.2

Summary of the Effects of Various Transfers of Genetic Information	
Kind of Transfer	**Effects**
Transformation	Transfers less than 1 percent of cell's DNA. Requires competence factor. Changes certain characteristics of an organism, depending on which genes are transferred.
Transduction	Transfer is effected by a bacteriophage.
Specialized	Only genes near the prophage are transferred to another bacterium.
Generalized	Fragments of host bacterial DNA of variable length and number are packed into the head of a virus.
Conjugation	Transfer is effected by a plasmid.
F^+	A single plasmid is transferred.
Hfr	An initiating segment of a plasmid and a linear sequence of bacterial DNA that follows the initiating segment are transferred.
F'	A plasmid and whatever bacterial genes adhere to it when it leaves a bacterium are transferred.

6. Some plasmids contain genes for catabolic enzymes. Generally, plasmids carry genes that code for functions not essential for cell growth; the chromosome carries the genes that code for essential functions.

RESISTANCE PLASMIDS

Resistance plasmids, also known as *R plasmids* or *R factors*, were discovered when it was noted that some enteric bacteria—bacteria found in the digestive tract—had acquired resistance to several commonly used antibiotics. We don't know how resistance plasmids arise, but we know that they are not induced by antibiotics. This has been demonstrated by the observation that cultures kept in storage from a time prior to the use of antibiotics exhibited antibiotic resistance on first exposure to the drugs. However, antibiotics contribute to the survival of strains that contain resistance plasmids. That is, when a population of organisms containing both resistant and nonresistant organisms is exposed to an antibiotic, the resistant organisms will survive and multiply, whereas the nonresistant ones will be killed. The resistant organisms are thus said to be *selected* to survive. Such selection is a major force in evolutionary change, as Charles Darwin realized.

According to Darwin, all living organisms are subject to *natural selection*, the survival of organisms on the basis of their ability to adapt to their environment. After studying many different kinds of plants and animals, Darwin drew two important conclusions. First, living organisms have certain heritable—that is, genetic—characteristics that help them adapt to their environment. Second, when environmental conditions change, those organisms with characteristics that allow them to adapt to the new environment will survive and reproduce. Organisms lacking such characteristics will perish and leave no offspring. A change in environmental conditions does not directly cause organisms to change. It merely provides a test of their ability to adapt. Only the organisms that can carry out their life processes under the new conditions will survive.

Resistance plasmids (**Figure 8.11**) generally contain two components: a **resistance transfer factor** (**RTF**) and one or more **resistance (R) genes**. The DNA in an RTF is similar to that in F plasmids. The RTF implements transfer by conjugation of the whole resistance plasmid and is essential for the transfer of resistance from one organism to another. Each R gene carries information that confers resistance to a specific antibiotic or to a toxic metal. For antibiotic resistance, such genes usually direct synthesis of an enzyme that inactivates the antibiotic. Some resistance plasmids carry R genes for resistance to four widely used antibiotics: sulfanilamide, chloramphenicol, tetracycline, and streptomycin. Transfer of such a plasmid to any recipient confers resistance to all four antibiotics. Other resistance plasmids carry genes for resistance to one or more of these antibiotics. And a few plasmids carry genes for resistance to as many as 10 antibiotics.

The transfer of resistance plasmids from resistant to nonresistant organisms is rapid, so large numbers of previously nonresistant organisms can acquire resistance quickly. Furthermore, transfer of resistance plasmids occurs not only within a species, but also between closely related genera such as *Escherichia, Klebsiella, Salmonella, Serratia, Shigella,* and *Yersinia*. Transfer has even been observed between less closely related genera. Transfer of resistance plasmids is of great medical

Antibiotic resistance can be acquired either by mutation or by gene transfer.

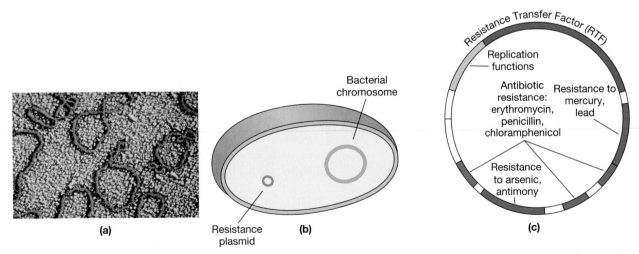

Figure 8.11 Resistance plasmids. **(a)** Resistance plasmids (magnified 268,000X). *(K. G. Murti/Visuals Unlimited)* **(b)** These circular pieces of DNA are much smaller than a bacterial chromosome. **(c)** A typical resistance plasmid can carry genes for resistance to various antibiotics and to inorganic toxic substances, sometimes used in disinfectants. The resistance transfer factor includes genes needed for the plasmid to undergo conjugation.

significance because it accounts for increasingly large populations of resistant organisms and reduces the effective use of antibiotics.

As scientists accumulate information on plasmids and how they confer antibiotic resistance, they become more concerned about the development of resistant strains and their potential danger to public health. As we shall see in Chapter 13, penicillin-resistant strains of *Neisseria gonorrhoeae*, *Haemophilus influenzae*, and some species of *Staphylococcus* already exist. Other antibiotics must now be used to treat the diseases caused by those strains, and the day may come when no antibiotic will effectively treat them. The more frequently antibiotics are used, the greater the selection is for resistant strains. Therefore, it is extremely important to identify the antibiotic to which an organism is most sensitive before using any antibiotic to treat a disease.

When antibiotics kill off good as well as harmful bacteria, this leaves "vacuums" into which other microbes can expand their populations, often causing side effects such as diarrhea. It would be better to stop the harmful activities of pathogens, and just leave them, and their neighbors, in place. A new way to do this involves ridding pathogens of the plasmids which carry the genes for toxin production and/or antibiotic resistance. As of January 2007, we have a new method to do this—**displacins**. These are pieces of DNA isolated from soil bacteria and attached to usually harmless *E. coli* bacteria. When let loose in a population of bacteria, the displacins move from the *E. coli* into the pathogens where they literally displace plasmids carrying harmful genes, leaving the former pathogenes now harmless.

CLOSE UP

Doing It the Hard Way

Geneticists receive respect these days for their research and discoveries, so you might be surprised to see geneticist Barbara McClintock's tiny Cold Spring Harbor laboratory. You might also be surprised to see a tennis racket, a pair of skates, an ironing board, and a hot plate among the Petri dishes and the Bunsen burners. Barbara McClintock worked and lived in tiny laboratory rooms before the days of research grants, huge laboratories, and teams of researchers, in a time when scientific research was considered beyond most women's abilities. Consequently, her mid-1940s reports on transposons were not accepted by most of her scientific peers until the 1970s, when genetic research verified her insights. But a life of being shunned by the scientific community did not stop her from establishing herself among the great geneticists of the twentieth century; her work, completed over 50 years ago, is highly esteemed and relevant today.

TRANSPOSONS

In addition to being transferred on resistance plamids by conjugation, R genes also can move from one plasmid to another in a cell or even become inserted in the chromosome. The ability of a genetic sequence to move from one location to another is called **transposition**. Such a mobile genetic sequence is called a **transposable element**. The simplest type of transposable element, an *insertion sequence*, contains a gene that codes for an enzyme (transposase) needed to transpose the insertion sequence; that gene is flanked on either side by a sequence of 9 to 41 nucleotides called *inverted repeats*. Transposable elements replicate only when in a plasmid or in a chromosome. During transposition, the insertion sequence is copied by the transposase and cellular enzymes. The copy randomly inserts itself into the bacterial chromosome or into another plasmid; the original insertion element remains in its original position. However, the ability to move among plasmids or to a chromosome greatly increases the ways in which a transposable element can affect the genetic makeup of a cell. The coding sequence or regulatory regions of any gene into which a transposable element inserts itself can be disrupted. Transposable elements are known to cause mutations and may be responsible for some spontaneous mutations.

A **transposon** (tranz-pos'on) is a transposable element that contains the genes for transposition, and one or more other genes as well (**Figure 8.12**). Typically, these other genes are for toxin production, or else are R genes, conferring resistance to antibiotics such as tetracycline, chloramphenicol, or ampicillin. Thus, a transposon can move R genes from one plasmid to another or to the bacterial chromosome. Viruses and plasmids can even move the transposon to a different cell, even one of a different species. Such movement can occur between prokaryotes and eukaryotes. The transposition of a transposon may disrupt gene function, depending on where it inserts. Most, however, insert between genes, rather than within a gene. Horticulturists are constantly on the look-out for new flower colors. Some of these are caused by transposons.

In 1983 Barbara McClintock won the Nobel Prize for her work on transposons, using corn. Transposons were next found in microorganisms and are now considered a universal phenomenon. Transposition is a relatively rare event and is not easily detected in eukaryotes. It is easier to detect in bacteria because researchers can work with large populations that can be tested more easily for particular characteristics.

BACTERIOCINOGENS

In 1925 the Belgian scientist André Gratia observed that some strains of *E. coli* release a protein that inhibits growth of other strains of the same organism. This allows them to compete more successfully for food and space against the other strains. About 20 such proteins,

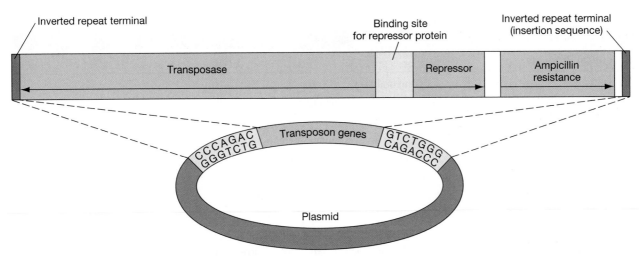

Figure 8.12 Transposons. A typical transposon is bounded by inverted repeat terminals—DNA segments with base sequences that are identical when read in opposite directions on different strands. A gene coding for a transposase enzyme that cuts DNA at these insertion sequences allows the transposon to cut itself into and out of plasmids and chromosomes. There is also a gene for a repressor protein that can keep the transposase gene from being transcribed.

called **colicins** (ko'leh-sinz), have been identified in *E. coli*, and similar proteins have been identified in many other bacteria. All these growth-inhibiting proteins are now called **bacteriocins** (bak-ter"e-o'sinz). Typically, bacteriocins inhibit growth only in other strains of the same species or in closely related species.

Bacteriocin production is directed by a plasmid called a **bacteriocinogen** (bak-ter"e-o-sin'o-jen). Although in most situations bacteriocinogens are repressed, in some cases the plasmid escapes repression and causes synthesis of its bacteriocin. Ultraviolet radiation can induce the formation and release of bacteriocin. When a bacteriocin is released, it can have a very potent effect on susceptible cells; one molecule of bacteriocin can kill a bacterium.

The mechanisms of action of bacteriocins are quite variable. Some enter a bacterial cell and destroy DNA. Others arrest protein synthesis by disrupting the molecular structure of ribosomes required for protein synthesis. Still others act on cell membranes by inhibiting active transport or by increasing membrane permeability to ions.

✓**CHECKLIST**

1. What do the designations F^+, F^-, F', and Hfr mean?
2. Are resistance plasmids induced by antibiotics?
3. How are resistance genes moved by transposons?
4. List several ways that insertion of a transposon could interfere with functioning of the *lac* operon (Figure 7.14).
5. What are displacins? What do they do?

BIOTECHNOLOGY

Bacteria That Don't Deserve a Bad Rap

In the news these days, we hear stories about bacteria causing outbreaks of food poisoning. But did you know that bacteria can actually keep food from spoiling? Naturally occurring proteins from certain bacteria, called bacteriocins, can control the growth of bacteria strains responsible for food spoilage. The U.S. Food and Drug Administration (FDA) has already approved the use of one bacteriocin, nisin, in pasteurized cheese; it inhibits the growth of *Clostridium botulinum*, a microbial strain that causes food poisoning. Scientists are researching other bacteriocins that kill nearly all of the bacterial cells associated with lean and fat beef tissues. We may soon have bacteria to thank for milk and meat products that are safer and stay fresh longer.

▌▌▌ GENETIC ENGINEERING

Genetic engineering refers to the purposeful manipulation of genetic material to alter the characteristics of an organism in a desired way (Table 8.3). Various methods of genetic manipulation enable microbial geneticists to create new combinations of genetic material in microbes. The transfer of genes between different members of the same species occurs in nature and has been done in the laboratory for several decades. Lederberg's experiment (Figure 8.6) is one example of such a technique. The transfer of genes between different species also is now possible. Here we will discuss five techniques of genetic engineering: *genetic fusion, protoplast fusion, gene amplification, recombinant DNA technology*, and the creation of *hybridomas*.

TABLE 8.3

Some Products and Applications of Genetic Engineering	
Pharmaceutical Products	**Use**
Human insulin	Treat diabetes
Human growth hormone	Prevent pituitary dwarfism
Blood clotting factor VIII	Treat hemophilia
Erythropoietin	Treat anemia; stimulate formation of new red blood cells
Alpha-, beta-, and gamma-interferon	Treat cancer and viral disease
Tumor necrosis factor	Disintegrate cancer cells
Interleukin-2	Treat cancer and immunodeficiencies
Tissue plasminogen activator	Treat heart attacks, dissolve clots
Taxol	Treat ovarian and breast cancers
Bone growth factor	Heal bone fractures, treat osteoporosis, stimulate bone growth
Epidermal growth factor	Heal wounds
Monoclonal antibodies	Diagnose and treat diseases
Hepatitis A and B vaccines	Prevent hepatitis
AIDS subunit vaccine (in clinical trials)	Incomplete virus vaccine
Human hemoglobin	Blood substitute in emergencies (produced in gene-altered pigs)
Antibiotics	Inhibit or kill microbial growth (increase yields by gene amplification)
Genetic Studies	
DNA and RNA probes	Identify organisms, diseases, genetic defects in fetuses and adults
Gene therapy	Insert missing gene, or replace defective gene, in adults or in egg and sperm; treat cystic fibrosis
Gene libraries	Understand gene structure and function, relatedness of organisms, Human Genome Project
Industrial Applications	
Oil-eating recombinant bacteria	Clean up oil spills, remove oil residue from empty tankers
Pollutant/toxic materials-degrading recombinant bacteria	Clean up contaminated sites
Enzymes, vitamins, amino acids, industrial chemicals	Various uses (yield increased by gene amplification in producing microbes)
Agricultural Applications	
Frostban bacteria (*Pseudomonas syringae*)	Prevent frost damage to strawberry crops
Breeding new types of plants and animals	Provide food, decoration, other uses
Herbicide-resistant crop plants	Allow crop plants to survive weeding done by spraying with herbicides; only crop plants survive
Viruses used as insecticides	Infect and kill insect pests

GENETIC FUSION

Genetic fusion allows transposition of genes from one location on a chromosome to another. It can also involve deletion of a DNA segment, resulting in the coupling of portions of two operons. For example, suppose that the *gal* operon, which regulates galactose use, and the *bio* operon, which regulates biotin synthesis, lie adjacent to each other on a chromosome **(Figure 8.13)**. Deletion of the control genes of the *bio* operon and subsequent coupling of the operons would constitute genetic fusion. Such fusion would allow the genes that control the use of galactose to control the entire operon, including the making of the enzymes involved in biotin synthesis.

The major application of genetic fusion within a species, as just described, are in research studies on the properties of microbes. However, the techniques developed for genetic fusion experiments have been extended and modified in the development of other kinds of genetic engineering.

One application of genetic fusion involves *Pseudomonas syringae*, a bacterium that grows on plants.

Figure 8.13 Genetic fusion. A possible example in which the deletion of a part of a chromosome causes two different adjacent operons to be joined together. The control mechanisms of the first operon will now govern the expression of the genes that were originally part of the second operon.

Genetically altered strains have been developed that increase the resistance of plants, such as potatoes and strawberries, to frost damage. Strains of this bacterium, which occur naturally on the leaves of plants, produce a protein that forms a nucleus for the formation of ice crystals. The ice crystals damage the plants by causing cracks in the cells and leaves. By removing part of the gene that produces the "ice crystal" protein, scientists have engineered strains of *P. syringae* that cannot make the protein. When organisms of this strain are sprayed on the leaves of plants, they crowd out the naturally occurring strain. The treated plants then become resistant to frost damage at temperatures as low as −5°C. (See box at end of this chapter.)

PROTOPLAST FUSION

A **protoplast** is an organism with its cell wall removed. **Protoplast fusion** (Figure 8.14) is accomplished by enzymatically removing the cell walls of organisms of two strains and mixing the resulting protoplasts. This allows fusion of the cells and their genetic material; that is, material from one strain recombines with that from the other strain before new cell walls are produced. Although genetic recombination occurs in nature in about one in a million cells, it occurs in protoplast fusion in as many as one in five cells. Thus, protoplast fusion simply speeds up a process that occurs in a very limited way in nature.

By mixing two strains, each of which has a desirable characteristic, new strains that have both characteristics can be produced. For example, a slow-growing strain that produces large quantities of a desired substance can be mixed with a fast-growing, poor producer. After protoplast fusion, some organisms will probably be fast-growing, good producers of the substance. Other organisms that turn out to be slow-growing, poor producers are discarded. Alternatively, two good producers can be

mixed to obtain a super producer. This has been done with two strains of the bacterium *Nocardia lactamdurans*, which produce the antibiotic cephalomycin. The new strains produce 10 to 15 percent more antibiotic than the best of the parent strains.

Protoplast fusion works best between strains of the same species. It has been accomplished in molds, however, between two species of the same genus (*Aspergillus nidulans* and *A. rugulosus*) and even between two genera of yeasts (*Candida* and *Endomycopsis*).

Microbiologists are exploring possible applications of protoplast fusion. It offers great promise for the future as procedures are refined and useful strains are developed. Also, both organisms' entire genomes, not just individual genes, are transferred, thus avoiding difficult maneuvers like transferring correct promoters.

Figure 8.14 Protoplast fusion. A colorized TEM of two tobacco plant leaf cells (magnified 775X) undergoing protoplast fusion. Protoplast fusion involves the use of enzymes to digest away the cell walls of organisms from two different strains. When placed together, the cells fuse and develop a new cell wall around the hybrid cell that contains the genes of both organisms. (*Dr. Jeremy Burgess/Photo Researchers, Inc.*)

GENE AMPLIFICATION

Gene amplification is a process by which plasmids, or in some cases bacteriophages, are induced to reproduce within cells at a rapid rate. If the genes required for the production of a substance are in the plasmids or can be moved to them, increasing the number of plasmids will increase production of the substance by the host cells.

Most bacteria and many fungi, including those that produce antibiotics, contain plasmids. Such plasmids, which often carry genes for antibiotic synthesis, provide many opportunities for using gene amplification to increase antibiotic yields. Even when genes concerned with antibiotic production are in the chromosome, scientists can transfer them to plasmids. Increased reproduction of plasmids would then greatly increase the number of copies of genes that act in antibiotic synthesis. This, in turn, would significantly increase the amount of antibiotic such cells could produce.

The possible applications of gene amplification are not limited to increasing antibiotic production. In fact, gene amplification may turn out to be even more effective in increasing production of substances that are synthesized by somewhat simpler pathways. These substances include enzymes and other products such as amino acids, vitamins, and nucleotides.

Rapid reproduction of bacteriophages already can be used to make the amino acid tryptophan. Bacteriophages carrying the *trp* operon (genes that control synthesis of enzymes to make tryptophan) of *E. coli* are induced to reproduce rapidly. Thus, cells containing large numbers of copies of the *trp* operon synthesize large quantities of the enzymes. Subsequent analysis of such cells has shown that half the intracellular proteins are enzymes for tryptophan synthesis.

RECOMBINANT DNA TECHNOLOGY

One of the most useful of all techniques of genetic engineering is the production of **recombinant DNA**—DNA that contains information from two different species of organisms. If these genes integrate permanently into the egg or sperm cells so that the genes can be transferred to offspring, the resulting organism is said to be a **transgenic**, or *recombinant*, organism. Making recombinant DNA involves three processes:

1. The manipulation of DNA *in vitro*—that is, outside cells. (*In vitro* literally means "in glass.")
2. The recombination of another organism's DNA with bacterial DNA in a phage or a plasmid.
3. The *cloning*, or production of many genetically identical progeny, of phages or plasmids that carry foreign DNA.

These processes were first carried out in 1972 by Paul Berg and A. D. Kaiser, who inserted other prokaryotic DNA into bacteria, and then by S. N. Cohen and Herbert Boyer, who inserted eukaryotic DNA into bacteria.

DNA from either prokaryotic or eukaryotic cells is removed from the cells and cut into small segments. The donor DNA segments are then incorporated into a **vector**, or self-replicating carrier such as a phage or a plasmid (**Figure 8.15**). First, a restriction endonuclease (an enzyme that cuts DNA at specific nucleotide sequences) is used to cut double-stranded DNA in the vector and donor DNA. The cuts leave overlapping ends. A particular restriction endonuclease always produces the same complementary ends. Then, donor DNA is incorporated into the vector by an enzyme called a *ligase*, which reunites the ends of nucleotide chains. Thus, the vector contains all the original DNA plus a new segment of donor DNA.

Once this new segment of DNA is inserted into the vector, it can be introduced into cells such as *E. coli* that have been rendered competent by heating (annealing) in a solution of calcium chloride or by **electroporation**. This technique uses a brief electrical pulse to produce temporary pores in the cell membrane through which the vector can pass. As the *E. coli* cells divide, the vectors in them also are reproduced by cloning. Such vector-containing cells can be identified, grown, and lysed, and the vectors containing a specific cloned segment of donor DNA can be retrieved. In other cases, large quantities of the protein product expressed by the donor gene can be obtained.

Restriction endonucleases are often just called **restriction enzymes**. Hundreds of kinds of restriction enzymes exist in various bacterial cells. There they protect the bacteria against bacteriophage infection by cutting up the foreign phage DNA into tiny segments. This ability to limit phage growth is called *restriction*. The bacteria protect their own DNA by adding methyl groups (during DNA synthesis) to sites where their own restriction enzymes would otherwise cut across. Each type of restriction enzyme recognizes a particular sequence of four to eight base pairs of DNA, which it then "snips" (cuts) across, creating what is called a **restriction fragment**. Enzymes that make a straight, or "blunt," cut across both strands of DNA create DNA pieces that are difficult to splice onto another DNA strand. Other DNA pieces that are cut at staggered sites (Figure 8.15) have a short, single-stranded piece sticking out at each end. These have been nicknamed "sticky" ends because they easily combine with complementary single-stranded portions of target DNA molecules from *any* other organism. Ligase enzymes then seal the short pieces into a chromosome or a plasmid.

Scientists have isolated and purified hundreds of bacterial restriction enzymes, each with a known target.

Bacteria, genetically engineered to be bioluminescent, can be inoculated into mice. Their transport and related disease can be followed by monitoring the glow.

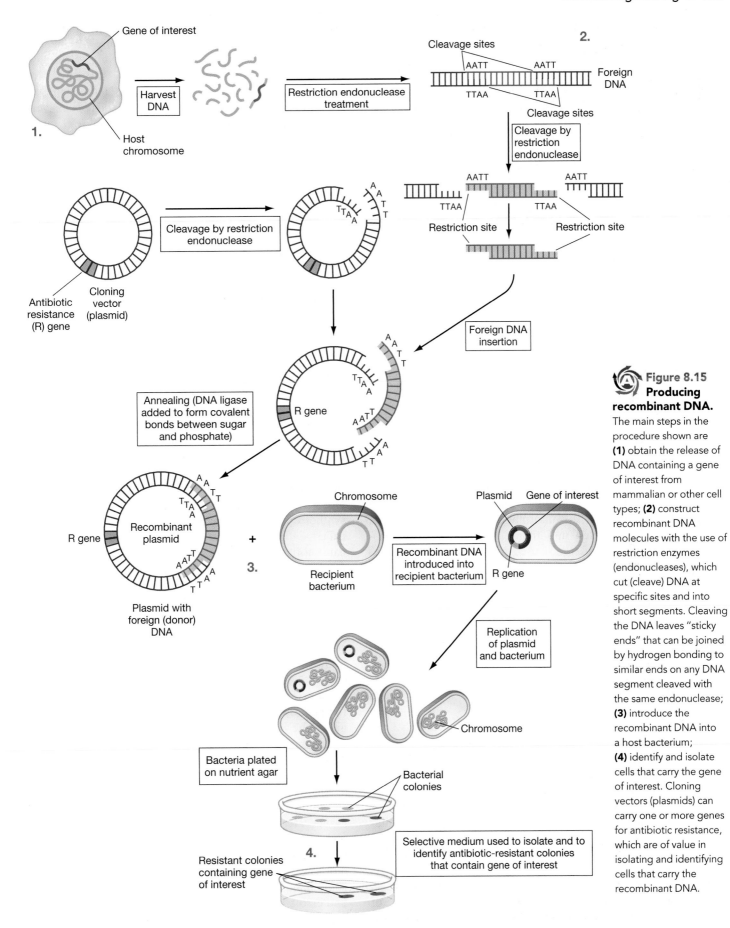

Figure 8.15
Producing recombinant DNA. The main steps in the procedure shown are (1) obtain the release of DNA containing a gene of interest from mammalian or other cell types; (2) construct recombinant DNA molecules with the use of restriction enzymes (endonucleases), which cut (cleave) DNA at specific sites and into short segments. Cleaving the DNA leaves "sticky ends" that can be joined by hydrogen bonding to similar ends on any DNA segment cleaved with the same endonuclease; (3) introduce the recombinant DNA into a host bacterium; (4) identify and isolate cells that carry the gene of interest. Cloning vectors (plasmids) can carry one or more genes for antibiotic resistance, which are of value in isolating and identifying cells that carry the recombinant DNA.

Using these as tools, they can now remove and/or splice in specific genes at exactly chosen sites on chromosomes or plasmids. This is how the human insulin gene was cut out of a human chromosome and inserted into a bacterial plasmid. Restriction enzymes are named for the organism from which they have been isolated, using the first letter of the genus name and the first two letters of the species designation, plus the number of the endonuclease, i.e., the order in which it was discovered. Thus, EcoRI is the first endonuclease found in *Escherichia coli* type-R, and HpaI is from *Haemophilus parainfluenzae*.

DNA sequences can be slightly different between members of the same species. This leads to what are called **restriction fragment length polymorphisms** (**RFLPs**); polymorphic (*poly*, "many," *morphos*, "shape, size"). When DNA from different individuals within a species is cut with the same restriction enzymes, restriction fragments will have different lengths due to deletions or insertions of DNA between the sites where the restriction enzyme cuts. RFLPs can be used to determine the ancestry of an individual, identify DNA from a specific individual, determine the location of genes responsible for genetic diseases, or identify new inserted genes or DNA sequences.

Medical Applications of Recombinant DNA

One of the most medically significant applications of recombinant DNA technology is the modification of bacterial cells to make substances useful to humans. To make bacterial cells produce human proteins, a human DNA gene with the information for synthesizing the protein is inserted into the vector. Interferon, a substance used to treat certain viral infections and cancers (Chapter 16), and the hormone insulin were among the first products made with recombinant DNA. Human growth hormone now can be made that way, and new products—vaccines, blood coagulation proteins for people with hemophilia, and enzymes such as cholesterol oxidase to diagnose disorders in cholesterol metabolism—have been developed. Many of the products in Table 8.3 are the result of recombinant DNA technology.

Recombinant insulin received FDA approval in 1982. The production of recombinant insulin is a $500-million-a-year business, making it one of the top 200 drugs sold.

The use of recombinant DNA technology to make substances useful to humans makes certain treatments potentially safer, cheaper, and available to more patients. For example, prior to the use of recombinant DNA to manufacture human insulin, the insulin for diabetic patients came exclusively from slaughtered cattle and pigs. Some patients develop allergies to such insulin, and the number of patients requiring insulin is increasing. Making nonallergenic human insulin and increasing the insulin supply are two important benefits of insulin production by recombinant DNA technology.

Most type 1 diabetic patients in the United States now use a preparation containing genetically engineered insulin.

Likewise, prior to the manufacture of human growth hormone by recombinant DNA technology, the hormone was obtained from the pituitary glands of cadavers at autopsy, and several cadavers were needed to obtain a single dose with which to treat children who have the congenital condition of pituitary dwarfism. This treatment was suspended in April 1985 in both the United States and the United Kingdom due to reports of several cases of the fatal, degenerative, neurological Creutzfeldt-Jakob disease (similar to mad cow disease). But later in 1985, the FDA approved a genetically engineered form of the hormone that produces the same effects as the human cadaver hormone. In addition to correcting congenital disorders, this product of recombinant DNA technology might prove useful in the treatment of delayed wound or fracture healing and in the metabolic problems associated with aging.

The manufacture of certain blood coagulation proteins by recombinant DNA technology makes these substances more readily available to individuals with hemophilia or other blood disorders. It also assures that the recipient will not acquire AIDS or hepatitis B from a contaminated blood product.

Recombinant DNA is being used to make vaccines more economically and in lager quantities than before. In this application, some microorganisms are used to combat the disease-causing capacity of others. Genes that direct the synthesis of specific substances, called *antigens*, from a disease-causing bacterium, virus, or protozoan parasite are inserted into another organism. The organism then makes a pure antigen. When the antigen is introduced into a human, the human immune system makes another specific substance, called an *antibody*, which takes part in the body's defense against the disease-causing organism (Chapter 17).

Hepatitis B vaccine, licensed in 1981, was the first vaccine for human use to be produced by recombinant DNA technology.

Recombinant DNA procedures for making vaccines for hepatitis A and B and influenza are already available. The vaccines not only are cheaper than conventional ones but also are purer and more specific, and they cause fewer undesirable side effects. Vaccines have proved to be highly specific and extremely effective against hepatitis A and B.

Many other applications of recombinant DNA techniques are being developed. An especially important one is the diagnosis of genetic defects in a fetus, which can be done by studying enzymes in fetal cells from amniotic fluid. Such defects are detected by using recombinant DNA with a known nucleotide sequence to find errors in the nucleotide sequence in fetal DNA segments. Such errors in fetal DNA denote genetic defects

BIOTECHNOLOGY

Need a Blood Transfusion? Call on a Genetically Altered Pig

Researchers soon expect to get government approval of a blood-substitute product composed primarily of human hemoglobin that is produced by transgenic (genetically altered) pigs. The biotechnology firm DNX, located in Princeton, New Jersey, began the project by injecting thousands of copies of two human hemoglobin genes into 1-day-old pig embryos that had been removed from their mothers' uteri. The embryos were then implanted into a second pig's uterus to grow until delivery. Two days after birth, the piglets were tested to see if they produced human hemoglobin along with pig hemoglobin—that is, if they were transgenic. Only about 0.5 percent of such transfers succeed.

It costs $50,000 to $75,000 to make one transgenic animal. DNX was initially successful in making three such pigs. Then the company bred the one transgenic male with more than 1,000 regular females and interbred the offspring of these crosses. Meanwhile the company was sold twice. The current owner, Baxter Healthcare, now has hundreds of transgenic pigs. Such crossing has continued for many generations. The human hemoglobin genes still function perfectly in the altered pigs, allaying fears that the genes would mutate or disintegrate in their new environment. The transgenic pigs currently produce blood containing more than 50 percent human hemoglobin.

To obtain the blood substitute, the pigs are bled and the red blood cells are ruptured. Pure human hemoglobin is separated from hybrid human/pig and pig hemoglobin by a multistep purification process that uses all available forms of chromatography—to safeguard against reliance on a single method that might fail and retain impurities.

The substitute product has several advantages over actual human blood:

1. It has a storage life of months instead of weeks.
2. Because naked hemoglobin does not stimulate the immune system to act against it, as do intact red blood cells containing hemoglobin, it can be transfused into anyone without the need for blood typing and matching.

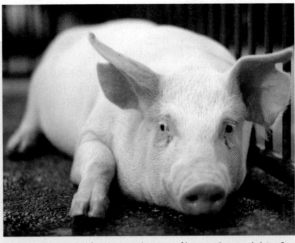

This pig has genes for the production of human hemoglobin. Such pigs will be bred and bled to collect human hemoglobin to save people's lives by transfusion. This is an example of biotechnological "pharming" of molecules. *(Mark Lyons/Lyons Photography, Inc.)*

3. It can ensure safety from human pathogens (including the AIDS virus) that might now contaminate human blood.
4. Because of the oxygen-carrying capacity of hemoglobin, it can serve as an immediate source of oxygen in earthquakes, in accidents, or in times of war, thereby enabling the injured to survive the trip to the hospital.

Human blood costs have recently risen to $300 to $500 per unit. Eventually, the blood substitute may cost the same as, or even less than, a unit of blood. Storage will be cheaper, and blood-typing costs would be eliminated.

One drawback to this procedure is that, once transfused, naked hemoglobin lasts only hours or days instead of 6 months, but this might be long enough to treat emergency cases. Another problem with this product is possible contamination with pig molecules or pig pathogens if purification processes fail. One issue that does *not* appear to be a problem is the use of this product by Jewish people who, for religious reasons, do not eat pork. The director of the Rabbinical Council of America has said that there would probably be no religious objection. By Judaic law, pigs may be used for purposes other than eating (such as sources of heart-valve replacements and insulin), and kosher laws are suspended in cases of life or death.

that can be responsible for absent or defective enzymes. Application of these techniques could greatly improve prenatal diagnosis of many genetic defects. Ultimately, as techniques for preparing recombinant DNA in animal cells improve, it may become possible to insert a missing gene or to replace a defective one in human cells (*gene therapy*). In fact, insertion of functional genes in appropriate cells may have cured the genetic disease known as severe combined immunodeficiency disease (Chapter 18). Insertion of such a gene in a defective gamete (egg or sperm) might prevent offspring from inheriting a genetic disease.

Forensic applications of DNA technology are rapidly coming into use in the courtroom, especially amplification of DNA by PCR, and RFLP analysis. In paternity cases, for example, experts can now determine with about 99 percent certainty that a given man is the father of a particular child, on the basis of DNA comparison (◄p. 204). Rapists and murderers can also be identified by the "DNA fingerprints" they leave behind at the scene of the crime in the form of semen, blood, hair, or tissue under their victim's fingernails. Sufficient DNA can even be collected for analysis from the rim of a used drinking glass or from the sweatband of a hat.

Industrial Applications of Recombinant DNA

Fermentation processes used in making wine, antibiotics, and other substances might be greatly improved by the use of recombinant DNA. For example, addition of genes for the synthesis of amylase to the yeast *Saccharomyces* could allow these organisms to produce alcohol from starch. Malting of grain to make beer would be unnecessary, and wines could be made from juices containing starches instead of sugars. Still other applications might include degradation of cellulose and lignin (plant materials often wasted), manufacture of fuels, clean-up of environmental pollutants, and leaching of metals from low-grade ores. Strains of *Pseudomonas putida*, already known to degrade different components of oil, might be engineered so that one strain degrades all components. Industrial leaching, or extraction, of metals from copper and uranium ores is already carried out by certain bacteria of the genus *Thiobacillus*. If these organisms could be made more resistant to heat and to the toxicity of the metals that they leach, the extraction process could be greatly accelerated.

Agricultural Applications of Recombinant DNA

Some bacteria are being engineered to control insects that destroy crops. The Monsanto Company has recently modified the genetic makeup of a strain of *Pseudomonas fluorescens*, which colonizes the roots of corn. This bacterium has been induced to carry genetic information inserted into it from *Bacillus thuringiensis*, allowing *P. fluorescens* to synthesize a protein that kills insects (**Figure 8.16**). Toxin made by *B. thuringiensis* has been extracted and used for many years as an insecticide. Now the pseudomonads, applied to the surface of corn seeds, can make the toxin as they grow around the roots of the corn.

Figure 8.16 SEM photo (118,750X) of crystals of a substance toxic to many insects. The genes for production of this toxin are being taken from *Bacillus thuringiensis*, which produces it naturally, and are being incorporated into other organisms through genetic engineering techniques. Imagine the benefits of crop plants that have built-in pesticide: no expense for chemical pesticides, no danger during application, no buildup in soil or water, and no entry or magnification of pesticides in the food chain. However, there may be a price to pay. Scientists have reported instances of resistance to Bt toxin in diamond back moths and pink bollworm. (*J. R. Adams/Science VU/Visuals Unlimited*)

BIOTECHNOLOGY

Virus with a Scorpion's Sting

Viruses that cause convulsions, paralysis, and death probably don't sound like viruses you want to keep around. Farmers, though, are glad for the geneticists who created a virus with these characteristics. Under ordinary circumstances, caterpillars that devour valuable crop plants are sometimes infected by a naturally occurring virus harmless to other organisms. The caterpillars develop a long illness that culminates in their climbing to the top of a plant and bursting, releasing a shower of infectious viral particles. But during their long illness, they still eat a lot of the crop. Genetic engineers have inserted the gene for scorpion venom into this virus. Caterpillars with the recombinant virus manufacture scorpion venom and quickly develop convulsions, paralysis, and death, all without stopping for a bite to eat. Don't worry—any altered virus that remains is destroyed in sunlight, and no chemical pesticide residues enter the food chain.

as well as they have in greenhouses, they could replace the use of chemical insecticides to control black cutworm and probably other insect larvae that damage crops With further research, other bacteria normally present on crops might be modified to control additional pests. Some optimistic scientists believe chemical pesticides may be phased out in favor of these safer and cheaper methods of pest control.

Also under development are genetically engineered seeds for crop plants that have high yield and other desirable characteristics and that will resist herbicides that kill weeds. The Environmental Protection Agency (EPA) has begun steps to approve growth of genetically altered crops to build up supplies of seeds that will produce their own insecticides. Potatoes, corn, and cotton have received genes for the production of insect-killing toxin from *B. thuringiensis*. Their seeds will kill the Colorado potato beetle, the European corn borer and other mothlike insects, and the cotton bollworm, pink bollworm, and tobacco budworm, respectively. Farmers will be able to buy seeds that will solve many cultivation problems. Attempts have also been made to introduce nitrogen-fixing genes into nonleguminous plants. (We will discuss nitrogen fixation in Chapter 25.) This work has been successful in some plants but not yet in any important crop plants. If it can be extended to crop plants, many agricultural crops could be made to satisfy their nitrogen needs and thrive without commercial fertilizers, which are expensive and tend to pollute groundwater. This would be especially beneficial in some developing nations, where famine is an ever-present threat, and money for expensive fertilizers is not available. However, concern over getting uniform dosages keep this idea from being used currently.

SEM

Figure 8.17 **A hybridoma cell.** A colorized SEM (11,875X) of a hybridoma, a single cell fused from two other cells. Hybridomas are often made by fusing an antibody-producing plasma cell and a cancer cell. The latter keeps the culture dividing and growing indefinitely; the former causes the hybridoma to produce pure antibody against the antigen to which it was sensitized. *(Dr. Jeremy Burgess/Photo Researchers, Inc.)*

HYBRIDOMAS

Along with the study of genetic recombinations in microorganisms came studies of such combinations in higher organisms. The first combination useful in industrial microbiology was the fusion of a myeloma (bone marrow cancer) cell with an antibody-producing white blood cell. A fusion of two such cells is called a **hybridoma** (hi-brid-o′ma), or hybrid cell **(Figure 8.17)**. This particular hybridoma can be grown in the laboratory, and it produces pure specific antibodies, called **monoclonal** (mon-o-klo′nal) **antibodies**, against any antigen to which the white blood cell was previously sensitized. Prior to the production of hybridomas and their monoclonal antibodies, no source of pure antibodies existed. Now many different kinds of monoclonal antibodies are produced commercially, and they represent a major advance in immunology. The production and uses of monoclonal antibodies are discussed in more detail in Chapter 17.

The ability to produce hybridomas may in the future lead to other important advances. Recently, agricultural scientists used this technique to fuse cells from a commercial potato plant with cells from an extremely rare wild strain. The wild strain was selected because it contains the gene for production of a natural insect repellent. The researchers were able to grow the fused cells into a plant that has the necessary commercial properties but also synthesizes the insect repellent in its leaves.

WEIGHING THE RISKS AND BENEFITS OF RECOMBINANT DNA

Despite the many potential benefits of recombinant DNA research, at first some scientists working with it were concerned about its hazards. They feared that some recombinants might prove to be new and especially virulent pathogens for which humans would have no natural defenses and no effective treatments. In 1974 they called for a moratorium on certain experiments until the hazards could be assessed. From this assessment emerged the idea of biological containment—the practice of making recombinant DNA only in organisms with mutations that prevent them from surviving outside the laboratory.

In 1981 the constraints on recombinant DNA research were relaxed because of the following observations:

1. No illnesses in laboratory workers could be traced to recombinants.
2. The strain of *E. coli* used in the experiments failed to infect humans who voluntarily received large doses.
3. Incorporation of mammalian genes into *E. coli* was observed in nature, and these genes invariably impaired the organism's ability to adapt to the environment. This suggested that if laboratory organisms did escape, they probably would not survive in the natural environment.
4. Mutants of *E. coli* containing recombinant DNA were subject to control by accepted sanitary practices.

Most scientists now agree that, as currently practiced, recombinant DNA techniques offer significant benefits and exceedingly small risks to humans.

✓**CHECKLIST**

1. Define genetic engineering.
2. What is recombinant DNA?
3. What three processes are involved in making recombinant DNA?
4. What is a restriction enzyme?

BIOTECHNOLOGY

Redesigning Bacteria

Many people are surprised to learn that frost is partly a microbiological phenomenon—certain bacteria are largely responsible for the formation of ice crystals. These "ice-plus bacteria" produce a protein that has the same structure as that of water molecules in an ice crystal, and water molecules tend to align themselves on it. As the ice crystals grow, they enter the spaces inside the leaf or flower, causing frost injury. This damage makes the plant susceptible to disease caused by *Pseudomonas syringae*, the predominant frost-causing strain.

Chemicals can control the bacteria that cause frost to form. But such chemicals are expensive, harm the environment, are of limited effectiveness, and may lead to diseases. These chemicals kill beneficial bacteria as well as harmful ones.

One of the microbiologists who is working on this problem is Trevor Suslow, director of microbial pesticides at the DNA Plant Technology Corporation in Oakland, California. The goal of Suslow's team in controlling frost-causing bacteria was to leave the beneficial bacteria untouched and remove just the ice-plus strains. The strategy was to locate or develop "ice-minus bacteria" to compete successfully with ice-plus strains.

Suslow's team focused on altering ice-plus bacteria so that they could no longer produce the ice-building protein. Using genetic engineering techniques, the team located the gene that controls this production and transferred it to *Escherichia coli* cells. With the aid of restriction enzymes, about one-third of the gene was moved. This was enough not only to inactivate the gene but also to ensure that no mutation could ever restore it to functional form. Next, the team cloned the defective gene, put it on a plasmid, and reinserted the plasmid into *P. syringae*. A plasmid and a chromosome in the same cell can exchange similar genes that are present on both. So Suslow's team looked for bacteria in which this natural process (called homologous recombination) had taken place, leaving the bacterial cells with a nonfunctional copy of the ice-forming gene in its chromosome. These were ice-minus

Spraying strawberry plants in Frostban field test in Brentwood, California (*Courtesy Trevor V. Suslow, University of California, Davis*)

bacteria. Deletion mutations of this kind occur constantly in nature. But it's more accurate, predictable, and cost effective to perform the genetic surgery in the laboratory.

Suslow and his team conducted several hundred tests of the ice-minus strain, which they called Frostban. Then it was time to do field trials. Suslow obtained the necessary permits and approvals for field testing but was not prepared for the commotion such tests would cause. Protests came from several quarters. It didn't matter that Frostban was safe. Some people objected to the genetic engineering of a living organism—any organism. Frostban became the symbol in their fight against a future full of recombinant DNA.

Environmentalists were concerned about the destruction of the ecological balance. Local farmers had more immediate concerns. Some simply weren't sure what these bacteria were and what they might do. Others worried about potential risks to their own crops from a product that offered them no benefit. Research showing that there was virtually no likelihood that Frostban would spread in this way didn't convince the farmers.

Suslow and his team were plagued by an increasing number of lawsuits and even vandalism. Discontinuing the field trials, they went back to the laboratory to isolate natural ice-minus bacteria—those in which the ice-building gene is already missing. Although such organisms have the same effect as the bioengineered ones, they're more acceptable to many people because they occur naturally.

BIOTECHNOLOGY

Should We Genetically Engineer Pets?

These zebra danio tropical aquarium fish are the United States' first officially sanctioned genetically modified pet! Thanks to the insertion of a gene from a sea coral, they glow red and blue, especially under ultraviolet light. Originally produced in a laboratory in Singapore, the fish were supposed to glow when swimming in polluted water—a kind of living indicator system. However, once they started glowing, it couldn't be stopped. Since then, aquarium lovers have been paying $5 apiece for them.

The U.S. Food and Drug Administration (FDA) decided in December of 2003 that these fish did not need to be regulated, thereby allowing them to be sold. A green-glowing genetically engineered medaka fish wasn't even subjected to FDA review and is also being sold. However, not everyone is happy about this. They ask: Is it ethical to alter the genetic make-up of an

Genetically altered fish
(©AP/Wide World Photos)

Normal zebra fish
(Paul Zahl/NG
Image Collection)

animal for no reason but our pleasure? How bizarre can alterations be allowed? What if modified animals escape into the wild and out-compete native species? (Zebra fish and medaka cannot live in cold waters.) What are your thoughts on this matter?

BIOTECHNOLOGY

Statement on Genetically Modified Organisms

Excerpt from Statement of the American Society for Microbiology on Genetically Modified Organisms

This statement released on 17 July 2000 outlines the Society's views on a much-debated topic:

In recent months, public understanding of biotechnology has been challenged by controversy concerning genetically modified organisms. The public has been confronted with charges and counter charges regarding the risks and benefits associated with using biotechnology to produce quality food in quantity. Since biotechnology enables well-characterized genes to be transferred from one organism to another with greater precision and predictability than is possible using traditional breeding procedures, ASM is sufficiently convinced to assure the public that plant varieties and products created with biotechnology have the potential of improved nutrition, better taste, and longer shelf-life.

Nothing in life is totally free of risk. However, to minimize risk it is important to rely on fact rather than on fear, and ASM is not aware of any acceptable evidence that food produced with biotechnology and subject to FDA oversight constitutes high risk or is unsafe. Rather, plant varieties created with biotechnology are grown more efficiently and economically than traditional crops. This eventually should result in a more nutritious product at less cost to the consumer as well as to reduced pesticide use and greater environmental protection. Those who resist the advance of biotechnology must address how otherwise to feed and care for the health of a rapidly growing global population forecast to increase at a rate of nearly 90 million people per year. However, a continued expression of public concern at the current level should be understood by federal agencies as reason to support more research, and to improve the quality and public accessibility of information on the regulation of products of biotechnology....

Source: David Pramer, "Statement of the American Society for Microbiology on Genetically Modified Organisms," *ASM News* 66 (2000):590–591. The statement was developed with an ASM ad hoc committee.

▌ RETRACING OUR STEPS

▌▌ THE TYPES AND SIGNIFICANCE OF GENE TRANSFER

- **Gene transfer** refers to the movement of genetic information between organisms. **Vertical gene transfer** passes genes from parent to offspring. **Lateral gene transfer** passes genes to other cells in the same generation. It occurs in bacteria by transformation, transduction, and conjugation.
- Gene transfer is significant because it increases genetic diversity within a population, thereby increasing the likelihood that some members of the population will survive environmental changes.

▌▌ TRANSFORMATION

THE DISCOVERY OF TRANSFORMATION

- Bacterial **transformation** was discovered in 1928 by Griffith, who showed that a mixed culture of live rough and heat-killed smooth pneumococci could produce live smooth pneumococci capable of killing mice.
- Avery later showed that a capsular polysaccharide was responsible for virulence and that DNA was the substance responsible for transformation. Watson and Crick determined the structure of DNA, which led to studies showing that a cell's genetic information is encoded in its nucleic acids.

THE MECHANISM OF TRANSFORMATION

- Transformation involves the release of naked DNA fragments and their uptake by other cells at a certain stage in their growth cycle: (1) Uptake of DNA requires a protein called **competence factor** to make recipient cells ready to bind DNA. (2) Endonucleases cut double-stranded DNA into units; the strands separate, and only one strand is transferred. (3) Ultimately, donor DNA is spliced into recipient DNA. Leftover recipient DNA is broken down, so a cell's total DNA remains constant.

THE SIGNIFICANCE OF TRANSFORMATION

- Transformation is significant because (1) it contributes to genetic diversity; (2) it can be used to introduce DNA into an organism, observe its effects, and study gene locations; (3) it can be used to create recombinant DNA.

▌▌ TRANSDUCTION

THE DISCOVERY OF TRANSDUCTION

- In **transduction**, genetic material is carried by a **bacteriophage (phage)**.

THE MECHANISMS OF TRANSDUCTION

- Phages can be virulent or temperate. (1) **Virulent phages** destroy a host cell's DNA, direct synthesis of phage particles, and cause lysis of the host cell in the lytic cycle. (2) **Temperate phages** can replicate themselves as a **prophage**—part of a bacterial chromosome—or eventually produce new phage particles and lyse the host cell. Persistence of the phage in the cell without the destruction of the host cell is called **lysogeny**.
- Prophage can be incorporated into the bacterial chromosome, or it can exist as a plasmid, a piece of extrachromosomal DNA. Cells that contain a prophage are called **lysogenic** cells because they have the potential to enter the **lytic cycle**.
- Transduction can be specialized or generalized. (1) In **specialized transduction**, the phage is incorporated into the chromosome and can transfer only genes adjacent to the phage.

(2) In **generalized transduction**, the phage exists as a plasmid and can transfer any DNA fragment attached to it.

THE SIGNIFICANCE OF TRANSDUCTION

• Transduction is significant because it transfers genetic material and demonstrates a close evolutionary relationship between prophage and host cell DNA. Also, its persistence in a cell suggests a mechanism for the viral origins of cancer, and it provides a possible mechanism for studying gene linkage.

III CONJUGATION

THE DISCOVERY OF CONJUGATION

• In **conjugation** large quantities of DNA are transferred from one organism to another during contact between donor and recipient cells.

• Conjugation was discovered by Lederberg in 1946 when he observed that mixing strains of *E. coli* with different metabolic deficiencies allowed the cells to overcome deficiencies.

• **Plasmids** are extrachromosomal DNA molecules.

THE MECHANISMS OF CONJUGATION

• Three mechanisms of conjugation have been observed: (1) In the transfer of **F plasmids**, a piece of extrachromosomal DNA (a **plasmid**) is transferred. (2) In high-frequency recombinations, parts of F plasmids that have been incorporated into the chromosome (the **initiating segment**) are transferred along with adjacent bacterial genes. (3) An F plasmid incorporated into the chromosome and subsequently separated becomes an **F′ plasmid** and transfers chromosomal genes attached to it.

THE SIGNIFICANCE OF CONJUGATION

• The significance of conjugation is that it increases genetic diversity, it may represent an evolutionary stage between asexual and sexual reproduction, and it provides a means of mapping genes in bacterial chromosomes.

GENE TRANSFER MECHANISMS COMPARED

• Genetic transfer mechanisms differ in the quantity of DNA transferred.

III PLASMIDS

CHARACTERISTICS OF PLASMIDS

• Plasmids are circular, self-replicating, double-stranded extrachromosomal DNA that carry information that is usually not essential for cell growth.

RESISTANCE PLASMIDS

• **Resistance (R) plasmids** carry genetic information that confers resistance to various antibiotics and to certain heavy metals. They generally consist of a **resistance transfer factor (RTF)** and one or more **resistance (R) genes**. Such plasmids can be removed from a bacterium by **displacins**.

TRANSPOSONS

• R genes that move from one plasmid to another in a cell or become inserted in the chromosome are part of a **transposon** because they transpose, or change, their locations.

BACTERIOCINOGENS

• **Bacteriocinogens** are plasmids that produce **bacteriocins**, which are proteins that inhibit growth of other strains of the same species or closely related species.

III GENETIC ENGINEERING

• **Genetic engineering** is the manipulation of genetic material to alter the characteristics of an organism.

GENETIC FUSION

• **Genetic fusion** allows transposition of genes from one location on a chromosome to another, sometimes deleting a portion, thereby causing the joining of genes from two different operons.

PROTOPLAST FUSION

• **Protoplast fusion** combines **protoplasts** (organisms without cell walls) and allows mixing of genetic information.

GENE AMPLIFICATION

• **Gene amplification** involves the addition of plasmids to microorganisms to increase yield of useful substances.

RECOMBINANT DNA TECHNOLOGY

• **Recombinant DNA** is DNA produced when genes from one kind of organism are introduced into the genome of a different kind of organism. The resulting organism is a **transgenic**, or recombinant, organism.

• Recombinant DNA has proven especially useful in medicine, industry, and agriculture.

HYBRIDOMAS

• **Hybridomas** are genetic recombinations involving cells of higher organisms.

WEIGHING THE RISKS AND BENEFITS OF RECOMBINANT DNA

• When recombinant DNA techniques were first developed, scientists were concerned that virulent pathogens might be created, and they developed containment procedures. As research proceeded and no illnesses caused by recombinants were observed, most scientists came to believe that the benefits of recombinant DNA techniques outweigh the risks.

▌ TERMINOLOGY CHECK

bacteriocin (*p. 225*)
bacteriocinogen (*p. 225*)
bacteriophage (*p. 215*)
chromosome mapping
    (*p. 218*)
clone (*p. 219*)
colicin (*p. 225*)

competence factor (*p. 214*)
conjugation (*p. 218*)
displacin (*p. 224*)
electroporation (*p. 228*)
F⁺ cell (*p. 219*)
F⁻ cell (*p. 219*)
F pilus (*p. 219*)

F plasmid (*p. 219*)
F′ plasmid (*p. 221*)
gene amplification (*p. 228*)
gene transfer (*p. 212*)
generalized transduction
    (*p. 217*)
genetic engineering (*p. 225*)

genetic fusion (*p. 226*)
high frequency of
    recombination (Hfr)
    strain (*p. 219*)
hybridoma (*p. 233*)
initiating segment
    (*p. 219*)

lateral gene transfer
(*p. 212*)
lysis (*p. 216*)
lysogenic (*p. 216*)
lysogeny (*p. 216*)
lytic cycle (*p. 216*)
monoclonal antibody
(*p. 233*)
phage (*p. 215*)
plasmid (*p. 219*)
promiscuous (*p. 221*)

prophage (*p. 216*)
protoplast (*p. 227*)
protoplast fusion (*p. 227*)
recombinant DNA (*p. 228*)
recombination (*p. 212*)
resistance (R) gene (*p. 223*)
resistance plasmid (*p. 223*)
resistance transfer factor
(RTF) (*p. 223*)
restriction endonuclease
(*p. 228*)

restriction enzyme
(*p. 228*)
restriction fragment
(*p. 228*)
restriction fragment length
polymorphism (RFLP)
(*p. 230*)
specialized transduction
(*p. 217*)
temperate phage (*p. 216*)
transduction (*p. 215*)

transformation (*p. 212*)
transgenic (*p. 228*)
transposable element
(*p. 224*)
transposition (*p. 224*)
transposon (*p. 224*)
vector (*p. 228*)
vertical gene transfer
(*p. 212*)
virulent phage (*p. 215*)

▌ CLINICAL CASE STUDY

Imagine that you are a scientist working for the local health department. Your microbiology laboratory has recently received several cultures of different Gram-negative enteric bacteria, *Salmonella typhimurium*, *Salmonella enteriditis*, *Escherichia coli*, and *Shigella* species that are all resistant to the same four different antibiotics. What could explain how all these different bacteria acquired resistance to the same four antibiotics? What would you look for to confirm your hypothesis?

▌ CRITICAL THINKING QUESTIONS

1. Drug-resistent bacteria are known to arise as a result of antibiotics added to animal feed. Sometimes these bacteria are present in foods, such as ground beef, from animal sources. Recent research has confirmed that genes for drug resistance spread to other species of bacteria in the human intestine following the ingestion of resistant bacteria. What processes allow these genes to move from species to species of bacteria? What can be done to reduce the significance of this problem?

2. Using some of the genetic engineering tools that were described in this chapter, design a protocol for creating a recombinant plasmid.

3. Many controversies (e.g., genetically modified foods) have emerged and will emerge regarding recombinant DNA technology. List as many potential benefits of this technology and as many potential harmful results as you can think of. On balance, do you see more benefits or more risks as transgenic organisms become increasingly common?

▌ SELF-QUIZ

1. What are the main characteristics of gene transfer?
2. Match the following descriptions to their modes of lateral gene transfer:
   - B Uptake of naked DNA
   - C Virus involved
   - A F⁺, F⁻, Hfr
   - B Competence factor
   - A F pilus
   - (a) Conjugation
   - (b) Transformation
   - (c) Transduction

3. Temperate phages can carry out both specialized transduction where any bacterial gene can be transferred and generalized transduction where only specific genes are transferred. True or false? *False*

4. In the lysogenic cycle of bacteriophages which of the following is true?
   - (a) Lysogeny is the term applied to a temperate phage that does not replicate itself independently and does not lyse the bacterial host cell.
   - (b) Phage DNA is incorporated into the host bacterium's DNA upon which time it is called a prophage.
   - (c) Temperate phages can be replicated either as a prophage along with bacterial chromosomal replication or can suddenly revert to the lytic cycle by replicating themselves and assembling into new phages.
   - (d) All the above are true.
   - (e) a and c only.

5. Which of the following is not true about conjugation?
   - (a) In conjugation large quantities of DNA are transferred during contact between donor and recipient cells.
   - (b) The number of genes transferred depends only on whether the donor cell is F⁺ or F′.
   - (c) Upon mixing strains of *E. coli* with different metabolic deficiencies, Lederberg observed that the cells could overcome their deficiencies, leading to his discovery of conjugation.
   - (d) All of the above are not true.
   - (e) a and c.

6. What are the three mechanisms of conjugation and how do they differ? *F plasmid F′ plasmid Hfr plasmid*

7. When the F plasmid is incorporated into the chromosome of F$^+$ bacteria, these cells are called:
   (a) F$^+$
   (b) High-frequency recombinants
   (c) F$^-$
   (d) F′
   (e) Prophage

8. Which of the following would inhibit transformation?
   (a) Presence of an exonuclease that degrades DNA
   (b) Vigorous shaking of a bacterial culture
   (c) Keeping the donor and recipient bacteria separated from each other
   (d) Addition of bacteriophage to a culture
   (e) None of the above would inhibit transformation

9. Match the following significant descriptions to their mode of gene transfer:
   A — Contributes to genetic diversity
   C — Provides a means of mapping genes in bacterial chromosomes
   B — Suggests a mechanism for viral origins of cancer
   C — May represent an evolutionary stage between asexual and sexual reproduction
   A — Can be used to create recombinant DNA
   B — Demonstrates a close evolutionary relationship between prophage and host cell DNA

   (a) Transformation
   (b) Transduction
   (c) Conjugation

10. Which of the following is characteristic of a plasmid?
    (a) Plasmids generally contain genes that are essential to the recipient's survival.
    (b) Plasmids can carry genes for antibiotic resistance and bacteriocins that inhibit the growth of other competing bacteria.
    (c) Some plasmids can carry genes for catabolic enzymes and virulence factors.
    (d) A plasmid is a circular, double-stranded extra-chromosomal piece of DNA that is self-replicating.
    (e) b, c, and d.

11. A genetic sequence that can move from one location to another within a cell is known as a:
    (a) Bacteriocin
    (b) Plasmid
    (c) Transposon
    (d) Hybridoma
    (e) Transducible element

12. Recombinant DNA is:
    (a) DNA that is produced when genes from one kind of organism are introduced by lateral transfer into the genome of another kind of organism
    (b) DNA that is destroyed by endonucleases
    (c) DNA amplified by the polymerase chain reaction
    (d) DNA that is rich in the nitrogenous base guanine
    (e) All of the above

13. A bacterium can acquire the ability to make a new enzyme or toxin naturally by:
    (a) Mutation
    (b) Transformation
    (c) Conjugation
    (d) All of the above
    (e) Only b and c

14. A bacterium that has had its cell wall removed is called:
    (a) A protoplast
    (b) A mycoplasma
    (c) Gram-positive
    (d) Gram-negative
    (e) A transposable element

15. Match the following genetic engineering terms to their descriptions:
    B — Gene amplification
    A — Genetic engineering
    D — Protoplast fusion
    C — Restriction endonucleases
    F — Genetic fusion
    E — Transgenic

    (a) Manipulation of genetic material to alter an organism's characteristics
    (b) The addition of plasmids to microorganisms in order to increase the yield of useful gene products
    (c) Enzymes that recognize specific sequences of 4 to 8 base pairs of DNA where it then cuts across to create a restriction fragment
    (d) Combines organisms without cell walls, allowing them to mix their genetic information
    (e) Results when a new DNA is produced by taking genes from one kind of organism and introducing them into the genome of another different organism
    (f) Allows the transposition and joining of genes from two separate genes that were originally located at different positions on the chromosome

16. In the laboratory, cells can be rendered ___ by the use of cold calcium chloride or electroporation.
    (a) Transposable
    (b) Competent
    (c) Conjugated
    (d) Recombinated
    (e) Transducible

17. Biotechnology and genetic engineering or molecular genetics can be used to make all of the following except:
    (a) Vaccines
    (b) Human hormones
    (c) Drugs
    (d) Life
    (e) Insulin

18. Formation of frost on agricultural plants is, in part, due to microorganisms. True or false?

    TRUE

19. Hybridomas are formed by fusing antibody-producing white blood cells with:
(a) Viruses
(b) Bacteria
(c) Myeloma cells
(d) Red blood cells
(e) Bone marrow cells

20. Individual hybridoma cells produce only one type, or _____ antibodies.
(a) Monoclonal
(b) Polyclonal
(c) IgE
(d) IgA
(e) Wild-type

21. Phages that destroy host cell DNA, direct synthesis of phage particles, and cause lysis of the host cell are called:
(a) Temperate phages
(b) Prophages
(c) Lysogenic phages
(d) Virulent phages
(e) Transductions

22. Which of the following statements concerning recombinant DNA technology is false?
(a) Thus far, no illnesses in laboratory workers have been traced to genetic recombinants.
(b) Production of large amounts of proteins such as insulin and human growth hormone has been made possible using recombinant DNA technology.
(c) Recombinant DNA technology offers specific benefits to the scientific, medical, and general population.
(d) Mutant strains of bacteria produced by genetic recombination are often unable to survive in the natural environment.
(e) Recombinant DNA technology provides a high degree of risk to the health of the general population.

23. The first vaccine for human use produced using recombinant DNA technology was the:
(a) Hepatitis B vaccine
(b) MMR vaccine
(c) AIDS vaccine
(d) Polio vaccine
(e) Hepatitis A vaccine

24. Which of the following is NOT a method by which bacteria naturally alter or exchange their genetic information?
(a) Transduction
(b) Transposition
(c) Transformation
(d) Protoplast fusion
(e) Conjugation

25. Identify processes (a) through (e) in the following diagram:

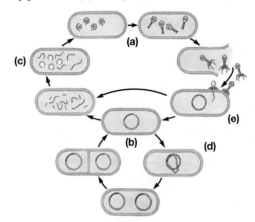

(a) Lytic cycle
(b) Lysogenic cycle
(c) Empty phage heads and pieces of DNA are assembled
(d) Phage and DNA are replicated
(e) Phage is resorbed by receptor site on cell wall, and inserts DNA.

▌ EXPLORATIONS ON THE WEB

http://www.wiley.com/college/black

If you think you've mastered this chapter, there's more to challenge you on the web. Go to the companion web site to fine-tune your understanding of the chapter concepts and discover answers to the questions posed below.

1. What do genetic engineers really do? What techniques do they use?

2. If bacteria only reproduce asexually, how can they transfer genes?

9

An Introduction to Taxonomy: The Bacteria

Come with me...

When you sip wine or eat cheese, do you ever wonder about the microorganisms necessary for making those products? Where do they come from? How are they stored? They might be kept at ATCC (American Type Culture Collection) in Manassas, Virginia. ATCC is a global nonprofit organization that preserves, authenticates, and distributes microorganisms and other biological materials for the scientific community. In addition to ensuring that high-quality microbial cultures are available for research and development, ATCC serves as a safe deposit location for industrial cultures owned by commercial organizations including winemakers, brewers, and cheese makers. ATCC stores these valuable cultures under a multi-layered security system in large liquid nitrogen tanks at very low temperatures (see photo). Manufacturers of wine, cheese, or other products requiring microbes can safely preserve their cultures at ATCC until they need to replenish their stocks. Other cultures stored at ATCC are available for use in research and development projects in many fields. ATCC scientists ensure that cultures used by researchers are correctly identified and the strain's genetic and biochemical properties are maintained.

 Video related to this topic is available within WileyPLUS.

Courtesy ATCC

Humans appear to have an innate need to name things. In many primitive societies, a person who knows the true name of an object or of another person is believed to have power over that object or person. Naming helps us to understand our world and to communicate with others about it.

TAXONOMY: THE SCIENCE OF CLASSIFICATION

In science, accurate and standardized names are essential. All chemists must mean the same thing when they talk about an element or a compound; physicists must agree on terms when they discuss matter or energy; and biologists must agree on the names of organisms, be they tigers or bacteria.

Faced with the great number and diversity of organisms, biologists use the characteristics of different organisms to describe specific forms of life and to identify new ones. The grouping of related organisms together is the basis of *classification*. The most obvious reasons for classification are (1) to establish the criteria for identifying organisms, (2) to arrange related organisms into groups, and (3) to provide important information on how organisms evolved. **Taxonomy** is the science of classification. It provides an orderly basis for the naming of organisms and for placing organisms into a category, or **taxon** (plural: *taxa*).

Another important aspect of taxonomy is that it makes use of and makes sense of the fundamental concepts of unity and diversity among living things. Organisms classified in any particular group have certain common characteristics—that is, they have unity with respect to these characteristics. For example, humans walk upright and have a well-developed brain; *Escherichia coli* cells are rod-shaped and have a Gram-negative cell wall. The organisms within taxonomic groups exhibit diversity as well. Even members of the same species display variations in size, shape, and other characteristics. Humans vary in height, weight, hair and eye color, and facial features. Certain kinds of bacteria vary somewhat in shape and in their ability to form specific structures, such as endospores.

A basic principle of taxonomy is that members of higher-level groups share fewer characteristics than those in lower-level groups. Like all other vertebrates, humans have backbones, but humans share fewer characteristics with fish and birds than with other mammals. Likewise, nearly all bacteria have a cell wall, but in some the wall is Gram-positive and in others it is Gram-negative.

LINNAEUS, THE FATHER OF TAXONOMY

The eighteenth-century Swedish botanist Carolus Linnaeus is credited with founding the science of taxonomy (**Figure 9.1**). He originated **binomial nomenclature**, the system that is still used today to name all living things. In the binomial,

Figure 9.1 Carolus Linnaeus (1707–1778). Linnaeus is known as the father of taxonomy. He is shown here in the cross-country skiing outfit he wore to collect specimens in Lapland. The curled boot toes held his skis onto his feet. *(Granger Collection)*

or "two-name," system, the first name designates the **genus** (plural: *genera*) of an organism, and its first letter is capitalized. The second name is the **specific epithet**, and it is not capitalized even when derived from the name of the person who discovered it. Together the genus and specific epithet identify the **species** to which the organism belongs. Both words are italicized in print but underlined when handwritten. When there is no danger of confusion, the genus name may be abbreviated to a single letter. Thus, *Escherichia coli* is often written *E. coli*, and humans (*Homo sapiens*) may be identified as *H. sapiens*.

The name of an organism often tells something about it, such as its shape, where it is found, what nutrients it uses, who discovered it, or what disease it causes. Some examples of names and their meanings are shown in **Table 9.1**.

The members of a species generally have several common characteristics that distinguish that species from all other species. As a rule, members of the species cannot be divided into

Why taxonomy? Common names are confusing. Passer domesticus is the English sparrow in America, house sparrow in England, gorrion in Spain, musch in Holland, and hussparf in Sweden.

TABLE 9.1

The Meaning of the Names of Some Microorganisms	
Name of Microorganism	Meaning of Name
Entamoeba histolytica	*Ent*, intestinal; *amoebae*, shape and means of movement; *histo*, tissue; *lytic*, lysing, or digesting tissue
Escherichia coli	Named after Theodor Escherich in 1888; found in the colon
Haemophilus ducreyi	*Hemo*, blood; *phil*, love; named after Augusto Ducrey in 1889
Neisseria gonorrhoeae	Named after Albert L. Neisser in 1879; causes gonorrhea
Saccharomyces cerevisiae	*Saccharo*, sugar; *myco*, mold; *cerevisia*, beer or ale
Staphylococcus aureus	*Staphylo*, cluster; *kokkus*, berry; *aureus*, golden
Lactococcus lactis	*Lacto*, milk; *kokkus*, berry
Shigella etousae	Named after Kiyoshi Shiga in 1898; European Theater of Operations of the U.S. Army (final *e* gives proper Latin ending)

About 4,400 animals, and over 7,700 plants still retain the names Linnaeus gave them. An "L." after a species name indicates Linnaeus named it.

one pure culture of a species differ from the organisms in another pure culture of the same species, the organisms in each culture are designated as strains. A **strain** is a subgroup of a species with one or more characteristics that distinguish it from other subgroups of the same species. Each strain is identified by a name, number, or letter that follows the specific epithet. For example, *E. coli* strain K12 has been extensively studied because of its plasmids and other genetic characteristics, and *E. coli* strain 0157:H7 causes hemorrhagic inflammation of the colon in humans.

In addition to introducing the binomial system of nomenclature, Linnaeus also established a hierarchy of taxonomic ranks: species, genus, family, order, class, phylum or division, and kingdom. At the highest level, Linnaeus divided all living things into two *kingdoms*—plant and animal. In his taxonomic hierarchy, much of which is still used today, each organism is assigned a species name, and species of very similar organisms are grouped into a genus. As we proceed up the hierarchy, several similar genera are grouped to form a *family*, several families to form an *order*, and so on to the top of the hierarchy. Some hierarchies today have additional levels, such as *subphyla*. Also, it has become accepted practice to refer to the first categories within the animal kingdom as *phyla* and to those within other kingdoms (we now have five) as *divisions*. Recently, the five kingdoms have been grouped together into three *domains*, a new category even higher than kingdom. Domains will be discussed later in the chapter. The classifications of a human, a dog, a wolf, and a bacterium are shown in **Figure 9.2**.

significantly different groups on the basis of a particular characteristic, but there are exceptions to this rule. Sometimes members of a species are divided on the basis of a small but permanent genetic difference, such as a need for a particular nutrient, resistance to a certain antibiotic, or the presence of a particular antigen. When organisms in

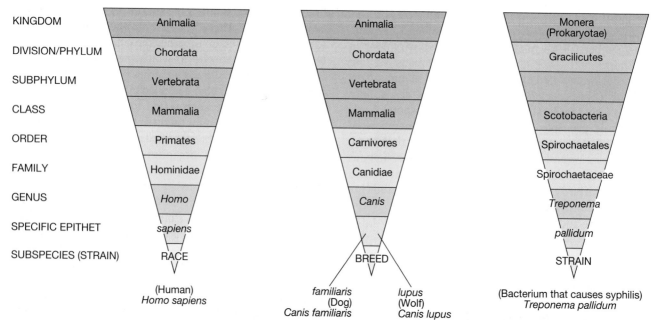

Figure 9.2 Classification of a human, a dog, a wolf, and a bacterium.

USING A TAXONOMIC KEY

Biologists often use a taxonomic *key* to identify organisms according to their characteristics. The most common kind is a **dichotomous key**, which has paired statements describing characteristics of organisms. Paired statements present an "either-or" choice, so that only one statement is true. Each statement is followed by directions to go to another pair of statements until the name of the organism finally appears. **Figure 9.3** is a dichotomous key that will identify each of the four most common U.S. coins: quarters, dimes, nickels, and pennies. Read statements 1a and 1b, and decide which statement applies to a given coin. Look at the number to the right of the statement; it tells you which pair of statements to look at next. Continue in this manner until you reach a group designation. If you have followed the key carefully, that designation will name the coin.

Of course, you don't need a taxonomic key to identify something as simple and as familiar as coins. But identifying all the many kinds of bacteria in the world is a more difficult task. Major groups of bacteria can be identified with the key in **Figure 9.4**. More detailed keys use staining reactions, metabolic reactions (fermentation of particular sugars or release of different gases), growth at different temperatures, properties of colonies on solid media, and similar characteristics of cultures. By proceeding step by step through the key, one should be able to identify an unknown organism, or even a strain, if the key is sufficiently detailed.

PROBLEMS IN TAXONOMY

Among the aims of a taxonomic system are organizing knowledge about living things and establishing standard names for organisms so that we can communicate about them. Ideally, we would like to classify organisms according to their **phylogenetic**, or evolutionary, relationships, but this is not always easy. Evolution occurs continuously and at a relatively rapid rate in microorganisms, and our knowledge of the evolutionary history of organisms is incomplete. Taxonomy must change with evolutionary

| 1a | Gram-positive | Go to 2 |
| 1b | Not Gram-positive | Go to 3 |

| 2a | Cells spherical in shape | Gram-positive cocci |
| 2b | Cells not spherical in shape | Go to 4 |

| 3a | Gram-negative | Go to 5 |
| 3b | Not Gram-negative (lack cell wall) | Mycoplasma |

| 4a | Cells rod-shaped | Gram-positive bacilli |
| 4b | Cells not rod-shaped | Go to 6 |

| 5a | Cells spherical in shape | Gram-negative cocci |
| 5b | Cells not spherical in shape | Go to 7 |

| 6a | Cells club-shaped | Corynebacteria |
| 6b | Cells variable in shape | Propionibacteria |

| 7a | Cells rod-shaped | Gram-negative bacilli |
| 7b | Cells not rod-shaped | Go to 8 |

| 8a | Cells helical with several turns | Spirochetes |
| 8b | Cells comma-shaped | Vibrioids |

Figure 9.4   A dichotomous key for classifying major groups of bacteria.

changes and new knowledge. *It is far more important to have a taxonomic system that reflects our current knowledge than to have a system that never changes.*

Creating a taxonomic system that provides an organized overview of all living things and how they are related to each other poses certain problems. Two such problems arise at opposite ends of the taxonomic hierarchy: (1) deciding what constitutes a species, and (2) deciding what constitutes a kingdom or in which domain a kingdom belongs. In the first case, taxonomists try to decide how much diversity can be tolerated within the unity of a species. In the second, taxonomists try to decide how to sort the diverse characteristics of living things into categories that reflect fundamental differences of evolutionary significance. In most advanced organisms, such as plants and animals, species that reproduce sexually are distinguished primarily by their reproductive capabilities. A male and a female of the same species are capable of DNA transfer through mating and producing fertile offspring, whereas members of different species ordinarily either cannot mate successfully or will have sterile offspring. *Morphology* (structural characteristics) and geographic distribution also are considered in defining species.

In bacteria, such criteria normally cannot be used in defining a species, primarily because lateral gene transfer (genetic recombination) among bacteria has been

| 1a | Smooth-edged | Go to 2 |
| 1b | Rough-edged | Go to 3 |

| 2a | Silver-colored | Nickel |
| 2b | Copper-colored | Penny |

| 3a | Large (about 1 in. diameter) | Quarter |
| 3b | Small (about 3/4 in. diameter) | Dime |

Figure 9.3   A dichotomous key for classifying typical U.S. coins. Why would the word *"flat"* not be useful in this key?

very common in evolution, but morphological differences are minor. A bacterial species is defined by the similarities found among its members. Properties such as biochemical reactions, chemical composition, cellular structures, genetic characteristics, and immunological features are used in defining a bacterial species. Identifying a species and determining its limits present the most challenging aspects of biological classification—for any type of organism.

Before taxonomists turned their attention to microorganisms, the two-kingdom system of plants and animals worked reasonably well. Anyone can tell plants from animals—for example, trees from dogs. Plants make their own food but cannot move, and animals move but cannot make their own food. Simple enough, or is it? In this scheme, how do you classify *Euglena*, a mobile microorganism that makes its own food? How would you classify jellyfishes and sponges, which are motile or immotile depending on their stage of life? And how do you classify colorless fungi that neither move nor make their own food? Finally, how do you classify slime molds, organisms that can be unicellular or multicellular and mobile or immobile? Obviously, many organisms pose a number of problems when one tries to use a two-kingdom system.

DEVELOPMENTS SINCE LINNAEUS'S TIME

The problem of classifying microorganisms was first addressed by the German biologist Ernst H. Haeckel in 1866 when he created a third kingdom, the Protista. He included among the protists all "simple" forms of life such as bacteria, many algae, protozoa, and multicellular fungi and sponges. Haeckel's original term, Protista, is still used in taxonomic schemes today, but it is now limited mainly to unicellular eukaryotic organisms.

Classification of bacteria has posed taxonomic problems over the centuries and still does. Until recently, many taxonomists regarded bacteria as small plants that lacked chlorophyll. As late as 1957, the seventh edition of *Bergey's Manual of Determinative Bacteriology*, a work devoted to the identification of bacteria, considered bacteria to be unicellular plants. Changes in this viewpoint came as the tools to study bacteria were developed. First, light microscopy and staining techniques were used to describe the basic structure of cells. Second, electron microscopy was used to study the ultrastructure of cells. And third, biochemical techniques were used to study chemical composition and chemical reactions in cells. One of the most important discoveries from these various studies was that DNA looked and behaved differently during cell division in bacteria than in cells whose DNA is organized into chromosomes within a nucleus.

Studies of the structure and function of cells also led to the recognition of two general patterns of cellular organization, prokaryotic and eukaryotic. Basing taxonomy on these two different patterns of cellular organization was proposed as early as 1937. Various taxonomists such as H. F. Copeland, R. Y. Stanier, C. B. van Niel, and R. H. Whittaker, working in the late 1950s, placed bacteria in a separate kingdom of anucleate (lacking a cell nucleus) organisms rather than with organisms that have true nuclei. In 1962, Stanier and van Niel stated, "The distinctive property of bacteria is the prokaryotic nature of their cells."

In 1956, Lynn Margulis and H. F. Copeland proposed a scheme of classifying prokaryotes and eukaryotes by the following four-kingdom system of classification:

1. Monera: all prokaryotes, including true bacteria and blue-green algae.
2. Protoctista: all eukaryotic algae, protozoa, and fungi.
3. Plantae: all green plants.
4. Animalia: all animals derived from a zygote, a cell formed by the union of an egg and a sperm.

These taxonomists also proposed that evolution from prokaryotic to eukaryotic life forms had taken place by endosymbiosis (◀Chapter 4, p. 104).

R. H. Whittaker felt that endosymbiosis could not account for all the differences between prokaryotes and eukaryotes. He also felt that a taxonomic system should give more consideration to the methods organisms use to obtain nourishment. Autotrophic nutrition by photosynthesis and heterotrophic nutrition by the ingestion of substances from other organisms had been considered in earlier taxonomies. Absorption as a sole means of acquiring nutrients had been overlooked. To Whittaker, fungi, which acquire nutrients solely by absorption, were sufficiently different from plants to justify placing them in a different kingdom. Also, fungi have certain reproductive processes not shared with any other organisms. Consequently, Whittaker proposed a taxonomic system in 1969 that separated the Protoctista into two kingdoms—Protista (pro-tis'tah) and Fungi—but retained the Monera, Plantae, and Animalia. Finally, through refinements of Whittaker's system by several taxonomists over the past few decades, the five-kingdom system was created.

THE FIVE-KINGDOM CLASSIFICATION SYSTEM

Before we discuss the five-kingdom classification system and how it applies to microorganisms, we must emphasize that all living organisms, regardless of the kingdom to which they are assigned, display certain characteristics that define the unity of life. All organisms are composed of cells, and all carry out certain functions, such as obtaining nutrients and getting rid of wastes. The cell is the basic structural and functional unit of all living things. The fact that viruses are not cells is one reason they are not considered to be living organisms. All cells are bounded by a

FUNGI

PLANTAE

ANIMALIA

PROTISTA

MONERA

Figure 9.5 **The five-kingdom system of classification.**

cell or plasma membrane, carry genetic information in DNA, and have ribosomes where proteins are made. All cells also contain the same kinds of organic compounds—proteins, lipids, nucleic acids, and carbohydrates. They also selectively transport material between their cytoplasm and their environment. Thus, although organisms may be classified in very diverse taxonomic groups, their cells have many similarities in structure and function.

No single classification system is completely accepted by all biologists. One of the most widely accepted is the **five-kingdom system (Figure 9.5).** A major advantage of this system is the clarity with which it deals with

microorganisms. It places all **prokaryotes**, microorganisms that lack a cell nucleus, in the kingdom Monera (Prokaryotae) (◀Chapter 4, p. 80). It places most unicellular **eukaryotes**, organisms whose cells contain a distinct nucleus, in the kingdom Protista. (Margulis proposed a very similar five-kingdom system in 1982, but she referred to the kingdom of simple eukaryotes as Protoctista instead of Protista.) The five-kingdom system also places fungi in the separate kingdom Fungi.

The properties and members of each of the five kingdoms are described below and summarized in **Table 9.2.** A more detailed classification of bacteria is provided in Appendix B

KINGDOM MONERA

The kingdom **Monera** (mo-ner′ah) is also called the kingdom **Prokaryotae**, as suggested by the French marine biologist Edouard Chatton in 1937. It consists of all prokaryotic organisms, including the eubacteria ("true bacteria"), the cyanobacteria, and the archaeobacteria **(Figure 9.6).**

All monerans are unicellular; they lack true nuclei and generally lack membrane-enclosed organelles. Their DNA has little or no protein associated with it. Reproduction in the kingdom Monera occurs mainly by binary fission. Of all monerans, the **eubacteria** (u′bak-ter″e-ah) are of greatest concern in the health sciences and will be considered in detail in several chapters of this book.

The **cyanobacteria** (si′an-o-bak-ter′e-ah), formerly known as blue-green algae, are of special importance in the balance of nature. They are photosynthetic, typically unicellular organisms, although cells may sometimes be connected to form threadlike filaments. Being autotrophs,

TABLE 9.2

The Five-Kingdom System of Classification					
	Monera (Prokaryotae)	Protista	Fungi	Plantae	Animalia
Cell type	Prokaryotic	Eukaryotic	Eukaryotic	Eukaryotic	Eukaryotic
Cell organization	Unicellular; occasionally grouped	Unicellular; occasionally multicellular	Unicellular or multicellular	Multicellular	Multicellular
Cell wall	Present in most	Present in some, absent in others	Present	Present	Absent
Nutrition	Absorption, some photosynthetic, some chemosynthetic	Ingestion or absorption, some photosynthetic	Absorption	Absorptive, photosynthetic	Ingestion; occasionally in some parasites by absorption
Reproduction	Asexual, usually by binary fission	Mostly asexual, occasionally both sexual and asexual	Both sexual and asexual, often involving a complex life cycle	Both sexual and asexual	Primarily sexual

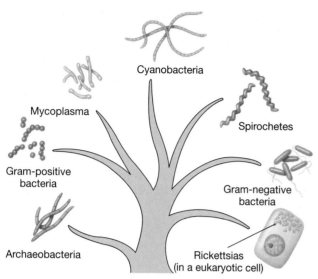

Figure 9.6 Some typical monerans. Monerans are prokaryotic organisms without a cell nucleus and other internal, membrane-enclosed structures.

Figure 9.7 Archaeobacteria: extremophiles that can exploit the unusual habitat of a "black smoker" vent. Living at deep ocean vents, where hot sulfurous volcanic gases are released from the Earth's interior, archaeobacteria survive in one of the most extreme environments known. This vent is located in the Mid-Atlantic Ridge, 3,100 meters below sea level—beneath a tremendous pressure of water, and at a temperature of 360°C. The bacteria obtain their energy from the sulfur compounds. *(Images courtesy of JAMSTEC)*

cyanobacteria do not invade other organisms, so they pose no health threat to humans, except for toxins (poisons) some release into water.

Cyanobacteria grow in a great variety of habitats, including anaerobic ones, where they often serve as food sources for more complex heterotrophic organisms. Some "fix" atmospheric nitrogen, converting it to nitrogenous compounds that algae and other organisms can use. Certain cyanobacteria also thrive in nutrient-rich water and are responsible for algal blooms—a thick layer of algae on the surface of water that prevents light from penetrating to the water below. Such blooms release toxic substances that can give the water an objectionable odor and even harm fish and livestock that drink the water.

Archaea (ar-kee-uh) surviving today are primitive prokaryotes adapted to extreme environments. The methanogens reduce carbon-containing compounds to the gas methane. The extreme halophiles live in excessively salty environments, and the thermoacidophiles live in hot acidic environments, such as volcanic vents in the ocean floor **(Figure 9.7)**. There, some species of bacteria form symbiotic relationships with organisms such as giant tubeworms (up to 2 meters tall). These worms lack a mouth, gut, or anus—how do they get fed? Chemolithotrophic archaeal bacteria, living inside the tubeworms, have metabolisms that fix inorganic sources (CO_3^-, HCO_3^-) into organic carbon sources via the same enzymes utilized in the Calvin cycle of certain autotrophs. The tubeworms are then able to use the organic carbon sources in cellular processes.

What do the tubeworms do for the bacteria? The tubeworms have well-vascularized plumes that trap O_2 and H_2S from the thermal vents and transport these substances to the chemolithotrophs. The bacteria use the O_2 and H_2S in their life-sustaining energy reactions,

providing nutrients to their ecosystem. Believed to be of very ancient origin, archaea have been found to differ from eubacteria in several distinctive ways, including the structure of their cell wall and the structure of their RNA polymerase. These organisms will be discussed in greater detail later in the chapter.

KINGDOM PROTISTA

Although the modern protist group is very diverse, it contains fewer kinds of organisms than when first defined by Haeckel. All organisms now classified in the kingdom **Protista** (Figure 9.8) are eukaryotic. Most are unicellular,

CLOSE UP

Going Where None Have Gone Before

If you were a microorganism, you'd love to be able to grow in places where competing microorganisms couldn't survive. That's what the Archaea have going for them. At the time of their discovery in 1977, the archaea were already thought to be quite odd. They lived in brines five times as salty as the oceans, geothermal environments that would cook other organisms to a crisp, and anaerobic habitats where even trace amounts of oxygen couldn't be found. Now they have been proven to be even odder. *Pyrolobus fumarii* holds the current record for life at high temperatures, growing at temperatures as high as a scalding 113°C. Antarctic archaea thrive at −1.8°C. Archaea have also been found in rice paddies, terrestrial soils, freshwater lake sediments, and even winery by-products.

CLOSE UP

Viroids and Prions

Science changes. For a long time, everyone thought that viruses were the smallest infectious agents. But particles even smaller than viruses have recently been discovered, and some appear to serve as infectious agents. They include viroids (vi′roid′s) and prions (pre′onz). A viroid is simply a fragment of RNA. The viroid that causes potato spindle tuber disease contains only 359 bases—enough information to specify the location of only 119 amino acids if all the bases function as codons. That's one-tenth the amount of nucleic acid found in the smallest viruses! Prions, or *proteinaceous infections particles*, are only one-tenth the size of a virus and consist of a protein molecule that folds incorrectly as a result of mutation. These particles, which are self-replicating, are responsible for some mysterious brain infections in humans, as well as mad cow disease in cattle.

invaded and by the diseases they caused. As more was learned about viruses, the early concept of "one virus, one disease" used in classification was found to be invalid for many viruses. Today viruses are classified by chemical and physical characteristics such as the type and arrangement of their nucleic acids, their shape (cubical or tubular), the symmetry of the protein coat that surrounds the nucleic acid, and the presence or absence of such things as a membrane covering (called an envelope), enzymes, tail structures, or lipids (**Figure 9.16**). These groupings reflect only common characteristics and are not intended to represent evolutionary relationships. A classification of viruses is presented in Appendix B

The study of viruses, or *virology*, is extremely important in any microbiology course for two reasons: (1) Virology is a recognized branch of microbiology, and techniques to study viruses are derived from microbiological techniques; and (2) viruses are of concern to health scientists because many cause diseases in humans, other animals, plants, and even microorganisms.

✓ CHECKLIST

1. What is the difference between a taxon and taxonomy?
2. What is the difference between species and specific epithet?
3. What is meant by a system of taxonomy that is phylogenetic? Why does such a system change frequently?
4. What is the difference between a kingdom and a domain? Name the five kingdoms, the three domains, and the types of organisms contained in each.
5. Where do viruses, viroids, and prions fit in today's taxonomy?

THE SEARCH FOR EVOLUTIONARY RELATIONSHIPS

Many biologists are interested in how living things evolved and how they are related to one another. In fact, most people have some curiosity about how life originated and gave rise to the diverse assortment of living things we see today. Although the details of the search for evolutionary relationships are of interest mainly to taxonomists, they are of some significance to health scientists. For example, many of the biochemical properties used to establish evolutionary relationships also can be used in identifying microorganisms. Whether it be a symbiotic association (e.g., between nitrogen-fixed bacteria and legumes) or a relationship between an infectious agent and its host, evolutionary relationships generally evolve together. Knowledge of such evolution is useful in order to understand the circumstances under which one organism becomes capable of infecting another, sometimes resulting in a symbiotic relationship and other times in the disease process.

The shattering of the long-held belief that all bacteria have a single circular chromosome (**Table 9.4**) has raised many questions. For example, how did multiple chromosomes, some of which are linear, come into being? And lacking mitosis, how is it assured that each daughter cell will receive the correct number and kinds of chromosomes?

First, let us remember the definition of chromosome. Plasmids contain genes which are needed only occasionally and are not essential for continuous use. If a large plasmid (*megaplasmid*) acquires a collection of "housekeeping" genes which are needed for daily life, it is then elevated to the status of chromosome. Confusingly enough, genes and plasmids can be acquired either vertically or by horizontal transfer. And, transposons can relocate genes from chromosomes into plasmids. Or, a chromosome could break, releasing a self-replicating portion of its

	DNA viruses	RNA viruses
Enveloped	Herpesvirus	Retrovirus
Naked (no envelope)	Adenovirus	Picornavirus

Figure 9.16 Some categories of viruses.

TABLE 9.4

Some Bacteria Having Two Chromosomes		
Organism	Major Chromosome Size in Kilobases (kb) (1 kb = 1,000 bases)	Minor Chromosome Size in kb (1 kb = 1,000 bases)
Agrobacterium rhizogenes	4,000	2,700
Agrobacterium tumefaciens	3,000	2,100 (linear)
Rhizobium galegae	5,850	1,200
Rhizobium loti	5,500	1,200
Sinorhizobium meliloti	3,400	1,700
Brucella suis (biovar 3)	3,100	None
Brucella suis (biovar 2 and 4)	1,850	1,350
Brucella ovis	2,100	1,150
Brucella melitensis	2,100	1,150
Brucella abortus	2,100	1,150
Ochrobactrum intermedium	2,700	1,900
Rhodobacter sphaeroides	3,046	914
Deinococcus radiodurans	2,649	412

genome into the cytoplasm. These are all ways that an ancestor with a single chromosome can develop a second chromosome.

On the other hand, genomic studies of close bacterial relatives suggest that, in some cases, the ancestral organism had two chromosomes which eventually fused to become one. In fact, some units accepted uncritically as being plasmids, because of their small size, may actually contain essential genes and be small chromosomes. Some plasmidless species' genomes reveal plasmid-type virulence gene sequences located in their single chromosomes, most likely having arrived there by horizontal fusion.

Within separate strains (biovars) of a single species, such as *Brucella suis*, the genome may exist as one or as two chromosomes, without conferring any obvious advantage to either biovar. This implies that having one versus two chromosomes has no evolutionary impact, at least in this species. However, when duplicate genes are found on both of the two chromosomes within a cell, they may have slightly different products (due to mutations) which are regulated differently. This could be an advantage. Meanwhile, what do we call such cells? They are not haploid or monoploid, but are not fully diploid either, as only some genes are duplicated. The term *mesoploid* has been suggested.

Right now there are more questions than answers. The correct apportioning of multiple chromosomes is not yet understood. Some researchers think it is aided by a mitosis-like process which is not yet well-identified, but relies on hypothetical microfilaments in the cytoplasm. We do know, however, that cells not receiving both kinds of chromosomes continue on for awhile, but eventually die. Maybe there is no "system" for assuring correct distribution, and those who are unlucky just die.

An extreme view of all this gene-swapping and reorganization of genomes is one that regards the entire bacterial universe as a single, huge, superorganism which possesses a network-like structure. A pool of genetic information is accessible to all bacterial cells by means of vertical and horizontal traffic, and is in continuous movement from one part of the superorganism to another. Indeed, for a long time scientists thought that genetic recombination among bacteria was extremely rare, and that mutation was the main driving force of evolution. However, we now must rethink this in view of the far greater frequency of horizontal gene transfer.

Eukaryotic genes enter this pool especially via intracellular endosymbiosis. Bacterial genes are horizontally transferred into host cell chromosomes, which in turn donate some of their genes to the bacterium. Eventually some essential genes of each wind up in the other's genome and both are then incapable of independent existence. Their symbiosis has become compulsory. Pathogenic bacteria sometimes use their pili to insert virulence gene sequences into eukaryotic cells. Parts of these virulence sequences of bacterial origin are not found integrated into eukaryotic chromosomes. So, perhaps we ought not to think of just all bacteria, but of all life, as being one huge superorganism, transferring genetic material among its parts through a network-like structure, rather than being limited to a vertical clonal descent.

SPECIAL METHODS NEEDED FOR PROKARYOTES

The taxonomy of most eukaryotes is based on morphology (structural characteristics) of living organisms, genetic features, and on knowledge of their evolutionary

Figure 9.17 Stromatolites. (a) Mats of cyanobacteria growing as stromatolites in shallow seawater off western Australia. These formations are 1,000–2,000 years old. *(Francois Gohier/Photo Researchers, Inc.)* **(b)** A cross-section through fossil stromatolites from Montana, showing horizontal layers of bacterial growth. *(Martin G.Miller/Visuals Unlimited)* **(c)** Filamentous cyanobacteria *(Paleolyngbya)* from the Lakhanda Formation in eastern Siberia. These microfossils date from the late Precambrian period and are approximately 950 million years old. *(Courtesy J. William Schopf, UCLA)*

relationships from fossil records. However, morphology and fossil records provide little information about prokaryotes. For one thing, prokaryotes have left few fossil records. As mentioned earlier, stromatolites, fossilized mats of prokaryotes, have been found mainly at sites where the environment millions of years ago allowed the deposition of dense layers of bacteria (**Figures 9.17a and b**). Stromatolites have provided much of our knowledge of the origin of the Archaea. Unfortunately, most bacteria do not form such mats, so most ancestral prokaryotes have disappeared without a trace.

Some rocks containing fossils of individual cells of cyanobacteria have been discovered (**Figure 9.17c**), but they have failed to reveal much information about the organisms. Moreover, prokaryotes have few structural characteristics, and these characteristics are subject to rapid change when the environment changes. Large organisms tend to require a fairly long period of time to reproduce, but prokaryotes reproduce rapidly. Assuming the same number of mutations per generation, organisms that reproduce most rapidly will accumulate a greater number of mutations over a given period of time. Because of this rapid mutational change rate, it is far more difficult to show the relationship between fossilized forms of prokaryotes and current organisms.

Chemical clues found in a manganese mine in South Africa suggest that bacteria, algae, or other simple organisms inhabited the soil there 2 billion years ago.

Because morphology and evolution are of little use in classifying prokaryotes, metabolic reactions, genetic relatedness, and other specialized properties have been used instead. Health scientists use these properties to identify infectious prokaryotes in the laboratory, but such identification does not necessarily reflect evolutionary relationships among the organisms. The methods described next are of use in exploring evolutionary relationships. Although the methods are particularly appropriate for eukaryotes, they can be used for prokaryotes as well.

NUMERICAL TAXONOMY

Numerical taxonomy is based on the idea that increasing the number of characteristics of organisms that we observe, increases the accuracy with which we can detect similarities among them. If the characteristics are genetically determined, the more characteristics two organisms share, the closer their evolutionary relationship. Although the idea of numerical taxonomy was developed before computers were available, computers allow us to compare large numbers of organisms rapidly and according to many different characteristics. In a simple example of numerical taxonomy, each characteristic is assigned a value of 1 if present and 0 if not present. Characteristics such as reaction to Gram staining, oxygen requirements, presence or absence of a capsule, properties of nucleic acids and proteins, and the presence or absence of particular enzymes and chemical reactions can be evaluated. Organisms are

Take another look

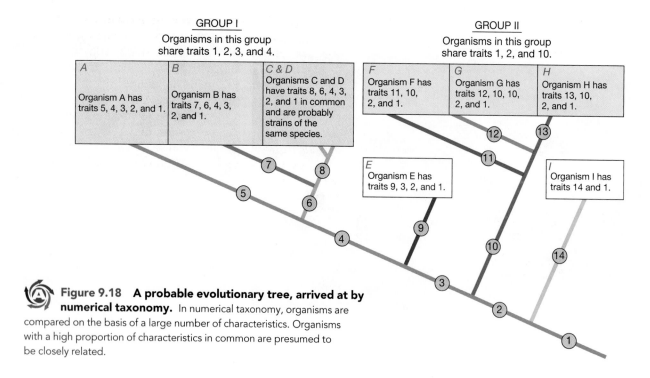

GROUP I
Organisms in this group
share traits 1, 2, 3, and 4.

GROUP II
Organisms in this group
share traits 1, 2, and 10.

A	B	C & D
Organism A has traits 5, 4, 3, 2, and 1.	Organism B has traits 7, 6, 4, 3, 2, and 1.	Organisms C and D have traits 8, 6, 4, 3, 2, and 1 in common and are probably strains of the same species.

F	G	H
Organism F has traits 11, 10, 2, and 1.	Organism G has traits 12, 10, 10, 2, and 1.	Organism H has traits 13, 10, 2, and 1.

E
Organism E has traits 9, 3, 2, and 1.

I
Organism I has traits 14 and 1.

Figure 9.18 A probable evolutionary tree, arrived at by numerical taxonomy. In numerical taxonomy, organisms are compared on the basis of a large number of characteristics. Organisms with a high proportion of characteristics in common are presumed to be closely related.

then compared, and patterns of similarities and differences are detected **(Figure 9.18)** With the use of numerical taxonomy, no single characteristic is used to arbitrarily divide all organisms into groups. If two organisms match on 90% or more of the characteristics studied, they are presumed to belong to the same species. Computerized numerical taxonomy offers great promise for improving our understanding of relationships among all organisms.

GENETIC HOMOLOGY

The discovery of the structure of DNA by James Watson and Francis Crick in 1953 provided new knowledge that was quickly applied by taxonomists, especially those studying taxonomic relationships and the evolution of eukaryotes. These scientists began to study the **genetic homology**, or the similarity of DNA, among organisms. Ideally one could just sequence the entire genome of every organism and compare them all to each other. This, however, is not a practical option at this time. It takes a lot of hard work and time to sequence just one genome—refer back to the Chapter 7 opening website interview with Dr. Karen Nelson. Several faster and easier techniques for estimating genetic homology are available. Similarities in DNA can be studied directly by determining the base composition of the DNA, by sequencing the bases in portions of DNA or RNA, and by using DNA hybridization. Because an organism's proteins are determined by its DNA, similarities in DNA can be studied indirectly by preparing *protein profiles* and by analyzing amino acid sequences in proteins.

Base Composition

Organisms can be grouped by comparing the relative percentages of bases present in the DNA of their cells. DNA contains four bases, abbreviated as A (adenine), T (thymine), G (guanine), and C (cytosine) (◀Chapter 2, p. 46). Base pairing occurs only between A and T and between G and C. In making base comparisons, we determine the total amount of G and C in a sample of DNA and express it as a percentage of total DNA. By subtracting this percentage from 100, we get the percentage of total A and T in the sample. For example, if the DNA is 60% G–C, then it is 40% A–T. The base composition of an organism is generally stated in terms of the percentage of guanine plus cytosine and is referred to as the G–C content. Base composition only determines the total amount of each nucleotide base present; it does not give any indication of the sequence of these bases.

Studies of base composition have shown that the G–C content varies from 23 to 75% in bacteria. These studies also have shown that certain species of bacteria, such as *Clostridium tetani* and *Staphylococcus aureus*, have very similar DNA compositions, but that *Pseudomonas aeruginosa* has a very different DNA composition. Thus, *C. tetani* and *S. aureus* are probably more closely related to each other than either is to *P. aeruginosa*. Similar percentages of bases do not in themselves prove that the organisms are closely related, because the *sequence* of bases may be quite different. (Human beings and *Bacillus subtilis*, for example, have nearly identical G–C percentages.) We can say, however, that if the percentages in two organisms are quite different, they are not likely to be closely related.

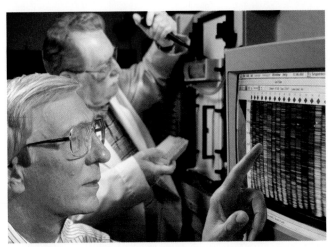

Figure 9.19 A DNA sequencer. Automated systems can identify the sequence of nucleotide bases in a piece of DNA. *(Keith Weller/USDA)*

DNA and RNA Sequencing

Automated equipment for identifying the base sequences in DNA or RNA is now available at reasonable cost **(Figure 9.19)**. It is therefore easier than before to search a culture for base sequences known to be unique to certain species. Using PCR techniques and a DNA synthesizer, one can produce a large number of **probes**, single-stranded DNA fragments that have sequences complementary to those being sought (◄Chapter 7, p. 204). A fluorescent dye or a radioactive tag (an indicator molecule) can be attached to the probe. When the probe finds its target DNA, it complementarily binds to it and does not wash off when rinsed. The specimen is then examined for fluorescing dye or for radioactivity. The presence or absence of the unique DNA sequence helps in identification of the specimen.

DNA Hybridization

In **DNA hybridization**, the double strands of DNA of each of two organisms are split apart, and the split strands from the two organisms are allowed to combine **(Figure 9.20)**. The strands from different organisms will **anneal** (bond to each other) by base pairing—A with T and G with C. The amount of annealing is directly proportional to the quantity of identical base sequences in the two DNAs. A high degree of homology (similarity) exists when both organisms have long, identical sequences of bases. Close DNA homology indicates that the two organisms are closely related and that they probably evolved from a common ancestor. A small degree of homology indicates that the organisms are not very closely related. Ancestors of such organisms probably diverged from each other thousands of centuries ago and have since evolved along separate lines.

You can use complementary DNA hybridization to detect the presence of DNA sequences from infectious viral agents.

Protein Profiles and Amino Acid Sequences

Every protein molecule consists of a specific sequence of amino acids and has a particular shape with an assortment of surface charges. Modern laboratory methods allow cells or organisms to be compared according to these properties of their proteins. Although variations in proteins among cells make these techniques difficult to apply to multicellular organisms, they are quite helpful in studying unicellular organisms.

A **protein profile** is a laboratory-prepared pattern of the proteins found in a cell **(Figure 9.21a)**. Because a cell's proteins are the products of its genes, the cells of each species synthesize a unique array of proteins—as

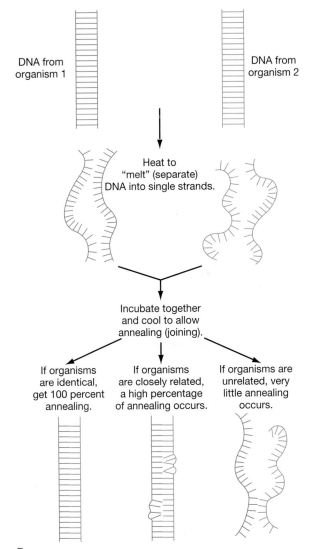

Figure 9.20 DNA hybridization. Strands of DNA are separated, and individual strands from two different organisms are allowed to anneal (join by hydrogen bonding at sites where there are many complementary base pairs). The degree of annealing reflects the degree of relatedness between the organisms, based on the assumption that annealing takes place only where genes, or parts of genes, are identical.

distinctive as a fingerprint is for humans. Analysis of the profiles of one or more proteins of different bacterial species provides a reasonable basis for comparisons.

Protein profiles are produced by the **polyacrylamide gel electrophoresis (PAGE)** method, which separates proteins on the basis of molecular size **(Figure 9.21b)**. In this method, samples of protein obtained from lysed cells are dissolved in a detergent and poured into wells (depressions) of a thin slab of polyacrylamide gel. The slab is inserted into a buffer-filled chamber. An electric current is then passed through the gel for a period of time. The current causes the protein molecules to migrate to the opposite end of the gel. Large protein molecules migrate more slowly than do smaller ones. After the smallest proteins have migrated, the current is turned off, and the gel slab is then removed. Next, it is stained so that the various proteins show up as separate stained bands in the gel slab (◄Chapter 18, p. 567).

Each band in the profile from one kind of cell represents a different protein in that cell. Bands at the same location in profiles from different kinds of cells indicate that the same protein is present in the different cells.

Determination of amino acid sequences in proteins also identifies similarities and differences among organisms. Certain proteins, such as cytochromes, which contribute to oxidative metabolism in many organisms, are commonly used to study amino acid sequences. The amino acid sequences in the same kind of protein from several organisms are determined. As with DNA hybridization, the extent of matching sequences of amino acids in the proteins indicates the relatedness of the organisms.

The proteins an organism contains are determined directly by the information in that organism's DNA. Thus, both protein profiles and determinations of amino acid sequences are as significant a measure of the relatedness of organisms as are DNA homologies. All are also related to the evolutionary history of the organisms.

OTHER TECHNIQUES

Other techniques for studying evolutionary relatedness include determining properties of ribosomes, immunological reactions, and phage typing.

Properties of Ribosomes

Ribosomes serve as sites of protein synthesis in both prokaryotic and eukaryotic cells. RNA in ribosomes can

Step 1: Gel solution is prepared and poured into the electrophoresis apparatus.

Step 2: Cells are broken apart, and a drop of sample solution is placed on the gel.

Step 3: Electric current is run through the gel to separate proteins into bands.

Step 4: Gel is stained to make bands of protein visible.

(a)

(b)

Figure 9.21 Separation of proteins. **(a)** Protein profiles, which provide a "fingerprint" of the proteins present in particular cells, can be used to compare different organisms to determine their degree of relatedness. *(Cytographics, Inc./Visuals Unlimited)* **(b)** The PAGE process.

Since ribosomal RNAs are functionally constant in bacteria, we may find drugs that target specific bacterial 16S or rRNA sequences to treat different infections.

be separated into several types according to the size of the RNA units. A particular RNA unit, the 16S rRNA component, has proven especially useful in studying evolutionary relationships for several reasons. Because rRNA molecules are easily rendered nonfunctional by even slight alterations in their genetic structure, mutations are rarely tolerated, and therefore ribosomes have evolved very slowly. It is the degree of similarity in 16S rRNA sequences between two organisms that indicate their evolutionary relatedness. If the nucleotide sequences of 16S rRNA molecules from two types of organism are very similar, those organisms are likely to be quite close evolutionarily. Although direct sequencing of 16S rRNA is used to show evolutionary relationships between species, newer methods, such as PCR, are beginning to replace it. The PCR technique, which is being used to amplify rRNA genes, requires less cell material and is more rapid and convenient for large studies than is direct rRNA sequencing.

Immunological Reactions

Immunological reactions also are used to identify and study surface structures and the composition of microorganisms, as explained in Chapter 17 As we shall see, one highly specific and sensitive technique involves proteins called *monoclonal antibodies*. Monoclonal antibodies can be created so that they will bind to a specific protein, usually a protein found on a cell surface. If the antibodies bind to the surfaces of more than one kind of organism, the organisms have that protein in common. This technique promises to be particularly useful in identifying specific biochemical properties of microorganisms. In turn, identification of such properties will be extremely useful in determining taxonomic relationships.

Phage Typing

Phage typing involves the use of bacteriophages, viruses that attack bacteria, to determine similarities among different bacteria. A separate agar plate is inoculated for each bacterium being studied. A sterile cotton swab or bent glass rod is used to spread the inoculum over the agar surface. After incubation, a *lawn*, or continuous sheet, of *confluent* bacterial growth will be produced. At the time of swabbing the plate, the underside of the plate is marked with numbered squares so that drops of known phages can be spotted onto specific zones of the

There are 10 times more kinds of phages than kinds of bacteria.

plate and later identified. After a suitable incubation period, as the lawn grows up, zones of lysis (*plaques*) appear in the bacterial lawn **(Figure 9.22)**. Because receptor sites for bacteriophages are highly specific, certain strains of a species of bacterium are attacked only by particular types of phages. By observing which phages cause holes in the lawn, researchers can identify the strain. Strains lysed by the same phages are presumed to be more closely related than strains that show different patterns of lysis by phages.

THE SIGNIFICANCE OF FINDINGS

The main significance of methods of determining evolutionary relationships is that these methods can be used to group closely related organisms and to separate them from less closely related ones. When groups of closely related organisms are identified, it is presumed that they probably had a common ancestor and that small differences among them have arisen by *divergent evolution*. **Divergent evolution** occurs as certain subgroups of a species with common ancestors undergo sufficient mutation to be identified as separate species.

Within the Bacteria, an early divergence gave rise to two important subgroups, the Gram-positive bacteria and the Gram-negative ones. Subsequent divergence within each group has given rise to many modern species of bacteria. Among the Gram-negative bacteria, the purple nonsulfur bacteria gave rise to modern bacteria that inhabit animal digestive tracts.

Figure 9.22 Phage typing. Receptor sites for bacteriophages are highly specific; certain strains of a species of bacterium are attacked only by particular types of phages. Clear sites (plaques) are left when phages have killed bacterial cells. On the basis of which phages have attacked a bacterial culture, one can determine which strain of that bacterial species is present. (*Dr. Edward J. Bottone, Mount Sinal School of Medicine*)

BACTERIAL TAXONOMY AND NOMENCLATURE

CRITERIA FOR CLASSIFYING BACTERIA

Most macroscopic organisms can preliminarily be classified according to observable structural characteristics. But it is more difficult to classify microscopic organisms, especially bacteria, because many of them have similar structures. Separating them according to cell shape, size, and arrangement does not produce a very useful classification system. Nor does the presence of specific structures such as flagella, endospores, or capsules allow identification of particular species. Therefore, other criteria must be used. Staining reactions, especially the Gram stain, were among the first properties other than morphology to be used to classify bacteria. Other properties now in use include features related to growth, nutritional requirements, physiology, biochemistry, genetics, and molecular analysis. These features include properties of DNA and proteins. Important criteria used in classifying bacteria are summarized in **Table 9.5**, and biochemical tests used in classifying and identifying them are described in **Table 9.6**.

By using various classification criteria, we can identify an organism as belonging to a particular genus and species. For bacteria, a species is regarded as a collection of strains that share many common features and differ significantly from other strains. A bacterial *strain* consists of descendants of a single isolation in pure culture. Bacteriologists designate one strain of a species as the **type strain**. Usually this is the first strain described. It is the name-bearer of the species and is preserved in one or more type culture collections. The American Type Culture Collection (ATCC), a nonprofit scientific organization established in 1925, collects, preserves, and distributes authenticated type cultures of microorganisms. Refer back to the opening vignette and video tour at the beginning of this chapter. Many important research studies dealing with classification, identification, and the industrial uses of microorganisms would be seriously hampered without the services of the ATCC.

For many strains of bacteria, scientists are able to determine that they are members of a particular species.

TABLE 9.5

Criteria for Classifying Bacteria		
Criteria	**Examples**	**Uses**
Morphology	Size and shape of cells; arrangements in pairs, clusters, or filaments; presence of flagella, pili, endospores, capsules	Primary distinction of genera and sometimes species
Staining	Gram-positive, Gram-negative, acid-fast	Separates eubacteria into divisions
Growth	Characteristics in liquid and solid cultures, colony morphology, development of pigment	Distinguish species and genera
Nutrition	Autotrophic, heterotrophic, fermentative with different products; energy sources, carbon sources, nitrogen sources, needs for special nutrients	Distinguish species, genera, and higher groups
Physiology	Temperature (optimum and range); pH (optimum and range), oxygen requirements, salt requirements, osmotic tolerance, antibiotic sensitivities and resistances	Distinguish species, genera, and higher groups
Biochemistry	Nature of cellular components such as cell wall, RNA molecules, ribosomes, storage inclusions, pigments, antigens; biochemical tests	Distinguish species, genera, and higher groups
Genetics	Percentage of DNA bases (G + C ratio); DNA hybridization	Determine relatedness within genera and families
Serology	Slide agglutination, fluorescent-labeled antibodies	Distinguish strains and some species
Phage typing	Susceptibility to a group of bacteriophages	Identification and distinguishing of strains
Sequence of bases in rRNA	rRNA sequencing	Determine relatedness among all living things
Protein profiles	Separate proteins by two-dimensional PAGE (electrophoresis)	Distinguish strains

TABLE 9.6

Specific Biochemical Tests Sometimes Used in Identifying and Classifying Bacteria

Biochemical Test	Nature of Test
Sugar fermentation	Organism is inoculated into a medium containing a specific sugar; growth and end products of fermentation, including gases, are noted. Anaerobic fermentations can be detected by inoculating organisms via a "stab" culture into solid medium.
Gelatin liquefaction	Organism is inoculated (stabbed) into a solid medium containing gelatin; liquefaction at room temperature or inability to resolidify at refrigerator temperature indicates the presence of proteolytic (protein-digesting) enzymes.
Starch hydrolysis	Organism is inoculated onto an agar medium containing starch; after the plate is flooded with Gram's iodine, clear areas around colonies indicate the presence of starch-digesting enzymes.
Litmus milk	Organism is inoculated into litmus milk medium (10% powdered skim milk plus litmus indicator); characteristic changes such as alteration of pH to acid or alkaline, denaturation of the protein casein (curdling), and gas production can be used to help identify specific organisms.
Catalase	Hydrogen peroxide (H_2O_2) is poured over heavy growth of an organism on an agar slant; release of O_2 gas bubbles indicates the presence of catalase, which oxidizes H_2O_2 to H_2O and O_2.
Oxidase	Two or three drops (or a disk) of an oxidase test reagent are added to an organism growing on an agar plate; a color change of the test reagent to blue, purple, or black indicates the presence of cytochrome oxidase.
Citrate utilization	Organism is inoculated into citrate agar medium in which citrate is the sole carbon source; an indicator in the medium changes color if citrate is metabolized; use of citrate indicates the presence of the permease complex that transports citrate into the cell.
Hydrogen sulfide	Organism is inoculated into peptone iron medium; formation of black iron sulfide indicates the organism produces hydrogen sulfide (H_2S).
Indole production	Organism is inoculated into a medium containing the amino acid tryptophan; production of indole, a nitrogenous breakdown product of tryptophan, indicates the presence of a set of enzymes that convert tryptophan to indole.
Nitrate reduction	Organism is inoculated into a medium containing nitrate (NO_3^-); presence of nitrite (NO_2^-) indicates that the organism has the enzyme nitrate reductase; absence of nitrite indicates either absence of nitrate reductase or presence of nitrite reductase (which reduces nitrite to N_2 or NH_3).
Methyl red	Organism is cultured in MR-VP broth; a methyl red indicator is added; presence of acid causes an indicator color change (red).
Voges-Proskauer	Organism is cultured in MR-VP broth; alpha naphthol and KOH-creatine are added; presence of the enzyme cytochrome oxidase causes color change in an indicator (rose color).
Phenylalanine deaminase	Organism is inoculated into a medium containing phenylalanine and ferric ions; formation of phenylpyruvate and its reaction with ferric ions produces a color change that demonstrates the presence of the enzyme phenylalanine deaminase.
Urease	Organism is inoculated into a medium containing urea; production of ammonia, usually detected by an indicator for alkaline pH, indicates the presence of the enzyme urease.
Specific nutrient	Organism is inoculated into a medium containing a specific nutrient, such as a particular amino acid (e.g., cysteine) or vitamin (e.g., niacin); growth of an organism that fails to grow in media lacking the specific nutrient can be used to identify some auxotrophs.

For other strains, however, difficult judgments must be made to decide whether the strain belongs to an existing species or differs sufficiently to be defined as a separate species. In recent years, similarities of DNA and proteins among organisms have proved a reliable means of assigning a strain to an existing species or establishing the basis for a new species.

Curiously, assigning bacterial genera to higher taxonomic levels—families, orders, classes, and divisions (or phyla)—can be even more difficult than organizing species and strains *within* genera. Many macroscopic organisms are classified by establishing their evolutionary relationships to other organisms from fossil records. Efforts are being made to classify bacteria by evolutionary relationships, too, but these efforts are hampered by the incompleteness of the fossil record and by the limited information gleaned from what fossils have been found. Even a complete fossil record might supply only morphological information and would thus be inadequate for determining evolutionary relationships.

THE HISTORY AND SIGNIFICANCE OF *BERGEY'S MANUAL*

The accepted reference on the identification of bacteria is commonly referred to as *Bergey's Manual*. The first edition of *Bergey's Manual of Determinative Bacteriology* was published in 1923 by the American Society for Microbiology; David H. Bergey **(Figure 9.23)** was chairperson of its editorial board. Since then, eight editions, an abridged version, and several supplements have been published. The *determinative information* (information used to identify bacteria) was collected into a single volume, the ninth edition of *Bergey's Manual of Determinative Bacteriology*, which was published in 1994. *Bergey's Manual* has become an internationally recognized reference for bacterial taxonomy. It has also served as a reliable standby for medical workers interested in identifying the causative agents of infections.

A four-volume first edition of *Bergey's Manual of Systematic Bacteriology* was published between 1984 and 1989, having a much broader scope. It provided descriptions and photographs of species, tests to distinguish among genera and species, DNA relatedness among organisms, and various numerical taxonomy studies **(Figure 9.24)**.

However, it is important to remember that, in their current state, both *Bergey's Manuals* do *not* present an accurate picture of evolutionary relationships among bacteria. Rather, they are practical groupings of bacteria that make it easy to identify them. We did not yet have enough information to draw a complete evolutionary tree for bacteria.

The five-volume second edition of *Bergey's Manual of Systematic Bacteriology* (Appendix B) represents a major departure from the first edition, as well as from the eighth and ninth editions of *Bergey's Manual of Determinative Bacteriology*. It is based on a phylogenetic (evolutionary) framework, rather than on a nonevolutionary grouping by phenotype. Sequencing of 16S rDNA has

Figure 9.23   David H. Bergey, the originator of *Bergey's Manuals*. The first *Bergey's Manual* was published in 1923. In 1936 he set up an educational trust to which all rights and royalties from the *Manuals* would be transferred for preparing, editing, and publishing future editions, as well as providing funds for research to clarify problems arising in the process. This nonprofit trust ensures that *Bergey's Manual* will be a self-perpetuating publication. *(Courtesy American Society for Microbiology Archives)*

provided guidance in this, but it is still very much a "work in progress."

PROBLEMS ASSOCIATED WITH BACTERIAL TAXONOMY

Despite the tremendous effort spent in classifying bacteria, the plight of bacteriologists looking at the taxonomy of bacteria might be described as follows: Those looking from the top level down can propose at least plausible divisions of the prokaryotes. Those bacteriologists looking from the bottom up can establish strains, species, and genera and can sometimes assign bacteria to higher-level groups. But too little is known about evolutionary relationships to establish clearly defined taxonomic classes and orders for many bacteria.

Less than 0.5% of the estimated 2 to 3 billion microbial species have been identified, let alone classified properly!

The difficulties of classifying bacteria are greatly magnified as one proceeds with total genome sequencing and discovery of more and more examples of lateral gene transfer.

BACTERIAL NOMENCLATURE

Despite all the taxonomic problems, there is an established nomenclature for bacteria. *Bacterial nomenclature* refers to the naming of species according to internationally agreed-upon rules. Both taxonomy and nomenclature are subject to change as new information is obtained. Organisms are sometimes moved from one category to another, and their official names are sometimes changed. For example, the bacterium that causes tularemia, a fever acquired by handling infected rabbits, was for many years called *Pasturella tularensis*. Its genus name was changed to *Francisella* after DNA hybridization studies revealed that hybridization between its DNA and that of *Pasteurella* species did not occur. It does, however, have a 78% match with the DNA of *Francisella novicida*. When considering specific orders and families, we must remember that such names have consistent endings: orders always end in *-ales* and families in *-aceae*.

BACTERIA

Some groups of bacteria, such as **Rickettsiae** and **Chlamydiae**, contain rather unusual organisms. These two groups are obligate intracellular parasites; that is, they can only grow inside living cells. Chlamydiae have an interesting and complex life cycle **(Figure 9.25)** rather than dividing by binary fission, as do Rickettsiae and most other bacteria. **Mycoplasmas** lack cell walls and form colonies that look like eggs fried sunny-side up. They have sterols in their cell membranes that give them great flexibility of shape (pleomorphism; ◄Chapter 4, p. 81). Also interesting are the **Ureaplasmas**, also with unusual cell walls and/or cell membranes. **Table 9.7** compares these groups with more typical bacteria and viruses.

BACTERIAL TAXONOMY AND YOU

As a beginning student, you will doubtless find it difficult to remember many characteristics of specific microorganisms that we cover in this course. A four-volume set of *Bergey's Manual*, First Edition, weighs about 21 pounds and costs about $400—not something you could carry to class and back. However, you can use the endpapers, located inside the front and back covers of this textbook. If you wish to find out whether a given organism is Gram-positive or Gram-negative, its shape, the disease(s) it causes, and so on, look it up by name in the back endpapers. Microorganisms are grouped as bacteria, viruses, fungi, and parasites (protozoa and

Figure 9.24 Samples from *Bergey's Manuals*. Shown here are three representative pages reproduced from the five-volume set of *Bergey's Manual of Systematic Bacteriology*. (top) Major groups of prokaryotes and their relatedness to one another *(Courtesy Bergey's Manual of Systematic Bacteriology)*. (middle) Map of evolutionary distances between various Archaea and Bacteria based on DNA sequences *(Courtesy Bergey's Manual of Systematic Bacteriology)*. (bottom) Life cycle of *Solentia*, an unusual bacterium that forms pseudofilaments composed of downward-opening envelopes *(Courtesy Bergey's Manual of Systematic Bacteriology)*.

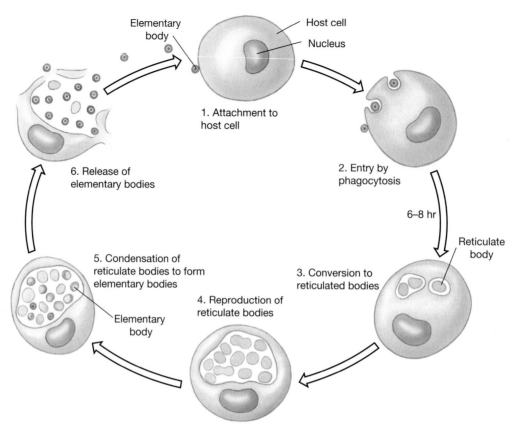

Figure 9.25 The life cycle of a chlamydia. (Step 1) Small, dark elementary bodies (the only infectious stage of the chlamydial life cycle) attach to a host cell and **(2)** enter by phagocytosis. **(3)** The elementary bodies, enclosed within membrane-enclosed vacuoles, lose their thick walls and enlarge to become reticulate bodies. **(4)** Reticulate bodies reproduce by binary fission, rapidly filling the cell. **(5)** They condense to form infectious elementary bodies, which **(6)** are then released by lysis and are free to attach to a new host cell.

helminths). Or if you are discussing a disease but can't remember which organism(s) cause it, look in the front endpapers under the name of the disease (again grouped as bacterial, viral, fungal, and parasitic diseases), and you will find the organism's name and some of its characteristics. Page numbers are given to direct you to further information. Thumbing through the book or index can be frustrating when you need some little piece of information. But flipping to the cover of your book is easy, and we encourage you to do so often, until the information gradually becomes more and more familiar.

✔ **CHECKLIST**

1. What is a type strain and a type culture collection? Why is such a collection essential to researchers?

2. Why was much of the first edition of *Bergey's Manual* not phylogenetically arranged?

3. What types of information are contained in *Bergey's Manual of Determinative Bacteriology*, compared to *Bergey's Manual of Systematic Bacteriology*?

TABLE 9.7

Characteristics of Typical Bacteria, Rickettsiae, Chlamydiae, Mycoplasmas, Ureaplasmas, and Viruses						
Characteristic	**Typical Bacteria**	**Rickettsiae**	**Chlamydiae**	**Mycoplasmas**	**Ureaplasmas**	**Viruses**
Cell wall	Yes	Yes	Yes	No	Sometimes	No
Grow only in cells	No	Yes	Yes	No	No	Yes
Require sterols	No	No	No	Sometimes	Yes	No
Contain DNA and RNA	Yes	Yes	Yes	Yes	Yes	No
Have metabolic system	Yes	Yes	Yes	Yes	Yes	No

CLOSE UP

The Discovery of New Organisms

Are there any new worlds to discover, or new creatures in them? Yes! In recent years, scientists have discovered living organisms in such diverse environments as submarine hot vents, inside volcanoes, and in deep oil wells. In 1990 a joint U.S. and Soviet team discovered hot vents in fresh water for the first time, complete with an associated community of archaeobacteria, worms, sponges, and other organisms.

The vents lie more than 400 meters deep in a most unusual Russian lake, Lake Baikal, which is the deepest lake in the world and holds the greatest quantity of fresh water in the world. Located in central Asia, in Siberia, it lies in a pocket between two continental plates. Asia was formed as a solid mass when several plates collided one after another and remained together. The area of Lake Baikal is being pulled apart, forming a rift valley and eventually a new ocean. This region is comparable to the sea-floor spreading centers (ridges) of the Pacific Ocean, where other vent communities have been found. In both locations, hot materials from deep inside the Earth are emerging. Lake Baikal is a unique treasure for studying the evolution of life and microbial forms. Most lakes are only thousands of years old, but Lake Baikal may be 25 million years old. Microbes similar to the early evolving stages of life may still exist in its depths.

What lives inside a volcano? Studies following the 1980 volcanic eruptions of Mount St. Helens have raised some interesting questions. Archaeobacteria, which had previously been known from the deep-sea volcanic vents (black smokers) located 2,200 meters below the sea surface, have been found living on and in Mount St. Helens at temperatures of 100°C. Where did they come from? Some scientists think that they may have been present deep inside the volcano. For that matter, where do the archaeobacteria in the submarine vents come from? Do they point out a linkage between terrestrial and submarine volcanic activity? We tend to think of life as being present only on the *surface* of the Earth, but perhaps there is a whole different range of life that we know nothing about, deep *within* the crust of the Earth. Daily, more and more evidence accumulates in favor of the idea of "continuous crustal culture."

Cores taken from the deepest oil wells being bored into the Earth reveal archaeobacteria at sites not connected with volcanic activities. Ancient bacteria from the early stages of our planet's development may still be colonizing the hot anaerobic interior of the Earth—places with conditions that resemble those formerly on the Earth's surface.

Mount St. Helens, shown here erupting in July 1980, is home to archaeobacteria. *(David Weintraub/Photo Researchers, Inc.)*

Various ecological problems have sent out scientists from universities, government, and industry to hunt for new microbes with properties that make the organisms useful in cleaning up the environment. Scientists from the Woods Hole Oceanographic Institution in Massachusetts took their search to a depth of more than 1,800 meters in the Gulf of California, where they have discovered anaerobic bacteria that can degrade naphthalene and possibly other hydrocarbons that might be found in oil spills. Sites in need of bioremediation often lack oxygen, making it impossible to utilize aerobic organisms for cleanup—hence the hunt in deep anaerobic environments. General Electric has also found an anaerobic bacterium that it plans to use to destroy polychlorinated biphenyls (PCBs), industrial by-product chemicals that accumulate in animal tissue and cause damage, including cancer and birth defects.

A new bacterium, initially referred to as GS-15, but now named *Geobacter metallireducens*, was discovered in the Potomac River by U.S. Geological Survey (USGS) scientists. It changes iron from one form to another. It seems, however, that these bacteria can just as easily feed on uranium, getting twice as much energy in the process and transforming the uranium into an insoluble precipitate. The USGS team has used GS-15 to remove uranium from contaminated well and irrigation water found in much of the western United States and at uranium mining, processing, and nuclear waste sites. It also can clean up underground petroleum spills.

There are many new microbes yet to be discovered. In addition to the naturally occurring species, new ones will be designed by scientists using genetic engineering techniques—or possibly even discovered on another planet! All these species will need to be classified and named. Clearly, *Bergey's Manual* will never be "finished."

▌ RETRACING OUR STEPS

▌ TAXONOMY: THE SCIENCE OF CLASSIFICATION

• Organisms are named according to their characteristics, where they are found, who discovered them, or what disease they cause. **Taxonomy** is the science of classification, and each category is a **taxon**.

LINNAEUS, THE FATHER OF TAXONOMY

• Linnaeus developed the system of **binomial nomenclature**, a two-name identification system for each living organism.

• The **genus** and **specific epithet** of each organism identify the **species** to which it belongs.

- Linnaeus also established the hierarchy of taxonomy and classified organisms into two kingdoms, Plantae and Animalia.

III USING A TAXONOMIC KEY

- A **dichotomous key** consists of a series of paired statements presented as either-or choices that describe characteristics of organisms. By selecting appropriate statements to progress through the key, one can classify organisms and, if the key is sufficiently detailed, identify them by genus and species.

PROBLEMS IN TAXONOMY

- Ideally, organisms should be classified by their **phylogenetic**, or evolutionary, relationships.
- Problems in taxonomy include the rapid pace of evolutionary change in microorganisms and the difficulty in deciding what constitutes a kingdom and what constitutes a species.

DEVELOPMENTS SINCE LINNAEUS'S TIME

- Since Linnaeus's time, several taxonomists have proposed three- and four-kingdom systems on the basis of various fundamental characteristics of living things. Whittaker proposed a five-kingdom system in 1969.
- Since 1925, *Bergey's Manual of Determinative Biology* has served as an important tool in identifying bacteria.

III THE FIVE-KINGDOM CLASSIFICATION SYSTEM

- The kingdoms of the **five-kingdom system** are **Monera** (**Prokaryotae**), **Protista**, **Fungi**, **Plantae**, and **Animalia**. The characteristics of members of each kingdom are summarized in Table 9.2.

KINGDOM MONERA

- All monerans are unicellular **prokaryotes**: They generally lack organelles, have no true nuclei, and their DNA has little or no protein associated with it.
- The **cyanobacteria** are photosynthetic monerans of great ecological importance.

KINGDOM PROTISTA

- The protists are a diverse group of mostly unicellular **eukaryotes**.

KINGDOM FUNGI

- The fungi include some unicellular and many multicellular organisms that obtain nutrients solely by absorption.

KINGDOM PLANTAE

- Most plants live on land and contain chlorophyll in organelles called chloroplasts.

KINGDOM ANIMALIA

- All animals are derived from zygotes; most are macroscopic.

III THE THREE-DOMAIN CLASSIFICATION SYSTEM

- The three **Domains** are higher than the category of kingdom. They include: **Bacteria**, **Archaea**, and **Eukarya**. Their characteristics are summarized in Table 9.3.
- The concept of a Universal Common Ancestor with a linear *tree of life* has now been replaced by a **shrub of life** with many roots, due to lateral gene transfer.

DOMAIN BACTERIA

- All Bacteria are unicellular prokaryotes and include the eubacteria ("true bacteria").

DOMAIN ARCHAEA

- All Archaea are unicellular prokaryotes, having a cell wall made of materials other than peptidoglycan.

DOMAIN EUKARYA

- All are eukaryotic cells, having a true nucleus.

III CLASSIFICATION OF VIRUSES

- **Viruses**, acellular infectious agents that share only a few characteristics with living organisms, are not included in any of the five kingdoms. Viruses are classified by their nucleic acids, chemical composition, and morphology.

III THE SEARCH FOR EVOLUTIONARY RELATIONSHIPS

SPECIAL METHODS NEEDED FOR PROKARYOTES

- Special methods are needed for determining evolutionary relationships among prokaryotes because they have few morphological characteristics and have left only a sparse fossil record.
- Several methods, including numerical taxonomy and genetic homology, are currently used to determine evolutionary relationships among organisms.

NUMERICAL TAXONOMY

- In **numerical taxonomy**, organisms are compared on the basis of a large number of characteristics and grouped according to the percentage of shared characteristics.

GENETIC HOMOLOGY

- **Genetic homology** is the similarity of DNA among different organisms, which provides a measure of their relatedness. Several techniques that determine genetic homology are available.
- The relative percentages of bases in DNA are a measure of relatedness. The base composition of DNA is determined, and G–C percentages are compared among organisms.
- Base sequences can be identified by DNA **probes**.
- In **DNA hybridization**, the degree of matching between strands of DNA is compared among organisms.
- **Protein profiles**, made by **polyacrylamide gel electrophoresis (PAGE)**, are used to indicate whether the same proteins are present in different organisms.
- The amino acid sequences of related organisms are similar, so the determination of amino acid sequences is another measure of relatedness.

OTHER TECHNIQUES

- Other methods make use of properties of ribosomes, immunological reactions, and **phage typing**.

THE SIGNIFICANCE OF FINDINGS

Evolutionary relationships can be used to group closely related organisms. Small differences among organisms descended from a common ancestor arise by **divergent evolution**. An early divergence gave rise to the two major subgroups of eubacteria, the Gram-positive bacteria and the Gram-negative ones.

III BACTERIAL TAXONOMY AND NOMENCLATURE

CRITERIA FOR CLASSIFYING BACTERIA

• The criteria used for classifying bacteria are summarized in Table 9.4. These criteria can be used to classify bacteria into species and even into strains within species.

• For many species a particular strain is designated as the **type strain**, which is preserved in a type culture collection.

THE HISTORY AND SIGNIFICANCE OF *BERGEY'S MANUAL*

• *Bergey's Manual of Determinative Bacteriology* was first published in 1923 and has been revised several times; a ninth edition was published in 1994.

• *Bergey's Manual of Systematic Bacteriology* (a four-volume set) provides definitive information on the identification and classification of bacteria.

PROBLEMS ASSOCIATED WITH BACTERIAL TAXONOMY

• Taxonomists do not agree on how members of the kingdom Prokaryotae (Monera) should be divided. Many species of bacteria have been grouped into genera and some into families. Four *divisions* (the equivalent of phyla) have been established. Much information is needed to determine evolutionary relationships and establish classes and orders.

BACTERIA

• A complete listing of the sections is provided in Appendix B.

• Groups of important bacteria include the spirochetes, mycoplasmas, rickettsiae, chlamydiae, mycobacteria, and cyanobacteria.

BACTERIAL TAXONOMY AND YOU

• Use the endpapers at the beginning and end of the book to familiarize yourself with taxonomic information.

▌ TERMINOLOGY CHECK

Animalia **(p. 247)**
anneal **(p. 257)**
Archaea **(p. 246)**
Bacteria **(p. 249)**
binomial nomenclature **(p. 241)**
Chlamydiae **(p. 263)**
cyanobacteria **(p. 245)**
dichotomous key **(p. 243)**
divergent evolution **(p. 259)**
DNA hybridization **(p. 257)**
domain **(p. 248)**
eubacteria **(p. 245)**

Eukarya **(p. 249)**
eukaryote **(p. 245)**
extreme halophile **(p. 252)**
extreme thermoacidophile **(p. 252)**
five-kingdom system **(p. 245)**
Fungi **(p. 247)**
genetic homology **(p. 256)**
genus **(p. 241)**
methanogen **(p. 252)**
Monera **(p. 245)**
Mycoplasmas **(p. 263)**

numerical taxonomy **(p. 255)**
phage typing **(p. 259)**
phylogenetic **(p. 243)**
Plantae **(p. 247)**
polyacrylamide gel electrophoresis (PAGE) **(p. 258)**
probe **(p. 257)**
Prokaryotae **(p. 245)**
prokaryote **(p. 245)**
protein profile **(p. 257)**
Protista **(p. 246)**

Rickettsiae **(p. 263)**
shrub of life **(p. 251)**
species **(p. 241)**
specific epithet **(p. 241)**
strain **(p. 242)**
stromatolite **(p. 248)**
taxon **(p. 241)**
taxonomy **(p. 241)**
type strain **(p. 260)**
Ureaplasmas **(p. 263)**
virus **(p. 252)**

▌ CLINICAL CASE STUDY

The following is a true story, though the name has been changed. It is an example of why it is critically important to identify organisms that are causing disease: Dr. Overland had just retired from teaching; she was a diabetic and went to the hospital to have a diagnostic procedure done, called an angiogram. This involves injecting dye in the circulatory system to analyze the function of the heart. A few days after the procedure Dr. Overland began to be feverish. She went to her doctor, who put her in the hospital and began treating her with an antibiotic to kill bacteria. Blood samples were collected to begin identification of the organism that was causing the disease process. The organism turned out not to be a bacterium but a yeastlike fungus, *Candida albicans*. Dr. Overland died before an antimicrobial could be used to kill the *C. albicans*. Why did it take the hospital so long to identify the pathogen?

▌ CRITICAL THINKING QUESTIONS

1. Before Linnaeus created binomial nomenclature, in which scientific names are uniform worldwide, the same organism had different names in different parts of the world. Is there any reason why that would have presented problems to scientists?

2. What is a species? How is this defined in organisms that do not reproduce sexually?

3. A series of DNA hybridization experiments were performed in which the DNA of two given organisms were separated into single stands. Then the two organisms' single-stranded DNA was incubated together, and the percentages that hybridized (combined with that of the other species) were determined. From the data given, which two species are probably most closely related?

Species	Percentage Hybridization
A and B	46
A and C	58
B and C	75

▋ SELF-QUIZ

1. Taxonomy is the science of ~~Classification~~. Organisms can be named according to their ~~characteristics~~, where they are found, ~~who~~ who discovered them, or what ~~disease~~ they cause.

2. Who invented binomial nomenclature (a two-name identification system) and is considered the father of taxonomy?
   (a) Frederick Griffith
   (b) R. H. Whittaker
   (c) Carolus Linnaeus
   (d) Lynn Margulis
   (e) H. F. Copeland

3. Using binomial nomenclature, the first name designates the genus while the second name designates the:
   (a) Specific epithet
   (b) Order
   (c) Kingdom
   (d) Phylum
   (e) Family

4. Which of the following is not a characteristic of a dichotomous key?
   (a) The dichotomous key has a "quartet" of statements describing the characteristics of organisms.
   (b) Each statement is followed by directions to go to another pair of statements until the name of the organism is identified.
   (c) It is the most commom type of key used by biologists to identify organisms based on their characteristics.
   (d) Paired statements present an "either-or" choice so that only one statement is true.
   (e) a and b.

5. When writing the name of a species one should have:
   (a) Two names—genus and specific epithet
   (b) The first letter of the genus name capitalized and the second name all in lowercase letters
   (c) Two names that are both either italicized or underlined
   (d) All of the above

6. Members of a species can sometimes be subdivided into subgroups called:
   (a) Orders
   (b) Genera
   (c) Strains
   (d) Families
   (e) Kingdoms

7. What are the five kingdoms in the five-kingdom system? List the characteristics of each.

8. All of the following statements are true about the three-domain classification system EXCEPT:
   (a) The three domains are Bacteria, Archaea, and Eukarya.
   (b) Lateral gene transfer has forced us to rethink our Domain model from a "tree of life" to a "shrub of life."
   (c) Domains are higher than the category of Kingdoms.
   (d) Compared to Bacteria, Archaea inhabit the same environs and have the same amount of peptidoglycan their cell walls.
   (e) All of the above are true.

9. What type of bacteria are also called blue-green algae?
   (a) Paramecium
   (b) Archaea
   (c) Eubacteria
   (d) Protists
   (e) Cyanobacteria

10. All of the following pertain to archaeobacteria EXCEPT:
    (a) They include microbes that live in hot acidic environments.
    (b) All are strict anaerobes.
    (c) They include microbes that live in extremely salty environments.
    (d) All lack peptidoglycan in their cell walls.
    (e) They include microbes that reduce carbon to methane gas.

11. Members of the kingdom Protista differ from members of the kingdom Monera mainly due to the presence of:
    (a) RNA
    (b) Ribosomes
    (c) Cell wall
    (d) DNA
    (e) Membrane-bound nucleus

12. Viruses are acellular infectious agents that contain nucleic acid and protein. What type of nucleic acid can they have?

13. Extreme halophiles grow in conditions containing high:
    (a) Nitrogen
    (b) Temperature
    (c) Amounts of Methane
    (d) Amounts of Oxygen
    (e) Amounts of Salt

14. An organism that contains 36% G–C will also contain:
    (a) 36% A–T
    (b) 36% A + 64%T
    (c) 64% A + 36%T
    (d) 64% A–T
    (e) 36% A + 36%T

15. Because prokaryotes have few morphological characteristics and as a group have a sparse fossil record, they are difficult to group in terms of evolutionary relationships. Match the special contemporary methods used today that help determine evolutionary relationships to their descriptions:

_e_ Genetic homology
_d_ Phage typing
_a_ Protein profiling
_f_ Numerical taxonomy
_b_ DNA hybridization
_c_ G–C content

(a) Polyarylamide gel electrophoresis is used to resolve whether the same proteins are present in different organisms

(b) Double-stranded DNA from two organisms are split apart with the spilt strands being allowed to combine; the degree of matching gives an idea of the amount of genetic homology between different organisms

(c) The relative percentages of bases of DNA are a measure of relatedness between two different organisms

(d) Employs the use of bacteriophages to determine similarities among different bacteria

(e) Similarity of DNA among different organisms provides a measure of their relatedness

(f) A large number of characteristics are compared and grouped according to the percentage of shared characteristics

16. DNA encoding which of the following cell components would most likely be conserved throughout evolution of an organism?
(a) Flagella
(b) Ribosomes
(c) Antibiotic resistance
(d) Antigenic proteins
(e) Membrane proteins

17. Which of the following would be the most specific method for classifying bacteria?
(a) DNA analysis
(b) Phage typing
(c) Morphology
(d) Size
(e) Capsules

18. What life-cycle characteristics do the Chlamydiae and Rickettsiae share?
(a) Both lack DNA that can be replicated.
(b) Both groups are not medically relevant in man.
(c) Both groups are obligate intracellular parasites.
(d) All of the above.
(e) None of the above.

19. _Viroids_ are smaller than viruses, contain only a fragment of _RNA_, and contain members that are infectious. Proteinaceous infectious particles or _prions_ are one-tenth the size of viruses, contain an incorrectly folded mutant protein molecule that is self-replicating, and are responsible for causing _brain_ infections in man and cattle.

20. Which of the following references would be most beneficial for identifying an unknown bacterial isolate?
(a) Encyclopedia
(b) Dictionary
(c) Bergey's Manual of Systematic Bacteriology
(d) Microbiology textbook
(e) Your microbiology professor

21. Other techniques useful in determining evolutionary relationships among prokaryotes are:
(a) Determining the amino acid sequences of two different organisms and comparing them
(b) Comparing ribosomal sequences and sizes between two different organisms
(c) Comparing the amplified ribosomal RNA gene products between two different organisms using PCR technology
(d) Immunological reactions employing monoclonal antibodies to determine the presence or absence of prokaryotic surface structures and biochemical enzymes.
(e) All of the above

22. Which of the following characteristics is used to classify viruses?
(a) Type and arrangement of nucleic acids
(b) Capsid shape
(c) Presence or absence of an envelope
(d) Presence or absence of tail structures
(e) All of these

23. Match the following with their respective descriptions:
_C_ Animalia
_D_ Plantae
_A_ Protista
_E_ Monera
_B_ Fungi

(a) Usually unicellular eukaryotes
(b) Unicellular and multicellular absorptive heterotrophs
(c) Multicellular ingestive heterotrophs
(d) Multicellular and photosynthetic
(e) Unicellular prokaryotes

24. Members of the extreme thermoacidophile group of Archaea must withstand extremes of temperature and pH. Which major obstacle have they been able to overcome in their evolution?
(a) Being able to survive in a high-saline environment
(b) Being able to withstand denaturation and/or inactivation of their enzymes
(c) Generating the enzymes necessary for aerobic respiration
(d) a and c
(e) None of the above

25. In the name _Mycobacterium tuberculosis_, what is the genus name, specific epithet, and species name of this organism?

Genus= mycobacterium
Specific epithet= tuberculosis
Species name= mycobacterium tuberculosis

26. In the following diagram of the life cycle of *Chlamydia trachomatis*, identify numbered stages 1–6 and parts (a)–(d).

LIFE CYCLE OF A CHLAMYDIA

1.) Attachment to host
2.) Entry by phagocytosis
3.) Conversion to reticulate bodies
4.) Reproduction of reticulate bodies
5.) Condensation of reticulate bodies
6.) Release of elementary bodies

(A.) Host cell
(B.) Nucleus
(C.) Reticulate body
(D.) Elementary body

▮ EXPLORATIONS ON THE WEB

http://www.wiley.com/college/black

If you think you've mastered this chapter, there's more to challenge you on the web. Go to the companion web site to fine-tune your understanding of the chapter concepts and discover answers to the questions posed below.

1. Members of genera belonging to the Enterobacteriaceae family cause such diseases as meningitis, bacillary dysentery, typhoid, and food poisoning. So, if you are going to get infected with an enteric bacteria, do you want it to be able to ferment lactose?

2. Which Gram-negative, nonfermenter rod is responsible for many nosocomial infections, especially in immunocompromised individuals, burn victims, and individuals on respirators or with indwelling catheters?

3. Learn more about why scientists now believe there are three rather than two evolutionary lineages.

Come with me...

Worldwide <u>273,500</u> women die of cervical cancer each year, <u>3900</u> in the United States alone. How many of them knew that this was a sexually transmitted disease, caused by the human papillomovirus (HPV)? This virus is found in 99.7% of all cervical cancer tissues. But now there is a new vaccine, Gardasil™, and a second one soon to be on the market, which can prevent cervical cancer! There are over 100 strains of HPV, 13 of which cause 99% of all cervical cancer. Other strains may cause genital warts (see photo in ◄Chapter 19, Figure 19.15, p. 587). Approximately 20 million people in the United States are currently infected with HPV. Eighty percent of sexually active women will be infected by age 50. Fortunately about 90% of these infections will spontaneously cure and do no harm.

©AP/Wide World Photos

Those strains that cause warts do not cause cervical cancer—it's the "silent" infections that cause no symptoms, but which become chronic long-lasting infections that can cause cancer. There is no cure for HPV infections—only prevention, which brings us to the Gardasil™ vaccine.

Gardasil™ is targeted against the 2 strains of HPV that cause 70% of all cervical cancer, and the 2 strains that cause 30% of genital warts. It is over 99% effective at this. It will not cure an existing infection, but can keep you from getting the 4 targeted strains on top of what you already have. Vaccine supply is limited, and so is being urged for use in girls who have not yet had sex, with ages 9 to 26 suggested. Three shots, 2 months apart over 6 months are necessary. Each shot now costs $120, and not all insurers cover this cost.

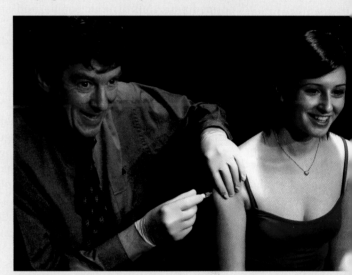

©AP/Wide World Photos

Some states have sought to require vaccinating all young girls, preferably before 6th grade. Some parents have raised an outcry against this. Reasons include fear that the vaccine could be damaging, or it is too expensive, or won't last very long, or that girls vaccinated will feel "protected" and thus go out and indulge in more sexual activity than they would if unvaccinated. But what if those girls are raped, or later have unfaithful husbands? Go to the website to see some news clips. What are your feelings, and those of the rest of your class? This is microbiology happening right now!

 Video related to this topic is available within WileyPLUS.

⫴ GENERAL CHARACTERISTICS OF VIRUSES
What Are Viruses? / Components of Viruses / Sizes and Shapes / Host Range and Specificity of Viruses / Origins of Viruses

⫴ CLASSIFICATION OF VIRUSES
RNA Viruses / DNA Viruses

⫴ EMERGING VIRUSES

⫴ VIRAL REPLICATION
General Characteristics of Replication / Replication of Bacteriophages / Lysogeny / Replication of Animal Viruses / Latent Viral Infections

⫴ CULTURING OF ANIMAL VIRUSES
Development of Culturing Methods / Types of Cell Cultures

⫴ VIRUSES AND TERATOGENESIS

⫴ VIRUSLIKE AGENTS: SATELLITES, VIROIDS, AND PRIONS
Satellites / Delta Hepatitis / Viroids / Prions

⫴ VIRUSES AND CANCER

⫴ HUMAN CANCER VIRUSES
How Cancer Viruses Cause Cancer / Oncogenes

Throughout human history, viral epidemics have caused us to become more aware of the impact microbes have on our lives and on the course of history. In the previous century, we need only look back at the influenza pandemic of 1918 and 1919, polio outbreaks in the 1940s and 1950, Marburg disease in the 1960s and the related Ebola hemorrhagic disease in the 1970s and 1990, or HIV in the 1980s and 1990s, to be reminded that virus infections have a significant impact on our society. In fact, we need not look back even that far. In 1996, 1997, and 2000, our world experienced outbreaks of hepatitis, dengue fever, yellow fever, Lassa fever, chicken influenza, and polio, plus West Nile fever in 2002 and SARS in 2003. Now we are worried, and preparing for, what could imminently be a pandemic outbreak of avian (bird) flu.

The media are full of both popularized and scientific writings on the "new" or "reemerging" viruses. What factors are contributing to their increased impact? Are they truly "new" viruses—or are they recurring forms of old viruses? What factors have led to their "reemergence"? We have seen the eradication of smallpox and read reports that polio is close to eradication in much of the world (in spite of recent outbreaks in the Mediterranean and Russia). Hantavirus, dengue fever, and yellow fever are said to be reemerging. Could smallpox also reappear? (What secret stocks of smallpox virus are held around the world for possible terrorist attacks of biological warfare?) Influenza is an annual occurrence, but we haven't seen a major pandemic for many years. Might we once again see a pandemic of the magnitude of that in 1918, which killed half a million Americans in just 10 months? It is likely that SARS will erupt again, as we now know that it can be carried by domestic animals such as housecats and ferrets as well as many species of wild animals. Hopefully we will have a vaccine ready before then. Some forms of cancer are definitely caused by viruses—viruses that we know are transmitted person to person. What are your chances of "catching" cancer?

These are just a few of the questions surrounding our knowledge of viral infections today. Current research continues to provide us with a better understanding of viral structure, viral replication, and viral diseases. As we learn more about how viral organisms function, it will increase our ability to develop new methods to control and/or eradicate viral infections. It will also extend our understanding of what the fundamental nature of life is—an important theme for all biologists.

This chapter examines the structure and behavior of viruses and viruslike agents. By the end of this chapter, you will have a better understanding of and appreciation for one of nature's tiniest-sized, but most dangerous, groups of microbes. Indeed, the name *virus* itself comes from the Latin word meaning "poison."

GENERAL CHARACTERISTICS OF VIRUSES

WHAT ARE VIRUSES?

Viruses are infectious agents that are too small to be seen with a light microscope and that are not cells. They have no cell nucleus, organelles, or cytoplasm. When they invade susceptible host cells, viruses display some properties of living organisms and so appear to be on the borderline between living and nonliving. Viruses can *replicate*, or multiply, only inside a living host cell. As such they are called **obligate intracellular parasites**, a distinction they share with chlamydias and rickettsias. We may have to reconsider the traditional definition of viruses following the announcement in December 1991 by E. Wimmer, A. Molla, and A. Paul that they successfully grew entire polioviruses in test tubes containing ground up human cells, but no live cells. RNA from polioviruses was added to the cell-free extract, and about 5 hours later complete new virus particles began to appear. This work has been replicated by many other groups, but has not yet been duplicated with other viruses.

Now in November 2003, Dr. Craig Venter (already famous for his role in completing the Human Genome Project) led a team at the Institute for Biological Energy Alternatives in creating a brand new, man-made virus. They didn't even make the parts themselves. They ordered them from various commercial companies. Then they put the over 5,000 DNA building blocks plus proteins together, creating a bacteriophage. They hope to eventually design genetically modified organisms that can eat carbon dioxide and clean the environment. Some environmentalists are very upset, fearing that such organisms could run amok.

Viruses differ from cells in important ways. Whereas prokaryotic and eukaryotic cells contain both DNA and RNA, individual virus particles contain only one kind of nucleic acid—either DNA or RNA but never both. Cells grow and divide, but viruses do neither. Viral replication requires that a virus particle infect a cell and program the host cell's machinery to synthesize the components required for the assembly of new virus particles. The infected cell may produce hundreds to thousands of new viruses and then usually dies. Tissue damage as a result of cell death accounts for the destructive effects seen in many viral diseases.

COMPONENTS OF VIRUSES

Typical viral components are shown in **Figure 10.1**. These components are a nucleic acid core and a surrounding protein coat called a **capsid**. In addition, some viruses have a surrounding lipid bilayer membrane called an **envelope**. A complete virus particle, including its envelope, if it has one, is called a **virion**.

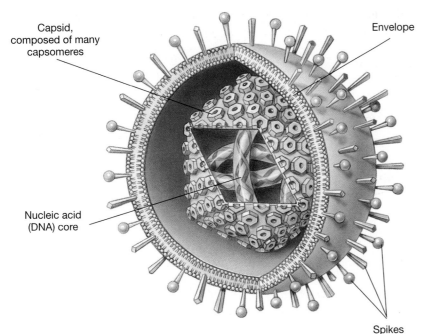

Capsid, composed of many capsomeres

Envelope

Nucleic acid (DNA) core

Spikes

Figure 10.1 **The components of an animal virus (a herpesvirus).**

Nucleic Acids

The British immunologist Peter Medawar, who shared the 1960 Nobel Prize in physiology or medicine, once described viruses as "a piece of bad news wrapped up in protein." The nucleic acid is the "bad news" because viruses use their **genome**, their genetic information ◀(Chapter 7), to replicate themselves in host cells. The result is often a disruption of host cellular activities or death of the host. Viral genomes consist of either DNA or RNA. Viral replication depends on the expression of the viral genome for the formation of viral proteins and the replication of new viral genomes within the infected host cell. Viral nucleic acid can be single-stranded or double-stranded, and linear, circular, or segmented (existing as several fragments). All genetic information in RNA viruses is carried by RNA. RNA genomes occur only in viruses and a viruslike agent called the viroid.

Capsids

The nucleic acid of an individual virion is in most cases enclosed within a capsid that protects it and determines the shape of the virus. Capsids also play a key role in the attachment of some viruses to host cells. Each capsid is composed of protein subunits called **capsomeres** (Figure 10.1). In some viruses, the proteins found in the capsomere are of a single type. In other viruses several different proteins may be present. The number of proteins and the arrangement of viral capsomeres are characteristic of specific viruses and thus can be useful in virus identification and classification.

Envelopes

Enveloped viruses have a typical bilayer membrane outside their capsids. Such viruses acquire their envelope after they are assembled in a host cell as they *bud*, or move through, one or several membranes. A virion's **nucleocapsid** comprises the viral genome together with the capsid. Viruses with only a nucleocapsid and no envelope are known as **naked**, or nonenveloped, viruses. The composition of an envelope generally is determined by the viral nucleic acid and by the substances derived from host membranes. Combinations of lipids, proteins, and carbohydrates make up most envelopes. Depending on the virus, projections referred to as **spikes** (Figure 10.1) may or may not extend from the viral envelope. These surface projections are **glycoproteins** that serve to attach virions to specific receptor sites on susceptible host cell surfaces. In certain viruses the possession of spikes causes various types of red blood cells to clump, or *hemagglutinate*—a property that is useful in viral identification.

What advantages might envelopes have for viruses? Because envelopes are acquired from and are therefore similar to host cell membranes, viruses may be "hidden" from attack by the host's immune system. Also, envelopes help viruses infect new cells by fusion of the envelope with the host's cell or plasma membrane. Conversely, enveloped viruses are damaged easily. Environmental conditions that destroy membranes—increased temperature, freezing and thawing, pH below 6 or above 8, lipid solvents, and some chemical disinfectants such as chlorine, hydrogen peroxide, and phenol—will also destroy the envelope. Naked viruses generally are more resistant to such environmental conditions.

SIZES AND SHAPES

Most viruses are too small to be seen with a light microscope, but **Figure 10.2**

Host cells are often 1,000 times the volume of a virus that is infecting it.

Figure 10.2 **Viral sizes and shapes.** Variations in shapes and sizes of viruses compared with a bacterial cell, an animal cell, and a eukaryotic ribosome.

shows that they have a range of sizes. The largest are or-thopoxviruses, which are about 240 nm by 300 nm—the size of the smallest bacteria or 1/10 the size of a red blood cell. The complex bacteriophages are about 65 nm by 200 nm. Among the smallest viruses known are the enteroviruses, which are less than 30 nm in diameter. But as Figure 10.2 shows, most viruses are quite small when compared with bacteria or eukaryotic cells. To put things in perspective, consider that typical ribosomes are about 25–30 nm in diameter.

Although some viruses are variable in shape, most viruses have a specific shape that is determined by the capsomeres or the envelope. Figure 10.2 shows several examples of viral symmetry. A *helical* capsid consists of a ribbonlike protein that forms a spiral around the nucleic acid. The tobacco mosaic virus is a helical virus (◀Chapter 1, p. 16). *Polyhedral* viruses are many-sided. The picornaviruses (pi-kor′na-vi″rus-ez) and the human adenoviruses (ad′e-no-vi″rus-ez) are polyhedral viruses. One of the most common polyhedral capsid shapes is the icosahedron; *icosahedral* viruses have 20 triangular faces. A *complex* capsid is a combination of helical and icosahe-dral shapes, and some viruses have a *bullet-shaped* capsid.

Most viruses with envelopes have a somewhat sphe-rical shape. For example, the herpesvirus shown in Figure 10.1 has a polyhedral capsid and an envelope. The filo-viruses (e.g., Ebola and Marburg) are threadlike in shape.

The poxviruses and many bacterial viruses are called **complex viruses** because they have a more elaborate coat or capsid (Figure 10.2). Many **bacteriophages**, or viruses that infect bacteria, have a complex shape that in-corporates specialized structures such as heads, tails, and tail fibers (Figure 10.2). Like spikes, the tail fibers are used by virions to attach to host bacteria. Other specialized bac-teriophage structures are used to infect the bacterial cells.

HOST RANGE AND SPECIFICITY OF VIRUSES

Although viruses are quite small and differ from one another structurally and in their replication strategies,

TRY IT

Another Evil of Tobacco

Keep smokers away from your tomato plants. Cigarette tobacco always contains some tobacco mosaic virus—enough to start an infection in tomato plants when carried there by smokers' hands or cigarette butts. Try an experiment: Is water in which cigarette tobacco has been soaked able to transmit tobacco mosaic disease? Dry tobacco? Cigarette smoke? Washed versus unwashed smokers' fingers? Some varieties of tomato plants are bred to resist tobacco mosaic virus infection.

APPLICATIONS

Plant Viruses

Besides the specificity shown by some viruses for bacteria and humans, other viruses are specific to and infect plants. Most viruses enter plant cells through damaged areas of the cell wall and spread through cytoplasmic connections called *plasmodesmata*.

Because plant viruses cause serious crop losses, much research has been done on them. The tobacco mosaic virus infects tobacco plants. Other plant viruses, which have either DNA or RNA genomes, infect various ornamental plants, including carnations and tulips. Food crops are not immune to viral infections. Lettuce, potatoes, beets, cucumbers, tomatoes, beans, corn, cauliflower, and turnips are all subject to infection by specific plant viruses.

The beautiful streaks in these tulips are caused by a viral infection. Unfortunately, the infection (which can spread from plant to plant) also weakens the tulips somewhat. Therefore plant breeders have developed varieties whose streaks are genetically produced. *(Science VU Wayside/Visuals Unlimited)*

Insects are known to cause serious crop losses because of their voracious eating habits. But many insects carry and transmit plant viruses as well. By damaging plants as they eat, insects provide an excellent infection mechanism for the plant viruses they harbor. Researchers now hope to control some crop-destroying insects by infecting them with specific insect viruses.

they are capable of infecting all forms of life (hosts). The **host range** of a virus refers to the spectrum of hosts that a virus can infect. Different viruses can infect bacteria, fungi, algae, protozoa, plants, vertebrates, or even invertebrates. However, most viruses are limited to only one host and to only specific cells and/or tissues of that host. Polioviruses, for example, can be grown in the laboratory in monkey kidney cells but have never been observed to cause a natural infection in any animals other than humans. In contrast, the rabies virus can attack the central nervous system of many warm-blooded animals. The host range of the rabies virus is much more extensive than that of polioviruses.

Viral specificity, another important property of viruses, refers to the specific kinds of cells a virus can infect. For example, certain papillomaviruses, which cause warts, are so specific in their replication strategy that they infect only skin cells. In contrast, cytomegaloviruses, known for their lethal effects, attack cells of the salivary glands, gastrointestinal tract, liver, lungs, and other organs. They can also cross the placenta and attack fetal tissues, especially those of the central nervous system. The discovery that a given virus can cause varying symptoms in several different body systems made the "one virus, one disease" concept untenable.

Viral specificity is determined mainly by whether a virus can attach to a cell. Attachment depends on the presence of specific receptor sites on the surfaces of host cells and on specific attachment structures on viral capsids or envelopes. Specificity is also affected by whether appropriate host enzymes and other proteins the virus needs in order to replicate are available inside the cell. Finally, specificity is affected by whether replicated viruses can be released from the cell to spread the infection to other cells.

Pig organs used for human transplants can carry pig viruses capable of infecting human cells. Since 1998, all pig organs, such as heart valves, must be tested and certified virus-free before they can be given to humans.

ORIGINS OF VIRUSES

Viruses are clearly quite different from cellular microbes. Free viruses are incapable of reproduction—they must infect host cells, uncoat their genetic material, and then use the host's machinery to copy or transcribe the viral genetic material. Thus, some debate remains as to whether viruses are living or are nonliving

CLOSE UP

Make Sure Lunch Can't Fight Back!

Some parasitic wasps lay their eggs in or on caterpillars. The eggs hatch into hungry wormlike larvae, who then begin eating their caterpillar hosts—nonessential parts first so as to keep lunch alive and fresh as long as possible. Meanwhile, the caterpillar responds to this by mounting an immune response designed to kill the wasp larvae. Larvae infected with the polydnavirus transmit an infection with this virus that suppresses the caterpillar's immune system. Thus, lunch is unable to fight back, the larvae finish eating, growing, and emerge as adults. Larvae that are not infected with the virus are losers to the caterpillar's immune defenses.

If you were a female wasp of this type, would you rather carry the polydnavirus in your digestive tract or in your ovaries? Why? Do you suppose most such wasps are, or are not, infected with polydnavirus? Why?

Many of the genes a human fetus acquires from its father are foreign to its mother. Some scientists think that before placental mammals could evolve, some ancestor(s) had to be infected by an immunosuppressive virus. Some women who suffer "chronic miscarriages" may have immune systems unable to tolerate the father's foreign genes.

chemical aggregates. Because viruses cannot reproduce or metabolize or perform metabolic functions on their own, some scientists say that they are not living. Other scientists claim that because viruses have the genetic information for replication, and this information is active after infection, they are living. Much of the genetic regulation of viral genes is similar to the regulation of host genes. In addition, viruses use the host's ribosomes for viral replication metabolism.

At present, we cannot definitively say whether viruses are living or nonliving. But we can ask, What are the origins of viruses? We do not know that either. There are probably several different ways in which viruses arose. In fact, they may appear and disappear continuously through time on our planet. However, because viruses cannot replicate without a host cell, it is likely that viruses were not present before primitive cells evolved. One hypothesis proposes that viruses and cellular organisms evolved together, with both viruses and cells originating from self-replicating molecules present in the precellular world. Another idea, sometimes referred to as reverse evolution, is that viruses were once cells that lost all cell functions, retaining only that information to replicate themselves by using another cell's metabolic machinery. A third hypothesis proposes that viruses evolved within the cells they infect, possibly from plasmids, the independently replicating DNA molecules found in many bacterial cells (◀Chapter 8, p. 219) or from retrotransposons (◀Chapter 8, p. 224). Plasmids are self-replicating and occur in both DNA and RNA forms. They do not, however, have genes to make capsids. In fact, it has been proposed that plasmids evolved from viroids. As some viroids moved from cell to cell, the viroid RNA may have picked up several pieces of genetic information, including the information for making a protein coat. Indeed, viruses, viroids, plasmids, and transposons all are agents of evolution through lateral gene transfer (◀Chapter 8, p. 212). Viruses that insert themselves into egg or sperm producing cells will be passed on from generation to generation, becoming a permanent addition to that species' genome.

In trying to understand the origins of viruses, virologists have uncovered some nucleotide sequence relationships common to certain viruses. On the basis of this information, these viruses have been placed into families with similar nucleotide sequences and genetic organization. However, they may have had different origins. It may be possible to predict the potential disease effects of newly discovered viruses by analyzing the nucleotide sequences of their genomes and comparing them with sequences found in other, known viruses.

✓CHECKLIST

1. Why are host cells necessary for viral replication?
2. Distinguish between capsid and capsomere.
3. Distinguish between naked and enveloped viruses.

▌▌▌ CLASSIFICATION OF VIRUSES

▌▌▌ Before they knew much about the structure or chemical properties of viruses, virologists classified viruses by the type of host infected or by the type of host structures infected. Thus, viruses have been classified as bacterial viruses (bacteriophages), plant viruses, or animal viruses. And animal viruses are grouped by the tissues they attack as *dermotropic* if they infected the skin, *neurotropic* if they infected nerve tissue, *viscerotropic* if they infected organs of the digestive tract, or *pneumotropic* if they infected the respiratory system.

As more was learned about the structure of viruses at the biochemical and molecular levels, classification of viruses came to be based on the type and structure of their nucleic acids, method of replication, host range, and other chemical and physical characteristics. And as more viruses were discovered (today, over 40,000 strains of viruses exist in collections worldwide), conflicting classification systems came into use, resulting in much confusion and some bad feelings. The need for a single, universal taxonomic scheme for viruses led to the establishment in 1966 of the International Committee on Taxonomy of Viruses (ICTV). This committee, which meets every 4 years, establishes the rules for classifying viruses. Virus classification is summarized in ◀Appendix B.

Virology, as a scientific field of study, is only about 100 years old.

Because viruses are so different from cellular organisms, it is difficult to classify them according to typical taxonomic categories—kingdom, phylum, and the like. The family has been the highest taxonomic category used by the ICTV. Viral genera also have been established, but most are new and slow to gain acceptance. Despite advances in classification, the problems of defining and naming *viral species*—a group of viruses that share the same genome and the same relationships with organisms—and of distinguishing between viral species and viral strains have not yet been resolved completely. Currently, the ICTV requires that the English common name, rather than a Latinized binomial term, be used to designate a viral species. For example, the formal taxonomic designations for the rabies virus would be as follows: family: Rhabdoviridae; genus: *Lyssavirus*; species: rabies virus. For the HIV virus the taxonomic designation is, family: Retroviridae; genus: *Lentivirus*; species: human immunodeficiency virus (HIV).

The names of specific viruses often consist of a group name and a number, such as HIV-1 or HIV-2. Virus families are often distinguished first on the basis of their nucleic acid type, capsid symmetry (shape), envelope, and size (**Tables 10.1** and **10.2**). The ICTV has assigned more than 5000 member viruses to 108 families and 203 genera, plus 30 genera which have not yet been assigned to families. Many of the virus families contain viruses that

TABLE 10.1

Classification of Major Groups of RNA Viruses That Cause Human Diseases

Family	Envelope and Capsid Shape	Typical Size (nm)	Example (Genus or Species)	Infection or Disease	
(+) Sense RNA Viruses					
Picornaviridae (1 copy)	Naked, polyhedral	18–30	*Enterovirus* *Rhinovirus* *Hepatovirus*	Polio Common cold Hepatitis A	
Togaviridae (1 copy)	Enveloped, polyhedral	40–90	Rubella Virus Equine encephalitis virus	Rubella (German measles) Equine encephalitis	
Flaviviridae (1 copy)	Enveloped, polyhedral	40–90	*Flavivirus*	Yellow fever	
Retroviridae (2 copies)	Enveloped, spherical	100	HTLV-I HIV	Adult leukemia, tumors AIDS	
(−) Sense RNA Viruses					
Paramyxoviridae (1 copy)	Enveloped, helical	150–200	*Morbillivirus*	Measles	
Rhabdoviridae (1 copy)	Enveloped, helical	70–180	*Lyssavirus*	Rabies	
Orthomyxoviridae (1 copy in 8 segments)	Enveloped, helical	100–200	*Influenzavirus*	Influenza A and B	
Filoviridae (1 copy)	Enveloped, filamentous	80	*Filovirus*	Marburg, Ebola	
Bunyaviridae (1 copy in 3 segments)	Enveloped, spherical	90–120	*Hantavirus*	Respiratory distress, hemorrhagic fevers	
Double-Stranded RNA Viruses					
Reoviridae (1 copy in 10–12 segments)	Naked, polyhedral	70	*Rotavirus*	Respiratory and gastrointestinal infections	

TABLE 10.2

Classification of Major Groups of DNA Viruses That Cause Human Diseases

Family	Envelope and Capsid Shape	Typical Size (nm)	Example (Genus or Species)	Infection or Disease	
Double-Stranded DNA Viruses					
Adenoviridae (linear DNA)	Naked, polyhedral	75	Human adenoviruses	Respiratory infections	
Herpesviridae (linear DNA)	Enveloped, polyhedral	120–200	*Simplexvirus* *Varicellovirus*	Oral and genital herpes Chickenpox, shingles	
Poxviridae (linear DNA)	Enveloped, complex shape	230 × 270	*Orthopoxvirus*	Smallpox, cowpox	
Papovaviridae (circular DNA)	Naked, polyhedral	45–55	Human papilloma-viruses	Warts, cervical and penile cancers	
Hepadnaviridae	Enveloped, polyhedral	40–45	Hepatitis B virus	Hepatitis B	
Single-Stranded DNA Viruses					
Parvoviridae (linear DNA)	Naked, polyhedral	22	B19	Fifth disease (erythema infectiosum) in children	

CLOSE UP

Naming Viruses

Although the ICTV approves all virus names, virologists often are creative in naming newly discovered viruses. As the following family names show: The Picornaviridae received their name from the fact that they are extremely small viruses (*piccolo*, Italian for "very small") and contain *RNA* as their genetic information. The Retroviridae have RNA as their genome and use it to direct the synthesis of DNA, reversing (*retro*, Latin for "backward") the usual direction of transcription. The Parvoviridae are very small viruses (*parvus*, Latin for "small"), whereas the Togaviridae received their name from someone who must have thought that the envelope of a togavirus resembled a *toga*, or cloak. And the arboviruses are a collection of a**r**thropod**bo**rne viruses (including togaviruses, flaviviruses, bunyaviruses, and arenaviruses). Can you see how the Coronaviridae got their name (*corona*, Latin for "crown")?

When it became possible to culture viruses, some were isolated that could not be linked to a known disease and that did not cause disease in laboratory animals. These viruses were dubbed "orphans." Thus, the Reoviridae are **r**espiratory **e**nteric **o**rphan viruses. Although the ICTV does not allow a person's name to be used in a virus name, geographical locations can be used. The Bunyaviridae received their name from *Bunya*mwere, Uganda, where they were first discovered.

Can you see why this virus is called a coronavirus? SARS is caused by a coronavirus (112,059X). (*Dr. Steve Patterson/Photo Researchers, Inc.*) [TEM]

cause important infections of humans and some other animals. Additional families contain viruses that infect only other animals, plants, fungi, algae, or bacteria.

Nucleic Acid Classification

Major groups of viruses are distinguished first by their nucleic acid content as either RNA or DNA viruses. Subsequent subdivisions are based largely on other properties of nucleic acids. The RNA viruses can be single-stranded (*ssRNA*) or double-stranded (*dsRNA*), although most are single-stranded (Table 10.1). Because most eukaryotic cells do not have the enzymes to copy

viral RNA molecules, the RNA viruses must either carry the enzymes or have the genes for those enzymes as part of their genome. Table 10.1 identifies two types of single-stranded RNA viruses—positive sense and negative sense RNA viruses. Many ssRNA viruses contain **positive (+) sense RNA**, meaning that during an infection the RNA acts like mRNA and can be translated by the host's ribosomes. Other ssRNA viruses have **negative (−) sense RNA**. In such viruses the RNA acts as a template during transcription to make a complementary (+) sense mRNA after a host cell has been entered (◀Chapter 7, p. 189). This strand is translated by host ribosomes. In order to perform the transcription step, (−) sense RNA viruses must carry an RNA polymerase within the virion.

Like RNA viruses, DNA viruses can also occur in single-stranded or double-stranded form (Table 10.2). For example, the human adenoviruses, responsible for some common colds, and the herpesviruses are double-stranded DNA (*dsDNA*) viruses. Only one single-stranded DNA (*ssDNA*) virus is currently known to produce human disease.

With this background, let us briefly examine several families of animal RNA and DNA viruses.

RNA VIRUSES

General Properties of RNA Viruses

The different families of RNA viruses are distinguished from one another by their nucleic acid content, their capsid shape, and the presence or absence of an envelope (Table 10.1; **Figure 10.3**). Most families of RNA viruses contain either one (+) sense RNA or one (−) sense RNA molecule. However, some RNA viruses are placed in separate families if the RNA exists as two complete copies of (+) sense RNA or contains small segments of (−) sense RNA. Finally, one family has segmented dsRNA.

Important Groups of RNA Viruses

Picornaviridae. The **picornaviruses** are very small (30 nm in diameter), naked, polyhedral, (+) sense RNA viruses. They include more than 150 species that cause disease in humans. After infection these viruses quickly interrupt all functions of DNA and RNA in the host cell. The Picornaviridae are divided into several groups, including the genera *Enterovirus*, *Hepatovirus*, and *Rhinovirus*.

Enteroviruses (*entero*, Greek for "intestine") include the polioviruses (Figure 10.3a). These viruses are resistant to many chemical substances and can replicate in and pass through the digestive tract unharmed. Unless inactivated by host defense mechanisms, the viruses invade the blood and lymph, spreading throughout the body but especially into the nervous system. Poor sanitation increases the numbers of enteroviruses, and human overcrowding helps their spread. As a result of early and frequent exposure, most children living in such conditions acquire infection in infancy, when paralysis is unlikely to occur, and symptoms are mainly flulike. It is the older

(a)   EM

(b)   TEM

(c)   TEM

(d)   EM

(e)   TEM

Figure 10.3   False-color electron micrographs of representative RNA viruses. **(a)** Picornaviruses (polioviruses; 71,500X); *(Omikron/Photo Researchers)* **(b)** retroviruses (oncovirus; 42,500X); *(CNRI/Custom Medical Stock Photo, Inc.)* **(c)** rhabdoviruses (rabies virus; 164,121X); *(Tektoff-RM/CNRI/Photo Researchers)* **(d)** orthomyxoviruses (influenza viruses (186,098X)); *(Herbert Wagner/Phototake)* **(e)** reoviruses (respiratory viruses (780,411X)). *(K. G. Murti/Visuals Unlimited)*

> *During the 1940s, in New York City, for every case of paralytic polio, 100 more people had symptomless cases of polio infection.*

child and adult that usually develop paralysis. Thus, paralytic polio epidemics are uncommon in developing nations.

Poor sanitation is also responsible for the spread of certain **hepatoviruses** (*hepato*, Greek for "liver"). The hepatitis A virus, for instance, is transmitted via the fecaloral route, with disease arising from the ingestion of contaminated food or water. The major organ infected is the liver.

The genus *Rhinovirus* (*rhino*, Greek for "nose"), which includes more than 100 types of human **rhinoviruses**, is one of the genera of viruses responsible for the

> *Rhinoviruses, causes of the common cold, can survive on household objects for up to 3 days.*

common cold. Human rhinoviruses do not cause digestive tract diseases because they cannot survive the acidic conditions in the stomach. Instead, they enter the body through the mucous membranes of the nasal passages and replicate in the epithelial cells of the upper respiratory tract. Much has recently been learned about the capsids of rhinoviruses **(Figure 10.4)**. Virologists have

Figure 10.4   Cold virus. Computer-generated model of a human rhinovirus, cause of the common cold. The colors represent different capsomeres of the capsid. *(Courtesy Michael G. Rossmann, Purdue University)*

discovered that these capsids attach to only a few receptors in the nasal mucous membranes. Thus, it may be possible in the future to prevent colds by designing chemicals that cover these receptors so that rhinoviruses cannot attach.

Togaviridae. The **togaviruses** are small, enveloped, polyhedral, (+) sense RNA viruses that multiply in the cytoplasm of many mammalian and arthropod host cells. Togaviruses known as arthropodborne viruses are transmitted by mosquitoes and cause several kinds of encephalitis (plural: *encephalitides*) in humans and in horses. The rubella virus, which causes German measles (rubella), is in this family but is not transmitted by arthropods; rather, it is spread person to person.

Flaviviridae. The **flaviviruses** are enveloped, polyhedral, (+) sense RNA viruses that are transmitted by mosquitoes and ticks. The viruses produce a variety of encephalitides or fevers in humans. The yellow fever virus is a flavivirus that causes a hemorrhagic fever—in which blood vessels in the skin, mucous membranes, and internal organs bleed uncontrollably. Hepatitis C infection is also caused by a flavivirus.

Retroviridae. The **retroviruses** are enveloped viruses that have two complete copies of (+) sense RNA (Figure 10.3b). They also contain the enzyme **reverse transcriptase**, which uses the viral RNA to form a complementary strand of DNA, which is then replicated to form a dsDNA. This reaction is exactly the reverse of the typical transcription step (DNA → RNA) in protein synthesis. For virus replication to continue, the newly formed DNA must be transcribed into viral RNA that will function as mRNA for viral protein synthesis and be incorporated into new virions. To do so, the DNA must first migrate to the host cell nucleus and become incorporated into chromosomes of host cells. Such integrated viral DNA is known as a **provirus**. Retroviruses cause tumors and leukemia in rodents and birds, as well as in humans. The human retroviruses invade immune defense cells called *T lymphocytes* and are referred to as <u>h</u>uman <u>T</u> cell <u>l</u>eukemia <u>v</u>iruses (HTLV). Both HTLV-1 and HTLV-2 are associated with malignancies (leukemia and other tumors), whereas the <u>h</u>uman <u>i</u>mmunodeficiency <u>v</u>irus (HIV-1 and HIV-2 strains) causes <u>a</u>cquired <u>i</u>mmune <u>d</u>eficiency <u>s</u>yndrome (AIDS). AIDS is discussed in ◄Chapter 18.

Paramyxoviridae. The **paramyxoviruses** (*para*, Latin for "near"; *myxo*, Greek for "mucus") are medium-sized, enveloped, (−) sense RNA viruses, with a helical nucleocapsid. Different genera of paramyxoviruses are responsible for mumps, measles, viral pneumonia, and bronchitis in children and mild upper respiratory infections in young adults.

Rhabdoviridae. Another (−) sense RNA virus group, the **rhabdoviruses** (*rhabdo*, Greek for "rod"), consists of medium-sized, enveloped viruses. Although these viruses have an envelope, the capsid is helical and makes the

viruses nearly rod- or bullet-shaped (Figure 10.3c). Rhabdoviridae virions contain an RNA-dependent RNA polymerase that uses the (−) sense strand to form a (+) sense strand. The newly produced strand serves as mRNA and as a template for the synthesis of new viral RNA. Human rabies almost always results from a bite by a rabid animal that is carrying the rabies virus. Rhabdoviruses also infect other vertebrates, invertebrates, or plants. The Lago virus, which produces disease in bats, and the Mokolo virus, which infects shrews in Africa, are closely related to rabies viruses.

> *Rabies is an ancient disease. It was recognized in Egypt before 2300 B.C. and in ancient Greece, where it was well described by Aristotle.*

Orthomyxoviridae. The **orthomyxoviruses** (*ortho*, Greek for "straight") are medium-sized, enveloped, (−)

PUBLIC HEALTH

What If It Comes to Your House?

Ebola outbreaks have occurred in Zaire and Sudan—hundreds of cases each time. The death rate was 88% in Zaire and 51% in Sudan. Again and again, Ebola comes back. There is no cure. Antibiotics do not help. There is no vaccine. What if it spreads to where you live?

One day you wake up with a throbbing headache. Then your back starts to ache. Fever, nausea, and vomiting follow. Patches of blood begin to show through your skin. Your blood cannot clot normally—hence the name "hemorrhagic fever" for the Ebola virus infection. Now bleeding is occurring in your brain, kidneys, and respiratory and digestive tracts. You vomit great quantities of blood from one end, while at the other end blood and tissue pour from your rectum, Death may be only hours away!

The surprising thing is that some people actually do survive an Ebola infection. In fact, nearly one-fifth of the population of rural Central Africa have antibodies against Ebola virus, meaning that they have survived their exposure to it.

Where has this virus come from? We do not know in what organisms it has been quietly living in remote jungle areas. But, as people push further into these areas, they come in contact with unknown viruses. As the virus "jumps species" to a new host, it usually causes severe disease. It had probably reached a quieter relationship with its original host, but it has not had time to yet adapt to this new human host. It is an "emerging" disease. (*Top photo: Gilbert Liz/Corbis Sygma/ Corbis; bottom: Barry Dowsett/ Photo Researchers, Inc.*)

sense RNA viruses that vary in shape from spherical to helical (Figure 10.3d). Their genome is segmented into eight pieces. Like the paramyxoviruses, orthomyxoviruses have an affinity for mucus. Influenza virus A, with which we are all too familiar, is an orthomyxovirus that also infects birds, swine, horses, and whales. Influenza virus B appears to be specific to humans.

Filoviridae. The **filoviruses** are enveloped, filamentous, single (−) sense RNA viruses. These viruses can be transmitted from person to person by close contact with blood, semen, or other secretions and by contaminated needles. The filoviruses include the viruses responsible for Marburg and Ebola diseases, which are hemorrhagic fevers.

Bunyaviridae. The **bunyaviruses** also are enveloped, (−) sense RNA viruses whose genome has three segments. Bunyaviruses can be transmitted by arthropods, but rodents typically are the principal host. The most recently recognized member of the Bunyaviridae is the hantavirus responsible for hantavirus pulmonary syndrome (HPS). Other forms of the genus *Hantavirus* cause hemorrhagic fevers.

Arenaviridae. Like the bunyaviruses, the **arenaviruses** are enveloped, (−) sense RNA viruses, but their genome has only two segments. Arenaviruses are carried by rodents. Human infections occur via aerosols, exposure to infectious urine or feces, or rat bites. Argentinean and Bolivian hemorrhagic fevers and Lassa fever are arenavirus infections.

Reoviridae. The **reoviruses** have a naked, polyhedral capsid (Figure 10.3e). They are medium-sized dsRNA

Ingestion of only 10 rotavirus particles is sufficient to cause infection and diarrhea.

viruses. They replicate in the cytoplasm and form distinctive inclusions that stain with eosin. Reoviruses include the orthoreoviruses, orbiviruses, and rotaviruses. The rotaviruses are the most common cause of severe diarrhea in infants and in young children under age 2. They also are responsible for minor upper respiratory and gastrointestinal infections in adults. The other reoviruses infect other animals.

DNA VIRUSES

General Properties of DNA Viruses

Like the RNA viruses, the animal DNA viruses are grouped into families according to their DNA organization (Table 10.2; **Figure 10.5**). The dsDNA viruses are further separated into families on the basis of the shape of their DNA (linear or circular), their capsid shape, and the presence or absence of an envelope. Only one family of viruses has ssDNA.

Important Groups of DNA Viruses

Adenoviridae. The **adenoviruses** (*adeno*, Greek for "gland") are medium-sized, naked viruses with linear dsDNA. First identified in adenoid tissue, they are highly resistant to chemical agents and are stable from pH 5 to 9 and from 36°C to 47°C. Freezing causes little loss of infectivity. More than 80 different types of adenoviruses have been identified, many being responsible for human respiratory disease. Adenovirus types 40 and 41 cause 10 to 30% of all cases of severe diarrhea in babies and young children. Only half the children carrying the virus in their throat actually become ill.

Diseases caused by adenoviruses are generally acute (that is, have sudden onset and short duration). Soon after entering the body, the virus appears in the blood and a measles-like rash may develop. Sources of adenoviruses are respiratory secretions and feces from infected persons.

Herpesviridae. The **herpesviruses** (*herpes*, Greek for "creeping") are relatively large, enveloped viruses with linear dsDNA (Figure 10.5a). Herpesviruses are widely distributed in nature, and most animals are infected with

PUBLIC HEALTH

New Virus Diseases in Animals

Several viruses have been identified as the cause of disease in animals other than humans. A retrovirus found in cats, called feline immunodeficiency virus (FIV), is very similar to the human immunodeficiency virus (HIV). Approximately 1 to 3% of randomly tested U.S. cats are infected. FIV infects the cat's lymph nodes but causes no immediate symptoms. However, like HIV, FIV gradually attacks the immune system over 3 to 6 years, leading to frequent infections of the mouth, skin, and respiratory systems. Loss of immune defenses also leads to diarrhea, weight loss, pneumonia, fever, and neurologic disease. Unlike HIV, FIV is not transmitted sexually; nor can it be transmitted to humans or other animals. Transmission between cats is usually through a bite. Veterinarians believe that the virus, like HIV, has been around for decades, and there is no cure.

Several other viruses also are responsible for recent disease outbreaks in animals. A virus closely related to FIV has been found in zoo lions and tigers and in wild panthers. *Caliciviruses*, members of a family similar to the picornaviruses, have been identified as the cause of disease outbreaks in swine, sea lions, and cats. These viruses cause a skin eruption similar to food-and-mouth disease in cattle. In swine, viral infection causes weight loss, fever, and the formation of skin blisters on the feet, snout, and tongue.

Between December 1993 and January 1995, more than 100 African lions in Tanzania died from an outbreak of canine distemper. The canine distemper virus, which is very unusual in large cats but common in dogs and wolves, probably was transmitted to the lions by domestic dogs that live near the Tanzanian game parks. African researchers believe the virus might spread to other animals, including leopards and wild dogs. In the U.S., many wildlife refuges prohibit you from bringing your dog into the park, for fear he will spread viruses to the wildlife.

Figure 10.5   False-color electron micrograph of representative DNA viruses (arrows point to one virion). **(a)** Herpesviruses (pink spheres within the cell (148,924X)); *(Centers for Disease Control/Photo Researchers, Inc.)* **(b)** papovaviruses (human papillomaviruses; 61,100X); *(CNRI/Photo Researchers, Inc.)* **(c)** parvoviruses (147,000X). *(Central Veterinary Laboratory, Weybridge, England/Photo Researchers, Inc.)*

one or more of the 100 types discovered. These viruses cause a broad spectrum of diseases, which are summarized in **Table 10.3.**

The core of the virion contains proteins around which the DNA is coiled. In cells infected with herpesviruses, the viral dsDNA can exist as a provirus. Therefore, a universal property of herpesviruses is **latency**, the ability to remain in host cells, usually in neurons, for long periods and to retain the ability to replicate. For example, a child who has recovered from chickenpox (varicella) will still have the virus in a latent form. Years or decades later the virus may be reactivated as a result of stress and/or physical factors. This adult disease, which can be quite painful and debilitating, is called *shingles* (zoster). Fluids oozing from the vesicles carry the varicella-zoster virus, which can then cause chickenpox in

persons who have not previously been infected with the virus. Of the more than 100 genes found in herpesviruses, 11 are known to be involved in latency.

Poxviridae.   The **poxviruses**, another group of enveloped, linear dsDNA viruses, are the largest and most complex of all viruses. They are widely distributed in nature; nearly every animal species can be infected by a form of poxvirus. The human poxviruses (orthopoxviruses) are large, enveloped, brick-shaped viruses 250 to 450 nm long and 160 to 260 nm wide. These viruses multiply in specialized portions of the host cell cytoplasm called *viroplasm*, where they can cause skin lesions typical of smallpox, molluscum contagiosum, and cowpox. Other poxviruses, such as monkeypox, can infect humans who have close contact with infected animals.

TABLE 10.3

Herpesviruses That Cause Human Disease			
Genus	**Virus Type**	**Infection or Disease**	**Additional Information**
Simplexvirus	Herpes simplex type 1	Oral herpes (sometimes genital and neonatal herpes), encephalitis	p. 298
	Herpes simplex type 2	Genital and neonatal herpes (sometimes oral herpes), meningoencephalitis	pp. 627–629
Varicellovirus	Varicella-zoster	Chickenpox (varicella) and shingles (zoster)	pp. 574, 583–584
Cytomegalovirus	Cytomegaloviruses (salivary gland virus)	Acute febrile illness; infections in AIDS patients, transplant recipients, and others with reduced immune system function; a leading cause of birth defects	pp. 298, 632–634
Roseolovirus	Roseola infantum (formerly called herpesvirus 6)	Exanthema subitum (roseola infantum), a common disease of infancy, featuring rash and fever	p. 583
Lymphocryptovirus	Epstein-Barr virus	Infectious mononucleosis and Burkitt's lymphoma (cancer of the jaw seen mainly in African children); also linked to Hodgkin's disease (cancer of lymphocytes) and B cell lymphomas, and to nasopharyngeal cancer in Asians	pp. 302, 740–741
Human herpesvirus 8	Kaposi's sarcoma virus	Kaposi's sarcoma linked to AIDS	p. 558

The smallpox virus represents the first human pathogen purposely eradicated from the face of the Earth.

Papovaviridae.

The **papovaviruses** (pa-po′va-vi″rus-ez) are named for three related viruses, the _pa_pilloma, _po_lyoma, and _va_cuolating viruses. These are small, naked, polyhedral dsDNA viruses that replicate in the nuclei of their host cells. Papovaviruses are widely distributed in nature; more than 25 human papillomaviruses and 2 human polyomaviruses have been found. Papillomaviruses are frequently found in host cell nuclei without being integrated into host DNA (Figure 10.5b); polyomaviruses are nearly always integrated as a provirus. The papillomaviruses cause both benign and malignant warts in humans, and about 13 papillomavirus strains are associated with cervical cancer. The most thoroughly studied vacuolating virus is simian virus 40 (SV-40). This virus has been used by virologists to study the mechanisms of viral replication, integration, and oncogenesis (the development of cancerous cells).

Hepadnaviridae.

The **hepadnaviruses** (he′pa-dee-en-ay-vi″rus-ez) are small, enveloped, mostly dsDNA (partially ssDNA) viruses. Their name comes from the infection of the liver—_hepa_titis—by a _DNA_ virus. The hepadnaviruses can cause chronic (that is, of long and continued duration) liver infections in humans and other animals, including ducks. In humans, the hepatitis B virus causes hepatitis B, which can progress to liver cancer. We will discuss in ◀Chapter 22 other forms of hepatitis that are caused by viruses.

Parvoviridae.

The **parvoviruses** are small, naked, linear ssDNA viruses (Figure 10.5c). Their genetic information is so limited that they must enlist the aid of an unrelated helper virus or a dividing host cell to replicate. Three genera, _Dependovirus_, _Parvovirus_, and _Erythrovirus_, have been identified in vertebrates. The dependoviruses are often called adeno-associated viruses because to replicate more virions they require coinfection with adenoviruses (or herpesviruses). No known human disease is associated with this genus. Members of the genus _Parvovirus_ can cause disease in rats, mice, swine, cats, and dogs. Rat parvovirus causes congenital defects in the unborn. Canine parvovirus is responsible for severe and sometimes fatal gastroenteritis in dogs and puppies. The only known parvovirus to infect humans (predominantly children) is _Erythrovirus_, which is also called B19. This virus, identified in 1974, is responsible for "fifth disease" (erythema infectiosum). It is so named because it was the fifth disease listed as a classical rash-associated childhood disease. It came behind measles (rubeola), scarlet fever, rubella, and a fourth rash producer no longer seen. B19 causes a deep red rash on children's cheeks and ears and both a rash and arthritis in adults. The B19 virus can cross the placenta and damage blood-forming cells in the fetus, leading to anemia, heart failure, and even fetal death.

✓ **CHECKLIST**

1. On what basis are viruses classified today?
2. Distinguish between positive (+) and negative (−) sense RNA.
3. What is the difference between dsDNA and ssDNA?

▌▌▌ EMERGING VIRUSES

Viruses have been infecting humans for thousands of years, and the diseases they cause have been responsible for millions of deaths. Microbiologists believe that many recent, unexpected viral diseases have been caused by **emerging viruses**—viruses that were previously _endemic_ (low levels of infection in localized areas) or had "crossed species barriers"—that is, expanded their host range to other species. For example, although the poliovirus has been endemic since ancient times, only since 1900 have _pandemics_ (high levels of infection worldwide) resulted from this virus, with numerous annual outbreaks. Why the increase in disease?

The poliovirus has not mutated over the centuries into a more pathogenic form. Rather, virologists and epidemiologists suggest that the urban populations that developed after the Industrial Revolution provided an ideal environment for viral spread. The large numbers of nonimmune people who had emigrated from nonendemic areas were exposed to immune people who carried the poliovirus. Thus, polio spread rapidly—and with deadly results. Only through the development of polio vaccines in the 1950s was the epidemic halted. Although there still are areas where polio is endemic, the World Health Organization (WHO) hoped to eradicate the poliovirus through vaccination programs by the year 2006. In March 2006, Egypt met the requirement of 3 years without any polio outbreaks needed to declare a country as being "polio-free". However, several other countries still have polio outbreaks, and so, the goal of eliminating polio will not be reached for years to come. But, like smallpox, once polio is eradicated, it will be gone forever, as it occurs only in humans, and has no reservoirs.

Other endemic viral diseases that are transmitted among humans, such as measles, also are recurring due to changes in human population densities and travel. But several viral diseases involve other animals that act as _reservoirs_ (a "healthy" organism harboring an infectious agent that is available to infect another host) or _vectors_ (carriers) for a virus. In these cases the virus could cross species barriers if the vector transmitted the virus from a reservoir species to humans. For example, prior to 1930, yellow fever was thought to be carried solely by one species of mosquito, _Aedes aegypti_. By controlling that mosquito in urban areas (through vaccination and spraying with DDT), the disease could be controlled. However, in the late 1950s, an outbreak of yellow fever occurred that was not carried by _A. aegypti_. Rather, jungle (sylvan) yellow fever was carried by another mosquito, of the genus *Haemagogus*. In the jungle these mosquitoes transmit the virus among monkeys that live high in the forest canopy. Forest clearing through tree cutting had brought the *Haemogogus* mosquito from the treetops to the forest floor, where it passed the virus to people involved in timber cutting and agricultural activities.

This yellow fever situation represents an excellent example of how viral diseases can remain endemic to those parts of the world where an insect vector lives. However, in tropical areas where once-uninhabited lands are being converted for use in agriculture and farming, contact with the insects (and the viruses they carry) is inevitable.

Of the more than 500 known arboviruses, about 80 cause disease in humans; of these, 20 are considered emerging viruses. The most dangerous are the yellow fever virus, which has been endemic for more than a century—and is reemeging—and the Dengue fever viruses, which are moving further north as global warming progresses. It is now entering the United States. Both are carried by mosquitoes.

Many virologists believe that a similar event may have occurred in the case of HIV. Retroviruses similar to HIV exist in domesticated cats (feline immunodeficiency virus, FIV) and in monkeys (simian immunodeficiency virus, SIV). It is possible that a mutated form of SIV crossed over to humans from contact with an infected monkey. However, antibodies to SIV have been found in humans. Therefore, SIV itself may initially have infected humans and later mutated. Natural selection could have favored these mutations because they were better adapted than SIV to the new human host.

In the United States one recent emerging virus is the hantavirus, which is transmitted from rodent feces and urine to humans. The virus, which causes hantavirus pulmonary syndrome (HPS), struck first in New Mexico in May 1993. Although we do not know how the virus got into the rodent population, genetic analysis and folklore suggest that the virus has been endemic in rodents for years. We do know that conditions were right for an explosion in the rodent population, guaranteeing more contact between rodent feces and humans. Other hantaviruses are known to cause hemorrhagic fever in millions of people worldwide. The first, called the Hantaan virus, was isolated in Korea in 1978 but is believed to have been causing disease since the 1930s (◄Chapter 21, p. 666).

New strains of influenza virus are particularly worrisome. If a host cell is simultaneously infected by two different influenza viruses, e.g., one human and one animal, they can "swap" parts of their genomes, thereby creating a new mutant-type virus (**Figure 10.6**). This virus may be so greatly altered, with perhaps a human flu virus core covered by a chicken, duck, or pig-type capsid, that neither host's immune system will be able to recognize or attack it. Rapid multiplication in the host can then lead to serious disease, or even death. This is what happened in the great swine flu pandemic

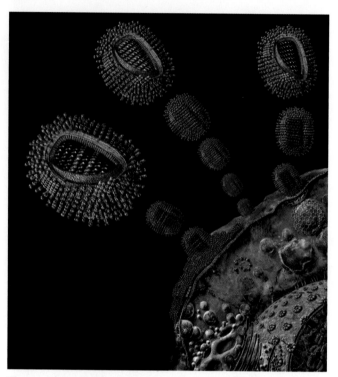

Figure 10.6   Production of a new strain of influenza virus. Two different viral types have infected the same cell (purple genome at lower left and orange genome at lower middle). During reproduction, portions of genomes can be "swapped" between the two strains, resulting in a new recombinant mutant virus (upper middle, purple and orange). This new strain could have the potential to spread rapidly with lethal results. It may even be able to attack a wider range of hosts. (*©Russell Kightley, http://www.rkm.com.au*)

of 1918 which killed 20 to 40 million people worldwide (see ◄Chapter 21, p. 660). Recently, virus recovered from exhumed bodies of victims was analyzed and found to be a combination of genetic material from both human and pig flu viruses. The gene for the hemagglutinin spike had sequences at its beginning and end from a human flu virus, while the middle sequences came from a pig flu virus. Therefore, health authorities became very alarmed in 1997, 1999, and 2003 when "chicken flu" (also know as bird flu or avian flu) broke out in Hong Kong, and people in contact with sick birds began coming down with a severe flu. Since 1998, a new flu virus of triple origin (human, duck, and pig) has been circulating in the U.S., which also worries us.

The avian flu virus is present in the feces of infected chickens, ducks, geese, other poultry, and migratory waterfowl. Bits of feces are easily spread into the air by flapping feathers and scratching feet. In 1997, 18 people were hospitalized; 6 died—a one-third mortality rate! The entire poultry stock of Hong Kong, 1.4 million birds, were killed to halt spread of the disease.

The 1999 outbreak again occurred in Hong Kong, but in 2003 it spread to 10 neighboring countries: Thailand, Cambodia, Indonesia, Japan, Laos, Vietnam, China, South Korea, Pakistan, and Taiwan (**Table 10.4**). Tens of

TABLE 10.4

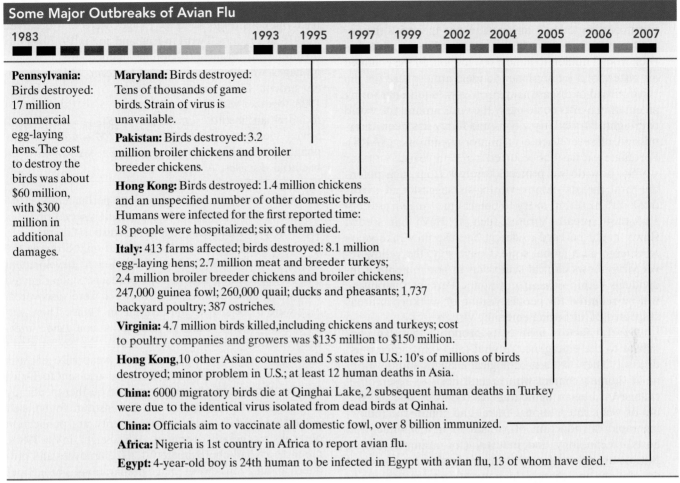

Some Major Outbreaks of Avian Flu										
1983	1993	1995	1997	1999	2002	2004	2005	2006	2007

Pennsylvania: Birds destroyed: 17 million commercial egg-laying hens. The cost to destroy the birds was about $60 million, with $300 million in additional damages.

Maryland: Birds destroyed: Tens of thousands of game birds. Strain of virus is unavailable.

Pakistan: Birds destroyed: 3.2 million broiler chickens and broiler breeder chickens.

Hong Kong: Birds destroyed: 1.4 million chickens and an unspecified number of other domestic birds. Humans were infected for the first reported time: 18 people were hospitalized; six of them died.

Italy: 413 farms affected; birds destroyed: 8.1 million egg-laying hens; 2.7 million meat and breeder turkeys; 2.4 million broiler breeder chickens and broiler chickens; 247,000 guinea fowl; 260,000 quail; ducks and pheasants; 1,737 backyard poultry; 387 ostriches.

Virginia: 4.7 million birds killed, including chickens and turkeys; cost to poultry companies and growers was $135 million to $150 million.

Hong Kong, 10 other Asian countries and 5 states in U.S.: 10's of millions of birds destroyed; minor problem in U.S.; at least 12 human deaths in Asia.

China: 6000 migratory birds die at Qinghai Lake, 2 subsequent human deaths in Turkey were due to the identical virus isolated from dead birds at Qinhai.

China: Officials aim to vaccinate all domestic fowl, over 8 billion immunized.

Africa: Nigeria is 1st country in Africa to report avian flu.

Egypt: 4-year-old boy is 24th human to be infected in Egypt with avian flu, 13 of whom have died.

Source: Diseases of Poultry, 11th Edition, edited by W. M. Saif, Iowa State Press 2003; Virginia state officials; Maryland Department of Agriculture

millions of birds were killed (**Figure 10.7**). Sequencing of the flu genome has indicated that the virus is still entirely avian. No "swapping" has yet occurred. But as long as humans and poultry remain in close contact, it is probably only a matter of time before it does. It has been recommended that live poultry markets be prohibited, and that birds be killed before being shipped to the city. Thus far it appears that most cases of disease are caused by close contact with birds, and the ability to transmit by human to human contact is extremely limited. The worry is that gene "swapping" might produce mutants that are easily passed between humans, leading to an uncontrolled pandemic. Properly cooked chicken and eggs have not been shown to cause disease. But, migratory birds leaving

Figure 10.7 **(a)** In 2003, an outbreak of "chicken flu" necessitated killing tens of millions of birds. (*©AP/Wide World Photos*) **(b)** Close contact between humans and birds can easily lead to coinfection by human and avian flu strains. Recombination may produce very dangerous mutants. (*EPA/Corbis*)

(a)

(b)

Hong Kong nest in Siberia and North Korea where they may add to the global spread of avian flu.

One of the disease transmission mechanisms of great concern today is air travel. Since the 1950s the annual number of international air travel passengers has risen from 2 million to almost 600 million. In the confined air system of a jet airplane (an ideal atmosphere for the rapid spread of disease), a person or mosquito harboring an emerging virus could carry the virus around the world overnight. Particularly worrisome today has been transmission of severe acute respiratory syndrome (SARS). Fortunately it has not occurred in recent years.

So, how do we protect ourselves from such potential viral threats? Many virologists suggest that "virus outposts" be set up to try to detect emerging viruses before they spread. Viruses such as HIV that spread slowly might be hard to detect because they take years to emerge on a global scale. Conversely, the yellow fever virus shows clinical symptoms in nonimmunized individuals within days, if not hours. Perhaps quarantine will be required for people visiting or working in areas suspected of harboring emerging viruses.

Several factors have been proposed that can contribute to the emergence of viral (and other infectious) diseases. These include ecological changes and development (human contact with natural hosts or reservoirs), changes in human demographics (typically due to famine or war), international travel and commerce (allowing rapid introduction of viruses to new habitats and hosts), technology and industry (for example, during processing and rendering of infected animals), microbial adaptations and change (high mutation rates or shifts in genetic composition), and environmental changes (such as extension of vector ranges by global warming).

VIRAL REPLICATION

GENERAL CHARACTERISTICS OF REPLICATION

In general, viruses go through the following five steps in their **replication cycles** to produce more virions:

1. **Adsorption**, the attachment of viruses to host cells.
2. **Penetration**, the entry of virions (or their genome) into host cells.
3. **Synthesis**, the synthesis of new nucleic acid molecules, capsid proteins, and other viral components within host cells while using the metabolic machinery of those cells.
4. **Maturation**, the assembly of newly synthesized viral components into complete virions.
5. **Release**, the departure of new virions from host cells. Release generally, but not always, kills (lyses) host cells.

REPLICATION OF BACTERIOPHAGES

Bacteriophages, or simply *phages*, are viruses that infect bacterial cells **(Figure 10.8)**. Phages were first observed in 1915 by Frederic Twort in England and in 1917 by Felix d'Herelle in France. d'Herelle named them bacteriophages, which means "eaters of bacteria." d'Herelle was an ardent Communist, and in 1923, together with Giorgi Eliava, founded an institute in Tbilisi, Soviet Georgia for the study of phages and **phage therapy** of bacterial diseases. A cottage was built for him on the institute

Seawater can hold 100 million bacteriophages per ml.

grounds, and he intended to live there permanently. However, after Eliava was executed by Stalin's secret police in 1937, d'Herelle left, never again to return to Georgia.

Meanwhile, the institute continued on, becoming the largest institution in the world devoted to development and production of phage therapy products. Stalin directed that when antibiotic resistant bacteria were discovered, they were to be sent to the institute in Tbilisi. There, experts would isolate strains of phages that could cure infections caused by these organisms. Later, the Soviet secret germ warfare program proceeded to weaponize bacteria such as anthrax. Samples of these were also sent to Tbilisi for identification of phages that would be therapeutically active against these "superbugs". The institute found itself embedded in KGB (secret police) security, and publication of results was greatly restricted. At its height, in the 1980s, the institute had approximately 1,200 employees and produced about 2 tons/day of phage preparations. Throughout the Soviet Union, phage therapy was preferred to the use

Figure 10.8 *Escherichia coli* being attacked by bacteriophages. *(Eye of Science/Photo Researchers, Inc.)*

Figure 10.9 Georgian scientist, Dr. Leila Kalandarishvili, working with d'Herelle's great grandson, Dr. Hubert Mazure, in Australia on phage therapy. *(Virginia Young/Newspix)*

of antibiotics. Bacteriophages are highly specific, attacking only the targeted bacteria, and leaving potentially beneficial bacteria that normally inhabit the human digestive tract and other locations alive. They also are cheap, effective in small doses, and rarely cause side effects. A typical treatment was 10 tablets or an aerosol spray; and recovery could be as rapid as 1 or 2 days away. The Polish microbiologist Stefan Slopek and his colleagues successfully used phages to treat 138 patients with long-term antibiotic-resistant bacterial infections. All patients benefited from the treatment, and 88% were completely cured. These results were published in English, in the 1980s, the first available to the Western world in decades.

With the discovery of antibiotics in the 1940s, plus the Soviet Russian secrecy, western medicine had turned away from phage therapy. Eli Lilly, which had been producing seven phage preparations in the United States, ceased production. Phage therapy continued only in the Soviet Union and its republics. Then with the collapse of the Soviet Union in 1992 and new independence for the Republic of Georgia, funding for the institute in Tbilisi dried up. Georgian scientists had to look to the West for money. And the West, plagued by increasing dangers from antibiotic-resistant strains, has begun to look to the Georgians. Today, d'Herelle's great grandson, Dr. Hubert Mazure, is working with the Georgian scientist Dr. Leila Kalandarishvili in Tbilisi and in Australia (**Figure 10.9**), helping to bring this needed therapy to us. With major pharmaceutical firms withdrawing from the very costly race to discover new antibiotics, we may soon have no choice about using phage therapy.

On a personal note: During one of my trips to Russia, I received phage therapy for an infection and was back on my feet in 48 hours. Phages replicating at the rate of 100's of new viruses per burst of a bacterial cell can rapidly outnumber the bacterial population which in the same time period can only double its numbers by binary fission. When the target bacterial population is gone, the remaining phages cannot reproduce, and are removed over a period of several days by the reticuloendothelial system.

Bacterial resistance to a given phage can occur, but within usually a few days researchers can quickly develop a new phage preparation that does work. However, most common phage preparations are "cocktails" of 20 to 50 different phage strains, thus improving the odds of success (◀Chapter 1, p. 22).

In the West, phages have been studied in great detail because it is much easier to manipulate bacterial cells and their viruses in the laboratory than to work with viruses that have multicellular hosts. In fact, work with phages provided us with the beginnings of modern molecular biology.

Properties of Bacteriophages

Like other viruses, bacteriophages can have their genetic information in the form of either double-stranded or single-stranded RNA or DNA. They can be relatively simple or complex in structure. To understand phage replication, we will examine the *T-even phages*. These phages, designated T2, T4, and T6 (T stands for "type"), are complex but well-studied naked phages that have dsDNA as their genetic material. The most widely studied is the T4 phage, an obligate parasite of the common enteric bacterium *Escherichia coli*. T4 has a distinctly shaped capsid made of a head, collar, and tail (**Figure 10.10; Table 10.5**). The DNA is packaged in the polyhedral head, which is attached to a helical tail.

Replication of T-Even Phages

Infection with and replication of new T4 phages occurs in the series of steps illustrated in **Figure 10.11**.

Adsorption. If T4 phages collide in the correct orientation with host cells, the phages will attach to, or adsorb onto, the host cell surface. Adsorption is a chemical attraction; it requires specific protein recognition factors found in the phage tail fibers that bind to specific receptor sites on the host cells. The fibers bend and allow the pins to touch the cell surface. Although many phages, including T4, attach to the cell wall, other phages can adsorb to flagella or to pili.

Penetration. The enzyme *lysozyme*, which is present within phage tails, weakens the bacterial cell wall. When the tail sheath contracts, the hollow tube (core) in the tail is

TABLE 10.5

The Functions of Bacteriophage Structural Components	
Component	**Function**
Genome	Carries the genetic information necessary for replication of new phage particles
Tail sheath	Retracts so that the genome can move from the head into the host cell's cytoplasm
Plate and tail fibers	Attach phage to specific receptor sites on the cell wall of a susceptible host bacterium

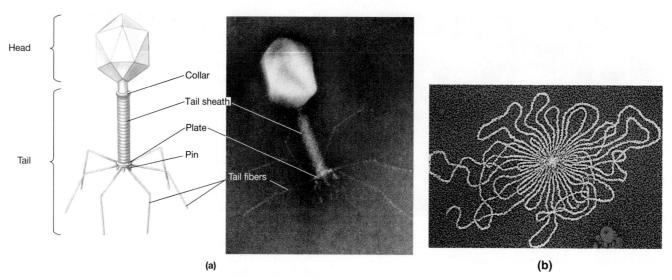

Head

Collar

Tail sheath

Plate

Pin

Tail fibers

Tail

(a)

(b)

Figure 10.10   Bacteriophages. **(a)** Structure and electron micrograph of a T-even (T4) bacteriophage (191,500X). *(Courtesy Robley C. Williams, Jr., Vanderbilt University)* **(b)** DNA normally is packaged into the phage head. Osmotic lysis has released the DNA from this phage, showing the large amount of DNA that must be packaged into a phage (or into an animal or plant virus (72,038X)). *(Omikron/ Photo Researchers, Inc.)*

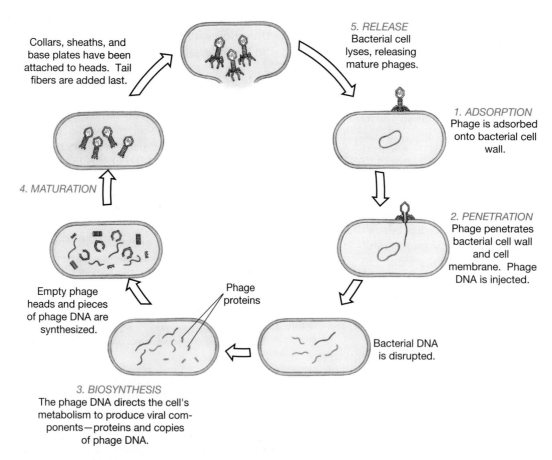

Collars, sheaths, and base plates have been attached to heads. Tail fibers are added last.

5. RELEASE
Bacterial cell lyses, releasing mature phages.

1. ADSORPTION
Phage is adsorbed onto bacterial cell wall.

4. MATURATION

2. PENETRATION
Phage penetrates bacterial cell wall and cell membrane. Phage DNA is injected.

Empty phage heads and pieces of phage DNA are synthesized.

Phage proteins

Bacterial DNA is disrupted.

3. BIOSYNTHESIS
The phage DNA directs the cell's metabolism to produce viral components—proteins and copies of phage DNA.

 Figure 10.11   Replication of a virulent bacteriophage. A virulent phage undergoes a lytic cycle to produce new phage particles within a bacterial cell. Cell lysis releases new phage particles that can infect more bacteria.

forced to penetrate the weakened cell wall and come into contact with the bacterial cell membrane. The viral DNA then moves from the head through the tube into the bacterial cell. It is not clear whether the DNA is introduced directly into the cytoplasm; according to recent evidence, T4 phages introduce their DNA into the periplasmic space, between the cell membrane and the cell wall. Either way, the phage capsid remains outside the bacterium.

Synthesis. Viral genomes, consisting of only a few thousand to 250,000 nucleotides, are too small to contain all the genetic information to replicate themselves. Therefore, they must use the biosynthetic machinery present in host cells. Once the phage DNA enters the host cell, phage genes take control of the host cell's metabolic machinery. Usually, the bacterial DNA is disrupted so that the nucleotides of hydrolyzed nucleic acids can be used as building blocks for new phage. Phage DNA is transcribed to mRNA, using the host cell's machinery. The mRNA, translated on host ribosomes, then directs the synthesis of capsid proteins and viral enzymes. Some of these enzymes are DNA polymerases that replicate the phage DNA. Thus, phage infection directs the host cell to make only viral products—that is, viral DNA and viral proteins.

Maturation. The head of a T4 phage is assembled in the host cell cytoplasm from newly synthesized capsid proteins. Then, a viral dsDNA molecule is packed into each head. At the same time, phage tails are assembled from newly formed base plates, sheaths, and collars. When the head is properly packed with DNA, each head is attached to a tail. Only after heads and tails are attached are the tail fibers added to form mature, infective phages.

Release. The enzyme lysozyme, which is coded for by a phage gene, breaks down the cell wall, allowing viruses to escape. In the process the bacterial host cell is lysed. Thus, phages such as T4 are called **virulent (lytic) phages** because they lyse and destroy the bacteria they infect (◄Chapter 8, p. 215). The released phages can now infect more susceptible bacteria, starting the infection process all over again. Such infections by virulent phages represent a **lytic cycle** of infection.

The time from adsorption to release is called the **burst time**; it varies from 20 to 40 minutes for different phages. The number of new virions released from each bacterial host represents the **viral yield**, or **burst size**. In phages such as T4, anywhere from 50 to 200 new phages may be released from one infected bacterium.

Phage Growth and the Estimation of Phage Numbers

Like bacterial growth, viral growth (biosynthesis and maturation) can be described by a **replication curve**, which generally is based on observations of phage-infected bacteria in laboratory cultures **(Figure 10.12)**. The replication curve of a phage includes an **eclipse period**, which spans from penetration through biosynthesis. During the eclipse period, mature virions cannot be detected in host cells. The **latent period** spans from penetration up to the point of phage release. As Figure 10.10 shows, the latent period is longer than—and includes—the eclipse period. The number of viruses per infected host cell rises after the eclipse period and eventually levels off.

If you had a phage suspension in a test tube, how could you determine the number of viruses in the tube? Phages cannot be seen with the light microscope, and it is not feasible to count such viruses from electron micrographs. Therefore, virologists and microbiologists use a different approach to estimate phage number. The viral assay method used is called a **plaque assay**. To perform a plaque assay, virologists start with a suspension of phages. Serial dilutions, like those described for bacteria, are prepared (◄Chapter 6, p. 151). A sample of each dilution is inoculated onto a plate containing a susceptible **bacterial lawn**—a layer of bacteria. Ideally,

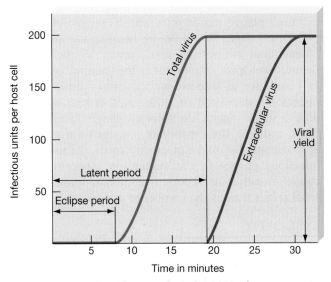

Figure 10.12 Growth curve for a bacteriophage. The eclipse period represents the time after penetration through the biosynthesis of mature phages. The latent period represents the time after penetration through the release of mature phages. The number of viruses per infected cell is the viral yield, or burst size.

Figure 10.13 Plaque assay. The number of bacteriophages in a sample is assayed by spreading the sample out over a "lawn" of solid bacterial growth. When the phages replicate and destroy the bacterial cells, they leave a clear spot, called a plaque, in the lawn. The number of plaques corresponds roughly to the number of phages that were initially present in the sample. Different kinds of phages produce plaques of different size or shape when replicating in the same bacterial species—in this case, *Escherichia coli*. The upper-left plate was inoculated with T2 phage; the upper-right plate, with T4 phage; and the lower plate, with lambda phage. *(Bruce Iverson/Bruce Iverson Photomicrography)*

virologists want a dilution that will permit only one phage to infect one bacterial cell. As a result of infection, new phages are produced from each infected bacterial cell, lysing the cell. These phages then infect surrounding susceptible cells and lyse them. After incubation and several rounds of lysis, the bacterial lawn shows clear areas called **plaques (Figure 10.13)**. Plaques represent areas where viruses have lysed host cells. In other parts of the bacterial lawn, uninfected bacteria multiply rapidly and produce a turbid growth layer.

Each plaque should represent the progeny from one infectious phage. Therefore, by counting the number of plaques and multiplying that number by the dilution factor, virologists can estimate the number of phages in a milliliter of suspension. Sometimes, however, two phages are deposited so close together that they produce a single plaque. And not all phages are infective. Thus, counting the number of plaques on a plate will approximate, but may not exactly equal, the number of infectious phages in the suspension. Therefore, such counts usually are reported as **plaque-forming units** (pfu) rather than as the number of phages.

LYSOGENY

General Properties of Lysogeny

The bacteriophages we have been discussing, virulent phages, destroy their host cells. **Temperate phages** do not always undergo a lytic cycle. The majority of the time they will exhibit **lysogeny**, a stable, long-term relationship between the phage and its host in which the phage nucleic acid becomes incorporated into the host nucleic acid.

Such participating bacteria are called *lysogenic cells.* One of the most widely studied lysogenic phages is the lambda (λ) phage of *Escherichia coli* **(Figure 10.14)**. Lambda phages attach to bacterial cells and insert their linear DNA into the bacterial cytoplasm **(Figure 10.15)**. However, once in the cytoplasm, the phage DNA circularizes and then integrates into the circular bacterial chromosome at a specific location. This viral DNA within the bacterial chromosome is called a **prophage**. The combination of a bacterium and a temperate phage is called a **lysogen**.

Insertion of a lambda phage into a bacterium alters the genetic characteristics of the bacterium. Two genes present in the prophage produce proteins that repress virus replication. The prophage also contains another gene that provides "immunity" to infection by another phage of the same type. This process, called **lysogenic conversion**, prevents the adsorption or biosynthesis of phages of the type whose DNA is already carried by the lysogen. The gene responsible for such immunity does not protect the lysogen against infection by a different type of temperate phage or by a virulent phage.

Lysogenic conversion can be of medical significance because the toxic effects of some bacterial infections are

TRY IT

Find Your Own Killer Phage

Obtain a sample likely to contain bacteriophages. The original samples used when phages were discovered were sewage-contaminated river waters. Manure is a rich source. Centrifuge to remove large particles and macroscopic organisms. Use a membrane filter system, such as those made by Millipore, having a proper pore size to remove bacteria but not viruses. You should now have a suspension of phage particles. However, it may be so concentrated that you might need to make several 1:10 dilutions before you find one that will give you a countable number of plaques.

Add one drop of the original extracted sample, or of one of its dilutions to a few milliliters of a fresh broth culture of the bacterium you want to study. Mix well. Spread 0.1 ml of the mixture over the surface of a plate of nutrient agar. Incubate at 37°C for several hours. The plate should then have a "lawn" of bacteria with clear spots (= plaques) where bacteria are not growing. Each plaque contains a population of bacteriophages that lyse that particular bacterium. With a sterile inoculating loop, remove a sample from one of the plaques. Add it to a turbid, growing broth culture of the same bacterium. Incubate the broth, examining it frequently. After a few hours it should change from turbid to clear as the phages lyse the bacteria. You now have a rich stock culture of phage. Multiplying the number of plaques by the dilution factor will give you an estimate of the total number of phages present in the original sample.

As a variation, try isolating a phage that will lyse cyanobacteria and clear up undesired algal growth.

Figure 10.14 False-color TEM of the temperate lambda phage (85,680X). This virus infects the bacterium *Escherichia coli*. (M. Wurtz, Biozentrum/Photo Researchers, Inc.)

caused by the prophages they contain. For example, the bacteria *Corynebacterium diphtheriae* and *Clostridium botulinum* contain prophages that have a gene that codes for the production of a toxin. The conversion from non-toxin production to toxin production is largely responsible for the tissue damage that occurs in diphtheria and botulism, respectively. Without the prophages, the bacteria do not cause disease.

Once established as a prophage, the virus can remain dormant for a long time. Each time a bacterium divides, the prophage is copied and is part of the bacterial chromosome in the progeny bacteria. Thus, this period of bacterial growth with a prophage represents a **lysogenic cycle** (Figure 10.15). However, either spontaneously or in response to some outside stimulation, the prophage can become active and initiate a typical lytic cycle. This process, called **induction**, may be due to a lack of nutrients for bacterial growth or the presence of chemicals toxic to the lysogen. The provirus seems to sense that "living" conditions are deteriorating and that it is time to find a new home. Through induction, the provirus removes itself from the bacterial chromosome. The phage DNA then codes for viral proteins to assemble new temperate phages in a manner similar to that

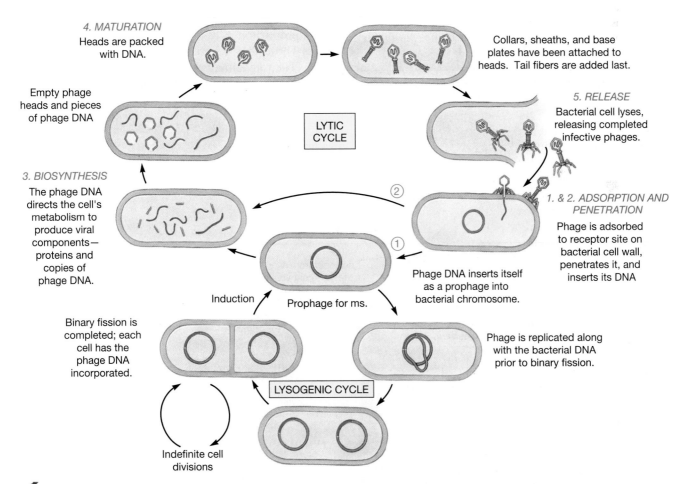

4. MATURATION
Heads are packed with DNA.

Collars, sheaths, and base plates have been attached to heads. Tail fibers are added last.

Empty phage heads and pieces of phage DNA

5. RELEASE
Bacterial cell lyses, releasing completed infective phages.

LYTIC CYCLE

3. BIOSYNTHESIS
The phage DNA directs the cell's metabolism to produce viral components—proteins and copies of phage DNA.

1. & 2. ADSORPTION AND PENETRATION
Phage is adsorbed to receptor site on bacterial cell wall, penetrates it, and inserts its DNA

Phage DNA inserts itself as a prophage into bacterial chromosome.

Induction

Prophage for ms.

Binary fission is completed; each cell has the phage DNA incorporated.

Phage is replicated along with the bacterial DNA prior to binary fission.

LYSOGENIC CYCLE

Indefinite cell divisions

Figure 10.15 Replication of a temperate bacteriophage. Following adsorption and penetration, the virus undergoes prophage formation. In the lysogenic cycle, temperate phages can exist harmlessly as a prophage within the host cell for long periods of time. Each time the bacterial chromosome is replicated, the prophage also is replicated; all daughter bacterial cells are "infected" with the prophage. Induction involves either a spontaneous or an environmentally induced excision of the prophage from the bacterial chromosome. A typical lytic cycle, involving biosynthesis and maturation, occurs, and new temperate phages are released.

TABLE 10.6

Comparison of Bacteriophage and Animal Virus Replication

Stage	Bacteriophage	Animal Virus
Attachment sites	Attachment of tail fibers to cell wall proteins	Attachment of spikes, capsid or envelope to plasma membrane proteins
Penetration	Injection of viral nucleic acid through bacterial cell wall	Endocytosis or fusion
Uncoating	None needed	Enzymatic digestion of viral proteins
Synthesis	In cytoplasm	In cytoplasm (RNA viruses) or nucleus (DNA viruses)
	Bacterial synthesis ceased	Host cell synthesis ceased
	Viral DNA or RNA replicated, formation of viral mRNA	Viral DNA or RNA replicated, formation of viral RNA
	Viral components synthesized	Viral components synthesized
Maturation	Addition of collar, sheath, base plate, and tail fibers to viral nucleic acid-containing head	Insertion of viral nucleic acid into capsid
Release	Host cell lysis	Budding (enveloped viruses), cell rupture (nonenvelopedviruses)
Chronic infection	Lysogeny	Latency, chronic infection, cancer

used by lytic phages. As a result, new temperate phages mature and are released through cell lysis.

The French microbiologist André Lwoff first described lysogeny in 1950. He also discovered that only a small proportion of lysogens produce phages at any one time. Those that do are lysed as a result of phage release. The remaining lysogens do not undergo induction and, due to lysogenic conversion, remain protected

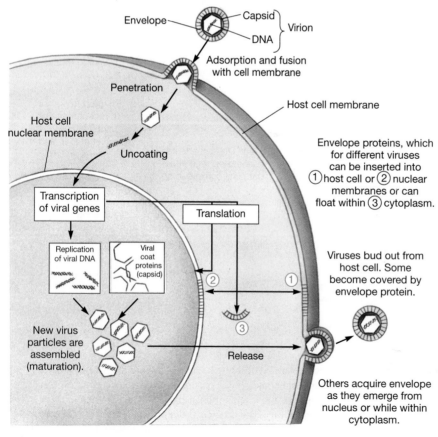

Figure 10.16 Replication of an enveloped dsDNA animal virus. Shown is a herpesvirus.

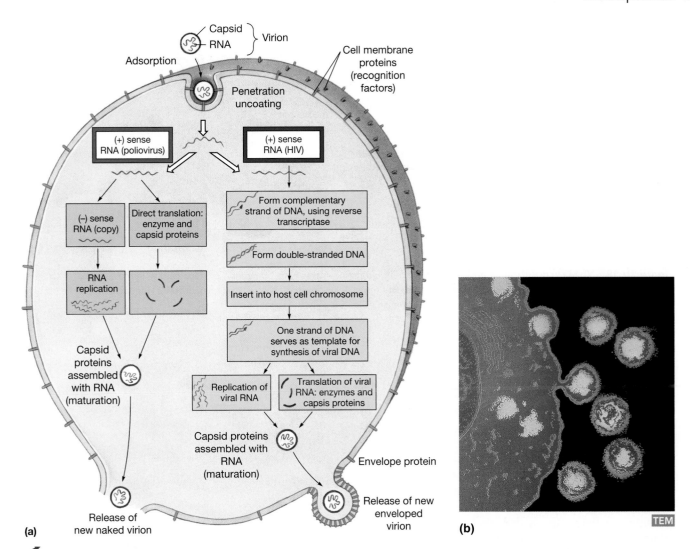

Figure 10.17 Replication of RNA viruses. (a) Two of the replication mechanisms used by different (+) sense RNA animal viruses. *Left:* In the poliovirus the viral (+) sense RNA serves as mRNA—it is translated immediately to produce proteins needed for reproduction of the virus. A (−) sense RNA copy is then made, which serves as a template for the production of more viral (+) sense RNA molecules. Mature polioviruses lyse the cell during release. *Right:* In HIV each (+) sense RNA, copied with the help of reverse transcriptase, forms an ssDNA, which serves as template for the synthesis of the complementary strand. The dsDNA is then inserted into the host chromosome, where it can remain for some time. When virus replication occurs, one strand of the DNA becomes the template for the synthesis of viral (+) sense RNA molecules. Mature HIV particles usually do not lyse the cell but rather bud off the cell surrounded by an envelope. **(b)** HIV viruses that are budding from a T-4 lymphocyte (84,777X). *(Chris Bjornberg/Photo Researchers)*

from infection by phages of the same type. In 1965 Lwoff shared the Nobel Prize in physiology or medicine with François Jacob and Jacques Monod.

The majority of bacteriophages undergo lysogeny. The reason may have to do with replication. Remember, virulent phages can move from host to host only by forming new phages that are released from one cell and that infect another. In contrast, a lysogenic cycle allows temperate phages to "infect" more bacteria without forming new bacteriophages. As a result of binary fission, a copy of the phage DNA is distributed to each new bacterial cell.

REPLICATION OF ANIMAL VIRUSES

Like other viruses, animal viruses invade and replicate in animal cells by the processes of adsorption, penetration, synthesis, maturation, and release. However, animal viruses perform these processes in ways that differ from those employed by bacteriophages—and in a number of different ways among themselves (Table 10.6). A complete replication cycle for an animal DNA virus is summarized in **Figure 10.16**, and two mechanisms for the replication of (+) sense RNA animal viruses are summarized in **Figure 10.17**.

Figure 10.18 Viral recognition of an animal host cell. (a) Rhinoviruses have "canyons," or depressions, in the capsid that attach to specific membrane proteins on the host cell membrane. (b) HIV has specific envelope spikes (viral glycoproteins) that attach to a membrane protein receptor on the surface of specific host immune defense cells.

Adsorption

As we have seen, bacteriophages have specialized structures for attaching to bacterial cell walls. Although animal cells lack cell walls, animal viruses have ways of attaching to host cells. Specificity involves a combination of virus and host cell recognition.

Naked viruses have attachment sites (proteins) on the surfaces of their capsids that bind to corresponding sites on appropriate host cells. For example, virologists have shown that rhinoviruses have "canyons," or depressions, in their capsids that bind to a specific membrane protein normally involved with cell adhesion (**Figure 10.18a**). Conversely, enveloped viruses, such as HIV, have spikes that recognize, in part, a membrane protein receptor on the surface of certain specific immune defense cells (**Figure 10.18b**).

Penetration

Penetration follows quite quickly after adsorption of the virion to the host's plasma membrane. Unlike bacteriophages, animal viruses do not have a mechanism for injecting their nucleic acid into host cells. Thus, both the nucleic acid and the capsid usually penetrate animal host cells. Most naked viruses enter the cell by endocytosis, in which virions are captured by pitlike regions on the surface of the cell and enter the cytoplasm within a membranous vesicle (**Figure 10.19**). Enveloped viruses may fuse their envelope with the host's plasma membrane or enter by endocytosis. In the latter case, the envelope fuses with the vesicle membrane.

Once the animal virus enters the host cell's cytoplasm, the viral genome must be separated from its protein coat (released) through a process called **uncoating**. Naked viruses are uncoated by proteolytic enzymes from host cells or from the viruses themselves. The uncoating of viruses such as the poxviruses is completed

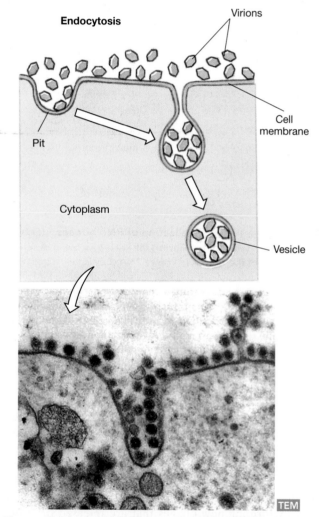

Figure 10.19 Animal virus penetration of host cells. Many naked virions adhere to the cell surface and become trapped in pits of the cell membrane. These pits invaginate to form separate cytoplasmic vesicles. In the electron micrograph, coronaviruses are being taken into the cytoplasm of a host cell (magnification unknown). *(Centers for Disease Control and Prevention CDC)*

by a specific enzyme that is encoded by viral DNA and formed soon after infection. Polioviruses begin uncoating even before penetration is complete.

Synthesis

The synthesis of new genetic material and proteins depends on the nature of the infecting virus.

Synthesis in DNA Animal Viruses.

Generally, DNA animal viruses replicate their DNA in the host cell nucleus with the aid of viral enzymes and synthesize their capsid and other proteins in the cytoplasm by using host cell enzymes. The new viral proteins move to the nucleus, where they combine with the new viral DNA to form virions (Figure 10.16). This pattern is typical of adenoviruses, hepadnaviruses, herpesviruses, and papovaviruses. Poxviruses are the only exception; their parts are synthesized in the host cell's cytoplasm.

In dsDNA viruses, replication proceeds in a complex series of steps designated as *early* and *late* transcription and translation. The early events take place before the synthesis of viral DNA and result in the production of the enzymes and other proteins necessary for viral DNA replication. The late events occur after the synthesis of viral DNA and result in the production of structural proteins needed for building new capsids. Compared with bacteriophage replication, synthesis in animal virus replication can take much longer. The capsids of the herpesviruses, for example, contain so many proteins that their synthesis requires 8 to 16 hours.

Some viruses, such as the adenoviruses, contain only ssDNA. Before viral replication can be initiated, the viral DNA must be copied, forming a dsDNA viral genome.

Synthesis in RNA Animal Viruses.

Synthesis in RNA animal viruses takes place in a greater variety of ways than is found in DNA animal viruses. In RNA viruses such as the picornaviruses, the (+) sense RNA acts as mRNA, and viral proteins are made immediately after penetration and uncoating (Figure 10.17). The nucleus of the host cell is not involved. Viral proteins also play key roles in the synthesis of these viruses. One protein inhibits synthetic activities of the host cell. For synthesis, an enzyme uses the (+) sense RNA as a template to make a (−) sense RNA. This (−) sense RNA in turn acts as a template RNA to replicate many (+) sense RNA molecules for virion formation.

In the retroviruses, such as HIV, the two copies of (+) sense RNA do not act as mRNA. Rather, they are transcribed into ssDNA with the help of reverse transcriptase (Figure 10.17). The ssDNA then is replicated through complementary base pairing to make dsDNA molecules. Once in the cell nucleus, this molecule inserts itself as a provirus into a host cell chromosome. The provirus can remain there for an indefinite period of time. When infected cells divide, the provirus is replicated along with the rest of the host chromosome. Thus, the viral genetic information is passed to progeny host cells.

Unlike prophages, however, the provirus cannot be excised. If an event occurs that activates the provirus, its genes are expressed; that is, the genes are used to make viral mRNA, which directs synthesis of viral proteins. Full-length (+) sense RNA molecules also are transcribed from the prophage. Two copies of the (+) sense RNA are packaged into each virion.

In (−) sense RNA animal viruses, such as the viruses causing measles and influenza A, a packaged transcriptase uses the (−) sense RNA to make (+) sense RNA molecules (mRNA). Prior to assembly, new (−) sense RNA is made from (+) sense RNA templates. The process is essentially the same regardless of whether the viral RNA is in one segment (measles) or in many segments (influenza A).

In the reoviruses, the dsRNA codes for several viral proteins. Each strand of the dsRNA acts as a template for its partner. Like DNA replication, RNA replication is semiconservative, so the molecules produced have one strand of old RNA and one strand of new RNA. These viruses have a double-walled capsid that is never completely removed, and replication takes place within the capsid.

Maturation

Once an abundance of viral nucleic acid, enzymes, and other proteins have been synthesized, assembly of components into complete virions starts. This step constitutes maturation or assembly of progeny viruses. The cellular site of maturation varies depending on the virus type. For example, human adenovirus nucleocapsids are assembled in the cell nucleus (Figure 10.16), whereas viruses such as HIV are assembled at the inner surface of the host cell's plasma membrane. The poxviruses, polioviruses, and picornaviruses are assembled in the cytoplasm.

Maturation of enveloped viruses is a longer and more complex process than that of most bacteriophages. As we have seen, both the infecting virus and nucleic acids and enzymes made in the host cell participate in synthesizing components. Among the components destined for the progeny viruses, the proteins and glycoproteins are coded by the viral genome; envelope lipids and glycoproteins are synthesized by host cell enzymes and are present in the host plasma membrane. If the virus is to have an envelope, the virion is not complete until it buds through a host membrane—either the nuclear, endoplasmic reticulum, Golgi, or plasma membrane—depending on the specific virus (Figure 10.16).

Release

The budding of new virions through a membrane may or may not kill the host cell. Human adenoviruses, for example, bud from the host cell in a controlled manner. This *shedding* of new virions does not lyse the host cells. Other types of animal viruses kill the host cell. When an infected animal cell is filled with progeny virions, the plasma membrane lyses and the progeny are released. Lysis of cells often produces the clinical symptoms of the infection or disease. The herpesviruses that cause cold sores and the poxviruses destroy skin cells as a result of virion release. And the polioviruses destroy nerve cells during the release process.

LATENT VIRAL INFECTIONS

Many individuals experience the reoccurrence of skin eruptions commonly called cold sores or fever blisters. These are caused by the herpes simplex virus, a member of the herpesviruses. As we saw earlier, these are dsDNA viruses that can exhibit a lytic cycle. They can also remain latent within the cells of the host organism throughout the individual's life—not in the skin cells we associate them with, but in the nerve cells. When activated, whether by a cold or fever or by stress or immunosuppression, they once again replicate resulting in cell lysis.

The ability to become latent is held by all of the herpesviruses. Another herpesvirus, the one that's responsible for chickenpox, can also remain dormant within the central nervous system. When it becomes activated, usually due to changes in cell-mediated immunity, the virus causes a rash to form along the nerve where it lay latent. This reactivation is known as shingles. Many individuals carry these viruses throughout their lives, never exhibiting any symptoms.

✔**CHECKLIST**

> **1.** List in correct order the five steps in viral replication.
>
> **2.** How do these five stages differ between bacteriophages and animal viruses?
>
> **3.** Compare lysogeny with the lytic cycle in bacteriophages.

▐▐▐ CULTURING OF ANIMAL VIRUSES

DEVELOPMENT OF CULTURING METHODS

Initially, if a virologist wanted to study viruses, the viruses had to be grown in whole animals. This made it difficult to observe specific effects of the viruses at the cellular level. In the 1930s virologists discovered that embryonated (intact, fertilized) chicken eggs could be used to grow herpesviruses, poxviruses, and influenza viruses. Although the chick embryo is simpler in organization than a whole mouse or rabbit, it is still a complex organism. The use of embryos did not completely solve the problem of studying cellular effects caused by

viruses. Another problem was that bacteria also grow well in embryos, and the effects of viruses often could not be determined accurately in bacterially contaminated embryos. Virology progressed slowly during these years until techniques for growing viruses in cultures improved.

Two discoveries greatly enhanced the usefulness of cell cultures for virologists and other scientists. First, the discovery and use of antibiotics made it possible to prevent bacterial contamination. Second, biologists found that proteolytic enzymes, particularly trypsin, can free animal cells from the surrounding tissues without injuring the freed cells. After the cells are washed, they are counted and then dispensed into plastic flasks, tubes, Petri dishes, or roller bottles (**Figure 10.20**). Cells in such suspensions will attach to the plastic surface, multiply, and spread to form sheets one cell thick, called **monolayers**. These monolayers can be subcultured. **Subculturing** is the process by which cells from an existing culture are transferred to new containers with fresh nutrient media. A large number of separate subcultures can be made from a single tissue sample, thereby assuring a reasonably homogeneous set of cultures with which to study viral effects.

The term **tissue culture** remains in widespread usage to describe the preceding technique, although the term **cell culture** is perhaps more accurate. Today, the majority of cultured cells are in the form of monolayers grown from enzymatically dispersed cells. With a wide variety of cell cultures available and with antibiotics to control contamination, virology entered its "Golden Age." In the 1950s and 1960s, more than 400 viruses were isolated and characterized. Although new viruses are still being discovered, emphasis is now on

Figure 10.20 A view from the end of a bottle lined with spiral plastic coils. One way to increase cell density is by increasing the surface area to which cells can attach. The bottle is rotated slowly at about 5 rev/h, so that a small volume of culture fluid can be used. Cells tolerate being out of the culture fluid for short periods. *(Science Source/Photo Researchers, Inc.)*

characterizing the viruses in more detail and on determining the precise steps in viral infection and viral replication.

Plaque assays similar to those used to study phages can be used for animal viruses. For example, cultures of susceptible human cells are grown in cell monolayers and then inoculated with viruses. If the viruses lyse cells, several rounds of infection will produce plaques.

TYPES OF CELL CULTURES

Three basic types of cell cultures are widely used in clinical and research virology: (1) primary cell cultures, (2) diploid fibroblast strains, and (3) continuous cell lines. **Primary cell cultures** come directly from the animal and are not subcultured. The younger the source animal, the longer the cells will survive in culture. They typically consist of a mixture of cell types, such as muscle and epithelial cells. Although such cells usually do not divide more than a few times, they support growth of a wide variety of viruses.

If primary cell cultures are repeatedly subcultured, one cell type will become dominant, and the culture is called a **cell strain**. In cell strains all the cells are genetically identical to one another. They can be subcultured for several generations with only a very small likelihood that changes in the cells themselves will interfere with the determination of viral effects.

Among the most widely used cell strains are **diploid fibroblast strains**. *Fibroblasts* are immature cells that produce collagen and other fibers as well as the substance of connective tissues, such as the dermis of the skin. Derived from fetal tissues, these strains retain the fetal capacity for rapid, repeated cell division. Such strains support growth of a wide range of viruses and are usually free of contaminating viruses often found in cell strains from mature animals. For this reason they are used in making viral vaccines.

The third type of cell culture in extensive use is the continuous cell line. A **continuous cell line** consists of cells that will reproduce for an extended number of generations. The most famous of such cultures is the HeLa cell line, which has been maintained and grown in culture since 1951 and has been used by many researchers worldwide. The original cells of the HeLa cell line came from a woman with cervical cancer and are named from the first letters of her name. In fact, many of the early continuous cell lines used malignant cells because of their capacity for rapid growth. Such immortal cell lines grow in the laboratory without aging, divide rapidly and repeatedly, and have simpler nutritional needs than normal cells. The HeLa cell line, for example, contains two viral genes necessary for its own immortality. Immortal cell lines are heteroploid (have different numbers of chromosomes) and are therefore genetically diverse.

Figure 10.21 Viral culture in eggs. Some viruses, such as influenza viruses, are grown in embryonated chicken eggs. *(Account Phototake/Phototake)*

Cell cultures have largely replaced animals and embryonated eggs for studies in animal virology. Yet, the embryonated chicken egg remains one of the best host systems for influenza A viruses (**Figure 10.21**). In addition, young albino Swiss mice are still used to culture *arboviruses* (*ar*thropod-*bo*rne viruses), and other mammalian cell lines—as well as mosquito cell lines—have been used for some time.

The Cytopathic Effect

The visible effect viruses have on cells is called the **cytopathic effect (CPE)**. Cells in culture show several common effects, including changes in cell shape and detachment from adjacent cells or the culture container (**Figure 10.22**). However, CPE can be so distinctive

PUBLIC HEALTH

Can a Caterpillar Make Your Next Flu Vaccine?

Allergic to vaccines made in eggs? Cheer up, a faster new method using insect cell culture may soon replace the egg. An insect virus (baculovirus), which is grown in insect cell culture, has been genetically modified to produce flu virus proteins. These proteins stimulate the human immune system and have been used to produce a flu vaccine which has been shown to be safe and 100% effective at appropriate dosages.

More importantly, large quantities of these flu proteins are produced more rapidly than in the egg procedure, which takes 6 months. Public health authorities are very anxious to be able to rapidly produce large quantities of vaccine effective against new flu strains such as the avian flu. Further clinical studies are underway.

(a) SEM **(b)** SEM

Figure 10.22 Viral transformation of cells. (a) Normal and **(b)** transformed (malignant) cells in culture (both 8,171X). Such transformation is an example of a cytopathic effect (CPE) caused by infection with the Rous sarcoma virus (RSV). In the transformed state, the cells become rounded and do not adhere to the culture container. *(G. Steven Martin/Visuals Unlimited)*

that an experienced virologist often can use it to make a preliminary identification of the infecting virus. For example, human adenoviruses and herpesviruses cause infected cells to swell because of fluid accumulation, whereas picornaviruses arrest cell functions when they enter and lyse cells when they leave. The paramyxoviruses cause adjacent cells in culture to fuse, forming giant, multinucleate cells called **syncytia** (sin-sish′e-a; singular: *syncytium*). Syncytia can contain 4 to 100 nuclei in a common cytoplasm. Another type of CPE produced by some viruses is *transformation*: the conversion of normal cells into malignant ones, which we discuss later in this chapter.

Some slow-growing viruses, such as cytomegaloviruses, rubella virus, and some adenoviruses, may not produce obvious CPE for 1 to 4 weeks.

VIRUSES AND TERATOGENESIS

Teratogenesis is the induction of defects during embryonic development. A **teratogen** (ter′a-to-jen) is a drug or other agent that induces such defects. Certain viruses are known to act as teratogens and can be transmitted across the placenta and infect the fetus. The earlier in pregnancy the embryo is infected, the more extensive the damage is likely to be. During the early stages of embryologic development, when an organ or body system may be represented by only a few cells, viral damage to those cells can interfere with the development of that organ or body system. Viral infections occurring later in development may damage fewer cells and thus have a proportionately smaller effect. That is because, by then, the total cell population in the fetus has greatly increased, and each organ or body system consists of thousands of cells.

Three human viruses—cytomegalovirus (CMV), herpes simplex virus (HSV) types 1 and 2, and rubella—account for a large number of teratogenic effects. Cytomegalovirus (CMV) infections are found in about 1% of live births; of those, about 1 in 10 eventually die from the CMV infection. Most of the defects are neurological, and the children have varying degrees of mental retardation. Some also have enlarged spleens, liver damage, and jaundice. HSV infections usually are acquired at or shortly after birth. Infections acquired before birth are rare. In cases of disseminated infections (those that spread through the body), some infants die and survivors have permanent damage to the eyes and central nervous system.

Cytomegalovirus infection during pregnancy is now the major viral cause of congenital abnormalities in the newborn.

Rubella virus infections in the mother during the first 4 months of pregnancy are most likely to result in fetal defects referred to as the "rubella syndrome." These defects include deafness, damage to other sense organs, heart and other circulatory defects, and mental retardation. The degree of impairment is highly variable. Some children adapt to their disabilities and live productive lives; in other cases the fetus is so impaired that death and natural abortion occur. Congenital rubella is discussed in ◀Chapter 19, p. 581.

A series of blood tests often referred to as the **TORCH series** is sometimes used to identify possibly teratogenic diseases in pregnant women and newborn infants. These tests detect antibodies made against _T_oxoplasma (a protozoan), _o_ther disease-causing viruses (usually including the hepatitis B virus and the varicella, or chickenpox, virus), _r_ubella virus, _C_MV, and _H_SV. All of these diseases can be transmitted to the fetus via the placenta. Intrauterine diseases other than those tested for in the TORCH series (e.g., syphilis and HIV) may also exist in a newborn. Therefore, passing the TORCH tests does not guarantee a healthy baby.

VIRUSLIKE AGENTS: SATELLITES, VIROIDS, AND PRIONS

Viruses represent the smallest microbes that, in most cases, have the genetic information to produce new virions in a host cell. The exceptions are those viruses that do not have all their own genetic information to produce new virions. We mentioned earlier that viruses such as the dependoviruses (family Parvoviridae) must use a helper virus to supply the necessary components to produce more virions. However, there are even smaller infectious agents that can cause disease: satellites, viroids, and prions.

SATELLITES

Satellites are small, single-stranded RNA molecules, usually 500 to 2,000 nucleotides in length, which lack genes required for their replication. However, in the presence of a helper virus, they can replicate. There are two types: **satellite viruses** and the **satellite nucleic acids** (also known as **virusoids**). They are called satellites because their reproduction "revolves around" a helper virus.

Satellite viruses are not defective versions of their helper viruses, in that they may have lost pieces or rearranged parts of the helper virus genome. The helper virus is not their parent. The two viruses are totally unrelated. The satellite is defective in being unable to replicate alone. It does, however, have genes coding for the capsid which covers it, in contrast to satellite nucleic acids (virusoids) which are covered by a capsid coded for by their helper virus.

Most satellites are associated with plant viruses. Unlike animal viruses, plant viruses very often have genomes split into several segments, each of which is encapsulated separately, and all of which collectively constitute the virus. Transmission from animal host to host, or cell to cell within a host, does not seem able to pass on complete sets of multiple particles. One exception appears to be the delta hepatitis virus, which does infect only humans, and appears to be a kind of hybrid between a satellite and a

viroid (to be discussed in the next section). The origins of viroids and satellites remain unclear.

DELTA HEPATITIS

Hepatitis delta virus (HDV) was discovered in the mid-1970s. Sequencing of its genome reveals similarity to the viroid and to virusoid RNAs that infect plants. HDV initially was thought to be part of the hepatitis B virus (HBV) because it was never found without the presence of hepatitis B infection. However, it was not present in all cases of hepatitis B, only in especially severe cases, having a death rate 10 times higher than when only B was present. By 1980 it was found to be a separate, defective pathogen that required coinfection with the hepatitis B virus in order to replicate. It can be prevented by vaccinating against HBV, because it cannot infect without its helper virus. HDV has the smallest genome of any known animal virus, with a length of a mere 1,679 to 1,683 nucleotides. In contrast, HBV has 3,000 to 3,300. HDV lacks a well-defined capsid of its own and is surrounded by the portion of the HBV that codes for the surface antigen (formerly called the Australian antigen) of HBV. It is primarily transmitted by blood and blood products. While present worldwide, delta hepatitis is especially frequent (over 60% infection rate) in parts of the Amazon basin, Central Africa, and the Middle East.

VIROIDS

In 1971 the plant pathologist T. O. Diener described a new type of infectious agent. He was studying potato tuber spindle disease, which was thought to be caused by a virus. However, no virions could be detected. Rather, Diener discovered molecules of RNA in the nuclei of diseased plant cells (**Figure 10.23a**). He proposed the concept of a **viroid**, an infectious RNA particle smaller than a virus. Since then viroids have been found to differ from viruses in six ways:

1. Each viroid consists of a single circular RNA molecule of low molecular weight, 246 to 399 nucleotides in length.
2. Viroids exist inside cells, usually inside of nucleoli, as particles of RNA without capsids or envelopes.
3. Unlike viruses such as the parvoviruses, viroids do not require a helper virus.
4. Viroid RNA does not produce proteins.
5. Unlike virus RNA, which may be copied in the host cell's cytoplasm or nucleus, viroid RNA is always copied in the host cell nucleus.
6. Viroid particles are not apparent in infected tissues without the use of special techniques to identify nucleotide sequences in the RNA.

(a)

(b)

Figure 10.23   Viroids and their effects.   (a) Viroid particles that cause potato spindle tuber disease (shown as yellow rods in this artist's rendition of an electron micrograph) are very short pieces of RNA containing only 300 to 400 nucleotides. The much larger (blue and purple) strand is DNA from a T7 bacteriophage. Such comparisons make it easy to see how viroids were overlooked for many years. *(Reprinted from Agricultural Research, vol. 37, no. 5 (May 1989), p. 4, published by the Agricultural Research Service of the USDA)* **(b)** The tomato plant on the left is normal; the one on the right is infected with a viroid causing tomato apical stunt disease. *(Courtesy United States Department of Agriculture)*

Viroids must disrupt host cell metabolism in some way, but because no protein products are produced, it is not clear how viroids and their RNA cause disease. They may interfere with the cell's ability to process mRNA molecules. Without mature mRNA molecules, proteins cannot be synthesized. If so, cell metabolism would be so disturbed that cell death could result. Although some viroids cause no apparent effect or only mild pathogenic effects in the host, other viroids are known to cause several lethal plant diseases, such as potato spindle tuber disease, chrysanthemum stunt disease, cucumber pale fruit disease, and tomato apical stunt disease **(Figure 10.23b)**. None of these diseases was recognized before 1922 despite centuries of intense cultivation of these crops. Several other diseases have been recognized recently. Some scientists believe that while isolated plants may have contained viroids for an unknown number of years, modern agricultural methods, such as growing large numbers of the same plant in close association and the use of machinery for harvesting, may have allowed viroid diseases to spread, allowing observers to recognize them. The viroid may have entered crop plants from unknown wild plants, an idea supported by the observation that the first viroid-infected crop plants appear along the margins of fields that abut wild plots. Viroids can even be transmitted through seeds, or by aphids. No viroid is presently known to infect animals, but there is no reason to suppose they cannot.

At least two hypotheses have been proposed to account for the origin of viroids. One suggests that they originated early in precellular evolution when the primary genetic material probably consisted of RNA. A second suggests that they are relatively new infectious agents that represent the most extreme example of parasitism.

PRIONS

Take another look

In the 1920s several cases of a slow but progressive dementing illness in humans were observed independently by Hans Gerhard Creutzfeldt and Alfons Maria Jakob. The disease, now called *Creutzfeldt–Jakob disease* (CJD), is characterized by mental degeneration, loss of motor function, and eventual death. Since that time several similar neurological degenerative diseases have been described ◀(Chapter 24). One is *kuru*, which caused loss of voluntary motor control and eventual death of natives of New Guinea. These deaths were attributed to an infective agent transmitted as a result of cannibalism. In other animals, *scrapie* in sheep and *bovine spongiform encephalopathy* (BSE)—commonly called mad cow disease—in dairy cattle have been observed to cause slow loss of neuronal function that leads to death. Consuming infected cattle has led to human cases termed *new-variant CJD*. In 2003 infected cattle were found in the United States and Canada. Turn to ◀Figure 24.14, p. 772, to see the holes in a section of brain that give it the name "spongiform." In the western part of the United States, some herds of deer and elk have spongiform encephalopathies. Unfortunately, some hunters have died from prion infection, presumably acquired during the skinning and butchering of these animals. Mice are also somehow involved.

Although some recent research into these similar diseases indicates that they may be caused by viruses, other evidence points to a different type of infective agent. The infective agent may be an exceedingly small *pro*teinaceous *in*fectious particle. In 1982, Stanley Prusiner proposed that such an infectious particle be called a **prion** (pre′on). In 1987, Prusiner received both the Nobel Prize in medicine and the Columbia University Louisa Gross Horwitz Prize for his work with prions. Prions have the following characteristics:

TABLE 10.7

Comparison of Viruses, Viroids, and Prions			
	Virus	**Viroid**	**Prion**
Nucleic acid	+ (ssDNA, dsDNA, ssRNA, or dsRNA)	+ (ssRNA)	−
Presence of capsid or envelope	+	−	−
Presence of protein	+	−	+
Need for helper viruses	+/− (Needed by some of the smaller viruses such as the parvoviruses)		
Viewed by	Electron microscopy	Nucleotide sequence identification	Host cell damage
Affected by heat and protein denaturing agents	+	−	−
Affected by radiation of enzymes that digest DNA or RNA	+	+	−
Host	Bacteria, animals, or plants	Plants	Mammals

1. Prions are resistant to inactivation by heating to 90°C, which will inactivate viruses.

2. Prion infection is not sensitive to radiation treatment that damages virus genomes.

3. Prions are not destroyed by enzymes that digest DNA or RNA.

4. Prions are sensitive to protein denaturing agents, such as phenol and urea.

5. Prions have direct pairing of amino acids.

See Table 10.7 for a comparison of virus, viroid, and prion characteristics.

Prusiner's research and that of others suggest that prions are normal proteins that become folded incorrectly, possibly as a result of a mutation (Figure 10.24). The harmless, normal proteins are found on the plasma membrane of many mammalian cells, especially brain cells. The prion proteins (*PrP*) are thought to stick together inside cells, forming small fibers, or fibrils. Because the fibrils cannot be organized in the plasma membrane correctly, such aggregations eventually kill the cell.

The most urgent question is to determine how a prion-caused disease spreads. Researchers believe that prions cause other copies of the normal protein to fold improperly. In an outbreak of mad cow disease in Britain early in the 1990s, the infectious prion originally came from a protein supplement in the feed. This supplement included by-products from scrapie-infected sheep! In fact, experiments show that when mice are injected with prion extracts, disease results. Prions have been shown to move easily from one species to another (Figure 10.25). When inoculated, a given prion can infect many different species. Prions do not appear to always be species-specific. Recent reports from Switzerland have demonstrated reversal of the spongiform condition of the brain in mice, early in infection, when prion protein is prevented from accumulating in neurons. Adjacent glial (non-neuronal) cells were packed with prion protein, but the mice remained normal. For more information on this and on prion-associated diseases, see ◀Chapter 24, pp. 769–773.

Figure 10.24 **Protein structure model of the two forms of the prion protein (PrP).** The protein helices are represented as spiral ribbons: **(a)** harmless form; **(b)** harmful form.

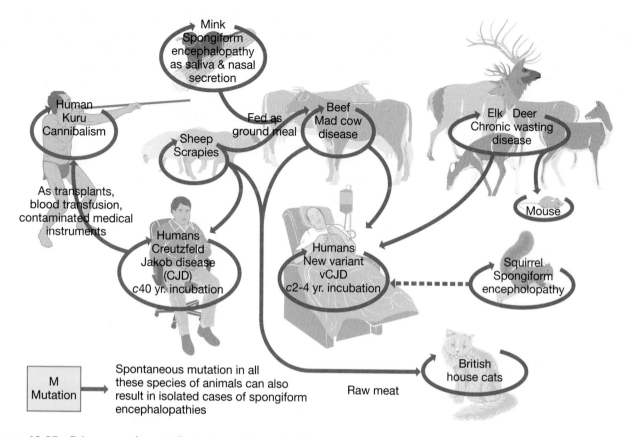

Figure 10.25 **Prion-caused spongiform encephalopathy diseases.** These occur in many species, used can be transmitted from one to another. Spontaneous mutations also produce some number of cases each year without involving transmission from another animal. Various zoo animals have acquired this disease when fed raw meat. See Chapter 24, pages 769–773 for more information.

✓**CHECKLIST**

1. What are the three most important types of cell culture used for growing viruses?
2. How are cell lines made immortal?
3. Define cytopathic effect (CPE), syncytia, and transformation.
4. Give two examples of viral teratogenesis.
5. What are satellites? How do their two types differ?
6. Compare viruses, viroids, and prions.

▍▍ VIRUSES AND CANCER

Cancer is known as a set of diseases that perturb the normal behavior and functioning of cells. We can define **cancer** as an uncontrolled, invasive growth of abnormal cells—in other words, cancer cells divide repeatedly. In many cases they cannot stop dividing; the result is a **neoplasm**, or localized accumulation of cells known as a **tumor**. A neoplasm can be **benign**—a noncancerous growth. But if the cells invade and interfere with the functioning of surrounding normal tissue, the tumor is **malignant**. Malignant tumors and their cells can **metastasize**, or spread, to other tissues in the body.

That viruses could cause some cancers in animals was discovered in 1911 by F. Peyton Rous. He showed that certain *sarcomas* (neoplasms of connective tissue) in chickens were caused by a virus, named the *Rous sarcoma virus* (RSV). Therefore, it was not surprising to discover that viruses can be associated with cancer in humans as well. Although most human cancers arise from genetic mutations, cellular damages from environmental chemicals, or both, can also cause cancer. Epidemiologists estimate that about 15% of human cancers arise from viral infections.

▍▍ HUMAN CANCER VIRUSES

After many years of research and testing, we now know of at least six viruses that are associated with human cancers. There probably are many more yet to be identified.

The Epstein-Barr virus (EBV) perhaps is the best understood of the human cancer viruses. This DNA virus is a herpesvirus that was first discovered in African children suffering from Burkitt's lymphoma, a malignant tumor that causes swelling and eventual destruction of the jaw (◄Figure 23.19, p. 741). In fact, evidence points to three other tumors also associated with EBV.

Several of the human papillomaviruses (HPV) have shown a strong correlation with some human cancers. Although some of these DNA viruses cause only benign warts, other types (HPV-8 and HPV-16) lead to a *carcinoma* (neoplasm of epithelial tissues) of the uterine cervix. Literally 99.7% of all cases of cervical cancer are caused by HPV, and are sexually transmitted. Another potential cancer-causing DNA virus is hepatitis B virus (HBV). It causes inflammation of the liver and leads to 80% of all liver cancer. *Kaposi's sarcoma*, a cancer of the endothelial cells of the blood vessels or lymphatic system, is associated with human herpesvirus 8.

The major human cancer viruses discovered so far are dsDNA viruses. However, some (+) sense RNA viruses, specifically the retroviruses, are also associated with cancers; for example, HTLV-I causes *adult T cell leukemia/lymphoma*.

HOW CANCER VIRUSES CAUSE CANCER

Like bacteriophages, some animal viruses that infect animal cells often cause cell death through cell lysis. Other animal viruses can infect cells and form proviruses. In some cases these infections result in physical and genetic changes to the host cells—the CPE discussed earlier. For example, RSV causes cells in culture to detach themselves from the culture flask and round up (Figure 10.22b). In the case of **DNA tumor viruses**, which can exist as proviruses, the major CPE is the uncontrollable division of the infected cells. This process, called **neoplastic transformation**, is typical of DNA tumor viruses. Many insert all or part of their DNA at random sites into the host DNA. However, only a few of these viral genes are necessary for transformation.

The papillomaviruses (family Papovaviridae) that cause human cancers infect cells, but their viral DNA remains free in the cytoplasm of the host (**Figure 10.26**). A few genes of the papillomavirus are active so that the virus can replicate with each cell division. Should the viral DNA accidentally integrate into the host cell DNA, unregulated replication of viral proteins can occur. These proteins cause host cells to divide uncontrollably. Some of these viral proteins block the effects of tumor-suppressor genes, which prevent uncontrolled cell divisions. Without the products of these genes, the host experiences uncontrolled cell divisions—and a tumor develops.

Many of the retroviruses are **RNA tumor viruses**. Recall that retroviruses use their own reverse transcriptase to transcribe (+) sense RNA into DNA that then integrates as a provirus into the host chromosome. The provirus of HTLV-I codes for proteins that transform the host cells into neoplastic ones. Infection also leads to the production of new virions by budding, which does not kill the infected cell. Thus, RNA tumor viruses can continue to infect other uninfected cells or sex cells. In the latter case, the presence of virus particles ensures transmission of virions to offspring.

ONCOGENES

The proteins produced by tumor viruses that cause uncontrolled host cell division come from segments of DNA called **oncogenes** (*onco*, Greek for "mass"). In DNA tumor-causing viruses, not only do oncogenes cause a neoplasm but they also contain the information for synthesizing viral proteins needed for viral replication. The oncogenes in RNA tumor viruses are quite different. Virologists and cell biologists have shown that

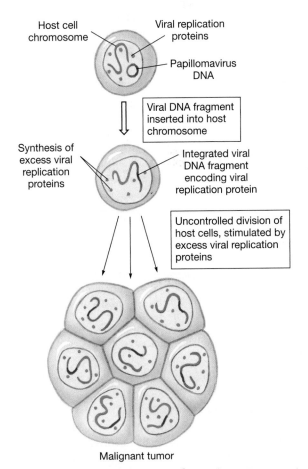

Figure 10.26   Malignant tumor formation. This particular tumor is caused by a papillomavirus (DNA tumor virus). Integration of the provirus causes synthesis of viral replication proteins that promote host cell divisions, leading to cancer.

some RNA tumor viruses pick up "extra" genes from normal host cells during viral replication. These genes, which are similar to oncogenes, are called proto-oncogenes. A **proto-oncogene** is a normal gene that, when under the control of a virus, can cause uncontrolled cell division; that is, it can act as an oncogene. Such oncogenes carried by these viruses are not needed for virus replication.

Many oncogenes have been discovered in oncogenic viruses, and most code for information leading to unlimited cell divisions. Such oncogenes are mutant genes containing deletions or substitutions (◀Chapter 7, p. 196). These mutations cause structural changes in the proteins for which the genes code. Such oncogenes work in one of two ways: (1) The product of the oncogene can disrupt normal cell function, leading to cell divisions. (2) The oncogene is controlled by viral regulators near the site of their integration into the host cell's chromosome. These regulators "turn on" the gene so that normal protein is made—but in excessive amounts or at the wrong time in the host cell's life. Again, excessive cell divisions occur. The discovery of oncogenes in viruses has had a major impact on our understanding of cancer. Although there is still much to be learned about cancer in humans, perhaps in the future effective antiviral drugs will prevent virus-induced cancers. Inhibitory RNA (RNAi) can also be used to turn off specific genes.

Following genomic sequencing it is now possible, using advanced technology called "microarrays," to determine which genes are turned on in a piece of normal tissue compared to those turned on in cancerous tissue. One can also see which drugs turn off specific genes. This should lead to rapid advances in cancer research.

▮ RETRACING OUR STEPS

▮▮▮ GENERAL CHARACTERISTICS OF VIRUSES

WHAT ARE VIRUSES?
- **Viruses** are submicroscopic **obligate intracellular parasites**—they replicate only inside a living host cell.

COMPONENTS OF VIRUSES
- Viruses consist of a nucleic acid core and a protein **capsid**. Some viruses also have a membranous **envelope**.
- Viral genetic information is contained in either DNA or RNA—not both.
- Capsids are made of subunits called **capsomeres**.
- A viral capsid and genome form a **nucleocapsid**. Such viruses are called **naked viruses**; those with a nucleocapsid surrounded by an envelope are **enveloped viruses**.

SIZES AND SHAPES
- Viruses have polyhedral, helical, binal, bullet, or complex shapes and vary in size from 20 to 300 nm in diameter.

HOST RANGE AND SPECIFICITY OF VIRUSES
- Viruses vary in **host range** and **viral specificity**. Many viruses infect a specific kind of cell in a single host species; others infect several kinds of cells, several hosts, or both.

ORIGINS OF VIRUSES
- Viruses have arisen and probably continue to arise, by multiple origins.
- Viruses act as agents of evolution by their participation in lateral gene transfer.

LATENT VIRAL INFECTIONS
- All herpesviruses have the ability to become latent, remaining dormant. Activation usually involves changes in cell-mediated immunity.

▮▮▮ CLASSIFICATION OF VIRUSES
- Viruses are classified by the nucleic acid (DNA or RNA) they contain, other chemical and physical properties, their mode of replication, shape, and host range. Some of these characteristics are summarized in Tables 10.1 and 10.2.
- Similar viruses are grouped into genera, and genera are grouped into families. Viruses that share the same genome and relationships with organisms generally constitute a viral species.

RNA VIRUSES
- Among the (+) sense RNA virus families are the Picornaviridae, which include the poliovirus, the hepatitis A virus, and the rhinoviruses; the Togaviridae, which include the virus that causes rubella; the Flaviviridae, which include the yellow fever virus; and the Retroviridae, which cause some cancers and AIDS. The (−) sense RNA viruses include the Paramyxoviridae, which cause measles, mumps, and several respiratory disorders; the Rhabdoviridae, one of which causes rabies; the Orthomyxoviridae, which include the influenza viruses; the Filoviridae, which cause Marburg and Ebola diseases; the Arenaviridae, which cause Lassa fever; and the Bunyaviridae, one of which causes hantavirus pulmonary syndrome. The double-stranded (ds) RNA virus families include the Reoviridae, which cause a variety of upper-respiratory and gastrointestinal infections.

DNA VIRUSES
- The dsDNA virus families include the Adenoviridae, some of which cause respiratory infections; the Herpesviridae, which cause oral and genital herpes, chickenpox, shingles, and infectious mononucleosis; the Poxviridae, which cause smallpox and similar, milder infections. The Papovaviridae cause warts; some papovaviruses are associated with certain cancers. The Hepadnaviridae cause human hepatitis B infection, and the Parvoviridae cause relatively rare human infections.

▮▮▮ EMERGING VIRUSES
- Many emerging diseases are caused by viruses which had been endemic at low levels in localized areas, but which have "jumped" species and acquired a new host range and spread;

sometimes due to human activities as well, e.g., colonizing previously uninhabited jungles.

VIRAL REPLICATION

GENERAL CHARACTERISTICS OF REPLICATION

• Viruses generally go through five steps in the replication process: **adsorption**, **penetration**, **synthesis**, **maturation**, and **release**. These steps are somewhat different in bacteriophages and animal viruses.

REPLICATION OF BACTERIOPHAGES

• **Bacteriophage** replication has been thoroughly studied in T-even phages, which are **virulent phages**.
• **Phage therapy** may replace antibiotics.
• T-even phages have recognition factors that attach to specific receptors on bacterial cell walls during adsorption. Enzymes weaken the bacterial wall so viral nucleic acid can penetrate it.
• During biosynthesis, viral DNA directs the making of viral components.
• In the maturation stage, the viral components are assembled into complete virions.
• Release, the final stage, is facilitated by the enzyme lysozyme. **Burst time** is the time from adsorption to release of progeny virions; **burst size** is the number of phage progeny released from one host cell.
• The growth curve of a phage includes an **eclipse period** (the time following penetration through biosynthesis) and a **latent period** (the time after penetration up to release).
• The number of phages produced in an infection can be determined by counting the number of **plaques** produced on a plate of virus-infected bacteria (**plaque assay**). Each plaque represents a **plaque-forming unit**.
• Phages carrying out these stages of replication, leading to host cell destruction, represent a **lytic cycle** of infection.

LYSOGENY

• **Lysogeny**, a stable, long-term relationship between certain phages and host bacteria, occurs in **temperate phages**. Temperate phage DNA can exist as a **prophage** or revert through **induction** to a lytic cycle.
• Prophages such as the lambda (λ) phage insert into a bacterial chromosome at a specific location.

REPLICATION OF ANIMAL VIRUSES

• On the surface of some viruses, proteins are used for attachment to host plasma membranes during adsorption; animal viruses thus gain entry into the cell. **Uncoating** (loss of the capsid) occurs at the plasma membrane or in the cytoplasm.
• Synthesis and maturation differ in DNA and RNA viruses. In most DNA viruses, DNA is synthesized in an orderly sequence in the nucleus, and proteins are synthesized in the cytoplasm of the host cell. In RNA viruses, RNA can act as a template for protein synthesis, for making mRNA, or for making DNA by reverse transcription. Virions are assembled in the cell; sometimes viral DNA is incorporated as a **provirus** into the host cell chromosome.
• Release can occur through direct lysis of the host cell or by budding through the host membrane.

CULTURING OF ANIMAL VIRUSES

DEVELOPMENT OF CULTURING METHODS

• The discovery of antibiotics to prevent bacterial contamination of chicken embryos and **cell cultures** and the use of trypsin to separate cells in culture systems into **monolayers** provided an important impetus to the study of virology.

TYPES OF CELL CULTURES

• **Primary cell cultures** come directly from animals and are not subcultured.
• All the cells in a **cell strain**, which are derived from subcultured primary cell cultures, are very similar. **Diploid fibroblast strains** from primary cultures of fetal tissues produce stable cultures that can be maintained for years; they are used to produce vaccines.
• **Continuous cell lines**, usually derived from cancer cells, grow in the laboratory without aging, can divide repeatedly, have greatly reduced nutritional needs, and display heteroploidy.
• The visible effects that viruses produce in infected host cells are collectively called the **cytopathic effect (CPE)**.

VIRUSES AND TERATOGENESIS

• A **teratogen** is an agent that induces defects during embryonic development.
• Viruses can act as teratogens by crossing the placenta and infecting embryonic cells. The earlier in pregnancy an infection occurs, the more extensive damage is likely to be.
• The rubella virus can be responsible for the death of fetuses and severe birth defects in others; cytomegaloviruses and occasionally herpesviruses also act as teratogens.

VIRUSLIKE AGENTS: SATELLITES, VIROIDS, AND PRIONS

SATELLITES

• **Satellites** are small RNA molecules unable to replicate without an unrelated helper virus. There are two types: **satellite viruses** which code for their own capsid protein, and **satellite nucleic acids** (= **virusoids**) whose helper virus encodes their capsid. Most satellites are associated with plant viruses.

VIROIDS

• **Viroids** are very different from viruses; each viroid is solely a small RNA molecule.
• Viroids may cause plant diseases by interfering with mRNA processing.

PRIONS

• **Prions** are infectious particles made of protein. Research indicates that prions are normal proteins that become folded incorrectly.
• Prions cause neurological degenerative diseases, including Creutzfeldt-Jakob disease, kuru, scrapie, mad cow disease, and chronic wasting disease.

VIRUSES AND CANCER

• **Cancer** is generally an uncontrolled and/or invasive growth of abnormal cells.

III HUMAN CANCER VIRUSES

• **Tumors**, or **neoplasms**, are **benign** (noncancerous) or **malignant** (cancerous). Malignant tumors spread by **metastasis**.

• Several animal viruses are thought to cause some forms of cancer, including the Epstein-Barr virus, certain human papillomaviruses, the hepatitis B virus, and some retroviruses, such as HTLV-1.

HOW CANCER VIRUSES CAUSE CANCER

• **DNA tumor viruses** contain viral genes whose protein products disrupt the activities of normal host cell proteins that control cell division.

• **RNA tumor viruses** contain viral genes used for **neoplastic transformation** and viral replication.

ONCOGENES

• **Oncogenes** are viral genes that cause host cells to divide uncontrollably.

• **Proto-oncogenes** are normal genes that, when under the control of a virus, act as oncogenes causing uncontrolled cell division.

• Oncogenes in RNA tumor viruses produce proteins in excessive amounts or produce proteins at the wrong times. In either case, infected host cells start uncontrolled cell division.

▌ TERMINOLOGY CHECK

adenovirus *(p. 281)*
adsorption *(p. 287)*
arenavirus *(p. 281)*
bacterial lawn *(p. 289)*
bacteriophage *(p. 274)*
benign *(p. 302)*
bunyavirus *(p. 281)*
burst size *(p. 289)*
burst time *(p. 289)*
cancer *(p. 302)*
capsid *(p. 272)*
capsomere *(p. 273)*
cell culture *(p. 296)*
cell strain *(p. 297)*
complex virus *(p. 274)*
continuous cell line *(p. 297)*
cytopathic effect (CPE)
  (p. 297)
diploid fibroblast strain *(p. 297)*
DNA tumor virus *(p. 303)*
eclipse period *(p. 289)*
emerging virus *(p. 283)*
enterovirus *(p. 278)*
envelope *(p. 272)*
enveloped virus *(p. 273)*
filoviruses *(p. 281)*
flavivirus *(p. 280)*

genome *(p. 273)*
glycoprotein *(p. 273)*
hepadnavirus *(p. 283)*
Hepatitis delta virus (HDV)
  (p. 299)
hepatovirus *(p. 279)*
herpesvirus *(p. 281)*
host range *(p. 275)*
induction *(p. 291)*
latency *(p. 282)*
latent period *(p. 289)*
lysogen *(p. 290)*
lysogenic conversion *(p. 290)*
lysogenic cycle *(p. 291)*
lysogeny *(p. 290)*
lytic cycle *(p. 289)*
lytic phage *(p. 289)*
malignant *(p. 302)*
maturation *(p. 289)*
metastasize *(p. 302)*
monolayer *(p. 296)*
naked virus *(p. 273)*
negative (−) sense RNA
  (p. 278)
neoplasm *(p. 302)*
neoplastic transformation
  (p. 303)

nucleocapsid *(p. 273)*
obligate intracellular
  parasite *(p. 272)*
oncogene *(p. 303)*
orthomyxovirus *(p. 280)*
papovavirus *(p. 283)*
paramyxovirus *(p. 280)*
parvovirus *(p. 283)*
penetration *(p. 287)*
phage therapy *(p. 286)*
picornavirus *(p. 278)*
plaque *(p. 290)*
plaque assay *(p. 289)*
plaque-forming unit *(p. 290)*
positive (+) sense RNA
  (p. 278)
poxvirus *(p. 282)*
primary cell culture *(p. 297)*
prion *(p. 300)*
prophage *(p. 290)*
proto-oncogene *(p. 304)*
provirus *(p. 280)*
release *(p. 289)*
reovirus *(p. 281)*
replication curve *(p. 289)*
replication cycle *(p. 286)*
retrovirus *(p. 280)*

reverse transcriptase *(p. 280)*
rhabdovirus *(p. 280)*
rhinovirus *(p. 279)*
RNA tumor virus *(p. 303)*
satellite virus *(p. 299)*
satellite nucleic acid *(p. 299)*
spike *(p. 273)*
subculturing *(p. 296)*
syncytia *(p. 298)*
synthesis *(p. 289)*
temperate phage *(p. 290)*
teratogen *(p. 298)*
teratogenesis *(p. 298)*
tissue culture *(p. 296)*
togavirus *(p. 280)*
TORCH series *(p. 299)*
tumor *(p. 302)*
uncoating *(p. 294)*
viral specificity *(p. 275)*
viral yield *(p. 289)*
virion *(p. 272)*
viroid *(p. 299)*
virulent (lytic) phage
  (p. 289)
virus *(p. 272)*
virusoid *(p. 299)*

▌ CLINICAL CASE STUDY

Can Koch's Postulates be fulfilled for all viral pathogens? What steps might present difficulties for viral pathogens? For some scientists Koch's Postulates still have not been sufficiently fulfilled to say that HIV causes the disease syndrome of AIDS. Visit this web site http://www.niaid.nih.gov/publications/hivaids/12.htm and explore more about this issue.

▌ CRITICAL THINKING QUESTIONS

1. One might expect that viruses, being so simple, would be quite easy to destroy. Yet many of the disinfectants, antiseptics, and antibiotics that effectively destroy bacteria fail to destroy viruses. How can that be?

2. The study of viruses has been greatly advanced through the development of modern cell culturing techniques.

What two discoveries led to improvements in cell culturing techniques?

3. It was once stated that "the death of the host is a result as harmful to the virus's future as to that of the host itself." Explain the significance of this statement.

▌ SELF-QUIZ

1. Viruses range in size from the smallest known _____ group, which are less than 30 nm, to the largest, which are approximately _____ nm, or about the size of the smallest bacteria.

2. Match the following viral structures to their descriptions:

___Capsid   (a) Surrounding lipid bilayer membrane
___Virion   (b) Complete virus particle, including
___Spike   envelope if it has one
___Envelope   (c) Surrounding protein coat
___Naked virus (d) Projection made of glycoprotein that
___Nucleocapsid   serves to attach virions to specific re-
               ceptor sites
           (e) Virion's genome together with capsid
           (f) Virus with a nucleocapsid but no
               envelope

3. A chemical component that is found in all viruses is:
(a) Protein   (d) RNA
(b) Lipid   (e) Glycoprotein
(c) DNA

4. A common polyhedral capsid shape of viruses is a:
(a) Pentagon   (d) Pyramid
(b) Cube   (e) Sphere
(c) Icosahedron

5. Viruses are capable of infecting all life forms; some can even infect members of different kingdoms but most are limited to only one host and to only specific cells and/or tissues of that host. True or false?

6. Enteroviruses differ from rhinoviruses mainly in their:
(a) Type of nucleic acid
(b) Size
(c) Capsid shape
(d) Ability to survive acidic conditions
(e) Strandedness

7. Which of the following properties do viruses have in common with the bacterial section containing Rickettsiae and Chlamydiae?
(a) They are both the same size.
(b) They both have RNA strands for their genomes.
(c) They are both obligate intracellular parasites.
(d) They both contain enzymes for glucose metabolism.
(e) None of the above.

8. Viruses that can remain latent (usually in neurons) for many years are most likely:
(a) Togaviruses   (d) Rhinoviruses
(b) Herpesviruses (e) Retroviruses
(c) Enteroviruses

9. What type of viruses contain the enzyme lysozyme to aid in their infection?
(a) Bacteriophages (d) Fungal viruses
(b) Animal viruses   (e) Human viruses
(c) Plant viruses

10. Viruses that infect bacteria are called:
(a) Satellites   (d) Bacteriophages
(b) Bacteriocins   (e) Bacterioviruses
(c) Delta hepatitis

11. Bacteriophages are readily counted by the process of:
(a) Immunoassays   (d) Tissue cell culture
(b) ELISA   (e) Electron microscopy
(c) Plaque assays

12. A type of cell culture that can reproduce for an extended number of generations and is used to support viral replication is a:
(a) Primary cell culture (d) Diploid fibroblast cell
(b) Continuous cell line (e) Connective tissue
(c) Cell strain

13. Which of the following is not a DNA virus?
(a) Adenovirus   (d) Herpesvirus
(b) Poxvirus   (e) Orthomyxovirus
(c) Papovavirus

14. All of the following are true about retroviruses EXCEPT?
(a) Retroviruses cause tumors and leukemia in rodents, birds, and humans.
(b) Retroviruses cause acquired immune deficiency syndrome (AIDS) in humans.
(c) Retroviruses contain reverse transcriptase to form a complementary strand of DNA which is then replicated to form double-stranded DNA (dsDNA).
(d) dsDNA must migrate to the cell nucleus and integrate into the chromosomes of the host whereby it becomes a provirus.
(e) Retroviruses have two complete copies of (−) sense RNA.

15. Match the following general replication steps to their description and place them in order:

Step #
___, ___ Release   (a) Host metabolic machinery is
___, ___ Adsorption   used to produce new nucleic acid
___, ___ Maturation   molecules, capsid proteins, and
___, ___ Penetration   other viral components
___, ___ Synthesis   (b) Entry of virion genome into the
                host cell
           (c) Attachment of viruses to host cell
           (d) Departure of new virions from host cell, generally with lysis of host cell
           (e) Assembly of newly synthesized viral components into complete virions

16. Bacteriophages that can enter into stable, long-term relationships with their hosts are called:
(a) Lytic phages   (d) Lazy phages
(b) Defective phages (e) Temperate phages
(c) Virulent phages

17. The positive (+) strand RNA of certain viruses does not act as a message but becomes converted into DNA and integrated into the host cellular DNA. These viruses are:
(a) Rhinoviruses   (d) Reoviruses
(b) Enteroviruses   (e) Picornaviruses
(c) Retroviruses

18. The replication of animal viruses differs from the replication of bacteriophages in what way?

(a) Once in the host cell, animal viruses undergo a process of "uncoating" whereby the viral genome is separated from its protein coat by proteolytic enzymes; the viral genome in a bacteriophage is ready to go once injected into the bacterial host cell.

(b) Compared to bacteriophage replication, synthesis in animal virus replication can take much longer.

(c) In the penetration stage, bacteriophages produce lysozyme to weaken the bacterial cell wall and inject their DNA through the tail core into the bacterial cell, whereas animal viruses either fuse their envelope with the host's plasma membrane or enter by endocytosis.

(d) The maturation stage in animal enveloped viruses is longer than that of bacteriophage replication.

(e) All of the above are ways in which animal viruses differ from bacteriophage replication.

19. In what way do animal viruses differ from each other?

(a) All animal viruses are RNA viruses.

(b) The RNA in RNA viruses can have different functions depending on the type of virus, whereby it can be used as a template for protein synthesis, mRNA production, or DNA production.

(c) Depending on the type of virus, release can occur through either lysis or by budding through the host membrane.

(d) DNA viruses differ from RNA viruses in that the genomes of DNA viruses always incorporate as a provirus into the host cell's chromosome.

(e) Both b and c.

20. The development of modern cell culturing techniques has greatly advanced our understanding of viral biology. Match the following cell culture terminology with their descriptions:

D Continuous cell line

C Primary cell culture

e Monolayer

f Subculturing

a Cytopathic effect

b Diploid fibroblast strain

(a) Visible effects viruses produce in infected host cells.

(b) Immature, fetally derived cell type that rapidly divides numerous times and supports a wide range of viruses

(c) Cells that come directly from the animal, have very few cell divisions, but support the growth of a wide variety of viruses

(d) Immortalized cells that reproduce for an extended number of generations

(e) Trypsinized and washed cells that attach to plastic surfaces where they multiply and spread to form sheets one cell thick

(f) Cells from an existing cell culture are transferred to new containers with fresh nutrient media

21. Creutzfeldt-Jakob disease (CJD), kuru, scrapie, and mad cow disease are caused by:

(a) Viroids

(b) Retroviruses

(c) DNA viruses

(d) Prions

(e) RNA viruses

22. The human virus that has been associated with Burkett's lymphoma (a malignant tumor of the jaw) is:

(a) Cytomegalovirus

(b) Human papilloma virus

(c) Retroviruses

(d) Epstein-Barr virus

(e) Enterovirus

23. Match the following viruslike agents to their descriptions:

B Prion

A Satellite

D Virusoid

C Satellite viruses

f Viroids

e Delta hepatitis virus

(a) Small, single-stranded RNA virus lacking genes required for its replication, and needing a helper virus

(b) Infectious, incorrectly folded protein

(c) Code for their own capsid protein

(d) Helper virus codes for its capsid

(e) Similar to viroids and virusoids, a defective pathogen requiring the presence of hepatitis B virus for its replication

(f) Infectious RNA particle smaller than a virus

24. Viruses that can induce defects during embryonic development (teratogenesis) in human are:

(a) Herpes simplex virus types I and II

(b) Rubella

(c) Rhinovirus

(d) Cytomegalovirus

(e) a, b, and d

25. Which is NOT true regarding viruses and cancer?

(a) An estimated 15% of human cancers arise from viral infections.

(b) Cancers can be caused by both RNA tumor viruses and DNA tumor viruses.

(c) Oncogenes are made up of DNA.

(d) All neoplasms are malignant.

(e) Examples of human cancers believed to arise from viral infections include: Kaposi's sarcoma, adult T cell leukemia/lymphoma, and cervical cancer.

26. Identify parts (a) through (c) of the following phage growth curve, and describe what happens in each phase.

▌ EXPLORATIONS ON THE WEB

If you think you've mastered this chapter, there's more to challenge you on the web. Go to the companion web site to fine-tune your understanding of the chapter concepts and discover answers to the questions posed below.

1. What strange rituals did certain Fore tribes of Papua, New Guinea practice that resulted in more than 2,500 cases of Kuru disease, an occurrence rate of at least 1% of the population, and why were most of the cases found in women and children?

2. How could Creutzfeldt-Jakob disease be an inherited disorder?

3. How might bacteriophage be used to treat disease or prevent disease.

Eukaryotic Microorganisms and Parasites

Come with me...
Look at What's on Me! Get it OFF!!!

Ticks cannot see or hear. Instead, organs at the tips of their front legs detect heat, carbon dioxide, and vibrations to help them find hosts. During an average 2- to 4-year life span, a female tick can survive on as little as three big blood meals. If not interfered with, the meal can last for about 1 week. Males take several smaller meals. Chemicals in the ticks' saliva prevent itching at the bite site, thus allowing them to feed for a long time before being discovered.

Aha! So, once you have found that feeding tick, what is the best way to get rid of it? The mouthparts (proboscis) are held firmly embedded in the skin by little hooks. If you cover the tick with chemicals such as nail polish remover or petroleum jelly that prevent it from breathing, the tick will struggle and force microbe-containing saliva out into the bite. Using narrow forceps, grasp the tick just behind where it enters the skin. Do not squash or crush the main body of the tick, as this will also force microbe-laden fluids into the bite. Pull back slowly and carefully until the entire proboscis is out. Broken pieces left behind in the bite may fester and become infected. Save the tick for identification, perhaps in a small amount of rubbing alcohol. Flush extra ticks down the toilet, or burn them. They will just crawl out of the trash can if you don't! Crushing them with your fingernails will release microbes onto your fingers. And hard-shelled ticks are difficult to crush!

(© Anthony Bannister/Gallo Images/Corbis)

 Video related to this topic is available within WileyPLUS.

In our survey of microbes, we have devoted significant attention to bacteria of the kingdom Monera and to viruses. However, some members of eukaryotic kingdoms are also of interest to microbiologists, ecologists, and health scientists. The kingdoms Protista and Fungi contain large numbers of microscopic species, some of which supply food and antibiotics, and some of which cause disease. The kingdom Animalia contains helminths that cause disease and arthropods that cause or transmit diseases. Studying the microscopic eukaryotes, as well as the helminths and arthropods, constitutes a significant part of a health scientist's training. Unless health scientists take a course in parasitology, their only opportunity to learn about helminths and arthropods is in conjunction with the study of microscopic infectious agents.

PRINCIPLES OF PARASITOLOGY

A **parasite** is an organism that lives at the expense of another organism, called the **host**. Parasites vary in the degree of damage they inflict on their hosts. Although some cause little harm, others cause moderate to severe damage. Parasites that cause disease are called **pathogens**. **Parasitology** is the study of parasites.

Although few people realize it, among all living forms, there are probably more parasitic than nonparasitic organisms. Many of these parasites are microscopic throughout their life cycle or at some stage of it. Historically, in the development of the science of biology, parasitology came to refer to the study of protozoa, helminths, and arthropods that live at the expense of other organisms. We will use the term *parasite* to refer to these organisms. Strictly speaking, bacteria and viruses that live at the expense of their hosts also are parasites.

The manner in which parasites affect their hosts differs in some respects from that described in earlier chapters for bacteria and viruses. Special terms also are used to describe parasites and their effects. This introduction to parasitology will make discussions of parasites here and in later chapters more meaningful.

THE SIGNIFICANCE OF PARASITISM

Parasites have been a scourge throughout human history. In fact, even with modern technology to treat and control parasitic diseases, there are more parasitic infections than there are living humans. It has been estimated that among the 60 million people dying each year, fully one-fourth die of parasitic infections or their complications.

Parasites play an important, though negative, role in the worldwide economy. For example, less than half the world's cultivable land is under cultivation, primarily because parasites endemic to (always present in) those lands prevent humans and domesticated animals from inhabiting some of them. As the world population increases, and the need for food with it, cultivation of such lands will become more important. In some inhabited regions, many people are near starvation and severely debilitated by parasites. Furthermore, parasitic infections in wild and domestic animals provide sources of human infection and cause debilitation and death among the animals, thus preventing the raising of cattle and other animals for food. Given the many human problems created by parasites, all citizens—and especially health scientists—need to understand the problems associated with the control and treatment of parasitic diseases.

PARASITES IN RELATION TO THEIR HOSTS

Parasites can be divided into **ectoparasites**, such as ticks and lice, which live on the surface of other organisms, and **endoparasites**, such as some protozoa and worms, which live within the bodies of other organisms. Most parasites are **obligate parasites**: They must spend at least some of their life cycle in or on a host. For example, the protozoan that causes malaria invades red blood cells. A few parasites are **facultative parasites**: They normally are free-living, such as some soil fungi, but they can obtain nutrients from a host, as many fungi do when they cause skin infections. Hosts that are invaded by parasites usually lack effective defenses against them, so such diseases can be serious and sometimes fatal.

A single tapeworm can live for 30 to 35 years. Its pear-shaped head is about 1 to 2 mm in diameter, and it may reach a length of 10 meters.

Parasites are also categorized according to the duration of their association with their hosts. **Permanent parasites**, such as tapeworms, remain in or on a host once they have invaded it. **Temporary parasites**, such as many biting insects, feed on and then leave their hosts. **Accidental parasites** invade an organism other than their normal host. Ticks that ordinarily attach to dogs or to wild animals sometimes attach to humans; the ticks are then accidental parasites. **Hyperparasitism** refers to a parasite itself having parasites. Some mosquitoes, which are temporary parasites, harbor the malaria parasite or other parasites. Such insects serve as **vectors**, or agents of transmission, of many human parasitic diseases.

An organism that transfers a parasite to a new host is a vector. A vector in which the parasite goes through part of its life cycle is a **biological vector**. The malaria mosquito is both a host and a biological vector. A **mechanical vector** is a vector in which the parasite does not go through any part of its life cycle during transit. Flies that carry parasite eggs, bacteria, or viruses from feces to human food are mechanical vectors.

Hosts are classified as **definitive hosts** if they harbor a parasite while it reproduces sexually; they are said to be **intermediate hosts** if they harbor the parasite during some other developmental stages. The mosquito is the definitive host for the malaria parasite because that parasite reproduces sexually in the mosquito; the human is an intermediate host, even though humans suffer greater damage from the parasite. **Reservoir hosts** are infected organisms that make parasites available for transmission to other hosts. Reservoir hosts for human parasitic diseases typically are wild or domestic animals. **Host specificity** refers to the range of different hosts in which a parasite can mature. Some parasites are quite host specific—they mature in only one host. The malaria parasite matures primarily in *Anopheles* mosquitoes. Other parasites can mature in many different hosts. The worm that causes trichinosis can mature in almost any warm-blooded animal, but the parasite is most often acquired by humans from pigs through the consumption of inadequately cooked, contaminated pork.

Over thousands of years of evolution, parasites tend to become less injurious to their hosts. Such an arrangement preserves the host so that the parasites are guaranteed a continuous supply of nutrients. A parasite that destroys its host also destroys its own means of support. The adjustment of parasites and hosts to each other is closely related to the host's defense mechanisms. Many parasites have one or more of the following mechanisms for evading host defense mechanisms:

1. *Encystment*, the formation of an outer covering that protects against unfavorable environmental conditions. These resistant cyst stages also sometimes provide a site for internal reorganization of the organism and cell division, help attach a parasite to a host, or serve to transmit a parasite from one host to another.

2. Changing the parasite's surface antigens (molecules that elicit immunity) faster than the host can make new antibodies (molecules that recognize and attack antigens).

3. Causing the host's immune system to make antibodies that cannot react with the parasite's antigens.

4. Invading host cells, where the parasites are out of reach of host defense mechanisms.

When parasites successfully evade host defenses, they can cause several kinds of damage. All parasites rob their hosts of nutrients. Some take such a large share of nutrients or damage so much surface area of the host's intestines that the host receives too little nourishment. Many parasites cause significant trauma to host tissues. They cause open sores on the skin, destroy cells in tissues and organs, clog and damage blood vessels, and may even cause internal hemorrhages. Parasites that do not evade defense mechanisms sometimes trigger severe inflammatory and immunological reactions. For example, treatment to rid human hosts of some worm infections effectively kills the worms, but toxins from the dead worms cause more tissue damage than do the living parasites. The dog heartworm, *Dirofilaria immitis*, perforates the heart wall and leaves holes in the heart when the worms die and decay. Therefore, it is important for a veterinarian to test all dogs for the presence of heartworms before administering preventive heartworm medication.

A hallmark of many parasites is their reproductive capability. Parasitism, although an easy life once the parasite is established, is a hazardous existence during transfers from one host to another. For example, many parasites that leave the human body through feces die from desiccation (drying out) before they reach another host. If several hosts are required to complete the life cycle, the hazards are greatly multiplied. Consequently, many parasites have exceptional reproductive capacities. Some parasites, such as certain protozoa, undergo **schizogony** (skiz-og'one), or multiple fission, in which one cell gives rise to many cells, all of which are infective. Others, such as various worms, produce large numbers of eggs. Some worms are **hermaphroditic**—that is, one organism has both male and female reproductive systems and both are functional. In fact, certain worms, such as tapeworms, lack a digestive tract and consist almost exclusively of reproductive systems.

PROTISTS

CHARACTERISTICS OF PROTISTS

The **protists**, members of the kingdom Protista, are a diverse assortment of organisms that share certain common characteristics. Protists are unicellular (though sometimes colonial), eukaryotic organisms with cells that have true nuclei and membrane-enclosed organelles. Although most protists are microscopic, they vary in diameter from 5 µm to 5 mm.

> *Protozoans are so tiny that some develop in the salivary glands of insects.*

THE IMPORTANCE OF PROTISTS

Protists have captured the fancy of biologists since Leeuwenhoek made his first microscopes. In fact, most

TRY IT

No More Pond Scum

A recent English discovery may keep your farm or garden ponds free of summertime algal scum. One hundred pounds of barley hay floated in a 1-acre pond (depth of water doesn't matter) will combat algae and keep it clear. A bale generally weighs about 25 pounds, costs $3–5, and lasts about 90 days. It should be wrapped in netting or chicken wire, and it may need empty, plastic container floats to keep it from sinking. Birds and turtles like to sun on it, and fish hide under it. Afterward it makes good compost.

But how does it accomplish its miracle? Possibly it acts like a giant teabag, releasing inhibitory chemicals into the water. What is the source of these chemicals? Could microbes be at work? Try to design experiments to investigate this phenomenon.

of the "animalcules" he observed were protists. Like Leeuwenhoek, many people find protists inherently interesting, and biologists have learned much about life processes from protists.

Protists also are important to humans for other reasons. For instance, they are a key part of food chains. Autotrophic protists capture energy from sunlight. Some heterotrophic protists ingest autotrophs and other heterotrophs. Others decompose, or digest, dead organic matter, which then can be recycled to living organisms. Protists also serve as food for higher-level consumers. Ultimately, some energy originally captured by protists reaches humans. For example, energy from the sun is transferred to protists, protists are eaten by oysters, and the oysters are eaten by humans.

Protists can be economically beneficial or detrimental. Certain protists have **tests**, or shells, of calcium carbonate. Carbonate shells deposited in great numbers by such protists that lived in ancient oceans formed the white cliffs of Dover, England and the limestone used in building the pyramids of Egypt. Because different test-forming protists gained prominence during different geological eras, the identification of the protists in rock layers helps determine the age of the rocks. Certain test-forming protists tend to occur in rock layers near petroleum deposits, so geologists looking for oil are pleased to find them. Some autotrophic protists produce toxins that do not harm the oysters that eat the protists, but the accumulated toxins can cause disease or even death in people who subsequently eat the oysters. Oyster beds infected with such protists can cause great economic losses to oyster harvesters. Other autotrophic protists multiply very rapidly in abundant inorganic nutrients and form a "bloom," a thick layer of organisms over a body of water. This process, called **eutrophication** (u″tro-fi-ka′shun), blocks sunlight, killing plants beneath the bloom and causing fish to starve. Microbes that decompose dead plants and animals use large quantities of oxygen, and the lack of oxygen leads to more deaths. Together these events result in great economic losses in the fishing industry.

Finally, some protists are parasitic. They cause debilitation in large numbers of people and sometimes death, especially in poor countries that lack the resources to eradicate those protists. Parasitic diseases caused by protozoa include amoebic dysentery, malaria, sleeping sickness, leishmaniasis, and toxoplasmosis. Together, these diseases account for severe losses in human productivity, incalculable human misery, and many deaths.

PUBLIC HEALTH

Red Tides

Certain species of *Gonyaulax*, *Pfiesteria piscicida*, and some other dinoflagellates produce toxins. When these marine organisms appear seasonally in large numbers, they cause a bloom known as a *red tide*. The toxins accumulate in the bodies of shellfish such as oysters and clams that feed on the protists. Although the toxin does not harm the shellfish, it causes paralytic shellfish poisoning in some fish, and in humans who eat the infected shellfish. Even animals as large as dolphins have been killed in large numbers by this toxin. Inhaling air that contains small quantities of the toxin can irritate respiratory membranes, so sensitive individuals should avoid the sea and its products during red tides.

(Bill Bachman/Photo Researchers)

In the last 30 years, the number of red tides worldwide has increased significantly—possibly linked to increased pollution.

CLASSIFICATION OF PROTISTS

Like all groups of living things, the protists display great variation, which provides a basis for dividing the kingdom Protista into sections and phyla. However, taxonomists do not agree about how these classifications should be made. We can accomplish our main purpose of illustrating diversity and avoid taxonomic problems by grouping protists according to the kingdom of macroscopic organisms they most resemble (**Table 11.1**). Thus, we speak of protists that resemble plants (**Figure 11.1**), protists that resemble fungi (**Figure 11.2**), and protists that resemble animals (**Figure 11.3**).

The Plantlike Protists

The plantlike protists, or algae, have chloroplasts and carry on photosynthesis. They are found in moist, sunny environments. Most have cell walls and one or two

TABLE 11.1

Properties of Protists		
Group	**Characteristics**	**Examples**
Plantlike protists	Have chloroplasts; live in moist, sunny environments	Euglenoids, diatoms, and dinoflagellates
Funguslike protists	Most are saprophytes; may be unicellular or multicellular	Water molds; plasmodial and cellular slime molds
Animal-like protists	Heterotrophs; most are unicellular, most are free-living, but some are commensals or parasites	Mastigophorans, sarcodines, apicomplexans, and ciliates

(a) SEM

(b) SEM

(c) SEM

Figure 11.1 **Representative algae, or plantlike protists.**
(a) *Euglena*, a euglenoid (895X). *(Carolina Biological Supply Company/Phototake)* **(b)** The diatom *Campylodiscus hibernicus* (250X). *(Andrew Syred/Photo Researchers, Inc.)* **(c)** *Gonyaulax*, a dinoflagellate that causes red tides (9,605X). *(David M. Phillips/ Visuals Unlimited)*

flagella, which allow them to move. The **euglenoids** (u-gle′noidz) usually have a single flagellum and a pigmented eyespot called a *stigma*. The stigma may orient flagellar movement so that the organism moves toward light. A typical euglenoid, *Euglena gracilis* **(Figure 11.1a)**, has an elongated, cigar-shaped, flexible body. Instead of a cell wall, it has a **pellicle**, or outer membranous cover. Euglenoids usually reproduce by binary fission. Most live in fresh water, but a few are found in the soil.

APPLICATIONS

Keep That Head on Your Beer!

Taxonomists disagree on how to classify eukaryotic algae. They are sometimes classified as protists, sometimes as plants, and sometimes divided between the protist and plant kingdoms. Eukaryotic algae should not be confused with blue-green algae (now called cyanobacteria), which are prokaryotes.

Although both eukaryotic and prokaryotic algae are important as producers in many environments, they are generally not of medical significance. (However, cyanobacteria of the genus *Prototheca*, which have lost their chlorophyll, have been reported to cause skin lesions.) Agar, which is of great importance in the microbiology laboratory, is a product extracted from smaller seaweeds (red algae). Some eukaryotic algae, such as kelps (brown algae), are used as food and in the manufacture of products such as cheese spreads, toothpaste, and mayonnaise, to which they add smoothness and spreadability. They also enable beer to retain a foamy "head."

Another group of plantlike protists have other pigments in addition to chlorophyll. These protists usually have cell walls surrounded by a loosely attached, secreted test that contains silicon or calcium carbonate. Most reproduce by binary fission. They include the **diatoms** (di′ah-tomz), which lack flagella **(Figure 11.1b)**, and several other groups, which have flagella and are distinguished by their yellow and brown pigments. Diatoms are an especially numerous group and are important as producers in both freshwater and marine environments. Fossil deposits of diatoms, known as diatomaceous earth, are used as filtering agents and abrasives in various industries.

The **dinoflagellates** (di′no-flaj″el-atz) are plantlike protists that usually have two flagella—one extending behind the organism like a tail, and the other lying in a transverse groove **(Figure 11.1c)**. They are small organisms that may or may not have a cell wall. Some have a *theca*, a tightly affixed, secreted layer that typically contains cellulose. Cellulose is an uncommon substance in protists, although it is abundant in plants. Whereas most dinoflagellates have chlorophyll and are capable of carrying on photosynthesis, others are colorless and feed on organic matter. Several dinoflagellates exhibit bioluminescence. The photosynthetic dinoflagellates are second only to the diatoms as producers (photosynthesizers) in marine environments.

The Funguslike Protists

The funguslike protists, or water molds and slime molds, have some characteristics of fungi and some of animals.

It's Not Just Fish That Are Damaged

Alone in the lab, late at night, searching for the cause of the deaths of over 1 billion fish in North Carolina estuaries, some scientists entered into their own horror stories. They were struck by toxins produced by *Pfiesteria piscicida*, a dinoflagellate whose numbers can suddenly skyrocket during an algal bloom. One toxin is water-borne, the other is airborne. By a twist of fate, the air from the laboratory where *Pfiesteria* was being grown was recycled by the air ducts into the scientists' next-door office.

This toxin wields a devastating blow to the immune system and higher level brain functions. As it starts, the eyes blur, there's difficulty catching one's breath, then hours of nausea and vomiting are followed by unwarranted fits of rage. After coming out of a narcotic-like delirium, the mind seems to be going, and bleeding lesions appear on the skin. The afflicted can't remember their name, can't form a sentence, can't read. Five to seven years later, nerve damage persists. Shortly after exposure to the toxin, 20 to 40% of the immune system is destroyed and remains impaired for years. *Pfiesteria* is now grown in closed containers, and scientists breathe through air packs. Local watermen tell stories of similar problems after driving a boat through areas of fish kill. Fish coming into contact with the toxins float belly-up within 30 seconds, flap disorientedly, and are dead within 1 minute.

(Courtesy Center for Applied Aquatic Ecology)

During the winter, *Pfiesteria* lives by eating algae, from which it takes intact chloroplasts, and then uses them to do photosynthesis when the supply of algae drops. A new term coined to describe this use of chloroplasts is "kleptochloroplasts," literally "stolen chloroplasts." When manure is dumped or runs off into embayed coastal waters, the excess nutrients cause an algal bloom, which attracts fish. The scent of fish feces causes the *Pfiesteria* to rise toward the water surface, changing en route from a peaceful amoeboid shape into one with "claws" that tear fish apart as the *Pfiesteria* feeds on the fish shreds. The attack usually begins at the fish's anus. *Pfiesteria* is the ultimate "shape-changer"—more than 20 stages and shapes in its life cycle are known to exist.

Water Molds. The **water molds** and related protists that cause mildew—the **Oomycota**—are sometimes classified as fungi. These molds, mildews, and plant blights produce flagellated spores, called *zoospores*, during asexual reproduction and large motile gametes during sexual reproduction. The most prominent phase of their life cycle consists of diploid cells from the union of gametes. These protists live freely in fresh water or as plant parasites; they cause such diseases as downy mildew on grapes and sugar beets and late blight in potatoes. A member of the

Oomycota was responsible for the Irish potato famine in the 1840s. With a few exceptions, water molds are not medically significant to humans. They do, however, cause disease in fish and other aquatic organisms.

Slime Molds. **Slime molds** are commonly found as glistening, viscous masses of slime on rotting logs; they also live in other decaying matter or in soil. Most slime molds are **saprophytes** (sap′ro-fitz), or organisms that feed on dead or decaying matter. A few are parasites of algae, fungi, or flowering plants, but not of humans. Slime molds occur as plasmodial slime molds and as cellular slime molds.

Plasmodial slime molds (Figure 11.2a) form a multinucleate, amoeboid mass called a **plasmodium**, which moves about slowly and phagocytizes dead matter. Sometimes a plasmodium stops moving and forms *fruiting bodies*. Each fruiting body develops *sporangia*, sacs that produce spores. When spores are released, they germinate into flagellated gametes. Two gametes fuse, lose their flagella, and form a new plasmodium. As a plasmodium feeds and grows, it can also divide and produce new plasmodia directly.

The **cellular slime molds (Figure 11.2b)** produce pseudoplasmodia, fruiting bodies, and spores with characteristics that are quite different from those of

(a)

(b)

Figure 11.2 Representative funguslike protists. **(a)** A plasmodial slime mold of the genus *Hemitrichia* on a decaying log. *(Dwight Kuhn Photography)* **(b)** Pseudoplasmodia of a cellular slime mold, *Dictyostelium discoideum* (93,583X). *(Cabisco/Visuals Unlimited)*

Figure 11.3 Representative protozoa, or animal-like protists. **(a)** *Trichonympha*, a mastigophoran, an endosymbiont from a termite gut. Particles seen inside the body are ingested wood particles (324X). *(Eric Grave/Photo Researchers, Inc.)* **(b)** *Amoeba proteus* (445X), a sarcodine, free-living inhabitant of ponds. *(Michael Abbey/Visuals Unlimited)* **(c)** *Plasmodium vivax* (inside red blood cells), an apicomplexan, one of the parasites that causes malaria (1,081X). *(Arthur M. Siegelman/Visuals Unlimited)* **(d)** *Paramecium caudatum* (171X), a ciliate. *(Michael Abbey/Visuals Unlimited)*

plasmodial slime molds. A **pseudoplasmodium** is a slightly motile aggregation of cells. It produces fruiting bodies, which in turn produce spores. The spores germinate into amoeboid phagocytic cells that divide repeatedly, producing more independent amoeboid cells. Depletion of the food supply causes the cells to aggregate into loosely organized new pseudoplasmodia.

The Animal-Like Protists

The animal-like protists, or **protozoa**, are heterotrophic, mostly unicellular organisms, but a few form colonies. Most are free-living. Some are **commensals**, which live in or on other organisms without harming them, and a few are parasites. The parasitic protozoa are of particular interest in the health sciences. Many protozoa live in watery environments and encyst when conditions are not favorable. Some protozoa are protected by a tough outer pellicle. Many are motile and are further classified on the basis of their means of locomotion (**Figure 11.3**). The protozoa that you will encounter in this book belong to the groups

Mastigophora, Sarcodina, Apicomplexa (also known as Sporozoa), or Ciliata (also known as Ciliophora).

Mastigophorans. The **mastigophorans** (mas″ti-gof′or-anz) have flagella. A few species are free-living in either fresh or salt water, but most live in symbiotic relationships with plants or animals. The symbiont *Trichonympha* (**Figure 11.3a**) lives in the termite gut and contributes enzymes that digest cellulose. Mastigophorans that parasitize humans include members of the genera *Trypanosoma*, *Leishmania*, *Giardia*, and *Trichomonas*. Trypanosomes cause African sleeping sickness, leishmanias cause skin lesions or systemic disease with fever, giardias cause diarrhea, and trichomonads cause vaginal inflammation. Leishmanias have been particularly a problem to troops in Iraq.

Amebozoa. The **amebozoa** (formerly called sarcodines) move by means of pseudopodia (**Figure 11.3b**) (◄Chapter 4, p. 103). A few amebozoa have flagella at some stage in their life cycle. They feed mainly on other microorganisms, including other protozoa and small algae. The amebozoa include foraminiferans and radiolarians, which have shells and are found mainly in marine environments, and amoebas, which have no shells and are typically parasites.

Numerous species of amoebas are capable of inhabiting the human intestinal tract. Most form cysts that help them withstand adverse conditions. The more commonly observed genera—*Entamoeba*, *Dientamoeba*, *Endolimax*, and *Iodamoeba*—cause amoebic dysenteries of varying degrees of severity. *Entamoeba gingivalis* is found in the mouth. *Dientamoeba fragilis*, which is unusual in that it has two nuclei and does not form cysts, is found in the large intestine of about 4% of the human population. Its means of transmission is unknown. Although usually considered a commensal, it can cause chronic, mild diarrhea.

Apicomplexans. The **apicomplexans** (or sporozoans) are parasitic and immobile (**Figure 11.3c**). Enzymes present in groups (complexes) of organelles at the tips (apices) of their cells digest their way into host cells, giving the group the name Apicomplexa. These parasites usually have complex life cycles. An important example is the life cycle of the malaria parasite, *Plasmodium*, which requires both a human and a mosquito host (**Figure 11.4**). (Do not

confuse this apicomplexan with the plasmodium form of slime molds.) The parasites, which are present as **sporozoites** (spo-ro-zo'itz) in the salivary glands of an infected mosquito, enter human blood through the mosquito's bite. The sporozoites migrate to the liver and become **merozoites** (meh-ro-zo'itz). After about 10 days, they emerge into the blood, invade red blood cells, and become **trophozoites** (tro-fo-zo'itz). Trophozoites reproduce asexually, producing many more merozoites, which are released into the blood by the rupture of red blood cells. Multiplication and release of merozoites is repeated several times during a bout of malaria. Some merozoites enter the sexual reproductive phase and become **gametocytes**, or male and female sex cells. When a mosquito takes a blood meal from an infected human, it also takes in gametocytes, most of which mature and unite to form zygotes in the lining of the mosquito's stomach. Zygotes pass through the stomach wall and produce sporozoites, which eventually make their way to the salivary glands.

> *People lacking the Duffy blood-group protein are resistant to Plasmodium vivax malaria; 90% of West African blocks and 60% of African-Americans lack this molecule.*

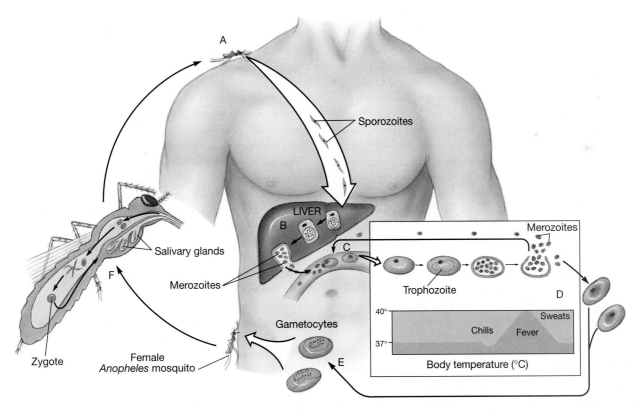

Figure 11.4 The life cycle of the malaria parasite *Plasmodium*. (**A**) Female *Anopheles* mosquito transmits sporozoites from its salivary glands when it bites a human. The sporozoites travel in human blood to the liver. (**B**) In the liver, the sporozoites multiply and become merozoites, which are shed into the bloodstream when liver cells rupture. (**C**) The merozoites enter red blood cells and become trophozoites, which feed and eventually form many more merozoites. (**D**) Merozoites are released by the rupture of the red blood cells, accompanied by chills, high fever (40°C), and sweating. They can then infect other red blood cells. (**E**) After several such asexual cycles, gametocytes (sexual stages) are produced. (**F**) Upon ingestion by a mosquito, the gametocytes form a zygote, which gives rise to more infective sporozoites in the salivary glands. These can then infect other people.

PUBLIC HEALTH

The War Against Malaria: Missteps and Milestones

Malaria, caused by protozoa of the genus *Plasmodium*, is one of the most serious parasitic infections that afflict humans. Despite massive efforts to control the spread of this disease, in any given year it strikes up to 500 million people and claims the lives of 1.5 to 3 million a year, many of them children. Malaria is an ancient human scourge. Written records on Egyptian papyrus from 1500 B.C. describe a disease with high, intermittent fever that must have been malaria. They refer to the use of tree oils as mosquito repellents—although not until some 30 centuries later was the mosquito proven to be the vector (or carrier) of the disease. In some low-lying ancient cities, nearly entire populations succumbed to the disease. Only the wealthy could afford to "head for the hills" in summer to escape the heat, mosquitoes, and fevers. Many of the Crusaders of medieval times also died of malaria, and slave trading in later years contributed significantly to its spread. The one bright spot in this dismal history was the discovery in the sixteenth century that malaria could be treated with quinine, a drug from the *Chinchona* tree.

A connection between swamps and fevers had long been recognized but was misinterpreted. Most early investigations of malaria's cause focused on air—indeed, the disease was named for the "bad air" (*mal*, bad) thought to be responsible—and water. The association between the disease and mosquitoes suggested in Egyptian records was ignored. By the 1870s, with the advent of the germ theory, some scientists came to believe that malaria was caused by a bacterium, which they named *Bacillus malariae*.

Alphonse Laveran, a French army physician in North Africa, was not convinced that the malaria organism had been found. With unstained preparations and a poor-quality, low-power microscope, he continued to search for it. He eventually found what we now know to be male sex cells of the malarial parasite in human blood. Most scientists rejected Laveran's findings in favor of the *Bacillus* theory, until he demonstrated the male sex cells to Pasteur in 1884. Only then did the scientific community accept that protozoa of the genus *Plasmodium* cause malaria.

The Italian physiologist Camillo Golgi added to Laveran's work in 1885 by identifying several species of *Plasmodium*. And in 1891 Russian researchers developed the Romanovsky staining procedure (methylene blue and eosin) for malarial blood smears. With slight modifications, it is still used.

Although the causative agent of malaria had been found, its transmission was not yet understood. Ronald Ross, a medical officer in India, spent years searching in his spare time for proof that malaria parasites are carried by mosquitoes. Eventually, he found the parasites in *Anopheles* mosquitoes, though he never fully explained the mode of transmission. His efforts were neither appreciated nor supported by his superiors, but in 1902 he was awarded the Nobel Prize in physiology or medicine.

Once the vector for malaria had been identified, researchers turned to controlling transmission of the disease by controlling vector populations. Credit for developing the first effective mosquito control measures is given to the American physician William Crawford Gorgas, chief medical officer for sanitation during the building of the Panama Canal early in the twentieth century. By draining wet areas and instituting the use of mosquito netting, Gorgas significantly reduced the incidence of both malaria and yellow fever among canal workers.

In the 1930s and 1940s, advances in malarial treatment and control led many people to believe that malaria was no longer a threat. However, when World War I soldiers thought to have been cured with quinine treatment suffered relapses after they had left malaria-infested areas, researchers began a new round of investigations. This time, they searched for sites in which the parasites became sequestered in human tissues. In 1938, S. P. James and P. Tate discovered the parasite outside red blood cells in certain birds; in 1948 H. C. Shortt and P. C. C. Garnham made a similar discovery of *P. vivax* in humans. The parasite is now known to disappear from blood circulation, out of the reach of drugs, by invading cells of the liver and other tissues.

The war against malaria continues, but with limited success. Massive insecticide (DDT) spraying in the 1960s appeared to have eradicated the disease from many regions of the world, but DDT-resistant mosquitoes soon emerged. Incidence of the disease increased, sometimes to epidemic proportions. Some strains of the parasite also became resistant to chloroquine, one of the best drugs for treating malaria.

Another factor contributing to malaria's recent comeback may be environmental changes in developing countries. In the late 1950s, for instance, the inhabitants of Kenya's Karo Plain turned from subsistence farming of maize and cattle grazing to a cash-producing rice culture. Rice must grow in wet conditions, so the dry plains were flooded and the cattle were banished from the area. But along with rice, the Karo Plain soon supported *Anopheles* mosquitoes. Attracted by the water and the increased humidity, malaria-carrying mosquitoes came to outnumber nonmalaria-carrying mosquitoes 2 to 1. And because the favorite host of the *Anopheles* mosquitoes—cattle—had been removed from the fields, the mosquitoes turned to humans for their meals. The result of all these environmental changes was a frightening increase in malaria incidence. In fact, global warming is predicted to bring about the spread of *P. falciparum*, the species that causes the most life-threatening form of malaria. Computer models predict it will spread to the eastern United States and Canada, to most of Europe, and to Australia.

Several new methods for controlling malaria are currently being studied. However, a successful vaccine still has not been developed. That means that health education and affordable replacements for chloroquine are paramount in controlling malaria. Mosquito nets impregnated with mosquito-killing chemicals could also be effective. In Gambia, childhood mortality from malaria fell by an astonishing 63% after the introduction of these nets.

Several species of *Plasmodium* cause malaria, and each displays variations in the life cycle just described and in the particular species of mosquito that serves as a suitable host. Another apicomplexan, *Toxoplasma gondii*, causes lymphatic infections and blindness in adults and severe neurological damage to the fetuses of infected pregnant women. It has also recently been implicated as a possible cause of schizophrenia. Contact with infected domestic cats and their feces, consumption of contaminated raw meat, and failure to wash one's hands after handling such meat are means of transmitting the parasite. See Chapter 23 for a discussion of the *T. gondii* life cycle and more information.

Ciliates. The largest group of protozoans, the **ciliates**, have cilia over most of their surfaces. Cilia have a basal body near their origin that anchors them in the cytoplasm and enables them to extend from the surface of the cell. Cilia allow the organisms to move, and in some genera, such as *Paramecium* (**Figure 11.3d**), cilia assist in food gathering. *Balantidium coli*, the only ciliate that parasitizes humans, causes dysentery.

Ciliates have several highly specialized structures. Most ciliates have a well-developed contractile vacuole, which regulates cell fluids. Some have a strengthened pellicle. Others have **trichocysts**, tentacles that can be used to capture prey, or long stalks by which they attach themselves to surfaces. Ciliates also undergo **conjugation**. Unlike bacterial conjugation, in which one organism receives genetic information from another, conjugation in ciliates allows exchange of genetic information between two organisms.

✓CHECKLIST

1. How do parasites differ from predators?
2. If a parasite has recently invaded a new population, is it more likely to cause severe or mild symptoms and effects?
3. What is a "bloom" of protists? Why does it occur? What can its effects be?

FUNGI

CHARACTERISTICS OF FUNGI

Fungi, studied in the specialized field of **mycology**, are a diverse group of heterotrophs. Many are saprophytes that digest dead organic matter and organic wastes. Some are parasites that obtain nutrients from the tissues of other organisms. Most fungi, such as molds and mushrooms, are multicellular, but yeasts are unicellular.

The body of a fungus is called a **thallus**. The thallus of most multicellular fungi consists of a **mycelium** (my-se′le-um), a loosely organized mass of threadlike structures called **hyphae** (hy′fe; singular: *hypha*; **Figure 11.5**). The mycelium is embedded in decaying organic matter, soil, or the tissue of a living organism. Mycelial cells release

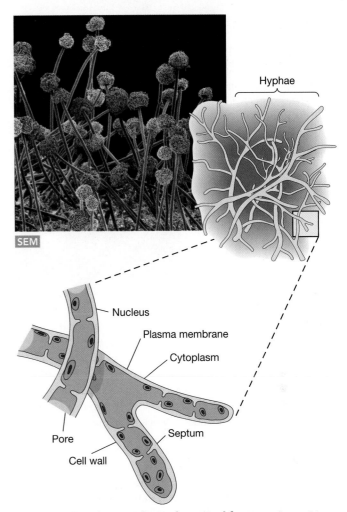

Figure 11.5   The mycelium of a typical fungus. The mold *Aspergillus niger* (85X) consists of filamentous hyphae, the cells of which can be multinucleate and separated by pore-containing septae. *(David Scharf/Peter Arnold, Inc.)*

enzymes that digest the *substratum* (the surface on which the fungus grows) and absorb small nutrient molecules. The cell walls of a few fungi contain cellulose, but those of most fungi contain **chitin** (ki′tin), a polysaccharide also found in the exoskeletons (outer coverings) of arthropods such as ticks and spiders. All fungi have lysosomal enzymes that digest damaged cells and help parasitic fungi to invade hosts. Many fungi synthesize and store granules of the nutrient polysaccharide glycogen. Some fungi, such as yeasts, are known to have plasmids. These plasmids can be used to clone foreign genes into the yeast cells, a technique of great use in genetic engineering (◀Chapter 8, p. 212).

The hyphal cells of most fungi have one or two nuclei, and many hyphal cells are separated by cross-walls called **septa** (singular: *septum*). Pores in septa allow both cytoplasm and nuclei to pass between cells. Some fungi have septa with so many pores that they are sievelike,

Phytophtora infestans, the fungus that caused the great Irish potato famine, led to the death of 1 million Irish, and to emigration of another 2 million.

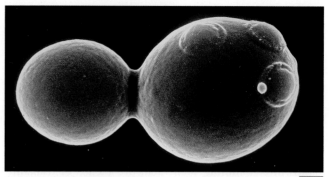

Figure 11.6 Budding yeast. Circular scars seen on the surface of the cell on the right represent sites of previous budding (6,160X). After 20 to 30 divisions, scars cover the cell surface and it cannot divide again. *(J. Forsdyke/Photo Researchers, Inc.)* `SEM`

THE IMPORTANCE OF FUNGI

In ecosystems, fungi are important decomposers. In the health sciences, they are important as facultative parasites—they can obtain nutrients from nonliving organic matter or from living organisms. Fungi are never obligate parasites because all fungi can obtain nutrients from dead organisms. Even when fungi parasitize living organisms, they kill cells and obtain nutrients as saprophytes. Nearly every form of life is parasitized by some type of fungus. Some fungi produce antibiotics that inhibit the growth of or kill bacteria. Parasitic fungi vary in the damage they inflict. Fungi such as those that cause athlete's foot are nearly always present on the skin and rarely cause severe damage. However, the fungus that causes histoplasmosis can spread through the lymphatic system to cause fever, anemia, and death.

Saprophytic fungi are beneficial as decomposers and as producers of antibiotics. The digestive activities of such fungi provide nutrients not only for the fungi themselves but for other organisms, too. The carbon and nitrogen compounds they release from dead organisms contribute significantly to the recycling of substances in ecosystems. Fungi are essential for decomposing lignins and other woody substances. Some fungi excrete metabolic wastes that are toxic to other organisms, especially soil microorganisms. In the soil, the production of such toxins, which are antibiotics, is called **antibiosis**. These toxins presumably help the species that produce them compete and survive. The antibiotics, when extracted and purified, are used to treat human infections (Chapter 13).

Parasitic fungi can be destructive when they invade other organisms. These fungi have three requirements for invasion: (1) proximity to the host, (2) the ability to penetrate the host, and (3) the ability to digest and absorb nutrients from host cells. Many fungi reach their hosts by producing spores that are carried by wind or water. Other fungi arrive on the bodies of insects or other animals. For example, wood-boring insects spread spores of the fungal Dutch elm disease **(Figure 11.9)** throughout North America

and a few lack septa entirely. Certain fungi with a single septal pore have an organelle called a *Woronin* (weh-ro′nin) *body*. When a hyphal cell ages or is damaged, the Woronin body moves to and blocks the pore so that materials from the damaged cell cannot enter a healthy cell.

Many fungi reproduce both sexually and asexually, but a few have only asexual reproduction. Asexual reproduction always involves mitotic cell division, which in yeast occurs by budding **(Figure 11.6)**. Sexual reproduction occurs in several ways. In one way, haploid gametes unite, and their cytoplasm mingles in a process called **plasmogamy** (plaz-mog′am-e). However, if the nuclei fail to unite, a **dikaryotic** ("two-nucleus") cell forms; it can persist for several cell divisions. Eventually, the nuclei fuse in a process called **karyogamy** (kar″-e-og′am-e) to produce a diploid cell. Such cells or their progeny later produce new haploid cells. Some fungi also can reproduce sexually during dikaryotic (diploid) phases of their life cycle. Fungi usually go through haploid, dikaryotic, and diploid phases in their life cycle **(Figure 11.7)**.

Fungi can produce spores both sexually and asexually, and spores can have one or several nuclei **(Figure 11.8)**. Typically, aquatic fungi produce motile spores with flagella, and terrestrial fungi produce spores with thick protective walls. Germinating spores produce either single cells or germ tubes. *Germ tubes* are filamentous structures that break through weakened spore walls and develop into hyphae.

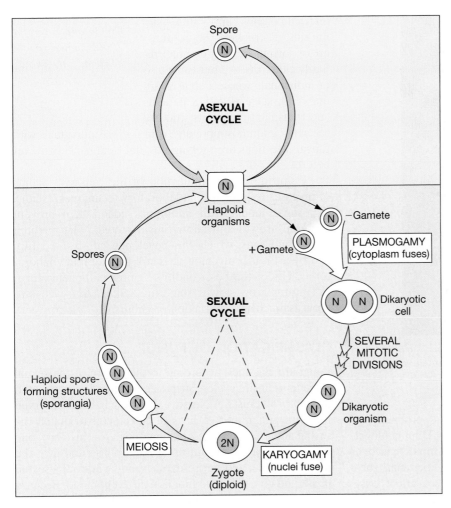

Spore

ASEXUAL CYCLE

N
Haploid organisms

N —Gamete

PLASMOGAMY (cytoplasm fuses)

+Gamete N

Spores N

SEXUAL CYCLE

N N Dikaryotic cell

SEVERAL MITOTIC DIVISIONS

N N Dikaryotic organism

Haploid spore-forming structures (sporangia)

MEIOSIS

2N
Zygote (diploid)

KARYOGAMY (nuclei fuse)

Figure 11.7 One method of sexual reproduction in fungi. Haploid organisms may maintain themselves by asexual spore formation (beige background) or budding. Alternatively (blue background), they may produce gametes that initially undergo plasmogamy (fuse their cytoplasmic portions). After several mitotic divisions of the still-separate nuclei, the two nuclei undergo karyogamy (fuse their nuclei) to form a diploid zygote. The zygote then undergoes meiosis to return to the haploid state and produces reproductive spores.

in the decades following World War I, killing almost all elm trees in some parts of the United States. Fungi penetrate plant cells by forming hyphal pegs that press on and push through cell walls. How fungi penetrate animal cells, which lack cell walls, is not fully understood, but lysosomes apparently play an important role. Once fungi have entered cells, they digest cell components and absorb

nutrients. As cells die, the fungus invades adjacent cells and continues to digest and absorb nutrients.

Fungal parasites in plants cause diseases such as wilts, mildews, blights, rusts, and smuts and thereby produce extensive crop damage and economic losses. Fungal infections of domestic birds and mammals are also responsible for extensive economic losses. Those fungi that invade

(a) SEM

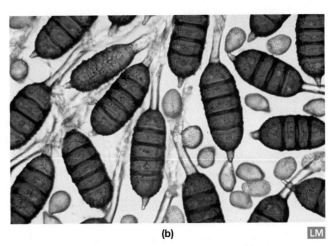

(b) LM

Figure 11.8 The formation of asexual spores (conidiospores). **(a)** Brushlike clusters of chains of spores (1,400X) of the fungus *Penicillium.* *(Andrew Syred/Photo Researchers, Inc.)* **(b)** Spores of the rose rust fungus *Phragmidium* (1,000X). *(Bruce Iverson/Visuals Unlimited)*

Figure 11.9   Dutch elm disease. American elms (*Ulmus americana*) killed by Dutch elm disease. *(Richard Thorn/Visuals Unlimited)*

humans cause human suffering, decreased productivity, and sometimes long-term medical expenses. Human fungal diseases, or **mycoses**, often are caused by more than one organism. Mycoses can be classified as superficial, subcutaneous, or systemic. *Superficial* diseases affect only keratinized tissue in the skin, hair, and nails. *Subcutaneous* diseases affect skin layers beneath keratinized tissue and can spread to lymph vessels. *Systemic* diseases invade internal organs and cause significant destruction. Some fungi are opportunistic; they do not ordinarily cause disease but can do so in individuals whose defenses are impaired, such as AIDS patients and transplant recipients who are receiving immunosuppressive drugs. More individuals with fungal infections are seeking hospital treatment than ever before.

Of the approximately 70,000 known species of fungi, only about 300 are pathogenic to humans.

Culturing and identification of the causative agents of mycoses require special laboratory techniques. Acidic, high-sugar media with antibiotics added help prevent bacterial growth and allow fungal growth. The medium Sabouraud agar, which was developed nearly a century ago by a French mycologist, is still used in many laboratories. Under the best conditions, most pathogenic fungi in cultures grow slowly; some may take 2 to 4 weeks to grow as much as bacteria do in 24 hours.

CLASSIFICATION OF FUNGI

Fungi are classified according to the nature of the sexual stage in their life cycles. Such classification is complicated by two problems: (1) No sexual cycle has been observed for some fungi, and (2) it is often difficult to match the sexual and asexual stages of some fungi. For instance, one researcher may work out an asexual phase and give the fungus a name; another researcher may work out a sexual phase and give the same fungus a different name. Because the relationship between the sexual and asexual phases is not always apparent, a particular species of fungi may have two names until someone discovers that the two

CLOSE UP
Fungi and Orchids

When explorers first brought orchids from South America back to England during the nineteenth century, the English were delighted to have such handsome specimens in their conservatories. However, they suffered great disappointment when the plants failed to thrive, no matter how carefully they were potted in fresh soil and new pots. It took the English several

Orchid blooming, showing roots. *(Paul Edmondson/Age Fotostock America, Inc.)*

Photomicrogrph of mycorrhizal fungi growing in association with fungal roots. *(©The School of Biological Sciences, University of Sydney.)*

years of experimentation, and perhaps the fortuitous importation of a few orchids in their native medium, to learn to grow orchids out of their natural environment. Eventually, it was discovered that orchids require certain fungi to thrive. These fungi form symbiotic associations with the orchid roots; such associations are called *mycorrhizae* (my″ko-ri′ze). When medium in which orchids had grown was used to pot fresh specimens, the orchids and the fungi formed mycorrhizae, and both thrived.

Figure 11.10   Dimorphism in fungi. (a) Hyphae of *Mucor* (667X). *(Courtesy Michael E. Oriowski, Louisiana State University)* **(b)** Yeast form of *Mucor* (667X). *(Courtesy Michael E. Oriowski, Louisiana State University)*

Figure 11.11   The black bread mold, *Rhizopus nigricans.* Sexual zygospores (black, spiny structures) are the result of the joining and fusion of genetic materials at the tips of special hyphal side branches. The zygospores germinate to produce a sporangium that, in turn, produces many asexual spores (377X). *(Bruce Iverson/Photo Researchers, Inc.)*

phases occur in the same organism. For example, a fungal cause of athlete's foot is called *Trichophyton* when it reproduces asexually, but is named *Arthroderma* when it reproduces sexually. Another problem is that many fungi look quite different when growing in tissues (yeastlike) and when growing in their natural habitats (filamentous). The ability of an organism to alter its structure when it changes habitats is called **dimorphism** (di-mor′fizm) **(Figure 11.10)**. Dimorphism in fungi has complicated the problem of identifying causative agents in fungal diseases. We will consider bread molds, sac fungi, club fungi, and the so-called Fungi Imperfecti, which are believed to have lost their sexual cycle **(Table 11.2)**.

Bread Molds

The **bread molds**, **Zygomycota**, or conjugation fungi, have complex mycelia composed of hyphae (lacking septa) with chitinous walls. The black bread mold, *Rhizopus* **(Figure 11.11)**, has hyphae that grow rapidly

TABLE 11.2

Properties of Fungi			
Phylum	**Common Name**	**Characteristics**	**Examples**
Zygomycota	Bread molds	Display conjugation	*Rhizopus* and other bread molds
Ascomycota	Sac fungi	Produce asci and ascospores during sexual reproduction	*Neurospora, Penicillium, Saccharomyces,* and other yeasts; *Candida, Trichophyton,* and several other human pathogens
Basidiomycota	Club fungi	Produce basidia and basidiospores	*Amanita* and other mushrooms; *Claviceps* (which produces ergot); *Cryptococcus*
Deuteromycota	Fungi Imperfecti	Sexual stage nonexistent or unknown	Soil organisms; various human pathogens

along a surface and into the substratum. Some bread-mold hyphae produce spores that are easily carried by air currents. When the spores reach an appropriate substratum, they germinate to produce new hyphae. Sometimes short branches of the hyphae of two different strains, called plus and minus strains, grow together. This joining of hyphae gave rise to the name conjugation fungi. Chemical attractants are involved in attracting hyphae to each other. Multinucleate cells form where the hyphae join, and many pairs of plus and minus nuclei fuse to form zygotes. Each zygote is enclosed in a **zygospore**, a thick-walled, resistant structure that also produces spores. Genetic information in zygospores comes from two strains, whereas that in hyphal spores comes from a single strain.

Although bread molds interest mycologists and frustrate bacteriologists whose cultures they contaminate, they usually do not cause human disease. *Rhizopus*, however, is an opportunistic human pathogen; it is especially

dangerous to people with diabetes mellitus that is not well controlled.

Sac Fungi

The **sac fungi** are a diverse group, containing over 30,000 species, including yeasts. Sac fungi have chitin in their cell walls and produce no flagellated spores. With the exception of some yeasts, which do not form hyphae, the hyphae of sac fungi have septa with a central pore. These fungi are properly called **Ascomycota** (as′ko-mi-ko″ta); unlike other fungi, they produce a saclike **ascus** (plural: *asci*) during sexual reproduction **(Figure 11.12)**. Yeasts are included among the ascomycetes, even though most yeasts have no known sexual stage. In species that reproduce both sexually and asexually, the asexual phase forms spores called **conidia** at the ends of modified hyphae. In the sexual phase, one strain has a large *ascogonium*, and an adjacent strain has a smaller *antheridium*. These structures fuse, their nuclei mingle,

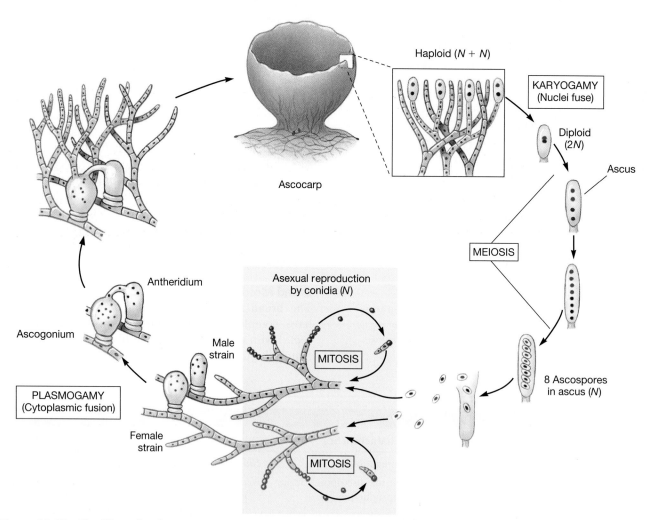

Figure 11.12 The life cycle of an ascomycete. In the asexual phase, spores called conidia are formed at the tips of modified hyphae. In the sexual phase, the mycelium that produces conidia also forms the gamete-producing structures, antheridia (male) and ascogonia (female). After cytoplasmic fusion of those structures occurs, dikaryotic hyphal cells develop and interweave into an ascocarp, where saclike asci grow. In each ascus, the dikaryotic nuclei fuse to form a zygote, and the zygote nucleus divides into eight nuclei. From them, eight ascospores form and are forcefully released.

and hyphal cells with dikaryotic nuclei grow from the fused mass. Eventually, dikaryotic nuclei fuse to form a zygote, and the zygote nucleus divides to form eight nuclei in each ascus. Each ascus forms eight **ascospores**, sometimes releasing them forcefully.

Several sac fungi are of interest in microbiology. *Neurospora* is significant because studies of its ascospores have provided important genetic information. *Penicillium notatum* produces the antibiotic penicillin; *P. roquefortii* and *P. camemberti* are responsible for the color, texture, and flavor of Roquefort and Camembert cheeses. Yeasts, especially those of the genus *Saccharomyces*, release carbon dioxide and alcohol as metabolic products of fermentation and are used to leaven bread and to make alcohol in beer and wine (Chapter 26). A number of sac fungi are human pathogens. *Candida albicans* causes vaginal yeast infections. *Trichophyton* is associated with athlete's foot, and *Aspergillus* with opportunistic respiratory infections. Species of *Blastomyces* and *Histoplasma* cause respiratory infections and can spread throughout the body.

Club Fungi

The **club fungi** include mushrooms, toadstools, rusts, and smuts. The rusts and smuts parasitize plants and cause significant crop damage. In addition to having hyphae aggregated to form mycelia, the club fungi have club-shaped sexual structures called **basidia**, from which the name **Basidiomycota** is derived (**Figure 11.13**). In a typical basidiomycete life cycle, sexual spores called **basidiospores** germinate to form septate mycelia, and cells of mycelia unite into dikaryotic forms. The dikaryotic mycelium grows and produces basidia, which in turn produce basidiospores. Come with me to visit a mushroom farm in the chapter 26 opener. Some mushrooms, such as *Amanita*, produce toxins that can be lethal to humans. *Claviceps purpurea*, a parasite of rye, produces the toxic substance ergot. This substance can be used in small quantities to treat migraine headaches and induce uterine contractions, but in larger quantities it can kill (Chapter 22). The yeast *Cryptococcus* causes opportunistic respiratory infections, which can be fatal if they spread to the central nervous system, causing

meningitis and brain infection. This organism is increasingly being seen in AIDS patients.

Fungi Imperfecti

The **Fungi Imperfecti**, or **Deuteromycota**, are called "imperfect" because no sexual stage has been observed in their life cycles. Without information on the sexual cycle, taxonomists cannot assign them to a taxonomic group. However, by their vegetative characteristics and the production of asexual spores, most of these fungi seem to belong with the sac fungi. Many of the Fungi Imperfecti have recently been placed in other phyla and given new genus names. We have kept the older designations, however, because the new ones are not yet familiar or widely used in clinical work. **Anamorphic** names refer to asexual life cycle stages, whereas **teleomorph** names refer to sexual stages.

TRY IT

Spore Prints

Mushroom identification requires knowledge about the spores of your unknown specimen. Some keys for mushroom identification are arranged according to spore color. How do you obtain such information? Try this simple method.

First, collect fresh mushrooms whose caps are just opening or that are fully open. Cut off the stem flush with the bottom of the cap. Place the cap, gill side down, on a piece of paper. Leave it undisturbed overnight or until dry. Gently lift the cap and see the sunburstlike pattern of spores that have been shed from the surfaces of the gills. If you have two mushrooms of the same variety, use a dark piece of paper under one and white paper under the other before the drying process, because you won't know what color spores to expect. Spores may range in color from black to white, tan, or even pink.

(Dwight R. Kuhn/Dwight Kuhn Photography)

(a)

(b)

Figure 11.13 Mushroom spores. (a) The gills on the bottom of a mushroom (*Leucoagaricus naucinus*) cap have microscopic, club-shaped structures called basidia. *(Grant Heilman Photography)* **(b)** Each basidium (of *Psilocybe mexicana* (225X) produces four balloonlike structures called basidiospores. LM *(S. Flegler/Visuals Unlimited)*

1. Distinguish between thallus, mycelium, and hyphae.

2. Explain dimorphism, how it arises, and the problems it may cause.

3. Why are the Fungi Imperfecti considered to be "imperfect"?

HELMINTHS

CHARACTERISTICS OF HELMINTHS

Helminths, or worms, are bilaterally symmetrical—that is, they have left and right halves that are mirror images. A helminth also has a head and tail end, and its tissues are differentiated into three distinct tissue layers: ectoderm, mesoderm, and endoderm. Helminths that parasitize humans include flatworms and roundworms (Table 11.3).

Flatworms

Flatworms (Platyhelminthes) are primitive worms usually no more than 1 mm thick, but some, such as large tapeworms, can be as long as 10 m. Flatworms lack a **coelom** (se′lom), a cavity that lies between the digestive tract and the body wall in higher animals. Most flatworms have a simple digestive tract with a single opening, but some parasitic flatworms, the tapeworms, have lost their digestive tracts. Most flatworms are hermaphroditic, each individual having both male and female reproductive systems. They have an aggregation of neurons in the head end, representing an early stage in the evolution of a brain. Flatworms lack circulatory systems, and most absorb nutrients and oxygen through their body walls.

More than 15,000 species of flatworms have been identified. They include free-living, mostly aquatic organisms such as *planarians* and two classes of parasitic organisms, the **flukes** (*trematodes*) and the **tapeworms** (*cestodes*). Both parasitic groups have highly specialized reproductive systems and suckers or hooks by which they attach to their host. The flukes can be internal or external parasites. *Fasciola hepatica* and several other flukes parasitize humans. Tapeworms parasitize the small intestine of animals almost exclusively, but occasionally occur in the eye or brain. The beef tapeworm,

🔍 CLOSE UP

Are Fungi the Biggest and Oldest Organisms on Earth?

Weighing about 100 tons (more than a blue whale) and extending through nearly 40 acres of soil in a forest near Crystal Falls in upper Michigan is a gigantic individual of the fungus *Armillaria bulbosa*. Between 1,500 and 10,000 years ago, most likely at the end of the last Ice Age, a single pair of compatible spores blew in from parent mushrooms, germinated, and mated. They began growth that continues today. The fungus grows primarily under the soil, so it is usually not visible to the casual observer. The hyphae of its mycelium probe through the soil, seeking woody debris to decompose and recycle. Experimental measurements of its growth rate through soil enabled scientists to estimate the time required to reach its present size.

DNA analysis of 12 genes from the organism's fruiting structures—commonly called *button* or *honey mushrooms*—and its stringlike underground colonizing structures—called *rhizomorphs*—revealed the huge fungus to be a giant clone. All parts of the clone are identical in genetic composition. Although there are minor breaks in its continuity, it is still regarded as a single individual.

Despite the massive size of this fungus, its discoverers, Myron Smith and James Anderson of the University of Toronto and Johann Bruhn of Michigan Technological University, predicted that it might not be the largest organism of its kind. Writing in the journal *Nature* in April 1992, they explained that they found the fungus in a mixed forest, containing many kinds of trees. In a single-type forest such as a large stand of birch or aspen, a fungus with a preference for that type of tree could reach even greater size. This one, however, has probably reached its maximum size, as it collides with competing fungi along its borders.

The scientists' prediction quickly proved to be prophetic. About a month after the *Nature* article was published, two forest pathologists—Ken Russell of the State Department of Natural Resources and Terry Shaw of the U.S. Forest Service—announced that they had been studying an even larger fungus near Mount Adams in southwestern Washington.

Mushrooms (Armillaria bulbosa)
(Courtesy Johann N. Bruhn, University of Missouri)

This organism, an individual of *Armillaria ostoyae*, covers 1,500 acres (about 2.5 square miles), making it almost 40 times as large as the Michigan fungus. The Washington fungus grows in a region populated largely by a single type of tree—in this case, pine—and therefore enjoys a vast source of nourishment. Although the Washington fungus dwarfs its Michigan counterpart in size, it is actually younger, having an estimated age of 400 to 1,000 years. Thus, the Michigan fungus retains the title "oldest" (at least for now) but not "largest."

Will scientists eventually discover fungi that are even bigger than the Washington fungus? Most likely, yes. In fact, in an interview, Shaw referred to an *A. ostoyae* in Oregon that might be larger than the one he discovered in Washington. And still bigger ones may remain to be found. The search for the "biggest and oldest" promises to be an exciting episode in the field of microbiology.

TABLE 11.3

Properties of Helminths		
Group	**Characteristics**	**Examples**
Flatworms (Platyhelminthes)	Worms live in or on hosts.	*Taenia* and other tapeworms are internal parasites; flukes can be internal or external parasites.
Roundworms (nematodes)	Most worms live in the intestine or circulatory system of hosts.	Hookworms, pinworms, and several other roundworms live in intestines or lymph system.

Taenia saginata, and several other worms parasitize humans.

Roundworms

Roundworms, or **nematodes**, share many characteristics with the flatworms, but they have a **pseudocoelom**, a primitive, fluid-filled body cavity that lacks the complete lining found in higher animals. The roundworms have cylindrical bodies with tapered ends and are covered with a thick, protective cuticle. They vary in length from less than 1 mm to more than 1 m. Contractions of strong muscles in the body wall exert pressure on the fluid in the pseudocoelom and stiffen the body. Pointed ends and stiff bodies allow roundworms to move through soil and tissues easily. Roundworm females are larger than males. Breeding is enhanced by chemical attractants released by females that attract males. Females can lay as many as 200,000 eggs per day. The large number of eggs, well protected by hard shells, ensures that some will survive and reproduce.

Over 80,000 species of roundworms have been described. They occur free living in soil, fresh water, and salt water and as parasites in every plant and animal species ever studied. A single acre of soil can contain billions of roundworms. Many parasitize insects and plants; only a relatively small number of species infect humans, but they cause significant debilitation, suffering, and death. Most roundworms that parasitize humans, such as hookworms and pinworms, live mainly in the intestinal tract, but a few, such as *Wuchereria*, have larval forms that live in blood or lymph. The effects of roundworms on humans were first recorded in ancient Chinese writings and have been noted by nearly every civilization since then. (For an account of modern American experiences with sushi and other forms of raw fish, which may harbor roundworms, see the box on sushi in Chapter 22.)

PARASITIC HELMINTHS

We will concern ourselves only with parasitic helminths and consider four groups: flukes, tapeworms, adult roundworms of the intestine, and roundworm larvae **(Figure 11.14)**. Because helminths have complex life cycles

| (a) LM | (b) LM | (c) LM | (d) LM |

Figure 11.14   Representative helminths. **(a)** *Clonorchis sinensis*, the Chinese liver fluke, stained to show internal organs. It infests the gallbladder, bile ducts, and pancreatic ducts, where it causes biliary cirrhosis and jaundice (59X). *(John D. Cunningham/Visuals Unlimited)* **(b)** Head (scolex) of a tapeworm (220X). The hooked spines and suckers are used for attachment to intestinal surfaces. *(G. Shih & R. Kessel/Visuals Unlimited)* **(c)** Mouth of the Old World hookworm *Ancylostoma duodenale* (59X). The muscular pharynx of this roundworm pumps blood from the intestinal lining of its host. *(Fred Marsik/Science VU/Visuals Unlimited)* **(d)** The microfilarial (miniature larval) stage of the heartworm *Dirofilaria immitis*, in a sample of dog blood (370X), is transmitted by mosquito bites. The larger stages live inside the heart and perforate its walls. *(George J. Wilder/Visuals Unlimited)*

related to their ability to cause diseases, we consider a typical life cycle for each group.

Flukes

Two types of fluke infections occur in humans. One involves tissue flukes, which attach to the bile ducts, lungs, or other tissues; the other involves blood flukes, which are found in blood in some stages of their life cycle. Tissue flukes that parasitize humans include the lung fluke, *Paragonimus westermani*, and the liver flukes, *Clonorchis sinensis* (Figure 11.14a) and *Fasciola hepatica*. Blood flukes include various species of the genus *Schistosoma*.

Parasitic flukes have a complex life cycle (Figure 11.15), often involving several hosts. The fusion of male and female gametes produces fertilized eggs that become encased in tough shells during their passage through the female fluke's uterus. The eggs pass from the host with the feces. When the eggs reach water, they hatch into free-swimming forms called **miracidia** (mi″ra-sid′e-ah). The miracidia penetrate a snail or other molluskan host, become **sporocysts**, and migrate to the host's digestive gland. The cells inside the sporocysts typically divide by mitosis to form **rediae** (re′de-e). Rediae, in turn, give rise to free-swimming **cercariae** (ser-ka′re-e), which escape from the mollusk into water. Using enzymes to burrow through exposed skin, cercariae penetrate another host (often an arthropod) and then encyst as **metacercariae**. When this host is eaten by the definitive host, the metacercariae excyst and develop into mature flukes in the host's intestine.

Tapeworms

Tapeworms consist of a **scolex** (sko′lex), or head end (Figure 11.14b), with suckers that attach to the intestinal wall, and a long chain of hermaphroditic **proglottids** (pro-glot′tidz), body components that contain mainly reproductive organs of both sexes. New proglottids develop behind the scolex, mature, and fertilize themselves. Old ones disintegrate and release eggs at the rear end. Among

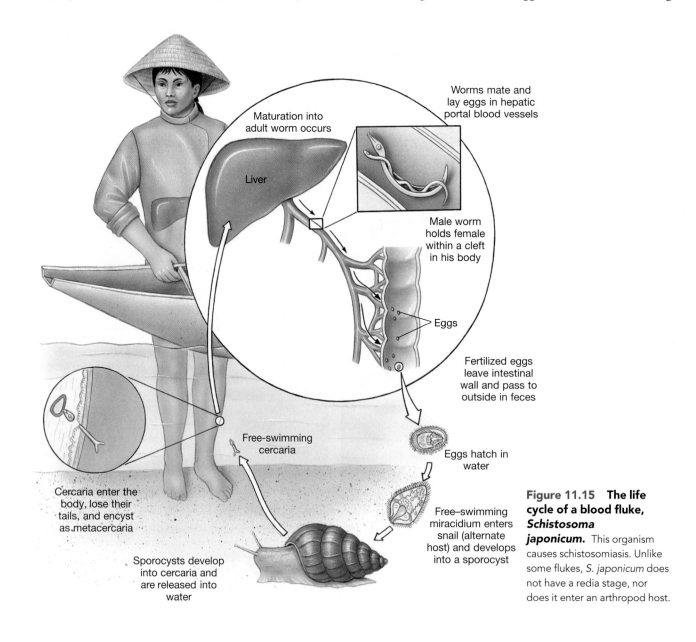

Maturation into adult worm occurs

Liver

Worms mate and lay eggs in hepatic portal blood vessels

Male worm holds female within a cleft in his body

Eggs

Fertilized eggs leave intestinal wall and pass to outside in feces

Free-swimming cercaria

Cercaria enter the body, lose their tails, and encyst as metacercaria

Eggs hatch in water

Free-swimming miracidium enters snail (alternate host) and develops into a sporocyst

Sporocysts develop into cercaria and are released into water

Figure 11.15   The life cycle of a blood fluke, *Schistosoma japonicum*. This organism causes schistosomiasis. Unlike some flukes, *S. japonicum* does not have a redia stage, nor does it enter an arthropod host.

the tapeworms that can infect humans are beef and pork tapeworms that are species of *Taenia*, dwarf and rat tapeworms that are species of *Hymenolepis*, the hydatid worm *Echinococcus*, the dog tapeworm *Dipylidium*, and the broad fish tapeworm *Diphyllobothrium*.

Although different species display minor variations, the life cycle of tapeworms **(Figure 11.16)** usually includes the following stages: Embryos develop inside eggs and are released from proglottids; the proglottids and eggs leave the host's body with the feces. When another animal ingests vegetation or water contaminated with eggs, the eggs hatch into larvae, which invade the intestinal wall and can migrate to other tissues. A larva can develop into a **cysticercus** (sis-ti-ser′kus), or bladder worm, or it can form a cyst. A cysticercus can remain in the intestinal wall or migrate through blood vessels to other organs. A cyst can enlarge and develop many tapeworm heads within it, becoming a **hydatid** (hi-da′tid) **cyst** (Chapter 22). If an animal eats flesh containing such a cyst, each scolex can develop into a new tapeworm.

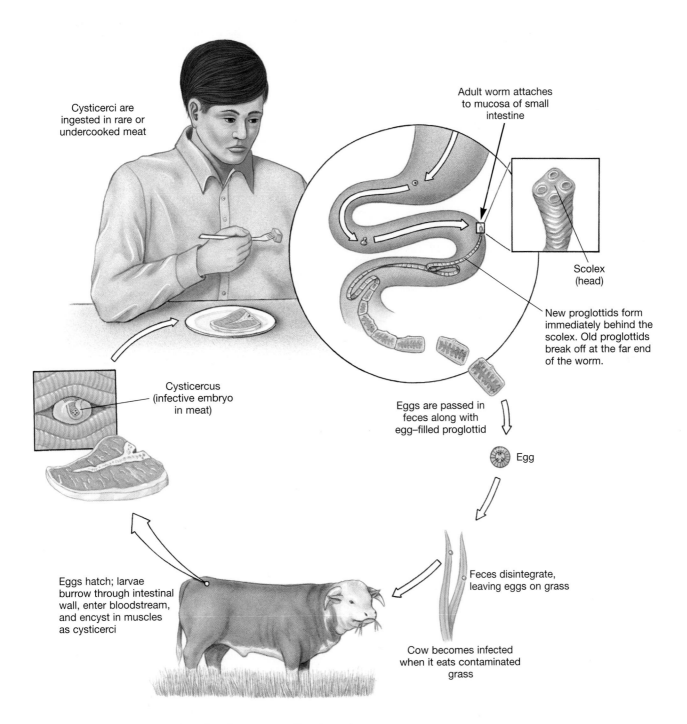

Cysticerci are ingested in rare or undercooked meat

Adult worm attaches to mucosa of small intestine

Scolex (head)

New proglottids form immediately behind the scolex. Old proglottids break off at the far end of the worm.

Cysticercus (infective embryo in meat)

Eggs are passed in feces along with egg–filled proglottid

Egg

Feces disintegrate, leaving eggs on grass

Eggs hatch; larvae burrow through intestinal wall, enter bloodstream, and encyst in muscles as cysticerci

Cow becomes infected when it eats contaminated grass

Figure 11.16 **The life cycle of the beef tapeworm *Taenia saginata*.**

Adult Roundworms

Most roundworms that parasitize humans live much of their life cycle in the digestive tract. They usually enter the body by ingestion with food or water, but some, such as the hookworm, penetrate the skin. These helminths include the pork roundworm *Trichinella spiralis*, the common roundworm *Ascaris lumbricoides*, the guinea worm *Dracunculus medinensis*, the pinworm *Enterobius vermicularis*, and the hookworm, *Ancylostoma duodenale* (**Figure 11.14c**) and *Necator americanus*.

Adult birds may temporarily act as host to Trichinella spiralis, but the microbe does not encyst in fowl muscle nor in that of any cold-blooded animal.

The life cycles of intestinal roundworms show considerable variation. We use the life cycle of *Trichinella spiralis* as an example (**Figure 11.17**). These worms enter humans as encysted larvae in the muscle of infected pigs when undercooked pork is eaten. The cyst walls are digested with the meat, and the larvae are released into the intestine. They mature sexually in about 2 days and then mate. Females burrow into the intestinal wall and produce eggs that hatch inside the adult worm and emerge as larvae. The larvae migrate to lymph vessels and are carried to the blood. From the blood, the larvae burrow into muscles and encyst. These cysts can remain in muscles for years. The same processes occur in the pigs themselves, so cysts are present in their tissues.

Roundworm Larvae

Whereas most roundworms cause much of their tissue damage as adults in the intestine, some cause their damage mainly as larvae in other tissues. These roundworms include *Wuchereria bancrofti*, which lives in lymphatic tissue and causes elephantiasis; *Loa loa*, which infects the eyes and eye

Dracunuculus in the United States! A different species infects raccoons and is sometimes seen as long white "strings" hanging from racoons' wrists as they wash their food in a pond.

Human ingests cyst in undercooked pork

Stomach acid dissolves cyst cover, releasing worms

Worms mate

Females release larvae into lacteals and vessels

Blood vessel

Larvae migrate to muscle and encyst

Encysted worm in pork

Undercooked pork in garbage

Figure 11.17 The life cycle of the roundworm *Trichinella spiralis.* This roundworm causes trichinosis.

membranes; *Onchocerca volvulus*, the cause of river blindness, which infects both the skin and eyes; and *Dracunculus medinensis* (Guinea worm), whose life cycle and symptoms are shown in **Figure 11.18**. Eradication of the Guinea worm is the special focus of the Carter Foundation of Atlanta, Georgia. Former President Jimmy Carter describes the life cycle and symptoms of this disease in an interview which begins here, and continues on the website for this chapter.

MR. CARTER: Guinea worm disease is contracted by drinking from ponds, step wells, cisterns, and other sources of stagnant water that have been contaminated by the worm larvae. Guinea worm, *Dracunculus medinensis*, affects only humans, and it actually uses its human host to further its life cycle. Contaminated water contains water fleas that have eaten immature Guinea worm larvae. The larvae escape when the digestive juices in the person's stomach kill the flea. The larvae penetrate the stomach wall, wander around the abdomen, mature in a few months, and mate, after which the male worms die. It is only the female worm that grows to 2 or 3 feet in length and, about a year later, secretes a toxin that causes a blister on the skin. When the blister ruptures, usually when the infected part of the body is immersed in cool water, the worm starts to emerge. This process can take 30 to 100 days before the worm finally finishes making its way out of the body. When an infected person enters the village pond or watering hole, the worm discharges hundreds of thousands of tiny larvae into the water, beginning the cycle again.

DR. HOPKINS: A small incision made before the emergence blister is raised allows the worm to be wound out gradually, wrapped around a stick. This may have been the origin of the symbol of the medical profession, the caduceus, a serpent coiled around a stick. Many scholars think the Guinea worm is the "fiery serpent" of the Bible. It takes several weeks of daily gentle winding to complete the removal of a worm. If the worm breaks and dies, it will decompose inside the host, causing festering and infection. If a portion of it retracts into the tissues, it can carry tetanus spores back with it, leading to fatal tetanus disease. The local practice in some countries of putting cow dung on the wound makes tetanus especially common. In Upper Volta and Nigeria, Guinea worm is the third leading cause of acquiring tetanus. Other types of microorganisms can also enter the wound, and secondary infections are frequent, even if tetanus is avoided. If the worm emerges near a major

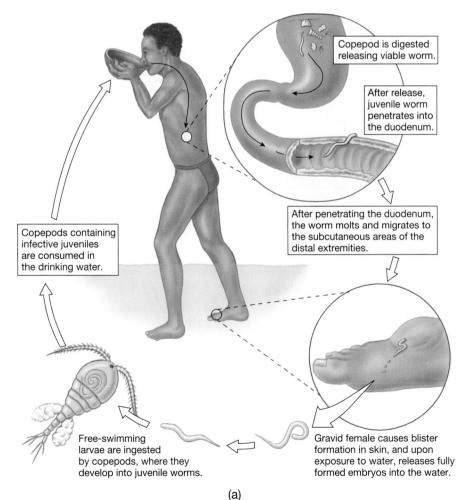

Copepod is digested releasing viable worm.

After release, juvenile worm penetrates into the duodenum.

Copepods containing infective juveniles are consumed in the drinking water.

After penetrating the duodenum, the worm molts and migrates to the subcutaneous areas of the distal extremities.

Free-swimming larvae are ingested by copepods, where they develop into juvenile worms.

Gravid female causes blister formation in skin, and upon exposure to water, releases fully formed embryos into the water.

(a)

(b)

Figure 11.18 *Dracunculus medinensis* **(Guinea worm).** **(a)** Life cycle. **(b)** Female Guinea worm emerging from a blister on the foot of a victim. *(Courtesy of the Carter Center)* joint, permanent scarring leads to stiff and crippled joints. One man died of starvation when a worm came out under his tongue and he couldn't eat. Although most worms emerge from the lower limbs, they can be found anywhere: scrotum, scalp, chest, face.

This interview continues on the Web. Go to www.wiley.com/college/black.

The life cycles of roundworms that parasitize humans as larvae also require a mosquito host (**Figure 11.19**). These worms enter the human body as immature larvae called **microfilariae** (mi″kro-fi-lar′e-e) with the bite of an infected mosquito. The microfilariae migrate through the tissues to lymph glands and ducts and mature and mate as they migrate. Females produce large numbers of new microfilariae, which enter the blood (**Figure 11.14d**), usually at night. The microfilariae are ingested by mosquitoes as they bite infected humans. Any one of

Over 1 billion people in 73 countries are at risk of contracting elephantiasis, and there are over 120 million people already infected.

several species of mosquitoes can serve as host. When the microfilariae reach the midgut of the mosquito, they penetrate its wall and migrate first to the thoracic muscles and then to the mosquito's mouthparts. There they can be transferred to a new human host, where the cycle is repeated.

Adult Loa loa worms migrate through subcutaneous tissue at the rate of 1 inch in 2 minutes. They are particularly troublesome while crossing the bridge of the nose.

✓**CHECKLIST**

1. How do flukes differ from tapeworms?

2. Describe the steps in the life cycle of a tapeworm, including cysticercus and hydatid cyst stages.

3. What are microfilariae? How are they usually transmitted? What do they become?

4. What are the definitive hosts in the life cycles of tapeworms and of schistosomes?

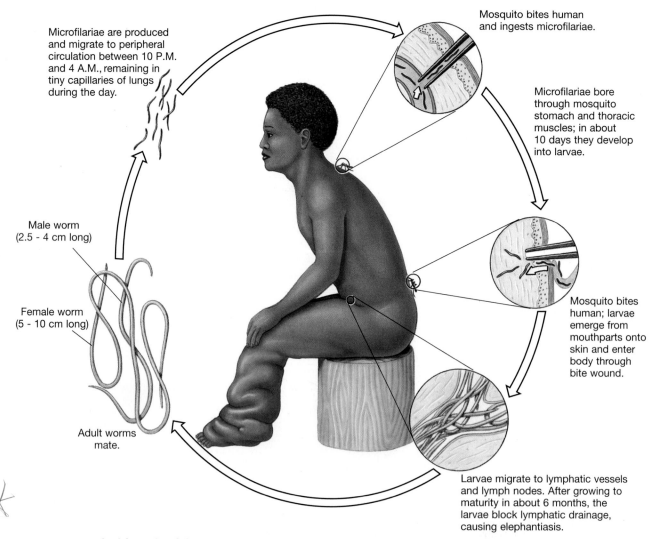

Microfilariae are produced and migrate to peripheral circulation between 10 P.M. and 4 A.M., remaining in tiny capillaries of lungs during the day.

Mosquito bites human and ingests microfilariae.

Microfilariae bore through mosquito stomach and thoracic muscles; in about 10 days they develop into larvae.

Male worm (2.5 - 4 cm long)

Female worm (5 - 10 cm long)

Mosquito bites human; larvae emerge from mouthparts onto skin and enter body through bite wound.

Adult worms mate.

Larvae migrate to lymphatic vessels and lymph nodes. After growing to maturity in about 6 months, the larvae block lymphatic drainage, causing elephantiasis.

Figure 11.19 **The life cycle of the roundworm *Wuchereria bancrofti*.** This roundworm produces microfilariae and causes elephantiasis (a chronic edema; see Figure 23.4), especially of the legs and scrotum.

ARTHROPODS

CHARACTERISTICS OF ARTHROPODS

Arthropods constitute the largest group of living organisms; as many as 80% of all animal species belong to the phylum Arthropoda. Arthropods are characterized by jointed chitinous exoskeletons, segmented bodies, and jointed appendages associated with some or all of the segments. The name arthropod is derived from *arthros*, joint, and *podos*, foot. The exoskeleton both protects the organism and provides sites for the attachment of muscles. These organisms have a true coelom, which is filled with fluid that supplies nutrients, as blood does in higher organisms. Arthropods have a small brain and an extensive network of nerves. Various groups have different structures that extract oxygen from air or from aquatic environments. The sexes are distinct in arthropods, and females lay many eggs. Arthropods are found in nearly all environments—free-living in soil, on vegetation, in fresh and salt water, and as parasites on many plants and animals.

CLASSIFICATION OF ARTHROPODS

Certain members of three subgroups (classes) of arthropods, the arachnids, insects, and crustaceans (Table 11.4),

TABLE 11.4

Properties of Three Classes of Arthropods		
Identifying Group	Characteristic	Examples
Arachnids	Have eight legs	Spiders, scorpions, ticks, mites
Insects	Have six legs	Lice, fleas, flies, mosquitoes, true bugs
Crustaceans	A pair of appendages on each body segment	Crabs, crayfish, copepods

are important either as parasites or as disease vectors (Figure 11.20). The diseases transmitted by arthropods are summarized in Table 11.5.

Arachnids

Arachnids have two body regions—a cephalothorax and an abdomen—four pairs of legs, and mouthparts that are used in capturing and tearing apart prey. They include spiders, scorpions, ticks, and mites. Spider bites and scorpion stings can produce localized inflammation and tissue death, and their toxins can produce severe systemic effects. Ticks

Figure 11.20 Representative arthropods that are parasitic or can serve as disease vectors. (a) A wood tick, *Dermacentor andersoni*. (*L. West/Photo Researchers, Inc.*) (b) False-color SEM of the pubic louse, *Phthirus pubis*, also known as a crab louse, clinging to a human pubic hair (55X). The lice suck blood, feeding about five times a day. (*Cath Wadforth/Photo Researchers, Inc.*) (c) The housefly, *Musca domestica* (4X), can carry microbes on its body. (*Runk Schoenberger/Grant Heilman Photography*) (d) The *Aedes* mosquito. (*Courtesy Centers for Disease Control and Prevention CDC*) (e) A flea, *Ctentocephalidis canis* (33X). (*A. M. Siegelman/Visuals Unlimited*)

TABLE 11.5

Diseases Transmitted by Arthropods			
Disease	**Causative Agents**	**Principal Vectors**	**Endemic Areas**
Plague	*Yersinia pestis*	Fleas	Only sporadic in modern times; reservoir of infection maintained in rodents
Tularemia	*Francisella tularensis*	Fleas and ticks	Western United States
Salmonellosis	*Salmonella* species	Flies	Worldwide
Lyme disease	*Borrelia burgdorferi*	Ticks	Parts of United States, Australia, and Europe
Relapsing fever	*Borrelia* species	Ticks and lice	Rocky Mountains and Pacific Coast of United States; many tropical and subtropical regions
Typhus fever	*Rickettsia prowazekii*	Lice	Asia, North Africa, and Central and South America
Tick-borne typhus fever	*Rickettsia conorii*	Ticks	Mediterranean area; parts of Africa, Asia, and Australia
Scrub typhus fever	*Rickettsia tsutsugamushi*	Mites	Asia and Australia
Murine typhus fever	*Rickettsia typhi*	Fleas	Tropical and subtropical regions
Rocky Mountain spotted fever	*Rickettsia rickettsii*	Ticks	United States, Canada, Mexico, and parts of South America
Q fever	*Coxiella burnetii*	Ticks and mites	Worldwide
Trench fever	*Rochalimaea quintana*	Lice	Known only in fighting armies
Viral encephalitis	Togaviruses	Mosquitoes	Worldwide but varies by virus and vector
Yellow fever	Togavirus	Mosquitoes	Tropics and subtropics
Dengue fever	Togavirus	Mosquitoes	India, Far East, Hawaii, Caribbean Islands, and Africa
Sandfly fever	A virus, probably of bunyavirus family	Female sandfly	Mediterranean region, India, and parts of South America
Colorado tick fever	An orbivirus	Ticks	Western United States
Tick-borne encephalitis	Various viruses	Ticks	Europe and Asia
African sleeping sickness	Trypanosomes	Tsetse fly	Africa
Chagas' disease	*Trypanosoma cruzi*	True bug	South America
Kala azar and other leishmaniases	*Leishmania* species	Sandfly	Tropical and subtropical regions
Malaria	*Plasmodium* species	Mosquitoes	Tropical and subtropical regions

and mites are external parasites on many animals; some also serve as vectors of infectious agents.

Infected ticks transmit several human diseases. Certain species of *Ixodes* carry viruses that cause encephalitis and the spirochete *Borrelia burgdorferi*, which causes Lyme disease. The common tick *Dermacentor andesoni*, which can cause tick paralysis, can also carry the viruses that cause encephalitis and Colorado tick fever, the rickettsiae that cause Rocky Mountain spotted fever, and the bacterium that causes tularemia. Several species of *Amblyoma* ticks also carry the Rocky Mountain spotted fever rickettsiae, and *Ornithodorus* ticks transmit the spirochete responsible for relapsing fever. Mites serve as vectors for the rickettsial disease scrub typhus and Q fever.

Recent studies in the United States show that 24% of people being treated for one tickborne disease are actually also infected with a second or third such disease. These infections are therefore called *polymicrobial*. Treatment that kills one of the microbes may leave the patient suffering from the other undiagnosed diseases.

Insects

Insects have three body regions—head, thorax, and abdomen—three pairs of legs, and highly specialized mouthparts. Some insects have specialized mouthparts for piercing skin and for sucking blood, and can inflict painful bites. Insects that can serve as vectors of disease include all lice and fleas and certain flies, mosquitoes, and true bugs, such as bedbugs and reduviid bugs. Although we often refer to all insects as "bugs," entomologists—scientists who study insects—use the term *true bug* to refer to certain insects that typically have thick, waxy wings and sucking, rather than biting, mouthparts.

The crab, or pubic, louse dies in 2 days without food. Both sexes suck blood.

The body louse is the main vector for the rickettsiae that cause typhus and trench fevers and a spirochete that causes relapsing fever. (This spirochete is a different species of *Borrelia* from the one carried by ticks.) Epidemics of all louseborne diseases usually occur under

Head and body lice interbreed, producing fertile offspring.

crowded, unsanitary conditions. All louseborne disease agents enter the body when louse feces are scratched into bite wounds.

The human flea, *Pulex irritans*, lives on other hosts and can transmit plague. However, fleas that normally parasitize rats and other rodents are more likely to transmit plague to humans. This bacterial disease still occurs in the United States in individuals who have had contact with wild rodents and their fleas.

Several kinds of flies feed on humans and serve as vectors for various diseases. The common housefly, *Musca domestica*, is not part of the life cycle of any pathogens, yet it is an important carrier of any pathogens found in feces. This fly is attracted to both human food and human excreta, and it leaves a trail of bacteria, vomit, and feces wherever it goes. Other insects, such as blackflies, serve as vectors for *Onchocerca volvulus*, which causes river blindness. Sandflies serve as vectors for leishmanias, for bacteria that cause bartonellosis, and for viruses that cause sandfly fever and several other diseases. Tsetse flies are vectors for trypanosomes that cause African sleeping sickness, and deer flies are vectors for the worm that causes loaiasis. Eye gnats, which look like tiny houseflies, may be responsible for transmission of bacterial conjunctivitis and the spirochete that causes yaws.

Many species of mosquitoes serve as vectors for diseases. *Culex pipiens*, a common mosquito, breeds in any water and feeds at night. It is a vector for *Wuchereria*. Another mosquito, *C. tarsalis*, breeds in water in sunny locations and also feeds at night. It is a vector for viruses that cause western equine encephalitis (WEE) and St. Louis encephalitis. Although WEE most often causes severe illness in horses, it also can cause severe encephalitis in children and a milder disease, with fever and central nervous system infections, in adults. (The latter form is sometimes called sleeping sickness, but it should not be confused with African sleeping sickness.) Many species of *Aedes* play a role in human discomfort and disease. *Aedes aegypti* is a vector of a variety of viral diseases, including dengue fever (breakbone fever), yellow fever, and epidemic hemorrhagic fever. Several species of *Anopheles* serve as vectors for malaria. They have a variety of breeding habits, and thus control of them requires the application of several different eradication methods.

Several species of reduviid bugs transmit the parasite that causes Chagas' disease, which is a leading cause of cardiovascular disorders in Central and South America. Bedbugs cause dermatitis and may be responsible for spreading one kind of hepatitis, a liver infection.

Crustaceans

Crustaceans are generally aquatic arthropods that typically have a pair of appendages associated with each segment. Appendages include mouthparts, claws, walking legs, and appendages that aid in swimming or in copu-

Figure 11.21 Embryo Japanese angelshark infected with parasite copepods. Long abdomened adult female copepods suck blood as ectoparasites, but while living inside the shark uterus as endosymbionts—an endosymbiotic octoparasite! *(Photo courtesy of George W. Benz, from K. Nagasawa, et al., The Journal of Parasitology, Vol. 84, No. 6, pp. 1218–1330 Dec. 1998)*

lation. Crustaceans that are hosts for disease agents that infect humans include some crayfish, crabs, and smaller crustaceans called copepods. Guinea worms are transmitted by copepods.

One very unusual copepod, *Trebius shiinoi*, is both an ectoparasite and an endosymbiont at the same time (**Figure 11.21**). Adult female copepods live inside the uterus of the Japanese angelshark, *Squatina japonica*, thus qualifying as endosymbionts, while sucking blood from the surfaces of shark embryos developing inside the uterus, and thus acting as ectoparasites. The world of parasitology is filled with amazing examples of biological flexibility!

✓ CHECKLIST

1. How do arthropods differ from other parasites?
2. Name the three classes or subgroups of arthropods that are associated with human disease.

∎ RETRACING OUR STEPS

∎∎∎ PRINCIPLES OF PARASITOLOGY

- A **parasite** is an organism that lives at the expense of another organism, the **host**. **Pathogens** are parasites that cause disease.
- **Parasitology** is the study of parasites, which typically include protozoa, helminths, and arthropods.

THE SIGNIFICANCE OF PARASITISM

- Parasites are responsible for much disease and death of humans, plants, and animals and for extensive economic losses.

PARASITES IN RELATION TO THEIR HOSTS

- Parasites can live on or in hosts. Parasites can be **obligate** or **facultative** and **permanent**, **temporary**, or **accidental**. **Vectors** are agents of parasite transmission.
- Parasites reproduce sexually in **definitive hosts** and spend other life stages in **intermediate hosts**. **Reservoir hosts** can transmit parasites to humans.
- **Host specificity** refers to the number of different hosts in which a parasite can mature.
- Over time parasites become more adapted to and less destructive of their hosts. Most parasites have mechanisms to evade host defenses and exceptionally adept reproductive capacities.

∎∎∎ PROTISTS

CHARACTERISTICS OF PROTISTS

- **Protists** are eukaryotic, and most are unicellular. They can be autotrophic or heterotrophic, and some are parasitic.

THE IMPORTANCE OF PROTISTS

- Protists are important in food chains as producers and decomposers; they can be economically beneficial or detrimental.

CLASSIFICATION OF PROTISTS

- Protists include plantlike organisms (such as **euglenoids**, **diatoms**, and **dinoflagellates**), funguslike organisms (the **water molds** and **slime molds**), and animal-like organisms (the **protozoa**, such as **mastigophorans**, **amebozoa**, **apicomplexans**, and **ciliates**). The groups of protists are summarized in Table 11.1.
- **Saprophytes** are organisms that feed on dead matter.

∎∎∎ FUNGI

CHARACTERISTICS OF FUNGI

- **Fungi** are saprophytes or parasites that generally have a **mycelium**, a loosely organized mass consisting of threadlike

hyphae. Most fungi reproduce both sexually and asexually, and their sexual stages are used to classify them.

THE IMPORTANCE OF FUNGI

- Fungi are important as decomposers in ecosystems and as parasites in the health sciences.

CLASSIFICATION OF FUNGI

- Fungi include **bread molds**, **sac fungi**, **club fungi**, and the **Fungi Imperfecti**, which cannot be classified in another group because either they lack a sexual stage or none has yet been identified. The groups of fungi are summarized in Table 11.2.

∎∎∎ HELMINTHS

CHARACTERISTICS OF HELMINTHS

- **Helminths**, or worms, are bilaterally symmetrical and have head and tail ends and differentiated tissue layers.

PARASITIC HELMINTHS

- Only two groups of helminths, the flatworms and the roundworms (nematodes), contain parasitic species.
- **Flatworms** lack a **coelom**, have a simple digestive tract with one opening, and are **hermaphroditic**. They include **tapeworms** and **flukes**.
- **Roundworms** have a **pseudocoelom**, separate sexes, and a cylindrical body. They include hookworms, pinworms, and other parasites of the intestinal tract and lymphatics.

∎∎∎ ARTHROPODS

CHARACTERISTICS OF ARTHROPODS

- **Arthropods** have jointed chitinous exoskeletons, segmented bodies, and jointed appendages.

CLASSIFICATION OF ARTHROPODS

- Parasitic and vector arthropods include some arachnids and insects; a few crustacea also serve as intermediate hosts for human parasites. Arthropod vectors of disease are summarized in Table 11.5.
- **Arachnids** have eight legs; they include scorpions, spiders, ticks, and mites.
- **Insects** have six legs; they include lice, fleas, flies, mosquitoes, and true bugs.
- **Crustaceans** are generally aquatic arthropods, typically with a pair of appendages on each segment; they include crayfish, crabs, and copepods.

∎ TERMINOLOGY CHECK

▌ CLINICAL CASE STUDY

George has recently returned from a trip to a family farm where they raise pigs and vegetables. They had a wonderful meal of roast pork and fresh vegetables. He had some diarrhea shortly after returning from his trip. A few weeks later he is experiencing severe muscle and joint pain, nausea, and fever.

What is the most probable etiologic agent? Where would this agent be found in George? What would it look like? What group of organisms is the pathogen in? How did George get infected?

▌ CRITICAL THINKING QUESTIONS

1. Cultures and other lab tests have confirmed that your patient has a life-threatening respiratory infection caused by a species of fungus that normally lives on a decaying organism. What are some underlying causes for this opportunistic infection that you would want to test for?

2. Protists classified as autotrophic are traditionally thought to be beneficial to the environment with their capacity for harnessing energy from sunlight, serving as the base for food chains, and recycling decomposed or dead organic matter. Can you think of a few examples in which they are detrimental to the environment?

3. Do you think that a big dose of penicillin would kill anything discussed in this chapter? Explain.

▌ SELF-QUIZ

1. Parasites that must spend at least some of their life cycle in or on a host are called _____ parasites whereas parasites that can either live on a host or freely are called _____ parasites.

2. Parasites may damage their host's body by:
   (a) Taking nutrients from the host
   (b) Clogging and damaging blood vessels
   (c) Triggering inflammatory responses
   (d) Causing internal hemorrhages
   (e) All of these

3. Match the following (more than one may apply):

   ___ Lice
   ___ Tapeworm
   ___ Biting mosquito
   ___ Housefly walking
       on manure
   ___ Ringworm fungus

   (a) Ecotoparasite
   (b) Endoparasite
   (c) Facultative parasite
   (d) Permanent parasite
   (e) Temporary parasite
   (f) Biological vector
   (g) Mechanical vector

4. All of the following are mechanisms used by parasites to evade host defenses EXCEPT:
   (a) Parasite causes host's immune system to make antibodies that cannot react with the parasite's antigens.
   (b) Encrystment.
   (c) Parasite kills the host.
   (d) Parasite changes its surface antigen faster than the host can make new antibodies.
   (e) Parasite invades host cells where it is out of the reach of host defense mechanisms.

5. Match the following parasitic terms to their definitions:
   ___ Accidental parasite
   ___ Host specificity
   ___ Intermediate host
   ___ Reservoir host
   ___ Definitive host
   ___ Obligate parasite

   (a) Host that harbors a parasite while it reproduces sexually
   (b) Parasite that invades an organism other than its normal host
   (c) Must spend at least some of its life cycle on or in host
   (d) Harbors a parasite during any part of its developmental stage except the sexual reproductive stage
   (e) Range of different hosts in which a parasite can mature
   (f) Infected organisms that make parasites available for transmission to other hosts

6. Which of the following is a reproductive method parasites use to overcome their death during transfer one host to another?
   (a) Schizogony (multiple cell fission)
   (b) Hermaphrodite (both sexes found within one organism)
   (c) Production of large numbers of eggs
   (d) a, b, and c
   (e) a and c

7. Group the following protists according to their characteristics:
   ___ Slime mold
   ___ Ciliate
   ___ Dinoflagellate
   ___ Apicomplexan
   ___ Euglenoid
   ___ Diatom
   ___ Mastigophoran
   ___ Amebozoa

   (a) Plantlike, contain chloroplasts, and live in moist, sunny environments
   (b) Funguslike, most are saprophytes, and can be uni- or multicellular
   (c) Animal-like, most are unicellular and free-living, and some are commensals or parasites

8. Most slime molds are _____ or organisms that feed on dead or decaying matter, _____ _____ _____ produce fruiting bodies, spores, and pseudoplasmodia. A _____ is a slightly mobile, loose aggregation of cells formed due to lack of nutrients. The _____ _____ _____ form multinucleate, amoeboid masses called plasmodia that move about slowly and phagocytize dead matter.

9. With which of the following is a dinoflagellate associated?
   (a) Most usually have flagella and can carry out photosynthesis.
   (b) Some produce toxins that accumulate in the bodies of shellfish.
   (c) Blooms of dinoflagellates are known as the "red tide."
   (d) Inhalation of air that contains small quantities of dinoflagellate toxin can cause respiratory membrane irritation in sensitive individuals.
   (e) All of the above are characteristics associated with dinoflagellates.

10. The protozoan group so named for digestive enzyme complexes located in organelles at the tips (apices) of their cells are known as:
    (a) Ciliates
    (b) Apicomplexans
    (c) Mastigophorans
    (d) Sarcodines
    (e) Water molds

11. All of the following are general characteristics of Fungi EXCEPT:
    (a) Many fungi reproduce both sexually and asexually via budding.
    (b) The body or thallus of multicellular fungi consists of a mycelium, which is a loosely organized mass of threadlike structures called hyphae that are used for embedding within and digestion of decaying organic matter, living tissue, or soil.
    (c) All hyphal cells of all fungi are separated by crosswalls called septa.
    (d) Saprophytic fungi are beneficial as decomposers and producers of antibiotics.
    (e) Parasite fungi are medically, economically, and environmentally important because of their ability to produce diseases in plants, animals, and man.

12. An important yet diverse group of fungi that includes those responsible for the color, texture, and flavor of some cheeses, the leavening of bread, the alcohol content of beer and wines, and the producer of pencillium are the:
    (a) Ascomycetes
    (b) Basidiomycetes
    (c) Zygomycetes
    (d) Deuteromycetes
    (e) Lichens

13. Fungi are not classified in the Kingdom Plantae primarily because they:
    (a) Have unicellular and multicellular forms
    (b) Are prokaryotes
    (c) Are heterotrophs
    (d) Need high moisture
    (e) Reproduce sexually

14. Match the following microorganisms and their descriptions:
    ___ Sac fungi
    ___ Water molds
    ___ Bread molds
    ___ Club fungi
    ___ Dimorphic fungi

    (a) Produce zygospores
    (b) Produce ascospores in an ascus
    (c) Produce basidiospores
    (d) Produce motile sexual and asexual spores
    (e) Exhibit yeastlike growth at 37°C and moldlike growth at 25°C

15. The symbiotic association between fungi and the roots of plants is called:
    (a) Hyphae
    (b) Plasmogamy
    (c) Mycorrhizae
    (d) Karyogamy
    (e) Dimorphism

16. Fungi Imperfecti or Deuteromycota:
(a) Do not have complete hyphae
(b) Do not form hyphae
(c) Have no observed sexual stage
(d) Cannot form conidia
(e) Only form antheridia

17. Which of the following is mismatched?
(a) Woronin body/organelle that blocks septal pore to protect healthy cells
(b) Basidiospores/sexual spores produced on club-shaped basidia
(c) Ergot/product of the fungus *Claviceps purpura* that is used to treat migraines
(d) Ascospores/single, thick-walled sexual spore produced between two hyphae
(e) Conidia/asexual spores at the ends of hyphae

18. The parasitic helminths that are most likely to be found in bile ducts, lungs, and blood are the:
(a) Tapeworms
(b) Roundworms
(c) Flatworms
(d) Flukes
(e) Heartworms

19. Match the following parasites and their descriptions:
___ *Wuchereria bancrofti*
___ *Taenia* species
___ *Trichinella spiralis*
___ *Enterobius vermicularis*
___ *Fasciola hepatica*

(a) Pinworms that infect the intestines
(b) Enter through digestive tract and then remain as cysts in human muscle
(c) Beef and pork tapeworms
(d) Microfilariae live in lymphatic tissue and cause elephantiasis
(e) Liver flukes

20. What portion of the tapeworm is responsible for attachment?
(a) Coelom
(b) Neck
(c) Proglottid
(d) Scolex
(e) Cuticle

21. The spirochete that causes Lyme disease is most likely transmitted by:
(a) Mosquitoes
(b) Flies
(c) Ticks
(d) Lice
(e) Roaches

22. Match the following diseases transmitted by arthropods to their causative agent and principal vector:
___ Yellow fever
___ Rocky Mountain spotted fever
___ African sleeping sickness
___ Dengue fever
___ Q fever
___ Lyme disease

(a) *Rickettsia rickettsii*
(b) *Borrelia burgdorferi*
(c) *Coxiella burnetti*
(d) *Francisella tularensis*
(e) Trypanosomes
(f) Togavirus

(1) Lice
(2) Mosquitoes
(3) Ticks
(4) Tsetse fly
(5) Fleas
(6) Mites

23. The unicellular, eukaryotic organisms with a true nucleus and organelles that are membrane-bound are classified as:
(a) Arthropods
(b) Lichens
(c) Flukes
(d) Protista
(e) Zygomycetes

24. Parasites that have a jointed, chitinous exoskeleton with segmented bodies and jointed appendages would be classified as:
(a) Protists
(b) Fungi
(c) Helminths
(d) Arthropods
(e) Protists

25. Match the following parasites with their characteristics
___ Arachnids
___ Crustaceans
___ Insects

(a) A scolex with proglottids
(b) Hyphae with conidiospores
(c) A pair of appendages on each body segment
(d) Have six legs
(e) Have eight legs

26. In the following diagram of a tapeworm, identify parts (a) and (b), the oldest proglottids, and the newest proglottids.

(a) _____

(b) _____

∎ EXPLORATIONS ON THE WEB
http://www.wiley.com/college/black

If you think you've mastered this chapter, there's more to challenge you on the web. Go to the companion web site to fine-tune your understanding of the chapter concepts and discover answers to the questions posed below.

1. The mortality rate due to malaria is estimated to be at over 1 million deaths each year and, because of global warming, the disease is spreading. In 1997, 2,364 cases of malaria were reported in the United Kingdom. How is malaria getting to England?

2. Dinoflagellates have their own "burglar alarm": they bioluminesce. How do dinoflagellates bioluminesce, and why are they called fire plants?

Sterilization and Disinfection

Come with me . . .

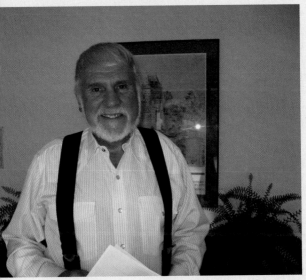

So you've just had your ears pierced. That bottle of antiseptic they sold you to put on your ears until they heal—does it really kill microbes? Those chemicals that they add to your local swimming pool—do they really keep the water safe? What about the disinfectants they use in the hospital—can they really kill those most virulent microbes from the sickest patients? Spores? Tuberculosis bacilli? What about a bioterrorist attack—what can we rely on to kill anthrax spores? The endoscope they used to examine someone else's bowel, and are now going to reuse on you—did the chemical sterilant really kill all the microbes? Wow! Sterilization and disinfection are really important.

Meet Dr. Norman Miner, head of MicroChem Laboratory, located in Euless, Texas, just outside Dallas. I travelled there to see how his laboratory tests antimicrobial products and certi-

Dr. Norman Miner. Courtesy MicroChem Laboratory, Euless, TX

fies them as effective or not. I first met him at an ASM (American Society for Microbiology) Conference on Biodefense where he presented a formula for a cheap, easy sterilant that every American household can have on hand. It even kills the oldest anthrax spores, which are the toughest to kill. Keep this recipe handy:

1 gallon water
1 cup of bleach
1 cup of vinegar

Mix in a well-ventilated area and use within 8 hours. Apply for 20 minutes, and then rinse to remove the chlorine.

 Video related to this topic is available within WileyPLUS.

Aspergillis niger growing on ceramic tiles in a mildew fungistatic test. Courtesy MicroChem Laboratory, Euless, TX

⫼ Principles of Sterilization and Disinfection

The Control of Microbial Growth

⫼ Chemical Antimicrobial Agents

The Potency of Chemical Agents / Evaluating the Effectiveness of Chemical Agents / Disinfectant Selection / Mechanisms of Action of Chemical Agents / Specific Chemical Antimicrobial Agents

⫼ Physical Antimicrobial Agents

Principles and Applications of Heat Killing / Dry Heat, Moist Heat, and Pasteurization / Refrigeration, Freezing, Drying, and Freeze-Drying / Radiation / Sonic and Ultrasonic Waves / Filtration / Osmotic Pressure

Do you like spicy foods? Perhaps you won't like the original reasons for their popularity. Before modern methods of food preservation, such as canning and refrigeration, were available, control of microbial growth in foods was a difficult problem. Inevitably after a short while, food began to take on the "off" flavors of spoilage. Spices were used to mask these unpleasant tastes. Some spices were also effective as preservatives. The antimicrobial effects of garlic have long been known. Fortunately, we need not eat spoiled food today, and we can use spices solely to enhance our enjoyment of safely preserved foods.

Medical care, especially in the operating room, is also safer today. As we have seen from the work of Ignaz Semmelweis and Joseph Lister, careful washing and the use of chemical agents are effective in controlling many infectious microorganisms (◄Chapter 1, p. 14). In this chapter, we will consider the properties of various chemical and physical agents used to control microorganisms in laboratories, in medical facilities, and in homes. Go to the website for this chapter to read about the specialized career of hospital infection control practitioner. This is a board-certified career open to both nurses and biology majors. The web essay also details specific methods of sterilization and disinfection used in hospitals.

PRINCIPLES OF STERILIZATION AND DISINFECTION

Sterilization is the killing or removal of all microorganisms in a material or on an object. There are no degrees of sterility—**sterility** means that there are *no* living organisms in or on a material. When properly carried

BIOTECHNOLOGY

Microbes in Space

Never a problem for Han Solo, but in the real world of space travel, bacteria carried by astronauts are of real concern. Some of the problems include infectious diseases, allergy to microbial metabolites, and microbial deterioration of structural materials. Preventing microbial problems inside space vehicles requires limiting the routes of infectious disease and the use of an efficient wastewater recovery system. Although sterility is not possible, most of the familiar routes of disease transmission are present within space vehicles: water, food, aerosols, and environmental surfaces. Providing potable water for extended space travel has been a challenge from the very beginning of the space program. For the past 15 years various prototype wastewater recovery systems have been under development at NASA's (National Aeronautics and Space Administration) Marshall Space Flight Center. Using these systems, potable water is produced when humidity condensate, hygiene water, urine, and fuel cell water are collected and treated to remove microbes and chemical contaminants.

out, sterilization procedures ensure that even highly resistant bacterial endospores and fungal spores are killed. Much of the controversy regarding spontaneous generation in the nineteenth century resulted from the failure to kill resistant cells in materials that were thought to be sterile. In contrast with sterilization, **disinfection** means reducing the number of pathogenic organisms on objects or in materials so that they pose no threat of disease.

Agents called **disinfectants** are typically applied to inanimate objects, and agents called **antiseptics** are applied to living tissue. A few agents are suitable as both disinfectants and antiseptics, although most disinfectants are too harsh for use on delicate skin tissue. *Antibiotics*, though often applied to skin, are considered separately in ◄Chapter 13. Terms related to sterilization and disinfection are defined in **Table 12.1**.

THE CONTROL OF MICROBIAL GROWTH

As explained in the discussion of the growth curves in ◄Chapter 6 (p. 149), both the growth and death of microorganisms occur at logarithmic rates. Here we are concerned with the death rate and the effects on it of antimicrobial agents—substances that kill microbes or inhibit their growth.

Organisms treated with antimicrobial agents obey the same laws regarding death rates as those declining in numbers from natural causes. We will illustrate this principle with heat as the agent because its effects have been the most thoroughly studied. When heat is applied to a material, the death rate of the organisms in or on it remains logarithmic but is greatly accelerated. Heat acts as an antimicrobial agent. If 20% of the organisms die in the first minute, 20% of those remaining alive will die in the second minute, and so on. If, at a different temperature, 30% die in the first minute, 30% of the remaining ones will die in the second minute, and so on. From these observations we can derive the principle that *a definite proportion of the organisms die in a given time interval.*

Consider now what happens when the number of live organisms that remain becomes small—100, for example, At a death rate of 30% per minute, 70 will remain after 1 minute, 49 after 2 minutes, 34 after 3 minutes, and only 1 after 12 minutes. Soon the probability of finding even a single live organism becomes very small. Most laboratories say a sample is sterile if the probability is no greater than one chance in a million of finding a live organism.

The total number of organisms present when disinfection is begun affects the length of time required to eliminate them. We can state a second principle: *The fewer organisms present, the shorter the time needed to achieve sterility.* Thoroughly cleaning objects before attempting to sterilize them is a practical application of this principle. Clearing objects of tissue debris and blood is also important because such organic matter impairs the effectiveness of many chemical agents.

Different antimicrobial agents affect various species of bacteria and their endospores differently. Furthermore, any

TABLE 12.1

Terms Related to Sterilization and Disinfection	
Term	**Definition**
Sterilization	The killing or removal of all microorganisms in a material or on an object.
Disinfection	The reduction of the number of pathogenic microorganisms to the point where they pose no danger of disease.
Antiseptic	A chemical agent that can safely be used externally on living tissue to destroy microorganisms or to inhibit their growth.
Disinfectant	A chemical agent used on inanimate objects to destroy microorganisms. Most disinfectants do not kill spores.
Sanitizer	A chemical agent typically used on food-handling equipment and eating utensils to reduce bacterial numbers so as to meet public health standards. Sanitization may simply refer to thorough washing with only soap or detergent.
Bacteriostatic agent	An agent that inhibits the growth of bacteria.
Germicide	An agent capable of killing microbes rapidly; some such agents effectively kill certain microorganisms but only inhibit the growth of others.
Bactericide	An agent that kills bacteria. Most such agents do not kill spores.
Viricide	An agent that inactivates viruses.
Fungicide	An agent that kills fungi.
Sporocide	An agent that kills bacterial endospores or fungal spores.

given species may be more susceptible to an antimicrobial agent at one phase of growth than at another. The most susceptible phase for most organisms is the logarithmic growth phase, because during that phase many enzymes are actively carrying out synthetic reactions, and interfering with even a single enzyme might kill the organism. From these observations, we can state a third principle: *Microorganisms differ in their susceptibility to antimicrobial agents.*

CHEMICAL ANTIMICROBIAL AGENTS

THE POTENCY OF CHEMICAL AGENTS

The potency, or effectiveness, of a chemical antimicrobial agent is affected by time, temperature, pH, and concentration. The death rate of organisms is affected by the length of time the organisms are exposed to the antimicrobial agent, as was explained earlier for heat. Thus, adequate time should always be allowed for an agent to kill the maximum number of organisms. The death rate of organisms subjected to a chemical agent is accelerated by increasing the temperature. Increasing temperature by 10°C roughly doubles the rate of chemical reactions and thereby increases the potency of the chemical agent. Acidic or alkaline pH can increase or decrease the agent's potency. A pH that increases the degree of ionization of a chemical agent often increases its ability to penetrate a cell. Such a pH also can alter the contents of the cell itself. Finally, increasing concentration may increase the effects of most antimicrobial chemical agents. High concentrations may be **bactericidal** (killing), whereas lower concentrations may be **bacteriostatic** (growth inhibiting).

Both ethyl and isopropyl alcohol are exceptions to the rule about increasing concentrations. They have long been believed to be more potent at 70% than at higher concentrations, although they are also effective at up to 99% concentration. Some water must be present for alcohols to disinfect because they act by coagulating (permanently denaturing) proteins, and water is needed for the coagulation reactions. Also, a 70% alcohol-water mixture penetrates more deeply than pure alcohol into most materials to be disinfected.

EVALUATING THE EFFECTIVENESS OF CHEMICAL AGENTS

Many factors affect the potency of chemical antimicrobial agents, so evaluation of effectiveness is difficult. No entirely satisfactory method is available. However, we need some way to compare the effectiveness of disinfecting agents, especially as new ones come on the market. Should you believe the salesman when he tells you his is better? Ask him what its phenol coefficients are.

The Phenol Coefficient
Since Lister introduced *phenol* (carbolic acid) as a disinfectant in 1867, it has been the standard disinfectant to which other disinfectants are compared under the same conditions. The result of this comparison is called the **phenol coefficient**. Two organisms, *Salmonella typhi*, a pathogen of the digestive system, and *Staphylococcus aureus*, a common wound pathogen, are typically used to determine phenol coefficients. A disinfectant with a phenol coefficient of 1.0 has the same effectiveness as phenol. A coefficient less than 1.0 means that the disinfectant is less effective than phenol; a coefficient greater than 1.0 means that it is more effective. Phenol coefficients are reported separately for the different test organisms **(Table 12.2)**. Lysol, for instance, has a coefficient of 5.0

TABLE 12.2

Phenol Coefficients of Various Chemical Agents		
Chemical Agent	*Staphylococcus aureus*	*Salmonella typhi*
Phenol	1.0	1.0
Chloramine	133.0	100.0
Cresols	2.3	2.3
Ethyl alcohol	6.3	6.3
Formalin	0.3	0.7
Hydrogen peroxide	—	0.01
Lysol	5.0	3.2
Mercury chloride	100.0	143.0
Tincture of iodine	6.3	5.8

(a)

(b)

Figure 12.1 **The filter paper method of evaluating disinfectants and antiseptics.** The difference in response of **(a)** *Staphylococcus aureus* (Gram-positive) *(Jack M. Bostrack/Visuals Unlimited)* and **(b)** *Escherichia coli* (Gram-negative) to several common chemical agents is shown here. In both cases, the greatest inhibition of growth is seen near the top of the Petri dish surrounding the merthiolate-soaked paper disk. The various disks were soaked in mercurochrome, iodine, Signal® mouthwash, or Listerine® mouthwash before being placed on the surface of the nutrient medium, which had first been confluently inoculated with one of the test organisms. *(Jack M. Bostrack/Visuals Unlimited)*

against *Staphylococcus aureus* but only 3.2 when used on *Salmonella typhi*, whereas ethyl alcohol has a value of 6.3 against both.

The phenol coefficient can be determined by the following steps. Prepare several dilutions of a chemical agent, and place the same volume of each in different test tubes. Prepare an identical set of test tubes, using phenol dilutions. Put both sets of tubes in a 20°C water bath for at least 5 minutes to ensure that the contents of all tubes are at the same temperature. Transfer 0.5 ml of a culture of a standard test organism to each tube. After 5, 10, and 15 minutes, use a sterile loop to transfer a specific volume of liquid from each tube into a separate tube of nutrient broth, and incubate the tubes. After 48 hours, check cultures for cloudiness, and find the smallest concentration (highest dilution) of the agent that killed all organisms in 10 minutes but not in 5 minutes. Find the ratio of this dilution to the dilution of phenol that has the same effect. For example, if a 1:1000 dilution of a chemical agent has the same effect as a 1:100 dilution of phenol, the phenol coefficient of that agent is 10 (1,000/100). If you performed this test on a new disinfectant and obtained these results, you would have found a very good disinfectant! The phenol coefficient provides an acceptable means of evaluating the effectiveness of chemical agents derived from phenol, but it is less acceptable for other agents. Another problem is that the materials on or in which organisms are found may affect the usefulness of a chemical agent by complexing with it or inactivating it. These effects are not reflected in the phenol coefficient number.

The Filter Paper Method

The **filter paper method** of evaluating a chemical agent is simpler than determining a phenol coefficient. It uses small filter paper disks, each soaked with a different chemical agent. The disks are placed on the surface of an agar plate that has been inoculated with a test organism. A different plate is used for each test organism. After incubation, a chemical agent that inhibits growth of a test organism is identified by a clear area around the disk where the

bacteria have been killed (**Figure 12.1**). Note: What is effective against one organism may have little or no effect on the others. Will the chemical agent having the widest zone of inhibition around it be the most effective to use? It may not be. Organic matter such as blood, feces, or vomitus can interfere with its action. Also, some chemical agents may just have molecules that are able to travel faster or farther through agar than the other agents tested did.

The Use-Dilution Test

A third way of evaluating chemical agents, the **use-dilution test**, uses standard preparations of certain test bacteria. A broth culture of one of these bacteria is coated onto small stainless steel cylinders and allowed to dry. Each cylinder is then dipped into one of several dilutions of the chemical agent for 10 minutes, removed, rinsed with water, and placed into a tube of broth. The tubes are incubated and then observed for the presence or absence of growth. Agents that prevent growth at the greatest dilutions are considered the most effective. Many microbiologists feel that this measurement is more meaningful than the phenol coefficient.

DISINFECTANT SELECTION

Several qualities should be considered in deciding which disinfectant to use. An ideal disinfectant should:

1. Be fast acting even in the presence of organic substances, such as those in body fluids.

2. Be effective against all types of infectious agents without destroying tissues or acting as a poison if ingested.

3. Easily penetrate material to be disinfected without damaging or discoloring the material.

4. Be easy to prepare and stable even when exposed to light, heat, or other environmental factors.

5. Be inexpensive and easy to obtain and use.

6. Not have an unpleasant odor.

No disinfectant is likely to satisfy all these criteria, so the agent that meets the greatest number of criteria for the task at hand is chosen.

In practice, many agents are tested in a wide range of situations and are recommended for use where they are most effective. Thus, some agents are selected for sanitizing kitchen equipment and eating utensils, whereas other agents are chosen for rendering pathogenic cultures harmless. Furthermore, certain agents can be used in dilute concentration on the skin and in stronger concentration on inanimate objects.

✓ **CHECKLIST**

1. Are there degrees of sterility? Why or why not?

2. Is it worth paying extra for a "sanitized" item? Explain.

3. Disinfectant A has a phenol coefficient of 0.5; disinfectant B has one of 5.0. How do these two compare to phenol?

MECHANISMS OF ACTION OF CHEMICAL AGENTS

Chemical antimicrobial agents kill microorganisms by participating in one or more chemical reactions that damage cell components. Although the kinds of reactions are almost as numerous as the agents, agents can be grouped by whether they affect proteins, membranes, or other cell components.

Reactions That Affect Proteins

Much of a cell is made of protein, and all its enzymes are proteins. Alteration of protein structure is called *denaturation* (◄Chapter 2, p. 44, and Figure 2.18, p. 44). In denaturation, hydrogen and disulfide bonds are disrupted, and the functional shape of the protein molecule is destroyed. Any agent that denatures proteins prevents them from carrying out their normal functions. When treated with mild heat or with some dilute acids, alkalis, or other agents, for a short time, proteins are temporarily denatured. After the agent is removed, some proteins can regain their normal structure. However, most antimicrobial agents are used in a strong enough concentration over a sufficient length of time to denature proteins permanently. Permanent denaturation of a microorganism's proteins kills the organism. Denaturation is bactericidal if it permanently alters the protein so that the protein's normal state cannot be restored. Denaturation is bacteriostatic if it temporarily alters the protein, and the normal structure can be recovered **(Figure 12.2)**.

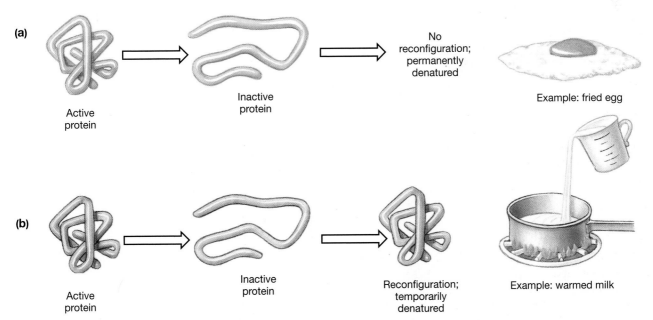

(a) Active protein → Inactive protein → No reconfiguration; permanently denatured — Example: fried egg

(b) Active protein → Inactive protein → Reconfiguration; temporarily denatured — Example: warmed milk

Figure 12.2 Denaturing proteins. (a) A permanently denatured protein, like that of a fried egg, cannot return to its original configuration. **(b)** A temporarily denatured protein, like that in warmed milk, can refold into its original configuration. The protein structure of milk that has been warmed is recovered when the milk is cooled.

Reactions that denature proteins include hydrolysis, oxidation, and the attachment of atoms or chemical groups. (Recall that hydrolysis is the breaking down of a molecule by the addition of water and that oxidation is the addition of oxygen to, or the removal of hydrogen from, a molecule (◄Chapter 2, p. 37). Acids, such as boric acid, and strong alkalis destroy protein by hydrolyzing it. Oxidizing agents (electron acceptors), such as hydrogen peroxide and potassium permanganate, oxidize disulfide linkages (—S—S—) or sulfhydryl groups (—SH). Agents that contain halogens—the elements chlorine, fluorine, bromine, and iodine—also sometimes act as oxidizing agents. Heavy metals, such as mercury and silver, attach to sulfhydryl groups. Alkylating agents, which contain methyl (—CH₃) or similar groups, donate these groups to proteins. Formaldehyde and some dyes are alkylating agents. Halogens can be substituted for hydrogen in carboxyl (—COOH), sulfhydryl, amino (—NH₂), and alcohol (—OH) groups. All these reactions can kill microorganisms.

Reactions That Affect Membranes

Membranes contain proteins and so can be altered by all the preceding reactions. Membranes also contain lipid and thus can be disrupted by substances that dissolve lipids. **Surfactants** (sur-fak'tantz) are soluble compounds that reduce surface tension, just as soaps and detergents break up grease particles in dishwater **(Figure 12.3)**. Surfactants include alcohols, detergents, and *quaternary ammonium compounds*, such as benzalkonium chloride, which dissolve lipids. Phenols, which are

Sponges should not be used in the microbiology lab, as soap does not prevent bacterial growth. Sponges spread "bacterial broth" on surfaces. Use paper towels!

alcohols, dissolve lipids and also denature proteins. Detergent solutions, also called **wetting agents**, are often used with other chemical agents to help the agent penetrate fatty substances. Although detergent solutions themselves usually do not kill microorganisms, they do help get rid of lipids and other organic materials so that antimicrobial agents can reach the organisms.

Reactions That Affect Other Cell Components

Other cell components affected by chemical agents include nucleic acids and energy-producing systems. Alkylating agents can replace hydrogen on amino or alcohol groups in nucleic aids. Certain dyes, such as crystal violet, interfere with cell wall formation. Some substances, such as lactic acid and propionic acid (end products of fermentation), inhibit fermentation and thus prevent energy production in certain bacteria, molds, and some other organisms.

Reactions That Affect Viruses

Like many cellular microorganisms, viruses can cause infections and must be controlled. Control of viruses requires that they be inactivated—that is, rendered permanently incapable of infecting or replicating in cells. Inactivation can be effected by destroying either the viruses' nucleic acid or their proteins.

Alkylating agents, such as ethylene oxide, nitrous acid, and hydroxylamine, act as chemical mutagens—they alter DNA or RNA. If the alteration prevents the DNA or RNA from directing the synthesis of new viral particles, the alkylating agents are effective inactivators. Detergents, alcohols, and other agents that denature proteins act on bacteria and viruses in the same way. Certain dyes, such as acridine orange and methylene blue, render viruses

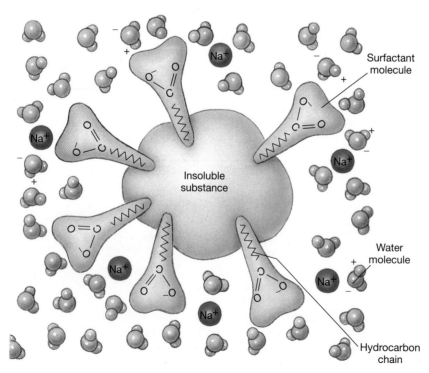

Figure 12.3   The action of a surfactant.
Here the surfactant molecule has ionized into sodium ions and long hydrocarbon chains, whose zigzagged, covalently bonded tails are able to enter an insoluble substance such as grease. The other end of these molecules has a carboxyl group with a negatively charged oxygen. These negative charges attract the positively charged sides of water molecules, thereby making the attached insoluble substance soluble in the water so the substance can be washed away.

Figure 12.4 Structural formulas of some important disinfectants.

susceptible to inactivation when exposed to visible light. This process disrupts the structure of the viral nucleic acid.

Viruses sometimes remain infective even after their proteins are denatured, so methods used to rid materials of bacteria may not be as successful with infectious viruses. Also, use of an agent that does not inactivate viruses can lead to laboratory-acquired infections.

SPECIFIC CHEMICAL ANTIMICROBIAL AGENTS

Now that we have considered general principles of sterilization and disinfection and the kinds of reactions caused by such agents, we can look at some specific agents and their applications. The structural formulas of some of the most important compounds discussed are shown in **Figure 12.4**.

Soaps and Detergents

Soaps and detergents remove microbes, oily substances, and dirt. Mechanical scrubbing greatly enhances their action. In fact, vigorous hand washing is one of the easiest and cheapest means of preventing the spread of disease among patients in hospitals, in medical and dental offices, among employees and patrons in food establishments, and among family members. Unlike surgical scrubs, germicidal soaps usually are not significantly better disinfectants than ordinary soaps.

Soaps contain alkali and sodium and will kill many species of *Streptococcus*, *Micrococcus*, and *Neisseria* and will destroy influenza viruses. Many pathogens that survive washing with soap can be killed by a disinfectant applied after washing. A common practice after washing and rinsing hands and inanimate objects is to apply a 70% alcohol solution. Even these measures do not necessarily rid hands of all pathogens. Consequently, disposable gloves are used

In public restrooms, only 68% of people are observed to wash their hands after using the toilet.

where there is a risk that health care workers may become infected or may transmit pathogens to other patients.

Detergents, when used in weak concentrations in wash water, allow the water to penetrate into all crevices and cause dirt and microorganisms to be lifted out and washed away. Detergents are said to be *cationic* if they are positively charged and *anionic* if they are negatively charged. Cationic detergents are used to sanitize food utensils. Although not effective in killing endospores, they do inactivate some viruses. Anionic detergents are used for laundering clothes and as household cleaning agents. They are less effective sanitizing agents than cationic detergents, probably because the negative charges on bacterial cell walls repel them.

Some bacterial spores can survive 20 years sitting in 70% ethyl alcohol.

PUBLIC HEALTH

Soap and Sanitation

Washing and drying clothing in modern public laundry facilities is generally a safe practice because the clothing is almost disinfected if the water temperature is high enough. Soaps, detergents, and bleaches kill many bacteria and inactivate many viruses. Agitation of the clothes in the washer provides good mechanical scrubbing. Many microbes that survive this action are killed by heat in the dryer. The use of bar soap in a public washroom is not such a safe practice. The soap may be a source of infectious agents. In a study of 84 samples of bar soap taken from public washrooms, every sample contained microorganisms. More than 100 strains of bacteria and fungi were isolated from the soap samples, and some of the organisms were potential pathogens. Many restaurants and other establishments have installed soap dispensers for this very reason. In fact, many jurisdictions have made the use of bar soap in such facilities illegal.

Figure 12.5 Heavy metals inhibit bacterial growth. The inhibitory effects of silver ions can be seen as clear zones in which no growth has occurred around the silver charm (which has been pushed aside) and silver dime. The nonsilver coin (a copper penny) has not inhibited growth of the organisms as effectively as the silver objects have. *(Centers for Disease Control and Prevention CDC)*

Many cationic detergents are **quaternary ammonium compounds**, or **quats**, which have four organic groups attached to a nitrogen atom. The ammonium ion (NH_4) has four hydrogens, each of which can be replaced by an organic group binding to the central nitrogen atom. *Quat* is the abbreviation for the Latin *quattuor* meaning "four." A variety of quats are available as disinfecting agents; their chemical structures vary according to their organic groups. One problem with quats is that their effectiveness is decreased in the presence of soap, calcium or magnesium ions, or porous substances such as gauze. An even more serious problem with these agents is that they support the growth of some bacteria of the genus *Pseudomonas* rather than killing them. Zephiran (benzalkonium chloride) was once widely used as a skin antiseptic. It is no longer recommended because it is less effective than originally thought and is subject to the same problems as other quats. It is still very often found in "ear-piercing care" kits. Quats are now often mixed with another agent to overcome some of these problems and to increase their effectiveness. Zephiran dissolved in alcohol kills about twice as many microbes in the same time as does a water (aqueous) solution of the same amount of Zephiran. Mouthwashes that foam when shaken usually contain a quat.

Acids and Alkalis

Soap is a mild alkali, and its alkaline properties help destroy microbes. A number of organic acids lower the pH of materials sufficiently to inhibit fermentation. Several are used as food preservatives. Lactic and propionic acids retard mold growth in breads and other products. Benzoic acid and several of its derivatives are used to prevent fungal growth in soft drinks, ketchup, and margarine. Sorbic acid and sorbates are used to prevent fungal growth in cheeses and a variety of other foods. Boric acid, formerly used as an eyewash, is no longer recommended because of its toxicity.

Heavy Metals

Heavy metals used in chemical agents include selenium, mercury, copper, and silver. Even tiny quantities of such metals can be very effective in inhibiting bacterial growth (**Figure 12.5**). Silver nitrate was once widely used to prevent gonococcal infection in newborn infants. A few drops of silver nitrate solution were placed in the baby's eyes at the time of delivery to protect against infection by gonococci entering the eyes during passage through the birth canal. For a time, many hospitals replaced silver nitrate with antibiotics such as erythromycin. However, the development of antibiotic-resistant strains of gonococci has led some localities to require the use of silver nitrate, to which gonococci do not develop resistance.

Organic mercury compounds, such as merthiolate and mercurochrome, are used to disinfect surface skin wounds. Such agents kill most bacteria in the vegetative state but do not kill spores. They are not effective against *Mycobacterium*. Merthiolate is generally prepared as a **tincture** (tingk′chur), that is, dissolved in alcohol. The alcohol in a tincture may have a greater germicidal action than the heavy metal compound. Thimerosal, another organic mercury compound, can be used to disinfect skin and instruments and as a preservative for vaccines. Phenylmercuric nitrate and mercuric naphthenate inhibit both bacteria and fungi and are used as laboratory disinfectants.

Selenium sulfide kills fungi, including spores. Preparations containing selenium are commonly used to treat fungal skin infections. Shampoos that contain selenium are effective in controlling dandruff. Dandruff, a crusting and flaking of the scalp, is often, though not always, caused by fungi. Mites sometimes play a role.

Copper sulfate is used to control algal growth. Although algal growth usually is not a direct medical problem, it is a problem in maintaining water quality in heating and air-conditioning systems and outdoor swimming pools. (The Environmental Protection Agency, however, is evaluating copper sulfate as an environmental hazard.)

Halogens

Hypochlorous acid, formed by the addition of chlorine to water, effectively controls microorganisms in drinking water and swimming pools. It is the active ingredient in household bleach and is used to disinfect food utensils and dairy equipment. It is effective in killing bacteria and inactivating many viruses. However, chlorine itself is easily inactivated by the presence of organic materials. That is why a substance such as copper sulfate is used to control algal growth in water to be purified with chlorine.

Iodine also is an effective antimicrobial agent. However, it should not be used on persons known to have an allergy to iodine. Often seafood allergies are triggered by iodine in the seafood. Tincture of iodine was one of the first skin antiseptics to come into use. Now *iodophors*, slow-release compounds in which the iodine is combined with organic molecules, are more commonly used. In such preparations, the organic molecules act as surfactants. Betadine and Isodine are used for surgical scrubs and on skin where an incision will be made. These compounds take several minutes to act and do not sterilize the skin. Betadine in concentrations of 3 to 5% destroys fungi, amoebas, and viruses, as well as most bacteria, but it does not destroy bacterial endospores. Contamination of Betadine with *Pseudomonas cepacia* has been reported.

Bromine is sometimes used in the form of gaseous methyl bromide to fumigate soil that will be used in the propagation of bedding plants. It is also used in some pools and indoor hot tubs because it does not give off the strong odor that chlorine does.

Chloramine, a combination of chlorine and ammonia, is less effective than other chlorine compounds at killing microbes, but superior at eliminating taste and odor problems. It is used in wound cleansing, root canal therapy, and is often added to water treatment procedures. But beware! Its residues will kill fish in aquaria and ponds. However, commercial products are available to neutralize this effect.

Alcohols

When mixed with water, alcohols denature protein. They are also lipid solvents and dissolve membranes. Ethyl and isopropyl alcohols can be used as skin antiseptics. Isopropyl alcohol is more often used because of legal regulation of ethyl alcohol. It disinfects skin where injections will be made or blood drawn. Alcohol disinfects but does not sterilize skin because it evaporates quickly and stays in contact with microbes for only a few seconds. It also does not penetrate deeply enough into pores in the skin. It kills vegetative microorganisms on the skin surface but does not kill endospores, resistant cells, or cells deep in skin pores. Ten to 15 minutes immersion in 70% ethyl alcohol is usually sufficient to disinfect a thermometer.

Phenols

Phenol and phenol derivatives called *phenolics* disrupt cell membranes, denature proteins, and inactivate enzymes. They are used to disinfect surfaces and to destroy discarded cultures because their action is not impaired by organic materials. Amphyl, which contains amylphenol, destroys vegetative forms of bacteria and fungi and inactivates viruses. It can be used on skin, medical instruments, dishes, and furniture. When used on surfaces, it retains its antimicrobial action for several days. The orthophenylphenol in Lysol gives it similar properties. A mixture of phenol derivatives called *cresols* is found in creosote, a substance used to prevent the rotting of wooden posts, fences, railroad ties, and such. However, because creosote is irritating to skin and is a carcinogen, its use is limited. The addition of halogens to phenolic molecules usually increases their effectiveness. Hexachlorophene and dichlorophene, which are halogenated phenols, inhibit staphylococci and fungi, respectively, on the skin and elsewhere. Chlorhexidine gluconate (Hibiclens), which is chlorinated and similar in structure to hexachlorophene, is effective against a wide variety of microbes even in the presence of organic material. It is a good agent for surgical scrubs.

PUBLIC HEALTH

Hexachlorophene

Hexachlorophene is an excellent skin disinfectant. In a 3% solution, it kills staphylococci and most other Gram-positive organisms, and its residue on skin is strongly bacteriostatic. Because staphylococcal skin infections can spread easily among newborn babies in hospitals, this antiseptic was used extensively in the 1960s for daily bathing of infants. The unforeseen price paid for controlling infections was permanent brain damage in infants bathed in it over a period of time. Hexachlorophene is absorbed through the skin and travels in the blood to the brain. Baby powder containing hexachlorophene killed 40 babies in France in 1972. Available in the United States today only by prescription, hexachlorophene is used routinely, though very cautiously, in hospital neonatal units because it is still the most effective agent for preventing the spread of staphylococcal infections.

APPLICATIONS

Do All Bacteria Come Out in the Wash?

In September 2006, fresh spinach was removed from the U.S. grocery store shelves. The *Escherichia coli* O157:H7 strain, which causes hemorrhagic diarrhea and very serious illness, caused 139 infections, including one death, in 26 states and Canada. These cases were traced back to contaminated spinach. Americans spend $4.4 billion per year on packaged spinach and lettuce, 80% of which is for packaged product. These packages of cut greens are sold as "ready to eat," no further washing required. Clearly something went wrong.

What should have happened? The U.S. government requires processors to follow GMP (good manufacturing practices) regulations. However, these are very general, and companies have much leeway as to how they follow them. Three washings of the leaves are standard, but how well they are washed varies. Upon arrival, workers remove branches, clumps of dirt, and obviously substandard produce. From there the produce is agitated in vats of lightly chlorinated water, to remove field debris. The next washing uses more heavily chlorinated water, usually containing 15 to 20 ppm (parts per million) of free

chlorine, much higher than ordinary tap water (3 ppm). The third wash is really more of a rinse, to remove the smell of chlorine. The produce is then spun dry and bagged.

Chlorinated water kills 90 to 99% of microbes on greens when the washing the process is done well. In one recent 6-year period, 12 out of 36 processing plants inspected by the FDA (Food and Drug Administration) failed to meet chlorination standards. Some companies used no chlorine at all; others did not even monitor the chlorine levels; and at some places the workers did not understand the measurements; another had only 1.5 to 3.0 ppm. So, are you better off washing your produce at home? Perhaps! It depends on how clean your hands, sink, utensils, etc., are. Other factors include how badly contaminated your greens are. Were cows depositing manure across the road from where your greens were grown, giving them a heavier *E. coli* contamination? Most lettuce is cored in the fields as it is harvested. Dirt entering cut surfaces is more difficult to remove, and may even move further into the plant. What to do? Recognize that eating uncooked greens constitutes a risk.

(a)

(b)

Do All Bacteria Come Out in the Wash?. **(a)** *(Andy Washnik)* **(b)** *(David Muench/Corbis)*

Trichlosan, made of two joined phenol rings, has become very popular in consumer products such as antibacterial soaps, kitchen cutting boards, highchair trays, toys, hand lotions, etc. It is fairly effective against bacteria, but does poorly against viruses and fungi. Furthermore, bacteria can develop resistance to it.

Oxidizing Agents

Oxidizing agents disrupt disulfide bonds in proteins and thus disrupt the structure of membranes and proteins. Hydrogen peroxide (H_2O_2), which forms highly reactive

superoxide (O_2^-) is used to clean puncture wounds. When hydrogen peroxide breaks down into oxygen and water, the oxygen kills obligate anaerobes present in the wounds. Hydrogen peroxide is quickly inactivated by enzymes from injured tissues. It is also very effective at disinfecting contact lenses, but all traces of it must be removed before use, or it may cause eye irritation. A recently developed method of sterilization that uses vaporized hydrogen peroxide can now be used for small rooms or areas, such as glove boxes and transfer hoods **(Figure 12.6)**. Another oxidizing agent, potassium permanganate, is used to disinfect instruments and, in low concentrations, to clean skin.

Alkylating Agents

Alkylating agents disrupt the structure of both proteins and nucleic acids. Because they can disrupt nucleic acids, these agents may cause cancer and should not be used in situations where they might affect human cells. Formaldehyde, glutaraldehyde, and β-propiolactone are used in aqueous solutions. Ethylene oxide is used in gaseous form.

Formaldehyde inactivates viruses and toxins without destroying their antigenic properties. Glutaraldehyde kills all kinds of microorganisms, including spores, and sterilizes equipment exposed to it for 10 hours. Beta-propiolactone destroys hepatitis viruses, as well as most other microbes, but penetrates materials poorly. It is, however, used to inactivate viruses in vaccines.

Gaseous ethylene oxide has extraordinary penetrating power. Used at a concentration of 500 ml/l at 50°C for 4 hours, it sterilizes rubber goods, mattresses, plastics, and other materials destroyed by higher temperatures. Also, NASA has used ethylene oxide to sterilize space probes that might otherwise carry earth microbes to other planets. Special equipment used during ethylene oxide sterilization is shown in **Figure 12.7**. As will be explained when we discuss autoclaving, an *ampule* (a sealed glass container) of endospores should be processed with ethylene oxide sterilization to check the effectiveness of sterilization.

All articles sterilized with ethylene oxide must be well ventilated for 8 to 12 hours with sterile air to remove all traces of this toxic gas, which can cause burns if it reaches living tissues and is also highly explosive. After exposure to ethylene oxide, articles such as catheters, intravenous lines, in-line valves, and rubber tubing must be thoroughly flushed with sterile air. Both the toxicity and flammability of ethylene oxide can be reduced by using it in gas that contains 90% carbon dioxide. *It is exceedingly important that workers be protected from ethylene oxide vapors, which are toxic to skin, eyes, and mucous membranes and may also cause cancer.*

APPLICATIONS

No More Frogs in Formalin

Formalin, a 37% aqueous solution of formaldehyde, was for many years the standard material used to preserve laboratory specimens for dissection. But formaldehyde is toxic to tissues and may cause cancer, so it is rarely used today as a preservative. A variety of other preservatives are now used. Although these are less toxic to students performing dissections, they are also less effective for long-term preservation. Molds growing on the surface of specimens are now a common problem.

Dyes

The dye acridine, which interferes with cell replication by causing mutations in DNA (◀Chapter 7, p. 200), can be used to clean wounds. Methylene blue inhibits growth of some bacteria in cultures. Crystal violet (gentian violet) blocks cell wall synthesis, possibly by the same reaction that causes this dye to bind to cell wall material in Gram staining. It effectively inhibits growth of Gram-positive bacteria in cultures and in skin infections. It can be used to treat protozoan (*Trichomonas*) and yeast (*Candida albicans*) infections.

Other Agents

Certain plant oils have special antimicrobial uses. Thymol, derived from the herb thyme, is used as a preservative, and eugenol, derived from oil of cloves, is used in dentistry to disinfect cavities. A variety of other agents are used primarily as food preservatives. They include sulfites and sulfur dioxide, used to preserve dried fruits and molasses; sodium diacetate, used to retard mold in bread; and sodium nitrite, used to preserve cured meats and some cold cuts. Foods containing nitrites should be eaten in moderation because the nitrites are converted during digestion to substances that may cause cancer.

The properties of chemical antimicrobial agents are summarized in **Table 12.3**.

Figure 12.6 Hydrogen peroxide disinfection. Recently developed biodecontamination equipment uses vaporized hydrogen peroxide to sterilize small, sealable enclosures such as glove boxes, hoods, or transfer rooms. It is not sufficient, however, to sterilize a larger space such as an operating room. *(Photo of VHP(R) 1000 courtesy of STERIS Corporation)*

Figure 12.7 Ethylene oxide sterilization. This equipment must be used very carefully, as ethylene oxide is both explosive and carcinogenic. An aerator is used to remove all traces of the gas from sterilized material. *(Photo of Eagle (R) 3017 100% EC Sterllizer courtesy of STERIS Corporation)*

TABLE 12.3

Properties of Chemical Antimicrobial Agents		
Agent	**Actions**	**Uses**
Soaps and detergents	Lower surface tension, make microbes accessible to other agents	Hand washing, laundering, sanitizing kitchen and dairy equipment
Surfactants	Dissolve lipids, disrupt membranes, denature proteins, and inactivate enzymes in high concentrations; act as wetting agents in low concentrations	Cationic detergents are used to sanitize utensils; anionic detergents to launder clothes and clean household objects; quaternary ammonium compounds are sometimes used as antiseptics on skin.
Acids	Lower pH and denature proteins	Food preservation
Alkalis	Raise pH and denature proteins	Found in soaps
Heavy metals	Denature proteins	Silver nitrate is used to prevent gonococcal infections, mercury compounds to disinfect skin and inanimate objects, copper to inhibit algal growth, and selenium to inhibit fungal growth.
Halogens	Oxidize cell components in absence of organic matter	Chlorine is used to kill pathogens in water and to disinfect utensils; iodine compounds are used as skin antiseptics.
Alcohols	Denature proteins when mixed with water	Isopropyl alcohol is used to disinfect skin; ethylene glycol and propylene glycol can be used in aerosols.
Phenols	Disrupt membranes, denature proteins, and inactivate enzymes; not impaired by organic matter	Phenol is used to disinfect surfaces and destroy discarded cultures; amylphenol destroys vegetative organisms and inactivates viruses on skin and inanimate objects; chlorhexidine gluconate is especially effective as a surgical scrub.
Oxidizing agents	Disrupt disulfide bonds	Hydrogen peroxide is used to clean puncture wounds, potassium permanganate to disinfect instruments.
Alkylating agents	Disrupt structure of proteins and nucleic acids	Formaldehyde is used to inactivate viruses without destroying antigenic properties, glutaraldehyde to sterilize equipment, betapropiolactone to destroy hepatitis viruses, and ethylene oxide to sterilize inanimate objects that would be harmed by high temperatures.
Dyes	May interfere with replication or block cell wall synthesis	Acridine is used to clean wounds, crystal violet to treat some protozoan and fungal infections.

✓**CHECKLIST**

1. How does a surfactant act?

2. If bacteria can grow in soap, why do we use it to clean things?

3. Explain the antimicrobial actions of acridine, mercurochrome, and Lysol.

4. What are the drawbacks of using ethylene oxide?

PUBLIC HEALTH

Is Cleanliness Really All That Great?

As part of a class investigative project, a student found that while she did not use the disinfectant Trichlosan, it was present in her skin. She lived with someone who did use Trichlosan. Join us on the web site for this chapter to examine the debates about the use of Trichlosan. Do you want it on your cutting board?

PHYSICAL ANTIMICROBIAL AGENTS

For centuries, physical antimicrobial agents have been used to preserve food. Ancient Egyptians dried perishable foods to preserve them. Scandinavians made holes in the centers of pieces of dry, flat, crisp bread in order to hang them in the air of their homes during the winter; likewise they kept seed grains in a dry place. Otherwise, both flour and grains would have molded during the long and very moist winters. Europeans used heat in the food-canning process 50 years before Pasteur's work explained why heating prevented food from spoiling. Today, physical agents that destroy microorganisms are still used in food preservation and preparation. Such agents remain a crucial weapon in the prevention of infectious disease. Physical antimicrobial agents include various forms of heat, refrigeration, desiccation (drying), irradiation, and filtration.

PRINCIPLES AND APPLICATIONS OF HEAT KILLING

Heat is a preferred agent of sterilization for all materials not damaged by it. It rapidly penetrates thick materials not easily penetrated by chemical agents. Several measurements have been defined to quantify the killing power of heat. The **thermal death point** is the temperature that kills all the bacteria in a 24-hour-old broth culture at neutral pH in 10 minutes. The **thermal death time** is the time required to kill all the bacteria in a particular culture at a specified temperature. *"Take Another Look"* at the difference between thermal death point and thermal death time on the web. The **decimal reduction time**, also known as the **DRT** or **D value**, is the length of time needed to kill 90% of the organisms in a given population at a specified temperature. (The temperature is indicated by a subscript: $D_{80°C}$, for example.)

These measurements have practical significance in industry, as well as in the laboratory. For example, a food-processing technician wanting to sterilize a food as quickly as possible would determine the thermal death point of the most resistant organism that might be present in the food and would use that temperature. In another situation it might be preferable to make the food safe for human consumption by processing it at the lowest possible temperature. This could be important in processing foods containing proteins that would be denatured, thereby altering their flavor or consistency. The processor would then need to know the thermal death time at the desired temperature for the most resistant organism likely to be in the food. Commercial food canning is discussed in ◀Chapter 26 (p. 825). Some organisms may remain alive after canning, and thus commercial canned goods are not always sterile.

DRY HEAT, MOIST HEAT, AND PASTEURIZATION

Dry heat probably does most of its damage by oxidizing molecules. *Moist heat* destroys microorganisms mainly by denaturing proteins; the presence of water molecules helps disrupt the hydrogen bonds and other weak interactions that hold proteins in their three-dimensional shapes (Figure 2.18, p. 44). Moist heat may disrupt membrane lipids as well. Heat also inactivates many viruses, but those that can infect even after their protein coats are denatured require extreme heat treatment, such as steam under pressure, that will disrupt nucleic acids.

Dry Heat

Dry (oven) heat penetrates substances more slowly than moist (steam) heat. It is usually used to sterilize metal objects and glassware and is the only suitable means of sterilizing oils and powders **(Figure 12.8)**. Objects are sterilized by dry heat when subjected to 171°C for 1 hour, 160°C for 2 hours or longer, or 121°C for 16 hours or longer, depending on the volume.

Figure 12.8 Hot-air oven used for sterilizing metal and glass items. *(Ulrich Sapountis/Okapia/Photo Researchers, Inc.)*

An open flame is a form of dry heat used to sterilize inoculating loops and the mouths of culture tubes by incineration and to dry the inside of pipettes. When flaming objects in the laboratory, you must avoid the formation of floating ashes and **aerosols** (droplets released into the air). These substances can be a means of spreading infectious agents if the organisms in them are not killed by incineration, as intended. For this reason, specially designed loop incinerators with deep throats are often used for sterilizing inoculating loops.

It is impossible to sterilize skin except by burning!

Moist Heat

Moist heat, because of its penetrating properties, is a widely used physical agent. Boiling water destroys vegetative cells of most bacteria and fungi and inactivates some viruses, but it is not effective in killing all kinds of spores. The effectiveness of boiling can be increased by adding 2% sodium bicarbonate to the water. However, if water is heated under pressure, its boiling point is elevated, so temperatures above 100°C can be reached. This is normally accomplished by using an **autoclave** (aw'to-klav), as shown in **Figure 12.9**, in which a pressure of 15 lb/in.² above atmospheric pressure is maintained for 15 to 20 minutes, depending on the volume of the load. At this pressure, the temperature reaches 121°C, which is high enough to kill spores, as well as vegetative organisms, and to disrupt the structure of nucleic acids in viruses. In this procedure it is the increased temperature, and not the increased pressure, that kills microorganisms.

Sterilization by autoclaving is invariably successful if properly done and if two commonsense rules are followed:

Figure 12.9   A small countertop autoclave. *(Richard Hutchings/Photo Researchers, Inc.)*

Prions are highly resistant and must be sterilized by longer and higher temperature autoclaving (134°C for 18 minutes).

First, articles should be placed in the autoclave so that steam can easily penetrate them; second, air should be evacuated so that the chamber fills with steam. Wrapping objects in aluminum foil is not recommended because it may interfere with steam penetration. Steam circulates through an autoclave from a steam outlet to an air evacuation port (Figure 12.10). In preparing items for autoclaving, containers should be unsealed and articles should be wrapped in materials that allow steam penetration. Large packages of dressings and large flasks of media require extra time for heat to penetrate them. Likewise, packing many

Figure 12.10   The autoclave. Steam is heated in the jacket of an autoclave, enters the sterilization chamber through an opening at the upper rear, and is exhausted through a vent at the bottom front.

articles close together in an autoclave lengthens the processing time to as much as 60 minutes to ensure sterility. It is more efficient and safer to run two separate, uncrowded loads than one crowded one.

Several methods are available to ensure that autoclaving achieves sterility. Modern autoclaves have devices to maintain proper pressure and record internal temperature during operation. Regardless of the presence of such a device, the operator should check the pressure periodically and maintain the appropriate pressure. Tapes impregnated with a substance that causes the word "sterile" to appear when they have been exposed to an effective sterilization temperature can be placed on packages. These tapes are not fully reliable because they do not indicate how long appropriate conditions were maintained. Tapes or other sterilization indicators should be placed inside and near the center of large packages to determine whether heat penetrated them. This precaution is necessary because when an object is exposed to heat, its surface becomes hot much more quickly than its center. (When a large piece of meat is roasted, for example, the surface can be well done while the center remains rare.)

The Centers for Disease Control and Prevention recommends weekly autoclaving of a culture containing heat-resistant endospores, such as those of *Bacillus stearothermophilus*, to check autoclave performance. Endospore strips are commercially available to make this task easy (Figure 12.11). The spore strip and an ampule of medium are enclosed in a soft plastic vial. The vial is placed in the center of the material to be sterilized and is autoclaved. Then the inner ampule is broken, releasing the medium, and the whole container is incubated. If no growth appears in the autoclaved culture, sterilization is deemed effective.

In large laboratories and hospitals, where great quantities of materials must be sterilized, special autoclaves, called *prevacuum autoclaves*, are often used (Figure 12.12). The chamber is emptied of air as steam flows in, creating a partial vacuum. The steam enters and heats the chamber much more rapidly than it would without the vacuum, so the proper temperature is reached quickly. The total sterilization time is cut in half, and the costs of sterilization are greatly decreased.

Pasteurization

Pasteurization, a process invented by Pasteur to destroy organisms that caused wine to sour, does not achieve sterility. It does kill pathogens, especially *Salmonella* and *Mycobacterium*, that might be present in milk, other dairy products, and beer. *Mycobacterium* used to cause many cases of tuberculosis among children who drank raw milk. Milk is pasteurized by heating it to 71.6°C for at least 15 seconds in the *flash method* or by heating it to 62.9°C for 30 minutes in the *holding method*. Some years ago certain strains of bacteria of the genus *Listeria* were found in pasteurized milk and cheeses. This pathogen causes diarrhea and encephalitis and can lead to death in pregnant

Cap that allows steam to penetrate

Crushable glass ampule

Flexible plastic vial

Nutrient medium

Spore strip

This unit is placed in autoclave.

After autoclaving, flexible vial is squeezed to break ampule and to release medium onto spore strip.

Incubation

If medium remains clear, spores were killed, and sterility was produced.

If medium becomes cloudy, spores were not killed, and sterility was not produced.

Figure 12.11 Checking for sterility. To check if an autoclave is operating properly, a commercially prepared spore test ampule is placed in the autoclave and run with the rest of the load. Afterward, the vial is crushed to release medium onto a strip containing spores. If the load was truly sterilized, the spores will have been killed, and growth will not occur in the medium. Sometimes an indicator dye is added to the medium, which will turn color if microbial growth occurs, due to the accumulation of acid by-products. This is faster than waiting for sufficient growth to turn the medium cloudy.

women. A few such infections have prompted questions about the need to revise standard procedures for pasteurization. However, finding these pathogens in pasteurized milk has not become a persistent problem, and no action has been taken, although back in 1950 the pasteurization temperature was raised to kill *Coxiella burnetii* bacteria found in milk.

Although most milk for sale in the United States is pasteurized fresh milk, sterile milk also is available. All evaporated or condensed canned milk is sterile, and some milk packaged in cardboard containers also is sterile. The canned milk is subjected to steam under pressure and has a "cooked" flavor. Sterilized milk in cardboard containers is widely available in Europe and can be found in some stores in the United States. It is subjected to a process that is similar to pasteurization but uses higher temperatures. It, too, has a "cooked" flavor but can be kept unrefrigerated as long as the container remains sealed. Such milk is often flavored with vanilla, strawberry, or chocolate. **Ultrahigh temperature (UHT) processing** raises the temperature from 74° to 140°C and then drops it back to 74°C in less than 5 seconds. A complex cooling process that keeps the milk from ever touching a surface hotter than itself prevents development of a "cooked" flavor. Some, but not all, small containers of coffee creamer are treated by this method.

REFRIGERATION, FREEZING, DRYING, AND FREEZE-DRYING

Cold temperature retards the growth of microorganisms by slowing the rate of enzyme-controlled reactions but does not kill many microbes. Heat is much more effective

Figure 12.12 A large automatic hospital autoclave. Recording charts keep records of the actual temperatures and pressures reached during the time of operation. These would alert the operator to some malfunctions. *(Courtesy Steris Corporation)*

APPLICATIONS

Yogurt

Certain foods such as yogurt are made by introducing into milk organisms such as lactobacilli that ferment the milk. The fermented foods are often heat-treated after initial pasteurization to kill the fermenting organisms and to increase the shelf-life of the products. The labels of such products indicate whether they contain live fermenting organisms. If you make yogurt at home, be sure to purchase a live-culture brand of yogurt to use as your starter.

APPLICATIONS

Home Canning

Home canning is done in an open water bath or in a pressure cooker. Food must be packed loosely in jars with ample fluid to carry heat to the center of the jar. Space must be left between the jars. Once adequately processed, canned foods will keep indefinitely. Jars of relish in the wreckage of the *Monitor*, an ironclad Civil War ship, were actually sterile after more than 100 years at the bottom of the Atlantic Ocean! The contents would have been edible if poisonous quantities of lead from the lids had not dissolved in the relish.

The water bath reaches a temperature of 100°C and is adequate for preventing spoilage of acidic foods such as fruits and tomatoes. Acid in these foods inhibits the germination of most spores, should some survive the boiling-water treatment. However, meats and alkaline vegetables, such as corn and beans, must be processed in a pressure cooker. Adding onions or green peppers to a jar of tomatoes increases the pH, so such mixtures also must be cooked under pressure. Because acid-tolerant spores do exist, home canning is safest when it is done by pressure cooking. All commercially canned foods are processed in pressurized equipment. A pressure cooker functions like an autoclave. Foods in jars are processed at least 15 minutes at 15 lb/in.^{2} pressure. Any spores that might be present in these foods are killed, so the food is sterile.

Failure to process alkaline foods at the high temperature reached in a pressure cooker can lead to the accumulation of toxin produced by still-living *Clostridium botulinum* bacteria while the food is stored. Even a tiny amount of this toxin can be lethal. Any home-canned or commercially canned foods should

(Courtesy Jacquelyn G. Black)

be discarded if they have an unpleasant odor or if the lids of the containers bulge, because this indicates that gas is being produced by living organisms inside the container. Unfortunately, toxins can be present even when the lids don't bulge and there is no noticeable odor, so great care must be used in home canning to be sure that adequate pressure is maintained for a long enough time. As a safety precaution, after a jar of home-canned food is opened, 15 to 20 minutes of vigorous boiling should destroy any botulism toxin. However, the best rule to follow is "When in doubt, throw it out."

---

than cold at killing microorganisms. *Refrigeration* is used to prevent food spoilage. *Freezing*, *drying*, and *freeze-drying* are used to preserve both foods and microorganisms, but these methods do not achieve sterilization. Freezing for several days, however, will probably kill most worm parasites found in meat.

APPLICATIONS

Home Freezing

Home freezing of foods is probably a more common practice today than home canning. Before freezing fresh fruits and vegetables, they should be blanched, or immersed in boiling water for about a minute. Blanching helps kill microorganisms on the foods, but its main purpose is to denature enzymes in the foods that can cause discoloration or changes in texture even at freezer temperatures. The foods should then be cooled quickly in cold water and placed in clean containers. Finally, they should be placed in the coldest part of the home freezer with space around the containers so that they freeze as quickly as possible.

Refrigeration

Many fresh foods can be prevented from spoiling by keeping them at 5°C (ordinary refrigerator temperature). However, storage should be limited to a few days because some bacteria and molds continue to grow at this temperature. To convince yourself of this, recall some of the strange things you have found growing on leftovers in the back of your refrigerator. In rare instances strains of *Clostridium botulinum* have been found growing and producing lethal toxins in a refrigerator when the organisms were deep within a container of food, where anaerobic conditions exist.

Freezing

Freezing at −20°C is used to preserve foods in homes and in the food industry. Although freezing does not sterilize foods, it does significantly slow the rate of chemical reactions so that microorganisms do not cause food to spoil. Frozen foods should not be thawed and refrozen. Repeated freezing and thawing of foods causes large ice crystals to form in the foods during slow freezing. Cell membranes in the foods are ruptured, and nutrients leak out. The texture of foods is thus altered, and they become

Figure 12.13 Preservation by drying. Sun drying is an ancient means of preventing the growth of microorganisms. These grapes will remain edible as raisins because microbes need more water than remains inside the dried fruit. *(Link/Visuals Unlimited)*

less palatable. It also allows bacteria to multiply while food is thawed, making the food more susceptible to bacterial degradation.

Freezing can be used to preserve microorganisms, but this requires a much lower temperature than that used for food preservation. Microorganisms are usually suspended in glycerol or protein to prevent the formation of large ice crystals (which could puncture cells), cooled with solid carbon dioxide (dry ice) to a temperature of −78°C, and then held there. Alternatively, they can be placed in liquid nitrogen and cooled to −180°C.

Drying

Drying can be used to preserve foods because the absence of water inhibits the action of enzymes. Many foods, including peas, beans, raisins, and other fruits, are often preserved by drying (**Figure 12.13**). Yeast used in baking also can be preserved by drying. Endospores present on such foods can survive drying, but they do not produce toxins. Dried pepperoni sausage and smoked fish retain enough moisture for microorganisms to grow. Because smoked fish is not cooked, eating it poses a risk of infection. Sealing such fish in plastic bags creates conditions that allow anaerobes such as *Clostridium botulinum* to grow.

Drying also naturally minimizes the spread of infectious agents. Some bacteria, such as *Treponema pallidum*, which causes syphilis, are extremely sensitive to drying and die almost immediately on a dry surface; thus they can be prevented from spreading by keeping toilet seats and other bathroom fixtures dry. Drying of laundry in dryers or in the sunshine also destroys pathogens.

Freeze-Drying

Freeze-drying, or **lyophilization** (li-of″i-li-za′shun), is the drying of a material from the frozen state (**Figure 12.14**). This process is used in the manufacture of some brands of instant coffee; freeze-dried instant coffee has a more natural flavor than other kinds. Microbiologists use lyophilization for long-term preservation rather than for destruction of cultures of microorganisms. Organisms in vials are rapidly frozen in alcohol and dry ice or in liquid nitrogen, are then subjected to a high vacuum to remove all the water while in the frozen state, and finally are sealed under a vacuum. Rapid freezing allows only very tiny ice crystals to form in cells, so the organisms survive this process. Organisms so treated can be kept alive for years, stored under vacuum in the freeze-dried state.

RADIATION

Four general types of radiation—ultraviolet light, ionizing radiation, microwave radiation, and strong visible light (under certain circumstances)—can be used to control

Figure 12.14 Freeze-drying (lyophilization) equipment.
(a) A stoppering tray dryer, in which the trays supply the heat needed to remove moisture. After completion of lyophilization (about 24 hours), the device automatically stoppers the vials. *(FTS Systems, Inc., photo by Gary Gold Photography)* **(b)** A manifold dryer, in which prefrozen samples in vials of different sizes can be attached via ports to the dryer, which supplies a vacuum to remove water. The sample will dry in 4 to 20 hours, depending on its initial thickness. Unlike the tray dryer, this device allows samples to be added and removed. *(FTS Systems, Inc., photo by Gary Gold Photography)*

(a)

(b)

microorganisms and to preserve foods. Refer to the electromagnetic spectrum (◄Figure 3.4) to review their relative wavelengths and positions along the spectrum.

Ultraviolet Light

Ultraviolet (UV) light consists of light of wavelengths between 40 and 390 nm, but wavelengths in the 200-nm range are most effective in killing microorganisms by damaging DNA and proteins. Ultraviolet light is absorbed by the purine and pyrimidine bases of nucleic acids. Such absorption can permanently destroy these important molecules. Ultraviolet light is especially effective in inactivating viruses. However, it kills far fewer bacteria than one might expect because of DNA repair mechanisms. Once DNA is repaired, new molecules of RNA and protein can be synthesized to replace the damaged molecules (◄Chapter 7, p. 201). Lying in soil and exposed to sunlight for decades, endospores are resistant to UV damage because of a small protein that binds to their DNA. This changes the geometry of the DNA by untwisting it slightly, thereby making it resistant to the effects of UV irradiation.

Ultraviolet light is of limited use because it does not penetrate glass, cloth, paper, or most other materials, and it does not go around corners or under lab benches. It does penetrate air, effectively reducing the number of airborne microorganisms and killing them on surfaces in operating rooms and rooms that will contain caged animals (Figure 12.15). Ultraviolet lights lose effectiveness over time and should be monitored often. To help sanitize the air without irradiating humans, UV lights can be turned on when the rooms are not in use. Exposure to UV light can cause burns, as anyone who has had a sunburn knows, and can also damage the eyes; years of skin exposure can lead to skin cancer. Hanging laundry outdoors on bright, sunny days takes advantage of the UV light present in sunlight. Although the quantity of UV rays in sunlight is small, these rays may help kill bacteria on clothing, especially diapers.

In some communities, UV light is replacing chlorine in sewage treatment. When chlorine-treated sewage effluent is discharged into streams or other bodies of water, carcinogenic compounds form and may enter the food chain. The cost of removing chlorine before discharging treated effluent could add more than $100 per year to the sewage bills of the average American family, and very few sewage plants do this. Running the sewage effluent under UV light before discharging it can destroy microorganisms without altering the odor, pH, or chemical composition of the water and without forming carcinogenic compounds.

Ionizing Radiation

X rays, which have wavelengths of 0.1 to 40 nm, and gamma rays, which have even shorter wavelengths, are forms of *ionizing radiation*, so named because it can dislodge electrons from atoms, creating ions. (Longer wavelengths are forms of *nonionizing radiation*.) These forms of radiation also kill microorganisms and viruses. Many bacteria are killed by absorbing 0.3 to 0.4 millirads of radiation; polioviruses are inactivated by absorbing 3.8 millirads. A **rad** is a unit of radiation energy absorbed per gram of tissue; a millirad is one-thousandth of a rad. Humans usually do not become ill from radiation unless they are subjected to doses greater than 50 rads.

Irradiation is used by hospitals to sterilize food for immune-compromised patients.

Ionizing radiation damages DNA and produces peroxides, which act as powerful oxidizing agents in cells. This radiation can also kill or cause mutations in human cells if it reaches them. It is used to sterilize plastic laboratory and medical equipment and pharmaceutical products. It can be used to prevent spoilage in seafoods by doses of 100 to 250 kilorads, in meats and poultry by doses of 50 to 100 kilorads, and in fruits by doses of 200 to 300 kilorads. (One kilorad equals 1,000 rads.) Many consumers in the United States reject irradiated foods for fear of receiving radiation, but such foods are quite safe—free of both pathogens and radiation. In Europe, milk and other foods are often irradiated to achieve sterility.

The bacterium, Deinococcus radiodurans, is remarkable for its ability to survive more than 1,000 times the amount of radiation that would kill a human. It is being studied as a candidate for use in the bioremediation of sites contaminated by radioactive materials.

Microwave Radiation

Microwave radiation, in contrast with gamma, X-ray, and UV radiation, falls at the long-wavelength

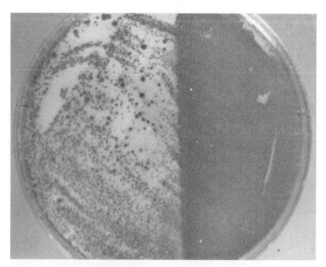

Figure 12.15 Ultraviolet radiation. The effects of UV exposure can be seen in this Petri plate of *Serratia marcescens*; the left side was exposed to UV rays while the right side was shielded. Most of the organisms on the left side have been killed. *(Grant Heilman Photography)*

Kitchen sponges should be run through the dishwasher to clean them or a wet sponge can be microwaved. But do not microwave a dry sponge, as it can burst into flames.

end of the electromagnetic spectrum. It has wavelengths of approximately 1 mm to 1 m, a range that includes television and police radar. Microwave oven frequencies are tuned to match energy levels in water molecules. In the liquid state, water molecules quickly absorb the microwave energy and then release it to surrounding materials as heat. Thus, materials that do not contain water, such as plates made of paper, china, or plastic, remain cool while the moist food on them becomes heated. For this reason the home microwave cannot be used to sterilize items such as bandages and glassware. Conduction of energy in metals leads to problems such as sparking, which makes most metallic items also unsuitable for microwave sterilization. Moreover, bacterial endospores, which contain almost no water, are not destroyed by microwaves. However, a specialized microwave oven has recently become available that can be used to sterilize media in just 10 minutes **(Figure 12.16)**. It has 12 pressure vessels, each of which holds 100 ml of medium. Microwave energy increases the pressure of the medium inside the vessels until sterilizing temperatures are reached.

Caution should be observed in cooking foods in the home microwave oven. Geometry and differences in density of the food being cooked can cause certain regions to become hotter than others, sometimes leaving very cold spots. Consequently, to cook foods thoroughly in a microwave oven, it is necessary to rotate the items either mechanically or by hand. For example, pork roasts must be turned frequently and cooked thoroughly to kill any cysts of the pork roundworm, *Trichinella* (◄Chapter 22). Failure to kill such cysts could lead to the disease trichinosis, in which cysts of the worm become embedded in human muscles and other tissues. All experimentally infected pork roasts, when microwaved without rotation, showed live worms remaining in some portion at the end of the standard cooking time.

Strong Visible Light

Sunlight has been known for years to have a bactericidal effect, but the effect is due primarily to UV rays in the sunlight. Strong visible light, which contains light of wavelengths from 400 to 700 nm (violet to red light), can have direct bactericidal effects by oxidizing light-sensitive molecules such as riboflavin and porphyrins (components of oxidative enzymes) in bacteria. For that reason, bacterial cultures should not be exposed to strong light during laboratory manipulations. The fluorescent dyes eosin and methylene blue can denature proteins in the presence of strong light because they absorb energy and cause oxidation of proteins and nucleic acids. The combination of a dye and strong light can be used to rid materials of both bacteria and viruses.

SONIC AND ULTRASONIC WAVES

Sonic, or sound, waves in the audible range can destroy bacteria if the waves are of sufficient intensity. Ultrasonic waves, or waves with frequencies above 15,000 cycles per second, can cause bacteria to cavitate. **Cavitation** (kav″i-ta′shun) is the formation of a partial vacuum in a liquid—in this case, the fluid cytoplasm in the bacterial cell. Bacteria so treated disintegrate, and their proteins are denatured. Enzymes used in detergents are obtained by cavitating the bacterium *Bacillus subtilis*. The disruption of cells by sound waves is called **sonication** (son″i-ka′-shun). Neither sonic nor ultrasonic waves are a practical means of sterilization. We mention them here because they are useful in fragmenting cells to study membranes, ribosomes, enzymes, and other components.

FILTRATION

Filtration is the passage of a material through a filter, or straining device. Sterilization by filtration requires filters with exceedingly small pores. Filtration has been used since Pasteur's time to separate bacteria from media and to sterilize materials that would be destroyed by heat. Over the years, filters have been made of porcelain, asbestos, diatomaceous earth, and sintered glass (glass that has been heated without melting). *Membrane filters* **(Figure 12.17)**, thin disks with pores that prevent the passage of anything larger than the pore size, are widely used today. They are usually made of nitrocellulose and have the great advantage that they can be manufactured with specific pore sizes from 25 μm to less than 0.025 μm. Particles filtered by various pore sizes are summarized in **Table 12.4**.

Membrane filters have certain advantages and disadvantages. Except for those with the smallest pore sizes, membrane filters are relatively inexpensive, do not

Figure 12.16 Microwave sterilization. The MikroClave® system is specifically designed for rapid sterilization of microbiological media and solutions. Using microwave energy, it can sterilize 1.2 liters of media in 6.5 minutes, or 100 ml in 45 seconds. Agar need not be boiled prior to sterilization. *(CEM Corporation)*

Figure 12.17 Sterilization by filtration. (a) Various types of membrane filters are available to sterilize large or small quantities of liquids. Some can be vacuum-filtered, ensuring that what is forced into the bottle or flask will be sterile. *(Courtesy Millpore Corporation, Billerca, Massachusetts)* **(b)** Scanning electron micrograph of *Staphylococcus epidermidis* cells (19,944X) trapped on the surface of a 0.22 *μ*m Millipore membrane filter. Membrane pore size can be selected to allow viruses, but not bacteria, to pass through or to prevent both from passing. *(Courtesy Millpore Corporation, Billerca, Massachusetts)*

clog easily, and can filter large volumes of fluid reasonably rapidly. They can be autoclaved or purchased already sterilized. A disadvantage of membrane filters is that many of them allow viruses and some mycoplasmas to pass through. Other disadvantages are that they may absorb relatively large amounts of the filtrate and may introduce metallic ions into the filtrate.

Membrane filters are used to sterilize materials likely to be damaged by heat sterilization. These materials include media, special nutrients that might be added to media, and pharmaceutical products such as drugs, sera,

and vitamins. Some filters can be attached to syringes so that materials can be forced through them relatively quickly. Filtration can also be used instead of pasteurization in the manufacture of beer. When using filters to sterilize materials, it is important to select a filter pore size that will prevent any infectious agent from passing into the product.

In the manufacture of vaccines that require the presence of live viruses, it is important to select a filter pore size that will allow viruses to pass through the filter but prevent bacteria from doing so. By selecting a filter with a proper pore size, scientists can separate polioviruses from the fluid and debris in tissue cultures in which they were grown. This procedure simplifies the manufacture of polio vaccine. Cellulose acetate filters with extremely tiny pores are now available and are capable of removing many viruses (although not the very smallest) from liquids. However, these filters are expensive and clog easily.

Membrane filters used to trap bacteria from air and water samples can be transferred directly to agar plates, and the quantity of bacteria in the sample can be determined. Alternatively, the filters can be transferred from one medium to another, so organisms with different nutrient requirements can be detected. Filtration is also used to remove microorganisms and other small particles from public water supplies and in sewage treatment facilities. This technique, however, cannot sterilize; it merely reduces contamination.

High-efficiency particulate air (HEPA) filters are used in the ventilation systems of areas where microbial control is especially important, such as in operating rooms, burn units, and laminar flow transfer hoods in

TABLE 12.4

Pore Sizes of Membrane Filters and Particles That Pass Through Them	
Pore Size (in *μ*m)	**Particles That Pass Through Them**
10	Erythrocytes, yeast cells, bacteria, viruses, molecules
5	Yeast cells, bacteria, viruses, molecules
3	Some yeast cells, bacteria, viruses, molecules
1.2	Most bacteria, viruses, molecules
0.45	A few bacteria, viruses, molecules
0.22	Viruses, molecules
0.10	Medium-sized to small viruses, molecules
0.05	Small viruses, molecules
0.025	Only the very smallest viruses, molecules
Ultrafilter	Small molecules

laboratories. HEPA filters also capture organisms released in rooms occupied by patients with tuberculosis or in laboratories where especially dangerous microbes are studied, such as the maximum containment units shown in ◀Figure 15.14. These filters remove almost all organisms larger than 0.3 μm in diameter. Used filters are soaked in formalin before they are disposed of. Most rooms for patients with tuberculosis have an outer "hallway" outside the main door to the room. Negative air pressure inside the room should cause air from outside the room to be sucked into it whenever the door is opened. However, tests have shown that some air still does escape from the room; hence the need for the little containment "hallway" outside. Just remember to always put on your mask *before* you enter the little "hallway," as it will have some TB germs in it!

OSMOTIC PRESSURE

High concentrations of salt, sugar, or other substances create a hyperosmotic medium, which draws water from microorganisms by osmosis (◀Chapter 4, p. 108). **Plasmolysis** (plaz-mol′i-sis), or loss of water, severely interferes with cell function and eventually leads to cell death. The use of sugar in jellies, jams, and syrups or salt solutions in curing meat and making pickles plasmolyzes most organisms present and prevents growth of new organisms. A few halophilic organisms, however, thrive in these conditions and cause spoilage, especially of pickles, and some fungi can live on the surface of jams.

Properties of physical antimicrobial agents are summarized in **Table 12.5**.

✓ CHECKLIST

1. Filtering water through paper does not sterilize the water. How, then, can membrane filters sterilize water?

2. Can a home microwave oven be used to sterilize items? Why or why not?

3. Are ice cubes safe sources of water in areas with poor water supplies?

4. Are pasteurized products sterilized?

TABLE 12.5

Properties of Physical Antimicrobial Agents

Agent	Action	Use
Dry heat	Denatures proteins	Oven heat used to sterilize glassware and metal objects; open flame used to incinerate microorganisms.
Moist heat	Denatures proteins	Autoclaving sterilizes media, bandages, and many kinds of hospital and laboratory equipment not damaged by heat and moisture; pressure cooking sterilizes canned foods.
Pasteurization	Denatures proteins	Kills pathogens in milk, dairy products, and beer.
Refrigeration	Slows the rate of enzyme-controlled reactions	Used to keep fresh foods for a few days; does not kill most microorganisms.
Freezing	Greatly slows the rate of most enzyme-controlled reactions	Used to keep fresh foods for several months; does not kill microorganisms; used with glycerol to preserve microorganisms.
Drying	Inhibits enzymes	Used to preserve some fruits and vegetables; sometimes used with smoke to preserve sausages and fish.
Freeze-drying	Dehydration inhibits enzymes	Used to manufacture some instant coffees; used to preserve microorganisms for years.
Ultraviolet light	Denatures proteins and nucleic acids	Used to reduce the number of microorganisms in air in operating rooms, animal rooms, and where cultures are transferred.
Ionizing radiation	Denatures proteins and nucleic acids	Used to sterilize plastics and pharmaceutical products and to preserve foods.
Microwave radiation	Absorbs water molecules, then releases microwave energy to surroundings as heat	Cannot be used reliably to destroy microbes except in special media-sterilizing equipment.
Strong visible light	Oxidation of light-sensitive materials	Can be used with dyes to destroy bacteria and viruses; may help sanitize clothing.
Sonic and ultrasonic waves	Cause cavitation	Not a practical means of killing microorganisms but useful in fractionating and studying cell components.
Filtration membranes	Mechanically removes microbes	Used to sterilize media, pharmaceutical products, and vitamins, in manufacturing vaccines, and in sampling microbes in air and water.
Osmotic pressure	Removes water from microbes	Used to prevent spoilage of foods such as pickles and jellies.

▍ RETRACING OUR STEPS

III PRINCIPLES OF STERILIZATION AND DISINFECTION

- **Sterilization** refers to the killing or removal of all organisms in any material or on any object.
- **Disinfection** refers to the reduction in numbers of pathogenic organisms on objects or in materials so that the organisms no longer pose a disease threat.
- Important terms related to sterilization and disinfection are defined in Table 12.1.

THE CONTROL OF MICROBIAL GROWTH

- Because of the logarithmic death rate of microorganisms, a definite proportion of organisms die in a given time interval.
- The fewer organisms present, the less time will be needed to achieve sterility.
- Microorganisms differ in their susceptibility to antimicrobial agents.

III CHEMICAL ANTIMICROBIAL AGENTS

THE POTENCY OF CHEMICAL AGENTS

- The potency, or effectiveness, of a chemical antimicrobial agent is affected by time, temperature, pH, and concentration of the agent.
- Potency increases with the length of time organisms are exposed to the agent, increased temperature, acidic or alkaline pH, and usually increased concentration of the agent.

EVALUATING THE EFFECTIVENESS OF CHEMICAL AGENTS

- Evaluation of effectiveness is difficult, and no entirely satisfactory method is available.
- For agents similar to phenol, the **phenol coefficient** is determined; it is the ratio of the dilution of the agent to the dilution of phenol that will kill all organisms in 10 minutes but not in 5 minutes.

DISINFECTANT SELECTION

- A variety of criteria are considered in selecting a **disinfectant**. In practice, most chemical agents are tested in various situations and used in situations where they produce satisfactory results.

MECHANISMS OF ACTION OF CHEMICAL AGENTS

- Actions of chemical antimicrobial agents can be grouped according to their effects on proteins, cell membranes, and other cell components.
- Reactions that alter proteins include hydrolysis, oxidation, and attachment of atoms or chemical groups to protein molecules. Such reactions denature proteins, rendering them nonfunctional.
- Membranes can be disrupted by agents that denature proteins and by **surfactants**, which reduce surface tension and dissolve lipids.
- Reactions of other chemical agents damage nucleic acids and energy-producing systems. Damage to nucleic acids is an important means of inactivating viruses.

SPECIFIC CHEMICAL ANTIMICROBIAL AGENTS

- Soaps and detergents aid in the removal of microbes, oils, and dirt but do not sterilize.

- Acids are commonly used as food preservatives; alkali in soap helps destroy microorganisms.
- Among the agents containing heavy metals, silver nitrate is used to kill gonococci, and mercury-containing compounds are used to disinfect instruments and skin.
- Among the agents containing halogens, chlorine is used to kill pathogens in water, and iodine is a major ingredient in several skin disinfectants.
- Alcohols are used to disinfect skin.
- Phenol derivatives can be used on skin, instruments, dishes, and furniture, and to destroy discarded cultures; they work well in the presence of organic materials.
- Oxidizing agents are particularly useful in disinfecting puncture wounds.
- Alkylating agents can be used to disinfect or to sterilize a variety of materials, but all are carcinogens.
- Some dyes, plant oils, sulfur-containing substances, and nitrates can be used as disinfectants or food preservatives.

III PHYSICAL ANTIMICROBIAL AGENTS

PRINCIPLES AND APPLICATIONS OF HEAT KILLING

- Heat destroys microorganisms by denaturing protein, by melting lipids, and, when open flame is used, by incineration.

DRY HEAT, MOIST HEAT, AND PASTEURIZATION

- Dry heat is used to sterilize metal objects and glassware.
- Flame is used to sterilize inoculating loops and the mouths of culture tubes.
- The **autoclave**, which uses moist heat under pressure, is a common instrument for sterilization and is very effective when proper procedures are followed.
- **Pasteurization** kills most pathogens in milk, other dairy products, and beer but does not sterilize.

REFRIGERATION, FREEZING, DRYING, AND FREEZE-DRYING

- Refrigeration, freezing, drying, and freeze-drying can be used to retard the growth of microorganisms.
- **Lyophilization**, drying in the frozen state, can be used for long-term preservation of live microorganisms.

RADIATION

- Radiation used to control microorganisms includes ultraviolet light, ionizing radiation, and sometimes microwaves and strong sunlight.

SONIC AND ULTRASONIC WAVES

- Sonic and ultrasonic waves can kill microorganisms but are used mostly for **sonication**, the disruption of cells by sound waves.

FILTRATION

- **Filtration** can be used to sterilize substances that are destroyed by heat, to separate viruses, and to collect microorganisms from air and water samples.

OSMOTIC PRESSURE

- High concentrations of sugar or salt create osmotic pressure that results in the **plasmolysis** of cells (causes them to lose water) and prevents growth of microorganisms in highly sweetened or salted foods.

▍ TERMINOLOGY CHECK

aerosol *(p. 353)*

antiseptic *(p. 342)*

autoclave *(p. 353)*

bactericidal *(p. 343)*

bacteriostatic *(p. 343)*

cavitation *(p. 359)*

decimal reduction time (DRT) *(p. 353)*

disinfectant *(p. 342)*

disinfection *(p. 342)*

DRT or D value *(p. 353)*

filter paper method *(p. 344)*

filtration *(p. 359)*

lyophilization *(p. 357)*

pasteurization *(p. 354)*

phenol coefficient *(p. 343)*

plasmolysis *(p. 361)*

quaternary ammonium compound (quat) *(p. 348)*

rad *(p. 358)*

sonication *(p. 359)*

sterility *(p. 342)*

sterilization *(p. 342)*

surfactant *(p. 346)*

thermal death point *(p. 353)*

thermal death time *(p. 353)*

tincture *(p. 348)*

ultrahigh temperature (UHT) processing *(p. 353)*

use-dilution test *(p. 344)*

wetting agent *(p. 346)*

▍ CLINICAL CASE STUDY

The following is a true story. A nursing student spent the day working with a nurse preceptor. At the end of the day, when meeting with their nursing school coordinator, the student remarked that the nurse had only washed her hands once during the entire day. The nurse had changed bed pans and treated patients! The nurse had used gloves on occasion. The student nurse was in a quandary as to how to deal with the situation. Visit the following CDC web site to learn more about how hand washing continues to be an issue in health care. Can health practitioners wash their hands too much? What new methods are replacing soap and water in hand washing? (http://www.cdc.gov/od/oc/media/pressrel/fs021025.htm)

▍ CRITICAL THINKING QUESTIONS

1. Should pasteurized milk be a sterile product? Why?

2. Would ultraviolet rays, X-rays, or gamma radiation be a good choice of sterilization method for destroying prions (such as those that cause "mad cow disease")? Explain.

3. A technician was testing a new disinfectant to determine its phenol coefficient and obtained the following results. (a) What is the phenol coefficient? (b) Is this likely to be a good disinfectant?

Exposure Time	Phenol Dilution				New Disinfectant Dilution			
	1:100	1:110	1:120	1:130	1:50	1:60	1:70	1:80
5 min	+	+	+	−	+	+	−	−
10 min	+	+	−	−	+	−	−	−

▍ SELF-QUIZ

1. Match the following types of antimicrobials with their actions:

D Bacteriostatic   (a) Kills microbes

A Germicidal   (b) Inactivates viruses

B Viricidal   (c) Kills bacteria

e E Sporicidal   (d) Stops bacterial growth

f E Fungicidal   (e) Kills bacterial endospores and fungal spores

C Bacteriocidal   (f) Kills yeasts and molds

2. Which term is used to describe the reduction in numbers of pathogenic organisms on objects or in materials so that they do not pose a disease threat?

(a) Sanitization

(b) Sterilization

(c) Disinfection

(d) Decontamination

(e) Lyophilization

3. Suppose you spilled two cultures of *Salmonella typhimurium* (each containing 100,000 cells) on your lab bench. You immediately applied the same disinfectant to both cultures at the same time. One culture had been freshly grown for 36 hours and the other culture is two weeks old. If the killing rate of the disinfectant is 90% per minute, do you think that microbes in both cultures will be completely killed after 6 minutes? Why or why not? *Probably not the 36 hr culture is in logarithmic phase and is*

4. When something is sterilized, there are levels or degrees which are reached for that object's or material's sterility. True or false?

susceptable to being killed, while the 2 week old culture is not and would take longer to being killed.

5. Which of the following is true of the phenol coefficient test?

(a) Uses *Salmonella typhi* and *Staphylococcus aureus*

(b) Uses phenol as the standard chemical against which other chemicals are compared

(c) If a chemical has a phenol coefficient less than 1.0, it is less effective than phenol

(d) It is particularly reliable for chemicals derived from phenol

(e) All of these

6. The potency or effectiveness of chemical antimicrobial agents can be enhanced by increases in exposure time,

temperature, and concentration of the agent. Which of the following is (are) true about potency of an antimicrobial agent and an increase in either its acidity or basicity?
(a) Acidic or alkaline pH can increase or decrease the agent's potency.
(b) The pH at which the least ionization occurs for the antimicrobial agent is most effective in killing cells.
(c) The pH at which the greatest ionization occurs for the antimicrobial agent is most effective in killing cells.
(d) When an antimicrobial agent reaches its optimal pH, it is more likely to penetrate cells and disrupt their contents.
(e) a, c, and d

7. Match the following chemical agents to their mechanism of action in damaging microbial cell components:
B Surfactant   (a) Protein denaturation
C Alkylating agents   (b) Membrane lipid disruption
A Oxidation agents   (c) Nucleic acid alteration
B Detergents   (d) Cell wall formation
A Hydrolyzing agents
A Heavy metals
D Crystal violet dye

8. Refrigeration, freezing, drying, and freeze-drying _Harad_ the growth of microorganisms but does not usually _Kill_ them. Dry heat, moist heat under pressure, and open flame are much more effective in killing microbes by _denaturing_ protein, melting _lipids_, and incineration.

9. The pasteurization process does which of the following in milk?
(a) It kills all microbes.
(b) It inactivates viruses.
(c) It kills all bacterial spores.
(d) It kills microbial pathogens that might be present in milk.
(e) It sterilizes milk.

10. The advantage of UV-radiation disinfection is that it readily penetrates through most samples. True or false?

11. Which of the following are reasons why UV light might be expected to be less effective in killing bacteria?
(a) UV light cannot penetrate glass, cloth, paper, or most materials under which microbes might be located.
(b) UV light can penetrate air.
(c) Small DNA-binding proteins in bacterial spores make the DNA resistant to UV light damage.
(d) UV light sources gain intensity over time.
(e) UV light kills fewer bacteria than expected because of their DNA repair mechanisms.

12. Gamma rays and X-rays are effective in killing microorganisms because they:
(a) Dislodge electrons from atoms, creating ions
(b) Damage DNA
(c) Produce powerful oxidizing agents (peroxides)
(d) All of these
(e) None of these

13. Quaternary ammonium compounds (quats) are a type of:
(a) Soap
(b) Alkylating agent
(c) Detergent

(d) Phenolic substance
(e) Basic solution

14. The active antimicrobial ingredient in bleach is:
(a) Phenol
(b) Hydrochloride
(c) Hypochlorite
(d) Iodine
(e) Bromide

15. Match the following chemical antimicrobial agents to their uses in combating microbes:
C,D Phenol derivatives   (a) Food preservation
C Iodine   (b) Puncture wound disinfection
C Alcohols   (c) Skin disinfection
A Acids   (d) Instrument disinfection
E Chlorine   (e) Water disinfection
B Oxidizing agents
A Nitrates

16. Heat-sensitive materials (rubber and plastic) and bulky materials (mattresses) can be sterilized using:
(a) Dry heat
(b) Autoclaving
(c) UV radiation
(d) Gaseous ethylene oxide
(e) None of these

17. In the process of autoclaving it is the increased temperature and not the increased pressure that kills all microbes, including spores and the nucleic acids of viruses. True or False?

18. The minimum time used for sterilization by autoclaving is:
(a) 5 minutes
(b) 15 minutes
(c) 45 minutes
(d) 1 hour
(e) 2 hours

19. Which of the following is a limitation of the autoclave?
(a) Length of time
(b) Ability to inactivate viruses
(c) Ability to kill endospores
(d) Use with heat-sensitive materials
(e) Use with glassware

20. The recommended method for testing that an autoclave has truly sterilized a load uses:
(a) *Mycobacterium tuberculosis*
(b) Influenza virus
(c) *Staphylococcus aureus*
(d) Bacteriophages
(e) *Bacillus stearothermophilus*

21. Microwave ovens will only heat materials that contain:
(a) Protein
(b) Water
(c) Lipid
(d) Metals, such as calcium or iron
(e) Food containing substrates

22. What does a researcher have to do in order to kill all bacteria in a liquid without damaging heat-labile proteins in the solution?
(a) Pass the liquid through a 0.5-μm filter
(b) Autoclave the solution

(c) Pass the liquid through a 0.22-μm filter
(d) Boil the solution
(e) Lower the pH of the solution

23. How does the presence of high concentrations of salt or sugars in food prevent growth of microorganisms?
   (a) It sets up a hypotonic environment causing the cells to lyse.
   (b) It sets up an isotonic environment in which cells die.
   (c) It sets up an environmental high in osmotic pressure resulting in cellular plasmolysis (water loss).
   (d) a and b.
   (e) None of the above.

24. How does the process of lyophilization work in order to preserve microorganisms?
   (a) Rapidly frozen organisms in vials are subjected to a vacuum instrument that removes water from them and seals the vials under vacuum.
   (b) The process allows large ice crystals to form inside the cell ensuring their preservation.
   (c) The rapid freeze-thaw cycles allow for rapid cryopreservation.
   (d) The process works by removing both water and contaminating organisms under vacuum.
   (e) All of the above.

25. Which of the following affects the elimination of bacteria from an object?
   (a) Number of bacteria present
   (b) Temperature
   (c) pH
   (d) Presence of organic matter
   (e) All of the above

26. Explain how antimicrobial agents A through D in the following diagram compare in effectiveness against Gram-positive and Gram-negative bacteria. The control filter paper was soaked in sterile water.

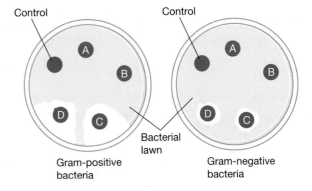

Gram-positive bacteria

Gram-negative bacteria

▌ EXPLORATIONS ON THE WEB

http://www.wiley.com/college/black

If you think you've mastered this chapter, there's more to challenge you on the web. Go to the companion web site to fine-tune your understanding of the chapter concepts and discover answers to the questions posed below.

1. Would you drink from a mountain stream or an unfamiliar water source? Probably not, but RVs do it every time they hook up to an unknown water source.

2. If a little *Clostridium botulinum* ended up in your canned peaches, would you still eat them? Why or why not?

Antimicrobial Therapy

Come with me...

Have you ever gone snorkeling on a coral reef? Then you know how absolutely gorgeous and exciting these places are. Even a TV special on reefs conveys the wonder of a reef and its inhabitants. However, worldwide, reefs are dying off. In 2000, at the International Coral Reef Symposium it was estimated that 27% of all reefs had been lost and that another 32% could be lost in the next 20 to 30 years.

Many factors are involved in reef death, e.g., disease of corals caused by human pathogens from sewage releases. However, the closest correlation is with raised water temperature. Even a 1 to 2 degree Celsius rise is enough to bleach and then kill coral. Global warming will only increase

(Secret Sea Visions/Peter Arnold, Inc.)

such deaths. New findings point to a possible explanation of how the higher temperature does its damage. Corals growing at normal temperatures produce antibiotics that keep out pathogens. Two degrees of elevation can cause production of antibiotics to cease, and in come pathogens like *Vibrio shiloi*, with death following quickly.

 Video related to this topic is available within WileyPLUS.

As you lie in your sickbed, suffering from some infectious disease, how reassuring it is to be able to reach over, swallow some capsules, and look forward to getting well soon. But people have not always been able to do this. Over the ages, anxious parents have watched their children die of fevers, diarrhea, infected wounds, and other maladies. Whole villages have been wiped out by plagues. During the middle of the fourteenth century, more than a quarter of the population of Europe died of the Black Death (bubonic plague) in just a few years. More people have died of infection in wartime than from swords or bullets. Until relatively recently, the only defenses against infectious diseases were such things as herbal teas and poultices (soaks), or fleeing the disease area. Modern and effective weapons against microbes—antibiotics and sulfa drugs—did not become available until the twentieth century.

Think back through your life. Have there been times when you might have died had it not been for antimicrobial drugs? At what age would you have died? Which of your family members might not be alive now? Happily, we live in a better time with respect to illnesses and deaths caused by infectious organisms. In the United States the life expectancy of a baby born in 1850 was less than 40 years; in 1900, about 50 years; and in 2006, over 77.6 years. Infectious diseases claimed the lives of about 1 in every 100 U.S. residents per year as late as 1900 but only about 1 in every 300 in 2000. Although antimicrobial agents still don't save all patients, they have drastically lowered the death rate from infectious disease. A period of increased infectious diseases could return, however, if patients and the medical community fail to protect the effectiveness of antimicrobial agents. As many pathogens develop resistance to available antimicrobial drugs, our ability to fight infectious diseases is dwindling.

ANTIMICROBIAL CHEMOTHERAPY

The term **chemotherapy** was coined by the German medical researcher Paul Ehrlich to describe the use of chemical substances to kill pathogenic organisms without injuring the host. Today, chemotherapy refers to the use of chemical substances to treat various aspects of disease—aspirin for headache and inflammation, drugs to regulate heart function, and agents to rid the body of malignant cells. With this broad modern definition of chemotherapy, we describe a **chemotherapeutic agent** as any chemical substance used in medical practice. Such agents also are referred to as **drugs**.

In microbiology we are concerned with **antimicrobial agents**, a special group of chemotherapeutic agents used to treat diseases caused by microbes. Thus, in modern terms, an antimicrobial agent is synonymous with a chemotherapeutic agent as Erhlich originally defined it. In this chapter we consider a variety of antimicrobial agents and a few agents used to treat helminth infections.

Antibiosis literally means "against life." In the 1940s Selman Waksman, the discoverer of streptomycin, defined an **antibiotic** as "a chemical substance produced by microorganisms which has the capacity to inhibit the growth of bacteria and even destroy bacteria and other microorganisms in dilute solution." In contrast, agents synthesized in the laboratory are called **synthetic drugs**. Some antimicrobial agents are synthesized by chemically modifying a substance from a microorganism. More often a synthetic precursor different from the natural one is supplied to a microorganism, which then completes synthesis of the antibiotic. Antimicrobial agents made partly by laboratory synthesis and partly by microorganisms are called **semisynthetic drugs**.

THE HISTORY OF CHEMOTHERAPY

Throughout history humans have attempted to alleviate suffering by treating disease—often by taking concoctions of plant substances. Although ancient Egyptians used moldy bread to treat wounds, they had no knowledge of the antibiotics it contained. Extracts of willow bark, now known to contain a compound closely related to aspirin, were used to alleviate pain. Parts of the foxglove plant were used to treat heart disease in the sixteenth century, although the active ingredient, digitalis, had not been identified. Likewise, quinine-containing extracts from the cinchona tree were used to treat malaria.

Despite their reputation for using rituals irrelevant to the cure of disease, traditional healers of primitive societies, especially in the tropics, are quite knowledgeable about medicinal properties of plants. Their knowledge has been passed down from generation to generation.

APPLICATIONS

When Doctors Learned to Cure

"Explanation was the real business of medicine. What the ill patient and his family wanted most was to know the name of the illness, and then if possible, what had caused it, and finally, most important of all, how it was likely to turn out.... It gradually dawned on us that we didn't know much that was really useful, that we could do nothing to change the course of the great majority of the diseases we were so busy analyzing.... Then came the explosive news of sulfanilamide, and the start of the real revolution in medicine. I remember with astonishment when the first cases of pneumococcal and streptococcal septicemia were treated in Boston in 1937. The phenomenon was almost beyond belief. Here were moribund patients, who would surely have died without treatment, improving in their appearance within a matter of hours of being given the medicine and feeling entirely well within the next day or so."
—Lewis Thomas, 1983

Because these healers are disappearing, pharmaceutical companies are attempting to learn from them and make written records of their treatments, as well as to test the plants they use.

In Western civilization, the first systematic attempt to find specific chemical substances to treat infectious disease was made by Paul Ehrlich (◄Chapter 1, p. 17). Although his discovery in 1910 of Salvarsan to treat syphilis was one of great therapeutic benefit, even more important were the concepts he developed in the new science of chemotherapy. He was interested in the mechanisms by which chemical substances bind to microorganisms and to animal tissues. His studies of chemicals that bind to tissues led to histological (tissue) stains that are still used today.

The next advances in chemotherapy were the nearly concurrent development of sulfa drugs and antibiotics. In 1935 Gerhard Domagk discovered that prontosil, a red dye, inhibits growth of many Gram-positive bacteria. The following year Ernest Fourneau found that the antimicrobial activity was due to the sulfanilamide portion of the prontosil molecule. These discoveries stimulated the development of a group of substances called *sulfonamides*, or *sulfa drugs*. As the number of sulfa drugs grew, it became possible to use them to attack directly a variety of pathogens. However, the usefulness of sulfa drugs is limited. They do not attack all pathogens, and they sometimes cause kidney damage and allergies. But they have saved many lives and continue to do so today.

Alexander Fleming (◄Chapter 1, p. 18) reasoned that the ability of the mold *Penicillium* to inhibit growth of microorganisms might be exploited. This idea led him to identify the inhibitory agent and name it *penicillin*. In 1928 Fleming had observed the contamination of his bacterial cultures with this fungus many times, as had many other microbiologists. However, instead of grumbling about another contaminated culture and tossing it out, Fleming saw the tremendous potential in this accidental finding. If only the substance (penicillin) could be extracted and collected in large quantities, it could be used to combat infection.

Fleming's idea did not come to fruition until the early 1940s, when Ernst Chain and Howard Florey finally isolated penicillin and worked with other researchers to develop methods of mass production. Such mass production occurred during World War II and saved the lives of many people whose wounds became infected. Supplies of the drug were limited, however, and it was not readily available to civilians until after the war. Following the war, research proceeded rapidly, and new antibiotics were discovered one after another.

The introduction of penicillin and sulfonamides in the 1930s can be said to mark the beginning of modern medicine. As the medical writer Lewis Thomas said, "Doctors could now *cure* disease, and this was astonishing, most of all to the doctors themselves." For an idea of how complicated and expensive it is to develop, test,

and get a new drug licensed, go to the web site interview with pharmacist Dan Albrant.

Take another look

GENERAL PROPERTIES OF ANTIMICROBIAL AGENTS

Antimicrobial agents share certain common properties. We can learn much about how these agents work and why they sometimes do not work by considering such properties as selective toxicity, spectrum of activity, mode of action, side effects, and resistance of microorganisms to them.

SELECTIVE TOXICITY

Some chemical substances with antimicrobial properties are too toxic to be taken internally and are used only for topical application—application to the skin's surface. For internal use, an antimicrobial drug must have **selective toxicity**—that is, it must harm the microbes without causing significant damage to the host. Some drugs, such as penicillin, have a wide range between the **toxic dosage level**, which causes host damage, and the **therapeutic dosage level**, which successfully eliminates the pathogenic organism if the level is maintained over a period of time. The relationship between an agent's toxicity to the body and its toxicity to an infectious agent is expressed in terms of its chemotherapeutic index. For any particular agent, the **chemotherapeutic index** is defined as the maximum tolerable dose per kilogram of body weight, divided by the minimum dose per kilogram of body weight, that will cure the disease. Thus, an agent with a chemotherapeutic index of 8 would be more effective and less toxic to the patient than an agent with a chemotherapeutic index of 1.

For drugs such as those containing arsenic, mercury, and antimony, the dosage must be calculated very precisely because these substances are highly toxic to human and other animal hosts, as well as to pathogens. Treatment of worm infections is especially difficult because what damages the parasite will also damage the host. In contrast, bacterial pathogens often can be treated by interfering with metabolic pathways not shared by the host. For example, penicillin interferes with cell wall synthesis; it is not toxic to human cells, which lack walls, though some patients are allergic to it.

THE SPECTRUM OF ACTIVITY

The range of different microbes against which an antimicrobial agent acts is called its **spectrum of activity**. Those agents that are effective against a great number of microorganisms from a wide range of taxonomic groups, including both Gram-positive and Gram-negative bacteria, are said to have a **broad spectrum** of activity. Those that are effective against only a small number of microorganisms or a single taxonomic group have a **narrow**

Figure 13.1 **The spectrum of antibiotic activity.** Broad-spectrum drugs, such as tetracycline, affect a variety of different organisms. Narrow-spectrum drugs, such as isoniazid, affect only a few specific types of organisms.

spectrum of activity (**Figure 13.1**). Some common antibiotics are classified according to their spectrum of activity in **Table 13.1**.

A broad-spectrum drug is especially useful when a patient is seriously ill with an infection caused by an unidentified organism. Using such a drug increases the chance that the organism will be susceptible to it. However, if the identity of the organism is known, a narrow-spectrum drug should be used. Using such a drug minimizes the destruction of the host's *microflora*, or *normal flora*—the indigenous microbes that naturally occur in or on the host—that sometimes compete with and help destroy infectious organisms. The use of narrow-spectrum

drugs also decreases the likelihood that organisms will develop drug resistance.

MODES OF ACTION

Like other medicines, antimicrobial agents are sometimes used simply because they work, without our always knowing how they work. Many people's lives have been saved by medicines whose actions at the cellular level have never been understood. However, it is always desirable to know the mode of action of an agent. With that knowledge, effects of actions on patients can be better monitored and controlled, and ways of improving them may be found.

Antimicrobial drugs generally act on an important microbial structure or function that usually differs from its counterpart in animals. This difference is exploited in exerting a *bactericidal*, or killing, effect or a *bacteriostatic*, or growth-inhibiting, effect on bacteria while having minimal effects on host cells (◄Chapter 12, p. 343). However, the host's immune systems or phagocytic defenses must still complete the elimination of the invading microbes.

Five different modes of action of antimicrobials are discussed here: (1) inhibition of cell wall synthesis, (2) disruption of cell membrane function, (3) inhibition of protein synthesis, (4) inhibition of nucleic acid synthesis, and (5) action as antimetabolites (**Figure 13.2**).

Inhibition of Cell Wall Synthesis

Many bacterial and fungal cells have rigid external cell walls, whereas animal cells lack cell walls. Consequently, inhibiting cell wall synthesis selectively damages bacterial and fungal cells. Bacterial cells, especially Grampositive ones, have a high internal osmotic pressure.

TABLE 13.1

The Spectrum of Activity of Selected Antimicrobial Agents		
Organisms Affected	**Broad-Spectrum Agents**[a]	**Narrow-Spectrum Agents**
Bacteroides and other anaerobes	Cephalosporins	Lincomycin Clindamycin
Yeasts	Chloramphenicol	Nystatin
Gram-positive bacteria	Gentamicin	Penicillin G
	Ampicillin	Erythromycin
Gram-negative bacteria	Kanamycin	Polymyxins
Streptococci and some Gram-negative bacteria	Tetracyclines	Streptomycin
Staphylococci and some clostridia	Tetracyclines	Vancomycin

[a]Broad-spectrum agents affect most bacteria.

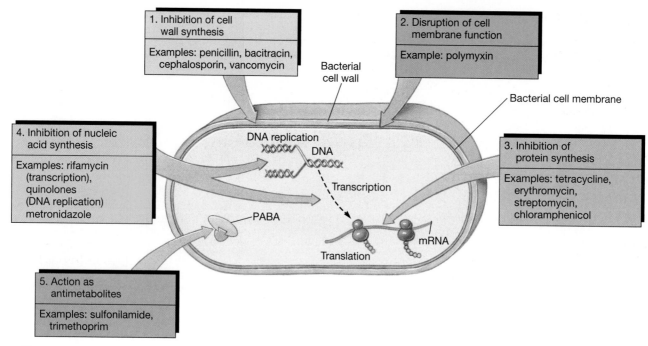

Figure 13.2 **Modes of action.** Five major modes of action by which drugs exert their antimicrobial effects on bacterial cells.

Without a normal, sturdy cell wall, these cells burst when subjected to the low osmotic pressure of body fluids (◀Chapter 4, p. 109). Antibiotics such as penicillin and cephalosporin contain a chemical structure called a *β-lactam ring*, which attaches to the enzymes that cross-link peptidoglycans (◀Chapter 4, p. 84). By interfering with the cross-linking of tetrapeptides, these antibiotics prevent cell wall synthesis **(Figure 13.3)**. Fungi and Archaea, whose cell walls lack peptidoglycan, are unaffected by these antibiotics.

Disruption of Cell Membrane Function

All cells are bounded by a membrane. Although the membranes of all cells are quite similar, those of bacteria and fungi differ sufficiently from those of animal cells to allow selective action of antimicrobial agents. Certain polypeptide antibiotics, such as polymyxins, act as detergents and distort bacterial cell membranes, probably by binding to phospholipids in the membrane. (With this distortion, the membrane is no longer regulated by membrane proteins, and the cytoplasm and cell substances are lost.) These antibiotics are especially effective against Gram-negative bacteria, which have an outer membrane rich in phospholipids (◀Chapter 4, p. 85). Polyene antibiotics, such as amphotericin B, bind to particular sterols, present in the membranes of fungal (and animal) cells. Thus, polymyxins do not act on fungi, and polyenes do not act on bacteria.

Inhibition of Protein Synthesis

In all cells, protein synthesis requires not only the information stored in DNA, plus several kinds of RNA, but also ribosomes. Differences between bacterial (70S) and animal (80S) ribosomes allow antimicrobial agents to attack bacterial cells without significantly damaging animal cells—that is, with selective toxicity. Aminoglycoside antibiotics, such as streptomycin, derive their name from the amino acids and glycosidic bonds they contain. They act on the 30S portion of bacterial ribosomes by interfering with the accurate reading (translation) of the mRNA message—that is, the incorporation of the correct amino acids (◀Chapter 7, p. 187). Chloramphenicol and erythromycin act on the 50S portion of bacterial ribosomes, inhibiting the formation of the growing polypeptide. Because animal cell ribosomes consist of 60S and 40S subunits, these antibiotics have little effect on host cells. (Mitochondria, however, which have 70S ribosomes, can be affected by such drugs.)

Inhibition of Nucleic Acid Synthesis

Differences between the enzymes used by bacterial and animal cells to synthesize nucleic acids provide a means for selective action of antimicrobial agents. Antibiotics of the rifamycin family bind to a bacterial RNA polymerase and inhibit RNA synthesis (◀Chapter 7, p. 186).

Action as Antimetabolites

The normal metabolic processes of microbial cells involve a series of intermediate compounds called *metabolites* that are essential for cellular growth and survival. **Antimetabolites** are substances that affect the utilization of metabolites and therefore prevent a cell from carrying out necessary metabolic reactions. Antimetabolites function in two ways: (1) by competitively inhibiting enzymes

(a) SEM

(b) SEM

Figure 13.3   Inhibition of cell wall synthesis by penicillin. Scanning electron micrograph of bacteria **(a)** before and **(b)** after exposure to penicillin (magnified 785X). Notice the distortion of the cell shape due to the disruption by penicillin of the cross-linking tetrapeptides in the peptidoglycan layer of the cell wall. (Both photos: From Victor Lorran, Some Effects of Subinhibitory Concentrations of Penicillin on the Structure and Division of Staphylococci, Antimicrobial Agents and Chemotherapy 7, 886, 1975.)

(a)

(b)   (c)

Figure 13.4   Competitive inhibition. **(a)** *Para*-aminobenzoic acid (PABA), a metabolite required by many bacteria. **(b)** Sulfanilamide, a sulfa drug. **(c)** *Para*-aminosalicylic acid (PAS). Sulfanilamide and PAS act as competitive inhibitors to PABA. Notice the similarity in the structures of the three compounds.

and (2) by being erroneously incorporated into important molecules such as nucleic acids. Antimetabolites are structurally similar to normal metabolites. The actions of antimetabolites are sometimes called **molecular mimicry** because they mimic, or imitate, the normal molecule, preventing a reaction from occurring or causing it to go awry.

In competitive inhibition an enzymatic reaction is inhibited by a substrate that binds to the enzyme's active site but cannot react (◀Chapter 5, p. 122). While this competing substrate occupies the active site, the enzyme is unable to function, and metabolism will slow or even cease if enough enzyme molecules are inhibited. Consider sulfanilamide and *para*-aminosalicylic acid (PAS), which are chemically very similar to *para*-aminobenzoic acid (PABA) **(Figure 13.4)**. They competitively inhibit an enzyme that acts on PABA. Many bacteria require PABA in order to make folic acid, which they use in

synthesizing nucleic acids and other metabolic products. When sulfanilamide or PAS instead of PABA is bound to the enzyme, the bacterium cannot make folic acid. Animal cells lack the enzymes to make the folic acid and must obtain it from their diets; thus their metabolism is not disturbed by these competitive inhibitors.

Antimetabolites such as the purine analog vidarabine and the pyrimidine analog idoxuridine are erroneously incorporated into nucleic acids. These molecules are very similar to the normal purines and pyrimidines of nucleic acids **(Figure 13.5)**. When incorporated into a nucleic

BIOTECHNOLOGY

Pharmacy of the Future: Antisense Drugs

The drugs of the future will do more than treat symptoms—they will attack the disease-causing genes. Many biotechnology companies, which have invested their future in the development of these therapies, are using synthetic nucleic acids such as RNAi (interference RNA) to prevent the expression of specific genes involved in AIDS, cancer, and inflammatory diseases. It was thought, because of high target specificity, that antisense drugs would produce few toxic side effects. However, toxicological studies have shown that antisense drugs can produce decreased blood cell counts in various rodents and extreme hypotension in monkeys. There are three basic types of antisense drugs. Classic antisense compounds are small, gene-specific oligonucleotides that bind to complementary regions on mRNA and prevent protein translation. The second type of antisense drugs uses ribozymes, enzymes made of RNA, to destroy specific mRNAs bound to antisense oligonucleotides. The third class of oligonucleotide antisense drugs targets specific tissues or organs.

acid, they garble the information that it encodes because they cannot form the correct base pairs during replication and transcription. Purine and pyrimidine analogs are generally as toxic to animal cells as to microbes because all cells use the same purines and pyrimidines to make nucleotides. These agents are most useful in treating viral infections, because viruses incorporate analogs more rapidly than do cells and are more severely damaged.

KINDS OF SIDE EFFECTS

The side effects of antimicrobial agents on infected persons (hosts) fall into three general categories: (1) toxicity, (2) allergy, and (3) disruption of normal microflora. The development of resistance to antibiotics can also be thought of as a side effect on the microorganisms. As is explained later, resistance produces infections that can be difficult to treat.

Toxicity

By their selective toxicity and modes of action, antimicrobial agents kill microbes without seriously harming host cells. However, some antimicrobials do exert toxic effects on the patients receiving them. The toxic effects of antimicrobial agents are discussed later in connection with specific agents.

Allergy

An *allergy* is a condition in which the body's immune system responds to a foreign substance, usually a protein. For example, breakdown products of penicillins combine with proteins in body fluids to form a molecule that the body treats as a foreign substance. Allergic reactions can be limited to mild skin rashes and itching, or they can be life-threatening. One kind of life-threatening allergic reaction, called *anaphylactic shock* (◄Chapter 18), occurs when an individual is subjected to a foreign substance to which his or her body has already become sensitized—that is, a substance to which the individual has been exposed and has developed antibodies against.

Disruption of Normal Microflora

Antimicrobial agents, especially broad-spectrum antibiotics, may exert their adverse effects not only on pathogens but also on indigenous microflora—the microorganisms that normally inhabit the skin and the digestive, respiratory, and urogenital tracts. When these microflora are disturbed, other organisms not susceptible to the antimicrobial agent, such as *Candida* yeast, invade the unoccupied areas and multiply rapidly. Invasion by replacement microflora is called **superinfection**. Superinfections are difficult to treat because they are susceptible to few antibiotics.

Although short-term use of penicillins generally does not severely disrupt normal microflora, oral ampicillin sometimes allows overgrowth of toxin-producing *Clostridia*. Long-term use of penicillin or aminoglycosides can abolish natural microflora and allow colonization of the gut with resistant Gram-negative bacteria and fungi such as *Candida*. Live-culture yogurt (which contains lactobacilli), or a preparation called Lactinex (which contains normal microflora) can be given to counteract the effects of antibiotics. Oral and vaginal superinfections with species of *Candida* yeasts are common after prolonged use of antimicrobial agents such as cephalosporins, tetracyclines, and chloramphenicol. An old remedy for this problem is douching with dilute suspensions of plain, live-culture yogurt. The risk of serious superinfections is greatest in hospitalized patients receiving broad-spectrum antibiotics, for two reasons. First, patients often are debilitated and less able to resist infection. Second, they are in an environment in which drug-resistant pathogens are prevalent.

THE RESISTANCE OF MICROORGANISMS

Resistance of a microorganism to an antibiotic means that a microorganism formerly susceptible to the action of the antibiotic is no longer affected by it. An important factor in the development of drug-resistant strains of microorganisms is that many antibiotics are bacteriostatic rather than bactericidal. Unfortunately, the most resilient microbes evade defenses (◄Chapter 14) and are likely to develop resistance to the antibiotic.

Figure 13.5 Base analogs. Nucleic acid bases and their analogs: molecules are so similar in structure that they can be incorporated in place of the correct molecule, thus acting as antimetabolites. **(a)** Basic structure of a purine. **(b)** The purine analog vidarabine. **(c)** Basic structure of a pyrimidine. **(d)** The pyrimidine analog idoxuridine.

(a)   (b)   (c)   (d)

For the past 55 years antibiotics have been used in animal feeds, not only to prevent disease, but also to promote live-stock growth. Antibiotics are used in animal feed at a rate of 2 to 50 grams per ton for improved growth and as high as 50 to 200 grams per ton of feed when specific diseases are being targeted. After animals have been fed antibiotics for long periods of time, they shed resistant bacteria in their feces. Transfer of these bacteria from animals to humans occurs when people, working with the animals, are infected on farms or in slaughterhouses. The Food and Drug Administration says that antibiotics, especially fluoroquinolones, are a "significant cause" of resistant *Campylobacter* bacterial infections of the digestive tract. Most are acquired by eating antibiotic-fed chicken. Such infections are rising rapidly, with 9,000 cases in 1999 increasing to 11,000 in 2000. Banning use of antibiotics as growth promoters in livestock would add $5 to $10 per person to the American family's meat bill. But think of the medical costs it would save—not to mention human suffering and deaths!

How Resistance Is Acquired

Microorganisms generally acquire antibiotic resistance by genetic changes, but sometimes they do so by nongenetic mechanisms. Nongenetic resistance occurs when microorganisms such as those that cause tuberculosis persist in the tissues out of reach of antimicrobial agents. If the sequestered microorganisms start to multiply and release their progeny, the progeny are still susceptible to the antibiotic. This type of resistance might more properly be called *evasion*. Another type of nongenetic resistance occurs when certain strains of bacteria temporarily change to L forms that lack most of their cell walls (◄Appendix B). For several generations, while the cell wall is lacking, these organisms are resistant to antibiotics that act on cell walls. However, when they revert to producing cell walls, they again become susceptible to the antibiotics.

Genetic resistance to antimicrobial agents develops from genetic changes followed by natural selection (**Figure 13.6**; ◄Chapter 8, p. 223). For example, in most bacterial populations, mutations occur spontaneously at a rate of about 1 per 10 million to 10 billion organisms. Bacteria reproduce so rapidly that billions of organisms can be produced in a short period of time, and among them there will always be a few mutants. If a mutant happens to be resistant to an antimicrobial agent in the environment, that mutant and its progeny will be most likely to survive, whereas the nonresistant organisms will die. After a few generations, most survivors will be resistant to the antimicrobial agent. Antibiotics do *not* induce mutations, but they can create environments that favor the survival of mutant resistant organisms.

Genetic resistance in bacteria, where it is best understood, can be due to changes in the bacterial chromosome

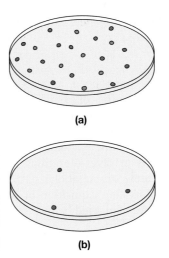

Figure 13.6 A method of detecting genetic resistance. (a) A mixed population of bacteria of varying resistance to a new antibiotic is present.
(b) Antibiotic is added to the Petri plate. Only those organisms with sufficient resistance will survive. Introduction of the antibiotic represents a change in the environment, but it does not create the resistant organisms— they were already there.

or to the acquisition of extrachromosomal DNA, usually in plasmids. (The mechanisms by which genetic changes occur were described in ◄Chapters 7 and 8.) **Chromosomal resistance** is due to a mutation in chromosomal DNA and will usually be effective only against a single type of antibiotic. Such mutations often alter the DNA that directs the synthesis of ribosomal proteins. **Extrachromosomal resistance** is usually due to the presence of particular kinds of **resistance (R) plasmids**, or **R factors** (◄Chapter 8, p. 223). How R plasmids originated is unknown, but they were first discovered in *Shigella* in Japan in 1959. Since that time many different R plasmids have been identified. Some R plasmids carry as many as six or seven genes, each of which confers resistance to a different antibiotic. R plasmids can also be transferred from one strain or species of bacteria to another. Most transfers occur by transduction (the transfer of plasmid DNA in a bacteriophage), and some occur by conjugation (◄Chapter 8, pp. 215, 218). Genes transferred by bacteriophages are responsible for the devastating

There is a concern that astronauts going up and staying for long periods of time in space may have difficulty treating their bacterial infections due to increased antibiotic resistance. Bacteria isolated from Russian cosmonauts showed an increased resistance to antibiotics after a space flight when compared to bacteria isolated before the mission. It has been since found that bacteria growing in zero gravity environments grow faster and have thicker cell walls. Dr. James Jorgensen, University of Texas Health Sciences Center, says the thick cell wall makes it more difficult for antibiotics to penetrate the microbe—thus contributing to the loss of antibiotic sensitivity. In addition there are many places within the confines of the space vehicle that permit the rapid spread of resistance genes. Future space-lab experiments are planned to determine if microbes do become more resistant to antibiotics as a result of space travel.

effects of MRSA (methicillin-resistant *Staphylococcus aureus*).

Mechanisms of Resistance

Five mechanisms of resistance have been identified, each of which involves the alteration of a different microbial structure. One involves the alteration of the target to which antimicrobial agents bind, a process that generally is caused by a mutation in the bacterial chromosome. The other mechanisms involve alterations in membrane permeability, enzymes, or metabolic pathways, which usually are caused by the acquisition of R plasmids. The five mechanisms are explained below:

1. *Alteration of Targets.* This mechanism usually affects bacterial ribosomes. The mutation alters the DNA such that the protein produced or target is modified. Antimicrobial agents can no longer bind to the target. Resistance to erythromycin, rifamycin, and antimetabolites has developed by this mechanism.

2. *Alteration of Membrane Permeability.* This mechanism occurs when new genetic information changes the nature of proteins in the membrane. Such alterations change a membrane transport system or pores in the membrane, so an antimicrobial agent can no longer cross the membrane. In bacteria, resistance to tetracyclines, quinolones, and some aminoglycosides has occurred by this mechanism. The presence of penicillin or cephalosporin can partially overcome such resistance because these agents interfere with cell wall synthesis.

3. *Development of Enzymes.* This common cause of resistance can destroy or inactivate antimicrobial agents. One enzyme of this type is β-lactamase. Several β-lactamases exist in various bacteria; they are capable of breaking the β-lactam ring in penicillins and some cephalosporins. Similar enzymes that can destroy various aminoglycosides and chloramphenicol have been found in certain Gram-negative bacteria.

4. *Alteration of an Enzyme.* This mechanism allows a formerly inhibited reaction to occur. It is exemplified by a mechanism found among certain sulfonamide-resistant bacteria. These organisms have developed an enzyme that has a very high affinity for PABA and a very low affinity for sulfonamide. Consequently, even in the presence of sulfonamide, the enzyme works well enough to allow the bacterium to function.

5. *Alteration of a Metabolic Pathway.* This mechanism bypasses a reaction inhibited by an antimicrobial agent that occurs in other sulfonamide-resistant bacteria. These organisms have acquired the ability to use ready-made folic acid from their environment and no longer need to make it from PABA.

APPLICATIONS

Microbial Resistance

Microbes resistant to disinfectants, antibiotics, or antiseptics are not created but are already present in microbial populations. It is the inappropriate use of these compounds that tends to select for and therefore promote the growth of resistant organisms in our hospitals and homes. Through the biological mechanisms of microbial competition and genetic exchange, resistant organisms can become the predominant organisms. When antimicrobial compounds are used, they must be used in concentrations that are strong enough to kill the most insensitive organisms present in the target population. If this situation has not been achieved, then the less sensitive bacteria, unencumbered by competition, are able to thrive.

First-Line, Second-Line, and Third-Line Drugs

As a strain of microorganism acquires resistance to a drug, another drug must be found to treat resistant infections effectively. If resistance to a second drug develops, a third drug is needed, and so on. Drugs used to treat gonorrhea illustrate this point. Before the 1930s no effective treatment was available for gonorrhea. But then sulfonamides were found to cure the disease. After a few years, sulfonamide-resistant strains developed, but penicillin was soon available as a "second-line" drug. Over several decades, penicillin-resistant strains developed but were combatted with very large doses of penicillin. By the 1970s some strains of gonococci developed the ability to produce a β-lactamase enzyme, which completely counteracted the effects of penicillin (**Figure 13.7**). "Third-line" spectinomycin was used. As spectinomycin-resistant strains started to appear, forcing physicians to resort to "fourth-line" drugs, we have to wonder whether the development of new drugs can go on indefinitely.

Figure 13.7 The effect of β-lactamase on penicillin. Numerous bacteria (staphylococci, streptococci, and gonococci) produce this enzyme, which inactivates penicillin. The enzyme can be transmitted by plasmids. Cephalosporins, although similar in action to penicillin, have a different cyclic ring structure and are more resistant to the effects of the enzyme.

The overall cost of antibiotic resistance in the United States is estimated between $350 million and $35 billion annually.

Drug-resistant organisms have most frequently been encountered in hospitals, where seriously ill patients with lowered resistance to infections serve as convenient hosts. However, more and more resistant organisms are being isolated from infections among the general population, and the risk of acquiring a drug-resistant infection is increasing for everyone. Moreover, many organisms are resistant to multiple antibiotics. Infections with such organisms are particularly difficult to treat. A new use for genetic probes will be to look for resistance genes in organisms, in order to avoid delays in effective treatment.

Cross-Resistance

Cross-resistance is resistance to two or more similar antimicrobial agents via a common mechanism. The action of β-lactamases provides a good example of cross-resistance.

In many instances, an enzyme that will break down one β-lactam antibiotic also will break down several other β-lactam antibiotics. The presence of such an enzyme would give a microorganism resistance to all the antibiotics it can break down.

Limiting Drug Resistance

Although, as we have seen, drug resistance is not induced by antibiotics, it is fostered by environments that contain antibiotics. The progress of microbes in acquiring resistance can be thwarted in three ways. First, high levels of an antibiotic can be maintained in the bodies of patients long enough to kill all pathogens, including resistant mutants, or to inhibit them so that body defenses can kill them. This is the reason your doctor admonishes you to be sure to take all of an antibiotic prescription and not to stop taking it once you begin to feel better. The development of resistance when medication is discontinued before all pathogens are killed is illustrated in **Figure 13.8**.

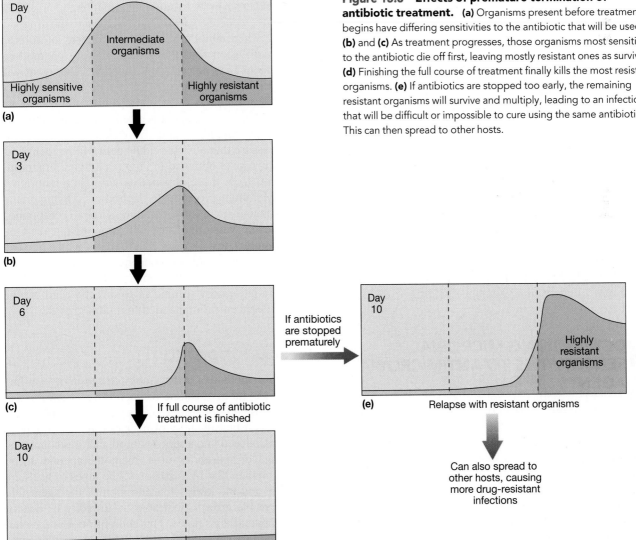

Figure 13.8 Effects of premature termination of antibiotic treatment. **(a)** Organisms present before treatment begins have differing sensitivities to the antibiotic that will be used. **(b)** and **(c)** As treatment progresses, those organisms most sensitive to the antibiotic die off first, leaving mostly resistant ones as survivors. **(d)** Finishing the full course of treatment finally kills the most resistant organisms. **(e)** If antibiotics are stopped too early, the remaining resistant organisms will survive and multiply, leading to an infection that will be difficult or impossible to cure using the same antibiotic. This can then spread to other hosts.

According to the CDC, nearly 50 million of the 150 million outpatient prescriptions for antibiotics are unnecessary.

Second, two antibiotics can be administered simultaneously so that they can exert an additive effect called **synergism**. For example, when streptomycin and penicillin are combined in therapy, the damage to the cell wall caused by the penicillin allows better penetration by streptomycin. A variation on this principle is the use of one agent to destroy the resistance of microbes to another agent. When clavulanic acid and a penicillin called amoxicillin are given together (Augmentin), the clavulanic acid binds tightly to β-lactamases and prevents them from inactivating the amoxicillin. However, some drugs are less effective when used in combination than when used alone. This decreased effect, called **antagonism**, can be observed when bacteriostatic drugs such as tetracyclines, which inhibit growth, are combined with bactericidal penicillins, which require growth to be effective.

Third, antibiotics can be restricted to essential uses only. For example, most physicians do not prescribe antibiotics for colds and other viral diseases, except in the case of patients at high risk of secondary bacterial infections, because such diseases do not respond to antibiotics. Restrictions on antibiotic use would be particularly valuable in hospitals, where microbes "just waiting to acquire resistance" lurk in antibiotic-filled environments. In addition, the use of antibiotics in animal feeds could be banned; see the box "Antibiotic Resistance: Drugs in Animal Feeds," p. 373.

✓**CHECKLIST**

1. Are all antimicrobial compounds properly called antibiotics? Why or why not?

2. When would you choose a narrow-spectrum, rather than a broad-spectrum, antibiotic? Why?

3. What is a superinfection? How is one acquired?

4. If exposure to antibiotics does not cause drug-resistant mutations to occur, why do we see more drug-resistant strains today?

DETERMINING MICROBIAL SENSITIVITIES TO ANTIMICROBIAL AGENTS

Microorganisms vary in their susceptibility to different chemotherapeutic agents, and susceptibilities can change over time. Ideally, the appropriate antibiotic to treat any particular infection should be determined before any antibiotics are given. Sometimes an appropriate agent can be prescribed as soon as the causative organism is identified from a laboratory culture. Often tests are needed to show which antibiotic kills the organism. Several methods—disk diffusion, dilution, and automated methods—are available to do this.

THE DISK DIFFUSION METHOD

In the **disk diffusion method**, or **Kirby-Bauer method**, a standard quantity of the causative organism is uniformly spread over an agar plate. Then several filter paper disks impregnated with specific concentrations of selected chemotherapeutic agents are placed on the agar surface (**Figure 13.9a**). Finally, the culture with the antibiotic disks is incubated.

During incubation, each chemotherapeutic agent diffuses out from the disk in all directions. Agents with lower molecular weights diffuse faster than those with higher molecular weights. Clear areas, called **zones of inhibition**, appear on the agar around disks where the agents inhibit the organism. The size of a zone of inhibition is not necessarily a measure of the degree of inhibition because of differences in the diffusion rates of chemotherapeutic agents. An agent of large molecular size might be a powerful inhibitor even though it might diffuse only a small distance and produce a small zone of inhibition. Standard measurements of zone diameters for particular media, quantities of organisms, and drug concentrations have been established and correlated to zone diameters in order to determine whether the organisms are *sensitive, moderately sensitive*, or *resistant* to the drug.

Even when inhibition has been properly interpreted in a disk diffusion test, the most inhibitory chemotherapeutic agent may not cure an infection. The agent will probably inhibit the causative organism, but it may not kill sufficient numbers of the organism to control the infection. A bactericidal agent is often needed to eliminate an infectious organism, and the disk diffusion method does not assure that a bactericidal agent will be identified. Moreover, results obtained *in vivo* (in a living organism) often differ from those obtained *in vitro* (in a laboratory vessel). Metabolic processes in the body of a living organism may inactivate or inhibit an antimicrobial compound.

A newer version of the diffusion test, called an **E(epsilometer) test** (**Figure 13.10**) uses a plastic strip containing a gradient of concentration of antibiotic. Printed on the strips are concentration values which allows the laboratory technician to directly read off the minimum concentration needed to inhibit growth.

THE DILUTION METHOD

The **dilution method** of testing antibiotic sensitivity was first performed in tubes of culture broth; it is now performed in shallow wells on standardized plates (**Figure 13.9b**). In this method a constant quantity of microbial inoculum (specimen) is introduced into a series of broth cultures containing decreasing concentrations of a chemotherapeutic agent. After incubation (for 16 to 20 hours) the tubes or wells are examined, and the lowest concentration of the agent that prevents visible growth (indicated by turbidity or dots of growing organisms) is noted. This concentration is the **minimum inhibitory concentration** (**MIC**) for a particular agent acting on a specific microorganism. This test can be done for

(a)

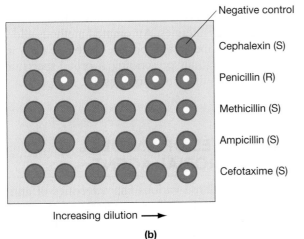

Negative control

Cephalexin (S)

Penicillin (R)

Methicillin (S)

Ampicillin (S)

Cefotaxime (S)

Increasing dilution ⟶

(b)

Figure 13.9 The disk diffusion (Kirby-Bauer) method of determining microbial sensitivities to various antibiotics. **(a)** A Petri plate is prestreaked on an agar medium with the organism to be tested. Paper disks, each containing a measured amount of a particular antibiotic, are placed firmly in contact with the medium and diffuse outward. After the plate is incubated, a clear area of no growth around a disk (a zone of inhibition) represents inhibition of the test organism by the antibiotic. Nonclear areas indicate resistance to that antibiotic. The largest zone of inhibition does not always indicate the most effective antibiotic, as different molecules do not diffuse at the same speed into the medium. Also, some drugs do not behave the same way in living organisms as they do on agar. However, the diameters of zones of inhibition, when compared with measured standards, help indicate whether an organism is sensitive or resistant to a drug. *(Science VU/Miles/Visuals Unlimited)* **(b)** Minimal inhibitory concentration (MIC) microbial susceptibility testing. A standardized microdilution plate with shallow wells that contain increasing dilutions (decreasing concentrations) of selected antibiotics in a broth is inoculated with a test bacterium. The plate is incubated; the lowest concentration that prevents growth (dots in well) is the MIC. The test bacterium on this plate is sensitive (S) to all the antibiotics except penicillin (R). The negative control well contains only broth.

Figure 13.10 An E (epsilometer) test, which determines antibiotic sensitivity and estimates MIC (minimal inhibitory concentration). A plastic strip containing an increaseing gradient of a given antibiotic is placed on the surface of a Petri dish which has been swabbed with the bacterial organism of interest. A zone of inhibition of growth around the strip indicates sensitivity of the organism to that specific antibiotic. The point at which inhibition begins (arrow) indicates the MIC for that antibiotic, and can be read off the printed scale. *(Courtesy AB Biodisk)*

several agents simultaneously by using several sets of tubes or wells, but it is time-consuming and therefore expensive.

Finding an inhibitory agent by the dilution method does no more to prove that it will kill the infectious organism in the patient than finding one by the disk diffusion method. However, the dilution method allows a second test to distinguish between bactericidal agents, which kill microorganisms, and bacteriostatic agents, which merely inhibit their growth. Samples from tubes that show no growth but that might contain inhibited organisms can be used to inoculate broth that contains no chemotherapeutic agent. In this test, the lowest concentration of the chemotherapeutic agent that yields no growth following this second inoculation, or *subculturing*, is the **minimum bactericidal concentration (MBC)**. Thus, both an effective chemotherapeutic agent and an appropriate concentration to control an infection can be determined. That concentration should be maintained at the sites of infection because it is the minimum concentration that will cure the disease.

SERUM KILLING POWER

Still another method of determining the effectiveness of a chemotherapeutic agent is to measure its **serum killing power**. This test is performed by obtaining a sample of a patient's blood while the patient is receiving an antibiotic. A bacterial suspension is added to a known quantity of the patient's **serum** (blood plasma minus the clotting factors). Growth (turbidity) in the serum after incubation means that the antibiotic is ineffective. Inhibition of growth suggests that the drug is working, and more quantitative determinations can be made to identify the lowest concentration that still provides serum killing power.

AUTOMATED METHODS

Automated methods (**Figure 13.11**) are now available to identify pathogenic organisms and to determine which antimicrobial agents will effectively combat them. One such method uses prepared trays with small wells into which a measured quantity of inoculum is automatically dispensed. As many as 36 organisms from different patients can be inoculated onto the same tray. Trays containing several kinds of media suitable for identifying members of different groups of organisms— such as Gram-positive bacteria, Gram-negative bacteria,

(a)

(b)

Figure 13.11 **An automated system for identifying microorganisms and determining their sensitivity to various antimicrobial agents.** **(a)** A sample containing the organism(s) is automatically inoculated into wells on a thin plastic tray, each containing a specific chemical reagent. *(Courtesy bioMerieux Vitek, Inc.)* **(b)** Tests are carried out in an incubation chamber, and the results are read and recorded by computer. Trays for a wide variety of different identification and antimicrobial sensitivity tests are available. *(Courtesy bioMerieux Vitek, Inc.)*

anaerobic bacteria, and yeasts—are available. Trays are also available to determine the sensitivity of organisms to a variety of antimicrobial agents.

The trays are inserted into a machine that measures microbial growth. Some machines do this by using a beam of light to measure turbidity. Others use media containing radioactive carbon. Organisms growing on such media release radioactive carbon dioxide into the air, and a sampling device automatically detects it. Machines vary in their degree of automation and the speed with which results become available. Some require technicians to perform some steps; others provide a computerized printout of results that is relayed to the patient's chart. Some machines provide results in 3 to 6 hours, and most provide them overnight, although slow-growing organisms may require 48 hours.

Automated methods make laboratory identification of organisms and their sensitivities to antimicrobials more efficient and less expensive. Once the results of laboratory tests are available, the physician can then choose an appropriate drug on the basis of the nature of the pathogen, the location of the infection, and other factors such as the patient's allergies. Automated methods allow physicians to prescribe an appropriate antibiotic early in an infection rather than prescribing a broad-spectrum antibiotic while awaiting laboratory results.

ATTRIBUTES OF AN IDEAL ANTIMICROBIAL AGENT

Having considered various characteristics of antimicrobial agents and methods of determining microbial sensitivities to them, we can now list the characteristics of an ideal antimicrobial agent:

1. *Solubility in Body Fluids.* Agents must dissolve in body fluids to be transported in the body and reach the infectious organisms. Even agents used topically must dissolve in the fluids of injured tissue to be effective; however, they must not bind too tightly to serum proteins.

2. *Selective Toxicity.* Agents must be more toxic to microorganisms than to host cells. Ideally, a great difference should exist between the low concentration that is toxic to microorganisms and the concentration that damages host cells.

3. *Toxicity not Easily Altered.* The agent should maintain a standard toxicity and not be made more or less toxic by interactions with foods, other drugs, or abnormal conditions such as diabetes and kidney disease in the host.

4. *Nonallergenic.* The agent should not elicit an allergic reaction in the host.

5. *Stability: maintenance of a constant, therapeutic concentration in blood and tissue fluids.* The agent

TRY IT
Be a Discoverer!

Soil microbes are good sources of antibiotics. Pharmaceutical companies pay people to bring back soil samples from all over the world in hopes of finding new and useful antibiotic-producing microbes. Old cemeteries, wildlife refuges, swamps, and airfield landing strips have all been sampled. Try collecting soil samples from some places that interest you. Using aseptic technique, swab the surface of a Petri plate of nutrient agar with a broth culture of an organism such as *Escherichia coli*. Make confluent strokes so that the entire surface is inoculated evenly, like a lawn. You could try different organisms on different plates. Then sprinkle tiny bits of your soil sample over each inoculated plate. Incubate the plates or leave them at room temperature, and examine them daily. Do clear areas develop around some of the colonies that have grown from soil particles? If so, these are areas where natural antibiotics have diffused out into the agar and prevented the background "lawn" bacteria from growing. Congratulations, you've found some antibiotics!

notatum. The discovery in the 1950s that certain strains of *Staphylococcus aureus* are resistant to penicillin provided the impetus to develop semisynthetic penicillins. The first of these was *methicillin*, which is effective against penicillin-resistant organisms because it is not broken down by β-lactamase enzymes. Other semisynthetic penicillins, including *nafcillin, oxacillin, ampicillin, amoxicillin, carbenicillin,* and *ticarcillin,* emerged in rapid succession. Each is synthesized by adding a particular side chain to a penicillin nucleus (**Figure 13.12**). Both the natural and semisynthetic penicillins are bactericidal.

Penicillin G, the most frequently used natural penicillin, is administered *parenterally*—that is, by some means other than through the gut, such as intramuscularly or intravenously. When administered orally, most of it is broken down by stomach acids. Penicillin is rapidly absorbed into the blood, reaches its

> *Some bacteriophages burrow holes in bacterial cell walls, thereby canceling the attempts of bacteria to pump out antibiotics. Use of phage along with antibiotics can reduce the needed concentration of antibiotics by 50 times.*

should be sufficiently stable in body fluids to have therapeutic activity over many hours; it should be degraded and excreted slowly.

6. *Resistance by Microorganisms not Easily Acquired.* There should be few, if any, microorganisms with resistance to the agent.

7. *Long Shelf-life.* The agent should retain its therapeutic properties over a long period of time with a minimum of special procedures such as refrigeration or shielding from light.

8. *Reasonable Cost.* The agent should be affordable to patients who need it.

Many antimicrobial agents meet these criteria reasonably well. But few, if any, meet all the criteria for the ideal antimicrobial agents. As long as better drugs might be found, the search for them will continue.

▋▋ ANTIBACTERIAL AGENTS

Most antimicrobial agents are *antibacterial agents,* so we will start our "catalog" of antimicrobial agents with them, keeping in mind that some are effective against other microbes as well. Antibacterial agents can be categorized in several ways; we have chosen to use their modes of action. Another way of grouping antibiotics is by the microorganism that produces them (**Table 13.2**).

INHIBITORS OF CELL WALL SYNTHESIS
Penicillins
Natural **penicillins**, such as *penicillin G* and *penicillin V,* are extracted from cultures of the mold *Penicillium*

TABLE 13.2

Selected Microbes That Serve as Sources of Antibiotics	
Microbe	**Antibiotic**
Fungi	
Cephalosporium species	Cephalosporins
Penicillium griseofulvum	Griseofulvin
Penicillium notatum and *P. chrysogenum*	Penicillin
Streptomycetes	
Streptomyces nodosus	Amphotericin B
Streptomyces venezuelae	Chloramphenicol
Streptomyces erythreus	Erythromycin
Streptomyces avermitilis	Ivermectin
Streptomyces griseus	Streptomycin
Streptomyces kanamyceticus	Kanamycin
Streptomyces fradiae	Neomycin
Streptomyces noursei	Nystatin
Streptomyces mediterranei	Rifampsin
Streptomyces aureofaciens	Tetracycline
Streptomyces orientalis	Vancomycin
Streptomyces antibioticus	Vidarabine
Actinomycetes	
Micromonospora species	Gentamicin
Other Bacteria	
Bacillus licheniformis	Bacitracin
Bacillus polymyxa	Polymyxins
Bacillus brevis	Tyrocidin

Penicillin

β-Lactam ring

Basic structure

Cephalosporin

(or O)

Basic structure

(R = sites for side-chain attachments)

Natural penicillins

Penicillin G
(Gram-positive cocci)

Penicillin V
(acid resistant)

Semisynthetic penicillins

Methicillin
(penicillinase resistant)

Ampicillin
(broad spectrum,
acid resistant)

Carbenicillin
(broad spectrum)

Amoxicillin
(broad spectrum)

1st Generation

Cephalothin
(Gram-positive cocci)

Cefazolin
(Gram-positive cocci)

Cephalexin
(broad spectrum)

2nd Generation

Cefuroxime
(broad spectrum)

Cefaclor
(broad spectrum)

3rd Generation

Cephtazidime
(broad spectrum)

Ceftriaxone
(broad spectrum)

4th Generation

Cefepime
(broad spectrum)

Figure 13.12 Penicillins. A comparison of the penicillin and cephalosporin molecules with β-lactam rings (blue). Cephalosporin differs slightly in the attached ring (red) and has two sites for side-chain attachments (purple) rather than one, as on the penicillin molecule.

In the laboratory, penicillin is sometimes added to mixed cultures to keep undesired species from overgrowing slower-growing species of archaea who, lacking peptidoglycan, are not sensitive to penicillin.

maximum concentration, and is excreted unless it is combined with an agent such as procaine, which slows excretion and prolongs activity.

Penicillin G is the drug of choice in treating infections caused by streptococci, meningococci, pneumococci, spirochetes, clostridia, and aerobic Gram-positive rods. It is also suitable for treating infections caused by a few strains of staphylococci and gonococci that are not resistant to it. Because it retains activity in urine, it is suitable for treating some urinary tract infections. Infections caused by organisms resistant to penicillin G can be treated with semisynthetics such as nafcillin, oxacillin, ampicillin, or amoxicillin. Carbenicillin and ticarcillin are especially useful in treating *Pseudomonas* infections. Allergy to penicillin is rare among children but occurs in 1 to 5% of adults. Penicillins are generally nontoxic, but large doses can have toxic effects on the kidneys, liver, and central nervous system.

In addition to their use as treatment for infections, penicillins also are used prophylactically—that is, to *prevent* infection. For example, patients with heart defects (in particular, malformed or artificial valves) or heart disease are especially susceptible to endocarditis, an inflammation of the lining of the heart, caused by a bacterial infection. Organisms tend to attack surfaces of damaged valves. To prevent such infections, susceptible patients often receive penicillin before surgery or dental procedures (even cleanings) that could release bacteria into the bloodstream.

Cephalosporins

Natural **cephalosporins** (sefa-lo-spor′-inz), derived from several species of the fungus *Cephalosporium*, have limited antimicrobial action. Their discovery led to the development of a large number of bactericidal, semisynthetic derivatives of natural cephalosporin C. The nucleus of a cephalosporin is quite similar to that of penicillin; both contain β-lactam rings (Figure 13.12). Semisynthetic cephalosporins, like semisynthetic penicillins, differ in the nature of their side chains. Frequently used cephalospirins include *cephalexin* (Keflex), *cephradine*, and *cefadroxil*, all of which are fairly well absorbed from the gut and therefore can be administered orally. Other cephalosporins, such as *cephalothin* (Keflin), *cephapirin*, and *cefazolin*, must be administered parenterally, usually into muscles or veins.

Although cephalosporins usually are not the first drug considered in the treatment of an infection, they are frequently used when allergy or toxicity prevents the use of other drugs. But because cephalosporins are structurally similar to penicillin, some patients who are allergic to penicillin may also be sensitive to the cephalosporins.

APPLICATIONS

Disarm and Kill

Sometimes a β-lactamase inhibitor such as clavulanic acid is added to a penicillin family antibiotic. Augmentin is such a preparation, composed of clavulanic acid plus ampicillin. The β-lactamase is destroyed by the clavulanic acid, allowing the ampicillin's β-lactam ring to remain intact and the ampicillin to kill the bacteria.

Nevertheless, cephalosporins account for one-fourth to one-third of pharmacy expenditures in U.S. hospitals, mainly because they have a fairly wide spectrum of activity, rarely cause serious side effects, and can be used prophylactically in surgical patients. Unfortunately, they are often used when a less expensive and narrower-spectrum agent would be just as effective.

The development of new varieties of cephalosporins seems to be a race against the ability of bacteria to acquire resistance to older varieties. When organisms became resistant to early "first-generation" cephalosporins, new, "second-generation" cephalosporins, including *cefuroxime* and *cefaclor*, were produced (Figure 13.12). Now "third-generation" cephalosporins, such as *ceftriaxone* and *cephtazidime*, and "fourth-generation" cefepime, are used against organisms resistant to older drugs. These drugs are especially effective (for now) in dealing with hospital-acquired infections resistant to many antibiotics. They are being tried in patients with AIDS and other immunodeficiencies. (Do not confuse these with second- and third-line drugs, described earlier, which are not derivatives of one another.)

Adverse effects from cephalosporins tend to be local reactions, such as irritation at the injection site or nausea, vomiting, and diarrhea when the drug is administered orally. Four to fifteen percent of patients allergic to penicillin also are allergic to cephalosporins. Moreover, newer cephalosporins have little effect on Gram-positive organisms, which can cause superinfections during the treatment of Gram-negative infections.

Other Antibacterial Agents That Act on Cell Walls

Carbapenems (kar′ba-pen-emz) represent a new group of bactericidal antibiotics with two-part structures. *Primaxin*, a typical carbapenem, consists of a β-lactam antibiotic (*imipenem*) that interferes with cell wall synthesis and *cilastatin sodium*, a compound that prevents degradation of the drug in the kidneys. As a group, the carbapenems have an extremely broad spectrum of activity.

Bacitracin, a small bactericidal polypeptide derived from the bacterium *Bacillus licheniformis*, is used only on lesions and wounds of the skin or mucous membranes because it is poorly absorbed and toxic to the kidneys. *Vancomycin* is a large, complex molecule produced by the soil actinomycete *Streptomyces orientalis*. It is too large a molecule to pass through pores in the outer membrane of Gram-negative cell walls, and is therefore not effective against most Gram-negative bacteria. It can be used to treat infections caused by methicillin-resistant staphylococci and enterococci. It is also the drug of choice against antibiotic-induced pseudomembranous colitis (enteritis with the formation of false membranes in stool). Because it is poorly absorbed through the gastrointestinal tract, it must be administered intravenously. Vancomycin is fairly toxic, causing hearing loss and kidney damage, especially in older patients, if the drug is not monitored carefully.

DISRUPTERS OF CELL MEMBRANES
Polymyxins

Five **polymyxins**, designated A, B, C, D, and E, have been obtained from the soil bacterium *Bacillus polymyxa*. Polymyxins B and E are the most common clinically. They are usually applied topically, often with bacitracin, to treat skin infections caused by Gram-negative bacteria such as *Pseudomonas*. Used internally, polymyxins can cause numbness in the extremities, serious kidney damage, and respiratory arrest. They are administered by injection when the patient is hospitalized and kidney function can be monitored.

APPLICATIONS

The Role of *Clostridium difficile* in Antibiotic-Associated Intestinal Disease

Antibiotic therapy sometimes has its dark side. Today, *Clostridium difficile* is one of the most important intestinal bacterial pathogens in the developed world, in terms of prevalence and severity of disease. Yet *C. difficile* causes intestinal disease only in patients to whom antibiotics have been administered. (A few exceptions to this rule existed in the preantibiotic era, and no completely convincing explanation is available to explain these cases.)

This pathogen is a Gram-positive anaerobic rod. *C. difficile* was first described in 1935, but it was not definitively associated with disease until 1978. During the 1950s, especially, *Staphylococcus aureus* was blamed for what are now known to have been *C. difficile* infections. *C. difficile* is now understood to be a major nosocomial pathogen and is currently the only organism recognized as a common cause of antibiotic-associated colitis. *C. difficile* is found in 15 to 25% of patients with antibiotic-associated diarrhea; 50 to 75% of patients with colitis; and more than 90% of patients with pseudomembranous colitis (PMC). In individuals with PMC, a fibrous pseudomembrane covers the mucosa of the colon due to fibrin-containing fluid that collects there.

The organism is rarely found in healthy people except for newborns, among whom it occurs frequently but without harm. These newborns can act as reservoirs for *C. difficile* and can easily spread it to others in hospitals or at home. It disappears as babies approach 6 to 12 months of age. Older children rarely develop *C. difficile* problems, despite frequent use of antibiotics. Older adults are most likely to develop disease, but this age factor has not yet been explained. Also, PMC can develop in patients who have undergone abdominal or intestinal surgery and in those whose gastrointestinal microbiota has been changed by the use of antibiotics.

C. difficile is a moderately strict anaerobe. It is widely distributed in nature, being found in soil and in feces of animals such as cows and horses. Once it has established itself in an environment, *C. difficile* is difficult to remove. It may persist in hospital wards or nursing homes for months or years despite vigorous efforts at eradication. Organisms have been recovered from the floor, bedpans, linens, and even walls of rooms of infected patients.

Clindamycin is the antibiotic most frequently associated with *C. difficile* infections, followed by ampicillin and cephalosporins. These drugs are often administered to patients who will undergo abdominal surgery, in order to decrease normal microflora. The *C. difficile* organisms are usually very susceptible to these antibiotics but survive by forming endospores. After the drug is discontinued, they regenerate and can overgrow the intestinal tract, but they do not invade its tissues.

Virtually all strains of *C. difficile* produce two toxins, but the strains vary widely in the amount of toxin produced. Toxin A, often called the enterotoxin, has a cytotoxic (cell-killing) effect and is responsible for most of the symptoms observed. Toxin B, called the cytotoxin, is considerably more cytotoxic but is not active in the gastrointestinal tract. The mechanism of action of these toxins is still not completely known.

Symptoms of this colitis include abdominal cramps, diarrhea, fever (up to 106°F, or 41.1°C), electrolyte imbalance, toxic megacolon (an acute dilation of the colon), and even perforation of the colon. The antibiotic vancomycin is the single most effective treatment and is almost 100% effective when perforation has not occurred. Fever drops within 24 to 48 hours, and normal bowel action returns in 5 to 7 days. However, relapse is frequent due to the sporulating forms. Vancomycin costs four times as much as an equivalent weight of gold. Cost of the treatment ranges from $200 to $600. Metronidazole is also used due to its antianaerobic properties.

Endoscopic observation of pseudomembranes in the colon is considered diagnostic for a severe *C. difficile* disease. Confirmation usually requires isolating the organism from stool specimens and identifying the toxins by enzyme-linked immunosorbent assay (ELISA) and tissue culture assay. Toxin A is so potent that one molecule is enough to cause the change seen in cultured cells.

INHIBITORS OF PROTEIN SYNTHESIS

Aminoglycosides

Aminoglycosides are obtained from various species of the genera *Streptomyces* and *Micromonospora*. The first, **streptomycin**, was discovered in the 1940s and was effective against a variety of bacteria. Since then, many bacteria have become resistant to it. Moreover, streptomycin can damage kidneys and the inner ear, sometimes causing permanent ringing in the ears and dizziness. Consequently, this compound is now used only in special situations and generally in combination with other drugs. For example, it can be used with tetracyclines to treat plague and tularemia and with isoniazid and rifampin to treat tuberculosis.

Other aminoglycosides, such as *neomycin, kanamycin, amikacin, gentamicin, tobramycin,* and *netilmicin,* also have special uses and display varying degrees of toxicity to the kidneys and inner ear. At lower, less toxic doses, aminoglycosides tend to be bacteriostatic. They are usually administered intramuscularly or intravenously because they are poorly absorbed when given orally.

An important property of aminoglycosides is their ability to act synergistically with other drugs—an aminoglycoside and another drug together often control an infection better than either could alone. For example, gentamicin and penicillin or ampicillin are effective against penicillin-resistant streptococci. In other synergistic actions, gentamicin or tobramycin work with carbenicillin or ticarcillin to control *Pseudomonas* infections, especially in burn patients, and aminoglycosides work with cephalosporins to control *Klebsiella* infections.

Other applications of aminoglycosides include the treatment of bone and joint infections, peritonitis (inflammation of the lining of the abdominal cavity), pelvic abscesses, and many hospital-acquired infections. In bone and joint infections, gentamicin and tobramycin are especially useful because they can penetrate joint cavities. Because peritonitis and pelvic abscesses are severe and are often caused by a mixture of enterococci and anaerobic bacteria, aminoglycoside treatment is usually started before the organisms are identified. Amikacin is especially effective in treating hospital-acquired infections resistant to other drugs. It should not be used in less demanding situations lest organisms become resistant to it, too.

Aminoglycosides often damage kidney cells, causing protein to be excreted in the urine, and prolonged use can kill kidney cells. These effects are most pronounced in older patients and those with preexisting kidney disease. Some aminoglycosides damage the eighth cranial nerve: streptomycin causes dizziness and disturbances in balance, and neomycin causes hearing loss.

Tetracyclines

Several **tetracyclines** are obtained from species of *Streptomyces*, in which they were originally discovered. Commonly used tetracyclines include tetracycline itself, *chlortetracycline* (Aureomycin), and *oxytetracycline* (Terramycin). Newer, semisynthetic tetracyclines include *minocycline* (Minocin) and *doxycycline* (Vibramycin). All are bacteriostatic at normal doses, are readily absorbed from the digestive tract, and become widely distributed in tissues and body fluids (with the exception of cerebrospinal fluid). These drugs easily enter the cytoplasm of host cells, making them especially useful at killing intracellularly infecting bacteria.

The fact that tetracyclines have the widest spectrum of activity of any antibiotics is a two-edged sword. They are effective against many Gram-positive and Gram-negative bacterial infections and are suitable for treating rickettsial, chlamydial, mycoplasmal, and some fungal infections. But because they have such a wide sprectrum of activity, they destroy the normal intestinal microflora and often produce severe gastrointestinal disorders. Recalcitrant superinfections of tetracycline-resistant *Proteus, Pseudomonas,* and *Staphylococcus,* as well as yeast infections, also can result.

Tetracyclines can cause a variety of mild to severe toxic effects. Nausea and diarrhea are common, and extreme sensitivity to light is sometimes seen. The drug can also cause pustules to form on the skin. Effects on the liver and kidneys are more serious. Liver damage can be fatal, especially in patients with severe infections or during pregnancy. Kidney damage can lead to acidosis (low blood pH) and to the excretion of protein and glucose. Anemia can occur, but it is rare. Tetracycline can also interfere with the effectiveness of birth control pills.

Staining of the teeth **(Figure 13.13)** occurs when children under 5 years of age receive tetracycline or when their mothers received it during the last half of their pregnancy. Both deciduous teeth (baby teeth) and permanent teeth will be mottled because the buds of both types of teeth form before birth. Tetracycline taken during pregnancy also can lead to abnormal bone formation in the fetal skull and a permanent abnormal skull shape. The ability of calcium ions to form a complex with tetracycline is responsible for its effects on bones and teeth. Because this reaction destroys the antibiotic effect of the drug, patients should not consume milk or other dairy products

Figure 13.13 **Staining of teeth caused by tetracycline.** If the condition results from ingestion of the antibiotic during pregnancy, both the deciduous (baby) and permanent teeth will be affected, as both sets of tooth buds are forming in the fetus at that time. (*SPL/Custom Medical Stock Photo, Inc.*)

APPLICATIONS

Antibiotics and Acne

Low doses of tetracyclines and erythromycin suppress skin bacteria, mostly *Propionibacterium acnes*, and reduce the release of microbial lipases, which contribute to skin inflammation. This therapy is used to treat acne, but its effectiveness has not been proved. Studies intended to assess antibiotic effectiveness in acne therapy have been useless because they lack suitable controls, fail to characterize adequately the type and severity of cases, and employ other concurrent therapies. Some studies have shown that low doses of many antibiotics can lead to the appearance of antibiotic-resistant strains. Are the benefits of antibiotics to acne patients worth the risk of promoting development of resistant organisms?

with the drug or for a few hours after taking it. Some greens, such as collard greens, are also very high in calcium and should be avoided when taking tetracyclines.

Chloramphenicol

Chloramphenicol, originally obtained from cultures of *Streptomyces venezuelae*, is now fully synthesized in the laboratory. Like tetracyclines it is bacteriostatic, is rapidly absorbed from the digestive tract, is widely distributed in tissues, and has a broad spectrum of activity. It is used to treat typhoid fever, infections due to penicillin-resistant strains of meningococci and *Haemophilus influenzae*, brain abscesses, and severe rickettsial infections.

Chloramphenicol damages bone marrow in two ways. It causes a dose-related, reversible aplastic anemia, in which bone marrow cells produce too few erythrocytes and sometimes too few leukocytes and platelets as well. Terminating use of the drug usually allows the bone marrow to recover normal function. It also causes a non–dose-related, permanent aplastic anemia due to destruction of bone marrow. Aplastic anemia appears days to months after treatment is discontinued and is most common in newborns. Unless a successful bone marrow transplant can be performed, aplastic anemia is usually fatal. It is seen in only one in 25,000 to 40,000 patients treated with chloramphenicol. Long-term use of chloramphenicol can cause inflammation of the optic and other nerves, confusion, delirium, and mild to severe gastrointestinal symptoms. Since chloramphenicol is sometimes prescribed or sold without prescription in countries outside the United States, you should always be careful to know the identity of antibiotics that you acquire abroad. In the United States, chloramphenicol is a drug of last choice when other effective agents exist.

Other Antibacterial Agents That Affect Protein Synthesis

Macrolides. **Erythromycin** (e-rith″ro-mi′sin), a commonly used **macrolide** (large-ring compound), is produced by several strains of *Streptomyces erythreus*. Erythromycin exerts a bacteriostatic effect, is readily absorbed, and reaches most tissues and body fluids (with the exception of cerebrospinal fluid). It is recommended for infections caused by streptococci, pneumococci, and corynebacteria but is also effective against *Mycoplasma* and some *Chlamydia* and *Campylobacter* infections. Erythromycin is most valuable in treating infections caused by penicillin-resistant organisms or in patients allergic to penicillin. Unfortunately, resistance to erythromycin often emerges during treatment. Dual antibiotic treatment—erythromycin and some other drug—is often used on patients with a pneumonia-like disease that might be Legionnaires' disease. Several antibiotics combat other pneumonias, but erythromycin is the only common antibiotic that will combat Legionnaires' disease. Erythromycin is one of the least toxic of commonly used antibiotics. Mild gastrointestinal disturbances are seen in 2 to 3% of patients receiving it. Two newer and more frequently used erythromycin relatives are *azithromycin* (Zithromax) and clarithromycin (Biaxin).

Lincosamides. *Lincomycin* is produced by *Streptomyces lincolnensis*, and *clindamycin* is a semisynthetic derivative that is more completely absorbed and less toxic than lincomycin. Both drugs, which are collectively called lincosamides, exert a bacteriostatic effect. Lincomycin can be used to treat a variety of infections but is not significantly better than other widely used antibiotics, and organisms quickly become resistant to it. Clindamycin is effective against *Bacteroides* and other anaerobes, except *Clostridium difficile*, which often becomes established as a superinfection during clindamycin therapy. Toxins from *C. difficile* can cause a severe, and sometimes fatal, colitis (inflammation of the large intestine) unless diagnosed early and treated with oral vancomycin.

INHIBITORS OF NUCLEIC ACID SYNTHESIS

Rifampin

From among the **rifamycins** produced by *Streptomyces mediterranei*, only the semisynthetic *rifampin* is currently used. Easily absorbed from the digestive tract except when taken directly after a meal, it reaches all tissues and body fluids. Rifampin blocks RNA transcription. Although it is bactericidal and has a wide spectrum of activity, it is approved in the United States only for treating tuberculosis and eliminating meningococci from the nasopharynx of carriers.

Rifampin can cause liver damage but usually does so only when excessive doses are given to patients with preexisting liver disease. It is unusual among antibiotics in its ability to interact with other drugs, and possibilities of such interactions should be considered before the drug is given. Taking rifampin concurrently with oral contraceptives has been implicated in an increased risk

Red Man Syndrome

Rifampin has been shown to cause the so-called red man syndrome. In this disorder, which occurs with high doses of the antibiotic, colored metabolic products of the drug accumulate in the body and are eliminated through sweat glands. It is characterized by bright orange or red urine, saliva, and tears, as well as skin that looks like a boiled lobster's. The red skin secretions can be washed away, but liver damage caused by the drug is only slowly repaired.

(Courtesy Bill Slue, New York University School of Medicine)

of pregnancy and menstrual disorders. Dosages of anticoagulants must be increased while a patient is taking rifampin to achieve the same degree of reduction in blood clotting. Finally, drug addicts who are receiving methadone sometimes suffer withdrawal symptoms if they are given rifampin without an increase in methadone dosage. One explanation for these diverse effects is that rifampin stimulates the liver to produce greater quantities of enzymes that are involved in the metabolism of a variety of drugs.

Quinolones

Quinolones, a new group of synthetic bactericidal analogs of *nalidixic acid*, are effective against many Gram-positive and Gram-negative bacteria. Quinolones' mode of action is to inhibit bacterial DNA synthesis by blocking DNA gyrase, the enzyme that unwinds the DNA double helix preparatory to its replication. *Norfloxacin, ciprofloxacin* (Cipro), and *enoxacin* are examples of this group of antibiotics. They are especially effective in the treatment of traveler's diarrhea and in urinary tract infections caused by multipley resistant organisms.

A recent advance has produced a hybrid class of antibiotics. One of these, a quinolone-cephalosporin combination, is currently being tested. When the β-lactamase enzymes act on the cephalosporin component, the quinolone is released from the hybrid molecule and is available to kill the cephalosporin-resistant organisms. The use of such a dual-acting synergistic antibiotic may also prevent or delay development of antibiotic resistance in organisms.

ANTIMETABOLITES AND OTHER ANTIBACTERIAL AGENTS

Sulfonamides

The **sulfonamides**, or *sulfa drugs*, are a large group of entirely synthetic, bacteriostatic agents. Many are derived from *sulfanilamide* (sul-fa-nil′a-mid), one of the first sulfonamides (Figure 13.4b). In general, orally administered sulfonamides are readily absorbed and become widely distributed in tissues and body fluids. They act by blocking the synthesis of folic acid, which is needed to make the nitrogenous bases of DNA. Sulfonamides have now been largely replaced by antibiotics because antibiotics are more specific in their actions and less toxic than sulfonamides.

When sulfonamides first came into use in the 1930s, they frequently led to kidney damage. Newer forms of these drugs usually do not damage kidneys, but they do occasionally produce nausea and skin rashes. Certain sulfonamides are still used to suppress intestinal microflora prior to colon surgery. They also are used to treat some kinds of meningitis because they enter cerebrospinal fluid more easily than do antibiotics. *Cotrimoxazole* (Septra), a combination of *sulfamethoxazole* and *trimethoprim*, is used to treat urinary tract infections and a few other infections. Cotrimoxazole is the primary drug of choice to control *Pneumocystis* pneumonia, a common fungal complication of AIDS patients. Unfortunately, both drugs are toxic to bone marrow and may cause nausea and skin rashes.

Isoniazid

Isoniazid (i-so-ni′a-zid) is an antimetabolite for two vitamins—nicotinamide (niacin) and pyridoxal (vitamin B_6). It binds to and inactivates the enzyme that converts the vitamins to useful molecules. This bacteriostatic, synthetic agent, which has little effect on most bacteria, is effective against the mycobacterium that causes tuberculosis. Isoniazid is completely absorbed from the digestive tract and reaches all tissues and body fluids, where it must first be activated by catalase (a host cell enzyme). Destruction of catalase, thus preventing activation of isoniazid, is one mechanism whereby mycobacteria can develop resistance to the drug. Once activated, isoniazid changes the acid-fastness by interfering with synthesis of mycolic acid, a component of mycobacteria cell walls. Because the mycobacteria present in any such infection usually include some isoniazid-resistant organisms, isoniazid usually is given with another two or three agents such as rifampin or ethambutol (discussed in the following section). Isoniazid kills the rapidly dividing bacilli; the other agents kill slow or dormant bacilli. Dietary supplements of nicotinamide and pyridoxal also should be given with isoniazid.

Ethambutol

The synthetic agent **ethambutol** is effective against certain strains of mycobacteria that do not respond to isoniazid. Ethambutol is well absorbed and reaches all tissues and body fluids. However, mycobacteria acquire resistance to it fairly rapidly, so it is used with other drugs such as isoniazid and rifampin. Its method of action is still unknown.

Nitrofurans

Nitrofurans (ni″tro-fyu′ranz) are antibacterial drugs that enter susceptible cells and apparently damage sensitive microbial respiratory systems. Several hundred nitrofurans have been synthesized since the first one was made in 1930. Only a few of these are currently used. Oral doses of *nitrofurantoin* (Furadantin) are bacteriostatic in low doses, easily absorbed, and quickly metabolized. This drug is especially

Agent	Used to Treat	Common Method of Administration*	Side Effects
Agents that inhibit cell wall synthesis			
Penicillin (natural)	Wide variety of infections, mostly of Gram-positive bacteria	IM, O	Relatively few side effects, but allergies do occur
Penicillin (semisynthetic)	Infections resistant to natural penicillin	O, IV	Same as natural penicillin
Cephalosporins	Wide variety of infections when allergy or toxicity makes other agents unsuitable	IV, IM, O	Relatively nontoxic but can lead to superinfections
Carbapenems	Mixed infections, nosocomial infections, infections of unknown etiology	IV	Allergic reactions, superinfections, seizures, gastrointestinal disturbances
Bacitracin	Skin infections (topical application)	T	Internal use toxic to kidneys

Imipenem (a carbapenem)

Bacitracin

Agents that interfere with cell membrane function			
Polymyxins	Skin infections (topical application, with bacitracin)	T, IV	Internal use highly toxic
Tyrocidins	Skin infections caused by Gram-positive cocci (topical application)	T, IV	Internal use highly toxic

Polymyxin B

Tyrocidin

Antimetabolites and other agents			
Sulfonamides	Some kinds of meningitis and to suppress intestinal flora before colon surgery	O, IV	Early forms caused kidney damage, but ones now in use do not
Isoniazid	Tuberculosis (used with ethambutol)	O	May cause pyridoxine deficiency
Ethambutol	Tuberculosis (used with isoniazid)	O	
Nitrofurantoin	Urinary tract infections	O	Nausea and vomiting

Sulfanilamide (a sulfonamide)

Isoniazid

Ethambutol

Nitrofurantoin

* IM = intramuscular   O = oral
IV = intravenous   T = topical

Figure 13.14 Selected antibacterial drugs.

Agent	Used to Treat	Common Method of Administration*	Side Effects
Agents that inhibit protein synthesis			
Streptomycin	Tuberculosis (used with isoniazid and rifampin)	IM, O	Damages kidneys and inner ear
Gentamicin and other aminoglycosides	Antibiotic-resistant and hospital-acquired infections (used synergistically with other drugs)	IM, T (burns)	Varying degrees of kidney and inner ear damage
Tetracyclines	A broad spectrum of bacterial infections and some fungal infections	O	Stain teeth; cause gastrointestinal symptoms; can lead to super-infections
Chloramphenicol	A broad spectrum of bacterial infections, brain abscesses, and penicillin-resistant infections	O	Can damage bone marrow and cause aplastic anemia
Erythromycin	Gram-positive bacterial infections, some penicillin-resistant infections, and Legionnaires' disease	O	One of the least toxic of commonly used antibiotics

Agents that inhibit nucleic acid synthesis			
Rifampin	Tuberculosis and to eliminate meningococci from the nasopharynx	O	Bright orange or red urine, saliva, tears, and skin; liver damage; many disorders when used with other agents
Quinolones	Urinary tract infections, traveller's diarrhea; effective against many drug-resistant organisms	O	Nausea; headaches and other nervous system disturbances

Figure 13.14 *Continued.*

useful in treating acute and chronic urinary infections. The low incidence of resistance to it makes it an ideal prophylactic agent to prevent recurrences. Unfortunately, 10% of patients experience nausea and vomiting as a side effect and must then be treated with an antibiotic instead.

The chemical structures, uses, and side effects of antibacterial agents are summarized in **Figure 13.14**.

✓ **CHECKLIST**

1. Why is the antibiotic with the largest zone of inhibition not necessarily the best one to use with a given patient?

2. What harmful side effects can occur when taking tetracycline? What things should be avoided while taking this drug?

3. What is a β-lactam ring? Where is it found? What importance does it have?

4. What organisms are affected by isoniazid and ethambutol? Are these two compounds antibiotics? What is isoniazid's mode of action?

▐▐▐ ANTIFUNGAL AGENTS

Antifungal agents are being used with greater frequency because of the emergence of resistant strains and an increase in the number of immunosuppressed patients, especially those with AIDS. Because fungi are eukaryotes and thus similar to human cells, antifungal treatment often causes toxic side effects. At less toxic levels, many systemic fungal infections are slow to respond. Furthermore, laboratory tests are not available to determine appropriate susceptibility and therapeutic levels. Despite these difficulties, numerous effective drugs are now becoming available, many without prescription.

Imidazoles and Triazoles

The **imidazoles** (im″id-az′olz) and *triazoles* comprise a large group of related synthetic fungicides. Several agents, including *clotrimazole, ketoconazole, miconazole,* and *fluconazole,* are currently in use; many are available without prescription. The imidazoles and triazoles appear to affect fungal plasma membranes by disrupting the synthesis of membrane sterols. All these agents are used topically in creams and solutions to control fungal skin infections (dermatomycoses) and *Candida* yeast infections of the skin, nails, mouth, and vagina. Ketoconazole has also been given orally to treat systemic fungal infections, especially when other antifungal agents have not been effective. Some patients, however, have experienced mild to severe skin irritations with the topical agents. Furthermore, potentially severe drug interactions may occur, especially with certain antihistamines and immunosuppressants.

Polyenes

The **polyene** family of antibiotics consists of antifungal agents that contain at least two double bonds. Amphotericin B and nystatin are two of the most common polyene antibiotics.

Amphotericin B. The fungicidal antibiotic *amphotericin* (am″fo-ter′i-sin) *B* (Fungizone) is derived from *Streptomyces nodosus.* This drug binds to plasma membrane ergosterol (a crystallizing sterol) found in fungi and some algae and protozoa but not in human cells. Amphotericin B increases membrane permeability such that glucose, potassium, and other essential substances leak from the cell. The drug is poorly absorbed from the digestive tract and so is given intravenously. Even then, only 10% of the dose given is found in the blood. Excretion persists for up to 3 weeks after treatment is discontinued, but it is not known where the drug is sequestered in the meantime.

Amphotericin B is the drug of choice in treating most systemic fungal infections, especially cryptococcosis, coccidioidomycosis, and aspergillosis. Although fungi are not known to develop resistance to this agent, side effects are numerous and sometimes severe. They include abnormal skin sensations, fever and chills, nausea and vomiting, headache, depression, kidney damage, anemia, abnormal heart rhythms, and even blindness. Because some of the fungal infections are fatal without treatment, patients, especially those who are immunocompromised or have AIDS, have little choice but to risk these unfortunate side effects.

Nystatin. The polyene antibiotic *nystatin* (Mycostatin) is produced by *Streptomyces noursei.* This drug has the same mode of action as amphotericin B but is also effective topically in the treatment of *Candida* yeast infections. Because it is not absorbed through the intestinal wall, it can be given orally to treat fungal superinfections in the intestine, which often occur after long-term treatment with antibiotics. Nystatin was named for the New York State Health Department, where it was discovered.

Griseofulvin

Griseofulvin (gris″e-o-ful′vin), originally derived from *Penicillium griseofulvum,* is used primarily for superficial fungal infections. This fungistatic drug is incorporated into new cells that replace infected cells; it interferes with fungal growth, probably by impairing the mitotic spindle apparatus used in cell division. Although griseofulvin (Fulvicin) is poorly absorbed from the intestinal tract, it is given orally and appears to reach the target tissues through perspiration. It is ineffective against bacteria and most systemic fungal agents but is very useful topically in treating fungal infections of the skin, hair, and nails. Most infections are cured within 4 weeks, but recalcitrant infections associated with fingernails and toenails may persist even after a year of treatment. Reactions to griseofulvin are usually limited to mild headaches but can include gastrointestinal disturbances, especially when prolonged treatment is required. It is also one of the antibiotics suspected, but not proven, to reduce effectiveness of birth control pills.

Other Antifungal Agents

Flucytosine is a synthetic drug used in treating infections caused by *Candida* and several other fungi. This fluorinated pyrimidine is transformed in the body to fluorouracil, an analog of uracil, and thereby interferes with nucleic acid and protein syntheses. The drug can be given orally and is easily absorbed, but 90% of the amount given is found unchanged in the urine within 24 hours. Because it is less toxic and causes fewer side effects than amphotericin B, flucytosine should be given instead of amphotericin B whenever possible.

Tolnaftate (Tinactin) is a common topical fungicide that is readily available without prescription. Although its mode of action is still not clear, it is effective in the treatment of various skin infections, including athlete's foot and jock itch.

Terbinafine (Lamisil), a relatively new fungicide, has been approved for topical use in skin infections and cutaneous candidiasis. Because it is absorbed directly through the skin, it reaches therapeutic levels in much less time than do orally administered agents such as griseofulvin.

▋▋▋ ANTIVIRAL AGENTS

Until recent years no chemotherapeutic agents effective against viruses were available. One reason for the difficulty in finding such agents is that the agent must act on viruses within cells without severely affecting the host cells. Currently available *antiviral agents* inhibit some phase of viral replication, but they do not kill the viruses.

Purine and Pyrimidine Analogs

Several purine and pyrimidine analogs are effective antiviral agents. All cause the virus to incorporate erroneous information (the analog) into a nucleic acid and thereby interfere with the replication of viruses (◄Chapter 7, p. 199). The drugs include idoxuridine, vidarabine, ribavirin, acyclovir, ganciclovir, and azidothymidine (AZT).

Idoxuridine and *trifluridine*, both analogs of thymine, are administered in eye drops to treat inflammation of the cornea caused by a herpesvirus. They should not be used internally because they suppress bone marrow.

Vidarabine (ARA-A), an analog of adenine, has been used effectively to treat viral encephalitis, an inflammation of the brain caused by herpesviruses and by cytomegaloviruses. It is not effective against cytomegalovirus infections acquired before birth. Vidarabine is less toxic than either idoxuridine or cytarabine, but it sometimes causes gastrointestinal disturbances.

Ribavirin (Virazole), a synthetic nucleotide analog of guanine, blocks replication of certain viruses. In an aerosol spray, it can combat influenza viruses; in an ointment, it can help to heal herpes lesions. Although it has low toxicity, it can induce birth defects and should not be given to pregnant women. It has been found to be effective against hantaviruses, such as those that caused the deadly outbreak of respiratory disease on the Navajo reservation in the Four Corners region of the American Southwest in 1993 (◄Chapter 21, p. 667). Ribavirin has shown activity against a wide variety of unrelated viruses, raising hopes of finding a broad-spectrum antiviral agent.

Acyclovir (Zovirax), an analog of guanine, is much more rapidly incorporated into virus-infected cells than into normal cells. Thus, it is less toxic than other analogs. It can be applied topically or given orally or intravenously. It is especially effective in reducing pain and promoting healing of primary lesions in a new case of genital herpes. It is given prophylactically to reduce the frequency and severity of recurrent lesions, which appear periodically after a first attack. It does not, however, prevent the establishment of latent viruses in nerve cells. Acyclovir is more effective than vidarabine against herpes encephalitis and neonatal herpes, an infection acquired at birth, but is not effective against other herpesviruses.

Ganciclovir is an analog of guanine similar to acyclovir. The drug is active against several kinds of herpesvirus infections, particularly cytomegalovirus eye infections in patients with AIDS.

Zidovudine (AZT) interferes with reverse transcriptase making DNA from RNA. It is used in treating AIDS.

Amantadine

The tricyclic amine **amantadine** prevents influenza A viruses from penetrating cells. Given orally, it is readily absorbed and can be used from a few days before to a week after exposure to influenza A viruses to reduce the incidence and severity of symptoms. Unfortunately, it causes insomnia and ataxia (inability to coordinate voluntary movements), especially in elderly patients, who also are often severely affected by influenza infections. *Rimantadine*, a drug similar to amantadine, may be effective against a wider variety of viruses and may be less toxic as well.

The Treatment of AIDS

Several agents are being tested for the treatment of AIDS. New information about AIDS, its complications, and its treatment is becoming available with great rapidity. We consider AIDS, agents used to treat it, and ramifications for health care workers in Chapter 18.

Interferons and Immunoenhancers

Cells infected with viruses produce one or more proteins collectively referred to as *interferons* (Chapter 16). When released, these proteins induce neighboring cells to produce antiviral proteins, which prevent these cells from becoming infected. Thus, interferons represent a natural defense against viral infection. Some interferons are currently being genetically engineered and tested as antiviral agents. Some positive results have been obtained in controlling chronic viral hepatitis and warts and arresting virus-related cancers, such as Kaposi's sarcoma.

Because cells produce interferons naturally, a possible way to combat viruses is to induce cells to produce interferons. Synthetic double-stranded RNA has been shown to increase the quantity of interferon in the blood. Experiments with one such substance in virus-infected monkeys have shown sufficient increase in interferon to prevent viral replication.

APPLICATIONS

Drug-Resistant Viruses

Evidence is accumulating that viruses, like bacteria, can develop resistance to chemotherapeutic agents. Herpesviruses and cytomegaloviruses with resistance to acyclovir have been observed in AIDS patients. Some laboratory strains of the virus that causes AIDS have become resistant to azidothymidine (AZT), the most effective drug currently available to treat the disease. Resistance to chemotherapeutic agents is a greater problem in viruses than in bacteria because so few antiviral agents are available. When a bacterium becomes resistant to one antibiotic, another usually can be found to which the bacterium is susceptible. Unfortunately, this is not the case with viruses, and we must hope that biotechnology can help us battle drug-resistant viruses.

Two other agents, *levamisole* and *inosiplex*, appear to stimulate the immune system to resist viral and other infections. Both seem to stimulate activity of leukocytes called T lymphocytes rather than to stimulate interferon release. Levamisole appears to be effective prophylactically in reducing the incidence and severity of chronic upper respiratory infections, which are probably viral in nature. It also reduces symptoms of autoimmune disorders such as rheumatoid arthritis, in which the body reacts against its own tissues. Inosiplex has a more specific action; it stimulates the immune system to resist infection with certain viruses that cause colds and influenza.

Although efforts to improve antiviral therapies by enhancing natural defenses have been somewhat successful, none is yet in widespread use. More research is needed to identify or synthesize effective agents, to determine how they act, and to discover how they can be most effectively used.

ANTIPROTOZOAN AGENTS

Although many protozoa are free-living organisms, a few are parasitic in humans. The parasite that causes malaria invades red blood cells and causes the patient to suffer alternating fever and chills. Other protozoan parasites cause intestinal or urinary tract infections. Several *antiprotozoan agents* have been found that are successful in controlling or even curing most protozoan infections, but some have rather unpleasant side effects.

Quinine

Quinine, from the bark of the cinchona tree (native to Peru and Bolivia, but now cultivated exclusively in Indonesia), was used for centuries to treat malaria. One of the first chemotherapeutic agents to come into widespread use, it is now used only to treat malaria caused by strains of the parasite resistant to other drugs.

Chloroquine and Primaquine

Currently the most widely used antimalarial agents are the synthetic agents **chloroquine** (Aralen) and **primaquine**. Chloroquine appears to interfere with protein synthesis, especially in red blood cells, which it enters more readily than it does other cells. The drug may concentrate in vacuoles within the parasite and prevent it from metabolizing hemoglobin. Chloroquine is used to combat active infections. The malarial parasite persists in red blood cells and can cause relapses when it multiplies and is released into blood plasma. A combination of chloroquine and primaquine can be used prophylactically to protect people who visit or work in regions of the world where malaria occurs and are thus at risk of becoming infected. However, the drugs must be taken both before and after a malarial zone is entered. A newer prophylactic agent, *mefloquine* (Lariam), has proved effective against resistant strains.

In some parts of the world chloroquine has been useless against resistant strains of malaria and has not been used for many years. Today some strains have again become sensitive to chloroquine, and it is coming back into use.

Metronidazole

The synthetic imidazole **metronidazole** (met-ro-ni′da-zol) causes breakage of DNA strands. It is effective in treating *Trichomonas* infections, which typically cause a vaginal discharge and itching. It also is effective against intestinal infections caused by parasitic amoebas and *Giardia*. Although metronidazole (Flagyl) controls these infections, it does not prevent overgrowth of *Candida* yeast infections. It also can cause birth defects and cancer and can be passed to infants in breast milk. Metronidazole sometimes causes an unusual side effect called "black hairy tongue," or "brown furry tongue," because it breaks down hemoglobin and leaves deposits in papillae (small projections) on the surface of the tongue (Figure 13.15).

Figure 13.15 Black hairy tongue, a reaction to the drug metronidazole (Flagyl). The papilli on the tongue's surface become elongated and filled with breakdown products of hemoglobin, which darken the tongue. *(Barts Medical Library/Phototake)*

Other Antiprotozoan Agents

A variety of other organic compounds have been found effective in treating certain infections caused by protozoa. *Pyrimethamine* (pir-i-meth′a-men) interferes with the synthesis of folic acid, which pathogenic protozoa need in greater quantities than do host cells. It is used with sulfanilamide to treat some protozoan infections such as toxoplasmosis. Pyrimethamine (Daraprim) can also be used prophylactically to prevent malaria.

Suramin sodium, a sulfur-containing compound, can be given intravenously to treat African sleeping sickness (trypanosomiasis) and other trypanosome infections. *Nifurtimox*, a nitrofuran, is used against the trypanosomes that cause Chagas' disease. Arsenic and antimony compounds, although very toxic, have been used with some success against stubborn amoebic infections and leishmanias. *Pentamidine isethionate* is used to treat African trypanosomiasis and as a drug of second choice for *Pneumocystis* pneumonia, a fungal complication of AIDS.

▌ ANTIHELMINTHIC AGENTS

Various helminths can infect humans. A variety of *antihelminthic agents* are available to help rid the body of these unwelcome parasites.

Niclosamide

Niclosamide interferes with carbohydrate metabolism, thereby causing a parasite to release large quantities of lactic acid. This drug may also inactivate products made by the worm to resist digestion by host proteolytic enzymes. It is effective mainly in the treatment of tapeworm infections.

Mebendazole

The imidazole **mebendazole** (Vermox) blocks the uptake of glucose by parasitic roundworms. It is useful in treating whipworm, pinworm, and hookworm infections. However, it can damage a fetus and thus should not be given to pregnant women.

Other Antihelminthic Agents

Piperazine (Antepar), a simple organic compound, is a powerful neurotoxin that paralyzes body-wall muscles of roundworms and is useful in treating *Ascaris* and pinworm infections. Although piperazine exerts its effect on worms in the intestine, if absorbed it can reach the human nervous system and cause convulsions, especially in children.

The compound *ivermectin*, originally developed for the treatment of parasitic nematodes in horses (and widely used to prevent heartworm infections in dogs), has been found to be extremely effective against *Onchocerca volvulus* in humans. Infection with this roundworm, widespread in many parts of Africa, causes a progressive loss of sight known as onchocerciasis, or river blindness.

Figure 13.16 provides the chemical structures, uses, and side effects of antifungal, antiviral, antiprotozoan, and antihelminthic agents.

✓ **CHECKLIST**

1. Do most antifungal agents also kill bacteria, or vice versa?

2. What difficulties result from the fact that parasitic protozoans and helminths have many of the same biochemical pathways as humans?

▌ SPECIAL PROBLEMS WITH DRUG-RESISTANT HOSPITAL INFECTIONS

As soon as antibacterial agents became available, resistant organisms began to appear. One of the first successes in treating bacterial infections was the use of sulfanilamide to treat infections caused by hemolytic streptococci. It was then discovered that sulfadiazine is useful in preventing recurrent streptococcal infections of rheumatic fever. Strains of streptococci resistant to sulfonamides soon emerged. Epidemics (mostly in military installations during World War II) caused by resistant strains led to many deaths. These epidemics were brought under control when penicillin became available, but soon penicillin-resistant streptococci were seen.

This chain of events has been repeated again and again. As new antibiotics were developed, strains of streptococci resistant to many of them evolved. Similar events led to the emergence of antibiotic-resistant strains of many other organisms, including staphylococci, gonococci, *Salmonella*, *Neisseria*, and especially *Pseudomonas*. *Pseudomonas* infections are now a major problem in hospitals. Many of these organisms are now resistant to several different antibiotics, and new resistant strains are constantly being encountered.

Why are resistant organisms found more often in hospitalized patients than among outpatients? This question can be answered by looking at the hospital environment and the patients likely to be hospitalized. First, despite efforts to maintain sanitary conditions, a hospital provides an environment where sick people live in close proximity and where many different kinds of infectious agents are constantly present and are easily spread. Second, hospitalized patients tend to be more severely ill than outpatients; many have lowered resistance to infection because of their illnesses or because they have received immunosuppressant drugs. Finally, and most importantly, hospitals typically make intensive use of a variety of antibiotics. Because many infections are being treated and different antibiotics are used, organisms resistant to one or more of the antibiotics are likely to emerge. The resistant strains can readily spread among patients.

Agent	Used to Treat	Common Method of Administration*	Side Effects
Antifungal agents			
Clotrimazole	Skin and nail infections	O	Skin irritation
Miconazole	Skin infections and systemic infections resistant to other agents	T, IV	Severe itching, nausea, fever, thrombophlebitis
Amphotericin B	Systemic infections	IV	Fever, chills, nausea, vomiting, anemia, kidney damage, blindness
Nystatin	*Candida* yeast infections, intestinal superinfections	T	
Griseofulvin	Infections of skin, hair, and nails	T, O	Mild headaches, nerve inflammations, gastrointestinal disturbances
Flucytosine	*Candida* and some systemic infections	O	Less toxic than many fungal agents

Agent	Used to Treat	Common Method of Administration*	Side Effects
Antihelminthic agents			
Niclosamide	Tapeworm infections	O	Irritation of gut
Piperazine	Pinworm and *Ascaris* infections	O	Can cause convulsions in children
Mebendazole	Whipworm, pinworm, and hookworm infections	O	Can damage fetus if given to pregnant women
Ivermectin	*Onchocerca volvulus* infections (cause of river blindness), heartworm infections in animals	O	Minimal

* IM = intramuscular   O = oral
IV = intravenous   T = topical

Figure 13.16 **Selected antifungal, antihelminthic, antiviral, and antiprotozoan drugs.**

Agent	Used to Treat	Common Method of Administration*	Side Effects
Antiviral agents			
Idoxuridine	Corneal infections	T	Suppresses bone marrow
Ganciclovir	CMV eye infections in AIDS	IV	Suppresses bone marrow
Vidarabine	Viral encephalitis	T, IV	Less toxic than other antiviral agents
Ribavirin	Herpes lesions (topical application), influenza (in aerosol)	T	Can cause birth defects if given to pregnant women
Acyclovir	Herpesvirus infections; lessens severity of symptoms	IV, O, T	Less toxic than other analogs
Amantadine	Infections of influenza A viruses from entering cells (preventive)	O	Insomnia and ataxia
AZT	AIDS	O	Can suppress bone marrow, nausea

Idoxuridine   Ganciclovir   Vidarabine   Ribavirin

Amantadine   Acyclovir   Azidothymidine (AZT)

Agent	Used to Treat	Common Method of Administration*	Side Effects
Antiprotozoan agents			
Quinine	Malaria resistant to other agents	O	
Chloroquine	Malaria	O	Headache, itching
Primaquine	With chloroquine to prevent relapse of malaria	O	Slight nausea and abdominal pain
Pyrimethamine	Various protozoan infections	O	Large doses damage bone marrow
Metronidazole	*Trichomonas, Giardia,* and amoebic infections	O, IV, T	Black hairy tongue

Quinine   Chloroquine   Primaquine

Pyrimethamine   Metronidazole

Figure 13.16 *continued.*

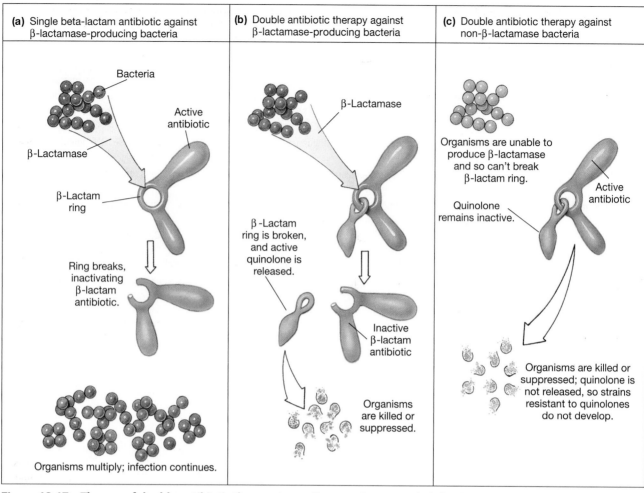

(a) Single beta-lactam antibiotic against β-lactamase-producing bacteria

Bacteria

Active antibiotic

β-Lactamase

β-Lactam ring

Ring breaks, inactivating β-lactam antibiotic.

Organisms multiply; infection continues.

(b) Double antibiotic therapy against β-lactamase-producing bacteria

β-Lactamase

β-Lactam ring is broken, and active quinolone is released.

Inactive β-lactam antibiotic

Organisms are killed or suppressed.

(c) Double antibiotic therapy against non-β-lactamase bacteria

Organisms are unable to produce β-lactamase and so can't break β-lactam ring.

Active antibiotic

Quinolone remains inactive.

Organisms are killed or suppressed; quinolone is not released, so strains resistant to quinolones do not develop.

Figure 13.17 **The use of double-antibiotic therapy to eradicate resistant-strain infections.**

Treatment of resistant infections creates a vicious cycle. If an antibiotic can be found to which an organism is susceptible, that drug can be used to treat the infection. However, some strains of the organism that are resistant to the new antibiotic may then proliferate and require treatment with another new drug. A recurrent cycle in which new antibiotics are used and the organisms subsequently develop resistance to them is established.

Preventing infections caused by antibiotic-resistant strains of microorganisms is a difficult task, but several guidelines should be followed. First, the use of antibiotics should be limited to situations in which the patient is unlikely to recover without antibiotic treatment. Second, sensitivity tests should be done, and patients should receive only an antibiotic to which the organism is known to be sensitive. Third, when antibiotics are used, they should be continued until the organism is completely eradicated from the patient's body. Double antibiotic use, as described earlier under quinolones, is especially helpful **(Figure 13.17)**. Finally, any patient with an infectious disease should be isolated from other patients.

▌ RETRACING OUR STEPS

▌▌▌ ANTIMICROBIAL CHEMOTHERAPY

- **Chemotherapy** is the use of any chemical agent in the treatment of disease.
- A **chemotherapeutic agent**, or **drug**, is any chemical agent used in medical practice.
- An **antimicrobial agent** is a chemical agent used to treat a disease caused by a microbe.

- An **antibiotic** is a chemical substance produced by microorganisms that inhibits the growth of or destroys other microorganisms.
- A **synthetic drug** is one made in the laboratory.
- A **semisynthetic drug** is one made partly by microorganisms and partly by laboratory synthesis.

▌ THE HISTORY OF CHEMOTHERAPY

- The first chemotherapeutic agents were concoctions from plant materials used by primitive societies.
- Paul Ehrlich's search for the "magic bullet" was the first systematic attempt to find chemotherapeutic agents. Subsequent events included the development of sulfa drugs, penicillin, and many other antibiotics.

▌ GENERAL PROPERTIES OF ANTIMICROBIAL AGENTS

SELECTIVE TOXICITY

- **Selective toxicity** is the property of antimicrobial agents that allows them to exert greater toxic effects on microbes than on the host.
- The **therapeutic dosage level** of an antimicrobial agent is the concentration over a period of time required to eliminate a pathogen.
- The **chemotherapeutic index** is a measure of the toxicity of an agent to the body relative to its toxicity for an infectious organism.

THE SPECTRUM OF ACTIVITY

- The **spectrum of activity** of an antimicrobial agent refers to the variety of microorganisms sensitive to the agent. A **broad-spectrum** agent attacks many different organisms. A **narrow-spectrum** agent attacks only a few different organisms.

MODES OF ACTION

- Agents that kill bacteria are bactericidal; those that inhibit bacterial growth are bacteriostatic.
- Agents that inhibit cell wall synthesis allow the membrane of the affected microbe to rupture and release the cell contents.
- Agents that disrupt membrane function dissolve the membrane or interfere with the movement of substances into or out of cells.
- Agents that inhibit protein synthesis prevent the growth of microbes by disrupting ribosomes or otherwise interfering with the process of translation.
- Agents that inhibit nucleic acid synthesis interfere with synthesis of RNA (transcription) or DNA (replication) or disrupt the information these molecules contain.
- Agents that act as **antimetabolites** affect normal metabolites by competitively inhibiting microbial enzymes or by being erroneously incorporated into important molecules such as nuclei acids.

KINDS OF SIDE EFFECTS

- Side effects of antimicrobial agents on the host include toxicity, allergy, and disruption of normal microflora.
- Allergic reactions to antimicrobial agents occur when the body reacts to the agent as a foreign substance.
- Many antimicrobial agents attack not only the infectious organism but also normal microflora. **Superinfections** with new pathogens can occur when the defensive capacity of normal microbiota is destroyed.

THE RESISTANCE OF MICROORGANISMS

- **Resistance** to an antibiotic means that a microorganism formerly susceptible to the action of an antibiotic is no longer affected by it.

- Nongenetic resistance occurs when microorganisms are sequestered from antibiotics or undergo a temporary change, such as the loss of their cell walls, that renders them nonsusceptible to antibiotic action.
- Genetic resistance occurs when organisms survive exposure to an antibiotic because of their genetic capacity to avoid damage by the antibiotic. As susceptible organisms die, the resistant survivors multiply unchecked and increase in numbers.
- **Chromosomal resistance** is due to a mutation in microbial DNA; **extrachromosomal resistance** is due to **resistance (R) plasmids**, or **R factors**.
- Mechanisms of resistance include alterations of receptors, cell membranes, enzymes, or metabolic pathways.
- **Cross-resistance** is resistance against two or more similar antimicrobial agents.
- Drug resistance can be minimized by (1) continuing treatment with an appropriate antibiotic at therapeutic dosage level until all the disease-causing organisms are destroyed; (2) using two antibiotics that exert **synergism**, an additive effect; and (3) using antibiotics only when absolutely necessary.

▌ DETERMINING MICROBIAL SENSITIVITIES TO ANTIMICROBIAL AGENTS

- Sensitivity of microbes to chemotherapeutic agents is determined by exposing them to the agents in laboratory cultures.

THE DISK DIFFUSION METHOD

- In the **disk diffusion (Kirby-Bauer) method**, antibiotic-impregnated filter paper disks are placed on agar plates inoculated with a lawn of the test organism. Sensitivities to the drugs are determined by comparing the size of clear zones around the disks to a table of standard measurements.

THE DILUTION METHOD

- In the **dilution method**, a constant inoculum is placed into broth cultures or wells with differing known quantities of chemotherapeutic agents. The **minimum inhibitory concentration (MIC)** of the agent is the lowest concentration in which no growth of the organism is observed. The **minimum bactericidal concentration (MBC)** of the agent is the lowest concentration in which subculturing of broth yields no growth.

SERUM KILLING POWER

- In the **serum killing power** method, a bacterial suspension is added to a patient's **serum** drawn while the patient is receiving an antibiotic, and it is noted whether the organisms are killed.

AUTOMATED METHODS

- Automated methods allow rapid identification of microorganisms and determination of their sensitivities to antimicrobial agents.

▌ ATTRIBUTES OF AN IDEAL ANTIMICROBIAL AGENT

- An ideal antimicrobial agent is soluble in body fluids, selectively toxic, and nonallergenic; can be maintained at a constant therapeutic concentration in blood and body fluids; is unlikely to elicit resistance; has a long shelf-life; and is reasonable in cost.

III ANTIBACTERIAL AGENTS

• Antibacterial agents inhibit cell wall synthesis, disrupt cell membrane functions, inhibit protein synthesis, inhibit nucleic acid synthesis, or act by some other means to kill bacteria.

III ANTIFUNGAL AGENTS

• Antifungal agents increase plasma membrane permeability, interfere with nucleic acid synthesis, or otherwise impair cell functions.

III ANTIVIRAL AGENTS

• Antiviral agents have been difficult to find because they must damage intracellular viruses without severely damaging host cells.
• Most antiviral agents are analogs of purines or pyrimidines.
• Interferon is released by virus-infected cells and stimulates neighboring cells to produce antiviral proteins. Interferons are being made by genetic engineering and being tested in treatment of viral infections and cancer.

III ANTIPROTOZOAN AGENTS

• Some antiprotozoan agents interfere with protein synthesis or folic acid synthesis. The mechanism of action of others is not well understood.

III ANTIHELMINTHIC AGENTS

• Antihelminthic agents interfere with carbohydrate metabolism or act as neurotoxins.

III SPECIAL PROBLEMS WITH DRUG-RESISTANT HOSPITAL INFECTIONS

• Resistant hospital infections are due largely to intensive use of a variety of antibiotics, which fosters the growth of resistant strains. Treatment and prevention of such infections are extremely difficult.

■ TERMINOLOGY CHECK

amantadine *(p. 389)*
aminoglycoside *(p. 383)*
antagonism *(p. 376)*
antibiosis *(p. 367)*
antibiotic *(p. 367)*
antimetabolite *(p. 370)*
antimicrobial agent *(p. 367)*
broad spectrum *(p. 368)*
carbapenem *(p. 381)*
cephalosporin *(p. 381)*
chemotherapeutic agent *(p. 367)*
chemotherapeutic index *(p. 368)*
chemotherapy *(p. 367)*
chloramphenicol *(p. 384)*
chloroquine *(p. 390)*
chromosomal resistance *(p. 373)*

cross-resistance *(p. 375)*
dilution method *(p. 376)*
disk diffusion method *(p. 376)*
drug *(p. 367)*
E (epsilometer) test *(p. 376)*
erythromycin *(p. 384)*
ethambutol *(p. 385)*
extrachromosomal resistance *(p. 373)*
griseofulvin *(p. 388)*
imidazole *(p. 388)*
isoniazid *(p. 385)*
Kirby-Bauer method *(p. 376)*
macrolide *(p. 384)*
mebendazole *(p. 391)*
metronidazole *(p. 390)*

minimum bactericidal concentration (MBC) *(p. 377)*
minimum inhibitory concentration (MIC) *(p. 376)*
molecular mimicry *(p. 371)*
narrow-spectrum *(pp. 368–369)*
niclosamide *(p. 391)*
nitrofuran *(p. 385)*
penicillin *(p. 379)*
polyene *(p. 388)*
polymyxin *(p. 382)*
primaquine *(p. 390)*
quinine *(p. 390)*
quinolone *(p. 385)*
resistance *(p. 372)*
resistance (R) plasmid *(p. 373)*

R factor *(p. 373)*
rifamycin *(p. 384)*
selective toxicity *(p. 368)*
semisynthetic drug *(p. 367)*
serum *(p. 377)*
serum killing power *(p. 377)*
spectrum of activity *(p. 368)*
streptomycin *(p. 383)*
sulfonamide *(p. 385)*
superinfection *(p. 372)*
synergism *(p. 376)*
synthetic drug *(p. 367)*
tetracycline *(p. 383)*
therapeutic dosage level *(p. 368)*
toxic dosage level *(p. 368)*
zone of inhibition *(p. 376)*

■ CLINICAL CASE STUDY

You are taking an introductory microbiology class and you cannot afford to get sick. An exam is scheduled for next week and you really need to be able to study. You tell your friend that you think you are coming down with a cold. Your friend says that he has some antibiotics left over from an illness he had a year ago. He would be glad to give them to you for you to take. Based on what you have learned from the text, what is wrong with what your friend said? To learn more about these issues visit the following web site: www.tufts.edu/med/apua/ncqa.pdf.

■ CRITICAL THINKING QUESTIONS

1. Is it safe to say that a specific chemical is either poisonous or not poisonous? Explain.

2. Why do you think that some microorganisms produce antibiotics?

3. Discuss the advantages and disadvantages of using more than one drug simultaneously for treatment.

▌ SELF-QUIZ

1. Match the following antimicrobial chemotherapy terms to their descriptions:

   C Synthetic drug
   E Antimicrobial agent
   B Chemotherapy
   A Semisynthetic drug
   D Antibiotic
   F Chemotherapeutic agent

   (a) Made partly by micro-organisms and partly synthetically in the lab
   (b) Use of any chemical agents in the treatment of disease
   (c) Made synthetically in the laboratory
   (d) Microbial-produced chemical that inhibits growth of or kills other microorganisms
   (e) Chemical agent used to treat a disease caused by microbes
   (f) Any chemical agent used in medical practice

2. How has chemotherapy developed since Ehrlich's time?
   development of sulfa drugs

3. Why do scientists study soil and water microorganism when searching for new antibiotics?
   many novel fungi and bacteria live there

4. What are the differences between bacteriostatic and bactericidal disinfectants?
   BS only prevent growth/bactericidal kills microbes

5. Match the following antibiotics to their modes of action:

   E Sulfanilamide
   C Erythromycin
   A Penicillin
   D Rifamycin
   B Polymyxin
   E Purine analog Vidarabine
   C Streptomycin

   (a) Inhibition of cell wall synthesis
   (b) Disruption of cell membrane function
   (c) Inhibition of protein synthesis
   (d) Inhibition of nucleic acid synthesis
   (e) Antimetabolite

6. An antibiotic that contains a β-lactam ring in its structure is:
   (a) Bacitracin
   (b) Streptomycin
   (c) Polymyxin
   (d) Penicillin
   (e) Tetracycline

7. Cephalosporins resemble which antibiotic in their mode of action and their structure?
   (a) Penicillin
   (b) Bacitracin
   (c) Streptomycin
   (d) Polymyxin
   (e) Tetracycline

8. Which of the following is not a property of an antimicrobial agent?
   (a) Must have a favorable chemotherapeutic index
   (b) Must be selectively toxic
   (c) Must be effective against viral diseases
   (d) Must have a known concentration over a period of time required to eliminate a pathogen
   (e) b and c

9. A narrow-spectrum agent attacks a ___ *few* different microorganisms while a ___-___ *broad spectrum* agent attacks many different microorganisms.

10. Penicillin is specific for bacteria because it:
    (a) Inhibits cell wall synthesis
    (b) Inhibits protein synthesis
    (c) Injures the plasma membrane
    (d) Inhibits nucleic acid synthesis
    (e) All of the above

11. Doctors prescribe synergistic drug combinations to treat bacterial infections. The purpose of such treatment is to:
    (a) Change the bacteria with cell walls to L forms lacking cell walls
    (b) Reduce the treatment time of the disease
    (c) Prevent microorganisms from acquiring drug resistance
    (d) Reduce the toxic side effects of the antibiotics
    (e) Use lower doses of antibiotics

12. An antibiotic that has a broad spectrum of activity but may cause aplastic anemia is:
    (a) Streptomycin
    (b) Cephalosporin
    (c) Penicillin
    (d) Bacitracin
    (e) Chloramphenicol

13. Which of the following is not a way in which antifungal drugs are effective?
    (a) They interfere with nucleic acid synthesis.
    (b) They increase plasma membrane permeability causing excessive leakiness of essential substances.
    (c) They can impair the mitotic spindle apparatus.
    (d) They can induce mycorrhizae production.
    (e) None of the above.

14. All of the following can be side effects of antimicrobial agents EXCEPT:
    (a) "Superinfections" can occur with new pathogens when defensive capacity of normal flora is destroyed.
    (b) Host toxicity.
    (c) Disruption of normal microflora in host.
    (d) Host allergic reaction.
    (e) Host "superimmunity."

15. The target for quinolones is:
    (a) RNA transcription
    (b) DNA replication
    (c) Protein synthesis
    (d) Cell wall formation
    (e) Membrane structure

16. Antimetabolites that block folic acid synthesis are:
    (a) Penicillins
    (b) Aminoglycosides
    (c) Cephalosporins
    (d) Sulfonamides
    (e) None of these

17. An antimetabolite that is effective against the *Mycobacterium* that causes tuberculosis is:
    (a) Sulfanilamide
    (b) Isoniazid
    (c) Bacitracin
    (d) Naladixic acid
    (e) Polymyxin A

18. The drug of choice for treating systemic fungal infections is:
    (a) Nystatin
    (b) Griseofulvin
    (c) Flucytosine
    (d) Amphotericin B
    (e) Tolnaftate

19. Which of the following is not a way to determine microbial sensitivity to an antimicrobial agent?
    (a) Dilution method

(b) Serum killing power

(c) Automated inoculating trays each containing a specific antimicrobial agent

(d) Antagonism

(e) Kirby-Bauer disk diffusion

20. What is interferon?

(a) An antifungal drug that inhibits topical fungal infections

(b) Secreted proteins elicited from virus-infected cells that induce neighboring cells to produce antiviral proteins that prevent these cells from becoming infected

(c) Secreted proteins elicited from virus-infected cells that protect the same cell from further infection

(d) Secreted proteins elicited from healthy cells that induce neighboring cells to produce antiviral proteins that prevent these cells from becoming infected

(e) a and c

21. Match the following antiviral drugs and the viral infections they are used to treat:

D Acyclovir
B Ganciclovir
C AZT
E Idoxuridine
A Ribavirin

(a) Hantavirus plus a wide variety of unrelated viruses

(b) Cytomegalovirus eye infections

(c) HIV

(d) Herpesvirus infections of the genitals

(e) Herpesvirus infections of the eyes

22. Chloroquine and primaquine are the agents most widely used to treat:

(a) Malaria

(b) Tuberculosis

(c) Lyme disease

(d) Legionnaires' disease

(e Thrush

23. Which of the following is a reason why helminthic and protozoan diseases are difficult to treat?

(a) They have a thick protective epidermis.

(b) They are hermaphroditic.

(c) They are prokaryotes.

(d) They have many biochemical pathways in common with man.

(e) Their cells are structurally different from human cells.

24. How do antihelminthic agents work?

(a) They suppress carbohydrate metabolism or act as neurotoxins.

(b) They suppress their immune systems.

(c) They suppress ribosomal synthesis.

(d) They suppress transcription and translation.

(e) They suppress mitosis.

25. Match the following drug resistance terms to their descriptions:

C Chromosomal resistance

F Nongenetic resistance

E Synergism

B Extrachromosomal resistance

A Cross-resistance

D Resistance

(a) Resistance against two or more similar antimicrobial agents

(b) Resistance due to resistance (R) plasmids

(c) Resistance due to microbial DNA

(d) Microbe formerly susceptible to antibiotic action that is no longer affected by it

(e) An additive effect of two antibiotics

(f) Microbes that are either sequestered from antibiotic or lose their cell walls rendering them nonsusceptible to antibiotic action

26. Indicate in the boxes the activity inhibited by the antibiotics and list some antibiotics which exhibit that activity.

▍ EXPLORATIONS ON THE WEB

If you think you've mastered this chapter, there's more to challenge you on the web. Go to the companion web site to fine-tune your understanding of the chapter concepts and discover answers to the questions posed below.

1. Choosing the right antibiotic provides the best chance of success and the least likelihood of toxicity.

2. Fleming's principle of limited sloppiness contributed to his discovery of penicillin. Fleming abhorred a tidy, meticulous lab; he left culture dishes lying around for weeks and would often discover interesting things in them, among them the mold *Penicillium*.

3. How do molds kill bacteria?

14 Host-Microbe Relationships and Disease Processes

Come with me . . .

After having spent the morning exploring your exotic surroundings, you are absolutely starving! Luckily, there's a market in the center of town.

Would you like to shop here? In many parts of the world there's no choice about the cleanliness of your local market. Will flies, rodents, and lack of refrigeration spread disease? You bet they will! If you think the melon and peppers look bad, you should see the stalls of meat and fish!

In the face of all this obvious contamination, why isn't everyone sick, dying, or dead? Well, some of them are sick, and the average life expectancy in developing countries is not very high. Probability of infection is rather like the two-pan balances on which the merchants were weighing out the food: host defenses on one side and microbial virulence on the other. When the balance shifts in favor of the microbes, infection occurs. Do you risk it or keep looking, hoping your hunger will pass? Come with me to see how you would feel about shopping or eating in third world countries.

Courtesy Jacquelyn G. Black

 Video related to this topic is available within WileyPLUS.

▐▐▐ HOST-MICROBE RELATIONSHIPS

Symbiosis / Contamination, Infection, and Disease / Pathogens, Pathogenicity, and Virulence / Normal (Indigenous) Microflora

▐▐▐ KOCH'S POSTULATES

▐▐▐ KINDS OF DISEASES

Infectious and Noninfectious Diseases / Classification of Diseases / Communicable and Noncommunicable Diseases

▐▐▐ THE DISEASE PROCESS

How Microbes Cause Disease / Signs, Symptoms, and Syndromes / Types of Infectious Disease / Stages of an Infectious Disease

▐▐▐ INFECTIOUS DISEASES—PAST, PRESENT, AND FUTURE

How is it that every now and then, no matter how careful you are, you "catch" an infectious disease? You become ill. With or without antimicrobial agents, you generally recover from the disease. In the process, you *may* develop *immunity*—that is, if you are exposed to the disease agent at another time, you may be protected from contracting the disease again.

Recall from ◀Chapter 11 (p. 311) that a **pathogen** is a parasite capable of causing disease in a host. The ability of a pathogen to cause a disease in you depends on whether the pathogen or you—the host—wins the battle. Pathogens have certain invasive capabilities, and you have a variety of defenses. For example, in many countries the measles virus is present in a portion of the population at all times. Those who are infected release the virus, and it makes its way into the tissues of susceptible individuals. There, the virus can overcome defenses, invade tissues, and cause disease. But some individuals do not become infected.

If a virus does make its way into your tissues, your immune-system defenses may destroy it before it can cause disease. You could become immune to further exposures without actually becoming sick. Even when your first defenses fail and the disease occurs, you may develop immunity and will not be susceptible to the disease on subsequent exposures.

To begin the study of host-microbe interactions, we will look at a variety of relationships between host and microbe and see how some of these relationships result in disease. We will then characterize diseases and look at the disease process brought on by pathogens.

HOST-MICROBE RELATIONSHIPS

▐ Microorganisms display a variety of complex relationships with other microorganisms and with larger forms of life that serve as hosts for them. A **host** is any organism that harbors another organism.

SYMBIOSIS

Symbiosis is an association between two (or more) species. Meaning "living together," the term *symbiosis* encompasses a spectrum of relationships. These include *mutualism*, *commensalism*, and *parasitism*.

At one end of the spectrum is **mutualism** (Table 14.1), in which both members of the association living

TABLE 14.1

The Spectrum of Symbiotic Associations		
Relationship	Effect on Species A	Effect on Species B
Mutualism	+	+
Parasitism	+	−
Commensalism	+	0
Antagonism	−	−

together benefit from the relationship (**Figure 14.1**). For example, the ability of termites to digest wood, or cellulose, depends on protozoa they harbor in their intestines. These protozoa and other microbes, such as bacteria and fungi, secrete into the intestine enzymes that digest the chewed wood. The protozoa themselves gain a safe, stable environment in which to live. For the termites, the relationship is obligatory; they would starve without their protozoan partners. Similarly, large numbers of *Escherichia coli* live in the large intestine of humans. These bacteria release useful products such as vitamin K, which we use to make certain blood-clotting factors. Although the relationship is not obligatory, *E. coli* does make a modest contribution toward satisfying our need for vitamin K. The bacteria, in turn, get a favorable environment in which to live and obtain nutrients.

At the other end of the spectrum is **parasitism**, in which one organism, the parasite, benefits from the relationship, whereas the other organism, the host, is harmed by it (**Figure 14.2**). By this broad definition of the term **parasite**, bacteria, viruses, protozoa, fungi, and helminths are parasites. (Some biologists use "parasite" to refer only to protozoa, helminths, and arthropods that live

Figure 14.1 Many of the bacteria on human skin are mutualistic. (41,616X) However, most of these organisms are commensals, which indirectly benefit us by competing with harmful organisms for nutrients and preventing those organisms from finding a site to attach to and invade tissue. *(David M. Phillips/Visuals Unlimited)* [SEM]

Figure 14.2 **Parasite infestation.** Female pinworms leaving the anus of a 5-year-old child to lay eggs on the adjacent skin. *(Photo by Martin Weber, MD., reproduced from The New England Journal of Medicine, vol. 328, no. 13, pg. 927 © 1993 by the Massachusetts Medical Society)*

The word "parasite" comes from the Greek parasitos, meaning "one who eats at another's table."

on or in their host.) Parasitism encompasses a wide range of relationships, from those in which the host sustains only slight harm to those in which the host is killed. Some parasites obtain comfortable living arrangements by causing only modest harm to their host. Other parasites kill their hosts, thereby rendering themselves homeless (◄Chapter 11, p. 311). The most successful parasites are those that maintain their own life processes without severely damaging their hosts.

Somewhere in the middle lies **commensalism**, in which two species live together in a relationship such that one benefits and the other one neither benefits nor is harmed. For example, many microorganisms live on our skin surfaces and utilize metabolic products secreted from pores in the skin. Because those products are released whether or not they are used by microorganisms, the microorganisms benefit, and ordinarily we are neither benefited nor harmed.

The line between commensalism and mutualism is not always clear. By taking up space and utilizing nutrients, microbes that show mutualistic or commensalistic behavior may prevent colonization of the skin by other, potentially harmful, disease-causing microbes—a phenomenon known as *microbial competition*. Hence these symbiotic relationships confer an indirect benefit on the host.

There is also a fine line between parasitism and commensalism. In healthy hosts, many microbes of the large intestine form harmless associations, simply feeding off digested food materials. But a 'harmless' microbe could act as a parasite if it gains access to a part of the body where it would not normally exist. The situation in which

both species harm each other without either benefiting is called *antagonism*.

CONTAMINATION, INFECTION, AND DISEASE

Contamination, infection, and disease can be viewed as a sequence of conditions in which the severity of the effects microorganisms have on their hosts increases. **Contamination** means that the microorganisms are present. Inanimate objects and the surfaces of skin and mucous membranes can be contaminated with a wide variety of microorganisms. Commensals do no harm, but parasites have the capacity to invade tissues. **Infection** refers to the multiplication of any parasitic organism within or on the host's body. (Sometimes the term **infestation** is used to refer to the presence of larger parasites, such as worms or arthropods, in or on the body.) If an infection disrupts the normal functioning of the host, disease occurs. **Disease** is a disturbance in the state of health wherein the body cannot carry out all its normal functions.

Both infection and disease result from interactions between parasites and their hosts. Sometimes an infection produces no observable effect on the host even though organisms have invaded tissues. More often an infection produces observable disturbances in the host's state of health; that is, disease occurs. When an infection causes disease, the effects of the disease range from mild to severe.

Let us look at some examples to understand the differences among contamination, infection, and disease. A health care worker who fails to follow aseptic procedures while dressing a skin wound contaminates her hands with staphylococci. However, after she finishes her task, she washes her hands properly and suffers no ill effects. Although her hands were contaminated, she did not develop an infection. Another worker performing the same task on another patient fails to wash his hands properly after treating the patient, and the organisms gain entrance to the body and infect a small cut. Soon the skin around the cut becomes reddened for a day or so. This worker was contaminated and infected. In a similar situation, a third worker develops a reddened area on her skin; she ignores it and in a few days has a large boil. This worker has experienced contamination, infection, and disease.

Disease, or illness, is characterized by changes in the host that interfere with normal function. These changes can be mild, severe but reversible, or irreversible. For example, if you become infected with one of the viruses that cause the common cold, you may have just a runny nose for a few days. Or you may have a severe cold with a sore throat, cough, fever, and headache, but the disease runs its course in a week or so without any permanent effects. The changes in your state of health are reversible. But if you develop trachoma, a bacterial infection of the eye, without treatment scarring of the cornea can occur, leading to permanent vision impairment and sometimes to blindness. Likewise, if you fail to get proper treatment for streptococcal infections, you might suffer irreversible damage to your heart or kidneys.

PATHOGENS, PATHOGENICITY, AND VIRULENCE

Pathogens vary in their abilities to disrupt the state of an individual's health—that is, they display different degrees of pathogenicity. **Pathogenicity** is the capacity to produce disease. An organism's pathogenicity depends on its ability to invade a host, multiply in the host, and avoid being damaged by the host's defenses. Some disease agents, such as *Mycobacterium tuberculosis*, frequently cause disease upon entering a susceptible host. Other agents, such as *Staphylococcus epidermidis*, cause disease only in rare instances and usually only in hosts with poor defenses. Most infectious agents exhibit a degree of pathogenicity between these extremes.

An important factor in pathogenicity is the number of infectious organisms that enter the body. If only a small number enter, the host's defenses may be able to eliminate the organisms before they can cause disease. If a large number enter, they may overwhelm the host's defenses and cause disease. Other organisms are so highly infectious that *Shigella*, for example, needs only 10 organisms to be ingested to cause a very nasty case of dysentery.

Virulence refers to the intensity of the disease produced by pathogens, and it varies among different microbial species. For example, *Bacillus cereus* causes mild gastroenteritis, whereas the rabies virus causes neurological damage that is nearly always fatal. Virulence also varies among members of the same species of pathogen. For example, organisms freshly discharged from an infected individual tend to be more virulent than those from a carrier, who characteristically shows no signs of disease. The virulence of a pathogen can increase by **animal passage**, the rapid transfer of the pathogen through animals of a species susceptible to infection by that pathogen. As one animal becomes diseased, organisms released from that animal are passed to a healthy animal, which then also gets sick. If this sequence is repeated two or three times, each newly infected animal suffers a more serious case of the disease than the one before it. Presumably the microbe becomes better able to damage the host with each animal passage. Sometimes an infectious disease spreads through human populations in this fashion, and an epidemic of the disease results. Influenza epidemics often proceed in this manner; the first people to become infected have a mild illness, but those infected later have a much more severe form of the disease. This process does not continue forever; the microbe reaches the height of its virulence, and the exposed population acquires immunity.

The virulence of a pathogen can be decreased by **attenuation**, the weakening of the disease-producing ability of the pathogen. Attenuation can be achieved by repeated subculturing on laboratory media or by transposal of virulence. **Transposal of virulence** is a laboratory technique in which a pathogen is passed from its normal host to a new host species and then passed sequentially through many individuals of the new host species. Eventually, the pathogen adapts so completely to the new host that it is no longer virulent for the original host. In other words, virulence has been transposed to another organism. Pasteur made use of transposal of virulence in preparing rabies vaccines. By repeated passage through rabbits, the virus eventually became harmless to humans and was safe to use in a human vaccine. We will see in ◄Chapter 17 that attenuation is an important step in the production of some vaccines in use today—for example, mumps and measles.

TABLE 14.2

Major Normal Microflora (Unless Otherwise Noted, Bacteria) of the Human Body	
Skin	**Intestine**
Staphylococcus epidermidis*	*Staphylococcus epidermidis**
Staphylococcus aureus	*Staphylococcus aureus*
Lactobacillus species	*Streptococcus mitis**
Propionibacterium acnes*	*Enterococcus* species*
Pityrosporon ovale (fungus)*	*Lactobacillus* species*
Mouth	*Clostridium* species*
Streptococcus salivarius*	*Eubacterium limosum**
Streptococcus pneumoniae	*Bifidobacterium bifidum**
Streptococcus mitis*	*Actinomyces bifidus*
Streptococcus sanguis	*Escherichia coli**
Streptococcus mutans	*Enterobacter* species*
Staphylococcus epidermidis*	*Klebsiella* species
Staphylococcus aureus	*Proteus* species
Moraxella catarrhalis	*Pseudomonas aeruginosa*
Veillonella alcalescens*	*Bacteroides* species*
Lactobacillus species*	*Fusobacterium* species
Klebsiella species	*Treponema denticola*
Haemophilus influenzae*	*Endolimax nana* (protozoan)
Fusobacterium nucleatum*	*Giardia intestinalis* (protozoan)
Treponema denticola*	
Candida albicans (fungus)*	**Urogenital Tract**
Entamoeba gingivalis (protozoan)*	*Streptococcus mitis**
Trichomonas tenax (protozoan)*	*Streptococcus* species*
	Staphylococcus epidermidis*
Upper Respiratory Tract	*Lactobacillus* species*
Staphylococcus epidermidis*	*Clostridium* species
Staphylococcus aureus	*Actinomyces bifidus*
Streptococcus mitis*	*Candida albicans* (fungus)*
Streptococcus pneumoniae	*Trichomonas vaginalis* (protozoan)
Moraxella catarrhalis	
Lactobacillus species	
Haemophilus influenzae	

*Well-established associations.

NORMAL (INDIGENOUS) MICROFLORA

As we have described, microorganisms found in various symbiotic associations with humans do not necessarily cause disease. An adult human body consists of approximately 10^{13} (10 trillion) eukaryotic cells. It harbors an additional 10^{14} (100 trillion) prokaryotic and eukaryotic microorganisms on the skin surface, on mucous membranes, and in the passageways of the digestive, respiratory, and reproductive systems. Thus, there are 10 times more microbial cells on or in the human body than there are cells making up the body!

Before birth, a fetus exists in a sterile environment, unless the mother is infected by microbes which can cross the placenta, e.g., German measles (rubella). During passage through the birth canal, the fetus acquires certain microorganisms that may become permanently or temporarily associated with it. Organisms that live on or in the body but do not cause disease are referred to collectively as **normal microflora**, or *normal microbiota* **(Table 14.2)**. Many such organisms have well-established associations with humans. Most organisms among the normal microflora are commensals—they obtain nutrients from host secretions, waste substances found on the surfaces of skin and mucous membranes. Two categories of organisms can be distinguished: *resident microflora* and *transient microflora*.

The **resident microflora (Figure 14.3)** comprise microbes that are always present on or in the human body.

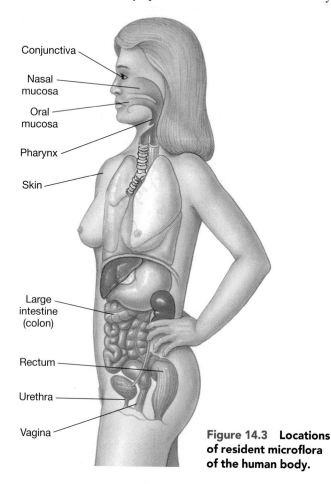

Figure 14.3 Locations of resident microflora of the human body.

Conjunctiva
Nasal mucosa
Oral mucosa
Pharynx
Skin
Large intestine (colon)
Rectum
Urethra
Vagina

They are found on the skin and conjunctiva, in the mouth, nose, and throat, in the large intestine, and in passageways of the urinary and reproductive systems, especially near their openings. In each of these body regions, resident microflora are adapted to prevailing conditions. The mouth and the lower part of the large intestine provide warm, moist conditions and ample nutrients. Mucous membranes of the nose, throat, urethra, and vagina also provide warm, moist conditions, although nutrients are in shorter supply. The skin provides ample nutrients but is cooler and less moist.

Other regions of the body lack resident microflora either because these regions provide conditions unsuitable for microorganisms, are protected by host defenses, or are inaccessible to microorganisms **(Table 14.3)**. For example, conditions in the stomach are too acidic to permit survival of microflora. Under normal conditions the nervous system is inaccessible to microbes. Blood has no resident microflora because it is relatively inaccessible, and host defense mechanisms normally destroy microorganisms before they become established.

Transient microflora are microorganisms that can be present under certain conditions in any of the locations

TABLE 14.3

Body Tissues, Organs, and Fluids That Are Normally Microbe-Free	
Internal Tissues and Organs	**Body Fluids**
Middle and inner ear	Blood
Sinuses	Cerebrospinal fluid
Internal eye	Saliva prior to secretion
Bone marrow	Urine in kidneys and in bladder
Muscles	Semen prior to entry into the urethra
Glands	
Organs	
Circulatory system	
Brain and spinal cord	
Ovaries and testes	

where resident microflora are found. They persist for hours to months, but only as long as the necessary conditions are met. Transient microflora appear on mucous membranes when greater than normal quantities of nutrients are available or on the skin when it is warmer and more moist than usual. Even pathogens can be transient microflora. For example, suppose that you come in contact with a child infected with measles, and some of the viruses enter your nose and throat. You had measles years ago and are immune to the disease, so your body's defenses prevent the viruses from invading cells. But you harbor the viruses as transients for a short time.

Among the resident and transient microflora are some species of organisms that do not usually cause disease but can do so under certain conditions. These organisms are called **opportunists** because they take advantage of particular opportunities to cause disease. Conditions that create opportunities for such organisms include:

1. *Failure of the Host's Normal Defenses.* Individuals with weakened immune defenses are said to be **immunocompromised**. Factors such as advanced malnutrition, the presence of another disease, advanced or very young age, treatment with radiation or immunosuppressive drugs, and physical or mental stress can lead to this state. The failure of host defenses in AIDS patients, for example, allows several different opportunistic infections to develop.

2. *Introduction of the Organisms into Unusual Body Sites.* The bacterium *Escherichia coli* is a normal resident of the human large intestine, but it can cause disease if it gains entrance to unusual sites such as the urinary tract, surgical wounds, or burns.

3. *Disturbances in the Normal Microflora.* Thriving populations of normal microflora compete with pathogenic organisms and in some instances actively combat their growth, an effect known as **microbial antagonism**. The normal microflora interfere with the growth of pathogens by competing for and depleting nutrients the pathogens need or by producing substances that create environments in which the pathogens cannot grow. As we saw in ◀Chapter 13 (p. 372), antibiotics sometimes destroy or disturb the normal microflora as they bring a pathogen under control. This disturbance allows other potential pathogens, such as yeasts that are not harmed by the antibiotic, to thrive in the absence of their antagonists, the normal microflora.

Although in later chapters we will focus on microorganisms that cause human disease, we must not lose sight of the importance of the many nonpathogenic microorganisms associated with the human body. In addition, we must remember that disease can result from disturbances in the normal ecological balance between resident populations and the host.

KOCH'S POSTULATES

The work of Robert Koch and the role of his Postulates in relating causative agents to specific diseases was described briefly in ◀Chapter 1 (p. 12). Now we can use our understanding of infection and disease to look at those postulates more carefully. For example, we now know that infection with an organism does not necessarily indicate that disease is present. With that knowledge, we can better appreciate the need for all four of **Koch's Postulates** to be satisfied in order to prove that a specific organism is the causative agent of a particular disease:

1. The specific causative agent must be observed in every case of a disease.

2. The agent must be isolated from a diseased host and must be grown in pure culture.

3. When the agent from the pure culture is inoculated into healthy, but susceptible, experimental hosts, the agent must cause the same disease.

4. The agent must be reisolated from the inoculated, diseased experimental host and identified as being identical to the original specific causative agent.

It is relatively easy today to demonstrate that each postulate is met for a variety of diseases caused by bacteria **(Figure 14.4)**. Some bacteria, however, are difficult to culture because they have fastidious nutritional requirements or other special needs for growth. For example, although the causative agent of syphilis, *Treponema pallidum*, has been known for many years, it has not been successfully grown on artificial media. Moreover, parasites such as viruses and rickettsias cannot be grown in artificial media and must instead be grown in living cells. For some agents that cause disease in humans, no other host has been found. Consequently, in such cases inoculation into a susceptible host is impossible unless human volunteers can be found. There are obvious ethical problems associated with

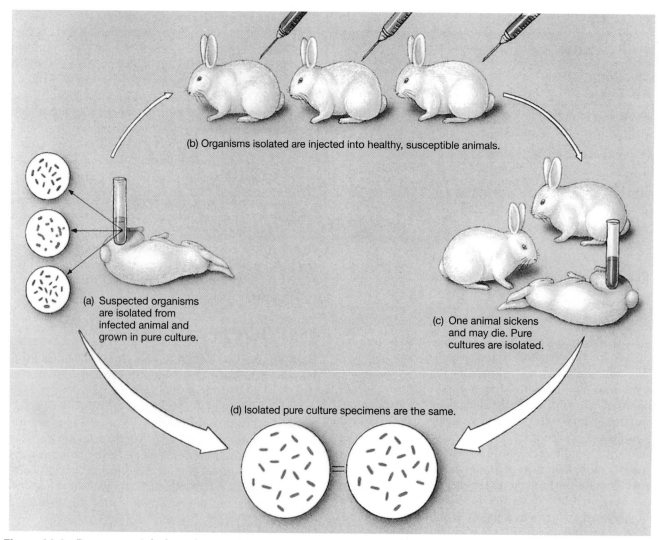

(b) Organisms isolated are injected into healthy, susceptible animals.

(a) Suspected organisms are isolated from infected animal and grown in pure culture.

(c) One animal sickens and may die. Pure cultures are isolated.

(d) Isolated pure culture specimens are the same.

Figure 14.4 Demonstration that a bacterial disease satisfies Koch's Postulates.

inoculating humans with infectious agents, even if volunteers might be available.

KINDS OF DISEASES

Human diseases are caused by infectious agents, structural or functional genetic defects, environmental factors, or any combination of these causes.

INFECTIOUS AND NONINFECTIOUS DISEASES

Infectious diseases are diseases caused by infectious agents such as bacteria, viruses, fungi, protozoa, and helminths. Chapters 19 through 24 of this text are devoted to discussions of particular infectious agents and the diseases they cause. **Noninfectious diseases** are caused by any factor other than infectious organisms.

CLASSIFICATION OF DISEASES

Classification of diseases as infectious or noninfectious gives a very limited view of human disease. The following scheme for classifying diseases provides a more comprehensive view. More importantly, it shows that infectious agents can interact with other factors in causing disease.

1. *Inherited diseases* are caused by errors in genetic information. The resulting developmental disorders may be caused by abnormalities in the number and distribution of chromosomes or by the interaction of genetic and environmental factors. Although inherited diseases have a noninfectious cause, some are associated with microbial activities. Sickle-cell anemia weakens patients and makes them more susceptible to infectious diseases. However, sickle-cell patients

PUBLIC HEALTH

Armadillo: Culture Vessel for Leprosy

The organism that causes Hansen's disease (leprosy) is difficult to culture. Many different methods had been tested and found unsatisfactory until someone tried inoculating the organism into the footpads of the nine-banded armadillo. There it grows very well; in fact, the organism multiplies faster there than in human tissues. When the organism does infect humans, it can have an incubation period of up to 30 years before disease symptoms appear. Before the armadillo was used to culture the organism, Koch's third Postulate could not be satisfied. No one wanted to hold out an arm and say, "Here, try to give me leprosy." Also, 30 years was a long time to wait to determine the results of such an experiment. The use of the armadillo has made it possible to confirm Koch's Postulates for *Mycobacterium leprae* as the causative agent of leprosy. This bacterium, seen by Armauer Hansen in 1878, was one of the first infectious agents to be identified and associated with a disease but one of the last to satisfy Koch's Postulates. The armadillo was chosen as an experimental host after naturally occurring leprosy infections were found in the armadillo populations of Texas and Louisiana.

DNA studies have shown the strains of *M. leprae* in the armadillo to be identical with those infecting humans. Cases of humans having acquired infection from armadillos have been confirmed, and we now designate armadillos as a reservoir for the disease in the U.S. Southwest. Cases of leprosy in African chimpanzees and mangabey monkeys have also been found. Injection of *M. leprae* collected from armadillos into mangabeys has caused the monkeys to develop leprosy.

(William J. Weber/Visuals Unlimited)

or carriers of the defect tend to be resistant to malaria. The abnormal hemoglobin S of sickle-cell patients gives up stored oxygen. With less oxygen, red blood cells change into a sickle shape and are removed by the spleen. Malaria-causing parasites that enter red blood cells make the cells sickle and are thereby killed before they complete their life cycle.

2. *Congenital diseases* are structural and functional defects present at birth, caused by drugs, excessive X-ray exposure, or certain infections. When a mother has a rubella (German measles) or a syphilis infection, the infectious agent may cross the placenta and cause congenital defects. Some medicines, such as the antiwrinkle drug retinoid-A and the antibiotic tetracycline, may cause congenital defects when taken by pregnant women.

3. *Degenerative diseases* are disorders that develop in one or more body systems as aging occurs. Patients with degenerative diseases such as emphysema or impaired kidney function are susceptible to infections. Conversely, infectious agents can cause tissue damage that leads to degenerative disease, as occurs in bacterial endocarditis, rheumatic heart disease, and some kidney diseases.

4. *Nutritional deficiency diseases* lower resistance to infectious diseases and contribute to the severity of infections. For example, the bacterium that causes diphtheria (*Corynebacterium diphtheriae*) produces more toxin in people with iron deficiencies than in those with normal amounts of iron. Poor nutrition also increases the severity of measles and contributes to deaths from the disease. Nutritional deficiencies can themselves develop from the action of, for example, helminths that severely damage the intestinal lining.

5. *Endocrine diseases* are due to excesses or deficiencies of hormones. Viral infection has been linked to pancreatic damage that leads to insulin-dependent diabetes.

6. *Mental disease* can be caused by a variety of factors, including those of an emotional, or psychogenic (si-ko-jen′-ik), nature as well as certain infections. For example, psychological stress may give rise to several gastrointestinal disorders, skin irritations, and even breathing difficulties. Mental disease can also result from brain infections such as in cases of neurosyphilis and the prion-caused Creutzfeldt-Jakob disease.

7. *Immunological diseases* such as allergies, autoimmune diseases, and immunodeficiencies are caused by malfunction of the immune system; AIDS is a consequence of a viral infection and destruction of certain cells of the immune system.

8. *Neoplastic diseases* involve abnormal cell growth that leads to the formation of various types of generally harmless or cancerous growths or tumors. Causes of such diseases include chemicals, physical agents such a various forms of radiation, and microorganisms, especially viruses. Papillomaviruses, which are known to cause warts, have been associated with the development of cervical cancer, and other viruses are known to cause tumorous growths in plants (◄Chapter 10, p. 302).

9. *Iatrogenic* (i-at″ro-jen′ik) *diseases* (*iatros*, Greek for "physician") are caused by medical procedures and/or treatments. Examples include surgical errors, drug reactions, and infections acquired from hospital treatment. The latter are called *nosocomial infections*. For example, *Staphylococcus aureus* is a common bacterium associated with surgical wound infections. Nosocomial infections are discussed in Chapter 15.

10. *Idiopathic* (id″e-o-path′ik) *diseases* are diseases whose cause is unknown. Some researchers believe that Alzheimer's disease, which causes mental deterioration, has an infectious basis.

COMMUNICABLE AND NONCOMMUNICABLE DISEASES

Some infectious diseases can be spread from one host to another and are said to be **communicable infectious diseases**. Some are more easily spread than others. Rubeola (red measles) and rubella are highly communicable, or **contagious diseases**, especially among young children. Vaccines protect children in developed countries, but nearly all unimmunized children in developing nations still get these diseases. Influenza is highly communicable among adults, especially the elderly. Gonorrhea and genital herpes infections are easily spread among unprotected sexual partners. Although also communicable, certain other diseases such as *Klebsiella* pneumonia are less contagious. Some diseases that normally affect other animals are transmissible to humans (Chapter 15), whereas diseases such as Hansen's disease (leprosy) can also be transmitted from humans to other animals.

Noncommunicable infectious diseases are not spread from one host to another. You cannot "catch" a noncommunicable disease from another person. Such diseases may result from (1) infections caused by an individual's normal microflora, such as an inflammation of the abdominal cavity lining following rupture of the appendix; (2) poisoning following the ingestion of preformed toxins, such as staphylococcal enterotoxin, a common cause of food poisoning; and (3) infections caused by certain organisms found in the environment, such as tetanus, a bacterial infection resulting from spores in the soil gaining access to a wound. Other noncommunicable infectious diseases, such as legionellosis, a form of pneumonia, can spread through contaminated air-conditioning systems.

▌▌▌ THE DISEASE PROCESS
▌▌▌ HOW MICROBES CAUSE DISEASE

Microorganisms act in certain ways that allow them to cause disease. These actions include gaining access to the host, adhering to and colonizing cell surfaces, invading tissues, and producing toxins and other harmful metabolic products. However, host defense mechanisms tend to thwart the actions of microorganisms. The occurrence of a disease depends on whether the pathogen or the host wins the battle; if it is a draw, a chronic, long-lasting disease may result.

Most of the pathogens considered in this text are prokaryotic microorganisms and viruses, which together account for the majority of human disease agents. However, several eukaryotes, such as fungi, protozoa, and multicellular parasites (mostly worms), display pathogenicity (◄Chapter 11, p. 325). Eukaryotic pathogens can be present in a host without causing disease signs or symptoms, or they can cause severe disease. The extent of damage caused by these pathogens, like that caused by prokaryotic infectious agents, is determined by the properties of the pathogens and by the host's response to them.

How Bacteria Cause Disease

Bacterial pathogens often have special structures or physiological characteristics that improve the chances of successful host invasion and infection. **Virulence factors** are structural or physiological characteristics that help organisms cause infection and disease. These factors include structures such as pili for adhesion to cells and tissues, enzymes that either help in evading host defenses or protect the organism from host defenses, and toxins that can directly cause disease.

> *The syphilis spirochete uses the tail end of its body to "hook" onto host cells.*

Direct Actions of Bacteria. Bacteria can enter the body by penetrating the skin or mucous membranes, by sexual transmission, by being ingested with food, by being inhaled in aerosols, or by transmission on a *fomite* (any inanimate object contaminated with an infectious agent). If the bacteria are immediately swept out of the body in urine or feces or by coughing or sneezing, they cannot initiate an infection.

A critical point in the production of bacterial disease is the organism's **adherence**, or attachment, to a host cell's surface. The occurrence of certain infections depends in part on the interaction between host plasma membranes and bacterial adherence factors. **Adhesins** are proteins or glycoproteins found on attachment pili (fimbriae) and capsules (◄Chapter 4, p. 96). Most adhesins that have been identified permit the pathogen to adhere only to receptors on membranes of certain cells or tissues (**Table 14.4**). For

TABLE 14.4

Examples of Adhesive Virulence Factors		
Bacterium	**Disease**	**Adhesion Mechanism**
Upper Respiratory Tract		
Mycoplasma pneumoniae	Atypical pneumonia	Adhesin on cell surface adheres to receptor in respiratory lining
Neisseria meningitidis	Meningitis	Adhesin on pili
Streptococcus pneumoniae	Pneumonia	Surface adhesins attach to carbohydrate on respiratory lining
Mouth		
Streptococcus mutans	Dental caries	Capsule attaches to tooth enamel
Intestinal Tract		
Shigella species	Dysentery	Unknown mechanism for attachment to intestinal lining
Escherichia coli	Diarrhea	Adhesins on pili attach to receptor on intestinal lining
Campylobacter jejuni	Diarrhea	Adhesins on flagella attach to intestinal lining
Vibrio cholerae	Cholera	Adhesins on flagella bind to receptors on intestinal lining
Urogenital Tract		
Treponema pallidum	Syphilis	Bacterial protein attaches to cells
Neisseria gonorrhoeae	Gonorrhea	Adhesins on pili attach to lining of genital tract

example, an adhesin on attachment pili of certain strains of *Escherichia coli* attaches to receptors on certain host epithelial cells. (Host leukocytes also have receptors for this adhesin, so the same adhesin that helps the bacterium attach may also help the host destroy it.) However, very often the capsules and attachment pili are also antiphagocytic structures. It is difficult for phagocytic cells to engulf bacteria that have capsules or attachment pili, so these structures make excellent virulence factors.

Some viruses gain entry to cells by mimicking substances of use to the host cell: rabies virus mimics the neurotransmitter acetylcholine.

Attachment to a host cell surface is not enough to cause an infection. The microbes must also be able to colonize the cell's surface or to penetrate it. **Colonization** refers to the growth of microorganisms on epithelial surfaces, such as skin or mucous membranes or other host tissues. For colonization to occur after adherence, the pathogens must survive and reproduce despite host defense mechanisms. For example, pathogenic bacteria on the skin's surface must withstand environmental conditions and bacteriostatic skin secretions. Those on respiratory membranes must escape the action of mucus and cilia. Those on the lining of portions of the digestive tract must withstand peristaltic movements, mucus, digestive enzymes, and acid. *Chlamydia pneumoniae*

biofilms that colonize coronary arteries have been implicated in heart disease (◄Chapter 23, p. 721).

Only a few pathogens cause disease by colonizing surfaces; most have additional virulence factors that enable the pathogen to invade tissues. The degree of **invasiveness** of a pathogen—its ability to invade and grow in host tissues—is related to the virulence factors the pathogen possesses and determines the severity of disease it produces. Some bacteria, such as pneumococci and other streptococci, release digestive enzymes that allow them to invade tissues rapidly and cause severe illnesses. Streptococci produce the enzyme **hyaluronidase**, or *spreading factor*. This enzyme digests hyaluronic acid, a gluelike substance that helps hold the cells of certain tissues together **(Figure 14.5a)**. Digestion of hyaluronic acid allows streptococci to pass between epithelial cells and invade deeper tissues. Some strains of *Streptococcus pyogenes* can cause a rapidly moving disintegration of tissues (necrotizing fasciitis) that can invade at the rate of 1 inch per hour!

In some cases the same pathogen can display varying degrees of invasiveness and pathogenicity in different tissues. Both bubonic plague and pneumonic plague are caused by the bacterium *Yersinia pestis*. In bubonic plague, the organisms enter the body by means of a flea bite, migrate through the blood, and infect many organs and tissues. Untreated, this disease has a mortality rate of about 55%. As a victim of pneumonic plague coughs or

(a)

1. Invasive pathogens reach epithelial surface.

2. Pathogens produce hyaluronidase.

3. Pathogens invade deeper tissues.

(b)

1. Pathogens produce coagulase.

2. Blood clot forms around pathogens.

3. Pathogens produce streptokinase, dissolving clot and releasing bacteria.

Figure 14.5 Enzymatic virulence factors help bacteria invade tissues and evade host defenses. **(a)** Hyaluronidase dissolves the "cement" that holds together the cells that line the intestinal tract. Bacteria that produce hyaluronidase can then invade deeper cells within the intestinal tissues. **(b)** Coagulase triggers blood plasma clotting, allowing bacteria protection from immune defenses. Streptokinase dissolves blood clots. Bacteria trapped within a clot can free themselves and spread the infection by producing streptokinase.

sneezes, the bacteria are spread by aerosols to other individuals. *Yersinia pestis* can cause a severe infection of the lungs with a mortality rate as high as 98% (◄Chapter 15, p. 433).

Most bacteria that invade tissues damage cells and are found around cells. Thus, enzymes that contribute to tissue damage are another important virulence factor. **Coagulase** is a bacterial enzyme that accelerates the coagulation (clotting) of blood. When blood plasma, the fluid portion of blood, leaks out of vessels into tissues, coagulase causes the plasma to clot. *Staphylococcus aureus* produces coagulase to aid in infection **(Figure 14.5b)**. Coagulase is a two-edged sword: It keeps organisms from spreading but also helps wall them off from immune defenses that might otherwise destroy them. Conversely, the bacterial enzyme **streptokinase** dissolves blood clots. Pathogens trapped in blood clots free themselves to spread to other tissues by secreting these virulence factors.

Some bacterial pathogens actually enter cells. The rickettsias, chlamydias, and a few other pathogens must invade cells to grow, reproduce, and produce disease. In other situations, organisms that can survive within host phagocytic cells not only escape destruction by the phagocytes but also obtain free transportation to deeper body tissues. Such organisms include *Mycobacterium tuberculosis* and *Neisseria gonorrhoeae*.

Bacterial Toxins. A **toxin** is any substance that is poisonous to other organisms. Some bacteria produce toxins, which are synthesized inside bacterial cells and are classified according to how they are released. **Exotoxins** are soluble substances secreted into host tissues. **Endotoxins** are part of the cell wall and are released into host tissues—sometimes in large quantities—from Gram-negative bacteria, often when the bacteria die or divide (◄Chapter 4, p. 86). Giving antibiotics that kill such bacteria can release sufficient toxin to cause the patient to die of severely reduced blood pressure (*endotoxic shock*). Let us look at some of the properties and effects of endotoxins and exotoxins **(Table 14.5)**.

Relatively weak (except in large doses), endotoxins are produced by certain Gram-negative bacteria. All endotoxins consist of lipopolysaccharide (LPS) complexes, the components of which vary among genera. They are relatively stable molecules that do not display affinities for particular tissues. Bacterial endotoxins have nonspecific

TABLE 14.5

Properties of Toxins		
Property	Exotoxins	Endotoxins
Organisms producing	Almost all Gram-positive; some Gram-negative	Almost all Gram-negative
Location in cell	Extracellular, excreted into medium	Bound within bacterial cell wall; released upon death of bacterium
Chemical nature	Mostly polypeptides	Lipopolysaccharide complex
Stability	Unstable; denatured above 60°C and by ultraviolet light	Relatively stable; can withstand several hours above 60°C
Toxicity	Among the most powerful toxins known (some are 100 to 1 million times as strong as strychnine)	Weak, but can be fatal in relatively large doses
Effect on tissues	Highly specific; some act as neurotoxins or cardiac muscle toxins	Nonspecific; ache-all-over systemic effects or local site reactions
Fever production	Little or no fever	Rapid rise in temperature to high fever
Antigenicity	Strong; stimulates antibody production and immunity	Weak; recovery from disease often does not produce immunity
Toxoid conversion and use	By treatment with heat or chemicals; toxoid used to immunize against toxin	Cannot be converted to toxoid; cannot be used to immunize
Examples	Botulism, gas gangrene, tetanus, diphtheria, staphylococcal food poisoning, cholera, enterotoxins, plague	Salmonellosis, tularemia, endotoxic shock

Diphtheria toxin molecules have a molecular weight of 62,000, contain 535 amino acids, and are not produced when iron is plentiful.

effects such as fever or a sudden drop in blood pressure. They also cause tissue damage in diseases such as typhoid fever and epidemic meningitis (an inflammation of membranes that cover the brain and spinal cord).

Exotoxins are more powerful toxins produced by several Gram-positive and a few Gram-negative bacteria. Most are polypeptides, which are denatured by heat, ultraviolet light, and chemicals such as formaldehyde. Species of *Clostridium, Bacillus, Staphylococcus, Streptococcus,* and several other bacteria produce exotoxins.

Some exotoxins are enzymes. **Hemolysins** were first discovered in cultures of bacteria grown on blood agar plates. The action of these exotoxins is to lyse (rupture) red blood cells. Two kinds of hemolysins were identified from bacteria grown on blood agar plates. **Alpha-hemolysins** (α-hemolysin) hemolyze blood cells, partially break down hemoglobin, and produce a greenish ring around colonies; **β-hemolysins** also hemolyze blood cells but completely break down hemoglobin and leave a clear ring around colonies **(Figure 14.6)**. Streptococci and staphylococci produce different hemolysins that are helpful in identifying them in laboratory cultures. There is no evidence that red blood cell lysis plays a role in the disease syndrome. Rather, the hemolysins release iron from the hemoglobin molecules in the red blood

Hemolysins also attack cells other than red blood cells, but are most easily visualized in blood agar.

cells. Iron is a critical element for growth of all cells, both host and microbe. But there is very little free iron within the human body. Most of it is bound in a form such as hemoglobin, and the microbe must enzymatically release it. Bacteria that can produce hemolysins can grow better than those that do not produce these enzymes. Especially in the staphylococci, the hemolysins can damage other types of cells as well. Alpha-hemolysin damages smooth muscle and kills skin cells.

Virulence factors called **leukocidins** are exotoxins produced by many bacteria, including the streptococci and staphylococci. These toxins damage or destroy certain kinds of white blood cells called *neutrophils* and *macrophages.* Leukocidins are most effective when released by microbes that have been engulfed by a neurophil. Because of the action of leukocidins, the number of white blood cells decreases in certain diseases, although most infections are characterized by an elevated white cell count. A similar substance, called **leukostatin,** interferes with the ability of leukocytes to engulf microorganisms that secrete the exotoxin.

Leukocidins release cell-killing enzymes from lysosomes.

In the preceding examples, the spreading of exotoxins by blood from the site of infection is called **toxemia.** But some diseases caused by microbes are due not to infection and invasion of tissues by pathogens but instead to the ingestion of preformed toxins made by pathogens. For example, botulism food poisoning strikes within hours of ingesting food that contains a significant amount of toxin produced by *Clostridium botulinum*—too short a

(a)

(b)

Figure 14.6 Types of hemolysis. (a) Alpha, or partial, hemolysis of red blood cells results in a greenish zone around colonies of *Streptococcus pneumoniae* grown on blood agar. *(L. M. Pope & D. R. Grote, University of Texas, Austin/Biological Photo Service)* **(b)** *Nocardia* colonies release β-hemolysins, which produce complete breakdown of hemoglobin, causing clear zones to form around colonies grown on blood agar. *(Courtesy ARUP Laboratories)*

APPLICATIONS

Clinical Use of Botulinum Toxin

The powerful effects of the neurotoxins produced by *Clostridium botulinum*, best known as a cause of lethal food poisoning, have been harnessed to help victims of dystonia. *Dystonia* refers to a group of neurologic disorders characterized by abnormal, sustained involuntary movements, often twisting, that are of unknown origin. In one form, blepharospasm, the patient's eyes remain tightly closed at all times. Physicians inject small quantities of botulinum toxin (trade name, Oculinum A) at several sites around each eye. The toxin blocks nerve impulses to muscles, thereby relieving the spasms of the eyelids [Photos **(a)** and **(b)**]. Injections are needed every 2 to 3 months. Some people receive such treatment for 4 to 5 years without problems; others develop antibodies (immune defenses) against the toxin after many injections. Because there are several different types of the toxin, it is hoped that patients can switch to a different form once they develop antibodies against one form. A round of injections can cost from $400 to $1,800.

A cosmetic use of botulinum toxin (Botox), rapidly gaining in favor, is removal of wrinkles, especially "frown" wrinkles in the center of the forehead.

(a)

(b)

(Courtesy Albert W. Biglan, M.D., University of Pittsburgh School of Medicine)

The forehead muscles are injected, become paralyzed in a relaxed state for several months, and cannot pull the skin together into deep creases [Photos **(c)** and **(d)**].

Botox can be used to stop severe underarm sweating. It is now also a great help to people suffering from urinary incontinence—and far less drastic than surgery. Botox is injected into muscles around the bladder and sphincters, a process taking about 15 min. It must be repeated approximately every 6 months.

Other dystonias being helped by botulinum toxin include oromandibular dystonia, in which the patient's jaws are clenched so tightly that the jaw bones may break, causing midfacial collapse. Eating and speaking are difficult, and some patients starve to death. Vocal cord spasms, causing a cracked tremulous voice, and "stenographers' cramp," causing the middle finger to extend rigidly, are also being treated experimentally with the toxin.

Oculinum A is licensed for treating adults with strabismus (cross-eye, lazy eye). Small amounts are injected into the overcontracted eye muscle, which then relaxes and lengthens. The antagonistic muscles on the other side of the eye contract to take up the slack, and the eye can then look straight ahead.

Botulinum toxin is a potential weapon for bioterrorism, and work is underway on an antibotulinum vaccine. So, don't get too enamoured of your botox shots. Once you've had the vaccine, the Botox shots will no longer smooth away your wrinkles.

(c)

(d)

(Courtesy Dr. Alastair Carruthers)

One milligram of botulinum toxin can kill more than 1 million guinea pigs.

time for the microbe to invade tissues and cause disease. The toxins accumulate during the storage of an improperly sterilized jar or can of food and have an immediate and often ultimately lethal effect on the consumer. Diseases that result from the ingestion of a toxin are termed **intoxications** rather than infections.

Many exotoxins have a special attraction for particular tissues. **Neurotoxins**, such as the botulism and tetanus toxins, are exotoxins that act on tissues of the nervous system to prevent muscle contraction (botulism) or muscle relaxation (tetanus). **Enterotoxins**, such as the toxin that causes cholera, are exotoxins that act on tissues of

the gut. Many exotoxins can act as *antigens*, foreign substances against which the immune system reacts. Antigenic exotoxins inactivated by treatment with chemical substances such as formaldehyde are called *toxoids*. A **toxoid** (*-oid* is Latin for "like") is an altered toxin that has lost its ability to cause harm but that retains antigenicity. Toxoids can be used to stimulate the development of immunity without causing disease. For example, when you get a tetanus booster shot, you are receiving the tetanus toxoid. It stimulates your body to produce immunity so that if you are exposed to active tetanus toxin through a cut or puncture of the skin, you will not get tetanus. The effects of bacterial exotoxins in human disease are summarized in Table 14.6. The specific effects of such diseases are discussed in later chapters.

TABLE 14.6

Effects of Exotoxins

Bacterium	Name of Toxin or Disease	Action of Toxin	Host Symptoms
Bacillus anthracis	Anthrax (cytotoxin)	Increases vascular permeability	Hemorrhage and pulmonary edema
Bacillus cereus	Enterotoxin	Causes excessive loss of water and electrolytes	Diarrhea
Clostridium botulinum	Botulism (eight serological types; neurotoxins)	Blocks release of acetylcholine at nerve endings	Respiratory paralysis, double vision
Clostridium perfringens	Gas gangrene (α-toxin, a hemolysin)	Breaks down lecithin in cell membranes	Cell and tissue destruction
	Food poisoning (enterotoxin)	Causes excessive loss of water and electrolytes	Diarrhea
Clostridium tetani	Tetanus (lockjaw) (neurotoxin)	Inhibits antagonists of motor neurons of brain; 1 nanogram can kill 2 tons of cells	Violent skeletal muscle spasms, respiratory failure
Corynebacterium diphtheriae	Diphtheria; produced by virus-infected (cytotoxin) bacteria	Inhibits protein synthesis	Heart damage can cause death weeks after apparent recovery
Escherichia coli	Traveler's diarrhea (enterotoxin)	Causes excessive loss of water and electrolytes	Diarrhea
Escherichia coli	O157;H7 (enterotoxin)	Hemolytic uremic syndrome	Destroys intestinal lining and causes hemorrhages in kidney
			Bleeding and kidney hemorrhage and failure
Pseudomonas aeruginosa	Various infections (exotoxin A)	Inhibits protein synthesis	Lethal, necrotizing lesions
Shigella dysenteriae	Bacillary dysentery (enterotoxin)	Cytotoxic effects; as potent as botulinum toxin	Diarrhea, causes paralysis in rabbits from spinal cord hemorrhage and edema
Staphylococcus aureus	Food poisoning (enterotoxin)	Stimulates brain center that causes vomiting	Vomiting
	Scalded skin syndrome (exfoliatin)	Causes intradermal separation of cells	Redness and sloughing of skin
Streptococcus pyogenes	Scarlet fever (erythrogenic, or red-producing toxin)	Causes vasodilation	Maculopapular (slightly raised, discolored) lesions
Vibrio cholerae	Cholera (enterotoxin)	Causes excessive loss of water (up to 30 liters/day) and electrolytes	Diarrhea; can kill within hours

(a) ▪LM▪   (b) ▪LM▪

Figure 14.7   An example of the cytopathic effect (CPE). **(a)** Uninfected mouse cells (magnification unknown); **(b)** the same cells 24 hours after infection with vesicular stomatitis virus (magnification unknown). A large number have died, and many others have rounded up into abnormal shapes. *(both photos: Gail W. T. Wertz, University of Alabama Medical School/ Biological Photo Service)*

How Viruses Cause Disease

Viruses can replicate only after they have attached to cells and then penetrated specific host cells. In tissue culture systems, once inside a cell, viruses cause observable changes collectively called the **cytopathic effect (CPE)** (◄Chapter 10, p. 297). CPE can be cytocidal when the viruses kill the cell and noncytocidal when they do not. Cytocidal viruses can kill cells by causing enzymes from cellular lysosomes to be released or by diverting the host cell's synthetic processes, thereby stopping the synthesis of host proteins and other essential macromolecules. CPE can be observed in laboratory tissue cultures with a compound microscope **(Figure 14.7)**. CPE can be so distinctive that an experienced clinical virologist can make a tentative identification by looking at infected cells through the microscope, even though further tests are needed to confirm the identification **(Table 14.7)**.

Many viruses produce pathogenic effects in host cells. These include *inclusion bodies*, which consist of nucleic acids and proteins not yet assembled into viruses, masses of viruses, or remnants of viruses. Rabiesviruses make inclusion bodies that are so distinctive they can be used to diagnose rabies. Retroviruses and oncoviruses integrate into host chromosomes and can remain in cells indefinitely, sometimes leading to the expression of their antigens on host-cell surfaces. Influenza and parainfluenza viruses produce hemagglutinins, which cause agglutination, or clumping together, of erythrocytes. This feature is of value in laboratory testing.

Viral infections can be productive or abortive. A **productive infection** occurs when viruses enter a cell and produce infectious offspring. An **abortive infection** occurs when viruses enter a cell but are unable to express all their genes to make infectious offspring. Productive infections vary in the degree of damage they cause, depending on the kind and number of cells the virus invades. An enterovirus, such as a human rotavirus or human adenovirus, that infects the gut can destroy millions of intestinal epithelial cells. Because these cells are rapidly replaced, the infection causes temporary, though sometimes severe, symptoms such as diarrhea, but no permanent damage. However, a poliovirus that infects motor neurons of the central nervous system can destroy these cells. Destroyed neurons cannot be replaced, so permanent paralysis may result. Human papillomaviruses that cause warts are limited to cells in localized areas. In contrast, the measles virus replicates and spreads throughout the body and may damage many tissues.

Latent viral infections are characteristic of herpesviruses. For example, chickenpox infections occur during childhood and usually are brought under control by host immune defenses. However, the virus may retreat into the nervous system and remain inactive, or latent. Later in life, factors such as stress, other infections, or fever can reactivate the virus. A weakened immune system allows the virus to multiply. Whatever the cause, the disease appears as shingles.

Persistent viral infections involve a continued production of viruses over many months or years. The hepatitis B virus (HBV) infects the liver in such a chronic fashion that there may be no outward signs of an infection. However, such persistent infections can lead to cirrhosis of the liver or even liver cancer.

How Fungi, Protozoa, and Helminths Cause Disease

In addition to bacteria and viruses, infectious diseases can also be caused by eukaryotes—specifically fungi, protozoa, and helminths. Even a few algae produce neurotoxins, and one alga (*Prototheca*) directly invades skin cells.

TABLE 14.7

Examples of Changes (Cytopathic Effects) to Virus-Infected Cells	
Virus Family	**Cytopathic Effect**
Adenoviridae	Cells swell
Herpesviridae	Cells swell
Picornaviridae	Cells swell and lyse
Paramyxoviridae	Cell membranes fuse, and up to 100 nuclei accumulate in a newly formed giant cell
Rhabdoviridae (rabies)	Inclusion bodies called Negri bodies form (site of viral replication or the accumulation of viral antigens)
Orthomyxoviridae	Produce hemagglutinins that cause erythrocytes to agglutinate, or clump together

Most fungal diseases result from fungal spores that are inhaled or from fungal cells and/or spores that enter cells through a cut or wound. Fungi damage host tissues by releasing enzymes that attack cells. As the first cells are killed, the fungi progressively digest and invade adjacent cells. Some fungi also release toxins or cause allergic reactions in the host. Certain fungi that parasitize plants produce *mycotoxins*, which cause disease if ingested by humans. Ergot, from a fungus that grows on rye, and aflatoxins, highly carcinogenic compounds that can be found in grains, cereals, and even peanut butter made from moldy peanuts, are mycotoxins (see Chapter 22).

Pathogenic protozoans and helminths cause human disease in several ways. Some protozoans, including those that cause malaria, invade and reproduce in red blood cells (◄Chapter 11, p. 317). The protozoan *Giardia intestinalis* attaches to tissues and ingests cells and tissue fluids of the host. *Giardia's* virulence factor is an *adhesive disk* by which it attaches to cells that line the small intestine (**Figure 14.8**). While burrowing into the tissue, the parasite uses its flagella to expel tissue fluids. This process creates so strong a suction that the parasite is not disturbed by peristaltic contractions.

Most helminths are extracellular parasites, inhabiting the intestines or other body tissues. However, some will destroy tissue as they migrate through the body. Many release toxic waste products and antigens in their excretions that often cause allergic reactions in the host. Humans are especially allergic to helminths. Some people will even have reactions to alcohol or formalin in which *Ascaris* worms have been stored. The outer surface of many helminths is quite tough and resistant to immune attacks.

✓ **CHECKLIST**

1. Give several examples of noninfectious diseases.
2. Can a disease be both infectious and noncommunicable?
3. Compare and contrast coagulase and streptokinase.
4. Compare and contrast infection and intoxication.

SIGNS, SYMPTOMS, AND SYNDROMES

Most diseases are recognized by their signs and symptoms. A **sign** is a characteristic of a disease that can be observed by examining the patient. Signs of disease include such things as swelling, redness, rashes, coughing, pus formation, runny nose, fever, vomiting, and diarrhea. A **symptom** is a characteristic of a disease that can be observed or felt only by the patient. Symptoms include such things as pain, shortness of breath, nausea, sore throat, headache, and malaise (discomfort).

A **syndrome** is a combination of signs and symptoms that occur together and are indicative of a particular disease or abnormal condition. For example, most infectious diseases cause the body to mount an acute inflammatory response. This response, which is discussed in Chapter 16, is characterized by a syndrome of fever, malaise, swollen lymph nodes, and **leukocytosis** (an increase in the numbers of white blood cells circulating in the blood).

In addition to the inflammatory response, many infectious diseases cause other signs and symptoms. Infections of the gut called enteric infections often cause nausea, vomiting, and diarrhea. Upper respiratory infections are usually characterized by coughing, sneezing, sore throat, and runny nose. Unfortunately, the signs and symptoms of diseases caused by different pathogens may be too similar to allow a specific diagnosis to be made. Thus, laboratory tests to identify infectious agents are an important component of modern medicine.

Even after recovery, some diseases leave aftereffects, called **sequelae** (se-kwel′e; singular: *sequela*). Bacterial infections of heart valves often cause permanent valve damage, and poliovirus infections leave permanent paralysis.

TYPES OF INFECTIOUS DISEASE

Infectious diseases vary in duration, location in the body, and other attributes. Several important terms, summarized in **Table 14.8**, are used to describe these attributes.

(a)   SEM   **(b)**   SEM

Figure 14.8 **Giardia intestinalis.** **(a)** *Giardia intestinalis* (21,150X) *(Jerome Paulin/Visuals Unlimited).* **(b)** The suction forces of *Giardia's* adhesive disk are so strong that they leave markings behind on the intestinal surface after they drop off. White arrowheads indicate where the suction cup attached, and white arrows point to marks left by the flagella. *(Courtesy Stanley L. Erlandsen, University of Minnesota School of Medicine, Minneapolis)*

TABLE 14.8

Terms Used to Describe Infections	
Term	Characteristic of Infection
Acute disease	Disease in which symptoms develop rapidly and that runs its course quickly
Chronic disease	Disease in which symptoms develop slowly and disease is slow to disappear
Subacute disease	Disease with symptoms intermediate between acute and chronic
Latent disease	Disease in which symptoms appear and/or reappear long after infection
Local infection	Infection confined to a small region of the body, such as a boil or bladder infection
Focal infection	Infection in a confined region from which pathogens travel to other regions of the body, such as an abscessed tooth or infected sinuses
Systemic infection	Infection in which the pathogen is spread throughout the body, often by traveling through blood or lymph
Septicemia	Presence and multiplication of pathogens in blood
Bacteremia	Presence but not multiplication of bacteria in blood
Viremia	Presence but not multiplication of viruses in blood
Toxemia	Presence of toxins in blood
Sapremia	Presence of metabolic products of saprophytes in blood
Primary infection	Infection in a previously healthy person
Secondary infection	Infection that immediately follows a primary infection
Superinfection	Secondary infection that is usually caused by an agent resistant to the treatment for the primary infection
Mixed infection	Infection caused by two or more pathogens
Inapparent infection	Infection that fails to produce full set of signs and symptoms

An **acute disease** develops rapidly and runs its course quickly. Measles and colds are examples of acute diseases. A **chronic disease** develops more slowly than an acute disease, is usually less severe, and persists for a long, indeterminate period. Tuberculosis and Hansen's disease (leprosy) are chronic diseases. A **subacute disease** is intermediate between an acute and a chronic disease. Gingivitis, or gum disease, can exist as a subacute disease. A **latent disease** is characterized by periods of inactivity either before signs and symptoms appear or between attacks. The herpes simplex virus and several other viral infections produce latent disease.

A **local infection** is confined to a specific area of the body. Boils and bladder infections are local infections. A **focal infection** is confined to a specific area, but pathogens from it, or their toxins, can spread to other areas. Abscessed teeth and sinus infections are focal infections. A **systemic infection**, or *generalized infection*, affects most of the body, and the pathogens are widely distributed in many tissues. Typhoid fever is a systemic infection. When focal infections spread, they become systemic infections. For example, organisms from an abscessed tooth can enter the bloodstream and be carried to other tissues, including the kidneys. The organisms can then infect the kidneys and other parts of the urinary tract.

Pathogens can be present in the blood with or without multiplying there. In **septicemia**, once known as blood poisoning, pathogens are present in and multiply in the blood. In **bacteremia** and **viremia**, bacteria and viruses, respectively, are transported in the blood but do not multiply in transit. Such spread of organisms often occurs in cases of injury, such as a cut, abrasion, or even teeth cleaning. As we have seen, some pathogens release toxins into the blood; the presence of toxins in blood is called *toxemia*. Saprophytes feed on dead tissues. Fungi behave as parasites when they destroy cells and as

APPLICATIONS

To Squeeze or Not to Squeeze

There is good cause for the common warning not to squeeze pimples and boils. Left alone, the body's defense mechanisms ordinarily confine these lesions to the skin. However, squeezing them can disperse microorganisms into the blood, producing bacteremia, which can lead to septicemia—a far worse condition. In septicemia, the organisms are spread throughout the body and can cause severe infections.

saprophytes when they feed on them or on other dead or decaying matter. They release metabolic products into the blood, thereby causing a condition called **sapremia**.

A **primary infection** is an initial infection in a previously healthy person. Most primary infections are acute. A **secondary infection** follows a primary infection, especially in individuals weakened by the primary infection. A person who catches the common cold as a primary infection, for instance, might come down with a middle-ear infection as a secondary infection. A **super-infection** (Chapter 13, p. 372) is a secondary infection that results from the destruction of normal microflora and often follows the use of broad-spectrum antibiotics. Although many infections are caused by a single pathogen, **mixed infections** are caused by several species of organisms present at the same time. Dental caries and periodontal disease are due to mixed bacterial infections. An **inapparent**, or **subclinical, infection** is one that fails to produce the full range of signs and symptoms either because too few organisms are present or because host defenses effectively combat the pathogens. Yet such mild infections are able to stimulate the immune system to protect against future infections. Sometimes people think they have never had a disease, and fail to come down with it despite repeated exposures. An inapparent case may have left them fully protected. People with inapparent infections, such as carriers of the hepatitis B virus, can spread the disease to others.

STAGES OF AN INFECTIOUS DISEASE

At one time or another, all of us have suffered from infectious diseases, such as the common cold, for which there is no cure. We simply must let the disease "run its course." Most diseases caused by infectious agents have a fairly standard course, or series of stages. These stages include the *incubation period*, the *prodromal phase*, the *invasive phase* (which includes the *acme*), the *decline phase*, and the *convalescence period* (**Figure 14.9**). Even when treatment is available to eliminate the pathogen, the disease process usually passes through most of the stages. Treatment commonly lessens the severity of symptoms because pathogens can no longer multiply. It shortens the duration of the disease and the time required for recovery. The signs and symptoms associated with the stages of an infectious disease and the resulting tissue damage caused by the pathogen are summarized in **Table 14.9**.

The Incubation Period

The **incubation period** for an infectious disease is the time between infection and the appearance of signs and symptoms. Although the infected person is not aware of the presence of an infectious agent, he or she can spread the disease to others. Each infectious disease has a typical incubation period (**Figure 14.10**). The length of the incubation period is determined by the properties of the pathogen and the response of the host to the organism.

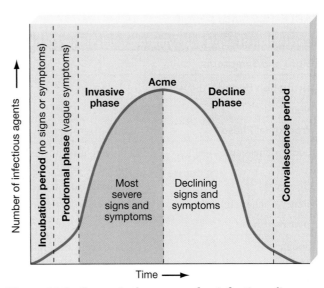

Figure 14.9   Stages in the course of an infectious disease.

Properties that affect the incubation period include the nature of the organism, its virulence, how many organisms enter the body, and where they enter in relation to the tissues they affect. For example, if large numbers of an extremely virulent strain of *Shigella* quickly reach the intestine, profuse diarrhea can appear in a day. In contrast, if only small numbers of a less virulent strain enter the digestive tract with a large quantity of food, the disease will develop more slowly. In fact, host defenses might be able to destroy the small number of organisms such that the disease will not occur at all. In the course of a lifetime, we undoubtedly have many more exposures than infections and more infections than we have overt diseases. As we will see in ◄Chapter 16, host defenses frequently attack pathogens as they start to invade tissues, thus averting potential diseases.

The Prodromal Phase

The **prodromal phase** of disease is a short period during which nonspecific, often mild, symptoms such as malaise and headache sometimes appear. A **prodrome** (*prodromos*, Greek for "forerunner") is a symptom indicating the onset of a disease. You wake up one morning feeling bad, and you know you're coming down with something, but you don't know yet whether you will break out in spots, start to cough, develop a sore throat, or experience other signs or symptoms. Many diseases lack a prodromal phase and begin with a sudden onset of symptoms such as fever and chills. During the prodromal phase, infected individuals are contagious and can spread the disease to others.

The Invasive Phase

The **invasive phase** is the period during which the individual experiences the typical signs and symptoms of the disease. These may include fever, nausea, headache, rash, and swollen lymph nodes. During this phase, the

TABLE 14.9

Correlation of Signs and Symptoms with Tissue Damage	
Signs and Symptoms	**Probable Nature of Tissue Damage**
Incubation Period	
None	None
Prodromal Phase	
Local redness and swelling	Pathogen has damaged tissue at site of invasion and caused release of chemicals that dilate blood vessels (redness) and allow fluid from blood to enter tissues (swelling).
Headache	Chemicals from tissue injury dilate blood vessels in the brain.
General aches and pains	Chemicals from tissue injury stimulate pain receptors in joints and muscles.
Invasive Phase	
Cough	Mucosal cells of respiratory tract have been damaged by pathogens; excess mucus is released, and neural centers in the brain elicit coughing to remove mucus.
Sore throat	Lymphatic tissue of the pharynx is swollen and inflamed by substances released by pathogens and by leukocytes.
Fever	Leukocytes release pyrogens that reset the body's thermostat and cause temperature to rise.
Swollen lymph nodes	Leukocytes release other substances that stimulate cell division and fluid accumulation in lymph nodes; lymph nodes themselves release substances that flow to and affect other lymph nodes; some pathogens multiply in lymph nodes.
Skin rashes	Leukocytes release substances that damage capillaries and allow small hemorrhages; some pathogens invade skin cells and cause pox, vesicles, and other skin lesions.
Nasal congestion	Nasal mucosal cells have been damaged by pathogens (usually viruses) that release fluids and increase mucous secretions.
Pain at specific sites (earache, local pain at a wound site)	Substances from pathogens or leukocytes have stimulated pain receptors; messages are relayed to the brain, where they are interpreted as pain.
Nausea	Toxins from pathogens have stimulated neural centers; you interpret the stimuli as nausea.
Vomiting	Toxins in food have stimulated the brain's vomiting center; vomiting helps rid the body of toxins.
Diarrhea	Toxins in food cause fluids to enter the digestive tract; some pathogens directly injure the intestinal epithelium; both toxins and pathogens stimulate peristalsis; frequent watery stools result.
Acme	
All signs and symptoms are at peak intensity	Full development of all signs and symptoms.
Decline Phase	
Signs and symptoms subside	Host defense mechanisms (and treatment, if applicable) have contributed to overcoming the pathogen.
Convalescence Period	
Patient regains strength	Tissue repair occurs; substances that caused signs and symptoms are no longer released.

time when the signs and symptoms reach their greatest intensity is known as the **acme**. During the acme, pathogens invade and damage tissues. In some diseases, such as some kinds of meningitis, this phase is referred to as being **fulminating** (*fulmen*, Latin for "lightning"), or sudden and severe. In other diseases, such as hepatitis B, it can be persistent or chronic or appear gradually with inapparent symptoms. A period of chills followed by fever marks the acme of many diseases. As signs and symptoms appear, the form the infection will take

becomes clear. Individuals at this critical stage are still contagious. The battle between pathogens and host defenses is at its height. A pathogen victory could lead to severe impairment of body function; if treatment is not available or provided in time, death can result.

Fever is an important component of the acme of many diseases. Certain pathogens produce substances called **pyrogens** that act on a center in the hypothalamus sometimes referred to as the body's "thermostat." Pyrogens set the thermostat at a higher-than-normal

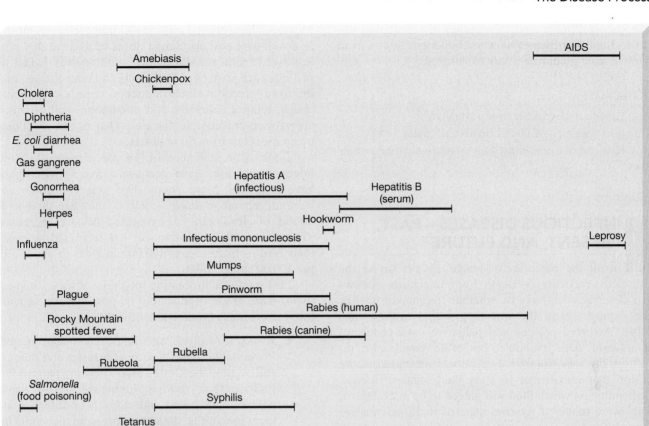

Figure 14.10 Incubation periods of selected infectious diseases. (Note that the time axis is not drawn to scale.)

temperature. The body responds with involuntary muscle contraction that generates heat and constriction (narrowing) of blood vessels in the skin to prevent heat loss. Because our bodies function at lower temperatures than the newly set temperature, we feel cold and have chills at this stage. We shiver and get "goose bumps" as muscles contract involuntarily.

As the effects of the pyrogens diminish, the thermostat is reset to the normal, lower temperature, and the body responds to reach and maintain that temperature. This response includes sweating and dilation (widening) of blood vessels in the skin to increase heat loss. Because our bodies are warmer than the newly set temperature, we feel hot and say that we have fever. Our skin gets moist from sweat as more blood circulates near the skin surface. In many infectious diseases, repeated episodes of pyrogen release occur, thereby accounting for bouts of fever and chills. A high fever that has risen rapidly will generally break suddenly by "*crisis*," whereas a low fever that arose gradually will probably return gradually to normal by "*lysis*."

The Decline Phase

As symptoms begin to subside, the disease enters the **decline phase**—the period of illness during which host defenses and the effects of treatment finally overcome the pathogen. The body's thermostat and other body activities gradually return to normal. Secondary infections may occur during this phase.

The Convalescence Period

During the **convalescence period**, tissues are repaired, healing takes place, and the body regains strength and recovers. Individuals no longer have disease symptoms. In some diseases, however, especially those in which scabs form over lesions, persons recovering from the disease can still transmit pathogens to others. Effects remaining after the disease has ended are called sequelae (e.g., pits and scarring following smallpox or chickenpox). Permanent heart or kidney damage can be due to streptococcal infections. Sometimes the sequelae are worse than the disease, as when blindness results from corneal scarring during a case of shingles.

✓ **CHECKLIST**

1. How does a sign differ from a symptom?
2. How does sapremia differ from bacteremia?
3. How does the prodromal stage differ from the incubation period?

INFECTIOUS DISEASES—PAST, PRESENT, AND FUTURE

In all the centuries of human history up to the twentieth, recovery or death from infectious diseases was determined largely by whether the human host or the pathogen won the war they waged against each other. Assorted potions and palliative (pain-reducing) treatments were available, but none could cure infectious diseases. Sometimes treatment was based on the notion that imbalances in body fluids caused disease. Depending on which fluid was judged to be in excess, efforts were made to remove some of it. Blood was removed by opening a vein or by applying blood-sucking leeches to the patient's skin. In eighteenth-century Europe, patients were bled until they lost consciousness. In 1774, when King Louis XV of France came down with smallpox, his desperate physicians bled him for 3 days in a row, each time removing "four large basinfuls of blood." In other cases, harsh laxatives were given to rid the body of excess bile. In most instances, these treatments failed to rid the patient of infectious agents, which at the time were unknown. At best, such treatments probably reduced suffering by hastening death.

Even after microorganisms came to be recognized as agents of disease, many years of painstaking research were required to relate specific diseases with the agents that caused them. More tedious research was needed to find antimicrobial agents that could cure diseases and to develop vaccines that could prevent them. The effects of these medical advances are clearly reflected in changes in the death rate in the United States (**Figure 14.11a**). That death rate has decreased from 1,560 per 100,000 people in 1900 to 505 per 100,000 people in the 1990s (Figure 14.11b). The greatest single factor in the decrease was the control of infectious diseases by better treatment or by immunization. Better sanitation has also helped. Figure 14.11a shows that in 1900 the proportion of the population dying from infectious diseases was 27 times the current rate. Deaths from typhoid fever, syphilis, and childhood diseases (measles, whooping cough, diphtheria) have nearly been eliminated, and deaths from pneumonia, influenza, and tuberculosis have been greatly reduced.

On a global scale, however, we have failed to eliminate diseases with technology that is already available.

Measles kills nearly 1.5 million children each year, mostly in developing nations. Single doses of vaccine that cost less than 12 cents could save most of these lives. In fact, as of 1995, each year 14 million children under the age of 5 die from infectious diseases such as measles, whooping cough, tetanus, diarrhea, and pneumonia—all of which are vaccine-preventable diseases. That is, one child dies every 2 seconds from these diseases.

Look again at Figure 14.11b. Because deaths from infectious diseases have decreased, average life spans have increased. More people live long enough to develop degenerative diseases, and as they age their likelihood of developing a malignant disease increases. Thus, the death rate for cancer has increased by more than 240%—from 55 per 100,000 in 1900 to about 133 per 100,000 in the 1990s.

Past successes in treating infectious diseases suggest that disease eradication should be possible. But at least four factors make eradication difficult:

1. *Available Medical Expertise is Not Always Applied.* Preventable diseases such as measles and mumps still occur in the United States because parents fail to have their children immunized. Also, some people, both young and old, fail to obtain treatment for curable diseases, a problem that could be solved by improved access to health care.

2. *Infectious Agents Are Often Highly Adaptable.* Many strains of microorganisms have developed resistance to several of the available antibiotics. The use of antibiotics has prevented so many deaths. However, the misuse and/or the overuse of antibiotics through the years has contributed to the development of mutant, drug-resistant bacterial strains. Treating diseases caused by such microbes presents a challenge that will not disappear or be solved quickly.

3. *Previously Unknown or Rare Diseases Become Significant as a Result of Changes in Human Activities and/or Social Conditions.* The epidemic of legionellosis that marred the festivities of a bicentennial celebration in 1976 was eventually found to be caused by a microorganism that was not commonly known but had existed and occasionally caused disease in the past. However, this time it was spread through a hotel air-conditioning system—something that could not have happened before air conditioning was invented. In the early 1980s many cases of toxic shock syndrome (TSS) suddenly appeared. This disease was shown to be caused by a staphylococcal toxin that usually reached the blood from organisms growing in certain rough, high-absorbency tampons used by women during menstruation. TSS was very rare before such tampons were invented; since then they have been modified by the manufacturer so that they no longer pose a health threat.

4. *Immigration and International Travel and Commerce Introduce New or Recurrent Strains of the Pathogen.* The high influx of legal and illegal immigrants often brings specific disease, such as tuberculosis, into areas where the disease had been eradicated. And the ease of international airline travel helps reintroduce previously eradicated diseases.

By the mid-twentieth century, microbiologists and public health officials thought that the use of antibiotics and vaccines would eliminate infectious disease. Although this dream has been realized for smallpox and probably will be true for polio shortly after the turn of the century, many infectious diseases remain a serious health threat. Tuberculosis and cholera represent diseases again on the rise, mostly due to antibiotic resistance and the absence of sanitary control measures. Other diseases have arisen from new infectious agents; AIDS, caused by HIV, is perhaps the most prominent. Scientists believe that HIV arose as a mutant strain in another animal species—an African monkey—and has "jumped species" to humans. Such jumps have probably occurred many times before, and with limited mobility of the human population, the virus would have disappeared without escaping into the rest of the world. However, the infection is slow to develop, and so unsuspecting carriers of the disease now take it with them, from one part of the world to another, thanks to the ease of international air travel.

In 2002 the Institute of Medicine stated that emerging infectious diseases in the United States should be taken seriously and that public health agencies were not prepared to deal with possible epidemics. In fact, AIDS might be the best example of our unpreparedness—and an indication of other diseases yet to come.

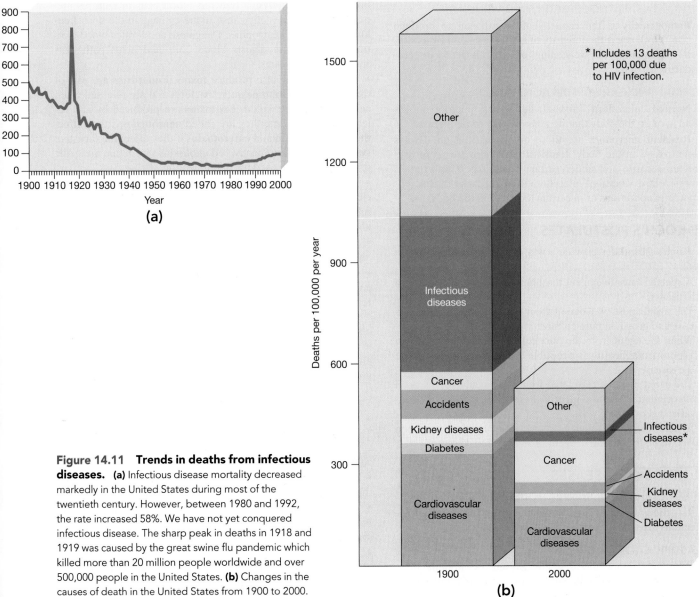

Figure 14.11 Trends in deaths from infectious diseases. **(a)** Infectious disease mortality decreased markedly in the United States during most of the twentieth century. However, between 1980 and 1992, the rate increased 58%. We have not yet conquered infectious disease. The sharp peak in deaths in 1918 and 1919 was caused by the great swine flu pandemic which killed more than 20 million people worldwide and over 500,000 people in the United States. **(b)** Changes in the causes of death in the United States from 1900 to 2000.

▌ RETRACING OUR STEPS

- A **pathogen** is a parasite capable of causing disease.

▌ HOST-MICROBE RELATIONSHIPS

- A **host** is an organism that harbors another organism.

SYMBIOSIS

- **Symbiosis** means "living together" and includes **commensalism**, in which one organism benefits and the other neither benefits nor is harmed; **mutualism**, in which both organisms benefit; and **parasitism**, in which one organism (the **parasite**) benefits and the other (the host) is harmed.

CONTAMINATION, INFECTION, AND DISEASE

- **Contamination** refers to the presence of microorganisms. In **infection**, pathogens invade the body; in **disease**, pathogens or other factors disturb the state of health such that the body cannot perform its normal functions. **Infestation** refers to the presence of worms or arthropods in or on the body.

PATHOGENS, PATHOGENICITY, AND VIRULENCE

- **Pathogenicity** is the capacity of a pathogen to produce disease. **Virulence** is the intensity of a disease caused by a pathogen. **Attenuation** is weakening of a pathogen's disease-producing capacity.

NORMAL (INDIGENOUS) MICROFLORA

- **Normal microflora** (normal flora) are microorganisms found in or on the body that do not normally cause disease.
- **Resident microflora** are those organisms that are always present on or in the body; **transient microflora** are those present temporarily and under certain conditions. **Opportunists** are resident or transient microflora that can cause disease under certain conditions or in certain locations in the body.

▌ KOCH'S POSTULATES

- **Koch's Postulates** provide a way to link a pathogen with a disease:
1. A specific causative agent must be observed in every case of a disease.
2. The agent must be isolated from a host displaying the disease and grown in pure culture.
3. When the agent from the pure culture is inoculated into an experimental healthy, susceptible host, the agent must cause the disease.
4. The agent must be reisolated from the inoculated, diseased experimental host and identified as the original specific causative agent.
- When Koch's Postulates are met, an organism has been proved to be the causative agent of an infectious disease.

▌ KINDS OF DISEASES

INFECTIOUS AND NONINFECTIOUS DISEASES

- **Infectious diseases** are caused by infectious agents; **noninfectious diseases** are caused by other factors.

CLASSIFICATION OF DISEASES

- Although many diseases are caused by noninfectious agents, some of these diseases may be associated with an infectious agent.

COMMUNICABLE AND NONCOMMUNICABLE DISEASES

- A **communicable**, or **contagious**, **infectious disease** can be spread from one host to another. A **noncommunicable infectious disease** cannot be spread from host to host and may be acquired from soil, water, or contaminated foods.

▌ THE DISEASE PROCESS

HOW MICROBES CAUSE DISEASE

- Many microbes have **virulence factors** that enable the establishment of infections. These factors include adhesion molecules, enzymes, and toxins.
- Bacteria cause disease by **adhering** to a host, by **colonizing** and/or invading host tissues, and sometimes by invading cells. The ability of a pathogen to invade and grow in host tissues, called **invasiveness**, is related to particular virulence factors. **Hyaluronidase** helps bacteria invade tissues.
- Bacteria release other substances, most of which damage host tissues. **Hemolysins** lyse red blood cells in cultures and may or may not directly cause tissue damage in the host. **Leukocidins** destroy neutrophils. **Coagulase** accelerates blood clotting. **Streptokinase** digests blood clots and helps pathogens spread to body tissues.
- Many bacteria also produce **toxins**. **Endotoxins** are part of the cell wall of Gram-negative bacteria and are released when cells divide or are killed. **Exotoxins** are produced by and released from bacteria; they are called **neurotoxins** if they affect the nervous system and **enterotoxins** if they affect the digestive system. **Toxoids** are inactivated exotoxins that retain antigenic properties and are used for immunization.
- Viruses damage cells and produce a variety of observable changes called the **cytopathic effect (CPE)**. A **productive infection** leads to the release of virus progeny, whereas an **abortive infection** does not produce infectious progeny.
- Pathogenic fungi can invade and progressively digest cells, and some produce toxins.
- Protozoa and helminths damage tissues by ingesting cells and tissue fluids, releasing toxic wastes, and causing allergic reactions.

SIGNS, SYMPTOMS, AND SYNDROMES

- A **sign** is an observable effect of a disease. A **symptom** is an effect of a disease felt by the infected person. A **syndrome** is a group of signs and symptoms that occur together.

TYPES OF INFECTIOUS DISEASE

- Terms used to describe types of diseases are defined in Table **14.8**.

STAGES OF AN INFECTIOUS DISEASE

- The **incubation period** is the time between infection and the appearance of signs and symptoms of a disease.
- The **prodromal phase** is the stage during which pathogens begin to invade tissues; it is marked by early nonspecific symptoms.
- The **invasive phase** is the period during which the individual experiences the typical signs and symptoms of the disease. During this phase the signs and symptoms reach their greatest intensity at the **acme**.

- The **decline phase** is the stage during which host defenses overcome pathogens; signs and symptoms subside during this phase, and secondary infections may occur.

- The **convalescence period** is the stage during which tissue damage is repaired and the patient regains strength. Recovering individuals may still transmit pathogens to others.

▌ TERMINOLOGY CHECK

abortive infection *(p. 414)*
acme *(p. 418)*
acute disease *(p. 416)*
adherence *(p. 408)*
adhesin *(p. 408)*
alpha (α) hemolysin *(p. 411)*
animal passage *(p. 403)*
attenuation *(p. 403)*
bacteremia *(p. 416)*
beta (β) hemolysin *(p. 411)*
chronic disease *(p. 416)*
coagulase *(p. 410)*
colonization *(p. 409)*
commensalism *(p. 402)*
communicable infectious
   disease *(p. 408)*
contagious disease *(p. 408)*
contamination *(p. 402)*
convalescence period
   (p. 419)
cytopathic effect (CPE)
   (p. 414)
decline phase *(p. 419)*
disease *(p. 402)*

endotoxin *(p. 410)*
enterotoxin *(p. 413)*
exotoxin *(p. 410)*
focal infection *(p. 416)*
fulminating *(p. 418)*
hemolysin *(p. 411)*
host *(p. 401)*
hyaluronidase *(p. 409)*
immunocompromised
   (p. 405)
inapparent infection *(p. 417)*
incubation period *(p. 417)*
infection *(p. 402)*
infectious disease *(p. 406)*
infestation *(p. 402)*
intoxication *(p. 413)*
invasiveness *(p. 409)*
invasive phase *(p. 417)*
Koch's Postulates *(p. 405)*
latent disease *(p. 416)*
latent viral infection *(p. 414)*
leukocidin *(p. 411)*
leukocytosis *(p. 415)*
leukostatin *(p. 411)*

local infection *(p. 416)*
microbial antagonism
   (p. 405)
mixed infection *(p. 417)*
mutualism *(p. 401)*
neurotoxin *(p. 413)*
noncommunicable infectious
   disease *(p. 408)*
noninfectious disease
   (p. 406)
normal microflora *(p. 404)*
opportunist *(p. 405)*
parasite *(p. 401)*
parasitism *(p. 401)*
pathogen *(p. 401)*
pathogenicity *(p. 403)*
persistent viral infection
   (p. 414)
primary infection *(p. 417)*
prodromal phase *(p. 417)*
prodrome *(p. 417)*
productive infection *(p. 404)*
pyrogen *(p. 418)*
resident microflora *(p. 404)*

sapremia *(p. 417)*
secondary infection
   (p. 417)
septicemia *(p. 416)*
sequela *(p. 415)*
sign *(p. 415)*
streptokinase *(p. 410)*
subacute disease *(p. 416)*
subclinical infection
   (p. 417)
superinfection *(p. 417)*
symbiosis *(p. 401)*
symptom *(p. 415)*
syndrome *(p. 415)*
systemic infection *(p. 416)*
toxemia *(p. 411)*
toxin *(p. 410)*
toxoid *(p. 413)*
transient microflora *(p. 404)*
transposal of virulence
   (p. 403)
viremia *(p. 416)*
virulence *(p. 403)*
virulence factor *(p. 408)*

▌ CLINICAL CASE STUDY

A microbiology student clad in shorts overturns his motorcycle on a patch of gravel. He sustains considerable abrasions and deep lacerations of the right leg. Emergency-room personnel have great difficulty trying to remove embedded gravel and dirt. Many areas of infection develop, and dead tissue can be seen at the sites of abrasion. Are his infections most likely due to pathogens or opportunists? Why is it difficult to determine?

▌ CRITICAL THINKING QUESTIONS

1. Performing step 3 of Koch's Postulates presents an ethical problem when it involves a life-threatening human disease. Can you suggest any alternatives to infecting a person with an organism that might prove fatal, while still accomplishing the goal of proving whether that organism causes the disease in question?

2. Infectious disease mortality decreased significantly in the twentieth century compared to all the time preceding this period. This recent success in treating infectious diseases has led us to believe that eradication of disease is possible. However, between 1980 and 1992 the rate increased 58% (Figure 14.11a). What factors have changed in modern times that make this goal harder to achieve?

3. Can you think of ways in which a species of microorganism that does not normally cause disease but inhabits the environs of resident and transient microflora can cause disease under certain circumstances?

▮ SELF-QUIZ

1. Match the following host-microbe relationship terms to their descriptions:

   C Parasitism
   E Pathogen
   A Symbiosis
   C Commensalisms
   D Host
   B Mutualism

   (a) An association between two or more species.
   (b) Both members of the association benefit from the relationship.
   (c) Only one organism benefits from the relationship.
   (d) Any organism that harbors another organism.
   (e) A parasite capable of causing disease in a host.
   (f) Two species living together in a relationship such that one benefits and the other neither benefits nor is harmed.

2. Contamination, infection, and disease are a sequence of conditions in which the severity of the effects microorganisms have on their hosts increases. True or false?

3. Transposal of virulence and attenuation are two techniques that are useful in the production of:
   (a) Antibiotics   (d) Virulent organisms
   (b) Antiseptics   (e) Vaccines
   (c) Pathogenic organisms

4. Endotoxins are associated with Gram-negative bacteria and are part of their cell _____ and are released when the cell _____/___ while exotoxins are produced and released by Gram-positive and some Gram-negative bacteria and are called _____ if they affect the nervous system and _____ if they affect the digestive system.
   (a) Cycle; shrivels/enlarges; hemolysins; leukocidins
   (b) Membrane; shrivels/enlarges; leukocidins; hemolysins
   (c) Walls; divides/dies; neurotoxins; enterotoxins
   (d) Flagella; divides/dies; enterotoxins; neurotoxins
   (e) Cycle divides/dies; hemolysins; neurotoxins

5. The best descriptive term for resident microflora is:
   (a) Parasites   (d) Commensals
   (b) Pathogens   (e) Mutualists
   (c) Infestations

6. Which of the following is not true about the transient microflora?
   (a) They are present under the same exact conditions, in any of the locations where resident microflora are found.
   (b) Pathogens can be transient microflora.
   (c) They occur in mucous membranes when greater than normal qualities of nutrients are available.
   (d) They occur on skin when it is warmer and more moist than usual.
   (e) They persist only as long as the necessary conditions are met.

7. Which of the following diseases can be directly caused by microorganisms?
   (a) Inherited disease
   (b) Congenital disease
   (c) Neoplastic disease
   (d) b and c
   (e) a and b

8. An iatrogentic disease in a patient caused by *Staphylococcus aureus*–contaminated surgical instruments would be known as a _____ _____.
   (a) Zoonotic invasion
   (b) Transient contamination
   (c) Invasive malignancy
   (d) Subclinical infection
   (e) Nosocomial infectionr

9. All of the following are virulence factors EXCEPT:
   (a) Adhesive pili
   (b) Enzymes that aid in evasion of host defenses
   (c) Enzymes that aid in direct protection of microbe from host defenses
   (d) Toxins
   (e) Chloroplasts

10. Which of the following is not true about the virulence factor coagulase?
    (a) It keeps microorganisms from spreading.
    (b) It increases the likelihood of exposure to host immune defenses.
    (c) It accelerates coagulation or clotting of host blood.
    (d) Streptokinase can counteract the effects of coagulase.
    (e) All of the above are true.

11. A communicable disease cannot be spread from host to host and may be acquired from soil, water, or contaminated food while a noncommunicable disease is a contagious disease that can be spread from one host to another. True or false?

12. An example of a latent disease is:
    (a) Chicken pox/shingles
    (b) Tuberculosis
    (c) Cold/flu
    (d) Gum disease
    (e) Leprosy

13. Which of the following would be a septicemia?
    (a) Bacteremia   (d) Viremia
    (b) Focal infection   (e) Systemic infection
    (c) Local infection

14. The presence of a few, nonmultiplying, bacteria in the blood is termed:
    (a) Viremia   (d) Toxemia
    (b) Septicemia   (e) Secondary infection
    (c) Bacteremia

15. Which of the following infectious disease stages is mismatched?
    (a) Convalescent period—tissue damage is repaired and patient strength returns
    (b) Decline phase—host defenses are overwhelmed by pathogen
    (c) Incubation period—time between infection and onset of signs and symptoms
    (d) Invasive phase—individual experiences typical signs and symptoms of disease
    (e) Prodromal phase—pathogens begin tissue invasion; marked by nonspecific symptoms

16. Viral damages to cells produce observable changes called the _____ effect. Viral infections that lead to release of viral progeny are known as _____ infections while those resulting in no infectious progeny are known as _____ infections.

(a) Prodromal; reproductive; chronic
(b) Productive; prodromal; cytopathic
(c) Abortive; cytopathic; productive
(d) Cytopathic; productive; abortive
(e) Cytopathic; prodromal; productive

17. A positive antibody test for HIV would be a _____ of disease.
    (a) Symptom   (c) Sign   (e) Sequela
    (b) Syndrome   (d) Virulence

18. Which of the following is not a condition of Koch's Postulates?
    (a) Isolate the causative agent of a disease.
    (b) Cultivate the microbe in the lab.
    (c) Inoculate a test animal to observe the disease.
    (d) Grow the organism in pure culture.
    (e) Produce a vaccine.

19. A laboratory bench with bacteria spilled on it could be correctly referred to as what?
    (a) Infected   (c) Infested   (e) Inflamed
    (b) Contaminated   (d) Diseased

20. A _____ is an observable effect of a disease while a _____ is an effect of a disease felt by the infected person. A _____ is a group of signs and symptoms that occur together.
    (a) Syndrome; sign; symptom
    (b) Syndrome; symptom; sign
    (c) Sign; symptom; syndrome
    (d) Symptom; sign; syndrome
    (e) Sign; syndrome; symptom

21. Which of the following is mismatched?
    (a) Abortive infection—infection leads to abortion in pregnant individuals
    (b) Inapparent infection—too few organisms present to produce typical signs and symptoms
    (c) Local infection—confined to specific area of body
    (d) Mixed infection—more than one type of organism is responsible for disease process
    (e) Productive infection—virus is produced from an infected host cell

22. The presence of *Staphylococci* on healthy skin helps to prevent pathogenic bacteria from colonizing and causing disease. This is an example of:
    (a) Virulence   (d) Opportunism
    (b) Pathogenicity   (e) Microbial antagonism
    (c) Antibiosis

23. _____ are soluble substances secreted from bacteria into host tissues, whereas _____ are part of the bacterial cell wall and enter host tissues during division or after cell death.
    (a) Exotoxins/endotoxins
    (b) Endotoxins/exotoxins
    (c) Lipopolysaccharides/proteins
    (d) Polysaccharides/porins
    (e) Toxoids/metatoxins

24. Botulinum toxin is an example of a(n):
    (a) Endotoxin   (d) Hemolysin
    (b) Lipopolysaccharide   (e) Exotoxin
    (c) Carbohydrate

25. Latent viral infections are brought under control by the use of drugs. True or false?

26. Trace the course of a disease in the accompanying graph. Identify stages (a) through (f), and relate each to signs and symptoms and to activities of a pathogen.

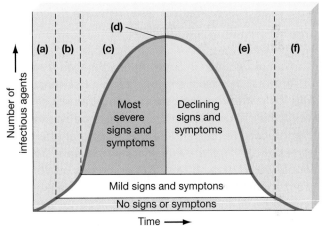

(a) Incubation period
(b) Prodromal phase
(c) Illness phase
(d) The acme
(e) Decline phase
(f) convalescence period

▌ EXPLORATIONS ON THE WEB

http://www.wiley.com/college/black

If you think you've mastered this chapter, there's more to challenge you on the web. Go to the companion web site to fine-tune your understanding of the chapter concepts and discover answers to the questions posed below.

1. Koch's Postulates developed in 1890 are still used today to show that an organism is the etiologic agent of a disease. While most scientists for some time have believed that HIV is the etiologic agent of AIDS, it is only recently that Koch's Postulates have been fulfilled. The third Postulate requires the organism be introduced into a healthy susceptible host and then cause the disease. How could this Postulate be fulfilled?

2. You are not alone. Your body houses about 10^{12} bacteria on the skin, 10^{10} in the mouth, and 10^{14} in the gastrointestinal tract.

3. In the eighteenth century, smallpox was a killer disease, as widespread as cancer or heart disease in the twentieth century. However, in 1980, the World Health Assembly officially declared "the world and its peoples" free from endemic smallpox. How was this infectious disease eliminated?

Epidemiology and Nosocomial Infections

Come with me . . .

What do you know about witch doctors beyond Hollywood's outdated imagery? The photo here shows a witch doctor from Southern Nigeria. The centuries of knowledge that he and other healers hold regarding local diseases and use of local plants to treat diseases is in danger of dying out. They have the distilled knowledge of countless years of treating illness, and in recent years more Western scientists are becoming apprenticed to these healers to help save this information.

One old shaman in South America began teaching an American husband and wife team of scientists his skills, only after they agreed to spend the rest of their lives in his village. No member of his tribe was willing to apprentice to him, and he was afraid that after they learned everything, the Americans would leave, and his people would have no care. The Americans publish the information and present it at meetings, but always return to the village. Many preindustrial peoples attributed diseases to "evil spirits" or "bad air"—an approach that differs from that of industrial societies. However, often their keen observations of nature give the clues needed to find the source of a disease. For example, Navajo medicine men noted an unusually high production of piñon pine nuts the year of the hantavirus outbreak in the desert Southwest. Rodents feeding on the nuts spread the virus in their urine.

Paul Almasy/Corbis Images

 Video related to this topic is available within WileyPLUS.

Up to this point in our study of host-microbe interactions, we have examined the characteristics of pathogens that lead to infectious diseases, and we have seen how the disease process occurs in individuals. But people with infectious diseases are members of a population; they acquire infectious diseases and transmit them within a population. Therefore, to further our understanding of such diseases, we must consider their effects on populations, including hospital populations. Unfortunately the high hopes of U.S. Surgeon General William H. Stewart, who testified before the U.S. Congress in 1967 that "the war against infectious diseases had been won," were premature.

EPIDEMIOLOGY

WHAT IS EPIDEMIOLOGY?

Epidemiology (ep″i-de-me-ol′o-je) is the study of factors and mechanisms involved in the frequency and spread of diseases and other health-related problems within populations of humans, other animals, or plants. The term is derived from the Greek words *epidemios*, meaning "among the people," and *logos*, meaning "study." Although **epidemiologists**, scientists who study epidemiology, may consider such health-related problems as automobile accidents, lead poisoning, or cigarette smoking, we will limit our discussion to factors and mechanisms that concern the transmission of infectious diseases as well as their **etiology** (e-te-ol′o-je), or cause, in a population. We will also see how epidemiologists use this information to design ways to control and prevent the spread of infectious agents.

Epidemiology is a branch of microbiology because many diseases that concern epidemiologists, such as AIDS, tuberculosis, and malaria, are caused by microbes. Parasitic worm diseases, such as hookworm disease and ascariasis, also are studied by epidemiologists. Epidemiology includes relationships among pathogens, their hosts, and the environment. It is related to public health because it provides information and methods used to understand and control the spread of disease within the human population. Bioterrorism is a special concern of epidemiologists. Agricultural and environmental scientists often are concerned with the epidemiology of animal and plant diseases. When such diseases can be transmitted to humans, they, too, become a public health problem.

In the United States, milk made available to other families from breast milk banks is first tested for HIV.

In following diseases and their spread, epidemiologists are especially interested in the frequencies of diseases within populations.

Incidence and Prevalence Rates

The **incidence** of a disease is the number of *new* cases contracted within a set population during a specific period of time (usually expressed as new cases per 100,000

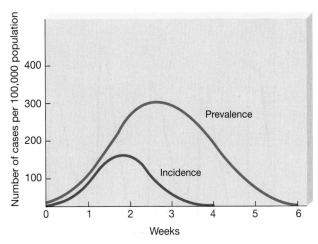

Figure 15.1 Incidence and prevalence rates. A hypothetical disease lasting 4 weeks illustrates the difference between the incidence and prevalence numbers.

people per year). The **prevalence** of a disease is the *total number* of people infected within the population at any time. Prevalence includes both old and newly diagnosed cases. For example, if epidemiologists conduct weekly surveys regarding a disease that lasts four weeks, an individual infected in week one could be counted as many as four times in a prevalence study but only once in an incidence survey. Thus, incidence data are reliable indicators of the spread of a disease—a drop in incidence suggests a reduction in the spread of the disease. In contrast, prevalence data measure how seriously and how long the disease is affecting a population (**Figure 15.1**).

Morbidity and Mortality Frequencies

Frequencies also are expressed as proportions of the total population. The **morbidity rate** is the number of individuals affected by a disease during a set period in relation to the total number in the population. It usually is expressed as the number of cases per 100,000 people per year. The **mortality rate** is the number of deaths due to a disease in a population during a specific period in relation to the total population. It is expressed as deaths per 100,000 people per year.

DISEASES IN POPULATIONS

When studying the frequency of diseases in populations, epidemiologists must consider the geographic areas affected and the degree of harm the diseases cause in the population. On the basis of their findings, they classify diseases as *endemic*, *epidemic*, *pandemic*, or *sporadic*.

An infectious disease agent is **endemic** if it is present continually in the population of a particular geographic area but both the number of reported cases and the severity of the disease remain too low to constitute a public health problem. For example, mumps is endemic in the entire United States, and valley fever is endemic in the southwestern United States. Chickenpox is an endemic

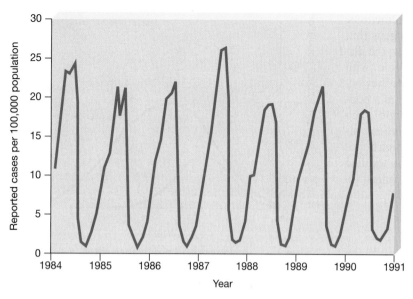

Figure 15.2 **The incidence rate of chickenpox in the United States.** Chickenpox, an endemic disease, shows obvious seasonal variation, with most cases occurring during the spring. (Chickenpox is no longer reportable to the CDC. Current data are from voluntary reporting, and while they show a drop since licensing of the chickenpox vaccine, only 14 states have consistently reported, and the data are not reliable.)

disease with seasonal variation; that is, many more cases are seen from late winter through spring than at other times (**Figure 15.2**). Endemic diseases also can vary in incidence in different parts of the endemic region.

An **epidemic** arises when a disease suddenly has a higher-than-normal incidence in a population. Then the morbidity rate or the mortality rate or both become high enough to pose a public health problem. Endemic diseases can give rise to epidemics, especially when a particularly virulent strain of a pathogen appears or when most of a population lacks immunity. For example, St. Louis encephalitis, a viral inflammation of the brain, reached epidemic proportions in the United States in 1975 (**Figure 15.3**). It arose due to the presence of both a large nonimmune bird population that carried the virus and a large mosquito population that transferred the virus from birds to humans, a situation mirrored today with West Nile Fever virus.

With the breakup of the former Soviet Union into the Russian Federation, epidemic diphtheria temporarily emerged. After 1990, overcrowding due to population

migrations resulted in a dramatic increase in the diphtheria incidence rate (**Figure 15.4**). In 1995, 50,319 cases were reported, with 1,746 deaths. There was also a cultural reluctance to immunize children who were not healthy and robust, as it was considered dangerous for ill or weak children to receive vaccine. World health officials considered the expanding diphtheria epidemic an international health emergency. Diphtheria cases arising in the Russian Federation have been reported in eastern and northern Europe. Health officials outlined a strategy in early 1995 to ensure that all NIS children are vaccinated against diphtheria. Since then, case rates have returned to a low rate, reaching 771 cases for the year 2000. In contrast, however, there was only 1 case in the United States in 2003.

A **pandemic** occurs when an epidemic spreads worldwide. In 1918 the swine flu reached pandemic proportions (as we shall see in ◀Chapter 21). Cholera has been responsible for seven pandemics over the centuries. Its spread through the Americas during the epidemic of 1991–1992 is shown in **Figure 15.5**. In the summer of 2006 there were over 60,000 cases in Sudan and Angola.

Figure 15.3 **The incidence rate of St. Louis encephalitis in the United States.** This viral disease showed a major outbreak during late 1975.

Figure 15.4 Diphtheria cases in the Russian Federation, 1965–2000. The number of reported cases increased from 1991 to 1994 because of insufficient health precautions as a result of the breakup of the Soviet Union. Since then, vaccination efforts have resumed and the incidence has fallen back to more normal levels. However, travellers are still advised to get diphtheria and polio booster shots before visiting these countries. (*Source:* World Health Organization.)

A **sporadic disease** occurs in a random and unpredictable manner, involving several isolated cases that pose no great threat to the population as a whole. Eastern equine encephalitis (EEE) is a sporadic disease in the Americas. **Figure 15.6** contrasts the sporadic nature of EEE with endemic California encephalitis (CE) in certain parts of the United States and epidemic western equine encephalitis (WEE) in the Americas.

Figure 15.5 The spread of cholera. Cholera spread through South and Central America during the period 1991–1992, beginning in Peru in January 1991. It then moved into Colombia and Ecuador. By late 1992 the epidemic had spread to Venezuela, Bolivia, Chile, and Brazil in South America and to Guatemala, Honduras, Panama, Nicaragua, and El Salvador in Central America. (*Map from Morbidity and Mortality Weekly Report*)

The nature and spread of epidemics can vary according to the source of the pathogen and how it reaches susceptible hosts. A **common-source outbreak** is an epidemic that arises from contact with contaminated substances. The term *outbreak* does not evoke the fear that the word *epidemic* can. The spread of a common-source outbreak can typically be traced to a water supply contaminated with fecal material or to improperly handled food. Many individuals become ill quite suddenly. For example, on a cruise in 1994 from San Pedro, California, to Ensenada, Mexico, 586 of 1,589 passengers acquired gastrointestinal illness due to *Shigella flexneri*. Such outbreaks subside quickly once the source of infection is eradicated.

A **propagated epidemic** arises from direct person-to-person contacts (horizontal transmission). The pathogen moves from infected people to uninfected but susceptible individuals. In a propagated epidemic, the number of cases rises and falls more slowly, and the pathogen is more difficult to eliminate, than in a common-source outbreak. Differences between common-source outbreaks and propagated epidemics are illustrated in **Figure 15.7**.

EPIDEMIOLOGIC STUDIES

Collecting frequency data and drawing conclusions are the foundation of any **epidemiologic study**. The British physician John Snow made what may have been the first epidemiologic study. In 1854, Snow investigated the cause of a cholera epidemic then sweeping through London. He eventually traced the source of the epidemic to the Broad Street pump in Golden Square (**Figure 15.8**). He proved that people became infected by drinking water contaminated with human feces.

Since Snow's landmark study, many other investigators have conducted epidemiologic studies to learn more about the spread of disease in populations. Such studies can be *descriptive*, *analytical*, or *experimental*.

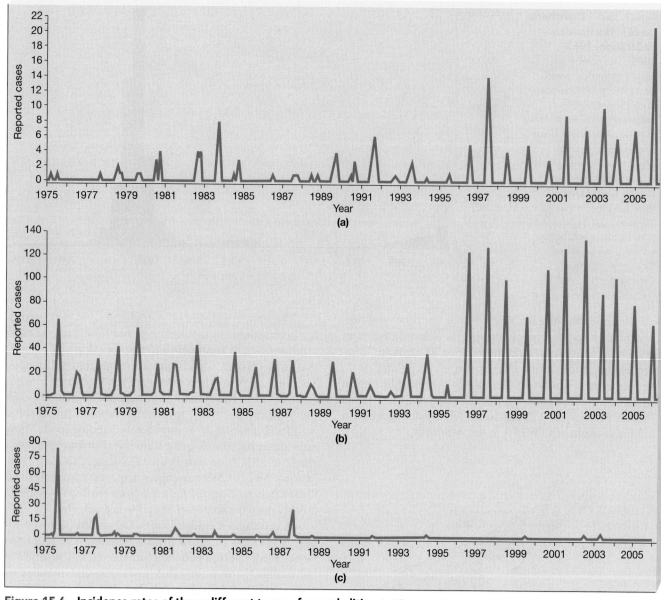

Figure 15.6 Incidence rates of three different types of encephalitis. **(a)** The sporadic pattern of eastern equine encephalitis is compared with the **(b)** endemic pattern of California encephalitis and **(c)** epidemic outbreaks of western equine encephalitis.

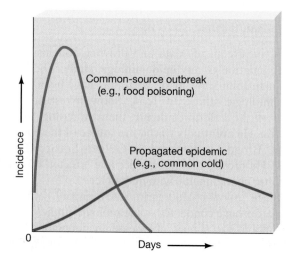

Descriptive Studies

A **descriptive study** is concerned with the physical aspects of an existing disease and disease spread. Such a study records (1) the number of cases of a disease, (2) those segments of the population that were affected, and (3) the locations and time period of the cases. The age, gender, race, marital status, socioeconomic status, and occupation of each patient also are recorded. From careful analysis of data accumulated from several studies, epidemiologists can determine whether people of a

Figure 15.7 Differences in incidence patterns of common-source outbreaks and propagated epidemics. In common-source outbreaks, all cases occur within a fairly short time period after exposure to the single source and then stop, whereas in propagated epidemics, new cases are continually seen.

Figure 15.8 The first epidemiologic study. In 1854, John Snow, a London physician, recorded the locations of cholera cases in the city. He found that they were clustered around the Broad Street pump, which supplied water to the nearby area. He tracked the epidemic's source to contamination of the pump by sewage. When the handle of the pump was removed, the outbreak subsided. Snow's study was the first modern, systematic, scientific epidemiologic study.

⊗ Pumps   .·. Deaths from Cholera

certain age group, males or females, or members of a certain race are particularly susceptible to the disease. Data on marital status or sexual behavior can help show whether the disease is transmitted sexually. Data on socioeconomic status might show that a disease is most easily transmitted among the undernourished or among those living in substandard conditions. Noting the occupations of infected individuals can help epidemiologists trace diseases to certain factories, slaughterhouses, or hide-processing plants. If most cases are among veterinarians, for example, the disease probably is transmitted by the animals they handle.

The geographic distribution of cases is also important, as Snow's study showed. Some of Snow's modern counterparts have traced disease outbreaks to contaminated water supplies, to restaurants where workers are infected with viral hepatitis, and to areas where particular infectious agents thrive.

Multiple sclerosis (MS) is a disease in which one's own immune system attacks the myelin sheath surrounding nerve cells in the spinal cord and brain. It eventually leads to a loss of muscle control and paralysis. Epidemiologically, most cases of MS are found in clusters, in temperate climates, and among Caucasians. In the United States, the incidence of MS increases significantly northward from Mississippi to Minnesota. However, an important determining factor seems to be where a person spent the first 14 years of life. For example, for an individual who grew up in a northern region with a high incidence of MS, moving south later would not reduce the likelihood of acquiring the disease. Various studies have led some investigators to propose that an infectious virus may contribute to the development of MS. Studies of the Faeroe Islands, north of Great Britain, showed them to be free of MS until British troops were stationed there during World War II. Since then, there have been several cycles of MS cases. Only people living on those islands where troops lived have developed MS, with one exception—a case in the house next to the dock where British troops changed ships, but never lived on that island. Thus, genetic factors coupled with an infection by one or more specific microbes increases the likelihood that a person will acquire MS later in life. (Chapter 18 discusses MS and other autoimmune disorders.)

Finally, the time period over which cases appear and the season of the year are important considerations in descriptive studies. To study the role of time in an epidemic, epidemiologists define an **index case** as the first case of the disease to be identified. As we have seen, common-source outbreaks can be distinguished from propagated epidemics by the rate at which the number of

APPLICATIONS

City Tales

"Hamburg persisted in postponing costly improvements to its water supply. . . . [It] drew its water from the Elbe without special treatment. Adjacent lay the town of Altona . . . where a solicitous government installed a water filtration plant. In 1892, when cholera broke out in Hamburg, it ran down one side of the street dividing the two cities and spared the other completely. . . . A more clearcut demonstration of the importance of the water supply in defining where the disease struck could not have been devised. Doubters were silenced; and cholera has, in fact, never returned to European cities, thanks to systematic purification of urban water supplies from bacteriological contamination."

—William H. McNeill, 1976

cases increases and the time required for the epidemic to subside. The season of the year in which epidemics occur may help identify the causative agent. Arthropod-borne infections usually occur in relatively warm weather, and certain respiratory infections usually occur in cold weather, when people often are crowded together indoors. The seasonal nature of encephalitis, which is most prevalent in the fall, is shown in Figure 15.6.

Analytical Studies

An **analytical study** focuses on establishing cause-and-effect relationships in the occurrence of diseases in populations. Such studies can be retrospective or prospective. A *retrospective* study takes into account factors that preceded an epidemic. For example, the investigator might ask patients where they had been and what they had done in the month or so before they became ill. The patients are then compared with a *control group*—individuals in the same population who are unaffected by the disease. Thus, if most patients had hiked in a certain wooded area, had contact with horses, or shared another common activity in which the control group did not participate, that activity might provide a clue to the source of the infection. Several investigations of this sort are described in Berton Roueché's fascinating book *The Medical Detectives* (New York: Truman Talley Books/Plume).

A *prospective* study considers factors that occur as an epidemic spreads. Which children in a population get chickenpox, at what age, and under what living conditions, for instance, are factors used to determine susceptibility and resistance to infection. As the 1993 hantavirus outbreak spread in the U.S. Southwest, epidemiologists had to determine what was causing the disease and how to change living conditions so that the infectious agent would stop spreading.

Experimental Studies

An **experimental study** designs experiments to test a hypothesis, often about the value of a particular treatment.

Such studies are limited to animals or to humans, in which participants are not subjected to harm. For example, an investigator might test the hypothesis that a particular treatment will be effective in controlling a disease for which no accepted cure is available. One group (the experimental group) from a population receives the treatment, and another group (the control group) receives a placebo. A **placebo** (pla-se′-bo) is a nonmedicinal substance that has no effect on the recipient but that the recipient *believes* is a treatment. From the results of the study, the investigator could learn whether the new treatment was effective.

However, in some circumstances control groups might not be used. In the early days of AIDS research, drug treatments were carried out without a control group because the U.S. government considered it unethical to allow control subjects to be deprived of potentially life-saving treatment.

Infectious diseases threaten human populations only when the diseases can be spread or transmitted. Factors important to the spread of disease or of infectious agents include (1) reservoirs of infection, (2) portals by which organisms enter and leave the body, and (3) mechanisms of transmission. In the next section we'll look at each of these factors in more detail.

✓ CHECKLIST

1. Distinguish between morbidity and mortality; between incidence and prevalence; between endemic, epidemic, and pandemic.

2. How do common-source outbreaks differ from propagated epidemics?

RESERVOIRS OF INFECTION

Most pathogens infecting humans cannot survive outside the body of a host long enough to serve as a source of infection. Therefore, sites in which organisms can persist and maintain their ability to infect are essential for new human infections to occur. Such sites are called **reservoirs of infection**. Examples are humans, other animals (including insects), plants, and certain nonliving materials, such as water and soil.

Human Reservoirs

Humans with active infections are important reservoirs because they can easily transmit organisms to other humans. **Carriers**, individuals who harbor an infectious agent without having any observable clinical signs or symptoms, also are important reservoirs. You may want at this time to skip ahead to ◀Chapter 22, p. 687, to read the story of a notorious carrier, Typhoid Mary. Thus, a disease may be transmitted by a person (or animal) with a **subclinical**, or **inapparent**, **infection**—an infection with signs and symptoms too

Children attending day care centers have 2-3 times greater risk of acquiring an infectious disease compared to children kept in the home environment.

mild to be recognized, except by special tests. Many cases of whooping cough (pertussis) in adults, for example, are never diagnosed; yet such *"healthy"* carriers harbor and transmit infectious agents. Adult cases do not have the characteristic "whoop." If you have ever had a really bad cough for weeks on end, you may have had whooping cough, and been spreading it. Infectious diseases are said to be *communicable* if they can be transmitted during the incubation period (before symptoms are apparent) and during recovery from the disease. A *chronic carrier* is a reservoir of infection for a long time after he or she has recovered from a disease. An *intermittent carrier* periodically releases infectious organisms.

Depending on the disease, carriers can discharge organisms from the mouth or nose, in urine, or in feces. Diseases commonly spread by carriers include diphtheria, typhoid fever, amoebic and bacillary dysenteries, hepatitis, streptococcal infections, polio, and pneumonia.

APPLICATIONS

What's in Dust?

Household dust typically contains an amazing assortment of microbes, spores, a few larger organisms, dandruff and other debris from the human body, and other nonliving materials. Spores of *Clostridium perfringens*, which causes gas gangrene in deep wounds, have been found in air conditioner filters. Many different genera of fungi, including *Penicillium*, *Rhizopus* (bread mold), and *Aspergillus*, which can cause swimmer's ear, have been found in dust. Bacterial endospores, fungal spores, plant pollens, insect parts, and hundreds of mites (small organisms related to spiders; see photo) also abound in dust. Mites feed on skin particles sloughed from our body surfaces; fortunately they do not feed on living skin. Dust also contains large quantities of human and animal hair, and sometimes nail clippings as well, all of which have microorganisms on their surfaces. Finally, dust contains nonliving particles from sources as diverse as flaking paint and meteorites.

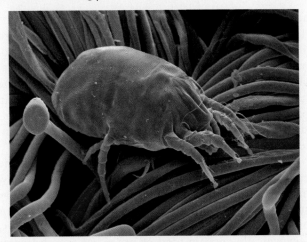

(Andrew Syred/Photo Researchers, Inc.)

International jet travel can increase the risk of introducing infectious agents, for example, cholera, from reservoirs in one region to populations visiting from another region.

Animal Reservoirs

About 150 pathogenic microorganisms can infect both humans and some other animals. In such instances, the animals can serve as reservoirs of infection for humans. Animals that are physiologically similar to humans are most likely to serve as reservoirs for human infections. Therefore, monkeys are important reservoirs for malaria, yellow fever, and many other human infections. Once humans are infected, they, too, can serve as reservoirs for the infections.

Diseases that can be transmitted under natural conditions to humans from other vertebrate animals are called **zoonoses** (zo-o-no'ses; singular: *zoonosis*; Greek *zoon*, "animal," and *nosos*, "disease"). Selected zoonoses are summarized in Table 15.1. Of these diseases, rabies is perhaps the greatest threat in the United States because of the severity of the disease and because both domestic pets and wild animals can serve as reservoirs for the rabies virus. In the United States, there were only 3 cases of rabies in humans during 2003 and one in 2005, but 5,451 cases reported in animals. Where vaccination of dogs and cats is widespread, humans are more likely to acquire rabies from wild animals in which the disease is endemic, such as skunks, raccoons, bats, and foxes. In 1990 to 2005, of 48 cases, 33 were bat-associated.

The larger the animal reservoir, both in the number of species and in the total number of susceptible animals, the more unlikely it is that a disease can be eradicated. This is especially true if the reservoir contains wild animals in which a disease is epidemic. It is impossible to find all the infected animals and to control the disease among them. Even today *Yersinia pestis*, the bacterium responsible for the plague, persists among gophers, ground squirrels, and other wild rodents in the American West and occasionally causes human cases.

Humans, their pets, and other domestic animals similarly serve as reservoirs of infection for wild animals. Distemper, an infectious viral disease in dogs, has spread to and killed many black-footed ferrets. This animal, already an endangered species, is now in even greater jeopardy. Thus, pets and domestic animals should not be allowed to enter wildlife refuges.

Nonliving Reservoirs

Soil and water can serve as reservoirs for pathogens. Soil, for example, is the natural environment of several bacterial species. *Clostridium tetani* (the cause of tetanus) and *C. botulinum* (the cause of botulism) are found everywhere, but especially where animal fecal matter is used as fertilizer. They are part of the normal intestinal microflora of cattle, horses, and some humans. Many fungi, including the organism that causes valley fever, also are common soil inhabitants. Often soil fungi can invade human tissues and cause ringworm, other skin diseases, or systemic infections. Water contaminated by

TABLE 15.1

Selected Zoonoses (with Emphasis on Those That Occur in Pets)		
Disease	**Animals Infected**	**Modes of Transmission**
Bacterial Diseases		
Avian tuberculosis	Birds	Respiratory aerosols
Anthrax	Domestic animals, including dogs and cats	Direct contact with animals, contaminated soil, and hides; ingestion of contaminated milk or meat; inhalation of spores
Brucellosis (undulant fever)	Domestic animals	Direct contact with infected tissues; ingestion of milk from infected animals
Bubonic plague	Rodents	Fleas
Lyme disease	Deer, field mice	Ticks
Leptospirosis	Primarily dogs; also pigs, cows, sheep, rodents, and other wild animals	Direct contact with urine, infected tissues, and contaminated water
Cat scratch fever	Cats	Scratches, bites, and licking
Psittacosis	Parrots, parakeets, and other birds	Respiratory aerosols
Relapsing fever	Rodents	Ticks and lice
Rocky Mountain spotted fever	Dogs, rodents, and other wild animals	Ticks
Salmonellosis	Dogs, cats, poultry, turtles, and rats	Ingestion of contaminated food or water
Viral Diseases		
Equine encephalitis (several varieties)	Horses, birds, and other domestic animals	Mosquitoes
Rabies	Dogs, cats, bats, skunks, and wolves	Bites, infectious saliva in wounds, and aerosols
Lassa fever, hantavirus pulmonary syndrome, hemorrhagic fevers	Rodents	Urine
Fungal Diseases		
Histoplasmosis	Birds	Aerosols of dried infected feces
Ringworm (several varieties)	Cats, dogs, and other domestic animals	Direct contact
Parasitic Diseases		
African sleeping sickness	Wild game animals	Tsetse flies
Tapeworms	Cattle, swine, rodents	Ingestion of cysts in meat or via proglottids in feces
Toxoplasmosis	Cats, birds, rodents, and domestic animals	Aerosols, contaminated food and water, and placental transfer

human or animal feces can contain a variety of pathogens, most of which cause gastrointestinal diseases. Improperly prepared or stored food also can serve as a temporary non-living reservoir of disease. Poorly cooked contaminated meats can be a source of infection with *Salmonella* species and a variety of helminths. Failure to refrigerate foods can lead to growth of microorganisms and to the production of toxins that cause food poisoning. Even with proper refrigeration, helminth larvae remain infectious unless the foods that contain them are cooked thoroughly.

PORTALS OF ENTRY

To cause an infection, a microorganism must enter body tissues. The sites at which microorganisms can enter the body are called **portals of entry**. Common portals of entry include the skin and the mucous membranes of the digestive, respiratory, and urogenital systems (**Figure 15.9**). Although intact skin usually prevents the entry of microorganisms, some enter through the ducts of sweat glands, mammary glands, or through hair follicles. Some fungi invade cells on the skin's surface, and a few can pass on to other tissues. Larvae of some parasitic worms such as the hookworm can bore through the skin to enter other tissues.

Openings to the outside of the body, such as the ears, nose, mouth, eyes, anus, urethra, and vagina, allow microbes to enter. Organisms that infect the respiratory system typically enter in inhaled air, on dust particles, or in airborne droplets. Those that infect the digestive system typically enter in food or water but can also enter from contaminated fingers. As a result of fluid discharge

Figure 15.9 Portals of entry for human pathogens.
Besides those portals common to females and males, additional
portals in females include the placenta and the milk ducts of the
mammary glands.

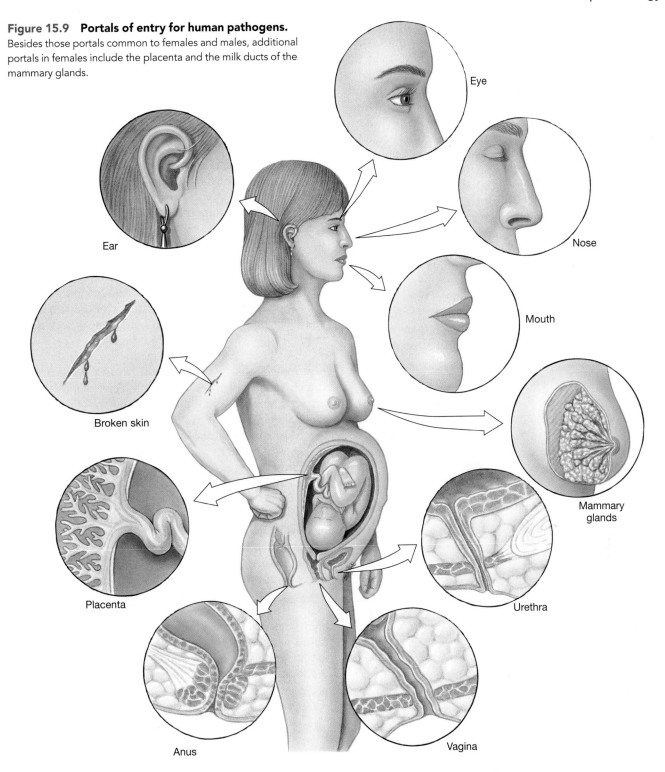

Eye

Nose

Mouth

Ear

Broken skin

Mammary
glands

Placenta

Urethra

Anus

Vagina

from mucous membranes, sexual intercourse provides a portal of entry for urogenital system infections. Some infectious microbes also travel through the urethra and vagina from skin surfaces.

Sometimes microorganisms are introduced directly into the damaged tissues created by bites, burns, injections, and accidental or surgical wounds. *Pseudomonas aeruginosa* infections are especially common in hospitals among burn and surgical patients. Insect bites are a

common portal of entry for a variety of parasitic protozoan and helminthic diseases carried by insects.

Finally, a few infectious organisms, mostly viruses, can cross the placenta from an infected mother and cause an infection in the fetus. Congenital infectious diseases such as cytomegalovirus infection, toxoplasmosis, syphilis, AIDS, and rubella (German measles) are acquired in this way.

Some organisms can enter the body through only a single portal. Others can enter through any of several portals, and their pathogenicity may depend on the portal of entry. For example, many pathogens that cause diseases of the digestive tract cause no disease at all if they happen to enter the respiratory tract. Similarly, most pathogens that cause respiratory diseases do not infect the skin or the tissues of the digestive tract. However, a few organisms cause illnesses no matter where they enter the body—but quite different illnesses, depending on which portal they enter. For example, the plague bacterium (*Yersinia pestis*) causes *bubonic plague*, which has a mortality of about 50% if untreated, and is acquired through a flea bite. But when inhaled into the lungs, the same organism causes *pneumonic plague*, which has a mortality of close to 100%.

Even when a pathogen enters the body, it may not reach an appropriate site to cause an infection. Almost everyone unknowingly carries *Klebsiella pneumoniae* in the pharynx at some time during winter. Why, then, do so few of us come down with pneumonia? The reason is that we do not get pneumonia of the pharynx. The organism must enter the lungs before it can cause disease.

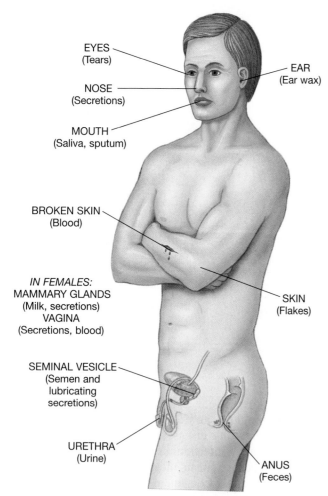

Figure 15.10   Portals of exit for human pathogens.
Besides those portals common to males and females, additional portals in males include the seminal vesicles and in females the mammary glands and vagina.

PORTALS OF EXIT

How infectious agents get out of their hosts is important to the spread of diseases. The sites where organisms leave the body are called **portals of exit** (**Figure 15.10**).

Generally, pathogens exit with body fluids or feces. Respiratory pathogens exit through the nose or mouth in fluids expelled during coughing, sneezing, or speaking. Saliva from dogs, cats, insects, and other animals can transmit infectious organisms. Pathogens of the gastrointestinal tract exit with fecal material. Some of these pathogens are helminth eggs, which are exceedingly resistant to drying and other environmental conditions. Urine and, in males, semen from the urethra carry urogenital pathogens. Semen is an important, though sometimes ignored, means by which pathogens, especially viruses, exit the body. For example, the AIDS virus (HIV) can be carried from the body by white blood cells and sperm present in semen, as can the hepatitis B and C viruses. Hepatitis B and C have now been added to the list of sexually transmitted diseases.

Blood from patients also sometimes contains infectious organisms, such as HIV or hepatitis viruses. As such, blood can be a source of infection for health care workers or others rendering aid to an injured person. Another person's blood should always be considered potentially infectious.

Milk is a significant portal of exit. Pasteurization of milk was important in stopping the spread of tuberculosis (TB) from cows to humans, as was tuberculin testing which removed infected animals from the herd. Breast milk transmits HIV to infants of infected mothers. If the mother has been infected before becoming pregnant, the transmission rate is 15%. If the mother acquires her HIV infection late in pregnancy or during breast feeding, the rate of infant infection jumps to 29%, presumably due to higher rates of viremia in the newly infected mother. In poor countries where infant formula is unavailable or not affordable, many of the babies lucky enough to escape being infected during pregnancy and birth (about a 50% risk) are unlucky enough to be infected by their mother's milk.

A contaminated whirlpool spa, displayed in a large home improvement center resulted in 15 cases of Legionnaires' disease and 2 deaths.

MODES OF DISEASE TRANSMISSION

For new cases of infectious diseases to occur, pathogens must be transmitted from a reservoir or portal of exit to a portal of entry. Transmission can occur by several modes, which we have grouped into three categories: contact transmission, transmission by vehicles, and transmission by vectors. **Figure 15.11** presents an overview of these transmission modes.

Contact Transmission
Contact transmission can be direct, indirect, or by droplets. **Direct contact transmission** requires body contact

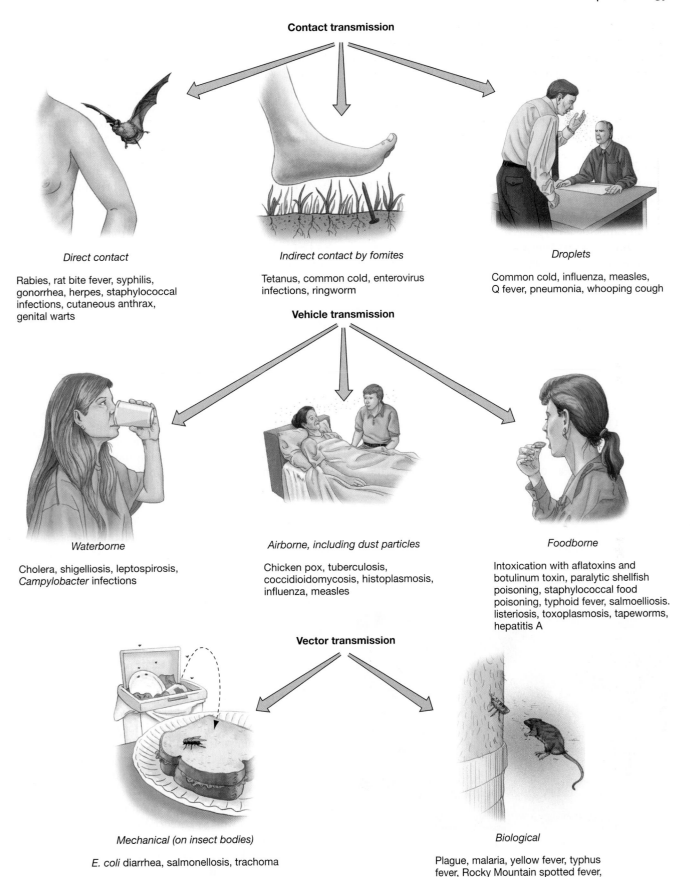

Contact transmission

Direct contact

Rabies, rat bite fever, syphilis, gonorrhea, herpes, staphylococcal infections, cutaneous anthrax, genital warts

Indirect contact by fomites

Tetanus, common cold, enterovirus infections, ringworm

Droplets

Common cold, influenza, measles, Q fever, pneumonia, whooping cough

Vehicle transmission

Waterborne

Cholera, shigelliosis, leptospirosis, *Campylobacter* infections

Airborne, including dust particles

Chicken pox, tuberculosis, coccidioidomycosis, histoplasmosis, influenza, measles

Foodborne

Intoxication with aflatoxins and botulinum toxin, paralytic shellfish poisoning, staphylococcal food poisoning, typhoid fever, salmoelliosis. listeriosis, toxoplasmosis, tapeworms, hepatitis A

Vector transmission

Mechanical (on insect bodies)

E. coli diarrhea, salmonellosis, trachoma

Biological

Plague, malaria, yellow fever, typhus fever, Rocky Mountain spotted fever, Chagas' disease, Lyme disease

Figure 15.11   Modes of disease transmission.

between individuals. Such transmission can be horizontal or vertical. In **horizontal transmission**, individuals pass on pathogens by shaking hands, kissing, touching sores, or having sexual contact. Pathogens can also be spread from one part of the body to another through unhygienic practices. For example, touching genital herpes lesions and then touching other parts of the body, such as the eyes, can spread the infection. Pathogens from fecal matter also can be spread by unwashed hands to the mouth; this is **direct fecal-oral transmission**. In **vertical transmission**, pathogens are passed from parent to offspring in an egg or sperm, across the placenta, in breast milk, or in the birth canal (as can happen with syphilis and gonorrhea).

Indirect contact transmission occurs through **fomites** (fó-mi-tez), nonliving objects that can harbor and transmit an infectious agent. Examples of fomites include soiled handkerchiefs, dishes, eating utensils, doorknobs, toys, bar soap, and money (U.S. paper currency is treated with an antimicrobial agent that reduces transmission of microorganisms).

Droplet transmission, a third kind of contact transmission, occurs when a person coughs, sneezes, or speaks near others **(Figure 15.12)**. **Droplet nuclei** consist of dried mucus, which protects microorganisms embedded in it. These particles can be inhaled directly, can collect on the floor with dust particles, or can become airborne. Droplets that travel less than 1 meter from a sneeze or cough to a host are not classified as airborne.

Transmission by Vehicles

A **vehicle** is a nonliving carrier of an infectious agent from its reservoir to a susceptible host. Common vehicles include water, air, and food. Blood, other body fluids, and intravenous fluids also can serve as vehicles of disease transmission.

Figure 15.12 Droplet transmission. Backlighting reveals a multitude of droplets from nose and mouth during a sneeze. Such dispersal is most important within a radius of about 1 m; however, the smallest particles can be dispersed much farther and kept aloft by air currents. Even a surgical mask will not prevent spread of all droplets. *(Lester V. Bergman/Corbis Images)*

Waterborne Transmission. Although waterborne pathogens do not grow in pure water, some survive transit in water with small quantities of nutrients or in water polluted with fertilizer. Waterborne pathogens usually thrive in and are transmitted in water contaminated with untreated or inadequately treated sewage. Such **indirect fecal-oral transmission** occurs when pathogens from feces of one organism infect another organism. Pathogens have been isolated from public water supplies, semiprivate water supplies (camps, parks, and hotels that have their own water systems), and private water supplies (springs and wells). Polioviruses, enteroviruses, *Giardia*, and *Cryptosporidium* as well as several bacteria are waterborne microbes that infect the digestive system and cause gastrointestinal symptoms. Waterborne infections can be prevented by proper treatment of water and sewage (Chapter 25), although enteroviruses are especially difficult to eradicate from water.

Airborne Transmission. Airborne microorganisms are mainly transients from soil, water, plants, or animals. They do not grow in air, but some reach new hosts through air despite dryness, temperature extremes, and ultraviolet radiation. In fact, dry air actually enhances transmission of many viruses. Pathogens are said to be airborne if they travel more than 1 m through the air. Both airborne pathogens and those suspended in droplets have the best chance of reaching new hosts when people are crowded together indoors. Increased incidence of airborne infections is associated with nearly sealed modern buildings in which temperatures are controlled with heating and air-conditioning systems, and little fresh air enters.

> Global warming may result in tropical diseases such as malaria moving into new regions of the world as their arthropod vectors spread.

Airborne pathogens fall to the floor and combine with dust particles or become suspended in aerosols. An **aerosol** is a cloud of tiny water droplets or fine solid particles suspended in air. Microorganisms in aerosols need not come directly from humans; they can also come from dust particles stirred by dry mopping, changing bedding, or even changing clothing. In the microbiology laboratory, flaming a transfer loop full of bacteria can disperse microorganisms into aerosols.

> A sneeze can blow air out at 100 to 200 miles per hour, releasing over 15,000 viruses per sneeze.

Dust particles can harbor many pathogens. Bacteria with sturdy cell walls, such as staphylococci and streptococci, can survive for several months in dust particles. Naked viruses as well as bacterial and fungal spores can survive for even longer periods.

Hospitalized patients are at great risk of getting airborne diseases because they often have lowered resistance and because former patients may have left pathogens deposited in dust particles. Cleaning floors with a wet mop, wiping surfaces with a damp cloth, and

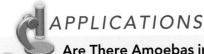

Are There Amoebas in Your Emergency Eyewash Station?

Emergency eyewash stations are intended to supply large volumes of clean water in case of a chemical accident. The American National Standards Institute recommends weekly flushing to keep them clean. However, often water stands in them for long periods of time at room temperature—ideal conditions for growth of biofilms of bacteria and fungi. These microbes can serve as food for water-dwelling amoebas, such as *Acanthamoeba*, which can cause serious eye infections, leading to blindness. Many species of amoebas supply growth factor and an intracellular home for *Legionella pneumophila*, the cause of Legionnaires' disease. Just 100 cells spread by airborne water droplets can cause infection. *Pseudomonas*, a dangerous pathogen often found, can cause considerable tissue destruction and is difficult to treat.

In a 1995 study of 30 dual-spray eyewash units connected to building plumbing, 60% of the units contained more than one genus of amoebas. Regular flushing could not eliminate them. What would you expect in eyewash bottles that are not connected to plumbing for tap water?

carefully unfolding bed linens and towels help reduce aerosols. Masks and special clothing are used in operating rooms, burn wards, and other areas where patients are at greatest risk of infection. Some hospitals also use ultraviolet lights and special air flow devices to prevent exposure of patients to airborne pathogens.

Foodborne Transmission. Pathogens are most likely to be transmitted in foods that are inspected improperly, processed unsanitarily, cooked incompletely, or refrigerated poorly. As with waterborne pathogens, foodborne pathogens are most likely to produce gastrointestinal symptoms.

Transmission by Vectors

As you learned in ◄Chapter 11 (p. 311), **vectors** are living organisms that transmit disease to humans. Most vectors are arthropods such as ticks, flies, fleas, lice, and mosquitoes. However, the mechanism of vector transmission can be mechanical or biological.

Mechanical Vectors. Insects act as *mechanical vectors* when they transmit pathogens passively on their feet and body parts. Houseflies and other insects, for example, frequently feed on animal and, if available, human fecal matter. If they then move on to feed on human fare, they can deposit pathogens in the process. Disease transmission by mechanical vectors does not require that the pathogen multiply on or in the vector.

This method of disease transmission can be prevented simply by keeping these vectors out of areas where food is prepared and eaten. The fly that walked across dog feces in the park should not be allowed to walk across your picnic potato salad. The use of screened areas to keep insects out also reduces disease transmission by mechanical

vectors. Unfortunately, in some poverty-ridden areas of the world, screens are lacking on windows—even those that open into hospital operating rooms!

Biological Vectors. Insects act as *biological vectors* when they transmit pathogens actively; that is, the infectious agent must complete part of its life cycle in the vector before the insect can transmit the infective form of the microbe. Compared with direct transmission through animal bites, the transmission of zoonoses through vectors is much more common. In most vector-transmitted diseases,

"Dog Germs!"

When Snoopy kisses Lucy, she always screams, "Dog germs!" However, Snoopy is the one who should be worried. A human mouth is more likely to contain pathogens than is a dog's mouth. Dog saliva is acidic and a less hospitable environment for microorganisms.

Although kissing probably does little harm, swimming with man's best friend may be more dangerous for humans. *Leptospira* spirochetes ordinarily infect animal kidneys and are shed during urination. One man discovered when he went swimming in a river with his dog ahead of him that this can lead to human leptospirosis. This disease is characterized by fever, headache, and kidney damage. An infected dog is a hazard in a crowded swimming pool. Although vigilant chlorination of pools and vaccination of pets against leptospirosis might prevent them from transmitting the disease, it is best not to allow pets in pools.

A variation on this occurred when some teenagers drove their swamp buggy through water contaminated with urine from leptospirosis-infected deer. Droplets splashed up by the tires apparently infected the passengers.

(Courtesy Jacquelyn G. Black)

such as malaria and schistosomiasis, a biological vector is the host for some phase of the life cycle of the pathogen. Control of zoonoses transmitted by biological vectors can often be achieved by controlling or eradicating the vectors. The spraying of standing water with oil kills many insect larvae. Spraying breeding grounds with pesticides also can be an effective control, at least until the vectors become resistant to the pesticides.

Special Problems in Disease Transmission

Transmission of disease by carriers poses special epidemiologic problems because carriers are often difficult to identify. The carriers themselves usually do not know they are carriers and sometimes cause sudden outbreaks of disease. Depending on the pathogen they carry, carriers can transmit disease by direct or indirect contact or through vehicles such as water, air, or food; they can even be a source of pathogens for vectors.

Over 20 million deaths each year are due to infectious disease.

Another special transmission problem arises with people who have *sexually transmitted diseases (STDs)*. Such diseases are most often transmitted by direct sexual contact, including kissing, but some can be transmitted by oral or anal sex. STDs present epidemiologic problems because infected individuals sometimes have contact with multiple sexual partners. In fact, the incidences of AIDS, genital herpes, genital warts, syphilis, and *Chlamydia* infections are rapidly increasing.

Zoonoses are another epidemiologic problem. They can be transmitted by direct contact, as when humans get rabies from the bite of an infected domestic or wild animal. An oral vaccine for rabies has been developed, to be administered to wildlife.

DISEASE CYCLES

Many diseases occur in cycles. For years or even decades, only a few cases are seen, but then many cases suddenly appear in epidemic or pandemic proportions. Let's look at one example. Bubonic plague—or the Black Death, as it was called—has occurred in pandemic outbreaks followed by recurrent cycles for centuries. Between A.D. 543 and 548, the disease spread from India or Africa, through Egypt, to Constantinople (now Istanbul, Turkey), where it killed 200,000 people in only 4 months. The disease quickly spread via fleas on rats aboard ships that were traveling to Europe and the Mediterranean basin. About a half century later it appeared in China, with equally devastating results. After the initial outbreaks, it occurred in cycles of 10 to 24 years over the next two centuries.

All of Europe breathed a sigh of relief for about the next 500 years, which were plague-free. But in 1346, a second pandemic, worse than the first, afflicted North Africa, the Middle East, and most of Europe. Nearly one-third of the population of Europe died, and in many cities three-fourths of the population lost their lives to the dreaded disease. Then cyclic recurrences claimed more lives in epidemics of the seventeenth century in England and the eighteenth century in France. Near the beginning of the twentieth century, a pandemic killed more than a million people in India and spread to many parts of the world, including San Francisco.

Thus, cyclic diseases pose special epidemiologic problems. Epidemiologists still cannot predict when one will break out and reach epidemic proportions. It is difficult to be prepared to treat sudden, large increases in the incidence of a disease and nearly impossible to persuade people to be immunized against a disease they have never seen.

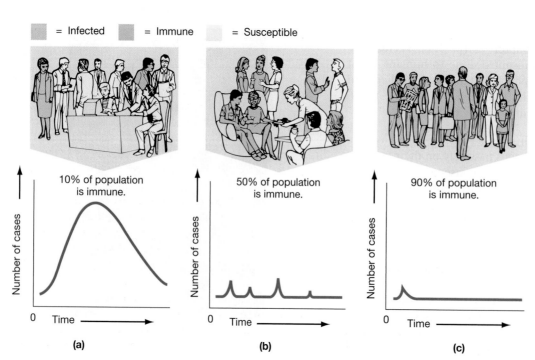

■ = Infected   ■ = Immune   ■ = Susceptible

10% of population is immune. *(a)*

50% of population is immune. *(b)*

90% of population is immune. *(c)*

Number of cases — Time

Figure 15.13 Herd immunity. (a) When the percentage of immune persons in a population is low, it is likely that susceptible individuals will be exposed to the disease. **(b** and **c)** As the percentage of immune persons increases, it becomes increasingly less likely that a susceptible individual will be exposed to the disease.

APPLICATIONS

Plague in the 1990s

In August 1994, outbreaks of bubonic and pneumonic plague occurred in parts of India where the bacterium was epidemic in rats. Widespread panic erupted, and health care workers worldwide became deeply concerned that the disease could become pandemic. By October 1994 the number of suspected cases of bubonic and pneumonic plague reported in India had reached 693, with 56 deaths. Due to rapid identification and control measures, the epidemic subsided.

With jet travel, it is easy for a disease to become pandemic almost overnight. Thus, the reports from India prompted U.S. health authorities to develop an emergency plan for detecting and managing any suspected cases arriving by plane. In fact, 11 air travelers arriving in New York, suspected of having the plague, were later found to be suffering from other diseases. This time we were lucky. The available vaccine is not completely effective, and many rats have become resistant to the best available rat poisons.

HERD IMMUNITY

An important factor in cyclic disease is **herd immunity** (or *group immunity*), which is the proportion of individuals in a community or population who are immune to a particular disease. If herd immunity is high—that is, if most of the individuals in a population are immune to a disease—then the disease can spread only among the small number of susceptible individuals in the population (**Figure 15.13**). Even when a member of the population becomes infected, the likelihood that that person will transmit the disease to others is small. Thus, a sufficiently high herd immunity protects the entire population, including its susceptible members.

It is easy to see, then, why public health officials want to maintain the highest possible herd immunity, especially against common cyclic diseases. They encourage parents to have children immunized against measles and other communicable diseases. In many cities in the United States, children are required to be immunized against measles before they can start school. As a result, about 95% of elementary school-age children are immune to measles. Although high school and college students who have neither had measles nor received vaccine are protected by herd immunity, many school systems and some colleges and universities are requiring immunization against measles for all students, whatever their age.

The loss of herd immunity could lead to a reemergence of a disease. The rise in diphtheria incidence in the Russian Federation portion of the former Soviet Union is due in part to a lack of childhood vaccinations. Herd immunity in these populations has dwindled. Some people fear that herd immunity to smallpox is bound to drop now that vaccinations have ceased as a result of the eradication of the smallpox virus. If the smallpox virus were to appear again, it could be devastating—few people would be immune to the disease. Smallpox vaccine also cross-protects against the monkeypox virus. Human herd immunity to monkeypox is also dwindling—a problem in some areas.

✓ CHECKLIST

1. Give examples of a nonliving reservoir and a nonliving vehicle of transmission.
2. Distinguish between biological and mechanical vectors.
3. What is herd immunity?

CONTROLLING DISEASE TRANSMISSION

Several methods are currently available for full or partial control of communicable diseases. They include *isolation*, *quarantine*, *immunization*, and *vector control*.

In **isolation**, a patient with a communicable disease is prevented from having contact with the general population. Isolation generally is accomplished in a hospital. There appropriate procedures can be carried out to reduce the spread of disease among susceptible individuals and to prevent the spread of disease in the general population. In all, there are seven categories of isolation (**Table 15.2**). Strict isolation makes use of all available procedures to prevent transmission of organisms or virulent infections to medical personnel and visitors. Even medical researchers working with highly pathogenic strains of microbes must use high levels of isolation and special laboratories equipped to contain such agents (**Figure 15.14**).

Quarantine is the separation of "healthy" human or animal carriers from the general population when they have been exposed to a communicable disease. Quarantine prevents spread of the disease during its incubation period. Although it is one of the oldest methods of controlling communicable diseases, it is now used mainly for serious diseases such as cholera and yellow fever. Quarantine differs from isolation in two ways: (1) It is applied to healthy people who were exposed to a disease during the incubation period, and (2) it pertains to limiting the movements of such people and not necessarily to precautions during treatment. Quarantine is rarely used today because it is very difficult to carry out. To ensure that no infected persons spread a disease, everyone who had been exposed to it would have to be quarantined for the disease's incubation period. This would mean, for example, that all travelers returning to the United States from a region of the world where cholera is endemic would have to be quarantined for 3 days in accommodations provided at airports and seaports—not likely to be a popular idea.

Large-scale *immunization* programs are an extremely effective means of controlling communicable diseases for which safe vaccines are available. Such programs greatly increase herd immunity and thus greatly decrease human suffering and deaths from infectious diseases. In the United States, immunizations have nearly eradicated polio, measles, mumps, diphtheria, and

TABLE 15.2

A Summary of Important Isolation Procedures						
Isolation Categories						
Strict	**Contact**	**Respiratory**	**Tuberculosis**	**Enteric Precautions**	**Drainage/ Secretion Precautions**	**Blood and Body Fluid Precautions**
Visitors must check in at nursing station before entering patient's room						
Yes	Yes	Yes	Yes	Yes	Yes	Yes
Hands must be washed on entering and on leaving patient's room						
Yes	Yes	Yes	Yes	Yes	Yes	Yes
Gowns must be worn by personnel and visitors						
Yes	Yes	No	Yes	Only for direct patient contact	Only if soiling is likely	Only if soiling is likely
Masks must be worn by personnel and visitors						
Yes	Yes	Unless not susceptible to disease	Only if coughing	No	No	No
Gloves must be worn by personnel and visitors						
Yes	Only for direct patient contact	No	No	Only for direct contact with patient or feces	Only for direct contact with lesion site	Only for direct contact with lesion site
Private room required with door closed						
Yes	Yes	Yes	Yes	Only for children	No	If hygiene is poor
Examples of Diseases						
Pneumonic plague, rabies, diphtheria, disseminated herpes, zoster, Lassa fever, chickenpox, draining *Staphylococcus aureus* wounds	Severe noninfected dermatitis, noninfected burns	Measles, mumps, rubella, pertussis	Pulmonary tuberculosis	Typhoid fever, cholera, salmonellosis, shigellosis, hepatitis	Bubonic plague, gas gangrene localized herpes puerperal sepsis	AIDS, hepatitis B

whooping cough. Unfortunately, as the incidence of these diseases becomes very small, people become complacent about getting immunized. Such complacency can lead to a sufficient decrease in herd immunity, resulting in outbreaks of vaccine-preventable diseases.

Vector control is an effective means of controlling infectious diseases if the vector, such as an insect or rodent, can be identified and its habitat, breeding habits,

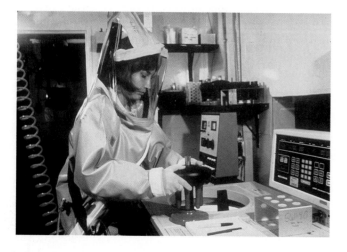

Figure 15.14 Biosafety level 4 lab. Scientists who work with very dangerous, and often easily disseminated, microbes do so in an isolation lab. This laboratory worker is in the highest level of isolation, a biosafety level 4 lab. Extreme precautions, such as the use of special ventilation systems and the wearing of "space suits," are required of personnel to avoid contact with microbes and to prevent the escape of the microbes from the facility. *(Centers for Disease Control and Prevention CDC).*

APPLICATIONS

Yummy Vaccines

The Oral Vaccination Program uses aerial distribution of an oral vaccine to control rabies in wildlife. A dose of vaccine is inserted in the hollow center of a bait composed of dog food or fish meal. The vaccine is a recombinant vaccine that has been shown to be safe and effective in almost 60 species of mammals and birds. The vaccine does not cause rabies in humans or animals and produces a protective immune response against the infection. To control rabies in red foxes, a massive bait drop was conducted over several years in Ontario, Canada. The bait program finally broke the rabies cycle in red foxes, and to date only four cases of rabies have been found. According to Ontario public health officials, if the trend continues, red fox rabies will be eliminated early this decade. Similar programs, targeted for coyotes and gray foxes, are ongoing in the United States.

and feeding behavior determined. Places where a vector lives and breeds can be treated with insecticides or rodenticides. Window screens, mosquito netting, insect repellents, and other barriers can be used to protect humans from becoming victims of the bites of feeding vectors. Unfortunately, vectors have their own defenses. Some escape or become resistant to pesticides or make their way through barriers.

Malaria control in the United States was accomplished mainly by the control of mosquito vectors. Consequently, Americans have little herd immunity to malaria because most have never had the disease. Since the 1940s new cases of malaria in the United States have occurred mainly among people infected in other countries (**Figure 15.15**). As long as infected individuals do not reintroduce the parasite into the mosquito population, a malaria epidemic in the United States is most unlikely despite low herd immunity. In recent years, however, laborers from

areas where malaria still is common have brought the parasite northward, and the disease has become endemic in some parts of California. In 2006 there were 1,245 cases of malaria in the United States.

Although communicable diseases are theoretically preventable, some still have a high incidence in every human population. In countries with relatively high living standards, the common cold and many sexually transmitted diseases occur with great frequency. In countries with lower standards of living—especially those in the tropics—the prevalence of malaria and a variety of other diseases, including some nearly eradicated in other countries, is extremely high. And certainly AIDS has become a worldwide threat that now extends beyond the special high-risk groups with which it was once associated.

PUBLIC HEALTH ORGANIZATIONS

In the United States and many other countries, the importance of controlling infectious diseases and reducing other health hazards has led to the creation of public health agencies. City and county health departments provide immunizations, inspect restaurants and food stores, and work with other local agencies to ensure that water and sewage are treated properly. State health departments deal with problems that extend beyond cities and counties. They often do laboratory tests, such as identifying rabies in animals, and hepatitis and toxins in water.

The Centers for Disease Control and Prevention

In the United States, the federal government operates the U.S. Public Health Service (USPHS), which has several branches. Of these branches, the *Centers for Disease Control and Prevention (CDC)* in Atlanta, Georgia (**Figure 15.16**), has major responsibilities for the control and prevention of infectious diseases and other

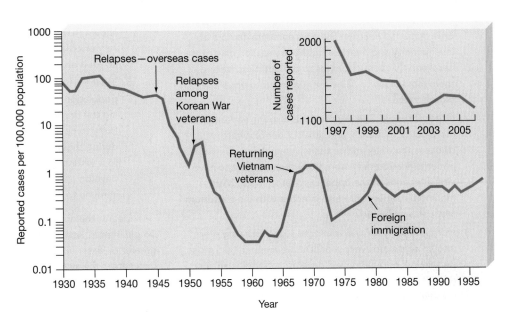

Figure 15.15 The incidence rate of malaria, as related to sources of infection into the United States. Also shown are recent total numbers of cases reported to CDC.

preventable conditions. Among the microbiology-related activities of the CDC are the following:

1. providing guidelines for occupational health and safety, quarantines, tropical medicine, co-operative activities with national agencies in other countries and with international agencies, and public health education;

2. making recommendations to the medical community about the use of antibiotics, especially for the treatment of diseases caused by antibiotic-resistant organisms;

3. storing infrequently used drugs and providing them to physicians who encounter patients with tropical parasitic diseases and other diseases rarely seen in the United States;

4. making recommendations regarding the administration of vaccines—which should be used, who should receive them, and at what ages.

PUBLIC HEALTH

The Meningitis Belt

Like many other diseases discussed in ◀Chapter 16, meningococcal meningitis, which is caused by *Neisseria meningitidis*, occurs in roughly 5- to 12-year cycles. During the 1960s, an estimated 3 million cases occurred in China. In mid-April of 1988, patients were entering the hospital in N'Djamena, the capital city of the north central African nation Chad, at a rate of 250 per day. Out in the countryside, thousands of others who were unable to reach medical facilities suffered and died uncounted. In 1989, 40,000 cases were diagnosed in an Ethiopian epidemic. In 1996, Africa had the largest outbreak or epidemic meningitis in history: 250,000 cases, 25,000 of which were fatal. Yet the United States does not have such epidemics. Looking specifically at Africa, what epidemiological factors cause meningitis epidemics to sweep across a broad belt of central Africa (see the map) in these cycles?

In considering such multiyear cycles, epidemiologists also must consider the seasonal cycles in Africa. Each year, whether an epidemic year or one of the years in between, meningitis outbreaks occur only during the dry season months of January through June. Then, as soon as the rains begin, the number of new cases drops to zero—only to climb again at the start of the dry season. However, it can be shown that the causative meningococci are transmitted year-round. How can the seasonality of meningitis outbreaks be explained?

1. *Environmental Factors* The environmental factors of low humidity and heat during the dry season may cause nasal and throat membranes to dry out, thus making them more likely to allow meningococci to enter the bloodstream.

2. *Other Diseases* The dry season is also the season of more colds, influenza, and other upper respiratory diseases that may add to tissue vulnerability. Meningitis patients are 23 times more likely than average to have viral infections of the upper respiratory tract. Many meningitis patients are also infected with the bacterium *Mycoplasma hominis*.

3. *Herd Immunity* Why doesn't the seasonal swing erupt into an epidemic every year? During an epidemic, most people in a population will develop antibodies against the prevalent strain of causative bacteria. Each year, however, new children are born, and they lack such immunity. Eventually the herd immunity drops so low as to allow another epidemic to occur.

4. *Strain Virulence* During the years between the epidemics, mutant strains will arise. If one of these new strains is more virulent than previous strains, it could initiate a new epidemic. Studies of epidemics have shown that usually only one strain of meningococci is responsible for any one epidemic. So virulence of strain is an important factor.

5. *Timing* The timing of entry of the mutant strain into the population is also crucial. If the mutant stain enters during the dry season, those who are exposed to it are likely to develop meningitis. But if a new strain enters during the rainy season, people will become exposed to it without developing the disease. Instead, they will develop antibodies against the new strain and will be immune to it when the following dry season begins.

Clearly, the epidemiology of meningitis epidemics is not a simple matter. Several factors are at work, and each may have a greater or lesser effect on an outbreak of a given epidemic. Unfortunately, the available meningitis vaccine confers only short-term immunity; it does not last from one epidemic to the next. However, it can be used during an epidemic to halt its spread.

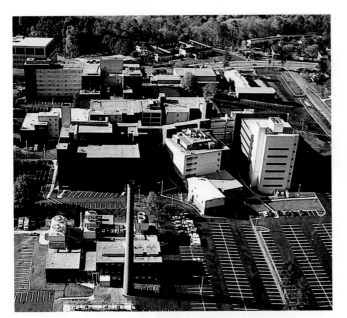

Figure 15.16 CDC headquarters in Atlanta, Georgia.
(Courtesy Centers for Disease Control and Prevention).

The CDC carries out epidemiologic studies, which are published as the *Morbidity and Mortality Weekly Report (MMWR)*. This publication provides statistics for specific diseases in various parts of the United States and the world (Figure 15.17). Other CDC periodical reports are *Recommendations and Reports* and *Surveillance Summaries*, which provide in-depth coverage of specific issues, many related to infectious diseases.

The World Health Organization

The *World Health Organization (WHO)* is an international agency based in Geneva, Switzerland, that coordinates and sets up programs to improve health in more than 100 member countries. Its basic objective is that all peoples attain the highest possible level of health. Specific activities are carried out by six regional organizations in Africa, the Eastern Mediterranean, Europe, Southeast Asia, the Western Pacific, and the Americas. WHO works closely with the United Nations on population control, management of food supplies, and various other scientific and educational activities.

WHO sets health standards for international disease control; helps developing nations establish effective control and immunization programs; collects, analyzes, and distributes health data; and maintains surveillance of potential epidemics (published in WHO's *Weekly Epidemiological Record*). It also provides training and research programs for health personnel and information for individuals (Figure 15.18). The agency has helped more than 100 countries in immunizing against diphtheria, measles, whooping cough, poliomyelitis, tetanus, and tuberculosis and hopes eventually to eradicate measles worldwide. It conducts research and training to combat widespread tropical diseases such as leprosy, malaria, and several diseases caused by helminths.

Take another look

WHO has been instrumental in coordinating the eradication of smallpox worldwide.

NOTIFIABLE DISEASES

Cooperation among state and national health organizations in the United States has led to the establishment of a list of **notifiable diseases**, which are infectious diseases that are potentially harmful to the public's health and must be reported by physicians. As of 2003, 58 infectious diseases were listed as notifiable at the national level (Table 15.3). On the basis of CDC suggestions, each year the Council of State and Territorial Epidemiologists (CSTE) adds diseases to, or deletes diseases from, the list. If a specific disease shows a decline in incidence, it may be removed from the list.

Although reporting infectious diseases is mandatory only at the state level, the reporting of notifiable diseases at the national level is intended to accomplish two things: (1) to ensure that public health officials learn of diseases that jeopardize the health of populations, and (2) to provide consistency and uniformity in the reporting of those

Figure 15.17 The *Morbidity and Mortality Weekly Report*, issued weekly by the CDC. It reports new and unusual cases and trends. It also lists, by state, the number of cases of notifiable diseases recorded that week, the number recorded for the same week in the previous year, and the cumulative totals. It is available online at www.cdc.gov. *(Centers for Disease Control and Prevention CDC)*

(a)

(b)

(c)

(d)

Figure 15.18 Some typical activities of the World Health Organization. (a) Eye tests for the prevention of blindness are given high in the Andes Mountains of Peru. *(Courtesy D. Espinoza, World Health Organization)* (b) A medical assistant travels by horseback to treat patients in remote villages of China. *(Courtesy Chang Hogen, World Health Organization)* (c) Ambulance boat transporting a patient to a hospital, Myanmar. *(Courtesy Ko San Win, Pan American Health Organization/World Health Organization)* (d) Health education classes, Guatemala. *(Courtesy Carlos Gaggero, Pan American Health Organization/World Health Organization).*

TABLE 15.3

Infectious Diseases Designated as Notifiable at the National Level—United States, 2003			
AIDS	Giardiasis	Malaria	Smallpox
Anthrax	Gonorrhea	Measles (rubeola)	Streptococcal disease, invasive, Group A
Arbovirus infection	*Haemophilus influenzae*	Meningococcal infection	
Botulism	Hantavirus, pulmonary syndrome*	Mumps	Streptococcal pneumonia, drug-resistant invasive disease
Brucellosis*		Pertussis	
Chancroid	Hemolytic uremic syndrome, postdiarrheal	Plague	Streptoccal toxic-shock syndrome
Chlamydia trachomatis, genital infection	Hepatitis A	Poliomyelitis	Syphilis
Cholera	Hepatitis B	Psittacosis	Syphilis, congenital
Coccidiomycosis (regional)	Hepatitis C	Q fever	Tetanus
Congenital rubella syndrome	Hepatitis, unspecified	Rabies, animal	Toxic shock syndrome
Cryptosporidiosis	HIV infection, adult	Rabies, human	Trichinosis
Cyclosporiasis	HIV infection, pediatric	Rocky Mountain spotted fever (typhus fever, tickborne)	Tuberculosis
Diphtheria	Influenza	Rubella	Tularemia*
Encephalitis: Eastern equine St. Louis, West Nile, Western equine	Legionellosis	Salmonellosis	Typhoid fever
	Leprosy (Hansen's disease)	SARS-associated coronavirus disease	Vancomycin-resistant *Staphylococcus aureus*
Ehrlichiosis	Listeriosis	Shigellosis	Varicella (chickenpox, shingles)
Escherichia coli O157:H7	Lyme disease		Yellow fever

*Not notifiable in all states.
Source: Morbidity and Mortality Weekly Report.

diseases. Various kinds of information about notifiable diseases in the United States are available from the CDC; samples of this information are provided in **Tables 15.4A** and **15.4B**. It is sobering to realize that the majority of deaths from infectious disease could have been prevented by simple, low-cost means **(Figure 15.19)**.

NOSOCOMIAL INFECTIONS

A **nosocomial** (nos-o-ko′me-al) **infection** is an infection acquired in a hospital or other medical facility. The term *nosocomial* is derived from the Greek words *nosos*, meaning "disease," and *komeo*, meaning "to take care of." Nosocomial diseases are acquired during medical treatment. Although many such infections occur in patients, infections acquired at work by staff members also are considered nosocomial infections.

Among patients admitted to American hospitals each year, about 2 million (10%) acquire an infection that increases the risk of death, the duration of the hospital stay, and the cost of treatment by close to $5 billion per year in hospital costs alone. Of these, over 90,000 people per year die of their nosocomial infections. Curiously, nosocomial infections are largely a product of advances in medical treatment. Intravenous, urinary, and other catheters; invasive diagnostic tests; and complex surgical procedures increase the likelihood that pathogens will enter the body. Intensive use of antibiotics contributes to the development of resistant strains of pathogens. And therapies to reduce the chances of rejection of transplanted organs impair the immune response to pathogens. Despite the risk of nosocomial infections, the medical treatments now available save far more patients than are lost to such infections.

THE EPIDEMIOLOGY OF NOSOCOMIAL INFECTIONS

The epidemiology of nosocomial infections, like the epidemiology of diseases acquired in the community, considers sources of infection, modes of transmission, susceptibility to infection, and prevention and control. In addition, it focuses on medical procedures that increase the risk of infection, the sites at which infections often occur, and the correlation between procedures and sites of infection.

Sources of Infection

Nosocomial infections can be exogenous or endogenous. **Exogenous infections** are caused by organisms that enter the patient from the environment. The organisms can come from other patients, staff members, or visitors. They can also be passed on by insects (ants, roaches, flies) from fomites (toilet, trash can) to patients. Other inanimate objects, such as equipment used in respiratory or intravenous therapy, catheters, bathroom fixtures and soap,

and water systems, also can be a source of exogenous infections. Some nosocomial infections have even been traced to disinfectants such as quaternary ammonium compounds, to which certain organisms are resistant. **Endogenous infections** are caused by opportunists among the patient's own normal microflora. Opportunists are most likely to cause infection if the patient has lowered resistance or if normal microflora that compete with pathogens have been eliminated by antibiotics.

A small group of organisms, including *Escherichia coli*, *Enterococcus* species, *Staphylococcus aureus*, and *Pseudomonas* species, are responsible for about half of all nosocomial infections **(Figure 15.20)**. These organisms are particularly likely to cause such infections because they are ubiquitous (present everywhere) and can survive outside the body for long periods. In addition, some strains of these organisms are resistant to many antibiotics; methicillin- and vancomycin-resistant staphylococci and carbapenem-resistant *Pseudomonas aeruginosa* are especially problematic.

Susceptibility and Transmission

Compared with the general population, patients in hospitals are much more susceptible to infection; that is, they are **compromised hosts**. Many patients have breaks in the skin from lesions, wounds (surgical and accidental), or bed sores. Some also have breaks in mucous membranes that line the digestive, respiratory, urinary, or reproductive system. The lack of intact skin and mucous membranes provides easy access for infectious organisms. Also, most patients are debilitated; their resistance to infectious organisms is lower than normal. Patients undergoing organ transplants receive immunosuppressant drugs, and patients with AIDS and other disorders of the immune system also have reduced resistance. Factors that contribute to host resistance are discussed in Chapters 16 and 17.

Theoretically, nosocomial infections can be transmitted by all modes of transmission that occur in the community. However, direct person-to-person transmission between an infected patient, staff member, or visitor and noninfected patients; indirect transmission through equipment, supplies, and hospital procedures; and transmission through air are most common in hospitals **(Figure 15.21)**. Some organisms can be transmitted by more than one route.

Universal Precautions

In 1988, the CDC, concerned about the possibilities that the AIDS virus would be transmitted in the health care setting, issued guidelines to reduce the risks. These guidelines, called **Universal Precautions**, are summarized in Appendix D. Some hospitals and other medical facilities choose to exercise even more caution than the CDC recommends. A shortened list of these precautions is given in **Table 15.5**. Universal Precautions apply to *all* patients, not just those infected with the viruses that cause AIDS or hepatitis B—hence the term *universal*.

TABLE 15.4A

Provisional Cases of Selected Notifiable Diseases, United States, Weeks Ending December 23, 2006, and December 24, 2005 (51st Week)*

Reporting area	AIDS		Chlamydia†			Gonorrhea			Lyme Disease			Tuberculosis		
	Cum. 2005‡	Cum. 2004	Current Week	Cum. 2006	Cum. 2005	Current Week	Cum. 2006	Cum. 2005	Current Week	Cum. 2006	Cum. 2005	Current Quarter	Cum. 2007	Cum. 2006
UNITED STATES	30,568	40,725	8,135	934,337	941,275	2,277	324,809	325,457	114	16,835	21,211	2,008	10,878	13,250
NEW ENGLAND	1,141	1,327	644	33,192	32,222	113	5,625	5,816	24	2,903	3,977	33	275	445
Connecticut	423	638	183	9,937	9,825	46	2,363	2,509	16	1,687	1,057	16	81	95
Maine	19	48	—	2,189	2,214	—	127	140	6	287	247	5	13	17
Massachusetts	561	451	421	15,183	14,235	66	2,409	2,503	—	33	2,333	—	142	274
New Hampshire	26	42	34	1,989	1,813	1	179	174	—	558	250	5	14	4
Rhode Island	105	132	—	2,851	3,200	—	484	431	—	235	37	2	16	47
Vermont	7	16	6	1,043	935	—	63	59	2	103	53	5	9	8
MID-ATLANTIC	6,597	10,045	1,580	117,717	116,683	328	31,526	33,647	67	9,462	12,039	508	2,016	2,096
New Jersey	956	1,766	—	16,110	18,852	—	4,580	5,632	—	1,918	3,345	67	437	485
New York (Upstate)	891	1,986	695	24,749	23,586	100	6,128	6,861	40	4,025	4,015	64	239	302
New York City	3,522	4,875	561	37,789	38,053	128	9,561	10,240	—	164	399	278	1,002	984
Pennsylvania	1,228	1,418	324	39,069	36,192	100	11,257	10,914	27	3,355	4,271	99	338	325
E. N. CENTRAL	2,929	3,217	804	152,330	161,041	320	62,642	65,604		1,444	1,730	254	1,078	1,319
Illinois	1,504	1,476	—	49,558	49,744	—	18,912	19,705	—	—	127	124	493	589
Indiana	348	389	—	18,820	19,756	—	8,285	7,970	—	21	30	18	111	146
Michigan	439	612	804	35,129	28,842	320	15,104	11,804	—	54	61	50	179	246
Ohio	518	585	—	30,717	42,527	—	14,124	20,337	—	42	57	45	226	260
Wisconsin	120	155	—	18,106	20,172	—	6,217	5,788	—	1,327	1,455	17	69	78
W. N. CENTRAL	690	801	180	57,208	57,675	41	18,113	18,382	1	845	944	108	465	490
Iowa	72	63	—	7,894	7,231	—	1,760	1,584	—	87	91	7	31	55
Kansas	94	113	79	7,029	7,150	23	1,984	2,481	—	5	3	12	97	61
Minnesota	176	216	—	11,022	12,035	—	2,844	3,437	1	729	829	51	206	199
Missouri	299	323	43	21,891	22,047	12	9,672	9,298	—	12	15	31	101	108
Nebraska	27	58	—	5,191	4,929	—	1,368	1,113	—	11	4	4	18	45
North Dakota	9	17	—	1,570	1,624	—	120	126	—	—	126	—	—	6
South Dakota	13	11	58	2,613	2,659	6	365	343	—	1	2	3	12	15
S. ATLANTIC	9,183	12,273	1,743	183,226	172,481	685	81,550	76,594	22	1,909	2,262	454	2,294	2,880
Delaware	134	157	81	3,551	3,343	31	1,462	895	—	465	643	2	30	25
District of Columbia	474	990	—	2,805	3,649	—	1,825	2,111	—	59	8	9	61	52
Florida	3,963	5,596	876	48,030	42,402	418	22,706	19,754	2	59	46	153	876	1,094
Georgia	1,701	1,508	—	32,856	31,420	109	16,504	14,846	—	7	6	56	442	504
Maryland	1,370	1,449	280	17,733	18,056	—	6,461	6,949	9	942	1,223	23	183	283
North Carolina	636	1,126	—	32,609	30,768	—	16,625	14,786	1	30	46	105	336	329
South Carolina	413	745	—	18,983	18,137	—	8,545	8,437	—	18	21	29	77	211
Virginia	441	613	506	23,634	21,828	127	6,457	8,056	10	315	252	72	268	355

Reporting area	AIDS Cum. 2005‡	AIDS Cum. 2004	Chlamydia† Current Week	Chlamydia† Cum. 2006	Chlamydia† Cum. 2005	Gonorrhea Current Week	Gonorrhea Cum. 2006	Gonorrhea Cum. 2005	Lyme Disease Current Week	Lyme Disease Cum. 2006	Lyme Disease Cum. 2005	Tuberculosis Current Quarter	Tuberculosis Cum. 2007	Tuberculosis Cum. 2006
West Virginia	51	89	—	3,025	2,878	—	966	758	—	14	17	5	21	26
E. S. CENTRAL	1,546	1,884	535	72,150	68,486	204	29,230	27,604	—	36	36	142	615	742
Alabama	385	464	63	20,408	16,721	34	9,389	9,245	—	16	3	51	196	216
Kentucky	198	217	—	8,854	8,165	—	3,250	2,871	—	7	5	22	81	124
Mississippi	288	429	—	18,341	20,756	—	7,241	6,989	—	1	—	35	117	103
Tennessee	675	774	472	24,547	22,844	170	9,350	8,498	—	12	28	34	221	299
W. S. CENTRAL	3,543	4,684	378	104,636	107,396	181	45,547	43,956	—	19	77	47	1,283	1,793
Arkansas	173	185	—	7,764	8,353	—	4,046	4,421	—	—	5	21	103	114
Louisiana	650	1,004	4	12,115	16,836	6	7,646	9,332	—	—	3	—	—	—
Oklahoma	229	195	374	12,659	11,248	175	4,797	4,454	—	—	—	26	134	144
Texas	2,491	3,300	—	72,098	70,959	—	29,058	25,739	—	19	69	—	1,048	1,535
MOUNTAIN	1,172	1,393	457	49,711	62,176	81	11,266	13,445	—	27	21	82	389	595
Arizona	473	508	457	18,692	20,816	81	4,604	4,842	—	7	8	64	240	281
Colorado	260	294	—	5,480	15,220	—	2,067	3,153	—	1	—	8	36	101
Idaho	15	19	—	2,333	2,713	—	139	116	—	7	2	—	—	23
Montana	15	7	—	2,459	2,205	—	186	144	—	—	—	—	—	10
Nevada	236	303	—	5,222	7,295	—	1,653	2,877	—	3	3	45	45	112
New Mexico	115	182	—	9,402	8,256	—	1,667	1,522	—	2	3	31	31	39
Utah	55	64	—	4,815	4,509	—	834	708	—	6	2	9	34	29
Wyoming	3	16	—	1,308	1,162	—	116	83	—	1	3	1	3	—
PACIFIC	3,767	5,101	1,814	164,167	163,115	324	39,310	40,409	—	190	125	380	2,463	2,890
Alaska	25	48	—	3,844	4,203	—	530	581	—	3	4	14	58	59
California	3,105	4,274	1,425	129,332	126,485	239	32,496	33,620	—	169	90	288	2,044	2,360
Hawaii	92	135	—	4,983	5,427	—	836	1,010	N	N	N	18	110	112
Oregon	193	277	—	8,508	8,850	—	1,293	1,535	—	15	21	—	—	103
Washington	352	367	389	17,400	18,150	85	4,155	3,563	—	3	10	60	251	256
American Samoa	U	U	U	U	U	U	U	U	U	U	U	U	U	—
C.N.M.I	2	U	U	U	U	U	U	U	U	U	U	U	—	U
Guam	2	1	—	—	841	—	—	96	—	—	—	—	—	63
Puerto Rico	814	636	135	4,569	3,922	11	274	357	N	N	N	—	79	113
U.S. Virgin Islands	10	19	—	178	196	—	30	45	—	—	—	—	—	—

N:Not notifiable. U:Unavailable. —:No reported cases. C.N.M.I.:Commonwealth of Northern Mariana Islands. Cum.: Cumulative years-to-date counts. *Individual cases can be reported through both the National Electronic Telecommunications System for Surveillance (NETSS) and the Public Health Laboratory Information System (PHLIS). †Chlamydia refers to genital infections caused by *C. trachomatis*. Totals reported to the Division of STD Prevention, NCHSTP. ‡Updated monthly from reports to the Division of HIV/AIDS Prevention—Surveillance and Epidemiology, National Center for HIV, STD, and TB Prevention. Last update January 6, 2006; 2006 data not being released at this time. *Source:* CDC, *Morbidity and Mortality Weekly Report.* Provisional Cases of Selected Notifiable Diseases, United States, weeks ending December 23, 2006, and December 24, 2005. MMWR 55 (51 & 52), 1/2/07.

TABLE 15.4B

Provisional Cases of Infrequently Reported Notifiable Diseases (less than 1,000 cases reported during preceding year), United States, Week Ending December 23, 2006 (51st Week)*

Diseases	Current Week	Cum. 2006	5-Year Weekly Average[†]	2005	2004	2003	2002	2001	States Reporting Cases During Current Week (No.)
Anthrax	—	1	—	—	—	—	2	23	
Botulism:									
Foodborne	2	15	1	19	16	20	28	39	CA (2)
Infant	1	82	2	90	87	76	69	97	WA (1)
Other (wound & unspecified)	—	46	1	33	30	33	21	19	
Brucellosis	—	106	3	122	114	104	125	136	
Chancroid	—	28	1	17	30	54	67	38	
Cholera	—	6	0	8	5	2	2	3	
Cyclosporiasis[§]	1	117	1	716	171	75	156	147	FL (1)
Diphtheria	—	—	—	—	—	1	1	2	
Domestic arboviral diseases[§¶]:									
California serogroup	—	7	1	80	112	108	164	128	
Eastern equine	—	—	0	21	8	14	10	9	
Powassan	—	—	0	1	1	—	1	N	
St. Louis	—	3	0	13	12	41	28	79	
Western equine	—	—	—	—	—	—	—	—	
Ehrlichiosis[¶]:									
Human granulocytic	3	446	24	790	537	362	511	261	NY (3)
Human monocytic	4	403	10	521	338	321	216	142	NY (1), NC (3)
Human (other & unspecified)	1	176	1	122	59	44	23	6	NC (1)
Haemophilus influenzae.**									
Invasive disease (age <5 yrs)[§]									
Serotype b	—	8	1	9	19	32	34	—	
Nonserotype b	1	81	5	135	135	117	144	—	CT (1)
Unknown serotype	4	204	4	217	177	227	153	—	NY (1), PA (1), FL (2)
Hansen disease[§]	—	69	3	88	105	95	96	79	
Hantavirus pulmonary syndrome[¶]	—	31	1	29	24	26	19	8	
Hemolytic uremic syndrome, postdiarrheal[¶]	4	241	6	221	200	178	216	202	NC (1), TX (2), CA (1)
Hepatitis C viral, acute	7	752	41	751	713	1,102	1,835	3,976	MI (2), MN (1), MO (1), KS (1), MD (1), CA (1)
HIV infection, pediatric (age <13 yrs)[§††]	7	52	5	380	436	504	420	543	NY (2), MD (2), FL (1), TX (1), WA (1)
Influenza-associated pediatric mortality[§§]	1	41	0	45	—	N	N	N	OH (1)
Listeriosis	7	715	15	892	753	696	665	613	NY (2), MD (2), FL (1), TX (1), WA (1)
Measles[¶¶]	—	45	1	66	37	56	44	116	
Meningococcal disease, invasive***:									
A, C, Y, & W-135	4	218	7	297	—	—	—	—	IA (1), FL (1), WA (2)
Serogroup B	2	130	5	157	—	—	—	—	FL (1), WA (1)
Other serogroup	1	24	0	27	—	—	—	—	FL (1)

Diseases	Current Week	Cum. 2006	5-Year Weekly Average[†]	Total Cases Reported for Previous Years 2005	2004	2003	2002	2001	States Reporting Cases During Current Week (No.)
Mumps	17	299	7	314	258	231	270	266	NY (1), MN (5), KS (4), MD (1), VA (1), WV (2), CA (3)
Plague	—	16	0	8	3	1	2	2	
Poliomyelitis, paralytic	—	—	—	1	—	—	—	—	
Psittacosis[¶]	—	20	0	19	12	12	18	25	
Q fever[¶]	—	162	2	139	70	71	61	26	
Rabies, human	—	2	—	2	7	2	3	1	
Rubella	—	8	0	11	10	7	18	23	
Rubella, congenital syndrome	—	1	0	1	—	1	1	3	
SARS-CoV[§***]	—	—	—	—	—	8	N	N	
Smallpox[§]	—	—	—	—	—	—	—	—	
Streptococcal toxic-shock syndrome[§]	—	87	3	129	132	161	118	77	
Streptococcus pneumoniae,[†] invasive disease (age <5 yrs)	15	1,079	28	1,257	1,162	845	513	498	NY (4), MI (1), MN (2), AR (2), OK (2), TX (2), AZ (2)
Syphilis, congenital (age <1 yr)	2	268	9	361	353	413	412	441	AZ (2)
Tetanus	1	22	1	27	34	20	25	37	PA (1)
Toxic-shock syndrome (other than streptococcal)[†]	1	96	3	96	95	133	109	127	CA (1)
Trichinellosis	—	11	0	19	5	6	14	22	
Tularemia[†]	—	85	3	154	134	129	90	129	
Typhoid fever	4	261	7	324	322	356	321	368	WA (2), CA (2)
Vancomycin-intermediate Staphylococcus aureus[¶]	—	3	—	2	—	N	N	N	
Vancomycin-resistant Staphylococcus aureus[¶]	—	—	—	3	1	—	—	1	
Yellow fever	—	—	0	—	—	—	1	—	

—: No reported cases.   N: Not notifiable.   Cum. Cumulative year-to-date counts.   *Incidence data for reporting year 2006 are provisional, whereas data for 2001, 2002, 2003, 2004, and 2005 are finalized. †Calculated by summing the incidence courts for the current week the two weeks preceding the current week, and the two weeks following the current week, for a total of 5 preceding years. Additional information is available at http://www.coc.gov./epo/dphsi/phs/files/5yearweeklyaverage.pdf §Not notifiable in all states. ¶Includes both neuroinvasive and non-neuroinvasive. Updated weekly from reports to the Division of Vector-Borne infectious Diseases. National Center for Zoonotic, Vector-Borne, and Enteric Diseases (proposed) ArboNET Surveillance). **Data for *H. influenzae* (all ages, all serotypes) are available in Table II. ††Updated monthly from reports to the Division of HIV/AIDS Prevention, National Center for HIV/AIDS, Viral Hepatitis, STD, and TB Prevention (proposed). Implementation of HIV reporting influences the number of cases reported. Pediatric HIV data will not be updated monthly for the remainder of this year due to upgrading of the national HIV/AIDS surveillance data management system. Data for HIV/AIDS are available in Table IV quarterly. §§Updated weekly from reports to the Influenza Division, National Center for Immunization and Respiratory Diseases (proposed). ¶¶No measles cases were reported for the current week. ***Data for meningococcal disease (all serogroups and unknown serogroups) are available in Table II. †††Updated weekly from reports to the Division of Viral and Rickettisial Diseases, National Center for Zoonotic, Vector-Borne, and Enteric Diseases (proposed).

Preventable deaths

It is estimated that the majority of deaths from infectious diseases can be prevented with existing, cost-effective strategies.

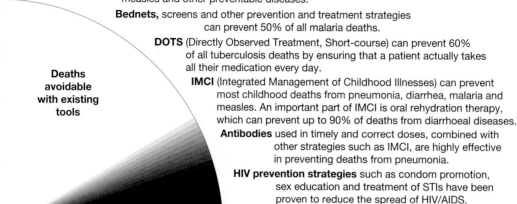

Childhood vaccinations have proven extremely effective in reducing deaths from measles and other preventable diseases.

Bednets, screens and other prevention and treatment strategies can prevent 50% of all malaria deaths.

DOTS (Directly Observed Treatment, Short-course) can prevent 60% of all tuberculosis deaths by ensuring that a patient actually takes all their medication every day.

IMCI (Integrated Management of Childhood Illnesses) can prevent most childhood deaths from pneumonia, diarrhea, malaria and measles. An important part of IMCI is oral rehydration therapy, which can prevent up to 90% of deaths from diarrhoeal diseases.

Antibodies used in timely and correct doses, combined with other strategies such as IMCI, are highly effective in preventing deaths from pneumonia.

HIV prevention strategies such as condom promotion, sex education and treatment of STIs have been proven to reduce the spread of HIV/AIDS.

Deaths avoidable with existing tools

Deaths unavoidable without new tools

© World Health Organization

Figure 15.19 **Preventable deaths.**

Universal Precautions apply to the following body fluids: blood, semen, and vaginal, tissue, cerebrospinal, synovial (joint cavity), pleural, peritoneal, pericardial, and amniotic fluids. The CDC has stated that Universal Precautions do not apply to feces, nasal secretions, sputum, sweat, tears, urine, and vomitus, as long as these do not contain visible blood. This is not to imply that no viruses are present in these fluids but rather that the risk of transmission is either very low or unproved. For example, studies have shown that HIV is present in all tested samples of saliva from AIDS patients. However, the levels are so low that it takes PCR techniques to detect them (◀Chapter 7, p. 203). AIDS patients are often infected with other disease organisms that may be present in these fluids. These organisms include tuberculosis bacilli in sputum, bacteria such as *Salmonella* and *Shigella*, *Cryptosporidium*

protozoa in feces, and herpesvirus in oral secretions. Therefore, some health care facilities require their employees to use Universal Precautions with all body fluids.

Equipment and Procedures That Contribute to Infection

Surgical procedures and the use of equipment such as catheters and respiratory devices are major contributors to nosocomial infections. The smallest abrasion can provide a site of entry for infectious agents. Infections can arise from a contaminated catheter, inadequate cleaning of the site of catheter insertion, or the movement of organisms from leaky connections. In addition, tubing, joints, containers of fluids, and the fluids themselves also can be contaminated.

All surgical procedures expose internal body parts to air, instruments, surgeons, and other operating room personnel, all of which can be contaminated. These procedures can also allow the patient's own microflora to enter sites where they can produce infection. For example, bacteria that cause pneumonia can reach the lungs from the pharynx during surgery.

Respiratory devices, including nebulizer jets, that administer oxygen or air and medications to expand passageways in the lungs provide a means for disseminating microorganisms deep into the lungs. Organisms can grow in the reservoir pans of both cold mist and warm steam humidifiers, and the organisms can be dispersed in an aerosol as the machines operate. Therefore, all respiratory equipment should be disinfected or sterilized daily and, if not disposable, should be disinfected before being moved from one patient to the next.

Other devices and procedures account for smaller, but significant, numbers of nosocomial infections. For

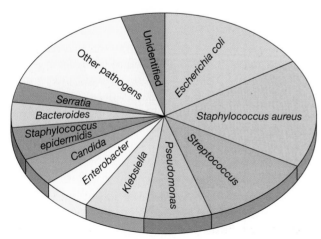

Figure 15.20 **Common causative agents of nosocomial infections.**

Figure 15.21 Some common modes of transmission of nosocomial infections.

example, hemodialysis, a procedure for removing wastes from blood, provides several ways to introduce microorganisms into the body **(Figure 15.22)**. Devices used to monitor blood pressure in the heart or major vessels or cerebrospinal fluid pressure have tubing extending outside the body. Such devices can be contaminated or can allow introduction of organisms from the patient or from the environment. Improperly cleaned gynecological instruments can transmit disease from one patient to another. Endoscopes, which are introduced through body openings and used to examine the linings of organs such as the bladder, large intestine, stomach, and respiratory passageways, are difficult to sterilize and thus can transfer microorganisms from one patient to another.

Another important factor that contributes to nosocomial infections is the intensive use of antibiotics, especially in hospital settings. How antibiotics contribute to the development of antibiotic-resistant pathogens and how these pathogens contribute to nosocomial infections were discussed in ◄Chapter 13.

Sites of Infection

The sites of nosocomial infections, in order from most to least common, are as follows: urinary tract, surgical wounds, respiratory tract, skin (especially burns), blood (bacteremia), gastrointestinal tract, and central nervous system **(Figure 15.23)**.

PREVENTING AND CONTROLLING NOSOCOMIAL INFECTIONS

The problem of nosocomial infections is widely recognized, and nearly all hospitals now have infection-control programs. In fact, to maintain accreditation by the American Hospital Association, hospitals must have programs that include surveillance of nosocomial

TABLE 15.5

Some Important Universal Precautions and Recommendations from CDC

1. Wear gloves and gowns if soiling of hands, exposed skin, or clothing with blood or body fluids is *likely*.

2. Wear masks *and* protective eyewear or chin-length plastic faceshields whenever splashing or splattering of blood or body fluids is *likely*. A mask alone is not sufficient.

3. Wash hands before and after patient contact, and after removal of gloves. Change gloves between *each* patient.

4. Use disposable mouthpiece/airway for cardiopulmonary resuscitation.

5. Discard contaminated needles and other sharp items *immediately* into a *nearby*, special puncture-proof container. Needles must *not* be bent, clipped, or recapped.

6. Clean spills of blood or contaminated fluids by (1) putting on gloves and any other barriers needed, (2) wiping up with disposable towels, (3) washing with soap and water, and (4) disinfecting with a 1:10 solution of household bleach and water. Allow it to stand on surface for at least 10 minutes. Bleach solution should not have been prepared more than 24 hours beforehand.

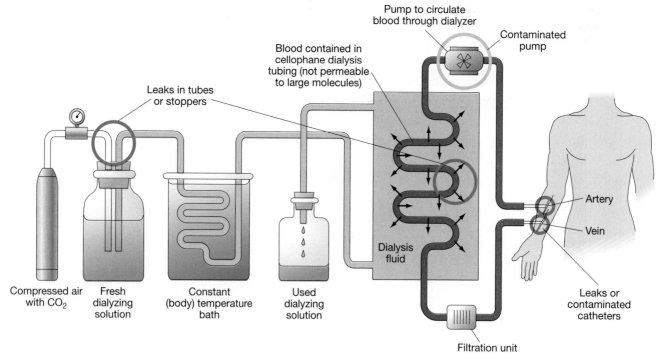

Figure 15.22 Possible sites of contamination in hemodialysis equipment.

APPLICATIONS

Infection Control in Dentistry

Infection control in the dental suite is designed to protect both patients and health professionals. Safe practice requires using latex gloves, eye protection, a mask to prevent inhalation of aerosols (which are used extensively with modern drills), and the use of sterile equipment. Between patients, all items that could be contaminated with oral secretions or blood must be disinfected. For example, the gloved hand of the dentist leaves the patient's mouth, moves up to the light to adjust it, pushes away the rim of the instrument tray, and then reaches for the drill handle, leaving organisms behind on all these sites. It is not enough for the dentist merely to change gloves between patients—the clean glove might pick up organisms left on the equipment and transport them into the mouth of the next patient. Many dentists now use disposable plastic covers on light adjustment handles and other sites, which can be changed between patients.

In September 1990, Florida dentist Dr. David Acer died of AIDS. Five of his patients have since been found to have the exact same strain of AIDS virus as Dr. Acer and are presumed to have contracted it from him in the dental office. None of the five had any other risk factors. Officials at CDC are at a loss to explain the method of possible transmission, but current suspicion falls on the drill handle. A University of Georgia study has shown that internal parts of the hollow handle, which holds drill bits,

polishing brushes, and other tools, can become coated with blood, saliva, and tooth fragments. The American Dental Association recommends heat sterilization of drill handles between patients, but this is inconvenient, and shortens the life span of expensive equipment, and requires having more handles to use with each patient; thus, some dentists do not want

(Larry Mulvehill/Photo Researchers, Inc.)

to do so. Although the drill handles may be chemically disinfected, this is not sufficient to sterilize against the AIDS virus. Use of gloves, face shields, and masks by all dental personnel is essential if the spread of infection is to be avoided, but other measures are also necessary.

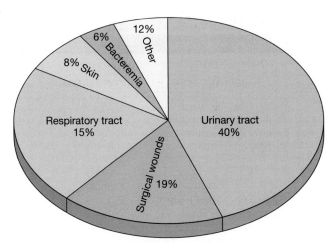

Figure 15.23   Relative frequencies of sites of nosocomial infections.

✓ **CHECKLIST**

1. List the Universal Precautions.
2. What is meant by a nosocomial infection?

infections in both patients and staff, a microbiology laboratory, isolation procedures, accepted procedures for the use of catheters and other instruments, general sanitation procedures, and a nosocomial disease education program for staff members. Most hospitals have an infection-control specialist to manage such a program.

Several techniques are available to prevent the introduction and spread of nosocomial infections. Hand washing is the single most important technique. Physicians, nurses, and other staff members who wash their hands thoroughly with soap and water between patient contacts can greatly reduce the risk of spreading diseases among patients. Scrupulous care in obtaining sterile equipment and maintaining its sterility while in use are also important. In addition, the use of gloves when handling infectious materials such as dressings and bedpans and when drawing blood prevents the spread of infections. And, as mentioned earlier, it is important to prevent insect infestations, as flies, ants, or roaches can easily spread an infectious agent.

Other techniques are needed to reduce the development of antibiotic-resistant pathogens. Routine use of antimicrobial agents to prevent infections has turned out to be a misguided effort because it contributes to the development of resistant organisms. Therefore, some hospitals maintain surveillance of antibiotic use. Antibiotics are given for known infections but are given prophylactically (as preventive measures) only in special situations. Prophylactic antibiotics are justified in surgical procedures such as those involving the intestinal tract and repair of traumatic injuries, in which the surgical field is invariably contaminated with potential pathogens. They also are justified in immunosuppressed and excessively debilitated patients, whose natural defense mechanisms may fail.

If all the known techniques for preventing nosocomial infections were practiced rigorously, the incidence of such infections probably could be reduced to half the present level.

BIOTERRORISM

It is sad that there needs to be a section on bioterrorism in this chapter. While most epidemiologists are busy investigating normal ways in which diseases are spread, others are needed to study the ways in which we can protect ourselves from the deliberate spread of disease.

However, the use of microbes in biological warfare is not something new. In ancient times, infected corpses were flung over the walls of besieged cities, or dumped into wells and other water sources. Gladiators sometimes thrust their swords and tridents into rotting cadavers before entering the arena. This ensured the death of an opponent from even minor cuts, should he kill them first.

In more recent times, in 1763 before the American revolution, colonial British military officers gave blankets containing smallpox scabs to native Americans with specific intent to kill them. During World War II, American and British scientists developed and produced 1000's of anthrax bombs by 1944, to drop on Hitler's Germany. These bombs were tested but never actually used in warfare. In April 1979, a Soviet anthrax factory in the Ural Mountains, in a city then called Sverdlovsk but now returned to its earlier name of Yekaterinburg, had a tragic accidental leak which caused an anthrax epidemic. Death claimed 68 of the victims. Complete details of this event were found in *Anthrax, the Investigation of a Deadly Outbreak*, by Jeanne Guillemin, 1999, University of California Press.

In 1984, in Oregon, members of a religious cult known as the Rajneeshees caused outbreaks of *Salmonella typhimurium* food poisoning by sprinkling organisms on the salad bars of ten local restaurants, a nursing home, and even the local medical center. They even gave the investigating judge a glass of water containing *Salmonella*, in hopes of killing him. More details of this and other attacks can be found in *Germs*, by Judith Miller, et, al., 2001, Simon & Schuster.

Then in 2001–2002, terrorist anthrax attacks along the east coast of the U.S., including government buildings in Washington, D.C. brought bioterrorism into the public mind, and awareness of its dangers. **Table 15.6** summarizes information about currently feared agents of bioterrorism. Ricin is not microbial, but rather is an extract from the castor bean, *Ricinus communis*. Preparations for homeland defense against bioterrorism have created more questions than answers. Resumption of civilian smallpox vaccination raised questions of safety. During January 24 to December 31, 2003,

TABLE 15.6—See Table 15.6 Key on p. 458

Agents of Bioterrorism

Agent	Incubation	Signs/Symptoms	Diagnostic Tests	Precautions
Anthrax	1–5 d	Fever, malaise, fatigue, cough, mild chest discomfort, cold/flu-like symptoms. Improvement 2–3 days. Abrupt onset resp. distress, shock, CXR: widened mediastinum.	Nasal, resp. culture, FA, PCR, blood–Gram stain, culture, PCR, serum–Ag ELISA	Standard
Botulism	1–5 d	Cranial nerve palsy–ptosis, blurred vision, dry mouth, dysphagia, dysphonia, descending flaccid paralysis w/gen weakness, resp. fail.	Nasal swab & resp.: PCR, Ag ELISA, serum toxin assays, blood & stool cultures	Standard
Brucellosis	5–60 d	Fever, headache, myalgias, arthralgias, back pain, sweats, depression, mental status changes, Osteoarticular findings.	Nasal swab, resp. secretions culture, PCR, blood culture & PCR, bone marrow cult., serology: agglutination	Contact, if lesions are draining
Cholera	4 h–5 d	Vomiting, malaise, headache (early onset), abdominal cramps, diarrhea.	Stool culture	Standard
Plague	2–3 d	*Pneumonic*—High fever, headache, malaise, cough, hemoptyosis, dyspnea, strider, cyanosis, GI symptoms present *Bubonic*—High fever, malaise, painful lymph nodes (buboes) progressing to shock thrombosis DIC or pneumonic form.	Nasal swab & resp. culture, FA, PCR, blood & CSF Gram stain & cult. lymph node—Wright or Giemsa stain, Ag-ELISA, culture	Pneumonic droplet
Q Fever	10–40 d	Fever, cough, pleuritic chest pain.	Nasal swab, sputum, blood: culture & PCR, serum—serology: ELISA, IFA	Standard
Ricin	4–6 h	Fever, chest tightness cough, dyspnea, nausea arthralgia. Airway necrosis, pul. edema, resp. distress	Nasal & resp. secretions—PCR, serum ELISA.	Standard
Smallpox (Variola)	7–17 d	Malaise, fever vomiting, headache, backache; 2–3 days later lesions appear—macules, papules, pustules (face/extremities).	Nasal/resp. PCR, viral cult., serum viral cult., skin lesions/scrapings PCR, viral cult., micro.	Airborne
Staphylococcal enterotoxin B	1–6 h	Fever, chills, HA, myalgia, nonprod. cough, SOB, retrosternal CP. Ingest. N/V, diarrhea. Sepsis, shock.	Nasal swab & resp. PCR, Ag ELISA, serum & urine Ag ELISA	Standard
Tularemia	2–10 d	HA, fever, malaise. Ulceroglandular: local ulcer and regional lymphadenopathy. Typhoidal: substernal CP, prostration, wt. loss.	Nasal swab & resp. cult., FA, PCR, blood culture, PCR, lymph node FA, serology–agglutination.	Standard
Tricothecene mycotoxicosis	2–4 h	Skin pain, pruritis, vesicles, necrosis, hemoptyosis, sev. ataxis, death.	Nasal swab & resp. FA, Serum & tissue toxin detection.	Contact w/decontam., Standard
Venezuelan Equine Encephalitis	2–6 d	Febrile w/encephalitis, rigors, sev. HA, photophobia, myalgias, N/V, cough sore throat, diarrhea.	Nasal swab & resp. RT, PCR, viral cult., serum PCR, ELISA or hemogglutination inhibition.	Standard

Chemotherapy	Chemoprophylaxis	Vaccine	Comments/(Human to Human Trans.?)
• Cipro 400 mg IV Q 8–12 h • Doxycycline 200 mg IV, then, 100 mg IV Q 12 h • PCN 2 million units IV Q 2 h • Streptomycin 30 mg/kg IM QD OR Gent. Child/Preg: Cipro, PCN, Doxycycline 3rd choice	• Cipro 500 mg po bid × 4 wk. If not vaccinated give vaccine. • Doxycycline 100 mg po bid + vaccine	Bioport vaccine 0.5 ml SC Q 1, 2, 4 wk, 6, 12, 18 mo. & annually	(No)
• DOD Heptavalent Equine—Desperciated antitoxin for serotypes (A–G IND) 1 vial IV	Pentavalent toxid vaccine Types (A–E)	DOD Heptavalent Equine Toxoid for Serotypes A–E (IND) 0.5 ml deep SC @ 0, 2, & 12 wk & annually	Skin test before vaccine. (NO)
• Doxycycline 200 mg/d po + Rifampin 600 mg, 900 mg/d po × 6 wks. • Ofloxacin 400 mg + Rifampin 600–900 mg qd × 6 wks	Doxycycline & Rifampin × 3 wk	No vaccine available	(NO)
• Tetracycline 500 mg Q 6 h × 3* • Doxycycline 300 mg once - 100 mg Q 12 h × 3 d* • Ciprofloxacin 500 mg Q 12 h × 3 d	None	Wyeth-Ayerst Vaccine 2 doses 0.5 ml IM or SC @ 7–30 d, then booster Q 6 mo	Quinolones for resistance. (RARE)
• Streptomycin 30 mg/kg/d IM In 2 divided doses × 10 days (or Gent.) • Doxycycline 200 mg IV, then 100 mg IV bid × 10–14 d • Chloramphenicol 1 g IV qid × 10–14 days (*) Child: 1st choice-Strep or Gent., Doxycyc., Cipro Preg. 1st choice Gent. then Doxycyc., Cipro	• Doxycycline 100 mg po bid × 7 d or duration of exposure • Ciprofloxacin 500 mg po bid × 7 d • Doxycycline 100 mg po bid × 7 • Tetracycline 500 mg po Q × 7 d	Vaccine no longer available	Chloramphenicol for plague meningitis. (YES)
• Tetracycline 500 mg po Q 6 h × 5–7 days • Doxycycline 100 mg Q 12 h × 5–7 days	• Tetracycline start 8–12 d post exp. × 5 d • Doxycycline start 8–12 postexp. × 5 d	Not available	(RARE)
• Inhalation, supportive therapy • GI gastric lavage, w/superactivated charcoal, carthartics	None	No vaccine	(NO)
Supportive care	Vaccinia immune globulin 0.6 ml/kg IM (within 3 d of exposure best w/ in 24 h)	Wyeth calf lymph vaccinia vaccine (licensed) 1 dose by scarification	Pre & post exp. vaccine if > 3 yr since last vaccine (YES)
Ventilatory support for inhalation exposure	None	No vaccine	(NO)
• Streptomycin 30 mg/kg IM divided bid 10–14 d • Gentamycin 3–5 mg/kg/d IV × 10–14 d	• Doxycycline 100 mg po bid × 14 d • Tetracycline 500 mg po Q × 14 d	IND—Live attenuated vaccine; one dose by scarification	(NO)
Decontamination of skin w/ soap and H$_2$O (*gown, glove)	Decontam. clothing/skin w/ soap/H$_2$O	No vaccine	(NO)
Supportive therapy: analgesics and anticonvulsants	N/A	VEE DOD TC-83 attenuated vac. (IND) 0.5 ml × 1 VEE DOD C-84 0.5 ml SC up to 3 doses	(LOW)

Figure 15.24 Training for bioterrorism. Persons exposed to possible germ warfare organisms will be isolated in special transport equipment to be brought to facilities where medical personnel wearing protective suits and respirator packs will treat them. *(Karen Kasmauski/National Geographic Society).*

smallpox vaccine was given to 39,213 civilian health workers. Nonserious adverse reactions were reported in 712 vaccinated persons—rash, fever, pain, headache, and fatigue. Serious adverse events, including 2 cases of myocardial infarction (heart attack), affected 97 persons. Among 578,286 military personnel who were vaccinated, contact transfer of vaccina virus to other people occurred in 30 cases—12 spouses, 8 adult intimate contacts, 8 adult friends, and 2 children in the same household, the majority (89%) of cases being uncomplicated infections of the skin; 2 (11%) involved the eye. There were no cases of transfer between health care workers and their patients in either direction.

However, transfer to health care workers is of concern in other cases. Personnel are conducting training sessions (**Figure 15.24**), and emergency plans are being made. Let us hope we will not have to use them.

Not high in public awareness is the threat of agricultural bioterrorism, affecting our food supply and economy. One study estimated that minimum cost of an outbreak of foot-and-mouth disease, confined to California and lasting only a few months, as reaching $13 billion dollars! Exports of cattle, sheep, and/or pigs would plummet as other countries refused to accept them. And think of the fears of American families wondering if their food was safe to eat. Also plant crops could easily be infected, especially since their pathogens generally don't infect humans, making it easy for terrorists to handle them. However, experts feel that animals, rather than plants, are the more likely targets. High-risk animal pathogens include those causing: foot-and-mouth disease, avian influenza, hog cholera, Newcastle disease of birds, and Rinderpest virus. The U.S. government is just now beginning to make plans to combat agroterrorism.

▌ RETRACING OUR STEPS

▌▌▌ EPIDEMIOLOGY

WHAT IS EPIDEMIOLOGY?

• **Epidemiology** is the study of factors and mechanisms in the spread of diseases in a population.

• In describing infectious diseases, **epidemiologists** use **incidence** to refer to the number of new cases in a specific period, **prevalence** to refer to the number of people infected at any one time, **morbidity rate** to indicate the number of cases as a proportion of the population, and **mortality rate** to indicate the number of deaths as a proportion of the population.

DISEASES IN POPULATIONS

• In a **sporadic disease**, several isolated cases appear in a population. In an **epidemic** disease, many cases appear in a population, and patients are sufficiently harmed to create a public health problem. Diseases that spread by **common-source outbreaks** originate with a single contaminated substance, such as a water supply. In **propagated epidemics**, diseases are spread by person-to-person contact. A **pandemic** disease is an epidemic disease that has spread over an exceptionally wide geographic area or several geographic areas.

EPIDEMIOLOGIC STUDIES

• The purpose of **epidemiologic studies** is to learn more about the spread of diseases in populations and how to control those diseases.
• The methods of epidemiologic studies are **descriptive, analytical** (retrospective and prospective), and **experimental**.

RESERVOIRS OF INFECTION

• **Reservoirs of infection** include humans, other animals, and nonliving sources from which infectious diseases can be transmitted.
• Among human reservoirs, **carriers** often transmit diseases. They are intermittent carriers if they periodically release pathogens.
• Diseases in animal reservoirs can be transmitted by direct contact with animals or by vectors. Diseases that can be naturally transmitted from animals to humans are called **zoonoses**.
• Diseases in nonliving reservoirs are transmitted by water, soil, or wastes.

PORTALS OF ENTRY

• **Portals of entry** include skin, mucous membranes that line various body systems, tissues, and the placenta.

PORTALS OF EXIT

• **Portals of exit** include the nose, ear, mouth, skin, and openings from which products of the digestive, urinary, and reproductive systems are released. Organisms usually are in body fluids or feces.

MODES OF DISEASE TRANSMISSION

• Transmission can be by contact, vehicle, or vector. **Direct contact transmission** includes person-to-person **horizontal transmission** and **vertical transmission** from parent to offspring. **Indirect contact transmission** occurs through **fomites** (inanimate objects) and by droplets. **Vehicles** of transmission include water, air, and food. **Vectors** of transmission are usually arthropods, which can transmit disease mechanically or biologically.
• Transmission by carriers, transmission of STDs, and transmission of zoonoses pose special epidemiologic problems.

DISEASE CYCLES

• Some diseases occur in cycles—there are just a few cases for several years, and then many cases suddenly appear.

HERD IMMUNITY

• **Herd immunity**, or group immunity, refers to immunity enjoyed by a large proportion of a population that reduces disease transmission among nonimmune individuals.
• A drop in herd immunity can lead to the sudden appearance of cases of a cyclic disease.

CONTROLLING DISEASE TRANSMISSION

• Methods used to control communicable diseases include isolation, quarantine, immunization, and vector control.
• **Isolation** procedures are summarized in Table 15.2. **Quarantine** is rarely used but can prevent exposed individuals from infecting others. Active immunization prevents many infections. Vector control is effective where vectors can be identified and eradicated.

PUBLIC HEALTH ORGANIZATIONS

• Public health organizations exist at city, county, state, federal, and world levels. They help establish and maintain health standards, cooperate in the control of infectious diseases, collect and disseminate information, and assist with professional and public education.

NOTIFIABLE DISEASE

• **Notifiable diseases** are listed in Table 15.3.

❚❚❚ NOSOCOMIAL INFECTIONS

• A **nosocomial infection** is an infection acquired in a hospital or other medical facility.

THE EPIDEMIOLOGY OF NOSOCOMIAL INFECTIONS

• Nosocomial infections can be **exogenous** (caused by external organisms) or **endogenous** (caused by opportunists in normal microflora). About half are caused by only four types of pathogens, of which many strains are antibiotic resistant.
• Host susceptibility is an important factor in the development of nosocomial infections.
• Medical equipment and procedures, including surgery, are often responsible for infections.
• Modes of nosocomial infections are illustrated in Figure 15.21.

PREVENTING AND CONTROLLING NOSOCOMIAL INFECTIONS

• Most hospitals have an extensive infection control program. Hand washing, use of gloves, scrupulous attention to maintaining sanitary conditions and sterility where possible, and surveillance of antibiotic use and other hospital procedures help reduce infections.
• Nosocomial infections could be reduced by half if all known procedures were carefully followed in all medical facilities at all times.

❚❚❚ BIOTERRORISM

• Bioterrorism is unfortunately not a new phenomenon. Table 15.6 lists the major agents of bioterrorism and their characteristics.

▌ TERMINOLOGY CHECK

aerosol **(p. 438)**
analytical study **(p. 430)**
carrier **(p. 432)**
common-source outbreak
   (p. 429)
compromised host **(p. 447)**
contact transmission **(p. 436)**
descriptive study **(p. 430)**
direct contact transmission
   (p. 436)
direct fecal-oral transmission
   (p. 438)
droplet nucleus **(p. 438)**
droplet transmission **(p. 438)**

endemic **(p. 427)**
endogenous infection
   (p. 447)
epidemic **(p. 428)**
epidemiologic study **(p. 429)**
epidemiologist **(p. 427)**
epidemiology **(p. 427)**
etiology **(p. 427)**
exogenous infection **(p. 447)**
experimental study **(p. 430)**
fomite **(p. 438)**
herd immunity **(p. 441)**
horizontal transmission
   (p. 438)

inapparent infection **(p. 432)**
incidence **(p. 427)**
index case **(p. 430)**
indirect contact transmission
   (p. 438)
indirect fecal-oral
   transmission **(p. 438)**
isolation **(p. 441)**
morbidity rate **(p. 427)**
mortality rate **(p. 427)**
nosocomial infection **(p. 447)**
notifiable disease **(p. 445)**
pandemic **(p. 428)**
placebo **(p. 432)**

portal of entry **(p. 434)**
portal of exit **(p. 436)**
prevalence **(p. 427)**
propagated epidemic **(p. 429)**
quarantine **(p. 441)**
reservoir of infection **(p. 432)**
sporadic disease **(p. 429)**
subclinical infection **(p. 432)**
Universal Precautions
   (p. 447)
vector **(p. 439)**
vehicle **(p. 438)**
vertical transmission **(p. 438)**
zoonosis **(p. 433)**

▌ CLINICAL CASE STUDY

Shari is the infection control officer at a county hospital. The hospital is building a new wing and has asked Shari and the infection control committee to make some suggestions that will help reduce the incidence of nosocomial infections. Visit the following web site to explore some simple ways to reduce the incidence of nosocomial infections: What recommendations should Shari make? http://www.healthtransformation.net/cases/index.asp?ID=14.

▌ CRITICAL THINKING QUESTIONS

1. If a particular disease occurs in humans in occasional, isolated, sporadic cases, but most of the time no one is ill, what does this suggest about the likely reservoir of such a disease?

2. Cyclic 5- to 12-year epidemics of meningococcal meningitis occur in the "Meningitis Belt" of Africa from the causative agent *Neisseria meningitidis*. What epidemiological factors can explain the cyclic nature and geographical location of meningococcal meningitis in this area of Africa?

3. Despite their best efforts, hospitals seem to be unable to reduce their nosocomial infection rate to zero. What are some reasons why there will always be some nosocomial infections?

▌ SELF-QUIZ

1. What is epidemiology?

2. Quarantine is used to prevent a patient with a communicable disease from having contact with the general population. True or false?

3. Match the following terms:
- ___ Persons in a population who become clinically ill during a specified period of time
- ___ The total lumber of sick individuals in a population at a particular time.
- ___ The colonization and growth of an infectious agent in a host
- ___ The number of new cases of a disease identified in a population during a defined period of time
- ___ The number of deaths within a population during a specified period of time

  (a) Incidence rate
  (b) Morbidity rate
  (c) Mortality rate
  (d) Prevalence rate
  (e) None of the above

4. The incidence rate can indicate whether there is an increase or decrease in the spread of a disease while the prevalence rate measures how seriously or long the disease is affecting a population. True or false?

5. An infectious disease agent that is continually present in a population located in a specific geographical location but has both the number of reported cases and the severity of the disease too low to constitute a public health problem is known as:
  (a) Epidemic
  (b) Endemic
  (c) Sporadic
  (d) Pandemic
  (e) Index case

6. What is the term used when a disease with a higher-than-normal incidence within a population that poses a public health problem suddenly spreads worldwide?
  (a) Pandemic
  (b) Epidemic
  (c) Sporadic
  (d) Endemic
  (e) Prevalent

7. What is the term used when the morbidity and/or mortality rate in a population becomes high enough to pose a public health problem?
   (a) Pandemic
   (b) Epidemic
   (c) Sporadic
   (d) Endemic
   (e) Prevalent

8. An epidemiological study focusing on cause-and-effect relationships in the occurrence of a disease in a population and in which factors preceding an epidemic are considered is known as an _____ _____ study.
   (a) Experimental prospective
   (b) Experimental retrospective
   (c) Analytical prospective
   (d) Analytical retrospective
   (e) Descriptive prospective

   Analytical retrospective

9. An epidemiological study in which an investigator tests the hypothesis that a particular treatment will be effective in controlling a disease for which no accepted cure is available is known as a(n) _____ study.
   (a) Descriptive
   (b) Analytical
   (c) Experimental
   (d) Prospective
   (e) Retrospective

10. Because most pathogens cannot survive for extended periods of time outside the body they must persist within _____ in order to maintain their ability to infect humans.
    (a) Macrophages
    (b) Viruses
    (c) Reservoirs of infection
    (d) Endospores
    (e) Capsules

11. *Salmonella typhi* has the ability to persist within the gallbladder of humans while causing no clinical symptoms. The infected individual is still contagious, however, and would be considered a(n):
    (a) Pathogen   (d) Carrier
    (b) Nuisance   (e) Bacteriophage
    (c) Endemic

12. Vertical transmission refers to transmission of a pathogen from a(n):
    (a) Lower life form to higher life form
    (b) Younger species to older species
    (c) Parent to offspring before or during birth
    (d) Younger child to an older child
    (e) Parent to a child

13. A disease in which a person contracts rabies virus after interaction with an infected raccoon would be known as a _____ disease.
    (a) Fungal
    (b) Communicable
    (c) Zoonotic
    (d) Vector-borne
    (e) Harmless

14. All of the following are common portals of entry and EXCEPT:
    (a) Cut skin and respiratory tract (cough)
    (b) Mucous membrane of urogenital system and nose (sneeze)
    (c) Mucous membrane of digestive tract and milk
    (d) Mucous membrane of respiratory tract and feces.
    (e) Intact skin and eyes

15. All of the following could be considered as reservoirs of infection EXCEPT:
    (a) A lizard
    (b) Your cousin
    (c) A piece of notebook paper
    (d) The inside contents of a glass container containing nutrient broth that was just autoclaved for 15 minutes at 15 psi
    (e) Your pillow

16. Match the following modes of disease transmission to their object(s):
    C _Fomites_   (a) Horizontal
    C _Bar soap_   (b) Vertical
    A _Kissing_   (c) Indirect contact
    F _Speaking_   (d) Vehicle
    A _Hand shaking_   (e) Biological vector
    D _Food_   (f) Droplet
    G _Housefly_   (g) Mechanical vector
    D _Mother breast-feeding her infant_
    E _Anopheles mosquito_
    C _Stepping on a rusty nail_

17. Which of the following would not be considered a disease vector?
    (a) Ticks
    (b) Fleas
    (c) Handkerchief
    (d) Lice
    (e) Mosquitoes

18. Which of the following are ways in which bioterrorism could be exerted?
    (a) Farm animals (cows and chickens)
    (b) Agricultural plants (corn and wheat)
    (c) Subway system
    (d) Water reservoirs
    (e) All of the above.

19. Diseases that are potentially harmful to the public's health and must be reported by physicians are called:
    (a) Notifiable diseases   (d) HIV
    (b) Nosocomial infections   (e) Epidemics
    (c) Recordable diseases

20. Hospital-acquired infections are called:
    (a) Notifiable diseases
    (b) Nosocomial infections
    (c) Pathogens
    (d) Staphylococcal infections
    (e) Transient infections

(handwritten in margin: Healthy Compromised)

21. Which of the following would not be considered a compromised host?
(a) AIDS patient   (d) Transplant patient
(b) Healthy individual   (e) Burn patient
(c) Chemotherapy patient

22. All of the following can help prevent nosocomial infections except:
(a) Maintaining sanitary conditions
(b) Use of gloves
(c) Patients remaining in bed
(d) Hand washing
(e) Judicious use of antibiotics

23. The extensive use of antibiotics as well as gene transfer has led to more virulent, antibiotic-resistant strains of all of the following organisms *except*:
(a) *Streptococcus pneumoniae*
(b) *Staphylococcus aureus*
(c) Enterococci
(d) Influenza
(e) *Escherichia coli*

24. Universal Precautions apply to all of the following bodily fluids except:
(a) Tears
(b) Blood
(c) Semen
(d) Cerebrospinal fluid
(e) Amniotic fluid

25. Match the following with their respective descriptions:
(handwritten answers: B, D, A, C, E)
___ Zoonoses
___ Fomite
___ Droplet nuclei
___ Exogenous infection
___ Endogenous infection

(a) Dried mucus that contains potential pathogens
(b) Naturally occurring animal disease that may be transmitted to humans accidentally
(c) Caused by external microbes that enter a patient's body
(d) Inanimate object contaminated with pathogens
(e) Caused by opportunists in body's normal flora

26. Suppose that the health department of city A mounts a successful campaign to get children immunized against measles. Only 100 out of 10,000 children fail to receive the vaccine. Now suppose that city B, the same size as city A, has not carried out a successful measles vaccination program. Of the 10,000 children in city B, 5,000 had measles when the disease last struck the population.
(a) If a child with measles moves to city A, what is the chance that child will encounter a susceptible child?
(b) If a child with measles moves to city B, what is the chance that child will encounter a susceptible child?
(c) Comparing the two scenarios, which city has the higher herd immunity, and in which city is an infected child more likely to transmit the disease to a susceptible child?

■ EXPLORATIONS ON THE WEB

http://www.wiley.com/college/black

If you think you've mastered this chapter, there's more to challenge you on the web. Go to the companion web site to fine-tune your understanding of the chapter concepts and discover answers to the questions posed below.

1. The bubonic plague struck and killed people with terrible speed. After 5 years the bubonic plague killed 25 million people—one-third of Europe's population. The Italian writer Boccaccio said its victims often "ate lunch with their friends and dinner with their ancestors in paradise." Discover more about the bubonic plague on the web site.

2. At its peak in 1918, the global flu epidemic was responsible for 11,000 deaths in 1 month. Learn more about this widespread infectious disease on the web site.

Innate Host Defenses

Patient suffering from advanced leprosy (Hansen's disease).
(©Kate Holt/eyevine/Zuma Press)

Come with me...

Sometimes, when you can't kill something that is harmful, the best thing to do is to wall it off. But if the wall gets too thick, too rigid, or just too many walls are needed, then your defense mechanism can wind up hurting you. In other words, things that your immune system does to try to protect you can sometimes be harmful. Granulomas are such an immune response.

A granuloma is a thick layer of cells around irritants such as chemicals, microbes, parasites, or even tissue damaged by trauma. A granuloma forms when the irritant can't be gotten rid of, e.g., Mycobacterium leprae bacteria which have been phagocytized by macrophages are difficult to kill because they divide so very slowly. A person with a strong immune response will form a granuloma around them typical of leprosy (now called Hansen's disease). This is what forms the disfiguring lumps and bumps. These lack sensation due to nerve damage, allowing infections to go unnoticed. At the same time, bone is resorbed and eventually infected fingers, toes, nose, and other tissues are lost.

Come with me to find out about other kinds of granulomas and their effects.

 Video related to this topic is available within WileyPLUS.

▌▌▌ INNATE AND ADAPTIVE HOST DEFENSES
▌▌▌ PHYSICAL BARRIERS
▌▌▌ CHEMICAL BARRIERS
▌▌▌ CELLULAR DEFENSES
Defensive Cells / Phagocytes / The Process of Phagocytosis / Extracellular Killing / The Lymphatic System

▌▌▌ INFLAMMATION
Characteristics of Inflammation / The Acute Inflammatory Process / Repair and Regeneration / Chronic Inflammation
▌▌▌ FEVER
▌▌▌ MOLECULAR DEFENSES
Interferon / Complement / Acute Phase Response

We can look at infectious disease as a battle between the power of infectious agents to invade and damage the body and the body's powers to resist such invasions. In Chapters 14 and 15 we considered how infectious agents enter and damage the body and how they leave the body and spread through populations. In the next three chapters we consider how the body resists invasion by infectious agents.

We begin this chapter by distinguishing between adaptive and innate defenses. Until recently these were called **specific** and **nonspecific defenses**. As the nonspecific defenses were studied, it became apparent that they involved very specific interactions but did not require a previous exposure to be active, hence the term *innate defense*. Then we will look at the innate defense mechanisms in more detail to see how they function in protecting the body against infectious agents.

INNATE AND ADAPTIVE HOST DEFENSES

With potential pathogens ever present, why do we rarely succumb to them in illness or death? The answer is that our bodies have defenses for resisting the attack of many dangerous organisms. Only when our resistance fails do we become susceptible to infection by pathogens.

Host defenses that produce resistance can be adaptive or innate. **Adaptive defenses** respond to particular agents called *antigens*. Viruses and pathogenic bacteria have molecules in or on them which serve as antigens. Adaptive defenses then respond to these antigens by producing protein *antibodies*. The human body is capable of making millions of different antibodies, each effective against a particular antigen. Adaptive responses also involve the activation of the *lymphocytes*, specific cells of the body's immune system. These antibody and cellular responses are more effective against succeeding invasions

by the same pathogen than against initial invasions thanks to memory cells. Chapter 17 focuses on these and other adaptive defenses of the immune system.

In the case of many threats to an individual's well being, adaptive defenses do not need to be called on because the body is adequately protected by its **innate defenses**—those that act against any type of invading agent. Often such defenses perform their function before adaptive body defense mechanisms are activated. However, the innate system's action is necessary to activate the adaptive system responses. Innate defenses include the following:

1. *Physical barriers*, such as the skin and mucous membranes and the chemicals they secrete.
2. *Chemical barriers*, including antimicrobial substances in body fluids such as saliva, mucus, gastric juices, and the iron limitation mechanisms.
3. *Cellular defenses*, consisting of certain cells that engulf (phagocytize) invading microorganisms.
4. *Inflammation*, the reddening, swelling, and temperature increases in tissues at sites of infection.
5. *Fever*, the elevation of body temperature to kill invading agents and/or inactivate their toxic products.
6. *Molecular defenses*, such as interferon and complement, that destroy or impede invading microbes.

The physical and certain chemical barriers operate to prevent pathogens from entering the body. The other innate defenses (cellular defenses, inflammation, fever, and molecular defenses) act to destroy pathogens or inactivate the toxic products that have gained entry or to prevent the pathogens from damaging additional tissues. Overactivity of the innate responses, however, can cause diseases such as autoimmune problems of lupus, rheumatoid arthritis, and others (◄Chapter 17). Underactivity will leave the host open to overwhelming infection (sepsis) leading to death. A delicate balance is needed. The innate defenses serve as the body's *first lines of defense* against pathogens. The adaptive defenses represent the *second lines of defense*. Let's look at each of the innate defenses now; we will discuss the adaptive defenses in Chapter 17.

APPLICATIONS

Take Two, Not Twenty-Two

Do you know someone who is a chronic aspirin or ibuprofen user? These days most people use the "harmless" painkillers freely. But, these little pills can have deadly effects. The problem is that aspirin, ibuprofen, and acetaminophen aren't specific enough. Their beneficial effects come from their ability to permanently block an enzyme that promotes inflammation, pain, and fevers. Unfortunately, the drugs are even more effective at permanently inhibiting a related enzyme that is necessary for the health of the stomach and kidneys. Aspirin also disrupts the body's acid-base balance, which can lead to whole organs—the kidneys, the liver, and the brain—shutting down forever, depending on the amount ingested. Patients can also have seizures and develop heart arrhythmias.

PHYSICAL BARRIERS

The skin and mucous membranes protect your body and internal organs from injury and infectious agents. These two physical barriers are made of cells that line the body surfaces and secrete chemicals, making the surfaces hard to penetrate and inhospitable to pathogens. The **skin**, for example, not only is exposed directly to microorganisms and toxic substances but also is subject to objects that

A natural antibiotic, human beta-defensin-2, lurks on the human skin and, when induced, can kill pathogens by punching holes in the bacterial membranes.

touch, abrade, and tear it. Sunlight, heat, cold, and chemicals can damage the skin. Cuts, scratches, insect and animal bites, burns, and other wounds can disrupt the continuity of the skin and make it vulnerable to infection.

Besides the skin, a **mucous membrane**, or *mucosa*, covers those tissues and organs of the body cavity that are exposed to the exterior. Mucous membranes, therefore, are another physical barrier that makes it difficult for pathogens to invade internal body systems.

The hairs and mucus of the nasal and respiratory system present mechanical barriers to invading microbes. But so do the physical reflex flushing activities of coughing and sneezing. Vomiting and diarrhea similarly act to flush harmful microbes and their chemical products from the digestive tract. Tears and saliva also flush bacteria from the eyes and mouth. Likewise, urinary flow is important in removing microbes that enter the urinary tract. Urinary tract infections are especially common among those unable to empty their bladder completely or frequently enough.

CHEMICAL BARRIERS

There are a number of chemical barriers that control microbial growth. The sweat glands of the skin produce a watery-salty liquid. The high salt content of sweat inhibits many bacteria from growing. Both sweat and the sebum produced by sebaceous glands in the skin produce secretions with an acid pH that inhibits the growth of many bacteria. The very acidic pH of the stomach is a major innate

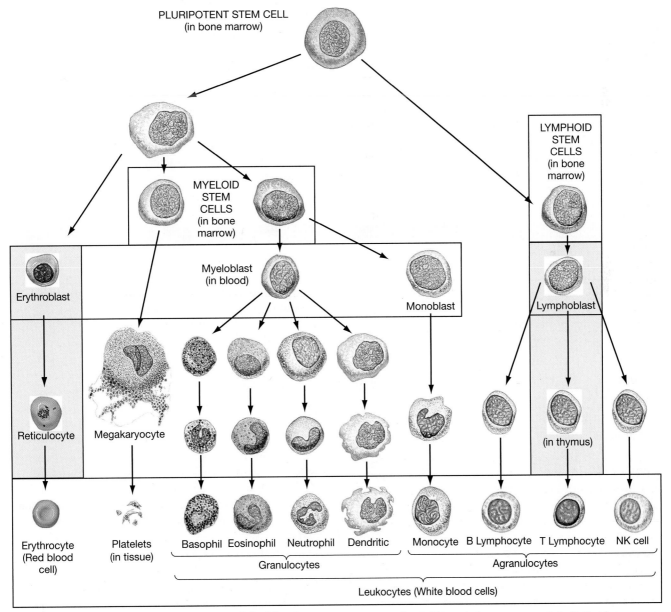

Figure 16.1 Formed (cellular) elements of the blood. These elements are derived from pluripotent stem cells (cells that form an endless supply of blood cells) in the bone marrow. The myeloid stem cells differentiate into several kinds of leukocytes, called granulocytes and agranulocytes. Lymphoid stem cells differentiate into B lymphocytes (B cells), T lymphocytes (T cells), and natural killer cells (NK cells).

defense against intestinal pathogens. Lysozyme, an enzyme present in tears, saliva, and mucus, cleaves the covalent linkage between the sugars in peptidoglycan; hence Gram-positive bacteria are particularly susceptible to killing by this enzyme (◀Chapter 19, p. 577) Transferrin, a protein present in the blood plasma, binds any free iron that is present in the blood. Bacteria require iron as a cofactor for some enzymes. The binding of iron by transferrin inhibits the growth of bacteria in the bloodstream. A similar protein, lactoferrin, present in saliva, mucus, and milk, also binds iron inhibiting bacterial growth. Small peptides called *defensins*, present in mucus and extracellular fluids, are a group of molecules that can kill pathogens by forming pores in their membranes, or inhibit growth by other mechanisms.

CELLULAR DEFENSES

Although the physical defense barriers do an excellent job of keeping microbes out of our bodies, we constantly suffer minor breaches of the physical defense barriers. A paper cut, the cracking of dry skin, or even brushing our teeth may temporarily breach the physical defenses and allow some microbes to enter the blood or connective tissue. However, we survive these daily attacks because ever-present cellular defenses can kill invading microbes or remove them from the blood or tissues.

When the skin is broken by any kind of trauma, microorganisms from the environment may enter the wound. Blood flowing out of the wound helps remove the microorganisms. Subsequent constriction of ruptured blood vessels and the clotting of blood help seal off the injured area until more permanent repair can occur. Still, if microorganisms enter blood through cuts in the skin or abrasions in mucous membranes, cellular defense mechanisms come into play.

DEFENSIVE CELLS

Cellular defense mechanisms use special-purpose cells found in the blood and other tissues of the body. Blood consists of about 60% liquid called **plasma** and 40% **formed elements** (cells and cell fragments). Formed elements include **erythrocytes** (red blood cells), **platelets**, and **leukocytes** (white blood cells) (**Figure 16.1** and **Table 16.1**). All are derived from *pluripotent stem cells*, cells that form a continuous supply of blood cells, in the bone marrow. Platelets, which are short-lived fragments of large cells called *megakaryocytes*, are important components of the blood-clotting mechanism.

Leukocytes are defensive cells that are important to both adaptive and innate host defenses. These cells are divided into two groups—granulocytes and agranulocytes—according to their cell characteristics and staining patterns with specific dyes.

TABLE 16.1

Formed Elements of the Blood in Healthy Adults			
Element	Normal Numbers (per microliter*)	Life Span	Functions
Erythrocytes		120 days	Transport oxygen gas from lungs to tissues; transport carbon dioxide gas from tissues to lungs
Adult male	4.6 to 6.2 million		
Adult female	4.2 to 5.4 million		
Newborn	5.0 to 5.1 million		
Leukocytes	5,000 to 9,000	Hours to days	
Granulocytes			
Dendritic cells			Phagocytic, antigen presentation in lymph node
Neutrophils	50–70% of total leukocytes		Phagocytic; contain oxidative chemicals to kill internalized microbes
Eosinophils	1–5% of total leukocytes		Release defensive chemicals to damage parasites (worms); phagocytic
Basophils	0.1% of total leukocytes		Release histamine and other chemicals during inflammation; responsible for allergic symptoms
Agranulocytes			
Monocytes	2–8% of total leukocytes		In tissues, develop into macrophages, which are phagocytic
Lymphocytes	20–50% of total leukocytes	Days to weeks	Essential to specific host immune defenses; antibody production
Platelets	250,000 to 300,000	5–9 days	Blood clotting

*1 microliter (μl) = 1 mm^3 = 1/1,000,000 liter.

Phlegm, Anyone?

Remember that thick, viscous mucus you coughed up last time you had a cold? Pretty gross stuff. And even grosser when you think of the tons of microorganisms your body had trapped with it. With barriers like that, how did those flu organisms manage to infect your respiratory tract in the first place? Some organisms, unfortunately, have evolved ways to get through this mucus barrier. For example, the influenza virus has a surface molecule that allows it to firmly attach itself to cells in the mucous membrane. Cilia can't sweep the attached virus out. As another example, the organism that causes gonorrhea has surface molecules that allow it to bind to mucous membrane cells in the urogential tract. With ingenious microorganisms like these, thank goodness your body has other defenses that lie in wait to attack any organisms that make it past your body's physical barriers.

Granulocytes

Granulocytes have granular cytoplasm and an irregularly shaped, lobed nucleus. They are derived from *myeloid stem cells* in the bone marrow (*myelos* is Greek for "marrow"). Granulocytes include basophils, mast cells, eosinophils, and neutrophils, which are distinguished from one another by the shape of their cell nuclei and by their staining reactions with specific dyes. **Basophils** release *histamine*, a chemical that helps initiate the inflammatory response. **Mast cells**, which are prevalent in connective tissue and alongside blood vessels, also release histamine and are associated with allergies. **Eosinophils** (e-o-sin′o-fils) are present in large numbers during allergic reactions (Chapter 18) and worm infections. These cells may also detoxify foreign substances and help turn off inflammatory reactions. **Neutrophils**, also called *polymorphonuclear leukocytes* (PMNLs), guard blood, skin, and mucous membranes against infection. These cells are phagocytic and respond quickly wherever tissue injury has occurred. **Dendritic cells** (DC) are cells with long membrane extensions that resemble the dendrites of nerve cells, hence their name. These cells are phagocytic and, as we will see in Chapter 17, are involved in initiating the adaptive defense response.

Agranulocytes

Agranulocytes lack granular cytoplasm and have round nuclei. These cells include monocytes and lymphocytes. **Monocytes** are derived from myeloid stem cells, whereas **lymphocytes** are derived from *lymphoid stem cells*, again in the bone marrow. The lymphocytes contribute to adaptive host immunity. They circulate in the blood and are found in large numbers in the lymph nodes, spleen, thymus, and tonsils.

The combined mass of all of the lymphocytes in your body is approximately equal to the mass of your brain or liver.

Neutrophils and monocytes are exceedingly important components of innate host defenses. They are phagocytic cells, or *phagocytes*.

PHAGOCYTES

Phagocytes are cells that literally eat (*phago*, Greek for "eating"; *cyte*, Greek for "cell") or engulf other materials. They patrol, or circulate through the body, destroying dead cells and cellular debris that must be removed constantly from the body as cells die and are replaced. Phagocytes also guard the skin and mucous membranes against invasion by microorganisms. Being present in many tissues, these cells first attack microbes and other foreign material at portals of entry, such as wounds in skin or mucous membranes. If some microbes escape destruction at the portal of entry and enter deeper tissues, phagocytes circulating in blood or lymph mount a second attack on them.

The neutrophils are released from the bone marrow continuously to maintain a stable circulating population. An adult has about 50 billion circulating neutrophils at all times. If an infection occurs, they are usually first on the scene because they migrate quickly to the site of infection. Being avid phagocytes, they are best at inactivating bacteria and other small particles. They are not capable of cell division and are "programmed" to die after only 1 or 2 days.

Neutrophils are released into the blood from the bone marrow, circulate for 7 to 10 hours, and then migrate into the tissues, where they have a 3-day life span.

The monocytes migrate from the bone marrow into the blood. When these cells move from blood into tissues, they go through a series of cellular changes, maturing into macrophages. **Macrophages** are "big eaters" (*macro*, Greek for "big") that destroy not only microorganisms but also larger particles, such as debris left from neutrophils that have died after ingesting bacteria. Although macrophages take longer than neutrophils to reach an infection site, they arrive in larger numbers.

Macrophages can be fixed or wandering. *Fixed macrophages* remain stationary in tissues and are given different names, depending on the tissue in which they reside (**Table 16.2**). *Wandering macrophages*, like the

TABLE 16.2

Names of Fixed Macrophages in Various Tissues	
Name of Macrophage	**Tissue**
Alveolar macrophage (dust cell)	Lung
Histiocyte	Connective tissue
Kupffer cell	Liver
Microglial cell	Neural tissue
Osteoclast	Bone
Sinusoidal lining cell	Spleen

SEM

Figure 16.2   False-color SEM of a macrophage moving over a surface (5,375X). The macrophage has spread out from its normal spherical shape and is using its ruffly cytoplasm to move itself and to engulf particles. Macrophages clear the lungs of dust, pollen, bacteria, and some components of tobacco smoke. *(SPL/Custom Medical Stock Photo, Inc.)*

neutrophils, circulate in the blood, moving into tissues when microbes and other foreign material are present **(Figure 16.2)**. Unlike neutrophils, macrophages can live for months or years. As we will see in ◄Chapter 17, besides having a nonspecific role in host defenses, macrophages also are critical to specific host defenses.

THE PROCESS OF PHAGOCYTOSIS

Phagocytes digest and generally destroy invading microbes and foreign particles by a process called **phagocytosis** ◄(Chapter 4) or by a combination of immune reactions and phagocytosis (p. 104). If an infection occurs, neutrophils and macrophages use this four-step process to destroy the invading microorganisms. The phagocytic cells must (1) find, (2) adhere to, (3) ingest, and (4) digest the microorganisms.

Fixed macrophages are named to reflect their tissue location: in the liver, they are called Kupffer cells; in the connective tissues, histiocytes; in the lung, alveoloar macrophages.

Chemotaxis

Phagocytes in tissues first must recognize the invading microorganisms. This is accomplished by receptors, called **toll-like receptors (TLRs)**, on the phagocytic cells that recognize molecular patterns unique to the pathogen, such as peptidoglycan, lipopolysaccharide, flagellin proteins, zymosan from yeast, and many other pathogen-specific molecules. Macrophages and dendritic cells can distinguish between Gram-negative and Gram-positive bacteria and between bacteria versus viral pathogens. They can then tailor the subsequent response to deal best with that type of pathogen. There are 10 TLRs now known in humans and over 200 in plants. Each is targeted at recognizing some particular bacterial, viral, or fungal component which is essential to the existence of that microbe; e.g., TLR 4 recognizes the lipopolysaccharide component of Gram-negative cell walls (◄Chapter 4, p. 85); TLRs 3,7, and 8 recognize the nucleic acids of viruses; TLR 5 recognizes a protein in bacterial flagella. They are called toll-like because they are closely related to the *toll* gene in fruitflies, which orients body parts properly. Files with defective *toll* genes have mixed-up, or weird-looking, bodies. *Toll* is the German word for weird. Both the infectious agents and the damaged tissues also release specific chemical substances to which monocytes and macrophages are attracted. In addition, basophils and mast cells release histamine, and phagocytes already at the infection site release chemicals called **cytokines** (si'to-kinz). These chemicals are a diverse group of small soluble proteins that have specific roles in host defenses, including the activation of cells involved in the inflammatory response. **Chemokines** are a class of cytokines that attract additional phagocytes to the site of the infection. Phagocytes make their way to this site by **chemotaxis**, the movement of cells toward a chemical stimulus (◄Chapter 4, p. 95). We will discuss cytokines in more depth in Chapter 17.

Some pathogens can escape phagocytes by interfering with chemotaxis. For example, most strains of the bacterium that causes gonorrhea (*Neisseria gonorrhoeae*) remain in the urogenital tract, but some strains escape local cellular defenses and enter the blood. Microbiologists believe that the invasive strains fail to release the chemical attractants that bring phagocytes to the infection site.

Adherence and Ingestion

Following chemotaxis and the arrival of phagocytes at the infection site, the infectious agents become attached to the plasma membranes of phagocytic cells. The ability of the phagocyte cell membrane to bind to specific molecules on the surface of the microbe is called **adherence**.

Complex antigens (substances that the body identifies as foreign), such as whole bacteria or viruses, tend to adhere well to phagocytes and are readily ingested.

A fundamental requirement for many pathogenic bacteria is to escape phagocytosis. The most common means by which bacteria avoid this defense mechanism is an *antiphagocytic capsule*. The capsules present on bacteria responsible for pneumococcal pneumonia (*Streptococcus pneumoniae*) and childhood meningitis (*Haemophilus influenzae*) make adherence difficult for phagocytes. The cell walls of the bacterium responsible for rheumatic fever (*Streptococcus pyogenes*) contain molecules of *M protein*, which interferes with adherence.

To overcome such resistance to adherence, the host's nonspecific defenses can make microbes more susceptible to phagocytosis. If microbes are first coated

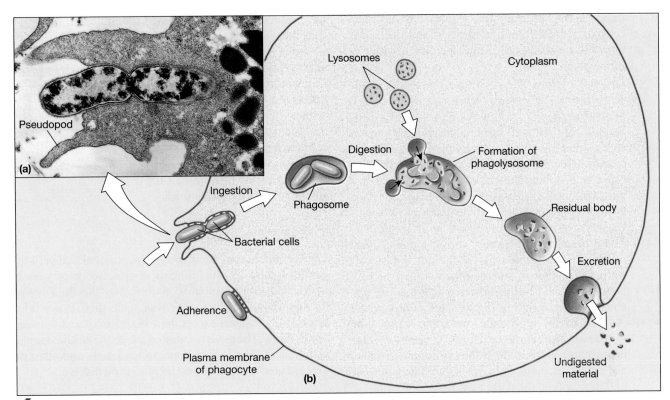

Figure 16.3 **Phagocytosis of two bacterial cells by a neutrophil.** **(a)** Extensions of cytoplasm, called pseudopodia, surround the bacteria. Fusion of the pseudopodia forms a cytoplasmic vacuole, called a phagosome, containing the bacteria (magnification unknown). *(Courtesy Dorothy F. Bainton, M.D., University of California at San Francisco)* **(b)** Phagocytes find their way to a site of infection by means of chemotaxis. Phagocytes, including macrophages and neutrophils, have proteins in their plasma membranes to which a bacterium adheres. The bacterium is then ingested into the cytoplasm of the phagocyte as a phagosome, which fuses with lysosomes to form a phagolysosome. The bacterium is digested, and any undigested material within the residual body is excreted from the cell.

with antibodies, or with proteins of the *complement system* (to be discussed later in this chapter), phagocytes have a much easier time binding to the microbes. Because both these mechanisms represent molecular defenses, we will discuss them later in this chapter.

Once captured, phagocytes rapidly ingest (engulf) the microbe. The cell membrane of the phagocyte forms fingerlike extensions, called *pseudopodia*, that surround the microbe (**Figure 16.3a**). These pseudopodia then fuse, enclosing the microbe within a cytoplasmic vacuole called a **phagosome** (**Figure 16.3b**).

Digestion

Phagocytic cells have several mechanisms for digesting and destroying ingested microbes. One mechanism uses the *lysosomes* found in the phagocyte's cytoplasm (◀Chapter 4, p. 102). These organelles, which contain digestive enzymes and small proteins called *defensins*, fuse with the phagosome membrane, forming a **phagolysosome** (Figure 16.3b). (More than 30 different types of antimicrobial enzymes have been identified with lysosomes.) In this way the digestive enzymes and defensins are released into the phagolysosome. The defensins eat holes in the cell membranes of microbes, allowing lysosomal enzymes to digest almost any biological molecule they contact. Thus, lysosomal enzymes rapidly (within 20 minutes) destroy the

microbes, breaking them into small molecules (amino acids, sugars, fatty acids) that the phagocyte can use as building blocks for its own metabolic and energy needs.

Macrophages can also use other metabolic products to kill ingested microbes. These phagocytic cells use oxygen to form hydrogen peroxide (H_2O_2), nitric oxide (NO), superoxide ions (O_2^-) and hypochlorite ions (OCl^-). (Hypochlorite is the ingredient in household bleach that accounts for its antimicrobial action.) All these molecules are effective in damaging plasma membranes of the ingested pathogens.

Once the microbes have been destroyed, there may be some indigestible material left over. Such material remains in the phagolysosome, which now is called a *residual body*. The phagocyte transports the residual body to the plasma membrane, where the waste is excreted (Figure 16.3b).

Just as some microbes interfere with chemotaxis and others avoid adherence, some microbes have developed mechanisms to prevent their destruction within a phagolysosome. In fact, a few pathogens even multiply within phagocytes. Some microbes resist digestion by phagocytes in one of three ways:

1. Some bacteria, such as those that cause the plague (*Yersinia pestis*), produce capsules that are not vulnerable to destruction by macrophages. If

these bacteria are engulfed by macrophages, their capsule protects them from lysosomal digestion, allowing the bacteria to multiply, even within a macrophage.

2. Other bacteria—such as those that cause Hansen's disease, or leprosy (*Mycobacterium leprae*), and tuberculosis (*M. tuberculosis*)—and the protozoan that causes leishmaniasis (*Leishmania* species) can resist digestion by phagocytes. In the case of *Mycobacterium*, each engulfed bacillus resides in a membrane-enclosed, fluid-filled compartment called a *parasitophorous vacuole* (PV). No lysosomal enzyme activity is associated with the PVs as they do not fuse with lysosomes. These organisms' resistance to lysosomal activity is due to the complexity of their acid-fast cell walls (◄Chapter 4, p. 88), which consist of wax D and mycolic acids. Lysosomal enzymes are unable to react with and digest these components. As the bacilli reproduce, new PVs arise. For *Leishmania* infections, each PV contains several protozoan cells. Although the lysosomal enzymes are active in these PVs, microbiologists do not understand how the pathogens resist digestion.

3. Still other microbes produce toxins that kill phagocytes by causing the release of the phagocyte's own lysosomal enzymes into its cytoplasm. Examples of such toxins are **leukocidin**, released by bacteria such as staphylococci, and **streptolysin**, released by streptococci.

Thus, some pathogens survive phagocytosis and can even be spread throughout the body in the phagocytes that attempt to destroy them. Because macrophages can live for months, they can provide pathogens with a long-term, stable environment in which they can multiply out of the reach of other host defense mechanisms.

EXTRACELLULAR KILLING

The phagocytic process described previously represents *intracellular killing*—that is, the microbe is degraded within a defense cell. However, other microbes, such as viruses and parasitic worms, are destroyed without being ingested by a defensive cell; they are destroyed *extracellularly* by products secreted by defensin cells.

Neutrophils and macrophages are too small to engulf a large parasite such as a worm (helminth). Therefore, another leukocyte, the eosinophil, takes the leading role in defending the body. Although eosinophils can be phagocytic, they are best suited for excreting toxic enzymes such as *major basic protein* (MBP) that can damage or perforate a worm's body. Once such parasites are destroyed, macrophages can engulf the parasite fragments.

Viruses must get inside cells to multiply (◄Chapter 1, p. 5). Therefore, host defenses must eliminate such infectious agents before they can reproduce in the cells they have infected. The leukocytes responsible for killing intracellular viruses are **natural killer (NK) cells**. NK cells are a type of lymphocyte whose activity is greatly increased by exposure to interferons and cytokines. Although the exact mechanism of recognition is not known, NK cells probably recognize specific glycoproteins on the cell surface of virus-infected cells. Such recognition does not lead to phagocytosis; rather, the NK cells secrete cytotoxic proteins that trigger the death of the infected cell.

> *In humans, Chediak-Higashi syndrome is associated with an absence of natural killer cells and with an increased incidence of lymphomas.*

THE LYMPHATIC SYSTEM

The **lymphatic system**, which is closely associated with the cardiovascular system, consists of a network of vessels, nodes and other lymphatic tissues, and the fluid *lymph* (**Figure 16.4**). The lymphatic system has three major functions: It (1) collects excess fluid from the spaces between body cells, (2) transports digested fats to the cardiovascular system, and (3) provides many of the innate and adaptive defense mechanisms against infection and disease.

Lymphatic Circulation

The process of draining excess fluid from the spaces between cells starts with the *lymphatic capillaries* found throughout the body. These capillaries, which are slightly larger in diameter than blood capillaries, collect the excess fluid and plasma proteins that leak from the blood into the spaces between cells. Once in the lymphatic capillaries, this fluid is called **lymph**. Lymphatic capillaries join to form larger **lymphatic vessels**. As fluid moves through the vessels, it passes through **lymph nodes**. Finally, the lymph is returned to the venous blood via the *right* and *left lymphatic ducts*, which drain the fluids into the right and left subclavian veins. There is no mechanism to move or pump lymphatic fluid. Hence, the flow of lymph depends on skeletal muscle contractions, which squeeze the vessels, forcing the lymph toward the lymphatic ducts. Throughout the lymphatic system, there are one-way valves to prevent backflow of lymph.

Lymphoid Organs

Specific organs of the lymphatic system are essential in the body's defense against infectious agents and cancers. These organs include the lymph nodes, thymus, and spleen. Although all lymphatic organs contain numerous lymphocytes, these cells originate in bone marrow and are released into blood and lymph. They live from weeks to years, becoming dispersed to various lymphatic organs or remaining in the blood and lymph. In humans most lymphocytes are either *B lymphocytes* (B cells) or *T lymphocytes* (T cells). B cells differentiate in the bone marrow itself and migrate to the lymph nodes and spleen. Immature T cells from the bone marrow migrate to the thymus, where they mature; they then migrate to

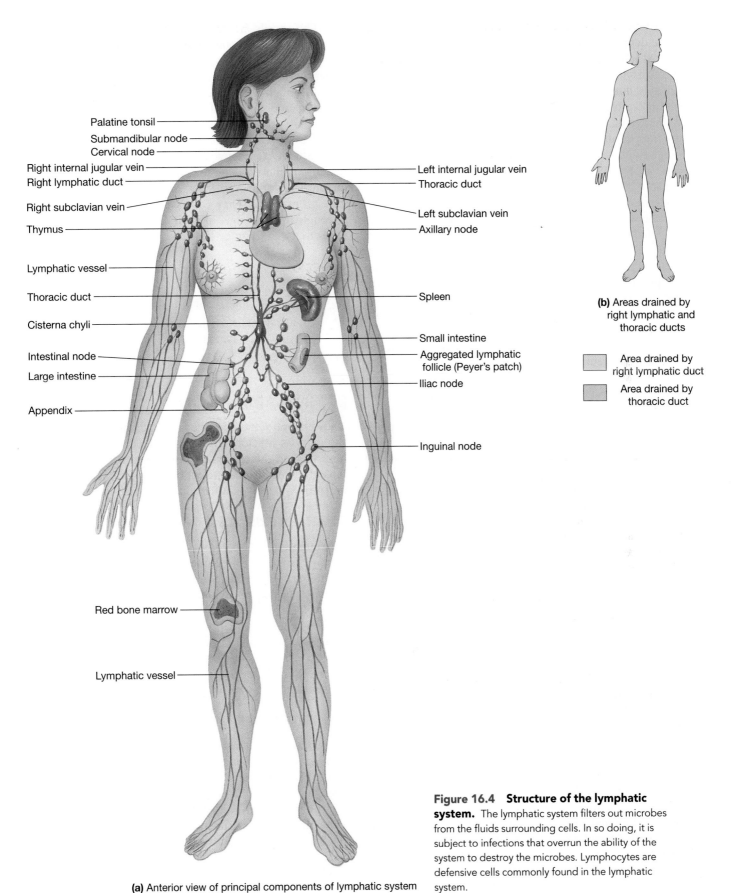

Palatine tonsil
Submandibular node
Cervical node
Right internal jugular vein
Right lymphatic duct
Right subclavian vein
Thymus
Lymphatic vessel
Thoracic duct
Cisterna chyli
Intestinal node
Large intestine
Appendix
Red bone marrow
Lymphatic vessel

Left internal jugular vein
Thoracic duct
Left subclavian vein
Axillary node
Spleen
Small intestine
Aggregated lymphatic follicle (Peyer's patch)
Iliac node
Inguinal node

(b) Areas drained by right lymphatic and thoracic ducts

Area drained by right lymphatic duct

Area drained by thoracic duct

(a) Anterior view of principal components of lymphatic system

Figure 16.4 Structure of the lymphatic system. The lymphatic system filters out microbes from the fluids surrounding cells. In so doing, it is subject to infections that overrun the ability of the system to destroy the microbes. Lymphocytes are defensive cells commonly found in the lymphatic system.

the lymph nodes or spleen. We will discuss these cells in more depth in Chapter 17.

At intervals along the lymphatic vessels, lymph flows through lymph nodes distributed throughout the body. They are most numerous in the thoracic (chest) region, neck, armpits, and groin. The lymph nodes filter out foreign material in the lymph. Most foreign agents passing through a node are trapped and destroyed by the defensive cells present.

Lymph nodes occur in small groups, each group covered in a network of connective tissue fibers called a **capsule (Figure 16.5)**. Lymph moves through a lymph node in one direction. Lymph first enters **sinuses**, wide passageways lined with phagocytic cells, in the outer cortex of the lymph node. The *outer cortex* houses large aggregations of B lymphocytes. The lymph then passes through the *deep cortex*, where T lymphocytes exist. The lymph moves through the inner region of a lymph node, the *medulla*, which contains B lymphocytes, macrophages, and plasma cells. Finally, lymph moves through sinuses in the medulla and leaves the lymph node.

This filtration of the lymph is important when an infection has occurred. For example, if a bacterial infection occurs, the bacteria that are not destroyed at the site of the infection may be carried to the lymph nodes. As the lymph passes through the nodes, a majority of the bacteria are removed. Macrophages and other phagocytic cells, especially dendritic cells, in the nodes bind to and phagocytize the bacterial cells, thereby initiating an adaptive immune response (Chapter 17).

The **thymus gland** is a multilobed lymphatic organ located beneath the sternum (breastbone) (Figure 16.4). It is present at birth, grows until puberty, then atrophies (shrinks) and is mostly replaced by fat and connective tissue by adulthood. Around the time of birth, the thymus begins to process lymphocytes and releases them into the blood as T cells. T cells play several roles in immunity: they regulate the development of B cells into antibody-producing cells, and subpopulations of T cells can kill virus-infected cells directly.

The **spleen**, located in the upper left quadrant of the abdominal cavity, is the largest of the lymphatic organs (Figure 16.4). Anatomically, the spleen is similar to the lymph nodes. It is encapsulated, lobed, and well supplied with blood and lymphatic vessels. Although it does not filter material, its sinusoids contain many phagocytes that engulf and digest worn-out erythrocytes and microorganisms. It also contains B cells and T cells.

Other Lymphoid Tissues

Earlier, we mentioned the lymphoid masses found in the ileum of the small intestine. Called Peyer's patches, these are **lymphoid nodules**, unencapsulated areas filled with lymphocytes. Collectively, the tissues of lymphoid nodules are referred to as **gut-associated lymphatic tissue (GALT)**, which are major sites of antibody production against mucosal pathogens. Similar nodules are found in the respiratory system, urinary tract, and appendix.

The **tonsils** are another site for the aggregation of lymphocytes. Although these tissues are not essential for fighting infections, they do contribute to immune defenses, as they contain B cells and T cells.

Although lymphatic tissues contain cells that phagocytize microorganisms, if these cells encounter more pathogens than they can destroy, the lymphatic tissues can become sites of infection. Thus, swollen lymph nodes and tonsillitis are common signs of many infectious diseases.

In summary, lymphoid tissues contribute to innate defenses by phagocytizing microorganisms and other foreign material. They contribute to adaptive immunity through the activities of their B and T cells, which we will discuss in Chapter 17.

✓ **CHECKLIST**

1. How do innate and adaptive defenses differ?
2. List six categories of innate defenses.
3. List and describe the steps in phagocytosis.
4. What are NK cells and how do they function?
5. What are the parts and functions of the lymphatic system?

▐▐▐ INFLAMMATION

Do you remember the last time you cut yourself? If the cut was not too serious, the bleeding soon stopped. You washed the cut and put on a bandage. A few hours later the area around the cut became warm, red, swollen, and perhaps even painful. It had become *inflamed*.

CHARACTERISTICS OF INFLAMMATION

Inflammation is the body's defensive response to tissue damage from microbial infection. It is also a response to mechanical injury (cuts and abrasions), heat and electricity (burns), ultraviolet light (sunburn), chemicals (phenols, acids, and alkalis), and allergies. But whatever the cause of inflammation, it is characterized by *cardinal signs* or symptoms: (1) calor—an increase in temperature, (2) rubor—redness, (3) tumor—swelling, and (4) dolor—pain at the infected or injured site. What happens in the inflammatory process, and why?

THE ACUTE INFLAMMATORY PROCESS

The duration of inflammation can be either acute (short-term) or chronic (long-term). In **acute inflammation**, the battle between microbes (or other agents of inflammation) and host defenses usually is won by the host. In an infection, acute inflammation functions to (1) kill invading microbes, (2) clear away tissue debris, and (3) repair injured tissue. Let's look at acute inflammation more closely. **Figure 16.6** illustrates the steps described next.

As phagocytic cells accumulate at the site of inflammation and begin to ingest bacteria, they release lytic enzymes, which can damage nearby healthy cells.

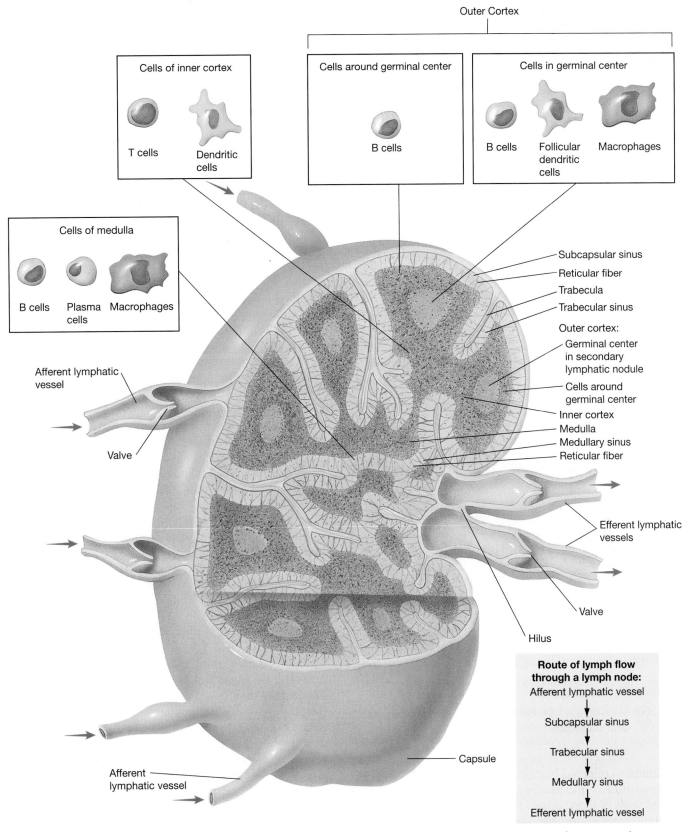

Figure 16.5 Structure of a lymph node. Lymph nodes are centers for removing microbes. These tissues contain phagocytes and lymphocytes. Swollen lymph nodes are usually an indication of a serious infection.

Epithelium   1. Cut allows bacteria to get beneath surface of skin.

2. Damaged cells release histamine and bradykinin.

3. Capillaries dilate (vasodilation), bringing more blood to the tissue. Skin becomes reddened and warmer.

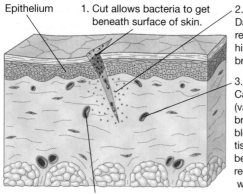

4. Capillaries become more permeable, allowing fluids to accumulate and cause swelling (edema).

5. Blood clotting occurs, and scab forms.

6. Bacteria multiply in cut.

7. Phagocytes enter tissue by moving through the walls of blood vessels (diapedesis).

8. Phagocytic cells are attracted to bacteria and tissue debris (chemotaxis) and engulf them.

9. Larger blood vessels dilate, further increasing blood supply to tissue and adding to heat and redness.

10. As dead cells and debris are removed, epithelial cells proliferate and begin to grow under the scab.

11. Scar tissue (connective tissue) replaces cells that replace themselves.

 Figure 16.6 Steps in the process of inflammation and subsequent healing.

When cells are damaged, the chemical substance histamine is released from basophils and mast cells. **Histamine** diffuses into nearby capillaries and venules, causing the walls of these vessels to dilate (**vasodilation**) and become more permeable. Dilation increases the amount of blood flowing to the damaged area, and it causes the skin around wounds to become red and warm to the touch. Because the vessel walls are more permeable, fluids leave the blood and accumulate around the injured cells, causing **edema** (swelling). The blood delivers clotting factors,

nutrients, and other substances to the injured area and removes wastes and some excess fluids. It also brings macrophages, which release cytokines. Some cytokines are chemokines and attract other phagocytes, and another cytokine, called *tumor necrosis factor alpha* (TNF-α) additionally causes vasodilation and edema.

All kinds of tissue injury—burns, cuts, infections, insect bites, allergies—cause histamine release. In conjunction with its effects on blood vessels, histamine also causes the red, watery eyes and runny nose of hay fever and the breathing difficulties in certain allergies. The drugs called **antihistamines** alleviate such symptoms by blocking the released histamine from reaching its receptors on target organs.

The fluid that enters the injured tissue carries the chemical components of the blood-clotting mechanism. If the injury has caused bleeding, platelets and clotting factors, such as fibrin, stop the bleeding by forming a blood clot in the injured blood vessel. Because clotting takes place near the injury, it greatly reduces fluid movement around damaged cells and walls off the injured area from the rest of the body. Pain associated with tissue injury is thought to be due to the release of **bradykinin**, a small peptide, at the injured site. How bradykinin stimulates pain receptors in the skin is unknown, but cellular regulators called **prostaglandins** seem to intensify bradykinin's effect.

Aspirin relieves pain by inhibiting prostaglandin synthesis.

Inflamed tissues also stimulate **leukocytosis**, an increase in the number of leukocytes in the blood. To do this, the damaged cells release cytokines that trigger the production and infiltration of more leukocytes. Within an hour after the inflammatory process begins, phagocytes start to arrive at the injured or infected site. For example, neutrophils pass out of the blood by squeezing between endothelial cells lining the vessel walls. This process, called **diapedesis** (di-a-pe-de′sis), allows neutrophils to congregate in tissue fluids at the injured region.

As we discussed earlier, when phagocytes reach an infected area, they attempt to engulf the invading microbes by phagocytosis. In that process many of the phagocytes themselves die. The accumulation of dead phagocytes, injured or damaged cells, the remains of ingested organisms, and other tissue debris forms the white or yellow fluid called **pus**. Many bacteria, such as *Streptococcus pyogenes*, cause pus formation because of their ability to produce leukocidins that destroy phagocytes. Viruses lack this activity and do not cause pus formation. Pus continues to form until the infection or tissue damage has been brought under control. An accumulation of pus in a cavity hollowed out by tissue damage is called an **abscess**. Boils and pimples are common kinds of abscesses.

Although the inflammatory process is usually beneficial, it can sometimes be harmful. For example, inflammation can cause swelling (edema) of the membranes (meninges) surrounding the brain or spinal cord, leading to brain damage. Swelling, which delivers phagocytes to injured tissue, can also interfere with breathing if it constricts the airways in the lung. Moreover, vasodilation delivers more oxygen and nutrients to injured tissues. Ordinarily this is of greater benefit to host cells than to pathogens, but sometimes it helps the pathogens thrive as well. Even though rapid clotting and the walling off of an injured area prevents pathogens from spreading, it can also prevent natural defenses and antibiotics from reaching the pathogens. Boils must be lanced before therapeutic drugs can reach them. Attempting to suppress the inflammatory process also can be harmful. Such attempts can allow boils to form when natural defenses might otherwise destroy the bacteria.

In summary, cellular defense mechanisms usually prevent an infection from spreading or from getting worse. However, sometimes these innate defense mechanisms are overwhelmed by sheer numbers of microbes or are inhibited by virulence factors that the microbes possess. The pathogens can then invade other parts of the body. For bacterial infections, medical intervention with antibiotics may inhibit microbial growth in injured tissue and reduce the chance of an infection spreading. Despite such measures, however, infections do spread. In Chapter 17 we will describe the mechanisms by which various lymphocytes act as agents of adaptive host immune defenses that help overcome an initial infection and prevent future infections by the same microbe.

REPAIR AND REGENERATION

During the entire inflammatory reaction, the healing process is also underway. Once the inflammatory reaction has subsided and most of the debris has been cleared away, healing accelerates. Capillaries grow into the blood clot, and **fibroblasts**, connective tissue cells, replace the destroyed tissue as the clot dissolves. The fragile, reddish, grainy tissue seen at the cut site consists of capillaries and fibroblasts called **granulation tissue**. As granulation tissue accumulates fibroblasts and fibers, it replaces nerve and muscle tissues that cannot be regenerated. New epidermis replaces the part destroyed. In the digestive tract and other organs lined with epithelium, an injured lining can similarly be replaced. Although scar tissue is not as elastic as the original tissue, it does provide a strong durable "patch" that allows the remaining normal tissue to function.

Several factors affect the healing process. The tissues of young people heal more rapidly than those of older people. The reason is that the cells of the young divide more quickly, their bodies are generally in a better nutritional state, and their blood circulation is more efficient. As you might guess from the many contributions of blood to healing, good circulation is extremely important. Certain vitamins also are important in the healing process. Vitamin A is essential for the division

of epithelial cells, and vitamin C is essential for the production of collagen and other components of connective tissue. Vitamin K is required for blood clotting, and vitamin E also may promote healing and reduce the amount of scar tissue formed.

CHRONIC INFLAMMATION

Sometimes an acute inflammation becomes a **chronic inflammation**, in which neither the agent of inflammation nor the host is a decisive winner of the battle. Rather, the agent causing the inflammation continues to produce tissue damage as the phagocytic cells and other host defenses attempt to destroy or at least confine the region of inflammation. In the process, pus may be formed continuously. Such chronic inflammation can persist for years.

Because the cause of inflammation is not destroyed, host defenses attempt to limit or confine the agent so that it cannot spread to surrounding tissue. For example, **granulomatous inflammation** results in granulomas. A **granuloma** is a pocket of tissue that surrounds and walls off the inflammatory agent. The central region of a granuloma contains epithelial cells and macrophages; the latter may fuse to form giant, multinucleate cells. Collagen fibers, which help wall off the inflammatory agent, and lymphocytes surround the core. Granulomas associated with a specific disease are sometimes given special names—for example, *gummas* (syphilis; ◀Figure 20.14, p. 622), *lepromas* (Hansen's disease; ◀Figure 24.5, p. 762), and *tubercles* (tuberculosis; ◀Figure 21.13, p. 656).

Tubercles usually contain necrotic (dead) tissue in the central region of the granuloma. As long as necrotic tissue is present, the inflammatory response will persist. If only a small quantity of necrotic tissue is present, the lesions sometimes become hardened as calcium is deposited in them. Calcified lesions are common in tuberculosis patients. When an anti-inflammatory drug such as cortisone is given, the organisms isolated in tubercles may be liberated and signs and symptoms of tuberculosis reappear (secondary tuberculosis).

▌▌▌ FEVER

A rise in temperature in infected or injured tissue is one sign of a local inflammatory reaction. **Fever**, a systemic increase in body temperature, often accompanies inflammation. Fever was first studied in 1868, when the German physician Carl Wunderlich devised a method to measure body temperature. He placed a foot-long thermometer in the armpit of his patients and left it in place for 30 minutes! Using this cumbersome technique, he could record human body temperatures during *febrile* (feverish) illnesses.

Normal body temperature is about 37°C (98.6°F), although individual variations in normal temperature

APPLICATIONS

Sweat It Out, Grandma

When you're in bed with a fever, it's hard to believe that fevers aren't just annoying side effects of being sick. They are actually important in fighting off infections. That's bad news for Grandma and Grandpa, since elderly people have trouble generating fevers. But a researcher at the University of Delaware in Newark found that sick geriatric rats, which also have problems developing fevers, benefited from living in rooms heated to 100°C. That doesn't necessarily mean that humans will benefit from such high temperatures, but if further studies show that they do, then cranking up the thermostat may help Grandma and Grandpa fight off the flu and other infections.

within the range 36.1° to 37.5°C (97.0° to 99.5°F) are not uncommon. Fever is defined clinically as an oral temperature above 37.8°C (100.5°F) or rectal temperature of 38.4°C (101.5°F). Fever accompanying infectious diseases rarely exceeds 40°C (104.5°F); if it reaches 43°C (109.4°F), death usually results.

Body temperature is maintained within a narrow range by a temperature-regulating center in the *hypothalamus*, a part of the brain. Fever occurs when the temperature established for this mechanism is reset and raised to a higher temperature. Fever can be caused by many pathogens, by certain immunological processes (such as reactions to vaccines), and by nearly any kind of tissue injury, even heart attacks. Most often, fever is caused by a substance called a **pyrogen** (*pyro*, Greek for "fire") (◀Chapter 14, p. 418). **Exogenous pyrogens** include exotoxins and endotoxins from infectious agents. These toxins cause fever by stimulating the release of an **endogenous pyrogen** from macrophages. The endogenous pyrogen is yet another cytokine, called *interleukin-1* (IL-1), that circulates via the blood to the hypothalamus, where it causes certain neurons to secrete prostaglandins. The prostaglandins then reset the hypothalamus thermostat at a higher temperature, which then causes the body temperature to begin rising within 20 minutes. In such situations, body temperature is still regulated, but the body's "thermostat" is reset at a higher temperature. (The sensation of chills that sometimes accompanies a fever was described in ◀Chapter 14, p. 418.)

Fever has several beneficial roles: (1) It raises the body temperature above the optimum temperature for growth of many pathogens. This slows their rate of growth, reducing the number of microorganisms to be combated. (2) At the higher temperatures of fever, some microbial enzymes or toxins may be inactivated. (3) Fever can heighten the level of immune responses by increasing the rate of chemical reactions in the body. This results in a faster rate at which the body's defense mechanisms attack pathogens, shortening the course of the infection. (4) Phagocytosis is enhanced. (5) The production of antiviral interferon is increased. (6) Breakdown of lysosomes is heightened, causing death of infected cells and the microbes inside of them. (7) Fever makes a patient feel ill. In this condition the patient is more likely to rest, preventing further damage to the body and allowing energy to be used to fight the infection.

In an infection, cells also release **leukocyte-endogenous mediator (LEM)**. Besides helping to elevate body temperature, LEM decreases the amount of iron absorbed from the digestive tract and increases the rate at which it is moved to iron storage deposits. Thus, LEM lowers the plasma iron concentration. Without adequate iron, growth of microorganisms is slowed (◀Chapter 6, p. 162).

Our current knowledge of the importance of fever has changed the clinical approach to this symptom. In the past, *antipyretics*—fever-reducing drugs such as aspirin—were given almost routinely to reduce fever caused by infections. For the beneficial effects cited above, many physicians now recommend allowing fevers to run their course. Evidence shows that medication can delay recovery. However, if a fever goes above 40°C or if the patient has a disorder that might be worsened by fever, antipyretics are still used. In fact, untreated extreme fever increases the metabolic rate by 20%, makes the heart work harder, increases water loss, alters electrolyte concentrations, and can cause convulsions, especially in children. Thus, patients with severe heart disease or fluid and electrolyte imbalances, as well as children subject to convulsions, usually receive antipyretics.

✓ CHECKLIST

1. List the cardinal signs or symptoms of inflammation.
2. What is the role of histamine in the inflammatory process?
3. Define *diapedesis*, *pus*, *edema*, *granuloma*, and *pyrogen*.
4. List four benefits of fever.

▌▌▌ MOLECULAR DEFENSES

Along with cellular defenses, inflammation, and fever, molecular defenses represent another formidable innate defense barrier. These molecular defenses involve the actions of *interferon* and *complement*.

INTERFERON

As early as the 1930s, scientists observed that infection by one virus prevented for a time infection by another virus. Then, in 1957, a small, soluble protein was discovered that was responsible for this viral interference. This protein, called **interferon** (in-ter-fer′on), "interfered" with virion replication in other cells. Such a molecule suggested to virologists that they might have the "magic bullet" for viral infections, similar to the antibiotics used to treat bacterial infections. As we will see, such hope has dwindled somewhat.

TABLE 16.3

Properties of Type I and Type II Human Interferons

Class	Cell Source	Subtypes	Stimulated By	Effects
Type I				
Alpha-interferon (INF-α)	Leukocytes	20	Viruses	Production of antiviral proteins in neighboring cells
Beta-interferon (INF-β)	Fibroblasts	1	Viruses	Same as INF-α
Type II				
Gamma-interferon (INF-γ)	T lymphocytes and NK cells	1	Viruses and other antigens	Activates tumor destruction and killing of infected cells

Take another look

Efforts to purify interferon led to the discovery that many different subtypes of interferon exist in different animal species, and that those produced by one species may be ineffective in other species. For example, interferon produced in a chicken is useful in protecting other chicken cells from viral infection. But chicken interferon is of no use in preventing viral infections in mice or in humans. Different interferons also exist in different tissues of the same animal. In humans there are three groups of interferons, called alpha (α), beta (β), and gamma (γ) (Table 16.3). Analysis of the protein structure and function show *α-interferon* and *β-interferon* to be similar, so they are placed together as *type I interferons*. *Gamma-interferon* is different structurally and functionally and represents the only known *type II interferon*.

Many researchers have tried to determine how these interferons act. The synthesis of α-interferon and β-

Interferons are usually species-specific but virus nonspecific.

interferon occurs after a virus infects a cell (**Figure 16.7**). These interferons do not interfere directly with viral replication. Rather, after viral infection, the cell synthesizes and secretes minute amounts of interferon. The interferon then diffuses to adjacent, uninfected cells and binds to their surfaces. Binding stimulates those cells to transcribe specific genes into mRNA molecules, which are then translated to produce many new proteins, most of them enzymes. Together these enzymes are called **antiviral proteins** (AVPs). Although viruses still infect cells possessing the AVPs, many of the proteins interfere with virus replication.

The AVPs are specifically effective against RNA viruses. Recall from ◄Chapter 10 (p. 278) that all RNA viruses must either produce dsRNA (Reoviridae) or go through a dsRNA stage during replication of (−) sense or (+) sense RNA. Two of the AVPs digest mRNA and limit translation of viral mRNA. The result is that the

Interferons are produced and released in response to viral infections, double-stranded RNA, endotoxins, and many parasitic organisms.

Figure 16.7 **The mechanism by which interferons α and β act.**

AVPs prevent the formation of new viral nucleic acid and capsid proteins. The infected cell that initially produced the interferon is thus surrounded by cells that can resist the replication of viruses, limiting viral spread.

Gamma-interferon also can block virus replication by AVP synthesis. However, lymphocytes and NK cells do not have to be infected with a virus to synthesize γ-interferon. Rather, it is produced in uninfected lymphocytes and NK cells that are sensitive to specific foreign antigens (viruses, bacteria, tumor cells) present in the body. The exact role of γ-interferon is unclear, but it is known to enhance the activities of lymphocytes, NK cells, and macrophages—the cells needed to attack microbes and tumors. It also enhances adaptive immunity by increasing antigen presentation (Chapter 17). Gamma-interferon (along with tumor necrosis factor-α, or TNF-α) also helps infected macrophages rid themselves of pathogens. For example, we mentioned earlier that macrophages can become infected with *Mycobacterium* bacilli. Such infected macrophages can be activated by γ-interferon and TNF-α, which bind to infected macrophages. New bactericidal activity is thereby triggered within the macrophage, usually leading to death of the bacteria and the restoration of normal macrophage function.

Therapeutic Uses of Interferon

Besides having the ability to block virus replication, interferons can also stimulate adaptive immune defenses. Therefore, interferons provide a potential therapy for viral infections and tumors. Unfortunately, infected animal cells produce very small quantities of interferons. However, today *recombinant interferon* (rINF) can be produced more cheaply and abundantly by using recombinant DNA techniques (◄Chapter 8, p. 228). Manufacture of recombinant interferon starts with the isolation and copying of the interferon gene and its insertion into plasmids. When recombinant plasmids are mixed with appropriate bacterial or yeast cells, some cells will take up the gene-containing plasmid and thereby acquire the human interferon gene. By growing these bacterial or yeast cells in very large vats and extracting the interferon that they produce, pharmaceutical companies can produce relatively significant quantities of recombinant interferon.

The ability to produce recombinant interferons spurred research on therapeutic applications for these proteins. In 1986, α-interferon was approved by the FDA for treating hairy cell leukemia, a very rare blood cancer. Since then, interferons have been approved for treatment of several other viral diseases, including genital warts and cancer. However, in most cases interferon is a treatment, not a cure. Patients must remain on the drug throughout their lives. With hairy cell leukemia, for example, removal of the drug results in a recurrence of the disease in 90% of the patients. For hepatitis C virus infection, treatment must be given 3 times a week for 6 months. Even so, if the patient is taken off treatment, the disease will reappear after 6 months in 70% of the cases.

Other studies have looked at the value of interferons to treat cancer. Tests on one form of bone cancer show that after most of the cancerous tissue is removed by surgery or destroyed by radiation, interferon therapy will reduce the incidence of metastasis (spread). How interferon stops metastasis is not known. Some cancers are the result of viral infections. Perhaps interferon interferes with viral replication. In addition to bone cancer, interferon is now used to treat renal cell carcinoma, kidney cancer, melanoma, multiple myeloma, carcinoid tumors, and some lymphomas. Interferon therapy could also prevent growth of the cancer cells through their destruction by macrophages and NK cells.

The therapeutic use of interferons has some drawbacks. When rINF is injected, it does not remain stable for very long in the body. This makes delivery of the interferons to the site of infection difficult. Recent research has led to the development of rINF that is chemically altered and remains active in the body longer. Injection of interferon (especially α-interferon) also has side effects, including fatigue, nausea, headache, vomiting, weight loss, and nervous system disorders. Whereas fever normally increases interferon production, which helps the body fight viral infections, the injection of interferon *produces* fever as a side effect. High doses can cause toxicity to the liver, kidneys, heart, and bone marrow.

Moreover, some microbes have developed resistance to interferons. Although some DNA viruses, such as the poxviruses, stimulate interferon synthesis, the human adenoviruses have resistance mechanisms to combat antiviral protein activity. In addition, the hepatitis B virus often fails to stimulate adequate interferon production in infected cells.

The therapeutic usefulness of interferon is clearly not the viral magic bullet that was originally envisioned. Nevertheless, interferons are being used to treat life-threatening viral infections and cancers.

COMPLEMENT

Complement, or the **complement system**, refers to a set of more than 20 large regulatory proteins that play a key role in host defense. They are produced by the liver and circulate in plasma in an inactive form. These proteins account for about 10% (by weight) of all plasma proteins. When complement was discovered, it was believed to be a single substance that "complemented," or completed, certain immunological reactions. Although complement can be activated by immune reactions, its effects are nonspecific—it exerts the same defensive effects regardless of which microorganism has invaded the body.

The general functions of the complement system are to (1) enhance phagocytosis by phagocytes; (2) lyse microorganisms, bacteria, and enveloped viruses directly; and (3) generate peptide fragments that regulate inflammation and immune responses. Furthermore, complement goes to work as soon as an invading microbe is detected; the system makes up an effective innate host defense long before adaptive host immune defenses are mobilized.

The complement system works as a cascade. A **cascade** is a set of reactions that amplify some effect—that is, more product is formed in the second reaction than in the first, still more in the third, and so on. Of the 20 different serum proteins so far identified in the complement system, 13 participate in the cascade itself, and 7 activate or inhibit reactions in the cascade.

Complement Function

Two pathways have been identified in the sequence of reactions carried out by the complement system. They are called the **classical pathway** and the **alternative pathway**, or *properdin pathway* (**Figure 16.8a**). The classical pathway begins when antibodies bind to antigens, such as microbes, and involves complement proteins C1, C4, and C2 (*C* stands for complement). The alternative pathway is activated by contact between complement proteins and polysaccharides at the pathogen surface. Complement proteins called factor B, factor D, and factor P (*properdin*) replace C1, C4, and C2 in the initial steps. However, the components of both pathways activate reactions involving C3 through C9.

The numbers attached to the complement cascade refer to their order of discovery, not the sequence in which they act.

Figure 16.8 The complement system. **(a)** Classical and alternative pathways of the complement cascade. Although the two pathways are initiated in different ways, they combine to activate the complement system. **(b)** Activation of the classical complement pathway. In this cascade each complement protein activates the next one in the pathway. The action of C3b is critical for opsonization and, along with C5b, for formation of membrane attack complexes. C4a, C3a, and C5a also are important to inflammation and phagocyte chemotaxis. (IgG is a class of antibodies that we will discuss in Chapter 17.)

(a)

(b)

Consequently, the effects of the complement systems are the same regardless of the pathway by which C3 is produced. However, the alternative pathway is activated even earlier in an infection than is the classical pathway.

The contributions of the complement system to innate defenses depend on C3, a key protein in the system. Once C3 is formed, it immediately splits into C3a and C3b, which then participate in three kinds of molecular defenses: opsonization, inflammation, and membrane attack complexes **(Figure 16.8b)**.

Opsonization. Earlier, we mentioned that some bacteria with capsules or surface proteins (M proteins) can prevent phagocytes from adhering to them. The complement system can counteract these defenses, making possible a more efficient elimination of such bacteria. First, special antibodies called **opsonins** bind to and coat the surface of the infectious agent. C1 binds to these antibodies, initiating the cascade. C1 causes the cleavage of C4 into C4a and C4b. C4b and C1 then cause C2 to split into C2a and C2b. The C4bC2a complex in turn leads to the splitting of C3 into C3a and C3b. C3b then binds to the surface of the microbe. Complement receptors on the plasma membrane of phagocytes recognize the C3b molecules; this recognition stimulates phagocytosis. This process, initiated by opsonins, is called **opsonization**, or *immune adherence*.

Inflammation. The complement system is also potent in initiating and enhancing inflammation. C3a, C4a, and C5a enhance the acute inflammatory reaction by stimulating chemotaxis and thus phagocytosis. These three complement proteins also adhere to the membranes of basophils and mast cells, causing them to release histamine and other substances that increase the permeability of blood vessels.

Membrane Attack Complexes. Another defense triggered by C3b is cell lysis. By a process called **immune cytolysis**, complement proteins produce lesions in the cell membranes of microorganisms and other types of cells. These lesions cause cellular contents to leak out. To cause immune cytolysis, C3b initiates the splitting of C5 into C5a and C5b. C5b then binds C6 and C7, forming a C5bC6C7 complex. This protein complex is hydrophobic (◀Chapter 4, p. 90) and inserts into the microbial cell membrane. C8 then binds to C5b in the membrane. Each C5bC6C7C8 complex causes the assembly in the cell membrane of up to 15 C9 molecules **(Figure 16.9)**. By extending all the way through the cell membrane, these proteins form a pore and constitute the **membrane attack complex (MAC)**. The MAC is responsible for the direct lysis of invading microorganisms. Importantly, host plasma

The completed MAC has a tubular form and a functional pore size of 70 to 100 angstroms (1 angstrom = 10⁻¹⁰ meter).

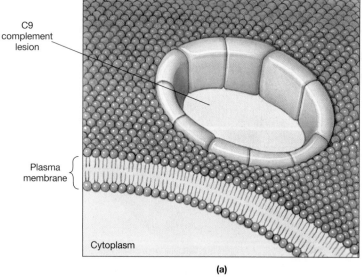

C9 complement lesion

Plasma membrane

Cytoplasm

(a)

(b)

(c)

Figure 16.9 Complement lesions in cell membranes. (a) Complement lyses a bacterial cell by creating a membrane attack complex (lesion) consisting of 10 to 15 molecules of C9. These protein molecules form a hole in the cell membrane through which the cytoplasmic contents leak out. **(b)** An EM showing the holes formed in red blood cell membranes by C9 (magnification unknown). *(From Sucharit Bhakdi et al., "Functions and relevance of the terminal complement sequence," Blut, vol. 60, p. 311, 1990. Reproduced by permission of Springer-Verlag New York, Inc.)* **(c)** Side view of complement lesion (MAC), 2,240,000X. The shorter arrows point to the edge of the cell membrane. The longer arrows point to the MAC itself, which consists of a cylinder with a central channel penetrating the cell membrane. This channel causes the flow of ions into and out of the cell to be unbalanced and results in lysis. Evidence suggests that the complement lesion consists almost entirely of C9. *(Courtesy Robert Dourmashkin, St. Bart's and Royal London School of Medicine).*

TEM

APPLICATIONS

Development of the Immune System

Can all organisms defend themselves against attacks by infectious microbes? For vertebrates, the answer is yes. As we saw in this chapter, they have nonspecific defense mechanisms, and as we will see in Chapter 17, they also have well-developed specific immune defenses.

Defenses against infection are not limited to animals. Other biological kingdoms also have host defense mechanisms, usually of a chemical nature. Plants, for example, produce chemical defenses that can wall off areas damaged or infected by bacteria or fungi. In fact, an important determinant of how well a given strain of plant can resist infection after pruning or damage is its chemical and physical defensive abilities. Many fungi are plant pathogens, getting their nutrients by parasitizing certain tissues within the plant. To infect a plant, the fungus must penetrate the plant cell (◀Chapter 11, p. 321). During infection, the plant cells produce enzymes that release carbohydrate molecules from the fungal cell walls. These fragments of fungal wall, called *elicitors*, trigger an immunological-like response by the plant. Elicitors cause the plant to produce lipidlike chemicals called *phytoalexins*. Phytoalexins inhibit fungal growth by restricting the infection to a small portion of the plant tissue (see the photo). Plant biotechnologists are trying to "breed" this response into other types of plants that are sensitive to fungal invasion.

Invertebrates also have nonspecific defenses for fending off invaders. Phagocytosis is important to invertebrates in obtaining food, but it is also necessary for preventing sedentary organisms permanently fixed to a surface, and living where space is limited, from being overgrown by neighbors. So, phagocytosis is used to defend one's territory. In animals lacking a cardiovascular system, amoebocytes wander through the body, engulfing foreign matter and damaged or aged cells. When your white blood cells phagocytize a bacterium, they are using an ancient mechanism preserved and transformed from simpler life forms.

Opsonization is also observed in invertebrates, made possible by complement-like components of body fluids. For example, fluids in the body cavity of sea urchins share many characteristics with human complement proteins. In fact, complement proteins, like phagocytosis, probably were derived from these early versions in invertebrates. Secretion of antimicrobial enzymes is another means of defense present even in simple protozoa. Thus, nonspecific defense processes, such as phagocytosis and opsonization, are often called a *primitive characteristic* because most animals have these ancient mechanisms.

Almost all invertebrates also can reject grafts of foreign tissue. Vertebrates reject such grafts more vigorously on a second encounter, but invertebrates do not; in fact, the second rejection may be slower than the first. Because invertebrates lack these memory responses, the presence of such specific immune defenses in vertebrates is considered an *advanced characteristic*. These defenses include the B cells, T cells, and antibodies.

Although immune defenses involving the production of specific antibodies are found in all types of fish, the swiftest and most complex immune responses are found in mammals and birds. Birds have a saclike structure, the *bursa of Fabricius*, that is not present in mammals and probably represents a higher state of evolution of the immune system. In chickens, immature B cells in the bone marrow migrate to the bursa of Fabricius. There they are stimulated to mature rapidly and are capable of recognizing foreign substances. In mammals, B cells originate and mature more slowly in the bone marrow. Thus, immune system development culminates in the two-part system of B cells and T cells. In Chapter 17 we will investigate this achievement of specific host defenses.

Experimentally damaged areas of tree trunk are walled off in trees that survive attack, thus keeping infection from spreading throughout the entire tree. *(Courtesy Agricultural Research Service, United States Department of Agriculture)*

membranes contain proteins that protect against MAC lysis. These proteins prevent damage by preventing the binding of activated complement proteins to host cells. The MAC forms the basis of *complement fixation*, a laboratory test used to detect antibodies against any one of many microbial antigens. That test is described in ◀Chapter 18.

A great advantage of the complement system to host defenses is that once it is activated, the reaction cascade occurs rapidly. A very small quantity of an activating substance (microbe) can activate a few molecules of C1. They, in turn, activate large quantities of C3; one C4b2a molecule can split 1,000 molecules of C3 into C3a and C3b. Thus, sufficient quantities of C3b are quickly available to cause opsonization and inflammation and to produce membrane attack complexes.

Unfortunately, complement activity can be impaired by the absence of one or more of its protein components. Impaired complement activity makes the host more vulnerable to various diseases **(Table 16.4)**, most of which are acquired or congenital. Acquired diseases result from temporary depletion of a complement protein; they subside when cells again synthesize the protein. Congenital complement deficiencies are due to genetic defects that prevent the synthesis of one or more complement components.

The most significant effect of complement deficiencies is the lack of resistance to infection. Deficiencies

TABLE 16.4

Disease States Related to Complement Deficiencies	
Disease State	**Complement Deficiencies**
Severe recurrent infections	C3
Recurrent infections of lesser severity	C1, C2, C5
Systemic lupus erythematosus (a bodywide) immunologic disease)	C1, C2, C4, C5, C8
Glomerulonephritis (an immunological disease of the kidneys)	C1, C8
Gonococcal infections	C6, C8
Meningococcal infections	C6

PHYSICAL BARRIERS	Hair, Secretions, Epithelium
Prevent approach and deny access to pathogens	
PHAGOCYTES	Fixed macrophage, Neutrophil, Free macrophage, Eosinophil, Monocyte
Remove debris and pathogens	
EXTRACELLULAR KILLING	Natural killer cell, Abnormal cell, Lysed abnormal cell
Destroys abnormal cells	
INFLAMMATORY RESPONSE	1. Blood flow increased 2. Phagocytes activated 3. Capillary permeability increased 4. Complement activated 5. Clotting reaction walls off region 6. Regional temperature increased 7. Specific defenses activated
Multiple effects	
FEVER	Body temperature rises above 37°C in response to pyrogens
Mobilizes defenses, accelerates repairs, inhibits pathogens	
INTERFERONS	Released by activated lymphocytes and macrophages and by virus-infected cells
Increase resistance of cells to infection, slow the spread of disease	
COMPLEMENT SYSTEM	Complement, Lysed pathogen
Attacks and breaks down cell walls, attracts phagocytes, stimulates inflammation	

Figure 16.10 A summary of the body's nonspecific defenses.

in several complement components have been observed. The greatest degree of impaired complement function occurs with a deficiency of C3—which is not surprising, because C3 is the key component in the system. In individuals with C3 deficiencies, chemotaxis, opsonization, and cell lysis are all impaired. Such individuals are especially subject to infection by pyogenic bacteria. A deficiency in MAC components (C5–C9) is associated with recurrent infections, especially by *Neisseria* species. Complement deficiencies are less important in defenses against viruses, although some viruses, such as the Epstein-Barr virus, use complement receptors to invade cells.

ACUTE PHASE RESPONSE

Observations of acutely ill patients have led to the characterization of the **acute phase response**, a response to acute illness that involves increased production of specific blood proteins called **acute phase proteins**. In an acute phase response, pathogen ingestion by macrophages stimulates the synthesis and secretion of several cytokines. One, called *interleukin-6* (IL-6), travels through the blood and causes the liver to synthesize and secrete the acute phase proteins into the blood. Thus, acute phase proteins form a nonspecific host defense mechanism distinct from both the inflammatory response and host-specific immune defenses. This mechanism appears to recognize foreign substances before the immune system defenses do and acts early in the inflammatory process, before antibodies are produced.

The best understood acute phase proteins are *C-reactive protein* (CRP) and *mannose-binding protein* (MBP). All humans studied thus far have the capacity to produce CRP and MBP. CRP recognizes and binds to phospholipids, and MBP to mannose sugars, in cell membranes of many bacteria and the plasma membranes of fungi. Once bound, these acute phase proteins act like an opsonin: They activate the complement system and immune cytolysis and stimulate phagocyte chemotaxis. If we knew how to enhance CRP and MBP activity, effective therapies could be developed to combat many bacterial and fungal infections.

In summary, the innate defense mechanisms operate regardless of the nature of the invading agent. They constitute the body's first line of defense against pathogens, whereas the adaptive defense mechanisms (Chapter 17) constitute the second line of defense. **Figure 16.10** reviews the major categories of innate defenses.

✓ CHECKLIST

1. What are interferons? How and where are they produced?

2. How might interferons be used to treat disease?

3. Describe the complement system, including the classical and alternative (properdin) pathways.

4. What are the results of activating the complement cascade?

5. What are the functions of acute phase proteins?

▌ RETRACING OUR STEPS

▌ INNATE AND ADAPTIVE HOST DEFENSES

- **Innate defenses** operate regardless of the kind of invading agent; they form a first line of defense that is often effective even before specific defenses are activated.
- **Adaptive defenses** respond to particular invading agents; provided by the immune system, they form a second line of defense against pathogens.

▌ PHYSICAL BARRIERS

- Skin and mucous membranes act as physical barriers to penetration and secrete chemicals inhospitable to pathogens.
- **Mucous membranes** consist of a thin layer of cells that secrete mucus.

▌ CELLULAR DEFENSES

DEFENSIVE CELLS

- **Formed elements**, found in blood but derived from bone marrow, provide a cellular defense barrier to infection.
- Defensive cells include **granulocytes (basophils, mast cells, eosinophils**, and **neutrophils)** and **agranulocytes (monocytes and lymphocytes)**.

PHAGOCYTES

- A **phagocyte** is a cell that ingests and digests foreign substances.
- Phagocytic cells include neutrophils in the blood and in injured tissues, monocytes in the blood, and fixed and wandering **macrophages**.

THE PROCESS OF PHAGOCYTOSIS

- The process of **phagocytosis** occurs as follows: (1) Invading microorganisms are located by **chemotaxis**, which is aided by the release of **cytokines** by phagocytes. (2) Ingestion occurs as the phagocyte surrounds and ingests a microbe or other foreign substance into a **phagosome**. (3) Digestion occurs as lysosomes surround a vacuole and release their enzymes into it, forming a **phagolysosome**. Enzymes and defensins break down the contents of the phagolysosome and produce substances toxic to microbes.
- Some microbes resist phagocytosis by producing capsules or specific proteins, preventing release of lysosomal enzymes, and by producing toxins (**leukocidin** and **streptolysin**).

EXTRACELLULAR KILLING

- Eosinophils defend against parasitic worm infections by secreting cytotoxic enzymes.
- **Natural killer** (NK) **cells** secrete products that kill virus-infected cells and certain cancer cells.

THE LYMPHATIC SYSTEM

- The **lymphatic system** consists of a network of **lymphatic vessels**, **lymph nodes** and **lymphoid nodules**, the **thymus gland**, the **spleen**, and **lymph**.
- All lymphatic tissues that filter blood and lymph are susceptible to infection by pathogens they filter when the pathogens overwhelm defenses.
- Nonspecific defenses consist of the actions of phagocytic cells.

▌ INFLAMMATION

CHARACTERISTICS OF INFLAMMATION

- **Inflammation** is the body's response to tissue damage. It is characterized by localized increased temperature, redness, swelling, and pain.

THE ACUTE INFLAMMATORY PROCESS

- **Acute inflammation** is initiated by **histamine** released by damaged tissues, which dilates and increases permeability of blood vessels (**vasodilation**). Activation of cytokines also contributes to initiation of inflammation.
- Dilation of blood vessels accounts for redness and increased tissue temperature; increased permeability accounts for **edema** (swelling).
- Tissue injury also initiates the blood-clotting mechanism.
- **Bradykinin** stimulates pain receptors; **prostaglandins** intensify its effect.
- Inflamed tissues also stimulate an increase in the number of leukocytes in the blood (**leukocytosis**) by releasing cytokines that trigger leukocyte production. Neutrophils and macrophages migrate from the blood to the site of injury (**diapedesis**).
- Leukocytes and macrophages phagocytize microbes and tissue debris.

REPAIR AND REGENERATION

- Repair and regeneration occur as capillaries grow into the site of injury and fibroblasts replace the dissolving blood clot. The resulting **granulation tissue** is strengthened by connective tissue fibers (from **fibroblasts**) and the overgrowth of epithelial cells.

CHRONIC INFLAMMATION

- **Chronic inflammation** is a persistent inflammation in which the inflammatory agent continues to cause tissue injury as host defenses fail to overcome the agent completely.
- **Granulomatous inflammation** is a chronic inflammation in which monocytes, lymphocytes, and macrophages surround necrotic tissue to form a **granuloma**.

▌ FEVER

- **Fever** is an increase in body temperature caused by **pyrogens**, which increase the setting (thermostat) of the temperature-regulating center in the hypothalamus.
- **Exogenous pyrogens** (usually pathogens and their toxins) come from outside the body and stimulate a cytokine that acts as an **endogenous pyrogen**.
- Fever and the chemicals associated with it augment the immune response and inhibit the growth of microorganisms by lowering plasma iron concentrations. Fever also increases the rate of chemical reactions, raises the temperature above the optimum growth rate for some pathogens, and makes the patient feel ill (thereby lowering activity); phagocytosis is enhanced; production of interferon is increased, and breakdown of lysosomes is heightened, causing death of infected cells and the microbes inside of them.
- Antipyretics are recommended only for high fevers and for patients with disorders that would be exacerbated by fever.

▌ MOLECULAR DEFENSES

INTERFERON

• **Interferons** are proteins that act nonspecifically to cause cell killing or to stimulate cells to produce **antiviral proteins**.

• Interferon can be made by recombinant DNA technology and has proved to be therapeutic for certain malignancies; other therapeutic applications are being studied.

COMPLEMENT

• **Complement** refers to a set of blood proteins that, when activated, produce a **cascade** of protein reactions. The **complement system** can be activated by the **classical pathway** or the **alternative pathway**.

• Action of the complement system is rapid and nonspecific. It promotes opsonization, inflammation, and immune cytolysis

through the formation of **membrane attack complexes** (MACs). In **opsonization**, invading agents are coated with **opsonins** (antibodies) and C3b complement protein, making the invaders recognizable to phagocytes. In **immune cytolysis**, complement proteins produce lesions on invaders' plasma membranes that cause cell lysis.

• Deficiencies in complement reduce resistance to infection.

ACUTE PHASE RESPONSE

• Acutely ill patients increase production of certain blood proteins (**acute phase proteins**). These substances are distinct from those involved in the inflammatory response and act quickly, before antibodies can be made. Such proteins initiate or accelerate inflammation, activate complement, and stimulate chemotaxis of phagocytes.

▌ TERMINOLOGY CHECK

abscess (*p. 474*)
acute inflammation (*p. 472*)
acute phase protein (*p. 482*)
acute phase response
  (*p. 482*)
adaptive defense (*p. 464*)
adherence (*p. 468*)
agranulocyte (*p. 467*)
alternative pathway (*p. 479*)
antihistamine (*p. 474*)
antiviral protein (*p. 477*)
basophil (*p. 467*)
bradykinin (*p. 474*)
capsule (*p. 472*)
cascade (*p. 479*)
chemokine (*p. 468*)
chemotaxis (*p. 468*)
chronic inflammation
  (*p. 475*)
classical pathway (*p. 479*)
complement (*p. 478*)
complement system (*p. 478*)

cytokine (*p. 468*)
dendritic cell (*p. 467*)
diapedesis (*p. 474*)
edema (*p. 474*)
endogenous pyrogen
  (*p. 476*)
eosinophil (*p. 467*)
erythrocyte (*p. 466*)
exogenous pyrogen (*p. 476*)
fever (*p. 475*)
fibroblast (*p. 475*)
formed element (*p. 466*)
granulation tissue (*p. 475*)
granulocyte (*p. 467*)
granuloma (*p. 475*)
granulomatous inflammation
  (*p. 475*)
gut-associated lymphatic
  tissue (GALT) (*p. 472*)
histamine (*p. 474*)
immune cytolysis (*p. 480*)
inflammation (*p. 472*)

innate defense (*p. 464*)
interferon (*p. 476*)
leukocidin (*p. 470*)
leukocyte (*p. 466*)
leukocyte-endogenous
  mediator (*p. 476*)
leukocytosis (*p. 474*)
lymph (*p. 470*)
lymphatic system (*p. 470*)
lymphatic vessel (*p. 470*)
lymph node (*p. 470*)
lymphocyte (*p. 467*)
lymphoid nodule (*p. 472*)
macrophage (*p. 467*)
mast cell (*p. 467*)
membrane attack complex
  (MAC) (*p. 480*)
monocyte (*p. 467*)
mucous membrane (*p. 464*)
natural killer (NK) cell
  (*p. 470*)
neutrophil (*p. 467*)

nonspecific defense (*p. 464*)
opsonin (*p. 480*)
opsonization (*p. 480*)
phagocyte (*p. 467*)
phagocytosis (*p. 468*)
phagolysosome (*p. 469*)
phagosome (*p. 469*)
plasma (*p. 466*)
platelet (*p. 466*)
prostaglandin (*p. 474*)
pus (*p. 474*)
pyrogen (*p. 476*)
sinus (*p. 472*)
skin (*p. 464*)
specific defenses (*p. 464*)
spleen (*p. 472*)
streptolysin (*p. 470*)
thymus gland (*p. 472*)
toll-like receptors (TLRs)
  (*p. 468*)
tonsil (*p. 472*)
vasodilation (*p. 474*)

▌ CLINICAL CASE STUDY

Patients with cystic fibrosis (a genetic disorder) produce thick secretions that do not drain easily from the respiratory passages. The buildup of such secretions leads to inflammation and the replacement of damaged cells with connective tissue that blocks those respiratory passages. Frequent infections result from impairment of which innate defense mechanism?

▌ CRITICAL THINKING QUESTIONS

1. Which of your body's nonspecific host defenses would help fight a pathogen entering your body through each of the following portals? (a) A small cut on your hand; (b) inhalation into your lungs; (c) ingestion with contaminated food.

2. Although the inflammatory process is beneficial in most cases, it can sometimes be harmful. In what ways can you think of where this is the case?

3. Is it a good idea to take steps to reduce a moderate fever? Why?

▌ SELF-QUIZ

1. Match each of the following innate defense mechanisms with its associated structure or body fluid:

    _C_ Lysozyme
    _D_ Very acidic pH
    _B_ Sebum and fatty acids
    _A_ Low pH, flushing action of urine
    _F_ Mucociliary escalator
    _E_ Phagocytes

    (a) Urogenital tract
    (b) Skin
    (c) Tears and saliva
    (d) Stomach
    (e) Lower respiratory tract
    (f) Bronchial tubes

2. Which of the following is true about adaptive immunity?
    (a) It is generally the first line of defense against invading agents.
    (b) It is a specific defense against foreign bodies or antigens (bacteria and viruses). The antigen activates lymphocytes, which in turn produce antibodies capable of fighting against the specific antigen.
    (c) It is a general defense that acts against any type of invading agent.
    (d) The antibody and cellular responses are more effective against succeeding invasions by the same pathogen than against initial invasions.
    (e) b and d.

3. Healthy skin is our number one line of defense against bacterial infections. True or false?

4. Match the following to their type of immunity:
    _B_ Lymphocytes
    _A_ Neutrophils
    _A_ Granulocytes
    _B_ Antibodies
    _A_ Mucous membrane
    _A_ Phagocyte

    (a) Innate immunity
    (b) Adaptive immunity

5. Which of the following is not a function of the lymphatic system?
    (a) Collects excess fluid from the spaces between body cells
    (b) Provides many of the nonspecific defense mechanisms
    (c) Transports digested fats to the cardiovascular system
    (d) Sequestration of iron
    (e) Provides many of the specific defense mechanisms

6. Inflammation is influenced by histamine, which is released by:
    (a) Eosinophils
    (b) Erythrocytes
    (c) Platelets
    (d) Basophils
    (e) Leukocytes

7. Describe what occurs in each step of the process of phagocytosis.

8. What is immune cytolysis and how is it related to the membrane attack complex (MAC)?

9. One of the common defense mechanisms pathogenic bacteria have to avoid phagocytosis is the presence of:
    (a) Pili
    (b) A cell membrane
    (c) Peptidoglycan
    (d) A capsule
    (e) Endospore formation

10. Beside capsule formation, microbes can resist phagocytosis by which of the following methods?
    (a) Interfering with chemotaxis.
    (b) Production of toxins such as leukocidin and streptolysin that cause the release of the phagocyte's own lysosomal enzymes into its cytoplasm, killing them.
    (c) Some microbes take up residence within macrophages and are protected from lysosomes and their contents by formation of parasitophorous vacuoles (PVs).
    (d) Avoidance of adherence to macrophages.
    (e) All of the above.

11. Interferon was at first thought to be the viral magic bullet; however, it has been found to have which of the following drawbacks?
    (a) In most cases, administration of interferon is only a treatment and not a cure of viral diseases such as genital warts and cancer.
    (b) Recombinant interferon can be made in large quantities and is relatively cheap.
    (c) Recombinant interferon is unstable and does not remain long in the body and some microbes have developed resistance to it.
    (d) Injection of interferon can produce side effects including fever and organ toxicity.
    (e) a, c, and d.

12. Match the following white blood cells with their respective description:
    _C_ Neutrophils
    _A_ Eosinophils
    _D_ Lymphocytes
    _B_ Monocytes

    (a) Released in large numbers during allergic responses
    (b) Agranular phagocytes
    (c) Most numerous leukocyte
    (d) Participate in specific defenses

13. Opsonization is a special type of innate molecular defense that works together with the complement system. Opsonins play an integral role in this defense and are specialized antibodies that bind to and coat the surfaces of which type of pathogen?
    (a) Acid-fast mycobacteria
    (b) Bacteria that produce metachromatic granules
    (c) Endospores
    (d) Capsule or surface protein producing bacteria
    (e) All of the above

14. An organelle found in phagocytic cells that contains ingested microbes, digestive enzymes, and small proteins called defensins is a:
    (a) Lysosome
    (b) Phagolysosome
    (c) Phagosome
    (d) Pseudopodium
    (e) None of these

15. Large parasites, such as helminths, are most likely attacked by:
    (a) Basophils
    (b) Erythrocytes
    (c) Platelets
    (d) Neutrophils
    (e) Eosinophils

16. Cells secreting cytotoxic proteins that trigger the death of virus infected cells are known as:
    (a) Basophils
    (b) Platelets
    (c) B-lymphocytes
    (d) Natural killer cells
    (e) Neutrophils

17. The largest lymphatic organ in the body that can digest "wornout" erythrocytes is the:
    (a) Thymus
    (b) Liver
    (c) Spleen
    (d) Pancreas
    (e) Tonsils

18. Match the following terms of inflammation to their descriptions:

    D Pyrogen
    E Chronic inflammation
    F Leukocytosis
    B Acute inflammation
    C Edema
    A Bradykinin

    (a) Small peptide released at injured site that is responsible for pain sensation.
    (b) Short-term inflammation that kills invading microbes, clears tissue debris, and repairs tissue injury
    (c) Fluid accumulation around injured cells causing swelling
    (d) Fever-causing substance
    (e) Long-term inflammation that attempts to destroy and/or confine the region of inflammation
    (f) Damaged cells release cytokines that trigger the production and infiltration of leukocytes to the inflammation site

19. Gummas, lepromas, and tubercules are all examples of pockets of tissue that surround and wall off areas of infection and inflammation that are called:
    (a) Peyer's patches
    (b) Phagolysosomes
    (c) Granulomas
    (d) Phagosomes
    (e) All of these, depending on the tissue involved

20. The use of an anti-inflammatory drug such as cortisone to treat chronic inflammation can result in a disease occurring due to the inflammatory agent. True or false?

21. What does leukocyte endogenous factor (LEF) do?
    (a) Aids blood clotting
    (b) Lowers plasma iron concentrations, slowing growth of microorganisms
    (c) Elevates the body temperature
    (d) b and c
    (e) a and c

22. Cells enter an antiviral state and produce antiviral proteins (AVPs) in response to the presence of:
    (a) Antigen
    (b) Lipopolysacharide
    (c) Specific antibody
    (d) Interferon
    (e) Complement

23. Which of the following is not true about the complement system?
    (a) It is a set of more than 20 proteins that play a key role in host defense by specifically acting in different ways toward different microorganisms.
    (b) Its general functions include enhancing phagocytosis by phagocytes, lysing microbes and enveloped viruses directly, and generating peptide fragments that regulate inflammation and immune responses.
    (c) It is a fast-acting innate host defense that works in a cascade.
    (d) There are two pathways (classical and alternative), with the former beginning when antibodies bind to microbes which trigger C1, C4, and C2 complement proteins, and the latter activated by contact between complement protein factors B, D, P and polysaccharides at the pathogen surface.
    (e) The effects of both pathways are the same.

24. Put the following events of the acute phase response in order:
    (a) 5 The acute phase proteins can now activate the complement system and immune cytolysis and stimulate phagocyte chemotaxis.
    (b) 3 C-reactive protein recognizes and binds to phospholipids and mannose-binding protein to mannose sugars, in cell membranes of many bacteria and the plasma membrane of fungi.
    (c) 2 Interleukin-6 reaches the liver via the bloodstream where it causes the liver to synthesize and secrete the acute phase proteins (C-reactive and mannose-binding proteins) into the blood.
    (d) 4 Once bound, the acute phase proteins act like opsonins.
    (e) 1 Macrophage ingestion of microbe stimulates synthesis and secretion of interleukin-6.

25. All of the following are true about interferon EXCEPT:
    (a) Its function is a form of innate defense.
    (b) Viral infection of a cell triggers synthesis and secretion of interferon.
    (c) Interferon prevents further viral replication in surrounding cells by binding to their surfaces, triggering production of antiviral proteins that interfere with virus replication.
    (d) Interferons can stimulate adaptive immune defenses.
    (e) All viruses are sensitive to the antimicrobial actions of interferons.

26. In the following diagram, identify the major steps in the phagocytic process. Describe what happens in each step.

(a) _____

(b) _____

(c) _____

(d) _____

(e) _____

▌ EXPLORATIONS ON THE WEB

http://www.wiley.com/college/black

If you think you've mastered this chapter, there's more to challenge you on the web. Go to the companion web site to fine-tune your understanding of the chapter concepts and discover answers to the questions posed below.

1. Phagocytes, also known as cell eaters, are large white cells that engulf and digest marauding microorganisms. Find out more about phagocytes.

2. Do you have allergies? If so blame your mast cells! Discover how mast cells produce cytokines that enhance your immune response.

Immunology I: Basic Principles of Adaptive Immunity and Immunization

17

Come with me . . .

Here come the Cossack soldiers! They've galloped up and surrounded the village already. Run, hide, try to save yourself! If they catch you, they will VACCINATE you!

Sobbing, the little girl who would become my grandmother was dragged off. It was about 1900 in Lithuania. A well-meaning tsar had ordered that all his people should be vaccinated against smallpox. In the village square a soldier took out his knife, made a star-shaped series of cuts into her upper arm, poured vaccine into them, and let her go. Then he cleaned off the bloody knife on the sole of his boot, and hollered, "Next!" For the next three months, little Tekla lay in her bed, flushed with fever, pus pouring out of her arm, dripping off her elbow. Half the other people in the village had died. When she recovered, she vowed never to have another vaccination again. Nor did she want her children or grandchildren vaccinated. She told me in hushed tones of the evils and deaths associated with it. Such "folk memories" of vaccine plans gone wrong have caused resistance to receiving vaccinations in many parts of the world, especially in developing nations.

© Bettmann/Corbis

 Video related to this topic is available within WileyPLUS.

When you were very young you probably received a variety of immunizations against diphtheria, tetanus, whooping cough, polio, and possibly measles, German measles, and mumps as well. Your parents or grandparents, however, probably became immune to both kinds of measles and to mumps by acquiring and then recovering from these diseases. Either being immunized or having a disease may confer specific immunity to the organism that causes that disease. As we saw in the last chapter, innate host defenses protect the host against infections in a general way. This chapter will show how adaptive host defenses and immunization protect the host against particular infectious agents. The next chapter will examine disorders of the immune system, such as allergies, AIDS, autoimmune diseases, and the tests used in studying them.

▐▐▐ IMMUNOLOGY AND IMMUNITY

The word *immune* literally means "free from burden." Used in a general sense, **immunity** refers to the ability of an organism to recognize and defend itself against infectious agents. *Susceptibility*, the opposite of immunity, is the vulnerability of the host to harm by infectious agents.

As we said in the last chapter, host organisms have many general defenses against invading infectious organisms, regardless of what type of organism invades (◄Chapter 16, p. 464). Immunity produced by such defenses is called *innate immunity*. In contrast, *adaptive immunity* is the ability of a host to mount a defense against particular infectious agents by physiological responses *specific to that infectious agent*.

Immunology is the study of adaptive immunity and how the immune system responds to specific infectious agents and toxins. The **immune system** consists of various cells, especially lymphocytes, and organs such as the thymus gland, that help provide the host with specific immunity to infectious agents (◄Chapter 16, p. 470).

▐▐▐ TYPES OF IMMUNITY

Innate immunity, also called **genetic immunity**, exists because of genetically determined characteristics. One kind of innate immunity is **species immunity**, which is common to all members of a species. For example, all humans have immunity to many infectious agents that cause disease in pets and domestic animals, and animals have similar immunity to some human diseases. Humans do not have the appropriate receptor sites and will not become infected with canine distemper no matter how much contact they have with infected puppies. *Mycobacterium avium* causes tuberculosis in birds, but rarely in humans with normal immune systems. (It does often infect people with AIDS.) Some diseases appear only in a few species. Gonococci infect humans and monkeys but usually not other species. *Bacillis anthracis* causes anthrax in all mammals and some birds but not in many other animals.

As discussed in ◄Chapter 16, innate immunity also includes the ability of an organism to recognize pathogens. Phagocytes and macrophages are activated in the innate immune response by unique molecules on pathogens, such as peptidoglycan, lipopolysaccharide, and zymosan of yeast. Receptors on the surface of phagocytic cells, called pattern recognition receptors (PRRs), or toll-like receptors (named for a protein receptor first discovered in fruit flies) bind to the pathogen-unique molecules.

ADAPTIVE IMMUNITY

In contrast to innate immunity, **adaptive** (also called **acquired) immunity** is immunity obtained in some manner other than by heredity. It can be naturally acquired or artificially acquired. **Naturally acquired adaptive immunity** is most often obtained by having a specific disease. During the course of the disease, the immune system responds to molecules called *antigens* on invading infectious agents. It activates cells called T cells, produces molecules called *antibodies*, and initiates other specific defenses that protect against future invasions by the same agent. Immunity also can be naturally acquired from antibodies transferred to a fetus across the placenta or to an infant in colostrum and breast milk. **Colostrum** (ko-los'trum) is the first fluid secreted by the mammary glands after childbirth. Although deficient in many nutrients found in milk, colostrum contains large quantities of antibodies that cross the intestinal mucosa and enter the infant's blood. However, they only protect for a short time and then disappear.

In contrast, **artificially acquired adaptive immunity** is obtained by receiving an antigen by the injection of vaccine or immune serum that produces immunity. Sticking needles full of vaccine or serum into people is not a natural process. Thus, the immunity produced is artificially acquired.

ACTIVE AND PASSIVE IMMUNITY

Regardless of whether immunity is naturally or artificially acquired, it can be active or passive. **Active immunity** is created when the person's own immune system activates T cells, or produces antibodies or other defenses against an infectious agent. It can last a lifetime or for a period of weeks, months, or years, depending on how long the antibodies persist. **Naturally acquired active immunity** is produced when a person is exposed to an infectious agent. **Artificially acquired active immunity** is produced when a person is exposed to a vaccine containing live, weakened, or dead organisms or their toxins. In both types of active immunity, the host's own immune system responds specifically to defend the body against an antigen. Furthermore, the immune system generally "remembers" the antigen to which it has responded and will mount another response any time it again encounters the same antigen.

IMMUNITY

INNATE (inborn) Genetic factors

ACQUIRED

ACTIVE Own antibodies

PASSIVE Ready-made antibodies

NATURAL Exposure to infectious agent

ARTIFICIAL Immunization

NATURAL Maternal antibodies

ARTIFICIAL Antibodies from other sources

Figure 17.1 **The various types of immunity.** Nonspecific immunity is largely innate or inborn, whereas specific immunity is acquired.

Passive immunity is created when ready-made antibodies are introduced into the body. This immunity is passive because the host's own immune system does not make antibodies. **Naturally acquired passive immunity** is produced when antibodies made by a mother's immune system are transferred to her offspring. New mothers are encouraged to breast-feed for a few days even if they are not planning to continue so that their infants obtain antibodies from colostrum. **Artificially acquired passive immunity** is produced when antibodies made by other hosts are introduced into a new host. For example, a person who is bitten by a rattlesnake may receive a snake antivenin injection. Antivenins are antibodies produced in another animal, such as horses or rabbits. In this kind of immunity, the host's immune system is not stimulated to respond. Ready-made antibodies and the immunity they confer persist for a few weeks to a few months and are destroyed by the host; the host's immune system cannot make new ones.

Relationships among the various types of immunity are shown in **Figure 17.1**. The properties of each type of immunity are summarized in **Table 17.1**.

CHARACTERISTICS OF THE IMMUNE SYSTEM

ANTIGENS AND ANTIBODIES

Actions of the immune system are triggered by antigens. An **antigen** is a substance the body identifies as foreign and toward which it mounts an immune response—often it is also referred to as an immunogen. Most antigens are large protein molecules with complex structures and molecular weights greater than 10,000. Some antigens are polysaccharides, and a few are glycoproteins (carbohydrate and protein) or nucleoproteins (nucleic acid and protein). Proteins usually have greater antigenic

TABLE 17.1

Characteristics of Types of Immunity

Characteristic	Kind of Immunity		
	Innate	Actively Acquired Adaptive	Passively Acquired Adaptive
Agent	Genetic and physiological factors	Antibodies elicited by antigens	Ready-made antibodies
Source of antibodies	None	Person immunized	Plasma of other, such as mother
How elicited	Genetic expression	*Natural:*	*Natural:*
		by having disease	by receiving antibodies across placenta or in colostrum
		Artificial:	
		by receiving vaccine	*Artificial:*
			by receiving injection of gamma globulin or immune serum
Time to develop immunity	Always present	5 to 14 days after receiving antigen	Immediately after receiving antibodies
Duration of immunity	Lifetime	Months to lifetime	Days to weeks

Stem cells can be used to replace dead or damaged tissues. Once implanted in a particular site, they will differentiate into functional tissue of the type ordinarily found in that site. Stem cells have been used to rebuild muscle in heart walls damaged by a heart attack, to rebuild bone lost through trauma, and to cure Parkinson's disease by replacing dead brain cells with live ones that produce dopamine. Stem cell research has been the subject of much controversy. At first, the only source of stem cells was from aborted fetuses, and many countries banned research and treatment using such cells. Later it was discovered that adults have previously unknown supplies of stem cells (e.g., in bone marrow) and that their own stem cells can be used for treatment.

(immunogenic) strength because they have a more complex structure than polysaccharides. Large, complex proteins can have several **epitopes**, or **antigenic determinants**, areas on the molecule to which antibodies can bind.

Antigens are found on the surface of viruses and all cells, including bacteria, other microorganisms, and human cells. The exact chemical structure of each of a cell's antigens is determined by genetic information in its DNA. Bacteria can have antigens on capsules, cell walls, and even flagella. Many microorganisms have several different antigens somewhere on their surface. Determining how the human body responds to these different antigenic determinants is important in making effective vaccines. As we shall see, antigens on the surfaces of red blood cells determine blood types, and antigens on other cells determine whether a tissue translated from another person will be rejected.

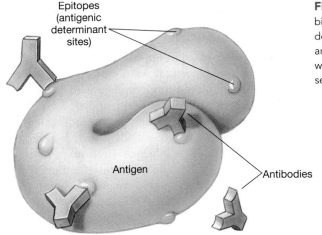

Epitopes (antigenic determinant sites)

Antigen

Antibodies

(a)

Figure 17.2 A typical antigen-antibody reaction. (a) Antibodies bind to specific chemical groups or structures, called epitopes, or antigenic determinants. (b) A Gram-negative bacterial pathogen may have several antigens, or immunogens (for example, for flagella, pili, and cell wall), each with particular epitopes. Large, complex protein molecules may have several different antigenic determinants.

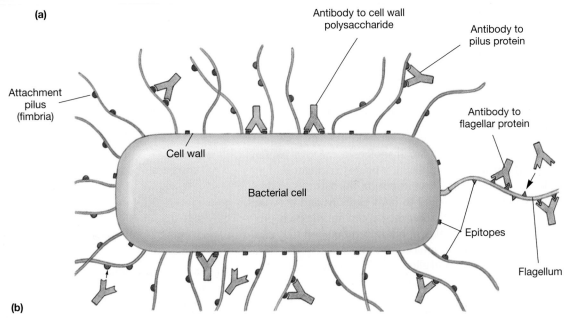

Antibody to cell wall polysaccharide

Antibody to pilus protein

Attachment pilus (fimbria)

Antibody to flagellar protein

Cell wall

Bacterial cell

Epitopes

Flagellum

(b)

Figure 17.3 Differentiation of stem cells into B cells and T cells. The occurs in the bone marrow and thymus respectively. The mature lymphocytes then migrate to lymphoid tissues such as the lymphiodes

In some instances a small molecule called a **hapten** (hap'ten) can act as an antigen if it binds to a larger protein molecule. Haptens act as epitopes on the surfaces of proteins. Sometimes they bind to body proteins and provoke an immune response. Neither the hapten nor the body protein alone acts as an antigen, but in combination they can. For example, penicillin molecules can act as haptens, bind to protein molecules, and elicit an allergic reaction, which is really a hypersensitivity reaction of the immune system.

One of the most significant responses of the immune system to any foreign substance is to produce antiantigen

proteins, or antibodies. An **antibody** is a protein produced in response to an antigen that is capable of binding specifically to the antigen. Each kind of antibody binds to a specific antigenic determinant. Such binding may or may not contribute to inactivation of the antigen. A typical antigen-antibody reaction is shown diagrammatically in **Figure 17.2**.

In discussing concentrations of antigens and antibodies, immunologists often refer to titers. A **titer** (ti′ter) is the quantity of a substance needed to produce a given reaction. For example, an antibody titer is the quantity required to bind to and neutralize a particular quantity of an antigen.

CELLS AND TISSUES OF THE IMMUNE SYSTEM

Specific immune responses are carried out by lymphocytes, which develop from stem cells as do other white blood cells, red blood cells, and platelets. Early in embryonic development undifferentiated stem cells, from regions in the yolk sac called primitive blood islands, proliferate. Later, they migrate into the body through the umbilical cord to various sites, where they differentiate into specific types of lymphocytes.

Differentiation of stem cells into lymphocytes is influenced by other tissues of the immune system **(Figure 17.3)**. Lymphocytes that are processed and mature in tissue, referred to as bursal-equivalent tissue, become **B lymphocytes**, or **B cells**. Differentiation of B cells was first observed in birds, where they are processed in an organ called the *bursa of Fabricius* (Fabris′e-us) **(Figure 17.4)**. Although no site equivalent to the bursa of Fabricius has been identified, B cells are produced in humans. This differentiation takes place in bone marrow where B cells differentiate. Functional B cells are found in all lymphoid tissues—lymph nodes,

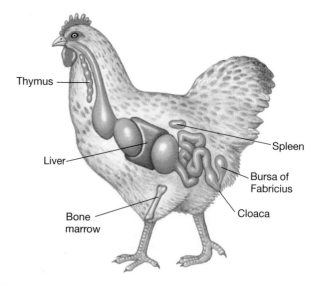

Figure 17.4 The bursa of Fabricius. In chickens this is where B cells develop. It is a pouch located off the cloaca, a chamber into which waste and reproductive materials empty. (Some other organs of importance to the immune system are also shown.)

TABLE 17.2

Proportions of B and T Lymphocytes in Human Lymphoid Tissues[a]		
Lymphoid Tissue	% B cells	% T cells
Peyer's patches and nodules in digestive trace	60	25
Spleen	45	45
Lymph nodes	20	70
Blood	10	75
Thymus	1	99

[a] Where percentages do not add to 100, some lymphocytes are undifferentiated. Based on data from E. J. Moticka. In R. F. Boyd and J. J. Marr (eds.), 1980. *Medical Microbiology*. New York: Little, Brown.

spleen, tonsils, adenoids, and gut-associated lymphoid tissues (GALT), which are lymphoid tissues in the digestive tract, including the appendix and Peyer's patches of the small intestine. B cells account for about one-tenth of the lymphocytes circulating in the blood.

Other stem cells migrate to the thymus, where they undergo differentiation into thymus-derived cells called **T lymphocytes**, or **T cells**. In adulthood, when the thymus becomes less active, differentiation of T cells still occurs in the thymus but at lower frequency. T cells are found in all tissues that contain B cells and account for about three-fourths of the lymphocytes circulating in the blood. The distribution of B and T cells in lymphatic tissues is summarized in **Table 17.2**. Subsequent differentiation of T cells produces four different kinds of cells: (1) cytotoxic (killer) T cells, (2) delayed-hypersensitivity T cells, (3) helper T cells, and (4) regulatory T cells. After differentiation these T cells migrate among lymphatic tissues and the blood.

A few lymphocytes that cannot be identified as either B cells or T cells are found in tissues and circulating in blood. These include the so-called **natural killer cells (NK cells)**, which nonspecifically kill cancer cells and cells infected with viruses, without having to utilize the specific immune responses. They "naturally" kill cells by releasing various cytotoxic molecules, some of which create holes in the target cell's membrane, leading to lysis. Other molecules enter the target cell and fragment its nuclear DNA, causing **apoptosis** (programmed cell death). NK cells are also affected by interferons.

DUAL NATURE OF THE IMMUNE SYSTEM

Lymphocytes give rise to two major types of immune responses, humoral immunity and cell-mediated immunity. However, the presence of a foreign substance in the body often triggers both kinds of responses.

Humoral (hu'mor-al) **immunity** is carried out by antibodies circulating in the blood. When stimulated by an antigen, B lymphocytes initiate a process that leads to the release of antibodies. Humoral immunity is most effective in defending the body against foreign substances

Humoral Immune Responses: What's in a Name?

The term *humoral* in humoral immune response is derived from the word *humor* (from *umor*, Latin for "liquid"). It originally referred to the four basic body fluids, or "humors"—blood, phlegm, yellow bile, and black bile—which ancient physicians believed must be present in proper proportions for an individual to enjoy good health. If any of these fluids were out of balance, a person was said to be "in bad humor" and likely to be diseased. Because this type of acquired immunity involves antibodies that circulate in the blood fluid, "humoral immune response" seemed logical.

outside of cells, such as bacterial toxins, bacteria, and viruses before these agents enter cells.

Cell-mediated immunity is carried out by T cells. It occurs at the cellular level, especially in situations where antigens are embedded in cell membranes or are inside host cells and are thus inaccessible to antibodies. It is most effective in clearing the body of virus-infected cells, but it also may participate in defending against fungi and other eukaryotic parasites, cancer, and foreign tissues, such as transplanted organs.

GENERAL PROPERTIES OF IMMUNE RESPONSES

Both humoral and cell-mediated responses have certain common attributes that enable them to confer immunity: (1) recognition of self versus nonself, (2) specificity, (3) heterogeneity, and (4) memory. We will look at each in some detail.

Recognition of Self Versus Nonself

For the immune system to respond to foreign substances, it must distinguish between host tissues and substances that are foreign to the host. Immunologists refer to normal host substances as **self** and foreign substances as **nonself**. The **clonal** (klo'nal) **selection hypothesis (Figure 17.5)**, first proposed by Frank Macfarlane Burnet in the 1950s, explains one way in which the immune system might distinguish self from nonself. According to this hypothesis, embryos contain many different lymphocytes, each genetically programmed to recognize a particular antigen and make antibodies to destroy it. If a lymphocyte encounters and recognizes that antigen after development is complete, it divides repeatedly to produce a clone, a group of identical progeny cells that make the same antibody. If, during development in the bone marrow (B cells) or thymus (T cells), it encounters its programmed antigen as part of a normal host substance (self), the lymphocyte is somehow destroyed or inactivated **(Figure 17.6)**. This mechanism removes lymphocytes that can destroy host tissues and thereby

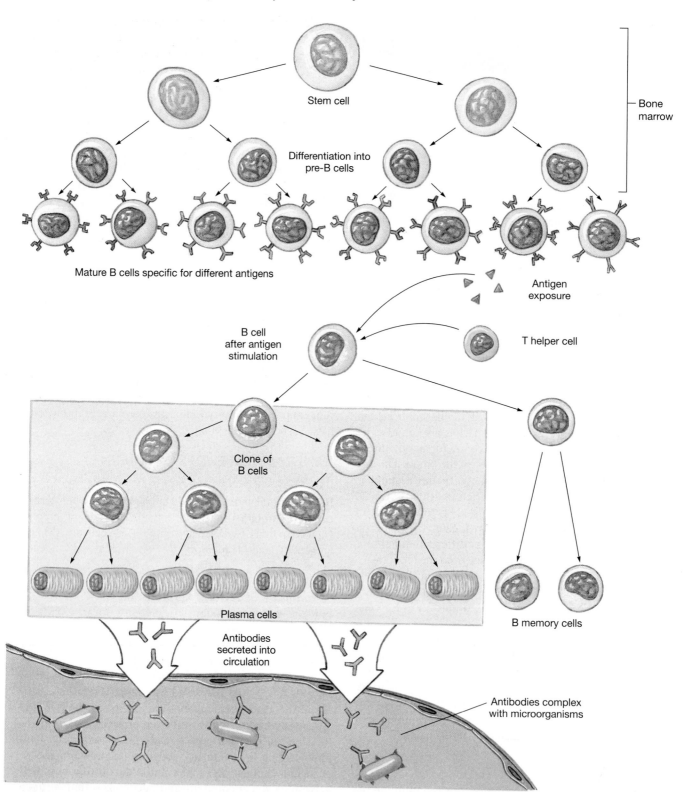

Figure 17.5 Clonal selection hypothesis. According to this theory, one of many B cells responds to a particular antigen and begins to divide, thereby producing a large population of identical B cells (a clone). All cells of such a clone produce the same antibody against the original epitope. B memory cells are also produced.

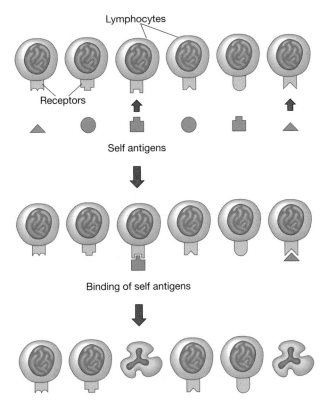

Lymphocytes

Receptors

Self antigens

Binding of self antigens

Clonal deletion of lymphocytes that have receptors for self

Figure 17.6 Clonal deletion. This process, which takes place in the bone marrow and thymus, removes those lymphocytes that have receptors for self antigens. When lymphocytes bind to self antigens, clonal deletion occurs; that is, those lymphocytes die as a result of condensation and disintegration of cell nuclei. Lymphocytes lacking self receptors survive.

creates **tolerance** for self. It also selects for survival lymphocytes that will protect the host from foreign antigens.

Tolerance also can be acquired by irradiation during cancer treatment or the administration of immunosuppressant drugs to prevent rejection of transplanted organs. The host loses the ability to detect and respond to foreign antigens in transplanted organs, but then also fails to respond to infectious organisms.

Specificity

By the time the immune system fully matures at age 2 to 3, it can recognize a vast number of foreign substances as nonself. Furthermore, it reacts in a different way to each foreign substance. This property of the immune system is called **specificity**. Due to specificity, each reaction is directed toward a specific foreign antigen, and the response to one antigen generally has no effect on other antigens. However, **cross-reactions**, reactions of a particular antibody with very similar antigens, can occur. For example, certain microorganisms, such as the bacterium that causes

Specificity of each T and B cell is determined by random gene rearrangements that occur during maturation in the bone marrow before it ever contacts an antigen.

syphilis, have the same haptens as some human cells, such as heart muscle cells, although the carrier molecules are quite different. This allows antibodies against this particular hapten to react with these otherwise vastly different cells. Cross-reactions also occur between strains of bacteria. For example, if three strains of pneumococci can cause pneumonia, and if each produces a particular antigen, A, B, or C, a person who has recovered from an infection with strain A has anti-A antibodies. The person then may also have some resistance to strains B and C because anti-A antibodies cross-react (that is, they react with antigens B and C).

Diversity

The ability of the immune system to respond specifically allows it to attack particular antigens. But in a lifetime, the human body encounters countless numbers of different foreign antigens. The property of **diversity** refers to the ability of the immune system to produce many different kinds of antibodies and T cell receptors, each of which reacts with a different epitope (antigenic determinant). When a bacterium or other foreign agent has more than one kind of antigenic determinant, the immune system may make a different antibody against each. And it is capable of producing antibodies even against foreign substances, such as newly synthesized molecules never before encountered by any immune system. Exposure to antigen is not necessary for diversity of antibody and T cell receptors. Lab animals raised in a germ-free environment still produce B cells and T cells with receptors specific for various antigens to which the animals have not been exposed. It is estimated that B cells have the ability to form antibodies to over 1 billion different epitopes or antigens.

Memory

In addition to its ability to respond specifically to a heterogeneous assortment of antigens, the immune system

APPLICATIONS

Kill That Virus! Not Me!

In a lethal meningitis in mice, the brain is covered with pus composed entirely of mouse lymphocytes that are produced in response to the virus. However, damage to the brain is due to the lymphocytes rather than to the virus. In mice infected with the virus before birth, the maturing immune system learns to recognize the virus as "self" and does not attack it. In the absence of an immune response, the virus invades all tissues but does no harm. However, if the mice subsequently receive transplants of normal lymphoid tissue, which has not acquired such tolerance, the virus elicits an immune response. Lymphocytes from the transplanted tissue then invade and damage the brain. (We will encounter other instances of diseases caused by the body's defenses rather than by the invading organism in the next chapter.)

TABLE 17.3

Main Attributes of Specific Immunity	
Attribute	**Description**
Recognition of self versus nonself	The ability of the immune system to tolerate host tissues while recognizing and destroying foreign substances, probably due to the destruction (deletion) of clones of lymphocytes during embryonic development
Specificity	The ability of the immune system to react in a different and particular way to each foreign substance
Heterogeneity	The ability of the immune system to respond in a specific way to a great variety of different foreign antigens
Memory	The ability of the immune system to recognize and quickly respond to foreign substances to which it has previously responded

also has the property of **memory**—that is, it can recognize substances it has previously encountered. Memory allows the immune system to respond rapidly to defend the body against an antigen to which it has previously reacted. In addition to producing antibodies during its first reaction to the antigen, the immune system also makes **memory cells** that stand ready for years or decades to quickly initiate antibody production. Consequently, the immune system responds to second and subsequent exposures to an antigen much more rapidly than to the first exposure. This prompt response due to "recall" by memory cells is called an **anamnestic** (secondary) **response**. The attributes of **specific immunity** are summarized in Table 17.3. With these attributes in mind, we will now look in more detail at the two kinds of specific immunity, humoral and cell-mediated.

✓ CHECKLIST

1. Distinguish between active and passive immunity. Give examples of each.
2. Distinguish between innate and acquired immunity. Give examples of each.
3. What are the differences between antigen, epitope, and hapten?
4. Distinguish between cellular and humoral immunity.

▌▌▌ HUMORAL IMMUNITY

Humoral immunity depends first on the ability of B lymphocytes to recognize specific antigens and second on their ability to initiate responses that protect the body against foreign agents. In most instances the antigens are on the surfaces of infectious organisms or are toxins produced by microbes. The most common response is the production of antibodies that will inactivate an antigen and lead to destruction of infectious organisms.

Each kind of B cell carries its specific antibody on its membrane and can bind immediately to a specific antigen. The binding of an antigen **sensitizes**, or activates, the B cell and causes it to divide many times. Some of the progeny are memory cells, but most are plasma cells. **Plasma cells** are large lymphocytes that synthesize and release many antibodies like those on their membranes. While it is active, a single plasma cell can produce as many as 2,000 antibodies per second!

After the B cell has bound antigen to antibody, it takes both into the cell where it "processes" the antigen by breaking it into short fragments which bind to a major histocompatibility complex II (MHCII) molecule on the surface of the B cell. This is called *presenting* the antigen. Macrophages and dendritic cells also present antigens in this way. T cells recognize the antigen plus MHCII, and become activated to produce interleukin 2 (IL-2). The direct contact of a T helper cell with the antigen-presenting B cell stimulates the B cell to proliferate further and to form B memory cells. Without T helper cell contact, no B memory cells are formed. How T cells carry out their functions will be explained later in this chapter.

PROPERTIES OF ANTIBODIES (IMMUNOGLOBULINS)

The basic units of antibodies, or **immunoglobulins (Ig)**, are Y-shaped protein molecules composed of four polypeptide chains—two identical **light (L) chains** and two identical **heavy (H) chains (Figure 17.7)**. The single Y-shaped molecule is called a monomer. The chains, which are held together by disulfide bonds, have constant regions and variable regions. The chemical structure of the *constant regions* determines the particular class that an immunoglobulin belongs to, as described next. The *variable regions* of each chain have a particular shape and charge that enable the molecule to bind a particular antigen. Each of the millions of different immunoglobulins has its own unique pair of identical antigen-binding sites formed from the variable regions at the ends of the L and H chains. These binding sites are identical to the receptors in the membrane of the parent B cell. In fact, the first immunoglobulins made by B cells are inserted into their membranes to form the receptors. When the B cells form plasma cells, they continue to make the same immunoglobulins. When an antibody is cleaved with the enzyme papain at the hinge region, two *Fab* (antibody binding fragment) pieces and one *Fc* (crystallizable fragment) piece result. The Fab fragment binds to the epitope. The Fc region formed by parts of the H chains in the tail of the Y has a site that can bind to and activate complement, participate in allergic reactions, and combine with phagocytes in opsonization.

Heavy chains

Antigen-
binding site

Light chain

Variable regions
of light and
heavy chains

Fab fragment

Carbohydrate

Complement
binding site

Disulfide bond

Constant regions
of light and
heavy chains

Fc fragment

Site of bonding
to macrophages

Ability to cross
placenta

(a)

(b)

Figure 17.7 **Antibody structure.** **(a)** The basic structure of the most abundant antibody (immunoglobulin) molecule in serum contains two heavy and two light chains, joined by disulfide bonds to form a Y shape. The upper ends of the Y, consisting of variable regions in both the light and heavy chains, differ from antibody to antibody. These variable regions form the two antigen-binding sites (part of the Fab fragment), which are responsible for the specificity of the antibody. The remaining part of the molecule consists of constant regions that are similar in all antibodies of a particular class. The Fc fragment determines the role each antibody plays in the body's immune responses. **(b)** A computer model of antibody structure. The two light chains are depicted in green, one heavy chain in red, and the other heavy chain in blue. (*R. Feldman/Visuals Unlimited*)

Classes of Immunoglobulins

Five classes of immunoglobulins have been identified in humans and other higher vertebrates (**Table 17.4**). Each class has a particular kind of constant region, which gives that class its distinguishing properties. The five classes are IgG, IgA, IgM, IgE, and IgD (**Figure 17.8**).

IgG, the main class of antibodies found in the blood accounts for as much as 20 percent of all plasma proteins. IgG is produced in largest quantities during a secondary response. The antigen-binding sites of IgG attach to antigens on microorganisms, and their tissue-binding sites attach to receptors on phagocytic cells. Thus, as a microorganism is surrounded by IgG, a phagocytic cell is brought into position to engulf the organism. The tail section of the H chains also activates complement. Complement, as

There are different subclasses of IgG molecules, distinguished from one another by subtle amino acid differences, affecting their biological activities.

explained in ◀Chapter 16, consists of proteins that lyse microorganisms and attract and stimulate phagocytes.

IgG is the only immunoglobulin that can cross the placenta from mother to fetus and provide antibody protection for it. IgG is also found in milk.

IgA occurs in small amounts in blood and in larger amounts in body secretions such as tears, milk, saliva, and mucus and attached to the linings of the digestive, respiratory, and genitourinary systems. IgA is secreted into the blood, transported through epithelial cells that line these tracts, and either released in secretions or attached to linings by tissue-binding sites. In blood, IgA consists of a single unit of two H and two L chains, but small amounts of dimers, trimers, and tetramers (2, 3, and 4 joined monomers) are present. Secretory IgA, which consists of two monomer units held together by a J chain (joining chain), has an attached **secretory component**, which protects the IgA from proteolytic (protein-splitting)

Each day, humans secrete 5-15 g of secretory IgA into their mucous secretions.

APPLICATIONS

How B Cells Build Diverse Antibodies

How can B cells make antibodies to almost any foreign antigen or foreign substance with which they come in contact? The key to such diversity lies in the immunoglobulin genes within each B cell. When B cells are formed in the bone marrow, each cell randomly pieces together different segments of its antibody genes.

In the embryo, the relatively few gene segments that code for the constant region of each light and heavy chain are not adjacent to the hundreds of gene segments that code for the variable regions. Let's look at how a light chain is built.

Light chains are formed when the DNA that separates a particular variable (V) segment from a constant (C) segment is removed, and the two gene segments are joined by a junction (J) segment. The now-joined segments form one continuous DNA sequence that represents the functional light-chain gene. Heavy chains are formed in a similar manner. Following transcription and translation, light-chain polypeptides are produced. These can be combined with heavy-chain polypeptides to form the functional antibody molecule. Thus, the diversity of antibody-binding sites comes from the random combinations of variable gene segments that join with constant gene segments to form the light and heavy chains.

TABLE 17.4

Properties of Antibodies

Property	Class of Immunoglobulin				
	IgG	IgM	IgA	IgE	IgD
Number of units	1	5	1 or 2	1	1
Activation of complement	Yes	Yes, strongly	Yes, by alternative pathway	No	No
Crosses placenta	Yes	No	No	No	No
Binds of phagocytes	Yes	No	No	No	No
Binds of lymphocytes	Yes	Yes	Yes	Yes	No
Binds of mast cells and basophils	No	No	No	Yes	No
Half-life (days) in serum	21	5–10	6	2	3
Percentage of total blood antibodies in serum	75–85	5–10	10	0.005	0.2
Location	Serum, extravascular, and across placenta	Serum and B cell membrane	Transport across epithelium	Serum and extracellular	B cell membrane

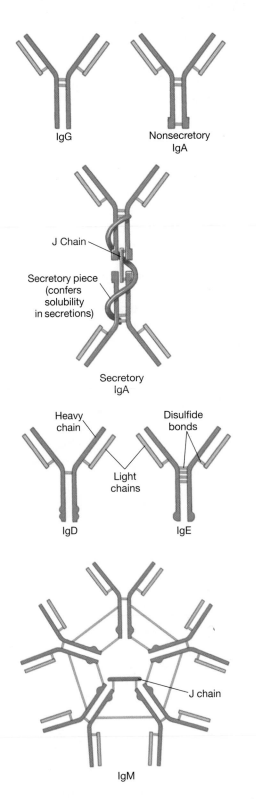

Figure 17.8 The structures of the different classes of antibodies.

enzymes and facilitates its transport. Mucosal surfaces, such as in the respiratory, urogenital, and digestive systems, are major sites for invasion by pathogens. The main function of IgA is to bind antigens on microorganisms before they invade tissues. It also activates complement,

which helps to kill the microorganisms. IgA does not cross the placenta, but it is abundant in colostrum where it helps protect infants from intestinal pathogens.

IgM is found as a monomer on the surface of B cells and is secreted as a pentamer by plasma cells. It is the first antibody secreted into the blood during the early stages of a primary response. IgM consists of five units connected by their tails to a J chain and so has 10 peripheral antigen-binding sites. As IgM binds to antigens, it also activates complement and causes microorganisms to clump together. These actions probably account for the initial effects the immune system has on infectious agents. It is also the first antibody formed in life, being synthesized by the fetus. In addition, it is the antibody of the inherited ABO blood types. Because of its size, IgM (M stands for macromolecule) is unable to cross the placenta and mostly stays inside blood vessels. High levels of IgM indicate recent infection or exposure to antigen.

IgE (also called *reagin*) has a special affinity for receptors on the plasma membranes of basophils in the blood or mast cells in the tissues. It binds to these cells by tissue-binding sites, leaving antigen-binding sites free to bind antigens to which humans can develop allergies, such as drugs, pollens, and certain foods. When IgE binds antigens, the associated basophils or mast cells secrete various substances, such as histamine, which produces allergy symptoms. IgE plays a damaging role in the development of allergies to such agents as drugs, pollens, and certain foods. Asthma and hay fever are common allergic diseases discussed in ◄Chapter 18. Levels of IgE are elevated in patients with allergies and in those harboring worm parasites. IgE is found mainly in body fluids and skin and is rare in blood. It has an extremely low concentration in serum.

Like IgM, **IgD** is found mainly on B-cell membranes and is rarely secreted. Although it can bind to antigens, its function is unknown. It may help initiate immune responses and some allergic reactions. In addition, IgD levels rise in some autoimmune conditions.

In discussing concentrations of antigens and antibodies, immunologists often refer to titers. A titer (ti′ter) is the quantity (concentration) of a substance present in a specific volume of body fluid. For example, during an infection, an individual's antibody titer (the concentration of antibody in the serum) normally increases. An increasing antibody titer serves as an indication of an immune response by the body.

PRIMARY AND SECONDARY RESPONSES

In humoral immunity the **primary response** to an antigen occurs when the antigen is first recognized by host B cells. After recognizing the antigen, B cells divide to form plasma cells, which begin to synthesize antibodies. In a few days, antibodies begin to appear in the blood plasma, and they increase in concentration over a period of 1 to 10 weeks. The first antibodies are IgM, which can bind to foreign substances directly. Cytokines trigger proliferating B cells to switch from making plasma cells that produce IgM to plasma cells that produce IgG. As IgM production wanes, IgG production accelerates, but eventually, it, too, wanes. The concentrations of both IgM and IgG can become so low as to be undetectable in plasma samples. However, the B cells that have proliferated and formed memory cells persist in lymphoid tissues. They do not participate in the initial response, but they retain their ability to recognize a particular antigen. They can survive without dividing for many months to many years.

When an antigen recognized by memory cells enters the blood, a **secondary response** occurs. The presence of memory cells (which are present in greater numbers than the original clone of B cells) makes the secondary response much faster than the primary response. Some memory cells divide rapidly, producing plasma cells, and others proliferate and form more memory cells. Plasma cells quickly synthesize and release large quantities of antibodies. In the secondary response, as in the primary response, IgM is produced before IgG. However, IgM is produced in smaller quantities over a shorter period, and IgG is produced sooner and in much larger quantities than in the primary response. Thus, the secondary response is characterized by a rapid increase in antibodies, most of which are IgG. The primary and secondary responses are compared in **Figure 17.9**.

The primary response of B cells can occur by two mechanisms. B cells can be activated by binding antigen, proliferating, and forming plasma cells. T helper (T_H) cells are not required for this response. These antigens are called **T-independent antigens**. This response usually only produces IgM antibody and no B memory cells are formed. For most antigens B cell activation requires contact with T_H cells activated by the same antigen. These are called **T-dependent antigens**. In this response the B cell becomes an antigen-presenting cell and makes contact with the T_H cell activating it. The activated T_H cell then secretes lymphokines that further activate the B cell causing it to differentiate and proliferate, producing B memory cells and plasma cells, and to undergo class switching so that IgG antibodies are produced (Figure 17.9).

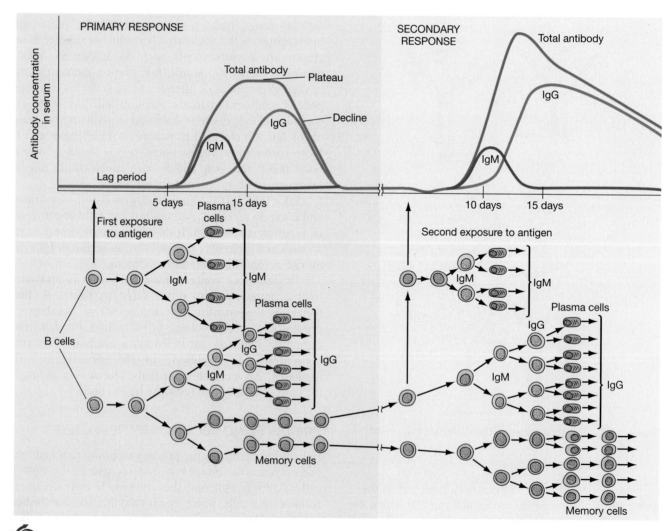

Figure 17.9 **Primary and secondary responses to an antigen.** This shows the correlation of antibody concentrations with the activities of B cells. Cytokines trigger the class switching from IgM to IgG.

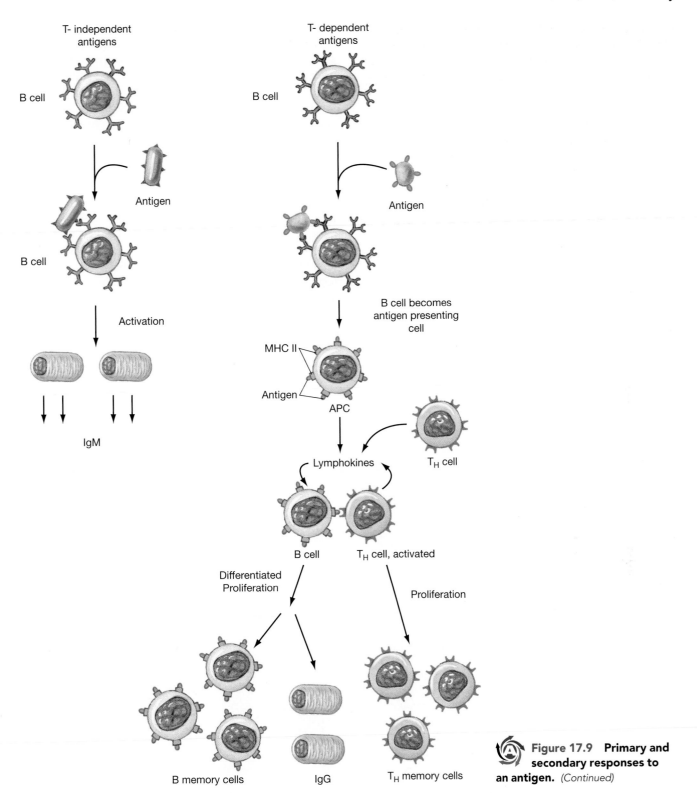

Figure 17.9 **Primary and secondary responses to an antigen.** *(Continued)*

KINDS OF ANTIGEN-ANTIBODY REACTIONS

The antigen-antibody reactions of humoral immunity are most useful in defending the body against bacterial infections, but they also neutralize toxins and viruses that have not yet invaded cells. The defensive capability of humoral immunity depends on recognizing antigens associated with pathogens.

For bacteria to colonize surfaces or for viruses to infect cells, these agents first must adhere to surfaces. IgA antibodies in tears, nasal secretions, saliva, and other fluids react with antigens on the microbes. They coat bacteria and viruses and prevent them from adhering to mucosal surfaces.

Microbes that escape IgA invade tissues and encounter IgE in lymph nodes and mucosal tissues. Gut-associated lymphoid tissue releases large quantities of IgE, which bind to mast cells; these cells then release histamine and other substances that initiate and accelerate the inflammatory process. Included in this process is the delivery of IgG and complement to the injured tissue.

Microbes that have reached lymphoid tissue without being recognized by B cells are acted on by macrophages and presented to B cells. B cells then bind the antigens and produce antibodies, usually with the aid of helper T cells. Antibodies binding with antigens on the surfaces of microbes form antigen-antibody complexes.

The formation of antigen-antibody complexes is an important component of the inactivation of infectious agents because it is the first step in removing such agents from the body. However, the means of inactivation varies according to the nature of the antigen and the kind of antibody with which it reacts. Inactivation can be accomplished by such processes as agglutination, opsonization, activation of complement, cell lysis, and neutralization. These reactions occur naturally in the body and can be made to occur in the laboratory. Here we will describe reactions chiefly as they relate to destruction of pathogens. We will discuss their laboratory applications more fully in ◀Chapter 18.

Because bacterial cells are relatively large particles, the particles that result from antigen-antibody reactions also are large. Such reactions result in **agglutination** (ag-lu-tin-a′shun), or the sticking together of microbes. IgM produces strong, and IgG produces weak, agglutination reactions with certain bacterial cells. Agglutination reactions produce results that are visible to the unaided eye and can be used as the basis of laboratory tests to detect the presence of antibodies or antigens. Some antibodies act as opsonins (◀Chapter 16, p. 480). That is, they neutralize toxins and coat microbes so that they can be phagocytized, a process called opsonization.

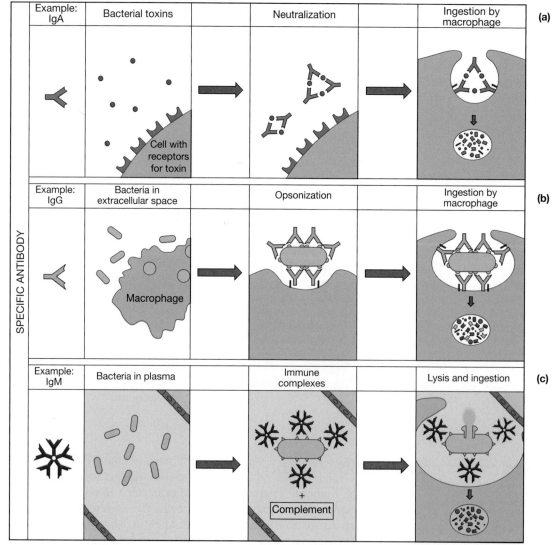

Figure 17.10 Antibodies produced by humoral immune responses eliminate foreign agents in three ways. (a) Neutralization of pathogens and toxins by IgA or IgG, **(b)** opsonization of bacteria by IgG, and **(c)** cell lysis initiated by IgM or IgG immune complexes allows for the formation of membrane attack complexes involving complement proteins.

Figure 17.11 Summary of humoral immunity.

Complement is an important component in inactivating infectious agents, as was discussed in ◀Chapter 16. Both IgG and IgM are powerful activators of the complement system; IgA is less powerful. Sometimes, antibodies, especially IgM, directly lyse cell membranes of infectious agents without the aid of complement.

Bacterial toxins, being small molecules secreted from the cell, usually are inactivated simply by the formation of antigen-antibody complexes, or **neutralization**. IgG is the main neutralizer of bacterial toxins. Neutralization effectively stops the toxin from doing further damage to the host. It does not destroy the organisms that produce the toxin—antibiotics are needed to prevent persisting organisms from continuing to produce toxin. Viruses, too, can be inactivated by neutralization (**Figure 17.10a**); IgM, IgG, and IgA are all effective neutralizers of viruses. Those viruses that have an envelope may then be lysed by complement (**Figure 17.10b**).

We have now considered the major characteristics of humoral immunity—how B cells are activated, how antibodies are produced, and how they function. These processes are summarized in **Figure 17.11**.

✓**CHECKLIST**

1. List the five types of immunoglobulins. Compare their structures and properties.
2. How do primary and secondary responses differ?
3. What is agglutination? Neutralization?

MONOCLONAL ANTIBODIES

Monoclonal antibodies are antibodies produced in the laboratory by a clone of cultured cells that make one specific antibody. In one method of making monoclonal antibodies, myeloma cells (malignant cells of the immune system) are mixed with sensitized lymphocytes. The malignant cells are used because they will keep dividing indefinitely. The lymphocytes are used because each makes a particular antibody. When the two cell types are mixed in cultures, they can be made to fuse with one another to make a cell called a *hybridoma* (**Figure 17.12**; ◀Chapter 8, p. 233). Hybridomas, which contain genetic information from each original cell, divide indefinitely, all the while producing large quantities of antibody. Which antibody a given hybridoma produces is determined by the antigen to which the lymphocytes were sensitized before their progeny were mixed with myeloma cells.

Generally, when a population of lymphocytes is exposed to an antigen, many different clones of B cells will proliferate, each making a different antibody. Many different hybridomas will therefore be produced by this technique. If one specific antibody is wanted, tests must be used to find which hybridomas are synthesizing that antibody, and those cells are then cloned.

Although monoclonal antibodies were first produced in 1975 as research tools, scientists were quick to recognize their practical uses. With experience, techniques for making monoclonal antibodies have improved. Culture media in which hybridomas thrive and produce large quantities of antibodies have been developed, and methods to grow hybridomas in large vat cultures in commercial laboratories are now available.

Theoretically, a monoclonal antibody can be produced for any antigen, provided lymphocytes sensitized to it can be obtained. Large numbers of hybridomas are now available, each producing a specific antibody. In addition to being used in research, many are produced commercially for use in diagnostic tests and in therapy. It is estimated that by the year 2010, the market value of monoclonal antibodies will reach $30.3 billion.

Several diagnostic procedures that use monoclonal antibodies are now available. Generally, these procedures are quicker and more accurate than previously used procedures. For example, a monoclonal antibody can be used to detect pregnancy only 10 days after conception. Other monoclonal antibodies allow rapid diagnosis of hepatitis, influenza, and herpes virus and chlamydial infections. Diagnostic tests for other infectious diseases and allergies are being developed at a rapid rate, and progress is being made in using monoclonal antibodies to diagnose various kinds of cancer. Some of the cancers for which monoclonal antibodies are currently used to monitor treatment or for diagnosis include prostate cancer, colorectal cancer, testicular cancer, thyroid cancer, lymphomas, myelomas, and small cell lung cancer.

Several monoclonal antibodies are being used to treat various cancers such as non-Hodgkin's lymphoma

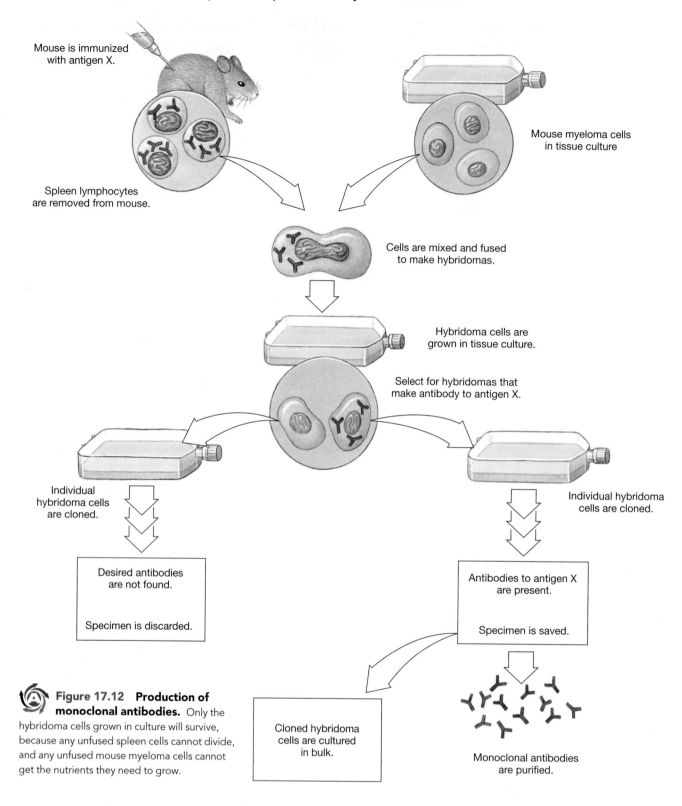

Mouse is immunized with antigen X.

Spleen lymphocytes are removed from mouse.

Mouse myeloma cells in tissue culture

Cells are mixed and fused to make hybridomas.

Hybridoma cells are grown in tissue culture.

Select for hybridomas that make antibody to antigen X.

Individual hybridoma cells are cloned.

Individual hybridoma cells are cloned.

Desired antibodies are not found.

Specimen is discarded.

Antibodies to antigen X are present.

Specimen is saved.

Cloned hybridoma cells are cultured in bulk.

Monoclonal antibodies are purified.

Figure 17.12 Production of monoclonal antibodies. Only the hybridoma cells grown in culture will survive, because any unfused spleen cells cannot divide, and any unfused mouse myeloma cells cannot get the nutrients they need to grow.

and breast cancer. These methods first require preparing antibodies to infectious agents or malignant cells. Then an appropriate drug or radioactive substance must be attached to the antibodies. If such antibodies are given to a patient, they carry the toxic substance directly to the cells that bear the appropriate antigen. The great advantage of therapeutic monoclonal antibodies is that they selectively damage infected or malignant cells without damaging normal cells. Monoclonal antibodies are also being used to prevent respiratory syncytial virus infections in children, prevent acute kidney transplant rejection, and to treat rheumatoid arthritis.

Monoclonal antibodies against tumor antigens have been tried in a few cancer patients. Unfortunately, the patients often displayed allergic reactions to myeloma proteins that accompany the antibodies. Researchers are now producing "humanized" monoclonal antibodies, which will kill malignant cells without causing allergic reactions in the patients that receive them. Humanized monoclonal antibodies, put together by genetic engineers, have a human constant region, plus a variable region made up of human and mouse portions. Diphtheria exotoxin delivered to cancer cells by monoclonal antibodies is being tried as a therapy for cancer.

▌▌▌ CELL-MEDIATED IMMUNITY

In contrast to humoral immunity, which involves B cells and immunoglobulins, cell-mediated immunity involves the direct actions of T cells. In cell-mediated immunity, T cells interact directly with other cells that display foreign antigens. These interactions clear the body of viruses and other pathogens that have invaded host cells. They also account for rejection of tumor cells, some allergic reactions, and immunological responses to transplanted tissues.

The cell-mediated immune response involves the differentiation and actions of different types of T cells and the production of chemical mediators called **cytokines** (lymphokines, interleukins). Much recent research has been devoted to determining the characteristics, origins, and functions of T cells, including the functions of secreted cytokines. Much more research is needed to fully understand cell-mediated immunity. What follows is a brief discussion of our current knowledge.

T cells, as noted earlier, are processed by the thymus. T cells differ from B cells in that they do not make antibodies. However, they do have a particular cell membrane receptor protein that corresponds to the antibodies of B cells and other receptor proteins as well.

THE CELL-MEDIATED IMMUNE REACTION

Cell-mediated immunity involves the response of T lymphocytes. Unlike B lymphocytes, T cells cannot be activated directly by antigen. The cell-mediated response requires presentation of the antigen on the surface of cells along with major histocompatibility complex (MHC) proteins ◄(Chapter 18). MHC proteins allow cells to recognize each other. There are two classes of MHC proteins. All nucleated cells have MHCI proteins on their surface. **Antigen-presenting cells** also have MHCII on their surface. The cell-mediated immune reaction typically begins with the processing of an antigen—usually one associated with a pathogenic organism—by dendritic cells, B cells, or macrophages. When macrophages and dendritic cells phagocytize pathogens, they ingest and degrade the pathogen. Pieces of the pathogen, peptides, are then transported to the surface of the macrophage or dendritic cell. Then they insert some of the pieces of the pathogen's antigen

molecules into their own cell membranes. This constitutes processing the antigen. The peptide is bound to the surface of the cell by MHCII proteins. When a macrophage presents the antigen to T cells that have the proper antigen receptor, the antigen and receptor bind. T cells cannot be activated without an appropriate MHC. T helper (T_H) cells are activated by antigen presented by MHCII, antigen-presenting cells. Cytotoxic (killer) T (T_C) cells are activated by antigen presented by MHCI, typically cells infected with virus, intracellular bacterial pathogens, transformed cancer cells, or foreign tissues, such as an organ transplant. Once activated, T_H cells can stimulate other T and B cells, as well as phagocytes. These reactions are summarized in **Figure 17.13**.

Binding with macrophages or B cells causes T cells to divide and differentiate into different types of T cells, including memory cells (**Figure 17.14**). Each cell is sensitized to the antigen that initiated the process, and each type has a different function in cell-mediated immune reactions. Some cells act directly and others release leukotrienes or cytokines, which are chemical substances that trigger certain immunologic reactions. The reactions of cell-mediated immunity are summarized in **Figure 17.15**. Refer to the figure as you read about the functions of different kinds of T cells.

Macrophages that have processed an antigen secrete the lymphokine interleukin-1 (IL-1), which activates **T helper (T_H) cells**. T_H cells, in turn, secrete lymphokines such as interleukin-2 (IL-2) and gamma interferon. IL-1 from macrophages and IL-2 from T_H cells activate other T cells, **delayed hypersensitivity T (T_D) cells**, and **cytotoxic (killer) T (T_C) cells**. T_C cells can be recognized by a CD8 glycoprotein on their cell membrane. Also, IL-1, IL-2, and gamma interferon together cause undifferentiated cells to become natural killer (NK) cells.

At the same time that these cells are differentiating, some T memory cells also are being formed. As in humoral immunity, the persistence of memory cells in cell-mediated immunity allows the body to recognize antigens to which T cells have previously reacted and to mount more rapid subsequent responses.

As we noted in the discussion of humoral immunity, T_H cells stimulate the growth and differentiation of B cells. Other regulatory cells, and the disappearance of foreign antigen as the immune response proceeds, apparently help to prevent both humoral and cell-mediated immune processes from getting out of hand.

Activated T_D cells also release various lymphokines. These include:

1. Macrophage chemotactic factor, which helps macrophages to find microbes.

2. Macrophage activating factor, which stimulates phagocytic activity.

3. Migration inhibiting factor, which prevents macrophages from leaving sites of infection.

4. Macrophage aggregation factor, which causes macrophages to congregate at such sites.

(a) ACTIVATION OF HELPER T CELLS

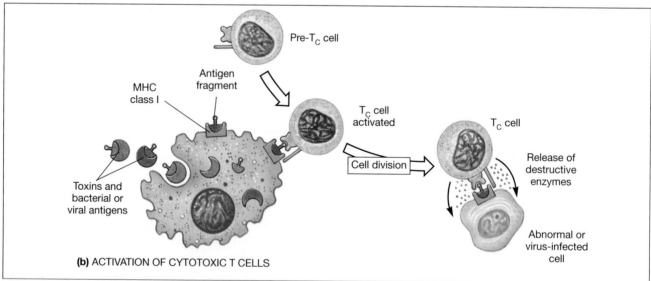

(b) ACTIVATION OF CYTOTOXIC T CELLS

Figure 17.13 The reactions in cell-mediated immunity. **(a)** The macrophage has processed an antigen and inserted an antigen (peptide) fragment into its plasma membrane as an MHC class II molecule. T_H cells have receptors that recognize the peptide fragment on MHC class II. Binding causes the T_H cells to become activated. The activated T cells then differentiate into either T_H1 cells or T_H2 cells. T_H1 cells activate infected macrophages to destroy internal bacterial infections. T_H2 cells activate B cells (humoral immune responses) by binding to MHC class II peptide presented by the B cells. **(b)** Presenting the same peptide fragment on MHC class I to T_C cells activates cells to attack infected cells, especially abnormal or virus-infected cells.

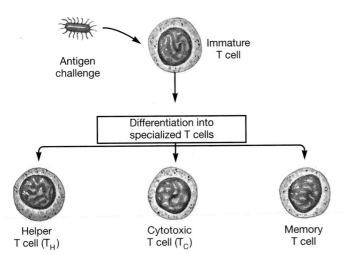

Figure 17.14 **Types of T cells.** After T cells are challenged by antigens, the cells differentiate into one of several types of functioning T cells.

T_D cells also participate in delayed hypersensitivity, a kind of allergic reaction explained in ◄Chapter 18.

T_C cells and NK cells kill infected host cells. When pathogens have evaded humoral immunity and established themselves inside cells, they can cause long-term infections unless the infected cells are destroyed by cell-mediated immunity. An agent that infects T cells is especially devastating because it destroys the very cells that might have combated the infection. AIDS is just such a disease. The AIDS virus invades T_H cells, prevents them from carrying out their normal immunological functions, and eventually kills them. The lack of T_H cells impairs both humoral and cell-mediated immune responses, including the destruction of malignant cells. Thus, because of extensive destruction of T_H cells, AIDS patients are susceptible to a host of opportunistic infections and to various malignancies.

HOW KILLER CELLS KILL

Recent research shows that T_C cells and NK cells kill other cells by making a lethal protein and firing it at target cells. Eosinophils have a similar protein, which they may use to kill certain helminths and other parasites. But lethal proteins are not the sole property of hosts' defensive cells—the amoebae that cause amoebic dysentery and some other parasites and fungi also have them. Learning more about these lethal proteins may one day enable us to treat amoebic dysentery and other parasitic

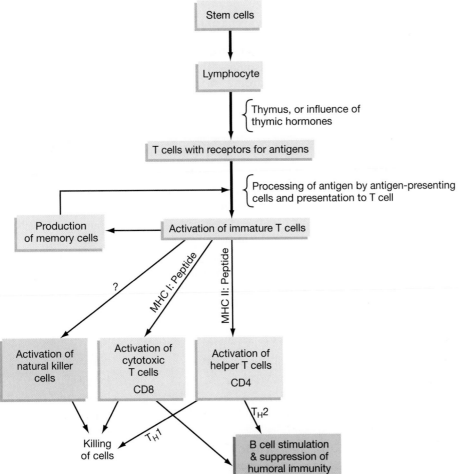

Figure 17.15 **Summary of cell-mediated immunity.** (CD stands for "cluster of differentiation.")

diseases by blocking the action of these proteins or to treat AIDS and malignant diseases by enhancing their actions.

Cytotoxic T cells act mainly on virally infected cells, whereas NK cells act mainly on tumor cells, cells of transplanted tissues, and possibly on cells infected with intracellular agents such as rickettsias and chlamydias. Each kind of killer cell acts by a different mechanism. Cytotoxic T cells bind to antigens presented by macrophages and then attack virus-infected cells. In contrast, NK cells bind directly to malignant or other target cells without the help of macrophages. If a target cell lacks certain proteins (major histocompatibility complex—MHC, to be discussed further in ◀Chapter 18), the NK cell will automatically attack and kill it.

Both kinds of killer cells contain granules of a lethal protein, **perforin**, which is released when they bind to a target cell. Perforin bores holes in the target cell membranes so that essential molecules leak out and the cells die. This process is similar to the action of complement. By killing infected cells while they are few in number and before new virus particles are released from them, cytotoxic T cells prevent the spread of infection—but at the expense of destroying host cells. Similarly, NK cells destroy malignant cells before they have a chance to multiply. Both kinds of killer cells can withdraw from cells they have damaged and move on to other target cells.

The discovery of such an efficient mechanism for killing cells raises two important questions: What prevents perforin from killing adjacent uninfected cells, and what prevents it from attacking the membranes of the killer cells themselves? Perforin doesn't kill adjacent cells because it is effective only when secreted at the binding site between the killer and target cells. Why perforin doesn't attack killer-cell membranes is not known, but it has been suggested that killer cells produce a protein, called protectin, that inactivates perforin.

THE ROLE OF ACTIVATED MACROPHAGES

Some bacteria, such as those that cause tuberculosis, Hansen's disease (leprosy), listeriosis, and brucellosis, can continue to grow even after they have been engulfed by macrophages. T_D cells combat such infections by releasing the lymphokine macrophage activating factor. This factor causes macrophages to increase production of toxic hydrogen peroxide, along with enzymes that attack the phagocytized organisms and accelerate the inflammatory response. Organisms that survive these defenses are walled off in granulomas.

We have now completed the discussion of cell-mediated immunity—how it is initiated and how its effects are produced. These processes are summarized in Figure 17.15. The various functions of B and T cells are summarized in **Table 17.5**. Comparing Figures 17.11 and 17.15 and studying Table 17.5 will serve to highlight similarities and differences in humoral and cell-mediated immunity and to provide an overview of specific immunity.

SUPERANTIGENS

Superantigens such as staphylococcal toxins that cause food poisoning, toxic shock syndrome, and scalded skin syndrome ◀(Chapter 19) or streptococcal toxins

TABLE 17.5

Characteristics of B cells, T cells, and Macrophages			
Characteristic	**B cells**	**T cells**	**Macrophages**
Site of production	Bursal-equivalent tissues	Thymus or under thymic hormones	
Type of immunity	Humoral	Cell-mediated, and assisting humoral	Humoral and cell-mediated
Subpopulations	Plasma cells and memory cells	Cytotoxic, helper, suppressor, delayed hypersensitivity, and memory cells	Fixed and wandering
Presence of surface antibodies	Yes	No	No
Presence of foreign surface antigens	No	No	Yes
Presence of receptors for antigens	Yes	Yes	No
Life span	Some long, most short	Long and short	Long
Secretory product	Antibodies	Cytokines	Interleukin-1
Distribution (% leukocytes)			
Peripheral blood	15–30	55–75	2–12
Lymph nodes	20	75	5
Bone marrow	75	10	10–15
Thymus	10	75	10

responsible for "flesh-eating," necrotizing fasciitis ◄(Chapter 19) are able to simultaneously bind to the MHCII molecule and the T cell receptor molecule on the surface of T cells. Binding to the receptor molecule does not involve specificity for the receptor site. The superantigen binds to T cells with different specificities; it is polyclonal. Up to 5% of the T cell population can react with a single superantigen. This activates T cells at up to 100 times the normal rate to bind to the macrophages and T helper cells then secrete immense quantities of interleukin-2 (IL-2). Instead of staying in the local area, the excess IL-2 gets into the bloodstream and is transported around the body, where it causes nausea, vomiting, fever, malaise, and symptoms of shock (e.g., as in toxic shock syndrome). The greatest number of T cells that are activated in this way do not have a use in fighting the causative infection. Many T cells, of all sorts, respond simultaneously to superantigens. They replicate furiously. Many of them die as a result, leaving the immune system deficient in those types of cells, thus leaving the host open to even more infections.

Superantigens may also play a role in autoimmune diseases. Not all T cells that recognize self are deleted in clonal deletion. However, these are usually so few in number that they do not cause disease. When these multiply excessively, the host tissues are attacked—a disorder called autoimmunity, which will be discussed in ◄Chapter 18. It is possible that autoimmune diseases such as rheumatoid arthritis and multiple sclerosis may be caused by superantigens.

MUCOSAL IMMUNE SYSTEM

The mucosal immune system (MALT = mucosal associated lymphoid tissue) is the largest component of the immune system and a major site of entry for pathogens. It consists of the entire gastrointestinal tract, urogenital tract, respiratory tract, and mammary glands. In humans it is typically more than 400 square meters of mucosa! A part of this system is the gut-associated lymphoid tissue (GALT) consisting of appendix, Peyer's patches of the small intestine, tonsils, and adenoids. The MALT system is partially separated from the systemic immune system. The immune response to pathogens on the epithelial surface of the mucosa has some characteristics different than the immune response to pathogens in the blood and lymph. In the gut M cells are interspersed between epithelial cells. These cells do not have microvilli on their surface. They take up antigens from the gut by endocytosis and release the antigens to antigen-presenting cells, such as dendritic cells beneath them. Previously unactivated lymphocytes that become activated by the antigen-presenting cells are transported to other mucosal surfaces by entering the blood, via lymph nodes that drain from the intestinal region (mesenteric lymph nodes) or thoracic region. Enteric pathogens cause an inflammatory response that activates antigen-presenting cells beneath M cells and increases response of lymphocytes to

antigens from the pathogen. Oral tolerance to food antigens prevents an immune response to food in the gut. The primary immunoglobulin on mucosal surfaces, breast milk, and colostrum is IgA. Recall that this immunoglobulin can be secreted across epithelial cells (p. 497). Colostrum has very high levels of IgA (50 mg/ml compared to 2.5 mg/ml in adult serum) for the first 4 days after birth. Interestingly the uterus, a part of the urogenital tract is a **privileged site**, that is, it is isolated from the adaptive immune system. Other privileged sites are the anterior chamber of the eye and testes.

FACTORS THAT MODIFY IMMUNE RESPONSES

The host defenses of young, healthy, human adults living in an unpolluted environment are capable of preventing nearly all infectious diseases. However, a variety of disorders, injuries, medical treatments, environmental factors, and even age can affect resistance to infectious diseases. An individual with reduced resistance is called a **compromised host**.

In the beginning of this chapter we noted that humans are genetically immune to some diseases. It has also been found that different races have different degrees of resistance and susceptibility to various diseases. When black and white military personnel live under the same conditions (same barracks, food, exercise regime, and such), blacks still develop TB at a higher rate.

Age also affects immune responses. In general, the very young and the elderly are most susceptible to infections, and young adults are least susceptible. The young are susceptible because the immune system is not fully developed until age 2 or 3. Infants can, however, produce some IgM shortly after birth, and they receive maternal IgG passively. The elderly are susceptible to infections and malignancies because the immune system, and especially cell-mediated response, is one of the first to decline in function during the aging process. Thus, it makes sense to take special precautions against unnecessarily exposing infants and elderly people to infectious agents. And it also makes sense to obtain recommended immunizations during infancy and early childhood.

Even seasonal patterns affect the immune system. For example, T cells have a yearly cycle, falling to their lowest level in June. People with Hodgkin's disease are most often diagnosed in spring, leading some researchers to believe that there is a link between the two cycles. AIDS patients with low T cell counts are 11 times more likely to develop Hodgkin's disease.

Genetic and age factors that modify immunity are beyond our control, but we have some control over diet and environment. Let's see how these factors contribute to resistance—or the lack of it.

An adequate diet, especially adequate protein and vitamin intake, is essential for maintaining healthy intact skin and mucous membranes and phagocytic activity. It is likewise important for lymphocyte production and antibody synthesis. Poor nutrition and poor inflammatory response of alcoholics and drug addicts greatly lower their resistance to infection. In the elderly an inadequate diet can further weaken a declining immunological response.

Regular moderate exercise such as 45 minutes of brisk walking, 5 days per week can produce a 20 percent increase in antibody level, which occurs during the exercise and for about 1 hour afterwards. Natural killer cell activity is also increased. However, excessive exercise such as running more than 20 miles per week, depresses the immune system. Marathoners who ran at their fastest pace for 3 hours experienced a drop in natural killer cell activity of more than 30 percent for about 6 hours. Long-distance runners are more vulnerable to infection, especially of the upper respiratory tract, for about 12 to 24 hours after a race, experiencing six times the rate of illness after a race compared with trained runners who did not race.

Pregnancy is a time when cell-mediated immunity decreases significantly. During a 1957 epidemic of influenza A in New York City, 50% of women of childbearing age who died were pregnant—even though they accounted for only 7% of women in that age group. No impairment of humoral immunity is observed during pregnancy. You may recall that IgG is the only immunoglobulin that can cross the placenta providing some specific immunity directly to the developing fetus.

Volunteers kept up to 3:00 A.M. one night suffered a 50 percent drop in natural killer cells. After a good night's rest the next evening, their NK cell count returned to normal.

Traumatic injuries lower resistance at the same time that they provide easier access to tissue for microbes. Tissue repair competes with immune processes because both require extensive protein synthesis. When normal systems that flush away microbes, such as tears, urinary excretions, and mucous secretions, are impaired by injuries, pathogens have easier access to tissues. Antibiotics destroy commensals that sometimes compete with pathogens. Impaired defenses and use of antibiotics allow opportunistic infections to become established.

Environmental factors such as pollution and exposure to radiation also lower resistance to infection. Air pollutants, including those in tobacco smoke, damage respiratory membranes and reduce their ability to remove foreign substances. They also depress the activities of phagocytes. Excessive exposure to radioactive substances damages cells, including cells of the immune system. These factors can be compounded by induced and inherited immunological disorders. Immunosuppressant drugs used to prevent rejection of transplanted tissue impair the functions of lymphocytes and some phagocytes. Diseases such as AIDS destroy T cells. Finally, genetic defects in the immune system itself can result in the absence of B cells, T cells, or both. How these disorders impair immunity is discussed in ◄Chapter 18.

◄Chapter 18.

✓CHECKLIST

1. What are monoclonal antibodies, and how are they produced? What are their uses?

2. Distinguish among the various kinds of T cells and their functions.

3. What are NK cells? How do they function?

▐▐▐ IMMUNIZATION

Throughout the world each year nearly 3.5 million children, most of them under 5 years old, die of three infectious diseases for which immunization is available. About 2 million die of measles, 800,000 die of tetanus, and 600,000 die of whooping cough. Another 4 million die of various kinds of diarrhea, against which some immunization is possible. Most of these deaths occur in underdeveloped countries.

These statistics dramatize three important facts about immunization. First, immunization can prevent significant numbers of deaths. Second, methods of immunization are not yet available for some infectious diseases, such as certain diarrheas. Most organisms that cause diarrhea exert their effects in the digestive tract, where antibodies and other immune defenses cannot reach

About 80 percent of the world's children are immunized against measles, diphtheria, pertussis, tetanus, tuberculosis and polio.

them. Finally, much greater effort is needed to make immunizations available in underdeveloped countries.

ACTIVE IMMUNIZATION

To develop active immunity, as noted earlier, the immune system must be induced to recognize and destroy infectious agents whenever they are encountered. **Active immunization** is the process of inducing active immunity. It can be conferred by administering vaccines or toxoids. A **vaccine** is a substance that contains an antigen to which the immune system responds. Antigens can be derived from living but attenuated (weakened) organisms, dead organisms, or parts of organisms. A **toxoid** is an inactivated toxin that is no longer harmful, but retains its antigenic properties.

Principles of Active Immunization

Regardless of the nature of the immunizing substance, the mechanism of active immunization is essentially the same. When the vaccine or toxoid is administered, the immune system recognizes it as foreign and produces antibodies, or sometimes cytotoxic T cells, and memory cells. This immune response is the same as the one that occurs during the course of a disease. The disease itself does not occur either because whole organisms are not

PUBLIC HEALTH

Where Can a Bacterium Hide from the Immune System?

Why is it so difficult to make vaccines against some bacteria? Some bacteria have really good hiding places—inside of phagocytes! Antibodies and complement proteins can't get at them there. But wait, aren't phagocytes supposed to be part of your immune system, programmed to kill the bacteria they ingest? How can some bacteria safely hide inside phagocytes? These intracellular bacteria have many very different evasion mechanisms, which differ from one species to another. These are very dangerous pathogens!

One trick is to cover yourself with impenetrable molecules. e.g., the waxy substances in *Mycobacterium* cell walls, or the smooth capsular materials of some *Salmonella*. Then the enzymes and toxins inside of the phagolysosome cannot reach and kill the bacteria. Or, it is possible that not all the bacteria inside one phagocyte will die. Those few that survive will multiply inside the cell and keep the infection going.

Another survival mechanism is to keep the phagosome and lysosome from fusing to make a phagolysosome filled with killing molecules. *Legionella* bacteria somehow prevent this fusion. The phagosome instead fuses with other cellular organelles such as mitochondria or the rough endoplasmic reticulum. *Legionella* bacteria divide to fill the phagocytes with more and more bacteria until the cell dies, ruptures open, and releases all those bacteria into the alveoli of the lung.

A third way to survive phagocytosis is to escape from the phagosome before it fuses with the lysosome. *Listeria* bacteria, which cause meningitis and septicemia, produce molecules that eat away the membrane covering the phagosome. Once safely outside in the cytoplasm, the bacteria have free access to all the cell's nutrients while staying hidden from antibodies and complement.

Some examples of intracellular bacteria are: *Brucella abortus, Chlamydia trachomatis, Escherichia coli* (enterohemorrhagic strains), *Legionella pneumophilla, Listeria monocytogenes, Mycobacterium tuberculosis, Mycobacterium leprae, Rickettsia rickettsiae, Salmonella typhi, Shigella dysenteriae,* and *Yersinia pestis.*

---

used or because they have been sufficiently weakened to have lost their virulence. In other words, vaccines retain important antigenic properties but lack the ability to cause disease. In fact, organisms in vaccines sometimes do multiply in the host, but without producing disease symptoms. Similarly, toxoids retain antigenic properties but cannot exert their toxic effects.

An important factor in the longevity of immunity from active immunization is the nature of the immunizing substance. In most instances, vaccines made with live organisms confer longer-lasting immunity than those made with dead organisms, parts of organisms, or toxoids. For example, measles (both rubella and rubeola) vaccines and oral poliomyelitis vaccine, which contain live viruses, usually confer lifelong immunity. Intramuscular polio vaccine, which contains killed viruses, and typhoid fever vaccine, which contains dead bacteria, confer immunity that lasts 3 to 5 years. Tetanus and diphtheria toxoids confer immunity of about 10 years' duration.

Because immunity is not always lifelong, "booster shots" are often needed to maintain it. As we have noted, the first dose of a vaccine or toxoid stimulates a primary immune response analogous to that during the course of a disease. Subsequent doses stimulate a secondary immune response analogous to that following exposure to an organism to which immunity has already developed. Thus, booster shots boost immunity by greatly increasing the number of antibodies. This increases the length of time sufficient antibodies are available to prevent disease.

The route of administration of a vaccine can affect the quality of immunity. Compared with injecting vaccines into muscle, immunity is more durable when oral vaccines are used against gastrointestinal infections and nasal aerosols are used against respiratory infections.

Vaccines and toxoids, especially those containing live organisms, must be properly stored to retain their effectiveness. Some require refrigeration, and serious failures of measles immunization have resulted from inadequate refrigeration. Others must be used within a certain number of hours or days after a vial of vaccine is opened. Thus, clinics offer some immunizations only on selected days. Immunizations that require expensive vaccines and that are in low demand may be given one day per week or even less frequently.

In general, active immunization cannot be used to prevent a disease after a person has been exposed. This is because the time required for immunity to develop is greater than the incubation period of the disease. Rabies immunization is an exception to this rule. Because rabies typically has a long incubation period, active immunization can be used with some hope that immunity will develop before the rabiesvirus reaches the brain. The farther the virus must travel to reach the brain, the greater the chance of effective immunization. So, a bite on the ankle received while kicking off a rabid animal may be less hazardous than one on the trunk or neck. Smallpox is another example of a disease that can be prevented by active immunization after exposure. As we shall see later, passive immunization sometimes is used to prevent or lessen the

PUBLIC HEALTH

Vaccines Against Addiction?

Have you struggled unsuccessfully to give up smoking, or watched patients who cannot kick their cocaine habit? Help may be on the way! Celtic Pharma has anti-nicotine and anti-cocaine vaccines in clinical trials. Promising results have currently led them to continue with larger studies. The vaccines are designed to raise anti-nicotine or anti-cocaine antibodies. These antibodies bind with nicotine or cocaine in the patient's bloodstream, forming a complex which is too big to pass the blood–brain barrier. This prevents the pleasure or high associated with use of the drug. Eventually the patient realizes that it is pointless to continue using a drug which gives them no pleasurable results, and they quit.

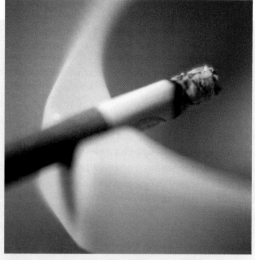

(© Cristina Pedrazzini/Photo Researchers, Inc.)

(Robin Nelsen/PhotoEdit)

The patient receives a series of injections over a 12-week period. Injecting assures patient compliance, unlike when people fail to use their anti-smoking patches. A cocaine derivative (succinyl norcaine) or a nicotine derivative (nicotinic butyric acid) is complexed with the B protein subunit of the cholera toxin (already in use as a cholera vaccine), and is adsorbed onto an aluminum hydroxide gel adjuvant, and then suspended in buffered saline. How long the vaccine will be effective, and whether periodic booster shots will be needed, is being worked out now.

Tobacco use is the second major cause of death in the world, and the heartbreak of cocaine addiction is unmeasurable. Let's hope that immunology can defeat both of these problems.

---

severity of diseases after exposure to them. Several vaccines and toxoids have been licensed for general use in the United Sates; their properties are summarized in **Table 17.6**. Many more are available for people with special needs, such as foreign travel, or for experimental purposes. Properties of some special use vaccines are given in **Table 17.7**.

Recommended Immunizations

Tables **17–8A** and **17–8B** list the vaccines currently recommended in the United States for routine immunization of normal infants, children, and adults. The **DTaP vaccine** contains diphtheria toxoid, acellular pertussis (whooping cough) bacteria, and tetanus toxoid. Although most children did tolerate the old whole cell

pertussis vaccine, a few suffered severe complications, as described shortly in connection with hazards of vaccines. There were two types of **poliomyelitis vaccine** generally used in the United States. One (the Sabin) contained three different types of live polioviruses. The other (the Salk) contains killed virus. Today only the inactivated Salk vaccine is used in the United States, as the Sabin vaccine can cause paralytic polio in recipients or in persons having close contact with someone recently vaccinated. Vaccines with similar antigens are available for administration orally or intramuscularly. **MMR vaccine** contains live rubella, rubeola, and mumps viruses. This vaccine can be used to immunize against all three diseases simultaneously, or separate vaccines can be given for each disease.

TABLE 17.6

Properties of Materials Available for Active Immunization

Disease	Nature of Material	Route of Administration	Use and Comments	Duration of Effectiveness
Anthrax	Cellular proteins	SC	2 and 4 weeks, 6, 12, 18 months, annual booster	Unknown
Bacterial meningitis	Polysaccharide-protein conjugate	IM	2, 4, 6, and 12–15 months; 75% effective	14–34 years
Cholera	Killed bacteria	SC, IM, ID	2 doses a week or more apart; 50% effective; may be required for travel	6 months
Diphtheria	Toxoid	IM	3 doses 4 weeks apart plus boosters; 90% effective	10 years
Hepatitis A	Inactivated virus	IM	2 doses, 2nd 6–18 months after 1st, varies with mfgr.	10 years
Hepatitis B	Viral antigen	IM	2 doses 4 weeks apart, booster in 6 months	About 5 years
Influenza (viral)	Inactivated virus	IM	1 or 2 doses, depending on type of virus; recommended for high-risk patients and medical personnel; 75% effective	1–3 years
Lobar pneumonia	Polysaccharide	SC, IM	1 dose before chemotherapy	5–7 years
Measles (rubeola)	Live virus	SC	1 dose at 15 months, revaccination around age 12; 95% effective; may prevent disease if given within 48 hours of exposure	Lifelong
Meningococcal meningitis	Polysaccharide	SC	1 dose, recommended during epidemics and for high-risk patients	Lifelong if given after age 2
Mumps	Live virus	SC	1 dose given after age 1; 95% effective	Lifelong
Pertussis (whooping cough)	Acellular proteins	IM	Same as diphtheria	10 years
Plague	Killed bacteria	IM	3 doses 4 weeks apart; for travel to some parts of the world	6 months
Poliomyelitis	Killed virus	IM	2, 4, and 6–18 months; booster at 4–6 years	Lifelong
Rabies	Killed virus	IM	2 doses 1 week apart with third dose in 2 weeks; 80% effective; used after probable exposure	2 year
Rubella	Live virus	SC	1 dose at 15 months; some recommend second dose at 12 years	Lifelong
Smallpox	Live vaccinia virus	ID	1 dose; 90% effective; used only by laboratory workers exposed to poxviruses, and military	3 years
Tetanus	Toxoid	IM	3 doses 4 weeks apart plus boosters	10 years
Tuberculosis	Attenuated bacteria	ID, SC	1 dose for inadequately treated patients and high-risk groups	Lifelong?
Typhoid fever	Live virus	O	4 doses 2 days apart; 70% effective; recommended for travel, epidemics, and carriers	5 years
Yellow fever	Live virus	SC	1 dose; recommended for travel to endemic areas	10 years

SC—subcutaneous; IM—intramuscular; ID—intradermal; O—oral.

TABLE 17.7

Selected Examples of Materials for Special Immunization and Experimentation		
Infectious Agent	**Nature of Material**	**Uses**
Adenovirus	Live virus	Military recruits
Bacillus anthracis	Antigen extract	Handlers of animals and hides
Campylobacter	Attenuated bacteria	Experimentation
Vibrio cholerae	Toxoids of *Escherichia coli* and *Vibrio cholerae*	Experimental oral administration to obtain more effective immunization
Cytomegalovirus	Live virus	Experimentation, may produce latent infections
Equine encephalitis virus	Live inactivated viruses	Laboratory workers and experimentation

TABLE 17.8A

Recommended Immunization Schedule—United States 2007
For Persons Aged 0–6 Years

Vaccine	Birth	1 month	2 months	4 months	6 months	12 months	15 months	18 months	19–23 months	2–3 years	4–6 years
Hepatitis B[1]	HepB	HepB		see footnote 1		HepB			HepB Series		
Rotavirus[2]			Rota	Rota	Rota						
Diphtheria, tetanus, pertussis[3]			DTaP	DTaP	DTaP		DTaP				DTaP
Haemophilus influenzae type b[4]			Hib	Hib	*Hib*[4]	Hib		Hib			
Pneumococcal[5]			PCV	PCV	PCV	PCV				PCV PPV	
Inactivated poliovirus			IPV	IPV		IPV					IPV
Influenza[6]						Influenza (yearly)					
Measles, mumps, rubella[7]						MMR					MMR
Varicella[8]						Varicella					Varicella
Hepatitis A[9]						HepA (2 doses)				HepA series	
Meningococcal[10]										MPSV4	

For Persons Aged 7–18 Years

Vaccine	7–10 years	11–12 years	13–14 years	15 years	16–18 years
Tetanus, diphtheria, pertussis[11]	See footnote 11	Tdap	Tdap		
Human papillomavirus[12]	See footnote 12	HPV (3 doses)	HPV series		
Meningococcal[13]	MPSV4	MCV4	MCV4[13]		
			MCV4		
Pneumococcal[14]	PPV				
Influenza[15]	Influenza (yearly)				
Hepatitis A[16]	HepA series				
Hepatitis B[17]	HepB series				
Inactivated poliovirus[18]	IPV series				
Measles, mumps, rubella[19]	MMR series				
Varicella[20]	Varicella series				

This schedule indicates the recommended ages for routine administration of currently licensed childhood vaccines, as of December 1, 2006. Additional information is available at http://www.cdc.gov/nip/recs/child-schedule.htm. Any dose not administered at the recommended age should be administered at any subsequent visit, when indicated and feasible. Additional vaccines may be licensed and recommended during the year. Licensed combination vaccines may be used whenever any components of the combination are indicated and other components of the vaccine are not contraindicated and if approved by the Food and Drug Administration for that dose of the series. Providers should consult the respective Advisory Committee on Immunization Practices statement for detailed recommendations. Clinically significant adverse events that follow immunization should be reported to the Vaccine Adverse Event Reporting System (VAERS). Guidance about how to obtain and complete a VAERS form is available at http://www.vaers,hhs.gov or by telephone, 800-822-7967.

    Range of recommended ages

    Catch-up Immunization

    Certain high-risk groups

1. **Hepatitis B vaccine (HepB).** (*Minimum age: birth*)
   At Birth:
   - Administer monovalent HepB to all newborns before hospital discharge.
   - If mother is hepatitis surface antigen (HBsAg)-positive, administer HepB and 0.5 ml. of hepatitis B immune globulin (HBIG) within 12 hours of birth.
   - If mother's HBsAg status is unknown, administer HepB within 12 hours of birth. Determine the HBsAg status as soon as possible and if HBsAg-positive, administer HBIG (no later than age 1 week).
   - If mother is HBsAg-negative, the birth dose can only be delayed with physician's order and mother's negative HBsAg laboratory report documented in the infant's medical record.

   After the Birth Dose:
   - The HepB series should be completed with either monovalent HepB or a combination vaccine containing HepB. The second dose should be administered at age 1–2 months. The final dose should be administered at age ≥24 weeks. Infants born to HBsAg-positive mothers should be tested for HBsAg and antibody to HBsAg after completion of ≥3 doses of a licensed HepB series, at age 9–18 months (generally at the next well-child visit).

   4-Month Dose:
   - It is permissible to administer 4 doses of HepB when combination vaccines are administered after the birth dose. If monovalent HepB is used for doses after the birth dose, a dose at age 4 months is not needed.

2. **Rotavirus vaccine (Rota).** (*Minimum age: 6 weeks*)
   - Administer the first dose at age 6–12 weeks. Do not start the series later than age 12 weeks.
   - Administer the final dose in the series by age 32 weeks. Do not administer a dose later than age 32 weeks.
   - Data on safety and efficacy outside of these age ranges are insufficient.

3. **Diphtheria and tetanus toxoids and acallular pertussis vaccine (DTaP).** (*Minimum age: 6 weeks*)
   - The fourth dose of DTaP may be administered as early as age 12 months, provided 6 months have elapsed since the third dose.
   - Administer the final dose in the series at age 4–6 years.

4. ***Haemophilus influenzae* type b conjugate vaccine (Hib).** (*Minimum age: 6 weeks*)
   - If PRP-OMP (PedvaxHIB® or ComVax® [Merck]) is administered at ages 2 and 4 months, a dose at age 6 months is not required.
   - TriHiBit® (DTaP/Hib) combination products should not be used for primary immunization but can be used as boosters following any Hib vaccine in children aged ≥ 12 months.

5. **Pneumococcal vaccine.** (*Minimum age: 6 weeks for pneumococcal conjugate vaccine [PCV]; 2 years for pneumococcal polysaccharide vaccine [PPV]*)
   - Administer PCV at ages 24–59 months in certain high-risk groups. Administer PPV to children ages ≥2 years in certain high-risk groups. See *MMWR* 2000; 49(No. RR-9): 1–35.

6. **Influenza vaccine.** (*Minimum age: 6 months for trivalent inactivated influenze vaccine [TIV]; 5 years for live, attenuated influenza vaccine [LAIV]*)
   - All children aged 6–59 months and close contacts of all children aged 0–59 months are recommended to receive influenza vaccine.
   - Influenza vaccine is recommended annually for children aged ≥59 months with certain risk factors, health-care workers, and other persons (including household members) in close contact with persons in groups at high risk. See MMWR 2006; 55(No. RR-10): 1–41.
   - For healthy persons aged 5–49 years, LAIV may be used as an alternative to TIV.
   - Children receiving TIV should receive 0.25 mL if aged 6–35 months or 0.5 mL if aged ≥3 years.
   - Children aged <9 years who are receiving influenza vaccine for the first time should receive 2 doses (separated by ≥4 weeks for TIV and ≥6 weeks for LAIV).

7. **Measles, mumps, and rubella vaccine (MMR).** (*Minimum age: 12 months*)
   - Administer the second dose of MMR at age 4–6 years. MMR may be administered before age 4–6 years, provided ≥4 weeks have elapsed since the first dose and both doses are administered at age ≥12 months.

8. **Varicella vaccine.** (*Minimum age: 12 months*)
   - Administer the second dose of varicella vaccine at age 4–6 years. Varicella vaccine may be administered before age 4–6 years, provided that ≥3 months have elapsed since the first dose and both doses are administered at age ≥12 months. If second dose was administered ≥28 days following the first dose, the second dose does not need to be repeated.

9. **Hepatitis A vaccine (HepA).** (*Minimum age: 12 months*)
   - HepA is recommended for all children aged 1 year (i.e., aged 12–23 months). The 2 doses in the series should be administered at least 6 months apart.
   - Children not fully vaccinated by age 2 years can be vaccinated at subsequent visits.
   - HepA is recommended for certain other groups of children, including in areas where vaccination programs target older children. See *MMWR* 2006; 55(No. RR-7): 1–23.

10. **Meningococcal polysaccharide vaccine (MPSV4).** (*Minimum age: 2 years*)

- Administer MPSV4 to children aged 2–10 years with terminal complement deficiencies or anatomic or functional asplenia and certain other high-risk groups. See *MMWR* 2005; 54(No. RR-7): 1–21.

11. Tetanus and diphtheria toxoids and acellular pertussis vaccine (Tdap). *(Minimum age: 10 years for BOOSTRIX® and 11 years for ADACEL℠)*
- Administer at age 11–12 years for those who have completed the recommended childhood DTP/DTap vaccination series and have not received a tetanus and diphtheria toxoid vaccine (Td) booster dose.
- Adolescents aged 13–18 years who missed the 11–12 year Td/Tdap booster dose should also receive a single dose of Tdap if they have completed the recommended childhood DTP/DTaP vaccination series.

12. Human papillomavirus vaccine (HPV). *(Minimum age: 9 years)*
- Administer the first dose of the HPV vaccine series to females at age 11–12 years.
- Administer the second dose 2 months after the first dose and the third dose 6 months after the first dose.
- Administer the HPV vaccine series to females at age 13–18 years if not previously vaccinated.

13. Meningococcal vaccine. *(Minimum age: 11 years for meningococcal conjugate vaccine [MCV4]; 2 years for meningococcal polysaccharide vaccine [MPSV4])*
- Administer MCV4 at age 11–12 years and to previously unvaccinated adolescents at high school entry (at approximately age 15 years).
- Administer MCV4 to previously unvaccinated college freshmen living in dormitories; MPSV4 is an acceptable alternative.
- Vaccination against invasive meningococcal disease is recommended for children and adolescents aged ≥2 years with terminal complement deficiencies or anatomic or functional asplenia and certain other high-risk groups. See *MMWR* 2005;54(No. RR-7):1–21. Use MPSV4 for children aged 2–10 years and MCV4 or MPSV4 for older children.

14. Pneumococcal polysaccharide vaccine (PPV). *(Minimum age: 2 years)*
- Administer for certain high-risk groups. See *MMWR* 1997;46(No. RR-8):1–24, and *MMWR* 2000;49(No. RR-9):1–35.

15. Influenza vaccine. *(Minimum age: 6 months for trivalent inactivated influenza vaccine [TIV]; 5 years for live, attenuated influenza vaccine [LAIV])*

- Influenza vaccine is recommended annually for persons with certain risk factors, health-care workers, and other persons (including household members) in close contact with persons in groups at high risk. See *MMWR* 2006;55 (No. RR-10):1–41.
- For healthy persons aged 5–49 years, LAIV may be used as an alternative to TIV.
- Children aged <9 years who are receiving influenza vaccine for the first time should receive 2 doses (separated by ≥4 weeks for TIV and ≥6 weeks for LAIV).

16. Hepatitis A vaccine (HepA). *(Minimum age: 12 months)*
- The 2 doses in the series should be administered at least 6 months apart.
- HepA is recommended for certain other groups of children, including in areas where vaccination programs target older children. See *MMWR* 20006;55 (No. RR-7): 1–23.

17. Hepatitis B vaccine (HepB). *(Minimum age: birth)*
- Administer the 3-dose series to those who were not previously vaccinated.
- A 2-dose series of Recombivax HB® is licensed for children aged 11–15 years.

18. Inactivated poliovirus vaccine (IPV). *(Minimum age: 6 weeks)*
- For children who received an all-IPV or all-oral poliovirus (OPV) series, a fourth dose is not necessary if the third dose was administered at age ≥4 years.
- If both OPV and IPV were administered as part of a series, a total of 4 doses should be administered, regardless of the child's current age.

19. Measles, mumps, and rubella vaccine (MMR). *(Minimum age: 12 months)*
- If not previously vaccinated, administer 2 doses of MMR during any visit, with ≥4 weeks between the doses.

20. Varicella vaccine. *(Minimum age: 12 months)*
- Administer 2 doses of varicella vaccine to persons without evidence of immunity.
- Administer 2 doses of varicella vaccine to persons aged <13 years at least 3 months apart. Do not repeat the second dose, if administered ≥28 days after the first dose.
- Administer 2 doses of varicella vaccine to persons aged ≥13 years at least 4 weeks apart.

Approved by the Advisory Committee on Immunization Practices the American Academy of Pediatrics, and the American Academy of Family Physicians.

The recommended age for administering vaccines varies. DTaP and polio vaccines can be administered effectively as early as 2 months of age. However, MMR vaccine is not recommended before 12 months of age. When it is given to younger infants, the quality of immunity that results usually is not sufficient to protect against infection, probably because of immaturity of the immune system. Vaccines against *Haemophilus influenzae*, type b (Hib), first became available in 1985 but didn't work well in children under 2 years old. The Food and Drug Administration (FDA) has now approved several new **Hib vaccines** for younger children that could prevent about 10,000 cases per year of meningitis, which kills about 500 children and leaves thousands of survivors mentally retarded, deaf, or otherwise neurologically damaged. Immunizations against Hib are scheduled at 2, 4, 6, and 12 to 15 months. Children between 15 months and 5 years need only one shot. Immunization is not recommended for those over 5 years old because nearly all children have contracted a Hib infection by then and have thus developed natural active immunity. When given early, the Hib vaccine can be combined with DTaP. Since you were born, influenza, hepatitis A and B, pneumococcal, rotavirus, meningococcal, varicella (chicken pox for children, shingles for adults), and **HPV** (human papilloma virus, the cause of 99% of cervical cancer) have been added to the lists of recommended vaccines. Do you need to catch up on any of these?

TABLE 17.8B

Recommended Adult Immunization Schedule, United States, October 2006–September 2007
By Vaccine and Age Group*

Vaccine	19–49 years	50–64 years	≥65 years
Tetanus, diphtheria pertussis (Td/Tdap)[1,†]	1-dose Td booster every 10 yrs Substitute 1 dose of Tdap for Td		
Human papillomavirus (HPV)[2]	3 doses (females)		
Measles, mumps, rubella (MMR)[3,†]	1 or 2 doses	1 dose	
Varicella[4,†]	2 doses (0, 4–8 wks)	2 doses (0, 4–8 wks)	
Influenza[5,†]	1 dose annually	1 dose annually	
Pneumococcal (polysaccharide)[6,7]	1–2 doses		1 dose
Hepatitis A[8,†]	2 doses (0, 6–12 mos, or 0, 6–18 mos)		
Hepatitis B[9,†]	3 doses (0, 1–2, 4–6 mos)		
Meningococcal[10]	1 or more doses		

By Vaccine and Medical and Other Indications

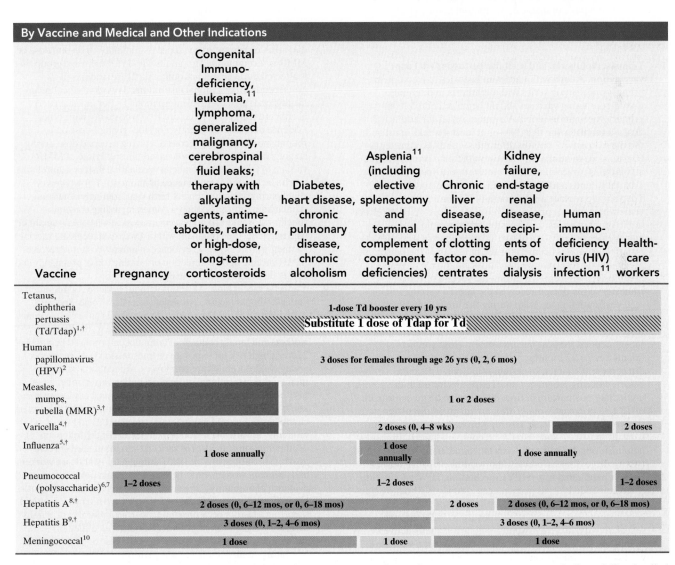

Vaccine	Pregnancy	Congenital Immuno-deficiency, leukemia,[11] lymphoma, generalized malignancy, cerebrospinal fluid leaks; therapy with alkylating agents, antime-tabolites, radiation, or high-dose, long-term corticosteroids	Diabetes, heart disease, chronic pulmonary disease, chronic alcoholism	Asplenia[11] (including elective splenectomy and terminal complement component deficiencies)	Chronic liver disease, recipients of clotting factor con-centrates	Kidney failure, end-stage renal disease, recipi-ents of hemo-dialysis	Human immuno-deficiency virus (HIV) infection[11]	Health-care workers
Tetanus, diphtheria pertussis (Td/Tdap)[1,†]	1-dose Td booster every 10 yrs Substitute 1 dose of Tdap for Td							
Human papillomavirus (HPV)[2]	3 doses for females through age 26 yrs (0, 2, 6 mos)							
Measles, mumps, rubella (MMR)[3,†]			1 or 2 doses					
Varicella[4,†]			2 doses (0, 4–8 wks)					2 doses
Influenza[5,†]	1 dose annually			1 dose annually	1 dose annually			
Pneumococcal (polysaccharide)[6,7]	1–2 doses		1–2 doses					1–2 doses
Hepatitis A[8,†]	2 doses (0, 6–12 mos, or 0, 6–18 mos)				2 doses	2 doses (0, 6–12 mos, or 0, 6–18 mos)		
Hepatitis B[9,†]	3 doses (0, 1–2, 4–6 mos)				3 doses (0, 1–2, 4–6 mos)			
Meningococcal[10]	1 dose			1 dose	1 dose			

* This schedule indicates the recommended age groups and medical indications for routine administration of currently licensed vaccines for persons aged ≥19 years, as of October 1, 2006. Licensed combination vaccines may be used whenever any components of the combination are indicated and when the vaccine's other components are not contraindicated. For detailed recommendations on all vaccines, including those used primarily for travelers or that are issued during the year, consult the manufacturers' package inserts and the complete statements from the Advisory Committee on Immunization Practices

(www.cdc.gov/nip/publications/acip-list.htm). Report all clinically significant postvaccination reactions to the Vaccine Adverse Event Reporting System [VAERS]. Reporting forms and instructions on filing a VAERS report are available at www.vaers.hhs.gov or by telephone, 800-822-7967. Information on how to file a Vaccine Injury Compensation Program claim is available at www.hrsa.gov/vaccine compensator or by telephone, 800-338-2382. To file a claim for vaccine injury, contact the U.S. Court of Federal Claims, 717 Madison Place, N.W., Washington, D.C. 20005; telephone, 202-357-6403. Additional information about the vaccines in this schedule and contraindications for vaccination is also available at www.cdc.gov/nip or from the CDC-INFO Contact Center at 800-CDC-INFO (800-232-4830) in English and Spanish, 24 hours a day, 7 days a week.

† Covered by the Vaccine Injury Compensation Program. NOTE: These recommendations must be read with the footnotes.

For all persons in this category who meet the age requirements and who lack evidence of immunity (e.g., lack documentation of vaccination or have no evidence of price infection)

Recommended if same other risk factor is present (e.g., as the basis of medical, occupational, lifestyle, or other indications)

Contraindicated

1. **Tetanus, diphtheria, and acellular pertussis (Td/Tdap) vaccination.** Adults with uncertain histories of a complete primary vaccination series with diphtheria and tetanus toxoid-containing vaccines should begin or complete a primary vaccination series. A primary series for adults is 3 doses; administer the first 2 doses at least 4 weeks apart and the third dose 6–12 months after the second. Administer a booster dose to adults who have completed a primary series and if the last vaccination was received ≥10 years previously. Tdap or tetanus and diphtheria (Td) vaccine may be used; Tdap should replace a single dose of Td for adults aged <65 years who have not previously received a dose of Tdap (either in the primary series, as a booster, or for wound management). Only one of two Tdap products (Adacel® [sanofi Pasteur]) is licensed for use in adults. If the person is pregnant and received the last Td vaccination ≥10 years previously, administer Td during the second or third trimester; if the person received the last Td vaccination in <10 years, administer Tdap during the immediate postpartum period. A one-time administration of 1 dose of Tdap with an interval as short as 2 years from a previous Td vaccination is recommended for postpartum women, close contacts of infants aged <12 months, and all healthcare workers with direct patient contact. In certain situations, Td can be deferred during pregnancy and Tdap substituted in the immediate postpartum period, or Tdap can be given instead of Td to a pregnant woman after an informed discussion with the woman (see www.cdc.gov/nip/publications/acip-list.htm). Consult the ACIP statement for recommendations for administering Td as prophylaxis in wound management (www.cdc.gov/mmwr/preview/mmwrhtml/00041645.htm).

2. **Human papillomavirus (HPV) vaccination.** HPV vaccination is recommended for all women aged ≤26 years who have not completed the vaccine series. Ideally, vaccine should be administered before potential exposure to HPV through sexual activity, however, women who are sexually active should still be vaccinated. Sexually active women who have not been infected with any of the HPV vaccine types receive the full benefit of the vaccination. Vaccination is less beneficial for women who have already been infected with one or more of the four HPV vaccine types. A complete series consists of 3 doses. The second dose should be administered 2 months after the first dose; the third dose should be administered 6 months after the first dose. Vaccination is not recommended during pregnancy. If a women is found to be pregnant after initiating the vaccination series, the remainder of the 3-dose regimen should be delayed until after completion of the pregnancy.

3. **Measles, mumps, rubella (MMR) vaccination.** *Measles component*; adults born before 1957 can be considered immune to measles. Adults born during or after 1957 should receive ≥1 dose of MMR unless they have a medical contraindication, documentation of ≥1 dose, history of measles based on healthcare provider diagnosis, or laboratory evidence of immunity. A second dose of MMR is recommended for adults who 1) have been recently exposed to measles or in an outbreak setting; 2) have been previously vaccinated with killed measles vaccine; 3) have been vaccinated with an unknown type of measles vaccine during 1963–1967; 4) are students in postsecondary educational institutions; 5) work in a healthcare facility; or 6) plan to travel internationally. Withhold MMR or other measles-containing vaccines from HIV-infected persons with severe immunosuppression.

Mumps component: adults born before 1957 can generally be considered immune to mumps. Adults born during or after 1957 should receive 1 dose of MMR unless they have a medical contraindication, history of mumps based on healthcare provider diagnosis, or laboratory evidence of immunity. A second dose of MMR is recommended for adults who 1) are in an age group that is affected during a mumps outbreak; 2) are students in postsecondary educational institutions; 3) work in a healthcare facility; or 4) plan to travel internationally. For unvaccinated healthcare workers born before 1957 who do not have other evidence of mumps immunity, consider giving 1 dose on a routine basis and strongly consider giving a second dose during an outbreak. *Rubella component:* administer I dose of MMR vaccine to women whose rubella vaccination history is unreliable or who lack laboratory evidence of immunity. For women of childbearing age, regarding of birth year, routinely determine rubella immunity and counsel women regarding congenital rubella syndrome. Do not vaccinate women who are pregnant or who might become pregnant within 4 weeks of receiving vaccine. Women who do not have evidence of immunity should receive MMR vaccine upon completion or termination of pregnancy and before discharge from the healthcare facility.

4. **Varicella vaccination.** All adults without evidence of immunity to varicella should receive 2 doses of varicella vaccine. Special consideration should be given to those who 1) have close contact with persons at high risk for severe disease (e.g., healthcare workers and family contacts of immunocompromised persons) or 2) are at high risk for exposure or transmission (e.g., teachers of young children; child care employees; residents and staff members of institutional settings, including correctional institutions; college students; military personnel; adolescents and adults living in households with children; nonpregnant women of childbearing age; and international travelers). Evidence of immunity to varicella in adults includes any of the following: 1) documentation of 2 doses of varicella vaccine at least 4 weeks apart; 2) U.S.-born before 1980 (although for healthcare workers and pregnant women, birth before 1980 should not be considered evidence of immunity); 3) history of varicella based on diagnosis or verification of varicella by a healthcare provider (for a patient reporting a history of or presenting with an atypical case, a mild case, or both, healthcare providers should seek either an epidemiologic link with a typical varicella case or evidence of laboratory confirmation, if it was performed at the time of acute disease); 4) history of herpes zoster based on healthcare provider diagnosis; or 5) laboratory evidence of immunity or laboratory confirmation of disease. Do not vaccinate women who are pregnant or might become pregnant within 4 weeks of receiving the vaccine. Assess pregnant women for evidence of varicella

immunity. Women who do not have evidence of immunity should receive dose 1 of varicella vaccine upon completion or termination of pregnancy and before discharge from the healthcare facility. Dose 2 should be administered 4–8 weeks after dose 1.

5. **Influenza vaccination.** *Medical indications:* chronic disorders of the cardiovascular or pulmonary systems, including asthma; chronic metabolic diseases, including diabetes mellitus, renal dysfunction, hemoglobinopathies, or immunosuppression (including immunosuppression caused by medications or HIV); any condition that compromises respiratory function or the handling of respiratory secretions or that can increase the risk of aspiration (e.g., cognitive dysfunction, spinal cord injury, or seizure disorder or other neuromuscular disorder); and pregnancy during the influenza season. No data exist on the risk for severe or complicated influenza disease among persons with asplenia; however, influenza is a risk factor for secondary bacterial infections that can cause severe disease among persons with asplenia. *Occupational indications:* healthcare workers and employees of long-term-care and assisted living facilities. *Other indications:* residents of nursing homes and other long-term-care and assisted living facilities; persons likely to transmit influenza to persons at high risk (e.g., in-home household contacts and caregivers of children aged 0–59 months or persons of all ages with high-risk conditions); and anyone who would like to be vaccinated. Healthy, nonpregnant persons aged 5–49 years without high-risk medical conditions who are not contacts of severely immunocompromised persons in special care units can receive either intranasally administered influenza vaccine (RuMist®) or inactivated vaccine. Other persons should receive the inactivated vaccine.

6. **Pneumococcal polysaccharide vaccination.** *Medical indications:* chronic disorders of the pulmonary system (excluding asthma); cardiovascular diseases; diabetes mellitus; chronic liver diseases, including liver disease as a result of alcohol abuse (e.g., cirrhosis); chronic renal failure or nephrotic syndrome; functional or anatomic asplenia (e.g., sickle cell disease or splenectomy [if elective splenectomy is planned, vaccinate at least 2 weeks before surgery]); immunosuppressive conditions (e.g., congenital immunodeficiency, HIV infection [vaccinate as close to diagnosis as possible when CD4 cell counts are highest], leukernia, lymphoma, multiple myeloma, Hodgkin disease, generalized malignancy, or organ or bone marrow transplantation); chemotherapy with alkylating agents, antimetabolites, or high-dose, long-term corticosteroids; and cochlear implants. *Other indications:* Alaska Natives and certain American Indian populations and residents of nursing homes or other long-term-care facilities.

7. **Revaccination with pneumococcal polysaccharide vaccine.** One-time revaccination after 5 years for persons with chronic renal failure or nephrotic syndrome; functional or anatornic aspleria (e.g., sickle cell disease or splenectomy); immunosuppressive conditions (e.g., congenital immunodeficiency, HIV infection, leukernia, lymphoma, multiple myeloma, Hodgkin disease, generalized malignancy, or organ or bone marrow transplantation); or chemotherapy with alkylating agents, antimetabolites, or high-dose, long-term corticosteroids. For persons aged ≥65 years, one-time revaccination if they were vaccinated ≥5 years previously and were aged <65 years at the time of primary vaccination.

8. **Hepatitis A vaccination.** *Medical indications:* persons with chronic liver disease and persons who receive clotting factor concentrates. *Behavioral indications:* men who have sex with men and persons who use illegal drugs. *Occupational indications:* persons working with hepatitis A virus (HAV)–infected primates or with HAV in a research laboratory setting. *Other indications:* persons traveling to or working in countries that have high or intermediate endemicity of hepatitis A (list of countries is available at www.cdc.gov/travel/diseases.html) and any person who would like to obtain immunity. Current vaccines should be administered in a 2-dose schedule at either 0 and 6–12 months, or 0 and 6–18 months. If the combined hepatitis A and hepatitis B vaccine is used, administer 3 doses at 0, 1, and 6 months.

9. **Hepatitis B vaccination.** *Medical indications:* persons with end-stage renal disease, including patients-receiving hemodialysis; persons seeking evaluation or treatment for a sexually transmitted disease (STD); persons with HIV infection; persons with chronic liver disease; and persons who receive clotting factor concentrates. *Occupational indications:* healthcare workers and public safety workers who are exposed to blood or other potentially infectious body fluids. *Behavioral indications:* sexually active persons who are not in a long-term mutually monogamous relationship (i.e., persons with >1 sex partner during the previous 6 months); current or recent injection-drug users; and men who have sex with men. *Other indications:* household contacts and sex partners of persons with chronic hepatitis B virus (HBV) infection; clients and staff members of institutions for persons with developmental disabilities; all clients of STD clinics; international travelers to countries with high or intermediate prevalence of chronic HBV infection (a list of countries is available at www.cdc.gov/travel/diseases.html); and any adult seeking protection from HBV infection. Settings where hepatitis B vaccination is recommended for all adults: STD treatment facilities; HIV testing and treatment facilities; facilities providing drug-abuse treatment and prevention services; healthcare settings providing services for injection-drug users or men who have sex with men; correctional facilities; end-stage renal disease programs and facilities for chronic hemodialysis patients; and institutions and nonresidential daycare facilities for persons with developmental disabilities. *Special formulation indications:* for adult patients receiving hemodialysis and other immunocompromised adults, 1 dose of 40 μg/ml. (Recombivax HB®) or 2 doses of 20 μg/mL (Engerix-B®).

10. **Meningococcal vaccination.** *Medical indications:* adults with anatomic or functional asplenia, or terminal complement component deficiencies. *Other indications:* first-year college students living in dormitories; microbiologists who are routinely exposed to isolates of *Neissena maningitidis*; military recruits; and persons who travel to or live in countries in which meningococcal disease is hyperendemic or epidemic (e.g., the "meningitis belt" of sub-Saharan African during the dry season [December–June]), particularly if their contact with local populations will be prolonged. Vaccination is required by the government of Saudi Arabis for all travelers to Mecca during the annual Hajj. Meningococcal conjugate vaccine is preferred for adults with any of the preceding indications who are aged ≤55 years, although meningococcal polysaccharide vaccine (MPSV4) is an acceptable alternative. Revaccination after 5 years might be indicated for adults previously vaccinated with MPSV4 who remain at high risk for infection (e.g., persons residing in areas in which disease is epidemic).

11. **Selected conditions for which *Haemophilus influenzae* type b (Hib) vaccine may be used.** Hib conjugate vaccines are licensed for children aged 6 weeks–71 months. No efficacy data are available on which to base a recommendation concerning use of Hib vaccine for older children and adults with the chronic conditions associated with an increased risk for Hib disease. However, studies suggest good immunogenicity in patients who have sickle cell disease, leukernia, or HIV infection or who have had splenectomies; administering vaccine to these patients is not contraindicated.

Approved by the Advisory Committee on Immunization Practices, the American College of Obstetricians and Gynecologists, the American Academy of Family Physicians, and the American College of Physicians.

Figure 17.16 Vaccination mark from inoculation with BCG vaccine. This vaccine is used in some countries to immunize against tuberculosis. It is introduced under the skin by a cluster of tines that leaves a permanent, raised mark. *(Science Photo Library/ Photo Researchers, Inc.)*

Immunization recommendations vary between developed and underdeveloped countries. Most developed countries use approximately the same immunizations as are recommended in the United States. However, several countries use BCG (Bacille Calmette-Guerin) **(Figure 17.16)** vaccine to protect against tuberculosis. The World Health Organization (WHO) has administered it to over 150,000 people in various countries in the past several decades. It is not used in the United States because serious controversy existed about its safety and efficacy when it was first developed. Now that it has been proven safe and effective, the incidence of tuberculosis is not sufficient to warrant widespread use. WHO recommends immunization at earlier ages in underdeveloped countries to combat serious contagious diseases: BCG and oral polio vaccines at birth; DTaP and polio vaccines at 6, 10, and 14 weeks, and measles vaccine at 9 months.

BCG vaccine is given, usually right after birth, to 75% of all children worldwide—but not in the United States.

HAZARDS OF VACCINES

There are overwhelming benefits of using vaccines to prevent serious infectious diseases in populations. However, vaccines also pose hazards that must be weighed when deciding whether they should be administered to entire populations, to certain individuals, or not at all. And, of course, the prevalence and severity of diseases also must be considered in such decisions.

Active immunization often causes fever, malaise, and soreness at the site of injection. Thus, patients already suffering from fever and malaise should not receive immunization because a worsening of their condition might be erroneously attributed to the vaccine. More importantly, the patient's immune system, overburdened by the existing infection, may be unable to mount an adequate response to the antigen in the vaccine. Particular reactions are associated with certain immunizations. For example, joint pain can be caused by rubella vaccine, and convulsions by pertussis vaccine. Allergic reactions sometimes follow the use of influenza and other vaccines that contain egg protein or vaccines that contain antibiotics as preservatives. However, reactions occur in only a small proportion of vaccine recipients and are generally less severe than the disease. An exceedingly small number of vaccine recipients die or suffer permanent damage from vaccines (see the Polio Vaccine Controversy box in ◀Chapter 24, p. 770). The FDA maintains a Vaccine Adverse Event Reporting System to ensure that vaccine hazards are not overlooked.

Live vaccines pose particular hazards to pregnant women, patients with immunological deficiencies, and patients receiving immunosuppressants such as radiation or corticosteroid drugs. In the case of pregnant women, live viruses sometimes cross the placenta and infect the fetus, whose immune system is immature. They also can cause birth defects. In immunodeficient or immunosuppressed patients, the attenuated virus sometimes has sufficient virulence to cause disease. Therefore, patients who test positive for the AIDS virus should not receive live-virus vaccines. This presents a problem for routine immunization of infants, some of whom, unknown to health care personnel, could have become infected with AIDS before birth. It also means that U.S. military and State Department employees and their accompanying families who are being posted abroad must be tested for AIDS to determine whether they can receive the required live-virus vaccines.

PASSIVE IMMUNIZATION

To induce passive immunity, ready-made antibodies are introduced into an unprotected individual. Because antibodies are found in the serum portion of the blood, these products are often called **antisera**. Although passive immunity is produced quickly, it is only temporary. It lasts only as long as there is a sufficiently high titer of circulating antibodies in the body. **Passive immunization** is established by administering a preparation such as gamma globulin, hyperimmune serum, or an antitoxin that contains large numbers of ready-made antibodies. However, the specificity and degree of this form of immunization depend on the antibody type and concentration used.

Immune serum globulin, formerly called **gamma globulin**, consists of pooled gamma globulin fractions (the portion of serum that contains antibodies) from many individuals. This kind of gamma globulin typically contains sufficient antibodies to provide passive immunity to a number of common diseases, such as mumps, measles, and hepatitis A.

If the donors are specially selected, gamma globulins that have high titers of specific kinds of antibodies can be prepared. Such preparations are often called **hyperimmune sera**, or *convalescent sera*. For example, gamma globulin from persons recovering from mumps or from recent recipients of mumps vaccine contains especially high titers of antimumps antibodies. Similar sera can be collected from donors with high titers of antibodies to other diseases. Hyperimmune sera can also be manufactured by introducing particular antigens—for instance, tetanus toxin—into another animal, such as a horse, and subsequently collecting the antibodies from the animal's serum.

Immunoglobulin in cows' colstrum has been used in humans to successfully treat myasthenia gravis, multiple sclerosis, systemic lupus, and rheumatoid arthritis, among other diseases.

Antitoxins are antibodies against specific toxins, such as those that cause botulism, diphtheria, or tetanus. Passive immunization against tetanus toxins can also be achieved by using tetanus immune globulin, a gamma globulin that contains antibodies against tetanus toxin. The properties of currently available materials used to produce passive immunity are summarized in **Table 17.9**.

Passive immunization gives immediate immunity to a nonimmune person who is exposed to a disease, or it at least lessens the severity of the disease process. Usually, a vaccine is given after the passive immunizing preparation to provide active immunity. Before the advent of antibiotics, passive immunization was frequently used to prevent or lessen the severity of several kinds of pneumonia and a variety of other infectious diseases. With respect to infectious diseases, the most common current use of passive immunity is to protect people with contaminated wounds against tetanus toxin. Although the incidence of exposure to diphtheria and botulism is lower than that of tetanus, passive immunization can also be used against these diseases.

Passive immunization is also used to counteract the effects of snake and spider bites and to prevent damage to fetuses from certain immunological reactions. Antivenins, or antibodies to the venom of certain poisonous snakes and the black widow spider, are given as emergency treatments. They react with venom molecules that have not already bound to tissues. Thus, the sooner after a bite the antivenin is administered, the more effective it is in counteracting the effects of the venom.

A fetal immunological reaction can occur when a mother with Rh-negative blood carries her second Rh-positive fetus. As is explained in more detail in ◄Chapter 18, the mother becomes sensitized to the Rh-positive red blood cells when her first Rh-positive child is born. The mother's immune system would subsequently make anti-Rh antibodies, which will harm the second Rh-positive child. To prevent this from happening, anti-Rh antibodies are given to the mother within 72 hours after the birth of the first child and after subsequent births, miscarriages, or abortions. Like the antivenins, the anti-Rh antibodies bind to Rh-positive red blood cells, so the cells are destroyed before the mother's immune system can make antibodies to them.

As with active immunization, passive immunization poses some hazards. The most common hazards are allergic reactions. Some antitoxins contain proteins from other animals as a result of their manufacture in eggs or horses. They are particularly likely to cause allergic reactions, especially when the patient receives them for the second time. Thus, vaccines of human origin are safer, at least with respect to the risk of allergic reactions. Allergic reactions to large IgG molecules also can occur if gamma globulins or hyperimmune sera are

TABLE 17.9

Properties of Materials Available for Passive Immunization	
Material	**Uses**
Human gamma globulin	To prevent recurrent infections in patients with deficiencies in humoral immunity and to prevent or lessen disease symptoms after exposure of nonimmune persons to measles or hepatitis A
Specific Gamma Globulins	
Varicella-zoster immune globulin	To prevent chickenpox in high-risk children; must be given within 4 days of exposure
Hepatitis-B immune globulin	To prevent hepatitis B after exposure (via blood or needles) and to prevent spread of the disease from mothers to newborns
Mumps immune globulin	May prevent orchitis (inflammation of testes) in adult males exposed to mumps
Pertussis immune globulin	To reduce severity of disease and mortality in children under 3 or in debilitated children
Rabies immune globulin	To prevent rabies after a bite from a possibly rabid animal; applied to the bite if possible and administered intramuscularly
Tetanus immune globulin	To prevent tetanus after injury in nonimmune patients
Vaccinia immune globulin	To halt progress of disease in immunodeficient patients who develop a progressive vaccinia (cowpox) infection

accidentally given intravenously instead of by their normal intramuscular route. A new immune globulin IV, containing smaller molecules, can be given safely by the intravenous route.

Another hazard, or at least detriment, of passive immunity is that giving ready-made antibodies can interfere with a host's ability to produce its own antibodies. One way this might occur is by ready-made antibodies binding to antigens and preventing them from stimulating the host's immune system. Thus, maternal antibodies, while they protect infants from some infections, may prevent the infant's immune system from making its own antibodies.

FUTURE OF IMMUNIZATION

Immunologists continue to search for new vaccines. Effective vaccines should meet five criteria: (1) Vaccines developed should be protective against the disease for which they were designed. (2) The vaccines must be safe and not have adverse side effects. (3) Protection should be sustained, providing long-term protection from infection. (4) Vaccines should generate neutralizing antibodies or protective T cells to vaccine antigens. (5) Vaccines should be practical in terms of stability and use.

Whole-cell killed vaccines (first-generation vaccines) sometimes produce unwanted side effects due to extraneous cellular materials. Therefore, efforts are made to identify and obtain cellular subunits that contain only the purified immunogenic portion of a microorganism that will produce immunity. *Subunit vaccines* (second-generation vaccines) are safer than *attenuated vaccines*, which use live organisms that are treated to eliminate their virulence. There is always the possibility that the organisms may revert to a virulent state. Many researchers consider attenuated vaccines too risky in the hunt for an AIDS vaccine, but they are functional for some other diseases, as is the BCG vaccine for tuberculosis. In general, live organisms produce higher and longer-lasting immunity than do nonliving organisms. *Recombinant DNA vaccines* (third-generation vaccines) are being produced by inserting the genes for specific antigens into the genomes of nonvirulent organisms. The hepatitis B viral antigens have been cloned in yeast cells, extracted and purified before use, to form a safe and very effective vaccine. Rabies virus antigen has been inserted into vaccinia (cowpox) virus and is being tested as a means for controlling rabies in wild animal populations such as raccoons.

Many vaccine researchers believe that by the year 2025 most Americans will be immunized routinely against some 30 diseases, including AIDS; genital herpes; influenza; hepatitis A, B, C, and E; and chickenpox and shingles. These immunizations will be administered in three stages: infancy and early childhood, for childhood diseases; prior to puberty, for certain sexually transmitted diseases; and adulthood, for influenza and shingles.

✓ CHECKLIST

1. What is the BCG vaccine? Why is it not used in the United States?
2. What are some hazards of vaccines?
3. What is gamma globulin? What is immune globulin? How are they prepared?
4. How are subunit vaccines prepared? Why are they better than whole-cell killed vaccines?

IMMUNITY TO VARIOUS KINDS OF PATHOGENS

This chapter has focused on the basic principles of immunity that apply to all pathogens. However, you may find it helpful to note the ways in which the immune system responds to various kinds of pathogens.

BACTERIA

As we saw in ◀Chapter 16 (p. 464), innate defenses such as skin, mucous membranes, and gastric secretions prevent many bacteria from entering host tissues. When bacteria do infect a host, certain immune responses alter the invading organisms so that they can be phagocytized. Once plasma cells produce specific antibodies, the antibodies can interfere with any of several steps in bacterial invasion. They can attach to pili and capsules, preventing bacterial attachment to cell surfaces. Antibodies can work with complement to opsonize bacteria for later phagocytosis or lysis by other cells of the immune system. Or they can neutralize bacterial toxins or inactivate bacterial enzymes.

VIRUSES

Viruses infect by invading cells—usually first attacking cells that line body passages. Then they directly invade target organs such as the lungs or travel in the blood (viremia) to target organs or organ systems such as the liver or nervous system. Polioviruses invade cells that line the digestive tract, but they also can enter nerve endings.

Immune responses can combat viral infections at any of these locations. Interferons, secretory IgA, and some IgG antibodies act at the surface-lining cells and prevent or minimize entry of viruses. IgG and IgM act in the blood to neutralize viruses directly or to promote their destruction by complement. Finally, cytotoxins and cellular immunity via T_C cells and NK cells are especially important in clearing the body of cells infected with viruses. The mechanisms by which the immune system combats viral infections are summarized in **Figure 17.17**.

Besides adaptive immune responses to a viral infection, many innate responses can limit infection. Fever is an important defense against viruses. Several viruses, such as influenza, parainfluenza, and rhinoviruses, are temperature-sensitive. They replicate in the lining cells of the respiratory tract, which normally has a temperature between

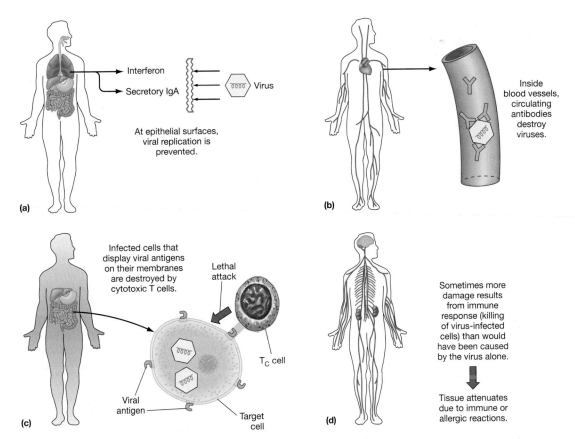

Figure 17.17   How the immune system combats viruses.

33° and 35°C—lower than the normal body temperature of 37°C because the cells are cooled as atmospheric air moves over their moist surfaces. When a person has a fever of even 1° to 2°C, the ability of the virus to replicate is reduced. Another benefit of fever in resisting viral infection is that temperature increases cause an increase in interferon production (◀Chapter 16, p. 476).

FUNGI

Certain fungal infections progress through a tissue as fungal cells invade and destroy one cell after another. Immunity to fungi is poorly understood, but it appears to be mainly cell-mediated. Fungal skin infections probably are combated by IgA antibodies and T_H cells, which release certain cytokines that activate macrophages. The macrophages, in turn, engulf and digest fungi. Commensal fungi apparently are kept in their place by cell-mediated responses. Supportive evidence comes from studies of individuals with impaired T cell functions. Such persons are extremely likely to become infected with opportunistic fungi, such as *Candida albicans*.

PROTOZOA AND HELMINTHS

Protozoa and helminths are quite dissimilar in size and complexity, but many use similar methods of invading the body. Host defenses against them also are similar, except that allergic reactions to helminths can be severe enough to cause more damage to host cells than to the

disease agent. Antigens on the surface of the roundworm *Ascaris* are potent inducers of IgE type allergic reactions. Individuals with an *Ascaris* allergy can absorb enough antigens through the skin to cause a severe allergic reaction, even just by coming into contact with fluids in which the worms have been preserved. The large quantities of IgE produced coat the surface of the worm, leading to its death. Some researchers believe the IgE evolved primarily as a defense against helminths. However, this does not always work out well.

Parasitic protozoa and helminths interact with their hosts in ways that do not endanger host survival and thereby ensure parasite survival. These pathogens cause chronic, debilitating diseases that usually are not immediately life-threatening. Most of these parasites are relatively large and difficult to phagocytize. Large helminths, such as heartworms in dogs, can block blood vessels and cause sudden death. When attacked by phagocytes, some helminths release toxins that significantly damage the host. Medical intervention also can be hazardous. Giving drugs to kill some helminths causes them to release large quantities of toxic decay products. Thus, once some worm infections have been acquired, coexistence with the parasites may be the best course of action.

Many parasites have complex life cycles with more than one host, and some infect animals that serve as reservoirs for human infection (◀Chapter 11, p. 311). Thus, they stand ready to take advantage of appropriate conditions in various hosts. Each life cycle stage of some

APPLICATIONS

Cancer and Immunology

Cancer is one of the leading causes of death in developed countries, and its incidence as a cause of death in developing nations is on the rise. Cancer cells can arise by mutation of a normal cell in response to chemical carcinogens, radiation, or viruses; by expression of previously repressed human oncogenes (tumor-producing genes) within cells (◄Chapter 10, p. 303); or by a combination of these factors.

Regardless of the means by which they are produced, many cancer cells have certain plasma membrane antigens not found in normal cells. These antigens are an ideal target for destruction by the immune system. According to the theory of *immune surveillance*, T_C cells and NK cells recognize and destroy these abnormal cells before they develop into cancers. If this theory is correct, each of us has within our body many different potentially malignant cells, but we develop cancer only if the T_C cells or NK cells fail to identify and destroy the mutated cells. Unfortunately, many tumors and cancers show little regard for control by the immune system.

One way in which T_C cells might fail to recognize malignant cells is by the actions of antigens on the malignant cells themselves. The antigens stimulate the formation of antibodies, which bind to the antigens without damaging the malignant cells. Although this binding does not occur before the antigens sensitize T_C cells, it does block T_C cells from attacking the malignant cells. Other tumors can avoid immune system attacks by lacking abnormal antigens or by losing MHC class I molecules through mutation. Such cells escape immune detection by T_C cells but, as described in this chapter, are still attacked by NK cells.

However, the notion that cancer cells can be destroyed by immune reactions has led researchers to work on developing immunotoxins and cancer vaccines. As discussed earlier in this chapter, an immunotoxin is a monoclonal antibody with an anticancer drug, a microbial toxin such as diphtheria toxin, or a radioactive substance attached to it. The antibody is designed to bind to a specific cancer cell antigen; the attached substance is selected for its ability to destroy the cancer cell. Such immunotoxins are expected to seek and destroy specific cells—the cancer cells that have the appropriate antigen. Cancer vaccines containing one or more cancer cell antigens are also being developed. Some of these vaccines induce immunity to specific cancer cells. Others contain antigens frequently found on certain kinds of cancer cells.

A major difficulty in making immunotoxins and cancer vaccines is that many different kinds of antigens are found on cancer cells from various patients. To be effective against an existing cancer, an immunotoxin or vaccine must be specific for the antigens on the cells of that specific cancer. Similarly, to immunize against common forms of cancer, vaccines need to generate antibodies or T_C cells that will destroy cancer cells should they develop. Another difficulty is that some malignant cells produce immunosuppressive cytokines, such as transforming growth factor-α (TGF-α), which stimulate tumor growth and may interfere with T_H1 cell activity. Immunotoxins and vaccines need to get powerful enough to overcome this inhibition. Finally, some antigens are found on both malignant and normal cells. Great care must be taken to develop immunotoxins and vaccines that will react only with malignant cells.

In October 1991 the first attempt was made to eliminate cancer by immunizing patients against their own tumors. Cells from a malignant melanoma tumor were removed, genes for tumor necrosis (killing) factor were added to the cancer cells' genomes, and the cells were reinjected (infused) into the patient. There, it is hoped, the genetically altered cells will secrete enough of this immune system factor to change the regulation of the immune system so as to overcome its tolerance of the tumor and cause it to begin attacking the cancerous cells. To date, there have been few successes.

Meanwhile, immunization already offers us the ability to prevent about 80% of liver cancer cases. Liver cancer is among the most prevalent kinds of cancer. One form of liver cancer, primary hepatocellular carcinoma, accounts for 80 to 90% of all cases. It now appears that this form is associated with infection by hepatitis B virus early in life and especially with being a carrier of the virus. Not only has hepatitis B viral DNA been found in chromosomes of cancerous liver cells but also the intact virus has been isolated from cancerous liver cells. The incidence of liver cancer is highest in regions of Africa and Asia where the incidence of hepatitis B infections also is high. Immunization against the hepatitis B virus would protect recipients of the vaccine against infection, prevent them from becoming carriers, and thus protect them against liver cancer.

Colorized SEM of a small T lymphocyte attacking two large tumor cells (5,700X). (Dr. Andreis Liepins/Photo Researchers, Inc.)

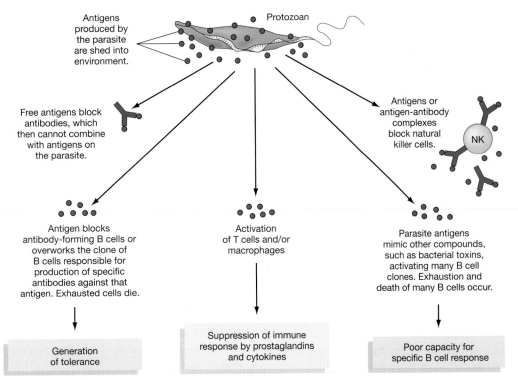

Figure 17.18 **How antigens of parasitic protozoans thwart the immune system.**

parasites can have several surface antigens. Although hosts may produce antibodies against such antigens, the ability of certain parasites to change them provides a way to thwart host defenses. For example, the malaria parasite induces host antibody formation before it invades host cells. While the protozoan multiplies inside cells, it produces different antigens, so the original antibodies formed by the host are not effective against the new parasite progeny when they are released.

The host's immune system also combats parasitic protozoa and the helminths by cell-mediated processes. Although T_C cells usually are not effective against such parasites, some T cells release cytokines, such as IL-3, that activate macrophages. These macrophages can attack malarial parasites and several kinds of worms, including blood flukes. Other cytokines, such as IL-5, enhance the ability of eosinophils to combat worm infections.

Parasitic protozoa and helminths have available to them an assortment of mechanisms that thwart immune responses. These mechanisms include the following:

1. Some protozoa protect themselves by invading cells, forming protective cases called cysts at certain life cycle stages, or otherwise becoming inaccessible to host defenses.

2. Some protozoa avoid immune recognition by changing their surface antigens (antigenic variation) regularly or with each reproductive cycle.

3. Some parasites suppress the host's immune responses by releasing toxins that damage lymphocytes, enzymes that inactivate IgG, or soluble antigens that thwart the immune system in a variety of ways **(Figure 17.18)**.

4. Some intracellular protozoa suppress the action of phagocytes by inhibiting fusion of lysosomes with vacuoles, by resisting digestion by lysosomal enzymes, or by impairing oxidative metabolism.

Given the various methods protozoa and helminths have to evade and thwart immune responses, it is not surprising that they cause chronic debilitating infections.

▌ RETRACING OUR STEPS

▌▌▌ IMMUNOLOGY AND IMMUNITY

- **Immunology** is the study of **immunity**, which refers to the capacity to recognize and defend against infectious agents and other foreign substances.

- **Susceptibility** is vulnerability to infectious agents. Immunity is **innate** when it acts against any infectious agent and **adaptive** when it acts against a particular infectious agent. **Immunology** is the study of specific immunity. The **immune**

system is the body system that provides the host with specific immunity to particular infectious agents.

▌ TYPES OF IMMUNITY

INNATE IMMUNITY

* **Innate immunity** provides nonspecific, hereditary defense and protection against many pathogens without prior exposure.

ADAPTIVE IMMUNITY

* **Acquired adaptive immunity** provides specific, nonhereditary defense and protection after exposure to a specific pathogen.

ACTIVE AND PASSIVE IMMUNITY

* In **active immunity**, an individual's own immune system makes antibodies.
* In **passive immunity**, ready-made antibodies are introduced into the body.
* Types of immunity are illustrated in Figure 17.1.

▌ CHARACTERISTICS OF THE IMMUNE SYSTEM

ANTIGENS AND ANTIBODIES

* An **antigen** is a foreign substance that can elicit a specific immune response. Most antigens are proteins, but some are polysaccharides, nucleoproteins, or glycoproteins.
* Each antigen has several **epitopes**, or **antigenic determinants**.
* An **antibody**, or **immunoglobulin**, is a protein produced in response to the presence of an antigen. Antibodies bind to epitopes on the antigen.

CELLS AND TISSUES OF THE IMMUNE SYSTEM

* Lymphocytes develop from lymphoid stem cells in the bone marrow. Lymphocytes differentiate into **B cells** in the bone marrow or into **T cells** in the thymus

DUAL NATURE OF THE IMMUNE SYSTEM

* The dual roles of the immune system consist of **humoral immunity**, which is carried out mainly by B cells and plasma cells, and **cell-mediated immunity**, which is carried out mainly by certain T cells.

GENERAL PROPERTIES OF IMMUNE RESPONSES

* Immune responses distinguish **self** from **nonself**.
* According to the **clonal selection theory**, B cells recognize specific epitopes on antigens according to the particular antibody present on the B cell plasma membrane. When a B cell detects an antigen with which it can react, it binds to the antigen, engulfs it, processes it, displays a peptide fragment as MHC class II to T_H2 cells, and divides many times. A **clone** of genetically identical B cells, which differentiate into many plasma cells and some **memory cells**, is produced.
* **Specificity** refers to the ability of immune responses to respond to and distinguish among different antigens and epitopes.
* **Diversity** refers to the ability of immune responses to produce many different antibodies and cell substances on the basis of the different antigens they encounter.
* **Immunological memory** refers to the ability of T and B cells to recognize substances to which the immune system has previously responded.

▌ HUMORAL IMMUNITY

* B cells are selected to respond to specific antigens in accordance with the particular antibody present on the B cell membrane prior to encountering an antigen.
* When a B cell detects an antigen with which it can react, it binds to the antigen and divides many times to produce a clone of many plasma cells and some memory cells.
* Many B cells require helper T cells to proliferate and differentiate into both plasma and memory B cells.
* **Plasma cells** synthesize and release large numbers of antibodies.
* **Memory cells** remain in lymphoid tissue ready to respond to subsequent exposure to the same antigen.

PROPERTIES OF ANTIBODIES (IMMUNOGLOBULINS)

* Structurally, antibodies consist of two **heavy** and two **light** polypeptide **chains**, each of which has a variable region capable of reacting with a specific antigen.
* Properties of particular kinds of antibodies (**immunoglobulins**) are summarized in Table 17.4.

PRIMARY AND SECONDARY RESPONSES

* **Primary responses** are the immune system's first encounter with foreign antigens.
* Memory cells remain in lymphoid tissue, ready to respond to subsequent exposure to the same antigen.
* **Secondary responses** bring fast and efficient destruction of antigens recognized by B and T memory cells. Primary and secondary responses are summarized in Table 17.5.

KINDS OF ANTIGEN-ANTIBODY REACTIONS

* Humoral immunity is most effective against bacteria, which are destroyed by **agglutination** (clumping) or lysed by complement after opsonization or directly by IgMs or neutralized. Toxins and some viruses can be inactivated by antibody **neutralization**.

▌ MONOCLONAL ANTIBODIES

* **Monoclonal antibodies** are antibodies produced in the laboratory from a clone of cultured cells that make one specific antibody to one specific epitope.
* Specific monoclonal antibodies can be used in some diagnostic tests, and methods to use them in treating infectious diseases and cancer are being developed.

▌ CELL-MEDIATED IMMUNITY

* Cell-mediated immunity concerns the direct actions of certain T cells that defend the body against viral infections and reject tumors and transplanted tissues.
* T cells do not make antibodies but instead have membrane receptors for antigens; these receptors bind to MHC molecules displaying foreign peptide fragments.
* Cell-mediated immune responses involve differentiation and activation of several kinds of T cells and the secretion of cytokines.

THE CELL-MEDIATED IMMUNE REACTION

* Processed antigens on MHC class II molecules bind with T cell receptors. Next, IL-1 secreted from macrophages and IL-2 secreted from T cells activate the T cells, which can then differentiate into T_H1 cells and T_H2 cells.

- Certain pathogenic bacteria can grow in macrophages after they have been phagocytized. T_H1 cells can release γ-interferon, a cytokine that causes such infected macrophages to become resensitized to other cytokines.
- AIDS destroys T_H cells, thereby impairing both humoral and cell-mediated immunity.

HOW KILLER CELLS KILL
- T_C and NK cells destroy target cells by releasing the lethal protein **perforin**.

THE ROLE OF ACTIVATED MACROPHAGES
- Certain pathogenic bacteria can grow in macrophages after phagocytosis. The lymphokine macrophage activating factor helps stimulate antimicrobial processes so that macrophages can kill the pathogens.
- When macrophages fail to kill pathogens, the pathogens are walled off in granulomas.

III FACTORS THAT MODIFY IMMUNE RESPONSES
- Host defenses in healthy adults in an unpolluted environment prevent nearly all infectious diseases. Individuals with reduced resistance are called **compromised hosts**.
- Factors that reduce host resistance include very young or old age, stress, seasonal patterns, poor nutrition, traumatic injury, pollution, and radiation. Complement deficiencies, immunosuppressants, infections such as HIV, and genetic defects impair immune system function.

III IMMUNIZATION

ACTIVE IMMUNIZATION
- **Active immunization** induces the same response as the one that occurs during a disease. Immunization challenges the immune system to develop specific defenses and memory cells.
- Active immunization is conferred by **vaccines** and **toxoids**. Vaccines can be made from live, attenuated organisms, dead organisms, parts of organisms, or a toxoid. Toxoids are made by inactivating toxins.
- The recommended immunizations for healthy infants, children, and adults in the United States are summarized in Tables 17.8A and 17.8B.
- The benefits of active immunization against life-threatening diseases nearly always outweigh the hazards. Reactions to

vaccines can cause serious side effects, but their incidence is lower than the incidences of the diseases themselves.

PASSIVE IMMUNIZATION
- **Passive immunization** occurs by the same mechanism as natural passive transfer of antibodies.
- Passive immunity is conferred by **antisera** such as **immune serum globulin (gamma globulin)**, **hyperimmune** or **convalescent sera**, and **antitoxins**.
- The benefits of passive immunization are limited to providing only temporary protection; the side effects are mainly allergic in nature.

FUTURE OF IMMUNIZATION
- **Subunit vaccines** produce fewer side effects than **whole-cell killed vaccines** and offer greater safety than do **attenuated vaccines**.
- **Recombinant DNA vaccines** contain genes for antigens of pathogens inserted into nonpathogenic organisms' genomes and are very safe.

III IMMUNITY TO VARIOUS KINDS OF PATHOGENS

BACTERIA
- Antibodies produced by plasma cells are the chief immunological defense against bacterial antigens. Most immune responses to bacteria serve to promote phagocytosis of the invading cells.

VIRUSES
- Viral infection is combatted by nonspecific defenses, interferon, and antibodies. In addition, the T_C cells of cell-mediated responses and NK cells are important in destroying virus-infected cells.

FUNGI
- Immune response to fungi involve IgA antibodies and are primarily cell-mediated.

PROTOZOA AND HELMINTHS
- Immune responses to parasitic protozoa and helminths are largely cell-mediated. T cells release cytokines that activate macrophages and attract other leukocytes. Allergic reactions to helminths can be more damaging to the host than to the parasite.

▌ TERMINOLOGY CHECK

aquired immunity (*p. 489*)

active immunity (*p. 489*)

active immunization (*p. 510*)

adaptive immunity (*p. 489*)

agglutination (*p. 502*)

anamnestic response (*p. 496*)

antibody (*p. 492*)

antigen (*p. 490*)

antigenic determinant (*p. 491*)

antigen-presenting cell (*p. 505*)

antiserum (*p. 520*)

antitoxin (*p. 521*)

apoptosis (*p. 493*)

artificially acquired active immunity (*p. 489*)

artificially acquired adaptive immunity (*p. 489*)

artificially acquired passive immunity (*p. 490*)

B cell (*p. 492*)

B lymphocyte (*p. 492*)

cell-mediated immunity (*p. 493*)

clonal selection hypothesis (*p. 493*)

colostrum (*p. 489*)

compromised host (*p. 509*)

cross-reaction (*p. 495*)

cytokines (*p. 505*)

cytotoxic (killer) T (T_C) cell (*p. 505*)

delayed hypersensitivity T (T_D) cells (*p. 505*)

diversity (*p. 495*)

DTaP vaccine (*p. 512*)

epitope (*p. 491*)

gamma globulin (*p. 520*)

genetic immunity (*p. 489*)

hapten (*p. 492*)

heavy (H) chain *(p. 496)*
Hib vaccine *(p. 516)*
HPV vaccine *(p. 516)*
humoral immunity *(p. 493)*
hyperimmune serum
  (p. 521)
IgA *(p. 497)*
IgD *(p. 499)*
IgE *(p. 499)*
IgG *(p. 497)*
IgM *(p. 499)*
immune serum globulin
  (p. 520)
immune system *(p. 489)*
immunity *(p. 489)*
immunoglobulin (Ig) *(p. 496)*

immunology *(p. 489)*
innate immunity *(p. 489)*
light (L) chain *(p. 496)*
memory *(p. 496)*
memory cell *(p. 496)*
MMR vaccine *(p. 513)*
monoclonal antibody
  (p. 503)
natural killer (NK) cells
  (p. 493)
naturally acquired active
  immunity *(p. 489)*
naturally acquired adaptive
  immunity *(p. 489)*
naturally acquired passive
  immunity *(p. 490)*

neutralization *(p. 503)*
nonself *(p. 493)*
passive immunity *(p. 490)*
passive immunization
  (p. 520)
perforin *(p. 508)*
plasma cell *(p. 496)*
poliomyelitis vaccine
  (p. 512)
primary response *(p. 499)*
privileged site *(p. 509)*
secondary response *(p. 500)*
secretory component
  (p. 497)
self *(p. 493)*
sensitize *(p. 496)*

species immunity *(p. 489)*
specific immunity *(p. 496)*
specificity *(p. 495)*
superantigen *(p. 508)*
T cell *(p. 493)*
T-dependent antigen
  (p. 500)
T helper (T_H) cell *(p. 505)*
T-independent antigen
  (p. 500)
titer *(p. 492)*
T lymphocyte *(p. 493)*
tolerance *(p. 495)*
toxoid *(p. 510)*
vaccine *(p. 510)*

▌ CLINICAL CASE STUDY

Your next-door neighbor has recently been diagnosed with cancer, a lymphoma, and is undergoing chemotherapy. He has been instructed not to eat anything that has not been cooked.

No raw vegetables or meat, including fish. He is not sure why this is. He knows you are taking a microbiology class and asks you to explain why. What can you tell him?

▌ CRITICAL THINKING QUESTIONS

1. How would you respond to parents who wish to avoid (a) all vaccines for their infant or (b) pertussis vaccine for their infant?

2. According to the clonal selection hypothesis, embryos contain many different lymphocytes, each genetically programmed to recognize a particular antigen and make antibodies to destroy it. If a lymphocyte encounters and recognizes that antigen after development is complete, it divides repeatedly to produce a clone or group of identical progeny cells that make the same antibody. If a developing fetus encountered a low-grade infection with a virus, survived, and then encountered a massive infection by the same virus 20 years later, what do you think would happen to this individual?

3. The parents of a 2-month-old infant delighted in taking him with them on their frequent trips to shopping malls. One

of the grandmothers suggested that they leave him home, as it was flu season and the malls were full of coughing and sneezing people. The mother immediately responded that she was breast-feeding the baby, and with all those antibodies from her, he was surely protected against any diseases in the mall. Later as she sat at the pediatrician's office holding her sick baby on her lap, she expressed disbelief that he could have caught the flu. If you were the doctor, what fallacies in this mother's beliefs could you explain to her?

4. Mutations in a liver cell have resulted in a deletion of two genes, one being involved in down-regulating uncontrolled cellular growth and the other being the MHCl gene. What do you think will happen to this cell and why?

▌ SELF-QUIZ

1. What is the difference between naturally acquired adaptive immunity and artificially acquired adaptive immunity?

2. What is the difference between active and passive immunity?

3. An epitope is an antigenic determinant, but is a hapten an epitope? Why or why not?

4. Which of the following is NOT true about T cells?
  (a) T cells develop from lymphoid stem cells in the bone marrow and mature in the thymus.
  (b) Cell-mediated immunity is primarily carried out by T cells.
  (c) Subsequent differentiation of T cells produces cytotoxic (killer) T cells, delayed-hypersensitivity T cells, helper T cells, and regulatory T cells.

  (d) Natural killer cells (NK) are exclusively differentiated T cells.
  (e) T cells act in situations where antigens are embedded in cells membranes or are inside host cells and thus are inaccessible to antibodies.

5. Which of the following is NOT true about B cells?
  (a) B cells develop from lymphoid stem cells in the bone marrow and mature in the bone marrow.
  (b) Humoral immunity is mediated by B cells and their antibodies.
  (c) Functional B cells are found in all tissues except lymphoid tissues (lymph nodes, spleen, adenoids, and gut-associated tissues).

(d) B cells produce and release antibodies when stimulated by an antigen.

(e) B cells are most effective in defending the body against foreign substances outside of cells, such as bacterial toxins, bacteria, and viruses before they enter cells.

6. Which of the following immune cells/molecules are most effective at destroying intracellular pathogens?
   (a) T_H cells   (d) B cells
   (b) Antibodies   (e) Complement
   (c) T_C cells

7. There are five classes of antibodies or immunoglobulins. Match the following antibody classes to their descriptions:

   ___ IgG
   ___ IgA
   ___ IgM
   ___ IgI
   ___ IgD

   (a) The "allergy" antibody that attaches to basophils and mast cells with their tissue-binding sites that in turn cause them to release substances that produce allergy symptoms when allergens such as pollen or certain foods are encountered

   (b) It is rarely secreted, being found mainly on B cell membranes, and their function is unknown.

   (c) Secreted as a pentamer, it is the first antibody secreted during the early stages of a primary response and is the antibody of the inherited ABO blood types.

   (d) The main antibody class that attaches to microbes with their antigen-binding sites and phagocytic cells through their tissue-binding sites, allowing engulfment of the microbe by the phagocytic cell.

   (e) Occurs in large amounts in body secretions and attaches to linings of the respiratory, digestive, and genitorurinary systems where it prevents microbes from invading tissues.

8. A living microbe with reduced virulence that is used for vaccination is considered:
   (a) A toxoid   (d) Dormant
   (b) Virulent   (e) Complement
   (c) Denatured

9. B cells that produce and release large amounts of antibody are called:
   (a) Memory cells   (d) Basophils
   (b) Plasma cells   (e) Killer cells
   (c) Neutrophils

10. The constant regions of an antibody determine its particular _____ it belongs to while its _____ regions impart its specificity and enable the molecule to bind a particular antigen.

11. Which of the following is NOT a way in which an antibody can destroy a bacterial cell?
    (a) Agglutination
    (b) Lysis by opsonization and complement fixation
    (c) Neutralization

(d) Direct lysis
(e) a and d

12. B cells are activated by:
    (a) Complement   (d) Antibody
    (b) Interferon   (e) Memory cells
    (c) Antigen

13. The best definition of an antigen is:
    (a) A foreign molecule in the body
    (b) A chemical that elicits antibody production and binds to that antibody
    (c) A molecule that binds to antibody
    (d) A pathogen
    (e) An enzyme that activates B cells

14. Fusion between a plasma cell and a tumor cell creates a
    (a) Myeloma   (d) Natural killer cell
    (b) Lymphoblast   (e) Lymphoma
    (c) Hybridoma

15. Monoclonal antibodies recognize a single:
    (a) Antigen   (d) Bacterium
    (b) Epitope   (e) Virus
    (c) B cell

16. Cell-mediated immunity is carried out by _____, while humoral immunity is mainly carried out by _____.
    (a) B cells/T cells   (d) Epitopes/antigens
    (b) T cells/B cells   (e) Antibodies/phagocytes
    (c) Antibodies/antigens

17. The ability of the immune system to recognize self antigens versus nonself antigens is an example of:
    (a) Specific immunity   (d) Tolerance
    (b) Cell-mediated immunity   (e) Antigenic immunity
    (c) Humoral immunity

18. Which is NOT true of memory cells?
    (a) They are part of a secondary response.
    (b) They are part of an anamnestic.
    (c) They can survive without dividing for many months to many years.
    (d) They produce more IgG than IgM.
    (e) They are natural killer cells.

19. Put the following steps in order for cell-mediated immune reactions:
    (a) Differentiated T cells include T helper, delayed hypersensitivity, cytotoxic, and memory T cells that all have different immunological functions depending on the antigen presented.
    (b) Antigen-presenting cells (macrophages and dendritic cells) phagocytize pathogens, ingesting and degrading them into pieces which are transported to the surface of the cell.
    (c) T cells bearing the corresponding receptor for the presented antigen bind to it and become activated only if the appropriate MHC is also present.
    (d) Some pieces of the pathogen's antigens are processed by inserting them into the antigen-presenting cell's membrane and are held in place by class II major histocompatibility complex (MHCII) proteins.

(e) Activated T cells are stimulated to divide and differentiate into different types of T cells, including memory cells.

20. Some pathogenic bacteria can grow in macrophages after phagocytosis. Which of the following is a way in which such macrophages can either kill or confine the invading microbes?
    (a) Become stimulated by the secreted lymphokine macrophage activating factor
    (b) Wall off the macrophage-ingested pathogens in granulomas
    (c) Present both MHCI and MHCII complexes
    (d) a and b.
    (e) All of the above

21. Individuals with reduced resistance to infectious diseases can have their immune systems compromised by all of the following factors EXCEPT:
    (a) Pollution or radiation
    (b) Very young or old age
    (c) Complement deficiencies or genetic defects
    (d) Poor nutrition and stress
    (e) All of the above

22. An antigen that overstimulates the immune system by bonding nonspecifically to MHC on antigen-presenting cells is termed a:
    (a) Nonspecific antigen   (d) Toxic shock syndrome
    (b) Superantigen   (e) Super necrotic
    (c) Epitope

23. A vaccine that activates both cell-mediated and humoral immunity would most likely be:
    (a) Heat-killed   (d) Live, attenuated
    (b) A toxoid   (e) A virion
    (c) Short-lived immunity

24. A patient with a _____ titer of antibodies has a greater protection against infection than a patient with a _____ titer.
    (a) High/low   (b) Low/high

25. A vaccine contains an antigen to which the immune system responds. Which of the following is not a possible component of a vaccine?
    (a) Dead pathogen
    (b) Toxoids
    (c) Live, weakened pathogen
    (d) Parts of a pathogen
    (e) c and d

26. In this diagram identify the various regions of an antibody molecule. What part(s) of the molecule do antigen bind to? What part(s) of the molecule does complement bind to?

Antibody

❚ EXPLORATIONS ON THE WEB

http://www.wiley.com/college/black

If you think you've mastered this chapter, there's more to challenge you on the web. Go to the companion web site to fine-tune your understanding of the chapter concepts and discover answers to the questions posed below.

1. Watch a short animation demonstrating how macrophages, T cells, and B cells interact to produce antibodies.

2. Find out how B cells are turned into antibody-producing factories called plasma cells.

3. How does our body make all these different antibodies that protect us from the millions of bacteria and viruses we encounter throughout our lifetime?

Immunology II: Immunological Disorders and Tests

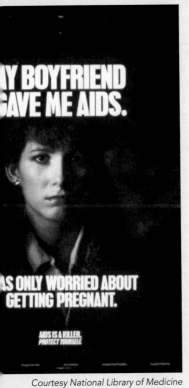

Courtesy National Library of Medicine

Come with me . . .

What do you know about the current state of HIV/AIDS? Take this True-False quiz:

_____ *The rate of HIV infection on college campuses is about 10 times higher than in the general population.*

_____ *15 million children under the age of 15 were orphaned by AIDS by the end of 2003.*

_____ *The vast majority of African AIDS cases are among heterosexuals who do not use injection drugs.*

_____ *HIV infection rates in the United States are estimated by CDC as being:*

22% cases are in black men	*32% cases are in black women*
17% cases are in white men	*8% cases are in white women*

_____ *Worldwide, one-half of newly infected people 25 years of age or younger will be dead before their 35th birthday.*

_____ *AIDS now kills 2.9 million people per year, worldwide. Tuberculosis now kills over 2.7 million people per year, worldwide. Malaria now kills over 1 million people per year, worldwide. Heart disease now kills 17 million people per year, worldwide.*

Sad to say, all these statements are true. Test yourself further online with more questions. Come with me to Africa to see AIDS at its worst.

 Video related to this topic is available within WileyPLUS.

||| OVERVIEW OF IMMUNOLOGICAL DISORDERS
Hypersensitivity / Immunodeficiency

||| IMMEDIATE (TYPE I) HYPERSENSITIVITY
Allergen / Mechanism of Immediate Hypersensitivity / Localized Anaphylaxis / Generalized Anaphylaxis / Genetic Factors in Allergy / Treatment of Allergies

||| CYTOTOXIC (TYPE II) HYPERSENSITIVITY
Mechanism of Cytotoxic Reactions / Examples of Cytotoxic Reactions

||| IMMUNE COMPLEX (TYPE III) HYPERSENSITIVITY
Mechanism of Immune Complex Disorders / Examples of Immune Complex Disorders

||| CELL-MEDIATED (TYPE IV) HYPERSENSITIVITY
Mechanism of Cell-Mediated Reactions / Examples of Cell-Mediated Disorders

||| AUTOIMMUNE DISORDERS
Autoimmunization / Examples of Autoimmune Disorders

||| TRANSPLANTATION
Histocompatibility Antigens / Transplant Rejection / Tolerance of the Fetus During Pregnancy / Immunosuppression

||| DRUG REACTIONS

||| IMMUNODEFICIENCY DISEASES
Primary Immunodeficiency Diseases / Secondary (or Acquired) Immunodeficiency Diseases

||| IMMUNOLOGICAL TESTS
The Precipitin Test / Agglutination Reactions / Tagged Antibody Tests

In ◀Chapter 17 we emphasized how specific immune responses defend the body against harmful substances. However, such responses are not always beneficial. Sometimes the humoral or cell-mediated responses react in ways that are physiologically unpleasant or even life-threatening. Perhaps you or someone you know gets a runny nose and watery eyes every time hay fever season rolls around. Maybe you know of people who have other allergies, who have had adverse reactions to a blood transfusion, or who suffer from more severe immunological disorders—like AIDS.

In this chapter we will learn more about the immune system by examining the ways in which the immune system goes awry and reacts inappropriately or inadequately. We also will look at some of the laboratory and clinical methods used to detect and measure immune reactions.

OVERVIEW OF IMMUNOLOGICAL DISORDERS

An **immunological disorder** is a condition that results from an inappropriate or inadequate immune response. Most inappropriate responses involve some type of hypersensitivity, whereas inadequate responses are due to an immunodeficiency.

HYPERSENSITIVITY

In **hypersensitivity**, or **allergy**, the immune system reacts in an exaggerated or inappropriate way to a foreign substance. Such responses can be thought of as "too much of a good thing"—the immune system responds to a harmless foreign agent by doing harm instead of protecting the body. Although allergy is another name for hypersensitivity, many disorders that people call "allergies" are not due to immunological reactions. These disorders include toxic responses to drugs, digestive upsets from nonallergic responses to foods, and emotional disturbances.

There are four types of hypersensitivity: (1) immediate hypersensitivity (Type I); (2) cytotoxic hypersensitivity (Type II); (3) immune complex hypersensitivity (Type III);

and (4) cell-mediated, or delayed, hypersensitivity (Type IV). The type that develops depends on which components of the immune response are involved and on how quickly the reaction develops. **Immediate (Type I) hypersensitivity**, or *anaphylaxis*, results from a prior exposure to a foreign substance called an *allergen*, an antigen that evokes a hypersensitivity response. Allergies to pollen, foods, and insect stings are examples of immediate hypersensitivity. **Cytotoxic (Type II) hypersensitivity** is elicited by antigens on cells, especially red blood cells, that the immune system recognizes as foreign. This reaction occurs when a patient receives the wrong blood type during a transfusion. **Immune complex (Type III) hypersensitivity** is elicited by antigens in vaccines, on microorganisms, or on a person's own cells. Large antigen-antibody complexes form, precipitate on blood vessel walls, and cause tissue injury within hours. **Cell-mediated (Type IV), or delayed, hypersensitivity** is triggered by exposure to foreign substances from the environment (such as poison ivy), infectious disease agents, transplanted tissues, and the body's own tissues and cells. *Delayed hypersensitivity T cells* react with the foreign cells or substances, causing in some cases extensive tissue destruction.

Autoimmune disorders represent a form of hypersensitivity in which the body's immune system responds to its own tissues as if they were foreign. Antibodies or T cells attack self-antigens.

IMMUNODEFICIENCY

In **immunodeficiency** the immune system responds inadequately to an antigen, either because of inborn or acquired defects in B cells or T cells. The weak responses in immunodeficiency disorders are "too little of a good thing." They do no direct harm, but they leave the individual susceptible to infections, which can be severe and even life-threatening.

Immunodeficiencies can be primary or secondary. **Primary immunodeficiencies** are genetic or developmental defects in which the person lacks T cells or B cells or has defective ones. **Secondary immunodeficiencies** result from damage to T cells or B cells after they have developed normally. These disorders can be caused by malignancies, malnutrition, infections such as AIDS, or drugs that suppress the immune system.

Let us now look more closely at the immunological disorders of hypersensitivity and immunodeficiency.

APPLICATIONS

Mother's Milk Is Best

Food allergies are less common in breast-fed infants than in bottle-fed infants for two reasons: Breast-fed infants are not subjected to potential allergens from cow's milk, and some components in the mother's milk may help to seal the newborn's immature intestinal lining against entry of allergens. In contrast, bottle-fed infants are subjected to foreign proteins early in life, and their intestinal lining may remain more permeable to allergens throughout life.

IMMEDIATE (TYPE I) HYPERSENSITIVITY

Immediate (Type I) hypersensitivity, or *anaphylactic hypersensitivity*, typically produces an immediate response upon exposure to an allergy-inducing antigen (also known as an *allergen*). Nonallergic persons do not respond to such antigens. **Anaphylaxis** (an'a-fi-lak'sis) (Greek: *ana*, "against," *phylaxis*, "protection") is an

APPLICATIONS

"Knockout" Kitties

Do you love cats, but have allergies to them? Relief may be in sight. Transgenic Pets, a small biotechnology company in Syracuse, NY, expects to soon be selling a genetically modified cat that will not cause allergies. Sixty to 90% of all cat allergies are due to just one protein called Fel dl. It is thought to help keep the cat's skin moist, but may have other as yet undiscovered functions. The gene for Fel dl was isolated and sequenced years ago. "Knocking out" (deleting) this gene will prevent production of the allergenic protein, and hopefully not cause any harm to the cat. "Knockout" mice are routinely produced and studied in laboratories. The same techniques will be used to produce "knockout" cats. Once a supply of both sexes is produced, the cats will be allowed to reproduce in the normal way, but will be neutered before sale, so as to protect the company's investment. They plan to sell the cats for $750 to $1000 each, and feel that owning such a cat will be safer than taking allergy medication. Dog lovers are not so lucky. Several allergens are involved in dog allergies.

immediate, exaggerated allergic reaction to antigens. The term *anaphylaxis* refers to detrimental effects to the host caused by an inappropriate immune response. These effects are the opposite of *prophylaxis*, the preventive effects generated by an immune response.

Early investigators discovered that a substance they called **reagin** (re-a'jin) was responsible for this type of hypersensitivity. We now know that reagin consists of IgE antibodies. However, the term *reagin* is still used in some allergy literature.

Anaphylaxis is the harmful result of IgE antibodies made in response to allergens. It can be local or generalized (systemic). **Localized anaphylaxis** appears as reddening of the skin, watery eyes, hives, asthma, and digestive disturbances. **Generalized anaphylaxis** appears as a systemic life-threatening reaction such as airway constriction or anaphylactic shock, a generalized condition resulting from a sudden extreme drop in blood pressure.

ALLERGEN

Immediate hypersensitivity results from two or more exposures to an allergen. An **allergen** is an ordinarily harmless foreign substance (typically a protein or a chemical bound to a protein) that can cause an exaggerated immunological response. The first exposure to the allergen produces no visible signs or symptoms. Allergens include airborne substances such as pollen, household dust, molds, and dander—tiny particles from hair, feathers, or skin. Household dust commonly contains nearly microscopic mites and their fecal pellets. Other allergens include venoms from insect stings, antibiotics, certain foods, sulfites, and foreign substances found in vaccines and in diagnostic or therapeutic materials. Allergens can be introduced into the body by inhalation, ingestion, or injection (Table 18.1).

MECHANISM OF IMMEDIATE HYPERSENSITIVITY

The typical sequence of events involved with the mechanism of Type I hypersensitivity is *sensitization*, which involves the *production of IgE antibodies (antiallergens)*, *allergen-IgE reactions*, and *local and systemic effects* of those reactions. Such reactions occur only in individuals who have previously been exposed to an allergen. In **sensitization**, the initial exposure to an allergen, B cells are activated (Figure 18.1a). The B cells differentiate into plasma cells, which produce IgE antibodies against the specific allergen (Figure 18.1b). The IgE antibodies attach by their Fc tails to the surface of mast cells in the respiratory and gastrointestinal tracts and to basophils in the blood (Figure 18.1c). Attachment leaves the antigen (allergen) binding sites of the IgE antibodies free to react with the same allergen upon future exposure. This sequence of sensitization

TABLE 18.1

Common Allergens		
Ingested	**Inhaled**	**Injected**
Animal proteins, especially from milk and eggs	Cocaine Dander	Antibiotics, especially cephalosporins and penicillins Heroin
Aspirin	Dust (household)	Hormones (adrenocorticotropic hormone and animal insulin)
Fruits	Face powder	Insect venoms (from bees, hornets, wasps, and yellow jackets)
Grains	Insecticides	Snake venoms (from vipers and cobras)
Hormone preparations	Mites and their feces	Spider venoms, especially from black widow and brown recluse
Nuts	Pollen (from grass, trees, and weeds)	
Penicillin	Spores (fungal and bacterial)	
Seafood		

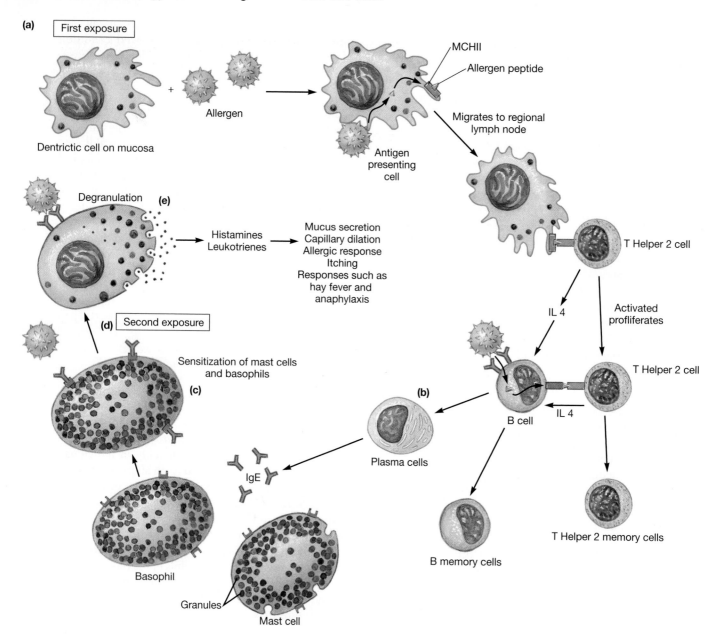

(a) First exposure

Dentrictic cell on mucosa

Allergen

MCHII

Allergen peptide

Antigen presenting cell

Migrates to regional lymph node

T Helper 2 cell

IL 4

Activated profliferates

T Helper 2 cell

IL 4

T Helper 2 memory cells

B cell

Plasma cells

B memory cells

IgE

Degranulation **(e)**

Histamines
Leukotrienes

Mucus secretion
Capillary dilation
Allergic response
Itching
Responses such as
hay fever and
anaphylaxis

(d) Second exposure

Sensitization of mast cells and basophils

(c)

(b)

Basophil

Granules

Mast cell

Figure 18.1 **The mechanism of immediate (Type I) hypersensitivity, or anaphylactic hypersensitivity.** In a first exposure, **(a)** allergen binds to B cells and is presented as allergen fragments on the surface of macrophages. Allergen fragment presentation activates T_H cells, which activate B cells. **(b)** B cells develop into the plasma cells that secrete IgE antibody. **(c)** IgE binds by its Fc tail to basophils and mast cells. In a second or later exposure, **(d)** allergen binds to sensitized mast cells and basophils, cross-linking IgE molecules. **(e)** This cross-linking stimulates degranulation of histamine and other mediators that cause the symptoms of allergies.

steps does not occur in all people. Why some people become sensitized to normally innocuous substances whereas others do not is poorly understood.

The sensitized mast cells and basophils now are primed to produce a massive chemical response to a second exposure from the same allergen. Although the *sensitizing* (first) *dose* of an allergen can be fairly high, the *triggering* or *eliciting* (subsequent) *dose* that causes the hypersensitive symptoms can be quite small. When a second or subsequent encounter occurs with the same allergen, the allergen attaches to sensitized mast cells and basophils, cross-linking

the IgE antibodies (**Figure 18.1d**). Cross-linking causes **degranulation**, the rapid release of *preformed mediators* (chemical substances that induce allergic responses) from cytoplasmic granules in mast cells and basophils (**Figure 18.1e**). **Histamine** is the main preformed mediator in humans. It dilates capillaries, thereby making them more permeable. It also causes

Theophylline, which is commonly administered orally or through inhalers to asthmatics, blocks an enzyme that would otherwise lead to degranulation.

TABLE 18.2

Some Mediators of Immediate Hypersensitivity and Their Effects	
Mediator	**Effects**
Preformed Mediators	
Histamine	Vascular dilation and increased capillary permeability, bronchial smooth muscle contraction, edema of mucosal tissues, secretion of mucus, and itching
Neutrophil and eosinophil chemotactic factors	Attraction of neutrophils, eosinophils, and other leukocytes to the site of an allergic reaction
Reaction Mediators	
Leukotrienes (SRS-A)	Prolonged bronchial smooth muscle contraction, increased capillary permeability, edema of mucosal tissues, and secretion of mucus
Prostaglandin D_2	Formation of minute blood clots, bronchial smooth muscle contraction, and capillary dilation

bronchial smooth muscle to contract, increases mucus secretion, and stimulates nerve endings that cause pain and itching.

Prostaglandins and **leukotrienes** are *reaction mediators* (chemical substances that control responses) that are also synthesized and released from mast cells after degranulation has occurred. Prostaglandin D_2 is a cellular messenger molecule produced in mast cells and basophils that also causes constriction of bronchial smooth muscle. *Slow-reacting substance of anaphylaxis* (SRS-A) is another mediator that causes a slow, long-lasting airway constriction in animals, SRS-A consists of three leukotriene mediators. These leukotrienes are 100 to 1000 times as potent as histamines and prostaglandin D_2 in causing prolonged airway constriction. Like histamine, leukotrienes and prostaglandin D_2 dilate and increase the permeability of capillaries, increase thick mucus secretion, and stimulate nerve endings that cause pain and itching. Preformed and reaction mediators and their effects are summarized in **Table 18.2**.

It is a second or later allergen exposure of a sensitized person that produces allergic signs and symptoms. Such hypersensitive responses by mast cells can be triggered by nonallergic factors, too. Emotional stress and temperature extremes often cause mediator release without the involvement of any IgE or allergen. Cold air can injure the cell membranes of mast cells lining the airway, which then release mediators that trigger asthma.

LOCALIZED ANAPHYLAXIS

Atopy (at'o-pe), which literally means "out of place," refers to localized allergic reactions. Atopic immune reactions occur first at the site where the allergen enters the body. If the allergen enters the skin, it causes a *wheal and flare reaction*, characterized by redness, swelling, and itching **(Figure 18.2)**. If the allergen is inhaled, mucous membranes of the respiratory tract become inflamed, and the patient has a runny nose and watery eyes. If the

(a)

(b)

Figure 18.2 Allergy testing. (a) Possible allergens are placed on prongs and introduced under the patient's skin. *(Southern Illinois University Biomed/Custom Medical Stock Photo, Inc.)* **(b)** If an individual is hypersensitive, a wheal (white raised area) and flare (reddened area) soon become visible on the skin. Such testing is usually done on the extremities to keep any hypersensitive reaction away from major organs. *(VU/Southern Illinois University/Visuals Unlimited)*.

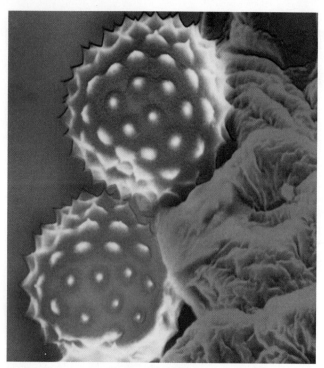

Figure 18.3 False-color SEM of ragweed (*Ambrosia*) pollen (1,619X). One of several pollen causes of hay fever. In early spring the culprits are primarily tree pollens such as oak, elm, birch (especially in Europe), and box elder. In late spring and early summer, grass pollens plus those of some broad-leaved plants are most likely to be involved. In late summer and early autumn, the chief allergens are ragweeds, saltbush, and Russian thistle pollens. (*Ralph C. Eagle/Photo Researchers, Inc.*)

allergen is ingested, mucous membranes of the digestive tract become inflamed, and the patient may have abdominal pain and diarrhea. Some ingested allergens, such as foods and drugs, also cause skin rashes.

Hay fever, or *seasonal allergic rhinitis*, is a common kind of atopy. More than 20 million Americans suffer from the typical signs and symptoms of watery eyes, sneezing, nasal congestion, and sometimes shortness of breath. First described in 1819 as resulting from exposure to newly mown hay, hay fever is now known to result from exposure to airborne pollen—tree pollens in spring, grass pollens in summer, and ragweed pollen (Figure 18.3) in fall. Some plants, such as goldenrod and roses, have long been blamed for hay fever but are innocent because their pollens are too heavy to be airborne for any great distance. It is the far less conspicuous green ragweed flowers that cause much of the misery for hay fever sufferers. On occasion, severe allergic rhinitis (inflammation of the nasal surfaces) can progress to sinus infections, middle ear problems, and temporary hearing loss. Although they share many symptoms, hay fever can be distinguished from the common cold by the increased numbers of eosinophils in nasal secretions. Finding elevated numbers of eosinophils in blood also suggests allergy (or infection with helminths).

GENERALIZED ANAPHYLAXIS

Some anaphylactic reactions are generalized, severe, and immediately life-threatening. In a sensitized person, a generalized reaction begins with sudden reddening of skin, intense itching, and hives, especially over the face, chest, and palms of hands. The disorder can then progress to respiratory anaphylaxis or anaphylactic shock.

In **respiratory anaphylaxis** the airways become severely constricted and filled with mucus secretions, and the allergic individual may die of suffocation. More than 4000 Americans die each year from respiratory anaphylaxis. Another 15 million Americans suffer from **asthma**, which is often caused by inhaled or ingested allergens, emotional stress, aspirin, or cold, dry air. Asthma can also be caused by hypersensitivity to endogenous microorganisms. For example, some patients become sensitized to *Moraxella catarrhalis*, a normal bacterial resident of respiratory mucous membranes.

In **anaphylactic shock** blood vessels suddenly dilate and become more permeable, causing an abrupt and life-threatening drop in blood pressure. Insect bites and stings are a common cause of anaphylactic shock in people sensitized to insect venoms (Figure 18.4a).

(a)

(b)

Figure 18.4 Honeybee sting allergy. (a) An eye has swollen shut but the bee sting has not caused the airway to close, as happens in more severe reactions. (*Scott Camazine/Photo Researchers, Inc.*) (b) Anaphylactic kit for emergency use, showing syringe loaded with epinephrine. Many individuals with several insect-sting allergies carry such kits with them at all times. They are available only by prescription. (*William C. Ober*).

Generalized anaphylaxis must be treated immediately. Unless epinephrine (adrenaline) is administered immediately, death can occur. Epinephrine acts by relaxing smooth muscle of respiratory passageways and constricting blood vessels. People who are sensitized to insect venom often carry an emergency anaphylactic kit (**Figure 18.4b**). The kit contains a tourniquet, benadryl (antihistamine) tablets, and a syringe containing two doses of epinephrine. Having such a kit on hand could easily mean the difference between life and death because of the rapid onset of life-threatening symptoms in patients who already have had anaphylactic reactions.

GENETIC FACTORS IN ALLERGY

In the United States, 50 million people have some kind of allergy. In many cases, genetic factors are thought to contribute to the development of allergies. Although different family members typically have different allergies (one person may suffer from asthma and another from dust allergies), all will have high levels of IgE antibodies. At least 60% of children with atopy have a family history of asthma or hay fever, and half these children later develop other allergies. Thus, allergy probably has a genetic basis, possibly in properties of membranes or the performance of various cells involved in immune responses, such as phagocytes. Normal membranes screen out all but the tiniest microorganisms and virtually all potential allergens. Membranes of allergic individuals, however, are more permeable to larger particles such as pollen grains. Even when allergens pass through membranes, phagocytic cells usually engulf them in normal individuals but sometimes fail to do so completely in allergic individuals.

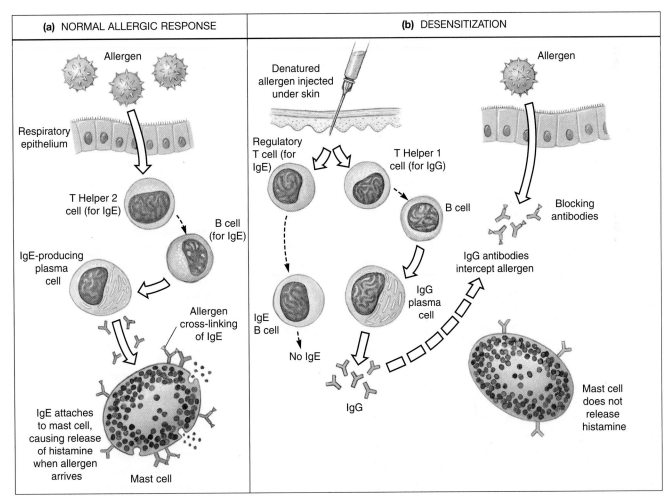

Figure 18.5   A proposed mechanism of action for desensitization allergy shots (hyposensitization).   (a) In a normal allergic response, natural exposure to an allergen causes helper T cells to stimulate those B cells that mature into plasma cells to make IgE antibodies. After binding to mast cells, a second allergen exposure causes degranulation. **(b)** Desensitization involves the injection of denatured allergen. Such shots may lead to tolerance, preventing B cells from maturing into plasma cells to make IgE antibodies. Exposure to the allergen also may activate those B cells that mature into the plasma cells that make IgG (blocking) antibody. Such IgG antibodies can bind to incoming allergen before it reaches the IgE molecules attached to mast cells. Complexing of allergen with these attached IgE molecules would cause the mast cells to degranulate and release histamine, so blocking this step is the key to preventing allergic response.

TREATMENT OF ALLERGIES

One approach to dealing with allergies is to avoid contact with the specific allergen. People with food allergies should not eat a food to which they have had a hypersensitivity reaction. Whatever the allergen, a subsequent exposure generally will trigger degranulation by mast cells **(Figure 18.5a)**. **Desensitization** (hyposensitization) is the only currently available treatment intended to cure an allergy. If denatured allergen is injected subcutaneously ("allergy shots"), it may induce a state of tolerance, preventing the activation of those B cells that mature into IgE-secreting plasma cells **(Figure 18.5b)**. Also, by receiving such injections with gradually increasing doses of the allergen, the patient may produce IgG antibodies, called **blocking antibodies**, against the allergen. Upon reexposure to an allergen, the blocking antibodies combine with the allergen before the allergen has a chance to react with IgE, so mast cells do not release mediators. The number of suppressor T cells sensitized to the allergen also increase significantly during desensitization. Thus, increases in IgG and decreases in IgE may work together to make the patient less sensitive to the allergen.

Desensitization has been very successful against insect venoms and drug allergies such as allergies to penicillin. Unfortunately, desensitization does not alleviate the signs and symptoms of many allergies, such as hay fever. In addition, the treatment itself can cause anaphylactic shock because the injections contain the very substance to which the patient is allergic. Patients must remain in the physician's office for 20 to 30 minutes after the injection so that emergency treatment will be available if a generalized anaphylactic reaction occurs.

Allergy injections are about 65 to 75% effective in individuals whose allergies are caused by inhaled allergens.

Other allergy treatments alleviate symptoms but do not cure the disorder. Antihistamines counteract the swelling and redness due to histamine but are not effective against SRS-A of asthmatic conditions, which involve constriction of the airways, whereas antiinflammatory agents such as corticosteroids suppress the inflammatory response. Two new antiallergy medications act by inhibiting leukotriene production. Better therapeutic methods of treating allergies are greatly needed. As we learn more about the properties of leukotrienes and IgE antibodies, perhaps this need can be met.

APPLICATIONS

The Origins of Allergies

Allergies account for almost 10% of all visits to physicians' offices in the United States. Why does the immune system often react violently to nonharmful substances? And why are some people allergic, whereas others are not? Research has shown that some people who lack the ability to produce IgE antibodies are prone to lung and sinus infections. Also, people who lack the ability to produce IgG or IgM antibodies often produce IgE antibodies to bacterial infections. These observations suggest that the IgE antibody may play a necessary role in immunity, besides causing allergies.

IgE is known to help fight infections by parasitic helminths (◀Chapter 17, p. 499). IgE antibodies also may protect against ectoparasites (ticks, chiggers, fleas). The American biologist Margie Profet believes that IgE antibodies also are a backup system to protect against the ingestion of toxins.

In the book *Why We Get Sick: The New Science of Darwinian Medicine*, the authors, physician Randolph Nesse and evolutionist George Williams, suggest that many allergies that exist today were not common 150 years ago. They say that hay fever was almost nonexistent in England in the early 1800s and rare in Japan even as recently as 1950. Yet today, about 10% of the Japanese have hay fever. Nesse and Williams suggest that modern comforts may be to blame. Well-insulated houses with thick carpeting are excellent breeding grounds for dust mites and catchalls for allergens such as pollen and mold spores. In fact, studies show that infants raised in homes relatively clean of allergens develop far fewer allergies than do infants raised in the average home. However, these ideas still do not explain fully why some people have allergies whereas others in the same household or family do not. Just to confuse things further, the data from a recent study suggest that early exposure to potential allergens may in fact decrease the incidence of allergic responses. Clearly there is still much to learn about the immune system and allergies.

Because about 15% of Americans suffer from allergies, it stands to reason that any one person would have about a 15% chance of being allergic to some allergen. In atopic families, if one parent is atopic, a child has a 25% chance of having an allergy. If both parents are atopic, the chance jumps to about 50%. Thus, some allergies can run in families and may, in part, have a genetic basis. Profet, Nesse, and Williams also suggest that if someone is simultaneously exposed, for example, to a plant toxin and an allergen, the immune system may respond to the toxin by producing IgE antibodies. In this attack, the immune system sees the allergen as "part of the toxin" and so reacts to it as well. Immune cells remain sensitized, and future exposures of the allergen alone will trigger an IgE response even though the toxin is not present.

Unfortunately, no matter what the causes or origins of allergies, we must suffer through them with only our allergy medications to help us along.

CYTOTOXIC (TYPE II) HYPERSENSITIVITY

In cytotoxic (Type II) hypersensitivity, specific antibodies react with cell surface antigens interpreted as foreign by the immune system, leading to phagocytosis, killer cell activity, or complement-mediated lysis. The cells to which the antibodies are attached, as well as surrounding tissues, are damaged because of the resulting inflammatory response. Antigens that initiate cytotoxic hypersensitivity typically enter the body in mismatched blood transfusions or during delivery of an Rh-positive infant to an Rh-negative mother.

MECHANISM OF CYTOTOXIC REACTIONS

When an antigen on a plasma membrane is first recognized as foreign, B cells become sensitized and stand ready for antibody production upon a subsequent antigen exposure. During subsequent exposures with the surface antigen, antibodies bind to the antigen and activate complement. Phagocytic cells, such as macrophages and neutrophils, are attracted to the site. The mechanisms of Type II hypersensitivity appear to be responsible for the tissue damage in cases of rheumatic fever following a streptococcal infection, in certain viral diseases, in transfusion reactions, and in hemolytic disease of the newborn (mother-infant Rh incompatibility).

EXAMPLES OF CYTOTOXIC REACTIONS

Cytotoxic reactions typical of Type II hypersensitivities are exemplified by mismatched blood transfusions and by hemolytic disease of the newborn.

Transfusion Reactions

Normal human red blood cells have genetically determined surface antigens (blood group systems) that form the basis for the different blood types. A **transfusion reaction** can occur when matching antigens and antibodies are present in the patient's blood at the same time. Such reactions can be triggered by any of the blood group antigens. We will focus on antigens A and B, which determine the **ABO blood group system**. As Table 18.3 shows, four blood types—A, B, AB, and O—are named according to whether red blood cells have antigen A, antigen B, both A and B antigens, or neither antigen. Normally, a person's serum has no IgM antibodies against the antigens present on his or her own red blood cells. However, if a sensitized patient receives red blood cells with a different blood cell antigen during a blood transfusion, IgM antibodies cause a Type II hypersensitivity reaction against the foreign antigen. The

In the United States alone, a transfusion takes place about every 3 seconds—some 12 million units of blood are used every year.

TABLE 18.3

Properties of the ABO Blood Group System		
Blood Type	**Antigens on Erythrocytes**	**Antibodies in Serum**
A	A	Anti-B
B	B	Anti-A
AB	A and B	Neither anti-A nor anti-B
O	Neither A nor B	Anti-A and anti-B

foreign red blood cells are agglutinated (clumped), complement is activated, and hemolysis (rupture of blood cells) occurs within the blood vessels **(Figure 18.6)**. Symptoms of a transfusion reaction include fever, low blood pressure, back and chest pain, nausea, and vomiting. Transfusion reactions can usually be prevented by careful cross-matching of donor and recipient blood group antigens so that the correct blood type can be selected for transfusion **(Figure 18.7)**.

Transfusion reactions to other erythrocyte antigens, such as Rh (Rhesus), also occur. However, they usually

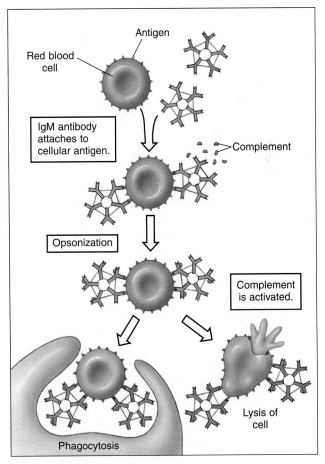

Figure 18.6 The mechanisms of cytotoxic (Type II) hypersensitivity. Mismatched red blood cell antigen usually is bound to IgM. Complement is activated and results in either subsequent phagocytosis or lysis of the red blood cells.

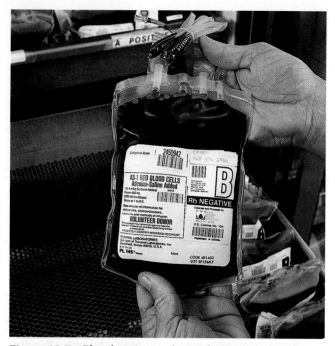

Figure 18.7 Blood typing and transfusions. Careful blood typing and matching of donor and recipient blood prevent most transfusion reactions. Persons with type AB blood can safely receive a transfusion of any of the four major blood types. Persons with type O blood can safely donate blood to recipients of any blood type. (Can you explain why? Refer to Table 18.3.) *(Larry Mulvehill/Photo Researchers, Inc.)*

are less serious than reactions to foreign A or B antigens because the antigen molecules are less numerous.

Hemolytic Disease of the Newborn

Another example of a cytotoxic reaction is **hemolytic disease of the newborn**, or *erythroblastosis fetalis*. In addition to the ABO blood group, red blood cells can have **Rh antigens**, so named because the antigens were discovered first in Rhesus monkeys. Blood with Rh antigens on red blood cells is designated Rh-positive; red blood cells lacking Rh antigens are designated Rh-negative. Anti-Rh antibodies normally are not present in the serum of either Rh-positive or Rh-negative blood. Consequently, sensitization is necessary for an Rh antigen-antibody reaction.

Sensitization typically occurs when an Rh-negative woman carries an Rh-positive fetus, which inherited this blood type from its father. The fetal Rh antigen rarely enters the mother's circulation during pregnancy but can leak across the placenta during delivery, miscarriage, or abortion **(Figure 18.8a)**. The Rh-negative mother's immune system then becomes sensitized to the Rh antigen and can produce anti-Rh antibodies if it again encounters the Rh antigen.

Because sensitization usually occurs at delivery, the first Rh-positive child of an Rh-negative mother rarely suffers from hemolytic disease. But when a sensitized

Rh-negative mother carries a second or subsequent Rh-positive fetus, the mother's anti-Rh antibodies cross the placenta and cause a Type II hypersensitivity reaction in the fetus **(Figure 18.8b)**. If this occurs, fetal red blood cells agglutinate, complement is activated, and the red blood cells are destroyed. The result is hemolytic disease of the newborn. The baby is born with an enlarged liver and spleen caused by efforts of these organs to eliminate damaged red blood cells **(Figures 18.8c, d)**. Such babies exhibit the yellow skin color of jaundice due to excessive bilirubin—a product of the breakdown of red blood cells—in their blood.

Hemolytic disease of the newborn can be prevented by giving Rh-negative mothers intramuscular injections of anti-Rh IgG antibodies (Rhogam) within 72 hours after delivery. The antibodies presumably bind to Rh antigens on the fetal red blood cells that have leaked into the mother's blood. These anti-Rh antibodies destroy the fetal red blood cells before they can act to sensitize her immune system. It is essential to treat all Rh-negative women after delivery, miscarriage, or abortion in case the fetus may have been Rh-positive. Today, anti-Rh antibodies are often administered to Rh-negative women during pregnancy as well. Such treatment at 3 and 5 months prevents sensitization to the fetus in case fetal antigens leak into the mother's circulation, which can result from hard coughing or sneezing. Before anti-Rh antibody (Rhogam) was given preventively, hemolytic disease of the newborn occurred in about 0.5% of all pregnancies; 12% of these terminated in stillbirths.

IMMUNE COMPLEX (TYPE III) HYPERSENSITIVITY

Immune complex (Type III) hypersensitivity results from the formation of antigen-antibody complexes. Under normal circumstances these large immune complexes are engulfed and destroyed by phagocytic cells.

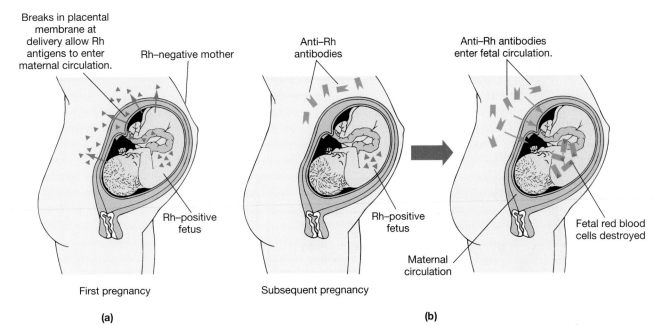

Breaks in placental membrane at delivery allow Rh antigens to enter maternal circulation.

Rh–negative mother

Rh–positive fetus

First pregnancy

Anti–Rh antibodies

Rh–positive fetus

Subsequent pregnancy

Anti–Rh antibodies enter fetal circulation.

Fetal red blood cells destroyed

Maternal circulation

(a)

(b)

Figure 18.8 Cause and effect of hemolytic disease of the newborn. (a) The stage is set for an Rh-incompatibility pregnancy when the mother is Rh⁻ and the fetus is Rh⁺ (which is usually the case if the father is Rh⁺). **(b)** Rh antigens may cross the placenta and enter the mother's bloodstream before or at delivery. She responds by making anti-Rh antibodies, which also can cross the placenta. Even if antibody production is not stimulated until delivery, the resulting antibodies will persist in the mother's circulation and attack the red blood cells of any subsequent Rh⁺ fetus. To prevent this situation, Rhogam (anti-Rh antibody) is injected into the mother early in the pregnancy, immediately after delivery, and in cases of miscarriage or abortion. Rhogam reduces exposure to the antigen and thus lessens anti-Rh antibody production. **(c)** Child affected by hemolytic disease caused by Rh incompatibility. *(From Edith Potter, Rh. Chicago: Year Book Medical Publishers, 1947.)* **(d)** The liver is greatly enlarged. *(From Edith Potter, Rh. Chicago: Year Book Medical Publishers, 1947.)*

(c)

(d)

Hypersensitivity occurs when antigen-antibody complexes persist or are continuously formed.

MECHANISM OF IMMUNE COMPLEX DISORDERS

Like anaphylactic and cytotoxic reactions, immune complex disorders also are initiated after sensitization. Upon subsequent exposure to the sensitizing antigen, specific IgG antibodies combine with the antigen in the blood to form an *immune complex* and activate complement (Figure 18.9). Antibodies bind to live cells or parts of damaged cells in blood vessel walls and other tissues. Normally, large immune complexes are removed by phagocytosis in the liver and spleen. However, immune complexes are often quite small and fail to bind tightly to Kupffer cells in the liver, thereby escaping elimination

from the blood, and are deposited in organs, tissues, or joints. Such antigen-antibody complexes and complement, in turn, cause basophils and mast cells to release histamine and other mediators of allergic reactions, with the effects described earlier. Phagocytes attracted chemotactically to these sites of activity release hydrolytic enzymes, causing tissue damage that is acute but can become chronic if the antigen remains for long periods of time.

EXAMPLES OF IMMUNE COMPLEX DISORDERS

We will illustrate immune complex disorders with two phenomena, the systemic serum sickness and the localized Arthus reaction. Other disorders that involve immune complexes, such as rheumatoid arthritis and

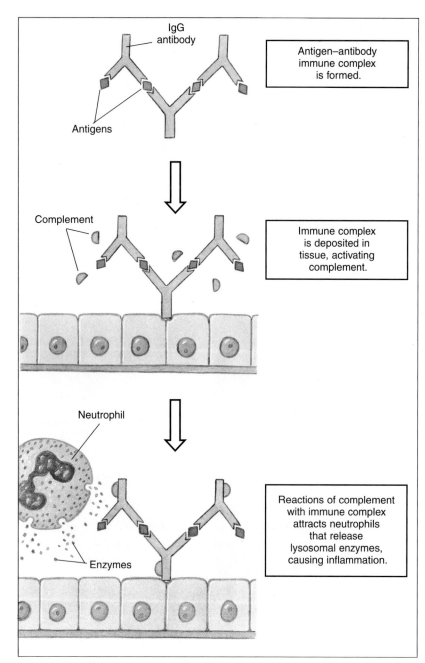

IgG antibody

Antigens

Antigen–antibody immune complex is formed.

Complement

Immune complex is deposited in tissue, activating complement.

Neutrophil

Reactions of complement with immune complex attracts neutrophils that release lysosomal enzymes, causing inflammation.

Enzymes

 Figure 18.9 The mechanism of immune complex (Type III) hypersensitivity. Immune complexes are formed when antigen is introduced into a previously sensitized individual. When the resulting immune complex is deposited, it activates complement, producing fever, itching, rash or hemorrhagic areas, joint pain, and acute inflammation. On a systemic basis, this can cause serum sickness.

systemic lupus erythematosus, are discussed later in connection with autoimmune diseases because the antibodies involved in these disorders react with the person's own tissues. Acute glomerulonephritis following certain streptococcal infections is another immune complex disease, in which the glomeruli of the kidneys can be severely damaged (◀Chapter 20).

Serum sickness was frequently seen in the preantibiotic era when large doses of antitoxin sera were used to immunize people passively against infectious diseases such as diphtheria. Diphtheria toxin given to horses caused them to make antibodies against the toxin. A patient then received the horse serum, which contained not only antidiphtheria toxin antibody but also horse

proteins. A sensitized patient's immune system would make sufficient antibodies against these horse proteins to form immune complexes consisting of human antibody reacting against horse serum protein, upon second exposure. These immune complexes, which are removed slowly by phagocytic cells, would attach to the glomeruli of the kidneys. The filtration capacity of the glomeruli was thereby impaired, causing proteins and blood cells to be excreted in the urine. Immune complexes are also deposited in joints and in skin blood vessels.

People with serum sickness usually have fever, enlarged lymph nodes, decreased numbers of circulating leukocytes, and swelling at the injection site. Most people recover from serum sickness as the complexes

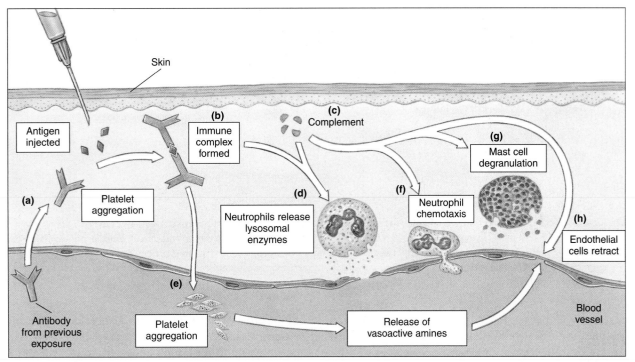

Figure 18.10 **The mechanism that produces an Arthus reaction and hemorrhagic areas.** **(a)** In severe cases, injection of horse protein antigens leads to **(b)** immune complex formation. **(c)** In association with complement, **(d)** neutrophils release lysosomal enzymes that damage the blood vessel wall. **(e)** Immune complexes also trigger platelet aggregation that can obstruct blood flow. **(f)** Complement also attracts more neutrophils to the site and **(g)** causes mast cell degranulation. **(h)** Finally, platelet and complement trigger endothelial retraction. Tissue death can occur if cells are cut off from blood vessel flow.

eventually are cleared from the blood and tissue repair occurs in the glomeruli. However, the disorder became chronic in many diphtheria patients because they received horse serum daily over the course of the disease.

Today, serum sickness is rare and usually is due to second exposure to a foreign substance in a biological product, such as horse serum in a vaccine preparation. An advantage of vaccines made by genetic engineering is that they do not contain foreign substances. When the

Figure 18.11 **Arthus reaction.** The patient shows an extensive area of hemorrhagic damage to buttocks that will result in tissue necrosis and sloughing. *(Reproduced by permission from F.H. Top, Sr., Communicable and Infectious Diseases, 6th ed. St. Louis, Mosby-Year Book, Inc. 1968).*

use of any biological product is contemplated, the patient should be tested for sensitivity first. A small quantity of the product should be given intradermally (within the skin) or intravenously. A wheal and flare reaction or a drop of 20 points or more in blood pressure following intravenous injection indicates hypersensitivity, and the product should not be given.

The **Arthus reaction**, named after Arthus, who discovered it in 1903, is a local reaction seen in the skin after subcutaneous (under the skin) or intradermal injection of an antigenic substance. It occurs in people who already have large quantities of antibodies (mainly IgG) to the antigen. In 4 to 10 hours, edema and hemorrhage develop around the injection site as immune complexes and complement trigger cell damage and platelet aggregation **(Figure 18.10)**. In severe reactions, tiny clots obstruct blood vessels, and cells normally nourished by the blocked vessels die **(Figure 18.11)**. In rare cases, injection of antigen may not occur. In "pigeon fancier's lung," the antigen is protein inhaled via dried pigeon feces, which triggers Arthus reactions in the lungs.

CELL-MEDIATED (TYPE IV) HYPERSENSITIVITY

Cell-mediated (Type IV) hypersensitivity is also called **delayed hypersensitivity** because reactions take more than 12 hours to develop. These reactions are

Antigens

Sensitized T$_H$1 cell

APC presents antigen fragments

Cytokines

Activated APC (macrophage)

Inflammatory action, mediator release

Eczema, swelling, granulomatous lesions

Figure 18.12   The mechanism of cell-mediated, or delayed (Type IV), hypersensitivity. This type of reaction is mediated by T cells rather than by B cells as in Types I, II, and III. T cells that have become sensitized to a particular antigen release cytokines on subsequent contact with the same antigen fragment. These cytokines cause inflammatory reactions that attract macrophages to the site. By degranulation, APCs release mediators that add to the inflammatory response. Contact dermatitis and poison ivy rash are examples of cell-mediated hypersensitivity.

mediated by T cells—specifically, a type of T$_H$1 cell [sometimes called a **delayed hypersensitivity T (T$_{DH}$) cell**]—not by antibodies.

MECHANISM OF CELL-MEDIATED REACTIONS

Cell-mediated hypersensitivity occurs as follows. On first exposure, antigen molecules bind to antigen-presenting cells that present antigen fragments to T$_H$1 (inflammatory T) cells (◀Chapter 17, p. 505). When APCs again present the same antigen during a second, later exposure, the sensitized T$_H$1 cells release various cytokines, including γ-interferon and migration inhibiting factor (MIF). Gamma-interferon stimulates macrophages to ingest the antigens. If the antigens are on microorganisms, the macrophages usually, but not always, kill the microorganisms. MIF prevents migration of macrophages, so they remain localized at the site of the hypersensitivity reaction. Other cytokines are presumed to cause the hypersensitivity reaction itself. Such reactions account for patches of raw, reddened skin in eczema, swelling, and granulomatous lesions. These processes are summarized in **Figure 18.12**.

EXAMPLES OF CELL-MEDIATED DISORDERS

Three common examples of delayed hypersensitivity—contact dermatitis, tuberculin hypersensitivity, and granulomatous hypersensitivity—illustrate the diversity of cell-mediated reactions.

Contact dermatitis occurs in sensitized individuals on second or subsequent exposure to allergens such as oils from poison ivy, rubber, certain metals, dyes, soaps, cosmetics, some plastics, topical medications, and other substances **(Table 18.4)**. Unlike Type I hypersensitivities, Type IV hypersensitivities do not appear to run in families. Molecules too small to cause immune reactions pass through the skin, where they become antigenic by

binding to normal proteins on Langerhans cells of the epidermis. These cells, which carry MHC class II antigens, migrate to the lymph nodes, where they act as antigen presenting cells to T$_H$1 cells. Within 4 to 8 hours after the next exposure, a hypersensitivity reaction begins, and eczema occurs within 48 hours.

Urushiol (u′ru-she-ol), an oil from the poison ivy plant, is a major cause of contact dermatitis in the United States **(Figure 18.13)**. Most people get poison ivy from direct contact with leaves or other plant parts, but some get it by inhaling smoke from burning brush that contains poison ivy plants. Poison ivy is particularly severe when oil droplets come in contact with respiratory membranes. Sensitivity to poison ivy can develop at any age, even among people who have come in contact with it for years without reacting to it. One way to minimize a reaction to

TABLE 18.4

Selected Contact Allergens	
Allergen	**Common Sources of Contact**
Benzocaine	Topical anesthetic
Chromium	Jewelry, watches, chrome-tanned leather, cement
Formaldehyde	Facial tissues, nail hardeners, synthetic fabrics
Latex	Surgical or examination gloves
Nickel	Jewelry, watches, objects made of stainless steel and white gold
Mercaptobenzothiazole	Rubber goods
Methapyrilene	Topical antihistamine
Merthiolate	Topical antiseptic
Neomycin	Topical antibiotic
Oleoresin	Oil from poison ivy and similar plants

(a)   (b)

Figure 18.13 Poison ivy (Type IV) hypersensitivity.
(a) Poison ivy (*Toxicodendron radicans*), showing the leaves with their characteristic three leaflets. Poison ivy vines also contain the irritating oil urushiol, so it is important to be able to recognize them in winter, when leaves may not be present. *(Ed Reschke/Peter Arnold, Inc.)* **(b)** Poison ivy dermatitis, showing fluid-filled vesicles. *(Beckman/Custom Medical Stock Photo, Inc.)*

Figure 18.14 A positive tuberculin skin test reaction. The raised area of induration should be observed and measured after 48 to 72 hours. A positive reaction will measure 5 mm or more; 2 mm or smaller is negative; 3 and 4 mm are considered doubtful. *(National Medical Slide/Custom Medical Stock Photo, Inc.)*

poison ivy is to wash exposed areas thoroughly with strong soap or detergent within minutes of contact, before much oil has penetrated the skin and bound chemically to skin cells. Once a person is sensitized, an exceedingly small quantity of oil will elicit a reaction upon the next exposure. Scratching lesions does not spread the oil, but it can lead to infections. Cashews and mangoes contain substances chemically similar to urushiol, and some people display delayed hypersensitivity (including digestive disturbances) to these substances.

Tuberculin hypersensitivity occurs in sensitized individuals when they are exposed to *tuberculin*, an antigenic lipoprotein from the tubercle bacillus *Mycobacterium tuberculosis*. Similar antigens from the bacterium that causes leprosy (*Mycobacterium leprae*) and the protozoan that causes leishmaniasis (*Leishmania tropica*) produce

similar reactions in sensitized individuals. The antigen activates T_H1 cells, which in turn release cytokines that cause large numbers of lymphocytes, monocytes, and macrophages to infiltrate the dermis. The normally soft tissues of the skin then form a raised, hard, sometimes red region called an **induration (Figure 18.14)**. In a **tuberculin skin test**, a purified protein derivative (PPD) from *Mycobacterium tuberculosis* is injected subcutaneously. If a person has been exposed to the bacterium or had received the BCG vaccine, an induration will form within 48 hours. The diameter and elevation of the induration, not its redness, indicate whether further tests are needed.

Granulomatous hypersensitivity, the most serious of the cell-mediated hypersensitivities, usually occurs when macrophages have engulfed pathogens but have failed to kill them. Inside the macrophages the protected pathogens survive and sometimes continue to divide. T_H1 cells sensitized to an antigen of the pathogen elicit the hypersensitivity reaction, attracting several cell types to the skin or lung. A granuloma in the skin (leproma) or lung (tubercle) develops. This kind of hypersensitivity is the most delayed of all, appearing 4 weeks or more after exposure to the antigen. Such persistent and chronic antigenic stimuli are also typical of the bacterial disease listeriosis, as well as many fungal and helminthic infections.

Characteristics of the four types of hypersensitivity are summarized in **Table 18.5**.

APPLICATIONS

Poison Ivy? But It Doesn't Grow Here!

Contact dermatitis similar to poison ivy was seen in American military personnel in Japan after World War II. Lesions appeared on elbows, forearms, and in a horseshoe shape on the buttocks and thighs. Knowing that poison ivy did not grow in Japan, the medical staff was puzzled. The puzzle was solved when scientists discovered that oils from a Japanese plant containing a small quantity of urushiol were used to manufacture lacquer. Although the quantity of urushiol in lacquer was not sufficient to sensitize Japanese people, it did elicit an allergic response in Americans previously sensitized to poison ivy. Resting their arms on countertops explained lesions on elbows and forearms; contact with toilet seats explained the horseshoe-shaped lesions.

✓CHECKLIST

1. List the four types of hypersensitivities and their names. What are the main mediators for each?

2. What are the signs and symptoms of anaphylaxis?

3. How does hemolytic disease of the newborn arise? How can it be prevented?

TABLE 18.5

	Characteristics of the Types of Hypersensitivity			
	Type I	**Type II**	**Type III**	**Type IV**
Characteristic	Immediate	Cytotoxic	Immune complex	Cell-mediated
Main Mediators	IgE	IgG, IgM	IgG, IgM	T cells
Other Mediators	Mast cells, basophils, histamine, prostaglandins, leukotrienes	Complement	Complement, inflammatory factors, eosinophils, neutrophils	Lymphokines, macrophages
Antigen	Soluble or particulate	On cell surfaces	Soluble or particulate	On cell surfaces
Reaction Time	Seconds to 30 minutes	Variable, usually hours	3 to 8 hours	24 hours to 4 or more weeks
Nature of Reaction	Local wheal and flare, airway restriction anaphylactic shock	Clumping of erythrocytes, cell destruction	Acute inflammation effects	Cell-mediated cell destruction
Therapy	Desensitization, antihistamines, steroids	Steroids	Steroids	Steroids

AUTOIMMUNE DISORDERS

Autoimmune disorders occur when individuals become hypersensitive to specific antigens on cells or tissues of their own bodies, despite mechanisms that ordinarily create tolerance to those self antigens. The antigens elicit an immune response in which **autoantibodies**, antibodies against one's own tissues, are produced. An autoimmune response can be T-cell-mediated as well. These disorders are characterized by cell destruction in various types of hypersensitivity reactions. Although autoimmune disorders arise from a response

TABLE 18.6

The Spectrum of Autoimmune Disorders		
Disorder	**Organ(s) or Tissues Affected**	**Autoantibody Target**
Organ-Specific Disorders		
Addison's disease	Adrenal glands	Adrenal gland proteins
Autoimmune hemolytic anemia	Erythrocytes	Red blood cell membrane proteins
Glomerulonephritis	Kidneys	Streptococcal cross-reactivity with kidney
Graves' disease	Thyroid gland	Thyroid-stimulating hormone receptor
Hashimoto's thyroiditis	Thyroid gland	Thyroglobulin
Idiopathic thrombocytopenic purpura	Blood platelets	Platelet glycoproteins
Juvenile diabetes	Pancreas	Beta cells and insulin
Myasthenia gravis	Skeletal muscles	Acetylcholine receptor
Pernicious anemia	Stomach	Vitamin B_{12} binding site
Postvaccine/postinfection encephalomyelitis	Myelin	Measles cross-reactivity with myelin
Premature menopause	Ovaries	Corpus luteum
Rheumatic fever	Heart	Streptococcal cross-reactivity with heart
Spontaneous male infertility	Testes	Spermatozoa
Ulcerative colitis	Colon	Colon cells
Systemic (Disseminated) Disorders		
Goodpasture's syndrome	Basement membranes	Basement membrane
Polymyositis/dermatomyositis	Muscles and skin	Cell nuclei
Rheumatoid arthritis	Joints	Cell nuclei, gamma globulins
Scleroderma	Connective tissues	Nucleoli
Sjögren's syndrome	Lacrimal and salivary glands	Cell nuclei
Systemic lupus erythematosus	Many tissues	Cell nuclei, histones

to a self antigen, they range over a wide spectrum—from those that affect a single organ or tissue (organ-specific) to those that are systemic, affecting many organs and tissues (Table 18.6).

AUTOIMMUNIZATION

Autoimmunization is the process by which hypersensitivity to "self" develops. Such a response is usually sustained and long-lasting and can cause long-term tissue damage. Immunologists are beginning to understand this process better. Several different mechanisms of autoimmunity probably exist:

1. *Genetic factors* may predispose a person toward autoimmune disorders. For example, the children of a parent who has autoantibodies to a single organ are likely to develop autoantibodies to the same or to a different single organ. As we shall see later, individuals who have genes for certain histocompatibility antigens are at greater than normal risk of developing particular autoimmune disorders.

2. In addition to predisposing genetic factors, **antigenic**, or *molecular*, **mimicry** can occur. T_H cells might attack tissue antigens that are similar to antigens of some pathogens. Some children who suffer rheumatic fever (caused by *Streptococcus pyogenes*) develop rheumatic heart disease later in life. For some reason the immune system of such individuals "sees" the heart valve tissue as similar to certain streptococcal antigens and attacks the heart valves.

3. The thymus is critical to the normal development of T cells. Aside from the T_H cells that recognize nonself antigens, T_H cells that recognize self antigens can exist if *clonal deletion* fails to remove these self-reactive T cells (◄Chapter 17, p. 495). If they survive and proliferate, they can attack self antigens and trigger B cell activity with antibody production. Antigens hidden in tissues and lacking contact with B or T cells during immune system development or clonal deletion could be released through physical injury. These antigens then will be perceived as foreign by the immune system (see the box "Sympathetic Blindness").

4. Mutations might give rise to aberrant proteins to which B cells react, producing plasma cells that make autoantibodies.

5. Viral components inserted into host cell membranes might act as antigens, or virus-antibody complexes might be deposited in tissues.

6. The sympathetic nervous system, which along with the parasympathetic system controls internal body functions, helps regulate the immune system. When the sympathetic nervous system is damaged, the number of regulatory T cells decreases.

EXAMPLES OF AUTOIMMUNE DISORDERS

Autoimmune disorders usually are chronic inflammatory disorders with symptoms that can alternately worsen and lessen. They affect about 6% of all humans and can affect one organ or many.

We now look at three other examples to illustrate the diversity of such disorders.

Myasthenia Gravis

Myasthenia gravis is an autoimmune disease that afflicts approximately 25,000 Americans, or 3 of every 10,000 people. It affects primarily women in their 20s and 30s and men in their 40s and 50s. The disease usually affects skeletal muscles of the limbs and those involved in eye movements, speech, and swallowing. The principal symptoms of the disease are progressive weakness and muscle fatigue. Eyelid drooping and double vision are common.

For muscles to contract normally, neurons secrete the neurohormone acetylcholine across the gap (junction) between neuron and muscle (**Figure 18.15**). When

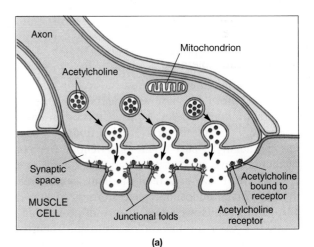

(a)

(b)

Figure 18.15 Myasthenia gravis. This disorder involves a loss of acetylcholine receptors from the neuromuscular junction. **(a)** A normally functioning neuromuscular junction has many acetylcholine receptors that bind acetylcholine. **(b)** Myasthenic patients have significantly fewer receptors for acetylcholine.

APPLICATIONS

Sympathetic Blindness

Lens proteins are normally confined within the lens capsule of the eye. Because they are never exposed to lymphocytes during development, the immune system never acquires tolerance for them. Sometimes an eye injury allows these proteins to leak into the bloodstream, where they elicit an immune response. The antibodies formed in this way then attack proteins in the undamaged eye. Because circulation (which transports antibodies) tends to be better in that eye, the immune response in the healthy eye can be more intense than that in the injured eye. This phenomenon sometimes leads to sympathetic blindness, or loss of sight in the uninjured eye.

(a)

(b)

Figure 18.16 Rheumatoid arthritis. Joint inflammation is typical in people suffering from rheumatoid arthritis **(b)**. In many cases, joint inflammation and destruction are so severe that they result in misshapen digits. In this gamma-ray photograph **(a)** swollen joints appear as bright spots. *(top: CNRI/Photo Researchers, Inc.; bottom: Custom Medical Stock Photo).*

acetylcholine receptors on the muscle cells bind acetylcholine, contraction of the muscle occurs. Evidence suggests that in people with myasthenia gravis, muscle contraction is prevented by IgG autoantibodies that either block the acetylcholine receptor or cause a reduction in the number of acetylcholine receptors (**Figure 18.15b**). In fact, most myasthenic patients have only 30 to 50% as many receptors as do unaffected persons.

Although myasthenia gravis is one of the best-understood autoimmune diseases, the reason why autoantibodies are formed is not well understood. One possibility is that autoantibodies are triggered by an immune response to an infectious virus or bacterium that has antigens mimicking a part of the acetylcholine receptor. Myasthenia gravis was once considered a fatal or disabling disease. Today myasthenic patients can be treated with drugs or immunosuppressive steroids so that they can live full lives. However, no means of preventing autoantibodies from forming or of removing them once they have formed is available. Most people with myasthenia gravis have tumors (benign and sometimes malignant) of the thymus gland. Surgical removal of the thymus sometimes results in cure, and even in patients without tumors, removal improves symptoms in more than half the patients.

In spite of muscle weakness, many women with myasthenia gravis have children. Their babies have temporary muscular weakness; they are like little rag dolls for the first few weeks of life. Small numbers of autoantibodies from the mother probably cross the placenta, affecting the fetus. Apparently the immune complexes do not permanently damage the fetal neurons because the babies soon have normal muscle function.

Rheumatoid Arthritis

In contrast with myasthenia gravis, which affects single organs, **rheumatoid arthritis (RA)** affects mainly the joints of the hands and feet, although it can extend to other tissues. Joints on opposite sides of the body are usually equally affected, in pairs. Of all forms of arthritis, RA is most likely to lead to crippling disabilities and to develop early in life (between the ages of 30 and 40). It is one of the most common autoimmune diseases, affecting about 2 million Americans. It is two to three times more prevalent in women than in men.

RA is characterized by inflammation and destruction of cartilage in the joints, often causing deformities in the fingers (**Figure 18.16**). Despite continued research, the cause of RA is unknown. Some researchers believe that an infectious microbe (mycoplasma or virus) is the cause, leading to antigenic mimicry and ultimately to an attack on self antigens. Others believe that a self antigen recognized by the immune system as

APPLICATIONS

New Treatment for Rheumatoid Arthritis

Recent research indicates that a protein called tumor necrosis factor-α (TNF-α) can trigger immune responses that attack bone and cartilage in joints, leading to RA. Centocor, a biotechnology company in Pennsylvania, has produced a genetically engineered antibody that binds to TNF-α so that it cannot trigger an immune response. In one study, 80% of patients receiving the antibody reported greatly reduced symptoms compared with patients who received a placebo. These antibodies do not survive for long periods in the body, so additional doses of the antibody were needed to prevent a recurrence of the disease.

(a)

(b)

Figure 18.17 Lupus erythematosus. The characteristic butterfly-shaped rash of systemic lupus erythematosus appears **(a)** red in fair-skinned people, *(Ken Greer/ Visuals Unlimited)* **(b)** but white in dark-skinned people. *(Ken Greer/ Visuals Unlimited)*.

foreign is involved. Whatever the stimulus, T_H1 cells recognize a self antigen together with MHC present in the joint. The interaction of T_H1 cells with the antigen leads to release of cytokines that initiate a local inflammation in the joint. This attracts polymorphonuclear leukocytes and macrophages whose activities damage the cartilage in the joint. These activities may include release of degrading enzymes from lysosomes (◄Chapter 4, p. 102). People with RA also have a T_H2 cell-dependent B-cell response to the Fc portion of IgG. The formation of IgM:IgG immune complexes also causes damage in the joint. These autoantibodies, called **rheumatoid factors**, are used as a diagnostic test for RA. All these enzymes, factors, and cells increase the inflammatory response, leading to swollen, painful joints.

Although no cure exists for RA, treatment can alleviate symptoms. Hydrocortisone lessens inflammation and reduces joint damage, but long-term use weakens bones and causes such undesirable side effects as a reduction of normal immune responses. Aspirin decreases inflammation and reduces pain with fewer side effects. Physical therapy is used to keep joints movable. In severe cases, surgical replacement of damaged joints can restore movement.

Systemic Lupus Erythematosus

About 200,000 Americans suffer from **systemic lupus erythematosus (SLE)**, a systemic autoimmune disease. The name is derived from the reddened skin rash (erythematose) that resembles a wolf's mask (*lupus* is Latin for "wolf"). The butterfly-shaped rash appears over the nose and cheeks of about 30% of SLE patients **(Figure 18.17)**, and gets worse in sunlight. SLE occurs 10 to 20 times as often in women as in men, with 80% of cases occurring in women during their reproductive years. African Americans and Asians are affected more than people of other races.

In SLE, autoantibodies (IgG, IgM, IgA) are made primarily against components of DNA but can also be made against blood cells, neurons, and other tissues. As the normal dying process of cells (skin, intestinal, kidney) occurs, anti-DNA antibodies attack the remnants of these cells. Immune complexes are deposited between the dermis and epidermis and in blood vessels, joints, glomeruli of the kidneys, and the central nervous system. They cause inflammation and interfere with normal functions at these sites.

Inflammation of blood vessels, heart valves, and joints are common effects of interference. Arthritis is the most common clinical characteristic of SLE; a patchy skin rash on the upper trunk and extremities is a common skin manifestation. This rash often is precipitated by exposure to sunlight. Most SLE patients eventually die from kidney failure as glomeruli fail to remove wastes from the blood. Among individuals with SLE, men tend to have a nonsystemic *discoid* form of the disease. It produces disk-shaped skin lesions and is less serious than the systemic form in its side effects.

SLE cannot be cured. Treatment depends on individual disease characteristics. It can include antipyretics to control fever, corticosteroids to reduce inflammation, and immunosuppressant drugs to present or decrease further autoimmune reactions.

▌▌▌TRANSPLANTATION

▌▌▌**Transplantation** is the transfer of tissue, called **graft tissue**, from one site to another. An **autograft** involves the grafting of tissue from one part of the body to another—for example, the use of skin from a patient's chest to help repair burn damage on a leg. A graft between genetically identical individuals (identical twins in humans, or members of highly inbred animal strains) is called an **isograft** (*iso*, Greek for "equal"). A graft between two people who are not genetically identical is termed an **allograft** (*allo*, Greek for "different"). Most organ transplants fall into this category. A transplant between individuals of different animal species is known as a **xenograft** (*xeno*, Greek for "foreign").

Early transplantation experiments involved the grafting of skin from one animal to another of the same species. The grafts appeared healthy at first but in a few days to a few weeks became inflamed and fell off. First thought to be due to infection, this reaction, called **transplant rejection**, is now known to be due to the destruction of the grafted tissue by the recipient's (that is, the host's) immune system. This process, which depends on T cells, also accounts for rejection of most organ transplants in humans. Transplants recognized as nonself are rejected.

A much less common transplantation effect is **graft-versus-host (GVH) disease**, in which the transplanted tissue contains immunocompetent T cells that launch a cell-mediated response against the recipient's tissues. This response occurs most often when immunodeficient patients receive bone marrow transplants and, of course, cannot reject the graft tissues that are rejecting their new host. Host cells then come to the reaction site, attracted by cytokines released by the donor's T cells. At the site, host cells are responsible for most of the tissue destruction. This then causes enlargement of liver, spleen, and lymph nodes, anemia, diarrhea, weight loss, and in severe cases, death of the graft recipient. GVH was more common before immunosuppressive drugs were introduced to block the immunological responses.

HISTOCOMPATIBILITY ANTIGENS

All human cells, and those of every other vertebrate, have a set of self antigens called **histocompatibility antigens** (*histo*, Latin for "tissue"). The genes producing these molecules are called the **major histocompatibility complex (MHC)**. Only identical twins have exactly the same MHC molecules, but all family members have a mixture of similar and different MHC molecules. These antigens are located on the surfaces of cells, including the cells of kidneys, hearts, and other commonly transplanted organs. If donor and recipient histocompatibility antigens are different, as they probably would be when donors and recipients are not related, recipient T cells recognize these cells as foreign and destroy the donor tissue.

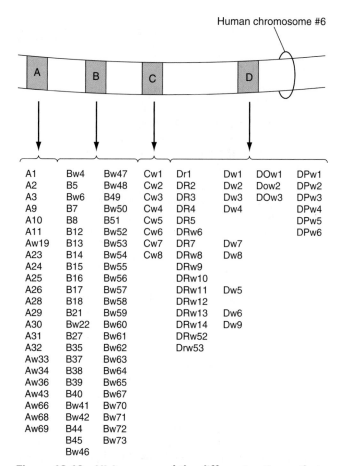

Figure 18.18   HLA genes and the different antigens that can be produced at each site along the human chromosome 6.

To try to prevent allograft rejection, it is necessary to determine if the graft has histocompatibility antigens not found in the recipient. Like red blood cell antigens, histocompatibility antigens can be identified by laboratory tests so that donor and recipient tissues can be as closely matched as possible. Such tests are one of several methods of *tissue typing*, or testing the compatibility of donor and recipient tissues. Because the first studies in humans involved antibody reaction with leukocytes, the MHC molecules on human cells are called **human leukocyte antigens (HLAs)**. Human HLAs are determined by a set of genes located on chromosome 6. They are designated A, B, C, and D (**Figure 18.18**). The information in each gene specifies a particular antigen. For example, HLA-B is so highly variable that there are alleles for 51 different antigens. Overall, about 120 different antigens are recognized in humans and produced from the HLA genes, resulting in a very high degree of genetic variability between individual people with regard to tissue types.

For tissue typing, it would be impossible to do a complete typing of all HLAs present in an individual. However, the HLA-DR antigens are known to generate the strongest rejection reactions. Therefore, the

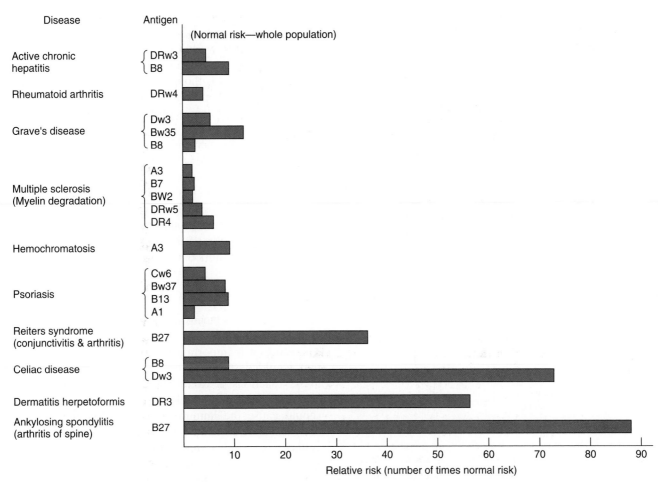

Figure 18.19 Correlations between specific HLAs and increased risk of developing certain diseases. Many of these are autoimmune conditions.

tissues of prospective transplant recipients are typed as to which of the 20 HLA-DR antigens are present. When a donor organ becomes available, it also is typed. It is transplanted into the recipient whose antigens most nearly match. This procedure reduces the chances of rejection. Identical twins are the best match for allografts because all their HLA antigens are the same, but siblings of the same parents can have some matching HLA antigens in common. The presence of certain HLA antigens is associated with a higher-than-normal risk of developing a particular disease (**Figure 18.19**), and many of these diseases are autoimmune disorders.

TRANSPLANT REJECTION

Like other immune reactions, transplant rejection displays specificity and memory (◄Chapter 17, p. 495). Rejection usually is associated with mismatched HLA-DR antigens. Certain cells that present antigens to phagocytes increase the likelihood of rejection. The fact that HLA-DR antigens are found on T cells and macrophages that carry out rejection reactions may explain why these antigens are so important in graft rejection.

T cells are responsible for rejection of grafts of solid tissue, such as kidney, heart, skin, or other organs. In animal experiments, allografts are retained by animals that lack T cells and are rejected by those lacking B cells. More specifically, T_H2 cells lead to rejection (**Figure 18.20**). These cells help stimulate cytotoxic T (T_C) cells, which reject the transplant through cell-mediated cytotoxicity. The T_H2 cells can also activate B cells to produce plasma cells and antibodies that cause rejection through lytic damage. Macrophages that are activated by T_H1 cells secrete inflammatory mediator and cause cytotoxic damage to the transplant. Natural killer cells can also act in transplant rejection.

The time required for rejection to occur varies from minutes to months. *Hyperacute rejection*, which is a cytotoxic hypersensitivity reaction, occurs when the recipient is already sensitized at the time the graft is done. For example, in kidney transplants, in which the graft is immediately supplied with host blood, extensive tissue destruction occurs within minutes to hours. (Corneal transplants, however, are not rejected because the cornea lacks blood vessels, and antibodies cannot reach them.) *Accelerated rejection* takes several days because it requires cells to reach the graft. *Acute rejection* occurs

Figure 18.20 Transplant rejection. A combination of both cell-mediated and humoral immune reactions is responsible for transplant rejection. T_H1 (inflammatory T) cells activate macrophages, which produce inflammatory mediators. T_H2 cells trigger both T_C and B cell activation. B cell activation leads to the production of plasma cells that synthesize antibodies, including anti–HLA-DR. Inflammatory mediators, T_C cell-mediated toxicity, and antibodies, along with complement, bring about transplant rejection.

in days to weeks, requiring T cell sensitization after transplantation. Rejection that begins months to years after transplantation represents a *chronic rejection*. This slow process is typical of cardiac and kidney transplants in which an interaction between the immune system and the transplant leads to eventual dysfunction of the transplant.

TOLERANCE OF THE FETUS DURING PREGNANCY

Given that half the genes of a fetus did not come from its mother, and that therefore a goodly number of them are surely foreign to her, how is it that the mother does not reject and abort this "non-self" fetus? In some cases of chronically miscarrying women, this may be what they *are* doing. But if the human race is to continue, somehow the fetus must be tolerated. We know from Rh-incompatibility pregnancies that mothers *are* able to make antibodies against foreign fetal proteins when they invade her bloodstream. Why are cytotoxic T cells and NK cells not produced and sent against the fetus? While not

understanding the situation fully, we can say that the fetus occupies an "*immunologically privileged site,*" with multiple factors at work. Cells on the surface and interior of the fetal portion of the placenta do not express MHC molecules. And, it is thought by some, that certain HLA molecules prevent maternal NK cells from killing fetal cells. Alpha-fetoprotein, a protein produced by the fetus, has been shown to have immunosuppressive properties. Cytokines, complement inhibitors, and who knows what else may all play a role in keeping the fetus safe.

Curiously, women whose tissue types are most closely alike with their husbands' suffer more miscarriages and infertility. It is thought that "foreignness" of the sperm may trigger maternal production of blocking antibodies that protect the fetus. If the cells are too much alike, not enough blocking antibodies are produced.

IMMUNOSUPPRESSION

When a patient is facing an organ transplant, the donor's HLA antigens are probably not a complete match. Therefore, it is important to prevent immune reactions that would destroy the organ. The minimizing of immune reactions is called **immunosuppression**. Ideally, immunosuppression should be as specific as possible—it should cause the immune system to tolerate only the antigens in transplanted tissue and allow the immune system to continue to respond to infectious agents.

In practice, radiation or cytotoxic drugs, both of which impair immune responses, are used to minimize rejection reactions. *Radiation* (X-rays) of lymphoid tissues suppresses the immune system, preventing rejection. Radiation

APPLICATIONS

Disguising Tissues Aids in Transplants

Transplanted tissues are rejected because the recipient's immune system recognizes them as foreign. However, scientists have succeeded in disguising foreign cells by covering HLAs that act as antigens and trigger rejection. Ordinarily the binding of antibodies to antigens on cells starts a process that leads to death of the cells. The researchers modified the antibodies such that although the antibodies fit tightly to foreign cell surface proteins, they did not destroy the cells. Human pancreatic cells with HLA proteins thus "covered" were transplanted into mice. The mouse immune system ignored the human cells, allowing them to live and produce insulin for more than 6 months. Further research may eventually provide transplants to treat diabetics. An advantage of using disguised cells is that immunosuppressants now used with transplants will be unnecessary. These drugs, sometimes taken for the rest of the patients' life, leave patients highly vulnerable to infections and often cause other undesirable side effects, including a predisposition to cancer.

also destroys other lymphoid functions, including the ability of the immune system to recognize infectious microbes. **Cytotoxic drugs**, such as azathioprine and methotrexate, damage many kinds of cells. But because they interfere with DNA synthesis, these drugs cause most damage to rapidly dividing cells. Because B cells and T cells divide rapidly after sensitization, the drugs exert a somewhat selective effect on the immune system.

Radiation and cytotoxic drugs impair T cell responses to infections. In contrast, the fungus-derived peptide cyclosporine A (CsA) suppresses, but does not kill, T cells, and it does not affect B cells. It is particularly useful in preventing transplant rejection: It allows T cells to regain function after the drug is stopped, and it does not reduce resistance to infections provided by B cells. The use of immunosuppressive drugs, especially CsA, has greatly increased the success rate of organ transplants. However, CsA may increase the transplant recipient's risk of developing cancer.

DRUG REACTIONS

Most drug molecules are too small to act as allergens. If a drug combines with a protein, however, the protein-drug complex sometimes can induce hypersensitivity. All four types of hypersensitivity have been observed in drug reactions.

Type I hypersensitivity can be caused by various drugs. Most reactions are localized, but generalized anaphylactic reactions sometimes occur, especially when drugs are given by injection. Orally administered drugs are less likely to cause hypersensitivity reactions because they are absorbed more slowly. Hypersensitivity reactions require prior sensitization and depend on the production of IgE antibodies. Although penicillin is one of the safest drugs in use, 5 to 10% of people receiving it repeatedly become sensitized. Of those sensitized, about 1% develop generalized anaphylactic reactions, which account for about 300 deaths per year in the United States.

Type II hypersensitivity (**Figure 18.21**) can occur when the drug binds to a plasma membrane directly; when it binds to a plasma protein, and the complex binds to a plasma membrane; or when it alters a plasma membrane in such a way that cellular antigens trigger autoantibody production. All such reactions involve IgG or IgM and complement. Their targets—erythrocytes, leukocytes, or platelets—are destroyed by complement-dependent cell lysis. Many antibiotics, sulfonamides, quinidine, and methyldopa elicit Type II reactions.

Type III hypersensitivity appears as serum sickness and can be caused by any drug that participates in the formation of immune complexes. Symptoms appear several days after administration, when sufficient quantities of immune complexes have accumulated to activate the complement system. A few patients sensitized to penicillin develop serum sickness.

Type IV hypersensitivity usually occurs as contact dermatitis after topical application of drugs. Antibiotics, antihistamines, local anesthetics, and additives such as lanolin are frequent agents of Type IV reactions. Medical personnel who handle drugs sometimes develop Type IV hypersensitivities.

IMMUNODEFICIENCY DISEASES

Immunodeficiency diseases arise from an absence or a deficiency of active lymphocytes, NK cells, or phagocytes, the presence of defective lymphocytes or

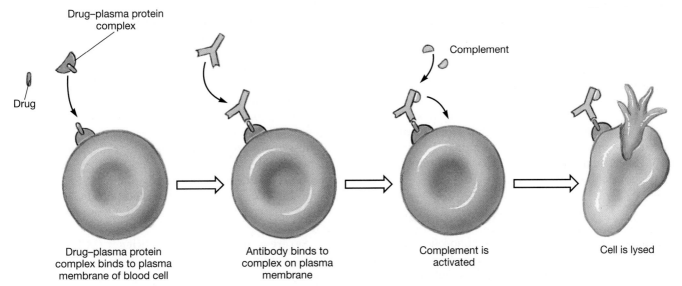

Figure 18.21 Drug reactions based on Type II hypersensitivity. A drug (or one of the products of its metabolism in the body) may bind to the plasma membrane of a blood cell, bind to a blood (plasma) protein to form a complex that binds to a plasma membrane, or alter a plasma membrane protein. Autoantibodies—IgG or IgM—are produced and then bind to the complex and activate complement to lyse the cell.

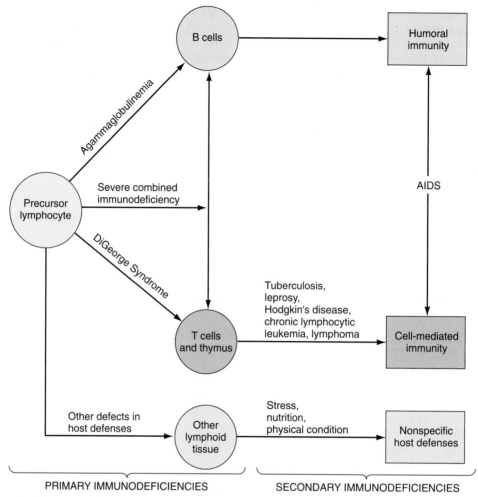

Figure 18.22 **Kinds of immunodeficiencies.**

phagocytes; or the destruction of lymphocytes. Such diseases invariably lead to impaired and inadequate immunity. **Primary immunodeficiency diseases** are caused by genetic defects in embryological development, such as failure of the thymus gland or Peyer's patches to develop normally. The result is a lack of T cells or B cells, or defective T and B cells. **Secondary immunodeficiency diseases** can be caused by (1) infectious agents, such as those responsible for leprosy, tuberculosis, measles, and AIDS; (2) malignancies, such as Hodgkin's disease or multiple myeloma; or (3) immunosuppressants, some chemotherapeutic drugs, certain antibiotics, and radiation. Such agents damage T cells or B cells after the cells have developed normally (**Figure 18.22**).

PRIMARY IMMUNODEFICIENCY DISEASES

Agammaglobulinemia, the first immunodeficiency disease to be understood, is a B cell deficiency. It occurs primarily in male infants in which B cells, and therefore antibodies, are absent. After maternal antibodies are lost by about 9 months of age, affected infants develop severe infections because they cannot produce IgM, IgA, IgD, and IgE

antibodies and produce only small amounts of IgG. Agammaglobulinemia is treated with massive doses of immune serum (gamma) globulin to replace missing antibodies and with antibiotics to prevent infections.

DiGeorge syndrome results from a deficiency of T cells, probably caused by an agent that interferes with embryological development of the thymus gland. Cell-mediated immunity is impaired, so viral diseases pose a greater-than-usual threat. Although B cells are normal, their activation requires T_H cell activation (Chapter ◄17, p. 500). Therefore, humoral immunity also is affected, because there are no functional T_H2 cells. Mice lacking a thymus, known as "nude" mice (**Figure 18.23**), are bred and raised in germ-free environments for research purposes. They are used in the study of DiGeorge syndrome, as well as in other areas of immunology and genetics.

Severe combined immunodeficiency (SCID) is particularly debilitating because both B and T cells are absent. SCID can have several genetic origins. For example, stem cells in the bone marrow that normally give rise to lymphocytes fail to develop properly because of a defective gene for IL-2, for the enzyme adenosine

Figure 18.23 **A nude mouse.** These animals lack a thymus gland, as well as their fur. They are delivered by cesarean section by sterile technique and must be kept in a germ-free environment all their lives, because they completely lack T cells. They are the equivalent of human cases of DiGeorge syndrome. Researchers use them in many types of studies of the immune system. (*National Institutes of Health/Photo Researchers, Inc.*)

deaminase (ADA), or for MHC molecules. An infant who inherits this condition is doomed to die within the first few years of life unless he or she is kept in a germ-free environment until satisfactory treatment can be devised. (See the box "The Boy in the Bubble.")

Bone marrow transplants can be effective in SCID patients if a compatible donor (usually a brother or sister) is found. If the transplant is not compatible, the transplanted lymphocytes respond immunologically to antigens in the recipient's tissues. This is another example of GVH disease, and it can be lethal.

Gene therapy, which attempts to replace a defective gene with a functional, therapeutic copy of the gene, has been used to treat SCID and has shown spectacular results. A few children have had their bone marrow cells removed and "infected' with a reproductively deficient retrovirus carrying the missing gene for ADA (which is essential to cell maturation). The cells were then returned to their bodies. Although the patients who received the infected bone marrow transplant must periodically have additional transplants, all of the children are living normal lives.

SECONDARY (OR ACQUIRED) IMMUNODEFICIENCY DISEASES

Immunodeficiency diseases are not always inherited; sometimes they are *acquired* as a result of infections, malignancies, autoimmune diseases, or other conditions. For example, congenital rubella infections can decrease T cell function and antibody production to the extent that infants fail to respond to vaccines. Once patients develop immunodeficiencies, they may suffer from chronic or frequent recurrent infections.

Among malignant diseases that produce immunodeficiencies, those of lymphoid tissues suppress T cell function, and those of bone marrow suppress both T cell function and antibody production. Autoimmune diseases, some kidney disorders, severe burns, malnutrition or starvation, and anesthesia also can cause temporary or permanent immunodeficiencies.

Acquired Immune Deficiency Syndrome (AIDS)

Certainly the most well-known secondary immunodeficiency is **acquired immune deficiency syndrome (AIDS)**, an infectious disease caused by the **human immunodeficiency virus (HIV)** which belongs in the family Lentiviridae. AIDS can be caused by at least two different types of human immunodeficiency viruses, designated HIV-1 and HIV-2. Most cases of AIDS in the United States, Canada, and Europe are caused by HIV-1. HIV-2, which is most common in certain parts of West Africa, may be less virulent. Both forms are tested for in screening the United States blood supply.

TABLE 18.7

Infections Frequently Found in AIDS Patients	
Pathogen	**Disease**
Bacteria	
Mycobacterium tuberculosis	Tuberculosis
Mycobacterium avium-intracellulare	Disseminated tuberculosis
Legionella pneumophila	Pneumonia
Salmonella species	Gastrointestinal disease
Viruses	
Herpes simplex	Skin and mucous membrane lesions, pneumonia
Cytomegalovirus	Encephalitis, pneumonia, gastroenteritis, fevers
Epstein-Barr	Oral hairy leukoplakia, possibly lymphoma
Varicella-zoster	Chickenpox, shingles
Fungi	
Pneumocystis carinii	*Pneumocystis carinii* pneumonia
Candida albicans	Mucous membrane and esophagus infections (thrush)
Cryptococcus neoformans	Meningitis, kidney disease
Histoplasma capsulatum	Pneumonia, disseminated infections, fevers
Other opportunistic fungi	Varies with opportunist
Protozoa	
Toxoplasma gondii	Encephalitis
Cryptosporidium species	Severe diarrhea

Recent studies based on DNA sequencing shows that HIV-2 is very closely related to the simian immunodeficiency virus (SIV) found in African Sooty Mangabey monkeys—so similar that HIV-2 is a mutated version of the same virus as SIV. However, HIV-1 differs significantly enough that it may have separated from the HIV-2/SIV evolutionary tree earlier. Concensus is that HIV-1 has evolved sometime in the last 100 years from the chimpanzee version of SIV. The oldest known case of AIDS in Europe was seen in a Danish surgeon who had worked in Zaire. She died in 1976. Cases occurring in Africa earlier in the 20th century would probably have been overlooked, given the lesser numbers of cases and the state of health care existing then.

Early evidence for the origin of HIV comes from studies of human blood stored in England and Zaire since 1959, in which HIV-1 antibodies were found. The virus may have existed in relatively isolated regions, perhaps in Central Africa, for decades. Migration of rural people to rapidly growing cities, where population density was much higher and sexual contact was more casual and more frequent, could have brought about a great increase in the number of infected individuals. The expansion of international travel in recent years could rapidly spread the virus to many other parts of the world. The virus probably made multiple entries into the United States before becoming established.

The virus gradually but relentlessly destroys the immune system. The lack of a functional immune system leaves the body open to a variety of malignancies and opportunistic infections, most of which are rarely seen among people who are not suffering from advanced HIV disease or AIDS. These complications—either alone or in combination—eventually prove fatal (Table 18.7).

HIV specifically targets and damages T_H cells, macrophages, dendritic cells, and Langerhans cells which have a CD4 molecule on their surface. The virus binds to the CD4 molecule and another protein, either CXCR4 on T cells or CCR5 on macrophages. The envelope of the virus fuses with the cell membrane leaving virus proteins on the surface of the infected cell that induce cell fusion to neighboring cells. Thus, HIV can infect another cell without having to be released from an infected cell. Dendritic cells and macrophages can acquire HIV on mucosa surfaces and then migrate to lymph nodes where T_H cells are infected. Macrophages that have phagocytized HIV from dead or dying tissue are impaired but usually do not die, they become reservoirs of HIV—in fact more virus is stored in macrophages than in T cells. HIV infected macrophages can deliver HIV to various organs of the body including the brain and lungs. Only 4% of HIV is in the blood—96% is in lymph nodes, intestines, and brain. The infectious cycle for the virus was described in ◀Chapter 10 (p. 291).

After a person is infected with HIV an enormous battle ensues between HIV and the immune system. Initially large quantities of the virus are produced which results in symptoms such as fever, fatigue, weight loss, diarrhea, and body aches. As immune cells become activated, antibody from B cells and T_C cells destroy large numbers of HIV. As virus-infected cells are destroyed by T_C cells, more immune cells replace those killed. Each day as HIV disease progresses, an estimated 1 billion virus particles are produced and destroyed as 2 billion immune cells are replaced! Although these early battles result in a draw, the virus eventually wins the war. As the years pass, it becomes more difficult to replace T_H cells and other immune cells. HIV is a retrovirus that uses an error-prone reverse transcriptase to make a DNA copy of its RNA genome. These errors result in a high mutation rate. Variations occur in the proteins on the surface of the virus so that antibodies can no longer recognize HIV. Eventually the immune system simply cannot keep up the fight.

Cells without the CD4 marker were formerly thought to be immune to the AIDS virus. Now it seems that cells with a CD8 marker begin to make CD4 when stimulated.

Without activated T_H cells and macrophages, the immune system cannot "see" infectious microbes. Because T_H cells are greatly diminished in number, B cells are not stimulated to form plasma cells, which produce antibodies to combat infections. (The anti-HIV antibodies detected early in the course of infection are made before T_H cell populations become too depleted to stimulate B cells.) Similarly, cytokines are produced in amounts insufficient to activate macrophages and T_C cells. A drop in T_H cell count can be used to predict the onset of disease symptoms. A normal T_H count is 800 to 1200 per μl of blood. If the count remains above 400, 8% of those infected will develop AIDS symptoms within 18 months. With a T_H count of 200, 33% progress to AIDS; if the count is below 100, 58% will develop AIDS within 18 months.

Progression of HIV Disease and AIDS

The sequence of events in HIV disease has now been established in some detail. The progression depends heavily on how much of the virus a person is exposed to (the *viral burden*) and how often the exposure is repeated. The CDC classification is based on the absence or presence of certain signs and symptoms and includes laboratory test results. Thus, persons categorized as being in groups 1 to 3 have HIV disease. Persons in group 4 have been diagnosed as having AIDS (**Figure 18.24**). Although diseases such as hairy leukoplakia (a white

Figure 18.24 CDC classification of HIV disease and AIDS. The progression of an HIV infection to AIDS is shown, along with diseases and dysfunctions that can occur.

lesion appearing on the tongue) are diagnostic of most AIDS patients, other opportunistic diseases or cancers vary among AIDS patients. For example, latent viral infections such as those caused by herpes simplex and cytomegalovirus, normally kept in check by the immune system, flare up and create a variety of disease symptoms. Severe diarrhea can be caused by opportunistic pathogens, including several species of *Cryptosporidium* (◀Chapter 22); encephalitis can be caused by *Toxoplasma gondii* (◀Chapter 24); and yeast infections can be caused by *Candida albicans* (◀Chapter 19). Pneumonia produced by the fungus *Pneumocystis carinii* is common to 80% of AIDS patients prior to their death. If patients do not first die of another opportunistic infection, about 50% will develop respiratory disease—caused either by *Mycobacterium tuberculosis* or by *Mycobacterium avium-intracellulare*. Overall, 88% of AIDS deaths result from an opportunistic infection.

Most AIDS patients develop malignancies not commonly found in the general population. A malignancy called **Kaposi's sarcoma**, caused by the human herpes virus 8, causes blood vessels to grow into tangled masses that are filled with blood and easily ruptured. In the skin and viscera, this sarcoma shows up as prominent pink or purplish spots (**Figure 18.25**). It can spread to the digestive tract, lungs, liver, spleen, and lymph nodes. However, there have been no reported cases of deaths among AIDS patients from Kaposi's sarcoma.

About 30% of individuals newly infected with HIV will progress to group 4 within 5 years if not treated. Without treatment, within 15 years 90% of HIV-infected patients will develop AIDS. Some of the drugs (especially protease inhibitors in combinations with other drugs—often called an AIDS cocktail) currently being used are capable of prolonging the life of AIDS patients, and so the death rate, but not the number of infected persons, is dropping. HAART (or Highly Active Antiretroviral Therapy), a combination of protease inhibitor and two nucleotide analogues which inhibits reverse transcriptase, has proven to be especially effective in reducing viral replication. HAART has increased the life expectancy of HIV-infected patients dramatically! However, it does do not cure the disease. Furthermore, if discontinued because of side effects or lack of funds, death usually follows swiftly—sometimes within a month!

Epidemiology of AIDS

AIDS has been called the epidemic of the century; certainly, few diseases have had such a dramatic impact. In 2006 an estimated one million Americans were living infected with HIV. At least 50,000 new infections occur each year; an estimated 4 or 5 people are infected per hour—although the actual number is likely to be higher. Amazingly, only about 25% of the 1 million individuals know they are infected! Between 1981 and the end of 2006 over 25 million people have died of AIDS, with 2.9 million in 2006 alone. AIDS has become the sixth leading cause of death among 25- to 44-year-olds. Fortunately, in the last several years, the rate of increase in incidence of new cases has slowed in the United States.

The AIDS pandemic is a global phenomenon. Worldwide, 70 to 80% of HIV infections are acquired from heterosexual contact. In developing nations 40% of pediatric infections are acquired via breast milk. AIDS has created over 15.2 million orphans worldwide. That is the equivalent of more than every child in America under the age of 5. World Health Organization (WHO) officials estimate that by the end of 2007, the worldwide number of people infected with HIV will exceed 40 million (**Figure 18.26**). But in many developing nations, AIDS cases also are probably undiagnosed or unreported. The region most seriously affected is sub-Saharan Africa, where over 24.7 million people are infected. In South Africa 600 people per day now die of AIDS and 1 in 4 people are dying of AIDS. One area where the virus is spreading rapidly is East Asia and the Pacific where cases went from 640,000 to 1.3 million between 2000 and 2003. Another 1.7 million people in Latin America and the Caribbean (250,000) are believed to be infected. Infection rates vary considerably by country. Botswana has the highest infection rate, with an estimated 38.8% of its adults now infected, a rate that has more than tripled since 1992 when the rate was 10%. Life expectancy at birth there is now 44 years instead of the 69 years it would be without AIDS. Swaziland (38.6%), Zimbabwe (33.7%), and Lesotho (31.5%) are close behind Botswana. The social and economic impacts in these countries are staggering. There is some good news amid all these statistics—rates of infection have slowed from 14 to 5% in Uganda following strong prevention campaigns.

Who Gets AIDS, and How

All available evidence suggests that it is virtually impossible to become infected with HIV through casual contact. Rather, a person becomes infected with the

Figure 18.25   Kaposi's sarcoma. This tumor of blood vessels is seen in AIDS patients as dark purple areas. *(St. Mary's Hospital Medical School/Photo Researchers, Inc.)*

Figure 18.26 **Estimated number of global HIV infections projected by the World Health Organization through 2006.** Of the 42 million HIV-infected persons, over 3.2 million of these are children. (*Source:* UNAIDS.).

AIDS virus only through intimate contact with the body fluids of an infected individual and by transmission from infected mother to fetus. The virus is most commonly transmitted through blood, semen, and vaginal secretions. It seems likely that, however it is transmitted, the virus must make contact with a break or abrasion in the skin or with mucous membranes to cause infection. For this reason, all practices that lead to an exchange of body fluids carry a risk of HIV infection. These include:

1. *Sexual contact with an infected individual.* All forms of sexual intercourse—heterosexual and homosexual, active and passive, vaginal, anal, and oral—carry the risk of HIV infection. Condoms can reduce but not eliminate transmission because they have a significant failure rate (17 to 54% in various studies), usually the result of improper use, and not all types of condoms are equally effective in blocking HIV. Natural skin condoms allow passage of the virus; latex condoms are much safer.

2. *The sharing of unsterilized needles by intravenous drug users.*

3. *Receipt of a blood transfusion or blood products contaminated with HIV.* HIV infection by blood transfusion caused many AIDS cases in the early 1980s. Many of the infected persons were hemophiliacs, who received injections of blood products to let their own blood clot properly. Transfusions are much less of a viral threat today, thanks to the testing of donated blood for HIV antibodies and to recombinant DNA technology. And, popular fears to the contrary, it is not possible to acquire the HIV virus by donating blood because new, sterile needles are used.

At least a quarter of the 2.5 million units of blood administered in Africa are not screened for the AIDS virus.

4. *Passage from an infected mother to an infant.* About 25% of infants carried by HIV-positive women become HIV infected. HIV transmission is possible while the fetus is in the uterus, during delivery, and through breast feeding. Preliminary studies indicate that more newborns are infected during delivery than before birth.

Treatment with AZT during the last few weeks of pregnancy reduces mother-to-fetus HIV transmission by half.

Health care personnel treating AIDS patients or HIV-infected patients are at risk of becoming infected.

The CDC has recommended the following precautions to minimize that risk (see also the Universal Precautions in ◄Chapter 15, p. 453).

1. *Wear gloves, masks, protective eyewear, and gowns* for touching blood, body fluids, mucous membranes, or skin lesions of patients and for procedures that might release droplets of body fluids. Discard these items, and wash hands immediately and thoroughly after seeing each patient. Like other medical personnel, dentists and their technicians should consider blood, saliva, and gingival fluids of all patients to be potentially infective and should use these procedures to prevent contact with such fluids.

2. *Avoid injury from needles* and other sharp objects, and discard them in puncture-proof containers.

3. *Use mouthpieces, resuscitation bags, or other ventilator devices* for emergency resuscitation.

4. Workers with skin lesions should *avoid direct patient care and handling of contaminated equipment.*

Keep in mind that almost any infectious disease (such as tuberculosis) that an HIV disease patient or AIDS patient may have contracted poses a danger to health care personnel. Of course, this problem is not limited to HIV infection; other types of infections can be a threat to health care workers.

What About an AIDS Vaccine?

The outlook for an AIDS vaccine is not promising. Scientists expect a period of trial and error lasting well beyond the year 2010 because research with HIV presents unusual problems. For one thing, HIV has a high mutation rate due to the imprecise operation of its reverse transcriptase. Thus, even if a successful vaccine against one strain of the virus is produced, another strain might not be affected by the vaccine, and new strains would likely develop. Outbreaks of influenza every few years are caused by similarly high mutation rates in the influenza virus.

Developing a vaccine poses more problems. Attenuated viruses cannot be used in a vaccine because they contain DNA that can be incorporated into the host's genome, possibly later giving rise to AIDS. Whole inactivated viruses, which have been used successfully in polio and influenza vaccines, are inappropriate for an AIDS vaccine. Such vaccines would, for HIV and other lentiviruses, predispose the host to severe infections. Recombinant viruses, such as HIV antigens on a vaccinia virus, are unacceptable because the host might develop disseminated vaccinia. Also, immunocompromised hosts may develop a variety of severe complications. Thus, many problems must be solved before we can have a safe and effective vaccine.

The Social Perspective: Economic, Legal, and Ethical Problems

AIDS and HIV disease will have an increasingly significant economic impact in the coming years. The yearly cost of medical care for a nonhospitalized AIDS patient in the United States can reach $36,000 or more, with drugs costing about $2,000 per month. On the basis of CDC estimates, the total cost for care of all U.S. AIDS patients in 2001 was almost $20 billion. If the care of AIDS patients is a burden in the United States, where average income exceeds $12,000 per person per year, imagine the catastrophic burden in many developing countries, where average annual income is less than $200 per person.

U.S. laws protect the confidentiality of medical information, including AIDS test results. The laws also provide equal opportunity with respect to employment, housing, and education, but many AIDS patients continue to encounter various kinds of discrimination. Protecting the rights of uninfected citizens and those of health professionals who treat AIDS patients also must be considered. Other legal issues concern the responsibility of HIV-infected individuals not to transmit the disease and the liabilities of distributors of blood products.

Many ethical issues are related to the legal issues. A major question is how the epidemic can be curtailed without infringing on individual freedoms. Another question weighs the moral obligation of health professionals to care for all patients against the risk of acquiring a fatal disease. Still other questions relate to allocation of scarce medical resources.

✓CHECKLIST

1. What is an autoimmune disease? What might cause one?

2. Summarize the causes, signs, and symptoms of myasthenia gravis, rheumatoid arthritis, and systemic lupus erythematosus.

3. How and why does transplant rejection occur?

4. Distinguish between primary and secondary immunodeficiency diseases. Give examples of each.

▌▌▌ IMMUNOLOGICAL TESTS

In ◄Chapter 17 we considered how certain immunological reactions—agglutination, cell lysis by complement and by IgM antibodies, and neutralization of viruses and toxins—kill pathogens. Now we will consider how those and other reactions are used as laboratory tests to detect and quantify antigens and antibodies. Such laboratory tests make up the branch of immunology called **serology**, so named because many of the tests are performed on serum samples. Today, some laboratory tests also use monoclonal antibodies derived from the culture fluid of animal tissue grown in culture (◄Chapter 17, p. 503). The tests and reactions described here represent a broad but incomplete sampling of laboratory and clinical

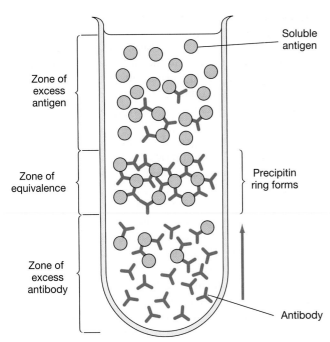

Figure 18.27 The precipitin test for antibodies. When IgG or IgM antibodies (which are soluble) react with soluble antigens, they quickly form small complexes, which over time combine into larger lattices that precipitate from the solution. Such precipitation occurs, however, only when there is an appropriate ratio of antigen to antibody. In the test, antibodies are placed in the bottom of a narrow tube. Soluble antigen is added, and the two are allowed to diffuse toward each other. Where the necessary concentration ratio is achieved (the zone of equivalence), precipitation takes place, visible as a hazy "precipitin ring" in the tube.

tests. The selection of the "right" test depends on the nature of the pathogen or disease being analyzed.

THE PRECIPITIN TEST

Historically, one of the first serologic tests to be developed was the **precipitin test (Figure 18.27)**, which can be used to detect antibodies or antigens. This test is based on a **precipitation reaction** in which antibodies called *precipitins* react with antigens, diffuse toward each other, and form a visible precipitate. During such reactions, antigen-antibody complexes form within seconds. Latticelike networks of these complexes, which are visually opaque, form minutes to hours later.

Many modifications have been made to the basic precipitin test to increase its sensitivity to detect specific antigen-antibody complexes. **Immunodiffusion tests** are based on the same principle as the precipitin test, but they are carried out in a thin layer of agar that has solidified on a glass slide. Immunodiffusion tests are used to determine if more than one antigen—and therefore more than one antibody—is present in a serum specimen. Small wells are made in the solidified agar, and the antigens and antibodies are placed in separate wells. Antigen-antibody complexes appear as detectable precipitation lines (after staining) in the agar between the wells. After diffusion

occurs, one or more bands of precipitation can be detected, each representing a different antigen-antibody complex **(Figure 18.28)**. The bands can be made more visible by washing the agar surface and applying a stain that colors the antigen-antibody complexes. An advantage of immunodiffusion tests is that in a single test medium, several antigens can be reacted with one kind of antibody or several kinds of antibodies with one antigen.

When serum samples contain several antigens, **immunoelectrophoresis** can be used to detect separate antigen-antibody complexes. The antigens are placed in a well on an agar-coated slide. An electric current then is passed through the gel. This process is called **electrophoresis**. During electrophoresis different antigen molecules migrate at different rates, depending on the size and electric charges of the molecules. Following electrophoresis, the antibody is placed in a trough made along one or both sides of the slide, and diffusion is allowed to occur. The results of immunoelectrophoresis are similar to those obtained in other immunodiffusion tests—precipitin bands form wherever matching antigen and antibody precipitate **(Figure 18.29)**. Thus, the advantage of immunoelectrophoresis is the ability to separate several antigens that might be present in a serum sample.

Immunoelectrophoresis and radial immunodiffusion are used routinely in large hospital clinical laboratories to

(a)

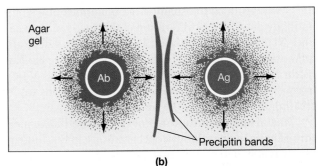

(b)

Figure 18.28 Immunodiffusion test. This test is a modification of the precipitin reaction. Wells are made in an agar gel and filled with test solutions of antigen (Ag) and antibody (Ab). **(a)** A single antibody and antigen diffuse outward from the wells, meet, react with each other, and precipitate. They form a line called a precipitin band, which is visualized by staining. **(b)** Two different antigen-antibody complexes diffuse at different rates, producing separate bands.

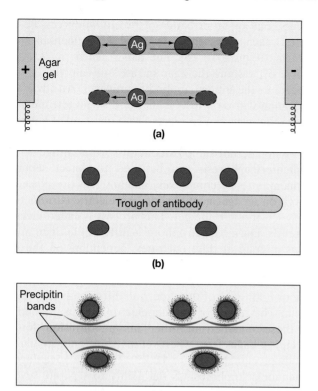

Figure 18.29 Immunoelectrophoresis. (a) Antigens placed in an agar gel are separated by means of an electric current. Positively charged molecules are drawn toward the negative pole; negatively charged ones move toward the positive pole. **(b)** A trough is cut into the agar between the wells and filled with antibody. **(c)** Curved precipitin bands form where antigens and antibodies diffuse to meet and react.

detect the presence of the various classes of immunoglobulins. IgG, IgM, and IgA are usually present in sufficient quantity to be detected by precipitin bands, but IgD and IgE are usually in too small a quantity to detect. Patients who do not make normal amounts of IgG, IgM, and IgA can be detected by this method. The method also allows clinicians to diagnose and monitor patients who have myeloma tumors (plasma cell tumors). These patients will produce unusually large quantities of a single antibody due to the proliferation of this single plasma cell line. Radial immunodiffusion is also used to detect and quantitate other proteins in the blood such as the clotting factor fibrinogen and complement components. Veterinary clinical laboratories also use these methods to detect immunoglobulins in the serum of animals.

Radial immunodiffusion provides a quantitative measure of antigen or antibody concentrations. In this test, antibody is added to molten agar, and the solution is allowed to solidify as a thin layer on a glass slide. Antigen samples of different concentrations are placed in wells made in the agar slide. After diffusion the antigen concentration is determined by measuring the diameter of the ring of precipitation around the antigen (**Figure 18.30**). Similarly, antibody concentrations can be determined by placing antibody samples of different concentrations in wells in a gel containing the antigen.

AGGLUTINATION REACTIONS

When antibodies react with antigens on cells, they can cause **agglutination**, or clumping together of the cells. One application of **agglutination reactions** is to determine whether the quantity of antibodies against a particular infectious agent in a patient's blood is increasing. The quantity of antibodies is called the **antibody titer**. It is reported as the reciprocal of the greatest serum dilution in which agglutination occurs. For example, an antiserum that agglutinated an agent's antigen at a dilution of 1;256 but not at 1;512 would be reported as an antibody titer of 256. An increase in the antibody titer over time indicates that the patient's immune system is attacking the agent. Diagnosis of the disease agent is possible when it can be shown that the patient's serum had no antibodies against the agent before the onset of disease or that a rise in titer occurred during the course of the disease. This production of antibodies in the serum resulting from infection (or immunization) is called **seroconversion**. The **tube agglutination test** measures antibody titers by comparing various dilutions of the patient's serum against the same known quantity of the antigen (cells).

Agglutination reactions are often used to diagnose disease due to an organism that is difficult to detect or

Figure 18.30 Radial immunodiffusion. (a) Wells cut into antibody-containing agar sheets are filled with antigen. The antigen diffuses outward, complexing with antibody as it goes. When the ratio of antigen to antibody is optimal, the complexes precipitate in a ring. **(b)** The diameter of the ring is proportional to the logarithm of the antigen concentration, which can be determined by reference to a standard curve. The concentration of an antibody can also be determined by placing it in a well cut into an antigen-containing agar sheet and comparing the size of the resulting ring with a standard curve for that antibody.

Figure 18.31 Hemagglutination. This test, used for matching blood types, is based on the agglutination reaction. In A, red blood cells were mixed with serum having antibodies to the antigens on the cells. The complex of cells and antibodies clumped together. In B, the serum added did not contain antibodies that recognize the blood type antigens on the cells, so no clumping occurred. *(George Whiteley/Photo Researchers, Inc.)*

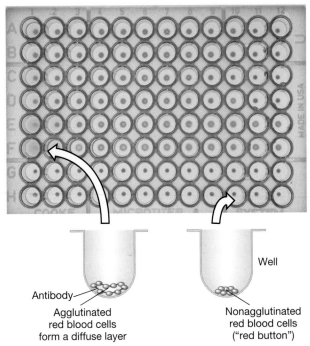

Figure 18.32 Microtiter plates. Hemagglutination tests often are carried out in microtiter plates. These plates contain 96 plastic wells, so many tests can be done simultaneously. When hemagglutination occurs, antibody reacts with antigen. All wells contain red blood cells. Positive hemagglutination results in a diffuse clumping of red blood cells (as in row F, column 1). Negative hemagglutination is seen as a clump of red blood cells ("red button") at the bottom of the wells (as in rows G and H). *(Southern Illinois University/Visuals Unlimited).*

grow directly in the clinical laboratory. These tests are used to detect antibodies to the three species of *Brucella* that cause brucellosis or undulant fever; *Franciscella tularensis*, the cause of tularemia; and antibodies to Epstein-Barr virus, the cause of mononucleosis.

Hemagglutination, or agglutination of red blood cells, is similar to the agglutination tests except the antigens are on the surface of red blood cells. Hemagglutination is used in blood typing **(Figure 18.31)**. In addition, hemagglutination tests can be used to detect viruses, such as those that cause measles and influenza. Such viruses bind to and cross-link red blood cells, causing **viral hemagglutination**. This process is inhibited by adding antibodies to the viruses. Because these antiviral antibodies bind to the viruses, the viruses cannot agglutinate red blood cells. Such inhibition is the basis of the **hemagglutination inhibition test**, which can be used to diagnose measles, influenza, and other viral diseases.

Today, these types of agglutination tests usually are performed in plastic *microtiter plates*. These plates contain 96 separate wells, so many tests can be done simultaneously. In the plate shown in **Figure 18.32**, antibody dilutions are added to the wells of the plate. Then, equal concentrations of red blood cells are added to each well. If sufficient antibody is present to agglutinate the red blood cells, the antibody-cell complexes sink into a diffuse layer at the bottom of the well. If the antibody titer is too low, the red blood cells settle and form a red "button" at the bottom of the well.

Earlier we saw that severe hemolytic disease of the newborn results from an Rh factor incompatibility between mother (Rh negative) and fetus (Rh positive). In a hemagglutination test, although anti-Rh antibodies will bind to Rh antigen on red blood cells, there are not sufficient Rh antigens to cause clumping with anti-Rh antibodies **(Figure 18.33a)**. The **Coomb's antiglobulin test** is designed to detect such antibodies. If red blood cells coated with anti-Rh antibody are treated with an antibody that recognizes the anti-Rh antibody, the antibody-antibody-cell

complexes will agglutinate **(Figure 18.33b)**. Thus, if a patient's serum contains Rh antibodies or if the red blood cells are Rh positive, agglutination will occur.

The body's natural defenses use complement to bind to antigen-antibody complexes, helping to destroy pathogens. This same ability is used in the laboratory or clinic

APPLICATIONS

Pink Means You're Pregnant

Did you know that some home pregnancy tests are modified agglutination assays? The tests are based on agglutination inhibition, a highly sensitive assay that can detect small quantities of an antigen. Latex particles are coated with human chorionic gonadotropin (HCG) and antibody to HCG. When a pregnant woman's HCG-containing urine comes in contact with these latex particles, the latex particles can't agglutinate, and so the absence of agglutination indicates pregnancy. And even if you never have to take a home pregnancy test, you still might be subject to an assay involving agglutination inhibition. Many jobs require prospective employees to take drug tests, and agglutination inhibition assays can be used to determine if an individual is using certain types of illegal drugs such as cocaine or heroin.

(a)

| Red blood cells | Anti-Rh antibodies | Cell-antibody complexes | Rabbit anti-human immunoglobulins | Agglutinated red blood cells |

(b)

Figure 18.33 The Coomb's antiglobulin test. **(a)** Anti-Rh antibodies are allowed to react with red blood cells. If Rh antigens are present on the blood cells, there are not enough of them to produce a hemagglutination reaction. **(b)** Therefore, anti-human antibodies prepared in rabbits are reacted with the red blood cell–antibody complexes. If Rh antigens are present on the red blood cells, hemagglutination will occur. A person with these red blood cells would be Rh-positive.

to detect very small quantities of antibodies. The **complement fixation test** is a multistep procedure that begins with the inactivation of complement from a patient's serum by heating. The serum is then diluted, and known quantities of nonhuman complement and the test antigen are added separately (**Figure 18.34a**). The antigen is specific to the antibody being sought. This mixture is incubated to allow the antigen to react with any antibody present. Next, an indicator system typically consisting of sheep red blood cells and antibody against those cells are

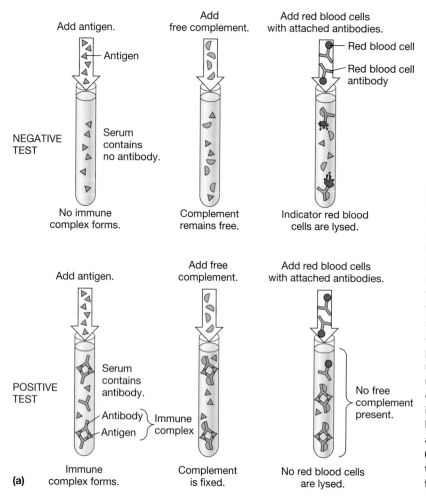

(a)

NEGATIVE TEST

Add antigen. — Antigen — Serum contains no antibody. — No immune complex forms.

Add free complement. — Complement remains free.

Add red blood cells with attached antibodies. — Red blood cell — Red blood cell antibody — Indicator red blood cells are lysed.

POSITIVE TEST

Add antigen. — Serum contains antibody. — Antibody } Immune — Antigen } complex — Immune complex forms.

Add free complement. — Complement is fixed.

Add red blood cells with attached antibodies. — No free complement present. — No red blood cells are lysed.

(b)

Figure 18.34 The complement fixation test for antibodies. **(a)** In the first step, the serum to be tested is diluted, and antigen to the antibody being sought is added. If the antibody is present, it reacts with the antigen and forms immune complexes. In the second step, free complement is added. If immune complexes have formed, the complement will interact with them and be fixed; if no immune complexes have formed, the complement will remain free. In the third step, sheep red blood cells with bound red blood cell antibody molecules are added. If free complement is present, it will lyse the red blood cells. This is a negative test result—which indicates that there was no antibody in the original serum. If all the complement has already been fixed by the earlier immune complex, the red blood cells will not be lysed. This is a positive test result—it indicates that antibody was present in the original serum specimen. **(b)** In a positive test, complement has been fixed, and the red blood cells will not be lysed. Instead, they form a characteristic red button in the bottom of the well. *(Leon J. LeBeau/Biological Photo Service/PO).*

added. If the antibody to the test antigen was present in the patient's serum, the antigen-antibody reaction will have fixed (combined with) the complement. Hence the blood cells will not be lysed, and the test will be positive, forming a red button of undamaged cells (**Figure 18.34b**). But if the antibody was not present, free complement remaining in the mixture will be fixed by the indicator system, resulting in the lysis of the cells. The test will be negative. The complement fixation test is used to diagnose such bacterial diseases as pertussis and gonorrhea and fungal diseases, including histoplasmosis. The Wassermann test for syphilis uses complement fixation.

Neutralization reactions can be used to detect bacterial toxins and antibodies to viruses. Immunity to diphtheria, which depends on the presence of diphtheria antitoxins (antibodies to diphtheria toxin), can be detected by the **Schick tests**. In this test a person is inoculated with a small quantity of diphtheria toxin. If the person is immune to the disease, diphtheria antitoxins (circulating in the blood) will neutralize the toxin, and no adverse reaction will occur. If the person is not immune and the antitoxin is not present, the toxin will cause tissue damage, detected as a swollen reddened area at the injection site after 48 hours.

Viral neutralization occurs when antibodies bind to viruses and neutralize them, or prevent them from infecting cells. In the laboratory or clinic, a patient's serum and a test virus are added to a cell culture or a chick embryo. If the serum contains antibodies to the virus, these antibodies will neutralize the virus and prevent the cells of the culture or the embryo from becoming infected.

Tagged Antibody Tests

The most sensitive immunological tests used to detect antibodies or antigens use antibodies that have a "molecular tag" that is easy to detect even at very low concentrations. In fact, the concentrations are so low that precipitation or agglutination does not occur.

Immunofluorescence makes use of antibodies (usually IgG) to which fluorescent dye molecules are bound (tagged) at the tail (Fc) ends of the antibodies. For example, IgG antibodies tagged with fluorescein isothiocyanate glow a bright yellow green when exposed to ultraviolet (UV) light. Fluorescent tagged antibodies can be used to detect antigens, other antibodies, or complement at their locations on cells or within tissues (**Figure 18.35**).

Figure 18.35   Immunofluorescence. Fluorescein is a fluorescent dye molecule that can be complexed with other molecules. When viewed with ultraviolet (UV) light under a fluorescence microscope, it will fluoresce, revealing the presence of the "tagged" molecule. To detect directly the presence of a specific antigen in a tissue, **(a)** a solution of fluorescein-tagged antibody to that antigen is prepared, added to cells or to a thin section of tissue, incubated, and then washed. Any dye-tagged antibody that has complexed with antigen in tissue will fluoresce when viewed by fluorescence microscopy. **(b)** In indirect testing the antibody to the antigen being sought is not itself tagged. Instead, its presence is detected by means of a fluorescein-tagged antibody (anti-antibody) to the original antibody. **(c)** Complement (protein C3) can be added to the tissue section along with the antibody, and a fluorescein-tagged antibody to one of the complement proteins can then be used to detect the presence of antigen-antibody complexes or complement attached to cells. **(d)** Immunofluorescent staining of influenza virus–infected lung cells (4,722X). The cells were treated with fluorescent dye–tagged antibodies against the specific virus. The nucleus of infected cells is bright yellow green, indicating the location of the viral antigen. (*Courtesy George A. Wistreich, East Los Angeles College*).

Because such cells or tissue samples can be examined with a fluorescence microscope, this technique is particularly useful in the research laboratory to locate cellular antigens and autoantibodies. A fluorescent tagged antibody that detects another antibody is known as an *anti-antibody*; one that detects complement is an *anti-complement antibody*. Immunofluorescence is helpful in diagnosing syphilis, gonorrhea, HIV infection, Legionnaires' disease, chlamydial cytomegalovirus, *Cryptosporidium* and fungal infections, immune complexes of IgA in renal biopsies, and the SARS virus, to name a few.

Fluorescence-Activated Cell Sorter (FACS)

Sometimes it is necessary to collect quantities of a particular cell type for study, for example, CD4 or CD8 T cells and their ratios to help assess progression of the disease in AIDS patients. This can be done under sterile conditions by using the **fluorescence-activated cell sorter (FACS)** (Figure 18.36), which is basically a modification of a machine called a flow cytometer. A series of droplets, each containing one cell, flows out of a nozzle. If the cell is fluorescent when struck by a laser beam of ultraviolet light, a detector will activate electrodes that give an electric charge to the droplet. Falling through an electromagnetic field, the charged droplets will be deflected into a separate container. As cells fall past the laser, they are sorted and counted by types.

Radioimmunoassay (RIA) also can be used to detect very small quantities (nanograms) of antigens and antibodies. To measure an antibody in a test specimen by RIA, a known antigen is placed in a saline (salt) solution and incubated in plastic well plates (**Figure 18.37a**). Some antigen molecules attach to the plastic; those that do not attach are washed away. The antibody being measured is added and allowed to bind with the antigen (**Figure 18.37b**). Then a radioactively tagged anti-antibody is applied (**Figure 18.37c**). After an incubation period, the excess unbound antibody is removed by washing. Radioactive material remaining in the well is measured with a radiation counter to determine the concentration of antibody in the test specimen.

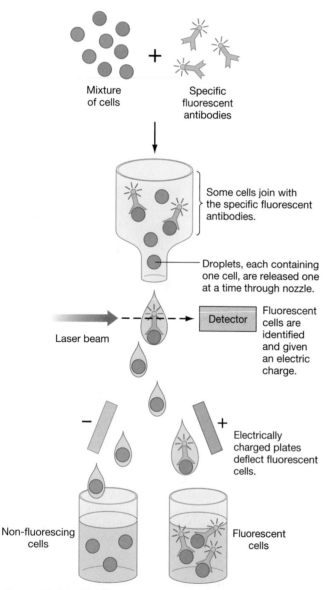

Figure 18.36   The fluorescence-activated cell sorter (FACS).

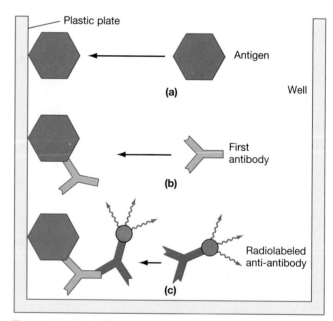

Figure 18.37   Radioimmunoassay (RIA) is used to detect very small quantities of antibody. (a) Antigen first is bound to a well of a plastic plate. After excess unbound antigen is washed out, the solution being tested for antibody is added and allowed to react. **(b)** If antibody is present, it reacts with the antigen. **(c)** After any unbound antibody is washed out, a radioactively labeled second antibody (anti-antibody) specific to the first antibody is added. The amount of bound radioactive anti-antibody present is measured; it is proportional to the concentration of the antibody in the original solution.

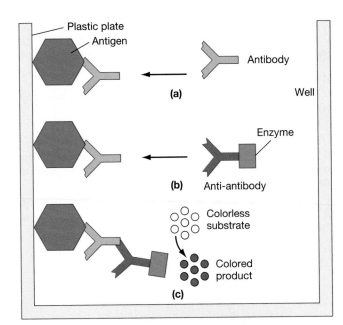

Figure 18.38 Enzyme-linked immunosorbent assay (ELISA) is a modification of RIA. **(a)** As in RIA, antigen bound to a well of a plastic plate reacts with the antibody being detected. **(b)** In one form of the ELISA, an anti-antibody is then added. However, rather than being radioactively labeled as in RIA, this antibody has a covalently attached enzyme. **(c)** A substrate specific to the enzyme is then added. If this enzyme has bound to the original antibody, it can catalyze a reaction, converting the colorless substrate into a color product. Such ELISA tests are done routinely as an initial test to detect HIV in blood samples.

Enzyme-linked immunosorbent assay (ELISA) is a modification of RIA in which the anti-antibody, instead of being radioactive, has an enzyme tag attached to it (**Figure 18.38a**). After the antibody being measured has reacted with the antigen, the anti-antibody–enzyme complex is added (**Figure 18.38b**). Finally, a substrate

that the enzyme converts to a colored product is applied (**Figure 18.38c**). The amount of colored product is proportional to the concentration of the antibody. RIA and ELISA are among the most widely used tests for antibodies or for antigens.

An important application of ELISA is the detection of HIV antibodies, generally within 6 weeks of infection. The test was developed to screen the U.S. blood supply and protect recipients of blood products from infection. ELISA is a sensitive test—so sensitive that the American Red Cross Blood Services reports that false-positive results (color product is detected when HIV antibodies are not present) occur at a rate of about 0.2%.

To confirm the presence of an HIV infection, a more expensive test, called **Western blotting**, can be performed on samples from an individual who has a positive ELISA test. In Western blotting, HIV proteins are isolated from the individual and are first separated in a gel by an electric current, similar to the procedure used in immunoelectrophoresis. The separated proteins are then transferred ("blotted") to cellulose filter paper. Next, serum from the individual is added to the blot. If HIV antibodies are present, they will react with the separated HIV proteins. Such antigen-antibody complexes can be visualized by the addition of an enzyme-labeled antihuman antibody. When an enzyme substrate is added, colored bands appear on the paper (**Figure 18.39**). Thus, Western blotting can determine the exact viral antigens to which the HIV antibodies are specific.

✓**CHECKLIST**

1. How do precipitin reactions differ from agglutination reactions?
2. What is meant by antibody titer?
3. Compare RIA, ELISA, and Western blotting tests.

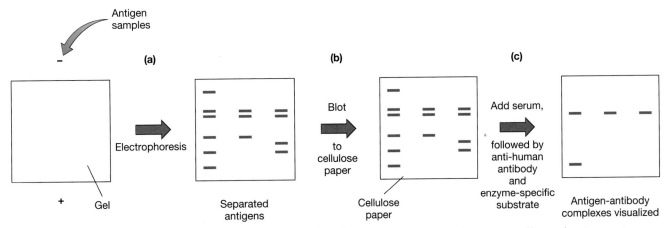

Figure 18.39 Western blotting test for HIV antigens in blood. **(a)** HIV antigens in a gel are separated by an electric current, forming bands of separate antigens. **(b)** The antigen bands are transferred (blotted) to cellulose paper. **(c)** HIV antibodies tagged with dye are added. Any antibodies recognizing specific HIV antigens attach to that antigen and form a visible band.

▌ RETRACING OUR STEPS

▐ OVERVIEW OF IMMUNOLOGICAL DISORDERS

- An **immunological disorder** results from an exaggerated or an inadequate immune response.

HYPERSENSITIVITY

- **Hypersensitivity**, or **allergy**, is due to an exaggerated reaction to an antigen.

IMMUNODEFICIENCY

- **Immunodeficiency** is due to an inadequate immune response. **Primary immunodeficiencies** are genetic or developmental defects resulting in lack of, or defective B or T cells, or both. **Secondary immunodeficiencies** result from damage to B or T cells after they have developed normally.

▐ IMMEDIATE (TYPE I) HYPERSENSITIVITY

- **Anaphylaxis** is due to harmful effects caused by an immediate, exaggerated immune response.

ALLERGENS

- An **allergen** is an ordinarily innocuous substance that can trigger a harmful immunological response in an allergen-sensitized person; common allergens are listed in Table 18.1.

MECHANISM OF IMMEDIATE HYPERSENSITIVITY

- The mechanism of **immediate hypersensitivity** is summarized in Figure 18.1; mediators of immediate hypersensitivity are summarized in Table 18.2.
- In the **sensitization** process, the affected individual produces IgE antibodies that attach to mast cells and basophils. A second or later exposure to the same allergen will trigger cellular **degranulation** with the release of **histamine** and other preformed mediators. The synthesis of reaction mediators also is responsible for symptoms of immediate hypersensitivity.

LOCALIZED ANAPHYLAXIS

- **Atopy** is a localized reaction to an allergen in which histamine and other mediators elicit wheal and flare reactions and other signs and symptoms of allergy.

GENERALIZED ANAPHYLAXIS

- **Generalized anaphylaxis** is a systemic reaction in which the airways are constricted (**respiratory anaphylaxis**) or blood pressure is greatly decreased (**anaphylactic shock**).

GENETIC FACTORS IN ALLERGY

- Genetic factors are thought to contribute to the development of allergies.

TREATMENT OF ALLERGIES

- Allergy is treated by **desensitization** (hyposensitization), as shown in Figure 18.5; symptoms are alleviated with antihistamines.

▐ CYTOTOXIC (TYPE II) HYPERSENSITIVITY

MECHANISM OF CYTOTOXIC REACTIONS

- The mechanism of **cytotoxic hypersensitivity** is summarized in Figure 18.6.

EXAMPLES OF CYTOTOXIC REACTIONS

- **Transfusion reactions** result in red blood cell **hemolysis** due to the production of antibodies to antigens on transfused red blood cells.
- **Hemolytic disease of the newborn** results when anti-Rh antibodies from a sensitized mother react with Rh antigens in an Rh positive fetus.

▐ IMMUNE COMPLEX (TYPE III) HYPERSENSITIVITY

MECHANISM OF IMMUNE COMPLEX DISORDERS

- The mechanism of **immune complex hypersensitivity** is summarized in Figure 18.9.

EXAMPLES OF IMMUNE COMPLEX DISORDERS

- **Serum sickness** occurs when foreign (sometimes animal) antigens in sera combine with antibody to form immune complexes, which are deposited in various tissues.
- The **Arthus reaction** is a local immune response to an antigenic substance (usually an injected substance) that causes edema and hemorrhage.

▐ CELL-MEDIATED (TYPE IV) HYPERSENSITIVITY

MECHANISM OF CELL-MEDIATED REACTIONS

- The mechanism of **cell-mediated (delayed) hypersensitivity** is summarized in Figure 18.12.

EXAMPLES OF CELL-MEDIATED DISORDERS

- **Contact dermatitis** occurs after second contact with poison ivy, metals, or another substance and usually appears as eczema.
- Other kinds of cell-mediated hypersensitivity include **tuberculin hypersensitivity** and **granulomatous hypersensitivity**.
- Characteristics of the four types of hypersensitivities are summarized in Table 18.5.

▐ AUTOIMMUNE DISORDERS

- **Autoimmune disorders** arise from a hypersensitivity to self antigens on cells and in tissues. Such diseases often produce **autoantibodies** to these self antigens. Autoimmune diseases can also be cell-mediated.

AUTOIMMUNIZATION

- **Autoimmunization** occurs when the immune system responds to a body component as if it were foreign. Genetic factor and **antigenic mimicry** are among the mechanisms that can cause autoimmune disease.
- Tissue damage from autoimmune disorders can be caused by cytotoxic, immune complex, and cell-mediated hypersensitivity reactions.

EXAMPLES OF AUTOIMMUNE DISORDERS

- Examples of autoimmune diseases include **myasthenia gravis**, **rheumatoid arthritis**, and **systemic lupus erythematosus**.

III TRANSPLANTATION

- **Transplantation** involves the transfer of **graft tissue** from one site to another on the same individual (**autograft**), between genetically identical individuals (**isograft**), from one individual to another nonidentical individual (**allograft**), or between different species (**xenograft**).

HISTOCOMPATIBILITY ANTIGENS

- Genetically determined **histocompatibility antigens** are found on the surface membranes of all cells. Some are correlated with increased risk of certain diseases.
- **Human leukocyte antigens** (HLAs) in graft tissue are a main cause of **transplant rejection**, but immunocompetent cells in bone marrow or other types of grafts sometimes destroy host tissue, as in **graft-versus-host (GVH) disease**.

TRANSPLANT REJECTION

- Transplant rejection is due to the presence of foreign HLAs. **Hyperacute rejection** is a Type II hypersensitivity, occurring in an already sensitized recipient. **Accelerated rejection** reactions are mainly cell-mediated (Type IV) hypersensitivities. **Chronic rejection** occurs over several months to years.

TOLERANCE OF THE FETUS DURING PREGNANCY

- The fetus occupies an "immunologically privileged site." Cells on the fetal part of the placenta do not express MHC molecules. Various molecules prevent maternal NK cells from killing fetal cells.
- "Foreignness" of sperm may trigger maternal production of blocking antibodies that protect the fetus.

IMMUNOSUPPRESSION

- **Immunosuppression** is a lowering of the responsiveness of the immune system to materials it recognizes as foreign. Immunosuppression is produced by **radiation** and by **cytotoxic drugs**. It minimizes transplant rejection but also can reduce the host's immune response to infectious agents.

III DRUG REACTIONS

- All four types of hypersensitivity reactions have been observed in immunological drug reactions.

III IMMUNODEFICIENCY DISEASES

- **Immunodeficiency diseases** arise from an absence of formed lymphocytes and other components of the immune system (**primary immunodeficiency**) or from the destruction of already formed lymphocytes (**secondary immunodeficiency**).

PRIMARY IMMUNODEFICIENCY DISEASES

- B cell deficiency, or **agammaglobulinemia**, leads to a lack of humoral immunity.
- T cell deficiency, such as **DiGeorge syndrome**, leads to a lack of cell-mediated and humoral immunity.
- Deficiencies of both B and T cells, or **severe combined immunodeficiency disease (SCID)**, leads to a lack of both humoral and cell-mediated immunity.

SECONDARY (OR ACQUIRED) IMMUNODEFICIENCY DISEASES

- Secondary immunodeficiencies are acquired through infections, malignancies, or autoimmune disorders.
- **Acquired immune deficiency syndrome (AIDS)** is caused by the **human immunodeficiency virus (HIV)**. HIV destroys T_H cells and eventually impairs all immune functions.
- HIV-infected individuals progress through a series of stages that lead to AIDS. Group 4 individuals suffer from opportunistic infections and malignancies such as **Kaposi's sarcoma**.
- The AIDS pandemic will have over 40 million people infected with HIV by the end of 2007.
- HIV infection can be acquired by: sexual contact, needle sharing, blood transfusions or products, and transfer across the placenta from mother to fetus.
- Health care personnel must practice Universal Precautions at *all* times.
- Vaccine development remains unpromising.
- Social, economic, legal, and ethical problems are becoming increasingly significant.

III IMMUNOLOGICAL TESTS

- **Serology** is the use of laboratory tests to detect antigens and antibodies.

THE PRECIPITIN TEST

- The **precipitin test** can be used to detect antibodies. Modifications of this test include **immunodiffusion**, **immunoelectrophoresis**, and **radial immunodiffusion**. All these tests rely on the formation of antigen-antibody complexes that precipitate from solutions or in agar gels.

AGGLUTINATION REACTIONS

- **Agglutination reactions** depend on the clumping of antigen and antibody combinations. Such tests can be used to determine one's **antibody titer** or if **seroconversion** has occurred.
- **Hemagglutination** includes the clumping of red blood cells by viruses and the binding of antibodies to specific antigen-bearing red blood cells. The **hemagglutination inhibition test** can be used to diagnose for measles or diseases caused by other viruses. **Coomb's antiglobulin test** is used to detect Rh antibodies.
- The **complement fixation test** indirectly detects antibodies in serum to antigens by determining whether complement combines (is fixed) with antigen-antibody complexes.
- **Neutralization reactions** can detect bacterial toxins and antibodies to certain viruses.

TAGGED ANTIBODY TESTS

- **Immunofluorescence** allows detection of products of immune reactions on cells or within tissues.
- **Fluorescence-activated cell sorter (FACS)** allows cells to be sorted, collected, and counted by type.
- Assays for antigens and antibodies can be performed with radioactive antibodies (**radioimmunoassay**) or antibodies containing enzymes coupled to them (**enzyme-linked immunosorbent assay, or ELISA**). **Western blotting** detects specific antibodies to specific antigens.

▌ TERMINOLOGY CHECK

ABO blood group system
(*p. 539*)

acquired immune deficiency
syndrome (AIDS) (*p. 555*)

agammaglobulinemia (*p. 554*)

agglutination (*p. 562*)

agglutination reaction (*p. 562*)

allergen (*p. 533*)

allergy (*p. 532*)

allograft (*p. 550*)

anaphylactic shock (*p. 536*)

anaphylaxis (*p. 532*)

antibody titer (*p. 561*)

antigenic mimicry (*p. 547*)

Arthus reaction (*p. 543*)

asthma (*p. 536*)

atopy (*p. 535*)

autoantibody (*p. 546*)

autograft (*p. 550*)

autoimmune disorder (*p. 546*)

autoimmunization (*p. 547*)

blocking antibody (*p. 538*)

cell-mediated (Type IV)
hypersensitivity (*p. 532*)

complement fixation test
(*p. 564*)

contact dermatitis (*p. 544*)

Coomb's antiglobulin test
(*p. 563*)

cytotoxic drug (*p. 553*)

cytotoxic (Type II)
hypersensitivity (*p. 532*)

degranulation (*p. 534*)

delayed (Type IV)
hypersensitivity (*p. 532*)

delayed hypersensitivity T
(T_{DH}) cell (*p. 544*)

desensitization (*p. 538*)

DiGeorge syndrome
(*p. 554*)

electrophoresis (*p. 561*)

enzyme-linked
immunoabsorbent assay
(ELISA) (*p. 567*)

fluorescence-activated cell
sorter (FACS) (*p. 566*)

generalized anaphylaxis
(*p. 533*)

graft tissue (*p. 550*)

graft-versus-host (GVH)
disease (*p. 550*)

granulomatous
hypersensitivity (*p. 545*)

hemagglutination (*p. 563*)

hemagglutination inhibition
test (*p. 563*)

hemolytic disease of the
newborn (*p. 540*)

histamine (*p. 534*)

histocompatibility antigen
(*p. 550*)

human immunodeficiency
virus (HIV) (*p. 555*)

human leukocyte antigen
(HLA) (*p. 550*)

hypersensitivity (*p. 532*)

immediate (Type I)
hypersensitivity (*p. 532*)

immune complex (Type III)
hypersensitivity (*p. 532*)

immunodeficiency (*p. 552*)

immunodeficiency disease
(*p. 553*)

immunodiffusion test
(*p. 561*)

immunoelectrophoresis
(*p. 561*)

immunofluorescence (*p. 565*)

immunological disorder
(*p. 532*)

immunosuppression (*p. 552*)

induration (*p. 545*)

isograft (*p. 550*)

Kaposi's sarcoma (*p. 558*)

leukotriene (*p. 535*)

localized anaphylaxis
(*p. 533*)

major histocompatibility
complex (MHC) (*p. 550*)

myasthenia gravis (*p. 547*)

neutralization reaction
(*p. 565*)

precipitation reaction
(*p. 561*)

precipitin test (*p. 561*)

primary immunodeficiency
(*p. 532*)

primary immunodeficiency
disease (*p. 554*)

prostaglandin (*p. 535*)

radial immunodiffusion
(*p. 562*)

radioimmunoassay (RIA)
(*p. 566*)

reagin (*p. 533*)

respiratory anaphylaxis
(*p. 536*)

Rh antigen (*p. 540*)

rheumatoid arthritis (RA)
(*p. 548*)

rheumatoid factor (*p. 549*)

Schick test (*p. 565*)

secondary immunodeficiency
(*p. 532*)

secondary immunodeficiency
disease (*p. 554*)

sensitization (*p. 533*)

seroconversion (*p. 561*)

serology (*p. 560*)

serum sickness (*p. 542*)

severe combined
immunodeficiency (SCID)
(*p. 554*)

systemic lupus erythematosus
(SLE) (*p. 549*)

transfusion reaction
(*p. 539*)

transplantation (*p. 550*)

transplant rejection (*p. 550*)

tube agglutination test
(*p. 562*)

tuberculin hypersensitivity
(*p. 545*)

tuberculin skin test (*p. 545*)

viral hemagglutination
(*p. 563*)

viral neutralization (*p. 565*)

Western blotting (*p. 567*)

xenograft (*p. 550*)

▌ CLINICAL CASE STUDY

A woman complains that whenever she goes to the dentist a few hours after her visit she experiences swelling of her face and mouth. What might be causing this reaction? What type of hypersensitivity reaction is this? What might her dentist do to prevent this problem? To learn more about this go to the following web site, http://www.shef.ac.uk/~arrp/case/case1-2.html.

▌ CRITICAL THINKING QUESTIONS

1. Why is it safer to state that a condition is "believed to be" an autoimmune disorder rather than just saying that it "is" an autoimmune disorder?

2. (a) If an Rh-negative mother carries three consecutive Rh-positive babies, why would only the first born survive and subsequent babies die?

(b) Can anything be done to prevent the deaths of subsequent babies?

3. As a health care worker, what would you do differently when handling the blood of someone you think might be infected with HIV as opposed to handling the blood of someone else?

■ SELF-QUIZ

1. Match the following immunological disorder terms to their descriptions:

___ Allergen
___ Primary immunodeficiency
___ Anaphylaxis
___ Atopy
___ Desensitization
___ Secondary immunodeficiency
___ Autoantibodies

(a) Antibodies against one's own tissues
(b) Injection of denatured allergens include tolerance and production of blocking IgG antibodies
(c) Localized reaction to allergen eliciting wheal and flare reaction and other signs and symptoms of allergy
(d) Results from damage to B or T cells after their normal development
(e) Genetic or developmental defects resulting in lack of or defective B or T cells or both
(f) Immediate, exaggerated allergic reaction to antigens
(g) Ordinary harmless foreign substance that can cause an exaggerated immunological response

2. Which of the following describes Type I immediate hypersensitivity?
(a) After initial sensitization to antigen, subsequent exposure to the sensitizing zntigen results in formation of antigen–IgG antibody complexes. These immune complexes deposit in tissues where they activate complement.
(b) Specific antibodies react with host cell surface antigens interpreted as foreign by the immune system, leading to phagocytosis, killer cell activity, or complement-mediated lysis. Typical sources of such antigens come form mismatched blood transfusions or Rh incompatibility between infant and mother.
(c) This rection is delayed and mediated by sensitized T_H1 cells that release cytokines responsible for the hypersensitivity reaction.
(d) Sensitization to allergen causes B cell activation resulting in production of anti-allergen IgE antibodies which attach to mast cells and basophils. Subsequent exposure to that allergen causes cross-linking of IgE antibodies resulting in deregulation of chemicals that induce allergic responses
(e) Both a and c.

3. Which of the following bind to mast cells and cross-link, resulting in degranulation and release of histamine?
(a) IgM
(b) IgA
(c) IgG
(d) Interleukins
(e) IgE

4. Desensitization to prevent a Type I allergic response is accomplished by stimulating the body to make:
(a) IgE antibodies and antigen presenting cells
(b) Antigen (allergen)
(c) Histamine
(d) IgG antibodies and regulatory T cells
(e) Antihistamine

5. A transfusion reaction ending in red blood cell hemolysis is due to an infusion of donor red cells carrying:
(a) Arthus reaction antigens
(b) Mismatched or different blood type antigen(s)
(c) Processed T cell antigens
(d) All of the above
(e) None of the above

6. An individual who got stung by a bee exhibited airway construction and an abrupt drop in blood pressure. Such an individual would have which of the following:
(a) Immune complex hypersentivity
(b) Antigenic mimicry
(c) Generalized anaphylaxis
(d) Atopy
(e) Arthus reaction

7. Type IV hypersensitivity is the only allergic reaction mediated by:
(a) NK cells
(b) Macrophages
(c) B cells
(d) Antibodies
(e) T cells

8. Production of autoantibodies may be due to:
(a) Emergence of mutant clones of B cells
(b) Production of antibodies against sequestered (hidden) tissues
(c) Genetic factors
(d) All are possible
(e) None of these

9. The formation of foreign antigen and antibody complexes in serum with subsequent deposition in the tissues is known as _____ while a local immune response to an antigenic substance that causes edema and hemorrhaging is known as _____. Both are examples of _____ disorders.

10. Match the following autoimmune disorders to their autoantibody targets and tissues affected:

___ Rheumatic fever
___ Rheumatoid arthritis
___ Ulcerative colitis
___ Myasthenia gravis
___ Scleroderma
___ Pernicious anemia
___ Systemic lupus erythematosus

(a) Cell nuclei and histones, many tissues
(b) Nucleoli, connective tissues
(c) Streptococcal cross-reactivity with heart, heart
(d) Vitamin B_{12} binding site, stomach
(e) Cell nuclei and gamma globulins, joints
(f) Colon cells, colon
(g) Acetylcholine receptor, skeletal muscle

11. Which of the following is NOT an example of Type IV cell-mediated hypersensitivity disorder?
(a) Anaphylactic shock
(b) Contact dermatitis

(c) Tuberculin hypersensitivity
(d) Granulomatous hypersensitivity
(e) Listeriosis

12. A tissue graft between two people who are not genetically identical is termed a:
(a) Isograft
(b) Homograft
(c) Endograft
(d) Enterograft
(e) Allograft

13. Major histocompatibility antigens (MHCS) are called human leukocyte antigens (HLAs) in humans and are a main cause of transplant rejection. True or false?

14. Graft-versus-host disease results when the recipient lacks or has a poor immune system, and the donor organ and recipient express different:
(a) HLA
(b) T cells
(c) Antibodies
(d) Autoantibodies
(e) Interleukins

15. Immunosuppression is a lowering of the responsiveness of the immune system to materials it recognizes as foreign and is produced by _____ and _____.
(a) Antibodies and foreign antigens
(b) NK cells and allergen desensitizers
(c) Radiation and cytotoxic drugs
(d) Hypersensitivity and transplant rejection
(e) a and d

16. Primary immunodeficiency diseases are caused by genetic defects in embryological development of lymphoid tissue resulting in a lack of T cells or B cells, or defective T and B cells. Which of the following diseases is a primary immunodeficiency disease?
(a) DiGeorge syndrome
(b) Severe combined immunodeficiency disease (SCID)
(c) Agammaglobulinemia
(d) All of the above
(e) None of the above

17. Human immunodeficiency virus (HIV) binds specifically to which immune cell marker?
(a) CD8
(b) MHC
(c) CDC
(d) CD4
(e) GP120

18. A person with AIDS will probably:
(a) Not make antibody
(b) Make a response to T-dependent antigens
(c) Make antibody to T-independent antigens
(d) Have large numbers of T-helper cells
(e) None of the above

19. HIV has a high mutation rate due to the imprecise operation of its:
(a) Viral membrane
(b) CD4 receptor
(c) Reverse transcriptase
(d) Protease
(e) Dismutase

20. Pregnancy tests detect the presence of which of the following?
(a) Rh
(b) Human chorionic gonadotropin (HCG)
(c) Fetal proteins
(d) Agglutination
(e) Depuration factor

21. Which of the following assays or devices depend on easily detectable "tagged" antibodies?
(a) Fluorescence-activated cell sorter (FACS)
(b) Enzyme-linked immunosorbent assay (ELISA)
(c) Radioimmunoassay (RIA)
(d) Western blot
(e) All of the above

22. All cells in the body display which class of major histocompatibility complex?
(a) Class I
(b) Class II

23. A transplant between individuals of different animal species is termed a(n):
(a) Allograft
(b) Isograft
(c) Enterograft
(d) Endograft
(e) Xenograft

24. Match the following immunological tests to their descriptions:

___ Immunoelectrophoresis
___ Hemagglutination inhibition
___ Neutralization reaction
___ Complement fixation test
___ Precipitin test
___ Coomb's antiglobulin test

(a) Viral cross-linking of red blood cells is inhibited by antiviral antibodies
(b) Used for detection of bacterial toxins and antibodies to viruses.
(c) Electric current through an agarose gel is used to detect separate antigen-antibody complexes based on their size and charge
(d) Used to detect antibodies; applications include immunodiffusion, immunoelectrophoresis, and radial diffusion
(e) Anti-human antibodies are used to detect low titers of anti-Rh antibodies bound to Rh antigens on red blood cells
(f) Indirectly detects antibodies in serum to antigens by determining whether complement combines (is fixed) with antigen-antibody complexes

25. Two tests that are used to detect the presence of HIV infection are:
(a) Agglutination and neutralization reactions
(b) Complement fixation and immunofluorescence tests
(c) Radioimmunoassay and immunofluorescence test

(d) Enzyme-linked immunosorbent assay (ELISA) and Western blotting

(e) Hemagglutination and Coomb's antiglobulin tests

26. Diagram _____ represents an immunodiffusion test in which only one antigen-antibody complex is present; in diagram _____, two different antigen-antibody complexes are present.

(a)

(b)

▌ EXPLORATIONS ON THE WEB

http://www.wiley.com/college/black

If you think you've mastered this chapter, there's more to challenge you on the web. Go to the companion web site to fine-tune your understanding of the chapter concepts and discover answers to the questions posed below.

1. They lurk under our beds and inside our sofas and carpets. Microscopic dust mites are everywhere and may well be responsible for your sneezing and allergies or even more serious symptoms like difficulty breathing or asthma attacks. Learn more at the web site.

2. If my sibling and I have the same blood group, then why can't I give her one of my kidneys? Find out why at the web site.

Diseases of the Skin and Eyes; Wounds and Bites

[zoster vaccine live (Oka/Merck)]
ZOSTAVAX®
STORE FROZEN

1 Single-dose Vial (0.65

No. 4309 | 0.

DILUEN
& CO., INC
VACCINES
Water)
PRESERVAT

Manuf. and Dist. by:
MERCK & CO., INC.
Whitehouse Station, NJ 088

(Courtesy Merck & Company)

Come with me ...

Sister Peter was the head librarian. How she loved being there where she could read all the latest books as they came in, and reread her old favorites. Then, one day she began to have itching, shooting and stabbing pains in the skin across her face. It worsened. Small blisters broke out. The painful rash ran across her eyes. It was shingles, a reactivation of the chickenpox virus which had been sitting in an inactive, latent, state over the many years since she had had a childhood case of chickenpox. Now elderly, Sister Peter's immune system had weakened, no longer able to keep the virus suppressed. The virus was moving out inside nerve fibers, causing the horrible pain and damaging the nerves. When it was over, Sister Peter was blind, never to read her beloved books again, nor to see the sky or friends' faces.

Mrs. Klein had a different experience. The stabbing pains shot through her hands, like knives constantly slashing through her fingers, alternating with electrical shocks. But when the rash left her, the pains didn't. They continued undiminished for the next 10 years—the rest of her life. How to tie your shoelaces, write, use a knife and fork, or even just sleep with such pain? It is called postherpetic neuralgia; herpetic because the chickenpox virus belongs in the herpes family of viruses. There is no good treatment for postherpetic neuralgia. It can last for months or years. Some patients unable to stand the relentless excruciating pains, year after year, eventually commit suicide.

A generation later, Mrs. Klein's daughter developed shingles around the waist, and had to switch to wearing loose dresses for the rest of her life. Now, granddaughter Elizabeth has a chance to escape such pains. There is a shingles vaccine (page 585). Come with me to the National Institutes of Health (NIH) where they are doing research on combating postherpetic neuralgia.

 Video related to this topic is available within WileyPLUS.

(P. Marazzi/Photo Researchers, Inc.)

Having considered the general principles of microbiology in the first five units of this text, we are now ready to apply those principles to understanding infectious diseases in humans. As you study diseases, keep in mind the information on normal microflora and disease processes in ◀Chapter 14, and on epidemiology in ◀Chapter 15.

THE SKIN, MUCOUS MEMBRANES, AND EYES

THE SKIN

The skin is the largest single organ of the body. The physical barrier it presents to most microbes is an important nonspecific defense. The surface consists of a thin **epidermis** and a thicker, underlying layer, the **dermis** (Figure 19.1). Lacking blood vessels, the epidermis is nourished by nutrients that diffuse from blood vessels in the dermis. The epidermis has several layers of dead *epithelial* (ep-i-the'le-al) *cells* that function as an excellent barrier against injury and infection of deeper layers of the body. Cells next to the dermis divide throughout the life of an individual and migrate toward the surface, where old cells are sloughed off. Thus the epidermis is renewed every 15 to 30 days. These cells contain a water-proofing protein called **keratin**, which prevents water-soluble substances from entering the body. In addition, thick skin on the palms of the hands and soles of the feet reduces the likelihood that the skin barrier will be broken. *Calluses* represent thickened areas subject to constant wear. Thus, the intact skin surface prevents pathogenic organisms and other foreign substances from entering the body.

Not only is the skin a physical barrier to infection, its surface is inhabited by a variety of normal microflora. They occupy so much of the surface that it is difficult for pathogens to find a site to attach to and colonize (◀Chapter 14, p. 404).

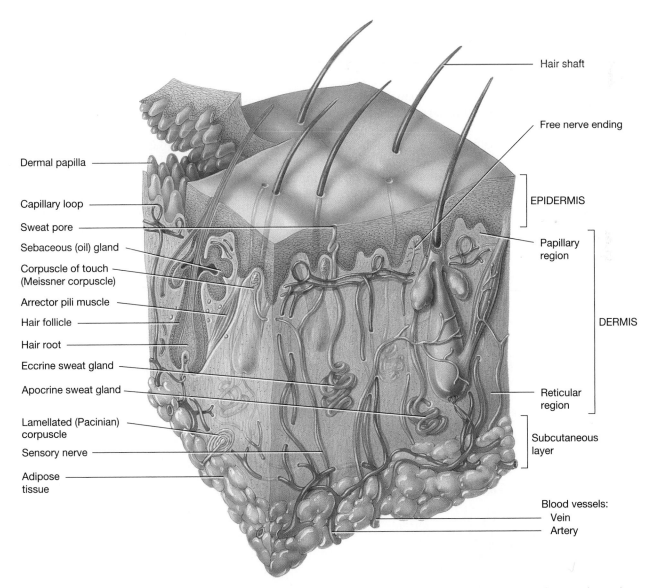

Figure 19.1 The skin. Consisting of the epidermis and underlying dermis, the skin forms a mechanical barrier to infection. Chemicals secreted onto the skin also retard pathogen attack and infection.

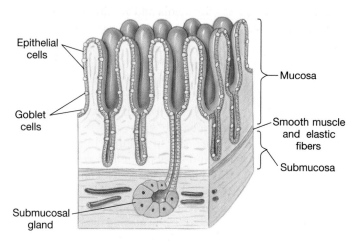

Figure 19.2 Mucous membranes. Much like the skin, mucous membranes consist of epithelial cells that line tissues. The secretion of mucus by goblet cells (special epithelial cells) and glands traps and clears away many potential pathogens and other foreign material.

The skin generates antimicrobial substances that make it an even more inhospitable environment. Embedded in the dermis, the durable part of the skin, are sebaceous (se-ba'shus) glands and sweat glands. Most **sebaceous glands** produce an oily secretion called **sebum** (se'bum), which consists mainly of organic acids and other lipids. Although these secretions provide some nutrients for the normal microflora, the acidic nature of the sebum helps maintain an acidic skin pH that discourages pathogen growth. The quantity of sebum secretions increases at puberty, contributing to the development of acne in some persons. **Sweat** (sudoriferous) **glands** are distributed over the body and empty a watery secretion through pores in the skin. In the armpits and groin, these glands secrete organic

substances in the sweat that lower the skin pH, again inhibiting the growth of most pathogens. The high salt concentration in sweat also inhibits many microorganisms.

MUCOUS MEMBRANES

Mucous membranes line those tissues and organs that open to the exterior of the body, especially those of the respiratory, digestive, and urogenital systems. Like the skin, mucous membranes have a thin epidermal layer and a deeper connective tissue layer (**Figure 19.2**). The cells of the epithelial layer block the penetration of microbes and secrete **mucus**, a thick but watery secretion of glycoproteins and electrolytes. Mucus, produced by *goblet cells* and certain glands, forms a protective layer over the epidermal cells, preventing drying and cracking of the mucous membrane. Mucus also traps pathogens before they can establish an infection. The pH of mucus (e.g., in the vagina) and lysozyme in mucus are good innate immune defense mechanisms. Additionally, secretory IgA in mucus provides adaptive immune defense.

THE EYES

Being exposed to the atmosphere, the eyes contact millions of microbes every day, but have no normal microflora. The eyes have several external protective structures to lessen the chances of an infection. In fact, these structures work so well that deep eye infections are exceedingly rare. Each eye is protected physically by eyelids, eyelashes, the **conjunctiva** (a mucous membrane covering the inner surface of each eyelid and the anterior region of each eye), and the tough **cornea** (the transparent part of the eyeball exposed to the environment) (**Figure 19.3**).

Anterior view of the lacrimal apparatus

Figure 19.3 Structures of the eye. Tears produced over the eye surface contain antibacterial molecules that inhibit potential infections and drain into the nasal cavity. Most eye infections involve the eyelids, conjunctiva, or cornea.

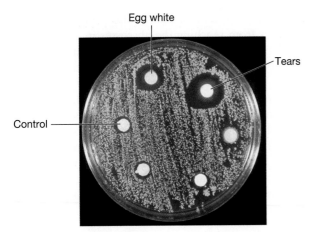

Figure 19.4   Lysozyme in tears. Lysis of *Micrococcus luteus* bacteria by lysozyme enzymes obtained from tears and from egg white. The control disks have been dipped in sterile distilled water. *(Jack M. Bostrack/Visuals Unlimited)*

Eyelashes and eyelids help prevent foreign objects from reaching the cornea.

The eyes also produce antimicrobial substances. Each eye has a **lacrimal gland**, which secretes a lacrimal fluid (tears) that continuously flushes the cornea, keeping it moist as the liquid carries away any microorganisms present. *Lacrimal canals* drain tears from the surface of the eyeball through the *nasolacrimal duct* and into the *nasal cavity*. Tears contain *lysozyme*, an enzyme that breaks down bacterial cell walls. It is especially effective in killing Gram-positive organisms, which have a thick peptidoglycan layer (**Figure 19.4**; ◀Chapter 4, p. 88). Lysozyme has no effect on viruses. Tears, like other body secretions, also contain specific defensive chemicals. Glandular cells in the conjunctiva add a mucous substance to the tears. This substance may help trap microorganisms and eliminate them from the eye.

NORMAL MICROFLORA OF THE SKIN

A huge population of normal microflora colonizes the nearly 2 square meters of skin covering the average human adult. Conditions differ widely at various locations—for example, armpits versus forehead—leading to populations of different compositions in different locations. (See the box "But I Keep Myself So Clean.") Moist areas will support larger populations; oils and sweat provide nutrients. Oily compounds are metabolized by some microbes, yielding fatty acid by-products that accumulate and give an acidic pH (5.5) to certain parts of the skin. This is a first-line, innate defense against infection, since most other microbes are unable to live or grow well at this pH. However, many of these fatty acid compounds have an unpleasant smell and are responsible for body and foot odors (◀Chapter 14, p. 404, Bloodhounds).

Sweat is rich in salt (sodium chloride). The accumulation of salt on the skin is another first-line, innate defense. Relatively few microbes are salt-tolerant. Staphylococci,

however, grow well on salty media and are thus one of the predominant organisms on skin. At least a dozen different species of *Staphylococcus* live on our skin.

A third innate line of defense involves the rapid and continuous shedding of skin cells. As younger cells rise up toward the skin surface, they fill with keratin. By the time they reach the top, the older cells are completely filled with keratin and have died. As they slough off, they carry away the microbes that have colonized them. Keratin-degrading fungi living on the skin sometimes, especially under moist conditions, will multiply too rapidly and cause infections such as athlete's foot.

The majority of organisms living on the skin are Gram-positive organisms such as *Staphylococcus* and *Micrococcus*, and coryneform bacteria (e.g., *Corynebacterium* and *Proprionibacterium acnes*).

One unsuspected inhabitant is the mite *Demodex folliculorum*, which dwells inside the openings of oil glands and down in hair follicles. You might raise your eyebrows to discover this population living in your eyebrows and eyelashes.

TRY IT

But I Keep Myself So Clean!

Think of your skin as a vast continent on which different habitats exist: forests, deserts, swamps, and so on. Not all places are equally hospitable to microbial life. Why not try an unofficial "census" of your skin? Which sites are the deserts, and which are teeming with microbial inhabitants?

Using sterile technique, press a standard-sized piece of sterile velvet or Velcro against a portion of your skin. Then press it against the surface of a Petri dish containing nutrient agar or tryptic soy agar (much as you would when using a stamp pad). Incubate at 37°C for 48 hours, and count the number of colonies growing on the agar. Comparing colony counts grown from standard-sized samples taken at different locations on the skin should reveal your "hot spots" for microbial growth.

Here are some ideas for sampling sites:

- Forehead, nose (inner versus outer surfaces).
- Cheek, armpit, middle of back, forearm.
- Between toes, upper versus lower surfaces of foot.

Other comparisons:

- Male versus female subjects: a shaved cheek compared to an unbearded cheek.
- Armpits: one side with deodorant, the other without.
- Areas of skin with and without makeup (use different sides of one person's face).
- Hands before and after washing. (You should see more colonies after washing; can you think of why?)
- Creases or folds of skin versus adjacent smooth areas.

DISEASES OF THE SKIN

Your skin covers your body and accounts for 15% of your body weight. It provides an effective barrier to invasion by most microbes except when it is damaged (◀Chapter 16, p. 464). Very few microbes can penetrate unbroken skin; however, mucous membranes are more easily entered. Here we will consider the kinds of skin diseases that occur when the skin surface fails to prevent microbial invasion.

BACTERIAL SKIN DISEASES

Many bacteria are found among the normal microflora of the skin. Ordinarily they are kept from invading tissues by an intact skin surface and innate defense mechanisms of the skin. Bacterial and other skin infections usually arise from a failure of these defenses. They are generally diagnosed by appearance and by clinical history.

Staphylococcal Infections

Folliculitis and Other Skin Lesions.   Everyone has had a pimple at some time; most likely it was caused by *Staphylococcus aureus*, the most pathogenic of the staphylococci. Staphylococcal skin infections are exceedingly common because the organisms are nearly always present on the skin. Strains of staphylococci colonize the skin and upper respiratory tract of infants within 24 hours of birth. (Half of all adults and virtually all children are nasal carriers of *S. aureus*.) Infection occurs when these organisms invade the skin through a hair follicle, producing **folliculitis** (fol-lik″u-li′tis), also referred to as **pimples** or **pustules**. An infection at the base of an eyelash is called a **sty**. A larger, deeper, pus-filled infection is an **abscess**; an exterior abscess is known as a **furuncle** (fu′rung-kl), or **boil (Figure 19.5)**. An estimated 1.5 million Americans have such infections annually. Further spread of infection, particularly on the neck and upper back, creates a massive lesion called a **carbuncle** (kar′bung-kl). Encapsulation of abscesses prevents them from shedding

Boils can become larger than ping-pong balls and are extremely painful.

organisms into the blood, but it also prevents circulating antibiotics from reaching the abscesses in effective quantities. Thus, in addition to antibiotic treatment, it is usually necessary to lance and drain abscesses surgically.

Staphylococcal infections are easily transmitted. Asymptomatic carriers, hospital personnel, and hospital visitors often spread staphylococci via the skin as well as by nasal droplets and fomites. Staphylococci usually cause infection in older patients only when foreign bodies such as catheters or splinters are present. Although 5 million organisms must be injected into the skin to cause infection, only 100 are needed if they are soaked into a suture and tied into the skin.

Scalded Skin Syndrome.   **Scalded skin syndrome** is caused by certain exotoxin-producing strains of *S. aureus*. Two different exotoxins, both called *exfoliatins*, are known. The genes for one are carried on the bacterial chromosome, whereas the genes for the other are located on a plasmid. An individual staphylococcal strain may carry genes for one, both, or neither of these exotoxins. The exotoxins are called exfoliatins because they travel through the bloodstream to sites far from the initial infection, causing the upper skin layers to separate and peel off in leaflike (foliar) sheets.

Scalded skin syndrome is most common in infants but can occur in adults as well, especially late in the disease called toxic shock syndrome. It begins with a slight reddened area, often around the mouth. Within 24 to 48 hours, it spreads to form large, soft, easily ruptured vesicles over the whole body. The skin over vesicles and adjacent reddened areas peels, leaving large, wet, scalded-looking areas **(Figure 19.6)**. The lesions dry and scale, and the skin returns to normal in 7 to 10 days. High fever is generally present. Bacteremia is common and can lead to septicemia and death within 36 hours. Exotoxins are highly antigenic (◀Table 14.5, page 411). They

Figure 19.5   A furuncle. This deep, pus-filled lesion is caused by *Staphylococcus aureus*. (Ken Greer/Visuals Unlimited)

Figure 19.6   Scalded skin syndrome in an infant. This infection is caused by *Staphylococcus aureus*. Reddened areas of skin peel off, leaving wet, scalded-looking areas. (*Kenneth Greer/Visuals Unlimited*)

stimulate the production of antibodies that prevent recurrence of this syndrome—but only if reinfection occurs by the same *S. aureus* strain.

Streptococcal Infections

Scarlet Fever.   **Scarlet fever**, sometimes called scarlatina, is caused by *Streptococcus pyogenes*, which also causes the familiar strep throat. Strains of the organism that cause scarlet fever have been infected by a temperate phage that enables them to produce an erythrogenic ("red-producing") toxin that causes the scarlet fever rash. Patients who have previously been exposed to the toxin, and thus have antibodies that can neutralize it, can develop strep throat without the scarlet fever rash. However, such people can still spread scarlet fever to others. Three different erythrogenic toxins have been identified. A person can develop scarlet fever once from each toxin. These are also called streptococcal pyrogenic exotoxins.

PUBLIC HEALTH

Flesh-Eating Bacteria

It was like a real-life *X-Files* episode in June 1994 in England and then again in 1998 in Texas. Newspapers and television sounded alerts after people began to die as a strain of Group A *Streptococcus*, called M1T1 strep, ate away at their flesh at the fearful rate of several inches per hour. After killing the initially infected tissue, the bacteria lived on the remains of the dead flesh, churning out toxins that diffused out of the dead area to kill more tissue. Antibiotics were useless against the bacteria because the circulatory system could not function in the dead tissue. The only hope was for the infected areas to be surgically removed, even if it meant amputating a limb.

Australian investigators have recently found that this fierce virulence is due to a gene brought in by a bacteriophage which infected the strep bacteria. This gene allows the bacteria to take plasminogen protein from the human bloodstream, and coat it on their surface. When activated, the protein becomes a protease which digests the proteins of cells and tissues, eating away flesh. The same gene also produces an enzyme that prevents the bacteria from being trapped and killed by neutrophil white blood cells.

(Ken Greer/Visuals Unlimited)

In the United States most scarlet fever organisms currently have low virulence. In past decades, when strains were more virulent, it was a much-feared killer. However, even low-virulence strains can cause serious complications, such as glomerulonephritis or rheumatic fever. The use of penicillin has appreciably lowered the mortality rate. Convalescent carriers can shed infective organisms from the nasopharynx for weeks or months after recovery. Fomites also are an important source of streptococcal infections.

Erysipelas.   **Erysipelas** (er″i-sip′e-las; from the Greek *erythros*, for red, and *pella*, for skin), also called **St. Anthony's fire**, has been known for over 2,000 years and is caused by hemolytic streptococci. Before antibiotics became available, erysipelas often occurred after wounds and surgery and sometimes after very minor abrasions. Mortality was high. Today, it rarely occurs, and mortality is low. The disease begins as a small, bright, raised, rubbery lesion at the site of entry. Lesions spread as streptococci grow at lesion margins, producing toxic products and enzymes such as hyaluronidase (◄Chapter 14, p. 410). Lesions are so sharply defined that they appear to be painted on (**Figure 19.7**). The organisms spread through lymphatics and can cause septicemia, abscesses, pneumonia, endocarditis, arthritis, and death if untreated. Curiously, erysipelas tends to recur at old sites. Instead of developing immunity, patients acquire greater susceptibility to future attacks.

Pyoderma and Impetigo

Pyoderma, a pus-producing skin infection, is caused by staphylococci, streptococci, and corynebacteria, singly or in combination. **Impetigo** (im-pe-ti′go), a highly contagious

Figure 19.7   Erysipelas. *(©ISM/Phototake)*

Figure 19.8 Impetigo. This highly contagious infection is caused by staphylococci, streptococci, or both together. *(Kenneth Greer/Visuals Unlimited)*

pyoderma, is caused by staphylococci, streptococci, or both (**Figure 19.8**). Fluid from early pustules usually contains streptococci, whereas fluid from later lesions contains both. Streptococcal strains that cause skin infections usually differ from those that cause strep throat. Impetigo occurs almost exclusively in children; why adults are not susceptible is unknown. Easily transmitted on hands, toys, and furniture, impetigo can rapidly spread through a day care center. Impetigo rarely produces fever, and it is easily treated with penicillin. Lesions usually heal without scarring, but skin can be discolored for several weeks, and pigment can be permanently lost.

Acne

Acne (acne vulgaris) affects more than 80% of teenagers and many adults, too. It is most often the result of male sex hormones that stimulate sebaceous glands to increase in size and secrete more sebum. Acne occurs in both males and females because the hormones are produced by the adrenal glands as well as by the testes. Microorganisms feed on sebum, and ducts of the glands and surrounding tissues become inflamed. "Blackheads" are a mild form of acne in which hair follicles and sebaceous glands become plugged with sebum and keratin. As the surface oxidizes, it takes on a dark or "black" appearance. In most severe cases (cystic acne) the plugged ducts become inflamed, rupture, and release secretions. Bacteria, especially *Propionibacterium acnes*, infect the area and cause more inflammation, more tissue destruction, and scarring. Such lesions can be widely distributed over the body, and some become encysted in connective tissue.

Acne is treated with frequent cleansing of the skin and topical ointments to reduce the risk of infection. Sometimes acne sufferers are advised to avoid fatty foods, but a connection between diet and acne is not well established. Dermatologists often prescribe oral antibiotics such as tetracycline in low doses to control bacterial infections in the lesions. However, continuous use of

antibiotics depletes natural intestinal flora and can contribute to the development of antibiotic-resistant strains of bacteria (◄Chapter 13, p. 373). The drug Accutane, derived from a molecule related to vitamin A, is now used to treat severe and persistent acne. It seems to inhibit sebum production for several months after treatment is stopped, but it can cause serious side effects, such as intestinal bleeding. If taken by a pregnant woman, even for just a few days, it can seriously damage the fetus. In most cases acne disappears or decreases in severity as the body adjusts to the hormonal changes of puberty and as the functioning of the sebaceous glands stabilizes.

Burn Infections

Severe burns destroy much of the body's protective covering and provide ideal conditions for infection. Burn infections, which are usually nosocomial, account for 80% of deaths among burn patients. *Pseudomonas aeruginosa* is the prime cause of life-threatening burn infections, but *Serratia marcescens* and species of *Providencia* also often infect burns. Many strains of these Gram-negative bacilli are antibiotic-resistant.

The thick crust or scab that forms over a severe burn is called **eschar** (es'kar). Bacteria growing in or on eschar pose no great threat, but those growing beneath it

PUBLIC HEALTH

What Goes Round in Whirlpool Baths?

Whirlpool baths can be quite dangerous if not properly cleaned and disinfected daily. Warm water, plus skin flakes, oils, and secretions makes a perfect breeding ground for microbes. The aerosol from a bubbling whirlpool can contain millions of bacteria per cubic meter of air. *Pseudomonas aeruginosa*, a frequent colonizer, is very difficult to eliminate—it can be back in the water just hours after disinfection. Often it grows attached to the sides of the whirlpool tub or inside the pipes connected to it. Disinfection may kill all the organisms in the water, but when the same bacteria attach to a surface and form a biofilm, they are much more resistant to the disinfecting chemicals.

A person in a whirlpool bath who vigorously rubs his or her eyes could cause a corneal abrasion and thereby become susceptible to bacterial eye infection. *Pseudomonas* infection can cause severe scarring of the cornea and distorted vision, or even blindness. Once such an infection is started, you can lose an eye almost overnight. Also, inhaling aerosols can cause *Pseudomonas* pneumonia, which has a very poor prognosis.

Many people break out in a body rash in whirlpool baths because of a hypersensitivity reaction to being immersed in water filled with *Pseudomonas*. And *Pseudomonas* isn't the only microbe to worry about in or on whirlpools, as two elderly widows discovered. They developed herpes lesions on the backs of their thighs after sitting on the moist rim of a hot tub where a previous occupant with a herpes outbreak had been seated.

cause severe local infections and can move into the blood. Delivering antibiotics to infections under eschar is difficult because the eschar lacks blood vessels. Topical antimicrobials can seep through the eschar; removal of some of the eschar by a surgical scraping technique called **débridement** (da-bred-maw′ or de-bred′ment) helps them to reach infection sites.

Prevention of burn infections is difficult even when patients are isolated within hospital burn units. Having lost skin, patients have lost the benefit of the movement of white blood cells to sites of infection in the skin. Such patients also have fluid and electrolyte deficiencies because of seepage from burned tissue. Finally, their appetites are depressed at a time when extensive tissue repair increases metabolic needs.

Burn infections are as difficult to diagnose as they are to treat. Early signs of infection can consist of only a mild loss of appetite or increased fatigue. Definitive diagnosis is made by finding more than 10,000 bacteria per gram of eschar. *Pseudomonas aeruginosa* infection should be suspected when a greenish discoloration appears at the burn site and cultures have a grapelike odor. This bacterium produces tissue-killing toxins that erode skin. It is extremely resistant to antimicrobial drugs and has been found growing in surgical scrub solutions.

Go to the chapter website to see an essay describing a visit to one of the premier burn units in the world, the U.S. Army Burn Unit in San Antonio, Texas, for more information on burn treatment.

Take another look

✓ CHECKLIST

1. What are some innate lines of defense provided by the skin?
2. Why are *Staphylococcus* species so numerous on skin?
3. What are some skin diseases caused by *Staphylococcus*?
4. What is débridement? Why is it necessary in burn cases?
5. What are some skin diseases caused by *Streptococcus*?

VIRAL SKIN DISEASES

Rubella

The Disease.   **Rubella**, or **German measles**, is the mildest of several human viral diseases that cause **exanthema** (ex-an-the′mah), or skin rash. A rash, the main symptom of rubella, appears first on the trunk 16 to 21 days after infection, but the virus (a togavirus) spreads in the blood and other tissues before the rash appears. Infected adult women often suffer from temporary arthritis and arthralgia (joint pain) from dissemination of the virus to joint membranes. These complications are seen less frequently in adult men.

Congenital rubella syndrome results from infection of a developing embryo across the placenta. When a woman becomes infected with rubella during the first 8 weeks of pregnancy, severe damage to the embryo's organ systems is likely because they are developing. After the eighteenth week of pregnancy, damage is rare. The spread of rubella viruses in the infant kills many cells, persistently infects other cells, reduces the rate of cell division, and causes chromosomal abnormalities. Many infants are stillborn, and those that survive may suffer from deafness, heart abnormalities, liver disorders, and low birth weight. Worldwide, approximately 100,000 cases per year occur.

To prevent congenital rubella syndrome, women should have their rubella immunity checked before trying to get pregnant.

Incidence and Transmission.   Prior to the development of a rubella vaccine, nearly all humans caught rubella at some time, but many cases were not detected. Half the cases in young children and up to 90% of those in young adults are not recognized. In the United States the incidence of rubella was nearly 30 cases per 100,000 people in 1969, when a vaccine was licensed. Rubella, including congenital rubella, has now been nearly eliminated (**Figure 19.9**), with only 8 cases reported in total in 2006.

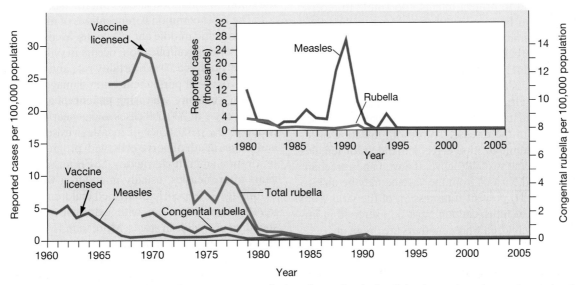

Figure 19.9   The U.S. incidence of rubella (German measles) and measles (rubeola).   The number of cases dropped rapidly after vaccines were licensed during the 1960s. Failure to vaccinate, however, led to temporary increases in measles cases. (*Source:* CDC.).

PUBLIC HEALTH

Measles and More

Did you ever think that you could catch measles from your dog? You can't, since humans are the only known hosts of measles, but distemper in dogs is caused by a virus with antigens similar to those of the measles virus. Did you know that measles probably couldn't have existed until 2500 B.C.? Measles can spread only if there is a continuous chain between infected and susceptible people, since the virus doesn't display latency, cannot survive for long outside the body, no other organism survives as a reservoir, and immunity from infection is lifelong.

A world population of at least 300,000 was therefore required to perpetuate the virus, and this population size was not reached until about 2500 B.C. Measles probably appeared as a human pathogen after that time—possibly as a mutant of a virus like the one that causes distemper.

Figure 19.10   Measles (rubeola). This shows the typical skin rash, plus white Koplik's spots in the mouth on the inner surface of the cheek. *(From R.T.D. Emond & H.A.K. Rowland,* A Colour Atlas of Infectious Diseases, *2nd ed, Wolfe Publishing/Mosby-Year Book, Inc.)*

Transmission is mainly by nasal secretions shortly before, and for about a week following, the appearance of a rash. Many infected individuals do not have a rash and transmit the virus without knowing it. Rubella is highly contagious, especially by direct contact among children aged 5 to 14. Infants infected before birth are rubella carriers; they excrete viruses and expose the hospital staff and visitors, including pregnant women, to the disease.

Diagnosis. Rubella can be positively diagnosed by a variety of laboratory tests. Determination of a four-fold rise in rubella-specific IgM antibody levels is particularly useful in identifying newborn carriers and assessing immunity of pregnant women exposed to rubella.

Immunity and Prevention. The currently available rubella vaccine, part of a combined attenuated viral vaccine preparation (MMR), is the only means of preventing rubella. Children should receive it both to protect themselves from measles and to protect infants, too young to receive the vaccine, from being exposed to a source of infection. High levels of immunity (◄Chapter 15) must be maintained, or outbreaks will occur (p. 441). The vaccine produces lower antibody levels than does infection, and immunity probably is not as long-lasting as that produced by infection. Also, viruses appear in the nasopharynx a few weeks after immunization, and some individuals experience mild symptoms of the disease. To prevent infection of fetuses, a second immunization is recommended for females before they become sexually active. If a woman is pregnant when immunized, viruses from the vaccine may be able to infect the fetus. To prevent transmission of the virus from child to mother to fetus, caution must be exercised in immunizing young children whose mothers are pregnant.

Measles

The Disease. **Measles**, or **rubeola**, is a febrile (accompanied by fever) disease with a rash caused by the rubeola virus. The paramyxovirus invades lymphatic tissue and blood. After the virus enters the body through the nose, mouth, or conjunctiva, symptoms appear in 9 to 11 days in children and in 21 days in adults. **Koplik's spots**, white spots with central bluish specks **(Figure 19.10)**, appear on the upper lip and cheek mucosa 2 or 3 days before other symptoms such as fever, conjunctivitis, and cough. These symptoms persist for 3 or 4 days and sometimes progressively worsen. They are followed by a rash, which spreads during a 3- to 4-day period from the forehead to the upper extremities, trunk, and lower extremities and disappears in the same order several days later. Rubeola can be distinguished from rubella on the basis of the rash. Rubella produces a flat pink rash, whereas the rubeola rash is red and raised. The rash is caused by the reaction of T cells with virus-infected cells in small blood vessels. Without such reactions, as in patients with defective cell-mediated immunity, no rash appears, but the virus is free to invade other organs (◄Chapter 18, p. 554). If the virus invades the lungs, kidneys, or brain, the common childhood disease often is fatal.

The most common complications of measles are upper respiratory and middle ear infections. **Measles encephalitis**, a more serious complication, occurs in only 1 or 2 patients per 1,000 but has a 30% mortality rate and leaves one-third of survivors with permanent brain damage. Another complication, **subacute sclerosing panencephalitis (SSPE)**, occurs in only 1 in 200,000 cases and is nearly always fatal. It is due to the persistence of measles viruses in brain tissue and causes death of nerve cells, with progressive mental deterioration and muscle rigidity. SSPE manifests itself 6 to 8 years after measles, usually in children who had measles before age 3. In poorly nourished children, measles causes intestinal inflammation with extensive protein loss and shedding of viruses in stools. More than 15% of children infected with measles in developing countries die of the disease or its complications. Much of the mortality is now known to result from long-lasting interleukin-12–mediated immunosuppression. In 2000, measles was considered to be the fifth leading killer of children, causing over 777,000 deaths worldwide. A concerted program of immunization is now lowering these sad statistics.

Incidence and Transmission. The measles virus is highly contagious; its portal of entry is the respiratory tract. A susceptible person has a 99% chance of infection if exposed directly to someone who is releasing the virus while coughing and sneezing. Before widespread immunization, most children had measles before they were 10 years old. In populations such as that of the United States, where periodic epidemics have occurred and most children are well nourished, the disease is serious but rarely fatal. In populations lacking immunity from periodic epidemics or in malnourished children, measles is a killer. In 1875, when measles was first introduced into Fiji, 30% of the population died of the disease.

Diagnosis and Treatment. Measles is diagnosed by its symptoms. Treatment is limited to alleviating symptoms and dealing with complications. Secondary bacterial infections can be effectively treated with antibiotics.

Immunity and Prevention. In a nonimmunized population, epidemics of numerous cases over a 3- or 4-week period generally occur every 2 to 5 years over wide regions. Measles vaccine now prevents such epidemics in many developed countries. In the United States, mandatory measles vaccination has greatly reduced the incidence of the disease (see Figure 19.9). Because the MMR vaccine containing attenuated viruses of measles, mumps, and rubella is generally used, the incidence of all three diseases has decreased simultaneously (◀Chapter 17, p. 512). Eradication of a disease requires that about 90% of the population be immune. Religious groups that reject immunization, immigrants who don't understand its importance, and apathetic parents who don't have their children immunized make eradication difficult. Compounding these problems have been federal government cutbacks in funding used to purchase vaccine for clinics. Hence, some poor children in the United States are going without vaccine.

Immunity acquired from having measles is lifelong. Recent epidemics among college students immunized as children suggest that immunity from the vaccine may not be lifelong or that the children were immunized before 15 months of age. A few cases of SSPE have been attributed to the vaccine, but the incidence of SSPE is much lower than it was prior to the use of the vaccine.

Roseola

Roseola (rose-colored) is a disease of infants and small children caused by the human herpesvirus 6 (HHV-6), which was first identified in 1988. An older name is exanthem subitum (exanthem=rash; subitum=sudden). The disease begins with an abrupt high fever lasting 3 to 5 days, sometimes accompanied by brief convulsions. The rose-colored rash appears following the sudden drop in fever. Within a day or two, the rash disappears. The virus, however, is not gone. It has begun a lifelong latent infection of T cells. Immunity to another outbreak is permanent. The virus replicates in salivary gland cells, and is shed in the saliva of most adults. Transmission is by saliva.

Chickenpox and Shingles

One Virus, Two Diseases. The **varicella-zoster virus (VZV)**, a herpesvirus, causes both **chickenpox** (varicella) and **shingles** (zoster). The same virus can be isolated from lesions of either disease. Chickenpox is a highly contagious disease that causes skin lesions and usually occurs in children. Before vaccine was available, almost everyone had chickenpox by the age of 30. There were probably over 4 million cases per year in the United States alone. Vaccination has now reduced that number by over 80%. Immunity lasts for about 20 years. Shingles is a sporadic disease that appears most frequently in older and immunocompromised individuals. A congenital varicella syndrome also is known to occur in about 5% of the U.S. population.

Chickenpox does not infect chickens nor is it caused by a poxvirus. A herpesvirus is its cause.

In chickenpox, the virus enters the upper respiratory tract and conjunctiva and replicates at the site of entry. New viruses are carried in blood to various tissues, where they replicate several more times. Release of these viruses causes fever and malaise. In 14 to 16 days after exposure, small, irregular, rose-colored skin lesions appear. The fluid in them becomes cloudy, and they dry and crust over in a few days. The lesions appear in cyclic crops over 2 to 4 days as the viruses go through cycles of replication. They start on the scalp and trunk and spread to the face and limbs, sometimes to the mouth, throat, and vagina, and occasionally to the respiratory and gastrointestinal tracts. The lesions are important portals of entry for secondary infections, especially with *Staphylococcus aureus*.

Chickenpox, although sometimes thought of as a mild childhood disease, can be fatal. The viruses invade and damage cells that line small blood vessels and lymphatics. Circulating blood clots and hemorrhages from damaged blood vessels are common. Death from varicella pneumonia is due to extensive blood vessel damage in the lungs and the accumulation of erythrocytes and leukocytes in alveoli. Cells in the liver, spleen, and other organs also die because of damage to blood vessels within them.

In shingles, painful lesions like those of chickenpox usually are confined to a single region supplied by a particular nerve **(Figure 19.11a)**. Such eruptions arise from latent viruses acquired during a prior case of chickenpox. During the latent period, these viruses reside in ganglia in the cranium and near the spine. When reactivated, the viruses spread from a ganglion along the pathway of its associated nerve(s). Pain and burning and prickling of the skin occur before lesions appear. The viruses damage nerve endings, cause intense inflammation, and produce clusters of skin lesions indistinguishable from chickenpox

(a)

(b)

Figure 19.11 Shingles. Shingles lesions usually result from varicella-zoster virus infections acquired during exposure to childhood chickenpox. The virus can remain latent within the body for many years before being reactivated in adulthood. **(a)** Vesicles commonly form a belt around the chest or hips, following the pathway of a nerve. *(Barts Medical Library/Phototake)* **(b)** The small, yellow vesicles dry up and heal by scabbing, but they can be excruciatingly painful and itchy. *(N.M. Hauprich/Photo Researchers, Inc.)*

lesions (Figure 19.11b). Symptoms range from mild itching to continuous, severe pain and can include headache, fever, and malaise. Lesions often appear on the trunk in a girdlelike pattern (*zoster*, "girdle") but can also infect the face and eyes. Shingles is most severe in individuals with malignancies or immune disorders. In such patients, lesions may cover wide areas of the skin and sometimes spread to internal organs, where they can be fatal.

The latent virus is activated when cell-mediated immunity drops below a critical, minimal level, as can occur in lymphatic cancers, spinal cord trauma, heavy-metal poisoning, or immunosuppression. In other cases, no cause of viral reactivation can be identified. Release of newly replicated viruses increases antibody production, but the antibodies may fail to stop viral replication. Recovery from shingles usually is complete. Second and third cases do occur, and they depend on the degree of development of cell-mediated immunity and local interferon production. Chronic shingles is found in the immunocompromised, including AIDS patients. New vesicles constantly erupt, whereas old ones fail to heal, which can be very debilitating.

Incidence and Transmission.

Chickenpox is endemic in industrialized societies in the temperate zone, and its incidence is highest in March and April. The primary infection usually occurs between the ages of 5 and 9. In general, chickenpox in adults who have not had it as children is more severe than in children. Shingles also is age-related, with most cases appearing in people over 45 years of age.

Pregnant women who have never had chickenpox and have not been vaccinated should avoid exposure to children who have chickenpox and to people who have shingles.

Infection can be spread by respiratory secretions and contact with moist lesions but not from crusted lesions. Children experiencing a mild case, with only a few lesions and no other symptoms, often spread the disease. In rare cases, adults with partial immunity can contract shingles from exposure to children with chickenpox.

Susceptible children, however, can easily contract chickenpox from exposure to adults with shingles.

Diagnosis and Treatment.

Chickenpox is usually diagnosed by a history of exposure and the nature of lesions, although a rapid laboratory test has recently become available. Shingles may be impossible to distinguish from other herpes lesions without laboratory tests. Treatment is limited to relieving symptoms, but aspirin should not be given to children because of the risk of Reye's syndrome. Acyclovir, once hoped to be useful in early stages of the disease to reduce its severity, has not proven successful. Antiviral agents are being tested on infections in immunosuppressed patients and those with disseminated disease. When given within the first 2 to 3 days after shingles symptoms appear, the anti-viral, Valtrex, may ease the course of the disease. Nerve pain is sometimes helped by neurontin, or by a lidocaine patch.

Immunity and Prevention.

Having chickenpox as a child confers lifelong immunity in most cases. Recurrences (in the form of shingles) are seen only in individuals with low concentrations of VZV antibodies and waning cell-mediated immunity. Chickenpox was the last of the common communicable childhood diseases for which no vaccine was widely available—until 1995, when a vaccine was finally licensed for use in the United States. An attenuated varicella vaccine, developed in Japan, prevents chickenpox when administered as late as 72 hours after exposure. It was used overseas for all children. The U.S. Food and Drug Administration approved this vaccine for general use in 1995. Because the mechanism of latency and reactivation is not well understood, however, immunologists had some reservations about administering vaccines containing viruses that can become latent. Cases have been found in which shingles has been caused by reactivation of the virus in the vaccine. Nevertheless, the vaccine is considered relatively safe and effective.

In 2007, Zostavax, a new vaccine against shingles, was introduced. It basically is a stronger version of the childhood chickenpox vaccine. It acts as a booster shot. Previously adults were exposed to children with chickenpox. These exposures to the natural virus acted as "mini booster shots" which helped to maintain adult immunity. Today, there are very few cases of chickenpox left to do that job. Zostavax reduces the incidence of shingles by 50%, and greatly reduces the seriousness and pain level of those who do still develop shingles.

Other Pox Diseases

Smallpox. In 1980, the World Health Organization (WHO) officially proclaimed that **smallpox** had been eradicated worldwide. This announcement marked the end of centuries of sickness and death from this disease.

The disease first appeared sometime after 10,000 B.C. in a small agricultural settlement in Asia or Africa. The mummy of Ramses V, who died in Egypt in 1160 B.C., has smallpox scars on the face, neck, shoulders, and arms. Smallpox ravaged villages in India and China for centuries, and an epidemic occurred in Syria in A.D. 302. The Persian physician Rhazez clearly described the disease in A.D. 900.

Crusaders returning from the Middle East brought smallpox to Europe in the twelfth century. The Spaniards carried the virus to the West Indies in 1507, and Cortez's army introduced it into Mexico in 1520. In each instance, the disease was introduced into a population that lacked immunity, and its spread was rampant. As many as 3.5 million Native Americans may have died from smallpox, and by the eighteenth century more than half the inhabitants of Boston had become infected. Slave traders introduced smallpox into Central Africa in the sixteenth and seventeenth centuries, and immigrants from India brought it to South Africa in 1713. The disease reached Australia in 1789.

The idea of immunization itself originated in Asia with the technique of variolation to prevent smallpox. In variolation, threads saturated with fluid from smallpox lesions were introduced into a scratch or dangled in the sleeve of a nonimmune person. This practice immunized some people, but it also started epidemics because live, virulent virus was used. Lady Mary Wortley Montague, wife of the British Ambassador to Turkey, introduced the practice to England in 1717. Despite high morbidity (illness) and even mortality (death) from this technique, General George Washington ordered all his troops variolated in 1777. By 1792, 97% of the population of Boston had been variolated.

Smallpox was given that name in contrast to syphilis, which was known as the "Great Pox."

Smallpox immunization was greatly improved by the English physician Edward Jenner, who noticed that milkmaids with cowpox scars never became infected with smallpox. It is now known that immunity to the less

Figure 19.12   One of the last cases of smallpox. Lesions are most numerous on the face and arms. As the disease progresses, the blisterlike lesions become opaque and eventually form crusts that fall off. Scarring is common. *(Courtesy Centers for Disease Control and Prevention CDC)*

severe cowpox confers immunity to smallpox as well. In 1796, Jenner inoculated an 8-year-old boy with cowpox and 6 weeks later inoculated him with smallpox. As Jenner had expected, the boy remained healthy. In the United States in 1799, the physician Benjamin Waterhouse introduced the cowpox vaccine by vaccinating his own children and then exposing them to smallpox.

Although it was well known by then that smallpox could be prevented by vaccination with cowpoxvirus, in 1947 an unimmunized traveler from Mexico infected 12 people in New York City, and 2 of them died. The last 8 cases of smallpox in the United States occurred in the Rio Grande Valley in 1949.

Smallpox was still endemic in 33 countries in 1967 when the WHO established its immunization campaign; in 1977 only a single natural case of smallpox occurred worldwide (**Figure 19.12**). When smallpox viruses escaped via air ducts from a laboratory in Birmingham, England, in 1978, an unimmunized medical photographer became infected and died. Her mother suffered a mild case of the disease and recovered. The director of the laboratory from which the virus escaped committed suicide. After this disaster, smallpox stock cultures have theoretically been maintained in only two maximum

Figure 19.13 **Progression of smallpox disease.** *(Courtesy World Health Organization)*

containment laboratories—one in the Atlanta division of the CDC, and the other in the Research Institute for Viral Preparations in Moscow. However, smallpox surveillance continues. It is now known that the Russians grew great quantities of smallpox virus for possible use in biological warfare, and that stocks from this supply have probably reached other Soviet-allied countries. Discussions of destroying the "last two" remaining stocks of smallpox virus have since fallen by the wayside.

The smallpox virus enters the throat and respiratory tract. During a 12-day incubation period, it infects phagocytic cells and later blood cells. The infection spreads to skin cells, causing pus-filled vesicles. Acute systemic symptoms begin with fever, backache, and headache. Vesicles appear first in the mouth and throat. They then rapidly spread to the face, forearms, hands, and finally the trunk and legs. Vesicles become opaque and pustular and encrust within about 2 weeks. Death is most likely to occur 10 to 16 days after onset of the first symptoms **(Figure 19.13)**.

Scrapings from lesions are used to differentiate smallpox from leukemia, chickenpox, and syphilis. In a nonimmune population, smallpox is highly contagious; in a largely immunized population, it spreads very slowly.

But now that smallpox has been eradicated worldwide, complications of vaccination are far more life-threatening than the disease itself.

Cowpox. **Cowpox**, caused by the vaccinia virus, causes lesions (similar to a smallpox vaccination) at abrasion sites, inflammation of lymph nodes, and fever. Vaccinia viruses can also cause a progressive disease, with numerous lesions and symptoms more like those of smallpox, that can be treated with the drug methisazone. Cattle appear to transmit the disease to humans.

Not only was the vaccinia virus used by Jenner to immunize against smallpox but also it was the first animal virus to be obtained in sufficient quantity for chemical and physical analysis (◄Chapter 1, p. 15). The modern vaccinia virus may have become attenuated by several centuries of passage on calves' skin, a procedure used in making smallpox vaccine, and is no longer exactly the same as the original cowpox virus.

Monkeypox. **Monkeypox** is sometimes mistaken for smallpox, as the lesions and death rates are very similar. Usually it occurs in western and central Africa, especially in Zaire and the Congo. Smallpox and monkeypox are both orthopoxviruses, so it is not surprising that smallpox vaccination protects against both. With discontinuation of smallpox vaccination, we are now seeing outbreaks of monkeypox among zoo monkeys (which is how it got its name), and most recently among Americans who purchased African giant rats at pet shops. These outbreaks are an unlooked for consequence of cessation of smallpox vaccination, and can be a serious threat in themselves. Some people suggest that smallpox immunization be resumed in areas where monkeypox is endemic.

Molluscum Contagiosum. The **molluscum contagiosum** virus is unusual for several reasons. First, it differs immunologically from both orthopoxviruses and parapoxviruses, the two major groups of poxviruses. Second, it elicits only a slight immune response. Third, although infected cells cease to synthesize DNA, the virus induces neighboring uninfected cells to divide rapidly. Thus, this virus may be intermediate between viruses that cause specific diseases and those that induce tumors. Molluscum contagiosum affects only humans and is distributed worldwide. It causes pearly white to light pink, painless, tumorlike growths scattered over the skin **(Figure 19.14)**. The disease usually affects children and young adults, and it can persist for years. It is acquired by personal contact or from items such as gym equipment and swimming pools. Treatment generally involves the removal of growths by chemicals or by localized freezing.

Warts

Human **warts**, or **papillomas**, are caused by **human papillomaviruses (HPV)**. HPV specifically attack skin and mucous membranes. Warts grow freely in many sites in the

(a)

(b)

Figure 19.15 Warts. Warts are caused by human papillomaviruses. **(a)** Genital warts around the vagina, and *(Kenneth Greer/Visuals Unlimited)* **(b)** dermal warts. *(Kenneth Greer/Visuals Unlimited)*

Figure 19.14 Molluscum contagiosum. Tumorlike growths develop in cases of this viral infection. *(Courtesy George A. Wistrelch, East Los Angeles College)*

body—the skin, the genital and respiratory tracts, and the oral cavity. Viral infection lasts a lifetime. Even when warts disappear or are removed, the virus remains in surrounding tissue, and warts may recur or form malignant tumors.

The Nature of Warts. Warts vary in appearance, area of occurrence, and pathogenicity. Some are barely visible and self-limiting—that is, they do not grow or spread—and others, such as laryngeal warts, are larger but benign. A few warts are malignant. **Genital warts**, also known as condylomata acuminata (kon″di-lo-ma′ta a-kum″i-na′ta) (**Figure 19.15a**), for example, generally do not become malignant. But other strains of the same virus, that do not

produce visible warts (remaining as chronic "invisible" infections), *do* cause 99 % of all cases of cervical cancer. Cervical cancer has now been added to the list of sexually transmitted diseases. Refer back to ◄Chapter 10, p. 271, and forward to ◄Chapter 20, p. 632, for information on Gardasil©, a new vaccine against 4 strains of HPV, which reduces the risk of gettings cervical cancer by 70%. Gardasil is currently recommended for use in girls aged 9 to 26 years old. It is given as a series of 3 shots which cost about $360, plus office visits. Warts of all types grow larger than normal in people with AIDS or other immunodeficiencies.

Transmission. Papillomaviruses are transmitted by direct contact, usually between humans, or by fomites. **Dermal warts (Figure 19.15b)** form when the virus enters the skin or mucous membranes through abrasions. Genital warts are sexually transmitted, and *juvenile onset laryngeal warts* are acquired during passage through an infected birth canal. The incubation period varies from 1 week to 1 month for dermal warts and 8 to 20 months for genital warts. Genital warts and warts acquired at birth will be discussed in ◄Chapter 20, (p. 631).

PUBLIC HEALTH

Mind Over Body

Imagine going to the doctor to get rid of a wart and being told that you need to see a hypnotist for treatment. According to Lewis Thomas, warts "can be ordered off the skin by hypnotic suggestion." In a controlled study, 14 patients with multiple warts were hypnotized, and the suggestion was made that all warts on one side of the body would go away. Within a few weeks, this happened to 9 patients, including someone who "got mixed up and destroyed the warts on the wrong side." But if hypnosis doesn't work for you, you could just try waiting your wart out. Thomas also said that warts "inexplicably and often very abruptly . . . come to the end of their lives and vanish without a trace."

APPLICATIONS

Algal Infections

Most algae manufacture their own food and are not parasitic, but some strains of the alga *Prototheca* have lost their chlorophyll and survive by parasitizing other organisms. Found in water and moist soil, they enter the body through skin wounds. By 1987, 45 cases of protothecosis had been reported, 2 from cleaning home aquariums. Protothecosis was first observed on the foot of a rice farmer, and most subsequent cases have occurred on legs or hands. In immunodeficient patients, the parasite can invade the digestive tract or peritoneal cavity. A few skin infections have responded to oral potassium iodide or intravenous amphotericin B and tetracycline therapy, but no satisfactory treatment has been found for others.

Dermal Warts. Epithelial cells become infected and proliferate to form dermal warts, which have distinct boundaries and remain above the basement membrane between the epidermis and the dermis. Children and young adults are more likely than older people to have dermal warts. Only a few warts are present at any one time, and most regress in less than 2 years. Removing one wart often causes regression of all others. In spontaneous regression, all warts usually disappear at the same time. Regression probably is an immunologic phenomenon.

Diagnosis and Treatment. Warts can be distinguished by immunological tests and microscopic examination of tissues. Enzyme immunoassay and immunofluorescent antibody tests can detect about three-fourths of the cases in which viruses are found microscopically. These tests sometimes fail because certain papillomas, especially genital and laryngeal papillomas and those progressing toward malignancy, produce only small quantities of antigens.

Available treatments for various kinds of warts are not entirely satisfactory. The most widely used treatment, cryotherapy, involves the freezing of tissue with liquid carbon dioxide or liquid nitrogen and excision of the infected tissue. Caustic chemical agents such as podophylin, salicylic acid, and glutaraldehyde; surgery; antimetabolites such as 5-fluorouracil; and interferon to block viruses also are used to get rid of warts. Recurrences are still common.

✓ CHECKLIST

1. What damages occur in congenital rubella syndrome?
2. What precautions must be observed in rubella immunization?
3. Distinguish between rubella and rubeola.
4. How can one virus cause both chickenpox and shingles?

FUNGAL SKIN DISEASES

The fungi that invade keratinized tissue are called **dermatophytes** (dur'ma'to-fitz), and fungal skin diseases are called **dermatomycoses** (dur″ma-to-mi-ko'ses). These diseases can be caused by any one of several organisms, mainly from three genera: *Epidermophyton*, *Microsporum*, and *Trichophyton* (Table 19.1). They cause various types of ringworm and attack the skin, nails, and hair.

Fungi that invade subcutaneous tissues live freely in soil or on decaying vegetation and can be found in bird droppings and as airborne spores. These fungi enter tissue through a wound and sometimes spread to lymph vessels. Subcutaneous fungal infections usually spread slowly and insidiously; response to treatment is likewise slow.

Ringworm

Ringworm occurs in several forms (including athlete's foot, discussed next) and is highly contagious. For example, ringworm of the scalp is easily acquired at hairstyling establishments if strict sanitary practices are not followed. Ringworm involves the skin, hair, and nails, and most forms are named according to where they are found. **Tinea corporis** (body ringworm) causes ringlike lesions with a central scaly area. (The shape of these lesions is what originally gave rise to the misleading name "ringworm.") **Tinea cruris** (groin ringworm, or "jock itch") occurs in skin folds in the pubic region (Figure 19.16). **Tinea unguium** (ringworm of the nails) causes hardening and discoloration of fingernails and toenails. In **tinea capitis** (scalp ringworm), hyphae grow down into hair follicles and often leave circular patterns of baldness. **Tinea barbae** (barber's itch) causes similar lesions in the beard.

None of the dermatomycoses result in severe diseases, and lesions usually do not invade other tissues; but they

TABLE 19.1

Dermatomycoses and Common Organisms Found in Lesions					
Dermatomycosis	Epidermophyton floccosum	Microsporum canis	Trichophyton mentagrophytes	Trichophyton rubrum	Trichophyton tonsurans
Body ringworm	X	X	X	X	X
Nail ringworm	X			X	
Groin ringworm	X		X	X	
Scalp ringworm		X	X		X
Beard ringworm			X	X	
Athlete's foot	X		X	X	

Figure 19.16 Tinea cruris, ringworm of the groin (jock itch). This infection is caused by the fungus *Trichophyton*. *(Everett S. Beneke/Visuals Unlimited)*

are unsightly, itchy, and persistent. The causative agents grow well at skin temperature, which is slightly below body temperature. Tissue damage caused by dermatophytes can allow secondary bacterial infections to develop.

Diagnosis and Treatment. Diagnosis of ringworm can be made by microscopic examination of scrapings from lesions, but observation of the skin itself is often sufficient. Although fungi in tissues generally do not form spores, those in laboratory cultures often do, and workers must be especially careful not to become infected from escaping spores. Large numbers of people were infected when spores from a mistakenly opened Petri plate in a university laboratory were dispersed by the building's ventilation system.

Treatment generally consists of removing all dead epithelial tissues and applying a topical antifungal ointment. If lesions are widespread or difficult to treat topically, as when they infect nailbeds, griseofulvin is administered orally. Prevention of ringworm requires avoiding contaminated objects and spores.

Athlete's Foot. In **athlete's foot**, or **tinea pedis**, hyphae invade the skin between the toes and cause dry, scaly lesions. Fluid-filled lesions develop on moist, sweaty feet. Subsequently, the skin cracks and peels, and a secondary bacterial infection leads to itchy, soggy, white areas between the toes. Athlete's foot, a form of ringworm, results from an ecological imbalance between normal flora and host defenses. The fungi that cause athlete's foot are widespread in the environment; they infect tissues when body defenses fail to repel them. Prevention of athlete's foot depends on maintaining healthy, clean, dry feet that can resist the opportunism of fungi. Antifungal agents, such as miconazole, also are effective in most cases.

Subcutaneous Fungal Infections

Sporotrichosis. **Sporotrichosis** is caused by *Sporothrix schenckii*, which usually enters the body from plants, especially sphagnum moss and rose and barberry thorns. The disease can also be acquired from other humans, dogs, cats, horses, and rodents. It is most common in the midwestern United States, especially in the Mississippi Valley. A lesion appears first as a nodular mass at the site of a minor wound. The mass ulcerates, becomes chronic, granulomatous, and pus-filled, and can spread easily to lymphatic vessels. In rare instances it disseminates to internal organs, especially the lungs. Diagnosis is made by culturing specimens of pus or tissue taken from lesions. The cutaneous and lymphatic forms can be treated with potassium iodide; disseminated infections require amphotericin B. People who work with plants or soil should cover injured skin areas to protect themselves from exposure to contaminated materials.

Blastomycosis. North American **blastomycosis** (Figure 19.17) is caused by the fungus *Blastomyces dermatitidis*, which is most common in soil of the central and southeastern United States. The fungus enters the body

(a)   (b)

Figure 19.17 Blastomycosis lesions. **(a)** Before treatment. *(Courtesy Julius Kane and Michael R. McGinnis, University of Texas Medical Branch at Galveston)* **(b)** after treatment with antifungal drugs. *(Courtesy Julius Kane and Michael R. McGinnis, University of Texas Medical Branch at Galveston)* through the lungs or wounds, where it causes disfiguring granulomatous, pus-producing lesions and multiple abscesses in skin and subcutaneous tissue. This condition, called **blastomycetic dermatitis**, for unknown reasons affects chiefly men in their 30s and 40s. In some cases, the fungus travels in the blood and invades internal organs, causing **systemic blastomycosis**. The lungs can be infected directly by the inhalation of spores, and organisms can travel from the lungs to infect other tissues. The disease usually causes relatively mild respiratory symptoms, fever, and general malaise. It can be diagnosed by finding budding yeast cells in sputum or pus, and is treated with amphotericin B or the less toxic hydroxystilbamidine.

Opportunistic Fungal Infections

Certain fungi, such as some yeasts and black molds, can invade the tissues of humans with impaired resistance. Two yeasts, *Candida albicans* and *Cryptococcus neoformans*, and certain molds, such as *Aspergillus fumigatus*, *A. niger*, and the zygomycetes *Mucor* and *Rhizopus*, account for many opportunistic fungal diseases.

Candidiasis. *Candida albicans*, an oval, budding yeast, is present among the normal flora of the digestive and urogenital tracts of humans. In debilitated individuals it can cause **candidiasis** (kan″di-di′uh-sis), or **moniliasis**, in one or several tissues. Superficial candidiasis appears as **thrush (Figure 19.18a)**, milky patches of inflammation on oral mucous membranes, especially in infants, diabetics, debilitated patients, and those receiving prolonged antibiotic therapy (◀Chapter 13, p. 372). The disease appears as **vaginitis** when vaginal secretions contain large amounts of sugar, as occurs during pregnancy, when oral contraceptives are used, when diabetes is poorly controlled, or when women wear tight synthetic undergarments, which promote warmth and moisture and thus favor yeast growth. Some strains of *Candida* may be

sexually transmitted. Cannery workers whose hands are in water for long periods sometimes develop skin and nail lesions **(Figure 19.18b)**. *Candida* can invade the lungs, kidneys, and heart or be carried in the blood, where it causes a severe toxic reaction. Candidiasis, the most common nosocomial fungal infection, is seen in patients with diseases such as tuberculosis, leukemia, and AIDS.

Finding budding cells in lesions, sputum, or exudates confirms the diagnosis of candidiasis. Various antifungal drugs are used to treat it. *Candida* is ubiquitous; infections can be prevented in large part by preventing debilitating conditions.

Aspergillosis. **Aspergillosis** in humans can be caused by various species of *Aspergillus*, but especially *A. fumigatus*. These fungi grow on decaying vegetation. *Aspergillus* initially invades wounds, burns, the cornea, or the external ear, where it thrives in earwax and can ulcerate the eardrum. In immunosuppressed patients, it can cause severe pneumonia. Diagnosis is made by finding characteristic hyphal fragments in tissue biopsies. Antifungal agents are only modestly successful in treatment. Because the mold is so ubiquitous that exposure is inevitable, prevention depends mainly on host defenses. Inhalation of spores by sensitized individuals can give rise to severe allergy symptoms as well.

> *Humans aren't the only ones affected by Aspergillus. Rusty splotches on the pages of old books are left by paper-digesting organisms from this genus.*

Recently, *Aspergillus* has been found to cause a wasting disease of sea fans growing on Caribbean coral reefs. Thus, *Aspergillus* is versatile enough to even invade marine environments.

Zygomycoses. Certain zygomycetes of the genera *Mucor* and *Rhizopus* can infect susceptible humans, such as untreated diabetics, and cause **zygomycoses**. Once

(a)

(b)

Figure 19.18 Candidiasis. (a) *Candida* infections of the oral cavity (thrush), seen as white patches, are a common complication of AIDS, diabetes, and prolonged antibiotic therapy. *(Kenneth Greer/Visuals Unlimited)* **(b)** *Candida* infections of the nails are very difficult to eradicate. *(Project Masters, Inc./The Bergman Collection)*

Following hurricane Katrina, and its flooding, many buildings in New Orleans are so badly filled with fungi that they must be pulled down as being unfit for human occupation.

established, the fungus invades the lungs, central nervous system, and tissues of the eye orbit and can be rapidly fatal, presumably because it evades body defenses. It can be diagnosed by finding broad hyphae in the lumens and walls of blood vessels. Antifungal drugs may or may not be effective in treating zygomycoses.

OTHER SKIN DISEASES

Madura foot, or **maduromycosis**, occurs mainly in the tropics. It is caused by a variety of soil organisms, including fungi of the genus *Madurella* and filamentous actinomycetes, such as *Actinomadura*, *Nocardia*, *Streptomyces*, and *Actinomyces*. Organisms enter the body through breaks in the skin, especially in people who do not wear shoes. Initial pus-filled lesions spread and form connected lesions that eventually become chronic and granulomatous. Untreated, the organisms invade muscle and bone, and the foot becomes massively enlarged **(Figure 19.19)**. Madura foot is diagnosed by finding white, yellow, red, or black granules of intertwined hyphae in pus. So-called sulfur granules in pus are yellow hyphae and not sulfur. Unless antibiotic therapy begins early in the infection and is sufficiently prolonged, amputation may be necessary. Keeping soil particles out of wounds prevents this disease.

Skin reactions to cercariae (larvae) of several helminth species of schistosomes cause **swimmer's itch**. These cercariae parasitize birds, domestic animals, and primates and can burrow into human skin. Immune reactions cause reddening and itching and usually prevent cercariae from reaching the blood and causing schistosomiasis (◀Chapter 11, p. 328). Swimmer's itch occurs throughout the United States but is especially common in the Great Lakes area.

Dracunculiasis (drah-kun"kew-li'a-sis) is caused by a parasitic helminth called a guinea worm. This disease is discussed in Chapter 11

Diseases of the skin are summarized in **Table 19.2**.

DISEASES OF THE EYES

Like skin diseases, eye diseases frequently result from pathogens that enter from the environment, and they cause damage to an exterior portion of our bodies. For that reason we discuss them here.

BACTERIAL EYE DISEASES

Ophthalmia Neonatorum
Ophthalmia neonatorum, or conjunctivitis of the newborn, is a pyogenic (pus-forming) infection of the eyes caused by organisms such as *Neisseria gonorrhoeae* and *Chlamydia trachomatis*. Organisms present in the birth canal enter the eyes as a baby is born. The resulting infection can cause **keratitis**, an inflammation of the cornea, which can progress to perforation and destruction of the cornea and blindness. Early in the 1900s, 20 to 40 % of children in European institutions for the blind had suffered from this disease. In developing countries the disease is still prevalent, but the true incidence is unknown because high infant mortality makes it difficult to estimate the number of original infections. Although infections are most common in newborns **(Figure 19.20)**, adults can transfer organisms by hands or by fomites from genitals to their eyes.

Penicillin was once the treatment of choice, but resistant strains have developed and tetracycline is now used. Tetracycline has the advantage of also being effective against chlamydias and other organisms, with which mothers may also be infected. Preventive measures have almost eradicated ophthalmia neonatorum in developed countries. A few drops of 1% silver nitrate solution in

(a)

(b)

Figure 19.19 Madura foot. This disease can be caused by true fungi or by actinomycetes, funguslike bacteria. Lesions filled with pus and pathogens may distort the foot to the point at which amputation is necessary. **(a)** Bottom view of foot *(Ed Rottinger/Custom Medical Stock Photo, Inc.)*; **(b)** top view of foot. *(Ed Rottinger/Custom Medical Stock Photo, Inc.)*

TABLE 19.2

A Summary of Diseases of the Skin

Disease	Agent	Characteristics
Bacterial Skin Diseases		
Folliculitis	*Staphylococcus aureus*	Skin abscess; encapsulated, so not reached by antibiotics
Scalded skin syndrome	*S. aureus*	Vesicular lesions over entire skin surface, fever; most common in infants
Scarlet fever	*Streptococcus pyogenes*	Sore throat, fever, rash caused by toxin; can lead to rheumatic fever and other complications
Erysipelas	*S. pyogenes*	Skin lesions spread to systemic infection; rare today, but common and fatal before antibiotics were available
Pyoderma and impetigo	Staphylococci, streptococci	Skin lesions, usually in children; easily spread by hands and fomites
Acne	*Propionibacterium acnes*	Skin lesions caused by excess of male sex hormones; infection is secondary; common in teenagers
Burn infections	*Pseudomonas aeruginosa* and other bacteria	Growth of bacteria under eschar, often a nosocomial infection; difficult to diagnose and treat; causative agents typically antibiotic-resistant
Viral Skin Diseases		
Rubella	Rubella virus	Mild disease with maculopapular exanthema (discolored, pimply rash); infection early in pregnancy can lead to congenital rubella; vaccine has greatly reduced incidence
Measles	Rubeola virus	Severe disease with fever, conjunctivitis, cough, and rash; encephalitis is a complication; occurs mainly in children; vaccine has greatly reduced incidence
Roseola	Human herpesvirus 6	Sudden fever, followed by rose-colored rash; virus shed in saliva
Chickenpox	Varicella-zoster virus	Generalized macular (discolored) skin lesions
Shingles	Varicella-zoster virus	Pain and skin lesions, usually on trunk; occurs in adults with diminished immunity; susceptible children exposed to cases of shingles can develop chickenpox
Smallpox	Smallpox virus	Eradicated by immunization as a human disease
Other pox diseases	Other poxviruses	Clear or bluish vesicles on skin surfaces; human infections are rare
Warts	Human papillomaviruses	Dermal warts are self-limiting; malignant warts occur in immunologic deficiencies
Fungal Skin Diseases		
Dermatomycoses	Dermatophytes	Dry, scaly lesions on various parts of the skin; difficult to treat
Sporotrichosis	*Sporothrix schenckii*	Granulomatous, pus-filled lesions; sometimes disseminates to lungs and other organs
Blastomycosis	*Blastomyces dermatitidis*	Granulomatous, pus-filled lesions that develop in lungs and wounds; sometimes disseminates to other organs
Candidiasis	*Candida albicans*	Patchy inflammation of mucous membranes of the mouth (thrush) or vagina (vaginitis); disseminated nosocomial infections occur in immunodeficient patients
Aspergillosis	*Aspergillus* species	Wound infection in immunodeficient patients; also infects burns, cornea, and external ear
Zygomycosis	*Mucor* and *Rhizopus* species	Occurs mainly with untreated diabetes; begins in blood vessels and can rapidly disseminate
Other Skin Diseases		
Madura foot	Various soil fungi and actinomycetes	Initial lesions spread and become chronic and granulomatous; can require amputation
Swimmer's itch	Cercariae of schistosomes	Itching due to cercariae burrowing into skin; immunological reaction prevents their spread
Dracunculiasis	*Dracunculus medinensis*	Larvae ingested in crustaceans in contaminated water migrate to skin and emerge through lesion; juveniles cause severe allergic reactions

Figure 19.20 Ophthalamia neonatorum. This gonococcal infection of the eye is most often seen in newborns, who acquire it as they pass through an infected birth canal. Adults, however, can also transfer bacteria to the eye from the genitals. (*National Medical Slide/ Custom Medical Stock Photo, Inc.*)

the eyes immediately after birth kills gonococci, but it is not effective against chlamydias and can irritate the eyes. Antibiotics such as penicillin, tetracycline, and erythromycin are effective against most, but not all, known bacterial causes of ophthalmia neonatorum.

Bacterial Conjunctivitis
Bacterial conjunctivitis, or pinkeye, is an inflammation of the conjunctiva caused by organisms such as *Staphylococcus aureus*, *Streptococcus pneumoniae*, *Neisseria gonorrhoeae*, *Pseudomonas* species, and *Haemophilus influenzae* biogroup *aegyptius*. The BPF clone of the latter causes Brazilian purpuric fever, in which young children initially have conjunctivitis and then develop potentially fatal septicemia with extensive hemorrhaging and symptoms of meningitis. Bacterial conjunctivitis is extremely contagious, especially among children, and

can spread rapidly through schools and day-care centers. Children rub itchy, runny eyes and transfer organisms to their playmates. In warm weather, gnats attracted to the moisture of tears may pick up organisms on their feet and transfer them to other persons' eyes. Topically applied sulfonamide ointment is an effective treatment. Children should not return to school until their infection is completely eliminated.

Trachoma
Trachoma (Figure 19.21), from the Greek word meaning "pebbled," or "rough," is caused by specific strains of *Chlamydia trachomatis*. The disease is marked by severely swollen conjunctiva with a pebbled appearance. Scarring of the eyelids causes the eyelashes to point inward, which leads to scarring and destruction of the cornea and eventually to blindness. Trachoma is the leading cause of preventable blindness worldwide; 500 million people have trachoma, and over 20 million are already blind from it. Recent research has shown that many, if not most, of these cases of blindness are actually due to secondary bacterial infections developing in the trachoma-infected tissues, rather than due to the trachoma alone. Although uncommon in the United States, except among Native Americans in the Southwest, the disease is widespread in parts of Asia, Africa, and South America, sometimes affecting 90% of the population. Flies are important mechanical vectors, and close mother-child contact facilitates human transfer.

Trachoma was described on Egyptian papyri in 1500 B.C.

VIRAL EYE DISEASES

Epidemic Keratoconjunctivitis
Epidemic keratoconjunctivitis (EKC), caused by an adenovirus, is sometimes called *shipyard eye*; workers are

TRY IT

What Grows in Your Health and Beauty Aids?

Are the shampoos, contact lens fluids, mascara, cold cream, and aerosol spray cans in your house sterile? Or are you smearing and spraying microorganisms on yourself and into your environment? A product may have passed the quality control tests at the factory, but after months on a store shelf, the antimicrobial "preservatives" may no longer be effective. Single-dose containers such as the bottles of shampoo in a hotel room are not required to have the preservatives needed in products that will have repeated reuse, with reinoculation of organisms into the product each time. With aerosol cans, check both the liquid inside and the nozzle. The product may be sterile, but a contaminated nozzle can spray organisms into the air. Does the storage case for your contact lenses have a biofilm of organisms on its inner sides?

Figure 19.21 Trachoma. This bacterial disease, caused by *Chlamydia trachomatis*, is the leading cause of preventable blindness in the world, affecting some 500 million people. Notice the pebbled appearance of the vastly swollen conjunctiva. (*Science VU/Visuals Unlimited*)

Save Your Eyes

When soft contact lenses became popular in the 1970s, fungal and bacterial eye infections became common among wearers. Fungi are not part of the normal microflora of the eye. Some contact lens users developed fungal keratitis (corneal ulcers) because the fungus grew in their hydrogel lens and then attacked the eye surface. Fungi of the genus *Fusarium* are usually the cause. People who complain that they can't get the spots off their contact lenses may not realize that those little spots are fungal colonies growing in the lens material itself. This problem typically results from poor hygiene and improper disinfection.

Fungal eye infections are a special problem for two reasons. First, we don't have an extensive arsenal of effective drugs against fungi, as we do against bacteria. Second, fungal infections are slow to develop, so people may not pay much attention to them at first.

Conversely, bacterial eye infections, such as those caused by *Pseudomonas*, can develop so rapidly that a person can suffer serious eye damage within 24 hours. The careless use of mascara is one common way of getting a corneal ulcer, usually caused by *Pseudomonas*. Within 24 to 48 hours after the initial use of health and beauty aids such as mascara, the material in the container is well colonized by *Pseudomonas* and other organisms. Scratching an eye with a contaminated mascara applicator introduces microbes into the cornea, allowing an infection to start. The person often suffers severe pain and redness within 3 to 4 hours after the injury and seeks treatment, unlike the victim of the slow, insidious attack of fungi.

Cosmetics, especially mascara, should be discarded after 3 to 6 months of use. Preservatives in them do not last forever, and bacteria multiply rapidly. The big "one-year-supply" jar of a cosmetic may not be a bargain after all.

frequently infected from dust particles in the environment. After 8 to 10 days' incubation, the conjunctiva become inflamed, and eyelid edema, pain, tearing, and sensitivity to light follow. Within 2 days the infection spreads to the corneal epithelium and sometimes to deeper corneal tissue. Clouding of the cornea can last up to 2 years. EKC can be nosocomially acquired in eye clinics and ophthalmologists' offices.

Acute Hemorrhagic Conjunctivitis

Another viral eye disease, **acute hemorrhagic conjunctivitis (AHC)**, caused by an enterovirus, appeared in 1969 in Ghana. Serological studies showed it was not prevalent anywhere in the world before then. First seen in the United States in 1981, the disease occurs chiefly in warm, humid climates under conditions of crowding and poor hygiene. AHC causes severe eye pain, abnormal sensitivity to light, blurred vision, hemorrhage under conjunctival membranes, and sometimes transient inflammation of the cornea. Onset is sudden, and recovery usually is complete in 10 days. A rare complication is a paralysis resembling poliomyelitis.

PARASITIC EYE DISEASES

Onchocerciasis (River Blindness)

The filarial (threadlike) larvae of the roundworm *Onchocerca volvulus* cause **onchocerciasis** (ong"ko-ser-si'a-sis), or **river blindness**, in many parts of Africa and Central America (**Figure 19.22**). Adult worms and microfilariae (small larvae) accumulate in the skin in nodules (◀Chapter 11, p. 332). When a blackfly bites an infected individual, it ingests microfilariae, which mature in the fly and move to its mouthparts. When the fly bites again, infectious microfilariae enter the new host's skin and invade various tissues, including the eyes. In many small villages where the people depend on water from blackfly-infested rivers, nearly all inhabitants who live past middle age are blind.

Adult worms cause skin nodules that can abscess. Microfilariae cause skin depigmentation and severe dermatitis via immunological responses to live filariae or toxins from dead ones. The worst tissue damage occurs as the worms invade the cornea and other parts of the eye. Over several years they cause eye blood vessels to become fibrous, and total blindness ensues by about age 40.

Onchocerciasis can be diagnosed by finding microfilariae in thin skin samples or adult worms visible through the skin. The drug ivermectin kills adult worms quickly and microfilariae over several weeks. Drugs are available to kill microfilariae rapidly, but toxins from very many dying microfilariae can cause anaphylactic shock. Onchocerciasis could be prevented by eliminating blackflies, which congregate along rivers. DDT destroys some flies, but it allows resistant ones to survive, and it accumulates in the environment.

Figure 19.22 Onchocerciasis, or river blindness. This disease is caused by the roundworm *Onchocerca volvulus*, whose microfilarial stages are transmitted by the bite of the blackfly. In some African villages, nearly all adults are blind and must be led by children, the only ones in the village who can still see, although they are already infected with the worms that will eventually blind them as well. *(Bettmann/Corbis)*

PUBLIC HEALTH

To Bathe or Not to Bathe

The historic baths in Bath, England, have been closed. Why? Because contamination of water sources by an amoeba, *Acanthamoeba*, led to a fatal amoebic brain infection in one visitor. But Bath is not the only place where *Acanthamoeba* is found. In the mid-1980s, 24 cases of *Acanthamoeba* keratitis were reported to the CDC, 20 of which were contact lens wearers. Since then, over 100 cases have been reported. The amoeba that causes this condition is found in brackish water, seawater, and hot tubs, as well as contaminated contact lens cleaning solution. It can also travel via airborne dust. The protozoan infection causes severe eye pain and destruction of corneal epithelium. Some patients have been successfully treated with ketoconazole or miconazole, but others have required corneal transplants.

The small size of the Pygmies of Uganda is due to onchocerciasis infections. When pregnant women are infected by *O. volvulus* (and most of them are), the parasite damages the pituitary gland of their fetuses. The result is dwarfism from a deficiency of growth hormone. A forest-dwelling strain of *O. volvulus* does not cause blindness but leads to severe itching, as well as psychological distress, in more than 9 million infected Africans. Treatment with ivermectin began in 1996.

Loaiasis

The eye worm *Loa loa*, a filarial worm endemic to African rainforests, is transmitted to humans by deer flies. Adult worms live in subcutaneous tissues and eyes (**Figure 19.23**); microfilariae appear in peripheral blood in the daytime and concentrate in the lungs at night. Deer flies feed during the day and acquire microfilariae from infected humans. The worms develop in the flies, migrate to their mouthparts, and are transferred to other humans when the flies feed again. Microfilariae migrate through subcutaneous tissue, leaving a trail of inflammation, and often settle in the cornea and conjunctiva. Although they usually do not cause blindness, the shock of finding a worm over an inch long in one's eye is undoubtedly a traumatic experience! In fact, the name *Loa loa* is believed to be derived from a voodoo term meaning "the rising of a devil within you."

Loaiasis (lo′ah-i″a-sis) is diagnosed by finding microfilariae in blood or by finding worms in the skin or eyes. It is treated by excising adult worms or by the use of suramin or other drugs to eradicate microfilariae. Control could be achieved by eradicating deer flies, but this is an exceedingly difficult task.

Diseases of the eyes are summarized in **Table 19.3**.

✓**CHECKLIST**

1. Why is it incorrect to say that you "catch" an athlete's foot infection?

2. How is river blindness transmitted?

3. What is thrush, and why is it especially common in AIDS patients?

4. Distinguish between ophthalmia neonatorium, bacterial conjunctivitis, and trachoma.

▐▐▐ WOUNDS AND BITES

Intact skin protects against most infectious disease agents, but we have seen that some are able to penetrate unbroken skin and that mucous membranes are not always effective barriers. In other cases, wounds and bites break the protective barrier of skin, allowing organisms to cause disease. (We discuss rabies, probably the most well-known, bite-related disease, in Chapter 23 because it is so closely associated with the nervous system.)

WOUND INFECTIONS

Gas Gangrene

Associated with deep wounds, **gas gangrene** often is a mixed infection caused by two or more species of *Clostridium*, especially *C. perfringens* (found in 80 to 90% of cases), *C. novyi*, and *C. septicum*. Spores of these obligately anaerobic organisms are introduced by injuries or surgery into tissues where circulation is impaired and dead and anaerobic tissue is present. In body regions where oxygen concentration is low, spores germinate, multiply, and produce toxins and enzymes such as collagenases, proteases, and lipases that kill other host cells and extend the anaerobic environment.

The onset of gas gangrene occurs suddenly 12 to 48 hours after injury. As the organisms grow and ferment muscle carbohydrates, they produce gas, mainly hydrogen,

Figure 19.23 The removal of a *Loa loa* worm from a patient's eye. *(Science VU/Armed Forces Institute of Pathology/Visuals Unlimited)*

TABLE 19.3

A Summary of Diseases of the Eyes

Disease	Agent	Characteristics
Bacterial Eye Diseases		
Ophthalmia neonatorum	*Neisseria gonorrhoeae*	Infection acquired during passage through birth canal causes corneal lesions and can lead to blindness; silver nitrate or antibiotics have nearly eradicated it in developed countries
Bacterial conjunctivitis	*Haemophilus influenzae* biogroup *aegyptius*, *Staphylococcus aureus*, *Streptococcus pneumoniae*, *Neisseria gonorrhoeae*, *Pseudomonas* species	Highly contagious inflammation of the conjunctiva in young children
Trachoma	*Chlamydia trachomatis*	Infection and destruction of cornea and conjunctiva; cause of preventable blindness
Keratitis	Bacteria, viruses, and fungi	Ulceration of the cornea; occurs mainly in immunodeficient and debilitated patients
Viral Eye Diseases		
Epidemic keratoconjunctivitis	Adenovirus	Inflammation of the conjunctiva that spreads to the cornea; transmitted in dust particles; also nosocomial
Acute hemorrhagic conjunctivitis	Enterovirus	Severe pain and hemorrhage under conjunctiva; highly contagious under crowded, unsanitary conditions
Parasitic Eye Diseases		
Onchocerciasis	*Onchocerca volvulus*	Microfilariae enter skin through blackfly bite and invade eyes and other tissues; causes dermatitis and blindness; occurs in tropics
Loaiasis	*Loa loa*	Microfilariae enter skin through deer fly bite; cause inflammation of conjunctiva and cornea

and gas bubbles distort and destroy tissue (**Figure 19.24**). Such tissue is called **crepitant** (rattling) **tissue**. The bubbles audibly "snap, crackle, and pop" when the patient is repositioned. Foul odor is such a prominent feature of gas gangrene that medical personnel can diagnose it without even entering the patient's room. High fever, shock, massive

Figure 19.24 **Gas gangrene.** In this photograph, the disease has blackened the toes of an infected foot. *(Science VU/Visuals Unlimited)*

tissue destruction, and blackening of skin accompany this rapidly spreading disease. If left untreated, death occurs swiftly. Diagnosis usually is made on the basis of clinical findings, and treatment is begun before laboratory test results are available. Penicillin is administered, and dead tissue is removed or limbs are amputated. Gas gangrene often follows illegal abortions performed under unsanitary conditions; it usually necessitates a hysterectomy. It is also more common in diabetics with high blood sugar. The increased sugar levels provide an adequate carbohydrate supply in muscles for clostridial organisms to ferment. *C. perfringens* is sometimes present in bile and can occasionally cause gas gangrene of abdominal muscles following gallbladder or bile duct surgery, especially if bile is spilled.

The use of *hyperbaric* (high-pressure) oxygen chambers to treat gas gangrene is somewhat controversial. Patients are placed in chambers containing 100% oxygen at 3 atmospheres pressure for 90 minutes two or three times a day. The actual mechanism by which high-pressure oxygen aids recovery is not known, but presumably it kills or inhibits obligate anaerobes. Gas gangrene can be prevented by adequate cleansing of wounds, delaying closing of wounds, and providing drainage when they are closed. No vaccine is available.

OTHER ANAEROBIC INFECTIONS

In addition to clostridia, certain non-spore-forming anaerobes are associated with some infections. *Bacteroides* and *Fusobacterium* species are normally present in the digestive tract, and *Bacteroides* accounts for nearly half of human fecal mass. *Fusobacterium* sometimes causes oral infections. If introduced into the abdominal cavity, genital region, or deep wounds by surgery or human bites, it causes infections there, too. Such infections are difficult to treat because the organisms are resistant to many antibiotics. Abscesses caused by anaerobic organisms must be drained surgically, and appropriate antibiotics used as supportive therapy.

Cat Scratch Fever

Cats are mechanical vectors of **cat scratch fever**, a disease caused by two different organisms: *Afipia felis*, a Gram-negative bacillus with a single flagellum, and, more commonly, the rickettsia *Bartonella (Rochalimaea) henselae*. The organisms are found mainly in capillary walls or in microabscesses. A cat scratch lesion resembling Kaposi's sarcoma has been found in AIDS patients. Another cat scratch lesion found in AIDS patients, as well as in other immunocompromised patients, is bacillary peliosis, in which blood-filled cavities form in bone marrow, liver, and spleen. Presumably cats acquire the organisms from the environment and carry them on claws and in the mouth. Over 40% of cats, especially kittens, carry the infective organisms without themselves being sick. When cats scratch, bite, or lick, they transmit the organisms to humans. Cat fleas may also play a role in transmission. After 3 to 10 days, a pustule appears at the site of entry **(Figure 19.25)**, and the patient has a mild fever, headache, sore throat, swollen glands, and conjunctivitis for a few weeks. Diagnosis is based on clinical findings and a history of contact with cats. Treatment with the antibiotics tetracycline or doxycycline is effective against cases caused by *Bartonella (Rochalimaea)* whereas cases caused by *Afipia* generally do not respond, making only symptomatic relief possible. No vaccine is available, and the only means of prevention is to avoid contact with cats. Over 25,000 cases per year occur in the United States.

Figure 19.25   Cat scratch fever disease lesion. A pus-filled nodule appears at the site of the scratch or bite, usually within 21 days. *(Kenneth E. Greer/Visuals Unlimited)*

Rat Bite Fever

One type of **rat bite fever** is caused by *Streptobacillus moniliformis*, which is presents in the nose and throat of about half of all wild and laboratory rats. However, only about 10 % of people bitten by rats develop rat bite fever. Most cases result from bites of wild rats; half are reported in children under 12 living in overcrowded, unsanitary conditions. It can also result from bites and scratches of mice, squirrels, dogs, and cats.

Rat bite fever begins as a localized inflammation at a bite site that heals promptly. In 1 to 3 days, headache begins and new lesions appear elsewhere, especially on palms and soles. Fever is intermittent. By the distribution and appearance of the rash, the disease sometimes is mistaken for Rocky Mountain spotted fever. An arthritis that develops can permanently damage joints.

Another form of rat bite fever, **spirillar fever**, is caused by *Spirillum minor*. First described in Japan as *sodoku* and now known to exist worldwide, this disease is still poorly understood. The initial bite heals easily, but 7 to 21 days later it flares up and occasionally forms an open ulcer. Chills, fever, and inflamed lymph nodes accompany a red or dark purple rash that spreads out from the wound site. After 3 to 5 days symptoms subside, but they can return after a few days, weeks, months, or even years.

Diagnosis of both forms of rat bite fever is made by dark-field examination of *exudates* (oozing fluids). The disease is treated with streptomycin or penicillin; without treatment the mortality rate is about 10 %. Technicians bitten by rodents should disinfect the bite site, seek medical treatment, and be alert for rat bite fever symptoms.

Other Bite Infections

Pasteurella multocida is transmitted by cat or dog bites and scratches. It is part of the normal flora of many wild and domestic animals' mouth, nasopharynx, and gastrointestinal tracts. Infections are reddened, diffuse cellulitis, usually localized in the soft tissue adjacent to the bite, and often appear within 24 hours. Virulence factors include production of an endotoxin, and a capsule that helps prevent phagocytosis. Penicillin is the drug of choice—something unusual for a Gram-negative bacillus.

Eikenella corrodens is part of the normal flora of the human mouth (often associated with chronic periodontal disease) and gastrointestinal tract. It causes opportunistic infections of human bites, and to fists damaged by hitting teeth. Treatment is with antibiotics such as penicillins, quinolones, and cephalosporins—although abscesses may need to be lanced and drained first.

ARTHROPOD BITES AND DISEASES

A variety of arthropods, including ticks, mites, and insects, cause human disease either directly or as vectors of pathogens. Many arthropods are ectoparasites that feed on blood. Their bites can be painful, and their toxins can lead to the development of anaphylactic shock. Here we consider the direct effects of bite injuries caused by arthropods; the diseases they transmit are discussed in other chapters.

Tick Paralysis

As ectoparasites, ticks attach themselves to the skin of a host, where they cause both local and systemic effects. The local effect is a mild inflammation at the bite site. Systemic effects usually occur when any of several species of hard-bodied ticks attach to the back of the neck, near the base of the skull, and feed for several days. This allows deep diffusion of anticoagulants and toxins secreted into the bite via the tick's saliva. The anticoagulant prevents the host's blood from clotting while the tick feeds on it. The toxins can cause **tick paralysis**, especially in children. Although the exact chemical composition of these toxins is not known, they appear to be produced by the ovaries of the ticks. They cause fever and paralysis, which affects first the limbs and eventually respiration, speech, and swallowing. Removal of the ticks when symptoms first appear prevents permanent damage. Failure to remove ticks can lead to death by cardiac or respiratory arrest. Both humans and livestock can be affected.

Although many other methods have been popularized over the years, the best way to extract a tick is to pull it straight out of the skin with tweezers.

Chigger Dermatitis

The term *mites* refers not to a particular species but to an assortment of species. As ectoparasites, adult mites attach to a host long enough to obtain a blood meal and then usually drop off. Chiggers, the larvae of certain species of *Trombicula* mites, burrow into the skin and release proteolytic enzymes that cause host tissue at the bite site to harden into a tube. The chiggers then insert mouthparts into the tube and feed on blood. They cause itching and inflammation in most people and can cause a violent allergic reaction called **chigger dermatitis** in sensitive individuals. Attempts to "smother" a chigger in its tube by painting nail polish, for example, over the itchy bite are misguided. Chiggers are especially prevalent along the southeastern coast of the United States. Unlike some South American chiggers, those found in the United States cease feeding and drop to the ground hours before the itching begins.

Figure 19.26 Scabies. *(©Ken Greer/Visuals Unlimited)*

Scabies and House Dust Allergy

Scabies, or **sarcoptic mange**, is caused by the itch mite *Sarcoptes scabiei*. By the time intense itching appears, the lesions usually are quite widespread **(Figure 19.26)**. Scratching lesions and causing them to bleed provides an opportunity for secondary bacterial infections. Scabies is spread by close human contact and can be transmitted by sexual activities. Outbreaks are especially problematic in hospitals and nursing homes. Disinfection of linens and strict isolation are necessary to prevent spread of the infestation.

House dust mites are ubiquitous and present another problem. We all inhale airborne mites or their excrement. Although this phenomenon is unappealing, it causes disease only in individuals with an allergy to house dust. Two other mites are human commensals: *Demodex folliculorum* lives in hair follicles, and *D. brevis* in sebaceous glands. The incidence of these mites in humans increases with age from 20 % in young adults to nearly 100 % in the elderly.

Flea Bites

The sand flea, *Tunga penetrans*, is also sometimes called a chigger because it burrows into the skin and lays its eggs. Sand fleas cause extreme itching, inflammation, and pain and provide sites for secondary infections, including infection with tetanus spores. Surgical removal and sterilization of the wound is used to treat sand flea infections. Fleas in homes and on pets are controlled by insecticides with residual effects for days to weeks after application. Wearing shoes and avoiding sandy beaches help protect against sand fleas. Although all fleas encountered by humans are a nuisance, only those that carry infectious agents are a public health hazard.

Figure 19.27 Screwworm fly larva (*Cochliomyia hominivorax*) on a steer. The eggs laid near the eyes have hatched into maggots that are now eating the eye and tunneling into the head. *(Courtesy Ann Czapiewski, Animal and Plant Health Inspection Service, USDA)*

APPLICATIONS

Screwworm Eradication

An ingenious method of controlling screwworm maggots, which do great harm to cattle, was developed in the 1930s by Edward Knipling of the U.S. Department of Agriculture. He captured and irradiated male screwworm flies with radioactive cobalt, making them incapable of fertilizing eggs. If females, which mate only once, mate with sterile males, they lay infertile eggs, so far fewer new flies are produced. Irradiated males must be released frequently to maintain a high percentage of sterile males in the population. Another control method uses insects, especially wasp larvae, that are parasitic on other insects. Farmers sometimes purposely release parasitic wasps to infect crop-destroying insects and thereby reduce crop damage. Both the use of sterilized males and the release of parasitic insects utilize natural habits of insects; neither contributes to environmental pollution, as most pesticides do.

Pediculosis

Common expressions such as "nit-picking," "going over with a fine-toothed comb," "lousy," and "nitty-gritty" attest to the association of lice with humans over the ages. To stay alive, lice must remain on their hosts for all but very short periods of their life. They glue their eggs (nits) to fibers of clothing and hair. Two varieties of the louse *Pediculus humanus* parasitize humans. One lives mainly on the body and on clothing in temperate climates where clothing usually is worn; the other lives on hair in any climate. **Pediculosis**, or lice infestation, results in reddened areas at bites, dermatitis, and itching. A lymph exudate from bites provides an ideal medium for secondary fungal infections, especially in the hair. The crab louse, *Phthirus pubis* (◀Figure 11.19b), clings to skin more tightly than does the body louse and causes intense itching at bites, especially in the pubic area. It is transmitted between humans by close physical contact, but the louse itself is not known to transmit other diseases. Insecticides can be used to eradicate lice, but sanitary conditions and good personal hygiene must be maintained to prevent their return.

Other Insect Bites

Blackflies inflict vicious wounds, and sensitive individuals who are bitten get **blackfly fever**, characterized by an inflammatory reaction, nausea, and headache. Bloodsucking flies, such as the tsetse fly and the deer fly, are related to horse flies. All inflict painful bites and sometimes cause anemia in domestic animals.

Myiasis (mi-i′a-sis) is an infection caused by maggots (fly larvae). Wild and domestic animals are susceptible to myiasis in wounds. Human myiasis can occur when larvae of the botfly, a greenish, metallic-looking fly, penetrate mucous membranes and small wounds. The larvae of more than 20 species of flies, including the common housefly, can inhabit the human intestine. One such species is the botfly, aptly named *Gastrophilus intestinalis*. The Congo floor maggot, the only known bloodsucking larva, sucks human blood. In the daytime it hides in soil and debris or beneath dirt floors; at night it seeks blood meals from hosts that are usually asleep.

Screwworm fly larvae enter cattle via wounds or openings such as the ears **(Figure 19.27)**. The maggots tunnel beneath the animals' skin, causing great pain and opening avenues for infection. One steer may have over 100 maggots tunneling in its head, and the poor beast will run frantically, barely stopping to eat or drink. Ranchers lost billions of dollars before the screwworm was eradicated in North America. It continues to plague other parts of the world today.

Many mosquitoes, bedbugs, and bloodsucking insects feed on humans and leave painful, itchy bites. Bedbugs hide in crevices during the day and come out at night to feed on their sleeping victims. Inflammation at bites results from an allergic reaction to the bug's saliva. Large numbers of bites can lead to anemia, especially in children. Mosquitoes and many other insects can be eradicated by the application of insecticides that have long-term residual effects. They can be kept out of dwellings by the use of tight housing construction and solid roofs instead of thatch, and by household cleanliness.

Wound and bite infections are summarized in **Table 19.4.**

✓ CHECKLIST

1. What is crepitant tissue?

2. What is a "nit"?

3. What economic impact does the screwworm have in North America today?

TABLE 19.4

A Summary of Wound and Bite Infections		
Disease	**Agent**	**Characteristics**
Wound Infections		
Gas gangrene	*Clostridium perfringens* and other species	Deep wound infections with gas production in anaerobic tissue; tissue necrosis and death can result if not treated promptly
Cat scratch fever	*Afipia felis* and *Bartonella (Rochalimaea) henselae*	Pustules at scratch site, fever, and conjunctivitis
Rat bite fever	*Streptobacillus moniliformis*	Inflammation at bite site, dissemination of lesions, intermittent fever
Spirillar fever	*Spirillum minor*	Inflammation at rat bite site heals; later reinflammation, fever, and rash
Pasteurella multocida infection	*Pasteurella multocida*	Inflammation at bite site
Eikenella corrodens infection	*Eikenella corrodens*	Inflammation at bite site
Athropod Bites and Infections		
Tick paralysis	Various ticks	Toxins introduced with tick bite cause fever and ascending motor paralysis
Chigger dermatitis	*Trombicula* mites	Larvae burrow into skin and cause itching and inflammation; can cause violent allergic reactions
Scabies	*Sarcoptes scabiei*	Widespread lesions with intense itching; other mites cause house dust allergy when their feces are inhaled by allergic individuals
Flea bites	*Tunga penetrans*	Itching and inflammation from adult females in skin
Pediculosis	*Pediculus humanus*	Inflammation at louse bite sites and itching; *Phthirus pubis* (pubic louse) found in pubic areas
Other Insect Bites		
Blackfly fever	Blackfly	Bites cause severe inflammatory reaction in sensitive individuals
Myiasis	Fly larvae	Maggots infect wounds in animals; Congo floor maggot sucks human blood; screwworms injure cattle
Mosquito and other bites	Mosquitoes, some flies, bedbugs	Painful, itchy bites; several insects serve as disease vectors

█ RETRACING OUR STEPS

• The agents and characteristics of the diseases discussed in this chapter are summarized in Tables 19.2, 19.3, and 19.4. Information in those tables is not repeated in this summary.

III THE SKIN, MUCOUS MEMBRANES, AND EYES

THE SKIN

• Skin consists of an outer **epidermis** and an inner **dermis**.

• Skin, a nonspecific defense, is a physical barrier that secretes antimicrobial chemicals, including acidic and salty secretions, waterproofing **keratin**, and **sebum**.

• Skin infections occur anywhere the skin is broken, in ducts of **sebaceous** and **sweat** (sudoriferous) **glands**, in hair follicles, and sometimes on unbroken skin.

MUCOUS MEMBRANES

• Mucous membranes consist of a thin layer of cells that secrete **mucus**. Pathogens can become trapped in mucus.

THE EYES

• Protective structures associated with the eyes include the eyelids, eyelashes, **conjunctiva**, **cornea**, and **lacrimal glands**.

• **Lysozyme** in tears is microbicidal. Eyes do not have a normal microflora.

NORMAL MICROFLORA OF THE SKIN

• Large populations of normal microflora colonize the skin, differing widely at various locations. The majority are Gram-positive organisms such as *Staphylococcus* and *Micrococcus*, and coryneform bacteria.

• Metabolic by-products of normal microflora produce an acid pH that is inhospitable to most pathogens. Salt in sweat also inhibits growth of most microbes.

• Continuous shedding of skin cells carries away microbes that have colonized them.

▌▌▌ DISEASES OF THE SKIN

BACTERIAL SKIN DISEASES

• Bacterial skin diseases are usually transmitted by direct contact, droplets, or fomites.

• Most such diseases (including **scalded skin syndrome, scarlet fever, erysipelas,** and **pyoderma**) can be treated with penicillin or other antibiotics. Treating **acne** with antibiotics may contribute to the development of antibiotic-resistant organisms. Burn infections often are nosocomial and caused by antibiotic-resistant organisms.

VIRAL SKIN DISEASES

• **Rubella (German measles), measles (rubeola),** and **chickenpox** viruses are usually transmitted via nasal secretions. Pox diseases (including **smallpox, cowpox,** and **molluscum contagiosum**) and **warts (papillomas)** are usually transmitted by direct contact.

• Treatment of viral skin diseases can usually relieve some symptoms but not cure the disease; warts can be excised or treated chemically.

FUNGAL SKIN DISEASES

• Fungal skin diseases are called **dermatomycoses.** They include various types of **ringworm,** including **athlete's foot (tinea pedis), sporotrichosis, blastomycosis, candidiasis (moniliasis), aspergillosis,** and **zygomycoses.**

• Subcutaneous fungal infections often persist despite treatment with topical fungicides. Systemic infections are even more difficult to eradicate.

• Because of the difficulty in treating fungal skin infections, prevention by maintaining healthy skin and avoiding contamination of wounds is especially important.

OTHER SKIN DISEASES

• Other skin diseases include those caused by both fungi and bacteria (such as **Madura foot**) and by parasitic helminths (**swimmer's itch** and **dracunculiasis**).

▌▌▌ DISEASES OF THE EYES

BACTERIAL EYE DISEASES

• Bacterial eye diseases include **ophthalmia neonatorum, bacterial conjunctivitis** (pinkeye), and **trachoma.** They are transmitted by direct contact, by fomites, by insect vectors, and (in certain infections) by contamination of infants during delivery.

• Most bacterial eye diseases are treated with antibiotics and prevented by good sanitation.

VIRAL EYE DISEASES

• Viral eye diseases, such as **epidemic keratoconjunctivitis** and **acute hemorrhagic conjunctivitis,** are transmitted by dust particles or direct contact. No effective treatments are available, but good sanitation can help prevent such diseases.

PARASITIC EYE DISEASES

• Controlling insect vectors reduces transmission of parasitic helminthic eye diseases, which include **onchocerciasis (river blindness)** and **loaiasis.**

▌▌▌ WOUNDS AND BITES

WOUND INFECTIONS

• **Gas gangrene** and other anaerobic infections can be prevented by careful cleaning and draining of deep wounds; they are treated with penicillin, antitoxins, and sometimes hyperbaric oxygen.

OTHER ANAEROBIC INFECTIONS

• **Cat scratch** and **rat bite fevers** can be prevented by avoiding injuries from cats and rats, respectively. Laboratory workers should be aware of the dangers of infections from rat bites.

• *Pasteurella multocida* and *Eikenella corrodens* also infect bites.

ARTHROPOD BITES AND DISEASES

• Arthropod bites cause diseases such as **tick paralysis, chigger dermatitis, scabies (sarcoptic mange), pediculosis,** and **myiasis.** Such diseases can be prevented by good sanitation and hygiene and protecting the skin from bites.

▌ TERMINOLOGY CHECK

abscess (*p. 578*)

acne (*p. 580*)

acute hemorrhagic
   conjunctivitis (AHC)
   (*p. 594*)

aspergillosis (*p. 590*)

athlete's foot (*p. 589*)

bacterial conjunctivitis
   (*p. 593*)

blackfly fever (*p. 599*)

blastomycetic dermatitis
   (*p. 590*)

blastomycosis (*p. 589*)

boil (*p. 578*)

candidiasis (*p. 590*)

carbuncle (*p. 578*)

cat scratch fever (*p. 597*)

chickenpox (*p. 583*)

chigger dermatitis (*p. 598*)

congenital rubella
   syndrome (*p. 581*)

conjunctiva (*p. 576*)

cornea (*p. 576*)

cowpox (*p. 586*)

crepitant tissue (*p. 596*)

débridement (*p. 581*)

dermal wart (*p. 587*)

dermatomycosis (*p. 588*)

dermatophyte (*p. 588*)

dermis (*p. 575*)

dracunculiasis (*p. 591*)

epidemic keratoconjunctivitis
   (EKC) (*p. 593*)

epidermis (*p. 575*)

erysipelas (*p. 579*)

eschar (*p. 580*)

exanthema (*p. 581*)

folliculitis (*p. 578*)

furuncle (*p. 578*)

gas gangrene (*p. 595*)

genital wart (*p. 587*)

German measles (*p. 581*)

human papillomavirus (HPV)
   (*p. 586*)

impetigo (*p. 579*)

keratin (*p. 575*)

keratitis (*p. 591*)

Koplik's spot (*p. 582*)

lacrimal gland (*p. 577*)

loaiasis (*p. 595*)

Madura foot (*p. 591*)

maduromycosis (*p. 591*)

measles (*p. 582*)

measles encephalitis (*p. 582*)

molluscum contagiosum
   (*p. 586*)

moniliasis (*p. 590*)

monkeypox (*p. 586*)

mucous membrane (*p. 576*)

mucus (*p. 576*)

myiasis (*p. 599*)
onchocerciasis (*p. 594*)
ophthalmia neonatorum
    (*p. 591*)
papilloma (*p. 586*)
pediculosis (*p. 599*)
pimple (*p. 578*)
pustule (*p. 578*)
pyoderma (*p. 579*)
rat bite fever (*p. 597*)
ringworm (*p. 588*)
river blindness (*p. 594*)
roseola (*p. 583*)

rubella (*p. 581*)
rubeola (*p. 582*)
St. Anthony's fire (*p. 579*)
sarcoptic mange (*p. 598*)
scabies (*p. 598*)
scalded skin syndrome
    (*p. 578*)
scarlet fever (*p. 579*)
sebaceous gland (*p. 576*)
sebum (*p. 576*)
shingles (*p. 583*)
smallpox (*p. 585*)
spirillar fever (*p. 597*)

sporotrichosis (*p. 589*)
sty (*p. 578*)
subacute sclerosing
    panencephalitis (SSPE)
    (*p. 582*)
sweat gland (*p. 576*)
swimmer's itch (*p. 591*)
systemic blastomycosis
    (*p. 590*)
thrush (*p. 590*)
tick paralysis (*p. 598*)
tinea barbae (*p. 588*)
tinea capitis (*p. 588*)

tinea corporis (*p. 588*)
tinea cruris (*p. 588*)
tinea pedis (*p. 589*)
tinea unguium (*p. 588*)
trachoma (*p. 593*)
vaginitis (*p. 590*)
varicella-zoster virus
    (VZV) (*p. 583*)
wart (*p. 586*)
zygomycosis (*p. 590*)

▌ CLINICAL CASE STUDY

John has a cat that sleeps with him. He has recently had recurring ringworm infections. His doctor has asked him if he has pets. Why do you think the doctor is asking this question?

What are the signs and symptoms of ringworm? Visit the following web site to learn more:http://www.cdc.gov/healthypets/diseases/ringworm.htm

▌ CRITICAL THINKING QUESTIONS

1. A skin disease caused by the molluscum contagiosum virus is unusual in that skin lesions take on the form of tumor-like growths. What properties of this virus might explain this phenomenon?

2. Why does *Candida albicans* sometimes cause disease when a person takes large amounts of antibiotics?

3. Smallpox, once a major killer, was the first disease to be eradicated from the entire world. What factors allowed the eradication of this disease?

▌ SELF-QUIZ

1. All of the following characteristics of skin help make it a good defensive barrier against pathogenic microbes EXCEPT:
(a) Springy elasticity
(b) Saltiness due to production of sweat from sweat glands
(c) Acidic pH due to oily compound (sebum) metabolism
(d) Normal microflora of skin
(e) Constant sloughing off of old dead epithelial skin cells

2. The majority of microorganisms comprising the normal microflora inhabiting the skin surface are:
(a) Gram-positive fungi of the single-celled yeasts
(b) Gram-positive bacteria of the *Staphylococcus, Micrococcus, Corynebacterium*, and *Proprionibacterium* genera
(c) Gram-negative bacteria of the Enterobacteriaceae family
(d) Gram-negative bacteria of the *Neisseria, Pseudomonas*, and *Salmonella* genera
(e) All of the above

3. Humans have protective devices against microbial invasion or infection of the eyes and mucous membranes. Match the following types of protection to their description:
___ Lysozyme   (a) Physical barrier of the eye
___ Eyelashes   (b) Physical barrier of mucous
___ Eyelid   membranes
___ Mucus   (c) Glandular secretion of the eye
___ IgA   (d) Glandular secretion of the mucous
___ Conjunctiva   membranes
___ Tears

4. Scalded skin syndrome is caused by exotoxins (exfoliatins) produced by:
(a) *Streptococcus pyogenes*   (d) *Pseudomonas aeruginosa*
(b) *Staphylococcus aureus*   (e) *Demodex folliculorum*
(c) *Propionibacterium acnes*

5. Pimples and scalded skin syndrome are caused by _____ whereas scarlet fever and erysipelas are caused by _____:

but pyoderma and impetigo could be caused by
_____.
(a) *Streptococcus, Staphylococcus aureus*; neither
(b) *Pseudomonas aeruginosa*; *Streptococcus*; both
(c) *Staphylococcus aureus*; *Streptococcus*; both
(d) *Staphylococcus aureus*; *Pseudomonas*; neither
(e) None of the above

6. The pus-producing skin infection (pyoderma) can be caused by:
(a) Staphylococci
(b) Streptococci
(c) Corynebacteria
(d) A combination of staphylococci, streptococci, and corynebacteria
(e) All of these

7. Which of the following is NOT a characteristic of impetigo?
(a) Common in children
(b) Can be caused by *Staphylococcus aureus*
(c) Can be caused by *Streptococcus pyogenes*
(d) Is highly contagious
(e) Can be caused by pseudomonads

8. Which of the following is NOT true about acne?
(a) It is most commonly the result of male sex hormones that stimulate increased sebaceous gland growth and secretion of sebum.
(b) *Propionibacterium acnes* in particular can turn a mild case of acne into a severe case by causing increased and widespread inflammation, tissue destruction, and scarring.
(c) Common treatments are frequent cleansing of the skin and topical ointments that reduce the risk of infection.
(d) Low doses of oral antibiotics are often prescribed to prevent bacterial infections of lesions and are without side effects.
(e) In most cases, acne disappears or decreases in severity on its own as the body adjusts to the hormonal changes of puberty and sebaceous gland function stabilizes.

9. Which of the following often cause nosocomial infections in burn patients?
(a) *Streptococcus pyogenes*   (d) *Staphylococcus aureus*
(b) *Serratia marcescens*   (e) b and c
(c) *Pseudomonas aeruginosa*

10. Match the following viral skin diseases with their descriptions:
___ Smallpox   (a) Mildest viral disease causing skin rash
___ Cowpox   by togavirus and is prevented by
___ Rubella   administration of attenuated virus in
    (German   the combined MMR vaccine
    measles)   (b) Paramyxovirus infection causing fever
___ Rubeola   and rash with "Koplik's spots"
    (measles)   (c) Highly contagious disease causing skin
___ Papilloma   lesions brought on by the varicella-
    (warts)   zoster herpesvirus and can reoccur
___Chickenpox   later in life causing shingles
                (d) Eradicated disease that brought on
                    sickness and death for centuries and
                    was found to be preventable by cow-
                    pox immunization by Edward Jenner
                (e) Transmission of the vaccinia virus from
                    cattle to humans causes this disease
                (f) Human papilloma virus (HPV) infec-
                    tion of skin and mucous membranes;
                    also responsible for 99% of all cervical
                    cancer cases

11. The bacterial eye diseases (ophthalmia neonatorum, bacterial conjunctivitis, and trachoma) can be transmitted by which of the following methods:
(a) Fomites   (d) Insect vectors
(b) The process of birth   (e) All of these above
(c) Direct contact

12. Gas gangrene is most likely associated with infection with:
(a) *Staphylococcus aureus*   (d) *Neisseria gonorrhoeae*
(b) *Clostridium perfringenes*   (e) *Pseudomonas*
(c) *Streptococcus pneumoniae*   *aeruginosa*

13. Hyperbaric oxygen may be useful in treating infections caused by:
(a) Gram-positive bacteria   (d) Yeast
(b) Gram-negative bacteria   (e) Viruses
(c) Anaerobic bacteria

14. The leading cause of preventable blindness in the world is caused by:
(a) *Chlamydia trachomatis*   (d) *Staphylococcus aureus*
(b) *Haemophilus influenzae* (e) *Streptococcus pneumoniae*
(c) *Neisseria gonorrhoeae*

15. Chickenpox is a disease of chickens and is caused by the poxvirus, cowpox is a disease of cattle, and smallpox is a human disease that commonly affects only children. True or false?

16. Antibiotic is used to treat the eyes of newborn infants when:
(a) The mother has gonorrhea
(b) The mother has genital herpes
(c) *Neisseria gonorrhoeae* is isolated from the newborn's eyes
(d) Always
(e) The mother has a history of multiple sex partners

17. The virus used to immunize individuals against smallpox causes the disease:
(a) Chickenpox   (d) Molluscum contagiosum
(b) Measles   (e) Warts
(c) Cowpox

18. A latent herpesvirus infection of cranial or spinal ganglia that can erupt along the pathway of the associated nerve(s) causing painful skin lesions is known as:
(a) Smallpox   (d) Shingles
(b) Chickenpox   (e) Warts
(c) Cowpox

19. Which of the following fungal diseases result from an opportunistic fungal infection?
(a) Zygomycoses   (d) All of the above
(b) Aspergillosis (e) None of the above
(c) Candidasis

20. Ringworm infections by *Tinea* spp. cause infections of the groin, nails, scalp, and beard. The term "ringworm":
(a) Refers to the small circular worms that cause these infections
(b) Refers to all disease-causing fungi
(c) Is misleading and gets its name from the shape of the ringlike lesions produced in *Tinea* spp. infections
(d) Is misleading and gets its name from the small circular shaped fungi
(e) b and d

21. A superficial infection (thrush) can be caused by *Candidia albicans*. This organism is a type of:
(a) Filamentous fungi
(b) Yeast
(c) Gram-positive bacteria
(d) Gram-negative bacteria
(e) Virus

22. Match the following etiological agents with their respective syndromes:

___ Adenovirus
___ *Sarcoptes scabiei*
___ *Chlamydia trachomatis*
___ *Neisseria gonorrhoeae*
___ *Pasteurella multocida*
___ Enterovirus
___ *Candida albicans*

(a) Vaginal yeast infection
(b) Acute hemorrhagic conjunctivitis
(c) Scabies
(d) Ophthalmia neonatorum
(e) Trachoma
(f) Cat or dog bites and scratches
(g) Epidemic keratoconjunctivitis

23. Pediculosis is an infestation of body lice that causes intense itching. True or false?

24. Match the following diseases with their respective etiological agent

___ Swimmer's itch
___ Cat scratch fever
___ Loaiasis
___ River blindness
___ Myiasis

(a) *Afipia* and *Bartonella*
(b) Nematode larvae from blackfly bites
(c) Filarial worm
(d) Schistosome larvae
(e) Screwworm fly larvae

25. Describe the harmful potential effects of the excessive use of alcohol-containing hand sanitizers.

26. Identify the various parts of the skin

d. _____

a. _____

e. _____

f. _____

b. _____

g. _____

h. _____

i. _____

c. _____

▮ EXPLORATIONS ON THE WEB

If you think you've mastered this chapter, there's more to challenge you on the web. Go to the companion web site to fine-tune your understanding of the chapter concepts and discover answers to the questions posed below.

1. Since the early 20th century *Staphylococcus* has been classified as among the deadliest of all disease-causing organisms. In fact, certain strains of *Staphylococcus* produce toxins that cause your skin to loosen and slough off. Then why do 20 to 30% of healthy people have staph on their skin or inside their nostrils? Find out why at the web site.

2. Edward Jenner is given credit for developing the first vaccine against smallpox, but the practice of inoculation with smallpox was first introduced in Britain by Lady Mary Wortley Montagu. Discover more at the web site.

20 Urogenital and Sexually Transmitted Diseases

Come with me . . .

When nature calls, we must answer—but where? We've all surely had to hold it at one time or another until we could get to a socially acceptable place to urinate. However, young children, pets, and wild animals are not too particular about where they urinate. Remember, though, that urine is a portal of exit for disease organisms.

Animals can, and often do, carry Leptospira bacteria in their urinary tracts. Deer, mice, rabbits, cattle, everyone urinates on the ground, and the rain washes residual Leptospira bacteria into rivers, streams, lakes, and ponds. Swim in these waters with your mouth open and you could acquire leptospirosis—a debilitating, difficult to diagnose, sometimes even fatal, disease of the kidneys. In chlorinated pools and larger bodies of water you're most likely safe, but think of leptospirosis the next time you let Fido jump in and out of the wading pool with the kids. One man found out the hard way that it's best not to let Fido swim upstream of you in a river, either. Fido evidently urinated in the water just ahead of his master and "shared" his leptospires. Better yet, get Fido and all other household pets vaccinated against leptospirosis. And whoever has to clean up the puppy puddles should be very diligent about keeping their own hands clean, remembering that leptospires can enter through skin and by contact with mucous membranes.

(Kavaler/Art Resource)

 Video related to this topic is available within WileyPLUS.

III COMPONENTS OF THE UROGENITAL SYSTEM

The Urinary System / The Female Reproductive System / The Male Reproductive System / Normal Microflora of the Urogenital System

III UROGENITAL DISEASES USUALLY NOT TRANSMITTED SEXUALLY

Bacterial Urogenital Diseases / Parasitic Urogenital Diseases

III SEXUALLY TRANSMITTED DISEASES

Acquired Immune Deficiency Syndrome (AIDS) / Bacterial Sexually Transmitted Diseases / Viral Sexually Transmitted Diseases

(a)

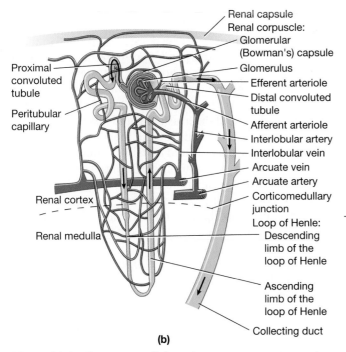

(b)

Figure 20.1 **Structures of the urinary system.** **(a)** Frontal view, including a blowup of the kidney. Infections of the urinary system often occur near the opening of the urethra, which serves as a portal of entry. The flow of urine out of the system tends to prevent many potential infections. **(b)** The structure of a nephron.

In this chapter we discuss diseases of the urogenital system, including sexually transmitted diseases. As with the other chapters on diseases, keep in mind what you have learned about normal microflora (◀Chapter 14, p. 404), disease processes (◀Chapter 14, p. 408), and body defenses (Chapters ◀16, 17, and 18). In studying urogenital diseases, it is important to recall that the urinary and reproductive (genital) systems are closely associated. Infections in one system easily spread to the other.

COMPONENTS OF THE UROGENITAL SYSTEM

The **urogenital system** includes the organs of both the urinary system and the reproductive system. Thus, in considering the urogenital system we will look at the structures and sites of infection in the urinary system, the female reproductive system, and the male reproductive system.

THE URINARY SYSTEM

The **urinary system** consists of paired kidneys and ureters, the urinary bladder, and the urethra (**Figure 20.1a**). The **kidneys** regulate the composition of body fluids and remove nitrogenous and certain other wastes from the body. To accomplish this, each kidney contains about 1 million functional units called **nephrons** (**Figure 20.1b**). In a nephron the fluid part of the blood is filtered from the **glomerulus**, a coiled cluster of capillaries, to the kidney tubules. Starting in the *renal cortex*, nephrons remove solutes, including wastes, and water from the blood. As these materials pass through the tubules known as the *renal medulla*, essential materials such as water, salts, and sugars pass back into the blood. **Urine**, the waste remaining in the kidney tubules, passes through collecting

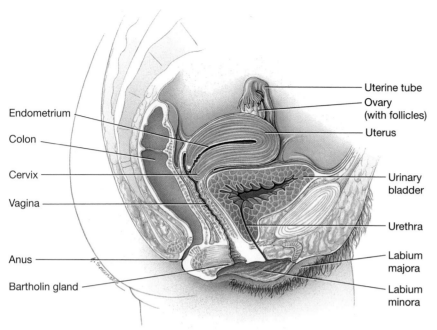

Endometrium

Colon

Cervix

Vagina

Anus

Bartholin gland

Uterine tube

Ovary (with follicles)

Uterus

Urinary bladder

Urethra

Labium majora

Labium minora

Figure 20.2 Structures of the female reproductive system. The portal of entry for the reproductive tract (vagina) is separate from the urinary tract (urethra). Chemical defenses, including cervical mucus, have antibacterial actions. In addition, secretions moving from the uterine (fallopian) tubes through the uterus and cervix to the vagina make microbial infections less likely.

ducts to the **ureter** of each kidney. The ureters carry urine, which is normally free of microbes, to the **urinary bladder**, where it is stored until released through the **urethra** during micturition (urination). **Urinalysis**, the laboratory analysis of urine specimens, can reveal imbalances in pH or water concentration, the presence of substances such as glucose or proteins, and other conditions associated with infections, metabolic disorders, and other diseases.

THE FEMALE REPRODUCTIVE SYSTEM

The **female reproductive system** consists of the ovaries, uterine tubes, uterus, vagina, and external genitalia (**Figure 20.2**). The paired **ovaries** contain cellular aggregations called **ovarian follicles**, each containing an *ovum* (plural: *ova*), or egg, and surrounding epithelial tissue. During a woman's reproductive years, an ovum capable of being fertilized is released once each month. The **uterine tubes** receive ova and convey them to the uterus. Fertilization usually occurs in the uterine tubes. The **uterus** is a pear-shaped organ in which a fertilized ovum develops. It is lined with a mucous membrane called the **endometrium**, the outer portion of which is sloughed during menstruation. The **vagina**, also lined with mucous membrane, extends from the **cervix** (an opening at the narrow, lower portion of the uterus) to the outside of the body. It allows passage of menstrual flow, receives sperm during intercourse, and forms part of the birth canal. The female external genitalia include the sexually sensitive *clitoris*, two pairs of *labia* (skin folds), and the mucous-secreting **Bartholin glands**. Because they nourish offspring, **mammary glands** (breasts) are considered part of the female reproductive system. These modified sweat glands develop at puberty and contain gland cells, embedded in fat, that produce milk and ducts that carry the milk to the nipple.

THE MALE REPRODUCTIVE SYSTEM

The **male reproductive system** consists of the testes, ducts, specific glands, and the penis (**Figure 20.3**). The **testes** (singular: *testis*) secrete the hormone testosterone into the bloodstream and produce sperm, which are conveyed through a series of ducts to the urethra. Secretions from **seminal vesicles** and the **prostate gland** mix with sperm to form **semen**. Other glands secrete mucus that lubricates the urethra. The **penis** is used to deliver semen into the female reproductive tract during sexual intercourse.

Sites of normal microflora in the urogenital system also are common sites of infections, partly because of their proximity to the rectum and partly because they provide warm, moist conditions suitable for microbial growth. The normal microflora of the urogenital system are found mainly at or near the external opening of the urethra of both sexes and the vagina of females. Many infectious diseases of the urogenital system are sexually transmitted, including genital herpes, gonorrhea, syphilis, and nongonoccocal urethritis. Other parts of

APPLICATIONS

Blue Breasts

The female breast, a part of the reproductive system, sometimes becomes infected. Mastitis (breast infection), usually caused by *Staphylococcus aureus*, is most likely to occur in nursing mothers. Appropriate antibiotic treatment generally gives prompt results with no sequelae. The disease is not limited to humans. In rabbits, for example, mastitis (known as "blue breasts") is common in nursing females and may be treated with penicillin.

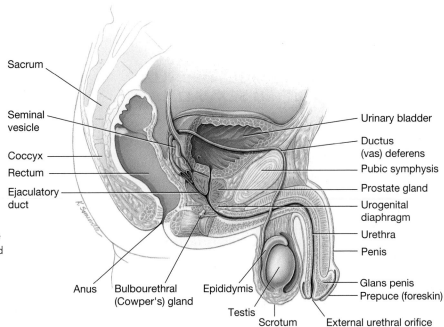

Figure 20.3 Structures of the male reproductive system. Because males have a single portal of entry for the reproductive and urinary systems (the urethra), infections of either system can occur. But chemical secretions containing lysozyme and spermine provide a flow that makes colonization by pathogens more difficult.

the urogenital system generally remain sterile, but a breakdown in defense mechanisms can lead to infection.

Defense mechanisms in the urogenital system are numerous. Normal microflora compete with opportunists and pathogens for nutrients and space and prevent them from causing disease. Urinary *sphincters* (muscles that close openings) act as mechanical barriers to microbes and also help prevent the backflow of urine. The flow of \urine through the urethra and of mucus through both the urethra and vagina helps wash away microbes. The low pH within both the urethra and, during reproductive years, the vagina prevents invasion by pathogens. Semen contains lysozyme and spermine, which help destroy invading pathogens.

Tiny nanobacteria, the size of viruses, live in urine, where by precipitating calcium and other minerals around themselves, they induce formation of kidney stones.

Despite its defenses, the urogenital tract is poorly protected against sexually transmitted diseases. The organisms that cause gonorrhea, syphilis, and nongonoccocal urethritis tolerate acidic conditions and successfully compete with natural microflora of the urogenital tract.

NORMAL MICROFLORA OF THE UROGENITAL SYSTEM

In healthy individuals, all parts of the urinary tract except the portion of the urethra closest to the urethral opening, are sterile. This colonization of the urethral end ensures that even "clean catch" urine specimens will contain bacteria flushed out of the urethra during voiding. *Escherichia coli* and *Lactobacillus* are the most common organisms, and can reach 100,000 bacteria/ml of urine. However, if urine is collected by direct puncture (suprapubic) of the

bladder with sterile needle and syringe, the urine is typically sterile in a healthy person.

Acid pH and high salt and urea concentrations retard the growth of bacteria in urine. However, urine specimens should be delivered to the lab promptly and refrigerated during any delays, as microbes will multiply rapidly in urine left sitting out at room temperature. When this happens, you typically see large numbers of several species of bacteria. A true urinary tract infection most often has large numbers of just a single organism.

Mycobacterium smegmatis (an acid-fast staining bacillus) lives on the external genitalia of both males and females. It is especially found under the foreskin of the penis of uncircumsized males, where it grows in collected secretions called *smegma*. If cleansing beneath the foreskin is inadequate when a voided urine specimen is collected, large numbers of *Mycobacterium smegmatis* bacteria will enter the specimen, where they might be confused with *Mycobacterium tuberculosis* bacilli found in the urine of people with tuberculosis of the kidneys.

In males, other than the last third of the urethra, the genital tract does not have a normal microflora and is sterile. The situation in the female reproductive tract is considerably more complex. There, hormones play a large role. During the child-bearing years, lactobacilli are the most numerous bacteria in the vagina, feeding on glycogen present in vaginal cells. Fermentation of glycogen produces lactic acid and a vaginal pH of about 4.7. This constitutes an innate line of defense, as most microbes other than lactobacilli cannot survive in this acidic environment. During childhood and after menopause, glycogen is absent from vaginal wall cells, and

Strains of Candida albicans that change from a single cell to a filamentous form seem to be more likely to cause infection than strains that remain in one form.

streptotocci and staphylococci rather than lactobacilli are the dominant organisms in those alkaline conditions.

UROGENITAL DISEASES USUALLY NOT TRANSMITTED SEXUALLY

BACTERIAL UROGENITAL DISEASES

Urinary Tract Infections

Urinary tract infections (UTIs) are among the most common of all infections seen in clinical practice. Second only to respiratory infections, they account for over 8.3 million office visits and 300,000 hospital stays per year in the United States. UTIs cause **urethritis** (u-re-thri′tis), or inflammation of the urethra, and **cystitis** (sis-ti′tis), or inflammation of the bladder. Because infection easily spreads from the urethra to the bladder, most infections are properly named **urethrocystitis** (u-re″thro-sis-ti′tis). Infectious agents reach the bladder more easily through the short (4-cm) female urethra than through the longer (20-cm) male urethra. Thus, women are affected some 40 to 50 times as often as men; by the age of 30, 20% of all women have acquired a UTI. In males the prostate gland is closely associated with the urethra and bladder, so **prostatitis** (pros-ta-ti′tis), or inflammation of the prostate gland, often accompanies UTIs.

Each year one out of five women develops **dysuria** (dis-yur′-e-ah), or pain and burning on urination indicative of urethral infection. One-fourth of these infections will develop into chronic cystitis, which will plague the unlucky victims intermittently for years. Symptoms of cystitis include continued dysuria, frequent and urgent urination, and sometimes pus in the urine. Elderly women are prone to UTIs, and up to 12% of some groups suffer chronically **(Table 20.1)**.

It is common for a person with a urinary infection to complain that, despite the urge to urinate, only a small amount of urine is passed.

A major cause of UTIs is incomplete emptying of the bladder during urination. Retained urine serves as a reservoir for microbial growth, thereby fostering infection. Any factor that interferes with the flow of urine and the complete emptying of the bladder can therefore predispose an individual to UTIs. The bladder is

TABLE 20.1

Urinary Tract Infections by Age and Sex		
Age	Female	Male
First 4 months of life	0.7%	1.3%
Four months to 5 years	4.5%	0.5%
Five years to 60 years	1–4%	<0.1%
Over 60 years	12%	1–4%
Very elderly	up to 30%	up to 30%

sometimes compressed by a "sagging" uterus or by the expansion of the uterus in pregnancy. Pregnancy can also cause decreased flow of urine through the ureters. Even the ring of a diaphragm can exert enough pressure on the bladder or ureters to interfere with voiding. In men the prostate tends to enlarge with age and constrict the urethra. Finally, the problem may be behavioral rather than mechanical: Some people simply do not visit the bathroom often enough. It is important for both females and males to empty the bladder frequently and completely. People with various kinds of paralysis who cannot void completely tend to have frequent UTIs.

UTIs originating in one area often spread throughout the urinary tract by "ascending" or "descending." Infections usually begin in the lower urethra and can ascend to cause inflammation of the kidneys, or **pyelonephritis** (pi″e-lo-ne-fri′tis). Less often, infections begin in the kidneys and descend to the urethra. Although UTIs do not occur more frequently during pregnancy, they usually are more serious because they are likely to ascend. Among pregnant women who have bacteria present in the urine, 40% develop pyelonephritis if the infection is not treated promptly. Descending UTIs originate outside the urinary tract. Organisms that enter the bloodstream from a focal infection, such as an abscessed tooth, can be filtered by the kidneys and cause a single infection or a chronic infection. For this reason it is often suggested that persons with chronic or frequent UTIs visit their dentist to see if some undiagnosed gum infection might be the source.

Escherichia coli is the causative agent in 80% of UTIs, but other enteric bacteria from feces such as *Proteus mirabilis* and *Klebsiella pneumoniae* can also cause such infections. Poor hygiene, such as wiping from back to front with toilet tissue, especially in females, can introduce fecal organisms into the urethra. It is important to

A newly discovered adhesin (a protein that sticks to a surface molecule), FimH, helps E. coli cells stick to the walls of the bladder, producing bladder infections.

teach children good toilet habits and the reasons for them. When *Chlamydia* or *Ureaplasma* are responsible for infection, they usually are sexually transmitted and cause non-gonococcal urethritis, which is discussed later.

Between 35 and 40% of all nosocomial infections are UTIs. Outpatients have a 1% chance of developing a UTI following a single catheterization, whereas hospitalized patients have a 10% chance. A great many patients with an indwelling catheter develop a UTI within the first week of use as organisms from skin, the lower urethra, or the catheter colonize the urethra. People with permanent catheters because of paralysis fight a never-ending battle against UTIs. *Staphylococcus epidermidis* and *S. saprophyticus* often cause infections in patients with indwelling catheters. *Pseudomonas aeruginosa* commonly causes infections that follow the use of instruments to examine the urinary tract. However, *E. coli* causes nearly half of nosocomial UTIs, and *Proteus mirabilis* causes about 13% of them.

UTIs are diagnosed by identifying organisms in urine cultures (**Figure 20.4**). Normal urine in the bladder is sterile, but urine is inevitably contaminated by bacteria as it passes through the lower part of the urethra. Even a clean-catch, midstream urine specimen will contain 10,000 to 100,000 organisms per milliliter. Low numbers of organisms do not necessarily rule out infection; in pyelonephritis and acute prostatitis, organisms sometimes enter the urine only in small numbers. Generally, however, only a single type of pathogen will be present at any one time. Finding several species in urine almost always means that the specimen was contaminated and the test should be repeated.

UTIs are treated with antibiotics, such as amoxicillin, trimethoprim, and quinolones, or with sulfonamides, according to susceptibilities of their causative agents. Prompt treatment helps prevent the spread of infection. UTIs can be prevented by good personal hygiene and frequent, complete emptying of the bladder.

Prostatitis

The symptoms of prostatitis are urgent and frequent urination, low fever, back pain, and sometimes muscle and joint pain. Most men have had at least one prostate infection by age 40. *Escherichia coli* is the cause of 80% of the cases, but it is still uncertain how the bacteria reach the prostate. Four routes of infection are possible: (1) by ascent through the urethra, (2) by back-flow of contaminated urine, (3) by passage of fecal organisms from the rectum through lymphatics and to the prostate, and (4) by descent of bloodborne organisms. Although uncommon, chronic prostatitis is a major cause of persistent UTIs in males, and it can cause infertility. Acute prostatitis usually responds well to appropriate antibiotic therapy without leaving sequelae.

Pyelonephritis

Pyelonephritis, an inflammation of the kidney, is usually caused by the backup of urine and consequent ascent of microorganisms. Urine backup can be caused by a number of factors, including lower urinary tract blockage or anatomical defects. Young children in particular often have imperfectly formed urinary tract valves that do not prevent urine from backing up. *Escherichia coli* causes 90% of outpatient cases and 36% of those in hospitalized patients, but yeasts such as *Candida* are occasionally responsible for the infection.

Pyelonephritis—and any other UTI—can be asymptomatic. When present, the symptoms are indistinguishable from those of cystitis, except that sometimes chills and fever occur. Dilute urine is another common finding, leading to frequent urination and **nocturia** (nok-tu're-ah), or nighttime urination. Patients must be carefully evaluated to identify underlying, predisposing conditions such as kidney stones or other blockages, which need to be relieved. Pyelonephritis is more difficult to treat than lower UTIs, but nitrofurantoin, sulfonamides, trimethoprime, ampicillin, gentamycin, ciprofloxacin, and quinolones usually are effective. Often these are started intravenously. During kidney failure or impaired kidney function, drugs must be used with care to prevent toxic accumulations.

Figure 20.4 Analysis of a urine sample. In this test, the color produced by a diagnostic reagent determines the presence of substances such as protein and nitrate, which can indicate bacterial growth, as in a urinary tract infection. Bacteria often produce nitrates while incubating in the bladder. Thus, a first morning urine specimen would contain the highest concentration of nitrates. A positive nitrate test indicates a urinary tract infection. However some bacteria further metabolize nitrates, so a negative test for nitrates does not preclude the presence of bacterial infection. (*Jon Meyer/Custom Medical Stock Photo, Inc.*)

Glomerulonephritis

Glomerulonephritis (glom-er″u-lo-nef-ri′tis), or *Bright's disease*, causes inflammation and damage to the glomeruli of the kidneys. It is an immune complex disease that sometimes follows a streptococcal or viral infection. As with rheumatic fever (rheumatogenic strains of *Streptococcus pyogenes*; ◀Chapter 23, p. 720), nephrogenic strains of *S. pyogenes* contain components in their cell walls that, when processed by the immune system, resemble tissue components present in glomerular tissue. This initiates the production of antibodies that cannot distinguish between bacterial cell wall and human glomerular components, resulting in antigen-antibody complexes. Antigen-antibody complexes are filtered out in the kidney, complement is activated, and there is an inflammatory response of glomerular capillaries **(Figure 20.5)**. The inflamed vessels leak blood and protein into the urine. Attraction and movement of phagocytic cells out of blood vessels, plus the release of hydrolytic enzymes in an attempt to engulf the newly formed complexes, further add to the damage of the glomeruli of the kidneys.

Because of the risk of glomerulonephritis, organisms from throat and other infections that might be streptococcal (◀Chapter 21, p. 644) should be cultured and appropriate antibiotics given. Although most people recover from glomerulonephritis and heal completely within 3 to 12 months, some have permanent residual kidney damage, and a few die.

Leptospirosis

Leptospirosis is caused by the spirochete *Leptospira interrogans* **(Figure 20.6)**. It is a zoonosis usually acquired by humans through contact with contaminated urine, directly or in water or soil. Dogs, cats, and many wild mammals carry these spirochetes (recall the "Dog Germs" box in ◀Chapter 15, p. 439). In some parts of the world, more than 50% of the rats are carriers of *Leptospira*. The bacteria live within the convoluted tubules of

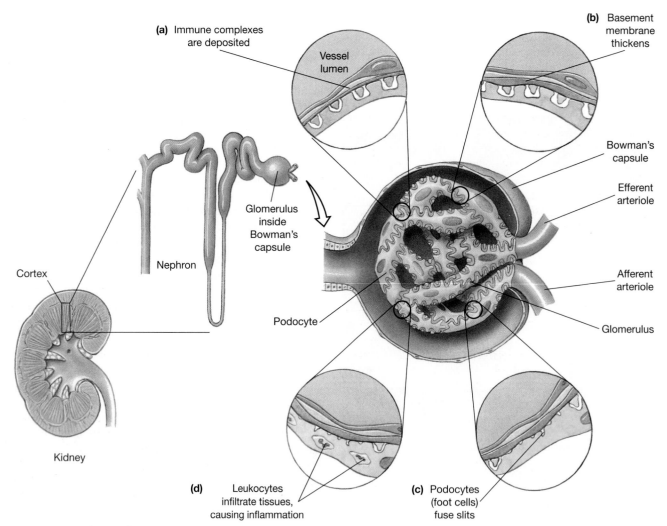

Figure 20.5 **Glomerulonephritis.** This disease occurs when **(a)** immune complexes are deposited on the filtering surfaces of the glomerulus, which is located inside Bowman's capsule in the cortex of the kidney; **(b)** basement membranes of cells thicken; **(c)** podocytes (foot cells) in the glomerulus, which normally act as filters, fuse their slits; and **(d)** leukocytes are attracted into the tissues. Kidney function is diminished, sometimes to the point of kidney failure.

Figure 20.6 A colorized TEM of *Leptospira interrogans*. Normally found in animals, this spirochete sometimes causes liver infections and jaundice in humans (11,000X). *(CNRI//Photo Researchers, Inc.)*

the kidneys and are shed into the urine. Often rain washes them from streets and the soil into natural bodies of water. They die quickly in brackish or acidic water but can survive for 3 months or longer in neutral or slightly basic water. Ponds or rivers bordering pastures where animals graze are especially likely to be contaminated.

In 1995 in Nicaragua, localized flooding led to an outbreak of leptospirosis in which 400 people were sickened, 150 hospitalized, and 40 died of pulmonary hemorrhage. In 1996, 9 American tourists returning from white-water rafting in Costa Rica developed leptospirosis. It appears that leptospirosis may be a more important cause of disease in the tropics than was previously recognized, and can be seen as an emerging disease. In 1997, in Illinois, entrants in a triathlon athletic competition suffered an outbreak.

The organisms enter the body through mucous membranes of the eyes, nose, or mouth or through skin abrasions. These organisms have a hook at the end, and can also burrow into the soft portions of the skin of the patient—usually soles of the feet or palms of the hands. Parents of small children often become infected from contact with pets. The children beg for a puppy, promising to take care of it, but the adults frequently clean up after the pets and become infected with leptospires. The highest incidence in the United States occurs in middle age women. Dogs, being ignorant of the niceties of toilet training, often account for a cluster infection that affects all who shared a swimming or wading pool with the dogs.

After an incubation period of 10 to 12 days, leptospirosis usually occurs as a febrile and otherwise nonspecific illness. In most cases, an uneventful recovery takes place in 2 to 3 weeks. However, 5 to 30% of the untreated cases result in death. A particularly virulent form of the infection, *Weil's syndrome*, is characterized by jaundice and significant liver damage.

Diagnosis can be made by direct microscopic examination of blood, but it is often not even considered because of the nonspecific symptoms and low incidence of leptospirosis. Culturing leptospires from blood or urine requires special media and extended incubation times (1 to 2 weeks). Many cases are asymptomatic, so it is difficult to know the true incidence of the disease. Generally, fewer than 100 cases per year are reported in the United States.

Leptospires are susceptible to almost any antibiotic if it is given in the first 2 or 3 days of the disease, but not after the fourth day. Patients have long-term immunity to the particular strain of *Leptospira* that infected them but not to any other strains. Leptospirosis can be prevented by vaccinating pets and avoiding swimming in or contact with contaminated water.

Bacterial Vaginitis

Vaginitis, or vaginal infection, usually is caused by opportunistic organisms that multiply when the normal vaginal microflora are disturbed by antibiotics or other factors. Predisposing conditions include diabetes mellitus, pregnancy, use of contraceptive pills, menopause, and conditions that result in imbalances of estrogen and progesterone, all of which change the pH and sugar concentration in the vagina. Several organisms account for a share each of vaginitis cases or at least serve as a marker for disruption of the vaginal flora. The bacterium *Gardnerella vaginalis*, in combination with anaerobic bacteria, accounts for about one-third of the cases. *Mobiluncus* is seen in other infections; it may be a separate organism or a clinical variant of *G. vaginalis*. The protozoan *Trichomonas vaginalis*, which can be opportunistic but usually is transmitted sexually, accounts for about one-fifth of the cases. The fungus *Candida albicans* causes most other cases.

Predisposing conditions to vaginitis include douching too often, tampons irritating the vaginal walls, tight slacks, and panties or panty hose without a cotton crotch.

Gardnerella vaginalis. This tiny Gram-negative bacillus or coccobacillus is present in the normal urogenital tract in 20 to 40% of healthy women. The normal vaginal pH is 3.8 to 4.4 in women of reproductive age and near neutral in young girls and elderly women. When the vaginal pH reaches 5 to 6, *Gardnerella vaginalis* interacts with anaerobic bacteria such as *Bacteroides* and *Peptostreptococcus* to cause vaginitis. None of these organisms alone produces disease. Because different anaerobes interact with *Gardnerella*, this kind of vaginitis sometimes is called *nonspecific vaginitis*. *Gardnerella* vaginitis produces a frothy, fishy-smelling vaginal discharge. The discharge, although usually small in volume, contains millions of organisms. Males occasionally get **balantitis** (bal-an-ti′-tis), an infection of the penis that corresponds to female vaginitis. Lesions appear on the penis after sexual contact with a woman who has vaginitis.

Diagnosis can be made when wet mounts of the discharge exhibit "clue cells," vaginal epithelial cells covered with tiny rods or coccobacilli (**Figure 20.7**). Metronidazole (Flagyl) suppresses vaginitis by eradicating the anaerobes necessary for continuance of the disease but allows normal lactobacilli to repopulate the vagina. This effect supports the notion that *Gardnerella* vaginitis requires an association with anaerobes. Ampicillin and tetracycline are also sometimes used in treatment. Unflavored (plain), live-culture yogurt used as a douche effectively replaces lactobacilli of normal microflora killed by antibiotic treatment.

Toxic Shock Syndrome

Infection with certain toxigenic strains of *Staphylococcus aureus* can produce **toxic shock syndrome (TSS)**. Prior to 1977 only 2 to 5 cases were reported each year, but then the incidence rose suddenly, reaching a peak of

(a)

(b)

Figure 20.7 Bacterial vaginitis. (a) A "clue cell" of *Gardnerella* infection in a vaginal smear. Notice the many bacteria clinging to the surface of the clue cell (507X). *(Science Photo Library/Photo Researchers, Inc.)* **(b)** Normal vaginal epithelial cells show no thick layer of bacteria clinging to their surfaces (1,014X). *(Dr. E. Walker/Photo Researchers, Inc.)*

APPLICATIONS

Know Your Yeast Infection

You've probably seen countless commercials and ads for vaginal yeast infection treatments. But how much do you actually know about yeast infections? In fungal vaginitis, most often caused by *Candida albicans*, lesions are raised gray or white patches surrounded by red areas, and the discharge is scanty, thick, and curdlike. Oddly, *Candida albicans* is a relatively common organism, found among the natural microflora of 20% of nonpregnant women and 30% of pregnant ones. It only becomes a problem when it causes opportunistic infections, especially in patients with AIDS or uncontrolled diabetes or in those being treated with antibiotics. Diabetes should therefore be regulated and antibiotic therapy stopped so that *Candida* vaginitis can be effectively treated with Nystatin or imidazole, the main ingredients in most of those yeast infection creams and pills you hear about.

about 320 cases in September 1980. Since 1986 between 412 and 102 cases per year have been reported. The sudden rise was associated with new superabsorbent, but abrasive, tampons, which were left in the vagina for longer than usual periods of time. The tampons caused small tears in the vaginal wall and provided appropriate conditions for bacteria to multiply.

Between 5 to 15% of women have *S. aureus* among their vaginal microflora, but only a small fraction of these strains cause TSS. Originally most cases occurred in menstruating women, but men, children, and postmenopausal women with focal infections of *S. aureus* such as boils and furuncles now account for most cases of TSS. Organisms enter the blood or grow in accumulated menstrual flow in tampons. They produce *exotoxin C*, which enhances the effects of an endotoxin; how these toxins exert their effects is not clearly understood. Clinical manifestations include fever, low blood pressure

APPLICATIONS

WARNING: Important Information About Toxic Shock Syndrome (TSS)

If you use tampons frequently, you are probably used to throwing out the insert that is found in practically every box. But have you ever read the warning? If you did, you'd know that the warning signs for TSS are sudden fever (usually 102°F or more), vomiting, diarrhea, fainting or near fainting when standing up, dizziness, or a rash that looks like a sunburn. You'd also know that the reported risks are higher for women under 30 years of age and that the incidence of TSS is estimated to be between 1 and 17 cases of TSS per 100,000 menstruating women and girls per year. You can reduce your risk of getting TSS during your period by alternating tampon use with pads, by selecting the minimum absorbency needed to control your menstrual flow, and by asking your doctor about any other questions you might have.

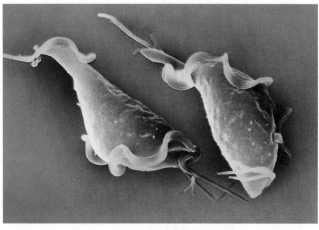

Figure 20.8 *Trichomonas vaginalis.* Notice the characteristic undulating membranes and flagella as seen with the scanning electron microscope (534X). *(David M. Phillips/Photo Researchers, Inc.)*

(shock), and a red rash, particularly on the trunk, that later peels. Immediate treatment with nafcillin has held the mortality rate to 3%. Deaths, when they occur, usually are due to shock. Recurrence is a frequent possibility, especially during subsequent menstrual cycles. Antibiotics can be given prophylactically to prevent recurrences, but women who have had TSS can reduce their risk of recurrence by ceasing to use tampons.

Although infrequent changing of superabsorbent tampons accounted for most cases of TSS, the contraceptive sponge caused some, and for this reason was withdrawn from the market.

PARASITIC UROGENITAL DISEASES

Trichomoniasis

Although **trichomoniasis** is transmitted primarily by sexual intercourse, it is discussed here because of its similarity to other kinds of vaginitis and because children (and rarely adults) can be infected from contaminated linens and toilet seats. There are an estimated 7.4 million cases of *Trichomonas* infections in the United States each year. At least three species of protozoa of the genus *Trichomonas* can parasitize humans, but only *T. vaginalis* causes trichomoniasis. The others are commensals; *T. hominis* is found in the intestine, and *T. tenax* in the mouth. *T. vaginalis* is a large flagellate with four anterior flagella and an undulating membrane **(Figure 20.8)**. It infects urogenital tract surfaces in both males and females and feeds on bacteria and cell secretions. Because the optimum pH for the organism is 5.5 to 6.0, it infects the vagina only when vaginal secretions have an abnormal pH. The symptoms of trichomoniasis are intense itching and a copious white discharge, especially in females, having the consistency of raw egg white.

Trichomoniasis is diagnosed by microscopic examination of smears of vaginal or urethral secretions and

treated with metronidazole (Flagyl) and restoration of normal vaginal pH in women. Flagyl cannot be used during pregnancy because it causes abortions, but it is important to get rid of the infection before delivery to prevent infecting the infant. A vinegar douche usually is effective.

The effects of different species of *Trichomonas* illustrate the extreme variation in the degree of damage parasites can do. *T. hominis* and *T. tenax* are considered commensals. However, *T. foetus* causes serious genital infections in cattle. It is the leading cause of spontaneous abortion in cows and, in fact, is responsible for losses to U.S. cattle breeders of nearly $1 million per year. Unfortunately outbreaks in 2004 of *Brucella abortus*, a bacterial infection also known as contagious abortion, if not controlled soon may replace *Trichomonas* as the number one cause. In 2000, while searching for birds infected with West Nile virus, the cause of an emerging disease of the brain, wildlife researchers found that flocks of doves were dying from *Trichomonas*, rather than from West Nile virus.

✓CHECKLIST

1. Why are urinary tract infections more common in women than in men?

2. Distinguish between the four routes of infection in prostatitis.

3. Why is kidney tissue damaged in glomerulonephritis, since bacteria are not directly infecting the kidneys?

4. What are "clue cells"?

▌▌▌ SEXUALLY TRANSMITTED DISEASES

▌▌▌ **Sexually transmitted diseases (STDs)** have become an increasingly serious public health problem in recent years, in part because of changing sexual behaviors. In addition, some causative agents are becoming resistant to antibiotics, and no vaccines have been developed to control any STDs. Consequently, the only means of preventing STDs is to avoid exposure to them.

ACQUIRED IMMUNE DEFICIENCY SYNDROME (AIDS)

Although AIDS is in many cases an STD, it is not transmitted exclusively by sexual contact. We discussed it in ◄Chapter 18 with disorders of the immune system (p. 555).

BACTERIAL SEXUALLY TRANSMITTED DISEASES
Gonorrhea

The term **gonorrhea** means "flow of seed" and was coined in A.D. 130 by the Greek physician Galen, who mistook pus for semen. By the thirteenth century, the venereal (from Venus, goddess of love in Roman mythology) transfer of this disease was known. But it was not until the mid-nineteenth century that gonorrhea was recognized as a specific disease; until then it was thought to be an early symptom of syphilis. The causative organism, *Neisseria gonorrhoeae*, was first described by the German physician Albert Neisser in 1879. It is a Gram-negative, spherical or oval diplococcus with flattened adjacent sides and resembles a pair of coffee beans facing each other **(Figure 20.9)**.

Drying kills the organisms in 1 to 2 hours, but they can survive for several hours on fomites. Cases in which *Neisseria* survived improper laundering in the hospital have been documented; in dried masses of pus, the bacteria can survive for 6 to 7 weeks! So although these organisms are ordinarily thought of as being very fragile, some are quite robust.

The infectivity of *Neisseria* is related in several ways to its pili. Attachment pili (fimbriae) enable gonococci

to attach to epithelial cells that line the urinary tract so that they are not swept out with the passage of urine. (In fact, sloughing of epithelial cells is one of the body's defenses against such infections.) These bacteria also use their pili to attach to sperm; conceivably, swimming sperm could carry gonococci into the upper part of the reproductive tract. Strains lacking pili usually are nonvirulent. The search for a vaccine against gonococcal pili is under way, but success has not yet been achieved.

Gonococci produce an endotoxin that damages the mucosa in fallopian tubes and releases enzymes such as proteases and phospholipases that may be important in pathogenesis. They also produce an extracellular protease that cleaves IgA, the immunoglobulin present in secretions (◄Chapter 17, p. 497). Gonococci adhere to neutrophils and are phagocytized by them (◄Chapter 16, p. 468). Phagocytosis kills some of the bacteria, but survivors multiply inside polymorphonuclear leukocytes (PMNLs). A typical wet mount of a urethral discharge will show PMNLs with diplococci in their cytoplasm (Figure 20.9). Gonococci also obtain iron for their own metabolic needs from the iron transport protein transferrin.

The Disease. Gonorrhoea is the second most commonly reported notifiable disease in the United States. Humans are the only natural hosts for gonococci. Gonorrhea is transmitted by carriers who either have no symptoms or have ignored them. As many as 40% of males and 60 to 80% of females remain asymptomatic after infection and can act as carriers for 5 to 15 years. Very few organisms are required to establish an infection; half of a group of males given a urethral inoculum of only 1,000 organisms developed gonorrhea. Following a single sexual exposure to an infected individual, about one-third of males become infected. After repeated exposures to the same infected person, three-quarters of males eventually become infected. Of males who develop symptoms, 95%

(a)

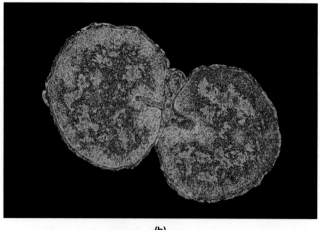

(b)

Figure 20.9 Diplococci of *Neisseria gonorrhoeae*. **(a)** As small, dark purple dots inside the cytoplasm of leukocytes in a urethral smear (4,807X), and *(A. M. Siegelman/Visuals Unlimited)* **(b)** magnified 96,051X by TEM, showing the internal structure of the pair. *(Dr. David M. Phillips/Photo Researchers, Inc.)*

(a)

(b)

Figure 20.10 **Symptoms of gonorrhea.** Symptoms sometimes include **(a)** in males, a urethral "drip" of pus and *(Kenneth Greer/Visuals Unlimited)* **(b)** an infrequent complication, gonococcally caused arthritis. *(Courtesy Centers for Disease Control and Prevention CDC)*

have pus dripping from the urethra within 14 days, and many develop symptoms sooner **(Figure 20.10a)**. The typical incubation period is 2 to 7 days.

Both the contraceptive pill and intrauterine devices (IUDs) have contributed to the epidemic of gonorrhea now seen in the Western world **(Figure 20.11)**. Their introduction in the 1960s led to increased sexual freedom and decreased use of condoms and spermicides. Women using the contraceptive pill have a 98% chance of being infected upon exposure because the pill alters vaginal conditions in favor of gonococcal growth. Although condoms and spermicides are less reliable contraceptives, they do offer some protection from gonorrhea. Gonorrhea spreads into the endometrial cavity and fallopian tubes two to nine times

The highest rate of gonorrhea in the United States is among women aged 15-19, with 624.7 of cases per 100,000. Men of the same age average 261.2 cases per 100,000.

faster in women who use IUDs than in those who do not use them. The exposure of blood vessels during menstruation allows bacteria to enter the circulatory system and thus facilitates the development of bacteremia. Gonorrhea lesions also have been found to make infection with HIV easier.

Although gonorrhea is thought of as a venereal disease (VD), it also affects other parts of the body (Figure 20.10b). Pharyngeal infections, which develop in 5% of those exposed by oral sex, are most common in women and homosexual men. The patient may develop a sore throat, but most pharyngeal cases are asymptomatic. The infected tissues act as a focal source for bacteremia. Anorectal infection, especially common in homosexual men, occurs in the last 5 to 10 cm of the rectum. It can be either asymptomatic or painful, with constipation, pus, and rectal bleeding. Women with vaginal gonorrhea also often have anorectal infection. Such infection can occur in the absence of anorectal intercourse due to contamination of the anal opening by organism-laden vaginal discharges. Medical personnel doing internal pelvic examinations can avoid spreading the infection by using a different gloved finger for anal examination than was inserted into the vagina. Pharyngeal and anorectal infections are also seen in children who are victims of sexual abuse.

The urethra is the most common site of gonorrhea infections in males. The most common site of infection in females is the cervix, followed by the urethra, anal canal, and pharynx. Skene's glands in the urethra and Bartholin's glands near the vaginal outlet also can be infected in females. As many as half of infected females develop **pelvic inflammatory disease (PID)**, in which the infection spreads throughout the pelvic cavity. Studies done in Sweden showed that sterility often follows PID because of tubal occlusion by scarring. The rate of sterility increases with the number of infections—13% after one infection, 35% after two infections, and 75% after three infections.

Disseminated infections, which occur in 1 to 3% of cases, produce bacteremia, fever, joint pain, endocarditis, and skin lesions that can be pustular, hemorrhagic, or necrotic. When organisms reach the joints, they can cause arthritis. Gonococcal arthritis is now the most common joint infection in people 16 to 50 years old. Another complication of gonorrhea is infection of the lymphatics that drain the pelvis. Scarring produces tight, inflexible tissue that immobilizes pelvic organs into the condition known as "frozen pelvis."

Transfer of organisms by contaminated hands, or fomites such as towels, can result in eye infections. If untreated, severe scarring of the cornea and blindness can result. Newborns can acquire ophthalmia neonatorum during passage through the birth canal of an infected mother (◄Chapter 19, p. 593). Pus that accumulates behind swollen eyelids can spurt forth with great pressure and infect health care workers. Such infections have occurred despite immediate treatment.

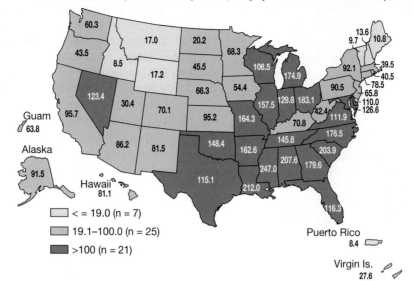

GONORRHEA. Reported cases per 100,000 population — United States, 2005

☐ <= 19.0 (n = 7)
☐ 19.1–100.0 (n = 25)
■ >100 (n = 21)

Note: The total rate of gonorrhea for the United States and outlying areas (Guam, Puerto Rico and Virgin Islands) was 114.2 per 100,000 population. The Healthy People 2010 target is 19.0 cases per 100,000 population.

(a)

(b)

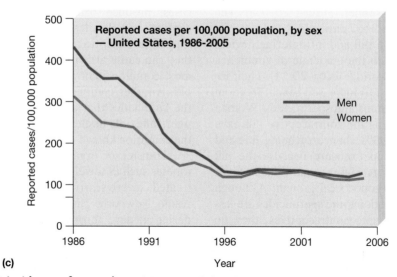

(c)

Figure 20.11 **The U.S. incidence of gonorrhea.** (a) By state; (b) by ethnicity; (c) by sex. (*Source:* CDC)

Age affects the site and type of gonococcal infection. During the first year of life, infection usually results from accidental contamination of the eye or vagina by an adult. Between 1 year and puberty, gonorrhea usually occurs as vulvovaginitis in girls who have been sexually molested. A girl's vaginal epithelium has less keratin and is more susceptible to infection than a woman's. The girls exhibit pain on urination, vulvar and perianal soreness, discomfort on defecation, and a yellowish-green discharge from vaginal and urethral openings. Both boys and girls can develop a pus-filled anal discharge indicative of anorectal gonorrhea as a result of sexual abuse. Failure of a hospital to launder sheets properly led to an outbreak of vulvovaginitis in female pediatric patients who acquired it from sitting on the sheets.

Diagnosis, Treatment, and Prevention. Diagnosis is made by identifying *N. gonorrhoeae* using molecular probes, since culturing it is not easy. Being a rather fastidious organism, it requires high humidity and ambient carbon dioxide for growth. Temperatures of 35° to 37°C and pH of 7.2 to 7.6 are optimal. Many species of *Neisseria* exist. A positive result on a screening test may not be due to *N. gonorrhoeae*; confirmatory tests should always be done. Samples from patients suspected of having *Neisseria* infections are inoculated from a swab directly into a special medium for transport and incubation in a laboratory (◄Chapter 6, p. 169).

Sulfonamides were the first agents found to treat gonococcal infections. As some strains became resistant to sulfonamides, penicillin became available. At first, penicillin G was effective in very small amounts, but now much larger doses must be used. Some strains of gonococci are entirely resistant to penicillin because they produce the enzyme β-lactamase, which enables them to break down this antibiotic. The CDC now recommends a single intramuscular (125- or 250-mg) dose of the cephalosporin ceftriaxone, ciprofloxacin, ofloxacin, or levofloxacin, *plus* azithromycin or doxycycline. Most frequently used is the combination of ceftriaxone plus azithromycin. Increases in resistance have made it very difficult to combat gonorrhea. Organisms typically acquire antibiotic resistance by means of their sex pili. One wonders if it will always be possible to find another antibiotic to combat resistant strains. The CDC, as of April

PUBLIC HEALTH

Finished Before It Started

Here in the United States, we made careful use of the then relatively new quinolone antibiotics to treat gonorrhea. We didn't want to selectively breed resistant strains, so we restricted their use to special cases. Unfortunately, the quinolones have been used so frequently in the Philippines that quinolone-resistant strains of *Neisseria gonorrheae* have already emerged there. If these strains make their way to the United States, the quinolones will be obsolete for treating gonorrhea. **Update**: These strains arrived here in 1991, and sadly in April 2007, the CDC declared most of the quinolone drugs were no longer useful for treating gonorrhea. Such resistance is 5 times more frequent among strains cultured from men who have sex with men.

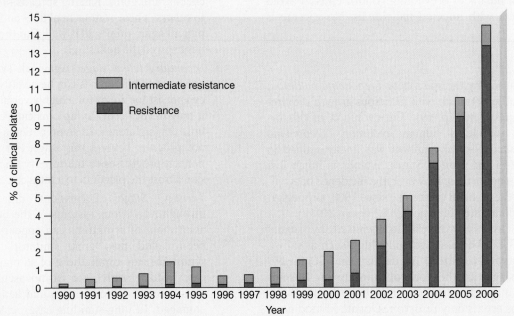

Percentage of *Neisseria gonorrhea* isolates with intermediate resistance or resistance to ciprofloxacin, by year. From Gonococcal Isolate Surveillance Project, United States, 1990–2006, CDC. *(Kavaler/Art Resource)*

2007, no longer recommends fluoroquinolones (i.e., cipro-floxacin, ofloxacin, or levofloxacin) for treatment of gonor-rhea, or associated conditions such as pelvic inflammatory disease (PID). Only one class of drugs, the cephalosporins, is still recommended and available. For persons with peni-cillin or cephalosporin allergies, a single intromuscular dose of spectinomycin 2g, is recommended. However, spec-tinomycin is not currently available in the United States. The CDC will post bulletins on its website when this drug does become available. Both cefixime and spectinomycin are insufficient for treatment of pharyngeal infections, as these are very difficult to eradicate. The CDC also rec-ommends a single oral dose of cefixime 400 mg. Un-fortunately 400-mg tablets are not currently available, only an injectable suspension. The CDC is also cautioning against widespread use of azithromycin, hoping to delay development of widespread resistance to it, as has hap-pened with Cipro and other fluoroquinolones. The CDC and the world Health Organization feel that 95% of cases treated with recommended procedures should be cured. Once resistant strains reach 5% of a microbial population, this is the threshold for changing recommendations. Strains resistant to fluoroquinolones now exceed 15%.

Gonorrhea patients often have other STDs. A 7-day course of treatment with oral doxycycline or a single dose of azithromycin has the advantage of killing *Chlamydia*, which may concurrently be present. Forty-five percent of patients having gonorrhea also test positive for *Chla-mydia*. Follow-up cultures should be done 7 to 15 days after completion of treatment to be sure the infection is cured. Additional follow-up cultures are also advised at 6 weeks, as 15% of women who are negative at 7 to 14 days are positive again at 6 weeks—possibly because of reinfection. All sexual partners must be treated.

The best means of preventing gonorrhea is to avoid sexual contact with infected individuals. No vaccine is available.

Syphilis

Syphilis is caused by the spirochete *Treponema pallidum*, an active motile organism with fastidious growth require-ments (◀Figure 3.30b, p. 69). The evolution of this or-ganism has paralleled human evolution. *Treponema pallidum* eluded discovery until it was finally stained in 1905. Today in the United States, syphilis is much less common than gonorrhea. However, the incidence of syphi-lis has, on average, been on the rise since 1960, although it has been declining steadily since 1990 (Figure 20.12).

The disease is ordinarily transmitted by sexual means, but it can be passed in body fluids such as saliva. It thereby creates a hazard for dentists and dental hy-gienists. Kissing is another form of transmission. It is not transmitted in food, water, or air or by arthropod vec-tors. Humans are its only reservoir. Donated blood need not be screened for syphilis because any spirochetes present are killed when the blood is refrigerated.

The Disease. A typical case of syphilis progresses as follows:

1. *Incubation stage:* Over a period of 2 to 6 weeks after entering the body, the organisms multiply and spread throughout the body.

2. *Primary stage:* On average, about 3 weeks after in-fection an inflammatory response at the original entry site causes formation of a **chancre** (shang-ker), a hard, painless nondischarging lesion about ½-inch in diameter. One or more primary chancres usually develop on the genitals but can develop on lips or hands (**Figure 20.13**). In fe-males, chancres on the cervix or another internal location sometimes escape detection. The patient often is embarrassed to seek medical attention for a lesion in a genital location and hopes that it will just "go away." And go away it does after about 4 to 6 weeks, without leaving any scarring. The pa-tient thinks all is well, but the disease has merely entered the next stage.

3. *Primary latent period:* All external signs of the disease disappear, but blood tests diagnostic for syphilis are positive, and the spirochete is spread-ing through the circulation.

4. *Secondary stage:* Symptoms can appear, disappear, and reappear over a period of up to 5 years, during which the patient is highly contagious. These symptoms include a copper-colored rash, particu-larly on the palms of the hands and the soles of the feet (**Figure 20.14a**), and various pustular rashes and skin eruptions. In females, the cervix usually has lesions. Painful, whitish mucous patches swarming with spirochetes appear on the tongue, cheeks, and gums. Kissing spreads the spirochetes to others. These lesions heal uneventfully, and the patient again thinks all is well. But the disease now has entered the next stage.

5. *Secondary latent stage:* Again all symptoms dis-appear, and blood tests can be negative. This stage can persist for life or for a highly variable period, or it may never occur. Symptoms can reoccur at any time during latency. In some patients syphilis does not progress beyond this stage, but in many pa-tients it progresses to the tertiary stage. Transmis-sion across the placenta to a fetus can also occur.

6. *Tertiary Stage:* Permanent damage occurs throughout various systems of the body. A wide assortment of symptoms can appear; syphilis has been called the "great imitator" because its symptoms can mimic those of so many other dis-eases. Most involve the cardiovascular and ner-vous systems. Blood vessels and heart valves are damaged. In long-standing cases, calcium depos-its in heart valves can be so extensive as to be visible on a chest X-ray. Neurological damage,

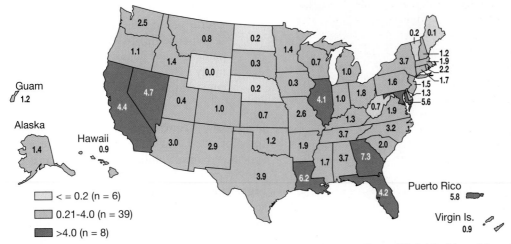

SYPHILIS, PRIMARY AND SECONDARY. Reported cases per 100,000 population — United States, 2005

Note: The total rate of P&S syphilis for the United States and outlying areas (Guam, Puerto Rico and Virgin Islands) was 3.0 per 100,000 population. The Healthy People 2010 target is 0.2 cases per 100,000 population.

(a)

(b)

(c)

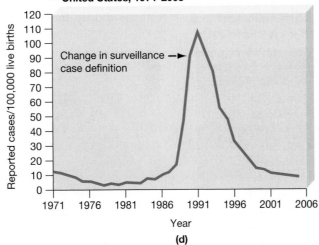

SYPHILIS, CONGENITAL. Reported cases per 100,000 live births among infants aged <1 year — United States, 1971-2005

(d)

Figure 20.12   The U.S. incidence of syphilis.   (a) By state. **(b)** By ethnicity. **(c)** By sex. **(d)** Congenital syphilis rates. As more women contract syphilis, transmission to the fetus becomes more common. (*Source:* CDC)

Figure 20.13   Primary chancres of syphilis.   (a) A genital site (penis); **(b)** an extragenital site (the face). *(Both Photos: Courtesy Centers for Disease Control and Prevention)*

called **neurosyphilis**, can include thickening of the meninges; ataxia (a-tax′e-ah), or an unsteady gait or inability to walk; and paresis (par′e-sis), or paralysis and insanity. These symptoms often are due to formation of granulomatous inflammations

called **gummas** (gum′ahz) (◀Chapter 16, p. 475). Internal gummas typically destroy neural tissue, whereas external gummas destroy skin tissue (**Figure 20.14b**). Mental illness accompanies neural damage. In the preantibiotic era, as many as half the beds in mental hospitals were occupied by patients with tertiary syphilis.

Diagnosis, Treatment, and Prevention. Diagnostic tests include DNA analysis of tissue for gene sequences specific to the syphilis organism, fluorescent antibody, and treponemal immobilization tests. Actively motile organisms can be observed under the dark-field microscope while specific antibody against *Treponema pallidum* is added. Immobilization of organisms by the antibody is 98% confirmatory for syphilis. Other blood tests such as VDRL (Venereal Disease Research Laboratory) and Wassermann tests have a high frequency of false positives, as they are based on the detection of tissue damage. A severe case of influenza, a myocardial infarction, or an autoimmune disease can cause sufficient damage to elicit a false positive reaction. Therefore, screening tests such as the VDRL must always be followed up by a confirmatory test, such as the fluorescent antibody test. Also, the VDRL test requires extra-clean glassware and very high attention to detail. A newer, much easier test, the rapid plasma reagin (RPR) test, makes use of disposable cardboard reaction cards, and is replacing the VDRL test.

Syphilis usually is treated with benzathine penicillin G. The longer the patient has had syphilis, the more important are continued treatment and testing to ensure that organisms have been eradicated. No vaccine is available, and recovery does not confer immunity.

Congenital Syphilis.   **Congenital syphilis** occurs when treponemes cross the placenta from mother to baby. At birth or shortly thereafter, the infant may show such

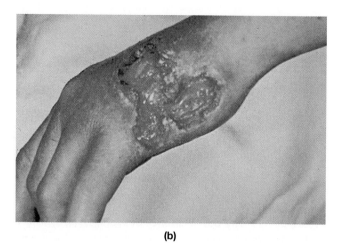

Figure 20.14   Signs of secondary and tertiary syphilis.   (a) A typical papular rash (secondary stage); *(Southern Illinois University/ Visuals Unlimited)* **(b)** a gumma (tertiary stage). *(Science VU/Visuals Unlimited)*

signs as notched incisors, or Hutchinson's teeth (**Figure 20.15a**), a perforated palate, saber shins (in which the shin bone projects sharply on the front of the leg) (Figure 20.15b), an aged-looking face with saddle-nose (a flat, saddle-shaped nose) (Figure 20.15c), and a nasal discharge. Women should be tested for syphilis prior to becoming pregnant, and as part of their prenatal testing if they are already pregnant, to prevent such congenital infections.

Chancroid

Chancroid (shang′kroid), called soft chancre to distinguish it from the hard, painless chancre of syphilis, is caused by *Haemophilus ducreyi*. Named for Augusto Ducrey, the Italian dermatologist who first observed it in skin lesions in 1889, the organism is a small Gram-negative rod that occurs in strands.

Relatively rare in the United States, with less than 106 cases reported per year, chancroid is seen most frequently in developing countries of Africa, the Caribbean, and Southeast Asia. Its worldwide incidence is believed to be greater than that of either gonorrhea or syphilis. Most cases in the United States occur in immigrants and a few in military personnel. In 1982 the CDC reported a sudden increase in the number of cases in southern California—from an average of 29 per year to over 400 per year. Of these, 90% were in Hispanic men, most of whom had recently immigrated from Mexico.

The Disease. Chancroid begins with the appearance of soft, painful lesions called chancres, which bleed easily, on the genitals 3 to 5 days after sexual exposure. They often occur on the labia and clitoris in females and on the penis in males. However, the infection can be present without apparent lesions, the only symptom being a burning sensation after urination. Chancres also can occur on the tongue and lips. Regardless of their location, chancres are extremely infective. Medical personnel sometimes acquire lesions on the hands merely from contact with chancres. In about one-third of the patients, chancroid spreads to the groin, where it forms enlarged masses of lymphatic tissue called *buboes*. These appear about 1 week after infection, swell to great size, and can break through the skin, discharging pus to the surface (**Figure 20.16**).

Diagnosis and Treatment. Chancroid is diagnosed by identifying the organism in scrapings from a lesion or in fluids from a bubo. Patients with chancroid also often have syphilis and other STDs. Thus, a patient with a positive diagnosis for one STD should be tested for other STDs. Untreated lesions can persist for months. Often the disease resolves on its own. Infection does not confer permanent immunity, and the disease can be acquired again and again. It is treated with antibiotics such as tetracycline, erythromycin, sulfanilamide, or a combination of trimethoprim and sulfamethoxazole. With treatment

(a)

(b)

(c)

Figure 20.15 Signs of congenital syphilis. **(a)** Hutchinson's teeth, showing notched central incisors. *(Science VU/Visuals Unlimited)* **(b)** Saber shin, displaying arching of the anterior edge of the tibia. *(Courtesy Kenneth E. Greer, M.D., University of Virginia Health Sciences Center)* **(c)** Saddle-nose, which causes a snuffling type of breathing. *(Science VU/Visuals Unlimited)*

APPLICATIONS

Homo sapiens and *Treponema pallidum*: An Evolutionary Partnership

Historically, four different diseases—syphilis, pinta, yaws, and bejel—have been caused by the same pathogen, *Treponema pallidum*. DNA studies show the causative organisms to be extremely similar for all four diseases. Only recently has the causative agent of pinta evolved significantly from *T. pallidum*; it has been given a separate species name, *T. carateum*. Researchers C. J. Hackett and Ellis Hudson suggest that the treponemes have evolved in accordance with changing human lifestyles.

The first pathogen infected a wild animal and occasionally infected people as a zoonosis. As such, it caused the set of symptoms known as *pinta*, a mild skin disease that does not reach deeper tissues. Its red, slate blue, and white lesions are disfiguring but not life-threatening. Found in the tropics, it is passed from person to person through direct skin contact. Warm skin, moist with perspiration, provides an ideal environment for the treponeme.

Around 10,000 B.C., probably in Africa, the pathogen mutated and gave rise to *yaws*, a disease that thrived in the somewhat more densely settled human villages. The highest incidence of yaws is still in tropical Africa and in Latin America. Typical symptoms are deep, oozing skin lesions that can penetrate underlying tissues and erode bone but that do not reach the viscera. Most victims are children 4 to 10 years old.

By 7000 B.C., people carrying the treponeme moved northward. The climate was cooler and drier, required more clothing, and made transmission via skin contact more difficult. The pathogen survived best in warm, moist areas such as the mouth, groin, and armpits. Transfer to a new host was most likely via oral contact such as kissing or by sharing bits of food or eating from the same utensils. Today the disease is known as nonvenereal syphilis, or *bejel*. The strain responsible for bejel is still more invasive; it causes granulomatous lesions like those of syphilis and attacks tissues of the skeletal, cardiovascular, and nervous systems. It is often transferred by young children, because older ones are generally past the infectious stage.

Bejel is almost always acquired prior to puberty and is not sexually transmitted. It is never transmmitted congenitally, presumably because the mother's infection becomes nontransmissible before she is old enough to become pregnant. Bejel epidemics occurred during the fifteenth century in Africa and in the seventeenth and eighteenth centuries in Scotland and Norway. Today the worldwide incidence of bejel is higher than that of venereal syphilis.

As standards of living improved and people covered more of their bodies, the surest means of transmitting the treponeme was via sexual activity and warm, moist reproductive tracts. The pathogen, again evolving in accordance with changing human lifestyles, came to cause venereal syphilis, in which the primary chancre appears to be analogous to yaws lesions. The disease is more severe than bejel because it penetrates organs to a greater and more devastating extent.

There is further evidence for the evolution of treponemes with human culture. First, none of the four diseases can be found in the same geographic area as any other. Second, infection with yaws prevents concurrent syphilis. Third, people with yaws who move to cooler climates lose their yaws symptoms and develop those of bejel instead. Evidently, the treponeme is extremely adaptive to its environment. Indeed, syphilis itself seems to have evolved considerably over the past five centuries. Accounts that date from the 1490s describe the disease as killing its victims in a matter of weeks. Today it is far less virulent; the incubation period alone is longer than that, and people with syphilis may survive for decades.

If the theory of parallel evolution of pathogen and humans is correct, historians who blamed Columbus's sailors for bringing the organism back from the New World were mistaken. The fact that the disease appeared in Europe at about the same time as Columbus's voyages is probably just a coincidence. The treponeme most likely arrived in Europe with slaves from Africa before the discovery of America, eventually evolving to the point at which it could cause venereal syphilis. Evidence for this hypothesis has been found in Native American burial grounds in New England. The remains of individuals who died before the arrival of settlers from the Old World show no trace of syphilis. Signs of syphilis appear first in the skeletons of young native women buried at the time of early European settlement.

(a)   (b)

(a) The rash of pinta. *(Courtesy Centers for Disease Control and Prevention CDC)* **(b)** Lesions caused by yaws. The bone damage can also be similar to that of saber shin produced by syphilis (Figure 20.15). *(Science VU/Visuals Unlimited)*

Figure 20.16 **Lesions of chancroid on the penis.** A draining bubo is in the adjacent groin area. Chancroid is caused by *Haemophilus ducreyi. (Courtesy Centers for Disease Control and Prevention CDC)*

lesions heal rapidly, but they often leave deep scars with much tissue destruction.

Nongonococcal Urethritis

As the name suggests, **nongonococcal urethritis (NGU)** is a gonorrhea-like STD caused by organisms other than gonococci. Most cases are caused by *Chlamydia trachomatis*, but some are caused by mycoplasmas. The prevalence of chlamydial infections is greater than that of any other STDs and is increasing dramatically (**Figure 20.17**).

 Chlamydial Infections. *Chlamydia trachomatis* is a tiny spherical bacterium with a complex intracellular life cycle (◀Chapter 9, p. 264). In addition to causing NGU, subspecies strains of *C. trachomatis* cause a wide range of disorders, including conjunctivitis and lymphogranuloma venereum (discussed later). The CDC has estimated that 3 to 5 million Americans contract chlamydial NGU each year. The subspecies that causes inclusion conjunctivitis in newborns also causes 30 to 50% of NGU cases in males, as well as 30 to 50% of vulvovaginitis cases and half of cervicitis (inflammation of the cervix) cases in women. The large numbers of humans who are infected and who have organisms in their body secretions makes transmission of chlamydial infections especially easy.

Following an incubation period of 1 to 3 weeks, symptoms of NGU similar to but milder than gonorrhea begin. A scanty, watery urethral discharge is observed, especially after passage of first morning urine. Sometimes this is accompanied by tingling sensations in the penis. Many chlamydial NGU infections are asymptomatic, and fortunately most chlamydial STDs produce no significant complications or aftereffects. However, inflammation of the epididymis (the tube through which sperm pass from the testis) can lead to sterility.

Pelvic inflammatory disease, which can be caused by more than 20 different infectious agents, is a common complication of NGU as well as of gonorrhea. Chlamydial PID increases the risk of sterility and ectopic pregnancy, a pregnancy in which the embryo begins development outside the uterus (for example, in a fallopian tube or the peritoneal cavity). Surveys of pregnant women reveal that 11% have *Chlamydia* in the cervix; postpartum fever is common among infected women. Their infants can develop neonatal chlamydial pneumonia, a rarely fatal disease that accounts for 30% of all pneumonias in children under 6 months of age. Because the condition typically does not develop until 2 or 3 months after delivery, its connection with cervical chlamydial infection is often overlooked.

It may be difficult to detect chlamydia; about 80% of women and 10% of men with the disease have no symptoms.

Chlamydial infections are difficult to control. Infants can become infected while passing through the birth canal of an infected mother. Silver nitrate used to prevent gonorrheal ophthalmia neonatorum does not protect against *Chlamydia*, but erythromycin protects against both. In venereal disease clinics, penicillin used to treat syphilis and gonorrhea fails to eliminate chlamydial

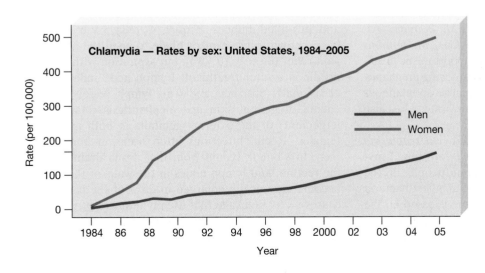

Figure 20.17 **Chlamydia infection rates (per 100,000 population) by sex, in the United States, 1984–2005.**

No News May Not Be Good News

If you've got a chlamydial infection, be prepared to hear your doctor tell you that laboratory tests proved negative for bacteria. Chlamydias can't be detected by standard culturing methods. In fact, before 1983, only 200 laboratories in the United Sates could culture them. Newly developed diagnostic tests such as enzyme immunoassays and monoclonal antibody tests have helped; they've actually revealed a previously unsuspected epidemic of chlamydial infections. Chlamydial infections are now known to be one of the most prevalent diseases in American society.

infections. The recent rise in chlamydial infections could be stopped with tetracycline and sulfa drugs if all sexual partners were treated.

Adult **inclusion conjunctivitis** can result from self-inoculation with *C. trachomatis* from genitals via fingers or towels; it is especially common in sexually active young adults. It closely resembles trachoma (◀Chapter 19, p. 593). Prior to the widespread use of chlorine in pools, it was called "swimming pool conjunctivitis." Reinfection can occur, but whether by inadequate immune response or by infection with another of eight different infectious strains is unknown.

Each year about 75,000 infants acquire chlamydial **inclusion blennorrhea** (blen-or-e'ah), a name derived from the Greek for "mucus flow." This usually benign conjunctivitis begins as a pus-filled mucous discharge 7 to 12 days after delivery and subsides with erythromycin treatment or spontaneously after a few weeks or months. If it persists, it becomes indistinguishable from childhood trachoma and can lead to blindness.

Mycoplasmal Infections. NGU also can be caused by *Mycoplasma hominis*, which is frequently among the normal urogenital microflora, especially in women. Mycoplasmas, which have no cell wall, apparently infect by fusing their cell membranes with the host cell membranes. Such infections are very common; more than half of normal adults have antibodies to *M. hominis*. Although most often associated with NGU, *M. hominis* sometimes causes PID in women and opportunistic urethritis in men. Mycoplasmas on the cervix during pregnancy probably colonize the placenta and cause spontaneous abortions, premature births, and low birthrate. They also foster ectopic pregnancies.

Yet another causative agent of NGU is *Ureaplasma urealyticum*, formerly called T-strain (T for "tiny") mycoplasma. Between 1 and 2.5 million people in the United States are infected. One of the smallest bacteria known to cause human disease, its name is based on the fact that it requires a 10% urea medium to grow. It is one of the first bacteria to have had its genome completely sequenced. Of patients seen in venereal disease

clinics, 50 to 80% carry *U. urealyticum* in addition to other sexually transmitted pathogens. The organism accounts for more than half of all infections that make couples infertile. Low sperm counts and poor sperm mobility have been observed in males; the organisms bind tightly to sperm and can be transmitted by them to sexual partners. They are a major cause of fetal death, recurrent miscarriage, prematurity, and low birth weight—itself a leading cause of neonatal death.

Diagnosis is made by culture from urethral and vaginal discharges and from placental surfaces. When both members of an infected couple are treated, pregnancy is achieved in 60% of cases, compared with a success rate of only 5% in untreated couples infected with this organism. Because mycoplasmas lack a cell wall, penicillin is not effective against them. Therefore, people being treated with penicillin for other STDs will not be cured of NGU. Tetracycline is most often used because it also controls *Chlamydia*. The 15% of *Mycoplasma* strains resistant to tetracycline can be treated with erythromycin and spectinomycin.

Lymphogranuloma Venereum

Another STD, **lymphogranuloma venereum (LGV)**, is common in tropical and subtropical regions. Although only about 250 cases per year occur in the United States, mostly in southeastern states, as many as 10,000 cases are treated annually in a single Ethiopian clinic. This nonreportable disease is about 20 times more common in males than in females.

The Disease. The causative agent of LGV was identified in 1940 as a highly invasive strain of *Chlamydia trachomatis*. Within 7 to 12 days after contact with the organism, lesions appear at the site of infection—usually the genitals, but sometimes the oral cavity. In most cases the lesions rupture and heal without scarring. Other symptoms include fever, malaise, headache, nausea and vomiting, and a skin rash. One week to two months later, the organisms invade the lymphatic system and cause regional lymph nodes to become enlarged, painful, and pus-filled buboes (**Figure 20.18a**). Victims have been known to use a razor blade to open the buboes to gain relief (**Figure 20.18b**), but aspiration with a sterile needle is a safer treatment. Lymph node inflammation occasionally obstructs and scars lymph vessels, causing edema of genital skin and elephantiasis (massive enlargement) of the external genitalia in both males and females. Rectal infections often occur in homosexual men. In women, lymph from the vagina drains toward the rectum, and lymph nodes in the walls of the rectum become chronically enlarged in 25% of the cases. This causes rectal blockage that usually requires surgery. Untreated cases produce a bloody, pus-filled anal discharge and can eventually lead to rectal perforation. Organisms transferred from the hands to the eyes can cause

(a)

(b)

Figure 20.18 Bilateral buboes of lymphogranuloma venereum, caused by *Chlamydia trachomatis*. **(a)** Early in development; *(Science VU/Visuals Unlimited)* **(b)** after reaching such size that the lesion was opened for drainage. *(Courtesy Centers for Disease Control and Prevention CDC)*

conjunctivitis. In rare instances the disease progresses to meningitis, arthritis, and pericarditis.

Spontaneous "cures" sometimes represent latent infections. Latency can be prolonged, as is evidenced by males who infect sex partners many years after their own initial infection. The genital tracts and rectums of chronically infected, but at times asymptomatic, persons serve as reservoirs of infection.

Diagnosis and Treatment. LGV is diagnosed by finding chlamydias as inclusions stained with iodine in pus from lymph nodes. Serological tests are available, but they frequently give false positive results. Doxycycline is the drug of choice for treating LGV. Enlarged lymph nodes can take 4 to 6 weeks to subside even after successful antibiotic therapy.

Granuloma Inguinale

Granuloma inguinale, or donovanosis, is caused by the small Gram-negative encapsulated rod *Klebsiella granulomatis*, formerly known as *Calymmatobacterium granulomatis*. The disease is uncommon in the United States, with about 50 cases reported per year, mostly among

male homosexuals. It is common in India, western coastal Africa, South Pacific islands, and some South American countries, from which it is occasionally brought to the United States. The epidemiology of granuloma inguinale is not completely understood. Some cases appear to be sexually transmitted and others not. Even in genital cases, infectivity is low and many partners of infected individuals do not get it.

Granuloma inguinale appears as irregularly shaped, painless ulcers on or around the genitals 9 to 50 days after intercourse. Fever is absent. Ulcers can be spread to other body regions by contaminated fingers. As ulcers heal, skin pigmentation is lost. Without treatment, tissue damage can be extensive. Diagnosis from scrapings of lesions is confirmed by finding large mononuclear cells called **Donovan bodies (Figure 20.19)**. Antibiotics such as ampicillin, tetracycline, erythromycin, and gentamicin offer effective treatment.

✓CHECKLIST

1. Which body sites can be affected by gonorrhea?
2. Does one develop immunity following a case of gonorrhea? After a case of syphilis?
3. What is a buboe?
4. What is PID? What can cause it?

VIRAL SEXUALLY TRANSMITTED DISEASES

Herpesvirus Infections

Two closely related herpesviruses cause disease in humans (◀Chapter 10, p. 282). **Herpes simplex virus**

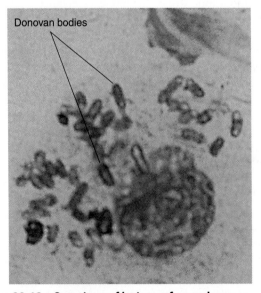

Donovan bodies

Figure 20.19 Scrapings of lesions of granuloma inguinale. Caused by *Klebsiella granulomatis*, these lesions reveal encapsulated bacterial cells inside the much larger macrophages. The bacterial cells resemble closed safety pins and are called Donovan bodies. *(Courtesy Centers for Disease Control and Prevention CDC)*

type 1 (HSV-1) typically causes fever blisters (cold sores), and **herpes simplex virus type 2 (HSV-2)**, sometimes called *herpes hominis virus*, typically causes **genital herpes**. Both HSV-1 and HSV-2 have an incubation period of 4 to 10 days, cause the same kind of lesions, and have been isolated from skin and mucous membranes of oral and genital lesions. Among oral infections, 90% are caused by HSV-1 and 10% by HSV-2. Among genital infections, 85% are caused by HSV-2 and 15% by HSV-1. The presence of HSV-1 in the genital region and HSV-2 in the oral region is most commonly due to the practice of oral sex. Genital herpes is by far the most common and most severe of the herpes simplex viral infections.

Initial HSV-1 or HSV-2 infection can be asymptomatic, especially in children, or cause localized lesions with or without symptoms of acute infection. Most adults have antibodies to herpesviruses, but only 10 to 15% have experienced symptoms. In both HSV-1 and HSV-2 infections, vesicles form under keratinized cells and fill with fluid from virus-damaged cells, particles of cell debris, and inflammatory cells. Vesicles are painful, but they heal completely in 2 to 3 weeks without scarring unless there is a secondary bacterial infection. Adjacent lymph nodes enlarge and are sometimes tender.

Latency (◀Chapter 10, p. 282) is a hallmark of herpes infections. More than 80% of the adult population worldwide harbors these viruses, but only a small proportion experience recurrent infections. Within 2 weeks of an active infection, the viruses travel via sensory neurons to ganglia **(Figure 20.20)**. Within the ganglia they replicate slowly or not at all. They can reactivate spontaneously or be activated by fever, ultraviolet radiation, stress, hormone imbalance, menstrual bleeding, a change in the immune system, or trauma. The infection can spread to and kill cells in the adrenal glands, liver, spleen, and lungs. In fatal herpes encephalitis, soft, discolored lesions appear in both gray and white matter of the brain.

Figure 20.20 Herpes simplex viruses are permanent "houseguests" once acquired. After an occurrence of a lesion, they migrate to ganglia, from which they travel back to the site of the original lesion at the time of the next outbreak. An infrequent exception to this pattern sometimes occurs at the trigeminal ganglion, at which three branches of the trigeminal (fifth cranial) nerve join. Virus from a lesion on the lower lip could migrate out along the ophthalmic branch on a subsequent episode, causing the eye to become affected, or even migrate backward into the brain, causing meningitis and brain damage. Why this sometimes happens is not known; fortunately, it is very rare.

PUBLIC HEALTH

Herpes Singles Club

Are you single? Do you have genital herpes? Then you might want to join a "herpes singles club." Some people with herpes join these clubs in hopes of being able to enjoy sexual activity with other members without inflicting the disease on uninfected people. Although this may prevent the spread of herpes infections to "outsiders," it can also give herpes sufferers a false sense of security. Because several different strains of both HSV-1 and HSV-2 exist, people can infect each other with different strains of HSV. Each new strain can cause a new primary infection and very painful lesions.

After reactivation, the virus moves along the nerve axon to the epithelial cells. There it replicates, causing recurrent lesions. These lesions, which always recur in exactly the same place as the original infection, are smaller, shed fewer viruses, contain more inflammatory cells, and heal more rapidly than primary lesions.

Successive recurrences usually become milder until they finally cease. While the virus is in a neuron, neither humoral nor cellular immunity can combat it. Once the virus reaches target epithelial cells and starts to replicate, antibodies can neutralize the viruses, and T cells can eliminate virus-infected cells. These immunological processes that make it difficult to isolate viruses from vesicular fluid in recurrent lesions also reduce their severity and duration. Recurrences can be limited to one or two episodes or can appear periodically for the life of the patient but typically occur five to seven times. Even if recurrences cease, the viruses remain latent in ganglia, and long-dormant viruses can be reactivated by severe stress, trauma, or impaired immune function (as, for example, in AIDS patients).

"HSV shedders," people who shed viruses while remaining asymptomatic, pose a significant problem. As many as 1 in 200 women have been shown to shed viruses even when they have no observable lesions and in some cases no knowledge of ever having had a herpes infection. Such women pose a serious threat to any infant they might bear.

Genital Herpes. Genital herpes infections usually are acquired after the onset of sexual activity because the virus is transmitted mainly by sexual contact. However, the virus can survive for short periods of time in moist areas such as hot tubs. Changes in sexual practices, especially an increase in oral sex, have increased the incidence of HSV-1 in genital lesions and HSV-2 in oral lesions. More than 20 million Americans now have genital herpes, and half a million new cases are seen each year.

In females the vesicles appear on the mucous membranes of the labia, vagina, and cervix. Ulcerations sometimes spread over the vulva and can even appear on the thighs. In males tiny vesicles appear on the penis and foreskin and are accompanied by urethritis and a watery discharge **(Figure 20.21)**. The prostate gland and seminal vesicles also can be affected. Both sexes experience intense pain and itching at the sites of lesions and swelling of lymph nodes in the groin.

A person infected with herpesvirus is contagious any time viruses are being shed. Shedding always occurs when active lesions are present and usually starts a few days before lesions appear. It can occur continuously even when lesions are not present. Therefore, abstention from sexual contacts when lesions are present does not always prevent spread of the disease. In recent years promiscuous sexual practices and ignorance of, or lack of concern about, transmitting the disease have greatly increased the number of cases of genital herpes. This incurable disease has now become one of the most common STDs.

Women infected with genital herpes may be subject to three other serious problems. First, the incidence of miscarriages among women with genital herpes is higher than that for uninfected women. Second, when infected women become pregnant, the infant must be delivered by

Figure 20.21 Penile herpes lesions. (CNRI/Phototake)

Caesarean section. Finally, infected women have an increased risk of becoming infected with the AIDS virus, as lesions provide open pathways of entrance for the virus.

Neonatal Herpes. Neonatal herpes (Figure 20.22) can appear at birth or up to 3 weeks after birth. Babies most often become infected by delivery through a birth canal contaminated with HSV-2, but they can also become infected through contaminated equipment and hospital procedures. In rare instances, infants are infected *in utero*. Because neonates are highly susceptible to HSV infections, they should not be cared for by individuals with such infections. Infected mothers who must care for their infants must scrupulously follow sanitary procedures.

At diagnosis, two-thirds of infected infants have skin vesicles; the others already have disseminated infection with neural or visceral lesions. Skin infections disseminate in 70% of infected infants. Infants with

Figure 20.22 Neonatal herpes. This type of herpes can be acquired when the baby passes through the birth canal of an infected mother. This could be avoided by Caesarean section delivery. Sometimes lesions are so extensive that they cover almost the entire skin surface. Such children either suffer profound brain damage or do not survive. (*Courtesy Centers for Disease Control and Prevention*)

disseminated infections display poor appetite, vomiting, diarrhea, respiratory difficulties, and hypoactivity. Some also have neurological symptoms, jaundice, and eye disorders. Neonates with disseminated infections deteriorate rapidly and usually die within 10 days. The few that survive usually have central nervous system and eye damage. Occasionally infants have only a few vesicles, but latent viruses may later cause significant damage. Early diagnosis and treatment of neonatal herpesvirus infections are essential for survival and to reduce the likelihood of neurological damage.

Other Herpes Simplex Infections. A variety of manifestations of herpesvirus infections have been seen. Most are thought to be caused by HSV-1, but HSV-2 may be responsible for some. They include gingivostomatitis, herpes labialis, keratoconjunctivitis, herpes meningoencephalitis, herpes pneumonia, eczema herpeticum, traumatic herpes, herpes gladiatorium, and whitlows. HSV-1 is spread by contaminated secretions and through contact with lesions and usually is acquired in childhood from relatives, nurses, or other children. The incidence of the virus is especially high within families, hospitals, and other institutions. Most humans are infected with HSV-1 in the first 18 months of life. Many of these infections are inapparent, and the first apparent lesions result from reactivations.

Gingivostomatitis (jin″gi-vo-sto″mah-ti′tis), lesions of the mucous membranes of the mouth, is most common in children 1 to 3 years old. After an incubation period of 2 to 20 days, small vesicles appear around the mouth over a 7-day period. Each vesicle accumulates fluid, scabs over, and heals in 2 to 3 weeks. Recurrent lesions typically occur in the form of **herpes labialis** (la′be-ah-lis), or fever blisters on the lips. If the eyes instead of the mouth are the site of initial infection, vesicles appear on the cornea and eyelids, causing **keratoconjunctivitis** (ker″a-to-kon-junk-ti-vi′tis).

The most serious manifestation of HSV infection is *herpes meningoencephalitis*, which can follow a generalized herpes infection in a newborn, child, or adult. How the virus crosses the blood-brain barrier to enter the central nervous system has recently been discovered. The virus remains latent in the ganglion of the trigeminal nerve, which supplies parts of the face and mouth, until some unknown factor reactivates it. Then, instead of moving outward to epithelial sites such as the tongue or lip, it ascends along the nerve into the brain. The disease has a rapid onset, with fever, headache, meningeal irritation, convulsions, and altered reflexes. In the middle-aged and elderly, meningoencephalitis can appear without any preceding symptoms. The patient experiences increasing confusion, hallucinations, and sometimes seizures. Most patients die in 8 to 10 days; survivors usually have permanent neurological damage.

Herpes pneumonia is rare. It usually is seen only in burn patients, alcoholics, and patients with AIDS or

other immunodeficiencies. **Eczema herpeticum** is a generalized eruption caused by entry of the virus through the skin. **Traumatic herpes** occurs when the virus enters traumatized skin in the area of a burn or other injury. **Herpes gladiatorium** occurs in skin injuries of wrestlers. A **whitlow (Figure 20.23)** is a herpetic lesion on a finger that can result from exposure to oral, ocular, and probably genital herpes lesions. Dental technicians, nurses, and other medical personnel should therefore use rubber gloves when treating patients with herpes lesions. A person with a whitlow can spread the infection to the mouth, eyes, and genital areas or to other persons.

Diagnosis, Treatment, and Prognosis. Herpesviruses are most easily isolated from vesicular fluid and cells from the base of a lesion and can be grown in the laboratory in a variety of cell types. The cytopathic effect of the viruses appears in the culture quickly. The time to obtain a diagnosis has been greatly decreased by rapid immunological tests. Speedy diagnosis is especially important for women in labor or near delivery, as a Caesarean delivery can prevent exposure of the infant to the virus.

Figure 20.23 Herpetic whitlow. This type of herpesvirus infection is a very painful infection of the finger, which can be spread to other areas of the body, for example, by rubbing one's eyes. Health care workers should protect themselves by wearing latex gloves. *(Ken Greer/Visuals Unlimited)*

In recent years several drugs have been used with varying degrees of success in treating HSV-1 and HSV-2 infections. Trifluorothymidine is effective in treating ocular herpes. Although acyclovir fails to prevent recurrences reliably, it does prevent the spread of lesions, decrease viral shedding, and shorten healing time. Best results are obtained with acyclovir when it is used at the beginning of a primary infection. The use of acyclovir, Ara-A, and vidarabine in various combinations early in the infection has increased survival rates in herpes meningoencephalitis and neonatal herpes. The prognosis for alleviating genital herpes symptoms is reasonably good. Most patients have few if any recurrences if they take acyclovir daily but may have a recurrence if they stop taking it. No treatment can eradicate latent viruses.

Immunity and Prevention. Although immunological knowledge of herpesviruses is sufficient to make a vaccine, such a vaccine is not yet available. Even if a vaccine were available and administered to all who lack HSV antibodies, it would take years to eradicate the disease. Large numbers of people already harbor latent viruses, so a vaccine to inactivate the viral genetic information responsible for *latency* is needed. A vaccine that elicits antibodies like those from natural infections would probably not be effective; such antibodies apparently have no effect on latent viruses and cannot alone prevent recurrences.

The best available means of preventing HSV infections is to avoid contact with individuals with HSV-1 or HSV-2 lesions. If all infected individuals were to refrain from sexual activity, especially when they have active lesions, some new cases of genital herpes could be prevented. If pregnant women would advise their obstetricians of herpes infections, their infants could be protected from exposure during delivery. Even these precautions will not prevent the spread of genital herpes. Most infected individuals shed viruses a few days before lesions appear, and some shed viruses continuously without having any lesions.

Genital Warts

Condylomas (kon-dil-o'mahz), or **genital warts**, caused by the human papillomavirus (HPV), most often occur in the sexually promiscuous, young adult population. The incidence of genital warts has increased rapidly in recent years to the extent that such warts are now among the most prevalent STDs. Approximately 20 million people in the United States are currently infected with HPV. At least half of all sexually active men and women become infected at some point in their life, with at least 80% of women being infected by age 50. Most will not have any symptoms and will clear the infection on their own. Currently in the United States, the incidence rate is 20% among young women. Two-thirds of the sexual partners of infected individuals also develop warts. Condoms are not as useful in prevention of HPV infection as they are against other STDs. The warts can be papillary or flat. In males the warts appear on the penis, anus, and perineum; in females they are on the vagina, cervix, perineum, and anus.

Like dermal warts (◄Chapter 19, p. 588), genital warts cause irritation and sometimes intense itching. They can persist or regress spontaneously. Genital warts often become infected with bacteria, and those that persist for many years can transform into malignant growths. Some victims suffer psychological damage from the presence of warts. Warts temporarily increase in number and size during pregnancy but decrease after delivery. Infants can be infected during delivery. The number of warts increases in immunosuppressed patients and in those with AIDS and other immunological deficiencies. (Treatment of warts was discussed in ◄Chapter 19, p. 587).

Infection with some HPV strains can result in visible warts. Other strains cause "silent" infections without wart formation. Either type of infection is transient in 90% of cases, with a mean duration of 8 months. Of the 10% that become established infections, one-tenth (i.e., 1% of all infected) will become malignant, causing cancer of the cervix, penis, or anus. One study found that 99.7% of 932 cervical cancers were infected with HPV. Externally visible (vulvar) wart strains (e.g., types 6 and 11) generally do not cause cancer. Those strains causing either cervical warts or silent infections (e.g., types 16, 18, 8, etc.) can cause malignancy when a nontransient infection develops due to a defective cell-mediated immune response in an individual. Thirteen strains cause 99% of all HPV cancers.

Incubation lasts 1 to 8 months, with immune response occurring between 3 and 9 months. A strong immune system can suppress and eliminate HPV infection during that time, preventing development of cancer. HPV does not have a viremic stage in which it would circulate through the body; instead it stays in the genital region. The virus numbers peak during 3 to 6 months and usually disappear by 9 months. Infection with one strain does not cross-protect against other strains.

Cervical cancer ranks number 9 among cancers of American women, but in developing countries such as Mexico and Brazil it is a major killer of women. Pap smears in the United States frequently identify abnormal cells from the cervix early enough so that surgical intervention is successful. In countries where Pap smears are not given regularly, by the time cancer is detected, it is too late to save the woman. Pap smears do not detect HPV virus, only abnormal cells, and that perhaps only 70% of the time, depending on the technician's skill. A newly developed DNA test (Hybrid Capture 2ᵂ) detects all 13 types of HPV that cause cancer, and costs $40 to $70. It appears to have no false negatives. Sixty percent of women with a positive result show some abnormality by 4 to 5 years after testing. It is far more accurate than the Pap test, but is just now coming into use. Vaccine research has produced Gardasil©, a vaccine which protects against HPV strains 16 and 18, the causative agents of 70% of all cases of cervical cancer. It is given in a series of 3 shots over 6 months, and costs a total of $360. For more information refer back to the opening vignette for ◄Chapter 10, p. 271. A second vaccine which offers even more protection will soon be on the market. Because the vaccines are not 100% protective, women will still need to have regular Pap smear tests done. Meanwhile education efforts must be increased to make all women aware that cervical cancer is just as much an STD as are syphilis and gonorrhea, and that a DNA test is available to tell them if they are infected, and a vaccine to keep them from getting infected.

Laryngeal Papillomas

Laryngeal papillomas (pap-il-lo′maz) are benign growths that can be dangerous if they block the airway. Hoarseness, voice changes, and respiratory distress occur when the airway becomes obstructed. Children are more likely to have laryngeal papillomas than adults. Surgical removal, sometimes every 2 to 4 weeks, is the only treatment for these obstructive growths. Also, there is danger of spreading the virus to the lungs during surgery. Laryngeal papillomas are usually caused by HPV-6 and HPV-11, which are thought to infect infants during birth to women with active genital warts.

Cytomegalovirus Infections

The **cytomegaloviruses (CMVs)** constitute a widespread and diverse group of herpesviruses. In general, each strain of CMV is capable of infecting a single species. The majority of human CMV infections occur in older children and adults and go unnoticed because they do not produce clinically apparent symptoms. An estimated 80% of U.S. adults carry the virus. When there are symptoms, they include malaise, myalgia, protracted fever, abnormal liver function, and lymph node inflammation without swelling. Symptoms are more severe in patients with AIDS and other immunodeficiencies.

Initially the virus can be recovered from the oropharynx; viremia with many virus-infected neutrophils can last for months. The viruses replicate and have low pathogenicity but are excreted intermittently over many months, during which time they can infect others. Viruses are shed in all body fluids—saliva, blood, semen, breast milk—but they are found most often and in the largest quantities in urine even a year or more after infection. In a symptomatic primary attack of CMV, cell-mediated immunity is depressed, and the ratio of T-helper to T-suppressor cells is reversed. However, the immune system returns to normal during convalescence. In subclinical cases cell-mediated immunity remains active. Large quantities of long-lasting antibodies are produced in response to

Studies show that CMV spreads to 1 in 5 nonimmune parents who have infected children.

CMV infection, but they do not prevent viral shedding. The virus is usually spread through long, close contact with children who are shedding the virus, but it can be spread by blood transfusions, organ transplants, and sexual intercourse.

Fetal and Infant CMV Infections. In fetuses and infants, CMV infections can be life threatening because the virus disseminates widely to various organs. Fetuses become infected by viruses that cross the placenta from infected mothers. Unlike rubella infection, maternal CMV infections are rarely detected, so the risk to the fetus is unknown. Maternal antibodies also can cross the placenta and inactivate small quantities of virus. Maternal hormones suppress CMV, but their effect diminishes as the pregnancy progresses. Both fetuses and neonates are dependent on maternal defenses against CMV because their own immune systems are too immature to mount a successful defense. Tests for maternal antibody levels are available. Young children are the most likely source of infection to nonimmune women.

In severe CMV infections in which the fetus has been infected with large numbers of viruses (about 4,000 cases per year in the United States), intrauterine growth retardation and severe brain damage can occur **(Figure 20.24)**. Many babies have CMV-infected cells in the inner ear and hearing loss due to nerve damage. Some have jaundice with liver damage, and others have impaired vision. Less severe infections (another 4,500–6,000 cases per year) cause damage to certain brain areas and mild central nervous system disorders with or without damage to hearing or sight. Mortality can be as high as 30%.

CMV infections contracted after birth generally cause fewer permanent defects than those contracted before birth. However, these infections can cause severe

Figure 20.24 A baby with birth defects due to congenitally acquired cytomegalovirus. The virus, sometimes called salivary gland virus, can cause grave damage to people who have impaired immune systems. *(Courtesy Centers for Disease Control and Prevention)*.

illness. Among babies infected with CMV, 5% have typical, generalized *cytomegalic inclusion disease (CID)*; another 5% have atypical, less generalized infections; the rest have subclinical but chronic infections. Many infants with CID infections have significant mental or sensory disorders. These include an abnormally small brain accompanied by intellectual deficits and inflammation of the eyes with impaired vision. In subclinical cases

TABLE 20.2

A Summary of Urogenital and Sexually Transmitted Diseases

Disease	Agent(s)	Characteristics
Bacterial Urogenital Diseases		
Urinary tract infections	*Escherichia coli, Proteus mirabilis,* and other bacterial species	Dysuria; sometimes leads to chronic cystitis; often ascend or descend in urinary tract
Prostatitis	*E. coli* and other bacterial species	Dysuria, urgent and frequent urination, low fever, back pain; can cause infertility
Pyelonephritis	*E. coli* and other bacterial species, sometimes the yeast *Candida*	Inflammation of pelvis of kidney, often caused by urinary tract blockage; dysuria, nocturia, sometimes fever
Glomerulonephritis	Streptococcal or viral infections from other sites	Deposition of immune complexes causes inflammation of glomeruli; can cause permanent kidney damage
Leptospirosis	*Leptospira interrogans*	Fever, nonspecific symptoms; can lead to Weil's syndrome with jaundice and liver damage
Bacterial vaginitis	*Gardnerella vaginalis* with anaerobes	Frothy, fishy-smelling discharge; pain and inflammation
Toxic shock syndrome	*Staphylococcus aureus*	Toxins reach blood and cause fever, rash, and shock that can lead to death
Parasitic Urogenital Disease		
Trichomoniasis	*Trichomonas vaginalis*	Intense itching, copious white discharge
Bacterial Sexually Transmitted Diseases		
Gonorrhea	*Neisseria gonorrhoeae*	Infectious organisms release endotoxin that damages mucosa; pus-filled discharge; can cause PID and infect other systems
Syphilis	*Treponema pallidum*	Chancre develops in primary stage; mucous membrane lesions and rash occur in secondary stage; permanent cardiovascular and neurological damage often occur in tertiary stage
Chancroid	*Haemophilus ducreyi*	Painful, bleeding lesions on genitals; often enlarged lymphatic buboes
Nongonococcal urethritis	*Chlamydia trachomatis* and mycoplasmas	Scanty, watery urethral discharge, inflammation, sometimes sterility; can cause neonatal infections and fetal death
Lymphogranuloma venereum	*Chlamydia trachomatis*	Genital lesions, fever, malaise, headache, nausea, vomiting, skin rash; lymph nodes become pus-filled buboes
Granuloma inguinale	*Klebsiella granulomatis*	Painful ulcers on genitals and other sites; loss of skin pigmentation as ulcers heal
Viral Sexually Transmitted Diseases		
Herpes simplex infections	Herpes simplex viruses	Fever blisters usually caused by HSV-1; genital herpes usually caused by HSV-2 (both are latent viruses); recurrent, painful vesicular lesions; neonatal herpes; and a variety of other manifestations
Genital warts	Human papillomaviruses	Warts on genitals, vagina, and cervix; irritation and sometimes intense itching; causes 99% of all cervical carcinoma
Cytomegalovirus infections	Cytomegaloviruses	Often asymptomatic but severe in fetuses, neonates, and immunodeficient patients; malaise, myalgia, fever, inflamed lymph nodes, neural damage, and death in fetuses and neonates

the prognosis is better, but 10% will have deafness or other sensory problems. Some will experience intellectual and behavioral disorders later. Babies who acquire CMV infections from blood transfusions have a gray pallor and symptoms like those of CID. Pneumonia, respiratory deterioration, and death may follow because the infants receiving transfusions usually are premature and debilitated before transfusions are given. Moreover, CMV infections sometime are not diagnosed in the presence of many other disorders. Finally, CMV sometimes causes pneumonia in infants 1 to 6 months of age in the presence of other infections, such as *Chlamydia trachomatis* and *Ureaplasma urealyticum*.

Disseminated CMV. In severe disseminated disease, the virus can be present in many organs. When the kidneys become infected, immune complexes are deposited in the glomeruli, but renal dysfunction is rare except in renal transplants. Subclinical hepatitis occurs in both children and adults. Lung infections are common in infants and immunosuppressed adults. Brain involvement is rare except in fetuses.

CMV is most virulent when present as a primary infection in immunosuppressed patients. Between 1 and 4 months after an organ transplant, the patient develops symptoms like those of infectious mononucleosis, with prolonged fever and enlargement of the liver and spleen. Pneumonia is frequent and severe in bone marrow transplant patients, and mortality can be as high as 40%. Hepatitis is usually reversible, but damage to the eyes is likely to be progressive and irreversible.

Diagnosis, Treatment, and Prevention. Definitive diagnosis of CMV infections requires identification of the virus from clinical specimens, a procedure that can take up to 6 weeks. New, faster techniques use monoclonal antibodies to detect viral antigens. It is anticipated that nucleic acid probes, which make use of the base sequences unique to CMV nucleic acids, will allow identification of the virus in clinical specimens in less than 24 hours.

No effective treatment for CMV infections in infants is available. The prognosis is poor when infections occur in fetuses because the damage is done before a diagnosis is made in the neonate. Interferon and hyperimmune gamma globulin given before and after organ transplantation reduce the incidence and severity of CMV infections in transplant patients. When blood transfusions are required, donor and recipient should be matched for CMV antigens to minimize introduction of CMV to which the recipient is not immune. Also, donors serologically negative for CMV should be used with infants, especially premature infants, to avoid the possibility of transmitting CMV through a transfusion. No effective vaccine exists.

The agents and characteristics of the diseases discussed in this chapter are summarized in **Table 20.2**.

✓**CHECKLIST**

1. How can genital herpes viral infections be acquired?
2. How can neonatal herpes infections be avoided?
3. Why is cervical cancer now listed as a sexually transmitted disease?

∎ RETRACING OUR STEPS

-   The agents and characteristics of the **sexually transmitted diseases (STDs)** discussed in this chapter are summarized in Table 20.2. Information in that table is not repeated in this summary.

∎∎∎ COMPONENTS OF THE UROGENITAL SYSTEM

-   The **urogenital system** includes the **urinary system** and the reproductive system. The urinary system consists of the **kidneys, ureters, urinary bladder,** and **urethra**. The **female reproductive system** consists of the **ovaries, uterine tubes, uterus, vagina,** external genitalia, and **mammary glands**. The **male reproductive system** consists of the **testes**, a system of ducts, glands, and the **penis**.
-   Urogenital infections are most common near the openings of the vagina and urethra.
-   Nonspecific defenses include urinary sphincters, cleansing action of the outflow of urine, acidity of mucous membranes, and competition from normal microflora.

NORMAL MICROFLORA OF THE UROGENITAL SYSTEM

-   Normally all parts of the urinary tract, except for the portion of the urethra nearest its opening, are sterile.

-   In males, other than the last third of the urethra, the genital tract lacks a normal microflora and is sterile.
-   In females, during childbearing years, lactobacilli predominate in the vagina producing an acid pH inhibitory to growth of most other types of microbes. During childhood and after menopause, streptococci and staphylococci predominate in the alkaline conditions of the vagina.

∎∎∎ UROGENITAL DISEASES USUALLY NOT TRANSMITTED SEXUALLY

BACTERIAL UROGENITAL DISEASES

-   **Urinary tract infections (UTIs)**, which are exceedingly common, can ascend or descend to spread through the urogenital system. They are diagnosed from urine cultures, treated with various antibiotics, and can be prevented by good personal hygiene and complete emptying of the bladder.
-   **Prostatitis** can result from the spread of a urinary tract infection and usually can be treated with antibiotics.
-   **Pyelonephritis** also can result from the spread of a urinary tract infection; it can be difficult to treat because renal failure allows drugs to accumulate to toxic levels.

- **Glomerulonephritis** usually results from throat and other infections in which immune complexes deposit in glomeruli. Prompt treatment of other infections minimizes the risk of glomerulonephritis.
- **Leptospirosis** is acquired through contact with contaminated urine and is diagnosed by microscopic examination of blood. It can be treated with antibiotics and prevented by vaccinating pets, which are usually responsible for transmitting the disease to humans.
- Bacterial and other forms of **vaginitis** occur when normal vaginal microflora are disturbed. *Gardnerella vaginitis* is diagnosed by the presence of clue cells. It is most often treated with metronidazole to eradicate anaerobes and allow the restoration of normal microflora.
- **Toxic shock syndrome (TSS)** most often arises from the use of superabsorbent tampons. It should be treated promptly with nafcillin and can be prevented by avoiding such tampons.

PARASITIC UROGENITAL DISEASES

- **Trichomoniasis** usually is transmitted sexually, is diagnosed from smears of secretions, and is treated with metronidazole.

III SEXUALLY TRANSMITTED DISEASES

BACTERIAL SEXUALLY TRANSMITTED DISEASES

- **Gonorrhea** is diagnosed by use of molecular probes and is treated with a combination of antibiotics, usually ceftriaxone plus azithromycin.
- **Syphilis** is diagnosed by immunological tests and is treated with benzathine penicillin G.

- **Chancroid** occurs mainly in developing countries. It is diagnosed by finding organisms from lesions or buboes and is treated with tetracycline and other antibiotics.
- The incidence of **nongonococcal urethritis (NGU)** is increasing dramatically. It can be diagnosed by culturing samples from discharges and is treated with erythromycin, tetracycline, or other antibiotics, depending on the susceptibility of the causative agent.
- **Lymphogranuloma venereum (LGV)** occurs mainly in tropical and subtropical regions. It is diagnosed by finding chlamydias as inclusions in pus and is treated with tetracycline or another antibiotic.
- **Granuloma inguinale** is common in India and several other countries. It is diagnosed by finding **Donovan bodies** in scrapings from lesions and can be treated with a variety of antibiotics.
- All the preceding diseases are usually transmitted by sexual contact and can be prevented by avoiding such contacts. No vaccines are available.

VIRAL SEXUALLY TRANSMITTED DISEASES

- **Genital herpes** is the most severe of herpes simplex virus infections. It is diagnosed by finding viruses in vesicular fluids or by immunological tests and can be treated, but not cured, with acyclovir and other antiviral agents.
- **Genital warts** must be distinguished from cervical dysplasia and carcinoma; the virus is responsible for over 99% of all cases of cervical cancer; treatment of warts was discussed in ◀Chapter 19.
- **Cytomegalovirus (CMV)** infections are of greatest significance in fetuses, newborns, and the immunosuppressed. Rapid diagnosis can be made by monoclonal antibody tests, but no effective treatment is available.

▌ TERMINOLOGY CHECK

balantitis *(p. 614)*
Bartholin gland *(p. 608)*
cervix *(p. 608)*
chancre *(p. 620)*
chancroid *(p. 623)*
condyloma *(p. 631)*
congenital syphilis *(p. 622)*
cystitis *(p. 610)*
cytomegalovirus (CMV) *(p. 632)*
Donovan body *(p. 627)*
dysuria *(p. 610)*
eczema herpeticum *(p. 630)*
endometrium *(p. 608)*
female reproductive system *(p. 608)*
genital herpes *(p. 628)*
genital wart *(p. 631)*
gingivostomatitis *(p. 630)*
glomerulonephritis *(p. 612)*
glomerulus *(p. 607)*
gonorrhea *(p. 616)*

granuloma inguinale *(p. 627)*
gumma *(p. 622)*
herpes gladiatorium *(p. 630)*
herpes labialis *(p. 630)*
herpes pneumonia *(p. 630)*
herpes simplex virus type 1 (HSV-1) *(p. 628)*
herpes simplex virus type 2 (HSV-2) *(p. 628)*
inclusion blennorhea *(p. 626)*
inclusion conjunctivitis *(p. 626)*
keratoconjunctivitis *(p. 630)*
kidney *(p. 607)*
laryngeal papilloma *(p. 632)*
leptospirosis *(p. 612)*
lymphogranuloma venereum (LGV) *(p. 626)*
male reproductive system *(p. 608)*
mammary gland *(p. 608)*

neonatal herpes *(p. 629)*
nephron *(p. 607)*
neurosyphilis *(p. 622)*
nocturia *(p. 611)*
nongonococcal urethritis (NGU) *(p. 625)*
ovarian follicle *(p. 608)*
ovary *(p. 608)*
pelvic inflammatory disease (PID) *(p. 617)*
penis *(p. 608)*
prostate gland *(p. 608)*
prostatitis *(p. 610)*
pyelonephritis *(p. 610)*
semen *(p. 608)*
seminal vesicle *(p. 608)*
sexually transmitted disease (STD) *(p. 615)*
syphilis *(p. 620)*
testis *(p. 608)*
toxic shock syndrome (TSS) *(p. 614)*

traumatic herpes *(p. 630)*
trichomoniasis *(p. 615)*
ureter *(p. 608)*
urethra *(p. 608)*
urethritis *(p. 610)*
urethrocystitis *(p. 610)*
urinalysis *(p. 608)*
urinary bladder *(p. 608)*
urinary system *(p. 607)*
urinary tract infection (UTI) *(p. 610)*
urine *(p. 607)*
urogenital system *(p. 607)*
uterine tube *(p. 608)*
uterus *(p. 608)*
vagina *(p. 608)*
vaginitis *(p. 613)*
whitlow *(p. 630)*

▌ CLINICAL CASE STUDY

A man whose wife had died of cervical cancer remarried. Ten years later his second wife also died of cervical cancer. Is there any possibility that this was not merely a sad coincidence? Go to the following web site to learn more.http://www.reproline.jhu.edu/english/3cc/3refman/cxca_hpv1.htm

▌ CRITICAL THINKING QUESTIONS

1. Currently we have enough immunological knowledge about herpesviruses to make a vaccine; however, there is none available. What is the primary reason for this?

2. Gonorrhea and genital chlamydia can have similar symptoms. If a patient came in with those symptoms, how would you determine whether he or she was infected with gonorrhea, chlamydia, both, or neither?

3. While the incidence of many other infectious diseases has declined sharply in the United States in the past 50 years, STDs remain a major problem. Can you identify some causes of this discrepancy?

▌ SELF-QUIZ

1. Urogenital infections are most common near the openings of the vagina and urethra. However, which of the following nonspecific defenses help prevent infection?
   (a) Cleansing action of urine outflow
   (b) Urinary sphincters
   (c) Acidic pH of mucous membranes
   (d) Competition from normal microflora
   (e) All of the above.
2. The causative agent of 80% of urinary tract infections is:
   (a) *Staphylococcus aureus*
   (b) *Escherichia coli*
   (c) *Proteus mirabilis*
   (d) *Staphylococcus epidermidis*
   (e) *Pseudomonas aeruginosa*
3. Which of the following is NOT true about the normal microflora found in the urogenital system?
   (a) Females of childbearing age normally bear vaginal lactobacilli that produce an acidic pH that inhibits the growth of most other types of microbes.
   (b) Females during childhood and after menopause normally bear a predominance of vaginal streptococci and staphylococci that keep an acidic pH.
   (c) In males, except for the last third of the urethra, the genital tract lacks a normal microflora and is sterile.
   (d) In both sexes, all parts of the urinary tract are normally sterile except for the portion of the urethra nearest its opening.
   (e) All of the above are true.
4. Urinary tract infections (UTI) caused by bacteria are common and can ascend or descend to spread through the urogenital system. Match the following resulting UTI-related diseases to their descriptions:
   ___ Glomerulonephritis
   ___ Toxic shock syndrome
   ___ Pyelonephritis
   ___ Leptospirosis
   ___ Prostatitis
   ___ Vaginitis

   (a) Opportunistic *Gardnerella vaginalis* in combination with anaerobic bacteria commonly cause this vaginal infection that can be diagnosed with "clue cells"
   (b) Associated with superabsorbent tampons left in too long, allowed for the multiplication of strains of *Staphylococcus aureus* producing exotoxin C
   (c) Infection of the prostate in men, commonly caused by *Escherichia coli*
   (d) An immune complex disease of the glomeruli of the kidney tipped-off by nephrogenic strains of *Streptococcus pyogenes*
   (e) An inflammation of the kidney usually caused by the backup of urine with consequent ascent of *Escherichia coli*
   (f) A zoonotic infection by *Leptospira interrogans* usually acquired by direct contact with contaminated dog, cat, or wild animal urine or in water or soil

5. Originally, toxic shock syndrome (TSS) occurred mostly in menstruating women. Currently, in which of the following groups do we see the most TSS cases?
   (a) Children
   (b) Men
   (c) Postmenopausal women
   (d) All of the above
   (e) None of the above

6. Urinary tract infections in pregnant women are more serious because they are more likely to:
   (a) Cause an infection of the fetus
   (b) Develop into pyelonephritis
   (c) Cause frequent urination
   (d) Cause decreased flow of urine
   (e) They are not more serious

7. Toxic shock syndrome is caused by a superantigen produced by which organism?
   - (a) *Salmonella enterica*
   - (b) *Escherichia coli*
   - (c) *Chlamydia trachomatis*
   - (d) *Neisseria gonorrhoeae*
   - (e) *Staphylococcus aureus*

8. Trichomoniasis is caused by what type of organism?
   - (a) Protozoan
   - (b) Bacterium
   - (c) Virus
   - (d) Prion
   - (e) Helminth

9. Although *Trichomonas vaginalis* causes trichomoniasis in humans, which other *Trichomonas* species is responsible for the leading cause of spontaneous abortions in cows and economic losses of up to $1 million per year for U.S. cattle breeders?
   - (a) *Trichomonas foetus*
   - (b) *Trichomonas hominis*
   - (c) *Trichomonas tenax*
   - (d) *Brucella abortus*
   - (e) b and c

10. Which of the following is NOT a way in which *Neisseria gonorrhoeae* infects or causes disease in the genitourinary tract of humans?
    - (a) It uses attachment pili.
    - (b) It produces an exfoliating exotoxin.
    - (c) It produces a damaging endotoxin that damages the fallopian tube mucosa.
    - (d) It produces a protease that cleaves IgA.
    - (e) It often survives phagocytosis by neutrophils and can survive and multiply within them.

11. Match the stages of syphilis to their descriptions:
    - ___ Incubation stage
    - ___ Primary stage
    - ___ Primary latent period
    - ___ Secondary stage
    - ___ Secondary latent stage
    - ___ Tertiary stage

    - (a) Disease of multiple body systems occur including "neurosyphilis" with formations of granulomatous inflammation called "gummas."
    - (b) All symptoms disappear again. Can persist for life, variable periods, or not at all. In some patients syphilis does not progress beyond this stage.
    - (c) Highly contagious stage in which symptoms can appear, disappear, and reappear over a period of up to 5 years. Copper-colored rashes are common.
    - (d) External signs of disease disappear but blood tests are still positive since the spirochetes are spreading through the circulation.
    - (e) Inflammatory "chancres" usually develop at the primary entry site 3 weeks later after initial infection. Chancres disappear 4–6 weeks later but disease has progressed to next stage.

    - (f) Organisms are multiplying and spreading throughout the body over a period lasting 2–6 weeks after entering body.

12. Which of the following is true of the chancre observed during primary syphilis?
    - (a) It disappears and leaves a scar.
    - (b) It lasts throughout all stages of syphilis.
    - (c) It is usually found on the face.
    - (d) It is a hard, painless, lesion.
    - (e) It is painful and purulent.

13. Which of the following diseases is caused by *Haemophilus ducreyi* and is often characterized by signs and symptoms of the "soft chancre" and "buboes" (enlarged masses of lymphatic tissue)?
    - (a) Syphilis
    - (b) Gonorrhea
    - (c) Chancroid
    - (d) Pelvic inflammatory disease
    - (e) Lymphogranuloma venereum

14. The treatment of choice for syphilis is:
    - (a) Vancomycin
    - (b) Sulfonamides
    - (c) Penicillin
    - (d) Acyclovir
    - (e) Protease inhibitors

15. Which of the following species causes lymphogranuloma venereum and most cases of nongonococcal urethritis (NGU)?
    - (a) *Mycoplasma hominis*
    - (b) *Ureaplasma urealyticum*
    - (c) *Klebsiella granulomatis*
    - (d) *Chlamydia trachomatis*
    - (e) *Treponema pallidum*

16. The disease caused by *Klebsiella granulomatis* and diagnosed by finding "Donovan bodies" (large mononuclear cells) from lesion scrapings is known as:
    - (a) Lymphogranuloma venereum
    - (b) Granuloma inguinale
    - (c) Mycoplasmosis
    - (d) Inclusion conjunctivitis
    - (e) a and b

17. Which of the following are associated with pelvic inflammatory disease?
    - (a) *Neisseria gonorrhoeae*
    - (b) *Escherichia coli*
    - (c) *Chlamydia trachomatis*
    - (d) Both a and c
    - (e) a, b, and c

18. Which of the following microbes lack a cell wall, is one of the smallest bacteria known to cause human disease, requires 10% urea to grow in, and accounts for more than half of all infections that make couples infertile?
    - (a) *Mycoplasma hominis*
    - (b) *Chlamydia trachomatis*
    - (c) *Herpes hominis* virus
    - (d) a and c
    - (e) *Ureaplasma urealyticum*

19. Herpes simplex virus type 1 typically causes fever blisters (cold sores) and herpes simplex virus type 2 typically causes genital herpes. True or false?

20. Herpesviruses usually establish a latent infection in which part of the body?
 (a) Nerve ganglia   (d) Intestine
 (b) Gallbladder   (e) Eyes
 (c) Stomach

21. Genital warts are caused by:
 (a) *Treponema pallidum*
 (b) Herpesviruses
 (c) *Chlamydia trachomatis*
 (d) Human papillomaviruses
 (e) *Neisseria gonorrhoeae*

22. What percentage of U.S. adults are believed to be carrying CMV?
 (a) 10%   (d) 80%
 (b) 20%   (e) 100%
 (c) 50%

23. A fetus is protected from CMV during gestation by:
 (a) Antibiotics
 (b) Its own immune system
 (c) Maternal immune defenses
 (d) Antivirals
 (e) Early diagnosis and cure

24. The human papillomavirus is thought to play a role in what disease other than genital warts?
 (a) Diarrhea   (d) Deafness
 (b) Blindness   (e) Cervical cancer
 (c) Arthritis

25. Cytomegaloviruses are a widespread and diverse group of:
 (a) Cytoviruses
 (b) Retroviruses
 (c) Poxviruses
 (d) Herpesviruses
 (e) Mycoplasmas

26. Complete the following table to describe how each disease is diagnosed and treated.

Disease	Mode of Diagnosis	Treatment
Gonorrhea		
Syphilis		
Chancroid		
Lymphogranuloma venereum		
Granuloma inguinale		

▌ EXPLORATIONS ON THE WEB   http://www.wiley.com/college/black

If you think you've mastered this chapter, there's more to challenge you on the web. Go to the companion web site to fine-tune your understanding of the chapter concepts and discover answers to the questions posed below.

1. Did you know that herpesviruses are multipotential, having the ability to cause more than one kind of disease including infectious mononucleosis, malignancies, and mental retardation? Find out more at the web site.

2. Urinary tract infections are responsible for over 8 million doctor visits per year; only respiratory infections are more common. Many women have recurrent urinary tract infections. It is estimated that nearly 20% of women who have had a urinary tract infection will have another, and 30% of those will have yet another, and around 80% of those will continue to have recurrences. Discover why at the web site.

Diseases of the Respiratory System

Linda Stannard/Photo Researchers, Inc.

Come with me...

Around the world people are watching for avian flu (◄Chapter 10, pp. 284–286). Thus far it has not burst out in a large epidemic of person-to-person spreading. But if it does, what are we in the United States going to do? Government officials feel that once such an epidemic does break out elsewhere, we have approximately 3 months before it reaches the United States. We have too many ports of entry to be able to keep it out. Nor can we keep out wild migratory birds that will bring it in. So, what will we do? The best advice is to "prepare to shelter in place." Stockpile food, medicine, and other necessities. You may need to remain fairly isolated for 3 months.

What about the vaccine? It has been found to be only 45% effective on adults under age 65. There is not enough vaccine, nor will there be, for everyone. The government has contracted for its manufacture and will keep it locked in government warehouses. None will be for sale or available privately. Another problem is that the vaccine must be given as a two-dose series, 2 months apart. If it is not released for use until an area has begun having an outbreak, many people will die before they can be vaccinated. And then only 45% will be protected by the vaccine.

 Video related to this topic is available within WileyPLUS.

The human respiratory system can be infected by various bacteria, viruses, fungi, and at least one helminth. Whether respiratory infections become established depends on host-microbe relationships (◄Chapter 14) and the condition of the respiratory system and its nonspecific defenses (◄Chapter 16). Respiratory infections are divided into upper respiratory infections, including ear infections, and lower respiratory infections.

COMPONENTS OF THE RESPIRATORY SYSTEM

The **respiratory system** consists of the **upper respiratory tract**—consisting of the nasal cavity, pharynx (far′ingks), larynx (lar′ingks), trachea (tra′ke-a), and

PUBLIC HEALTH

Elegant Tuberculosis

Flashbulbs, music, wisps of fabric as the top designers display their new season of clothing on toothpick-thin supermodels. What does this have to do with the photo you see here? This scene from a movie version of the opera *La Boheme* reflects how fashionable it was to be languorous and pale in the latter part of the nineteenth century. Much of the art of this period shows the same pale, lethargic women, reclining across sofas, with the bright pink cheeks that marked tuberculosis.

But why was the tuberculosis look admired? It was believed then that highly intellectual and creative individuals were especially susceptible to tuberculosis. The merits of poets, artists, and musicians were validated by their deaths in unheated attics. Death from tuberculosis was also a prominent theme in paintings. Indeed, tuberculosis became a fashionable disease and set the standards of beauty. Could that still be echoing in the emaciated look of many of today's high-fashion models?

(Photofest)

bronchi (brong′ki)—and the **lower respiratory tract**—composed of the lungs **(Figure 21.1)**. This entire system is lined with moist epithelium. However, in the upper respiratory tract this epithelium contains mucus-secreting cells and is covered with cilia.

THE UPPER RESPIRATORY TRACT

The respiratory system moves oxygen from the atmosphere to the blood and removes carbon dioxide from the blood to the atmosphere. Every breath of air contains millions of microorganisms and other suspended particles. Some microorganisms and particles breathed in are removed by hairs and mucus as the air passes through the **nasal cavity**. If microbes enter the **nasal sinuses**, which are hollow cavities lined with mucous membrane within the skull, sinus infections can occur. Air with any remaining microbes and particles then passes through the **pharynx** (throat), a common passageway for the respiratory and digestive systems. The auditory (Eustachian) tube connects the pharynx with the middle ear.

From the pharynx, air and any remaining microbes pass through a series of rigid-walled tubes, the larynx and the trachea. The **larynx** (voice box) contains the vocal cords, which produce sound when they vibrate. The *epiglottis* is a flap of tissue that prevents food and fluids from entering the larynx. The **trachea** (windpipe) branches into primary **bronchi** (singular: *bronchus*), all of which are lined with cilia.

The upper respiratory tract contains a variety of normal microflora that help prevent infection by pathogens that may be inhaled. In addition, mucus from the membranes that line the nasal cavity and pharynx traps microorganisms and most particles of debris, preventing them from passing beyond the pharynx. Mucus also contains lysozyme. Coughing and sneezing mechanically agitate mucus, increasing exposure of microorganisms to mucus and helping to expel them.

The beating of cilia generally serves to move a cell through its environment (◄Chapter 4, p. 103). In the nasal cavity and bronchi, however, cilia extend from the epithelial cells. Because these cells are anchored in place, they function instead to trap and move microbes and particles near the cell surfaces. There mucus with debris trapped in it is moved up into the pharynx. This mechanism, the **mucociliary escalator**, allows materials in the bronchi to be lifted to the pharynx and to be spit out or swallowed. Nevertheless, the mucous membranes of the upper respiratory system are common sites of infection. Such infections often spread to the sinuses, the middle ear, and even the lower respiratory tract.

THE LOWER RESPIRATORY TRACT

From the bronchi, air passes into the lungs. Secondary bronchi divide into smaller **bronchioles**, forming a branching structure known as the *bronchial tree*. This complex branching arrangement greatly increases the surface area

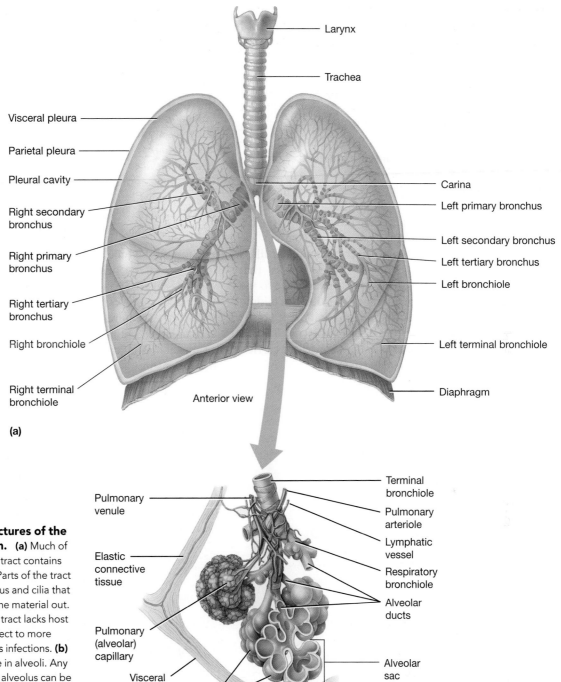

Figure 21.1 Structures of the respiratory system. (a) Much of the upper respiratory tract contains resident microbiota. Parts of the tract are covered with mucus and cilia that trap and move airborne material out. The lower respiratory tract lacks host microflora and is subject to more severe and dangerous infections. **(b)** Bronchioles terminate in alveoli. Any microbes entering an alveolus can be destroyed by alveolar macrophages (not shown).

exposed to oxygen flowing into, and carbon dioxide flowing out of, the lungs. Air passing through the terminal bronchioles then enters the **respiratory bronchioles (Figure 21.1b)**. These microscopic channels end in a series of saclike **alveoli** (singular: *alveolus*), which form clusters (nodules). It is in the alveoli that gas exchange occurs. There oxygen diffuses into the blood, and carbon dioxide from the blood diffuses into the alveoli. The bronchial tree, alveoli, blood vessels, and lymphatic vessels form the bulk of the lungs. The surfaces of the lungs and the cavities they occupy are covered by a membrane called the **pleura**,

which secretes *serous fluid*, a watery fluid that lubricates some tissues.

Mucus from the lining of the bronchial tree also traps foreign materials that have passed beyond the pharynx. If the upper respiratory tract defense mechanisms fail and microorganisms get into bronchi and bronchioles, alveolar macrophages help remove them. Only when the numbers of organisms exceed the capacity of the phagocytes to destroy them, or when an infection reaches the lungs by way of blood or lymph vessels, does a lower respiratory infection occur.

THE EARS

Like the eyes, the ears are exposed to the environment and thus are subject to microbial attack. The ears contain physical protective structures to prevent infection. Each ear is divided into the outer ear, the middle ear, and the inner ear (**Figure 21.2**). The *outer ear* has a flaplike **pinna**, or auricle (commonly called the ear), covered with skin, and an **auditory canal**, which is lined with skin that has many small hairs and many ceruminous glands. The **ceruminous** (se-ru′mi-nus) **glands** are modified sebaceous glands that secrete **cerumen** (earwax). Both hairs and wax help trap microorganisms and other foreign objects and prevent them from entering the auditory canal. Nevertheless, this canal can become infected with fungi.

The **tympanic membrane**, or *eardrum*, separates the outer and middle ears. The *middle ear* is a small, air-filled cavity containing the small bones, called *ossicles*, that transmit sound waves from the tympanic membrane to the inner ear. If the tympanic membrane is intact, most middle ear infections arise from microorganisms that move up the *auditory (Eustachian) tube* from the nasopharynx. The *inner ear* converts sound waves into nerve impulses carried by the vestibulocochlear cranial nerve (VIII).

Outer and middle ear infections are relatively common—especially in children, because their auditory tubes are shorter and wider, facilitating the transmission of microbes. Because of the relative inaccessibility of the inner ear to pathogens, inner ear infections are rare. However, if such an infection does occur, it can easily spread to the **mastoid area**, a bony projection of the skull located behind and below the auditory canal. There is only a thin bony separation between mastoid and brain. Should the mastoid area become infected, therefore—a condition known as *mastoiditis*—the results can be quite serious, as there is danger that the infection will reach the brain.

NORMAL MICROFLORA OF THE RESPIRATORY SYSTEM

Because of the functions of the respiratory system, its surfaces cannot be lined with a tough, keratinized, thick, multiple-layered type of tissue such as we have on our skin. Gas exchange, along with warming and humidifying the air we breathe, requires a very thin, moist, delicate layer of cells in close contact with the circulatory system. In some cases, such as alveoli, only a single flat epithelial cell lies pressed against a capillary. Microbial colonization of such

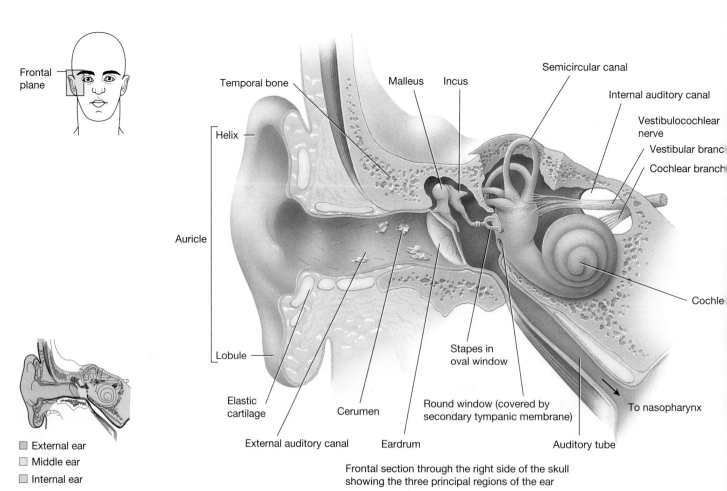

□ External ear
□ Middle ear
□ Internal ear

Frontal plane

Temporal bone

Malleus   Incus

Semicircular canal

Internal auditory canal

Vestibulocochlear nerve

Vestibular branc

Cochlear branch

Helix

Auricle

Lobule

Elastic cartilage

Cerumen

External auditory canal

Stapes in oval window

Round window (covered by secondary tympanic membrane)

Eardrum

Cochle

To nasopharynx

Auditory tube

Frontal section through the right side of the skull showing the three principal regions of the ear

Figure 21.2 Structures of the ear. The outer ear is protected from foreign material by ear hairs and earwax. Nevertheless, some ear infections involve the outer ear. Infections of the middle ear usually arise from infections moving up the auditory (Eustachian) tube.

surfaces would hinder gas exchange and very likely damage the delicate cells. So it is no surprise that the upper respiratory tract has many protective mechanisms built into it to prevent colonization of the lower respiratory tract.

Healthy lungs are usually sterile. In the trachea (below the level of the larynx), bronchi, and larger bronchioles, there likewise is no normal microflora. Organisms found there are merely transient microbes that have been inhaled. They are removed by the mucociliary escalator. Cigarette smoking and inhalation of toxic fumes decrease the activity of the mucociliary escalator. Failure to clear organisms can then lead to infection.

Within the alveoli, macrophages (◄Chapter 16, p. 467) engulf particles and microbes. These macrophages can also move up the mucociliary escalator to the throat. However, not all microbes are trapped in the mucus of the middle portion of the respiratory tract. Some, such as *Mycobacterium tuberculosis* and *Coccidioides immitis*, are able to remain airborne through the trachea and bronchi, and thus reach the alveoli. There, macrophages may be unable to destroy them. In fact, *M. tuberculosis* can survive and multiply inside of macrophages eventually causing infection within the alveoli. Conditions that impair the coughing reflex, or that prevent the epiglottis from fitting tightly over the glottis, also predispose to lower respiratory infection. Most infections of the larynx, trachea, and bronchi are viral in nature.

The pharynx has a normal microflora similar to that of the mouth. Aspiration of secretions from this area can cause infections such as pneumonia lower in the respiratory tract. The pharynx has its own set of defenses, a ring of lymphoid tissues, such as the tonsils, which are an important part of the immune system (◄Chapter 16). Still, potential pathogens, such as *Klebsiella pneumoniae*, *Streptococcus pneumoniae*, *Staphylococcus aureus*, and species of *Haemophilus*, are able to survive in and colonize the pharynx. Their presence does not necessarily indicate an infection. For example, you do not develop pneumonia of the pharynx—the organisms must successfully invade the bronchi and/or lungs in order to cause pneumonia.

The upper respiratory tract forms the first line of defense against infection. These areas have a normal microflora similar to that of skin, with *Staphylococcus epidermidis* and corynebacteria being most numerous. *Staphylococcus aureus* is also often found here, especially in the anterior nostrils. About one-third of the healthy population carries *S. aureus* at any given time—half permanently, and half transiently. Nasal carriers of *S. aureus* can easily spread it to other individuals, especially newborns who are not yet fully colonized. A nasal carrier of a highly virulent strain of *S. aureus* working in a neonatal unit presents a very real hazard, as most newborns are colonized by a single strain of *S. aureus* from their environment within the first 24 hours after delivery. However, it has been shown that deliberate colonization of the infant's nasal mucosa with a fairly harmless strain of *S. aureus* during this first 24-hour period prevents subsequent colonization with a different (potentially more virulent) strain of *S. aureus*.

Hairs lining the anterior nostrils prevent the passage of larger particles. But they are also important because they cause air to swirl and eddy, thereby slowing it down and forcing particulate matter to settle out onto mucus. The mucus is propelled toward the pharynx at a rate of 0.5 to 1.0 cm per minute, where it forms a postnasal drip. Swallowed into the stomach, trapped microbes are destroyed. The nasal passages have projections and plates of bone that, like the hairs, cause turbulence and slowing of air flow.

All in all, the defenses of the respiratory system are quite efficient, considering that the average adult male inhales an estimated average of eight microbes per minute at rest, and more while breathing harder, yet develops only a few respiratory infections each year.

APPLICATIONS

Alcoholics and Pneumonia

The opening to the larynx is covered by the epiglottis, a flap of tissue that prevents food and fluids from entering the larynx. When the epiglottis fails to operate properly, fluids can be drawn or sucked into the lungs. Such failure in alcoholics often leads to pneumonia. When alcoholics are "sleeping off" drunken episodes, the epiglottis gapes and allows pharyngeal secretions to be drawn into the lungs with breathing. If pneumonia-causing organisms are present in those secretions, they can gain access to and infect the lungs. Once in the lungs, it is unlikely that the microbes will be destroyed because alcoholics have "lazy leukocytes." In other words, their macrophages are less efficient in destroying invading microbes than are those of nonalcoholics.

DISEASES OF THE UPPER RESPIRATORY TRACT

BACTERIAL UPPER RESPIRATORY DISEASES

Bacterial infections of the upper respiratory tract are exceedingly common. They are easily acquired through the inhalation of droplet nuclei from infected persons or carriers, especially in winter, when people are crowded indoors in poorly ventilated areas.

It's more likely that you will catch a cold when you are around more infected people, but the more friendly and supportive people you associate with, the greater your immune system resistance will be.

Pharyngitis and Related Infections

Pharyngitis, or sore throat, is an infection of the pharynx. It is frequently caused by a virus but is sometimes bacterial in origin. **Laryngitis** is an infection of the larynx, often with loss of voice. **Epiglottitis**, an infection of the epiglottis, can close the airway and cause suffocation.

Croup, a viral infection found in children, also involves the larynx and epiglottis. Difficult breathing, spasms of the larynx, and membrane formation may occur. When an infection reaches the sinus cavities, bronchi, or tonsils, the conditions are referred to as **sinusitis**, **bronchitis**, and **tonsilitis**, respectively. When an infection spreads to the lungs, it is no longer an upper respiratory disease and is called *pneumonia*.

Streptococcal Pharyngitis.

Less than 10% of cases of pharyngitis are caused by the group A *β*-hemolytic *Streptococcus pyogenes*. This infection, familiarly known as *strep throat*, is most common in children 5 to 15 years old, but is often seen in adults as well. It is acquired by inhaling droplet nuclei from active cases or healthy carriers. Dogs and other family pets also can be carriers. The detection and elimination of carrier states is often difficult in recurrent or cluster outbreaks. Contaminated food, milk, and water also can spread the disease, so it is important that infected persons and carriers not handle food. In strep throat, the throat typically becomes inflamed, and the adenoids and lymph nodes in the neck swell. The tonsils become quite tender and develop white, pus-filled lesions **(Figure 21.3)**. Onset is usually abrupt, with chills, headache, acute throat soreness, especially upon swallowing, and often nausea and vomiting. Fever is generally high. The absence of cough and nasal discharge helps distinguish strep throat from the common cold. Strains of *S. pyogenes* infected with a specific toxin-producing temperature phage cause scarlet fever along with pharyngitis (◄Chapter 19, p. 579).

With a strep test now available, it takes a physician about 15 minutes to detect a streptococcal infection, instead of the 24 hours or more required for a culture.

Diagnosis is made by positive throat culture. A rapid, enzyme-labeled antibody screening test, using a throat swab, can be completed in a physician's office within minutes. Immediate treatment is important. If treatment is delayed, *S. pyogenes* can interact with the immune system and give rise to rheumatic fever (◄Chapter 23), which occurs in 3% of untreated cases. For this reason treatment with penicillin or one of its derivatives is often begun even before culture results are available.

Laryngitis and Epiglottitis.

Laryngitis can be caused by bacteria such as *Haemophilus influenzae* and *Streptococcus pneumoniae*, by viruses alone, or by a combination of bacteria and viruses. Acute epiglottitis was almost invariably caused by *H. influenzae*, but today, widespread immunization against *H. influenzae* has resulted in most cases being of viral origin. Inflammation of the tissues rapidly closes the airway, causing difficulty in breathing or even death.

You should suspect epiglottitis when swallowing is extremely painful (causing drooling), when speech is muffled, and when breathing becomes difficult.

Sinusitis.

More than half the cases of sinusitis are caused by *Streptococcus pneumoniae*, *Moraxella catarrhalis*, or *H. influenzae*, but some cases are caused by *Staphylococcus aureus* or *Streptococcus pyogenes*. Swelling of sinus cavity linings slows or prevents drainage and leads to pressure and severe pain. When drainage remains impeded, mucus accumulates and fosters bacterial growth. Secretions consisting of mucus, bacteria, and phagocytic cells then collect in the sinuses. Chronic sinusitis can permanently damage sinus linings and cause *polyps* (smooth, pendulous growths) to form. Anaerobic bacterial infections, such as with *Bacteroides*, are most likely to be the cause of chronic sinusitis. Roots of the upper teeth are very close to the maxillary sinuses, and 5 to 10% of these sinus infections are of dental

Every year, over $1.5 billion worth of "sinus" medicine is purchased in the United States for the symptoms of sinus disease.

(a)   (b)

Figure 21.3   Strep throat. This common form of pharyngitis is characterized by **(a)** enlarged and reddened adenoids at the sides of the throat *(Biophoto Associates/Photo Researchers, Inc.)* and **(b)** white, pus-filled lesions on tonsils. *(Southern Illinois University/Peter Arnold, Inc.)*

origin. Applying moist heat over the sinuses, instilling drops of vasoconstrictors, such as ephedrine, into the nasal passages, humidifying the air, and holding the head in a position to promote drainage help alleviate symptoms. Treatment with antibiotics such as penicillin and medications to lessen discomfort may be necessary. Swimmers and divers often suffer from sinusitis if water is forced into the sinuses. Such people can prevent water from entering their sinuses by using nose clips or by exhaling as they submerge and continuing to exhale while their head is underwater.

Bronchitis. Bronchitis involves the bronchi and bronchioles but does not extend into the alveoli. About 15% of the general population has chronic bronchitis. It is most common in older people and is linked to smoking, air pollution, inhalation of coal dust, cotton lint, and other particles, and heredity. Patients cough up sputum containing mucus, organisms, and phagocytic cells. Common causative agents include *Streptococcus pneumoniae*, *Mycoplasma pneumoniae*, and various species of *Haemophilus*, *Moraxella*, *Streptococcus*, and *Staphylococcus*. Infections can spread to the alveoli of the lung and cause pneumonia. Diagnosis is made from sputum cultures. By the time an infected individual seeks medical attention, respiratory membranes may have been permanently damaged. Antibiotic treatment can halt further deterioration but cannot reverse the damage already done. Eventually severe shortness of breath develops.

Diphtheria

Although fewer than 5 cases per year are now seen in the United States, **diphtheria** was once a feared killer. A century ago, 30 to 50% of patients died; most deaths occurred by suffocation in children under age 4. Diphtheria is still a problem today in the former Soviet Union, where the disease recently made a return to epidemic proportions but has since dropped. Since 1990, close to 200,000 cases have occurred there, with over 5,000 deaths resulting (Chapter ◀15, p. 429).

It is thought that George Washington may have died from diphtheria.

Sequelae, or adverse signs that follow a disease, are common in diphtheria. Myocarditis, an inflammation of the heart muscle, and polyneuritis, an inflammation of several nerves, account for deaths even after apparent recovery. Significant cardiac abnormalities occur in 20% of patients. Neurologic problems, including paralysis, can occur after particularly severe cases.

Causative Agent. Diphtheria is caused by strains of *Corynebacterium diphtheriae* infected with a prophage that carries an exotoxin-producing gene (◀Chapter 10, p. 290). Club-shaped cells of this Gram-positive rod grow side by side in palisades (like the upright logs used to fortify old forts). These cells contain metachromatic granules of phosphates, which turn reddish when stained

Figure 21.4 *Corynebacterium diphtheriae*, **the cause of diphtheria.** This stained photomicrograph shows metachromatic granules of phosphates that have stained a deep reddish blue (2,801X). *(G. W. Willis, M.D./Visuals Unlimited)*

with methylene blue **(Figure 21.4)**. **Diphtheroids** are corynebacteria that are found in or on such body sites as the nose, throat, nasopharynx, urinary tract, and skin. These organisms differ from *C. diphtheriae* by not being toxin-producers. On inoculation into a guinea pig, toxin-producing *C. diphtheriae* strains cause disease, but diphtheroids do not. Some laboratories also use a gel diffusion test to detect toxin-producing strains.

To produce toxin, the bacterium must be infected by an appropriate strain of bacteriophage in the lysogenic prophage state (◀Chapter 10, p. 290). In other words, the bacteriophage DNA must be integrated into the bacterial chromosome, where its toxin-producing genes are expressed. "Curing" the bacterium of its phage infection eliminates its ability to produce toxin. The phage-infected cell also fails to produce toxin unless the patient's blood iron concentration drops to a critically low level as in anemia. The toxin inhibits protein synthesis, leading to cell death. PCR techniques can be used to detect the toxin gene, for rapid diagnosis.

The Disease. Humans are the only natural hosts for *C. diphtheriae*. Theoretically, diphtheria could therefore be eradicated by worldwide immunization. Diphtheria usually is spread by droplets of respiratory secretions. Infection usually begins in the pharynx 2 to 4 days after exposure. The organism, damaged epithelial cells, fibrin, and blood cells combine to form a **pseudomembrane (Figure 21.5)**. Although it is not a true membrane, removing it leaves a raw, bleeding surface that is soon covered by another pseudomembrane. The pseudomembrane can block the airway and cause suffocation. The organisms very rarely invade deeper tissues or spread to other sites, but the extremely potent toxin spreads throughout the body and kills cells by interfering with protein synthesis. The heart, kidneys, and nervous system are most susceptible.

Diphtheria organisms are sometimes found in unusual sites. They can invade the nasal cavities, which have

Figure 21.5 A pseudomembrane characteristic of diphtheria. This is not a true membrane; rather, it adheres to the underlying tissue. If torn off, it will leave a raw, bloody surface and will eventually reform. However, it must be removed when it blocks the airway. *(Science VU/Visuals Unlimited).*

APPLICATIONS

Diphtheria and the "Desert Fox"

German general Erwin Rommel, the "Desert Fox," led the North African tank warfare during World War II. He was always seen in newsreels with a handkerchief pressed to his nose. This was not because of desert dust, but because of nasal diphtheria. He periodically returned to Berlin for treatment, leaving his troops without his brilliant and charismatic leadership. In fact, the Allies may have defeated the Nazis in North Africa because of the battles lost due to troops who were disheartened in his absence. In Berlin, Rommel joined other generals in a plot to assassinate Hitler. The plot failed; Rommel's role was discovered. As Gestapo agents were taking him to headquarters, he died of a heart attack. This could have been the result of a heart weakened by years of exposure to diphtheria toxin. Had Rommel never contracted diphtheria, would the outcome of World War II have changed?

relatively few blood vessels. There they cause a milder disease because less toxin enters the blood. They also can invade the skin in *cutaneous diphtheria* and cause tissue injury. Only a small amount of toxin is absorbed by the blood. Cutaneous diphtheria is a tropical disease associated with poor hygiene; it is rare in the United States.

Treatment and Prevention. Diphtheria is treated by administering both antitoxin to counteract toxin and antibiotics such as erythromycin or clindamycin to kill the organisms. The disease can be prevented with DTP (*diphtheria, tetanus, pertussis*) vaccine, which contains diphtheria toxoid, tetanus toxoid, and killed *Bordetella pertussis* and is administered in a series of shots beginning at age 2 months. Boosters are needed throughout childhood and adulthood **(Figure 21.6)**. In the past, when

diphtheria was endemic in the United States, most of the adult population maintained or even acquired immunity by continual exposure to small quantities of the bacterium. Such exposures acted as natural booster shots. Now that diphtheria is no longer endemic, immunity depends on the vaccine. Without revaccination, adults in countries with good childhood immunization programs eventually become susceptible again, due to waning immunity. Because diphtheria is not well controlled elsewhere in the world, especially in the former Soviet Union, vaccination must be continued in the United States to prevent epidemics should the disease spread from other places. Reimmunization of adults every 10 years could be aided by administration of diphtheria toxoid and tetanus toxoid together, rather than tetanus toxoid alone after injuries.

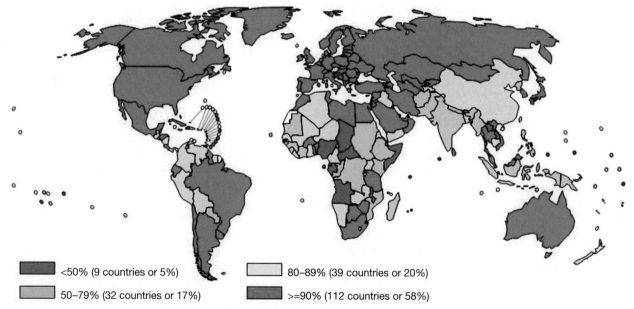

■ <50% (9 countries or 5%)	▨ 80–89% (39 countries or 20%)
▨ 50–79% (32 countries or 17%)	▨ >=90% (112 countries or 58%)

Figure 21.6 Immunization coverage with DTP3 vaccines in infants, 2005. Source: WHO/UNICEF coverage estimates, 1980–2005, as of August 2006 192 WHO Member States.

Ear Infections

Ear infections commonly occur as **otitis media** in the middle ear and as **otitis externa** in the external auditory canal. Because otitis media usually produces a puslike exudate, it often is referred to as otitis media with effusion (OME). *Streptococcus pneumoniae*, *S. pyogenes*, and *Haemophilus influenzae* account for about half of acute cases. Species of various anaerobes are often responsible for chronic cases, although inadequately treated *Streptococcus* and *Haemophilus* can become chronic cases. Otitis externa usually is caused by *Staphylococcus aureus* or *Pseudomonas aeruginosa*. Pseudomonad infections are common in swimmers because the organisms are highly resistant to chlorine.

OME infections arise from the passage of pharyngeal organisms through the Eustachian tube. Fever and earache, which arise from pressure the pus creates in the middle ear, usually are present, but some cases are asymptomatic. The disease is treated with antibiotics, usually penicillin. It is important to continue treatment until all organisms are eradicated to prevent complications. Even after successful therapy, sterile fluid can remain in the middle ear, impairing vibration of the tiny bones of hearing there and decreasing sound transmission. Tubes are sometimes inserted to prevent fluid accumulation and repeated middle ear infections (**Figure 21.7**). If the impairment occurs during speech development, as it often does, speech can be adversely affected. Some children thought to be inattentive or defiant really cannot hear. School can be a nightmare for such children. As children grow older, the Eustachian tube changes shape and develops an angle that prevents most organisms from reaching the middle ear—much to the relief of the children and their parents.

VIRAL UPPER RESPIRATORY DISEASES

The Common Cold

The common cold, or **coryza**, probably causes more misery and loss of work hours than any other infectious disease, but it is not life-threatening. Although exact statistics are not available on the number of infections per year, Americans lose more than 200 million days of work and school per year because of colds. Economically, colds are a bane to employers, who must deal with lost work time, and a boon to manufacturers and retailers of cold remedies. Cold viruses are ubiquitous and present year round, but most infections occur in early fall or early spring. After an incubation period of 2 to 4 days, signs and symptoms such as sneezing, inflammation of mucous membranes, excessive mucus secretion, and airway obstruction appear. Sore throat, malaise, headache, cough, and occasionally tracheobronchitis occur. Pus and blood may be present in nasal secretions. The illness lasts about 1 week. The severity of symptoms is directly correlated with the quantity of viruses released from infected epithelial cells of the upper respiratory tract. Viral cold infections can also predispose an individual to secondary bacterial infections such as pneumonia, sinusitis, bronchitis, and otitis media. These infections add time, misery, and additional cost to the original cold. More serious diseases such as whooping cough and respiratory syncytial virus infections can be mistaken early

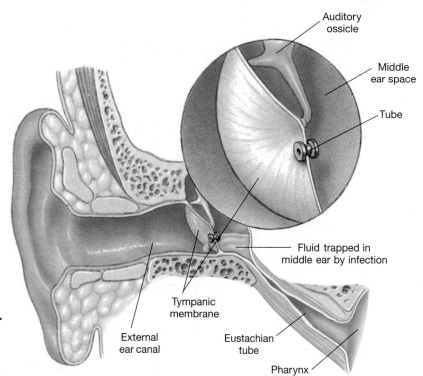

Figure 21.7 Treating middle ear infections.
Tubes are placed through the tympanic membrane (eardrum) to promote drainage. This can be done in the physician's office. The phlange on the rear end of the tube is designed to hold it in place.

on for colds, so correct diagnosis and subsequent treatment are delayed.

Causative Agents.   Different cold viruses predominate during different seasons. In fall and spring, rhinoviruses are in the majority. Parainfluenza virus is present all year but peaks in late summer. In mid-December the coronaviruses appear. Adenoviruses are present at a low level all year. About 200 different viruses can cause colds. Often it is not a virus at all that causes many upper and lower respiratory infections, although they are reported back as "viral." Actually they are caused by *Chlamydophila pneumonia*, a very difficult to grow bacterium. Perhaps as many as half of all "viral" respiratory infections could really be treated with antibiotics.

Rhinoviruses are the most common cause of colds, accounting for about half of all cases. They are resistant to antibiotics, chemotherapeutics, and disinfectants but are quickly inactivated by acidic conditions. They grow best at 33° to 34°C and thus replicate in the epithelium of the upper respiratory tract, where air movements lower the tissue temperature. Rhinoviruses can be isolated from nasal secretions and throat washings and identified by their sensitivity to low pH and resistance to ether and chloroform, a unique combination of traits among viruses. At least 113 different rhinoviruses have been identified, all with different antigens. Natural immunity is short-lived, and no effective vaccine has been developed. Even if a person becomes immune to some rhinoviruses, there are always others to cause another cold.

The second most common cause of colds is another group of ubiquitous viruses, the **coronaviruses**. Coronaviruses have clublike projections that give them a halo (*corona*, Latin for "halo" or "crown"). The projections are responsible for attaching the virus to host cells, as well as stimulating the immune system to produce antibodies against the virus. In addition to causing colds, these viruses also cause acute respiratory distress, in some cases mild pneumonia, and acute gastroenteritis. They infect the epithelium of both the respiratory and digestive tracts. In the digestive tract they reduce absorptive capacity and cause diarrhea, dehydration, and electrolyte imbalances.

Transmission.   Cold viruses are more often spread by fomites than by close contact with infected persons. Blowing the nose and handling used tissues contaminates the fingers, so that anything touched becomes contaminated. Conscientiously using tissues once, discarding them immediately in covered containers, and thoroughly washing hands after each nose wipe can significantly reduce the spread of rhinoviruses.

Diagnosis and Treatment.   Most people diagnose and treat their own colds with remedies that alleviate some symptoms. Over-the-counter antihistamines are reasonably effective in counteracting inflammatory reactions as the body attempts to defend itself against the viruses.

Experiments with certain kinds of human interferon have shown potential in blocking or limiting rhinovirus infections, but they must be given with another agent that helps the interferon reach epithelial cells beneath a thick blanket of mucus. A quantity of alpha interferon sufficient to block infection causes irritation of membranes and bleeding. Different combinations of interferons and other factors are being explored to control rhinovirus infections.

Parainfluenza

Parainfluenza is characterized by rhinitis (nasal inflammation), pharyngitis, bronchitis, and sometimes pneumonia, mainly in children. **Parainfluenza viruses** (paramyxoviruses; ◀Chapter 10, p. 280) initially attack the mucous membranes of the nose and throat. In very mild cases, symptoms may be inapparent. When symptoms do appear, the first symptoms are cough and hoarseness for 2 or 3 days, harsh breathing sounds, and a red throat. Symptoms can progress to a barking cough and a high-pitched, noisy respiration called *stridor* (stri'dor). Recovery is usually quick—within a few days.

Of the four parainfluenza viruses capable of infecting humans, two can cause croup. **Croup** is defined as any acute obstruction of the larynx and can be caused by a variety of infectious agents, including parainfluenza viruses. Both the larynx and the epiglottis become swollen and inflamed; the high-pitched barking cough of croup, which sounds like a seal's bark, results from partial closure of these structures. Increased humidity from a cool-mist vaporizer or from a hot shower helps to relieve croup symptoms.

By age 10, most children, whether or not they have had recognizable illness, have antibodies to all four parainfluenza viruses. Thus, the incidence of infection is

APPLICATIONS

A Cure for the Common Cold?

Why is there a vaccine against polio but not against the common cold? There are only three distinct forms (serotypes) of wild poliovirus. Once a separate vaccine was made against each serotype, the job was done. However, there are at least 113 different rhinoviruses; we would need 113 vaccines to attack the rhinoviruses, and then more for the coronaviruses and adenoviruses that also cause colds. Why are there only three types of poliovirus? Polioviruses must survive the stomach's acid conditions in order to establish an infection. Mutations in their capsid structure almost always leave them vulnerable to low (acidic) pH, so mutated forms do not succeed. Rhinoviruses are spared this harrowing route. Capsid changes due to mutations enable them to evade the upper respiratory tract's defenses easily, so some mutated forms succeed. The result? There are too many versions of cold viruses for vaccine researchers to keep up with.

TABLE 21.1

A Summary of Diseases of the Upper Respiratory Tract		
Disease	Agent(s)	Characteristics
Bacterial Upper Respiratory Diseases		
Pharyngitis	*Streptococcus pyogenes*	Inflammation of the throat; fever without cough or nasal discharge
Laryngitis and epiglottitis	*Haemophilus influenzae, Streptococcus pneumoniae, Moraxella*	Inflammation of the larynx and epiglottis, often with loss of voice
Sinusitis	*H. influenzae, S. pneumoniae, S. pyogenes, Staphylococcus aureus*	Inflammation of sinus cavities, sometimes with severe pain
Bronchitis	*Streptococcus pneumoniae, Mycoplasma pneumoniae*, and others	Inflammation of bronchi and bronchioles with a mucopurulent (mucus- and pus-filled) cough; shortness of breath in chronic cases
Diphtheria	*Corynebacterium diphtheriae*	Inflammation of the pharynx with pseudomembrane and systemic effects of toxin
Otitis externa	*Staphylococcus aureus, Pseudomonas aeruginosa*	Inflammation of external ear canal; common in swimmers
Otitis media	*Streptococcus pneumoniae, S. pyogenes, Haemophilus influenzae*	Pus-filled infection of the middle ear with pressure and pain
Viral Upper Respiratory Diseases		
Common cold	Rhinoviruses, coronaviruses	Sore throat, malaise, head and cough
Parainfluenza	Parainfluenza viruses	Nasal inflammation, pharyngitis, bronchitis, croup, sometimes pneumonia

very high, although the incidence of clinically apparent disease is much lower. Epidemics and smaller outbreaks of parainfluenza infections occur primarily in fall but also sometimes in early spring after the "flu" season. The viruses are spread by direct contact or by large droplets. The causative viruses can be inactivated by drying, increased temperature, and most disinfectants, so they do not remain long on surfaces or in the environment. Resistance to infection comes from secretory IgA that defends mucous membranes against infection, and not from blood-borne IgG (◄Chapter 17, p. 497). Reinfection with parainfluenza viruses is rare, so secretory immunoglobulins must create effective immunity. However, efforts to make a vaccine for parainfluenza viruses have not been successful.

Diseases of the upper respiratory tract are summarized in **Table 21.1**.

CHECKLIST

1. Of what importance is bacteriophage infection in *Corynebacterium diphtheriae*
2. How do tubes in the tympanic membrane help to prevent frequent middle ear infections?
3. How is the common cold usually transmitted?
4. What is the most common cause of pharyngitis?
5. Why is it important to know when a pharyngitis is due to a virus or *Streptococcus pyogenes*? How can this be determined?

DISEASES OF THE LOWER RESPIRATORY TRACT

BACTERIAL LOWER RESPIRATORY DISEASES

Among the bacterial diseases of the lower respiratory tract are two of the great killer infections of history: pneumonia and tuberculosis. The advent of antibiotic therapy brought these diseases under control to a considerable extent. Both are making comebacks today as a result of the spread of AIDS and of treatment with immunosuppressive drugs for transplant patients and with antiinflammatory agents for autoimmune disorders, such as rheumatoid arthritis and multiple sclerosis. Lowered resistance, overcrowding, chronic diseases, aging, and other immunosuppressive factors also contribute to the severity of the problem. Chronic lower respiratory diseases now rank fourth among the 10 leading causes of death in the U.S., following heart disease, cancer, and stroke.

Whooping Cough

Whooping cough, also called **pertussis**, is a highly contagious disease known only in humans. The word *pertussis* means "violent cough," and the Chinese call it the "cough of 100 days." Although distributed worldwide, strains found in the United States are less virulent than most. However, jet travel could bring virulent strains from North Africa or other parts of the world at any time. Whooping cough is a major health problem in developing nations, where lack of immunization allows 80% of those exposed to contract the disease. In the United States, concern and negative publicity regarding vaccine safety discouraged some parents from having very young children immunized. As a result, incidence of the disease more than doubled during the 1980s and reached nearly 8,400 in 2003 (**Figure 21.8a**). The disease tends to occur sporadically, especially in infants or young children; 50% of cases occur in the first year of life (**Figure 21.8b**).

Before the development of the vaccine, nearly every child got whooping cough. Adults who contract it today either were not vaccinated, or their immunity has declined. Immunity wanes 5 to 10 years after vaccination. Partial immunity lessens the severity of the disease. Many adults may fail to display the characteristic "whooping" sound and so are misdiagnosed. Such people serve as a reservoir to spread the infection. Pertussis is considered to be the least well controlled reportable bacterial vaccine-preventable disease in the United States.

Causative Agent. *Bordetella pertussis*, a small, aerobic, encapsulated, Gram-negative coccobacillus first isolated in 1906, is the usual causative agent of whooping cough. Only about 5% of the cases are due to *Bordetella parapertussis* and *B. bronchiseptica*, which usually produce a milder disease. *B. bronchiseptica* is a normal resident of canine respiratory tracts, where it sometimes causes "kennel cough."

Susceptible people become infected by inhaling respiratory droplets. The organisms colonize cilia lining the respiratory tract. Only active cases of whooping cough are known to shed organisms; carriers of the disease are unknown. *Bordetella pertussis* does not invade tissues or enter the blood, but it does produce several substances that contribute to its virulence. It produces an endotoxin, an exotoxin, and *hemagglutinins* (hem′ah-gloo′ti-ninz), surface antigens that help it attach to the cilia of epithelial cells in the upper respiratory tract. There, the toxins destroy the ciliated epithelial cells.

*Introduction of universal pediatric diphteria and tetanus toxoids and whole-cell pertussis vaccine.
SOURCE: 1950–2006, CDC, National Notifiable Diseases Survaillance System, and 1922–1949, passive reports to the Public Health Service

(a)

*Of 7,580 cases, 25 were reported with unknown age.

(b)

Figure 21.8   The U.S. incidence of whooping cough.
(a) By year. **(b)** by age group. **(c)** The technique of culturing for whooping cough. A swab attached to a thin, flexible wire is inserted through the nostrils, and the patient is asked to cough several times. (*Source:* CDC.)

(c)

APPLICATIONS

The Whooping Cough Fashion Statement

In the past, nannies in Manhattan's Central Park were often seen wearing white uniforms covered with blood from holding infants who suffered paroxysms of coughing. The infants also produced lots of aerosol droplets containing infectious organisms. Perhaps you can't remember ever seeing a case of whooping cough, but ask older family members or friends if they remember seeing people with whooping cough or if they had it themselves.

Those cells are then sloughed, leaving a surface of non-ciliated cells, which in turn allows mucus to accumulate within the respiratory airway.

The Disease.   After an incubation period of 7 to 10 days, the disease progresses through three stages: *catarrhal*, *paroxysmal*, and *convalescent*. The **catarrhal stage** is characterized by fever, sneezing, vomiting, and a mild, dry, persistent cough. A week or two later the **paroxysmal** (par-oks-iz′mal) (or *intensifying*) **stage** begins as mucus and masses of bacteria fill the airway and immobilize the cilia. Strong, sticky, ropelike strings of mucus in the airway elicit violent, paroxysmal coughing. Failure to keep the airway open leads to **cyanosis** (si-an-o′sis), or a bluing of the skin, because too little oxygen gets to the blood. Keeping the airway open is especially difficult in infants; airway blockage accounts for the high death rate in patients under 1 year of age. Straining to draw in air gives the characteristic loud "whooping" sound. Coughing sieges occur several times a day and cause exhaustion. Sometimes coughing is so severe that it causes hemorrhage, convulsions, and rib fractures. Vomiting that usually follows a coughing siege leads to dehydration, nutrient deficiency, and electrolyte imbalance—all especially dangerous in infants.

After a paroxysmal stage, usually of 1 to 6 weeks' duration but sometimes longer, the patient enters the **convalescent stage**. Milder coughing can continue for several months before it eventually subsides. Secondary infections with other organisms are common at this stage.

Diagnosis and Treatment.   Whooping cough is diagnosed by obtaining organisms from the posterior nasal passages (**Figure 21.8c**) and culturing them on freshly made charcoal–blood agar medium, which has largely replaced the classic potato–blood–glycerol agar (Bordet-Gengou) medium. Because penicillin does not kill *B. pertussis*, the antibiotic is often added to the medium to suppress the growth of other organisms. Once typical colonies appear, fluorescent antibody stain can be used for identification. Whooping cough treatment includes antitoxin given early in the disease to combat toxin, and erythromycin to shorten the paroxysmal, or second, stage (◀Chapter 17, p. 521). The antibiotic cannot entirely eliminate that stage, but it does reduce the number of viable organisms being shed. Supportive measures such as suctioning, rehydration, oxygen therapy, and attention to nutrition and electrolyte balance are very important, as is appropriate treatment of secondary infections.

Prevention.   Pertussis whole-cell vaccine has saved many lives, but it has not been entirely safe. In the United States it lowered the number of cases from 227,319 in 1938 to 3,590 in 1994, but each year vaccine reactions caused 5 to 20 children to die and another 50 to suffer permanent brain damage. Yet the number of children hurt by the vaccine was far smaller than the number who would die without it. Recombinant DNA technology was being used for the development of acellular vaccines. Because they contain only bacterial proteins, such preparations reduce or even eliminate the side effects associated with previous preparations. Acellular pertussis vaccines are now approved for use as all five doses. Previously it was used only as fourth and fifth doses. The designation *aP* indicates "acellular Pertussis" vaccine. Vaccination begins with the DTaP series at 2 months of age, as soon as the immune system is capable of responding to antigens in the vaccine. Early immunization is important because pertussis antibodies do not cross the placenta, so infants have no passive immunity to whooping cough. Immunity gradually decreases after vaccination. However, immunization of unvaccinated children over 7 years of age or adults was not recommended due to the possible risks associated with the older vaccine. This means that a large part of the teenage and adult population is actually susceptible to disease, but mistakenly feel they are protected by their early vaccinations. Incidence has been rising in this group.

Recovery from the disease confers immunity, but not for a lifetime. Second cases are known, especially in adulthood, but they are usually milder than the first. The United States is fortunate right now in having low-virulence strains of *B. pertussis*. African strains are far more virulent and would probably cause massive outbreaks if they were to become established here outside that continent.

APPLICATIONS

DTP Vaccine Liability

What happens when a child is harmed rather than helped by a vaccine? Deaths and injuries from pertussis and other vaccines have, in fact, resulted in numerous lawsuits against the manufacturers. After several years of deliberations, the National Vaccine Compensation Act was passed by Congress and went into effect January 1, 1988. This law established a trust fund and a minimum compensation of $250,000 per child and requires parents who accept this compensation to sign away their right to pursue further court action. So each time you pay for a DTP injection or an MMR (measles, mumps, rubella) injection, your money goes into a trust fund—a fund that might provide some comfort to the parents of affected children.

Classic Pneumonia

In the United States, more deaths result from pneumonia than from any other infectious disease. **Pneumonia**, an inflammation of lung tissue, can be caused by bacteria, viruses, fungi, certain helminths, chemicals, radiation, and some allergies. Infectious forms of the disease develop when pathogens that are able to evade upper respiratory defenses are inhaled. Such organisms initiate the disease process by colonizing the upper respiratory tract and then entering the lower respiratory tract accidentally during a deep breath or suppressed cough, or by means of a large amount of mucus. Several bacteria are known causes of pneumonia, including *Streptococcus pneumoniae*, also known as the pneumococcus; *Staphylococcus aureus*; *Klebsiella pneumoniae*; and *Mycoplasma pneumoniae*, a recently recognized cause of human respiratory infections that is responsible for about 10% of community-acquired cases of pneumonia. *Pseudomonas aeruginosa* is another respiratory pathogen. It causes pneumonia in immunocompromised persons and hospitalized patients. Once the organism gains access to damaged respiratory epithelium, it attaches by means of its pili and multiplies (Figure 21.9).

Classification of Pneumonias. Pneumonias are classified by site of infection as lobar or bronchial. **Lobar pneumonia** affects one or more of the five major lobes of the lungs. It is a serious primary disease that affects about 500,000 people in the United States annually. *S. pneumoniae* causes 95% of all cases. However, back in 1881 when *K. pneumoniae* was first discovered, it was thought to be the primary cause. Fibrin deposits are characteristic of lobar pneumonia. When they solidify, they cause **consolidation**, or blockage of air spaces. **Pleurisy**, inflammation of the pleural membranes that causes painful breathing, often accompanies lobar pneumonia.

Bronchial pneumonia begins in the bronchi and can spread into the surrounding tissues toward the alveoli in a patchy manner. Although also most likely to be caused by pneumococci, bronchial pneumonia differs from lobar pneumonia in two ways: (1) It often appears as a secondary infection after a primary infection such as a viral influenza or heart disease, or another lung disease, and (2) it lacks the plentiful fibrin deposits of lobar pneumonia. Bronchial pneumonia can follow exposure to chemicals or aspiration of vomitus or other fluids. Infants sometimes aspirate amniotic fluid during birth, especially during delivery by Caesarean section. Bronchial pneumonia is common in elderly and debilitated patients. In fact, it is sometimes called the "old man's friend" because it spares an elderly person a possibly longer, more painful death from some other cause.

Transmission. Both lobar and bronchial pneumonia are transmitted by respiratory droplets and, in the winter, by carriers—including health care workers—who have had contact with pneumonia patients. As much as 60% of a population can be carriers in closely associated groups, such as personnel on military bases or children in preschools.

The Disease. After a few days of mild upper respiratory symptoms, the onset of pneumococcal pneumonia is sudden. The infected person suffers violent chills and high fever (up to 106°F). Chest pain, cough, and sputum containing blood, mucus, and pus follow. Fever may end 5 to 10 days after onset when untreated or within 24 hours after antibiotics are given.

Pneumococcal pneumonia is the fourth leading cause of death in the U.S., and the only infectious disease among the top 10 causes of death. With immediate and adequate antibiotic therapy, the mortality rate is 5%; without treatment, it is 30%. This compares favorably with *Klebsiella* pneumonia, which despite best treatment still has a mortality rate of 50%. *Klebsiella* causes an extremely severe pneumonia that can lead to chronic ulcerative lesions in the lungs and extensive destruction of lung tissue. Failure to seek medical attention promptly keeps the overall mortality rate for pneumonia in the United States at 25%. Conditions that predispose toward pneumonia include old age, chilling, drugs, anesthesia, alcoholism, and a variety of disease states.

Diagnosis, Treatment, and Prevention. Diagnosis of pneumonia is based on clinical observations, X-rays, or sputum culture. *Klebsiella* pneumonia is usually treated with cephalosporins. Penicillin is the drug of choice for treatment of pneumococcal pneumonia, however, high rates of resistance are now being seen in many communities. Therefore, a third-generation cephalosporin, or a fluoroquinolone such as levofloxacin or gatifloxacin, is often used. Following recovery, immunity exists for a

Figure 21.9 Pneumonia-causing bacteria. SEM
This SEM shows *Pseudomonas aeruginosa* (see arrow) attaching to human respiratory epithelium (magnification unknown). *(Courtesy Dr. S. Girod de Bentzmann)*

few months only against the particular serotype that caused the infection. Thus, a patient can develop case after case of pneumonia from infection with other serotypes or other species. More than 85 serotypes are recognized. Of these, only 23 are responsible for 85% of pneumococcal disease in the United States. Artificial immunity can be induced with the polyvalent vaccine Pneumovax. This vaccine contains 23 serotype antigens, so it protects against 80% of pneumococcal pneumonias in all age groups except children under 2 years of age. Immunization is especially recommended for the elderly and at-risk populations. Revaccination every 10 years is currently recommended, although in 2003, CDC noted that in patients over 65 years of age, immunity will most likely drop below protective levels somewhere between 5 to 7 years after vaccination. In some cases it may not last past 2 years. Prevnar, an anti-*Streptococcus pneumoniae* vaccine introduced in February 2000, can be given to young children to prevent pneumonia and ear infections. Ear infections cause 27 million doctors' visits per year in the United States. However, the 4-dose Prevnar series costs $232. Still, its manufacturer sold $461 million in its first year of use. There is, however, some concern that Prevnar could be linked to a rise in insulin-dependent diabetes among recipients.

Mycoplasma Pneumonia

One of the tiniest bacterial pathogens known, *Mycoplasma pneumoniae* ordinarily causes mild, and sometimes inapparent, upper respiratory tract infections. In 3 to 10% of infections it causes **primary atypical pneumonia**, or *Mycoplasma* pneumonia, usually a mild pneumonia with an insidious onset. The disease is said to be atypical because the symptoms are different from those of classic pneumonia. Some patients have no signs or symptoms related to their respiratory tract—only fever and malaise. Patients often remain ambulatory, so the disease is sometimes called **walking pneumonia**. The mortality rate is less than 0.1%. It is unusual in preschool children and is most common among young people 5 to 19 years old, although it is found in all age groups. *Mycoplasma pneumoniae* may cause up to 20% of all noninfluenza community-acquired pneumonias.

Transmission is by respiratory secretions in droplet form, and onset of symptoms follows an incubation period of 12 to 14 days. Fever lasts 8 to 10 days and gradually declines along with the cough and chest pain. Unusual features of *Mycoplasma* pneumonia are that alveoli decrease in size by inward swelling of the alveolar walls and that the alveoli do not fill with fluid.

Diagnosis can be made by isolating *M. pneumoniae* from sputum or from a nasopharyngeal swab. This process takes 2 to 3 weeks due to the slow growth rate of the organisms (**Figure 21.10**). Serologic tests such as indirect immunofluorescence, latex agglutination, and ELISA are also of value during the early stages of the disease. Commercially available DNA probes are also used and

(a) SEM

(b) SEM

Figure 21.10 Mycoplasma. (a) These cells lack a cell wall and assume irregular shapes (151,000X). **(b)** A cell infected by numerous *Mycoplasma* cells (45,357X). *(Both photos: David M. Phillips/Visuals Unlimited)*

produce results similar to those of culture. However, treatment is usually based on clinical symptoms.

Azithromycin or a fluroquinolone are the drugs of choice. Penicillin has no effect because *Mycoplasma* lack cell walls (the site of penicillin's antibacterial action). Even untreated cases have a favorable prognosis. No vaccine is currently available, so prevention requires avoidance of contact with infected persons and their secretions.

Legionnaires' Disease

In 1976 many war veterans attending a convention in Philadelphia became victims of a mysterious ailment that became known as **Legionnaires' disease**. After 29 deaths and much frantic investigation, the previously unidentified causative organism, *Legionella pneumophila*, was finally isolated (**Figure 21.11**). Researchers at CDC later found antibodies to *L. pneumophila* in frozen blood samples from unidentified outbreaks that occurred decades ago. One wonders how the organism was overlooked for such a long time. However, it is sufficiently different from previously classified bacteria that a new genus had to be created for it.

Legionella pneumophila is a weakly Gram-negative, strictly aerobic bacillus with fastidious nutritional

Figure 21.11 *Legionella pneumophila*, **the cause of Legionnaires' disease (53,361X).** *(CDC/Photo Researchers, Inc.)*

requirements. It does not ferment sugars and has an obscure life cycle. More than 20 *Legionella* species have been identified. Most are free-living in soil or water and do not ordinarily cause disease. However, some strains live as intracellular parasites of amoebas of many species such as *Acanthamoeba*, *Naegleria*, *Hartmanella*, and *Echinamoeba*. Some of these amoebas colonize cooling towers, shower heads, and other wet areas. Legionellosis is transmitted when organisms growing in soil or water become airborne and enter the patient's lungs as an aerosol. Person-to-person transmission has never been documented. Air conditioners, ornamental fountains, produce sprayers in grocery stores, humidifiers, and vaporizers in patient rooms have been implicated in the spread of the disease. Such devices should be regularly disinfected. Once inhaled, the *Legionella* organisms are taken up into amoeboid phagocytes by a spirally progressing form of phagocytosis. They thrive inside the acidic conditions of the phagolysosome, multiplying and eventually rupturing the cell (◀Chapter 16, p. 469). They adapt well to living inside amoeboid white blood cells.

One family was even infected during a visit to a local home improvement store that had a working, bubbling hot tub on display.

After an incubation period of 2 to 10 days, Legionnaires' disease appears with fever, chills, headache, diarrhea, vomiting, fluid in the lungs, pain in the chest and abdomen, and, less frequently, profuse sweating and mental disorders. When death occurs, it is usually due to shock and kidney failure. In nonpneumonic legionellosis, after about 48 hours' incubation, the patient suffers 2 to 5 days of flulike symptoms without infiltration of the lungs.

Pontiac fever is a mild legionellosis named for an outbreak in 1968 in Pontiac, Michigan that affected 144 people—95% of the employees of the Health Department. None died; all recovered within 3 to 4 days.

Culture, along with a urine antigen test for *L. pneumophila* serogroup 1 (the most common serogroup) are used to diagnose *Legionella* infections. Azithromycin, a fluoroquinolone, or erythromycin are used to treat them; most other antibiotics have no effect. Because it is difficult to distinguish Legionnaires' disease from other pneumonias, erythromycin is typically given along with a second antibiotic such as penicillin to treat any pneumonia-like disease.

Control of *Legionella*-associated infections depends on maintaining adequate chlorine levels in all potable water sources, cooling towers, and other reservoirs of potable water when not in use. Periodic cleaning of surfaces in air conditioners, humidifiers, and similar equipment is valuable in reducing the incidence of outbreaks, especially in hospitals.

Tuberculosis

Tuberculosis, or **TB** (formerly called *consumption*), has plagued humankind since ancient times, as indicated by skeletal damage in 3,000-year-old Egyptian mummies and earlier human remains. It remains a massive global health problem today. One-third of the world's population have tuberculosis. Each year 3 million die, and 10 million new cases arise (**Figure 21.12**).

Incidence in the United States. The incidence of TB in the general U.S. population decreased by nearly 6% per year from 1953, when uniform national reporting began, to 1986. Temporary increases in the 1980s were caused by Asian and Haitian refugees, many of whom became infected in crowded, unsanitary camps and escape boats. Disproportionately large numbers of resident non–white groups—blacks, Eskimos, Native Americans, and Hispanics—suffer from TB, sometimes in epidemic proportions (**Figure 21.13**). In 1986 the incidence increased by 2.6%. Many new cases occurred among AIDS patients; others appeared to represent reactivations of old

Before 1900, approximately one-third of adults died of TB before reaching old age.

![microscope icon] **APPLICATIONS**

The Prowler Who's Hard to Stop

Our cells are pretty impressive at killing intruders. For example, most of our cells have little sacs of digestive enzymes called lysosomes. These enzymes can break down carbohydrates, proteins, nucleic acids, and some lipids, making them pretty effective at killing just about anything foreign that enters the cell. But some bacteria can get into our cells and survive anyway. How do they do it? *Legionella pneumophila*, which causes Legionnaires' pneumonia, invades macrophages. Somehow, the bacterium alters its host so that the host's lysosome can't release its destructive enzymes. That's why antibiotics are necessary—our cells can't always kill the bacterium on their own.

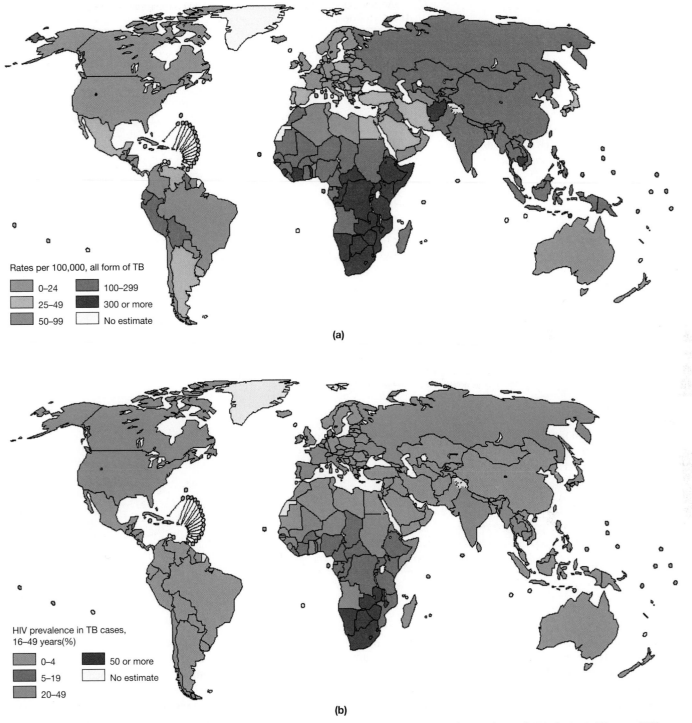

Figure 21.12 **Worldwide tuberculosis.** (a) Estimated TB Incidence Rates per 100,000, 2003; (b) prevalence of HIV infection in TB cases, 2003.

infections triggered by immunodeficiency, crowding, stress, and use of antiinflammatory drugs. In 1993 the number of new cases reported per month rose from 778 in January to 5,130 in December. The increase continues, and people are afraid that tuberculosis may return and grow into an even bigger and more frightening plague than AIDS. Tuberculosis is an airborne disease to which anyone could be exposed, whereas behavior modification can lessen much of one's chances of exposure to the HIV virus.

Causative Agents

The causative agents of tuberculosis are members of the genus *Mycobacterium*, with *M. tuberculosis* causing the vast majority of cases (Table 21.2). *Mycobacterium tuberculosis* was discovered by Robert Koch in 1882, when the disease was called the "White Plague" of Europe. Certain other agents, referred to as atypical mycobacteria, also cause tuberculosis, especially *M. avium-intracellulare* complex (MAC) in AIDS patients. These

TUBERCULOSIS. Reported cases per 100,000 population — United States and U.S. territories, 2006

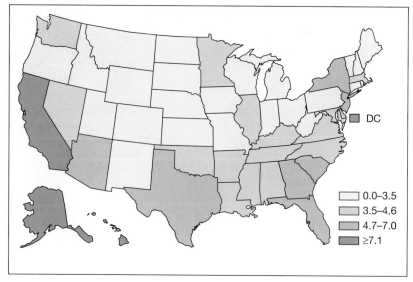

Figure 21.13 The U.S. incidence of tuberculosis, 2006. **(a)** By state; **(b)** by place of birth; **(c)** by ethnicity. (*Source:* CDC.).

	0.0–3.5
	3.5–4.6
	4.7–7.0
	≥7.1

(a)

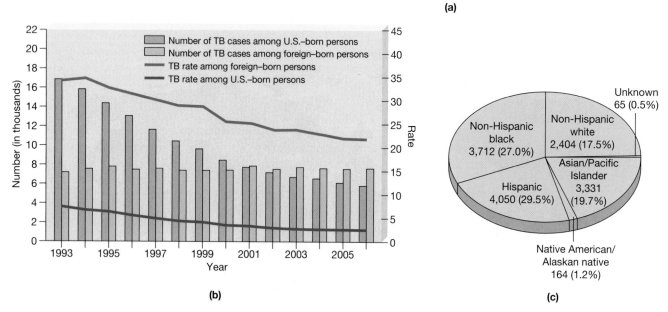

(b)

(c)

infections are often acquired by ingestion and are spread throughout all organs of the body by the bloodstream. The symptoms are therefore not primarily respiratory. All mycobacteria are straight or slightly curved rods that stain acid-fast, as illustrated in ◀Figure 3.32 (p. 72).

Certain properties of mycobacteria are closely associated with their role in tuberculosis. Waxes and long-chain mycolic acids in mycobacterial cell walls make mycobacteria difficult to Gram stain, contribute to environmental survival of these organisms, and protect them from some host defenses. Being obligate aerobes sensitive to slight decreases in oxygen concentration, mycobacteria grow best in the apical, or upper portions of the lungs, which are the most highly oxygenated. Pathogenic

In June 1998 microbiologists from the Pasteur Institute unveiled the complete DNA sequence of Mycobacterium tuberculosis, revealing more than 4,000 genes.

mycobacteria have an exceedingly long generation time (12 to 18 hours, compared with 20 to 30 minutes for most bacteria), which accounts for the long time (up to 8 weeks) it takes to produce a visible colony on laboratory media. Mycobacteria are highly resistant to drying and can remain viable for 6 to 8 months in dried sputum, a property that contributes to public health problems. They are, however, quite sensitive to direct sunlight.

The Disease. Tuberculosis is acquired by the inhalation of droplet nuclei of respiratory secretions or particles of dry sputum containing tubercle bacilli. Young children and elderly people are particularly at risk, so screening school, day-care, and nursing home workers for tuberculosis is important. After organisms are inhaled, they multiply very slowly *inside* white blood cells that have phagocytized them. They elicit a host response that

TABLE 21.2

Mycobacteria That Cause Human Disease

Species	Disease
Mycobacterium tuberculosis	Tuberculosis
M. avium-intracellulare (MAC complex)	Tuberculosis-like disease in humans, transmitted from birds and swine
M. bovis	Tuberculosis, transmitted from cattle; could be transmitted from nonhuman primates
M. fortuitum complex	Wound infections, indwelling catheter infections
M. kansasii	Tuberculosis-like disease
M. leprae	Hansen's disease (leprosy)
M. marinum	Cutaneous lesions in humans, tuberculosis in fish
M. ulcerans	Ulcerative lesions

includes neutrophil infiltration and fluid accumulation within the alveoli of the lungs. The organisms eventually rupture and destroy the neutrophils. Later, macrophages and lymphocytes move into the area. Alveolar macrophages also phagocytize the living tubercle bacilli, which are again able to multiply within and destroy their new hosts. Rupture of the dead phagocytes releases infective organisms. No toxins are produced. As additional cells are infected, an acute inflammatory response occurs. A large quantity of fluid is released, especially in lung tissue,

where it produces pneumonia-like symptoms. Lesions sometimes heal, but more often they produce massive tissue necrosis or solidify to become chronic granulomas, or **tubercles (Figure 21.14)**. Tubercles consist of central accumulations of enlarged macrophages, multinucleate Langerhans giant cells containing tubercle bacilli, peripheral lymphocytes, macrophages, and newly formed connective tissue. The central portion of the tubercle undergoes destruction, typically giving it a cheesy, or **caseous**, appearance. Some organisms gain access to the lymphatic and circulatory systems. About 3 to 4 weeks after exposure, delayed hypersensitivity and cell-mediated immunity develop. On occasion, however, the host fails to respond immunologically. Uncontrolled multiplication of tubercle bacilli then takes place in the lungs, resulting in the formation of numerous tubercles. Organisms are subsequently spread by the circulatory system to other body tissues and organisms. This condition is called **miliary tuberculosis**, named after the small lesions that form that resemble millet seeds. Lesions can be walled off from the rest of the lungs by encapsulation when the host has sufficient resistance. Lesions near blood vessels can perforate the vessels and cause hemorrhage, which leads to bloody sputum, a major symptom of tuberculosis.

Granulomas can keep viable organisms walled off for decades. When the immune system becomes impaired by age or other infections, granulomas can open and the disease can be reactivated. People with AIDS often have tuberculosis, either from new infections or reactivation of old ones. Except for initial infections among immigrants, AIDS patients, and residents of certain large cities such as New York and Washington, D.C., most tuberculosis

(a)

(b)

Figure 21.14 Tuberculosis X-ray photos. (a) In this normal chest X-ray, the faint white lines are arteries and other blood vessels. The heart is visible as a white bulge in the lower right quadrant. *(Biophoto Associates/Photo Researchers)* **(b)** An advanced case of pulmonary tuberculosis; white patches (arrows) indicate areas of disease. These patches, or tubercles, may contain live *Mycobacterium tuberculosis*. Lung tissue and function in these lesions are permanently destroyed. *(Biophoto Associates/Photo Researchers, Inc.)*

Relief Could Cause Disease

If you were suffering from inflamed arthritic joints, you might not be happy to hear that your physician wouldn't let you have cortisone. But there would be a reason for caution. When cortisone is used to treat the arthritis of older patients, the drug also affects the granulomatous lesions of tuberculosis; it actually reactivates tuberculosis organisms as it counteracts the factors that keep the lesions walled off. Bringing relief to the aching joints of arthritic patients can activate hidden tuberculosis lesions, and then these patients can go on to infect others, particularly children. So listen to your doctor when he or she says that cortisone may not be the best option for arthritis relief!

cases in the United States are reactivations rather than first-time infections.

Tuberculosis of the bones can cause extensive erosion, especially in the spine (**Figure 21.15**). The urogenital tract, meninges, lymphatic system, and peritoneum also are prone to develop extrapulmonary tuberculosis. In **disseminated tuberculosis**, now frequently seen in AIDS patients, infected cells become casts: As the tubercle bacillus multiples in a cell, the cell's organelles and membranes are destroyed, and a clump of microorganisms in the shape, or cast, of the former cell remains. When this process occurs in the intestine, a replica of the intestine consisting of live *Mycobacterium* can be seen.

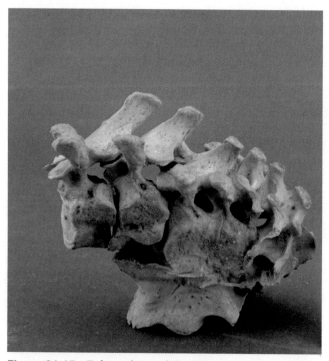

Figure 21.15 Tuberculosis of the spinal column. Notice that several of the lower vertebrae have fused as a result of infection damage. (*Courtesy National Museum of Health and Medicine, Armed Forces Institute of Pathology*).

Although primary tuberculosis is nowadays most likely to affect the lungs, it can also affect the digestive tract. Before milk pasteurization became widespread, primary digestive tract infections were often seen, especially in children. Most were caused by *Mycobacterium bovis* transmitted in raw milk or other dairy products. By law, dairy cattle are now regularly screened for tuberculosis, and infected animals are destroyed. However, some cows escape testing, so drinking raw, unpasteurized milk is a health risk. Today tuberculosis of the digestive tract usually occurs as a secondary infection site, as when pathogen-laden sputum coughed up from a primary lung infection is swallowed.

Diagnosis, Treatment, and Prevention. Tuberculosis can be diagnosed by sputum culture, but because the organisms grow very slowly, cultures must be kept for at least 8 weeks before they are declared negative. Genetic probes are also of value in identification of mycobacteria. Chest X-rays fail to reveal lesions outside the lungs and detect only relatively large ones within them. Therefore, screening is now done by skin tests rather than by X-ray examinations. In a skin test, a small quantity of *purified protein derivative (PPD)*, a protein from tuberculosis bacilli, is injected intracutaneously and examined 48 to 72 hours later for an *induration*, or a raised, not necessarily reddened bump. Induration is a delayed hypersensitivity reaction to the PPD. A positive skin test indicates previous exposure and some degree of immune response. It does not indicate that the person now has or has ever had tuberculosis—the immune response may have prevented infection or eliminated or walled off organisms. In fact, health care workers, teachers, and others who have frequent skin tests may eventually develop a positive response due to sensitization to the test material itself.

Treatment is with isoniazid and rifapin for at least 1 year. Many strains of *Mycobacterium*, however (particularly atypical species that are often found in AIDS patients but are rare in healthy individuals), are now resistant to isoniazid. Such strains must be treated with a "second-line" or sometimes even a "third-line" drug. If treatment is successful, sputum cultures become negative for mycobacteria within 3 weeks. Before the development of isoniazid, patients were often sent to tuberculosis sanatoriums, where cold, fresh air was believed to have curative effects. Patients were bundled up and left outdoors for most of the day. In reality, the rest and proper nutrition were what really helped some patients survive.

Tuberculosis can be prevented by vaccination with attenuated organisms in the vaccine BCG, or bacillus of Calmette and Guérin. The French bacteriologsits Albert Calmette and Camille Guérin developed the vaccine at the Pasteur Institute in Paris in the early 1900s. BCG immunization is widely practiced in parts of the world where the prevalence of tuberculosis is high (◀Chapter 17, p. 520). However, the BCG vaccine has not been approved for use in the United States.

Psittacosis and Ornithosis

In the early 1900s, **psittacosis** (sit″ah-ko′sis), or *parrot fever*, a respiratory disease associated with psittacine birds, such as parrots and parakeets, was discovered. Today at least 130 different species of birds, including ducks, chickens, and turkeys, are found to carry a form of this disease. Because most of these birds are not psittacines, their disease is referred to as **ornithosis** (from *ornis*, Greek for "bird"). Wild and domestic birds can be infected, often without showing symptoms. However, stresses such as overcrowding, chilling, or shipping to pet shops can activate the disease. Both forms of the disease are caused by *Chlamydophila psittaci* and are spread by direct contact, infectious nasal droplets, and feces. Organisms can be found in every organ of an infected bird. The birds have diarrhea and a mucopurulent discharge from the nose and mouth.

Humans usually acquire the disease from birds. Poultry-plant workers are especially susceptible. Human-to-human transmission also has been documented, and medical personnel caring for patients can contract the disease. Organisms are inhaled and spread systemically to the lungs and reticuloendothelial system, especially Küpffer's cells. Most cases of the disease are mild and self-limiting, but some patients develop a serious pneumonia. After an incubation time of 1 to 2 weeks, onset of symptoms is sudden, with sore throat, coughing, difficulty in breathing, headache, fever, and chills. This disease is difficult to distinguish from other pneumonias on a clinical basis. Definitive diagnosis is made by inoculation into tissue culture. However, the organism is so highly infective that this should be undertaken only by very experienced technicians, in specially equipped labs.

With tetracycline treatment, the mortality rate from ornithosis pneumonia is about 5%; in untreated cases, it is around 20%. No vaccine is available, so prevention involves strict quarantine regulations for the birds entering the United States.

Q Fever

Q fever was first described in Queensland, Australia. The *Q* stands not for "Queensland" but for "query," because determining which organism caused it remained a question for a long time. Q fever is now known to be caused by *Coxiella burnetii*, an organism included among the rickettsias. Unlike other rickettsias, this organism survives long periods outside cells and can be transmitted aerially as well as by ticks. For a long time it was a mystery how Q fever organisms can survive 7 to 9 months in wool, 6 months in dried blood, and over 2 years in water or skim milk. The mystery now appears to have been solved. *C. burnetii* has two forms, called large- and small-cell variants **(Figure 21.16)**. Electron microscope studies suggest that a type of endospore may be formed at one end of the large-cell variant, thereby giving resistance uncharacteristic of other rickettsias.

Once *C. burnetii* cells with endospore-like bodies inside them have been inhaled, they are phagocytized by host cells. *Coxiella* grow in phagolysosomes of host cells.

There they display the unusual property of responding to the acidic conditions by increasing their metabolism and multiplying rapidly until they almost fill infected cells. Eventually the cells rupture, releasing new groups of pathogenic bacteria. These pathogens, when spread by the bloodstream, then infect other body cells. There appears to be no exit from the human body.

Coxiella burnetii exist all over the world, especially in cattle- and sheep-raising areas. Wild animals and domestic sheep and cattle are the normal hosts of *C. burnetii*. The bacteria are transmitted via tick bites, feces, and genital secretions of infected animals. Humans become infected by

Small cell forms — Large cell forms

Figure 21.16 *Coxiella burnetii*. Small and large cell forms (16,008X). The cell walls of the large forms have less peptidoglycan with no cross-links—hence the diversity of cell shapes. Notice the endospore-like body within the large cell form, which is probably responsible for the relative resistance of the organism. *(Reproduced from T.F. McCaul & J.C. Williams, Journal of Bacteriology 147:1063–1076, Figure 1b, with permission of the American Society for Microbiology).*

inhaling aerosol droplets from infected domestic animals, which usually do not appear to be ill. Farmers become infected while attending a cow giving birth or miscarrying if the placenta is laden with organisms. (In Canada recently a dozen card players all contracted Q fever after an infected pet cat had a litter of kittens in the same room.) Slaughterhouse and tannery workers become infected by inhaling dried tick feces from the hides of animals. Because of its resistance to drying, *Coxiella* can remain viable in the environment for long periods of time. Transmission to humans also occurs by the ingestion of milk from contaminated animals. In areas such as Los Angeles, an estimated 10% of the cows shed organisms into their milk. Rates are lower in some other areas. Flash pasteurization eliminates this hazard (◀Chapter 12, p. 354). In fact, flash pasteurization was designed to kill Q fever organisms without giving the milk a "cooked" flavor.

Symptoms of Q fever—chills, fever, headache, malaise, and severe sweats—are very similar to those of primary atypical pneumonia. The incubation period is 18 to 20 days. Diagnosis is made by serologic testing or by direct immunofluorescent antibody staining. Treatment is with antibiotics such as tetracycline or fluoroquinolone. Lifelong immunity generally follows an attack of Q fever; a vaccine is available for workers with occupational exposure.

Untreated or inadequately treated cases of Q fever can go into long periods of remission, sometimes lasting years. Cortisone treatment can reactivate it. In chronic cases, endocarditis and heart valve infections are sometimes seen. Endocarditis is invariably fatal, as the organisms do not respond to antibiotic therapy. Special strains appear to be genetically adapted to infect the heart and are very resistant, causing the most serious disease. Other genetic strains cause milder forms of disease.

Nocardiosis

Nocardiosis is characterized by tissue lesions and abscesses. It is caused by the aerobic, acid-fast staining, filamentous bacterium *Nocardia asteroides* (Figure 21.17). The organisms are found in soil and water. Disease is usually initiated by inhalation of *N. asteroides*. The bacterium was first identified in 1888 as the causative agent of a disease in cattle

Figure 21.17
Nocardia asteroides.
This clearly shows the branching, filamentous shapes of the bacteria (1,400X). (*R.J. Hawley & T. Imaeda/Visuals Unlimited*).

known as *farcy*. In humans the primary infection site is the lungs in about three-fourths of all cases, but the disease can also originate in the skin and in other organs. The disease usually occurs in immunosuppressed patients. Human-to-human transmission is rare.

The genus Nocardia was named after Edmond Nocard (1850-1908), a French bacteriologist and veterinary pathologist.

Mortality rates as high as 85% have been reported. However, with early diagnosis and aggressive treatment with sulfonamides in combination with trimethoprim, mortality can be kept below 50%. Diagnosis is based on finding the organism in sputum or other specimens and on culture results.

✓**CHECKLIST**

1. What dangers are associated with pertussis vaccine?
2. How does bronchial pneumonia differ from lobar pneumonia?
3. What sanitary measures can be taken to prevent Legionnaires' disease?
4. How is AIDS related to TB?
5. What organism causes walking or atypical pneumonia?
6. How is it that the fastidious organism *Legionella pneumophilia* can survive inside a phagolysosome?

VIRAL LOWER RESPIRATORY DISEASES

Influenza

Influenza is the last remaining great plague from the past. Since Hippocrates described an influenza-like outbreak in 412 B.C., repeated influenza epidemics and pandemics have been recorded, even in recent times (**Figure 21.18**). One of the greatest killers of all time was the pandemic of swine flu (also known as Spanish flu) of 1918–1919, when 20 to 40 million people died (**Figure 21.19**). The causative virus was unusually virulent, and the crowded, unsanitary conditions created by World War I increased the spread of the virus.

Causative Agents. Influenza is caused by **orthomyxoviruses**. These RNA viruses have an envelope surface antigen hemagglutinin that is responsible for their infectivity. It attaches specifically to a receptor on erythrocytes and other host cells. Some influenza viruses have an enzyme called *neuraminidase* (nur″am-in′i-das), which helps the virus penetrate the mucus layer protecting the respiratory epithelium (**Figure 21.20**). The enzyme also plays a role in the budding of new virus particles from infected cells. These properties allow identification of influenza viruses in the laboratory.

On the basis of their nucleoprotein antigens, three major influenza virus serotypes are recognized: types A, B, and C. Influenza viruses have a tendency to undergo **antigenic variations** (*changeability*), or mutations that affect viral antigens. Thus, immunity developed through infection with one influenza virus is often insufficient to prevent infection by a variant.

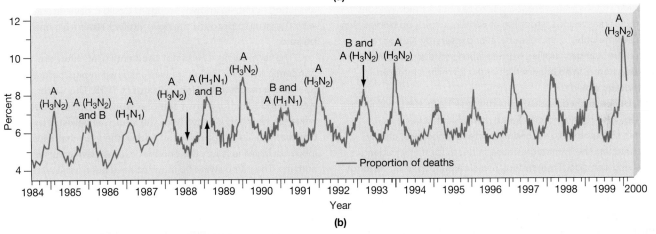

Figure 21.18 The U.S. incidence of influenza. **(a)** Since 1900; **(b)** recently (as a proportion of all deaths in 121 cities). Influenza varieties are labeled by type and by hemagglutinin (H) and neuraminidase (N) numbers. (*Source*: CDC.).

Figure 21.19 Swine flu. During the height of the great flu pandemic of 1918, the gymnasium of Iowa State University was temporarily converted into a hospital ward. (*Courtesy Iowa State University Archives*)

APPLICATIONS

The Great Flu Pandemic of 1918

Where could he be? He should have been home an hour ago! In normal times, this wouldn't have been so alarming—but those weren't normal times. Was his body one of those piled up like cordwood on a street corner downtown?

Such fears raced through the minds of frantic families during the great Spanish flu pandemic of 1918. Seemingly healthy people dropped dead without warning. Undertakers couldn't meet the demands of a pandemic that killed half a million Americans in only 10 months. (The U.S. Civil War claimed about that many lives over 4 years.) Philadelphia alone recorded 528 bodies piled up, awaiting burial, in just 1 day. Numerous others lay undiscovered. Rescue teams went door to door, seeking victims who were too sick to call out for aid. Wagons made regular rounds, stopping at corners to pick up bodies that families, neighbors, storekeepers, or passers-by had carried there. No wonder families worried when someone was late getting home. Who knew which of the 25 million infected Americans would die?

Unlike many diseases, influenza took its heaviest toll among young, healthy individuals. Some died of high fever; others succumbed to secondary bacterial infections, chiefly pneumonia. The immediate cause of death for many patients was lung damage. Tremendous pressure built up until the lungs degenerated completely (on autopsy, they typically had the consistency of pudding.) If the crushing pressure was relieved, the patient had a chance of survival. One physician called to the bedside of a teenage girl could do little but comfort the parents. Suddenly a stream of blood shot from her nose and soaked the trio. Projectile hemorrhaging had reduced the pressure in her lungs, and she lived. But people who survived cases with high fever often developed Parkinson's disease. We now associate Parkinson's disease with the elderly. But after the 1918 pandemic, it affected all ages, including many small children.

The cause of this grim experience was something too tiny for the researchers of 1918 to see—the swine flu virus. What they did see with light microscopes were secondary bacterial invaders. (One bacillus was named *Haemophilus influenzae* in the mistaken belief that it was the cause of the pandemic.) Indeed, in one New York City family, all six members were found to be infected with different species of bacteria. No wonder the medical world was confused. Attempts at vaccine production proved futile. Few people even guessed at the existence of the virus.

Where had this virus come from? A new type of influenza had appeared in France in April 1918 as American and European troops fought during the final months of World War I. From France the virus traveled to Spain. That summer it spread throughout Europe, China, and West Africa. By then, the English-speaking countries had dubbed it the "Spanish lady." The Lady entered Boston in August and crossed the United

States within a month. Around the world she went, causing a pandemic. By the time it had subsided in 1920, more than 25 million had died. The country hardest hit was India, where death wiped out the entire population increase of a decade. In some parts of the world, nearly half the population died. There is some evidence that the virus took nonhuman victims as well—baboons in South Africa, for example.

It is not certain what brought on this deadly kind of flu. The influenza virus, with a long history as a human pathogen, is one of the most rapidly mutating viruses known. But the natural mutation process may have had human help. Mustard gas, which was widely used in trench warfare in France, is a powerful alkylating agent—a known mutagen. Could it have interacted with the virus to produce the new, virulent strain? No one can be sure.

The term *swine flu* was not used until years after the pandemic was over. Influenza was unknown in pigs before the Spanish flu reached the United States in 1918. Thereafter, swine were found to suffer from a complex disease involving both an influenza virus and a bacterium. The virus was believed to be the one that caused the Spanish flu. Today, the virus that causes influenza in swine is not the one that causes "swine flu" in humans. The name swine flu may not have been a mistake, however, for the rapidly evolving virus strains are likely to have mutated quite a bit since 1918.

Is the virus that caused the great pandemic still a threat? In 1976, the U.S. government mounted a massive immunization program against swine flu. Flu viruses almost always march westward around the globe. By looking at the viruses that have caused problems to the east, U.S. public health officials decide which viral strains to incorporate into next year's vaccine. So when swine flu was detected in Asia in 1975, it sent chills up and down medical spines. Not only could millions of deaths occur—there were also the sequelae. How many Parkinson's cases would be survivors of a 1976 pandemic?

Nervously the government exhorted all to be immunized. Then came a disturbing development. Some recipients of swine flu vaccine had subsequently developed *Guillain-Barré syndrome*, resulting in complete paralysis. (Although most persons recover completely, a small percentage die of the paralysis.) Which would be worse: swine flu or paralysis? Lines at vaccine clinics shortened. With fewer people immunized, what would happen when the swine flu arrived? The medical community waited. But swine flu never came. Was it a reprieve or just a postponement? No one can say.

Meanwhile, frozen bodies of swine flu victims buried in permafrost in Alaskan and Scandinavian cemeteries have been exhumed under very strict isolation conditions—whole villages were even evacuated. PCR and DNA sequencing studies of the recovered swine flu virus are now being compared to today's strains.

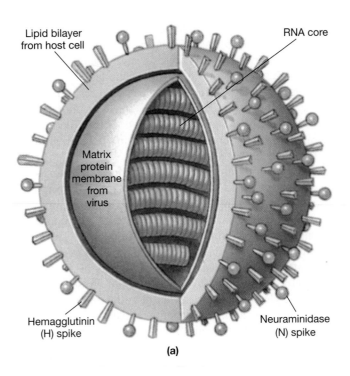

Lipid bilayer from host cell

RNA core

Matrix protein membrane from virus

Hemagglutinin (H) spike

Neuraminidase (N) spike

(a)

(b)

Figure 21.20 **The influenza virus.** **(a)** The virus shows tinin and neuraminidase spikes on its outer surface and an RNA core. **(b)** A colorized TEM of influenza virus type A (85,243X). *(Dr. Dennis Kunkel/Phototake).*

Influenza virus type A, first isolated in 1933, causes epidemics and pandemics at irregular intervals because it periodically undergoes antigenic variation. Secondary bacterial pneumonias are the most important and serious complications of influenza virus A infections. Various species of birds and mammals are infected with strains of influenza A and may serve as reservoirs of infection.

Influenza B viruses also undergo antigenic changes, but less extensively and at a slower rate than do influenza A viruses. Epidemics caused by influenza B viruses are limited geographically and tend to center around schools and other institutions. These viruses are found only in humans.

Influenza C viruses are structurally different from the A and B viruses. Infections with C viruses are rarely recognized. When the disease is recognized, it is typically limited to the children in a single family or single classroom. The low infectivity of C viruses, which lack neuraminidase, suggests that this enzyme may enhance infectivity in viruses that have it.

Antigenic Variation. Antigenic variation in influenza viruses occurs by two processes, *antigenic drift* and *antigenic shift*. **Antigenic drift** results from mutations in genes that code for hemagglutinin and neuraminidase. Such mutations change the configuration of the part of the antigen molecule that stimulates the production of, and combines with, specific antibodies. Thus, antibodies formed against the parental hemagglutinin or neuraminidase are less effective in inhibiting the mutated forms of these viral components of viral offspring. If sufficient antigenic drift occurs naturally, an influenza epidemic will follow. **Antigenic shift** results from gene reassortment, possibly after two different viruses infect the same cell, for example, bird and human influenza

viruses infect a pig cell and exchange large segments of their genomes. It represents more dramatic changes; the viral strains that emerge are significantly different antigenically from previously known strains. Thus, antibodies formed against one type of hemagglutinin are not protective against another type. Antigenic shift generally precedes a major pandemic. It is fortunate that antigenic shift is rare because the strains it produces cause severe pandemics. Six have occurred over the last 100 years or so—in 1890, 1900, 1918, 1957, 1968, and 1977. See the discussion of emerging influenza viruses in ◄Chapter 10, page 283, including "chicken flu."

PUBLIC HEALTH

What's in Flu Vaccine?

There are so many different strains of the flu virus, due to antigenic variation. How does the CDC decide which to include in each year's vaccine? Each year beginning in February, the CDC looks at which strains are most prevalent and what alterations they have undergone in order to select the strains to be used in the coming flu season's vaccine. For example, the vaccine for the 2006–2007 season contained three strains: influenza A/Wisconsin/67/2005 (H3N2)-like, B/Malaysia/2506/2004-like, and A/New Caledonia/20/99-like (H_1N_1) antigens. These strains were named for the geographic location where they were first isolated, a number assigned by the laboratory that isolated them, the year of isolation, and—for type A viruses—the numbers of their envelope antigens, hemagglutinin (H), and neuraminidase (N). The vaccines are thoroughly evaluated for safety and effectiveness before they are given to humans. If you want to avoid the flu each year, the CDC will help you.

Immune responses vary according to antigenic changes. It is as if a contest occurs between the virus's ability to elude the immune system and the immune system's ability to recognize and inactivate the virus. The antigenic change in most viruses is small, and existing antibodies prevent the viruses from causing infections. The viruses cause severe disease in a few people who lack antibodies, mild disease in others who have some effective antibodies, and no disease in others. The diverse assortment of antibodies that hosts acquire from exposure to several different viral strains tends to protect them from new strains. For a virus to cause an epidemic, it must have undergone sufficient antigenic change to elude most available antibodies.

Influenza viruses replicate in the cytoplasm of nucleated host cells but not in erythrocytes, which lack a nucleus. The effects of influenza viruses on cells they infect are noteworthy. The viruses acquire their envelopes by budding through host cell membranes and can do so without immediately killing the host cell (◄Chapter 10, p. 295). A host cell typically produces thousands of viruses per minute over many hours before the cell's macromolecules are exhausted and the cell dies.

The Disease. Influenza, like many other viral respiratory infections, is a relatively superficial infection. The viruses enter the body through inhalation of virus-containing droplets or indirect contact with infectious respiratory secretions. Immediate invasion of the oropharyngeal epithelial lining results. The invading viruses multiply and spread quickly to other portions of the respiratory tract, including mucus-secreting and ciliated epithelial cells. First the cilia are destroyed; then the cells are damaged. Severely damaged cells die and are sloughed. Although host cells begin regeneration immediately, it takes 10 days or more to restore an intact ciliated epithelium. Loss of the mucociliary escalator,

ordinarily a major host defense, allows bacterial invasion and enhanced adherence of bacteria to virus-infected cells (p. 640). Impaired phagocytosis and accumulation of fluid in the lungs add to the risk of secondary bacterial infections, especially pneumonia. Death can result from influenza alone, secondary bacterial infection alone, or a combination. Although antibiotics reduce the risk of death from bacterial infections, they have no effect on viral infections (◄Chapter 13, p. 389).

In a nonimmune person, disease signs and symptoms begin to appear 36 to 48 hours after infection. Fever, malaise, and muscle soreness are the most common symptoms; cough, nasal discharge, sore throat, and gastroenteritis (usually only in children) also occur frequently. The fever usually lasts about 3 days. As systemic symptoms decrease, respiratory symptoms increase. The severity of the disease is directly proportional to the quantity of viruses released from cells. Release of viruses starts as the first malaise appears and peaks about a day before maximum fever and interferon production occur. Viral shedding usually ceases within 8 days of initial exposure. Although the acute phase of the illness is over in about a week, fatigue, cough, and weakness may persist over several weeks.

Incidence and Transmission. In northern temperate zones, influenza appears in late November or early December and disappears in April. In most years the greatest number of cases are seen between January and mid-March. In any one region influenza has a 5- to 7-week period of high prevalence. Indoor crowding, poor air circulation, and dry air expedite the spread of the virus. Schools furnish almost ideal conditions for influenza transmission and usually are the focal points of outbreaks.

Grain dust harbors bacterial endotoxins that can cause lung inflammation in both farm animals and agricultural workers.

TABLE 21.3

Differences Among a Common Cold, Influenza, and Pneumonia			
Symptoms	**Cold**	**Influenza**	**Pneumonia**
Fever	Rare	Characteristic high (100.4°–104°F) sudden onset, lasts 3 to 4 days	May or may not be high
Headache	Occasional	Prominent	Occasional
General aches and pains	Slight	Usual; often quite severe	Occasionally quite severe
Fatigue and weakness	Quite mild	Extreme; can last up to a month	May occur, depending on type
Exhaustion	Never	May occur early and prominent	May occur, depending on type
Runny, stuffy nose	Common	Sometimes	Not characteristic
Sneezing	Usual	Sometimes	Not characteristic
Sore throat	Common	Sometimes	Not characteristic
Chest discomfort, cough	Mild to moderate; hacking cough	Can become severe	Frequent and may be severe
Complications	Sinus and ear infections	Bronchitis, pneumonia; can be life-threatening	Widespread infections of other organs; can be life-threatening, especially in elderly and debilitated persons

People often claim to have the "flu," however, when they are actually suffering from another condition such as an allergy or a bad cold (Table 21.3). Gastrointestinal problems, often called "stomach flu," are probably not flu at all but are most likely due to another viral infection.

Diagnosis and Treatment. The best specimens for isolation of viruses are throat swabs taken as early in the illness as possible. The viruses can be cultured in embryonated chick eggs and various cell lines and identified by hemagglutination inhibition and immunofluorescent antibody tests.

Moderately effective treatment is now becoming available for influenza. The drug amantadine blocks influenza A virus replication, probably by interfering with uncoating (◄Chapter 13, p. 389). If given soon after the onset of symptoms, amantadine is effective in shortening the disease's course and reducing the severity. It is useful for short-term protection of selected individuals but impractical to use for the duration of an influenza epidemic unless patients are especially compromised. Furthermore, it prevents users from developing antibodies to the virus and has some unpleasant side effects. Rimantadine is more effective and less toxic (◄Chapter 13, p. 389). Ribavirin has been demonstrated to inactivate both A and B viruses.

Immunity and Prevention. Because of influenza viruses' ability to mutate, annual immunization is recommended, especially for high-risk persons, such as those with chronic conditions. The killed virus vaccine provides effective protection against influenza, and annual immunization increases the diversity of the recipient's antibodies—which may be of use in some future flu outbreak. Even less frequent immunization can provide some degree of protection. Although immunized individuals sometimes become infected with flu viruses, they generally have a milder disease of shorter duration than do nonimmunized individuals. Furthermore, they shed fewer viruses over a shorter duration of time. Thus, immunization of some people tends to reduce the spread of the disease to other people. A new type of

Figure 21.21   Flu Mist® vaccine is inhaled, rather than injected. *(Courtesy Wyeth-MedImmune, Inc.)*

immunization, first made available in 2003, is Flu Mist®, which can be inhaled rather than injected (Figure 21.21).

A human gene that confers resistance to influenza was identified at the University of California in Santa Barbara. This gene is turned on by interferon. It produces a specific protein, called the Mx protein, which prevents the virus from making viral RNA and proteins.

SARS (Severe Acute Respiratory Syndrome)

Quietly, in November 2002, people in China began falling sick of a new respiratory disease. However, Chinese authorities said nothing of this until February 2004, when they reported 305 cases of "atypical pneumonia." These were subsequently called SARS (severe acute respiratory syndrome) and are caused by a coronavirus (SARS-CoV, Figure 21.22). The disease spread rapidly, including to other countries. In March, the World Health Organization (WHO) issued a global SARS alert. In May, China threatened to execute or jail for life anyone found breaking SARS quarantine orders. By the time the epidemic was declared over in July 2003, it had spread to 29 countries, had killed 774, and infected 8,098 people. Of the

APPLICATIONS

Flu Shots May Help Prevent Heart Attacks

Some viruses may cause inflammation of blood vessels, which in turn causes plaque deposits inside the vessels to bleed. This leads to heart attacks. A recent study showed that in a group of people who had received flu shots, there were 67% fewer cases of angina and second heart attacks. In the year 2000, the CDC lowered the recommended age group for receiving flu shots from "over 65" to "over 55" so as to include more of the age group likely to experience heart attacks.

Figure 21.22   SARS virus (253,467X). *(Sercomi/Photo Researchers, Inc.)*

dead, 350 were in China, none in the U.S. The Chinese government killed over 10,000 civet cats, a weasel-like animal sold in wild-game markets and served as a delicacy in many restaurants. Other mammals, including domestic house cats, have been shown to carry the virus.

Symptoms of SARS begin with a high fever, dry cough, shortness of breath, difficulty breathing, and most often X-rays indicating pneumonia. Less often, headache, muscular stiffness, confusion, rash, and diarrhea may also occur. The virus is spread by close contact with an infected person, usually by exhaled or coughed aerosol droplets. People touching contaminated objects or surfaces, who then touched their nose, mouth, or eyes became infected, as did those kissing or hugging.

The causative coronavirus is a relative of the common cold virus, and is not a type of influenza. Vaccine development is underway, with human trials scheduled to begin in Fall 2004. Thus far animal trials look promising. It is expected that future outbreaks will occur, due to the large number of reservoir animals and the ease with which it is spread. But as of 2007, no new outbreaks have occurred.

Respiratory Syncytial Virus Infection

The **respiratory syncytial virus (RSV)** is the most common and most important and costly cause of lower respiratory tract infections in children under 1 year and especially in male infants 1 to 6 months old. The virus derives its name from the fact that it causes cells in cultures to fuse their plasma membranes, lose independent identity, and become multinucleate masses, or **syncytia** (sin-sish'ya; singular: *syncytium*). The disease, a form of **viral pneumonia**, begins with a 3- or 4-day period of fever as the virus infects the respiratory tract. Fever is followed by hyperventilation, hyperinflation, and infiltration on the lungs with fluid. If an infant with RSV is placed on a rapidly pulsing respirator, the infant can shed enough virus to infect an entire intensive care nursery. Major outbreaks of this nature generally result in fatalities. Viruses are released for weeks to months after infection, and reinfection is common. In older children and adults, RSV affects mainly the upper respiratory tract, and adults who carry the virus in their nose can be the source of nursery infections.

The incidence and transmission of RSV are quite similar to those for parainfluenza viruses, and simultaneous infection by both kinds of viruses is not uncommon. The virus can be identified by enzyme immunoassay performed on nasal secretions or by a direct immunofluorescent test on cells from such secretions. Ribavirin shortens the duration of the disease. There is no vaccine. The best means of prevention is to reduce an infant's exposure to crowds and other sources of infection.

Hantavirus Pulmonary Syndrome

In May and June 1993, 24 cases of severe respiratory illness among residents of the Four Corners area of the

southwestern United States were reported. Healthy adults suddenly became ill and died within a few hours. Fear raced through Arizona, Colorado, New Mexico, and parts of the Navajo reservations where deaths had occurred. Investigation identified a previously unreported hantavirus as the cause of the disease, later named **hantavirus pulmonary syndrome (HPS)**. The virus was identified by PCR techniques and RNA sequencing (◄Chapter 7, p. 204) and was shown to be distinct from all other known hantaviruses. Hantavirus illness had been known elsewhere in the world for years (◄Chapter 23), but with renal and hemorrhagic symptoms—not pulmonary illness. The new virus was tentatively named Muerto Canyon virus, after the area from which it was isolated. After complaints from residents of the area, however, the name Sin Nombre ("No-Name") virus was proposed and has been adopted.

The fatality rate (60%) has been more than 10 times higher than for other hantaviruses. During 1994, other cases of HPS were found throughout the United States—in Florida, Louisiana, Indiana, and Rhode Island. In each case, a different, new hantavirus was found to be the cause. Each virus was carried by rodents that appeared healthy but shed virus in urine, feces, and saliva. Sin Nombre virus is carried by the deer mouse, *Peromyscus maniculatus*. The Florida variety of the virus is carried by the cotton rat, *Sigmodon hispidus*; the vector for the Louisiana strain has not yet been identified. Virus from dried excreta becomes airborne and is inhaled. Exposure to rodent-contaminated areas, both indoors and outdoors, should be avoided. Ribavirin has been somewhat useful in treatment, but further studies are needed.

Acute Respiratory Disease

Acute respiratory disease (ARD) is a lower respiratory tract viral disease that ranges from mild to severe. Respiratory symptoms such as sore throat, cough, and other cold symptoms, fever, headache, and malaise are frequently seen. Severe viral pneumonia that lasts about 10 days sometimes occurs in ARD.

Cases of viral ARD have been seen in military training facilities in the United States and in Europe. Epidemics usually start 3 to 6 weeks after the start of training. The epidemic nature of the disease has been attributed to crowding of people from different geographic areas under stressful conditions. However, epidemics have not been observed in colleges and other institutions where conditions are similar, so other factors may be involved in such epidemics.

Adenoviruses cause about 5% of ARD cases in children under 5 years old. Usually the symptoms are mild and nonspecific—stuffy nose, cough, and nasal discharge. More severe symptoms such as tonsillitis, pharyngitis, bronchitis, bronchiolitis, croup, and often conjunctivitis and abdominal pain may also appear. Adenovirus pneumonia accounts for about 10% of all childhood pneumonia and is occasionally fatal.

FUNGAL RESPIRATORY DISEASES

Compared with bacteria and viruses, fungi are much less frequent causes of respiratory diseases. Most fungal infections are seen in immunodeficient and debilitated patients. Blastomycosis, usually a skin disease, can also cause mild respiratory symptoms (◀Chapter 19, p. 589).

Fungi can also grow inside of buildings, known as "sick" buildings, where they infect occupants or cause allergic reactions. One such mold, *Stachybotrys atra*, grows in houses that have continuous dampness or frequent flooding problems. Children are more susceptible to *S. atra* than are adults, and may develop potentially fatal cases of bleeding lungs. It is very difficult to remediate "sick" buildings—sometimes all that can be done is to tear them down.

Coccidioidomycosis

The soil fungus *Coccidioides immitis* causes **coccidioidomycosis** (kok″sid-i-oy′do-mi-ko″sis), or San Joaquin valley fever (**Figure 21.23**). This organism is found mainly in warm, arid regions of the southwestern United States and Mexico. Infection usually occurs as a result of inhaling dust particles laden with fungal spores. Dogs, cattle, sheep, and wild rodents deposit spores in feces, and the spores easily become airborne. Susceptible individuals have become infected merely from standing on the platform at a train station for a short period of time to get a breath of "fresh air." Travelers passing through dust storms in the Southwest often become infected and return to states where the disease is not expected. Recent increases in disease incidence in California are due in part to earthquake activity, which has produced excess dust. When spores are inhaled by susceptible individuals, an influenza-like illness results. Coccidioidomycosis is always highly infectious, and it can be self-limiting or progressive. In fewer than 1% of affected individuals, dissemination to the meninges or the bones occurs within 1 year of the initial infection. Dissemination is much more common in blacks than in whites.

Large spherules containing numerous small endospores (reproductive spores that are not ordinarily infectious—not to be confused with bacterial endospores) can be found in sputum, pus, spinal fluid, or biopsied tissue (**Figure 21.23b**). The fungus forms cottony, white mycelia (colonies) in culture. Various immunological tests are available to aid in diagnosis, but some give

(a)

(b) LM

Figure 21.23 Coccidioidomycosis. **(a)** The skin rash of coccidioidomycosis on face. *(Courtesy National Institute of Allergy and Infectious Diseases)* **(b)** A tissue section showing a spherule that contains many endospores (1,201X). *(Science VU/Visuals Unlimited)*

false-positive results in individuals who have antibodies against other fungal diseases. A skin test is also available to determine previous exposure to the fungus. The test is performed in the same manner as the tuberculosis skin test and measures the delayed hypersensitivity to antigens of *C. immitis*. Individuals exhibiting a positive skin test are immune to a second attack of the disease.

Coccidioidomycosis in its common form is an acute, self-limiting infection that generally does not require treatment. However, in cases of disseminated disease, therapy is needed. Amphotericin B is the most effective agent available. Newer polyenes, such as the azoles, show promise. If treatment fails, the disease is often fatal. Prevention is difficult, but reducing dust in endemic regions may be helpful. A vaccine is being developed for coccidioidomycosis.

Histoplasmosis

The soil fungus *Histoplasma capsulatum* causes **histoplasmosis**, or *Darling's disease*. The disease is endemic to the central and eastern United States but is found globally in major river valleys. The highest incidence is seen in the Mississippi and Ohio valleys, where 80% of the population shows immunological evidence of exposure to the fungus. *H. capsulatum* thrives in soil mixed with feces and especially in chicken houses and in caves containing bat guano (feces). Cave dust is responsible for so many cases that the disease is sometimes referred to as cave sickness.

The fungus enters the body by the inhalation of conidia (fungal spores) (◀Chapter 11, p. 324). These conidia are engulfed, but not killed, by macrophages and thus travel in macrophages throughout the body. Person-to-person spread does not occur. Although most infections do not produce disease symptoms, granulomatous lesions occur in the lungs and spleen of susceptible individuals. Inhaling large numbers of conidia can cause pulmonary infection resembling pneumonia. In a few individuals, especially the very young, very old, or those receiving immunosuppressants, *H. capsulatum* can spread to the spleen, liver, and lymph nodes. Anemia, high fever, and an enlarged spleen and liver are common in disseminated histoplasmosis, and death often results.

Diagnosis is made by microscopic identification of small ovoid cells of the organism inside infected human cells. When cultured at body temperature, the fungi look like budding yeasts; at room temperature, they form mycelia with hyphae and spores. An intradermal skin test is also available to determine previous exposure to *H. capsulatum*.

Supportive therapy is used for pulmonary histoplasmosis, and amphotericin B is sometimes effective in treating the disseminated disease. Humans can very easily become infected in environments such as chicken houses and caves, where the air is laden with spores. Some employers hire only individuals with positive histoplasmosis skin tests, and therefore immunity to histoplasmosis, to work in high-risk environments. Spraying infected soil and fecal deposits with 3% formaldehyde destroys some of the spores.

Cryptococcosis

Cryptococcosis is caused by *Filobasidiella* (formerly *Cryptococcus*) *neoformans*, a budding, encapsulated yeast. Organisms typically enter the body through the skin, nose, or mouth. Birds carry the fungi on their feet and beak. Although birds do not suffer from cryptococcosis, they do disseminate the opportunistic yeast. These organisms thrive on the nitrogenous waste creatinine, which is present in high concentration in bird feces.

Cryptococcosis is usually characterized by mild symptoms of respiratory infection, but it can become systemic if large quantities of the spores are inhaled by debilitated patients. *Filobasidiella neoformans* often

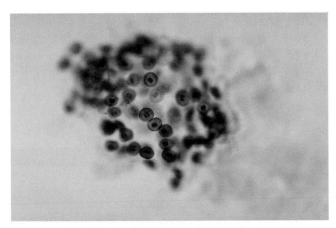

Figure 21.24 *Pneumocystis jiroveci* **in sputum.** The organism is a frequent cause of pneumonia in AIDS patients (353X). (G.W. Willis, M.D./Biological Photo Service)

spreads to the meninges, which become thickened and matted, and the organism can invade brain tissue. As with other fungal opportunists, cryptococcosis is increasing in incidence among AIDS patients.

Observing the organism in body fluids confirms the diagnosis. A latex agglutination test is also used to detect the presence of capsular material in body fluids. Flucytosine and aphotericin B can be used in combination to treat systemic disease. Because *F. neoformans* thrives in pigeon droppings, some degree of prevention can be achieved by reducing pigeon populations and decontaminating droppings with alkali.

Pneumocystis Pneumonia

Pneumocystis jiroveci, formerly *P. carinii* (**Figure 21.24**), was long thought to be a protozoan of the sporozoan group. It is now thought to be an opportunistic fungus. It invades cells of the lungs and causes alveolar septa to thicken and the epithelium to rupture. Then the parasites and a foamy exudate from cells collect in the alveoli. The disease, called ***Pneumocystis* pneumonia**, occurs in infants, the elderly, and the immunocompromised. The marked increase in the incidence of this disease in recent years is due mainly to its ability to infect persons with AIDS. The fungus can spread to other organs and cause extrapulmonary infections.

Diagnosis is made by finding organisms in biopsied lung tissue or bronchial lavage (washings from the bronchial tubes). Treatment of *Pneumocystis* infection generally involves the administration of a combination of trimethoprim and sulfamethoxazole or pentamidine. People with HIV or AIDS are treated with these drugs as a preventive measure.

Aspergillosis

Aspergillosis (◄Chapter 19, p. 590), called *farmer's lung disease* when it occurs in the lungs, is usually caused by *Aspergillus fumigatus* or *A. flavus*. Other species of *Aspergillus* and even other genera of fungi also can cause farmer's lung disease. Fungal spores inhaled from piles of rotting vegetation or compost may cause clinical allergy, such as asthma, or may produce invasive infection in the lower respiratory tract. Masses of fungal mycelium may grow large enough to be visible on X-rays as a *fungus ball*, or *aspergilloma*. These masses can obstruct gas exchange and cause death by asphyxiation. *Aspergillus* growing in the lungs may also serve as antigens that trigger chronic asthma. Amphotericin B is the drug of choice for invasive infections. Immunosuppressed, immunodeficient (AIDS), and diabetic patients are at higher than normal risk. Fungus balls are very difficult to treat, and surgery is sometimes needed to remove them.

PARASITIC RESPIRATORY DISEASES

The lung fluke *Paragonimus westermani* is found in many parts of Asia and the South Pacific (**Figure 21.25**). The fluke's life cycle starts when egg-laden feces are released into water; the eggs hatch and invade first a snail and then a crab or crayfish, in which the last larval form, or metacercariae, develop (◄Chapter 11, p. 328). When a human eats an infected shellfish, the metacercariae leave the crab or crayfish as it is digested in the human's small intestine. The larvae then bore through the intestine and embed in the abdominal wall temporarily. They soon leave it and penetrate the diaphragm and membranes around the lungs to reach the bronchioles. The larvae mature into adults and lay eggs in the

Literally every wild musk ox in Alaskan herds is host to several lung flukes, each about a foot long, and several inches thick. It takes your breath away just to think about it!.

Figure 21.25 The lung fluke *Paragonimus westermani*. This is an adult worm stained to show internal structures (13X). (A.M. Siegelman/Visuals Unlimited)

LM

bronchioles. When the host coughs, the eggs move into the pharynx, are swallowed, and exit the body in the feces. Infected humans have a chronic cough, bloody sputum, and difficult breathing. Diagnosis can be made by finding eggs in sputum or by any one of several immunologic tests. The drug praziquantel is effective in treating lung fluke infections. Infections can be avoided by cooking shellfish before eating them.

Diseases of the lower respiratory tract are summarized in **Table 21.4.**

TABLE 21.4

A Summary of Diseases of the Lower Respiratory Tract		
Disease	**Agent(s)**	**Characteristics**
Bacterial Lower Respiratory Diseases		
Whooping cough	*Bordetella pertussis*	Catarrhal stage with fever, sneezing, vomiting, and mild cough; paroxysmal stage with ropy mucus and violent cough; convalescent stage with mild cough
Classic pneumonia	*Streptococcus pneumoniae, Staphylococcus aureus, Klebsiella pneumoniae*	Inflammation of bronchi or alveoli of lungs with fluid accumulation and fever
Mycoplasma pneumonia	*Mycoplasma pneumoniae*	Mild inflammation of bronchi or alveoli
Legionnaires' disease	*Legionella pneumophila*	Inflammation of the lungs, fever, chills, headache, diarrhea, vomiting, and fluid in lungs
Tuberculosis	*Mycobacterium tuberculosis*	Tubercles in lungs and sometimes in other tissues; organisms can persist in walled-off lesions and be reactivated
Ornithosis	*Chlamydophila psittaci*	Pneumonia-like disease transmitted to humans by birds
Q fever	*Coxiella burnetii*	Disease similar to *Mycoplasma* pneumonia but transmitted by ticks, aerosols, and fomites.
Nocardiosis	*Nocardia asteroides*	Pneumonia-like disease seen in immunodeficient patients
Viral Lower Respiratory Diseases		
Influenza	Influenza viruses	Viruses subject to antigenic variation, with new strains causing epidemics; inflammation of oropharyngeal membranes, fever, malaise, muscle pain, cough, nasal discharge, and gastroenteritis
Respiratory syncytial virus infection	Respiratory syncytial virus	Febrile disease of the respiratory tract; can cause viral pneumonia
Hantavirus pulmonary syndrome	Hantaviruses	Fever, kidney abnormalities; in severe cases shock, bleeding, and pulmonary edema
Acute respiratory disease	Adenoviruses	Mild cough and nasal discharge; can cause viral pneumonia
Fungal Respiratory Diseases		
Coccidioidomycosis	*Coccidioides immitis*	Influenza-like illness; dissemination to meninges and bones can occur
Histoplasmosis	*Histoplasma capsulatum*	Granulomatous lesions in lungs and spleen in susceptible individuals; can cause pneumonia
Cryptococcosis	*Filobasidiella (Cryptococcus) neoformans*	Usually a mild pulmonary disease; pneumonia and dissemination to meninges can occur
Blastomycosis	*Blastomyces dermatitidis*	Usually a skin disease (see ◄Chapter 19); sometimes disseminated to lungs or acquired directly by inhalation; mild respiratory symptoms
Pneumocystis pneumonia	*Pneumocystis jiroveci*	Rupture of alveolar septa, foamy sputum; occurs mainly in immunodeficient patients
Aspergillosis	*Aspergillus* species	Allergic asthmatic response to inhalation of spores or invasive infection of lung; fungal balls can cause asphyxiation
Parasitic Respiratory Diseases		
Lung fluke infection	*Paragonimus westermani*	Larvae mature in bronchioles and cause chronic cough, bloody sputum, and difficulty breathing

▌ RETRACING OUR STEPS

• The agents and characteristics of the diseases discussed in this chapter are summarized in Tables 21.1 and 21.4. Information in these tables is not repeated in this summary.

▌ COMPONENTS OF THE RESPIRATORY SYSTEM

THE UPPER RESPIRATORY TRACT

• The **upper respiratory tract** consists of the nasal cavity, pharynx, larynx, trachea, and bronchi. It is lined with mucus-secreting epithelium and is covered with cilia.
• The **mucociliary escalator** is a nonspecific defense whereby microbes trapped in mucus are moved up to the pharynx.

THE LOWER RESPIRATORY TRACT

• The **lower respiratory tract** consists of the **bronchioles**, **respiratory bronchioles**, and **alveoli**, forming the lungs. The surfaces of the lungs and the cavities they occupy are covered by the **pleura**, which secretes **serous fluid**.

THE EARS

• The outer ear consists of a **pinna** and an **auditory canal**, lined with **ceruminous glands** that secrete cerumen (earwax).
• The **tympanic membrane** separates the outer and middle ears. The middle ear contains three ossicles that transmit sound waves to the inner ear, where the vestibulocochlear cranial nerve carries impulses to the brain.

NORMAL MICROFLORA OF THE RESPIRATORY SYSTEM

• Healthy lungs are usually sterile. Below the level of the larynx, the trachea, bronchi, and larger bronchioles have a transient rather than a normal resident microflora.
• The pharynx has a normal microflora similar to that of the mouth.
• The upper respiratory tract above the pharynx has a normal microflora similar to that of skin.

▌ DISEASES OF THE UPPER RESPIRATORY TRACT

BACTERIAL UPPER RESPIRATORY DISEASES

• Infections related to **pharyngitis**, or sore throat, include **laryngitis**, **epiglottitis**, **sinusitis**, and **bronchitis**. These diseases usually are transmitted by respiratory droplets. If severe enough, they can be treated with penicillin or other antibiotics.
• **Diphtheria**, no longer common in the United States, occurs only in humans. Both the organism and its toxin, produced by genes from a lysogenic prophage, contribute to disease signs and symptoms. A major disease sign is the formation of a **pseudomembrane** that can block the airway. Diphtheria is spread by respiratory droplets, treated with antitoxin and antibiotics such as erythromycin and clindamycin, and prevented by DTP vaccine.

• Ear infections occur in the middle (**otitis media**) and outer (**otitis externa**) ear. Organisms reach the middle ear via the Eustachian tube and usually can be eradicated by penicillin.

VIRAL UPPER RESPIRATORY DISEASES

• The common cold, **coryza**, is transmitted by fomites and aerosols. Treatment is limited to alleviating symptoms; no vaccine is available.
• **Parainfluenza** infections range from inapparent disease to severe **croup**. Most children develop antibodies to **parainfluenza viruses** by age 10.

▌ DISEASES OF THE LOWER RESPIRATORY TRACT

BACTERIAL LOWER RESPIRATORY DISEASES

• **Whooping cough**, or **pertussis**, is distributed worldwide, but immunization has decreased its incidence. It is transmitted by respiratory droplets and treated with antitoxin and erythromycin. Vaccine prevents the disease but may also cause some complications and deaths.
• Classic **pneumonia** can be **lobar** or **bronchial**; it is transmitted by respiratory droplets and carriers. *Klebsiella* pneumonia is more severe than pneumococcal pneumonia. Penicillin is the drug of choice for pneumococcal pneumonia, and vaccine is recommended for high-risk populations. *Mycoplasma* pneumonia is transmitted by respiratory droplets and is treated with erythromycin or tetracycline.
• **Legionnaires' disease** is transmitted via aerosols from contaminated water and treated with erythromycin, azithromycin, or a fluoroquinolone.
• **Tuberculosis (TB)** has been a major health problem worldwide for centuries, and its incidence in the United States is increasing. It is transmitted by respiratory droplets. Inactive but viable organisms can persist for years walled off in **tubercles**. Treatment with isoniazid is effective except for resistant organisms, which must be treated with "second- or third-line" drugs. A vaccine is available worldwide, but its use in the United States is limited to high-risk individuals.
• **Ornithosis** is transmitted to humans from infected birds. It is usually mild but can cause a serious pneumonia. The causative organism is dangerous to handle in the lab.
• **Q fever** is transmitted by ticks, aerosol droplets, and fomites. It is treated with tetracycline, and a vaccine is available for workers with occupational exposure.
• **Nocardiosis**, marked by lesions and abscesses, infects primarily the lungs but can infect the skin or other organs.

VIRAL LOWER RESPIRATORY DISEASES

• The viruses that cause **influenza** display **antigenic variation** (mutation that affects antigens). The disease occurs mainly from December through April. It is transmitted in crowded, poorly ventilated conditions and diagnosed by immunological tests. Vaccine can prevent the disease, but its effectiveness can be lessened by antigenic variation. Chicken flu is discussed in ◀Chapter 10.

- **SARS (severe acute respiratory syndrome)** is caused by a coronavirus. No vaccine is yet available.
- **Respiratory syncytial virus (RSV)** infections and **acute respiratory disease (ARD)** are important acute respiratory tract illnesses. RSV infections tend to be severe, especially in young children.
- **Hantavirus pulmonary syndrome (HPS)**, a severe infection, has a high mortality. The hantavirus infects primarily rodents and is found in their excreta.

FUNGAL RESPIRATORY DISEASES
- Opportunistic fungal infections occur mainly in immunocompromised and debilitated patients. These infections usually are transmitted by spores, and some can be treated with amphotericin B.
- In the United States, **coccidioidomycosis** occurs in warm, arid regions; **histoplasmosis** is endemic in eastern states, and **cryptococcosis** occurs wherever there are infected birds, especially pigeons.
- ***Pneumocystis* pneumonia**, an opportunistic fungal infection, is a common cause of death among AIDS patients.
- **Aspergillosis** is contracted by the inhalation of fungal spores and can involve the growth of large masses of fungal mycelia (fungus balls) that are visible on X-rays.

PARASITIC RESPIRATORY DISEASES
- Lung fluke infections occur mainly in Asia and the South Pacific where infected shellfish are eaten. The disease can be treated with drugs and prevented by cooking shellfish adequately.

▌ TERMINOLOGY CHECK

acute respiratory disease (ARD) *(p. 667)*
alveolus *(p. 641)*
antigenic drift *(p. 663)*
antigenic shift *(p. 663)*
antigenic variation *(p. 660)*
aspergillosis *(p. 669)*
auditory canal *(p. 642)*
bronchial pneumonia *(p. 652)*
bronchiole *(p. 640)*
bronchitis *(p. 644)*
bronchus *(p. 640)*
caseous *(p. 657)*
catarrhal stage *(p. 651)*
cerumen *(p. 642)*
ceruminous gland *(p. 642)*
coccidioidomycosis *(p. 667)*
consolidation *(p. 652)*
convalescent stage *(p. 651)*
coronavirus *(p. 648)*
coryza *(p. 647)*
croup *(p. 648)*

cryptococcosis *(p. 668)*
cyanosis *(p. 651)*
diphtheria *(p. 645)*
diphtheroid *(p. 645)*
disseminated tuberculosis *(p. 658)*
epiglottitis *(p. 643)*
hantavirus pulmonary syndrome (HPS) *(p. 666)*
histoplasmosis *(p. 668)*
influenza *(p. 660)*
laryngitis *(p. 643)*
larynx *(p. 640)*
Legionnaires' disease *(p. 653)*
lobar pneumonia *(p. 652)*
lower respiratory tract *(p. 640)*
mastoid area *(p. 642)*
miliary tuberculosis *(p. 657)*
mucociliary escalator *(p. 640)*
nasal cavity *(p. 640)*
nasal sinus *(p. 640)*

nocardiosis *(p. 660)*
ornithosis *(p. 659)*
orthomyxovirus *(p. 660)*
otitis externa *(p. 647)*
otitis media *(p. 647)*
parainfluenza *(p. 648)*
parainfluenza virus *(p. 648)*
paroxysmal stage *(p. 651)*
pertussis *(p. 649)*
pharyngitis *(p. 643)*
pharynx *(p. 640)*
pinna *(p. 642)*
pleura *(p. 641)*
pleurisy *(p. 652)*
Pneumocystis pneumonia *(p. 669)*
pneumonia *(p. 652)*
Pontiac fever *(p. 654)*
primary atypical pneumonia *(p. 653)*
pseudomembrane *(p. 645)*
psittacosis *(p. 659)*

Q fever *(p. 659)*
respiratory bronchiole *(p. 641)*
respiratory syncytial virus (RSV) *(p. 666)*
respiratory system *(p. 640)*
rhinovirus *(p. 648)*
SARS *(p. 665)*
sinusitis *(p. 644)*
syncytium *(p. 666)*
tonsilitis *(p. 644)*
trachea *(p. 640)*
tubercle *(p. 657)*
tuberculosis (TB) *(p. 654)*
tympanic membrane *(p. 642)*
upper respiratory tract *(p. 640)*
viral pneumonia *(p. 666)*
walking pneumonia *(p. 653)*
whooping cough *(p. 649)*

▌ CLINICAL CASE STUDY

Mildred had always done her grocery shopping at the grocery store across from the retirement community she lives in. Recently she was sick with flu-like symptoms: muscle aches and soreness, malaise, and then fever and chills. She recovered but she has heard about other people at the retirement community with similar symptoms, some of whom have ended up in the hospital and one of whom died. She has heard Legionnaires' disease mentioned. This week when she was at the grocery store she noticed that the vegetable section was closed for remodeling. She asked the check-out clerk about it. The clerk said the local health department had come out and done some testing. As a result, they were getting a new water spraying system put in.

Were Mildred's symptoms consistent with Legionnaires' disease? What diagnostic tests might have been done if she had gone to the doctor? Why do you think the grocery store was putting a new water spraying system. How do you think Mildred got infected with *Legionella pneumophila*? Visit the following web sites to learn more about actual outbreaks of Legionnaires' disease and ways to prevent infection.

Outbreak at automotive plant http://www.q-net.net.au/ ~legion/Legionnaires_Disease_Fords.html
www.ncbi.nlm.nih.gov/entrez/query.fcgi?cmd=Retrieve &db=PubMed&list_uids=11749746&dopt=Abstract

▪ CRITICAL THINKING QUESTIONS

1. What are some factors that make the respiratory system prone to infection?

2. In the United States, the incidence of tuberculosis decreased during the period of 1953 to 1986. What factors have contributed toward its increased incidence since then?

3. Following Spring Break 2001 in Acapulco, several hundred college students returned to school ill with pneumonia-like symptoms including fever, chest pain, and coughing. The symptoms ranged from mild to fairly severe. Their sputum contained yeast-like cells. Acapulco has a damp climate and organic soil. There had been considerable wind during the week. What do you think was wrong with these students?

▪ SELF-QUIZ

1. Rheumatic fever is caused by which organism?
   (a) *Corynebacterium diphtheriae*
   (b) *Staphylococcus epidermidis*
   (c) *Staphylococcus aureus*
   (d) *Escherichia coli*
   (e) *Streptococcus pyogenes*

2. Which of the following components of the respiratory system is normally free of microorganisms?
   (a) Lungs
   (b) Nasal cavity
   (c) Pharynx
   (d) Trachea
   (e) Bronchi

3. Which of the following is true about diphtheria?
   (a) It is caused by strains of *Corynebacterium diphtheriae* infected with a prophage that carries an exotoxin-producing gene.
   (b) A pseudomembrane composed of *Corynebacterium diphtheriae* cells, damaged epithelial cells, fibrin, and blood cells forms that can block the airway and cause suffocation.
   (c) The exotoxin spreads throughout the body damaging multiple organ system by interfering with cellular protein synthesis.
   (d) It can be prevented with DTP (diphtheria, tetanus, and pertussis) vaccine and treated with both antitoxin and antibiotics.
   (e) All of the above are true.

4. Which of the following is NOT true about strep throat?
   (a) It is caused by group A β-hemolytic *Streptococcus pyogenes*.
   (b) It can be acquired through droplet nuclei inhalation from active cases, healthy carriers, and family pets, as well as from contaminated food, milk, and water.
   (c) Active cases can have inflamed and sore throat, swollen adenoids and lymph nodes in neck, pus-filled tonsillar lesions, fever, chills, headache, nausea, and vomiting.
   (d) Coughing and nasal discharge are common hallmarks of strep throat.
   (e) If immediate treatment with an antibiotic is not given, the patient has a risk of also getting rheumatic fever.

5. All of the following are true about ear infections EXCEPT:
   (a) Middle ear infections (otitis media) often produce a puslike exudate (referred to as otitis media with effusion) due to infection by genera of *Streptococcus, Haemophilus*, or various species of anaerobes.
   (b) External auditory canal (otitis externa) infections are typically caused by *Staphylococcus aureus* or *Pseudomonas aeruginosa*.
   (c) After successful treatment of middle ear infections with antibiotics, sometimes tubes are inserted to prevent fluid accumulation, repeated infections, and hearing loss.
   (d) Pseudomonad infections are common in swimmers because these organisms are able to produce a soluble greenish-blue pigment.
   (e) Repeated ear infections in children often decrease markedly because as the child ages and the Eustachian tube changes shape and develops an angle that prevents most organisms from reaching the middle ear.

6. Why are colds difficult to treat and prevent?
   (a) Different cold viruses predominate during different seasons.
   (b) Because there are so many different types of rhino- and coronaviruses, each with different antigens, development and administration of vaccines are currently prohibitive.
   (c) Although some human interferons have been shown to block or limit rhinovirus infections, delivery to infection sites is difficult, and unwanted side effects can occur.
   (d) All of the above.
   (e) Both a and c.

7. Which of the following is NOT true about parainfluenza viruses?
   (a) Most infants by the age of 6 months have been exposed to and develop antibodies against all 4 parainfluenza viruses.
   (b) They cause rhinitis characterized by nasal, pharyngeal, and bronchial inflammation, primarily in children.
   (c) Parainfluenza virus infection can be prevented by vaccination.
   (d) They can cause noisy respiration (stridor) and acute obstruction of the larynx called croup.
   (e) The viruses are spread by direct contact or by large droplets.

8. Whooping cough is usually caused by _____ _____ and causes signs and symptoms of disease by its production of _____, _____, and _____.
   (a) *Mycobacterium tuberculosis*; pili, pigment, and exfoliants

(b) *Bordetella pertussis*; endotoxin, exotoxin, and hemagglutinins

(c) *Bordetella pertussis*; metachromatic granules, endospores, and fimbrae

(d) b and c

(e) None of the above

9. In whooping cough, patients experience violent coughing episodes during the _____ stage and can have bluing of the skin or _____ due to lack of oxygen from massive mucus and bacterial airway blockage.
   (a) Paroxysmal; cyanosis
   (b) Catarrhal; pigmentation
   (c) Convalescent; osmosis
   (d) Pneumotic; hydrolysis
   (e) Primary; emesis

10. The treatment of choice for whooping cough is:
    (a) Tetracycline
    (b) Penicillin
    (c) Antitoxin and erythromycin
    (d) Chloramphenicol
    (e) Live, attenuated *Bordetella pertussis*

11. Penicillins have no effect on *Mycoplasma pneumoniae* because:
    (a) Mycoplasmas are viruses
    (b) Mycoplasmas possess beta lactamases
    (c) Mycoplasmas are too small
    (d) Mycoplasmas are eukaryotes
    (e) Mycoplasmas lack cell walls

12. *Legionella pneumophila*, the causative agent of Legionnaires' disease, is transmitted by aerosol inhalation. Which of the following fomites have NOT been implicated in the spread of this disease?
    (a) Clothing
    (b) Air conditioners
    (c) Ornamental fountains
    (d) Humidifiers
    (e) Patient vaporizers

13. How many cases of tuberculosis are reported globally each year?
    (a) 10
    (b) 10,000
    (c) 100,000
    (d) 3 million
    (e) 10 million

14. Mycobacteria are difficult to Gram stain, and are termed "acid-fast" due to their:
    (a) Ability to survive in acidic conditions
    (b) Resistance to drying
    (c) Thick, waxy cell walls
    (d) Resistance to sunlight
    (e) Lack of a peptidoglycan layer

15. Which of the following is true about pathogenic mycobacteria?
    (a) They have a long generation time of 12 to 18 hours.
    (b) They are obligate anaerobes sensitive to small amounts of oxygen.
    (c) They are highly resistant to drying and can remain viable in dried sputum 6 to 8 months later.
    (d) Both a and b.
    (e) Both a and c.

16. Clinical symptoms of tuberculosis are primarily due to:
    (a) Host inflammatory response
    (b) Mycotoxins
    (c) Endotoxin
    (d) Mucus production
    (e) Exotoxin

17. Birds serving as reservoir hosts for *Chlamydophila psittaci* is the only way humans can contract ornithosis. True or false?

18. Which of the following is NOT true of *Mycobacterium tuberculosis*?
    (a) Can remain viable in dried sputum for 6 to 8 months
    (b) It is highly resistant to direct sunlight
    (c) Can grow inside macrophages
    (d) Can cause disease many years after initial infection
    (e) Transmission is by inhalation of respiratory secretions or dried sputum

19. What characteristic of *Coxiella burnetii*, the causative agent of Q fever, allows it to survive for up to 2 years?
    (a) Metachromatic granule
    (b) Exotoxin
    (c) Endospore
    (d) Hemagglutinin
    (e) Cilia

20. Diphtheroids are the strains of *Corynebacterium diphtheriae* that cause diphtheria. True or false?

21. Some strains of *Legionella pneumophila* live as endosymbionts of many amoebas in moist environments. True or false?

22. Antigenic shifts in influenza viruses are represented by dramatic changes in the viral antigens. It is likely they arise from rare events in which:
    (a) Two different influenza viruses infect a cell at the same time
    (b) The lysogenic conversion of two separate viruses
    (c) Conjugation of two viruses
    (d) Accumulation of mutations
    (e) All of the above

23. Hantavirus pulmonary syndrome is:
    (a) Associated with inhalation of dried feces and urine of carrier rodents
    (b) A syndrome that has only been reported in Africa and Latin America
    (c) Transmitted by the bite of a rodent carrying the virus
    (d) Attributed to rats but not mice
    (e) Preventable by vaccination

24. Match the following microorganisms to their perspective description:

    ___ *Corynebacterium diphtheriae*

    ___ *Streptococcus pneumoniae*

    ___ *Mycoplasma pneumoniae*

    ___ *Histoplasma capsulatum*

    ___ *Cryptococcus neoformans*

    (a) Infects the pharynx and produces a systemic toxin
    (b) Causes the majority of cases of lobar pneumonia
    (c) Causes primary atypical pneumonia
    (d) Fungus present in chicken-impacted soil and bat guano
    (e) Yeast that causes a mild respiratory infection that can spread to meninges

25. Match the following organisms to their staining reaction and shape:

___ *Streptococcus pyogenes*   (a) Gram-negative rod
___ *Haemophilus influenzae*   (b) Gram-positive rod
___ *Mycoplasma*   (c) Gram-negative
     pneumoniae   coccobacillus
___ *Legionella pneumophilia*   (d) Gram-negative coccus
___ *Corynebacterium*   (e) Gram-positive coccus
     diphtheriae   (f) Acid-fast rod
___ *Bordetella pertussis*   (g) No staining reaction
___ *Mycobacterium*
     tuberculosis

26. Using the accompanying diagram, name six diseases each of the upper and lower respiratory systems. Indicate the specific part of the respiratory system each is likely to affect, identify the infectious agent, and describe each as bacterial, viral, fungal, or parasitic.

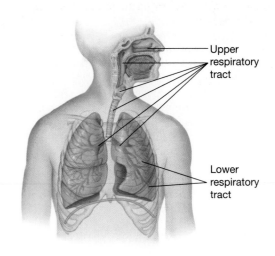

Upper respiratory tract

Lower respiratory tract

▮ EXPLORATIONS ON THE WEB   http://www.wiley.com/college/black

If you think you've mastered this chapter, there's more to challenge you on the web. Go to the website to fine-tune your understanding of the chapter concepts and discover answers to the questions posed below.

1. Read about the story of the worst epidemic the United States has ever known, which killed more Americans than all the combat deaths of this century combined, at the companion website.

2. Don't stand too close to that air conditioner, and stay out of the whirlpool spa, for either could be contaminated with the *Legionella* bacteria. Learn more at the companion website.

22 Oral and Gastrointestinal Diseases

Come with me . . .

Oh dear! Is this seasickness? My stomach is heaving right along with the waves! I saved so long for this trip, and this is the only vacation time I have off during the year—please, please, please don't let me be sick. All the hoping and praying didn't save 100s of cruise passengers from catching norovirus (formerly called "Norwalk-like-viruses") in recent years. This nasty virus causes acute onset vomiting, diarrhea with abdominal cramps, and nausea, sometimes with fever. Symptoms usually last 24 to 60 hours. No specific therapy exists, just rehydration and mainly over-the-counter anti-diarrhea medication. Antibiotics are useless against a virus. Recovery is usually complete without lasting effects. However, no immunity develops after an attack, and you can catch it again and again.

Monika Graff/Getty Images News and Sport Services

So how do cruise passengers catch this disease? It is often spread through food, ice, water, by person-to-person contact, by inhaled aerosoled droplets of vomitus, and contaminated surfaces. It is so contagious that as few as 10 viral particles may be enough to cause an infection. It is estimated that 23 million cases occur every year in the United States, and most are on land. But when you are one of 700 sick passengers on a cruise ship, that's little consolation.

Come with me to see how these ships are disinfected, and for hints on how to hopefully avoid getting sick on one.

 Video related to this topic is available within WileyPLUS.

We all have a wealth of firsthand experience with diseases of the oral cavity and gastrointestinal tract. We've had plaque removed from our teeth and cavities filled. We've had episodes of nausea, vomiting, and diarrhea—usually of short duration—after we have eaten contaminated food or have drunk impure water. Fortunately, for most of us these problems have been minor inconveniences rather than serious illnesses. In other parts of the world, however, especially where water and sewage treatment are inadequate, gastrointestinal infections are major problems. In this chapter we will consider diseases of the oral cavity and intestinal tract, both minor and serious.

Worldwide, foodborne diseases cause an estimated 1.5 billion cases per year of diarrhea in children under 5, with over 3 million of these resulting in death.

Like the preceding chapters dealing with diseases of body systems, this chapter presumes a knowledge of disease processes, normal and opportunistic microflora (◀Chapter 14), epidemiology (◀Chapter 15), host systems and host defenses (◀Chapter 16), and immunity (◀Chapters 17 and 18). It also assumes familiarity with the characteristics of bacteria, viruses, and eukaryotic microorganisms and helminths (◀Chapters 9 through 11).

▌▌▌ COMPONENTS OF THE DIGESTIVE SYSTEM

The **digestive system** consists of an elongated tube, or tract, that extends from the mouth to the rectum and includes accessory organs that help in digestion of food materials **(Figure 22.1)**. The digestive tract consists of various organs—mouth, pharynx, esophagus, stomach, and intestines—whereas the accessory organs include the teeth, salivary glands, liver, gallbladder, and pancreas. The digestive system has five major functions:

1. *Movement* of food materials through the tract.
2. *Secretion* of digestive juices and mucus to break down food materials.
3. *Digestion*, the process of breaking large food molecules into small molecules that can be passed from the digestive tract into the bloodstream.
4. *Absorption* of digested food substances into the blood or lymph.
5. *Elimination* of indigestible food materials and intestinal microflora.

Although food contains many microorganisms, most are killed by various defense mechanisms in the digestive tract. Throughout the digestive tract, **mucin** (myu'sin), a glycoprotein in mucus, coats bacteria and prevents their attaching to surfaces.

THE MOUTH

The mouth, or *oral cavity*, is lined with a mucous membrane and contains the tongue, teeth, and salivary

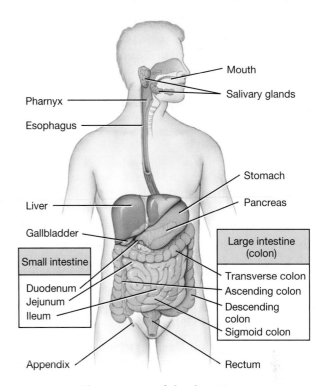

Figure 22.1 The structure of the digestive system.
Normal microflora of the digestive system prevent the easy growth of pathogens, and stomach acids and enzymes destroy most disease agents. Vomiting and diarrhea are two expulsion mechanisms used to rid the system of toxic materials, including bacterial toxins.

glands. The mouth is a portal of entry for microorganisms; a normal mouth contains more resident microbes than there are people on Earth.

Each tooth has a *crown* covered with **enamel** above the gum and a *root* covered with **cementum** below the gum **(Figure 22.2)**. Under these coverings is a porous substance called *dentin*, a central *pulp cavity*, and the *root canals*, where blood vessels and nerves are located. Each tooth is held in a tooth socket by fibers running from the cementum to the bone of the socket. Although enamel is the hardest substance in the body, it can be attacked by

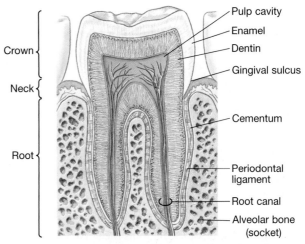

Figure 22.2 A section through a typical tooth and gum.

microbial-produced acids and enzymes. Microbes also can infect the gums, form pockets of infection between teeth and gums, and spread to the underlying bone.

The salivary glands secrete saliva, containing both antibodies that can coat bacteria and lysozyme that kills some bacteria. The salivary glands themselves, however, are subject to infection.

THE STOMACH

After food is chewed and mixed with saliva in the mouth, it passes through the pharynx and esophagus to the stomach. There the food mixes with hydrochloric acid and the enzyme pepsin, which together begin the digestion of proteins. The lining of the stomach is protected from the acid by viscous mucus. Only alcohol, aspirin, and some lipid-soluble drugs can cross the mucous barrier and be absorbed in the stomach.

THE SMALL INTESTINE

Partially digested food leaving the stomach enters the **small intestine** and is mixed immediately with secretions from the liver and pancreas. Liver cells secrete bile—a mixture of bile salts, cholesterol, and other lipids—that aids in the digestion of fats.

Materials digested in the small intestine enter the bloodstream and are carried directly to the liver via the portal vein. There is a variety of toxins in the foods we eat; the liver has the job of detoxifying these materials. In tissues such as the liver, capillaries are enlarged to form networks of vessels called **sinusoids**. These sinusoids are lined with phagocytic *Kupffer cells*, which remove any dead blood cells, bacteria, and toxins from the blood as it passes through the sinusoids. The pancreas releases hormones into the blood and digestive juices into ducts that empty into the small intestine. These digestive juices include enzymes that digest starch, proteins, lipids, and nucleic acids, along with bicarbonate that neutralizes the acidic materials from the stomach.

The small intestine has an extensive internal surface area (about one-tenth the area of a football field) due to folds in the wall—fingerlike projections called **villi** (singular: *villus*)—and folds called **microvilli** in membranes of mucosal cells. Villi contain blood and lymph vessels. Digestion is completed in the small intestine; simple sugars and amino acids are absorbed into blood vessels, and fats into lymph vessels. Enzymes in the small intestine kill most microorganisms and inactivate most viruses in food. Also, the lower end of the intestine, called the *ileum*, contains lymphoid tissue clustered into *Peyer's patches*. These patches contain lymphocytes and macrophages that protect the small intestine from microbial invasion from the large intestine.

THE LARGE INTESTINE

The **large intestine**, or *colon*, joins the small intestine near the appendix and ends at the rectum. What digestion occurs in the large intestine is done by the bacteria of the normal microflora. By-products of bacterial metabolism, such as amino acids, thiamine, riboflavin, and vitamins K and B_{12}, are absorbed but in quantities too small to satisfy nutritional needs. Much water is absorbed, and undigested food is converted to feces. **Feces**, which consist of about three-fourths water and one-fourth solids, are stored in the rectum until they are eliminated from the body.

The large intestine also contains many resident microbes that ultimately come from the mouth or from the anus, another portal of entry. However, the normal microflora of the large intestine compete with opportunists and pathogens and prevent their growth and reproduction under normal conditions. Gastrointestinal infections generally occur only when defense mechanisms have been overwhelmed by large numbers of a pathogenic organism. Even when an infection occurs, vomiting and diarrhea help rid the body of the pathogens and their toxins.

NORMAL MICROFLORA OF THE MOUTH AND DIGESTIVE SYSTEM

Unless a microbe entering the mouth can quickly attach to a surface, it will be washed down into the digestive tract by the flow of saliva. Patterns of salivary flow vary from person to person; some are highly efficient at rinsing food and microbes away, while others are inefficient. The development of tooth decay is greatly affected by the degree of efficiency.

The ecology of the oral microflora is poorly understood. Although over 400 species of microbes have been identified as living there, probably even more than that number of species have not yet been identified. Research has focused on those organisms associated with tooth decay and gum disease, but there are many other microbes living on the cells lining the cheek, on the tongue, palate, and floor of the mouth. These may contribute to bad breath.

The human esophagus does not appear to have a permanent, normal microflora. The stomach's acid pH usually prevents colonization by microbes. It and the first two-thirds of the small intestine contain very few microbes, mainly lactobacilli and streptococci that are merely passing through. In the last third of the small intestine, motility of contents is slower, and some microbes are able to colonize its surfaces. These are mainly Gram-negative, facultatively anaerobic bacteria, especially members of the family Enterobacteriaceae (e.g., *Escherichia coli*), plus obligate anaerobes (e.g., *Bacteroides* and *Clostridium*). In the large intestine, food may remain as long as 60 hours, allowing ample time for microbial colonization and replication. Feces are composed about 50% by weight and volume of bacteria. Most of these are species of *Bacteroides*. The microbes of the lower digestive tract utilize foodstuffs that passed through the upper digestive tract without being completely digested. One important by-product of their activities is the production of vitamin K, necessary for the proper clotting of blood.

DISEASES OF THE ORAL CAVITY

Although we would like to believe otherwise, our mouths are microbial breeding grounds. About 12 billion organisms live in a healthy human mouth. We pick up microbes in many ways—from food, kissing, and dirty fingers, to name a few—and the microbes thrive.

BACTERIAL DISEASES OF THE ORAL CAVITY

Dental Plaque

Dental plaque is a continuously formed coating of microorganisms and organic matter on tooth surfaces. Plaque formation, although not a disease itself, is the first step in tooth decay and gum disease. Scrupulous and frequent cleaning of teeth minimizes but does not entirely prevent plaque formation. Plaque begins to form within 24 hours after cleaning. Unless it is removed regularly, plaque can become so securely attached to teeth that it can no longer be removed by home methods. Professional cleaning removes plaque, but it begins to form again even before you get home from the dentist's office.

Plaque formation begins as positively charged proteins in saliva adhere to negatively charged enamel surfaces and form a *pellicle* (film) over the tooth surface. Cocci such as *Streptococcus mutans* and some filamentous bacteria among the normal oral flora microbiota attach to the newly formed pellicle. These organisms may hydrolyze sucrose (table sugar) to glucose and fructose, both of which can be metabolized for energy production and growth. Plaque-forming bacteria such as *S. mutans* and *S. sanguis* can also convert sucrose to polymers of glucose units (such as the polysaccharide dextran) that serve as bridges, holding together cells in plaque. Plaque consists of up to 30 different genera of bacteria and their products, such as dextran, saliva proteins, and minerals **(Figure 22.3)**. If plaque is not removed thoroughly and

regularly, streptococci, lactobacilli, and other acid-producing bacteria may accumulate within it in layers 300 to 500 cells thick. In some individuals, the total number of bacteria in their plaque alone may approach 10 billion! These organisms metabolize fructose and other sugars that diffuse into the plaque and set the stage for tooth decay. Plaque that accumulates near the gumline also offers protection to bacteria in the gingival (gum) crevices between the teeth and the gums. These bacteria include species of *Actinomyces*, *Veillonella*, *Fusobacterium*, and sometimes spirochetes, as well as the streptococci just noted. If plaque is allowed to accumulate, some crevices change into anaerobic pockets full of bacteria that can irritate the gums or destroy the bone in which the teeth are set. In some people, this gumline plaque mineralizes into *calculus (tartar)* that can also irritate the gum and contribute to inflammation and bleeding.

Dental Caries

The Disease. **Dental caries**, or tooth decay, is the chemical dissolution of enamel and deeper parts of teeth. Taking its name from the Latin *cariosus*, which means "rotten," caries is the most common infectious disease in developed countries where the diet contains relatively large amounts of refined sugar. Unchecked, the decay can proceed through enamel, into dentin, into the pulp cavity, and eventually can cause an abscess in the bone that supports the tooth. Sugars easily diffuse through plaque to bacteria embedded in it, but acids produced by bacterial fermentation fail to diffuse out. The acids gradually dissolve enamel, after which protein-digesting enzymes break down any remaining material.

In the 1960s, it was shown that tooth decay is a transmissible disease. Rats without caries developed them when caged with rats having caries. Think twice about who you kiss!

(a)

(b)

Figure 22.3 Dental plaque. (a) A variety of organisms that accumulate in plaque deposits (42,689X). *(Fred E. Hossler/Visuals Unlimited)* **(b)** Production of sticky dextran from sucrose enables bacteria to adhere to tooth surfaces, where the lactic acid produced in this process eats away tooth enamel and forms cavities. *(Courtesy Dr. Ross P. Karlin)*

APPLICATIONS

It's a Zoo in There

It's wet, wild, and full of interesting sights and sounds. Exotic animals, isolated from their native surroundings and diet, have dental problems they wouldn't have in the wild. According to Dr. Edward Shagam, orthodontist to humans and zoo dentist, the carnivores have most of the dental problems.

Instead of gnawing on coarse meat and bones that help keep their mouth healthy, animals are given mashed-up meat to limit complaints from offended zoo patrons. Just like their human counterparts, who are not able to floss after every meal, these animals suffer from gum disease and cavities. Other dental problems involve root canal therapies, due to broken and worn teeth, resulting from fighting and chewing on zoo enclosures.

The combination of sucrose and the action of *S. mutans* on it accounts for much tooth decay. Consequently, the more sucrose you eat and the more frequently you eat it, the greater your risk of dental caries. Saliva helps rinse sugars from the mouth, but its rinsing efficiency varies with flow rate and mouth shape. Sugar accumulates in poorly rinsed areas. Although starchy foods are only partially digested in the mouth, those that are sticky can adhere to tooth surfaces and remain on the teeth long enough for bacterial action to contribute to tooth decay. Sugar alcohols, such as sorbitol and xylitol used in "sugar-free" chewing gums, do not contribute to tooth decay because many bacteria cannot metabolize them. Bacteria that can utilize sorbitol, such as *S. mutans*, metabolize it slowly. They release acid at a rate that allows buffers in saliva to neutralize it, preventing a drop in pH in areas of plaque.

Treatment and Prevention. Dental caries are treated by removing decay and filling the cavity with *resins* (plastic materials) or with *amalgam* (a mixture of silver and other metals). Dental caries can be prevented or their incidence greatly reduced by limiting the intake of sugary and sticky foods, brushing regularly with plaque-removing toothpaste, and flossing between teeth. Vaccines to prevent dental caries are being developed against strains of *S. mutans*, which are most prevalent at decay sites. An injectable vaccine that elicits circulating IgG production has been successful in monkeys. An oral vaccine that causes secretory IgA production has been successful in rats. Neither vaccine has yet been adequately tested in humans. Stress may be a factor in the effectiveness of a vaccine because it appears to depress the immune system. This was noted when dental students were found to secrete less salivary IgA while taking examinations than while on summer vacation (◄Chapter 17, p. 497).

The use of **fluoride** has been the most significant factor in reducing tooth decay. It works by hardening the surface enamel of teeth. Fluoride reduces the solubility of tooth enamel by inhibiting demineralization and also enhances

remineralization. To understand how fluoride affects enamel, imagine the tooth surface as consisting of the ends of rods packed together like a fistful of pencils (**Figure 22.4**). No matter how tightly the rods are packed, channels exist between them. Acids produced by bacteria in plaque seep through channels and dissolve the enamel rods. As channels are enlarged, more acid enters and dissolves more enamel. Eventually, sufficient enamel is eroded to form cavities, in which dental caries can form. Fluoride fills the spaces between rods with a hard mineralized material that strengthens the tooth surface and prevents acid penetration. Fluoride may also inhibit certain enzymes that produce phosphates needed by bacteria to capture energy from nutrients. Without the phosphates, the bacteria die.

Numerous studies have shown fluoride to be safe and effective in the prevention of tooth decay. Added to city water supplies in a 1-part-per-million concentration, fluoride reduces tooth decay in children by as much as 60%. Because caries occur mainly in childhood and adolescence, a successful program to prevent childhood tooth decay is

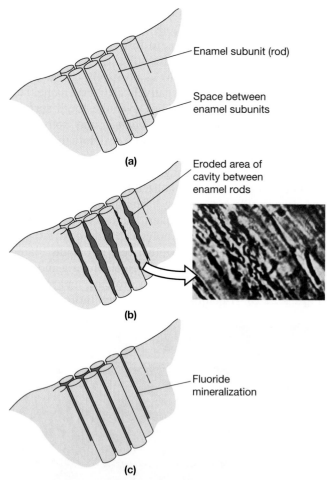

Figure 22.4 The effect of fluoride on tooth decay.
(a) Normal tooth structure; **(b)** dental caries forming as a result of bacterially produced acid seeping down between rods of enamel; *(From S. Rosen, Lenney, and O'Malley, "Dental caries in gnotobiotic rats," Journal of Dental Research, vol. 47, no. 3, p. 362 May–June 1968)* **(c)** mineralization by application of fluoride, which fills in the spaces between enamel rods, thereby preventing acid from seeping in.

Fluoride
treatment gel

Figure 22.5 **Fluoride treatment of teeth.** Visits to the dentist can be minimized by proper oral hygiene and the use of fluoride-containing toothpaste or oral rinse. Fluoride can also be applied in the dentist's office. *(Courtesy Dr. Ross P. Karlin)*

(b)

(b)

Figure 22.6 **Advanced periodontal disease.** **(a)** Severe gum inflammation, which can lead to *(Dr. R. Gottsegen/Peter Arnold, Inc.)* **(b)** loss of bone (see arrows) surrounding roots of teeth, causing loosening and eventual loss of teeth. *(Science VU/Max Listgarten/ Visuals Unlimited)*

of major importance. Yet nearly half the U.S. population obtains water from wells or from nonfluoridated community supplies. In some communities that lack fluoridated water, children receive fluoride tablets or gels at school. Ingestion of fluoride by children during tooth formation strengthens the entire tooth, whereas topical application affects only the surface of the tooth.

Although fluoride is most beneficial before age 20, it provides some benefits at any age, even when water is fluoridated and fluoride toothpastes and mouthwashes are used. Fluoride gel treatments should be given every 6 months from about age 4 to adulthood. Each time teeth are cleaned, a little surface is removed, and remineralization with topical fluoride helps restore the surface **(Figure 22.5)**.

Another means of preventing dental caries is sealing of the teeth with a type of resin that mechanically bonds to the tooth structure. Because the grinding surfaces of teeth have tiny pits and crevices that fill with plaque that cannot be removed with a toothbrush, these surfaces are more susceptible to decay than are smooth sides. Applying sealant to teeth provides nearly complete protection against decay as long as the sealant remains, which may be 10 years or longer. Applied at about age 6 or 7 after secondary grinding teeth have erupted, sealants will last through most of the decay-prone years. In the sealing process, teeth are first thoroughly cleaned and etched with acid; then the sealant is applied. Tiny cavities cease to grow in sealed teeth and eventually become sterile as the decay organisms die. Sealant can be applied only to teeth without fillings, but it can be used on adults as well as children.

Periodontal Disease

The Disease. When bacteria become trapped in gingival crevices, they cause both tooth decay and gum inflammation. **Periodontal disease** is a combination of gum inflammation and erosion of periodontal ligaments and the alveolar bone that supports teeth **(Figure 22.6)**. The disease is a chronic, generally painless infection that affects over 80% of teenagers and adults and is the major cause of tooth loss.

Plaque formation is the initiating event in periodontal disease. Organisms in gingival crevices are believed to produce endotoxins and acids that in turn produce an inflammatory response. This response breaks down the epithelial cells of the gums, and new groups of organisms replace previous crevice inhabitants as the disease progresses. If the process is not arrested, the gums recede and can become necrotic, and the teeth loosen as surrounding bone and ligaments are eroded and weakened.

About 33% of the U.S. population over the age of 65 have no teeth left.

In its mildest form, periodontal disease is called **gingivitis** (jin-jiv-i′tis), which affects only the gums. In its most severe form, gingivitis is called **acute necrotizing ulcerative gingivitis (ANUG)**, or trench mouth. The disease got its name because it was common among soldiers in World War I under stress "in the trenches." It is now common in young people, and stress seems to be an important factor in its development. ANUG responds to antibiotics, but more advanced forms of periodontal disease usually do not. Unchecked, periodontal disease can lead to chronic **periodontitis**, which affects the bone and tissue that support the teeth, as well as the gums. The loss of bone due to periodontitis is not reversible.

Periodontal disease occurs when plaque is allowed to accumulate, resulting in overgrowth of potentially virulent bacteria such as *Porphyromonas gingivalis*, *Actinobacillus actinomycetemcomitans*, *Prevotella intermedia*, *Bacteroides forsythus*, *Fusobacterium nucleatum*, and other Gram-negative rods. Many of the organisms living in the mouth, especially, spirochetes, have yet to be isolated, cultured, and identified. Certain other bacteria, such as *Streptococcus mitis*, *S. sanguis*, *Veillonella* species, and some *Actinomyces* species, may help control virulent bacterial populations. If plaque is not removed by brushing or flossing, calcium is deposited on plaque surfaces, forming a very rough, hard crust called **tartar**, or *calculus*. Tartar strongly binds to teeth surfaces. As more plaque accumulates, tartar forms a thick layer. The calculus contributes to the conditions that lead to bleeding of gums. As the condition worsens, pockets of inflammation form in the gum crevices. Measuring the depth of these pockets serves as an indication of the extent of periodontal disease.

A 1982 study of organisms associated with periodontal disease, which was done by Paul H. Keyes, illustrates a method of studying a complex microbial environment. Keyes obtained plaque from individuals with healthy gums, mildly inflamed gums, and severely diseased gums. When he examined them by phase-contrast microscopy, he found distinctly different populations of microorganisms. Plaque from healthy gums contains nonmotile filamentous bacteria, a few colonies of cocci, fewer than five leukocytes per microscopic field, and no amoebas or spirochetes. Plaque from mildly inflamed gums shows a greater variety of microbes. Nonmotile organisms form dense masses surrounded by clusters of motile spirochetes and rapidly spinning bacilli too numerous to count. Some spirochetes and leukocytes are present, but amoebas and trichomonads are absent. Plaque from severely diseased gums contains dense mats of nonmotile filaments, cocci, and numerous motile organisms. Extending from the dense mats are brushlike aggregations of spirochetes and flexible rods that move in rippling waves and migrate from one surface to another. Amoebas always are present, trichomonads sometimes are seen, and leukocytes are numerous.

More recent studies suggest that of 300 species of bacteria in the mouth, *Porphyromonas gingivalis* may be a specific cause of some cases of periodontal disease. Researchers at the University of Texas have succeeded in causing a burst of periodontal disease in monkeys given doses of the bacterium. The researchers have also had some success in treating the monkeys with rifampin.

Plaque organisms do not appear to penetrate gum tissue. Their effects are caused largely by the secretion of destructive enzymes, inflammation, and allergic reactions to bacterial products spreading from the plaque.

Treatment and Prevention. The treatment of chronic periodontal disease is a somewhat controversial subject in dentistry today, and patients vary in their responses to different treatments. Treatments include antimicrobial mouth rinses, brushing with a mixture of bicarbonate of soda and hydrogen peroxide, surgery to eliminate pockets, and antibiotic therapy in unresponsive or rapidly progressing cases. Chronic periodontal disease can be prevented or its onset delayed by daily thorough cleaning of teeth and, most important, frequent professional removal of plaque from the pockets. Once organisms erode gums and form pockets between the teeth and gums, infections must be kept under control.

VIRAL DISEASES OF THE ORAL CAVITY

Mumps

Mumps is caused by a paramyxovirus somewhat similar to the measles (rubeola) virus. The virus is transmitted by saliva or as aerosol droplets, enters via the oral cavity or via the respiratory tract, and invades cells of the oropharynx.

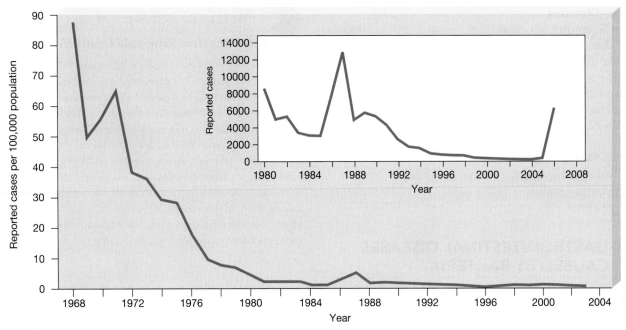

Figure 22.7 **The U.S. incidence of mumps.** Since a vaccine became available in 1967, the number of cases has sharply declined. (*Source:* CDC.)

After initially replicating in the upper respiratory tract, the virus travels in the blood to the salivary glands and sometimes to other glands and organs, such as the testes and the meninges. Swelling of the parotid glands appears 14 to 21 days after initial infection and can persist as long as 7 days. Viruses are released and the infected person is contagious for 7 days before the glands swell and for up to 9 days after swelling subsides. Mumps viruses are excreted in urine for up to 2 weeks after the onset of symptoms.

Humans are the only known hosts for the mumps virus. The virus is found worldwide, occurring especially in the spring. Infections are most common in children 6 to 10 years of age. Although up to 85% of an exposed susceptible population will be infected by mumps, 20 to 40% will have no symptoms. These subclinical cases, as well as those involving only one side are still able to produce lasting immunity. When the disease appears in postpubertal males, 20 to 30% develop **orchitis**, inflammation of the testes. Such infections are capable of causing sterility but

rarely do so. Other complications of mumps, regardless of the infected person's age or sex, include meningoencephalitis, eye and ear infections, and inflammation of glands other than the parotid, such as the ovaries and pancreas (where it may be a cause of juvenile onset diabetes). An effective vaccine for mumps contains viruses weakened by passage through a sequence of embryonated egg inoculations. The vaccine usually is administered in combination with measles and rubella vaccines—collectively known as the MMR vaccine. The incidence of mumps has generally decreased dramatically since the vaccine was licensed and put to use in 1967 **(Figure 22.7)**. The vaccine is recommended for all persons who were born after 1957 and have no immunity to the disease. Infants are immunized after 15 months of age.

In 2006, there was an outbreak of 6,339 cases of mumps in the United States. It was due to the same strain that caused 70,000 cases in the U.K., during 2004 to 2006, mostly in unvaccinated people aged 18 to 25.

TABLE 22.1

A Summary of Oral Diseases		
Disease	**Agent(s)**	**Characteristics**
Bacterial Diseases of the Oral Cavity		
Dental caries	*Streptococcus mutans* and other species	Erosion of tooth enamel and other structures by acids from microbial metabolism
Periodontal disease	Various bacteria, *Porphyromonas gingivalis*	Inflammation and destruction of gums, loosening of teeth, erosion of bone
Viral Disease of the Oral Cavity		
Mumps	Paramyxovirus	Inflammation and swelling of salivary glands and sometimes testes, epididymis, and other tissues

Other Diseases

Thrush, an oral infection by the yeast *Candida albicans*, was discussed in ◄Chapter 19 and illustrated in Figure 19.18a (p. 590). Herpes simplex virus infections, a cause of fever blisters and cold sores on the lips and in the mouth, are discussed in ◄Chapter 20.

Diseases of the oral cavity are summarized in Table 22.1.

✓CHECKLIST

1. How does fluoride help reduce dental decay?
2. Distinguish between plaque and tartar.

GASTROINTESTINAL DISEASES CAUSED BY BACTERIA

BACTERIAL FOOD POISONING

Food poisoning is caused by ingesting food contaminated with preformed toxins. It also can be caused by the ingestion of foods contaminated with pesticides, heavy metals, or other toxic substances. In food poisoning caused by microbial toxins, organisms that can continue to produce toxin may also be ingested with the toxins. However, tissue damage is due to action of the toxin, so most cases of microbial food poisoning are intoxications rather than infections. Because the toxin is preformed, the onset of symptoms in intoxication is more rapid than in infection. Food poisoning can be prevented by following proper food-handling procedures, described in ◄Chapter 1.

Bacteria that produce toxins responsible for food poisoning include *Campylobacter jejune*, *Staphylococcus aureus*, *Clostridium perfringens*, *C. botulinum*, and *Bacillus cereus*. Staphylococci usually enter food by way of infected food handlers. Other organisms that cause food poisoning are ubiquitous soil organisms found in water, feces, sewage, and nearly all foods. We will consider the nature of the food poisoning intoxication caused by each of these organisms.

Foodborne microbes trigger some 81 million cases of disease in the U.S. each year, according to the Government Accounting Office.

Staphylococcal Enterotoxicosis

Certain strains of *Staphylococcus aureus* cause food poisoning, or **enterotoxicosis** (en″ter-o-tox-i-ko′sis), by releasing certain **enterotoxins**—enterotoxins A or D—which are exotoxins that inflame the intestinal lining and inhibit water adsorption from the intestine. These enterotoxins also cause neural stimulation of the vomiting center of the brain. Because the organisms are relatively resistant to heat and drying, foods easily become contaminated with them from food handlers or from the environment. The organisms multiply and release toxin in uncooked or inadequately cooked foods, especially if the foods are unrefrigerated. Nearly any food can be

contaminated with *S. aureus*, but those with a starch or cream base are most likely candidates. Cream pies, dairy products, poultry products, and picnic foods such as potato salad are common culprits. Contamination is difficult to detect because it produces no change in the food's appearance, taste, or odor. Unlike most exotoxins, the toxin is heat-stable and withstands boiling for 30 minutes. Hence cooking foods can kill the organisms but does not destroy the toxin (◄Chapter 14, p. 413).

When food contaminated with *S. aureus* enterotoxin enters the intestine, the toxin acts directly. The organisms usually continue to produce toxin (but do not multiply). When it comes in contact with the mucosa, the toxin causes tissue damage only after it has entered the blood and has circulated back to the intestine. Symptoms such as abdominal pain, nausea, vomiting, and **diarrhea** (excessive frequency and looseness of bowel movements), but usually not fever, appear 1 to 6 hours after ingestion of contaminated food. The time required for symptoms to appear depends on how long it takes for absorption of a sufficient quantity of toxin to produce them. A food loaded with toxin elicits symptoms quickly; one with only a little toxin takes longer. Once elicited, symptoms usually last about 8 hours. For otherwise healthy adults, no treatment is required because the disease is self-limiting. It can be severe in infants, the elderly, and debilitated patients. Recovery from food poisoning does not confer immunity, as sufficient antibody is not produced. The best way of preventing food poisoning by *S. aureus* is to use sanitary food-handling procedures.

Other Kinds of Food Poisoning

An enterotoxin from *Clostridium perfringens* also causes food poisoning (◄Chapter 14, p. 413). The enterotoxin, which is released only during sporulation, is produced under anaerobic conditions, as when undercooked meats and gravies are kept warm for a while. The main symptom is diarrhea. Compared with *S. aureus* food poisoning, *C. perfringens* food poisoning takes longer to appear—8 to 24 hours after ingestion—and lasts longer (about 24 hours). It, too, is self-limiting and can be prevented by sanitary food handling.

Although *C. perfringens* enterotoxin causes food poisoning, the organism itself can infect tissues. Upon introduction of spores into a wound, germination to vegetative cells can result in gas gangrene (◄Chapter 19, p. 596) and anaerobic cellulitis (inflammation of connective tissue). Vegetative cells multiply and release toxins and enzymes that diffuse into surrounding healthy tissue, causing cellular destruction and death.

Botulism, which is caused by a neurotoxin produced by *Clostridium botulinum*, is acquired when toxin-contaminated food is eaten. Although it is a kind of food poisoning, it has little effect on the digestive system. Its effects on the nervous system are discussed in ◄Chapter 2.

Bacillus cereus secretes a toxin that acts as an emetic—that is, it induces vomiting. The incubation period and signs and symptoms resemble those of staphylococcal food poisoning. Symptoms occur less than 12 hours after ingestion and last only a short time. Such toxins are often found in contaminated rice or meat dishes. *B. cereus* exists as a saprophyte in water and soil.

Food poisoning by *Pseudomonas cocovenenans*, which occurs in Polynesia, is named **bongkrek disease**, after its association with the native coconut delicacy, bongkrek. The bacterium produces a potent, and often fatal, toxin that is frequently found in dishes prepared with coconut.

BACTERIAL ENTERITIS AND ENTERIC FEVERS

For most of us diarrhea is an unpleasant inconvenience, but it can be deadly. In 1900 in New York City the death rate among infants from diseases grouped together as diarrhea was 5,603 per 100,000. Although today the death rate is less than 60 per 100,000, the disease is still life-threatening, and death from diarrhea can occur in hours in infants.

Enteritis is an inflammation of the intestine. **Bacterial enteritis** is an intestinal infection, not an intoxication, as is food poisoning. The causative bacteria actually invade and damage the intestinal mucosa or deeper tissues. Enteritis that affects chiefly the small intestine usually causes diarrhea. When the large intestine is affected, the result is often called **dysentery**, a severe diarrhea that often contains large quantities of mucus and sometimes blood or even pus. Some pathogens spread through the body from the intestinal mucosa and cause systemic infections, as is the case with typhoid fever. Such infections are called **enteric fevers**.

Salmonellosis

Salmonellosis (sal″mo-nel-o′sis) is a common enteritis caused by some members of the genus *Salmonella*. The annual reported incidence of salmonellosis (excluding typhoid fever) in the United States climbed from 20,000 cases in the early 1970s to more than 65,000 cases in 1986

but has dropped to about 41,924 cases in 2006. Many more cases go unreported. Some health experts estimate the true prevalence to exceed 2 million. The classification of the salmonellas has recently been revised. The number of species within the genus *Salmonella* has been reduced to three: *Salmonella typhi*, *S. choleraesuis* (usually a swine pathogen), and *S. enteritidis*. About 2,000 strains of salmonellas have been identified by their surface antigens and grouped into **serovars** (se′ro-varz). Prior to 1972 each was named as a separate species. Most of the 2,000 strains have been consolidated into the species *S. enteritidis*. The former *S. typhimurium* is now considered to be a serovar (strain) of *S. enteritidis*. However, most investigators still refer to *S. typhimurium* rather than more properly to *S. enteritidis* serovar *typhimurium*. Identification of serovars is sometimes useful in tracing the source of disease outbreaks.

Salmonella species other than *S. typhi*, the cause of typhoid fever, can be found in the gastrointestinal tracts of many animals, including poultry, wild birds, and rodents. In these hosts the bacteria either cause obvious disease or are carried without producing harmful effects. The sale of live Easter chicks and baby turtles was banned in the U.S. after children playing with them became ill with salmonellosis. As many as 90% of pet reptiles carry *Salmonella*. It is very important never to clean the turtle's bowl in the kitchen sink. It is now known that chicken eggs can become infected if the hens laying them are infected. Salmonellas have also been traced to contaminated water and to food contaminated by carriers. The practice of confining poultry and pigs to small spaces while being reared for market makes feeding and caring for the animals more efficient, but it also facilitates the transmission of *Salmonella* and other pathogens among them.

Salmonella infection is generally associated with the ingestion of improperly prepared, previously contaminated food. Meat and dairy products are most likely candidates. Foods containing uncooked eggs can also be a source.

Signs and symptoms of salmonellosis include abdominal pain, fever, and diarrhea with blood and mucus. They

PUBLIC HEALTH

Homeless Bacteria Looking for a New Host

Would you believe that chicken farmers are spraying newly hatched chicks with bacteria? It's true. But the bacteria that they are spraying, known collectively as Preempt, consists of 29 types of living, nontoxic bacteria isolated from the gut of mature chickens. As the hatchlings preen, the chicks ingest the bacteria, seeding their intestines with beneficial microbes. These microbes cover niches in the intestines so quickly and heavily that intestinal pathogens find it all but impossible to move in. *Salmonella*, *Campylobacter*, *Listeria*, and the deadly *Escherichia coli* O157:H7 are going to have to find new animals to infect.

PUBLIC HEALTH

Mom Knows Best, but She Doesn't Always Know Everything

Mom probably said to wash fresh fruits and vegetables to rid them of germs. That's a great habit, but it takes more than simple washing to eliminate all of the bacteria. In May 1998, researcher Elizabeth Ehrenfeld deposited up to 100 million *Escherichia coli* or *Salmonella* on bacteria-free bean sprouts, then washed them three times in clean water. While this reduced the bacterial population, it still left up to 1 million microbes per gram of food—many thousands of times the number needed to sicken someone with a weakened immune system. What can you do to protect yourself? Researchers are looking into nontoxic mixes of vinegar and off-the-shelf hydrogen peroxide that will eliminate all *E. coli* organisms from foods. Others are considering other chlorine-free, antibacterial products that you dissolve in tap water and spray on foods to kill all germs. Some of the disinfectants are already being marketed.

appear 8 to 48 hours after ingestion of organisms and are associated with the organisms' invading the mucosa of both the small and large intestines. Fever probably is caused by endotoxins, toxins that are released from a cell only when it is lysed. In otherwise healthy adults, salmonellosis lasts 1 to 4 days and is self-limiting. Antibiotics usually are not given because they tend to induce carrier states and to contribute to the development of antibiotic-resistant strains. Infants and elderly or debilitated patients often have more severe and prolonged symptoms. In such cases, antibiotics may be prescribed.

Other serovars of *Salmonella* also cause disease. Because of their ability to invade intestinal tissue and enter the blood, *S. typhimurium* and *S. paratyphi* cause a somewhat more serious condition called **enterocolitis**, or enteric fever. Symptoms and bacteremia appear after an incubation period of 1 to 10 days. Enteric symptoms such as fever and chills can last 1 to 3 weeks. Chronic infections of the gallbladder and other tissues are not uncommon. A carrier state becomes established when, after the patient's recovery, organisms from chronically infected tissues continue to be excreted with feces. Broad-spectrum antibiotics rid carriers of organisms but can activate the disease in some carriers by upsetting the balance of intestinal microflora.

Prevention of salmonellosis and enterocolitis depends on the maintenance of sanitary water and food supplies and eradication of organisms from carriers. The organisms cannot be entirely eradicated because poultry and other animals serve as reservoirs, and no effective vaccine is available.

Typhoid Fever

Typhoid fever, one of the most serious of the epidemic enteric infections, is caused by *Salmonella typhi* (**Figure 22.8**). The disease is rare in places where good sanitation is practiced but is more common where faulty water

and sewage systems exist. Uncooked shellfish, raw fruit, or raw vegetables have often served as sources of the pathogen in outbreaks in the United States. In 1942 the total number of U.S. cases was 5,000. For most of the last 30 years, however, fewer than 500 cases per year have been reported in the United States. The organisms enter the body in food or water and invade the mucosa of the upper small intestine. From there they invade lymphoid tissues and are phagocytized and disseminated. The organisms multiply in the phagocytes, emerge, and continue to multiply intracellularly.

Bacteremia and septicemia occur at the same time as symptoms appear. During the first week, the patient suffers from headache, malaise, and fever, probably due to an endotoxin. During the second week, the patient's condition worsens. The organisms invade many tissues, including the intestinal mucosa, and are excreted in the stools. *Salmonella typhi* thrive and multiply in bile; organisms from the gallbladder reinfect the intestinal mucosa and lymphoid tissue such as Peyer's patches. Characteristic "rose spots" often appear on the trunk and abdomen for a few days. Abdominal distention and tenderness and enlargement of the spleen are common complaints, but diarrhea usually is absent. Unlike most other infections, in which leukocytes increase in number, leukocytes decrease in number in typhoid fever. Some patients become delirious and can suffer complications such as internal hemorrhage, perforation of the bowel, and pneumonia. Fluoroquinolones (e.g., Ciprofloxacin), a broad-spectrum cephalosporin, or chloramphenicol are the antibiotics of choice in treating typhoid fever, but some strains of *S. typhi* are resistant to chloramphenicol.

In the American Civil War, 81,360 Union Army soldiers died of dysentery and typhoid; 93,443 were killed in combat.

By the fourth week, symptoms subside, convalescence begins, and immunity develops. Cell-mediated

Figure 22.8 *Salmonella typhi*, the cause of typhoid fever. Notice the flagella, which have been made visible by the use of stain. They would not otherwise be visible when using a light microscope (5,000X). *(A. M. Siegelman/Visuals Unlimited)* [LM]

PUBLIC HEALTH

The Saga of Typhoid Mary

1901—Mary Mallon is working as a cook for a family. A visitor with typhoid fever arrives, and Mary develops typhoid fever. One month later the family laundress falls ill with typhoid.

1902—Mary changes employers. Two weeks later the laundress develops typhoid and is soon followed by six more members of the household. Mary quickly leaves.

1903–Now working as a cook in a household in Ithaca, New York, Mary is believed to have started a waterborne typhoid outbreak that spreads widely, killing 1,300 people. Mary quickly leaves.

1904—Mary moves on to a household on Long Island; within 3 weeks, four fellow servants develop typhoid.

1906—Moving again, Mary arrives at a new destination, and six people fall ill with typhoid, all within 1 week. One girl dies. Two weeks later, Mary—having been suspected as the cause—flees to employment with yet another family. After another 2 weeks, the family's laundress develops typhoid fever.

1907—Using a false name, Mary starts work as a cook for a family in New York City. Two months later, two household members develop typhoid; one dies. Mary leaves soon after the disease breaks out, but this time the New York City Health Department tracks her down. In March 1907 she refuses to co-operate with Health Department workers who want to test her stool for the presence of typhoid bacilli that would mark this otherwise healthy woman as a carrier of typhoid. Police arrive and forcibly remove her to a hospital on an island in the East River. Tests reveal typhoid bacilli in her stool specimens; presumably they are being shed from her bacilli-colonized gallbladder. No antibiotics are available at this time—nor will they be until the late 1940s. Surgical removal of Mary's gallbladder is the only way to end her carrier state. Not a believer in the germ theory of disease, Mary is certain that she will die under the surgeon's knife, all for foolish nonsense, and refuses to undergo the surgery. She is sentenced to involuntary isolation in the hospital until such time as she is no longer a menace to public health.

1907–1910—Mary's stools are cultured every few days. Some days there are no typhoid bacilli; other days huge numbers appear. Meanwhile, Mary has attained public celebrity status—imprisoned against her will, threatened by the surgeon's knife!

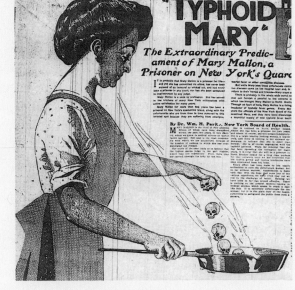

(Courtesy The New York Public Library)

1910—Public sympathy leads to Mary's release. She promises never again to work as a cook.

1915—Twenty-five cases of typhoid occur at the Sloane Hospital for Women in New York City. Eight die—mostly doctors and nurses. The hospital cook steps out "for a few minutes," never to return. She does not use the name Mary Mallon, but when the Health Department catches up with her, it is indeed Mary again, insisting on her right to be employed as a cook—the only job she likes.

It is back to the island hospital again for Mary. On arrival at the quarantine facilities, she offers to work in the kitchen—an offer authorities refuse. Mary steadfastly refuses a gallbladder operation and forcefully insists that she will work as a cook if she gets out. What to do? Mary remains quarantined there for the rest of her life.

1938—Mary Mallon dies at age 70 from a stroke.

Had you been a Health Department official, a judge, a lawyer, or someone else associated with Mary, what would you have done? Before antibiotics and vaccines were available, quarantine was a common practice.

---

immunity provides protection against future infection. Antibodies to surface antigens of the bacterium also are produced, but these are of more use in laboratory diagnosis than in protecting the patient against infection. The Widal test detects antibodies and is used to confirm the diagnosis of typhoid fever. A live, attenuated oral typhoid vaccine is now available. It stimulates cell-mediated, as well as humoral, immunity, including the production of secretory antibody (IgA), which aids in prevention of mucosal invasion (◀Chapter 17, p. 497). After the initial series of doses, a booster dose is needed every 3 years. Means of protection against typhoid fever include good sanitation and sewage disposal and proper hygiene.

Shigellosis

Shigellosis (shig″el-o′sis), or **bacillary dysentery**, can be caused by several serovars (se′-ro-var) of *Shigella* (**Figure 22.9**). They include *Shigella dysenteriae* (serovar A), *S. flexneri* (serovar B), *S. boydii* (serovar C), and *S. sonnei* (serovar D). The latter causes 80% of the cases in the

Figure 22.9 A colorized TEM of *Shigella*, the cause of shigellosis (bacillary dysentery). (43,313X) *(Dr. Kari Lounatmaa/Photo Researchers, Inc.)*

TEM

United States. Shigellosis was first described in the fourth century B.C. Although less invasive than salmonellosis, it spreads rapidly in overcrowded conditions with poor sanitation. Humans and other higher primates such as chimpanzees and gorillas are the only reservoirs of infection, but the organisms can persist in foods for up to a month.

The pathogens are spread by contaminated food, fingers, flies, feces, and fomites. Playing, bathing, and washing clothes in contaminated water play significant roles in transmitting *Shigella*. In areas of good sanitation, infection is usually acquired by poor handwashing.

Children between the ages of 1 to 10 are most susceptible to *Shigella*, which accounts for 15% of the infant diarrhea cases in the United States. For the past several decades, the total reported number of annual cases of shigellosis in the United States has ranged between 15,000 and 20,000 with the true number being estimated at 450,000 cases per year. Worldwide an estimated 150 million cases occur annually. Many recent outbreaks have been recorded in day-care centers. The ingestion of just 10 organisms can be sufficient to cause infection. Thus, even slight lapses in hygiene can allow easy spread of the disease. In developing nations it is a major cause of infant mortality, with one-third of all infant deaths due to dehydration from shigellosis and other enteric pathogens. Essential fluid-replacement treatment often is not available in developing countries.

Once contaminated food or water is ingested, *Shigella* survive the acidity of the stomach, pass through the small intestine, and attach to portions of it and the large intestine. The pathogens invade host cells and induce the cells to create special filaments for invading adjacent host cells. After an incubation period of 1 to 4 days, abdominal cramps, fever, and profuse diarrhea with blood and mucus suddenly appear. The severity of signs and symptoms ranges from most to least serious in the same order as the

serovar designations A to D. In the most serious cases, diarrhea can cause dangerous protein deficiencies—called *kwashiorkor* (kwash-e-or′kor)—and vitamin B_{12} deficiencies, which, together with loss of electrolytes, can result in neurological damage. Along with the fever-eliciting endotoxin found in all serovars, *S. dysenteriae* also produces Shiga toxin, which acts as a neurotoxin. The neurotoxin activity is believed to be responsible for the severity and relatively high fatality rate of *S. dysenteriae* disease due to convulsions and coma. All serovars cause ulceration and bleeding of the intestinal lining and sometimes deeper intestinal layers. Symptoms persist for 2 to 7 days and usually are self-limiting, but they can cause severe dehydration and fluid and electrolyte imbalances.

Specific diagnosis of shigellosis can be difficult because the organisms are very sensitive to acids in feces. The organisms die if fecal specimens are not maintained in a buffered transport medium. Viable specimens can be obtained by swabbing a bowel lesion during internal examination of the bowel. Stool specimens are inoculated into selective media. Distinctive colonies usually appear after 24 hours of incubation.

Treatment is necessary in children and debilitated patients. Restoring fluid and electrolytes is essential to recovery. Usually hydration fluids, with or without antibiotics, are used for this purpose. Fluoroquinolones or trimethoprim/sulfamethoxazole are used for treatment. Prevention is difficult because many people have inapparent infections. A carrier state, usually lasting less than 1 month, exists and accounts for many new cases by fecal-oral transmission. Any breakdown in sanitation can lead to transmission by way of feces, fingers, and flies. Immunity following recovery from shigellosis is transient. From limited experience in recent years, oral vaccines containing certain strains of *S. flexneri* and *S. sonnei* seem to be safe and effective.

Asiatic Cholera

Asiatic cholera, so named because of its high incidence in Asia, can affect people anywhere sanitation is poor and fecal contamination of water occurs. Worldwide, more than 100,000 cases are reported annually. In the United States, fewer than 10 cases are reported per year. Some of those are transmitted to humans from contaminated shellfish in Gulf Coast states, where mild winters do not kill off the causative vibrios, leaving waters contaminated year-round. Occasionally some immunosuppressed individuals die from *Vibrio vulnificus* and *V. parahemolyticus* infections, which are ordinarily rather mild and do not cause true cholera. In endemic regions, such as parts of Asia, 5 to 15% of patients die; when a seasonal epidemic occurs, as many as 75% of patients die.

Cholera epidemics are often dependent on weather events, but may be decreased by using satellites to detect resultant algal blooms, the food of cholera-carrying copepods.

Cholera Around the World—and On Our Doorstep

In 1892 the great German hygienist Max von Pettenkofer, in the presence of witnesses, raised a broth culture of *Vibrio cholerae* to his lips and swallowed approximately 1 billion organisms. His aim was to disprove Robert Koch's studies, done in India, showing that cholera is a contagious disease spread by polluted water. Von Pettenkofer firmly believed that cholera was neither contagious nor related to drinking water. He thought the disease was due instead to interactions of microbes and soil, with "soil factors" playing the greatest role. Control of the disease would depend on the removal of these "soil factors." Within a few days, von Pettenkofer developed a mild but genuine case of cholera, and *V. cholerae* organisms were recovered from his stool. However, he insisted that his was *not* a real case of cholera. So, a few days later, his assistant repeated the experiment, fell severely ill with cholera, but recovered. The culture used in these experiments was a weakly virulent strain supplied by the bacteriologist Georg Gaffky, who had guessed at what von Pettenkofer intended to do and did not want the 74-year-old man to die.

In February 1991 Peruvians saw a similar scene broadcast on their televisions. The president of the country, Alberto Fujimori, and his wife were filmed eating raw fish and proclaiming that this popular dish was safe to eat. At that time 45,000 Peruvians had become cholera victims, and at least 193 had died from the disease. To slow the epidemic, the Peruvian Health Ministry declared that raw fish was likely to be contaminated and therefore unsafe to eat. Physicians worldwide criticized President Fujimori for his foolhardy defense of the Peruvian fisheries industry. Fujimori neglected to make it clear to his viewers that the fish he and his wife ate had been caught far out to sea, away from the sewage-contaminated coastal and river waters where poorer Peruvians caught their fish. Ironically, the Peruvian minister of fisheries, Felix Canal, attempted the same demonstration and promptly came down with cholera.

In April 1991 eight people in New Jersey contracted cholera by eating crab meat illegally imported from Ecuador. And in September 1991, Alabama state health officials closed an oyster reef near Dauphin Island, in the Gulf of Mexico, that they said was teeming with the deadly strain of *V. cholerae* that was causing the outbreak in South America. Thankfully, health officials swiftly caught and managed these two outbreaks. But cholera can quickly veer out of control, causing untold suffering and death.

Cholera had existed on the Indian subcontinent for centuries before the first Europeans arrived. Portuguese explorers described it early in the sixteenth century, but it did not invade other areas until 1817. The world is now gripped by the seventh cholera pandemic to have swept the globe. The United States experienced the second, third, and fourth of these, beginning in 1832, 1849, and 1866, respectively. The fourth pandemic left more than 50,000 Americans dead, with outbreaks continuing until about 1878. In 1887 and 1892 boatloads of infected passengers arrived in New York City as part of the fifth pandemic. However, a drastic shift in knowledge and attitudes in the United States prevented its spread. During the 1832 outbreak, cholera was viewed as a punishment from God. Indeed, health authorities resisted pleas to clean up streets and water supplies; to do so, they felt, would be to oppose God's will. Pigs were allowed to wander the New York streets, and they left feces and garbage in their wake. In contrast, by 1866 most intelligent physicians realized that cholera was a contagious disease spread by filth and ignorance. During that pandemic, physicians and city officials learned a great deal about control measures. Implementing these measures prevented future pandemics from spreading extensively into the United States.

The current pandemic, caused by the E1 Tor strain of *V. cholerae*, began in Indonesia in 1958. From there it spread to Macao, Hong Kong, and the Philippines and reached eastern Europe by 1965. It is still rampant in parts of Africa and Asia. In 1989 sub-Saharan Africa accounted for three-quarters of all cases in the world. In the summer of 1994, tribal conflicts in the central African nation of Rwanda claimed more than 500,000 lives, and 3 million Rwandans fled to surrounding nations, including Zaire, Burundi, and Tanzania. There, the refugees set up primitive camps while they waited for the slaughter to subside. With no running water, no sanitation facilities, and little food, the refugees succumbed to cholera, dysentery, and malaria before worldwide relief efforts came to their aid. About 50,000 Rwandan refugees contracted cholera in the camps, and thousands died due to lack of treatment. With proper treatment, cholera victims can recover, but treatment requires clean water for victims to drink to replace lost fluids. In late July 1994 the United States and Canada joined the relief efforts by sending physicians, food, clean water, and portable latrines to the region.

Curbing the disease is a different matter where sanitation remains primitive over the long term. Health officials fear that cholera will establish itself as an endemic disease in South America and will bring death for decades to poor areas that lack proper sewage and water treatment systems. Carried by ship from Africa, cholera reached a port city north of Lima, Peru, in January 1991. It spread through the untreated drinking waters and via infected people and contaminated food to Lima. Within 3 months, some 175,000 people had become ill, and 1,258 had died. Because only about one-fourth of infected people show symptoms, vast numbers of Peruvians must have been infected. Meanwhile, cholera spread into nearby countries, including Colombia, Ecuador, Chile, Guatemala, Brazil, Mexico—and even the United States. Because the current vaccine protects only half of those who receive it, and then only for 6 months, the World Health Organization is not recommending mass vaccination campaigns.

Many South American countries have instituted educational programs about cholera transmission and offer practical solutions to the sanitation problems. For instance, the Peruvian government now uses television and radio to broadcast information about boiling water, proper handling of food, and the necessity of disposing of garbage and sewage away from drinking water supplies. The government has distributed millions of chlorination tablets to its citizens. In addition, physicians and nurses take special courses about the treatment of cholera, and food handlers are trained in workshops about how to handle food safely. As a result of this massive educational effort, many areas of Peru have witnessed a decrease in the number of cholera cases.

Overcoming ingrained habits, however, is not easy. The fight against cholera is also a battle against ignorance, but with proper education, perhaps this fight can be won.

Figure 22.10   A colorized SEM of *Vibrio cholerae*, the cause of Asiatic cholera. The bacterium is slightly curved and has a single polar flagellum. It is seen as paler pink, curved rods lying on darker red intestinal lining cells (7,620X) *(Veronika Burmeister/Visuals Unlimited).*

The causative organism, *Vibrio choleráe* (Figure 22.10), can survive outside the body in cool, alkaline water, especially if organic and/or fecal matter is present. When ingested, it invades the intestinal mucosa, multiplies, and releases a potent enterotoxin. The enterotoxin, known as *choleragen*, binds to epithelial cells of the small intestine and makes plasma membranes highly permeable to water. This action results in a significant secretion of fluids and chloride ions and the inhibition of sodium absorption. At this time the intestinal lining becomes shredded, causing numerous small, white flecks resembling rice grains to be passed in the feces. Infected individuals experience severe nausea, vomiting, abdominal pain, and diarrhea. Stools rapidly become clear, containing numerous mucus plugs, thus giving rise to the term "rice-water stools." As many as 22 liters of fluids and electrolytes can be lost per day, so all patients, regardless of age, are subject to severe dehydration. Special "cholera cots" made of canvas with a hole cut beneath the buttocks have been used. A bucket marked in liters is placed below the hole to measure the fluid lost so that it can be replaced. Most deaths are attributable to shock due to greatly reduced blood volume.

Fluid and electrolyte replacement is the most effective treatment of cholera. During the 1971 epidemic in India and Pakistan, medical personnel were able to save large numbers of victims of the disease largely because of the availability of replacement therapy. Treatment with doxycycline or tetracycline reduces the duration of the symptoms but does not eliminate the organism or the toxin. Recovery from the disease confers only temporary immunity. Many recovered patients remain in a carrier state and can infect others and reinfect themselves. The only available vaccine is not very effective and is not widely used. However, a toxoid-type vaccine is being tested. It also may be possible to develop a vaccine that makes use of secreted IgA against the cholera organism.

A strain of *V. cholerae* known as the E1 Tor biotype (named after the quarantine camp where it was first isolated) causes a form of cholera that is slower and more insidious in its onset than the classical form of the disease. Cholera outbreaks have often resulted in the imposition of quarantine and the halting of shipping from port cities. In the past, therefore, countries in South and Central America that were economically dependent on their exports would sometimes report that cholera was not present, only "El Tor." The two forms of the disease, however, are essentially the same. Even today countries attempt to avoid quarantine by refusing to declare cholera present, even in the face of numerous documented cases. Therefore, the WHO (World Health Organization) has recently announced that it will independently declare the presence of cholera in a country, and that shipping will follow their findings rather than waiting for a country to close down its own ports by admitting to the presence of cholera, rather than to "cholera-like diarrhea."

Vibriosis

An enteritis called **vibriosis** (vib-re-o'sis) is caused largely by *Vibrio parahaemolyticus*. Although the disease is most common in Japan, where raw fish is considered a delicacy, the organism is widely distributed in marine environments. In the United States, infections usually are acquired from contaminated fish and shellfish that have not been thoroughly cooked. Most outbreaks in the United States occur at outdoor festivities where crab and shrimp are served without thorough cooking and proper refrigeration. Some outbreaks have been traced to the eating of contaminated raw oysters. *Vibrio parahaemolyticus* also can infect skin wounds of people exposed to contaminated water. Once inside the intestine, the organisms colonize the mucosa and release an enterotoxin. Symptoms of nausea, vomiting, diarrhea, and abdominal pain appear about 12 hours after contaminated food or water has been ingested, and they last 2 to 5 days. The disease usually is not treated, and no vaccine is available.

Traveler's Diarrhea

Among the 250 million people who travel internationally each year, it has been estimated that over 100 million suffer from a self-limiting mild to severe diarrhea. Of these travelers, 30% are confined to bed, and another 40% are forced to curtail their activities. This disorder, officially called **traveler's diarrhea**, also has been called "Delhi belly," "Montezuma's revenge," and some even less attractive names.

The most common causes of traveler's diarrhea are pathogenic strains of *Escherichia coli*, which account for 40 to 70% of all cases. Strains differ with geographic location, so travelers are likely to be exposed to new strains. Some strains of *E. coli* are normal inhabitants of the human digestive tract, and only certain strains are capable of causing enteritis. **Enteroinvasive strains** have a plasmid with a gene coding for a particular surface antigen called

E. coli can survive easily for 2 months on a dry stainless-steel work surface, and 2 to 5 months in natural bodies of water.

K antigen, which enables the strains to attach to and invade mucosal cells. **Enterotoxigenic strains** contain a plasmid that enables them to make an enterotoxin. They attach to the mucosa by attachment pili or fimbriae. These organisms also cause numerous cases of infant diarrhea. Other causes of traveler's diarrhea include bacteria such as *Shigella*, *Salmonella*, *Campylobacter*, rotaviruses, and protozoa such as *Giardia* and *Entamoeba*. Jet lag and other stresses of travel do not cause diarrhea, but they can lower resistance to infection. Travelers can experience symptoms of diarrhea even when no pathogens are present. Such cases are due to unusual kinds and amounts of substances dissolved in water, which lead to gastrointestinal upsets.

Symptoms of traveler's diarrhea vary from mild to severe and include nausea, vomiting, diarrhea, bloating, malaise, and abdominal pain. A typical case causes four to five loose stools per day for 3 or 4 days. Fluid loss is greater with invasive strains than with toxigenic strains. The disease is especially hazardous in infants, who are subject to severe dehydration. Bottle-fed infants are much more likely to become infected than breast-fed infants. Before their normal microflora are sufficiently established to compete with the pathogens, newborns are especially at risk of acquiring pathogenic strains from hospital workers. After infancy, children most often acquire *E. coli* infections during travel in foreign countries, especially those with poor sanitation.

Travelers sometimes medicate themselves with antibiotics before and during a stay in a foreign country. This treatment is not recommended because antibiotics usually are not effective and because such use contributes to the development of antibiotic-resistant strains. A better practice is to keep antidiarrhea medicine available and to use it only after symptoms appear.

Traveler's diarrhea can persist for months or years as a postinfectious irritable bowel syndrome. It can also cause lactose intolerance by damaging intestinal lining cells that normally produce the enzyme lactase, which digests lactose (milk sugar). During bouts of diarrhea, patients should eliminate from their diet dairy products and other foods containing lactose. After a few weeks, small quantities of such foods can be reintroduced and gradually increased as long as they are well tolerated. In some cases normal lactose tolerance is lost forever. *Escherichia coli* and other bacteria, the protozoan *Giardia*, and certain helminths frequently cause lactose intolerance.

Escherichia coli has significance far beyond its ability to cause diarrhea. It is an important indicator organism because it is always present in water contaminated with fecal material. *Escherichia coli* is usually more numerous than other organisms and is easier to isolate. Finding *E. coli* in water indicates that any pathogens found in feces might also be present. **Enterohemorrhagic strains** of *E. coli* O157:H7 have caused deadly outbreaks

PUBLIC HEALTH

Watch Out for Those Kiddie Pools ... Maybe

Perhaps you heard about the June 1998 White Water park incident. A single child with diarrhea had an accident in the kiddie pool. The *E. coli* O157:H7 organisms in the child's fecal material spread through the water, leading to the infection of at least 13 other children. That wasn't an unusual incident—*E. coli* O157:H7 is estimated to cause 10,000 to 20,000 cases of infection in the United States annually, according to the CDC. So you might be glad to hear that a technique has been developed that detects even one cell of the disease-causing bacteria in just 4 hours.

of bloody diarrhea, massive recalls of food, and a change in how we cook our hamburgers—no more rare or pink meat. It began with two outbreaks in 1982, traced back to undercooked hamburgers in fast-food chain restaurants. Out of 732 cases, in four western U.S. states, four children died. In Japan, in 1996, more than 6,000 school children were affected. In the United Kingdom it is more likely to be acquired from lamb than from beef. The CDC now requires reporting all cases. Because many cases are never diagnosed, statistics are uncertain, but it is estimated that about 3,000 infections per year occur in the United States, with about about 30 deaths. Its prevalence in Europe, Asia, Africa, and South America is unknown, but it is commonly reported from Canada.

In December 1997, the Food and Drug Administration at last approved the use of radiation for eliminating harmful microorganisms such as E. coli from red meat.

Escherichia coli O157:H7 is found in a small number of healthy cattle's intestines, where it is then passed out in manure. Meat is not the only product that can be contaminated. Apples picked up from manure-laden soil underneath apple trees were pressed into a lethal apple juice. Visitors to an agricultural fair drank water from a manure-contaminated well and became ill. One contaminated carcass put through a meat grinder can contaminate tons of meat that follow it. In one case, in 1998, that amounted to 25 million pounds of hamburger that the U.S. Department of Agriculture ordered recalled by a Nebraska meat processor.

The serotype designation O157:H7 refers to the numbered identities of the somatic (O) and flagellar (H) antigens carried by that strain of *E. coli*. This strain produces one or both of two toxins called Shiga toxins 1 and 2, so named because of their identity with toxins produced by species of *Shigella*. Lateral gene transfer by plasmid to a harmless strain of *E. coli* could have turned it into a killer. Only about 100 cells are needed to start an infection, with an incubation period of 3 or 4 days. Onset begins with abdominal cramping, non-bloody diarrhea, which usually becomes bloody on the second or third day; plus vomiting in one-third of patients; and little or no fever. In some cases the diarrhea never becomes bloody. About 6% of infected patients, especially children and the elderly, develop kidney involvement leading to kidney failure called **hemorrhagic uremic syndrome (HUS)**. HUS is currently the most common cause of acute kidney failure in children in the United States. Multiple diagnostic tests are available. Treatment with antimicrobials does not seem to help.

Escherichia coli is also an extremely versatile opportunistic pathogen—it can infect any part of the body subject to fecal contamination, including the urinary and reproductive tracts and the abdominal cavity after perforation of the bowel. It is present in many bacteremias, causes septicemias, and can infect the gallbladder, meninges, surgical wounds, skin lesions, and lungs, especially in debilitated and immunodeficient patients.

Other Kinds of Bacterial Enteritis

Certain strains of *Campylobacter jejuni* and *C. fetus* can be found in food and water and are becoming increasingly associated with human gastroenteritis, especially in infants and in elderly or debilitated patients. Some strains of *C. fetus* cause infectious abortions in several kinds of domestic animals. Although these organisms apparently do not multiply in foods, they are transmitted passively in undercooked chicken, unpasteurized milk, and poultry held in unchlorinated water during processing. Improper cooking of contaminated poultry can lead to enteritis. In some areas *Campylobacter* surpasses *Salmonella* as the major cause of foodborne enteric disease. Most health departments are now testing for *Campylobacter* in food-handling areas and on equipment, and *Campylobacter* is now the most reported foodborne disease.

Campylobacter infections cause copious diarrhea, foul-smelling feces, fever, and abdominal pain. They also cause arthritis in 2 to 10% of infected children but rarely in adults. Because large quantities of fluid can be lost, dehydration and fluid and electrolyte imbalances are common among the populations most affected. The disease is treated with fluid and electrolyte replacement and sometimes with tetracycline and erythromycin.

Yersiniosis (yer-sin″e-o′sis), a severe enteritis, is caused by *Yersinia enterocolitica*. The disease is most common in western Europe, but some cases are seen in the United States. This free-living organism is found mainly in marine environments, but it can survive in many places. Infection can be acquired from water, milk, seafoods, fruits, and vegetables, even when they are refrigerated, because the organism grows more rapidly at refrigerator temperatures than at body temperature. The organism is most easily identified when cultured at 25°C, a temperature at which the organisms are motile and easily distinguished from other bacteria. Yersiniosis symptoms, which are related to release of an enterotoxin, are similar to other kinds of enteritis, but abdominal pain usually is more severe, and white blood cells increase in number. Yersiniosis sometimes is misdiagnosed as appendicitis because of the similarity of symptoms.

A new source of *Yersinia* infection has caused the CDC to issue a warning to preparers of the southern

U.S. delicacy chitterlings, or chitlins (pork intestines). Until the preparers have very carefully washed their hands, they should avoid touching children or anything used by children. Children should not be allowed to handle raw chitterlings. Fifteen children in Atlanta recently became ill, mainly from contact with preparers, whereas the adults remained well.

BACTERIAL INFECTIONS OF THE STOMACH, ESOPHAGUS, AND INTESTINES

Peptic Ulcer and Chronic Gastritis

Recent studies have revealed a bacterial cause of peptic ulcers and chronic gastritis and a probable cofactor of stomach cancer **(Figure 22.11)**. The organism, *Helicobacter pylori* (formerly called *Campylobacter pylori*) **(Figure 22.12)**, was first cultured in 1982 from gastric biopsy tissues by Barry Marshall and J. Robin Warren, who were awarded the Nobel prize in 2005 for their discoveries. It is able to survive the very acidic conditions of the stomach by generating ammonia from urea. The ammonia is believed to neutralize gastric acidity around the *Helicobacter* cells, thereby allowing the organisms to survive and reproduce. They colonize and multiply in the gastric mucosa directly above the epithelial cell layer of the stomach. In culture they are slow to grow and require microaerophilic conditions, plus an enriched medium.

Peptic ulcers are lesions of the mucous membranes lining the esophagus, stomach, or duodenum. The lesions are caused by the sloughing away of dead inflammatory tissue and exposure to acid; they eventually result in an excavation into the surface of the organ. Four million Americans suffer from ulcers each year. Approximately 10% of the population will suffer from an ulcer at some point in their lives. Ulcers are responsible for about 46,000 operations and 14,000 deaths per year.

Take another look

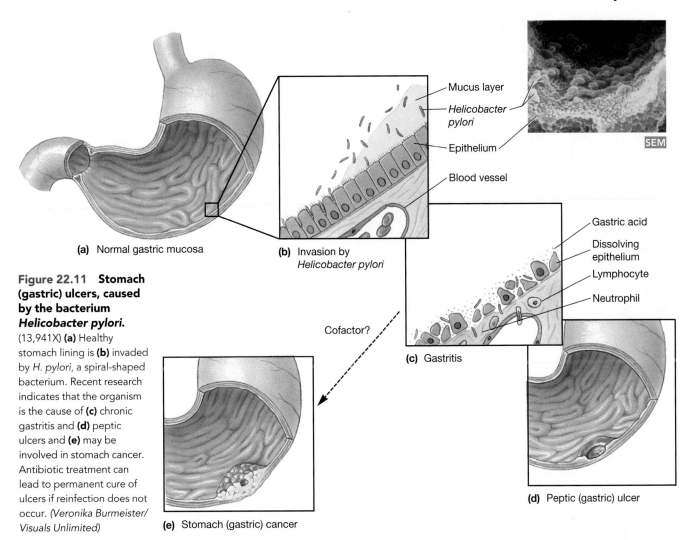

(a) Normal gastric mucosa

(b) Invasion by *Helicobacter pylori*

Mucus layer
Helicobacter pylori
Epithelium
Blood vessel

SEM

Gastric acid
Dissolving epithelium
Lymphocyte
Neutrophil

(c) Gastritis

Cofactor?

(e) Stomach (gastric) cancer

(d) Peptic (gastric) ulcer

Figure 22.11 Stomach (gastric) ulcers, caused by the bacterium *Helicobacter pylori*. (13,941X) **(a)** Healthy stomach lining is **(b)** invaded by *H. pylori*, a spiral-shaped bacterium. Recent research indicates that the organism is the cause of **(c)** chronic gastritis and **(d)** peptic ulcers and **(e)** may be involved in stomach cancer. Antibiotic treatment can lead to permanent cure of ulcers if reinfection does not occur. *(Veronika Burmeister/ Visuals Unlimited)*

Chronic gastritis (stomach inflammation) may be so mild as to cause no noticeable signs or symptoms, or it can produce pain and indigestion. It is observed in 70 to 95% of people who have peptic ulcers. Severe gastritis may lead to ulceration.

H. pylori is present in 95% of patients with duodenal ulcers and in 70% of those with gastric ulcers. In developed countries, it is uncommon for children to be infected, but the rate of infection increases by about 1% per year of age above 20, eventually reaching an average of 20 to 30% among U.S. adults. In developing countries, it is commonly found in children and reaches levels of 80% of the adult population. In Latin America, stomach cancer rates are the highest in the world, and so is the rate of *H. pylori* infection. Interestingly, U.S. Hispanics also have a 75% infection rate. In Japan, rates of stomach cancer and of *H. pylori* infection are dropping together. It may be that just having chronic inflammation of the stomach for a long period of time—which occurs when a person suffers *H. pylori* infection early in life—

The adhesin used by Helicobacter pylori to bind to the stomach lining has been identified. This may make a vaccine to prevent H. pylori's colonization possible.

predisposes toward stomach cancer. Thus, some scientists are unwilling to call *H. pylori* a definite cause of stomach cancer but feel it may be only a cofactor because most people infected with *H. pylori* never develop stomach cancer. This may be due to differences among strains, as *H. pylori* has a higher degree of genetic diversity than most human pathogens. However, among the various kinds of stomach cancer, the most common type (gastric intestinal) has an 89% correlation with *H. pylori* infection.

No one is certain yet of the route of infection or the portal of exit and no animal reservoir has been found. It is of great importance, however, if we find that we can prevent or cure ulcers by eliminating *H. pylori* infection. Currently available drugs such as Tagament only control, but cannot cure, ulcers. Experimental treatments with antibiotics have reported cure rates as high as 80 to 90%, but these drugs must be used very carefully. Patients treated only with drugs that suppress stomach acid had a relapse rate of 75 to 95% within 2 years. Researchers expect to have better tests, drugs, and treatment plans worked out in

Data linking chronic H. pylori infection to stomach cancer have led the International Agency for Research on Cancer to classify it as a Group 1 carcinogen.

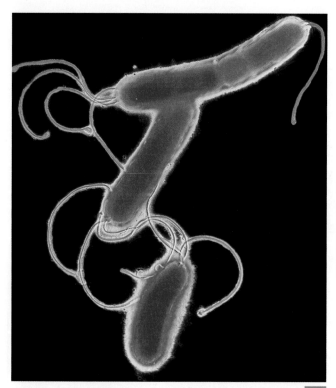

Figure 22.12 *Helicobacter pylori* (9,684X). Stomach `SEM` (gastric) ulcers are now known to be caused by this bacterium. The ulcers can be cured by killing the bacteria with antibiotics. Note the knob at the end of each flagellum. These make it possible for *H. pylori* to move through the thick mucus that coats the stomach lining. Other bacteria with normal flagella are unable to move forward and get trapped in the mucus. *(A. B. Dorselt/Photo Researchers, Inc.)*

a few years. Currently, over 90% of cases are cured by a 2-week administration of omeprazole (a proton pump inhibitor) plus one or more antibiotics (e.g., metronidazole, tetracycline, amoxicillin, or clarithromycin) and bismuth.

Approaches to *H. pylori* identification include direct detection of the organism in gastric biopsies, testing for the presence of the enzyme urease in biopsy tissue, and culturing specimens on special media. Serological tests for specific antibody against *H. pylori* have been developed and are undergoing evaluation.

Pseudomembranous Colitis

Pseudomembranous colitis, a condition characterized by the formation of a membrane covering on the mucosal surface of the colon, is caused by *Clostridium difficile*. The signs and symptoms, diagnosis, and treatment for *C. difficile* disease were discussed in ◀Chapter 13, p. 382.

BACTERIAL INFECTIONS OF THE GALLBLADDER AND BILIARY TRACT

The gallbladder and its associated ducts can also be sites of bacterial infection. The liver produces bile, which is stored in the gallbladder and slowly released into the small intestine. Organisms with lipid-containing

envelopes are destroyed by the action of bile, which breaks down lipids. Most enteric viruses, poliovirus, and hepatitis A virus (discussed in the next section) lack lipid envelopes and are protected from bile activity. Organisms such as the typhoid bacillus are so resistant to the action of bile that they can actually grow in the gallbladder itself. They are shed from the gallbladder into the intestine and are eliminated in feces. People carrying these organisms in the gallbladder are asymptomatic.

Gallstones, formed from crystals of cholesterol and calcium salts, can block the bile ducts, decreasing the flow of bile and predisposing the individual to inflammation of the gallbladder (*cholecystitis*) or biliary ducts (*cholangitis*). Distension of the gallbladder due to the accumulation of bile fosters the entry of microbes—most commonly *E. coli*—into the bloodstream, probably through minute tears in the gallbladder wall. Infection of the gallbladder can also ascend into the liver.

Intestinal bacteria have been linked to chronic gallstone formation. Antibiotics halt development of new stones until discontinued. Then stone formation resumes.

Diagnosis of gallbladder infections is based on clinical findings of recurring pains (*biliary colic*), nausea, vomiting, chills, fever, and often jaundice due to the absorption of blocked bile into the bloodstream. Despite prompt treatment with antibiotics, gallbladder infections cannot be cured unless the causative obstruction is removed, either by surgery or by the spontaneous passage of the gallstone.

Gastrointestinal diseases caused by bacteria are summarized in **Table 22.2**.

✓CHECKLIST

1. Distinguish between diarrhea and dysentery.
2. Is cholera a problem in the Western Hemisphere? Explain.
3. Why are travelers advised not to take antibiotics before and during a trip in order to avoid traveler's diarrhea?
4. What causes stomach (gastric) ulcers? How can they be cured?

TABLE 22.2

A Summary of Gastrointestinal Diseases Caused by Bacteria

Disease	Agent(s)	Characteristics
Bacterial Food Poisoning		
Staphylococcal enterotoxicosis	*Staphylococcus aureus*	Heat-stable enterotoxin causes tissue damage, abdominal pain, nausea, vomiting
Other kinds of food poisoning	*Clostridium perfringens, C. botulinum, Bacillus cereus*	Diarrhea and sometimes intestinal infection and gas gangrene
Bacterial Enteritis and Enteric Fevers		
Salmonellosis	*Salmonella typhimurium, S. enteritidis*	Abdominal pain, fever, diarrhea with blood and mucus from toxin; enterocolitis from invasion of organisms; chronic infections and carrier states occur
Typhoid fever	*Salmonella typhi*	Organisms invade mucosa and lymphatics, multiply in phagocytes and other tissues; high fever and "rose spots"; carrier state and life-threatening complications can occur
Shigellosis	*Shigella* strains	Organisms cause intestinal lesions and release toxins; symptoms include cramps, fever, profuse diarrhea with blood and mucus
Asiatic cholera	*Vibrio cholerae*	Organisms invade intestinal lining, release potent toxin that increases lining permeability; symptoms include nausea, vomiting, copious diarrhea, fluid imbalances
Vibriosis	*Vibrio parahaemolyticus*	Organisms colonize mucosa and release toxin; nausea, vomiting, diarrhea are self-limiting
Traveler's diarrhea	Pathogenic strains of *Escherichia coli*, other bacteria (also viruses and protozoa)	Organisms can invade mucosa and/or produce toxin, cause nausea, vomiting, diarrhea, bloating, malaise, abdominal pain; self-limiting except for postinfection complications; dehydration and death in infants
Other kinds of bacterial enteritis	*Campylobacter jejuni, C. fetus, Yersinia enterocolitica*	*Campylobacter* species cause enteritis in infants and debilitated patients; *Yersinia* releases a toxin that causes enteritis with pain resembling appendicitis
Bacterial Infections of the Upper Gastrointestinal Tract		
Ulcers, stomach cancer	*Helicobacter pylori*	Definite association with ulcers; probable cofactor in stomach cancer
Cholecystitis, cholangitis	Usually *E. coli*	Blockage of bile ducts by gallstones causes inflammation of gallbladder and bile ducts; accumulation of bile can cause infection to spread to bloodstream or liver
Pseudomembranous colitis	*Clostridium difficile*	Formation of a pseudomembrane on the mucosal surface of the colon; occurs in patients whose gastrointestinal microflora have been altered by the use of antibiotics

GASTROINTESTINAL DISEASES CAUSED BY OTHER PATHOGENS

VIRAL GASTROINTESTINAL DISEASES

Viral Enteritis

Rotavirus infection is a major cause of **viral enteritis** among infants and young children. Rotaviruses are transmitted by the fecal-oral route, replicate in the intestine, damage the intestinal epithelium, and cause a watery diarrhea within 48 hours. Rotavirus infection is a major cause of infant morbidity and mortality in developing countries, where 3 to 5 billion cases occur annually, with 5 to 10 million deaths in children under age 5. Worldwide, most children have been infected by age 4. Rotavirus infections account for a third of childhood deaths in some countries. Rotavirus infections also occur in infants and young children in developed countries, where the infections are often nosocomial. Special care should be used with hospitalized children to prevent such infections. The number of cases rises dramatically during the winter in the United States. This timing helps distinguish it from bacterial diarrheas.

Rotaviruses are reoviruses that replicate their double-stranded RNA in such great numbers (10^{10} viruses/g of stool) that they can easily be found by electron microscopy in fecal suspensions (**Figure 22.13**); no effort to concentrate the viruses is required. The capsid is double-layered, without an envelope, and covers 10 to 12 segments of the double-stranded RNA genome. Immunoelectron microscopy (immunologic techniques combined with electron microscopy) can be used to identify rotaviruses when antibodies from a patient's serum are reacted with virions in diagnostic specimens. In addition, many hospitals now use ELISA tests that detect rotaviruses in stool specimens. Although there is no specific treatment, restoring fluid and electrolyte balance is crucial and should be prompt. A vaccine was approved in the U.S., but proved dangerous and had to be recalled.

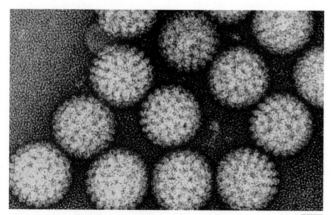

Figure 22.13 Rotaviruses. Seen here as large yellow spheres in fecal suspension (650,000X), they resemble little wheels and are a cause of diarrhea. *(K. G. Murti/Visuals Unlimited).*   `SEM`

APPLICATIONS

Another Plug for Breast-Feeding

If you're a female, and even if you're a male, you've probably heard more than you ever wanted to know about the benefits of breast-feeding. You probably know that breast milk allows antibodies to pass from the mother to her infant. But you might be interested to know that one of these antibodies acts specifically against rotavirus, the most common cause of diarrhea in infants, and that a complex carbohydrate in breast milk affords babies even more protection. How? The compound, called lactadherin, doesn't break down in the baby's stomach. Instead, it acts like it's one of the natural carbohydrates found in the child's intestines. The rotavirus attaches to the lactadherin, believing that it's anchoring itself in the gut, and then gets flushed out of the child's system as the lactadherin passes through.

Rotavirus infection in humans is important in medical practice in the United States. Antibodies to the virus have been identified in as many as 90% of groups of children tested, even though the infection may not have been identified at the time it occurred. Much more research is needed to determine how immunity is produced and how the disease might be controlled.

Enteritis can also be caused by viruses other than rotaviruses. Species of *Enterovirus* such as the echoviruses (*e*nteric *c*ytopathic *h*uman *o*rphan viruses) can cause mild gastrointestinal symptoms and damage intestinal cells. They also sometimes infect other tissues, and certain species can cause meningoencephalitis (inflammation of the brain and meninges).

Patients who have received bone marrow transplants are especially susceptible to infection with rotaviruses, enteroviruses, and the bacterium *Clostridium difficile*. As many as 55% of such patients succumb to these infections.

The *Norwalk virus* (named for a 1968 outbreak in Norwalk, Ohio) is responsible for nearly half of all U.S. outbreaks of acute infectious nonbacterial enteritis. Norwalk virus infection affects older children and adults more often than it affects preschoolers or infants. Outbreaks occur throughout the year and are common at schools, camps, and nursing homes and on cruise ships. The infection is the second most common cause of illness (after respiratory disease) among U.S. families and occurs worldwide. It is characterized by 1 to 2 days of diarrhea, vomiting, or both. Immunity does not follow an attack, which makes development of a vaccine unlikely. Careful sanitary practices are the best means of prevention.

Hepatitis

Hepatitis, an inflammation of the liver, usually is caused by viruses (**Figure 22.14** and **Table 22.3**). It also can be caused by an amoeba and various toxic chemicals. The most common viral hepatitis is **hepatitis A**, formerly called **infectious hepatitis**. It is caused by the hepatitis A

Figure 22.14 Hepatitis A virus (172,758X). *(Scott Camazine/Photo Researchers).*

virus (HAV), a single-stranded RNA virus usually transmitted by the fecal-oral route. **Hepatitis B**, formerly called **serum hepatitis**, is caused by the hepatitis B virus (HBV), a double-stranded DNA virus usually transmitted via blood. A third type of hepatitis is transmitted parenterally (by blood) and is probably caused by at least two viral agents. This type is diagnosed in the absence of HAV and HBV as **hepatitis C (HCV)**, formerly called *non-A, non-B (NANB) hepatitis* **(Figure 22.15)**. A fourth type of hepatitis, transmitted by the fecal-oral route and formerly called *non-A, non-B, non-C hepatitis*, has been separated out as **hepatitis E (HEV)**. An especially severe form of the disease **hepatitis D**, or **delta hepatitis**, is caused by the presence of both hepatitis D virus (HDV) and HBV. However, HDV alone does not cause disease; it cannot infect without HBV.

Hepatitis A. Hepatitis A occurs most often in children and young adults, especially in autumn and winter. It can occur in epidemics if a population is subjected to water or food, especially shellfish, contaminated with HAV. There are no animal reservoirs. Outbreaks due to contaminated food in fast-food restaurants have been on the rise.

Hepatitis A has an incubation period of 15 to 40 days and begins as an acute febrile illness. After entering the body through the mouth, the viruses (an RNA picornavirus) replicate in the gastrointestinal tract and spread through the blood to the liver, spleen, and kidneys.

TABLE 22.3

Comparison of Types of Viral Hepatitis					
Characteristic	Hepatitis A	Hepatitis B	Hepatitis C	Hepatitis D	Hepatitis E
Alternate names	Infectious hepatitis; epidemic hepatitis; short-term hepatitis	Serum hepatitis	Parenterally transmitted non-A, non-B, post-transfusion hepatitis	Delta hepatitis	Enterically transmitted non-A, non-B, non-C hepatitis
Agent	HAV RNA virus, Picornaviridae	HBV DNA virus, Hepadnaviridae	HCV At least two unclassified RNA viruses; ? Flavivirus, ? Togavirus	HDV Defective RNA virus; has hepatitis B capsid	HEV Unclassified RNA virus; ? Calcivirus
Transmission	Fecal-oral	Blood and other body fluids; crosses placenta with high frequency	Blood and blood products; occasionally crosses placenta	Blood, must coinfect or superinfect with hepatitis B; can cross the placenta	Fecal-oral; more common in adults than in children
Incubation period	15–40 days; average, 28 days	45–180 days; average, 90 days	Short, 2–4 weeks; long, 8–12 weeks	2–12 weeks	2–6 weeks
Severity of disease	Self-limiting; usually mild, rarely severe	Subclinical to severe; most recover completely	Subclinical to severe; most resolve spontaneously	Severe; high mortality rate	Moderate but high mortality in pregnant women
Carrier state	No	Yes, is associated with 80% of liver cancer	Yes, possible association with liver cancer	Yes	No
Chronic liver disease	No	Yes	Yes	Yes	No
Vaccines	Yes	Yes	No	No	No

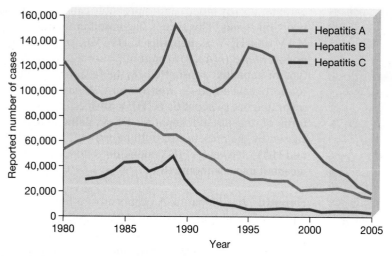

Figure 22.15 Hepatitis. Reported cases by year—U.S., 1980–2005. (*Source*: CDC).

* Hepatitis A vaccine was first licensed in 1995.
† Hepatitis B vaccine was first licensed in 1982.
§ An anti-HCV (hepatitis C virus) antibody test became available in May 1990.

In 2001, the hepatitis A rate was the lowest ever recorded. However, cyclic increases in hepatitis A have been observed approximately every 10 years, and thus rates could increase again. The incidence of hepatitis B continues to decline, but because of asymptomatic infections and underreporting, reported caes represent only a fraction of actual infections occurring (approximately 78,000 new infections in 2001). The trend in reported hepatitis C; non-A, non-B after 1990 is misleading because reported cases have included those based only on a positive laboratory test for the anti-HCV, and most of these cases represent chronic HCV infection.

Jaundice, a yellowing of the skin common in hepatitis, is caused by impaired liver function. The liver fails to rid the body of a yellow substance called **bilirubin**, which is a product of the breakdown of hemoglobin from red blood cells. Other symptoms of hepatitis are malaise, nausea, diarrhea, abdominal pain, and lack of appetite for a period of 2 days to 3 weeks. Probably over half the cases are asymptomatic. Chronic infections are rare, and recovery usually is complete and confers lifelong immunity. Immunological tests are available to detect hepatitis A viruses and host antibodies against them. There is no treatment for hepatitis other than alleviating symptoms. A vaccine for hepatitis A has been available since 1995. Gamma globulin injections are also used to provide temporary immunity. While hepatitis A has declined in all parts of the U.S., some areas retain high incidence **(Figure 22.16)**.

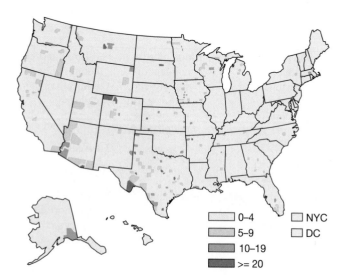

0–4
5–9
10–19
>= 20
NYC
DC

Figure 22.16 Hepatitis A. Reported cases per 100,000 population—U.S. and U.S. territories, 2003. (*Source*: CDC.)

Hepatitis A rates have declied in all regions of the United States, including the western states where rates have historically been higher than elsewhere in the country. Because hepatitis A rates vary from year to year with nationwide increases approximately every 10 years, further monitoring is needed to determine whether these rates wil remain low.

Hepatitis B. Hepatitis B occurs in people of all ages with about the same incidence throughout the year. It can be transmitted by intravenous or percutaneous (into the skin) injections, by anal/oral sexual practices (common among homosexual males), by contact with other virus-containing body secretions (including semen and breast milk), and by contaminated needles among intravenous drug users. Health care workers who have routine contact with patients' body fluids (especially blood) have a higher incidence of the disease than does the general community. Transmission via contaminated semen in artificial insemination has been documented.

Hepatitis B has an incubation period of 45 to 180 days, with an average of 90 days. The virus replicates in cells of the liver, lymphoid tissues, and blood-forming tissues. It can persist in the blood for years, thus creating a carrier state. The onset of symptoms is insidious, and fever is uncommon. Otherwise the symptoms are similar to those of hepatitis A, except that chronic, active hepatitis B frequently destroys liver cells.

Immunological methods are available to detect hepatitis B virus and the host's antibodies. Treatment relieves some symptoms but does not cure the disease. An effective vaccine became available in 1986, and government regulations require that it be provided by employers for health care workers who may have contact with blood or body fluids that might contain HBV. Sexually active teenagers and adults should also be vaccinated. Pediatricians recommend vaccination of all preteens, as parents cannot be sure of when their teenagers may become sexually active. Babies in the United States are now routinely immunized. However, most of the older population remains unprotected. The vaccine is safe. It is produced by recombinant DNA technology—the insertion of the appropriate hepatitis B virus genes into a plasmid. The plasmid

is then inserted into yeast cells, thereby avoiding all contact with human cells. This yeast-produced vaccine has been given to over 2 million Americans and is 95% effective. When given to hepatitis B–infected pregnant women, it reduces from 90 to 23% the number of infants who become carriers. This number can be reduced to 5% when gamma globulin is given along with the vaccine. Some 40% of carriers die of liver disease, and in some parts of the world nearly 90% of mothers are infected. Health officials in the United States currently recommend that all infants be vaccinated at birth. But such vaccination is expensive and is probably beyond the means of many developing countries, where it is most important.

The hepatitis B virus, a member of the Hepadnaviridae family, is unusually stable and resists drying and irradiation. Its double-stranded, circular DNA has a gap in one strand that may help it insert into liver cell DNA. Insertion of viral DNA into liver cell DNA, in turn, may contribute to liver cell carcinoma, a kind of cancer that occurs much more frequently in people who have had hepatitis B than in the general population.

Hepatitis C. Hepatitis C virus, the leading reason for liver transplants **(Figure 22.17)**, has been recovered from parenterally transmitted non-A, non-B hepatitis cases. Because hepatitis C disease has two different incubation periods—2 to 4 weeks and 8 to 12 weeks—some researchers believe that there may be two different causative agents, both RNA viruses: one a flavivirus, the other a togavirus. Hepatitis C can be distinguished from other kinds of hepatitis by the high blood concentration of a liver enzyme, alanine transferase. Various enzymes are released into the blood from damaged liver cells in all types of hepatitis, but this particular enzyme is unusually elevated in HCV hepatitis. Although the disease usually is

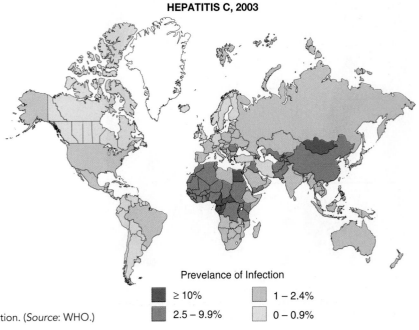

HEPATITIS C, 2003

Prevelance of Infection

- ≥ 10%
- 2.5 – 9.9%
- 1 – 2.4%
- 0 – 0.9%

Figure 22.17 **Hepatitis C.** Worldwide distribution. (*Source:* WHO.)

mild or even inapparent, the infection can be severe in compromised individuals and becomes chronic in about 80% of those infected. About 20% of chronic infections progress to cirrhosis and liver cancer. No vaccine is available, and immunity does not follow infection. Of the 170 million carriers of HCV worldwide, about 4 million live in the U.S.

Hepatitis D. Affecting over 15 million people worldwide, hepatitis D has an incubation period of 2 to 12 weeks. That period is shorter when HBV carriers are superinfected with HDV than when individuals are infected with both viruses at the same time. Sometimes called a defective virus, HDV alone fails to cause disease because it requires HBV antigens for replication. HDV and HBV together can result in death. Vaccination against HBV will prevent infection with HDV. There is no separate vaccine against HDV.

Hepatitis E. Transmitted through fecally contaminated water supplies, hepatitis E has caused large outbreaks in Asia and Africa. It is more common in adults than in children. The mortality rate is low (1%), except in pregnant women, for whom it is about 20%. Chronic cases do not occur. No vaccine is available, and immunity does not follow infection. One strain of HEV has been found infecting pigs as well as humans. The role of pigs in the epidemiology of HEV infection is yet to be determined. It is caused by an RNA virus, probably a calcivirus.

The hepatitis virus story has not ended. Recently three more viruses have been discovered that are associated with hepatitis-like disease. If these viruses turn out to be causative agents for true hepatitis cases, it has been proposed that they be named hepatitis F, G, and H viruses. Hepatitis G, curiously enough, seems to spread by both the fecal-oral route and the blood route. It is a single-stranded RNA virus, and can cause chronic cases and carriers. Some investigators believe that there may be dozens more viruses that can cause little-known forms of hepatitis.

PROTOZOAN GASTROINTESTINAL DISEASES

Giardiasis

The flagellated protozoan *Giardia intestinalis*, sometimes called *G. lamblia* (**Figure 22.18a**), was first observed by Leeuwenhoek in 1681 when he was studying organisms in his own stools. It is, however, a far older organism. Examination of *Giardia's* DNA has shown it to be the most primitive DNA of any eukaryote, very similar to that of older prokaryotes.

Giardia infects the small intestine of humans, especially children, and causes a disorder called **giardiasis** (je′ar-di′a-sis). After cysts ingested from fecal material pass through the stomach and small intestine, motile trophozoites are released in the colon (◄Chapter 11, p. 316). The parasite has an adhesive disk by which it attaches to the bowel wall (**Figure 22.18b**). In severe infections nearly every cell is covered by a parasite. It feeds mainly on mucus, forming cysts that are deposited in the mucus and passed intermittently in mucous stools. Symptoms of giardiasis include inflammation of the bowel, diarrhea, dehydration, and weight loss. Nutritional deficiencies are common in infected children because the parasites can occupy much of the intestinal absorptive area. Fat absorption is greatly reduced, and deficiencies

(a) SEM

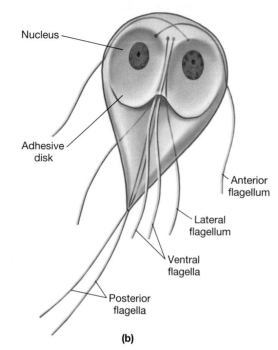

Nucleus

Adhesive disk

Anterior flagellum

Lateral flagellum

Ventral flagella

Posterior flagella

(b)

Figure 22.18 *Giardia intestinalis,* **a protozoan parasite that causes diarrhea.** **(a)** (35,600X). *(Jerome Paulin/Visuals Unlimited)* **(b)** The structure of *G. intestinalis.*

of fat-soluble vitamins are common. Diarrhea is copious and frothy from bacterial action on unabsorbed fats, but it is not bloody because the parasitic protozoa usually do not invade cells. Some infected persons experience severe joint inflammation and an itchy rash even before they have diarrhea. This *reactive arthritis of Giardia* fails to respond to anti-inflammatory drugs ordinarily used to treat arthritis. The arthritis disappears along with the diarrhea when antiprotozoan drugs are given.

Giardiasis is transmitted through food, water, and hands contaminated with fecal matter. It is occasionally transmitted from wild animals and is thus called "beaver fever" by backpackers and hunters in the western United States. Contaminated water supplies in Aspen, Colorado; Saint Petersburg, Russia; and probably many other places have caused large numbers of cases. In some child-care centers, up to 70% of the children are infected with the pathogen. *Giardia* cysts are not killed by ordinary sewage treatment and chlorination. Some localities in Pennsylvania are being forced to drink bottled water until they can afford to add sand filters to their treatment plants to trap *Giardia* cysts.

Diagnosis is made by microscopic examination and finding cysts of the protozoan in stools. *Giardia* is found in the mucous sheets that line the intestinal surface. These sheets are released every few days. Samples such as those taken from a bedpan should include some mucus. Because of the intermittent passing of cyst-containing mucus, daily stool samples for several days are needed to increase the likelihood of a positive diagnosis. Trophozoites are sometimes found in watery stools but once passed out are unable to encyst. As many as 14 billion parasites may be contained in a single diarrheic stool. Immnofluorescent antibody techniques and ELISA procedures are also used for diagnosis.

Metronidazole (Flagyl), furazolidine, and quinacrine (Atabrine) are used to treat giardiasis. The disease can be prevented by maintaining pure water supplies uncontaminated by human or animal wastes.

Amoebic Dysentery and Chronic Amebiasis

Amebiasis (am-e-bi′a-sis) is caused by *Entamoeba histolytica* (**Figure 22.19**), a major pathogenic amoeba. It was originally isolated from intestinal ulcers of a patient who had succumbed to severe diarrhea, used to infect a dog in which the same disease appeared, and recovered from the dog, thereby demonstrating Koch's Postulates (◀Chapter 14, p. 405). Ameobiasis can appear as a severe acute disease called **amoebic dysentery** or as **chronic amebiasis**, which can suddenly revert to the acute stage. Approximately 400 million people are infected worldwide, most with chronic amebiasis. The proportion of the infected population varies from 1% in Canada to 5% in the United States to 40% in tropical areas.

Humans become infected with the parasite by ingesting cysts in food or water contaminated with fecal matter. After ingestion and passage through the stomach and small intestine, cysts rupture and release amoeboid

Figure 22.19 *Entamoeba histolytica* **trophozoite.** LM
Active amoebic form (3,860X). *(Lauritz Jensen/Visuals Unlimited).*

trophozoites in the colon. Trophozoites reproduce asexually within the colon. There they feed on bacteria that make up the normal microflora of the large intestine. They may cause very little trouble to the host, or they may invade the intestinal mucosa, where they can live indefinitely. Once they have invaded the intestinal mucosa, the parasites multiply and cause significant ulceration. Sometimes their proteolytic enzymes digest deep into, or even through, the bowel wall. Thus, the protozoa sometimes enter blood vessels and travel to other tissues, or they allow bacteria in fecal material to enter the body cavity and cause peritonitis. Patients with amebiasis have abdominal tenderness, 30 or more bowel movements per day, and dehydration from excessive fluid loss. If the parasites invade liver and lung tissue, they can cause abscesses. Bacterial infection of lesions in any tissues can occur.

Because fecal material becomes dehydrated as it passes through the colon, *E. histolytica* trophozoites tend to encyst there. The cysts are passed with the feces. Cysts can survive up to 30 days in a cool, moist environment and are not killed by normal chlorine concentrations in water. The most common means of transmission is through the fecal-oral route. Flies and cockroaches can also be mechanical vectors. Infection can be acquired through sexual practices in which there is ingestion of fecal material.

Amoebic infections can be diagnosed by finding trophozoites or cysts in stools, but several stool samples on consecutive days may be needed to find them. Immunofluorescent antibody and ELISA procedures are also available for diagnosis. Metronidazole is widely used in treatment, even though it has been found to be mutagenic in bacteria and carcinogenic in rats. Antibiotics also are used to prevent or cure secondary bacterial infections. Such infections can be prevented by sanitary handling of water and food.

Balantidiasis

Balantidium coli (**Figure 22.20**) is the only ciliated protozoan that causes human disease. It is distributed worldwide, particularly in the tropics, but human infection is rare except in the Philippines. *Balantidium coli* is

(a) LM

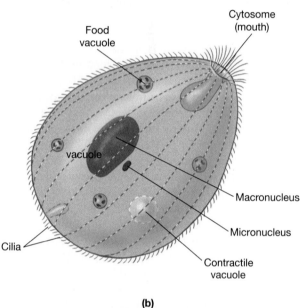

(b)

Figure 22.20 *Balantidium coli.* This very large, ciliated protozoan causes diarrhea. It is the only ciliate known to infect humans. **(a)** It is photographed here in a fecal smear (340X). *(Larry Jensen/Visuals Unlimited)* **(b)** The structure of *B. coli.*

transmitted by cysts in fecal matter. After being ingested, cysts rupture and release trophozoites that invade the walls of the large intestine, causing a dysentery known as **balantidiasis** (bal″an-tid-i′a-sis). Symptoms of the disease are similar to those of amoebic dysentery; as with amoebic dysentery, perforation of the intestine from balantidiasis can lead to fatal peritonitis.

Diagnosis is made by finding trophozoites or cysts in fecal specimens. Tetracycline or metronidazole are used to treat the disease, but some people remain carriers even after treatment. As with other organisms transmitted through fecal matter, infection can be prevented by good sanitation. Pigs serve as a reservoir of infection, so contact with their feces must be avoided.

Cryptosporidiosis
Protozoans of the genus *Cryptosporidium* commonly cause opportunistic infections worldwide, probably by fecal-oral transmission from kittens and puppies. These organisms live

in or under the membrane of cells lining the digestive and respiratory systems. After being swallowed, cysts burst in the intestine, releasing protozoa that invade intestinal cells or migrate to other tissues. In 1993 an outbreak of watery diarrhea affected some 403,000 residents of Milwaukee, Wisconsin; the outbreak was attributed to *Cryptosporidium* infection of one of the city's water treatment plants. In 2006 there were 5,140 cases in the United States. In immunocompetent individuals, the disease is self-limiting, but in immunosuppressed patients, it causes severe diarrhea—up to 25 bowel movements per day with a loss of as much as 17 liters of fluid. The majority of severe cases of **cryptosporidiosis** (krip″to-spor-id-e-o′sis) have been seen in AIDS patients. No effective treatment has been found.

> *Diarrhea hit 403,000 residents of Milwaukee, Wisconsin—4,400 badly enough to be hospitalized—in 1993 when Cryptosporidium contaminated the city water supply.*

Cyclosporiasis
The protozoan parasite, *Cyclospora cayentanenisis*, first attracted public attention in 1996 when 1,465 cases of infection, with severe diarrhea, resulted from consumption of raspberries imported into the United States from Guatemala. Other outbreaks followed rapidly in 1996 and 1997, involving mesclun lettuce and the herb, basil. However, the causative organism has been lurking in microbial literature under various other names since it was first described in 1979. It has been called a "coccidian-like body," a "cyanobacterium-like body," a "blue-green alga," and a "large form of *Cryptosporidium*." Finally, in 1993 it was given its present name. It is very closely related to species of *Eimeria* that cause similar disease in cattle, rats, and domestic fowl.

Cyclospora produces oocysts which are spread via the oral-fecal route. Perhaps only 10 to 100 oocysts are enough to initiate infection. After an incubation period of about 7 days, the disease begins with flu-like symptoms, watery diarrhea, bloating, anorexia, abdominal pain, weight loss, and extreme fatigue. The disease lasts at least 1 to 2 weeks, but more often averages 6 to 7 weeks, and often becomes relapsing. Diagnosis involves finding oocysts in the feces, a very difficult job. Detection by PCR has been hampered by inhibitory factors in the stool, plus difficulties in extracting DNA. Treatment is a 7-day course of trimethoprimsulfamethoxazole (Bactrim, Septra), traditional antiprotozoan drugs being of no help. Two theories about the "emergence" of this new disease have been advanced: (1) More virulent strains of *Cyclospora* have just recently arisen, or (2) *Cyclospora* has recently "jumped species" to humans, perhaps from livestock.

EFFECTS OF FUNGAL TOXINS

Fungi produce a large number of toxins, and most come from members of the genera *Aspergillus* and *Penicillium.* Their various effects on humans include loss of

Hold the Aflatoxin!

That peanut butter and jelly sandwich you packed for lunch might give you more than the sugary protein-filled energy rush you hoped for. Aflatoxin can be present in peanut butter if any moldy peanuts are used to make it. In one study, 7% of peanut butter samples tested contained aflatoxin. Similar toxins can be present in jellies. Even when the moldy top layer of jelly is discarded, some toxins may have diffused down into the jelly—in other words, don't scrape the top off and eat what's underneath! If, in addition, moldy bread is used, you may be eating a potentially highly carcinogenic sandwich for lunch.

(a)   (b)

Figure 22.22 Deadly toxin-producing mushrooms.
Amanita mushrooms kill by means of a toxin that inhibits RNA polymease. **(a)** *Amanita muscaria*, which was formerly used as an insecticide. It was sprinkled with sugar and set out to attract flies, which died after nibbling on the insecticide-laced sugar. *(W. Ormerod/Visuals Unlimited)* **(b)** *Amanita virosa*, commonly called the "destroying angel." *(Forest W. Buchanan/Visuals Unlimited)*

muscle coordination, tremors, and weight loss. Some are carcinogenic.

Aspergillus flavus and other aspergilli produce poisonous substances called **aflatoxins** (af″lah-tox′inz). Aflatoxins are the most potent carcinogens yet discovered. Although the effects of the toxins on humans are not fully understood, their presence in foodstuffs may cause cancer of the liver. The toxins reach humans in food made from mold-infested grain and peanuts.

Claviceps purpurea is a fungus that grows parasitically on rye and wheat **(Figure 22.21)**. **Ergot** (er′got)

is the name of the vegetative structure *C. purpurea* forms, as well as the name of the toxin and the disease it causes in grains. Although most strains of these grains grown in the United States are genetically resistant to the ergot fungus, many strains grown in other parts of the world are not. Ergot causes a variety of effects on humans. When the fungus is harvested with rye and incorporated into foodstuffs, it can cause **ergot poisoning**, or *ergotism*—hallucinations, high fever, convulsions, gangrene of the limbs, and ultimately death. The same substance that causes ergot poisoning in large quantities can be used therapeutically in small quantities. Drugs derived from ergot are used in carefully measured doses to control bleeding in childbirth, induce abortions, treat migraine headaches, and lower high blood pressure.

When Napoleon was in the Ukraine, he received bad grain and his horses became afflicted with ergot poisoning.

Mushroom toxins are found mainly in various species of *Amanita*, which are widely distributed around the world **(Figure 22.22)**. The mushroom toxins phallotoxin and amatoxin act on liver cells. They cause vomiting, diarrhea, and jaundice. Ingestion of a sufficient quantity of the toxin can be lethal or can so damage the liver that an organ transplant is necessary.

(a)   LM

Ergotamine

(b)

Figure 22.21 Ergot-producing fungus. **(a)** Micrograph of a cross-section through the parasitic fungus *Claviceps purpurea* (65X), which produces ergot. *(John D. Cunningham/Visuals Unlimited)* **(b)** The structure of ergotamine, a toxic alkaloid produced by the ergot-producing mold.

✓ CHECKLIST

1. Which types of hepatitis can be prevented by vaccines?
2. Who should receive hepatitis vaccines?
3. What is aflatoxin? Its source? Its effects?

PUBLIC HEALTH

Sushi

In Japan, several hundred people each year find out what it is like to ingest a live worm with their tasty snack. Typically they consume sushi at dinner and awake in the small hours of the morning with such agonizing pain that they are rushed to the hospital. There, examination of the stomach with a fiber-optic gastroscope shows *Anisakis* larvae penetrating the gastric or duodenal mucosa. In some cases, the damage by this round-worm is so severe that a portion of the stomach wall must be removed. In more fortunate cases, the worm larvae can be removed by forceps attached to the gastroscope. This infection is called anisakiasis (an"is-a-ki'a-sis). There is no treatment for it other than removal of the larvae.

What are the odds that a sushi dinner will give you anisa-kiasis in the United States? Only a half dozen or so cases have been reported in the United States; curiously, these cases tended to involve problems somewhat higher up in the digestive tract. A California man had eaten raw white sea bass 10 days before he felt a peculiar sensation at the back of his throat. Coughing, he reached into his mouth and pulled out a lively 8.5-cm-long worm. The other U.S. victims have similarly coughed up and removed their own worms—one as quickly as 4 hours after a meal of cod fillet. Still, considering the number of people in the United States who eat raw or undercooked fish, the total number of diagnosed cases of anisakiasis is so low that there is not much to worry about—so far!

(R. Marcialis/Photo Researchers, Inc.)

HELMINTH GASTROINTESTINAL DISEASES

A wide variety of helminths can parasitize the human intestinal tract, and some also invade other tissues. Although most are prevalent only in tropical regions, several are endemic to the United States. Health care workers in the United States should be alert to the possibility that patients may have acquired such parasites while traveling in tropical areas.

Fluke Infections

Foodborne fluke infections affect 40 million people worldwide. The sheep liver fluke *Fasciola hepatica* (**Figure 22.23**) is the most thoroughly studied of all flukes. It is found in humans in South America, Cuba, northern Africa, and some parts of Europe. Its intermediate host is a snail. Cercaria (larval forms) develop in the snail and upon their release mature in water and encyst as metacercaria on water vegetation. When humans eat such vegetation, especially watercress, the metacercaria are released in the large intestine, bore through the intestinal wall, and migrate to the liver. There they feed on blood, block bile ducts, and cause inflammation. Sometimes they migrate to the eyes, brain, or lungs. Adult flukes can be found in the bile ducts and the gallbladder. They live about 10 years. Infections can be diagnosed by finding eggs in stool specimens. Infected humans can be treated with bithionol and other anti-helminthic agents. Infection can be prevented by not eating water vegetation unless it has been cooked.

The Chinese liver fluke *Clonorchis sinensis* is widely distributed in Asia. As much as 80% of the population in rural areas is infected; some travelers and users of raw imported products also become infected. The life cycle is similar to *Fasciola's* except that a second intermediate host, typically a fish but sometimes a crustacean, is required. Metacercaria excyst (emerge from a cyst) in the duodenum and migrate to the liver, and adult flukes take up residence in bile ducts. They destroy bile duct epithelium, block ducts, and sometimes perforate and damage the liver. The incidence of liver cancer is unusually high in areas where fluke infections are high, but it is unknown

Figure 22.23 *Fasciola hepatica*, **the sheep liver fluke.** This specimen has been stained to reveal its internal structures (2.5X). *(Bruce Iverson/Photo Researchers, Inc.)*

whether the fluke is responsible. Finding eggs in feces confirms the diagnosis, but there is no effective treatment. The parasite could be eradicated by cooking fish and crustaceans. However, the cultural habit of eating raw fish and shellfish and the lack of fuel for cooking maintain infections.

Yet another fluke, *Fasciolopsis buski*, is common in pigs and humans in the Orient. It lives in the small intestine and causes chronic diarrhea and inflammation. If several flukes are present, they can cause obstruction, abscesses, and **verminous intoxication**, an allergic reaction to toxins in the flukes' metabolic wastes. Antihelminthic drugs can be used to rid the body of flukes. Human infection can be prevented by controlling snails, avoiding uncooked vegetation, and stopping the use of human excreta for fertilizer.

Tapeworm Infections

Human tapeworm infections can be caused by several species, most of which have worldwide distribution. Illustrations of the tapeworm life cycle were provided in ◄Chapter 11 (p. 328). Humans are most often infected by eating uncooked or poorly cooked, contaminated pork or beef. Tapeworm infections can also occur through contact with infected dogs or from eating infected raw fish.

The pork tapeworm *Taenia solium* **(Figure 22.24a)**, reaches a length of 2 to 7 m, and the beef tapeworm, *T. saginata* reaches a length of 5 to 25 m. These worms usually enter the body as larvae in raw or poorly cooked meat, especially pork. Viable larvae can develop into adult tapeworms. When adult tapeworms develop in the intestine, they absorb large quantities of nutrients and lead to malnutrition even when the person has an adequate diet. Long, ribbonlike worms may tangle up into a mass that blocks passage of materials through the intestine. Internal autoinfection can occur if eggs are released while the proglottids are still inside the intestine, each segment carrying about 60,000 eggs. The eggs can invade the bloodstream and spread to other body sites, most worrisomely to the central nervous system (CNS).

When humans ingest or are autoinfected by tapeworm eggs (or ova) instead of larvae **(Figure 22.24b)**, the egg coverings disintegrate in the small intestine, and the released larvae penetrate the intestinal wall and

(a) SEM

(b) LM

(c) LM

(d)

Figure 22.24 Tapeworms. (a) A colorized SEM of the scolex, or head, of *Taenia solium*, showing suckers and the double ring of 22 to 36 movable hooks, which aid in attaching to the intestinal lining of its host (224X). *(Stanley Flegler/Visuals Unlimited)* **(b)** Proglottids, or body segments, of *T. pisiformis* (2X). Small new ones grow behind the scolex; they increase in size as they age and move backward away from the scolex. The last row of proglottids is filled with mature eggs. *(Bruce Iverson/Photo Researchers)* **(c)** Protoscolices of *Echinococcus granulosus* (308X). Each round structure, the size of a grain of sand, contains the head of a tapeworm and can grow into a complete worm. As many as millions of such structures are found in a fluid-filled hydatid cyst, which can contain up to 15 liters of fluid. When even a more moderately sized cyst forms in the brain, it can do extensive damage. *(Larry Jensen/Visuals Unlimited)* **(d)** Hydatid cysts in brain. *(Courtesy Dr. Ana Flisser, Facultad de Medicina, National University of Mexico)* enter the blood. In 60 to 70 days, a larva migrates to various tissues and develops into a **cysticercus**, or bladder worm (◀Chapter 11, p. 329). The bladder worm consists of an oval white sac with the tapeworm head invaginated into it. In humans, cysticerci are frequently found in the brain, where they can reach a diameter of 6 cm. When pigs ingest tapeworm eggs, the cysticerci migrate to muscle and encyst. The pigs' defenses cause the encysted larvae to be surrounded by calcium deposits. In humans, calcification fails to stop the growth of cysticerci; if vital organs such as brain, heart, or lungs are involved, patients may suffer paralysis and convulsions. When a cysticercus dies, it releases toxins and usually causes a severe, or even fatal, allergic response. Human tapeworm infections can be prevented by the sanitary disposal of human wastes and by thorough cooking of meats and fish. Even freezing meats at −5°C for at least a week appears to kill the parasites.

Humans ingest eggs of *Echinococcus granulosus* tapeworms through contact with infected dogs, especially when the dogs lick the faces of small children. Eggs of this tapeworm are especially likely to produce cysts, called **hydatid cysts (Figure 22.24c)**, in vital tissues such as the liver, lungs, and brain (◀Chapter 11, p. 329). The cysts, which can contain hundreds of tiny immature worm heads and often reach the size of a grapefruit or larger, exert pressure on organs. If a cyst ruptures, as may happen during an attempt at surgical removal, it can release all these infective units. A ruptured cyst can also cause severe allergic reactions such as anaphylactic shock.

Humans can ingest *Hymenolepis nana* tapeworm eggs in cereals or other foods that contain parts of infected insects. Such tapeworms in the intestine, especially in children, can cause diarrhea, abdominal pain, and convulsions. Worldwide it is the most common tapeworm infecting humans, and its incidence is rising in the United States.

The broad fish tapeworm, *Diphyllobothrium latum*, is common in fish-eating carnivores. It reaches humans through ingestion of raw or poorly cooked, contaminated fish in sushi and other dishes. Fish tapeworm infections are common in Scandinavia, Russia, and the Baltic—approaching 100% infestation of the population in some areas—and have occurred in the Great Lakes area of the United States. The worm requires as intermediate hosts both a small crustacean and a fish to complete its life cycle. When humans ingest infected fish tissue, the worms coiled

PUBLIC HEALTH

With Immigration Comes Tapeworm Baggage

Taenia solium is being carried around the world inside human and pig hosts. It is most common in parts of Central and South America, Africa, and Asia (see map), where it is endemic in rural areas. A new significant public health problem is appearing in developed countries—CNS symptoms (headaches, seizures, and hydrocephalus) due to neurocysticercosis. The International League Against Epilepsy recently called it the main cause of epilepsy in the world. The pig tapeworm requires humans as its definitive host, where sexual reproduction takes place. Humans acquire the worms by eating cysticerci in undercooked pork. Once established in the human intestine, the worms shed great quantities of eggs in feces.

In some underdeveloped countries with poor hygiene, pigs running loose eat human feces, or are actually kept in "pig-sty privies," where they are purposefully fed human feces. This ensures infection of almost all pigs, some with as many as 2,500 cysticerci per kilogram of meat. Lacking education, and understanding of infection, the human owners think this is a fine arrangement: free, good nutrition for pigs, and clean-up of feces in a place without sewage treatment facilities. Subsequently, the sale of such pigs to areas free of *T. solium* has led to its establishment there, plus neurocysticercosis among the unsuspecting human populace. Such was the case when Balinese pigs were given to the native people of another island by the Indonesian government in an attempt to get them to accept Indonesian rule.

But one need not eat pork or come in contact with pigs in order to suffer from cysticercosis—all you need to do is swallow some *T. solium* eggs, possibly from the unwashed hands of a worm-infected food handler. Most U.S. cases are associated with Mexican immigrants or visitors to Mexico.

■ High or moderate prevalence
□ Low or moderate prevalence

up in the fish muscle reach and mature in the intestine. Adult worms attach to the intestine and begin producing eggs. While attached, the parasites absorb large quantities of vitamin B_{12} and impair the victim's ability to absorb the vitamin. Vitamin B_{12} deficiency, or pernicious anemia, from tapeworm infections is especially high in Finland.

Tapeworm infections are diagnosed by finding eggs or proglottids in feces. Infections can be treated with niclosamide and other antihelminthic agents. Prompt diagnosis and treatment are important to remove worms before they invade tissues beyond the intestine. Human infections could be completely prevented by avoiding raw meats and fish and infected dogs.

Trichinosis

Trichinosis (trik″in-o′sis) is caused by the small round-worm *Trichinella spiralis*, also sometimes called *trichina worm*. This parasite, unlike most, is more common in temperate than in tropical climates. (Almost all U.S. adults have antibodies to *Trichinella* and therefore carry around a few worms. But a few will not cause symptoms.) The parasite usually enters the digestive tract as encysted larvae **(Figure 22.25)** in poorly cooked pork, but infections have been traced to venison and meat from other game animals and to horse meat in France. In the intestine, cysts release larvae that develop into adults. The adults mate, the males then die, and the females produce living larvae before they, too, die. The larvae migrate through blood and lymph vessels to the liver, heart, lungs, and other tissues. When they reach skeletal muscles, especially eye, tongue, diaphragm, and chewing muscles, they form cysts. The formation of cysts in humans represents a dead-end for the worms, as they will not be eaten and passed on to another host. Encysted worms remain alive and infectious for years.

These parasites cause tissue damage as adults and as migrating and encysted larvae. The adult females penetrate the intestinal mucosa and release toxic wastes that produce symptoms similar to those of food poisoning. Wandering larvae damage blood vessels and any tissues they enter. Death can result from heart failure, kidney failure, respiratory disorders, or reactions to toxins. Encysted larvae cause muscle pain. Trichinosis is difficult to diagnose, but muscle biopsies and immunological tests are sometimes positive. Treatment is directed toward relieving symptoms because the disease cannot be cured. It can be prevented by eating only thoroughly cooked meat. Freezing does not necessarily kill encysted larvae, and microwave cooking is safe only if the internal temperature of the meat reaches 77°C. Microwave cooking depends on geometry. Because pieces of meat are irregularly shaped, they must be rotated during cooking to heat evenly. Studies of pork experimentally infected with *Trichinella* show survival of some worms whenever microwave cooking is done without rotation. By law, pigs in the United States must not be fed garbage or restaurant scraps unless it has been cooked first. This prevents raw pork scraps, containing live trichina worms, from being recycled back into pigs, which they will then infect. Therefore, it is not very likely that you will get trichinosis from U.S.-raised pork. The average number of cases diagnosed per year in the United States is less than 30 **(Figure 22.26)**.

Hookworm Infections

Hookworm is most often caused by one of two species of small (11 to 13 mm long) roundworms—*Ancylostoma duodenale* and *Necator americanus*. Although these parasites have a complex life cycle, that cycle can occur in a single host, and the host is often a human. Eggs in feces quickly hatch in moist soil. There the eggs release free-living larvae that feed on bacteria and organic debris, grow, molt, and become mature parasitic larvae. If these larvae reach the skin, typically of the feet or legs, they burrow through it to reach blood vessels that carry them to the heart and lungs. The larvae then penetrate lung tissue, and some are coughed up and swallowed. In the intestine the larvae burrow into villi and mature into adult worms. The adult worms mate and start the cycle over again. Worldwide, an estimated 500 million people are infested with hookworms.

As hookworm larvae burrow through the skin, host inflammatory reactions kill many of them, although bacterial infection of penetration sites causes **ground itch**. In the lungs the parasites cause many tiny hemorrhages, but they cause the greatest damage to the lining of the entire small intestine. They feed on blood and cause abdominal pain, loss of appetite, and protein and iron deficiencies, so people infected with hookworms often appear to be lazy. These effects are especially debilitating in people whose diets are barely adequate without the burden of worm infections.

Diagnosis is made by finding eggs or worms in feces, but samples must be concentrated to find them. Tetrachloroethylene is effective against *Necator* infections. It is inexpensive and easy to administer in mass treatment efforts. Bephenium hydroxynaphthalate (Alcopar), mebendazole, and several other drugs, although more expensive, kill both species of hookworms. Dietary supplements, especially iron, should be provided for all

Figure 22.25 *Trichinella spiralis.* This specimen is curled up as a cyst, embedded in striated muscle fibers (1,020X). *(James Solliday/Biological Photo Service)*

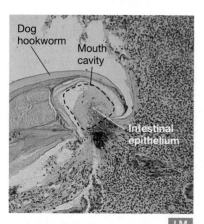

Figure 22.26 Trichinosis. Reported cases, by year—U.S., 1971–2006. (*Source:* CDC.)

In 2001, 22 cases of trichinosis were reported from seven states (Alaska, California, Illinois, New Jersey, Wisconsin, and Wyoming). The year 2006 was the eleventh consecutive year in which <25 new cases were reported.

The intestinal tracts of many cats dissected in anatomy labs are infested with Ascaris lumbricoides, whose eggs may still be viable.

hookworm patients. Hookworm is preventable through sanitary disposal of human wastes, but stopping the use of human excreta as fertilizer and getting uneducated people to use latrines can be difficult. Plantation workers often defecate repeatedly in areas near fields where they work. If infected, such individuals serve as a continuous source of larvae to themselves and to others.

In 1991 cases of human hookworm infection by the dog hookworm *Ancylostoma caninum* (**Figure 22.27**) were linked with intestinal problems such as diarrhea, abdominal pain, and weight loss. Larvae of other species of hookworm for which humans are not the normal host sometimes penetrate the skin and cause *cutaneous larva migrans*, or *creeping eruption*. Severe skin inflammation results from body defenses that prevent further migration of the parasites. Such infections often are acquired from infected cats and dogs and can be treated with thiabendazole.

Ascariasis

Ascaris lumbricoides (**Figure 22.28**) is a large roundworm 25 to 35 cm long that causes **ascariasis** (as″kar-i′a-sis). People become infected by ingesting food or water contaminated with *Ascaris* eggs. Once in the intestine, the eggs hatch, and larvae penetrate the intestinal wall and enter lymph vessels and venules. Although the larvae can invade and cause immunological reactions in almost any tissue, most move through the respiratory tract to the pharynx and are swallowed. Larvae move to the small intestine, mature, and begin to produce eggs. A single female worm produces an average of 200,000 eggs per day, with a total lifetime output of about 26 million eggs. A female lives 12 to 18 months. Eggs are especially resistant to acids and can develop in 2% formalin. They also resist drying, and people can be

Figure 22.28 *Ascaris lumbricoides,* **a large roundworm.** Females may reach a length of 35 cm and can produce several hundred eggs per day. The eggs are passed out in feces and can survive in soil for months or even years. These approximately 800 worms were all removed from the ileum of one child at autopsy. (*Courtesy Dr. Daniel H. Connor, Armed Forces Institute of Pathology*)

Figure 22.27 Dog hookworm, *Ancylostoma caninum.* This roundworm is about 1 cm long, attached to intestinal epithelium, where it sucks blood (250X). It has taken a large piece of intestine into its mouth. (Heavy infestations may result in anemia.) (*R. Calentine/ Visuals Unlimited*)

infected by airborne eggs. In some areas of the southern United States where the soil never freezes, 20 to 60% of children are infected, usually with 5 to 10 worms. Worldwide, 25% of the population is infested with *Ascaris* worms. Mebendazole and pyrantel are effective in treatment.

Ascaris worms cause three kinds of damage:

1. Larvae burrowing through lung tissue cause *Ascaris* pneumonitis, which involves hemorrhage, edema, and blockage of alveoli from worms, dead leukocytes, and tissue debris. If secondary bacterial pneumonia develops, it can be fatal.

2. Adult worms cause malnutrition, but because they feed mainly on the contents of the intestine, they do little damage to the mucosa. They also release toxic wastes that elicit allergic reactions. If sufficiently numerous, they cause intestinal blockage and sometimes perforation. The peritonitis that follows perforation is nearly always fatal.

3. Wandering worms cause abscesses in the liver and other organs and sometimes traumatize victims or doctors examining them for other disorders by crawling from body openings such as the nose or umbilicus.

Diagnosis is made by finding eggs or worms in feces. The adult worms can be eradicated from the body by piperazine, mebendazole, and several other drugs, but no treatment is available to rid the body of larvae. Piperazine temporarily relaxes the worm's muscles. Failing to move against the current, and to press closely against the wall of the intestine, allows peristalsis to carry the worms out through the anus. Infection is fully preventable by good sanitation and personal hygiene.

Toxocara species, another type of roundworm, ordinarily parasitize cats and dogs. It has been estimated that in the United States, 98% of puppies, including those from good kennels, are infected. Infection rates may be almost as high in dogs and cats of any age. *Visceral larva migrans* is the migration of larvae of these parasites in human tissues such as liver, lung, and brain, where they cause tissue damage and allergic reactions. The risk of such human infections can be minimized by worming pets periodically, disposing of pet wastes carefully, and keeping children's sandboxes inaccessible to pets.

Trichuriasis

Trichuriasis (trik″u-ri′a-sis) is caused by the **whipworm**, *Trichuris trichiura*, which is distributed nearly worldwide. It is estimated that nearly 300 million people are infected, some of them in the southeastern United States. For humans to be infected, human feces must be deposited on warm, moist soil in shady areas. Small children, who put dirty hands in their mouth, are particularly susceptible to infection. Eggs deposited with feces contain partially developed embryos. When the eggs are swallowed, they hatch. Juveniles crawl into enzyme-secreting glands of the intestine called crypts of Lieberkühn, where they develop.

They return to the intestinal lumen (central space), where they reach full maturity within 3 months of initial infection.

Adult whipworms damage the intestinal mucosa and feed on blood. They cause chronic bleeding, anemia, malnutrition, allergic reactions to toxins, and susceptibility to secondary bacterial infection. Infections can result in rectal bleeding in children. Infection is diagnosed by finding worm eggs in stools. The drug mebendazole is effective in ridding the intestine of parasites. Sanitary disposal of wastes is essential to prevent reinfection.

Strongyloidiasis

Strongyloidiasis (stron″jil-oi-di′a-sis) is caused by *Strongyloides stercoralis* (**Figure 22.29**). This parasite is unusual in that females produce eggs by *parthenogenesis*—that is, without fertilization by a male. In fact, no male of this species has ever been reliably identified in the parasitic stage. Adult females, about 2.2 mm long by 0.04 mm wide, attach to the small intestine, burrow into underlying layers, and release eggs containing noninfective larvae. Many eggs hatch in the intestine and are passed with the feces. In soil, larvae can become free-living adults or develop into infective larvae and penetrate the skin of new hosts. Infective larvae, which penetrate the skin, are carried by blood to the lungs. There they bore their way to the trachea, travel to the pharynx, and are swallowed. When they arrive in the small intestine, they develop into adults and restart the life cycle.

The *Strongyloides* larvae cause itching, swelling, and bleeding at penetration sites, which often become infected with bacteria. Migrating larvae cause immunological reactions in the host, but the reactions usually do not stop the parasites. Coughing and burning sensations in the chest

Figure 22.29 *Strongyloides stercoralis.* This is a roundworm parasite of the small intestine (675X). *(Arthur M. Slegelman/ Visuals Unlimited)*

LM

TABLE 22.4

A Summary of Gastrointestinal Diseases Caused by Pathogens Other Than Bacteria

Disease	Agent(s)	Characteristics
Viral Gastrointestinal Diseases		
Viral enteritis	Rotaviruses	Viral replication destroys intestinal epithelium; causes diarrhea and dehydration that can be fatal under age 5
Hepatitis A	Hepatitis A virus	Viral replication in intestinal and other cells causes malaise, nausea, diarrhea, abdominal pain, lack of appetite, fever, and jaundice; usually self-limiting; vaccine available
Hepatitis B	Hepatitis B virus	Viral replication and symptoms similar to hepatitis A, except onset is insidious and fever usually absent; chronic infections and carrier state occur; transmitted parenterally; vaccine available
Hepatitis C	Unknown, but may involve two agents	Mild symptoms, but disease often chronic
Hepatitis D	Hepatitis D and B viruses	Can result in fatal hepatitis
Hepatitis E	Hepatitis E virus	Moderate disease, but high mortality in pregnant women
Protozoan Gastrointestinal Diseases		
Giardiasis	*Giardia intestinalis*	Parasite attaches to intestinal wall; feeds on mucus and causes inflammation, diarrhea, dehydration, nutritional deficiencies, and sometimes reactive arthritis
Amoebic dysentery	*Entamoeba histolytica*	Parasites ulcerate mucosa and cause severe acute diarrhea, abdominal tenderness, and dehydration; parasites can live indefinitely in intestine, causing latent amebiasis
Balantidiasis	*Balantidium coli*	Organisms invade walls of intestine; cause dysentery and sometimes perforation and peritonitis
Cryptosporidiosis	*Cryptosporidium* species	Organisms live in or under mucosal cells and cause severe diarrhea in immunocompromised patients
Cyclosporiasis	*Cyclospora cayetanensis*	Dysentery, often relapsing—severe in immunocompromised patients
Helminth Gastrointestinal Diseases		
Fluke infections	*Fasciola hepatica, Clonorchis sinensis, Fasciolopsis buski*	Organisms excyst in intestine and migrate to liver, block ducts, and damage tissues; *F. buski* causes intestinal obstruction, abscesses, and verminous intoxication
Tapeworm infections	*Taenia solium, T. saginata, Echinococcus granulosis, Hymenolepis nana, Diphyllobothrium latum*	Most infections caused by ingesting encysted larvae, which mature and erode intestinal mucosa; ingestion of eggs allows larval forms to develop in humans as cysticerci or hydatid cysts that can damage the brain and other vital organs
Trichinosis	*Trichinella spiralis*	Larvae excyst in intestine, mature, and produce new larvae, which migrate to various tissues and encyst in muscles, where they cause pain; larvae cause tissue damage, and adults release toxins
Hookworm infections	*Ancylostoma duodenale, Necator americanus*	Larvae penetrate skin and migrate via blood vessels to heart and lungs; coughed-up larvae enter digestive tract and burrow into villi; mature worms feed on blood, cause abdominal pain, loss of appetite, and protein and iron deficiencies
Ascariasis	*Ascaris lumbricoides*	Eggs hatch in the intestine, and larvae cause immunological reactions in many tissues; adults cause malnutrition by feeding on nutrients in gut; wandering worms cause abscesses
Trichuriasis (whipworm)	*Trichuris trichiura*	Eggs hatch in the intestine, and juveniles invade crypts of Lieberkühn and mature; adults damage intestinal mucosa, causing chronic bleeding, anemia, malnutrition, and allergic reactions to toxins
Strongyloidiasis	*Strongyloides stercoralis*	Filariae penetrate the skin, bore to trachea, climb to pharynx, and are swallowed; mature worms bore into intestine and release embryos; cause inflammation, bleeding, and immunologic reactions at various sites, severe diarrhea and fluid loss
Pinworm infections	*Enterobius vermicularis*	Ingested eggs give rise to mature worms in the intestine; infection interferes with nutrition, especially in children

occur in lung infections, and burning and ulceration occur in intestinal infections. Secondary bacterial infection in any tissue can lead to septicemia. The diarrhea and fluid loss associated with intestinal parasites are severe and difficult to control even with electrolyte therapy. Consequently, patients often die of complications such as heart failure or paralysis of respiratory muscles. Immunosuppressed patients can die when worms are disseminated beyond the gut, into lungs and meninges.

Strongyloides infection in humans usually occurs when larvae are encountered in contaminated soil or water. Diagnosis is difficult because larvae can be found in fecal smears only in massive infections. Work is in progress to develop a reliable immunological test. The drugs thiabendazole and cambendazole have the greatest effect on the parasites with the fewest undesirable side effects.

Pinworm Infections

Pinworm infections are caused by the small roundworm *Enterobius vermicularis*. Like the hookworm, this parasite can complete its life cycle without an alternate host. In fact, humans are its only known host. It has the greatest geographic distribution of any human worm parasite in the world. While it is more common in lower socioeconomic groups, it can be found even in the seats of the mighty. An estimated 209 million people worldwide are infested with pinworms, 18 million of these in the United States and Canada. Adult pinworms attach to the epithelium of the large intestine and mate, and the females produce eggs. Egg-laden females, containing 11,000 to 15,000 eggs, migrate toward the anus during the night, release their eggs on the exterior of the anus, and then crawl back in. Lower temperature and aerobic environment are the factors that trigger egg laying. These eggs are easily transmitted

Eggs of the dog pinworm (which does not infect humans) can infect your dog when he stops to sniff other dogs' feces.

to other people by bedclothes, by debris under fingernails of those who scratch the itchy area around the anus, and even by inhalation of airborne eggs. Ingested eggs hatch in the small intestine and release larvae that mature and reproduce in the large intestine. Some few worms may travel upward to the stomach, esophagus, and nose. Others may reach the bladder, the uterus, and via the fallopian tubes the peritoneal cavity.

Although pinworm infection usually is not debilitating, it does cause considerable discomfort and can interfere with adequate rest and nutrition, especially in children. Infection with large numbers of the worms can cause the rectum to protrude from the body. Pinworm infections are diagnosed by finding eggs around the anus; at night or right after the host awakens, the pinworms can be picked up with the sticky side of transparent cellophane tape (translucent will not work) affixed to a wooden tongue depressor. If one member of a family has pinworms, all are presumed infected and are treated with piperazine or another antihelminthic agent. These agents are generally inexpensive and nontoxic. Bed linens, clothing, and towels should be washed and the house thoroughly cleaned at the time of treatment. The treatment and cleaning are repeated in 10 days. Despite these efforts, reinfection in families is very likely. In the absence of reinfection, the infection is self-limiting and will cease without treatment. Parents sometimes need to be reminded that this is a non-fatal infection (◀Figure 14.2, p. 402).

Gastrointestinal diseases caused by pathogens other than bacteria are summarized in **Table 22.4**.

✓**CHECKLIST**

1. Describe the life cycle of a tapeworm. Where do the stages of cysticercus and hydatid cysts fit in the life cycle?
2. Can you safely cook pork in a microwave oven?
3. Describe the technique used to diagnose a case of pinworms.

▌ RETRACING OUR STEPS

• The agents and characteristics of the diseases discussed in this chapter are summarized in Tables 22.1, 22.2, and 22.4. Information in those tables is not repeated here.

▌▌ COMPONENTS OF THE DIGESTIVE SYSTEM

• The **digestive system** is an elongated tube comprised of the mouth, pharynx, esophagus, stomach, and intestines, along with the accessory organs—teeth, salivary glands, liver, gallbladder, and pancreas.
• The five major functions of the digestive system are:
1. *movement* of food through the tract
2. *secretion* of digestive juices and mucus
3. *digestion* of food

4. *absorption* of digested food into the bloodstream
5. *elimination* of undigested food components and intestinal microflora
• Nonspecific defenses include mucus, stomach acid, Kupffer cells in liver sinusoids, patches of submucosal lymphatic tissue, and competition from normal flora.

NORMAL MICROFLORA OF THE MOUTH AND DIGESTIVE SYSTEM

• Perhaps a thousand species of microbes live within the human oral cavity, and over half of these have not yet been identified. Their ecology is poorly understood.
• The esophagus lacks a permanent normal microflora, as does the stomach. The first two-thirds of the small intestine

contains mainly transient microbes, but the last third's surface is colonized mainly by Enterobacteriaceae and obligate anaerobes. In the large intestine, feces are approximately 50% bacteria, mostly *Bacteroides*.

- Bacteria of the large intestine produce important metabolic by-products, such as vitamin K.

III DISEASES OF THE ORAL CAVITY

BACTERIAL DISEASES OF THE ORAL CAVITY

- **Dental plaque** is a continuously formed coating on teeth that consists of microorganisms in organic matter.
- **Dental caries** (tooth decay) is the chemical breakdown of enamel and deeper parts of the teeth. Caries can be prevented by good oral hygiene, use of **fluoride**, and application of sealants.
- **Periodontal disease** involves the gums, periodontal ligaments, and alveolar bone and can be prevented by prevention of plaque buildup. It is treated with plaque removal techniques, mouth rinses, peroxide-sodium bicarbonate, surgery, or antibiotics.

VIRAL DISEASES OF THE ORAL CAVITY

- **Mumps** is transmitted by saliva or as an aerosol and occurs worldwide, mainly in children. It can be prevented by a vaccine.

III GASTROINTESTINAL DISEASES CAUSED BY BACTERIA

BACTERIAL FOOD POISONING

- **Food poisoning** is caused by ingesting food containing pre-formed toxins. It can be prevented by sanitary handling and proper cooking and refrigeration of foods. Bacterial toxins in foods cause intoxications.
- Staphylococcal **enterotoxicosis** usually results from eating contaminated, poorly refrigerated foods, especially dairy and poultry products.
- Other kinds of food poisoning usually arise from eating undercooked, contaminated meats and gravies and rice.

BACTERIAL ENTERITIS AND ENTERIC FEVERS

- **Enteritis** is an inflammation of the intestine. **Enteric fever** is a systemic disease caused by pathogens that invade other tissues. All enteritises and enteric fevers are transmitted via the fecal-oral route and can be prevented by good sanitation.
- **Salmonellosis** is a self-limiting form of enteritis that is treated with antibiotics only in high-risk patients.
- **Typhoid fever** is an enteric infection caused by the bacterium *Salmonella typhi*. The disease is treated with chloramphenicol. Cell-mediated immunity follows infection. The available vaccine is limited in its effectiveness.
- **Shigellosis**, or **bacillary dysentery**, is treated with antibiotics, and recovery does not produce immunity.
- **Asiatic cholera**, common in Asia and other regions with poor sanitation, is treated with fluid and electrolyte replacement and tetracycline. Recovery can produce a long-lasting immunity, but the duration is unknown. Vaccines are not effective in cases of massive exposures.
- **Vibriosis** is a mild disease common where raw seafood is eaten.
- **Traveler's diarrhea** occurs in more than 1 million travelers each year. It is usually self-limiting but can have complications.
- *Escherichia coli* is an indicator of fecal contamination and an opportunistic pathogen.

BACTERIAL INFECTIONS OF THE STOMACH, ESOPHAGUS, AND INTESTINES

- *Helicobacter pylori* is considered to be the cause of peptic ulcers and chronic gastritis and a probable cofactor of stomach cancer.

BACTERIAL INFECTIONS OF THE GALLBLADDER AND BILIARY TRACT

- Bile destroys most organisms that have lipid envelopes. *Salmonella typhi* is resistant to bile; it can live in the gallbladder and be shed in feces without causing symptoms. Gallstones blocking bile ducts can cause infections of the gallbladder and ducts, usually by *E. coli*. Infection can spread to the bloodstream or ascend to the liver.

III GASTROINTESTINAL DISEASES CAUSED BY OTHER PATHOGENS

VIRAL GASTROINTESTINAL DISEASES

- Most viral gastrointestinal diseases arise from the ingestion of contaminated water or food and are transmitted via the fecal-oral route, but hepatitis B, C, and D are transmitted parenterally from contaminated blood and other body fluids.
- **Rotavirus** infections kill many children in developing countries.
- Treatment for viral **hepatitis** relieves symptoms. Gamma globulin gives temporary immunity to **hepatitis A**. Vaccines are available for both hepatitis A and **hepatitis B**. **Hepatitis D** can only infect in the presence of hepatitis B.

PROTOZOAN GASTROINTESTINAL DISEASES

- Protozoan gastrointestinal diseases arise from contaminated food and water via the fecal-oral route and can be prevented by good sanitation.
- Diagnosis usually is made by finding protozoan cysts in fecal matter. Most such diseases can be treated with antiprotozoan agents.
- **Giardiasis** is especially common in children. **Amoebic dysentery**, **chronic amebiasis**, and **balantidiasis** occur worldwide but mainly in tropical areas. **Cryptosporidiosis** occurs mainly in immunocompromised patients, but recent outbreaks have involved persons with normal immune systems.

EFFECTS OF FUNGAL TOXINS

- **Aflatoxins** are potent carcinogens produced by fungi of the genus *Aspergillus*; humans ingest them from moldy grain and peanuts.
- **Ergot poisoning** comes from eating grains contaminated with *Claviceps purpurea*. Small quantities of **ergot** toxin can be used medicinally.
- Mushroom toxins, which are associated mainly with species of *Amanita*, cause vomiting, diarrhea, jaundice, and hallucinations. In sufficient quantities, they are fatal.

HELMINTH GASTROINTESTINAL DISEASES

- Helminth gastrointestinal diseases are acquired mainly in tropical regions and include several kinds of fluke, roundworm, and tapeworm infections.
- Helminths that infect humans often have complex life cycles in which some animals, fish, vegetation, snails, and crustaceans may also serve as hosts.
- Most such diseases can be diagnosed by finding eggs in fecal specimens. They can be prevented by good sanitation—avoiding contaminated water and soil and thoroughly cooking foods that might be contaminated.

▌ TERMINOLOGY CHECK

acute necrotizing ulcerative
  gingivitis (ANUG)
  (*p. 681*)
aflatoxin (*p. 703*)
amebiasis (*p. 701*)
amoebic dysentery (*p.701*)
ascariasis (*p. 708*)
Asiatic cholera (*p. 688*)
bacillary dysentery (*p. 687*)
bacterial enteritis (*p. 685*)
balantidiasis (*p. 702*)
bilirubin (*p. 698*)
bongkrek disease (*p. 685*)
cementum (*p. 677*)
chronic amoebiasis (*p. 701*)
cryptosporidiosis (*p. 702*)
cysticercus (*p. 706*)
delta hepatitis (*p. 697*)
dental caries (*p. 679*)
dental plaque (*p. 679*)

diarrhea (*p. 684*)
digestive system (*p. 677*)
dysentery (*p. 685*)
enamel (*p. 677*)
enteric fever (*p. 685*)
enteritis (*p. 696*)
enterocolitis (*p. 686*)
enterohemorrhagic strain
  (*p. 691*)
enteroinvasive strain (*p. 690*)
enterotoxicosis (*p. 684*)
enterotoxigenic strain (*p. 691*)
enterotoxin (*p. 684*)
ergot (*p. 703*)
ergot poisoning (*p. 703*)
feces (*p. 678*)
fluoride (*p. 680*)
food poisoning (*p. 684*)
giardiasis (*p. 700*)
gingivitis (*p. 681*)

ground itch (*p. 707*)
hemorrhagic uremic
  syndrome (HUS) (*p. 692*)
hepatitis (*p. 696*)
hepatitis A (*p. 696*)
hepatitis B (*p. 697*)
hepatitis C (HCV) (*p. 697*)
hepatitis D (*p. 697*)
hepatitis E (HEV) (*p. 697*)
hookworm (*p. 707*)
hydatid cyst (*p. 706*)
infectious hepatitis (*p. 696*)
large intestine (*p. 678*)
microvillus (*p. 678*)
mucin (*p. 677*)
mumps (*p. 682*)
orchitis (*p. 683*)
periodontal disease (*p. 681*)
periodontitis (*p. 682*)
pinworm (*p. 711*)

rotavirus (*p. 696*)
salmonellosis (*p. 685*)
serovar (*p. 685*)
serum hepatitis (*p. 697*)
shigellosis (*p. 687*)
sinusoid (*p. 678*)
small intestine (*p. 678*)
strongyloidiasis (*p. 709*)
tartar (*p. 682*)
traveler's diarrhea (*p. 690*)
trichinosis (*p. 707*)
trichuriasis (*p. 709*)
typhoid fever (*p. 686*)
verminous intoxication
  (*p. 705*)
vibriosis (*p. 690*)
villus (*p. 678*)
viral enteritis (*p. 696*)
whipworm (*p. 709*)
yersiniosis (*p. 692*)

▌ CLINICAL CASE STUDY

A group of high school students from a military base in Germany went on a field trip to Russia and visited a soft drink bottling plant. They were all given free bottles of soft drink. After returning, students started coming to the base infirmary complaining of diarrhea, with frothy greasy stools, weight loss, and in some cases, fever or lactose intolerance. The students were diagnosed with giardiasis. The clinical laboratory at the military base found *Giardia intestinalis* in unopened soft drink bottles. What was the probable source of the *Giardia* found in the soft drink bottles? What could be done to prevent contamination of the soft drink? How would the clinical lab have diagnosed the cases of giardiasis? What is the minimum number of ingested cysts that may cause the disease? Visit the following web site to learn about this most common nonbacterial gastrointestinal pathogen. (FDA Bad Bug Book, available at: http://vm.cfsan.fda.gov/~mow/chap22.html.)

▌ CRITICAL THINKING QUESTIONS

1. Melany, a 59-year-old women with a history of heart disease, was placed on omeprazole (proton pump inhibitor), tetracycline, and bismuth to treat her chronic gastric ulcers. During a follow-up visit two weeks later she reported that she was pain free and did not feel limited by her ulcers anymore. Should Melany and her doctor be content with this outcome? Why or why not?

2. Typhoid fever and intestinal salmonellosis are caused by similar organisms. What are some important differences in the sources of these organisms and the diseases they produce?

3. A client has obvious symptoms of hepatitis: yellowing of her skin and eyes, anorexia, abdominal pain, enlarged liver, nausea, and diarrhea. Why is it important to make an accurate diagnosis of which type of hepatitis is present?

▌ SELF-QUIZ

1. Which of the following are incorrectly paired?
   (a) Oral cavity—colonized by hundreds of species of microbes
   (b) Esophagus—colonized by hundreds of species of microbes
   (c) Stomach—acid content kills many pathogens
   (d) Small intestine—first two-thirds colonized by transient microbes, last third mainly by Enterobacteriaceae and obligate anaerobes
   (e) Large intestine—colonized with many bacteria, many of which produce vitamin K

2. Which of the following is NOT true about dental plaque?
   (a) It is an intermittently formed coating of microbes and organic matter on tooth surfaces that serve as the first step in tooth decay and gum disease.
   (b) It begins as positively charged salivary proteins adhere to negatively charged enamel surfaces, forming a pellicle (film) over the tooth surface.
   (c) *Streptococcus mutans* attaches to the pellicle where it hydrolyzes sucrose to fructose and glucose, or polymers of glucose that serve as bridges, holding together cells in plaque.

(d) Plaque consists of up to 30 different genera of bacteria and their products, such as polymers of glucose (dextran), saliva proteins, and minerals.

(e) Plaque that is allowed to accumulate in crevices near the gumline can change into anaerobic pockets full of bacteria that can irritate the gums or destroy underlying bone.

3. Dental caries occurs when sugars diffuse through plaque to the embedded bacteria, which in turn ferment the sugars into organic _____ which cannot diffuse out of the area, resulting in a gradual _____ of the enamel.
   (a) Bases; precipitation
   (b) Bases; buildup
   (c) Acids; dissolving
   (d) Acids; osmosis
   (e) None of the above

4. All of the following are ways to prevent or treat dental caries EXCEPT:
   (a) Remove decay and fill cavities with resins or amalgams.
   (b) Limit the intake of sticky foods.
   (c) Brush with plaque-removing toothpaste and flossing between teeth.
   (d) Use fluoridated water, toothpastes, and mouthwashes.
   (e) Increase sugar intake to enhance increased salivary flow, which will wash away oral bacteria.

5. Periodontal disease occurs when bacteria become trapped in gingival crevices where they cause tooth decay, gum inflammation, and erosion of periodontal ligaments and alveolar bone that supports teeth. True or false?

6. Longstanding plaque that becomes calcified on its surface, forming a hard crust, is referred to as:
   (a) Crest
   (b) Calculus
   (c) Tartar
   (d) Colgate
   (e) b and c

7. A viral disease caused by paramyxoviruses that affect glands of the oral cavity, and sometimes other non–oral cavity glands and meninges, is called:
   (a) Measles
   (b) Mumps
   (c) Rubeola
   (d) AIDS
   (e) Thrush

8. Which of the following apply to *Staphylococcus aureus* enterotoxin?
   (a) It causes food poisoning.
   (b) It inflames the intestinal lining and inhibits intestinal water adsorption.
   (c) It is heat stable.
   (d) It stimulates the vomiting center in the brain.
   (e) All of the above.

9. What is the difference between bacterial food poisoning and bacterial enteritis?
   (a) Food poisoning is caused by an intestinal infection, whereas bacterial enteritis is caused by an enterotoxin.
   (b) Food poisoning is caused by bacterial pathogens, whereas bacterial enteritis is caused by members of the normal intestinal flora.
   (c) There is no difference, both are caused by enterotoxin.
   (d) Food poisoning is caused by an enterotoxin, whereas bacterial enteritis is caused by an intestinal infection.
   (e) There is no difference, both are caused by a bacterial infection of the intestine.

10. Match the following bacterial enteric diseases or terms to their descriptions:
    ___ Typhoid fever
    ___ Traveler's diarrhea
    ___ Vibriosis
    ___ Salmonellosis
    ___ Shigellosis
    ___ Asiatic cholera
    ___ Dysentery
    ___ Enteric fever

    (a) An enteritis caused by ingestion of raw or undercooked contaminated seafood
    (b) Systemic infection spread from intestinal mucosa
    (c) Enteritis of the large intestine causing diarrhea
    (d) Serious enteric fever caused by ingestion of *Salmonella typhi*–contaminated food or water
    (e) A self-limiting enteritis commonly caused by members of the genus *Salmonella* and associated with ingestion of improperly prepared, previously contaminated foods
    (f) Bacillary dysentery caused by serovars of *Shigella* that can produce fever-eliciting endotoxin and neural-acting Shiga toxin
    (g) Diarrhea caused most commonly by enteroinvasive or enterotoxigenic strains of *Escherichia coli*
    (h) Associated with fecal contamination of water, this enteritis causes massive electrolyte loss and diarrhea accompanied by severe nausea, vomiting, and abdominal pain

11. Which of the following organisms produces a potent toxin most similar to Shiga-toxin?
    (a) *Salmonella typhi*
    (b) *Campylobacter jejuni*
    (c) *Escherichia coli* O157:H7
    (d) *Clostridium botulinum*
    (e) *Clostridium perfringens*

12. Which of the following species of microbe is associated with being a good indicator of fecal contamination and an extremely versatile opportunistic pathogen?
    (a) *Escherichia coli*
    (b) *Campylobacter jejuni*
    (c) *Yesinia enterocolitica*
    (d) *Shigella dysenteriae*
    (e) *Salmonella typhi*

13. Complications arising following campylobacteriosis include which of the following?
    (a) Arthritis
    (b) Brain damage
    (c) Kidney damage
    (d) Paralysis
    (e) Blindness

14. From the following list, which is the BEST way to prevent outbreaks of Gram-negative gastroenteritis?
    (a) Cook foods hot and thoroughly
    (b) Salt foods
    (c) Add vinegar and spices to food
    (d) Freeze-dry foods
    (e) Refrigerate foods

15. *Helicobacter* is able to survive the acidic conditions of the stomach by producing which enzyme?
    (a) B-galactosidase
    (b) Coagulase
    (c) Lactase
    (d) Amylase
    (e) Urease

16. Which of the following are *incorrectly* matched?
    (a) Hamburger—*Escherichia coli* O157:H7 (EHEC)
    (b) Chickens and turkeys—*Salmonella enteritidis*
    (c) Cooked rice—*Bacillus cereus*
    (d) Potato salad—*Staphylococcus aureus*
    (e) None of the above are incorrectly matched

17. Which of the following cell wall components are most affected by the action of bile?
    (a) Lipids   (d) Porins
    (b) Proteins   (e) LPS
    (c) Carbohydrates

18. Rotavirus infection is a major cause of viral enteritis among infants and children. Which of the following is NOT true about rotaviruses?
    (a) They are transmitted by the fecal-oral route, damage intestinal epithelium, and cause a watery diarrhea.
    (b) They are a major cause of adult morbidity and mortality in developing countries.
    (c) They are reoviruses that replicate their double-stranded RNA in great numbers.
    (d) Breast-fed infants are protected by anti-rotavirus antibodies and the complex carbohydrate lactadherin.
    (e) They are not the only major cause of viral enteritis.

19. Which of the following hepatitis viruses would most likely be transmitted by contaminated water?
    (a) Hepatitis A   (e) Hepatitis E
    (b) Hepatitis B   (f) a and c
    (c) Hepatitis C   (g) a and e
    (d) Hepatitis D

20. Which of the following viruses requires a surrogate virus infection in order to cause disease?
    (a) Hepatitis B   (d) Rotavirus
    (b) Hepatitis D   (e) Echovirus
    (c) Norwalk virus

21. How many different hepatitis viruses have been identified to date?
    (a) 4   (d) 7
    (b) 5   (e) 8
    (c) 6

22. Which of the following protozoan parasites inhibits the adsorption of fats in the intestine?
    (a) *Giardia*   (d) *Amoeba*
    (b) *Cryptosporidium*   (e) Hepatitis A
    (c) Rotavirus

23. Aflatoxins are produced by a:
    (a) Bacterium   (d) Fungus
    (b) Virus   (e) Protozoan
    (c) Unknown organism

24. Which of the following infectious organisms is a prokaryotic organism?
    (a) *Cryptosporidium parvum*
    (b) *Aspergillus niger*
    (c) *Taenia solium*
    (d) *Salmonella typhi*
    (e) *Giardia lamblia*

25. Infection with which of the following can be diagnosed by finding its eggs in fecal specimens?
    (a) Rotavirus   (d) Hepatitis A
    (b) *Vibrio cholerae* (e) *Aspergillus niger*
    (c) Helminths

26. Identify the structures that are part of the digestive system. Indicate what mechanisms may be present to prevent disease at each structure.
    (a) _____
    (b) _____
    (c) _____
    (d) _____
    (e) _____
    (f) _____
    (g) _____
    (h) _____
    (i) _____
    (j) _____
    (k) _____
    (l) _____
    (m) _____
    (n) _____
    (o) _____
    (p) _____
    (q) _____

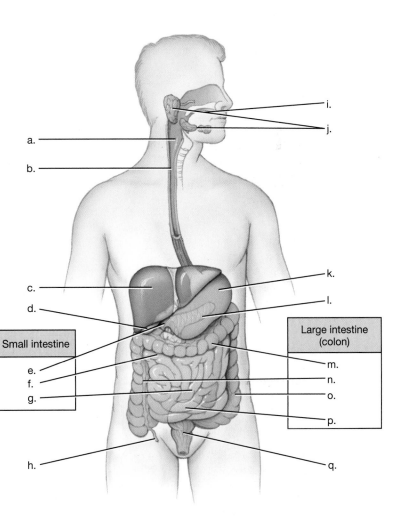

Small intestine

Large intestine (colon)

▌ **EXPLORATIONS ON THE WEB**   http://www.wiley.com/college/black

If you think you've mastered this chapter, there's more to challenge you on the web. Go to the companion web site to fine-tune your understanding of the chapter concepts and discover answers to the questions posed below.

1. Most diarrheal illness is self-limiting and people recover without treatment or seeing a physician. However, for some people diarrhea can be a life-threatening illness. How do you decide whether it is just an inconvenience or a serious life-threatening disease?

2. Imagine going to the dentist and getting vaccinated! No more cavities! Better yet the vaccine is sprayed into your nose, no needles! When might such a vaccine be available?

Cardiovascular, Lymphatic, and Systemic Diseases

David Schultz/Getty Images

Come with me . . .

"Home, home on the range. . . ." More than deer and antelope make their home on the range at Yellowstone National Park. You're also likely to find brucellosis roaming free among herds of bison. Brucellosis is a contagious bacterial disease that affects the circulatory system. It causes the reduction of meat and milk production, as well as abortion. Surveys show that half of the Yellowstone animals are infected with brucellosis, along with many elk.

This disease costs U.S. cattle producers over $30 million each year. When infected animals are found on cattle ranches, they are slaughtered, at a loss of about $1,000 per steer. Pregnant cows abort their fetuses. Vaccines have proven to be poorly effective. Perhaps this is a place where bacteriophage therapy can be tried. Until then, ranchers and park rangers will have to rely on all possible preventive measures. At $900/hour, they are using helicopters to try to herd bison back into the park. In the first 3 months of 2004, 278 of the park's 4,200 wild buffalo have been shot by National Park Service and Montana Department of Livestock employees. It is sad to see the killing of animals who once numbered 35 million in the mid-1800s, but by 1902 numbered only 23 individuals left! To discourage contact with ranch herds, some ranchers have established separate feeding sites for Park animals. Others simply keep their rifles handy.

 Video related to this topic is available within WileyPLUS.

||| THE CARDIOVASCULAR SYSTEM
The Heart and Blood Vessels / The Blood / Normal Microflora of the Cardiovascular System

||| CARDIOVASCULAR AND LYMPHATIC DISEASES
Bacterial Septicemias and Related Diseases / Helminthic Diseases of the Blood and Lymph

||| SYSTEMIC DISEASES
Bacterial Systemic Diseases / Rickettsial and Related Systemic Diseases / Viral Systemic Diseases / Protozoan Systemic Diseases

Diseases of the cardiovascular and lymphatic systems frequently affect several other systems because the infectious agents are easily disseminated through blood and lymph. Therefore, diseases that usually affect multiple systems are included in this chapter.

▮▮▮ THE CARDIOVASCULAR SYSTEM

The **cardiovascular system** consists of the heart, blood vessels, and blood (**Figure 23.1**). This system supplies oxygen and nutrients to all parts of the body and removes carbon dioxide and other wastes from them. It also carries hormones from endocrine glands to the appropriate tissues and cells and regulates the pH of the blood via buffers.

THE HEART AND BLOOD VESSELS

The heart lies between the lungs in a tough, membranous *pericardial sac* lubricated by serous fluid. Its wall consists of a thin internal *endocardium*, a thick, muscular *myocardium*, and an outer *epicardium*. The heart has four chambers—two *atria* and two *ventricles*, with valves that direct blood flow at the exits of each chamber. The right side of the heart pumps blood to the respiratory portion of the lungs, the left side to all other organs and tissues.

Blood leaving the heart circulates through a closed system of blood vessels and subsequently returns to the heart. Its flow is regulated so that all cells receive nutrients and get rid of wastes according to their needs. The blood vessels include *arteries* that receive blood from the heart, *arterioles* that branch from arteries; *capillaries* that branch from arterioles, *venules* that receive blood from capillaries; and *veins* that receive blood from venules and return to the heart. Capillary walls are composed of a single layer of cells that allows exchange of materials between the blood and the tissues. Leukocytes, such as neutrophils and macrophages, sometimes force their way between cells of the capillary walls (diapedesis) during infections.

THE BLOOD

The blood consists of plasma and formed elements (cells and cell fragments). Plasma is more than 90% water and contains proteins such as albumins, globulins, and fibrinogen. Certain globulins, the antibodies, are important in defending the body against infection, and fibrinogen is important in blood clotting. Plasma also contains **electrolytes** (ions such as Na^+, K^+, and Cl^-), gases (such as oxygen and carbon dioxide), nutrients, and waste products. In contrast with plasma, serum is the fluid that remains after both formed elements and clotting factors have been removed.

Besides leukocytes and platelets, formed elements of the blood include erythrocytes. Erythrocytes are the most abundant of the formed elements, accounting for 40 to 45% of the total blood volume. Erythrocyte volume is an important indicator of the oxygen-carrying capacity of the blood because erythrocytes contain the oxygen-binding molecule hemoglobin.

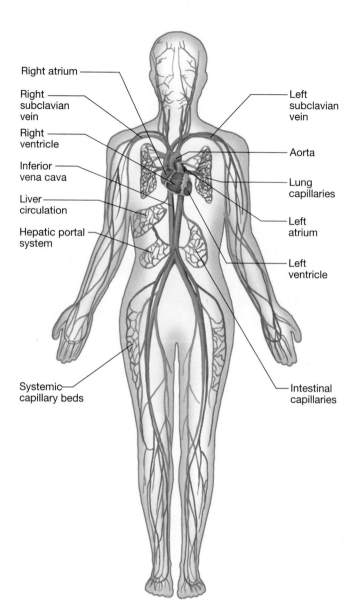

Figure 23.1 **The general structure of the cardiovascular system.** Oxygenated blood is shown in red, unoxygenated blood in blue. The cardiovascular system is normally sterile and contains no resident microbes.

NORMAL MICROFLORA OF THE CARDIOVASCULAR SYSTEM

The cardiovascular system does not have any normal resident microflora. Because it is a closed system, blood, blood vessels, and the heart should ordinarily be sterile. However, even in healthy individuals, microorganisms will occasionally enter the bloodstream, causing a transient **bacteremia** (◀Chapter 14, p. 416). If not killed or removed, the microbes begin to grow and multiply in the bloodstream, causing **septicemia** (blood poisoning). Points of entry may be through wounds (some as minor as vigorous brushing of your teeth) or by shedding from other areas of infection such as a boil, a pimple, or an

PUBLIC HEALTH

Man's Best Friend's Worst Enemy

Keep the mosquitoes away from Fido, or better yet give him preventive medication. The heartworm, *Dirofilia immitis*, is transmitted by mosquitoes. Once in the dog's bloodstream, the larvae migrate to the skin, mature, and move to the heart. There they mate and release microfilariae that become infective larvae only when ingested by a mosquito. Adult worms 15 to 30 cm long accumulate in the right side of the heart and are also found in the lung and liver. The unfortunate dog dies an early death from heart damage and circulatory damage.

The drug Ivermectin, when given in monthly doses, prevents heartworm infection. However, it can be administered only to dogs who are not already infected. If adult worms are present, the drug kills them. But disintegration of worm remains releases toxic substances and debris that can clog blood vessels while further weakening the wall of the heart where they had resided. *(From Howard Shiand, D.V.M., Journal of American Veterinary Medical Association 163: 981, Oct. 1973)*

Figure 23.2 Lymphangitis from an infected burn. The reddened streaks in the arm indicate the spread of organisms through lymph vessels, a symptom of septicemia. The old-fashioned name for such an infection was blood poisoning. *(Barts Medical Library/Phototake)*

abscessed tooth. Once in the bloodstream, organisms are eliminated by antibodies and/or phagocytosis (◄Chapters 16 and 17). Certain sites, such as heart valves, are especially susceptible to bacterial colonization and infection.

CARDIOVASCULAR AND LYMPHATIC DISEASES

BACTERIAL SEPTICEMIAS AND RELATED DISEASES

Septicemias

Before antibiotics, septicemia was often fatal; even with antibiotics, it is still not easy to treat. Gram-positive organisms such as *Staphylococcus aureus* and *Streptococcus pneumoniae* once commonly caused most septicemias. Today, broad-spectrum antibiotics have made septicemias from these organisms less frequent. However, other bacterial species—including *Pseudomonas aeruginosa*, *Bacteroides fragilis*, and species of *Klebsiella*, *Proteus*, *Enterobacter*, and *Serratia*—have come into play. These organisms cause **septic shock**, a life-threatening septicemia accompanied by low blood pressure and the collapse of blood vessels. Probably one-third of all septicemias are now Gram-negative septic shock, and 10% of all septicemias are caused by multiple organisms. Endotoxins produced by these organisms are directly responsible for shock. Antibiotics often worsen the situation; when they

kill organisms, the disintegrating organisms release larger quantities of endotoxin, causing more damage to the host's blood vessels, and blood pressure drops even further. A few cases of septic shock are caused by organisms that produce exotoxins.

Symptoms of septicemia include fever, shock, and **lymphangitis** (lim-fan-ji′tis), or red streaks due to inflamed lymphatic vessels beneath the skin **(Figure 23.2)**. One-third of all septicemia cases are nosocomial and appear within 24 hours after an invasive medical procedure has been performed. The transition from bacteremia to septicemia can be sudden or gradual. Therefore, hospital patients who have undergone invasive procedures should be carefully watched for signs of septicemia. Septicemia has a mortality of 50 to 70% and accounts for about 35,000 deaths per year in the United States alone.

Diagnosis of septicemia is made by culturing of blood, catheter tips, urine, or other sources of infection. In treating septicemias, blood pressure must be elevated and stabilized; then the infectious organisms must be eliminated by appropriate antibiotic therapy.

Puerperal Fever

Puerperal (pu-er′per-al) **fever**, also called *puerperal sepsis* or **childbed fever**, was a common cause of death before antibiotics became available (◄Chapter 1, p. 13). It is caused by Group A, β-hemolytic streptococci (*Streptococcus pyogenes*), which are normal vaginal and respiratory microflora. They can also be introduced during delivery by medical personnel. Streptococci pass through irritated uterine surfaces and invade the blood, giving rise to septicemia. Signs and symptoms of the disease are chills, fever, pelvic distention and tenderness, and a bloody vaginal discharge. Streptococci can be isolated from blood cultures to diagnose puerperal fever. Penicillin is effective except against resistant organisms, and mortality is low with prompt treatment, but recovery usually takes many weeks, and relapses are common.

Group B Streptococcal Disease

Group B streptococci (*Streptococcus agalactiae*) are the leading cause of neonatal sepsis and meningitis in the United States and Europe. They cause 1 to 3 cases per 1,000 births, with a mortality rate of about 50%. Over 30% of those who survive meningitis will suffer central nervous system damage. Group B streptococci are common in vaginal flora, being found in 10 to 30% of both pregnant and nonpregnant women. Rupture of membranes more than 12 hours before birth allows the organisms ample time to infect the baby. Most cases of infection become apparent within a few days of birth, with fever, respiratory distress, and lethargy. "Late-onset" cases, appearing 3 to 8 weeks after birth, most often feature meningitis. Pregnant women should be tested for Group B streptococci in their third trimester of pregnancy, hopefully before labor, and given ampicillin or immune globulin if colonization is found. Newborns of colonized mothers should also be given ampicillin for 7 to 10 days. If colonization is not detected until labor, intravenous penicillin G given to the mother at least 4 hours before delivery will result in protective levels of antibiotic in the baby's bloodstream during birth. This is a disease that should be almost totally preventable. A vaccine is currently being tested.

Rheumatic Fever

Rheumatic fever is a multisystem disorder following infection by β-hemolytic *Streptococcus pyogenes*. That rheumatic fever can follow such infections has been known for decades, but the mechanisms by which it occurs are not yet completely understood. Some form of genetic predisposition is suspected because a certain human leukocyte antigen (HLA) is present in 75% of rheumatic fever patients but in only 12% of the general population.

Most rheumatic fever patients are between the ages of 5 and 15. Onset of the disease usually occurs 2 to 3 weeks after a strep throat, but it can occur within 1 week or as late as 5 weeks after the initial infection. Strep throat symptoms have disappeared by the time the rheumatic fever symptoms begin. Classic signs and symptoms include fever, arthritis, and a rash. Evidence of damage to the mitral valve of the heart confirms a specific diagnosis of rheumatic fever. Weeks or months later subcutaneous nodules appear, especially near the elbows. Approximately 3% of untreated strep throat cases progress to rheumatic fever. Culturing of streptococci and serological tests are helpful in diagnosis, as is a previous history of streptococcal infection.

Heart damage in rheumatic fever results from immunological events, and is thus an autoimmune problem. Certain streptococcal strains have an antigen that is very similar to heart cell antigens. Antibodies that bind to one antigen will bind to the other; that is, the antibodies are *cross-reactive*. In the immune reaction, lymphocytes probably become sensitized to the antigen and attack heart tissue as well as the streptococci. The resulting heart damage can be fatal. Antibiotic therapy will not reverse existing damage, but it can prevent further damage and can be used to prevent recurrences. Once rheumatic fever occurs, mitral valve deformities contribute to eddies in blood flow that predispose toward bacterial colonization of heart valve surfaces. (This condition, *bacterial endocarditis*, is discussed next.) Individuals at risk of rheumatic fever should receive a prophylactic antibiotic—usually penicillin—before dental work or other invasive procedures to prevent possible streptococcal infection.

> *Rheumatic fever can cause aortic stenosis, a narrowing of the aortic valve opening, which causes the left ventricle of the heart to work harder.*

Prompt treatment of β-hemolytic *S. pyogenes* infections with antibiotics before cross-reactive antibodies can form is the only practical way to prevent rheumatic fever. No effective vaccine exists; vaccines produced so far elicit damaging antibodies, but this problem might someday be solved. Antiinflammatory drugs such as steroids and aspirin can lessen scarring of heart tissue.

Bacterial Endocarditis

Bacterial endocarditis (en'-do-kar-di"tis), or *infective endocarditis*, is a life-threatening infection and inflammation of the lining and valves of the heart. It can be subacute or acute. Two out of three patients have the subacute type, which manifests itself as fever, malaise, bacteremia, and regurgitating heart murmur usually lasting 2 weeks or more. It occurs primarily in people over age 45 who have a history of valvular disease from rheumatic fever or congenital defects. Many microbes, including fungi, can cause endocarditis, but most cases are due to bacteria, especially strains of *Streptococcus* or *Staphylococcus*, many of which are normal residents of the mouth or throat. Acute endocarditis is a rapidly progressive disease that destroys heart valves and causes death in a few days.

In bacterial endocarditis, organisms from another body site of infection are transported to the heart. A **vegetation** develops, in which exposed collagen fibers on damaged valvular surfaces trigger fibrin deposition (**Figure 23.3**). Transient bacteria attach to fibrin and form a bacteria-fibrin mass. Vegetations can interfere with the conduction of normal electric impulses across the heart. Vegetations also deform heart valves, decrease their flexibility, and prevent them from closing completely. Blood flows backward from ventricles into atria when the ventricles contract, decreasing the pumping efficiency of the heart. Congestive heart failure, an accumulation of fluids around the heart, is the most common complication and direct cause of death from bacterial endocarditis.

Bacterial endocarditis is diagnosed from blood cultures and is treated with penicillin or other antibiotics,

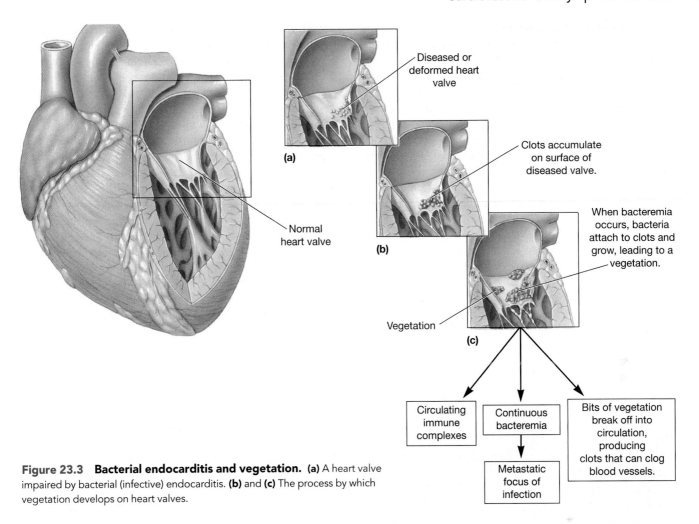

Figure 23.3 Bacterial endocarditis and vegetation. **(a)** A heart valve impaired by bacterial (infective) endocarditis. **(b)** and **(c)** The process by which vegetation develops on heart valves.

depending on the susceptibilities of causative organisms. In some cases surgical valve replacement is necessary. Untreated endocarditis results in death. Antibiotics cure about half of all patients; surgery cures another quarter, and one-quarter die. Deaths are most frequent among intravenous drug abusers and others with compromising conditions.

Myocarditis (mi′o-kar-di″tis), an inflammation of the heart muscle (myocardium), and **pericarditis** (per′e-kar-di″tis), an inflammation of the protective membrane around the heart (pericardial sac), also can be caused by microbial infections. Although most such infections are viral, *Staphylococcus aureus* causes 40% of all pericarditis cases. Untreated cases have a mortality rate of nearly 100%, whereas cases receiving appropriate treatment have a mortality rate between 20 and 40%.

Coronary artery disease and atherosclerosis (hardening of the arteries) have been linked to *Chlamydia pneumoniae*. Over 90% of plaque deposits in arteries contain *Chlamydia*. Exactly what this means is still a topic of vigorous debate. Adding to the confusion,

Marek's disease virus, a herpesvirus, has been shown to cause atherosclerosis in chickens.

✓CHECKLIST

1. Describe the normal flora of the cardiovascular system.

2. Why could treatment with antibiotics bring on cases of septic shock?

3. What does a vegetation found in the heart consist of? How and why does it form? What effects does it have?

4. Describe the two diseases of the cardiovascular system and the bacteria that cause them that are associated with childbirth.

HELMINTHIC DISEASES OF THE BLOOD AND LYMPH

Schistosomiasis

Three species of blood flukes of the genus *Schistosoma* cause **schistosomiasis** (shis″to-so-mi′a-sis), and each requires a particular snail as intermediate host to complete its life cycle (see Figure 11.15, ◀p. 328). The first of these helminths was identified by the German parasitologist Theodor Bilharz in the 1850s, so the disease is also called *bilharzia* (bil-har′ze-a). Bilharzia has been known since biblical times; some think Joshua's curse on Jericho was the placing of blood flukes in communal wells. In fact, the Egypt of the pharaohs was called "land of the menstruating men" by ancient writers because the prevalence of flukes made bloody urine so common. Eggs of these parasites have been found in the bladder walls of Egyptian mummies.

The World Health Organization (WHO) reports some 250 million cases of schistosomiasis worldwide. Incidence of the disease has increased significantly in Egypt since the building of the Aswan Dam in 1960 because collection of water behind the dam created exceptionally favorable conditions for snail hosts. *Schistosoma japonicum* is found in Asia. *S. haematobium* in Africa, and *S. mansoni* (**Figure 23.4**) in Africa, South America, and the Caribbean. The latter two species probably reached South America during slave trading days, but only *S. mansoni* found an appropriate snail host there. New species have been reported in Vietnam (for instance, *S. mekongi*) and in other geographic areas.

Humans become infected by free-swimming cercariae that have emerged from their snail hosts (**Figure 23.4b**). Cercariae penetrate the skin when humans wade in snail-infested waters, migrate to blood vessels, and are carried to the lungs and liver. The flukes mature and migrate to veins between the intestine and liver or sometimes the urinary bladder, where they mate and produce eggs, as many as 3,000 eggs per day. Adults have a special ability to coat themselves with host antigens, thereby evading the host's immune system. Some eggs become trapped in the tissues and cause inflammation; others penetrate the intestinal wall and are excreted in the feces. Cercariae cause dermatitis at penetration sites and tissue damage during migration. Metacercariae and adults migrate to and invade the liver (**Figure 23.4c**), where they cause cirrhosis. These flukes also can invade and damage other organs.

Schistosome eggs are highly antigenic, and allergic reactions to them are responsible for much of the damage caused by blood flukes. If the eggs are released near the spinal cord, the resulting inflammation can cause

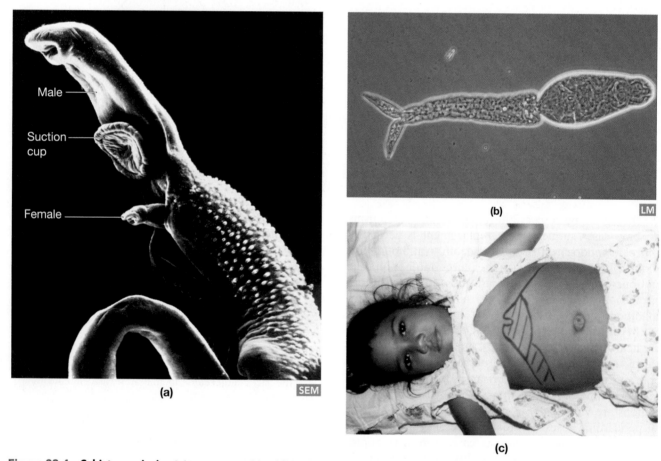

(a) SEM

(b) LM

(c)

Figure 23.4 Schistosomiasis. Schistosomes, or blood flukes, are trematodes that cause schistosomiasis in Africa, South America, and the Caribbean. **(a)** A mating pair of schistosomes. The larger, male worm holds the smaller, female worm in a groove of his body. The male attaches itself to the wall of a human blood vessel by means of the suction cup just below its head. The worms, thinner than a cotton fiber, are barely visible to the unaided eye. A mating pair can live in the human body for up to 10 years. (Magnification unknown.) *(Courtesy National Institute of Allergy and Infectious Diseases/National Institutes of Health)* **(b)** The cercarial stage (160X) leaves snails and penetrates the skin of humans wading or swimming in water. *(Sinclair Stammers/Photo Researchers, Inc.)* **(c)** A child suffering from schistosomiasis. The tracing delineates her enlarged liver, typical of this disease. *(Courtesy National Institute of Allergy and Infectious Diseases/National Institutes of Health)* neurological disorders. The eggs most often damage blood vessels, but which vessels are damaged depends on the species. *Schistosoma japonicum* severely damages blood vessels in the small intestine; *S. masoni*, those of the large intestine; and *S. haematobium*, those of the urinary bladder. Symptoms of schistosome bladder infections include pain on urination, bladder inflammation, and bloody urine.

Diagnosis can be made by finding eggs in feces or urine, but eggs may not be present in chronic cases, which can last for 20 to 30 years. Intradermal injection of schistosome antigen and measurement of the area of the wheal or a complement fixation test are good immunological methods of diagnosis (◀Chapter 18, p. 564). Until recently, toxic antimony compounds were used to treat the disease. Several new drugs, especially praziquantal, seem to be quite effective and less toxic. Some investigators, however, believe that praziquantal-resistant schistosomes are on the increase.

Schistosomiasis could be prevented entirely if human wastes were not emptied into rivers or if humans never waded in snail-infested waters. The practice of wading in the local river to wash the body after defecation or urination is an important means of transmission wherever the infection occurs. Chemical molluscicides have been used to reduce snail populations, but it is difficult to determine the proper concentrations under varying river conditions. Recent experiments using predatory snails to destroy cercariae-carrying snails have shown great promise. Finally, work toward developing a vaccine is underway. One new vaccine in testing does not prevent infection but protects against damage caused by schistosomes.

Filariasis

Filariasis (fil″ar-i′a-sis) can be caused by several different roundworms. *Wuchereria bancrofti* (see Figure 11.19, ◀p. 000) is a common cause of this tropical disease, for which WHO reports over 120 million cases worldwide; 30% in India and 30% in Africa. Adult worms are found in the lymph glands and ducts of humans. Females release embryos called *microfilariae*. They are present in peripheral blood vessels during the night and retreat to deep vessels, especially those of the lungs, during the day. Mosquitoes also are essential hosts in the life cycle of this parasite, and several species of night feeders among the genera *Culex*, *Aedes*, and *Anopheles* serve as hosts. When a mosquito bites an infected person, it ingests microfilariae that develop into larvae and migrate to the mosquito's mouthparts. When the mosquito bites again, the larvae are infected and can infect another person. They enter the blood, develop, and reproduce in the lymph glands and ducts, thereby completing the life cycle. Adult worms are responsible for inflammation in lymph ducts, fever,

Over 1 billion people in 73 countries are at risk of contracting elephantiasis, and there are over 120 million people already infected.

and eventual blockage of lymph channels in affected areas. Repeated infections over a period of years can lead to **elephantiasis** (el″ef-an-ti′a-sis), gross enlargement of limbs, the scrotum, and sometimes other body parts due to the blockage of lymph vessels and subsequent accumulation of fluid and connective tissue in those vessels **(Figure 23.5a)**.

(a)

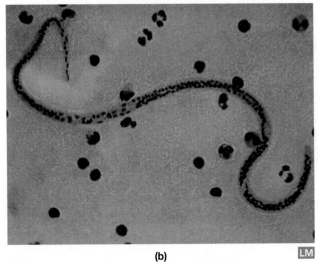

(b) LM

Figure 23.5 **Elephantiasis.** **(a)** Elephantiasis of the scrotum, caused by the roundworm *Wuchereria bancrofti*. Swelling results from blockage of the lymphatic system by adult worms. Another common site of elephantiasis is the leg. *(Phototake)* **(b)** The microfilaria stage of the life cycle (2,268X) is transmitted to humans by the bite of mosquitoes. *(Fred Marsik/Visuals Unlimited)*

A Good Deed

Elephantiasis is actually a lymphedema caused by the congestion of the lymphatic vessels by filarial worms. To end the scourge of lymphatic filariasis, the World Bank, the World Health Organization, and the health care company Glaxo SmithKline Beecham are cooperating on a program in which albendazole, the drug to treat the disease, will be donated free-of-charge to needy countries. Because it takes one yearly dose for 4 to 5 years to treat all people in infected areas, Glaxo SmithKline Beecham's donation commitment is several billion doses of albendazole. It is estimated that because one-fifth of the world's population is at risk of infection, it will take approximately 20 years to eliminate the disease. This generous gift of many billions of dollars of drugs will also reduce cases of hookworms and other roundworms, thus improving general health in affected countries.

Filariasis is diagnosed by finding microfilariae in thick blood smears **(Figure 23.5b)** made from blood samples taken at night or by an intradermal test (skin snip biopsies). The drugs diethylcarbamiazine (Hetrazan) and metronidazole are effective in treating the disease. Swollen limbs are wrapped in pressure bandages to force lymph from them; if distortion is not too great, nearly normal size can be regained. To control the disease, it would be necessary to treat all infected individuals and to eradicate the species of mosquitoes that carry the parasite. Limited progress has been made.

Cardiovascular and lymphatic diseases are summarized in **Table 23.1**.

SYSTEMIC DISEASES

BACTERIAL SYSTEMIC DISEASES

Anthrax

Anthrax is a zoonosis that affects mostly plant-eating animals, especially sheep, goats, and cattle. Meat-eating animals can acquire the disease by eating infected flesh or by inhaling anthrax spores, but the disease is not spread from live animal to animal. Each year, many thousands of animals worldwide have anthrax, but only 20,000 to 100,000 humans develop it, mostly in Africa, Asia, and Haiti. An anthrax outbreak at a secret germ warfare factory in Sverdlovsk (now called Yelkaterinburg) Russia in 1979 killed at least 77 people out of 88 cases. For a very detailed analysis of this outbreak, see the book by Jeanne Guillemin, *Anthrax: The Investigation of a Deadly Outbreak* (1999, University of California Press), which strips away lies intended to cover up this gruesome accident. Health authorities in the United States have made vigorous efforts to eradicate anthrax and to prevent its import from other countries. Prior to 2001, when letters containing anthrax were mailed to Washington, D.C., and New York City, and spores released in Florida, only 5 human cases had occurred in the United States since 1980, and no more than 6 per year had been reported since 1970 **(Figure 23.6)**. In 2006 one naturally acquired case of inhalation anthrax occurred in a man travelling with a dance group. He was a drummer and had made his own drums using anthrax infected goat hides from West Africa.

The Disease. The causative agent of anthrax, *Bacillus anthracis*, was discovered in 1877 by Robert Koch. The bacillus is a large, Gram-positive, facultatively anaerobic, endospore-forming rod. The endospores form only under aerobic conditions and are not found in tissues or

TABLE 23.1

A Summary of Cardiovascular and Lymphatic Diseases		
Disease	Agent(s)	Characteristics
Bacterial Septicemias and Related Diseases		
Septicemia	Various bacterial species	Septic shock due to endotoxins of causative agent(s), fever, lymphangitis
Puerperal fever	*Streptococcus pyogenes*	Organisms from uterus invade blood and cause septicemia, pelvic distention, bloody discharge
Rheumatic fever	*Streptococcus pyogenes*	Fever, arthritis, rash, mitral valve damage due to immunological reaction
Bacterial endocarditis	*Staphylococcus* or *Streptococcus* strains	Inflammation and vegetation of heart valves and lining, fever, malaise, bacteremia, heart murmur, congestive heart failure that can cause death
Parasitic Diseases of the Blood and Lymph		
Schistosomiasis	*Schistosoma haematobium, S. mansoni, S. japonicum*	Dermatitis from cercaria, cirrhosis of liver from eggs, allergic reactions to eggs, tissue damage in intestine and urinary bladder
Filariasis	*Wuchereria bancrofti*	Inflammation and blockage of lymph ducts, leading to elephantiasis, fever

ANTHRAX. Reported cases, by year – United States, 1951–2006

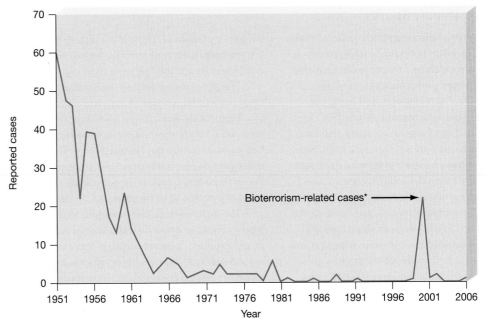

*One epizootic-associated cutaneous case was reported in 2001 from Texas.

In 2001, 22 anthrax cases (11 inhalation and 11 cutanous [four suspected, seven confirmed]) were associated with an unprecedented biological terrorism event. Five of the 11 international cases were fatal. Cases occurred among residents of seven states. In addition, one naturaly occurring case was reported from Texas. In 2006 one isolation case occurred associated with imported African hides. *Bacillus anthracis* remains a Class A bioterrorrism threat agent.

Figure 23.6   Anthrax, reported cases by year—U.S., 1951–2006.

circulating blood. But if an infected animal's blood is spilled during an autopsy, bacilli exposed to air rapidly form endospores. Veterinarians and farmers should be very careful to avoid contaminating soil or other materials, as endospores can remain viable over 60 years, perhaps even over 100 years, ruining a pasture for animal usage during those years.

Other than by bioterrorism, most human anthrax results from contact with endospores during occupational exposure on farms or in industries that handle wool, hides, meat, or bones. Respiratory anthrax, or "wool-sorters' disease," was such a problem in nineteenth-century England that legislation was enacted to protect textile workers from this occupational disease.

Human anthrax cases appear in 3 different clinical forms: 90% are cutaneous, 5% are respiratory, and 5% are intestinal. **Cutaneous anthrax** has a mortality rate of 10 to 20% when untreated but only 1% with adequate treatment. **Respiratory anthrax** is almost always fatal, regardless of treatment. **Intestinal anthrax** has a fatality rate of 25 to 50%. In addition, regardless of the initial site of infection, if bacteria enter the bloodstream causing

Bacillus anthracis produces a toxin that inhibits a protein kinase used in cell signaling. This may mean that protease inhibitors could be used against anthrax.

septicemia, this leads to meningitis (which is almost always fatal in 1 to 6 days) in about 5% of patients.

Cutaneous anthrax develops 2 to 5 days after endospores enter epithelial layers of the skin. Lesions 1 to 3 cm in diameter develop at the site of entry (**Figure 23.7**)

Figure 23.7   Cutaneous anthrax. Spores introduced into the skin through abrasions or cuts germinate. The newly emerged organisms multiply, release exotoxin locally, and invade adjacent tissue, where they produce extensive damage. *(Courtesy Centers for Disease Control)*

Prairie Heritage: Deadly Spores

For centuries, vast herds of bison roamed the Great Plains of the United States. The bison periodically suffered outbreaks of anthrax. Their dead carcasses rotted, leaving anthrax spores in the soil. Then came the great cattle drives of the 1800s, from Texas to Canada. Carcasses were left along the trails, adding more anthrax spores to the soil. By the 1930s, ranchers realized that the bodies of anthrax-infected animals must be dealt with in order to prevent soil contamination. Some ranchers still remember the huge bonfires of burning carcasses in the 1930s and 1940s. But by then the soil was contaminated with anthrax spores—which remain viable even today.

In dry years, wind blows away the topsoil, exposing spores from the older, deeper layers. In wet years, or along river and creek beds, erosion washes away topsoil, likewise exposing and spreading spores. From Texas to Canada, yearly anthrax immunization of cattle is largely routine. But after 10 or 15 years without an outbreak, ranchers may get a false sense of security and discontinue using the vaccine. That is what happened in South Dakota during the summer of 1993.

In August 1993, an unvaccinated herd of beef cattle in southeastern South Dakota began to sicken. On August 13, after three animals had died, a veterinarian was summoned. His diagnosis was pulmonary emphysema. On August 15, a second veterinarian arrived and confirmed that diagnosis. The rancher was told that it was safe to send the rest of the herd to slaughter. Soon 19 animals had died, 9 others went to slaughter immediately, and the rest were shipped to the stockyards to be sold.

At the slaughterhouses the animals' livers did not look healthy to inspectors. The rancher was notified of the possible diagnosis of anthrax. Laboratory testing on August 17 confirmed the presence of anthrax. An immediate order of quarantine was issued the following morning.

The slaughtered animals' remains were retrieved, and the meat plants and storage lockers had to be disinfected by steaming. Fortunately, none of the animals at the stockyards had been sold yet. All were shipped back to the ranch from which they had come and were held in quarantine. There they received anthrax vaccine plus antibiotics, and a booster shot 2 weeks later. The manure and bedding from the areas they had occupied in the stockyards had to be collected, hauled away, and sterilized.

The dead animals had been sent to a rendering plant, where carcasses or waste parts such as bones and fat are boiled down. Products from rendering plants may eventually be used in food products for humans or other animals. Thus, these 19 carcasses were a danger. Had the animals been milk cows, their milk could also have been contaminated.

State health inspectors and others who had handled the herd were given tetracycline to prevent the development of anthrax infection. Happily, no human cases occurred. But by then, 32 head of cattle had died. The last 13 carcasses were burned. Thirty days after the last death, the rest of the herd was released from quarantine. Nearby ranchers who had neglected to vaccinate their herds swiftly administered the vaccine.

South Dakota and the other Great Plains states have periodic outbreaks. It is part of life there. Health authorities must be ever vigilant against the prairie heritage of spores from the distant past. And, would-be terrorists wanting to grow cultures of anthrax have only to look to our own soils for their starting material. The spores are there.

and expand outward. Eventually, the center of the lesion becomes black and necrotic, the scab resembling a piece of coal; hence the name anthrax from the Greek word for coal. Finally, it heals but leaves a scar. Recovery probably confers some—but not total—immunity.

Symptoms of intestinal anthrax closely mimic those of food poisoning (◀Chapter 22, p. 684). Ulcerations may form anywhere along the digestive tract, from the mouth and esophagus on down, even in the appendix. These frequently cause septicemia and death. An outbreak in the United States resulted from consumption of imported cheese made from unpasteurized goat's milk. The cheese was served at a wine and cheese party to a group of physicians. Some of them traced the source of the infection to a particular cheese factory in France, which was closed as a result.

Pulmonary anthrax is the most deadly form, and is therefore the one chosen for bioterrorism. Under natural circumstances it is uncommon in humans, but common among grazing animals whose noses are close to the soil in which anthrax spores wait. Once inhaled into the lungs, the spores germinate in alveoli where they are phagocytized, but not killed, by macrophages. Eventually they kill the macrophages. It may take as long as 60 days before all spores have germinated. Therefore discontinuing antibiotics 10 or 30 days after a known or suspected exposure is a risky thing to do. Unfortunately, early symptoms of pulmonary anthrax resemble those of many common respiratory infections such as colds or the flu. Antibiotics given at this stage can still help. However, if exposure is not suspected, it is unlikely that antibiotics will be given now—later they will not save the patient. Then comes a period of false relief: symptoms ease, but bacteria enter the bloodstream leading to septicemia and death 2 or 3 days later from septic shock. The lymph nodes of the *mediastinum* (the central, vertical part of the chest) become greatly enlarged, and can be seen on X-ray as a distinct widening of the mediastinum. Person-to-person spread does not occur, as organisms are not coughed up from the bronchi.

Virulence factors include a capsule made of glutamic acid, whose genes are carried on one plasmid. A second plasmid carries the genes for three exotoxins: edema factor, lethal factor, and protective antigen. All three toxins plus the capsule must be present for disease to occur. Loss of either plasmid renders the bacterium avirulent. Edema factor combines with protective antigen to form *edema toxin* which causes swelling and prevents phagocytosis by macrophages. Lethal factor combines with protective antigen to form *lethal toxin* which causes macrophages to release tumor necrosis factor-α, and interleukin-1β, plus other inflammatory cytokines, and to die. The resulting toxemia from the exotoxins causes clots to form inside pulmonary capillaries and lymph nodes, causing mediastinal swelling which obstructs airways. The disease is nearly 100% fatal. The postal workers who developed pulmonary anthrax in the 2001 attacks, and who were saved by heroic efforts, have not regained their former health. They are still considerably impaired and are retired on disability.

Diagnosis, Treatment, and Prevention. Anthrax is diagnosed by culturing blood or examining smears from cutaneous lesions of patients with a history of possible exposure. Serologic and DNA tests are also used. Development of rapid diagnostic tests are of prime defense importance. The disease is treated with ciprofloxacin (Ciprot) as the drug of choice. However, many other antibiotics such as penicillin, doxycycline, erythromycin, and chloramphenicol can usually be used successfully. During the 2001 attacks, it was feared that the anthrax was genetically engineered to be resistant to penicillin and other antibiotics. Cipro was a newer drug and was therefore thought less likely to have had resistance to it genetically engineered into the anthrax strain. And so, Cipro became the drug of choice. Later, the anthrax strain was found to be susceptible to penicillin and other common antibiotics. However, while an antibiotic may kill the anthrax organisms, it does not inactivate the lethal toxins which they have produced. Thus, a patient on antibiotics can still die. A vaccine is available, which requires an initial series of 6 shots over an 18-month period followed by yearly booster shots. It can be given to workers with occupational exposure to anthrax, but industries still should maintain dust-free environments and provide respirators to prevent endospore inhalation. Worker education and on-site employee health services for prompt detection of infections also are important. Unimmunized visitors must be kept away from work areas. Clothing worn by workers should be sterilized and cleaned at the facility to prevent family members from becoming infected by handling it.

Animal immunization is an important means of prevention. Farmers must avoid using bone meal contaminated with anthrax spores and must dispose of infected animals by burying them in deep, lime-lined pits. Lime prevents earthworms from bringing anthrax endospores to the surface; incineration is used but must be done properly to avoid the spread of bits of contaminated carcass and spores by the wind. Veterinarians must be especially careful when working with infected animals or giving vaccinations because accidental vaccination of humans with a vaccine intended for animals can cause anthrax.

Plague

From 1937 through 1974, fewer than 10 cases per year of plague, a zoonosis, were reported in the United States, with no cases reported in some years. Then 20 cases were reported in 1975 and 40 in 1983, mostly in rural settings in the Rocky Mountain states (Figure 23.8). There, the disease is referred to as *sylvatic plague* because it is carried by wild rodents such as ground squirrels, chipmunks, and desert rats. The United States is fortunate in not suffering from *urban plague*, in which the European or city rat is the main host. Plague remains an endemic disease in certain other areas of the world, but the number of cases worldwide is much smaller than in the great pandemics mentioned in ◄Chapter 1 (p. 8).

The Disease. The causative agent of plague, *Yersinia pestis*, is a short Gram-negative rod. When approached by a phagocyte, the bacilli release proteins which prevent phagocytosis and that kill the phagocyte. They also reduce the inflammatory response of the host immune system. Other virulence genes destroy complement factors C3b and C5a, degrade fibrin clots thereby aiding spread of organisms, and help in absorption of iron from the host cells. All these virulence factors, plus others, help make plague a very lethal disease, and one of the most devastating in history. As a zoonosis, the disease is spread by infected rodents, especially rats, which transmit the organisms by animal-to-animal contact and occasionally to humans by flea bites. As infected rats die of plague, their body temperature drops, and their blood coagulates; hungry fleas jump to nearby sources of warmth and liquid blood. The new host usually is another rat, but in crowded, rat-infested living quarters or when contact is made with the carcass of a plague-infected animal, the next host can easily be a human.

PUBLIC HEALTH

Plague in the United States?

Here? Today? Yes! In fact, plague is quite prevalent in gophers, chipmunks, pack rats, prairie dogs, and ground squirrels in 15 western states. Indeed, plague is thought to have persisted for years among similar wild rodent populations of the Central Asian steppe before breaking out to decimate Europe as the "Black Death" in 1346. Should our rural (*sylvatic*) plague infect city rats, giving rise to *urban* plague, we may be in for big trouble. So far only sporadic cases of human plague have occurred in the American West, often among hunters and Native Americans on reservations. But we've been lucky. Some years ago, a pet store owner in Los Angeles trapped some desert rodents with the intention of selling them. Unfortunately for him—but perhaps fortunately for the community—he died of plague before he could sell any of the dangerous animals as pets.

PLAGUE. Reported cases among humans, by year – United States, 1971–2006

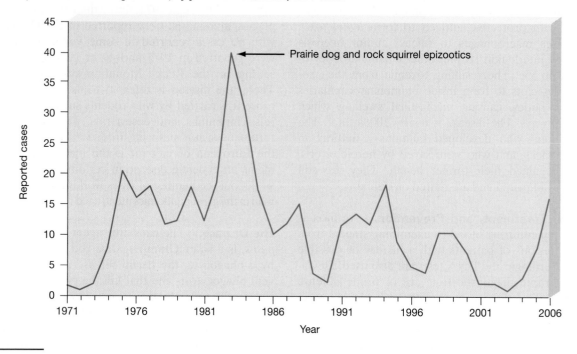

Only two laboratory-confirmed cases of human plague were identified from the United States in 2001. Utah and New Mexico each reported a single case, both of which were of the primary bubonic form of the disease. Both cases were acquired from the natural sources within known enzootic regions, and both patients recovered.

World Distribution of Plague, 2004

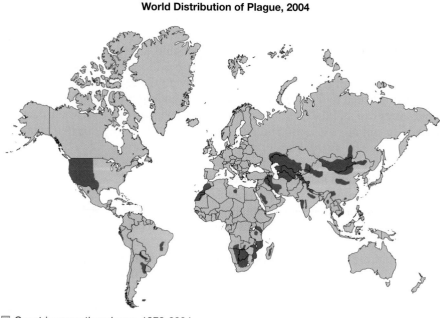

Figure 23.8 The incidence of plague.
(a) U.S. incidence, 1955–2006. (b) Worldwide regions of endemic plague. (*Source:* CDC.)

☐ Countries reporting plague, 1970-2004
■ Probable Sylvetic foci

The flea itself suffers from plague infection. The plague organisms ingested from a sick rat multiply and block the flea's digestive tract until food (blood meals) cannot pass through it. The flea gets hungrier, bites more ferociously, and infects new victims as it dispenses plague organisms with each bite. Eventually the flea dies, but this is little consolation to a new human victim, who has a 50 to 60% chance of dying if not treated.

Once inside the host, plague bacilli multiply and travel in lymphatics to lymph nodes, where they cause hemorrhages and immense lymph node enlargements called **bu-boes** (singular: *bubo*), especially in the groin and armpit

Figure 23.9 Buboes. In this drawing from a fourteenth-century Flemish manuscript, a physician lances a plague-caused bubo. A second patient awaits the lancing of his underarm bubo. *(Granger Collection)*

(Figure 23.9). Buboes are characteristic of **bubonic plague** and appear after an incubation time of 2 to 7 days. Hemorrhages turn the skin black—hence the name Black Death. Deaths from bubonic plague can be prevented with adequate, timely antibiotic treatment, otherwise death occurs within a few days after appearance of the buboes. If organisms move from the lymphatics into the circulatory system, **septicemic plague** develops. It is characterized by hemorrhage and necrosis in all parts of the body, meningitis, and pneumonia. This form of plague is invariably fatal despite the best modern care. **Pneumonic plague** occurs with lung involvement and can be spread when aerosol droplets from a coughing patient are inhaled. It, too, has a mortality rate approaching 100% despite excellent care. Medical personnel working with patients are more likely to acquire pneumonic than bubonic plague.

Diagnosis, Treatment, and Prevention. Plague can be diagnosed by fluorescent antibody tests or by the identification of *Yersinia pestis* from stained smears of sputum or from fluid aspirated from lymph nodes. It is treated with streptomycin, tetracycline, or both. Fortunately no drug-resistant strains have yet appeared.

Recovery from a case of plague confers lifetime immunity. A vaccine is available to protect medical personnel and the families of victims. However, during treatment of plague patients in the Vietnam War, it was discovered that workers, even those protected by immunization, can become pharyngeal carriers of plague organisms for a short period of time. Plague can be prevented by immunizing those who travel to endemic regions, controlling rat populations, and maintaining surveillance for infections in wild rodent populations. CDC surveys have found plague only among wild rural (usually desert) rodents, which are not likely to come in contact with urban populations. However, plague has moved eastward from the California coast, where it first arrived on a ship from China that docked at San Francisco in 1899. If the disease spreads to

PUBLIC HEALTH

Did Your Ancestors Survive Plague? What Good Might That Do You?

In 1665–1666, all residents of the small village of Eyam, located in Northern England, were shut up in their town for 1 year of quarantine; plague victims and healthy, all together. Food was left for them at the city limits. After 1 year, it was assumed that everyone would be dead, a sad thing for them but good in that it prevented spread of plague to neighboring towns. But there *were* survivors, some of whom had never gotten sick, others who did but recovered. Why? Might they have had some genetic difference that kept them alive? A grave digger who handled 100's of bodies and a woman who nursed and buried her husband and all six children in one week, never got sick themselves. Remarried to other survivors, their descendants have mostly remained in Eyam. DNA sequencing of today's descendants revealed that a high percentage of them have the delta 32 mutation in the CCR5 gene. This mutation prevents plague bacilli from entering macrophages and causing disease. Geneticists found that the frequency of delta 32 in the European population rocketed upward about 700 years ago—matching the time of arrival of plague in Europe. American descendants of European immigrants also have this mutation. African, Asian, and East Indian populations do not. Plague killed 60 to 75% of the European population (35 million people). Presence of two copies of delta 32 prevent you from ever getting plague; one copy means you can get sick, but have a fighting chance of recovery; no copies means infection and rapid death. Clearly delta 32 conferred a selective advantage then. But what about now? The AIDS virus invades in the same way that plague did. Two copies of delta 32 prevented 3,000 times the amount of HIV ordinarily needed from infecting macrophages. One copy means you are slow to become infected, and disease progresses more slowly. No copies, and you are truly vulnerable. About 3 million people in the U.S. and Britain have two copies, a wonderful legacy across 100s of years from their ancestors. What might your ancestors have left you?

urban rats, many of which are resistant to rat poisons, the risk to humans will increase.

Tularemia

The Disease. **Tularemia**, caused by *Francisella tularensis*, is a zoonosis found in more than 100 mammals—especially cottontail rabbits, muskrats, and rodents—and arthropod vectors, such as ticks and deer flies. In ticks the pathogen is incorporated in eggs as they leave the ovaries in **transovarian trasmission**—infection of eggs before they are fertilized—and thereby passes from one generation to the next. Although in about half of human cases the vector is never identified, tularemia is most often associated with cottontail rabbits; the number of cases reported always rises significantly during rabbit-hunting season.

Francisella tularensis, a small Gram-negative coccobacillus with worldwide distribution, was first isolated in

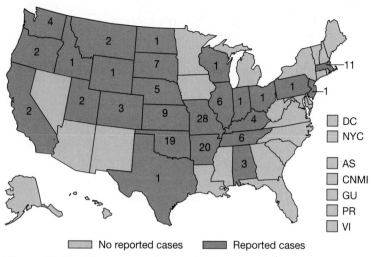

TULAREMIA. Reported cases – United States and U.S. territories, 2004

DC
NYC
AS
CNMI
GU
PR
VI

☐ No reported cases   ☐ Reported cases

Figure 23.10 **The U.S. incidence of tularemia, 2004.** (*Source:* CDC)

1911 from Tulare County, California, for which the species was named. The genus was named after Edward Francis, who did much of the early work on this organism. The annual incidence of tularemia in the United States has dropped from more than 2,000 cases in 1939 to fewer than 200 cases in recent years **(Figure 23.10)**. The disease is an occupational hazard for taxidermists.

Tularemia can be acquired in three ways. First, organisms usually enter through minor cuts, abrasions, or bites. Second, organisms can be inhaled, especially from aerosols formed during the skinning of infected animals. Third, organisms can be acquired by consumption of contaminated water or meat, resulting in an intestinal form of the disease. Swimming in a river near a colony of infected river rats and eating undercooked infected rabbit have been reported as sources of infection. Freezing, even for years, does not destroy the organisms.

Entry through the skin results in the **ulceroglandular** form of disease. After a 48-hour incubation period, symptoms begin with abrupt high fever of 40° to 41°C (104° to 106°F) with chills and shaking. If untreated, fever, severe headache, and buboes can last a month. An ulcer sometimes forms at the site of entry. The handling of animals and skins is most likely to cause ulcers on hands, whereas a bite by an arthropod vector is most likely to cause buboes (lymph node lesions) in the groin or armpit. The patient is initially disabled for 1 to 2 months and can have frequent relapses. The mortality is about 5% if left untreated. In the days of American pioneers, rabbit was a major part of the diet, and tularemia also must have been a major feature of pioneer life. Until the 1960s, when biological warfare was banned, *F. tularensis* was one organism studied by government scientists for possible use in biological warfare.

Bacteremia from lesions can lead to **typhoidal tularemia**, a septicemia that resembles typhoid fever.

Touching the eyes with contaminated hands can lead to conjunctivitis, but this happens in very few cases. Inhalation of organisms or their spread from blood produces a patchy bronchial pneumonia that leads to lung tissue necrosis and 30% mortality.

Diagnosis, Treatment, and Prevention. Diagnosis from blood cultures is difficult; the highly infectious organisms are hard to grow on ordinary laboratory media. Organisms live intracellularly in macrophages where they resist degradation. Only 50 organisms are sufficient to produce a human infection, regardless of the route of administration. Laboratory infections are easily acquired, so culturing of *Francisella tularensis* should be attempted only by very experienced personnel in isolation laboratories equipped with special air-flow hoods and other safety devices. Animal inoculations should be avoided. Agglutination tests are the standard methods of diagnosis (◄Chapter 18, p. 562). Streptomycin is the drug of choice for all forms of tularemia.

Prevention by eliminating the organism from wild reservoir populations is impractical. Therefore, one should avoid handling sickly animals, wear gloves when handling or skinning wild game, and, in tick-infested areas, wear protective clothing and frequently search clothing and skin for ticks. A vaccine exits, but it is not always protective and must be readministered every 3 to 5 years.

Brucellosis
The Disease. **Brucellosis**, also called **undulant fever**, **Bang's disease**, or **Malta fever**, is a zoonosis highly infective for humans. It is caused by several species of *Brucella*. *Brucella melitensis* was isolated in 1887 on the Mediterranean isle of Malta by Sir David Bruce. *Brucella* are small Gram-negative bacilli, each of which has a preferred host: *B. abortus*, cattle; *B. melitensis*, sheep and goats; *B. suis*, swine; and *B. canis*, dogs. In addition to preferred hosts, each species can infect several other hosts, including humans. The incidence of brucellosis in the United States has dropped sharply from more than 6,000 cases per year at the end of World War II to fewer than 200 cases per year since 1978, with 106 cases in 2006.

In Alaska, reindeer and caribou are known to be infected with brucellosis and serve as reservoir hosts. Eating their meat can cause human infection.

Brucella enter hosts through the digestive tract via contaminated dairy products and animal feed, the respiratory tract via aerosols, or the skin via contact with infected animals on farms or in slaughterhouses. Inside the host these facultative intracellular parasites multiply and move through the lymphatic system into the blood, where they cause an acute bacteremia within 1 to 6 weeks. Uncontrolled infections form granulomas (◄Chapter 16, p. 475) in the reticuloendothelial system. Brucellosis has a gradual onset with a daily fever cycle—high in the afternoon and low at night after profuse sweating. These episodes are caused by the release of bacteria from granulomas into the bloodstream. The spleen, lymph nodes, and liver can be enlarged, and jaundice can be present, but symptoms may be too mild to diagnose. The initial acute phase lasts from several weeks to 6 months. Recovery usually is spontaneous, but chronic aches and nervousness can develop. If attributed to a psychiatric disturbance, the disease may not be diagnosed or properly treated.

Diagnosis, Treatment, and Prevention. Brucellosis is diagnosed by serologic tests and treated with tetracycline. Streptomycin, gentamicin, or rifampin may be added in severe cases. Treatment must be prolonged because the organisms are well protected from antibiotics in the blood. When death does occur, it is usually from endocarditis. Brucellosis can be prevented by pasteurizing dairy products, immunizing animal herds, and providing education and protective clothing for workers with occupational exposure. No good human vaccine is available at present.

Cattle ranchers in the vicinity of Yellowstone National Park in Wyoming and in Montana face a danger from diseased bison who wander outside the park boundaries. Over 50% of the Yellowstone bison herd are infected with brucellosis. They can spread it to cattle when they mingle with them, especially during hard winters, when food may be more abundant on ranchers' lands. Hunters are allowed to kill bison who leave the park. Nearly 600 are shot in some years. The state of Montana has spent millions of dollars in the last decade to eradicate brucellosis from its herds. In cattle, *Brucella* multiples in the breast, placenta, uterus, and fetus, with its membranes, due to the presence of the carbohydrate mesoerythritol. This causes abortion of the fetus. Human uteri lack this sugar, which is preferred to glucose by many *Brucella* strains, and so are not subject to abortion. The epididymis also contains the sugar in animals but not humans.

Relapsing Fever

The Disease. Relapsing fever is an acute arthropodborne disease characterized by alternating fever and nonfever periods. Nearly a dozen species of the genus *Borrelia* are the causative agents **(Figure 23.11)**. Two different vectors transmit relapsing fever: soft ticks of the genus *Ornithodoros*, and human body and head lice of the genus *Pediculus*. Fifteen species of soft ticks are known to transmit relapsing fever. Louseborne infections are called **epidemic relapsing fever**, whereas tickborne infections are called **endemic relapsing fever**. Relapsing fever appeared in ancient Greece; since then outbreaks have occurred during times of war, poverty, and in institutions of extreme overcrowding that occur, for example, after natural

(a)

(b)

Figure 23.11 Relapsing fever. This disease is caused by various *Borrelia* species, such as **(a)** *Borrelia recurrentis* (among blood cells; 1,275X), carried by ticks *(Eric Grave/Phototake)* and by **(b)** lice such as the body louse (32X). *(National Medical Slide/Custom Medical Stock Photo, Inc.)*

disasters. During epidemics, mortality can reach 30% of untreated cases.

Borrelia are large, spiral, Gram-negative bacteria; their coarser, more irregular spirals and ease of staining distinguish them from *Treponema* (syphilis) spirochetes.

Transmission of relapsing fever varies with the vector involved. Lice must be crushed and their body contents scratched into the skin to transmit the disease. Each louse acquires the organisms by biting an infected host. Ticks transmit disease organisms in their salivary secretions when biting and by transovarian transmission. The ticks can survive for up to 5 years without a meal and still contain live infectious *Borrelia*. Eradication of these vectors is impossible. The ticks feed mainly at night for only about half an hour; victims such as campers or occupants of houses infested with tick-carrying rodents never realize they have been bitten. Lack of this important piece of clinical history makes diagnosis difficult.

Much of what we know about relapsing fever was learned before the antibiotic era, when syphilis was treated by inducing high fever. Patients were purposely infected with relapsing fever or with malaria; the high fever killed the syphilis organisms and left the patient with a more manageable disease. After an incubation period of 3 to 5 days, relapsing fever begins with a sudden onset of chills and high fever. Fever persists for 3 to 7 days and ends by crisis. About 16% of patients never experience a relapse, but most, after 7 to 10 days, have fever for 2 or 3 days. Typically, more fevers followed by intervals of relief occur—hence the name relapsing fever. The disease is particularly dangerous in pregnant women because the organisms can cross the placenta and infect the fetus.

Relapses are explained by changes in the organisms' antigens. During a febrile period, the body's immune response destroys most of the organisms. The few that remain have surface antigens the host's immune system fails to recognize. These organisms multiply during a relief period until they are numerous enough to cause a relapse. Each relapse represents a new population of organisms that have evaded the host's defense mechanisms.

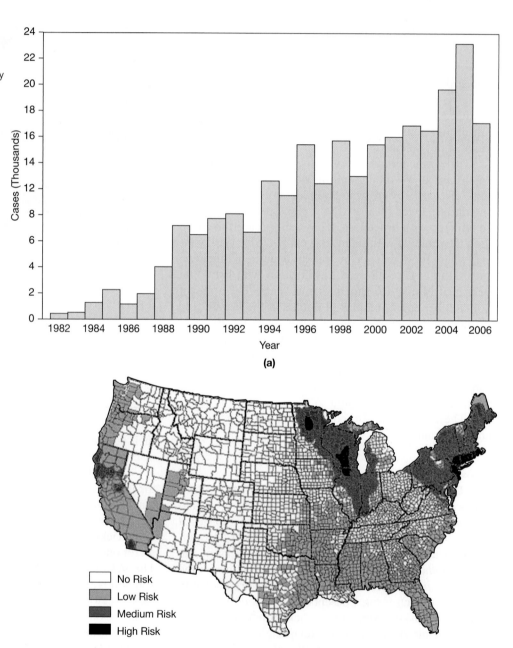

Figure 23.12 The U.S. incidence of Lyme disease, 2006. (a) What appears to be a dramatic spread of the disease may really be better diagnosis due to heightened awareness.
(b) However, the disease is also spreading in some areas. (*Source:* CDC)

APPLICATIONS

Outwitted by the Spirochetes Again

Some microorganisms are really "smart." They'd have to be in order to survive against our modern-day arsenal of medicines and vaccines and outwit our ever-watchful immune systems. If you look at pathogens that way, then the smartest ones are those that have been around the longest. That would make the spirochete that causes relapsing fever quite clever, since it has been around for thousands of years. The *Borrelia* spirochete is transmitted to humans by tick bites. Within the mammalian host, these spirochetes periodically replace one outer surface antigen with another, a process that presumably allows them to evade our immune systems. You can't help but give these microbes some respect.

Diagnosis, Treatment, and Prevention. Diagnosis is made by identifying the organisms in stained blood smears prepared during a rising fever phase. Tetracycline or chloramphenicol are used to treat relapsing fever. Immunity following recovery is usually short-lived. No vaccine is available, so prevention primarily is directed toward tick and louse control, as well as educating the public.

Lyme Disease

Take another look

The Disease. Alteration of ecosystems can give rise to new human diseases or to increased incidence and recognition of previously unidentified diseases (*emerging diseases*), as has been demonstrated in the case of **Lyme disease**. More Virginia white-tailed deer, which thrive along borders between forests and clearings and are a major reservoir of the disease agent, now inhabit the United States than when the Pilgrims landed. Settlers clearing fields created suitable habitats; as hunting of deer for food has declined, their populations have increased to record levels. With this increase has come Lyme disease, first described in 1974 by Allen Steere and his colleagues at Yale University and named after the Connecticut town where the earliest recognized cases occurred. The disease has now been identified on three continents and in more than 46 states in the United States. It is common, along with deer, from Cape Cod, Massachusetts, to Virginia and in Minnesota **(Figure 23.12)**.

In 1982 Willy Burgdorfer of the National Institutes of Health laboratory in Montana isolated and described the causative organism, *Borrelia burgdorferi*, a previously unknown spirochete. By 1985 Lyme disease was the most commonly reported tickborne disease in the United States. It is carried by *Ixodes scapularis*, the blacklegged tick, which feeds on deer and small mammals such as mice. Initially described as the deer tick because it was first found on deer, scientists have since shown that the deer tick is actually the blacklegged tick. The tick takes 3 blood meals during its 2-year life cycle **(Figure 23.13a)**. It ingests contaminated blood in one meal and transmits disease during a subsequent one.

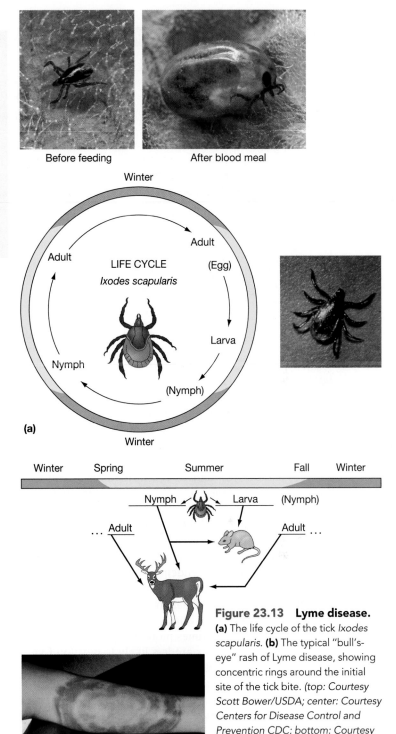

Before feeding After blood meal

Figure 23.13 Lyme disease.
(a) The life cycle of the tick *Ixodes scapularis*. **(b)** The typical "bull's-eye" rash of Lyme disease, showing concentric rings around the initial site of the tick bite. (*top: Courtesy Scott Bower/USDA; center: Courtesy Centers for Disease Control and Prevention CDC; bottom: Courtesy Allen C. Steere/www.invasive.org*)

Dogs, horses, and cows, as well as humans, can be infected. As the tick spreads to areas of high human density, the risk of tickborne infections increases.

Diagnosis, Treatment, and Prevention. Symptoms of Lyme disease vary, but most patients develop flulike symptoms shortly after being bitten by an infected tick. A bull's-eye rash at the site of the bite is also seen in about half of all

TABLE 23.2

A Summary of Bacterial Systemic Diseases		
Disease	**Agent(s)**	**Characteristics**
Anthrax	*Bacillus anthracis*	Cutaneous lesions become necrotic; respiratory infections always fatal; intestinal infections similar to food poisoning
Plague	*Yersinia pestis*	Bubonic plague causes buboes, or enlarged lymph nodes, and hemorrhages turn skin black; septicemic plague occurs when organisms invade blood and cause hemorrhage and necrosis in many tissues; pneumonic plague from inhalation of organisms causes pneumonia
Tularemia	*Franciscella tularensis*	Ulceroglandular form causes high fever, headache, buboes; bacteremia leads to typhoidal tularemia; inhalation leads to bronchopneumonia; relapses often occur
Brucellosis	*Brucella* species	Gradual onset of symptoms, cyclic fever, enlarged lymph nodes and liver, jaundice
Relapsing fever	*Borrelia recurrentis*	Several days of high fever, respites, and shorter periods of fever due to changes in organisms' antigens; can cross placenta
Lyme disease	*Borrelia burgdorferi*	Rash and flulike symptoms at onset, later arthritis and nerve and heart damage; can cross placenta

Doxycycline is a good inexpensive, broad-spectrum antibiotic that can be used to treat Lyme disease, bronchitis, and even malaria.

cases **(Figure 23.13b)**. Weeks or months later, other symptoms occur. Arthritis is the most common symptom, but loss of insulating myelin from nerve cells can cause symptoms resembling Alzheimer's disease and multiple sclerosis. Myocarditis occurs in some patients. Because most patients do not seek medical attention early, Lyme disease usually is diagnosed by clinical symptoms after arthritis appears. An antibody test is available. Lyme disease is treated with antibiotics such as doxycycline and amoxicillin, which are more effective the earlier they are administered in the course of the disease. Unfortunately, the earlier stages are often misdiagnosed. Furthermore, undiagnosed subclinical infections are common in endemic areas.

A vaccine for humans currently exists, but there have been numerous claims of damage due to the vaccine. A better vaccine to protect dogs from Lyme disease is also available. Dogs are six times more likely than humans to develop the disease. Cats are less likely to develop Lyme disease than dogs, possibly because of their continual grooming. Other animals such as cattle are also susceptible, as a Wisconsin dairy farmer discovered when three-quarters of his 60 dairy cows went lame from Lyme arthritic enlargements of leg joints, some nearly as large as basketballs. Only about 10% of infected people develop crippling Lyme arthritis. Researchers have found a genetic predisposition in those who do, with 89% having HLA-DR 4 or RLA-DR 2 tissue antigens (◀Chapter 18, p. 550). Antibiotics given in the first 6 weeks after infection generally prevent this chronic arthritis. After a certain point in the course of the infection, however, antibiotics are no longer able to prevent the development of arthritis in susceptible individuals.

Because no good vaccine for humans is available, Lyme disease can be controlled only by avoiding tick bites. Bites are difficult to avoid, however, because the tick—especially the immature form, which often carries the spirochete—is very tiny, the size of a poppy seed, and may go unnoticed. In tick-infested areas, a hat, long sleeves, and long pants should be worm, with socks pulled up over the bottom ends of the pants. Tick repellant can also be used. Prevention is a lot easier than cure. Once chronic Lyme disease is established, it can even cross the placenta and infect a fetus.

A drastic attempt to control the disease was tried on Great Island, Massachusetts, where an entire 52-member deer herd was killed. Putting antitick eartags on deer and mice, a difficult job, has not been very effective. Investigators have also tried saturating cotton balls with insecticide and leaving them around for the mice to take back to their nests, thereby killing off many of their ticks. Now, a European wasp that parasitizes the ticks is being released in tick-infected areas. This biological control method has given excellent results, almost exterminating the tick in some areas. The low incidence of Lyme disease on the U.S. West Coast may be due to a protein, toxic to ticks, found in the blood of lizards whose blood is the primary food of ticks in that area.

Bacterial systemic diseases are summarized in **Table 23.2**.

Symbiotic microbes and nematodes may be used in the future to control deer ticks, the arthropod vector responsible for the spread of Lyme disease.

✓**CHECKLIST**

1. Why would anthrax make such a good agent for biological warfare?

2. How does portal of entry affect plague symptoms and mortality rate?

3. Why is it important to seek early treatment for Lyme disease?

RICKETTSIAL AND RELATED SYSTEMIC DISEASES

Rickettsias are named after Howard T. Ricketts, who identified them as the causative agents of typhus and Rocky Mountain spotted fevers. Both he and another investigator, Baron von Prowazek, died of laboratory infections from these highly infectious organisms. Rickettsias are small, Gram-negative coccobacilli and are microscopic obligate intracellular parasites (◄Chapter 9, p. 264). Those that cause typhus fever grow in the cytoplasm of infected cells; those that cause spotted fever grow in both the nucleus and the cytoplasm. Rickettsias can be cultured only in cells. Embryonated eggs also are often used, but they should be handled only by very experienced technicians and only in specially equipped isolation laboratories. Despite improvements in laboratory design and techniques and the availability of vaccine, laboratory infections are still common, and occasionally fatal.

Before 1984 only 8 rickettsial diseases were known, but in the next 13 years, 7 new rickettsial diseases were discovered. They are now considered to be an emerging disease group. Rickettsial diseases have several properties in common. The organisms invade and damage the cells of blood vessel linings and cause the linings to leak. This leakage causes skin lesions and especially **petechiae** (pe-te′ke-e; singular: *petechia*), pinpoint-size hemorrhages most common in skin folds. It also causes necrosis in organs such as the brain and heart. Although each disease produces a particular kind of skin rash, all rickettsial diseases cause fever, headache, extreme weakness, and liver and spleen enlargement. Headaches can be very painful, causing the patient to appear confused.

Except for Q fever (◄Chapter 21, p. 659), rickettsial diseases are transmitted among vertebrate hosts by an arthropod vector. Humans often are accidental hosts of zoonoses, but they are the sole hosts of epidemic typhus fever and trench fever. Because of the hazards of culturing organisms, rickettsial diseases usually are diagnosed by clinical findings and serologic tests. The diseases are treated with tetracycline or chloramphenicol; even these antibiotics merely inhibit, but do not kill, rickettsias. Therapy must be prolonged until the body's defenses can overcome the infection. Often rickettsias are not totally eliminated; they can remain latent in lymph nodes and have been known to reactivate 20 years after initial infection.

Typhus Fever

Typhus fever occurs in a variety of forms, including *epidemic*, *endemic* (*murine*), and *scrub typhus*. *Brill-Zinsser disease* is a recurrent form of endemic typhus.

Epidemic typhus, also called *classic*, *European*, or *louseborne typhus*, is caused by *Rickettsia prowazekii*. The disease is most frequently seen during wars and other conditions of overcrowding and poor sanitation. In 1812, epidemic typhus helped drive Napoleon from Russia; more recently, during World War I, it infected over 30

APPLICATIONS

Microbes and War

"Soldiers have rarely won wars. They more often mop up after the barrage of epidemics. And typhus, with its brothers and sisters—plague, cholera, typhoid, dysentery—has decided more campaigns than Caesar, Hannibal, Napoleon, and all the…generals of history. The epidemics get the blame for defeat, the generals the credit for victory. It ought to be the other way 'round…." —Hans Zinsser, 1935.

million Russians and killed 3 million. The history of warfare is filled with instances in which typhus was the "commanding general." These and other examples of the effects of epidemic typhus on human affairs are well documented in Hans Zinsser's book *Rats, Lice, and History*. Only after the discovery of the pesticide DDT during World War II were typhus epidemics halted.

Epidemic typhus is transmitted by human body lice. After a louse feeds on an infected person, rickettsias multiply in its digestive tract and are shed in its feces. When a louse bites, it also defecates; infected lice deposit organisms next to a bite and themselves die of typhus in a few weeks. As victims scratch bites, they inoculate organisms into the wound. Lice become infected by biting an infected human. They abandon dead bodies or people with high fevers, moving to and infecting new hosts.

After about 12 days' incubation, onset of fever and headache is abrupt and is followed 6 or 7 days later by a rash on the trunk that spreads to the extremities but rarely affects palms or soles. Antibiotic therapy should be started immediately. Without treatment the disease typically lasts up to 3 weeks, and mortality ranges between 3 and 40%. The disease can be prevented by eradicating lice with insecticides and by maintaining hygienic living conditions. A vaccine is available. Recovery generally gives lifetime immunity, except when Brill-Zinser disease occurs.

Brill-Zinsser disease, or *recrudescent typhus*, is a recurrence of a typhus infection. It is named for Nathan Brill and Hans Zinsser, who studied it in the 1930s among New York City's Eastern European immigrants. Compared with first infections, this disease has milder symptoms, is shorter in duration, and often does not cause a skin rash. It is caused by reactivation of latent organisms harbored in lymph nodes, sometimes for years. Lice feeding on patients with Brill-Zinsser disease can transmit the organism and cause initial typhus infections in susceptible individuals. The disease can be prevented by preventing typhus itself and can be reduced in incidence by adequate antibiotic therapy of those already infected. Despite rigorous treatment, some victims of typhus still develop Brill-Zinsser disease later. It can be distinguished from epidemic typhus by the type of antibodies formed shortly after onset. Epidemic typhus first elicits IgM and then IgG antibodies, whereas Brill-Zinsser disease, being a secondary response, elicits primarily IgG antibodies (◄Chapter 17, p. 497).

Endemic typhus, or **murine typhus** (so named for its association with rats—*murine* refers to rats and mice) is a fleaborne typhus caused by *Rickettsia typhi*. It occurs in isolated pockets around the world, including southeastern and Gulf Coast states in the United States, especially Texas. Its incidence in the United States is fewer than 100 cases per year. Fleas from infected rats defecate while biting, infecting the humans they bite. The host rubs organisms into the bite wound or transfers them to mucous membranes, another portal of entry. After 10 to 14 days' incubation, onset of fever, chills, and a crushing headache is abrupt, followed by a rash in 3 to 5 days. The disease is self-limiting and lasts about 2 weeks if untreated. Mortality is about 2%.

Scrub typhus, or *tsutsugamushi* (soot″soo-ga-moosh′e) *disease*, is caused by *Rickettsia tsutsugamushi*. "Tsutsugamushi" is Japanese for "bad little bug." The "bug" that transmits this disease is a mite that feeds on rats in Japan,

Australia, and parts of Southeast Asia. Mites drop off rodent hosts and infect humans with their bites. Scrub typhus was a problem during World War II and the Vietnam War when soldiers crawled on their bellies through low scrub vegetation to avoid snipers. After 10 to 12 days' incubation, scrub typhus begins abruptly with fever, chills, and headache. Many patients develop sloughing lesions at the bite sites and later a generalized spotty rash. In untreated cases, the fatality rate can reach 50%, but with prompt antibiotic treatment, fatalities are rare. Although no vaccine is available, infections can be prevented by controlling mite populations.

Rocky Mountain Spotted Fever

Rocky Mountain spotted fever was first recognized around 1900 in Rocky Mountain states such as Idaho and Montana. However, its geographic distribution suggests that "Appalachian spotted fever" would be a more appropriate name

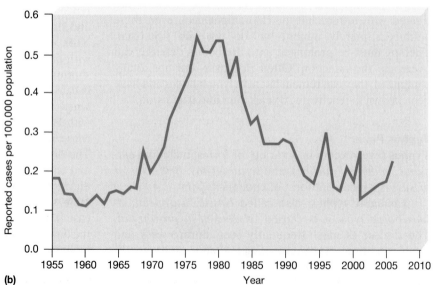

Figure 23.14 The U.S. incidence of Rocky Mountain spotted fever.
(a) By county, **(b)** By year. (*Source:* CDC.)

(a)

LM

(b)

Figure 23.15   Rocky Mountain spotted fever. **(a)** Vectors include the dog tick *Dermacentor variabilis* (20X). *(R. Calentine/ Visuals Unlimited)* **(b)** The characteristic rash of the disease on a patient's feet. *(Kenneth Greer/Visuals Unlimited)*

(Figure 23.14). The incidence has stayed below 1,000 cases per year since the 1930s but might increase as more people spend time in tick-infested areas. This disease is caused by *Rickettsia rickettsii* and is transmitted by ticks of the genus *Dermacentor* **(Figure 23.15a)**. After 3 to 4 days' incubation, onset of fever, headache, and weakness is abrupt, followed in 2 to 4 days by a rash **(Figure 23.15b)**. The rash begins on ankles and wrists, is prominent on palms and soles, and progresses toward the trunk—just the reverse of the progression in typhus. Spots are caused by blood leaking out of damaged blood vessels beneath the skin surface; they coalesce as blood leaks from many damage sites. Blood vessels in organs throughout the body are similarly damaged.

Strains vary considerably in virulence; likewise, mortality varies from 5 to 80%, with an average of 20% in untreated cases. Prompt antibiotic treatment keeps the mortality rate to between 5 and 10%. Rocky Mountain spotted fever can be prevented by wearing protective clothing and by vigilantly inspecting clothing and skin during visits to tick-infested areas. Inspecting children's hair is especially important. The only vaccine is not completely effective.

Worldwide, there are many other types of spotted fevers. Each is caused by a specific rickettsial species and is typically named for its location, such as "Siberian spotted fever."

Rickettsialpox

Rickettsialpox, caused by *Rickettsia akari*, was first discovered in New York City in 1946 and is now known to occur in Russia and Korea as well. It is carried by mites found on house mice. The disease is relatively mild, and its lesions resemble those of chickenpox. Because of misdiagnosis as chickenpox or other diseases, incidence and mortality data are unreliable, but no fatalities have been reported. It can be prevented by controlling rodents.

Trench Fever

Trench fever, also called *shinbone fever*, resembles epidemic typhus in that it is transmitted among humans by lice and is prevalent during wars and under unsanitary conditions. Stress probably is a predisposing factor. The causative agent is the bacterium *Bartonella (Rochalimaea) quintana*, classified with rickettsias even though it is not an obligate intracellular parasite. It can be cultured in artificial media and has worldwide distribution, but it rarely causes disease.

Trench fever was first seen in World War I among soldiers living in trenches and wearing the same clothing day after day. British "trench coats" were developed to protect troops who were without shelter during bad weather, but they were not entirely effective. Soldiers and their clothing, including the trench coats, became infested with body lice; exhausted soldiers crowded together in filth fell victim to the disease. After World War I the disease disappeared and did not reappear until World War II. Cases of trench fever have recently been found among the urban poor. The symptoms of trench fever are a 5-day fever and severe leg pain, but many soldiers have reported recurrent symptoms, including mental confusion and depression, as long as 19 years after infection. No vaccine is available, and prevention depends on control of lice.

Bartonellosis

Bartonellosis (bar″to-nel-o′sis) is caused by *Bartonella bacilliformis*, named for the Peruvian physician A. L. Barton, who first described it in 1901. The disease occurs in two forms: **Oroya fever**, or *Carrion's disease*, an acute fatal fever with severe anemia, and **verruga peruana** (ver-oo′gah per-oo-a′nah), a chronic, nonfatal skin disease **(Figure 23.16)**. Both are found only on the western slopes of the Andes in Peru, Ecuador, and Colombia— the habitat of the sandfly *Phlebotomus*, which transmits the organism. In 1885 Daniel Carrion, a Peruvian medical student, inoculated himself with material from a wartlike verruga peruana lesion to show a connection between it and Oroya fever. His death 39 days later from Oroya fever clearly demonstrated the connection.

After being transmitted to a human host by the bite of an infected sandfly, *B. bacilliformis* enters the blood and multiplies during an incubation period of a few weeks to 4 months. Little is known of the epidemiology

Figure 23.16   Lesions of verruga peruana. *(Courtesy Armed Forces Institute of Pathology)*

of bartonellosis, but humans appear to be the sole reservoir. Oroya fever is a severe, febrile, hemolytic anemia. Verruga peruana causes only skin lesions that persist for 1 month to 2 years but usually last for about 6 months. Lesions heal spontaneously but can recur. Oroya fever probably develops in people with no immunity, and verruga peruana occurs in those with partial immunity. Penicillin, tetracycline, or streptomycin can cure Oroya fever but not verruga peruana. No vaccine is available, and prevention depends on the control of sandflies.

Ehrlichiosis

Some recently identified human pathogens related to the rickettsias belong to the genera *Ehrlichia* and *Bartonella*. They are Gram-negative coccobacilli and are microscopic obligate intracellular parasites.

Ehrlichiosis, originally recognized as a disease in dogs, has now been found in humans, with 1,025 cases reported in 2006 in the United States. *Ehrlichia canis* and *E. chaffeensis*, the causative agents, are spread by dog ticks, as well as by ticks that spread Lyme disease. Clinically, the human form of the disease resembles other rickettsial diseases. Typical symptoms include fever, headache, hepatitis, and muscle pain. Blood abnormalities, such as a reduction of white blood cells, are common. The absence of a rash distinguishes ehrlichiosis from Rocky Mountain spotted fever. Diagnosis requires special serological tests, but finding typical ehrlichial intracytoplasmic inclusions in white blood cells may suggest the presence of *Ehrlichia*.

The characteristics of rickettsial diseases are summarized in **Table 23.3**.

Bacillary Angiomatosis

Bacillary angiomatosis is caused by *Bartonella henselae*, another rickettsial organism. The disease involves small blood vessels of the skin and internal organs. It is seen in persons with AIDS and other immunocompromised patients. Molecular probes and related techniques are used in diagnosis. The organism is also the cause of cat scratch fever (◀Chapter 19, p. 597).

TABLE 23.3

A Summary of Rickettsial Diseases

Disease	Causative Organism	Geographic Area of Prevalence	Arthropod Vector Reservoir	Vertebrate Reservoir
Typhus Group				
Epidemic (classic, European) typhus	*Rickettsia prowazekii*	Worldwide	Louse	Human
Brill-Zinsser disease (recrudescent typhus)	*R. prowazekii*	Worldwide	(Recurring infection)	Human
Endemic (murine) typhus	*R. typhi*	Worldwide, small scattered foci	Flea	Rodents
Scrub Typhus Group				
Scrub typhus (tsutsugamushi disease)	*R. tsutsugamushi*	Japan, Southwest Asia	Mite	Rat
Spotted Fever Group				
Rocky Mountain spotted fever	*R. rickettsii*	Western Hemisphere	Tick	Rodents, dogs
Rickettsialpox	*R. akari*	United States, Korea, Russia	Mite	House mouse
Trench fever	*Bartonella (Rochalimaea) quintana*	Worldwide, but disease only during wars	Louse	Human
Bartonellosis	*Bartonella bacilliformis*	Western slopes of the Andes	Sandfly	Humans only known host
Ehrlichiosis	*Ehrlichia canis, E. chaffeensis*	Southeastern and south central United States	Tick	Dogs, humans

✓CHECKLIST

✓CHECKLIST

1. How does Q fever differ from other rickettsial diseases?
2. How does *Bartonella (Rochalimaea) quintana* differ from other rickettsiae?

VIRAL SYSTEMIC DISEASES

Dengue Fever

Dengue (den′ga or den″ge) **fever**, first characterized in 1780 by the American physician Benjamin Rush in Philadelphia, has also been called *breakbone fever* because of the severe bone and joint pain it causes. Other symptoms include high fever, headache, loss of appetite, nausea, weakness, and in some cases a rash. The disease is self-limiting and runs its course in about 10 days. Four distinct immunological types of the dengue virus—an arbovirus of the family Flaviviridae—have been identified, and two of them have been correlated with disease symptoms. A first dengue fever infection produces the symptoms just noted. A second dengue fever infection with a different immunologic type of the virus causes a hemorrhagic form of the disease. The hemorrhage occurs while the virus is replicating in circulating lymphocytes. It involves an immune response made possible by prior infection. Other symptoms are rapid breathing and low blood pressure that may progress to shock. The shock is reversible if treatment is initiated promptly. Serologic tests are available to diagnose dengue fever, and a vaccine against one immunological type of the virus appears to confer immunity. Infection with one serotype does not provide cross-protection against the other serotypes, so one person can have dengue fever four times during his or her life.

Dengue fever is distributed worldwide in tropical areas, causing 500,000 to 3 million cases per year, with occasional episodes in the subtropics. It is endemic in 101 countries **(Figure 23.17)**. Currently South America and the Caribbean are experiencing serious outbreaks of dengue fever—enough to now make health authorities refer to it as an *emerging disease*. Its main vector is *Aedes aegypti*, although in some areas the Asian tiger mosquito, *A. albopictus*, can be important. Since 1985, health officials have been concerned about the arrival and spread of *A. albopictus* in the United States. The mosquito apparently arrived in used tire casings imported from Asia. In 1980 the first outbreak of dengue fever in the United States in about 40 years occurred. Since then, all four serotypes have arrived in the United States. The rapid spread and aggressive biting habits of *A. albopictus* could bring dengue fever to areas that previously had been safe. Mosquito control is the primary method of preventing the disease, as vaccines are not available for all dengue viral serotypes.

Yellow Fever

Yellow fever was first studied by Carlos Finlay and Walter Reed in the early 1900s when infection of workers threatened to disrupt the construction of the Panama Canal. Although adequate techniques to identify the

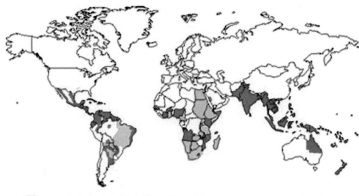

☐ Areas infested with *Aedes aegypti*

☐ Areas with *Aedes aegypti* and dengue epidemic activity

Figure 23.17   Dengue fever worldwide distribution, 2006.

causative flavivirus (◄Chapter 10, p. 280), an arbovirus, were not available, Finley and Reed identified the mosquito vector, *Aedes aegypti*, and instituted control measures that prevented transmission of the disease. The disease is now limited to tropical areas of Central and South America and Africa. Incidence is greatest in remote jungle areas, where monkeys serve as reservoirs of infection and carrier mosquitoes bite both monkeys and people. In recent decades the number of reported cases each year has varied from 12 to 304, but yellow fever incidence has been vastly underreported. The actual number of cases worldwide is probably 200,000 annually, 30,000 of whom die each year.

In 1802, Napoleon lost 22,000 of his 25,000 troops to yellow fever in Haiti while trying to stop a slave revolt.

Yellow fever causes fever, nausea, and vomiting, which coincide with viremia. Liver damage from viral replication in liver cells causes the jaundice for which the disease is named. The disease is of short duration: In less than a week, the patient has either died or is recovering. In most instances the fatality rate is about 5%, but in some epidemics it reaches 30%. Two strains of yellow fever viruses are used to produce vaccines. The

APPLICATIONS

More Than Those Slave Owners Bargained For

The slave trade brought more to the Americas than slaves—it brought the yellow fever mosquito, *Aedes aegypti*. The mosquitoes laid their eggs in the water cisterns of sailing ships and lived almost exclusively on the blood of their human shipmates after hatching. And once the boat docked in the Western Hemisphere, the mosquitoes promptly began spreading diseases: in particular, the devastating epidemics of yellow fever that didn't end until the dawn of the twentieth century, and dengue and dengue hemorrhagic fever in even more ferocious epidemics.

Dakar strain is scratched into the skin, whereas the 17D strain is administered subcutaneously. Both are effective in establishing immunity.

Infectious Mononucleosis

In 1962 Dennis Burkitt suggested that a virus caused the lymphoid malignancy now called *Burkitt's lymphoma*, found in children in East Africa. This herpesvirus, now called **Epstein-Barr virus** (**EBV**) or human herpesvirus number 4 **(Figure 23.18)**, is known to cause Burkitt's lymphoma, most cases of **infectious mononucleosis**, and oral hairy leukoplakia, a disease found among AIDS cases (see Table 10.3, ◄p. 282). EBV infects primarily human B lymphocytes. It replicates like most other herpesviruses and derives its envelope from the inner nuclear membrane of the host cell. The virus has an unusually large number of genes—more than 50 different proteins are produced by complete expression of the DNA in EBV.

EBV enters the body through the oropharynx. The virus first infects epithelial cells and eventually B cells. It establishes a persistent infection in which viruses are shed for months to years. The viruses invade such sites as lungs, bone marrow, and lymphoid organs, where they infect certain kinds of mature B lymphocytes. They penetrate B lymphocytes over a 12-hour period, and EBV replication begins within 6 hours after penetration. Viral DNA replicates much faster than cellular DNA. Viral DNA can exist as circular plasmids or can become integrated into the cellular DNA.

EBV exerts three significant effects on lymphocytes:

1. The virus acts on antibody-producing cells and elicits EBV antibodies.
2. Infection and transformation of viral genes to the cellular DNA (or presence of the viral chromosome as a free plasmid in the host cell) are complex events that occur primarily in B lymphocytes, which have receptors for EBV. The cells produce a variety of antigens, some of which are recognized by T cells. This recognition causes the T cells to proliferate, which accounts for the excess of lymphocytes seen in infectious mononucleosis.
3. Other antigens are induced on the surface of some infected B cells. They appear to play a role in some B-cell and T-cell interactions and may account for some symptoms of infectious mononucleosis.

The proliferation of EBV-infected lymphocytes is limited by cytotoxic T cells and cells that make humoral antibodies and complement (◄Chapter 17, p. 507). If these defenses fail to limit lymphocyte proliferation, uncontrolled B cell proliferation can lead to B cell cancer or Burkitt's lymphoma.

Infectious mononucleosis is an acute disease that affects many systems. Lymphatic tissues become inflamed, some liver cells become necrotic, and monocytes accumulate in liver sinusoids. In some cases myocarditis and glomerulonephritis are seen. The incubation period for the disease is from 30 to 50 days. Mild symptoms—headache, fatigue, and malaise—occur during the first 3 to 5 days of the disease and worsen as the disease progresses. About 80% of patients have a sore throat during the first week. The spleen is enlarged, and cells in lymphoid tissues in the oropharynx multiply. The tonsils are coated with a gray exudate, and the soft palate may be covered with petechiae. Secondary infection with β-hemolytic streptococci frequently occurs. Although the disease causes great discomfort and requires several weeks of recuperation, fatalities are rare and usually result from underlying immunological defects.

Diagnosis of EBV infections is complicated by the fact that the disease resembles cytomegalovirus infections, toxoplasmosis, and acute leukemia. The distinguishing symptoms of EBV infections are the concurrent sore throat, multiplication of lymphocytes, and the presence of antibodies against the antigens on sheep and human erythrocytes. Infectious mononucleosis is treated with bed rest and antibiotics for secondary infections. Ampicillin is not used because it causes a rash in infectious mononucleosis patients. The presence of IgG antibodies to viral capsid protein indicates a past infection, and their numbers furnish an index of immunity. Increasing numbers of IgM antibodies to the protein are evidence of a current infection. No vaccine is available.

In developing countries, the entire population has antibodies to EBV by age 1. Exposure to the virus in infancy produces mild symptoms or no symptoms and confers immunity to later infection. Where living standards are higher, a more severe disease is seen later in life. The incidence of infectious mononucleosis in the United States is highest among relatively affluent teenagers and young adults; 10 to 15% become infected. The affected age group and the large inoculum required to transmit the disease may be responsible for its being called the "kissing disease." Actually, infectious mononucleosis is

Figure 23.18 The Epstein-Barr virus. A colorized TEM `TEM` (magnification unknown) of the virus that causes infectious mononucleosis and, along with other factors, leads to Burkitt's lymphoma. *(Courtesy Centers for Disease Control)*

not highly contagious. About 15% of people who have recovered from the disease are shedding low levels of virus in their saliva at any one time. Those who have recently been infected will shed the virus continuously for about 18 months. About half of immunosuppressed individuals constantly shed large amounts of the virus. Once infected by EBV, even after symptoms have disappeared, the virus remains in B cells in a latent state as a permanent infection. Immunosuppression allows the virus to resume replication. In a person with a normal immune system, any reactivation of EBV is quickly halted.

Burkitt's Lymphoma.

Burkitt's lymphoma, a tumor of the jaw and viscera such as liver and spleen, is seen mainly in children (**Figure 23.19**). It occurs about 6 years after the primary infection with EBV, but genetic and environmental factors play a role in the development of the disease. The tumor frequently arises from a single cell. The immune system of affected individuals appears to be normal, but it is incapable of eliminating the tumor cells. Burkitt's lymphoma is found mainly in regions of Africa where malaria is endemic, and infection with malarial parasites may enhance growth of the virus or interfere with the immune response.

Other Effects.

Another tumor associated with EBV, which also shows a distinct geographic localization, is *nasopharyngeal carcinoma*, found most often in China and rarely in the Western Hemisphere. It is the most common tumor in the male population of southern China, comprising about 20% of all cancers. Genetic and cultural patterns predispose to this tumor. Chemicals consumed in salted fish, plus plant extracts (from the plant family *Euphorbiaceae*) used in Chinese traditional medicine, can induce replication of EBV.

Individuals with immunological defects are especially susceptible to developing lymphomas, presumably because they lack the necessary immune mechanisms to eliminate the malignant cells. Cyclosporin A, used to depress immunity in organ transplant patients, enhances lymphoid cell growth in donor organs. Organs from donors who have had EBV infections may contain EBV; immunosuppression from any cause, including AIDS, can release EBV and lead to mononucleosis or malignancy. Recipients of EBV-free organs will probably have latent EBV infections of their own, which can then proliferate due to the use of immunosuppressive drugs to prevent organ rejection. Blood transfusions can also transmit EBV.

Chronic Fatigue Syndrome.

Since 1985, researchers have been attempting to determine whether a *chronic EBV syndrome* exists. Patients report fever and persistent fatigue along with a variety of other nonspecific symptoms similar to those of infectious mononucleosis. Some, but not all, have EBV antibodies. Others have had measles or herpesvirus infections. Because a direct relationship between the symptoms and previous EBV infection has not been established, the illness has been renamed **chronic fatigue syndrome**. More study is needed to determine whether the syndrome is associated with one or more previous viral illnesses or whether it might be a psychological disorder. Some recent studies point to possible immune system defects. Human herpes virus number 6 (HHV6), the recently discovered cause of the childhood disease roseola, is also a candidate for causative agent of chronic fatigue syndrome.

Figure 23.19 Burkitt's lymphoma. (a) This condition is a form of cancer of the jaw that results from infection with the EBV; it is usually seen only in African children. *(Science VU/Visuals Unlimited)* **(b)** Internal view. *(Science VU/Visuals Unlimited)*

Other Viral Infections

Filovirus Fevers. **Filoviruses**, or filamentous viruses, display unusual variability in shape. Some are branched, others are fishhook- or U-shaped, and still others are circular. They contain negative-sense RNA in a helical capsid and vary in length from 130 to 4,000 mm (◀Chapter 10, p. 281). Two filoviruses have been associated with human disease. The **Ebola virus** caused outbreaks of hemorrhagic fever first in 1976, with a mortality of 88% in Zaire and 51% in Sudan. Nearly one-fifth of the population of rural areas of Central Africa have antibodies to Ebola. Transmission is person to person. A 1995 outbreak of Ebola virus in Zaire became world news; more than 200 cases were documented, with a mortality rate of about 75%. The *Marburg virus* was first recognized in Germany when technicians preparing monkey kidney cell cultures died of a hemorrhagic disease. Nosocomial Marburg virus infections have since been encountered with a mortality of about 25%. Hemorrhage into skin, mucous membranes, and internal organs, death of cells of the liver, lymph tissue, kidneys, and gonads, and brain edema also have been observed. The virus has been isolated directly from monkeys and from laboratory inoculation of guinea pigs.

Bunyavirus Fevers. Infections caused by bunyaviruses begin suddenly with fever and chills, headache, and muscle aches. Although usually not fatal and without permanent effects, the infections are temporarily incapacitating. When encephalitis occurs, it progresses slowly, either because the viruses replicate slowly in neural tissues or because certain viruses capable of replicating in neural tissue are selected. Rats, bats, and animals with hooves serve as reservoirs of infection. Tropical and temperate forest mosquitoes are vectors.

The *LaCrosse bunyavirus* has been identified in the northeastern and north-central United States. It causes a mild disease in adults but can cause seizures, convulsions, mental confusion, and paralysis in children. The *California encephalitis virus*, a bunyavirus initially isolated from mosquitoes in the San Joaquin Valley, has similar effects on humans.

Bunyaviruses called **phleboviruses** (because they are carried by the sandfly *Phlebotomus papatasii*) have been recovered from human infections. The **Rift Valley fever** virus causes epidemics and has unpredictable virulence. It causes sudden vomiting, joint pain, and slowing of the heart rate. One epidemic of Rift Valley fever in 1975 in central Africa infected thousands but left only 4 people dead. Only 2 years later the disease appeared in Egypt, where 200,000 cases and 598 deaths were recorded.

The *hantaviruses* are associated with hemorrhagic fevers, including Korean or epidemic hemorrhagic fever, and kidney disease. They are distributed widely over Eurasia and find reservoirs in rodents. They cause capillary leakage, hemorrhage, and cell death in the pituitary gland, heart, and kidneys. Kidney damage can be severe, and low blood pressure can proceed quickly to shock, from which one-third of patients die. Even more die if bleeding occurs in the gastrointestinal tract and the

PUBLIC HEALTH

Ebola Virus Scare

In April and May of 1995, an outbreak of the Ebola virus occurred in Zaire, Africa. The outbreak was traced to a forest worker who had become infected in December 1994. People infected with the virus develop symptoms of hemorrhagic fever, including fever and muscle aches. Most patients experience respiratory and kidney problems, abdominal pain, sore throat, and severe bleeding as well. Because the blood of Ebola patients cannot clot properly, bleeding occurs from needle injection sites, the digestive system, internal organs, and the skin. The virus is spread by direct contact with body fluids. Thus, outbreaks can be limited by isolating infected individuals, sterilizing needles and syringes, disposing of hospital waste and corpses properly, and using masks, gowns, boots, and gloves by hospital personnel—who are at special risk. Thousands of dollars worth of such supplies, as well as medicines and blood plasma, have been donated from around the world to limit the spread of the latest Ebola outbreak.

The reservoir for the Ebola virus is unknown. The first known outbreaks occurred in 1976 near the Ebola River (for which the virus was named) in Zaire and in western Sudan. That first Zaire outbreak resulted from Ebola-contaminated needles and syringes that were reused without sterilization. In fact, most of the outbreaks, including the one in 1995, were the result of poor medical conditions in hospitals.

As of mid-1998 there were four known Ebola viruses: subspecies Zaire, Sudan, Tai, and Reston. Ebola Zaire and Ebola Sudan can infect humans. Ebola Tai is the first Ebola virus known to be passed to humans from other animals—naturally infected chimpanzees of the Tai Forest in the Ivory Coast. Because infections of chimps, like humans, can lead to death, the original source of the virus is not the chimps. Ebola Reston, isolated from infected monkeys delivered to Reston, Virginia, in 1989, appears to be spread only among monkeys, but in this case aerosol transmission was documented. Since then, two more shipments of monkeys from the same supplier were found to have Ebola Reston. The U.S. government has now banned further import from that source.

For a factual account (as opposed to some sensational fictionized movies) of the Reston outbreak, read *"The Hot Zone"* by Richard Preston, Random House, 1994.

central nervous system or if fluid accumulates in the lungs. The major source of such infections is contact with infected rodents or their excreta. Some rodents shed viruses in feces and saliva for 30 days and into their urine for a year. Humans under 10 or over 60 years are rarely infected, perhaps because they are less likely to come in contact with infected materials.

Arenavirus Fevers. Like bunyaviruses, arenaviruses cause hemorrhagic fevers. Of these **Lassa fever** is perhaps the most widely known. It is an African disease that begins with pharyngeal lesions and proceeds to severe liver damage. The prognosis is poor in 20 to 30% of cases in which hemorrhage from mucous membranes occurs. Several other arenavirus infections, including **Bolivian hemorrhagic fever**, have been identified in humans, especially in Africa and South America.

Bolivian and other *South American hemorrhagic fevers* are multisystem diseases with insidious onset and progressive effects. The viruses attack lymphatic tissues and bone marrow and cause vascular damage, bleeding, and shock. However, death, which occurs in about 15% of cases, usually results from damage to the central nervous system. How the viruses affect the nervous system is not known.

Colorado Tick Fever. **Colorado tick fever** is caused by an **orbivirus** (a member of the Reoviridae) that is transmitted to humans by dog ticks from reservoir animals such as squirrels and chipmunks. High levels of viremia with infection of immature erythrocytes occur. The patient suffers from headache, backache, and fever, but recovery usually is complete.

Parvovirus Infections. Two parvovirus infections are now known to affect cats and dogs, respectively. **Feline panleukopenia virus (FPV)** causes severe disease with fever, decreased numbers of white blood cells, and enteritis in cats. The virus replicates in blood-forming and lymphoid tissues and secondarily invades the intestinal mucosa. In 1978 a new virus, the **canine parvovirus**, appeared and infected dogs in widespread geographic areas.

APPLICATIONS

No Good Choices

If you had a choice, which would you choose: Bolivian hemorrhagic fever or malaria? Probably neither. But this was a dilemma facing the Bolivians. For years, hemorrhagic fever existed in Bolivia without causing major problems. This situation changed when a campaign was initiated to control malaria by eradicating mosquitoes. Large quantities of pesticides were used, killing many village cats; in the absence of cats, the rodent population multiplied and brought arenavirus into human dwellings. This caused the incidence of hemorrhagic fever to rise to epidemic levels.

This virus first surfaced in North America, Europe, and Australia and rapidly spread worldwide. It causes severe vomiting and diarrhea in dogs of all ages and sudden death with myocarditis in puppies under 3 months old. When the canine parvovirus first appeared in an area, the death rate often exceeded 80%. Vaccines are now available for both feline panleukopenia and canine parvovirus.

Aplastic Crisis. A member of the genus *Erythrovirus*, also called B19, has been identified as the probable cause of **aplastic crisis**—a period in which erythrocyte production ceases—in sickle-cell anemia. The virus appears to replicate in rapidly dividing cells of the bone marrow. Afflicted children soon are in acute distress. Although normal erythrocytes remain functional in the blood for about 120 days, sickled cells survive only 10 to 15 days. Under such circumstances severe destruction of erythrocytes occurs. A child generally experiences only one such crisis, possibly because immunity to the virus develops. Another bit of evidence that aplastic crisis is infectious is that, although it occurs only in sickle-cell anemia patients, it does occur in 3- to 5-year cycles within particular communities.

Fifth Disease. The parvovirus B19 destroys the stem cells that give rise to red blood cells. This is not a major problem in healthy adults or children, but it is a serious danger to those who have chronic hemolytic anemias, such as sickle-cell anemia, and who therefore have difficulty maintaining normal levels of red blood cells. It is also a danger to the fetus if a pregnant woman contracts the virus. The virus can be transmitted across the placenta and cause the fetus to develop fatal anemia. The virus does not, however, cause birth defects. Immunodeficient patients cannot control replication of the virus and may develop chronic anemia. After B19 was recognized as the cause of aplastic crisis in sickle-cell anemia, it was also found to cause **fifth disease** (*erythema infectiosum*) in normal children. It is common in children of ages 5 to 14 and often goes unnoticed. Infected children have a bright red rash on the cheeks that may spread to the trunk and extremities (**Figure 23.20**). A low-grade fever may accompany the rash. Often the infection is totally asymptomatic. The virus appears to be spread via the respiratory route. The disease is self-limiting and confers lifetime immunity. The name "fifth disease" comes from a nineteenth-century list of childhood rash diseases. The first disease on this list is scarlet fever; the second is rubeola; the third, rubella; the fourth, epidemic pseudoscarlatina (a type of sepsis); and the fifth, erythema infectiosum.

Coxsackie Virus Infections. *Coxsackie viruses* have an affinity for the pericardium and myocardium. They can cause meningoencephalitis, diarrhea, rashes, pharyngitis, and liver disease. Epidemic muscle pain, diabetes, and inflammation of the pancreas, heart muscle, and the pericardium are associated with coxsackie B viruses. Coxsackie viruses are highly infectious and

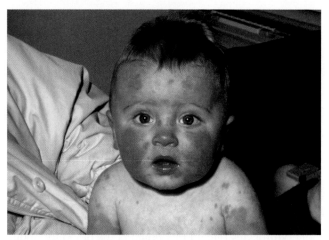

Figure 23.20 **Fifth disease caused by parvovirus B19.**
(H.C. Robinson/Photo Researchers, Inc.)

readily spread among family members and in institutions. Most infections probably arise from fecal-oral transmission, but because the viruses can be isolated from nasal secretions, infection also may occur by respiratory fomites. Coxsackie virus infections during pregnancy can cause congenital defects, but their incidence is much lower than that for rubella virus infections, and aborting the fetus is usually not recommended. No effective means are available for treatment, immunization, or prevention.

✓ **CHECKLIST**

1. If you have had infectious mononucleosis, do you still carry the virus? If so, what can it do?
2. What diseases are caused by B19?

PROTOZOAN SYSTEMIC DISEASES
Leishmaniasis
Three species of protozoa of the genus *Leishmania* can cause **leishmaniasis** (lish″man-i′a-sis) in humans (**Figure 23.21a**). The protozoa are transmitted by sandflies. When an infected sandfly bites, the parasites enter the blood of the host and are phagocytized by macrophages (◄Chapter 16, p. 469). The parasites multiply inside the macrophages, and new parasites are released when the macrophages rupture. Leishmaniasis is endemic to most tropical and subtropical countries, where appropriate species of sandfly vectors are available. Rodents can be reservoir hosts of the disease. Some 12 million cases worldwide are reported by WHO.

The Disease. *Leishmania donovani* causes **kala azar** (ka′lah ah-zar′; Hindi for "black poison"), or *visceral leishmaniasis*. Symptoms include high irregular fever, progressive weakness, wasting, and protrusion of the abdomen due to extensive liver and spleen enlargement. Extensive damage to the immune system results when parasites destroy large numbers of phagocytic cells. If

untreated, the disease is usually fatal in 2 to 3 years, and it can be fatal in 6 months in patients with impaired immunity and secondary infections.

Other leishmanias are more localized in their effects and are rarely fatal. *L. tropica* causes a cutaneous lesion, sometimes called an *oriental sore*, at the site of a sandfly bite. *L. braziliensis* causes skin and mucous membrane lesions and sometimes nasal and oral polyps (**Figure 23.21b**). Some parents, observing that people who had oriental sores rarely got kala azar, have purposely infected children with oriental sores on inconspicuous parts of the body to protect them from the more serious disease.

(a)   LM

(b)

Figure 23.21 **Leishmaniasis.** (a) *Leishmania donovani*, seen as small dots among cells (2,335X). (George A. Wistreich, East Los Angeles College) (b) A patient suffering from leishmaniasis, probably caused by *L. braziliensis*. (Science VU/Visuals Unlimited)

Distribution of Malaria, 2004

Figure 23.22 The worldwide incidence of malaria, 2004. *(Source: CDC)*

Diagnosis, Treatment, and Prevention.   Diagnosis is made by identifying the protozoa in blood smears in kala azar and from scrapings of skin and mucous membrane lesions. Antimony compounds are used to treat both kala azar and skin and mucous membrane lesions. However, such drugs are very toxic. Prevention depends mainly on controlling sandfly breeding and elimination of rodent reservoir infections.

Malaria

Several species of the protozoan *Plasmodium* are capable of causing **malaria**, one of the most severe of all parasitic diseases. Malaria is one of the world's greatest public health problems. It is endemic in most tropical areas **(Figure 23.22)**. The number of clinical cases has been estimated at up to 500 million cases worldwide, with 1.5 million to 3 million deaths—of which more than 1 million occur in children under age 5. Nearly all adults in Africa and India have been infected, and annual economic losses due to malaria exceed $1 billion in Africa alone. Drug-resistant strains are increasing rapidly, driving the death rates up. At one time malaria was thought to have been eradicated from the United States, but military personnel, travelers, and immigrants have carried the disease with them to the United States from endemic areas, causing 1,245 cases in 2006.

Members of the genus *Plasmodium* are amoeboid, intracellular protozoa that infect erythrocytes and other tissues. They are transmitted to humans through the bite of *Anopheles* mosquitoes. *Plasmodium* species have a complex life cycle (see Figure 11.4, ◄p. 317). At least four species, *P. vivax*, *P. malariae*, *P. ovale*, and *P. falciparum*, infect humans. They can be identified by their effects on, and in some cases by their appearance in, red blood cells and by the nature of the disease the parasites cause.

Certain individuals, especially West African blacks, and persons of Mediterranean origin, are protected from malaria by carrying the gene for sickle-cell anemia. The presence of two such genes causes sickle-cell anemia, but the presence of a single gene prevents malarial parasites from growing in erythrocytes. When the parasite enters the cell, it uses up oxygen. Under low-oxygen conditions, the cell sickles (takes on a distorted spiney or sickle shape). The spleen removes abnormally shaped blood cells, such as sickle-cells, usually destroying them before the malaria parasite has time to complete its life cycle, thus reducing the number of infected cells to a level where symptoms are minor or do not occur.

The Disease.   In the pathogenesis of malaria, sporozoites (◄Chapter 11, p. 317) enter the blood from the bite of an infected female mosquito. (Male mosquitoes are not equipped to take a blood meal.) The parasites disappear from the blood within an hour and invade cells of the liver and other organs. In about a week they begin releasing merozoites, which invade and reproduce in red blood cells as trophozoites **(Figure 23.23)**. At

PUBLIC HEALTH

Desert Storm Leishmaniasis

The soldiers who served our country in the Persian Gulf War in 1991 probably didn't expect to die at the hands of a microorganism, but several dozen cases of leishmaniasis were confirmed in U.S. military personnel involved in Desert Storm. The disease is fatal in 90% of untreated cases. Currently, available treatment lowers the mortality rate to 10%. Nevertheless, for several years, health officials banned blood transfusions and organ transplants from Gulf War veterans to prevent the spread of the disease. Today these concerns return for troops serving in Iraq and Afghanistan.

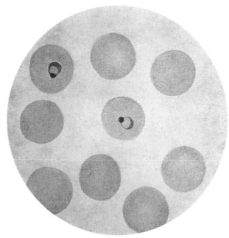

Figure 23.23   Plasmodium, the cause of LM
malaria. The "ring" stage of the malarial parasite *Plasmodium falciparum*, seen as dark, circular structures within red blood cells (1,500X). At this stage, merozoites have become trophozoites upon invading the host's red blood cells. *(Biophoto Associates/Photo Researchers, Inc.)*

intervals of 48 to 72 hours, depending on the infecting species of *Plasmodium*, blood cells rupture in a characteristic pattern and release more merozoites, which infect other red blood cells. The release of merozoites soon becomes synchronized and corresponds to intervals of high fever. Some merozoites become gametocytes, which can undergo sexual reproduction in mosquitoes, should they feed on the patient's blood. The average malarial parasites destroy 25 to 75% of the hemoglobin in an erythrocyte, degrading it in acidic food vacuoles, in order to satisfy their protein requirements. Even after the initial disease has subsided, patients are subject to relapses as dormant protozoa become activated, emerge from the liver, and initiate a new cycle of disease. Relapses do not occur after infections caused by *P. falciparum* because this species does not remain in the liver.

PUBLIC HEALTH

Anemia Protects Against Malaria

If you are suffering from anemia, some scientists might think that you are lucky. Low blood iron levels, which are characteristic of certain types of anemia, seem to boost the immune system's ability to fight malaria. Studies suggest that compounds that remove iron from the blood, called iron chelators, interfere with the DNA replication of the parasite that causes malaria, helping to clear the parasite from the body. These iron chelators also appear to boost levels of NO_3, leading to an immune response. But this immune response also causes fever and inflammation. That's not so lucky, and scientists agree that much more work must be done to develop an effective iron-chelating drug with only mild side effects.

Of the four species of malarial parasites, *P. falciparum* causes the most severe disease because it agglutinates red blood cells and obstructs blood vessels. Such obstruction causes tissue **ischemia**, or reduced blood flow with oxygen and nutrient deficiency and waste accumulation. This species can also cause malignant malaria—an especially virulent, rapidly fatal disease—and a condition called **blackwater fever**. In blackwater fever, large numbers of erythrocytes are lysed, probably because of the host's autoimmune reaction to the parasites. Products of hemoglobin breakdown cause jaundice and kidney damage. Pigments from hemoglobin blacken the urine and give the disease its name.

Diagnosis, Treatment, and Prevention. The main means of diagnosing malaria is by identifying the protozoa in red blood cells. The species of *Plasmodium* responsible for a particular infection can be identified by the distinctive appearance of erythrocytes invaded by the parasite. Chloroquine (Aralen) is the drug of choice for all forms of malaria in the acute stage. A serious problem in the treatment of malaria is that some strains, especially strains of *P. falciparum*, have become resistant to chloroquine. Drugs have recently been found that can be administered with chloroquine to overcome such resistance. This strategy has been tested in monkeys but not yet in humans. A traveler who will enter a malarial region can take chloroquine prophylactically for 2 weeks before entry, during the stay, and for 6 weeks after leaving the area. The drug suppresses clinical symptoms of malaria, but it does not necessarily prevent infection. In areas where malaria is resistant to chloroquine, Lariam™ is used. However, in some people, Lariam™ can cause psychiatric problems, possibly including suicide. Disease caused by *P. vivax* or *P. ovale* can appear months or years after a person leaves a malarial area, even when the suppressant drug has been taken. Primaquine is the drug of choice for eliminating parasites from the liver and other tissues if they have been infected.

Attempts to destroy mosquitoes that carry malaria have been an important component of malaria control efforts. In the early 1960s the pesticide DDT was used successfully to eradicate malaria-carrying mosquitoes from the United States. (But it was soon banned in the United States due to its toxicity.) DDT and other insecticides also have been tried in other areas, especially in Africa. Unfortunately, the region in which such mosquitoes thrive in Africa is so vast that no insecticide spraying program has been effective. Some of the mosquitoes, particularly those that carry *P. falciparum*, have now become resistant to DDT and probably to other pesticides. Thus, the use of insecticides has made malaria more deadly by increasing the proportion of mosquitoes that carry the more virulent parasite.

Researchers at the CDC and their colleagues at other institutions have recently developed a strain of *Anopheles gambiae* that is highly resistant to infection with malaria

Four new mosquito genes have been found: two that prevent development of malaria parasites inside a mosquito, and two that protect the parasite. Eliminating the protective proteins kills the parasite.

parasites. The researchers believe that the resistance is due to a relatively simple genetic change and that it might be possible to induce such resistance in natural vector populations. If large numbers of *A. gambiae*, the most important malaria vector in Africa, could be made resistant, transmission of malaria could be greatly reduced.

Another control effort is directed toward developing a malaria vaccine. One problem has been to identify the stage of the parasite responsible for triggering the immune response in humans. An antigen on the sporozoite has now been identified, and the gene for this antigen can be cloned, using recombinant DNA technology. Consequently, the antigen can be manufactured and used to make a vaccine. Thus, great strides have been made toward the development of an effective vaccine; the vaccine will hopefully be available in the near future. But even when a vaccine becomes available, administering it to the vast numbers of people living where malaria is endemic will be a monumental task. Mechanisms to distribute the vaccine and ways of gaining the people's cooperation will be needed. The cost of the vaccine in the quantities required will be another major problem.

Toxoplasmosis

Toxoplasma gondii **(Figure 23.24)** is a widely distributed protozoan that infects many warm-blooded animals—both domestic and wild. It is an intracellular parasite and can invade many tissues. Humans usually become infected through contact with the feces of domestic cats that forage for natural foods, especially infected rodents **(Figure 23.25)**. Cats that are kept inside

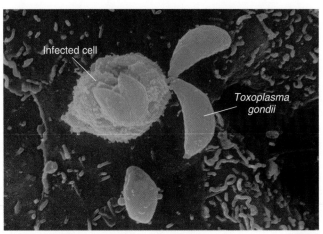

Figure 23.24 Toxoplasmosis. Crescent-shaped protozoans, *Toxoplasma gondii*, leaving an infected cell in which they have multiplied (7,684X). This organism can be a danger to immunocompromised patients and to pregnant women, in whom it can cause serious birth defects and miscarriages. *(Moredun Animal Health Ltd./Custom Medical Stock Photo, Inc.)*

and are fed only dry cat food, canned or cooked foods are unlikely to acquire the parasites. Another common means of transmission is the consumption of raw or undercooked contaminated meat. The French, who consume large amounts of steak tartare (raw ground beef), have the highest incidence of infection in the world.

The Disease. *Toxoplasma gondii* causes only mild lymph node inflammation in most humans. In the majority of cases, the infection is chronic, asymptomatic, and self-limiting. However, *T. gondii* can cause serious **toxoplasmosis**, especially in developing fetuses, newborn infants, and sometimes in young children. The organism can be transferred across the placenta of an infected mother to the fetus. There it causes serious congenital defects, including the accumulation of cerebrospinal fluid, abnormally small head, blindness, mental retardation, and disorders of movement. However, only half of infected newborns show symptoms at birth. Later, severe symptoms may occur from 3 months of age into early adulthood, especially blindness and mental retardation. It can also be responsible for stillbirths and spontaneous abortions. If infection occurs after birth, the symptoms are similar to, but less severe than, those seen in fetuses. In patients with severe immunosuppression, such as AIDS patients, the disease can appear as encephalitis and can also cause dermatologic problems.

A recent study indicates a possible connection between toxoplasmosis and schizophrenia. Over 50% of schizophrenics and their mothers test positive for toxoplasmosis—far more than the general population does. We know that Toxoplasma infection can alter behavior. Infected mice lose all fear of cats. Schizophrenia has a definite biological basis. If a person with no history of schizophrenia receives a unit of blood from a person exhibiting schizophrenia, the recipient will also exhibit symptoms for several hours.

APPLICATIONS

Using Malaria to Unite the World

Who would have thought that malaria would play a role in world politics? These days, that's exactly what it's doing. The World Health Organization has united the world in an effort to halve world deaths from malaria by 2010. The program, called Roll Back Malaria, involves the Multilateral Initiative on Malaria (MIM), which includes 37 African countries, the United States, Europe, and the African Initiative on Malaria. The coalition brings together governments, the World Bank, and other international agencies. And it's a good thing that we're uniting now; malaria already causes about 500 million cases of illness and up to 3 million deaths a year worldwide as the parasite develops increasing drug resistance and spreads into new regions such as Central Asia and eastern Europe. Bed nets, pesticides, and medications are available, but their distribution is failing.

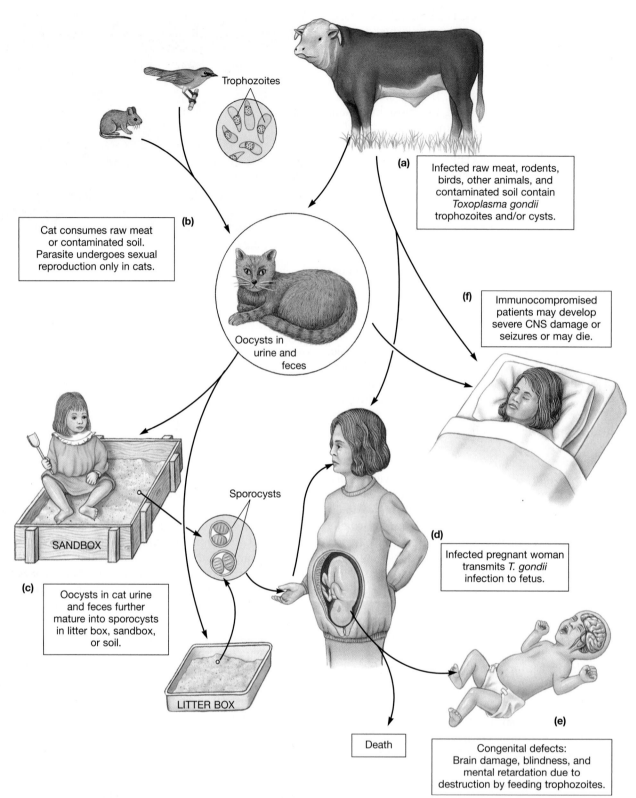

Figure 23.25 *Toxoplasma gondii* **cycle of natural transmission.** **(a)** Contaminated raw meat or soil containing trophozoites (feeding forms) of *T. gondii* are consumed. **(b)** In the intestinal tract of cats, the trophozoites undergo sexual reproduction. Oocysts are released in urine and feces and further mature into sporocysts in litter boxes, sandboxes, or soil. **(c)** If the trophozoites are consumed by older children or adults, a mild flulike syndrome may result. **(d)** Pregnant women suffer only mild or inapparent infection. **(e)** But they can transmit this infection to the fetus, which may develop congenital defects or die. **(f)** Immunocompromised individuals, who may already have the organism in their body, rapidly die of central nervous system destruction.

TABLE 23.4

Viral and Protozoan Systemic Diseases		
Disease	Agent(s)	Characteristics
Viral Systemic Diseases		
Dengue fever	Dengue fever virus	Severe bone and joint pain, high fever, headache, loss of appetite, weakness, sometimes a rash
Yellow fever	Yellow fever virus	Fever, anorexia, nausea, vomiting, liver damage, jaundice
Infectious mononucleosis	Epstein-Barr virus	Headache, fatigue, malaise, usually sore throat, secondary streptococcal infections common
Other viral fevers	Filoviruses, bunyaviruses, phleboviruses, arenaviruses, orbivirus, and coxsackie viruses	Some cause hemorrhagic fevers; some cause encephalitis, joint pain, slow heart rate, infection of erythrocytes, diarrhea, rashes, sore throat, liver disease, meningitis, inflammation of the heart and the sac around it
Protozoan Systemic Diseases		
Kala azar	*Leishmania donovani*	Visceral leishmaniasis with irregular fever, weakness, wasting, enlarged liver and spleen
Localized leishmaniasis	*L. tropica, L. braziliensis*	Oriental sore and skin and mucous membrane lesions
Malaria	*Plasmodium* species	Periods of high fever associated with release of parasites from red blood cells; relapse can occur; one species can cause malignant malaria and blackwater fever
Toxoplasmosis	*Toxoplasma gondii*	Mild lymph node inflammation in adults; can cross placenta and cause serious damage to nervous system of fetus; also causes damage in small children and immunosuppressed patients
Babesiosis	*Babesia microti*	High fever, headache, muscle pain, anemia, and jaundice; fatal in patients whose spleens have been removed

Diagnosis, Treatment, and Prevention. Toxoplasmosis can be diagnosed by finding the parasites in the blood, cerebrospinal fluid, or tissues, by animal inoculation with subsequent isolation of the organisms, or by indirect immunofluorescence tests (◀Chapter 18, p. 565). Pyrimethamine and trisulfapyridine are used in combination to treat toxoplasmosis, but no treatment can reverse permanent damage from prenatal infection. To prevent this disease, pregnant women should avoid contact with raw meat and cat feces. A cat may shed up to 10 million oocysts per day in its feces. These sporulate and become infectious in 1 to 5 days. Someone other than the pregnant woman should change the litter daily to prevent accumulation of infectious oocysts. In warm areas, the oocysts may remain infectious in moist soil for a year. Cats should be kept out of sandboxes where children play, especially if a child might carry the organism to a pregnant woman.

Babesiosis

Several species of the sporozoan *Babesia* can cause **babesiosis** (ba-be″se-o′sis). Cattle are affected with babesiosis caused by the tickborne protozoan *Babesia bigemina*, but *B. microti* is most often associated with human infections. The parasites enter the blood via bites of infected ticks and invade and multiply in red blood cells.

The Disease. Although many cases are asymptomatic, when symptoms appear, they usually begin with sudden high fever, headache, and muscle pain. Anemia and jaundice can occur as red blood cells are destroyed. The symptoms last for several weeks and are followed by a prolonged carrier state. If babesiosis occurs in a person who has undergone spleen removal, it is usually fatal in 5 to 8 days. This is because the lack of a spleen impairs the body's ability to break down defective red blood cells.

Diagnosis, Treatment, and Prevention. Diagnosis is made from blood smears, but the parasite can be confused with *Plasmodium falciparum*. Chloroquine is the drug of choice for treatment, and avoiding tick bites is the best means of protection.

The properties of nonbacterial systemic diseases are summarized in Table 23.4.

✓CHECKLIST

1. How does carrying a gene for sickle-cell anemia protect against malaria?
2. Over what period should travelers take antimalarial drugs? Why?
3. Who is at risk for toxoplasmosis?

▌ RETRACING OUR STEPS

- The agents and characteristics of the diseases discussed in this chapter are summarized in Table 23.1 through 23.4. Information in those tables is not repeated in this summary.

▌ THE CARDIOVASCULAR SYSTEM

- The **cardiovascular system** consists of the heart, an extensive system of blood vessels, and blood.
- Although the cardiovascular system is normally sterile, pathogens can be transported in blood (*bacteremia*), multiply in blood (*septicemia*), and infect the heart valves and pericardium.

▌ CARDIOVASCULAR AND LYMPHATIC DISEASES

BACTERIAL SEPTICEMIAS AND RELATED DISEASES

- **Septicemia**, or blood poisoning, involves multiplication of bacteria in the blood. Septicemias and related diseases are diagnosed by culturing appropriate samples and are treated with antibiotics.
- **Rheumatic fever** and **bacterial endocarditis** occur most often in patients who have had previous streptococcal infections.

HELMINTHIC DISEASES OF THE BLOOD AND LYMPH

- **Schistosomiasis** is acquired from schistosome larvae that penetrate the skin and is diagnosed by finding eggs in feces. It is treated with praziquantel and could be prevented by the eradication of infected snails or avoidance of snail-infested water.
- **Filariasis** is transmitted by mosquitoes and is diagnosed by finding microfilariae in blood. It is treated with diethylcarbamiazane or metronidazole and could be prevented if infected mosquitoes could be eradicated.

▌ SYSTEMIC DISEASES

BACTERIAL SYSTEMIC DISEASES

- **Anthrax** is acquired through contact with *Bacillus anthracis* spores from infected domestic animals or their hides. It is diagnosed by culturing blood or from smears from lesions and is treated with penicillin or tetracycline. The disease can be prevented by immunizing animals and humans with occupational exposure and by careful burial of infected animals.
- Plague has occurred in periodic epidemics since the Middle Ages, remains endemic in some regions, and is increasing in incidence in the United States. The form that is transmitted by fleas from infected rats is known as **bubonic plague**, which forms enlarged lymph nodes, called **buboes**. If the disease progresses to the circulatory system, it is called **septicemic plague**. Lung involvement results in **pneumonic plague**, which is contagious and transmitted by aerosols. Plague is diagnosed by stained smears and antibody tests and treated with streptomycin or tetracycline. It can be prevented by controlling rat populations and immunizing people who enter endemic areas.
- **Tularemia** is transmitted through the skin, inhaled, or ingested. It can be diagnosed by agglutination tests and treated with streptomycin. It is prevented by avoiding infected mammals and arthropods; the vaccine is short-lasting and not fully protective.
- **Brucellosis** is transmitted to humans through the skin from domestic animals by the ingestion of contaminated dairy products and by inhalation or ingestion and is diagnosed by serological tests. It is treated with prolonged antibiotic therapy and can be prevented by avoiding infected animals and contaminated fomites.
- **Relapsing fever** is transmitted by lice and ticks and can be diagnosed from blood smears. It is treated with tetracycline or chloramphenicol and can be prevented by avoiding or controlling ticks and lice.
- **Lyme disease** is transmitted by ticks from infected deer and other animals and is diagnosed by clinical signs and serological tests. It is treated with antibiotics and can be prevented by avoiding tick bites.

RICKETTSIAL AND RELATED SYSTEMIC DISEASES

- **Typhus fever** occurs in several forms. **Epidemic typhus**, transmitted by human body lice, usually occurs in unsanitary, overcrowded conditions. It has a high mortality rate unless treated with antibiotics.
- **Brill-Zinsser disease**, or recrudescent typhus, is a recurrence of a latent typhus infection. **Endemic**, or **murine**, **typhus** is carried by fleas, and **scrub typhus**, by mites from infected rats.
- **Rocky Mountain spotted fever**, carried by ticks, damages blood vessels. Strains of the causative rickettsia vary in virulence, and the mortality of the untreated infection can be high.
- **Rickettsialpox** is carried by mites that live on house mice. **Trench fever**, transmitted by lice, is prevalent under unsanitary conditions, most often among persons under stress. **Bartonellosis**, transmitted by sandflies, occurs in two forms: **Oroya fever**, an acute fever that causes life-threatening anemia, and **verruga peruana**, a self-limiting skin rash.
- Recently identified human pathogens resembling the rickettsias include *Ehrlichia canis* and *E. chaffeensis*, which cause **ehrlichiosis**, and *Bartonella henselae*, which causes **bacillary angiomatosis**.

VIRAL SYSTEMIC DISEASES

- **Dengue fever**, an arbovirus disease, can be diagnosed by serologic tests; a vaccine is available against one immunological type of dengue fever virus.
- **Yellow fever**, another arbovirus disease, is diagnosed by symptoms and can be prevented with vaccine.
- **Infectious mononucleosis** is caused by the **Epstein-Barr virus**, diagnosed by symptoms, and treated symptomatically and with antibiotics for secondary infections. In developing countries babies have mild symptoms and produce antibodies by age 1, but in developed countries patients are teenagers or young adults, who have a much more serious disease.
- **Chronic fatigue syndrome** has been linked with the Epstein-Barr virus, which is responsible for infectious mononucleosis, **Burkitt's lymphoma**, nasopharyngeal carcinoma, and oral hairy leukoplakia.
- Other viral infections include **filovirus** fevers (such as **Ebola virus** infection), bunyavirus fevers (such as **Rift Valley fever**), arenavirus fevers (such as **Lasssa fever** and **Bolivian hemorrhagic fever**), **Colorado tick fever**, feline panleukopenia virus and canine parvovirus infections, and **fifth disease**.

PROTOZOAN SYSTEMIC DISEASES

- **Leishmaniasis** occurs in tropical and subtropical countries where appropriate species of sandfly vectors are available. It is diagnosed from blood smears or scrapings from lesions and is treated with antimony. The disease could be prevented by controlling sandfly breeding and eliminating rodent reservoir infections.

• **Malaria** is one of the world's greatest public health problems; it kills a million people annually, most of them children. Cases in the United States come from people who have been in endemic areas. Malaria is transmitted by female *Anopheles* mosquitoes and is diagnosed by identifying protozoa in blood smears. Active disease is treated with chloroquine (except for resistant strains), and latent parasites are eliminated with primaquine. Research is in progress to find a way to control mosquitoes and to develop an effective vaccine.

• **Toxoplasmosis** is usually transmitted from feces of cats that have consumed infected rodents and by human ingestion of contaminated raw or undercooked meat. It is diagnosed by finding parasites in body fluids or tissue and is treated with pyrimetamine and trisulfapyridine. The disease can be prevented by avoiding contaminating materials.

• **Babesiosis** is transmitted by ticks and is diagnosed from blood smears. It is treated with chloroquine and can be prevented by avoiding contact with ticks.

▌ TERMINOLOGY CHECK

anthrax *(p. 724)*
aplastic crisis *(p. 743)*
babesiosis *(p. 749)*
bacillary angiomatosis *(p. 738)*
bacteremia *(p. 718)*
bacterial endocarditis *(p. 720)*
Bang's disease *(p. 730)*
bartonellosis *(p. 737)*
blackwater fever *(p. 746)*
Bolivian hemorrhagic fever *(p. 743)*
Brill–Zinsser disease *(p. 735)*
brucellosis *(p. 730)*
bubo *(p. 728)*
bubonic plague *(p. 729)*
Burkitt's lymphoma *(p. 741)*
canine parvovirus *(p. 743)*
cardiovascular system *(p. 718)*
childbed fever *(p. 719)*

chronic fatigue syndrome *(p. 741)*
Colorado tick fever *(p. 743)*
cutaneous anthrax *(p. 725)*
dengue fever *(p. 739)*
Ebola virus *(p. 742)*
ehrlichiosis *(p. 738)*
electrolyte *(p. 718)*
elephantiasis *(p. 723)*
endemic relapsing fever *(p. 731)*
endemic typhus *(p. 736)*
epidemic relapsing fever *(p. 731)*
epidemic typhus *(p. 735)*
Epstein–Barr virus (EBV) *(p. 740)*
feline panleukopenia virus (FPV) *(p. 743)*
fifth disease *(p. 743)*
filariasis *(p. 723)*
filovirus *(p. 742)*

infectious mononucleosis *(p. 740)*
intestinal anthrax *(p. 725)*
ischemia *(p. 746)*
kala azar *(p. 744)*
Lassa fever *(p. 743)*
leishmaniasis *(p. 744)*
Lyme disease *(p. 733)*
lymphangitis *(p. 719)*
malaria *(p. 745)*
Malta fever *(p. 730)*
murine typhus *(p. 736)*
myocarditis *(p. 721)*
orbivirus *(p. 743)*
Oroya fever *(p. 737)*
pericarditis *(p. 721)*
petechia *(p. 735)*
phlebovirus *(p. 742)*
pneumonic plague *(p. 729)*
puerperal fever *(p. 719)*
relapsing fever *(p. 731)*
respiratory anthrax *(p. 725)*

rheumatic fever *(p. 720)*
rickettsialpox *(p. 737)*
Rift Valley fever *(p. 742)*
Rocky Mountain spotted fever *(p. 736)*
schistosomiasis *(p. 721)*
scrub typhus *(p. 736)*
septicemia *(p. 718)*
septicemic plague *(p. 729)*
septic shock *(p. 719)*
toxoplasmosis *(p. 747)*
transovarian transmission *(p. 729)*
trench fever *(p. 737)*
tularemia *(p. 729)*
typhoidal tularemia *(p. 730)*
typhus fever *(p. 735)*
ulceroglandular *(p. 730)*
undulant fever *(p. 730)*
vegetation *(p. 720)*
verruga peruana *(p. 737)*
yellow fever *(p. 739)*

▌ CLINICAL CASE STUDY

Ruth was having her first baby. Thirty-six weeks into her pregnancy her doctor did an examination and had a culture done to see if she was colonized with group B streptococcus (*Streptococcus agalactiae*). The culture came back positive. Ruth is now in labor and her doctor is recommending that she be given antibiotics. Ruth is confused. She is not sick and her baby is not sick. Why does she need to be given an antibiotic? What might happen if she does not get the antibiotic? Visit the following web site to learn the answers to these questions and find out more about this important pathogen: Learn more about this disease and prevention at the following web site and other CDC site links. http://www.cdc.gov/groupbstrep/gbs/hospitalsguidelines.recommend.htm.

▌ CRITICAL THINKING QUESTIONS

1. List the virulence factors of *Bacillus anthracis* in anthrax. How do these factors work to cause death in the host?

2. Plague has been only sporadic in humans in the United States in recent years, but has a world history of producing massive epidemics. Do you think that the United States could ever have a plague epidemic?

3. Arthropod-transmitted diseases usually have rather specific vectors. Why can't a particular arthropod-transmitted disease be transmitted by a wider variety of arthropods?

4. Relapsing fever is caused by species of the genus *Borrelia*. What causes the victim to relapse after intervals of relief from fever in this disease?

▌ SELF-QUIZ

1. The normal microflora of the heart includes species of:
   - (a) Gram-positive bacteria
   - (b) Gram-negative bacteria
   - (c) Fungi
   - (d) Viruses
   - (e) None of these

2. Which of the following is commonly directly responsible for causing septic shock today?
   - (a) Bacterial hemagglutinins
   - (b) Bacterial endotoxins
   - (c) Bacterial exotoxins
   - (d) Fungal aflatoxins
   - (e) Bacterial neurotoxins

3. All of the following are symptoms of septic shock EXCEPT
   - (a) Shock
   - (b) Lymphangitis
   - (c) Chills
   - (d) Fever
   - (e) Collapsed blood vessels

4. _____ are the leading cause of neonatal sepsis and meningitis while _____ cause the multisystem disorder rheumatic fever.
   - (a) Group B streptococci (*Streptococcus agalactiae*); β-hemolytic *Streptococcus pyogenes*
   - (b) β-hemolytic *Staphylococcus pyogenes*; Group B streptococci (*Streptococcus agalactiae*)
   - (c) β-hemolytic *Staphylococcus aureus*; *Streptococcus pneumoniae*
   - (d) *Pseudomonas aeruginosa*; *Klebsiella pneumoniae*
   - (e) (b) and (c)

5. What is the most likely explanation of how *Streptococcus pyogenes* causes rheumatic fever?
   - (a) Strep throat infections migrate down to the heart.
   - (b) Endotoxin production causes septicemia.
   - (c) Vegetation followed by fibrin deposition.
   - (d) Cross-reactivity of an antibody for a *Streptococcus pyogenes* antigen to a heart cell antigen.
   - (e) It is an opportunistic infection triggered by coronary artery disease and atherosclerosis.

6. Bacterial endocarditis is an infection and inflammation of the lining and valves of the heart and occurs as a result of transient bacteria attaching to fibrin from exposed collagen fibers of damaged valvular surfaces. True or false?

7. Blood flukes causing schistosomiasis require a particular _____ as an intermediate host to complete their life cycle, while roundworms causing filariasis depend on a _____ host to complete their life cycle.
   - (a) mussel; black fly
   - (b) fish; tick
   - (c) virus; porcine
   - (d) metacercariae; tsetse fly
   - (e) snail, mosquito

8. An organism that can cause cutaneous infections and can be transmitted by endospores is:
   - (a) *Staphylococcus aureus*
   - (b) *Pseudomonas aeruginosa*
   - (c) *Yersinia pestis*
   - (d) *Bacillus anthracis*
   - (e) *Streptococcus pyogenes*

9. Which of the following statements is true of infection caused by *Yersinia pestis*?
   - (a) It can produce enlarged lymph nodes (buboes).
   - (b) It can infect the circulatory system, causing septicemia.
   - (c) It can cause pneumonia.
   - (d) It is called sylvatic plague in areas where it is carried by wild rodents.
   - (e) All of these are true.

10. A pathogen of both humans and cattle that causes undulant fever is a member of the genus:
    - (a) *Yersinia*
    - (b) *Borellia*
    - (c) *Brucella*
    - (d) *Pediculus*
    - (e) *Streptococcus*

11. Alternate periods with fever along with periods without fever are characteristics of relapsing fever caused by members of the genus:
    - (a) *Borellia*
    - (b) *Yersinia*
    - (c) *Brucella*
    - (d) *Pediculus*
    - (e) *Streptococcus*

12. Which of the following diseases is not a zoonosis?
    - (a) Relapsing fever
    - (b) Tularemia
    - (c) Anthrax
    - (d) Plague
    - (e) Brucellosis

13. Lyme disease is caused by the spirochete _____ _____ which is carried and transmitted by the blacklegged tick _____ _____.
    - (a) *Yersinia petis*; *Pulex irritans*
    - (b) *Borrelia recurrentis*; *Musca domestica*
    - (c) *Borrelia burgdorferi*; *Ixodes scapularis*
    - (d) *Rickettsia akari*; *Aedes aegypti*
    - (e) *Francisella tularensis*; *Culex pipiens*

14. You advise your friend who is spending the summer working for the U.S. Forest Service to be sure to tuck her pants in her boots, use insect repellent, and to check daily for ticks. You are concerned that she could get:
    - (a) Rocky Mountain spotted fever
    - (b) Plague
    - (c) Lyme disease
    - (d) a and c
    - (e) a, b, and c

15. A louseborne disease that has influenced the course of wars (epidemic typhus) is caused by:
    - (a) *Bartonella bacilliformis*
    - (b) *Rickettsia prowazekii*
    - (c) *Rickettsia akari*
    - (d) *Borrelia burgdorferi*
    - (e) *Yersinia pestis*

16. A possible symptom of Lyme disease is:
    - (a) A bull's-eye rash at the point of the insect bite
    - (b) Flulike symptoms
    - (c) Arthritis
    - (d) Symptoms resembling Alzheimer's disease
    - (e) All of these

17. Match the following Rickettsial and Rickettsial-related systemic diseases to their descriptions:
    - ___ Rickettsialpox
    - ___ Bartonellosis
    - ___ Trench fever
    - ___ Rocky mountain spotted fever
    - ___ Ehrlichiosis
    - ___ Bacillary angiomatosis
    - ___ Brill-Zinsser disease

    - (a) spread by dog and blacklegged ticks, this disease resembles rickettsial diseases and is caused by *Ehrlichia canis* or *E. chaffeensis*
    - (b) seen in immunocompromised patients, this disease involves the blood vessels of skin and internal organs
    - (c) occurs in two forms, Oroya fever and verruga peruana, caused by

Bartonella bacilliformis and transmitted by sandflies

(d) a recurrence of a typhus infection caused by reactivation of latent organisms harbored in lymph nodes

(e) resembles chickenpox and is caused by *Rickettsia akari* and transmitted by mites found on house mice

(f) caused by *Bartonella quintana*, it resembles epidemic typhus because it is transmitted by lice and is prevalent during wars and unsanitary conditions

(g) caused by *Rickettsia rickettsii* and transmitted by ticks of the *Dermacentor* genus, a characteristic rash forms on the palms and soles, as well as other places

18. Although the initial infections are usually self-limiting, a second infection with a different virus strain can lead to immunological reactions that produce hemorrhagic disease. This occurs in:
(a) Dengue fever
(b) Typhus fever
(c) Yellow fever
(d) Rift Valley fever
(e) All of these

19. Monkeys may serve as a reservoir for this disease that is found mainly in tropical areas of the Americas and Africa. It is:
(a) Typhus fever
(b) Dengue fever
(c) Rift Valley fever
(d) Yellow fever
(e) All of these

20. In humans, the causative agent of infectious mononucleosis (Epstein-Barr virus) primarily infects:
(a) Heart muscle cells
(b) Cells lining the alveoli
(c) B lymphocytes
(d) Sensory neurons
(e) Cells of the intestinal epithelium

21. A tumor of the jaw and viscera (Burkitt's lymphoma) occurs approximately 6 years after a primary infection with:
(a) Yellow fever virus
(b) Rift Valley fever virus
(c) Ebola virus
(d) Dengue virus
(e) Epstein-Barr virus

22. Filoviruses have an unusual shape that may be U- or fish-hook-shaped. One disease caused by a filovirus is:
(a) Yellow fever virus
(b) Rift Valley fever virus
(c) Epstein-Barr virus
(d) Ebola virus
(e) Dengue virus

23. Which of the following is true about fifth disease?
(a) It is caused by parvovirus B19 that destroys stem cells that give rise to red blood cells.
(b) It is common in children aged 5–14 and is not a concern unless the afflicted has chronic hemolytic anemia or an immunodeficiency.
(c) It gets its name from a nineteenth-century list of childhood rash diseases.
(d) b and c
(e) a, b, and c

24. Which of the following is NOT currently a way in which one could combat malaria?
(a) Take the drugs chloroquine and/or primaquine.
(b) Carry a single gene for the sickle-cell trait.
(c) Eradication of *Anopheles* species mosquitoes.
(d) Take a malarial vaccine.
(e) Wear protective clothing while in areas of endemic malaria.

25. Match the following protozoan systemic diseases with their mode of transmission and causative agent:
___ Leishmaniasis
___ Malaria
___ Babesiosis
___ Toxoplasmosis
(a) cat feces
(b) raw or undercooked meat
(c) *Anopheles* mosquito
(d) sandfly
1. *Toxoplasma gondii*
2. *Leishmania donovani*
3. *Plasmodium* species
4. *Babesia microti*
5. *Leishmania braziliensis*

26. Which of the following systemic diseases is one of the world's greatest public health problems, being endemic in most tropical areas, and with up to 500 million cases and 1.5 to 3 million deaths per year?
(a) Rheumatic fever
(b) Lyme disease
(c) Malaria
(d) Yellow fever
(e) Infectious mononucleosis

▌ EXPLORATIONS ON THE WEB http://www.wiley.com/college/black

If you think you've mastered this chapter, there's more to challenge you on the web. Go to the companion web site to fine-tune your understanding of the chapter concepts and discover answers to the questions posed below.

1. Did you know that Lyme disease is called the great imitator because its symptoms can mimic a variety of illnesses? Learn more about Lyme disease at the companion web site.

2. Major Walter Reed, U.S. Army Medical Corps, needs your help. It's June 25, 1900, and he has just been posted to Cuba. He has big trouble—hundreds of his men have died from yellow jack and it is his responsibility to find out what's causing it and put a stop to it. Find out how you can help Major Reed at the companion web site.

Come with me . . .

Do you like eating ethnic foods—Chinese, Mexican, middle-Eastern? Well how about some Alaskan Native American dishes: "fermented" whale or seal blubber (fat), "fermented" seal flippers, or "fermented" fish heads ("stinkheads") and fish eggs ("stink eggs")? Traditional ways of preparing these foods increase the probability of botulism poisoning. Alaskan Natives have the highest incidence of botulism in the world.

Whales and seals are often butchered on the beach where Clostridium botulinum spores are plentiful in the soil, as well as in the intestines of the animal. Fish gills also contain large numbers of these spores. Contamination of the food is certain to occur. Traditionally the food was then stored in a cool, shallow hole in the ground and left to "ferment" for a month or two. This softens bones, making calcium available, a nutrient not easily available in the traditional diet. This, however, is not a true fermentation as there are little or no carbohydrates in the food. Carbohydrate fermentation would produce acid, which would kill the C. botulinum spores. Originally the pit was lined with wood, leaves and animal skins. Today people use plastic bags or plastic buckets with tight snap-on lids. This now creates an anaerobic environment. And sometimes the containers are left out on top of the ground where the temperature is warmer. This warm anaerobic environment is just perfect for germination of C. botulinum spores, which then grow and produce toxin. By the time the food is consumed months later, large amounts of deadly toxin have accumulated in it. As a more rapid means of preparation, sometimes food is placed in a glass jar with the lid screwed on tightly, and is left next to the stove. The food is ready in about one week—a potentially lethal "fast food".

Come with me to learn more about botulism in Alaska as well as in the rest of the United States.

Doug Allan/npl/Minden Pictures, Inc.

 Video related to this topic is available within WileyPLUS.

As with cardiovascular and lymphatic diseases, the diseases of the nervous system also often affect other systems. This chapter assumes a knowledge of disease processes (◄Chapter 14) and host systems and host defenses (◄Chapters 16 and 17).

COMPONENTS OF THE NERVOUS SYSTEM

As you read this sentence, millions of neural signals allow you to understand words while you maintain an upright posture and carry on various internal processes, such as breathing. The ability of the nervous system to control body functions at the same time depends on many *neurons*, or nerve cells, working together. Structurally, the **nervous system** has two components: the central and peripheral nervous systems **(Figure 24.1)**. The **central nervous system (CNS)**, which consists of the brain and spinal cord, receives signals from and issues commands via the **peripheral nervous system (PNS)**, composed of nerves that supply all parts of the body. **Nerves** of the peripheral nervous system consist of nerve fibers that transmit sensory information and motor responses. Aggregations of these cell bodies in the PNS are called **ganglia** (singular: *ganglion*). The brain and spinal cord are covered and protected by membranes called **meninges** (me-nin′jez), which are sheets of connective tissues. Hollow chambers in the brain and spinal cord and spaces between meninges are filled with *cerebrospinal fluid*.

Like the cardiovascular system, the nervous system is ordinarily sterile and has no normal microflora. Bacterial and viral pathogens, however, can enter cerebrospinal fluid from the blood. The CNS is well protected by bone and meninges from invasion by pathogens. In addition, phagocytes of the nervous system called *microglial cells* can destroy invaders that reach the brain and spinal cord. The brain itself has special thick-walled capillaries without pores in their walls. These capillaries form the **blood-brain barrier**, which limits entry of substances into brain cells. Although it does protect the cells from microorganisms and toxic substances, the blood-brain barrier prevents the cells from receiving medications that easily reach other cells.

NORMAL MICROFLORA OF THE NERVOUS SYSTEM

The nervous system does not have a normal microflora.

DISEASES OF THE BRAIN AND MENINGES

BACTERIAL DISEASES OF THE BRAIN AND MENINGES

Bacterial Meningitis

Bacterial meningitis is an inflammation of the *meninges*, the membranes that cover the brain and spinal cord. This life-threatening disease can be caused by several kinds of bacteria, each of which has a prevalence that can be correlated with the age of the host **(Table 24.1)**. Meningitis causes *necrosis* (death of tissues in a given area), clogging of blood vessels, increased pressure within the skull from edema, decreased cerebrospinal fluid flow, and impaired central nervous system function. The early symptoms are headache, fever, and chills. In rare instances, seizures develop. Onset can be insidious or fulminating. Death occurs from shock and other serious complications within hours of the appearance of symptoms.

Most cases of meningitis are acute, but some are chronic. Acute meningitis is acquired from carriers or endogenous organisms. Organisms gain access to the meninges directly during surgery or trauma, or spread to them in blood from other infections such as pneumonia and otitis media. Host defenses in the arachnoid layer, one of the meninges, ordinarily combat bacteremia. If organisms overwhelm the defenses, however, meningitis results. Chronic meningitis occurs with extension of underlying diseases such as syphilis or tuberculosis. The bacteria causing these underlying diseases are slow growers; in such cases, onset of typical meningitis symptoms is insidious, occurring over a period of weeks.

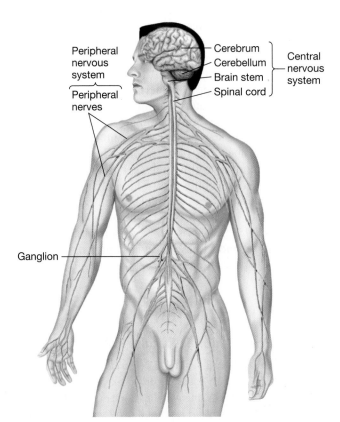

Figure 24.1 The structure of the nervous system. The nervous system normally is free of resident microbes. Infections typically affect the meninges (which cover the brain and spinal cord) or the sensory nerve ganglia.

Peripheral nervous system

Peripheral nerves

Ganglion

Cerebrum
Cerebellum
Brain stem
Spinal cord

Central nervous system

TABLE 24.1

Types of Bacterial Meningitis		
Age	**Most Frequent Causative Agents**	**Comments**
Newborn (0–2 months)	*Escherichia coli*, other Enterobacteriaceae, *Streptococcus* species	Average mortality about 50%; incidence 40–50/100,000 live births; maternal transmission
Preschool (2 months–5 years)	*Haemophilus influenzae*, type b; *Neisseria meningitidis*	Maximum incidence 6–8 months; overall incidence 180/100,000 children
Youth and young adult (5–40 years)	*Neisseria meningitidis*, *Streptococcus pneumoniae*	Sporadic or epidemic
Mature adult (over 40 years)	*Streptococcus pneumoniae*, *Staphylococcus* species	Sporadic

Meningitis is diagnosed by culturing cerebrospinal fluid. The fluid usually is turbid—sometimes so thick with pus that it is difficult to remove with a syringe. Antibiotic treatment varies with the causative organism. If tubercular meningitis is suspected, isoniazid therapy is called for immediately. Treatment for tubercular meningitis lasts a year or more, whereas some very rare cases of fungal meningitis may require several years of treatment.

Meningococcal Meningitis. The bacterium *Neisseria meningitidis* has caused 2,000 to 3,000 cases of meningitis per year for the past decade in the United States. Mortality is about 85% when the disease is untreated but only 1% with the best treatment. The 15% mortality rate in the United States probably reflects delay in seeking treatment. Death can occur in 12 to 48 hours if treatment is delayed. This disease was the leading cause of death from infectious disease among U.S. armed forces during World War II. Infants are the most susceptible group, followed by teens and college students between the ages of 15 and 24. Have *you* had your meningitis vaccine? More and more colleges are now requiring, or at least recommending it.

In meningococcal meningitis the organisms colonize the nasopharynx, spread to the blood, and make their way to the meninges, where they grow rapidly (**Figure 24.2**). In a complication called Waterhouse-Friderichsen syndrome, the meningococci invade all parts of the body (sepsis), and death occurs within hours from endotoxin shock. Meningococci produce 100 to 1,000 times more endotoxin than do other types of bacteria. Therefore symptoms develop very rapidly, and even a short delay in seeking treatment can be fatal. The immediate cause of death is usually clotting of blood followed by massive hemorrhage in the adrenal glands (located above the kidneys), leading to fatal deficiency of essential adrenal hormones. A lesser degree of hemorrhaging is sometimes seen in meningitis patients who develop a petechial skin rash which does not fade when pressed. However, many of these patients lose fingers, toes and even limbs, or may need large areas of skin grafts.

Penicillin is the drug of choice for treatment; the prevalence of resistant strains makes sulfonamides no longer useful. Third-generation cephalosporins and ampicillin are also used now. Vaccines are available against types A and C but are not effective against the most common type B meningococci. The type B bacteria are covered by molecules similar to ones found on human cells, and are thus not recognized by the human immune system. The risk of this disease can be decreased by prevention of overtiring and overcrowding. The number of inches required between bunks in military barracks is based on experience with meningococcal outbreaks. This is the only type of meningitis causing large outbreaks, e.g., in sub-Saharan Africa (◄Chapter 15, p. 444) where type A is most common, and 150,000 cases per year cause 16,000 deaths. One-third of secondary cases occur within 2 days of being exposed to the primary case.

Figure 24.2
Meningococci. These organisms are the cause of meningococcal meningitis. They have been phagocytized by white blood cells in a cerebrospinal fluid specimen. (*Courtesy Edward J. Bottone, Mount Sinai School of Medicine*)

APPLICATIONS

Post-Genomic Results Roll in Rapidly

Only 3 months after TIGR (The Institute for Genomic Research, Rockville, MD) released early draft versions of its genome sequences for the *N. meningitidis* bacterium, an Italian pharmaceutical company has identified gene sequences common to all strains, including type B. They have a single vaccine, designed to protect against all strains, already available. A quick payoff from genome research!

In some closed environments, such as military bases, dormitories, and day-care centers, 90% of the population can be carriers of meningococci, yet only 1 per 1,000 carriers develops the disease. Among members of patient households, 80% to 90% are carriers, compared with only 5% to 30% in the general public. Antibiotics can eliminate the carrier state.

Haemophilus Meningitis. Prior to development of a vaccine, about two-thirds of bacterial meningitis cases during the first year of life are caused by *Haemophilus influenzae* type B (hib). Today, meningococcal meningitis has replaced *Haemophilus* meningitis as the number one type seen in the United States. Among children, 30% to 50% are carriers of this organism; among adults, the figure is only 3%. Humans are exposed to *H. influenzae* early in life and rapidly acquire immunity, so the disease is rare in adults. Only 10% of children between ages 3 and 6 lack antibodies, and all those over age 6 have antibodies. Without treatment this disease is nearly always fatal. Even with treatment, one-third die. Of those who recover, 30% to 50% have serious mental retardation, and 5% are permanently institutionalized because of damage to the central nervous system. *Haemophilus* meningitis is the leading cause of mental retardation in the United States and worldwide. Hib vaccines are available, which all children age 5 and under should receive (◀Chapter 17, p. 516). These vaccines have dramatically reduced incidence of the disease.

Streptococcus Meningitis. Among adults *Streptococcus pneumoniae* is the most common cause of meningitis. Organisms generally spread via the blood from lung, sinus, mastoid, or ear infections. Mortality is 40%.

Listeriosis. Another kind of meningitis, **listeriosis**, is caused by *Listeria monocytogenes*, a small, Gram-positive bacillus that is widely distributed in nature.

APPLICATIONS

A Road Less Traveled

Why are brain infections hard to treat? Because the blood-brain barrier separates the brain from the rest of the body's cardiovascular system. This barrier allows only selective substances to pass through it. The body's own antibodies and complement proteins have a hard time passing from the blood into brain tissue to fight off infections. Antibiotics such as penicillin cannot easily pass through the blood-brain barrier, and if the meninges are inflamed, passage is easier. In most cases, average concentration of antibiotics (administered orally or intravenously) in the cerebrospinal fluid generally reaches only about 15% of their concentration in plasma. But all is not lost. Some other antibiotics, such as chloramphenicol and tetracycline, are lipid-soluble and diffuse easily across the blood-brain barrier. Although the road to the brain is less traveled, at least it is not abandoned.

Foodborne transmission by improperly processed milk, cheese, meat, and vegetables is the most common source of infection, surviving both high and low temperatures. It is sometimes acquired as a zoonosis and is particularly threatening to those with impaired immune systems. Although not an especially significant human disease for several decades, listeriosis is now a leading cause of infection in kidney transplant patients. In pregnant women the bacillus can cross the placenta, infect the fetus, and cause abortion, stillbirth, or neonatal death. Listeriosis is responsible for many cases of fetal damage, sometimes not appearing until several weeks after birth, with onset of meningitis.

Certain nontoxic bacteria that are also found in hot dogs produce bacteriocins that effectively kill the Listeria

There is now a new U.S.-approved bacteriophage product available, which contains 5 different strains of bacteriophage, each of which will kill one or more strains of *Listeria*. It is sprayed on cut surfaces of fruit such as melons sold prepackaged in bite-sized chunks. Consumer response will be interesting to observe. Already people, who do not understand that bacteriophages each kill only one specific species of bacteria, are raising an outcry that it is not safe to feed their children fruit covered with viruses. There are about 1 million bacteriophages in every milliliter of water or food that you already eat or drink. They are present in everyone's mouth, and the only things they harm are bacteria. We need to educate the public regarding bacteriophages.

Brain Abscesses

Microorganisms that cause **brain abscesses** reach the brain from head wounds or via blood from another site. As would be expected with wounds, multispecies infections are common, and anaerobes are as likely to be responsible as are aerobes. Most such abscesses occur in patients under age 40, but two age periods—birth to 20 years and 50 to 70 years—show peak incidences. The infection gradually grows in mass and compresses the brain. The masses can be detected by CAT (computerized axial tomography) scans or X-rays, and causative agents can be identified by serologic tests and by culturing cerebrospinal fluid. In very early stages, antibiotic treatment can be sufficient, but later surgical drainage or removal of abscesses is usually necessary. Abscesses in areas of the brain that control the heart or other vital organs cannot be treated surgically. Without treatment, half the patients die, but with the current best treatment, only 5% to 10% die.

VIRAL DISEASES OF THE BRAIN AND MENINGES

Viral Meningitis

Unlike bacterial meningitis, which is always fatal if untreated, **viral meningitis** is usually self-limiting and nonfatal. Enteroviruses account for approximately 40% of viral meningitis cases, and mumps virus for 15%. The causative virus in 30% of cases remains unidentified.

Figure 24.3 **Negri bodies.** These characteristic signs of [LM] rabies are shown here in the cerebellum of a human brain (1,768X). *(Science VU/Visuals Unlimited)*

Rabies

Rabies was described by Democritus in the fifth century B.C. and by Aristotle in the fourth century B.C. Pasteur made real progress in understanding the disease when he found evidence of the infectious agent in saliva, the central nervous system, and peripheral nerves. He attenuated the agent and proved that a suspension of it could be used to prevent rabies. In 1903 the Italian physician Adelchi Negri discovered inclusion bodies (◄Chapter 14, p. 414), called *Negri bodies*, or clusters of viruses, in neurons **(Figure 24.3)**. Negri bodies were used to diagnose rabies for more than 50 years until an immunofluorescent antibody test (IFAT) was developed in 1958. Still used today, the IFAT is so sensitive that after an animal suspected of having rabies bites a human, it can be killed immediately and its brain examined for rabies antigens. Prior to the availability of IFAT, such animals had to be held for 30 days, or until symptoms of rabies developed, before a search for Negri bodies could be made. Most biting animals are not rabid, so the IFAT has saved many people both the anguish of wondering if they have been exposed to rabies and the discomfort and risk of treatment.

The virus that causes rabies has a worldwide distribution. It infects all mammals exposed to it, so the possibilities for reservoir infections are almost limitless. Different types of rabies are found in different regions of the world. In almost all of Asia, Africa, Mexico, and Central and South America, rabies is endemic in dogs. In Canada, the United States, and Western Europe, wildlife rabies predominates, accounting for 90% of all cases, and dog rabies is controlled. The World Health Organization (WHO) lists 60 countries, including England, Australia, Japan, Sweden, and Spain, as rabies-free. This success is due to animal vaccination and quarantine programs. Unfortunately, in the United States, rabies has become a "reemerging disease." Cats rather than dogs are now the most common rabid domestic animals. The number of cases in raccoons now exceeds the number of dog cases 45 years ago. Cases of rabid skunks

have tripled in the same time period, and bat rabies has significantly increased. In Texas, coyote cases have gone from zero in 1987 to nearly 100 per year today. Along the eastern coast of the United States, a northward relocation of Florida raccoons to the Virginia/West Virginia border for hunting purposes in 1977 has spread the disease **(Figure 24.4)**. Health authorities are trying to prevent a westward spread of rabies by means of a "vaccine corridor." However, there has been a breakthrough of the barrier in Ohio. Mathematical modeling is being used to calculate where additional baited vaccine should be deposited to halt this spread.

Identifying rabid animals can be a problem. About half of all rabid dogs release viruses into saliva 3 to 6 days before they show symptoms of rabies. In contrast, 90% of rabid cats have viruses in their saliva about 1 day before they become symptomatic. Any change in the behavior of an animal can be a warning that it might be rabid. A "friendly" wild animal approaching people or a gentle family animal snapping without provocation sometimes indicates impending rabies symptoms. Small wild animals and some domestic ones, too, will bite if suddenly grabbed and held. Biting is their only defense in such a stressful situation. Much unnecessary anguish could be avoided if people, especially children, were taught to appreciate unfamiliar animals without touching them.

Even a healthy pet that has bitten someone should be confined and watched for 10 days to see if it develops symptoms of rabies.

Animals vary in their susceptibility to rabies, and susceptibility is directly correlated with the role of the animals in maintaining reservoir infections. Foxes, coyotes, skunks, raccoons, and bats are highly susceptible. No infected fox has ever been known to survive, whereas 10% to 20% of dogs and 30% to 40% of mongooses do survive. Bats are particularly dangerous because they are asymptomatic and shed viruses into their feces, urine, and saliva. Two explorers of bat-infested caves in Texas died of rabies. Dogs, cats, cattle, horses, and sheep are less susceptible. Whether a person becomes infected depends mainly on whether an animal is shedding viruses in its saliva at the time of the bite. Even animals later proved to have rabies may not have been shedding viruses at the time they inflicted bites.

The Disease. Rabies is caused by the **rabies virus**, an RNA-containing rhabdovirus (◄Figure 10.3c, p. 280). After entering the body through an animal bite or other break in the skin, the rabies virus first replicates in injured tissue for 1 to 4 days and then migrates to nerves, where it replicates slowly until it reaches the spinal cord. It progresses rapidly up the spinal cord to the brain by the flow of cytoplasm through axons. The time from infection to the appearance of symptoms varies from 13 days to 2 years but is usually between 20 and 60 days. The length of time required for symptoms to appear is proportional to

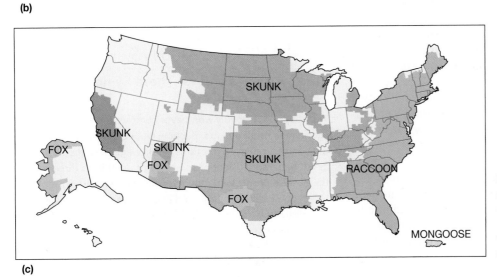

Figure 24.4 Spread of rabies.
(a) Changing distribution of raccoon rabies in the United States, 1947 to 1996. (b) Presence/absence of rabies worldwide, 2005. (c) Predominant species of wildlife rabies. (*Source: CDC.*)

APPLICATIONS

How Do Bats Survive Rabies?

Almost all the recent cases of rabies in the United States have been caused by bat-associated strains of rabies. Bats can develop furious type rabies and die shortly after being infected. However, some don't die for periods of up to 3 months. And many others become healthy carriers of rabies. During hibernation, the rabiesvirus is found localized in the brown fat of bats, where it remains without multiplying. When hibernation ends, and the bat's temperature rises, the virus begins to replicate and spread to other tissues, from which it may be transmitted to other hosts. And yes, the vampire bat can and does transmit rabies, especially to cattle (and a few humans) in southern climes.

the distance between the wound and the brain and is affected by the accessibility of nerve fibers. Thus, a bite on the face, which is well supplied with nerves and close to the brain, produces symptoms much more quickly than does a bite on the leg. The rabies virus has a predilection for nervous tissue but also infects salivary glands and the respiratory tract lining. There are even documented cases of transmission of rabies virus via corneal transplants.

It is the typically lengthy incubation period of rabies that allows postexposure immunization to be possible. Ordinarily there is sufficient time for the bitten individual to be vaccinated and to respond by making enough protective antibodies to prevent onset of the disease. Once symptoms have occurred, it is too late to vaccinate, and death usually follows quickly.

In humans the first symptoms are headache, fever, nausea, and partial paralysis near the bite site. These symptoms persist for 2 to 10 days and then worsen until the acute neurological phase of the disease ensues. The patient's gait becomes uncoordinated as paralysis becomes more general. Hydrophobia (fear of water) occurs as throat muscles undergo painful spasms, especially during swallowing. Aerophobia (fear of moving air) occurs because the skin is hypersensitive to any sensations. Confusion, hyperactivity, and hallucinations also occur. Within 10 to 14 days of the onset of symptoms, the patient typically goes into a coma and dies. Of all human patients who have suffered from clinical rabies, only two have been known to survive and make a complete recovery. Both had some degree of protection from an earlier immunization, and they were known to have been bitten by a rabid animal, so postexposure immunological treatment was started immediately.

Diagnosis, Treatment, and Prevention. A sample from a brain or skin biopsy can be stained by the IFAT to identify rabies virus antigen before the patient dies. Finding the antigen confirms the diagnosis, but failure to find it does not rule out rabies. Sometimes a diagnosis can be made before death by testing cerebrospinal fluid or serum for neutralizing antibodies, which increase 10 to 12 days after the symptoms appear.

The bite of a rabid animal is treated by first thoroughly cleaning it with soap and flushing it with large amounts of water. Hyperimmune rabies serum is introduced into and around the wound in hope of neutralizing viruses before they reach the nervous system, where they are beyond reach of the antibodies. Interferon also can be applied to the wound. A series of injections of vaccine is given to induce the production of neutralizing antibodies. The biting animal should be located and confined for examination by IFAT.

The best means of preventing rabies is to immunize pets, and such immunization is required in many countries. Attempts to reduce rabies in raccoons have been made by using vaccine in small sponges covered with food bait. When the raccoons eat the bait, they also ingest enough vaccine to prevent them from acquiring rabies. Preliminary studies conducted in the United States show promising results. When raccoons from a bait-treated wildlife area were trapped, 15 out of 16 survived a challenge dose of rabies virus. All 16 caught in the nonbaited control area died of the same challenge doses. It is not certain whether the animal that died in the experimental group ever ate any bait. It is estimated that a distribution of one bait pellet per acre, at a cost of $1 per pellet, could prove sufficient to reduce rabies dramatically in wild raccoons.

Rabies immunization is recommended for veterinarians and their staffs, hunters who may have contact with wild animals, and technicians who work with the virus. The first vaccine, and for many years the only one available, was developed by Pasteur. This vaccine contained viruses modified by 50 transfers, involving drying of infected spinal cords, from one rabbit to another. It was used in all cases of suspected rabies because the disease might develop while the patient waited for test results on the biting animal. The vaccine was given subcutaneously (into the skin) in 14 or more daily injections into the abdomen, where the thick tissue slows the rate of absorption. These injections had unpleasant to serious side effects, including severe abdominal pain, fatigue, fever,

APPLICATIONS

Watch Out for These Symptoms in Your Dog

Your pet dog probably won't ever contract rabies, thanks to vaccination requirements. But how would you know if your dog did contract rabies? Symptoms would begin with the dog acting as if it had a sore throat or something caught in its throat. As the disease progressed, the dog could become groggy or paralyzed (dumb rabies) or agitated and aggressive, biting anything that disturbs it (furious rabies). Throat muscle spasms, difficulty in swallowing, and drooling would also be indicative of rabies. Eventually, the dog would become apathetic and stuporous and slide into a final coma.

and sometimes rabies infection. A relatively new vaccine produced from viruses grown in human diploid fibroblast cultures elicits high levels of neutralizing antibodies with only a few injections and minimal side effects. The current vaccine is given intramuscularly on days 0, 2, 7, 14, and 28. In addition, hyperimmune globulin is placed deep in the wound and infiltrated around the wound.

Encephalitis

The Disease. **Encephalitis** is an inflammation of the brain caused by a variety of togaviruses or by a flavivirus. We will first consider the following four diseases, each of which is caused by a different virus: **eastern equine encephalitis (EEE)**, seen most often in the eastern United States; **western equine encephalitis (WEE)**, seen in the western United States (◀Figure 15.6, p. 430); **Venezuelan equine encephalitis (VEE)**, seen in Florida, Texas, Mexico, and much of South America; and **St. Louis encephalitis (SLE)**, seen from east to west in the central United States (◀Figure 15.3, p. 428). The equine varieties, caused by togaviruses, are so named because they infect horses more often than humans. The life cycles of these viruses generally involve transmission from a mosquito to a bird, back to a mosquito, and then to a horse, human, or other mammal, and finally back to a mosquito. The St. Louis variety, caused by a flavivirus, is so named because the first epidemic was identified in St. Louis in 1933. That virus appears to be transmitted mostly among English sparrows, mosquitoes, and humans.

The viruses, which are introduced into the body through bites of infected mosquitoes, first multiply in the skin and spread to lymph nodes. Viremia involving especially large numbers of viruses follows. In a few infections, the viruses invade the central nervous system, where they cause shrinkage and lysis of neurons. WEE appears every summer, and about a third of the cases occur in children under 1 year of age. Fever and headache are common symptoms, and convulsions sometimes occur. EEE is a much more serious disease; it causes a severe necrotizing infection of the brain. The disease is fatal in 50% to 80% of cases, and survivors often suffer permanent brain damage. Fortunately, because swamp birds are the major reservoir for the virus and swamp mosquitoes the major vectors, very few humans become infected. VEE is mainly a disease of horses; when it occurs in humans, it resembles influenza.

SLE occurs in late summer epidemics about every 10 years and causes the most severe symptoms in elderly patients. The illness starts with malaise, fever, and chills as consequences of viremia. Other common symptoms include anorexia, myalgia (muscle pain), sore throat, and drowsiness. In addition, certain patients have symptoms of a urinary tract infection, neurological disorders, altered states of consciousness, and convulsions. Complications can include secondary bacterial infections, blood clots in the lungs, and gastrointestinal hemorrhage. Most patients escape the complications and recover fully.

West Nile Fever

West Nile fever, an emerging disease new to the United States, is an old story elsewhere, e.g., along the Nile River and in Israel. There it is an endemic disease that causes outbreaks about once in every 10 years. Most deaths are among people in their 70's and 80's, plus some very young children. In 1999, probably carried in the blood of an imported bird, the West Nile arbovirus made its entrance into New York City—just like so many other immigrants before it, arriving in the United States. Carried by at least 43 species of mosquitoes, it has infected over 60 species of birds and countless other animals, plus humans. In 1999, 55 people in the New York area were sickened with encephalitis caused by West Nile virus, 7 of whom died. That year the virus appeared to have spread only within a 30-mile radius of its center. However, despite spraying efforts to control mosquitoes, in 2000 it spread more than 300 miles, from Virginia to Canada to Pennsylvania and to all of New England. It spread to the Mississippi Valley in 2001 and became coast-to-coast a year later. The 2006 incidence numbers are shown in **Figure 24.5**. Symptoms range from fever to neuroinvasive disease. Death is most likely in diabetics over the age of 65. Meanwhile, we have discovered the virus in horses, raccoons, bats, rabbits, and many other mammals, as well as in many birds. Mortality rate approaches 100% in crows and blue jays.

PUBLIC HEALTH

Chickens Stand Sentinel Guard Duty

Brave chickens across the country have been enlisted in the war against West Nile fever. They are placed in cages in various neighborhoods, awaiting the bites of mosquitoes. Immediately after being bitten by one that gives them West Nile virus, the chickens (who are immune to West Nile) begin forming antibodies against the virus. Public health workers regularly test the chickens for presence of these antibodies, obtaining a specimen by swabbing their mouths. A positive test alerts the local authorities to step up mosquito prevention efforts.

(Getty Images News and Sports Services)

2006 West Nile Virus Activity in the United States

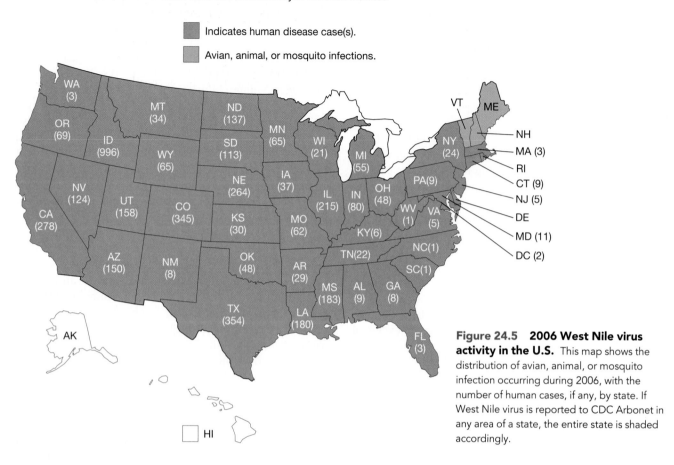

Figure 24.5   2006 West Nile virus activity in the U.S. This map shows the distribution of avian, animal, or mosquito infection occurring during 2006, with the number of human cases, if any, by state. If West Nile virus is reported to CDC Arbonet in any area of a state, the entire state is shaded accordingly.

Diagnosis, Treatment, and Prevention. Encephalitis sometimes can be diagnosed by isolating the causative agent from cell cultures or mice inoculated with blood or spinal fluid. Cultures can be negative when the disease is present because the viremic phase of the disease usually is over before the patient seeks medical attention. Serological methods can be used to identify antibodies at any time during and following the illness. Treatment only alleviates symptoms. Vaccines are available for immunizing horses, but they are rarely used in humans because of the danger of inducing a virulent form of the disease. Prevention by eradicating mosquito vectors is a more appropriate means of decreasing the already low incidence of encephalitis among humans.

Other Viral Diseases of the Brain and Meninges

Herpes Meningoencephalitis. Herpes simplex virus, which usually is responsible for cold sores, also can cause **herpes meningoencephalitis**. This disease often follows a generalized herpes infection in a newborn infant, child, or adult. The virus reaches the brain by ascending from the trigeminal ganglion. The disease has a rapid onset with fever and chills, headache, convulsions, and altered reflexes. In the middle-aged or elderly, meningoencephalitis causes confusion, loss of speech, hallucinations, and sometimes seizures. Most patients die in 8 to 10 days; survivors usually display neurological damage.

Polyomavirus Infections. Polyomaviruses enter the body through the respiratory or gastrointestinal tract. Initial replication takes place in the cells the virus first enters. The viremia that follows allows the viruses to reach target organs, particularly the kidneys, lungs, and brain. Polyomaviruses, which are papovaviruses, were first recognized in the 1960s as viral particles in the enlarged nuclei of oligodendrocytes. These are the cells that produce myelin, the lipoprotein that coats nerve fibers in the central nervous system. Infected oligodendrocytes are observed to surround areas that lack myelin in the brains of patients dying from **progressive multifocal leukoencephalopathy**.

One cause of progressive multifocal leukoencephalopathy is now known to be the JC virus, a polyomavirus named with the initials of a victim from whom it was isolated. Onset is insidious, with vision and speech impairment being the first signs. Typical symptoms of viral infections, such as fever and headache, are absent. Mental deterioration, limb paralysis, and blindness follow. Diagnosis is difficult because cerebrospinal fluid remains normal, and only nonspecific changes are seen in electroencephalograms. The JC virus infects and kills oligodendrocytes but does not affect neurons. On occasion a young patient develops this disease as a complication of a primary infection, but most cases result from reactivated latent viruses from childhood infections.

Other polyomaviruses have been isolated from various patients. The BK virus was isolated from the urine of a kidney-transplant patient. In one instance a 16-year-old boy with an immunodeficiency had BK viremia and developed kidney inflammation with viruses present in kidney cells. Irreversible kidney failure resulted. The BK virus also has been associated with, but not isolated from, respiratory illness and cystitis, an inflammation of the urinary bladder.

Half the children in the United States have antibodies to the JC virus by age 14; antibodies to the BK virus are found by age 4. Although JC and BK viruses apparently persist in most humans for years without causing disease, they sometimes reappear as complications of chronic diseases, immunodeficiencies, and disorders in which lymphocytes proliferate. Pregnancy, diabetes, organ transplantation, antitumor therapy, and immunodeficiency diseases, including AIDS, are among the conditions that can reactivate polyomaviruses. Many kidney-transplant patients, for example, excrete either BK or JC viruses, but the shedding of these viruses rarely has serious consequences. However, unchecked viral multiplication, which is particularly likely with T cell deficiencies, can sometimes occur and cause clinically apparent disease. Both JC and BK viruses are oncogenic in laboratory animals and possibly also in humans. No diagnostic tests for polyomaviruses are available for routine use; nor is any treatment available for the infections, even if they can be recognized.

✓ **CHECKLIST**

1. List five organisms that can cause meningitis.
2. Why are bats a particularly dangerous source of rabies infection?
3. How does encephalitis differ from meningitis?

APPLICATIONS

Amoebic Invaders

If you are an avid swimmer or an avid hot tubber, here's something you might want to know: two different genera of soil amoebas, *Naegleria* and *Acanthamoeba*, are opportunistic human pathogens you could encounter. *Naegleria fowleri* is usually seen in swimmers. The amoebas probably enter through the nasal passage and make their way up along nerves to the meninges. There, they cause meningoencephalitis. *Acanthamoeba polyphaga* accumulates on the water surface of the contaminated hot tubs when the tubs are covered. They are dispersed after the lid is removed and bathers begin to enter the water. These amoebas cause ulceration of the eyes and the skin; if they invade the central nervous system, death can occur within a few weeks due to meningoencephalitis.

OTHER DISEASES OF THE NERVOUS SYSTEM

BACTERIAL NERVE DISEASES

Hansen's Disease

Hansen's disease, the currently preferred name for **leprosy**, has been known since biblical times, when many things, even houses, were said to have leprosy. Many so-called leprosy cases were other skin diseases such as fungal and viral infections, and the houses probably had fungi growing on their walls.

The disease is rapidly disappearing, 10 million people having been cured over the last 15 years, but an estimated 2.5 million cases exist worldwide today, mainly in Asia, Africa, and South America **(Figure 24.6)**. Five countries account for 90% of the cases: Brazil, Madagascar, Mozambique, Tanzania, and Nepal. Hansen's disease also occurs in the United States, where a peak number of 361 cases were reported in 1985, mostly among immigrants from countries where Hansen's disease is endemic. Infected people sometimes show no symptoms when they enter the United States. No reliable test is available to disclose all these subclinical cases, although the **lepromin skin test**, similar to the tuberculin skin test for tuberculosis, detects some of them. Health care workers should watch for Hansen's disease among immigrants.

The Disease. The acid-fast bacillus *Mycobacterium leprae* is found in all cases of Hansen's disease. Although *M. leprae* was the first bacterium to be recognized as a human pathogen, demonstrating that it fulfills Koch's Postulates has been slow because the organism is difficult to grow in the laboratory. For one thing, it reproduces very slowly, having a 12-day division cycle. Recently developed methods for growing *M. leprae* in such diverse hosts as nine-banded armadillos (◄Chapter 14, p. 407), chimpanzees, mangabey monkeys, and mice have made the bacterium available for research. It may now be possible to develop a vaccine. PCR techniques allow *M. leprae* to be identified in samples of skin or nasal tissues, thus simplifying diagnosis.

M. leprae bacilli occur within cells in a characteristic arrangement, grouped together like a bundle of logs, and are covered by capsules.

Clinical forms of Hansen's disease vary along a spectrum from tuberculoid to lepromatous. In the **tuberculoid**, or anesthetic, form **(Figure 24.7a)**, areas of skin lose pigment and sensation. In the **lepromatous** (lep-ro'mat-us), or nodular, form **(Figure 24.7b)**, a granulomatous response (◄Chapter 16, p. 475) causes enlarged, disfiguring skin lesions called **lepromas**. Incubation time averages 2 to 5 years for the tuberculoid form and 9 to 12 years for the lepromatous form.

Mycobacterium leprae is the only bacterium known to destroy peripheral tissue; it also destroys skin and mucous membranes. The organism has a predilection for cooler parts of the human body, such as the nose, ears, and

(a)

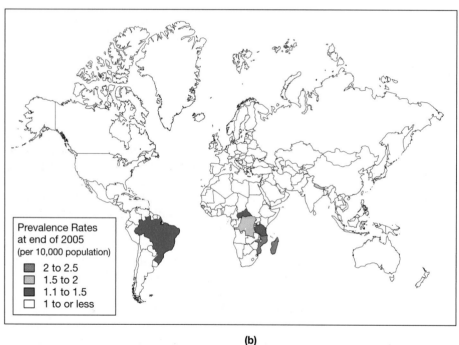

(b)

Figure 24.6 **The incidence of Hansen's disease (leprosy).** **(a)** In the United States; **(b)** worldwide prevalence rates 2005. (*Source:* CDC.)

fingers, but large numbers of organisms are seen throughout the body, except for the central nervous system. Continuous bacteremia of 1,000 organisms per milliliter of blood has been demonstrated in lepromatous cases. Large numbers of bacilli are shed in respiratory secretions and in pus discharged from lesions. Although Hansen's disease is not highly contagious, shedding of organisms probably transmits the disease to those with extensive, close contact with patients, such as children of infected parents.

As Hansen's disease progresses, it deforms hands and feet **(Figure 24.8)**. Severe lepromatous disease erodes bone: fingers and toes become needlelike, pits develop in the skull, nasal bones are destroyed, and teeth fall from the jaw as bone surrounding them is lost.

Surgery sometimes can restore the use of extremely crippled hands and feet. The National Hansen's Disease Center, a U.S. Public Health Service facility in Carville, Louisiana, has pioneered development of these special surgical techniques. In the past decade, it was recognized that changes in the feet of diabetic patients can be helped by these same surgical techniques. Thus, Carville now has an active teaching program for surgeons who will use knowledge gained from one of the most shunned diseases to help thousands of victims of diabetes.

Examination of ancient skeletons has provided insights into the epidemiology of Hansen's disease in past centuries. It is clear that the disease traveled from the Old World to the New World and that in the past, even

Figure 24.7 Hansen's disease: two extremes with gradations and combinations between. (a) In the tuberculoid, or anesthetic, form, areas of skin lose pigment and sensitivity. A pin may be stuck into these "anesthetized" areas and not be felt because of destruction of nerves and nerve endings. *(CNRI/Phototake)* **(b)** The nodular form is characterized by disfiguring granulomas called lepromas. *(Ken Greer/ Visuals Unlimited)*

(a)   **(b)**

allowing for misdiagnosis, its incidence in Europe was much higher than it is today. In England, lepers were shunned even in death, being buried in separate cemeteries. Examination of skeletons from these graves reveals that, by far, the majority of the people buried there did in fact have leprosy. Genetic factors may predispose toward resistance to Hansen's disease; as susceptible individuals have died, resistant individuals have made up a greater percentage of the population.

Diagnosis, Treatment, and Prevention. Hansen's disease is diagnosed by PCR or by finding the organism in acid-fast-stained smears and in scrapings from lesions or biopsies. The disease is treated with dapsone, clofazimine, and rifampin, but dapsone-resistant strains are beginning to appear. Treatment greatly reduces nodules of lepromatous disease, but it cannot restore lost tissue. Until recently, victims of Hansen's disease were isolated

Figure 24.8 A deformed "claw" hand of Hansen's disease. This deformity can be treated surgically in its early stages, thereby preventing crippling. *(NMSB/Custom Medical Stock Photo, Inc.)*

in special hospitals called leprosariums. Now the disease can be arrested, and the people can live nearly normal lives without infecting others in their community. (They must still sleep in separate bedrooms and use only their own linens and utensils, however, and they cannot live in a household where children are present.)

Immune responses in Hansen's disease are cell mediated and vary from strong to weak. Strong responses and distinctly positive skin tests are seen in patients with the less serious tuberculoid disease. Weak responses and negative skin tests are seen in patients with rapidly progressing lepromatous disease. However, test results may change over time from positive to negative and vice versa as immune response rises and falls. Lepromatous patients usually have adequate cell-mediated response to other antigens, so their lack of immunity is not due to generalized T cell absence or dysfunction. The absence of T cells in "nude" mice, which are hairless and lack a thymus, makes them suitable organisms for growing large quantities of organisms.

Vaccine is not available for Hansen's disease. Even if a vaccine became available now, its effectiveness would take years to determine because of the disease's long incubation period. Avoiding exposure and receiving prophylactic chemotherapy after exposure are the only means of prevention. However, a group called The Global Alliance for Leprosy Elimination expects to treat and cure all leprosy patients remaining in the world by distributing free drugs for this purpose!

Tetanus
Tetanus is caused by the obligately anaerobic, Gram-positive, spore-forming rod *Clostridium tetani* (Figure 24.9). The organism can be cultured in the laboratory only under strict anaerobic conditions. *Clostridium tetani* endospores are exceedingly resistant to drying, disinfectants, and heat.

(a) TEM

(b) TEM

Figure 24.9 Tetanus spores. (a) A TEM of *Clostridium tetani* bacillus with large, dark terminal endospores (105,800X). *(Dr. T. J. Beveridge/Biological Photo Service/PO)* **(b)** A freeze-etch TEM preparation, showing the rounded endospore inside a *Clostridium* bacillus (107,250X). *(Biological Photo Service)*

Kitasato isolated C. tetani, established it as the causative organism of tetanus, showed that it could not invade the bloodstream, and showed that the disease is an intoxication.

Boiling for 20 minutes does not kill them, and they can survive for years if not exposed to sunlight (◀Chapter 4, p. 92). Spores are found in all soils but especially in those enriched with manure. The organisms are part of the normal bowel microflora of horses and cattle and about 25% of humans. Therefore, handling bedpans, dirty diapers, or other objects contaminated with feces can transmit organisms to persons who have any breaks in their skin.

Since the development of tetanus vaccine in 1933, the incidence of tetanus in the United States has steadily dropped, with the annual number of cases remaining below 100 since 1975. The highest incidence is in older people, especially women. Vaccine was not available during their childhoods, and they did not receive it during military service, as men did. They remain susceptible to tetanus spores as they enjoy gardening in their retirement years. The elderly should be immunized for their protection. Get grandma a tetanus shot for her birthday!

The Disease. To cause tetanus, spores must be deposited deep in tissues, where oxygen is unavailable. This occurs in deep cuts and puncture wounds. Stepping on a rusty nail has a reputation for leading to tetanus, but it is tetanus endospores, not rust, that cause the disease—a shiny new nail can be just as dangerous if the spores are present. Making puncture wounds bleed helps flush tetanus spores and other organisms from them. Once inside the host, the noninvasive tetanus organisms stay at the wound site and release a powerful exotoxin; tetanus is a toxin-mediated disease. After 4 to 10 days' incubation, symptoms begin, with generalized muscle stiffness followed by spasms that affect every muscle. An arched back and clenched fists and jaws (hence the term *lockjaw*), are classic symptoms **(Figure 24.10)**. Spasms can be violent enough to break bones. Eventually, respiratory muscles become paralyzed, heart function is disturbed, and, with rare exceptions, the patient dies. Survivors experience a period of sore muscles but suffer no further sequelae. Before vaccine was available, many soldiers died from tetanus. On battlefields strewn with horses and manure, contamination of wounds with tetanus spores was inevitable. War-related cases were virtually eliminated by vaccinating soldiers; only 12 cases occurred during World War II.

Figure 24.10 A soldier dying of tetanus. Tetanus was a common cause of death in the days of cavalry troops. The extreme contraction of all muscles, from those of the face to those of the toes, is a classic symptom of the disease. *(Charles Bell (1774–1842), Oposthotunus. Reproduced by kind permission of The Royal College of Surgeons of Edinburgh)*

Treatment and Prevention. Tetanus toxoid vaccine given prior to injuries protects against the toxin. Antitoxin and antibiotics are given to nonimmunized patients when injuries are treated. Because antitoxin must be administered to inactivate the toxin before the immune system has time to become sensitized to it, infection treated in this way confers no immunity. Patients should receive toxoid immunization after they recover.

Tetanus neonatorum is acquired through the raw stump of the umbilical cord. In some societies, contaminated knives are used to cut umbilical cords after a baby is delivered, and mud is smeared on the cut end. In parts of some developing nations, 10% of deaths within a month of birth are due to neonatal tetanus.

Botulism

Botulism derives its name from the Latin word *botulus*, which means "sausage." It was coined at a time when the disease was often acquired from eating sausages. Botulism is caused by *Clostridium botulinum*, a spore-forming obligate anaerobe that releases a potent exotoxin (neurotoxin). (See the box "Clinical Use of Botulinum Toxin" in ◄Chapter 14, p. 412) The disease occurs in three forms: foodborne, infant, and wound. Foodborne botulism accounts for 90% of cases and is caused by ingestion of toxin, usually from improperly home-canned nonacid foods, especially green beans and green peppers (**Figure 24.11**). Thus, foodborne botulism is an intoxication; the organisms do not infect tissues. Infant botulism and wound botulism involve both infection and intoxication because the organisms grow in tissues and produce toxin.

Endospores of *C. botulinum* are more heat resistant than those of any other anaerobe; they withstand several hours at 100°C and 10 minutes at 120°C. They also are very resistant to freezing (down to −190°C) and irradiation. Found in most soils in the Northern Hemisphere, these endospores remain viable for long periods of time and enable the organism to withstand aerobic conditions. Endospores will germinate only in anaerobic conditions.

The ability of *C. botulinum* to form toxin depends on infection with a bacteriophage. This phage carries the information for botulism toxin production. If infected with an appropriate bacteriophage, *C. botulinum* produces one of eight different toxins, of which only four cause human disease. The other toxins cause disease in various other animals. If a strain of the bacillus is "cured" of its phage infection, it no longer produces toxin. If later infected with a different phage, the bacillus will produce another toxin.

Botulism toxin is the most potent toxin known—even more toxic than *Shigella* and tetanus toxins. As little as 0.000005 μg can kill a mouse! One ounce could kill the entire U.S. population. Originally thought to be an exotoxin, it is now known to be produced inside the cytoplasm and released only upon death and autolysis of the cell. It is activated by proteolytic enzymes, possibly including trypsin in the host's intestine. The toxin is colorless, odorless, and tasteless; people have died from a single taste of affected food. If endospores are not destroyed, they germinate in food during storage under anaerobic conditions and can release large quantities of toxin. Whereas endospores are highly heat resistant, the toxin can be inactivated by only a few minutes' boiling. Boiling home-canned foods vigorously before serving would eliminate most foodborne botulism.

The Disease. Botulism is a neuroparalytic disease with sudden onset and rapidly progressing paralysis. It ends in death from respiratory arrest if not treated promptly. The toxin acts at junctions between neurons and muscle cells and prevents the release of acetylcholine, the chemical that neurons release to cause muscle cells to contract. The toxin thus paralyzes muscles in a relaxed (flaccid) state—starting with small eye muscles, progressing to the larynx and pharynx, and on to the

To prevent botulism, follow the 2-40-140 rule. Don't eat meats, dressings, or salads that have been kept for more than 2 hours between 40°F (4.4°C) and 140°F (60°C).

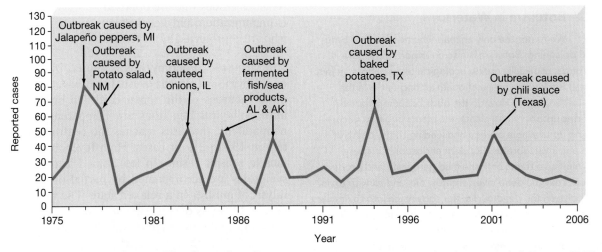

Figure 24.11 **The U.S. incidence of foodborne botulism.** The foods that caused certain outbreaks are included. (*Source:* CDC.)

respiratory muscles. This causes double vision, difficulty in speaking and swallowing, and difficulty in breathing. It causes no fever but can cause gastrointestinal disturbances. Although the toxin is an antigen, people who have recovered do not have antibodies, so the amount of antigen required to elicit antibodies must be greater than the lethal dose.

Diagnosis and Treatment.   Diagnosis is based on clinical symptoms and history, with confirmation later by demonstrating toxin in serum, feces, or food remains. Although the confirmation test takes 24 to 96 hours, treatment with a polyvalent antitoxin is started immediately. A polyvalent antitoxin is used to ensure effectiveness against all toxins that affect humans. Help in maintaining respiration is important and may be continued for up to 2 months. Antibiotics are of no use because foodborne botulism is due to preformed toxin and not to growth or organisms. With proper treatment, the mortality rate is less than 10%.

Infant botulism was first recognized in 1976, and incidence has ranged between 30 and 100 cases per year, mostly in California. The disease is associated with feeding honey to infants. Studies in California have shown that 10% of the jars of honey sold there contain botulism endospores. Endospores germinate and grow in the immature digestive tract of infants, probably because they lack appropriate competing microbiota. As toxin is absorbed, the infant becomes lethargic and loses the ability to suck and swallow, so the disease is often called "floppy baby" syndrome. Infant botulism usually occurs in infants under 6 months of age and rarely after 12 months. If parents would not give honey to children under 1 year, most cases could be prevented. The prognosis is excellent, and death is rare, but the child usually must remain hospitalized for several months.

APPLICATIONS

Botulism in Waterfowl

We're not the only animals who need to fear botulism food poisoning. Botulism is also an important disease in wetland birds, especially during ecological disturbances. In fact, the disease is a leading cause of death among ducks in the western United States. How do the birds become infected? After storms uproot marsh plants, the plants begin to decompose, using up available oxygen and leading to the death of small aquatic invertebrates. Botulism endospores in these animals' digestive tracts or in the mud germinate and produce toxin. Ducks eat the dead invertebrates, die, and fill with toxin. Flies lay eggs on the dead ducks; the eggs hatch into toxin-filled maggots. Healthy ducks eat the maggots, die, and provide a site for new fly eggs. The cycle is repeated until thousands of ducks have died and no more ducks are left at that site.

Wound botulism is the least common form of botulism; no more than one case per year has been seen in the United States since 1942. It occurs in deep, crushing wounds. Tissue damage impairs circulation and creates anaerobic conditions, so endospores germinate, multiply, and produce toxin. Toxin enters the blood and is distributed throughout the body. It reaches junctions between neurons and muscle cells about a week after injury and causes progressive paralysis. The mortality rate is about 25%.

✓**CHECKLIST**

1. Distinguish between tuberculoid and lepromatous forms of Hansen's disease.

2. Why is Hansen's disease no longer a common disease in Europe? Why is it more common in South America?

3. Why must patients who recover from tetanus be immunized against tetanus afterward? Why hasn't having a case of the disease made them immune?

4. What symptoms would you experience as you develop a case of botulism food poisoning?

VIRAL NERVE DISEASES

Poliomyelitis

Poliomyelitis is a very ancient disease; its effects are clearly depicted in Egyptian wall paintings thousands of years old. As recently as the early 1950s, it was a dreaded disease in the United States, with nearly 58,000 cases reported in the peak year of 1952. The coming of summer struck terror in the minds of parents, camps, swimming pools, and theaters were closed and the diagnosis of a case of paralytic polio in the community was cause for outright panic. In the United States today, those most likely to become infected are members of religious groups who are opposed to immunization, and illegal aliens who are unprotected by a vaccine.

Absence of animal reservoirs plus availability of effective vaccines have made health authorities choose polio as the next disease to eradicate totally from our planet. The Type 2 strain has already been totally eradicated as of the year 2000.

The Disease.   Poliomyelitis is caused by three strains of polioviruses (picornaviruses) that have an affinity for motor neurons of the spinal cord and brain. Although most poliovirus infections are inapparent or mild and nonparalytic, the virus reaches the central nervous system in 1% to 2% of cases. High fever, back pain, and muscle spasms result. In less than 1% of cases, these symptoms are accompanied by partial or complete paralysis of muscles in a relaxed state. The nature and degree of paralysis depend on which neurons in the spinal cord and brain are infected and how severely they are damaged, or if they are lysed. Any paralysis remaining after several months is permanent. The very old and the very young are likely to suffer paralysis as a result of poliovirus infection. Malnutrition, physical exhaustion, corticosteroids, radiation, and pregnancy can increase the severity of the disease.

Poliovirus infections in small children in impoverished areas may go undetected, whereas teenagers and young adults in affluent areas sometimes acquire severe, paralytic poliovirus infections. Good sanitation in the affluent areas reduces exposure and, therefore, natural immunity to the viruses.

Diagnosis, Treatment, and Prevention. Diagnosis of poliomyelitis is made by isolating the virus from pharyngeal swabs or feces, culturing it, and noting its cytopathic effects. Methods also are available for identifying antibodies to the virus in serum. Treatment alleviates symptoms, but patients with paralyzed breathing muscles must forever live in an "iron lung" **(Figure 24.12)**.

(a)

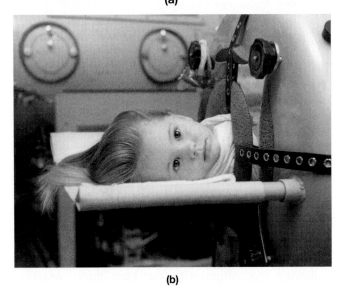

(b)

Figure 24.12 Polio. During polio epidemics (before 1955) in the United Sates, **(a)** row after row of iron lungs was filled with patients *(Courtesy March of Dimes Birth Defects Foundation)* **(b)** such as this 2-year-old girl. Patients sometimes remained in them for years, or until their death. *(Courtesy March of Dimes Birth Defects Foundation)*

Before vaccine became available in 1955, only nonspecific public health measures were available to prevent the spread of poliomyelitis. Schools, swimming pools, and other places where crowds, especially crowds of children, gathered were closed. Large quantities of insecticides were sprayed in the mistaken belief that insect bites somehow played a role in transmission of the disease. Transmission is now known to occur both by the fecal-oral route and from pharyngeal secretions, thus explaining the dangers of fecally contaminated swimming pools in summer. During the first few years that vaccine was available, it could not be made in sufficient quantities to immunize the whole population. Clinics were set up to immunize pregnant women and young children.

Vaccines. In 1955 the injectable Salk polio vaccine became available. It contained viruses that were inactivated by formalin at neutral pH but that still retained their antigenic properties. Unfortunately, before the technique was perfected, some batches of vaccine still contained infectious viruses. More than 200 cases of polio and 10 deaths resulted. In 1963 the oral Sabin vaccine, which contained attenuated live viruses, was introduced. In addition to ease of administration on a sugar cube or in a sugary liquid, this vaccine has the advantages of longer-lasting immunity and prevention of fecal-oral transmission by eliminating viruses in the gastrointestinal tract, where they multiply. Vaccine use has reduced polio incidence in the United States from about 29,000 cases in 1955 to 20 cases in 1969 in unimmunized and immunosuppressed individuals **(Figure 24.13a)**. In October 1995 CDC recommended a combination of polio vaccines to reduce the incidence of vaccine-related polio. (See the box on the next page.) Infants would receive shots of the inactivated vaccine at 2 and 4 months, then oral doses of the attenuated vaccine at 6 months and up to 18 months. The oral vaccine is needed because it is effective against "wild" polio virus that, although eradicated from the United States, could be brought in from other areas by travelers **(Figure 24.13b)**.

Postpolio syndrome is a condition in which people who survived polio, years before, suffer weakening or paralysis of muscles, which requires them again to use crutches and braces. It is neither infectious nor a recurrence of the disease. It is believed to be due to overuse of compensating muscles that have labored too hard for too many years and now cannot function properly.

PRION DISEASES OF THE NERVOUS SYSTEM

Since the 1920s, when **Creutzfeldt-Jakob disease (CJD)** was first investigated, several degenerative diseases of the nervous system have been identified. Although the existence of **prions** was not universally accepted, many researchers believe that such diseases are associated with prions (◄Chapter 10, p. 300). Stanley Prusiner received the 1997 Nobel Prize for his work on prions and disease.

APPLICATIONS

The Polio Vaccine Controversy

In the United States, the attenuated-virus vaccine developed by Albert Sabin is almost universally used. Some other countries, however, have returned to the inactivated-virus vaccine developed by Jonas Salk. Which is preferable? The oral attenuated version is no longer approved for use in the United States, but is still used elsewhere.

Both vaccines have advantages and disadvantages. The inactivated-virus vaccine can be incorporated with other pediatric vaccines, and at the appropriate dose it can be given to immunodeficient patients. However, it does not produce immunity in all recipients, and recipients may require booster shots. Moreover, virulent viruses are used to make the vaccine; a tragedy could occur if they are not completely inactivated.

The attenuated-virus vaccine induces immunity similar to that produced by a natural infection in that the vaccine causes antibodies to be produced in the gut (secretory IgA) as well as in the bloodstream. This reduces the possibility that people who are themselves immune can serve as reservoirs of infection by carrying the virus in their intestines and passing it on to others. Immunity develops quickly and may be lifelong. Oral administration requires less skill and is more acceptable than injection to many people. Finally, the oral vaccine remains potent without refrigeration for a longer time than does the injectable vaccine.

Unfortunately, the live virus used in the oral vaccine does occasionally mutate, and some mutants are virulent. The viruses multiply in gut tissue of the vaccinated individual and are shed in their feces. Thus, changing the diaper of a recently vaccinated child can pose a hazard, especially to family members or day-care workers and immunodeficient patients, depending on the degree of hygiene practiced. About six cases of polio are caused by the vaccine each year in the United States, and about that many are caused by close contact with these recently immunized children. Since 1980, a total of 143 out of 145 cases of paralytic polio have been associated with oral polio vaccine. The other two cases were classified as indeterminate. Liability costs are now included in the price of the oral vaccine to cover awards that may have to be made to victims of vaccine-induced polio. In the United States, these costs have made the oral vaccine more expensive than the injected vaccine, although the oral vaccine is cheaper to make and administer. In warm countries with endemic poliomyelitis and other viral infections, repeated administration has often failed to induce immunity, probably because immune responses being mounted against other viruses prevent an adequate immune response to the vaccine organisms. Finally, the attenuated-virus vaccine, like all vaccines that use live organisms, cannot be administered to immunosuppressed or immunocompromised persons.

Polio has been virtually eliminated in Israel through the use of a combination of inactivated-virus and attenuated-virus vaccines. This procedure shows promise for controlling polio in developing countries with warm climates, fecally contaminated water, and endemic polio. However, the cost of obtaining and administering both vaccines will delay implementation.

Some western U.S. herds of elk and deer have prion infections. The first hunters' cases of CJD in young men who butchered their kill have already resulted in several deaths. We have our own prion problem in this country.

These diseases are referred to collectively as **transmissible spongiform encephalopathies** because in damaging neurons they give brain tissue a "spongy" appearance **(Figure 24.14)**. They include *kuru*, CJD, and a special form of CJD called *Gerstmann-Strassler disease* in humans; *scrapie* in sheep and goats; *transmissible mink encephalopathy; chronic wasting disease* of elk and mule deer; and *"mad cow disease,"* or *bovine spongiform encephalopathy*, in British dairy cattle. In addition, in 1991, 29 cases of transmissible spongiform encephalopathies were reported in British cats, and 2 were reported in ostriches in the Berlin Zoo.

A prominent feature of all prion-associated diseases is the lack of any inflammatory response, which is a hallmark of other infectious diseases. There is, however, an increase in the size of *astrocytes*—the cells that regulate the passage of materials from the blood to neurons—throughout the central nervous system. These cells apparently produce large quantities of a filamentous protein called *amyloid*, which is characteristic of a variety of degenerative diseases of the nervous system, including Alzheimer's disease.

Like other infectious agents, prions are transmissible, although disease symptoms may not appear until years after initial infection. **Kuru**, which occurred mainly in New Guinea, is transmitted through small breaks in the skin. Why this disease appeared mainly in women was puzzling until it was discovered that New Guinean women prepared the bodies of the dead for cannibalistic consumption and smeared their own bodies with the raw flesh of the corpses. Prions in diseased tissues entered the blood of the women or the children playing at their feet, traveled to the brain, and eventually caused kuru. Men and adolescent boys live apart from the women and small children and so were spared infection with kuru.

D. Carleton Gajdusek, a U.S. investigator at the National Institutes of Health (NIH), won a Nobel Prize in 1976 for his work on kuru. He found that 1 to 15 years following inoculation with the kuru prion, onset of symptoms began with headaches, minor loss of coordination, and a tendency to giggle at inappropriate times. Three months later, victims needed crutches to walk or stand; a month after that, victims could not move except for spasms. At that point, the swallowing

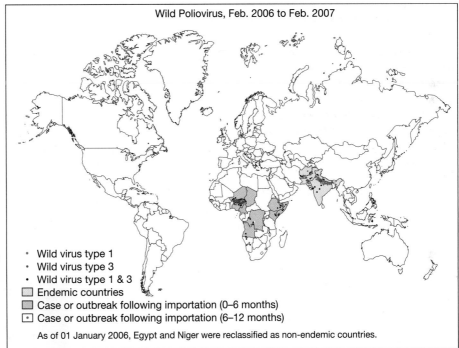

Figure 24.13 Incidence of polio.
(a) U.S. polio. The drastic drop since vaccines became available (first Salk, then Sabin) has not happened in Asia and Africa. (b) Thus, many people in 2007 still suffer from this preventable disease. In fact, more people die each year from vaccine-preventable diseases than die from AIDS. (*Source:* CDC.)

muscles no longer functioned, and malnutrition became a serious problem. Relatives would prechew food and massage it down the esophagus of victims (**Figure 24.15**), but within 1 year the patients died.

Some cases of kuru have been traced to inadvertent inoculation with prions present in corneal transplant tissue from patients having undiagnosed CJD, to dwarfed children who received human growth hormone injections made from cadaver pituitaries, and to silver electrodes implanted in the brain during surgical procedures following use in a CJD patient. The same electrodes, located 17 months later after numerous supposed sterilizations, were implanted into a chimpanzee's brain, where they caused CJD. These findings have caused some newer methods of instrument sterilization to be abandoned. Autoclaving at 121°C at 15 psi of pressure for 1 hour destroys infectivity, something that years of storage in formaldehyde fails to do.

In one experiment, Gajdusek and Paul Brown (also at NIH) buried scrapie-infected hamster brains in soil and left them undisturbed for 3 years. When the brain material was dug up, it was still infectious. Gajdusek and Brown suspect that infected material may retain its deadliness for more than a decade under such circumstances. Thus, sheep that graze in fields where contaminated carcasses have been placed have become infected with **scrapie (Figure 24.16)**. Marking of burial sites and reexamination of the accepted technique of adding corrosive quicklime to the bodies need to be done before eradication can be achieved.

(a)

(b)

Figure 24.14 **A prion-caused disease, Creutzfeldt-Jakob disease.** **(a)** A section through the cerebral cortex of a normal human brain (370X) reveals a solid structure *(Courtesy Frederick C. Skvara, M.D.)* whereas **(b)** a section through the brain of a patient with Creutzfeldt-Jakob disease (370X) shows many holes. It is clear why CJD is referred to as a subacute *spongiform* encephalopathy. *(Courtesy Frederick C. Skvara, M.D.)*

Figure 24.15 **Kuru.** Kuru victims who have reached the point where they must be fed prechewed food. Since cannibalistic rites have been stopped in New Guinea, the disease has disappeared there. *(Courtesy D. Carleton Gajdusek, M.D.)*

Figure 24.16 **Scrapie.** Sheep infected with scrapie will rub or scrape—sometimes until they are bloodied—against fences, poles, or trees. This fatal disease has no cure. *(Courtesy United States Department of Agriculture)*

In most cases of CJD, no source of prions has been identified. In the most thoroughly studied instance of Gerstmann-Strassler syndrome, CJD developed in every generation of a family line for over 100 years. In this form of CJD, a genetic mechanism seems to facilitate prion infection, but how this mechanism works is not clear. Perhaps the presence of a particular gene renders patients susceptible to infection by prions from outside the body, or perhaps the gene activates synthesis of prions within the body.

Mad cow disease (Figure 24.17) reached a peak rate in the early 1990s, with 400–500 deaths of British cattle per week, but it has now diminished, following

Figure 24.17 **Mad cow disease.** Carcasses of infected cattle are burned rather than buried. *(Nigel Dickinson/Peter Arnold, Inc.)*

slaughter of infected herds. It is thought to have existed in Britain since at least the late 1960s, and the number of cases began to mushroom in early 1987. At that time the method of rendering (boiling down) animal remains for livestock feed was changed so as to omit a solvent extraction step and to increase the number of sheep heads used (including the brains). Feeding of British ground animal meal to any kind of livestock has been banned in most European countries since the late 1990s. However, greed and ignorance led to widespread violation of these laws. In the year 2000 many countries in Europe were rocked by sudden rises in human and animal cases. There had been considerable debate as to whether prions can cross species lines in nature. But prions have clearly been transmitted from one species to another in laboratory trials. The British government has prohibited the addition of beef brains to hamburger, formerly a common practice. The British have also changed their butchering methods so that knife or saw blades do not pass through the spinal cord, thus preventing blades from contaminating edible parts. Many countries, including the United States, have banned the import of British beef, cattle, and beef products. But, we have been importing cattle from Canada where mad cow disease was discovered in May 2003. In December of 2003, a cow who had been born on a dairy farm in Alberta, Canada, was slaughtered in the State of Washington, and found to be positive for mad cow disease. This has brought about a cry for better safety measures in the United States. However, May 2004 found millions of tons of Canadian beef still being imported into the United States, in violation of the new laws which had been passed. A second case of mad cow disease has since been found in Texas, in a cow that was born in Texas. Spontaneous mutation does produce a few cases (◄Figure 10.25, p. 302)

A long ignored transmissible spongiform encephalopathy (TSE) called **chronic wasting disease** of elk and deer has been an American problem since its discovery in 1977 in wild deer in Colorado. It spreads more easily than BSE, probably through saliva, urine, feces or hair. The rapidity of spread through other states is alarming **(Figure 24.18)**. Ranch-raised elk from infected herds have inadvertently been shipped around the country, thereby spreading the disease to distant states. Since deer and elk often mingle with grazing cattle, we need to know if they can spread their prions to cattle. Also, hunters need to know if it is safe to butcher and eat deer or elk that they have shot. There is also the question of whether carnivores such as wolves and mountain lions who kill and eat deer or elk (especially the brain and central nervous system) could contract prion diseases. We may be sitting on a time bomb of our own making for having so long ignored these questions.

The possibility that slow-onset prions play a role in other neurodegenerative diseases such as Alzheimer's and Parkinson's diseases is being investigated.

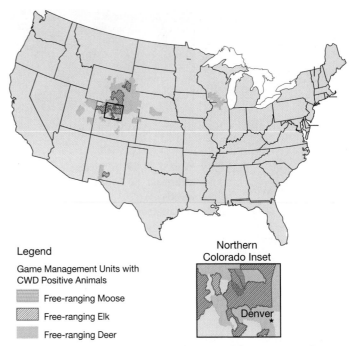

Figure 24.18 Chronic wasting disease of free-ranging deer and elk by county, U.S.

Alzheimer's disease, first described by Alois Alzheimer in 1907, now affects more than 2 million people in the United States—more than 5% of the population over age 65. Amyloid proteins arranged in structures called *neurofibrillary tangles*, or *plaques*, have been found at autopsy in the brains of Alzheimer's patients, but the holes found in kuru and other spongiform encephalopathies are missing. Whether prions play a role in the deposition of the proteins is not yet known. So far researchers have not been able to transmit Alzheimer's disease to laboratory animals, as has been done with kuru, CJD, and scrapie. Whether this is due to non-susceptibility of the animals or the absence of an infectious agent also is not yet known. However, bits of β-amyloid protein injected into rat brains have recently produced a disease similar to Alzheimer's in these rats. Genetic factors may be involved, at least in Alzheimer's disease and CJD, because multiple cases of both diseases have been seen in some families. Much remains to be learned about prions, neurodegenerative diseases, and the relationship between them.

PARASITIC DISEASES OF THE NERVOUS SYSTEM

African Sleeping Sickness

African sleeping sickness, or **trypanosomiasis** (tri-pan″o-so-mi′a-sis), is a disease of equatorial Africa caused by protozoan blood parasites of the genus *Trypanosoma*. Although 100 or more species of this parasite infect various vertebrates and invertebrates, two species, *T. brucei gambiense* and *T. brucei rhodesiense*, cause disease in

(b)

Figure 24.19 African sleeping sickness. **(a)** *Trypanosoma brucei gambiense* (1707X), seen in a blood sample, causes African sleeping sickness. *(A. M. Siegelman/Visuals Unlimited)* **(b)** Trypanosomes are spread by the bite of the tsetse fly. *(©ABPL ImaGE LIBRARY/Animals Animals/Earth Scenes)*

(a)

Both male and female tsetse flies bite and transmit disease. They bite only during daylight hours.

humans. Typically, trypanosomes have an undulating membrane and a flagellum **(Figure 24.19a)**, but in some stages they are shorter and lack a flagellum. For humans to get African sleeping sickness, they must be bitten by an infected tsetse (tset′se) fly **(Figure 24.19b)**. When a tsetse fly bites, it injects infectious trypanosomes, sometimes hundreds in a single bite, into the blood of its victim. The flies serve as vector and as hosts for part of the life cycle of trypanosomes. Although transmission usually is from one human to another via the fly, game animals serve as natural reservoirs for *T. brucei rhodesiense*.

The Disease. African sleeping sickness is a progressive disease characterized according to the tissue in which the parasites congregate during stages that affect first the blood, then the lymph nodes, and finally the central nervous system. Although the parasites do not actually invade cells, they can damage every tissue and organ in the body.

The bites of tsetse flies cause a local inflammatory reaction. After an incubation period of 2 to 23 days, fever appears initially for about a week while parasites are in the blood and at irregular intervals as the parasites are released from lymph nodes. Patients are able to work through the first and second stages, but they suffer from various symptoms—shortness of breath, cardiac pain, disturbed vision, anemia, and weakness—that become increasingly severe. Invasion of the nervous system by the parasites causes headache, apathy, tremors, and an uncoordinated, shuffling gait. Pain and stiffness in the neck and paralysis occur as the disease progresses. Eventually, the patient cannot be roused to eat, becomes emaciated, has convulsions, sleeps continuously, goes into a profound coma, and dies.

Infection with *T. brucei gambiense* produces a slowly progressive, chronic disease that, if untreated, lasts several years before central nervous system symptoms intensify and death occurs. Infection with *T. brucei rhodesiense* produces a more rapidly progressive disease; it is often fatal within a few months, before central nervous system damage is apparent.

Diagnosis, Treatment, and Prevention. Diagnosis of African sleeping sickness is made by finding the trypanosome parasite in the blood. Until recently,

APPLICATIONS

Nagana and HDL

In up to 25% of Africa, it is impossible to raise livestock because of the tsetse-fly-borne disease *nagana*. The causative agent, the flagellated parasite *Trypanosoma brucei brucei*, probably originated in Africa. It lives in the blood of large native grazing animals without harming them, presumably because a long association has allowed host and parasite to adapt to each other. But when humans migrated into the areas inhabited by tsetse flies, the domesticated livestock the humans brought with them were not able to survive infection with *T. brucei brucei*. When bitten by tsetse flies, humans are also infected by the parasite but do not become ill; the parasites rapidly disintegrate in human blood.

Which component of human blood protects against *T. brucei brucei*? The surprising answer involves a protein never before recognized as providing protection against infectious disease. The high-density lipoprotein (HDL) fraction of blood serum, known for its role in removing cholesterol from the

blood, contains a small subfraction that binds to hemoglobin released from dying red blood cells. This binding allows the iron in hemoglobin to be recycled within the body. When a tsetse fly bite releases *T. brucei brucei* into the human bloodstream, the protein-hemoglobin complex is taken in by the trypanosome and eventually enters the parasite's lysosomes. The acidic pH inside the lysosomes causes the protein-hemoglobin to undergo enzymatic changes that result in the release of free radicals. These extremely reactive molecules destroy the membranes of the lysosomes, whose contents—lysozyme enzymes—are expelled into the cytoplasm. There the enzymes digest the parasite from within.

Only humans, some apes, and Old World monkeys have the gene for this *trypanosome lytic factor* (TLF). However, it is hoped that genetic engineering techniques can be used to transfer the gene into livestock, thereby producing trypanosome-resistant breeds of cattle, for example.

arsenic-containing drugs were used to treat the disease, but they cause eye damage, and the parasites quickly become tolerant to them. Pentamidine, suramin, and melarsoprol are now used, usually in that order. If pentamidine, the least toxic drug, fails to combat the infection, more toxic ones are tried. Melarsoprol has the advantage of penetrating the blood-brain barrier, and can work wonders even late in the disease, if you survive its very high toxicity. Results of treatment with any drugs are generally more successful if treatment is begun before central nervous system involvement occurs. A combination of Berenil and nitroimidazole is being used to treat the disease after it has progressed to the central nervous system.

Preventing human infection is nearly impossible because of the wide range of tsetse flies (4.5 million square miles) and possibly because of the reservoir of infection in large game animals. Some control has been achieved by clearing brush where flies congregate and by applying aerial pesticides. Another significant means of reducing the tsetse fly population is to release irradiated male flies, which fail to produce viable sperm. The eggs of females that mate with them fail to develop.

Trypanosomes that cause African sleeping sickness have a special means of evading the host's defenses. The intermittent fever of the disease is directly correlated with increasing numbers of parasites in the blood. The unusual thing about these parasites is that each time parasites appear in the blood, they have a glycoprotein coat different from that of previously released parasites. By the time the immune system has developed antibodies to a trypanosome surface antigen, the trypanosome has a different surface antigen. This ability to alter antigens has thwarted efforts to produce a vaccine for African sleeping sickness. When trypanosomes first enter a human, they seem to be able to make only about 15 antigens; later they can make 100 or more. Researchers hope to produce a vaccine that will confer immunity against any antigen the trypanosome might have when it enters the human body. Such a vaccine would allow the body to attack the trypanosome before an infection becomes established.

Should a vaccine be developed, a means to mount a costly, massive vaccination program would be needed. Because available vaccines, such as those for measles and mumps, have not been administered to many of the children in the region endemic to African sleeping sickness, the prospect for mass immunizaton is not bright. More than 20,000 people per year are likely to continue to die of the disease.

Chagas' Disease

Chagas' (Cha′gahs) **disease**, named for the Brazilian physician Carlos Chagas, who first described it in 1909, is caused by *Trypanosoma cruzi*. The disease occurs sporadically in the southern United States and is endemic in Mexico, Central America, and much of South America.

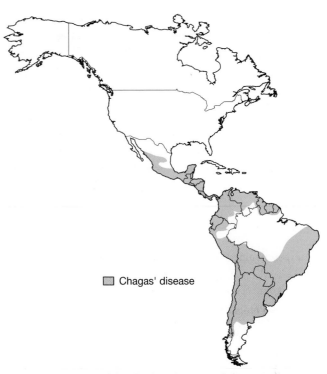

Figure 24.20   Worldwide distribution of Chagas' disease, 1998. (*Source:* CDC.)

☐ Chagas' disease

It affects more than 18 million people **(Figure 24.20)**. *T. cruzi* looks like the trypanosomes that cause African sleeping sickness. It is transmitted by several kinds of reduviid bugs that are hosts for the sexual phase of the trypanosome's life cycle (◄Chapter 11, p. 335). Each species of bug occupies a particular region, so the bugs that transmit Chagas' disease in Mexico are different from those that transmit it in South America. Reduviid bugs often bite near the eyes; they defecate as they bite, depositing infectious parasites on the skin. Humans

APPLICATIONS

Assassin Bugs

A few cases of Chagas' disease usually occur in the United States each year, along the Texas border. Fortunately the infective reduviid bugs (often called "kissing bugs" or "assassin bugs") don't live much farther north than that. However, there are other species of reduviids living in most of the United States. Could they also spread Chagas' disease? The trypanosomes live well in these other bugs. But, our northern bugs don't necessarily defecate when they bite. The two activities are linked in Mexican and South American reduviids. A northern bug could defecate at or near a bite site, just by accident. So there is a small likelihood of infection. A greater worry is global warming, which could extend the range of southern species well into the United States. And just think, there is no known cure for Chagas' disease.

TABLE 24.2

A Summary of Diseases of the Nervous System		
Disease	**Agent(s)**	**Characteristics**
Bacterial Diseases of the Brain and Meninges		
Bacterial meningitis	See Table 24.1	Tissue necrosis, brain edema, headache, fever, occasionally seizures
Listeriosis	*Listeria monocytogenes*	A kind of meningitis seen in fetuses and immunodeficient patients
Brain abscesses	Various anaerobes	Infection that grows in mass and compresses brain
Viral Diseases of the Brain and Meninges		
Rabies	Rabies virus	Invades nerves and brain; headache, fever, nausea, partial paralysis, coma, and death ensue unless patient has immunity
Encephalitis	Several encephalitis viruses	Shrinkage and lysis of neurons of the central nervous system; headache, fever, and sometimes brain necrosis and convulsions
Herpes meningoencephalitis	Herpesvirus	Fever, headache, meningeal irritation, convulsions, altered reflexes
Progressive multifocal leukoencephalopathy	Polyomavirus, JC virus	Infects oligodendrocytes in areas of brain lacking myelin; mental deterioration, limb paralysis, and blindness
Bacterial Nerve Diseases		
Hansen's disease	*Mycobacterium leprae*	Range of symptoms from loss of skin pigment and sensation to lepromas and erosion of skin and bone
Tetanus	*Clostridium tetani*	Toxin-mediated disease; muscle stiffness, spasms, paralysis of respiratory muscles, heart damage, and usually death
Botulism	*Clostridium botulinum*	Preformed toxin from food prevents release of acetylcholine; paralysis and death result unless treated promptly; in infants and wounds, endospores germinate and produce toxin
Viral Nerve Diseases		
Poliomyelitis	Several types of polioviruses	Fever, back pain, muscle spasms, partial or complete flaccid paralysis from destruction of motor neurons
Prion Diseases of the Nervous System		
Transmissible spongiform encephalopathies	Prions	Death of brain cells leaves holes, creating spongiform brain tissue; amyloid plaques form; long delay before symptoms appear; then spasms rapidly worsening to collapse; no cure
Parasitic Diseases of the Nervous System		
African sleeping sickness	*Trypanosoma brucei gambiense, T. brucei rhodesiense*	Fever, weakness, anemia, tremors, shuffling gait, apathy; as parasites invade nervous system, emaciation, convulsions, and coma ensue
Chagas' disease	*Trypanosoma cruzi*	Subcutaneous inflammation, damage to lymphatic tissues, muscle, and nerve ganglia; muscle pain and paralysis of intestinal, heart, and skeletal muscle

almost automatically rub such a bug bite and thereby transfer parasites to the eyes or the bite wound.

The Disease. Chagas' disease begins with subcutaneous inflammation around the bug bite. After 1 to 2 weeks, the parasites have made their way to lymph nodes, where they repeatedly divide and form aggregates called **pseudocysts**. Wherever the pseudocysts rupture, they cause inflammation and tissue necrosis. These parasites enter cells either by invasion or by phagocytosis and can damage lymphatic tissues, all kinds of muscle, and especially supporting tissues around nerve ganglia. Destruction of nerve ganglia in the heart

accounts for nearly three-fourths of deaths from heart disease among young adults in endemic areas.

Chagas' disease appears in an acute form and a chronic form. The acute disease, which is most common in children under 2, is characterized by severe anemia, muscle pain, and nervous disorders. In especially virulent acute disease, death can occur in 3 to 4 weeks, but many patients recover after several months of less virulent disease. The chronic disease, which is seen mainly in adults, probably arises from a childhood infection. It is a mild disease and is sometimes asymptomatic but often causes enlargement of various organs. Insidious damage to nerves can cause several severe effects. In the digestive tract, it slows or stops muscle contractions due to death of up to 85% of the neurons in the esophagus and 50% of those in the colon; in the heart, it can cause irregular heartbeat and accumulation of fluid around the heart; in the central nervous system, it can cause paralysis by destroying motor centers. *T. cruzi* also crosses the placenta, so chronically infected mothers often give birth to infants with severe acute disease. In some parts of the world, untested blood used for transfusion also spreads the disease.

Diagnosis, Treatment, and Prevention.   A new PCR technique allows rapid and easy diagnosis of Chagas' disease. The older method, still in use in some places, involved feeding patients' blood to animals, and was a long and difficult procedure. The parasites can be found in the blood during fever in acute cases. Small animals such as guinea pigs and mice can be inoculated with blood from patients and observed for disease symptoms. This technique is known as *xenodiagnosis. Xenos* is Greek for "strange" or "foreign"; in this context it refers to the use of an organism different from a human. Chronic Chagas' disease sometimes can be detected by allowing patients to be bitten by uninfected laboratory-reared reduviid bugs and examining the bugs in 2 to 4 weeks for trypanosomes, which develop in the bug's intestine.

Different agents in the body usually activate T cells of the immune system; in the presence of Trypanosoma cruzi, T cells are not activated by any of these agents.

Despite several ways to diagnose the disease, no effective treatment is available. Drugs used to treat other trypanosome infections are of no use because they fail to reach the parasites inside cells. Work is underway to develop new drugs and a vaccine, but until these are available, control of the reduviid vectors is the only means of reducing misery from this disease. Treating homes with insecticides offers some protection, but the bugs crawl into crevices in walls and thatched roofs and are difficult to eradicate.

The agents and characteristics of the diseases discussed in this chapter are summarized in **Table 24.2.**

✓CHECKLIST

1. Compare the live (oral) versus killed polio vaccines as to advantages and disadvantages.
2. How is African sleeping sickness transmitted?
3. What damage occurs in Chagas' disease?

▌ RETRACING OUR STEPS

• The agents and characteristics of the diseases discussed in this chapter are summarized in Table 24.2. Information in that table is not repeated in this summary.

▌ COMPONENTS OF THE NERVOUS SYSTEM

• The **central nervous system (CNS)** and the **peripheral nervous system (PNS)** are the two main components. The brain and spinal cord form the CNS; **nerves** reaching all parts of the body compose the PNS. Aggregations of nerve cell bodies in the PNS are called **ganglia. Meninges** cover the brain and spinal cord.
• The CNS is protected by special thick-walled capillaries without pores in their walls, which form the **blood-brain barrier**.
• The nervous system has no normal microflora.

▌ DISEASES OF THE BRAIN AND MENINGES

BACTERIAL DISEASES OF THE BRAIN AND MENINGES
• **Bacterial meningitis**, acquired from carriers or endogenous organisms, can usually be treated with penicillin. A vaccine is available to protect children from *Haemophilus* meningitis.
• **Listeriosis** can be transmitted by improperly processed dairy products and can cross the placenta. It is a great threat to immunodeficient patients.
• **Brain abscesses** arise from wounds or as secondary infections. Antibiotics are effective if given early in an infection, and surgery can be used later unless the abscess is in a vital area of the brain.

VIRAL DISEASES OF THE BRAIN AND MENINGES
• **Viral meningitis** is usually self-limiting and nonfatal.

• **Rabies** has a worldwide distribution except for a few rabies-free countries. It is difficult to control because of the large number of small mammals that serve as reservoirs.
• Rabies is diagnosed by IFAT and treated by thoroughly cleaning bite wounds and injecting hyperimmune rabies serum and giving vaccine. It can be prevented by immunizing pets and people at risk and by avoidance of contact with wild animals by those who are not immunized.
• **Encephalitis** is transmitted by mosquitoes, often from horses, and can sometimes be diagnosed by culturing blood or spinal fluid in cells or mice. Vaccine is available for horses.
• **Herpes meningoencephalitis** often follows a generalized herpes infection.

▌ OTHER DISEASES OF THE NERVOUS SYSTEM

BACTERIAL NERVE DISEASES
• **Hansen's disease (leprosy)** affects millions of people worldwide. Many patients are asymptomatic for years after infection, and no diagnostic test is available to identify such patients. Dapsone, clofazimine, and rifampin are used to treat Hansen's disease, but no vaccine is available.
• **Tetanus** typically follows introduction of spores into deep wounds. It is treated with antitoxin and antibodies. If all people received the available vaccine, tetanus could be prevented.
• **Botulism** is acquired by ingesting foods that contain a potent preformed neurotoxin and is treated with polyvalent

antitoxin. In **infant botulism** and **wound botulism**, spores germinate and produce toxin.

VIRAL NERVE DISEASES

• Prior to the development of vaccines in the 1950s, **poliomyelitis** was a common and dreaded disease. It can be diagnosed from culture and by immunological methods. Treatment merely alleviates symptoms and includes the use of an iron lung to maintain breathing when respiratory muscles are paralyzed.

• Both an injectable and an oral polio vaccine are available; each has advantages and disadvantages. Worldwide use of vaccine could eradicate poliomyelitis.

PRION DISEASES OF THE NERVOUS SYSTEM

• Following a long delay after inoculation, brain cells die, causing a spongiform appearance of the tissue. Amyloid plaques accumulate.

• Symptoms begin with spasms, which rapidly worsen to collapse, followed by death.

• Prions cause **transmissible spongiform encephalopathy** in humans and many animals. Prions can be transmitted across species lines.

• **Kuru** and **Creutzfeldt-Jakob disease (CJD)** affect humans, **scrapies** attacks sheep, and **mad cow disease** occurs in cattle.

PARASITIC DISEASES OF THE NERVOUS SYSTEM

• **African sleeping sickness** occurs in equatorial Africa and is transmitted by the tsetse fly. It is diagnosed by finding parasites in the blood and can be treated with pentamidine and other drugs.

• **Chagas' disease** occurs from the southern United States to all but the southernmost part of South America and is transmitted by various species of bugs. It is diagnosed by finding parasites in the blood and by xenodiagnosis. No effective treatment is available.

▌ TERMINOLOGY CHECK

African sleeping sickness
  (p. 773)
bacterial meningitis *(p. 755)*
blood-brain barrier *(p. 755)*
botulism *(p. 767)*
brain abscess *(p. 757)*
central nervous system (CNS)
  (p. 755)
Chagas' disease *(p. 775)*
chronic wasting disease
  (p. 773)
Creutzfeldt-Jakob disease
  (CJD) *(p. 769)*
eastern equine encephalitis
  (EEE) *(p. 761)*

encephalitis *(p. 761)*
ganglion *(p. 755)*
Hansen's disease *(p. 763)*
herpes meningoencephalitis
  (p. 762)
infant botulism *(p. 768)*
kuru *(p. 770)*
lepromatous *(p. 763)*
lepromin skin test *(p. 763)*
leprosy *(p. 763)*
listeriosis *(p. 757)*
mad cow disease *(p. 772)*
meninges *(p. 755)*
nerve *(p. 755)*
nervous system *(p. 755)*

peripheral nervous system
  (PNS) *(p. 755)*
poliomyelitis *(p. 768)*
prion *(p. 770)*
progressive multifocal
  leukoencephalopathy
  (p. 762)
pseudocyst *(p. 776)*
rabies *(p. 758)*
rabies virus *(p. 758)*
St. Louis encephalitis
  (SLE) *(p. 761)*
scrapie *(p. 771)*
tetanus *(p. 765)*

tetanus neonatorum
  (p. 767)
transmissible spongiform
  encephalopathies
  (p. 770)
trypanosomiasis *(p. 773)*
tuberculoid *(p. 763)*
Venezuelan equine
  encephalitis (VEE)
  (p. 761)
viral meningitis *(p. 757)*
western equine encephalitis
  (WEE) *(p. 761)*
West Nile fever *(p. 761)*
wound botulism *(p. 768)*

▌ CLINICAL CASE STUDY

Harry was driving to the university when he suddenly saw a dog writhing on the roadside, having just been hit by a car. Harry stopped, wrapped the dog in a blanket, and started driving to a veterinarian's office. While wrapping the dog, Harry was covered in its blood, and was bitten and scratched several times. The dog died en route to the animal hospital, so Harry disposed of it in a convenient dumpster. Some weeks later, Harry's roommate noticed strange behavior in Harry. He became reclusive, insisted on closing drapes over the window,

refused to have lights on in the room, stopped eating, was very thirsty but refused to drink water and other liquids, laid in bed, and made strange noises. Finally his roommate asked Student Health to come to the dorm and examine Harry. They did, and had to take him away forcibly to the hospital, where he died a few days later. What do you think the diagnosis was? How was the infection acquired? Think carefully before you stop to help an injured animal—this case is based on a true story.

▌ CRITICAL THINKING QUESTIONS

1. What is it about the brain that makes it hard to treat for bacterial meningitis? Why do you think lipid-soluble antibiotics like chloramphenicol and tetracycline work better in combating bacterial meningitis?

2. Rabies is among the few diseases where people can be immunized after the pathogen has already entered their body. What is different about rabies that makes this possible?

3. Long ago, people with Hansen's disease were involuntarily confined to facilities called leprosariums to prevent transmission to others. Once effective treatments became available, this approach was viewed as counterproductive in the efforts to eradicate Hansen's disease. Why might that be the case?

▮ SELF-QUIZ

1. Place the following signs and sypmtoms associated with bacterial meningitis in the appropriate category:

___ Earlier (a) Shock
___ Later   (b) Fever
              (c) Necrosis
              (d) Decreased cerebrospinal fluid flow
              (e) Clogging of blood vessels
              (f) Headache
              (g) Increased pressure within skull
              (h) Death
              (i) Impaired central nervous system function
              (j) Chills

2. Which of the following is *not* true of the nervous system?
(a) Consists of central and peripheral systems
(b) Central nervous system is composed of brain and spinal cord
(c) Ganglia are part of brain
(d) Meninges are membranes that cover brain and spinal cord
(e) Normally free of microbes

3. Which of the following bacterial meningitis diseases currently is the most common type in the U.S., with increased incidences in infants and in the 15- to 24-year-old age group, and has three types or strains of the infectious microbe?
(a) *Haemophilus* meningitis
(b) Listeriosis
(c) *Streptococcus* meningitis
(d) Meningococcal meningitis
(e) a and c

4. Which of the following is a common cause of meningitis in nonimmunized young children?
(a) *Streptococcus pneumoniae*
(b) *Escherichia coli*
(c) *Staphylococcus*
(d) *Haemophilus influenzae*
(e) None of the above

5. A complication of infection with this organism (Waterhouse-Friderichsen syndrome) can occur if the organism becomes widely distributed in the body, leading to endotoxin shock and death. The organism is:
(a) *Haemophilus influenzae* type A
(b) *Neisseria meningitidis*
(c) *Haemophilus influenzae* type B
(d) *Streptococcus pneumoniae*
(e) *Listeria monocytogenes*

6. Before development of a vaccine against this microbe, the disease it caused accounted for two-thirds of bacterial meningitis cases during the first year of life but is still the number one leading cause of mental retardation in patients who survive serious disease due to permanent central nervous system disorders. What is the microorganism?
(a) *Haemophilus influenzae* type B
(b) *Haemophilus influenzae* type A
(c) *Neisseria meningitidis*
(d) *Streptococcus pneumoniae*
(e) *Listeria monocytogenes*

7. Bacterial meningitis caused by _____ _____ is responsible for the leading cause of infection in kidney transplant patients, while _____ _____ is the leading cause among adults, and _____ _____ is the leading cause among one-year-olds in the United States.

8. Which of the following bacteria that may cause meningitis is Gram-positive and therefore does not cause endotoxin shock in infected individuals?
(a) *Escherichia coli*
(b) *Neisseria meningitidis*
(c) *Listeria monocytogenes*
(d) *Haemophilus influenzae* type B
(e) None of these

9. Avid swimmers and hot tub enthusiasts sometimes acquire an opportunistic infection caused by *Naegleria fowleri* that results in a meningoencephalitis. This organism is a (an):
(a) Protist
(b) Helminth
(c) Bacteria
(d) Amoeba
(e) a and d

10. Which of the following in NOT true about listeriosis?
(a) Foodborne transmission is the most common source of infection.
(b) It is a leading cause of bacterial meningitis today.
(c) It is responsible for many cases of fetal damage and death.
(d) It is sometimes acquired as a zoonosis.
(e) Individuals with impaired immune systems are highly susceptible to this disease.

11. Viral meningitis is usually always fatal if untreated, unlike bacterial meningitis, which is always self-limiting and non-fatal. True or false?

12. Which of the following is acid-fast and causes Hansen's disease?
(a) *Mycobacterium tuberculosis*
(b) *Clostridium tetani*
(c) *Neisseria meningitidis*
(d) *Haemophilus influenzae*
(e) *Mycobacterium leprae*

13. An immunofluorescent antibody test (IFAT) has replaced the older test for the presence of inclusions in neurons (Negri bodies) for the detection of infections caused by:
(a) *Neisseria meningitidis*
(b) Enteroviruses

(c) Mumps virus
(d) Rabies virus
(e) *Haemophilus influenzae*

14. The viruses that cause encephalitis are most likely to be:
    (a) Togaviruses
    (b) Enteroviruses
    (c) Mumps virus
    (d) Rabies virus
    (e) Hepatitis viruses

15. What do the following viral encephalitis diseases have in common: eastern equine encephalitis, western equine encephalitis, Venezuelan equine encephalitis, St. Louis encephalitis, West Nile fever?
    (a) They are all caused by the same virus.
    (b) They are have the horse as a common intermediate host.
    (c) They are all transmitted into the body through the bite of an infected mosquito.
    (d) Birds are never involved in the life cycle of the viruses that cause these diseases.
    (e) Both a and d.

16. The only bacterium known to damage peripheral nerves is:
    (a) *Naegleria fowleri*
    (b) *Mycobacterium leprae*
    (c) *Streptococcus pneumoniae*
    (d) *Neisseria meningitidis*
    (e) *Haemophilus influenzae*

17. All of the following are true about *Clostridium tetani*, the causative agent of tetanus, EXCEPT:
    (a) They are spore-forming rods.
    (b) They are part of the normal bowel microflora of horses, cattle, and humans.
    (c) They produce a powerful exotoxin that mediates the disease.
    (d) Their toxin elicits muscle stiffness, spasms, and paralysis in the victim.
    (e) They require oxygen to survive.

18. A vaccine that was developed in 1933 has proven effective in reducing the incidence of disease caused by:
    (a) *Clostridium botulinum*
    (b) *Mycobacterium leprae*
    (c) *Clostridium tetani*
    (d) *Streptococcus pneumoniae*
    (e) *Listeria monocytogenes*

19. The most common form of disease caused by *Clostridium botulinum* is:
    (a) Infant
    (b) Wound
    (c) Lung, following exposure to aerosols
    (d) Foodborne
    (e) None of these

20. The endospores of this organism are the most heat-resistant known; it is:

    (a) *Clostridium tetani*

(b) *Clostridium botulinum*
(c) *Mycobacterium leprae*
(d) *Streptococcus thermicos*
(e) *Listeria monocytogenes*

21. All of the following are true about *Clostridium botulinum*, the causative agent of botulism, EXCEPT:
    (a) Their ability to form toxin depends on infection with a bacteriophage.
    (b) They can produce one of eight different toxins, all of which cause human disease.
    (c) They produce the most potent toxin known.
    (d) Although their endospores are highly heat-resistant, their toxin can be inactivated by only a few minutes' boiling.
    (e) All of the above are true.

22. Match the following descriptions to parasitic diseases of the nervous system: _____ African sleeping sickness _____ Chagas' disease
    (a) *Trypanosoma cruzi*
    (b) *Trypanosoma brucei gambiense*
    (c) *Trypanosoma brucei rhodesiense*
    (d) Transmitted by defecating reduviid bugs
    (e) No treatment or cure
    (f) Treated with pentamidine, suramin, and melarsoprol
    (g) Transmitted by tsetse fly
    (h) Evasion of host defenses by constant altering of surface antigens
    (i) Evasion of host defenses by preventing activation of T cells

23. All of the following are true about poliomyelitis EXCEPT:
    (a) Most poliovirus infections are inapparent or mild and nonparalytic.
    (b) Polio can be caused by three strains of picornaviruses.
    (c) The Salk vaccine contains live, attenuated polioviruses while the Sabin vaccine contains formalin-inactivated polioviruses.
    (d) Transmission is by the fecal-oral route and pharyngeal secretions.
    (e) Poliomyelitis has afflicted man for thousands of years.

24. Which form of poliovirus vaccine is better at eliminating viruses in the gastrointestinal tract?
    (a) Live attenuated vaccine
    (b) Formalin-killed vaccine

25. A major difference between infections with prions and other agents is that infections with prions:
    (a) Do not lead to an inflammatory reponse
    (b) Are not transmissible
    (c) Do not cause an increase in the size of astrocytes
    (d) Are not fatal
    (e) All of these

26. Name a pathogen that can cause disease by infecting or affecting each part of the nervous system. How does the pathogen cause the disease or get to the infection site? Can the disease be prevented? Can the disease be treated?

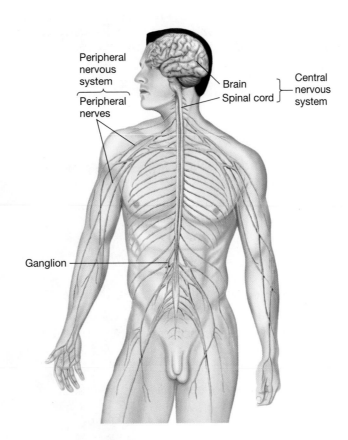

Peripheral nervous system

Peripheral nerves

Brain

Spinal cord

Central nervous system

Ganglion

▌ EXPLORATIONS ON THE WEB   http://www.wiley.com/college/black

If you think you've mastered this chapter, there's more to challenge you on the web. Go to the compansion web site to fine-tune your understanding of the chapter concepts and discover answers to the questions posed below.

1. Don't let Cujo bite you. Did you know that once symptoms of rabies appear there is no cure? Find out more at the web site.

2. The iron lung was not a medieval torture device, but instead a medical apparatus used to treat patients suffering from polio. Learn more at the web site.

3. Can you believe the advent of indoor plumbing and improvements in sanitation gave rise to increasing polio outbreaks? Discover why at the web site.

Come with me . . .

Have you been to Yellowstone National Park in Wyoming? Did you see the geysers, like Old Faithful, erupting, or areas of mud boiling with huge gas bubbles? These are some pretty impressive geothermal phenomena. But, did you know that there are also areas at the bottom of the ocean where heat is being released from the earth's interior in similar fashion? They're not as easy to get to as Yellowstone. You may need to go 8,500 feet below the ocean's surface. And once there, you may find plumes of water as hot as 330°C—that's 626° F! There may be "forests" of 50 or more black smoking chimneys in a vent field, some as tall as a 5-story building. The heat they release is enough to affect ocean circulation patterns. How can anything live down there? Yet these smoking vents are the most productive areas of our planet. Since they were first discovered in 1977, more than 500 new species of organisms, unique to the vents, have been found.

Come online with me to see video clips of vents erupting, diagrams explaining why they erupt, photos of vast numbers of animals and microbial mats, and to see some "white smoking vents."

 Video related to this topic is available within WileyPLUS.

Courtesy OAR/National Undersea Research Program (NURP); NOAA

FUNDAMENTALS OF ECOLOGY

Ecology is the study of the relationships among organisms and their environment. These relationships include interactions of organisms with physical features—the **abiotic factors**—of the environment and interactions of organisms with one another—the **biotic factors**—of the environment. An **ecosystem** comprises all the organisms in a given area together with the surrounding abiotic and biotic factors.

THE NATURE OF ECOSYSTEMS

Ecosystems are organized into various biological levels. The **biosphere** is the region of the Earth inhabited by living organisms. It consists of the *hydrosphere* (Earth's water supply), the *lithosphere* (the soil and rock that include the Earth's crust), and the *atmosphere* (the gaseous envelope surrounding Earth). A wide diversity of organisms exists within the biosphere. A terrestrial ecosystem, such as a desert, tundra, grassland, or tropical rain forest, is characterized by a particular climate, soil type, and organisms. The hydrosphere is divided into freshwater and marine ecosystems.

The organisms within an ecosystem live in communities. An ecological **community** consists of all the kinds of organisms that are present in a given environment. Microorganisms can be categorized as indigenous or nonindigenous to an environment. **Indigenous**, or *native*, **organisms** are always found in a given environment. They generally are able to adapt to normal seasonal changes or changes in the quantity of available nutrients in the environment. For example, *Spirillum volutans* is indigenous to stagnant water, various species of *Streptomycetes* are indigenous to soil, and *Escherichia coli* is indigenous to the human digestive tract. Regardless of variations in the environment (except for cataclysmic changes), an environment will always support the life of its indigenous organisms. **Nonindigenous organisms** are temporary inhabitants of an environment. They become numerous when growth conditions are favorable for them and disappear when conditions become unfavorable.

Communities are made up of *populations*, groups of organisms of the same species. In general, communities composed of many populations of organisms are more stable than those composed of only a few populations—that is, of only a few different species. The various species create a system of "checks and balances" such that the numbers of each species remain relatively constant.

The basic unit of the population is the individual organism. Organisms occupy a particular habitat and niche. The *habitat* is the physical location of the organism. Microorganisms are so small that they often occupy a **microenvironment**, a habitat in which the oxygen, nutrients, and light are stable, including the environment immediately surrounding the microbe. A particle of soil

Figure 25.1 **The flow of energy (E) and nutrients (N) in ecosystems.** Energy flows through the system (it is obtained continuously from the sun), whereas nutrients in the environment must be recycled for new life to continue.

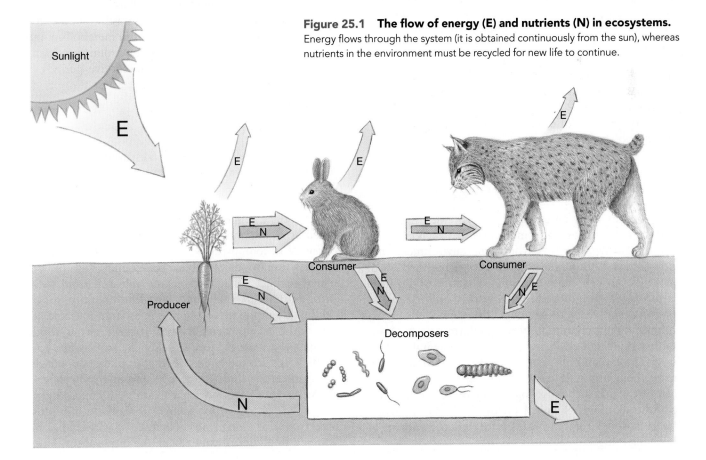

could be the microenvironment of a bacterium. This environment is more important to the bacterium than the more extensive *macroenvironment*. An organism's *niche* is the role it plays in the ecosystem—that is, its use of biotic and abiotic factors in its environment. Microbes may be *producers*, *consumers*, or *decomposers*. We will discuss each of these groups in the next section.

THE FLOW OF ENERGY IN ECOSYSTEMS

Energy is essential to life, and radiant energy from the sun is the ultimate source of energy for nearly all organisms in any ecosystem. (The chemolithotrophic bacteria that extract energy from inorganic compounds are an exception; ◀Chapter 5, p. 138). Organisms called **producers** (autotrophs) capture energy from the sun. They use this energy and various nutrients from soil or water to synthesize the substances they need to grow and to support their other activities. Energy stored in the bodies of producers is transferred through an ecosystem when **consumers** (heterotrophs) obtain nutrients by eating the producers or other consumers. **Decomposers** obtain energy by digesting dead bodies or wastes of producers and consumers. The decomposers release substances that producers can use as nutrients. The flow of energy and nutrients in an ecosystem is summarized in **Figure 25.1**.

Microorganisms can be producers, consumers, or decomposers in ecosystems. Producers include photosynthetic organisms among bacteria, cyanobacteria, protists, and eukaryotic algae. Although green plants are the primary producers on land, microorganisms fill this role in the ocean. Consumers include heterotrophic bacteria, protists, and microscopic fungi. (To the extent that viruses divert a cell's energy to the synthesis of new viruses, they too act as consumers.) Many microorganisms act as decomposers. In fact, they play a greater role in the decomposition of organic substances than larger organisms do.

▌▌▌ BIOGEOCHEMICAL CYCLES

As they carry on essential life processes, living organisms incorporate water molecules and carbon, nitrogen, and other elements from their environment into their bodies. Without decomposers to ensure the flow of nutrients through ecosystems, much matter would soon be incorporated into bodies and wastes, and life would soon become extinct. Although the supply of energy from sunlight is continuously renewed, the supply of water and the chemical elements that serve as nutrients is fixed. These materials must be recycled continuously to make them available to living organisms. The mechanisms by which such recycling occurs are referred to collectively as **biogeochemical cycles**; *bio* refers to living things, and *geo* refers to the Earth, the environment of the living things.

CLOSE UP

Bacteria Make the World Go 'Round

"The bacteria particularly ... are still more important than we. Omnipresent in infinite varieties ... they release the carbon and nitrogen held in the dead bodies of plants and animals which would—without bacteria and yeasts—remain locked up forever in useless combinations, removed forever as further sources of energy and synthesis. Incessantly busy in swamp and field, these minute benefactors release the frozen elements and return them to the common stock, so that they may pass through other cycles as parts of other living bodies. ... Without the bacteria to maintain the continuities of the cycles of carbon and nitrogen between plants and animals, all life would eventually cease. ... Without them, the physical world would become a storehouse of well-preserved specimens of its past flora and fauna ... useless for the nourishment of the bodies of posterity. ..."—Hans Zinsser, 1935

THE WATER CYCLE

The **water cycle**, or **hydrologic cycle (Figure 25.2)**, recycles water. Water reaches the Earth's surface as precipitation from the atmosphere. It enters living organisms during photosynthesis and by ingestion. It leaves them as a by-product of respiration, in wastes, and by evaporation from the surfaces of living things, such as by *transpiration* (water loss from pores in plant leaves). Like all other living things, microorganisms use water in metabolism, but they also live in water or very moist environments. Many form spores or cysts that help them survive periods of drought, but vegetative cells must have water.

THE CARBON CYCLE

In the **carbon cycle (Figure 25.3)**, carbon from atmospheric carbon dioxide (CO_2) enters producers during photosynthesis or chemosynthesis. Consumers obtain carbon compounds by eating producers, other consumers, or the remains of either. Carbon dioxide is returned to the atmosphere by respiration and by the actions of decomposers on the dead bodies and wastes of other organisms. Carbon compounds can be deposited in peat, coal, and oil and released from them during burning. A small but significant quantity of carbon dioxide in the atmosphere comes from volcanic activity and from the weathering of rocks, many of which contain the carbonate ion, CO_3^{2-}. The oceans and carbonate rocks are the largest reservoirs of carbon, but recycling of carbon through these reservoirs is very slow.

As we have seen in earlier chapters, all microorganisms require some carbon source to maintain life. Most carbon entering living things comes from carbon dioxide dissolved in bodies of water or in the atmosphere. Even

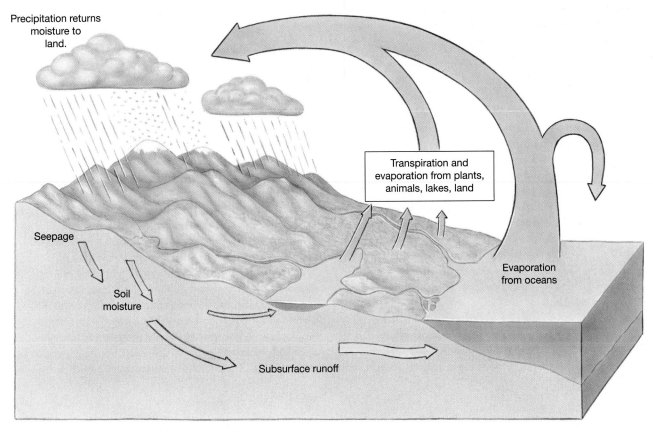

Figure 25.2 **The water cycle.**

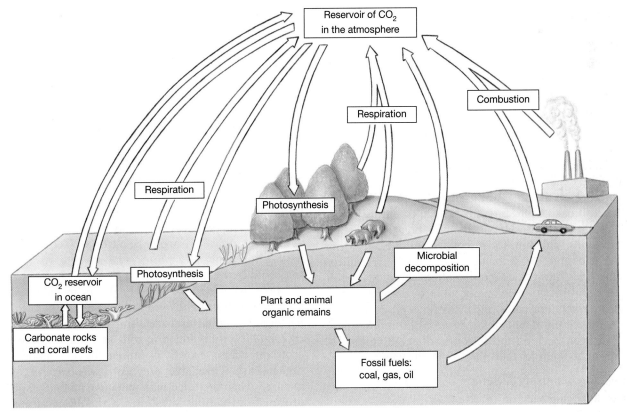

Figure 25.3 **The carbon cycle.**

APPLICATIONS

The Greenhouse Effect

Atmospheric carbon dioxide and water vapor form a blanket over the Earth's surface, creating a "greenhouse effect." These gases allow the sun's radiation to penetrate the atmosphere, thereby reaching the Earth's surface and warming it and the atmosphere. But they trap much of the infrared (heat) radiation produced by the warm surface, reflecting it back to Earth. Solar energy is thus captured within the "greenhouse." The overall effect is to reduce temperature variation and to raise the temperature near the Earth's surface.

Over the past century, human activities have released relatively large quantities of carbon dioxide from coal and oil combustion. In fact, the concentration of CO_2 in the atmosphere has increased by more than 10% since 1958. Yet forests, which absorb carbon dioxide and replenish the Earth's oxygen through photosynthesis, have been cut down and burned at an ever-increasing rate. Some scientists think these processes are causing a general warming trend, increasing average temperatures around the world and possibly changing the balance of organisms in ecosystems. This trend would make now-temperate regions too warm to grow wheat and other food crops, create droughts in other areas, and make new deserts. It might even melt polar ice, raising the level of oceans and flooding coastal cities.

During the first 5 months of 1998, global temperature increased by 0.5°C! Global warming might lead to an increased incidence of infectious diseases as the ranges of the vectors (mosquitoes, snails, flies, and such) carrying the diseases expand. Several scenarios have been developed by computer modeling. According to one model, a global temperature rise of just 3°C over the next century could result in 50 million to 80 million new malaria cases per year. Other diseases, including schistosomiasis, African trypanosomiasis, and dengue and yellow fevers, are also likely to spread with global warming.

Marine photosynthetic plankton (*phytoplankton*) utilize CO_2 and are thought to affect the Earth's temperature. (See the box "And You Thought It Was El Niño," ◀ p. 798). In June 1995, researchers reported that iron added in regular doses to part of the Pacific Ocean promoted extensive phytoplankton growth. The plankton absorbed huge amounts of CO_2 from the air. Scientists are quick to warn, however, against the prospect of seeding the oceans with iron to reduce global CO_2 levels. Iron inputs would alter the food chain and could increase levels of other greenhouse gases.

the carbon in sugars and starches ingested by consumers is derived from carbon dioxide. Because the atmosphere contains only a limited quantity of carbon dioxide (0.03%), recycling is essential for maintaining a continuous supply of atmospheric carbon dioxide.

THE NITROGEN CYCLE AND NITROGEN BACTERIA

In the **nitrogen cycle (Figure 25.4)**, nitrogen moves from the atmosphere through various organisms and back into the atmosphere. This cyclic flow depends not only on decomposers but also on various nitrogen bacteria.

Decomposers use several enzymes to break down proteins in dead organisms and their wastes, releasing nitrogen in much the same way as they release carbon. Proteinases break large protein molecules into smaller molecules. Peptidases break peptide bonds to release amino acids. Deaminases remove amino groups from amino acids and release ammonia. Eventually, free nitrogen gas finds it way back into the atmosphere. Many soil microorganisms produce one or more of these enzymes. Clostridia, actinomycetes, and many fungi produce extracellular proteinases that initiate protein decomposition.

Nitrogen bacteria fall into one of three categories according to the roles they play in the nitrogen cycle:

- Nitrogen-fixing bacteria
- Nitrifying bacteria
- Denitrifying bacteria

Nitrogen-Fixing Bacteria

Nitrogen fixation is the reduction of atmospheric nitrogen gas (N_2) to ammonia (NH_3). Organisms that can fix nitrogen are essential for maintaining a supply of physiologically usable nitrogen on earth. About 255 million metric tons of nitrogen is fixed annually—70% of it by nitrogen-fixing bacteria. Bacteria and cyanobacteria fix nitrogen in many different environments—from Antarctica to hot springs, acid bogs to salt flats, flooded lands to deserts, in marine and fresh water, and even in the gut of some organisms.

The energy for nitrogen fixation can come from fermentation, aerobic respiration, or photosynthesis. The various organisms that fix nitrogen live independently, in loose associations, or in intimate symbiosis. Regardless of the environment or the associations of the organisms, nitrogen-fixing bacteria must have a functional nitrogen-fixing enzyme called **nitrogenase**, a reducing agent that supplies hydrogen, as well as energy from ATP. In aerobic environments, nitrogen fixers must also have a mechanism to protect the oxygen-sensitive nitrogenase from inactivation.

Free-living aerobic nitrogen fixers include several species of genus *Azotobacter* and some methylotrophic bacteria. Cyanobacteria are also free-living nitrogen fixers. *Azotobacter* is found in soils and is a versatile heterotroph whose growth is limited by the amount of organic carbon available. *Methylotrophic bacteria* can fix nitrogen when provided with methane, methanol, or hydrogen from various substrates. Cyanobacteria can fix nitrogen by using hydrogen from hydrogen sulfide, so

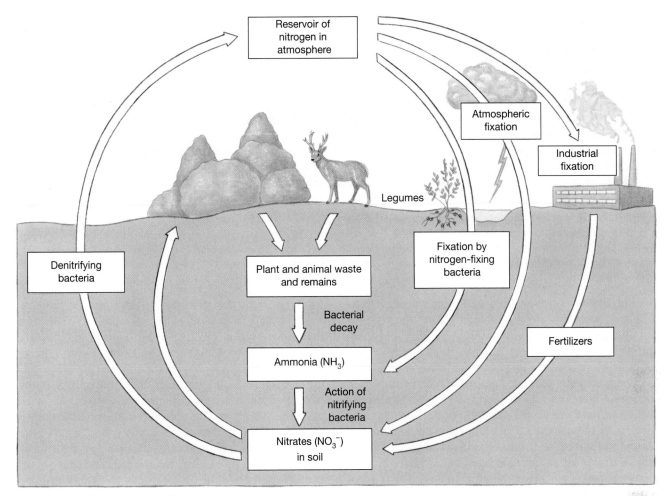

Figure 25.4   The nitrogen cycle.

they increase the availability of nitrogen in sulfurous environments.

Nitrogen-fixing facultative anaerobes include species of *Klebsiella, Enterobacter, Citrobacter,* and *Bacillus.* In addition, a number of obligate anaerobes, including photosynthetic Rhodospirillaceae and bacteria of the genera

TRY IT

Nitrogen-Fixing Bacteria in Root Nodules

It's easy to find and visualize nitrogen-fixing bacteria in root nodules. To do this, first dig up the root system from a leguminous plant such as beans or peas. Carefully wash away the dirt to expose the nodules. Crush a nodule on a clean glass slide. Add a small drop of water, and use the nodule pieces or an inoculating loop to spread the fluid out into a thin film. Remove the nodule pieces. Allow the slide to air-dry, and then quickly pass it through a Bunsen burner flame three or four times to heat-fix the organisms. Flood it with methylene blue stain for 1 minute. Rinse the slide with water, dry it, and examine it under the oil immersion lens. The bacilli you see are likely to belong to the genus *Rhizobium.*

Clostridium, Desulfovibrio, and *Desulfotomaculum,* also fix nitrogen. Several species of *Klebsiella* capable of fixing nitrogen are found in *rhizomes* (subsurface stems) of legumes such as peas and beans and in the intestines of humans and other animals. Nitrogen fixation has been observed in about 12% of *Klebsiella pneumoniae* organisms from patients. Various species of *Clostridium* are found in soils and muds. They use a variety of organic substances for energy and withstand unfavorable conditions as spores. They tolerate a range of pH from 4.5 to 8.5 but fix nitrogen best at pH 5.5 to 6.5. *Desulfovibrio* and *Desulfotomaculum,* anaerobic sulfate reducers that live in mud and soil sediments, fix nitrogen at pH 7 to 8.

Some nitrogen fixers are found in symbiotic association with other organisms, which provide them with organic carbon sources. For example, the cyanobacterium *Anabaena* (**Figure 25.5a**) is found in pores of the leaves of *Azolla,* a small water fern found in many parts of the world. The nitrogen-fixing *Anabaena* supplies the necessary nitrogen; the fern supplies substrates for energy capture in ATP and reductants for nitrogen fixation. Together, they fix 100 kg of nitrogen per hectare (about two football fields in area) per year and are used as "green manure" in rice cultivation in southeast Asia.

Figure 25.5 *Rhizobium* **bacteria and root nodules.** **(a)** SEM of the beadlike, filamentous *Anabaena azollae*, a nitrogen-fixing cyanobacterium that lives in a mutualistic relationship with the water fern *Azolla* (1,228X). *(David Hall/Photo Researchers)* **(b)** Nodules on the roots of a bean plant, resulting from invasion of the plant by nitrogen-fixing bacteria. *(John D. Cunningham/Visuals Unlimited)* **(c)** A cross section through a root of a leguminous plant shows densely packed *Rhizobium* bacteria inside the nodule (1227X). *(A.M. Siegelman/Visuals Unlimited)*

Rhizobium is the primary symbiotic nitrogen fixer. It lives in *nodules*, growths extending from the roots of certain plants, usually legumes (**Figure 25.5b** and **c**). In this symbiotic relationship, the plant benefits by receiving nitrogen in a usable form, and the bacteria benefit by receiving nutrients needed for growth. When paired with legumes, *Rhizobium* can fix 150 to 200 kg of nitrogen per hectare of land per year. In the absence of legumes, the bacteria fix only about 3.5 kg of nitrogen per hectare per year—less than 2% of that fixed in the symbiotic relationship. Farmers often mix nitrogen-fixing bacteria with seed peas and beans before planting to ensure that nitrogen fixation will be adequate for their crops to thrive.

The mechanisms by which *Rhizobium* establishes a symbiotic relationship with a legume has been the subject of many research studies. *Rhizobium* multiplies in the vicinity of legume roots, probably under the influence of root secretions. As the rhizobia increase in numbers, they release enzymes that digest cellulose and the substances that cement cellulose fibers together in the root cell walls. The rhizobia then change from their free-living rod shape and become spherical, flagellated cells called **swarmer cells**. These cells are thought to produce indoleacetic acid, a plant growth hormone that causes curling of root hairs. The swarmer cells then invade the root hairs and form hyphalike networks, killing some root cells and proliferating in others. Swarmer cells become large, irregularly shaped cells called **bacteroids**, which are tightly packed into root cells, probably under the influence of chemical substances in the plant cells. Accumulations of bacteroids in adjacent root cells form nodules on the plant roots.

Bacteroids contain the enzyme nitrogenase, which catalyzes the following reaction:

$$\underset{\text{nitrogen gas}}{N_2} + \underset{\text{hydrogen gas}}{3\,H_2} \xrightarrow{\textit{Rhizobium}} \underset{\text{ammonia}}{2\,NH_3}$$

However, this enzyme is inactivated by oxygen, so nitrogen fixation can occur only when oxygen is prevented from reaching the enzyme. Nitrogenase is protected from oxygen by a kind of hemoglobin, a red pigment that binds oxygen. This particular hemoglobin is synthesized only in root nodules containing bacteroids, because part of the genetic information for its synthesis is in the bacteroids and part is in the plant cells. Synthesis of nitrogenase is repressed in the presence of excess ammonium (NH_4^+) and derepressed in the presence of free nitrogen. Thus, fixation occurs only when "fixed" nitrogen is in short supply and free nitrogen is available.

Rhizobium species vary in their capacity to invade particular legumes and in their capacity to fix nitrogen once they have invaded. Some species cannot invade any legumes, whereas others invade only certain legumes. This invasive specificity is determined genetically (probably by a single gene or a group of closely related genes) and can be altered by genetic transformation (◄Chapter 8, p. 212). Such a transformation might enable a species of *Rhizobium* to invade a group of legumes it previously could not colonize.

> *Bacterial nitrogen-fixing genes added to major crop plants such as corn could make them self-fertilizing, eliminating costs and pollution problems.*

Although symbiotic nitrogen fixation occurs mainly in association between rhizobia and legumes, other such associations are known. The alder tree, which grows in soils low in nitrogen, has root nodules similar to those formed by rhizobia. These nodules contain nitrogen-fixing actinomycetes of the genus *Frankia*.

Nitrifying Bacteria

Nitrification is the process by which ammonia or ammonium ions are oxidized to nitrites or nitrates. Carried out by autotrophic bacteria, nitrification is an important part of the nitrogen cycle because it supplies plants with *nitrate* (NO_3^-) the form of nitrogen most usable in plant metabolism. Nitrification occurs in two steps. In each step, nitrogen is oxidized and energy is captured by the

Figure 25.6 A TEM of a microcolony of the nitrifying bacterium *Nitrosomonas*. (17,750X). *(Paul W. Johnson/Biological Photo Service)* TEM

bacteria that carry out the reaction. Various species of *Nitrosomonas* (**Figure 25.6**) and related genera—Gram-negative, rod-shaped bacteria—produce *nitrite* (NO_2^-) the reduced form of nitrates:

$$NH_4^+ \ + \ 1\tfrac{1}{2}O_2 \ \xrightarrow{\textit{Nitrosomonas}}$$

ammonium ion   oxygen gas

$$NO_2^- \ + \ H_2O \ + \ 2H^+ \ + \text{energy}$$

nitrite   water   hydrogen ions

Species of *Nitrobacter* and related genera, also Gram-negative rods, produce nitrates:

$$NO_2^- \ + \ \tfrac{1}{2}O_2 \ \xrightarrow{\textit{Nitrobacter}} \ NO_3^- \ + \ \text{energy}$$

nitrite   oxygen   nitrate

These bacteria use energy derived from the preceding reactions to reduce carbon dioxide in autotrophic metabolism. Because oxygen is required for the nitrification reactions, the reactions occur only in oxygenated water and soils. Furthermore, because nitrite is toxic to plants, it is essential that these reactions be carried out in sequence to provide nitrates and to prevent excessive accumulation of nitrites in soil.

Denitrifying Bacteria

Denitrification is the process by which nitrates are reduced to nitrous oxide (N_2O) or nitrogen gas (unbalanced equation):

$$NO_3^- \ \rightarrow \ NO_2^- \ \rightarrow \ N_2O \ \rightarrow \ N_2$$

nitrate   nitrate   nitrous oxide   nitrogen gas

Although this process does not occur to any significant degree in well-oxygenated soils, it does occur in oxygen-depleted, waterlogged soils. Most denitrification is performed by *Pseudomonas* species, but it can also be

accomplished by *Thiobacillus denitrificans, Micrococcus denitrificans,* and several species of *Serratia* and *Achromobacter.* These bacteria, although usually aerobic, use nitrate instead of oxygen as a hydrogen acceptor under anaerobic conditions. Another process that reduces soil nitrate is the reduction of nitrate to ammonia. Several anaerobic bacteria carry out this process (called *dissimilative nitrate reduction*) in a complex reaction that can be summarized as follows (unbalanced equation):

$$NO_3^- \ \rightarrow \ H_2 \ \rightarrow \ NH_3 \ \rightarrow \ N_2O$$

nitrate   hydrogen gas   ammonia   nitrous oxide

Although the latter process reduces the quantity of available nitrate, it retains the nitrogen in the soil in other forms.

Denitrification is a wasteful process because it removes nitrates from the soil and interferes with plant growth. It is responsible for significant losses of nitrogen from fertilizer applied to soils. Another unfortunate effect of denitrification is the production of nitrous oxide, which is converted to nitric oxide (NO) in the atmosphere. Nitric oxide, in turn, reacts with ozone (O_3) in the upper atmosphere. Ozone provides a barrier between living things on Earth and the sun's ultraviolet radiation. If enough ozone is destroyed, it no longer serves as an effective screen, and living things can be exposed to excessive ultraviolet radiation, which can cause cancer and mutations (◄Chapter 7, p. 202).

THE SULFUR CYCLE AND SULFUR BACTERIA

The **sulfur cycle** (**Figure 25.7**), which involves the movement of sulfur through an ecosystem, resembles the nitrogen cycle in several respects. Sulfhydryl (–SH) groups in proteins of dead organisms are converted to hydrogen sulfide (H_2S) by a variety of microorganisms. This

APPLICATIONS

Pig Poop Problems Prevented

Pigs are cute—a few at a time. But half a million being raised on a pig "factory farm" can cause problems. Each adult pig produces 50 to 60 lb of manure per day. Multiply by half a million, and that's a mighty big pile to deal with. Diluted with water and stored in a lagoon, the stench alone keeps neighbors awake nights gagging and vomiting. Nighttime is usually when the lagoon wastewater is pumped out over cropland, to act as fertilizer. Sometimes there are more nutrients than crop plants can take in, or it rains right after application. Then nutrients wash off into waterways causing pollution, algal blooms, fish kills, etc. Not much fun for swimmers either!

U.S. Department of Agriculture scientists are now adapting Japanese technology for treating municipal wastewater to our large-scale pig farms. *Nitrosomonas* and *Nitrobacter* bacteria are entrapped in polymer gel pellets about an eighth- to a quarter-inch wide. Ammonia from wastewater along with oxygen and carbon dioxide from aerators can enter and nourish the bacteria. Screens and filters keep the pellets from being pumped out with treated wastewater. Inside the pellets, bacteria break the ammonia down into nitrites, then into nitrates (nitrification). Then in the process of denitrification, they convert nitrates into free nitrogen gas (N_2). These bacteria are hard little workers: after 2 months of acclimation to higher and higher ammonia concentrations, they are ready and able to stay

on the job for 10 years or more! Flasks of pig wastewater show what a difference 12 hours of treatment can make: 675 parts per million nitrogen in the dark untreated flask versus less than 24 parts per million in the clear treated flask.

(Courtesy Agricultural Research Service, USDA)

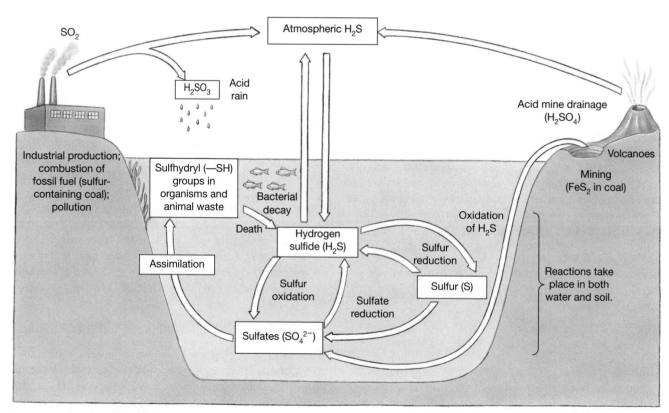

Figure 25.7   The sulfur cycle.

process is analogous to the release of ammonia from proteins in the nitrogen cycle. Hydrogen sulfide is toxic to living things and thus must be oxidized rapidly. Oxidation to elemental sulfur is followed by oxidation to sulfate (SO_4^{2-}), which is the form of sulfur most usable by both microorganisms and plants. This process is analogous to nitrification. The sulfur cycle is of special importance in aquatic environments, where sulfate is a common ion, especially in ocean water.

The various sulfur bacteria can be categorized according to their roles in the sulfur cycle. These roles include sulfate reduction, sulfur reduction, and sulfur oxidation.

Sulfate-Reducing Bacteria

Sulfate reduction is the reduction of sulfate (SO_4^{2-}) to hydrogen sulfide (H_2S). The sulfate-reducing bacteria are among the oldest life forms, probably more than 3 billion years old. They include the closely related genera *Desulfovibrio*, *Desulfomonas*, and *Desulfotomaculum*. In these bacteria, sulfate is the final electron acceptor in anaerobic oxidation, as oxygen is the final electron acceptor in aerobic oxidation. By reducing sulfate, these bacteria produce large amounts of hydrogen sulfide. However, for this process to occur, energy from ATP is required to phosphorylate the sulfate and convert it to ADP–SO_4. Then ADP–SO_4 can act as an electron acceptor and compete successfully for substrates.

Sulfate-reducing bacteria **(Figure 25.8)** are strict anaerobes. They are widely distributed and predominate in nearly all anaerobic environments. As discussed in ◀Chapter 6, p. 157, sulfate-reducing bacteria can be psychrophilic, mesophilic, thermophilic, or halophilic. The variety of organic carbon sources they can metabolize is limited. Most use lactate, pyruvate, fumarate, malate, or ethanol, and some can use glucose and citrate. End products of such metabolism are typically acetate and carbon dioxide. Some sulfate reducers use products derived from the anaerobic degradation of plant material by other organisms. *Desulfovibrio*, *Desulfomonas*, and *Desulfotomaculum* oxidize fatty acids and a variety of other organic acids.

Figure 25.8 **The sulfate-reducing bacterium** LM *Desulfotomaculum acetoxidans..* Notice the spherical spores and gas vacuoles (phase contrast, 5000X). *(F. Widdel/Visuals Unlimited)*

Sulfur-Reducing Bacteria

Sulfur reduction is the reduction of sulfur to hydrogen sulfide. Like sulfate-reducing bacteria, sulfur-reducing bacteria are anaerobes. They can use intracellular or extracellular sulfur as an electron acceptor in fermentation. The sulfur can be in elemental form or in disulfide bonds or organic molecules. This process provides energy for the organisms when sunlight is not available for photosynthesis.

Sulfur-Oxidizing Bacteria

Sulfur oxidation is the oxidation of various forms of sulfur to sulfate. *Thiobacillus* and similar bacteria oxidize hydrogen sulfide, ferrous sulfide, or elemental sulfur to sulfuric acid (H_2SO_4). When this acid ionizes, it greatly decreases the pH of the environment, sometimes lowering the pH to 1 or 2. Sulfur-oxidizing organisms are responsible for oxidizing ferrous sulfide in coal-mining wastes, and the acid they produce is extremely toxic to fish and other organisms in streams fed by such wastes.

OTHER BIOGEOCHEMICAL CYCLES

In addition to water, carbon, nitrogen, and sulfur—whose cycles we have examined—phosphorus and other elements also move through ecosystems cyclically. Any element (including trace elements) that appears in the cells of living organisms must be recycled to extract it from dead organisms and make it available to living ones. We will conclude our discussion of biogeochemical cycles with a brief description of the phosphorus cycle.

The **phosphorus cycle (Figure 25.9)** involves the movement of phosphorus among inorganic and organic forms. Soil microorganisms are active in the phosphorus cycle in at least two important ways: (1) They break down organic phosphates from decomposing organisms to inorganic phosphates, and (2) they convert inorganic phosphates to orthophosphate (PO_4^{3-}), a water-soluble nutrient used by both plant and microorganisms. These functions are particularly important because phosphorus is often the limiting nutrient in many environments.

THE DEEP HOT BIOSPHERE

Thomas Gold, in his controversial book (which is gaining in acceptance) *The Deep Hot Biosphere, the Myth of Fossil Fuels* (2001, Copernicus Books, an Imprint of Springer-Verlag), holds that the entire crust of the Earth, down to a depth of several miles, is inhabited by a culture of microbes. These bacteria deep inside the Earth feed on oil and methane gas deposits, which are an original part of the Earth, from the time it first coalesced out of planetary debris. Oil would no longer be viewed as a product formed by the compression and transformation of plant and animal debris. Primitive life began deep in the depths of the planet, and has only emerged onto its surfaces as the surfaces have cooled and modified. Today at the borderlands between deep and surface environments we find upwelling of chemicals

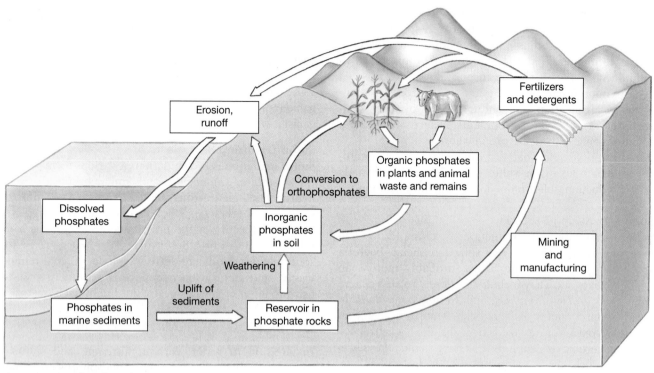

Figure 25.9 **The phosphorus cycle.**

and organisms, first seen in 1977 at a depth of 2.6 km, northeast of the Galapagos Islands, along the Pacific Rise. Later in this chapter we will discuss the hydrothermal black-smoking vents, cold seeps, and methane and sulfur bubbling caves, which are the areas of borderlands. See also the opening of this chapter.

✓**CHECKLIST**

1. Distinguish between autotrophs, heterotrophs, and decomposers.
2. What is meant by nitrogen fixation? Who does it? What is its importance to us?
3. What is the importance of denitrification?

▐▐▐ AIR

Having discussed briefly some fundamentals of ecology and of biogeochemical cycles, we will now explore different kinds of environments and the microorganisms they contain. We will begin with the air around us and then consider soil and water.

MICROORGANISMS FOUND IN AIR

Microorganisms do not grow in air, in part because it lacks the nutrients needed for metabolism and growth. However, spores are carried in air, and vegetative cells can be carried on dust particles and water droplets in air. The kinds and numbers of airborne microorganisms vary tremendously in different environments. Large numbers of many different kinds of microorganisms are present in

indoor air where humans are crowded together and building ventilation is poor. Small numbers have been detected at altitudes of 3,000 m.

Among the organisms found in air, mold spores are by far the most numerous, and the predominant genus is usually *Cladosporium*. Bacteria commonly found in air include both aerobic spore formers such as *Bacillus subtilis* and non–spore formers such as *Micrococcus* and *Sarcina*. Algae, protozoa, yeasts, and viruses also have been isolated from air. While coughing, sneezing, or even talking, infected humans can expel pathogens along with water droplets. Health care workers should handle wastes from patients carefully to avoid producing aerosols (tiny droplets that remain suspended for some time) of pathogens (◄Chapter 15, p. 438).

Determining the Microbial Content of Air

Airborne microbes can be detected by collecting those that happen to fall onto an agar plate or in a liquid medium. A special centrifuge-like air-sampling instrument provides a better measure of airborne microbes (**Figure 25.10a**).

METHODS FOR CONTROLLING MICROORGANISMS IN AIR

Chemical agents, radiation, filtration, and laminar airflow all can be used to control microorganisms in air. Certain chemical agents, such as triethylene glycol, resorcinol, and lactic acid dispersed as aerosols, kill many if not all microorganisms in room air. These agents are highly bactericidal, remain suspended long enough to act at normal room temperature and humidity, are nontoxic to humans, and do not damage or discolor objects in the room.

APPLICATIONS

Sick Buildings

The U.S. Environmental Protection Agency (EPA) estimates that 30 million to 75 million American workers are at risk of getting sick from the buildings in which they work. Since the late 1970s, indoor air specialists from the National Institute for Occupational Safety and Health (NIOSH) have investigated more than 1,000 instances of building-related illness. Some of the ailments can be fatal. Sick building syndrome (SBS) is a collection of symptoms including headache, dizziness, nausea, burning eyes, nosebleeds, and respiratory and skin problems. Symptoms vary among individuals.

How do you know when a building—instead of a cold, allergies, or another cause—is to blame for SBS symptoms? If your symptoms get worse the more time you spend in a building and improve when you are away from it, take a close look at the building. The ventilation system has been implicated in many cases of SBS. In many new "tight" buildings, occupants are dependent on a central ventilation system for the air they breathe. The system pulls in air from outside, runs it through filters, warms or cools it, and then delivers it throughout the building by a series of ducts. Once the air has circulated, return ducts channel some of it outside. Dirt, dust, insects, mold, mildew, particles of building materials, and other detritus are found on the system's filters. Whatever gets into the ventilation system gets into workers' lungs as well. New filters and proper cleaning of the system's equipment often correct ventilation problems.

Another possible cause of SBS is too little air. In the 1930s the American Society of Heating, Refrigeration, and Air Conditioning Engineers set 15 cubic feet of fresh outdoor air per person per minute as the ventilation standard. The current standard is 20 cubic feet—but it assumes that no more than seven people will occupy a 1,000-square-foot area. With more people, more air is needed. The "fleece factor," or pollution from "man-made mineral fibers" (MMMF), may also cause SBS. New carpets, padded partitions, curtains, and upholstery are treated with volatile compounds that are released into the air. Copy machines, paper shredders, laser printers, and blueprint copiers release vapors and particles that may be irritating. Finally, biological contaminants such as molds, fungi, and bacteria thrive in building ventilation systems and are spread as aerosols. In recent years, bacteria that collect in ventilation systems have proved fatal—for example, when *Legionella* species have caused Legionnaires' disease (◀Chapter 21, p. 648).

If you suspect that you are in a sick building, document your symptoms, check the ventilation system, inspect the air vents for dirt, check for exhaust vents around copy machines, and so on. Ask the maintenance engineers how much fresh outdoor air is circulating in cubic feet per minute per person. Once you have targeted hazards, convince someone to do something about them.

Ultraviolet (UV) radiation has little penetrating power and is bactericidal only when rays make direct contact with microorganisms in air (◀Chapter 12, p. 358). Thus, ultraviolet lamps must be carefully placed to ensure treatment of all air in a room. They are most useful in maintaining sterile conditions in rooms only sporadically occupied by humans. They can be turned off while technicians are performing sterile techniques and turned on again when the room is not in use. Humans entering a room while ultraviolet lights are on must wear protective clothing and special glasses to protect their eyes from burns. Ultraviolet lamps can also be

(a)

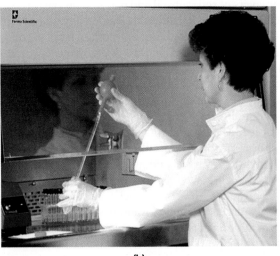

(b)

Figure 25.10 Measuring airborne microbes. **(a)** An air-sampling device used to measure airborne bacteria and fungi. *(Courtesy Graseby Andersen, Inc.)* **(b)** Technicians are protected from the airborne spread of organisms by the use of a laminar flow hood, which suctions air away from the opening and filters it before expelling it. *(Forma/Science VU/Visuals Unlimited)* installed in air ducts to reduce the number of micro-organisms entering a room through its ventilation system.

Air filtration involves passing air through fibrous materials, such as cotton or fiberglass. It is useful in industrial processes in which sterile air must be bubbled through large fermentation vats. Cellulose acetate filters can be installed in a *laminar airflow system* (**Figure 25.10b**) to remove microorganisms that may have escaped into the air underneath the laminar flow hood. The air is suctioned away from the opening, filtered, and then returned to the room.

SOIL

We might think of the soil we walk on as an inert substance, but nothing could be further from the truth. Soil, in fact, is teeming with microscopic and small macroscopic organisms, and it receives animal wastes and organic matter from dead organisms. Microorganisms act as decomposers to break down this organic matter into simple nutrients that can be used by plants and by the microbes themselves. (Animals, of course, obtain their nutrients from plants or from other animals.) Soil microorganisms are thus extremely important in recycling substances in ecosystems.

COMPONENTS OF SOIL

Soil is divided into a number of layers, or *soil horizons* (**Figure 25.11**). These horizons include the *topsoil*, *subsoil*, and *parent material*, which overlie bedrock. Soil contains inorganic and organic components. The inorganic components are rocks, minerals, water, and gases. The organic components are **humus** (nonliving organic matter) and living organisms. Soils differ greatly in the relative proportions of these components. Topsoil, the surface layer of soil, contains the greatest number of microorganisms because it is well supplied with oxygen and nutrients. Lower layers of soil (subsoil and parent material) and soils depleted of oxygen and nutrients contain fewer organisms.

The most abundant inorganic components of soil are pulverized rocks and minerals. *Weathering* of rocks—mechanical or chemical breakdown—releases minerals, and the elements that make up those minerals, into the soil. The most abundant elements in many soils are silicon, aluminum, and iron. Small quantities of other elements, such as calcium, potassium, magnesium, sodium, phosphorus, nitrogen, and sulfur, also are present in soil.

In addition to rocks and minerals, soil contains water and the gases carbon dioxide, oxygen, and nitrogen. Water molecules adhere to soil particles or are interspersed among them. The amount of water in soil is highly variable and depends on the climate, the quantity of recent rainfall, and the drainage of the soil. Gases are dispersed among soil particles or dissolved in water. The concentration of gases in the soil also varies with the metabolic activity of soil organisms. Compared with atmospheric air, soil is lower in oxygen and higher in carbon dioxide.

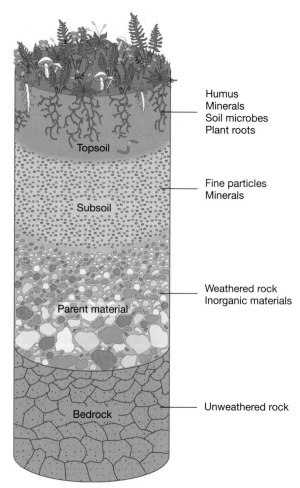

Figure 25.11 Soil horizons. A vertical segment of the main soil layers.

Humus is constantly changing as organisms die and as decomposers degrade complex molecules to simpler ones. Soils differ greatly in the amount of humus they contain. Most soils are 2% to 10% humus, but a peat bog can be 95% humus.

In addition to microorganisms, soil also contains the root systems of many plants, a variety of invertebrate animals (nematodes, earthworms, snails, slugs, insects, centipedes, millipedes, and spiders), and a few burrowing reptiles and mammals. Despite the great diversity of soil organisms, microorganisms are by far the most numerous both in total numbers and in the number of species present. As we have seen, microorganisms convert humus to nutrients usable by other organisms. They also make great demands on soil nutrients. When nutrients are abundant, the number of microbes increases rapidly. As nutrients are depleted, the number of microbes decreases accordingly.

MICROORGANISMS IN SOIL

All the major groups of microorganisms—bacteria, fungi, algae, and protists, as well as viruses—are present in soil, but bacteria are more numerous than all other

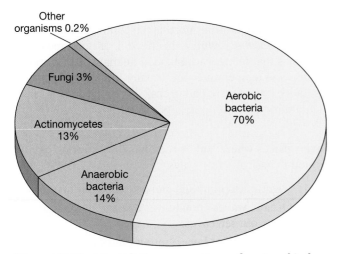

Figure 25.12 The relative proportions of various kinds of organisms found in soil. "Other organisms" include such things as algae, protists, and viruses.

kinds of microorganisms (**Figure 25.12**). Among the bacteria in soil are autotrophs, heterotrophs, aerobes, anaerobes, and depending on the soil temperature, mesophiles and thermophiles. In addition to nitrogen-fixing, nitrifying, and denitrifying bacteria, soil contains bacteria that digest special substances such as cellulose, protein, pectin, butyric acid, and urea.

Soil fungi are mostly molds. Both mycelia and spores are present mainly in topsoil, the aerobic surface layer of the soil. Fungi serve two functions in soil: They decompose plant tissues such as cellulose and lignin, and their mycelia form networks around soil particles, giving the soil a crumbly texture (◀Chapter 11, p. 320). In addition to molds, yeasts are abundant in soils in which grapes and other fruits are growing.

Small numbers of cyanobacteria, algae, protists, and viruses are found in most soils. Algae are found only on the soil surface, where they can carry on photosynthesis. In desert and other barren soils, algae contribute significantly to the accumulation of organic matter in the soil. Protists, mostly amoebas and flagellated protozoa, also are found in many soils. They feed on bacteria and may help control bacterial populations. Soil viruses infect mostly bacteria, but a few infect fungi, and quite a few infect plants. Very little is known about the identity or classification of many plant viruses.

APPLICATIONS

Fungal Confusion

The rust fungus, *Puccinia recondita*, causes millions of dollars of crop loss in wheat and rye each year. When its spores land on a leaf, they send out tiny "infection tubes" that enter pores (stomata) on the leaf surface. Hairy leaves developed by breeders "confuse" the fungus, and it dies before locating a stoma.

Viroids in soil are often spread by mechanical means and can cause serious plant diseases. Viruses that attack insects are common, most of them belonging to the families Baculoviridae and Iridoviridae. Animal viruses are not indigenous to soil but are often added to soil through human activities, such as manure spreading. The survival of viruses in soil varies with environmental conditions and the particular virus. It ranges from hours to years.

It is possible to use viruses for biocontrol of insect pests in soil. Viruses of an appropriate type are inoculated into soil. When all insect pests have been killed, the virus tends to disappear, offering no further risk—an attribute important for government approval. Caterpillar stages of the celery looper butterfly can be controlled by the addition of the *Anagrapha falcifera* multiple nuclear polyhedrosis virus to the soil.

Factors Affecting Soil Microorganisms

Soil microorganisms, like all other organisms, interact with their environment. Their growth is influenced both by abiotic factors and by other organisms. The microorganisms, in turn, affect the physical characteristics of soil and other organisms in the soil.

Abiotic factors in soil, as in any other environment, include moisture, oxygen concentration, pH, and temperature. The moisture and oxygen content of soil are closely related. Spaces among soil particles ordinarily contain both water and oxygen, and aerobic organisms thrive there. However, in waterlogged soils all the spaces are filled with water, so only anaerobic bacteria grow there.

Soil pH, which can vary from 2 to 9, is an important factor in determining which microorganisms will be present. Most soil bacteria have an optimum pH between 6 and 8, but some molds can grow at almost any soil pH. Molds thrive in highly acidic soils partly because of reduced competition from bacteria for available nutrients. Lime neutralizes acidic soil and increases the bacterial population. Fertilizer containing ammonium salts has two effects on soil: (1) It provides a nitrogen source for plants; and (2) when it is metabolized by certain bacteria, the bacteria release nitric acid, which decreases the soil pH and increases the mold population.

Soil temperature varies seasonally from below freezing to as high as 60°C at soil surfaces exposed to intense summer sunlight. Mesophilic and thermophilic bacteria are quite numerous in warm to hot soils, whereas cold-tolerant mesophiles (and not true psychrophiles) are present in cold soils. Most soil molds are mesophilic and are found mainly in soils of moderate temperature. Surprisingly, in 2003, soil of cold high mountain ranges in the United States was found to greatly increase in number of fungi during the winter.

Exceedingly wide variations exist in physical characteristics of soil and in the numbers and kinds of organisms it contains—even in soil samples taken only a few centimeters apart. Such observations led ecologists to develop the concept of microenvironments. The interactions among organisms and between the organisms and their environment can be quite different in different microenvironments, no matter how close together they are.

The Importance of Decomposers in Soil

Soil microorganisms that are decomposers are important in the carbon cycle because of their ability to decompose organic matter. The decomposition of complex organic substances from dead organisms is a stepwise process that requires the action of several kinds of microorganisms. The organic substances include cellulose, lignins, and pectins in the cell walls of plants; glycogen from animal tissues; and proteins and fats from both plants and animals. Cellulose is degraded by bacteria, especially those of the genus *Cytophaga*, and by various fungi. Lignins and pectins are partially digested by fungi, and the products of fungal action are further digested by bacteria. Protozoa and nematodes also can play a role in the degradation of lignins and pectins. Proteins are degraded to individual amino acids mainly by fungi, actinomycetes, and clostridia.

Under the anaerobic conditions of waterlogged soils in marshes and swamps, methane is the main carbon-containing product. It is produced by three genera of strictly anaerobic bacteria—*Methanococcus*, *Methanobacterium*, and *Methanosarcina*. In addition to degrading carbon compounds to methane, they also obtain energy by oxidizing hydrogen gas:

$$4H_2 \; + \; CO_2 \; \rightarrow \; CH_4 \; + \; 2H_2O$$
hydrogen   carbondioxide   methane   water
gas

In one way or another, organic substances are metabolized to carbon dioxide, water, and other small molecules. In fact, for each naturally occurring organic compound, there is one or more organisms that can decompose it. Thus, carbon is continuously recycled. However, certain organic compounds manufactured by humans resist the actions of microorganisms. Accumulations of these synthetic substances create environmental hazards.

Nitrogen enters the soil through the decomposition of proteins from dead organisms and through the actions of nitrogen-fixing organisms. In addition to protein decomposition, which introduces nitrogen into the soil, gaseous nitrogen is fixed both by free-living microorganisms and by symbiotic microorganisms associated with the roots of legumes, as previously described.

SOIL PATHOGENS

Soil pathogens are primarily plant pathogens, many of which have been discussed in earlier chapters. A few soil pathogens can affect humans and other animals. The main human pathogens found in soil belong to the genus *Clostridium* (◀Chapter 24, p. 765). All are anaerobic spore formers. *Clostridium tetani* causes tetanus and can be introduced easily into a puncture wound. *Clostridium botulinum* causes botulism. Its spores, found on many edible plants, can survive in incompletely processed foods to produce a deadly toxin. *Clostridium perfringens* causes gas gangrene in poorly cleaned wounds. Grazing animals can contract anthrax from spores of *Bacillus anthracis* in the soil. In fact, most soil organisms that infect warmblooded animals exist as spores because soil temperatures are usually too low to maintain vegetative cells of these pathogens.

CAVES

Caves, places where soil or rock has disappeared, can be formed in basically four ways:

1. Slightly acid rainwater can dissolve away limestone.
2. Ocean waves pounding against cliff bottoms erode out sea caves.
3. Streams of hot lava running out of a volcano cool on their outer surfaces and harden into a tube. Inside, the lava remains liquid, running out, and leaving empty tunnels called lava tubes.
4. But strangest of all the ways, is the just recently discovered process whereby microorganisms produce sulfuric acid that eats away rock. Five percent of all limestone caves are formed in this manner.

Take another look

The opening page of ◀Chapter 1 described Lechuguilla Cave, the deepest cave in the continental United States, which is located in the Carlsbad Caverns National Park, New Mexico. It and Cueva de Villa Luz (Cave of the Lighted House) in Mexico are caves dissolved by microbial-produced sulfuric acid. A film entitled "The Mysterious Life of Caves" was produced by NOVA, and was originally broadcast on PBS (Public Broadcasting System) on October 1, 2002. It features geomicrobiologists Penelope J. Boston and Diana Northrup visiting these and other caves, explaining the role of microbes, and some of the experiments that they have conducted in the caves. This film is now available for purchase. NOVA also has a web site describing this program, with a transcript, spectacular photos of Lechuguilla Cave, and interviews with the scientists. It also features links to many other web sites about caves. Please visit it at http://www.pbs.org/wgbh/nova/caves/about.html.

Carlsbad and Lechuguilla caverns are no longer actively forming. However, Cueva de Villa Luz is vigorously growing. Hydrogen sulfide (H_2S) gas bubbles up from the bottom of the cave, forming sulfuric acid when it reacts with water. Microbes eating oil deposits deep below the cave release the H_2S. The walls of the cave are covered with **snottites**, mucus-like strings of bacterial colonies (p. 1). These bacteria eat sulfur and drip sulfuric acid. The entire cave is like walking around in battery acid, so strong that it will eat away skin and clothing. Behind it, it leaves a telltale residue, beautiful crystals of gypsum (**Figure 25.13**). The bacteria living here are extremeophiles, possibly much like many of the bacteria that were present when life first emerged on Earth. As we explore other planets such as Mars, and the moons of Jupiter, we are looking for similar extremeophiles which may be living below their surfaces.

Figure 25.13 Chandelier Ballroom in Lechuguilla Cave. These gypsum chandeliers, up to 20 feet long, are thought to be the largest in the world. Note the size of the man in the red suit. *(Courtesy Sura Ballmann and Urs Widmer/Speleo Projects)*

▐▌▌ WATER

All aquatic ecosystems—freshwater, ocean water, even rainwater—contain microorganisms, as well as inorganic substances. Most of the organisms in such environments have been considered in earlier chapters. Here we will consider properties of the environments, interactions of microorganisms with their environment, transmission of human pathogens in water, and methods for maintaining safe water supplies.

FRESHWATER ENVIRONMENTS

Freshwater environments include surface water, such as lakes, ponds, rivers, and streams, and groundwater that runs through underground strata of rock. Although groundwater contains few microorganisms, surface water contains large numbers of many different kinds of microorganisms.

Ponds and lakes are divided vertically into zones. The shoreline, or *littoral zone*, is an area of shallow water near the shore where light penetrates to the bottom. The *limnetic zone* comprises the sunlit water away from the shore; resident microorganisms include algae and cyanobacteria. Between the limnetic zone and the lake sediment is the *profundal zone*. When organisms in the limnetic zone die, they sink to the profundal zone, where they provide nutrients for other organisms. The sediment, or *benthic zone*, is composed of organic debris and mud (**Figure 25.14**).

In aquatic environments, water temperatures vary from 0°C to nearly 100°C. Most microorganisms thrive in water at moderate temperatures. However, some thermophilic bacteria have been found in geysers at a water temperature above 90°C, and psychrophilic fungi and bacteria have been found in water at 0°C. The pH of

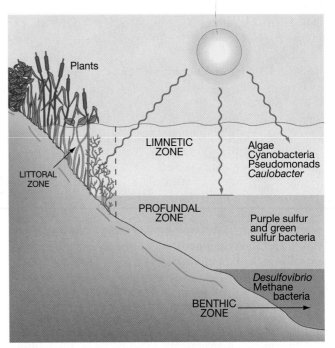

Figure 25.14 Zonation in a typical lake or pond. Organisms commonly found in each zone are listed at the right.

fresh water varies from 2 to 9. Although most microorganisms grow best in waters of nearly neutral pH, a few have been found in both extremely acidic and extremely alkaline waters. Most natural waters are rich in nutrients, but the quantities of various nutrients in different bodies of water vary considerably. Sometimes nutrients become so abundant that a "bloom," or sudden proliferation of organisms in a body of water, occurs (**Figure 25.15**); (See the box Red Tides in ◀Chapter 11, p. 313).

◆ CLOSE UP

The Nose of the Beholder

Thermal hot springs derive their heat from masses of magma that have pushed their way near to the Earth's surface. Natural hot springs, many used as health and bathing resorts, are home to thermophilic and metal-reducing bacteria. The water in thermal hot springs, which may emerge from rocks at temperatures well over 100°C, is rich in dissolved minerals. It is the hot water and high mineral content that many people feel is the source of the benefit of bathing in these natural spas. Thermal springs that are rich in dissolved sulfur oxides may support the growth of a range of sulfur-fixing bacteria. These bacteria, which include *Desulfovibrio* and *Desulfomonas*, reduce sulfate and produce hydrogen sulfide. Release of the hydrogen sulfide gas from the hot bubbling water smells like sewer gas or rotten eggs. However, in the context of a natural thermal hot spring, the odor is not unpleasant.

Figure 25.15 **Algal bloom.** An overabundance of nutrients has allowed an algal bloom to develop on this pond. Bubbles of oxygen can be seen collecting underneath some of the algae. However, algal blooms often lead to increased biological oxygen demand (BOD) as they die off and decompose, using up oxygen needed by other pond organisms such as fish. *(Kevin and Betty Collins/Visuals Unlimited)*

In aquatic environments, oxygen can be the limiting factor in the growth of microorganisms. Because of oxygen's low solubility in water, its concentration never exceeds 0.007 g per 100 g of water. When water contains large quantities of organic matter, decomposers rapidly deplete the oxygen supply as they oxidize the organic matter. Oxygen depletion is much more likely in standing water in lakes and ponds than in running water in rivers and streams because the movement of running water causes it to be continuously oxygenated.

Though the Earth's surface is 70% water, only 1% is available for human use. The remainder is unavailable in oceans, seas, and glaciers.

Yet another factor affecting microorganisms in aquatic environments is the depth to which the sunlight penetrates the water. Of minor importance in fresh water except for deep lakes, it is very important in the ocean. Photosynthetic organisms are limited to locations with adequate sunlight.

Which organisms are present depends on the temperature and pH of the water, dissolved minerals, the depth of sunlight penetration, and the quantity of nutrients in the water. Aerobic bacteria are found where oxygen supplies are adequate, whereas anaerobic bacteria are found in oxygen-depleted waters. Eukaryotic algae, cyanobacteria, and sulfur bacteria are limited to water that receives adequate sunlight. *Desulfovibrio* and methane bacteria are found in lake sediments. Protists are found in many different freshwater environments.

MARINE ENVIRONMENTS

The marine, or ocean, environment covers about 70% of the Earth's surface and is therefore larger than all other environments combined. Compared with fresh water, the ocean is much less variable in both temperature and pH. Except in the vicinity of volcanic vents in the seafloor, where water reaches 250°C (◀Chapter 9, p. 265), ocean water ranges from 30°C to 40°C at the surface near the equator to 0°C in polar regions and in the lowest depths. At any one location and depth, the temperature is nearly constant. The pH of ocean water ranges from nearly neutral to slightly alkaline (pH 6.5 to 8.3). This pH range is suitable for growth of many microorganisms, and sufficient carbon dioxide dissolves in the water to support photosynthetic organisms.

Ocean water, which is about seven times as salty as fresh water, displays a remarkably constant concentration of dissolved salts—3.3 to 3.7 g per 100 g of water. Thus, organisms that live in marine environments must be able to tolerate high salinity but need not tolerate variations in salinity.

Marine environments display a far greater range of hydrostatic pressure than do fresh waters. Hydrostatic pressure increases with depth at a rate of approximately 1 atm per 10 m, so the pressure at a depth of 1,000 m is 100 times that at the surface. A few microorganisms, including Archaea, have been isolated from Pacific Ocean trenches at depths greater than 1,000 m!

Other factors that vary with the depth of ocean water are penetration of sunlight and oxygen concentration. Sunlight of sufficient intensity to support photosynthesis penetrates ocean water to depths of only 50 to 125 m, depending on the season, latitude, and transparency of the water. Oxygen diffuses into surface water and is released by photosynthetic organisms in sunlit water. However, deep water lacks both sunlight and oxygen.

Nutrient concentrations vary in ocean water, depending on depth and proximity to the shore (**Figure 25.16**). Nutrients are most plentiful near the mouths of rivers and near the shore, where runoff from land enriches the water. However, ocean water is generally lower in phosphates and nitrates than fresh water. Nutrients in the waters of the open ocean are relatively dilute. Photosynthetic organisms near the surface serve as food for the heterotrophic organisms at the same or deeper levels. Decomposers are usually found in bottom sediments, where they release nutrients from dead organisms.

CLOSE UP

And You Thought It Was El Niño

Oceanic phytoplankton absorb large amounts of carbon dioxide as they perform photosynthesis. Atmospheric carbon dioxide is one of the greenhouse gases that trap heat and raise global temperature. Seasonal fluctuations of phytoplankton populations affect the Earth's temperature more than we had previously realized. Recent research has revealed that massive populations of marine viruses may be at the root of these processes. Earlier, the ocean was thought to contain relatively few viruses. Viral infections destroy phytoplankton and thus may help determine the Earth's weather.

Figure 25.16 Nutrients flow into the ocean with the influx of water from rivers. The light-colored area of water is silt-bearing, nutrient-rich river water. *(Bruce F. Molnia/Biological Photo Service)*

Large numbers of many different kinds of microorganisms live in the ocean. Yet, much remains to be learned about the exact numbers and the particular species living in various parts of the ocean. The primary producers of the ocean are photosynthetic microorganisms called *phytoplankton*. They are motile and contain oil droplets or other devices for buoyancy, allowing them to remain in sunlit waters. Phytoplankton include cyanobacteria, diatoms, dinoflagellates, chlamydomonads (kla″mi-do-mo′nadz), and a variety of other protists and eukaryotic algae **(Figure 25.17)**.

Many of the consumers of the ocean are heterotrophic bacteria. Which species inhabit a given region is determined by the temperature and pH of the water and by the available nutrients. Members of the genera *Pseudomonas*, *Vibrio*, *Achromobacter*, and *Flavobacterium* are common in ocean water.

There are over a million bacteria/ml of open ocean water.

Protozoa, especially radiolarians and foraminiferans, and a variety of fungi also feed on producers in the open ocean. However, most of these organisms inhabit an aquatic zone beneath the region of intense sunlight.

Between the stratum in which these consumers live and the bottom of the ocean is a relatively uninhabited stratum. In the sediments at the bottom of the ocean, the number of microorganisms again increases. The bottom dwellers are generally strict or facultative anaerobic decomposers. Many of them contribute significantly to the maintenance of biogeochemical cycles and produce such substances as ammonium ion, hydrogen sulfate, and nitrogen gas.

HYDROTHERMAL VENTS AND COLD SEEPS

At certain places along the bottom of the oceans, materials and/or heat leave the deep hot biosphere of the Earth's interior, and enter into the seawater. There, steep chemical and thermal gradients create specific extreme environments. "Black smokers" venting clouds of sulfur compounds, plus very high temperatures, up to 350°C, through tall chimneys are called **hydrothermal vents (Figure 25.18)**. Cooler ones bubbling up methane through patchy areas near continental margins are called **cold seeps**.

The chimneys shown in Figure 25.18 were removed from the seafloor for study in the Edifice Rex sulfide recovery project in 1997 and 1998. Inside the hollow tube of Finn, hydrothermal fluids were rapidly venting at a temperature of 302°C, while outside the seawater at its surfaces ranged in temperature from 2°C to 10°C. This is an almost 300°C temperature gradient across the mere 5 to 42 cm thick chimney wall. Inside the wall, various zones existed, supporting different communities of microbes. Microscopic examination showed bacteria attached to all mineral surfaces throughout the wall. DNA probes showed mixed communities of eubacteria and archaea near the cool outer walls, gradually changing to primarily archaea at the hot inner wall surfaces. Most of the organisms were unculturable and unidentifiable.

(a)   SEM

(b)   LM

Figure 25.17 Phytoplankton: producers of the ocean. They include organisms such as **(a)** the marine diatom *Isthmia nervosa* (625X) *(Manfred Kage/Peter Arnold. Inc.)* and **(b)** diatoms collected in a plankton-tow from Rhode Island Sound (100X). *(Paul W. Johnson/Biological Photo Service)*

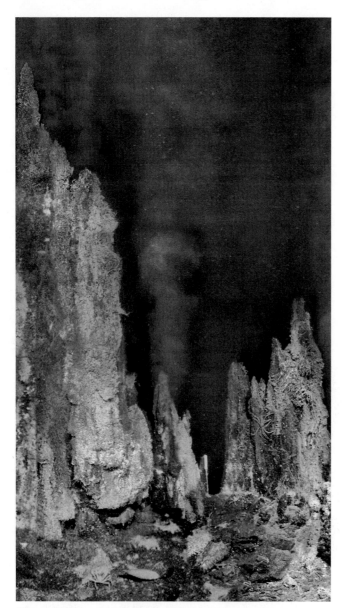

Figure 25.18 A hydrothermal vent chimney. Three sulfide chimneys (Finn, Roane, and Phang) from the Faulty Towers Complex of the Mothra Vent Field on the Juan de Fuca Ridge in the Pacific Northwest were removed and taken to a lab for analysis and study as part of the Edifice Rex project. Finn was venting 302°C fluids at the time of collection. *(Courtesy D. S. Kelley and J. R. Delaney, University of Washington)*

However, DNA sequences were closely related to the *Crenarchaeota* and *Euryarchaeota*, both kingdoms of archaea, and to the methanogens *Thermococcales* and *Archaeoglobales*. Thermophilic aerobic heterotrophic bacteria of the genera *Bacillus* and *Thermus* have also been found. Much basic work still remains to be done.

Cold seeps are common features found along continental margins (e.g., at Monterey Bay, California; the Eel River Basin, located on the continental slope off Northern California; and on the slope near the Florida Everglades). They have chemosynthetic biological communities such as clam beds and sulfide oxidizing mats and balls of bacteria associated with them. Groups of bacteria involved in anaerobic oxidation of methane are especially common. Layers of carbonated deposits are laid down at these sites, as a result of microbial metabolism.

Much has been discovered about these marine features in just over 25 years since the first hydrothermal chimney was discovered. Several web sites offer stunning "fly-by" videos of vents and seeps, as well as information about their communities. Try visiting http://oceanexplorer.noaa.gov/explorations/04fire/logs/april12/april12.html, or http://www.pmel.noaa.gov/vents/marianas/multimedia04.html. Move around to various sites from the vents home page.

WATER POLLUTION

Water is considered polluted if a substance or condition is present that renders the water useless for a particular purpose. Thus, the concept of water pollution is a relative one, depending on both the nature of the pollutants and the intended uses of the water. For example, human drinking water is considered polluted if it contains pathogens or toxic substances. Water that is too polluted for drinking may be safe for swimming; water that is too polluted for swimming may be acceptable for boating, industrial uses, or generating electrical power. The EPA has established drinking water standards and methods for water testing. Water that is fit for human consumption is termed **potable water**.

> *30,000 people die every day in developing countries due to lack of clean water (WHO).*

Pollutants

The major types of water pollutants are organic wastes such as sewage and animal manures, industrial wastes, oil, radioactive substances, sediments from soil erosion, and heat. Organic wastes suspended in water are usually decomposed by microorganisms, provided that the water contains sufficient oxygen for oxidation of the substances. The oxygen required for such decomposition is called **biological oxygen demand (BOD)**. When BOD is high, the water can be depleted of oxygen rapidly. Anaerobes increase in number while populations of aerobic decomposers dwindle, leaving behind large quantities of organic wastes. Organic wastes also can contain pathogenic organisms from among the bacteria, viruses, and protozoa.

Industrial wastes contain metals, minerals, inorganic and organic compounds, and some synthetic chemicals. The metals, minerals, and other inorganic substances can alter the pH and osmotic pressure of the water, and some are toxic to humans and other organisms. Synthetic chemicals can persist in water because most decomposers lack enzymes to degrade them. Oil is another important water pollutant. Radioactive substances released into water persist as a hazard to living organisms until they have undergone natural radioactive decay.

Soil particles, sand, and minerals from soil erosion enter water from agricultural, mining, and construction activities. Nitrates, phosphates, and other nutrients enter the water from detergents, fertilizers, and animal manures. Abundant nutrient enrichment of water, called **eutrophication**, leads to excessive growth of algae and other plants. Eventually the plants become so dense that sunlight cannot penetrate the water. Many algae and other plants die, leaving large quantities of dead organic matter in the water. The high BOD of this organic matter leads to oxygen depletion and the persistence of undecomposed matter in the water.

Even heat can act as a water pollutant when large quantities of heated water are released into rivers, lakes, or oceans. Increasing the temperature of water decreases the solubility of oxygen in it. The altered temperature and decreased oxygen supply significantly change the ecological balance of the aquatic environment.

The effects of water pollution are summarized in Table 25.1.

TABLE 25.1

The Effects of Water Pollution

Pollutant	Effects	Comment
Organic wastes (sewage, decaying plants, animal manures, wastes from food-processing plants, oil refineries, and leather, paper, and textile plants)	Increase biological oxygen demand of water	If adequate oxygen is available, these substances can be degraded by microorganisms usually present in water. If oxygen becomes depleted, decomposition is limited to what can be done by anaerobic decomposers. Water plants may be killed, and animals may be killed or caused to migrate.
Pathogenic organisms	Cause disease in humans who drink the water	Most bacteria are well controlled in public drinking water, but certain viruses, especially those that cause hepatitis, still cause human disease. More effective means of removing viruses during purification are needed.
Inorganic chemicals and minerals	Increase the salinity and acidity of water and render it toxic	Such chemicals should be removed during waste treatment. Heavy metals such as mercury, which are toxic to humans, should be prevented from entering water supplies.
Synthetic organic chemicals (herbicides, pesticides, detergents, plastics, wastes from industrial processes)	Can cause birth defects, cancer, neurological damage, and other illness	Because these substances are not biodegradable, chemical or physical means must be used to remove them during waste treatment. Many such substances become magnified (increased in concentration) as they are passed along food chains.
Plant nutrients	Cause excessive and sometimes uncontrolled growth of aquatic plants (eutrophication); impart undesirable odors and tastes to drinking water	Removal of excess phosphates and nitrates from water during waste treatment is costly and difficult.
Sediments from land erosion	Cause silting of waterways and destruction of hydroelectric equipment near dams; reduce light reaching plants in water and oxygen content of water	
Radioactive wastes	Can cause cancer, birth defects, radiation sickness when in large doses	Effects can be magnified through food chains. Because such wastes are difficult to remove from water, preventing them from reaching water is exceedingly important.
Heated water	Reduces oxygen solubility in water; alters habitats and kinds of organisms present; encourages growth of some aquatic life, but can decrease growth of desired organisms such as fish	

TABLE 25.2

Human Pathogens Transmitted in Water	
Organisms	**Diseases Caused**
Salmonella typhi	Typhoid fever
Other *Salmonella* species	Salmonellosis (gastroenteritis)
Shigella species	Shigellosis (bacillary dysentery)
Vibrio cholerae	Asiatic cholera
Vibrio parahaemolyticus	Gastroenteritis
Escherichia coli	Gastroenteritis
Yersinia enterocolitica	Gastroenteritis
Campylobacter fetus	Gastroenteritis
Legionella pneumophilia	Legionnaires' disease (pneumonia)
Hepatitis A virus	Hepatitis
Poliovirus	Poliomyelitis
Giardia intestinalis	Giardiasis
Balantidium coli	Balantidiasis
Entamoeba histolytica	Amoebic dysentery
Cryptosporidium parvum	Cryptosporidiosis (gastroenteritis)

Pathogens in Water

Human pathogens in water supplies usually come from contamination of the water with human feces. When water is contaminated with fecal material, any pathogens that leave the body through the feces—many bacteria and viruses and some protozoa—can be present. The most common pathogens transmitted in water are listed in Table 25.2. Water usually is tested for fecal contamination by isolating *Escherichia coli* from a water sample. *Escherichia coli* is called an **indicator organism** because, as *E. coli* is a natural inhabitant of the human digestive tract, its presence in water indicates that the water is contaminated with fecal material.

✓ CHECKLIST

1. What is the effect of UV radiation on organisms in air? In water?
2. Where are most organisms found in soil?
3. List four factors affecting soil microorganisms.
4. What is BOD? What do microbes have to do with this?

WATER PURIFICATION
Purification Procedures

Purification procedures for human drinking water are determined by the degree of purity of the water at its source. Water from deep wells or from reservoirs fed by clean mountain streams requires very little treatment to make it safe to drink. In contrast, water from rivers that contain industrial and animal wastes and even sewage from upstream towns requires extensive treatment before it is safe

APPLICATIONS
Just Don't Drink the Water

In the spring of 1993, the worst outbreak of waterborne disease in U.S. history occurred in Milwaukee. More than 400,000 cases of gastrointestinal illness and more than 100 deaths resulted from parasitic contamination of the drinking water supply. The city's water treatment facilities were not using the proper screens to exclude the waterborne parasite *Cryptosporidium parvum* from release into the drinking water. A more recent outbreak in Las Vegas has been linked to the deaths of 35 HIV-infected individuals.

Other contaminants include disease-producing microorganisms such as *C. parvum* and chemicals such as lead, nitrates, arsenic, and radon. Scientists have found that even chlorine reacts with decaying organic matter in water to form carcinogenic chemical by-products. Chief sources of water pollution include municipal and industrial wastes in landfills, residential septic systems, active and abandoned oil and gas wells, active and inactive coal and mineral mines, active and inactive underground storage tanks at gas stations, and a vast quantity of pesticides, fertilizers, and improperly disposed of motor oil.

to drink. Such water is first allowed to stand in a holding reservoir until some of the particulate matter settles out. Then alum (aluminum potassium sulfate) is added to cause **flocculation**, or precipitation of suspended colloids (◄Chapter 2, p. 33) such as clay. Many microorganisms are also removed from the water by flocculation.

Following flocculation treatment, water is filtered. **Filtration**, the passage of water through beds of sand, removes nearly all the remaining microorganisms. Charcoal can be used instead of sand for filtration; it has the advantage of removing organic chemicals that are not removed by sand. Finally, water is chlorinated. **Chlorination**, the addition of chlorine to water, readily kills bacteria but is less effective in destroying viruses and cysts of pathogenic protozoa. The amount of chlorine required to destroy most of the microorganisms is increased by the presence of organic matter in the water. Chlorine may combine with some organic molecules to form carcinogenic substances. Although current water chlorination procedures have not been proven to increase the risk of cancer in humans, the long-term effects of the interaction of chlorine and organic compounds are difficult to assess.

In the United States over 98% of the water supply systems that disinfect water use chlorine-based compounds as the disinfectant.

Tests for Water Purity

Water purity is usually tested by looking for **coliform bacteria**. The coliform bacteria, which include *E. coli*, are Gram-negative, non-spore-forming, aerobic or facultative anaerobic bacteria that ferment lactose and produce acid and gas. Most municipal water supplies are

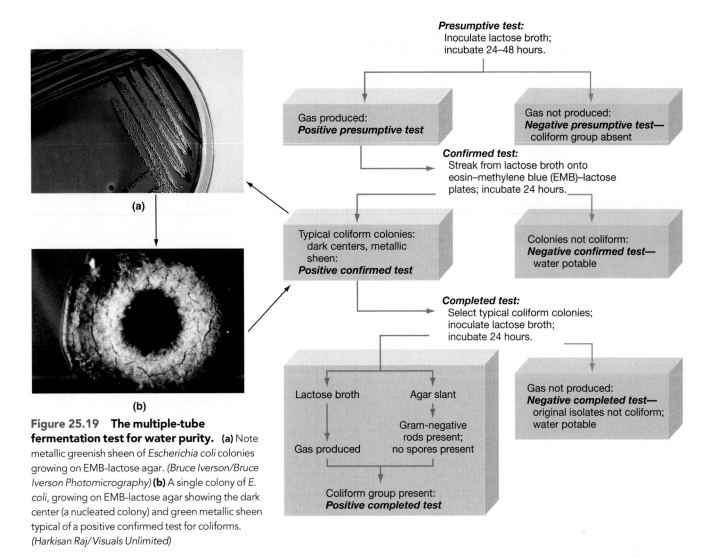

Figure 25.19 The multiple-tube fermentation test for water purity. (a) Note metallic greenish sheen of *Escherichia coli* colonies growing on EMB-lactose agar. *(Bruce Iverson/Bruce Iverson Photomicrography)* **(b)** A single colony of *E. coli*, growing on EMB-lactose agar showing the dark center (a nucleated colony) and green metallic sheen typical of a positive confirmed test for coliforms. *(Harkisan Raj/Visuals Unlimited)*

regularly tested for the presence of coliform bacteria. The presence of a significant number of coliforms is evidence that the water may not be safe for drinking. Three methods of testing for coliform bacteria currently in use are the multiple-tube fermentation method, the membrane filter method, and the ONPG and MUG test.

The **multiple-tube fermentation method (Figure 25.19)** involves three stages of testing: the presumptive test, the confirmed test, and the completed test. In the **presumptive test**, a water sample is used to inoculate lactose broth tubes. Each tube receives a water volume of 10 ml, 1 ml, or 0.1 ml. The tubes are incubated at 35°C and observed after 24 and 48 hours for evidence of gas production. Gas production provides presumptive evidence that coliform bacteria are present. In using the presumptive test, microbiologists analyzing water samples determine approximate numbers of organisms by means of the most probable number method (◀Chapter 6, p. 153).

Because certain noncoliform bacteria also produce gas, additional testing is necessary to confirm the presence of coliforms. In the **confirmed test**, samples from the highest dilution showing gas production are streaked onto eosin–methylene blue (EMB) agar plates. EMB prevents growth of Gram-positive organisms. Coliforms (which are Gram-negative) produce acid; under acidic conditions the eosin and methylene blue dyes are absorbed by the organisms of a colony. Thus, after 24-hour incubation, coliform colonies have dark centers and may also have a metallic greenish sheen. Observing such colonies confirms the presence of coliforms. In the **completed test**, organisms from dark colonies are used to inoculate lactose broth and agar slants. The production of acid and gas in the lactose broth and the microscopic identification of Gram-negative, non-spore-forming rods from slants constitute a positive completed test.

In the **membrane filter method (Figure 25.20)**, a 100-ml water sample is drawn through a sterile membrane filter that has pores about 0.45 μm in diameter. This membrane, which traps bacteria on its surface, is then incubated on the surface of a sterile absorbent pad previously saturated with an appropriate growth medium. After incubation, colonies form on the filter where bacteria were trapped during filtration. The presence of more than one colony per 100 ml of water indicates that

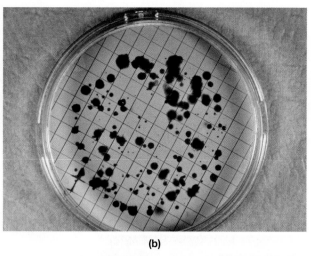

(a)

(b)

Figure 25.20   The membrane filter test for water purity. (a) Filtering water samples and trapping organisms on filter pads, which are then incubated. *(Leon J. LeBeau/Biological Photo Service)* (b) After incubation, the coliform colonies that appear are counted. *(Raymond B. Otero/Visuals Unlimited)*

the water may be unsafe for human consumption. If colonies are observed, additional tests can be performed to identify them specifically. The membrane filter method is much faster, and permits larger volumes of water to be tested, than the multiple-tube fermentation method.

The **ONPG and MUG test** relies on the ability of coliform bacteria to secrete enzymes that convert a substrate into a product that can be detected by a color change (◄Chapter 5, p. 120). In this test, a water sample is inoculated into a nutrient-broth tube containing the substrates ONPG (*O-n*itrophenyl-β-d-galactopyranoside) and MUG (4-*m*ethyl*u*mbelliferyl-β-d-glucuronide). After a suitable incubation period, if coliforms are present, they will have secreted the enzymes β-galactosidase and β-glucuronidase and other substances. Beta-galactosidase hydrolyzes ONPG into a yellow-colored product, whereas β-glucuronidase hydrolyzes MUG into a product that fluoresces blue when illuminated with ultraviolet light **(Figure 25.21)**. This test can be used in conjunction with the other tests for water purity.

The Easyphage™ test looks not for coliform bacteria themselves but for bacteriophage viruses that indicate the presence of coliforms **(Figure 25.22)**. The kit can be used for food as well as water testing.

On the basis of CDC estimates, 940,000 Americans become ill each year from contaminated water, and 900 die.

In addition to contamination with coliforms, drinking water can also contain other organisms sometimes referred to as "nuisance" organisms. Although these organisms do not produce human disease, they can affect the taste, color, or odor of the water. Some also can form insoluble precipitates inside water pipes. Nuisance organisms include sulfur bacteria, iron bacteria, slime-forming bacteria, and algae. Among the sulfur bacteria are *Desulfovibrio*, which produces hydrogen sulfide, and *Thiobacillus*, which produces sulfuric acid that can corrode pipes. Iron bacteria deposit insoluble iron compounds that can obstruct water flow in pipes. Eukaryotic algae, diatoms, and cyanobacteria reproduce rapidly in water exposed to sunlight (as in reservoirs). They can become so numerous that they clog filters used in the purification process. Identifying these various nuisance organisms in

Figure 25.22   Easyphage™. This test uses bacteriophages to indicate the presence of coliform bacteria in food or water testing. Presence of a plaque (hole) is a positive indicator of a specific bacterium which has been grown on the Petri dish. This means that this bacterium was present in the original food or water sample; otherwise bacteriophages which attack it would not be present. *(Courtesy Scientific Methods, Inc.)*

Figure 25.21   The ONPG and MUG test. A positive ONPG produces a yellow color; a positive MUG gives off a blue fluorescence on ultraviolet illumination. A sample without coliforms remains clear. *(Colibert photo courtesy of IDEXX Laboratories)*

water is a tedious task because different tests are required for each kind of organism. None of the tests are performed routinely but may be done when citizens complain of objectionable tastes, odors, or colors in water.

Water can also be contaminated with various organic substances, especially when water supplies are drawn from rivers that have received upstream effluents of industrial wastes. Although it is at least theoretically possible to detect these substances through chemical analyses, such tests are rarely performed.

▮▮▮ SEWAGE TREATMENT

▮▮ **Sewage** is used water and the wastes it contains. It is about 99.9% water and about 0.1% solid or dissolved wastes. These wastes include household wastes (human feces, detergents, grease, and anything else people put down the drain or garbage disposal unit), industrial wastes (acids and other chemical wastes and organic matter from food-processing plants), and wastes carried by rainwater that enters sewers.

Sewage treatment is a relatively modern practice. Until recent years, many large U.S. cities dumped untreated sewage into rivers and oceans; many cities along the Mediterranean Sea still do! When small amounts of sewage are dumped into fast-flowing, well-oxygenated rivers, the natural activities of decomposers in the river purify the water. However, large amounts of sewage overload the purification capacity of rivers. Cities downstream are then forced to take their water supplies from rivers that contain wastes of the cities upstream. Fortunately, most U.S. cities now have some form of sewage treatment.

Complete sewage treatment consists of three steps: primary, secondary, and tertiary treatment (**Figure 25.23**). In **primary treatment**, physical means are used to remove solid wastes from sewage. In **secondary treatment**, biological means (actions of decomposers) are used to remove solid wastes that remain after primary treatment. In **tertiary treatment**, chemical and physical means are used to produce an effluent of water pure enough to drink. Let us look at each of these processes in more detail.

Figure 25.23 **An overview of a sewage treatment plant.** Primary, secondary, and tertiary treatment facilities are shown.

Figure 25.24 Trickling filters. These filters are used in secondary sewage treatment. *(R.F. Ashley/Visuals Unlimited)*

PRIMARY TREATMENT

As raw sewage enters a sewage treatment plant, several physical processes are used to remove wastes in primary treatment. Screens remove large pieces of floating debris, and skimmers remove oily substances. Water is then directed through a series of sedimentation tanks, where small particles settle out. The solid matter removed by these procedures accounts for about half the total solid matter in sewage. Flocculating substances can be used to increase the amount of solids that settle out and thus the

(a)   TEM

(b)   LM

Figure 25.25 Two sheathed bacteria used in trickling filters. (a) *Sphaerotilus* (29,300X); *(Judith F.M. Hoeniger, University of/Biological Photo Service)* **(b)** *Beggiatoa* (400X). *(Paul W. Johnson/ Biological Photo Service)*

proportion of solids removed by primary treatment. Sludge is removed from the sedimentation tanks intermittently or continuously, depending on the design of the treatment plant.

SECONDARY TREATMENT

The effluent from primary treatment flows into secondary treatment systems. These systems are of two types: trickling filter systems and activated sludge systems. Both systems make use of the decomposing activity of aerobic microorganisms. The BOD is high in secondary treatment systems, so the systems provide for continuous oxygenation of the wastewater.

In a **trickling filter system (Figure 25.24)**, sewage is sprayed over a bed of rocks about 2 m deep. The individual rocks are 5 to 10 cm in diameter and are coated with a slimy film of aerobic organisms such as *Sphaerotilus* and *Beggiatoa* **(Figure 25.25)**. Spraying oxygenates the sewage so that the aerobes can decompose organic matter in it. Such a system is less efficient but less subject to operational problems than an activated sludge system. It removes about 80% of the organic matter in the water.

In an **activated sludge system**, the effluent from primary treatment is constantly agitated, aerated, and added to solid material remaining from earlier water treatment. This **sludge** contains large numbers of aerobic organisms that digest organic matter in wastewater. However, filamentous bacteria multiply rapidly in such systems and cause some of the sludge to float on the surface of the water instead of settling out. This phenomenon, called **bulking**, allows the floating matter to contaminate the effluent. The sheathed bacterium *Sphaerotilus* **(Figure 25.25a)**, which sometimes proliferates rapidly on decaying leaves in small streams and causes a bloom, can interfere with the operation of sewage systems in this way. Its filaments clog filters and create floating clumps of undigested organic matter.

Sludge from both primary and secondary treatments can be pumped into **sludge digesters**. Here, oxygen is virtually excluded, and anaerobic bacteria partially digest the sludge to simple organic molecules and the gases carbon dioxide and methane. The methane can be used for heating the digester and providing for other power needs of the treatment plant. Undigested matter can be dried and used as a soil conditioner or as landfill **(Figure 25.26)**.

TERTIARY TREATMENT

The effluent from secondary treatment contains only 5% to 20% of the original quantity of organic matter and can be discharged into flowing rivers without causing serious problems. However, this effluent can contain large quantities of phosphates and nitrates, which can increase the growth rate of plants in the river.

Figure 25.26 **Sludge digestion.** The injection of municipal sewage sludge into farmland is designed to return nutrients to the soil. But if treatment is not complete, it may also add pathogens to the soil. *(Larry Lefever/Grant Hellman Photography)*

Tertiary treatment is an extremely costly process that involves physical and chemical methods. Fine sand and charcoal are used in filtration. Various flocculating chemicals precipitate phosphates and particulate matter. Denitrifying bacteria convert nitrates to nitrogen gas. Finally, chlorine is used to destroy any remaining organisms. Water that has received tertiary treatment can be released into any body of water without danger of causing eutrophication. Such water is pure enough to be recycled into a domestic water supply. However, the chlorine-containing effluent, when released into streams and lakes, can react to produce carcinogenic compounds that may enter the food chain or be ingested directly by humans in their drinking water. It would be safer to remove the chlorine before releasing the effluent, but this is rarely done today, although the cost is not great. Ultraviolet lights are now replacing chlorination as the final treatment of effluent (◄Chapter 12, p. 358). They destroy microbes without adding carcinogens to our streams and waters. Likewise, especially in Europe, the treatment of effluent with ozone is

Figure 25.27 **Sewage disposal by means of a septic tank and drainage field system.** **(a)** Solid materials settle out as sludge, which undergoes microbial decay, while soluble materials continue to the drainage field. There **(b)** they seep out into the soil, which filters out bacteria and some viruses and binds phosphate. Soil bacteria decompose organic materials.

replacing chlorination. Ozone generators are simple and not very costly, and they do not add carcinogens to natural waterways.

SEPTIC TANKS

About 50 million rural families in the United States do not have access to city sewer connections or their treatment facilities. These homes rely on backyard **septic tank** systems (Figure 25.27). Homeowners must be careful not to flush or put materials such as poisons and grease down the drain, as these might kill the beneficial microbes in the septic tank that decompose sludge solids that accumulate there. This would necessitate immediate pumping of the tank by a vehicle known as the "honey wagon" to prevent sewage

from backing up into the house. Even with normal operation, it is occasionally necessary to pump the sludge from the tank and haul it to a sewage treatment plant.

Soluble components of the sewage continue out of the septic tank into the drainage (leaching) field. There they seep through perforated pipe, past a gravel bed, and into the soil, which filters out bacteria and some viruses and binds phosphate. Soil bacteria decompose organic materials. Drainage fields must be placed where they will not allow seepage into wells, a difficult problem on hills or in densely populated areas. Drainage fields cannot be used where the water table is too high or the soil is insufficiently permeable, such as in rocky areas. After 10 years or more, the average drainage field clogs up and can no longer be used.

BIOTECHNOLOGY

Putting Microbes to Work to Clean Our Environment

In 1991, manufacture of the East German auto, the Trabant, ceased. The two-cylinder, two-stroke engine couldn't compete with the Mercedes and BMWs of what was then West Germany. Imagine the pain, after perhaps 10 years of saving to buy one, of finding that what should have been a status symbol was now a costly dinosaur. Customers seeking to trade in their Trabants for a bigger car not only got no trade-in allowance but were charged $300 more to dispose of their old Trabant. Space in landfills is scarce and costly in Germany.

Biotechnology may soon have a solution to this type of problem by means of microbial *biodegradation*, or decomposition of man-made materials by microbial activity. Trabants are made mostly of plastic. They are so light that two people can lift one into a dumpster. This plastic has a great deal of agricultural waste (cellulose) from the former Soviet Union incorporated into it, which should make it easily biodegradable by microbes. Scientists are searching the soil of landfills for microbes that are degrading the plastic of Trabants already buried there. These organisms could be used to disintegrate the glut of unwanted Trabants more rapidly. One Berlin biotechnology company claims to have developed a bacterium that will eat a Trabant in 20 days, leaving only a small pile of compost.

Plastics such as soda bottles and disposable diapers are an ever-increasing component of already overcrowded landfills. Some companies are producing biodegradable plastics meant to be broken down by microbes over time. In fact, the manufacture of those plastics depends on bacteria. Poly-β-hydroxy-alkanes (PHAs) are biodegradable polyesters (polymers formed with ester bonds; ◀Chapter 2, p. 39) made by bacteria such as *Alcaligenes eutrophus*. These bacteria store PHAs in

The end of a dream. Dozens of discarded Trabant cars are stacked up in a Berlin junkyard. Scientists are looking for microbes that will eat the cars, which are mostly made of plastic. East Germans once saved for years to afford one of these "dream" cars. (©AP/Wide World Photos)

granules as a reserve of carbon. When carbon is readily available in the form of glucose, *A. eutrophus* produces large quantities of PHAs to store it for lean times. When carbon is scarce, the bacterium uses its own stores of carbon in the granules. Thus, this one bacterium can be used to manufacture PHAs—which are then formed into bottles, plastic bags, and such—as well as to degrade those products once their useful days have passed!

Although old Trabants and plastics are a nuisance, they aren't harmful, unlike contaminated soil and water. Dangerous chemical contaminants must be removed from soil and water.

BIOREMEDIATION

Bioremediation is a process that uses naturally occurring or genetically engineered microorganisms such as yeast, fungi, and bacteria to transform harmful substances into less toxic or nontoxic compounds. Microorganisms break down a variety of organic compounds in nature to obtain nutrients, carbon, and energy for growth and survival. Bioremediation promotes the growth of microorganisms to degrade contaminants by utilizing those contaminants as carbon and energy sources.

Bioremediation has been used since the late 1970s to degrade petroleum product and hydrocarbons. (◀See the box "Microbial Cleanup" in Chapter 5, p. 135.) In March 1989, the supertanker *Exxon Valdez* ran aground in Prince William Sound, Alaska. The leaking tanker flooded the shoreline with about 42 million liters (11 million gal) of crude oil. The gravel and sand beaches were saturated with oil more than half a meter thick. A massive cleanup program was organized. At first conventional cleanup techniques—such as booms, high-pressure hot-water sprays, skimmers, and manual scrubbers—were used. But the shore remained black and gooey because these methods could not remove all the oil from beneath rocks and within beach sediments.

Scientists from the EPA decided to use bioremediation to enhance the cleanup. They sprayed the beach with fertilizer (nutrients) to stimulate the growth of the native microorganisms and promote the use of the oil wastes as the microbes' carbon source. The areas sprayed with fertilizer were soon nearly clean of oil to a depth of about one-third of a meter, while untreated areas were still coated with sticky oil. Studies of one treated beach show that 60% of total hydrocarbons and 45% of polycyclic aromatic hydrocarbons (PAHs), which are potentially toxic, were degraded by bacteria within 3 months. The *Exxon Valdez* cleanup was a bioremediation success story.

Another recent bioremediation success story involves the natural oil-detoxification actions of microbes. In 1995, scientists studying the Persian Gulf region, where lakes of oil had been left after pipelines and oil wells were damaged in the 1991 Gulf War, concluded that a natural recovery of the region's plants was occurring. Samir Radwan and his colleagues in Kuwait found that the roots of wildflowers in the oil-soaked desert were healthy and oil-free. Cultures made of bacteria and fungi from the sand revealed several types of known oil-eating microorganisms, such as the bacterium *Arthrobacter*. The researchers believe they may have found a cheap, safe, and natural method for cleaning up oil spills on land—cultivating plants whose roots recruit microbial oil-eaters.

The applications for bioremediation are rapidly expanding. It has already proved to be a successful means of decomposing wastes in landfills. It may be suitable for cleaning up soil and groundwater contaminated by leaks from underground tanks storing petroleum, heating oil, and other materials. The wood-preservative industry also appears to be a promising area for bioremediation. Each year the United States uses 450,000 tons of creosote, an oily liquid that is distilled from coal tar and used as a wood preservative. Creosote sometimes leaks from its holding tanks and seeps into the soil and underlying groundwater. The white-rot fungus *Phanerochaete chrysosporium* has been shown to degrade pentachlorophenol, the main contaminant at wood-preserving sites. This superfungus can also destroy other toxic compounds in the soil, including dioxins, polychlorinated biphenyls (PCBs), and PAHs.

At present only naturally occurring microbes are used in bioremediation. Scientists are experimenting with genetic engineering techniques, however, to develop microorganisms for use at hazardous waste sites. Before these bioengineered organisms can be used in the field, the EPA requires that they undergo a safety review in accordance with the Toxic Substance Control Act to evaluate any possible risks to human health or to the environment.

There are both advantages and disadvantages to the use of bioremediation. *Advantages*: Bioremediation is an ecologically sound, "natural" process; it destroys target chemicals at the contamination site instead of transferring contaminants from one site to another; and the process is usually less expensive than other methods used for cleaning up hazardous wastes. *Disadvantages*: Using bioremediation often takes longer than other remedial methods such as excavation or incineration, and bioremediation techniques are not yet refined for sites with mixtures of contaminants. More research is needed to perfect this technology. Nevertheless, bioremediation holds enormous promise for the future. As scientists develop more practical uses for bioremediation, this technology will become more important in cleaning up and protecting the environment.

✓CHECKLIST

1. Define coliform bacteria. If they are not pathogens, why are they important?
2. Compare primary, secondary, and tertiary sewage treatment as to what they remove.
3. Why must you be very careful about what you flush down into a septic tank system?
4. What are some of the advantages and disadvantages of bioremediation?

▌ RETRACING OUR STEPS

▌▌▌ FUNDAMENTALS OF ECOLOGY

- **Ecology** is the study of relationships among organisms and their environment.

THE NATURE OF ECOSYSTEMS

- An **ecosystem** includes all the **biotic** and **abiotic factors** of an environment. It will always support life of **indigenous organisms** and sometimes supports **nonindigenous** ones. All the living organisms in an **ecosystem** make up a **community**.

THE FLOW OF ENERGY IN ECOSYSTEMS

- Energy in an ecosystem flows from the sun to **producers** to **consumers**. **Decomposers** obtain energy from digesting dead bodies and wastes of other organisms. The nutrients they release can be recycled.

▌▌▌ BIOGEOCHEMICAL CYCLES

- Although energy is continuously available, nutrients must be recycled from dead organisms and waste to make them available for other organisms.

THE WATER CYCLE

- The **water cycle** is summarized in Figure 25.2.

THE CARBON CYCLE

- The **carbon cycle** is summarized in Figure 25.3.

THE NITROGEN CYCLE AND NITROGEN BACTERIA

- The **nitrogen cycle** is summarized in Figure 25.4.
- **Nitrogen fixation** is the reduction of atmospheric nitrogen to ammonia. It is accomplished by certain free-living aerobes and anaerobes but mostly by *Rhizobium*.
- *Rhizobium* cells accumulate around legume roots and change to **swarmer cells**, which invade root cells and become bacteroids. **Nitrogenase** in **bacteroids** catalyzes nitrogen-fixation reactions.
- **Nitrification** is the conversion of ammonia to nitrites and nitrates; nitrites are formed by *Nitrosomonas*, and nitrates by *Nitrobacter*.
- **Denitrification** is the conversion of nitrates to nitrous oxide and nitrogen gas. It is accomplished by a variety of organisms, especially in waterlogged soils.

THE SULFUR CYCLE AND SULFUR BACTERIA

- The **sulfur cycle** is summarized in Figure 25.7.
- Various bacterial species carry out **sulfate reduction**, **sulfur reduction**, or **sulfur oxidation**.

OTHER BIOGEOCHEMICAL CYCLES

- The **phosphorus cycle** is summarized in Figure 25.9.
- All elements found in living organisms must be recycled.

THE DEEP HOT BIOSPHERE

- Bacteria have been found at the deepest depths ever drilled. It is probable that the inner areas of the Earth have a continuous, subcrustal culture of bacteria living there, consuming oil and gas originally present in the Earth.

▌▌▌ AIR

MICROORGANISMS FOUND IN AIR

- Microorganisms are transmitted by air but do not grow in air. Air is analyzed by exposing agar plates to it and by drawing air over the surface of agar or into a liquid medium and then studying the organisms found.

METHODS FOR CONTROLLING MICROORGANISMS IN AIR

- Microorganisms in air are controlled by chemical agents, radiation, filtration, and unidirectional airflow.

▌▌▌ SOIL

COMPONENTS OF SOIL

- Soil contains inorganic rocks, minerals, water, gases, organic matter (**humus**), and many microorganisms.

MICROORGANISMS IN SOIL

- Microbes of all major taxonomic groups are found in soil.
- Physical factors that affect soil microorganisms include moisture, oxygen concentrations, pH, and temperature.
- Organisms also alter the characteristics of their environment as they use nutrients and release wastes.
- Soil microorganisms are important as decomposers in the carbon cycle and in all phases of the nitrogen cycle.

SOIL PATHOGENS

- Soil pathogens affect mainly plants and insects.
- Various species of *Clostridium* are important human pathogens found in soil.

CAVES

- Microbes eating oil deposits deep underground release H_2S which forms sulfuric acid when it reacts with water. The acid eats out caves along cracks in the rocks.
- Other bacteria growing on the cave walls eat sulfur, forming sulfuric acid which drops from strings of mucus-like bacterial colonies, further enlarging the cave.

▌▌▌ WATER

FRESHWATER ENVIRONMENTS

- Freshwater environments are characterized by low salinity and variability in temperature, pH, and oxygen concentration.
- Microorganisms from all major taxonomic groups are found in freshwater environments. Bacteria—aerobes where oxygen is plentiful, and anaerobes where oxygen is depleted—are especially abundant.

MARINE ENVIRONMENTS

- Marine environments are characterized by high salinity, smaller variability in temperature, pH, and oxygen concentration. As depth increases, pressure increases and sunlight penetration decreases. Microorganisms from all major taxonomic groups are also found in marine environments. Photosynthetic organisms are found nearest the surface, heterotrophs in surface and lower strata, and decomposers in bottom sediments.

HYDROTHERMAL VENTS AND COLD SEEPS

• Steep chemical and thermal gradients create specific extreme environments inhabited by eubacteria and archaea, as well as higher forms of life, in very productive communities.

WATER POLLUTION

• Water is polluted if a substance or condition is present that renders the water useless for a particular purpose.
• Table 25.1 summarizes the effects of water pollution.
• Many human pathogens can be transmitted in water.
• Such pathogens are listed in Table 25.2.

WATER PURIFICATION

• Water purification involves **flocculation** of suspended matter, **filtration**, and **chlorination**.
• Tests for water purity are designed to detect **coliform bacteria**; they include the **multiple-tube fermentation** and **membrane filter methods** and the **ONPG and MUG test**, and the **Easyphage**® test for fecal indicator bacteriophages.

█ SEWAGE TREATMENT

• **Sewage** is used water and the wastes it contains.

PRIMARY TREATMENT

• **Primary treatment** is the physical removal of solid wastes.

SECONDARY TREATMENT

• **Secondary treatment** is the removal of organic matter by the action of aerobic bacteria.

TERTIARY TREATMENT

• **Tertiary treatment** is the removal of most organic matter, nitrates, phosphates, and any surviving microorganism by physical and chemical means.

SEPTIC TANKS

• Soil bacteria decompose soluble components of sewage in the drainage field of a **septic tank** system.

█ BIOREMEDIATION

• Naturally occurring or genetically engineered microbes transform harmful substances into less toxic or nontoxic compounds by using the contaminants as carbon and energy sources.

█ TERMINOLOGY CHECK

abiotic factor **(p. 783)**	completed test **(p. 803)**	membrane filter method **(p. 803)**	primary treatment **(p. 805)**
activated sludge system **(p. 806)**	confirmed test **(p. 803)**	microenvironment **(p. 783)**	producer **(p. 784)**
bacteroid **(p. 788)**	consumer **(p. 784)**	multiple-tube fermentation method **(p. 803)**	secondary treatment **(p. 805)**
biogeochemical cycle **(p. 784)**	decomposer **(p. 784)**	nitrification **(p. 788)**	septic tank **(p. 808)**
biological oxygen demand (BOD) **(p. 800)**	deep hot biosphere **(p. 791)**	nitrogenase **(p. 786)**	sewage **(p. 805)**
bioremediation **(p. 809)**	denitrification **(p. 789)**	nitrogen cycle **(p. 786)**	sludge **(p. 806)**
biosphere **(p. 783)**	ecology **(p. 783)**	nitrogen fixation **(p. 786)**	sludge digester **(p. 806)**
biotic factor **(p. 783)**	ecosystem **(p. 783)**	nonindigenous organism **(p. 783)**	snottite **(p. 796)**
bulking **(p. 806)**	eutrophication **(p. 801)**	ONPG and MUG test **(p. 804)**	sulfate reduction **(p. 791)**
carbon cycle **(p. 784)**	filtration **(p. 802)**	phosphorus cycle **(p. 791)**	sulfur cycle **(p. 789)**
chlorination **(p. 802)**	flocculation **(p. 802)**	potable water **(p. 800)**	sulfur oxidation **(p. 791)**
cold seep **(p. 799)**	humus **(p. 794)**	presumptive test **(p. 803)**	sulfur reduction **(p. 791)**
coliform bacterium **(p. 802)**	hydrologic cycle **(p. 784)**		swarmer cell **(p. 788)**
community **(p. 783)**	hydrothermal vent **(p. 799)**		tertiary treatment **(p. 805)**
	indicator organism **(p. 802)**		trickling filter system **(p. 806)**
	indigenous organism **(p. 783)**		water cycle **(p. 784)**

█ CLINICAL CASE STUDY

Helen developed red, itchy scaly patches on her scalp. When the doctor told her that she had tinea capitis, ringworm of the scalp, she was shocked. "But I don't know anyone with ringworm—I haven't been around anyone to catch it from. I lead a fairly isolated life. My main activity is gardening!" What could you tell Helen about her possible source of infection?

█ CRITICAL THINKING QUESTIONS

1. With so much advertising for various antibacterial products, many people have gotten the impression that all bacteria are harmful and that maybe the Earth would be better off if all bacteria were eliminated. What would life be like in a bacteria-free world?

2. Oceanic phytoplankton are the primary producers of the sea, providing nutrients for heterotrophic organisms. During their photosynthetic process, phytoplankton absorb large amounts of carbon dioxide (a major greenhouse gas), which in turn affects global temperatures. Phytoplankton populations undergo seasonal fluctuations, which in turn can affect global temperatures. What do you think is directly responsible for the seasonal fluctuation in phytoplankton populations and, as a result, indirectly for the Earth's weather?

3. Water is usually tested for safety by using fecal coliforms (primarily *Escherichia coli*) as an indicator of fecal contamination. The validity of this approach has recently been questioned. Do you know why?

4. How do microorganisms make caves?

▌SELF-QUIZ

1. Match the following ecological terms with their descriptions:
___ Abiotic factor
___ Biotic factor
___ Ecosystem
___ Indigenous organisms
___ Community
___ Biosphere
___ Ecology

(a) Study of the relationships among organisms and their environment
(b) The region of earth inhabited by living organisms
(c) Native organisms always found in a given environment
(d) All organisms within a given area together with surrounding biotic and abiotic factors
(e) The living organisms of the environment
(f) Consists of all the kinds of organisms present in a given environment
(g) Physical or nonliving factors of the environment

2. Match the following microorganisms to their corresponding ecological category. Some may have more than one.
_____ Producer _____ Consumer _____ Decomposer
(a) Heterotrophic bacteria
(b) Fungi
(c) Photosynthetic cyanobacteria
(d) Virus
(e) Chemoautotrophic bacteria
(f) Parasitic amoeba
(g) Bacteria found on a fallen, uprooted tree that has been there for one year

3. What percentage of the nitrogen fixed annually on Earth is fixed by the action of bacteria?
(a) Less than 10%   (d) 70%
(b) 20%   (e) 90%
(c) 50%

4. Place the following nitrogen cycle definitions/terms and microorganisms with their corresponding categories:
___ Nitrogen-fixing bacteria
___ Nitrifying bacteria
___ Denitrifying bacteria

1. Nitrates are reduced to nitrous oxide or nitrogen gas
2. Reduction of atmospheric nitrogen gas to ammonia
3. Process by which ammonia or ammonia ions are oxidized to nitrites or nitrates
4. Nitrogenase enzyme
(a) Species of the genus *Azotobacter*
(b) Autotrophic bacteria

(c) Species of the genera *Nitrosomonas* and *Nitrobacter*
(d) Cyanobacteria
(e) Symbiotic *Rhizobium*
(f) Species of the genus *Clostridium*
(g) Gram-negative rod-shaped bacteria
(h) Species of the genera *Pseudomonas*

5. The form of nitrogen that is most usable in plants is:
(a) Nitrate   (d) Ammonia
(b) Nitrite   (e) Ammonium ions
(c) Molecular nitrogen

6. Which step of the nitrogen cycle is considered to be wasteful and detrimental to the environment?
(a) Nitrogen fixation   (d) All of the above
(b) Nitrification   (e) None of the above
(c) Denitrification

7. Which form of sulfur is most usable by both microorganisms and plants?
(a) Sulfite   (d) Sulfate
(b) Hydrogen sulfide   (e) None of these
(c) Elemental sulfur

8. All sulfate-reducing microorganisms can be classified under the following headings EXCEPT:
(a) Psychrophilic   (d) Halophilic
(b) Mesophilic   (e) Aerobic
(c) Thermophilic

9. Which type of bacteria is responsible for producing extremely toxic sulfuric acid?
(a) *Thiobacillus*   (d) Sulfur-oxidizing
(b) Sulfate-reducing   (e) a and d
(c) Sulfur-reducing

10. Soil microorganisms involved in the phosphorus cycle are important in which of the following ways?
(a) They convert inorganic phosphates to water-soluble orthophosphate.
(b) They break down organic phosphates from decomposing organisms to inorganic phosphates.
(c) They convert ATP to ADP.
(d) Both a and b.
(e) Both b and c.

11. All of the following are true about soil microorganisms EXCEPT:
(a) They never affect or change the physical characteristics of their soil microenvironment.
(b) They are considered to be a part of the soil.

(c) All major microbial taxonomic groups can be found in soil (bacteria, fungi, algae, protists, viruses).

(d) Species of the genus *Clostridium* contain important human pathogens found in soil.

(e) They are important as decomposers in the carbon and nitrogen cycles.

12. Which of the following could be an effect of an excess of sewage running into a freshwater reservoir?
    (a) It could cause the water to become polluted and nonpotable.
    (b) It could cause an increase in the biological oxygen demand (BOD).
    (c) It could cause oxygen depletion, leading to an increase in the anaerobic microbial population.
    (d) Large quantities of organic wastes could be left behind, killing off many plants and animals or causing animals to migrate.
    (e) All of the above.

13. All of the following are true about releasing untreated sewage into a river EXCEPT:
    (a) It is a health hazard
    (b) It increases the BOD
    (c) It decreases the dissolved oxygen
    (d) It kills bacteria
    (e) None of the above

14. The microorganism that is mainly used as an indicator of fecal pollution in water is:
    (a) *Escherichia coli*
    (b) *Clostridium tetani*
    (c) *Clostridium botulinum*
    (d) Cyanobacteria
    (e) All of these can be used

15. Developing sewage treatment and clean water systems for drinking and washing could help to eradicate disease caused by all of the following organisms EXCEPT:
    (a) *Escherichia coli*, enterotoxigenic
    (b) *Vibrio cholerae*
    (c) *Giardia lamblia*
    (d) *Toxoplasma gondii*
    (e) None of the above

16. The worst outbreak of waterborne disease in the U.S. was caused by:
    (a) *Escherichia coli*
    (b) *Salmonella typhi*
    (c) *Cryptosporidium parvum*
    (d) *Clostridium botulinum*
    (e) None of these

17. Which of the following microorganisms are not derived from human pollution but can be found in water supply systems and are considered "nuisance" organisms?
    (a) Coliform bacteria
    (b) *Rhizobium*
    (c) *Azotobacter*
    (d) *Thiobacillus*
    (e) *Clostridium*

18. Biological treatment of sewage by microorganisms (mainly decomposers) would most likely occur at which stage of wastewater treatment?
    (a) Primary
    (b) Secondary
    (c) Tertiary
    (d) Advanced
    (e) All of these

19. Which of the following wastewater treatments is most likely to produce carcinogens as a by-product?
    (a) Chlorination
    (b) Ozonation
    (c) Ultraviolet (UV) light
    (d) Sand filtration
    (e) Carbon filtration

20. What are the consequences of having an excess of nutrients such as nitrates and phosphates running into a freshwater reservoir in the process of eutrophication?
    (a) It could lead to an excess of algal and plant growth.
    (b) It could lead to a decrease in sunlight penetration and an increase in BOD.
    (c) It could lead to a long-standing rich and diverse growth of more plants and animals.
    (d) Both a and b.
    (e) Both b and c.

21. Categorize the following as to whether they are advantages or disadvantages of bioremediation:
    ___ Advantages
    ___ Disadvantages
    (a) Lower costs
    (b) Longer time requirement
    (c) Destruction of target chemicals at the contamination site
    (d) Lack of refinement for dealing with mixed contaminants
    (e) Genetically engineered microbes

22. Which of the following microorganisms is the primary producer of oceans?
    (a) Heterotrophic bacteria
    (b) Phytoplankton
    (c) Thermophilic bacteria
    (d) Chemoheterotrophs
    (e) Psychrophiles

23. Which of the following is NOT a factor that can influence where and what types of microorganisms grow in a marine environment?
    (a) Depth penetration of sunlight and oxygen
    (b) Depth penetration and shoreline proximity of nutrients
    (c) Salinity
    (d) Hydrostatic pressure
    (e) Temperature

24. The most common microbiological contaminant of air is:
    (a) Spores from bacteria
    (b) Spores from molds
    (c) Fungal hyphae
    (d) Gram-positive bacteria
    (e) Gram-negative bacteria

25. Which of the following nutrients is often in limited supply in freshwater lakes and ponds?
    (a) Carbon
    (b) Nitrogen
    (c) Oxygen
    (d) Phosphorus
    (e) Minerals

26. Use the following diagram to explain how energy flows through an ecosystem. In doing so, identify parts (a) through (d).

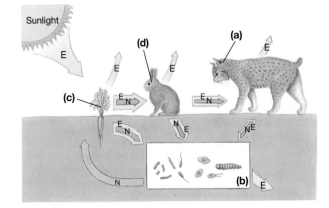

■ EXPLORATIONS ON THE WEB   http://www.wiley.com/college/black

If you think you've mastered this chapter, there's more to challenge you on the web. Go to the companion web site to fine-tune your understanding of the chapter concept and discover answers to the questions posed below.

1. Some microscopic organisms have the ability to remove harmful pollutants from wastewater. Discover how they can accomplish this at the web site.

2. Coral reefs are dying at an alarming rate, and if the rate continues, up to 70% of the reefs could be dead in 20 to 40 years. Who or what is the cause? Find out at the web site.

Come with me . . .

"Did you know that mushrooms are the number one cash crop in Pennsylvania?" asked Jim Agelucci, general manager of Phillips Mushroom Farms in Kennett Square, Pennsylvania, as we walked toward the mushroom growing sheds. I hadn't realized what a big industry it is, and it's one that definitely requires a good knowledge of microbiology. Over 735 million pounds of mushrooms are grown each year in the United States, 47% of these in Pennsylvania. And 44% of Pennsylvania mushrooms are grown right here in Chester County, the largest production area in the United States.

Americans prefer the white button mushrooms, Agaricus bisporus, even though the brown strain is larger and meatier and has better flavor, longer shelf life, and higher productivity (more crop per square foot). Genetic engineers are trying to combine the admirable qualities of the brown strains with the desired white color. In fact, almost all varieties of button mushrooms grown today are genetically engineered strains and may have an off-white color. Phillips is also experimenting with growing small quantities of specially mushrooms such as Portobello, Nameko, winecaps, and mitake. These are new to Americans and are appearing in gourmet restaurants and specialty produce stores. Once the cultivation methods are better worked out, production will increase, and more Americans will be able to enjoy these delicacies. Most of the specialty mushrooms must be cooked, although some can be eaten raw in salads. Come with me to the Phillips Mushroom Farm and see how complicated this really is!

 Video related to this topic is available within WileyPLUS.

(top: Stockdisc/Getty Images; bottom: Stockbyte/Getty Images)

▐▐▐ MICROORGANISMS FOUND IN FOOD

Grains / Fruits and Vegetables / Meats and Poultry / Fish and Shellfish / Milk / Other Edible Substances

▐▐▐ PREVENTING DISEASE TRANSMISSION AND FOOD SPOILAGE

Food Preservation / Drying and Lyophilization / Pasteurization of Milk / Standards for Food and Milk Production

▐▐▐ MICROORGANISMS AS FOOD AND IN FOOD PRODUCTION

Algae, Fungi, and Bacteria as Food / Food Production

▐▐▐ BEER, WINE, AND SPIRITS

▐▐▐ INDUSTRIAL AND PHARMACEUTICAL MICROBIOLOGY

Useful Metabolic Processes / Problems of Industrial Microbiology

▐▐▐ USEFUL ORGANIC PRODUCTS

Simple Organic Compounds / Antibiotics / Enzymes / Amino Acids / Other Biological Products

▐▐▐ MICROBIOLOGICAL MINING

▐▐▐ MICROBIOLOGICAL WASTE DISPOSAL

Throughout history, humans have used microorganisms. We have long found uses for microbes in making food and medicine, and today other industrial applications of microbes abound. And because biotechnology is advancing rapidly, the future of applied microbiology looks very bright. As we consider the many applications of microbiology—from food production to mining—we must also discuss food spoilage and methods of preservation.

❚❚❚ MICROORGANISMS FOUND IN FOOD

Anything that people eat or drink can be used as food by microorganisms, too (Figure 26.1). Most substances consumed by humans are derived from plants, which of course grow in soil, or from animals, which live in contact with soil and so have soil organisms on them. Although soil organisms usually are not human pathogens, many can cause food spoilage. The handling of foods from harvest or slaughter to human consumption provides many more opportunities for the foods to become contaminated with microorganisms. Unsanitary practices by food handlers and unsanitary working conditions frequently lead to contamination of foods with pathogens. Improper storage and preparation procedures at home and especially in restaurants can lead to further contamination with pathogens. Improper refrigeration of prepared foods is a major source of food poisoning.

Twenty percent of all emergency room visits are due to foodborne illness.

Globalization affects food safety. Fruits and vegetables imported from third world countries, where sanitation standards are much lower, can bring in disease and parasites. Are their workers infected with foodborne illnesses? Are there toilets? Handwashing facilities? Do the workers use them? Do we carefully and thoroughly wash the produce before we eat it? Those fresh berries in winter may carry hidden costs.

GRAINS

When properly harvested, the various kinds of edible grains, such as rye and wheat, are dry. Because of the lack of moisture, few microorganisms thrive on them. However, if stored under moist conditions, grains can easily become contaminated with molds and other microorganisms. Insects, birds, and rodents also transmit microbial contaminants to grains.

Raw grains contaminated with the mold *Claviceps purpurea*, known as *ergot*, cause ergot poisoning, or ergotism. Compounds produced by this fungus are hallucinogenic and can alter behavior or even be deadly if eaten. Certain species of *Aspergillus*, infecting peanuts and some grains, produce *aflatoxins*. These toxic compounds have proved to be potent mutagens and carcinogens (◀Chapter 22, p. 703).

Most grains are used in making breads and cereals. Humans have made breads for thousands of years, and some loaves of bread on display in the British Museum are 4,000 years old. Although the first instance of leavening bread with yeast probably occurred by accident, special strains of *Saccharomyces cerevisiae* that produce large amounts of carbon dioxide are now added to bread dough to make it rise. We will describe this process in more detail later.

Like raw grain, bread is susceptible to contamination and spoilage by various molds. *Rhizopus nigricans* is the most common bread mold, but several species of *Penicillium*, *Aspergillus*, and *Monilia* also thrive in bread. Contamination of bread with *M. sitophila*, a pink bread mold, is especially dreaded by bakers because it is almost impossible to eliminate from a bakery once it has become established. Rye bread is particularly likely to become contaminated with *Bacillus* species, which hydrolyze proteins and starch and give the bread a stringy texture. If not killed by baking, *Bacillus* spores germinate and cause rapid and extensive damage to freshly baked bread.

FRUITS AND VEGETABLES

Millions of commensal bacteria, especially *Pseudomonas fluorescens*, are found on the surfaces of fruits and vegetables. These foods also easily become contaminated with organisms from soil, animals, air, irrigation water, and equipment used to pick, transport, store, or process them. Pathogens such as *Salmonella*, *Shigella*, *Entamoeba histolytica*, *Ascaris*, and a variety of viruses can be transmitted on the surfaces of fruits and vegetables. However, the skins of most plant foods contain waxes and release antimicrobial substances,

Figure 26.1 Microbes in food. Microorganisms use the same foods that humans do, as can be seen from these oranges covered with mold. *(R. Calentine/Visuals Unlimited)*

both of which tend to prevent microbial invasion of internal tissues.

Cantaloupes present a special problem. Their netted surfaces develop biofilms of *Salmonella*, which grow into the fissures beneath the netting (**Figure 26.2**). Fimbriae and cellulose secretions attach the bacteria firmly and thickly to the surface of the cantaloupe rind. The upper layers of the biofilm protect bacteria located deeper

(a)

(b)

Figure 26.2 Scanning electron micrograph of a cantaloupe. (a) SEM (×2,500) of a cantaloupe rind showing attachment and initiation of biofilm formation by *Salmonella poona* cells inside the netting of inoculated cantaloupe. Cantaloupes were spot inoculated, allowed to dry for 2 h, dissected, and treated for SEM imaging. *(Courtesy Bassam A. Annous, USDA-ARS Food Safety Intervention Technologies Research Unit, Wyndmoor, PA)* **(b)** Dried for 72 h, dissected, and treated for SEM imaging. *(Courtesy Bassam A. Annous, USDA-ARS Food Safety Intervention Technologies Research Unit, Wyndmoor, PA)*

within the film from the effects of sanitizing solutions. About 5% of melons imported into the United States from Mexico have such layers of *Salmonella*. When you cut through the rind, you can drag bacteria over the cut surfaces of the flesh. Sanitizing the surface would prevent this.

Now, United States Department of Agriculture (USDA) scientist, Dr. Bassam Annous, has developed a surface pasteurization process that kills 99.999% of *Salmonella* on surfaces of artificially contaminated cantaloupe. Melons are immersed in 169°F water for 3 minutes (**Figure 26.3**). Each melon is then sealed in a plastic bag before being rapidly cooled in an ice-water bath. The bag prevents recontamination of the fruit. Moreover, shelf life is extended because natural spoilage organisms were also killed by the surface pasteurization process. All this without affecting the quality of the melon.

Dr. Annous has also been working with mangos, papayas, and tomatoes. Results look promising. However, surface pasteurization does not seem to be the answer for sanitizing leafy produce. It is more likely to destroy the product. For a discussion of how spinach and lettuce are currently processed, refer back to ◀Chapter 12, p. 350.

Certain vegetables are particularly vulnerable to microbial attack and spoilage. Leafy vegetables and potatoes are susceptible to bacterial soft rot by *Erwinia carotovora*. The fungus *Phytophthora infestans* caused the Irish potato famine of 1846. Biotechnology is being called upon to help prevent such crop damage. Attempts are under way, for example, to insert into soybeans a gene for disease resistance, found in wild mustard plants, that could yield pathogen-resistant soybean crops. Similarly, a gene that protects plants against the bacterium *Pseudomonas syringae* may soon be introduced into corn, bean, and tomato plants.

Fruits are likewise susceptible to spoilage by microbial action. Tomatoes, cucumbers, and melons can be damaged by the fungus *Fusarium*, which causes soft rot and cracking of tomato skins. Fruit flies pick up the fungus from infected tomatoes and transmit it to healthy ones as they deposit eggs in surface cracks. Other insects pierce tomatoes to feed on them and at the same time introduce *Rhizopus*, which breaks down pectin and can turn a tomato into a bag of water. Fresh fruit juices, because of their high sugar and acid content, provide an excellent medium for growth of molds, yeasts, and bacteria of the genera *Leuconostoc* and *Lactobacillus*. Grapes and berries are damaged by a wide variety of fungi, and large numbers of stone fruits, such as peaches, can be destroyed overnight by brown rot due to *Monilia fructicola*. *Penicillium*

Years ago, a small flock of diseased chickens raised on a family farm affected only a few people. Today "factory" farm flocks, numbering in the 1000's, can spread disease to far more people.

Figure 26.3 **Surface pasteurization of cantaloupes.** At the rear, cantaloupes are being lifted out of the bath following a 3 minute immersion at 169°F. *(Courtesy Bassam A. Annous, USDA-ARS Food Safety Intervention Technologies Research Unit, Wyndmoor, PA)*

expansum, which grows on apples, produces the toxin patulin, which can easily contaminate cider. Other *Penicillium* species produce blue and green mold on citrus fruits.

MEATS AND POULTRY

Meat animals arrive at slaughterhouses with numerous and varied microorganisms in the gut and feces, on hides and hoofs, and sometimes in tissues. At least 70 pathogens have been identified among these microbes. Nearly all animal carcasses in slaughterhouses in the United States are inspected by a veterinarian or a trained inspector, and those found to be diseased are condemned and discarded **(Figure 26.4)**. Inspection, however, cannot guarantee that meat is parasite-free. The inspector examines the carcass surface and the heart (where helminths are often concentrated) but cannot look inside

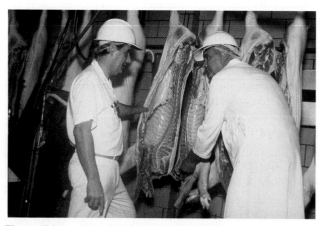

Figure 26.4 **Meat inspection at a slaughterhouse.** Inspection helps provide, but cannot guarantee, a safe supply of meat for human consumption. *(Courtesy United States Department of Agriculture)*

each cut of meat. Following the discovery, in December 2003, of the first case of BSE (bovine spongiform encephalopathy, or mad cow disease) in the United States, there has been a loud outcry for stricter laws for meat inspection (◀Chapter 24, p. 772). Some of the most common diseases identified in slaughterhouses are abscesses, pneumonia, septicemia, enteritis, toxemia, nephritis, and pericarditis. Lymphadenitis, an inflammation of the lymph nodes, is especially common in sheep and lambs.

Even after animals are slaughtered and the carcasses are hung in refrigerated rooms to age, microorganisms sometimes spoil the meat. Several molds grow on refrigerated meats, and *Cladosporium herbarum* can grow on frozen meats. Mycelia of *Rhizopus* and *Mucor* produce a fluffy, white growth referred to as "whiskers" on the surfaces of hanging carcasses. The bacterium *Pseudomonas mephitica* releases hydrogen sulfide and causes green discoloration on refrigerated meat under low oxygen conditions. Several species of *Clostridium* cause putrefaction called **bone stink** deep in the tissues of large carcasses.

Ground meats sometimes contain helminth eggs and always contain large numbers of lactobacilli and molds. In most areas butcher shops are required to maintain

PUBLIC HEALTH

Can You Trust Your Microwave?

I'm too tired to cook tonight—let's just nuke some chicken pot pies in the microwave. But beware! Pot pies are not "heat and serve" items that just need to be warmed up. They are uncooked, and need to be thoroughly cooked to destroy pathogens that may be lurking inside them. You wouldn't eat half-done chicken, and you'd better not eat a lukewarm pot pie.

In Fall 2007, over 175 people in 32 states found this out the hard way when they developed Salmonella infections associated with Banquet brand pot pies. One Minnesota family's 19 month-old-daughter became violently ill, passed out, had a seizure, ran a 104-degree fever and had continuous diarrhea that required 6 to 8 diaper changes per hour! She continued having diarrhea for another 6 weeks. Of the other 174 cases, 33 were hospitalized, but none died.

ConAgra initially blamed the consumers for not following directions properly, but then admitted that the directions were not sufficiently clear. After 3 days, Con Agra finally recalled their pot pies. Lawsuits have been filed over this delay. The directions specified microwaving on high for 4 minutes in medium- or high-wattage microwaves, or 6 minutes in low-wattage ovens. The Minnesota family had used 7 1/2 minutes. One scientist used a 1,000-watt microwave for 4 minutes, which heated the pie to 45 degrees. A safe temperature is 165 degrees. After 6 minutes the pie was 204 degrees near the top, but 127 deeper in the pie.

two separate meat-grinding machines—one for pork and one for other meats. This practice is encouraged because it is difficult to clean a machine thoroughly enough to eliminate any possibility of transmitting raw pork bits, which may carry the helminth *Trichinella spiralis*, the cause of trichinosis, to other meats (◀Chapter 11, p. 330). Lactobacilli in ground meats produce acids that retard the growth of enteric pathogens. Nevertheless, ground meats are subject to spoilage even when refrigerated and should be frozen if they are not to be used within a day or two. All meats, but especially ground meats, should be cooked thoroughly to kill pathogens.

More than 20 bacterial genera have been found on dressed poultry, and mishandling of poultry in restaurants accounts for many foodborne infections. Nearly half these infections have been traced to *Salmonella* and a fourth each to *Clostridium perfringens* and *Staphylococcus aureus*. Freezing fails to rid poultry of *Salmonella*. Pseudomonads and several other Gram-negative bacteria are common contaminants of poultry, where they cause slime and offensive odors.

APPLICATIONS

Beware of the Bad Egg!

Salmonella species are found in many animal reservoirs, especially poultry. Recent investigations of *Salmonella* in the poultry industry were not reassuring, and consumers can assume that poultry products will be heavily contaminated. What steps can you take to maneuver defensively in the sea of *Salmonella*?

First, assume that the shells of all eggs are contaminated, and wash your hands after handling them. (But do not wash the eggs—washing removes a protective surface coating that helps prevent microbes from entering the eggs.) *Salmonella* are part of the intestinal microflora of poultry, and eggs surely come in contact with feces, feathers, and contaminated surfaces. Cracked eggs cannot be sold for human consumption, but in some areas they can be sold as pet food. When organisms enter an egg, they find a superb, nutrient-rich culture medium in which they can rapidly multiply.

Does it make a difference whether you refrigerate eggs with the big end or the little end up? Big end up is the recommendation of food scientists. Why? The object is to keep the egg yolk and any embryo residing on it as close to the center of the egg as possible. This maximizes the distance an invading microbe would have to swim from the shell to the yolk. The egg white is filled with chemical hazards for the bacteria. Lysozymes attack the bacterial cell walls, killing many bacteria by rupturing their walls. Nutrients, vitamins, and metal ions of iron, copper, and zinc are wrapped tightly by proteins and other substances in the egg white and so are unavailable to the bacteria. Deprived of nutrients and chewed up by lysozymes, bacteria do not survive the swim; the embryo remains safe.

But why big end up? The natural tendency of the lipid-rich yolk is to rise, just as oil floats to the top of water. Bird eggs have two ropy cords (look for them next time you crack a raw egg) called chalaza, which act much like the ropes of a hammock to suspend the yolk from the shell lining. The larger chalaza is at the small end of the egg and can hold the yolk down better, preventing it from rising too close to the edge of the protective egg white.

Contaminated eggs being destroyed (*Courtesy United States Department of Agriculture*)

Additionally, when baking, control your urge to lick raw batter if you've added any eggs to it. If the batter contains eggs, even powdered eggs in mixes, it might contain *Salmonella*. Pieces of shell dropped into batter, even when promptly retrieved, can inoculate it with organisms. Also, raw eggs can be contaminated internally if laid by an infected hen or if the eggs have undetected cracks or have been immersed in water.

Finally, handle raw poultry carefully. Poultry is naturally contaminated with fecal organisms, and contamination is exacerbated by the industry's practice of holding poultry in water baths during certain phases of the plucking and evisceration processes. The water becomes a *Salmonella* broth, and the surfaces of the birds become uniformly contaminated. Thus, use poultry promptly and discard contaminated wrappings and juices carefully. If you have laid poultry on a countertop or cutting board, scrub those surfaces thoroughly with hot soapy water before placing other foods on them. And don't touch other foods until you have washed your hands thoroughly.

You might suppose that eggs, with their hard shells, would be free of microbial contamination. Most are, but the shells are porous, and pseudomonads and some other bacteria, as well as fungi such as *Penicillium*, *Cladosporium*, and *Sporotrichum*, grow on eggshells. These microbes can pass through pores in the shells to infect the inside of eggs. *Salmonella* also survives on eggshells and can enter broken eggs or be deposited with bits of shell in foods. Hens infected with *S. pullorum* lay infected eggs. Those of you who have ever cleaned the internal organs out of a chicken have no doubt seen that the shell is not laid down around an egg until it has traveled a goodly distance down the oviduct. The sperm fertilize the egg at the top of the oviduct where there is no shell. (Did you ever wonder how the sperm fertilizes a chicken egg?) Bacteria such as *Salmonella* can also get inside the egg before the shell is laid down. The shell does not have to be cracked for an egg to be filled with *Salmonella*! The CDC reports that 1 out of every 10,000 eggs has *Salmonella* inside the shell. Any pathogens on eggshells or in eggs can be transmitted to humans unless eggs and foods that contain them are cooked thoroughly. Eating raw eggs, such as in eggnog, is a calculated risk.

FISH AND SHELLFISH

Fresh fish abound with microorganisms. Several species of enteric bacteria and clostridia, enteroviruses, and parasitic worms are commonly found on or in fresh fish. Many of these organisms survive shipment of fish packed in crushed ice, especially if the fish are packed too tightly together or are pressed against the slats of contaminated crates.

Shellfish, such as oysters and clams, carry many of the same organisms as fish. Raw oysters typically carry *Salmonella typhimurium* and sometimes *Vibrio choleae*.

Clams are especially likely sources of human infection because they are filter feeders—that is, they obtain food by filtering water and extracting microbes. If clams are exposed to increased levels of sewage, red tides, and other sources of large numbers of pathogens or toxin producers, clamming may be prohibited until the numbers of organisms decrease. Scallops are less likely to transmit diseases to humans because only the muscular part of the organism, and not its digestive tract, is eaten.

Fish contaminated with Photobacterium phosphoreum will glow. Although there have been no reports of this bacterium causing illness, it does indicate that the seafood is not fresh.

Among crustaceans, shrimp are extremely likely to be contaminated. Some studies have shown that over half the breaded shrimp on the market contain in excess of 1 million bacteria (and more than 5,000 coliforms) per gram. Such high bacterial counts probably result from growth of bacteria during processing before the shrimp are frozen. Lobsters and crabs are even more perishable than shrimp and can carry a variety of enteric pathogens. Along the U.S. Gulf Coast, improperly cooked crabs have transmitted cholera. Crabs also carry *Clostridium botulinum* (◀Chapter 24, p. 767) and the pathogenic fungi *Cryptococcus* and *Candida*. Keep in mind, however, that the mere presence of microorganisms in seafoods—or in any other foods—does not necessarily mean that the foods are spoiled or contaminated with pathogens. In fact, *Lactobacillus bulgaricus*, which produces hydrogen peroxide, can be used to inhibit growth of other organisms found on seafood.

MILK

Modern mechanized milking and milk handling has greatly reduced the microbial content of raw milk (**Figure 26.5**). However, breeding dairy cattle for increased milk production has resulted in exceptionally large udders and teats that easily admit bacteria. The first few milliliters of milk drawn from such cows can contain as many as 15,000 bacteria per milliliter, whereas the last milk drawn is free of microorganisms. Most microorganisms in freshly drawn milk are *Staphylococcus epidermidis* and *Micrococcus*, but *Pseudomonas*, *Flavobacterium*, *Erwinia*, and some fungi also can be present.

Microorganisms have many opportunities to enter milk before it is consumed. Hand milking, as opposed to mechanical milking, allows organisms from the body of the cow to enter the milk. These include *Escherichia*

APPLICATIONS

Combating Lactose Intolerance

Lactose-intolerant people are not able to break down lactose into glucose and galactose due to the inability to produce the enzyme lactase. Unfortunately that process occurs in the intestines where lactose-positive organisms, in the process of digesting the sugar, produce gas and acid. Lactose-intolerant people experience gas, cramps, nausea, and diarrhea when they eat more lactose than is tolerable. Many people whose families originate from African, Asian, Near Eastern, and Mediterranean areas, as well as North and South American Indians, are lactose-intolerant. Because calcium is an important part of the diet, lactose-intolerant individuals need to eat non-dairy calcium-rich foods, take calcium supplements, or purchase lactose-reduced foods and lactase tablets.

coli, which give milk a fecal flavor, and *Acinetobacter johnsoni* (formerly known as *Alcaligenes viscolactis*), which is especially abundant during summer months and causes a viscous slime to form in milk. Storing, transporting, and processing milk allow for contamination with any organisms in containers and for growth of those already present. Infectious organisms in milk usually come either from infected cows or from unsanitary practices of milk handlers. Diseased cattle can transmit *Mycobacterium bovis* and *Brucella* species to their milk. Dairy herds are tested for tuberculosis (with infected animals removed from the herd) and vaccinated against brucellosis (undulant fever); hence the risk of transmitting these diseases to humans is small. *Staphylococcus aureus*, *Salmonella* species, and other enteric bacteria can enter milk through unsanitary handling. Bacteria and molds will grow even in dried milk, which is made from pasteurized liquid milk but is unsterilized, if the water content of the powder reaches 10%.

Figure 26.5   A merry-go-round milking parlor. Mechanized procedures for milking and milk handling have greatly reduced the microbial content of raw milk. *(Jane Latta/Photo Researchers)*

Some microorganisms, such as certain *Pseudomonas* species and some soil organisms, grow in refrigerated milk. These organisms are psychrophilic; although they normally grow at higher temperatures, they can grow at 5°C (refrigerator temperature). They also survive the concentration of chlorine normally used to purify drinking water.

Organisms that sour milk include *Streptococcus lactis* and species of *Lactobacillus*. When these microbes release enough lactic acid to bring the pH below 4.8, the milk proteins coagulate, and the milk is said to have soured. Souring of milk does not mean that the milk is unsafe for human consumption, but it does greatly alter the taste and appearance of the milk.

> *Lactobacillus acidophilus adds lactic acid to milk, making that milk digestible by lactose-intolerant people.*

Various microorganisms that are involved in the spoilage of food are summarized in **Table 26.1**.

OTHER EDIBLE SUBSTANCES

Humans consume sugar, spices, condiments, tea, coffee, and cocoa—all of which are subject to microbial contamination—in addition to major nutrients. Fresh, dry, refined sugar is sterile, but raw sugar-cane juice supports growth of fungi such as *Aspergillus*, *Saccharomyces*, and *Candida* and several species of bacteria, including *Bacillus* and *Micrococcus*. Most are removed by filtration, and the remainder are killed by heat during evaporation of the juice. Foods to which sugar is added are especially susceptible to spoilage because the sugar is an excellent nutrient for many organisms. *Bacillus stearothermophilus*, a facultative anaerobe that grows best at 55° to 60°C, can multiply rapidly during processing of foods. Another thermophilic anaerobe, *Clostridium thermosaccharolyticum*, is often responsible for the gas production that causes cans to bulge. Conversely, sugar in high concentration acts as a preservative. The high sugar concentration in such foods as jellies, jams, candies, and candied fruits creates sufficient osmotic pressure to inhibit microbial growth.

Maple trees are tapped, and sap is collected from them, in the early spring. The sap becomes increasingly contaminated as the weather becomes warmer. Organisms that thrive in maple sugar sap include species of *Leuconostoc*, *Pseudomonas*, and *Enterobacter*. Although these organisms may consume large quantities of sugar, they are killed as the sap is evaporated to syrup or sugar.

Honey can contain toxins if it is made of nectars from such plants as *Rhododendron* or *Datura*. It can also contain spores of *Clostridium botulinum*. Although the spores do not germinate in the honey, in infants they can germinate after the honey is ingested. Their toxins can cause "floppy baby syndrome." Generally, somewhere

TABLE 26.1

Microorganisms Involved in Food Spoilage		
Food	**Organism**	**Type of Spoilage**
Grains		
Bread	*Rhizopus nigricans*	Bread mold
	Penicillium species	
	Aspergillus species	
	Monilia sitophilia	
	Bacillus species	Protein and starch hydrolysis—stringy texture
Fruits and Vegetables		
Leafy vegetables	*Erwinia carotovora*	Bacterial soft rot
Potatoes	*Phytophthora infestans*	Potato blight
Tomatoes, melons	*Fusarium* species	Soft rot; cracking skin
Grapes, berries, "stone" fruit	*Monilia fructicola*	Brown rot
Apples	*Penicillium expansum*	Patulin, a toxic cider contaminant
Meat		
	Rhizopus species	"Whiskers," white fluffy growth
	Mucor species	
	Pseudomonas	Green discoloration on refrigerated meat
	Clostridium species	Bone stink
Poultry		
	Pseudomonads	Sliminess
	Penicillium	Eggshell contamination
	Cladosporium species	
	Sporotrichum species	
Milk		
	Acinetobacter johnsoni	Sliminess
	Streptococcus lactis	Sour milk
	Lactobacillus species	

between 70 and 100 cases of infant botulism are reported each year in the United States.

Spices have been used in food preservation and embalming for centuries and have thereby gained a

APPLICATIONS

Bon Bon Bombs

Candy is generally safe from microbial contamination because its high sugar concentration creates an osmotic pressure too high for most organisms to survive. Occasionally, the cream filling in a batch of chocolates becomes contaminated with *Salmonella* or *Clostridium*. Chocolate lovers have been surprised to discover that the gas produced by clostridia has caused their favorite chocolates to explode.

reputation as antimicrobials. This reputation is undeserved; spices more often mask odors of putrefaction than prevent spoilage. Leeuwenhoek, who first observed bacteria in spices, reported that water containing whole peppercorns teemed with them. Among the many microorganisms found in spices (Table 26.2), most are not pathogens. The small amounts of spices used in cooking are probably not a health hazard.

Condiments such as salad dressings, catsup, pickles, and mustard usually are markedly acidic. Although low pH prevents growth of many microorganisms, some molds are able to grow in these foods if they are not refrigerated.

Americans consume huge quantities of carbonated beverages and coffee and lesser quantities of tea and cocoa. The automated equipment in modern manufacturing plants can prepare carbonated beverages aseptically,

TABLE 26.2

Numbers of Bacteria in Spices					
Number of Microorganisms Per Gram in Dry Sample					
Type of Spice	**Total Aerobes**	**Coliforms**	**Yeasts and Molds**	**Aerobic Spores**	**Anaerobic Spores**
Bay leaves	520,000	0	3,300	9,200	<2
Cloves	3,000	0	18,005	<2	<2
Curry	<7,500,000	0	70	>240,000	>240,000
Marjoram	370,000	0	18,000	54,000	>24,000
Paprika	<5,500,000	600	2,300	>240,000	>620
Pepper	<2,000,000	0	15	>240,000	>24,000
Sage	6,800	0	10	7	>1,700
Thyme	1,900,000	0	11,000	160,000	>24,000
Turmeric	1,300,000	50	70	>110,000	>240,430

Source: Adapted from Karlson and Gunderson, 1965, as cited in *Food Technology 1986*, with permission from the Institute of Food Technologies.

but syrups can become contaminated with molds if there is a mechanical failure. Syrups sold in bulk to restaurants also can become contaminated. Freshly harvested coffee beans are subject to contamination by various molds and by microorganisms transmitted to them by

APPLICATIONS

Tea or Coffee for Britain

When the coffee plantations of Ceylon (now Sri Lanka) were destroyed in the 1860s by rust, the fields were replanted with tea. The British, who had depended on Ceylon for their coffee supplies, were forced to switch to tea. Thus, the lowly fungus played a major role in changing the British into a nation of tea drinkers. The Irish, 2 decades earlier, had not been so lucky. When a fungal disease destroyed the potato crop in the late 1840s, no replacement capable of feeding the population was available. A million people starved to death, and over a million emigrated.

Could such catastrophes occur today? The answer, unfortunately, is probably "Yes." Most of the food crops grown today are potentially vulnerable to diseases and pests because they are monotypes, or pure strains, lacking the genetic diversity that is typical of wild plants. An outbreak of plant disease can therefore spread very rapidly through an entire crop. Agricultural scientists are using genetic engineering techniques to develop strains that are resistant to diseases and insects—rust-resistant strains of coffee, for example (as well as varieties that are naturally low in caffeine). However, there is always the danger that such strains will turn out to be vulnerable to a new species of microorganism—either a mutation or one to which previously cultivated varieties had not been susceptible. This happened to the U.S. corn crop in the 1970s and resulted in large economic losses.

insects. The coffee rust *Hemileia vastatrix*, a fungus, has devastated coffee plantations in Asia and is now a serious problem in South America. Tea leaves allowed to become moist are susceptible to contamination by *Aspergillus* and *Penicillium* molds, which impart an unpleasant aroma to tea brewed from them.

Microorganisms are useful in preparing coffee and cacao beans for marketing. The bacterium *Erwinia dissolvens* is used to digest pectin in outer coverings of coffee beans. Other bacteria are used to dissolve the covering of cacao beans, the beans from which cocoa and chocolate are made. These beans are subsequently treated with fermenting bacteria. Yeasts, which turn bean pulp into alcohol, are essential to develop the flavor and aroma of chocolate as well.

✓ CHECKLIST

1. What organisms are most likely to contaminate milk?
2. What dangers does honey present?
3. What is bone stink?
4. What pathogenic bacteria are associated with eggs?

PREVENTING DISEASE TRANSMISSION AND FOOD SPOILAGE

Diseases acquired from food are due mainly to the direct effects of microorganisms or their toxins (Table 26.3), but they can result from microbial action on food substances. Industrialization has increased the spread of foodborne pathogens. Large processing plants provide opportunities for contamination of great quantities of food unless sanitation is strictly practiced. In institutions

TABLE 26.3

Pathogenic Organisms Transmitted in Food and Milk		
Organism	Disease	Vector
Staphylococcus aureus	Food poisoning	Infected food handlers, unrefrigerated foods, milk from infected cows
Clostridium perfringens	Food poisoning	Unrefrigerated foods
Bacillus cereus	Food poisoning	Unrefrigerated foods
Clostridium botulinum	Botulism	Inadequately processed canned goods
Salmonella species	Salmonellosis	Infected food handlers, poor sanitation, contaminated seafood
Shigella species	Shigellosis	Infected food handlers, poor sanitation
Enteropathogenic *Escherichia coli*	Traveler's diarrhea and other diseases	Infected food handlers (sometimes asymptomatic), poor sanitation, contaminated meat
Campylobacter	Gastroenteritis	Undercooked chicken and raw milk
Vibrio cholerae	Cholera	Poor sanitation
Vibrio parahaemolyticus	Asian food poisoning	Undercooked fish and shellfish
Listeria	Listeriosis	Inadequately processed milk
Hepatitis A virus	Hepatitis	Infected food handlers

that feed large numbers of people, contaminated food will cause many cases of disease. The increased popularity of convenience foods, especially fast foods, has also raised the risk of infection.

In addition to the enteric diseases described in ◀Chapter 19, several other diseases can be transmitted in food. *Klebsiella pneumoniae* is commonly found in the human digestive tract. Although it is thought of mainly as a pathogen of the respiratory system, it can cause diarrhea in infants, abscesses, and nosocomial wound and urinary tract infections. Tuberculosis can be transmitted by fomites in food, unpasteurized milk and cheese, and meats from infected animals.

Several diseases can be transmitted to humans who eat infected meat. They include anthrax, burcellosis, Q fever, and listeriosis. Because of meat inspection procedures, animal and meat handlers are exposed to these diseases much more often than are consumers. Adirondack disease, caused by *Yersinia enterocolitica*, is thought to be transmitted by eating infected meat, but it also may be transmitted through milk and water. The disease has many forms, including mild to severe gastroenteritis, arthritis, glomerulonephritis, a fatal typhoidlike septicemia, and fatal ileitis, an inflammation of the small intestine that can be mistaken for appendicitis. People have also become infected with *Erysipelothrix rhusiopathiae* by eating infected pork. The resulting disease—called erysipelas (er-i-sip′e-las) in animals and erysipeloid in humans—infects swine, sheep, and turkeys. It is most likely to infect farmers

Over half of France's 350 kinds of cheese are made from raw, unpasteurized milk. Immunocompromised people should not eat soft cheeses like Brie, Camembert, and blue-veined cheeses.

and packing plant workers through skin wounds. The organism can infect skin, joints, and the respiratory tract.

Viruses are frequently transmitted through food. Enteroviruses are often spread during unsanitary handling of food, especially by asymptomatic food handlers. Droplet infection of food can transmit echovirus and coxsackie respiratory infections. Poliomyelitis viruses can be transmitted through milk and other foods. And the viruses responsible for hepatitis A can be transmitted through shellfish from contaminated waters. The virus that causes lymphocytic choriomeningitis, a flulike illness, can be spread to foods by mice.

Milk is an ideal medium for the growth of many pathogens. In addition to toxin producers and pathogens found in other foods, milk can contain organisms from cows. These organisms include *Mycobacterium bovis*, *Brucella* species, *Listeria monocytogenes*, and *Coxiella burnetii*. Spore-producing organisms such as *Bacillus anthracis* enter milk from infected cows or from soil. But milk also contains certain antibacterial substances, including lysozyme, agglutinins, leukocytes, and lactenin. Lactenin is a combination of thiocyanate, lactoperoxidase, and hydrogen peroxide. It is also present in human milk and other body secretions and may help prevent enteric infections in newborns. Fermented milk contains bacteria, such as *Leuconostoc cremoris* that kill pathogens. A crucial factor in preventing spoilage and disease transmission in food and milk is cleanliness in handling. Other common-sense practices—prompt use of fresh foods, careful refrigeration, and prompt and adequate processing of foods to be stored—also help control disease transmission and spoilage.

FOOD PRESERVATION

Many procedures used to preserve foods are based on practices begun early in human civilization. The ability to maintain a stable year-round food supply was essential in allowing humans to abandon a nomadic lifestyle, which followed the food supply, for a more settled village lifestyle. These practices were most likely based on simple observations like the following: Grain that was kept dry did not turn moldy. Dried and salted foods remained edible over a long period of time. And milk allowed to sour or made into cheese was usable over a much longer period of time than was fresh milk. Modern methods of food and milk preservation still use some of these early methods, but they also use heat and cold and other specialized procedures.

Many of the methods of food preservation have been described under antimicrobial physical agents in ◀Chapter 12. They include canning using moist heat; refrigeration, freezing; lyophilization and drying; and the use of radiation. A number of chemical food additives also are used to retard spoilage.

Canning

The most common method of food preservation is **canning**—the use of moist heat under pressure. This method, analogous to laboratory autoclaving, is used to preserve fruits, vegetables, and meats in metal cans or glass jars (**Figure 26.6**). If properly carried out, canning

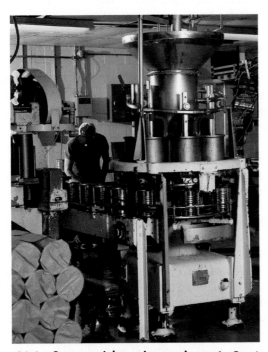

Figure 26.6 Commercial canning equipment. Canning destroys all microbes and their spores, thereby preventing food spoilage and disease transmission. (*Larry Lefever/Grant Heilman Photography*)

APPLICATIONS

Home Canning

Clostridium botulinum spores are present on most fresh food surfaces. The bacterial spores can be killed with a combination of acid and heat. *C. botulinum* does not grow or produce toxin below pH 4.5, and so is not a danger in highly acidic food. It may be possible to acidify certain foods by the addition of lemon juice, citric acid, or vinegar. Low-acid foods, not safely processed with boiling water, must be sterilized using heat and pressurized steam. All low-acid foods should be sterilized at temperatures between 240° to 250°F, attainable with a pressure canner operated at 10 to 15 psi. The time needed to sterilize the food depends on the kind of food being canned, the way it is packed in jars, and the size of the jars. In addition, the time for processing your home preserves will depend on the altitude where you live. Therefore it is important to carefully follow all directions.

destroys all harmful spoilage microorganisms, including most heat-resistant endospores; prevents spoilage; and averts any hazard of disease transmission. Foods so treated may remain edible for years.

Some thermophilic anaerobic endospores, such as those of *Bacillus stearothermophilus*, may remain alive even after commercial canning. Hence canned food should not be stored in a hot environment, such as a car trunk or hot attic. At elevated temperatures, the endospores can germinate, grow, and cause spoilage. Usually this produces gas, which causes the ends of cans to bulge so that the ends can be pressed up and down. Such spoilage also usually produces acid, which gives a sour flavor. Such changes are called **thermophilic anaerobic spoilage**. In some cases, however, spoilage due to growth of such spores does not cause cans to bulge with gas; such spoilage is called **flat sour spoilage**. Cans can also bulge due to **mesophilic spoilage**, which occurs when canning procedures were improperly followed or the seal has been broken. This type of spoilage can take place at room temperature, unlike flat sour and thermophilic spoilage, which occur only in properly processed and sealed cans that have been stored at high temperatures.

Because of the danger of botulism and other kinds of spoilage in improperly processed canned foods, people who do home canning should carefully follow instructions in a good, up-to-date canning handbook. The U.S. Department of Agriculture (USDA) now recommends that all low-acid foods be processed by pressure cooking. The USDA also recommends that jellies and jams, formerly packed hot and sealed with wax, be processed in a boiling water bath and sealed with lids like other canned goods to prevent accumulation of mold toxins. (See the box "Home Canning" in ◀Chapter 12, p. 356). *Any can that has bulging ends—whether produced at home or commercially—should be discarded.*

Although many condiments are heat processed, some of their properties help preserve them. Sugar in jellies and jams increases osmotic pressure and retards growth of microorganisms. (Saccharin does not have this effect, so artificially sweetened products may require more stringent precautions to prevent spoilage than do those with a high sugar content.) Likewise, the high acidity of pickles and other sour foods helps prevent microbial growth.

Refrigeration and Freezing

Refrigeration at temperatures slightly above freezing (about 4°C) is suitable for preserving foods for only a few days. It does not prevent growth of psychrophilic organisms that can cause food poisoning. Freezing, another common method of food preservation, involves storage of foods at temperatures below freezing (about −10°C in most home freezers). All kinds of foods can be preserved by freezing for several months and some for much longer periods of time. An advantage of freezing is that it preserves the natural flavor of foods better than does canning. However, freezing has two disadvantages: (1) It causes some foods, especially fruits and watery vegetables, to become somewhat soft upon thawing and detracts from their appearance. (2) Although freezing prevents growth of most microorganisms, it does not destroy them. As soon as food begins to thaw, microorganisms begin to grow. In fact, freezing and thawing of foods actually promotes growth of microorganisms. Ice crystals puncture cell and plasma membranes and cell walls, allowing nutrients to escape from foods. These nutrients are then readily available to support microbial growth. Consequently, it is important never to thaw and refreeze foods.

DRYING AND LYOPHILIZATION

Drying (dessication, or dehydration) is one of the oldest methods used for food preservation. A certain level of water activity is necessary for microbial growth. Ideally, if more than 90% of the water is removed, the food can be stored in that state. Drying stops microbial growth but does not kill all microorganisms in or on food. Foods can be dehydrated by natural means, such as sun drying, or by artificial means, by passing heated air over the food with controlled humidity. Addition of salt, high concentrations of sugar, or chemical preservatives that alter osmotic pressure and reduce water content are often included in the drying process.

Currently, **lyophilization** (freeze-drying) is employed in the food industry almost exclusively for the preparation of instant coffee and dry yeast for breadmaking. (The technique is also used to preserve bacterial cultures; see ◀Chapter 12, p. 357). Lyophilization involves drying frozen food in a vacuum. The process yields higher quality foods than are produced by ordinary drying methods.

Irradiation

Irradiation of food as a method of preservation is still new and quite controversial due to public concern about the hazards of radiation. There are two categories of radiation used to control microorganisms in food: non-ionizing and ionizing.

Ultraviolet (UV) radiation, a form of *nonionizing radiation*, is limited by poor penetrating power. The radiation wavelength and exposure time also determine the effectiveness of this method of preservation. UV radiation is effective as a sanitizing agent for food-processing equipment and other surfaces. Microwaves, another form of nonionizing radiation, are useful for food preparation, cooking, and processing but not for preservation. Microwaves do not kill microorganisms directly, but the heat generated during cooking can be microbicidal. Uneven heating can interfere with this antimicrobial activity. Therefore, frequent rotation of food is necessary during microwaving.

Ionizing radiation, such as gamma rays, has great penetrating ability and is microbicidal. This type of radiation can be utilized either before or after packaging, depending on the foodstuff. Gamma radiation from cobalt-60 or cesium-137 has been used in Japan and some European countries for food preservation for a number of years. The U.S. Food and Drug Administration (FDA) has declared such irradiation safe for preserving certain foods. Radiation has successfully been used to control microbial growth on fresh fish during transport to market and to kill insects in spices. It has also proved effective in reducing spoilage of fresh fruits and vegetables. Most recently the USDA proposed rules providing for the irradiation of fresh poultry. The public has much skepticism about irradiated foods. It should be emphasized that irradiation kills microorganisms but does not cause the food itself to become radioactive.

Chemical Additives

A large number of chemical compounds are added to various foods to kill microorganisms or retard their growth. A few examples and their uses are described here.

Organic acids, some of which occur naturally in some foods, lower the pH of foods enough to prevent growth of human pathogens and toxin-producing bacteria. Acids such as benzoic, sorbic, and propionic inhibit growth of yeast and other fungi in margarine, fruit juices, breads, and other baked goods.

Alkylating agents such as ethylene oxide and propylene oxide are used only in nuts and spices. Sulfur dioxide, which is most effective at an acidic pH, is used in the United States only to bleach dried fruits and to eliminate bacteria and undesired yeasts in wineries. Ozone, a highly reactive form of oxygen, is used to kill coliform bacteria in shellfish and to treat water used in beverages.

APPLICATIONS

The Great Irradiation Debate: Are Gamma Rays Safe for Food?

What method of food preservation allows you to store onions sprout-free for up to 3 months, keeps refrigerated strawberries firm and fresh for 3 weeks, and virtually eliminates *Salmonella* from poultry? Irradiation—but it may not be coming to a store near you. Despite its approval by the FDA and its endorsement by the World Health Organization, the American Medical Association, and even Julia Child, irradiation is practiced by only one major poultry company in the United States, and only one U.S. company is in the business of irradiating fruits and vegetables. Although advocates praise irradiation as an effective way to reduce bacterial and insect contamination of foods, critics charge that the practice may cause more harm than the organisms it destroys. Go to the website for this chapter to read more about the pros and cons, plus the actual methods of irradiation.

NON - IRRADIATED -

IRRADIATED - (0.2 M RAD)

STRAWBERRIES -
15 DAYS STORAGE 38°F (4°C)

(Courtesy International Atomic Energy Agency)

It has the advantage of leaving no residue. The disadvantages of ozone are that it tends to give foods a rancid taste by oxidizing fats, and it can damage molecules, especially lung lysozyme, if inhaled.

Sodium chloride, perhaps one of the first food additives, increases osmotic pressure in foods, keeping most microorganisms from growing. Salt used to cure meats is especially useful in preventing growth of clostridia deep in tissues, although fungi eventually grow on the surfaces of salted foods. Salt dehydrates bacteria and makes it difficult for them to take in water and nutrients. A recent discovery suggests that salt, in addition to increasing osmotic pressure, may create electric charges on the surface of meats, preventing bacteria from adhering to the surfaces.

Still other chemical additives have special uses. Halogen compounds, such as sodium hypochlorite, disinfect water and food surfaces. Gaseous chlorine prevents growth of microorganisms on food-processing equipment. Nitrates and nitrites suppress microbial growth in meats, especially ground meats and cold cuts. During cooking, however, they can be converted to nitrosamines, which are carcinogenic and toxic to the liver. We continue to use nitrates and nitrites because we have no good alternatives, particularly for sausage. The name *botulism* comes from the Latin word for sausage, so common was food poisoning from sausage in the days before the use of nitrites. Another reason nitrites are used is to keep colors bright, especially the red color of fresh meat. Carbon dioxide kills microorganisms, except for some fungi, in carbonated beverages. It also retards maturation of fruits and decreases spoilage during shipment. Finally, quaternary ammonium compounds (quats) can be used to sanitize many objects—utensils, cows' udders, fresh vegetables, and the surfaces of eggshells, which they do not penetrate.

Antibiotics

Antibiotics are added to foods and milk in some countries. In the United States only the anticlostridial agent *nisin*, a bacteriocin produced naturally during fermentation of milk by *Streptococcus lactis*, may be used. Antibiotic use in foods and milk is prohibited for the following good reasons:

- The antibiotics might be relied on instead of good sanitation.
- Pathogenic microorganisms might develop resistance to the antibiotics, so treatment of the diseases they cause would become difficult or impossible.
- Humans might be sensitized to the antibiotics and subsequently suffer allergic reactions.
- The antibiotics might interfere with the activities of microorganisms essential for fermenting milk and making cheese.

PASTEURIZATION OF MILK

Prevention of spoilage and disease transmission in milk begins with the maintenance of health in both dairy animals and in milk handlers. In the past, bovine tuberculosis was sometimes transmitted to humans from cows' milk. Many children were infected early in life and usually died by age 15. Establishment of mandatory testing of dairy herds for tuberculosis every 3 years (the time it takes an infection to progress to a transmissible stage) has greatly decreased incidence of the disease in the United States.

Milk is collected under clean, but not sterile, conditions and is usually subjected to pasteurization. Two methods of pasteurization are currently in use:

- In **high-temperature short-time (HTST) pasteurization**, or **flash pasteurization**, milk is heated to 71.6°C for at least 15 seconds.

- In **low-temperature long-time (LTLT) pasteurization**, or the **holding method**, milk is heated to 62.9°C for at least 30 minutes.

Both methods destroy vegetative cells of pathogens likely to be found in milk and decrease the number of organisms that can cause souring. Following pasteurization, milk is quickly cooled and refrigerated in sealed containers until it is used.

Milk can be preserved and kept safe to drink by means other than pasteurization. In Europe and increasingly in parts of the United States, milk may be sterilized by **ultra-high temperature (UHT) treatment** rather than simple pasteurization. UHT milk is heated to 87.8°C for 3 seconds. Such milk can be kept in sealed paper containers—called aseptic packaging—and stay unrefrigerated for about 6 months. Milk can also be preserved as canned condensed milk, which also is sterilized. It is reconstituted by adding an equal volume of water. Although sterilized milk is completely free of microorganisms, the heat treatment necessary to render it sterile alters its flavor.

Various chemical additives are sometimes used in milk. Addition of hydrogen peroxide to milk reduces the temperature required to destroy most pathogens. However, it fails to kill mycobacteria, and its use in milk has been banned in the United States for this reason.

STANDARDS FOR FOOD AND MILK PRODUCTION

Because food and milk production are carefully regulated by federal, state, and local laws, consumers are better protected in the United States than in many other

Figure 26.7 Testing for trichinosis. This new blood test kit can detect trichinosis infection in pigs. *(Tim McCabe/United States Department of Agriculture)*

countries. Despite regulations, some hazards remain because of the use of food additives or heat treatment. The FDA regulates inspection of meat and poultry, accurate labeling, and quality standards for products being shipped across state lines. Other similar regulations are imposed by many state and local agencies within their own jurisdictions. And the USDA is now proposing that meat and poultry be inspected by microscope—a much stricter testing requirement than has ever been in place (**Figure 26.7**). Federal and state courts are battling this out.

TABLE 26.4

Tests for Determining Milk Quality		
Test	**Description**	**Purpose and Significance**
Phosphatase test	Detects the presence of phosphatase, an enzyme destroyed during pasteurization.	To determine whether adequate heat was used during pasteurization. If active phosphatase remains, pathogens also might be present.
Reductase test	Indirectly measures the number of bacteria in milk. The rate at which methylene blue is reduced to its colorless form is directly proportional to the number of bacteria in a milk sample.	To estimate the number of bacteria in a milk sample. High-quality milk contains so few bacteria that a standard concentration of methylene blue will not be reduced in 5½ hours. Low-quality milk has so many bacteria that methylene blue is reduced in 2 hours or less.
Standard plate count	Directly measures viable bacteria. Diluted milk is mixed with nutrient agar and incubated 48 hours; colonies are counted, and the number of bacteria in original sample is calculated.	To determine the number of bacteria in a milk sample. The number per milliliter must not exceed 100,000 in raw milk before pooled with other milk or 20,000 after pasteurization.
Test for coliforms	Same as the test used for water (see ◀Chapter 25).	To determine whether coliforms are present. A positive coliform test indicates contamination with fecal material.
Test for pathogens	Detects the presence of pathogens. Methods depend on the suspected pathogens.	To identify pathogens. Usually not required but can help locate the source of infectious agents that may appear in milk.

Many food producers maintain quality checks on their own products. In canning factories, for example, counts of microorganisms in samples of foods are made during processing in an effort to minimize the number of organisms in the foods. Milk, because it is an extremely good growth medium for microorganisms, is subjected to several tests (Table 26.4). The use of these tests virtually guarantees high-quality milk to consumers.

MICROORGANISMS AS FOOD AND IN FOOD PRODUCTION

ALGAE, FUNGI, AND BACTERIA AS FOOD

Take another look

The rapid rise in world population is greatly increasing the demand for human food. Given present birthrates, the Earth's population is expected to double in about 40 years, from 5 billion to 10 billion. To put these large numbers in perspective, consider that the world population increases by about 156 people per minute, 225,000 per day, or 9 million (the population of New York City) every 40 days. Even now, 25,000 people in developing countries die of starvation every day, and many more suffer from malnutrition. Clearly, this situation calls for expanding the human food supply. Go online to see how mycorrhizal fungi are being raised for distribution onto farmland, or as an additive to seedling soil, as a means of improving harvests.

Among microorganisms, yeasts show great promise for increasing our food supplies. Yeasts are a good source of protein and vitamins, and they can be grown on a variety of waste materials—grain husks, corncobs, citrus rinds, paper, and sewage. Each kilogram of yeast introduced into one of these media can produce 100 kg of protein—1,000 times that obtained from 1 kg of soybeans and 100,000 times that obtained from 1 kg of beef. Yeast food products have been manufactured in Africa, Australia, Puerto Rico, Hawaii, Florida, and Wisconsin. Taiwan makes about 73,000 tons of yeast food per year, most of which is shipped to the United States for adding to processed foods. Dried yeast is sold as a nutritional supplement in health-food stores. Growing yeast on wastes could increase the food supply in the United States by 50 to 100%, but it has some drawbacks. Expensive equipment is required to begin production, and, more important, a way must be found to make people accept yeast as a desirable food. At present, yeast is used chiefly in animal feeds. Only small amounts of single-celled foodstuffs (yeast or algal) can be handled well by the human digestive tract. Large quantities of nucleic acid derivatives can aggravate gout; therefore, yeast must be processed to reduce their levels.

Algal culture is another promising avenue for increasing human food supplies (Figure 26.8). Algae such as *Scenedesmus* and *Chlorella* have been cultivated in Asia, Israel, Central America, several European countries, and even the western United States. Algae have also been used as ingredients in ice cream (as well as in nonfood consumer products such as diapers and cosmetics).

The use of algae as human food shortens the food chain. In other words, if humans eat algae directly rather than eating fish that have been nourished on algae, the algae will feed more people than will the fish. Approximately 100,000 kg of algae are required to make 1 kg of fish. Each acre of pond used to cultivate algae can produce 40 tons of dried algae—40 times the protein per acre from soybeans and 160 times that from beef. However, algal culture has so far proved economically feasible only in urban areas where large quantities of treated sewage on which to grow the algae are available. In addition to the problem of getting people to accept algal products as food, growing algae on sewage creates a potential health hazard because the products may contain viral pathogens.

Even some bacteria are used as food. The cyanobacterium *Spirulina* has been grown for centuries in alkaline lakes in Africa, Mexico, and by the Incas in Peru. The cyanobacteria are harvested, sun-dried, washed to remove sand, and made into cakes for human consumption. Dried *Spirulina* is about 65% protein, a very valuable food in many developing nations. An acre of *Spirulina* aquaculture could yield 100 times as much protein as the same acre planted in wheat, or 1,000 times as much as beef herded on an acre. Ancient Aztecs in Mexico ate *Spirulina*.

If technical difficulties could be overcome and the products made acceptable as human food, yeasts, algae, and some bacteria could increase the world food supply. But, at best, the use of microorganisms as food can buy only a little time to allow humans to control their own numbers.

Figure 26.8 Increasing food production by algal culture. In Asia, many red algae are grown in mariculture farms (agriculture in seawater). Nori, the dried sheets of pressed red algae used to wrap some sushi rolls, are produced in this way. *(Biophoto Associates/Photo Researchers Inc.)*

FOOD PRODUCTION

The use of microorganisms in making bread, cheese, and wine is as old as civilization itself. Long before any microorganisms were identified, milk was being made into cheese and fermented beverages, and bread was being leavened by microbes. In modern food production, specific organisms are purposely used to make a variety of foods.

Bread

In bread, yeast is used as a **leavening agent**—that is, to produce gas that makes the dough rise. A particular strain of *Saccharomyces cerevisiae* is added to a mixture of flour, water, salt, sugar, and shortening. The mixture is allowed to ferment at about 25°C for several hours. During fermentation, yeast cells produce a little alcohol and much carbon dioxide. As the carbon dioxide bubbles become trapped in the dough, they cause the dough to increase in bulk and acquire a lighter, finer texture. When the dough is baked, the alcohol and carbon dioxide are driven off. The bread becomes light and porous because of spaces created by carbon dioxide bubbles. Home bakers often use activated dry yeast, a product that is prepared by the lyophilization of yeast cells.

A genetically engineered strain of "fast-acting" yeast cuts the rising time of most breads in half.

Dairy Products

Microorganisms are used in making a wide variety of dairy products. Cultured buttermilk, popular in the United States, is made by adding *Streptococcus cremoris* to pasteurized skim milk and allowing fermentation to occur until the desired consistency, flavor, and acidity are reached. Other organisms—*Streptococcus lactis, S. diacetylactis* and *Leuconostoc citrovorum, L. cremoris,* or *L. dextranicum*—make buttermilk with different flavors because of variations in fermentation products. Sour cream is made by adding one of these organisms to cream. Yogurt is made by adding *Streptococcus thermophilus* and *Lactobacillus bulgaricus* to milk. These organisms release still other products, and so yogurt has a different texture and flavor.

Next time you buy yogurt "with active cultures," examine a sample with your microscope to see masses of rod-shaped Lactobacillus bulgaricus cells.

Fermented milk beverages have been made for centuries in various countries, especially Eastern European countries, by adding specific organisms or groups of organisms to milk. The products vary in acidity and alcohol content **(Table 26.5)**. Acidophilus milk is made by adding *Lactobacillus acidophilus* to sterile milk. Sterilization prevents uncontrolled fermentation by organisms that might already be present in nonsterilized milk. Bulgarian milk is made by *L. bulgaricus*; it is similar to buttermilk except that it is more acidic and

(Burke/Triolo Productions/Jupiter Images)

lacks the flavor imparted by the leuconostocs. The Balkan product kefir is made from the milk of cows, goats, or sheep, and the fermentation is usually carried out in goatskin bags. The Russian product koumiss is made from mare's milk. In kefir and koumiss, *Streptococcus*

TABLE 26.5

Fermented Milk Beverages	
Characteristics	**Beverages and Countries Where Made**
Less than 1% lactic acid	Sour cream, cultured buttermilk (United States)
	Filmjolk (Finland)
2–3% lactic acid	Yogurt (United States); called leben in Egypt, matzoon in Armenia, naja in Bulgaria, and dahi in India
	Tarho (Hungary)
	Kos (Albania)
	Fru-fru (Switzerland)
	Kaimac (Yugoslavia)
	Acidophilus milk (United States)
Alcoholic (1–3%)	Koumiss, kefir, and araka (former Soviet Union)
	Fuli and puma (Finland)
	Taette (Norway)
	Lang (Sweden)

lactis, L. bulgaricus, and yeasts are responsible for the production of lactic acid, alcohol, and other products. These products are usually made by continuous fermentation—fresh milk is added as fermented product is removed.

Cheeses

The first step in making almost any cheese is to add lactic acid bacteria and either **rennin** (an enzyme from calves' stomachs) or bacterial enzymes to milk. The bacteria sour the milk, and the enzymes coagulate the milk protein casein. The solid portion—the **curd**—is used to make cheese, and the liquid portion—the **whey**—is a waste product of this process (**Figure 26.9**). Sometimes lactic acid is extracted from whey. In separating the curd and the whey, different amounts of moisture are removed according to the kind of cheese being made. For soft cheeses, the whey is simply allowed to drain from the curd; for harder cheeses, heat and pressure are used to extract more moisture. Nearly all cheeses are salted. Salting helps remove water, prevents growth of undesired microorganisms, and contributes to the flavor of the cheese.

A few cheeses, such as cream cheese, cottage cheese, and ricotta, are unripened, but most cheeses are ripened. Ripening involves the action of microorganisms on the curd after it is pressed into a particular form. Soft cheeses are ripened by the action of microorganisms that occur naturally or are inoculated onto the surface of the pressed curd. Because the enzymes that ripen the cheese must diffuse from the surface of the center of the cheese, soft cheeses are relatively small. In contrast, hard cheeses are ripened by the action of microorganisms distributed through the curd. Because microbial action does not depend on diffusion, these cheeses can be quite large (**Figure 26.10**). Cheese is ripened by microorganisms in a cool, moist environment. Many modern factories have environmentally controlled rooms, but some still use natural caves similar to those where cheeses were first ripened.

Cheeses can be classified by their consistency (soft to hard), by the kind of microorganisms involved in the ripening process, and by the length of time required for ripening (**Table 26.6**). The ripening period is shorter for soft cheeses (1 to 5 months) than for hard cheeses (2 to 16 months).

Several microbial actions, such as decomposition of the curd and fermentation, occur during the ripening of cheeses. Prior to ripening, the curd consists of protein, lactose, and if the cheese is made from whole milk, fat. As microorganisms act on the curd, they first break down the lactose to lactic acid and other products, such as alcohols and volatile acids. Proteolytic enzymes break down protein, more extensively in soft than in hard cheese. *Brevibacterium linens* and the mold *Penicillium camemberti* are especially adept at releasing proteolytic enzymes. Lipase, particularly in *Penicillium roqueforti*, releases short-chain fatty acids such as butyric, caproic, and caprylic acids. These acids and their oxidation products contribute significantly to the flavor of the cheeses. The effects of fermentation in cheese ripening are most easily seen in Swiss cheese. Bacteria of the genus *Propionibacterium* ferment lactic acid and produce propionic acid, acetic acid, and carbon dioxide. The acids flavor the cheese, and the carbon dioxide, which becomes trapped in the curd, produces the characteristic holes in the cheese.

Other Products

A great variety of fermented foods and food products are made throughout the world; here we consider only a few. (Beers, wine, and spirits are discussed later.)

Vinegar. Vinegar is made from ethyl alcohol by the acetic acid bacterium, *Acetobacter aceti* (**Figure 26.11**), which oxidizes the alcohol to acetic acid. Commercially produced vinegar contains about 4% acetic acid. Cider vinegar is made from alcohol in fermented apple cider; wine vinegar is made from alcohol in wine.

Sauerkraut. Sauerkraut was first made in Europe in the sixteenth century. Bacteria naturally present on cabbage leaves act on shredded cabbage placed in layers in large crocks. Enough dry salt is placed between the layers to make a 2 to 3% salt solution as the salt draws

(a)

(b)

(c)

Figure 26.9   Making Gouda cheese. (a) Lactobacilli and the enzyme rennin are added to pasteurized milk. The bacteria sour the milk, and the rennin coagulates the milk protein casein. *(John Colwell/Grant Hellman Photography)* **(b)** The milk curdles into the solid curd and the liquid whey. Curds are drained and placed into "hoops," which are then squeezed in a press. *(John Colwell/Grant Hellman Photography)* **(c)** The pressed cheeses are removed from their hoops and floated in a tank of brine (salt solution). The high salt concentration extracts even more moisture from the cheese by osmosis (recall ◀Chapter 4, p. 108) and thus hardens it. The salt also adds flavor to the cheese and prevents the growth of undesired microorganisms. *(John Colwell/Grant Hellman Photography)*

water out of the cabbage. The cabbage is firmly packed and weighted down to create an anaerobic environment. Although many organisms cannot tolerate such an environment, anaerobic, halophilic species of *Lactobacillus* and *Leuconostoc* can carry out fermentation under these conditions. These microbes produce lactic acid, acetic acid, carbon dioxide, alcohol, and small amounts of other substances. After 2 to 4 weeks at room temperature, the cabbage is changed by fermentation to sauerkraut, which can be refrigerated until eaten or preserved by canning.

Pickles.   Pickles are made by essentially the same process as that used to make sauerkraut. Fresh cucumbers, either whole, sliced, or ground, are packed in brine and allowed to ferment for a few days to a few weeks. *Leuconostoc mesenteroides* is the major fermenting organism in low-salt brines (less than 5% salt), whereas species of

Pediococcus are more active in high-salt brines (5% or higher salt concentration). A problem in pickle making is to prevent a yeast film from forming on the surface of vats. Direct sunlight and ultraviolet light help solve this problem. After fermentation, vinegar and spices are added to sour pickles; sugar also is added to sweet pickles. Most pickles are pasteurized. Some pickles, such as sweet pickles and pickle relish, are made without fermentation. After the pickles have been soaked in brine for a few hours, they are seasoned with vinegar and spices, heat-processed, and sealed. The entire operation can be done in a day in any kitchen.

Olives.   Olives are treated with lye to hydrolyze oleuropein, a very bitter phenolic glucoside they contain that gives an undesirable flavor. Green olives are fermented in a 5 to 8% salt solution by *Leuconostoc mesenteroides* and *Lactobacillus plantarum*. They are packed in water

Figure 26.10 Ripening Gouda cheese. The ripening process involves the action of microorganisms on the curd after it is pressed into its form. A plastic coating is added to the cheese to keep it from drying out. The cheese is then placed on shelves in the aging room, where the microorganisms that are distributed throughout the curd will ripen it. Because gouda is a hard cheese, it must ripen for a relatively long period—3 months to a year—depending on the size of the wheel. *(John Colwell/Grant Heilman Photography)*

and pasteurized. Ripe olives are picked when reddish in color but not fully ripe. They are oxidized with tannins to blacken them, fermented in dilute (less than 5%) salt solution, packed in water in cans, and processed at 116°C for 60 minutes.

Poi. Poi, a common food in the South Pacific, is made from ground roots of the taro plant. A paste of ground roots is allowed to ferment through the action of a succession of naturally occurring organisms. Pseudomonads

Figure 26.11 Fermented foods: vinegar. The manufacture of vinegar makes use of the bacterium *Acetobacter aceti* (412X). This organism oxidizes ethyl alcohol to acetic acid. *(David M. Phillips/Visuals Unlimited)*

and coliforms begin the fermentation process, and lactobacilli later take over. Yeasts add alcohol to the mixture.

Soy Sauce. Soy sauce is made in a stepwise process (Figure 26.12). A salted mixture of crushed soybeans and wheat is treated with the mold *Aspergillus oryzae* to break down starch into fermentable glucose. The product, called *koji*, is mixed with an equal quantity of salt solution to make a mixture called moromi. Moromi is fermented for 8 to 12 months at low temperature and with occasional stirring. The fermenting organisms are mainly the bacterium *Pediococcus soyae* and the yeasts *Saccharomyces rouxii* and *Torulopsis* species. Lactic acid, other acids, and alcohol are produced. Upon completion of fermentation, the liquid and solid parts of the moromi are separated. The liquid part is bottled as soy sauce, and the solid part is sometimes used as animal food.

Soy Products. Other soy products include miso, tofu, and sufu. *Miso* is a fermented paste of soybeans made like soy sauce. *Tofu* is a soft curd of soybeans. It is

TABLE 26.6

Classification of Ripened Cheeses		
Consistency and Ripening Period	**Examples**	**Organisms Associated with Ripening**
Soft (1–2 months)	Limburger	*Streptococcus lactis, S. cremoris, Brevibacterium linens*
Soft (2–5 months)	Brie and Camembert	*S. lactis, S. cremoris, Penicillium camemberti,* and *P. candidum*
Semihard (1–8 months)	Muenster and brick	*S. lactis, S. cremoris, B. linens*
Semihard (2–12 months)	Roquefort and blue	*S. lactis, S. cremoris,* and *P. roqueforti* or *P. glaucum*
Hard (3–12 months)	Cheddar and Colby	*S. lactis, S. cremoris, S. durans,* and *Lactobacillus casei*
	Edam and Gouda	*S. lactis* and *S. cremoris*
	Gruyere and Swiss	*S. lactis, S. thermophilus, S. helveticus, Propionibacterium shermani,* or *L. bulgaricus* and *P. freudenreichii*
Hard (12–16 months)	Parmesan and Romano	*S. lactis, S. cremoris, S. thermophilus,* and *L. bulgaricus*

(a)

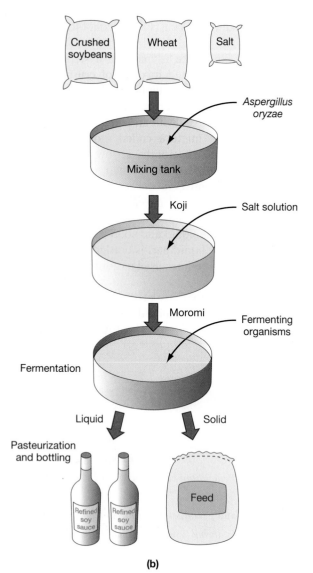

(b)

Figure 26.12 Making soy sauce. **(a)** The fermentation of soy and wheat is carried out in large steel tanks. *(Bill Franz Photography)* **(b)** A diagram of the soy fermentation process.

made from soybeans ground to make a milk, which is boiled to inactivate enzymes. The curd is precipitated with calcium or magnesium sulfate and pressed into a soft mass similar in consistency to soft cheese. *Sufu* is made by action of a fungus on soy curd. Cubes of curd are dipped in a mixture of salt and citric acid, inoculated with *Mucor*, and incubated until coated with mycelia. The fungus-coated cubes are aged 6 weeks in a rice-wine brine.

Fermented Meats. Microbes such as *Lactobacillus plantarum* and *Pediococcus cerevisiae* add flavor by fermenting meats such as salami, summer sausage, and Lebanon bologna. The heterolactic acid fermentation helps preserve the meat and also gives it a tangy flavor. Fungi such as *Penicillium* and *Aspergillus*, growing naturally on the surfaces of country hams, help to produce their distinctive flavor.

✓**CHECKLIST**

1. What is flat sour spoilage?
2. How does thawing frozen foods make them more susceptible to microbial food spoilage?
3. Why are there different methods of pasteurization?
4. Describe the process of cheese production.

▌▌▌ BEER, WINE, AND SPIRITS

Beer and wine are made by fermenting sugary juices; spirits, such as whiskey, gin, and rum, are made by fermenting juices and distilling the fermented product. **Distillation** separates alcohol and other volatile substances from solid and nonvolatile substances. Strains of *Saccharomyces* are the fermenters for all alcoholic beverages. Many different strains have been developed, each having distinctive characteristics. Both the organisms and how they are used are carefully guarded brewers' secrets.

To make beer, cereal grains (usually barley) are **malted** (partially germinated) to increase the concentration of starch-digesting enzymes that provide the sugar for fermentation **(Figure 26.13)**. Malted grain is crushed and mixed with hot water (about 65°C), producing **mash**. After a few hours, a liquid extract called **wort** is separated from the mix. *Hops* (flower cones from the hop plant) are added to the wort for flavoring, and the mixture is boiled to stop enzyme action and precipitate proteins. A strain of *Saccharomyces* is added. Fermentation produces ethyl alcohol, carbon dioxide, and other substances, including amyl and isoamyl alcohols and acetic and butyric acids, which add to the flavor of the beer. After fermentation, the yeast is removed, and the beer is filtered, pasteurized, and bottled.

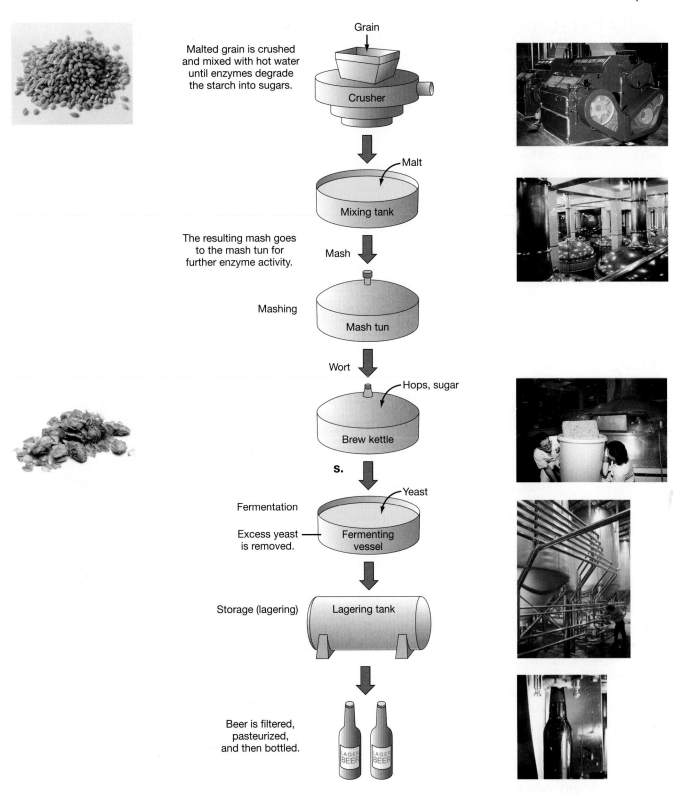

Grain

Malted grain is crushed and mixed with hot water until enzymes degrade the starch into sugars.

Crusher

Malt

Mixing tank

The resulting mash goes to the mash tun for further enzyme activity.

Mash

Mashing

Mash tun

Wort

Hops, sugar

Brew kettle

s.

Fermentation

Yeast

Excess yeast is removed.

Fermenting vessel

Storage (lagering)

Lagering tank

Beer is filtered, pasteurized, and then bottled.

LAGER BEER   LAGER BEER

Figure 26.13   The beer fermentation process.   *(left photos: Courtesy of Carolina Brewing Company; right top 3 photos: Photos and descriptions courtesy of Coors Brewing Company, Golden, Colorado; right bottom 2 photos: Courtesy of Carolina Brewing Company.)*

APPLICATIONS

What Is Not in Your Beer

Fermentation wastes from breweries contain significant quantities of vitamins, proteins, and carbohydrates. Releasing these wastes into streams is prohibited in some areas because it enriches the waters, causing an overgrowth of algae. Some breweries dispose of these wastes by drying them and marketing them as supplements to animal feed, or as the famous Australian product, Vegemite℗, beloved throughout the British empire. Its dark golden-colored spread is used on bread and to flavor stews, for example. It is very high in B vitamins. Multigallon tins of it were shipped to Australian troops during World War II for a little taste of home, plus vitamins.

Yeasts can also be grown specifically for the vitamins they produce. Certain *Candida* species can make and release into the medium up to 0.1 mg of riboflavin per gram of dry weight of yeast. This process has limited commercial value because the organisms are poisoned by traces of iron in the medium, and iron is nearly always present from equipment. Higher fungi, such as some that parasitize coffee and other plants, are not harmed by traces of iron and can be used commercially to produce riboflavin.

Saccharomyces uvarum produces ergosterol, a sterol that can be converted to vitamin D by UV radiation. This process is commercially feasible when carbon sources are adequate and vats are aerated.

(a)

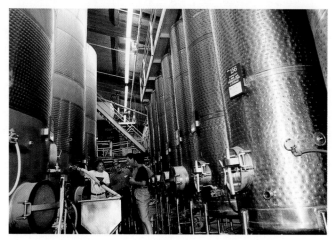

(b)

Most wine is made from juice extracted from grapes (**Figure 26.14**), although it can be made from any fruit—and even from nuts or dandelion blossoms. Juice is treated with sulfur dioxide to kill any wild yeasts that may already be present. Sugar and a strain of *Saccharomyces* are then added, and fermentation proceeds. Although ethyl alcohol is the main product of fermentation, other products similar to those in beer add to the flavor of the wine. In both beer and wine, the particular characteristics of the juice and the yeast strain determine the flavor of the final product. When fermentation is completed, liquid wine is siphoned to separate it from yeast sediment and, if necessary, cleared with agents such as charcoal to remove suspended particles. Finally, it is bottled and aged in a cool place.

Spirits are made from the fermentation of a variety of foods, including malted barley (Scotch whiskey), rye (rye whiskey, gin), corn (bourbon), wine or fruit juice (brandy), potatoes (vodka), and molasses (rum). After fermentation, distilling separates alcohol and other volatile substances that impart flavor from the solid and nonvolatile substances. Because of distillation, the alcohol content of spirits ranges from 40 to 50%—much higher than the typical 12% for wine and 6% for beer. (Wines do not contain more alcohol because when the alcohol concentration reaches 12 to 15%, it poisons the yeasts

(c)

Figure 26.14 The fermentation of wine. (a) The process begins with the rapid bubbling of the grape juice and yeast mixture *(Fred Lyon)*, which is **(b)** stored in two-story-high stainless steel fermentation vats until the fermentation process is complete. *(Sylvain Grandadam/Photo Researchers, Inc.)* **(c)** The wine is then transferred to wooden barrels, where it is aged, sometimes for many years. During this time, the flavor mellows and develops fully. *(Photo Researchers, Inc.)*

carrying out the fermentation. To produce fortified wines such as sherry and cognac, extra alcohol is added after fermentation.)

▌▌▌ INDUSTRIAL AND ▌▌▌ PHARMACEUTICAL MICROBIOLOGY

Industrial microbiology deals with the use of microorganisms to assist in the manufacture of useful products or to dispose of waste products. **Pharmaceutical microbiology** is a special branch of industrial microbiology concerned with the manufacture of products used in treating or preventing disease. Today many industrial and pharmaceutical processes make use of genetic engineering, as we saw in ◀Chapter 8 (p. 232).

Industrial microbiology, albeit in primitive form, had its beginnings more than 8,000 years ago when the Babylonians fermented grain to make beer. However, little was understood about fermentation until Pasteur studied the process in the nineteenth century. Over the next several decades, various researchers studied fermentation and its products, but their findings were largely ignored until shortages of materials for making explosives in World War I created a use for them. Both glycerol and acetone are used to make explosives and other materials. The Germans developed a process for making glycerol; the British used acetone-butanol fermentation by *Clostridium acetobutylicum* to make acetone. An important sidelight of the acetone-butanol fermentation was the development of techniques to maintain pure cultures in industrial fermentation vats.

The serendipitous discovery by Fleming in 1928 that *Penicillium notatum* kills *Staphylococcus aureus* was the beginning of the antibiotic industry (◀Chapter 13, p. 367). Today the manufacture of antibiotics is an immense component of pharmaceutical microbiology. Concurrent with the development of antibiotics was the development and industrial production of a variety of vaccines. The manufacture of antibiotics, vaccines, and many other pharmaceuticals all require pure-culture technology.

In recent years genetic engineering has been used to cause cells to synthesize products they otherwise would not make or to increase the yield of products they normally make. Genetic engineering may make it possible to program organisms to carry out specific industrial and pharmaceutical processes. Such processes might well be more efficient and more profitable than any others now available.

Today hundreds of different substances are manufactured with the aid of microorganisms. Various species of yeasts, molds, bacteria, and actinomycetes are used in manufacturing processes. The organisms themselves are sometimes useful, as when they can serve as a source of protein. Animal feed consisting of microorganisms is called **single-cell protein (SCP)**. Single-cell protein is an important high-yield, relatively inexpensive source of protein-rich food. More often the valuable substance is a product of microbial metabolism.

USEFUL METABOLIC PROCESSES

The production of complex molecules and metabolic end products in amounts that are commercially profitable requires the manipulation of microbial processes. In nature, microbes have regulatory mechanisms, such as induction and repression, that cause them to make substances only in the amounts needed (◀Chapter 7, p. 194). An important task in industry is to manipulate regulatory mechanisms so that the organisms continue producing large quantities of useful substances for humans. Industrial microbiologists accomplish this in several ways: (1) by altering nutrients available to the microbes, (2) by altering environmental conditions, (3) by isolating mutant microbes that produce excesses of a substance because of a defective regulatory mechanism, and (4) by using genetic engineering to program organisms to display particular synthetic capabilities. In some instances these efforts have been remarkably successful. The industrial strain of the mold *Ashbya gossypii* produces 20,000 times as much of the vitamin riboflavin as it uses. Industrial strains of *Propionibacterium shermanii* and *Pseudomonas denitrificans* produce 50,000 times as much cobalamin (vitamin B_{12}) as they use.

PROBLEMS OF INDUSTRIAL MICROBIOLOGY

It is one thing to cause a microbe to carry out a useful process in a test tube. It is quite another to adapt the process so that it will be profitable in a large-scale industrial setting. In the past, most industrial processes have been carried out in large fermentation vats. Many new processes simply do not work in large vats, so smaller vats must be used (**Figure 26.15**). Moreover, many of today's industrial microorganisms have been so extensively modified by selection of mutants or by genetic engineering that their products, which are useful to humans, may be useless or even toxic to the organisms.

Isolating and purifying the product, with or without killing the organisms, often presents technical difficulties. When the product remains within the cell, the plasma membrane must be disrupted to obtain the product. Membranes can be disrupted by spraying a medium containing the organisms through a nozzle under high pressure or by putting the organisms in alcohol or salt solutions that cause the product to form a precipitate. Product molecules from disrupted cells sometimes can be collected on resin beads of appropriate size and electric charge. Secreted products can be collected relatively easily, sometimes without killing the organisms. This can be done by using a **continuous reactor**, in which fresh medium is introduced on one side and medium containing

(a)

(b)

Figure 26.15   Scaling up for fermentation. Some fermentations do not progress satisfactorily **(a)** in large vats (at a Philadelphia beer-brewing plant) *(Joseph Nettis/Photo Researchers, Inc.)* and therefore must be conducted **(b)** in small-scale containers. *(Science VU/Visuals Unlimited)*

Figure 26.16   A continuous reactor. Fresh culture medium is introduced on one side, and medium containing product is withdrawn at the other side in a continuous process.

the product is withdrawn on the other side **(Figure 26.16)**. Continuous reactors, which operate under strictly controlled temperature and pH conditions, are used in many types of industrial fermentations.

USEFUL ORGANIC PRODUCTS

SIMPLE ORGANIC COMPOUNDS

Simple organic compounds such as solvents and organic acids can be manufactured with the aid of microorganisms. The solvents include ethanol (ethyl alcohol), butanol, acetone, and glycerol. The acids include acetic, lactic, and citric acids. Although microorganisms are not now regularly used to make these substances, microbial synthesis will become economically feasible as the cost of raw materials from petroleum increases, especially if organisms can be genetically programmed to increase their productivity.

Ethanol, an industrial chemical with annual sales of about $300 million, is used not only as a solvent but also in the manufacture of antifreeze, dyes, detergents, adhesives, pesticides, explosives, cosmetics, and pharmaceuticals. In addition, ethanol is used as a fuel, either alone or mixed with gasoline. Microbes are employed to produce ethanol for industrial purposes in much the same way they are used to make alcoholic beverages, but microbiologists are developing new methods. In one method, cellulose is extracted from wood, digested to sugars, and fermented by thermophilic clostridia. Because these organisms act at high temperatures, their metabolic rates are higher than those of other fermenters, and they produce alcohol at a more rapid rate. Also, because the effluent is already heated, less energy is needed to distill and purify the product. In another method, the bacterium *Zymomonas mobilis*, which ferments sugar twice as fast as yeast can, is used **(Figure 26.17)**.

The yeast *Pachysolen tannophilus* produces relatively large quantities of alcohol from five-carbon sugars. This is significant because grain products contain both five-carbon and six-carbon sugars. Thus, making alcohol with yeasts that use only six-carbon sugars fails to extract a significant amount of energy from the grain. Producing alcohol for fuel is currently almost a break-even activity; the energy obtained from the six-carbon sugars in the alcohol is approximately equal to the energy required to produce it. Extracting energy from both five- and six-carbon sugars would be much more economical.

Clostridium acetobutylicum acting on starch or *C. saccharoacetobutylicum* acting on sugar produces both butanol and acetone. Butanol is used in the manufacture of brake fluid, resins, and gasoline additives. Acetone is used mainly as a solvent.

Figure 26.17 Alcohol production. Equipment for the industrial production of ethanol, using *Zymomonas mobilis* as a catalyst. *(Courtesy Warren Gretz, National Renewable Energy Laboratory)*

Glycerol is made by adding sodium sulfite to a yeast fermentation. The sodium sulfite shifts the metabolic pathway so that glycerol, and not alcohol, is the main product. Glycerol is used as a lubricant and as an emollient in a variety of foods, toothpaste, cosmetics, and paper.

Of the organic acids made by microorganisms, acetic acid (vinegar) is used in the greatest volume in the manufacture of rubber, plastics, fabrics, insecticides, photographic materials, dyes, and pharmaceuticals. Acetic acid bacteria oxidize ethanol to make vinegar. But thermophilic bacteria can make it from cellulose, and *Acetobacterium woodii* and *Clostridium aceticum* can make it from hydrogen and carbon dioxide. Other acids made by microorganisms include lactic and citric acids. *Lactobacillus delbrueckii* metabolizes glucose to lactic acid, which is used to acidify foods, to make

synthetic textiles and plastics, and in electroplating. The mold *Aspergillus niger* makes citric acid with great efficiency when molasses is used as the fermentation substrate. Citric acid is widely used to acidify and improve the flavor of foods.

ANTIBIOTICS

The antibiotic industry came into being in the 1940s with the manufacture of penicillin (**Figure 26.18a**), and about 100 antibiotics have been manufactured in quantity since then. The market value of antibiotics worldwide now exceeds $5 billion annually.

Industrial microbiologists work diligently to find ways to cause organisms to make their particular antibiotic in large quantities. An effective way is to induce mutations and then screen the progeny to find strains that produce more antibiotic than the parent strain. Such methods, combined with improved fermentation procedures, have been quite successful. For example, a strain of *Penicillium chrysogenum* that once produced 60 mg of penicillin per liter of culture now produces 20 g/l. Genetic engineering may lead to more effective methods.

The reason only 2% of all known antibiotics have been marketed is that many antibiotics are either too toxic to use therapeutically or of no greater benefit than antibiotics already in use. Because pathogens continually develop resistance to antibiotics, industrial microbiologists not only seek new antibiotics but also try to make available antibiotics more effective. They look for ways to increase potency, improve therapeutic properties, and make antibiotics more resistant to inactivation by microorganisms. These efforts have led to the development of *semisynthetic antibiotics* (**Figure 26.18b**), which are made partly by microorganisms and partly by chemists (◀Chapter 13, p. 367). This collaboration is illustrated by the way chemists have modified β-lactam antibiotics (penicillins and cephalosporins), which some microorganisms destroy with

(a)

(b)

Figure 26.18 Penicillin production. **(a)** The amber-colored droplets seen on the surface of this culture of the mold *Penicillium notatum* are the antibiotic penicillin, which it has produced. *(A. McClenaghan/Science Photo Library/Photo Researchers, Inc.)* **(b)** Synthetic and semisynthetic antibiotics are now made in the laboratory for use alongside those produced by microorganisms. *(Dan McCoy/Rainbow)* the enzyme β-lactamase. Binding a molecule such as clavulanic acid to the β-lactam ring prevents the enzyme from inactivating the antibiotic. The antibiotic augmentin, for example, is a semisynthetic antibiotic consisting of amoxicillin and clavulanic acid.

ENZYMES

All enzymes used in industrial processes are synthesized by living organisms. With a few exceptions, such as the extraction of the meat tenderizer papain from papaya fruit, industrial enzymes are made by microorganisms. Enzymes are extracted from microorganisms rather than being synthesized in the laboratory because laboratory synthesis is as yet too complicated to be practical. Enzymes, like other proteins, consist of complex chains of amino acids in specific sequences. Organisms use genetic information to synthesize enzymes easily, but laboratory chemists find this a tedious and expensive process. Enzymes are especially useful in industrial processes because of their specificity. They act on a certain substrate and yield a certain product, thus minimizing problems of product purification.

Scientists are interested in the enzymes that allow extremophilic bacteria to function in harsh environments, because of their possible application to industrial environments.

The methods industrial microbiologists use to produce enzymes include screening for mutants and gene manipulation. Screening for mutants that produce large quantities of an enzyme has been an effective technique. Gene manipulation has been somewhat less effective. Compared with synthetic pathways for antibiotics, those for enzymes are simpler, and fewer genes are involved. Thus, gene manipulation shows great promise for increasing the yield of some enzymes and making industrial production of others feasible. Only about 200 of some 2,000 known enzymes are now produced commercially, so there is room for significant progress in this area.

Of the commercially available enzymes, proteases and amylases are produced in greatest quantities. Proteases, which degrade proteins and are added to detergents to increase cleaning power, are made industrially by molds of the genus *Aspergillus* and bacteria of the genus *Bacillus*. Amylases, which degrade starches into sugars, also are made by *Aspergillus* species. Other useful degradative enzymes are lipase from the yeast *Saccharomycopsis* and lactase from the mold *Trichoderma* and the yeast *Kluyveromyces*. Another important industrial enzyme is invertase (glucose isomerase) from *Saccharomyces*; it converts glucose to fructose, which is used as a sweetener in many processed foods.

Proteolytic Enzymes. Pancreatic enzymes were first used more than 70 years ago to remove blood stains from butchers' aprons without weakening the cloth. These enzymes were tried as laundry aids, but it was found that they were inactivated by soap. In the 1970s proteolytic enzymes from bacteria, which retain their activity in the presence of detergents and hot water, were added to detergents. When this was first tried, workers in the detergent factories developed respiratory problems and skin irritations. The enzymes were removed from the detergents when the illnesses were traced to airborne enzyme molecules. Today, enzymes are incorporated in coated granules that dissolve in the wash instead of in the workers' environment.

Proteolytic enzymes have also been added to drain cleaners, where they are especially useful in degrading hair, which often clogs bathroom drains. A drain cleaner with a lipase should do a good job on kitchen drains.

Enzymes in Paper Production. To make high-quality paper, much of the lignin, a coarse material in wood, is removed by expensive chemical means to produce a purer cellulose wood pulp. The fungus *Phanerochaete chrysosporium* secretes enzymes that digest both lignin and cellulose. If the enzymes can be separated and purified, one might be used selectively to digest lignin, leaving cellulose unchanged. If perfected, this process would provide a cheap way to prepare wood pulp for high-quality paper. A by-product of the research may benefit humans. Another lignin-digesting enzyme has been identified in a strain of *Streptomyces* bacteria. This enzyme modifies lignin into a molecule that enhances antibody production in mice. Might it someday do the same for humans?

AMINO ACIDS

Microbial production of amino acids has become a commercially successful industry. Twenty different amino acids are utilized by animals in the production of proteins. Eight of the 20 are *essential amino acids*—they cannot be synthesized by the animal and therefore must be provided in the diet. Over 30,000 tons of lysine, an essential amino acid, are manufactured by microbial fermentation annually. Lysine is produced by a mutant strain of *Corynebacterium glutamicum* in which the biosynthetic pathway has been altered to promote the production of large quantities of the amino acid. Lysine is added to animal feed as a supplement and sold in health-food stores for human consumption.

Glutamic acid, another product of microbial fermentation, is also produced by a mutant strain of *C. glutamicum*. This mutant contains a high level of the enzyme glutamic acid dehydrogenase, which increases the yield of glutamic acid. The bacterium is grown in a culture medium deficient in biotin, which results in the formation of "leaky" plasma membranes. Such membranes permit the excretion of glutamic acid from the cell. Glutamic acid is used to make the flavor enhancer monosodium glutamate (MSG). About 200,000 tons are produced annually. Other microbially synthesized amino acids include phenylalanine, aspartic acid, and tryptophan.

Figure 26.19 Hepatitis B vaccine is being produced by recombinant DNA technology. Technicians use a chromatography column to separate key proteins from batches of yeast cells. (Hank Morgan/Photo Researchers, Inc.)

OTHER BIOLOGICAL PRODUCTS

Vitamins, hormones, and single-cell proteins are the major categories of other biologically useful products of industrial microbiology. The ability of microorganisms to make vitamin B_{12} and riboflavin was cited earlier as an example of highly successful amplification of microbial synthesis. Vaccines, of course, are exceedingly useful products (**Figure 26.19**). The development of hepatitis vaccine was described in ◀Chapter 8 (p. 230).

Microorganisms also are used in the production of steroid hormones. The process used is called **bioconversion**, a reaction in which one compound is converted to another by enzymes in cells. The first application of bioconversion to hormone synthesis was the use of the mold *Rhizopus nigricans* to hydroxylate progesterone. This microbial step simplified the chemical synthesis of cortisone from bile

acids from 37 to 11 steps. It reduced the cost of cortisone from \$200 per gram to \$6 per gram. (Subsequent improvements in procedures have brought the price below \$0.70 per gram.) Other hormones that can now be produced industrially include insulin, human growth hormone, and somatostatin. They are made by recombinant DNA technology, using modified strains of *Escherichia coli*.

Single-cell proteins, as noted earlier, are whole organisms rich in protein. Currently used in animal feed, they may someday feed humans, too. An important advantage of single-cell protein is that it can be made from substances of less value than the protein produced. Certain *Candida* species make protein from paper pulp wastes, *Saccharomycopsis* makes it from petroleum wastes, and the bacterium *Methylophilus* makes it from methane or methanol.

▐▐▐ MICROBIOLOGICAL MINING

As the availability of mineral-rich ores decreases, methods are needed to extract minerals from less concentrated sources. This need spawned the new discipline known as **biohydrometallurgy**, the use of microbes to extract metals from ores. Copper and other metals originally were thought to be leached from the wastes of ore crushing as a result of an inorganic chemical reaction such as those reactions used to extract metals from ores. It was then discovered that this leaching is due to the action of *Thiobacillus ferrooxidans*. This chemolithotrophic acidophilic bacterium lives by oxidizing the sulfur that binds copper, zinc, lead, and uranium into their respective sulfide minerals, with a resultant release of the pure metal. Copper in low-grade ores is often present as copper sulfide. When acidic water is sprayed on such ore, *T. ferrooxidans* obtains energy as it uses oxygen from the atmosphere to oxidize the sulfur atoms

(a)

(b)

Figure 26.20 Microbial mining. Mining is made easier in many places by the activity of microorganisms. **(a)** Microbial leaching (extraction) helps concentrate copper from ore when conventional chemical approaches are not economically feasible. When acid solutions are added to dump surfaces, acidophilic bacteria such as *Thiobacillus ferrooxidans* quickly catalyze the oxidation of copper sulfate. (Corale L. Brierley/Visuals Unlimited) **(b)** Microorganisms in these ponds reclaim metals from low-grade ores. Ions generated by microbial activity include ferric ions (orange) and ferrous ions (green). (Corale L. Brierley/Visuals Unlimited) in sulfide ores to sulfate. The bacterium doesn't use the copper; it merely converts it to a water-soluble form that can be retrieved and used by humans (**Figure 26.20a**).

Other minerals also can be degraded by microbes. *T. ferrooxidans* releases iron from iron sulfide by the same process (**Figure 26.20b**). Combinations *of T. ferrooxidans* and a similar organism, *T. thiooxidans*, degrade some copper and iron ores more rapidly than either one does alone. Another combination of organisms, *Leptospirillum ferrooxidans* and *T. organoparus*, degrades pyrite (FeS$_2$) and chalcopyrite (CuFeS$_2$), although neither organism can degrade the minerals alone. Other bacteria can be used to mine uranium, and bacteria may eventually be used to remove arsenic, lead, zinc, cobalt, and gold. However, of late, fewer mining companies are actually using microbes in their processing.

MICROBIOLOGICAL WASTE DISPOSAL

Sewage-treatment plants (◄Chapter 25, p. 805) are prime examples of microbiological waste-disposal systems, but sewage disposal is a relatively simple problem compared with the problems associated with disposing of chemical pollutants and toxic wastes. Some wastes persist in the environment and contaminate water supplies for wildlife and humans. Bioremediation, the use of microorganisms to dispose of some chemical wastes, can help avert a monumental environmental disaster from accumulation of toxic wastes (◄Chapter 25, p. 809).

Three strains of microorganisms have been found to deactivate Arochlor 1260, one of the most highly toxic of polychlorinated biphenyl (PCB) compounds. Other organisms have been shown to detoxify chemical substances such as cyanide and dioxin and to degrade oil spilled in the ocean. Genetic engineering has been used to develop a bacterium capable of detoxifying the defoliant Agent Orange, and work is under way to modify bacteria to detoxify other toxins.

One of the problems in developing microorganisms that can degrade toxic substances is the limited information available on the genetic characteristics of microorganisms found in wastes. Many researchers in this area focus their efforts on organisms found in wastes because these organisms probably already have degradatory capabilities. The researchers think it should be somewhat easier to modify them to metabolize other wastes than to use better-known organisms that have no known capacity to degrade wastes.

✓**CHECKLIST**

1. What are the origins and use of single-cell protein?
2. List several enzymes and amino acids that are produced by industrial microbiology methods.
3. How do microbes help in mining and in waste disposal?

▌ RETRACING OUR STEPS

▌▌▌ MICROORGANISMS FOUND IN FOOD

- Anything that humans eat is food for microorganisms, too.
- Many organisms in food are commensals; some cause spoilage, and a few cause human disease.

GRAINS
- Grains stored in moist areas become contaminated with various molds.
- Flour is purposely inoculated with yeasts to make bread.

FRUITS AND VEGETABLES
- Fruits and vegetables are subject to soft rot and mold damage.

MEATS AND POULTRY
- Meats and poultry contain many kinds of microorganisms, some of which cause zoonoses. Poultry typically is contaminated with *Salmonella*, *Clostridium perfringens*, and *Staphylococcus aureus*.

FISH AND SHELLFISH
- Seafood can be contaminated with several kinds of bacteria and viruses.

MILK
- Milk can contain organisms from cows, milk handlers, and the environment.

OTHER EDIBLE SUBSTANCES
- Sugars support growth of various organisms, which are killed during refining.
- Spices contain large numbers of microorganisms; condiments may support mold growth.
- Syrups for carbonated beverages can become contaminated with molds.
- Tea, coffee, and cocoa are also subject to mold contamination if not kept dry.

▌▌▌ PREVENTING DISEASE TRANSMISSION AND FOOD SPOILAGE

- Common diseases transmitted in food and milk are listed in Table 26.3.
- Good sanitation reduces the chance of getting foodborne diseases. The most important factor in preventing spoilage and disease transmission in food and milk is cleanliness in handling.

FOOD PRESERVATION
- Methods of food preservation include **canning**, refrigeration, freezing, **lyophilization**, drying, ionizing radiation, and use of chemical additives.

PASTEURIZATION OF MILK
- Methods of preserving milk include pasteurization and sterilization.

STANDARDS FOR FOOD AND MILK PRODUCTION
- Certain standards for food and milk production are maintained by federal, state, and local laws. Tests for milk quality are summarized in Table 26.4.

III MICROORGANISMS AS FOOD AND IN FOOD PRODUCTION

ALGAE, FUNGI, AND BACTERIA AS FOOD
- Rapid human population growth has created a need for new food sources.
- Yeasts can be grown on a variety of wastes and are good sources of cheap protein and vitamins. Equipment to grow them is expensive, and some effort will be needed to persuade humans to consider them acceptable as food.
- Algae can be grown in lakes and on sewage and also are good sources of protein. Problems with growing them include the danger of viral contamination and lack of acceptability as food.

FOOD PRODUCTION
- Yeast is used to leaven bread.
- The fermentative capabilities of certain bacteria are used to make dairy products such as buttermilk, sour cream, yogurt, a variety of fermented beverages, and cheeses.
- In cheese making, the **whey** of milk is discarded, and microorganisms ferment the **curd** and impart flavor and texture to a cheese.
- Other foods produced by microbial fermentation include vinegar, sauerkraut, pickles, olives, poi, soy sauce, other soy products, salami, Lebanon bologna, and summer sausage.

III BEER, WINE, AND SPIRITS

- Beer is made from **malted** grains; hops are added, and the mixture is fermented.
- Wine is made by fermenting fruit juices.
- Spirits are made by fermenting various substances and distilling the products.

III INDUSTRIAL AND PHARMACEUTICAL MICROBIOLOGY

- **Industrial microbiology** deals with the use of microorganisms to assist in the manufacture of useful products or to dispose of waste products.
- **Pharmaceutical microbiology** deals with use of microorganisms in the manufacture of medically useful products.

USEFUL METABOLIC PROCESSES
- Modifications of microbial processes in industry include altering nutrients available to microbes, altering environmental conditions, isolating mutants that produce excesses of useful products, and modifying the organisms by genetic engineering.

PROBLEMS OF INDUSTRIAL MICROBIOLOGY
- One problem in industrial microbiology is the development of small-scale processes; other problems concern techniques for recovering products.

III USEFUL ORGANIC PRODUCTS

SIMPLE ORGANIC COMPOUNDS
- Microorganisms can make alcohols, acetone, glycerol, and organic acids. Microbial processes to make such products are likely to become more economically feasible in the future.

ANTIBIOTICS
- Antibiotics are derived from *Streptomyces*, *Penicillium*, *Cephalosporin*, and *Bacillus* species.
- Many antibiotics have a β-lactam ring; some semisynthetic antibiotics are made by modifying that ring so that microorganisms cannot degrade it.

ENZYMES
- Enzymes extracted from microorganisms include proteases, amylases, lactases, lipases, and invertase. They are used in detergents, drain cleaners, enrichment of food, and manufacture of paper.

AMINO ACIDS
- Several amino acids, such as lysine and glutamic acid, can be produced microbially.

OTHER BIOLOGICAL PRODUCTS
- Vitamins and hormones are usually made by manipulating organisms so that they produce excessive amounts of these useful products.
- **Single-cell proteins** consist of whole organisms rich in proteins; they are used mainly as animal feeds.

III MICROBIOLOGICAL MINING

- Microbes are currently used to extract copper from low-grade ores. Other minerals that can be extracted by microorganisms include iron, uranium, arsenic, lead, zinc, cobalt, and nickel.

III MICROBIOLOGICAL WASTE DISPOSAL

- Sewage-treatment plants (◄Chapter 25) use microorganisms in waste disposal.
- A few organisms have been found that degrade toxic wastes, and research is under way to identify or develop others.

I TERMINOLOGY CHECK

bioconversion (*p. 841*)
biohydrometallurgy (*p. 841*)
bone stink (*p. 818*)
canning (*p. 825*)
continuous reactor (*p. 837*)
curd (*p. 831*)
distillation (*p. 834*)
flash pasteurization (*p. 827*)

flat sour spoilage (*p. 825*)
high-temperature short-time (HTST) pasteurization (*p. 827*)
holding method (*p. 828*)
industrial microbiology (*p. 837*)
leavening agent (*p. 830*)

low-temperature long-time (LTLT) pasteurization (*p. 828*)
lyophilization (*p. 826*)
malted (*p. 834*)
mash (*p. 834*)
mesophilic spoilage (*p. 825*)
pharmaceutical microbiology (*p. 837*)

rennin (*p. 831*)
single-cell protein (SCP) (*p. 837*)
thermophilic anaerobic spoilage (*p. 825*)
ultra-high temperature (UHT) treatment (*p. 828*)
whey (*p. 831*)
wort (*p. 834*)

∎ CLINICAL CASE STUDY

A young married student became pregnant. Her husband told her that fatherhood was not for him, and he ran off, never to be located again. Finances were hard for the young mother; trying to work, finish college, and take care of her baby. She lived out in the country, and on her drive to school she passed a chicken farm that advertised "Pet Eggs" for sale, very cheaply. These were cracked eggs. In an effort to save money, she bought some, scrambled them, and fed them to her young son. He became very ill, nearly died, and survived only to be profoundly retarded for the rest of his life. How did this happen? What was the diagnosis? Would you buy those eggs for your pet? Unfortunately this is a true story.

∎ CRITICAL THINKING QUESTIONS

1. Suppose that your microbiology class is going to have a party where everything consumed is produced by microorganisms. What are some items you could take?

2. Industrial microbiology today has been a huge success in providing us (the consumer) with hundreds of different substances made with the use of microorganisms. However, industrial microbiology has not been without its problems. What are some of the difficulties this industry has had to overcome to be successful?

3. What are the safeguards one can take to protect against *Salmonella*-contaminated poultry, eggs, and meats?

∎ SELF-QUIZ

1. Match the following grains and breads to their most likely microbial contaminant/inhabitant and desired/detrimental by-product:

___ Leavened bread
___ Rye bread
___ Some grains and peanuts
___ Raw grains

1. *Claviceps purpurea*
2. *Aspergillus species*
3. *Bacillus* species
4. *Saccharomyces cerevisiae*

(a) Aflatoxins
(b) Hallucinogen—ergot poisoning
(c) Carbon dioxide
(d) Hydrolyzed protein and starch

2. Fruits and vegetables are both subjected to bacterial and fungal soft rot. How do plants protect themselves from this occurrence?
(a) They emit a low-pitched sonar wave to disrupt attachment of microbes.
(b) They release antimicrobial substances.
(c) The skins of plant foods contain protective waxes.
(d) All of the above.
(e) Both b and c.

3. The Irish potato famine of 1846 was caused by the fungus:
(a) *Erwinia carotovora*
(b) *Pseudomonas syringae*
(c) *Monilia sitophila*
(d) *Rhizopus nigricans*
(e) *Phytophthora infestans*

4. A green discoloration on refrigerated meat may be caused by the growth of:
(a) *Pseudomonas syringae*
(b) *Monilia sitophila*
(c) *Rhizopus nigricans*
(d) *Pseudomonas mephitica*
(e) *Pseudomonas fluorescens*

5. Approximately 50% of the foodborne outbreaks in restaurants that were associated with poultry were caused by:
(a) *Salmonella* spp.
(b) *Clostridium perfringens*
(c) *Staphylococcus aureus*
(d) *Escherichia coli*
(e) *Pseudomonas mephitica*

6. Hens may lay infected eggs if they are infected with:
(a) *Clostridium perfringens*
(b) *Staphylococcus aureus*
(c) *Escherichia coli*
(d) *Pseudomonas mephitica*
(e) *Salmonella pullorum*

7. Which of the following is a protective measure within egg whites that protects the embryo within from microbial contamination?
(a) Egg whites are filled with lysozyme that ruptures bacterial cell walls, killing them.
(b) Egg whites produce antimicrobial antibodies.
(c) Nutrients, vitamins, and minerals are tightly wrapped up by proteins and other substances in the egg white, rendering them unavailable for bacterial use.
(d) All of the above.
(e) None of the above.

8. What is it about clams that make them especially likely sources of human infection?
(a) They have sticky interiors that make microorganism attachments easy.
(b) Their ability to glow with *Photobacterium phosphoreum* attracts pathogenic species of microorganisms.
(c) They are best eaten raw.
(d) They are filter feeders, obtaining food by filtering water and extracting microbes.
(e) Their hard outer shell seals in and concentrates any contaminating microbes.

9. The most common microorganism in freshly drawn milk is:
   (a) *Staphylococcus epidermidis*
   (b) *Staphylococcus aureus*
   (c) *Mycobacterium bovis*
   (d) *Escherichia coli*
   (e) None of these

10. Match the following microorganisms that can be found in milk to their descriptions:
    ___ Can grow in refrigerated milk
    ___ Causes a fecal flavor in milk
    ___ Causes a viscous slime to form in milk
    ___ Cause milk to sour
    ___ Dairy herds are tested or vaccinated against these human pathogens
    ___ Present in freshly drawn milk

    (a) *Brucella* species
    (b) *Staphylococcus epidermidis*
    (c) *Pseudomonas* species
    (d) *Micrococcus*
    (e) *Streptococcus lactis*
    (f) *Lactobacillus* species
    (g) *Mycobacterium tuberculosis*
    (h) *Escherichia coli*
    (i) *Acinetobacter johnsoni*

11. Why do foods high in sugar concentration (jellies, jams, candies, candied fruits) resist microbial contamination?
    (a) Microbes do not have the necessary enzymes to break down the sugars.
    (b) High sugar concentrations create sufficient osmotic pressure that inhibits microbial growth.
    (c) Microbes prefer foods low in sugar.
    (d) Both a and b.
    (e) Both b and c.

12. At times, canned foods can be seen with the can bulging due to gas production within, especially when stored in a hot environment. Which of the following microorganisms can survive the canning process and cause bulging cans and food spoilage?
    (a) *Clostridium perfringens*
    (b) *Lactobacillus acidophilus*
    (c) *Pseudomonas aeruginosa*
    (d) *Bacillus stearothermophilus*
    (e) *Aspergillus oryzae*

13. Which of the following methods still in use today were developed by early civilizations to preserve food?
    (a) Canning
    (b) Drying
    (c) Acidic conditions
    (d) Lyophilization
    (e) b and c

14. In contrast to nonionizing, the use of ionizing radiation (such as gamma rays) to preserve food is discouraged because it can make food radioactive. True or false?

15. Which of the following is NOT a method of food preservation?
    (a) Repeated cycles of freezing and thawing.
    (b) Drying
    (c) Canning with moist heat under pressure
    (d) Lyophilization
    (e) Salting foods

16. Match the following chemical additives for food preservation to their uses or mode of action:
    ___ Sodium chloride
    ___ Ethylene and propylene oxides
    ___ Nitrates and nitrites
    ___ Organic acids
    ___ Quaternary ammonium compounds
    ___ Carbon dioxide

    (a) Lowers the pH, preventing microbial growth
    (b) Increases osmotic pressure
    (c) Sanitizes objects (utensils, cow udders, eggshell surfaces, etc.)
    (d) Kills microorganisms in carbonated beverages, retards maturation of fruits, and decreases spoilage during shipment
    (e) Inhibits microbial growth in ground meats and coldcuts
    (f) Alkylating agents used in nuts and spices to kill and inhibit microbes

17. Which of the following foods may not be produced by microbial action?
    (a) Sauerkraut
    (b) Vinegar
    (c) Pickles
    (d) Olives
    (e) Soy sauce

18. Which of the following type of microorganism shows promise as a food source because of its ability to grow on waste material and to provide a high amount of protein?
    (a) Algae
    (b) Fungi
    (c) Yeasts
    (d) Gram-positive bacteria
    (e) Gram-negative bacteria

19. *Spirulina* spp. have been used as a source of food. These microorganisms are a type of:
    (a) Cyanobacterium
    (b) Yeas
    (c) Fungus
    (d) Alga
    (e) None of these

20. Which of the following microorganisms can be added to milk to make a cultured dairy product?
    (a) *Streptococcus lactis*
    (b) *Streptococcus cremoris*
    (c) *Leuconostoc citrovorum*
    (d) *Lactobacillus bulgaricus*
    (e) All of these

21. What happens to curd during the process of cheese ripening?
    (a) Microorganisms carry out anabolic reactions.
    (b) Microorganisms oxidize glucose, assemble new polypeptide chains, and make fatty acids.
    (c) Microorganisms make alkaline products from the curd, giving the cheese its flavor.
    (d) Microorganisms ferment lactose and decompose curd by catabolizing proteins and fatty acids to form acids and their oxidation products that flavor the cheese.
    (e) None of the above.

22. *Lactobacillus planetarum* and *Pediococcus cerevisiae* are two microorganisms that ferment heterolactic acid, which aids in this food's preservation and tangy flavor.
(a) Tofu
(b) Pickles
(c) Cold cuts (salami, summer sausage, and Lebanon bologna)
(d) Sauerkraut
(e) Vinegar

23. Which of the following compounds can be made by the action of microorganisms?
(a) Ethanol
(b) Acetone
(c) Butanol
(d) Glycerol
(e) All of these

24. *Thiobacillus ferrooxidans* is useful in mining operations because of its ability to oxidize:
(a) Sulfur
(b) Copper
(c) Iron
(d) Metallic oxides
(e) Silver

25. What modifications can be made to microbial metabolic processes in order for them to produce desirable products in amounts that are commercially profitable?

(a) Modify organisms by genetic engineering.
(b) Alter environmental conditions.
(c) Alter nutrients available to microbes.
(d) Isolate mutants that produce large amounts of useful products.
(e) All of the above.

26. In each accompanying diagram, explain how organisms get into milk, and list the diseases they could cause.

(©Reuters/Corbis Images)

▍ EXPLORATIONS ON THE WEB

http://www.wiley.com/college/black

If you think you've mastered this chapter, there's more to challenge you on the web. Go to the companion web site to fine-tune your understanding of the chapter concepts and discover answers to the questions posed below.

1. Everyone knows that meat, fish, poultry, and vegetables have to be canned using a steam-pressure canner. So why are tomatoes the exception? Find out at the web site.

2. Should I eat my canned pears if they are pink and/or blue? How about mustard that turned green? Explore the answers to these questions at the web site.

3. Do irradiated peas glow, and is it still safe to eat my mutated pork chops? Discover why or why not at the web site.

4. Scientists want to zap my chocolate cake and ice cream with low-dose radiation to increase shelf life and kill dangerous bacteria, but will my cake still taste the same? Uncover the answer at the web site.

Metric System Measurements, Conversions, and Math Tools

Metric System Prefixes

pico (p) $= 10^{-12}$
nano (n) $= 10^{-9}$
micro (μ) $= 10^{-6}$
milli (m) $= 10^{-3}$
centi (c) $= 10^{-2}$
deci (d) $= 10^{-1}$
kilo (k) $= 10^{3}$

Length

1 kilometer (km) = 0.62 mile
1 meter (m) = 39.37 inches = 3.281 feet
1 meter = 100 centimeters = 1000 millimeters
1 centimeter (cm) = 10 millimeters = 0.394 inch
1 millimeter (mm) = 0.0394 inch
1 micrometer (μm) $= 10^{-6}$ meter
1 nanometer (nm) $= 10^{-9}$ meter
1 angstrom (Å) $= 10^{-10}$ meter

Volume

1 liter (l) = 1.057 quarts
1 liter = 1000 milliliters
1 milliliter (ml) = 1 cm^3 = 0.061 cubic inch
1 mm^3 $= 10^{-3}$ cm^3 $= 10^{-6}$ liter

Mass

1 kilogram (kg) = 1000 grams = 2.205 pounds
1 pound = 453.6 g
1 gram (g) = 1000 milligrams = 0.0353 ounce
1 ounce = 28.35 g
1 milligram (mg) $= 10^{-3}$ g
1 microgram (μg) $= 10^{-6}$ g

Temperature

degrees Fahrenheit ($^\circ$F) $= \frac{9}{5}(^\circ$C$) + 32$

degrees Celsius ($^\circ$C) $= \frac{5}{9}(^\circ$F$ - 32)$

0°C $= 32^\circ$F (freezing point of water)
100°C $= 212^\circ$F (boiling point of water)
37°C $= 98.6^\circ$F (normal body temperature)

Exponential (Scientific) Notation

Numbers that are either very large or very small are usually represented in *exponential* (scientific) *notation* as a number between 1 and 10 multiplied by a power of 10. In this kind of expression, the small raised number to the right of the 10 is the *exponent*.

Number	Exponential Form	Exponent
1,000,000	1×10^6	6
100,000	1×10^5	5
10,000	1×10^4	4
1,000	1×10^3	3
100	1×10^2	2
10	1×10^1	1
1		
0.1	1×10^{-1}	-1
0.01	1×10^{-2}	-2
0.001	1×10^{-3}	-3
0.0001	1×10^{-4}	-4
0.00001	1×10^{-5}	-5
0.000001	1×10^{-6}	-6
0.0000001	1×10^{-7}	-7

Numbers greater than 1 have *positive* exponents, which tell how many times a number must be *multiplied* by 10 to obtain the correct value. For example, the expression 5.2×10^3 means that 5.2 must be multiplied by 10 three times:

$$5.2 \times 10^3 = 5.2 \times 10 \times 10 \times 10 = 5.2 \times 1000 = 5200$$

In so doing, we move the decimal point three places to the right:

$$5\underset{1\ \ 2\ \ 3}{2\,0\,0.}$$

The value of a positive exponent indicates *how many places to the right the decimal point must be moved* to give the correct number in ordinary decimal notation.

Numbers less than 1 have *negative* exponents, which tell how many times a number must be *divided* by 10 (or multiplied by one-tenth) to obtain the correct value. Thus, the expression 3.7×10^{-2} means that 3.7 must be divided by 10 two times:

$$3.7 \times 10^{-2} = \frac{3.7}{10 \times 10} = \frac{3.7}{100} = 0.037$$

In so doing, we move the decimal point two places to the left:

$$0\underset{2\ \ 1}{.0\,3\,7}$$

The value of a negative exponent indicates *how many places to the left the decimal point must be moved* to give the correct number in ordinary decimal notation.

Converting Decimal Numbers to Exponential Notation

To convert a number greater than 1 from decimal notation to exponential notation, first move the decimal point to the *left* until only a single digit is to the left of the decimal point. The *positive* exponent needed for the exponential notation is the same as the *number of places the decimal point was moved*:

$$6\underset{5\ 4\ 3\ 2\ 1}{3\,5\,7\,8\,1.} = 6.35781 \times 10^5$$

To convert a number less than 1 from decimal notation to exponential notation, first move the decimal point to the *right* until a single *nonzero* digit is to the left of the decimal point. The *negative* exponent needed for exponential notation is the same as *the number of places the decimal point was moved*:

$$0.\underset{1\ 2\ 3\ 4}{0\,0\,0\,4}2\,6 = 4.26 \times 10^{-4}$$

Multiplying Exponential Numbers

To multiply two numbers in exponential form, *add* the exponents. For example:

$$
\begin{aligned}
(3.5 \times 10^3) \times (4.2 \times 10^4) &= 3.5 \times 4.2 \times 10^{(3+4)} \\
&= 14.7 \times 10^7 \\
&= 1.47 \times 10^8 = 1.5 \times 10^8 \\
&\qquad\qquad \textbf{(rounded off)}
\end{aligned}
$$

$$
\begin{aligned}
(5.2 \times 10^4) \times (4.6 \times 10^{-3}) &= 5.2 \times 4.6 \times 10^{[4+(-3)]} \\
&= 23.92 \times 10^1 \\
&= 2.392 \times 10^2 = 2.4 \times 10^2 \\
&\qquad\qquad \textbf{(rounded off)}
\end{aligned}
$$

Dividing Exponential Numbers

To divide two numbers in exponential form, *subtract* the exponents. For example:

$$
\frac{4.1 \times 10^4}{6.2 \times 10^6} = \frac{4.1}{6.2} \times 10^{(4-6)} = 0.6613 \times 10^{-2}
$$

$$
= 6.613 \times 10^{-3} = 6.6 \times 10^{-3}
$$
$$\textbf{(rounded off)}$$

$$
\frac{6.6 \times 10^3}{8.4 \times 10^{-2}} = \frac{6.6}{8.4} \times 10^{[3-(-2)]} = 0.7857 \times 10^5
$$

$$
= 7.857 \times 10^4 = 7.9 \times 10^4
$$
$$\textbf{(rounded off)}$$

pH Equation

The most convenient way to express the acidity or alkalinity of a solution is in terms of the *concentration of protons, or hydrogen ions* (H^+), present in the solution. Thus, to calculate the pH of a solution, you need to know the H^+ concentration—expressed as $[H^+]$—of that solution. Conversely, if you know the pH, you can determine $[H^+]$. A simple logarithmic equation can be used in either case.

$$\text{pH} = -\log_{10}[H^+]$$

In words, the pH of a solution is equal to the negative logarithm (to the base of 10) of the hydrogen ion concentration, where $[H^+]$ is given in moles per liter. For example, water normally has $[H^+] = 10^{-7}$ moles/liter. The pH of water then is $-\log_{10}[10^{-7}] = -(-7) = 7$. Your stomach "juices" have $[H^+] + 10^{-2}$ moles/liter. Therefore, the pH of your stomach juices is 2.

To calculate $[H^+]$ from the pH equation when you know the pH, consider a structure in an animal cell with pH = 5. For this structure $[H^+]$ must be 10^{-5} moles/liter. Antacids have a pH of 9, so for antacids $[H^+] = 10^{-9}$ moles/liter.

Classification of Bacteria and Viruses

CLASSIFICATION OF BACTERIA

Taxonomic Outline of the Procaryotic Genera
BERGEY'S MANUAL® OF SYSTEMATIC BACTERIOLOGY, 2nd Edition
May 2004, Springer-Verlag, New York.

George M. Garrity, Matthew Winters, and Denise B. Searles

Domain "*Archaea*"
Phylum AI. *Crenarchaeota phy. nov.*
  Class I. *Thermoprotei class. nov.*
    Order I. *Thermoproteales*
      Family I. *Thermoproteaceae*
        Genus I. *Thermoproteus*
        Genus II. *Caldivirga*
        Genus III. *Pyrobaculum*
        Genus IV. *Thermocladium*
        Genus V. *Vulcanisaeta*
      Family II. *Thermofilaceae*
        Genus I. *Thermofilum*
    Order II. *Caldisphaerales*
      Family I. *Caldisphaeraceae*
        Genus I. *Caldisphaera*
    Order III. *Desulfurococcales ord. nov.*
      Family I. *Desulfurococcaceae*
        Genus I. *Desulfurococcus*
        Genus II. *Acidolobus*
        Genus III. *Aeropyrum*
        Genus IV. *Ignicoccus*
        Genus V. *Staphylothermus*
        Genus VI. *Stetteria*
        Genus VII. *Sulfophobococcus*
        Genus VIII. *Thermodiscus gen. nov.*
        Genus IX. *Thermosphaera*
      Family II. *Pyrodictiaceae*
        Genus I. *Pyrodictium*
        Genus II. *Hyperthermus*
        Genus III. *Pyrolobus*
    Order IV. *Sulfolobales*
      Family I. *Sulfolobaceae*
        Genus I. *Sulfolobus*
        Genus II. *Acidianus*
        Genus III. *Metallosphaera*
        Genus IV. *Stygiolobus*
        Genus V. *Sulfurisphaera*
        Genus VI. *Sulfurococcus*
Phylum AII. *Euryarchaeota phy. nov.*
  Class I. *Methanobacteria class. nov.*
    Order I. *Methanobacteriales*
      Family I. *Methanobacteriaceae*
        Genus I. *Methanobacterium*
        Genus II. *Methanobrevibacter*
        Genus III. *Methanosphaera*

        Genus IV. *Methanothermobacter*
      Family II. *Methanothermaceae*
        Genus I. *Methanothermus*
  Class II. *Methanococci class. nov.*
    Order I. *Methanococcales*
      Family I. *Methanococcaceae*
        Genus I. *Methanococcus*
        Genus II. *Methanothermococcus gen. nov.*
      Family II. *Methanocaldococcaceae fam. nov.*
        Genus I. *Methanocaldococcus gen. nov.*
        Genus II. *Methanotorris gen. nov.*
  Class III. *Methanomicrobia*
    Order I. *Methanomicrobiales*
      Family I. *Methanomicrobiaceae*
        Genus I. *Methanomicrobium*
        Genus II. *Methanoculleus*
        Genus III. *Methanofollis*
        Genus IV. *Methanogenium*
        Genus V. *Methanolacinia*
        Genus VI. *Methanoplanus*
      Family II. *Methanocorpusculaceae*
        Genus I. *Methanocorpusculum*
      Family III. *Methanospirillaceae fam. nov.*
        Genus I. *Methanospirillum*
        Genus II. *Methanocalculus*
    Order II. *Methanosarcinales ord. nov.*
      Family I. *Methanosarcinaceae*
        Genus I. *Methanosarcina*
        Genus II. *Methanococcoides*
        Genus III. *Methanohalobium*
        Genus IV. *Methanohalophilus*
        Genus V. *Methanolobus*
        Genus VI. *Methanomethylovorans*
        Genus VII. *Methanomicrococcus*
        Genus VIII. *Methanosalsum gen. nov.*
      Family II. *Methanosaetaceae fam. nov.*
        Genus I. *Methanosaeta*
  Class IV. *Halobacteria class. nov.*

    Order I. *Halobacteriales*
      Family I. *Halobacteriaceae*
        Genus I. *Halobacterium*
        Genus II. *Haloarcula*
        Genus III. *Halobaculum*
        Genus IV. *Halobiforma*
        Genus V. *Halococcus*
        Genus VI. *Haloferax*
        Genus VII. *Halogeometricum*
        Genus VIII. *Halomicrobium*
        Genus IX. *Halorhabdus*
        Genus X. *Halorubrum*
        Genus XI. *Halosimplex*
        Genus XII. *Haloterrigena*
        Genus XIII. *Natrialba*
        Genus XIV. *Natrinema*
        Genus XV. *Natronobacterium*
        Genus XVI. *Natronococcus*
        Genus XVII. *Natronomonas*
        Genus XVIII. *Natronorubrum*
  Class V. *Thermoplasmata class. nov.*
    Order I. *Thermoplasmatales ord. nov.*
      Family I. *Thermoplasmataceae fam. nov.*
        Genus I. *Thermoplasma*
      Family II. *Picrophilaceae*
        Genus I. *Picrophilus*
      Family III: *Ferroplasmatacaea*
        Genus I. *Ferroplasma*
  Class VI. *Thermococci class. nov.*
    Order I. *Thermococcales*
      Family I. *Thermococcaceae*
        Genus I. *Thermococcus*
        Genus II. *Palaeococcus*
        Genus III. *Pyrococcus*
  Class VII. *Archaeoglobi class. nov.*
    Order I. *Archaeoglobales ord. nov.*
      Family I. *Archaeoglobaceae fam. nov.*
        Genus I. *Archaeoglobus*
        Genus II. *Ferroglobus*
        Genus III. *Geoglobus*
  Class VIII. *Methanopyri class. nov.*
    Order I. *Methanopyrales ord. nov.*
      Family I. *Methanopyraceae fam. nov.*
        Genus I. *Methanopyrus*

Domain "*Bacteria*"
Phylum BI. *Aquificae phy. nov.*
  Class I. *Aquificae class. nov.*
    Order I. *Aquificales ord. nov.*
      Family I. *Aquificaceae fam. nov.*
        Genus I. *Aquifex*
        Genus II. *Calderobacterium*
        Genus III. *Hydrogenobaculum*
        Genus IV. *Hydrogenobacter*
        Genus V. *Hydrogenothermus*
        Genus VI. *Persephonella*
        Genus VII.
        Sulfurihydrogenibium
        Genus VIII. *Thermocrinis*
      Genera incertae sedis
        Genus I. *Balnearium*
        Genus II. *Desulfurobacterium*
        Genus III. *Thermovibrio*
Phylum BII. *Thermotogae phy. nov.*
  Class I. *Thermotogae class. nov.*
    Order I. *Thermotogales ord. nov.*
      Family I. *Thermotogaceae fam.*
      nov.
        Genus I. *Thermotoga*
        Genus II. *Fervidobacterium*
        Genus III. *Geotoga*
        Genus IV. *Marinotoga*
        Genus V. *Petrotoga*
        Genus VI. *Thermosipho*
Phylum BIII. *Thermodesulfobacteria*
  Class I. *Thermodesulfobacteria*
    Order I. *Thermodesulfobacteriales*
      Family I. *Thermodesulfobacteriaceae*
        Genus I. *Thermodesulfobacterium*
        Genus II. *Theromodesulfatator*
Phylum BIV. *"Deinococcus-Thermus"*
  Class I. *Deinococci*
    Order I. *Deionococcales*
      Family I. *Deinococcaceae*
        Genus I. *Deinococcus*
    Order II. *Thermales ord. nov.*
      Family I. *Thermaceae fam. nov.*
        Genus I. *Thermus*
        Genus II. *Marinithermus*
        Genus III. *Meiothermus*
        Genus IV. *Oceanithermus*
        Genus V. *Vulcanithermus*
Phylum BV. *Chrysiogenetes phy. nov.*
  Class I. *Chrysiogenetes class. nov.*
    Order I. *Chrysiogenales ord. nov.*
      Family I. *Chrysiogenaceae fam. nov.*
        Genus I. *Chrysiogenes*
Phylum BVI. *Chloroflexi*
  Class I. *"Chloroflexi"*
    Order I. *"Chloroflexales"*
      Family I. *"Chloroflexaceae"*
        Genus I. *Chloroflexus*
        Genus II. *Chloronema*
        Genus III. *Heliothrix*
        Genus IV. *Roseiflexus*
      Family II. *Oscillochloridaceae*
        Genus I. *Oscillochloris (moved)*

    Order II. *"Herpetosiphonales"*
      Family I. *"Herpetosiphonaceae"*
        Genus I. *Herpetosiphon*
  Class II. *Anaerolineae*
    Order I. *Anaerolinaeles*
      Family I. *Anaerolinaeceae*
        Genus I. *Anaerolinea*
        Genus II. *Caldilinea*
Phylum BVII. *Thermomicrobia phy. nov.*
  Class I. *Thermomicrobia class. nov.*
    Order I. *Thermomicrobiales ord. nov.*
      Family I. *Thermomicrobiaceae fam.*
      nov.
        Genus I. *Thermomicrobium*
Phylum BVIII. *Nitrospira*
  Class I. *"Nitrospira"*
    Order I. *"Nitrospirales"*
      Family I. *"Nitrospiraceae"*
        Genus I. *Nitrospira*
        Genus II. *Leptospirillum*
        Genus III. *Magnetobacterium*
        Genus IV. *Thermodesulfovibrio*
Phylum BIX. *Deferribacteres phy. nov.*
  Class I. *Deferribacteres class. nov.*
    Order I. *Deferribacterales ord.*
    nov.
      Family I. *Deferribacteraceae fam.*
      nov.
        Genus I. *Deferribacter*
        Genus II. *Denitrovibrio*
        Genus III. *Flexistipes*
        Genus IV. *Geovibrio*
      Genera incertae sedis
        Genus I. *Synergistes*
        Genus II. *Caldithrix*
Phylum BX. *Cyanobacteria*
  Class I. *"Cyanobacteria"*
    Subsection I.
      Family I.
        Form genus I. *Chamaesiphon*
        Form genus II. *Chroococcus*
        Form genus III. *Cyanobacterium*
        Form genus IV. *Cyanobium*
        Form genus V. *Cyanothece*
        Form genus VI. *Dactylococcopsis*
        Form genus VII. *Gloeobacter*
        Form genus VIII. *Gloeocapsa*
        Form genus IX. *Gloeothece*
        Form genus X. *Microcystis*
        Form genus XI. *Prochlorococcus*
        Form genus XII. *Prochloron*
        Form genus XIII. *Synechococcus*
        Form genus XIV. *Synechocystis*
    Subsection II
      Family I.
        Form genus I. *Cyanocystis*
        Form genus II. *Dermocarpella*
        Form genus III. *Stanieria*
        Form genus IV. *Xenococcus*
      Family II.
        Form genus I. *Chroococcidiopsis*
        Form genus II. *Myxosarcina*

        Form genus III. *Pleurocapsa*
    Subsection III.
      Family I.
        Form genus I. *Arthrospira*
        Form genus II. *Borzia*
        Form genus III. *Crinalium*
        Form genus IV. *Geitlerinema*
        Genus V. *Halospirulina*
        Form genus VI. *Leptolyngbya*
        Form genus VII. *Limnothrix*
        Form genus VIII. *Lyngbya*
        Form genus IX. *Microcoleus*
        Form genus X. *Oscillatoria*
        Form genus XI. *Planktothrix*
        Form genus XII. *Prochlorothrix*
        Form genus XIII.
        Pseudanabaena
        Form genus XIV. *Spirulina*
        Form genus XV. *Starria*
        Form genus XVI. *Symploca*
        Genus XVII. *Trichodesmium*
        Form genus XVIII. *Tychonema*
    Subsection IV.
      Family I.
        Form genus I. *Anabaena*
        Form genus II. *Anabaenopsis*
        Form genus III. *Aphanizomenon*
        Form genus IV. *Cyanospira*
        Form genus V.
        Cylindrospermopsis
        Form genus VI.
        Cylindrospermum
        Form genus VII. *Nodularia*
        Form genus VIII. *Nostoc*
        Form genus IX. *Scytonema*
      Family II.
        Form genus I. *Calothrix*
        Form genus II. *Rivularia*
        Form genus III. *Tolypothrix*
    Subsection V.
      Family I.
        Form genus I. *Chlorogloeopsis*
        Form genus II. *Fischerella*
        Form genus III. *Geitleria*
        Form genus IV. *Iyengariella*
        Form genus V. *Nostochopsis*
        Form genus VI. *Stigonema*
Phylum BXI. *Chlorobi phy. nov.*
  Class I. *"Chlorobia"*
    Order I. *Chlorobiales*
      Family I. *Chlorobiaceae*
        Genus I. *Chlorobium*
        Genus II. *Ancalochloris*
        Genus III. *Chlorobaculum*
        Genus IV. *Chloroherpeton*
        Genus V. *Pelodictyon*
        Genus VI. *Prosthecochloris*
Phylum BXII. *Proteobacteria*
  Class I. *"Alphaproteobacteria"*
    Order I. *Rhodospirillales*
      Family I. *Rhodospirillaceae*
        Genus I. *Rhodospirillum*

Genus II. Azospirillum
Genus III. Inquilinus
Genus IV. Magnetospirillum
Genus V. Phaeospirillum
Genus VI. Rhodocista
Genus VII. Rhodospira
Genus VIII. Rhodovibrio
Genus IX. Roseospira
Genus X. Skermanella
Genus XI. Thalassospira
Genus XII. Tistrella
Family II. Acetobacteraceae
  Genus I. Acetobacter
  Genus II. Acidiphilium
  Genus III. Acidisphaera
  Genus IV. Acidocella
  Genus V. Acidomonas
  Genus VI. Asaia
  Genus VII. Craurococcus
  Genus VIII. Gluconacetobacter
  Genus IX. Gluconobacter
  Genus X. Kozaki
  Genus XI. Muricoccus
  Genus XII. Paracraurococcus
  Genus XIII. Rhodopila
  Genus XIV. Roseococcus
  Genus XV. Rubritepida
  Genus XVI. Stella
  Genus XVII. Teichococcus
  Genus XVIII. Zavarzinia
Order II. *Rickettsiales*
  Family I. Rickettsiaceae
    Genus I. Rickettsia
    Genus II. Orientia
  Family II. Anaplasmataceae
    Genus I. Anaplasma
    Genus II. Aegyptianella
    Genus III. Cowdria
    Genus IV. Ehrlichia
    Genus V. Neorickettsia
    Genus VI. Wolbachia
    Genus VII. Xenohaliotis
  Family III. "Holosporaceae"
    Genus I. Holospora
  Genera incertae sedis
    Genus I. Caedibacter
    Genus II. Lyticum
    Genus III. Odyssela
    Genus IV. Pseudocaedibacter
    Genus V. Symbiotes
    Genus VI. Tectibacter
Order III. *"Rhodobacterales"*
  Family I. "Rhodobacteraceae"
    Genus I. Rhodobacter
    Genus II. Ahrensia
    Genus III. Albidovulum
    Genus IV. Amaricoccus
    Genus V. Antarctobacter
    Genus VI. Gemmobacter
    Genus VII. Hirschia
    Genus VIII. Hyphomonas
    Genus IX. Jannaschia

Genus X. Ketogulonicigenium
Genus XI. Leisingera
Genus XII. Maricaulis
Genus XIII. Methylarcula
Genus XIV. Oceanicaulis
Genus XV. Octadecabacter
Genus XVI. Pannonibacter
Genus XVII. Paracoccus
Genus XVIII.
Pseudorhodobacter
Genus XIX. Rhodobaca
Genus XX. Rhodothalassium
Genus XXI. Rhodovulum
Genus XXII. Roseibium
Genus XXIII.
Roseinatronobacter
Genus XXIV. Roseivivax
Genus XXV. Roseobacter
Genus XXVI. Roseovarius
Genus XXVII. Rubrimonas
Genus XXVIII. Ruegeria
Genus XXIX. Sagittula
Genus XXX. Silicibacter
Genus XXXI. Staleya
Genus XXXII. Stappia
Genus XXXIII. Sulfitobacter
Order IV. *"Sphingomonadales"*
  Family I. Sphingomonadaceae
    Genus I. Sphingomonas
    Genus II. Blastomonas
    Genus III. Erythrobacter
    Genus IV. Erythromicrobium
    Genus V. Erythromonas
    Genus VI. Novosphingobium
    Genus VII. Porphyrobacter
    Genus VIII. Rhizomonas
    Genus IX. Sandaracinobacter
    Genus X. Sphingobium
    Genus XI. Sphingopyxis
    Genus XII. Zymomonas
Order V. *Caulobacterales*
  Family I. Caulobacteraceae
    Genus I. Caulobacter
    Genus II. Asticcacaulis
    Genus III. Brevundimonas
    Genus IV. Phenylobacterium
Order VI. *"Rhizobiales"*
  Family I. Rhizobiaceae
    Genus I. Rhizobium
    Genus II. Agrobacterium
    Genus III. Allorhizobium
    Genus IV. Carbophilus
    Genus V. Chelatobacter
    Genus VI. Ensifer
    Genus VII. Sinorhizobium
  Family II. Aurantimonadaceae
    Genus I. Aurantimonos
    Genus II. Fulvimarina
  Family III. Bartonellaceae
    Genus I. Bartonella
  Family IV. Brucellaceae
    Genus I. Brucella

Genus II. Mycoplana
Genus III. Ochrobactrum
Family V. "Phyllobacteriaceae"
  Genus I. Phyllobacterium
  Genus II. Aminobacter
  Genus III. Aquamicrobium
  Genus IV. Defluvibacter
  Genus V. "Candidatus
    Liberibacter"
  Genus VI. Mesorhizobium
  Genus VII. Nitratireductor
  Genus VIII. Pseudaminobacter
Family VI. "Methylocystaceae"
  Genus I. Methylocystis
  Genus II. Albibacter
  Genus III. Methylopila
  Genus IV. Methylosinus
  Genus V. Terasakiella
Family VII. "Beijerinckiaceae"
  Genus I. Beijerinckia
  Genus II. Chelatococcus
  Genus III. Methylocapsa
  Genus IV. Methylocella
Family VIII. "Bradyrhizobiaceae"
  Genus I. Bradyrhizobium
  Genus II. Afipia
  Genus III. Agromonas
  Genus IV. Blastobacter
  Genus V. Bosea
  Genus VI. Nitrobacter
  Genus VII. Oligotropha
  Genus VIII. Rhodoblastus
  Genus IX. Rhodopseudomonas
Family IX. Hyphomicrobiaceae
  Genus I. Hyphomicrobium
  Genus II. Ancalomicrobium
  Genus III. Ancylobacter
  Genus IV. Angulomicrobium
  Genus V. Aquabacter
  Genus VI. Azorhizobium
  Genus VII. Blastochloris
  Genus VIII. Devosia
  Genus IX. Dichotomicrobium
  Genus X. Filomicrobium
  Genus XI. Gemmiger
  Genus XII. Labrys
  Genus XIII. Methylorhabdus
  Genus XIV. Pedomicrobium
  Genus XV. Prosthecomicrobium
  Genus XVI. Rhodomicrobium
  Genus XVII. Rhodoplanes
  Genus XVIII. Seliberia
  Genus XIX. Starkeya
  Genus XX. Xanthobacter
Family X. "Methylobacteriaceae"
  Genus I. Methylobacterium
  Genus II. Microvirga
  Genus III. Protomonas
  Genus IV. Roseomonas
Family XI. "Rhodobiaceae"
  Genus I. Rhodobium
  Genus II. Roseospirillum

Class II. *"Betaproteobacteria"*
  Order I. *"Burkholderiales"*
    Family I. *"Burkholderiaceae"*
      Genus I. *Burkholderia*
      Genus II. *Cupriavidus*
      Genus III. *Lautropia*
      Genus IV. *Limnobacter*
      Genus V. *Pandoraea*
      Genus VI. *Paucimonas*
      Genus VII. *Polynucleobacter*
      Genus VIII. *Ralstonia*
      Genus IX. *Thermothrix*
      Genus X. *Wautersia*
    Family II. *"Oxalobacteraceae"*
      Genus I. *Oxalobacter*
      Genus II. *Duganella*
      Genus III. *Herbaspirillum*
      Genus IV. *Janthinobacterium*
      Genus V. *Massilia*
      Genus VI. *Oxalicibacterium*
      Genus VII. *Telluria*
    Family III. *Alcaligenaceae*
      Genus I. *Alcaligenes*
      Genus II. *Achromobacter*
      Genus III. *Bordetella*
      Genus IV. *Brackiella*
      Genus V. *Derxia*
      Genus VI. *Kersteria*
      Genus VII. *Oligella*
      Genus VIII. *Pelistega*
      Genus IX. *Pigmentiphaga*
      Genus X. *Sutterella*
      Genus XI. *Taylorella*
    Family IV. *Comamonadaceae*
      Genus I. *Comamonas*
      Genus II. *Acidovorax*
      Genus III. *Alicycliphilus*
      Genus IV. *Brachymonas*
      Genus V. *Caldimonas*
      Genus VI. *Delftia*
      Genus VII. *Diaphorobacter*
      Genus VIII. *Hydrogenophaga*
      Genus IX. *Hylemonella*
      Genus X. *Lampropedia*
      Genus XI. *Macromonas*
      Genus XII. *Ottowia*
      Genus XIII. *Polaromonas*
      Genus XIV. *Ramlibacter*
      Genus XV. *Rhodoferax*
      Genus XVI. *Variovorax*
      Genus XVII. *Xenophilus*
    Genera incertae sedis
      Genus I. *Aquabacterium*
      Genus II. *Ideonella*
      Genus III. *Leptothrix*
      Genus IV. *Roseateles*
      Genus V. *Rubrivivax*
      Genus VI. *Schlegelella*
      Genus VII. *Sphaerotilus*
      Genus VIII. *Tepidomonax*
      Genus IX. *Thiomonas*
      Genus X. *Xylophilus*

  Order II. *"Hydrogenophilales"*
    Family I. *"Hydrogenophilaceae"*
      Genus I. *Hydrogenophilus*
      Genus II. *Thiobacillus*
  Order III. *"Methylophilales"*
    Family I. *"Methylophilaceae"*
      Genus I. *Methylophilus*
      Genus II. *Methylobacillus*
      Genus III. *Methylovorus*
  Order IV. *"Neisseriales"*
    Family I. *Neisseriaceae*
      Genus I. *Neisseria*
      Genus II. *Alysiella*
      Genus III. *Aquaspirillum*
      Genus IV. *Chromobacterium*
      Genus V. *Eikenella*
      Genus VI. *Formivibrio*
      Genus VII. *Iodobacter*
      Genus VIII. *Kingella*
      Genus IX. *Laribacter*
      Genus X. *Microvirgula*
      Genus XI. *Morococcus*
      Genus XII. *Prolinoborus*
      Genus XIII. *Simonsiella*
      Genus XIV. *Vitreoscilla*
      Genus XV. *Vogesella*
  Order V. *"Nitrosomonadales"*
    Family I. *"Nitrosomonadaceae"*
      Genus I. *Nitrosomonas*
      Genus II. *Nitrosolobus*
      Genus III. *Nitrosospira*
    Family II. *Spirillaceae*
      Genus I. *Spirillum*
    Family III. *Gallionellaceae*
      Genus I. *Gallionella*
  Order VI. *"Rhodocyclales"*
    Family I. *"Rhodocyclaceae"*
      Genus I. *Rhodocyclus*
      Genus II. *Azoarcus*
      Genus III. *Azonexus*
      Genus IV. *Azospira*
      Genus V. *Azovibrio*
      Genus VI. *Dechloromonas*
      Genus VII. *Dechlorosoma*
      Genus VIII. *Ferribacterium*
      Genus IX. *Propionibacter*
      Genus X. *Propionivibrio*
      Genus XI. *Quadricoccus*
      Genus XII. *Sterolibacterium*
      Genus XIII. *Thauera*
      Genus XIV. *Zoogloea*
  Order VII. *Procabacteriales*
    Family I. *Procabacteriaceae*
      Genus I. *Procabacter*
Class III. *"Gammaproteobacteria"*
  Order I. *"Chromatiales"*
    Family I. *Chromatiaceae*
      Genus I. *Chromatium*
      Genus II. *Allochromatium*
      Genus III. *Amoebobacter*
      Genus IV. *Halochromatium*
      Genus V. *Isochromatium*

      Genus VI. *Lamprobacter*
      Genus VII. *Lamprocystis*
      Genus VIII. *Marichromatium*
      Genus IX. *Nitrosococcus*
      Genus X. *Pfennigia*
      Genus XI. *Rhabdochromatium*
      Genus XII. *Rheinheimera*
      Genus XIII. *Thermochromatium*
      Genus XIV. *Thioalkalicoccus*
      Genus XV. *Thiobaca*
      Genus XVI. *Thiocapsa*
      Genus XVII. *Thiococcus*
      Genus XVIII. *Thiocystis*
      Genus XIX. *Thiodictyon*
      Genus XX. *Thioflavicoccus*
      Genus XXI. *Thiohalocapsa*
      Genus XXII. *Thiolamprovum*
      Genus XXIII. *Thiopedia*
      Genus XXIV. *Thiorhodococcus*
      Genus XXV. *Thiorhodovibrio*
      Genus XXVI. *Thiospirillum*
    Family II. *Ectothiorhodospiraceae*
      Genus I. *Ectothiorhodospira*
      Genus II. *Alcalilimnicola*
      Genus III. *Alkalispirillum*
      Genus IV. *Arhodomonas*
      Genus V. *Halorhodospira*
      Genus VI. *Nitrococcus*
      Genus VII. *Thioalkalispira*
      Genus VIII. *Thioalkalivibrio*
      Genus IX. *Thiorhodospira*
    Family III. *Halothiobacillaceae*
      Genus I. *Halothiobacillus*
  Order II. *Acidithiobacillales*
    Family I. *Acidithiobacillaceae*
      Genus I. *Acidithiobacillus*
    Family II. *Thermithiobacillaceae*
      Genus I. *Thermithiobacillus*
  Order III. *"Xanthomonadales"*
    Family I. *"Xanthomonadaceae"*
      Genus I. *Xanthomonas*
      Genus II. *Frateuria*
      Genus III. *Fulvimonas*
      Genus IV. *Luteimonas*
      Genus V. *Lysobacter*
      Genus VI. *Nevskia*
      Genus VII. *Pseudoxanthomonas*
      Genus VIII. *Rhodanobacter*
      Genus IX. *Schineria*
      Genus X. *Stenotrophomonas*
      Genus XI. *Thermomonas*
      Genus XII. *Xylella*
  Order IV. *"Cardiobacteriales"*
    Family I. *Cardiobacteriaceae*
      Genus I. *Cardiobacterium*
      Genus II. *Dichelobacter*
      Genus III. *Suttonella*
  Order V. *"Thiotrichales"*
    Family I. *"Thiotrichaceae"*
      Genus I. *Thiothrix*
      Genus II. *Achromatium*
      Genus III. *Beggiatoa*

Genus IV. Leucothrix
Genus V. Thiobacterium
Genus VI. Thiomargarita
Genus VII. Thioploca
Genus VIII. Thiospira
Family II. Franciscellaceae
Genus I. Franciscella
Family III. "Piscirickettsiaceae"
Genus I. Piscirickettsia
Genus II. Cycloclasticus
Genus III. Hydrogenovibrio
Genus IV. Methylophaga
Genus V. Thioalkalimicrobium
Genus VI. Thiomicrospira
Order VI. "Legionellales"
Family I. Legionellaceae
Genus I. Legionella
Family II. "Coxiellaceae"
Genus I. Coxiella
Genus II. Aquicella
Genus III. Rickettsiella
Order VII. "Methylococcales"
Family I. Methylococcaceae
Genus I. Methylococcus
Genus II. Methylobacter
Genus III. Methylocaldum
Genus IV. Methylomicrobium
Genus V. Methylomonas
Genus VI. Methylosarcina
Genus VII. Methylosphaera
Order VIII. "Oceanspirillales"
Famliy I. "Oceanospirillaceae"
Genus I. Oceanospirillum
Genus II. Balneatrix
Genus III. Marinomonas
Genus IV. Marinospirillum
Genus V. Neptunomonas
Genus VI. Oceanobacter
Genus VII. Oleispira
Genus VIII. Pseudospirillum
Genus IX. Thalassolitus
Family II. Alcanivoraceae
Genus I. Alcanivorax
Genus II. Fundibacter
Family III. Hahellaceae
Genus I. Hahella
Genus II. Zooshikella
Family IV. Halomonadaceae
Genus I. Halomonas
Genus II. Carnimonas
Genus III. Chromohalobacter
Genus IV. Cobetia
Genus V. Deleya
Genus VI. Zymobacter
Family V. Oleiphilaceae
Genus I. Oleiphilus
Family VI. Saccharospirillaceae
Genus I. Saccharospirillum
Order IX. Pseudomonadales
Family I. Pseudomonadaceae
Genus I. Pseudomonas
Genus II. Azomonas

Genus III. Azotobacter
Genus IV. Cellvibrio
Genus V. Chryseomonas
Genus VI. Flaviomonas
Genus VII. Mesophilobacter
Genus VIII. Rhizobacter
Genus IX. Rugamonas
Genus X. Serpens
Family II. Moraxellaceae
Genus I. Moraxella
Genus II. Acinetobacter
Genus III. Psychrobacter
Family III. Incertae sedis
Genus I. Enhydrobacter
Order X. "Alteromonadales"
Family I. "Alteromonadaceae"
Genus I. Alteromonas
Genus II. Aestuariibacter
Genus III. Allishewanella
Genus IV. Colwellia
Genus V. Ferrimonas
Genus VI. Glaciecola
Genus VII. Idiomarina
Genus VIII. Marinobacter
Genus IX. Marinobacterium
Genus X. Microbulbifer
Genus XI. Moritella
Genus XII. Pseudoalteromonas
Genus XIII. Psychromonas
Genus XIV. Shewanella
Genus XV. Thalassomonas
Order XI. "Vibrionales"
Family I. Vibrionaceae
Genus I. Vibrio
Genus II. Allomonas
Genus III. Catenococcus
Genus IV. Enterovibrio
Genus V. Grimontia
Genus VI. Listonella
Genus VII. Photobacterium
Genus VIII. Salinivibrio
Order XII. "Aeromonadales"
Family I. Aeromonadaceae
Genus I. Aeromonas
Genus II. Oceanomonas
Genus III. Oceanisphaera
Genus IV. Tolumonas
Family II. Succinivibrionaeae
Genera incertae sedis
Genus I. Succinivibrio
Genus II. Anaerobiospirillum
Genus III. Ruminobacter
Genus IV. Succinimonas
Order XIII. "Enterobacteriales"
Family I. Enterobacteriaceae
Genus I. Escherichia
Genus II. Alterococcus
Genus III. Arsenophonus
Genus IV. Brenneria
Genus V. Buchnera
Genus VI. Budvicia
Genus VII. Buttiauxella

Genus VIII. Calymmatobacterium
Genus IX. Cedecea
Genus X. Citrobacter
Genus XI. Edwardsiella
Genus XII. Enterobacter
Genus XIII. Erwinia
Genus XIV. Ewingella
Genus XV. Hafnia
Genus XVI. Klebsiella
Genus XVII. Kluyvera
Genus XVIII. Leclercia
Genus XIX. Leminorella
Genus XX. Moellerella
Genus XXI. Morganella
Genus XXII. Obesumbacterium
Genus XXIII. Pantoea
Genus XXIV. Pectobacterium
Genus XXV. Phlomobacter
Genus XXVI. Photorhabdus
Genus XXVII. Plesiomonas
Genus XXVIII. Pragia
Genus XXIV. Proteus
Genus XXX. Providencia
Genus XXXI. Rahnella
Genus XXXII. Raoultella
Genus XXXIII. Saccharobacter
Genus XXXIV. Salmonella
Genus XXXV. Samsonia
Genus XXXVI. Serratia
Genus XXXVII. Shigella
Genus XXXVIII. Sodalis
Genus XXXIX. Tatumella
Genus XL. Trabulsiella
Genus XLI. Wigglesworthia
Genus XLII. Xenorhabdus
Genus XLIII. Yersinia
Genus XLIV. Yokenella
Order XIV. "Pasteurellales"
Family I. Pasteurellaceae
Genus I. Pasteurella
Genus II. Actinobacillus
Genus III. Gallibacterium
Genus IV. Haemophilus
Genus V. Lonepinella
Genus VI. Mannheimia
Genus VII. Phocoenobacter
Class IV. "Deltaproteobacteria"
Order I. "Desulfurellales"
Family I. "Desulfurellaceae"
Genus I. Desulfurella
Genus II. Hippea
Order II. "Desulfovibrionales"
Family I. "Desulfovibrionaceae"
Genus I. Desulfovibrio
Genus II. Bilophila
Genus III. Lawsonia
Family II. "Desulfomicrobiaceae"
Genus I. Desulfomicrobium
Family III. "Desullfohalobiaceae"
Genus I. Desulfohalobium
Genus II. Desulfomonas

Genus III. Desulfonatronovibrio
Genus IV. Desulfothermus
Family IV. "Desulfonatronumaceae"
Genus I. Desullfonatronum
Order III. *"Desulfobacterales"*
Family I. "Desulfobacteraceae"
Genus I. Desulfobacter
Genus II. Desulfotibacillum
Genus III. Desulfobacterium
Genus IV. Desulfobacula
Genus V. Desulfobotulus
Genus VI. Desulfocella
Genus VII. Desulfococcus
Genus VIII. Desulfofaba
Genus IX. Desulfofrigus
Genus X. Desulfomusa
Genus XI. Desulfonema
Genus XII. Desulforegula
Genus XIII. Desulfosarcina
Genus XIV. Desulfospira
Genus XV. Desulfotignum
Family II. "Desulfobulbaceae"
Genus I. Desulfobulbus
Genus II. Desulfocapsa
Genus III. Desulfofustis
Genus IV. Desulforhopalus
Genus V. Desulfotalea
Family III. "Nitrospinaceae"
Genus I. Nitrospina
Order IV. *"Desulfarcales"*
Family I. Desulfarculaceae
Genus I. Desulfarculas
Genus II. Desulfuromusa
Family II. "Geobacteraceae"
Genus I. Geobacter
Genus II. Tricholorobacter
Order V. *"Desulfuromonales"*
Family I. "Desulfuromonaceae"
Genus I. Desulfuromonas
Genus III. Malomonas
Genus IV. Pelobacter
Order VI. *"Syntrophobacterales"*
Family I. "Syntrophobacteraceae"
Genus I. Syntrophobacter
Genus II. Desulfacinum
Genus III. Desulforhabdus
Genus IV. Desulfovirga
Genus V. Thermodesulforhabdus
Family II. "Syntrophaceae"
Genus I. Syntrophus
Genus II. Desulfobacca
Genus III. Desulfomonile
Genus IV. Smithella
Order VII. *"Bdellovibrionales"*
Family I. "Bdellovibrionaceae"
Genus I. Bdellovibrio
Genus II. Bacteriovorax
Genus III. Micavibrio
Genus IV. Vampirovibrio
Order VIII. *Myxococcales*
Suborder I. *Cystobacterineae*
Family I. Cystobacteriaceae

Genus I. Cyctobacter
Genus II. Anaeromyxobacter
Genus III. Archangium
Genus IV. Hyalangium
Genus V. Melittangium
Genus VI. Stigmatella
Family II. Myxococcaceae
Genus I. Myxococcus
Genus II. Corallococcus
Genus III. Pyxicoccus
Suborder II. *Sorangineae*
Family I. Polyangiaceae
Genus I. Polyangium
Genus II. Byssophaga
Genus III. Chondromyces
Genus IV. Haploangium
Genus V. Jahnia
Genus VI. Sorangium
Suborder III. *Nannocystineae*
Family I. Nannocystaceae
Genus I. Nannocystis
Family II. Haliangiaceae
Genus I. Haliangium
Family III. Kofleriaceae
Genus I. Kofleria
Class V. *"Epsilonproteobacteria"*
Order I. *"Campylobacterales"*
Family I. Campylobacteraceae
Genus I. Campylobacter
Genus II. Arcobacter
Genus III. Dehalospirillum
Genus IV. Sulfurospirillum
Family II. "Helicobacteraceae"
Genus I. Helicobacter
Genus II. Sulfurimonas
Genus III. Thiovulum
Genus IV. Wolinella
Family III. Nautilaceae
Genus I. Nautila
Genus II. Caminibacter
Family IV. Hydrogenimonaceae
Genus I. Hydrogenimonas
Phylum BXIII. *Firmicutes*
Class I. *"Clostridia"*
Order I. *Clostridiales*
Family I. Clostridiaceae
Genus I. Clostridium
Genus II. Acetivibrio
Genus III. Acidaminobacter
Genus IV. Alkaliphilus
Genus V. Anaerobacter
Genus VI. Aerotruncus
Genus VII. Bryantella
Genus VIII. Caminicella
Genus IX. Caloramator
Genus X. Caloranaerobacter
Genus XI. Coprobacillus
Genus XII. Dorea
Genus XIII. Faecalibacterium
Genus XIV. Hespellia
Genus XV. Natronincola
Genus XVI. Oxobacter

Genus XVII. Parasporobacterium
Genus XVIII. Sarcina
Genus XIX. Soehngenia
Genus XX. Tepidibacter
Genus XXI. Thermobrachium
Genus XXII. Thermohalobacter
Genus XXIII. Tindallia
Family II. "Lachnospiraceae"
Genus I. Lachnospira
Genus II. Acetitomaculum
Genus III. Anaerofilum
Genus IV. Anaerostipes
Genus V. Butyrivibrio
Genus VI. Catenibacterium
Genus VII. Catonella
Genus VIII. Coprococcus
Genus IX. Johnsonella
Genus X. Lachnobacterium
Genus XI. Pseudobutyrivibrio
Genus XII. Roseburia
Genus XIII. Ruminococcus
Genus XIV. Shuttleworthia
Genus XV. Sporobacterium
Family III. "Peptostreptococcaceae"
Genus I. Peptostreptococcus
Genus II. Anaerococcus
Genus III. Filifactor
Genus IV. Finegoldia
Genus V. Fusibacter
Genus VI. Gallicola
Genus VII. Helcococcus
Genus VIII. Micromonas
Genus IX. Peptoniphilus
Genus X. Sedimentibacter
Genus XI. Sporanaerobacter
Genus XII. Tissierella
Family IV. "Eubacteriaceae"
Genus I. Eubacterium
Genus II. Acetobacterium
Genus III. Anaerovorax
Genus IV. Mogibacterium
Genus V. Pseudoramibacter
Family V. Peptococcaceae
Genus I. Peptococcus
Genus II. Carboxydothermus
Genus III. Dehalobacter
Genus IV. Desulfitobacterium
Genus V. Desulfonispora
Genus VI. Desulfosporosinus
Genus VII. Desulfotomaculum
Genus VIII. Pelotomaculum
Genus IX. Syntrophobotulus
Genus X. Thermoterrabacterium
Family VI. "Heliobacteriaceae"
Genus I. Heliobacterium
Genus II. Heliobacillus
Genus III. Heliophilum
Genus IV. Heliorestis
Family VII. "Acidaminococcaceae"
Genus I. Acidaminococcus
Genus II. Acetonema
Genus III. Alisonella

Genus IV. *Anacroarus*
Genus V. *Anaeroglobus*
Genus VI. *Anaeromusa*
Genus VII. *Anaerosinus*
Genus VIII. *Anaerovibrio*
Genus IX. *Centipeda*
Genus X. *Dendrosporobacter*
Genus XI. *Dialister*
Genus XII. *Megasphaera*
Genus XIII. *Mitsuokeller*
Genus XIV. *Papillibacter*
Genus XV. *Pectinatus*
Genus XVI. *Phascolarctobacterium*
Genus XVII. *Propionispira*
Genus XVIII. *Propionispora*
Genus XIX. *Quinella*
Genus XX. *Schwartzia*
Genus XXI. *Selenomonas*
Genus XXII. *Sporomusa*
Genus XXIII. *Succiniclasticum*
Genus XXIV. *Succinispira*
Genus XXV. *Veillonella*
Genus XXVI. *Zymophilus*
Family VIII. *Syntrophomonadaceae*
Genus I. *Syntrophomonas*
Genus II. *Acetogenium*
Genus III. *Aminobacterium*
Genus IV. *Aminomonas*
Genus V. *Anaerobaculum*
Genus VI. *Anaerobranca*
Genus VII. *Caldicellulosiruptor*
Genus VIII. *Carboxydocella*
Genus IX. *Dethiosulfovibrio*
Genus X. *Pelospora*
Genus XI. *Syntrophospora*
Genus XII. *Syntrophothermus*
Genus XIII. *Thermaerobacter*
Genus XIV. *Thermanaerovibrio*
Genus XV. *Thermohydrogenium*
Genus XVI. *Thermosyntropha*
Order II. *"Thermoanaerobacteriales"*
Family I. *"Thermoanaerobacteriaceae"*
Genus I. *Thermoanaerobacterium*
Genus II. *Ammonifex*
Genus III. *Caldanaerobacter*
Genus IV. *Carboxydobrachium*
Genus V. *Coprothermobacter*
Genus VI. *Gelria*
Genus VII. *Moorella*
Genus VIII. *Sporotomaculum*
Genus IX. *Thermacetogenium*
Genus X. *Thermoanaeromonas*
Genus XI. *Thermoanaerobacter*
Genus XII. *Theremoanaerobium*
Genus XIII. *Thermovenabulum*
Family II. *Thermodesulfobiaceae*
Genus I. *Thermodesulfobium*
Order III. *Haloanaerobiales*
Family I. *Haloanaerobiaceae*
Genus I. *Haloanaerobium*
Genus II. *Halocella*
Genus III. *Halothermothrix*

Family II. *Halobacteroidaceae*
Genus I. *Halobacteroides*
Genus II. *Acetohalobium*
Genus III. *Haloanaerobacter*
Genus IV. *Halonatronum*
Genus V. *Natrionella*
Genus VI. *Orenia*
Genus VII. *Selenihalanaerobacter*
Genus VIII. *Sporohalobacter*
Class II. *Mollicutes*
Order I. *Mycoplasmatales*
Family I. *Mycoplasmataceae*
Genus I. *Mycoplasma*
Genus II. *Eperythrozoon*
Genus III. *Haemobartonella*
Genus IV. *Ureaplasma*
Order II. *Entomoplasmatales*
Family I. *Entomoplasmataceae*
Genus I. *Entomoplasma*
Genus II. *Mesoplasma*
Family II. *Spiroplasmataceae*
Genus I. *Spiroplasma*
Order III. *Acholeplasmatales*
Family I. *Acholeplasmataceae*
Genus I. *Acholeplasma*
Genus II. *Phytoplasma*
Order IV. *Anaeroplasmatales*
Family I. *Anaeroplasmataceae*
Genus I. *Anaeroplasma*
Genus II. *Asteroleplasma*
Order V. *Incertae sedis*
Family I. *"Erysipelotrichaceae"*
Genus I. *Erysipelothrix*
Genus II. *Bulleidia*
Genus III. *Holdemania*
Genus IV. *Solobacterium*
Class III. *"Bacilli"*
Order I. *Bacillales*
Family I. *Bacillaceae*
Genus I. *Bacillus*
Genus II. *Amphibacillus*
Genus III. *Anoxybacillus*
Genus IV. *Exiguobacterium*
Genus V. *Filobacillus*
Genus VI. *Geobacillus*
Genus VII. *Gracilibacillus*
Genus VIII. *Halobacillus*
Genus IX. *Jeotgalibacillus*
Genus X. *Lentibacillus*
Genus XI. *Marinibacillus*
Genus XII. *Oceanobacillus*
Genus XIII. *Parapiobacillus*
Genus XIV. *Saccharococcus*
Genus XV. *Salibacillus*
Genus XVI. *Ureibacillus*
Genus XVII. *Virgibacillus*
Family II. *Alicyclobacillaceae*
Genus I. *Alicyclobacillus*
Genus II. *Pasteuria*
Genus III. *Sulfobacillus*
Family III. *Caryophanaceae*
Genus I. *Caryophanon*

Family IV. *"Listeriaceae"*
Genus I. *Listeria*
Genus II. *Brochothrix*
Family V. *"Paenibacillaceae"*
Genus I. *Paenibacillus*
Genus II. *Ammoniphilus*
Genus III. *Aneurinibacillus*
Genus IV. *Brevibacillus*
Genus V. *Oxalophagus*
Genus VI. *Thermicanus*
Genus VII. *Thermobacillus*
Family VI. *Planococcaceae*
Genus I. *Planococcus*
Genus II. *Filibacter*
Genus III. *Kurthia*
Genus IV. *Planomicrobium*
Genus V. *Sporosarcina*
Family VII. *"Sporolactobacillaceae"*
Genus I. *Sporolactobacillus*
Genus II. *Marinococcus*
Family VIII. *"Staphylococcaceae"*
Genus I. *Staphylococcus*
Genus II. *Gemella*
Genus III. *Jeotgalicoccus*
Genus IV. *Macrococcus*
Genus V. *Salinicoccus*
Family IX. *"Alicyclobacillaceae"*
Genus I. *Alicyclobacillus*
Genus II. *Pasteuria*
Genus III. *Sulfobacillus*
Family X. *"Thermoactinomycetaceae"*
Genus I. *Thermoactinomyces*
Family XI. *Turcibacteriaceae*
Genus I. *Turcibacter*
Order II. *"Lactobacillales"*
Family I. *Lactobacillaceae*
Genus I. *Lactobacillus*
Genus II. *Paralactobacillus*
Genus III. *Pediococcus*
Family II. *"Aerococcaceae"*
Genus I. *Aerococcus*
Genus II. *Abiotrophia*
Genus III. *Dolosicoccus*
Genus IV. *Eremococcus*
Genus V. *Facklamia*
Genus VI. *Globicatella*
Genus VII. *Ignavigranum*
Family III. *"Carnobacteriaceae"*
Genus I. *Carnobacterium*
Genus II. *Agitococcus*
Genus III. *Alkalibacterium*
Genus IV. *Allofustis*
Genus V. *Alloiococcus*
Genus VI. *Desemzia*
Genus VII. *Dolosigranulum*
Genus VIII. *Granulicatella*
Genus IX. *Isobaculum*
Genus X. *Lactosphaera*
Genus XI. *Marinilactibacillus*
Genus XII. *Trichococcus*
Family IV. *"Enterococcaceae"*
Genus I. *Enterococcus*

Genus II. Atopobacter
Genus III. Melissococcus
Genus IV. Tetragenococcus
Genus V. Vagococcus
Family V. "Leuconostocaceae"
Genus I. Leuconostoc
Genus II. Oenococcus
Genus III. Weissella
Family VI. Streptococcaceae
Genus I. Streptococcus
Genus II. Lactococcus
Family VII. Incertae sedis
Genus I. Acetoanaerobium
Genus II. Oscillospira
Genus III. Syntrophococcus
Phylum BXIV. *Actinobacteria phy. nov.*
Class I. *Actinobacteria*
Subclass I. *Acidimicrobidae*
Order I. *Acidimicrobiales*
Suborder I. *"Acidimicrobineae"*
Family I. Acidimicrobiaceae
Genus I. Acidimicrobium
Subclass II. *Rubrobacteridae*
Order I. *Rubrobacterales*
Suborder II. *"Rubrobacterineae"*
Family I. Rubrobacteraceae
Genus I. Rubrobacter
Genus II. Conexibacter
Genus III. Solirubrobacter
Genus IV. Thermoleophilum
Subclass III. *Coriobacteridae*
Order I. *Coriobacteriales*
Suborder III. *"Coriobacterineae"*
Family I. Coriobacteriaceae
Genus I. Coriobacterium
Genus II. Atopobium
Genus III. Collinsella
Genus IV. Cryptobacterium
Genus V. Denitrobacterium
Genus VI. Eggerthella
Genus VII. Olsenella
Genus VIII. Slackia
Subclass IV. *Sphaerobacteridae*
Order I. *Sphaerobacterales*
Suborder IV. *"Sphaerobacterineae"*
Family I. Sphaerobacteraceae
Genus I. Sphaerobacter
Subclass V. *Actinobacteridae*
Order I. *Actinomycetales*
Suborder V. *Actinomycineae*
Family I. Actinomycetaceae
Genus I. Actinomyces
Genus II. Actinobaculum
Genus III. Arcanobacterium
Genus IV. Mobiluncus
Genus V. Varibaculum
Suborder VI. *Micrococcineae*
Family I. Micrococcaceae
Genus I. Micrococcus
Genus II. Arthrobacter
Genus III. Citricoccus
Genus IV. Kocuria

Genus V. Nesterenkonia
Genus VI. Renibacterium
Genus VII. Rothia
Genus VIII. Stomatococcus
Genus IX. Yania
Family II. Bogoriellaceae
Genus I. Bogoriella (moved)
Family III. Rarobacteraceae
Genus I. Rarobacter (moved)
Family IV. Sanguibacteraceae
Genus I. Sanguibacter
Family V. Brevibacteriaceae
Genus I. Brevibacterium
Family VI. Cellulomonadaceae
Genus I. Cellulomonas
Genus II. Oerskovia
Genus III. Tropheryma
Family VII. Dermabacteraceae
Genus I. Dermabacter
Genus II. Brachybacterium
Family VIII. Dermatophilaceae
Genus I. Dermatophilus
Genus II. Kineosphaera
Family IX. Dermacoccaceae
Genus I. Dermacoccus
Genus II. Demetria
Genus III. Kytococcus
Family X. Intrasporangiaceae
Genus I. Intrasporangium
Genus II. Arsenicoccus
Genus III. Janibacter
Genus IV. Knoellia
Genus V. Ornithinicoccus
Genus VI. Ornithinimicrobium
Genus VII. Nostocoidia
Genus VIII. Terrabacter
Genus IX. Terracoccus
Genus X. Tetrasphaera
Family XI. Jonesiaceae
Genus I. Jonesia
Family XII. Microbacteriaceae
Genus I. Microbacterium
Genus II. Agreia
Genus III. Agrococcus
Genus IV. Agromyces
Genus V. Aureobacterium
Genus VI. Clavibacter
Genus VII. Cryobacterium
Genus VIII. Curtobacterium
Genus IX. Frigoribacterium
Genus X. Leifsonia
Genus XI. Leucobacter
Genus XII. Mycetocola
Genus XIII. Okibacterium
Genus XIV. Plantibacter
Genus XV. Rathayibacter
Genus XVI. Rhodoglobus
Genus XVII. Salinibacterium
Genus XVIII. Subtercola
Family XIII. "Beutenbergiaceae"
Genus I. Beutenbergia
Genus II. Georgenia

Genus III. Salana
Family VII. Promicromonosporaceae
Genus I. Promicromonospora
Genus II. Cellulosimicrobium
Genus III. Xylanibacterium
Genus IV. Xylanimonas
Suborder VII. *Corynebacterineae*
Family I. Corynebacteriaceae
Genus I. Corynebacterium
Family II. Dietziaceae
Genus I. Dietzia
Family III. Gordoniaceae
Genus I. Gordonia
Genus II. Skermania
Family IV. Mycobacteriaceae
Genus I. Mycobacterium
Family V. Nocardiaceae
Genus I. Nocardia
Genus II. Rhodococcus
Family VI. Tsukamurellaceae
Genus I. Tsukamurella
Family VII. "Williamsiaceae"
Genus I. Williamsia
Suborder VIII. *Micromonosporineae*
Family I. Micromonosporaceae
Genus I. Micromonospora
Genus II. Actinoplanes
Genus III. Asanoa
Genus IV. Catellatospora
Genus V. Catenuloplanes
Genus VI. Couchioplanes
Genus VII. Dactylosporangium
Genus VIII. Pilimelia
Genus IX. Spirilliplanes
Genus X. Verrucosispora
Genus XI. Virgisporangium
Suborder IX. *Propionibacterineae*
Family I. Propionibacteriaceae
Genus I. Propionibacterium
Genus II. Luteococcus
Genus III. Microlunatus
Genus IV. Propioniferax
Genus V. Propionimicrobium
Genus VI. Tessaracoccus
Family II. Nocardioidaceae
Genus I. Nocardioides
Genus II. Aeromicrobium
Genus III. Actinopolymorpha
Genus IV. Friedmanniella
Genus V. Hongia
Genus VI. Kribbella
Genus VII. Micropruina
Genus VIII. Marmoricola
Genus IX. Propionicimonas
Suborder X. *Pseudonocardineae*
Family I. Pseudonocardiaceae
Genus I. Pseudonocardia
Genus II. Actinoalloteichus
Genus III. Actinopolyspora
Genus IV. Amycolatopsis
Genus V. Croissiella
Genus VI. Kibdelosporangium

Genus VII. Kutzneria
Genus VIII. Prauserella
Genus IX. Saccharomonospora
Genus X. Saccharopolyspora
Genus XI. Streptoalloteichus
Genus XII. Thermobispora
Genus XIII. Thermocrispum
Family II. Actinosynnemataceae
Genus I. Actinosynnema
Genus II. Actinokineospora
Genus III. Lechevalieria
Genus IV. Lentzea
Genus V. Saccharothrix
Suborder XI. *Streptomycineae*
Family I. Streptomycetaceae
Genus I. Strepstomyces
Genus II. Kitasatospora
Genus III. Streptoverticillium
Suborder XII. *Streptosporangineae*
Family I. Streptosporangiaceae
Genus I. Streptosporangium
Genus II. Acrocarpospora
Genus III. Herbidospora
Genus IV. Microbispora
Genus V. Microtetraspora
Genus VI. Nonomuraea
Genus VII. Planobispora
Genus VIII. Planomonospora
Genus IX. Planopolyspora
Genus X. Planotetraspora
Family II. Nocardiopsaceae
Genus I. Nocardiopsis
Genus II. Streptomonospora
Genus III. Thermobifida
Family III. Thermomonosporaceae
Genus I. Thermomonospora
Genus II. Actinomadura
Genus III. Spirillospora
Suborder XIII. *Frankineae*
Family I. Frankiaceae
Genus I. Frankia
Family II. Geodermatophilaceae
Genus I. Geodermatophilus
Genus II. Blastococcus
Genus III. Modestobacter
Family III. Microsphaeraceae
Genus I. Microsphaera
Family IV. Sporichthyaceae
Genus I. Sporichthya
Family V. Acidothermaceae
Genus I. Acidothermus
Family VI. "Kineosporiaceae"
Genus I. Kineosporia
Genus II. Cryptosporangium
Genus III. Kineococcus
Suborder XIV. *Glycomycineae*
Family I. Glycomycetaceae
Genus I. Glycomyces
Order II. *Bifidobacteriales*
Family I. Bifidobacteriaceae
Genus I. Bifidobacterium
Genus II. Aeriscardovia

Genus III. Falcivibrio
Genus IV. Gardnerella
Genus V. Parascardovia
Genus VI. Scardovia
Family II. Unknown Affiliation
Genus I. Actinobispora
Genus II. Actinocorallia
Genus III. Excellospora
Genus IV. Pelczaria
Genus V. Turicella
Phylum BXV. *Planctomycetes phy. nov.*
Class I. *"Planctomycetacia"*
Order I. *Planctomycetales*
Family I. Planctomycetaceae
Genus I. Planctomyces
Genus II. Gemmata
Genus III. Isosphaera
Genus IV. Pirellula
Phylum BXVI. *Chlamydiae phy. nov.*
Class I. *"Chlamydiae"*
Order I. *Chlamydiales*
Family I. Chlamydiaceae
Genus I. Chlamydia
Genus II. Chlamydophila
Family II. Parachlamydiaceae
Genus I. Parachlamydia
Genus II. Neochlamydia
Family III. Simkaniaceae
Genus I. Simkania
Genus II. Rhabdochlamydia
Family IV. Waddliaceae
Genus I. Waddlia
Phylum BXVII. *Spirochaetes phy. nov.*
Class I. *"Spirochaetes"*
Order I. *Spirochaetales*
Family I. Spirochaetaceae
Genus I. Spirochaeta
Genus II. Borrelia
Genus III. Brevinema
Genus IV. Clevelandina
Genus V. Cristispira
Genus VI. Diplocalyx
Genus VII. Hollandina
Genus VIII. Pillotina
Genus IX. Treponema
Family II. "Serpulinaceae"
Genus I. Serpulina
Genus II. Brachyspira
Family III. Leptospiraceae
Genus I. Leptospira
Genus II. Leptonema
Phylum BXVIII. *Fibrobacteres*
Class I. *"Fibrobacteres"*
Order I. *"Fibrobacterales"*
Family I. "Fibrobacteraceae"
Genus I. Fibrobacter
Phylum BXIX. *Acidobacteria*
Class I. *"Acidobacteria"*
Order I. *"Acidobacteriales"*
Family I. "Acidobacteriaceae"
Genus I. Acidobacterium
Genus II. Geothrix

Genus III. Holophaga
Phylum BXX. *Bacteroidetes phy. nov.*
Class I. *"Bacteroidetes"*
Order I. *"Bacteroidales"*
Family I. Bacteroidaceae
Genus I. Bacteroides
Genus II. Acetofilamentum
Genus III. Acetomicrobium
Genus IV. Acetothermus
Genus V. Anaerophaga
Genus VI. Anaerorhabdus
Genus VII. Megamonas
Family II. "Rikenellaceae"
Genus I. Rikenella
Genus II. Alistepes
Genus III. Marinilabilia
Family III. "Porphyromonadaceae"
Genus I. Porphyromonas
Genus II. Dysgonomonas
Genus III. Tannerella
Family IV. "Prevotellaceae"
Genus I. Prevotella
Class II. *"Flavobacteria"*
Order I. *"Flavobacteriales"*
Family I. Flavobacteriaceae
Genus I. Flavobacterium
Genus II. Aequorivita
Genus III. Arenibacter
Genus IV. Bergeyella
Genus V. Capnocytophaga
Genus VI. Cellulophaga
Genus VII. Chryseobacterium
Genus VIII. Coenonia
Genus IX. Croceibacter
Genus X. Empedobacter
Genus XI. Gelidibacter
Genus XII. Gillisia
Genus XIII. Mesonia
Genus XIV. Miricauda
Genus XV. Myroides
Genus XVI. Ornithobacterium
Genus XVII. Polaribacter
Genus XVIII. Psychroflexus
Genus XIX. Psychroserpens
Genus XX. Riemerella
Genus XXI. Saligentibacter
Genus XXII. Tenacibaculum
Genus XXIII. Weeksella
Family II. "Blattabacteriaceae"
Genus I. Blattabacterium
Class III. *"Sphingobacteria"*
Order I. *"Sphingobacteriales"*
Family I. Sphingobacteriaceae
Genus I. Sphingobacterium
Genus II. Pedobacter
Family II. "Saprospiraceae"
Genus I. Saprospira
Genus II. Haliscomenobacter
Genus III. Lewinella
Family III. "Flexibacteraceae"
Genus I. Flexibacter
Genus II. Belliella

Genus III. Cyclobacterium
Genus IV. Cytophaga
Genus V. Dyadobacter
Genus VI. Flectobacillus
Genus VII. Hongiella
Genus VIII. Hymenobacter
Genus IX. Meniscus
Genus X. Microscilla
Genus XI. Reichenbachia
Genus XII. Runella
Genus XIII. Spirosoma
Genus XIV. Sporocytophaga
Family IV "Flammeovirgaceae"
  Genus I. Flammeovirga
  Genus II. Flexithrix
  Genus III. Persicobacter
  Genus IV. Thermonema
Family V. Crenotrichaceae
  Genus I. Crenothrix
  Genus II. Chitinophaga

Genus III. Rhodothermus
Genus IV. Salinibacter
Genus V. Toxothrix
Phylum BXXI. *Fusobacteria phy. nov.*
  Class I. *"Fusobacteria"*
    Order I. *"Fusobacteriales"*
      Family I. "Fusobacteriaceae"
        Genus I. Fusobacterium
        Genus II. Ilyobacter
        Genus III. Leptrotrichia
        Genus IV. Propionigenium
        Genus V. Sebaldella
        Genus VI. Streptobacillus
        Genus VII. Sneathia
      Family II. Incertae sedis
        Genus I. Cetobacterium
Phylum BXXII. *Verrucomicrobia*
  Class I. *Verrucomicrobiae*
    Order I. *Verrucomicrobiales*
      Family I. Verrucomicrobiaceae

Genus I. Verrucomicrobium
Genus II. Prosthecobacter
Family II. Opitutaceae
  Genus I. Opitutus
Family III. Victivallaceae
  Genus I. Victivallis
Family IV. Xiphinematobacteriaceae
  Genus I. Xiphinematobacter
Phylum BXXIII. *Dictyoglomi*
  Class I. *"Dictyoglomi"*
    Order I. *"Dictyoglomales"*
      Family I. "Dictyoglomaceae"
        Genus I. Dictyoglomus
Phylum BXXIV. *Gemmatimonadetes*
  Class I. *Gemmatimonadetes*
    Order I. *Gemmatimonadales*
      Family I. Gemmatimonadaceae
        Genus I. Gemmatimonas

CLASSIFICATION OF VIRUSES

The classification of viruses has undergone great change, as has bacterial taxonomy. Most viruses have not even been classified due to a lack of data concerning their reproduction and molecular biology. Estimates suggest that more than 30,000 viruses are being studied in laboratories and reference centers worldwide.

The classification and viral information presented here follows the outline given in Chapter 10 (Tables 10.1 and 10.2). Information also can be found in *Human Virology: A Text for Students of Medicine, Dentistry, and Microbiology* (L. Collier and J. Oxford, 1993, Oxford University Press), and in *Virology* (J. Levy, H. Fraenkel-Conrat, and R. Owens, 2d ed., 1994, Prentice-Hall).

The 21 families of viruses listed here are primarily those that infect vertebrates. Thus, these families represent only a small part of the 108 families and unassigned genera and more than 5000 viruses recognized in *Virus Taxonomy—Seventh Report of the International Committee on Taxonomy of Viruses*, van Regenmortal, et al. [eds.], 2000, Academic Press, San Diego CA.

1. Family: Picornaviridae

Genera:

Enterovirus (gastrointestinal viruses, poliovirus, coxsackie viruses A and B, echoviruses)
Hepatovirus (hepatitis A virus)
Cardiovirus (encephalomyocarditis virus of mice and other rodents)
Rhinovirus (upper respiratory tract viruses, common cold viruses)
Aphthovirus (foot-and-mouth disease virus)

Naked, polyhedral, positive-sense, ssRNA. Synthesis and maturation take place in the host cell cytoplasm. Viruses are released via cell lysis.

2. Family: Caliciviridae

Genus:

Calicivirus (Norwalk viruses and similar viruses causing gastroenteritis, hepatitis E virus)

Naked, polyhedral, positive-sense, ssRNA. Synthesis and maturation take place in the host cell cytoplasm. Viruses are released via cell lysis.

3. Family: Togaviridae

Genera:

Alphavirus (eastern, western, and Venezuelan equine encephalitis viruses, Semliki forest virus)
Rubivirus (rubella virus)
Arterivirus (equine arteritis virus, simian hemorrhagic fever virus)

Enveloped, polyhedral, positive-sense, ssRNA. Synthesis occurs in the host cell cytoplasm; maturation involves budding of nucleocapsids through the host cell plasma membrane. Viruses are released via cell lysis *(Arterivirus)*. Many replicate in arthropods and vertebrates.

4. Family: Flaviviridae

Genera:

Flavivirus (yellow-fever virus, dengue fever virus, St. Louis and Japanese encephalitis viruses, tickborne encephalitis virus)
Pestivirus (bovine diarrhea virus, hog cholera virus)
Hepacivirus (hepatitis C)

Enveloped, polyhedral, positive-sense, ssRNA. Synthesis occurs in the host cell cytoplasm; maturation involves budding through host cell endoplasmic reticulum and Golgi apparatus membranes. Most replicate in arthropods.

5. Family: Coronaviridae

Genus:

Coronavirus (common cold viruses, avian infectious bronchitis virus, feline infectious peritonitis virus, mouse hepatitis virus)

Enveloped, helical, positive-sense, ssRNA. Synthesis occurs in the host cell cytoplasm; maturation involves budding through membranes of the endoplasmic reticulum and Golgi apparatus. Viruses are released via cell lysis.

6. Family: Rhabdoviridae

Genera:

Vesiculovirus (vesicular stomatitis-like viruses)
Lyssavirus (rabies and rabieslike viruses)
Ephernerovirus (bovine ephemeral fever virus)

Enveloped, helical, negative-sense, ssRNA. Synthesis occurs in the host cell nucleus; maturation occurs via budding from the host cell plasma membrane. Many replicate in arthropods.

7. Family: Filoviridae

Genera:

Marburgvirus (Marburg; 23–88% lethal to humans)
Ebolavirus (Ebola; 50–90% lethal to humans)

Enveloped; long, filamentous forms, sometimes with branching, and sometimes U-shaped, 6-shaped, or circular; negative-sense, ssRNA. Synthesis occurs in the host cell cytoplasm; maturation involves budding from the host cell plasma membrane. Viruses are released via cell lysis. These viruses are "Biosafety Level 4" pathogens—they must be handled in the laboratory under maximum containment conditions.

8. Family: Paramyxoviridae

Genera:

Paramyxovirus (parainfluenza viruses 1–4, mumps virus, Newcastle disease virus)
Morbillivirus (measles and measleslike viruses, canine distemper virus)
Pneumovirus (respiratory syncytial virus)

Enveloped, helical, negative-sense, ssRNA. Synthesis occurs in the host cell cytoplasm; maturation involves budding through the host cell plasma membrane. Viruses are released via cell lysis. Morbilliviruses can cause persistent infections.

9. Family: Orthomyxoviridae

Genera:

Influenzavirus A and B (influenza viruses A and B)
Influenzavirus C (influenza C virus)

Enveloped, helical, negative-sense, ssRNA (eight segments). Synthesis occurs in the host cell nucleus; maturation takes place in the host cell cytoplasm. Viruses are released through budding from the host cell's plasma membrane. These viruses can reassort genes during mixed infections.

10. Family: Bunyaviridae

Genera:

Bunyavirus (Bunyamwera supergroup)
Phlebovirus (sandfly fever viruses)
Nairovirus (Nairobi sheep diseaselike viruses)
Uukuvirus (Uukuniemi-like viruses)
Hantavirus (hemorrhagic fever viruses, Korean hemorrhagic fever, Sin Nombre hantavirus)

Enveloped, spherical, negative-sense, ssRNA (three segments; *Phlebovirus* ambisense ssRNA). Synthesis occurs in the host cell cytoplasm; maturation occurs within the Golgi apparatus. Viruses are released via cell lysis. Closely related viruses can reassort genes during mixed infections.

11. Family: Arenaviridae

Genus:

Arenavirus (Lassa fever virus, lymphocytic choriomeningitis virus, Machupo virus, Junin virus)

Enveloped, helical, ambisense, ssRNA (two segments). Synthesis occurs in the host cell cytoplasm; maturation involves budding from the host cell plasma membrane. Virions contain ribosomes. The human pathogens Lassa, Machupo, and Junin viruses are "Biosafety Level 4" pathogens—they must be handled in the laboratory under maximum containment conditions.

12. Family: Reoviridae

Genera:

Orthoreovirus (reoviruses 1, 2, and 3)
Orbivirus (Orungo virus)

Rotavirus (human rotaviruses)
Cypovirus (cytoplasmic polyhidrosis viruses)
Coltivirus (Colorado tick fever virus)
Plant reovirus 1/3 (plant reoviruses subgroups 1, 2, and 3)

Each genus differs in morphology and physiochemical details. In general, virions are naked, polyhedral, dsRNA (10–12 segments). Synthesis and maturation take place in the host cell cytoplasm. Viruses are released via cell lysis. Virions contain ribosomes.

13. Famiy: Birnaviridae

Genus:

Birnavirus (infectious pancreatic necrosis virus of fish and infectious bursal disease virus of fowl)

Naked, polyhedral, dsRNA (two segments). Synthesis and maturation take place in the host cell cytoplasm. Viruses are released via cell lysis.

14. Family: Retroviridae

Genera:

MLV-related virus (spleen necrosis virus, mouse and feline leukemia viruses)
Betaretrovirus (mouse mammary tumor virus)
Type D (squirrel monkey retrovirus)
Alpharetrovirus (avian leukemia virus, Rous sarcoma virus)
HTLV-BLV group (human T cell leukemia virus HTLV-I, HTLV-II, bovine leukemia virus)
Spumavirus (the foamy viruses)
Lentivirus (human, feline, simian, and bovine immunodeficiency viruses)

Enveloped, spherical, negative-sense, ssRNA (two identical strands). Synthesis occurs in the host cell cytoplasm; maturation involves budding through the host cell plasma membrane. These viruses contain the enzyme reverse transcriptase. The retroviruses (except the *Spumavirus* and *Lentivirus* genera) represent the RNA tumor viruses, causing leukemias, carcinomas, and sarcomas.

15. Family: Hepadnaviridae

Genera:

Orthohepadnavirus (hepatitis B virus)
Avihepadnavirus (duck hepatitis virus)

Enveloped, polyhedral, partially dsDNA. Synthesis and maturation take place in the host cell nucleus. Surface antigen production occurs in the cytoplasm. Persistence is common and is associated with chronic disease and neoplasia.

16. Family: Parvoviridae

Genera:

Parvovirus (feline leukopenia virus, canine parvovirus)
Dependovirus (adeno-associated viruses)
Densovirus (insect parvoviruses)
Erythrovirus (human erythrovirus B19)

Naked, polyhedral, negative-sense, ssDNA *(Parvovirus)* or positive-sense and negative-sense, ssDNA (other genera). Synthesis and maturation occur in rapidly dividing host cells, specifically in the host cell nucleus. Viruses are released via cell lysis.

17. Family: Papovaviridae

Genera:

Papillomavirus (wart viruses, genital condylomas, DNA tumor viruses)
Polyomavirus (human polyoma-like viruses, SV-40)

Naked, polyhedral, dsDNA. Synthesis and maturation take place in the host cell nucleus. Viruses are released via cell lysis.

18. Family: Adenoviridae

Genera:

Mastadenovirus (human adenoviruses A–F, infectious canine hepatitis virus)
Aviadenovirus (avian adenoviruses)

Naked, polyhedral, dsDNA. Synthesis and maturation take place in the host cell nucleus. Viruses are released via cell lysis.

19. Family: Herpesviridae

Subfamily:
Alphaherpesvirinae

Genera:
Simplexvirus (herpes simplex viruses 1 and 2)
Varicellovirus (varicella-zoster virus)

Subfamily:
Betaherpesvirinae

Genera:
Cytomegalovirus (human cytomegalovirus)
Muromegalovirus (murine cytomegalovirus)

Subfamily:
Gammaherpesvirinae

Genera
Lymphocryptovirus (Epstein-Barr viruses)
Rhadinovirus (saimiri-ateles-like viruses)
Enveloped, polyhedral, dsDNA. Synthesis and maturation occur in the host cell nucleus, with budding through the nuclear envelope. Although most herpesviruses cause persistent infections, virions can be released by rupture of the host cell plasma membrane.

20. Family: Poxviridae

Subfamily:
Chordopoxvirinae

Genera:
Orthopoxvirus (vaccinia and variola viruses, cowpox virus)
Parapoxvirus (orf virus, pseudocowpox virus)
Avipoxvirus (fowlpox virus)
Capripoxvirus (sheep pox virus)
Leporipoxvirus (myxoma virus)
Suipoxvirus (swinepox virus)
Yatapoxvirus (yabapox virus and tanapox virus)
Molluscipoxvirus (molluscum contagiosum virus)

Subfamily:
Entomopoxvirinae

Genus:
Entomopoxvirus A/B/C (poxviruses of insects)
External envelope, large, brick-shaped (or ovoid), dsDNA. Synthesis and maturation take place in the portion of the host cell cytoplasm called viroplasm ("viral factories"). Viruses are released via cell lysis.

21. Family: Irdoviridae

Genera:
Iridovirus (small iridescent insect viruses)
Chloriridovirus (large iridescent insect viruses)
Ranavirus (frog viruses)
Lymphocystivirus (lymphocystis viruses of fish)
Enveloped (missing on some insect viruses), polyhedral, dsDNA. Synthesis occurs in both the host cell nucleus and cytoplasm. Most virions remain cell-associated.

Word Roots Commonly Encountered in Microbiology

a-, an-   not, without, absence   abiotic, not living; anaerobic, in the absence of air

acantho-   thorn or spinelike   *Acanthamoeba*, an amoeba with spinelike projections

actino-   having rays   *Actinomyces*, a bacterium forming colonies that look like sunbursts

aero-   air   aerobic, in the presence of air

agglutino-   clumping or sticking together   hemagglutinatin, clumping of blood cells

albo-   white   *Candida albicans*, a white fungus

amphi-   around, doubly, both   Amphitrichous describes flagella found at both ends of a bacterial cell.

ant-, anti-   against, versus   Antibacterial compounds kill bacteria.

archaeo-   ancient   Archaeobacteria are thought to resemble ancient forms of life.

arthro-   joint   arthritis, inflammation of joints

asco-   sac, bag   Ascospores are held in a saclike container, the ascus.

-ase   denotes enzyme   lipase, an enzyme attacking lipids

aureo-   golden   *Staphylococcus aureus* has gold-colored colonies.

auto-   self   autotrophs, self-feeding organisms

bacillo-   rod   bacillus, a rod-shaped bacterium

basid-   base, foundation   basidium, fungal cell bearing spores at its end

bio-   life   biology, the study of living things

blast-   bud   blastospore, spore formed by budding

bovi-   cow   *Mycobacterium bovis*, bacterium causing tuberculosis in cattle

brevi-   short   *Lactobacillus brevis*, a bacterium with short rod-shaped cells

butyr-   butter   Butyric acid gives rancid butter its unpleasant odor.

campylo-   curved   *Campylobacter*, a curved bacterium

carcino-   cancer   A carcinogen causes cancer.

caryo-, karyo-   center, kernel   Prokaryotic cells lack a true, discrete nucleus.

caseo-   cheese   caseous, cheeselike lesions

caul-   stalk, stem   *Caulobacter*, a stalked bacterium

ceph-, cephalo-   of the head or brain   encephalitis, inflammation of the brain

chlamydo-   cloaked hidden   *Chlamydia* are difficult bacteria to detect.

chloro-   green   chlorophyll, a green pigment

chromo-   colored   Metachromatic granules stain various colors within a cell.

chryso-   golden   *Streptomyces chryseus*, a bacterium forming golden colonies

-cide   to kill   Fungicide kills fungi.

co-, con-   with, together   congenital, existing from birth

cocc-   berry   *Streptococcus*, spherical bacteria in chains

coeno-   shared in common   coenocytic, many nuclei not separated by septa

col-, colo-   colon   coliform bacteria, found in the colon (large intestine)

conidio-   dust   conidia, tiny dustlike spores produced by fungi

coryne-   club   *Corynebacterium diphtheriae*, a club-shaped bacterium

-cul   little, tiny   molecule, a tiny mass

cut-, -cut   skin   cutaneous, of the skin

cyan-   blue   cyanobacteria, formerly called the blue-green algae

cyst-, -cyst   bladder   cystitis, inflammation of the urinary bladder

cyt-, -cyte   cell   leukocyte, white blood cell

de-   lack of removal   decolorize, to remove color

dermato-   skin   dermatitis, inflammation of the skin

di-, diplo-   two, double   diplococci, pairs of spherical cells

dys-   bad, faulty, painful   dysentery, a disease of the enteric system

ec-, ecto-, ex-   outside, outer   ectoparasite, found on the outside of the body

em-, en-   in, inside   encapsulated, inside a capsule

-emia   of the blood   pyemia, pus in the blood

endo-   inside   endospore, spore found inside a cell

entero-   intestine   enteric, bacteria found in the intestine

epi-   atop, over   epidemic, a disease spreading over an entire population at one time

erythro-   red   lupus erythematosus, disease with a red rash

etio-   cause   etiology, study of the causes of disease

eu-   true, good, normal   eukaryote, cell with a true nucleus

exo-   outside   exotoxin, toxin released outside of a cell

extra-   outside, beyond   extracellular, outside of a cell

fil-   thread   filament, thin chain of cells

flav-   yellow   flavivirus, cause of yellow fever

-fy   to become, make   solidify, to become solid

galacto-   milk   galactose, monosaccharide from milk sugar

gamet-   marriage   gamete, a reproductive cell, such as egg or sperm

gastro-   stomach   gastroenteritis, inflammation of the stomach and intestines

gel-   to stiffen, congeal   gelatinous, jellylike

gen-, -gen   to give rise to   pathogen, microbe that causes disease

-genesis   origin, development   pathogenesis, development of disease

germ, germin-   bud   germination, process of growing from a spore

-globulin   <u>protein</u>   immunoglobulins, proteins of the immune system

haem-, hem-   <u>blood</u>   hemagglutinatin, clumping of blood cells

halo-   <u>salt</u>   halophilic, organisms that thrive in salty environments

hepat-   <u>liver</u>   hepatitis, inflammation of the liver

herpes   <u>creeping</u>   herpes zoster, or shingles in which vesicles erupt sequentially along a nerve pathway

hetero-   <u>different, other</u>   heterotroph, organism deriving nutrition from other sources

histo-   <u>tissue</u>   histology, the study of tissues

homo-   <u>same</u>   homologous, having the same structure

hydro-   <u>water</u>   hydrologic cycle, water cycle

hyper-   <u>over, above</u>   hyperbaric oxygen, higher than atmospheric pressure oxygen

hypo-   <u>under, below</u>   hypodermic, going beneath the skin

im-, in-   <u>not</u>   insoluble, cannot be dissolved

inter-   <u>between</u>   intercellular, between cells

intra-   <u>inside</u>   intracellular, inside a cell

io-   <u>violet</u>   iodine, element that is purple in gaseous state

iso-   <u>same, equal</u>   isotonic, having the same osmotic pressure

-itis   <u>inflammation of</u>   meningitis, inflammation of the meninges

kin-   <u>moving</u>   kinetic energy, energy of movement

leuko-   <u>white</u>   leukocyte, white blood cell

lip-, lipo-   <u>fat, lipid</u>   lipoprotein, molecule having both fatty and proteinaceous parts

-logy, -ology   <u>study of</u>   microbiology, study of microbes

lopho-   <u>tuft</u>   lophotrichous, having a tuft or group of flagella

luc-, luci-   <u>light</u>   luciferase, enzyme that catalyzes a light-producing reaction

luteo-   <u>yellow</u>   *Micrococcus luteus*, bacterium producing yellow colonies

lys-, lysis   <u>slitting</u>   cytolysis, rupture of a cell

macro-   <u>large</u>   macroconidia, large spores

meningo-   <u>membrane</u>   meninges, membranes of the brain

meso-   <u>middle</u>   mesophile, organism growing best at medium temperatures

micro-   <u>small, tiny</u>   microbiology, study of tiny forms of life

mono-   <u>one, single</u>   monosaccharide, a single sugar unit

morph-   <u>shape, form</u>   pleiomorphic, having many different shapes

multi-   <u>many</u>   multicellular, having many cells

mur-   <u>wall</u>   muramic acid, a component of cell walls

muri-, mus-   <u>mouse</u>   murine, in or of mice

mut-, -mute   <u>to change</u>   mutagen, agent that causes genetic change

myc-, -myces   <u>fungus</u>   *Actinomyces*, a bacterium that resembles a fungus

myxo-   <u>slime, mucus</u>   myxomycetes, slime molds

necro-   <u>dead, corpse</u>   necrotizing toxin, causes death of tissue

nema-, -nema   <u>thread</u>   *Treponema*, nematode, threadlike organisms

nigr-   <u>black</u>   *Rhizopus nigricans*, a black mold

oculo-   <u>eye</u>   binocular, microscope with two eyepieces

-oid   <u>like, resembling</u>   toxoid, harmless molecule that resembles a toxin

-oma   <u>tumor</u>   carcinoma, tumor of epithelial cells

onco-   <u>mass, tumor</u>   oncogenes, genes that cause tumors

-osis   <u>condition of</u>   brucellosis, condition of being infected with *Brucella*

pan-   <u>all, universal</u>   pandemic, a disease affecting a large part of the world

para-   <u>beside, near, abnormal</u>   parainfluenza, a disease resembling influenza

patho-   <u>abnormal</u>   pathology, study of abnormal diseased states

peri-   <u>around</u>   peritrichous, flagella located all around an organism

phago-   <u>eating</u>   phagocytosis, cell eating by engulfing

-phile,   <u>loving, preferring</u>   capnophile, organism needing
philo-, -phil   higher than normal levels of carbon dioxide

-phob, -phobe   <u>hating, fearing</u>   hydrophobic, water-repelling

-phore   <u>bearing, carrying</u>   electrophoresis, technique in which ions are carried by an electric current

-phyte   <u>plant</u>   dermatophyte, fungus that attacks skin

pil-   <u>hair</u>   pilus, hairlike tube on bacterial surface

-plast   <u>formed part</u>   chloroplast, green body inside plant cell

pod-, -pod   <u>foot</u>   podocyte, foot cell of kidney

poly-   <u>many</u>   polyribosomes, many ribosomes on the same piece of messenger RNA

post-   <u>afterward, behind</u>   post-streptococcal glomerulonephritis, kidney damage following a streptococcal infection

pre-, pro-   <u>before, toward</u>   prepubertal, before puberty

pseudo-   <u>false</u>   pseudopod, projection resembling a foot, false foot

psychro-   <u>cold</u>   psychrophilic, preferring extreme cold

pyo-   <u>pus</u>   pyogenic, producing pus

pyro-   <u>fire, heat</u>   pyrogen, fever-producing compound

rhin-   <u>nose</u>   rhinitis, inflammation of nasal membranes

rhizo-   <u>root</u>   mycorrhiza, symbiotic growth of fungi and roots

rhodo-   <u>red</u>   *Rhodospirillum*, a large red spiral bacterium

-rrhea   <u>flow</u>   diarrhea, abnormal flow of liquid feces

rubri-   <u>red</u>   *Rhodospirillum rubrum*, a large red spiral bacterium

saccharo-   <u>sugar</u>   polysaccharide, many sugar units linked together

sapro-   <u>rotten, decaying</u>   saprophyte, organism living on dead matter

sarco-   <u>flesh</u>   sarcoma, tumor made up of muscle or connective tissue

schizo-   <u>to split</u>   schizogony, a type of fission in malarial parasites

-scope, -scopy   <u>to see, examine</u>   microscopy, use of the microscope to examine small things

sept-, septo-   <u>partition, wall</u>   septum, wall between cells

septi-   <u>rotting</u>   septic, exhibiting decomposition due to bacteria

soma-, -some   <u>body</u>   chromosome, colored body (when stained)

spiro-   <u>coil</u>   spirochete, spiral-shaped bacterium

sporo-   <u>spore</u>   sporocidal, spore killing

staphylo-   <u>in bunches, like grapes</u>   staphylococci, spherical bacteria growing in clusters

-stasis, stat- <u>stopping, not changing</u> bacteriostatic, able to stop the growth of bacteria

strepto- <u>twisted</u> *Streptobacillus*, twisted chains of bacilli

sub- <u>under, below</u> subclinical, signs and symptoms not clinically apparent

super- <u>above, more than</u> superficial mycosis, fungal infection of the surface tissues

sym-, syn- <u>together</u> symbiosis, living together

tact-, -taxis <u>touch</u> chemotaxis, orientation or movement in response to chemicals

tax-, taxon- <u>arrangement</u> taxonomy, the classification of organisms

thermo- <u>heat</u> thermophile, organism preferring or needing high temperatures

thio- <u>sulfur</u> *Thiobacillus*, organism that oxidizes hydrogen sulfide to sulfates

tox- <u>poison</u> toxin, a harmful compound

trans- <u>through, across</u> transduction, movement of genetic information from one cell to another

trich- <u>hair</u> monotrichous, having a single, hairlike flagellum

-troph <u>feeding, nutrition</u> phototroph, organism that makes its own food, using energy from light

uni- <u>one, singular</u> unicellular, composed of one cell

undul- <u>waving</u> undulant fever, disease in which fever rises and falls

vac-, vaccin- <u>cow</u> vaccine, disease-preventing product originally produced by inoculating it onto skin of calves

vacu- <u>empty</u> vacuole, empty-appearing structure in cytoplasm

vesic- <u>blister, bladder</u> vesicle, small blisterlike lesions

vitr- <u>glass</u> *in vitro*, grown in laboratory glassware

xantho- <u>yellow</u> *Xanthomonas oryzae*, bacterium producing yellow colonies

xeno- <u>strange, foreign</u> xenograft, graft from a different species

zoo- <u>animal</u> protozoan, first animal

zygo- <u>yoke, joining</u> zygote, fertilized egg

-zyme <u>ferment</u> enzymes, biological catalysts, some of which are involved in fermentation

Safety Precautions in the Handling of Clinical Specimens

Concern for maintaining safe conditions in school and hospital laboratories, in other work settings, and especially during patient interactions has led the federal government to formulate various regulations and recommendations. These are far too extensive to reproduce here in their entirety; several are nearly 200 pages long. As an introduction to the kinds of safety measures that should be taken, a few of the guidelines set forth in these publications are provided below. Some key references and sources are also listed, with the hope that they will stimulate the interested reader to investigate further.

In 1983 CDC published a document entitled "Guidelines for isolation precautions in hospitals" that contained a section headed "Blood and body fluid precautions." The recommendations in this section specified precautions to be observed regarding contact with the blood or body fluids of any patient known or suspected to be infected with bloodborne pathogens.

In August 1987 CDC published a document entitled "Recommendations for prevention of HIV transmission in health-care settings." In contrast with the 1983 document, the 1987 publication recommended that blood and body fluid precautions be consistently used for *all* patients regardless of their bloodborne infection status. These blood and body fluid precautions as they pertain to all patients are referred to as "Universal Blood and Body Fluid Precautions," or more simply as "Universal Precautions."

Following publication of this document, there were many requests for clarification—for example, as to which bodily fluids these precautions should apply. This led to a CDC publication on June 24, 1988, in *MMWR* entitled "Update: Universal precautions for prevention of transmission of human immunodeficiency virus, hepatitis B virus, and other bloodborne pathogens in health-care settings."

Copies of these two most recent reports (CDC August 1987 and June 1988) are available through the National AIDS Information Clearinghouse, P.O. Box 6003, Rockville, MD 20850.

In these two publications CDC makes the following recommendations regarding sharp instruments:

1. Take care to prevent injuries when using needles, scalpels, and other sharp instruments or devices; when handling sharp instruments after procedures; when cleaning used instruments; and when disposing of used needles. Do not recap used needles by hand; do not remove used needles from disposable syringes by hand; and do not bend, break, or otherwise manipulate used needles by hand. Place used disposable syringes and needles, scalpel blades, and other sharp items in puncture-resistant containers for disposal. Locate the puncture-resistant containers as close to the use area as is practical.

2. Use protective barriers to prevent exposure to blood, body fluids containing visible blood, and other fluids to which universal precautions apply. The type of protective barrier(s) should be appropriate for the procedure being performed and the type of exposure anticipated.

3. Immediately and thoroughly wash hands and other skin surfaces that are contaminated with blood, body fluids containing visible blood, or other body fluids to which universal precautions apply.

Glove use is recommended during phlebotomy (drawing blood samples) but cannot protect against penetrating injuries. Some institutions have relaxed recommendations for using gloves for phlebotomy procedures by skilled phlebotomists in settings where the prevalence of bloodborne pathogens is known to be very low (for example, volunteer blood-donation centers). Such institutions should periodically reevaluate their policy. Gloves should always be available to health care workers who wish to use them for phlebotomy. In addition, the following general guidelines apply:

1. Use gloves for performing phlebotomy when the health care worker has cuts, scratches, or other breaks in his/her skin.

2. Use gloves in situations where the health care worker judges that hand contamination with blood may occur, for example, when performing phlebotomy on an uncooperative patient.

3. Use gloves for performing finger and/or heel sticks on infants and children.

4. Use gloves when persons are receiving training in phlebotomy.

In March 1989 the Environmental Protection Agency (EPA) published a new set of "Standards for the tracking and management of medical waste," in part designed to prevent the deplorable pollution of our nation's beaches by medical wastes (*Federal Register*, March 24, 1989, pp. 12325–95). In May 1989 the Occupational Safety and Health Administration (OSHA) published a new set of rules for "Occupational exposure to bloodborne pathogens" (*Federal Register*, May 30, 1989, pp. 23041–139). Both publications provide very detailed information about procedures that must be followed.

Another useful and quite detailed publication, issued by CDC in February 1989 and reprinted in *MMWR* for June 23, 1989, is entitled "Guidelines for prevention of transmission of human immunodeficiency virus and hepatitis B virus to health-care and public-safety workers."

For those concerned primarily with the teaching laboratory, we recommend the publication "Handling infectious materials in the education setting" (G. Ballman, *American Clinical Laboratory*, July 1989, pp.10–11). Other publications of interest include *Biosafety in Microbiological and Biomedical Laboratories* (CDC, U.S. Department of Health and Human Services, Public Health Service, 1988), which describes the four biosafety levels classified by CDC and recommended safety precautions for each, and "Labeling of microbial risks" (C. Robinson and T. H. Hatfield, *Journal of College Science Teaching*, May 1995, pp. 407–9), which evaluates that classification system of biosafety levels.

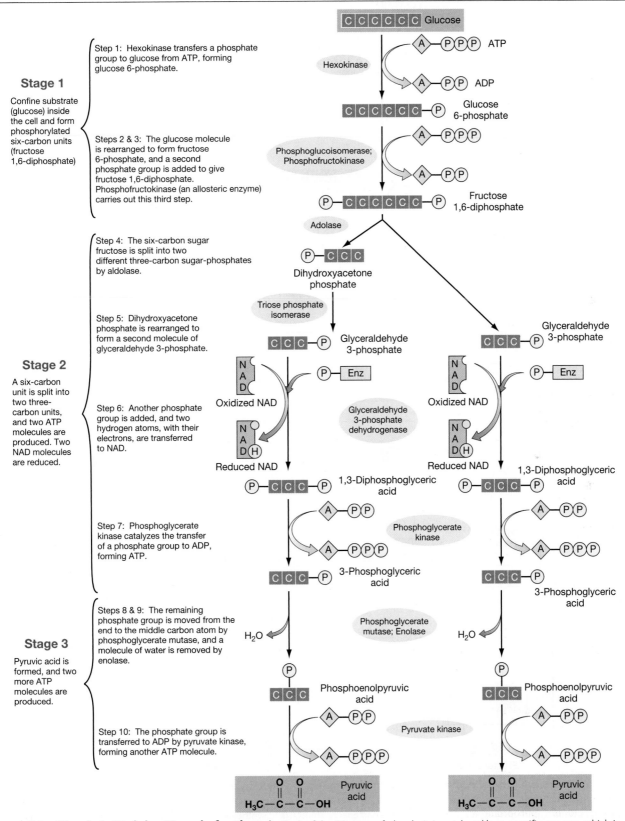

Figure E.1 Glycolysis (Embden-Meyerhof pathway). Each of the 10 steps of glycolysis is catalyzed by a specific enzyme, which is indicated in a purple oval. (Refer to Chapter 5 for an explanation; Figure 5.11 shows a simplified version of the process. ◀p. 125)

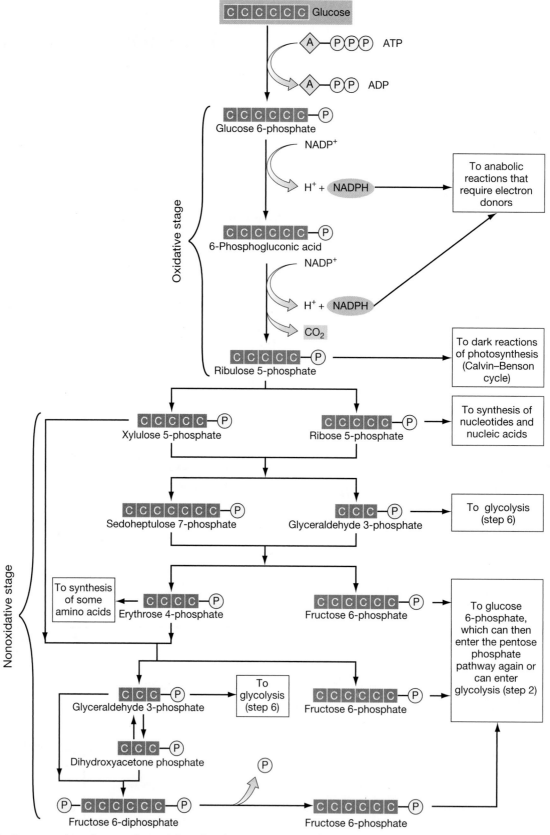

Figure E.2 Pentose phosphate pathway (phosphogluconate pathway). This metabolic pathway occurs with glycolysis. It provides an alternative pathway for the breakdown of glucose as well as pentoses (five-carbon sugars). This pathway plays three important roles: (1) It provides intermediate pentoses, especially ribose, that the bacterial cell must use to synthesize nucleic acids. (2) This pathway's intermediates can be used to synthesize some amino acids. (3) The pentose phosphate pathway reduced NADP to NADPH. This coenzyme, like NADH, is an electron carrier and thus is a source of reducing power. The fates of several intermediates are indicated. For clarity, the specific enzymes catalyzing these reactions and the structural formulas of substrates have been omitted. (Refer to Chapter 5 for an explanation of this pathway. ◀p. 126)

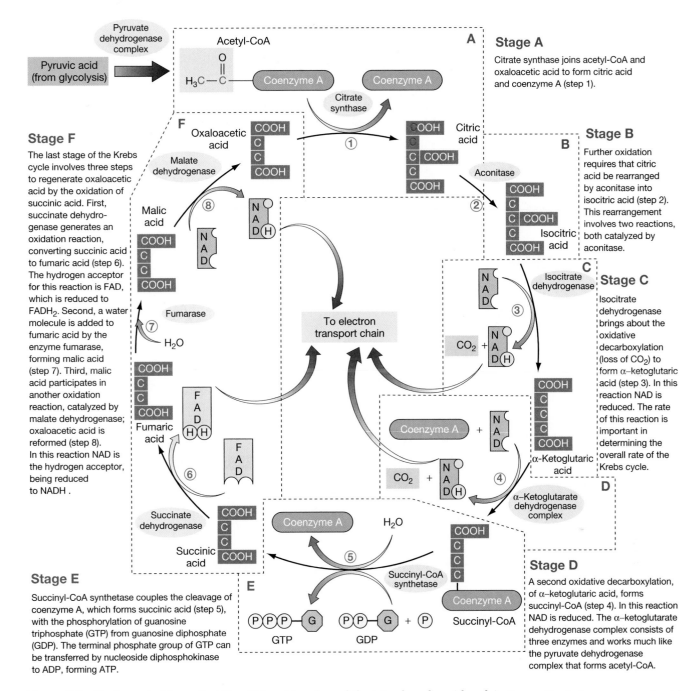

Pyruvate dehydrogenase complex

Pyruvic acid (from glycolysis)

Acetyl-CoA

H_3C-C (O double bond) — **Coenzyme A** → **Coenzyme A**

Citrate synthase ①

Stage A

Citrate synthase joins acetyl-CoA and oxaloacetic acid to form citric acid and coenzyme A (step 1).

Stage F

The last stage of the Krebs cycle involves three steps to regenerate oxaloacetic acid by the oxidation of succinic acid. First, succinate dehydrogenase generates an oxidation reaction, converting succinic acid to fumaric acid (step 6). The hydrogen acceptor for this reaction is FAD, which is reduced to FADH$_2$. Second, a water molecule is added to fumaric acid by the enzyme fumarase, forming malic acid (step 7). Third, malic acid participates in another oxidation reaction, catalyzed by malate dehydrogenase; oxaloacetic acid is reformed (step 8). In this reaction NAD is the hydrogen acceptor, being reduced to NADH .

Oxaloacetic acid COOH C C COOH

Malate dehydrogenase

Malic acid COOH C C COOH

NAD(H) ⑧ NAD

Fumarase ⑦ H$_2$O

Fumaric acid COOH C C COOH

FAD(H)(H) ⑥ FAD

Succinate dehydrogenase

Succinic acid COOH C C COOH

Citric acid COOH C C COOH C COOH

Aconitase ②

Stage B

Further oxidation requires that citric acid be rearranged by aconitase into isocitric acid (step 2). This rearrangement involves two reactions, both catalyzed by aconitase.

Isocitric acid COOH C C COOH C COOH

NAD ③ CO$_2$ + NAD(H)

Isocitrate dehydrogenase

Stage C

Isocitrate dehydrogenase brings about the oxidative decarboxylation (loss of CO$_2$) to form α–ketoglutaric acid (step 3). In this reaction NAD is reduced. The rate of this reaction is important in determining the overall rate of the Krebs cycle.

α-Ketoglutaric acid COOH C C C COOH

Coenzyme A + NAD CO$_2$ + NAD(H) ④

α–Ketoglutarate dehydrogenase complex

Stage D

A second oxidative decarboxylation, of α–ketoglutaric acid, forms succinyl-CoA (step 4). In this reaction NAD is reduced. The α–ketoglutarate dehydrogenase complex consists of three enzymes and works much like the pyruvate dehydrogenase complex that forms acetyl-CoA.

To electron transport chain

Coenzyme A H$_2$O ⑤ **Succinyl-CoA synthetase**

Succinyl-CoA COOH C C C **Coenzyme A**

Stage E

Succinyl-CoA synthetase couples the cleavage of coenzyme A, which forms succinic acid (step 5), with the phosphorylation of guanosine triphosphate (GTP) from guanosine diphosphate (GDP). The terminal phosphate group of GTP can be transferred by nucleoside diphosphokinase to ADP, forming ATP.

(P)(P)(P)–G **GTP**   (P)(P)–G + (P) **GDP**

Figure E.3   Krebs cycle (also called the citric acid cycle and the tricarboxylic acid cycle). The reaction that converts pyruvic acid to acetyl-CoA precedes the Krebs cycle (see Figure 5.16). This reaction is catalyzed by a pyruvate dehydrogenase complex, which contains three enzymes. Each of the eight steps of the Krebs cycle is also catalyzed by a specific enzyme, as indicated in a purple oval. (Refer to Chapter 5 for an explanation; Figure 5.17 shows a simplified version of the process. ◀p. 130)

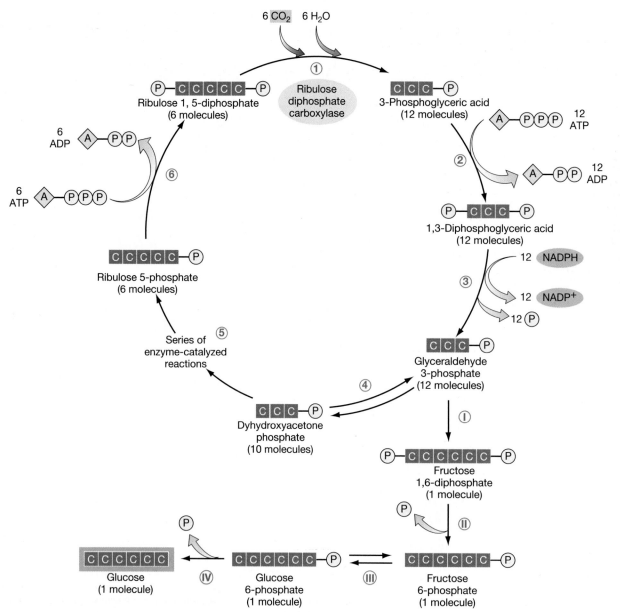

Figure E.4 Calvin-Benson cycle (dark reactions of photosynthesis). Each step of the Calvin-Benson cycle is catalyzed by a specific enzyme, which for simplicity is not shown. Steps 1 through 3 produce 12 three-carbon intermediates. These three steps are dependent on photophosphorylation products (ATP and NADPH). Two of every 12 three-carbon molecules undergo chemical reactions (steps 1 through IV) to produce a six-carbon glucose molecule. The other 10 three-carbon molecules are recycled (steps 4 through 6), forming 6 five-carbon molecules. These are phosphorylated by ATP to ribulose-1,5-diphosphate. Each of these five-carbon molecules then combines with a CO_2 molecule, starting the process once again. The enzyme catalyzing this step is ribulose diphosphate carboxylase, the most prevalent enzyme in the biological world. (Refer to Chapter 5 for an explanation of the process. ◀pp. 136–137)

abiotic factor A physical feature of the environment that interacts with organisms

ABO blood group system One of the blood typing systems that is based on the presence or absence of blood group antigens A and B on red blood cells

abortive infection Viral infection in which viruses enter a cell but are unable to express all of their genes to make infectious progeny

abscess An accumulation of pus in a cavity hollowed out by tissue damage

absorption Process in which light rays are neither passed through nor reflected off an object but are retained and either transformed to another form of energy or used in biological processes

accidental parasite A parasite that invades an organism other than its normal host

acid A substance that releases hydrogen ions when it is dissolved in water

acidic dye *See* **anionic dye**

acidophile An acid-loving organism that grows best in an environment with a pH of 4.0 to 5.4

acme (sometimes referred to as fulminating) During the illness phase of the disease process, the time of most intense signs and symptoms

acne Skin condition caused by bacterial infection of hair follicles and the ducts of sebaceous glands

acquired immune deficiency syndrome (AIDS) An infectious disease caused by the human immunodeficiency virus that destroys the individual's immune system

acquired immunity Immunity obtained in some manner other than by heredity

acridine derivative A chemical mutagen that can be inserted between bases of the DNA double helix, causing frameshift mutations

actinomycetes Gram-positive bacteria that tend to form filaments

activated sludge system Procedure in which the effluent from the primary stage of sewage treatment is agitated, aerated, and added to sludge containing aerobic organisms that digest organic matter

activation energy The energy required to start a chemical reaction

active immunity Immunity created when an organism's own immune system produces antibodies or other defenses against an agent recognized as foreign

active immunization Use of vaccines to control diseases by increasing herd immunity through stimulation of the immune response

active site Area on the surface of an enzyme to which its substrate binds

active transport Movement of molecules or ions across a membrane against a concentration gradient; requires expenditure of energy from ATP

acute disease A disease that develops rapidly and runs its course quickly

acute hemorrhagic conjunctivitis Eye disease caused by an enterovirus

acute inflammation The relatively short duration of inflammation during which time host defenses destroy invading microbes and repair tissue damage

acute necrotizing ulcerative gingivitis (**ANUG**; also called trench mouth) A severe form of periodontal disease

acute phase protein Protein, such as C-reactive protein or mannose-binding protein, that forms a nonspecific host-defense mechanism during an acute phase response

acute phase response A response to an acute illness that produces specific blood proteins called acute phase proteins

acute respiratory disease (ARD) Viral disease that occurs in epidemics with cold symptoms as well as fever, headache, and malaise; sometimes causes viral pneumonia

adaptive defenses Host defenses that produce resistance by responding to particular antigens, such as viruses and pathogenic bacteria

adaptive immunity The ability of a host to mount a defense against particular infectious agents by physiological responses specific to that infectious agent

adenovirus A medium-sized, naked DNA virus that is highly resistant to chemical agents and often causes respiratory infections or diarrhea

adherence The attachment of a microorganism to a host's cell surface

adhesin A protein or glycoprotein on attachment pili (fimbriae) or capsules that helps a microorganism attach to a host cell

adsorption The attachment of the virus to the host cell in the replication process

aerobe An organism that uses oxygen, including ones that must have oxygen

aerobic respiration Process in which aerobic organisms gain energy from the catabolism of organic molecules via the Krebs cycle and oxidative phosphorylation

aerosol A cloud of tiny liquid droplets suspended in air

aerotolerant anaerobe A bacterium that can survive in the presence of oxygen but does not use oxygen in its metabolism

aflatoxin Fungal toxin that is a potent carcinogen; found in food made from contaminated grain or peanuts infested with *Aspergillus flavus* and other aspergilli

African sleeping sickness (also called trypanosomiasis) Disease of equatorial Africa caused by protozoan blood parasites of the genus *Trypanosoma*

agammaglobulinemia Primary immunodeficiency disease caused by failure of B cells to develop, resulting in lack of antibodies

agar A polysaccharide extracted from certain marine algae and used to solidify medium for the growth of microorganisms

agar plate A plate of nutrient medium solidified with agar

agglutination The clumping together of the cells when antibodies react with antigens on cells

agglutination reaction A reaction of antibodies with antigens that results in agglutination, the clumping together of cells or other large particles

agranulocyte A leukocyte (monocyte or lymphocyte) that lacks granules in the cytoplasm and has rounded nuclei

AIDS (acquired immune deficiency syndrome) An infectious disease caused by the human immunodeficiency virus that destroys the individual's immune system

alcoholic fermentation Fermentation in which pyruvic acid is reduced to ethyl alcohol by electrons from reduced NAD (NADH)

algae (singular: *alga*) Photosynthesis, eukaryotic organisms in the kingdoms Protista and Plantae

alkaline (also called basic) Condition caused by an abundance of hydroxyl ions (OH^-) resulting in a pH of greater than 7.0

alkaliphile A base- (alkaline) loving organism that grows best in an environment with a pH of 7.0 to 11.5

alkylating agent A chemical mutagen that can add alkyl groups ($-CH_3$) to DNA bases, altering their shapes and causing errors in base pairing

allele The form of a gene that occupies the same place (locus) on the DNA molecule as another form but may carry different information for a trait

allergen An ordinarily innocuous foreign substance that can elicit an adverse immunological response in a sensitized person

allergy (also called hypersensitivity) When the immune system reacts in an exaggerated or inappropriate way to a foreign substance

allograft A graft of tissue between two organisms of the same species that are not genetically identical

allosteric site The site at which a noncompetitive inhibitor binds

alpha (α) hemolysin A type of enzyme that partially lyses red blood cells, leaving a greenish ring in the blood agar medium around the colonies

alpha (α) hemolysis Incomplete lysis of red blood cells by bacterial enzymes

alternative pathway One of the sequences of reactions in nonspecific host responses by which proteins of the complement system are activated

alveolus A saclike structure arranged in clusters at the ends of the respiratory bronchioles, having walls one cell layer thick, where gas exchange occurs

amantadine An antiviral agent that prevents penetration by influenza A viruses

amebozoa Major group of protozoans which move by pseudopodia and ingest food by phagocytosis, e.g., *Amoeba* species

Ames test Test used to determine whether a particular substance is mutagenic, based on its ability to induce mutations in auxotrophic bacteria

amino acid An organic acid containing an amino group and a carboxyl group, composing the building blocks of proteins

aminoglycoside An antimicrobial agent that blocks bacterial protein synthesis

amoebic dysentery Severe, acute form of amebiasis, caused by *Entamoeba histolytica*

amoeboid movement Movement by means of pseudopodia that occurs in cells without walls, such as amoebas and some white blood cells

amphibolic pathway A metabolic pathway that can yield either energy or building blocks for synthetic reactions

amphitrichous The presence of flagella at both ends of the bacterial cell

anabolic pathway A chain of chemical reactions in which energy is used to synthesize biologically important molecules

anabolism Chemical reactions in which energy is used to synthesize large molecules from simpler components (also called synthesis)

anaerobe An organism that does not use oxygen, including some organisms that are killed by exposure to oxygen

anaerobic respiration Respiration in which the final electron acceptor in the electron transport chain is an inorganic molecule other than oxygen (e.g., a sulfate, nitrate)

analytical study An epidemiological study that focuses on establishing cause-and-effect relationships in the occurrence of diseases in populations

anamnestic response (*See also* **secondary response**) Prompt immune response due to "recall" by memory cells

anamorphic Asexual part of the life cycle of a fungus.

anaphylactic shock Condition resulting from a sudden extreme drop in blood pressure caused by an allergic reaction

anaphylaxis An immediate, exaggerated allergic reaction to antigens, usually leading to detrimental effects

Ångstrom (Å) Unit of measurement equal to 0.0000000001 m, or 10^{-10} m. No longer officially recognized

Animalia The kingdom of organisms to which all animals belong

animal passage The rapid transfer of a pathogen through animals of a species susceptible to infection by the pathogen

anion A negatively charged ion

anionic dye (also called acidic dye) An ionic compound, used for staining bacteria, in which the negative ion imparts the color

anneal To bond or join the double strands of DNA by hydrogen bonding at sites where there are many complementary base pairs, used in reference to DNA hybridization

antagonism The decreased effect when two antibiotics are administered together

anthrax A zoonosis caused by *Bacillus anthracis* that exists in cutaneous, respiratory ("woolsorters disease"), or intestinal forms; transmitted by endospores

antibiosis The natural production of an antimicrobial agent by a bacterium or fungus

antibiotic A chemical substance produced by microorganisms that can inhibit the growth of or destroy other microorganisms

antibody (also called immunoglobulin) A protein produced in response to an antigen that is capable of binding specifically to that antigen

antibody titer The quantity of a specific antibody in an individual's blood, often measured by means of agglutination reactions

anticodon A three-base sequence in tRNA that is complementary to one of the mRNA codons, forming a link between each codon and the corresponding amino acid

antigen (also called immunogen) A substance that the body identifies as foreign and toward which it mounts an immune response

antigen binding site The site on the antibody to which the antigen (epitope) binds

antigen challenge Exposure to a foreign antigen

antigenic determinant *See* **epitope**

antigenic drift Process of antigenic variation that results from mutations in genes coding for hemagglutinin and neuraminidase

antigenic mimicry Self-antigen that is similar to an antigen on a pathogen

antigenic shift Process of antigenic variation probably caused by a reassortment of viral genes

antigenic presenting cell An immunological cell, such as a macrophage, dendritic cell, or B cell, that processes antigen fragments and presents peptide fragments from the antigen on its cell surface

antigenic variation Mutations of influenza viruses that occur by antigenic drift and antigenic shift

antigen-presenting cell Cells with MHC II proteins on their surface, in addition to MHC 1 proteins

antihistamine Drug that alleviates symptoms caused by histamine

antimetabolite A substance that prevents a cell from carrying out an important metabolic reaction

antimicrobial agent A chemotherapeutic agent used to treat diseases caused by microbes

antiparallel The opposite head-to-tail arrangement of the two strands in a DNA double helix

antiseptic A chemical agent that can be safely used externally on tissues to destroy microorganisms or to inhibit their growth

antiserum (plural: *antisera*) Serum that contains antibodies

antitoxin An antibody against a specific toxin

antiviral protein A protein induced by interferon that interferes with the replication of viruses

apicomplexan (also called sporozoan) A parasitic protozoan such as *Plasmodium*, that generally has a complex life cycle

aplastic crisis A period during which erythrocyte production ceases

apoenzyme The protein portion of an enzyme

apoptosis Genetically programmed cell death

arachnid An arthropod with two body regions, four pairs of legs, and mouth parts that are used in capturing and tearing apart prey

Archaea One of the three Domains of living things; all members are bacterial organisms lacking peptidoglycan in their cell walls and differing from eubacteria in many ways

arenavirus An enveloped RNA virus that causes Lassa fever and certain other hemorrhagic fevers

arthropod Makes up the largest group of living organisms, characterized by a jointed chitinous exoskeleton, segmented body, and jointed appendages associated with some or all of the segments

Arthus reaction A local reaction seen in the skin after subcutaneous or intradermal injection of an antigenic substance, an immune complex (type III) hypersensitivity

artificially acquired active immunity When an individual is exposed to a vaccine containing live, weakened, or dead organisms or their toxins, the host's own immune system responds specifically to defend the body (e.g., by making specific antibodies)

artificially acquired adaptive immunity When an individual's immune system is stimulated to react by some man-made process (e.g., given a vaccine or an immune serum)

artificially acquired passive immunity When antibodies made by other hosts are introduced into a new host (e.g., via mother's milk or shots of gamma globulin)

ascariasis Disease caused by a large roundworm, *Ascaris lumbricoides*, acquired by ingestion of food or water contaminated with eggs

Ascomycota *See* **sac fungus**

ascospore One of the eight sexual spores produced in each ascus of a sac fungus

ascus (plural: *asci*) Saclike structures produced by sac fungi during sexual reproduction

aseptic technique A set of procedures used to minimize chances that cultures will be contaminated by organisms from the environment

Asiatic cholera Severe gastrointestinal disease caused by *Vibrio cholerae*; common in areas of poor sanitation and fecal contamination of water

aspergillosis (also called farmer's lung disease) Skin infection caused by various species of *Aspergillus*, which can cause severe pneumonia in immunosuppressed patients

asthma Respiratory anaphylaxis caused by inhaled or ingested allergens or by hypersensitivity to endogenous microorganisms

athlete's foot (also called tinea pedis) A form of ringworm in which hyphae invade the skin between the toes, causing dry, scaly lesions

atom The smallest chemical unit of matter

atomic force microscope (AFM) Advanced member of the family of scanning tunneling microscopes, allowing 3-dimensional views of structures from atomic size to about 1 μm

atomic number The number of protons in an atom of a particular element

atomic weight The sum of the number of protons and neutrons in an atom

atopy Localized allergic reactions that occur first at the site where an allergen enters the body

atrichous A bacterial cell without flagella

attachment pilus (also called fimbria) Type of pilus that helps bacteria adhere to surfaces

attenuation (1) A genetic control mechanism that terminates transcription of an operon prematurely when the gene products are not needed. (2) The weakening of the disease-producing ability of an organism

auditory canal Part of the outer ear lined with skin that contains many small hairs and ceruminous glands

autoantibody An antibody against one's own tissues

autoclave An instrument for sterilization by means of moist heat under pressure

autograft A graft of tissue from one part of the body to another

autoimmune disorder An immune disorder in which individuals are hypersensitive to antigens on cells of their own bodies

autoimmunization The process by which hypersensitivity to "self" develops; it occurs when the immune system responds to a body component as if it were foreign

autotrophs Nutritionally deficient mutants that have lost the ability to synthesize a particular enzyme

autotrophy "Self-feeding"—the use of CO_2 as a source of carbon atoms for the synthesis of biomolecules

auxotroph An organism that uses carbon dioxide gas to synthesize organic molecules

auxotrophic mutant An organism that has lost the ability to synthesize one or more metabolically important enzymes through mutation, therefore it requires special substances in its growth medium

axial filament (also called endoflagellum) A subsurface filament attached near the ends of the cytoplasmic cylinder of spirochetes that causes the spirochete body to rotate like a corkscrew

babesiosis A protozoan disease caused by the apicomplexan *Babesia microti* and other species of *Babesia*

bacillary angiomatosis A disease of the small blood vessels of the skin and internal organs caused by the rickettsial organism *Bartonella hensalae*

bacillary dysentery *See* **shigellosis**

bacillus (plural: *bacilli*) A rodlike bacterium

bacteremia An infection in which bacteria are transported in the blood but do not multiply in transit

bacteria (singular: *bacterium*) All prokaryotic organisms

Bacteria When spelled with a capital B, it is the name of one of the three domains of living things; all members are bacterial

bacterial conjunctivitis (also called pinkeye) A highly contagious inflammation of the conjunctiva caused by various bacterial species

bacterial endocarditis (also called infective endocarditis) A life-threatening infection and inflammation of the lining and valves of the heart

bacterial enteritis An intestinal infection caused by bacterial invasion of intestinal mucosa or deeper tissues

bacterial lawn A uniform layer of bacteria grown on the agar surface in a Petri dish

bacterial meningitis An inflammation of the meninges, the membranes that cover the brain and spinal cord by any one of several bacterial species

bactericidal Referring to an agent that kills bacteria

bacteriocin A protein released by some bacteria that inhibits the growth of other strains of the same or closely related species

bacteriocinogen A plasmid that directs production of a bacteriocin

bacteriophage (also called phage) A virus that infects bacteria

bacteriostatic Referring to an agent that inhibits the growth of bacteria

bacteroid Irregularly shaped cell usually found in tight packets that develop from *Rhizobium* swarmer cells and form nodules in the roots of leguminous plants

balantidiasis Type of dysentery caused by the ciliated protozoan *Balantidium coli*

balantitis An infection of the penis

Bang's disease, also called brucellosis, undulant fever, or Malta fever A zoonosis caused by several species of *Brucella*, highly infective for humans. It is caused by several species of *Brucella*

barophile An organism that lives under high hydrostatic pressure

Bartholin gland A mucus-secreting gland of the female external genitalia

bartonellosis Rickettsial disease, caused by *Bartonella bacilliformis*, that occurs in two forms (*see also* **Oroya fever** and **verruga peruana**)

base A substance that absorbs hydrogen ions or donates hydroxyl ions

base analog A chemical mutagen similar in molecular structure to one of the nitrogenous bases found in DNA that causes point mutations

basic dye *See* **cationic dye**

Basidiomycota *See* **club fungus**

basidiospore A sexual spore of the club fungi

basidium (plural: *basidia*) A clublike structure in club fungi bearing four external spores on short, slender stalks

basophil A leukocyte that migrates into tissues and helps initiate the inflammatory response by secreting histamine

B cell *See* **B lymphocyte**

benign Not harmful

beta (β) hemolysin A type of enzyme that completely lyses red blood cells, leaving a clear ring in the blood agar medium around the colonies

beta (β) hemolysis Complete lysis of red blood cells by bacterial enzymes

beta oxidation A metabolic pathway that breaks down fatty acids into 2-carbon pieces

bilirubin A yellow substance, the product of the breakdown of hemoglobin from red blood cells

binary fission Process in which a bacterial cell duplicates its components and divides into two cells

binocular Referring to a light microscope having two eyepieces (oculars)

binomial nomenclature The system of taxonomy developed by Linnaeus in which each organism is assigned a genus and specific epithet

biochemistry The branch of organic chemistry that studies the chemical reactions of living systems

bioconversion A reaction in which one compound is converted to another by enzymes in cells

biogeochemical cycle Mechanism by which water and elements that serve as nutrients are recycled

biohydrometallurgy The use of microbes to extract metals from ores

biological oxygen demand (BOD) The oxygen required to degrade organic wastes suspended in water

biological vector An organism that actively transmits pathogens that complete part of their life cycle within the organism

bioremediation A process that uses naturally occurring or genetically engineered microorganisms to transform harmful substances into less toxic or nontoxic compounds

biosphere The region of the earth inhabited by living organisms

biotic factor An organism in the biosphere

blackfly fever Illness resulting from bites by blackflies, characterized by an inflammatory reaction, nausea, and headache

blackwater fever Malaria caused by *Plasmodium falciparum* that results in jaundice and kidney damage

blastomycetic dermatitis Fungal skin disease caused by *Blastomyces dermatitidis*; characterized by disfiguring, granulomatous, pus-producing lesions

blastomycosis Fungal skin disease caused by *Blastomyces dermatitidis* that enters the body through wounds

blocking antibody IgG antibody, elicited in allergy patients by increasing doses of allergen, that complexes with allergen before it can react with IgE antibody

blood agar Type of medium containing sheep blood, used to identify organisms that cause hemolysis, or breakdown of red blood cells

blood–brain barrier Formation in the brain of special thick-walled capillaries without pores in their walls that limit entry of substances into brain cells

B lymphocyte (also called B cell) A lymphocyte that is produced in and matures in bursal-equivalent tissue, it gives rise to antibody-producing plasma cells

body tube Microscope part that conveys an image from the objective to the eyepiece

boil *See* **furuncle**

Bolivian hemorrhagic fever A multisystem disease caused by an arenavirus with insidious onset and progressive effects

bone stink Putrefaction deep in the tissues of large carcasses that is caused by several species of *Clostridium*

bongkrek disease Type of food poisoning caused by *Pseudomonas cocovenenans*, named for a native Polynesian coconut dish

botulism Disease caused by *Clostridium botulinum*. The most common form, food-borne botulism, results from ingestion of preformed toxin and is, therefore, an intoxication rather than an infection

bradykinin Small peptide thought to cause the pain associated with tissue injury

brain abscess A pus-filled cavity caused by microorganisms reaching the brain from head wounds or via blood from another site

bread mold (also called Zygomycota or conjugation fungus) A fungus with complex mycelia composed of aseptate hyphae with chitinous cross walls

bright-field illumination Illumination produced by the passage of visible light through the condenser of a light microscope

Brill-Zinsser disease (also called recrudescent typhus) A recurrence of an epidemic typhus infection caused by reactivation of latent organisms harbored in the lymph nodes

broad spectrum Referrring to the range of activity of an antimicrobial agent that attacks a wide variety of microorganisms

bronchial pneumonia Type of pneumonia that begins in the bronchi and can spread through surrounding tissue toward the alveoli

bronchiole A finer subdivision of the air-conveying bronchi

bronchitis An infection of the bronchi

bronchus (plural: *bronchi*) A subdivision of the trachea that conveys air to and from the lungs

brucellosis (also called undulant fever and Malta fever) A zoonosis highly infective for humans, caused by any of several species of *Brucella*

bubo Enlargement of infected lymph nodes, especially in the groin and armpit, due to accumulation of pus; characteristic of bubonic plague and other diseases

bubonic plague A bacterial disease, caused by *Yersinia pestis* and transmitted by flea bites, that spreads in the blood and lymphatic system

budding Process that occurs in yeast and a few bacteria in which a small new cell develops from the surface of an existing cell

bulking Phenomenon in which filamentous bacteria multiply, causing sludge to float on the surface of water rather than settling out

bunyavirus An enveloped RNA virus that causes some forms of respiratory distress and hemorrhagic fever

Burkitt's lymphoma A tumor of the jaw, seen mainly in African children; caused by the Epstein-Barr virus

burst size (also called viral yield) The number of new virions released in the replication process

burst time The time from absorption to release of phages (in the replication process)

cancer An uncontrolled, invasive growth of abnormal cells

candidiasis (also called moniliasis) A yeast infection caused by *Candida albicans* that appears as thrush (in the mouth) or vaginitis

canine parvovirus A parvovirus that causes severe disease in dogs

canning The use of moist heat under pressure to preserve food

capillary A blood vessel that branches from an arteriole

capnophile An organism that prefers carbon dioxide gas for growth

capsid The protein coating of a virus, which protects the nucleic acid core from the environment and usually determines the shape of the virus

capsomere A protein aggregate that makes up a viral capsid

capsule (1) A protective structure outside the cell wall, secreted by the organism. (2) A network of connective fibers covering organs such as the lymph nodes

carbapenem A bactericidal antibiotic that acts on bacterial cell walls

carbohydrate A compound composed of carbon, hydrogen, and oxygen that serves as the main source of energy for most living things

carbon cycle Process by which carbon from atmospheric carbon dioxide enters living and nonliving things and is recycled through them

carbuncle A massive pus-filled lesion resulting from an infection, particularly of the neck and upper back

carcinogen A cancer-producing substance

cardiovascular system Body system that supplies oxygen and nutrients to all parts of the body and removes carbon dioxide and other wastes from them

carrier An individual who harbors an infectious agent without having observable clinical signs or symptoms

cascade A set of reactions in which magnification of effect occurs, as in the complement system

casein hydrolysate A substance derived from milk protein that contains many amino acids; used to enrich certain media

caseous Characterizing lesions with a "cheesy" appearance that form in lung tissue of patients with tuberculosis

catabolic pathway A chain of chemical reactions that capture energy by breaking down large molecules into simpler components

catabolism The chemical breakdown of molecules in which energy is released

catabolite repression Process by which the presence of a preferred nutrient (often glucose) represses the genes coding for enzymes used to metabolize some alternative nutrient

catalase An enzyme that converts hydrogen peroxide to water and molecular oxygen

catarrhal stage Stage of whooping cough characterized by fever, sneezing, vomiting, and a mild, dry persistent cough

cation A positively charged ion

cationic dye (also called basic dye) An ionic compound, used for staining bacteria, in which the positive ion imparts the color

cat scratch fever A disease caused by *Afipia felis* or, more commonly, *Bartonella (Rochalimaea) henselae* and transmitted in cat scratches and bites

cavitation The formation of a cavity inside the cytoplasm of a cell

cell culture A culture in the form of a monolayer from dispersed cells and continuous cultures of cell suspensions

cell-mediated immune response The immune response to an antigen carried out at the cellular level by T cells

cell-mediated immunity The immune response involving the direct action of T cells to activate B cells or to destroy microbe-infected cells, tumor cells, or transplanted cells (organ transplants)

cell-mediated (Type IV) hypersensitivity (also called delayed hypersensitivity) Type of allergy elicited by foreign substances from the environment, infectious agents, transplanted tissues, and the body's own malignant cells; mediated by T cells

cell membrane (also called plasma membrane) A selectively permeable lipoprotein bilayer that forms the boundary between a bacterial cell's cytoplasm and its environment

cell nucleus A distinct organelle enclosed by a nuclear envelope and containing nucleoplasm, nucleoli, and (typically paired) chromosomes

cell strain Dominant cell type resulting from subculturing

cell theory Theory formulated by Schleiden and Schwann that cells are the fundamental units of all living things

cellular slime mold Funguslike protist consisting of amoeboid, phagocytic cells that aggregate to form a pseudoplasmodium

cell wall Outer layer of most bacterial, algal, fungal, and plant cells that maintains the shape of the cell

cementum The hard, bony covering of the tooth below the gumline

central nervous system The brain and spinal cord

cephalosporin An antibacterial agent that inhibits cell wall synthesis

cercaria A free-swimming fluke larva that emerges from the snail or mollusk host

cerumen Earwax

ceruminous gland A modified sebaceous gland that secretes cerumen

cervix An opening at the narrow lower portion of the uterus

Chagas' disease Disease caused by *Trypanosoma cruzi* that occurs in the southern United States and is endemic to Mexico; transmitted by several kinds of reduviid bugs

chancre A hard, painless, nondischarging lesion; a symptom of primary stage syphilis

chancroid Sexually transmitted disease caused by *Haemophilus ducreyi* that causes soft, painful skin lesions on the genitals, which bleed easily

chemical bond The interaction of electrons in atoms that form a molecule

chemical equilibrium A steady state in which there is no net change in the concentrations of substrates or products

chemically nondefined medium *See* **complex medium**

chemiosmosis Process of energy capture in which a proton gradient is created by means of electron transport and then used to drive the synthesis of ATP

chemoautotroph An autotroph that obtains energy by oxidizing simple inorganic substances such as sulfides and nitrites

chemoheterotroph A heterotroph that obtains energy from breaking down ready-made organic molecules

chemokines A class of cytokines that attract additional phagocytes to the site of the infection

chemolithotroph *See* **chemoautotroph**

chemostat A device for maintaining the logarithmic growth of a culture by the continuous addition of fresh medium

chemotaxis A nonrandom movement of an organism toward or away from a chemical

chemotherapeutic agent (also called drug) Any chemical substance used to treat disease

chemotherapeutic index The maximum tolerable dose of a particular drug per kilogram body weight divided by the minimum dose per kilogram body weight that will cure the disease

chemotherapy The use of chemical substances to treat various aspects of disease

chickenpox A highly contagious disease, characterized by skin lesions, caused by the varicella-zoster herpesvirus; usually occurs in children

chigger dermatitis A violent allergic reaction caused by chiggers, the larvae of *Trombicula* mites

childbed fever *See* **puerperal fever**

chitin A polysaccharide found in the cell walls of most fungi and the exoskeletons of arthropods

Chlamydiae Tiny, nonmotile, spherical bacteria; all are obligate intracellular parasites with a complex life cycle

chloramphenicol A bacteriostatic agent that inhibits protein synthesis

chlorination The addition of chlorine to water to kill bacteria

chloroplast A chlorophyll-containing organelle found in eukaryotic cells that carry out photosynthesis

chloroquine An antiprotozoan agent effective against the malaria parasite

chocolate agar Type of medium made with heated blood, so named because it turns a chocolate brown color

chromatin The appearance of chromosomes as fine threads in cells

chromatophore The internal membranes of photosynthetic bacteria and cyanobacteria

chromosomal resistance Drug resistance of a microorganism due to a mutation in chromosomal DNA

chromosome A structure that contains the DNA of organisms

chromosome mapping The identification of the sequence of genes in a chromosome

chronic amebiasis Chronic infection caused by the protozoan *Entamoeba histolytica*

chronic disease A disease that develops more slowly than an acute disease, is usually less severe, and persists for a long, indeterminate period

chronic fatigue syndrome (previously called chronic EBV syndrome) Disease of uncertain origin similar to mononucleosis with symptoms including persistent fatigue and fever

chronic inflammation A condition in which there is a persistent, indecisive standoff between an inflammatory agent and the phagocytic cells and other host defenses attempting to destroy it

chronic wasting disease Spongiform encephalopathy of deer and elk, caused by prions

ciliate A protozoan that moves by means of cilia that cover most of its surface

cilium (plural: *cilia*) A short cellular projection used for movement that beats in coordinated waves

citric acid cycle *See* **Krebs cycle**

classical pathway One of the two sequences of reactions by which proteins of the complement system are activated

clonal deletion The process in which the binding of lymphocytes to self antigens triggers a genetically programmed destruction of those lymphocytes

clonal selection hypothesis Theory that explains how exposure to an antigen stimulates a lymphocyte capable of making antibodies against that particular antigen to proliferate, giving rise to a clone of identical antibody-producing cells

clone A group of genetically identical cells descending from a single parent cell

club fungus (also called Basidiomycota) A fungus, including mushrooms, toadstools, rusts, and smuts, that produces spores on basidia

cluster of differentiation marker An antigen found on the cell surface of B and T cells that can be used to distinguish the cells from one another

coagulase A bacterially produced enzyme that accelerates the coagulation (clotting) of blood

coarse adjustment Focusing mechanism of a microscope that rapidly changes the distance between the objective lens and the specimen

coccidioidomycosis (also called valley fever) Fungal respiratory disease caused by the soil fungus *Coccidioides immitis*

coccus (plural: *cocci*) A spherical bacterium

codon A sequence of three bases in mRNA that specifies a particular amino acid in the translation process

coelom The body cavity between the digestive tract and body wall in higher animals

coenzyme An organic molecule bound to or loosely associated with an enzyme

cofactor An inorganic ion necessary for the function of an enzyme

cold seeps Cool chemical and thermal gradients along the bottom of oceans bubbling up methane through patchy areas near continental margins

colicin A protein released by some strains of *Escherichia coli* that inhibits growth of other strains of the same organism

coliform bacterium Gram-negative, non-spore-forming, aerobic or facultatively anaerobic bacterium that ferments lactose and produces acid and gas; significant numbers may indicate water pollution

colloid A mixture formed by particles too large to form a true solution dispersed in a liquid

colonization Growth of microorganisms on epithelial surfaces such as skin or mucous membranes

colony A group of descendants of an original cell

colony-forming unit (CFU) Live bacterial cell which can give rise to a colony

Colorado tick fever Disease caused by an orbivirus carried by dog ticks, characterized by headache, backache, and fever

colostrum The protein-rich fluid secreted by the mammary glands just after childbirth, prior to the appearance of breast milk

commensal An organism that lives in or on another organism without harming it and that benefits from the relationship

commensalism A symbiotic relationship in which one organism benefits and the other one neither benefits nor is harmed by the relationship

common-source outbreak An epidemic that arises from contact with contaminated substances

communicable infectious disease (also called contagious disease) Infectious disease that can be spread from one host to another

community All the kinds of organisms present in an environment

competence factor A protein released into the medium that facilitates the uptake of DNA into a bacterial cell

competitive inhibitor A molecule similar in structure to a substrate that competes with that substrate by binding to the active site

complement (also called contagious disease) A set of more than 20 large regulatory proteins that circulate in plasma and when activated form a nonspecific defense mechanism against many different micro-organisms

complementary base pairing Hydrogen bonding between adenine and thymine (or uracil) bases or between guanine and cytosine bases

complement fixation test A complex serologic test used to detect small quantities of antibodies

complement system *See* **complement**

completed test The final test for coliforms in multiple-tube fermentation in which organisms from colonies grown on eosin methylene blue agar are used to inoculate broth and agar slants

complex medium (also called chemically nondefined medium) A growth medium that contains certain reasonably well-defined materials but that varies slightly in chemical composition from batch to batch

complex virus A virus, such as bacteriophage or poxvirus, that has an envelope or specialized structures

compound A chemical substance made up of atoms of two or more elements

compound light microscope A light microscope with more than one lens

compromised host An individual with reduced resistance, being more susceptible to infection

conclusions End result of an analysis of experimental results

condenser Device in a microscope that converges light beams so that they will pass through the specimen

condyloma *See* **genital wart**

confirmed test Second stage of testing for coliforms in multiple-tube fermentation in which samples from the highest dilution showing gas production are streaked into eosin methylene blue agar

confocal microscopy Uses beams of ultraviolet laser light to excite fluorescent chemical dye molecules into emitting (returning) light

congenital rubella syndrome Complication of German measles causing death or damage to a developing embryo infected by virus that crosses the placenta

congenital syphilis Syphilis passed to a fetus when treponemes cross the placenta from mother to child before birth

conidium (plural: *conidia*) A small, asexual, aerial spore organized into chains in some bacteria and fungi

conjugation (1) The transfer of genetic information from one bacterial cell to another by means of conjugation pili. (2) The exchange of information between two ciliates (protists)

conjugation pilus (also called sex pilus) A type of pilus that attaches two bacteria together and provides a means for the exchange of genetic material

conjunctiva Mucous membranes of the eye

consolidation Blockage of air spaces as a result of fibrin deposits in lobar pneumonia

constitutive enzyme An enzyme that is synthesized continuously regardless of the nutrients available to the organism

consumer (also called heterotroph) An organism that obtains nutrients by eating producers or other consumers

contact dermatitis Cell-mediated (type IV) hypersensitivity disorder that occurs in sensitized individuals on second exposure of the skin to allergens

contact transmission A mode of disease transmission effected directly, indirectly, or by droplets

contagious disease *See* **communicable infectious disease**

contamination The presence of microorganisms on inanimate objects or surfaces of the skin and mucous membranes

continuous cell line Cell culture consisting of cells that can be propagated over many generations

continuous reactor A device used in industrial and pharmaceutical microbiology to isolate and purify a microbial product often without killing the organism

control variable A factor that is prevented from changing during an experiment

convalescent period or stage The stage of an infectious disease during which tissues are repaired, healing takes place, and the body regains strength and recovers

Coomb's antiglobulin test An immunological test designed to detect anti-Rh antibodies

core The living part of an endospore

cornea The transparent part of the eyeball exposed to the environment

coronavirus Virus with clublike projections that causes colds and acute upper respiratory distress

cortex A laminated layer of peptidoglycan between the membranes of the endospore septum

corynebacteria Club-shaped, irregular, non-pore-forming, Gram-positive rods

coryza The common cold

countable number A number of colonies on an agar plate small enough so that one can clearly distinguish and count them (30 to 300 per plate)

covalent bond A bond between atoms created by the sharing of pairs of electrons

cowpox Disease caused by the vaccinia virus and characterized by lesions, inflammation of lymph nodes, and fever; virus is used to make vaccine against smallpox and monkeypox

crepitant tissue Distorted tissue caused by gas bubbles in gas gangrene

Creutzfeldt-Jakob disease (CJD) A transmissible spongiform encephalopathy of the human brain caused by prions

crista (plural: cristae) A fold of the inner mitochondrial membrane

cross-reaction Immune reaction of a single antibody with different antigens that are similar in structure

cross-resistance Resistance against two or more similar antimicrobial agents through a common mechanism

croup Acute obstruction of the larynx that produces a characteristic high-pitched, barking cough

crustacean A usually aquatic arthropod that has a pair of appendages associated with each body segment

cryptococcosis Fungal respiratory disease caused by a budding, encapsulated yeast, *Filobasidiella neoformans*

cryptosporidiosis Disease caused by protozoans of the genus *Cryptosporidium*, common in AIDS patients

curd The solid portion of milk resulting from bacterial enzyme addition and used to make cheese

cutaneous anthrax Infection by *Bacillus anthracis* that appears on the surface of the skin 2 to 5 days after endospores enter epithelial layers of the skin

cyanobacteria Photosynthetic, prokaryotic, typically unicellular organisms that are members of the kingdom Monera

cyanosis Bluish skin characteristic of oxygen-poor blood

cyclic photophosphorylation Pathway in which excited electrons from chlorophyll are used to generate ATP without the splitting of water or reduction of NADP

cyst A spherical, thick-walled cell that resembles an endospore, formed by certain bacteria

cysticercus (also called bladder worm) An oval white sac with a tapeworm head invaginated into it

cystitis Inflammation of the bladder

cytochrome An electron carrier functioning in the electron transport chain; heme protein

cytokine One of a diverse group of soluble proteins that have specific roles in host defenses

cytomegalovirus (CMV) One of a widespread and diverse group of herpesviruses that often produces no symptoms in normal adults but can severely affect AIDS patients and congenitally infected children

cytopathic effect (CPE) The visible effect viruses have on cells

cytoplasm The semifluid substance inside a cell, excluding, in eukaryotes, the cell nucleus

cytoplasmic streaming Process by which cytoplasm flows from one part of a eukaryotic cell to another

cytoskeleton A network of protein fibers that support, give rigidity and shape to a eukaryotic cell, and provide for cell movements

cytotoxic drug A drug that interferes with DNA synthesis, used to suppress the immune system and prevent the rejection of transplants

cytotoxic T cell (Tc) Lymphocyte that destroys virus-infected cells

cytotoxic (Type II) hypersensitivity Type of allergy elicited by antigens on cells, especially red blood cells, that the immune system treats as foreign

cytotoxin Toxin produced by cytotoxic cells that kills infected host cells

dark-field illumination In light microscopy, the light that is reflected from an object rather than passing through it, resulting in a bright image on a dark background

dark reactions light-independent (also called carbon fixation) Part of photosynthesis in which carbon dioxide gas is reduced by electrons from reduced NADP (NADPH) to form various carbohydrate molecules, chiefly glucose

dark repair Mechanism for repair of damaged DNA by several enzymes that do not require light for activation; they excise defective nucleotide sequences and replace them with DNA complementary to the unaltered DNA strand

daughter cell One of the two identical products of cell division

deaminating agent A chemical mutagen that can remove an amino group ($-NH_2$) from a nitrogenous base, causing a point mutation

death phase *See* **decline phase**

débridement Surgical scraping to remove the thick crust or scab that forms over burnt tissue (eschar)

decimal reduction time (DRT; also called D value) The length of time needed to kill 90% of the organisms in a given population at a specified temperature

decline phase (1) The fourth of four major phases of the bacterial growth curve in which cells lose their ability to divide (due to less supportive conditions in the medium) and thus die (also called **death phase**). (2) In the stages of a disease, the period during which the host defenses finally overcome the pathogen and symptoms begin to subside

decomposer Organism that obtains energy by digesting dead bodies or wastes of producers and consumers

deep hot biosphere Theory that the entire crust of the Earth, down to a depth of several miles, is inhabited by a culture of microbes that feed on oil and methane gas deposits that are an original part of the Earth

defined synthetic medium A synthetic medium that contains known specific kinds and amounts of chemical substances

definitive host An organism that harbors the adult, sexually reproducing form of a parasite

degranulation Release of histamine and other preformed mediators of allergic reactions by sensitized mast cells and basophils after a second encounter with an allergen

dehydration synthesis A chemical reaction that builds complex organic molecules

delayed (Type IV) hypersensitivity *See* **cell-mediated (Type IV) hypersensitivity**

delayed hypersensitivity (T_D) cells Those T cells (inflammatory T_H1) that produce lymphokines in cell-mediated (Type IV) hypersensitivity reactions

deletion The removal of one or more nitrogenous bases from DNA, usually producing a frameshift mutation

delta hepatitis *See* **hepatitis D**

denaturation The disruption of hydrogen bonds and other weak forces that maintain the structure of a globular protein, resulting in the loss of its biological activity

dendritic cells Cells with long membrane extensions that resemble the dendrites of nerve cells

dengue fever (also called breakbone fever) Viral systemic disease that causes severe bone and joint pain

denitrification The process by which nitrates are reduced to nitrous oxide or nitrogen gas

dental caries (also called tooth decay) The erosion of enamel and deeper parts of teeth

dental plaque A continuously formed coating of microorganisms and organic matter on tooth enamel

deoxyribonucleic acid (DNA) Nucleic acid that carries hereditary information from one generation to the next

dermal wart Wart resulting from the viral infection of epithelial cells

dermatomycosis A fungal skin disease

dermatophyte A fungus that invades keratinized tissue of the skin and nails

dermis The thick inner layer of the skin

descriptive study An epidemiologic study that notes the number of cases of a disease, which segments of the population are affected, where the cases have occurred, and over what time period

desensitization Treatment designed to cure allergies by means of injections with gradually increasing doses of allergen

Deuteromycota *See* **Fungi Imperfecti**

diapedesis The process in which leukocytes pass out of blood into inflamed tissues by squeezing between cells of capillary walls

diarrhea Excessive frequency and looseness of bowel movements

diatom An alga or plantlike protist that lacks flagella and has a glasslike outer shell

dichotomous key Taxonomic key used to identify organisms; composed of paired (either-or) statements describing characteristics

differential medium A growth medium with a constituent that causes an observable change (in color or pH) in the medium when a particular chemical reaction occurs, making it possible to distinguish between organisms

differential stain Use of two or more dyes to differentiate among bacterial species or to distinguish various structures of an organism; for example, the Gram stain

diffraction Phenomenon in which light waves, as they pass through a small opening, are broken up into bands of different wavelengths

DiGeorge syndrome Primary immunodeficiency disease caused by failure of the thymus to develop properly, resulting in a deficiency of T cells

digestive system The body system that converts ingested food into material suitable for the liberation of energy or for assimilation into body tissues

digital microscopy Has built-in digital camera and preloaded software

dikaryotic Referring to fungal cells within hyphae that have two nuclei, produced by plasmogamy in which the nuclei have not united

dilution method A method of testing antibiotic sensitivity in which organisms are incubated in a series of tubes containing known quantities of a chemotherapeutic agent

dimer Two adjacent pyrimidines bonded together in a DNA strand, usually as a result of exposure to ultraviolet rays

dimorphism The ability of an organism to alter its structure when it changes habitats

dinoflagellate An alga or plantlike protist, usually with two flagella

diphtheria A severe upper respiratory disease caused by *Corynebacterium diphtheriae*; can produce subsequent myocarditis and polyneuritis

diphtheroid Organism found in normal throat cultures that fails to produce exotoxin but is otherwise indistinguishable from diphtheria-causing organisms

dipicolinic acid Acid found in the core of an endospore that contributes to its heat resistance

diplo- Prefix that indicates that a bacteria divides in one plane and produces cells in pairs

diploid A eukaryotic cell that has paired sets of chromosomes

diploid fibroblast strain A culture derived from fetal tissues that retains fetal capacity for rapid, repeated cell division

direct contact transmission Mode of disease transmission requiring person-to-person body contact

direct fecal-oral transmission Direct contact transmission of disease in which pathogens from fecal matter are spread by unwashed hands to the mouth

direct microscopic count A method of measuring bacterial growth by counting cells in a known volume of medium that fills a specially calibrated counting chamber on a microscope slide

disaccharide A carbohydrate formed by the joining of two monosaccharides

disease A disturbance in the state of health wherein the body cannot carry out all its normal functions. (*See also* **epidemiology** and **infectious disease**)

disinfectant A chemical agent used on inanimate objects to destroy microorganisms

disinfection Reducing the number of pathogenic organisms on objects or in materials so that they pose no threat of disease

disk diffusion method (also called Kirby-Bauer method) A method used to determine microbial sensitivity to antimicrobial agents in which antibiotic disks are placed on an inoculated Petri dish, incubated, and observed for inhibition of growth

displacin A molecule that displaces (removes) gene(s) from a chromosome

disseminated tuberculosis Type of tuberculosis spread throughout body; now seen in AIDS patients, usually caused by *Mycobacterium avium-intercellulare*

distillation The separation of alcohol and other volatile substances from solid and nonvolatile substances

divergent evolution Process in which descendants of a common ancestor species undergo sufficient change to be identified as separate species

diversity The ability of the immune system to produce many different kinds of antibodies and T cell receptors, each of which reacts with a different epitope (antigenic determinant)

DNA hybridization Process in which the double strands of DNA of each of two organisms are split apart and the split strands from the two organisms are allowed to combine

DNA polymerase An enzyme that moves along behind each replication fork, synthesizing new DNA strands complementary to the original ones

DNA replication Formation of new DNA molecules

DNA tumor virus An animal virus capable of causing tumors

domain A new taxonomic category above the kingdom level, consisting of the Archaea, Bacteria, and Eukarya

Donovan body A large mononuclear cell found in scrapings of lesions that confirms the presence of granuloma inguinale

DPT vaccine Diphtheria, killed whole cell pertussis, and tetanus vaccine

dracunculiasis Skin disease caused by a parasitic helminth, the guinea worm *Dracunculus medinensis*

droplet nucleus A particle consisting of dried mucus in which microorganisms are embedded

droplet transmission Contact transmission of disease through small liquid droplets

drug *See* **chemotherapeutic agent**

DTaP vaccine Diphtheria, tetanus, and acellular pertussis vaccine

D value *See* **decimal reduction time**

dyad A set of paired chromosomes in eukaryotic cells that are prepared to divide by mitosis or meiosis

dysentery A severe diarrhea that often contains mucus and sometimes blood or pus

dysuria Pain and burning on urination

eastern equine encephalitis Type of viral encephalitis seen most often in the eastern United States; infects horses more frequently than humans

Ebola virus A filovirus that causes hemorrhagic fevers

eclipse period Period during which viruses have absorbed to and penetrated host cells but cannot yet be detected in cells

ecology The study of relationships among organisms and their environment

ecosystem All the biotic and abiotic components of an environment

ectoparasite A parasite that lives on the surface of another organism

eczema herpeticum A generalized eruption caused by entry of the herpesvirus through the skin; often fatal

edema An accumulation of fluid in tissues that causes swelling

E (epsilometer) test A newer version of the diffusion test that uses a plastic strip containing a gradient of concentration of antibiotic to determine antibiotic sensitivity

and estimate MIC (minimal inhibitory concentration)

ehrlichiosis A tick-borne disease found in dogs and humans and caused by *Ehrlichia canis* and *E. chaffeensis*

electrolyte A substance that is ionizable in solution

electron A negatively charged subatomic particle that moves around the nucleus of an atom

electron acceptor An oxidizing agent in a chemical reaction

electron donor A reducing agent in a chemical reaction

electron micrograph A "photograph" of an image taken with an electron microscope

electron microscope Microscope that uses a beam of electrons rather than a beam of light and electromagnets instead of glass lenses to produce an image

electron transport Process in which pairs of electrons are transferred between cytochromes and other compounds

electron transport chain (also called respiratory chain) A series of compounds that pass electrons to oxygen (the final electron acceptor)

electrophoresis Process used to separate large molecules such as antigens or proteins by passing an electrical current through a sample on a gel

electroporation A brief electric pulse produces temporary pores in the cell membrane, allowing entrance of vectors carrying foreign DNA

element Matter composed of one kind of atom

elementary body An infectious stage in the life cycle of chlamydias

elephantiasis Gross enlargement of limbs, scrotum, and sometimes other body parts from accumulation of fluid due to blockage of lymph ducts by the helminth *Wuchereria bancrofti*

emerging virus Viruses that were previously endemic (low levels of infection in localized areas) or had "cross-species barriers" and expanded their host range to other species

enamel The hard substance covering the crown of a tooth

encephalitis An inflammation of the brain caused by a variety of viruses or bacteria

endemic Referring to a disease that is constantly present in a specific population

endemic relapsing fever Tick-borne cases of relapsing fever caused by several species of *Borrelia*

endemic typhus (also called murine typhus) A flea-borne typhus caused by *Rickettsia typhi*

endergonic Requiring energy for a chemical reaction

endocytosis Process in which vesicles form by invagination of the plasma membrane to move substances into eukaryotic cells

endoenzyme An enzyme that acts within the cell producing it

endoflagellum *See* **axial filament**

endogenous infection An infection caused by opportunistic microorganisms already present in the body

endogenous pyrogen Pyrogen secreted mainly by monocytes and macrophages that circulates to the hypothalamus and causes an increase in body temperature

endometrium The mucous membrane lining the uterus

endoparasite A parasite that lives within the body of another organism

endoplasmic reticulum An extensive system of membranes that form tubes and plates in the cytoplasm of eukaryotic cells; involved in synthesis and transport of proteins and lipids

endospore A resistant, dormant structure, formed inside some bacteria, such as *Bacillus* and *Clostridium*, that can survive adverse conditions

endospore septum A cell membrane without a cell wall that grows around the core of an endospore

endosymbiotic theory Holds that the organelles of eukaryotic cells arose from prokaryotes that came to live, in a symbiotic relationship, inside the eukaryote-to-be cell

endotoxin (also called lipopolysaccharide) A toxin incorporated in Gram-negative bacterial cell walls and released when the bacterium dies

end-product inhibition *See* **feedback inhibition**

enrichment medium A medium that contains special nutrients that allow growth of a particular organism

enteric bacteria Members of the family Enterobacteriaceae, many of which are intestinal; small facultatively anaerobic Gram-negative rods with peritrichous flagella

enteric fever Systemic infection, such as typhoid fever, spread throughout the body from the intestinal mucosa

enteritis An inflammation of the intestine

enterocolitis Disease caused by *Salmonella typhimurium* and *S. paratyphi* that invade intestinal tissue and produce bacteremia

enterohemorrhagic strain of *Escherichia coli* One that causes bloody diarrhea and is often fatal; often from contaminated food

enteroinvasive strain Strain of *Escherichia coli* with a plasmid-borne gene for a surface antigen (K antigen) that enables it to attach to and invade mucosal cells

enterotoxicosis *See* **food poisoning**

enterotoxigenic strain Strain of *Escherichia coli* carrying a plasmid that enables it to make an enterotoxin

enterotoxin An exotoxin that acts on tissues of the gut

enterovirus One of the three major groups of picornaviruses that can infect nerve and muscle cells, the respiratory tract lining, and skin

envelope A bilayer membrane found outside the capsid of some viruses, acquired as the virus buds through one of the host's membrane

enveloped virus A virus with a bilayer membrane outside its capsid

enzyme A protein catalyst that controls the rate of chemical reactions in cells

enzyme induction A mechanism whereby the genes coding for enzymes needed to metabolize a particular nutrient are activated by the presence of that nutrient

enzyme-linked immunosorbent assay (ELISA) Modification of radioimmunoassay in which the anti-antibody, instead of being radioactive, is attached to an enzyme that causes a color change in its substrate

enzyme repression Mechanism by which the presence of a particular metabolite represses the genes coding for enzymes used in its synthesis

enzyme-substrate complex A loose association of an enzyme with its substrate

eosinophil A leukocyte present in large numbers during allergic reactions and worm infections

epidemic Referring to a disease that has a higher than normal incidence in a population over a relatively short period of time

epidemic keratoconjunctivitis (EKC) (sometimes called shipyard eye) Eye disease caused by an adenovirus

epidemic relapsing fever Louseborne cases of relapsing fever caused by several species of *Borrelia*

epidemic typhus (also called classic, European, or louseborne typhus) Louseborne rickettsial disease caused by *Rickettsia prowazekii*, seen most frequently in conditions of overcrowding and poor sanitation

epidemiologic study A study conducted in order to learn more about the spread of a disease in a population

epidemiologist A scientist who studies epidemiology

epidemiology The study of factors and mechanisms involved in the spread of disease within a population

epidermis The thin outer layer of the skin

epiglottitis An infection of the epiglottis

epitope (also called antigenic determinant) An area on an antigen molecule to which antibodies bind

Epstein-Barr virus (EBV) Virus that causes infectious mononucleosis and Burkitt's lymphoma

ergot Toxin produced by *Claviceps purpurea*, a parasite fungus of rye and wheat that causes ergot poisoning when ingested by humans

ergot poisoning Disease caused by ingestion of ergot, the toxin produced by *Claviceps purpurea*, a fungus of rye and wheat

erysipelas (also called St. Anthony's fire) Infection caused by hemolytic streptococci that spreads through lymphatics, resulting in septicemia and other diseases

erythrocyte A red blood cell

erythromycin An antibacterial agent that has a bacteriostatic effect on protein synthesis

eschar The thick crust or scab that forms over a severe burn

ethambutol An antibacterial agent effective against certain strains of mycobacteria

etiology The assignment or study of causes and origins of a disease

eubacteria True bacteria

euglenoid An alga or plantlike protist, usually with a single flagella and a pigmented eyespot (stigma)

Eukarya One of the three domains of living things; all members are eukaryotic

eukaryote An organism composed of eukaryotic cells

eukaryotic cell A cell that has a distinct cell nucleus and other membrane-bound structures

eutrophication The nutrient enrichment of water from detergents, fertilizers, and animal manures, which causes overgrowth of algae and subsequent depletion of oxygen

exanthema A skin rash

exergonic Releasing energy from a chemical reaction

exocytosis Process by which vesicles inside a eukaryotic cell fuse with the plasma membrane and release their contents from the eukaryotic cell

exoenzyme (also called **extracellular enzyme**) An enzyme that is synthesized in a cell but crosses the cell membrane to act in the periplasmic space or the cell's immediate environment

exogenous infection An infection caused by microorganisms that enter the body from the environment

exogenous pyrogen Exotoxins and endotoxins from infectious agents that cause fever by stimulating the release of an endogenous pyrogen

exon The region of a gene (or mRNA) that codes for a protein in eukaryotic cells

exonuclease An enzyme that removes segments of DNA

exosporium A lipid-protein membrane formed outside the coat of some endospores by the mother cell

exotoxin A soluble toxin secreted by microbes into their surroundings, including host tissues

experimental study An epidemiological study designed to test a hypothesis about an outbreak of disease, often about the value of a particular treatment

experimental variable The factor that is purposely changed in an experiment

exponential rate (also called logarithmic rate) The rate of growth in a bacterial culture characterized by doubling of the population in a fixed interval of time

extracellular enzyme *See* **exoenzyme**

extrachromosomal resistance Drug resistance of a microorganism due to the presence of resistance (R) plasmids

extreme halophile Archaea that grow in highly saline environments such as the Great Salt Lake, the Dead Sea, salt evaporation ponds, and the surfaces of salt-preserved foods

extreme thermoacidophile Organism requiring very hot and acidic environment; usually belonging to the domain Archaea.

Fab fragment That portion of an antibody that contains an antigen-binding site

facilitated diffusion Diffusion (down a concentration gradient) across a membrane (from an area of higher concentration to lower concentration) with the assistance of a carrier molecule, but not requiring ATP

facultative Able to tolerate the presence or absence of a particular environmental condition

facultative anaerobe A bacterium that carries on aerobic metabolism when oxygen is present but shifts to anaerobic metabolism when oxygen is absent

facultative parasite A parasite that can live either on a host or freely

facultative psychrophile An organism that grows best at temperatures below 20°C but can also grow at temperatures above 20°C

facultative thermophile An organism that can grow both above and below 37°C

FAD Flavin adenine dinucleotide, a coenzyme that carries hydrogen atoms and electrons

fastidious Referring to microorganisms that have special nutritional needs that are difficult to meet in the laboratory

fat A complex organic molecule formed from glycerol and one or more fatty acids

fatty acid A long chain of carbon atoms and their associated hydrogens with a carboxyl group at one end

F⁻ cell Cell lacking the F plasmid; called recipient or female cell

F⁺ cell Cell having an F plasmid; called donor or male cell

Fc fragment The tail region of an antibody that may contain sites for macrophage and complement binding

feces Solid waste produced in the large intestine and stored in the rectum until eliminated from the body

feedback inhibition (also called end-product inhibition) Regulation of a metabolic pathway by the concentration of one of its intermediates or, typically, its end product, which inhibits an enzyme in the pathway

feline panleukopenia virus (FPV) A parvovirus that causes severe disease in cats

female reproductive system The host system consisting of the ovaries, uterine tubes, uterus, vagina, and external genitalia

fermentation Anaerobic metabolism of the pyruvic acid produced in glycolysis

fever A body temperature that is abnormally high

fibroblast A new connective tissue cell that replaces fibrin as a blood clot dissolves, forming granulation tissue

fifth disease (also called erythema infectiosum) A normal disease in children caused by the *Erythrovirus* called B19; characterized by a bright red rash on the cheeks and a low-grade fever

filariasis Disease of the blood and lymph caused by any of several different roundworms carried by mosquitos

filovirus A filamentous virus that displays unusual variability in shape. Two filoviruses, the Ebola virus and the Marburg virus, have been associated with human disease

filter paper disk method Method of evaluating the antimicrobial properties of a chemical agent using filter paper disks placed on an inoculated agar plate

filtration (1) A method of estimating the size of bacterial populations in which a known volume of air or water is drawn through a filter with pores too small to allow passage of bacteria. (2) A method of sterilization that uses a membrane filter to separate bacteria from growth media. (3) The filtering of water through beds of sand to remove most of the remaining microorganisms after flocculation in water treatment plants

fimbria *See* **attachment pilus**

fine adjustment Focusing mechanism of a microscope that very slowly changes the distance between the objective lens and the specimen

five-kingdom system System of classifying organisms into one of five kingdoms: Monera (Prokaryotae), Protista, Fungi, Plantae, and Animalia

flagellar staining A technique for observing flagella by coating the surfaces of flagella with a dye or a metal such as silver

flagellum (plural: *flagella*) A long, thin, helical appendage of certain cells that provides a means of locomotion

flash pasteurization *See* **high-temperature short-time pasteurization**

flat sour spoilage Spoilage due to the growth of spores that does not cause cans to bulge with gas

flatworm (also called Platyhelminthes) A primitive, unsegmented, hermaphroditic often parasitic worm

flavivirus A small, enveloped, (+) sense RNA virus that causes a variety of encephalitides, including yellow fever

flavoprotein An electron carrier in oxidative phosphorylation

flocculation The addition of alum to cause precipitation of suspended colloids, such as clay, in the water purification process

fluctuation test A test to determine that resistance to chemical substances occurs spontaneously rather than being induced

fluid-mosaic model Current model of membrane structure in which proteins are dispersed in a phospholipid bilayer

fluke A flatworm with a complex life cycle; can be an internal or external parasite

fluorescence Emission of light of one color when irradiated with another, shorter wavelength of light

fluorescence-activated cell sorter (FACS) A machine that collects quantities of a particular cell type under sterile conditions for study

fluorescence microscopy Use of ultraviolet light in a microscope to excite molecules so that they release light of different colors

fluorescent antibody staining Procedure in fluorescence microscopy that uses a fluorochrome attached to antibodies to detect the presence of an antigen

fluoride Chemical that helps in reducing tooth decay by poisoning bacterial enzymes and hardening the surface enamel of teeth

focal infection An infection confined to a specific area from which pathogens can spread to other areas

folliculitis (also called pimple or pustule) Local infection produced when hair follicles are invaded by pathogenic bacteria

fomite A nonliving substance capable of transmitting disease, such as clothing, dishes, or paper money

food poisoning (also called enterotoxicosis) A gastrointestinal disease caused by ingestion of foods contaminated with preformed toxins or other toxic substances

formed elements Cells and cell fragments comprising about 40% of the blood

F pilus A bridge formed from an F1 cell to an F2 cell for conjugation

F plasmid Fertility plasmid containing genes directing synthesis of proteins that form an F pilus (sex pilus, or conjugation pilus)

F′ plasmid An F plasmid that has been imprecisely separated from the bacterial chromosome so that it carries a fragment of the bacterial chromosome

frameshift mutation Mutation resulting from the deletion or insertion of one or more bases

freeze-etching Technique in which water is evaporated under vacuum from the freeze-fractured surface of a specimen before the observation with electron microscopy

freeze-fracturing Technique in which a cell is first frozen and then broken with a knife so that the fracture reveals structures inside the cell when observed by electron microscopy

fulminating *See* **acme**

functional group Part of a molecule that generally participates in chemical reactions as a unit and gives the molecule some of its chemical properties

Fungi (singular: *fungus*) The kingdom of nonphotosynthetic, eukaryotic organisms that absorb nutrients from their environment

Fungi Imperfecti (also called Deuteromycota) Group of fungi termed "imperfect" because no sexual stage has been observed in their life cycles

furuncle (also called a boil) A large, deep, pus-filled infection

gamete A male or female reproductive cell

gametocyte A male or female sex cell

gamma globulin *See* **immune serum globulin**

ganglion An aggregation of neuron cell bodies

gas gangrene A deep wound infection, destructive of tissue, often caused by a combination of two or more species of *Clostridium*

gene A linear sequence of DNA nucleotides that form a functional unit within a chromosome or plasmid

gene amplification A technique of genetic engineering in which plasmids or bacteriophages carrying a specific gene are induced to reproduce at a rapid rate within host cells

generalized anaphylaxis *See* **anaphylactic shock**

generalized transduction Type of transduction in which a fragment of DNA from the degraded chromosome of an infected bacteria cell is accidentally incorporated into a new phage particle during viral replication and thereby transferred to another bacterial cell

generation time Time required for a population of organisms to double in number

genetic code The one-to-one relationship between each codon and a specific amino acid

genetic engineering The use of various techniques to purposefully manipulate genetic material to alter the characteristics of an organism in a desired way

genetic fusion A technique of genetic engineering that allows transposition of genes from one location on a chromosome to another location; the coupling of genes from two different operons

genetic homology The similarity of DNA base sequences among organisms

genetic immunity Inborn or innate immunity

genetics The science of heredity, including the structure and regulation of genes and how these genes are passed between generations

gene transfer Movement of genetic information between organisms by transformation, transduction, or conjugation

genital herpes *See* **herpes simplex virus type 2**

genital wart (also called condyloma) An often malignant wart associated with sexually transmitted viral disease having a very high association rate with cervical cancer

genome The genetic information in an organism or virus

genotype The genetic information contained in the DNA of an organism

genus A taxon consisting of one or more species; the first name of an organism in the binomial system of nomenclature; for example, *Escherichia* in *Escherichia coli*

German measles *See* **rubella**

germination The start of the process of development of a spore or an endospore

germ theory of disease Theory that microorganisms (germs) can invade other organisms and cause disease

giardiasis A gastrointestinal disorder caused by the flagellated protozoan *Giardia intestinalis*

gingivitis The mildest form of periodontal disease, characterized by inflammation of the gums

gingivostomatitis Lesions of the mucous membranes of the mouth

glomerulonephritis (also called Bright's disease) Inflammation of and damage to the glomeruli of the kidneys

glomerulus A coiled cluster of capillaries in the nephron

glycocalyx Term used to refer to all substances containing polysaccharides found external to the cell wall

glycolysis An anaerobic metabolic pathway used to break down glucose into pyruvic acid while producing some ATP

glycoprotein A long, spikelike molecule made of carbohydrate and protein that projects beyond the surface of a cell or viral envelope; some viral glycoproteins attach the virus to receptor sites on host cells, while others aid fusion of viral and cellular membranes

glycosidic bond A covalent bond between two monosaccharides

Golgi apparatus An organelle in eukaryotic cells that receives, modifies, and transports substances coming from the endoplasmic reticulum

gonorrhea A sexually transmitted disease caused by *Neisseria gonorrhoeae*

graft tissue Tissue that is tranplanted from one site to another

graft-versus-host (GVH) disease Disease in which host antigens elicit an immunological response from graft cells that destroys host tissue

gram molecular weight *See* **mole**

Gram stain A differential stain that uses crystal violet, iodine, alcohol, and safranin to differentiate bacteria. Gram-positive bacteria stain dark purple; Gram-negative ones stain pink/red

granulation tissue Fragile, reddish, grainy tissue made up of capillaries and fibroblasts that appears with the healing of an injury

granule An inclusion that is not bounded by a membrane and contains compacted substances that do not dissolve in the cytoplasm

granulocyte A leukocyte (basophil, mast cell, eosinophil, neutrophil) with granular cytoplasm and irregularly shaped, lobed nuclei

granuloma In a chronic inflammation, a collection of epithelial cells, macrophages, lymphocytes, and collagen fibers

granuloma inguinale (also called donovanosis) A sexually transmitted dis-ease caused by *Calymmatobacterium granulomatis*

granulomatous hypersensitivity Cell-mediated hypersensitivity reaction that occurs when macrophages have engulfed pathogens but have failed to kill them

granulomatous inflammation A special kind of chronic inflammation characterized by the presence of granulomas

granzyme A cytotoxin produced by cytotoxic T cells that help kill infected host cells

griseofulvin An antifungal agent that interferes with fungal growth

ground itch Bacterial infection of sites of penetration by hookworms

group translocation An active transport process in bacteria that chemically modifies substance so it cannot diffuse out of the cell

growth curve The different growth periods of a bacterial or phage population

gumma A granulomatous inflammation, symptomatic of syphilis, that destroys tissue

gut-associated lymphatic tissue (GALT) Collective name for the tissues of lymphoid nodules, especially those in the digestive, respiratory, and urogenital tracts; main site of antibody production

halobacteria One of the groups of the archaeobacteria that live in very concentrated salt environments

halophile A salt-loving organism that requires moderate to large concentrations of salt

hanging drop A special type of wet mount often used with dark-field illumination to study motility of organisms

Hansen's disease The preferred name for leprosy; caused by *Mycobacterium leprae*, it exhibits various clinical forms ranging from tuberculoid to lepromatous

hantavirus pulmonary syndrome (HPS) The "Sin Nombre" hantavirus responsible for severe respiratory illness

haploid A eukaryotic cell that contains a single, unpaired set of chromosomes

hapten A small molecule that can act as an antigenic determinant when combined with a larger molecule

heat fixation Technique in which air-dried smears are passed through an open flame so that organisms are killed, adhere better to the slide, and take up dye more easily

heavy chain (H chain) Larger of the two identical pairs of chains comprising immunoglobulin molecules

helminth A worm, with bilateral symmetry; includes the roundworms and flatworms

helper T cell (T_H) Lymphocytes that stimulate other immune cells such as B cells and macrophages

hemagglutination Agglutination (clumping) of red blood cells; used in blood typing

hemagglutination inhibition test Serologic test used to diagnose measles, influenza, and other viral diseases, based on the ability of antibodies to viruses to prevent viral hemagglutination

hemoglobin The oxygen-binding compound found in erythrocytes

hemolysin An enzyme that lyses red blood cells

hemolysis The lysis of red blood cells

hemolytic disease of the newborn (also called erythroblastosis fetalis) Disease in which a baby is born with enlarged liver and spleen caused by efforts of these organs to destroy red blood cells damaged by maternal antibodies; mother is Rh-negative and baby is Rh-positive

hemorrhagic uremic syndrome (HUS) Infection with O157:H7 strain of *Escherichia coli* causing kidney damage and bleeding in the urinary tract

hepadnavirus A small, enveloped DNA virus with circular DNA; one such virus causes hepatitis B

hepatitis An inflammation of the liver, usually caused by viruses but sometimes by an amoeba or various toxic chemicals

hepatitis A (formerly called infectious hepatitis) Common form of viral hepatitis caused by a single-stranded RNA virus transmitted by the fecal-oral route

hepatitis B (formerly called serum hepatitis) Type of hepatitis caused by a double-stranded DNA virus usually transmitted in blood or semen

hepatitis C (formerly called non-A, non-B hepatitis) Type of hepatitis distinguished by a high level of the liver enzyme alanine transferase; usually mild or inapparent infection but can be severe in compromised individuals

hepatitis D (also called delta hepatitis) Severe type of hepatitis caused by presence of both hepatitis D and hepatitis B viruses; hepatitis D virus is an incomplete virus and cannot replicate without presence of hepatitis B virus as a helper

hepatitis E Type of hepatitis transmitted through fecally contaminated water supplies

hepatovirus One of three major groups of picornaviruses that can infect nerve and is responsible for causing hepatitis A

herd immunity (also called group immunity) The proportion of individuals in a population who are immune to a particular disease

heredity The transmission of genetic traits from an organism to its progeny

hermaphroditic Having both male and female reproductive systems in one organism

herpes gladiatorium Herpesvirus infection that occurs in skin injuries of wrestlers; transmitted by contact or on mats

herpes labialis Fever blisters (cold sores) on lips

herpes meningoencephalitis A serious disease caused by herpesvirus that can cause permanent neurological damage or death and that sometimes follows a generalized herpes infection or ascends from the trigeminal ganglion

herpes pneumonia A rare form of herpes infection seen in burn patients, AIDS patients, and alcoholics

herpes simplex virus type 1 (HSV-1) A virus that most frequently causes fever blisters (cold sores) and other lesions of the oral cavity, and less often causes genital lesions

herpes simplex virus type 2 (HSV-2; sometimes called herpes hominis virus) A virus that typically causes genital herpes, but which can also cause oral lesions

herpesvirus A relatively large, enveloped DNA virus that can remain latent in host cells for long periods of time

heterogeneity The ability of the immune system to produce many different kinds of antibodies, each specific for a different antigenic determinant

heterotroph An organism that uses compounds to produce biomolecules

heterotrophy "Other-feeding," the use of carbon atoms from organic compounds for the synthesis of biomolecules

Hib vaccine Vaccine against *Haemophilus influenzae* b

high-energy bond A chemical bond that releases energy when hydrolyzed; the energy can be used to transfer the hydrolyzed product to another compound

high frequency of recombination (Hfr) strain A strain of F$^+$ bacteria in which the F plasmid is incorporated into the bacterial chromosome

high-temperature short-time (HTST) pasteurization (also called flash pasteurization) Process in which milk is heated to 71.6°C for at least 15 seconds

histamine Amine released by basophils and tissues in allergic reactions

histocompatibility antigen An antigen found in the membranes of all human cells that is unique in all individuals except identical twins

histone A protein that contributes directly to the structure of eukaryotic chromosomes

histoplasmosis (also called Darling's disease) Fungal respiratory disease endemic to the central and eastern United States, caused by the soil fungus *Histoplasma capsulatum*

holding method *See* **low-temperature long-time pasteurization**

holoenzyme A functional enzyme consisting of an apoenzyme and a coenzyme or cofactor

homolactic acid fermentation A pathway in which pyruvic acid is directly converted to lactic acid using electrons from reduced NAD (NADH)

hookworm A disease caused by two species of small roundworms, *Ancylostoma duodenale* and *Necator americanus*, whose larvae burrow through skin and feet, enter the blood vessels, and penetrate lung and intestinal tissue

horizontal transmission Direct contact transmission of disease in which pathogens are usually passed by handshaking, kissing, contact with sores, or sexual contact

host Any organism that harbors another organism

host range The different types of organisms that a microbe can infect

host specificity The range of different hosts in which a parasite can mature

HPV vaccine Vaccine against cervical cancer, 99% of which are caused by the human papilloma virus (HPV)

human immunodeficiency virus (HIV) One of the retroviruses that is responsible for AIDS

human leukocyte antigen (HLA) A lymphocyte antigen used in laboratory tests to determine compatibility of donor and recipient tissues for transplants

human papillomavirus (HPV) Virus that attacks skin and mucous membranes, causing papillomas or warts

humoral immune response A response to foreign antigens carried out by antibodies circulating in the blood

humoral immunity The immune response most effective in defending the body against bacteria, bacterial toxins, and viruses that have not entered cells

humus The nonliving organic components of soil

hyaluronidase (also called spreading factor) A bacterially produced enzyme that digests hyaluronic acid, which helps hold the cells of certain tissues together, thereby making tissues more accessible to microbes

hybridoma A hybrid cell resulting from the fusion of a cancer cell with another cell, usually an antibody-producing white blood cell

hydatid cyst An enlarged cyst containing many tapeworm heads

hydrogen bond A relatively weak attraction between a hydrogen atom carrying a partial positive charge and an oxygen or nitrogen atom carrying a partial negative charge

hydrologic cycle *See* **water cycle**

hydrolysis A chemical reaction that produces simpler products from more complex organic molecules

hydrophilic Water-loving

hydrophobic Water-repelling

hydrostatic pressure Pressure exerted by standing water

hydrothermal vents "Black smokers" venting clouds of sulfur compounds, plus very high temperatures, up to 350°C, through tall chimneys at certain places along the bottom of the oceans

hyperimmune serum (also called convalescent serum) A preparation of immune serum globulins having high titers of specific kinds of antibodies

hyperparasitism The phenomenon of a parasite itself having parasites

hypersensitivity (also called allergy) Disorder in which the immune system reacts inappropriately, usually by responding to an antigen it normally ignores

hypertonic solution A solution containing a concentration of dissolved material greater than that within a cell

hypha (plural: *hyphae*) A long, threadlike structure of cells in fungi or actinomycetes

hypothesis A tentative explanation for an observed condition or event

hypotonic solution A solution containing a concentration of dissolved material lower than that within a cell

IgA Class of antibody found in the blood and secretions

IgD Class of antibody found on the surface of B cells and rarely secreted

IgE Class of antibody that binds to receptors on basophils in the blood or mast cells in the tissues; responsible for allergic or immediate (Type I) hypersensitivity reactions

IgG The main class of antibodies found in the blood; produced in largest quantities during secondary response

IgM The first class of antibody secreted into the blood during the early stages of a primary immune response (a rosette of five immunoglobulin molecules) or found on the surface of B cells (a single immunoglobulin molecule)

illness phase In an infectious disease, the period during which the individual experiences the typical signs and symptoms of the disease

imidazole An antifungal agent that disrupts fungal plasma membranes

immediate (Type I) hypersensitivity (also called anaphylactic hypersensitivity) Response to a foreign substance (allergen) resulting from prior exposure to the allergen

immersion oil Substance used to avoid refraction at a glass-air interface when examining objects through a microscope

immune complex An antigen-antibody complex that is normally eliminated by phagocytic cells

immune complex disorder (also called immune complex [Type III] hypersensitivity) A disorder caused by antigen-antibody complexes that precipitate in the blood and injure tissues; elicited by antigens in vaccines, on microorganisms, or on a person's own cells

immune complex (Type III) hypersensitivity An exaggerated or inappropriate reaction by the immune system to a foreign substance, elicited by antigens in vaccines, on microorganisms, or on a person's own cells

immune cytolysis Process in which the membrane attack complex of complement produces lesions on cell membranes through which the contents of the bacterial cells leak out

immune serum globulin (also called gamma globulin) A pooled sample of antibody-containing fractions of serum from many individuals

immune system Body system that provides the host organism with specific immunity to infectious agents

immunity The ability of an organism to defend itself against infectious agents

immunization Stimulation of the immune system to recognize and destroy infectious agents whenever they are encountered

immunocompromised Referring to an individual whose immune defenses are weakened due to fighting another infectious disease, or because of an immunodeficiency disease or an immunosuppressive agent

immunodeficiency Inborn or acquired defects in lymphocytes (B or T cells)

immunodeficiency disease A disease of impaired immunity caused by lack of lymphocytes, defective lymphocytes, or destructive lymphocytes

immunodiffusion test A serologic test similar to the precipitin test but carried out in agar gel medium

immunoelectrophoresis Serologic test in which antigens are first separated by gel electrophoresis and then allowed to react with antibody placed in a trough in the gel

immunofluorescence Referring to the use of antibodies to which a fluorescent substance is bound and used to detect antigens, other antibodies, or complement within tissues

immunogen *See* **antigen**

immunogenic Something that is a potent stimulator of antibody production and defense cell activity

immunoglobulin (Ig) (also called antibody) The class of protective proteins produced by the immune system in response to a particular epitope

immunological disorder Disorder that results from an inappropriate or inadequate immune system

immunological memory The ability of the immune system to recognize substances it has previously encountered

immunology The study of specific immunity and how the immune system responds to specific infectious agents

immunosuppression Minimizing of immune reactions using radiation or cytotoxic drugs

impetigo A highly contagious pyoderma caused by staphylococci, streptococci, or both

inapparent infection (also called subclinical infection) An infection that fails to produce symptoms, either because too few organisms are present or because host defenses effectively combat the pathogens

incidence rate The number of new cases of a particular disease per 100,000 population seen in a specific period of time

inclusion A granule or vesicle found in the cytoplasm of a bacterial cell

inclusion blennorrhea A mild chlamydial infection of the eyes in infants

inclusion body (1) An aggregation of reticulate bodies within chlamydias. (2) A form of cytopathic effect consisting of viral components, masses of viruses, or remnants of viruses

inclusion conjunctivitis A chlamydial infection that can result from self-inoculation with *Chlamydia trachomatis*

incubation period In the stages of an infectious disease, the time between infection and the appearance of signs and symptoms

index case The first case of a disease to be identified

index of refraction A measure of the amount that light rays bend when passing from one medium to another

indicator organism An organism such as *Escherichia coli* whose presence indicates the contamination of water by fecal matter

indigenous organism (also called native organism) An organism native to a given environment

indirect contact transmission Transmission of disease through fomites

indirect fecal-oral transmission Transmission of disease in which pathogens from feces of one organism infect another organism

induced mutation A mutation produced by agents called mutagens that increase the mutation rate

inducer A substance that binds to and inactivates a repressor protein

inducible enzyme An enzyme coded for by a gene that is sometimes active and sometimes inactive

induction The stimulation of a temperate phage (prophage) to excise itself from the host chromosome and initiate a lytic cycle of replication

induration A raised, hard, red region on the skin resulting from tuberculin hypersensitivity

industrial microbiology Branch of microbiology concerned with the use of microorganisms to assist in the manufacture of useful products or disposal of waste products

infant botulism (also called "floppy baby" syndrome) Form of botulism in infants associated with ingestion of honey

infection The multiplication of a parasite organism, usually microscopic, within or upon the host's body

infectious disease Disease caused by infectious agents (bacteria, viruses, fungi, protozoa, and helminths)

infectious hepatitis *See* **hepatitis A**

infectious mononucleosis An acute disease that affects many systems, caused by the Epstein-Barr virus

infestation The presence of helminths (worms) or arthropods in or on a living host

inflammation The body's defensive response to tissue damage caused by microbial infection

influenza Viral respiratory infection caused by orthomyxoviruses that appears as epidemics

initiating segment That part of the F plasmid that is transferred to the recipient cell in conjugation with an Hfr bacterium

innate defenses Nonspecific host defenses that act against any type of invading agent. They include physical barriers, chemical barriers, cellular defenses, inflammation, fever, and molecular defenses

innate immunity Immunity to infection that exists in an organism because of genetically determined characteristics

insect An arthropod with three body regions, three pairs of legs, and highly specialized mouthparts

insertion The addition of one or more bases to DNA, usually producing a frameshift mutation

interferon A small protein often released from virus-infected cells that binds to adjacent uninfected cells, causing them to produce antiviral proteins that interfere with viral replication

interleukin A cytokine produced by leukocytes

intermediate host An organism that harbors a sexually immature stage of a parasite

intestinal anthrax Infection by *Bacillus anthracis* that appears in the intestine. If bacteria enter the bloodstream, causing septicemia, this leads to meningitis

intoxication The ingestion of a microbial toxin that leads to a disease

intron (also called intervening region) Region of a gene (or mRNA) in eukaryotic cells that does not code for a protein

invasiveness The ability of a microorganism to take up residence in a host

invasive stage (or phase) Disease spreads into body from site of entry causing symptoms to appear

ion An electrically charged atom produced when an atom gains or loses one or more electrons

ionic bond A chemical bond between atoms resulting from attraction of ions with opposite charges

iris diaphragm Adjustable device in a microscope that controls the amount of light passing through the specimen

ischemia Reduced blood flow to tissues with oxygen and nutrient deficiency and waste accumulation

isograft A graft of tissue between genetically identical individuals

isolation Situation in which a patient with a communicable disease is prevented from contact with the general population

isomer An alternative form of a molecule having the same molecular formula but different structure

isoniazid An antimetabolite that is bacteriostatic against the tuberculosis-causing mycobacterium

isotonic Fluid containing the same concentration of dissolved materials as is in a cell; causes no change in cell volume

isotope An atom of a particular element that contains a different number of neutrons

kala azar Visceral leishmaniasis caused by *Leishmania donovani*

Kaposi's sarcoma A malignancy often found in AIDS patients in which blood vessels grow into tangled masses that are filled with blood and easily ruptured

karyogamy Process by which nuclei fuse to produce a diploid cell

keratin A waterproofing protein found in epidermal cells

keratitis An inflammation of the cornea

keratoconjunctivitis Condition in which vesicles appear in the cornea and eyelids

kidney One of a pair of organs responsible for the formation of urine

Kirby-Bauer method *See* **disk diffusion method**

Koch's Postulates Four postulates formulated by Robert Koch in the nineteenth century; used to prove that a particular organism causes a particular disease

Koplik's spots Red spots with central bluish specks that appear on the upper lip mucosa in early stages of measles

Krebs cycle (also called tricarboxylic acid cycle and the citric acid cycle) A sequence of enzyme-catalyzed chemical reactions that metabolizes 2-carbon units called acetyl groups to CO_2 and H_2O

Kupffer cells Phagocytic cells that remove foreign matter from the blood as it passes through sinusoids

kuru Transmissible spongiform encephalopathy disease of the human brain, caused by prions, associated with cannibalism and tissue/organ transplants

lacrimal gland Tear-producing gland of the eye

lactobacilli Type of regular, nonsporing, Gram-positive rods found in many foods; used in production of cheeses, yogurt, sourdough, and other fermented foods

lagging strand The new strand of DNA formed in short, discontinuous DNA segments during DNA replication

lag phase First of four major phases of the bacterial growth curve, in which organisms grow in size but do not increase in number

large intestine The lower area of the intestine that absorbs water and converts undigested food into feces

laryngeal papilloma Benign growth caused by herpesvirus that can be dangerous if such papillomas block the airway; infants are often infected during birth by mothers having genital warts

laryngitis An infection of the larynx, often with loss of voice

larynx The voicebox

Lassa fever Hemorrhagic fever, caused by arenaviruses, that begins with pharyngeal lesions and proceeds to severe liver damage

latency The ability of a virus to remain in host cells for long periods of time while retaining the ability to replicate

latent disease A disease characterized by periods of inactivity either before symptoms appear or between attacks

latent period Period of a bacteriophage growth curve that spans the time from penetration through biosynthesis

latent viral infection An infection typical of herpesviruses in which an infection in childhood that is brought under control later in life is reactivated

lateral gene transfer Genes pass from one organism to another within the same generation

leading strand The new strand of DNA formed as a continuous strand during DNA replication

leavening agent An agent, such as yeast, that produces gas to make dough rise

Legionnaires' disease Disease caused by *Legionella pneumophila*, transmitted by airborne bacteria

legionellas The causative bacterial agent in Legionnaires' disease, *Legionella pneumophila*

leishmaniasis A parasitic systemic disease caused by three species of protozoa of the genus *Leishmania* and transmitted by sandflies

leproma An enlarged, disfiguring skin lesion that occurs in the lepromatous form of Hansen's disease

lepromatous Referring to the nodular form of Hansen's disease (leprosy) in which a granulomatous response causes enlarged, disfiguring skin lesions called lepromas

lepromin skin test Test used to detect Hansen's disease (leprosy); similar to the tuberculin test

leprosy *See* **Hansen's disease**

leptospirosis A zoonosis caused by the spirochete *Leptospira interrogans*, which enters the body through mucous membranes or skin abrasions

leukocidin An exotoxin produced by many bacteria, including the streptococci and staphylococci, that kills phagocytes

leukocyte A white blood cell

leukocyte-endogenous mediator (LEM) A substance that helps raise the body temperature while decreasing iron absorption (increasing iron storage)

leukocytosis An increase in the number of white blood cells (leukocytes) circulating in the blood

leukostatin An exotoxin that interferes with the ability of leukocytes to engulf microorganisms that release the toxin

leukotriene A reaction mediator released from mast cells after degranulation that causes prolonged airway constriction, dilation, and increased permeability of capillaries, increased thick mucous secretion, and stimulation of nerve endings that cause pain and itching

L-forms Irregularly shaped naturally occurring bacteria with defective cell walls

ligase An enzyme that joins together DNA segments

light chain (L chain) Smaller of the two identical pairs of chains constituting immunoglobulin molecules

light-dependent (light) reactions The part of photosynthesis in which light energy is used to excite electrons from chlorophyll, which are then used to generate ATP and NADPH

light independent (dark) reactions (also called carbon fixation) Part of photosynthesis in which carbon dioxide gas is reduced by electrons from reduced NADP (NADPH) to form various carbohydrate molecules, chiefly glucose

light microscopy The use of any type of microscope that uses visible light to make specimens observable

light repair (also called photoreactivation) Repair of DNA dimers by a light-activated enzyme

lipid One of a group of complex, water-insoluble compounds

lipid A Toxic substance found in the cell wall of a Gram-negative bacteria

lipopolysaccharide (also called endotoxin) Part of the outer layer of the cell wall in Gram-negative bacteria

listeriosis A type of meningitis caused by *Listeria monocytogenes* that is especially threatening to those with impaired immune systems

loaiasis Tropical eye disease caused by the filarial worm *Loa loa*

lobar pneumonia Type of pneumonia that affects one or more of the five major lobes of the lungs

local infection An infection confined to a specific area of the body

localized anaphylaxis An immediate (Type I) hypersensitivity restricted to only some tissues/organs resulting in, e.g., reddening of the skin, watery eyes, hives, etc.

locus The location of a gene on a chromosome

logarithmic rate *See* **exponential rate**

log phase Second of four major phases of the bacterial growth curve, in which cells divide at an exponential or logarithmic rate

lophotrichous Having two or more flagella at one or both ends of a bacterial cell

lower respiratory tract Thin-walled bronchioles and alveoli where gas exchange occurs

low-temperature long-time (LTLT) pasteurization (also called holding method) Procedure in which milk is heated to 62.9°C for at least 30 minutes

luminescence Process in which absorbed light rays are reemitted at longer wavelengths

Lyme disease Disease caused by *Borrelia burgdorferi*, carried by the deer tick

lymph The excess fluid and plasma proteins lost through capillary walls that is found in the lymphatic capillaries

lymphangitis Symptom of septicemia in which red streaks due to inflamed lymphatics appear beneath the skin

lymphatic system Body system, closely associated with the cardiovascular system, that transports lymph in lymphatic vessels through body tissues and organs; performs important functions in host defenses and specific immunity

lymphatic vessel Vessel that returns lymph to the blood circulatory system

lymphilization The drying of a material from the frozen state; freeze-drying

lymph node An encapsulated globular structure located along the routes of the lymphatic vessels that helps clear the lymph of microorganisms

lymphocyte A leukocyte (white blood cell) found in large numbers in lymphoid tissues that contribute to specific immunity

lymphogranuloma venereum A sexually transmitted disease, caused by *Chlamydia trachomatis*, that attacks the lymphatic system

lymphoid nodule A small, unencapsulated aggregation of lymphatic tissue that develops in many tissues, especially the digestive, respiratory, and urogenital tracts, collectively called gut-associated lymphatic tissue (GALT); they are the body's main sites of antibody production

lymphoid stem cell A cell in the bone marrow from which lymphocytes develop

lymphokine A cytokine secreted by T cells when they encounter an antigen

lyophilization Freeze drying, a means of preservation of cultures

lysis The destruction of a cell by the rupture of a cell or plasma membrane, resulting in the loss of cytoplasm

lysogen The combination of a bacterium and a temperate phage

lysogenic Pertaining to a bacterial cell in the state of lysogeny

lysogenic conversion The ability of a prophage to prevent additional infections of the same cell by the same type of phage; also the conversion of a non-toxin-producing bacterium into a toxin-producing one by a temperate phage

lysogeny The ability of temperate bacteriophages to persist in a bacterium by the integration of the viral DNA into the host chromosome and without the replication of new viruses or cell lysis

lysosome A small membrane-bound organelle in animal cells that contains digestive enzymes

lytic cycle The sequence of events in which a bacteriophage infects a bacterial cell, replicates, and eventually causes lysis of the cell

lytic phage *See* **virulent phage**

macrolide A large-ring compound, such as erythromycin, that is antibacterial by affecting protein synthesis

macrophage Ravenously phagocytic leukocytes found in tissues

mad cow disease Transmissible spongiform encephalopathy disease of the brain of cattle, caused by prions

Madura foot (also called maduromycosis) Tropical disease caused by a variety of soil organisms (fungi and actinomycetes) that often enter the skin through bare feet

maduromycosis *See* **Madura foot**

magnetosome Membranous vesicles where magnetite (Fe_3O_4) synthesized by magnetotactic bacteria is stored. They are nearly constant in size and are oriented in parallel chains like a string of tiny magnets

major histocompatibility complex (MHC) A group of cell surface proteins that are essential to immune recognition reactions

malaria A severe parasitic disease caused by several species of the protozoan *Plasmodium* and transmitted by mosquitos

male reproductive system The host system consisting of the testes, ducts, specific glands, and the penis

malignant Relating to a tumor that is cancerous

Malta fever *See* **brucellosis**

malted Referring to cereal grains that are partially germinated to increase the concentration of starch-digesting enzymes

mammary gland A modified sweat gland that produces milk and ducts that carry milk to the nipple

mash Malted grain that is crushed and mixed with hot water

mast cell A leukocyte that releases histamine during an allergic response

mastigophoran A flagellate protozoan such as *Giardia*

mastoid area Portion of the temporal bone prominent behind the ear opening

matrix Fluid-filled inner portion of a mitochondrion

maturation The process by which complete virions are assembled from newly synthesized components in the replication process

measles (also called rubeola) A febrile disease with rash caused by the rubeola virus, which invades lymphatic tissue and blood

measles encephalitis A serious complication of measles that leaves many survivors with permanent brain damage

mebendazole An antihelminthic agent that blocks glucose uptake by parasitic roundworms

mechanical stage Attachment to a microscope stage that holds the slide and allows precise control in moving the slide

mechanical vector A vector in which the parasite does not complete any part of its life cycle during transit

medium A mixture of nutritional substances on or in which microorganisms grow

megakaryocyte Large cell normally present in bone marrow that gives rise to platelets

meiosis Division process in eukaryotic cells that reduces the chromosome number in half

membrane attack complex (MAC) A set of proteins in the complement system that lyses invading bacteria by producing lesions in their cell membranes

membrane filter method Method of testing for coliform bacteria in water in which bacteria are filtered through a membrane and then incubated on the membrane surface in growth medium

memory cell Long-lived B or T lymphocyte that can carry out an anamnestic or secondary response

meninges Three layers of membrane that protect the brain and spinal cord

merozoite A malaria trophozoite found in infected red blood or liver cells

mesophile An organism that grows best at temperatures between 25° and 40°C, including most bacteria

mesophilic spoilage Spoilage due to improper canning procedures or because the seal has been broken

messenger RNA (mRNA) A type of RNA that carries the information from DNA to dictate the arrangement of amino acids in a protein

metabolic pathway A chain of chemical reactions in which the product of one reaction serves as the substrate for the next

metabolism The sum of all chemical processes carried out by living organisms

metacercaria The postcercarial encysted stage in the development of a fluke, prior to transfer to the final host

metachromasia Property of exhibiting a variety of colors when stained with a simple stain

metachromatic granule (also called volutin) A polyphosphate granule that exhibits metachromasia

metastasize Relating to the spread of malignant tumors to other body tissues

methanogens One of the groups of the Archaeobacteria that produce methane gas

metronidazole An antiprotozoan agent effective against *Trichomonas* infections

microaerophile A bacterium that grows best in the presence of a small amount of free oxygen

microbe *See* **microorganism**

microbial antagonism The ability of normal microbiota to compete with pathogenic organisms and in some instances to effectively combat their growth

microbial growth Increase in the number of cells, due to cell division

microbiology The study of microorganisms

micrococci Aerobes or facultative anaerobes that form irregular clusters by dividing in two or more planes

microenvironment A habitat in which the oxygen, nutrients, and light are stable, including the environment immediately surrounding the microbe

microfilament A protein fiber that makes up part of the cytoskeleton in eukaryotic cells

microfilaria An immature microscopic roundworm larva

micrometer (μm) Unit of measure equal to 0.000001 m or 10^{-6} m; formerly called a micron (μ)

microorganism (also called microbe) Organism studied with a microscope; includes the viruses

microscopy The technology for making very small things visible to the unaided eye

microtubule A protein tubule that forms the structure of cilia, flagella, and part of the cytoskeleton in eukaryotic cells

microvillus (plural: *microvilli*) A minute projection from the surface of an animal cell

miliary tuberculosis Type of tuberculosis that invades all tissues, producing tiny lesions

minimum bactericidal concentration (MBC) The lowest concentration of an antimicrobial agent that kills microorganisms, as indicated by absence of growth following subculturing in the dilution method

minimum inhibitory concentration (MIC) The lowest concentration of an antimicrobial agent that prevents growth in the dilution method of determining antibiotic sensitivity

miracidium Ciliated, free-swimming first-stage fluke larva that emerges from an egg

mitochondrion An organelle in eukaryotic cells that carries out oxidative reactions that capture energy

mitosis Process by which the cell nucleus in a eukaryotic cell divides to form identical daughter nuclei

mixed infection An infection caused by several species of organisms present at the same time

mixture Two or more substances combined in any proportion and not chemically bound

MMR vaccine Measles, mumps, and rubella vaccine

mole (also called gram molecular weight) The weight of a substance in grams equal to the sum of the atomic weights of the atoms in a molecule of the substance

molecular mimicry Imitation of the behavior of a normal molecule by an antimetabolite

molecule Two or more atoms chemically bonded together

molluscum contagiosum A viral infection characterized by flesh-colored, painless lesions

Monera (also called Prokaryotae) The kingdom of prokaryotic organisms that are unicellular and lack a true cell nucleus

moniliasis *See* **candidiasis**

monkeypox An orthopoxvirus, usually occurring in western and central Africa, especially in Zaire and the Congo, and sometimes mistaken for smallpox, as the lesions and death rates are very similar

monoclonal antibody A single, pure antibody produced in the laboratory by a clone of cultured hybridoma cells

monocular Refers to a light microscope having one eyepiece (ocular)

monocyte A ravenously phagocytic leukocyte, called a macrophage after it migrates into tissues

monolayer A suspension of cells that attach to plastic or glass surfaces as a sheet one cell layer thick

monosaccharide A simple carbohydrate, consisting of a carbon chain or ring with several alcohol groups and either an aldehyde or ketone group

monotrichous A bacterial cell with a single flagellum

morbidity rate The number of persons contracting a specific disease in relation to the total population (cases per 100,000)

mordant A chemical that helps a stain adhere to the cell or cell structure

mortality rate The number of deaths from a specific disease in relation to the total population

most probable number (MPN) A statistical method of measuring bacterial growth, used when samples contain too few organisms to give reliable measures by the plate-count method

mother cell (also called parent cell) A cell that has approximately doubled in size and is about to divide into two daughter cells

mucin A glycoprotein in mucus that coats bacteria and prevents their attaching to surfaces

mucociliary escalator Mechanism involving ciliated cells that allows materials in the bronchi, trapped in mucus, to be lifted to the pharynx and spit out or swallowed

mucous membrane (also called mucosa) A covering over those tissues and organs of the body cavity that are exposed to the exterior

mucus A thick but watery secretion of glycoproteins and electrolytes secreted by the mucous membranes

multiple-tube fermentation method Three-step method of testing for coliform bacteria in drinking water

mumps Disease caused by a paramyxovirus that is transmitted by saliva and invades cells of the oropharynx

murine typhus *See* **endemic typhus**

mutagen An agent that increases the rate of mutations

mutation A permanent alteration in an organism's DNA

mutualism A form of symbiosis in which two organisms of different species live in a relationship that benefits both of them

myasthenia gravis Autoimmune disease specific to skeletal muscle, especially muscles of the limbs and those involved in eye movements, speech, and swallowing

mycelium (plural: *mycelia*) In fungi, a mass of long, threadlike structures (hyphae) that branch and intertwine

mycobacteria Slender, acid-fast rods, often filamentous; include organisms that cause tuberculosis, leprosy, and chronic infections

mycology The study of fungi

mycoplasmas Very small bacteria with cell membranes, RNA and DNA, but no cell walls

mycosis (plural: *mycoses*) A disease caused by a fungus

myiasis An infestation caused by maggots (fly larvae)

myocarditis An inflammation of the heart muscle

NAD Nicotinamide dinucleotide, a coenzyme that carries hydrogen atoms and electrons

naked virus A virus that lacks an envelope

nanometer (nm) Unit of measure equal to 0.00000000.1 m or 10^{-9} m; formerly called a millimicron (mμ)

narrow spectrum The range of activity of an antimicrobial agent that attacks only a few kinds of microorganisms

nasal cavity Part of the upper respiratory tract where air is warmed and particles are removed by hairs as they pass through

nasal sinus A hollow cavity within the skull that is lined with mucous membrane

natural killer (NK) cell A lymphocyte that can destroy virus-infected cells, malignant tumor cells, and cells of transplanted tissues

naturally acquired active immunity When an individual is exposed to an infectious agent, often having the disease, and their own immune system responds in a protective way

naturally acquired adaptive immunity Defense against a specific disease is acquired sometime after birth, without the intervention or use of man-made products such as vaccines or gamma globulin

naturally acquired passive immunity When antibodies made by another individual are given to a host (e.g., in mother's milk), without intervention by man

negative (−) sense RNA An RNA strand made up of bases complementary to those of a positive (+) sense RNA

negative staining Technique of staining the background around a specimen, leaving the specimen clear and unstained

nematode *See* **roundworm**

neonatal herpes Infection in infants, usually with HSV-2, most often acquired during passage through a birth canal contaminated with the virus

neoplasm A localized tumor

neoplastic transformation The uncontrollable division of host cells caused by infection with a DNA tumor virus

nephron A functional unit of the kidney in which fluid from the blood is filtered

nerve A bundle of neuron fibers that relays sensory and motor signals throughout the body

nervous system The body system, comprising the brain, spinal cord, and nerves, that coordinates the body's activities in relation to the environment

neuron A conducting nerve cell

neurosyphilis Neurological damage, including thickening of the meninges, ataxia, paralysis, and insanity, that results from syphilis

neurotoxin A toxin that acts on nervous system tissues

neutral Referring to a solution with a pH of 7.0

neutralization Inactivation of microbes or their toxins through the formation of antigen-antibody complexes

neutralization reaction An immunological test used to detect bacterial toxins and antibodies to viruses

neutron An uncharged subatomic particle in the nucleus of an atom

neutrophil (also called polymorphonuclear leukocyte, PMNL) A phagocytic leukocyte

neutrophile An organism that grows best in an environment with a pH of 5.4 to 8.5

niclosamide An antihelminthic agent that interferes with carbohydrate metabolism

nitrification The process by which ammonia or ammonium ions are oxidized to nitrites or nitrates

nitrofuran An antibacterial drug that damages cellular respiratory systems

nitrogenase Enzyme in nitrogen-fixing bacteroids that catalyzes the reaction of nitrogen gas and hydrogen gas to form ammonia

nitrogen cycle Process by which nitrogen moves from the atmosphere through various organisms and back into the atmosphere

nitrogen fixation The reduction of atmospheric nitrogen gas to ammonia

nocardioforms Gram-positive, nonmotile, pleomorphic, aerobic bacteria, often filamentous and acid-fast; include some skin and respiratory pathogens

nocardiosis Respiratory disease characterized by tissue lesions and abscesses; caused by the filamentous bacterium *Nocardia asteroides*

nocturia Nighttime urination, often a result of urinary tract infections

Nomarski microscopy Differential interference contrast microscopy; utilizes differences in refractive index to visualize structures, producing a nearly three-dimensional image

noncommunicable infectious disease Disease caused by infectious agents but not spread from one host to another

noncompetitive inhibitor A molecule that attaches to an enzyme at an allosteric site (a site other than the active site), distorting the shape of the active site so that the enzyme can no longer function

noncyclic photoreduction The photosynthetic pathway in which excited electrons from chlorophyll are used to generate ATP and reduce NADP with the splitting of water molecules

nongonococcal urethritis (NGU) A gonorrhea-like sexually transmitted disease most often caused by *Chlamydia trachomatis* and mycoplasmas

nonindigenous organism An organism temporarily found in a given environment

noninfectious disease Disease caused by any factor other than infectious agents

nonself Antigens recognized as foreign by an organism

nonsense codon (also called terminator codon) A set of three bases in a gene (or mRNA) that does not code for an amino acid

nonspecific defenses Those host defenses against pathogens that operate regardless of the invading agent

nonspecific immunity Produced by general defenses, such as skin, lysozyme, and complement, that protect against many different kinds of organisms rather than a specific one or two

nonsynchronous growth Natural pattern of growth during the log phase in which every cell in a culture divides at some point during the generation time, but not simultaneously

normal microflora Microorganisms that live on or in the body but do not usually cause disease (also called normal flora)

nosocomial infection An infection acquired in a hospital or other medical facility

notifiable disease A disease that a physician is required to report to public health officials

nuclear envelope The double membrane surrounding the cell nucleus in a eukaryotic cell

nuclear pore An opening in the nuclear envelope that allows for the transport of materials between nucleus and cytoplasm

nuclear region (also called nucleoid) Central location of DNA, RNA, and some proteins in bacteria; not a true nucleus

nucleic acids Long polymers of nucleotides that encode genetic information and direct protein synthesis

nucleocapsid The nucleic acid and capsid of a virus

nucleoid *See* **nuclear region**

nucleolus (plural: *nucleoli*) Area in the nucleus of a eukaryotic cell that contains RNA and serves as the site for the assembly of ribosomes

nucleoplasm The semifluid portion of the cell nucleus in eukaryotic cells that is surrounded by the nuclear envelope

nucleotide An organic compound consisting of a nitrogenous base, a five-carbon sugar, and one or more phosphate groups

null cells Undifferentiated cells that cannot be identified as either B cells or T cells; include the natural killer (NK) cells

numerical aperature The widest cone of light that can enter a lens

numerical taxonomy Comparison of organisms based on quantitative assessment of a large number of characteristics

nutritional complexity The number of nutrients an organism must obtain to grow

nutritional factor One factor that influences both the kind of organisms found in an environment and their growth

objective lens Lens in a microscope closest to the specimen that creates an enlarged image of the object viewed

obligate Requiring a particular environmental condition

obligate aerobe A bacterium that must have free oxygen to grow

obligate anaerobe A bacterium that is killed by free oxygen

obligate intracellular parasite An organism or virus that can live or multiply only inside a living host cell

obligate parasite A parasite that must spend some or all of its life cycle in or on a host

obligate psychrophile An organism that cannot grow at temperatures above 20°C

obligate thermophile An organism that can grow only at temperatures above 37°C

ocular lens Lens in the microscope that further magnifies the image created by the objective lens

ocular micrometer A glass disk with an inscribed scale that is placed inside the eyepiece of a microscope; used to measure the actual size of an object being viewed

Okazaki fragment One of the short, discontinuous DNA segments formed on the lagging strand during DNA replication

onchocerciasis (also known as river blindness) An eye disease caused by the filarial larvae of the nematode *Onchocerca volvulus*, transmitted by blackflies; common in Africa and Central America

oncogene A cancer-causing gene

ONPG and MUG test Water purity test that relies on the ability of coliform bacteria to secrete enzymes that convert a substrate into a product that can be detected by a color change

Oomycota *See* **water mold**

operon A sequence of closely associated genes that includes both structural genes and regulatory sites that control transcription

opportunist A species of resident or transient microbiota that does not ordinarily cause disease but can do so under certain conditions

opsonin An antibody that promotes phagocytosis when bound to the surface of a microorganism

opsonization The process by which microorganisms are rendered more attractive to phagocytes by being coated with antibodies (opsonins) and C3b complement protein (also called immune adherence)

opthalmia neonatorum Pyrogenic infection of the eyes caused by *Neisseria gonorrhoeae* (also known as conjunctivitis of the newborn)

optical microscope *See* **compound light microscope**

optimum pH The pH at which microorganisms grow best

orbivirus Type of virus that causes Colorado tick fever

orchitis Inflammation of the testes; a symptom of mumps in postpubertal males

organelle An internal membrane-enclosed structure found in eukaryotic cells

organic chemistry The study of compounds that contain carbon

ornithosis Disease with pneumonia-like symptoms, caused by *Chlamydia psittaci* and acquired from birds (previously called psittacosis and parrot fever)

Oroya fever (also called Carrion's disease) One form of bartonellosis; an acute fatal fever with severe anemia

orthomyxovirus A medium-sized, enveloped RNA virus that varies in shape from spherical to filamentous and has an affinity for mucus

osmosis A special type of diffusion in which water molecules move from an area of higher concentration to one of lower concentration across a selectively permeable membrane

osmotic pressure The pressure required to prevent the net flow of water molecules by osmosis

otitis externa Infection of the external ear canal

otitis media Infection of the middle ear

outer membrane A bilayer membrane, forming part of the cell wall of Gram-negative bacteria

ovarian follicle An aggregation of cells in the ovary containing an ovum

ovary In the female, one of a pair of glands that produce ovarian follicles, which contain an ovum and hormone-secreting cells

oxidation The loss of electrons and hydrogen atoms

oxidative phosphorylation Process in which the energy of electrons is captured in high-energy bonds as phosphate groups combine with ADP to form ATP

pandemic An epidemic that has become worldwide

papilloma *See* **wart**

papovavirus A small, naked DNA virus that causes both benign and malignant warts in humans; some types cause cervical cancer

parainfluenza Viral disease characterized by nasal inflammation, pharyngitis, bronchitis, and sometimes pneumonia, mainly in children

parainfluenza virus Virus that initially attacks the mucous membranes of the nose and throat

paramyxovirus A medium-sized, enveloped RNA virus that has an affinity for mucus

parasite An organism that lives in or on, and at the expense of, another orgnaism, the host

parasitism A symbiotic relationship in which one organism, the parasite, benefits from the relationship, whereas the other organism, the host, is harmed by it

parasitology The study of parasites

parfocal For a microscope, remaining in approximate focus when minor focus adjustments are made

paroxysmal stage Stage of whooping cough in which mucus and masses of bacteria fill the airway, causing violent coughing

parvovirus A small, naked DNA virus

passive immunity Immunity created when ready-made antibodies are introduced into, rather than created by, an organism

passive immunization The process of inducing immunity by introducing ready-made antibodies into a host

***Pasteurella-Haemophilus* group** Very small Gram-negative bacilli and coccobacilli that lack flagella and are nutritionally fastidious

pasteurization Mild heating to destroy pathogens and other organisms that cause spoilage

pathogen Any organism capable of causing disease in its host

pathogenicity The capacity to produce disease

pediculosis Lice infestation, resulting in reddened areas at bites, dermatitis, and itch

pellicle (1) A thin layer of bacteria adhering to the air-water interface of a broth culture by their attachment pili. (2) A strengthened plasma membrane of a protozoan cell. (3) Film over the surface of a tooth at the beginning of plaque formation

pelvic inflammatory disease (PID) An infection of the pelvic cavity in females, caused by any of several organisms including *Neisseria gonorrhoeae* and *Chlamydia*

penetration The entry of the virus (or its nucleic acid) into the host cell in the replication process

penicillin An antibacterial agent that inhibits cell wall synthesis

penis Part of the male reproductive system used to deliver semen to the female reproductive tract during sexual intercourse

peptide bond A covalent bond joining the amino group of one amino acid and the carboxyl group of another amino acid

peptidoglycan (also called murein) A structural polymer in the bacterial cell wall that forms a supporting net

peptococci Anaerobes that form pairs, tetrads, or irregular clusters; they lack both catalase and the enzyme to ferment lactic acid

peptone A product of enzyme digestion of proteins that contains many small peptides; a common ingredient of a complex medium

perforin A cytotoxin produced by cytotoxic T cells that bores holes in the plasma membrane of infected host cells

pericarditis An inflammation of the protective membrane around the heart

periodontal disease A combination of gum inflammation, decay of cementum, and erosion of periodontal ligaments and bone that support teeth

periodontitis A chronic periodontal disease that affects the bone and tissue that support the teeth and gums

peripheral nervous system All nerves outside the central nervous system

periplasm Those substances (enzymes, transport proteins) located in the periplasmic space of Gram-negative bacteria or in the older cell wall of Gram-positive bacteria

periplasmic enzyme An exoenzyme produced by Gram-negative organisms, which acts in the periplasmic space

periplasmic space The space between the cell membrane and the outer membrane in Gram-negative bacteria that is filled with periplasm

peritrichous Having flagella distributed all over the surface of a bacterial cell

permanent parasite A parasite that remains in or on a host once it has invaded the host

permease An enzyme complex involved in active transport through the cell membrane

peroxisome An organelle filled with enzymes that in animal cells oxidate amino acids and in plant cells oxidize fats

persistent viral infection The continued production of viruses within the host over many months or years

pertussis *See* **whooping cough**

petechia (plural: *petechiae*) A pinpoint-size hemorrhage, most common in skin folds, that often occurs in rickettsial diseases

pH A means of expressing the hydrogen-ion concentration, and thus the acidity, of a solution

phage *See* **bacteriophage**

phage (bacteriophage) therapy The use of highly specific viruses that attack only the targeted bacteria and leave potentially beneficial bacteria that normally inhabit the human digestive tract and other locations alive

phage typing Use of bacteriophages to determine similarities or differences among different bacteria

phagocyte A cell that ingests and digests foreign particles

phagocytosis Ingestion of solids into cells by means of the formation of vacuoles

phagolysosome A structure resulting from the fusion of lysosomes and a phagosome

phagosome A vacuole that forms around a microbe within the phagocyte that engulfed it

pharmaceutical microbiology A special branch of industrial microbiology concerned with the manufacture of products used in treating or preventing disease

pharyngitis An infection of the pharynx, usually caused by a virus but sometimes bacterial in origin; a sore throat

pharynx The throat, a common passageway for the respiratory and digestive systems with tubes connecting to the middle ear

phase-contrast microscopy Use of a microscope having a condenser that accentuates small differences in the refractive index of various structures within the cell

phenol coefficient A numerical expression for the effectiveness of a disinfectant relative to that of phenol

phenotype The specific observable characteristics displayed by an organism

phlebovirus Bunyavirus that is carried by the sandfly *Phlebotomus papatsii*

phospholipid A lipid composed of glycerol, two fatty acids, and a polar head group; found in all membranes

phosphorescence Continued emission of light by an object when light rays no longer strike it

phosphorus cycle The cyclic movement of phosphorus between inorganic and organic forms

phosphorylation The addition of a phosphate group to a molecule, often from ATP; generally increasing the molecule's energy

phosphotransferase system A mechanism that uses energy from phosphoenolpyruvate to move sugar molecules into cells by active transport

photoautotroph An autotroph that obtains energy from light

photoheterotroph A heterotroph that obtains energy from light

photolysis Process in which light energy is used to split water molecules into protons, electrons, and oxygen molecule

photoreactivation *See* **light repair**

photosynthesis The capture of energy from light and use of this energy to manufacture carbohydrates from carbon dioxide

phototaxis A nonrandom movement of an organism toward or away from light

phylogenetic Pertaining to evolutionary relationships

physical factor Factor in the environment, such as temperature, moisture, pressure, or radiation, that influences the kinds of organisms found and their growth

picornavirus A small, naked RNA virus; different genera are responsible for polio, the common cold, and hepatitis

pilus (plural: *pili*) A tiny hollow projection used to attach bacteria to surfaces (attachment pilus) or for conjugation (conjugation pilus)

pimple *See* **folliculitis**

pinna Flaplike external structure of the ear

pinworm A small roundworm, *Enterobius vermicularis*, that causes gastrointestinal disease

placebo An unmedicated, usually harmless substance given to a recipient as a substitute for or to test the efficacy of a medication or treatment

Plantae The kingdom of organisms to which all plants belong

plaque A clean area in a bacterial lawn culture where viruses have lysed cells

plaque assay A viral assay used to determine viral yield by culturing viruses on a bacterial lawn and counting plaques

plaque-forming unit A plaque counted on a bacterial lawn that gives only an approximate number of phages present, because a given plaque may have been due to more than one phage

plasma Liquid portion of the blood, excluding the formed elements

plasma cell A large lymphocyte differentiated from a B cell that synthesizes and releases antibodies like those on the B cell surface

plasma membrane (also called cell membrane) A selectively permeable lipoprotein bilayer that forms the boundary between the cytoplasm of a eukaryotic cell and its environment

plasmid (also called extrachromosomal DNA) A small, circular, independently replicating piece of DNA in a cell that is not part of its chromosome and can be transferred to another cell

plasmodial slime mold Funguslike protist consisting of a multinucleate amoeboid mass, or plasmodium, that moves about slowly and phagocytizes dead matter

plasmodium A multinucleate mass of cytoplasm that forms one of the stages in the life cycle of a plasmodial slime mold

plasmogamy Sexual reproduction in fungi in which haploid gametes unite and their cytoplasm mingles

plasmolysis Shrinking of a cell, with separation of the cell membrane from the cell wall, resulting from loss of water in a hypertonic solution

platelet A short-lived fragment of large cells called megakaryocytes, important component of the blood-clotting mechanism

pleomorphism Phenomenon in which bacteria vary widely in form, even within a single culture under optimal conditions

pleura Serous membrane covering the surfaces of the lungs and the cavities they occupy

pleurisy Inflammation of pleural membranes that causes painful breathing; often accompanies lobar pneumonia

***Pneumocystis* pneumonia** A fungal respiratory disease caused by *Pneumocystis carinii*

pneumonia An inflammation of lung tissue caused by bacteria, viruses, or fungi

pneumonic plague Usually fatal form of plague transmitted by aerosol droplets from a coughing patient

point mutation Mutation in which one base is substituted for another at a specific location in a gene

polar compound A molecule with an unequal distribution of charge due to an unequal sharing of electrons between atoms

poliomyelitis Disease caused by any of several strains of polioviruses that attack motor neurons of the spinal cord and brain

polyacrylamide gel electrophoresis (PAGE) A technique for separating proteins from a cell based on their molecular size

polyene An antifungal agent that increases membrane permeability

polymer A long chain of repeating subunits

polymerase chain reaction (PCR) A technique that rapidly produces a billion or more identical copies of a DNA fragment without needing a cell

polymyxin An antibacterial agent that disrupts the cell membrane

polynucleotide A chain of many nucleotides

polypeptide A chain of many amino acids

polyribosome (also called polysome) A long chain of ribosomes attached at different points along an mRNA molecule

polysaccharide A carbohydrate formed when many monosaccharides are linked together by glycosidic bonds

Pontiac fever A mild variety of legionellosis

porin A protein in the outer membrane of Gram-negative bacteria that nonselectively transports polar molecules into the periplasmic space

portal of entry A site at which microorganisms can gain access to body tissues

portal of exit A site at which microorganisms can leave the body

positive chemotaxis Movement of an organism toward a chemical

positive (+) sense RNA An RNA strand that encodes information for making proteins needed by a virus

potable water   Water that is fit for human consumption

pour plate   A plate containing separate colonies and used to prepare a pure culture

pour plate method   Method used to prepare pure cultures using serial dilutions, each of which is mixed with melted agar and poured into a sterile Petri plate

poxvirus   DNA virus that is the largest and most complex of all viruses

precipitation reaction   Immunological test in which antibodies called precipitants react with antigens to form latticelike networks of molecules that precipitate from solution

precipitin test   Immunological test used to detect antibodies that is based on the precipitation reaction

prediction   The expected outcome if a hypothesis is correct

preserved culture   A culture in which organisms are maintained in a dormant state

presumptive test   First stage of testing in multiple-tube fermentation in which gas production in lactose broth provides presumptive evidence that coliform bacteria are present

prevalence rate   The number of people infected with a particular disease at any one time

primaquine   An antiprotozoan agent that interferes with a protein synthesis

primary atypical pneumonia   (also called mycoplasma pneumonia and walking pneumonia) A mild form of pneumonia with insidious onset

primary cell culture   A culture that comes directly from an animal and is not subcultured

primary immunodeficiency disease   A genetic or developmental defect in which T cells or B cells are lacking or nonfunctional

primary infection   An initial infection in a previously healthy person

primary response   Humoral immune response that occurs when an antigen is first recognized by host B cells

primary structure   The specific sequence of amino acids in a polypeptide chain

primary treatment   Physical treatment to remove solid wastes from sewage

prion   An exceedingly small infectious particle consisting of protein without any nucleic acid

privileged site   An area of the body that is isolated from the adaptive immune system, such as the uterus, the testes, and the anterior chamber of the eye

probe   A single-stranded DNA fragment that has a sequence of bases that can be used to identify complementary DNA base sequences

prodromal phase   In an infectious disease, the short period during which nonspecific symptoms such as malaise and headache sometimes appear

prodrome   A symptom indicating the onset of a disease

producer   (also called autotroph) Organism that captures energy from the sun and synthesizes food

product   The material resulting from an enzymatic reaction

productive infection   Viral infection in which viruses enter a cell and produce infectious progeny

proglottid   One of the segments of a tapeworm, containing the reproductive organs

progressive multifocal leukoencephalopathy   Disease caused by the JC polyomavirus with symptoms including mental deterioration, limb paralysis, and blindness

Prokaryotae   Alternate name for kingdom Monera, consisting of all prokaryotic organisms including the eubacteria, the cyanobacteria, and the archaeobacteria

prokaryote   Microorganism that lacks a cell nucleus and membrane-enclosed internal structures; all bacteria in the kingdom Monera (Prokaryotae) are prokaryotes

prokaryotic cell   A cell that lacks a cell nucleus; includes all bacteria

promiscuous   Plasmids that are self-transmissible (have genes for the formation of an F pilus) that transfer into species other than their own kind

propagated epidemic   An epidemic that arises from person-to-person contacts

prophage   The DNA of a lysogenic phage that has integrated into the host cell chromosome

propionibacteria   Pleomorphic, irregular, nonsporing, Gram-positive rods

prostaglandin   A reaction mediator that acts as a cellular regulator, often intensifying pain

prostate gland   Gland located at the beginning of the male urethra whose milky fluid discharge forms a component of semen

prostatitis   Inflammation of the prostate gland

protein   A polymer of amino acids joined by peptide bonds

protein profile   A technique for visualizing the proteins contained in a cell; obtained by the use of polyacrylamide gel electrophoresis

protist   A unicellular eukaryotic organism that is a member of the kingdom Protista

Protista   The kingdom of organisms that are unicellular but contain internal organelles typical of the eukaryotes

proton   A positively charged subatomic particle located in the nucleus of an atom

proto-oncogene   A normal gene that can cause cancer in uncontrolled situations; often the normal gene comes under the control of a virus

protoplast   A Gram-positive bacterium from which the cell wall has been removed

protoplast fusion   A technique of genetic engineering in which genetic material is combined by removing the cell walls of two different types of cells and allowing the resulting protoplasts to fuse

prototroph   A normal, nonmutant organism (also called wild type)

protozoa   (singular: *protozoan*) Single-celled, microscopic, animal-like protists in the kingdom Protista

provirus   Viral DNA that is incorporated into a host-cell chromosome

pseudocoelom   A primitive body cavity, typical of nematodes, that lacks the complete lining found in higher animals

pseudocyst   An aggregate of trypanosome protozoa that forms in lymph nodes in Chagas' disease

pseudomembrane   A combination of bacilli, damaged epithelial cells, fibrin, and blood cells resulting from infection with diphtheria that can block the airway, causing suffocation

pseudomonads   Aerobic motile rods with polar flagella

pseudoplasmodium   A multicellular mass composed of individual cellular slime mold cells that have aggregated

pseudopodium   A temporary footlike projection of cytoplasm associated with amoeboid movement

psittacosis   *See* **ornithosis**

psychrophile   A cold-loving organism that grows best at temperatures of 15° to 20°C

puerperal fever   (also called childbed fever or puerperal sepsis) Disease caused by β-hemolytic streptococci, which are normal vaginal and respiratory microbiota that can be introduced during child delivery by medical personnel

pure culture   A culture that contains only a single species of organism

purine   The nucleic acid bases adenine and guanine

pus   Fluid formed by the accumulation of dead phagocytes, the materials they have ingested, and tissue debris

pustule   *See* **folliculitis**

pyelonephritis   Inflammation of the kidneys

pyoderma   A pus-producing skin infection caused by staphylococci, streptococci, and corynebacteria, singly or in combination

pyrimidine   Any of the nucleic acid bases thymine, cytosine, and uracil

pyrimidine dimer   Two adjacent pyrimidines (two thymines, two cytosines, or thymine and cytosine) bonded together in a DNA strand, caused when ultraviolet rays strike DNA

pyrogen A substance that acts on the hypothalamus to set the body's "thermostat" to a higher-than-normal temperature

Q fever Pneumonia-like disease caused by *Coxiella burnetii*, a rickettsia that survives long periods outside cells and can be transmitted aerially as well as by ticks

quarantine The separation of humans or animals from the general population when they have a communicable disease or have been exposed to one

quaternary ammonium compound (quat) A cationic detergent that has four organic groups attached to a nitrogen atom

quaternary structure The three-dimensional structure of a protein molecule formed by the association of two or more polypeptide chains

quinine An antiprotozoan agent used to treat malaria

quinolone A bactericidal agent that inhibits DNA replication

quinone (also called coenzyme Q) A nonprotein, lipid-soluble electron carrier in oxidative phosphorylation

rabies A viral disease that affects the brain and nervous system with symptoms including hydrophobia and aerophobia; transmitted by animal bites

rabies virus An RNA-containing rhabdovirus that is transmitted through animal bites

rad A unit of radiation energy absorbed per gram of tissue

radial immunodiffusion Serological test used to provide a quantitative measure of antigen or antibody concentration by measuring the diameter of the ring of precipitation around an antigen

radiation Light rays, such as X-rays and ultraviolet rays, that can act as mutagens

radioimmunoassay (RIA) Technique that uses a radioactive anti-antibody to detect very small quantities of antigens or antibodies

radioisotope Isotope with unstable nuclei that tends to emit subatomic particles and radiation

rat bite fever A disease caused by *Streptobacillus moniliformis* transmitted by bites from wild and laboratory rats

reactant Substance that takes part in a chemical (enzymatic) reaction

reagin Older name for immunoglobulin E (IgE); very important in allergies

recombinant DNA DNA combined from two different species by restriction enzymes and ligases

recombination The combining of DNA from two different cells, resulting in a recombinant cell

redia The development stage of the fluke immediately following the sporocyst stage

reduction The gain of electrons and hydrogen atoms

reference culture A preserved culture used to maintain an organism with its characteristics as originally defined

reflection The bouncing of light off an object

refraction The bending of light as it passes from one medium to another medium of different density

regulator gene Gene that controls the expression of structural genes of an operon through the synthesis of a repressor protein

regulatory site The promoter and operator regions of an operon

relapsing fever Disease caused by various species of *Borrelia*, most commonly by *B. recurrentis*; transmitted by lice

release The exit from the host cell of new virions, which usually kills the host cell

rennin An enzyme from calves' stomachs used in cheese manufacture

reovirus A medium-sized RNA virus that has a double capsid with no envelope; causes upper respiratory and gastrointestinal infections in humans

replica plating A technique used to transfer colonies from one medium to another

replication Process by which an organism or structure (especially a DNA molecule) duplicates itself

replication curve A description of viral growth (biosynthesis and maturation) based on observations of phage-infected bacteria in laboratory cultures

replication cycle The series of steps of virus replication in a host cell

replication fork A site at which the two strands of the DNA double helix separate during replication and new complementary DNA strands form

repressor In an operon it is the protein that binds to the operator, thereby preventing transcription of adjacent genes

repressor protein Substance produced by host cells that keeps a virus in an inactive state and prevents the infection of the cell by another phage

reservoir host An infected organism that makes parasites available for transmission to other hosts

reservoir of infection Site where microorganisms can persist and maintain their ability to infect

resident microflora Species of microorganisms that are always present on or in an organism

resistance The ability of a microorganism to remain unharmed by an antimicrobial agent

resistance (R) gene A component of a resistance plasmid that confers resistance to a specific antibiotic or to a toxic metal

resistance (R) plasmid (also called R factor) A plasmid that carries genes that provide resistance to various antibiotics or toxic metals

resistance transfer factor (RTF) A component of a resistance plasmid that implements transfer by conjugation of the plasmid

resolution The ability of an optical device to show two items as separate and discrete entities rather than a fuzzily overlapped image

resolving power A numerical measure of the resolution of an optical instrument

respiratory anaphylaxis Life-threatening allergy in which airways become constricted and filled with mucous secretions

respiratory anthrax Also known as "woolsorters' disease," an infection by *Bacillus anthracis* that affects the respiratory tract. It is almost always fatal

respiratory bronchiole Microscopic channel in the lower respiratory system that ends in a series of alveoli

respiratory syncytial virus (RSV) Cause of lower respiratory infections affecting children under 1 year old; causes cells in culture to fuse their plasma membranes and become multinucleate masses (syncytia)

respiratory system Body system that moves oxygen from the atmosphere to the blood and removes carbon dioxide and other wastes from the blood

restriction endonuclease An enzyme that cuts DNA at precise base sequences

restriction enzyme Another term for restriction endonuclease

restriction fragment length polymorphism (RFLP) A short piece of DNA snipped out by restriction enzymes

reticulate body An intracellular stage in the life cycle of chlamydias

retrovirus An enveloped RNA virus that uses its own reverse transcriptase to transcribe its RNA into DNA in the cytoplasm of the host cell

reverse transcription An enzyme found in retroviruses that copies RNA into DNA

R factor *See* **resistance (R) plasmid**

R group An organic chemical group attached to the central carbon atom in an amino acid

rhabdovirus A rod-shaped, enveloped RNA virus that infects insects, fish, various other animals, and some plants

Rh antigen   An antigen found on some red blood cells; discovered in the cells of Rhesus monkeys

rheumatic fever   A multisystem disorder following infection by β-hemolytic *Streptococcus pyogenes* that can cause heart damage

rheumatoid arthritis   Autoimmune disease that affects mainly the joints but can extend to other tissues

rheumatoid factor   IgM found in the blood of patients with rheumatoid arthritis and their relatives

rhinovirus   A virus that replicates in cells of the upper respiratory tract and causes the common cold

ribonucleic acid (RNA)   Nucleic acid that carries information from DNA to sites where proteins are manufactured in cells and that directs and participates in the assembly of proteins

ribosomal RNA (rRNA)   A type of RNA that, together with specific proteins, makes up the ribosomes

ribosome   Site for protein synthesis consisting of RNA and protein, located in the cytoplasm

Rickettsiae   Small, nonmotile, Gram-negative organisms; obligate intracellular parasites of mammalian and arthropod cells

rickettsialpox   Mild rickettsial disease with symptoms resembling those of chickenpox; caused by *Rickettsia akari* and carried by mites found on house mice

rifamycin   An antibacterial agent that inhibits ribonucleic acid (RNA) synthesis

Rift Valley fever   Disease caused by bunyaviruses that occurs in epidemics

ringworm   A highly contagious fungal skin disease that can cause ringlike lesions

river blindness   *See* **onchocerciasis**

RNA polymerase   An enzyme that binds to one strand of exposed DNA during transcription and catalyzes the synthesis of RNA from the DNA template

RNA primer   Molecule to which a polymerase can attach, to begin DNA replication

RNA tumor virus   Any retrovirus that causes tumors and cancer

Rocky Mountain spotted fever   Disease caused by *Rickettsia rickettsia* and transmitted by ticks

roseola   Disease of infants and small children caused by the human herpesvirus 6 (HHV-6). An older name is exanthem subitum

rotavirus   Virus transmitted by the fecal-oral route that replicates in the intestine, causing diarrhea and enteritis

roundworm   (also called nematode) A worm with a long, cylindrical, unsegmented body and a heavy cuticle

rubella   (also called German measles) Viral disease characterized by a skin rash; can cause severe congenital damage

rubeola   *See* **measles**

rule of octets   Principle that an element is chemically stable if it contains eight electrons in its outer shell

sac fungus   (also called Ascomycota) A member of a diverse group of fungi that produces saclike asci during sexual reproduction

St. Anthony's fire   *See* **erysipelas**

St. Louis encephalitis   Type of viral encephalitis most often seen in humans in the central United States

salmonellosis   A common enteritis characterized by abdominal pain, fever, and diarrhea with blood and mucus; caused by *Salmonella* species

sapremia   A condition caused when saprophytes release metabolic products into the blood

saprophyte   An organism that feeds on dead or decaying organic matter

sarcina   A group of eight cocci in a cubicle packet

sarcoptic mange   *See* **scabies**

SARS (severe acute respiratory syndrome)   Caused by a coronavirus (SARSCoV), a relative of the common cold virus, and not a type of influenza; transmitted by mammals, including civet cats

satellite nucleic acids   (also known as virusoids) Small, single-stranded RNA molecules that lack genes required for their replication. They require a helper virus (or "satellite") to replicate

satellite virus   Small, single-stranded RNA molecules, usually 500 to 2,000 nucleotides in length, which lack genes required for their replication. They require a helper ("satellite") to replicate

saturated fatty acid   A fatty acid containing only carbon-hydrogen single bonds

scabies   (also called sarcoptic mange) Highly contagious skin disease caused by the itch mite *Sarcoptes scabiei*

scalded skin syndrome   Infection caused by staphylococci consisting of large, soft vesicles over the whole body

scanning electron microscope (SEM)   Type of electron microscope used to study the surfaces of specimen

scanning tunneling microscope (STM)   Also called scanning probe microscope; type of microscope in which electrons tunnel into each other's clouds, can show individual molecules, live specimens, and work underwater

scarlet fever   (sometimes called scarlatina) Infection caused by *Streptococcus pyogenes* that produces an erythrogenic toxin

Schaeffer-Fulton spore staining   A differential stain used to make endospores easier to visualize

Schick test   Test to determine immunity to diphtheria

schistosomiasis   (also called bilharzia) Disease of the blood and lymph caused by blood fluke of the genus *Schistosoma*

schizogony   Multiple fission, in which one cell gives rise to many cells

scolex   Head end of a tapeworm, with suckers and sometimes hooks that attach to the intestinal wall

scrapie   Transmissible spongiform encephalopathy disease of the brain of sheep, causing extreme itching so that the sheep repeatedly scrape themselves against posts, trees, etc.

scrub typhus   (also called tsutsugamushi disease) A typhus caused by *Rickettsia tsutsugamushi*; transmitted by mites that feed on rats

sebaceous gland   Epidermal structure, associated with hair follicles, that secretes an oily substance called sebum

sebum   Oily substance secreted by the sebaceous glands

secondary immunodeficiency disease   Result of damage to T cells or B cells after they have developed normally

secondary infection   Infection that follows a primary infection, especially in patients weakened by the primary infection

secondary response   The folding or coiling of a polypeptide chain into a particular pattern, such as a helix or pleated sheet

secondary structure   The folding or coiling of a polypeptide chain into a particular pattern, such as a helix or pleated sheet

secondary treatment   Treatment of sewage by biological means to remove remaining solid wastes after primary treatment

secretory piece   A part of the IgA antibody that protects the immunoglobulin from degradation and helps in the secretion of the antibody

secretory vesicle   Small membrane-enclosed structure that stores substances coming from the Golgi apparatus

selectively permeable   Able to prevent the passage of certain specific molecules and ions while allowing others through

selective medium   A medium that encourages growth of some organisms and suppresses growth of others

selective toxicity   The ability of an antimicrobial agent to harm microbes without causing significant damage to the host

self   Molecules that are not recognized as antigenic or foreign by an organism

semen   The male fluid discharge at the time of ejaculation, containing sperm and various glandular and other secretions

semiconservative replication Replication in which a new DNA double helix is synthesized from one strand of parent DNA and one strand of new DNA

seminal vesicle A saclike structure whose secretions form a component of semen

semisynthetic drug An antimicrobial agent made partly by laboratory synthesis and partly by microorganisms

sense codon A set of three DNA (or mRNA) bases that code for an amino acid

sensitization Initial exposure to an antigen, which causes the host to mount an immune response against it

septicemia (also called blood poisoning) An infection caused by rapid multiplication of pathogens in the blood

septicemic plague Fatal form of plague that occurs when bubonic plague bacteria move from the lymphatics to the circulatory system

septic shock A life-threatening septicemia with low blood pressure and blood-vessel collapse, caused by endotoxins

septic tank An underground tank for receiving sewage, where solid materials settle out as sludge, which must be pumped periodically

septum (plural: *septa*) A cross-wall separating two fungal cells

sequela (plural: *sequelae*) The aftereffect of a disease; after recovery from it

serial dilution A method of measurement in which sucessive 1:10 dilutions are made from the original sample

seroconversion The identification of a specific antibody in serum as a result of an infection

serology The branch of immunology dealing with laboratory tests to detect antigens and antibodies

serovar Strain; a subspecies category

serum The liquid part of blood after cells and clotting factors have been removed

serum hepatitis *See* **hepatitis B**

serum killing power Test used to determine effectiveness of an antimicrobial agent in which a bacterial suspension is added to the serum of a patient who is receiving an antibiotic and incubated

serum sickness Immune complex disorder that occurs when foreign antigens in sera cause immune complexes to be deposited in tissues

severe combined immunodeficiency (SCID) Primary immunodeficiency disease caused by failure of stem cells to develop properly, resulting in deficiency of both B and T cells

sewage Used water and the wastes it contains

sexually transmitted disease (STD) An infectious disease spread by sexual activities

shadow casting The coating of electron microscopy specimens with a heavy metal, such as gold or palladium, to create a three-dimensional effect

shigellosis (also called bacillary dysentery) Gastrointestinal disease caused by several strains of *Shigella* that invade intestinal lining cells

shinbone fever *See* **trench fever**

shingles Sporadic disease caused by reactivation of varicella-zoster herpesvirus that appears most frequently in older and immunocompromised individuals

shrub of life A diagram that represents our current understanding of the early evolution of life. There are many roots rather than a single ancestral line, and the branches criss-cross and merge again and again

sign A disease characteristic that can be observed by examining the patient, such as swelling or redness

simple diffusion The net movement of particles from a region of higher to one of lower concentration; does not require energy from a cell

simple stain A single dye used to reveal basic cell shapes and arrangements

single-cell protein (SCP) Animal feed consisting of microorganisms

sinus A large passageway in tissues, lined with phagocytic cells

sinusitis An infection of the sinus cavities

sinusoid An enlarged capillary

skin The largest single organ of the body that presents a physical barrier to infection by microorganisms

slime layer A thin protective structure loosely bound to the cell wall that protects the cell against drying, helps trap nutrients, and sometimes binds cells together

slime mold A funguslike protist

sludge Solid matter remaining from water treatment that contains aerobic organisms that digest organic matter

sludge digester Large fermentation tank in which sludge is digested by anaerobic bacteria into simple organic molecules, carbon dioxide, and methane gas

small intestine The upper area of the intestine where digestion is completed

smallpox A formerly worldwide and serious viral disease that has now been eradicated

smear A thin layer of liquid specimen spread out on a microscopic slide

snottite Mucus-like strings of bacterial colonies growing on the walls of caves that were created by microbial-produced sulfuric acid that dissolves rock. These bacteria eat sulfur and drip sulfuric acid

solute The substance dissolved in a solvent to form a solution

solution A mixture of two or more substances in which the molecules are evenly distributed and will not separate out on standing

solvent The medium in which substances are dissolved to form a solution

sonication The disruption of cells by sound waves

specialized transduction Type of transduction in which the bacterial DNA transduced is limited to one or a few genes lying adjacent to a prophage that are accidentally included when the prophage is excised from the bacterial chromosome

species A group of organisms with many common characteristics; the narrowest taxon

species immunity Innate or inborn genetic immunity

specific defense A host defense that operates in response to a particular invading pathogen

specific epithet The second name of an organism in the binomial system of nomenclature, following that of the genus—for example, *coli* in *Escherichia coli*

specific immunity Defense against a particular microbe

specificity (1) The property of an enzyme that allows it to accept only certain substrates and catalyze only one particular reaction. (2) The property of a virus that restricts it to certain specific types of host cells. (3) The ability of the immune system to mount a unique immune response to each antigen it encounters.

spectrum of activity Refers to the range of different microbes against which an antimicrobial agent is effective

spheroplast A Gram-negative bacterium that lacks the cell wall but has not lysed

spike A glycoprotein projection that extends from the viral capsid or envelope and is used to attach to or fuse with host cells

spindle apparatus A system of microtubules in the cytoplasm of a eukaryotic cell that guides the movement of chromosomes during mitosis and meiosis

spirillar fever A form of rat bite fever, caused by *Spirillum minor*, first described as *sodoku in* Japan

spirillum (plural: *spirilla*) A flexible, wavy-shaped bacterium

spirochete Corkscrew-shaped motile bacterium

spleen The largest lymphatic organ; acts as a blood filter

spontaneous generation The theory that living organisms can arise from nonliving things

spontaneous mutation A mutation that occurs in the absence of any agent known to cause changes in DNA; usually caused by errors during DNA replication

sporadic disease A disease that is limited to a small number of isolated cases posing no great threat to a large population

spore A resistant reproductive structure formed by fungi and actinomycetes; different from a bacterial endospore

spore coat A keratinlike protein material that is laid down around the cortex of an endospore by the mother cell

sporocyst Larval form of a fluke that develops in the body of its snail or mollusk host

sporotrichosis Fungal skin disease caused by *Sporothrix schenckii* that often enters the body from plants

sporozoite A malaria trophozoite present in the salivary glands of infected mosquitoes

sporulation The formation of spores, such as endospores

spread plate method A technique used to prepare pure cultures by placing a diluted sample of cells on the surface of an agar plate and then spreading the sample evenly over the surface

stain (also called dye) A molecule that can bind to a structure and give it color

standard bacterial growth curve A graph plotting the number of bacteria versus time and showing the phases of bacterial growth

staphylo- Prefix that indicates a group of bacteria cells arranged in grapelike clusters, created by random division planes

start codon The first codon in a molecule of mRNA which begins the sequence of amino acids in protein synthesis; in bacteria it always codes for methionine

stationary phase The third of four major phases of the bacterial growth curve in which new cells are produced at the same rate that old cells die, leaving the number of live cells constant

sterility The state in which there are no living organisms in or on a material

sterilization The killing or removal of all microorganisms in a material or on an object

steroid A lipid having a four-ring structure, includes cholesterol, steroid hormones, and vitamin D

stock culture A reserve culture used to store an isolated organism in pure condition for use in the laboratory

stop codon (also called terminator codon) The last codon to be translated in a molecule of mRNA, causing the ribosome to release from the mRNA

strain A subgroup of a species with one or more characteristics that distinguish it from other subgroups of that species

streak plate method Method used to prepare pure cultures in which bacteria are lightly spread over the surface of agar plates, resulting in isolated colonies

strepto- Prefix that indicates a group of bacteria cells arranged in chains, created by division in one plane

streptococci Aerotolerant anaerobes that form pairs, tetrads, or chains by dividing in one or two planes; most lack the enzyme catalase

streptokinase A bacterially produced enzyme that digests (dissolves) blood clots

streptolysin Toxin produced by streptococci that kills phagocytes

streptomycetes Gram-positive, filamentous, sporing, soil-dwelling bacteria; producer of many antibiotics

streptomycin An antibacterial agent that blocks protein synthesis

stroma The fluid-filled inner portion of a chloroplast

stromatolite Live or fossilized layered mats of photosynthetic prokaryotes associated with warm lagoons or hot springs

strongyloidiasis Parasitic disease caused by the roundworm *Stongyloides stercoralis* and a few closely related species

structural gene A gene that carries information for the synthesis of a specific polypeptide

structural protein A protein that contributes to the structure of cells, cell parts, and membranes

sty An infection at the base of an eyelash

subacute disease A disease that is intermediate between an acute and a chronic disease

subacute sclerosing panencephalitis (SSPE) A complication of measles, nearly always fatal, that is due to the persistence of measles viruses in brain tissue

subclinical infection *See* **inapparent infection**

subculturing The process by which cells from an existing culture are transferred to fresh medium in new containers

substrate (1) The substance on which an enzyme acts. (2) A surface or food source on which a cell can grow or a spore can germinate

sulfate reduction The reduction of sulfate ions to hydrogen sulfide

sulfonamide (also called sulfa drug) A synthetic, bacteriostatic agent that blocks the synthesis of folic acid

sulfur cycle The cyclic movement of sulfur through an ecosystem

sulfur oxidation The oxidation of various forms of sulfur to sulfate

sulfur reduction The reduction of elemental sulfur to hydrogen sulfide

superantigens Powerful antigens, such as bacterial toxins, that activate large numbers of T cells, causing a large immune response that can cause diseases such as toxic shock

superinfection A secondary infection from the removal of normal microbiota, allowing colonization by pathogenic, and often antibiotic-resistant, microbes

superoxide A highly reactive form of oxygen that kills obligate anaerobes

superoxide dismutase An enzyme that converts superoxide to molecular oxygen and hydrogen peroxide

suppressor T cell (T_s) Possibly a type of cytotoxic or helper T cells that inhibits immune responses

surface tension A phenomenon in which the surface of water behaves like a thin, invisible, elastic membrane

surfactant A substance that reduces surface tension

susceptibility The vulnerability of an organism to harm by infectious agents

swarmer cell Spherical, flagellated *Rhizobium* cell that invades the root hairs of leguminous plants, eventually to form nodules

sweat gland Epidermal structure that empties a watery secretion through pores in the skin

swimmer's itch Skin reaction to cercariae of some species of the helminth *Schistosoma*

symbiosis The living together of two different kinds of organisms

symptom A disease characteristic that can be observed or felt only by the patient, such as pain or nausea

synchronous growth Hypothetical pattern of growth during the log phase in which all the cells in a culture divide at the same time

syncytium (plural: *syncytia*) A multinucleate mass in a cell culture, for example, that caused by the repiratory syncytial virus

syndrome A combination of signs and symptoms that occur together

synergism Referring to an inhibitory effect produced by two antibiotics working together that is greater than either can achieve alone

synthesis The step of viral replication during which new nucleic acids and viral proteins are made

synthetic drug An antimicrobial agent synthesized chemically in the laboratory

synthetic medium A growth medium prepared in the laboratory from materials of precise or reasonably well-defined composition

syphilis A sexually transmitted disease, caused by the spirochete *Treponema pallidum*, characterized by a chancre at the site of entry and often eventual neurological damage

systemic blastomycosis Disease resulting from invasion by *Blastomyces dermatitides* of internal organs, especially the lungs

systemic infection (also called generalized infection) An infection that affects the entire body

systemic lupus erythematosus A widely disseminated, systemic autoimmune disease resulting from production of antibodies against DNA and other body components

tapeworm Flatworm that lives in the adult stage as a parasite in the small intestine of animals

tartar Calcium deposition on dental plaque forming a very rough, hard crust

taxon (plural: *taxa*) A category used in classification, such as species, genus, order, family

taxonomy The science of classification

T cell *See* **T lymphocyte**

T-dependent antigen Antigen requiring helper T cell (T$_H$2) activity to activate B cells

teichoic acid A polymer attached to peptidoglycan in Gram-positive cell walls

teleomorphic Sexual part of the life cycle of a fungus

temperature phage A bacteriophage that does not cause a virulent infection; rather its DNA is incorporated into the host cell chromosome, as a prophage, and replicated with the chromosome

template DNA used as a pattern for the synthesis of a new nucleotide polymer in replication or transcription

temporary parasite A parasite that feeds on and then leaves its host (such as a biting insect)

teratogen An agent that induces defects during embryonic development

teratogenesis The induction of defects during embryonic development

terminator *See* **stop codon**

terminator codon (also called nonsense codon or stop codon) A codon that signals the end of the information for a particular protein

tertiary structure The folding of a protein molecule into globular shapes

tertiary treatment Chemical and physical treatment of sewage to produce an effluent of water pure enough to drink

test A shell made of calcium carbonate and common to some protists

testis (plural: *testes*) One of a pair of male reproductive glands that produce testosterone and sperm

tetanus (also called lockjaw) Disease caused by *Clostridium tetani* in which

muscle stiffness progresses to eventual paralysis and death

tetanus neonatorum Type of tetanus acquired through the raw stump of the umbilical cord

tetracycline An antibacterial agent that inhibits protein synthesis

tetrad Cuboidal groups of four cocci

thallus The body of a fungus

theca A tightly affixed, secreted outer layer of dinoflagellates that often contains cellulose

T helper (T$_H$) cell Type of T cell which works together with B cells to produce antibodies

therapeutic dosage level Level of drug dosage that successfully eliminates a pathogenic organism if maintained over a period of time

thermal death point The temperature that kills all the bacteria in a 24-hour-old broth culture at neutral pH in 10 minutes

thermal death time The time required to kill all the bacteria in a particular culture at a specified temperature

thermoacidophile A member of one of the groups of the archaeobacteria that live in extremely hot, acidic environments

thermophile A heat-loving organism that grows best at temperatures from 50 to 60°C

thermophilic anaerobic spoilage Spoilage due to endospore germination and growth in which gas and acid are produced, making cans bulge

thrush Milky patches of inflammation on oral mucous membranes; a symptom of candidiasis, caused by *Candida albicans*

thylakoid An internal membrane of chloroplasts that contains chlorophyll

thymus gland Multilobed lymphatic organ located beneath the sternum that processes lymphocytes into T cells

tick paralysis A disease characterized by fever and paralysis due to anticoagulants and toxins secreted into a tick's bite via the ectoparasite's saliva

tincture An alcoholic solution

T-independent antigen Antigen not requiring helper T cell (T$_H$2) activity to activate B cells

tinea barbae Barber's itch; a type of ringworm that causes lesions in the beard

tinea capitis Scalp ringworm, a form of ringworm in which hyphae grow in hair follicles, often leaving circular patterns of baldness

tinea corporis Body ringworm, a form of ringworm that causes ringlike lesions with a central scaly area

tinea cruris (also called jock itch) Groin ringworm, a form of ringworm that occurs in skin folds in the pubic region

tinea pedis *See* **athlete's foot**

tinea unguium A form of ringworm that causes hardening and discoloration of fingernails and toenails

tissue culture Culture made from a single tissue, assuring a reasonably homogenous set of cultures in which to test the effects of a virus or to culture an organism

titer The quantity of a substance needed to produce a given reaction

T lymphocyte (also called T cell) Thymus-derived cell of the immune system and agent of cellular immune responses

togavirus A small, enveloped RNA virus that multiplies in many mammaliam and arthropod cells

tolerance A state in which antigens no longer elicit an immune response

toll-like receptors (TLRs) Molecules on phagocytes that recognize pathogens

tonsil Lymphoid tissue that contributes immune defenses in the form of B cells and T cells

tonsilitis A bacterial infection of the tonsils

TORCH series A group of blood tests used to identify teratogenic diseases in pregnant women and newborn infants

total magnification Obtained by multiplying the magnifying power of the objective lens by the magnifying power of the ocular lens

toxemia The presence and spread of exotoxins in the blood

toxic dosage level Amount of a drug necessary to cause host damage

toxic shock syndrome (TSS) Condition caused by infection with certain toxigenic strains of *Staphylococcus aureus*; often associated with the use of superabsorbent but abrasive tampons

toxin Any substance that is poisonous to other organisms

toxoid An exotoxin inactivated by chemical treatment but which retains its antigenicity and therefore can be used to immunize against the toxin

toxoplasmosis Disease caused by the protozoan *Toxoplasma gondii* that can cause congenital defects in newborns

trace element Minerals, such as copper, iron, zinc, and cobalt ions, that are required in minute amounts for growth

trachea The windpipe

trachoma Eye disease caused by *Chlamydia trachomatis* that can result in blindness

transcription The synthesis of RNA from a DNA template

transduction The transfer of genetic material from one bacterium to another by a bacteriophage

transfer RNA (tRNA) Type of RNA that transfers amino acids from the cytoplasm to the ribosomes for placement in a protein molecule

transformation A change in an organism's characteristics through the transfer of naked DNA

transfusion reaction Reaction that occurs when matching antigens and antibodies are present in the blood at the same time

transgenic State of permanently changing an organism's characteristics by integrating foreign DNA (genes) into the organism

transient microflora Microorganisms that may be present in or on an organism under certain conditions and for certain lengths of time at sites where resident microbiota are found

translation The synthesis of protein from information in mRNA

transmissible spongiform encephalopathies Prion-caused diseases resulting in brain tissue developing multiple holes such that it resembles a sponge, includes Creutzfeldt-Jakob disease, mad cow disease, kuru, scrapie, chronic wasting disease and others

transmission The passage of light through an object

transmission electron microscope (TEM) Type of electron microscope used to study internal structures of cells; very thin slices of specimens are used

transovarian transmission Passing of pathogen from one generation of ticks to the next as eggs leave the ovaries

transplantation The moving of tissue from one site to another

transplant rejection Destruction of grafted tissue or of a tranplanted organ by the host immune system

transposable element A mobile genetic sequence that can move from one plasmid to another plasmid or chromosome

transposal of virulence A laboratory technique in which a pathogen is passed from its normal host sequentially through many individual members of a new host species, resulting in a lessening or even total loss of its virulence in the original host

transposition The process whereby certain genetic sequences in bacteria or eukaryotes can move from one location to another

transposon A mobile genetic sequence that contains the genes for transposition as well as one or more other genes not related to transposition

traumatic herpes Type of herpes infection in which the virus enters traumatized skin in the area of a burn or other injury

traveler's diarrhea Gastrointestinal disorder generally caused by pathogenic strains of *Escherichia coli*

trench fever (also called shinbone fever) Rickettsial disease, caused by *Rochalimaea*

quintana, resembling epidemic typhus in that it is transmitted by lice and is prevalent during wars and under unsanitary conditions

treponemes Spirochetes belonging to the genus *Treponema*

triacylglycerol A molecule formed from three fatty acids bonded to glycerol

tricarboxylic acid cycle *See* **Krebs cycle**

trichinosis A disease caused by a small nematode, *Trichinella spiralis*, that enters the digestive tract as encysted larvae in poorly cooked meat, usually pork

trichocyst Tentaclelike structure on ciliates for catching prey (for attachment)

trichomoniasis A parasitic urogenital disease, transmitted primarily by sexual intercourse, that causes intense itching and a copious white discharge, especially in females

trichuriasis Parasitic disease caused by the whipworm, *Trichuris trichiura*, that damages intestinal mucosa and causes chronic bleeding

trickling filter system Procedure in which sewage is spread over a bed of rocks coated with aerobic organisms that decompose the organic matter in it

trophozoite Vegetative form of a protozoan such as *Plasmodium*

trypanosomiasis *See* **African sleeping sickness**

tube agglutination test Serologic test that measures antibody titers by comparing various dilutions of the patient's serum against known quantities of an antigen

tubercle A solidified lesion or chronic granuloma that forms in the lungs in patients with tuberculosis

tuberculin hypersensitivity Cell-mediated hypersensitivity reaction that occurs in sensitized individuals when they are exposed to tuberculin

tuberculin skin test An immunological test for tuberculosis in which a purified protein derivative from the *Mycobacterium tuberculosis* is injected subcutaneously, resulting in an induration if there was previous exposure to the bacterium

tuberculoid Referring to the anesthetic form of Hansen's disease (leprosy) in which areas of skin lose pigment and sensation

tuberculosis Disease caused mainly by *Mycobacterium tuberculosis*

tularemia Zoonosis caused by *Franciscella tularensis*, most often associated with cottontail rabbits

tumor An uncontrolled division of cells, often caused by viral infection

turbidity A cloudy appearance in a culture tube indicating the presence of organisms

tympanic membrane (also called the eardrum) Membrane separating the outer and middle ear

type strain Original reference strain of a bacterial species, descendants of a single isolation in pure culture

typhoidal tularemia Septicemia that resembles typhoid fever, caused by bacteremia from tularemia lesions

typhoid fever An epidemic enteric infection caused by *Salmonella typhi*; uncommon in areas with good sanitation

typhus fever Rickettsial disease that occurs in a variety of forms including epidemic, endemic (murine), and scrub typhus

tyrocidin An antibacterial agent that disrupts cell membranes

ulceroglandular Referrring to the form of tularemia caused by entry of *Franciscella tularensis* through the skin and characterized by ulcers on the skin and enlarged regional lymph nodes

ultra-high temperature (UHT) processing A method of sterilizing milk and dairy products by raising the temperature to 87.8°C for 3 seconds

uncoating Process in which protein coats of animal viruses that have entered cells are removed by proteolytic enzymes

undulant fever *See* **brucellosis**

Universal Precautions A set of guidelines established by the CDC to reduce the risks of disease transmission in hospital and medical laboratory settings

unsaturated fatty acid A fatty acid that contains at least one double bond between adjacent carbon atoms

upper respiratory tract The nasal cavity, pharynx, larynx, trachea, bronchi, and larger bronchioles

Ureaplasmas Bacteria with unusual cell walls, require sterols as a nutrient

ureter Tube that carries urine from the kidney to the urinary bladder

urethra Tube through which urine passes from the bladder to the outside during micturition (urination)

urethritis Inflammation of the urethra

urethrocystitis Common term used to describe urinary tract infections involving the urethra and the bladder

urinalysis The laboratory analysis of urine specimens

urinary bladder Storage area for urine

urinary system Body system that regulates the composition of body fluids and removes nitrogenous and other wastes from the body

urinary tract infection (UTI) A bacterial urogenital infection that causes urethritis or cystitis

urine Waste collected in the kidney tubules

urogenital system Body system that (1) regulates the composition of body fluids and removes certain wastes from the body and

(2) enables the body to participate in sexual reproduction

use-dilution test A method of evaluating the antimicrobial properties of a chemical agent using standard preparations of certain test bacteria

uterine tube (also called Fallopian tubes or oviducts) A tube that conveys ova from the ovaries to the uterus

uterus The pear-shaped organ in which a fertilized ovum implants and develops

vaccine A substance that contains an antigen to which the immune system responds. (*See also* **immunization**; specific types of vaccines)

vacuole A membrane-bound structure that stores materials such as food or gas in the cytoplasm or eukaryotic cells

vagina The female genital canal, extending from the cervix to the outside of the body

vaginitis Vaginal infection, often caused by opportunistic organisms that multiply when the normal vaginal microflora are disturbed by antibiotics or other factors

variable Anything that can change in an experiment

varicella-zoster virus (VZV) A herpesvirus that causes both chickenpox and shingles

vasodilation Dilation of the capillary and venule walls during an acute inflammation

vector (1) A self-replicating carrier of DNA; usually a plasmid, bacteriophage, or eukaryotic virus. (2) An organism that transmits a disease-causing organism from one host to another

vegetation A growth that forms on damaged heart valve surfaces in bacterial endocarditis; exposed collagen fibers elicit fibrin deposits, and transient bacteria attach to the fibrin

vegetative cell A cell that is actively metabolizing nutrients

vehicle A nonliving carrier of an infectious agent from its reservoir to a susceptible host

Venezuelan equine encephalitis Type of viral encephalitis seen in Florida, Texas, Mexico, and South America; infects horses more frequently than humans

verminous intoxication An allergic reaction to toxins in the metabolic wastes of liver flukes

verruga peruana One form of bartonellosis; a chronic nonfatal skin disease

vertical gene transfer Genes pass from parents to offspring

vertical transmission Direct contact transmission of disease in which pathogens are passed from parent to offspring in an egg or sperm, across the placenta, or while traversing the birth canal

vesicle A membrane-bound inclusion in cells

vibrio A comma-shaped bacterium

vibriosis An enteritis caused by *Vibrio parahaemolyticus*, acquired from eating contaminated fish and shellfish that have not been thoroughly cooked

villus (plural: *villi*) A multicellular projection from the surface of a mucous membrane, functioning in absorption

viral enteritis Gastrointestinal disease caused by rotaviruses, characterized by diarrhea

viral hemagglutination Hemagglutination caused by binding of viruses, such as those that cause measles and influenza, to red blood cells

viral meningitis Usually self-limiting and nonfatal form of meningitis

viral neutralization The binding of antibodies to viruses, which is used in an immunological test to determine if a patient's serum contains viruses

viral pneumonia Disease caused by viruses such as respiratory syncytial virus

viral specificity Refers to the specific types of cells within an organism that a virus can infect

viral yield *See* **burst size**

viremia An infection in which viruses are transported in the blood but do not multiply in transit

viridans group A group of streptococci that often infect the valves and lining of the heart and cause incomplete (alpha) hemolysis of red blood cells in laboratory cultures

virion A complete virus particle, including its envelope if it has one

viroid An infectious RNA particle, smaller than a virus and lacking a capsid, that causes various plant diseases

virulence The degree of intensity of the disease produced by a pathogen

virulence factor A structural or physiological characteristic that helps a pathogen cause infection and disease

virulent phage (also called lytic phage) A bacteriophage that enters the lytic cycle when it infects a bacterial cell, causing eventual lysis and death of the host cell

virus A submicroscopic, parasitic, acellular microorganism composed of a nucleic acid (DNA or RNA) core inside a protein coat

virusoid (also known as satellite nucleic acids) Small, single-stranded RNA molecules, usually 500 to 2,000 nucleotides in length, which lack genes required for their replication. They require a helper (satellite) virus to replicate

visceral larva migrans The migration of larvae of *Toxocara* species in human tissues, where they cause damage and allergic reactions

vitamin A substance required for growth that the organism cannot make

volutin (also called metachromatic granule) Polyphosphate granules

walking pneumonia *See* **primary atypical pneumonia**

wandering macrophages Phagocytic cells that circulate in the blood or move into tissues when microbes and other foreign material are present

wart (also called papilloma) A growth on the skin and mucous membranes caused by infection with human papillomaviruses

water cycle (also called the hydrologic cycle) Process by which water is recycled through precipitation, ingestion by organisms, respiration, and evaporation

water mold (also called Oomycota) A fungus-like protist that produces flagellated asexual spores (zoospores) and large, motile gametes

wavelength The distance between successive crests or troughs of a light wave

Western blotting A technique used to transfer and identify proteins

western equine encephalitis Type of viral encephalitis seen most often in the western United States; infects horses more frequently than humans

West Nile fever Emerging viral disease new to U.S., transmitted by mosquitoes, causing seizures and encephalitis, lethal to crows

wet mount Microscopy technique in which a drop of fluid containing organisms (often living) is placed on a slide

wetting agent A detergent solution often used with other chemical agents to penetrate fatty substances

whey The liquid portion (waste product) of milk resulting from bacterial enzyme addition

whipworm *Trichuris trichiura*, a worm that causes trichuriasis infestation of the intestine

whitlow A herpetic lesion on a finger that can result from exposure to oral, ocular, and probably genital herpes

whooping cough (also called pertussis) A highly contagious respiratory disease caused primarily by *Bordetella pertussis*

wort The liquid extract from mash

wound botulism Rare form of botulism that occurs in deep wounds when tissue damage impairs circulation and creates anaerobic conditions in which *Clostridium botulinum* can multiply

xenograft A graft between individuals of different species

yeast extract Substance from yeast containing vitamins, coenzymes, and nucleosides; used to enrich media

yellow fever Viral systemic disease found in tropical areas, carried by the mosquito *Aedes aegypti*

yersiniosis Severe enteritis caused by *Yersinia enterocolitica*

Ziehl-Neelsen acid-fast stain A differential stain for organisms that are not decolorized by acid in alcohol, such as the bacteria that cause Hansen's disease (leprosy) and tuberculosis

zone of inhibition A clear area that appears on agar in the disk diffusion method, indicating where the agent has inhibited growth of the organism

zoonosis (plural: *zoonoses*) A disease that can be transmitted from animals to humans

zygomycosis Disease in which certain fungi of the genera *Mucor* and *Rhizopus* invade lungs, the central nervous system, and tissues of the eye orbit

Zygomycota *See* **bread mold**

zygospore In bread molds, a thick-walled, resistant, spore-producing structure enclosing a zygote

zygote A cell formed by the union of gametes (egg and sperm)

Clinical Case Study Answers

CHAPTER 1 The cause (etiology) of AIDS was shown to be the human immunodeficiency virus (HIV). Many tests had to be done to rule out bacteria, fungi, etc. as the causative agent. Some early observations were that (at the beginning of the epidemic in the U.S.) gay men in the San Francisco, California area were the most frequently infected group.

CHAPTER 2 Lipids are not very soluble in water. They are covalently bonded and therefore lack the electrical charge that is needed to be water soluble. Stains used in the microbiology lab are dissolved in water, which has difficulty penetrating waxy lipids. Antibiotics also find it difficult to penetrate lipids. Leprosy and TB bacteria also live inside host cells which makes it difficult for antibiotics to reach them.

CHAPTER 3 He should make 3 slides of the bacterium cultured from the patient. To each slide he should add a drop of one of the 3 antibodies that react, each with a different one of the 3 probable organisms suspected. Using a fluorescent (UV) microscope he should examine each slide. If the fluorescent dye labeled antibodies attach to the bacteria on the slide, the bacteria will glow a fluorescent color and indicate a positive test for that organism. Failure to react with the bacteria will indicate a negative test.

CHAPTER 4 Cold sores are caused by herpes viruses. Viruses are acellular, and do not have cell membranes that could be disrupted by polymyxin.

CHAPTER 5 Enzymes are proteins that will begin to denature at temperatures above 40°C, thereby losing their activity. Death and/or irreversible brain damage can occur at about 109°F (43°C). Some children also will have convulsions with high fevers.

CHAPTER 6 Ammonia is a base that can neutralize the stomach acid that is around the bacterium.

CHAPTER 7 You could tell Cathy that not all strains of a particular virus are identical. Viruses do mutate. The strain of herpes virus that she has probably has had a mutation in its thymidine kinase which protects it from the action of acyclovir, unlike the strain with which Mary is infected.

CHAPTER 8 Conjugation and transfer of a resistance transfer factor (RTF). Look for the same plasmid in the different bacteria.

CHAPTER 9 Microorganisms usually cannot be identified by simply looking at them. A series of tests must be done. Some of these take time, especially when the organism must be cultured first. By the time the pathogen was identified as a fungus, it was too late to successfully treat the infection. Antibiotics that kill bacteria often do not kill fungi. Had they immediately begun treatment with an antifungal antibiotic, the infection would not have progressed as far, and she might have been saved.

CHAPTER 10 Yes, Koch's postulates can be fulfilled, though it can be more difficult since viruses must be grown in cells. This means other cells would always be present. Scientists can still isolate pure cultures of viruses in tissue culture cells.

CHAPTER 11 The most probable infectious agent is *Trichinella spiralis*. The worm will most likely be found in the muscles. It will look like a coiled worm in a cyst. It is a nematode or roundworm. George was infected from eating undercooked pork meat.

CHAPTER 12 Yes, washing hands with soap and water removes much of normal flora and oils from the surface of the skin. This provides a niche for other organisms to then live in and can also lead to infections of the clinician's hands. To avoid this, some hospitals and clinics are now using a disinfectant hand cream that will kill bacteria but not remove the natural oils of the skin.

CHAPTER 13 There are several issues here. It is too early in the disease process. It is not clear whether you are having problems with an allergy, viral infection, or bacterial infection. Antibiotics will not help with an allergy or viral infection. Using antibiotics when they are not needed can lead to an increase in the percentage of antibiotic resistance among the bacteria in your body as the sensitive ones die off. Even if you have a bacterial infection, the antibiotic your friend has may not be the right one for your infection. Also, the dosage and number of doses may not be right for the infection you have. Finally, you could help your friend by pointing out that they should always take all the antibiotic they are given for an infection. By not taking all of the antibiotic prescribed they run the risk of selecting for antibiotic-resistant bacteria. Having so many bacteria causing disease in their body, their innate and adaptive immune systems may not be able to kill the remaining bacteria present. This could lead to a relapse of their disease, only this time they may have a larger population of antibiotic-resistant bacteria.

CHAPTER 14 Most likely, the motorcyclist's infections are due to opportunists, because the abraded skin and lacerations opened up the body to possible infection. However, because the gravel and dirt can contain a great variety of microbes, it is always possible that some could be transients that could establish a disease regardless of the availability of the open wound.

CHAPTER 15 Install sinks at entrance to patient rooms for hospital staff to wash their hands. Simple solution: Hospital staff wash hands between each patient. Provide alcohol-based waterless gel disinfectant for use in place of hand washing. Use filters in patient rooms and have patient rooms dedicated for use with patients with communicable diseases such as chickenpox and tuberculosis.

CHAPTER 16 The mucociliary escalator system allows materials in the bronchi to be lifted to the pharynx and to be spit out or swallowed. Any infection such as whooping cough, pneumococcal pneumonia, pseudomonas sepsis, or chronic ulcerative bronchitis that is normally controlled by the proper function of this system could cause severe infections in these patients.

CHAPTER 17 Chemotherapy often decreases the number of T and B cells in the body. Hence, individuals undergoing chemotherapy are more susceptible to infections. Raw foods are more likely to have bacteria and viruses on them. Bacteria and viruses are more likely to be killed by cooking.

CHAPTER 18 The reaction is being caused by the latex gloves that her dentist wears. The type of hypersensitivity described is contact dermatitis. To prevent the problem, the dentist should wear gloves made of polyvinyl chloride.

CHAPTER 19 Many animals including cats and dogs also get ringworm and can be the source of the infection. In cats, a pink nose may be the only sign of ringworm. Ringworm on the scalp usually makes a bald patch of scaly skin. On other parts of the skin there can be a ring-shaped rash that is reddish and may be itchy. The rash can be dry and scaly or wet and crusty.

CHAPTER 20 It may not be a coincidence. The man may have been infected with human papillomavirus (HPV), which causes 99% of all cases of cervical carcinoma.

CHAPTER 21 Yes, Mildred's symptoms are consistent with legionellosis. If she had seen a doctor the tests would include direct fluorescent antibody microscopy, ELISA, and PCR with genetic probes of a sputum specimen. The bacterium can live in water and the probable source of the infection was the sprayed water in the vegetable section of the grocery store.

CHAPTER 22 The probable source of contamination was at the bottling plant from the contaminated water used to make the soda or to wash the bottles. The use of properly disinfected water would prevent contamination. The clinical lab would probably have looked for trophozoites or cysts in stool samples or culture the organism. The minimum number of cysts required to cause disease (infectious dose) is one or more cysts, unlike bacterial pathogens that usually require hundreds to thousands of microbes.

CHAPTER 23 The *Streptococcus agalactiae* could infect Ruth's baby or her at birth when the integrity of the skin on both mother and child is broken and then cause disease. The antibiotic will decrease the numbers of organism present and also help control organisms that might get into the bloodstream. If no antibiotic is administered the baby could develop sepsis, pneumonia, and meningitis; the mother could develop sepsis or an infection of the soft tissues. Prior to birth stillbirth could occur.

CHAPTER 24 Harry died of rabies, acquired when the dog bit him. Often a rabid animal is disoriented and may run out in front of a car. A rabid fox was killed when it was hit by a bus, as it was running toward the bus stop where my mother was waiting.

CHAPTER 25 Fungal spores capable of causing ringworm exist in soil, where they metabolize dead organic matter. Raking and digging in her yard could get some spores airborne and into Helen's hair. Even just standing outside on a windy day could deposit spores on her head.

CHAPTER 26 *Salmonella* bacteria must have entered through the cracks and multiplied inside the eggs producing lots of endotoxin which caused the illness and brain damage. If the eggs were not cooked thoroughly, *Salmonella* organisms would also have multiplied inside the boy, and produced more toxin. The farmer should not have been selling those eggs at all; not for humans or pets!

Critical Thinking Questions Answers

Note: For many of these questions, there is no single correct answer. You may come up with a great answer that no one has ever thought of before.

CHAPTER 1

1. Today, a new vaccine or drug is first tested for safety using animal hosts or, if no suitable animal host can be found, tests may be performed using cultures of human cells or tissues. Then safety testing is conducted using small numbers of people at first, then larger numbers. Testing for effectiveness similarly begins with animal hosts when possible, followed by testing with gradually expanding numbers of human subjects. If the disease is life-threatening, rather than injecting the disease agent into human subjects, people are immunized and followed over a period of time while they go about their lives and may be naturally exposed to the disease agent. The infection rate in the immunized individuals is then compared to that of a similar, nonimmunized control group.

2. If toxin-producing bacteria in an infection were killed or became dormant (as in endospore production) but their toxin remained behind in the body causing signs and symptoms of the disease, one would have no living infectious agent to isolate and inoculate. Isolation of spores and injection into a healthy subject may reproduce the disease only if the right conditions exists for dormant spores to germinate into the living bacteria.

An opportunistic infection by *Pneumocystis carinii* is often responsible for the indirect killing of AIDS patients due to their weakened immune cell function from the primary infection by the Human Immunodeficiency Virus. If one were to isolate *Pneumocystis carinii* (thinking this was the primary cause of infection) and inject it into a healthy individual (AIDS-free) it might not produce the pneumonia that could kill this individual if he or she was able to mount a vigorous immune response.

3. Angelina Hess suggested to Robert Koch that he use agar (a thickener she used in cooking) to firm up his bacteriological media, which would enable him to isolate bacterial colonies. In 1879, Louis Pasteur's assistant accidentally inoculated some chickens with an old chicken cholera culture. They did not get sick and later when the same chickens were inoculated with a new fresh cholera culture, they remained healthy. Pasteur realized that this "mistake" led to the chickens becoming immunized against chicken cholera. In 1928, Fleming observed that contaminating *Penicillium* mold in a *Staphylococcus* culture prevented bacterial growth adjacent to the *Penicillium*. As a result, Fleming recognized the potential of this observation for fighting bacterial infections in humans.

4. The reason we know so much about the observations of van Leeuwenhoek is that he sent letters to the Royal Academy of Science in London describing his observations. These letters were written in enough detail that others could understand what he was observing and some of the significance of that work. By publishing the results of their research scientists make those results available to many other scientists to examine. This allows scientists to discuss the results, make criticisms, improve the observations, and build on the observations of others to further the understanding of the biological world, and develop practical applications.

5. (a) Pharmacogenomics, proteomics, functional genomics, regulatory genomics, evolutionary genomics. (b) By comparing bacterial genomes to the human genome, we could exploit genes found in the former and not the latter, or vice versa, by selectively targeting the gene or its protein product in order to effect killing of just the prokaryotic cell.

6. (e)

CHAPTER 2

1. Water is able to dissolve many different compounds, making it an ideal basis for the cytoplasm of all cells. The powerful attraction of water molecules to each other allows a thin film of water to cover membranes and keep them moist. Also, water can gain or lose a lot of heat with a relatively small temperature change, helping living things maintain an ideal temperature for their functions. Finally, water participates in many of the chemical reactions that are essential to life. So could life exist on a water-free planet? Certainly not life in any form that would seem familiar to us.

2. With its valence of 4, each carbon atom can bond to up to 4 other atoms, including other carbon atoms, allowing the formation of extremely large and/or complex molecules. Also, the ability of carbon to form either single or double covalent bonds increases the diversity of compounds possible. Perhaps somewhere in the universe, living systems have developed that are based on some other element, but once again, such life forms would seem very different from any on Earth.

3. The use of chemical treatments and heat on hair follicles denature the α-keratin protein by disrupting its secondary, tertiary, and quaternary structure, allowing one to mold the hair follicles into a different shape.

4. The mistake or mutation in the DNA sequence may allow the gene to code for a different amino acid sequence, thus changing the protein product. The new protein product may be useful, for example if it confers antibiotic resistance upon the bacterium, or harmful if it changes an essential protein into another protein, preventing it from doing its normal function, as in an enzyme involved in glucose metabolism.

5. Figure 10.24, p. 301 shows two different configurations (folding patterns) of the same protein. Somehow, by a method as yet unknown, the abnormal form serves as a template to fold normal versions into the abnormal form.

CHAPTER 3

1. The Gram stain technique has timed steps. Initiating the primary crystal violet stain without washing, mordant treatment, and secondary staining would expose the microbes to the primary stain for too long, most probably skewing the results. Gram staining a smear prepared from cultures aged over 48 hours often results in "Gram variable" results or cells that do not react distinctively to this stain.

2. Both organisms on Craig's slide appeared red. Iodine is necessary for the proper bonding of the crystal violet into the Gram-positive organism, so when he decolorized, both organisms lost their violet color.

3. Heat-fixed specimens: *Advantages:* Kills pathogenic microorganisms, causes microorganisms to adhere to slide, alters microorganisms so they can more readily accept stains (dyes). *Disadvantages:* Wet slides passed through a flame can boil and destroy microbes on the slide. Too little heat-fixation may cause

the microbes to not stick to the slide resulting in their being washed off in subsequent steps. Any remaining live, non-heat-fixed cells will stain poorly. Too much heat-fixation can incinerate the microbes, resulting in distorted cells and/or cellular debris. Some structures such as bacterial capsules are destroyed by heat-fixing. **Wet Mount specimens:** *Advantages:* Allows one to view undistorted, living, motile microbes. Wet mounts also allows one to view the size and shape of individual organisms as well as any characteristic arrangement or groupings of bacterial cells. *Disadvantages:* Specimens cannot be preserved and must be viewed within the hour. Specimens cannot be stained for enhanced viewing.

CHAPTER 4

1. The essence of the difference between prokaryotes and eukaryotes is that prokaryotes lack a clearly defined nucleus and other membrane-bound internal structures.

2. Human cells have no cell walls, and so are not affected by drugs that block the growth of the cell wall.

3. Currently, a wealth of evidence taken from prokaryotic endosymbionts living in eukaryotes exists that support the endosymbiotic theory. Some eukaryotes lacking mitochondria and living in low oxygen environments form a symbiotic relationship with prokaryotes which serve as surrogate mitochondria. Dr. Lynn Margulis has proposed that eukaryotic flagellae and cilia originated from symbiotic associations of motile spirochete bacteria with nonphotosynthetic protists. There is also evidence from giant tube worms living near hydrothermal vents deep in the ocean that form a prokaryotic endosymbiosis with bacteria colonizing their internal tissues. The worms lack mouths, anuses, and digestive tracts yet are provided energy from the bacteria metabolizing the hydrogen sulfide spewing from the hot vents.

CHAPTER 5

1. One can start the separation process with the use of differential and selective media that might exploit differences in fermentation metabolism between the two. For example, growing *Klebsiella pneumoniae* and *Staphylococcus aureus* on high salt mannitol sugar plates containing a pH indicator, would allow one to separate and identify *Staphylococcus aureus* due to this species' ability to grow in a high salt environment and being able to ferment mannitol (*Klebsiella pneumoniae* cannot do either). One could also test the end products of fermentation from the growth of an isolated colony to help identify it. For example, you could carry out the Voges-Proskauer test for acetoin which is an intermediate in butanediol fermentation in *Klebsiella pneumoniae*.

2. (a) Evolution and conditions at the time have determined the sequence of different types of metabolism. For example, billions of years ago there was very little atmospheric oxygen available so microorganisms at the time had to evolve a way to capture energy in its absence and keep glycolysis going. The solution was to evolve the fermentation pathway which would recycle the limited amount of NAD needed for glycolysis by passing the electrons of reduced NAD off to other molecules. Later, when earth attained an atmosphere rich in oxygen, enzymes involved in the aerobic respiration pathway evolved to take advantage of the new conditions. Another example is the purple and green bacteria capable of photosynthesis. They were probably the first to evolve photosynthesis because of the relatively anaerobic environment on earth billions of years ago. They evolved an H2S splitting system for the liberation of electrons due to the high availability of this compound at the time. Later, when earth attained an atmosphere richer in oxygen, water splitting enzymes involved in the light reactions of photosynthesis evolved to take advantage of the new conditions. (b) There are so many different types of environments on earth that helped to drive the evolution of so many different types of metabolism. Many different types of metabolism evolved to exploit the particular environmental conditions. Ultimately, genetics and Darwin's "survival of the fittest" played a huge role in this. Had only one type of metabolism evolved, those species containing it would have become extinct due to the changing conditions on earth.

3. Many effective drugs have been developed by studying the metabolic pathways of microorganisms in search for enzymes possessed by pathogens, but not by ourselves. Once such an enzyme is found, it may be possible to synthesize molecules that are so similar to the enzyme's substrate that they block the active site on the enzyme or attach elsewhere on the enzyme, changing the shape of its active site.

CHAPTER 6

1. By 3:00 p.m., the original 100 bacteria would have increased to 1,638,400. By 5:00 p.m., there would have been 18 generations of bacteria.

2. No organism can continue to reproduce at its full potential indefinitely. Factors such as exhaustion of food supply and accumulation of toxic metabolic wastes limit the population ultimately achieved.

3. Most likely, the original culture was not in nonsynchronized growth. This would lead us to believe that some of the population of microbes were in the tail end of logarithmic growth. As some of these latter microbes were transferred over to the new media, fresh nutrients allowed them to continue their growth.

CHAPTER 7

1. In organisms with only one set of genes, such as bacteria, there are no such things as dominant and recessive genes. If an organism has a certain gene it can be expressed. While every human carries thousands of "hidden" recessive genes, this is not true in bacteria.

2. The longevity of individual microbes at this time was probably very short because of the DNA damage induced by the UV light. The rate of evolution was most likely accelerated by the UV light in favor of those populations of microbes that developed DNA repair mechanisms. Otherwise microbial populations would have become extinct because they would not have been able to make the proteins needed for their survival due to the induced mutations in their DNA by the UV light.

3. The polymerase chain reaction (PCR) is a method that can be used to confirm the presence of DNA from the tubercle bacillus (*Mycobacterium tuberculosis*) in just a few hours. PCR is a technique that allows one to rapidly produce or amplify a billion copies of a segment of DNA. As long as at least one copy of DNA can be isolated from the sample, one can theoretically amplify large quantities of a segment from this DNA which is used for easy analysis. PCR works by designing a set of complementary oligonucleotides based on a known sequence segment of target DNA (in this case the sequence from *Mycobacterium tuberculosis*). The complementary oligonucleotide pair primes off the DNA sample, and is exponentially amplified in repeated thermal cycles of heat denaturation, amplification, and priming with a heat-stable DNA polymerase. The amplified DNA can then either be directly run on an electrophoretic gel along with known standards or can first be cut with restriction endonucleases and then run on the gel for identification / verification.

CHAPTER 8

1. Drug-resistant genes move with ease from one microorganism to another through processes of recombination such as conjugation, transformation, and transduction. The risk of newly drug-resistant pathogens reaching humans from animal sources can be reduced by minimizing the use of antibiotics in animal feeds, following strict sanitary precautions when slaughtering and butchering meat animals, and by adequately cooking all foods from animal sources.

2. Using some of the genetic engineering tools that were described in this chapter, design a protocol for creating a recombinant plasmid.

I. Isolate bacterial plasmid DNA containing a single recognition sequence for a specific restriction endonuclease, for example the restriction enzyme EcoRI.

II. Isolate total genomic DNA from a mouse. The mouse gene for growth hormone (mGH) is flanked on either side by DNA sequences that are also recognized by the specific restriction endonuclease EcoRI. There are no other EcoRI sites within the gene.

III. Use EcoRI to cut or digest I and II above separately.

IV. Run the restriction fragments generated from III above on an electrophoretic gel to separate, isolate, and purify the cut plasmid and mGH gene flanked by cut EcoRI ends.

V. Mix the two DNAs (EcoRI cut plasmid and mGH gene) together in a test tube that contains DNA ligase. This enzyme will ligate or splice together the ends of the mGH gene into the digested EcoRI site in the plasmid. You now have a recombinant plasmid.

VI. One can then purify the recombinant plasmid and use it to transform competent bacteria, which will uptake the recombinant plasmid. Growing the bacteria en masse will also replicate the recombinant plasmid DNA which you can isolate from the bacteria to use as you wish.

3. Of course, the answer to this question depends a lot on some of your personal values and experiences. Anyone who owes their survival to genetically engineered insulin, for example, is likely to be enthusiastic over transgenic organisms. Similarly, anyone concerned with feeding the world's ever-increasing human population is likely to praise genetically modified plant varieties.

CHAPTER 9

1. One of the oldest traditions in science is publishing and otherwise sharing information. In this way, scientists don't have to repeat all the research that has already been done. It would have been very difficult for science to move ahead if there had been no uniform system of naming organisms. How would scientists in different places have been able to share information if the same organism had a different name in every region of the world? Biology would have progressed very slowly!

2. (a) Species is a group whose members are capable of inter-breeding with one another but incapable of producing viable and fertile offspring with members of other species. (b) In advanced organisms (plants and animals) species that reproduce sexually are distinguished by their reproductive capabilities. Morphological (structural) characteristics and geographic distribution are also considered in defining a species. However, this definition of species is hard to fit to the prokaryotes as lateral gene transfer (genetic recombination) is relatively common in evolution and morphological differences are minor. As a result, bacterial species is defined by the similarities found among its members. Properties such as biochemical reactions, chemical composition, cellular structures, genetic characteristics, and immunological features are used in defining a bacterial species.

3. Species B and C are most closely related because they share the highest percentage of DNA homology.

CHAPTER 10

1. The extreme simplicity of viruses eliminates most of the vulnerable features found in cellular organisms such as bacteria. Antibacterial products commonly attack cell walls, plasma membranes, and cytoplasm, none of which are present in viruses.

2. First, the discovery and use of antibiotics made it possible to prevent bacterial contamination of eukaryotic cell cultures, preventing their overshadowing of cellular effects caused by the virus. Second, biologists found that proteolytic enzymes, especially trypsin, could free animal cells from their surrounding tissues without injuring the freed cells. These cells can then be washed, isolated, and dispensed into plastic culture dishes containing nutrient media, where they will attach, multiply, and form monolayers.

3. The most successful pathogens are the ones that do not kill the host. Remember, viruses depend on their host's cellular machinery to replicate themselves. If a virus kills its host, in essence it is killing itself unless it can infect another host. A virus that is able to infect a host and replicate itself in the host, without killing the host, will be able to continue its replicative cycle again and again as long as the host is able to recover from the infection.

CHAPTER 11

1. When decay organisms infect a client, you would want to order tests to identify HIV infection, diabetes, and leukemia or other cancers. You would also want to explore the possibility of alcoholism or other chemical dependency, which can impair body defenses against pathogens.

2. Some autotrophic protists produce toxins. Oysters feed on these protists with no ill effects, however the toxins get concentrated within their bodies so when humans eat oysters they can become quite ill or even die. Oyster beds infected with such protists can cause great economic losses to oyster harvesters. Other autotrophic protists rapidly multiply when abundant inorganic nutrients are available, forming a "bloom," a thick layer of organisms over a body of water. This "eutrophication" process blocks sunlight, killing plants beneath the bloom, which in turn cause fish to starve. Microbes that decompose dead organic matter in bodies of water can also consume large quantities of oxygen which can cause massive fish kills, due to the lack of oxygen.

3. Penicillin is quite specific against bacteria, affecting the growth of their cell wall. The organisms described in this chapter either have no cell wall at all or a cell wall that is very different in chemical composition from the bacterial cell wall. Thus, nothing mentioned in this chapter would be controlled by penicillin.

CHAPTER 12

1. In this era of paranoid fear of microorganisms, many people expect their foods and beverages to be completely free of bacteria. The fact is that pasteurization is a selective process, using only enough heat to kill certain "target" organisms, either pathogens or spoilage organisms. Thus, a pasteurized product is not a sterile product and does not need to be sterile. The public needs to be educated to the fact that most microorganisms are not harmful and, in many cases, are beneficial.

2. No. Ultraviolet rays, X-rays, and gamma rays all kill pathogens by damaging their nucleic acids. Prions have no nucleic acid so are quite resistant to these radiant energy forms.

3. (a) The phenol coefficient is the ratio of the dilution of the test agent to the dilution of phenol that will kill all organisms in 10 minutes but not in 5 minutes. Highest dilution of new disinfectant at 10 minutes = 1:50; Highest dilution of phenol at 10 minutes = 1:110. Therefore 50/110 = 0.45 phenol coefficient. (b) No. The phenol coefficient of the new disinfectant being less than 1 makes it less effective than phenol.

CHAPTER 13

1. The identical chemical can be deadly to one species and harmless to another. This is the concept of selective toxicity that is basic to today's antimicrobial therapy (though this has not always been the case). The drugs we use today affect some aspect of the physiology or anatomy of the pathogen that is not a part of the host. Common examples include drugs that attack cell walls (we have none) or inhibit an enzyme that we don't have.

2. Antibiotic-forming microorganisms typically live in the soil or other locations where there are abundant microorganisms. Even one gram of soil contains dozens of species of bacteria and fungi, all competing for the available food. The organism that releases an antibiotic into its habitat gains a definite competitive advantage over the others.

3. The advantages for using more than one drug to treat a bacterial infection are: 1) synergism or the additive effect of two or more antibiotics. Such an example is the quinolone–cephalosporin combination, which will kill cephalosporin-resistant microbes. By prescribing a "back-up" antibiotic acting in synergy such as the above example, one may prevent or delay development of antibiotic resistance in organisms.

The disadvantages for using more than one drug to treat a bacterial infection are the development of multiple drug-resistant strains of bacteria, rendering many antibiotics ineffective. Also, a patient's risk of side-effects can increase due to the increased number of administered antibiotics. Adverse drug-interaction could also occur in patient's taking different drugs for different reasons.

CHAPTER 14

1. The first approach to performing step 3 of Koch's Postulates for a life-threatening human disease would be to look for a susceptible animal host for the organism. Another approach would be to just infect isolated human cells or tissues, rather than a person, and see if the same pathological changes developed in the cells or tissues that had been observed in diseased people.

2. (a) Available medical treatments such as antibiotics and vaccines, preventative measures, and sanitization efforts have not been applied. Examples of this are parents who fail to have their children immunized, lack of access to healthcare facilities, and poor economic status. (b) Infectious agents are highly adaptable. Many strains of microorganisms have developed resistance to multiple antibiotics that are currently available. Misuse/overuse of antibiotics have contributed toward this problem. (c) Changes in social conditions and/or human activity have allowed previously unknown or rare diseases to become significant. An example of this is modern air conditioning technology and Legionellosis. (d) Immigration and international travel and commerce have quickly introduced new or recurrent strains of pathogens. Immigrants infected with *Mycobacterium tuberculosis* bring tuberculosis disease into areas where it has been previously eradicated. Developmental encroachment on land that has not been disturbed for a long period of time has also contributed to this problem. Ebola virus in Africa is an example of this.

3. Individuals with weakened immune defenses due to malnutrition, presence of another disease, advanced or very young age, treatment with radiation or immunosuppressive drugs, and physical or mental stress can lead to this state. The failure of host defenses in AIDS patients can allow several different opportunistic infections to develop, for example.

The introduction of a bacterial resident from its normal environment (tissue) of the body into an unusual site can cause an opportunistic infection. An example is *E. coli* (a normal resident of the large intestine) may gain entrance to the urinary tract, surgical wounds, or burns.

Disturbances in the normal microflora may contribute toward an opportunistic infection. An example of this is when an individual is treated with an antibiotic to bring one pathogen under control, it also disturbs or kills the normal microflora (releasing a check on other potential pathogens), and in so doing allows another pathogen to gain a foothold and cause a different disease.

CHAPTER 15

1. If a disease is sporadic in humans, its reservoir is very likely not humans. Most human sporadic diseases have animal reservoirs; the pathogens of a few reside in soil or water.

2. *Environmental factors:* Seasonal cycles help to explain this. Epidemic outbreaks only occur during the dry season, a time during which nasal and throat membranes are likely to dry out, making such individuals more susceptible to meningococci entering their blood. During the wet season, cases fall to zero because of the high humidity and heat.

Other diseases: More colds, influenza, and other respiratory diseases occur during the dry season, making such individuals more susceptible to meningitis.

Herd immunity: During an epidemic in a given year, most people develop antibodies against the prevalent strain of *Neisseria meningitides*. However, each year new children are born that lack this immunity. Eventually the herd immunity drops so low as to allow another epidemic to occur.

Strain virulence: Between epidemics, mutant strains arise. If one of these strains is more virulent than previous strains, it could initiate a new epidemic.

Timing: The timing of entry of the mutant strain into the population is also crucial. If it enters during the dry season, those exposed to it are likely to contract the disease. If it enters during the wet season, people will become exposed to it without contracting the disease and instead develop antibodies against this new strain and will be immune to it when the following dry season begins.

3. There are many reasons why there will always be some nosocomial infections. Foremost among these is the extremely precarious health of many hospitalized individuals. Because of old age, cancer therapies, antibiotic usage, and immune deficiency disorders, many patients have almost no ability to fight pathogens. Everyone carries potentially pathogenic organisms at all times. Every patient admitted brings along pathogens; every hospital worker and every visitor similarly carries a huge population of potentially troublesome organisms. Also, because of the intensive use of antibiotics and various forms of radiation, hospitals are the source of some of the most highly drug-resistant pathogens.

CHAPTER 16

1. (a) A small cut on your hand would initiate inflammation, bringing additional phagocytes to the damaged area; the phagocytes would engulf any pathogens present. (b) Pathogens inhaled into your lungs would usually be trapped on the mucus layer and carried by the respiratory cilia up to the pharynx where they would be swallowed and destroyed by stomach acid. (c) Pathogens ingested with contaminated food would encounter powerful stomach acid and digestive enzymes. Unfortunately, some kinds of pathogens would survive and initiate digestive infections. Then they would be attacked by phagocytes.

2. Inflammation can cause swelling in places of the body where it can cause harm. Examples of this are the membranous meninges surrounding the brain or spinal cord causing brain damage and in the lungs where it can cause obstructed breathing. The vasodilation that occurs in inflammation brings in additional oxygen and nutrients to the area which in most cases benefits the host cells but sometimes it causes the pathogens to thrive as well. The walling off of pathogens prevents them from spreading but this can also prevent the host's natural defenses and antibiotics from reaching them. Suppression of the inflammatory response can also allow formation of boils when natural defenses might otherwise destroy the bacteria.

3. No, it is not a good idea to take aspirin or in other ways reduce a moderate fever. Fever helps fight pathogens; it speeds up our own body defenses and slows down the growth of pathogens. Only when fever becomes high is it advisable to act to reduce it.

CHAPTER 17

1. (a) The diseases can be more deadly or damaging than the vaccines. (b) We now have a new acellular pertussis vaccine which does not cause the damage that the older whole-cell pertussis vaccines of a few years ago did. It is much safer.

2. Most likely, this individual would get quite sick and maybe even die due to lack of an immunological response. Remember, this individual encountered the same type of virus during development. During this time a lymphocyte that recognizes its specific antigen is either destroyed or inactivated. This is a necessary process that creates "tolerance" preventing the individual from mounting an immune response against self antigens later in life. However, viral exposure during development of this individual (although the virus technically was a nonself antigen), resulted in tolerance, eliminating those specific lymphocytes that could have destroyed the virus.

3. Not all types of antibodies present in a mother's bloodstream are able to be secreted into her milk; e.g., pertussis antibodies are absent. Also, it is unlikely that the mother has had every last disease that people in the mall have (e.g., flu strains change every year). Therefore, she will not have antibodies against every disease that her baby is exposed to. Furthermore, she may have had a disease so long ago that her antibody supply could have declined to a nonprotective level.

4. The deletion of the gene responsible for restraining the uncontrolled cell growth will result in a cancerous liver cell. The fact that it does not contain an MHCI antigen also protects it from the effects of cytotoxic T cell killing. Therefore, it has a good chance of cloning itself and causing a malignancy unless the NK cell can keep it in check as NK cells do not need MHCI complexes and can nonspecifically kill cancer cells.

CHAPTER 18

1. There is always some possibility that what appears to be an autoimmune disorder is in fact the body's normal response to an unrecognized infection by an intracellular parasite. (You will recall that your body deals with viral and other intracellular infections by destroying the infected cells.)

2. (a) Some red blood cells bearing the Rh antigen of the first born must have leaked across the placenta during delivery thereby "sensitizing" the Rh-negative mom. When mom carries a subsequent Rh-positive fetus, her anti-Rh antibodies can cross the placenta, causing a type II hypersensitivity reaction in the fetus resulting in hemolytic disease of the newborn or erythroblastosis fetalis. (b) Intramuscular injection of anti-Rh IgG antibodies (Rhogam) to Rh-negative mothers can protect subsequent Rh-positive babies by binding to Rh antigens on the fetal red blood cells that have leaked into the mother's bloodstream. The anti-Rh antibodies destroy the fetal red blood cells before they can sensitize the mother's immune system.

3. NOTHING! All blood and body fluids are handled exactly the same. You have no way to know whose blood carries HIV and/or hepatitis B or C viruses.

CHAPTER 19

1. Although the molluscum contagiosum virus is classified within the Poxviridae family, it differs immunologically from both orthopoxviruses and parapoxviruses. It also elicits only a slight immune response and although infected cells cease to synthesize DNA, the virus induces neighboring uninfected cells to divide rapidly. This suggests that it has undergone genetic mutation (acquiring and/or changing existing genes) that enable it to be immunologically distinguishable from other Poxviridae family members as well as giving it the properties of tumor induction.

2. *Candida albicans* is present among the normal flora of the digestive and urogenital tracts of humans. The normal microflora compete with pathogenic organisms such as *Candida albicans*, keeping them in check, and in some instances actively combat their growth. If there is a disturbance in the normal microflora due to antibiotic killing of bacteria, then the normal checks and balances on *Candida albicans* can be removed, allowing it to gain a foothold, and thrive, causing disease. This is so because antibiotics do not affect fungi.

3. Smallpox was eradicated because, first of all, it has no non-human reservoir (it would probably be impossible to eradicate a pathogen that could live in soil, water, or wild animals). Secondly, an effective vaccine was available. And finally, the World Health Organization assisted impoverished developing countries with their immunization programs. The concept of herd immunity allows the eradication of a disease even though not every individual is immunized. Polio and measles are now nearing world eradication.

CHAPTER 20

1. A vaccine that elicits antibodies against common immunological epitopes of the herpesvirus would probably not be effective because a large portion of the human population already harbors *latent* herpesvirus. So a vaccine to inactivate the viral genetic information responsible for latency is needed.

2. First, you would keep in mind that some people will be infected with both gonorrhea and chlamydia; if you detect one, don't forget to check for the other one as well. Gonorrhea is detected by Gram-stained slides, cultures, and antibody-based or DNA technology lab tests. Chlamydia are invisible on stained slides and don't grow on culture media, so they are identified by fluorescent antibody tests, other antibody-based tests, or DNA technology.

3. Of course, there is no simple answer to this question, but here are some factors that experts have identified as responsible for the high incidence of STDs. You will probably have identified additional good answers. (a) People often become sexually active at a younger age and may have more sexual partners than they did 50 years ago. (b) Birth control pills were not yet available 50 years ago. Their use can reduce motivation for using condoms and definitely increases the vaginal susceptibility to gonorrhea and other STDs. (c) A large percentage of the cases of STDs are in very young people who may not be well informed about the transmission, prevention, and symptoms of these diseases. (d) Other than for HIV, there has not been much political backing for research dollars for STDs. Many less common diseases receive better research funding. (e) Some people practice denial when it comes to STDs: "It couldn't happen to me." (f) Many people are reluctant to bring up the issue of STDs with a potential sex partner, either out of shyness, lack of communication skills, or not wanting to spoil the romance of the moment.

CHAPTER 21

1. The human respiratory system contains a huge surface area of delicate membranous tissue, as large as a tennis court by some estimates. Each of us inhales a huge amount of air each day and every breath we inhale contains microorganisms, some of which are potentially infectious. Fortunately, at least for nonsmokers, there are effective mechanisms, such as cilia, for sweeping these microorganisms out of our lungs.

2. Temporary increases in the 1980s were caused by Asian and Haitian refugees who became infected in crowded, unsanitary camps and escape boats. Also, many new cases occurred among AIDS patients due to opportunistic infections caused by immunocompromised immune systems and reactivations of old infections triggered by immunodeficiency, crowding, stress, and use of anti-inflammatory drugs.

3. These students had histoplasmosis. Had they been a much younger or much older group, their illness might have been more severe including dissemination to other organs.

CHAPTER 22

1. The fact that Melanie's gastric ulcers were cured by this treatment suggests that she had been chronically infected with *Helicobacter pylori*. Beyond this, chronic *Helicobacter pylori* infection has been associated with a high incidence of stomach cancer and heart disease, so she should be checked out for the former and followed closely for the latter to see if her heart disease symptoms improve.

2. Typhoid fever has humans as its only reservoir, while intestinal salmonellosis usually has animal reservoirs (human to human transmission is possible). Typhoid fever is always systemic, while intestinal salmonellosis is usually confined to the intestine (some cases do become systemic). Typhoid fever seldom causes diarrhea, which is a leading symptom of intestinal salmonellosis.

3. Most hepatitis cases in the United States are hepatitis A, B, or C. Proper diagnosis may help prevent further transmission of hepatitis. Hepatitis A carries the risk of transmission to others primarily through fecal contamination of food or water; someone with hepatitis A should not be working with food. Hepatitis B and C are transmitted through blood and other body fluids, and can be sexually transmitted. Hepatitis A does not cause chronic infection; once a person recovers he or she will have no further risk of liver damage from that virus. Both hepatitis B and C commonly persist and are chronic infections, carrying the possibility of future life-threatening illness. Both require long-term monitoring and possible injections of interferon to reduce the risks of cirrhosis, liver failure, and liver cancer.

CHAPTER 23

1. *Bacillus anthracis* virulence factors: (a) Glutamic acid capsule, whose genes are carried on one plasmid; (b) a second plasmid carries the genes for three exotoxins: edema factor, lethal factor, and protective antigen. All *Bacillus anthracis* virulence factors must be present for disease to occur. Edema factor combines with protective antigen to form edema toxin which causes swelling and prevents phagocytosis by macrophages. Lethal factor combines with protective antigen to form lethal toxin which causes macrophages to release tumor necrosis factor-α and interleukin-1β, plus other inflammatory cytokines, and to die. The resulting toxemia from the exotoxins causes clots to form inside pulmonary capillaries and lymph nodes, causing mediastinal swelling which obstructs airways.

2. Given its reservoir of rodents and given the abundance of rodents in the United States, there would seem to be some possibility of an epidemic at some time. Neighborhoods do sometimes become overrun with rats and pathogens do sometimes mutate into more virulent forms. A huge epidemic with thousands of deaths seems unlikely now that the transmission of plague is clearly understood and most parts of the United States are served by effective Public Health Departments. The first few cases of plague would bring in a small army of public health workers to combat the rodents and fleas and to treat infected human cases before person-to-person transmission occurred.

3. Arthropod-transmitted diseases usually have specific vectors because in most cases the pathogen undergoes a critical part of its life cycle within the vector. The physiology of the pathogen is so intimately associated with the anatomy and physiology of the vector that a substitute arthropod cannot support the development of the pathogen.

4. The relapses are caused by changes in the organisms' antigens. During a febrile period, the body's immune response kills most of the organisms. The few that remain have surface antigens the host's immune system fails to recognize, so these

organisms multiply during a relief period until they are numerous enough to cause a relapse.

CHAPTER 24

1. (a) The blood-brain barrier (BBB) separates the brain from the rest of the body's circulatory system. There are thick-walled capillaries in the brain that have no pores, which limits the easy passage of substances and cells. The barrier allows only selective substances to pass through it. Even the body's own antibodies and complement proteins have a hard time passing through it. (b) These antibiotics, being lipid-soluble, diffuse much more easily through the lipid bilayer membranes of the capillary cells making up the BBB.

2. Rabies has an unusually long incubation period as the virus travels slowly up through the nervous system on its way to the brain. This usually allows enough time to confer active immunity by means of a series of five vaccine injections.

3. From time to time there are proposals to involuntarily confine people with some infectious disease. This is usually viewed as counterproductive in the fight against that disease, especially when effective treatment is available. History has shown that when people face involuntary confinement they will hide the fact that they are infected, thereby not receiving the treatments that could arrest or cure their illness. Additionally, there are ethical questions about involuntary confinement because of infectious disease.

CHAPTER 25

1. We wouldn't like a bacteria-free world. In the first place we would be pretty hungry. Everything we eat is either a plant or an animal that depends on plants for its growth. And plants depend on soil bacteria to produce the nitrate nitrogen that plants require for growth. Many natural cycles, such as the nitrogen cycle, the carbon cycle, and the sulfur cycle have bacteria-dependent stages. And there might not be much oxygen to breathe in a bacteria-free world. Much of our atmospheric oxygen comes as a by-product of photosynthesis by cyanobacteria (formerly called blue-green algae).

2. Research has revealed that there are massive populations of marine viruses, some of which infect and destroy phytoplankton.

3. One problem with using fecal coliform bacteria as an indicator of fecal contamination in water is that some viruses and protozoa of fecal origin live longer in water than do fecal coliforms. Water with a negative coliform test could still carry pathogens. We can expect to see some changes in how water is tested in coming years.

4. Some microorganisms are capable of producing hydrogen sulfide (H_2S) by consuming oil deposits below a cave. The H_2S gas then bubbles up and reacts with water to form sulfuric acid. Sulfuric acid in turn eats out caves along cracks in the rocks above. The walls of caves are also covered with snottites (mucus-like strings of bacterial colonies). These bacteria eat sulfur and drip sulfuric acid which also helps in etching out the cave.

CHAPTER 26

1. Don't worry, you'll have plenty to eat. You could take bread (product of yeast), cheese (bacteria and/or fungi), yogurt (bacteria), pickles and sauerkraut (bacteria), hard salami (bacteria and fungi), and black olives (bacteria). And don't forget the beer or wine, produced by yeast.

2. To meet commercial demands, large-scale fermentation vats were used in the past; however many new processes do not work efficiently in large vats, so smaller vats must be used. The extensive selection of mutants and creation of genetically modified microorganisms has produced products useful to humans but these may be useless or even toxic to the microbe. Isolation and purification of the product with or without killing the organisms often present technical difficulties. When the product remains within the cell, the plasma membrane must be disrupted to obtain the product but this kills the organism. Secreted products can be collected easily and sometimes without killing the organisms. The use of a continuous reactor can achieve this.

3. Poultry should be used promptly and contaminated wrappings and juices carefully discarded. Any countertop, cutting board, and utensils that have come into contact with poultry should be scrubbed thoroughly with hot soapy water before other foods come into contact with them. Also, wash your hands thoroughly before touching other foods to prevent cross-contamination of them. Cook poultry thoroughly to kill off the *Salmonella*.

Assume that the shells of all eggs are contaminated with *Salmonella*. Wash your hands after handling them but do not wash the eggs. Doing so will remove a protective surface coating that helps prevent microbes from entering the eggs. Refrigerate your eggs with the big end up. Doing so keeps the egg yolk and any embryo residing within as close to the center as possible. This maximizes the distance an invading microbe would have to travel from the shell to the yolk. Never select a cracked or broken egg for consumption. Ingesting raw eggs, such as in egg nog or cake batter, is a calculated risk. One should thoroughly cook eggs or foods that contain them in order to kill off the *Salmonella*.

Meats should be refrigerated promptly and frozen if they are not going to be used within a day or two. All meats should be cooked thoroughly to kill off the pathogen.

4. *Salmonella* are very resistant to drying, surviving months of total dryness. Also, this outbreak reminds us that *Salmonella* can contaminate just about every kind of food and beverage, including those not from animal sources. Even orange juice has been the source of major outbreaks of *Salmonella*.

Self-Quiz Answers

CHAPTER 1

1. True   **2.** False

3. (c)   **4.** (b)

5. They are important to study because of their relation to human health (the diseases they produce). Some are directly useful for their production of antibiotics, digestive enzymes in ruminants (breakdown of cellulose), and in the food industry (mushrooms, pickles, sauerkraut, yogurt, dairy products, beer, wine, and bread.) Microbes are important to the human environment, being the first and last links in the complex web of life (photosynthetic and chemosynthetic microbes are able to respectively capture the sun's or mineral's energy, utilizing it to make molecules that can be used by other organisms as food when they ingest them). Microbes are also responsible for the decomposition of dead organic matter, returning minerals and inorganic matter to the soil for plants to use in making protoplasm. In turn, man can eat these plants directly or indirectly by eating the herbivores (plant eaters) who feed on them. Some microbes are also beneficial to man in that they can degrade industrial waste products into less toxic or harmless chemicals. Microbes are invaluable study subjects because of their versatility in research (being able to be genetically engineered to produce products useful to man such as insulin and human growth hormone). The study of microbes also gives us insight into the life processes in all life forms.

6. False

7. (c) Algae; (f) Bacteria; (a) Fungi; (e) Protozoa; (b) Viruses; (d) Helminthes

8. False. Worms have microscopic stages in their life cycles that can cause disease, and the arthropods (as represented by the tick) can transmit these stages, as well as other disease-causing microbes.

9. Etiology is the assignment or study of causes and origins of a disease. Epidemiology is the study of factors and mechanisms involved in the spread of disease within a population.

10. (b)   **11.** (a)

12. (e)

13. The development of high-quality lenses by Leeuwenhoek made it possible to observe microorganisms and later to formulate the cell theory.

14. (d)   **15.** (d)

16. (c)   **17.** (d)

18. (a)

19. (c), (a), (d), (b)

20. Lister and Semmelweis contributed to improved sanitation in medicine by applying the germ theory and using aseptic technique.

21. I, (c), 3; II, (a), 4; III, (d), 1; IV, (b), 2

22. Beijerinck was the first to characterize viruses. Fleming discovered penicillin. Metchnikoff identified the role of phagocytosis in immune defenses.

23. (a) Viruses are smaller than bacteria. Viruses are considered to be on the "borderline" of the living and non-living; bacteria are living. Viruses are acellular, bacteria are cellular. (b) As early as the 1920s, bacteriophage therapy, which employs the use of certain viruses that attack and kill specific bacteria, has been successfully used to fight disease-causing bacteria.

24. True

25. (d) James Watson & Francis Crick; (e) Reed and colleagues; (f) Avery, McCarty, & McLeod; (b) Beadle & Tatum; (c) Selman Waksman; (a) Frederick Griffith

26. (a) Heating the broth and the neck of the flask to boiling would kill all the vegetative cells. (Note: had resistant endospores of bacteria been present in the broth, the boiling would not have destroyed them.) Boiling also forced out any remaining air, thus removing any dust-laden bacteria. (b) Cooling the broth slowly allowed air to return to the flask. The "swan-neck" bend in the flask trapped any bacteria and dust that would have entered had the neck been straight. The broth thus remains free of microbial growth. (c) Tipping of the flask allowed some of the sterile broth to contact the dust and the microbes present in the bend of the neck. Returning the flask upright allowed the contamination to reach the broth in the flask and recontaminate the flask.

CHAPTER 2

1. (c)

2. <u>A</u>tom: the smallest unit of any element that retains the properties of that element; <u>E</u>lement: matter composed of one kind of atom; <u>M</u>olecule: two or more atoms chemically combined; <u>C</u>ompound: two or more different kinds of atoms chemically combined.
M, C H_2O; E, M O_2; M, C   salt; A, E   sulfur; M, C CH_4; A, E   sodium; M, C   glucose; E, M H_2; A, E   chlorine

3. Ions are atoms that have gained or lost one or more electrons.

4. Protons, neutral, electrons, nucleus, orbits, protons and neutrons, number

5. An isotope of the same element contains a different number of neutrons.

6. (c) Solute; (d) Mixture; (a) Solution; (b) Solvent

7. (e)

8. (a) An acid is a hydrogen ion (H^+) or proton donor. The pH scale is a logarithmic scale that measures the (H^+) concentration. This means the concentration of (H^+) changes by a factor of 10 for each unit of the scale. A solution with a pH of 7 is neutral. As a solution goes from pH 7 to pH 0, it is increasing its acidity. (b) pH 3. (c) pH 13. (d) The production of sulfuric acid by *Thiobacillus thioparis* would produce a pH of 3 or lower within the marble stone environment.

9. (e)

10. (a) Monosaccharides are carbon chains or rings with attached alcohol groups plus one other functional group

such as an aldehyde or ketone group.
(b) Disaccharides are formed from two monosaccharides that are connected by the removal of water and the formation of a glycosidic bond (sugar alcohol/sugar linkage). Polysaccharides are formed from multiple (greater than two) glycosidic linkages of monosaccharides.

11. (b)

12. Living organisms use carbohydrates as a direct source of energy (glucose), energy storage (glycogen), cellular structure and support (cellulose), cell marking and signaling, and in the genetic code (nucleic acids).

13. Lipids constitute a chemically diverse group of substances that include fats, phospholipids, and steroids. Simple lipids (fats) are composed of a three-carbon glycerol and one or more fatty acids (a fatty acid consists of a long chain of carbon atoms with associated hydrogen atoms and a carboxyl group at one end of the chain). Many fats are used as a source of energy.

Some lipids contain one or more other molecules in addition to fatty acids and glycerol. One example of this is phospholipids, which differ from fats by the substitution of phosphoric acid for one of the fatty acids. Phospholipids are an integral structural component of cell membranes.

Steroids are lipids with a characteristic four-ring structure that have different chemical groups attached to the rings depending on the specific steroid. An example of a steroid is cholesterol. Steroids play many roles including cellular membrane composition, cell signaling (hormones), and in metabolic reactions as vitamins.

14. (e)   15. (d)

16. The primary structure of a protein consists of the

specific sequence of amino acids in a polypeptide chain and is maintained by peptide bonds. The secondary structure of a protein consists of the folding or coiling of amino acid chains into a particular pattern, such as an α-helix or β-pleated sheet and is maintained by hydrogen bonding between different amino acids. Tertiary structure consists of further bending and folding of the protein molecule into globular shapes or fibrous thread-like strands. Quaternary structure is formed by the association of several separate tertiary-structured polypeptide chains. Tertiary and quaternary structures are maintained by disulfide linkages, hydrogen bonds, and other forces between R groups of amino acids.

17. (e)   18. True
19. (c)   20. (a)
21. (d) Adenine; (a) Thymine; (c) Phosphate; (b) Ribose; (d) Nucleotide; (a) Deoxynucleotide; (b) Uracil; (c) Guanine; (d) Nitrogenous base
22. Adenine / uracil or thymine; Cytosine / guanine; Guanine / cytosine; Uracil / adenine; Thymine / adenine.
23. (d) Polysaccharides; (e) Polypeptide; (c) Fat; (b) DNA; (c) Steroids
24. (d)   25. (d)
26. This structure shows two amino acids joined together by a peptide bond, making it a dipeptide. (a) Amino group; (b) Peptide bond—remove the singly bonded oxygen; (c) Side group of amino acid 2; (d) Carboxyl group

CHAPTER 3

1. Width: $0.5\ \mu m \times 10^{-6} m/\mu m \times 10^9\ nm/1\ m = 5 \times 10^2\ nm$.
Length: $15\ \mu m \times 10^{-6} m/\mu m \times 10^3\ mm/1\ m = 1.5 \times 10^{-2}\ mm$
2. (e)   3. (c)
4. (a) Resolution is the ability to see two items as separate and discrete units and is

closely related to the wavelength of light used.
(b) Resolution is important to microscopy because without it, two separate objects would appear as one. Light of a wavelength short enough to pass between two objects is necessary to resolve them. If the wavelength of light is too long to pass between two separate objects then we would see them as one.

5. (a) **a**. reflection; (b) **b**. refraction; (c) **c**. transmission; (d) **d**. absorption.
Absorption occurs when light rays neither pass through nor bounce off an object but are taken up by the object. Reflection occurs when light rays strike an object and bounce back off the object into our eyes. Transmission occurs when light rays pass through an object. Refraction is the bending of light as it passes from one medium to another of different density.

6. (c)

7. (c) Fluorescence; (e) Diffraction; (a) Immersion oil; (d) Phosphorescence; (b) Luminescence

8. (b)

9. (d) Phase-contrast; (g) Dark-field; (f) Bright-field; (i) Transmission electron; (b) Confocal; (e) Scanning electron; (c) Fluorescence; (a) Nomarski

10. A simple stain makes use of a single dye and reveals basic cell shapes and cell arrangements. A differential stain makes use of two or more dyes and distinguishes between two kinds of organisms or between two different parts of an organism.

11. (c)   12. (b)
13. (c)
14. Cationic, acidic, positively
15. (f)   16. (d)
17. (e)   18. (c)
19. (e)
20. False. Although special stains can offer valuable

information about microorganisms, they are usually not enough to permit identification. Many species look identical under the microscope because there are only a limited number of shapes, arrangements, and staining reactions but thousands of kinds of bacteria.

21. (b)
22. See p. 68, Table 3.2
23. (d)
24. (c) Phase contrast microscopy; (d) Fluorescent microscopy; (b) Transmission electron microscopy; (a) Bright-field microscopy; (e) Dark-field microscopy; (f) Scanning electron microscopy
25. See p. 61
26. For functions, see p. 59
(a) Eyepiece (ocular); (b) Arm; (c) focus adjustment; (d) Illuminator; (e) condenser; (f) stage; (g) objectives
27. (b)

CHAPTER 4

1. (c)
2. (a) P Single chromosome; (b) E Membrane-bound nucleus; (c) B Fluid-mosaic membrane; (d) N Viruses; (e) B 70S ribosomes; (f) E Endoplasmic reticulum; (g) E Respiratory enzymes in mitochondria; (h) E Mitosis; (i) P Peptidoglycan in cell wall; (j) E Cilia; (k) E 80S ribosomes; (l) E Chloroplasts; (m) E "9+2" microtubule arrangement in flagella; (n) P Bacteria (o) B Can have extra chromosomal DNA; (p) E Meiosis
3. (c) Coccus; (a) Bacillus; (d) Spirillum; (f) Vibrio; (b) Staph; (e) Tetrad
4. True
5. Bacterial cell membranes contain phospholipids and proteins. Phospholipids in the membrane are in a fluid state and proteins are dispersed among the lipid molecules in the membrane, forming a mosaic pattern.

Membrane phospholipids form a bilayer with the charged phosphate ends of the lipid molecules located at the inner and outer membrane surfaces where they are hydrophilic (water-loving) and can interact with the watery environment. The fatty acid ends of the phospholipids extend inward from both charged phosphate surfaces where they are tucked away from water and hydrophobic (water-hating), forming a barrier between the cell and its environment.

Interspersed among the phospholipid molecules are different protein molecules (the mosaic), some of which extend through the entire membrane, forming pores or channels that selectively let materials into or out of the cell and other proteins that are embedded in or loosely attached to the inner or outer surface membrane providing different functions (enzymatic, signaling, structural).

6. (a), (c), (d), (e)
7. (b)   **8.** (e)
9. (e)
10. (d) Phototaxis; (f) Flagellum; (b) Conjugation pili; (g) Slime layer; (h) Chemotaxis; (c) Glycocalyx; (a) Axial filament; (e) Capsule
11. See p. 109, Figure 4.31
12. (d)   **13.** (d)
14. Cilia have the same basic 9 + 2 arrangement of microtubules found in flagella. Cilia allow faster movement of an organism compared to flagella, and in some cells they can propel fluids, dissolved particles, bacteria, mucus, and so on past the cell. This function is especially important in host defenses against disease. Pseudopodia are temporary projections of cytoplasm in cells without walls. This renders the cell capable of amoeboid movement.

Flagella are found in many Eukaryotes (protozoans, algae, mammals-sperm cells) and in prokaryotes (many species of bacteria). Cilia are found in eukaryotes (protozoans and mammals—cells lining the respiratory tract). Pseudopods are found in eukaryotes (amoebas and mammals—white blood cells).

15. (c)
16. (c) Cell wall; (b) Lipopolysaccharide; (c) Flagellum; (d) Cilia; (a) Teichoic acid
17. (d)   **18.** True
19. Cell membrane/ mitochondria
20. Chloroplast characteristics include: chloroplasts are found in eukaryotic plant and algal cells only, contain the pigment chlorophyll, are membrane-enclosed, and contain DNA that is replicated independently of the cell in which they function. The thylakoid membranes inside of the chloroplast contain chlorophyll pigment and the enzymes needed to harness and capture the energy of sunlight for its use in manufacturing a useful form of chemical energy (photosynthesis).
21. Positive, spheroplasts, L-forms
22. (e) Cytoskeleton; (b) Lysosomes; (a) Smooth endoplasmic reticulum; (c) Rough endoplasmic reticulum; (d) Nucleus
23. (a)
24. (c) Facilitated diffusion; (a) Osmosis; (g) Simple diffusion; (b) Passive transport; (f) Active transport; (e) Hypotonic solution; (f) Isotonic solution; (d) Hypertonic solution
25. (b)
26. (a) flagellum; (b) inclusion; (c) ribosomes; (d) pilus (fimbria); (e) chromosome; (f) capsule or slime layer; (g) cell wall; (h) cell membrane; (i) cytoplasm; (j) plasmid; For functions see p. 111–112.

CHAPTER 5

1. (c)
2. (c) Photoautotrophs; (a) Chemoautotrophs; (b) Photoheterotrophs; (d) Chemoheterotrophs
3. (b) Many microorganisms in this group are infectious; (a) Many microorganisms in this group can carry out photosynthesis; (a) Members of this group usually do not cause disease; (b) Members of this group carry out the same metabolic processes as man; (b) Members of this group break down organic compounds to obtain energy; (a) Members of this group synthesize organic compounds to obtain energy.
4. (c) Oxidation; (a) Catabolic reaction; (d) Anabolic reaction; (e) Reduction; (b) Phosphorylation
5. (b) Cellular enzymes decrease or lower the AE required for the reaction to proceed.
6. Metabolism is the sum of all chemical processes carried out by living organisms. A subdivision of metabolism is anabolism which are reactions that require energy to synthesize complex molecules from simpler ones. Catabolism is the other subdivision of metabolism and involve reactions that release energy by breaking complex molecules into simpler ones, which in turn, can be reused as building blocks.
7. (e), (g) Membrane transport; (f) Bioluminescence; (b), (d) Movement; (a), (c) Biosynthesis
8. True.   **9.** (a) and (d).
10. True   **11.** (a)
12. (e)   **13.** (e)
14. (b)   **15.** (a)
16. (e)   **17.** (c)
18. (a)
19. No. There are the pentose phosphate and Entner-Doudoroff pathways.
20. Fats are hydrolyzed to 3 fatty acids and glycerol. The glycerol is metabolized in glycolysis. Fatty acids are

broken down into two-carbon pieces by a metabolic pathway called beta oxidation. This results in the formation of acetyl-CoA which enters the Krebs cycle and is oxidized to obtain additional energy.

21. (b) Oxidative phosphorylation; (d) Chemiosmosis; (c) Flavoproteins, cytochromes, & quinones; (a) Electron transport
22. (c)   23. (b)
24. (b)
25. (b) Chemiosmosis; (a) Glycolysis; (f) Electron transport chain; (c) Fermentation; (d) Phytosynthesis; (e) Krebs cycle
26. (a) Apoenzyme; (b) Cofactor; (c) Substrate; (d) Active site; (e) Coenzyme; (f) Allosteric site; (g) Holoenzyme

CHAPTER 6

1. True   2. False
3. (c)
4. (b) Decline / death phase; (c) Stationary phase; (e) Lag phase; (f) Chemostat; (d) Log phase; (a) Medium
5. (b)
6. (a) 64. (b) Most likely around 64. The freezing would prevent growth. During the 30-minute post freezing period the cells are thawing out and probably would not approach a logarithmic phase during this time. Freezing can also kill cells.
7. (b)   8. (e)
9. (c)
10. High concentrations of dissolved substances exert sufficient osmotic pressure to kill or inhibit microbial growth.
11. (e)
12. plasmolysis / catalase / breaks / hydrostatic
13. (d)
14. (e) Aerotolerant anaerobe; (b) Obligate aerobe; (c) Capnophile; (d) Microaerophile; (e) Facultative anaerobe; (a) Obligate anaerobe

15. (e)   16. (e)
17. (c)   18. (c)
19. False   20. (a)
21. (e)
22. The streak plate method involves the spreading out of bacteria across a sterile, solid surface such as an agar plate. It is done to isolate a single pure colony so it can be picked up to inoculate a pure culture, avoiding cross contamination with other possible species of microbes on the plate. The pour plate method involves serial dilution of a culture of microbes and transferring a measured volume of the dilutions to melted agar, mixing, and plating. In this way pure isolated and separate colonies grow out and in the agar plate. These colonies can be picked and transferred to another plate or medium to obtain a pure culture.
23. (e)   24. (e)
25. (c)   26. b; c; a; b; d

CHAPTER 7

1. (c) Heredity; (a) Chromosome; (e) Phenotype; (f) Gene; (g) Alleles; (b) Mutation; (d) Genotype
2. (a)   3. (b)
4. (c)
5. (c) Semiconservative replication; (d) Anti-codon; (f) Translation; (a) Replication fork; (b) Transcription; (e) Okazaki fragment
6. mRNA = UACGUCAUC; tRNA anticodons = ATG, CAG, UAG; amino acids = methionine, glutamine, stop; There is a terminator (nonsense) codon = AUC.
7. (d)   8. (b)
9. (b)
10. In their evolution, bacteria and all other organisms have developed mechanisms to turn reactions on and off in accordance with their needs. All cells try to limit their waste of energy and materials. Control mechanisms have evolved to regulate metabolic activity so as to

produce only what is needed, not to squander energy, and not to excessively produce wasteful amounts of enzymes and materials.
11. The mechanisms that control metabolism either regulate pre-existing enzyme activity directly or regulate the synthesis of an enzyme by turning on or off the corresponding gene that codes for the particular enzyme.
12. (e) Enzyme repression; (f) Feedback inhibition; (a) Catabolite repression; (c) Enzyme induction; (d) Repressor; (b) Operon
13. (d) Inducer; (e) Place where repressor binds to shut off operon; (f) Substance that binds to promoter site to start transcription; (d) Combines with repressor to keep operon "on"; (c) Z, Y, A; (a) May be located some distance from the operon and is not under control of the promoter; (g) Protein that binds to operator preventing transcription of structural genes
14. (b)   15. (b)
16. (e)   17. (e)
18. (b)   19. (c)
20. (c)   21. (e)
22. (a) Microorganisms are useful in the study of mutations because of their short generation times and their relatively low expense in maintaining large populations of mutant organisms. (b) A spontaneous mutation is a random mutation that occurs in the absence of any agent known to cause changes in DNA. Induced mutations are produced by agents called mutagens that increase the mutation rate above the spontaneous mutation rate. (c) The fluctuation test demonstrates that resistance to chemical substances occurs spontaneously rather than being induced. Replica plating does the same as the fluctuation test but in addition allows one to isolate mutant bacterial

colonies without exposing them to the substance to which they are resistant. (d) The Ames test is used to detect the ability of auxotrophic bacteria to revert back to their original synthetic ability. The test is used for screening chemicals which may be mutagens; this may indicate their potential for being carcinogenic in man.
23. light repair; enzyme – controlled
24. (d)   25. (d)
26. (a) Frameshift (deletion) would result in a significant change in reading of message. (b) Point (base substitution) would result in a significant change in reading and in the creation of a stop signal. (c) Frameshift (insertion of two bases) would result in a significant change in reading of message.

CHAPTER 8

1. Gene transfer is the movement of genetic information between organisms. In most eukaryotes, it is an essential part of the organism's life cycle and usually occurs by sexual reproduction. When genes pass from parents to offspring, it is called vertical gene transfer. Sexual reproduction is usually what we think of as vertical gene transfer. However, in bacteria vertical gene transfer takes place asexually by binary fission. Bacteria also do horizontal or lateral gene transfer when they pass genes to other microbes of their same generation. Bacteria have three main mechanisms of lateral gene transfer: transformation, transduction, and conjugation. Gene transfer is significant because it increases the genetic diversity of organisms, which increases the likelihood that some will adapt to changing environmental conditions and not die out. This diversity leads to evolutionary changes.

2. (b) Uptake of naked DNA; (c) Virus involved; (a) F⁺, F⁻, Hfr; (b) Competence factor; (a) F pilus

3. False   4. (d)

5. (b)

6. The three mechanisms of conjugation are **F** plasmid transfer, **Hfr**, and **F'**. In F plasmid transfer a whole extrachromosomal piece of DNA (plasmid) is transferred from **F⁺**, sex pilus bearing donor cells to **F⁻**, recipient cells. The recipient cell now becomes **F⁺**.

    In the high-frequency recombination (**Hfr**) mechanism, only parts of F plasmids that have been incorporated into the bacterial chromosome (initiating segment) are transferred along with adjacent bacterial genes. The recipient cell does not become **F⁺**, as only part of the F plasmid is transferred.

    In the **F'** mechanism, an F plasmid incorporated into the chromosome and bearing one or more bacterial genes from the previous donor subsequently separates from the chromosome and is completely transferred along with a fragment of the new chromosome (and bearing one or more genes) to the recipient **F⁻** cells. Such a transferred plasmid is known as a **F'** plasmid.

7. (b)   8. False

9. (a), (b), (c) Contributes to genetic diversity; (c) Provides a means of mapping genes in bacterial chromosomes; (b) Suggests a mechanism for viral origins of cancer; (c) May represent an evolutionary stage between asexual and sexual reproduction; (a) Can be used to create recombinant DNA; (b) Demonstrates a close evolutionary relationship between prophage and host cell DNA

10. (e)   11. (c)

12. (a)   13. (e)

14. (a)

15. (b) Gene Amplification; (a) Genetic Engineering; (d) Protoplast fusion; (c) Restriction Endonucleases; (f) Genetic Fusion; (e) Transgenic

16. (b)   17. (d)

18. True   19. (c)

20. (a)   21. (d)

22. (e)   23. (d)

24. (d)

25. (a) Lytic cycle; (b) Lysogenic cycle; (c) Empty phage heads and pieces of phage DNA are assembled; (d) Phage is replicated along with bacterial DNA; (e) Phage is adsorbed to receptor site on bacterial cell wall, penetrates it, and inserts DNA

CHAPTER 9

1. classification / characteristics / who / disease

2. (c)   3. (a)

4. (a)   5. (a), (b), (d)

6. (c)

7. Refer to Table 9.2.

8. (d)   9. (e)

10. (d)   11. (e)

12. Depending on the Family, viral genomes can be single-stranded RNA, double-stranded RNA, single-stranded DNA, or double-stranded DNA.

13. (e)   14. (d)

15. (e) Genetic homology; (d) Phage typing; (a) Protein profiling; (f) Numerical taxonomy; (b) DNA hybridization; (c) G–C content

16. (b)   17. (a)

18. (c)

19. Viroids / RNA / prions / brain

20. (c)   21. (e)

22. (e)

23. (c) Animalia; (d) Plantae; (a) Protista; (e) Monera; (b) Fungi

24. (b)

25. Genus = *Mycobacterium*, Specific epithet = *tuberculosis*, Species name = *Mycobacterium tuberculosis*

26. (1) Attachment to host cell; (2) Entry by phagocytosis; (3) Conversion to reticulated bodies; (4) Reproduction of reticulate bodies; (5) Condensation of reticulate bodies; (6) Release of elementary bodies; (a) Host cell; (b) Nucleus; (c) Reticulate body; (d) Elementary body.

CHAPTER 10

1. enterovirus / 300

2. (c) Capsid; (b) Virion; (d) Spike; (a) Envelope; (f) Naked virus; (e) Nucleocapsid

3. (a)   4. (c)

5. True   6. (d)

7. (c)   8. (b)

9. (a)   10. (d)

11. (c)   12. (b)

13. (e)   14. (e)

15. (d), 5, Release; (c), 1, Adsorption; (e), 4, Maturation; (b), 2, Penetration; (a), 3, Synthesis

16. (e)   17. (c)

18. (e)   19. (e)

20. (d) Continuous cell line; (c) Primary cell culture; (e) Monolayer; (f) Subculturing; (a) Cytopathic effect; (b) Diploid fibroblast strain

21. (d)   22. (d)

23. (b) Prion; (a) Satellite; (d) Virusoid; (c) Satellite viruses; (f) Viroids; (e) Hepatitis delta virus

24. (e)   25. (d)

26. Phage growth curve: (a) Latent period; time from penetration through maturation. (b) Eclipse period; time from penetration through biosynthesis. (c) Viral yield (burst size); number of phage progeny released from one host cell.

CHAPTER 11

1. obligate / facultative

2. (e)

3. (a), (e), (f) Lice; (b), (d) Tapeworm; (a), (e), (f) Biting mosquito; (g) housefly walking on manure; (a), (c) Ringworm fungus

4. (c)

5. (b) Accidental parasite; (e) Host specificity; (d) Intermediate host; (f) Reservoir host; (a) Definitive host; (c) Obligate parasite

6. (d)

7. (b) Slime mold; (c) Ciliate; (a) Dinoflagellate; (c) Apicomplexan; (a) Euglenoid; (a) Diatom; (c) Mastigophoran; (c) Sarcodine

8. saprophytes / Cellular slime molds / pseudoplasmodium / plasmodial slime molds

9. (e)   10. (b)

11. (c)   12. (a)

13. (c)

14. (b) Sac fungi; (d) Water molds; (a) Bread molds; (c) Club fungi; (e) Dimorphic fungi

15. (c)   16. (c)

17. (d)   18. (d)

19. (d) *Wucheria bancrofti*; (c) *Taenia* species; (b) *Trichinella spiralis*; (a) *Enterobius vermicularis*; (e) *Fasciola hepatica*

20. (d)   21. (c)

22. (f), 2 Yellow Fever; (a), 3 Rocky Mountain Spotted Fever; (e), 4 African Sleeping sickness; (f), 2 Dengue Fever; (d), 3, 5 Tularemia; (c), 3, 6 Q Fever; (b), 3 Lyme Disease

23. (d)   24. (d)

25. Arachnids (e); Crustacean (c); Insects (d)

26. (a) Scolex; (b) Proglottids. The oldest proglottids are at the end of the tapeworm. The newest ones are nearest the germinal center next to the scolex.

CHAPTER 12

1. (d) Bacteriostatic; (a) Germicidal; (b) Viricidal; (e) Sporicidal; (f) Fungicidal; (c) Bacteriocidal

2. (c)

3. Probably not. The 36-hour culture is probably in the logarithmic phase of growth (a time of maximum susceptibility to antimicrobial agents) and would be killed within 6 minutes. However, the 2-week-old culture is not in the logarithmic growth phase and therefore not as

susceptible to the same anti-microbial agent. As a result, it would probably take longer for all the cells to be killed in this culture.

4. False. There are no degrees of sterility. When properly carried out, sterilization procedures ensure that even highly resistant bacterial endospores and fungal spores are killed.

5. (e)   6. (e)

7. (b) Surfactant; (a), (c) Alkylating agents; (a) Oxidation agents; (b) Detergents; (a) Hydrolyzing agents; (a) Heavy metals; (d) Crystal violet dye

8. retard / kill / denaturing / lipids

9. (d)   10. False
11. (a)   12. (d)
13. (b)   14. (c)
15. (c), (d) Phenol derivatives; (c) Iodine; (c) Alcohols; (a) Acids; (e) Chlorine; (b) Oxidizing agents; (a) Nitrates
16. (d)   17. True
18. (b)   19. (d)
20. (e)   21. (b)
22. (c)   23. (c)
24. (a)   25. (e)

26. Disinfectant A has some inhibitory effect on Gram-positive bacteria but no effect on Gram-negative bacteria. Disinfectant B has no effect on either type of bacterium. Disinfectants C and D appear to be very effective on Gram-positive bacteria but only slightly affect Gram-negative bacteria.

CHAPTER 13

1. (c) Synthetic drug; (e) Antimicrobial agent; (b) Chemotherapy; (a) Semisynthetic drug; (d) Antibiotic; (f) Chemotherapeutic agent

2. Development of chemotherapy since Ehrlich's time first started with the concurrent development of sulfa drugs by Gerhard Domagk, with prontosil inhibition of Gram-positive bacteria, and

antibiotics by Alexander Fleming with *Penicillium* mold, in the 1930s. Ernst Chain and Howard Florey in the 1940s helped to isolate and mass produce penicillin. After World War II, many new antibiotics were discovered and marketed.

3. Many novel fungi and bacteria live there.

4. Bacteriostatic disinfectants only prevent growth, whereas bactericidal kill microbes.

5. (e) Sulfanilamide; (c) Erythromycin; (a) Penicillin; (d) Rifamycin; (b) Polymyxin; (d), (e) Purine analog vidarabine; (c) Streptomycin

6. (d)   7. (a)
8. (c)
9. few / broad-spectrum
10. (a)   11. (c)
12. (e)   13. (d)
14. (e)   15. (b)
16. (d)   17. (b)
18. (d)   19. (d)
20. (b)
21. (d) Acyclovir; (b) Ganciclovir; (c) AZT; (e) Idoxuridine; (a) Ribavirin
22. (a)   23. (d)
24. (a)
25. (c) Chromosomal resistance; (f) Nongenetic resistance; (e) Synergism; (b) Extrachromosomal resistance; (a) Cross-resistance; (d) Resistance
26. Box 1: Inhibition of cell wall synthesis; Antibiotics: penicillin, bacitracin, cephalosporin, vancomycin. Box 2: Disruption of cell membranes function; Antibiotics: polymyxin. Box 3: Inhibition of protein synthesis; Antibiotics: tetracycline, erythromycin, streptomycin, chloramphenicol. Box 4: Inhibition of nucleic acid synthesis; Antibiotics: rifamycin (transcription), quinolones (DNA replication), metronidazole. Box 5: Action as antimetabolites; Antibiotics: sulfonilamide, trimethoprim.

CHAPTER 14

1. (c) Parasitism; (e) Pathogen; (a) Symbiosis; (f) Commensalism; (d) Host; (b) Mutualism
2. True.   3. (e)
4. (c)   5.
6. (a)   7. (d)
8. (e)   9. (e)
10. (b)   11. False.
12. (a)   13. (e)
14. (c)   15. (b)
16. (d)   17. (c)
18. (e)   19. (b)
20. (c)   21. (a)
22. (e)   23. (a)
24. (e)   25. False
26. (a) The incubation period is the time between infection and the appearance of signs and symptoms; (b) The prodromal phase is the period during which pathogens invade tissues; it is marked by early nonspecific symptoms; (c) The illness phase is the period during which the individual experiences the typical signs and symptoms of the disease. During this phase, the signs and symptoms reach their greatest intensity at the acme; (d) The acme is the time when the signs and symptoms reach their greatest intensity; (e) The decline phase is the stage during which signs and symptoms subside and the host defenses overcome the pathogens; (f) The convalescence period is the stage during which tissue damage is repaired and the patient regains strength.

CHAPTER 15

1. Epidemiology is the study of factors and mechanisms in the spread of diseases in a population.
2. False
3. (b) Persons in a population who become clinically ill during a specified period of time; (d) The total number of people infected in a population at a particular time; (e) The colonization and growth of an infectious agent in a host; (a) The

number of new cases of a disease identified in a population during a defined period of time; (c) The number of deaths within a population during a specified period of time

4. True   5. (b)
6. (a)   7. (b)
8. (d)   9. (c)
10. (c)   11. (d)
12. (c)   13. (c)
14. (e)   15. (d)
16. (c) Fomites; (c) Bar soap; (a) Kissing; (f) Speaking; (a) Hand shaking; (d) Food; (g) Housefly; (b) Mother breast-feeding her infant; (e) *Anopheles* mosquito; (c) Stepping on a rusty nail
17. (c)   18. (e)
19. (a)   20. (b)
21. (b)   22. (c)
23. (d)   24. (a)
25. (b) Zoonoses; (d) Fomite; (a) Droplet nuclei; (c) Exogenous infection; (e) Endogenous infection
26. (a) If a child with measles moves to city A, what is the chance that child will encounter a susceptible child? 100 non-immunized children divided by 10,000 total children = 1/100 or a 1% chance that a child with measles coming to city A will encounter a susceptible child.

   (b) If a child with measles moves to city B, what is the chance that child will encounter a susceptible child? 5000 children in city B have had measles and are presumed to now be immune from the last outbreak. This leaves 5000 susceptible children out of a total of 10,000 children or 1/2 (50%) chance that a child with measles coming to city B will encounter a susceptible child.

   (c) Comparing the two scenarios, which city has the higher herd immunity, and in which city is an infected child more likely to transmit the disease to a susceptible child? City A has the higher herd immunity since only

1% of their children are susceptible to measles versus 50% in city B. Since city B has the lower herd immunity, a measles-bearing child is more likely to transmit the disease here.

CHAPTER 16

1. (c) Lysozyme; (d) Very acidic pH; (b) Sebum, and fatty acids; (a) Low pH, flushing action of urine; (f) Mucociliary escalator; (e) Phagocytes
2. (e) 3. True
4. (b) Lymphocytes; (a) Neutrophils; (a) Granulocytes; (b) Antibodies; (a) Mucous membrane; (a) Phagocyte
5. (d) 6. (d)
7. (a) Phagocytes first recognize invading microbes by their pattern recognition receptors (PRPs) that recognize molecular patterns unique to the pathogen. (b) In addition, infectious agents and damaged tissues at the infection site both release specific chemical substances that attract phagocytes. At the infection site phagocytes release cytokines, which are a diverse group of soluble proteins, each having a specific role in host defenses including activation of cells involved in the inflammatory response. (c) Chemokines (a class of cytokine) attract additional phagocytes to the infection site via chemotaxis (the movement of a cell toward a chemical stimulus). (d) The phagocyte's cellular membrane must then bind to specific molecules on the surface of the microbes in a process called "adherence." (e) Once captured, the phagocytic membrane extends outward forming pseudopodia that surround the microbe. The pseudopodia then fuse, engulfing the microbe within a cytoplasmic vacuole called a phagosome. (f) Once ingested, phagocytes can digest the microbe

in several ways. One way is through the fusion of lysosomes with the phagosome forming a phagolysosome. The lysosome brings many types of antimicrobial digestive enzymes to kill the microbe. Phagocytes can also use oxidative methods to kill ingested microbes. (g) Any indigestible material remains in the phagolysosome (now called a residual body) which is then transported to the plasma membrane and is excreted as waste.

8. Immune cytolysis is another complement C3b mediated immune response that causes complement proteins to produce lesions in the cell membranes of microorganisms and other types of cells. The lesions cause cellular contents to leak out. C3b initiates splitting of C5 into C5a and C5b. C5b then binds C6 and C7, forming a hydrophobic C5bC6C7 complex that inserts into the microbial cell membrane. C8 then binds to C5b in the membrane, forming a C5bC6C7C8. This complex causes multiple C9 molecules to assemble completely through the cell membrane, forming a pore. This C9 pore constitutes the membrane attack complex (MAC) and is responsible for the direct lysis of invading microbes.
9. (d) 10. (e)
11. (e)
12. (c) Neutrophils; (a) Eosinophils; (d) Lymphocytes; (b) Monocytes
13. (d) 14. (b)
15. (e) 16. (d)
17. (c)
18. (d) Pyrogen; (e) Chronic inflammation; (f) Leukocytosis; (b) Acute inflammation; (c) Edema; (a) Bradykinin
19. (c) 20. True
21. (d) 22. (d)
23. (a)
24. 5 The acute phase proteins can now activate the

complement system and immune cytolysis and stimulate phagocyte chemotaxis. 3 C-reactive protein recognizes and binds to phospholipids and mannose-binding protein to mannose sugars, in cell membranes of many bacteria and the plasma membrane of fungi. 2 Interleukin-6 reaches the liver via the bloodstream where it causes the liver to synthesize and secrete the acute phase proteins (C-reactive and mannose-binding proteins) into the blood. 4 Once bound, the acute phase proteins act like opsonins. 1 Macrophage ingestion of microbe stimulates synthesis and secretion of interleukin-6.
25. (e)
26. The steps are as follows: (a) Invading microorganisms are located by chemotaxis. The phagocyte cell membrane adheres to the surface of the microbe. (b) Ingestion occurs as the phagocyte surrounds and ingests a microbe or other foreign substance into a phagosome. (c) Digestion occurs as lysosomes surround a vacuole and release their enzymes into it; enzymes break down the contents of the phagolysosome and produce substances toxic to the microbe. (d) Any indigestible material remains in the phagolysosome, which is now called a residual body. (e) The phagocyte transports the residual body to the plasma membrane, where the waste is excreted.

CHAPTER 17

1. Naturally acquired adaptive immunity is often obtained from having a specific disease and results in T cell activation and antibody production against antigens on invading infectious microbes. A fetus who receives antibodies from its mother

across the placenta or in colostrums or breast milk is also considered to have naturally acquired immunity. In contrast, artificially acquired adaptive immunity is obtained by receiving an antigen by the injection of vaccine or immune serum that produces immunity.

2. Active immunity is created when a person's own immune system activates T cells or produces antibodies or other defenses against an infectious agent. It can last a lifetime or for a period of weeks, months, or years, depending on how long the antibodies persist. Two types of active immunity exist. They are naturally acquired or artificially acquired. The former is produced when a person is exposed to an infectious agent. The latter is produced when a person is exposed to a vaccine containing live, weakened, or dead organisms or their toxins. In both types of active immunity, the host's own immune system responds specifically to defend the body against an antigen. The immune system also "remembers" the antigen to which it has responded and will mount another vigorous response any time it encounters that same antigen again.

   Passive immunity is created when ready-made antibodies are introduced into the body. The immunity is passive because the host's own immune system does not make antibodies. There are two types of passive immunity. 1) Naturally acquired passive immunity whereby antibodies made by a mother's immune system is transferred to her offspring. 2) Artificially acquired passive immunity is produced when antibodies made by other hosts are introduced into a new host. Rabbit antibodies against rattlesnake venom called antivenom is injected into a person who has been bitten by a rattlesnake in order to protect him from the effects of the snake venom. This type of immunity is temporary because the host's immune system is not stimulated to respond and the antivenom is eventually destroyed by the host.

3. A hapten is a small molecule that alone cannot elicit an immune response. It can act as an antigen if it binds to a larger protein molecule. Here the hapten acts as an epitope on the surface of the protein. Together, they can elicit an immune response.

4. (d) 5. (c)
6. (c)
7. (d) IgG; (e) IgA; (c) IgM; (a) IgE; (b) IgD
8. (e) 9. (b)
10. class / variable
11. (c) 12. (c)
13. (b)
14. The sensitized lymphocytes are used because each makes a particular, specific antibody. The malignant cells are used because they will keep dividing indefinitely allowing the hybridomas to be able to be cloned and cultured in the laboratory.
15. (b) 16. (b)
17. (d) 18. (e)
19. (b), (d), (c), (e), (a)
20. (d) 21. (e)
22. (b) 23. (d)
24. (a) 25. (b)
26. Antigen binds to the Fab portion of the antibody molecule. Complement binds to the Fc portion of the antibody molecule.

CHAPTER 18
1. (g) Allergen; (e) Primary immunodeficiency; (f) Anaphylaxis; (c) Atopy; (b) Desensitization; (d) Secondary immunodeficiency; (a) Autoantibodies
2. (d) 3. (e)

4. (a) 5. (b)
6. (c) 7. (e)
8. (d)
9. serum sickness / Arthus reaction / immune complex
10. (c) Rheumatic fever; (e) Rheumatoid arthritis; (f) Ulcerative colitis; (g) Myasthenia gravis; (b) Scleroderma; (d) Pernicious anemia; (a) Systemic lupus erythematosus
11. (a) 12. (e)
13. True 14. (a)
15. (c) 16. (d)
17. (d) 18. (c)
19. (c) 20. (b)
21. (e) 22. (a)
23. (e)
24. (c) Immunoelectrophoresis; (a) Hemagglutination inhibition; (b) Neutralization reaction; (f) Complement fixation test; (d) Precipitin test; (e) Coomb's antiglobulin test
25. (d) 26. (b)/(a)

CHAPTER 19
1. (a) 2. (b)
3. (c) Lysozyme; (a) Eyelashes; (a) Eyelid; (b), (d) Mucus; (d) IgA; (a) Conjunctiva; (a), (c) Tears
4. (b) 5. (c)
6. (e) 7. (e)
8. (d) 9. (e)
10. (d) Smallpox; (e) Cowpox; (a) Rubella (German measles); (b) Rubeola (measles); (f) Papilloma (warts); (c) Chickenpox
11. (e) 12. (b)
13. (c) 14. (a)
15. False 16. (d)
17. (c) 18. (d)
19. (d) 20. (c)
21. (b)
22. (g) Adenovirus; (c) *Sarcoptes scabiei*; (d), (e) *Chlamydia trachomatis*; (d) *Neisseria gonorrhoeae*; (f) *Pasteurella multocida*; (b) Enterovirus; (a) *Candida albicans*
23. True
24. (d) Swimmer's itch; (a) Cat scratch fever; (c) Loaiasis; (b) River blindness; (e) Myiasis

25. Normal flora can be disturbed, allowing pathogens to overgrow, causing infection
26. (a) Epidermis; (b) Dermis; (c) Subcutaneous layer; (d) Hair shaft; (e) Sebaceous gland; (f) Nerve; (g) Sweat duct; (h) Hair follicle; (i) Sweat gland; (j) Blood vessels; (k) Fat

CHAPTER 20
1. (e) 2. (b)
3. (b)
4. (d) Glomerulonephritis; (b) Toxic shock syndrome; (e) Pyelonephritis; (f) Leptospirosis; (c) Prostatitis; (a) Vaginitis
5. (d) 6. (b)
7. (e) 8. (a)
9. (a) 10. (b)
11. (f) Incubation stage; (e) Primary stage; (d) Primary latent period; (c) Secondary stage; (b) Secondary latent stage; (a) Tertiary stage
12. (d) 13. (c)
14. (c) 15. (d)
16. (b) 17. (d)
18. (e) 19. True
20. (a) 21. (d)
22. (d) 23. (c)
24. (e) 25. (d)
26.

Disease	Mode of Diagnosis	Treatment
Gonorrhea	Culture tests	Penicillin/broad-spectrum drugs
Syphilis	Dark-field microscopy/immunology	Penicillin/broad-spectrum drugs
Chancroid	Observe bacteria in lesions	Tetracyclines
Lymphogranuloma venereum	Find inclusions in pus	Tetracyclines
Granuloma inguinale	Observe Donovan bodies	Broad-spectrum drugs

CHAPTER 21
1. (e) 2. (a)
3. (e) 4. (d)
5. (e) 6. (d)
7. (c) 8. (b)
9. (a) 10. (c)
11. (e) 12. (a)
13. (e) 14. (c)
15. (e) 16. (a)
17. False 18. (b)
19. (c) 20. False
21. True 22. (a)
23. (a)
24. (a) *Corynebacterium diphtheriae*; (b) *Streptococcus pneumoniae*; (c) *Mycoplasma pneumoniae*; (d) *Histoplasma capsulatum*; (e) *Cryptococcus neoformans*
25. (e) *Streptococcus pyogenes*; (c) *Haemophilus influenzae*; (g) *Mycoplasma pneumoniae*; (a) *Legionella pneumophilia*; (b) *Corynebacterium diphtheriae*; (c) *Bordetella pertussis*; (f) *Mycobacterium tuberculosis*
26. Upper respiratory tract; see Table 21.1; lower respiratory tract; see Table 21.4

CHAPTER 22
1. (b) 2. (a)
3. (c) 4. (e)
5. True 6. (e)
7. (b) 8. (e)
9. (d)
10. (d) Typhoid fever; (g) Traveler's diarrhea; (a) Vibriosis; (e) Salmonellosis; (f) Shigellosis; (h) Asiatic cholera; (c) Dysentery; (b) Enteric fever
11. (c) 12. (a)
13. (a) 14. (a)
15. (e) 16. (e)
17. (a) 18. (b)
19. (g) 20. (b)
21. (e) 22. (a)
23. (d) 24. (e)
25. (c)
26. (a) Pharynx; (b) Esophagus; (c) Liver; (d) Gallbladder; (e) Duodenum; (f) Jejunum; (g) Ileum; (h) Appendix; (i) Mouth; (j) Salivary glands, (k) Stomach; (l) Pancreas; (m) Transverse colon;

(n) Ascending colon;
(o) Descending colon;
(p) Sigmoid colon;
(q) Rectum.

All of the gastro-intestinal system is lined by epithelial cells that form epithelial tissue which blocks pathogens from entering the bloodstream and other tissues.

- *Mouth:* Salivary glands secrete mucus that contains lysozyme that kills bacteria and antibodies that react with bacteria
- *Stomach:* Low pH
- *Small intestine:* Peristalsis, liver detoxifies toxins, mucus
- *Large intestine:* Normal microbiota

CHAPTER 23

1. (e)	**2.** (b)
3. (c)	**4.** (a)
5. (d)	**6.** True
7. (e)	**8.** (d)
9. (e)	**10.** (c)
11. (a)	**12.** (a)
13. (c)	**14.** (d)
15. (b)	**16.** (e)

17. (e) Rickettsialpox; (c) Bartonellosis; (f) Trench fever; (g) Rocky mountain spotted fever; (a) Ehrlichiosis; (b) Bacillary angiomatosis; (d) Brill-Zinsser disease

18. (a)	**19.** (d)
20. (c)	**21.** (e)
22. (d)	**23.** (e)
24. (d)	

25. (d), 2, 5, Leishmaniasis; (c), 3, Malaria; (a), (b), 4, Babesiosis; (a), (b), 1, Toxoplasmosis

26. (c)

CHAPTER 24

1. (b), (f), (j), Earlier; (a), (c), (d), (e), (g), (h), (i), Later

2. (c)	**3.** (d)
4. (d)	**5.** (b)
6. (a)	

7. *Listeria monocytogenes / Streptococcus pneumoniae / Neisseria meningitidis*

8. (c)	**9.** (e)
10. (b)	**11.** False
12. (e)	**13.** (d)
14. (b)	**15.** (c)
16. (a)	**17.** (e)
18. (c)	**19.** (d)
20. (d)	**21.** (b)

22. (b), (c), (f), (g), (h), African sleeping sickness; (a), (d), (e), (i), Chagas' disease

23. (c)	**24.** (a)
25. (a)	

26. There are many possible answers. Below are some examples:

Brain:

(a) *Bacterial meningitis:* unknown how it gets to site, causes disease by growing in cerebrospinal fluid, making toxins and products that cause disease. Bacterial meningitis due to *Haemophilus influenzae* can be prevented by vaccine. Bacterial meningitis is treated with antibiotics.

(b) *Rabies:* bite delivers virus to bloodstream and then travels to brain, immunize pets and other animals, immunize people who handle wild animals. Disease can be treated with vaccine after exposure.

(c) *Polio:* pathogen travels from intestine to blood to brain, vaccine is available to prevent disease. Treatment consists of supportive therapy for damaged muscles, e.g., braces, iron-lung machine.

(d) *West Nile fever:* mosquito bite, no vaccine, supportive measures only, no cure.

Spinal cord:

(a) *bacterial meningitis:* see Brain

(b) *polio:* see Brain

Ganglion:

(a) *Chagas' disease:*, bite from infected reduviid bugs and feces of bug enters bloodstream or by rubbing feces into eye from a bite then to bloodstream. *Trypanosoma cruzi* infects heart nerve ganglia, no good drug treatment, prevention involves removing bugs with insecticides.

Peripheral nerves:

(a) *Botulism:* eat preformed toxin in food, crosses intestine and travels to peripheral nerves where it interferes with neuromuscular junction and causes flaccid paralysis. Prevention consists of cooking foods thoroughly and canning foods properly. Treatment consists of supportive therapy and administration of antitoxin.

CHAPTER 25

1. (g) Abiotic factor; (e) Biotic factor; (d) Ecosystem; (c) Indigenous organisms; (f) Community; (b) Biosphere; (a) Ecology

2. (c), (e), Producer; (a), (b), (d), (f), Consumer; (b), (g), Decomposer

3. (d)

4. 2, 4, (a), (d), (e), (f) Nitrogen-fixing bacteria 3, (b), (c), (g) Nitrifying bacteria 1, (h) Denitrifying bacteria

5. (a)	**6.** (c)
7. (d)	**8.** (e)
9. (e)	**10.** (d)
11. (a)	**12.** (e)
13. (d)	**14.** (a)
15. (d)	**16.** (c)
17. (d)	**18.** (b)
19. (a)	**20.** (d)

21. (a), (c), (e) Advantages; (b),(d), (e) Disadvantages

22. (b)	**23.** (c)
24. (b)	**25.** (c)

26. Energy in an ecosystem flows from the sun to producers (c), to consumers (d and a), and then to decomposers (b) who obtain energy by digesting the dead bodies and wastes of other organisms.

CHAPTER 26

1. 4, (c), Leavened bread; 3, (d), Rye bread; 2, (a), Some grains and peanuts; 1, (b), Raw grains

2. (e)	**3.** (e)
4. True	**5.** (a)
6. (e)	**7.** (a), (c)
8. (d)	**9.** (a)

10. (c) Can grow in refrigerated milk; (h) Causes a fecal flavor in milk; (i) Causes a viscous slime to form in milk; (e), (f) Cause milk to sour; (a), (g) Dairy herds are tested or vaccinated against these human pathogens; (b), (d) Present in freshly drawn milk

11. (b)	**12.** (d)
13. (e)	**14.** False
15. (a)	

16. (b) Sodium chloride; (f) Ethylene and propylene oxides; (e) Nitrates and nitrites; (a) Organic acids; (c) Quaternary ammonium compounds; (d) Carbon dioxide

17. (c)	**18.** (c)
19. (a)	**20.** (e)
21. (d)	**22.** (c)
23. (e)	**24.** (a)
25. (e)	

26. Milk can become contaminated from the cows and milk handlers (a), from the packaging plant (b), and even while in the refrigerator (c). Diseases include salmonellosis, listeriosis, tuberculosis, brucellosis. Q fever, and others.

Index

Index

Pathogens and the Diseases They Cause

VIRUSES

Virus	Group Family	Disease	Page
adenovirus	Adenoviridae	acute upper & lower respiratory tract distress, pharyngitis, pneumonia, follicular conjunctivitis, epidemic keratoconjunctivitis	277, 281 593–594 647–649 652–653
arenavirus	Arenaviridae	Bolivian hemorrhagic fever	281, 743
	Arenaviridae	Lassa fever	281, 743
bunyavirus	Bunyaviridae	encephalitis	283, 742
canine parvovirus	Paroviridae	severe vomiting & diarrhea	743
Colorado tick fever	Reoviridae	encephalitis	743
coronavirus	Coronaviridae	colds, GI disturbances	665–666
coxsackle	Picornaviridae	common cold syndrome & pharyngitis; severe systemic illness of newborn; muscle pain & damage; diabetes; meningoencephalitis	743–744
cytomegalovirus	Herpesviridae	mononucleosis, congenital cytomegalic inclusion disease, severe birth defects	282, 632–634
dengue	Flaviviridae	dengue fever (break-bone fever)	334, 739
Eastern equine encephalitis	Togaviridae	encephalitis	429, 761
Ebola	Filoviridae	hemorrhagic fever	742
enterovirus	Picornaviridae	acute hemorrhagic conjunctivitis	278–279
Epstein-Barr	Herpesviridae	Burkitt's lymphoma, infectious mononucleosis, nasopharyngeal carcinoma	282, 740
erythrovirus (B19)	Parvoviridae	aplastic crisis in sickle cell anemia, fifth disease (erythema infectiosum)	283, 743
feline panleukopenia	Parvoviridae	decreased number of white blood cells with fever	743
Hantaan	Bunyaviridae	Korean hemorrhagic fever	281
hantavirus	Bunyaviridae	hantavirus pulmonary syndrome	281, 284, 666
hepatitis A	Picornaviridae	infectious hepatitis	696–698
hepatitis B	Hepadnaviridae	serum hepatitis	699
hepatitis C	?	hepatitis C (non-A, non-B)	699–700
hepatitis D	?	hepatitis D (delta hepatitis)	700
hepatitis E	?	hepatitis E (enterically transmitted non-A, non-B, non-C)	700
herpes simplex type 1	Herpesviridae	oral herpes, gingivostomatitis, herpes labialis (cold sores), keratoconjunctivitis, herpetic whitlow	281–282, 627
herpes simplex type 2	Herpesviridae	genital herpes, oral & whitlow	282, 627–628
herpesvirus	Herpesviridae	meningoencephalitis	282, 630
human immunodeficiency (HIV)	Retorviridae	HIV disease, AIDS	555–560
human papillomavirus	Papovaviridae	common warts (papillomas), genital warts (condylomas); associated with cervical cancer	271, 586–588, 592
influenza	Orthomyxoviridae	influenza (flu)	280–281, 660–661
Marburg	Filoviridae	hemorrhagic fever	742
measles	Paramyxoviridae	rubeola, sometimes subacute sclerosing panencephalitis (SSPE)	582–583
monkeypoxvirus	Orthopoxviridae	monkeypox	586
parainfluenza	Paramyxoviridae	rhinitis, pharyngitis, bronchitis, pneumonia, croup	280, 648–649, 683
paramyxovirus (mumps)	Paramyxoviridae	mumps	280, 682–684
poliovirus	Picornaviridae	poliomyelitis	278, 768–770
polyomavirus: BK	Papovaviridae	associated with renal transplant infection, immunosuppressed patients	762–763
polyomavirus: JC	Papovaviridae	mild respiratory illness	762–763
poxvirus group (unclassified)	?	molluscum contagiosum	282, 586
rabies	Rhabdoviridae	rabies	280, 434, 758–760
respiratory syncytial	Paramyxoviridae	pneumonia in children under age 1, upper respiratory infection in older children & adults	666–667 652–653
rhinovirus	Picornaviridae	common cold	279, 648
Rift Valley fever	Bunyaviridae	fever & hemorrhage	742
rotavirus	Reoviridae	enteritis	696–697
rubella	Togaviridae	German measles, 3-day measles	581–583
St. Louis encephalitis	Flaviviridae	encephalitis	428, 761
smallpoxvirus	Orthopoxviridae	smallpox	585–586
varicella-zoster	Herpesviridae	chickenpox, shingles	282, 583–585
Venezuelan equine encephalitis	Togaviridae	encephalitis	456, 761
Western equine encephalitis	Togaviridae	encephalitis	761
yellow fever	Flaviviridae	yellow fever	277, 280, 334, 739

FUNGI

Organism	Disease	Page
Aspergillus sp.	aspergillosis, pneumonia in compromised patients, skin infections in burn patients, corneal & external ear infections	590
Blastomyces dermatitidis	blastomycosis	589–590
Candida albicans	candidiasis	590
Calviceps purpurea	ergot poisoning	816
Coccidioides immitis	coccidiodomycosis (valley fever)	667–668
Epidermophyton sp.	ringworm (tinea)	588
Filobasidiella neoformans	cryptococcosis	668–669
Histoplasma capsulatum	histoplasmosis	668
Microsporum sp.	ringworm (tinea)	588
Mucor sp.	zygomycosis	590–591
Pneumocystis carinii	Pneumocystis pneumonia	669
Rhizopus sp.	zygomycosis	590–592
Sporothrix schenckii	sporotrichosis	589
Trichophyton sp.	ringworm (tinea)	588

BACTERIA—ALSO SEE APPENDIX B

Organism	Gram Stain*	Basic Morphology	Diseases	Page	Organism	Gram Stain*	Basic Morphology	Diseases	Page
Actinomadura sp.	+	rod, some filamentous forms	Madura foot (maduromycosis)	591	*Haemophilus influenzae*	−	coccobacillus, some strains form capsules	meningitis in children under 5, epiglottitis, ear infections, pneumonia in elderly or compromised patients	224, 644
Actinomyces israelii	+	filamentous, diptheroid, & coccal	actinomycosis, mouth, & other lesions	591	*Helicobacter pylori*	−	curved rod	chronic gastritis, peptic ulcer	692–694
Afipia felis	−	rod	cat scratch fever	597	*Klebsiella pneumoniae*	−	rod, encapsulated	pneumonia, infant diarrhea, urinary tract infections	128, 223, 408, 652-3, 787, 824
Bacillus anthracis	+	rod, encapsulated	anthrax	413, 724–25, 796	*Legionella pneumophilia*	−	coccoid rod	Legionnaires' disease (pneumonia)	653
Bacillus cereus	+	rod, encapsulated	food poisoning	684	*Leptospiria interrogans*	−	spiral	leptospirosis	612–13
Bacteroides sp.	−	small rod	mouth lesions, septicemia, abscesses, Vincent's angina	719	*Listeria monocytogenes*	+	rod	listeriosis, meningitis, abortion	757, 776
Bartonella bacilliformis	−	curved or coccoid	Oroya fever (systemic form), verruga peruana (cutaneous form)	737	*Mycobacterium avium*	A-F	rod	chronic pulmonary disease, opportunistic infections in immunosuppressed patients	655
Bartonella henselae	NA	coccobacillus	cat scratch fever	597, 738	*Mycobacterium leprae*	A-F	rod	Hansen's disease (leprosy)	407, 763–765
Bordetella pertusssis	−	coccobacillus	whooping cough	649–651	*Mycobacterium tuberculosis*	A-F	rod, branching forms	tuberculosis	654–658
Borrelia burgdorferi	−	spiral	Lyme disease	334, 733–4	*Mycoplasma pneumoniae*	NA	too small to be visualized by light microscope	primary atypical bacterial pneumonia	645, 653
Borrelia recurrentis	−	large spiral	epidemic relapsing fever	731	*Neisseria gonorrhoeae*	−	cocci in pairs	gonorrhea, ophthalmia neonatorum, meningitis, arthritis, keratitis	616–620
Brucella sp.	−	coccobacillus	brucellosis (undulant fever or Malta fever)	730–731	*Neisseria meningitidis*	−	cocci in pairs; capsules formed in young cells	meningitis, Waterhouse-Friderichson syndrome	444, 756
Calymmatobacterium granulomatis	−	rod, encapsulated	granuloma inguinale (donovanosis)	627–628	*Nocardia* sp.	+	rod, some filamentous forms	nocardiosis, Madura foot (maduromycosis)	660
Campylobacter sp.	−	rod	gastroenteritis	684	*Porphyromonas gingivalis*	−	rod	periodontal disease	682
Chlamydia psittaci	NA	coccoid, very tiny	ornithosis (psittacosis)	659	*Propionibacterium acnes*	+	rod	acne	384, 580
Chlamydia trachomatis	NA	coccoid, very tiny	conjunctivitis, trachoma, genital tract infection (nongonococcal urethritis), infant pneumonitis, lymphogranuloma venereum	591–593, 625–626	*Providencia stuartii*	−	rod	urinary tract infections, wound infections	580
Clostridium botulinum	+	rod	food poisoning (botulism), wound infections, infant botulism	413, 684–685 767–8, 796, 818, 821	*Pseudomonas aeruginosa*	−	rod	urinary tract infections, skin lesions, eye & ear infections, septicemia in immunocompromised patients	413, 580–81, 611, 719
Clostridium difficile	+	rod	pseudomembranous colitis	382, 694					
Clostridium perfringens	+	rod	gas gangrene, food poisoning	413, 595–6, 684, 819					
Clostridium tetani	+	rod	tetanus	413, 765–766	*Rickettsia akari*	NA	coccobacillus	rickettsialpox	737
Corynebacterium diptheriae	+	rod, club-shaped, pleomorphic, forms palisades	diptheria: pharyngeal, laryngeal & cutaneous	413, 614–615	*Rickettsia prowazekii*	NA	coccobacillus	epidemic typhus, Brill-Zinsser disease	736
Coxiella burnetii	NA	coccobacillus	Q fever pneumonia	659–60	*Rickettsia rickettsii*	NA	coccobacillus	Rocky Mountain spotted fever	736–37
Escherichia coli	−	rod	urinary tract infections, "traveler's diarrhea," nosocomial infections	84, 180, 216, 223, 413, 690	*Rickettsia tsutsugamushi*	NA	coccobacillus	tsutsugamushi fever	334, 736
					Rickettsia typhi	NA	coccobacillus	endemic or murine typhus	736
					Rochalimaea quintana	NA	coccobacillus	trench fever	737
Francisella tularensis	−	small rod (coccobacillus)	tularemia	334, 729–730	*Salmonella enteritidis,* *S. paratyphi,* *S. typhimurium*	−	rod	salmonellosis (food poisoning)	685–686, 816
Gardnerella vaginalis	−	small rod	bacterial vaginitis (nonspecific), urethritis	613	*Salmonella typhi*	−	rod	typhoid fever	686–687
Haemophilus aegyptius	−	coccobacillus	bacterial conjunctivitis	593	*Serratia marcescens*	−	rod	urinary tract infections, hospital epidemics, septicemia, peritonitis, arthritis, pneumonia	719
Haemophilus ducreyi	−	rod	chancroid	624–25					

BACTERIA *(Concluded)*—ALSO SEE APPENDIX B

Organism	Gram Stain*	Basic Morphology	Diseases	Page	Organism	Gram Stain*	Basic Morphology	Diseases	Page
Shigella boydii, S. dysenteriae, S. flexneri, S. sonnei	−	rod, generally single	shigellosis (bacterial (dysentery)	413, 687–688	*Streptomyces* sp.	+	rod, some filamentous forms	Madura foot (madfuromycosis)	384, 388
Spirillum minor	−	spiral	rat bite fever	597	*Treponema pallidum*	−	spiral	acquired & congenital syphilus	620–624
Staphylococcus aureus	+	cocci in clusters	skin lesions, abscesses, boils, scalded skin syndrome, impetigo, toxic shock syndrome, food poisoning, pericarditis	84, 128, 413, 578, 614, 643, 684, 819	*Ureaplasma urealyticum*	NA	very small rod	nongonococcal urethritis; may be responsible for low sperm counts	626
					Vibrio cholerae	−	comma-shaped rod	Asiatic cholera	13, 413, 689
Staphylococcus epidermidis	+	cocci in clusters	skin lesions, contamination of prosthesis	128, 611	*Vibrio parahaemolyticus*	−	rod	food poisoning	690
					Yersinia enterocolitica	−	rod	yersiniosis (enteritis)	692, 334
Streptobacillus moniliformis	−	rodlike, often pleomorphic	rat bite fever	597	*Yersinia pestis*	−	short, thick rod; exhibits bipolar staining	bubonic plague, septicemic plague, pneumonic plague	433, 436, 727–729
Streptococcus mutans	+	cocci in chains	dental caries, possibly plaque, subacute endocarditis	679–681					
Streptococcus pneumaniae	+	cocci in pairs, encapsulated	bacterial pneumonis, otitis media, meningitis, sinusitis	652–653, 757					
Streptococcus pyogenes	+	cocci in chains	pharyngitis, skin lesions, impetigo, scarlet fever, erysipelas, puerperal fever, rheumatic fever, glomerulonephritis	413, 579, 644, 719–720					

*Key to Gram stain
− = Gram-negative
+ = Gram-positive
A-F = acid fast
NA = not applicable

PARASITES

Organism	Type	Disease	Page	Organism	Type	Disease	Page
Ancylostoma duodenale (hookworm)	roundworm	Old World hookworm disease	707	*Necator americanus* (hookworm)	roundworm	New World hookworm disease	707
Ascaris lumbricoides	roundworm	ascariasis	330, 708–709	*Onchocerca volvulus*	roundworm	onchocerciasis (river blindness)	331, 335, 594–595
Babesia microti	protozoan	babesiosis	749	*Paragonimus westermani* (liver/lung fluke)	flatworm	paragonimiasis	327, 669–670
Balantidium coli	protozoan	balantidiasis	701–702				
Clonorchis sinensis (Chinese liver fluke)	flatworm	clonorchiasis	704				
Cryptosporidium sp.	protozoan	cryptosporidiosis	702	*Pediculus humanus*	louse	pediculosis (lice infestation)	599
Diphyllobothrium latum (broad fish tapeworm)	flatworm	diphyllobothriasis	706	*Phthirus pubis*	louse	"crabs" (pubic lice)	599
Dirofilaria immitis (heartworm)	roundworm	heartworm disease (filariasis)	719	*Plasmodium* sp.	protozoan	malaria	317–318, 745–747
Dracunculus medinensis (Guinea) worm)	roundworm	dracunculiasis	330, 591	*Sarcoptes scabiei*	mite	scabies	598
				Schistosoma sp.	flatworm	swimmer's itch	591
				Schistosoma sp.	flatworm	schistosomiasis	328
Echinococcus glanulosus	flatworm	echinococcosis	706	*Strongyloides stercoralis*	roundworm	strongyloidiasis	709–711
Entamoeba histolytica	protozoan	amoebic dysentery	701	*Taenia saginata* (beef tapeworm)	flatworm	taeniasis	326–327
Enterobius vermicularis (pinworm)	roundworm	pinworm infestation	328, 701	*Taenia solium* (pork tapeworm)	flatworm	taeniasis	705–707
Fasciola hepatica (sheep liver fluke)	flatworm	fascioliasis	326, 328, 704	*Toxocara* sp.	roundworm	visceral larva migrans	709
Fasciolopsis buski (Chinese liver fluke)	flatworm	fasciolopsiasis	705, 710	*Toxoplasma gondii*	protozoan	toxoplasmosis	747–749
Giardia intestinalis	protozoan	giardiasis	700	*Trichinella spiralis*	roundworm	trichninosis	330, 707
Hymenolepsis nana (dwarf tapeworm)	flatworm	hymenolepiasis	706	*Trichomonas vaginalis*	protozoan	trichomoniasis	615
Leishmania *L. braziliensis*	protozoan	leishmaniasis cutaneous and mucous membrane infection	744	*Trichuris trichiura* (whipworm)	roundworm	trichuriasis	709
				Trombicula sp.	mite	chigger dermatitis	598
				Trypanosoma brucei gambiense and *T. brucei rhodesiense*	protozoan	trypanosomiasis (African sleeping sickness)	773–775
L. donovani		kala azar	744	*Trypanosoma cruzi*	protozoan	Chagas' disease	334, 775–776
L. tropica		oriental sore (cutaneous)	744	*Tunga penetrans*	sandflea	chigger infestation	598
				Wuchereria bancrofti	roundworm	elephantiasis	331–332, 723
Loa loa	roundworm	loaiasis	330, 595				

The tables of viral and fungal pathogans appear on the back of the facing page.